GeoRef Thesaurus

Sixth Edition

Barbara A. Goodman
Editor

Published by the
AMERICAN GEOLOGICAL INSTITUTE

Earlier Editions
GeoRef Thesaurus and Guide to Indexing, First Edition
Copyright ©1977 American Geological Institute
Second Edition, 1978
Third Edition, 1981
Fourth Edition, 1986
Fifth Edition, 1989

GeoRef Thesaurus, Sixth Edition
Copyright ©1992 American Geological Institute
All rights reserved

International Standard Book Number 0-922152-17-9
Printed in the United States of America

American Geological Institute
4220 King Street
Alexandria, Virginia 22302
703/379-2480

Price: $95

Contents

Acknowledgement iv

Introduction v
   Source and Development of the Vocabulary v
   Changes in this Edition vi
   Indexing vii
   Searching vii
   Autoposting viii
   Singular and Plural Terms viii
   Term Relationships viii
   Notes for Terms x
   Term Validation xi
   Alphabetization xi

Sources xii

Abbreviations xiv

Thesaurus, A-Z 1

Hierarchies and Other Lists 725
   List A   Level-one terms 725
   List B   Area sets 737
   List C   Commodities 739
   List D   Elements 742
   List E   Geologic age (stratigraphic) terms 743
   List F   Fossils 749
   List G   Meteorites 754
   List H   Igneous rocks 755
   List I   Sedimentary rocks 757
   List J   Metamorphic rocks 759
   List K   Sedimentary structures 761
   List L   Minerals 762
   List M   Soils 771
   List N   Sediments 774
   List O   Geographic terms 774
   List P   Fields of interest 782
   List R   Rock Units 787

Maps & Charts 800
   U. S. Physiographic Map 800
   Alaskan Subdivisions 801
   Atlantic Ocean 802
   Pacific Ocean 803

Acknowledgements

The 6th edition of the Thesaurus was prepared at AGI by Barbara Goodman, Editor/Indexer, GeoRef in consultation with Sharon Tahirkheli, Chief Editor, GeoRef and John Mulvihill, Director, GeoRef. In addition the Vocabulary Task Force (VTF), a subcommittee of the GeoRef Advisory Committee, assisted in the preparation. The VTF reviewed many new Term proposals and contributed ideas and expert advice on the other changes. Most of the work of the VTF was done away from AGI on long lists of proposals generated at AGI. In addition one meeting of the VTF was held at AGI to resolve specific differences and discuss general problems and proposals. Joseph Stables and Cecile Lethem , GeoRef editor/indexers helped incorporate some changes which affected thousands of Terms. Lawrence Berg and Mike Cacic handled aspects of the computer programming, and Kay Yost and Leigh Sutherland provided needed expertise with Ventura Publisher.

Members of the Vocabulary Task Force:

> Marilyn Stark (Chair)
> U.S.Geological Survey, Denver,
> Colorado
>
> Dena Fracolli Hanson
> Fort Worth, Texas
>
> Anne Krum
> Shell Western Exploration and
> Production, Inc., Houston, Texas
>
> Dorothy McGarry
> University of California, Los Angeles
>
> Patricia Sheahan
> Konsult International Inc., Willowdale,
> Ontario, Canada

Cover photograph is an electron micrograph of a Coccolithus huxleyi (Lohmann) (magnification 58,600), from the Pleistocene, in the Caribbean from the May-June 1966 Geotimes, Vol. 10, no. 9, article by William H. Hay, University of Illinois.

Introduction

This sixth edition of the GeoRef *Thesaurus* contains more than 27,000 valid and invalid Terms, of which 3,600 are newly added. In addition, 2,300 terms from the previous edition were modified. The Thesaurus is a guide to the Index Terms used in GeoRef. It includes Term relationships, usage notes, dates of addition, indexing rules, guidelines for searching, and lists of systematic and other Terms.

The body of the Thesaurus is organized along the lines of most information retrieval thesauri. The American National Standard, *Guidelines for Thesaurus Construction and Use* (Z39.19-1980) has generally been followed herein.

Source and Development of the Vocabulary

The first edition of the Thesaurus was published in 1977. At that time GeoRef had been in its eleventh year of production at AGI and was using an indexing scheme based on earlier geology bibliographies. Impetus and funding to produce the 1977 Thesaurus came from petroleum companies which had begun to search GeoRef on ORBIT in 1973. The 1977 Thesaurus was based on a Term frequency list consisting of terms used in GeoRef from 1967-1976.

Before the 1977 Thesaurus, GeoRef indexers worked from rules in a Guide to Indexing. The lists in the back of the Thesaurus were derived from that Guide.

When the Thesaurus began to be used in 1977, the indexer was to consult it but was also supposed to continue to use the terminology in the source document for Terms not found in the Thesaurus. This practice continues, and assures that fresh terminology, the natural language of geology, will continue to appear in GeoRef without the substantial delay involved in formally adding a Term and its appearance in the Thesaurus.

This practice also means that variant forms of Terms will have been used in GeoRef up to the date of entry of a Term in the Thesaurus.

The date a Term was entered in the Thesaurus is indicated in parentheses following the term.

In the body of this sixth edition, the first two editions are combined under the year 1978, because Terms introduced at that time could only be distinguished by manually comparing the two editions. The vast majority of those Terms, especially the geographic terms were introduced in the 1977 first edition.

Addition to GeoRef of older bibliographies from which the GeoRef Index Term vocabulary was derived, was completed in 1987. When these old bibliographies were added, the Index Terms were included in GeoRef as originally published. These are a second major source of variant terms in GeoRef.

Presently the GeoRef database consists of four segments or subfiles:

(N) *Bibliography of North American Geology* (publications from 1785-1970)
(G) *Geophysical Abstracts* (published from 1966-1971)
(E) *Bibliography and Index of Geology Exclusive of North America* (published from 1933-1968)
(B) *Bibliography and Index of Geology* (published from 1969 to date)

The Index Term vocabulary in all of the above probably derived from that in the early volumes of the *Bibliography of North American Geology*, edited by John M. Nickles and published by the U.S. Geological Survey. But many changes have occurred since the first volume appeared in 1923.

Some of the Terms used in the pre-1967 bibliographies have been noted in the Thesaurus. In particular, the Entry Terms (level 1 Terms) have been cross-referenced in the body of the Thesaurus and gathered into List A in the back of the Thesaurus. But the bulk of the Terms from these old bibliographies are not accounted for in the Thesaurus.

The procedure followed in adding Terms to the Thesaurus begins with the compilation of a list of candidate Terms consisting of all non-valid Terms used since the previous edition. For this edition, a list of non-valid Terms used from October 1987 to March 1991 was compiled. Each Term was then given a type such as "physiographic" and "formation." Terms became Thesaurus candidate terms depending on type and frequency of use. Each candidate Term was then searched to determine how and how often it occurred in GeoRef, and a way of handling it was proposed by the Thesaurus editor. These proposals were discussed with the Vocabulary Task Force. In this 6th edition, 3,600 new Terms were added.

Given the three to four years between editions of the Thesaurus, and the requirement of at least ten postings by the time a Term is added, we have a good idea of the final form it will take in the literature. This is particularly significant for stratigraphic

Terms, which commonly have several variants, such as Great Smoky Group, Great Smoky Conglomerate, Great Smoky Formation and Great Smoky Quartzite. By examining the frequencies of such Terms over the years, it is usually possible to choose that form which has become generally accepted in the geological literature as the new Thesaurus Term.

As explained in this section, natural language - the terminology found in the geological literature - has been used in indexing for GeoRef and continues to be used therein. Consequently, in searching GeoRef, to retrieve all papers on a concept before that concept was added as a Term in the Thesaurus, it is necessary to search any Term's Used Fors and other variants found through an "expand," "neighbor," or "browse" on the Term (use "expand" in STN and DIALOG, "neighbor" in ORBIT, "scan" in OCLC, and "browse" in CAN/OLE). To call attention to this, the following cautionary note has been added at the foot of each page spread in the Thesaurus:

> A Term is controlled as of the year shown (see p. vi). Before then variants of the Term may occur in GeoRef.

Changes in this Edition

New Terms

The number of new Terms added to the Thesaurus in this edition reflects the volume of place names, names of geologic formations, and similar proper nouns encountered in the geological literature of the world. But this number also reflects our decision to relate cities and political divisions of appreciable size outside the United States to their countries as of this edition, e.g. Paris became Paris France. 2,200 of the total of 3,600 new Terms were altered city, district, province, etc. Terms which were valid prior to 1993.

Provinces were used as the basic unit in Canada, e.g. Perth Ontario; the traditional basic divisions in Great Britain were used, e.g. London England. This allowed autoposting to terms such as Perth Ontario and Perth Scotland.

Simplified BT-NT Relationships

The four classes of broader term relationships previously used in the Thesaurus have been simplified to one class which is shown as BT in the text. For each term in this edition, all of its broader terms are listed as BTs.

In keeping with previous editions, physiographic terms are not generally used as part of political hierarchies, i.e. one would not expect Plaquemines Parish Louisiana to autopost Gulf Coastal Plain. The indexer is expected to add the physiographic Terms relevant to each article.

An effort was made in this edition to autopost hierarchies for mountains where they could be established.

Specific Hierarchy Changes

A new hierarchy was established for sedimentation.

The hierarchies for chemical elements, minerals and Phanerozoic do not autopost those terms.

North America only autoposts to those terms which may not be placed in one country, e.g.

```
Rocky Mountains        U.S. Rocky Mountains
  BT North America       BT1 United States
                         BT2 Rocky Mountains
```

Most former geographic areas of the USSR now autopost Europe or Asia instead of the former USSR.

In the United States, individual states no longer autopost general regional terms such as Western U.S.

Quadrangles have been eliminated as controlled terms. They will continue to be indexed where they occur. Users are encouraged to search quadrangles and coordinates as supplements to a basic search, especially in Alaska (see map in Maps and Charts section).

Hierarchies which were updated for this edition include the Primates and Reptiles in List F and the igneous rocks in List H. The latter were compared with the recent classification produced by the International Union of Geological Sciences Subcommission on the Systematics of Igneous Rocks and adjusted where current usage conforms to the new system.

Used For

In this edition the Prior Term (PT) relationship used in the 5th edition for previously valid terms has been dropped, and all prior terms have been included in the Used For (UF) relationship.

Indexer Note and Thesaurus-based Cross References

Three-level set building instructions have been removed from Indexer Notes. The sets are no longer needed for the printed index of the Bibliography and Index of Geology. Starting in 1993 the printed index of the B.I.G. will consist of single-level entry points and, for heavily-posted terms, subdivisions based on categories and/or area terms. There will be extensive cross references in this index which will come from the BT-NT, UF and SA relationships in this Thesaurus.

Indexing

Certain specific Terms are routinely added in indexing, even if they have not occurred enough in GeoRef to have been added to the Thesaurus. These include quadrangle names, local geographic names, U.S. counties, rock formations, minerals, and fossils. Indexing for a paper on an area in the United States, for example, includes all counties in the area, unless there are more than 10. The same applies to North American rock formations (up to 10 per citation), and to new fossils mentioned in taxonomic papers (again, the maximum of 10 applies). If the limit of 10 is exceeded in a paper, Broader Terms are used instead. Frequent use makes such specific Terms candidates for the Thesaurus, and many have been added to this edition.

In GeoRef indexing, Terms for cities refer not only to the area within the city limits, but also to the surrounding region. Terms for materials in GeoRef, for example, "iron oxides" and "isotopes", almost always indicate naturally occurring materials. When a paper discusses tests done on man-made minerals, the Term "synthetic materials" is indexed.

In GeoRef, for many materials with economic value, there is a Term for the material in the economic context, i.e. as a commodity, and a different Term for the material in a mineralogical, petrological or geochemical context. Diatomaceous earth, for example, is the GeoRef Term for the sedimentary rock. The Term for the same rock, when discussed as a commodity, is diatomite. These commodity Terms are grouped in List C in the back of the Thesaurus.

In this Thesaurus, the number of Terms beginning with "northern", "eastern", "northwestern", "south-central", etc., is limited. However, other such Terms are often used informally in the indexing, especially in combination with state names in the United States and province names in Canada, e.g. southern Kansas, south-central Ontario. They are used with country names outside the U.S. and Canada, except for Great Britain, e.g. eastern Malawi, northeastern England. Before 1976, directional descriptors were not linked to their geographic referrants, e.g. DESCRIPTORS: England, southern; petrology; metamorphic rocks . . .

Another type of directional is formal and valid but not used in hierarchies. Examples of these are: Southwest England, Central Swiss Alps. These generally describe natural geologic regions and are usually indexed only when the author uses the Term.

Searching

The Thesaurus provides a historical record of the Index Term vocabulary of GeoRef from 1978 to date. All valid terms are dated by year. Many of the terms dated 1978 were in use for years prior to then, but we have not attempted to reconstruct their history before 1978, which was the date of the second edition of the Thesaurus, as explained in the previous section, Source and Development of the Vocabulary.

Since there was no Thesaurus before 1978, variant forms of Terms are likely to have been used then. But for each segment of GeoRef, those Terms used on level 1 and level 2 were controlled. A list of level 1 Terms (List A) appears in the back of the Thesaurus and most level 1 Terms which are no longer valid have been cross referenced in the Thesaurus.

When searching GeoRef, be aware that the form of an Index Term is "guaranteed" only from 1978 or the year given in its note. You can depend on the Term to have been used in the form shown as of that year. To search for it before that year, usually a search in the basic index on the portion of the Term common to its uncontrolled variants is best. For example:

Spacelab Program (1982)

To retrieve citations entered into GeoRef before 1982, search "Spacelab" in the basic index. This will retrieve "Spacelab," "Spacelab Program" and other variants.

Stratigraphic formations are particularly fluid entities. Since any author can propose a new formal name with proper documentation, the searcher is cautioned to search every possible variant, e.g. Smith Formation, Smith Sandstone, Smith Group, etc., even "after" the year GeoRef established a valid Term. Indexers are encouraged to use the GeoRef Term. However, in some cases it may not fit, most often because of lithologic (sand versus siltstone) variation. In a few cases, the same geographic name occurs in unrelated locations, e.g. Windermere System in Canada and the Windermere Group in England. The searcher will have to be aware of this possibility. In most of these cases, the ages will also be different.

The basic geographic unit of the world is the country name. In the United States, Canada and Australia the states and provinces are used. Very few directionals, e.g. eastern, central have been formalized and used in hierarchies as valid Terms. Some examples are East Africa and Central Europe. These are displayed in the back in List O and in the hierarchies of their narrow terms. They were not necessarily autoposted prior to the 6th edition. However, directionals are indexed "extensively" as terms. For a further discussion of use of geographic directionals, see Indexing above.

The user is encouraged to search Used For Terms even after the Term was established, as they may have been added after the establishment of the Term.

Autoposting

Whenever Terms are used by an indexer, their Broader Terms are added in the record through a computer lookup. For example, whenever an indexer uses Atchison County Kansas the Terms Kansas and United States are added as index Terms in the record. This is referred to as autoposting. Kansas, and United States are Broader Terms of the Term Atchison County Kansas.

In previous editions some classes of BTs were autoposted, some were not. In this sixth edition of the Thesaurus, all BTs of a term are autoposted, except for North America as noted under Specific Hierarchy Changes above.

Singular and Plural Terms

In deciding whether an Index Term should be singular or plural in GeoRef, the guideline followed is that individual entities - C-14, bayerite, etc., are singular and groups of things - rocks, minerals, etc., are plural. Significant exceptions may still be found in Rock types.

Term Relationships

In the GeoRef Thesaurus, the following relationships may occur for a given Term:

Geographic Coordinates (CO) - For a geographic Term, the rectangular area covered by the Term, expressed in latitudes and longitudes.

Used for (UF) - A synonym or alternative form of the Term which may have been used prior to the time the Term was adopted.

Broader Term (BT) - A Broader Term refers to a group of which the Term is a member or an area in which the Term is a smaller area. Beginning in 1993, all Broader Terms are autoposted.

Narrower Term (NT) - A Narrower Term refers to specific member of a group which is represented by the Term or an area within a larger area which is represented by the Term.

See Also (SA) - A valid Term which is related to the Term in some way other than the above relationships.

a. Geographic Coordinates (CO)

In use since September 1977, Geographic Coordinates have a fixed length of 30 characters. The Coordinates define a rectangular geographic area using latitudes and longitudes. For example, for Brazos County Texas, lat. 30°20'-30°58'N and long. 96°5'-96°40'W, the coordinates are:

Brazos County Texas
CO N302000N305800
W0960500W0964000

In indexing, coordinates have been assigned at the discretion of the indexer for the principal area or areas which are studied in a document. They are not automatically added for each area Term in a citation. Also coordinates are found only in the portion of Subfile B, from September 1977 on.

For the other subfiles, including all the backfile citations, there are no coordinates and searches for geographic locations are limited to Index Terms. It is advisable to use coordinates to supplement a search on geographic Index Terms.

It is the responsibility of the indexer to assign coordinates for the area studied where enough information is provided to define the area. If the area corresponds to an Index Term, the indexer uses the coordinates for that Term in the Thesaurus. Otherwise, the indexer must derive the coordinates from information in the paper or by consulting an atlas.

c. Used For (UF)

Used For relationships direct the Thesaurus user from synonyms and alternate forms to valid Terms.

In this connection, it is well to note that GeoRef does not have a separate field in its records for invalid Terms. Instead, all Index Terms in a GeoRef record are in the same field. Invalid Terms are periodically analyzed. Some are added to the Thesaurus as valid Terms; others become cross references to valid Terms, either as Used for Terms or as entry points in See references; and others not used often enough to be considered are not put in the Thesaurus either as valid or invalid.

When a Term has been added to the Thesaurus, its form and usage are established. But during the years prior to being added, synonyms and variants of the Term may have been used in GeoRef. These can be factored into a search by including Used For Terms as alternates in the search, and by doing an "expand" or "neighbor" or "browse" on a Term and including alternate forms in the search. Used Fors have been included from synonyms and alternate forms. In the latter case, if an alternate form could be found through an "expand" or "neighbor" or "browse" on the Term, it may not appear as a Used For. Names of formations, for example, commonly have several variants in the literature. There may not be a Used For in the Thesaurus from each of these variants to the valid Term if all forms begin with the same word and thus could be found in an "expand" or "neighbor" or "browse." But there will be Used Fors from Terms which would not sort alphabetically with the valid Term. Thus, there may not be a Used For from PCB to PCBs, but there will be a Used For from polychlorinated biphenyls to PCBs. Examples of Used Fors are:

muscovite
  UF potash mica

praseodymium
  UF Pr

greenschist
  UF prasinite

Holocene
  UF Post-glacial

Brunswick Germany
  UF Braunschweig Germany

neutron probe
  UF probe, neutron

For every Used For there is a corresponding Use entry in the Thesaurus from the invalid Term to the valid Term. For example:

bornite
  UF erubescite

erubescite
  use bornite

Some of the Used Fors were valid terms, e.g. Abyssinia and spectrometry and for one reason or another have been replaced by other Terms. These will have Scope Notes indicating they were once valid.

The Used For references under a Term help define the Term by showing concepts which are considered to be synonymous in GeoRef.

d. <u>Broader Term (BT)</u>

Most Broader Terms represent groups of which the Term is a member. For example:

peridot
  BT olivine group
  BT nesosilicates
  BT orthosilicates
  BT silicates

Broader Terms are listed in order from the most specific to the most general. Thus in the above example, peridot is a member of the olivine group, the olivine group is a member of the nesosilicates, etc.

Two notable exceptions to this member-group relationship are geographic locations and ages. For geographic locations, a part-whole relationship prevails. For example:

Minas Gerais Brazil
  State in E central Brazil.
  BT Brazil
  BT South America

Brazil
  BT South America

Specific lakes, rivers, faults and mountains which are wholly within a country have that country as a Broader Term. Stratigraphic Terms such as formations and groups have ages as Broader Terms. And ages have part-whole relationships. For example:

Lockatong Formation
  BT Upper Triassic
  BT Triassic
  BT Mesozoic

Senonian
  BT Upper Cretaceous
  BT Cretaceous
  BT Mesozoic

All Broader Terms of a Term are shown under that Term. Before the 6th edition, the geographic locations within a country stopped at the country level. The country which displayed its broader terms was always added where relevant. North America, however, is not autoposted to narrower terms of a country.

A few Terms have multiple sets of Broader Terms. These sets are shown as BT1, BT2, etc. For example:

C-14/C-12 (1993)
  BT1 radioactive isotopes
  BT1 isotopes
  BT2 stable isotopes
  BT2 isotopes
  BT3 carbon

In this example, the immediate broader terms are the first encountered in each hierarchy: BT1 radioactive isotopes, BT2 stable isotopes, and BT3 carbon.

e. <u>Narrower Term (NT)</u>

These are Terms for specific members of the groups represented by the Term. For example:

gymnosperms
  NT Bennettitales
  NT Caytoniales
  NT Coniferae
  NT Coniferales
    etc.

In this example Bennettitales is a member of the group gymnosperms, as are Caytoniales, Coniferae, Coniferales and the other Narrower Terms of gymnosperms.

For any Term, only its Narrower Terms on the next level down are listed. If a Narrower Term itself has Narrower Terms, these will not appear under the Term. Thus, in the above example,

Pachypteris, a Narrower Term of Caytoniales, will not appear under gymnosperms.

Terms which have multiple sets of Broader Terms are displayed as Narrower Terms under each of the immediate broader terms. In the example above, the Term, C-14/C-12 appears as an NT under radioactive isotopes, stable isotopes and carbon.

Narrower Terms are arranged in alphabetical order.

An important exception to the member-group relationship of Narrower Terms is geographic locations in which the Narrower Terms of a Term are in a part-whole relationship. For example:

> **Missouri**
> NT Atchison County Missouri
> NT Barton County Missouri
> NT Benton County Missouri
> NT Boone County Missouri
>   etc.

Another exception is for ages where stratigraphic Terms appear under appropriate ages in Narrower Terms and age Terms appear in Narrower Terms under ages of which they are parts. For example:

> **Devonian**
> NT Acadian Phase
> NT Ackley Granite
> NT Barre Granite
> NT Cedar City Formation
>   etc.
>
> **Paleozoic**
> NT Antler Orogeny
> NT Arbuckle Group
> . . . .
> NT Devonian
>   etc.

f. <u>See Also (SA)</u>

The See Also is used to indicate relationships other than Used For, Broader Term and Narrower Term. Examples of See Also relationships are:

> **Gallup Sandstone**
> SA Mesaverde Group
> SA New Mexico
>
> **geochronology**
> SA absolute age
>
> **periodicity**
> SA earthquakes
> SA frequency
> SA Milankovitch theory
> SA orogeny

Notes for Terms

In addition to the above relationships, Terms have explanatory notes. There are three kinds, year, general notes and Indexer Notes (IN).

a. <u>Year</u>

Year Term became valid - The year in parenthesis following each Term is the year the Term was introduced in the Thesaurus. Terms with the year 1978, the year of the first edition of the Thesaurus, may have been valid in earlier years as well.

b. <u>General notes</u>

These immediately follow Terms and are not prefaced by a caption. Valid Terms are underlined in the notes.

- "Also search" - An "also search" statement is included in the Note when it is not possible to suggest searching an alternative Term through a UF relationship. These notes are used when the user needs to search on a combination of Terms rather than a single Term. This applies to all U.S. counties and cities, for example:

> **Bonneville County Idaho**
> Before 1989, also search Bonneville County AND Idaho.

In this context, note that it is usually advisable to also search relevant Used For Terms and Terms with similar spelling found through an "expand" or "neighbor" or "browse", since these may have been used in the years before the Term was introduced (see Used For above).

- Geographical Term locations - Brief notes on the location of many geographical Terms are given.
- Scope notes - These define the use of the Term in GeoRef. For example:

> **hydrocarbons**
> Used for small amounts of hydrocarbons in rocks or sediments. Also used for pollution studies of petroleum products. For pollution studies, before 1993, also search petroleum and pollution. For economic deposits, see petroleum; natural gas; bitumens; asphalt; oil sands; oil shale.

- Autoposting notes - Some broader terms contain notes regarding narrower terms which began autoposting after both terms were introduced.

> **halides (1978)**
> Autoposting of this term to fluoborates and fluosilicates began in 1981.

c. <u>Indexer Note (IN)</u>

These notes tell how a Term is to be used in both the printed index and in GeoRef.

The phrase "includes use" which occurs frequently in the notes is for examples of current and significant GeoRef usage. It means "this is an important use" but not "this is the only use."

Term Validation

The Thesaurus is used in the production of GeoRef to validate the Terms used to index new citations. As each Index Term is typed, it is compared to a list of valid Thesaurus Terms. If the Term is found in the validation file, it is accepted. If not, and the Term is correctly spelled, the invalid Term is accepted for the citation and a record of it is saved. Such invalid Terms become candidate Terms for the next edition of the Thesaurus.

Alphabetization

Terms are sorted word-by-word rather than letter-by-letter. Specifically:

- The sort is on letters, numbers, spaces and virgules(/).
- A space sorts before a virgule or slash (/); a virgule sorts before a letter; a letter sorts before a number.
- Hyphens and open parentheses sort as if they were spaces.
- All characters, including punctuation marks, other than letters, numbers, hyphens, virgules and the open parenthesis, are squeezed out in the sort, i.e., treated as if they did not exist.
- Acronyms sort as words.

Sources

The following publications are consulted for identifying and determining term hierarchies for new GeoRef index terms. Publications are listed according to subject headings and within subject headings, are listed from most to least frequently consulted.

General

Bates, Robert L. and Jackson, Julia A. (eds.), Glossary of Geology, 3rd ed. (Am. Geol. Inst., Alexandria, VA, 1987)
Parker, Sybil P., (ed.), McGraw-Hill Encyclopedia of the Geological Sciences, 2nd ed. (McGraw-Hill, New York, NY, 1988)
Bates, Robert L. and Jackson, Julia A. (eds.), Glossary of Geology, 2nd ed. (Am. Geol. Inst., Falls Church, VA, 1980)

Geography

Seltzer, L.E. (ed.), Columbia Lippincott Gazetteer of the World (Columbia Univ. Press, New York, NY, 1966)
Webster's New Geographic Dictionary (G & C Merriam Co., Springfield, MA, 1988)
Hammond Citation World Atlas (Hammond Inc., Maplewood, NJ, 1992)
Hammond Ambassador World Atlas (Hammond Inc., Maplewood, NJ, 1977)
The Times Atlas of the World, Comprehensive Edition, 6th ed. (Times Books, New York, NY, 1980)
The Times Index-Gazetteer of the World (Houghton Mifflin Co., Boston, MA, 1966)
Geographic Names Data Base, U.S. Geological Survey
U.S. Dept. of the Interior, Board on Geographic Names, Official Standard Names

Meteorites

Graham, A.L., Bevan, A.W.R. and Hutchison, R., Catalogue of Meteorites, 4th ed. (Univ. Arizona Press, Tucson, AZ, 1985)
Hey, Max H., Catalogue of Meteorites (British Museum, London, 1966)
Wasson, John T., Meteorites; Classification and Properties (Springer-Verlag, Berlin, 1974)

Mineralogy (Silicates)

Strunz, H., Mineralogische Tabellen, 6th ed. (Akad. Verlag., Leipzig, 1977)

Mineralogy

Nickel, Earnest H. and Nichols, Monte C., Mineral Reference Manual (Van Nostrand and Reinhold, New York, 1991)

Mineralogy (Nonsilicates)

Fleischer, Michael and Mandarino, Joseph A., Glossary of Mineral Species 1991 (Mineral. Record. Inc., Tucson, AZ)

Paleontology, Invertebrate

Moore, R.C. (ed.), Treatise on Invertebrate Paleontology, 1st ed. [2nd ed. with Teichert, C., (ed). for Part E (Archaeocyatha); Part V (Graptolithina); and Part W, Supplement 1 (Trace Fossils). Revised with Robison, R.A., (ed.), for Part G (Bryozoa). Robison, R.A., (ed.) for Part W, Supplement 2 (Conodonta).] (Geol. Soc. Am./ Univ. Kansas Press)

Paleontology, Vertebrate

Romer, A.S., Vertebrate Paleontology, 3rd ed. (Univ. Chicago Press, 1974)
Gregory, J.T.; Bacskai, J.A. Shkurkin, G.V.; Winans, M.C. and Riuschev, B.H., Bibliography of Fossil Vertebrates 1940-(series published by Geological Society of America 1940-1973, American Geological Institute, 1981-83, and Society of Vertebrate Paleontology, 1984-
Weishampel, Peter Dodson and Osmolska, Halszka, The Dinosauria (Univ. of Calif. Press, Berkeley, 1990)
Benton, M. J., The Phylogeny and Classification of the Tetrapods (Clarendon Press, Oxford, 1988)

Igneous and Metamorphic Rocks

Tomkeieff, S. I., Dictionary of Petrology (John Wiley & Sons, Chichester, 1983)
Le Maitre, R. W., editor, with others, A Classification of Igneous Rocks and Glossary of Terms, Recommendations of the International Union of Geological Sciences Subcommission on the Systematics of Igneous Rocks (Blackwell Scientific Publications, Oxford, 1989)

Sedimentary Rocks

Pettijohn, F.J., Sedimentary rocks, 2nd ed. (Harper & Row, New York, NY, 1957)

Stratigraphy

Lexique Stratigraphique International (Cent. Natl. Rech. Sci., Paris, 1950s-1960s [for United States, 1967 ed.]) For American names, the Lexique follows U.S. Geological Survey Bulletin 1200, Lexicon of Geologic Names of the United States.

Haq, B.U. and van Eysinga, F.W.B., Geological Time Table, 4th ed. (Elsevier Sci. Publ. Co., Amsterdam, 1987)

van Eysinga, F.W.B. (compiler), Geological Time Table, 3rd ed. (Elsevier Sci. Publ. Co., Amsterdam, 1975)

Fabre, J. (ed.), Lexique Stratigraphique International, Nouvelle Serie no 1; Afrique de l'ouest; Introduction geologique et termes stratigraphique (Pergamon Press, New York, NY, 1983)

Lexicon of Geologic Names of the United States (U.S. Geological Survey Bulletins 896, 1056-A, 1200, 1350, 1520, 1564, 1565, U.S. Government Printing Office, Washington, DC)

Luttrell, G. N., Hubert, M. L. and Murdock, C. R., GEONAMES; Data Base of Geologic Names of the United States through 1988, U.S. Geological Survey, Reston, 1990)

Abbreviations

Entries for index terms listed in the main body of the Thesaurus contain abbreviations, most of which indicate relationships between terms. They are as follows:

IN	Indexer Note
CO	Geographic Coordinates
UF	Used for
BT	Broader Term, of a Term with one hierarchy
BT1,BT2, etc.	Broader Term, of a Term with multiple hierarchies
NT	Narrower Term
SA	See Also

For a detailed explanation of the above, see the Introduction.

A

A-type granites (1993)
Anorogenic granites. General term used for rift zone and interior stable continental plate granites, which often have an alkaline composition.
BT granites
BT plutonic rocks
BT igneous rocks
SA alkalic composition
SA I-type granites
SA rift zones
SA S-type granites

aa
use aa lava

aa lava (1978)
Before 1978, also search aa AND lava.
UF aa
BT lava
SA pahoehoe
SA volcanism

AABW
use Antarctic bottom water

Aachen
No longer a valid term for GeoRef. See Aachen Germany.

Aachen Germany (1993)
City near the intersection of the Belgium and Netherlands borders.
BT North Rhine-Westphalia Germany
BT Germany
BT Central Europe
BT Europe

Aaland
use Aland

Aalenian (1978)
Europe.
BT Middle Jurassic
BT Jurassic
BT Mesozoic
SA Dogger
SA Lower Jurassic

AAPG (1985)
Acronym.
UF American Association of Petroleum Geologists
SA associations

Aar Massif (1978)
BT Switzerland
BT Central Europe
BT Europe

Aar Valley (1978)
River valley in central and N Switzerland.
UF Aare Valley
BT Switzerland
BT Central Europe
BT Europe

Aare Valley
use Aar Valley

Aargau
No longer a valid term for GeoRef. As of 1993, see Aargau Switzerland.

Aargau Switzerland (1993)
Canton in N Switzerland.
BT Switzerland
BT Central Europe
BT Europe

Abakan
No longer a valid term for GeoRef. As of 1993, see Abakan Russian Federation.

Abakan Russian Federation (1993)
Town on Yenisei River SW of Krasnoyarsk in Khakass Autonomous Oblast. Before 1993, also search Abakan. This term has multiple hierarchies.
BT1 Russian Federation
BT1 Commonwealth of Independent States
BT2 Asia

abandoned mines (1985)
UF closed mines
BT mines
SA land subsidence
SA mining geology
SA reclamation

Abashiri
No longer a valid term for GeoRef. See Abashiri Japan.

Abashiri Japan (1993)
City and sub-prefecture in NE Hokkaido. Before 1993, also search Abashiri or Abasiri.
UF Abasiri Japan
BT Hokkaido
BT Japan
BT Far East
BT Asia
NT Kitami Japan
SA Kitami Basin

Abasiri Japan
use Abashiri Japan

Abean Orogeny
use Hercynian Orogeny

Abee
No longer a valid term for GeoRef. See Abee Meteorite.

Abee Meteorite (1985)
Impact at Abee, E central Alberta, Canada. Before 1985, also search Abee AND meteorites.
BT enstatite chondrites
BT chondrites
BT stony meteorites
BT meteorites
SA Alberta
SA Canada

Aberdeen
No longer a valid term for GeoRef as of 1993. See Aberdeen Scotland.

Aberdeen Scotland (1993)
City on North Sea. Before 1993, also search Aberdeen and Scotland.
BT Aberdeenshire Scotland
BT Grampian region Scotland
BT Scotland
BT Great Britain
BT United Kingdom
BT Western Europe
BT Europe

Aberdeenshire
No longer a valid term for GeoRef as of 1993.
use Aberdeenshire Scotland

Aberdeenshire Scotland (1993)
Former county in NE Scotland.
CO N565000N574300 W0014500W0034800
UF Aberdeenshire
BT Grampian region Scotland
BT Scotland
BT Great Britain
BT United Kingdom
BT Western Europe
BT Europe
NT Aberdeen Scotland

Aberfeldy (1989)
Metal ore and barite mining region in Perthshire, Tayside region, central Scotland. As of 1993, use Aberfeldy Scotland for the town. After 1992, see Aberfeldy Field for the oil field in Lloydminster, W Saskatchewan.
BT Perthshire Scotland
BT Scotland
BT Great Britain
BT United Kingdom
BT Western Europe
BT Europe
SA barite deposits
SA lead-zinc deposits

Aberfeldy Field (1993)
Oil field in Lloydminster, W Saskatchewan. Before 1993, also search Aberfeldy AND Saskatchewan.
BT Saskatchewan
BT Western Canada
BT Canada
SA oil and gas fields

Aberystwyth
No longer a valid term for GeoRef as of 1993. See Aberystwyth Wales.

Aberystwyth Grits (1978)
Upper Llandoverian. Forms part of the Ystwyth Stage. N and central Wales and NW England.
BT Silurian
BT Paleozoic
SA England
SA United Kingdom
SA Wales

Aberystwyth Wales (1993)
Town on Saint Georges Channel in Cardiganshire (Dyfed). Before 1993, also search Aberystwyth.
BT Cardiganshire Wales
BT Dyfed Wales
BT Wales
BT Great Britain
BT United Kingdom
BT Western Europe
BT Europe

Abies (1989)
Genus.
BT Pinaceae
BT Coniferales
BT gymnosperms
BT Spermatophyta
BT Plantae

Abilene
Valid through 1988. Search in combination with state term. After 1988, use specific city-state term.

Abilene Kansas (1989)
City in E central Kansas. Before 1989, search Abilene AND Kansas.
CO N385500N385500 W0971400W0971400
BT Dickinson County Kansas
BT Kansas
BT United States

Abilene Texas (1989)
City in NW central Texas. Before 1989, search Abilene AND Texas.
CO N322700N322700 W0994500W0994500
BT Taylor County Texas
BT Texas
BT United States

Abitibi
No longer a valid term for GeoRef. As of 1993, see Abitibi County Quebec.

Abitibi Belt (1981)
Structural subprovince in SW Quebec and E Ontario.
CO N460000N503000 W0735000W0850000
BT Superior Province
BT Canadian Shield
BT North America
SA Blake River Group
SA Ontario
SA Quebec
SA Sigma Mine
SA Timiskaming Group

Abitibi County Quebec (1993)
County on James Bay in W Quebec. Before 1993, also search Abitibi or Abitibi County in conjunction with Quebec.
BT Quebec
BT Eastern Canada
BT Canada
NT Chibougamau Quebec
NT Val d'Or Quebec

Abkhasia Georgian Republic
use Abkhazia Georgian Republic

Abkhazia
No longer a valid term for GeoRef. As of 1993, see Abkhazia Georgian Republic.

Abkhazia Georgian Republic (1993)
Former Abkhaz Autonomous Soviet Socialist Republic. Before 1993, also search Abkhasia.
CO N422000N432000 E0421500E0400000
UF Abkhasia Georgian Republic
BT Georgian Republic
BT Europe
NT Kodor River basin

ablation (1978)
SA debris cones
SA glacial geology
SA glaciers
SA mass balance
SA sedimentation
SA wind erosion
SA wind transport

Abo
Use Abo Finland for the city in Finland. Use Abo Formation for the Lower Permian unit in New Mexico.

Abo Finland (1993)
Port city in SW Finland. Before 1993, search Turku AND Finland.
UF Turku Finland
BT Finland
BT Scandinavia
BT Western Europe
BT Europe

Abo Formation (1985)
Lower Permian, New Mexico.
BT Lower Permian
BT Permian
BT Paleozoic
SA Leonardian
SA New Mexico
SA Wolfcampian

abrasion (1978)
UF mechanical erosion
SA detritus
SA erosion
SA glaciation

SA grinding
SA planation
SA ventifacts

abrasives (1978)
General.
IN When used as a commodity term index also industrial minerals.
SA carborundum
SA corundum deposits
SA diamonds
SA diatomite
SA garnet deposits
SA industrial minerals
SA pumice deposits
SA silica

Abruzzi
No longer a valid term for GeoRef. See Abruzzi Italy.

Abruzzi Italy (1993)
Autonomous region on the Adriatic Sea.
CO N414000N425500
  E0144800E0130000
BT Italy
BT Southern Europe
BT Europe
NT Chieti Italy
NT Marsica

Absaroka Fault (1989)
Western Overthrust Belt of Idaho, Utah and Wyoming.
IN Index states as applicable.
BT United States
SA Idaho
SA Utah
SA Western Overthrust Belt
SA Wyoming

Absaroka Range (1978)
Range of the Rocky Mountains. This term has multiple hierarchies.
IN Index states and counties as applicable.
CO N433000N450000
  W1090000W1101500
BT1 U. S. Rocky Mountains
BT1 Rocky Mountains
BT2 U. S. Rocky Mountains
BT2 United States
NT Beartooth Mountains
SA Montana
SA Wyoming

Absaroka Supergroup (1985)
Montana and NW Wyoming.
BT Eocene
BT Paleogene
BT Tertiary
BT Cenozoic
SA Montana
SA Wyoming

Absaroka Thrust
Use Absaroka Fault or Absaroka thrust sheet if applicable.

absarokite (1978)
BT basalts
BT volcanic rocks
BT igneous rocks

absolute age (1978)
Used for isotopic (radiometric or radiogenic) dating. For non-isotopic dating, see geochronology. Before 1971, also search absolute dating.
IN Index dates or specific methods if applicable. Include type of material.
UF absolute dating
UF actual age (absolute age)
SA accelerator mass spectroscopy
SA age

SA Ar/Ar
SA C-14
SA charcoal
SA cosmochronology
SA dates
SA diffusion
SA geochemistry
SA geochronology
SA He-4/He-3
SA He/He
SA I/Xe
SA Io/Th
SA Io/U
SA isochrons
SA isotopes
SA K/Ar
SA Kr/Kr
SA new methods
SA orogeny
SA overprinting
SA Pb-210
SA Pb/Pb
SA Pb/Th
SA radioactive decay
SA Rb/Sr
SA Re/Os
SA relative age
SA Sm/Nd
SA Sr/Sr
SA standard materials
SA Th/Th
SA Th/U
SA tritium
SA U/He
SA U/Pb
SA U/Th/Pb
SA uranium disequilibrium
SA whole rock

absolute age, dates
Not a valid GeoRef index term after 1971. Was used on level 1 in subfiles E, G, N, T and B.

absolute age, methods
Not a valid GeoRef index term after 1971. Was used on lovel 1 in subfiles E, G, N, T and B.

absolute dating
Not a valid term for GeoRef. Before 1971, included use in subfiles E, G, N and B.
use absolute age

absorbent materials (1981)
SA industrial minerals
SA materials

absorption (1978)
SA adsorption
SA atomic absorption
SA atomic absorption spectra
SA emission spectroscopy
SA geochemistry
SA sorption
SA spectroscopy
SA wave absorption
SA wettability

absorption and scattering
No longer a valid term for GeoRef. Usually out-of-scope. Used with aeronomy or aurora. As of 1993, see absorption or wave dispersion if applicable.

absorption spectra
use atomic absorption spectra

absorption spectrophotometry
Not a valid GeoRef index term after 1970. Was used on level 1 in subfile N.

absorption spectroscopy
Not a valid index term. Use absorption with spectroscopy.

Abu Dhabi (1978)

Sheikdom. One of federation of 7 states at S end of Persian Gulf.
BT United Arab Emirates
BT Arabian Peninsula
BT Asia

Abukuma Mountains (1978)
Mostly in Fukushima Prefecture in E central Honshu.
UF Abukuma Plateau
BT Honshu
BT Japan
BT Far East
BT Asia
SA Fukushima Japan
SA Futaba Group
SA Ibaraki Japan
SA Miyagi Japan

Abukuma Plateau
use Abukuma Mountains

abundance
No longer a valid term for GeoRef. From 1981 through 1992, used only in the chemical sense. See geochemistry.

abyssal cones
use submarine fans

abyssal environment
As of 1981, no longer a valid term for GeoRef.
use deep-sea environment

abyssal fans
use submarine fans

abyssal hills (1993)
UF submarine hills
SA bottom features
SA deep-sea environment
SA submarine dunes
SA submarine environment

abyssal plains (1978)
SA bottom features
SA continental rise
SA ocean basins
SA ocean floors
SA oceanography
SA plains

Abyssinia
No longer a valid GeoRef index term. Before 1945, was used on level 1 in subfile E.
use Ethiopia

Abyssinian Rift valley
use Ethiopian Rift

abyssolith
use batholiths

Ac
use actinium

academic institutions (1989)
Reserved for any educational institution.
UF colleges
UF universities
SA associations
SA Birmingham University
SA Cambridge University
SA Colorado School of Mines
SA education
SA institutions
SA Lamont-Doherty Geological Observatory
SA Leningrad Mining Institute
SA Moscow University
SA museums
SA Ohio State University
SA organization
SA Pennsylvania State University
SA Scripps Institution of Oceanography
SA University of Arizona
SA University of Bonn

SA University of California
SA University of Louvain
SA University of Lund
SA University of Michigan
SA University of Pennsylvania
SA University of Rome
SA University of Tokyo
SA University of Wisconsin
SA Woods Hole Oceanographic Institution

Acadian (1978)
Provincial series, Canada.
BT Middle Cambrian
BT Cambrian
BT Paleozoic
NT Lancara Formation
SA Acadian Phase
SA Canada

Acadian Orogeny
use Acadian Phase

Acadian Phase (1978)
Use Acadian for the age. Before 1978, also search Acadian Orogeny. Also search orogeny AND Acadian.
UF Acadian Orogeny
BT Devonian
BT Paleozoic
SA Acadian
SA Antler Orogeny
SA orogeny
SA tectonics

Acantharina (1978)
Suborder. As of 1990, microfossils and Protista are autoposted to this term. This term has multiple hierarchies.
BT1 Porulosida
BT1 Radiolaria
BT1 Protista
BT1 Invertebrata
BT2 Porulosida
BT2 Radiolaria
BT2 Protista
BT2 microfossils

acanthite (1978)
BT sulfides
SA argentite
SA silver ores

Acanthodes (1978)
Genus.
BT Acanthodii
BT Pisces
BT Vertebrata
BT Chordata

acanthodians (1989)
As of 1981, used as the common name for Acanthodii.
BT fish
BT vertebrates
SA Acanthodii
SA biostratigraphy

Acanthodii (1978)
Subclass. Before 1981, also search acanthodians. Autoposting of this term to Acanthodes began in 1981.
BT Pisces
BT Vertebrata
BT Chordata
NT Acanthodes
SA acanthodians
SA Osteichthyes
SA Placodermi

acanthopores (1978)
Skeletal structures in stenolaemate bryozoans.
SA Bryozoa

Acceglio
No longer a valid term for GeoRef. See Acceglio Italy.

Acceglio Italy (1993)
Village in NW Italy.
BT Piemonte Italy
BT Italy
BT Southern Europe
BT Europe

acceleration (1993)
General term.
SA accelerograms
SA accelerometers
SA motions
SA seismic response
SA velocity

accelerator mass spectroscopy (1993)
Particle accelerator and mass spectrometer. May be used in dating using cosmogenic isotopes. Before 1993, also search AMS.
BT mass spectroscopy
BT spectroscopy
SA absolute age
SA Al-26
SA Be-10
SA C-14
SA chemical analysis
SA instruments
SA Si-32

accelerograms (1978)
Used for data. For the instrument, use accelerometers. Also search acceleration.
SA acceleration
SA accelerometers
SA earthquakes
SA engineering geology
SA seismology

accelerographs
 use accelerometers

accelerometers (1981)
Used for the instrument. Use accelerograms for data.
UF accelerographs
SA acceleration
SA accelerograms
SA engineering geology
SA instruments
SA seismology

accessory minerals (1978)
Before 1978, search minerals AND accessory.
SA heavy minerals
SA minerals
SA petrology

Accomac County
 use Accomack County Virginia

Accomack County
Valid through 1988. Search in combination with state term. After 1988, use specific city-state term.

Accomack County Virginia (1989)
On the Delmarva Peninsula. Before 1989, search Accomack County AND Virginia.
CO N372500N380300
    W0751500W0755700
UF Accomac County
BT Virginia
BT United States
NT Wachapreague Inlet

Accra
No longer a valid term for GeoRef. See Accra Ghana.

Accra Ghana (1993)
City on the Gulf of Guinea.
BT Ghana
BT West Africa
BT Africa

accreting plate boundary (1993)
UF divergent plate boundaries
SA accretion
SA passive margins
SA plate boundaries
SA plate divergence
SA plate tectonics
SA rifting
SA sea-floor spreading

accretion (1978)
Term used for sedimentation through 1977. After 1977, term includes use as genetic concept for the Moon and planets.
SA accreting plate boundary
SA deposition
SA Moon
SA planetesimals
SA planetology
SA plate divergence
SA rifting
SA sedimentation
SA tectonics

accretionary prisms
 use accretionary wedges

accretionary wedges (1989)
UF accretionary prisms
SA plate tectonics
SA tectonics
SA terranes
SA underplating

accumulation
As of 1981, no longer a valid term for GeoRef. Before 1981, used as a general term.

accumulation, petroleum
 use petroleum accumulation

accuracy (1978)
Used as a general term. Also search precision.
SA calibration
SA corrections
SA efficiency
SA errors
SA precision
SA reliability

Acer (1978)
Genus.
BT Dicotyledoneae
BT angiosperms
BT Spermatophyta
BT Plantae

acetate (1989)
Compound derived from acetic acid.
BT organic materials

acetylene (1989)
BT alkynes
BT aliphatic hydrocarbons
BT hydrocarbons
BT organic materials

Acheulean
 use Acheulian

Acheulian (1978)
Archaeological classification.
UF Acheulean
UF Acheullian
BT Paleolithic
BT Stone Age
BT Cenozoic
SA archaeology

Acheullian
 use Acheulian

achondrites (1978)
Autoposting of this term began in 1985.
BT stony meteorites
BT meteorites
NT ALHA 77005
NT ALHA 81005
NT angrite
NT chassignite
NT chladnite
NT diogenite
NT EETA 79001
NT eucrite
NT howardite
NT nakhlite
NT shergottite
NT SNC Meteorites
NT ureilite
SA chondrites
SA Elephant Moraine Meteorites
SA iron meteorites
SA stony irons

acid mine drainage (1978)
Before 1978, also search drainage AND mines.
SA acids
SA drainage
SA environmental geology
SA mine drainage
SA mines
SA pollution
SA sulfuric acid

acid precipitation
 use acid rain

acid rain (1985)
UF acid precipitation
UF acid rainfall
BT atmospheric precipitation
SA acidification
SA pH
SA pollution
SA rainfall

acid rainfall
 use acid rain

acid sulfate soils (1985)
BT soils

acidic
A valid index term through 1977.

acidic composition (1978)
Before 1978, also search acidic. Before 1985, also search silicic.
UF silicic composition
SA acidic magmas
SA acids
SA composition
SA igneous rocks
SA pH

acidic magmas (1981)
Autoposting of magmas to this term began in 1989.
BT magmas
SA acidic composition

acidification (1985)
SA acid rain
SA petroleum engineering
SA tertiary recovery

acidity
 use pH

acids (1978)
SA acid mine drainage
SA acidic composition
SA alanine
SA amino acids
SA carboxylic acids
SA compounds
SA DNA
SA fatty acids
SA fulvic acids
SA humic acids
SA hydrochloric acid

SA hydrofluoric acid
SA inorganic acids
SA organic acids
SA pH
SA sulfuric acid

Ackley Granite (1989)
S Newfoundland.
BT Devonian
BT Paleozoic
SA Newfoundland Island

acmite (1978)
BT clinopyroxene
BT pyroxene group
BT chain silicates
BT silicates
SA aegirine

Acoela (1981)
BT Gastropoda
BT Mollusca
BT Invertebrata

Aconcagua Chile
 use Valparaiso Chile

Aconcagua Province
Former province. No longer a valid term for GeoRef. As of 1993, use Valparaiso Chile.

acoustic emissions
 use acoustical emissions

acoustic logging
 use acoustical logging

acoustic methods
 use acoustical methods

acoustic surveys
 use acoustical surveys

acoustic waves
 use acoustical waves

acoustical
A valid index term through 1977.

acoustical emissions (1985)
UF acoustic emissions
UF emissions, acoustical
SA acoustical waves
SA avalanches
SA earthquakes
SA precursors
SA rock mechanics
SA seismology

acoustical exploration
Not a valid GeoRef index term after 1970. Was used on level 1 in subfile N. Now use acoustical logging, acoustical methods or acoustical surveys.

acoustical logging (1978)
Not a valid index term from 1975 to 1977. Before 1978, also search well-logging AND acoustic.
UF acoustic logging
UF logging, acoustical
BT well-logging
SA acoustical methods
SA acoustical surveys
SA borehole televiewers
SA impedance
SA seismic logging
SA tube waves

acoustical methods (1978)
Also search acoustic.
UF acoustic methods
BT geophysical methods
SA acoustical logging
SA acoustical surveys
SA air guns
SA borehole televiewers
SA deep-tow methods
SA echo sounding
SA geophones

SA GLORIA
SA Green function
SA hydrophones
SA impedance
SA Seabeam
SA side-scanning methods
SA sonar methods
SA ultrasonic methods

acoustical properties (1978)
SA physical properties
SA properties
SA seismology

acoustical surveys (1978)
Also search acoustic.
UF acoustic surveys
BT geophysical surveys
BT surveys
SA acoustical logging
SA acoustical methods
SA GLORIA
SA Seabeam
SA sonar methods
SA ultrasonic methods

acoustical waves (1978)
Also search acoustic AND waves; acoustical AND waves.
UF acoustic waves
UF sonic waves
UF sound waves
SA acoustical emissions
SA elastic waves
SA impedance
SA seismology
SA waves

Acqui
No longer a valid term for GeoRef. See Acqui Italy.

Acqui Italy (1993)
Town in NW Italy.
BT Piemonte Italy
BT Italy
BT Southern Europe
BT Europe
SA Alessandria Italy

acquisition, data
use data acquisition

Acre
No longer a valid term for GeoRef. See Acre Brazil.

Acre Brazil (1993)
State in W Brazil.
CO S110000S070000
 W0663000W0740000
BT Brazil
BT South America

Acreodi (1981)
Autoposting of Condylarthra, Eutheria and Theria to this term began in 1989.
BT Condylarthra
BT Eutheria
BT Theria
BT Mammalia
BT Tetrapoda
BT Vertebrata
BT Chordata

acritarch flora (1981)
Common name for acritarchs. Autoposting of palynomorphs and microfossils to this term began in 1989.
BT palynomorphs
BT microfossils
SA acritarchs
SA biostratigraphy

acritarchs (1978)
Hystrichosphaeridae is included here and under Dinoflagellata. Autoposting of microfossils to this term began in 1989.
BT palynomorphs
BT microfossils
NT Baltisphaeridium
NT Hystrichosphaeridae
SA acritarch flora
SA Dinoflagellata

Acropora cervicornis (1989)
Species.
BT Scleractinia
BT Zoantharia
BT Anthozoa
BT Coelenterata
BT Invertebrata

Acropora palmata (1985)
Species. Autoposting of Zoantharia to this term began in 1989.
BT Scleractinia
BT Zoantharia
BT Anthozoa
BT Coelenterata
BT Invertebrata

Actiniaria (1978)
Order. Autoposting of Zoantharia to this term began in 1989.
BT Zoantharia
BT Anthozoa
BT Coelenterata
BT Invertebrata

actinides (1989)
Series of heavy, radioactive, metallic elements with atomic numbers 89 to 103.
BT metals
NT actinium
NT americium
NT berkelium
NT californium
NT curium
NT einsteinium
NT fermium
NT lawrencium
NT mendelevium
NT neptunium
NT nobelium
NT plutonium
NT protactinium
NT thorium
NT uranium

Actinistia (1981)
Suborder. Autoposting of Crossopterygii to this term began in 1989.
UF Coelacanthini
BT Crossopterygii
BT Sarcopterygii
BT Osteichthyes
BT Pisces
BT Vertebrata
BT Chordata

actinium (1978)
Autoposting of actinides and metals to this term began in 1989.
UF Ac
BT actinides
BT metals

Actinodontida (1981)
UF Modiomorphoida
BT Bivalvia
BT Mollusca
BT Invertebrata

actinolite (1978)
Autoposting of clinoamphibole to this term began in 1981.
BT clinoamphibole
BT amphibole group
BT chain silicates
BT silicates
SA actinolite facies
SA asbestos

actinolite facies (1978)
BT facies
SA actinolite
SA metamorphic rocks

Actinopterygii (1978)
Subclass. Holostei, Teleostei and Cyprinidae are narrower terms as of 1981. Autoposting of this term began in 1978.
BT Osteichthyes
BT Pisces
BT Vertebrata
BT Chordata
NT Chondrostei
NT Holostei
NT Teleostei

action, frost
use frost action

activation analysis (1978)
Including field applications and instruments. Also search activation.
UF radioactivation analysis
SA analysis
SA chemical analysis
SA isotopes
SA neutron activation analysis

activation energy (1978)
SA energy
SA particles

active faults (1978)
BT faults
SA seismic gaps

active layer (1978)
UF annually thawed layer
UF layer, active
UF mollisol
SA permafrost
SA soils
SA taliks

active margins (1985)
SA continental margin
SA plate boundaries
SA plate convergence
SA plate tectonics
SA plates
SA subduction zones
SA underplating

active tectonics
use neotectonics

activity (1978)
SA geochemistry

activity, igneous
use igneous activity

actual age (absolute age)
use absolute age

actualism
use uniformitarianism

Acungui Group (1978)
BT Precambrian
SA Brazil
SA Parana Brazil

Ada County
Valid through 1988. Search in combination with state term. After 1988, use specific county-state term.

Ada County Idaho (1989)
SW Idaho. Before 1989, also search Ada County AND Idaho.
CO N430700N434700
 W1155800W1163300
BT Idaho
BT United States
NT Boise Idaho

Adair County Kentucky (1989)
S Kentucky. Before 1989, search Adair County AND Kentucky.
CO N365500N371800
 W0850300W0853100
BT Kentucky
BT United States
NT Columbia Kentucky

Adak Island (1978)
In central part of Aleutian Islands SW of Alaska Peninsula.
BT Aleutian Islands
BT Southwestern Alaska
BT Alaska
BT United States

Adalia Turkey
use Antalya Turkey

Adamawa (1978)
Region in West Africa, now part of N central Cameroon and E Nigeria.
IN Index Cameroon and/or Nigeria as applicable.
BT West Africa
BT Africa
SA Cameroon
SA Nigeria

adamellite (1978)
BT granites
BT plutonic rocks
BT igneous rocks

Adamello Massif (1978)
In Rhaetian Alps in N Italy. This term has multiple hierarchies.
BT1 Lombardy Italy
BT1 Italy
BT1 Southern Europe
BT1 Europe
BT2 Rhaetian Alps
BT2 Central Alps
BT2 Alps
BT2 Europe

adamite (1978)
BT arsenates

Adamow Mine (1978)
Lignite mine in W central Poland. Also search Adamow.
BT Poznan Poland
BT Poland
BT Central Europe
BT Europe
SA lignite
SA mines

Adams County
Valid through 1988. Search in combination with state term. After 1988, use specific county-state term.

Adams County Colorado (1989)
N central Colorado. Before 1989, also search Adams County AND Colorado.
CO N394500N400000
 W1033800W1050400
BT Colorado
BT United States
NT Rocky Mountain Arsenal
SA Wattenberg Field

Adams County Idaho (1989)
W Idaho. Before 1989, also search Adams County AND Idaho.
CO N442500N451800
 W1160800W1165300
BT Idaho
BT United States
SA Seven Devils Mountains

Adams County Illinois (1989)
W Illinois. Before 1989, also search Adams County AND Illinois.
CO N394500N401200
 W0905500W0913500
BT Illinois
BT United States

Adams County Indiana (1989)
  E Indiana. Before 1989, also search Adams County AND Indiana.
  CO N403500N405500
    W0844800W0850500
  BT Indiana
  BT United States

Adams County Iowa (1989)
  SW Iowa. Before 1989, also search Adams County AND Iowa.
  CO N405300N411000
    W0942800W0945600
  BT Iowa
  BT United States

Adams County Mississippi (1989)
  SW Mississippi. Before 1989, also search Adams County AND Mississippi.
  CO N311000N314500
    W0911200W0914000
  BT Mississippi
  BT United States

Adams County Nebraska (1989)
  S Nebraska. Before 1989, also search Adams County AND Nebraska.
  CO N402000N404200
    W0982200W0984200
  BT Nebraska
  BT United States

Adams County North Dakota (1989)
  SW North Dakota. Before 1989, also search Adams County AND North Dakota.
  CO N455700N461700
    W1020000W1030000
  BT North Dakota
  BT United States

Adams County Ohio (1989)
  S Ohio. Before 1989, also search Adams County AND Ohio.
  CO N383500N390400
    W0831700W0834300
  BT Ohio
  BT United States

Adams County Pennsylvania (1989)
  S Pennsylvania. Before 1989, also search Adams County AND Pennsylvania.
  CO N394000N403500
    W0765600W0772800
  BT Pennsylvania
  BT United States

Adams County Washington (1989)
  SE Washington. Before 1989, also search Adams County AND Washington.
  CO N464300N471700
    W1175700W1192500
  BT Washington
  BT United States

Adams County Wisconsin (1989)
  Central Wisconsin. Before 1989, also search Adams County AND Wisconsin.
  CO N433800N441500
    W0893700W0900200
  BT Wisconsin
  BT United States

Adana
  No longer a valid term for GeoRef. As of 1993 see Adana Turkey.

Adana Turkey (1993)
  Province in S Anatolia. Also a city.
  BT Turkey
  BT Middle East
  BT Asia
  SA Anatolia

adaptation (1978)
  SA biologic evolution
  SA biotopes
  SA ecology
  SA paleoecology
  SA paleontology

adaptive filters (1981)
  SA filters

Adavale Basin (1978)
  S central Queensland.
  BT Queensland Australia
  BT Australia
  BT Australasia

Addis Ababa
  No longer a valid term for GeoRef. See Addis Ababa Ethiopia.

Addis Ababa Ethiopia (1993)
  City in central Ethiopia.
  BT Ethiopia
  BT East Africa
  BT Africa

Addison County
  Valid through 1988. Search in combination with state term. After 1988, use specific county-state term.

Addison County Vermont (1989)
  W Vermont. Before 1989, also search Addison County AND Vermont.
  CO N434500N441800
    W0724400W0732500
  BT Vermont
  BT United States

addresses
  No longer a valid GeoRef index term. Before 1971, was used on level 1 in subfiles E and N. See symposia.

Adelaide
  No longer a valid term for GeoRef. See Adelaide Australia.

Adelaide Australia (1993)
  City on Gulf Saint Vincent in SE South Australia.
  BT South Australia
  BT Australia
  BT Australasia

Adelaide Geosyncline (1978)
  Also search Adelaide.
  BT South Australia
  BT Australia
  BT Australasia
  SA geosynclines
  SA Stuart Shelf

Adelaide Island (1978)
  Off W coast of Antarctic Peninsula S of Cape Horn.
  BT Antarctica

Adelaidean (1978)
  SE Australia. Middle and upper Proterozoic. Autoposting of Precambrian to this term ended in 1989. As of 1989, Proterozoic is autoposted to this term. As of 1990, upper Precambrian and Precambrian are autoposted to this term.
  UF Adelaidean system
  BT Proterozoic
  BT upper Precambrian
  BT Precambrian

Adelaidean system
  use Adelaidean

Adelie Coast (1978)
  In the French Sector on the Indian Ocean side S of Australia.
  UF Adelie Land
  BT Antarctica

Adelie Land
  use Adelie Coast

Aden
  No longer a valid term for GeoRef. See Aden Yemen.

Aden protectorate
  No longer a valid GeoRef index term. Before 1969, was used on level 1 in subfile E.
  use Aden Yemen

Aden Yemen (1993)
  City on Gulf of Aden. Former British Colony. Before 1969, also search Aden protectorate. Before 1993, also search Aden and Southern Yemen.
  UF Aden protectorate
  BT Yemen
  BT Arabian Peninsula
  BT Asia

Adetognathus (1989)
  Genus. As of 1990, microfossils is autoposted to this term.
  BT Conodonta
  BT microfossils

adiabatic demagnetization (1981)
  BT demagnetization
  SA geophysics
  SA magnetization
  SA paleomagnetism
  SA remanent magnetization

Adirondack Anorthosite (1978)
  NE New York.
  BT Precambrian
  SA New York

Adirondack Mountains (1978)
  NE New York.
  CO N430000N443000
    W0733000W0750000
  UF Adirondacks
  BT New York
  BT United States
  SA Appalachians
  SA Blue Mountain Lake
  SA Canadian Shield
  SA Goodnow earthquake 1983
  SA Marcy Massif

Adirondacks
  use Adirondack Mountains

Admire
  Valid through 1988. Search in combination with state term. After 1988, use specific city-state term.

Admire Kansas (1989)
  Village in E central Kansas. Before 1989, search Admire AND Kansas.
  CO N383800N383800
    W0960600W0960600
  BT Lyon County Kansas
  BT Kansas
  BT United States

Adour Basin (1978)
  River basin in extreme SW France. Also search Adour.
  BT France
  BT Western Europe
  BT Europe
  SA Gers France
  SA Landes France

Adrar (1978)
  Interior region in W Mauritania near Western Sahara border.
  BT Mauritania
  BT West Africa
  BT Africa

Adrar des Iforas (1989)
  Region in E Mali and S Algeria.
  IN Index countries as applicable.
  BT Africa
  SA Algeria
  SA Mali
  SA Sahara

Adrastea Satellite (1989)
  One of the satellites of Jupiter. Before 1989, also search Adrastea AND Jupiter.
  SA Jupiter
  SA satellites
  SA Voyager Program

Adriatic Coast
  use Adriatic region

Adriatic region (1978)
  IN Index countries as applicable.
  UF Adriatic Coast
  BT Europe
  SA Albania
  SA Croatia
  SA Italy
  SA Slovenia
  SA Yugoslavia

Adriatic Sea (1978)
  Between Italy on the W, and Albania and Yugoslavia on the E.
  CO N395000N454500
    E0200000E0121500
  BT East Mediterranean
  BT Mediterranean Sea

adsorption (1978)
  SA absorption
  SA chromatography
  SA clay mineralogy
  SA desorption
  SA geochemistry
  SA Langmuir equation
  SA processes
  SA solution
  SA sorption
  SA wettability

adularia (1978)
  BT alkali feldspar
  BT feldspar group
  BT framework silicates
  BT silicates
  SA aluminosilicates
  SA K-feldspar
  SA orthoclase

advection (1993)
  SA convection
  SA diffusion
  SA geochemistry
  SA processes
  SA solution transport

Adzhar Georgian Republic
  use Adzharistan Georgian Republic

Adzharia Georgian Republic
  use Adzharistan Georgian Republic

Adzharistan
  No longer a valid term for GeoRef. As of 1993, see Adzharistan Georgian Republic.

Adzharistan Georgian Republic (1993)
  Former Adzhar Autonomous Soviet Socialist Republic in SW Georgian Republic. Before 1978, also search Adzhar or Adzharia.
  CO N410000N411500

E0423000E0411500
UF Adzhar Georgian Republic
UF Adzharia Georgian Republic
BT Georgian Republic
BT Europe
NT Batumi Georgian Republic

Aegean Islands (1978)
In 1981, broader term changed from Greece to Mediterranean region.
BT Mediterranean region
NT Greek Aegean Islands
SA Turkish Aegean region

Aegean Sea (1978)
Between Greece on the W and Turkey on the E.
CO N360000N410000
E0282000E0230000
BT East Mediterranean
BT Mediterranean Sea
SA Greek Aegean Islands
SA Kalamata earthquake 1986

aegerine
use aegirine

Aegina (1978)
Island in Saronic Gulf. After 1993, use Aegina Greece for the town.
BT Sterea Ellas
BT Greece
BT Southern Europe
BT Europe

aegirine (1978)
UF aegerine
BT clinopyroxene
BT pyroxene group
BT chain silicates
BT silicates
SA acmite

aenigmatite (1978)
UF enigmatite
BT chain silicates
BT silicates
SA rhonite

Aeolian Islands
use Lipari Islands

aeolianite
use eolianite

aeolotropy
use anisotropy

aerial photographs
use aerial photography

aerial photography (1978)
Also search aerial.
UF aerial photographs
SA cartography
SA geomorphology
SA landform description
SA maps
SA mosaics
SA photogeology
SA photogrammetry
SA photography
SA pixels
SA remote sensing
SA space photography

aerobic environment (1978)
Before 1978, search aerobic.
SA anaerobic environment
SA depositional environment
SA environment
SA paleoecology

aeromagnetic maps (1983)
BT maps
SA airborne methods
SA geophysical survey maps
SA geophysical surveys
SA magnetic survey maps
SA magnetic surveys

aeromagnetic surveys
A valid term through 1973. Use magnetic surveys and airborne methods. Also search aeromagnetic.

aeronomy
As of 1993, no longer a valid term. Usually out-of-scope for GeoRef.

aerosols (1978)
SA air-sea interface
SA ash
SA atmosphere
SA convection
SA meteorology
SA particles
SA radioactive tracers
SA turbidity

aeschynite (1978)
UF aschynite
UF eschinite
UF eschynite
BT niobates
BT oxides

Afar (1978)
Large desert region mostly in Ethiopia.
IN Index Djibouti and/or Ethiopia as applicable.
UF Danakil
UF Dankalia
BT Africa
SA Afar Depression
SA Djibouti
SA Ethiopia
SA Hadar
SA Hadar Formation

Afar Depression (1978)
Part of Afar lying between the "Danakil Alps" and the Ethiopian highlands in NE Ethiopia. Section of the Great Rift Valley.
UF Afar Rift
UF Danakil Depression
BT Ethiopia
BT East Africa
BT Africa
SA Afar
SA East African Rift
SA Great Rift Valley

Afar Rift
use Afar Depression

Afars and Issas
Not a valid GeoRef index term after 1971. Was used on level 1 in subfile G.
use Djibouti

Afars and Issas Territory
A valid index term through 1977.
use Djibouti

affinities (1978)
Also search geochemical affinities if applicable.
SA Invertebrata
SA mineral resources
SA Plantae
SA Vertebrata

afforestation
use vegetation

Afghan-Tadzhik Basin
use Afghan-Tadzhik Depression

Afghan-Tadzhik Depression (1978)
Also search Afghan-Tadzhik.
IN Index Afghanistan and/or Tadzhikistan as applicable.
UF Afghan-Tadzhik Basin
BT Asia
SA Afghanistan
SA Tadzhik Depression

SA Tadzhikistan

Afghanistan (1978)
CO N290000N381500
E0750000E0600000
BT Indian Peninsula
BT Asia
NT Altimur Mountains
NT Badakhshan Afghanistan
NT Bamian Afghanistan
NT Dasht-i-Nawar
NT Ghazni Afghanistan
NT Kabul Afghanistan
NT Logar Afghanistan
NT Paktia Afghanistan
NT Wardak Afghanistan
SA Afghan-Tadzhik Depression
SA Amu Darya
SA Amu Darya Basin
SA Badakhshan
SA Hindu Kush
SA Kohistan
SA Murgab Basin
SA Turkestan

Africa (1978)
To retrieve all documents, individual countries and physiographic regions should also be searched (see list O). Autoposting of this term began in 1978.
CO S350000N370000
E0510000W0180000
NT Adrar des Iforas
NT Afar
NT African Platform
NT Blue Nile
NT Cap Blanc
NT Cape Verde Islands
NT Central Africa
NT Chad Basin
NT Congo Basin
NT East Africa
NT East African Lakes
NT East African Rift
NT Gregory Rift
NT Kalahari Desert
NT Kasai River
NT Lebombo Mountains
NT Libyan Desert
NT Limpopo Basin
NT Limpopo Belt
NT Madagascar
NT Nile River
NT Nile Valley
NT North Africa
NT Nubia
NT Nubian Shield
NT Red Sea Hills
NT Reguibat Ridge
NT Sahara
NT Sahel
NT Senegal River
NT Southern Africa
NT Tanezrouft
NT Umba
NT Volta Basin
NT West Africa
NT West African Shield
NT Zambezi Valley
SA African Plate
SA Dwyka Formation
SA Gondwana
SA Karroo Supergroup
SA Katangan Orogeny
SA Nubian Sandstone
SA Tethys

African Plate (1978)
Includes the continent of Africa, adjoining areas of the Atlantic and Indian oceans, Madagascar, and much of E Mediterranean Sea.
SA Africa
SA plate tectonics
SA plates

African Platform (1978)
A vast platform of complex Precambrian and Paleozoic rocks, partly covered by Mesozoic and Tertiary sedimentary rocks, which underlies the entire continent.
BT Africa

aftershocks (1978)
SA earthquakes
SA focus
SA foreshocks
SA main shocks
SA seismology

Aftonian (1978)
Interglacial interval. North America.
BT lower Pleistocene
BT Pleistocene
BT Quaternary
BT Cenozoic

Ag
use silver

Agades
No longer a valid term for GeoRef. As of 1993 see Agades Niger.

Agades Niger (1993)
Town in W central Niger. Before 1993 also search Agades or Agadez.
UF Agadez Niger
BT Niger
BT West Africa
BT Africa

Agadez Niger
use Agades Niger

Agarak (1978)
Mining region in central Armenia. This term has multiple hierarchies.
BT1 Armenia
BT1 Commonwealth of Independent States
BT2 Armenia
BT2 Europe

agate (1978)
Variety of quartz.
BT silica minerals
BT framework silicates
BT silicates
SA amygdules
SA chalcedony
SA gems
SA quartz

age (1978)
SA absolute age
SA exposure age
SA ground water
SA maturity
SA ocean basins
SA relative age

age determination
Not a valid term for GeoRef. For absolute age determination, use absolute age in conjunction with dates. See relative age for specific methods.

Ager Formation (1978)
Catalonia and the Pyrenees.
BT lower Eocene
BT Eocene
BT Paleogene
BT Tertiary
BT Cenozoic
SA Spain

agglomerate (1978)
SA breccia
SA conglomerate
SA igneous rocks
SA pyroclastics

SA sedimentary rocks
agglutinates (1978)
SA fines
SA Moon
SA particles
aggradation (1978)
SA geomorphology
SA processes
aggregate (1978)
IN Index as a commodity term with construction materials.
UF lightweight aggregate
SA construction materials
SA gravel deposits
SA pumice deposits
SA rocks
SA sand
SA sands
aggregation
A valid term through 1975. After 1975, use sedimentation.
AGI (1989)
Acronym.
UF American Geological Institute
SA associations
Agly Massif (1978)
S France.
BT Pyrenees-Orientales France
BT France
BT Western Europe
BT Europe
agmatite (1978)
BT migmatites
BT metamorphic rocks
Agnatha (1978)
Class. Autoposting of this term began in 1978.
BT Vertebrata
BT Chordata
NT Heterostraci
NT Ostracodermi
SA agnathans
SA Cyclostomata
SA Pisces
agnathans (1989)
Used as the common name for Agnatha.
IN Index for all non-paleontologic studies of fossils.
BT vertebrates
NT ostracoderms
SA Agnatha
SA biostratigraphy
Agnes (1978)
A storm in 1972 on the East Coast of United States.
SA hurricanes
Agnostida (1978)
BT Trilobita
BT Trilobitomorpha
BT Arthropoda
BT Invertebrata
Agnotozoic
use Proterozoic
agpaite (1978)
BT syenites
BT plutonic rocks
BT igneous rocks
Agricola Lake (1978)
BT Northwest Territories
BT Western Canada
BT Canada
agricultural waste (1978)
Before 1978, search agricultural.
UF waste, agricultural
SA agriculture
SA waste disposal

agriculture (1978)
SA agricultural waste
SA channelization
SA fertilization
SA fertilizers
SA land use
SA soils
Agrokipia Deposit (1993)
Iron-zinc-copper sulfide stockwork deposit. Before 1993, also search Agrokipia and commodity.
BT Cyprus
BT Middle East
BT Asia
SA base metals
SA copper ores
SA iron ores
SA Troodos Massif
SA Troodos Ophiolite
SA zinc ores
Agto
No longer a valid term for GeoRef. As of 1993 see Agto Greenland.
Agto Greenland (1993)
Fishing settlement on small island in Davis Strait. Before 1993 also search Agto AND Greenland.
BT Greenland
BT Arctic region
AGU (1989)
Acronym.
UF American Geophysical Union
SA associations
Agua Blanca Fault (1978)
IN Index North America or states as applicable.
SA Baja California
SA California
SA Mexico
SA North America
Aguacate Group (1989)
Miocene-Pliocene. Costa Rica and Mexico.
IN Index ages as applicable.
BT Neogene
BT Tertiary
BT Cenozoic
SA Costa Rica
SA Mexico
SA Miocene
SA Pliocene
Aguja Formation (1989)
W Texas.
BT Gulfian
BT Upper Cretaceous
BT Cretaceous
BT Mesozoic
SA Texas
Agulhas Bank (1978)
South of Cape of Good Hope.
BT Indian Ocean
SA Table Mountain Group
Ahaggar (1978)
Volcanic upland in SE Algeria.
CO N194000N264800
    E0110000E0005900
UF Ahaggar Mountains
UF Hoggar Mountains
BT Algeria
BT North Africa
BT Africa
SA Sahara
Ahaggar Mountains
use Ahaggar
ahermatypic taxa (1978)
Before 1978, search ahermatypic.
UF taxa, ahermatypic
SA Coelenterata
SA hermatypic taxa

Ahnet (1978)
Region in S central Algeria.
BT Algeria
BT North Africa
BT Africa
AI
Not a valid term for GeoRef. Sometimes used as the abbreviation for artificial intelligence.
Aichi
No longer a valid term for GeoRef. See Aichi Japan.
Aichi Japan (1993)
Prefecture in central Honshu. Capital is Nagoya. Before 1993, also search Aichi AND Japan.
BT Honshu
BT Japan
BT Far East
BT Asia
NT Atsumi Peninsula
NT Shonai River
SA Chubu Japan
SA Ise Bay
SA Kiso Mountains
SA Mikawa
SA Yahagi River
aid projects (1981)
SA economic agreements
SA economics
SA international cooperation
Aigoual Massif (1978)
In Cevennes Mountains in S France. This term has multiple hierarchies.
BT1 France
BT1 Western Europe
BT1 Europe
BT2 Cevennes
BT2 France
BT2 Western Europe
BT2 Europe
Aiguilles Rouges (1978)
Alpine range in SE France. This term has multiple hierarchies.
BT1 Alps
BT1 Europe
BT2 France
BT2 Western Europe
BT2 Europe
SA French Alps
Aiken
Valid through 1988. Search in combination with state term. After 1988, use specific city-state term.
Aiken County
Valid through 1988. Search in combination with state term. After 1988, use specific county-state term.
Aiken County South Carolina (1989)
W South Carolina. Before 1989, also search Aiken County AND South Carolina.
CO N331000N335200
    W0811200W0820300
BT South Carolina
BT United States
NT Aiken South Carolina
SA Savannah River Plant
Aiken South Carolina (1989)
City in W South Carolina. Before 1989, search Aiken AND South Carolina.
CO N333400N333400
    W0814400W0814400
BT Aiken County South Carolina
BT South Carolina
BT United States

aikinite (1978)
BT sulfides
Ain
No longer a valid term for GeoRef. See Ain France.
Ain France (1993)
Department in E France.
CO N453500N463000
    E0061000E0044000
BT France
BT Western Europe
BT Europe
SA Jura Mountains
air (1978)
SA atmosphere
SA background level
SA pollution
air guns (1985)
UF airguns
SA acoustical methods
SA geophysical methods
SA instruments
SA seismic methods
air-sea interface (1978)
UF interface, air-sea
SA aerosols
SA atmosphere
SA boundary interactions
SA interfaces
SA meteorology
SA oceanography
Aira Caldera (1978)
Kagoshima Prefecture, Kyushu, S Japan.
BT Kagoshima Japan
BT Kyushu
BT Japan
BT Far East
BT Asia
SA Ontake
SA Sakura-jima
airborne
A valid index term through 1977. After 1977, use airborne methods.
airborne methods (1978)
Before 1978, also search airborne.
SA aeromagnetic maps
SA geophysical methods
SA geophysical surveys
SA ground methods
SA magnetic methods
SA magnetic surveys
SA multispectral scanner
SA remote sensing
SA SAR
SA SLAR
SA thematic mapper
SA total-field methods
SA vertical-gradient methods
airfields (1981)
SA engineering geology
SA environmental geology
SA land use
airglow
As of 1981, no longer a valid term for GeoRef.
airguns
use air guns
Airy waves (1978)
BT elastic waves
SA seismology
SA surface waves
SA waves
Aisne
No longer a valid term for GeoRef. See Aisne France.
Aisne France (1993)

Department in N France.
CO N484500N501000
 E0041000E0010000
BT France
BT Western Europe
BT Europe
SA Oise River valley
SA Somme River valley

Aix-en-Provence
No longer a valid term for GeoRef.
use Aix-en-Provence France

Aix-en-Provence France (1993)
City in SE France.
UF Aix-en-Provence
BT Bouches-du-Rhone France
BT France
BT Western Europe
BT Europe

Aizu Basin (1978)
Fukushima Prefecture, N central Honshu. Also search Aizu.
SA Fukushima Japan
SA Honshu

Ajay River (1978)
IN Index Indian states as applicable.
BT India
BT Indian Peninsula
BT Asia
SA Bihar India
SA West Bengal India

Ajman (1978)
Sheikdom. One of federation of 7 states at S end of Persian Gulf.
BT United Arab Emirates
BT Arabian Peninsula
BT Asia

Ajmer
No longer a valid term for GeoRef. See Ajmer India.

Ajmer India (1993)
City and district in central Rajasthan, NW India.
BT Rajasthan India
BT India
BT Indian Peninsula
BT Asia

akaganeite (1978)
BT oxides
SA iron oxides

Akaishi Mountains (1989)
Central Honshu, Japan.
SA Fossa Magna
SA Honshu
SA Japanese Alps
SA Nagano Japan
SA Shizuoka Japan
SA Yamanashi Japan

Akbastau
No longer a valid term for GeoRef. As of 1993, see Akbastau Kazakhstan.

Akbastau Kazakhstan (1993)
Village NE of Chimkent in S central Kazakhstan. This term has multiple hierarchies.
BT1 Kazakhstan
BT1 Central Asia
BT1 Asia
BT2 Kazakhstan
BT2 Commonwealth of Independent States

Akchagylian (1978)
Europe.
BT upper Pliocene
BT Pliocene
BT Neogene
BT Tertiary

BT Cenozoic
SA upper Tertiary

Akchatau
No longer a valid term for GeoRef. As of 1993, see Akchatau Kazakhstan.

Akchatau Kazakhstan (1993)
Town in SE Karaganda oblast in E central Kazakhstan. This term has multiple hierarchies.
IN Index oblast as applicable.
BT1 Kazakhstan
BT1 Central Asia
BT1 Asia
BT2 Kazakhstan
BT2 Commonwealth of Independent States
SA Karaganda Kazakhstan

Akenobe Mine (1978)
Polymetallic ores. Hyogo Prefecture, S Honshu.
BT Hyogo Japan
BT Honshu
BT Japan
BT Far East
BT Asia
SA mines
SA polymetallic ores

Akera (1978)
River in SW Azerbaidzhan. This term has multiple hierarchies.
BT1 Azerbaidzhan
BT1 Europe
BT2 Azerbaidzhan
BT2 Commonwealth of Independent States

akermanite (1978)
BT melilite group
BT sorosilicates
BT orthosilicates
BT silicates

Akhaltsikhe
No longer a valid term for GeoRef. As of 1993, see Akhaltsikhe Georgian Republic.

Akhaltsikhe Georgian Republic (1993)
City in S Georgian Republic. Before 1993, also search Akhaltsikh.
BT Georgian Republic
BT Europe

Akhtala
No longer a valid term for GeoRef. As of 1993, see Akhtala Armenia.

Akhtala Armenia (1993)
Town in N Armenia. This term has multiple hierarchies.
BT1 Armenia
BT1 Commonwealth of Independent States
BT2 Armenia
BT2 Europe

Akita
No longer a valid term for GeoRef. See Akita Japan or Akita City Japan.

Akita City Japan (1993)
City on Japan Sea in N Honshu. Before 1993, also search Akita.
BT Akita Japan
BT Honshu
BT Japan
BT Far East
BT Asia

Akita Japan (1993)

Prefecture including the city in N Honshu. Before 1993, also search Akita or Akita Prefecture AND Japan.
UF Akita Prefecture
BT Honshu
BT Japan
BT Far East
BT Asia
NT Akita City Japan
NT Hanaoka Mine
NT Hanawa Mine
NT Hokuroku Japan
NT Kosaka Mine
NT Odate Japan
NT Oga Peninsula
NT Shakanai Mine
SA Chokai
SA Koma-ga-take
SA Nagano City Japan
SA Tamagawa
SA Tohoku

Akita Prefecture
No longer a valid term for GeoRef.
use Akita Japan

Akiyoshi (1978)
Limestone region in extreme SW Honshu.
BT Honshu
BT Japan
BT Far East
BT Asia

Akiyoshi Limestone (1978)
SW Honshu.
BT Permian
BT Paleozoic
SA Honshu
SA Japan

Akjoujt
No longer a valid term for GeoRef. See Akjoujt Mauritania.

Akjoujt Mauritania (1993)
Village in W central Mauritania. Before 1993 also search Akjoujt and Mauritania.
BT Mauritania
BT West Africa
BT Africa

Akron
Valid through 1988. Search in combination with state term. After 1988, use specific city-state term.

Akron Ohio (1989)
City in NE Ohio. Before 1989, search Akron AND Ohio.
CO N410400N410400
 W0813100W0813100
BT Summit County Ohio
BT Ohio
BT United States

Aksu
No longer a valid term for GeoRef. As of 1993, see Aksu Kazakhstan.

Aksu Kazakhstan (1993)
Village in Taldy Kurgan oblast, Kazakhstan. Before 1993, Aksu may also have been used for town in Xinjiang China. That town is now called Wensu. This term has multiple hierarchies.
BT1 Kazakhstan
BT1 Central Asia
BT1 Asia
BT2 Kazakhstan
BT2 Commonwealth of Independent States

Aktyubinsk
No longer a valid term for GeoRef. As of 1993, see Aktyubinsk Kazakhstan.

Aktyubinsk Kazakhstan (1993)
Oblast and city in W central Kazakhstan. This term has multiple hierarchies.
BT1 Kazakhstan
BT1 Central Asia
BT1 Asia
BT2 Kazakhstan
BT2 Commonwealth of Independent States
NT Emba Field
NT Emba Kazakhstan
NT Irgiz Kazakhstan
NT Zhamanshin Crater
SA irghizite
SA Mugodzhar Hills

Akzhal
No longer a valid term for GeoRef. As of 1993, see Akzhal Kazakhstan.

Akzhal Kazakhstan (1993)
Town in E Semipalatinsk oblast in E Kazakhstan. This term has multiple hierarchies.
BT1 Semipalatinsk Kazakhstan
BT1 Kazakhstan
BT1 Central Asia
BT1 Asia
BT2 Semipalatinsk Kazakhstan
BT2 Kazakhstan
BT2 Commonwealth of Independent States

Al
use aluminum

Al-26 (1978)
Autoposting of broader terms began in 1989. This term has multiple hierarchies.
BT1 radioactive isotopes
BT1 isotopes
BT2 aluminum
BT2 metals
SA accelerator mass spectroscopy
SA Al-27/Al-26

Al-26/Al-27
use Al-27/Al-26

Al-27 (1989)
This term has multiple hierarchies.
BT1 stable isotopes
BT1 isotopes
BT2 aluminum
BT2 metals
SA Al-27/Al-26

Al-27/Al-26 (1985)
Autoposting of broader terms began in 1989. This term has multiple hierarchies.
UF Al-26/Al-27
BT1 radioactive isotopes
BT1 isotopes
BT2 stable isotopes
BT2 isotopes
BT3 aluminum
BT3 metals
SA Al-26
SA Al-27
SA isotope ratios

Al-Kufrah
use Kufra Basin

Al-Qusayr Egypt
use Kosseir Egypt

Al-Quseir Egypt
use Kosseir Egypt

Ala-Kul Lake
   use Alakol

Alabama   (1978)
   Autoposting of this term began in 1978.
   CO N300000N350000
         W0850000W0883000
   BT United States
   NT Alabama River
   NT Baldwin County Alabama
   NT Blount County Alabama
   NT Butler County Alabama
   NT Calhoun County Alabama
   NT Chambers County Alabama
   NT Cherokee County Alabama
   NT Chilton County Alabama
   NT Choctaw County Alabama
   NT Clarke County Alabama
   NT Clay County Alabama
   NT Cleburne County Alabama
   NT Coffee County Alabama
   NT Conecuh County Alabama
   NT Coosa County Alabama
   NT Dallas County Alabama
   NT De Kalb County Alabama
   NT Elmore County Alabama
   NT Escambia County Alabama
   NT Etowah County Alabama
   NT Fayette County Alabama
   NT Franklin County Alabama
   NT Greene County Alabama
   NT Henry County Alabama
   NT Houston County Alabama
   NT Jackson County Alabama
   NT Jefferson County Alabama
   NT Lamar County Alabama
   NT Lauderdale County Alabama
   NT Lawrence County Alabama
   NT Lee County Alabama
   NT Limestone County Alabama
   NT Lowndes County Alabama
   NT Macon County Alabama
   NT Madison County Alabama
   NT Marion County Alabama
   NT Marshall County Alabama
   NT Mobile Bay
   NT Mobile County Alabama
   NT Monroe County Alabama
   NT Montgomery County Alabama
   NT Morgan County Alabama
   NT Perry County Alabama
   NT Pickens County Alabama
   NT Pike County Alabama
   NT Randolph County Alabama
   NT Russell County Alabama
   NT Saint Clair County Alabama
   NT Shelby County Alabama
   NT Sumter County Alabama
   NT Talladega Front
   NT Tallapoosa County Alabama
   NT Tuscaloosa County Alabama
   NT Walker County Alabama
   NT Warrior coal field
   NT Washington County Alabama
   SA Appalachian Basin
   SA Appalachian Plateau
   SA Bangor Limestone
   SA Black Warrior Basin
   SA Bluffport Marl Member
   SA Blufftown Formation
   SA Brevard Zone
   SA Bucatunna Formation
   SA Byram Formation
   SA Catahoula Formation
   SA Chattahoochee River
   SA Chattanooga Shale
   SA Chesterian
   SA Chickamauga Group
   SA Chipola Formation
   SA Citronelle Formation
   SA Claiborne Group
   SA Clayton Formation
   SA Coker Formation
   SA Conasauga Group
   SA Cumberland Plateau
   SA Demopolis Chalk
   SA Eutaw Formation
   SA Fernvale Formation
   SA Fort Payne Formation
   SA Glendon Limestone
   SA Golconda Formation
   SA Gulf Coastal Plain
   SA Hartselle Sandstone
   SA Hatchetigbee Formation
   SA Haynesville Formation
   SA Highland Rim
   SA Hillabee Chlorite Schist
   SA Jackson Group
   SA Lenoir Limestone
   SA Lisbon Formation
   SA Marianna Limestone
   SA Martin Lake
   SA Mary Lee Coal
   SA Midway Group
   SA Mississippi Embayment
   SA Mississippi Sound
   SA Monteagle Limestone
   SA Moodys Branch Formation
   SA Mooreville Chalk
   SA Naheola Formation
   SA Nanafalia Formation
   SA Nashville Dome
   SA New River Formation
   SA Oak Hill Member
   SA Ocala Group
   SA Parkwood Formation
   SA Pennington Formation
   SA Piedmont
   SA Porters Creek Formation
   SA Pottsville Group
   SA Red Mountain
   SA Red Mountain Formation
   SA Ripley Formation
   SA Rome Formation
   SA Saint Louis Limestone
   SA Sainte Genevieve Limestone
   SA Selma Group
   SA Shady Dolomite
   SA Shubuta Member
   SA Smackover Formation
   SA Talladega Group
   SA Tallahatta Formation
   SA Tennessee River
   SA Tennessee Valley
   SA Tombigbee River
   SA Tuscaloosa Formation
   SA Valley and Ridge Province
   SA Vicksburg Group
   SA Warsaw Formation
   SA Wiggins Arch
   SA Yazoo Clay

Alabama River   (1978)
   Central and SW Alabama.
   BT Alabama
   BT United States

alabandine
   use alabandite

alabandite   (1978)
   UF alabandine
   UF manganblende
   BT sulfides
   SA manganese minerals

alabaster   (1978)
   UF onyx marble
   BT sulfates
   SA aragonite
   SA calcite
   SA gypsum

Alachua County
   Valid through 1988. Search in combination with state term. After 1988, use specific county-state term.

Alachua County Florida   (1989)
   N Florida. Before 1989, also search Alachua County AND Florida.
   CO N292600N295600
         W0820500W0824000
   BT Florida
   BT United States

Alacran Reef   (1978)
   Off north Yucatan coast.
   BT Gulf of Mexico
   BT North American Atlantic
   BT North Atlantic
   BT Atlantic Ocean

Alae Crater   (1978)
   On Kilauea Volcano on Hawaii Island. Autoposting of broader terms to this term began in 1989. This term has multiple hierarchies.
   BT1 Hawaii Island
   BT1 Hawaii County Hawaii
   BT1 Hawaii
   BT1 United States
   BT2 Hawaii Island
   BT2 Hawaii County Hawaii
   BT2 Hawaii
   BT2 Polynesia
   BT2 Oceania
   BT3 Hawaii Island
   BT3 Hawaii County Hawaii
   BT3 Hawaii
   BT3 East Pacific Ocean Islands
   SA Kilauea

Alagoas
   No longer a valid term for GeoRef. See Alagoas Brazil.

Alagoas Brazil   (1993)
   State in NE Brazil.
   CO S103000S084500
         W0350000W0382000
   BT Brazil
   BT South America
   SA Sao Francisco Basin
   SA Sergipe-Alagoas Basin

Alai Range   (1978)
   One of the W ranges of the Tien Shan in SW Kyrgyzstan. Before 1978, also search Alai.
   IN Index former Soviet republics as applicable.
   UF Alay Range
   BT Tien Shan
   BT Asia
   NT Hissar Range
   SA Darvaza Range
   SA Kyrgyzstan
   SA Tadzhikistan
   SA Zeravshan Range

Alaia Turkey
   use Alanya Turkey

Alais
   Not a valid term for GeoRef. See Ales France.

Alaiye Turkey
   use Alanya Turkey

Alakol   (1978)
   Lake in Taldy Kurgan Oblast E of Lake Balkhash in E Kazakhstan.
   IN Index Kazakhstan if applicable.
   UF Ala-Kul Lake
   SA Kazakhstan

Alameda County
   Valid through 1988. Search in combination with state term. After 1988, use specific county-state term.

Alameda County California   (1989)
   On San Francisco Bay, W California. Before 1989, also search Alameda County AND California.
   CO N373000N375000
         W1213000W1222500
   BT California
   BT United States
   NT Berkeley California
   NT Hayward California
   SA Greenville Fault
   SA San Francisco Bay region

Alaminos Canyon   (1978)
   BT Gulf of Mexico
   BT North American Atlantic
   BT North Atlantic
   BT Atlantic Ocean

Alamosa County
   Valid through 1988. Search in combination with state term. After 1988, use specific county-state term.

Alamosa County Colorado   (1989)
   S Colorado. Before 1989, also search Alamosa County AND Colorado.
   CO N371800N374200
         W1052800W1060200
   BT Colorado
   BT United States
   SA San Luis Valley

Alan Hills Meteorites
   use Allan Hills Meteorites

Aland   (1978)
   Islands belonging to Finland.
   UF Aaland
   UF Aland Archipelago
   UF Aland Islands
   BT Baltic Sea
   BT European Atlantic
   BT North Atlantic
   BT Atlantic Ocean

Aland Archipelago
   use Aland

Aland Islands
   use Aland

Alandroal
   No longer a valid term for GeoRef. As of 1993 see Alandroal Portugal.

Alandroal Portugal   (1993)
   Town in S central Portugal. Before 1993 also search Alandroal AND Portugal.
   BT Evora Portugal
   BT Portugal
   BT Iberian Peninsula
   BT Southern Europe
   BT Europe

alanine   (1989)
   BT amino acids
   BT organic materials
   SA acids
   SA geochemistry
   SA organic acids

Alanya
   No longer a valid term for GeoRef. As of 1993 see Alanya Turkey.

Alanya Turkey   (1993)
   Town on the Gulf of Adalia in S Anatolia. Before 1993 also search Alaia or Alaiye or Alaya.
   UF Alaia Turkey
   UF Alaiye Turkey
   UF Alaya Turkey
   BT Turkey
   BT Middle East
   BT Asia
   SA Anatolia

Alaska (1978)
Autoposting of this term began in 1978. As of 1993, divided into six divisions based on U. S. Geological Survey quadrangles. See Maps and Charts section for divisions and quadrangle names.
CO N510000N720000
      W1300000E1730000
BT United States
NT Alaska Range
NT Alexander Terrane
NT Barrow Alaska
NT Delta River
NT East-Central Alaska
NT Northern Alaska
NT Prince William Terrane
NT Southeastern Alaska
NT Southern Alaska
NT Southwestern Alaska
NT Tracy Arm
NT Trans-Alaska Pipeline
NT West-Central Alaska
NT Yakutat Terrane
NT Yukon-Koyukuk Basin
NT Yukon-Tanana Upland
SA Alaska earthquake 1964
SA Andreanof Islands earthquake 1986
SA Arctic Coastal Plain
SA Arctic region
SA Bear Mountain
SA Bootlegger Cove Clay
SA Border Ranges Fault
SA Bowser Formation
SA Chitistone Pass
SA Coast Mountains
SA Coast plutonic complex
SA Coast Ranges
SA Colville Group
SA Colville River
SA Continental Offshore Stratigraphic Test
SA Cook Inlet
SA Crazy Mountains
SA Denali Fault
SA Endicott Group
SA Ghost Rocks Formation
SA Gubik Formation
SA Hazelton Group
SA Hemlock Conglomerate
SA Ivishak Formation
SA Kaltag Fault
SA Kanayut Conglomerate
SA Kayak Island
SA Kekiktuk Conglomerate
SA Kenai Group
SA Kingak Shale
SA Kodiak Formation
SA Kuskokwim Group
SA Lisburne Group
SA McHugh Complex
SA Meade Basin
SA Montague Island
SA Naknek Formation
SA Nanushuk Group
SA Nikolai Greenstone
SA Norton Basin
SA Old Crow Tephra
SA Orca Group
SA Pebble Shale
SA Prince of Wales Island
SA Rampart Group
SA Rat Island
SA Road River Formation
SA Rocky Mountains
SA Sadlerochit Group
SA Sag River Sandstone
SA Saint Elias Mountains
SA Saint George Island
SA Saint Paul Island
SA Shublik Formation
SA Simpson Field
SA TACT
SA Talkeetna Formation
SA Tanana River
SA Tintina Fault
SA Torok Formation
SA Tyonek Formation
SA Valdez Group
SA White Mountain
SA White River
SA Wrangellia
SA Yakataga Formation
SA Yukon River
SA Yukon-Tanana Terrane

Alaska earthquake 1964 (1989)
S Alaska.
BT earthquakes
SA Alaska
SA Aleutian Trench

Alaska Panhandle (1993)
Before 1993, also search Panhandle AND Alaska.
BT Southeastern Alaska
BT Alaska
BT United States
SA Coast Mountains

Alaska Peninsula (1978)
SW Alaska.
BT Southwestern Alaska
BT Alaska
BT United States
SA Katmai
SA Pavlof
SA Valley of Ten Thousand Smokes

Alaska Range (1978)
S central Alaska.
BT Alaska
BT United States
SA North American Cordillera

alaskite (1978)
BT granites
BT plutonic rocks
BT igneous rocks

Alava
No longer a valid term for GeoRef. As of 1993, see Alava Spain.

Alava Spain (1993)
One of the Basque Provinces in N Spain. Before 1993, also search Alava and Spain.
CO N422900N431200
      W0021400W0031500
BT Basque Provinces Spain
BT Spain
BT Iberian Peninsula
BT Southern Europe
BT Europe
NT Vitoria Spain

Alaverdi
No longer a valid term for GeoRef. As of 1993, see Alaverdi Armenia.

Alaverdi Armenia (1993)
City in N Armenia. Before 1978, also search Allaverdy. This term has multiple hierarchies.
UF Allaverdy Armenia
BT1 Armenia
BT1 Commonwealth of Independent States
BT2 Armenia
BT2 Europe

Alay Range
use Alai Range

Alaya Turkey
use Alanya Turkey

Alazani (1978)
Mining and oil producing area in E Georgian Republic.
BT Georgian Republic

BT Europe

Alba Patera (1989)
Volcano on Mars.
BT Mars

Alba-Iulia
No longer a valid term for GeoRef. As of 1993 see Alba-Iulia Romania.

Alba-Iulia Romania (1993)
Town in Alba County, SW Transylvania. Before 1993 also search Alba-Iulia AND Romania.
BT Transylvania
BT Romania
BT Southern Europe
BT Europe

Albacete
No longer a valid term for GeoRef. As of 1993, see Albacete Spain.

Albacete Spain (1993)
Refers only to the province in SE Spain. Before 1993, also search Albacete and Spain.
CO N380200N392700
      W0005500W0025500
BT Murcia region
BT Spain
BT Iberian Peninsula
BT Southern Europe
BT Europe

Alban Hills (1978)
Part of the Lower Apennines SE of Rome. This term has multiple hierarchies.
BT1 Latium Italy
BT1 Italy
BT1 Southern Europe
BT1 Europe
BT2 Apennines
BT2 Italy
BT2 Southern Europe
BT2 Europe

Albania (1978)
CO N393500N424000
      E0210500E0191500
BT Southern Europe
BT Europe
NT Albanides
SA Adriatic region
SA Balkan Peninsula
SA Dinaric Alps
SA Ionian Zone
SA Mediterranean region
SA Montenegro earthquake 1979

Albanides (1978)
Mountain area in W and SW Albania. This term has multiple hierarchies.
BT1 Alps
BT1 Europe
BT2 Albania
BT2 Southern Europe
BT2 Europe

Albany
Valid through 1988. Search in combination with state term. After 1988, use specific city-state term.

Albany County
Valid through 1988. Search in combination with state term. After 1988, use specific county-state term.

Albany County New York (1989)
E New York. Before 1989, also search Albany County AND New York.
CO N422300N424800
      W0734100W0775900
BT New York

BT United States
NT Albany New York

Albany County Wyoming (1989)
SE Wyoming. Before 1989, also search Albany County AND Wyoming.
CO N410000N422300
      W1051700W1062000
BT Wyoming
BT United States

Albany New York (1989)
City in E New York. Before 1989, search Albany AND New York.
CO N424000N424400
      W0734900W0734900
BT Albany County New York
BT New York
BT United States

Albatross Cordillera
use East Pacific Rise

albedo (1978)
Before 1993, also used with ionosphere and magnetosphere.
IN May be used with ionosphere for neutron albedo, magnetic albedo, and albedo of electromagnetic waves; or with cosmic rays
UF Bond albedo
SA interplanetary space
SA Moon
SA reflectance
SA remote sensing

Albemarle County
Valid through 1988. Search in combination with state term. After 1988, use specific county-state term.

Albemarle County Virginia (1989)
Central Virginia. Before 1989, also search Albemarle County AND Virginia.
CO N374200N381700
      W0781200W0785200
BT Virginia
BT United States

Albert Canyon
No longer a valid term for GeoRef. See Albert Canyon British Columbia.

Albert Canyon British Columbia (1993)
Village between Mount Revelstone and Glacier National Park in SE British Columbia.
BT British Columbia
BT Western Canada
BT Canada

Albert Formation (1989)
New Brunswick, Canada.
BT Carboniferous
BT Paleozoic
SA New Brunswick

Alberta (1978)
CO N490000N600000
      W1100000W1200000
BT Western Canada
BT Canada
NT Athabasca Glacier
NT Athabasca Oil Sands
NT Athabasca River
NT Banff Alberta
NT Banff National Park
NT Calgary Alberta
NT Dinosaur Provincial Park
NT Drumheller Alberta
NT Edmonton Alberta
NT Fort McMurray Alberta
NT Jasper National Park

NT Leduc Alberta
NT Medicine Hat Alberta
NT Peace River Arch
NT Red Deer River
NT Red Deer River valley
NT Smoky River
NT Wabasca Alberta
SA Abee Meteorite
SA Alberta Basin
SA Altyn Limestone
SA Banff Formation
SA Bearpaw Formation
SA Belly River Formation
SA Blairmore Group
SA Bluesky Formation
SA Bow River valley
SA Bruderheim Meteorite
SA Cadomin Formation
SA Canadian Cordillera
SA Canadian Rocky Mountains
SA Canadian Shield
SA Cardium Formation
SA Churchill Province
SA Clearwater Formation
SA Clearwater River
SA Cold Lake
SA Columbia Icefield
SA Disturbed Belt
SA Edmonton Formation
SA Elk Lake
SA Elk Point Basin
SA Elk Point Group
SA Elmworth Field
SA Fernie Formation
SA Gates Formation
SA Gething Formation
SA Gog Group
SA Great Plains
SA Grosmont Formation
SA Hector Formation
SA Horseshoe Canyon Formation
SA Innisfree Meteorite
SA Ishbel Group
SA Judith River Formation
SA Keg River Formation
SA Kicking Horse River valley
SA Kootenay Formation
SA Lake Athabasca
SA Lake Louise
SA Lewis thrust fault
SA Mannville Group
SA Mazama Ash
SA McMurray Formation
SA Miette Group
SA Milk River Formation
SA Missouri River basin
SA Mist Mountain Formation
SA Moosebar Formation
SA Mount Erebus
SA Nisku Formation
SA North Saskatchewan River
SA Oldman Formation
SA Palliser Formation
SA Paskapoo Formation
SA Peace River
SA Peace River Formation
SA Purcell System
SA Ranger Canyon Formation
SA Rundle Group
SA Slave Point Formation
SA Souris River basin
SA South Saskatchewan River
SA Spirit River Formation
SA Sturgeon Lake
SA Sullivan Mine
SA Swan Hills
SA Swan Hills Formation
SA Sweetgrass Arch
SA Viking Formation
SA Waterways Formation
SA Western Interior Seaway
SA Western Overthrust Belt

Alberta Basin (1989)
Structural basin in Alberta, Canada.
IN Index Alberta as applicable.
SA Alberta

Albian (1978)
Europe. Above Aptian, below Cenomanian. Autoposting of this term began in 1978.
BT Lower Cretaceous
BT Cretaceous
BT Mesozoic
NT upper Albian
SA Peace River Formation

Albion Range (1978)
Cassia County in S Idaho.
BT Idaho
BT United States

albite (1978)
BT plagioclase
BT feldspar group
BT framework silicates
BT silicates
SA andesine
SA granites
SA peristerite

albitite (1978)
BT syenites
BT plutonic rocks
BT igneous rocks

albitization (1978)
BT metasomatism
SA autometamorphism
SA processes
SA spilitization

albitophyre (1978)
As of 1993, this term is considered obsolete.
BT syenites
BT plutonic rocks
BT igneous rocks
SA keratophyre

Alboran Sea (1978)
Between S Spain and Morocco in W Mediterranean Sea.
CO N350500N364500
W0010000W0053000
BT West Mediterranean
BT Mediterranean Sea
SA El Alboran Island

Alborz Mountains
use Elburz

Albuquerque
As of 1985, no longer a valid term for GeoRef. From 1985 through 1988, Albuquerque region was used.
use Albuquerque New Mexico

Albuquerque Basin (1989)
Colorado Plateau of New Mexico and adjacent areas of Texas and Colorado.
IN Index states as applicable.
BT United States
SA Colorado
SA Colorado Plateau
SA New Mexico
SA Texas

Albuquerque New Mexico (1989)
City in central New Mexico. From 1985 to 1988, also search Albuquerque region. Before 1985, also search Albuquerque.
CO N350500N350500
W1063800W1063800
UF Albuquerque
UF Albuquerque region
BT Bernalillo County New Mexico
BT New Mexico
BT United States

Albuquerque region
Valid from 1985 through 1988.
use Albuquerque New Mexico

alcohols (1985)
BT organic materials
NT sterols
SA phenols

Alcona County Michigan (1989)
NE Michigan.
CO N443000N445000
W0831800W0835200
BT Michigan Lower Peninsula
BT Michigan
BT United States
NT Glennie Michigan

Alcoy
No longer a valid term for GeoRef. As of 1993, see Alcoy Spain.

Alcoy Spain (1993)
City in SE Spain. Before 1993, also search Alcoy and Spain.
BT Alicante Spain
BT Valencia region
BT Spain
BT Iberian Peninsula
BT Southern Europe
BT Europe

Alcudia Valley (1978)
S central Spain.
BT Ciudad Real Spain
BT New Castile Spain
BT Castile Spain
BT Spain
BT Iberian Peninsula
BT Southern Europe
BT Europe

Aldabra Island (1978)
Atoll 250 miles NW of Malagasy Republic.
CO S092400S092400
E0462000E0462000
BT Seychelles
BT Indian Ocean Islands

Aldan
No longer a valid term for GeoRef. As of 1993, see Aldan Russian Federation for the city.

Aldan Plateau (1978)
SE Yakutia. This term has multiple hierarchies.
CO N562000N581000
E1303000E1230000
BT1 Russian Federation
BT1 Commonwealth of Independent States
BT2 Asia
SA Yakutia Russian Federation

Aldan River (1978)
SE Yakutia. Flows into the Lena River. This term has multiple hierarchies.
BT1 Russian Federation
BT1 Commonwealth of Independent States
BT2 Asia
SA Yakutia Russian Federation

Aldan Russian Federation (1993)
City in SE Yakutia. Before 1993, also search Aldan AND Russian Republic. This term has multiple hierarchies.
BT1 Yakutia Russian Federation
BT1 Russian Federation
BT1 Commonwealth of Independent States
BT2 Yakutia Russian Federation
BT2 Asia

Aldan Shield (1978)
The Aldan Plateau area of SE Yakutia. Part of greater Angara Shield. This term has multiple hierarchies.
CO N553000N613000
E1370000E1163000
BT1 Siberian Platform
BT1 Russian Federation
BT1 Commonwealth of Independent States
BT2 Siberian Platform
BT2 Asia
SA Yakutia Russian Federation
SA Yudomian

Aldanian (1978)
Europe.
BT Lower Cambrian
BT Cambrian
BT Paleozoic

Aldridge Formation (1989)
In Belt Supergroup.
BT middle Proterozoic
BT Proterozoic
BT upper Precambrian
BT Precambrian
SA Belt Supergroup
SA British Columbia
SA Idaho
SA Montana
SA Washington

Aleksinac (1978)
Town in E Serbia.
UF Alexinats
UF Alexinatz
BT Serbia
BT Yugoslavia
BT Southern Europe
BT Europe

Aleksod (1978)
Region in Ahaggar in SE Algeria.
BT Algeria
BT North Africa
BT Africa

Alentejo (1978)
Former province in central and S Portugal.
BT Portugal
BT Iberian Peninsula
BT Southern Europe
BT Europe

Ales
No longer a valid term for GeoRef. As of 1993, see Ales France.

Ales France (1993)
City in S France. Before 1978, also search Alais.
BT Gard France
BT France
BT Western Europe
BT Europe

Alessandria
No longer a valid term for GeoRef. See Alessandria Italy.

Alessandria Italy (1993)
City and province in NW Italy.
BT Piemonte Italy
BT Italy
BT Southern Europe
BT Europe
SA Acqui Italy

Aleutian Arc
use Aleutian Islands

Aleutian Islands (1978)
Chain of volcanic islands extending in great curve from Alaska Peninsula in SW Alaska.
CO N510000N553000
W1610000E1720000

UF Aleutian Arc
UF Aleutians
BT Southwestern Alaska
BT Alaska
BT United States
NT Adak Island
NT Amchitka Island
NT Rat Islands
NT Shemya Island
NT Shumagin Islands
NT Unalaska Island
SA Aleutian Trench
SA Andreanof Islands earthquake 1986
SA Rat Island

Aleutian Ridge (1978)
Along S side of Aleutian chain.
BT Pacific Ocean
SA Aleutian Trench

Aleutian Trench (1978)
S of Aleutian Ridge and extending along the entire Aleutian chain.
BT Pacific Ocean
SA Alaska earthquake 1964
SA Aleutian Islands
SA Aleutian Ridge
SA DSDP Site 183
SA DSDP Site 186
SA Leg 19

Aleutians
use Aleutian Islands

Alexander Archipelago (1989)
SE Alaska.
CO N544000N582500
    W1305000W1364000
BT Southeastern Alaska
BT Alaska
BT United States
NT Chichagof Island
SA Alexander Terrane
SA Sitka Sound

Alexander Island (1978)
Large island off W coast of Antarctic Peninsula S of Cape Horn.
BT Antarctica

Alexander Terrane (1989)
Comprised of the Alexander Archipelago, SE Alaska.
BT Alaska
BT United States
SA Alexander Archipelago
SA Southeastern Alaska

Alexandria
No longer a valid term for GeoRef. See Alexandria Egypt.

Alexandria Egypt (1993)
Governate. Also a city on the Mediterranean Sea.
BT Egypt
BT North Africa
BT Africa

Alexandrian (1978)
North American provincial series. Lower Silurian (above Cincinnatian of Ordovician, below Niagaran).
UF Alexandrian Series
BT Lower Silurian
BT Silurian
BT Paleozoic

Alexandrian Series
use Alexandrian

alexandrite (1978)
Variety of chrysoberyl.
BT oxides
SA chrysoberyl

Alexinats
use Aleksinac

Alexinatz
use Aleksinac

Alfisols (1978)
BT soils
SA soil group

Alfold (1978)
Plain in central and E Hungary.
CO N455500N482700
    E0225500E0184500
UF Great Alfold
UF Great Hungarian Plain
UF Hungarian Basin
UF Hungarian Great Plain
UF Hungarian Plain
BT Hungary
BT Central Europe
BT Europe

Alfred Wegener
use Wegener, Alfred

algae (1978)
Autoposting of this term to Pyrrhophyta and Receptaculitaceae began in 1981. This term has multiple hierarchies.
BT1 thallophytes
BT1 Plantae
BT2 microfossils
NT calcareous algae
NT Chlorophyta
NT Chrysophyta
NT Coccolithophoraceae
NT Collenia
NT Cyanophyta
NT diatoms
NT Epiphyton
NT Microcodium
NT nannofossils
NT Phaeophyta
NT Pyrrhophyta
NT Rhodophyta
SA algal biscuits
SA algal flora
SA algal mats
SA algal mounds
SA algal structures
SA bioherms
SA endolithic taxa
SA fungi
SA lichens
SA microorganisms
SA phytoplankton
SA plankton
SA Protista
SA reef builders
SA stromatolites
SA zooxanthellae

algal banks (1978)
BT algal structures
BT biogenic structures
BT sedimentary structures
SA banks
SA carbonate banks

algal biscuits (1978)
UF lake biscuits
UF water biscuits
BT algal structures
BT biogenic structures
BT sedimentary structures
SA algae
SA girvanella

algal flora (1981)
Common name for algae. This term has multiple hierarchies.
IN Index for all non-paleontologic studies of fossils.
BT1 plants
BT2 microfossils
NT charophytes
NT coccoliths
NT desmids
NT diatom flora
NT nannofossils

SA algae
SA biostratigraphy
SA encrustations
SA stromatolites

algal limestone (1978)
Common in early Paleozoic, especially in the Cambro-Ordovician of the Appalachian region of the United States. Before 1978, also search algal AND limestone.
BT limestone
BT carbonate rocks
BT sedimentary rocks

algal mats (1978)
BT algal structures
BT biogenic structures
BT sedimentary structures
SA algae
SA stromatolites

algal mounds (1978)
UF mounds, algal
BT algal structures
BT biogenic structures
BT sedimentary structures
SA algae
SA calcilutite
SA Epiphyton
SA limestone

algal structures (1981)
BT biogenic structures
BT sedimentary structures
NT algal banks
NT algal biscuits
NT algal mats
NT algal mounds
SA algae
SA thrombolites

Algarve (1978)
Region in extreme S Portugal. Former province.
BT Portugal
BT Iberian Peninsula
BT Southern Europe
BT Europe

Algau Alps
use Allgau Alps

Alger Algeria
use Algiers Algeria

Algeria (1978)
CO N190000N370000
    E0120000W0090000
BT North Africa
BT Africa
NT Ahaggar
NT Ahnet
NT Aleksod
NT Algiers Algeria
NT Annaba Algeria
NT Aures
NT Babor Range
NT Bechar Algeria
NT Constantine Algeria
NT Dellys Algeria
NT Gour Oumelalen
NT Great Kabylia
NT Guelma Algeria
NT Hassi Messaoud Field
NT Hodna Basin
NT Kabylia
NT Laghouat
NT Mouydir
NT Oran Algeria
NT Ougarta Algeria
NT Taourirt Algeria
NT Tassili n'Ajjer
NT Tell
NT Tindouf Algeria
NT Tindouf Basin
SA Adrar des Iforas
SA Atlas Mountains
SA El Asnam earthquake 1980

SA Mediterranean region
SA Sahara
SA Tanezrouft

Algerian Plain
use Balearic Basin

Algiers
No longer a valid term for GeoRef after 1981. See Algiers Algeria.

Algiers Algeria (1993)
Department, N Algeria. Before 1993, also search Alger or Algiers region.
UF Alger Algeria
BT Algeria
BT North Africa
BT Africa

Algiers region
No longer a valid term for GeoRef. Used as large department of French Algeria. As of 1993, see Algeria for departments.

alginite (1985)
BT macerals
SA organic residues

Algoma
As of 1985, no longer a valid term for GeoRef.
use Algoma District Ontario

Algoma District
No longer a valid term for GeoRef. As of 1993, see Algoma District Ontario.

Algoma District Ontario (1993)
On Lake Superior. Before 1993, also search Algoma District, District of Algoma or Algoma in conjunction with Ontario.
CO N461000N494000
    W0824500W0852000
UF Algoma
UF District of Algoma
BT Ontario
BT Eastern Canada
BT Canada
NT Blind River Ontario
NT Echo Bay Ontario
NT Elliot Lake Ontario
NT Mamainse Point
NT Quirke Lake
NT Sault Sainte Marie Ontario
NT Wawa Ontario
SA Michipicoten Island

algoma-type (1989)
SA iron formations
SA iron ores
SA mineral deposits, genesis

Algonkian (1978)
Autoposting of Precambrian to this term ended in 1989. As of 1989, Proterozoic is autoposted to this term. As of 1990, upper Precambrian and Precambrian are autoposted to this term.
BT Proterozoic
BT upper Precambrian
BT Precambrian
NT Baraboo Quartzite

algorithms (1978)
SA computer programs
SA data processing
SA equations
SA Kalman filters
SA mathematical geology
SA mathematical methods

ALHA 77005 (1989)
One of the Allan Hills Meteorites. This term has multiple hierarchies.
UF Allan Hills 77005

BT1 Allan Hills Meteorites
BT1 meteorites
BT2 achondrites
BT2 stony meteorites
BT2 meteorites
SA Antarctica
SA Victoria Land

ALHA 81005 (1989)
One of the Allan Hills Meteorites. This term has multiple hierarchies.
UF Allan Hills 81005
BT1 Allan Hills Meteorites
BT1 meteorites
BT2 achondrites
BT2 stony meteorites
BT2 meteorites
SA Antarctica
SA Victoria Land

Alicante
No longer a valid term for GeoRef. As of 1993, see Alicante City Spain.

Alicante City Spain (1993)
Refers to only the city in SE Spain. Before 1993, also search Alicante and Spain.
BT Alicante Spain
BT Valencia region
BT Spain
BT Iberian Peninsula
BT Southern Europe
BT Europe

Alicante Province
No longer a valid term for GeoRef. As of 1993, see Alicante Spain.

Alicante Spain (1993)
Refers only to the province in SE Spain. From 1981-1992, also search Alicante Province and Spain. Before 1981, also search Alicante and Spain. For the city, see Alicante City and Spain.
CO N375200N385300
    E0001500W0010500
BT Valencia region
BT Spain
BT Iberian Peninsula
BT Southern Europe
BT Europe
NT Alcoy Spain
NT Alicante City Spain

Alice Arm
No longer a valid term for GeoRef. See Alice Arm British Columbia.

Alice Arm British Columbia (1993)
Village on Observatory Inlet near Alaska border NE of Prince Rupert.
BT British Columbia
BT Western Canada
BT Canada

Alice Springs
No longer a valid term for GeoRef. See Alice Springs Australia.

Alice Springs Australia (1993)
Town in S central Northern Territory.
BT Northern Territory Australia
BT Australia
BT Australasia

aliphatic hydrocarbons (1985)
BT hydrocarbons
BT organic materials
NT alkanes
NT alkenes
NT alkynes
SA aromatic hydrocarbons

Alisitos Formation (1989)
Baja California, NW Mexico. S California.
BT Cretaceous
BT Mesozoic
SA Baja California
SA California
SA Mexico

Alismidae (1981)
BT Monocotyledoneae
BT angiosperms
BT Spermatophyta
BT Plantae

alkali amphibole
use alkalic amphibole

alkali basalt
As of 1989, no longer a valid term for GeoRef. From 1981 to 1988, alkali basalts, basalts and volcanic rocks were autoposted to this term.
use alkali basalts

alkali basalt family
As of 1981, no longer a valid term for GeoRef. Use alkali basalts or trachybasalts as applicable. From 1978-1980, was autoposted to the rocks that were classified under it.

alkali basalts (1981)
Before 1981, search alkali basalt family. From 1978-1980, alkali basalt family was autoposted to the rocks that were classified under that term. Before 1989, also search alkali basalt. Autoposting of this term to alkali basalt ended in 1989.
UF alkali basalt
UF alkalic basalt
UF alkaline basalt
BT basalts
BT volcanic rocks
BT igneous rocks
NT alkali olivine basalt
NT ankaratrite
NT basanite
NT crinanite
NT hawaiite
NT mugearite
NT nepheline basalt
NT spilite
NT tephrite
NT trachybasalts
NT trachydolerite
SA leucitite
SA melilitite
SA nephelinite
SA ocean-island basalts

alkali diorites (1981)
BT diorites
BT plutonic rocks
BT igneous rocks

alkali earths
use alkaline earth metals

alkali feldspar (1978)
Autoposting of this term began in 1985.
BT feldspar group
BT framework silicates
BT silicates
NT adularia
NT amazonite
NT anorthoclase
NT antiperthite
NT cryptoperthite
NT K-feldspar
NT microcline
NT microperthite
NT moonstone
NT orthoclase
NT perthite
NT sanidine

alkali gabbro family
As of 1981, no longer a valid term for GeoRef. From 1978-1980, was autoposted to the rocks that were classified under it.
use alkali gabbros

alkali gabbros (1981)
Before 1981, search alkali gabbro family. From 1978-1980, alkali gabbro family was autoposted to the rocks that were classified under that term.
UF alkali gabbro family
BT gabbros
BT plutonic rocks
BT igneous rocks
NT essexite
NT ijolite
NT leucitite
NT melteigite
NT nephelinite
NT olivine nephelinite
NT teschenite
NT urtite
SA foyaite
SA shonkinite

alkali granite
A valid term for GeoRef from 1981 to 1988. From 1981 to 1988, alkali granites was autoposted to this term.
use alkali granites

alkali granites (1981)
Before 1981, search alkalic granite. Before 1989, also search alkali granite. Autoposting of this term to alkali granite ended in 1989.
UF alkali granite
UF alkalic granite
BT granites
BT plutonic rocks
BT igneous rocks

alkali metals (1978)
Metallic elements from group IA of the periodic table. Autoposting of this term to specific alkali metals began in 1989. Autoposting of metals to this term began in 1989.
UF alkaline metals
BT metals
NT cesium
NT francium
NT lithium
NT potassium
NT rubidium
NT sodium
SA precious metals

alkali olivine basalt (1978)
Autoposting of alkali basalts to this term began in 1989.
BT alkali basalts
BT basalts
BT volcanic rocks
BT igneous rocks
SA olivine basalt

alkali syenite
As of 1989, no longer a valid term for GeoRef. From 1981 to 1988, alkali syenites and syenites were autoposted to this term.
use alkali syenites

alkali syenites (1981)
Before 1989, search alkali syenite. Autoposting of this term to alkali syenite ended in 1989.
UF alkali syenite
UF alkalic syenite
BT syenites
BT plutonic rocks
BT igneous rocks

alkalic
Valid index term through 1977.

alkalic amphibole (1981)
Before 1981, also search alkali amphibole.
UF alkali amphibole
BT amphibole group
BT chain silicates
BT silicates

alkalic basalt
use alkali basalts

alkalic composition (1978)
Before 1978, also search alkalic AND igneous rocks.
SA A-type granites
SA calc-alkalic composition
SA composition
SA feldspathoid rocks
SA igneous rocks

alkalic granite
As of 1981, no longer a valid term for GeoRef.
use alkali granites

alkalic pyroxene (1981)
BT pyroxene group
BT chain silicates
BT silicates

alkalic syenite
use alkali syenites

alkaline basalt
use alkali basalts

alkaline earth elements
use alkaline earth metals

alkaline earth metals (1989)
Metallic elements from Group IIA of the periodic table.
UF alkali earths
UF alkaline earth elements
BT metals
NT barium
NT beryllium
NT calcium
NT magnesium
NT radium
NT strontium

alkaline metals
use alkali metals

alkalinity (1978)
SA buffers
SA geochemistry
SA pH
SA sodium hydroxide

alkanes (1985)
BT aliphatic hydrocarbons
BT hydrocarbons
BT organic materials
NT butane
NT ethane
NT methane
NT n-alkanes
NT paraffins
NT pentane
NT phytane
NT pristane
NT propane
SA biomarkers
SA saturated hydrocarbons

alkenes (1985)
BT aliphatic hydrocarbons
BT hydrocarbons
BT organic materials
NT ethylene

alkynes (1989)
BT aliphatic hydrocarbons
BT hydrocarbons
BT organic materials
NT acetylene

Allan Hills (1981)
Victoria Land, Antarctica.
CO S770000S763000
E1600000E1590000
BT Victoria Land
BT Antarctica

Allan Hills 77005
use ALHA 77005

Allan Hills 81005
use ALHA 81005

Allan Hills Meteorites (1981)
Due to numbering of these meteorites, such as Allan Hills 76-01 and Allan Hills 77005, also truncate when searching. Due to abbreviations for these meteorites, such as ALH 82130 and ALHA 77219, also search ALH or ALHA in combination with number. Before 1981, also search Allan Hills AND meteorites. Autoposting of this term to individual meteorites began in 1989.
UF Alan Hills Meteorites
BT meteorites
NT ALHA 77005
NT ALHA 81005
SA Antarctica
SA Victoria Land

allanite (1978)
UF orthite
BT epidote group
BT sorosilicates
BT orthosilicates
BT silicates

Allarechenskiy (1978)
Ore field in the Kola Peninsula in extreme NW European Russia. This term has multiple hierarchies.
UF Allarechenskoye
BT1 Russian Federation
BT1 Commonwealth of Independent States
BT2 Europe
SA Kola Peninsula

Allarechenskoye
use Allarechenskiy

Allaverdy Armenia
use Alaverdi Armenia

Allegany County
Valid through 1988. Search in combination with state term. After 1988, use specific county-state term.

Allegany County Maryland (1989)
W Maryland. Before 1989, also search Allegany County AND Maryland.
CO N392600N394330
W0781900W0790400
BT Maryland
BT United States
NT Cumberland Maryland

Allegany County New York (1989)
W New York. Before 1989, also search Allegany County AND New York.
CO N420000N423200
W0775100W0781900
BT New York
BT United States
SA Genesee River

Alleghanian
use Allegheny Group

Alleghany County
Valid through 1988. Search in combination with state term. After 1988, use specific county-state term.

Alleghany County North Carolina (1989)
NW North Carolina. Before 1989, also search Alleghany County AND North Carolina.
CO N362100N363400
W0805600W0812200
BT North Carolina
BT United States

Alleghany County Virginia (1989)
W Virginia. Before 1989, also search Alleghany County AND Virginia.
CO N373700N375800
W0793600W0801900
BT Virginia
BT United States

Alleghany Mountains
use Allegheny Mountains

Alleghany Orogeny (1993)
An event which deformed the rocks of the Valley and Ridge Province, and those of the adjacent Allegheny Plateau in the Central Appalachians and Southern Appalachians. Pennsylvanian to Triassic? Before 1978, also search Allegheny or Alleghany AND orogeny.
IN Index ages as applicable.
UF Alleghenian Orogeny
UF Allegheny Orogeny
SA orogeny
SA Pennsylvanian
SA Permian
SA tectonics
SA Triassic

Alleghenian
use Allegheny Group

Alleghenian Orogeny
use Alleghany Orogeny

Alleghenies
use Allegheny Mountains

Allegheny County
Valid through 1988. Search in combination with state term. After 1988, use specific county-state term.

Allegheny County Pennsylvania (1989)
W Pennsylvania. Before 1989, also search Allegheny County AND Pennsylvania.
CO N401200N404000
W0794200W0802200
BT Pennsylvania
BT United States
NT Pittsburgh Pennsylvania

Allegheny Front (1978)
Eastern slope of the Allegheny Mountains.
IN Index states as applicable.
BT United States
SA Allegheny Mountains
SA Maryland
SA Pennsylvania
SA Virginia
SA West Virginia

Allegheny Group (1978)
Subdivided into three formations: Clarion, Kittanning, and Freeport.
E Kentucky, W Maryland, E Ohio, W Pennsylvania, W Virginia, and West Virginia. Above Pottsville Group; below Conemaugh Group.
UF Alleghanian
UF Alleghenian
BT Middle Pennsylvanian
BT Pennsylvanian
BT Carboniferous
BT Paleozoic
SA Freeport Formation
SA Kentucky
SA Kittanning Formation
SA Maryland
SA Ohio
SA Pennsylvania
SA Virginia
SA West Virginia

Allegheny Mountains (1978)
Ranges of the Appalachians constituting a part of the Allegheny Plateau. As of 1993, Appalachians is a broader term. This term has multiple hierarchies.
IN Index states as applicable.
UF Alleghany Mountains
UF Alleghenies
BT1 Appalachians
BT2 United States
SA Allegheny Front
SA Allegheny Plateau
SA Appalachian Plateau
SA Maryland
SA Pennsylvania
SA Shenandoah Valley
SA Virginia
SA West Virginia

Allegheny Orogeny
As of 1993, no longer a valid term for GeoRef.
use Alleghany Orogeny

Allegheny Plateau (1978)
W section of the Appalachians extending from Cumberland Plateau on S to Mohawk Valley in New York on N and including the Allegheny Mountains.
IN Index states as applicable.
BT United States
SA Allegheny Mountains
SA Appalachian Plateau
SA Appalachians
SA Kentucky
SA New York
SA Ohio
SA Pennsylvania
SA Virginia
SA West Virginia

Allen County
Valid through 1988. Search in combination with state term. After 1988, use specific county-state term.

Allen County Kansas (1989)
SE Kansas. Before 1989, also search Allen County AND Kansas.
CO N374300N380300
W0950500W0953000
BT Kansas
BT United States

Allen County Kentucky (1989)
S Kentucky. Before 1989, also search Allen County AND Kentucky.
CO N363700N365500
W0855600W0861900
BT Kentucky
BT United States

Allen County Ohio (1989)
W Ohio. Before 1989, also search Allen County AND Ohio.
CO N403800N405500
W0835300W0842400
BT Ohio
BT United States

Allende Meteorite (1981)
Impact in SE Chihuahua. Before 1981, also search Allende AND meteorites.
BT carbonaceous chondrites
BT chondrites
BT stony meteorites
BT meteorites
SA Chihuahua Mexico

Allerod (1978)
Europe. An interval of late-glacial time following the Older Dryas and preceding the Younger Dryas. Autoposting of upper Pleistocene and Pleistocene to this term began in 1989.
BT Weichselian
BT upper Pleistocene
BT Pleistocene
BT Quaternary
BT Cenozoic

allevardite
use rectorite

Allgaeu Alps
use Allgau Alps

Allgau Alps (1978)
This term has multiple hierarchies.
IN Index Bavaria Germany and/or Tyrol Austria as applicable.
CO N471000N473000
E0105000E0100000
UF Algau Alps
UF Allgaeu Alps
BT1 Alps
BT1 Europe
BT2 Central Europe
BT2 Europe
SA Austria
SA Bavaria Germany
SA Germany
SA Tyrol Austria

Allier
No longer a valid term for GeoRef. See Allier France.

Allier France (1993)
Department in central France.
CO N455000N464500
E0040000E0023000
BT France
BT Western Europe
BT Europe
SA Limagne

Alligator Ridge Deposit
use Alligator Ridge Mine

Alligator Ridge Mine (1989)
E Nevada.
UF Alligator Ridge Deposit
BT White Pine County Nevada
BT Nevada
BT United States
SA gold ores
SA mines

Alligator Rivers Field (1993)
Arnhem Land, N Northern Territory. The East Alligator, South Alligator, and West Alligator rivers debouch into Van Diemen Gulf in the Alligator Rivers region. Uranium deposits occur in the Ranger, Koongarra, Rum Jungle, Nabarlek, and Jabiluka mining areas. Before 1993, also search Alligator River and commodity. Also search South Alligator, East Alligator, and West Alligator if applicable.
BT Northern Territory Australia
BT Australia

BT Australasia
SA Jabiluka Australia
SA mines
SA Pine Creek Geosyncline
SA Ranger Mine
SA uranium ores

allochthons (1978)
SA autochthons
SA faults
SA klippen
SA nappes
SA overthrust faults
SA structural geology
SA tectonics
SA windows

alloclasite (1978)
BT sulfides

Allogromiina (1978)
As of 1990, microfossils and Protista are autoposted to this term. This term has multiple hierarchies.
BT1 foraminifera
BT1 Protista
BT1 Invertebrata
BT2 foraminifera
BT2 Protista
BT2 microfossils

allometry (1978)
SA fossils

allophane (1978)
BT clay minerals
BT sheet silicates
BT silicates

alloys (1978)
Before 1981, also search native elements and alloys. As of 1981, minerals which formerly were classified under native elements and alloys are included here or under native elements. Autoposting of this term began in 1981.
NT awaruite
NT carbides
NT electrum
NT hedleyite
NT josephinite
NT kamacite
NT niggliite
NT nitrides
NT phosphides
NT plessite
NT silicides
NT taenite
NT tetrataenite
SA annealing
SA base metals
SA construction materials
SA metallurgy
SA metals
SA minerals
SA native elements

alluvial cones
use alluvial fans

alluvial deposits
use alluvium

alluvial fans (1978)
Autoposting of fluvial features to this term began in 1989.
UF alluvial cones
UF detrital fan
UF dry delta
UF talus fan
BT fluvial features
NT fan deltas
SA Alluvial soils
SA alluvium
SA cones
SA debris cones
SA deposition
SA geomorphology

alluvial plains (1978)
Autoposting of fluvial features to this term began in 1989.
UF river plain
BT fluvial features
SA deposition
SA geomorphology
SA paleogeography
SA plains

Alluvial soils (1978)
BT soils
SA alluvial fans
SA alluvium
SA soil group

alluvion
use alluvium

alluvium (1978)
UF alluvial deposits
UF alluvion
BT clastic sediments
BT sediments
SA alluvial fans
SA Alluvial soils
SA alluvium aquifers
SA clay
SA colluvium
SA eluvium
SA floodplains
SA gravel
SA sand
SA silt
SA soil mechanics
SA soils

alluvium aquifers (1982)
BT aquifers
SA alluvium
SA ground water

Alma Dag
use Amanos Mountains

Alma District
No longer a valid term for GeoRef.
use Alma mining district

Alma mining district (1993)
Mining district in central Colorado. Before 1993, also search Alma District.
UF Alma District
BT Park County Colorado
BT Colorado
BT United States
SA metal ores
SA mines

Alma River (1978)
In Crimea. Also search Alma. This term has multiple hierarchies.
BT1 Crimea Ukraine
BT1 Ukraine
BT1 Europe
BT2 Crimea Ukraine
BT2 Ukraine
BT2 Commonwealth of Independent States

Alma-Ata
No longer a valid term for GeoRef. As of 1993, see Alma-Ata Kazakhstan.

Alma-Ata Kazakhstan (1993)
City in SE Kazakhstan. Before 1978, also search Vernyi or Vyernyi. This term has multiple hierarchies.
UF Vernyi Kazakhstan
UF Vyernyi Kazakhstan
BT1 Kazakhstan
BT1 Central Asia
BT1 Asia
BT2 Kazakhstan
BT2 Commonwealth of Independent States

Almaden
No longer a valid term for GeoRef. As of 1993, see Almaden Spain.

Almaden Spain (1993)
Village in S central Spain. Before 1993, also search Almaden and Spain.
BT Ciudad Real Spain
BT New Castile Spain
BT Castile Spain
BT Spain
BT Iberian Peninsula
BT Southern Europe
BT Europe

Almalyk
No longer a valid term for GeoRef. As of 1993, see Almalyk Uzbekistan.

Almalyk Uzbekistan (1993)
Town SE of Tashkent. This term has multiple hierarchies.
BT1 Uzbekistan
BT1 Asia
BT2 Uzbekistan
BT2 Commonwealth of Independent States

almandine (1978)
UF almandite
BT garnet group
BT nesosilicates
BT orthosilicates
BT silicates

almandite
use almandine

Almas Valley basin (1978)
River basin in W central Transylvania.
BT Transylvania
BT Romania
BT Southern Europe
BT Europe

Almeria
No longer a valid term for GeoRef. As of 1993, see Almeria City Spain.

Almeria City Spain (1993)
Refers only to the city. Before 1993, also search Almeria and Spain. On the Mediterranean Sea.
BT Almeria Spain
BT Andalusia Spain
BT Spain
BT Iberian Peninsula
BT Southern Europe
BT Europe

Almeria Province
No longer a valid term for GeoRef. As of 1993, see Almeria Spain.

Almeria Spain (1993)
Refers only to the province. From 1981-1992, also search Almeria Province and Spain. Before 1981, also search Almeria and Spain. In SE Spain. For the city, see Almeria City Spain.
CO N364000N375700
W0013500W0031000
BT Andalusia Spain
BT Spain
BT Iberian Peninsula
BT Southern Europe
BT Europe
NT Almeria City Spain
NT Cerro del Hoyazo Spain
NT Sierra de Gador
NT Sierra de los Filabres
NT Velez Rubio Spain

almeriite
use natroalunite

Almond Formation (1978)
In Mesaverde Group. Overlies Ericson Sandstone and underlies Lewis Shale. SW Wyoming.
BT Upper Cretaceous
BT Cretaceous
BT Mesozoic
SA Mesaverde Group
SA Wyoming

Almora
No longer a valid term for GeoRef. See Almora India.

Almora India (1993)
Town and district in N Uttar Pradesh, N India.
BT Uttar Pradesh India
BT India
BT Indian Peninsula
BT Asia

Alno (1978)
Island in Gulf of Bothnia.
CO N623000N623000
E0173000E0173000
BT Sweden
BT Scandinavia
BT Western Europe
BT Europe
SA Gulf of Bothnia

alnoite (1985)
BT lamprophyres
BT plutonic rocks
BT igneous rocks

Alnus (1978)
Genus.
BT Dicotyledoneae
BT angiosperms
BT Spermatophyta
BT Plantae

Alpago Basin (1978)
Belluno Province in N Veneto.
BT Belluno Italy
BT Veneto Italy
BT Italy
BT Southern Europe
BT Europe

Alpena County
Valid through 1988. Search in combination with state term. After 1988, use specific county-state term.

Alpena County Michigan (1989)
On Lake Huron in NE Michigan. Before 1989, also search Alpena County AND Michigan.
CO N445000N451200
W0831000W0835200
BT Michigan Lower Peninsula
BT Michigan
BT United States

Alpenvorland (1981)
High plain at N edge of Alps between Jura and Bohemian Massifs.
CO N471500N490400
E0133000E0083500
BT Germany
BT Central Europe
BT Europe
SA Baden-Wurttemberg Germany
SA Bavaria Germany

Alpes-de-Haute Provence
No longer a valid term for GeoRef.
use Alpes-de-Haute Provence France

Alpes-de-Haute Provence France (1993)

Alpes-de-Haute-Provence
Department in SE France. Before 1991, also search Alpes-de-Haute-Provence. Before 1977, also search Basses-Alpes.
CO N433000N450000
E0070000E0053000
UF Alpes-de-Haute Provence
UF Alpes-de-Haute-Provence
UF Basses-Alpes
BT France
BT Western Europe
BT Europe
NT Castellane France
NT Digne France
SA Eoulx Basin
SA Provence
SA Provence Alps
SA Verdon Valley

Alpes-de-Haute-Provence
No longer a valid term for GeoRef. Valid 1977-1990.
use Alpes-de-Haute Provence France

Alpes-Maritimes
No longer a valid term for GeoRef.
use Alpes-Maritimes France

Alpes-Maritimes France (1993)
Department in extreme SE France.
CO N433000N442000
E0074500E0064000
UF Alpes-Maritimes
BT France
BT Western Europe
BT Europe
NT Antibes France
NT Nice France
NT Saint-Vallier France
SA Esterel
SA Maritime Alps
SA Provence
SA Provence Alps

alpha activation analysis
Not a valid GeoRef index term after 1970. Was used on level 1 in subfile N. Now use alpha-ray spectroscopy, or activation analysis.

alpha chalcocite
use digenite

Alpha Cordillera (1978)
Undersea feature N and NW of Queen Elizabeth Islands of Northwest Territories, Canada.
UF Alpha Ridge
BT Arctic Ocean
SA CESAR

alpha rays (1978)
Use when referring to the rays themselves. For application or methodology, use alpha-ray spectroscopy.
UF rays, alpha
SA alpha-ray spectroscopy
SA beta rays
SA gamma rays
SA radioactivity
SA spectroscopy
SA X-rays

Alpha Ridge
use Alpha Cordillera

alpha-ray spectroscopy (1978)
Before 1978, also search alpha ray AND spectroscopy.
BT spectroscopy
SA alpha rays
SA analysis
SA chemical analysis
SA gamma-ray spectroscopy
SA radioactivity

Alpine
No longer a valid term for GeoRef. After 1977, use Alpine Orogeny.

alpine
No longer a valid term for GeoRef. After 1977, use alpine environment.

Alpine County
Valid through 1988. Search in combination with state term. After 1988, use specific county-state term.

Alpine County California (1989)
On Nevada border in E central California. Before 1989, also search Alpine County AND California.
CO N382000N385500
W1193000W1201000
BT California
BT United States

alpine environment (1978)
Before 1978, also search alpine.
SA Alps
SA climate
SA depositional environment
SA ecology
SA environment
SA mountains
SA subalpine environment

Alpine Fault (1978)
IN Index country and islands as applicable.
SA New Zealand
SA South Island

Alpine Foreland
use Prealps

Alpine Geosyncline (1978)
Before 1978, search geosynclines AND Alpine.
SA Alpine Orogeny
SA geosynclines

Alpine Orogeny (1978)
Before 1978, also search orogeny AND Alpine.
IN Index ages as applicable.
UF Alpine structure
SA Alpine Geosyncline
SA alpine-type
SA Cenozoic
SA Mesozoic
SA orogeny
SA tectonics

Alpine structure
use Alpine Orogeny

alpine-type (1978)
SA Alpine Orogeny
SA Alps
SA ecology
SA mineral deposits, genesis
SA tectonics

Alps (1978)
Great mountain system of S central Europe. Autoposting of this term to Bavarian Alps, Central Swiss Alps, French Alps, North Austrian Alps, South Austrian Alps, and Swiss Alps began in 1981.
IN Index countries as applicable.
BT Europe
NT Aiguilles Rouges
NT Albanides
NT Allgau Alps
NT Central Alps
NT Central Austrian Alps
NT Eastern Alps
NT French Alps
NT Gailtal Alps
NT Hochschwab
NT Lechtal Alps
NT Limestone Alps
NT North Austrian Alps
NT Piedmont Alps
NT Prealps
NT Sau Alps
NT South Austrian Alps
NT Swiss Alps
NT Western Alps
SA alpine environment
SA alpine-type
SA Austria
SA Brianconnais Zone
SA France
SA Germany
SA Hellenides
SA Italy
SA San Giorgio Mountain
SA Sesia-Lanzo Zone
SA Slovenia
SA Switzerland
SA Tavetsch
SA Yugoslavia

Alps, Australian
use Southern Alps

Alps, Japanese
use Japanese Alps

Alpujarras (1978)
Mountainous area in S Spain.
BT Spain
BT Iberian Peninsula
BT Southern Europe
BT Europe

Alsace (1978)
Region and former province in NE France.
BT France
BT Western Europe
BT Europe
SA Bas-Rhin France
SA Haut-Rhin France

Alster River (1985)
Schleswig-Holstein and Lower Saxony. Joins the Elbe River at Hamburg.
BT Germany
BT Central Europe
BT Europe
SA Hamburg Germany
SA Lower Saxony Germany
SA Schleswig-Holstein Germany

Alston Block (1989)
Cumbria, NW England.
CO N540000N553000
W0010000W0033000
BT Cumbria England
BT England
BT Great Britain
BT United Kingdom
BT Western Europe
BT Europe

alstonite (1978)
UF bromlite
BT carbonates

Alta
No longer a valid term for GeoRef. As of 1993 see Alta Norway.

Alta Norway (1993)
Village in extreme N Norway. Before 1993 also search Alta AND Norway.
BT Finnmark Norway
BT Norway
BT Scandinavia
BT Western Europe
BT Europe

Altai
No longer a valid term for GeoRef. As of 1993, see Altai Russian Federation for the kray.

Altai Mountains (1978)
Mountain system at meeting of borders of Altai Kray, Kazakhstan, Mongolia, and Xinjiang China.
IN Index countries and provinces and/or republics as applicable.
CO N443000N500000
E1030000E0870000
UF Altay Mountains
BT Asia
NT Chuya Alps
NT Gorny Altai
NT Kalba Range
NT Kuznetsk Alatau
NT Mongolian Altai
SA Altai-Sayan region
SA China
SA Kazakhstan
SA Mongolia
SA Siberian fold belt
SA Xinjiang China

Altai Russian Federation (1993)
Administrative Territory or Kray on the Kazakhstan and Mongolian borders. Before 1978, also search Altai. Before 1993, also search Altai. This term has multiple hierarchies.
UF Altay Russian Federation
BT1 Russian Federation
BT1 Commonwealth of Independent States
BT2 Asia
NT Chuya Basin
NT Gorny Altai
NT Zmeinogorsk Russian Federation
SA Chuya Alps
SA Rudny Altai
SA Salair Ridge

Altai-Sayan region (1978)
Region including adjoining Sayan and Altai Mountains. Also search Altai-Sayan.
IN Index Mongolia and/or Russian Federation as applicable.
CO N500000N550000
E1000000E0800000
BT Asia
SA Altai Mountains
SA Eastern Sayan
SA Mongolia
SA Russian Federation
SA Sayan
SA Western Sayan

altaite (1978)
Autoposting of sulfides to this term ended in 1989.
BT tellurides

Altamaha River (1978)
SE Georgia.
BT Georgia
BT United States

Altan Teeli
No longer a valid term for GeoRef. As of 1993 see Altan Teeli Mongolia.

Altan Teeli Mongolia (1993)
Village in W Mongolia. Before 1993 also search Altan Teeli AND Mongolia.
UF Altan Teli
UF Dzereg
BT Mongolia
BT Far East
BT Asia

Altan Teli
use Altan Teeli Mongolia

Altay Mountains
use Altai Mountains

Altay Russian Federation
use Altai Russian Federation

Altenberg
No longer a valid term for GeoRef. See Altenberg Germany.

Altenberg Germany (1993)
Town near Czechoslovak border in Saxony, E Germany.
BT Saxony Germany
BT Germany
BT Central Europe
BT Europe

alteration (1978)
Formerly a general term appropriate to a large number of topics. As of 1976, use was restricted to any process having to do with changes in temperature and pressure, i.e. diagenesis, hydrothermal alteration, metasomatism, weathering, etc. The term is used to denote the whole spectrum of chemical and physical change.
SA alunitization
SA amphibolitization
SA argillization
SA autometamorphism
SA changes
SA diagenesis
SA fenitization
SA hydrothermal alteration
SA hydrothermal processes
SA igneous rocks
SA illitization
SA kaolinization
SA leaching
SA metasomatic rocks
SA metasomatism
SA microclinization
SA muscovitization
SA palagonitization
SA processes
SA scapolitization
SA serpentinization
SA thermal alteration
SA tourmalinization
SA wall-rock alteration
SA weathering

alternating field demagnetization (1989)
UF alternating-field demagnetization
BT demagnetization
SA electrical field
SA experimental studies
SA geophysics
SA magnetic field
SA paleomagnetism

alternating-field demagnetization
use alternating field demagnetization

altimetry (1985)
SA altitude
SA atmospheric pressure
SA cartography
SA geodesy
SA Geosat
SA topography

Altimur Mountains (1978)
E Afghanistan.
BT Afghanistan
BT Indian Peninsula
BT Asia

Altin Tepe
use Altin-Tepe

Altin-Tepe (1978)
Ore bearing region. in 1981.
UF Altin Tepe
BT Romania
BT Southern Europe
BT Europe
SA Romanian Dobruja

Altiplano (1993)
High intermontane plateau region in the Andes between the Western Cordilleras and Eastern Cordilleras. Extends from NW Argentina through W Brazil to SE Peru. Also refers to other regions.
IN Index mountains and regions as applicable.
SA Andes
SA Argentina
SA Bolivia
SA Lake Titicaca
SA Mexico
SA Peru
SA Spain

altitude (1978)
Used as a general term.
UF elevation
SA altimetry
SA latitude

Alto Adige (1978)
Along with Trento comprises Trentino-Alto Adige autonomous region in NE Italy.
BT Trentino-Alto Adige Italy
BT Italy
BT Southern Europe
BT Europe
SA Trentino

Alto Alentejo (1978)
Former province in central Portugal.
BT Portugal
BT Iberian Peninsula
BT Southern Europe
BT Europe

Altonian (1978)
Provincial series, New Zealand.
BT lower Miocene
BT Miocene
BT Neogene
BT Tertiary
BT Cenozoic

Altoona
Valid through 1988. Search in combination with state term. After 1988, use specific city-state term.

Altoona Pennsylvania (1989)
City in W central Pennsylvania. Before 1989, search Altoona AND Pennsylvania.
CO N403200N403200
    W0782300W0782300
BT Blair County Pennsylvania
BT Pennsylvania
BT United States

Altyn Limestone (1978)
In Ravalli Group of Belt Supergroup. SW Alberta, SE British Columbia and NW Montana.
BT middle Proterozoic
BT Proterozoic
BT upper Precambrian
BT Precambrian
SA Alberta
SA Belt Supergroup
SA British Columbia
SA Montana
SA Ravalli Group

Altyn-Topkan
No longer a valid term for GeoRef. As of 1993, see Altyn-Topkan Tadzhkistan.

Altyn-Topkan Tadzhkistan (1993)
Village in N Tadzhikistan. This term has multiple hierarchies.
BT1 Tadzhikistan
BT1 Asia
BT2 Tadzhikistan
BT2 Commonwealth of Independent States

alum (1978)
UF potash alum
UF potassium alum
BT sulfates

alum rock
use alunite

alumina (1978)
UF aluminum oxide
SA aluminum
SA corundum
SA geochemistry
SA oxides

aluminosilicates (1978)
Autoposting of this term began in 1978.
BT silicates
NT maskelynite
SA adularia
SA amazonite
SA bavenite
SA berthierine
SA bytownite
SA carpholite
SA chloritoid
SA clinozoisite
SA cookeite
SA cordierite
SA corrensite
SA cymrite
SA epidote
SA euclase
SA eucryptite
SA glaucophane
SA hydrobiotite
SA hydrosodalite
SA kaliophilite
SA kalsilite
SA lawsonite
SA leifite
SA milarite
SA mordenite
SA offretite
SA osumilite
SA petalite
SA prehnite
SA pumpellyite
SA sapphirine
SA spodumene
SA stilpnomelane
SA sudoite

aluminum (1978)
Chemical element. As of 1981, use aluminum ores for aluminum as a commodity. Autoposting of metals to this term began in 1989.
UF Al
BT metals
NT Al-26
NT Al-27
NT Al-27/Al-26
SA alumina
SA aluminum ores
SA bauxite
SA lithophile elements

aluminum ores (1981)
Before 1981, also search aluminum AND (deposit OR deposits OR ore OR ores OR economic) in the basic index. Autoposting of metal ores to this term began in 1985.
BT metal ores
SA aluminum
SA bauxite
SA boehmite
SA gibbsite

aluminum oxide
use alumina

alumite
use alunite

alumohydrocalcite (1978)
BT carbonates

alumstone
use alunite

alunite (1978)
UF alum rock
UF alumite
UF alumstone
BT sulfates
SA alunitization
SA jarosite
SA potash
SA Red Mountain

alunitization (1989)
BT metasomatism
SA alteration
SA alunite
SA natroalunite
SA processes

Alveolina (1978)
Autoposting of Miliolacea to this term began in 1989. Autoposting of microfossils and Protista to this term began in 1990. This term has multiple hierarchies.
BT1 Alveolinellidae
BT1 Miliolacea
BT1 Miliolina
BT1 foraminifera
BT1 Protista
BT1 Invertebrata
BT2 Alveolinellidae
BT2 Miliolacea
BT2 Miliolina
BT2 foraminifera
BT2 Protista
BT2 microfossils

Alveolinellidae (1978)
Autoposting of Miliolacea to this term began in 1989. Autoposting of microfossils and Protista to this term began in 1990. This term has multiple hierarchies.
BT1 Miliolacea
BT1 Miliolina
BT1 foraminifera
BT1 Protista
BT1 Invertebrata
BT2 Miliolacea
BT2 Miliolina
BT2 foraminifera
BT2 Protista
BT2 microfossils
NT Alveolina

Alwar
No longer a valid term for GeoRef. As of 1993, for the city in NE Rajasthan, see Alwar India. Formerly a princely state. Also search Alwur.

Alwar India (1993)
City in NE Rajasthan. Before 1978, also search Alwur.
UF Alwur India
BT Rajasthan India
BT India
BT Indian Peninsula
BT Asia

Alwur India
use Alwar India

Am
use americium

Am-241 (1985)
Autoposting of broader terms began in 1989. This term has multiple hierarchies.
BT1 radioactive isotopes
BT1 isotopes
BT2 americium
BT2 actinides
BT2 metals

Amadeus Basin (1978)
SW Northern Territory and W Western Australia.
IN Index state and/or territory as applicable.
CO S260000S230000 E1340000E1290000
BT Australia
BT Australasia
SA Northern Territory Australia
SA Western Australia

Amador County
Valid through 1988. Search in combination with state term. After 1988, use specific county-state term.

Amador County California (1989)
E California. Before 1989, also search Amador County AND California.
CO N381500N384500 W1200500W1210100
BT California
BT United States

Amagase
No longer a valid term for GeoRef. See Amagase Japan.

Amagase Japan (1993)
Village S of Osaka.
BT Kyoto Japan
BT Honshu
BT Japan
BT Far East
BT Asia

Amalthea Satellite (1989)
One of the satellites of Jupiter. Before 1989, also search Amalthea AND Jupiter.
SA icy satellites
SA Jupiter
SA satellites
SA Voyager Program

Amami
use Oshima

Amami Gunto
use Oshima

Amami-O-shima (1978)
Largest island in Oshima (Amami Gunto) island group, SW of Kyushu. This term has multiple hierarchies.
BT1 Oshima
BT1 Ryukyu Islands
BT1 Japan
BT1 Far East
BT1 Asia
BT2 Oshima
BT2 Kagoshima Japan
BT2 Kyushu
BT2 Japan
BT2 Far East
BT2 Asia
SA O-shima
SA Oshima Peninsula

Amanos Mountains (1978)
Along SE Gulf of Alexandretta in SE Anatolia.
UF Alma Dag
UF Gavur Mountains
BT Turkey
BT Middle East
BT Asia

Amapa
No longer a valid term for GeoRef. See Amapa Brazil.

Amapa Brazil (1993)
Federal Territory on French Guiana border.
CO S011000N043000 W0495000W0550000
BT Brazil
BT South America
NT Serra do Navio Brazil

Amargosa Desert (1978)
IN Index California and/or Nevada as applicable.
BT United States
SA California
SA Nevada

amargosite
use bentonite

Amarjola (1978)
Region in the Rajmahal Hills in E Bihar, E India.
SA Bihar India
SA Rajmahal Hills

Amasra
No longer a valid term for GeoRef. As of 1993 see Amasra Turkey.

Amasra Basin (1978)
Coal basin in NW Anatolia.
BT Turkey
BT Middle East
BT Asia
SA Anatolia
SA coal fields

Amasra Turkey (1993)
Town on the Black Sea in NW Anatolia. Before 1993 also search Amasra AND Turkey.
BT Zonguldak Turkey
BT Turkey
BT Middle East
BT Asia

Amazon Basin (1978)
River drainage basin. Also search Amazon.
IN Index countries as applicable.
UF Amazon region
BT South America
SA Bolivia
SA Brazil
SA Colombia
SA Ecuador
SA Peru
SA Venezuela

Amazon region
use Amazon Basin

Amazon River (1978)
Largest river in the world if measured by volume of water carried. Rises in 2 major headstreams in the Peruvian Andes flowing N and then E across N Brazil into the Atlantic Ocean. Also search Amazon.
IN Index countries as applicable.
UF Rio Amazonas
BT South America
SA Brazil
SA Colombia
SA Marajo
SA Peru

Amazonas
As of 1993, no longer a valid term for GeoRef. See Amazonas Brazil, Amazonas Colombia, Amazonas Peru or Amazonas Venezuela. For Rio Amazonas, use Amazon River.

Amazonas Brazil (1993)
State in NW Brazil. Before 1993, search Amazonas in combination with Brazil.
BT Brazil
BT South America
SA Orinoco River

Amazonas Colombia (1993)
Commissary in SE Colombia. Before 1993, search Amazonas AND Colombia.
BT Colombia
BT South America
SA Orinoco River

Amazonas Peru (1993)
Department in N Peru. Before 1993, search Amazonas AND Peru.
BT Peru
BT South America

Amazonas Venezuela (1993)
Territory in S Venezuela. Before 1993, search Amazonas AND Venezuela.
BT Venezuela
BT South America
SA Orinoco River
SA Rio Negro

Amazonian (1993)
Youngest system of three in the Martian stratigraphic column.
SA Mars

Amazonian Shield
use Brazilian Shield

amazonite (1978)
UF amazonstone
BT alkali feldspar
BT feldspar group
BT framework silicates
BT silicates
SA aluminosilicates
SA microcline

amazonstone
use amazonite

Amba Dongar (1978)
Region in Gujarat, W India.
BT Gujarat India
BT India
BT Indian Peninsula
BT Asia

amber (1978)
UF bernstein
BT organic compounds
SA resins

amber mica
use phlogopite

Ambin Massif (1978)
SE France.
BT Savoie France
BT France
BT Western Europe
BT Europe

amblygonite (1978)
UF hebronite
BT phosphates
SA lithium
SA montebrasite

Amblypoda (1978)
Order. Autoposting of this term to Dinocerata, Pantodonta, Pyrotheria and Xenungulata began in 1989. Autoposting of Eutheria and Theria to this term began in 1989.
BT Eutheria
BT Theria
BT Mammalia
BT Tetrapoda
BT Vertebrata
BT Chordata
NT Dinocerata
NT Pantodonta
NT Pyrotheria
NT Xenungulata
SA Ungulata

Ambre Mountain (1978)
N Madagascar. This term has multiple hierarchies.
BT1 Madagascar
BT1 Indian Ocean Islands
BT2 Madagascar
BT2 Africa

Ambrosia Lake (1978)
BT New Mexico
BT United States

Ambrosia Lake District
No longer a valid term for GeoRef.
use Ambrosia Lake mining district

Ambrosia Lake mining district (1993)
Uranium-mining district in New Mexico. Before 1993, also search Ambrosia Lake District.
UF Ambrosia Lake District
BT McKinley County New Mexico
BT New Mexico
BT United States
SA Grants mineral belt
SA mines
SA uranium ores

Amchitka Island (1978)
In center of Aleutian Islands SW of Alaska Peninsula. Also search Amchitka.
BT Aleutian Islands
BT Southwestern Alaska
BT Alaska
BT United States
NT Cannikin Alaska

AMCOR (1985)
Acronym.
UF Atlantic Margin Coring Project
SA Atlantic Coastal Plain
SA continental margin
SA continental shelf
SA continental slope
SA cores
SA marine drilling
SA marine sediments
SA North American Atlantic
SA oceanography
SA offshore
SA soil mechanics
SA U. S. Geological Survey
SA United States

Ameki Formation (1989)
BT Eocene
BT Paleogene
BT Tertiary
BT Cenozoic
SA Nigeria

Amelia (1978)
Mineral bearing region in Amelia County in S central Virginia.
IN Index county and state.
SA Virginia

America (1981)
Includes all of the Americas.
SA Central America
SA Latin America
SA North America
SA South America

SA West Indies

American Association of Petroleum Geologists
use AAPG

American Geological Institute
use AGI

American Geophysical Union
use AGU

American River (1985)
Central California. Joins Sacramento River at Sacramento.
BT California
BT United States

American Samoa
use Samoa

americium (1978)
Autoposting of actinides and metals to this term began in 1989.
UF Am
BT actinides
BT metals
NT Am-241

Americus Limestone Member (1978)
Of Foraker Limestone. E Kansas, SE Nebraska and central N Oklahoma.
BT Permian
BT Paleozoic
SA Kansas
SA Nebraska
SA Oklahoma

Ames Limestone (1978)
Of Conemaugh Group. E Kentucky, SW Maryland, E Ohio, SW Pennsylvania, and N West Virginia.
BT Upper Pennsylvanian
BT Pennsylvanian
BT Carboniferous
BT Paleozoic
SA Conemaugh Group
SA Kentucky
SA Maryland
SA Pennsylvania
SA West Virginia

amethyst (1978)
Variety of quartz.
BT silica minerals
BT framework silicates
BT silicates
SA corundum
SA gems
SA quartz

Amiidae (1978)
Family. Autoposting of Halecostomi to this term ended in 1989.
BT Holostei
BT Actinopterygii
BT Osteichthyes
BT Pisces
BT Vertebrata
BT Chordata

amino acids (1978)
Autoposting of this term began in 1978.
BT organic materials
NT alanine
NT aspartic acid
NT glycine
NT isoleucine
SA acids
SA fatty acids
SA organic acids
SA peptides
SA proteins
SA racemization

Amisk Group (1978)
Considered as oldest rocks in Flin Flon Belt. Overlain by Missi Group.
BT Archean
BT Precambrian
SA Manitoba
SA Missi Group

Amisk Lake (1978)
Near Manitoba border in NE Saskatchewan.
BT Saskatchewan
BT Western Canada
BT Canada

Amite River (1978)
Rises in SW Mississippi and flows S and E into Lake Maurepas in SE Louisiana.
IN Index states as applicable.
BT United States
SA Louisiana
SA Mississippi

Amitsoq Gneiss (1989)
W Greenland.
BT upper Archean
BT Archean
BT Precambrian
SA Greenland

Ammobaculites (1978)
Genus. As of 1990, microfossils and Protista are autoposted to this term. This term has multiple hierarchies.
BT1 Lituolacea
BT1 Textulariina
BT1 foraminifera
BT1 Protista
BT1 Invertebrata
BT2 Lituolacea
BT2 Textulariina
BT2 foraminifera
BT2 Protista
BT2 microfossils

Ammodiscacea (1978)
As of 1990, microfossils and Protista are autoposted to this term. This term has multiple hierarchies.
BT1 Textulariina
BT1 foraminifera
BT1 Protista
BT1 Invertebrata
BT2 Textulariina
BT2 foraminifera
BT2 Protista
BT2 microfossils
NT Ammodiscidae
NT Astrorhizidae

Ammodiscidae (1978)
Family. As of 1990, microfossils and Protista are autoposted to this term. This term has multiple hierarchies.
BT1 Ammodiscacea
BT1 Textulariina
BT1 foraminifera
BT1 Protista
BT1 Invertebrata
BT2 Ammodiscacea
BT2 Textulariina
BT2 foraminifera
BT2 Protista
BT2 microfossils

Ammonia (1978)
Genus. As of 1990, microfossils and Protista are autoposted to this term. This term has multiple hierarchies.
BT1 Rotaliacea
BT1 Rotaliina
BT1 foraminifera
BT1 Protista
BT1 Invertebrata
BT2 Rotaliacea
BT2 Rotaliina
BT2 foraminifera
BT2 Protista
BT2 microfossils
NT Ammonia beccarii
SA ammonia compound

Ammonia beccarii (1978)
Species. As of 1990, microfossils and Protista are autoposted to this term. This term has multiple hierarchies.
BT1 Ammonia
BT1 Rotaliacea
BT1 Rotaliina
BT1 foraminifera
BT1 Protista
BT1 Invertebrata
BT2 Ammonia
BT2 Rotaliacea
BT2 Rotaliina
BT2 foraminifera
BT2 Protista
BT2 microfossils

ammonia compound (1978)
Use to distinguish from fossil genus, Ammonia. Before 1978, also search ammonia.
SA Ammonia
SA ammonium
SA ammonium ion
SA geochemistry

Ammonites (1978)
BT Ammonoidea
BT Tetrabranchiata
BT Cephalopoda
BT Mollusca
BT Invertebrata
SA aptychi
SA Ceratites

ammonium (1978)
SA ammonia compound
SA geochemistry

ammonium ion (1985)
SA ammonia compound
SA geochemistry
SA ions

Ammonoidea (1978)
Autoposting of this term began in 1978.
BT Tetrabranchiata
BT Cephalopoda
BT Mollusca
BT Invertebrata
NT Ammonites
NT Anarcestida
NT Bactritida
NT Baculites
NT Bouleiceras
NT Ceratitida
NT Clymeniida
NT Dactylioceratidae
NT Desmoceratida
NT Goniatitida
NT Hildoceratacea
NT Lytoceratida
NT Perisphinctida
NT Phylloceratida
NT Prolecanitida
NT Psiloceratida
NT Scaphites
SA ammonoids

ammonoids (1978)
As of 1981, restricted to use as the common name for Ammonoidea. Autoposting of cephalopods and mollusks to this term began in 1989.
BT cephalopods
BT mollusks
BT invertebrates

SA Ammonoidea
SA biostratigraphy

Ammonoosuc Volcanics (1985)
NW New Hampshire, N central Massachusetts, and Vermont.
BT Middle Ordovician
BT Ordovician
BT Paleozoic
SA Massachusetts
SA New Hampshire
SA Vermont

amorphous materials (1978)
General term for all materials not crystalline. Descriptive, not geologic term. Before 1978, also search amorphous AND materials.
SA materials
SA opal-A

amosite (1978)
Commercial term. Autoposting of amphibole group to this term ended in 1985 and began again in 1989.
BT clinoamphibole
BT amphibole group
BT chain silicates
BT silicates
SA anthophyllite
SA asbestos
SA gedrite

Amparo
No longer a valid term for GeoRef. See Amparo Brazil.

Amparo Brazil (1993)
City in E Sao Paulo State.
BT Sao Paulo Brazil
BT Brazil
BT South America

Amphibia (1978)
Class. Autoposting of this term began in 1978.
BT Tetrapoda
BT Vertebrata
BT Chordata
NT Labyrinthodontia
NT Lepospondyli
NT Lissamphibia
SA amphibians

amphibians (1981)
Common name for Amphibia.
IN Index for all non-paleontologic studies of fossils.
BT tetrapods
BT vertebrates
SA Amphibia
SA biostratigraphy

amphibole
As of 1981, no longer a valid term for GeoRef. Was used in subfiles E, G, N, T and B.
use amphibole group

amphibole group (1978)
Autoposting of this term began in 1978. Before 1978, also search amphibole.
UF amphibole
BT chain silicates
BT silicates
NT alkalic amphibole
NT clinoamphibole
NT orthoamphibole
SA amphibolitization
SA asbestos
SA jade

amphibolite
As of 1981, no longer a valid term for GeoRef. From 1978-1980, amphibolites was autoposted to this term.

use amphibolites

amphibolite facies (1978)
BT facies
SA epidote-amphibolite facies
SA metamorphic rocks

amphibolites (1978)
Before 1981, also search amphibolite. Autoposting of this term to amphibolite ended in 1981.
UF amphibolite
BT metamorphic rocks
NT orthoamphibolite
NT para-amphibolite
SA amphibolitization

amphibolitization (1989)
UF amphibolization
BT metasomatism
SA alteration
SA amphibole group
SA amphibolites
SA processes

amphibolization
use amphibolitization

Amphineura
Not a valid term for GeoRef. Use Aplacophora or Polyplacophora.

Amphipora (1989)
Genus.
BT Stromatoporoidea
BT Coelenterata
BT Invertebrata

Amphistegina (1978)
Genus. As of 1990, microfossils and Protista are autoposted to this term. This term has multiple hierarchies.
BT1 Orbitoidacea
BT1 Rotaliina
BT1 foraminifera
BT1 Protista
BT1 Invertebrata
BT2 Orbitoidacea
BT2 Rotaliina
BT2 foraminifera
BT2 Protista
BT2 microfossils

amplitude (1978)
Not to be used to indicate abundance. Strictly for use in seismology.
SA AVO methods
SA elastic waves
SA Hilbert transformations
SA seismic intensity
SA seismology

amplitude distortion (1981)
SA distortion
SA elastic waves
SA seismology

amplitude variation with offset
use AVO methods

amplitude-versus-offset analysis
use AVO methods

Ampurdan (1978)
Coastal plain in Catalonia.
BT Gerona Spain
BT Catalonia Spain
BT Spain
BT Iberian Peninsula
BT Southern Europe
BT Europe

AMS C-14 dating
Not a valid term for GeoRef. As of 1993, use C-14 and accelerator mass spectroscopy and absolute age.

Amsden Formation (1985)
In Montchauve Group. Montana, Wyoming and North Dakota.
BT Carboniferous
BT Paleozoic
SA Mississippian
SA Montana
SA North Dakota
SA Pennsylvanian
SA Wyoming

Amsterdam
No longer a valid term for GeoRef. As of 1993 see Amsterdam Netherlands.

Amsterdam Island (1978)
Dependency of Madagascar 2800 miles off SE African coast. In 1985, broader term changed from Indian Ocean to Indian Ocean Islands.
BT Indian Ocean Islands

Amsterdam Netherlands (1993)
City on Ijsselmeer (former Zuider Zee). North Holland. Before 1993 also search Amsterdam AND Netherlands.
BT Netherlands
BT Western Europe
BT Europe

Amu Darya (1978)
River which rises in Pamirs and flows into Aral Sea.
IN Index Afghanistan and former Soviet republics as applicable.
UF Oxus
BT Asia
SA Afghanistan
SA Amu Darya Basin
SA Surkhan Darya
SA Tadzhikistan
SA Turkmenia
SA Uzbekistan
SA Vakhsh

Amu Darya Basin (1989)
River basin of the Amu Darya.
IN Index former Soviet republics and/or Afghanistan as applicable.
UF Amu Darya Syneclise
BT Asia
SA Afghanistan
SA Amu Darya
SA Tadzhikistan
SA Turkmenia
SA Uzbekistan

Amu Darya Syneclise
use Amu Darya Basin

Amund Ringnes Island (1978)
One of the Queen Elizabeth Islands in Franklin District.
BT Northwest Territories
BT Western Canada
BT Canada
SA Queen Elizabeth Islands

Amur Basin (1978)
River basin covering an area including parts of Amur and Chita oblasts, and Khabarovsk region in the Russian Federation; parts of Inner Mongolia, and N China; and E Mongolia. Before 1981, also search Amur; Amur region.
IN Index Chinese provinces, Mongolia, or Russian Federation as applicable.
CO N470000N540000 E1410000E1210000
BT Asia
SA Amur region
SA Inner Mongolia China
SA Mongolia
SA Russian Far East
SA Russian Federation

Amur region (1981)
SE Russian Federation. Larger than the oblast. This term has multiple hierarchies.
CO N473000N553000 E1390000E1213000
BT1 Russian Federation
BT1 Commonwealth of Independent States
BT2 Asia
SA Amur Basin
SA Baikal-Amur Railroad region
SA Bureya Russian Federation
SA Khabarovsk Russian Federation
SA Stanovoy Range
SA Udokan Series
SA Zeya-Bureya Basin

Amur River (1978)
Formed at junction of the Shilka River and the Argun River. After serving as the boundary between N China and 2 oblasts of the Russian Federation, it flows NE across Khabarovsk region into the N end of the Tatar Strait. Also search Amur. In China, the river is also called the Hei-lung Chiang or Heilong Jiang.
IN Index Chinese provinces and/or Russian Federation as applicable.
BT Asia
SA Argun River
SA Heilongjiang China
SA Inner Mongolia China
SA Russian Federation

Amur Russian Federation (1993)
Oblast. Before 1993, also search Amur or Amur region. This term has multiple hierarchies.
BT1 Russian Federation
BT1 Commonwealth of Independent States
BT2 Asia
NT Bureya Russian Federation
NT Zeya-Bureya Basin
SA Russian Far East
SA Transbaikalia

amygdules (1989)
Includes use as a general term.
SA agate
SA igneous rocks
SA pebbles
SA secondary minerals
SA sediments

Anabar Anteclise
use Anabar Shield

Anabar Bay (1978)
On the Laptev Sea in NW Yakutia. Also search Anabar. This term has multiple hierarchies.
BT1 Yakutia Russian Federation
BT1 Russian Federation
BT1 Commonwealth of Independent States
BT2 Yakutia Russian Federation
BT2 Asia

Anabar Massif
use Anabar Shield

Anabar River (1978)
Flows into Anabar Bay in NW Yakutia. Also search Anabar. This term has multiple hierarchies.
BT1 Yakutia Russian Federation
BT1 Russian Federation
BT1 Commonwealth of Independent States
BT2 Yakutia Russian Federation
BT2 Asia

Anabar Shield (1978)
Part of the Siberian Platform in NW central Yakutia. Also search Anabar. This term has multiple hierarchies.
CO N650000N720000 E1230000E0960000
UF Anabar Anteclise
UF Anabar Massif
BT1 Yakutia Russian Federation
BT1 Russian Federation
BT1 Commonwealth of Independent States
BT2 Yakutia Russian Federation
BT2 Asia
BT3 Siberian Platform
BT3 Russian Federation
BT3 Commonwealth of Independent States
BT4 Siberian Platform
BT4 Asia
SA Udachnaya Pipe

Anacostia River basin (1978)
Also search Anacostia River.
IN Index Maryland and/or District of Columbia as applicable.
BT United States
SA District of Columbia
SA Maryland

Anadarko Basin (1978)
IN Index states as applicable.
BT United States
SA Arbuckle Anticline
SA Kansas
SA Oklahoma
SA Texas

Anadyr Basin (1978)
River basin in Chukchi National Okrug in NE Siberia. Also search Anadyr. This term has multiple hierarchies.
BT1 Russian Federation
BT1 Commonwealth of Independent States
BT2 Asia
SA Chukchi Peninsula

Anadyr Range (1978)
Extends SE from Chaun Bay into W Chukchi Peninsula in extreme NE Siberia. Also search Anadyr. This term has multiple hierarchies.
UF Chukchi Range
BT1 Russian Federation
BT1 Commonwealth of Independent States
BT2 Asia

anaerobic environment (1978)
Before 1978, search anaerobic.
UF anoxic environment
SA aerobic environment
SA anaerobic taxa
SA deep-water environment
SA depositional environment
SA environment
SA paleoecology
SA sapropel

anaerobic taxa (1978)
Before 1978, search anaerobic.
UF taxa, anaerobic
SA anaerobic environment
SA ecology
SA fossils
SA paleoecology

Anagalida (1981)
Family. Autoposting of Proteutheria, Insectivora, Eutheria and Theria to this term began in 1989.
BT Proteutheria
BT Insectivora

BT Eutheria
BT Theria
BT Mammalia
BT Tetrapoda
BT Vertebrata
BT Chordata

Anahuac Formation (1985)
Upper Oligocene or lower Miocene. Louisiana and Texas.
BT Tertiary
BT Cenozoic
SA Louisiana
SA lower Miocene
SA Texas
SA upper Oligocene

analbite (1978)
Autoposting of feldspar group to this term ended in 1985 and began again in 1989.
BT plagioclase
BT feldspar group
BT framework silicates
BT silicates

analcime (1978)
Autoposting of zeolite group to this term began in 1981.
UF analcite
BT zeolite group
BT framework silicates
BT silicates

analcite
  use analcime

analog filters (1981)
SA continuous filters
SA discrete filters
SA filters

analog simulation (1978)
Also search analog AND automatic data processing; analog AND data processing.
UF analog techniques
SA data processing
SA mathematical geology
SA models
SA simulation

analog techniques
As of 1981, no longer a valid term for GeoRef.
  use analog simulation

analyses
A valid term through 1974. After 1974, use analysis.

analysis (1978)
Before 1975, also search analyses.
IN For data index material name.
SA activation analysis
SA alpha-ray spectroscopy
SA atomic absorption
SA Auger spectroscopy
SA automated analysis
SA autoradiography
SA Backus-Gilbert analysis
SA Bayesian analysis
SA boundary element analysis
SA canonical analysis
SA chemical analysis
SA chemical composition
SA chromatography
SA cluster analysis
SA colorimetry
SA correspondence analysis
SA covariance analysis
SA differential thermal analysis
SA discriminant analysis
SA electrolytic analysis
SA electron diffraction analysis
SA electron paramagnetic resonance
SA electron probe
SA emission spectroscopy
SA environmental analysis
SA factor analysis
SA finite difference analysis
SA finite element analysis
SA flame photometry
SA fluid inclusions
SA Fourier analysis
SA frequency domain analysis
SA gamma-ray spectroscopy
SA Hilbert transformations
SA Hugoniot analysis
SA infrared spectroscopy
SA ion chromatography
SA ion probe
SA isotopes
SA laser methods
SA laser ranging
SA least-squares analysis
SA major-element analyses
SA microwave spectroscopy
SA minor-element analyses
SA modal analysis
SA Mossbauer spectroscopy
SA multispectral analysis
SA multivariate analysis
SA neutron activation analysis
SA neutron diffraction analysis
SA neutron probe
SA nuclear magnetic resonance
SA numerical analysis
SA optical spectroscopy
SA pattern recognition
SA polarography
SA pollen analysis
SA qualitative analysis
SA quantitative analysis
SA radio-frequency spectroscopy
SA Raman spectroscopy
SA regression analysis
SA sampling
SA shape analysis
SA spectral analysis
SA spectroscopy
SA spherical harmonic analysis
SA statistical analysis
SA structural analysis
SA systems analysis
SA thermal analysis
SA thermogravimetric analysis
SA thermomagnetic analysis
SA Thomson-Haskell analysis
SA time domain analysis
SA time series analysis
SA trace-element analyses
SA trend-surface analysis
SA ultraviolet spectroscopy
SA univariate analysis
SA vacuum fusion analysis
SA variance analysis
SA velocity analysis
SA wet methods
SA Wiener-Hopf analysis
SA X-ray analysis
SA X-ray diffraction analysis
SA X-ray fluorescence
SA X-ray spectroscopy

analytical
A valid index term through 1977. See analysis, analytical methods.

analytical data
A valid term through 1975. After 1975, use data.

analytical methods
As of 1982, no longer a valid term for GeoRef. A valid term from 1978-1981. Before 1978. also search analytical AND methods.

Anantapur
No longer a valid term for GeoRef. See Anantapur India.

Anantapur India (1993)
City and district in SW Andhra Pradesh, S India.
BT Andhra Pradesh India
BT India
BT Indian Peninsula
BT Asia

Anantnag (1978)
Town and district in S central Jammu and Kashmir, disputed territory between India and Pakistan. As of 1993, India is no longer a broader term.
IN Index Jammu and Kashmir; and India and/or Pakistan as applicable.
SA India
SA Jammu and Kashmir
SA Pakistan

Anapa
No longer a valid term for GeoRef. As of 1993, see Anapa Russian Federation.

Anapa Russian Federation (1993)
Town in Krasnodar Kray in the Northern Caucasus. Before 1993, also search Anapa. This term has multiple hierarchies.
BT1 Krasnodar Russian Federation
BT1 Russian Federation
BT1 Commonwealth of Independent States
BT2 Krasnodar Russian Federation
BT2 Europe

Anapsida (1978)
Subclass. Autoposting of this term to Proganosauria began in 1989.
BT Reptilia
BT Tetrapoda
BT Vertebrata
BT Chordata
NT Chelonia
NT Cotylosauria
NT Mesosauria

Anarak
No longer a valid term for GeoRef. See Anarak Iran.

Anarak Iran (1993)
Town in Esfahan, central Iran.
BT Iran
BT Middle East
BT Asia

Anarcestida (1981)
BT Ammonoidea
BT Tetrabranchiata
BT Cephalopoda
BT Mollusca
BT Invertebrata

Anasco Bay (1978)
On Mona Passage on the W coast.
BT Puerto Rico
BT Greater Antilles
BT Antilles
BT West Indies
BT Caribbean region

anastomosing streams
  use braided streams

anatase (1978)
UF octahedrite (mineral)
BT oxides
SA brookite
SA rutile

anatectite
  use anatexite

anatexis (1978)
SA assimilation
SA contamination
SA magmas
SA metamorphism
SA palingenesis

anatexite (1978)
From 1978-1980, metamorphic rocks was autoposted to this term. Autoposting of migmatites to this term began in 1981.
UF anatectite
BT migmatites
BT metamorphic rocks

Anatolia (1978)
In W and W central Turkey.
CO N360500N420200
   E0445000E0261000
BT Turkey
BT Middle East
BT Asia
SA Adana Turkey
SA Alanya Turkey
SA Amasra Basin
SA Ankara Turkey
SA Antalya Turkey
SA Elbistan Turkey
SA Eskisehir Turkey
SA Hatay Turkey
SA Kocaeli Turkey
SA Konya Basin
SA Konya Turkey
SA Mudurnu earthquake 1967
SA Tuz Golu

anatomy (1982)
SA bones
SA extremities
SA functional morphology
SA jaws
SA morphology
SA muscles
SA nervous system
SA paleontology
SA physiology
SA skeletons
SA skulls
SA spinal column
SA teeth

Ancenis
No longer a valid term for GeoRef. See Ancenis France.

Ancenis France (1993)
Town on the Loire River in W France.
BT Loire-Atlantique France
BT France
BT Western Europe
BT Europe

anchimetamorphism (1982)
BT metamorphism
SA ground water
SA weathering

Anchorage
Valid through 1988. Search in combination with state term. After 1988, use specific city-state term.

Anchorage Alaska (1989)
City in Cook Inlet in S Alaska. Before 1989, search Anchorage AND Alaska.
CO N611000N611000
   W1500000W1500000
UF Anchorage region
BT Southern Alaska
BT Alaska
BT United States
SA Cook Inlet

Anchorage region
  use Anchorage Alaska

anchors (1978)
SA engineering geology

ancient
A valid term through 1977 used with ice ages (i.e. After 1977, use ancient ice ages.

ancient ice ages (1978)
Before 1978, search ice ages AND ancient.
SA glacial geology

Ancona
No longer avalid term for GeoRef. See Ancona Italy.

Ancona Italy (1993)
City and province on the Adriatic Sea.
BT Marches Italy
BT Italy
BT Southern Europe
BT Europe

Ancylus Lake (1989)
Stage of Baltic Glaciation. W Finland and N Sweden.
IN Index countries as applicable.
BT lower Holocene
BT Holocene
BT Quaternary
BT Cenozoic
SA Baltic Glaciation
SA Baltic region
SA Baltic Sea
SA Finland
SA Sweden

Andalusia
No longer a valid term for GeoRef. As of 1993, see Andalusia Spain.

Andalusia Spain (1993)
Eight provinces in S and SW Spain. Provinces are considered narrower terms as of 1981. Before 1993, also search Andalusia and Spain.
CO N360000N385000
    W0004000W0073000
BT Spain
BT Iberian Peninsula
BT Southern Europe
BT Europe
NT Almeria Spain
NT Cadiz Spain
NT Cordoba Spain
NT Granada Spain
NT Huelva Spain
NT Jaen Spain
NT Malaga Spain
NT Serrania de Ronda
NT Seville Spain
SA Andalusian

Andalusian (1978)
BT Triassic
BT Mesozoic
SA Andalusia Spain
SA Betic Cordillera
SA Spain

andalusite (1978)
Autoposting of nesosilicates began in 1985. As of 1981, use andalusite deposits for andalusite as a commodity.
BT nesosilicates
BT orthosilicates
BT silicates
SA andalusite deposits
SA kyanite
SA sillimanite

andalusite deposits (1981)
Before 1981, search andalusite AND deposits.
IN May be used as a commodity term with ceramic materials.
SA andalusite
SA ceramic materials

Andaman Basin (1978)
Between the Andaman and the Nicobar islands.
BT Bay of Bengal
BT Indian Ocean

Andaman Islands (1978)
E Bay of Bengal, N of Nicobar Islands. Andaman and Nicobar Territory.
CO N103000N134500
    E0930000E0922000
BT Bengal Islands
BT India
BT Indian Peninsula
BT Asia
SA Neill Island

Andaman Sea (1978)
W of Burmese and Thai section of Malay Peninsula. Before 1981, Bay of Bengal was included as a broader term.
CO N050000N170000
    E0980000E0923000
BT Indian Ocean
SA Bay of Bengal

Andean Geosyncline (1978)
Before 1978, search geosynclines AND Andean.
SA Andean Orogeny
SA Andes
SA geosynclines

Andean Orogeny (1978)
Before 1978, search orogeny AND Andean.
SA Andean Geosyncline
SA Andes
SA orogeny
SA tectonics

Andepts (1985)
BT Inceptisols
BT soils

Anderson County
Valid through 1988. Search in combination with state term. After 1988, use specific county-state term.

Anderson County Kansas (1989)
E Kansas. Before 1989, also search Anderson County AND Kansas.
CO N380300N382300
    W0950400W0953000
BT Kansas
BT United States
NT Garnett Kansas

Anderson County Kentucky (1989)
Central Kentucky. Before 1989, also search Anderson County AND Kentucky.
CO N375200N380800
    W0845000W0851000
BT Kentucky
BT United States

Anderson County South Carolina (1989)
NW South Carolina. Before 1989, also search Anderson County AND South Carolina.
CO N341300N345000
    W0822000W0825900
BT South Carolina
BT United States

Anderson County Tennessee (1989)
E Tennessee. Before 1989, also search Anderson County AND Tennessee.
CO N355600N361600
    W0835700W0842700
BT Tennessee
BT United States
NT Walker Branch watershed

Anderson County Texas (1989)
E Texas. Before 1989, also search Anderson County AND Texas.
CO N313000N320500
    W0951400W0960300
BT Texas
BT United States

Andes (1978)
Mountain system extending along W coast from Venezuela and Colombia to Tierra del Fuego. Autoposting of this term began in 1978.
IN Index countries as applicable.
UF Andes Mountains
BT South America
NT Argentine Andes
NT Central Andes
NT Eastern Cordillera
NT Northern Andes
NT Patagonian Andes
NT Sierra de Perija
NT Southern Andes
NT Venezuelan Andes
NT Western Cordillera
SA Altiplano
SA Andean Geosyncline
SA Andean Orogeny
SA Argentina
SA Bolivia
SA Caribbean Mountain Range
SA Chile
SA Colombia
SA Ecuador
SA Peru
SA Venezuela

Andes Mountains
use Andes

andesine (1978)
BT plagioclase
BT feldspar group
BT framework silicates
BT silicates
SA albite
SA anorthite

andesite
As of 1989, no longer a valid term for GeoRef. From 1981 to 1988, andesites and volcanic rocks were autoposted to this term.
use andesites

andesite basalt
Not a valid term for GeoRef. See basalt.

andesite porphyry (1978)
BT andesites
BT volcanic rocks
BT igneous rocks
SA porphyry

andesite tuff (1978)
UF andesitic tuff
BT pyroclastics
BT volcanic rocks
BT igneous rocks
SA tuff

andesite-rhyolite family
As of 1981, no longer a valid term for GeoRef. Use andesites, trachyandesites, dacites, rhyodacites, or rhyolites as applicable. From 1978-1980, was autoposted to the rocks that were classified under it.

andesites (1981)
Before 1981, search andesite-rhyolite family. From 1978-1980, andesite-rhyolite family was autoposted to the rocks that were classified under that term. Before 1989, also search andesite. Autoposting of this term to andesite ended in 1989.
UF andesite
BT volcanic rocks
BT igneous rocks
NT andesite porphyry
NT boninite
NT pyroxene andesite
SA andesitic composition
SA propylite

andesitic composition (1978)
Before 1978, search andesitic AND composition.
SA andesites
SA composition
SA igneous rocks

andesitic tuff
use andesite tuff

Andhra Pradesh
No longer a valid term for GeoRef.
use Andhra Pradesh India

Andhra Pradesh India (1993)
State in central India. Before 1993, also search Andhra Pradesh.
CO N130000N200000
    E0845000E0764000
UF Andhra Pradesh
BT India
BT Indian Peninsula
BT Asia
NT Anantapur India
NT Cuddapah Basin
NT Cuddapah India
NT Guntur India
NT Hyderabad India
NT Khammam India
NT Kurnool India
NT Nellore India
NT Prakasam India
NT Rajahmundry India
NT Srikakulam India
NT Tirupati India
NT Visakhapatnam India
SA Aravalli System
SA Coromandel Coast
SA Cuddapah System
SA Deccan Plateau
SA Godavari River
SA Godavari Valley
SA Krishna
SA Krishna-Godavari Basin
SA Kurnool System
SA Nellore mica belt
SA Pranhita-Godavari Valley
SA Raghavapuram Shales

Ando soils
use Andosols

Andorra (1978)
A principality in E Pyrenees between France and Spain.
CO N422500N424000
    E0014500E0012500
BT Western Europe
BT Europe
SA Pyrenees

Andosols (1978)
UF Ando soils
BT soils
SA soil group

andradite (1978)
BT garnet group
BT nesosilicates
BT orthosilicates

BT silicates

Andreanof Islands earthquake 1986 (1989)
Epicenter near the Andreanof Islands, Aleutian Islands, SW Alaska.
BT earthquakes
SA Alaska
SA Aleutian Islands

Andrews County
Valid through 1988. Search in combination with state term. After 1988, use specific county-state term.

Andrews County Texas (1989)
W Texas. Before 1989, also search Andrews County AND Texas.
CO N320500N323000
W1021000W1030000
BT Texas
BT United States
SA Central Basin Platform

Andriamena
No longer a valid term for GeoRef. As of 1993 see Andriamena Madagascar.

Andriamena Madagascar (1993)
Village in N central Madagascar. Before 1993 also search Andriamena AND (Madagascar OR Malagasy Republic). This term has multiple hierarchies.
BT1 Madagascar
BT1 Indian Ocean Islands
BT2 Madagascar
BT2 Africa

Andros Island (1978)
Large island in W Bahamas.
CO N233000N250000
W0773000W0783000
BT Bahamas
BT West Indies
BT Caribbean region

Androscoggin County
Valid through 1988. Search in combination with state term. After 1988, use specific county-state term.

Androscoggin County Maine (1989)
SW Maine. Before 1989, also search Androscoggin County AND Maine.
CO N435200N443000
W0700000W0703000
BT Maine
BT United States

Andrychow (1978)
Borehole. After 1993, use Andrychow Poland for the town.
BT Poland
BT Central Europe
BT Europe
SA Andrychow Poland

Andrychow Poland (1993)
Town in S Poland. Before 1993 also search Andrychow AND Poland.
BT Bielsko Poland
BT Poland
BT Central Europe
BT Europe
SA Andrychow
SA Cracow Poland

Anegada Island (1978)
Northernmost island. British Virgin Islands.
BT British Virgin Islands

BT Virgin Islands
BT Lesser Antilles
BT Antilles
BT West Indies
BT Caribbean region

anelastic materials (1981)
Before 1981, also search anelastic media; anelastic.
UF anelastic media
SA materials
SA seismology

anelastic media
As of 1981, no longer a valid term for GeoRef.
use anelastic materials

anelasticity (1978)
SA brittle deformation
SA deformation
SA elasticity
SA seismology

Angara
use Angara River

Angara River (1978)
Flows from Lake Baikal into Yenisei River. Formerly called Upper Tunguska in its lower course. This term has multiple hierarchies.
UF Angara
UF Upper Tunguska
BT1 Russian Federation
BT1 Commonwealth of Independent States
BT2 Asia
SA Tunguska
SA Tunguska River

Angara-Lena Basin (1978)
Adjoining basin of two rivers in Irkutsk Oblast W and NW of Lake Baikal. This term has multiple hierarchies.
CO N520000N630000
E1150000E0950000
BT1 Siberian Platform
BT1 Russian Federation
BT1 Commonwealth of Independent States
BT2 Siberian Platform
BT2 Asia
SA Irkutsk Russian Federation
SA Lena Basin

angiosperm flora (1982)
Common name for angiosperms.
IN Index for all non-paleontologic studies of fossils.
BT plants
SA angiosperms
SA biostratigraphy

angiosperms (1978)
Autoposting of this term began in 1978.
BT Spermatophyta
BT Plantae
NT Dicotyledoneae
NT Monocotyledoneae
SA angiosperm flora
SA Florinites
SA fossil wood
SA gymnosperms
SA Pteropsida
SA seeds
SA sporangia

angle of dip
use dip

Anglesea
use Anglesey

Anglesey (1978)

Island in N Wales. Before 1978, also search Anglesea. As of April 1974, part of Gwynedd County. As of 1993, see Anglesey Wales for the former county.
CO N531000N533000
W0040000W0044000
UF Anglesea
BT Anglesey Wales
BT Wales
BT Great Britain
BT United Kingdom
BT Western Europe
BT Europe

Anglesey Wales (1993)
Former county in N Wales including the island. As April 1974, part of Gwynedd County. Before 1993, also search Anglesey or Anglesea.
CO N531000N533000
W0040000W0044000
BT Wales
BT Great Britain
BT United Kingdom
BT Western Europe
BT Europe
NT Anglesey

anglesite (1978)
BT sulfates

Anglian (1989)
W Europe.
BT Quaternary
BT Cenozoic

Angola (1978)
Peoples Republic of Angola.
CO S180000S060000
E0230000E0120000
UF Portuguese West Africa
BT Central Africa
BT Africa
NT Cabinda Angola
NT Catanda Angola
NT Cuanza Basin
NT Cuanza-Sul Angola
NT Luanda Angola
NT Mossamedes
NT Namibe Angola
SA Congo Basin
SA Congo River
SA Kasai River
SA Zambezi Valley

Angola abyssal plain
use Angola Basin

Angola Basin (1985)
SE Atlantic Ocean. As of 1990, Atlantic Ocean is autoposted to this term.
CO S300000S030000
E0080000W0080000
UF Angola abyssal plain
BT Southeast Atlantic
BT South Atlantic
BT Atlantic Ocean
SA DSDP Site 364
SA DSDP Site 365
SA DSDP Site 522
SA DSDP Site 523
SA Leg 40

Angora Turkey
use Ankara Turkey

Angouleme
No longer a valid term for GeoRef. See Angouleme France.

Angouleme France (1993)
City in W France.
BT Charente France
BT France
BT Western Europe
BT Europe

Angra dos Reis
No longer a valid term for GeoRef. See Angra dos Reis Brazil. Use Angra dos Reis Meteorite and achondrites if applicable.

Angra dos Reis Brazil (1993)
City in SW Rio de Janeiro.
BT Rio de Janeiro Brazil
BT Brazil
BT South America

Angren
No longer a valid term for GeoRef. As of 1993, see Angren Uzbekistan.

Angren Uzbekistan (1993)
City in Tashkent Oblast. This term has multiple hierarchies.
BT1 Tashkent Uzbekistan
BT1 Uzbekistan
BT1 Asia
BT2 Tashkent Uzbekistan
BT2 Uzbekistan
BT2 Commonwealth of Independent States

angrite (1981)
BT achondrites
BT stony meteorites
BT meteorites

angular momentum (1989)
Used as a general term, sometimes in relation to atmospheric circulation.
SA atmosphere
SA circulation
SA convection
SA Earth
SA motions

angular unconformities (1982)
BT unconformities

Angus Scotland (1993)
Former county on the North Sea. Before 1978, also search Forfar or Forfarshire. Became part of Tayside region Scotland in 1974.
CO N563000N570000
W0023000W0032500
UF Forfarshire
BT Scotland
BT Great Britain
BT United Kingdom
BT Western Europe
BT Europe

Anhui
No longer a valid term for GeoRef. See Anhui China.

Anhui China (1993)
Province. E China. Before 1985, also search Anhwei.
CO N293000N343500
E1192500E1150000
UF Anhwei China
BT China
BT Far East
BT Asia
NT Qianshan China
SA Dabie Mountains
SA Taiyuan Formation
SA Tancheng-Lujiang Fault
SA Tancheng-Lujiang fault zone
SA Tangshan China
SA Wufeng Formation
SA Yangtze Platform
SA Yishu Fault

Anhwei China
use Anhui China

anhydrite (1978)
As of 1981, use anhydrite deposits for anhydrite as a commodity.

IN   Refers to both the rock and the mineral.
UF   cube spar
BT   sulfates
SA   anhydrite deposits
SA   calcium sulfate
SA   cap rocks
SA   chemically precipitated rocks
SA   evaporites
SA   gypsum
SA   sedimentary rocks
SA   sediments

anhydrite deposits   (1981)
Before 1981, search anhydrite AND deposits.
SA   anhydrite
SA   evaporite deposits
SA   gypsum deposits

anhysteretic remanent magnetization   (1978)
Autoposting of magnetization to this term began in 1989.
BT   remanent magnetization
BT   magnetization
SA   paleomagnetism

anilite   (1978)
BT   sulfides

animals
As of 1981, no longer a valid term for GeoRef. Before 1981, used as a general term.

Animas Mountains   (1978)
Hidalgo County in extreme SW New Mexico.
BT   New Mexico
BT   United States
SA   Hidalgo County New Mexico

Animikie Group   (1978)
Includes Pokegama Quartzite, Biwabik Iron Formation Series, and Virginia Formation, N Michigan, NE Minnesota, Ontario, and N Wisconsin.
BT   Precambrian
SA   Biwabik Iron Formation
SA   Gunflint Iron Formation
SA   Michigan
SA   Minnesota
SA   Ontario
SA   Virginia Formation
SA   Wisconsin

anion exchange-spectrochemical analysis
Not a valid GeoRef index term after 1970. Was used on level 1 in subfile N.

anions   (1978)
SA   cations
SA   electrolytes
SA   geochemistry
SA   ion exchange
SA   ions

Anisian   (1978)
Europe. Above Scythian, below Ladinian.
BT   Middle Triassic
BT   Triassic
BT   Mesozoic

anisotropic materials   (1978)
Before 1978, search anisotropic.
SA   anisotropy
SA   materials

anisotropy   (1978)
UF   aeolotropy
SA   anisotropic materials
SA   magnetic susceptibility
SA   mechanical properties
SA   seismology

Anjou   (1978)
Old province in W France.
BT   France
BT   Western Europe
BT   Europe

Anjouan Island   (1978)
One of the French Comoro Islands in the Mozambique Channel. In 1985, broader term changed from Indian Ocean to Indian Ocean Islands.
UF   Johanna Island
BT   Comoro Islands
BT   Indian Ocean Islands

Ankara
No longer a valid term for GeoRef. As of 1993 see Ankara Turkey.

Ankara Turkey   (1993)
Province in W central Anatolia. Also a city formerly called Angora. Before 1993 also search Ankara AND Turkey.
UF   Angora Turkey
BT   Turkey
BT   Middle East
BT   Asia
SA   Anatolia

ankaramite   (1978)
BT   basalts
BT   volcanic rocks
BT   igneous rocks
SA   ultramafics

ankaratrite   (1978)
BT   alkali basalts
BT   basalts
BT   volcanic rocks
BT   igneous rocks
SA   nepheline basalt
SA   olivine nephelinite

ankerite   (1978)
UF   cleat spar
UF   ferroan dolomite
RT   carbonates
SA   dolomite

Ann Arbor
Valid through 1988. Search in combination with state term. After 1988, use specific city-state term.

Ann Arbor Michigan   (1989)
City in SE Michigan. Before 1989, search Ann Arbor AND Michigan.
CO   N421800N421800
     W0834300W0834300
BT   Washtenaw County Michigan
BT   Michigan Lower Peninsula
BT   Michigan
BT   United States

Annaba
No longer a valid term for GeoRef. As of 1993, see Annaba Algeria.

Annaba Algeria   (1993)
Department on the Mediterranean Sea. Also a city which was formerly called Bone.
BT   Algeria
BT   North Africa
BT   Africa

Annapolis Valley   (1978)
River valley in SW Nova Scotia. Also search Annapolis.
IN   Index Nova Scotia if applicable.
SA   Nova Scotia

Anne Arundel County
Valid through 1988. Search in combination with state term. After 1988, use specific county-state term.

Anne Arundel County Maryland   (1989)
Central Maryland. Before 1989, also search Anne Arundel County AND Maryland.
CO   N384300N391400
     W0762400W0765100
BT   Maryland
BT   United States

annealing   (1989)
Process for strengthening materials.
SA   alloys
SA   graphitization
SA   metal ores
SA   metallurgy

Annecy
No longer a valid term for GeoRef. See Annecy France.

Annecy France   (1993)
Town S of Lake Geneva in W Haute-Savoie.
BT   Haute-Savoie France
BT   France
BT   Western Europe
BT   Europe

Annelida   (1978)
Before 1981, Chaetopoda, Polychaetia, Myzostomia, Oligochaetia and Serpulidae were indexed under this term.
BT   worms
BT   Invertebrata
NT   Chaetopoda
SA   scolecodonts

Annonaceae   (1978)
Family.
BT   Dicotyledoneae
BT   angiosperms
BT   Spermatophyta
BT   Plantae

Annot Sandstone   (1978)
Priabonian (Bartonian). In the Alps of SE France.
BT   Eocene
BT   Paleogene
BT   Tertiary
BT   Cenozoic
SA   France

annual growth rings
use tree rings

annual meeting
As of 1981, no longer a valid term for GeoRef. See symposia.

annual report   (1978)
BT   report
SA   associations
SA   current research
SA   industry
SA   progress report
SA   survey organizations
SA   symposia

annual variations   (1985)
SA   magnetic field
SA   seasonal variations
SA   secular variations
SA   variations

annually thawed layer
use active layer

Anoka County
Valid through 1988. Search in combination with state term. After 1988, use specific county-state term.

Anoka County Minnesota   (1989)
E Minnesota. Before 1989, also search Anoka County AND Minnesota.
CO   N450200N452500
     W0930200W0933100
BT   Minnesota
BT   United States

anomalies   (1978)
SA   Bouguer anomalies
SA   electrical anomalies
SA   free-air anomalies
SA   geochemical anomalies
SA   gravity anomalies
SA   leakage anomalies
SA   magnetic anomalies
SA   regional anomalies
SA   residual anomalies
SA   thermal anomalies

Anomalinidae   (1978)
Family. As of 1990, microfossils and Protista are autoposted to this term. This term has multiple hierarchies.
BT1   Cassidulinacea
BT1   Rotaliina
BT1   foraminifera
BT1   Protista
BT1   Invertebrata
BT2   Cassidulinacea
BT2   Rotaliina
BT2   foraminifera
BT2   Protista
BT2   microfossils

anorthite   (1978)
BT   plagioclase
BT   feldspar group
BT   framework silicates
BT   silicates
SA   andesine

anorthoclase   (1978)
BT   alkali feldspar
BT   feldspar group
BT   framework silicates
BT   silicates
SA   orthoclase

anorthosite   (1978)
UF   plagioclasite
BT   gabbros
BT   plutonic rocks
BT   igneous rocks
SA   gabbroic anorthosite
SA   meta-anorthosite

anorthositic gabbro   (1978)
BT   gabbros
BT   plutonic rocks
BT   igneous rocks

anoxic environment
use anaerobic environment

Ansbach
No longer a valid term for GeoRef. See Ansbach Germany.

Ansbach Germany   (1993)
City in W Bavaria.
BT   Bavaria Germany
BT   Germany
BT   Central Europe
BT   Europe

Anshan
No longer a valid term for GeoRef. See Anshan China.

Anshan China   (1993)
NE China. Independent municipality located in Liaoning, S Manchuria.
BT   Liaoning China
BT   China
BT   Far East
BT   Asia

Anshan Group   (1989)
NE China.
BT   Precambrian
SA   China
SA   Jilin China

SA Liaoning China
SA Shandong China

Antalya
No longer a valid term for GeoRef.
As of 1993, see Antalya Turkey.

Antalya Complex (1989)
BT Mesozoic
SA Turkey

Antalya Turkey (1993)
Province in S Anatolia on Gulf of
Adalia. Also a city formerly called
Adalia.
UF Adalia Turkey
BT Turkey
BT Middle East
BT Asia
SA Anatolia
SA Taurus Mountains

Antananarivo Madagascar
(1993)
City on the Indian Ocean in E
Madagascar. Before 1993 also
search Tananarive OR An-
tananarivo. This term has multiple
hierarchies.
UF Tananarive Madagascar
BT1 Madagascar
BT1 Indian Ocean Islands
BT2 Madagascar
BT2 Africa

Antarctic bottom water (1985)
Cold, deep water mass. Circulates
counterclockwise (westward) at
the foot of the Antarctic continen-
tal shelf.
UF AABW
BT Antarctic Ocean
SA bottom water
SA Brazil Basin
SA DSDP Site 515

Antarctic Continent
use Antarctica

Antarctic ice sheet (1993)
BT Antarctica
SA Elephant Moraine

Antarctic meteorites
As of 1993, use Antarctica AND
meteorites.

Antarctic Ocean (1978)
An arbitrary definition, considered
the equivalent of Southern Ocean.
Waters about Antarctica are
sometimes called the Antarctic
Ocean, but actually are only the
parts of the Atlantic Ocean, Pa-
cific Ocean and Indian Ocean
south of approximately 55° S. Also
search Antarctic. Before 1981,
South Sandwich Islands was in-
cluded as a narrower term.
IN Index oceans as applicable.
UF Southern Ocean
NT Antarctic bottom water
NT Bellingshausen Sea
NT Bransfield Strait
NT Drake Passage
NT McMurdo Sound
NT Mid-Antarctic Ridge
NT Prydz Bay
NT Ross Sea
NT Scotia Sea
NT Weddell Sea
SA Antarctica
SA Atlantic Ocean
SA DSDP Site 266
SA DSDP Site 270
SA DSDP Site 271
SA DSDP Site 272
SA DSDP Site 273
SA DSDP Site 274

SA DSDP Site 322
SA DSDP Site 323
SA DSDP Site 325
SA Indian Ocean
SA Leg 28
SA Leg 29
SA Leg 35
SA Leg 36
SA Leg 113
SA Leg 114
SA Leg 119
SA Leg 120
SA ODP Site 689
SA ODP Site 690
SA ODP Site 691
SA ODP Site 692
SA ODP Site 693
SA ODP Site 694
SA ODP Site 695
SA ODP Site 696
SA ODP Site 697
SA ODP Site 738
SA ODP Site 739
SA ODP Site 741
SA ODP Site 743
SA ODP Site 744
SA ODP Site 745
SA ODP Site 746
SA ODP Site 748
SA ODP Site 749
SA ODP Site 750
SA ODP Site 751
SA Pacific Ocean
SA Pacific-Antarctic Ridge
SA South Sandwich Islands
SA subantarctic regions

Antarctic Peninsula (1978)
Peninsula jutting northward to-
ward Cape Horn.
UF Palmer Peninsula
BT Antarctica

Antarctic Plate (1978)
SA Antarctica
SA plate tectonics
SA plates

Antarctic Platform (1978)
Central area which constitutes a
high plateau. Also search Antarctic.
BT Antarctica

Antarctic Polar Cap (1981)
Before 1981, also search Antarc-
tica AND Polar Cap.
BT Antarctica
SA polar caps
SA South Pole

Antarctic region
Not a valid GeoRef index term
after 1971. Was used on level 1 in
subfile G. Now use Antarctica or
Antarctic Ocean as applicable.

Antarctica (1978)
Continent centering on the South
Pole. To retrieve all documents,
individual countries and physio-
graphic regions should also be
searched (see list O). Also search
Antarctic. Autoposting of this term
to Dry Valley Drilling Project
began in 1981. South Georgia,
Signy Island, South Sandwich Is-
lands and Kerguelen Islands are
narrower terms as of 1981.
CO S900000S610000
      W1800000E1800000
UF Antarctic Continent
NT Adelaide Island
NT Adelie Coast
NT Alexander Island
NT Antarctic ice sheet
NT Antarctic Peninsula
NT Antarctic Platform

NT Antarctic Polar Cap
NT Anvers Island
NT Beardmore Glacier
NT Byrd Station
NT Coalsack Bluff
NT Don Juan Pond
NT Dry Valley Drilling Project
NT Dufek Intrusion
NT East Antarctica
NT East Ongul Island
NT Elephant Moraine
NT Ellsworth Land
NT Ellsworth Mountains
NT Enderby Land
NT Filchner Ice Shelf
NT Fosdick Mountains
NT Graham Land
NT Horlick Mountains
NT Hut Point Peninsula
NT James Ross Island
NT Kerguelen Islands
NT Lutzow-Holm Bay
NT Marguerite Bay
NT Marie Byrd Land
NT McMurdo Ice Shelf
NT Mirnyy Station
NT Molodezhnaya Station
NT Ohio Range
NT Pensacola Mountains
NT Prince Charles Mountains
NT Queen Alexandra Range
NT Queen Maud Land
NT Queen Maud Range
NT Ronne Ice Shelf
NT Ross Dependency
NT Ross Ice Shelf
NT Ross Island
NT Scotia Sea Islands
NT Shackleton Glacier
NT Shackleton Range
NT Sor-Rondane Mountains
NT South Pole
NT Syowa Station
NT Taylor Glacier
NT Transantarctic Mountains
NT Vestfold Hills
NT Victoria Land
NT Victoria Land Basin
NT Vostok Station
NT West Antarctica
NT Wilkes Land
NT Wisconsin Range
NT Wright Valley
SA ALHA 77005
SA ALHA 81005
SA Allan Hills Meteorites
SA Antarctic Ocean
SA Antarctic Plate
SA Arthur Harbor
SA Beacon Supergroup
SA Beacon Valley
SA Bowers Supergroup
SA Circum-Antarctic region
SA EETA 79001
SA Elephant Moraine Meteorites
SA Falkland Islands
SA Ferrar Group
SA Fremouw Formation
SA Gondwana
SA Jones Mountains
SA Kirkpatrick Basalt
SA La Meseta Formation
SA Lewis Cliff Meteorites
SA Lopez de Bertodano Forma-
   tion
SA Mount Erebus
SA Napier Complex
SA Polar Continental Shelf
SA polar regions
SA Robertson Bay Group
SA Ross Formation
SA Seymour Island
SA Sobral Formation
SA subantarctic regions
SA Taylor Valley

SA Yamato Meteorites

Antelope Valley (1981)
W Mojave Desert, N of Los Ange-
les.
IN Index counties and Mojave
   Desert as applicable.
SA California
SA Mojave Desert

Antelope Valley Limestone
(1978)
Lower and Middle Ordovician. In
Pogonip Group. Central Nevada.
BT Ordovician
BT Paleozoic
SA Nevada
SA Pogonip Group

anthophyllite (1978)
BT orthoamphibole
BT amphibole group
BT chain silicates
BT silicates
SA amosite
SA asbestos
SA cummingtonite
SA gedrite

Anthozoa (1978)
Autoposting of this term to Actinia-
ria, Ceriantipatharia,
Corallimorpharia, Heterocorallia,
Hexactiniaria, Octocorallia and
Zoanthiniaria began in 1981.
BT Coelenterata
BT Invertebrata
NT Ceriantipatharia
NT Octocorallia
NT Zoantharia

anthracite (1978)
This term has multiple hierarchies.
BT1 coal
BT1 organic residues
BT2 coal
BT2 sedimentary rocks
SA lignite
SA organic materials

Anthraconaia (1978)
Genus. Autoposting of Pteriina to
this term began in 1989.
BT Pteriina
BT Bivalvia
BT Mollusca
BT Invertebrata

Anthraconauta (1978)
Genus. Autoposting of Pteriina to
this term began in 1989.
BT Pteriina
BT Bivalvia
BT Mollusca
BT Invertebrata

anthraxolite (1978)
A hard, black asphaltite often as-
sociated with oil shale. Autopost-
ing of broader terms to this term
began in 1989.
BT bitumens
BT organic materials
SA coal
SA oil shale

anthropogenic (activity)
use human activity

anthropology (1978)
SA fossil man
SA paleoindian

Anti-Atlas (1978)
Range of the Atlas Mountains SW
of High Atlas Mountains in SW
Morocco. This term has multiple
hierarchies.
CO N274000N320500
    W0020000W0131500

UF Anti-Atlas Mountains
BT1 Moroccan Atlas Mountains
BT1 Morocco
BT1 North Africa
BT1 Africa
BT2 Moroccan Atlas Mountains
BT2 Atlas Mountains
BT2 North Africa
BT2 Africa

Anti-Atlas Mountains
use Anti-Atlas

Antibes
No longer a valid term for GeoRef. See Antibes France.

Antibes France (1993)
Town on the Mediterranean Sea in extreme SE France.
BT Alpes-Maritimes France
BT France
BT Western Europe
BT Europe

anticlinal
A valid term through 1977. After 1977, use anticlines.

anticlines (1978)
Not a valid index term from 1971 to 1977. Before 1978, also search folds AND anticlinal.
BT folds
SA anticlinoria
SA antiform folds
SA arches
SA diapirs
SA domes
SA geanticlines
SA synclines
SA upright folds

anticlinoria (1978)
BT folds
SA anticlines
SA antiform folds
SA geanticlines
SA geosynclines
SA synclinoria
SA systems
SA upright folds

Anticosti Island (1978)
In Gulf of Saint Lawrence.
BT Quebec
BT Eastern Canada
BT Canada
SA Ellis Bay Formation

antidunes (1978)
BT bedding plane irregularities
BT sedimentary structures
SA dunes
SA flame structures
SA sand waves

Antietam Formation (1978)
In Chilhowee Group. Maryland, SE Pennsylvania, Virginia, West Virginia.
BT Lower Cambrian
BT Cambrian
BT Paleozoic
SA Chilhowee Group
SA Maryland
SA Pennsylvania
SA Virginia
SA West Virginia

antiform folds (1978)
Before 1978, also search folds AND antiform.
BT folds
SA anticlines
SA anticlinoria
SA synform folds
SA upright folds

Antigonish
No longer a valid term for GeoRef. As of 1993, see Antigonish Nova Scotia.

Antigonish County
No longer a valid term for GeoRef. As of 1993, see Antigonish County Nova Scotia.

Antigonish County Nova Scotia (1993)
N Nova Scotia. Before 1993, also search Antigonish County AND Nova Scotia.
CO N452200N455400 W0612600W0621500
BT Nova Scotia
BT Maritime Provinces
BT Eastern Canada
BT Canada
NT Antigonish Nova Scotia

Antigonish Nova Scotia (1993)
Town in Antigonish County, NE Nova Scotia. Before 1993, also search Antigonish and Nova Scotia.
BT Antigonish County Nova Scotia
BT Nova Scotia
BT Maritime Provinces
BT Eastern Canada
BT Canada

antigorite (1978)
BT serpentine group
BT sheet silicates
BT silicates

Antilles (1978)
As of 1993, West Indies and Caribbean region are broader terms.
IN Index Greater Antilles and/or Lesser Antilles if applicable.
BT West Indies
BT Caribbean region
NT Greater Antilles
NT Lesser Antilles
SA Caribbean Mountain Range

antimonates (1981)
Before 1981, search antimonates and antimonites.
SA minerals

antimonates and antimonites
As of 1981, no longer a valid term for GeoRef. Use antimonates or antimonites as applicable.

antimonides (1978)
Autoposting of sulfides to this term ended in 1989.
NT dyscrasite
SA minerals

antimonite
use stibnite

antimonites (1981)
Before 1981, search antimonates and antimonites.
SA minerals

antimony (1978)
Chemical element. As of 1981, use antimony ores for antimony as a commodity. Autoposting of metals to this term began in 1989.
UF Sb
UF stibium
BT metals
SA antimony ores
SA native elements

antimony ores (1981)
Before 1981, also search (antimony OR Sb) AND (deposit OR deposits OR ore OR ores OR economic) in the basic index. Autoposting of metal ores to this term began in 1985.
IN Commodity. See List C.
BT metal ores
SA antimony
SA Dachang Deposit
SA native elements
SA stibnite

Antioquia
No longer a valid term for GeoRef. See Antioquia Colombia.

Antioquia Colombia (1993)
Department including the city in NW Colombia.
BT Colombia
BT South America

antiperthite (1978)
Autoposting of alkali feldspar to this term began in 1989.
BT alkali feldspar
BT feldspar group
BT framework silicates
BT silicates
SA perthite

antitrust legislation (1981)
SA economics
SA legislation

Antler orogenic belt
use Antler Orogeny

Antler Orogeny (1978)
An orogeny which extensively deformed Paleozoic rocks of the Great Basin in Nevada. Named for relations in the Antler Peak Quadrangle near Battle Mountain. Before 1978, also search Antler; Antler orogenic belt.
IN Index ages as applicable.
UF Antler orogenic belt
BT Paleozoic
SA Acadian Phase
SA Devonian
SA Mississippian
SA Nevada
SA orogeny
SA tectonics

Antlers Sands (1978)
In Trinity Group. SW Oklahoma and NE Texas. This term has multiple hierarchies.
BT1 Comanchean
BT1 Cretaceous
BT1 Mesozoic
BT2 Lower Cretaceous
BT2 Cretaceous
BT2 Mesozoic
SA Oklahoma
SA Texas
SA Trinity Group

Antofagasta
No longer a valid term for GeoRef. See Antofagasta Chile.

Antofagasta Chile (1993)
Province including the seaport in N Chile.
BT Chile
BT South America
NT Chuquicamata Chile
NT Socompa
SA Atacama Desert

Antong Java Rise
use Ontong Java Plateau

Antrim
No longer a valid term for GeoRef. As of 1993, see Antrim Northern Ireland.

Antrim Northern Ireland (1993)
Traditional county in NE North Ireland. Also a town in S central Antrim County. Includes the smaller administrative region of the same name adopted after 1973.
CO N543000N551500 W0054500W0063000
BT Northern Ireland
BT United Kingdom
BT Western Europe
BT Europe
SA Giant's Causeway

Antrim Shale (1985)
Upper Devonian and Lower Mississippian.
IN Index ages as applicable.
BT upper Paleozoic
BT Paleozoic
SA Lower Mississippian
SA Michigan
SA Upper Devonian

Antwerp
No longer a valid term for GeoRef. See Antwerp Belgium.

Antwerp Belgium (1993)
Province in N Belgium including the city at the mouth of the Western Scheldt River.
BT Belgium
BT Western Europe
BT Europe
NT Malines Belgium

Anura (1978)
Order.
BT Lissamphibia
BT Amphibia
BT Tetrapoda
BT Vertebrata
BT Chordata

Anvers Island (1978)
Just off NW coast of the Antarctic Peninsula.
BT Antarctica

Anvil Points
No longer a valid term for GeoRef. See Anvil Points Mine.

Anvil Points Mine (1993)
NW Colorado. Before 1993, also search Anvil Points.
BT Garfield County Colorado
BT Colorado
BT United States
SA mines
SA oil shale

ANWR
use Arctic National Wildlife Refuge

Anza Desert State Park (1978)
Occupies most of E San Diego County. Also search Anza.
BT California
BT United States
SA San Diego County California

Anzoategui
No longer a valid term for GeoRef. As of 1993 see Anzoategui Venezuela.

Anzoategui Venezuela (1993)
State in N extending from the Caribbean Sea to the Orinoco River. Before 1993 also search Anzoategui AND Venezuela.
BT Venezuela
BT South America
NT Oficina
SA Orinoco River

Aomori
No longer a valid term for GeoRef. See Aomori Japan.

Aomori Japan (1993)

City and prefecture in N Honshu.
BT Honshu
BT Japan
BT Far East
BT Asia
NT Hachinohe Japan
NT Shimokita Peninsula
SA Iwaki
SA Tohoku

Aosta Valley Italy
use Valle d'Aosta Italy

Aouelloul (1978)
A meteor crater near Chinguetti in W central Mauritania.
UF Aouelloul Crater
BT Mauritania
BT West Africa
BT Africa

Aouelloul Crater
use Aouelloul

Apache County
Valid through 1988. Search in combination with state term. After 1988, use specific county-state term.

Apache County Arizona (1989)
NE Arizona. Before 1989, also search Apache County AND Arizona.
CO N332800N370000
    W1090400W1100000
BT Arizona
BT United States
SA Black Mesa
SA Petrified Forest National Park
SA Rio Puerco

Apache Group (1989)
Central Arizona.
BT middle Proterozoic
BT Proterozoic
BT upper Precambrian
BT Precambrian
SA Arizona

Apalachicola
Valid through 1988. Search in combination with state term. After 1988, use specific city-state term.

Apalachicola Bay (1989)
Sliver of the Gulf of Mexico located between Saint George Island and mainland border of NW Florida.
CO N293400N294900
    W0844500W0851500
BT Franklin County Florida
BT Florida
BT United States
SA Apalachicola Florida
SA Gulf of Mexico
SA Saint George Island

Apalachicola Florida (1989)
City at mouth of Apalachicola River on Apalachicola Bay in the panhandle of NW Florida. Before 1989, search Apalachicola AND Florida.
CO N294300N294300
    W0850100W0850100
BT Franklin County Florida
BT Florida
BT United States
SA Apalachicola Bay
SA Apalachicola River

Apalachicola River (1985)
Florida panhandle. Flows from Lake Seminole S to Apalachicola Bay, Gulf of Mexico.
BT Florida
BT United States
SA Apalachicola Florida

apatite (1978)
BT phosphates
SA carbonate apatite
SA chlorapatite
SA fluorapatite
SA francolite
SA heavy minerals
SA hydroxylapatite
SA phosphate deposits
SA pyromorphite

apatite ores
use phosphate deposits

Apennine Front (1978)
Linear outer slopes of the Apennines in Italy. Also search Apennine AND front. Also used for feature on the Moon.
IN Index Italy and Apennines; or Moon as applicable.
SA Apennines
SA Italy
SA Moon

Apennine Mountains
use Apennines

Apennine Range
use Apennines

Apennines (1978)
Mountain range extending the full length of the Italian peninsula. As of 1993, Apennines is autoposted to its narrower terms.
IN Also index San Marino if appropriate.
CO N380000N450000
    E0170000E0082000
UF Apennine Mountains
UF Apennine Range
BT Italy
BT Southern Europe
BT Europe
NT Alban Hills
NT Apuane Alps
NT Ligurian Apennines
NT Monte Amiata
NT Northern Apennines
NT Southern Apennines
SA Apennine Front
SA San Marino
SA Scaglia Formation

aphanitic texture (1978)
Before 1978, search aphanitic.
BT textures
SA igneous rocks

Aphebian (1978)
North America, lower Proterozoic. Autoposting of Precambrian to this term ended in 1989. As of 1989, lower Proterozoic and Proterozoic are autoposted to this term. As of 1990, upper Precambrian and Precambrian are autoposted to this term.
BT lower Proterozoic
BT Proterozoic
BT upper Precambrian
BT Precambrian
NT Hurwitz Group
SA middle Precambrian

Aphrodite Terra (1989)
Physiographic province on Venus.
BT Venus
SA Pioneer Program

Aplacophora (1978)
BT Mollusca
BT Invertebrata

aplite (1978)
BT granites
BT plutonic rocks
BT igneous rocks

apogranite (1978)
BT granites
BT plutonic rocks
BT igneous rocks

Apollo
As of 1981, use Apollo Program.

Apollo Program (1981)
Autoposting of this term to specific Apollo missions began in 1989. Before 1981, search Apollo.
NT Apollo 8
NT Apollo 9
NT Apollo 11
NT Apollo 12
NT Apollo 13
NT Apollo 14
NT Apollo 15
NT Apollo 16
NT Apollo 17
SA extraterrestrial geology
SA landing sites
SA Moon
SA planetology
SA remote sensing

Apollo 8 (1989)
BT Apollo Program
SA Moon
SA remote sensing

Apollo 9 (1978)
Autoposting of Apollo Program to this term began in 1989.
BT Apollo Program
SA Moon
SA remote sensing

Apollo 11 (1978)
Autoposting of Apollo Program to this term began in 1989.
BT Apollo Program
SA Moon
SA remote sensing

Apollo 12 (1978)
Autoposting of Apollo Program to this term began in 1989.
BT Apollo Program
SA Moon
SA remote sensing

Apollo 13 (1978)
Autoposting of Apollo Program to this term began in 1989.
BT Apollo Program
SA remote sensing

Apollo 14 (1978)
Autoposting of Apollo Program to this term began in 1989.
BT Apollo Program
SA Moon
SA remote sensing

Apollo 15 (1978)
Autoposting of Apollo Program to this term began in 1989.
BT Apollo Program
SA Moon
SA remote sensing

Apollo 16 (1978)
Autoposting of Apollo Program to this term began in 1989.
BT Apollo Program
SA Moon
SA remote sensing

Apollo 17 (1978)
Autoposting of Apollo Program to this term began in 1989.
BT Apollo Program
SA Aristarchus
SA Moon
SA remote sensing

Apollo-Soyuz Program (1981)
SA extraterrestrial geology
SA planetology

SA remote sensing

apophyllite (1978)
BT sheet silicates
BT silicates

Appalachian Basin (1978)
IN Index states or provinces as applicable.
BT North America
SA Alabama
SA Eastern Gas Shales Project
SA Kentucky
SA Maryland
SA New York
SA North Carolina
SA Ohio
SA Ontario
SA Pennsylvania
SA Quebec
SA Rome Trough
SA Tennessee
SA Virginia
SA West Virginia

Appalachian Mountains
As of 1973, no longer a valid term for GeoRef. Used in subfiles B and N.
use Appalachians

Appalachian Orogeny
use Appalachian Phase

Appalachian Phase (1978)
Before 1978, also search Appalachian structure; orogeny AND Appalachian; Appalachian Orogeny.
UF Appalachian Orogeny
UF Appalachian structure
BT Permian
BT Paleozoic
SA orogeny
SA tectonics

Appalachian Plateau (1978)
Westernmost part of the Appalachians including the Allegheny and Cumberland plateaus. Also search Appalachian AND plateau.
IN Index states as applicable.
BT Appalachians
BT North America
SA Alabama
SA Allegheny Mountains
SA Allegheny Plateau
SA Catskill Mountains
SA Cumberland Plateau
SA Kentucky
SA Maryland
SA Mohawk Valley
SA New York
SA Ohio
SA Pennsylvania
SA Pine Mountain Window
SA Tennessee
SA Virginia
SA West Virginia

Appalachian region
As of 1971, no longer a valid term for GeoRef. Used in subfiles N, B, and G.
use Appalachians

Appalachian structure
use Appalachian Phase

Appalachians (1978)
Mountain system of eastern North America extending from Canadian Maritime Provinces to Alabama. In 1985, Great Appalachian Valley, Valley and Ridge Province, Piedmont and Appalachian Plateau were added as narrower terms. Before 1973, also search Appalachian Mountains; Appalachian re-

gion. Autoposting of this term began in 1978.
IN   Index countries as applicable.
CO N330000N473000
    W0670000W0870000
UF Appalachian Mountains
UF Appalachian region
BT North America
NT Allegheny Mountains
NT Appalachian Plateau
NT Blue Ridge Mountains
NT Blue Ridge Province
NT Carolina slate belt
NT Catskill Mountains
NT Central Appalachians
NT Great Appalachian Valley
NT Northern Appalachians
NT Piedmont
NT Shawangunk Mountains
NT Southern Appalachians
NT Valley and Ridge Province
SA Adirondack Mountains
SA Allegheny Plateau
SA Avalon Terrane
SA Berkshire Hills
SA Brevard Zone
SA Canada
SA Chain Lakes Massif
SA Charlotte Belt
SA Cumberland Plateau
SA Eastern Gas Shales Project
SA Eastern Overthrust Belt
SA Great Smoky Mountains
SA Green Mountains
SA Hudson Highlands
SA Kings Mountain Belt
SA Kiokee Belt
SA Norumbega fault zone
SA Pulaski thrust sheet
SA Raleigh Belt
SA Reading Prong
SA South Mountain
SA Taconic Allochthon
SA United States
SA White Mountains

apparatus   (1978)
Do not use for instruments. Before 1981, included use under conodonts. Before 1971, was used for instruments.
SA Conodonta

apparent polar wandering (1989)
SA continental drift
SA paleolatitude
SA paleomagnetism
SA plate tectonics
SA polar wandering
SA pole positions

apparent resistivity   (1989)
SA conductivity
SA electrical logging
SA electrical methods
SA electrical surveys
SA geophysical methods
SA geophysical surveys
SA resistivity

appinite   (1978)
From 1978-1980, plutonic rocks was autoposted to this term. As of 1993, plutonic rocks is autoposted to this term.
BT plutonic rocks
BT igneous rocks
SA diorites
SA syenites

applications   (1978)
SA methods

Apsheron Peninsula   (1978)
Peninsula jutting into the Caspian Sea on E. Also search Apsheron. This term has multiple hierarchies.
CO N402000N404000
    E0503000E0493000
BT1 Azerbaidzhan
BT1 Europe
BT2 Azerbaidzhan
BT2 Commonwealth of Independent States
SA   Peschanyy
SA   Surakhany Azerbaidzhan

Apt
No longer a valid term for GeoRef. See Apt France.

Apt France   (1993)
Town in SE France.
BT Vaucluse France
BT France
BT Western Europe
BT Europe

Aptian   (1978)
Europe. Above Barremian below Albian.
BT Lower Cretaceous
BT Cretaceous
BT Mesozoic
SA Areado Formation
SA Gargasian
SA Zubair Formation

aptychi   (1978)
UF aptychus
SA Ammonites

aptychus
use aptychi

Apuane Alps   (1978)
Division of Etruscan Apennines in central Italy. This term has multiple hierarchies.
BT1 Tuscany Italy
BT1 Italy
BT1 Southern Europe
BT1 Europe
BT2 Apennines
BT2 Italy
BT2 Southern Europe
BT2 Europe

Apulia
No longer a valid term for GeoRef. See Apulia Italy.

Apulia Italy   (1993)
Autonomous region comprising approximately the southern third of east coast. Before 1978, also search Puglia.
CO N394500N415800
    E0183200E0145700
UF Puglia Italy
BT Italy
BT Southern Europe
BT Europe
NT Bari Italy
NT Brindisi Italy
NT Foggia Italy
NT Gargano
NT Lecce Italy
NT Murge
NT Otranto Italy
NT Salentina Peninsula
NT Taranto Italy

Apure Guarico Plain   (1981)
CO N061000N095000
    W0644000W0715500
BT Venezuela
BT South America

Apuseni Mountains   (1978)
A large mountain massif in WSW Transylvania. Also search Apuseni.
BT Romania
BT Southern Europe
BT Europe
SA Muntii Metalici

aquamarine   (1985)
Variety of beryl.
BT ring silicates
BT silicates
SA beryl
SA gems

aquatic environment   (1985)
SA deep-water environment
SA depositional environment
SA ecology
SA estuarine environment
SA fluvial environment
SA fresh-water environment
SA lacustrine environment
SA marine environment
SA paleoecology
SA sedimentation

aqueous solutions   (1978)
SA geochemistry
SA hydrochemistry
SA ion chromatography
SA solution
SA solutions
SA water

Aquia Formation   (1978)
In Pamunkey Group. Delaware, Maryland and E Virginia.
BT lower Eocene
BT Eocene
BT Paleogene
BT Tertiary
BT Cenozoic
SA Delaware
SA Maryland
SA Virginia

aquifer properties
A valid term through 1978. After 1978, see aquifers or specific property, e.g. transmissivity.

aquifers   (1978)
NT alluvium aquifers
NT confined aquifers
NT perched aquifers
NT shallow aquifers
NT surficial aquifers
NT unconfined aquifers
SA aquitards
SA artesian waters
SA Darcy's law
SA dispersivity
SA ground water
SA heat storage
SA hydrogeology
SA leaky aquifers
SA multiphase flow
SA multiple aquifers
SA RASA
SA recharge
SA shallow depth
SA storage coefficient
SA transmissivity
SA water balance
SA water resources
SA water supply
SA water yield

Aquitaine   (1978)
A region of SW France.
BT France
BT Western Europe
BT Europe

Aquitaine Basin   (1978)
Large lowland in SW bounded by Pyrenees on S, Central Massif on N and NE, and Bay of Biscay on W.
CO N431500N460000
    E0020000W0013000
BT France
BT Western Europe
BT Europe

Aquitanian   (1978)
Europe. Above Chattian (Oligocene), below Burdigalian.
BT lower Miocene
BT Miocene
BT Neogene
BT Tertiary
BT Cenozoic

aquitards   (1989)
SA aquifers
SA artesian waters
SA confined aquifers
SA ground water
SA leaky aquifers
SA permeability
SA seepage

Ar
use argon

Ar-36   (1978)
Autoposting of broader terms began in 1989. This term has multiple hierarchies.
BT1 stable isotopes
BT1 isotopes
BT2 argon
BT2 noble gases
SA   Ar-38/Ar-36
SA   Ar-40/Ar-36

Ar-36/Ar-38
use Ar-38/Ar-36

Ar-36/Ar-40
use Ar-40/Ar-36

Ar-37   (1978)
Artificial. Autoposting of broader terms began in 1989. This term has multiple hierarchies.
BT1 radioactive isotopes
BT1 isotopes
BT2 argon
BT2 noble gases

Ar-38   (1978)
Autoposting of broader terms began in 1989. This term has multiple hierarchies.
BT1 radioactive isotopes
BT1 isotopes
BT2 argon
BT2 noble gases
SA   Ar-38/Ar-36

Ar-38/Ar-36   (1985)
Autoposting of broader terms began in 1989. This term has multiple hierarchies.
UF   Ar-36/Ar-38
BT1 radioactive isotopes
BT1 isotopes
BT2 stable isotopes
BT2 isotopes
BT3 argon
BT3 noble gases
SA   Ar-36
SA   Ar-38
SA   isotope ratios

Ar-39   (1978)
Artificial. Autoposting of broader terms began in 1989. This term has multiple hierarchies.
BT1 radioactive isotopes
BT1 isotopes
BT2 argon
BT2 noble gases
SA   Ar-40/Ar-39

Ar-39/Ar-40
use Ar-40/Ar-39

Ar-40   (1978)
Autoposting of broader terms began in 1989. This term has multiple hierarchies.
BT1 stable isotopes
BT1 isotopes

BT2 argon
BT2 noble gases
SA Ar-40/Ar-36
SA Ar-40/Ar-39

Ar-40/Ar-36 (1978)
Autoposting of broader terms began in 1989. This term has multiple hierarchies.
UF Ar-36/Ar-40
BT1 stable isotopes
BT1 isotopes
BT2 argon
BT2 noble gases
SA Ar-36
SA Ar-40
SA isotope ratios

Ar-40/Ar-39 (1978)
Autoposting of broader terms began in 1989. This term has multiple hierarchies.
UF Ar-39/Ar-40
BT1 radioactive isotopes
BT1 isotopes
BT2 stable isotopes
BT2 isotopes
BT3 argon
BT3 noble gases
SA Ar-39
SA Ar-40
SA isotope ratios

Ar/Ar (1978)
Isotopic ratio used in age determination.
UF argon-argon
SA absolute age
SA argon
SA isotope ratios

Arab Formation (1989)
BT Upper Jurassic
BT Jurassic
BT Mesozoic
SA Saudi Arabia
SA United Arab Emirates

Arabia
 use Arabian Peninsula

Arabian Desert
 use Eastern Desert

Arabian Gulf
 use Persian Gulf

Arabian Peninsula (1978)
Between Red Sea on the W; and the Persian Gulf, Gulf of Oman, and Arabian Sea on the E. Autoposting of this term began in 1978.
IN Index countries as applicable.
CO N123000N304000
 E0594000E0343000
UF Arabia
BT Asia
NT Arabian Shield
NT Bahrain
NT Kuwait
NT Oman
NT Qatar
NT Saudi Arabia
NT Timna
NT United Arab Emirates
NT Yemen
SA Middle East
SA Red Sea Rift

Arabian Plate (1978)
SA plate tectonics
SA plates
SA Saudi Arabia

Arabian Ridge
 use Carlsberg Ridge

Arabian Sea (1978)
Between the Arabian Peninsula on the W and Pakistan and India in the E. Persian Gulf is considered a narrower term as of 1981. Autoposting of this term began in 1978.
CO S010000N303000
 E0770000E0473000
BT Indian Ocean
NT Gulf of Aden
NT Gulf of Cambay
NT Gulf of Oman
NT Persian Gulf
SA DSDP Site 219
SA DSDP Site 220
SA DSDP Site 233
SA Leg 23
SA Leg 24
SA Leg 117
SA ODP Site 720
SA ODP Site 721
SA ODP Site 722
SA ODP Site 723
SA ODP Site 724
SA ODP Site 725
SA ODP Site 726
SA ODP Site 727
SA ODP Site 728
SA ODP Site 729
SA ODP Site 730
SA ODP Site 731

Arabian Shield (1978)
NE part of the Ethiopian Shield.
IN Index countries as applicable.
BT Arabian Peninsula
BT Asia
SA Saudi Arabia
SA Yemen

Arabian-Indian Midoceanic Ridge
 use Carlsberg Ridge

Arabian-Indian Ridge
 use Carlsberg Ridge

Arabian-Nubian Shield
 Not a valid term for GeoRef. Use Arabian Shield or Nubian Shield or both as applicable.

Arabic (1978)
Used to indicate language of a catalog, dictionary, glossary or lexicon.
SA catalogs
SA dictionaries
SA glossaries
SA lexicons

Aracaju
 No longer a valid term for GeoRef. See Aracaju Brazil.

Aracaju Brazil (1993)
City on the Atlantic Ocean in NE Brazil.
BT Sergipe Brazil
BT Brazil
BT South America

Aracena
 No longer a valid term for GeoRef. As of 1993, see Aracena Spain.

Aracena Spain (1993)
Town near the Gulf of Cadiz. Before 1993, also search Aracena and Spain.
BT Huelva Spain
BT Andalusia Spain
BT Spain
BT Iberian Peninsula
BT Southern Europe
BT Europe

Arachnida (1978)
Class.
BT Chelicerata
BT Arthropoda

BT Invertebrata

Arad
 No longer a valid term for GeoRef. See Arad Romania.

Arad Romania (1993)
City and county near the Hungarian border. This term has multiple hierarchies.
BT1 Banat
BT1 Southern Europe
BT1 Europe
BT2 Romania
BT2 Southern Europe
BT2 Europe

Araeoscelidia (1981)
Order. Autoposting of Euryapsida to this term ended in 1993. As of 1993, Diapsida is autoposted to this term. Before 1993, Trilophosauria was a narrower term.
BT Diapsida
BT Reptilia
BT Tetrapoda
BT Vertebrata
BT Chordata
SA Trilophosauria

Arafura Sea (1978)
Sea between N Australia and Indonesia; 800 miles long by about 350 miles wide. As of 1990, Pacific Ocean is autoposted to this term.
BT North Australian Seas
BT West Pacific
BT Pacific Ocean
SA Australia
SA Indonesia
SA McArthur Basin

Aragats (1978)
A mountain in NW.
IN Index regions as applicable.
SA Armenia

Aragon
 No longer a valid term for GeoRef. As of 1993, see Aragon Spain.

Aragon Spain (1993)
Region and ancient kingdom in NE Spain. Provinces are considered narrower terms as of 1981. Before 1993, also search Aragon and Spain.
CO N395200N425700
 E0004500W0021000
BT Spain
BT Iberian Peninsula
BT Southern Europe
BT Europe
NT Huesca Spain
NT Saragossa Spain
NT Teruel Spain

aragonite (1978)
BT carbonates
SA alabaster
SA calcite
SA calcium carbonate
SA pearls
SA witherite

Aragua
 No longer a valid term for GeoRef. As of 1993 see Aragua Venezuela.

Aragua Venezuela (1993)
State in N Venezuela. Before 1993 also search Aragua AND Venezuela.
BT Venezuela
BT South America

Aragvi (1978)

River in N central Georgian Republic.
BT Georgian Republic
BT Europe

Aral region (1978)
Before 1978, also search Aral. This term has multiple hierarchies.
IN Index former Soviet republics as applicable.
UF Aral Sea region
BT1 Commonwealth of Independent States
BT2 Central Asia
BT2 Asia
SA Kazakhstan
SA Turkmenia
SA Uzbekistan

Aral Sea (1978)
Between Kazakhstan and Uzbekistan. Also search Aral. As of 1993, Central Asia is a broader term. This term has multiple hierarchies.
CO N440000N460000
 E0620000E0580000
BT1 Central Asia
BT1 Asia
BT2 Commonwealth of Independent States

Aral Sea region
 use Aral region

Arapahoe County
 Valid through 1988. Search in combination with state term. After 1988, use specific county-state term.

Arapahoe County Colorado (1989)
S and SE of Denver in N central Colorado. Before 1989, also search Arapahoe County AND Colorado.
CO N393300N394300
 W1034300W1050400
BT Colorado
BT United States

Arapien Shale (1985)
Central Utah.
BT Jurassic
BT Mesozoic
SA Utah

Ararat (1978)
Mountain near Iranian and Armenian borders.
BT Turkey
BT Middle East
BT Asia

Arauco
 No longer a valid term for GeoRef. See Arauco Chile.

Arauco Chile (1993)
Former province in S central Chile.
BT Chile
BT South America

Aravalli Group
 use Aravalli System

Aravalli Range (1978)
Central and S Rajasthan, NW India.
SA Khetri copper belt
SA Rajasthan India

Aravalli Supergroup
 use Aravalli System

Aravalli System (1978)
Includes Dabari, Matoon, and Udaipur formations.
UF Aravalli Group

UF Aravalli Supergroup
BT Archean
BT Precambrian
SA Andhra Pradesh India
SA Gujarat India
SA India
SA Rajasthan India

Araxa
No longer a valid term for GeoRef. See Araxa Brazil.

Araxa Brazil (1993)
Town in W Minas Gerais.
BT Minas Gerais Brazil
BT Brazil
BT South America

Araya Peninsula (1978)
W Sucre.
BT Sucre Venezuela
BT Venezuela
BT South America

Arbarastakh (1978)
Carbonatite complex in Yakutia.
IN Index region as applicable.
SA Yakutia Russian Federation

Arbuckle Anticline (1989)
Anadarko Basin, S Oklahoma.
BT Oklahoma
BT United States
SA Anadarko Basin
SA Arbuckle Mountains

Arbuckle Group (1978)
Upper Cambrian and Lower Ordovician. Subdivided into Fort Sill Limestone, Royer Marble, Signal Mountain Limestone, Chapman Ranch Dolomite, Wolf Creek Dolomite, Cool Creek Limestone, Alden Limestone, and West Spring Creek Formation. S Oklahoma.
IN Index ages as applicable.
BT Paleozoic
SA Lower Ordovician
SA Oklahoma
SA Upper Cambrian

Arbuckle Mountains (1978)
Carter and Murray counties, S central Oklahoma.
CO N340000N343000
W0964500W0973000
BT Oklahoma
BT United States
SA Arbuckle Anticline
SA Carter County Oklahoma
SA Murray County Oklahoma

Arcachon Basin (1978)
Inlet of Bay of Biscay in SW France.
BT Gironde France
BT France
BT Western Europe
BT Europe
SA Bay of Biscay

arch dams (1978)
UF archdams
BT dams

Archaean
use Archean

Archaediscidae (1978)
Family. As of 1990, microfossils and Protista are autoposted to this term. This term has multiple hierarchies.
BT1 Fusulinina
BT1 foraminifera
BT1 Protista
BT1 Invertebrata
BT2 Fusulinina
BT2 foraminifera
BT2 Protista

BT2 microfossils

Archaeocopida
use Archeocopida

Archaeocyatha (1978)
BT Invertebrata

Archaeogastropoda (1978)
Autoposting of this term to Bellerophontina began in 1989.
BT Gastropoda
BT Mollusca
BT Invertebrata
NT Bellerophontina

Archaeolithothamnium (1978)
As of 1990, microfossils and thallophytes are autoposted to this term. This term has multiple hierarchies.
BT1 Corallinaceae
BT1 Rhodophyta
BT1 algae
BT1 thallophytes
BT1 Plantae
BT2 Corallinaceae
BT2 Rhodophyta
BT2 algae
BT2 microfossils

archaeological sites (1978)
UF sites, archaeological
SA archaeology
SA artifacts
SA fossil man
SA middens

archaeology (1978)
Before 1971, also search archeology. Before 1993, also search geoarchaeology.
UF archeology
UF geoarchaeology
SA Acheulian
SA archaeological sites
SA artifacts
SA Aurignacian
SA Bronze Age
SA Chalcolithic
SA Clovis
SA fossil man
SA Iron Age
SA Levalloisian
SA Lubbock Lake
SA Magdalenian
SA Mesolithic
SA middens
SA Mousterian
SA Neolithic
SA ocher
SA paleoecology
SA Paleolithic
SA paleopathology
SA paleoseismicity
SA Stone Age
SA type localities
SA upper Paleolithic

Archaeomonadaceae (1978)
This term has multiple hierarchies.
BT1 ebridians
BT1 Protista
BT1 Invertebrata
BT2 ebridians
BT2 Protista
BT2 microfossils

Archaeopteris (1989)
Genus.
BT Filicopsida
BT pteridophytes
BT Plantae

Archaeopteryx (1978)
Genus.
BT Archaeornithes
BT Aves
BT Tetrapoda
BT Vertebrata

BT Chordata

Archaeornithes (1978)
Subclass. Autoposting of this term began in 1978.
BT Aves
BT Tetrapoda
BT Vertebrata
BT Chordata
NT Archaeopteryx
SA Neornithes

Archaeozoic
use Archean

Archangel Russian Federation
use Arkhangelsk Russian Federation

archdams
use arch dams

Archean (1978)
As of 1981, used for lower Precambrian. Autoposting of this term began in 1978.
UF Archaean
UF Archaeozoic
UF Archeozoic
UF lower Precambrian
BT Precambrian
NT Amisk Group
NT Aravalli System
NT Arunta Complex
NT Blake River Group
NT Blondeau Formation
NT Bulawayan Group
NT Dharwars
NT Gangpur Series
NT Gilman Formation
NT Hayes River Group
NT Iron Ore Group
NT J-M Reef
NT Kalgoorlie System
NT Krivoy Rog Series
NT lower Archean
NT Napier Complex
NT Nuk Gneiss
NT Opemisca Group
NT Peninsular Gneiss
NT Rice Lake Group
NT Roy Group
NT Sausar Series
NT Singhbhum Granite
NT Taihua Group
NT Timiskaming Group
NT upper Archean
NT Vermilion granitic complex
NT Warrawoona Group
NT Yellowknife Group
SA Closepet Granite
SA Kaapvaal Craton
SA Lewisian Complex
SA Moldanubian

Archeocopida (1978)
Order. As of 1990, microfossils, Crustacea, Mandibulata and Arthropoda are autoposted to this term. This term has multiple hierarchies.
UF Archaeocopida
BT1 Ostracoda
BT1 Crustacea
BT1 Mandibulata
BT1 Arthropoda
BT1 Invertebrata
BT2 Ostracoda
BT2 microfossils

archeology
No longer a valid term for GeoRef. Before 1971, was used in subfiles E, G, N and B.
use archaeology

Archeozoic
use Archean

arches (1978)

SA anticlines
SA arcuate structures
SA domes
SA folds
SA tectonics

Arches National Park (1981)
In SE Utah. Before 1981, also search Arches.
BT Utah
BT United States
SA national parks

Archie equation
use Archie's law

Archie's law (1989)
Equation expressing the relationship between resistivity of the geologic formation, resistivity of water contained within the pores of the rock, porosity of the rock, and amount of pore space occupied by water.
UF Archie equation
SA equations
SA mathematical methods
SA petroleum engineering
SA porosity
SA reservoirs
SA resistivity
SA well-logging

Archosauria (1978)
Subclass. Autoposting of this term began in 1978. As of 1993, Archosauromorpha is considered a broader term.
BT Diapsida
BT Reptilia
BT Tetrapoda
BT Vertebrata
BT Chordata
NT Crocodilia
NT Ornithischia
NT Pterosauria
NT Saurischia
NT Thecodontia

Archostemata (1978)
BT Insecta
BT Mandibulata
BT Arthropoda
BT Invertebrata

Archuleta County
Valid through 1988. Search in combination with state term. After 1988, use specific county-state term.

Archuleta County Colorado (1989)
On New Mexico border in SW Colorado. Before 1989, also search Archuleta County AND Colorado.
CO N370000N372500
W1063000W1073000
BT Colorado
BT United States

Arcina (1981)
Autoposting of Cyrtodontida to this term ended in 1989.
BT Bivalvia
BT Mollusca
BT Invertebrata
SA Cyrtodontida

arcs
As of 1993, no longer a valid term for GeoRef. Usually out-of-scope. Used with electrical field. See arcuate structures.

arcs, island
use island arcs

Arctic
Not a valid GeoRef index term after 1969. Was used on level 1 in subfiles G and N. Now use Arctic Ocean, Arctic region, or arctic environment.

Arctic America
Not a valid GeoRef index term after 1970. Was used on level 1 in subfile N. Now use Alaska, Yukon Territory or Northwest Territories as applicable. See also Arctic Archipelago.

Arctic Archipelago (1978)
Large group of islands in the Arctic Ocean nearly coextensive with Franklin District, Northwest Territories, Canada. Used as level 1 area term from 1977 to 1978.
UF Arctic Islands
BT Franklin District Northwest Territories
BT Northwest Territories
BT Western Canada
BT Canada
SA Arctic Coastal Plain
SA Arctic region
SA Baffin Island
SA Bathurst Island

Arctic Coastal Plain (1978)
IN Index countries or regions as applicable.
BT Arctic region
SA Alaska
SA Arctic Archipelago
SA Arctic National Wildlife Refuge
SA Greenland
SA Jan Mayen
SA North Slope
SA Northern Alaska
SA Northwest Territories
SA Russian Federation
SA Svalbard

arctic environment (1978)
Before 1978, also search arctic.
SA climate
SA depositional environment
SA ecology
SA environment
SA paleoclimatology
SA paleoecology

Arctic Islands
use Arctic Archipelago

Arctic National Wildlife Refuge (1989)
Arctic Coastal Plain of NE Alaska.
CO N693500N702000
W1420000W1470000
UF ANWR
BT Northern Alaska
BT Alaska
BT United States
SA Arctic Coastal Plain
SA Arctic region
SA Sadlerochit Mountains

Arctic Ocean (1978)
Extends from North Pole southward to approximately 70° N and is bounded by Alaska, Canada, Greenland, Norway, and the Russian Federation. Before 1981, Jan Mayen, Svalbard, Beerenberg and Bear Island were included as narrower terms. Norwegian Sea and White Sea are narrower terms as of 1981.
CO N700000N900000
W1800000E1800000
UF Arctic Sea
UF North Polar Sea
NT Alpha Cordillera
NT Barents Sea
NT Beaufort Sea
NT Canada Basin
NT Chukchi Sea
NT East Siberian Sea
NT Eurasia Basin
NT Fram Strait
NT Greenland Sea
NT Kara Sea
NT Laptev Sea
NT Lomonosov Ridge
NT Makarov Basin
NT Mendeleyev Ridge
NT Mid-Arctic Ocean Ridge
NT Mohns Ridge
NT Nares Strait
NT Norwegian Sea
NT White Sea
SA Arctic region
SA Baffin Bay
SA Barrow Alaska
SA Bear Island
SA Beerenberg
SA Bering Strait
SA CESAR
SA DSDP Site 336
SA DSDP Site 338
SA DSDP Site 343
SA DSDP Site 344
SA DSDP Site 345
SA DSDP Site 346
SA DSDP Site 348
SA Jan Mayen
SA Leg 38
SA Leg 104
SA Leg 105
SA ODP Site 642
SA ODP Site 643
SA ODP Site 644
SA ODP Site 645
SA ODP Site 646
SA ODP Site 647
SA Russian Federation
SA Svalbard

Arctic region (1978)
The Arctic Ocean and islands in it and adjacent to it plus mainland surfaces N of the Arctic circle. Jan Mayen and Svalbard are considered narrower terms as of 1981. Arctic Coastal Plain is considered a narrower term as of 1993. Autoposting of this term began in 1978.
IN Index Alaska, Arctic Archipelago, Arctic Ocean, Greenland, and countries as applicable.
NT Arctic Coastal Plain
NT Greenland
NT Jan Mayen
NT North Pole
NT Russian Arctic
NT Svalbard
SA Alaska
SA Arctic Archipelago
SA Arctic National Wildlife Refuge
SA Arctic Ocean
SA Bear Island
SA Canada
SA Finland
SA Hecla Hoek Formation
SA Iceland
SA Norway
SA polar regions
SA subarctic regions
SA Sweden

Arctic Sea
use Arctic Ocean

Arctocyonia (1981)
Autoposting of Condylarthra, Eutheria and Theria to this term began in 1989.
BT Condylarthra
BT Eutheria
BT Theria
BT Mammalia
BT Tetrapoda
BT Vertebrata
BT Chordata

arcuate faults (1978)
Before 1978, also search arcuate AND faults.
BT faults
SA arcuate structures
SA boundary faults
SA peripheral faults

arcuate structures (1993)
General. Use more specific term if applicable.
SA arches
SA arcuate faults
SA coronae
SA fold and thrust belts
SA greenstone belts
SA island arcs
SA ring structures

Ardara Pluton (1978)
An igneous intrusion in NW Ireland.
SA Donegal Ireland

Ardeche
No longer a valid term for GeoRef. See Ardeche France.

Ardeche France (1993)
Department in S France.
CO N441500N453000
E0050000E0034500
BT France
BT Western Europe
BT Europe
NT Largentiere France
NT Privas France
SA Cevennes
SA Juvinas Meteorite
SA Vivarais

Ardennes (1978)
Wooded plateau region.
IN Index countries as applicable.
CO N493000N503000
E0053000E0041000
UF Forest of Ardennes
BT Western Europe
BT Europe
SA Ardennes France
SA Givet France
SA Luxembourg
SA Luxembourg Belgium
SA Rocroi France
SA Sedan France
SA Stavelot Belgium

Ardennes Department
No longer a valid term for GeoRef.
use Ardennes France

Ardennes France (1993)
Department in NE France. Before 1993, also search Ardennes Department. Before 1981, also search Ardennes.
CO N491200N501000
E0052500E0040200
UF Ardennes Department
BT France
BT Western Europe
BT Europe
NT Givet France
NT Rocroi France
NT Sedan France
SA Ardennes

Ardmore Basin (1985)
S Oklahoma and N Texas.
IN Index states as applicable.
CO N334000N344000
W0960000W0990000
BT United States
SA Oklahoma
SA Texas

Ardnamurchan (1978)
Peninsula and parish in Argyll County, N of Mull Island in W Scotland. This term has multiple hierarchies.
BT1 Argyllshire Scotland
BT1 Scotland
BT1 Great Britain
BT1 United Kingdom
BT1 Western Europe
BT1 Europe
BT2 Highland region Scotland
BT2 Scotland
BT2 Great Britain
BT2 United Kingdom
BT2 Western Europe
BT2 Europe

arduinite
use mordenite

Areado Formation (1978)
BT Lower Cretaceous
BT Cretaceous
BT Mesozoic
SA Aptian
SA Brazil
SA Minas Gerais Brazil

areal geology (1978)
Used for entries that might properly be placed under three or more of the major discipline headings such as geomorphology, stratigraphy, structural geology.
SA bibliography
SA expeditions
SA explanatory text
SA geology
SA guidebook
SA maps
SA regional

areal studies (1978)
UF studies, areal
SA clay mineralogy
SA mineral data

areas described
No longer a valid GeoRef index term. Before 1971, was used on level 2 in subfiles E and N. See areal geology; areal studies.

Arecibo
No longer a valid term for GeoRef. As of 1993 see Arecibo Puerto Rico.

Arecibo Puerto Rico (1993)
City in S Puerto Rico on Caribbean Sea. Before 1993 also search Arecibo AND Puerto Rico.
BT Puerto Rico
BT Greater Antilles
BT Antilles
BT West Indies
BT Caribbean region

Arecidae (1981)
BT Monocotyledoneae
BT angiosperms
BT Spermatophyta
BT Plantae

arenaceous texture (1978)
Before 1978, also search arenaceous AND specific sediment type or sedimentary rock.
UF arenarious texture
UF psammitic texture
UF sabulous texture
UF sandy texture
BT textures
SA sand
SA sandstone

SA sedimentary rocks
SA sediments

Arenal (1978)
Volcano in NW Costa Rica.
BT Costa Rica
BT Central America

arenarious texture
use arenaceous texture

Arendal
No longer a valid term for GeoRef. As of 1993 see Arendal Norway.

Arendal Norway (1993)
A city on the Skagerrak. Before 1993 also search Arendal AND Norway.
BT Norway
BT Scandinavia
BT Western Europe
BT Europe

arendalite
use epidote

Arenicolites (1985)
BT ichnofossils
SA burrows
SA worms

Arenig (1978)
Two mountains in N Merionethshire (Gwynedd) in N Wales. One is called Arenig Fach and the other Arenig Fawr. For the age term, see Arenigian.
BT Merionethshire Wales
BT Wales
BT Great Britain
BT United Kingdom
BT Western Europe
BT Europe
SA Arenigian

Arenigian (1978)
Europe. Above Tremadocian, below Llanvirnian. Also search Arenig and Ordovician.
BT Lower Ordovician
BT Ordovician
BT Paleozoic
NT Ballantrae Complex
SA Arenig
SA Buchans Group

arenite (1978)
Autoposting of this term began in 1978.
BT clastic rocks
BT sedimentary rocks
NT quartz arenite
SA terrigenous materials

Arequipa
No longer a valid term for GeoRef. As of 1993 see Arequipa Peru.

Arequipa Peru (1993)
Department in S Peru. Also a city. Before 1993 also search Arequipa AND Peru.
BT Peru
BT South America

arfvedsonite (1978)
BT clinoamphibole
BT amphibole group
BT chain silicates
BT silicates

Argentera
No longer a valid term for GeoRef. See Argentera Italy.

Argentera Italy (1993)
Village in NW Italy.
BT Italy
BT Southern Europe

BT Europe
SA Piemonte Italy

Argentera Massif (1978)
S Piemonte.
BT Piemonte Italy
BT Italy
BT Southern Europe
BT Europe

Argentiere Glacier (1978)
In the Mount Blanc Massif of the Pennine Alps.
BT Haute-Savoie France
BT France
BT Western Europe
BT Europe
SA Mont Blanc
SA Pennine Alps

Argentina (1978)
CO S550000S220000
    W0533000W0730000
BT South America
NT Argentine Andes
NT Buenos Aires Argentina
NT Catamarca Argentina
NT Chaco Argentina
NT Chubut Argentina
NT Cordoba Argentina
NT Corrientes Argentina
NT Jujuy Argentina
NT La Pampa Argentina
NT La Rioja Argentina
NT Mendoza Argentina
NT Neuquen Argentina
NT Neuquen Basin
NT Pampas
NT Rio Negro Argentina
NT Salado Basin
NT Salta Argentina
NT San Juan Argentina
NT San Luis Argentina
NT Santa Cruz Argentina
NT Santa Fe Argentina
NT Santiago del Estero Argentina
NT Tucuman Argentina
SA Altiplano
SA Andes
SA Baquero Formation
SA Campo del Cielo Meteorite
SA Chaco
SA Chanares Formation
SA Colorado River
SA Paganzo Group
SA Parana Basin
SA Parana River
SA Patagonia
SA Patagonian Andes
SA Precordillera
SA Puncoviscana Formation
SA Rio de la Plata
SA San Juan River
SA Santa Cruz River
SA Staten Island
SA Tierra del Fuego
SA Yacoraite Formation

Argentine Andes (1981)
Region in Argentina. Autoposting of Argentina to this term began in 1989. Autoposting of South America to this term began in 1990. This term has multiple hierarchies.
CO S450000S213000
   W0644000W0734000
BT1 Andes
BT1 South America
BT2 Argentina
BT2 South America

Argentine Basin (1978)
E of Argentina in South American Atlantic. Autoposting of Atlantic Ocean began in 1990.
BT South American Atlantic
BT South Atlantic

BT Atlantic Ocean
SA Leg 39

argentite (1978)
UF argyrite
BT sulfides
SA acanthite
SA silver
SA silver ores

Arges River (1978)
In S Romania.
BT Romania
BT Southern Europe
BT Europe

argillaceous rocks
use argillaceous texture

argillaceous texture (1978)
Before 1978, also search argillaceous AND specific sediment type or sedimentary rock.
UF argillaceous rocks
BT textures
SA argillite
SA clay
SA clay mineralogy
SA clay minerals
SA marl
SA sedimentary rocks
SA sediments
SA shale

argillite (1978)
UF dropstone
BT clastic rocks
BT sedimentary rocks
SA argillaceous texture
SA argillization
SA terrigenous materials

argillization (1989)
BT metasomatism
SA alteration
SA argillite
SA clay minerals
SA feldspar group
SA processes

Argo abyssal plain (1978)
Eastern Indian Ocean.
BT Indian Ocean
SA DSDP Site 261
SA Leg 123
SA ODP Site 765

Argolis
No longer a valid term for GeoRef. As of 1993 see Argolis Greece.

Argolis Greece (1993)
Department in E Peloponnesus. Before 1993 also search Argolis AND Greece.
BT Peloponnesus Greece
BT Greece
BT Southern Europe
BT Europe

argon (1978)
Autoposting of noble gases to this term began in 1989.
UF Ar
BT noble gases
NT Ar-36
NT Ar-37
NT Ar-38
NT Ar-38/Ar-36
NT Ar-39
NT Ar-40
NT Ar-40/Ar-36
NT Ar-40/Ar-39
SA Ar/Ar
SA isotopes
SA K/Ar

argon-argon
use Ar/Ar

Argovian (1978)
Europe. Substage in Great Britain: Upper Jurassic (lower Lusitanian; above Oxfordian Stage, below Rauracian Substage).
BT Upper Jurassic
BT Jurassic
BT Mesozoic

Argun River (1978)
Rises in NE China and serves as part of boundary between NE China and Russian Federation. Joins with Shilka River to form the Amur River. Also a river in the Greater Caucasus, in Georgian Republic and Russian Federation. IN Index Europe or Asia and Chinese provinces or Russian Federation and/or Georgian Republic as applicable.
SA Amur River
SA Asia
SA Europe
SA Georgian Republic
SA Greater Caucasus
SA Russian Far East
SA Russian Federation
SA Shilka Valley
SA Transbaikalia

Argyll (County)
Not a valid term for GeoRef. See Argyllshire Scotland.

Argyllshire
No longer a valid term for GeoRef as of 1993.
use Argyllshire Scotland

Argyllshire Scotland (1993)
Former county in NW Scotland which in 1975 was split between Highland and Strathclyde regions. The latter has the major part. Before 1978, also search Argyll. IN Index region as applicable.
UF Argyllshire
BT Scotland
BT Great Britain
BT United Kingdom
BT Western Europe
BT Europe
NT Ardnamurchan
NT Islay
NT Kintyre
NT Mull Island
SA Firth of Clyde
SA Glen Coe
SA Highland region Scotland
SA Strathclyde region Scotland

argyrite
use argentite

Ariake Bay (1978)
On East China Sea on W coast of Kyushu. Also search Ariake Sea.
BT Kyushu
BT Japan
BT Far East
BT Asia

arid environment (1978)
Before 1978, also search arid; before 1975 search arid regions.
UF arid regions
SA barchans
SA climate
SA depositional environment
SA desert varnish
SA deserts
SA ecology
SA environment
SA geomorphology
SA land use
SA paleoclimatology
SA semi-arid environment
SA wadis

SA yardangs

arid regions
A valid term through 1974.
use arid environment

Arida-gawa (1993)
W Kii Peninsula S of Osaka, S Honshu. Before 1993, also search Arita River, Aritagawa, and Arida River.
UF Arita River
UF Aritagawa
BT Wakayama Japan
BT Honshu
BT Japan
BT Far East
BT Asia

Aridisols (1978)
BT soils
SA soil group

Ariege
No longer a valid term for GeoRef. See Ariege France.

Ariege France (1993)
Department in S France.
CO N423500N432500 E0021000E0005000
BT France
BT Western Europe
BT Europe
NT Castillon Massif
NT Querigut Massif
NT Saint-Girons France
SA Lherz
SA Salat Valley

ariegite (1978)
BT ultramafics
BT plutonic rocks
BT igneous rocks
SA websterite

Ariel Satellite (1989)
One of the satellites of Uranus. Before 1989, also search Ariel AND Uranus.
SA icy satellites
SA satellites
SA Uranus
SA Voyager Program

Aries Valley (1978)
River valley in W central Transylvania. Also search Aries.
BT Transylvania
BT Romania
BT Southern Europe
BT Europe

Arikaree Group (1978)
Comprises Sharps, Monroe Creek, and Harrison Formations. Colorado, S South Dakota, SE Montana, and SE Wyoming.
BT Miocene
BT Neogene
BT Tertiary
BT Cenozoic
SA Colorado
SA Montana
SA South Dakota
SA Wyoming

Arikareean (1978)
Provincial series, North America.
IN Index ages as applicable.
BT Tertiary
BT Cenozoic
SA Kirkwood Formation
SA lower Miocene
SA upper Oligocene

Aristarchus (1989)
Lunar crater near Vallis Schroteri in region between Ocean of Storms and Sea of Rains.
UF Aristarchus Plateau
BT Moon
SA Apollo 17
SA Ocean of Storms
SA Sea of Rains

Aristarchus Plateau
use Aristarchus

Arita River
No longer a valid term for GeoRef.
use Arida-gawa

Aritagawa
use Arida-gawa

Ariyalur
No longer a valid term for GeoRef. See Ariyalur India.

Ariyalur India (1993)
Town in S Tamil Nadu, S India.
BT Tamil Nadu India
BT India
BT Indian Peninsula
BT Asia

Ariyalur Stage (1978)
S India.
BT Upper Cretaceous
BT Cretaceous
BT Mesozoic
SA India

Arize Massif (1978)
N of the central Pyrenees.
BT France
BT Western Europe
BT Europe

Arizona (1978)
Autoposting of this term began in 1978.
CO N311500N370000 W1090000W1150000
BT United States
NT Apache County Arizona
NT Cochise County Arizona
NT Coconino County Arizona
NT Date Creek basin
NT Gila County Arizona
NT Graham County Arizona
NT Greenlee County Arizona
NT La Paz County Arizona
NT Maricopa County Arizona
NT Mogollon Plateau
NT Mogollon Rim
NT Mohave County Arizona
NT Naha test well
NT Navajo County Arizona
NT Petrified Forest National Park
NT Pima County Arizona
NT Pinal County Arizona
NT Rincon Mountains
NT San Carlos Indian Reservation
NT San Francisco Peaks
NT Santa Cruz County Arizona
NT Stanton's Cave
NT Tucson Basin
NT Yavapai County Arizona
NT Yuma County Arizona
SA Apache Group
SA Basin and Range Province
SA Bidahochi Formation
SA Bisbee Group
SA Black Mesa
SA Black Mesa Basin
SA Black Mountain
SA Bright Angel Shale
SA Brushy Basin Shale Member
SA Burro Canyon Formation
SA Canyon Diablo Meteorites
SA Carmel Formation
SA Chinle Formation
SA Chocolate Mountains
SA Chuar Group
SA Coconino Sandstone
SA Colorado Lineament
SA Colorado Plateau
SA Colorado River
SA Colorado River basin
SA Cutler Formation
SA El Paso Group
SA Entrada Sandstone
SA Esplanade Sandstone Member
SA Four Corners
SA Gila River
SA Glen Canyon Group
SA Grand Canyon
SA Hermosa Formation
SA Jupiter Member
SA Kaibab Formation
SA Kayenta Formation
SA Lake Mead
SA Lake Powell
SA Laramie Formation
SA Manakacha Formation
SA Mancos Shale
SA Marble Canyon
SA Martin Formation
SA Mesaverde Group
SA Moenave Formation
SA Moenkopi Formation
SA Mohave Mountains
SA Montoya Group
SA Morrison Formation
SA Muddy Creek Formation
SA Mural Limestone
SA Navajo Indian Reservation
SA Navajo Sandstone
SA O'Leary Peak
SA Orocopia Schist
SA Paradox Basin
SA Peach Springs Tuff
SA Pedregosa Basin
SA Petrified Forest Member
SA Pioche Shale
SA Redwall Limestone
SA Rio Puerco
SA Salt River
SA Salt River valley
SA Salt Wash Sandstone Member
SA San Francisco Mountains
SA San Juan Basin
SA San Pedro Valley
SA San Rafael Group
SA Santa Cruz River
SA Shinarump Member
SA Sierrita Mountains
SA Sonoran Desert
SA Sullivan Mine
SA Supai Formation
SA Superstition Mountains
SA Tapeats Sandstone
SA Todilto Formation
SA Toroweap Formation
SA University of Arizona
SA Unkar Group
SA Verde Formation
SA Verde Valley
SA Virgin River valley
SA Watahomigi Formation
SA Wescogame Formation
SA Westwater Canyon Sandstone Member
SA White Mountains
SA Wingate Sandstone

Arkansas (1978)
Autoposting of this term began in 1978.
CO N330000N363000 W0894000W0944000
BT United States
NT Benton County Arkansas
NT Boone County Arkansas
NT Calhoun County Arkansas
NT Carroll County Arkansas
NT Clark County Arkansas
NT Clay County Arkansas
NT Cleburne County Arkansas
NT Cleveland County Arkansas
NT Columbia County Arkansas
NT Crawford County Arkansas
NT Crittenden County Arkansas
NT Dallas County Arkansas
NT Franklin County Arkansas
NT Fulton County Arkansas
NT Garland County Arkansas
NT Grant County Arkansas
NT Greene County Arkansas
NT Hot Spring County Arkansas
NT Howard County Arkansas
NT Independence County Arkansas
NT Jackson County Arkansas
NT Jefferson County Arkansas
NT Johnson County Arkansas
NT Lafayette County Arkansas
NT Lawrence County Arkansas
NT Lee County Arkansas
NT Lincoln County Arkansas
NT Logan County Arkansas
NT Madison County Arkansas
NT Marion County Arkansas
NT Monroe County Arkansas
NT Montgomery County Arkansas
NT Nevada County Arkansas
NT Newton County Arkansas
NT Perry County Arkansas
NT Phillips County Arkansas
NT Pike County Arkansas
NT Polk County Arkansas
NT Pope County Arkansas
NT Prairie County Arkansas
NT Pulaski County Arkansas
NT Randolph County Arkansas
NT Saline County Arkansas
NT Scott County Arkansas
NT Sevier County Arkansas
NT Union County Arkansas
NT Washington County Arkansas
NT White County Arkansas
SA Arkansas Novaculite
SA Arkansas River
SA Arkansas River valley
SA Arkoma Basin
SA Atoka Formation
SA Benton Uplift
SA Bloyd Formation
SA Boggy Shale
SA Boone Formation
SA Bossier Formation
SA Brassfield Formation
SA Buckner Formation
SA Carrizo Sand
SA Chattanooga Shale
SA Collier Shale
SA Cotton Valley Group
SA Desmoinesian
SA Everton Formation
SA Fayetteville Formation
SA Fayetteville Meteorite
SA Fernvale Formation
SA Georgetown Formation
SA Hartshorne Sandstone
SA Haynesville Formation
SA Hosston Formation
SA Illinois River
SA Imo Formation
SA Jackfork Group
SA Johns Valley Formation
SA Louann Salt
SA Midcontinent
SA Midway Group
SA Mississippi Embayment
SA Mississippi River
SA Mississippi Valley
SA Morrow Formation
SA New Madrid region
SA Norphlet Formation
SA Ouachita Belt
SA Ouachita Mountains
SA Ozark Mountains
SA Paluxy Formation
SA Pitkin Limestone
SA Red River
SA Red River valley

Before then, variants of the term may occur in GeoRef.

SA Reelfoot Rift
SA Rodessa Formation
SA Saint Peter Sandstone
SA Saratoga Chalk
SA Savanna Formation
SA Schuler Formation
SA Sligo Formation
SA Smackover Formation
SA Smithville Formation
SA Sparta Sand
SA Spiro Sandstone
SA Stanley Group
SA Tennessee Sandstone
SA Trinity Group
SA Walker Creek Field
SA Washita Group
SA White River
SA Womble Shale
SA Woodbine Formation

Arkansas City
Valid through 1988. Search in combination with state term. After 1988, use specific city-state term.

Arkansas City Kansas (1989)
City in S Kansas. Before 1989, search Arkansas City AND Kansas.
CO N370300N370300
    W0970200W0970200
BT Cowley County Kansas
BT Kansas
BT United States

Arkansas Novaculite (1985)
Devonian and Mississippian. SW Arkansas and SE Oklahoma.
IN Index ages as applicable.
BT upper Paleozoic
BT Paleozoic
SA Arkansas
SA Devonian
SA Mississippian
SA Oklahoma

Arkansas River (1978)
S central U.S., flows 1, 450 mi. generally ESE from the Rocky Mountains of central Colorado, through Kansas, Oklahoma, and Arkansas, to Mississippi River N of Greenville, Mississippi. Also search Arkansas River valley.
IN Index states as applicable.
BT United States
SA Arkansas
SA Colorado
SA Kansas
SA Mississippi
SA Oklahoma

Arkansas River valley (1978)
IN Index states as applicable.
BT United States
SA Arkansas
SA Colorado
SA Kansas
SA Oklahoma

Arkhangelsk
No longer a valid term for GeoRef. As of 1993, see Arkhangelsk Russian Federation.

Arkhangelsk Russian Federation (1993)
City and oblast on the White Sea in NW Russian Federation. Before 1993, also search Archangelsk. Before 1978, also search Archangel. This term has multiple hierarchies.
UF Archangel Russian Federation
BT1 Russian Federation
BT1 Commonwealth of Independent States
BT2 Europe

NT Belaya Gora Russian Federation
NT Belozero Russian Federation
NT Franz Josef Land
NT Mezen Russian Federation
NT Nenets Russian Federation
NT Novaya Zemlya
NT Onega Russian Federation
NT Vetrenyy Ridge
SA Mezen River basin
SA Northern Dvina River

Arkoma Basin (1978)
IN Index states as applicable.
BT United States
SA Arkansas
SA Oklahoma
SA Spiro Sandstone

arkose (1978)
BT clastic rocks
BT sedimentary rocks
SA arkosic composition
SA meta-arkose
SA sandstone
SA subarkose
SA terrigenous materials

arkosic composition (1978)
Before 1978, also search arkosic AND specific sediment type or sedimentary rock.
SA arkose
SA composition
SA sedimentary rocks
SA sediments

Arlan (1978)
Oil bearing region in Bashkiria in S Urals.
IN Index region as applicable.
SA Bashkiria Russian Federation

Arlington County
Valid through 1988. Search in combination with state term. After 1988, use specific county-state term.

Arlington County Virginia (1989)
N Virginia. Along Potomac River opposite Washington, D.C. Before 1989, also search Arlington County AND Virginia.
CO N385000N385600
    W0770300W0771200
BT Virginia
BT United States

armalcolite (1978)
BT oxides
SA pseudobrookite

Armenia (1978)
Formerly the Armenian Soviet Socialist Republic. This term has multiple hierarchies.
CO N390000N410000
    E0460000E0430000
BT1 Commonwealth of Independent States
BT2 Europe
NT Agarak
NT Akhtala Armenia
NT Alaverdi Armenia
NT Arteni
NT Dastakert Armenia
NT Kadzharan Armenia
NT Kafan Armenia
NT Lake Sevan
NT Megri Armenia
NT Megrinskiy Pluton
NT Pambak Armenia
NT Razdan
NT Sevan Armenia
NT Shamlug Armenia
NT Sisian Armenia
NT Yerevan Armenia
NT Zod Armenia

SA Aragats
SA Caucasus
SA Lesser Caucasus
SA Oktemberyan Series
SA Shakhdag Range
SA Spitak earthquake 1988
SA Transcaucasia
SA Zangezur

Armenia earthquake 1988
use Spitak earthquake 1988

Armidale
No longer a valid term for GeoRef. See Armidale Australia.

Armidale Australia (1993)
Town in NE New South Wales.
BT New South Wales Australia
BT Australia
BT Australasia

armored mud balls (1978)
UF mud balls, armored
BT secondary structures
BT sedimentary structures
SA clay

Armorica
use Armorican Massif

Armorican Massif (1978)
NW France.
CO N463000N494500
    W0000000W0045000
UF Armorica
BT France
BT Western Europe
BT Europe

Armstrong County
Valid through 1988. Search in combination with state term. After 1988, use specific county-state term.

Armstrong County Pennsylvania (1989)
W Pennsylvania. Before 1989, also search Armstrong County AND Pennsylvania.
CO N403200N410900
    W0791200W0794200
BT Pennsylvania
BT United States

Armstrong County Texas (1989)
Extreme N Texas. Before 1989, also search Armstrong County AND Texas.
CO N344800N351400
    W1010700W1014100
BT Texas
BT United States

Arno River Basin (1978)
N central Tuscany.
BT Tuscany Italy
BT Italy
BT Southern Europe
BT Europe

aromatic hydrocarbons (1978)
Before 1978, search aromatic or aromatics AND hydrocarbons.
BT hydrocarbons
BT organic materials
NT benzene
NT polycyclic aromatic hydrocarbons
NT toluene
NT xylene
SA aliphatic hydrocarbons
SA aromatization
SA biomarkers

aromatization (1989)
SA aromatic hydrocarbons
SA hydrocarbons
SA petroleum
SA processes

SA production

Aroostook County
Valid through 1988. Search in combination with state term. After 1988, use specific county-state term.

Aroostook County Maine (1989)
N Maine. Before 1989, also search Aroostook County AND Maine.
CO N453500N473000
    W0674700W0700200
BT Maine
BT United States
NT Presque Isle Maine
SA Saint John River

Arosa Bay (1978)
On the Atlantic Ocean in W between provinces of La Coruna and Pontevedra.
BT Galicia Spain
BT Spain
BT Iberian Peninsula
BT Southern Europe
BT Europe
SA La Coruna Spain
SA Pontevedra Spain

Arran (1978)
Island in Firth of Clyde in SW Scotland. Before 1975, it was included in Buteshire.
IN Index Strathclyde region Scotland if applicable.
SA Strathclyde region Scotland

arrays (1978)
SA computer networks
SA electrical methods
SA geophysical methods
SA geophysical surveys
SA LASA
SA networks
SA NORESS
SA NORSAR
SA PASSCAL
SA Schlumberger methods
SA seismic networks

arrival time (1978)
SA dip moveout
SA earthquakes
SA elastic waves
SA moveout
SA normal moveout
SA ray tracing
SA raypaths
SA seismic migration
SA seismology

Arrow Canyon Range (1978)
BT Nevada
BT United States

arroyos (1978)
This term has multiple hierarchies.
BT1 fluvial features
BT2 erosion features
SA ephemeral streams
SA geomorphology
SA gullies
SA wadis

arsenates (1978)
Autoposting of this term began in 1978.
NT adamite
NT austinite
NT beudantite
NT conichalcite
NT haidingerite
NT legrandite
NT mimetite
NT pharmacolite
NT pharmacosiderite
NT scorodite
SA minerals

arsenic (1978)
  Chemical element. As of 1981, use arsenic ores for arsenic as a commodity. Autoposting of metals to this term began in 1989.
  UF As
  BT metals
  SA arsenic ores
  SA arsenopyrite
  SA chalcophile elements
  SA native elements

arsenic ores (1981)
  Before 1981, also search arsenic AND (deposit OR deposits OR ore OR ores OR economic) in the basic index. Autoposting of metal ores to this term began in 1985.
  IN Commodity. See List C.
  BT metal ores
  SA arsenic
  SA arsenopyrite
  SA native elements

arsenical pyrites
  use arsenopyrite

arsenides (1978)
  Autoposting of sulfides to this term ended in 1989. Autoposting of this term to gersdorffite began in 1985.
  NT arsenopyrite
  NT cobaltite
  NT gersdorffite
  NT lollingite
  NT maucherite
  NT niccolite
  NT pararammelsbergite
  NT rammelsbergite
  NT safflorite
  NT skutterudite
  NT sperrylite
  SA minerals

arsenites (1978)
  Autoposting of this term began in 1978.
  NT mixite
  SA minerals

arsenopyrite (1978)
  This term has multiple hierarchies.
  UF arsenical pyrites
  UF white pyrites
  BT1 arsenides
  BT2 sulfides
  SA arsenic
  SA arsenic ores
  SA lollingite

arsenosulfides (1978)
  BT sulfides

Arsia Mons (1985)
  Volcano on Mars.
  BT Mars
  SA Tharsis
  SA Tharsis Montes

Artemisia (1978)
  Genus.
  BT Dicotyledoneae
  BT angiosperms
  BT Spermatophyta
  BT Plantae
  SA miospores
  SA palynomorphs

Artemovsk
  No longer a valid term for GeoRef. See Artemovsk Ukraine.

Artemovsk Ukraine (1993)
  City in the Donets Basin. Before 1993, also search Artemovsk and Bakhmut in combination with Ukraine. This term has multiple hierarchies.
  UF Bakhmut Ukraine
  BT1 Ukraine
  BT1 Europe
  BT2 Ukraine
  BT2 Commonwealth of Independent States

Arteni (1978)
  Region in Armenia. This term has multiple hierarchies.
  BT1 Armenia
  BT1 Commonwealth of Independent States
  BT2 Armenia
  BT2 Europe

artesian waters (1978)
  Before 1976, also search artesian waters and wells.
  UF artesian waters and wells
  SA aquifers
  SA aquitards
  SA confined aquifers
  SA ground water
  SA hydrogeology
  SA water
  SA water resources

artesian waters and wells
  A valid term through 1971. Was used in subfiles E and N.
  use artesian waters

Arthrodira (1978)
  Order.
  BT Placodermi
  BT Pisces
  BT Vertebrata
  BT Chordata

Arthropoda (1978)
  Autoposting of this term began in 1978.
  BT Invertebrata
  NT Chelicerata
  NT Mandibulata
  NT Trilobitomorpha
  SA arthropods
  SA cuticles

arthropods (1985)
  Common name for Arthropoda.
  IN Index for all non-paleontologic studies of fossils.
  BT invertebrates
  NT crustaceans
  NT insects
  NT trilobites
  SA Arthropoda
  SA biostratigraphy
  SA segmentation

Arthur Harbor (1978)
  On Anvers Island. Off NW Antarctic Peninsula.
  IN Index regions as applicable.
  SA Antarctica

Articulata (1978)
  Autoposting of this term to Terebratulidae and Terebratulina began in 1981.
  BT Brachiopoda
  BT Invertebrata
  NT Dictyonellidina
  NT Orthida
  NT Pentamerida
  NT Rhynchonellida
  NT Spiriferida
  NT Strophomenida
  NT Terebratulida
  NT Thecideidina
  SA Inarticulata

Articulatae (1978)
  BT Sphenopsida
  BT pteridophytes
  BT Plantae
  SA Sphenopteris

artifacts (1978)
  Before 1993, also search tools AND archaeology.
  UF pottery
  SA archaeological sites
  SA archaeology
  SA Clovis
  SA fossil man

artificial
  A valid term through 1977. After 1977, use artificial recharge; artificial intelligence, or synthetic materials, if applicable.

artificial intelligence (1989)
  SA data processing
  SA expert systems

artificial islands (1993)
  Often used in sedimentation, shore protection, and reclamation.
  UF man-made islands
  SA barrier islands
  SA engineering geology
  SA marine installations
  SA reclamation
  SA sedimentation
  SA shorelines
  SA stabilization
  SA submarine installations

artificial minerals
  Not a valid index term for GeoRef.

artificial recharge (1978)
  Also search artificial AND recharge.
  SA fluid injection
  SA ground water
  SA injection
  SA natural recharge
  SA recharge
  SA reinjection wells

artinite (1978)
  BT carbonates

Artinskian (1978)
  Europe. Above Sakmarian, below Kungurian.
  BT Lower Permian
  BT Permian
  BT Paleozoic

Artiodactyla (1978)
  Order. Autoposting of Eutheria and Theria to this term began in 1989.
  BT Eutheria
  BT Theria
  BT Mammalia
  BT Tetrapoda
  BT Vertebrata
  BT Chordata
  NT Hippopotamus
  NT Ruminantia
  NT Suiformes
  SA Ungulata

Artois (1978)
  Historic region in extreme N France, located to the S and SW of Flanders.
  BT France
  BT Western Europe
  BT Europe

Arunachal Pradesh
  No longer a valid term for GeoRef.
  use Arunachal Pradesh India

Arunachal Pradesh India (1993)
  State India, NE of Assam.
  UF Arunachal Pradesh
  BT Northeastern India
  BT India
  BT Indian Peninsula
  BT Asia
  SA Himalayas

Arunta Block (1989)
  Northern Territory, central Australia.
  BT Northern Territory Australia
  BT Australia
  BT Australasia
  SA Arunta Complex

Arunta Complex (1978)
  BT Archean
  BT Precambrian
  SA Arunta Block
  SA Australia
  SA Northern Territory Australia

Arve Valley (1978)
  River valley.
  IN Index countries as applicable.
  BT Europe
  SA France
  SA Switzerland

Arvicolidae (1978)
  In some classifications, sometimes includes some genera from the family Cricetidae. Autoposting of Eutheria and Theria to this term began in 1989.
  BT Myomorpha
  BT Rodentia
  BT Eutheria
  BT Theria
  BT Mammalia
  BT Tetrapoda
  BT Vertebrata
  BT Chordata

As
  use arsenic

Asal Rift (1985)
  Rift zone, central Djibouti and Gulf of Tadjoura.
  IN Index Djibouti and/or Gulf of Tadjoura with Gulf of Aden as applicable.
  UF Asal-Ghoubbat Rift
  UF Asal-Ghoubbet Rift
  SA Djibouti
  SA Gulf of Tadjoura

Asal-Ghoubbat Rift
  use Asal Rift

Asal-Ghoubbet Rift
  use Asal Rift

Asama (1978)
  Active volcano on W Gumma and E Nagano prefecture border, 85 miles NW of Tokyo. Also a place in Nagano prefecture. Also search Mount Asama. After 1993, use Asama Japan for the place.
  IN Index prefecture or prefectures as applicable.
  UF Asama Volcano
  UF Asamayama
  UF Mount Asama
  BT Honshu
  BT Japan
  BT Far East
  BT Asia
  SA Gumma Japan
  SA Nagano Japan

Asama Volcano
  use Asama

Asamayama
  use Asama

Asaphidae (1978)
  Family.
  BT Ptychopariida
  BT Trilobita
  BT Trilobitomorpha
  BT Arthropoda
  BT Invertebrata

asbestos (1978)

asbestos deposits
As of 1981, use asbestos deposits for asbestos as a commodity.
BT silicates
SA actinolite
SA amosite
SA amphibole group
SA anthophyllite
SA asbestos deposits
SA chain silicates
SA chrysotile
SA sheet silicates

asbestos deposits (1981)
Before 1981, search asbestos AND deposits.
IN Commodity. See List C.
SA asbestos
SA chrysotile
SA insulation materials
SA Thetford Mines

Asbian (1989)
Europe.
BT Dinantian
BT Carboniferous
BT Paleozoic

asbolane
use asbolite

asbolite (1985)
UF asbolane
BT oxides
SA manganese oxides

Ascension Island (1978)
British possession 700 miles NW of Saint Helena. In 1985, broader term changed from Atlantic Ocean to Atlantic Ocean Islands.
BT Atlantic Ocean Islands

ascharite
use szaibelyite

Aschersleben
No longer a valid term for GeoRef. See Aschersleben Germany.

Aschersleben Germany (1993)
City in Saxony-Anhalt, E central Germany. Before 1993, Aschersleben was used with Halle Bezirk.
BT Saxony-Anhalt Germany
BT Germany
BT Central Europe
BT Europe

aschynite
use aeschynite

Ascoli Piceno
No longer a valid term for GeoRef. See Ascoli Piceno Italy.

Ascoli Piceno Italy (1993)
Town and province in S Marches.
BT Marches Italy
BT Italy
BT Southern Europe
BT Europe

Ascraeus Mons (1989)
Shield volcano on Mars.
BT Mars
SA Tharsis
SA Tharsis Montes

ASEAN (1981)
Acronym. In 1985, Brunei, Indonesia, Malaysia, Philippine Islands, Singapore and Thailand were members.
UF Association of South East Asian Nations
SA associations
SA Brunei
SA Indonesia
SA international cooperation
SA Malaysia

SA Philippine Islands
SA Singapore
SA Thailand

aseismic design (1981)
SA design
SA engineering geology
SA seismic response

aseismic margins (1981)
SA continental margin
SA passive margins
SA plate tectonics

aseismic ridges (1978)
Before 1978, search aseismic.
SA microcontinents
SA mid-ocean ridges

ash (1978)
As of 1978 term is used to indicate "artificial" ash.
UF fly ash
UF industrial ash
SA aerosols
SA volcanic ash

ash falls (1978)
SA fallout
SA pyroclastics
SA volcanic ash
SA volcaniclastics
SA volcanism
SA volcanoes
SA volcanology

ash flows (1978)
UF glowing avalanche
SA ash-flow tuff
SA clastic sediments
SA flows
SA fluidization
SA igneous rocks
SA ignimbrite
SA nuees ardentes
SA pyroclastic flows
SA pyroclastics
SA sediments
SA tuff
SA volcanic ash
SA volcanic rocks
SA volcanism
SA volcanology

Ash Hollow Formation (1978)
In Ogallala Formation. W Kansas, W Nebraska, South Dakota, and SE Wyoming.
BT Pliocene
BT Neogene
BT Tertiary
BT Cenozoic
SA Kansas
SA Nebraska
SA Ogallala Formation
SA South Dakota
SA Wyoming

ash-flow tuff (1978)
BT pyroclastics
BT volcanic rocks
BT igneous rocks
SA ash flows
SA ignimbrite
SA pyroclastic flows
SA tuff
SA volcanic ash

Ashe Formation (1978)
NW North Carolina, E Tennessee, and SW Virginia.
BT lower Paleozoic
BT Paleozoic
SA North Carolina
SA Tennessee
SA Virginia

Ashgillian (1978)
Europe. Upper Caradocian, below Llandoverian (Silurian).
BT Upper Ordovician
BT Ordovician
BT Paleozoic
NT Hirnantian
SA Bala

Ashio
No longer a valid term for GeoRef. See Ashio Japan.

Ashio Japan (1993)
Town on border of Tochigi and Gumma prefectures, central Honshu.
BT Honshu
BT Japan
BT Far East
BT Asia
SA Gumma Japan
SA Tochigi Japan

Ashio Mine (1993)
Before 1993, also search Ashio copper mines.
IN Index prefectures as applicable.
BT Honshu
BT Japan
BT Far East
BT Asia
SA copper ores
SA Gumma Japan
SA mines
SA Tochigi Japan

Ashkhabad
No longer a valid term for GeoRef. As of 1993, see Ashkhabad Turkmenia.

Ashkhabad Turkmenia (1993)
Oblast and city in S Turkmenia near the Iranian border. This term has multiple hierarchies.
BT1 Turkmenia
BT1 Asia
BT2 Turkmenia
BT2 Commonwealth of Independent States
NT Cheleken Turkmenia
NT Darvaza Turkmenia

Ashland County
Valid through 1988. Search in combination with state term. After 1988, use specific county-state term.

Ashland County Ohio (1989)
N central Ohio. Before 1989, also search Ashland County AND Ohio.
CO N403300N410500 W0820700W0822600
BT Ohio
BT United States

Ashland County Wisconsin (1989)
Extreme N Wisconsin. Before 1989, also search Ashland County AND Wisconsin.
CO N455800N464000 W0901800W0905500
BT Wisconsin
BT United States

Ashley River Fault (1989)
Charleston region, SE South Carolina.
BT Charleston County South Carolina
BT South Carolina
BT United States

Ashmore Meteorite (1981)
Impact in Gaines County, W Texas. Before 1981, also search Ashmore AND meteorites.
BT H chondrites

BT chondrites
BT stony meteorites
BT meteorites
SA Gaines County Texas
SA Texas

Ashtabula County
Valid through 1988. Search in combination with state term. After 1988, use specific county-state term.

Ashtabula County Ohio (1989)
NE Ohio. Before 1989, also search Ashtabula County AND Ohio.
CO N413000N415800 W0803100W0810000
BT Ohio
BT United States

ashtonite
use mordenite

Asia (1978)
To retrieve all documents, individual countries and physiographic regions should also be searched (see list O). Autoposting of Asia to Middle East and to Middle Eastern countries began in 1981. Before 1981, Malay Archipelago was included as a narrower term. As of 1993, Asia is autoposted to many parts of the former USSR and to the Altai Mountains.
NT Abakan Russian Federation
NT Afghan-Tadzhik Depression
NT Aldan Plateau
NT Aldan River
NT Altai Mountains
NT Altai Russian Federation
NT Altai-Sayan region
NT Amu Darya
NT Amu Darya Basin
NT Amur Basin
NT Amur region
NT Amur River
NT Amur Russian Federation
NT Anadyr Basin
NT Anadyr Range
NT Angara River
NT Arabian Peninsula
NT Baikal Mountains
NT Baikal rift zone
NT Baikal-Amur Railroad region
NT Brahmaputra River
NT Buryat Russian Federation
NT Central Asia
NT Chadobets Uplift
NT Chaya Massif
NT Chirchik River
NT Chita Russian Federation
NT Chukchi Peninsula
NT Chuya Alps
NT Dzhagdy Range
NT Dzhida River
NT Dzhugdzhur region
NT Dzhungarian Alatau
NT Euphrates River
NT Far East
NT Ganges River
NT Gobi Desert
NT Gornaya Shoriya
NT Himalayas
NT Hindu Kush
NT Ili Basin
NT Indian Peninsula
NT Indus River
NT Indus-Yarlung Zangbo suture zone
NT Irkutsk Basin
NT Irkutsk Russian Federation
NT Irtysh River
NT Ishim
NT Iya River

NT Kamchatka Russian Federation
NT Kamenka River
NT Kansk-Achinsk Basin
NT Karelia Russian Federation
NT Karymskaya Sopka
NT Kemerovo Russian Federation
NT Kerulen River
NT Khabarovsk Russian Federation
NT Khamar-Daban Range
NT Khanka Lake
NT Khatanga Basin
NT Kiya River
NT Kodar Range
NT Kolyma River
NT Kolyma River basin
NT Kolyma Uplift
NT Krasnoyarsk Russian Federation
NT Kuban River
NT Kuban Valley
NT Kulunda Steppe
NT Kura River
NT Kurama Range
NT Kuznetsk Basin
NT Kyrgyzstan
NT Lake Baikal
NT Lena Basin
NT Lena River
NT Lower Tunguska River
NT Magadan Russian Federation
NT Main Boundary Fault
NT Main Central Thrust
NT Mama River
NT Maya River basin
NT Maymecha
NT Maymecha-Kotuy
NT Megion Field
NT Middle East
NT Moneron Island
NT Moskva River
NT Murgab Basin
NT Naga Hills
NT Nayba River basin
NT Novosibirsk Russian Federation
NT Ob River
NT Ob-Irtysh Interfluve
NT Okhotsk region
NT Okhotsk-Chukchi
NT Okhotsk-Chukchi volcanic belt
NT Olekma
NT Olekma-Vitim Highlands
NT Omolon
NT Omolon Block
NT Omsk Russian Federation
NT Ossetia
NT Patom Plateau
NT Penzhina Bay
NT Primorye Russian Federation
NT Pskem Range
NT Rudny Altai
NT Russian Far East
NT Russian Pacific region
NT Sakhalin Russian Federation
NT Salair Ridge
NT Salym
NT Selenga River valley
NT Sette-Daban Range
NT Shoriya Mountains
NT Siberia
NT Siberian fold belt
NT Siberian Platform
NT Sikhote-Alin Range
NT Siwalik Range
NT Southeast Asia
NT Stanovoy Range
NT Stony Tunguska River
NT Strait of Malacca
NT Suchan Basin
NT Surkhan Darya
NT Surkhan Darya basin
NT Syr Darya
NT Tadzhikistan
NT Tannu-Ola Range
NT Tarbagatay Range
NT Tatar Strait
NT Tien Shan
NT Tomsk Russian Federation
NT Transbaikalia
NT Tunguska Basin
NT Tunguska River
NT Tunguska Syneclise
NT Turan
NT Turanian Platform
NT Turkestan
NT Turkmenia
NT Turukhan
NT Tyumen Russian Federation
NT Uchur River basin
NT Uda River
NT Udokan Mountains
NT Uzbekistan
NT Verkhoyansk region
NT Vilyuy River
NT Vilyuy River basin
NT Vindhyan Basin
NT Vitim
NT West Siberia
NT Wrangel Island
NT Yakutia Russian Federation
NT Yenisei Basin
NT Yenisei River
NT Yenisei-Khatanga basin
NT Yudoma
NT Zeya
SA Argun River
SA Beringia
SA Blyava
SA Commonwealth of Independent States
SA Elbrus
SA Eurasia
SA Malay Archipelago
SA Oka River
SA Pacific mobile belt
SA Phuket Group
SA Polar Continental Shelf
SA Russian Federation
SA Tethys
SA Yangtze Plate

Asia Minor
Not a valid GeoRef index term after 1969. Was used on level 1 in subfiles E and B.
use Turkey

Asiatic
A valid term through 1977 used in combination with USSR. After 1977, see Asiatic USSR.

Asiatic USSR
As of 1993, no longer a valid term for GeoRef. Included the Russian Republic E of the Urals and the republics of Soviet Central Asia: Kazakhstan, Kirghizia, Tadzhikistan, Turkmenia, and Uzbekistan. Before 1978, search Asiatic AND USSR. Now see Russian Federation, Kazakhstan, Kyrgyzstan, Tadzhikistan, Turkmenia, or Uzbekistan.

Askania gravimeters (1981)
BT gravimeters

Aso (1978)
Volcanic mountain with five peaks in central Kyushu. Before 1993, also used for town in Kumamoto Prefecture.
UF Aso Volcano
UF Aso-san
BT Kumamoto Japan
BT Kyushu
BT Japan
BT Far East
BT Asia
SA Aso Caldera

Aso Caldera (1978)
Aso, central Kyushu.
BT Kumamoto Japan
BT Kyushu
BT Japan
BT Far East
BT Asia
SA Aso

Aso Volcano
use Aso

Aso-san
use Aso

aspartic acid (1985)
BT amino acids
BT organic materials
SA organic acids

asperities (1989)
UF asperity
SA engineering geology
SA faults
SA fractures
SA rock mechanics
SA shear tests
SA three-dimensional models

asperity
use asperities

asphalt (1978)
UF asphaltene
UF asphaltenes
BT bitumens
BT organic materials
SA biomarkers

asphaltene
use asphalt

asphaltenes
use asphalt

Assam
No longer a valid term for GeoRef. See Assam India.

Assam India (1993)
State NE of Bangladesh in NE India. Territory was reorganized in early 1970s.
BT Northeastern India
BT India
BT Indian Peninsula
BT Asia
SA Garo Hills
SA Meghalaya India
SA Shillong India
SA Shillong Plateau

Asse Mine (1989)
Salt mine and radioactive waste disposal site. Lower Saxony, N central Germany.
IN Index regions as applicable.
UF Asse Salt Mine
BT Lower Saxony Germany
BT Germany
BT Central Europe
BT Europe
SA radioactive waste
SA waste disposal
SA waste disposal sites

Asse Salt Mine
use Asse Mine

Asselian (1978)
Europe. Stage. Below Sakmarian. Above Stephanian of Upper Carboniferous.
BT Permian
BT Paleozoic

assemblages (1978)
Used for assemblages of fossils. For assemblages of minerals, use mineral assemblages.
UF faunal assemblages
UF fossil assemblages
SA biofacies
SA biotopes
SA communities
SA faunal provinces
SA floral provinces
SA fossils
SA index fossils

assimilation (1978)
As of 1981, restricted to magmas. Not to be used as a general term.
SA anatexis
SA contamination
SA country rocks
SA emplacement
SA hybridization
SA inclusions
SA intrusions
SA magmas
SA palingenesis

Assiniboine River (1978)
S Canada; 590 mi. long. Rises in SE Saskatchewan, flows S and E across S Manitoba into the Red River of the North at Winnipeg.
IN Index provinces as applicable.
BT Western Canada
BT Canada
SA Assiniboine River valley
SA Manitoba
SA Saskatchewan

Assiniboine River valley (1978)
Important wheat growing region.
IN Index provinces as applicable.
UF Assiniboine Valley
BT Western Canada
BT Canada
SA Assiniboine River
SA Manitoba
SA Saskatchewan

Assiniboine Valley
use Assiniboine River valley

Association of South East Asian Nations
use ASEAN

associations (1978)
Reserved for organizations not included under academic institutions, government agencies, museums or survey organizations. Before 1976, also search meetings.
SA AAPG
SA academic institutions
SA AGI
SA AGU
SA annual report
SA ASEAN
SA bibliography
SA catalogs
SA collections
SA current research
SA education
SA Euratom
SA expeditions
SA geology
SA government agencies
SA GSA
SA IGCP
SA INQUA
SA institutions
SA international cooperation
SA International Geological Congress
SA International Monetary Fund
SA IUGS
SA JOIDES
SA London Metal Exchange
SA museums

SA organization
SA planning
SA programs
SA progress report
SA publications
SA report
SA research
SA survey organizations
SA symposia
SA UNCTAD
SA UNDP
SA UNIDO
SA United Nations
SA World Bank

Assynt (1989)
Sutherland, N Scotland.
UF Assynt region
BT Sutherland Scotland
BT Highland region Scotland
BT Scotland
BT Great Britain
BT United Kingdom
BT Western Europe
BT Europe

Assynt region
use Assynt

Assyntian Orogeny
use Assyntic Orogeny

Assyntic Orogeny (1978)
Orogenic event at the end of the Precambrian. Before 1978, search Assyntian.
UF Assyntian Orogeny
BT Precambrian
SA Cambrian
SA orogeny
SA tectonics

Astarte (1978)
Genus. Autoposting of Astartida to this term began in 1989.
BT Astartida
BT Bivalvia
BT Mollusca
BT Invertebrata

Astartida (1981)
Autoposting of this term to Astarte began in 1989.
BT Bivalvia
BT Mollusca
BT Invertebrata
NT Astarte

astatine (1989)
Chemical element. Autoposting of halogens to this term began in 1989.
UF At
BT halogens

Asteridae (1981)
BT Dicotyledoneae
BT angiosperms
BT Spermatophyta
BT Plantae

Asterocyclina (1978)
Genus. Autoposting of microfossils and Protista to this term began in 1990. This term has multiple hierarchies.
BT1 Orbitoidacea
BT1 Rotaliina
BT1 foraminifera
BT1 Protista
BT1 Invertebrata
BT2 Orbitoidacea
BT2 Rotaliina
BT2 foraminifera
BT2 Protista
BT2 microfossils

Asteroidea (1978)
BT Stelleroidea
BT Asterozoa

BT Echinodermata
BT Invertebrata

asteroids (1978)
UF minor planets
SA cosmic dust
SA cosmochemistry
SA interplanetary space
SA meteorites
SA meteoroids
SA Moon
SA parent bodies
SA planetology
SA planets
SA solar system
SA tektites

Asterozoa (1978)
Autoposting of this term began in 1978.
BT Echinodermata
BT Invertebrata
NT Stelleroidea

asthenosphere (1978)
UF zone of mobility
SA crust
SA Earth
SA elastic waves
SA isostasy
SA lithosphere
SA lower crust
SA magmas
SA mantle
SA Mohorovicic discontinuity
SA plate tectonics
SA tectonophysics
SA upper mantle

Asti
No longer a valid term for GeoRef. See Asti Italy.

Asti Italy (1993)
City and province in NW Italy.
BT Piemonte Italy
BT Italy
BT Southern Europe
BT Europe

Astian (1978)
Europe. Above the Piacenzian.
BT upper Pliocene
BT Pliocene
BT Neogene
BT Tertiary
BT Cenozoic

Astoria Canyon (1978)
Just off Astoria, Oregon and the mouth of the Columbia River.
BT Pacific Ocean
SA Oregon

Astoria Formation (1985)
Lower to middle Miocene. NW Oregon and SW Washington.
BT Miocene
BT Neogene
BT Tertiary
BT Cenozoic
SA Oregon
SA Washington

Astrakhan
No longer a valid term for GeoRef. As of 1993, see Astrakhan Russian Federation.

Astrakhan Arch (1985)
W Caspian Depression, near mouth of Volga River.
IN Index former Soviet republics as applicable.
BT Commonwealth of Independent States
SA Astrakhan Russian Federation
SA Caspian Depression
SA Kazakhstan
SA Russian Federation

SA Volga region

Astrakhan Russian Federation (1993)
Oblast and city on NW Caspian Sea. This term has multiple hierarchies.
BT1 Russian Federation
BT1 Commonwealth of Independent States
BT2 Europe
SA Astrakhan Arch

Astrapotheria (1978)
Order. Autoposting of Eutheria and Theria to this term began in 1989.
BT Eutheria
BT Theria
BT Mammalia
BT Tetrapoda
BT Vertebrata
BT Chordata

astroblemes (1978)
Autoposting of impact features to this term began in 1989. Autoposting of impact craters to this term began in 1993.
UF fossil meteorite craters
BT impact craters
BT impact features
SA craters
SA geomorphology
SA meteor craters

astronomy
As of 1993, no longer a valid term for GeoRef. Usually out-of-scope. Used with astrophysics and solar physics or planetology.

astrophyllite (1978)
BT sheet silicates
BT silicates

astrophysics and solar physics
As of 1993, no longer a valid term for GeoRef. Usually out-of-scope. Used for special bibliographies.

Astrorhizidae (1978)
Family. Autoposting of microfossils and Protista to this term began in 1990. This term has multiple hierarchies.
BT1 Ammodiscacea
BT1 Textulariina
BT1 foraminifera
BT1 Protista
BT1 Invertebrata
BT2 Ammodiscacea
BT2 Textulariina
BT2 foraminifera
BT2 Protista
BT2 microfossils

Asturian Arc (1981)
BT Spain
BT Iberian Peninsula
BT Southern Europe
BT Europe

Asturian Massif (1978)
In NW Spain.
BT Asturias Spain
BT Spain
BT Iberian Peninsula
BT Southern Europe
BT Europe

Asturian Orogenic Phase
use Asturian Orogeny

Asturian Orogeny (1978)
Before 1978, search Asturian AND orogeny.
UF Asturian Orogenic Phase
BT Carboniferous
BT Paleozoic
SA orogeny

SA tectonics

Asturias
No longer a valid term for GeoRef. As of 1993, see Asturias Spain.

Asturias Spain (1993)
Region and ancient kingdom in NW Spain. Before 1981, also search Oviedo. Asturian Massif, Aviles and La Caridad are considered narrower terms as of 1981. Before 1993, also search Asturias and Spain.
CO N425400N434000 W0043000W0071500
UF Oviedo Spain
BT Spain
BT Iberian Peninsula
BT Southern Europe
BT Europe
NT Asturian Massif
NT Aviles Spain
NT La Caridad Spain
SA Cantabrian Basin

Asturreta (1978)
Region in N Spain.
BT Navarra Spain
BT Spain
BT Iberian Peninsula
BT Southern Europe
BT Europe

Asu River Group (1989)
In the Benue Valley.
BT Cretaceous
BT Mesozoic
SA Benue Valley
SA Nigeria

Aswan (1978)
Governate in SE Egypt. Also a city on the Nile and name of dam. As of 1993, use Aswan Egypt for governate and city.
SA Nile River

Aswan Egypt (1993)
Governate. Also a city on the Mediterranean Sea.
BT Egypt
BT North Africa
BT Africa
NT Kom Ombo Egypt

asymmetric distribution (1982)
A type of statistical distribution.
SA distribution
SA statistical analysis
SA statistical distribution
SA symmetry

asymmetric folds (1978)
Before 1978, also search folds AND asymmetric.
BT folds
SA symmetric folds

asymmetry
use symmetry

At
use astatine

Ata Caldera (1978)
Volcanic caldera on southern Kyushu, Japan. Also a caldera on uninhabited island of Ata, S Tonga. Search in combination with Kyushu or with Tonga as applicable.
IN Index Kyushu or Tonga as applicable.
SA Japan
SA Kyushu
SA Polynesia
SA Tonga

Atacama
No longer a valid term for GeoRef. See Atacama Chile.

Atacama Chile (1993)
Province in N central Chile.
BT Chile
BT South America
NT Copiapo Chile
SA Atacama Desert
SA Copiapo

Atacama Desert (1978)
N central part of country.
IN Index provinces as applicable.
BT Chile
BT South America
SA Antofagasta Chile
SA Atacama Chile

atacamite (1978)
BT chlorides
BT halides
SA paratacamite

Atascosa County
Valid through 1988. Search in combination with state term. After 1988, use specific county-state term.

Atascosa County Texas (1989)
SW Texas. Before 1989, also search Atascosa County AND Texas.
CO N283800N291500
 W0981000W0985000
BT Texas
BT United States

Atasu
No longer a valid term for GeoRef. As of 1993, see Atasu Kazakhstan.

Atasu Kazakhstan (1993)
Village in central Kazakhstan, NW of Lake Balkhash. This term has multiple hierarchies.
BT1 Kazakhstan
BT1 Central Asia
BT1 Asia
BT2 Kazakhstan
BT2 Commonwealth of Independent States

ataxite (1981)
BT iron meteorites
BT meteorites
NT Campo del Cielo Meteorite
NT Warburton Meteorite

Ataxophragmiidae (1978)
Family. Autoposting of microfossils and Protista to this term began in 1990. This term has multiple hierarchies.
BT1 Lituolacea
BT1 Textulariina
BT1 foraminifera
BT1 Protista
BT1 Invertebrata
BT2 Lituolacea
BT2 Textulariina
BT2 foraminifera
BT2 Protista
BT2 microfossils

Atbay
 use Red Sea Hills

Atchafalaya Bay (1978)
On the Gulf of Mexico off Saint Mary and Terrebonne parishes.
BT Louisiana
BT United States
SA Belle Isle
SA Gulf Coastal Plain
SA Gulf of Mexico
SA Saint Mary Parish Louisiana
SA Terrebonne Parish Louisiana

Atchafalaya River (1985)
S central Louisiana. Flows from Grand Lake N to Red River.
BT Louisiana
BT United States

Atchison
Valid through 1988. Search in combination with state term. After 1988, use specific city-state term.

Atchison County
Valid through 1988. Search in combination with state term. After 1988, use specific county-state term.

Atchison County Kansas (1989)
NE Kansas. Before 1989, also search Atchison County AND Kansas.
CO N392600N394000
 W0945900W0953300
BT Kansas
BT United States
NT Atchison Kansas

Atchison County Missouri (1989)
Extreme NW Missouri. Before 1989, also search Atchison County AND Missouri.
CO N401600N403600
 W0951100W0954700
BT Missouri
BT United States

Atchison Kansas (1989)
City in NE Kansas. Before 1989, search Atchison AND Kansas.
CO N393300N393300
 W0951900W0951900
BT Atchison County Kansas
BT Kansas
BT United States

Atdabanian (1989)
Russian Federation, Canada.
BT Lower Cambrian
BT Cambrian
BT Paleozoic

atectonic processes (1981)
UF nontectonics
SA processes
SA sedimentation
SA tectonics

Atera Fault (1978)
Gifu Prefecture in central Honshu.
BT Gifu Japan
BT Honshu
BT Japan
BT Far East
BT Asia

Athabasca District (1981)
N Saskatchewan.
UF Athabaska Basin
BT Saskatchewan
BT Western Canada
BT Canada
SA Lake Athabasca
SA Midwest Lake
SA Midwest Lake Deposits

Athabasca Formation (1981)
Primarily an orthoquartzite. Outcrops over a large area S of Lake Athabasca in Churchill Province.
BT Proterozoic
BT upper Precambrian
BT Precambrian
SA Churchill Province
SA Saskatchewan

Athabasca Glacier (1978)
In Rocky Mountains between Jasper and Banff national parks near the British Columbia border.
UF Athabaska Glacier
BT Alberta
BT Western Canada
BT Canada
SA Canadian Rocky Mountains

Athabasca Oil Sands (1981)
In E Alberta. One of the largest petroleum reservoirs in the world. Oil is found in the McMurray, Clearwater, and Grand Rapids formations.
UF Athabaska Oil Sands
BT Alberta
BT Western Canada
BT Canada
SA McMurray Formation

Athabasca River (1978)
Flows into Lake Athabasca.
UF Athabaska River
BT Alberta
BT Western Canada
BT Canada
SA Clearwater River

Athabaska Basin
 use Athabasca District

Athabaska Glacier
 use Athabasca Glacier

Athabaska Oil Sands
 use Athabasca Oil Sands

Athabaska River
 use Athabasca River

Athens
No longer a valid term for GeoRef. Before 1989, included use for cities named Athens in the U.S.; now use specific city-state term or see Athens Greece.

Athens County
Valid through 1988. Search in combination with state term. After 1988, use specific county-state term.

Athens County Ohio (1989)
SE Ohio. Before 1989, also search Athens County AND Ohio.
CO N391200N393500
 W0814200W0821800
BT Ohio
BT United States
NT Athens Ohio

Athens Georgia (1989)
City in NE Georgia. Before 1989, search Athens AND Georgia.
CO N392000N392000
 W0820600W0820600
BT Clarke County Georgia
BT Georgia
BT United States

Athens Greece (1993)
City in central Greece and Euboea. Before 1993 also search Athens AND Greece.
BT Greece
BT Southern Europe
BT Europe

Athens Ohio (1989)
City in SE Ohio. Before 1989, search Athens AND Ohio.
CO N335700N335700
 W0832400W0832400
BT Athens County Ohio
BT Ohio
BT United States

Athgarh Sandstone (1978)
Sandstones of this formation are also grouped as part of Cuttack Stage. Part of Upper Gondwana sequence of the E coast.
BT Jurassic
BT Mesozoic
SA India
SA upper Gondwana System

Atikokan
No longer a valid term for GeoRef. As of 1993, see Atikokan Ontario.

Atikokan Ontario (1993)
Rainy River District, W Ontario. Before 1993, also search Atikokan AND Ontario.
BT Rainy River District Ontario
BT Ontario
BT Eastern Canada
BT Canada

Atlanta
Valid through 1988. Search in combination with state term. After 1988, use specific city-state term.

Atlanta Georgia (1989)
City in N central Georgia. Before 1989, search Atlanta AND Georgia.
CO N334500N334500
 W0842300W0842300
BT Fulton County Georgia
BT Georgia
BT United States

Atlantic (1978)
Used primarily in Europe for an interval of postglacial time following the Boreal and preceding the Subboreal. It corresponds to most of the Altithermal and the middle part of the Hypsithermal.
BT Holocene
BT Quaternary
BT Cenozoic

Atlantic Coast
Not a valid term for GeoRef. Search Atlantic Coast OR Atlantic Ocean in conjunction with appropriate country or state. See also Atlantic region.

Atlantic Coastal Plain (1978)
As of 1993, restricted to U.S.
IN Index states as applicable.
BT United States
NT Central Atlantic Coastal Plain
NT Southern Atlantic Coastal Plain
SA AMCOR
SA Baltimore Canyon
SA Carolina Bays
SA Chesapeake Bay
SA Connecticut
SA Delaware
SA Florida
SA Georgia
SA Lydonia Canyon
SA Maryland
SA Massachusetts
SA Middendorf Formation
SA Middle Atlantic Bight
SA New Jersey
SA New York
SA New York Bight
SA North Carolina
SA Outer Banks
SA Rhode Island
SA Salisbury Embayment
SA Snuggedy Swamp
SA South Carolina
SA Virginia

Before then, variants of the term may occur in GeoRef.

Atlantic County
  Valid through 1988. Search in combination with state term. After 1988, use specific county-state term.

Atlantic County New Jersey (1989)
  SE New Jersey. Before 1989, also search Atlantic County AND New Jersey.
  CO N391700N394400 W0741900W0745900
  BT New Jersey
  BT United States
  SA Great Bay

Atlantic Margin Coring Project
  use AMCOR

Atlantic Ocean (1978)
  Before 1981, many islands which are now included under Atlantic Ocean Islands were included as narrower terms of Atlantic Ocean. For several others, the change was made in 1985. Hudson Bay is a narrower term as of 1981. Autoposting of Atlantic Ocean to Caribbean Sea began in 1981.
  CO S550000N750000 E0200000W0800000
  NT Atlantis fracture zone
  NT Atlantis II
  NT Atlantis Seamount
  NT Barbados Ridge
  NT Barracuda Ridge
  NT Bay of Fundy
  NT Bermuda Platform
  NT Blake-Bahama Basin
  NT Blake-Bahama Outer Ridge
  NT Block Island Sound
  NT Celtic Sea
  NT Davis Strait
  NT East Atlantic
  NT Equatorial Atlantic
  NT Falkland Plateau
  NT Flemish Cap
  NT Georges Bank
  NT Gibbs fracture zone
  NT Great Bahama Bank
  NT Great Meteor Seamount
  NT Gulf of Maine
  NT Gulf Stream
  NT Hatteras abyssal plain
  NT Iceland-Faeroe Ridge
  NT Labrador Basin
  NT Long Island Sound
  NT Maury Channel
  NT Mid-Atlantic Ridge
  NT North Atlantic
  NT Northeast Atlantic
  NT Northwest Atlantic
  NT Puerto Rico Trench
  NT Reykjanes Ridge
  NT Rockall Bank
  NT Rockall Plateau
  NT Rockall Trough
  NT Romanche fracture zone
  NT Romanche Trench
  NT Sable Island Bank
  NT Sargasso Sea
  NT Scotia Ridge
  NT Scotian Shelf
  NT South Atlantic
  NT Southwest Atlantic
  NT Strait of Gibraltar
  NT Straits of Florida
  NT TAG hydrothermal field
  NT Trou Sans Fond
  NT West Atlantic
  NT Wilkinson Basin
  NT Wilmington Canyon
  NT Yucatan Channel
  SA Antarctic Ocean
  SA Atlantic Ocean Islands
  SA Atlantic region
  SA Baffin Bay
  SA Blake-Bahama Formation
  SA Central Graben
  SA Continental Offshore Stratigraphic Test
  SA DSDP Site 4
  SA DSDP Site 19
  SA DSDP Site 20
  SA DSDP Site 21
  SA DSDP Site 100
  SA DSDP Site 102
  SA DSDP Site 103
  SA DSDP Site 104
  SA DSDP Site 105
  SA DSDP Site 108
  SA DSDP Site 111
  SA DSDP Site 116
  SA DSDP Site 117
  SA DSDP Site 119
  SA DSDP Site 120
  SA DSDP Site 137
  SA DSDP Site 141
  SA DSDP Site 327
  SA DSDP Site 328
  SA DSDP Site 329
  SA DSDP Site 330
  SA DSDP Site 332
  SA DSDP Site 332B
  SA DSDP Site 333
  SA DSDP Site 334
  SA DSDP Site 335
  SA DSDP Site 354
  SA DSDP Site 355
  SA DSDP Site 356
  SA DSDP Site 357
  SA DSDP Site 360
  SA DSDP Site 361
  SA DSDP Site 362
  SA DSDP Site 363
  SA DSDP Site 364
  SA DSDP Site 365
  SA DSDP Site 366
  SA DSDP Site 367
  SA DSDP Site 368
  SA DSDP Site 369
  SA DSDP Site 370
  SA DSDP Site 382
  SA DSDP Site 384
  SA DSDP Site 385
  SA DSDP Site 386
  SA DSDP Site 387
  SA DSDP Site 390
  SA DSDP Site 391
  SA DSDP Site 392
  SA DSDP Site 395
  SA DSDP Site 396
  SA DSDP Site 397
  SA DSDP Site 398
  SA DSDP Site 400
  SA DSDP Site 401
  SA DSDP Site 402
  SA DSDP Site 403
  SA DSDP Site 404
  SA DSDP Site 405
  SA DSDP Site 406
  SA DSDP Site 407
  SA DSDP Site 408
  SA DSDP Site 410
  SA DSDP Site 416
  SA DSDP Site 417
  SA DSDP Site 418
  SA DSDP Site 511
  SA DSDP Site 512
  SA DSDP Site 513
  SA DSDP Site 514
  SA DSDP Site 515
  SA DSDP Site 516
  SA DSDP Site 517
  SA DSDP Site 518
  SA DSDP Site 519
  SA DSDP Site 520
  SA DSDP Site 521
  SA DSDP Site 522
  SA DSDP Site 523
  SA DSDP Site 524
  SA DSDP Site 525
  SA DSDP Site 526
  SA DSDP Site 527
  SA DSDP Site 528
  SA DSDP Site 529
  SA DSDP Site 530
  SA DSDP Site 531
  SA DSDP Site 532
  SA DSDP Site 533
  SA DSDP Site 534
  SA DSDP Site 541
  SA DSDP Site 542
  SA DSDP Site 543
  SA DSDP Site 544
  SA DSDP Site 545
  SA DSDP Site 546
  SA DSDP Site 547
  SA DSDP Site 548
  SA DSDP Site 549
  SA DSDP Site 550
  SA DSDP Site 551
  SA DSDP Site 552
  SA DSDP Site 553
  SA DSDP Site 554
  SA DSDP Site 555
  SA DSDP Site 556
  SA DSDP Site 558
  SA DSDP Site 561
  SA DSDP Site 562
  SA DSDP Site 563
  SA DSDP Site 564
  SA DSDP Site 603
  SA DSDP Site 604
  SA DSDP Site 605
  SA DSDP Site 606
  SA DSDP Site 607
  SA DSDP Site 608
  SA DSDP Site 609
  SA DSDP Site 610
  SA DSDP Site 611
  SA DSDP Site 612
  SA DSDP Site 613
  SA Faeroe Islands
  SA Falkland Islands
  SA FAMOUS
  SA Hare Bay
  SA Hatteras Formation
  SA HEBBLE
  SA Iapetus
  SA Iceland
  SA Iceland Research Drilling Project
  SA Jan Mayen
  SA Leg 1
  SA Leg 2
  SA Leg 3
  SA Leg 4
  SA Leg 11
  SA Leg 12
  SA Leg 13
  SA Leg 14
  SA Leg 36
  SA Leg 37
  SA Leg 38
  SA Leg 39
  SA Leg 40
  SA Leg 41
  SA Leg 43
  SA Leg 44
  SA Leg 45
  SA Leg 46
  SA Leg 47
  SA Leg 48
  SA Leg 49
  SA Leg 50
  SA Leg 51
  SA Leg 52
  SA Leg 53
  SA Leg 71
  SA Leg 72
  SA Leg 73
  SA Leg 74
  SA Leg 75
  SA Leg 76
  SA Leg 78A
  SA Leg 78B
  SA Leg 79
  SA Leg 80
  SA Leg 81
  SA Leg 82
  SA Leg 93
  SA Leg 94
  SA Leg 95
  SA Leg 101
  SA Leg 102
  SA Leg 103
  SA Leg 106
  SA Leg 108
  SA Leg 109
  SA Leg 110
  SA Leg 114
  SA Norwegian Sea
  SA ODP Site 626
  SA ODP Site 627
  SA ODP Site 628
  SA ODP Site 629
  SA ODP Site 630
  SA ODP Site 631
  SA ODP Site 632
  SA ODP Site 633
  SA ODP Site 634
  SA ODP Site 635
  SA ODP Site 637
  SA ODP Site 638
  SA ODP Site 639
  SA ODP Site 640
  SA ODP Site 641
  SA ODP Site 648
  SA ODP Site 649
  SA ODP Site 657
  SA ODP Site 658
  SA ODP Site 659
  SA ODP Site 660
  SA ODP Site 661
  SA ODP Site 662
  SA ODP Site 663
  SA ODP Site 664
  SA ODP Site 665
  SA ODP Site 666
  SA ODP Site 667
  SA ODP Site 668
  SA ODP Site 669
  SA ODP Site 670
  SA ODP Site 671
  SA ODP Site 672
  SA ODP Site 673
  SA ODP Site 674
  SA ODP Site 675
  SA ODP Site 676
  SA ODP Site 698
  SA ODP Site 699
  SA ODP Site 700
  SA ODP Site 701
  SA ODP Site 702
  SA ODP Site 703
  SA ODP Site 704
  SA Orgon III
  SA Scotia Sea
  SA Signy Island
  SA South Georgia
  SA South Sandwich Islands
  SA Weddell Sea

Atlantic Ocean Islands (1981)
  Before 1981, many islands in this group were included as narrower terms of Atlantic Ocean. For several others, the change was made in 1985. Cape Verde Islands was formerly a narrower term of Africa. As of 1993, Saint Pierre and Miquelon is a narrower term.
  NT Ascension Island
  NT Azores
  NT Bermuda
  NT Bouvet Island
  NT Canary Islands
  NT Cape Verde Islands
  NT Faeroe Islands

NT Falkland Islands
NT Madeira
NT Saint Helena
NT Saint Paul Rocks
NT Saint Pierre and Miquelon
NT Sao Tome e Principe
NT Shetland Islands
NT Tristan da Cunha
SA Atlantic Ocean
SA Iceland

Atlantic region (1976)
An artificial term used to indicate the coastal region immediately adjacent to the Atlantic Ocean. Includes land and the immediate littoral zone.
SA Atlantic Ocean
SA Laurentia

Atlantic-type margins
use passive margins

Atlantis (1981)
Legendary island or continent said to have been destroyed by a cataclysm.
SA popular geology

Atlantis fracture zone (1978)
Across the Mid-Atlantic Ridge in N Atlantic Ocean.
BT Atlantic Ocean
SA Mid-Atlantic Ridge

Atlantis II (1978)
Seamounts off U.S. E coast roughly midway between Nova Scotia and Bermuda.
BT Atlantic Ocean

Atlantis II Deep (1978)
BT Red Sea
BT Indian Ocean
SA DSDP Site 227

Atlantis Seamount (1981)
SW of the Azores. Before 1981, search Atlantis.
BT Atlantic Ocean

atlas (1978)
SA catalogs
SA glossaries
SA maps
SA publications
SA report

Atlas Mountains (1978)
Mountain system of NW and N Africa extending from SW Morocco to NE Tunisia. Also search Atlas AND appropriate area.
IN Index countries as applicable.
BT North Africa
BT Africa
NT Moroccan Atlas Mountains
SA Algeria
SA Morocco
SA Tunisia

Atlin
No longer a valid term for GeoRef. See Atlin British Columbia.

Atlin British Columbia (1993)
Village near Alaska border NE of Juneau in NW British Columbia.
BT British Columbia
BT Western Canada
BT Canada

atmosphere (1978)
For studies dealing with geochemical changes (short-term or during geologic time), observations related to interactions with phenomena of the Earth's surface, and other geophysical aspects. For changes in climate, see paleoclimatology. Autoposting of this term began in 1978.
SA aerosols
SA air
SA air-sea interface
SA angular momentum
SA atmospheric precipitation
SA atmospheric pressure
SA biosphere
SA boundary interactions
SA circulation
SA climate
SA convection
SA convection currents
SA cycles
SA degassing
SA Earth
SA environmental geology
SA fallout
SA geochemistry
SA geophysics
SA glacial geology
SA greenhouse effect
SA humidity
SA hydrology
SA hydrosphere
SA meteorology
SA Milankovitch theory
SA Moon
SA ozone
SA paleoatmosphere
SA paleoclimatology
SA planetology
SA pollution
SA sea water
SA storms
SA sulfur dioxide
SA volcanology

atmospheric precipitation (1978)
Before 1978, also search hydrology AND precipitation. This term began autoposting in 1993.
NT acid rain
NT snow
SA atmosphere
SA climate
SA drought
SA evapotranspiration
SA hydrologic cycle
SA hydrology
SA meteorology
SA moisture
SA precipitation
SA raindrops
SA rainfall
SA retention
SA water

atmospheric pressure (1978)
BT pressure
SA altimetry
SA atmosphere

Atoka County
Valid through 1988. Search in combination with state term. After 1988, use specific county-state term.

Atoka County Oklahoma (1989)
SE Oklahoma. Before 1989, also search Atoka County AND Oklahoma.
CO N341000N344100 W0954000W0962300
BT Oklahoma
BT United States

Atoka Formation (1978)
Sandstone members: Coody (Coata), Pope Chapel, Georges Fork, Dirty Creek, Webbers Falls, and Blackjack School. Subdivided, in Arkansas, to include Greenland Sandstone Member at base.
BT Atokan
BT Middle Pennsylvanian
BT Pennsylvanian
BT Carboniferous
BT Paleozoic
SA Arkansas
SA Oklahoma

Atokan (1978)
Provincial series, North America: lower Middle Pennsylvanian. Above Morrowan, below Desmoinesian.
BT Middle Pennsylvanian
BT Pennsylvanian
BT Carboniferous
BT Paleozoic
NT Atoka Formation
SA Manakacha Formation
SA Watahomigi Formation

atolls (1978)
BT reefs
SA bioherms
SA Coelenterata
SA islands
SA lagoons

atomic absorption (1978)
SA absorption
SA analysis
SA atomic absorption spectra
SA chemical analysis
SA spectroscopy

atomic absorption spectra (1985)
Used for data. For methodology, use atomic absorption. Before 1985, also search atomic absorption AND spectra.
UF absorption spectra
BT spectra
SA absorption
SA atomic absorption
SA spectroscopy

atomic energy
Not a valid GeoRef index term after 1970. Was used on level 1 in subfile N.
use nuclear energy

atomic packing (1978)
Before 1978, also search packing.
SA crystal structure
SA crystal systems
SA crystallography
SA mineralogy
SA packing

Atotsugawa Fault (1989)
Honshu, central Japan.
BT Honshu
BT Japan
BT Far East
BT Asia
SA Hida Mountains
SA Hokuriku

atrazine (1993)
Used for triazine pesticides. 2-chloro-4-ethylamine-6-isopropyl amino s-triazine.
SA herbicides
SA pollutants

Atrypa (1978)
Genus.
BT Atrypidae
BT Spiriferida
BT Articulata
BT Brachiopoda
BT Invertebrata

Atrypidae (1978)
Autoposting of this term began in 1978.
BT Spiriferida
BT Articulata
BT Brachiopoda
BT Invertebrata
NT Atrypa

Atsumi Peninsula (1978)
Pacific coast of Aichi Prefecture, S central Honshu.
BT Aichi Japan
BT Honshu
BT Japan
BT Far East
BT Asia

attapulgite
use palygorskite

attenuation (1978)
Before 1981, also search damping.
UF damping
SA elastic waves
SA near-field spectra
SA seismology
SA wave dispersion

Atterberg limits (1978)
UF consistency limits
UF limits, Atterberg
SA deformation
SA plasticity
SA rock mechanics
SA soil mechanics

Attica
No longer a valid term for GeoRef. As of 1993 see Attica Greece.

Attica Greece (1993)
Ancient division and state. Currently a department in E central Greece. Before 1993 also search Attica AND Greece.
BT Sterea Ellas
BT Greece
BT Southern Europe
BT Europe

Attock
No longer a valid term for GeoRef. See Attock Pakistan.

Attock Pakistan (1993)
Town in NW Punjab Province, W Pakistan.
BT Punjab Pakistan
BT Pakistan
BT Indian Peninsula
BT Asia

Aturia (1978)
Genus.
BT Nautiloidea
BT Tetrabranchiata
BT Cephalopoda
BT Mollusca
BT Invertebrata

Au
use gold

Aube
No longer a valid term for GeoRef. See Aube France.

Aube France (1993)
Department in N central France.
CO N475500N484500 E004500E00320000
BT France
BT Western Europe
BT Europe

Aubrac (1978)
Mountain range in S central France.
BT France
BT Western Europe
BT Europe

aubrite
use chladnite

Auckland
No longer a valid term for GeoRef. See Auckland New Zealand.

Auckland Islands (1978)
200 miles south of New Zealand. In 1985, broader term changed from Pacific Ocean to West Pacific Ocean Islands.
BT West Pacific Ocean Islands

Auckland New Zealand (1993)
City in N North Island.
IN Also Index North Island.
BT New Zealand
BT Australasia
SA North Island

Aude
No longer a valid term for GeoRef. See Aude France.

Aude France (1993)
CO N423000N434000 E0031500E0014500
BT France
BT Western Europe
BT Europe
NT Corbieres
NT Mouthoumet Massif
NT Narbonne France
NT Portel France
NT Quillan France
NT Salsigne Mine
SA Montagne Noire

audio-magnetotelluric methods
use audiomagnetotelluric methods

audiomagnetotelluric methods (1985)
UF audio-magnetotelluric methods
NT CSAMT methods
SA Earth-current methods
SA geophysical methods
SA geophysical surveys
SA magnetic methods
SA magnetotelluric methods
SA magnetotelluric surveys

augen gneiss (1978)
BT gneisses
BT metamorphic rocks

Auger effect
use Auger spectroscopy

Auger spectroscopy (1982)
UF Auger effect
UF autoionization
BT spectroscopy
SA analysis
SA chemical analysis

augite (1978)
BT clinopyroxene
BT pyroxene group
BT chain silicates
BT silicates
SA basalts
SA fassaite
SA omphacite
SA titanaugite

augitite (1978)
BT ultramafics
BT plutonic rocks
BT igneous rocks

Augusta
Valid through 1988. Search in combination with state term. After 1988, use specific city-state term.

Augusta County
Valid through 1988. Search in combination with state term. After 1988, use specific county-state term.

Augusta County Virginia (1989)
NW Virginia. Before 1989, also search Augusta County AND Virginia.
CO N375300N382800 W0784500W0793000
BT Virginia
BT United States
NT Waynesboro Virginia

Augusta Maine (1989)
City in S Maine. Before 1989, search Augusta AND Maine.
CO N441700N441700 W0694800W0694800
BT Kennebec County Maine
BT Maine
BT United States

Augustine (1978)
Volcano on Augustine Island in Kamishak Bay at mouth of Cook Inlet in S Alaska.
IN Index Southwestern Alaska if applicable.
UF Augustine Volcano
SA Cook Inlet
SA Southwestern Alaska

Augustine Volcano
use Augustine

Augustow
No longer a valid term for GeoRef. As of 1993 see Augustow Poland.

Augustow Poland (1993)
Town in NE Suwalki, NE Poland. Before 1993 also search Augustow AND Poland.
BT Suwalki Poland
BT Poland
BT Central Europe
BT Europe

aulacogens (1978)
SA grabens
SA tectonics
SA troughs

aureoles (1978)
UF contact aureole
UF contact zone
UF exomorphic zone
UF metamorphic aureoles
UF metamorphic zone
UF thermal aureole
SA contact metamorphism
SA country rocks
SA haloes
SA intrusions
SA metamorphism
SA zoning

Aures (1978)
Mountain massif in NE Algeria.
BT Algeria
BT North Africa
BT Africa

Aurignacian (1985)
Archaeological classification. Europe and N Africa.
BT Paleolithic
BT Stone Age
BT Cenozoic
SA archaeology
SA Pleistocene
SA Quaternary

aurora
As of 1993, no longer a valid term for GeoRef. Usually out-of-scope. Used as a level 1 term from 1978-1992.

auroral oval
As of 1993, no longer a valid term. Usually out-of-scope for GeoRef.

auroral zone
No longer a valid term. Usually out-of-scope for GeoRef. Before 1993, included auroral oval.

Austin
Valid through 1988. Search in combination with state term. After 1988, use specific city-state term.

Austin Chalk (1989)
Member of the Austin Group. E Texas and S Louisiana.
BT Gulfian
BT Upper Cretaceous
BT Cretaceous
BT Mesozoic
SA Austin Group
SA Louisiana
SA Texas

Austin Group (1978)
Comprises Ector Tongue, Bonham Clay, Blossom Sand, Brownstone Marl, Gober Tongue, Austin Chalk, and Burditt Marl. E Texas. Also search Austin Chalk.
BT Gulfian
BT Upper Cretaceous
BT Cretaceous
BT Mesozoic
SA Austin Chalk
SA Texas

Austin Texas (1989)
City in central Texas. Before 1989, search Austin AND Texas.
CO N311800N311800 W0974700W0974700
BT Travis County Texas
BT Texas
BT United States

austinite (1978)
BT arsenates

Austral Islands (1978)
Group in French Oceania 330 miles S of Society Islands. Also search Austral. In 1981, broader term changed from Pacific Ocean to French Polynesia and Polynesia.
UF Tubuai Islands
BT French Polynesia
BT Polynesia
BT Oceania
SA Pacific Ocean

Australasia (1978)
Islands of the South Pacific including Australia, New Zealand, and Papua New Guinea. Autoposting of this term to Australia and its states began in 1981. Before 1981, New Guinea was included as a narrower term.
IN Index countries as applicable.
NT Australia
NT New Zealand
NT Papua New Guinea
NT Tasman orogenic zone
SA New Guinea
SA Oceania
SA Solomon Islands

Australia (1978)
To retrieve all documents, individual states and physiographic regions should also be searched (see list O). Autoposting of this term began in 1978.
CO S440000S100000 E1540000E1130000
BT Australasia
NT Amadeus Basin
NT Carpentaria Basin
NT Great Artesian Basin
NT Joseph Bonaparte Gulf
NT Lachlan fold belt
NT Musgrave Ranges
NT New South Wales Australia
NT Northern Territory Australia
NT Otway Basin
NT Perth Australia
NT Queensland Australia
NT Snowy Mountains
NT South Australia
NT Tasman Geosyncline
NT Tasmania Australia
NT Victoria Australia
NT Western Australia
SA Arafura Sea
SA Arunta Complex
SA Australian Plate
SA Australian Shield
SA Baralaba Coal Measures
SA Bass Strait
SA Birkhead Formation
SA Bitter Springs Formation
SA Black Mountain
SA Brockman Iron Formation
SA Broken Hill Block
SA Broken River Formation
SA Bulldog Shale
SA Bulli Seam
SA Burra Group
SA Cahill Formation
SA Cooper Basin
SA Elliot Formation
SA Eucla Basin
SA Fortescue Group
SA Gambier Embayment
SA Georgina Basin
SA Giles Complex
SA Gondwana
SA Hamersley Group
SA Hawkesbury Sandstone
SA Heemskirk Granite
SA Henbury Meteorite
SA Illawarra Coal Measures
SA Indo-Australian Plate
SA Ipswich Coal Measures
SA Kombolgie Formation
SA Latrobe Group
SA Lilydale Limestone
SA McArthur Basin
SA Melville Island
SA Mersey Valley
SA Murchison Meteorite
SA Murray Basin
SA Murray River
SA Narrabeen Group
SA New England Batholith
SA New England Orogeny
SA Newcastle Coal Measures
SA Nullarbor Plain
SA Oceania
SA Officer Basin
SA Rocknest Formation
SA Rundle Group
SA Simpson Desert
SA Singleton Coal Measures
SA Stockton Formation
SA Tapley Hill Formation
SA Toolebuc Formation
SA Umberatana Group
SA Warburton Meteorite
SA Weekeroo Station Meteorite
SA Willyama Complex
SA Wilpena Group
SA Winton Formation
SA Wolf Creek Meteorite
SA Wonoka Formation

Australian Capital Territory (1978)
SE Australia, within state of New South Wales. Formerly called Federal Capital Territory.
BT New South Wales Australia
BT Australia
BT Australasia
NT Canberra Australia

Australian Plate (1993)
Pre-dates modern Indo-Australian Plate. Before 1993, also search Indian Plate and Australia.
SA Australia
SA Indo-Australian Plate
SA New Guinea
SA New Zealand
SA plate tectonics
SA plates

Australian Shield (1978)
SW and central Australia.
SA Australia

australite (1978)
BT tektites
SA meteorites

Australopithecinae (1978)
Subfamily. Autoposting of Hominidae, Eutheria and Theria to this term began in 1989. Autoposted to Australopithecus after 1992.
BT Hominidae
BT Primates
BT Eutheria
BT Theria
BT Mammalia
BT Tetrapoda
BT Vertebrata
BT Chordata
NT Australopithecus

Australopithecus (1978)
Genus. Autoposting of this term to Australopithecus afarensis began in 1989. Autoposting of Eutheria and Theria to this term began in 1989.
BT Australopithecinae
BT Hominidae
BT Primates
BT Eutheria
BT Theria
BT Mammalia
BT Tetrapoda
BT Vertebrata
BT Chordata
NT Australopithecus afarensis

Australopithecus afarensis (1985)
Species. Autoposting of Australopithecus, Eutheria and Theria to this term began in 1989.
BT Australopithecus
BT Australopithecinae
BT Hominidae
BT Primates
BT Eutheria
BT Theria
BT Mammalia
BT Tetrapoda
BT Vertebrata
BT Chordata

Austria (1978)
CO N461500N490000 E0171500E0093000
BT Central Europe
BT Europe
NT Austrian Vienna Basin
NT Burgenland Austria
NT Carinthia Austria
NT Central Austrian Alps
NT Enns Valley
NT Gailtal Alps
NT Hochschwab
NT Hohe Tauern
NT Koralpe Range
NT Limestone Alps
NT Lower Austria
NT North Austrian Alps
NT North Austrian Crystallines
NT North Austrian Molasse
NT Penninic Zone
NT Salzburg State Austria
NT Salzkammergut
NT South Austrian Alps
NT South Austrian Molasse
NT Styria Austria
NT Tauern Tunnel
NT Tauern Window
NT Totes Gebirge
NT Tyrol Austria
NT Upper Austria
NT Venediger Group
NT Vorarlberg Austria
NT Wechsel
SA Allgau Alps
SA Alps
SA Bohemian Massif
SA Central Alps
SA Danube River
SA Danube Valley
SA Eastern Alps
SA Gosau Formation
SA Hallstatt Limestone
SA Lake Neusiedler
SA Molasse Basin
SA Morava River valley
SA Pannonia
SA Pannonian Basin
SA Rhaetian Alps
SA Rhine Basin
SA Rhine River
SA Semmering
SA Vienna Basin
SA Wetterstein Limestone

Austrian Vienna Basin (1981)
As of 1990, Central Europe and Europe are autoposted to this term. This term has multiple hierarchies.
CO N474000N485000 E0171000E0154000
BT1 Austria
BT1 Central Europe
BT1 Europe
BT2 Vienna Basin
BT2 Central Europe
BT2 Europe

authigenesis (1978)
SA authigenic minerals
SA diagenesis
SA minerals
SA recrystallization
SA sedimentary rocks

authigenetic minerals
use authigenic minerals

authigenic minerals (1978)
Also search authigenic AND minerals.
UF authigenetic minerals
SA authigenesis
SA minerals

authors
No longer a valid GeoRef index term. Before 1971, was used on level 2 in subfile N. See bibliography; biography; geologists.

autochthons (1978)
SA allochthons
SA overthrust faults
SA tectonics
SA windows

autocorrelation (1978)
BT statistical analysis
SA crosscorrelation
SA mathematical geology

autoionization
use Auger spectroscopy

automated analysis (1985)
For chemical analysis and geophysical surveys in which data acquisition and analysis are computerized.
SA analysis
SA chemical analysis
SA data processing
SA geochemical methods
SA geophysical methods
SA geophysical surveys
SA remote sensing

automatic cartography
As of 1993, no longer a valid term for GeoRef.
use digital cartography

automatic data processing
As of 1985, no longer a valid GeoRef term.
use data processing

autometamorphism (1978)
Before 1981, also search autometasomatism.
UF autometasomatism
BT metamorphism
SA albitization
SA alteration
SA hydrothermal alteration
SA igneous rocks
SA mineral assemblages
SA processes

autometasomatism
As of 1981, no longer a valid term for GeoRef.
use autometamorphism

autoradiography (1978)
UF radiography
SA analysis
SA geophysical methods
SA radioactivity

autoregression (1985)
BT regression analysis
BT statistical analysis
SA geostatistics
SA mathematical geology

Autun
No longer a valid term for GeoRef. See Autun France.

Autun France (1993)
Town in E central part of country.
BT Saone-et-Loire France
BT France
BT Western Europe
BT Europe

Autunian (1978)
Europe. Above Stephanian, below Saxonian.
BT Lower Permian
BT Permian
BT Paleozoic
SA Rotliegendes

autunite (1978)
BT phosphates
SA uranium ores

Auvergne (1978)
Historical region in S central France.
CO N443000N461500 E0044500E0020000
BT France
BT Western Europe
BT Europe

Auversian (1978)
Europe. Above Lutetian, below Bartonian.
BT upper Eocene
BT Eocene
BT Paleogene
BT Tertiary
BT Cenozoic
SA Priabonian

Aux Vases Sandstone (1989)
SW Illinois, S Indiana and E Missouri.
BT Chesterian
BT Upper Mississippian
BT Mississippian
BT Carboniferous
BT Paleozoic
SA Illinois
SA Illinois Basin
SA Indiana
SA Missouri

Auxerre
No longer a valid term for GeoRef. See Auxerre France.

Auxerre France (1993)
City in NE central France.
BT Yonne France
BT France
BT Western Europe
BT Europe

Avacha (1978)
Active volcano on S Kamchatka Peninsula. This term has multiple hierarchies.
UF Avacha Volcano
BT1 Kamchatka Russian Federation
BT1 Russian Federation
BT1 Commonwealth of Independent States
BT2 Kamchatka Russian Federation
BT2 Asia
SA Kamchatka Peninsula

Avacha Volcano
use Avacha

avalanches (1978)
Autoposting of mass movements to this term began in 1989.
BT mass movements
SA acoustical emissions
SA debris avalanches
SA geologic hazards
SA geomorphology
SA landslides
SA slope stability

Avallon
No longer a valid term for GeoRef. See Avallon France.

Avallon France (1993)
Town in NE central France.
BT Yonne France
BT France
BT Western Europe
BT Europe

Avalon Peninsula (1978)
SW of Saint John's.
CO N463000N481000 W0523000W0541500
BT Newfoundland Island
BT Newfoundland
BT Eastern Canada
BT Canada

Avalon Terrane (1993)
Continental sliver that extends from the Avalon Peninsula of Newfoundland to N Nova Scotia, through coastal New Brunswick, Maine, New Hampshire, Massachusetts, Rhode Island, and E Connecticut. Some find evidence

that it extends to the Carolina slate belt.
IN   Index provinces and/or states as applicable.
BT   North America
SA   Appalachians
SA   Avalon Zone
SA   Carolina slate belt
SA   Connecticut
SA   Georgia
SA   Maine
SA   Massachusetts
SA   New Brunswick
SA   New Hampshire
SA   Newfoundland Island
SA   North Carolina
SA   Nova Scotia
SA   Rhode Island
SA   South Carolina

Avalon Zone   (1985)
Primarily in Newfoundland. Also known in S New Brunswick and Cape Breton Island, Nova Scotia.
IN   Index provinces as applicable.
BT   Eastern Canada
BT   Canada
SA   Avalon Terrane
SA   Avalonian Orogeny
SA   Cape Breton Island
SA   New Brunswick
SA   Newfoundland Island
SA   Nova Scotia

Avalonian Orogenic Phase
use Avalonian Orogeny

Avalonian Orogeny   (1978)
Used in connection with orogenic event near the end of Precambrian time along E border of North America. Named for the Avalon Peninsula in SE Newfoundland. Before 1978, search Avalonian AND orogeny.
UF   Avalonian Orogenic Phase
BT   Precambrian
SA   Avalon Zone
SA   orogeny

Avawatz Mountains   (1989)
S California.
BT   San Bernardino County California
BT   California
BT   United States

Avery Island   (1985)
Island surrounded by marsh, S Iberia Parish, Louisiana. Also a town. As of 1993, use Avery Island Louisiana for the town.
SA   Gulf Coastal Plain
SA   Louisiana
SA   Mississippi Delta

Aves   (1978)
Class. Autoposting of this term began in 1978.
BT   Tetrapoda
BT   Vertebrata
BT   Chordata
NT   Archaeornithes
NT   Neornithes
SA   birds
SA   eggs
SA   feathers

Aves Island   (1978)
Uninhabited Venezuelan island just East of Bonaire, Netherlands Antilles. Also search Aves in combination with Caribbean Sea.
UF   Bird Island
BT   Caribbean Sea
BT   North American Atlantic
BT   North Atlantic
BT   Atlantic Ocean

Aves Ridge   (1978)
S of Aves Island in W Caribbean Sea.
BT   Caribbean Sea
BT   North American Atlantic
BT   North Atlantic
BT   Atlantic Ocean
SA   Leg 15

Aveyron
No longer a valid term for GeoRef. See Aveyron France.

Aveyron France   (1993)
Department in S central France.
CO   N434000N450000 E0033000E0014500
BT   France
BT   Western Europe
BT   Europe
NT   Decazeville France
SA   Central Massif
SA   Rodez Trough
SA   Rouergue

Avila
No longer a valid term for GeoRef. As of 1993, see Avila City Spain.

Avila City Spain   (1993)
Refers to only the city in W central Spain. Before 1993, also search Avila and Spain.
BT   Avila Spain
BT   Old Castile Spain
BT   Castile Spain
BT   Spain
BT   Iberian Peninsula
BT   Southern Europe
BT   Europe

Avila Province
No longer a valid term for GeoRef. As of 1993, see Avila Spain.

Avila Spain   (1993)
Refers only to the province in W central Spain. From 1981-1992, also search Avila Province and Spain. Before 1981, also search Avila and Spain. For the city, see Avila City Spain.
CO   N400500N411000 W0041000W0054500
BT   Old Castile Spain
BT   Castile Spain
BT   Spain
BT   Iberian Peninsula
BT   Southern Europe
BT   Europe
NT   Avila City Spain

Aviles
No longer a valid term for GeoRef. As of 1993, see Aviles Spain.

Aviles Spain   (1993)
City on the Bay of Biscay. Before 1993, also search Aviles and Spain.
BT   Asturias Spain
BT   Spain
BT   Iberian Peninsula
BT   Southern Europe
BT   Europe

AVO methods   (1993)
Seismic technique frequently used in hydrocarbon exploration. Before 1993, also search offset and amplitude.
UF   amplitude variation with offset
UF   amplitude-versus-offset analysis
SA   amplitude
SA   geophones
SA   geophysical methods
SA   hydrocarbon indicators
SA   reflection methods
SA   seismic methods

Avon
No longer a valid term for GeoRef as of 1993. See Avon England.

Avon England   (1993)
County, SW England. Formed in 1975 from Gloucestershire and Somerset.
CO   N511500N513500 W0022000W0030500
BT   England
BT   Great Britain
BT   United Kingdom
BT   Western Europe
BT   Europe
NT   Bath England
NT   Bristol England
SA   Cotswold Hills
SA   Gloucestershire England
SA   Somerset England

Avonian   (1978)
Europe.
BT   Carboniferous
BT   Paleozoic
SA   Culm
SA   Dinantian

Awamoan   (1978)
Stage. New Zealand.
UF   Awamoan Stage
BT   lower Miocene
BT   Miocene
BT   Neogene
BT   Tertiary
BT   Cenozoic

Awamoan Stage
use Awamoan

awaruite   (1978)
BT   alloys
SA   iron
SA   nickel

Awash Valley   (1978)
River valley in E Ethiopia.
SA   Ethiopia

Awashima   (1978)
Island in Sea of Japan off Niigata Prefecture.
BT   Honshu
BT   Japan
BT   Far East
BT   Asia
SA   Niigata Japan

Axel Heiberg Island   (1978)
Largest of Sverdrup Islands of Franklin District.
CO   N780000N820000 W0850000W0970000
BT   Franklin District Northwest Territories
BT   Northwest Territories
BT   Western Canada
BT   Canada

axes, fold
use fold axes

Axial Seamount   (1989)
In Juan de Fuca Ridge, NE Pacific.
BT   North American Pacific
BT   Pacific Ocean
SA   Juan de Fuca Ridge

axial surface
use axial-plane structures

axial-plane structures   (1978)
Used for foliation or cleavage. Before 1978, also search axial-plane; axial plane; axial plane foliation; axial planes.
UF   axial surface

SA   cleavage
SA   deformation
SA   flow cleavage
SA   folds
SA   foliation
SA   schistosity
SA   slaty cleavage
SA   structural analysis

axinite   (1978)
BT   ring silicates
BT   silicates
SA   borosilicates

Ayacucho
No longer a valid term for GeoRef. As of 1993 see Ayacucho Peru.

Ayacucho Peru   (1993)
Department in SW Peru. Before 1993 also search Ayacucho AND Peru.
BT   Peru
BT   South America

Aycross Formation   (1985)
NW Wyoming.
BT   middle Eocene
BT   Eocene
BT   Paleogene
BT   Tertiary
BT   Cenozoic
SA   Wyoming

Ayrshire
No longer a valid term for GeoRef as of 1993.
use Ayrshire Scotland

Ayrshire Scotland   (1993)
Former county in Strathclyde region, SW Scotland.
CO   N550000N555000 W0033000W0050000
UF   Ayrshire
BT   Strathclyde region Scotland
BT   Scotland
BT   Great Britain
BT   United Kingdom
BT   Western Europe
BT   Europe
NT   Girvan Scotland
SA   Firth of Clyde

Azalea Field   (1978)
In Midland County, W Texas.
IN   Index county or region as applicable.
SA   Midland County Texas
SA   oil and gas fields

Azerbaidzhan   (1978)
Formerly the Azerbaidzhan Soviet Socialist Republic in Transcaucasia. When referring to Iran, use Azerbaijan. This term has multiple hierarchies.
CO   N383000N420000 E0500000E0450000
BT1   Europe
BT2   Commonwealth of Independent States
NT   Akera
NT   Apsheron Peninsula
NT   Baku Archipelago
NT   Baku Azerbaidzhan
NT   Bank Azerbaidzhan
NT   Dashkesan Azerbaidzhan
NT   Kedabek Azerbaidzhan
NT   Kirovabad Azerbaidzhan
NT   Kobystan
NT   Lenkoran Azerbaidzhan
NT   Mugan Steppe
NT   Nagorno-Karabakh Azerbaidzhan
NT   Nakhichevan Azerbaidzhan
NT   Neftechala Azerbaidzhan
NT   Neftyanyye Kamni

NT Ordubad Azerbaidzhan
NT Shemakha Azerbaidzhan
NT Surakhany Azerbaidzhan
SA Azerbaijan
SA Caspian Basin
SA Caucasus
SA Filizchay
SA Greater Caucasus
SA Indarch Meteorite
SA Kura Lowland
SA Kura River
SA Lesser Caucasus
SA Peschanyy
SA Shakhdag Range
SA Transcaucasia
SA Zangezur

Azerbaijan (1978)
Used when referring to Iran. For the former Soviet republic, use Azerbaidzhan. As of 1993, also use East Azerbaijan Iran or West Azerbaijan Iran to refer to provinces.
BT Iran
BT Middle East
BT Asia
SA Azerbaidzhan

Azof Russian Federation
use Azov Russian Federation

Azolla (1978)
Genus.
BT Filicopsida
BT pteridophytes
BT Plantae

Azores (1978)
Island group belonging to Portugal in E central North Atlantic Ocean. In 1981, broader term changed from Atlantic Ocean to Atlantic Ocean Islands.
CO N370000N400000
   W0250000W0310000
BT Atlantic Ocean Islands
NT Sao Miguel Island
NT Terceira Island
SA DSDP Site 556
SA DSDP Site 558
SA DSDP Site 561
SA DSDP Site 562
SA DSDP Site 563
SA DSDP Site 564
SA Leg 82

Azores-Gibraltar Ridge (1981)
As of 1990, Atlantic Ocean is autoposted to this term.
BT North Atlantic
BT Atlantic Ocean

Azov
No longer a valid term for GeoRef. As of 1993, see Azov Russian Federation.

Azov region (1978)
Includes Azov Sea, the Crimea, the southern Ukraine bordering Azov Sea, and western section of the Northern Caucasus. Also search Azov.
IN Index Azov Sea and former Soviet republics as applicable.
CO N443000N470000
   E0390000E0350000
UF Azov-Kuban Basin
BT Europe
SA Azov Sea
SA Russian Federation
SA Ukraine

Azov Russian Federation (1993)
Town near NE arm of Azov Sea in SW Rostov Oblast. Before 1993, also search Azov or Azof. Before 1978, Azov was also used to index Azov Sea. This term has multiple hierarchies.
UF Azof Russian Federation
BT1 Rostov Russian Federation
BT1 Russian Federation
BT1 Commonwealth of Independent States
BT2 Rostov Russian Federation
BT2 Europe
SA Azov Sea

Azov Sea (1978)
Shallow north arm of the Black Sea bounded on the SW by the Crimea, N by the Ukraine, and E by the Northern Caucasus. Also search Azov. This term has multiple hierarchies.
CO N450000N470000
   E0390000E0340000
UF Sea of Azof
UF Sea of Azov
BT1 Commonwealth of Independent States
BT2 Europe
SA Azov region
SA Azov Russian Federation
SA Black Sea

Azov-Kuban Basin
use Azov region

Aztec Sandstone (1989)
S California and SE Nevada.
BT Jurassic
BT Mesozoic
SA California
SA Mojave Desert
SA Nevada

azurite (1978)
BT carbonates
SA copper minerals
SA copper ores
SA smithsonite

B

B
use boron

b value
use b-values

b-values (1989)
UF b value
SA earthquakes
SA experimental studies
SA physical properties
SA seismology
SA shock waves

Ba
use barium

Babadag Lake basin (1978)
N Dobruja.
BT Romania
BT Southern Europe
BT Europe
SA Romanian Dobruja

babingtonite (1978)
BT chain silicates
BT silicates

Babor Range (1978)
Coastal range of the Little Atlas Mountains in N Algeria. Also search Babor.
UF Babors
BT Algeria
BT North Africa
BT Africa

Babors
use Babor Range

Baca Field (1989)
Geothermal field in Valles Caldera, N New Mexico.
UF Baca geothermal field
BT Sandoval County New Mexico
BT New Mexico
BT United States
SA geothermal fields
SA Valles Caldera

Baca Formation (1989)
Central New Mexico.
BT Eocene
BT Paleogene
BT Tertiary
BT Cenozoic
SA New Mexico

Baca geothermal field
use Baca Field

Bacau
No longer a valid term for GeoRef. As of 1993 see Bacau Romania.

Bacau Romania (1993)
City and county in E central Romania. Before 1993 also search Bacau AND Romania. This term has multiple hierarchies.
BT1 Moldavia
BT1 Europe
BT2 Romania
BT2 Southern Europe
BT2 Europe
SA Bicaz Valley

back-arc basins (1985)
BT basins
SA continental margin
SA fore-arc basins
SA forelands
SA island arcs
SA orogeny
SA plate tectonics
SA subduction zones
SA tectonics

backfill (1993)
Used for mining and construction.
SA construction
SA cut-and-fill mining
SA engineering geology
SA materials
SA mining geology
SA soil-structure interface

background level (1993)
General term. Also see background radiation. Before 1993, also search geochemical background if applicable.
SA air
SA background radiation
SA detection
SA geochemistry
SA ground water
SA isotopes
SA levels
SA pollution
SA soils

background radiation (1993)
Before 1993, also search background and radioactive isotopes or radioactive waste as applicable.
SA background level
SA levels
SA radiation
SA radioactive isotopes
SA radioactive waste

Backus-Gilbert analysis (1981)
SA analysis
SA seismology

bacteria (1978)
Includes schizomycetes. Also search bacterial. Autoposting of this term began in 1978.
UF schizomycetes
BT thallophytes
BT Plantae
NT Thiobacillus ferrooxidans
SA biology
SA biota
SA fungi
SA microorganisms
SA parasites
SA prokaryotes
SA Protista
SA soils

Bactritida (1981)
Sometimes classified under Nautiloidea, or as a separate order.
BT Ammonoidea
BT Tetrabranchiata
BT Cephalopoda
BT Mollusca
BT Invertebrata
SA Nautiloidea

Baculites (1978)
Genus.
BT Ammonoidea
BT Tetrabranchiata
BT Cephalopoda
BT Mollusca
BT Invertebrata

Bad Deutsch Altenburg
No longer a valid term for GeoRef.
use Bad Deutsch Altenburg Austria

Bad Deutsch Altenburg Austria (1993)
Village on the Danube in E Lower Austria.
UF Bad Deutsch Altenburg
BT Lower Austria
BT Austria
BT Central Europe
BT Europe

Bad Gastein
No longer a valid term for GeoRef. See Bad Gadstein Austria.

Bad Gastein Austria (1993)
Town in W central Austria.
BT Salzburg State Austria
BT Austria
BT Central Europe
BT Europe

Bad Pyrmont
No longer a valid term for GeoRef. See Bad Pyrmont Germany.

Bad Pyrmont Germany (1993)
Town in S Lower Saxony, central Germany.
BT Lower Saxony Germany
BT Germany
BT Central Europe
BT Europe

Badajoz
No longer a valid term for GeoRef. As of 1993, see Badajoz City Spain.

Badajoz City Spain (1993)
Refers to only the city in W Spain. Before 1993, also search Badajoz and Spain.
BT Badajoz Spain
BT Extremadura Spain
BT Spain
BT Iberian Peninsula
BT Southern Europe
BT Europe

Badajoz Province
No longer a valid term for GeoRef. As of 1993, see Badajoz Spain.

Badajoz Spain (1993)
Refers only to the province in W Spain on Portuguese border. Before 1993, also search Badajoz Province and Spain. Before 1981, also search Badajoz and Spain. For the city, see Badajoz City Spain.
CO N375700N393000
 W0044000W0072000
BT Extremadura Spain
BT Spain
BT Iberian Peninsula
BT Southern Europe
BT Europe
NT Badajoz City Spain

Badakhshan (1978)
Part of Gorno-Badakshan Autonomous Oblast in Tadzhikistan. Before 1993, also used for frontier province in NE Afghanistan. After 1992, use Badakhshan Afghanistan for the province. This term has multiple hierarchies.
BT1 Tadzhikistan
BT1 Asia
BT2 Tadzhikistan
BT2 Commonwealth of Independent States
SA Afghanistan
SA Shugnan Tadzhikistan

Badakhshan Afghanistan (1993)
Frontier province in NE Afghanistan. Before 1993, also search Badakhshan and Afghanistan.
BT Afghanistan
BT Indian Peninsula
BT Asia
SA Tadzhikistan

Badami Series (1978)
The Badami Series lies over the Kaladgi Series and is overlain by thick cover of Deccan Traps. Belgaum District in SW part of India.
BT Proterozoic
BT upper Precambrian
BT Precambrian
SA Deccan Traps
SA India
SA Karnataka India

baddeleyite (1978)
BT oxides
SA ceramic materials

Baden (1978)
Former German state.
BT Baden-Wurttemberg Germany
BT Germany
BT Central Europe
BT Europe

Baden-Baden
No longer a valid term for GeoRef. See Baden-Baden Germany.

Baden-Baden Germany (1993)
City, Baden-Wurttemberg, SW Germany. Before 1978, also search Baden.
BT Baden-Wurttemberg Germany
BT Germany
BT Central Europe
BT Europe

Baden-Wuerttemberg
use Baden-Wurttemberg Germany

Baden-Wurttemberg
No longer a valid term for GeoRef.
use Baden-Wurttemberg Germany

Baden-Wurttemberg Germany (1993)
State in SW Germany.
CO N473000N494800
 E0103000E0073000
UF Baden-Wuerttemberg
UF Baden-Wurttemberg
BT Germany
BT Central Europe
BT Europe
NT Baden
NT Baden-Baden Germany
NT Black Forest
NT Freiburg Germany
NT Hegau
NT Heidelberg Germany
NT Holzmaden region
NT Kaiserstuhl
NT Karlsruhe Germany
NT Katzenbuckel
NT Neckar River
NT Stuttgart Germany
NT Swabian Alb
NT Urach Germany
NT Weinheim Germany
NT Wurttemberg
NT Wutach Valley
SA Alpenvorland
SA Clara Mine
SA Franconia
SA Lake Constance
SA Main River
SA Odenwald
SA Palatinate
SA Swabia
SA Upper Rhine Valley

Badenian (1978)
Provincial series, Europe.
BT Miocene
BT Neogene
BT Tertiary
BT Cenozoic
SA lower Miocene
SA middle Miocene

Badkhyz (1978)
Steppe region in extreme S Turkmenia near Afghanistan and Iranian borders. This term has multiple hierarchies.
BT1 Turkmenia
BT1 Asia
BT2 Turkmenia
BT2 Commonwealth of Independent States

badlands (1978)
Generic term for a type of topography. Autoposting of erosion features to this term began in 1989.
BT erosion features
SA geomorphology

Badlands National Monument (1978)
Badlands area in Theodore Roosevelt National Memorial Park in SW South Dakota. Also search Badlands AND South Dakota.
BT South Dakota
BT United States
SA national monuments

Baffin Bay (1978)
Part of Arctic Ocean between Greenland and Baffin Island, and inlet of Laguna Madre, S Texas. Before 1981, broader term was Atlantic Ocean.
IN Index Arctic Ocean or Texas and counties as applicable.
SA Arctic Ocean
SA Atlantic Ocean
SA Kenedy County Texas
SA Kleberg County Texas
SA Leg 105
SA Nares Strait
SA Northwest Territories
SA ODP Site 645
SA Texas

Baffin Island (1978)
Largest and most easterly of the Canadian Arctic Archipelago in Franklin District.
IN Index Northwest Territories if applicable.
SA Arctic Archipelago
SA Barnes ice cap
SA Cape Dyer
SA Cumberland Peninsula
SA Franklin District Northwest Territories
SA Frobisher Bay
SA Keewatin District Northwest Territories
SA Narpaing Fjord
SA Northwest Territories

Bagalkot
No longer a valid term for GeoRef. See Bagalkot India.

Bagalkot India (1993)
Town in NW Karnataka, S India.
BT Karnataka India
BT India
BT Indian Peninsula
BT Asia

Bagh Beds (1978)
Includes Nimar Sandstone, Nodular Limestone, Deola Marl, Coralline Limestone.
BT Upper Cretaceous
BT Cretaceous
BT Mesozoic
SA Gujarat India
SA India
SA Madhya Pradesh India

Bahama Islands
use Bahamas

Bahama Platform
use Blake Plateau

Bahamas (1978)
A chain of islands SE of Florida and N of Cuba.
CO N204500N280000
 W0720000W0793000
UF Bahama Islands
BT West Indies
BT Caribbean region
NT Andros Island
NT Bimini
NT Tongue of the Ocean
SA Black Rock
SA Exuma Sound
SA Leg 101
SA Little Bahama Bank
SA Northeast Providence Channel
SA Northwest Providence Channel
SA ODP Site 631
SA ODP Site 632
SA ODP Site 633
SA ODP Site 634
SA ODP Site 635
SA San Salvador

Bahariya Oasis (1978)
In the Libyan Desert in W Egypt.
BT Egypt
BT North Africa
BT Africa
SA Libyan Desert

Bahia
No longer a valid term for GeoRef. See Bahia Brazil.

Bahia Brazil (1993)
State in E Brazil.
CO S182000S082000
 W0372000W0464000
BT Brazil
BT South America
NT Boquira Mine
NT Curaca River basin
NT Reconcavo Basin
NT Salvador Brazil
NT Tucano Basin
NT Uaua Brazil
SA Barreiras Formation
SA Brazilian Shield
SA Sao Francisco Basin
SA Sao Francisco Craton
SA Serra do Espinhaco

Bahrain (1978)
An archipelago in the W Persian Gulf.
CO N253000N263000
 E0505500E0501000
UF Bahrein
BT Arabian Peninsula
BT Asia

Bahrein
use Bahrain

Baia Mare
No longer a valid term for GeoRef. As of 1993 see Baia Mare Romania.

Baia Mare Romania (1993)
Town in Maramures County in N Transylvania. Also search Baia-Mare.
UF Baia-Mare
BT Maramures Romania
BT Transylvania
BT Romania
BT Southern Europe
BT Europe

Baia-Mare
use Baia Mare Romania

Baie of Islands Ophiolite
use Bay of Islands Ophiolite

Baikal (Lake)
use Lake Baikal

Baikal Mountains (1978)
West shore of Lake Baikal mostly in Irkutsk Oblast. Also search Baikal. This term has multiple hierarchies.
BT1 Russian Federation
BT1 Commonwealth of Independent States
BT2 Asia
SA Buryat Russian Federation
SA Irkutsk Russian Federation

Baikal region (1978)
Lake Baikal, Baikal Mountains, NE Irkutsk Oblast, and N Buryat Autonomous Republic. Also search Baikal. This term has multiple hierarchies.
CO N510000N610000
 E1183000E1040000
BT1 Siberian fold belt
BT1 Asia
BT2 Russian Federation
BT2 Commonwealth of Independent States
SA Baikal-Amur Railroad region
SA Buryat Russian Federation
SA Chaya Massif
SA Irkutsk Russian Federation
SA Lake Baikal
SA Transbaikalia

Baikal rift zone (1981)
S central Siberia. This term has multiple hierarchies.
CO N513000N580000
   E1200000E1000000
BT1 Russian Federation
BT1 Commonwealth of Independent States
BT2 Asia
SA Buryat Russian Federation
SA Irkutsk Russian Federation
SA Lake Baikal
SA Siberia

Baikal-Amur Railroad region (1985)
SE Russian Federation, from Lake Baikal to the Pacific. This term has multiple hierarchies.
BT1 Russian Federation
BT1 Commonwealth of Independent States
BT2 Asia
SA Amur region
SA Baikal region
SA Buryat Russian Federation
SA Chita Russian Federation
SA Russian Far East

Baikalian Orogenic Phase
   use Baikalian Phase

Baikalian Orogeny
   use Baikalian Phase

Baikalian Phase (1978)
Widely used throughout the Russian Federation for orogeny occurring during the Precambrian and Cambrian. Named after Lake Baikal in Siberia. Before 1978, search Baikalian AND orogeny; Baikalian Orogeny.
IN Index ages as applicable.
UF Baikalian Orogenic Phase
UF Baikalian Orogeny
SA Cambrian
SA orogeny
SA Precambrian
SA Russian Federation
SA Siberia

Bainbridge Formation (1978)
Characterized by two lithologically well-differentiated formations, Saint Clair below, and Moccasin Springs above. SW Illinois and E Missouri.
BT Niagaran
BT Middle Silurian
BT Silurian
BT Paleozoic
SA Illinois
SA Missouri

Bairdia (1978)
Genus. As of 1990, microfossils, Crustacea, Mandibulata, and Arthropoda are autoposted to this term. This term has multiple hierarchies.
BT1 Bairdiidae
BT1 Bairdiacea
BT1 Bairdiomorpha
BT1 Podocopida
BT1 Ostracoda
BT1 Crustacea
BT1 Mandibulata
BT1 Arthropoda
BT1 Invertebrata
BT2 Bairdiidae
BT2 Bairdiacea
BT2 Bairdiomorpha
BT2 Podocopida
BT2 Ostracoda
BT2 microfossils

Bairdiacea (1978)
As of 1990, microfossils, Crustacea, Mandibulata, and Arthropoda are autoposted to this term. This term has multiple hierarchies.
BT1 Bairdiomorpha
BT1 Podocopida
BT1 Ostracoda
BT1 Crustacea
BT1 Mandibulata
BT1 Arthropoda
BT1 Invertebrata
BT2 Bairdiomorpha
BT2 Podocopida
BT2 Ostracoda
BT2 microfossils
NT Bairdiidae

Bairdiidae (1978)
As of 1990, microfossils, Crustacea, Mandibulata, and Arthropoda are autoposted to this term. This term has multiple hierarchies.
BT1 Bairdiacea
BT1 Bairdiomorpha
BT1 Podocopida
BT1 Ostracoda
BT1 Crustacea
BT1 Mandibulata
BT1 Arthropoda
BT1 Invertebrata
BT2 Bairdiacea
BT2 Bairdiomorpha
BT2 Podocopida
BT2 Ostracoda
BT2 microfossils
NT Bairdia

Bairdiomorpha (1981)
Includes Bairdiacea. Autoposting of this term began in 1985. As of 1990, microfossils, Crustacea, Mandibulata, and Arthropoda are autoposted to this term. This term has multiple hierarchies.
BT1 Podocopida
BT1 Ostracoda
BT1 Crustacea
BT1 Mandibulata
BT1 Arthropoda
BT1 Invertebrata
BT2 Podocopida
BT2 Ostracoda
BT2 microfossils
NT Bairdiacea

Baita
No longer a valid term for GeoRef. As of 1993 see Baita Romania.

Baita Romania (1993)
Village in Maramures County, NW Romania. Before 1993 also search Baita AND Romania.
BT Maramures Romania
BT Transylvania
BT Romania
BT Southern Europe
BT Europe

Baixo Alentejo (1978)
A region in S Portugal.
BT Portugal
BT Iberian Peninsula
BT Southern Europe
BT Europe

Baja California (1978)
A peninsula. Northern half is state of Baja California, formerly Baja California Norte. Southern half is the territory of Baja California Sur.
CO N230000N323000
   W1090000W1170000
BT Mexico
SA Agua Blanca Fault
SA Alisitos Formation
SA Baja California Mexico
SA Baja California Sur Mexico
SA Cerro Prieto
SA Cerro Prieto Fault
SA Imperial Valley
SA Mexicali Valley
SA Ojo de Liebre Lagoon
SA Peninsular Ranges Batholith
SA Rosario Formation
SA Sierra de Juarez
SA Todos Santos Bay
SA Vizcaino Peninsula

Baja California Mexico (1993)
Used for the Mexican state in the northern part of the peninsula. Before 1993 also search Baja California Norte.
UF Baja California Norte
BT Mexico
NT Cerro Prieto
NT Mexicali Mexico
NT Ojo de Liebre Lagoon
NT Sierra de Juarez
NT Todos Santos Bay
SA Baja California
SA Baja California Sur Mexico
SA Colorado River
SA Colorado River delta
SA Peninsular Ranges Batholith
SA Santa Susana Formation

Baja California Norte
No longer a valid term for GeoRef.
use Baja California Mexico

Baja California Sur
No longer a valid term for GeoRef.
use Baja California Sur Mexico

Baja California Sur Mexico (1993)
Used for the Mexican territory in the southern part of the peninsula of Baja California. Before 1993 also search Baja California Sur.
CO N225000N280000
   W1091000W1150000
UF Baja California Sur
BT Mexico
NT Vizcaino Peninsula
SA Baja California
SA Baja California Mexico

Bajocian (1978)
Europe.
BT Middle Jurassic
BT Jurassic
BT Mesozoic
NT Brent Group
SA Dogger

Bakal
No longer a valid term for GeoRef. As of 1993, see Bakal Russian Federation.

Bakal Russian Federation (1993)
City in Chelyabinsk Oblast in Southern Urals. Before 1993, also search Bakal.
IN Index region as applicable.
BT Russian Federation
BT Commonwealth of Independent States
SA Chelyabinsk Russian Federation
SA Southern Urals

Baker County
Valid through 1988. Search in combination with state term. After 1988, use specific county-state term.

Baker County Florida (1989)
N Florida. Before 1989, also search Baker County AND Florida.
CO N300700N303100
   W0820200W0822500
BT Florida
BT United States
SA Okefenokee Swamp

Baker County Georgia (1989)
SW Georgia. Before 1989, also search Baker County AND Georgia.
CO N310500N312600
   W0840800W0843700
BT Georgia
BT United States

Baker County Oregon (1989)
On Powder River in NE Oregon. Before 1989, also search Baker County AND Oregon.
CO N441500N450500
   W1164900W1183300
BT Oregon
BT United States
SA Wallowa Mountains

Baker Lake (1985)
Lake in E central Keewatin District. As of 1993, use Baker Lake Northwest Territories for the town at W end of the lake.
IN Index Keewatin District or region as applicable.
SA Keewatin District Northwest Territories

Bakersfield
Valid through 1988. Search in combination with state term. After 1988, use specific city-state term.

Bakersfield California (1989)
City in S central California. Before 1989, search Bakersfield AND California.
CO N352500N352500
   W1190000W1190000
BT Kern County California
BT California
BT United States

Bakhchisarai
No longer a valid term for GeoRef. See Bakhchisarai Ukraine.

Bakhchisarai Ukraine (1993)
Town in S Crimea. Before 1993, also search Bakhchisarai or Bakhchisaray in combination with Ukraine. This term has multiple hierarchies.
UF Bakhchisaray Ukraine
BT1 Crimea Ukraine
BT1 Ukraine
BT1 Europe
BT2 Crimea Ukraine
BT2 Ukraine
BT2 Commonwealth of Independent States

Bakhchisaray Ukraine
use Bakhchisarai Ukraine

Bakhmut Ukraine
use Artemovsk Ukraine

Bakhtiari Formation (1978)
Comprises Upper and Lower Bakhtiari Formations.
BT Pliocene
BT Neogene
BT Tertiary
BT Cenozoic
SA Iran
SA Iraq

Bakken Formation (1985)

Upper Devonian? and Lower Mississippian. North Dakota, Montana, Manitoba and Saskatchewan.
 BT upper Paleozoic
 BT Paleozoic
 SA Lower Mississippian
 SA Manitoba
 SA Montana
 SA North Dakota
 SA Saskatchewan
 SA Upper Devonian

Bakony Mountains (1978)
 In NW central Hungary. Also search Bakony.
 UF Transdanubian Central Mountains
 BT Hungary
 BT Central Europe
 BT Europe

Baku
 No longer a valid term for GeoRef. As of 1993, see Baku Azerbaidzhan.

Baku Archipelago (1978)
 Also search Baku. This term has multiple hierarchies.
 BT1 Azerbaidzhan
 BT1 Europe
 BT2 Azerbaidzhan
 BT2 Commonwealth of Independent States

Baku Azerbaidzhan (1993)
 City on Apsheron Peninsula on Caspian Sea. This term has multiple hierarchies.
 BT1 Azerbaidzhan
 BT1 Europe
 BT2 Azerbaidzhan
 BT2 Commonwealth of Independent States

Bala (1978)
 England: Middle and Upper Ordovician. Comprises the Caradocian and Ashgillian. Before 1993, this term and Bala District were also used to refer to a Welsh town and to various other locations.
 BT Ordovician
 BT Paleozoic
 SA Ashgillian
 SA Caradocian
 SA Hirnantian
 SA Merionethshire Wales
 SA Middle Ordovician
 SA Upper Ordovician
 SA Wales

Balaghat
 No longer a valid term for GeoRef. See Balaghat India.

Balaghat India (1993)
 Town and district in central Madhya Pradesh, central India.
 BT Madhya Pradesh India
 BT India
 BT Indian Peninsula
 BT Asia

Balaklala Rhyolite (1989)
 N California. Overlies Copley Greenstone and interfingers with it in some areas.
 BT Middle Devonian
 BT Devonian
 BT Paleozoic
 SA California
 SA Copley Greenstone

balance, mass
 use mass balance

balance, water
 use water balance

Balasore
 No longer a valid term for GeoRef. See Balasore India.

Balasore India (1993)
 Town and district on Bay of Bengal in NE Orissa.
 BT Orissa India
 BT India
 BT Indian Peninsula
 BT Asia

Balaton (Lake)
 use Lake Balaton

Balaton region (1978)
 Area around Lake Balaton in W Hungary. Also search Balaton.
 BT Hungary
 BT Central Europe
 BT Europe
 SA Lake Balaton

Balcanoona
 No longer a valid term for GeoRef. See Balcanoona Australia.

Balcanoona Australia (1993)
 Village in E South Australia.
 CO S303300S303300
  E1392200E1392200
 BT South Australia
 BT Australia
 BT Australasia

Balcones Escarpment
 use Balcones fault zone

Balcones fault zone (1989)
 Central Texas.
 UF Balcones Escarpment
 UF Balcones-Ouachita Trend
 UF Ouachita-Balcones Trend
 BT Texas
 BT United States

Balcones-Ouachita Trend
 use Balcones fault zone

Bald Mountain (1978)
 Name of peaks which occur in several states: central Colorado; central and SW central Idaho; E central Oregon; W South Dakota; NE Utah; N Wyoming and W central Wyoming.
 IN Index states as applicable.
 SA Colorado
 SA Idaho
 SA Oregon
 SA South Dakota
 SA Utah
 SA Wyoming

Baldwin County
 Valid through 1988. Search in combination with state term. After 1988, use specific county-state term.

Baldwin County Alabama (1989)
 On Mobile Bay and Gulf of Mexico in SW Alabama. Before 1989, also search Baldwin County AND Alabama.
 CO N301300N311800
  W0872200W0880200
 BT Alabama
 BT United States

Baldwin County Georgia (1989)
 Central Georgia. Before 1989, also search Baldwin County AND Georgia.
 CO N325500N331100
  W0830400W0832200
 BT Georgia
 BT United States

Balearic abyssal plain
 use Balearic Basin

Balearic Basin (1978)
 Between Balearic Islands and Sardinia.
 UF Algerian Plain
 UF Balearic abyssal plain
 BT Mediterranean Sea
 SA DSDP Site 372

Balearic Islands (1978)
 Off E coast of Spain. Forms Spanish province of Baleares. In 1981, broader term changed from Mediterranean Sea to Spain. Autoposting of Mediterranean region, Southern Europe and Europe to this term began in 1989. This term has multiple hierarchies.
 CO N383500N400500
  E0042500E0010500
 BT1 Mediterranean region
 BT2 Spain
 BT2 Iberian Peninsula
 BT2 Southern Europe
 BT2 Europe
 NT Ibiza
 NT Majorca
 NT Minorca
 SA Mediterranean Sea

Balei
 No longer a valid term for GeoRef. As of 1993, see Balei Russian Federation.

Balei Russian Federation (1993)
 City in S central Chita Oblast E of Lake Baikal. Before 1993, also search Balei or Baley.
 IN Index regions as applicable.
 UF Baley Russian Federation
 BT Russian Federation
 BT Commonwealth of Independent States
 SA Chita Russian Federation

Baley Russian Federation
 use Balei Russian Federation

Bali (1978)
 Island off E end of Java. Lesser Sunda Islands is considered a broader term as of 1981. This term has multiple hierarchies.
 BT1 Lesser Sunda Islands
 BT1 Far East
 BT1 Asia
 BT2 Indonesia
 BT2 Far East
 BT2 Asia
 NT Mount Agung

Balkan Foreland (1978)
 A stable area marginal to the Balkan Mountains of central Bulgaria.
 BT Bulgaria
 BT Southern Europe
 BT Europe

Balkan Mountains (1978)
 Range extending E and W across central Bulgaria.
 CO N423500N440000
  E0280000E0221500
 UF Stara Planina
 BT Bulgaria
 BT Southern Europe
 BT Europe

Balkan Peninsula (1978)
 SE Europe between the Adriatic and Ionian seas on the W; the Mediterranean Sea on the S; and the Aegean and Black seas on the E. Before 1978, also search Balkans.
 UF Balkan States
 UF Balkans
 BT Europe
 SA Albania
 SA Bulgaria
 SA Greece
 SA Romania
 SA Turkey
 SA Yugoslavia

Balkan States
 use Balkan Peninsula

Balkans
 use Balkan Peninsula

Balkhan (1978)
 Two mountain ranges in Krasnovodsk Oblast in W: the Greater Balkhan Range and the Lesser Balkhan Range. This term has multiple hierarchies.
 BT1 Turkmenia
 BT1 Asia
 BT2 Turkmenia
 BT2 Commonwealth of Independent States

Balkhash
 No longer a valid term for GeoRef. As of 1993, see Balkhash Kazakhstan.

Balkhash (Lake)
 use Lake Balkhash

Balkhash Kazakhstan (1993)
 City on N shore of Lake Balkhash. This term has multiple hierarchies.
 BT1 Kazakhstan
 BT1 Central Asia
 BT1 Asia
 BT2 Kazakhstan
 BT2 Commonwealth of Independent States

Balkhash region (1978)
 Area around Lake Balkhash. Also search Balkhash. This term has multiple hierarchies.
 BT1 Kazakhstan
 BT1 Central Asia
 BT1 Asia
 BT2 Kazakhstan
 BT2 Commonwealth of Independent States
 SA Lake Balkhash

ball and pillow
 use ball-and-pillow

ball-and-pillow (1978)
 UF ball and pillow
 BT soft sediment deformation
 BT sedimentary structures
 SA pillow structure
 SA primary structures

Ballantrae Complex (1989)
 SW Scotland.
 BT Arenigian
 BT Lower Ordovician
 BT Ordovician
 BT Paleozoic
 SA Great Britain
 SA Scotland
 SA United Kingdom

balls, coal
 use coal balls

Balmat
 No longer a valid term for GeoRef. See Balmat-Edwards mining district.

Balmat-Edwards District
 No longer a valid term for GeoRef.
 use Balmat-Edwards mining district

Balmat-Edwards mining district (1993)

Mining district in N New York. Before 1981, also search Balmat; Edwards. Before 1993, also search Balmat-Edwards District.
CO N441500N442000
   W0751500W0752500
UF Balmat-Edwards District
BT Saint Lawrence County New York
BT New York
BT United States
SA mines

Baltic Basin
   use Baltic region

Baltic Coast
   use Baltic region

Baltic Glaciation (1978)
Comprises Baltic region plus Norway which was completely covered by Pleistocene to upper Holocene glaciation. Also search Baltic AND glaciation. Before 1993, also search Baltic Sea Stages. Includes Baltic ice lake, Yoldia Sea, Ancylus Lake, and Litorina Sea.
IN Index ages as applicable.
UF Baltic Sea Stages
BT upper Quaternary
BT Quaternary
BT Cenozoic
SA Ancylus Lake
SA Baltic ice lake
SA Baltic region
SA Baltic Sea
SA Denmark
SA Estonia
SA Germany
SA Kola Peninsula
SA Latvia
SA Lithuania
SA Litorina Sea
SA lower Holocene
SA Norway
SA Poland
SA Russian Federation
SA Sweden
SA upper Pleistocene
SA Yoldia Sea

Baltic ice lake (1993)
Upper Pleistocene. Stage of Baltic Glaciation which covered most of Scandinavia, Baltic Sea, Finland, and part of Kola Peninsula.
BT upper Pleistocene
BT Pleistocene
BT Quaternary
BT Cenozoic
SA Baltic Glaciation
SA Baltic region
SA Baltic Sea
SA Finland
SA Kola Peninsula
SA Norway
SA Scandinavia
SA Sweden

Baltic Plain (1981)
This term has multiple hierarchies.
IN Index countries as applicable.
CO N540000N600000
   E0313000E0190000
BT1 Russian Platform
BT2 Europe
SA Estonia
SA Latvia
SA Russian Federation

Baltic region (1978)
Also search Baltic. Eurasia was autoposted from 1981-1992. As of 1993, Baltic region is autoposted to Estonia, Latvia and Lithuania.

IN Index Baltic Sea and countries as applicable.
CO N520000N680000
   E0310000E0100000
UF Baltic Basin
UF Baltic Coast
BT Europe
NT Courland Spit
NT Estonia
NT Latvia
NT Lithuania
SA Ancylus Lake
SA Baltic Glaciation
SA Baltic ice lake
SA Baltic Sea
SA Denmark
SA Finland
SA Germany
SA Gulf of Riga
SA Kaliningrad Russian Federation
SA Litorina Sea
SA Peribaltic Syneclise
SA Russian Federation
SA Saint Petersburg Russian Federation
SA Sweden
SA Yoldia Sea

Baltic Sea (1978)
Enclosed by Denmark, Sweden, Finland, Russian Federation, Estonia, Latvia, Lithuania, Poland, and Germany.
CO N540000N660000
   E0300000E0100000
BT European Atlantic
BT North Atlantic
BT Atlantic Ocean
NT Aland
NT Gulf of Bothnia
NT Gulf of Finland
NT Gulf of Riga
NT Kiel Bay
SA Ancylus Lake
SA Baltic Glaciation
SA Baltic ice lake
SA Baltic region
SA Bornholm
SA Oresund
SA Saaremaa
SA Stevns Klint
SA Yoldia Sea

Baltic Sea Stages
   use Baltic Glaciation

Baltic Shield (1978)
Before 1993, also search Fennoscandian Shield.
IN Index countries as applicable.
UF Fennoscandian Shield
BT Europe
SA Denmark
SA Estonia
SA Fennoscandia
SA Finland
SA Ladoga Series
SA Lake Ladoga region
SA Latvia
SA Norway
SA Russian Federation
SA Sweden

Baltic Syneclise
   use Peribaltic Syneclise

Baltimore
Valid through 1988. Search in combination with state term. After 1988, use specific city-state term.

Baltimore Canyon (1985)
Submarine canyon off Delaware and New Jersey.
BT North American Atlantic
BT North Atlantic
BT Atlantic Ocean

SA Atlantic Coastal Plain
SA Baltimore Canyon Trough
SA Leg 95
SA submarine canyons

Baltimore Canyon Trough (1993)
Basin offshore E North America, extending from Cape Hatteras to Connecticut. In North American Atlantic.
BT North American Atlantic
BT North Atlantic
BT Atlantic Ocean
SA Baltimore Canyon

Baltimore County
Valid through 1988. Search in combination with state term. After 1988, use specific county-state term.

Baltimore County Maryland (1989)
Nearly surrounds city of Baltimore in N Maryland. Before 1989, also search Baltimore County AND Maryland.
CO N391200N394326
   W0762000W0765400
BT Maryland
BT United States
NT Baltimore Maryland

Baltimore Gneiss (1978)
Unconformable below Glenarm Series. W Maryland, SE Pennsylvania, and central W Virginia.
UF Baltimore Mafic Complex
BT Precambrian
SA Maryland
SA Pennsylvania
SA Virginia

Baltimore Mafic Complex
   use Baltimore Gneiss

Baltimore Maryland (1989)
City in but independent of Baltimore County, on Chesapeake Bay in N Maryland. Before 1989, search Baltimore AND Maryland.
CO N391800N391800
   W0763800W0763800
BT Baltimore County Maryland
BT Maryland
BT United States

Baltisphaeridium (1978)
Autoposting of microfossils to this term began in 1990.
BT acritarchs
BT palynomorphs
BT microfossils

Baluchistan
No longer a valid term for GeoRef. See Baluchistan Pakistan.

Baluchistan Pakistan (1993)
Province in W Pakistan.
BT Pakistan
BT Indian Peninsula
BT Asia
SA Sulaiman Range

Bambak Armenia
   use Pambak Armenia

Bamble
No longer a valid term for GeoRef. As of 1993, see Bamble Norway.

Bamble Norway (1993)
Village on the Skagerrak in S Telemark.
BT Telemark Norway
BT Norway
BT Scandinavia
BT Western Europe
BT Europe

Bambu
No longer a valid term for GeoRef. As of 1993 see Bambu Zaire.

Bambu Zaire (1993)
Village in NE Zaire. Before 1993 also search Bambu or Kilo-Mines.
UF Kilo Zaire
UF Kilo-Mines Zaire
BT Zaire
BT Central Africa
BT Africa

Bambui Group (1978)
BT Proterozoic
BT upper Precambrian
BT Precambrian
SA Brazil
SA Minas Gerais Brazil

Bamian
No longer a valid term for GeoRef. See Bamian Afghanistan.

Bamian Afghanistan (1993)
Use for the province in N central Afghanistan. Also a town. Before 1993, also search Bamian and Afghanistan.
BT Afghanistan
BT Indian Peninsula
BT Asia

Banat (1978)
Agricultural region. Part of Vojvodina in Serbia and the former Banat region of Romania. Also includes part of Hungary near Szeged.
IN Index countries as applicable.
UF Banat of Temesvar
BT Southern Europe
BT Europe
NT Arad Romania
NT Delinesti Romania
NT Mehadia Romania
NT Moldova Noua Romania
NT Resita Romania
NT Rusca Montana Romania
NT Sasca-Montana Romania
NT Severin
NT Svinita Romania
SA Getic Nappe
SA Hungary
SA Mehedinti Plateau
SA Romania
SA Semenic Mountains
SA Serbia
SA Vojvodina
SA Yugoslavia

Banat of Temesvar
   use Banat

Bancroft
No longer a valid term for GeoRef. As of 1993, see Bancroft Ontario.

Bancroft Ontario (1993)
Village between Ottowa and Georgian Bay in SE Ontario. Before 1993, also search Bancroft AND Ontario.
BT Hastings County Ontario
BT Ontario
BT Eastern Canada
BT Canada

Band-e-Amir
No longer a valid term for GeoRef. See Band-e-Amir Iran.

Band-e-Amir Iran (1993)
A settlement in S Iran.
BT Fars Iran
BT Iran
BT Middle East
BT Asia

Banda Arc (1993)

Island arc in the Banda Sea, Indonesian Seas region separated from Australian continental crust by the horseshoe-shaped Timor, Aru, and Seram troughs. Includes Sumba, Timor, Seram.
BT West Pacific
BT Pacific Ocean
SA Banda Sea
SA Indonesian Seas
SA Timor
SA Timor Trough

Banda Sea (1985)
N of Timor, E of Flores Sea, S of Moluccas and W of New Guinea. As of 1990, Pacific Ocean is autoposted to this term.
BT Indonesian Seas
BT West Pacific
BT Pacific Ocean
SA Banda Arc

Bandama River (1978)
Rises in N Ivory Coast and flows S into Gulf of Guinea.
BT Ivory Coast
BT West Africa
BT Africa

banded gneiss (1978)
BT gneisses
BT metamorphic rocks

Banded Gneissic Complex (1993)
Proterozoic. Rajasthan-Gujarat area, India.
BT Proterozoic
BT upper Precambrian
BT Precambrian
SA India

banded iron formations
 use iron formations

banded materials (1978)
Before 1978, also search banded.
SA materials

banded structures (1978)
Also search banded.
UF banding
SA igneous rocks
SA metamorphic rocks
SA petrology

Bandelier Tuff (1985)
N central New Mexico. Also search Bandelier Rhyolite Tuff.
BT Pleistocene
BT Quaternary
BT Cenozoic
SA New Mexico

banding
 use banded structures

Bandundu
 No longer a valid term for GeoRef. As of 1993 see Bandundu Zaire.

Bandundu Zaire (1993)
Province. E Zaire. Before 1993 also search Bandundu AND Zaire.
CO S081000S004000
   E0210000E0121000
BT Zaire
BT Central Africa
BT Africa

Banff
 No longer a valid term for GeoRef. See Banff Alberta.

Banff Alberta (1993)
Town in Banff National Park near the British Columbia border in SW Alberta.
BT Alberta
BT Western Canada

BT Canada

Banff Formation (1989)
BT Kinderhookian
BT Lower Mississippian
BT Mississippian
BT Carboniferous
BT Paleozoic
SA Alberta
SA British Columbia

Banff National Park (1978)
On the British Columbia border in SW Alberta.
BT Alberta
BT Western Canada
BT Canada
SA Columbia Icefield
SA Kicking Horse River valley
SA Lake Louise

Banffshire
 No longer a valid term for GeoRef as of 1993.
 use Banffshire Scotland

Banffshire Scotland (1993)
Former county on Moray Firth in NE Scotland. Before 1993, also search Banffshire.
UF Banffshire
BT Grampian region Scotland
BT Scotland
BT Great Britain
BT United Kingdom
BT Western Europe
BT Europe

Bangalore
 No longer a valid term for GeoRef. See Bangalore India.

Bangalore India (1993)
City and district in SE Karnataka, S India.
BT Karnataka India
BT India
BT Indian Peninsula
BT Asia

Bangka (1978)
An island between SE Sumatra and SW Borneo.
BT Indonesia
BT Far East
BT Asia
SA Sumatra

Bangkok
 No longer a valid term for GeoRef. As of 1993 see Bangkok Thailand.

Bangkok Thailand (1993)
City on the Gulf of Siam in S Thailand. Before 1993 also search Bangkok AND Thailand.
BT Thailand
BT Far East
BT Asia

Bangla Desh
 use Bangladesh

Bangladesh (1978)
Formerly East Pakistan. Before 1973, also search East Pakistan or Pakistan AND east.
CO N203000N263200
   E0924000E0880000
UF Bangla Desh
UF East Pakistan
BT Indian Peninsula
BT Asia
NT Magura Bangladesh
SA Bengal
SA Brahmaputra River
SA Ganges River
SA Indian Plate
SA Indo-Australian Plate
SA Rajmahal Series

SA Surma Group
SA Vindhyan Basin

Bangor Limestone (1978)
Includes Burgess Oolite and Rockwood Oolite, and Spout Spring Oolite near base. Alabama, central and E Tennessee, and NW Georgia.
BT Upper Mississippian
BT Mississippian
BT Carboniferous
BT Paleozoic
SA Alabama
SA Georgia
SA Tennessee

Banja Luka (1978)
Town in N Bosnia and Herzegovina.
BT Bosnia-Herzegovina
BT Yugoslavia
BT Southern Europe
BT Europe

Bank
 No longer a valid term for GeoRef. As of 1993, see Bank Azerbaidzhan.

Bank Azerbaidzhan (1993)
Town in SE Azerbaidzhan. This term has multiple hierarchies.
BT1 Azerbaidzhan
BT1 Europe
BT2 Azerbaidzhan
BT2 Commonwealth of Independent States

banks (1978)
BT biogenic structures
BT sedimentary structures
SA algal banks
SA carbonate banks
SA mud banks

Banks Island (1978)
W Franklin District.
IN Index Northwest Territories and district as applicable.
SA Franklin District Northwest Territories
SA Northwest Territories

Bankura
 No longer a valid term for GeoRef. See Bankura India.

Bankura India (1993)
Town and district in W West Bengal, E India.
BT West Bengal India
BT India
BT Indian Peninsula
BT Asia

Banning Fault (1989)
Southern California.
BT California
BT United States

Bannock Basin (1993)
Central Mediterranean Sea, WSW of Crete.
CO N341500N342400
   E0200800E0195900
BT East Mediterranean
BT Mediterranean Sea
SA Leg 13
SA Leg 42A

Bannock County
 Valid through 1988. Search in combination with state term. After 1988, use specific county-state term.

Bannock County Idaho (1989)
SE Idaho. Before 1989, also search Bannock County AND Idaho.

CO N421500N430200
   W1115200W1124300
BT Idaho
BT United States
SA Bannock Range

Bannock Range (1985)
SE Idaho.
IN Index counties as applicable.
CO N421000N425000
   W1121500W1124000
BT Idaho
BT United States
SA Bannock County Idaho
SA Oneida County Idaho

Banska Stiavnica (1978)
Town in S Slovakia.
BT Slovakia
BT Czechoslovakia
BT Central Europe
BT Europe

Baoha
 No longer a valid term for GeoRef. As of 1993 see Baoha Vietnam.

Baoha Vietnam (1993)
A town in N North Vietnam. Before 1993 also search Baoha AND Vietnam.
BT Vietnam
BT Far East
BT Asia

Baquero Formation (1978)
BT Lower Cretaceous
BT Cretaceous
BT Mesozoic
SA Argentina
SA Santa Cruz Argentina

Bar
 No longer a valid term for GeoRef. See Bar Ukraine.

Bar Ukraine (1993)
Town SW of Kiev in W Vinnitsa Oblast. This term has multiple hierarchies.
BT1 Ukraine
BT1 Europe
BT2 Ukraine
BT2 Commonwealth of Independent States

Barabash Suite (1978)
Primorye Kray in Russian Far East.
BT Upper Permian
BT Permian
BT Paleozoic
SA Primorye Russian Federation
SA Russian Federation

Baraboo
 Valid through 1988. Search in combination with state term. After 1988, use specific city-state term.

Baraboo Quartzite (1978)
Underlies Seeley Slate. Sauk and Columbia counties in S central Wisconsin.
BT Algonkian
BT Proterozoic
BT upper Precambrian
BT Precambrian
SA Wisconsin

Baraboo Wisconsin (1989)
City in S central Wisconsin. Before 1989, search Baraboo AND Wisconsin.
CO N432700N432700
   W0894500W0894500
BT Sauk County Wisconsin
BT Wisconsin
BT United States

Baraga County
Valid through 1988. Search in combination with state term. After 1988, use specific county-state term.

Baraga County Michigan (1989)
NW Michigan Upper Peninsula, N Michigan. Before 1989, also search Baraga County AND Michigan.
CO N462500N465800
     W0875800W0884500
BT Michigan Upper Peninsula
BT Michigan
BT United States

Baraga Group (1989)
N Michigan. Includes Goodrich Quartzite, Michigamme Formation (Bijiki iron formation, Clarksburg volcanics and Greenwood iron formation); Hemlock Formation, Fence River Formation and Badwater Greenstone.
BT Proterozoic
BT upper Precambrian
BT Precambrian
SA Hemlock Formation
SA Michigamme Formation
SA Michigan

Barail Group (1989)
In Assam and Meghalaya, NE India.
BT Oligocene
BT Paleogene
BT Tertiary
BT Cenozoic
SA India

Barakar
No longer a valid term for GeoRef. See Barakar India.

Barakar Formation
use Barakar Stage

Barakar India (1993)
Town in W West Bengal, E India.
BT West Bengal India
BT India
BT Indian Peninsula
BT Asia

Barakar Stage (1978)
Lowest stage of the Damuda Series. Named for Barakar River in Raniganj coalfields of West Bengal. In N, NE, and E India.
UF Barakar Formation
BT Lower Permian
BT Permian
BT Paleozoic
SA Damuda Series
SA India

Baralaba Coal Measures (1978)
Permo-Carboniferous.
IN   Index ages as applicable.
BT Paleozoic
SA Australia
SA Carboniferous
SA Permian
SA Queensland Australia

Baranya Mountains
use Mecsek Mountains

Baraolt
No longer a valid term for GeoRef. As of 1993 see Baraolt Romania.

Baraolt Basin (1978)
SE Transylvania. Also search Baraolt.
BT Transylvania
BT Romania
BT Southern Europe
BT Europe

Baraolt Romania (1993)
Town in SE Transylvania. Before 1993 also search Baraolt AND Romania.
BT Transylvania
BT Romania
BT Southern Europe
BT Europe

Barataria Bay (1978)
Inlet of Gulf of Mexico between Lafourche and Plaquemines parishes.
BT Louisiana
BT United States
SA Gulf Coastal Plain
SA Lafourche Parish Louisiana
SA Plaquemines Parish Louisiana

Barbados (1978)
Easternmost island of the Lesser Antilles.
CO N130000N132000
     W0592000W0594500
BT Lesser Antilles
BT Antilles
BT West Indies
BT Caribbean region

Barbados Ridge (1978)
Extends N and S of Barbados with the island in the center.
BT Atlantic Ocean
SA DSDP Site 541
SA DSDP Site 542
SA DSDP Site 543
SA Leg 78A

Barber County
Valid through 1988. Search in combination with state term. After 1988, use specific county-state term.

Barber County Kansas (1989)
On the Oklahoma border in S Kansas. Before 1989, also search Barber County AND Kansas.
CO N370000N372500
     W0982200W0990000
BT Kansas
BT United States

Barberton
No longer a valid term for GeoRef. As of 1993 see Barberton South Africa

Barberton greenstone belt (1993)
In Kaapvaal Craton, SE Transvaal and N Swaziland. Before 1993, also search Barberton and greenstone. Includes Onverwacht Group, Fig Tree Group, and Moodies Group.
BT Southern Africa
BT Africa
SA Barberton Mountain Land
SA Fig Tree Series
SA Kaapvaal Craton
SA Onverwacht Group
SA South Africa
SA Swaziland
SA Transvaal South Africa

Barberton Mountain Land (1978)
SE Transvaal.
BT Transvaal South Africa
BT South Africa
BT Southern Africa
BT Africa
SA Barberton greenstone belt

Barberton South Africa (1993)
Town near Swaziland border in E Transvaal. Before 1993 also search Barberton AND South Africa.
BT Transvaal South Africa
BT South Africa
BT Southern Africa
BT Africa

barbosalite (1978)
BT phosphates

Barcelona
No longer a valid term for GeoRef. As of 1993, see Barcelona City Spain.

Barcelona City Spain (1993)
Refers only to the city on the Mediterranean Sea. Before 1993, also search Barcelona and Spain.
BT Barcelona Spain
BT Catalonia Spain
BT Spain
BT Iberian Peninsula
BT Southern Europe
BT Europe

Barcelona Province
No longer a valid term for GeoRef. As of 1993, see Barcelona Spain.

Barcelona Spain (1993)
Refers only to the province on the Mediterranean Sea. From 1981-1993, also search Barcelona Province and Spain. Before 1981, also search Barcelona and Spain. For the city, see Barcelona City Spain.
CO N411000N422000
     E0025000E0012200
BT Catalonia Spain
BT Spain
BT Iberian Peninsula
BT Southern Europe
BT Europe
NT Barcelona City Spain
NT Vich Spain
SA Llobregat River basin
SA Ter River basin

barchans (1985)
This term has multiple hierarchies.
BT1 dunes
BT2 eolian features
SA arid environment
SA deserts
SA geomorphology
SA wind transport

Bardo Mountains (1978)
In SW Poland.
BT Wroclaw Poland
BT Poland
BT Central Europe
BT Europe

Barents Sea (1978)
Part of Arctic Ocean N of Norway and Russian Federation, and between Spitsbergen and Novaya Zemlya. Bear Island was considered a narrower term before 1981.
CO N700000N800000
     E0670000E0100000
BT Arctic Ocean
SA Bear Island

Bari
No longer a valid term for GeoRef. See Bari Italy.

Bari Italy (1993)
City and province on the Adriatic Sea.
BT Apulia Italy
BT Italy
BT Southern Europe
BT Europe

Barinas
No longer a valid term for GeoRef. As of 1993 see Barinas Venezuela.

Barinas Venezuela (1993)
State in SW Venezuela. Also a city. Before 1993 also search Barinas AND Venezuela.
BT Venezuela
BT South America

Baringo Lake
use Lake Baringo

barite (1978)
As of 1981, use barite deposits for barite as a commodity.
BT sulfates
SA barite deposits

barite deposits (1981)
Before 1981, search barite AND deposits.
IN   Commodity. See List C.
SA Aberfeldy
SA barite
SA Clara Mine
SA Fukazawa Mine
SA North Pole Deposit

barium (1978)
Autoposting of alkaline earth metals and metals to this term began in 1989.
UF Ba
BT alkaline earth metals
BT metals
SA barylite

barium feldspar (1981)
BT feldspar group
BT framework silicates
BT silicates
NT celsian
NT hyalophane
NT paracelsian
SA barium silicates

barium silicates (1985)
BT silicates
SA barium feldspar

Barmer
No longer a valid term for GeoRef. See Barmer India.

Barmer India (1993)
Town in W Rajasthan, W India.
BT Rajasthan India
BT India
BT Indian Peninsula
BT Asia

barnacles (1989)
Common name for crustaceans of the order Cirripedia.
BT crustaceans
BT arthropods
BT invertebrates
SA biostratigraphy
SA Cirripedia

Barnes ice cap (1978)
Central Baffin Island in Franklin District.
IN   Index Baffin Island and Franklin District if applicable.
SA Baffin Island
SA Franklin District Northwest Territories

Barnett Formation (1989)
Central Texas.
BT Mississippian
BT Carboniferous
BT Paleozoic
SA Texas

Barnstable County
  Valid through 1988. Search in combination with state term. After 1988, use specific county-state term.
Barnstable County Massachusetts (1989)
  Coextensive with Cape Cod in SE Massachusetts. Before 1989, also search Barnstable County AND Massachusetts.
  CO N413100N420500
      W0695500W0704200
  BT Massachusetts
  BT United States
  NT Cape Cod
  NT Woods Hole Massachusetts
  SA Buzzards Bay
Barnwell County
  Valid through 1988. Search in combination with state term. After 1988, use specific county-state term.
Barnwell County South Carolina (1989)
  W South Carolina. Before 1989, also search Barnwell County AND South Carolina.
  CO N330700N333000
      W0811300W0814700
  BT South Carolina
  BT United States
  SA Savannah River Plant
Barnwell Formation (1985)
  In Jackson Group. SW South Carolina and E Georgia.
  BT upper Eocene
  BT Eocene
  BT Paleogene
  BT Tertiary
  BT Cenozoic
  SA Georgia
  SA Jackson Group
  SA South Carolina
Baroda
  No longer a valid term for GeoRef. Former state in SE Gujarat, W India. For the city and district, see Baroda India.
Baroda India (1993)
  City and district in SE Gujarat, W India.
  BT Gujarat India
  BT India
  BT Indian Peninsula
  BT Asia
barometry, geologic
  use geologic barometry
Baron France (1993)
  SW Oise, France. Before 1993, also search Baron or Baron region and France.
  BT Oise France
  BT France
  BT Western Europe
  BT Europe
Baron region
  No longer a valid term for GeoRef. See Baron France if applicable.
Baronnies (1978)
  Hilly region in SE France.
  BT Drome France
  BT France
  BT Western Europe
  BT Europe
Barracuda Ridge (1978)
  In the Anegada Passage area between the Virgin Islands and the N Leeward Islands.

  BT Atlantic Ocean
  SA Caribbean Sea
Barrandian (1978)
  Stage.
  BT Middle Cambrian
  BT Cambrian
  BT Paleozoic
Barrandian area
  use Barrandian Basin
Barrandian Basin (1978)
  CO N492400N501500
      E0143000E0131000
  UF Barrandian area
  BT Czechoslovakia
  BT Central Europe
  BT Europe
Barre Granite (1978)
  An elongated, oval-shaped body that crops out in northwestern and west-central parts of East Barre Quadrangle, Washington County, Vermont.
  BT Devonian
  BT Paleozoic
  SA Vermont
Barreal
  No longer a valid term for GeoRef. See Barreal Argentina.
Barreal Argentina (1993)
  Town in SW San Juan Province.
  BT San Juan Argentina
  BT Argentina
  BT South America
Barreiras Formation (1978)
  BT Tertiary
  BT Cenozoic
  SA Bahia Brazil
  SA Brazil
  SA Espirito Santo Brazil
  SA Para Brazil
  SA Pernambuco Brazil
Barreirinhas Basin (1978)
  NE Maranhao.
  BT Maranhao Brazil
  BT Brazil
  BT South America
Barremian (1978)
  Europe. Above Hauterivian, below Aptian.
  BT Lower Cretaceous
  BT Cretaceous
  BT Mesozoic
  SA Speeton Clay
Barren Measures (1993)
  Middle Permian. Northeastern Indian coal fields. Underlies Raniganj Stage; overlies Barakar Formation.
  IN Index age or ages as applicable.
  SA India
  SA Middle Permian
barrier bars
  use longshore bars
barrier beaches (1985)
  Autoposting of shore features to this term began in 1989.
  BT shore features
  SA barrier islands
  SA beaches
  SA geomorphology
  SA longshore bars
  SA sand ridges
barrier islands (1978)
  Before 1981, also search barriers.
  IN Index shore features if applicable.
  SA artificial islands
  SA barrier beaches

  SA bars
  SA capes
  SA coastlines
  SA geomorphology
  SA islands
  SA littoral erosion
  SA longshore bars
  SA sand ridges
  SA shore features
  SA shorelines
  SA tombolos
barrier reefs (1978)
  BT reefs
  SA fringing reefs
  SA lagoons
barriers
  As of 1981, no longer a valid term for GeoRef. See barrier islands; longshore bars. As of 1993, also see disposal barriers.
Barrow
  Valid through 1988. Search in combination with state term. After 1988, use specific city-state term.
Barrow Alaska (1989)
  City about 12 miles S of Point Barrow on the Arctic Ocean. Before 1989, search Barrow AND Alaska. As of 1993, also use for the Eskimo village on Point Barrow headland, N Alaska. Also search Point Barrow AND Alaska or Point Barrow Alaska or Nuwuk AND Alaska.
  CO N761600N761600
      W1565000W1565000
  UF Point Barrow Alaska
  BT Alaska
  BT United States
  SA Arctic Ocean
  SA Barrow Field
  SA Point Barrow
Barrow Field (1989)
  Natural gas field. North Slope, N Alaska. Also in other locations.
  IN Index county or region as applicable.
  SA Barrow Alaska
  SA North Slope
  SA Northern Alaska
  SA oil and gas fields
Barrow Island (1978)
  Off NW coast of Western Australia.
  SA Western Australia
bars (1978)
  As of 1993, restricted to sedimentation. Autoposting of this term began in 1978.
  UF sand bars
  NT longshore bars
  NT point bars
  SA barrier islands
  SA bedding plane irregularities
  SA capes
  SA coastlines
  SA fluvial features
  SA geomorphology
  SA littoral erosion
  SA planar bedding structures
  SA sand ridges
  SA shore features
  SA tombolos
  SA veins
Barsakelmes (1978)
  Region in Ustyurt in SW Kazakhstan.
  IN Index Kazakhstan if applicable.
  SA Kazakhstan
  SA Ustyurt

Barstovian (1978)
  North America.
  BT Miocene
  BT Neogene
  BT Tertiary
  BT Cenozoic
Barstow
  Valid through 1988. Search in combination with state term. After 1988, use specific city-state term.
Barstow California (1989)
  City in S California. Before 1989, search Barstow AND California.
  CO N345500N345500
      W1170100W1170100
  BT San Bernardino County California
  BT California
  BT United States
Barstow Formation (1985)
  Middle and upper Miocene. S California.
  BT Miocene
  BT Neogene
  BT Tertiary
  BT Cenozoic
  SA California
Barth
  No longer a valid term for GeoRef. See Barth Germany.
Barth Germany (1993)
  Town on the Baltic Sea in Mecklenburg-Western Pomerania. Before 1993, Barth was used with Mecklenburg and with Rostock Bezirk.
  BT Mecklenburg-Western Pomerania Germany
  BT Germany
  BT Central Europe
  BT Europe
  SA Mecklenburg
Bartlesville Sand (1978)
  In Kansas and Oklahoma.
  BT Carboniferous
  BT Paleozoic
  SA Kansas
  SA Mississippian
  SA Oklahoma
  SA Pennsylvanian
Barton Beds (1978)
  Overlain by the lower Headon Beds and rest on the upper Bracklesham Beds. Divided into lower, middle, and upper Barton Beds. Hampshire County in S England. Also search Barton.
  BT Bartonian
  BT upper Eocene
  BT Eocene
  BT Paleogene
  BT Tertiary
  BT Cenozoic
  SA Bracklesham Beds
  SA England
  SA United Kingdom
Barton County
  Valid through 1988. Search in combination with state term. After 1988, use specific county-state term.
Barton County Kansas (1989)
  Central Kansas. Before 1989, also search Barton County AND Kansas.
  CO N381600N384100
      W0982800W0990200
  BT Kansas
  BT United States
Barton County Missouri (1989)

SW Missouri. Before 1989, also search Barton County AND Missouri.
CO N372200N374000
W0940400W0943700
BT Missouri
BT United States

Barton Springs (1989)
S central Texas.
IN Index counties or regions as applicable.
SA Hays County Texas
SA Texas
SA Travis County Texas

Bartonian (1978)
Europe. Above Auversian, below Ludian.
BT upper Eocene
BT Eocene
BT Paleogene
BT Tertiary
BT Cenozoic
NT Barton Beds
SA Priabonian

Bartow County Georgia (1989)
NW Georgia. Before 1989, search Bartow County AND Georgia.
CO N340400N342500
W0844000W0850300
BT Georgia
BT United States
NT Cartersville Georgia

Barwell Meteorite (1981)
Impact in Leicestershire, England. Before 1981, also search Barwell.
BT L chondrites
BT chondrites
BT stony meteorites
BT meteorites
SA England
SA Leicestershire England

barylite (1978)
BT ring silicates
BT silicates
SA barium

Barytherioidea (1981)
Suborder? Autoposting of Eutheria and Theria to this term began in 1989.
BT Proboscidea
BT Eutheria
BT Theria
BT Mammalia
BT Tetrapoda
BT Vertebrata
BT Chordata

Bas-Dauphine (1978)
Region. Part of the historic province of Dauphine in SE France. Associated with the Dauphine Alps.
BT France
BT Western Europe
BT Europe
SA Dauphine
SA Dauphine Alps

Bas-Rhin
No longer a valid term for GeoRef. See Bas-Rhin France.

Bas-Rhin France (1993)
Department in Alsace in extreme NE France.
CO N480500N490500
E0081200E0070000
BT France
BT Western Europe
BT Europe
NT Bouxwiller France
NT Strasbourg France
SA Alsace

SA Saar Basin

Bas-Zaire
use Lower Zaire

basalt
As of 1989, no longer a valid term for GeoRef. From 1981 to 1988, basalts and volcanic rocks were autoposted to this term.
use basalts

basalt family
As of 1981, no longer a valid term for GeoRef. Use basalts, alkali basalts, trachybasalts, or diabase as applicable. From 1978-1980, was autoposted to the rocks that were classified under it.

basalt flows (1993)
Before 1993, also search lava flows AND basalts.
SA basalts
SA flood basalts
SA lava flows

basalt glass
Not a valid term for GeoRef. Use volcanic glass AND basalts or basaltic composition.

basaltic
Not a valid term for GeoRef. Use basaltic composition or basaltic layer.

basaltic composition (1978)
Before 1978, also search basaltic.
SA basalts
SA composition
SA igneous rocks

basaltic domes
use shield volcanoes

basaltic layer (1978)
UF layer, basaltic
UF sima
SA crust
SA granitic layer
SA lower crust

basaltic rocks
Not a valid term for GeoRef. Use basalts or basaltic composition.

basalts (1981)
Before 1981, search basalt family. From 1978-1980, basalt family was autoposted to the rocks that were classified under that term. Before 1989, also search basalt. Also search basaltic rocks. Autoposting of this term to basalt ended in 1989.
UF basalt
BT volcanic rocks
BT igneous rocks
NT absarokite
NT alkali basalts
NT ankaramite
NT columnar basalt
NT flood basalts
NT melaphyre
NT mid-ocean ridge basalts
NT ocean-island basalts
NT oceanite
NT olivine basalt
NT olivine tholeiite
NT shoshonite
NT tholeiite
NT tholeiitic basalt
NT trap rocks
SA augite
SA basalt flows
SA basaltic composition
SA diabase
SA lava
SA metabasalt

SA picrite
SA picrite porphyry
SA pillow lava
SA pyrolite
SA quartz diabase
SA tholeiitic composition

basanite (1978)
BT alkali basalts
BT basalts
BT volcanic rocks
BT igneous rocks

basculating faults
use wrench faults

base exchange
use ion exchange

base maps (1989)
BT maps
SA topographic maps

base metals (1978)
Autoposting of metal ores to this term began in 1985.
IN See list C for use as a commodity term.
BT metal ores
SA Agrokipia Deposit
SA alloys
SA Broken Hill Mine
SA copper ores
SA lead ores
SA metals
SA Nasliden Mine
SA precious metals
SA zinc ores

base surges (1978)
Used in sedimentation and for pyroclastic flows.
SA nuees ardentes
SA pyroclastic flows
SA sedimentation
SA surges
SA volcanism

base-level peneplain
use peneplains

Basel
No longer a valid term for GeoRef. See Basel Switzerland.

Basel Switzerland (1993)
Canton composed of demicantons of Basel-Land and Basel-Stadt. Also a city. Also search Basle.
UF Basle Switzerland
BT Switzerland
BT Central Europe
BT Europe

baselevel
Not a valid GeoRef index term after 1970. Was used on level 1 in subfile N. Now use peneplains, erosion surfaces, or changes of level.

basement (1978)
UF basement rocks
UF basement structure
SA basement tectonics
SA crust
SA igneous rocks
SA metamorphic rocks
SA Mohorovicic discontinuity
SA sedimentary cover
SA shields
SA supracrustals
SA tectonics

basement rocks
use basement

basement structure
use basement

basement tectonics (1993)

Before 1993, also search basement AND tectonics.
BT tectonics
SA basement
SA metamorphic rocks
SA supracrustals

Bashkiria
As of 1993, no longer a valid term for GeoRef.
use Bashkiria Russian Federation

Bashkiria Russian Federation (1993)
Former Bashkir Autonomous Soviet Socialist Republic in SE Russian Federation. Before 1993, also search Bashkiria. This term has multiple hierarchies.
CO N520000N570000
E0600000E0530000
UF Bashkiria
BT1 Russian Federation
BT1 Commonwealth of Independent States
BT2 Europe
NT Sibay Russian Federation
NT Tuymazy Russian Federation
NT Uchaly Russian Federation
NT Ufa Russian Federation
SA Arlan
SA Ural region

Bashkirian (1978)
Europe.
UF Bashkirian Stage
BT Upper Carboniferous
BT Carboniferous
BT Paleozoic

Bashkirian Stage
use Bashkirian

BASIC (1985)
BT computer languages
SA computer programs
SA data processing

basic rocks
Not a valid term for GeoRef. Use mafic composition.

Basilicata
No longer a valid term for GeoRef. See Basilicata Italy.

Basilicata Italy (1993)
Autonomous region in S Italy.
CO N395500N411000
E0165000E0152200
BT Italy
BT Southern Europe
BT Europe
NT Matera Italy
NT Potenza Italy
SA Irpinia earthquake 1980
SA Lucania

Basin and Range
Not a valid GeoRef index term after 1970. Was used on level 1 in subfile N.
use Basin and Range Province

Basin and Range Province (1978)
Physiographic province in W and SW United States characterized by a series of tilted fault blocks forming longitudinal, asymmetric ridges of mountains and broad intervening basins. Before 1971, also search Basin and Range.
CO N290000N433000
W1023000W1220000
UF Basin and Range
BT United States
NT Great Basin
SA Arizona
SA Battle Mountain High

SA California
SA Idaho
SA Manzano Mountains
SA Mineral Mountains
SA Nevada
SA New Mexico
SA Nopah Range
SA Oregon
SA Pilot Range
SA Quitman Mountains
SA Rincon Mountains
SA Sevier Desert
SA Texas
SA Tijeras Canyon
SA Toquima Range
SA Utah
SA Whipple Mountains
SA Yucca Mountain

basin management (1982)
SA basins
SA conservation
SA drainage basins
SA ground water
SA hydrology
SA land use
SA management
SA reclamation
SA surface water
SA water management
SA water resources

Basin of Mexico
Not a valid term for GeoRef. See Sigsbee Deep or Valley of Mexico.

basin range structure (1978)
Also search basin range; basin-range.
UF basin-range structure
UF range, basin
SA basins
SA faults
SA tectonics

basin, structural
use basins

basin-range structure
use basin range structure

basins (1978)
Before 1971 also search basin, structural. Before 1972, was sometimes used in combination with structural (i.e. basins, structural).
UF basin, structural
UF structural basins
NT back-arc basins
NT fore-arc basins
NT foreland basins
NT intermontane basins
NT intracratonic basins
NT marginal basins
NT pull-apart basins
NT sedimentary basins
SA basin management
SA basin range structure
SA depressions
SA drainage basins
SA extension tectonics
SA folds
SA morphostructures
SA ocean basins
SA paleogeography
SA playas
SA sedimentation
SA tectonics

basins, drainage
use drainage basins

basins, structural
Not a valid GeoRef index term after 1971. Was used on level 1 in subfiles N, T, E and B. Now use basins.

Baskunchak Series (1978)
S Urals.
IN Index age as applicable.
SA Permian
SA Russian Federation
SA Triassic

Basle Switzerland
use Basel Switzerland

Basommatophora (1981)
BT Gastropoda
BT Mollusca
BT Invertebrata

Basque Provinces
As of 1993, no longer a valid term for GeoRef.
use Basque Provinces Spain

Basque Provinces Spain (1993)
On the Bay of Biscay near the French border. Before 1978, also search Basque. Provinces are considered narrower terms as of 1981.
IN Index provinces as applicable.
CO N422900N432800
  W0014500W0033300
UF Basque Provinces
BT Spain
BT Iberian Peninsula
BT Southern Europe
BT Europe
NT Alava Spain
NT Guipuzcoa Spain
NT Vizcaya Spain
SA Cantabrian Basin

Bass Basin
use Bass Strait

Bass Islands Dolomite (1989)
Michigan Basin.
BT Silurian
BT Paleozoic
SA Kentucky
SA Michigan
SA Michigan Basin
SA New York
SA Ohio
SA Ontario
SA Pennsylvania

Bass Strait (1978)
Large channel separating Victoria from Tasmania and the Indian Ocean from the Tasman Sea.
UF Bass Basin
SA Australia
SA Western Port
SA Wynyard Australia

bassanite (1978)
BT sulfates

Basses-Alpes
use Alpes-de-Haute Provence France

Basses-Pyrenees France
use Pyrenees-Atlantiques France

Bastar
No longer a valid term for GeoRef. Former Indian State. For the district in SE Madhya Pradesh, see Bastar India.

Bastar India (1993)
District in SE Madhya Pradesh, E central India.
BT Madhya Pradesh India
BT India
BT Indian Peninsula
BT Asia

bastnaesite (1978)
From 1978-1980, halides and carbonates were autoposted to this term. Autoposting of fluorides began in 1981. As of 1989, halides is autoposted to this term. This term has multiple hierarchies.
UF bastnasite
BT1 carbonates
BT2 fluorides
BT2 halides
SA rare earth deposits

bastnasite
use bastnaesite

Bastogne
No longer a valid term for GeoRef. See Bastogne Belgium.

Bastogne Belgium (1993)
Town in E Luxembourg Province, SE Belgium.
BT Luxembourg Belgium
BT Belgium
BT Western Europe
BT Europe

Bastrop County Texas (1993)
S central Texas. Before 1989, also search Bastrop County AND Texas.
CO N294500N302500
  W0965500W0974000
BT Texas
BT United States

Basutoland
use Lesotho

Bataan
No longer a valid term for GeoRef. As of 1993 see Bataan Philippine Islands.

Bataan Philippine Islands (1993)
Province, S Luzon, Philippine Islands. Before 1993 also search Bataan AND Philippine Islands.
BT Luzon
BT Philippine Islands
BT Far East
BT Asia

Batchawana Bay (1978)
On Lake Superior N of Sault Sainte Marie.
BT Ontario
BT Eastern Canada
BT Canada

Batenev Ridge (1978)
In the Kuznetsk Alatau of the Khakass Autonomous Oblast SW of Krasnoyarsk. This term has multiple hierarchies.
BT1 Kuznetsk Alatau
BT1 Russian Federation
BT1 Commonwealth of Independent States
BT2 Kuznetsk Alatau
BT2 West Siberia
BT2 Commonwealth of Independent States
BT3 Kuznetsk Alatau
BT3 West Siberia
BT3 Asia
BT4 Kuznetsk Alatau
BT4 Altai Mountains
BT4 Asia

Batesville
Valid through 1988. Search in combination with state term. After 1988, use specific city-state term.

Batesville Arkansas (1989)
City in NE Arkansas. Before 1989, search Batesville AND Arkansas.
CO N354500N354500
  W0913900W0913900
BT Independence County Arkansas
BT Arkansas
BT United States

Bath
No longer a valid term for GeoRef as of 1993. See Bath England.

Bath County
Valid through 1988. Search in combination with state term. After 1988, use specific county-state term.

Bath County Kentucky (1989)
NE Kentucky. Before 1989, also search Bath County AND Kentucky.
CO N375900N381800
  W0833000W0835800
BT Kentucky
BT United States

Bath County Virginia (1989)
W Virginia. Before 1989, also search Bath County AND Virginia.
CO N375100N381600
  W0792700W0800200
BT Virginia
BT United States

Bath England (1993)
City near Bristol in Somerset, SW England.
CO N512300N512300
  W0022200W0022200
BT Avon England
BT England
BT Great Britain
BT United Kingdom
BT Western Europe
BT Europe
SA Somerset England

batholiths (1978)
UF abyssoliths
UF central granite
UF intrusive mountain
BT intrusions
SA dikes
SA domes
SA igneous rocks
SA laccoliths
SA lopoliths
SA plutons
SA stocks

Bathonian (1978)
Europe. Above Bajocian, below Callovian.
BT Middle Jurassic
BT Jurassic
BT Mesozoic
NT Great Oolite Series
SA Dogger

Bathurst
No longer a valid term for GeoRef. As of 1993, see Bathurst New Brunswick or Bathurst Australia as applicable.

Bathurst Australia (1993)
City in New South Wales. Before 1993, also search Bathurst AND Australia.
BT New South Wales Australia
BT Australia
BT Australasia

Bathurst Island (1978)
Island in NW Franklin District, Northwest Territories, Canada. Also island in Northern Territory, Australia.
IN Index regions as applicable.
SA Arctic Archipelago
SA Franklin District Northwest Territories
SA Northern Territory Australia
SA Northwest Territories

Bathurst mining district (1993)

Gloucester County, New Brunswick.
  BT Gloucester County New Brunswick
  BT New Brunswick
  BT Maritime Provinces
  BT Eastern Canada
  BT Canada
  SA metal ores
  SA mines

Bathurst New Brunswick (1993)
  City, Gloucester County. Before 1993, also search Bathurst AND New Brunswick. Also see Bathurst mining district if applicable.
  BT Gloucester County New Brunswick
  BT New Brunswick
  BT Maritime Provinces
  BT Eastern Canada
  BT Canada

bathymetric features
  Use bottom features for ocean floors. Use bottom features and lakes for lakebed features.

bathymetric maps (1978)
  Before 1978, search maps AND bathymetric.
  BT maps
  SA bathymetry
  SA isobath maps
  SA ocean floors

bathymetric surveys
  use bathymetry

bathymetry (1978)
  UF bathymetric surveys
  SA bathymetric maps
  SA echo sounding
  SA geophysical methods
  SA geophysical surveys
  SA isobath maps
  SA marine geology
  SA ocean floors
  SA oceanography
  SA paleo-oceanography
  SA paleobathymetry
  SA Seabeam

Baton Rouge
  Valid through 1988. Search in combination with state term. After 1988, use specific city-state term.

Baton Rouge Louisiana (1989)
  City on the Mississippi River in SE central Louisiana. Before 1989, search Baton Rouge AND Louisiana.
  CO N303000N303000
      W0911000W0911000
  BT East Baton Rouge Parish Louisiana
  BT Louisiana
  BT United States

Battambang
  No longer a valid term for GeoRef. See Battambang Cambodia.

Battambang Cambodia (1993)
  Town in W Cambodia.
  BT Cambodia
  BT Far East
  BT Asia

Battery Point Formation (1989)
  Gaspe Peninsula. Above York River Formation; below Malbaie Formation.
  BT Lower Devonian
  BT Devonian
  BT Paleozoic
  SA Gaspe Peninsula
  SA Quebec

Battle Mountain (1989)
  Lander County, N central Nevada; Eagle County, N central Colorado.
  IN Index counties or regions as applicable.
  SA Colorado
  SA Eagle County Colorado
  SA Lander County Nevada
  SA Nevada

Battle Mountain High (1981)
  Heat flow anomaly in region including parts of N Nevada, SE Oregon, S Idaho, SW Montana, and NW Wyoming.
  IN Index states as applicable.
  BT United States
  SA Basin and Range Province
  SA Great Basin
  SA heat flow
  SA Idaho
  SA Montana
  SA Nevada
  SA Oregon
  SA Western U.S.
  SA Wyoming

Batumi
  No longer a valid term for GeoRef. As of 1993, see Batumi Georgian Republic.

Batumi Georgian Republic (1993)
  City on the Black Sea near the Turkish border. Before 1993, also search Batum.
  BT Adzharistan Georgian Republic
  BT Georgian Republic
  BT Europe

Bauchi Plateau
  use Jos Plateau

Bauer Deep (1978)
  SE Pacific Ocean, E of the East Pacific Rise.
  BT Pacific Ocean

Baumkirchen (1978)
  Region in Inn Valley.
  BT Tyrol Austria
  BT Austria
  BT Central Europe
  BT Europe
  SA Inn Valley

Bauru Formation (1978)
  Upper Cretaceous to lower Tertiary.
  IN Index ages as applicable.
  SA Brazil
  SA K-T boundary
  SA lower Tertiary
  SA Minas Gerais Brazil
  SA Sao Paulo Brazil
  SA Upper Cretaceous

bauxite (1978)
  Includes use both as a commodity term and a sedimentary rock.
  IN See List C for use as a commodity term.
  SA aluminum
  SA aluminum ores
  SA bauxitization
  SA clastic rocks
  SA clay minerals
  SA laterites
  SA sedimentary rocks

bauxitization (1978)
  SA bauxite
  SA laterites
  SA laterization
  SA mineral deposits, genesis
  SA processes
  SA weathering

Bavaria
  No longer a valid term for GeoRef. As of 1993, see Bavaria Germany.

Bavaria Germany (1993)
  State in SE Germany.
  CO N474000N503500
      E0134500E0093000
  BT Germany
  BT Central Europe
  BT Europe
  NT Ansbach Germany
  NT Bavarian Forest
  NT Bavarian Massif
  NT Bayreuth Germany
  NT Coburg Germany
  NT Eichstatt Germany
  NT Erlangen Germany
  NT Fichtelgebirge
  NT Franconian Forest
  NT Franconian Jura
  NT Hagendorf
  NT Kelheim Germany
  NT Middle Franconia
  NT Munchberg Gneiss Massif
  NT Nabburg Germany
  NT Neuburg Germany
  NT Nuremberg Germany
  NT Passau Germany
  NT Regensburg Germany
  NT Ries Crater
  NT Sandelzhausen Germany
  NT Solnhofen Germany
  NT Spessart
  NT Upper Bavaria Germany
  NT Upper Franconia Germany
  NT Upper Palatinate
  NT Weissenburg Germany
  SA Allgau Alps
  SA Alpenvorland
  SA Bavarian Alps
  SA Franconia
  SA Inn Valley
  SA Isar Valley
  SA KTB
  SA Lake Constance
  SA Main River
  SA Odenwald
  SA Palatinate
  SA Rhon Mountains
  SA Saale River
  SA Salzach River
  SA Schwarzenberg Germany
  SA Solnhofen Limestone
  SA Swabia

Bavarian Alps (1978)
  Range of the Central Alps along the Austro-German border. Autoposting of Alps to this term ended in 1989 and began again in 1993. This term has multiple hierarchies.
  IN Index name of country and province or state.
  CO N472200N475000
      E0130700E0104500
  BT1 Central Alps
  BT1 Alps
  BT1 Europe
  BT2 Central Europe
  BT2 Europe
  SA Bavaria Germany
  SA Tyrol Austria
  SA Upper Bavaria Germany

Bavarian Forest (1978)
  A subsidiary forested mountain range of the Bohemian Forest in E Bavaria.
  BT Bavaria Germany
  BT Germany
  BT Central Europe
  BT Europe

Bavarian Massif (1981)

  CO N483000N502000
      E0135000E0113500
  BT Bavaria Germany
  BT Germany
  BT Central Europe
  BT Europe

bavenite (1978)
  BT sheet silicates
  BT silicates
  SA aluminosilicates

Bay of Bengal (1978)
  Part of Indian Ocean between India and the W coast of Burma. Before 1981, Andaman Sea was considered a narrower term.
  CO N050000N230000
      E0940000E0800000
  BT Indian Ocean
  NT Andaman Basin
  SA Andaman Sea
  SA Bengal Fan
  SA Godavari River
  SA Krishna-Godavari Basin

Bay of Biscay (1978)
  Large inlet of the Atlantic Ocean between the W tip of Brittany and NW Spain. As of 1990, Atlantic Ocean is autoposted to this term. Before 1981, also search Gulf of Gascony.
  CO N432000N480000
      W0011000W0080000
  UF Biscay (Bay)
  UF Biscay Bay
  UF Gulf of Gascony
  BT European Atlantic
  BT North Atlantic
  BT Atlantic Ocean
  NT Cantabria Seamount
  SA Arcachon Basin
  SA Bay of Bourgneuf
  SA DSDP Site 119
  SA DSDP Site 402
  SA Gironde Estuary
  SA Groix
  SA Vilaine Bay

Bay of Bourgneuf (1978)
  On Bay of Biscay off Vendee and Loire-Atlantique departments.
  BT France
  BT Western Europe
  BT Europe
  SA Bay of Biscay
  SA Loire-Atlantique France
  SA Vendee France

Bay of Fundy (1978)
  Inlet between Maine and New Brunswick on the W and Nova Scotia on the E.
  BT Atlantic Ocean
  SA Cumberland Basin
  SA Minas Basin

Bay of Haifa
  use Haifa Bay

Bay of Islands (1978)
  Inlet of Gulf of Saint Lawrence off W coast.
  BT Newfoundland Island
  BT Newfoundland
  BT Eastern Canada
  BT Canada
  SA Gulf of Saint Lawrence

Bay of Islands Ophiolite (1989)
  Ordovician? W Newfoundland.
  UF Baie of Islands Ophiolite
  BT lower Paleozoic
  BT Paleozoic
  SA Newfoundland Island

Bay of Mont-Saint-Michel
  use Bay of Saint-Michel

Bay of Naples (1978)
Inlet of Tyrrhenian Sea off the city of Naples.
BT Campania Italy
BT Italy
BT Southern Europe
BT Europe
SA Sorrento Peninsula

Bay of Plenty (1978)
Large inlet off NE coast of North Island, New Zealand.
BT Pacific Ocean
SA Edgecumbe earthquake 1987
SA New Zealand
SA North Island

Bay of Saint Michel
use Bay of Saint-Michel

Bay of Saint-Michel (1978)
On English channel off SW corner of Normandy. Location of Mont Saint Michel.
UF Bay of Mont-Saint-Michel
UF Bay of Saint Michel
UF Mont-Saint-Michel Bay
BT France
BT Western Europe
BT Europe
SA English Channel
SA Manche France

Bay of the Seine (1978)
Inlet of the English Channel off Normandy between Cotentin Peninsula on W to mouth of Seine River on E.
BT France
BT Western Europe
BT Europe
SA English Channel

Bay Park
Valid through 1988. Search in combination with state term. After 1988, use specific city-state term.

Bay Park New York (1989)
Town on Long Island in SE New York. Before 1989, search Bay Park AND New York.
CO N403700N403800
W0733900W0734100
BT Nassau County New York
BT New York
BT United States
SA Long Island

Bayan-Nuurin-khotnor Basin (1981)
Part of the Bayan-khongor fault valley, central Mongolia. Between the Hangay Mountains to the NE and the South Hangay Upland to the SW. Contained a pluvial lake during the Pleistocene.
CO N463000N473000
E1010000E0990000
BT Mongolia
BT Far East
BT Asia
SA Hangay Mountains

bayerite (1978)
Often artificial.
BT oxides
SA gibbsite

Bayesian analysis (1985)
BT statistical analysis
SA analysis
SA data processing
SA discriminant analysis
SA geostatistics
SA mathematical geology
SA mathematical methods
SA mathematical models
SA variance analysis

Bayfield County
Valid through 1988. Search in combination with state term. After 1988, use specific county-state term.

Bayfield County Wisconsin (1989)
N Wisconsin. Before 1989, also search Bayfield County AND Wisconsin.
CO N461000N470000
W0904500W0913500
BT Wisconsin
BT United States

Baylor County
Valid through 1988. Search in combination with state term. After 1988, use specific county-state term.

Baylor County Texas (1989)
SW of Wichita Falls in N Texas. Before 1989, also search Baylor County AND Texas.
CO N332500N335300
W0985800W0993000
BT Texas
BT United States

Bayn Dzak (1978)
Region in the Gobi Desert.
BT Mongolia
BT Far East
BT Asia
SA Gobi Desert

Bayreuth
No longer a valid term for GeoRef. See Bayreuth Germany.

Bayreuth Germany (1993)
City in NE Bavaria.
BT Bavaria Germany
BT Germany
BT Central Europe
BT Europe

bays (1978)
BT shore features
SA coastlines
SA embayments
SA fjords
SA geomorphology
SA inlets
SA shorelines

Bazhenov Formation (1985)
W Siberia and Siberian Platform.
BT Jurassic
BT Mesozoic
SA Russian Federation
SA Siberian Platform
SA West Siberia

Be
use beryllium

Be-10 (1978)
Autoposting of broader terms began in 1989. This term has multiple hierarchies.
BT1 radioactive isotopes
BT1 isotopes
BT2 beryllium
BT2 alkaline earth metals
BT2 metals
SA accelerator mass spectroscopy
SA Be-10/Be-9

Be-10/Be-9 (1985)
Autoposting of broader terms began in 1989. This term has multiple hierarchies.
UF Be-9/Be-10
BT1 radioactive isotopes
BT1 isotopes
BT2 stable isotopes
BT2 isotopes
BT3 beryllium
BT3 alkaline earth metals
BT3 metals
SA Be-10
SA isotope ratios

Be-7 (1978)
Autoposting of broader terms began in 1989. This term has multiple hierarchies.
BT1 radioactive isotopes
BT1 isotopes
BT2 beryllium
BT2 alkaline earth metals
BT2 metals

Be-9/Be-10
use Be-10/Be-9

beach erosion
After 1981, not a valid term for GeoRef.
use littoral erosion

beach placers (1982)
BT placers
SA marine sediments
SA stream placers

beach ridges (1978)
BT shore features
SA beaches
SA cheniers
SA geomorphology
SA marine terraces
SA shorelines

beach rock
use beachrock

beaches (1978)
BT shore features
SA barrier beaches
SA beach ridges
SA berms
SA capes
SA chenier plains
SA coastal environment
SA coastlines
SA foredunes
SA geomorphology
SA lacustrine features
SA lagoons
SA landform evolution
SA littoral erosion
SA paleoecology
SA sedimentation
SA shorelines
SA spits

beachrock (1978)
UF beach rock
BT carbonate rocks
BT sedimentary rocks

Beacon Supergroup (1978)
In Queen Maud Range along SW Ross Ice Shelf and in Victoria Land W of Ross Sea, both in Ross Dependency S of New Zealand.
IN Index ages as applicable.
SA Antarctica
SA Carboniferous
SA Devonian
SA Jurassic
SA Permian
SA Queen Maud Range
SA Ross Dependency
SA Ross Ice Shelf
SA Triassic
SA Victoria Land

Beacon Valley (1978)
In S Victoria Land near SE Ross Sea in Ross Dependency S of New Zealand.
SA Antarctica

Bear Gulch Limestone Member (1978)
In Tyler Formation. Central N Montana.
BT Mississippian
BT Carboniferous
BT Paleozoic
SA Montana
SA Tyler Formation

Bear Island (1978)
240 miles N of Norway. Along with Spitsbergen forms Svalbard island group. In 1981, broader terms changed from Barents Sea and Arctic Ocean to Svalbard and Arctic region. Also occurs in other locations. Before 1993, also search Bjornoya if applicable.
UF Bjornoya
SA Arctic Ocean
SA Arctic region
SA Barents Sea
SA Svalbard

Bear Lake (1978)
IN Index counties or regions as applicable.
SA Idaho
SA Utah

Bear Mountain (1989)
Occurs in various locations.
IN Index appropriate regions as applicable.
SA Alaska
SA California
SA New Mexico
SA New York
SA South Dakota

Bear Province (1978)
Structural province of the Canadian Shield.
CO N630000N733000
W1060000W1240000
BT Canadian Shield
BT North America
SA Northwest Territories

Bear River basin (1978)
IN Index counties or regions as applicable.
SA Idaho
SA Utah

Bear River Formation (1985)
SW Wyoming, SE Idaho and NE Utah.
BT Lower Cretaceous
BT Cretaceous
BT Mesozoic
SA Idaho
SA Utah
SA Wyoming

Bear River Range (1978)
IN Index counties or regions as applicable.
SA Idaho
SA United States
SA Utah

Bear Valley (1978)
Central California.
IN Index counties or regions as applicable.
SA California

Bear-Slave Operation (1978)
Geochemical operation in Great Bear and Great Slave lakes region of Mackenzie District.
SA geochemical methods
SA Great Bear Lake
SA Great Slave Lake
SA Mackenzie District Northwest Territories
SA mineral exploration
SA Northwest Territories

Beardmore Glacier (1978)

At NW end of Queen Maud Range just off Ross Ice Shelf in Ross Dependency S of New Zealand.
BT Antarctica
SA Coalsack Bluff
SA Fremouw Formation
SA Lewis Cliff Meteorites
SA Queen Maud Range
SA Ross Dependency
SA Ross Ice Shelf

bearing capacity (1978)
SA California bearing ratio
SA compaction
SA foundations
SA physical properties
SA soil mechanics
SA stress

Bearn (1978)
An historical region of SW France.
BT France
BT Western Europe
BT Europe

Bearpaw Formation (1978)
In Montana Group. S Alberta; N, E, and S Montana; and Elk Basin region of central N Wyoming.
BT Upper Cretaceous
BT Cretaceous
BT Mesozoic
SA Alberta
SA Edmonton Formation
SA Montana
SA Montana Group
SA Two Medicine Formation
SA Wyoming

Beartooth Mountains (1978)
NE spur of Absaroka Range. This term has multiple hierarchies.
IN Index states as applicable.
CO N441000N455000
  W1090000W1110000
BT1 Absaroka Range
BT1 U. S. Rocky Mountains
BT1 Rocky Mountains
BT2 Absaroka Range
BT2 U. S. Rocky Mountains
BT2 United States
SA Montana
SA Wyoming

Beas River (1978)
One of the "Five Rivers" of the Punjab. Rises in Himachal Pradesh and flows into the Sutlej River.
IN Index states as applicable.
BT India
BT Indian Peninsula
BT Asia
SA Himachal Pradesh India
SA Punjab India

Beata Ridge (1978)
Off south central coast of Hispaniola.
BT Caribbean Sea
BT North American Atlantic
BT North Atlantic
BT Atlantic Ocean
SA Leg 15

Beatty
No longer a valid term for GeoRef. As of 1993, see Beatty Ontario.

Beatty Nevada (1989)
Town and site of underground nuclear explosions in S Nevada. Before 1989, search Beatty AND Nevada.
CO N365400N365400
  W1164500W1164500
BT Nye County Nevada
BT Nevada

Beatty Ontario (1993)
Used for the township in Cochrane District, N Ontario. Before 1993, also search Beatty AND Ontario. For town in Nevada, use specific city-state term.
BT Cochrane District Ontario
BT Ontario
BT Eastern Canada
BT Canada

Beauce (1978)
Ancient region in N central France.
IN Index departments as applicable.
BT France
BT Western Europe
BT Europe
SA Eure-et-Loir France
SA Loir-et-Cher France

Beaufort
Valid through 1988. Search in combination with state term. After 1988, use specific city-state term.

Beaufort County
Valid through 1988. Search in combination with state term. After 1988, use specific county-state term.

Beaufort County North Carolina (1989)
On the Atlantic coast in E North Carolina. Before 1989, also search Beaufort County AND North Carolina.
CO N351300N354300
  W0762800W0771200
BT North Carolina
BT United States
NT Lee Creek Mine
SA Pamlico River

Beaufort County South Carolina (1989)
On Atlantic coast in extreme S South Carolina. Before 1989, also search Beaufort County AND South Carolina.
CO N320700N324300
  W0802500W0810200
BT South Carolina
BT United States
SA Port Royal Sound

Beaufort Formation (1978)
Underlies Castle Hayne Limestone or Yorktown Formation, or locally, an unnamed middle(?) Miocene unit; overlies Upper Cretaceous Peedee Formation. In E North Carolina. Search in combination with area.
BT Paleocene
BT Paleogene
BT Tertiary
BT Cenozoic
SA Beaufort Group
SA North Carolina

Beaufort Group (1993)
In Karroo Supergroup. Permian and Triassic of South Africa. Before 1993, also search Beaufort Formation or Beaufort Series AND South Africa.
IN Index ages as applicable.
SA Beaufort Formation
SA Botswana
SA Karroo Supergroup
SA Namibia
SA Permian
SA South Africa
SA Triassic
SA Zimbabwe

Beaufort North Carolina (1989)
City in E North Carolina. Before 1989, search Beaufort AND North Carolina.
CO N344400N344400
  W0764100W0764100
BT Carteret County North Carolina
BT North Carolina
BT United States

Beaufort Sea (1978)
Between N Alaska and the Arctic Archipelago of Canada.
CO N690000N760000
  W1230000W1560000
BT Arctic Ocean

Beaumont
Valid through 1988. Search in combination with state term. After 1988, use specific city-state term.

Beaumont Clay (1985)
E Texas and SW Louisiana.
BT Pleistocene
BT Quaternary
BT Cenozoic
SA Louisiana
SA Texas

Beaumont Texas (1989)
City in E Texas. Before 1989, search Beaumont AND Texas.
CO N300400N300400
  W0940600W0940600
BT Jefferson County Texas
BT Texas
BT United States

Beaver County
Valid through 1988. Search in combination with state term. After 1988, use specific county-state term.

Beaver County Oklahoma (1989)
Extreme NW Oklahoma. Before 1989, also search Beaver County AND Oklahoma.
CO N363000N370000
  W1000000W1005500
BT Oklahoma
BT United States
SA Oklahoma Panhandle

Beaver County Pennsylvania (1989)
W Pennsylvania. Before 1989, also search Beaver County AND Pennsylvania.
CO N402800N405200
  W0800900W0803200
BT Pennsylvania
BT United States

Beaver County Utah (1989)
SW Utah. Before 1989, also search Beaver County AND Utah.
CO N381000N383700
  W1122000W1140300
BT Utah
BT United States
NT Mineral Mountains
NT Roosevelt Hot Springs KGRA
SA Tushar Mountains
SA Wah Wah Mountains

Beaver Creek (1978)
Two Beaver Creeks. One in NE and one in W central Wyoming.
IN Index state and counties as applicable.
SA Wyoming

Beaver Lake (1978)
Garden County in western panhandle.
IN Index counties or regions as applicable.
SA Garden County Nebraska

Beaver River (1978)
Extreme SE Yukon Territory.
IN Index territory or region as applicable.
SA Yukon Territory

Beaverhead County
Valid through 1988. Search in combination with state term. After 1988, use specific county-state term.

Beaverhead County Montana (1989)
Extreme SW Montana. Before 1989, also search Beaverhead County AND Montana.
CO N442300N455300
  W1112700W1135600
BT Montana
BT United States
NT Tendoy Range
SA Gravelly Range

Beaverhead Formation (1978)
Upper Cretaceous, Paleocene and Eocene. Unconformably overlies Kootenay Formation, Dinwoody Formation, Thaynes Formation, Phosphoria Formation, Quadrant Quartzite and Madison Group. E central Idaho and SW Montana.
IN Index ages as applicable.
SA Dinwoody Formation
SA Eocene
SA Idaho
SA K-T boundary
SA Madison Group
SA Montana
SA Paleocene
SA Phosphoria Formation
SA Thaynes Formation
SA Upper Cretaceous

Beaverhead Mountains (1978)
S part of Bitterroot Range.
IN Index Bitterroot Range and states as applicable.
SA Bitterroot Range
SA Idaho
SA Montana
SA U. S. Rocky Mountains
SA United States

Beaverlodge (1978)
Lake N of Lake Athabasca.
BT Saskatchewan
BT Western Canada
BT Canada

Bechar
No longer a valid term for GeoRef. See Bechar Algeria.

Bechar Algeria (1993)
Town about 440 miles SW of Algiers. Before 1993, search Bechar and Algeria.
BT Algeria
BT North Africa
BT Africa

Bechuanaland
Not a valid GeoRef index term after 1971. Was used on level 1 in subfile G.
use Botswana

Becraft Mountain (1978)
Columbia County in E New York.
IN Index county as applicable.
SA Columbia County New York

bed forms
   use bedforms

bed load
   use bedload

bed-load
   use bedload

Bedarieux
   No longer a valid term for GeoRef. See Bedarieux France.

Bedarieux France (1993)
   Town in S France.
   BT Herault France
   BT France
   BT Western Europe
   BT Europe

bedded volcano
   use stratovolcanoes

bedding (1978)
   BT planar bedding structures
   BT sedimentary structures
   SA bedding plane irregularities
   SA columnar joints
   SA cross-bedding
   SA flaser bedding
   SA fractures
   SA graded bedding
   SA laminations
   SA lineation
   SA massive bedding
   SA primary structures
   SA rhythmic bedding
   SA stratification
   SA strike

bedding faults (1978)
   Before 1978, search faults AND bedding.
   BT faults
   SA bedding joints
   SA duplexes

bedding joints (1981)
   SA bedding faults
   SA joints

bedding plane irregularities (1978)
   Autoposting of this term to shrinkage cracks ended in 1981.
   BT sedimentary structures
   NT antidunes
   NT chevron marks
   NT current markings
   NT dune structures
   NT flute casts
   NT fossil ice wedges
   NT frost features
   NT groove casts
   NT grooves
   NT megaripples
   NT mounds
   NT mud lumps
   NT mudcracks
   NT parting lineation
   NT ripple marks
   NT sand ridges
   NT sand waves
   NT scour casts
   NT scour marks
   NT striations
   NT tool marks
   SA bars
   SA bedding
   SA casts
   SA dunes
   SA imbrication
   SA load casts
   SA sand bodies
   SA sole marks

bedding structures, planar
   use planar bedding structures

bedding-plane slip
   use flexural-slip

Bedford County
   Valid through 1988. Search in combination with state term. After 1988, use specific county-state term.

Bedford County Pennsylvania (1989)
   S Pennsylvania. Before 1989, also search Bedford County AND Pennsylvania.
   CO N394300N401800
         W0780800W0784800
   BT Pennsylvania
   BT United States
   NT Woodbury Pennsylvania

Bedford County Tennessee (1989)
   Central Tennessee. Before 1989, also search Bedford County AND Tennessee.
   CO N352000N354200
         W0861500W0864000
   BT Tennessee
   BT United States

Bedford County Virginia (1989)
   SW central Virginia. Before 1989, also search Bedford County AND Virginia.
   CO N370200N373800
         W0791300W0795000
   BT Virginia
   BT United States

Bedford Shale (1978)
   Devonian and Mississippian(?): NE Kentucky, E Ohio, and SW Pennsylvania.
   BT Paleozoic
   SA Devonian
   SA Kentucky
   SA Mississippian
   SA Ohio
   SA Pennsylvania

Bedfordshire
   No longer a valid term for GeoRef as of 1993.
   use Bedfordshire England

Bedfordshire England (1993)
   County in SE England.
   CO N514500N522000
         W0001000W0004000
   UF Bedfordshire
   BT England
   BT Great Britain
   BT United Kingdom
   BT Western Europe
   BT Europe

bedforms (1978)
   UF bed forms
   SA fluvial features
   SA geomorphology
   SA hydrology
   SA lacustrine features
   SA rivers and streams

bedload (1978)
   UF bed load
   UF bed-load
   UF bottom load
   UF sediment budget
   UF sediment load
   SA sedimentation
   SA sediments
   SA stream transport
   SA streams

bedrock (1978)
   Used as a general term.
   SA outcrops
   SA rocks
   SA soils

beds, convoluted
   use convoluted beds

beds, key
   use key beds

beds, marker
   use marker beds

Beech Creek Limestone Member (1985)
   Of Golconda Formation. SW Indiana and N central Kentucky.
   BT Chesterian
   BT Upper Mississippian
   BT Mississippian
   BT Carboniferous
   BT Paleozoic
   SA Golconda Formation
   SA Indiana
   SA Kentucky

Beekmantown Group (1985)
   From SE Canada to Virginia.
   BT Lower Ordovician
   BT Ordovician
   BT Paleozoic
   SA Maryland
   SA New York
   SA Ontario
   SA Pennsylvania
   SA Quebec
   SA Tennessee
   SA Vermont
   SA Virginia

Beemerville
   Valid through 1988. Search in combination with state term. After 1988, use specific city-state term.

Beemerville New Jersey (1989)
   Village in extreme N New Jersey. Before 1989, search Beemerville AND New Jersey.
   CO N411300N411300
         W0744100W0744100
   BT Sussex County New Jersey
   BT New Jersey
   BT United States

Beerenberg (1978)
   Extinct volcano on Norway's Jan Mayen Island. In 1981, broader term changed from Arctic Ocean to Jan Mayen.
   BT Jan Mayen
   BT Arctic region
   SA Arctic Ocean

Behar India
   use Bihar India

behavior (1978)
   Used as a general term applicable to a variety of disciplines.
   SA biologic evolution
   SA ecology
   SA paleoecology
   SA properties
   SA radiation

beidellite (1978)
   BT clay minerals
   BT sheet silicates
   BT silicates
   SA montmorillonite

Beijing
   No longer a valid term for GeoRef. See Beijing China.

Beijing China (1993)
   Independent municipality and capital of China, including surrounding counties. Located in Hebei. Before 1985, also search Peking.
   CO N393000N410000
         E1173000E1155000
   BT Hebei China
   BT China
   BT Far East
   BT Asia
   SA Xingtai earthquake 1966

Beil Limestone Member (1978)
   Of Lecompton Limestone. SW Iowa, E Kansas, SE Nebraska.
   BT Virgilian
   BT Upper Pennsylvanian
   BT Pennsylvanian
   BT Carboniferous
   BT Paleozoic
   SA Iowa
   SA Kansas
   SA Lecompton Limestone
   SA Nebraska

Beira Baixa (1978)
   Former province in N central Portugal.
   BT Portugal
   BT Iberian Peninsula
   BT Southern Europe
   BT Europe

Beirut
   No longer a valid term for GeoRef. See Beirut Lebanon.

Beirut Lebanon (1993)
   City on the Mediterranean Sea. Before 1993 also search Beirut AND Lebanon.
   BT Lebanon
   BT Middle East
   BT Asia

Beius Basin (1978)
   Bihor County in W Transylvania.
   BT Transylvania
   BT Romania
   BT Southern Europe
   BT Europe

Bek-Budi Uzbekistan
   use Karshi Uzbekistan

Belarus (1993)
   Formerly the Byelorussian Soviet Socialist Republic. Before 1978, also search Belorussia and White Russia. Before 1993, search Byelorussia. This term has multiple hierarchies.
   CO N510000N560000
         E0330000E0240000
   UF Belorussia
   UF Byelorussia
   UF White Russia
   BT1 Europe
   BT2 Commonwealth of Independent States
   NT Brest Belarus
   NT Byelorussian Massif
   NT Kletno Belarus
   NT Mezhrechye Belarus
   NT Minsk Belarus
   NT Mogilev Belarus
   NT Orsha Belarus
   NT Ostashkovichi Belarus
   NT Rechitsa Belarus
   NT Starobin Belarus
   NT Ustron Belarus
   NT Vyshkovo Belarus
   NT Zhitkovichi Belarus
   SA Brest Basin
   SA Bug region
   SA Bug River
   SA Dnieper Basin
   SA Dnieper River
   SA Dvina River
   SA European Platform
   SA Neman River basin
   SA Polesye
   SA Pripet Basin
   SA Russian Plain

Belau (1993)

Island group, W Caroline Islands, Micronesia. Before 1993, also search Palau AND Pacific Ocean or Micronesia.
CO N065500N080500 E1345000E1341500
UF Palau
UF Pelew
BT Caroline Islands
BT Micronesia
BT Oceania

Belaya Gora
No longer a valid term for GeoRef. As of 1993, see Belaya Gora Russian Federation.

Belaya Gora Russian Federation (1993)
Village S of Arkhangelsk in Arkhangelsk Oblast in NW Russian Federation. This term has multiple hierarchies.
BT1 Arkhangelsk Russian Federation
BT1 Russian Federation
BT1 Commonwealth of Independent States
BT2 Arkhangelsk Russian Federation
BT2 Europe

Belchatow
No longer a valid term for GeoRef. As of 1993 see Belchatow Poland.

Belchatow Mine (1993)
Lignite mine, Piotrkow Province, central Poland. Before 1993, also search Belchatow.
BT Poland
BT Central Europe
BT Europe
SA lignite
SA Lodz Poland
SA mines

Belchatow Poland (1993)
Town in Piotrkow Province, central Poland. Use Belchatow Mine for the lignite mine. Before 1993 also search Belchatow AND Poland.
BT Poland
BT Central Europe
BT Europe
SA Lodz Poland

Belcher Islands (1978)
Group in SE Hudson Bay.
BT Northwest Territories
BT Western Canada
BT Canada
SA Hudson Bay

belemnites (1978)
As of 1981, restricted to use as the common name for Belemnitidae. Autoposting of mollusks and cephalopods to this term began in 1989.
BT cephalopods
BT mollusks
BT invertebrates
SA Belemnitidae
SA biostratigraphy

Belemnitidae (1978)
Family. Autoposting of this term to Belemnopsis began in 1985.
BT Belemnoidea
BT Dibranchiata
BT Cephalopoda
BT Mollusca
BT Invertebrata
NT Belemnopsis
SA belemnites

Belemnoidea (1978)
Autoposting of this term to Belemnitidae and Belemnopsis began in 1985.
BT Dibranchiata
BT Cephalopoda
BT Mollusca
BT Invertebrata
NT Belemnitidae

Belemnopsis (1978)
Genus.
BT Belemnitidae
BT Belemnoidea
BT Dibranchiata
BT Cephalopoda
BT Mollusca
BT Invertebrata

Belfast
No longer a valid term for GeoRef. After 1993, see Belfast Northern Ireland.

Belfast Northern Ireland (1993)
Traditional county in W Northern Ireland. Includes the smaller administrative region of the same name adopted after 1973.
BT Northern Ireland
BT United Kingdom
BT Western Europe
BT Europe

Belfort
No longer a valid term for GeoRef. See Belfort France.

Belfort France (1993)
Department. Territore de Belfort in NE France.
CO N472500N475000 E0071000E0064000
BT France
BT Western Europe
BT Europe

Belgaum
No longer a valid term for GeoRef. See Belgaum India.

Belgaum India (1993)
Town and district in W Karnataka, S India.
BT Karnataka India
BT India
BT Indian Peninsula
BT Asia

Belgian Congo
use Zaire

Belgium (1978)
CO N493000N513000 E0063000E0023000
BT Western Europe
BT Europe
NT Antwerp Belgium
NT Brabant Belgium
NT Brabant Massif
NT Hainaut Belgium
NT Liege Belgium
NT Limburg Belgium
NT Luxembourg Belgium
NT Namur Belgium
SA Boom Clay
SA Campine
SA European Platform
SA Grimmertingen
SA Liege earthquake 1983
SA Luxembourg Sandstone
SA Meuse River
SA Meuse Valley
SA North Sea Coast
SA North Sea region
SA Scheldt River
SA Stavelot-Venn Massif
SA University of Louvain

Belgorod
No longer a valid term for GeoRef. As of 1993, see Belgorod Russian Federation.

Belgorod Russian Federation (1993)
Oblast including the city N of Kharkov in SW Russian Federation. This term has multiple hierarchies.
BT1 Russian Federation
BT1 Commonwealth of Independent States
BT2 Europe

Belgrade (1978)
City on Danube River in N central Serbia.
UF Beograd
BT Serbia
BT Yugoslavia
BT Southern Europe
BT Europe

Beli Isvor Bulgaria
use Beli Izvor Bulgaria

Beli Izvor
No longer a valid term for GeoRef. See Beli Izvor Bulgaria.

Beli Izvor Bulgaria (1993)
Village in E Bulgaria. Also search Beli Izvor or Beli Isvor.
UF Beli Isvor Bulgaria
BT Bulgaria
BT Southern Europe
BT Europe

Belitung
use Billiton

Belize (1978)
Formerly British Honduras. Before 1975, also search British Honduras.
CO N155500N183000 W0872000W0892000
UF British Honduras
BT Central America
NT Maya Mountains

Beljak Austria
use Villach Austria

Bell Canyon Formation (1978)
In Delaware Mountain Group. Includes Hegler Limestone, Pinery Limestone, Radar Limestone, Lamar Limestone and McCombs Limestone members. S New Mexico, and W Texas.
BT Guadalupian
BT Permian
BT Paleozoic
SA Delaware Mountain Group
SA New Mexico
SA Texas

Bell County
Valid through 1988. Search in combination with state term. After 1988, use specific county-state term.

Bell County Kentucky (1989)
Extreme SE Kentucky. Before 1989, also search Bell County AND Kentucky.
CO N363600N365700 W0832600W0835600
BT Kentucky
BT United States

Bell County Texas (1989)
Central Texas. Before 1989, also search Bell County AND Texas.
CO N304700N311700 W0970500W0975500
BT Texas
BT United States
SA Edwards Aquifer

Bell Creek Field (1978)
Oil and gas field.
IN Index counties or regions as applicable.
SA Carter County Montana
SA Montana
SA oil and gas fields
SA Powder River County Montana

Bell Island (1978)
In Conception Bay W of Saint John's.
IN Index province or region as applicable.
SA Newfoundland Island

Bellary
No longer a valid term for GeoRef. See Bellary India.

Bellary India (1993)
Town in Karnataka, S central India.
BT Karnataka India
BT India
BT Indian Peninsula
BT Asia

Belle Isle (1978)
Island constituting a salt dome on N shore of Atchafalaya Bay in Saint Mary Parish in S Louisiana. Also island off Labrador in Newfoundland.
IN Index state and parish or Newfoundland as applicable.
SA Atchafalaya Bay
SA Newfoundland
SA Saint Mary Parish Louisiana

Belledonne
Not a valid term for GeoRef. See Belledonne Massif.

Belledonne Massif (1978)
Part of mountain range in SE France. Also search Belledonne.
UF Belledonne Mountain Range
UF Belledonne Range
BT France
BT Western Europe
BT Europe

Belledonne Mountain Range
use Belledonne Massif

Belledonne Range
use Belledonne Massif

Bellerophontina (1981)
Autoposting of Archaeogastropoda to this term began in 1989.
BT Archaeogastropoda
BT Gastropoda
BT Mollusca
BT Invertebrata

Bellingshausen Sea (1978)
Inlet of South Pacific Ocean extending westward from Alexander Island off the W coast of Antarctic Peninsula to Thurston Island.
CO S730000S660000 W0750000W0980000
BT Antarctic Ocean
SA Bransfield Strait

Belluno
No longer a valid term for GeoRef. See Belluno Italy.

Belluno Italy (1993)
Town and province in N Veneto.
BT Veneto Italy
BT Italy
BT Southern Europe
BT Europe

NT Alpago Basin

Belly River Formation (1978)
Has been considered equivalent to Two Medicine Formation and Virgelle Sandstone of Blackfoot Indian Reservation, N Montana, and Judith River Formation, Claggett Formation, and Eagle Sandstone of central Montana.
BT Upper Cretaceous
BT Cretaceous
BT Mesozoic
SA Alberta
SA Saskatchewan

Belomorsk
No longer a valid term for GeoRef. As of 1993, see Belomorsk Russian Federation.

Belomorsk Russian Federation (1993)
Town in former Karelian ASSR in NW Russian Federation. Before 1993, also search Belomorsk or Soroka. This term has multiple hierarchies.
UF Soroka Russian Federation
BT1 Karelia Russian Federation
BT1 Russian Federation
BT1 Commonwealth of Independent States
BT2 Karelia Russian Federation
BT2 Asia

Belorussia
use Belarus

Belozerka
No longer a valid term for GeoRef. See Belozerka Ukraine.

Belozerka Ukraine (1993)
Town in Kherson Oblast N of the Crimea. This term has multiple hierarchies.
BT1 Ukraine
BT1 Europe
BT2 Ukraine
BT2 Commonwealth of Independent States

Belozero
No longer a valid term for GeoRef. As of 1993, see Belozero Russian Federation.

Belozero Russian Federation (1993)
Village W of Arkhangelsk in Arkhangelsk Oblast. Before 1993, also search Belozero. This term has multiple hierarchies.
BT1 Arkhangelsk Russian Federation
BT1 Russian Federation
BT1 Commonwealth of Independent States
BT2 Arkhangelsk Russian Federation
BT2 Europe

Belsk
No longer a valid term for GeoRef. As of 1993 see Belsk Poland.

Belsk Poland (1993)
Town in in NE Poland. Before 1993 also search Belsk AND Poland.
UF Bielsk
UF Bielsk Podlaski
BT Bialystok Poland
BT Poland
BT Central Europe
BT Europe

Belt Basin (1978)
W Montana, N Idaho, NE Washington and S British Columbia.
BT North America
SA British Columbia
SA Idaho
SA Montana
SA Washington

Belt Supergroup (1978)
Comprises Piegan Group, Missoula Group, and Ravalli Group.
BT middle Proterozoic
BT Proterozoic
BT upper Precambrian
BT Precambrian
SA Aldridge Formation
SA Altyn Limestone
SA Bonner Formation
SA British Columbia
SA Disturbed Belt
SA Empire Formation
SA Helena Formation
SA Idaho
SA Missoula Group
SA Montana
SA Mount Shields Formation
SA Newland Limestone
SA Prichard Formation
SA Ravalli Group
SA Revett Quartzite
SA Saint Regis Formation
SA Shepard Formation
SA Snowslip Formation
SA Spokane Formation
SA Wallace Formation
SA Washington
SA Yellowjacket Formation

Beltana
No longer a valid term for GeoRef. See Beltana Australia.

Beltana Australia (1993)
Village and mining center in E central South Australia.
BT South Australia
BT Australia
BT Australasia

Beltrami County
Valid through 1988. Search in combination with state term. After 1988, use specific county-state term.

Beltrami County Minnesota (1989)
NW Minnesota. Before 1989, also search Beltrami County AND Minnesota.
CO N472500N483300
W0942500W0954000
BT Minnesota
BT United States
SA Red Lake

belts
A valid term through 1978. After 1978, see specific terms, e.g. fold belts, metamorphic belts, terranes, etc.

belts, fold
use fold belts

belts, greenstone
use greenstone belts

belts, metamorphic
use metamorphic belts

belts, mobile
use mobile belts

belts, volcanic
use volcanic belts

Bembridge Marls (1978)
Defined to include Bembridge Oyster Beds, Lower Bembridge Marls, and Upper Bembridge Marls. Isle of Wight in English Channel S of Southampton.
BT Oligocene
BT Paleogene
BT Tertiary
BT Cenozoic
SA England
SA Isle of Wight England
SA United Kingdom

benches (1978)
SA erosion features
SA fluvial features
SA geomorphology
SA littoral erosion
SA marine terraces
SA mesas
SA shore features
SA terraces
SA wave-cut platforms

beneficiation (1978)
Used for commodity production technology.
SA flotation
SA methods
SA mineral deposits, genesis
SA optimization
SA production
SA purification

Bengal (1978)
A region and former Indian province.
IN Index Bangladesh and Indian states as applicable.
UF Bengal Basin
SA Bangladesh
SA India
SA Tripura India
SA West Bengal India

Bengal Basin
use Bengal

Bengal Fan (1993)
Submarine fan extending from the Ganges-Brahmaputra delta system south and west approximately to the equator.
BT Indian Ocean
SA Bay of Bengal
SA Leg 116
SA ODP Site 717
SA ODP Site 718
SA ODP Site 719

Bengal Islands (1981)
Includes Andaman and Nicobar Islands.
CO N063000N135000
E0943000E0913000
BT India
BT Indian Peninsula
BT Asia
NT Andaman Islands
NT Nicobar Islands
SA Indian Ocean Islands

Benham
Valid through 1988. Search in combination with state term. After 1988, use specific city-state term.

Benham Kentucky (1989)
Village in SE Kentucky. Before 1989, search Benham AND Kentucky.
BT Harlan County Kentucky
BT Kentucky
BT United States

Beni Bouchera (1978)
Region in the Rif in NE Morocco.
BT Rif
BT Morocco
BT North Africa
BT Africa

Benin (1978)
Was established in 1960 as Dahomey. Before 1976, also search Dahomey.
CO N061000N123000
E0035000E0004000
UF Dahomey
BT West Africa
BT Africa
SA Imo Shale
SA Mali-Niger Syneclise
SA Niger River
SA Niger Valley

Benioff seismic zone
use Benioff zone

Benioff zone (1978)
UF Benioff seismic zone
UF Wadati zone
UF Wadati-Benioff Zone
UF zone, Benioff
SA lithosphere
SA plate tectonics
SA plates
SA sea-floor spreading
SA seismology
SA subduction
SA subduction zones
SA trenches

benmoreite (1989)
Silica-saturated to undersaturated igneous rock intermediate in composition between mugearite and trachyte.
BT volcanic rocks
BT igneous rocks
SA trachybasalts

Bennettitales (1978)
BT gymnosperms
BT Spermatophyta
BT Plantae
SA Ptilophyllum

Benoue Valley
use Benue Valley

Benson Mines (1978)
N New York. Also search Benson AND New York.
IN Index counties or regions as applicable.
SA iron ores
SA mines
SA Saint Lawrence County New York

benthic taxa
use benthonic taxa

benthonic
A valid index term through 1977. After 1977, use benthonic taxa.

benthonic taxa (1978)
Before 1978, search benthonic; benthic.
UF benthic taxa
UF taxa, benthonic
SA bottom features
SA ecology
SA fossils
SA marine geology
SA paleoecology

Benton County
Valid through 1988. Search in combination with state term. After 1988, use specific county-state term.

Benton County Arkansas (1989)
Extreme NW Arkansas. Before 1989, also search Benton County AND Arkansas.
CO N360600N362800
W0935300W0943700
BT Arkansas
BT United States

Benton County Indiana (1989)
W Indiana. Before 1989, also search Benton County AND Indiana.
CO N402800N404300 W0870600W0873200
BT Indiana
BT United States

Benton County Iowa (1989)
E central Iowa. Before 1989, also search Benton County AND Iowa.
CO N415200N421800 W0915000W0921800
BT Iowa
BT United States

Benton County Minnesota (1989)
Central Minnesota. Before 1989, also search Benton County AND Minnesota.
CO N453300N454800 W0934600W0942300
BT Minnesota
BT United States

Benton County Mississippi (1989)
N Mississippi. Before 1989, also search Benton County AND Mississippi.
CO N343500N345900 W0890200W0892200
BT Mississippi
BT United States

Benton County Missouri (1989)
Central Missouri. Before 1989, also search Benton County AND Missouri.
CO N380500N383300 W0930300W0933100
BT Missouri
BT United States

Benton County Oregon (1989)
W Oregon. Before 1989, also search Benton County AND Oregon.
CO N441700N444300 W1230500W1234900
BT Oregon
BT United States

Benton County Tennessee (1989)
W Tennessee. Before 1989, also search Benton County AND Tennessee.
CO N354800N362100 W0875500W0881300
BT Tennessee
BT United States

Benton County Washington (1989)
On Oregon line in S Washington. Before 1989, also search Benton County AND Washington.
CO N455100N464200 W1185700W1195200
BT Washington
BT United States
NT Richland Washington
SA Hanford Reservation
SA Rattlesnake Hills

Benton Formation (1978)
In Colorado Group. Comprising Graneros Shale, Greenhorn Limestone and Carlile Shale as members. Contains Codell Sandstone Member. E Colorado, South Dakota, Kansas, S Minnesota, SE Montana, Nebraska, NE New Mexico, E Wyoming.
BT Cretaceous
BT Mesozoic
SA Carlile Shale
SA Codell Sandstone Member
SA Colorado
SA Colorado Group
SA Graneros Shale
SA Greenhorn Limestone
SA Kansas
SA Minnesota
SA Montana
SA Nebraska
SA New Mexico
SA South Dakota
SA Wyoming

Benton Uplift (1985)
W Arkansas and SE Oklahoma.
IN Index states as applicable.
BT United States
SA Arkansas
SA Oklahoma
SA Ouachita Mountains
SA Ouachita Orogeny

bentonite (1978)
As of 1981, use bentonite deposits for bentonite as a commodity.
UF amargosite
UF mineral soap
UF soap clay
UF volcanic clay
BT clastic rocks
BT sedimentary rocks
SA bentonite deposits
SA clastic sediments
SA clay
SA clay mineralogy
SA clay minerals
SA metabentonite
SA pyroclastics
SA sheet silicates
SA terrigenous materials
SA tuff
SA volcanic ash

bentonite deposits (1981)
Before 1981, search bentonite AND deposits.
IN Commodity. See List C.
SA bentonite
SA clays

Benue Trough
use Benue Valley

Benue Valley (1978)
River valley.
IN Index countries as applicable.
UF Benoue Valley
UF Benue Trough
UF Binue Valley
BT West Africa
BT Africa
SA Asu River Group
SA Cameroon
SA Nigeria

Benxi Formation (1989)
In Henan and Shanxi provinces.
BT Carboniferous
BT Paleozoic
SA China
SA Henan China
SA Shanxi China

benzene (1985)
BT aromatic hydrocarbons
BT hydrocarbons
BT organic materials
SA petroleum

Beograd
use Belgrade

Beppu
No longer a valid term for GeoRef. See Beppu Japan.

Beppu Japan (1993)
City and geothermal area in NE Kyushu.
BT Oita Japan
BT Kyushu
BT Japan
BT Far East
BT Asia

beraunite (1978)
BT phosphates

Berchtesgaden
No longer a valid term for GeoRef. See Berchtesgaden Germany.

Berchtesgaden Germany (1993)
Town in extreme SE Bavaria.
BT Upper Bavaria Germany
BT Bavaria Germany
BT Germany
BT Central Europe
BT Europe

Berea Sandstone (1978)
Devonian or Mississippian. NE Kentucky, S Michigan, Ohio, W Pennsylvania, and N West Virginia.
BT Paleozoic
SA Devonian
SA Kentucky
SA Michigan
SA Mississippian
SA Ohio
SA Pennsylvania
SA West Virginia

Beregovo
No longer a valid term for GeoRef. See Beregovo Ukraine.

Beregovo Ukraine (1993)
Town in Transcarpathian Oblast in extreme SW Ukraine. This term has multiple hierarchies.
BT1 Transcarpathia Ukraine
BT1 Ukraine
BT1 Europe
BT2 Transcarpathia Ukraine
BT2 Ukraine
BT2 Commonwealth of Independent States

Beresford Lake (1978)
In SW Manitoba.
BT Manitoba
BT Western Canada
BT Canada

beresite (1978)
From 1978-1980, granite-granodiorite family was autoposted to this term. As of 1990, metasomatic rocks is autoposted to this term.
BT greisen
BT metasomatic rocks
BT metamorphic rocks
SA granites
SA granodiorites
SA hypabyssal rocks
SA quartz porphyry

berg crystal
use quartz crystal

Bergamo
No longer a valid term for GeoRef. See Bergamo Italy.

Bergamo Italy (1993)
Province and city in central Lombardy.
BT Lombardy Italy
BT Italy
BT Southern Europe
BT Europe

Bergell Massif (1978)
In SE Switzerland.
CO N461000N462000 E0094500E0093000
BT Switzerland
BT Central Europe

BT Europe

Bergen
No longer a valid term for GeoRef. As of 1993 see Bergen Norway.

Bergen County
Valid through 1988. Search in combination with state term. After 1988, use specific county-state term.

Bergen County New Jersey (1989)
NE New Jersey. Before 1989, also search Bergen County AND New Jersey.
CO N404500N410800 W0735300W0741700
BT New Jersey
BT United States

Bergen Norway (1993)
City on the North Sea in SW Norway. Before 1993 also search Bergen AND Norway.
BT Norway
BT Scandinavia
BT Western Europe
BT Europe

Bergisch Gladbach
No longer a valid term for GeoRef. See Bergisch Gladbach Germany.

Bergisch Gladbach Germany (1993)
City in SW central North Rhine-Westphalia.
BT North Rhine-Westphalia Germany
BT Germany
BT Central Europe
BT Europe

Bergisches Land (1978)
Region on right bank of the Rhine River.
BT North Rhine-Westphalia Germany
BT Germany
BT Central Europe
BT Europe
SA Rhine River

Bergstrasse (1978)
Name applied to the W slope of the Odenwald hills of SW Germany.
BT Hesse Germany
BT Germany
BT Central Europe
BT Europe
SA Odenwald

Berici Hills (1978)
Range of volcanic hills in Vicenza Province in W central Veneto.
BT Vicenza Italy
BT Veneto Italy
BT Italy
BT Southern Europe
BT Europe

Bering land bridge
As of 1981, no longer a valid term for GeoRef.
use Beringia

Bering Sea (1978)
Between NE Siberia and Alaska with the Aleutian Islands on the S and Bering Strait on the N.
CO N513000N653000 W1573000E1620000
BT West Pacific
BT Pacific Ocean
NT Bowers Ridge
NT Bristol Bay
NT Navarin Basin

NT Norton Sound
SA Bering Strait
SA Beringia
SA Kaltag Fault
SA Karaginskiy Island
SA Komandorski Islands
SA Leg 19
SA Norton Basin
SA Nunivak Island
SA Saint George Basin
SA Saint George Island
SA Saint Paul Island

Bering Strait (1981)
Connects Chukchi Sea and Bering Sea between Asia and North America.
CO N650000N670000
    W1670000W1700000
SA Arctic Ocean
SA Bering Sea
SA Beringia
SA Chukchi Sea
SA Pacific Ocean
SA West Pacific

Beringia (1981)
NE Siberia, NW North America, and intervening continental shelves. Early man entered North America via this region. Before 1981, also search Bering land bridge.
UF Bering land bridge
SA Asia
SA Bering Sea
SA Bering Strait
SA Chukchi Sea
SA North America
SA Siberia

Berkeley
Valid through 1988. Search in combination with state term. After 1988, use specific city-state term.

Berkeley California (1989)
City in W Central California, across the San Francisco Bay from San Francisco. Before 1989, search Berkeley AND California.
CO N375300N375300
    W1221700W1221700
BT Alameda County California
BT California
BT United States
SA San Francisco Bay

Berkeley County
Valid through 1988. Search in combination with state term. After 1988, use specific county-state term.

Berkeley County South Carolina (1989)
SE South Carolina. Before 1989, also search Berkeley County AND South Carolina.
CO N325000N332900
    W0792700W0802200
BT South Carolina
BT United States

Berkeley County West Virginia (1989)
In eastern panhandle of West Virginia. Before 1989, also search Berkeley County AND West Virginia.
CO N391700N393800
    W0775000W0781500
BT West Virginia
BT United States

berkelium (1989)
Chemical element.
UF Bk
BT actinides

BT metals

Berks County
Valid through 1988. Search in combination with state term. After 1988, use specific county-state term.

Berks County Pennsylvania (1989)
SE central Pennsylvania. Before 1989, also search Berks County AND Pennsylvania.
CO N401000N404000
    W0753300W0762500
BT Pennsylvania
BT United States
NT Reading Pennsylvania

Berkshire
No longer a valid term for GeoRef as of 1993.
use Berkshire England

Berkshire Anticlinorium
use Berkshire Hills

Berkshire County
Valid through 1988. Search in combination with state term. After 1988, use specific county-state term.

Berkshire County Massachusetts (1989)
W Massachusetts. Before 1989, also search Berkshire County AND Massachusetts.
CO N420300N424500
    W0725600W0733100
BT Massachusetts
BT United States
SA Berkshire Hills

Berkshire England (1993)
County in S England. As of 1975, some of its former territory is included in Oxfordshire.
CO N512000N514500
    W0003000W0014000
UF Berkshire
BT England
BT Great Britain
BT United Kingdom
BT Western Europe
BT Europe
SA Oxfordshire England

Berkshire Highlands
use Berkshire Hills

Berkshire Hills (1978)
Hills in Berkshire County in W Massachusetts.
IN Index counties or regions as applicable.
UF Berkshire Anticlinorium
UF Berkshire Highlands
UF Berkshire Massif
UF Berkshires
SA Appalachians
SA Berkshire County Massachusetts

Berkshire Massif
use Berkshire Hills

Berkshires
use Berkshire Hills

Berlin
No longer a valid term for GeoRef. See Berlin Germany.

Berlin Germany (1993)
City partitioned into East Berlin and West Berlin from 1945 to 1989.
BT Brandenburg Germany
BT Germany
BT Central Europe
BT Europe

SA North German Plain

berms (1989)
Type of bench or shelf. Term used in relation to engineering geology and geomorphology.
SA beaches
SA coastlines
SA embankments
SA engineering geology
SA erosion features
SA geomorphology
SA retaining walls
SA slope stability

Bermuda (1978)
British colony comprising over 300 coral islands about 640 miles ESE of Cape Hatteras. In 1981, broader term changed from Atlantic Ocean to Atlantic Ocean Islands. Before 1972, also search Bermuda Islands.
CO N321500N322500
    W0644000W0645000
UF Bermuda Island
UF Bermuda Islands
UF Bermudas
BT Atlantic Ocean Islands
NT Harrington Sound
SA Bermuda Platform
SA Bermuda Rise

Bermuda Island
use Bermuda

Bermuda Islands
Not a valid GeoRef index term after 1971. Was used on level 1 in subfile G.
use Bermuda

Bermuda Platform (1978)
Part of Bermuda Rise.
BT Atlantic Ocean
SA Bermuda
SA Bermuda Rise

Bermuda Rise (1978)
Runs in SW-NE direction to the west of Bermuda. Autoposting of Northwest Atlantic to this term began in 1989. Autoposting of Atlantic Ocean to this term began in 1990.
BT Northwest Atlantic
BT Atlantic Ocean
SA Bermuda
SA Bermuda Platform
SA DSDP Site 386
SA DSDP Site 387
SA DSDP Site 417
SA DSDP Site 418
SA Leg 43
SA Leg 51
SA Leg 52
SA Leg 53

Bermudas
use Bermuda

Bern
No longer a valid term for GeoRef, as of 1993, see Bern Switzerland.

Bern Switzerland (1993)
Canton in W central Switzerland. Also search Bern or Berne.
UF Berne Switzerland
BT Switzerland
BT Central Europe
BT Europe
NT Biel Lake
SA Bernese Alps
SA Grindelwald

Bernalillo County
Valid through 1988. Search in combination with state term. After 1988, use specific county-state term.

Bernalillo County New Mexico (1989)
Central New Mexico. Before 1989, also search Bernalillo County AND New Mexico.
CO N345200N351200
    W1061000W1071500
BT New Mexico
BT United States
NT Albuquerque New Mexico
NT Tijeras Canyon
SA Manzano Mountains
SA Rio Puerco

berndtite (1978)
BT sulfides

Berne Switzerland
use Bern Switzerland

Bernese Alps (1978)
N division of Central Alps in Bern and Valais cantons. This term has multiple hierarchies.
UF Bernese Oberland
BT1 Switzerland
BT1 Central Europe
BT1 Europe
BT2 Central Alps
BT2 Alps
BT2 Europe
SA Bern Switzerland
SA Grindelwald
SA Uri Switzerland
SA Valais Switzerland

Bernese Oberland
use Bernese Alps

Bernic Lake (1978)
BT Manitoba
BT Western Canada
BT Canada

bernstein
use amber

Berounka System (1981)
CO N492000N501500
    E0142500E0125000
BT Czechoslovakia
BT Central Europe
BT Europe

Berriasian (1978)
Europe. Lowermost Lower Cretaceous, above Portlandian (Jurassic), below Valanginian.
BT Lower Cretaceous
BT Cretaceous
BT Mesozoic
SA Neocomian

Berrien County
Valid through 1988. Search in combination with state term. After 1988, use specific county-state term.

Berrien County Georgia (1989)
S Georgia. Before 1989, also search Berrien County AND Georgia.
CO N310300N312800
    W0830200W0832400
BT Georgia
BT United States

Berrien County Michigan (1989)
Extreme SW Michigan. Before 1989, also search Berrien County AND Michigan.
CO N414500N421500
    W0861500W0865200
BT Michigan Lower Peninsula

BT Michigan
BT United States

berryite (1978)
Autoposting of sulfobismuthites began in 1989.
BT sulfobismuthites
BT sulfosalts

berthierine (1978)
BT sheet silicates
BT silicates
SA aluminosilicates
SA chamosite

berthonite
use bournonite

bertrandite (1978)
BT ring silicates
BT silicates

Berwick Formation (1989)
In Merrimack Group. Probably Ordovician and Silurian. SW Maine and SE New Hampshire.
IN Index ages as applicable.
BT lower Paleozoic
BT Paleozoic
SA Maine
SA Merrimack Group
SA New Hampshire
SA Ordovician
SA Silurian

Berwickshire
No longer a valid term for GeoRef as of 1993.
use Berwickshire Scotland

Berwickshire Scotland (1993)
Former county in SE Scotland. Since 1974, part of Border region Scotland.
UF Berwickshire
BT Scotland
BT Great Britain
BT United Kingdom
BT Western Europe
BT Europe

beryl (1978)
IN May be used as a mineral or commodity term.
BT ring silicates
BT silicates
SA aquamarine
SA beryllium
SA beryllium ores
SA emerald
SA gems

beryllium (1978)
Chemical element. As of 1981, use beryllium ores for beryllium as a commodity. Autoposting of alkaline earth metals and metals to this term began in 1989.
UF Be
BT alkaline earth metals
BT metals
NT Be-10
NT Be-10/Be-9
NT Be-7
SA beryl
SA beryllium ores

beryllium ores (1981)
Before 1981, also search beryllium AND (deposit OR deposits OR ore OR ores OR economic) in the basic index. Autoposting of metal ores to this term began in 1985.
IN Commodity. See List C.
BT metal ores
SA beryl
SA beryllium

Besancon
No longer a valid term for GeoRef.
See Besancon France.

Besancon France (1993)
City near Swiss border in E France.
BT Doubs France
BT France
BT Western Europe
BT Europe

Beshchady Mountains (1978)
Section of the Carpathians.
IN Index countries as applicable.
BT Carpathians
BT Europe
SA Czechoslovakia
SA Poland
SA Ukraine

Beskid Mountains (1978)
Both West and East Beskid mountains comprise mountain group of the Carpathians on NE border of Czechoslovakia extending into Poland. This term has multiple hierarchies.
IN Index countries as applicable.
CO N495000N494000
    E0222300E0180000
UF Beskids
BT1 Carpathians
BT1 Europe
BT2 Central Europe
BT2 Europe
NT Pieniny Mountains
SA Bielsko Poland
SA Nowy Sacz Poland
SA Polish Carpathians
SA Slovakia

Beskids
use Beskid Mountains

Bessarabian
No longer a valid term for GeoRef.
See Moldova.

besshi-type (1989)
SA copper ores
SA iron ores
SA mineral deposits, genesis
SA stratabound deposits

Bet-Pak-Dala (1978)
Desert steppe W of Lake Balkhash. This term has multiple hierarchies.
CO N450000N470000
    E0730000E0673000
UF Golodnaya Step
UF Hunger Steppe
BT1 Kazakhstan
BT1 Central Asia
BT1 Asia
BT2 Kazakhstan
BT2 Commonwealth of Independent States

beta particles
use beta rays

beta rays (1982)
Use when referring to the rays themselves.
UF beta particles
UF rays, beta
SA alpha rays
SA gamma rays
SA radioactivity
SA X-rays

Beta Regio (1993)
Low relief area above Venusian plains. Also search Beta AND Venus.
BT Venus

betafite (1978)
Autoposting of niobotantalates to this term began in 1989. This term has multiple hierarchies.
UF blomstrandite
UF ellsworthite
UF hatchettolite
BT1 niobotantalates
BT1 niobates
BT1 oxides
BT2 niobotantalates
BT2 tantalates
BT2 oxides
SA pyrochlore

betekhtinite (1978)
BT sulfides

Betic Cordillera (1978)
Extends along the N extremity of Andalusia between Guadiana and Guadalquivir rivers. It is a southern series of the Alpine System. Before 1978, also search Sierra Morena.
CO N380000N390000
    W0030000W0060000
UF Cordillera Betica
UF Cordillera Marianica
BT Spain
BT Iberian Peninsula
BT Southern Europe
BT Europe
NT Serrania de Ronda
NT Spanish Sierra Nevada
SA Andalusian
SA Betic Zone
SA Prebetic Zone
SA Subbetic Zone

Betic Zone (1978)
Used in association with the Betic Cordillera in SW Spain.
BT Spain
BT Iberian Peninsula
BT Southern Europe
BT Europe
SA Betic Cordillera

Betsiboka Basin (1978)
Around town of Majunga on Bombetoka Bay. Before 1978, also search Majunga; Majunga Basin. This term has multiple hierarchies.
UF Majunga Basin
BT1 Madagascar
BT1 Indian Ocean Islands
BT2 Madagascar
BT2 Africa

Betts Cove (1978)
BT Newfoundland Island
BT Newfoundland
BT Eastern Canada
BT Canada

Betts Cove Ophiolite (1989)
BT Ordovician
BT Paleozoic
SA Newfoundland Island

Betula (1978)
Genus.
BT Dicotyledoneae
BT angiosperms
BT Spermatophyta
BT Plantae

beudantite (1978)
This term has multiple hierarchies.
BT1 arsenates
BT2 sulfates

Bevier Coal (1978)
Bevier Formation comprises Bevier Coal with underlying gray clay and overlying black slate. SE Kansas.
BT Pennsylvanian
BT Carboniferous
BT Paleozoic
SA Desmoinesian
SA Kansas

Bewcastle
No longer a valid term for GeoRef as of 1993. See Bewcastle England.

Bewcastle England (1993)
Village and parish in N Cumberland County, NW England. Before 1993, also search Bewcastle AND England.
BT Cumberland England
BT Cumbria England
BT England
BT Great Britain
BT United Kingdom
BT Western Europe
BT Europe

Bexar County
Valid through 1988. Search in combination with state term. After 1988, use specific county-state term.

Bexar County Texas (1989)
SW central Texas. Before 1989, also search Bexar County AND Texas.
CO N290000N294400
    W0981000W0985000
BT Texas
BT United States
NT San Antonio Texas
SA Edwards Aquifer

Beyrichicopina (1981)
As of 1990, microfossils, Crustacea, Mandibulata, and Arthropoda are autoposted to this term. This term has multiple hierarchies.
BT1 Ostracoda
BT1 Crustacea
BT1 Mandibulata
BT1 Arthropoda
BT1 Invertebrata
BT2 Ostracoda
BT2 microfossils

Bhandara
No longer a valid term for GeoRef. See Bhandara India.

Bhandara India (1993)
Town in extreme NE Maharashtra, central India.
BT Maharashtra India
BT India
BT Indian Peninsula
BT Asia

Bhander Group (1978)
Includes upper Bhander Sandstone, Sirbu shales, Middle Bhander Sandstone and Ganurgarh shales. Named after range of hills N of Narmada Valley.
IN Index ages as applicable.
SA Cambrian
SA India
SA Madhya Pradesh India
SA Precambrian
SA Vindhyan

Bhilwara
No longer a valid term for GeoRef. See Bhilwara India.

Bhilwara India (1993)
Town in S central Rajasthan, NW India.
BT Rajasthan India
BT India
BT Indian Peninsula
BT Asia

BHTV
  use borehole televiewers
Bhuj
  No longer a valid term for GeoRef. See Bhuj India.
Bhuj India (1993)
  Town in Cutch, Gujarat, W India.
  BT Cutch India
  BT Gujarat India
  BT India
  BT Indian Peninsula
  BT Asia
Bhuj Series (1978)
  Consists of Zamia beds, Ptilophyllum beds, and Palmoxylon beds. Named after town of Bhuj in Cutch.
  BT Middle Cretaceous
  BT Cretaceous
  BT Mesozoic
  SA Gujarat India
  SA India
Bhutan (1978)
  Kingdom in E Himalayas between Tibet and NE India.
  CO N263000N283000
       E0921000E0884000
  BT Indian Peninsula
  BT Asia
  SA Himalayas
  SA Lesser Himalayas
Bi
  use bismuth
Bialowieza (1978)
  National park in E Poland.
  BT Poland
  BT Central Europe
  BT Europe
  SA national parks
Bialystok
  No longer a valid term for GeoRef. As of 1993 see Bialystok Poland.
Bialystok Poland (1993)
  Province in extreme NE Poland. Also a city. After 1993, use for the province. Before 1993 also search Bialystok AND Poland.
  BT Poland
  BT Central Europe
  BT Europe
  NT Belsk Poland
Biarritz
  No longer a valid term for GeoRef. See Biarritz France.
Biarritz France (1993)
  Town on the Bay of Biscay near Spanish border.
  BT Pyrenees-Atlantiques France
  BT France
  BT Western Europe
  BT Europe
Biarritzian (1978)
  Europe. Middle to upper Eocene.
  BT Eocene
  BT Paleogene
  BT Tertiary
  BT Cenozoic
Bibb County Georgia (1989)
  Central Georgia. Before 1989, search Bibb County AND Georgia.
  CO N323900N325700
       W0832800W0835300
  BT Georgia
  BT United States
  NT Macon Georgia
bibliographic review
  use bibliography

bibliographies
  use bibliography
bibliography (1978)
  Covers major bibliographies on topics, areas or persons.
  UF bibliographical review
  UF bibliographies
  SA areal geology
  SA associations
  SA biography
  SA catalogs
  SA geology
  SA GeoRef
  SA Geoscan
  SA glossaries
  SA lexicons
  SA libraries
  SA mineralogy
  SA monographs
  SA oceanography
  SA paleobotany
  SA paleontology
  SA petrology
  SA publications
  SA research
  SA review
  SA sedimentary petrology
  SA stratigraphy
  SA structural geology
  SA survey organizations
  SA tectonophysics
  SA volcanology
bicarbonate ion (1982)
  SA carbonate ion
  SA carbonates
  SA geochemistry
  SA ions
Bicaz Valley (1978)
  River valley in Bacau County. This term has multiple hierarchies.
  BT1 Moldavia
  BT1 Europe
  BT2 Romania
  BT2 Southern Europe
  BT2 Europe
  SA Bacau Romania
Bida
  No longer a valid term for GeoRef. As of 1993 see Bida Nigeria.
Bida Nigeria (1993)
  Region in W central Nigeria. Before 1993 also search Bida AND Nigeria.
  BT Nigeria
  BT West Africa
  BT Africa
Bidahochi Formation (1978)
  Divided into six members for mapping purposes. Fifth member informally referred to as White Cone Member. In NE Arizona.
  BT Pliocene
  BT Neogene
  BT Tertiary
  BT Cenozoic
  SA Arizona
  SA New Mexico
bideauxite (1978)
  This term has multiple hierarchies.
  BT1 chlorides
  BT1 halides
  BT2 fluorides
  BT2 halides
Biel Lake (1978)
  In NW Bern.
  UF Bienne Lake
  UF Lake of Bienne
  BT Bern Switzerland
  BT Switzerland
  BT Central Europe

  BT Europe
Biella
  No longer a valid term for GeoRef. See Biella Italy.
Biella Italy (1993)
  Town in Vercelli Province in NW Italy.
  BT Vercelli Italy
  BT Piemonte Italy
  BT Italy
  BT Southern Europe
  BT Europe
Bielsk
  use Belsk Poland
Bielsk Podlaski
  use Belsk Poland
Bielsko
  No longer a valid term for GeoRef. As of 1993 see Bielsko Poland.
Bielsko Poland (1993)
  As of 1989, used only for the province in S Poland. Before 1989, also used for the city of Bielsko-Biala. Before 1993 also search Bielsko AND Poland.
  BT Poland
  BT Central Europe
  BT Europe
  NT Andrychow Poland
  NT Bielsko-Biala Poland
  NT Cieszyn Poland
  NT Wadowice Poland
  SA Beskid Mountains
  SA Katowice Poland
Bielsko-Biala
  No longer a valid term for GeoRef. As of 1993 see Bielsko-Biala Poland.
Bielsko-Biala Poland (1993)
  Name of city which was combined with Biala in 1950 to form Bielsko-Biala in what was formerly part of Katowice Province. Before 1989, also search Bielsko. Before 1993 also search Bielsko-Biala AND Poland.
  BT Bielsko Poland
  BT Poland
  BT Central Europe
  BT Europe
  SA Katowice Poland
Bienne Lake
  use Biel Lake
BIFs
  Not a valid term for GeoRef. See iron formations.
Big Belt Mountains (1978)
  Range of Rocky Mountains in W central Montana.
  IN Index Rocky Mountains and Montana as applicable.
  SA Montana
  SA Rocky Mountains
Big Bend National Park (1978)
  In Brewster County on the Mexican border in SW Texas. Also search Big Bend.
  BT Brewster County Texas
  BT Texas
  BT United States
  SA Rattlesnake Mountain
Big Horn Basin
  use Bighorn Basin

Big Horn County
  Valid through 1988. Search in combination with state term. After 1988, use specific county-state term.
Big Horn County Montana (1989)
  S Montana. Before 1989, also search Big Horn County AND Montana.
  CO N450000N460400
       W1061900W1084400
  BT Montana
  BT United States
  SA Bighorn River
Big Horn County Wyoming (1989)
  N Wyoming. Before 1989, also search Big Horn County AND Wyoming.
  CO N440900N450000
       W1070800W1083900
  BT Wyoming
  BT United States
  SA Bighorn River
Big Horn Mountains
  use Bighorn Mountains
Big Maria Mountains (1985)
  SE California. Autoposting of broader terms to this term began in 1989.
  CO N334000N335500
       W1143000W1144500
  BT Riverside County California
  BT California
  BT United States
  SA Little Maria Mountains
Big Pine Key (1978)
  One of the principal islands in the Florida Keys.
  BT Florida Keys
  BT Monroe County Florida
  BT Florida
  BT United States
Big Snowy Group (1989)
  Central Montana. Includes Heath Formation.
  BT Carboniferous
  BT Paleozoic
  SA Heath Formation
  SA Idaho
  SA Montana
Big Snowy Mountains (1978)
  N central Montana.
  IN Index counties or regions and mountains as applicable.
  SA Montana
Bigby-Cannon Limestone (1978)
  Contemporary facies comprising this unit are Bigby Limestone and Cannon Limestone. Overlies Hermitage Formation and underlies Catheys Formation. Central Tennessee.
  BT Middle Ordovician
  BT Ordovician
  BT Paleozoic
  SA Tennessee
Bighorn Basin (1978)
  River basin.
  IN Index states and counties as applicable.
  UF Big Horn Basin
  SA Madison Aquifer
  SA Montana
  SA United States
  SA Wyoming
Bighorn Mountains (1978)
  Range of Rocky Mountains. This term has multiple hierarchies.

IN   Index states as applicable.
CO   N433000N453000
     W1063000W1083000
UF   Big Horn Mountains
BT1 U. S. Rocky Mountains
BT1 Rocky Mountains
BT2 U. S. Rocky Mountains
BT2 United States
SA   Montana
SA   Wyoming

Bighorn River   (1989)
S Montana and N Wyoming.
IN   Index states and counties as applicable.
SA   Big Horn County Montana
SA   Big Horn County Wyoming
SA   Fremont County Wyoming
SA   Hot Springs County Wyoming
SA   Montana
SA   Washakie County Wyoming
SA   Wyoming
SA   Yellowstone County Montana

Bihar
No longer a valid term for GeoRef.
See Bihar India.

Bihar India   (1993)
State in E India. Before 1978, also search Behar and India.
CO   N220000N273000
     E0882000E0831000
UF   Behar India
BT   India
BT   Indian Peninsula
BT   Asia
NT   Dhanbad India
NT   Gaya India
NT   Hazaribagh India
NT   Jharia India
NT   Karanpura India
NT   Karharbari India
NT   Monghyr India
NT   Palamau India
NT   Rajgir India
NT   Rajmahal India
NT   Ranchi India
NT   Singhbhum India
SA   Ajay River
SA   Amarjola
SA   Bihar mica belt
SA   Bokaro
SA   Chainpur Meteorite
SA   Damodar Valley
SA   Damuda Series
SA   Jharia coal field
SA   Karanpura coal field
SA   Karharbari Stage
SA   Kosi Basin
SA   Mosaboni Mines
SA   Panchet Series
SA   Rajmahal Hills
SA   Ramgarh coal field
SA   Semri Series
SA   Shergotty Meteorite
SA   Singhbhum Granite
SA   Singhbhum shear zone
SA   Son Valley

Bihar mica belt   (1993)
Extends WSW-ENE from Gurpa in Gaya district to near Simultala in Bhagalpur district, Bihar. Also passes through districts of Nawadah, Ghiridih, and Munger.
SA   Bihar India
SA   Gaya India
SA   Hazaribagh India

Bihar Mountains
use Bihor Mountains

Bihor
No longer a valid term for GeoRef.
As of 1993 see Bihor Romania.

Bihor Mountains   (1978)
In W central Transylvania.
UF   Bihar Mountains
BT   Transylvania
BT   Romania
BT   Southern Europe
BT   Europe
NT   Vladeasa Mountain

Bihor Romania   (1993)
County in W Transylvania. Before 1993 also search Bihor AND Romania.
BT   Transylvania
BT   Romania
BT   Southern Europe
BT   Europe
NT   Rosia Romania

Biikzhal   (1978)
Drill hole in Emba River region of Caspian Basin. This term has multiple hierarchies.
BT1 Kazakhstan
BT1 Central Asia
BT1 Asia
BT2 Kazakhstan
BT2 Commonwealth of Independent States

Bijapur
No longer a valid term for GeoRef.
See Bijapur India.

Bijapur India   (1993)
Town in NW Karnataka, S India.
BT   Karnataka India
BT   India
BT   Indian Peninsula
BT   Asia

Bijawar System   (1978)
Includes Majhauli Group, Bhitri Group, Lora Group, and Chanderdip Group.
BT   Precambrian
SA   India
SA   Madhya Pradesh India

Bikaner
No longer a valid term for GeoRef. Former state in NW Rajasthan.
See Bikaner India.

Bikaner India   (1993)
A city in NW Rajasthan, NW India.
BT   Rajasthan India
BT   India
BT   Indian Peninsula
BT   Asia

Bilaspur
No longer a valid term for GeoRef.
See Bilaspur India.

Bilaspur India   (1993)
City and district in S Himachal Pradesh. Also a city and district in E central Madhya Pradesh.
IN   Index states as applicable.
BT   India
BT   Indian Peninsula
BT   Asia
SA   Himachal Pradesh India
SA   Madhya Pradesh India

Bilbao
No longer a valid term for GeoRef.
As of 1993, see Bilbao Spain.

Bilbao Spain   (1993)
City on the Bay of Biscay in the Basque country, N Spain. Before 1993, also search Bilbao and Spain.
BT   Vizcaya Spain
BT   Basque Provinces Spain
BT   Spain
BT   Iberian Peninsula
BT   Southern Europe
BT   Europe

Billings
Valid through 1988. Search in combination with state term. After 1988, use specific city-state term.

Billings County
Valid through 1988. Search in combination with state term. After 1988, use specific county-state term.

Billings County North Dakota   (1989)
W North Dakota. Before 1989, also search Billings County AND North Dakota.
CO   N464000N472000
     W1030500W1034000
BT   North Dakota
BT   United States

Billings Montana   (1989)
City in S central Montana. Before 1989, search Billings AND Montana.
CO   N454700N454700
     W1083000W1083000
BT   Yellowstone County Montana
BT   Montana
BT   United States

Billiton   (1978)
Island between Sumatra and SW Borneo.
UF   Belitung
BT   Indonesia
BT   Far East
BT   Asia

Bimini   (1978)
Two small islands off S Florida coast.
UF   Bimini Islands
BT   Bahamas
BT   West Indies
BT   Caribbean region
SA   Northwest Providence Channel

Bimini Islands
use Bimini

Bingham
Valid through 1988. Search in combination with state term. After 1988, use specific city-state term.

Bingham County
Valid through 1988. Search in combination with state term. After 1988, use specific county-state term.

Bingham County Idaho   (1989)
SE Idaho. Before 1978, also search Bingham for the county. Before 1989, also search Bingham County AND Idaho.
CO   N425000N434000
     W1113500W1130000
BT   Idaho
BT   United States

Bingham Utah   (1989)
Mining town in N central Utah. Before 1989, search Bingham AND Utah.
CO   N403230N403230
     W1120850W1120850
BT   Salt Lake County Utah
BT   Utah
BT   United States

Binghamton
Valid through 1988. Search in combination with state term. After 1988, use specific city-state term.

Binghamton New York   (1989)
City on the Pennsylvania border in S New York. Before 1989, search Binghamton AND New York.
CO   N420600N420600
     W0755500W0755500
BT   Broome County New York
BT   New York
BT   United States

Binn Valley
use Binnental

Binnatal
use Binnental

Binnen Tal
use Binnental

Binnental   (1978)
Valley in E Valais.
UF   Binn Valley
UF   Binnatal
UF   Binnen Tal
BT   Valais Switzerland
BT   Switzerland
BT   Central Europe
BT   Europe

Binue Valley
use Benue Valley

biocalcarenite   (1978)
BT   limestone
BT   carbonate rocks
BT   sedimentary rocks
SA   calcarenite

biocenoses   (1978)
UF   biocenosis
UF   biocoenoses
UF   life assemblage
SA   biotopes
SA   communities
SA   paleontology
SA   thanatocenoses

biocenosis
use biocenoses

biochemical sedimentation   (1981)
BT   sedimentation

biochemistry   (1978)
Before 1971, also search paleobiochemistry.
UF   biogeochemistry
UF   paleobiochemistry
SA   biodegradation
SA   biology
SA   biomineralization
SA   diet
SA   fossils
SA   geochemistry
SA   organic materials
SA   paleobiology
SA   steroids

biochronology   (1993)
Geochronology studies using biological events or or paleontologic methods.
SA   biologic evolution
SA   biostratigraphy
SA   correlation
SA   extinction
SA   geochronology
SA   mass extinctions
SA   relative age

biocirculation
As of 1993, no longer a valid term for GeoRef. Usually out-of-scope. Used with ocean circulation.

bioclastic sedimentation   (1981)
BT   sedimentation

biocoenoses
use biocenoses

biodegradation   (1982)

    UF biodeterioration
    UF biological degradation
    SA biochemistry
    SA biogenic processes
    SA biogenic structures
    SA biology
    SA biomarkers
    SA biomass
    SA biorhexistasy
    SA biosphere
    SA biota
    SA bioturbation
    SA degradation
    SA weathering

biodeterioration
    use biodegradation

bioerosion
    Not a valid term for GeoRef. See biogenic effects, erosion.

biofacies (1978)
    UF biologic facies
    SA assemblages
    SA biostratigraphy
    SA biota
    SA biotopes
    SA ecology
    SA facies
    SA sedimentation
    SA sediments

biogenic
    A valid term through 1977. See biogenic structures; biogenic effects.

biogenic effects
    No longer a valid term for GeoRef.
    use biogenic processes

biogenic origin
    use biogenic processes

biogenic processes (1993)
    Between 1981 and 1992, also search biogenic effects or biogenic origin.
    UF biogenic effects
    UF biogenic origin
    SA biodegradation
    SA biogenic structures
    SA biorhexistasy
    SA biota
    SA bioturbation
    SA ecology
    SA effects

biogenic structures (1978)
    Autoposting of this term to ichnofossils ended in 1981. Autoposting of this term to borings, burrows, coprolites, tracks and trails ended in 1985.
    BT sedimentary structures
    NT algal structures
    NT banks
    NT bioherms
    NT bioturbation
    NT carbonate banks
    NT girvanella
    NT lebensspuren
    NT oncolites
    NT stromatactis
    NT stromatolites
    NT thrombolites
    SA biodegradation
    SA biogenic processes
    SA borings
    SA burrows
    SA coal
    SA coprolites
    SA fecal pellets
    SA hardground
    SA ichnofossils
    SA peat
    SA reefs
    SA tracks
    SA trails

biogeochemical cycles
    Not a valid term for GeoRef. See biochemistry and geochemical cycle.

biogeochemical methods (1978)
    Before 1972, also search biogeochemical prospecting; biogeochemistry; biogeochemical surveys.
    UF biogeochemical prospecting
    UF biogeochemical surveys
    BT geochemical methods
    SA geobotanical methods
    SA geochemistry
    SA methods
    SA mineral exploration

biogeochemical prospecting
    A valid term through 1971.
    use biogeochemical methods

biogeochemical surveys
    Not a valid GeoRef index term after 1970. Was used on level 1 in subfile N.
    use biogeochemical methods

biogeochemistry
    use biochemistry

biogeographic maps (1985)
    Before 1985, also search maps AND biogeography.
    BT maps
    SA biogeography
    SA paleontology

biogeography (1978)
    Used for descriptions of the geographic distribution of both fossil and modern organisms. Before 1971, also search paleobiogeography; zoogeography.
    IN Index with common fossil names and age where available.
    UF chorology
    UF paleobiogeography
    UF phytogeography
    UF zoogeography
    SA biogeographic maps
    SA biologic evolution
    SA Boreal
    SA colonization
    SA continental drift
    SA ecology
    SA endemic taxa
    SA faunal provinces
    SA floral provinces
    SA fossils
    SA geography
    SA land bridges
    SA paleobotany
    SA paleoecology
    SA paleogeography
    SA paleoindian
    SA paleontology
    SA palynology
    SA range
    SA stratigraphy
    SA vicariance

biographies
    No longer a valid term for GeoRef. In 1959, included use on level 1 in subfile N.
    use biography

biography (1978)
    IN Index also name of individual (last, first)
    UF biographies
    SA bibliography
    SA Darwin, Charles
    SA Gilbert, Grove Karl
    SA Hutton, James
    SA Lyell, Charles
    SA Schimper, Wilhelm-Philippe
    SA Vernadskiy, Vladimir Ivanovich
    SA Wegener, Alfred

bioherms (1978)
    As of 1984, use instead of reefs under sedimentary structures.
    UF organic mound
    BT biogenic structures
    BT sedimentary structures
    NT mud mounds
    SA algae
    SA atolls
    SA biostromes
    SA calcareous composition
    SA carbonate banks
    SA Coelenterata
    SA reefs

biologic evolution (1983)
    Before 1983, also search evolution AND appropriate paleontology category codes; evolution AND appropriate fossil name(s); evolution AND paleontology; evolution AND paleobotany.
    UF biological evolution
    SA adaptation
    SA behavior
    SA biochronology
    SA biogeography
    SA cladistics
    SA creationism
    SA Darwinism
    SA diachronism
    SA DNA
    SA evolution
    SA extinct taxa
    SA extinction
    SA gradualism
    SA life origin
    SA micropaleontology
    SA natural selection
    SA paleobotany
    SA paleoecology
    SA paleoindian
    SA paleontology
    SA phylogeny
    SA prokaryotes
    SA punctuated equilibria
    SA radiation
    SA speciation
    SA vicariance

biologic facies
    use biofacies

biological degradation
    use biodegradation

biological evolution
    use biologic evolution

biological markers
    use biomarkers

biological processes
    As of 1981, no longer a valid term for GeoRef. Use biology.

biological zones
    use biozones

biology (1978)
    Before 1981, also search biological processes.
    SA bacteria
    SA biochemistry
    SA biodegradation
    SA biosphere
    SA biotypes
    SA ecology
    SA epibiotism
    SA exobiology
    SA geomicrobiology
    SA paleobiology

biomarkers (1989)
    UF biological markers
    UF molecular fossils
    SA alkanes
    SA aromatic hydrocarbons
    SA asphalt
    SA biodegradation
    SA biostratigraphy
    SA chemical fossils
    SA geochemical indicators
    SA saturated hydrocarbons
    SA source rocks
    SA stratigraphy
    SA thermal maturity

biomass (1978)
    SA biodegradation
    SA ecology
    SA new energy sources
    SA trophic analysis

biometry (1978)
    SA cladistics
    SA fossils
    SA mathematical geology
    SA paleontology
    SA statistical analysis

biomicrite (1978)
    BT limestone
    BT carbonate rocks
    BT sedimentary rocks
    SA micrite

biomineralization (1993)
    Interaction between organism and environment to produce minerals often as body parts. Applies to many plants and animals.
    SA biochemistry
    SA bones
    SA fossilization
    SA processes
    SA shells
    SA taphonomy

biopelite
    use black shale

bioremediation (1993)
    BT remediation
    SA polluted water
    SA pollution
    SA reclamation
    SA soil treatment
    SA strip mining
    SA waste disposal
    SA water treatment

biorhexistasy (1981)
    SA biodegradation
    SA biogenic processes
    SA laterites
    SA soils
    SA weathering

biosparite (1978)
    BT limestone
    BT carbonate rocks
    BT sedimentary rocks

biosphere (1978)
    SA atmosphere
    SA biodegradation
    SA biology
    SA Earth
    SA hydrosphere
    SA IGBP
    SA lithosphere
    SA tectonosphere

biostratigraphy (1978)
    SA acanthodians
    SA acritarch flora
    SA agnathans
    SA algal flora
    SA ammonoids
    SA amphibians
    SA angiosperm flora
    SA arthropods
    SA barnacles
    SA belemnites

SA biochronology
SA biofacies
SA biomarkers
SA biotopes
SA biozones
SA birds
SA bivalves
SA brachiopods
SA bryozoans
SA cephalopods
SA charophytes
SA chitinozoans
SA coccoliths
SA conodonts
SA corals
SA crinoids
SA crustaceans
SA desmids
SA diachronism
SA diatom flora
SA dinoflagellates
SA dinosaurs
SA discoasters
SA echinoderms
SA echinoids
SA feathers
SA ferns
SA fish
SA foraminifers
SA fossils
SA fusulinids
SA gastropods
SA graptolites
SA gymnosperm flora
SA historical geology
SA ichthyoliths
SA index fossils
SA insects
SA invertebrates
SA K-T boundary
SA mammals
SA megaspores
SA miospores
SA mollusks
SA nannoconids
SA nannofossils
SA ostracoderms
SA ostracods
SA paleoecology
SA paleontology
SA palynomorphs
SA plants
SA radiolarians
SA range
SA reptiles
SA rodents
SA rudists
SA selachians
SA silicoflagellates
SA sponges
SA stratigraphy
SA stromatolites
SA stromatoporoids
SA tetrapods
SA trilobites
SA vertebrates

biostromes (1978)
SA bioherms
SA Coelenterata
SA reefs

biota (1978)
UF organisms
SA bacteria
SA biodegradation
SA biofacies
SA biogenic processes
SA ecology
SA factors
SA faunal studies
SA floral studies
SA living taxa
SA microorganisms
SA soils

biotite (1978)
BT mica group
BT sheet silicates
BT silicates
SA heavy minerals

biotite gneiss (1978)
BT gneisses
BT metamorphic rocks

biotite granite (1978)
BT granites
BT plutonic rocks
BT igneous rocks

biotite schist (1978)
BT schists
BT metamorphic rocks

biotopes (1982)
SA adaptation
SA assemblages
SA biocenoses
SA biofacies
SA biostratigraphy
SA communities
SA ecology
SA environment
SA faunal provinces
SA floral provinces
SA habitat
SA paleoecology
SA paleoenvironment

bioturbation (1978)
BT biogenic structures
BT sedimentary structures
SA biodegradation
SA biogenic processes
SA borings
SA Bouma sequence
SA burrows
SA sedimentation

biotypes (1978)
SA biology
SA ecology
SA paleoecology

biozones (1989)
UF biological zones
SA biostratigraphy
SA stratigraphic units
SA stratigraphy

bipedalism (1985)
SA functional morphology
SA locomotion
SA paleontology

Birch Creek
Valid through 1988. Search in combination with state term. After 1988, use specific city-state term.

Birch Creek Alaska (1989)
Village in E Alaska. Before 1989, search Birch Creek AND Alaska.
CO N661800N661800 W1455900W1455900
BT East-Central Alaska
BT Alaska
BT United States

Bird Island
use Aves Island

Bird River
No longer a valid term for GeoRef. See Bird River Manitoba.

Bird River Manitoba (1993)
Village in SE Manitoba.
BT Manitoba
BT Western Canada
BT Canada

Bird Spring Formation (1978)
Upper Mississippian, Pennsylvanian and Lower Permian. In Spring Mountain area, Nevada. Underlies Spring Mountain Formation and overlies Illipah Formation.
IN Index ages as applicable.
BT Paleozoic
SA California
SA Lower Permian
SA Nevada
SA Pennsylvanian
SA Upper Mississippian
SA Utah

Birdbear Formation (1989)
In Jefferson Group. Overlies Duperow Formation.
BT Upper Devonian
BT Devonian
BT Paleozoic
SA Jefferson Group
SA Montana
SA Nisku Formation
SA North Dakota
SA Saskatchewan
SA Williston Basin

birds (1981)
Common name for Aves.
IN Index for all non-paleontologic studies of fossils.
BT tetrapods
BT vertebrates
SA Aves
SA biostratigraphy
SA feathers

birefraction
use birefringence

birefringence (1978)
UF birefraction
UF double refraction
SA optical properties
SA polarization
SA refractive index

Birkhead Formation (1989)
Eromanga Basin, Queensland.
BT Jurassic
BT Mesozoic
SA Australia
SA Eromanga Basin
SA Queensland Australia

Birmingham
No longer a valid term for GeoRef. As of 1993, see Birmingham England. As of 1989, see Birmingham Alabama.

Birmingham Alabama (1989)
City in central Alabama. Before 1989, search Birmingham AND Alabama.
CO N333000N333000 W0865500W0865500
BT Jefferson County Alabama
BT Alabama
BT United States

Birmingham England (1993)
City in West Midlands, W central England. Before 1993, also search Birmingham AND England.
CO N523000N523000 W0015000W0015000
BT England
BT Great Britain
BT United Kingdom
BT Western Europe
BT Europe
SA Warwickshire England

Birmingham University (1978)
In Birmingham in W central England.
SA academic institutions
SA England

birnessite (1978)
BT oxides

BIRPS (1989)
Acronym.
UF British Institutes Reflection Profiling Syndicate
SA deep seismic sounding
SA elastic waves
SA reflection methods
SA seismic methods
SA seismic surveys

Birrimian (1989)
W Africa. As of 1990, Precambrian and upper Precambrian are autoposted to this term.
BT lower Proterozoic
BT Proterozoic
BT upper Precambrian
BT Precambrian

Biryusa (1978)
River SW of Tayshet in W Irkutsk Oblast.
BT Russian Federation
BT Commonwealth of Independent States
SA Irkutsk Russian Federation

Bisbee
Valid through 1988. Search in combination with state term. After 1988, use specific city-state term.

Bisbee Arizona (1989)
City in SE Arizona. Before 1989, search Bisbee AND Arizona.
CO N312700N312700 W1095500W1095500
BT Cochise County Arizona
BT Arizona
BT United States

Bisbee Group (1989)
SE Arizona, SW New Mexico. This term has multiple hierarchies.
BT1 Comanchean
BT1 Cretaceous
BT1 Mesozoic
BT2 Lower Cretaceous
BT2 Cretaceous
BT2 Mesozoic
SA Arizona
SA Mural Limestone
SA New Mexico

Biscay (Bay)
use Bay of Biscay

Biscay Bay
use Bay of Biscay

Biscayne Aquifer (1989)
SE Florida.
IN Index counties as applicable.
BT Florida
BT United States
SA Broward County Florida
SA Dade County Florida
SA Palm Beach County Florida

Biscayne Bay (1978)
S of Miami on E coast of Dade County.
BT Dade County Florida
BT Florida
BT United States

bischofite (1985)
BT chlorides
BT halides

Bishkek Kyrgyzstan (1993)
City in N Kyrgyzstan. Before 1993, also search Frunze or Pishpek. This term has multiple hierarchies.
UF Pishpek Kyrgyzstan
BT1 Kyrgyzstan
BT1 Asia
BT2 Kyrgyzstan

BT2 Commonwealth of Independent States

Bishop Tuff (1978)
E central California. Bishop Tuff was erupted a short time after Sherwin glacial stage.
BT Pleistocene
BT Quaternary
BT Cenozoic
SA California

Bismarck Archipelago (1978)
Volcanic group of islands N and NE of E New Guinea. Before 1972, also search Bismarck Islands.
UF Bismarck Islands
BT Papua New Guinea
BT Australasia
NT Rabaul Caldera
SA Melanesia
SA New Britain
SA New Ireland
SA Rat Island

Bismarck Islands
Not a valid GeoRef index term after 1971. Was used on level 1 in subfile G.
use Bismarck Archipelago

Bismarck Sea (1978)
Part of W Pacific Ocean NE of New Guinea and NW of New Britain.
BT West Pacific
BT Pacific Ocean

bismuth (1978)
Chemical element. As of 1981, use bismuth ores for bismuth as a commodity. Autoposting of metals to this term began in 1989.
UF Bi
BT metals
SA bismuth ores
SA heavy metals
SA native elements

bismuth ores (1981)
Before 1981, also search (bismuth OR Bi) AND (deposit OR deposits OR ore OR ores OR economic) in the basic index. Autoposting of metal ores to this term began in 1985.
IN Commodity. See List C.
BT metal ores
SA bismuth
SA bismuthinite
SA native elements

bismuthides (1978)
Autoposting of sulfides to this term ended in 1989. Michenerite is a narrower term as of 1981.
NT froodite
NT michenerite
SA minerals

bismuthinite (1978)
UF bismuthite
BT sulfides
SA bismuth ores

bismuthite
use bismuthinite

Bison (1978)
Genus. Autoposting of Eutheria and Theria to this term began in 1989.
BT Bovidae
BT Ruminantia
BT Artiodactyla
BT Eutheria
BT Theria
BT Mammalia
BT Tetrapoda

BT Vertebrata
BT Chordata
NT Bison latifrons
NT Bison occidentalis

Bison latifrons (1978)
Species. Autoposting of Eutheria and Theria to this term began in 1989.
BT Bison
BT Bovidae
BT Ruminantia
BT Artiodactyla
BT Eutheria
BT Theria
BT Mammalia
BT Tetrapoda
BT Vertebrata
BT Chordata

Bison occidentalis (1978)
Species. Autoposting of Eutheria and Theria to this term began in 1989.
BT Bison
BT Bovidae
BT Ruminantia
BT Artiodactyla
BT Eutheria
BT Theria
BT Mammalia
BT Tetrapoda
BT Vertebrata
BT Chordata

Bistrita
No longer a valid term for GeoRef. As of 1993 see Bistrita Romania.

Bistrita Mountains (1978)
Range of the Carpathians in NE Romania. This term has multiple hierarchies.
UF Bistritei Mountains
BT1 Carpathians
BT1 Europe
BT2 Romania
BT2 Southern Europe
BT2 Europe

Bistrita Romania (1993)
City in Bistrita-Nasaud County in central Transylvania. Before 1993 also search Bistrita AND Romania.
BT Transylvania
BT Romania
BT Southern Europe
BT Europe

Bistrita Valley (1978)
River valley in W central Moldavia. This term has multiple hierarchies.
BT1 Moldavia
BT1 Europe
BT2 Romania
BT2 Southern Europe
BT2 Europe

Bistritei Mountains
use Bistrita Mountains

Bitkov
No longer a valid term for GeoRef. See Bitkov Ukraine.

Bitkov Ukraine (1993)
In W Ukraine. Before 1993, also search Bitkov or Bitkow or Bytkov in combination with Ukraine. This term has multiple hierarchies.
UF Bitkow Ukraine
UF Bytkov Ukraine
BT1 Ukraine
BT1 Europe
BT2 Ukraine
BT2 Commonwealth of Independent States

Bitkow Ukraine
use Bitkov Ukraine

Bitlis
No longer a valid term for GeoRef. As of 1993 see Bitlis Turkey.

Bitlis Turkey (1993)
Town and province in SE Turkey. Before 1993 also search Bitlis AND Turkey.
BT Turkey
BT Middle East
BT Asia

bitter spar
use dolomite

Bitter Springs Formation (1978)
Underlies Middle Cambrian sediments and overlies Heavitree Quartzite. SW and S central Northern Territory.
BT Proterozoic
BT upper Precambrian
BT Precambrian
SA Australia
SA Northern Territory Australia

Bitterfeld
No longer a valid term for GeoRef. See Bitterfeld Germany.

Bitterfeld Germany (1993)
City in S Saxony-Anhalt, S central Germany. Before 1993, Bitterfeld was also used with Halle Bezirk.
BT Saxony-Anhalt Germany
BT Germany
BT Central Europe
BT Europe

Bitterroot Range (1978)
Range of the Rocky Mountains.
IN Index states and U. S. Rocky Mountains as applicable.
SA Beaverhead Mountains
SA Idaho
SA Idaho Batholith
SA Montana
SA Rocky Mountains
SA Selway-Bitterroot Wilderness
SA U. S. Rocky Mountains

bitumenite
use torbanite

bitumens (1978)
Autoposting of hydrocarbons to this term ended in 1989.
IN See List C for use as a commodity term.
BT organic materials
NT anthraxolite
NT asphalt
NT gilsonite
NT grahamite
SA kerogen
SA oil seeps
SA petroleum

bituminous coal (1978)
Before 1978, also search coal AND bituminous. This term has multiple hierarchies.
UF soft coal
BT1 coal
BT1 organic residues
BT2 coal
BT2 sedimentary rocks

bituminous rocks and sands
No longer a valid GeoRef index term. Before 1969, was used on level 1 in subfile E. See oil sands; oil shale; bitumens.

bituminous sands
Not a valid GeoRef index term after 1970. Was used on level 1 in subfile N.
use oil sands

bituminous shale
As of 1981, no longer a valid term for GeoRef. From 1978-1980, clastic rocks was autoposted to this term.
use black shale

bivalves (1981)
Common name for Bivalvia. Autoposting of mollusks to this term began in 1989.
UF pelecypods
BT mollusks
BT invertebrates
NT rudists
SA biostratigraphy
SA Bivalvia

Bivalvia (1978)
Class. Before 1976, also search Pelecypoda. Autoposting of this term began in 1978.
UF Pelecypoda
BT Mollusca
BT Invertebrata
NT Actinodontida
NT Arcina
NT Astartida
NT Carbonicola
NT Cardiidae
NT Carditida
NT Ctenodontida
NT Cyrtodontida
NT Didacna
NT Edmondia
NT Glycymeris
NT Gryphaea
NT Hiatella arctica
NT Hippuritacea
NT Limnocardium
NT Lucinidae
NT Mactra
NT Mercenaria
NT Myophoria
NT Mytilus
NT Nuculanidae
NT Nuculidae
NT Ostreacea
NT Pholadomyida
NT Praecardiida
NT Pteriina
NT Rudistae
NT Septibranchia
NT Solemyida
NT Tridacna
NT Trigoniidae
NT Unionidae
NT Venerida
SA bivalves

Biwa Lake
use Lake Biwa

Biwabik Iron Formation (1978)
In Animikie Group. Saint Louis County, NE Minnesota. Part of Mesabi Iron Range.
BT Precambrian
SA Animikie Group
SA Minnesota

bixbyite (1978)
UF partridgeite
BT oxides

Bjerkrem-Sogndal Massif (1978)
S Norway.
BT Norway
BT Scandinavia
BT Western Europe
BT Europe

Bjornoya
use Bear Island

Bjurbole Meteorite (1981)
Impact ENE of Helsinki. Before 1981, also search Bjurbole AND meteorites.
BT L chondrites
BT chondrites
BT stony meteorites
BT meteorites
SA Finland

Bk
use berkelium

Black Creek Formation (1989)
E South Carolina and W North Carolina.
BT Upper Cretaceous
BT Cretaceous
BT Mesozoic
SA North Carolina
SA South Carolina

Black earth
use Chernozems

Black Forest (1978)
Mountainous region in W Baden-Wurttemberg.
CO N473000N490000
    E0083000E0073000
UF Schwarzwald
BT Baden-Wurttemberg Germany
BT Germany
BT Central Europe
BT Europe

Black Hills (1978)
Group of mountains primarily in SW South Dakota. May be in other locations.
IN Index states and counties as applicable.
SA South Dakota
SA United States
SA Wyoming

black lead
use graphite

Black Mesa (1978)
Tableland region in NE Arizona.
IN Index counties or regions as applicable.
SA Apache County Arizona
SA Arizona
SA Navajo County Arizona

Black Mesa Basin (1978)
In NE Arizona.
IN Index counties or regions as applicable.
SA Arizona

Black Mountain (1989)
Occurs in various locations.
IN Index appropriate regions or states and U.S. counties as applicable.
SA Arizona
SA Australia
SA California
SA South Africa
SA Wales

Black Range (1978)
Extends N-S through parts of Grant and Sierra counties in SW New Mexico.
IN Index counties or state and mountains as applicable.
SA Grant County New Mexico
SA New Mexico
SA Sierra County New Mexico

Black River Group (1978)
Comprises Pamelia, Lowville, and Chaumont formations.
BT Middle Ordovician
BT Ordovician
BT Paleozoic
SA Blackriverian
SA New York
SA Pennsylvania

Black River-Matheson Program
use BRIM

Black Rock (1981)
W central Utah. Also search Black Rock Desert AND Utah for location in Utah. Also in the Bahamas.
IN Index states and specific regions as applicable.
SA Bahamas
SA Black Rock Desert
SA Utah

Black Rock Desert (1978)
In Pershing and Humboldt counties in NW Nevada.
IN Index counties and states as applicable.
SA Black Rock
SA Humboldt County Nevada
SA Nevada
SA Pershing County Nevada

Black Sea (1978)
Large inland sea bounded on the N and NE by the Central Asian republics, the Russian Federation, and Ukraine; S by Turkey, and W by Bulgaria and Romania. In 1981, broader term changed from Eurasia to East Mediterranean and Mediterranean Sea. As of 1993,
CO N410000N470000
    E0420000E0280000
BT East Mediterranean
BT Mediterranean Sea
SA Azov Sea
SA DSDP Site 379
SA DSDP Site 380
SA DSDP Site 381
SA Eurasia
SA Leg 42B

Black Sea Basin
use Black Sea region

Black Sea Coast
use Black Sea region

Black Sea Lowland
use Black Sea region

Black Sea region (1978)
IN Index countries as applicable.
CO N423000N473000
    E0400000E0300000
UF Black Sea Basin
UF Black Sea Coast
UF Black Sea Lowland
BT Europe
SA Bulgaria
SA Dobruja Basin
SA Eurasia
SA Moldova
SA Romania
SA Turkey
SA Ukraine

black shale (1978)
Before 1981, also search bituminous shale.
UF biopelite
UF bituminous shale
UF carbonaceous shale
BT clastic rocks
BT sedimentary rocks
SA carbonaceous composition
SA shale
SA terrigenous materials

black smokers (1993)
BT hydrothermal vents
SA bottom features
SA fumaroles
SA hydrothermal conditions
SA mid-ocean ridges
SA ocean floors
SA sulfides
SA volcanic features

Black Warrior Basin (1978)
Primarily in Alabama.
IN Index states as applicable.
BT United States
SA Alabama
SA Mary Lee Coal
SA Mississippi
SA Warrior coal field

Blackbird District
No longer a valid term for GeoRef.
use Blackbird mining district

Blackbird mining district (1993)
E central Idaho. Before 1993, also search Blackbird District.
CO N450500N451000
    W1142000W1142500
UF Blackbird District
BT Lemhi County Idaho
BT Idaho
BT United States
SA cobalt ores
SA copper ores
SA mines

Blackford County
Valid through 1988. Search in combination with state term. After 1988, use specific county-state term.

Blackford County Indiana (1989)
E Indiana. Before 1989, also search Blackford County AND Indiana.
CO N402200N403500
    W0851300W0852600
BT Indiana
BT United States

Blackhawk Formation (1978)
In Mesaverde Group. Six principal members are Desert, Grassy, Sunnyside, Kenilworth, Aberdeen, and Spring Canyon. Central E Utah.
BT Upper Cretaceous
BT Cretaceous
BT Mesozoic
SA Mesaverde Group
SA Utah

Blackleaf Formation (1993)
Lower Cretaceous. NW and central Montana. In Colorado Group. Members ascending are Flood Member, Taft Hill Member, Vaughn Member, and Bootlegger Member. Underlies Marias River Shale; overlies Kootenay Formation.
BT Lower Cretaceous
BT Cretaceous
BT Mesozoic
SA Colorado Group
SA Montana
SA Sweetgrass Arch

Blackriverian (1978)
North America. Lower Mohawkian, below Trentonian.
BT Middle Ordovician
BT Ordovician
BT Paleozoic
SA Black River Group

Blacksburg
Valid through 1988. Search in combination with state term. After 1988, use specific city-state term.

Blacksburg Virginia (1989)
Town in SW Virginia. Before 1989, search Blacksburg AND Virginia.
CO N371500N371500
    W0802500W0802500
BT Montgomery County Virginia
BT Virginia
BT United States

Blackstone Series (1989)
SE Massachusetts and N Rhode Island.
BT upper Proterozoic
BT Proterozoic
BT upper Precambrian
BT Precambrian
SA Massachusetts
SA Rhode Island

Blackwater (1978)
River and falls in Tucker County, E West Virginia.
IN Index county or region as applicable.
SA West Virginia

Blaine County
Valid through 1988. Search in combination with state term. After 1988, use specific county-state term.

Blaine County Idaho (1989)
S central Idaho. Before 1989, also search Blaine County AND Idaho.
CO N423500N435900
    W1130000W1150000
BT Idaho
BT United States
SA Craters of the Moon
SA Pioneer Mountains
SA Sawtooth Range

Blaine County Montana (1989)
N Montana. Before 1989, also search Blaine County AND Montana.
CO N474200N490000
    W1081500W1095000
BT Montana
BT United States

Blaine County Nebraska (1989)
Central Nebraska. Before 1989, also search Blaine County AND Nebraska.
CO N414300N421100
    W0994300W1001800
BT Nebraska
BT United States

Blaine County Oklahoma (1989)
W central Oklahoma. Before 1989, also search Blaine County AND Oklahoma.
CO N353300N361000
    W0981200W0983800
BT Oklahoma
BT United States

Blaine Formation (1978)
In Nippewalla, El Reno, or Pease River Group. W Oklahoma, central North and panhandle of Texas.
BT Permian
BT Paleozoic
SA Oklahoma
SA Texas

Blair County
Valid through 1988. Search in combination with state term. After 1988, use specific county-state term.

Blair County Pennsylvania (1989)

Central Pennsylvania. Before 1989, also search Blair County AND Pennsylvania.
CO N401500N404500
    W0780800W0783800
BT Pennsylvania
BT United States
NT Altoona Pennsylvania

Blair Formation (1989)
In Mesaverde Group. Underlies Rock Springs Formation and overlies Hilliard Shale. SW Wyoming and NE Utah.
BT Upper Cretaceous
BT Cretaceous
BT Mesozoic
SA Mesaverde Group
SA Utah
SA Wyoming

Blairmore Group (1978)
Includes the Gladstone Formation, Beaver Mines Formation, and Mill Creek Formation in S Alberta.
BT Cretaceous
BT Mesozoic
SA Alberta

Blake Basin
  use Blake-Bahama Basin

Blake Nose (1981)
NE-jutting spur of the Blake Plateau.
CO N293000N303000
    W0760000W0770000
BT Blake Plateau
BT Northwest Atlantic
BT Atlantic Ocean
SA DSDP Site 390
SA DSDP Site 392
SA Leg 44

Blake Outer Ridge
  use Blake-Bahama Outer Ridge

Blake Plateau (1978)
E of N Florida and N of Grand Bahama Island. As of 1990, Atlantic Ocean is autoposted to this term.
UF Bahama Platform
BT Northwest Atlantic
BT Atlantic Ocean
NT Blake Nose
SA Leg 101
SA ODP Site 627

Blake Ridge
  use Blake-Bahama Outer Ridge

Blake River Group (1985)
W Quebec and E Ontario.
IN Index provinces as applicable.
BT Archean
BT Precambrian
SA Abitibi Belt
SA Ontario
SA Quebec

Blake-Bahama Basin (1978)
Just to the E of the Blake Plateau E of North Florida.
UF Blake Basin
BT Atlantic Ocean
SA DSDP Site 391
SA Leg 44

Blake-Bahama Formation (1989)
North American Atlantic.
BT Lower Cretaceous
BT Cretaceous
BT Mesozoic
SA Atlantic Ocean
SA DSDP Site 603
SA Leg 93
SA North American Atlantic

Blake-Bahama Outer Ridge (1978)
E of the Blake-Bahama Basin which is E of the Blake Plateau off N Florida.
UF Blake Outer Ridge
UF Blake Ridge
BT Atlantic Ocean
SA DSDP Site 100
SA DSDP Site 102
SA DSDP Site 103
SA DSDP Site 104
SA Leg 44

Blancan (1978)
North America. Above Hemphillian; below Irvingtonian.
IN Index ages as applicable.
BT Cenozoic
SA Pleistocene
SA Pliocene
SA Texas

Blanco fracture zone (1978)
Off coast of Oregon. NE of E end of Mendocino Escarpment.
BT Pacific Ocean

Blanco River basin
  use Rio Blanco Basin

blasting (1978)
SA engineering geology
SA excavations
SA explosions
SA foundations

Blastoidea (1978)
BT Crinozoa
BT Echinodermata
BT Invertebrata

blastomylonite (1978)
BT mylonites
BT metamorphic rocks

Blattopteroida (1981)
BT Insecta
BT Mandibulata
BT Arthropoda
BT Invertebrata

Bleiberg
  No longer a valid term for GeoRef. See Bleiberg Austria.

Bleiberg Austria (1993)
Town in S Austria.
BT Carinthia Austria
BT Austria
BT Central Europe
BT Europe

Bleikvassli (1978)
Ore bearing region.
BT Nordland Norway
BT Norway
BT Scandinavia
BT Western Europe
BT Europe

Blekinge
  No longer a valid term for GeoRef. As of 1993 see Blekinge Sweden.

Blekinge Sweden (1993)
County on the Baltic Sea in extreme SE Sweden. Before 1993 also search Blekinge AND Sweden.
BT Sweden
BT Scandinavia
BT Western Europe
BT Europe

blind deposits (1981)
SA mineral deposits, genesis
SA mineral exploration

Blind River
  No longer a valid term for GeoRef. As of 1993, see Blind River Ontario.

Blind River Ontario (1993)
Town on Lake Huron in S Ontario. Before 1993, also search Blind River AND Ontario.
BT Algoma District Ontario
BT Ontario
BT Eastern Canada
BT Canada

bloating shale
  use shale

block
  A valid index term through 1977. Now use block structures.

block clay
  use melange

block faults
  use block structures

Block Island Sound (1978)
Between Washington County, Rhode Island and Block Island.
BT Atlantic Ocean
SA Rhode Island

block mountains
  use block structures

block structures (1978)
Restricted to tectonics. Before 1978, also search faults AND block; block faults; block structure; block mountains.
UF block faults
UF block mountains
UF fault blocks
SA faults
SA horsts
SA normal faults
SA tesserae

blomstrandite
  use betafite

Blondeau Formation (1989)
BT Archean
BT Precambrian
SA Quebec

Bloody Bluff Fault (1989)
E Massachusetts. Also search Bloody Bluff.
IN Index state as applicable.
SA Massachusetts

Bloomsburg Formation (1978)
In Cayuga Group. Central and S Pennsylvania, W Maryland, W Virginia, and N West Virginia.
BT Upper Silurian
BT Silurian
BT Paleozoic
SA Maryland
SA Pennsylvania
SA Virginia
SA West Virginia

Blount County
  Valid through 1988. Search in combination with state term. After 1988, use specific county-state term.

Blount County Alabama (1989)
N central Alabama. Before 1989, also search Blount County AND Alabama.
CO N334500N341800
    W0861800W0865600
BT Alabama
BT United States

Blount County Tennessee (1989)
E Tennessee. Before 1989, also search Blount County AND Tennessee.
CO N353000N355300
    W0834000W0841100
BT Tennessee
BT United States

Blow River (1978)
Flows into Mackenzie Bay in the extreme N Yukon Territory.
BT Yukon Territory
BT Western Canada
BT Canada

blowouts (1993)
Use sand boils if applicable.
SA drilling
SA dunes
SA engineering geology
SA erosion features
SA explosions
SA geomorphology
SA geopressure
SA geothermal systems
SA ground water
SA oil and gas fields
SA petroleum
SA petroleum engineering
SA sand bodies
SA sand boils
SA shore features

Bloyd Formation (1978)
Shale consists of (ascending) Brentwood Limestone Member, Woolsey Member and an unnamed shale division. NW Arkansas and E Oklahoma.
BT Morrowan
BT Lower Pennsylvanian
BT Pennsylvanian
BT Carboniferous
BT Paleozoic
SA Arkansas
SA Oklahoma

blue algae
  use Cyanophyta

blue chalcocite
  use digenite

Blue Glacier (1978)
Mount Olympus, Olympic National Park in NW Washington.
IN Index counties or regions as applicable.
SA Olympic Mountains
SA Washington

blue lead
  use galena

Blue Mountain (1989)
Occurs in various locations.
IN Index appropriate regions as applicable.
SA California
SA Idaho
SA New Mexico
SA Nova Scotia
SA Ontario
SA Oregon
SA Pennsylvania
SA Washington

Blue Mountain Lake (1978)
In the Adirondack Mountains.
IN Index counties or regions as applicable.
SA Adirondack Mountains
SA New York

Blue Mountains (1978)
In W Idaho, NE Oregon, and SE Washington, and also a range in E Jamaica.

IN Index states or Jamaica as applicable.
SA Idaho
SA Jamaica
SA Oregon
SA United States
SA Washington

Blue Nile (1978)
Right headstream of the Nile River.
IN Index countries as applicable.
BT Africa
SA Ethiopia
SA Nile River
SA Sudan

Blue Ridge Mountains (1978)
E and SE range of the Appalachians. Also search Blue Ridge. This term has multiple hierarchies.
BT1 Appalachians
BT2 United States
SA Georgia
SA Grandfather Mountain
SA North Carolina
SA Shenandoah Valley
SA South Mountain
SA Virginia
SA West Virginia

Blue Ridge Province (1978)
Embraces the Appalachians E of the Great Appalachian Valley from South Mountain, Pa., southward including the Piedmont. Composed of some of the most ancient rocks of the Appalachians. Also search Blue Ridge.
IN Index states as applicable.
BT Appalachians
BT North America
SA Georgia
SA Maryland
SA North Carolina
SA Pennsylvania
SA Piedmont
SA South Carolina
SA Virginia
SA West Virginia

blue-green algae
 use Cyanophyta

Bluebell Mine (1978)
E shore of Kootenay Lake in SE British Columbia.
IN Index province or region as applicable.
SA British Columbia
SA lead-zinc deposits
SA mines

blueschist (1978)
BT schists
BT metamorphic rocks
SA glaucophane schist

blueschist facies (1978)
BT facies
SA greenschist facies
SA metamorphic rocks

Bluesky Formation (1989)
BT Lower Cretaceous
BT Cretaceous
BT Mesozoic
SA Alberta
SA British Columbia

Bluff Formation (1993)
Oligocene, Cayman Islands. Also Oligocene-Miocene basalt in New Zealand. Also occurs as sandstone in the Paradox Basin, SW United States.
IN Index ages as applicable.
SA Cayman Islands
SA Miocene

SA New Zealand
SA Oligocene
SA Paradox Basin

Bluffport Marl Member (1989)
Of Demopolis Chalk. W Alabama and E Mississippi.
BT Upper Cretaceous
BT Cretaceous
BT Mesozoic
SA Alabama
SA Demopolis Chalk
SA Mississippi

bluffs (1989)
Cliff with steep, broad face.
SA cliffs
SA coastlines
SA erosion features
SA fluvial features
SA geomorphology
SA shore features
SA slopes

Blufftown Formation (1989)
In Selma Group. E Alabama and W Georgia.
BT Upper Cretaceous
BT Cretaceous
BT Mesozoic
SA Alabama
SA Georgia
SA Selma Group

Blyava (1978)
A section of the City of Mednogorsk in the S Urals near the Kazakh border.
BT Russian Federation
BT Commonwealth of Independent States
SA Asia

BM (NH)
 use British Museum

Bo Hai
 use Bohai Bay

Bochnia
No longer a valid term for GeoRef. As of 1993 see Bochnia Poland.

Bochnia Poland (1993)
A mining town with evaporite deposits in S Poland. Before 1993 also search Bochnia AND Poland.
BT Tarnow Poland
BT Poland
BT Central Europe
BT Europe

Bochum
No longer a valid term for GeoRef. See Bochum Germany.

Bochum Germany (1993)
City in the Ruhr Valley.
BT North Rhine-Westphalia Germany
BT Germany
BT Central Europe
BT Europe

Bodaibo
No longer a valid term for GeoRef. As of 1993, see Bodaibo Russian Federation.

Bodaibo Russian Federation (1993)
Town in NE Irkutsk Oblast. Before 1993, also search Bodaibo or Bodaybo.
IN Index regions as applicable.
UF Bodaybo Russian Federation
BT Russian Federation
BT Commonwealth of Independent States
SA Irkutsk Russian Federation

Bodaybo Russian Federation
 use Bodaibo Russian Federation

Bodega Head (1978)
Tip of peninsula sheltering Bodega Bay from the Pacific Ocean in Sonoma County.
BT Sonoma County California
BT California
BT United States

Bodensee
 use Lake Constance

bodies, ore
 use ore bodies

bodily tide
 use Earth tides

bodily waves
 use body waves

body waves (1978)
Autoposting of this term began in 1978.
UF bodily waves
BT elastic waves
NT P-waves
NT PcP-waves
NT PKiKP-waves
NT PKP-waves
NT PKS-waves
NT Pn-waves
NT PP-waves
NT PPP-waves
NT PPS-waves
NT PS-waves
NT S-waves
NT ScS-waves
NT SKP-waves
NT SKS-waves
NT Sn-waves
SA seismology
SA surface waves
SA waves

boehmite (1978)
BT oxides
SA aluminum ores
SA hoegbomite

Boeotia
No longer a valid term for GeoRef. As of 1993 see Boeotia Greece.

Boeotia Greece (1993)
A department in Central Greece. Before 1993 also search Boeotia AND Greece.
BT Sterea Ellas
BT Greece
BT Southern Europe
BT Europe

Boerzsoeny Mountains
 use Borzsony Mountains

Bog soils (1978)
BT soils
SA Histosols
SA Hydromorphic soils
SA Intrazonal soils
SA peat
SA soil group

Boggy Shale (1985)
In Krebs Group. Oklahoma and W Arkansas.
BT Desmoinesian
BT Middle Pennsylvanian
BT Pennsylvanian
BT Carboniferous
BT Paleozoic
SA Arkansas
SA Krebs Group
SA Oklahoma

Bogota
No longer a valid term for GeoRef. See Bogota Colombia.

Bogota Colombia (1993)
City in Cundinamarca Department, central Colombia.
BT Colombia
BT South America
SA Sabana de Bogota

Bogota Plateau
 use Sabana de Bogota

bogs (1978)
Autoposting of this term began in 1978.
NT muskegs
NT peat bogs
SA fluvial features
SA geomorphology
SA lacustrine features
SA marshes
SA swamps
SA wetlands

Boguszow
No longer a valid term for GeoRef. As of 1993 see Boguszow Poland.

Boguszow Poland (1993)
Province and city in SW Poland. After 1993, use for the province. Before 1993 also search Boguszow AND Poland.
BT Walbrzych Poland
BT Poland
BT Central Europe
BT Europe

Bohai Bay (1989)
Arm of the Yellow Sea bordered by Hebei, Liaoning, and Shandong.
UF Bo Hai
UF Gulf of Chihli
UF Po Hai
UF Po Hai Wan
BT Yellow Sea
BT West Pacific
BT Pacific Ocean
SA Hebei China
SA Huanghua Depression
SA Liaodong Peninsula
SA Liaoning China
SA Shandong China
SA Shandong Peninsula

bohdanowiczite (1978)
BT sulfides

Bohemia (1978)
Region in westernmost Czechoslovakia.
CO N483200N510400
 E0165000E0120600
BT Czech Republic
BT Czechoslovakia
BT Central Europe
BT Europe
NT Central Bohemian Pluton
NT Chvaletice
NT Jachymov
NT Karlovy Vary
NT Kladno
NT Kutna Hora
NT Pilsen Basin
NT Prague
NT Pribram
NT Slany
NT Sobotka
NT Sokolov
NT Sokolov Basin
NT Teplice
SA Bohemian Massif
SA Erzgebirge
SA Izera Mountains
SA Karkonosze Mountains
SA Plana
SA Sniеznik
SA Sudeten
SA Sudeten Mountains

SA Weisse Elster Basin
SA White Mountain

Bohemia-Moravia System
As of 1993, no longer a valid term for GeoRef.

Bohemian Basin (1981)
CO N494000N504500
    E0163000E0134000
BT Czechoslovakia
BT Central Europe
BT Europe

Bohemian Massif (1978)
A dissected quadrangular plateau mainly in Bohemia with parts in Austria, Germany and Poland.
IN Index region or country as applicable.
BT Central Europe
BT Europe
SA Austria
SA Bohemia
SA Central Bohemian Pluton
SA Czechoslovakia
SA Germany
SA KTB
SA Poland

Boise
Valid through 1988. Search in combination with state term. After 1988, use specific city-state term.

Boise County
Valid through 1988. Search in combination with state term. After 1988, use specific county-state term.

Boise County Idaho (1989)
W Idaho. Before 1989, also search Boise County AND Idaho.
CO N433000N442000
    W1145500W1162000
BT Idaho
BT United States

Boise Idaho (1989)
City in SW Idaho. Before 1989, search Boise AND Idaho.
CO N433800N433800
    W1161200W1161200
BT Ada County Idaho
BT Idaho
BT United States

Bojnice (1978)
Village in W central Slovakia.
BT Slovakia
BT Czechoslovakia
BT Central Europe
BT Europe

Bokaro (1978)
Coal field in central Bihar.
SA Bihar India
SA coal fields

Bol'shezemel'skaya Tundra
use Bolshezemelskaya Tundra

Bolangir
No longer a valid term for GeoRef. See Bolangir India.

Bolangir India (1993)
Town and district in W Orissa, E India.
BT Orissa India
BT India
BT Indian Peninsula
BT Asia

Bolboforma (1989)
Genus. This term has multiple hierarchies.
BT1 problematic microfossils
BT1 microfossils
BT2 problematic microfossils
BT2 problematic fossils

Boleslawiec
No longer a valid term for GeoRef. As of 1993 see Boleslawiec Poland.

Boleslawiec Poland (1993)
Town in Jelinia Gora Province, SW Poland. Before 1993 also search Boleslawic AND Poland.
BT Poland
BT Central Europe
BT Europe

Boliden
No longer a valid term for GeoRef. As of 1993 see Boliden Sweden.

Boliden Sweden (1993)
Village in N Sweden. Before 1993 also search Boliden AND Sweden.
BT Vasterbotten Sweden
BT Sweden
BT Scandinavia
BT Western Europe
BT Europe
SA Nasliden Mine

Bolinas Lagoon (1978)
Off Bolinas Bay in SW Marin County, W California.
BT Marin County California
BT California
BT United States

Bolivar
No longer a valid term for GeoRef. As of 1993, see Bolivar Colombia or Bolivar Venezuela if applicable. Also used in other locations.

Bolivar Colombia (1993)
Department in N Colombia. Before 1993, also search Bolivar AND Colombia.
BT Colombia
BT South America
NT Cartagena Colombia
SA Magdalena Delta

Bolivar Venezuela (1993)
State in SE Venezuela. Before 1993, also search Bolivar AND Venezuela.
BT Venezuela
BT South America
NT Imataca Complex
SA Orinoco River

Bolivia (1978)
CO S230000S093000
    W0583000W0693000
BT South America
NT Cochabamba Bolivia
NT La Paz Bolivia
NT Oruro Bolivia
NT Santa Cruz Bolivia
NT Sucre Bolivia
SA Altiplano
SA Amazon Basin
SA Andes
SA Eastern Cordillera
SA San Juan River

Bolivina (1978)
Genus. Autoposting of Bolivinitidae to this term began in 1989. Autoposting of microfossils and Protista to this term began in 1990. This term has multiple hierarchies.
BT1 Bolivinitidae
BT1 Buliminacea
BT1 Rotaliina
BT1 foraminifera
BT1 Protista
BT1 Invertebrata
BT2 Bolivinitidae
BT2 Buliminacea
BT2 Rotaliina
BT2 foraminifera
BT2 Protista
BT2 microfossils

Bolivinitidae (1978)
Autoposting of this term to Bolivina and Bolivinoides began in 1989. Autoposting of microfossils and Protista to this term began in 1990. This term has multiple hierarchies.
BT1 Buliminacea
BT1 Rotaliina
BT1 foraminifera
BT1 Protista
BT1 Invertebrata
BT2 Buliminacea
BT2 Rotaliina
BT2 foraminifera
BT2 Protista
BT2 microfossils
NT Bolivina
NT Bolivinoides

Bolivinoides (1978)
Genus. Autoposting of Bolivinitidae to this term began in 1989. Autoposting of microfossils and Protista to this term began in 1990. This term has multiple hierarchies.
BT1 Bolivinitidae
BT1 Buliminacea
BT1 Rotaliina
BT1 foraminifera
BT1 Protista
BT1 Invertebrata
BT2 Bolivinitidae
BT2 Buliminacea
BT2 Rotaliina
BT2 foraminifera
BT2 Protista
BT2 microfossils

Bologna
No longer a valid term for GeoRef. See Bologna Italy.

Bologna Italy (1993)
City and province in E central Emilia-Romagna.
BT Emilia-Romagna Italy
BT Italy
BT Southern Europe
BT Europe
NT Farneto Cave

Bolshezemelskaya Tundra (1978)
In Nenets National Okrug of Arkhangelsk Oblast. This term has multiple hierarchies.
UF Bol'shezemel'skaya Tundra
BT1 Nenets Russian Federation
BT1 Arkhangelsk Russian Federation
BT1 Russian Federation
BT1 Commonwealth of Independent States
BT2 Nenets Russian Federation
BT2 Arkhangelsk Russian Federation
BT2 Europe

Bolshoy Selerikan River (1985)
E Yakutia, Russian Federation. Joins Elgi River SW of Ust-Nera. This term has multiple hierarchies.
BT1 Yakutia Russian Federation
BT1 Russian Federation
BT1 Commonwealth of Independent States
BT2 Yakutia Russian Federation
BT2 Asia

Boltyshka Depression (1978)
Central Ukraine. Also search Boltyshka. This term has multiple hierarchies.
BT1 Ukraine
BT1 Europe
BT2 Ukraine
BT2 Commonwealth of Independent States

Bolzano
No longer a valid term for GeoRef. See Bolzano Italy.

Bolzano Italy (1993)
City and province in N central Trentino-Alto Adige.
BT Trentino-Alto Adige Italy
BT Italy
BT Southern Europe
BT Europe
NT Merano Italy

Bombay
No longer a valid term for GeoRef. Former state, W India. See Bombay India.

Bombay India (1993)
City on the Arabian Sea in W Maharashtra, W India.
BT Maharashtra India
BT India
BT Indian Peninsula
BT Asia

Bonaire (1978)
Island off coast of Venezuela, 30 miles E of Curacao.
BT Netherlands Antilles
BT Lesser Antilles
BT Antilles
BT West Indies
BT Caribbean region
SA Rincon Formation

Bonanza Group (1989)
Vancouver Island, SW British Columbia.
BT Jurassic
BT Mesozoic
SA British Columbia
SA Vancouver Island

Bonanza King Formation (1985)
Middle and Upper Cambrian. SE California and S Nevada.
BT Cambrian
BT Paleozoic
SA California
SA Nevada

Bonaparte Basin
use Bonaparte Gulf basin

Bonaparte Gulf basin (1978)
N of Joseph Bonaparte Gulf and W of Northern Territory.
UF Bonaparte Basin
BT Timor Sea
BT North Australian Seas
BT West Pacific
BT Pacific Ocean
SA Joseph Bonaparte Gulf

Bond albedo
use albedo

bonding (1978)
SA cell dimensions
SA chelation
SA crystal chemistry
SA crystal structure
SA ligands
SA order-disorder

Bondoc Peninsula (1978)
Between Ragay Gulf and Mompog Pass of S Quezon Province in S Luzon.
BT Luzon
BT Philippine Islands

BT Far East
BT Asia

bone beds (1981)
Before 1981, also search bone bed. From 1981-1984, <u>phosphate rocks</u> and <u>chemically precipitated rocks</u> were autoposted to this term.
BT sedimentary rocks
SA carbonate rocks
SA clastic rocks
SA phosphate rocks

Bone Spring Limestone (1989)
SE <u>New Mexico</u> and W <u>Texas</u>. Underlies <u>Delaware Mountain Group</u>; overlies <u>Hueco Limestone</u> or <u>Wolfcampian</u>, undifferentiated.
BT Leonardian
BT Lower Permian
BT Permian
BT Paleozoic
SA New Mexico
SA Texas

Bone Valley Formation (1978)
Unconformably overlain by <u>Pleistocene</u> terrace deposits ranging in <u>age</u> from Sunderland to Pamlico. S central <u>Florida</u>.
BT middle Pliocene
BT Pliocene
BT Neogene
BT Tertiary
BT Cenozoic
SA Florida

bones (1978)
SA anatomy
SA biomineralization
SA fossils
SA jaws
SA osteology
SA paleontology
SA paleopathology
SA skeletons
SA skulls
SA teeth
SA Vertebrata

Bong Range (1978)
In N <u>Liberia</u>.
BT Liberia
BT West Africa
BT Africa

Bonin Islands (1978)
Group of 27 volcanic <u>islands</u> 600 miles S of Tokyo. Part of Tokyo Prefecture. In 1985, broader term changed from <u>Pacific Ocean</u> to <u>West Pacific Ocean Islands</u>. Also search Ogasawara and Ogasawara Islands.
UF Ogasawara Islands
BT West Pacific Ocean Islands
SA Izu-Bonin Arc
SA Japan

boninite (1985)
BT andesites
BT volcanic rocks
BT igneous rocks

Bonner County
Valid through 1988. Search in combination with state term. After 1988, use specific county-state term.

Bonner County Idaho (1989)
N <u>Idaho</u>. Before 1989, also search Bonner County AND <u>Idaho</u>.
CO N475000N485000
    W1160500W1170500
BT Idaho
BT United States
SA Idaho Panhandle

Bonner Formation (1978)
In <u>Missoula Group</u>, <u>Precambrian (Belt Supergroup)</u> of W <u>Montana</u>.
BT middle Proterozoic
BT Proterozoic
BT upper Precambrian
BT Precambrian
SA Belt Supergroup
SA Missoula Group
SA Montana

Bonneterre Formation (1978)
Includes Tom Sauk <u>limestone</u> member at base.
BT Upper Cambrian
BT Cambrian
BT Paleozoic
SA Missouri

Bonneville County
Valid through 1988. Search in combination with state term. After 1988, use specific county-state term.

Bonneville County Idaho (1989)
On <u>Wyoming</u> border in SE <u>Idaho</u>. Before 1989, also search Bonneville County AND <u>Idaho</u>.
CO N430000N434000
    W1110000W1123500
BT Idaho
BT United States

Bonneville Salt Flats (1978)
Tooele County in NW <u>Utah</u>.
BT Tooele County Utah
BT Utah
BT United States

Book Cliffs (1985)
Escarpment in E <u>Utah</u> and W <u>Colorado</u>.
IN Index states and counties as applicable.
SA Carbon County Utah
SA Colorado
SA Garfield County Colorado
SA Grand County Colorado
SA Grand County Utah
SA Mesa County Colorado
SA Utah

book reviews (1978)
SA catalogs
SA monographs
SA publications
SA review
SA textbooks

books
use monographs

Boom Clay (1989)
BT Oligocene
BT Paleogene
BT Tertiary
BT Cenozoic
SA Belgium

Boone County
Valid through 1988. Search in combination with state term. After 1988, use specific county-state term.

Boone County Arkansas (1989)
N <u>Arkansas</u>. Before 1989, also search Boone County AND <u>Arkansas</u>.
CO N360700N362600
    W0925300W0931500
BT Arkansas
BT United States

Boone County Illinois (1989)
On <u>Wisconsin</u> line in N <u>Illinois</u>. Before 1989, also search Boone County AND <u>Illinois</u>.
CO N421000N423000
    W0884500W0885600
BT Illinois
BT United States

Boone County Indiana (1989)
Central <u>Indiana</u>. Before 1989, also search Boone County AND <u>Indiana</u>.
CO N395500N401200
    W0861500W0864300
BT Indiana
BT United States

Boone County Iowa (1989)
Central <u>Iowa</u>. Before 1989, also search Boone County AND <u>Iowa</u>.
CO N415200N421300
    W0934200W0940900
BT Iowa
BT United States

Boone County Kentucky (1989)
N <u>Kentucky</u>. Before 1989, also search Boone County AND <u>Kentucky</u>.
CO N384500N391000
    W0843600W0845400
BT Kentucky
BT United States

Boone County Missouri (1989)
On the <u>Missouri River</u> in central <u>Missouri</u>. Before 1989, also search Boone County AND <u>Missouri</u>.
CO N384000N391500
    W0920700W0923500
BT Missouri
BT United States
NT Columbia Missouri
SA Devils Icebox

Boone County Nebraska (1989)
E central <u>Nebraska</u>. Before 1989, also search Boone County AND <u>Nebraska</u>.
CO N412800N415500
    W0975000W0981800
BT Nebraska
BT United States

Boone County West Virginia (1989)
SW <u>West Virginia</u>. Before 1989, also search Boone County AND <u>West Virginia</u>.
CO N374600N381400
    W0812700W0815800
BT West Virginia
BT United States

Boone Formation (1993)
Lower and <u>Upper Mississippian</u>. N <u>Arkansas</u>, SW <u>Missouri</u>, and E <u>Oklahoma</u>.
IN Index <u>age</u> as applicable.
BT Mississippian
BT Carboniferous
BT Paleozoic
SA Arkansas
SA Missouri
SA Oklahoma
SA Osagian

Boothia Felix Peninsula
use Boothia Peninsula

Boothia Peninsula (1978)
N extremity of mainland <u>Canada</u> W of Gulf of Boothia in Franklin District.
CO N690000N720000
    W0920000W0963000
UF Boothia Felix Peninsula
BT Franklin District Northwest Territories
BT Northwest Territories
BT Western Canada
BT Canada
SA Somerset Island

Bootlegger Cove Clay (1989)
Separates the Knik from overlying Naptowne glacial deposits.
BT Pleistocene
BT Quaternary
BT Cenozoic
SA Alaska

Boquira Mine (1978)
Mine in Bahia, <u>Brazil</u>. Also search Boquira.
BT Bahia Brazil
BT Brazil
BT South America
SA metal ores
SA mines

Bor (1978)
Town in <u>Serbia</u>.
IN Index <u>Serbia</u> as applicable.
SA Serbia

boracite (1978)
From 1978-1980, <u>borates</u> and <u>halides</u> were autoposted to this term. Autoposting of <u>chlorides</u> began in 1981. As of 1989, <u>halides</u> is autoposted to this term. This term has multiple hierarchies.
BT1 borates
BT2 chlorides
BT2 halides

Borah Peak (1989)
Mountain in <u>Lost River Range</u>, central <u>Idaho</u>. This term has multiple hierarchies.
CO N440900N440900
    W1134700W1134700
UF Mount Borah
BT1 Custer County Idaho
BT1 Idaho
BT1 United States
BT2 Lost River Range
BT2 Idaho
BT2 United States

Borah Peak earthquake 1983 (1989)
Epicenter near <u>Borah Peak</u>, Custer County, central <u>Idaho</u>.
UF Mount Borah earthquake, 1983
BT earthquakes
SA Custer County Idaho
SA Idaho

borate deposits (1981)
Before 1981, search <u>borates</u> AND deposits.
IN May be used as a commodity term with <u>boron deposits</u>.
SA borates
SA boron deposits
SA evaporite deposits

borates (1978)
As of 1981, use <u>borate deposits</u> for borates as a commodity. Before 1981, <u>voltaite</u> was included as a narrower term. Autoposting of this term began in 1978.
NT boracite
NT borax
NT colemanite
NT fluoborite
NT hulsite
NT hydroboracite
NT kaliborite
NT kernite
NT kotoite
NT kurchatovite
NT kurnakovite
NT ludwigite
NT preobrazhenskite
NT sakhaite
NT solongoite

NT suanite
NT szaibelyite
NT ulexite
NT vimsite
NT vonsenite
SA borate deposits
SA minerals

borax (1978)
BT borates

Borax Lake (1978)
Located in San Bernardino County, S California. Also a lake in Harney County, Oregon.
IN Index states and counties as applicable.
SA California
SA Harney County Oregon
SA Oregon
SA San Bernardino County California

Borborema (1978)
A plateau in NE Brazil.
IN Index states as applicable.
BT Brazil
BT South America
SA Paraiba Brazil
SA Rio Grande do Norte Brazil

Bordeaux
No longer a valid term for GeoRef. See Bordeaux France.

Bordeaux France (1993)
City in SW France.
BT Gironde France
BT France
BT Western Europe
BT Europe

Bordelais (1978)
Region in SW France.
BT Gironde France
BT France
BT Western Europe
BT Europe

Borden Group (1978)
In Kentucky, comprises New Providence Formation, Brodhead Formation, Floyds Knob Formation, and Muldraugh Formation. In S Indiana, comprises New Providence Shale, Locust Point, Carwood, Floyds Knob, and Edwardsville formations. S Indiana and E central Kentucky.
BT Mississippian
BT Carboniferous
BT Paleozoic
SA Indiana
SA Kentucky

border faults
use peripheral faults

Border Ranges Fault (1989)
S Alaska.
IN Index regions as applicable.
SA Alaska
SA Southern Alaska

borderland, continental
use continental borderland

bore
use boreholes

Boreal (1978)
Used primarily in Europe for an interval of postglacial time following the Preboreal and preceding the Atlantic, during which the inferred climate was warm and dry.
BT Holocene
BT Quaternary
BT Cenozoic
SA biogeography
SA paleoclimatology

SA Preboreal

boreal environment (1978)
Before 1978, also search boreal; before 1981, also search boreal region. (Boreal is used as an age unit).
SA depositional environment
SA ecology
SA environment
SA paleoclimatology
SA paleoecology

boreal region
As of 1981, no longer a valid term for GeoRef. Use boreal environment.

borehole breakouts (1993)
Enlargement of borehole in direction of minimum stress.
UF breakouts
UF wellbore breakouts
SA boreholes
SA drilling
SA geopressure
SA petroleum engineering

borehole sections (1981)
SA boreholes
SA cross sections
SA well-logging

borehole televiewers (1993)
Well-logging instrument or technique using acoustic(sonar) beams. BHTV is a tradename.
UF BHTV
SA acoustical logging
SA acoustical methods
SA boreholes
SA geophysical methods
SA instruments
SA sonar methods
SA ultrasonic methods
SA video methods
SA well-logging

boreholes (1978)
Before 1976, also search drill holes.
UF bore
UF drill holes
SA borehole breakouts
SA borehole sections
SA borehole televiewers
SA cores
SA crosshole methods
SA cuttings
SA deep drilling
SA dipmeter logging
SA downhole methods
SA drilling
SA electrical logging
SA heat flow
SA horizontal drilling
SA Iceland Research Drilling Project
SA Illinois Deep Hole Project
SA KTB
SA marine drilling
SA measurement-while-drilling
SA three-dimensional methods
SA tube waves
SA well-logging
SA wells

borings (1978)
As of 1978, restricted to worm borings. For well-logging, use boreholes or cores. From 1981-1984, biogenic structures and sedimentary structures were autoposted to this term.
UF worm borings
SA biogenic structures
SA bioturbation
SA burrows
SA ichnofossils

SA lebensspuren
SA sedimentary structures

Borislav
No longer a valid term for GeoRef. See Borislav Ukraine.

Borislav Ukraine (1993)
City in Lvov Oblast in W Ukraine. This term has multiple hierarchies.
BT1 Lvov Ukraine
BT1 Ukraine
BT1 Europe
BT2 Lvov Ukraine
BT2 Ukraine
BT2 Commonwealth of Independent States

Borkut Deposit (1978)
Region of mineral deposits (including mercury ores) in Transcarpathian Oblast in W Ukraine. Also search Borkut. This term has multiple hierarchies.
BT1 Transcarpathia Ukraine
BT1 Ukraine
BT1 Europe
BT2 Transcarpathia Ukraine
BT2 Ukraine
BT2 Commonwealth of Independent States
SA mercury ores

Borneo (1978)
An island, the S two-thirds of which is known as Kalimantan, Indonesia, with the remainder comprising the two states of East Malaysia and Brunei. As of 1981, Malay Archipelago or Far East may be used as a broader term. Before 1981, Malay Archipelago was the broader term. Before 1981, Brunei was a narrower term. As of 1993, both Malay Archipelago and Far East are used as broader terms. This term has multiple hierarchies.
IN Index Brunei, Kalimantan, and the states of East Malaysia as applicable.
CO S040000N073000
    E1192000E1085000
BT1 Far East
BT1 Asia
BT2 Malay Archipelago
NT Brunei
NT East Malaysia
NT Kalimantan Indonesia
SA Indonesia
SA Malaysia

Borneo-Java Shelf
use Sunda Shelf

Bornholm (1978)
Island in W Baltic Sea S of Sweden.
UF Bornholm Island
BT Denmark
BT Scandinavia
BT Western Europe
BT Europe
SA Baltic Sea

Bornholm Island
use Bornholm

bornite (1978)
UF erubescite
UF horseflesh ore
UF purple copper ore
BT sulfides
SA copper
SA copper minerals
SA copper ores

boron (1978)

Chemical element. As of 1981, use boron deposits for boron as a commodity.
UF B
SA boron deposits
SA chemical elements
SA lithophile elements

boron deposits (1981)
Before 1981, search boron AND deposits.
SA borate deposits
SA boron

boronatrocalcite
use ulexite

borosilicates (1978)
BT silicates
SA axinite
SA dravite

Borrego Mountain (1978)
In S California.
BT California
BT United States
SA Borrego Mountain earthquake 1968

Borrego Mountain earthquake 1968 (1993)
Superstition Hills Fault, Imperial Valley, California.
BT earthquakes
SA Borrego Mountain
SA California
SA Imperial Valley

Borrowdale Volcanic Series (1978)
Llanvirnian and ? Llandeilian. Underlain by Skiddaw Slates and overlain by the Coniston Limestone Group. Lake District in NW England.
BT Ordovician
BT Paleozoic
SA England
SA Llandeilian
SA Llanvirnian

borrowing (1981)
SA economics

Borshchev
No longer a valid term for GeoRef. See Borshchev Ukraine.

Borshchev Ukraine (1993)
City in Ternopol Oblast in W Ukraine. Before 1993, also search Borshchev or Borshchov or Borszczow in combination with Ukraine. This term has multiple hierarchies.
UF Borshchov Ukraine
UF Borszczow Ukraine
BT1 Ukraine
BT1 Europe
BT2 Ukraine
BT2 Commonwealth of Independent States

Borshchov Ukraine
use Borshchev Ukraine

Borsod Basin (1978)
Coal basin in NE Hungary. Also search Borsod.
BT Hungary
BT Central Europe
BT Europe
SA coal fields

Borszczow Ukraine
use Borshchev Ukraine

Borzeta (1978)
Boreholes in W Carpathians.
BT Poland
BT Central Europe
BT Europe

Borzsony Mountains (1978)
Near Slovakian border in NW Nograd, Hungary.
UF Boerzsoeny Mountains
BT Hungary
BT Central Europe
BT Europe

Bos (1978)
Genus. Autoposting of Eutheria and Theria to this term began in 1989.
BT Bovidae
BT Ruminantia
BT Artiodactyla
BT Eutheria
BT Theria
BT Mammalia
BT Tetrapoda
BT Vertebrata
BT Chordata
NT Bos primigenius

Bos primigenius (1978)
Species. Autoposting of Eutheria and Theria to this term began in 1989.
BT Bos
BT Bovidae
BT Ruminantia
BT Artiodactyla
BT Eutheria
BT Theria
BT Mammalia
BT Tetrapoda
BT Vertebrata
BT Chordata

Bosano (1978)
Region in NW Sardinia.
BT Sardinia Italy
BT Italy
BT Southern Europe
BT Europe

Boschekul
No longer a valid term for GeoRef. As of 1993, see Boshchekul Kazakhstan.

Boshchekul Kazakhstan (1993)
Village in Pavlodar Oblast in NE Kazakhstan. This term has multiple hierarchies.
BT1 Pavlodar Kazakhstan
BT1 Kazakhstan
BT1 Central Asia
BT1 Asia
BT2 Pavlodar Kazakhstan
BT2 Kazakhstan
BT2 Commonwealth of Independent States

Bosnia (1978)
Region. Northern part of Republic of Bosnia and Herzegovina.
BT Bosnia-Herzegovina
BT Yugoslavia
BT Southern Europe
BT Europe
NT Sarajevo

Bosnia-Herzegovina (1981)
Region in W central Yugoslavia.
CO N423500N452000
    E0194000E0153800
BT Yugoslavia
BT Southern Europe
BT Europe
NT Banja Luka
NT Bosnia
NT Herzegovina
SA Neretva Valley

Boso Peninsula
use Chiba Peninsula

Bosphorus
use Bosporus

Bosporus (1978)
Strait connecting Black Sea with Sea of Marmara.
UF Bosphorus
BT Turkey
BT Middle East
BT Asia
SA Istanbul Turkey

Bossier Formation (1978)
In Cotton Valley Group. S Arkansas, W Louisiana, and E Texas.
BT Upper Jurassic
BT Jurassic
BT Mesozoic
SA Arkansas
SA Cotton Valley Group
SA Louisiana
SA Texas

Boston
Valid through 1988. Search in combination with state term. After 1988, use specific city-state term.

Boston Bay Group (1989)
E Massachusetts. Previously thought to be Devonian or Carboniferous in age.
IN Index ages as applicable.
SA Cambridge Argillite
SA Lower Cambrian
SA Massachusetts
SA Roxbury Conglomerate
SA upper Proterozoic

Boston Massachusetts (1989)
City on Massachusetts Bay in E Massachusetts. Before 1989, search Boston AND Massachusetts.
CO N422000N422000
    W0710500W0710500
BT Suffolk County Massachusetts
BT Massachusetts
BT United States

bostonite (1978)
BT syenites
BT plutonic rocks
BT igneous rocks

Bosumtwi Crater (1978)
S central Ghana.
BT Ghana
BT West Africa
BT Africa

botany, paleo-
use paleobotany

Botetourt County Virginia (1993)
W Virginia. Before 1989, also search Botetourt County AND Virginia.
CO N371900N374800
    W0793000W0800400
BT Virginia
BT United States

Bothnian Sea
use Gulf of Bothnia

Botryococcus (1978)
Autoposting of thallophytes and microfossils began in 1990. This term has multiple hierarchies.
BT1 Chlorophyta
BT1 algae
BT1 thallophytes
BT1 Plantae
BT2 Chlorophyta
BT2 algae
BT2 microfossils

Botswana (1978)
Before 1966 was British protectorate of Bechuanaland. Before 1972, also search Bechuanaland.
CO S270000S174000
    E0293000E0200000
UF Bechuanaland
BT Southern Africa
BT Africa
SA Beaufort Group
SA Dwyka Formation
SA Ecca Group
SA Kaapvaal Craton
SA Kalahari Desert
SA Karroo Supergroup
SA Limpopo Basin
SA Limpopo Belt
SA Pretoria Group
SA Ventersdorp Supergroup
SA Zambezi Valley

bottling (1981)
SA drinking water
SA potability
SA water
SA water quality
SA water resources

bottom currents (1978)
Papers are usually out-of-scope for GeoRef.
BT currents
SA ocean circulation

bottom features (1978)
UF features, bottom
UF submarine features
SA abyssal hills
SA abyssal plains
SA benthonic taxa
SA black smokers
SA deep-sea environment
SA furrows
SA gravity flows
SA hydrothermal vents
SA marginal basins
SA marine geology
SA microcontinents
SA mid-ocean ridges
SA ocean floors
SA oceanography
SA pockmarks
SA sand ridges
SA seamounts
SA submarine canyons
SA submarine dunes
SA submarine environment
SA submarine volcanoes
SA trenches
SA troughs

bottom load
use bedload

bottom water (1978)
SA Antarctic bottom water
SA North Atlantic Deep Water
SA oceanography
SA water

Botucatu Formation (1978)
IN Index ages as applicable.
SA Brazil
SA Mato Grosso Brazil
SA Paraguay
SA Permian
SA Sao Paulo Brazil
SA South America
SA Triassic

Bou Azzer (1978)
Locality in Marrakesh region in W central Morocco.
UF Bou-Azzer
BT Morocco
BT North Africa
BT Africa

Bou-Azzer
use Bou Azzer

Bouches-du-Rhone
No longer a valid term for GeoRef.
use Bouches-du-Rhone France

Bouches-du-Rhone France (1993)
CO N431500N435500
    E0054500E0041000
UF Bouches-du-Rhone
BT France
BT Western Europe
BT Europe
NT Aix-en-Provence France
NT Carry-le-Rouet
NT Marseilles France
NT Rhone Delta
NT Sainte-Victoire Mountain
SA Provence

boudinage (1978)
Before 1981, also search boudins.
UF boudins
UF pull apart structures
UF sausage structure
BT soft sediment deformation
BT sedimentary structures
SA deformation
SA lineation
SA melange
SA structural analysis

boudins
As of 1981, no longer a valid term for GeoRef.
use boudinage

Bougainville (1978)
Largest of the Solomon Islands. ESE of the Bismarck Archipelago in the SW Pacific. This term has multiple hierarchies.
BT1 Papua New Guinea
BT1 Australasia
BT2 Solomon Islands
BT2 Melanesia
BT2 Oceania

Bouguer
A valid term through 1976. After 1976, use Bouguer anomalies.

Bouguer anomalies (1978)
Also search Bouguer.
BT gravity anomalies
SA anomalies
SA free-air anomalies
SA geophysical methods
SA geophysical surveys

boulangerite (1978)
BT sulfantimonites
BT sulfosalts

Boulder
Valid through 1988. Search in combination with state term. After 1988, use specific city-state term.

Boulder Batholith (1978)
BT Montana
BT United States

boulder clay (1981)
BT clastic sediments
BT sediments
SA glacial geology
SA till

Boulder Colorado (1989)
City in N central Colorado. Before 1989, search Boulder AND Colorado.
CO N400200N400200
    W1051600W1051600
BT Boulder County Colorado
BT Colorado
BT United States

Boulder County
Valid through 1988. Search in combination with state term. After 1988, use specific county-state term.

Boulder County Colorado (1989)
N central Colorado. Before 1989, also search Boulder County AND Colorado.
CO N395000N402000
W1050500W1053500
BT Colorado
BT United States
NT Boulder Colorado
SA Rocky Mountain National Park

boulder trains (1985)
Autoposting of glacial features to this term began in 1989.
BT glacial features
SA boulders
SA geomorphology
SA glacial geology
SA glaciated terrains
SA till

boulders (1978)
Included use as sedimentary rock until 1977.
BT clastic sediments
BT sediments
SA boulder trains
SA cobbles
SA erosion features
SA erratics
SA geomorphology
SA glacial features
SA glacial geology
SA gravel
SA rock glaciers
SA sedimentary rocks
SA terrigenous materials

Bouleiceras (1978)
Genus.
BT Ammonoidea
BT Tetrabranchiata
BT Cephalopoda
BT Mollusca
BT Invertebrata

Boulogne
No longer a valid term for GeoRef. See Boulogne France.

Boulogne France (1993)
City on the English Channel, N France. Before 1993, also search Boulogne and France. Before 1978, also search Boulogne-sur-Mer.
UF Boulogne-sur-Mer France
BT Pas-de-Calais France
BT France
BT Western Europe
BT Europe

Boulogne-sur-Mer France
use Boulogne France

Boulonnais (1978)
Old district in extreme N France.
BT Pas-de-Calais France
BT France
BT Western Europe
BT Europe

Bouma cycle
use Bouma sequence

Bouma sequence (1989)
A succession consisting of five intervals, each characterized by a particular sedimentary structure, which forms a turbidite.
UF Bouma cycle
BT turbidity current structures
BT sedimentary structures
SA bioturbation
SA flow structures
SA graded bedding
SA laminations
SA ripple marks
SA turbidite

boundaries, stratigraphic
use stratigraphic boundary

boundary
No longer a valid term for GeoRef. Used for stratigraphic boundaries only after 1978. See stratigraphic boundary or plate boundaries.

boundary conditions (1985)
Use as a general term.
SA boundary interactions
SA convection
SA deformation
SA ground water
SA heat flow
SA rock mechanics
SA soil mechanics

Boundary County
Valid through 1988. Search in combination with state term. After 1988, use specific county-state term.

Boundary County Idaho (1989)
On Canadian border in N Idaho. Before 1989, also search Boundary County AND Idaho.
CO N480000N490000
W1160500W1170500
BT Idaho
BT United States
SA Idaho Panhandle
SA Purcell Mountains

boundary element analysis (1985)
UF boundary value problems
SA analysis
SA finite element analysis
SA mathematical models

boundary faults (1993)
UF marginal faults
BT faults
SA arcuate faults
SA decollement
SA peripheral faults

boundary interactions (1978)
UF interactions, boundary
SA air-sea interface
SA atmosphere
SA boundary conditions
SA meteorology

boundary layer (1978)
Before 1976, also search boundary layer processes.
UF boundary layer processes
UF layer, boundary
SA HEBBLE
SA ocean circulation
SA oceanography
SA turbulence

boundary layer processes
A valid term through 1975.
use boundary layer

boundary value problems
use boundary element analysis

boundstone (1978)
BT carbonate rocks
BT sedimentary rocks

Bourges
No longer a valid term for GeoRef. See Bourges France.

Bourges France (1993)
City in central France.
BT Cher France
BT France
BT Western Europe
BT Europe

Bourgogne
use Burgundy

bournonite (1978)
UF berthonite
UF endellionite
UF wheel ore
BT sulfantimonites
BT sulfosalts

Bouvet Island (1978)
Norwegian island 1600 miles SSW of Cape of Good Hope. In 1985, broader term changed from Atlantic Ocean to Atlantic Ocean Islands.
BT Atlantic Ocean Islands

Bouxwiller
No longer a valid term for GeoRef. See Bouxwiller France.

Bouxwiller France (1993)
Town in extreme NE France.
BT Bas-Rhin France
BT France
BT Western Europe
BT Europe

Bovidae (1978)
Family. Autoposting of Eutheria and Theria to this term began in 1989.
BT Ruminantia
BT Artiodactyla
BT Eutheria
BT Theria
BT Mammalia
BT Tetrapoda
BT Vertebrata
BT Chordata
NT Bison
NT Bos

Bow River valley (1993)
In S Alberta.
SA Alberta

bow shock waves
As of 1993, no longer a valid term. Usually out-of-scope for GeoRef. Before 1976, also search bow shock. Used with magnetosphere.

Bowen
No longer a valid term for GeoRef. See Bowen Australia.

Bowen Australia (1993)
Town on the Coral Sea in the S Great Barrier Reef area, E Queensland, NE Australia.
BT Queensland Australia
BT Australia
BT Australasia

Bowen Basin (1978)
In E Queensland. Also search Bowen.
IN Index state or region as applicable.
SA Queensland Australia
SA Springsure Australia

Bowers Ridge (1978)
N of the Aleutian Islands.
BT Bering Sea
BT West Pacific
BT Pacific Ocean

Bowers Supergroup (1989)
Cambrian to Ordovician. N Victoria Land.
IN Index ages as applicable.
BT Paleozoic
SA Antarctica
SA Cambrian
SA Ordovician
SA Victoria Land

Bowie Seamount (1978)
Just W of the Queen Charlotte Islands off central British Columbia.
BT Pacific Ocean

Bowser Basin (1989)
BT British Columbia
BT Western Canada
BT Canada

Bowser Formation (1978)
In Tuxedni Group, Middle(?) and Upper Jurassic. Central S Alaska.
BT Jurassic
BT Mesozoic
SA Alaska

Bowser Lake Group (1989)
BT Upper Jurassic
BT Jurassic
BT Mesozoic
SA British Columbia

Box Elder County
Valid through 1988. Search in combination with state term. After 1988, use specific county-state term.

Box Elder County Utah (1989)
Extreme NW Utah. Before 1989, also search Box Elder County AND Utah.
CO N410000N420000
W1115300W1140500
BT Utah
BT United States
SA Great Salt Lake
SA Moxa Arch
SA Pilot Range

Boyle Dolomite (1989)
Central Kentucky.
BT Middle Devonian
BT Devonian
BT Paleozoic
SA Kentucky

Boza Wola (1978)
Region in S Poland.
BT Poland
BT Central Europe
BT Europe

Br
use bromine

Braarudosphaeridae (1978)
Autoposting of thallophytes and microfossils to this term began in 1990. This term has multiple hierarchies.
BT1 nannofossils
BT1 algae
BT1 thallophytes
BT1 Plantae
BT2 nannofossils
BT2 algae
BT2 microfossils
BT3 nannofossils
BT3 algal flora
BT3 plants
BT4 nannofossils
BT4 algal flora
BT4 microfossils

Brabant
No longer a valid term for GeoRef. See Brabant Belgium.

Brabant Belgium (1993)
Province in central Belgium.
BT Belgium
BT Western Europe
BT Europe
NT Brussels Belgium
NT Louvain Belgium

Brabant Massif (1978)
Also search Brabant.
BT Belgium
BT Western Europe
BT Europe

Brachiopoda (1978)

Autoposting of this term began in 1978.
BT Invertebrata
NT Articulata
NT Inarticulata
SA brachiopods
SA reef builders

brachiopods (1981)
Common name for Brachiopoda.
IN Index for all non-paleontologic studies of fossils.
BT invertebrates
SA biostratigraphy
SA Brachiopoda

Brachiopterygii (1981)
BT Osteichthyes
BT Pisces
BT Vertebrata
BT Chordata

Brachyura (1978)
Suborder.
BT Malacostraca
BT Crustacea
BT Mandibulata
BT Arthropoda
BT Invertebrata

brackish water (1978)
Before 1978, also search brackish.
SA brackish-water environment
SA ecology
SA hydrochemistry
SA paleosalinity
SA salinity
SA salt lakes
SA salt water
SA sea water
SA water

brackish-water environment (1978)
Before 1978, also search brackish or brackish water.
SA brackish water
SA deep-water environment
SA depositional environment
SA ecology
SA environment
SA salinity
SA salt marshes
SA swamps
SA tidal flats

Bracklesham Beds (1989)
BT Paleogene
BT Tertiary
BT Cenozoic
SA Barton Beds
SA England

Brad
No longer a valid term for GeoRef. As of 1993 see Brad Romania.

Brad Romania (1993)
Town in Hunedoara County in W Romania. Before 1993 also search Brad AND Romania.
BT Transylvania
BT Romania
BT Southern Europe
BT Europe

Bradford County
Valid through 1988. Search in combination with state term. After 1988, use specific county-state term.

Bradford County Florida (1989)
N Florida. Before 1989, also search Bradford County AND Florida.
CO N294100N300700
W0820200W0822300
BT Florida

BT United States

Bradford County Pennsylvania (1989)
NE Pennsylvania. Before 1989, also search Bradford County AND Pennsylvania.
CO N413300N420000
W0760700W0765600
BT Pennsylvania
BT United States
SA Chemung River

Bradyodonti (1978)
BT Elasmobranchii
BT Chondrichthyes
BT Pisces
BT Vertebrata
BT Chordata

Braganca
No longer a valid term for GeoRef. As of 1993 see Braganca Portugal.

Braganca Portugal (1993)
District in NE Portugal. Also a city. Before 1993 also search Braganca AND Portugal.
BT Portugal
BT Iberian Peninsula
BT Southern Europe
BT Europe
SA Tras-os-Montes

Brahmaputra River (1978)
Rises in the Kailas Range of SW Xizang; it flows E across S Xizang, then S, SW, and again S through Bangladesh where it becomes the Jamuna River which merges with the Ganges River in the Ganges-Brahmaputra Delta on the Bay of Bengal.
IN Index Xizang and countries as applicable.
BT Asia
SA Bangladesh
SA China
SA India
SA Indian Peninsula
SA Xizang China

braided channels
use braided streams

braided streams (1978)
Autoposting of fluvial features to this term began in 1989.
UF anastomosing streams
UF braided channels
BT streams
BT fluvial features
SA channels
SA geomorphology
SA sedimentation

Braila
No longer a valid term for GeoRef. As of 1993 see Braila Romania.

Braila Romania (1993)
City and county in NE Walachia. Before 1993 also search Braila AND Romania.
BT Walachia
BT Romania
BT Southern Europe
BT Europe

Brallier Shale (1985)
In Susquehanna Group. Central Pennsylvania, Virginia and West Virginia.
BT Upper Devonian
BT Devonian
BT Paleozoic
SA Pennsylvania
SA Virginia
SA West Virginia

Bramsche
No longer a valid term for GeoRef. See Bramsche Germany.

Bramsche Germany (1993)
Town in SW Lower Saxony.
BT Lower Saxony Germany
BT Germany
BT Central Europe
BT Europe

Branchiopoda (1978)
As of 1981, used for Conchostraca. Before 1981, also search Conchostraca.
UF Conchostraca
BT Crustacea
BT Mandibulata
BT Arthropoda
BT Invertebrata

Brandenburg
No longer a valid term for GeoRef. See Brandenburg Germany.

Brandenburg Germany (1993)
State in NE Germany. Before 1993, used as region. Includes the town on the Havel River.
BT Germany
BT Central Europe
BT Europe
NT Berlin Germany
NT Cottbus Germany
NT Doberlug Germany
NT Potsdam Germany
SA Frankfurt Germany
SA North German Plain
SA Oder Valley

Brandenburg Stade (1978)
Europe. Before 1978, also search Brandenburg.
BT Weichselian
BT upper Pleistocene
BT Pleistocene
BT Quaternary
BT Cenozoic

Brandon (1978)
Region in Warwickshire, central England. After 1992, see Brandon Manitoba for city in SW Manitoba. Before 1989, also included use for the town in Vermont; now use specific city-state term when applicable.
IN Index England and counties as applicable.
SA England
SA Warwickshire England

Brandon Manitoba (1993)
City in SW Manitoba. Before 1993, search Brandon AND Manitoba.
BT Manitoba
BT Western Canada
BT Canada

Brandon Stade (1978)
Europe. Before 1978, also search Brandon.
BT Weichselian
BT upper Pleistocene
BT Pleistocene
BT Quaternary
BT Cenozoic

Brandon Vermont (1989)
Town in W Vermont. Before 1989, search Brandon AND Vermont.
CO N434700N434700
W0730500W0730500
BT Rutland County Vermont
BT Vermont
BT United States

brannerite (1978)
BT oxides

SA uranium ores

Bransfield Strait (1993)
Antarctic Ocean between South Shetland Islands and Palmer Peninsula, Antarctica. Connects Weddell Sea on the Atlantic Ocean side with Bellingshausen Sea, Pacific Ocean side.
CO S634500S611500
W0520000W0650000
BT Antarctic Ocean
SA Bellingshausen Sea
SA Weddell Sea

Brasov
No longer a valid term for GeoRef. As of 1993 see Brasov Romania.

Brasov Romania (1993)
City and county in SE Transylvania. From 1950 to 1960, the city was named Stalin; therefore, also search Stalin AND (Transylvania OR Romania) for the city. Before 1993 also search Brasov AND Romania.
BT Transylvania
BT Romania
BT Southern Europe
BT Europe
NT Hoghiz Romania

Brassfield Formation (1978)
Considered to consist of lower dolomitic limestone, including Whitfieldella bed, and Plum Creek Clay Member above. In central N Arkansas, S Indiana, SW Illinois, central Kentucky, SW Ohio, and S Tennessee.
BT Lower Silurian
BT Silurian
BT Paleozoic
SA Arkansas
SA Illinois
SA Indiana
SA Kentucky
SA Ohio
SA Tennessee

Bratislava (1978)
City on the Danube River in SW Slovakia.
BT Slovakia
BT Czechoslovakia
BT Central Europe
BT Europe

braunite (1978)
Autoposting of nesosilicates began in 1985.
BT nesosilicates
BT orthosilicates
BT silicates
SA manganese ores

Braunschweig Germany
use Brunswick Germany

Bravais lattice
use lattice

bravoite (1978)
BT sulfides
SA pyrite

Brawley Fault (1989)
S California.
UF Brawley seismic zone
BT Imperial County California
BT California
BT United States

Brawley seismic zone
use Brawley Fault

Brazil (1978)
CO S340000N051500
W0340000W0740000
BT South America

NT Acre Brazil
NT Alagoas Brazil
NT Amapa Brazil
NT Amazonas Brazil
NT Bahia Brazil
NT Borborema
NT Brazilian Shield
NT Ceara Brazil
NT Espirito Santo Brazil
NT Goias Brazil
NT Maranhao Brazil
NT Mato Grosso Brazil
NT Minas Gerais Brazil
NT Para Brazil
NT Paraiba Brazil
NT Parana Brazil
NT Pernambuco Brazil
NT Piaui Brazil
NT Recife-Joao Pessoa
NT Rio de Janeiro Brazil
NT Rio Grande do Norte Brazil
NT Rio Grande do Sul Brazil
NT Rondonia Brazil
NT Roraima Brazil
NT Santa Catarina Brazil
NT Sao Francisco Craton
NT Sao Paulo Brazil
NT Sergipe Brazil
NT Serra do Espinhaco
NT Serra do Mar
NT Tocantins River region
SA Acungui Group
SA Amazon Basin
SA Amazon River
SA Areado Formation
SA Bambui Group
SA Barreiras Formation
SA Bauru Formation
SA Botucatu Formation
SA Colorado River
SA Estrada Nova Formation
SA Guiana Basin
SA Guyana Shield
SA Irati Formation
SA Itarare Subgroup
SA Minas Series
SA Parana Basin
SA Parana River
SA Passa Dois Group
SA Pirabas Formation
SA RADAM
SA Rio Bonito Formation
SA Rio Negro
SA Rio Negro Meteorite
SA Roraima Formation
SA Santana Formation
SA Sao Francisco Basin
SA Serra Geral Formation
SA Tubarao Group

Brazil Basin (1989)
In South American Atlantic. As of 1990, Atlantic Ocean is autoposted to this term.
BT South American Atlantic
BT South Atlantic
BT Atlantic Ocean
SA Antarctic bottom water
SA DSDP Site 19
SA DSDP Site 20
SA DSDP Site 21
SA DSDP Site 355
SA DSDP Site 515
SA Leg 39
SA Leg 72
SA Vema Channel

Brazilian Cycle (1989)
Tectonic cycle.
SA cycles
SA tectonics

Brazilian Shield (1978)
The Brazilian uplands of N central, E central, and S Brazil.

IN Index states as applicable.
UF Amazonian Shield
BT Brazil
BT South America
SA Bahia Brazil
SA Goias Brazil
SA Maranhao Brazil
SA Mato Grosso Brazil
SA Minas Gerais Brazil
SA Para Brazil
SA Parana Brazil
SA Santa Catarina Brazil
SA Sao Paulo Brazil

Brazoria County
Valid through 1988. Search in combination with state term. After 1988, use specific county-state term.

Brazoria County Texas (1989)
On Gulf of Mexico in SE Texas. Before 1989, also search Brazoria County AND Texas.
CO N284000N294000
W0950000W0955000
BT Texas
BT United States
NT Pleasant Bayou

Brazos County
Valid through 1988. Search in combination with state term. After 1988, use specific county-state term.

Brazos County Texas (1989)
E central Texas. Before 1989, also search Brazos County AND Texas.
CO N302000N305800
W0960500W0964000
BT Texas
BT United States

Brazos River (1978)
Central Texas.
BT Texas
BT United States

Brazzaville
No longer a valid term for GeoRef. See Brazzaville Congo.

Brazzaville Congo (1993)
City on the Congo River across from Kinshasa.
BT Congo
BT Central Africa
BT Africa

breakers
use breaking waves

breaking strength
use fracture strength

breaking waves (1978)
Before 1978, also search waves AND breaking; breakers.
UF breakers
BT ocean waves
SA waves

breakouts
use borehole breakouts

breakwaters (1978)
UF water-break
BT marine installations
SA engineering geology
SA shorelines
SA tombolos

Breathitt County
Valid through 1988. Search in combination with state term. After 1988, use specific county-state term.

Breathitt County Kentucky (1989)

E central Kentucky. Before 1989, also search Breathitt County AND Kentucky.
CO N372000N374000
W0825700W0833400
BT Kentucky
BT United States

Breathitt Formation (1978)
In Pottsville Group. In SE Kentucky.
BT Middle Pennsylvanian
BT Pennsylvanian
BT Carboniferous
BT Paleozoic
SA Kentucky
SA Pottsville Group

breccia (1978)
As of 1981, use lunar breccia for breccia on the Moon. Autoposting of this term began in 1978.
UF rubblerock
BT clastic rocks
BT sedimentary rocks
NT tectonic breccia
NT volcanic breccia
SA agglomerate
SA breccia pipes
SA brecciation
SA conglomerate
SA faults
SA gouge
SA gravel
SA lunar breccia
SA meteor craters
SA microbreccia
SA sediments
SA shear zones
SA slickensides
SA structural analysis
SA suevite
SA tectonics
SA terrigenous materials

breccia pipes (1978)
BT pipes
BT intrusions
SA breccia
SA diatremes
SA plugs
SA volcanic breccia
SA volcanic rocks
SA volcanism
SA volcanoes

brecciation (1978)
SA breccia
SA processes

Brecknock (shire)
use Brecknockshire Wales

Brecknockshire
No longer a valid term for GeoRef as of 1993.
use Brecknockshire Wales

Brecknockshire Wales (1993)
Former county SE Wales. As of April 1974, part of Powys. Before 1993, also search Brecknockshire or Breconshire.
UF Brecknock (shire)
UF Brecknockshire
UF Breconshire Wales
BT Powys Wales
BT Wales
BT Great Britain
BT United Kingdom
BT Western Europe
BT Europe

Breconshire Wales
use Brecknockshire Wales

Breidamerkurjokull (1978)
Glacier.
BT Iceland
BT Western Europe

BT Europe

Bremen Germany (1993)
City state in NW Germany.
BT Lower Saxony Germany
BT Germany
BT Central Europe
BT Europe
SA Weser River

Brenham
Valid through 1988. Search in combination with state term. After 1988, use specific city-state term.

Brenham Texas (1989)
City in SE central Texas. Before 1989, search Brenham AND Texas.
CO N300900N300900
W0962400W0962400
BT Washington County Texas
BT Texas
BT United States

Brent Crater (1978)
Meteor crater at Brent in Algonquin Provincial Park E of Georgian Bay.
IN Index district or region as applicable.
SA Nipissing District Ontario
SA Ontario

Brent Group (1989)
BT Bajocian
BT Middle Jurassic
BT Jurassic
BT Mesozoic
SA North Sea

Brescia
No longer a valid term for GeoRef. See Brescia Italy.

Brescia Italy (1993)
City and province in E Lombardy.
BT Lombardy Italy
BT Italy
BT Southern Europe
BT Europe

Breslau Poland
use Wroclaw Poland

Brest
No longer a valid term for GeoRef. As of 1993, see Brest France and Brest Belarus.

Brest Basin (1978)
Named after city of Brest in SW Belarus. Also search Brest.
IN Index Poland and former Soviet republics as applicable.
BT Europe
SA Belarus
SA Poland
SA Ukraine

Brest Belarus (1993)
City in SW Belarus. Before 1993, also search Brest AND Byelorussia. This term has multiple hierarchies.
BT1 Belarus
BT1 Europe
BT2 Belarus
BT2 Commonwealth of Independent States

Brest France (1993)
City on the Atlantic in W France. Before 1993, also search Brest AND France.
BT Finistere France
BT France
BT Western Europe
BT Europe

Brevard County
Valid through 1988. Search in combination with state term. After 1988, use specific county-state term.

Brevard County Florida (1989)
On the Atlantic Ocean in central Florida. Before 1989, also search Brevard County AND Florida.
CO N275000N284800
    W0802500W0805800
BT Florida
BT United States
NT Cape Canaveral

Brevard fault zone
  use Brevard Zone

Brevard Zone (1978)
IN Index Appalachians and states as applicable.
UF Brevard fault zone
SA Alabama
SA Appalachians
SA Georgia
SA North Carolina
SA Southern Appalachians
SA United States

Brewster County
Valid through 1988. Search in combination with state term. After 1988, use specific county-state term.

Brewster County Texas (1989)
Extreme SW Texas. Before 1989, also search Brewster County AND Texas.
CO N290000N304000
    W1022000W1035000
BT Texas
BT United States
NT Big Bend National Park
NT Mule Ear Diatreme
NT Terlingua Texas
SA Christmas Mountains
SA Glass Mountains
SA Rattlesnake Mountain

Brezno (1978)
Town in central Slovakia.
BT Slovakia
BT Czechoslovakia
BT Central Europe
BT Europe

Briancon
No longer a valid term for GeoRef. See Briancon France.

Briancon France (1993)
Town in SE France.
BT Hautes-Alpes France
BT France
BT Western Europe
BT Europe

Briancon Zone
  use Brianconnais Zone

Brianconnais Zone (1978)
SE France and NW Italy.
IN Index countries as applicable.
UF Briancon Zone
UF Brianconnaise Zone
BT Europe
SA Alps
SA Cretaceous
SA Eocene
SA France
SA Italy
SA Jurassic
SA Mesozoic
SA Paleocene
SA Permian
SA Tertiary

Brianconnaise Zone
  use Brianconnais Zone

briartite (1978)
BT sulfides

brick clays
Not a valid term for GeoRef. Use clays.

Bridge Creek Limestone Member (1989)
Of Greenhorn Limestone. SE Colorado and W Kansas.
BT Upper Cretaceous
BT Cretaceous
BT Mesozoic
SA Colorado
SA Greenhorn Limestone
SA Kansas

Bridger Formation (1985)
Middle and upper? Eocene. SW Wyoming, NW Colorado and NE Utah.
BT Eocene
BT Paleogene
BT Tertiary
BT Cenozoic
SA Colorado
SA Utah
SA Wyoming

Bridger Range (1985)
SW Montana.
IN Index U. S. Rocky Mountains and counties as applicable.
SA Gallatin County Montana
SA Meagher County Montana
SA Montana
SA Park County Montana
SA Rocky Mountains
SA U. S. Rocky Mountains

Bridgerian (1978)
North America.
BT Eocene
BT Paleogene
BT Tertiary
BT Cenozoic

bridges (1978)
Used for the man-made structures. For the naturally occurring, archlike features, use natural curiosities.
SA engineering geology
SA foundations
SA geomorphology
SA highways
SA natural curiosities

brief report
  use report

Brigham Group (1989)
Lower and Middle Cambrian. SE Idaho, NE Utah and NE Washington.
BT Cambrian
BT Paleozoic
SA Idaho
SA Utah
SA Washington

Bright Angel Shale (1985)
N Arizona, SE California and S Nevada.
BT Middle Cambrian
BT Cambrian
BT Paleozoic
SA Arizona
SA California
SA Nevada

bright spots
Not a valid term for GeoRef. As of 1993, use hydrocarbon indicators if applicable.

BRIM (1989)
Acronym for program in Cochrane District, Ontario.
UF Black River-Matheson Program
SA Cochrane District Ontario
SA Ontario
SA programs

Brindisi
No longer a valid term for GeoRef. See Brindisi Italy.

Brindisi Italy (1993)
City and province on the S Adriatic Sea.
BT Apulia Italy
BT Italy
BT Southern Europe
BT Europe

brine
  use brines

brines (1978)
UF brine
SA bromine
SA bromine deposits
SA desalinization
SA geothermal energy
SA geothermal systems
SA hydrochemistry
SA iodine
SA iodine deposits
SA mineral waters
SA salt
SA salt lakes
SA salt water
SA salt-water intrusion
SA sea water
SA thermal waters

Brioverian (1978)
Europe. Autoposting of Precambrian to this term ended in 1989. As of 1989, upper Proterozoic and Proterozoic are autoposted to this term. Precambrian and upper Precambrian are autoposted to this term as of 1990.
BT upper Proterozoic
BT Proterozoic
BT upper Precambrian
BT Precambrian

Brisbane
No longer a valid term for GeoRef. See Brisbane Australia.

Brisbane Australia (1993)
City in extreme SE Queensland.
BT Queensland Australia
BT Australia
BT Australasia

Bristol
No longer a valid term for GeoRef as of 1993. See Bristol England.

Bristol Bay (1978)
Between SW Alaskan mainland and Alaska Peninsula.
BT Bering Sea
BT West Pacific
BT Pacific Ocean

Bristol Channel (1978)
Arm of Atlantic Ocean. In 1985, Great Britain was added as a broader term.
IN Index England and/or Wales.
BT Great Britain
BT United Kingdom
BT Western Europe
BT Europe
SA England
SA Swansea Bay
SA Wales

Bristol England (1993)
City on the Bristol Channel in SW central England. Before 1993, also search Bristol AND England.
BT Avon England
BT England
BT Great Britain
BT United Kingdom
BT Western Europe
BT Europe
SA Gloucestershire England

britholite (1978)
Autoposting of orthosilicates to this term began in 1989. This term has multiple hierarchies.
BT1 phosphates
BT2 orthosilicates
BT2 silicates

British Columbia (1978)
CO N490000N600000
    W1140000W1390000
BT Western Canada
BT Canada
NT Albert Canyon British Columbia
NT Alice Arm British Columbia
NT Atlin British Columbia
NT Bowser Basin
NT Cariboo Mountains
NT Crowsnest Pass
NT Dogtooth Mountains
NT Endako British Columbia
NT Fernie Basin
NT Guichon Creek Batholith
NT Hope British Columbia
NT Kamloops British Columbia
NT Kimberley British Columbia
NT Kootenay Lake
NT Manning Park
NT Omineca Belt
NT Omineca Mountains
NT Pinchi Lake
NT Prince Rupert British Columbia
NT Princeton British Columbia
NT Queen Charlotte Islands
NT Queen Charlotte Sound
NT Revelstoke British Columbia
NT Rossland British Columbia
NT Saanich Inlet
NT Salmo British Columbia
NT Selkirk Mountains
NT Sifton Basin
NT Skeena Mountains
NT Slocan mining camp
NT Smithers British Columbia
NT Sustut Basin
NT Tofino Basin
NT Topley Intrusions
NT Tulameen coal area
NT Vancouver British Columbia
NT Vancouver Island
NT Ware British Columbia
SA Aldridge Formation
SA Altyn Limestone
SA Banff Formation
SA Belt Basin
SA Belt Supergroup
SA Bluebell Mine
SA Bluesky Formation
SA Bonanza Group
SA Bowser Lake Group
SA Burgess Shale
SA Cache Creek
SA Cache Creek Group
SA Cadomin Formation
SA Canadian Cordillera
SA Canadian Rocky Mountains
SA Cardium Formation
SA Cassiar Mountains
SA Chilliwack Group
SA Clearwater Lake
SA Coast Mountains
SA Coast plutonic complex
SA Coast Ranges
SA Columbia Icefield

SA Columbia River
SA Columbia River basin
SA Copper Mountain
SA Eagle Bay Formation
SA Earn Group
SA Equity Mine
SA Fernie Formation
SA Fraser River
SA Fraser River delta
SA Gates Formation
SA Gething Formation
SA Glacier National Park
SA Gog Group
SA Graham Island
SA Harrison Lake
SA Hazelton Group
SA Horsethief Creek Group
SA Indian River
SA Ishbel Group
SA Juan de Fuca Strait
SA Karmutsen Group
SA Kicking Horse River valley
SA Kootenay Arc
SA Kootenay Formation
SA Kunga Group
SA Ladner Group
SA Lardeau Group
SA Leech River Fault
SA Liard River
SA Methow Basin
SA Miette Group
SA Mist Mountain Formation
SA Monashee Complex
SA Moosebar Formation
SA Nahanni Formation
SA Nanaimo Group
SA Nicola Group
SA Okanagan Valley
SA Okanogan Range
SA Pasayten Group
SA Peace River
SA Peace River Formation
SA Purcell Mountains
SA Purcell System
SA Queen Charlotte Basin
SA Queen Charlotte Fault
SA Quesnel Lake
SA Quesnellia Terrane
SA Ranger Canyon Formation
SA Road River Formation
SA Rocky Mountain Trench
SA Rundle Group
SA Shuksan Thrust
SA Shuswap Complex
SA Sicker Group
SA Skagit Valley
SA Slave Point Formation
SA Stikinia Terrane
SA Straight Creek Fault
SA Strait of Georgia
SA Sullivan Mine
SA Takla Group
SA Valhalla Complex
SA Western Interior Seaway
SA Western Overthrust Belt
SA Whitehorse Trough
SA Windermere System
SA Wrangellia
SA Yukon-Tanana Terrane

British Guiana
  use Guyana

British Honduras
  Not a valid index term for GeoRef since 1974.
  use Belize

British Institutes Reflection Profiling Syndicate
  use BIRPS

British Museum (1985)
  Natural history museum, located in London, England.
  UF BM (NH)
  SA England

SA museums

British North Borneo
  use Sabah Malaysia

British Somaliland
  No longer a valid GeoRef index term. Before 1969, was used on level 1 in subfile E.
  use Somali Republic

British Virgin Islands (1978)
  Virgin Islands excluding those of the United States. Before 1978, also search Virgin Islands for British Virgin Islands. Before 1981, West Indies was the broader term.
  BT Virgin Islands
  BT Lesser Antilles
  BT Antilles
  BT West Indies
  BT Caribbean region
  NT Anegada Island
  SA United Kingdom

Brittany (1978)
  Region and former province occupying peninsula between English Channel and Bay of Biscay.
  BT France
  BT Western Europe
  BT Europe
  SA Cotes-du-Nord France
  SA Finistere France
  SA Morbihan France

brittle deformation (1985)
  UF brittle fracture
  BT deformation
  SA anelasticity
  SA brittle materials
  SA brittleness
  SA cataclasis
  SA fractures
  SA structural analysis
  SA tectonics

brittle fracture
  use brittle deformation

brittle materials (1989)
  SA brittle deformation
  SA brittleness
  SA deformation
  SA materials
  SA rock mechanics

brittleness (1989)
  SA brittle deformation
  SA brittle materials
  SA deformation
  SA mechanical properties
  SA rigidity
  SA rock mechanics
  SA strain
  SA stress

Brive (1978)
  Basin in S central France.
  UF Brive-la-Gaillarde
  BT Correze France
  BT France
  BT Western Europe
  BT Europe

Brive-la-Gaillarde
  use Brive

Brno (1978)
  City in S central Moravia.
  BT Moravia
  BT Czech Republic
  BT Czechoslovakia
  BT Central Europe
  BT Europe

Broach
  No longer a valid term for GeoRef. See Broach India.

Broach India (1993)

City in SE Gujarat, W India.
  BT Gujarat India
  BT India
  BT Indian Peninsula
  BT Asia

broad band spectra
  use broad-band spectra

broad-band spectra (1993)
  UF broad band spectra
  UF broadband spectra
  SA electromagnetic methods
  SA frequency
  SA satellite methods
  SA spectra

broadband spectra
  use broad-band spectra

Broadlands (1978)
  Geothermal field. Central North Island, New Zealand.
  IN Index North Island and New Zealand as applicable.
  SA geothermal fields
  SA New Zealand
  SA North Island
  SA Taupo volcanic zone
  SA Wairakei

Broadwater County
  Valid through 1988. Search in combination with state term. After 1988, use specific county-state term.

Broadwater County Montana (1989)
  W central Montana. Before 1989, also search Broadwater County AND Montana.
  CO N454800N464800 W1110500W1114800
  BT Montana
  BT United States

brochantite (1978)
  BT sulfates

Brocken Massif (1978)
  In Harz Mountains in central Germany.
  BT Germany
  BT Central Europe
  BT Europe
  SA Harz Mountains
  SA Saxony-Anhalt Germany

Brockman Iron Formation (1978)
  Wittenoom Gorge area in NW Western Australia.
  BT Precambrian
  SA Australia
  SA Western Australia
  SA Wittenoom Gorge

Broegger Peninsula
  use Brogger Peninsula

Brogger Peninsula (1978)
  On the Island of Spitsbergen.
  UF Broegger Peninsula
  UF Broggerhalvoya
  BT Spitsbergen Island
  BT Spitsbergen
  BT Svalbard
  BT Arctic region

Broggerhalvoya
  use Brogger Peninsula

Broken Bay (1978)
  Inlet of Pacific Ocean.
  IN Index state or region as applicable.
  SA New South Wales Australia

Broken Hill (1978)

City and metals mining center in W New South Wales. As of 1993, use Broken Hill Australia for the city.
  IN Index state or region as applicable.
  SA Broken Hill Block
  SA metal ores
  SA New South Wales Australia

Broken Hill Block (1985)
  Structural unit, W New South Wales and E South Australia.
  IN Index state or region as applicable.
  SA Australia
  SA Broken Hill
  SA New South Wales Australia
  SA South Australia

Broken Hill Mine (1993)
  Metal ore mine in Broken Hill area, New South Wales. Also base metals mine in Aggeneys, Cape Province, South Africa. Also lead-zinc deposits in Kabwe, Zambia. Before 1993, also search Broken Hill and region and commodity.
  IN Index countries and commodities as applicable.
  SA base metals
  SA Cape Province South Africa
  SA lead-zinc deposits
  SA metal ores
  SA mines
  SA New South Wales Australia
  SA South Africa
  SA Zambia

Broken River Formation (1978)
  Silurian and Devonian. NE Queensland.
  IN Index ages as applicable.
  BT Paleozoic
  SA Australia
  SA Devonian
  SA Queensland Australia
  SA Silurian

bromellite (1978)
  BT oxides

Bromide Formation (1985)
  In Simpson Group.
  BT Middle Ordovician
  BT Ordovician
  BT Paleozoic
  SA Oklahoma
  SA Simpson Group

bromide ion (1993)
  Before 1978, search bromide.
  SA bromides
  SA bromine
  SA ions

bromides (1981)
  Before 1981, also search bromide.
  BT halides
  SA bromide ion

bromine (1978)
  Chemical element. As of 1981, use bromine deposits for bromine as a commodity. Autoposting of halogens to this term began in 1989.
  UF Br
  BT halogens
  SA brines
  SA bromide ion
  SA bromine deposits
  SA salt

bromine deposits (1981)
  Before 1981, search bromine AND deposits.

IN Commodity. See List C.
SA brines
SA bromine
SA salt

bromlite
use alstonite

Bronson Hill Anticlinorium
(1978)
In S New Hampshire and N central Massachusetts.
IN Index states as applicable.
BT United States
SA Massachusetts
SA New Hampshire

Bronze Age (1985)
Archaeological classification.
BT Cenozoic
SA archaeology
SA Holocene
SA Quaternary

bronzite (1978)
BT orthopyroxene
BT pyroxene group
BT chain silicates
BT silicates
SA enstatite

bronzitite (1978)
BT ultramafics
BT plutonic rocks
BT igneous rocks
SA pyroxenite

brookite (1978)
BT oxides
SA anatase
SA rutile

Brooks Range (1978)
Between Yukon River and Arctic Ocean in N central Alaska. At one time, was considered part of Rocky Mountains.
BT Northern Alaska
BT Alaska
BT United States
NT De Long Mountains
NT Endicott Mountains
NT Sadlerochit Mountains
SA Rocky Mountains

Broome County
Valid through 1988. Search in combination with state term. After 1988, use specific county-state term.

Broome County New York (1989)
S New York. Before 1989, also search Broome County AND New York.
CO N420000N422500
    W0752000W0761000
BT New York
BT United States
NT Binghamton New York

Brothers fault zone (1985)
Cascade Range, central Oregon.
IN Index counties as applicable.
BT Oregon
BT United States
SA Cascade Range

Broward County
Valid through 1988. Search in combination with state term. After 1988, use specific county-state term.

Broward County Florida (1989)
On the Atlantic Ocean in S Florida. Before 1989, also search Broward County AND Florida.
CO N255700N262100
    W0800400W0805400

BT Florida
BT United States
SA Biscayne Aquifer

brown algae
use Phaeophyta

brown coal
As of 1981, no longer a valid term for GeoRef. Use lignite. From 1978-1980, organic residues was autoposted to this term.

Brown County
Valid through 1988. Search in combination with state term. After 1988, use specific county-state term.

Brown County Illinois (1989)
W Illinois. Before 1989, also search Brown County AND Illinois.
CO N394900N400700
    W0903300W0905500
BT Illinois
BT United States
NT Versailles Illinois

Brown County Indiana (1989)
S central Indiana. Before 1989, also search Brown County AND Indiana.
CO N390300N392200
    W0860500W0862300
BT Indiana
BT United States

Brown County Kansas (1989)
NE Kansas. Before 1989, also search Brown County AND Kansas.
CO N394000N400000
    W0952000W0954800
BT Kansas
BT United States

Brown County Minnesota (1989)
S Minnesota. Before 1989, also search Brown County AND Minnesota.
CO N440800N443500
    W0942200W0950800
BT Minnesota
BT United States
NT New Ulm Minnesota

Brown County Wisconsin (1989)
E Wisconsin. Before 1989, also search Brown County AND Wisconsin.
CO N441500N444000
    W0874200W0881500
BT Wisconsin
BT United States
NT Green Bay Wisconsin

Brown forest soils (1978)
BT soils
SA Intrazonal soils
SA soil group

brown mica
use phlogopite

Brown soils (1978)
BT soils
SA soil group
SA Zonal soils

Browns Park Formation (1985)
Miocene. NE Utah, NW Colorado and S Wyoming. Autoposting of Neogene and Tertiary to this term began in 1989.
BT Neogene
BT Tertiary
BT Cenozoic
SA Colorado
SA Miocene
SA Pliocene

SA Utah
SA Wyoming

Browse Basin (1985)
Continental shelf off N Western Australia.
IN Index land or ocean regions as applicable.
CO S160000S120000
    E1260000E1190000
SA Indian Ocean
SA Timor Sea
SA Western Australia

Bruce Peninsula (1981)
Peninsula of SE Ontario, Canada. Extends N between Lake Huron and Georgian Bay.
CO N444000N451500
    W0805500W0814500
UF Saugeen Peninsula
BT Ontario
BT Eastern Canada
BT Canada
SA Niagara Escarpment

brucite (1978)
BT oxides

brucite marble (1978)
BT marbles
BT metamorphic rocks

Bruderheim
No longer a valid term for GeoRef. See Bruderheim Meteorite.

Bruderheim Meteorite (1993)
Olivine-hypersthene chondrite named for village NE of Edmonton in central Alberta. Before 1993, also search Bruderheim and Bruederheim.
BT L chondrites
BT chondrites
BT stony meteorites
BT meteorites
SA Alberta

Bruederheim
Not a valid term for GeoRef. See Bruderheim Meteorite.

Brule Formation (1978)
In White River Group. Middle and upper Oligocene. In NE Colorado, W Nebraska, W South Dakota, and E Wyoming.
BT Oligocene
BT Paleogene
BT Tertiary
BT Cenozoic
SA Colorado
SA Nebraska
SA South Dakota
SA White River Group
SA Wyoming

Brunei (1978)
Sultanate formerly under British protection which gained independence and joined the Commonwealth in 1984. Found in NE part of Borneo. In 1981, Far East became the broader term. Before 1981, Borneo and Malay Archipelago were broader terms. As of 1993, Borneo is a broader term. This term has multiple hierarchies.
CO N040000N051000
    E1153000E1140000
BT1 Borneo
BT1 Far East
BT1 Asia
BT2 Borneo
BT2 Malay Archipelago
SA ASEAN
SA East Malaysia

Brunhes Epoch (1978)

Geomagnetic epoch.
UF Brunhes Normal
BT Quaternary
BT Cenozoic
SA paleomagnetism
SA reversals

Brunhes Normal
use Brunhes Epoch

Brunswick
No longer a valid term for GeoRef. As of 1993, see Brunswick Germany for the city in Lower Saxony, Germany. As of 1989, for the city in Georgia, use Brunswick Georgia.

Brunswick County
Valid through 1988. Search in combination with state term. After 1988, use specific county-state term.

Brunswick County North Carolina (1989)
Before 1989, also search Brunswick County AND North Carolina.
BT North Carolina
BT United States

Brunswick County Virginia (1989)
S Virginia. Before 1989, also search Brunswick County AND Virginia.
CO N363300N370200
    W0773900W0780300
BT Virginia
BT United States

Brunswick Formation (1989)
In Newark Supergroup. New Jersey and SE Pennsylvania.
IN Index ages as applicable.
BT Mesozoic
SA Lower Jurassic
SA New Jersey
SA Newark Supergroup
SA Passaic Formation
SA Pennsylvania
SA Upper Triassic

Brunswick Georgia (1989)
City in SE Georgia. Before 1989, search Brunswick AND Georgia.
CO N310900N310900
    W0813000W0813000
BT Glynn County Georgia
BT Georgia
BT United States

Brunswick Germany (1993)
City in Lower Saxony, Germany. Before 1989, also search Brunswick AND Germany and Braunschweig.
CO N521500N521500
    E0103000E0103000
UF Braunschweig Germany
BT Lower Saxony Germany
BT Germany
BT Central Europe
BT Europe

brushite (1978)
BT phosphates

Brushy Basin Shale Member (1985)
Member of Morrison Formation. SE Utah, NE Arizona, SW Colorado and NW New Mexico.
BT Upper Jurassic
BT Jurassic
BT Mesozoic
SA Arizona
SA Colorado
SA Morrison Formation
SA New Mexico

SA Utah

Brussels
No longer a valid term for GeoRef. See Brussels Belgium.

Brussels Belgium (1993)
City in central Belgium. Search in conjunction with Belgium. Also search Bruxelles.
CO N505000N505000
   E0042100E0042100
UF Bruxelles Belgium
BT Brabant Belgium
BT Belgium
BT Western Europe
BT Europe

Bruxelles Belgium
use Brussels Belgium

Bryan County
Valid through 1988. Search in combination with state term. After 1988, use specific county-state term.

Bryan County Georgia (1989)
SE Georgia. Before 1989, also search Bryan County AND Georgia.
CO N313700N321400
   W0811000W0814600
BT Georgia
BT United States

Bryan County Oklahoma (1989)
S Oklahoma. Before 1989, also search Bryan County AND Oklahoma.
CO N334100N340900
   W0954500W0963300
BT Oklahoma
BT United States

Bryansk
No longer a valid term for GeoRef. As of 1993, see Bryansk Russian Federation.

Bryansk Russian Federation (1993)
City and oblast SW of Moscow in W Russian Federation. Before 1993, also search Bryansk. This term has multiple hierarchies.
BT1 Russian Federation
BT1 Commonwealth of Independent States
BT2 Europe

bryophytes (1978)
Autoposting of this term began in 1978.
UF mosses
BT Plantae
NT Hepaticae
NT Musci
SA mud mounds
SA thallophytes

Bryozoa (1978)
Phylum. Before 1989, also search Ectoprocta. Autoposting of this term began in 1978.
UF Ectoprocta
UF Polyzoa
BT Invertebrata
NT Cheilostomata
NT Cryptostomata
NT Ctenostomata
NT Cystoporata
NT Trepostomata
SA acanthopores
SA bryozoans
SA Cyclostomata
SA reef builders

bryozoans (1981)
Common name for Bryozoa.

IN Index for all non-paleontologic studies of fossils.
BT invertebrates
SA biostratigraphy
SA Bryozoa
SA encrustations

Bucatunna Formation (1989)
In Byram Formation of Vicksburg Group. SW Alabama, Florida and SE Mississippi.
BT middle Oligocene
BT Oligocene
BT Paleogene
BT Tertiary
BT Cenozoic
SA Alabama
SA Byram Formation
SA Florida
SA Mississippi
SA Vicksburg Group

Buccaneer Field (1985)
In Gulf of Mexico, off Texas.
CO N283000N290000
   W0943000W0950000
BT Gulf of Mexico
BT North American Atlantic
BT North Atlantic
BT Atlantic Ocean
SA oil and gas fields

Bucegi Mountains (1978)
Group in the E Transylvanian Alps in S central Romania.
BT Romania
BT Southern Europe
BT Europe
SA Sinaia Romania
SA Transylvanian Alps

Buchanan County
Valid through 1988. Search in combination with state term. After 1988, use specific county-state term.

Buchanan County Iowa (1989)
E Iowa. Before 1989, also search Buchanan County AND Iowa.
CO N421800N423900
   W0910700W0920500
BT Iowa
BT United States

Buchanan County Missouri (1989)
NW Missouri. Before 1989, also search Buchanan County AND Missouri.
CO N393300N394800
   W0943600W0950800
BT Missouri
BT United States

Buchanan County Virginia (1989)
SW Virginia. Before 1989, also search Buchanan County AND Virginia.
CO N370300N373300
   W0814500W0821800
BT Virginia
BT United States

Buchans Group (1989)
Lower and/or Middle Ordovician.
BT Ordovician
BT Paleozoic
SA Arenigian
SA Llanvirnian
SA Newfoundland Island

Buchans Newfoundland (1993)
Mining center. Central Newfoundland. Before 1993, also search Buchans or Buchans region AND Newfoundland.
UF Buchans region
BT Newfoundland Island

BT Newfoundland
BT Eastern Canada
BT Canada

Buchans region
No longer a valid term for GeoRef.
use Buchans Newfoundland

Bucharest
No longer a valid term for GeoRef. As of 1993 see Bucharest Romania.

Bucharest Romania (1993)
City and administrative district in S central Walachia. Before 1993 also search Bucharest AND Romania.
BT Walachia
BT Romania
BT Southern Europe
BT Europe

Buckingham County
Valid through 1988. Search in combination with state term. After 1988, use specific county-state term.

Buckingham County Virginia (1989)
Central Virginia. Before 1989, also search Buckingham County AND Virginia.
CO N372000N374800
   W0781500W0785100
BT Virginia
BT United States

Buckner Formation (1978)
Subsurface. Overlies Smackover Limestone; underlies Cotton Valley Group.
BT Upper Jurassic
BT Jurassic
BT Mesozoic
SA Arkansas
SA Louisiana
SA Mississippi
SA Texas

Bucks County
Valid through 1988. Search in combination with state term. After 1988, use specific county-state term.

Bucks County Pennsylvania (1989)
SE Pennsylvania. Before 1989, also search Bucks County AND Pennsylvania.
CO N400300N403800
   W0744000W0752800
BT Pennsylvania
BT United States

Buckskin Mountains (1985)
SW Arizona.
IN Index counties or regions as applicable.
SA Yuma County Arizona

Bucksport Formation (1989)
E Maine. Overlies Penobscot Formation; appears to be overlain and interbedded with unit referred to as Veazie Formation.
BT Paleozoic
SA Maine

Buda Limestone (1985)
In Washita Group. S Texas and N Mexico. Also search Buda Formation in combination with appropriate state or country name. This term has multiple hierarchies.
BT1 Comanchean
BT1 Cretaceous
BT1 Mesozoic

BT2 Upper Cretaceous
BT2 Cretaceous
BT2 Mesozoic
SA Mexico
SA Texas
SA Washita Group

Buda Marl (1989)
Also search Buda Formation AND Hungary.
BT Oligocene
BT Paleogene
BT Tertiary
BT Cenozoic
SA Hungary

Budapest
No longer a valid term for GeoRef. See Budapest Hungary.

Budapest Hungary (1993)
County including city of Budapest on both sides of the Danube River in N central Hungary. Contained within the county of Pest. Before 1993, also search Budapest or Budapest region and Hungary.
CO N472000N474500
   E0193000E0184500
UF Budapest region
BT Hungary
BT Central Europe
BT Europe

Budapest region
No longer a valid term for GeoRef.
use Budapest Hungary

Budennovsk Russian Federation
use Prikumsk Russian Federation

Buenos Aires
No longer a valid term for GeoRef. For the city or province in Argentina, see Buenos Aires City Argentina or Buenos Aires Argentina.

Buenos Aires Argentina (1993)
Province. Before 1993, also search Buenos Aires Province. Before 1978, also search Buenos Aires.
CO S410000S331500
   W0564000W0633000
UF Buenos Aires Province
BT Argentina
BT South America
NT Buenos Aires City Argentina
NT Sauce Grande River valley

Buenos Aires City Argentina (1993)
City on the Rio de la Plata coextensive with the independent Federal District or Distrito Federal of Argentina within the province, Buenos Aires Argentina. Before 1993, also search Buenos Aires or Federal District or Distrito Federal AND Argentina.
UF Federal District Argentina
BT Buenos Aires Argentina
BT Argentina
BT South America

Buenos Aires Province
No longer a valid term for GeoRef.
use Buenos Aires Argentina

buergerite (1978)
BT ring silicates
BT silicates
SA schorl
SA tourmaline

Buffalo River (1978)
E of the Tennessee River in W central Tennessee.

IN Index counties or regions as applicable.
SA Tennessee

buffering
use buffers

buffers (1985)
UF buffering
UF chemical buffers
SA alkalinity
SA geochemistry
SA pH
SA reagents

Bug region (1978)
Area including both the Southern Bug and the Western Bug rivers in Ukraine, Belarus, and E Poland.
IN Index Poland and/or Ukraine or Belarus as applicable.
BT Europe
SA Belarus
SA Poland
SA Ukraine

Bug River (1978)
The Southern Bug River in S Ukraine and the Western Bug River of Ukraine, Belarus and E Poland.
IN Index Poland and/or former Soviet republics as applicable.
BT Europe
SA Belarus
SA Poland
SA Ukraine

building stone (1978)
UF stone, building
BT construction materials
SA dimension stone
SA granite deposits
SA limestone deposits
SA marble deposits
SA sandstone deposits

buildings (1978)
SA foundations
SA raft foundations
SA structures

Bukantau (1978)
Mountains SE of Aral Sea in N Uzbekistan.
IN Index country as applicable.
SA Uzbekistan

Bukhara
No longer a valid term for GeoRef. As of 1993, see Bukhara Uzbekistan.

Bukhara Uzbekistan (1993)
Oblast and city near border of Turkmenia in S Uzbekistan. This term has multiple hierarchies.
BT1 Uzbekistan
BT1 Asia
BT2 Uzbekistan
BT2 Commonwealth of Independent States
NT Kermine Uzbekistan

Bukhara-Khiva (1978)
A region comprising neighboring former Khanates SE of the Aral Sea. This term has multiple hierarchies.
BT1 Uzbekistan
BT1 Asia
BT2 Uzbekistan
BT2 Commonwealth of Independent States

Bukk Mountains (1978)
Southern spur of the Carpathians in NE Hungary. This term has multiple hierarchies.
BT1 Hungary

BT1 Central Europe
BT1 Europe
BT2 Carpathians
BT2 Europe

Bukovina (1978)
Region in E central Europe.
IN Index Ukraine and/or Romania as applicable.
BT Europe
SA Romania
SA Ukraine

Bukusu (1978)
Carbonatite complex.
BT Uganda
BT East Africa
BT Africa

Bulawayan Group (1978)
Strong unconformities separate it from the overlying and underlying systems (Shamvaian and Sebakwian).
BT Archean
BT Precambrian
SA Zimbabwe

Bulgaria (1978)
CO N413000N441500
    E0284500E0221500
BT Southern Europe
BT Europe
NT Balkan Foreland
NT Balkan Mountains
NT Beli Izvor Bulgaria
NT Bulgarian Dobruja
NT Bulgarian Rhodope Mountains
NT Burgas Bulgaria
NT Central Bulgaria
NT Glavatsi
NT Isker River
NT Khaskovo Bulgaria
NT North Bulgarian Hills
NT Pernik coal basin
NT Pleven Bulgaria
NT Plovdiv Bulgaria
NT Rila Mountains
NT Ruse Bulgaria
NT Sakar Mountains
NT Sofia Bulgaria
NT Sredna Gora
NT Varna Bulgaria
NT Vidin Bulgaria
NT Vratsa Bulgaria
NT West Bulgarian Hills
SA Balkan Peninsula
SA Black Sea region
SA Danube Plain
SA Danube River
SA Danube Valley
SA Dobruja Basin
SA Istranca Mountains
SA Krajiste
SA Macedonia
SA Maritsa River
SA Moesia
SA Moesian Platform
SA Osogovo Mountains
SA Rhodope Mountains
SA Struma River valley
SA Vardar Zone

Bulgarian Dobruja (1981)
Autoposting of Bulgaria to this term began in 1989. As of 1990, Southern Europe and Europe are autoposted to this term. This term has multiple hierarchies.
CO N431100N440000
    E0284000E0273000
BT1 Bulgaria
BT1 Southern Europe
BT1 Europe
BT2 Dobruja Basin
BT2 Southern Europe
BT2 Europe

Bulgarian Rhodope Mountains (1981)
Autoposting of Bulgaria to this term began in 1989. Southern Europe and Europe are autoposted to this term as of 1990. This term has multiple hierarchies.
CO N411200N421800
    E0261500E0230500
BT1 Bulgaria
BT1 Southern Europe
BT1 Europe
BT2 Rhodope Mountains
BT2 Southern Europe
BT2 Europe
SA Madan Bulgaria
SA Rila Mountains

Bulimina (1978)
Genus. Autoposting of microfossils and Protista to this term began in 1990. This term has multiple hierarchies.
BT1 Buliminacea
BT1 Rotaliina
BT1 foraminifera
BT1 Protista
BT1 Invertebrata
BT2 Buliminacea
BT2 Rotaliina
BT2 foraminifera
BT2 Protista
BT2 microfossils

Buliminacea (1978)
Autoposting of microfossils and Protista to this term began in 1990. This term has multiple hierarchies.
BT1 Rotaliina
BT1 foraminifera
BT1 Protista
BT1 Invertebrata
BT2 Rotaliina
BT2 foraminifera
BT2 Protista
BT2 microfossils
NT Bolivinitidae
NT Bulimina
NT Gabonella
NT Uvigerinidae

bulk density (1989)
SA density
SA properties
SA soil mechanics
SA soils

bulk modulus (1978)
UF incompressibility modulus
UF modulus of incompressibility
UF modulus, bulk
UF volume elasticity
BT elastic constants
SA deformation
SA elasticity
SA shear modulus

Bull Lake Glaciation (1985)
Defined from Bull Lake Drift, E Wind River Range, Wyoming.
BT upper Quaternary
BT Quaternary
BT Cenozoic
SA Colorado
SA Idaho
SA Montana
SA Neoglacial
SA Pinedale Glaciation
SA Utah
SA Wyoming

Bulldog Shale (1989)
BT Cretaceous
BT Mesozoic
SA Australia
SA New South Wales Australia
SA Queensland Australia

SA South Australia

Bullfrog Member (1989)
Member of Crater Flat Tuff. Found at the Nevada Test Site in Nye County, S Nevada.
BT Miocene
BT Neogene
BT Tertiary
BT Cenozoic
SA Crater Flat Tuff
SA Nevada

Bulli Seam (1978)
Top of the Illawarra Coal Measures is marked by the Bulli Coal Seam. Coal seam in Sydney Basin in E New South Wales.
BT Permian
BT Paleozoic
SA Australia
SA Illawarra Coal Measures
SA New South Wales Australia

Bullion Creek Formation (1989)
In the upper Fort Union Formation.
BT Paleocene
BT Paleogene
BT Tertiary
BT Cenozoic
SA Fort Union Formation
SA North Dakota

Buncombe County
Valid through 1988. Search in combination with state term. After 1988, use specific county-state term.

Buncombe County North Carolina (1989)
W North Carolina. Before 1989, also search Buncombe County AND North Carolina.
CO N352500N355000
    W0821000W0825000
BT North Carolina
BT United States

Bundelkhand (1978)
Region S of the Jumna River.
IN Index states as applicable.
BT India
BT Indian Peninsula
BT Asia
SA Madhya Pradesh India
SA Uttar Pradesh India

Bundenbach (1978)
Area in Westerwald region in E central Germany.
BT Hesse Germany
BT Germany
BT Central Europe
BT Europe
SA Westerwald

Bundesrepublik Deutschland
Not a valid term for GeoRef. As of 1993, see Germany and also Mecklenburg-Western Pomerania Germany, Brandenburg Germany, Saxony-Anhalt Germany, Thuringia Germany, and Saxony Germany.

Bungonia Caves (1978)
In SE New South Wales.
BT New South Wales Australia
BT Australia
BT Australasia

Bunker Hill and Sullivan Mine
use Bunker Hill Mine

Bunker Hill Mine (1985)
Mine in N Idaho. Autoposting of broader terms to this term began in 1989.

CO N473000N474000
    W1160000W1162000
UF Bunker Hill and Sullivan Mine
BT Shoshone County Idaho
BT Idaho
BT United States
SA Coeur d'Alene mining district
SA lead ores
SA lead-zinc deposits
SA mines
SA zinc ores

Bunol
  No longer a valid term for GeoRef. As of 1993, see Bunol Spain.

Bunol Spain (1993)
  Town in E Spain. Before 1993, also search Bunol and Spain.
BT Valencia Spain
BT Valencia region
BT Spain
BT Iberian Peninsula
BT Southern Europe
BT Europe

Bunter (1978)
  Europe. Above Permian, below Muschelkalk.
UF Bunter Sandstone
UF Buntsandstein
BT Lower Triassic
BT Triassic
BT Mesozoic
NT Voltzia Sandstone

Bunter Sandstone
  use Bunter

Buntsandstein
  use Bunter

buoyancy (1985)
SA density
SA flotation
SA specific gravity

Burbank Sand (1978)
  Subsurface. Osage and Kay counties in N Oklahoma.
BT Carboniferous
BT Paleozoic
SA Kansas
SA Oklahoma

Burdekin Delta (1978)
  On Upstart Bay on Pacific Ocean in Great Barrier Reef area in E Queensland.
BT Queensland Australia
BT Australia
BT Australasia

Burdekin River (1978)
  Flows into Upstart Bay on Pacific Ocean in Great Barrier Reef area in E Queensland.
BT Queensland Australia
BT Australia
BT Australasia

Burdigalian (1978)
  Europe.
BT lower Miocene
BT Miocene
BT Neogene
BT Tertiary
BT Cenozoic

Burdwan
  No longer a valid term for GeoRef. See Burdwan India.

Burdwan India (1993)
  A city, district and division in central West Bengal, E India.
BT West Bengal India
BT India
BT Indian Peninsula
BT Asia

Bureau County
  Valid through 1988. Search in combination with state term. After 1988, use specific county-state term.

Bureau County Illinois (1989)
  N Illinois. Before 1989, also search Bureau County AND Illinois.
CO N411000N413500
    W0891000W0895200
BT Illinois
BT United States

Bureya
  No longer a valid term for GeoRef. As of 1993, see Bureya Russian Federation.

Bureya Russian Federation (1993)
  Town near the Manchurian border in SE Amur Oblast, SE Russian Federation. Before 1993, also search Bureya. This term has multiple hierarchies.
CO N480000N520000
    E1333000E1260000
BT1 Amur Russian Federation
BT1 Russian Federation
BT1 Commonwealth of Independent States
BT2 Amur Russian Federation
BT2 Asia
SA Amur region

Burgas
  No longer a valid term for GeoRef. See Burgas Bulgaria.

Burgas Bulgaria (1993)
  Province on the Black Sea in SE Bulgaria. Also a city.
BT Bulgaria
BT Southern Europe
BT Europe

Burgenland
  No longer a valid term for GeoRef.
  use Burgenland Austria

Burgenland Austria (1993)
  State on the Hungarian border in E Austria.
CO N464600N480800
    E0171000E0160000
UF Burgenland
BT Austria
BT Central Europe
BT Europe
SA Lake Neusiedler

Burgess Shale (1978)
BT Middle Cambrian
BT Cambrian
BT Paleozoic
SA British Columbia
SA Canada

Burgos
  No longer a valid term for GeoRef. As of 1993, see Burgos City Spain.

Burgos City Spain (1993)
  Refers to only the city in N central Spain. Before 1993, also search Burgos and Spain.
BT Burgos Spain
BT Old Castile Spain
BT Castile Spain
BT Spain
BT Iberian Peninsula
BT Southern Europe
BT Europe

Burgos Province
  No longer a valid term for GeoRef. As of 1993, see Burgos Spain.

Burgos Spain (1993)
  Refers only to the province in N central Spain. From 1981-1992, also search Burgos Province and Spain. Before 1981, also search Burgos and Spain. For the city, see Burgos City Spain.
CO N413000N431500
    W0023000W0042000
BT Old Castile Spain
BT Castile Spain
BT Spain
BT Iberian Peninsula
BT Southern Europe
BT Europe
NT Burgos City Spain

Burgundy (1978)
  Region and former province of E central and E France.
UF Bourgogne
BT France
BT Western Europe
BT Europe

Buriadia (1978)
BT Coniferales
BT gymnosperms
BT Spermatophyta
BT Plantae

burial (1993)
  Restricted to the natural process.
SA depth indicators
SA processes

burial depth
  Not a valid term for GeoRef. As of 1989, see depth, depth indicators, diagenesis. As of 1993, also see burial diagenesis.

burial diagenesis (1993)
  Increasing temperature, fluid pressure, and confining pressure characterize the environment where physical compaction results in expulsion of trapped, geochemically evolved pore fluids. Variable processes include fluid mixing, recycling, and geochemical evolution. Before 1993, also search burial and diagenesis.
BT diagenesis
SA depth indicators
SA late diagenesis

burial metamorphism (1978)
  Before 1978, also search metamorphism AND burial.
BT metamorphism
SA geosynclines
SA regional metamorphism

buried channels (1978)
  Autoposting of fluvial features to this term began in 1989.
BT fluvial features
SA buried features
SA buried valleys
SA channels
SA drift
SA geomorphology
SA glacial geology
SA paleogeography
SA paleorelief
SA planar bedding structures

buried features (1978)
  Before 1978, search buried.
UF features, buried
SA buried channels
SA buried valleys
SA paleokarst
SA paleorelief
SA Paleosols

buried valleys (1978)
SA buried channels
SA buried features
SA drift
SA geomorphology
SA glacial features
SA glacial geology
SA paleogeography
SA paleorelief
SA valleys

Burin Peninsula (1985)
S Newfoundland.
CO N470000N480000
    W0540000W0563000
BT Newfoundland Island
BT Newfoundland
BT Eastern Canada
BT Canada

Burke County
  Valid through 1988. Search in combination with state term. After 1988, use specific county-state term.

Burke County Georgia (1989)
  E Georgia. Before 1989, also search Burke County AND Georgia.
CO N325000N331900
    W0813300W0822000
BT Georgia
BT United States
NT Waynesboro Georgia

Burke County North Carolina (1989)
  W central North Carolina. Before 1989, also search Burke County AND North Carolina.
CO N353400N360000
    W0812200W0815900
BT North Carolina
BT United States

Burke County North Dakota (1989)
  NW North Dakota. Before 1989, also search Burke County AND North Dakota.
CO N483300N490000
    W1020200W1025700
BT North Dakota
BT United States

Burkina Faso (1985)
  Formerly Upper Volta. Former French protectorate which achieved independence in 1960. Before 1985, also search Upper Volta.
CO N092500N151000
    E0312000E0221000
UF Upper Volta
UF Voltaic Republic
BT West Africa
BT Africa
SA Sahel
SA Volta Basin

Burleson County
  Valid through 1988. Search in combination with state term. After 1988, use specific county-state term.

Burleson County Texas (1989)
  S central Texas. Before 1989, also search Burleson County AND Texas.
CO N301500N304500
    W0961500W0970000
BT Texas
BT United States

Burlington County
Valid through 1988. Search in combination with state term. After 1988, use specific county-state term.

Burlington County New Jersey (1989)
W central New Jersey. Before 1989, also search Burlington County AND New Jersey.
CO N393500N401200
W0742200W0750500
BT New Jersey
BT United States

Burlington Limestone (1985)
From Iowa to Kentucky.
BT Osagian
BT Lower Mississippian
BT Mississippian
BT Carboniferous
BT Paleozoic
SA Illinois
SA Iowa
SA Kentucky
SA Missouri

Burlington Peninsula (1981)
On N coast of Newfoundland.
CO N490000N501000
W0553000W0570000
BT Newfoundland Island
BT Newfoundland
BT Eastern Canada
BT Canada

Burma (1978)
Before 1993, also search Myanmar.
CO N090000N284500
E1014500E0920000
UF Myanmar
BT Far East
BT Asia
NT Kama Burma
NT Shan State Burma
SA Indochina
SA Naga Hills
SA Phuket Group
SA Siwalik System

Burnet County
Valid through 1988. Search in combination with state term. After 1988, use specific county-state term.

Burnet County Texas (1989)
Central Texas. Before 1989, also search Burnet County AND Texas.
CO N302500N310000
W0975200W0983000
BT Texas
BT United States

Burnie
No longer a valid term for GeoRef. See Burnie Australia.

Burnie Australia (1993)
Town on Bass Strait in N Tasmania.
BT Tasmania Australia
BT Australia
BT Australasia

Burono
use Vourinos

Burra Group (1989)
BT upper Proterozoic
BT Proterozoic
BT upper Precambrian
BT Precambrian
SA Australia
SA South Australia

Burro Canyon Formation (1985)
SW Colorado, NE Arizona, NW New Mexico and SE Utah.
BT Lower Cretaceous
BT Cretaceous
BT Mesozoic
SA Arizona
SA Colorado
SA New Mexico
SA Utah

Burro Mountain (1978)
IN Index counties or regions and mountains as applicable.
SA California

burrows (1978)
From 1978-1984, biogenic structures and sedimentary structures were autoposted to this term.
SA Arenicolites
SA biogenic structures
SA bioturbation
SA borings
SA ichnofossils
SA lebensspuren
SA sedimentary structures

bursts, rock
use rock bursts

Burundi (1978)
Formerly Urundi which was part of Belgian trust territory of Uranda-Urundi.
CO S043000S022000
E0305000E0285500
BT Central Africa
BT Africa
SA East African Rift
SA Lake Tanganyika
SA Nile River

Buryat
No longer a valid term for GeoRef. As of 1993, see Buryat Russian Federation.

Buryat Russian Federation (1993)
Former Buryat Autonomous Soviet Socialist Republic. S and SE of Lake Baikal. Before 1993, also search Buryat. Before 1978, also search Buryat-Mongol, Buryat-Mongolia, or Buryatia. This term has multiple hierarchies.
UF Buryat-Mongol
UF Buryat-Mongolia
UF Buryatia
BT1 Russian Federation
BT1 Commonwealth of Independent States
BT2 Asia
NT Mukhor-Tala Russian Federation
NT Vitim Plateau
SA Baikal Mountains
SA Baikal region
SA Baikal rift zone
SA Baikal-Amur Railroad region
SA Dzhida River
SA Khamar-Daban Range
SA Lake Baikal
SA Olekma-Vitim Highlands
SA Transbaikalia

Buryat-Mongol
use Buryat Russian Federation

Buryat-Mongolia
use Buryat Russian Federation

Buryatia
use Buryat Russian Federation

buserite (1993)
Problematic mineral, which some consider to be variant of todorokite.
BT oxides
SA manganese oxides
SA todorokite

Bushveld Complex (1978)
Igneous complex in E Transvaal. Also search Bushveld.
BT Transvaal South Africa
BT South Africa
BT Southern Africa
BT Africa
SA Merensky Reef

bustamite (1978)
BT chain silicates
BT silicates

butane (1985)
Autoposting of alkanes and aliphatic hydrocarbons to this term began in 1989.
BT alkanes
BT aliphatic hydrocarbons
BT hydrocarbons
BT organic materials

Butler County
Valid through 1988. Search in combination with state term. After 1988, use specific county-state term.

Butler County Alabama (1989)
S Alabama. Before 1989, also search Butler County AND Alabama.
CO N313000N315700
W0862400W0865400
BT Alabama
BT United States

Butler County Iowa (1989)
N central Iowa. Before 1989, also search Butler County AND Iowa.
CO N423400N425400
W0923300W0930200
BT Iowa
BT United States

Butler County Kansas (1989)
SE Kansas. Before 1989, also search Butler County AND Kansas.
CO N372800N380600
W0963100W0970900
BT Kansas
BT United States

Butler County Kentucky (1989)
W central Kentucky. Before 1989, also search Butler County AND Kentucky.
CO N370000N372400
W0862400W0865500
BT Kentucky
BT United States

Butler County Missouri (1989)
SE Missouri. Before 1989, also search Butler County AND Missouri.
CO N363000N365600
W0900800W0904100
BT Missouri
BT United States

Butler County Nebraska (1989)
E Nebraska. Before 1989, also search Butler County AND Nebraska.
CO N410300N412700
W0965400W0972300
BT Nebraska
BT United States

Butler County Ohio (1989)
Extreme SW Ohio. Before 1989, also search Butler County AND Ohio.
CO N391700N393600
W0842100W0845000
BT Ohio
BT United States
NT Oxford County Ohio
NT Oxford Ohio

Butler County Pennsylvania (1989)
W Pennsylvania. Before 1989, also search Butler County AND Pennsylvania.
CO N404000N411100
W0794200W0801100
BT Pennsylvania
BT United States

Butte
Valid through 1988. Search in combination with state term. After 1988, use specific city-state term.

Butte County
Valid through 1988. Search in combination with state term. After 1988, use specific county-state term.

Butte County California (1989)
N central California. Before 1989, also search Butte County AND California.
CO N392000N401000
W1210500W1220400
BT California
BT United States
NT Oroville California
NT Oroville Dam
SA Oroville earthquake 1975

Butte County Idaho (1989)
SE central Idaho. Before 1989, also search Butte County AND Idaho.
CO N431600N441400
W1124200W1134500
BT Idaho
BT United States
SA Craters of the Moon
SA Lemhi Range
SA Lost River Fault
SA Lost River Range

Butte County South Dakota (1989)
W South Dakota. Before 1989, also search Butte County AND South Dakota.
CO N443200N451300
W1025700W1040400
BT South Dakota
BT United States
SA Whitewood Creek

Butte District
As of 1993, no longer a valid term for GeoRef.
use Butte mining district

Butte mining district (1993)
Mining district in the Butte area in SW Montana. Before 1993, also search Butte District. Before 1978, also search Butte and Montana.
IN Index counties or regions as applicable.
UF Butte District
BT Montana
BT United States
SA mines

Butte Montana (1989)
City in SW Montana. Before 1989, search Butte and Montana.
CO N460000N460000
W1123100W1123100
BT Silver Bow County Montana
BT Montana
BT United States

Buzachi Peninsula (1985)

On E side of Caspian Sea. This term has multiple hierarchies.
CO N441000N454000
 E0531500E0511500
 BT1 Kazakhstan
 BT1 Central Asia
 BT1 Asia
 BT2 Kazakhstan
 BT2 Commonwealth of Independent States

Buzau
No longer a valid term for GeoRef. As of 1993 see Buzau Romania.

Buzau River (1978)
In NE Walachia.
 BT Walachia
 BT Romania
 BT Southern Europe
 BT Europe

Buzau Romania (1993)
City and county in NE Walachia. Before 1993 also search Buzau AND Romania.
 BT Walachia
 BT Romania
 BT Southern Europe
 BT Europe

Buzzards Bay (1978)
Inlet of Atlantic Ocean W of Cape Cod.
 IN Index counties as applicable.
 BT Massachusetts
 BT United States
 SA Barnstable County Massachusetts
 SA Plymouth County Massachusetts

Bydgoszcz
No longer a valid term for GeoRef. As of 1993 see Bydgoszcz Poland.

Bydgoszcz Poland (1993)
Province in N central Poland. Also a city. After 1993, use for the province. Before 1993 also search Bydgoszcz AND Poland.
 BT Poland
 BT Central Europe
 BT Europe
 NT Inowroclaw Poland
 NT Mogilno Poland
 SA Dobrzyn Poland
 SA Grudziadz Poland

Byelorussia
As of 1993, no longer a valid term for GeoRef.
 use Belarus

Byelorussian Massif (1978)
S part of Lithuanian-Byelorussian Upland. This term has multiple hierarchies.
 BT1 Belarus
 BT1 Europe
 BT2 Belarus
 BT2 Commonwealth of Independent States

Byram Formation (1989)
In Vicksburg Group. Mississippi, SW Alabama, NW Florida and Louisiana. Includes Bucatunna Formation and Glendon Limestone.
 BT middle Oligocene
 BT Oligocene
 BT Paleogene
 BT Tertiary
 BT Cenozoic
 SA Alabama
 SA Bucatunna Formation
 SA Florida
 SA Glendon Limestone
 SA Louisiana
 SA Mississippi
 SA Vicksburg Group

Byrd Station (1978)
In Marie Byrd Land SW of the Amundsen Sea on the Pacific Ocean side.
CO S795900S795900
 W1200100W1200100
 BT Antarctica

Bystrzyca (1978)
River in SW Poland.
 BT Wroclaw Poland
 BT Poland
 BT Central Europe
 BT Europe

Bytkov Ukraine
 use Bitkov Ukraine

Bytom
No longer a valid term for GeoRef. As of 1993 see Bytom Poland.

Bytom Poland (1993)
City in central Katowice, S Poland. Before 1993 also search Bytom AND Poland.
 BT Katowice Poland
 BT Poland
 BT Central Europe
 BT Europe

bytownite (1978)
 BT plagioclase
 BT feldspar group
 BT framework silicates
 BT silicates
 SA aluminosilicates

C

C
 use carbon

C-12 (1978)
Autoposting of broader terms began in 1989. This term has multiple hierarchies.
 UF carbon-12
 BT1 stable isotopes
 BT1 isotopes
 BT2 carbon
 SA C-13/C-12
 SA C-14/C-12

C-12/C-13
 use C-13/C-12

C-13 (1978)
Autoposting of broader terms began in 1989. This term has multiple hierarchies.
 UF carbon-13
 BT1 stable isotopes
 BT1 isotopes
 BT2 carbon
 SA C-13/C-12

C-13/C-12 (1978)
Autoposting of broader terms began in 1989. This term has multiple hierarchies.
 UF C-12/C-13
 BT1 stable isotopes
 BT1 isotopes
 BT2 carbon
 SA C-12
 SA C-13
 SA geologic thermometry
 SA isotope ratios
 SA paleoclimatology

C-14 (1978)
Isotope used in age determination. Autoposting of broader terms began in 1989. Before 1976, also search carbon-14. For absolute age use, also search radiocarbon dating before 1971. This term has multiple hierarchies.
 UF carbon-14
 UF radiocarbon dating
 BT1 radioactive isotopes
 BT1 isotopes
 BT2 carbon
 SA absolute age
 SA accelerator mass spectroscopy
 SA C-14/C-12

C-14/C-12 (1993)
As of 1993, not to be used for absolute age. This term has multiple hierarchies.
 BT1 radioactive isotopes
 BT1 isotopes
 BT2 stable isotopes
 BT2 isotopes
 BT3 carbon
 SA C-12
 SA C-14
 SA isotope ratios

Ca
 use calcium

Ca/Sr
 use Sr/Ca

Caballo Mountains (1989)
S New Mexico.
 IN Index state and county as applicable.
 SA New Mexico
 SA Sierra County New Mexico

Cabaniss Formation (1978)
In Cherokee Group, SE Kansas, W Missouri, central and NE Oklahoma.
 BT Desmoinesian
 BT Middle Pennsylvanian
 BT Pennsylvanian
 BT Carboniferous
 BT Paleozoic
 SA Cherokee Group
 SA Kansas
 SA Missouri
 SA Oklahoma

Cabarrus County
Valid through 1988. Search in combination with state term. After 1988, use specific county-state term.

Cabarrus County North Carolina (1989)
S central North Carolina. Before 1989, also search Cabarrus County AND North Carolina.
CO N351000N353100
 W0801500W0804800
 BT North Carolina
 BT United States

Cabinda
No longer a valid term for GeoRef. See Cabinda Angola.

Cabinda Angola (1993)
A district and an enclave N of the mouth of the Congo River. Before 1993, also search Cabinda or Kabinda AND Angola.
 UF Kabinda Angola
 BT Angola
 BT Central Africa
 BT Africa

Cabo Ortegal (1978)
Cape in N La Coruna Province, NW Spain.
 BT La Coruna Spain
 BT Galicia Spain
 BT Spain
 BT Iberian Peninsula
 BT Southern Europe
 BT Europe

Cabo Rojo
No longer a valid term for GeoRef. As of 1993 see Cabo Rojo Puerto Rico.

Cabo Rojo Puerto Rico (1993)
Town in SW Puerto Rico. Before 1993 also search Cabo Rojo AND Puerto Rico.
 BT Puerto Rico
 BT Greater Antilles
 BT Antilles
 BT West Indies
 BT Caribbean region

Cabrieres
No longer a valid term for GeoRef. See Cabrieres France.

Cabrieres France (1993)
Location in the Montagne Noire in S France.
 BT Herault France
 BT France
 BT Western Europe
 BT Europe

Caceres
No longer a valid term for GeoRef. As of 1993, see Caceres City Spain.

Caceres City Spain (1993)
Refers to only the city in W Spain. Before 1993, also search Caceres and Spain.
 BT Caceres Spain
 BT Extremadura Spain
 BT Spain
 BT Iberian Peninsula
 BT Southern Europe
 BT Europe

Caceres Province
No longer a valid term for GeoRef. As of 1993, see Caceres Spain.

Caceres Spain (1993)
Refers only to the province in W Spain on Portuguese border. From 1981-1992, also search Caceres Province and Spain. Before 1981, also search Caceres and Spain. For the city, see Caceres City Spain.
CO N390300N403000
 W0045800W0073000
 BT Extremadura Spain
 BT Spain
 BT Iberian Peninsula
 BT Southern Europe
 BT Europe
 NT Caceres City Spain

Cache County
Valid through 1988. Search in combination with state term. After 1988, use specific county-state term.

Cache County Utah (1989)
N Utah. Before 1989, also search Cache County AND Utah.
CO N412200N420000
 W1112200W1121000
 BT Utah
 BT United States

Cache Creek (1985)

Creek in Mackenzie Delta area, Yukon Territory and Northwest Territories. Also other creeks in S British Columbia, N California, NW Wyoming and SW Oklahoma.
IN Index states or Canadian provinces as applicable.
SA British Columbia
SA California
SA Northwest Territories
SA Oklahoma
SA Wyoming
SA Yukon Territory

Cache Creek Group (1989)
S British Columbia and San Juan Island, NW Washington.
IN Index ages as applicable.
SA British Columbia
SA Pennsylvanian
SA Permian
SA San Juan Islands
SA Triassic
SA Washington

Cache La Poudre River (1978)
N central Colorado.
BT Colorado
BT United States

Cache Valley (1978)
In several locations.
IN Index states as applicable.
SA Idaho
SA Illinois
SA Utah

cacoxenite (1978)
BT phosphates

Caddo County
Valid through 1988. Search in combination with state term. After 1988, use specific county-state term.

Caddo County Oklahoma (1989)
West central Oklahoma. Before 1989, also search Caddo County AND Oklahoma.
CO N345000N353500
W0980600W0983800
BT Oklahoma
BT United States
SA Meers Fault

Cadeby Formation (1989)
E England.
BT Upper Permian
BT Permian
BT Paleozoic
SA England
SA Great Britain

Cadiz
No longer a valid term for GeoRef. As of 1993, see Cadiz City Spain.

Cadiz City Spain (1993)
Refers only to the city in W Cadiz Province, SW Spain. Before 1993, also search Cadiz and Spain.
BT Cadiz Spain
BT Andalusia Spain
BT Spain
BT Iberian Peninsula
BT Southern Europe
BT Europe

Cadiz Province
No longer a valid term for GeoRef. As of 1993, see Cadiz Spain.

Cadiz Spain (1993)
Refers only to the province. In SW Spain, bordering the Atlantic Ocean. From 1981-1992, also search Cadiz Province and Spain. Before 1981, also search Cadiz

and Spain. For the city, see Cadiz City Spain.
CO N360000N370500
W0050500W0063000
BT Andalusia Spain
BT Spain
BT Iberian Peninsula
BT Southern Europe
BT Europe
NT Cadiz City Spain
NT Vejer de la Frontera Spain
SA Ceuta
SA Serrania de Ronda

cadmium (1978)
Chemical element. As of 1982, use cadmium ores for cadmium as a commodity. Autoposting of metals to this term began in 1989.
UF Cd
BT metals
SA cadmium ores
SA chalcophile elements
SA heavy metals

cadmium blende
use greenockite

cadmium ocher
use greenockite

cadmium ores (1982)
Before 1982, also search (cadmium OR Cd) AND (deposit OR deposits OR ore OR ores OR economic) in the basic index. Autoposting of metal ores to this term began in 1985.
BT metal ores
SA cadmium

Cadomian Orogeny (1978)
Before 1978, search Cadomian AND orogeny.
IN Index ages as applicable.
SA Cambrian
SA orogeny
SA Precambrian
SA tectonics

Cadomin Formation (1989)
SW Alberta and E British Columbia.
BT Lower Cretaceous
BT Cretaceous
BT Mesozoic
SA Alberta
SA British Columbia

Cady Mountains (1985)
SE California. Autoposting of broader terms to this term began in 1989.
BT San Bernardino County California
BT California
BT United States
SA Mojave Desert

Caen
No longer a valid term for GeoRef. See Caen France.

Caen France (1993)
City in Normandy, N Calvados, NE France.
BT Calvados France
BT France
BT Western Europe
BT Europe

Caernarvon County
Not a valid term for GeoRef as of 1993. See Caernarvonshire Wales.

Caernarvonshire Wales (1993)
Former county in NW Wales. As of April 1974, part of Gwynedd. Before 1993, also search

Caernarvon County or Caernarvonshire.
CO N524500N532000
W0034500W0044500
BT Wales
BT Great Britain
BT United Kingdom
BT Western Europe
BT Europe
NT Lleyn Peninsula
NT Snowdonia
SA Tremadoc Bay

caesium
use cesium

Cahill Formation (1989)
BT Proterozoic
BT upper Precambrian
BT Precambrian
SA Australia
SA Northern Territory Australia

CAI
Not a valid term for GeoRef. Acronym for many concepts including conodont alteration index; color alteration index; computer-aided instruction; computer-assisted instruction; calcium-aluminum inclusions. Use appropriate phrase as applicable.

Cainozoic
use Cenozoic

Cairngorm Mountains (1978)
Range of the Grampian Hills in highlands of NE central Scotland.
BT Scotland
BT Great Britain
BT United Kingdom
BT Western Europe
BT Europe
SA Grampian Highlands
SA Grampian region Scotland
SA Highland region Scotland

Cairo
No longer a valid term for GeoRef. See Cairo Egypt.

Cairo Egypt (1993)
City on the Nile River, coextensive with the governorate in NE Egypt.
BT Egypt
BT North Africa
BT Africa

Caithness
No longer a valid term for GeoRef as of 1993. See Caithness Scotland.

Caithness Scotland (1993)
Former county in NE Scotland. Before 1993, also search Caithness.
CO N581000N584000
W0030000W0035000
BT Highland region Scotland
BT Scotland
BT Great Britain
BT United Kingdom
BT Western Europe
BT Europe

Cajon Pass (1989)
Railroad and highway pass. S California. Extends from E San Gabriel Mountains to NW San Bernardino Mountains, connecting the Mojave Desert with Los Angeles Basin.
BT San Bernardino County California
BT California
BT United States
SA Los Angeles Basin
SA Mojave Desert

Calabria
No longer a valid term for GeoRef. See Calabria Italy.

Calabria Italy (1993)
An autonomous region in S Italy. The toe of the Italian boot.
CO N380000N401000
E0170000E0154500
BT Italy
BT Southern Europe
BT Europe
NT Catanzaro Italy
NT Cosenza Italy
NT Reggio di Calabria Italy
NT Sila Massif

Calabrian (1978)
Europe. Above Astian, below middle Pleistocene.
BT lower Pleistocene
BT Pleistocene
BT Quaternary
BT Cenozoic
SA Villafranchian

Calamites (1978)
Genus.
BT Equisetales
BT Sphenopsida
BT pteridophytes
BT Plantae

Calatayud-Teruel Basin (1978)
NE Spain.
IN Index provinces as applicable.
BT Spain
BT Iberian Peninsula
BT Southern Europe
BT Europe
SA Saragossa Spain
SA Teruel Spain

Calaveras County
Valid through 1988. Search in combination with state term. After 1988, use specific county-state term.

Calaveras County California (1989)
Central California. Before 1989, also search Calaveras County AND California.
CO N375000N383500
W1200100W1210000
BT California
BT United States

Calaveras Fault (1978)
Central California.
BT California
BT United States

Calaveras Formation (1985)
N California.
BT upper Paleozoic
BT Paleozoic
SA California
SA Permian

calaverite (1978)
Autoposting of sulfides to this term ended in 1989.
BT tellurides
SA gold ores

calc-alkalic
A valid term through 1977. After 1977, use calc-alkalic composition with igneous rocks.

calc-alkalic composition (1978)
SA alkalic composition
SA composition
SA I-type granites
SA igneous rocks

calc-schist (1978)
BT schists

BT metamorphic rocks
calc-silicate composition (1985)
SA composition
SA metamorphic rocks

calc-sinter
use travertine

calc-tufa
use tufa

calcarenite (1978)
BT limestone
BT carbonate rocks
BT sedimentary rocks
SA biocalcarenite

calcareous
Valid index term through 1977. After 1977, use calcareous composition.

calcareous algae (1978)
Autoposting of thallophytes and microfossils began in 1990. This term has multiple hierarchies.
BT1 algae
BT1 thallophytes
BT1 Plantae
BT2 algae
BT2 microfossils
SA stromatolites

calcareous clay
use marl

calcareous composition (1978)
Before 1978, also search calcareous AND specific sediment type or sedimentary rock.
SA bioherms
SA composition
SA ooze
SA sedimentary rocks
SA sediments

calcareous nannofossils
use nannofossils

calcareous ooze
Not a valid term for GeoRef. Use calcareous composition and ooze.

calcareous sinter
use travertine

Calcareous soils (1978)
Before 1978, also search calcareous.
BT soils
SA soil group

calcareous tufa
use tufa

Calcasieu Parish
Valid through 1988. Search in combination with state term. After 1988, use specific county-state term.

Calcasieu Parish Louisiana (1989)
On the Texas border in SW Louisiana. Before 1989, also search Calcasieu Parish AND Louisiana.
CO N300200N302900 W0925300W0934800
BT Louisiana
BT United States
SA Calcasieu River

Calcasieu River (1993)
SW Louisiana.
IN Index parishes as applicable.
BT Louisiana
BT United States
SA Calcasieu Parish Louisiana
SA Cameron Parish Louisiana
SA Jefferson Parish Louisiana
SA Rapides Parish Louisiana

calcic composition (1978)

Before 1978, also search igneous rocks AND calcic.
SA composition
SA igneous rocks

calcicrete
use calcrete

calcification (1978)
Term refers to fossils or soils.
SA diagenesis
SA fossilization
SA fossils
SA processes
SA soils

calcilutite (1978)
BT limestone
BT carbonate rocks
BT sedimentary rocks
SA algal mounds

calciphyre (1978)
BT marbles
BT metamorphic rocks

Calcisphaerulidae (1978)
Family.
UF Calcispherulidae
BT problematic fossils

Calcispherulidae
use Calcisphaerulidae

Calcispongea (1978)
Autoposting of this term began in 1978.
BT Porifera
BT Invertebrata
NT Sphinctozoa

calcite (1978)
As of 1981, use calcite deposits for calcite as a commodity.
BT carbonates
SA alabaster
SA aragonite
SA calcite deposits
SA calcitization
SA calcium carbonate
SA carbonate rocks
SA chalk
SA dolomite
SA Iceland spar
SA light minerals
SA limestone
SA magnesian calcite
SA travertine
SA tufa

calcite deposits (1981)
Before 1981, search calcite AND deposits.
IN Commodity. See List C.
SA calcite
SA marble deposits

calcitization (1978)
SA calcite
SA dedolomitization
SA diagenesis
SA dolomitization
SA processes

calcium (1978)
Autoposting of alkaline earth metals and metals to this term began in 1989.
UF Ca
BT alkaline earth metals
BT metals
NT Sr/Ca
SA calcium ion
SA lithophile elements

calcium carbonate (1978)
SA aragonite
SA calcite
SA carbonates
SA geochemistry
SA lime

SA limestone
calcium chloride (1985)
SA chlorides
SA geochemistry
SA halides

calcium ion (1985)
SA calcium
SA geochemistry
SA ions

calcium sulfate (1985)
SA anhydrite
SA geochemistry
SA gypsum
SA sulfates

calcium-aluminum inclusions (1989)
Found in meteorites. Sometimes abbreviated as CAI.
BT inclusions
SA meteorites

calcrete (1978)
UF calcicrete
BT carbonate rocks
BT sedimentary rocks
SA caliche
SA conglomerate
SA duricrust
SA soils
SA weathering crust

CALCRUST (1989)
Acronym. Seismic reflection program.
UF California Consortium for Crustal Studies
SA California
SA continental crust
SA geophysical methods
SA reflection methods
SA seismic methods
SA seismic surveys

calculation
A valid term through 1978. After 1978, see mathematical methods or mathematical models.

calculators
Use hand calculators if applicable.

Calcutta
No longer a valid term for GeoRef. See Calcutta India.

Calcutta India (1993)
City on the Hooghly River near Bay of Bengal in S West Bengal, E India.
BT West Bengal India
BT India
BT Indian Peninsula
BT Asia

calderas (1978)
BT volcanic features
SA cauldrons
SA crater lakes
SA craters
SA geomorphology
SA volcanism
SA volcanoes
SA volcanology

caldron subsidence
No longer a valid GeoRef index term. Before 1972, was used on level 2 in subfiles G and N. The variant form cauldron subsidence was also used. See cauldrons; calderas; subsidence.

Caldwell County
Valid through 1988. Search in combination with state term. After 1988, use specific county-state term.

Caldwell County Kentucky (1989)
W Kentucky. Before 1989, also search Caldwell County AND Kentucky.
CO N365600N365600 W0874000W0880700
BT Kentucky
BT United States

Caldwell County Missouri (1989)
NW Missouri. Before 1989, also search Caldwell County AND Missouri.
CO N393200N394800 W0934800W0941200
BT Missouri
BT United States

Caldwell County North Carolina (1989)
W central North Carolina. Before 1989, also search Caldwell County AND North Carolina.
CO N354700N360800 W0812000W0814800
BT North Carolina
BT United States

Caldwell County Texas (1989)
S central Texas. Before 1989, also search Caldwell County AND Texas.
CO N293800N300500 W0971800W0975400
BT Texas
BT United States

Caldwell Parish Louisiana (1989)
NE central Louisiana. Before 1989, search Caldwell Parish AND Louisiana.
BT Louisiana
BT United States
NT Columbia Louisiana

Caledonia County
Valid through 1988. Search in combination with state term. After 1988, use specific county-state term.

Caledonia County Vermont (1989)
NE Vermont. Before 1989, also search Caledonia County AND Vermont.
CO N441000N444500 W0715000W0722500
BT Vermont
BT United States

Caledonian
A valid term through 1977. After 1977, use Caledonian Orogeny.

Caledonian Geosyncline
As of 1981, no longer a valid term for GeoRef. Use Caledonian Orogeny. Term was introduced in 1978.

Caledonian Orogeny (1978)
Name commonly used for the early Paleozoic deformation in Europe which created an orogenic belt extending from Ireland and Scotland northwestward through Scandinavia. Before 1978, also search Caledonian. Before 1981, also search Caledonian Geosyncline.
BT Paleozoic
SA Caledonides
SA orogeny

Caledonides (1978)

Orogenic belt extending from Ireland and Scotland northwestward through Scandinavia, formed by the Caledonian Orogeny.
SA Caledonian Orogeny
SA Europe
SA Great Britain
SA Ireland
SA Koli Nappe
SA Scandinavia

Calera Limestone (1989)
In Franciscan Formation?
BT Cretaceous
BT Mesozoic
SA California
SA Franciscan Complex
SA Oregon

Calgary
No longer a valid term for GeoRef. See Calgary Alberta.

Calgary Alberta (1993)
City in SW Alberta.
BT Alberta
BT Western Canada
BT Canada

Calhoun County
Valid through 1988. Search in combination with state term. After 1988, use specific county-state term.

Calhoun County Alabama (1989)
E Alabama. Before 1989, also search Calhoun County AND Alabama.
CO N333200N335700
    W0853200W0861000
BT Alabama
BT United States

Calhoun County Arkansas (1989)
S Arkansas. Before 1989, also search Calhoun County AND Arkansas.
CO N331800N334700
    W0921900W0924700
BT Arkansas
BT United States

Calhoun County Florida (1989)
NW Florida. Before 1989, also search Calhoun County AND Florida.
CO N301100N304500
    W0850000W0852200
BT Florida
BT United States

Calhoun County Georgia (1989)
SW Georgia. Before 1989, also search Calhoun County AND Georgia.
CO N312500N313900
    W0842400W0844700
BT Georgia
BT United States

Calhoun County Illinois (1989)
W Illinois. Before 1989, also search Calhoun County AND Illinois.
CO N385300N392300
    W0902900W0905800
BT Illinois
BT United States

Calhoun County Iowa (1989)
Central Iowa. Before 1989, also search Calhoun County AND Iowa.
CO N421300N423300
    W0942300W0945500
BT Iowa
BT United States

Calhoun County Michigan (1989)
S Michigan. Before 1989, also search Calhoun County AND Michigan.
CO N420400N422400
    W0844200W0851700
BT Michigan Lower Peninsula
BT Michigan
BT United States

Calhoun County Mississippi (1989)
N central Mississippi. Before 1989, also search Calhoun County AND Mississippi.
CO N334200N341000
    W0890800W0893000
BT Mississippi
BT United States

Calhoun County South Carolina (1989)
Central South Carolina. Before 1989, also search Calhoun County AND South Carolina.
CO N332800N335300
    W0803000W0810300
BT South Carolina
BT United States

Calhoun County Texas (1989)
S Texas. Before 1989, also search Calhoun County AND Texas.
CO N280700N284500
    W0954300W0965500
BT Texas
BT United States
SA Matagorda Bay

Calhoun County West Virginia (1989)
W central West Virginia. Before 1989, also search Calhoun County AND West Virginia.
CO N383700N390300
    W0805900W0811700
BT West Virginia
BT United States

calibration (1978)
Used as a general term.
SA accuracy
SA instruments

caliche (1978)
BT carbonate rocks
BT sedimentary rocks
SA calcrete
SA duricrust
SA soils
SA weathering
SA weathering crust

Calico Mountains (1978)
Small range in Mojave Desert.
IN Index counties or regions and mountains as applicable.
SA California
SA Mojave Desert

Caliente Range (1978)
IN Index counties or regions as applicable.
SA California

California (1978)
Autoposting of this term began in 1978.
CO N323000N420000
    W1141500W1243000
BT United States
NT Alameda County California
NT Alpine County California
NT Amador County California
NT American River
NT Anza Desert State Park
NT Banning Fault
NT Borrego Mountain
NT Butte County California
NT Calaveras County California
NT Calaveras Fault
NT Central California
NT Channel Islands
NT Colorado Desert
NT Colusa County California
NT Contra Costa County California
NT Coyote Creek Fault
NT Cuyama Basin
NT Death Valley
NT Death Valley Fault
NT Del Norte County California
NT El Dorado County California
NT Elsinore Fault
NT Fresno County California
NT Gabilan Range
NT Garlock Fault
NT Golden Trout Wilderness
NT Greenville Fault
NT Hilton Creek Fault
NT Hosgri Fault
NT Humboldt County California
NT Imperial County California
NT Inyo County California
NT Kern County California
NT Kettleman Hills
NT Kings County California
NT Lake County California
NT Lassen County California
NT Lassen Volcanic National Park
NT Los Angeles Basin
NT Los Angeles County California
NT Madera County California
NT Marin County California
NT Mariposa County California
NT Melones Fault
NT Mendocino County California
NT Merced County California
NT Modoc County California
NT Modoc Plateau
NT Mojave Desert
NT Mono County California
NT Monterey Bay
NT Monterey County California
NT Mount Lyell
NT Napa County California
NT Nevada County California
NT Newport-Inglewood Fault
NT Nopah Range
NT Northern California
NT Orange County California
NT Pajaro Valley
NT Placer County California
NT Plumas County California
NT Rattlesnake Creek Terrane
NT Riverside County California
NT Sacramento County California
NT Salinian Block
NT Salton Sea
NT Salton Trough
NT San Andreas Fault
NT San Benito County California
NT San Bernardino County California
NT San Diego County California
NT San Emigdio Mountains
NT San Francisco Bay
NT San Francisco Bay region
NT San Francisco County California
NT San Gabriel Fault
NT San Gabriel Mountains
NT San Gregorio Fault
NT San Jacinto Mountains
NT San Joaquin County California
NT San Joaquin Valley
NT San Luis Obispo County California
NT San Mateo County California
NT Santa Ana Mountains
NT Santa Barbara Channel
NT Santa Barbara County California
NT Santa Clara County California
NT Santa Clara Valley
NT Santa Cruz County California
NT Santa Maria Basin
NT Santa Monica Mountains
NT Santa Rosa Mountains
NT Santa Ynez Mountains
NT Shasta County California
NT Sierra County California
NT Sierra Nevada Batholith
NT Siskiyou County California
NT Solano County California
NT Sonoma County California
NT Southern California
NT Southern California Batholith
NT Stanislaus County California
NT Sur fault zone
NT Sutter County California
NT Sylmar Fault
NT Tehachapi Mountains
NT Tehama County California
NT Temblor Range
NT The Geysers
NT Transverse Ranges
NT Trinity Complex
NT Trinity County California
NT Tulare County California
NT Tuolumne County California
NT Ventura County California
NT Yolla Bolly Terrane
NT Yolo County California
NT Yosemite National Park
NT Yuba County California
SA Agua Blanca Fault
SA Alisitos Formation
SA Amargosa Desert
SA Antelope Valley
SA Aztec Sandstone
SA Balaklala Rhyolite
SA Barstow Formation
SA Basin and Range Province
SA Bear Mountain
SA Bear Valley
SA Bird Spring Formation
SA Bishop Tuff
SA Black Mountain
SA Blue Mountain
SA Bonanza King Formation
SA Borax Lake
SA Borrego Mountain earthquake 1968
SA Bright Angel Shale
SA Burro Mountain
SA Cache Creek
SA Calaveras Formation
SA CALCRUST
SA Calera Limestone
SA Calico Mountains
SA Caliente Range
SA California Current
SA Capistrano Formation
SA Carrara Formation
SA Cascade Range
SA Castaic Formation
SA Catalina Schist
SA Central Valley
SA Cerro Prieto Fault
SA Chainman Shale
SA Chalfant Valley earthquake 1986
SA Chatsworth Formation
SA Chocolate Mountains
SA Clear Lake
SA Coalinga earthquake 1983
SA Coast Range Ophiolite
SA Coast Ranges
SA Colorado River
SA Colorado River basin
SA Condrey Mountain Schist
SA Copley Greenstone
SA Copper Mountain
SA Coyote Lake earthquake 1979
SA Diablo Range

Before then, variants of the term may occur in GeoRef.

SA Diamond Peak Formation
SA Dunderberg Shale
SA Eagle Lake
SA Eel River
SA El Centro earthquake 1940
SA El Paso Mountains
SA Ely Springs Dolomite
SA Eureka Quartzite
SA Feather River
SA Forbes Formation
SA Franciscan Complex
SA Furnace Creek
SA Galice Formation
SA Great Basin
SA Great Valley Sequence
SA Hat Creek Basalt
SA Hayfork Terrane
SA Hayward Fault
SA Hector Formation
SA Hidden Valley Dolomite
SA Holz Shale
SA Homestake Mine
SA Hornbrook Formation
SA Imperial Formation
SA Imperial Valley
SA Imperial Valley earthquake 1979
SA Johnnie Formation
SA Josephine Ophiolite
SA Josephine Peridotite
SA Kaibab Formation
SA Kettleman Hills earthquake 1985
SA Kingston Peak Formation
SA Klamath Mountains
SA Ladd Formation
SA Lake Lahontan
SA Lake Tahoe
SA Llajas Formation
SA Loma Prieta earthquake 1989
SA Long Valley
SA Lost Burro Formation
SA Mammoth Lakes earthquakes
SA Marca Shale Member
SA McCloud Limestone
SA Medicine Lake
SA Mexicali Valley
SA Mill Creek
SA Mission Valley Formation
SA Modelo Formation
SA Moenkopi Formation
SA Mohave Mountains
SA Mono Basin
SA Mono Lake
SA Monte Cristo Limestone
SA Monterey Formation
SA Moreno Formation
SA Morgan Hill earthquake 1984
SA Noonday Dolomite
SA Nopah Formation
SA North Palm Springs earthquake 1986
SA Orocopia Schist
SA Oroville earthquake 1975
SA Owens Valley
SA Pahrump Series
SA Parkfield earthquake 1966
SA Parkfield earthquakes
SA Peace Valley Beds
SA Peach Springs Tuff
SA Pelona Schist
SA Peninsular Ranges
SA Peninsular Ranges Batholith
SA Pico Formation
SA Pioche Shale
SA Pisgah Crater
SA Pogonip Group
SA Point Loma Formation
SA Poway Conglomerate
SA Puente Formation
SA Punchbowl Formation
SA Purisima Formation
SA Rand Mountains
SA Ridge Basin
SA Ridge Route Formation
SA Rincon Formation
SA Rose Canyon Formation
SA Sacramento Basin
SA Sacramento River
SA Sacramento Valley
SA Salinas Valley
SA San Bernardino Mountains
SA San Diego Formation
SA San Fernando earthquake 1971
SA San Fernando Valley
SA San Francisco earthquake 1906
SA San Francisco Peninsula
SA San Jacinto Fault
SA San Joaquin Basin
SA San Joaquin River
SA San Lorenzo Formation
SA San Nicolas Basin
SA San Onofre Breccia
SA San Pedro Basin
SA Santa Ana River
SA Santa Barbara Basin
SA Santa Barbara Formation
SA Santa Clara Formation
SA Santa Cruz Mountains
SA Santa Lucia Range
SA Santa Margarita Formation
SA Santa Maria Formation
SA Santa Susana Formation
SA Saugus Formation
SA Scripps Institution of Oceanography
SA Searles Lake
SA Sespe Formation
SA Shadow Mountains
SA Shoo Fly Complex
SA Sierra Madre
SA Sierra Nevada
SA Sisquoc Formation
SA Smartville Complex
SA Sonoran Desert
SA Stirling Quartzite
SA Supai Formation
SA Superstition Hills earthquake 1987
SA Table Mountain
SA Tapeats Sandstone
SA Tejon Formation
SA Temblor Formation
SA Trinity River
SA Trinity River basin
SA Truckee River
SA Tulare Formation
SA Twin Lakes
SA Umpqua Formation
SA University of California
SA Vaqueros Formation
SA Ventura Basin
SA White Mountain
SA White Mountains
SA Whittier Fault
SA Whittier Narrows earthquake 1987
SA Wildcat Group
SA Williams Formation
SA Wood Canyon Formation
SA Wyman Formation
SA Zabriskie Quartzite

California bearing ratio (1985)
SA bearing capacity
SA penetration tests
SA soil mechanics

California City
No longer a valid term for GeoRef. See California City Colombia.

California City Colombia (1993)
Before 1993, also search California City ANd Colombia. Before 1981, also search California AND Colombia.
BT Santander Colombia
BT Colombia
BT South America

California Consortium for Crustal Studies
use CALCRUST

California Continental Borderland
Not a valid term for GeoRef. As of 1989, use California in conjunction with continental borderland.

California Current (1989)
Branch of the Aleutian Current in NE Pacific. Flows S along coast of W North America from N48° to N23°, where it joins the North Equatorial Current.
BT Northeast Pacific
BT Pacific Ocean
SA California
SA East Pacific
SA Mexico
SA North Pacific
SA Pacific Coast
SA United States

californium (1978)
Autoposting of actinides and metals to this term began in 1989.
UF Cf
BT actinides
BT metals

Caliman (Mountains)
use Calimani Mountains

Calimani Mountains (1978)
Range in the Eastern Carpathians in E Transylvania. Also search Calimani. This term has multiple hierarchies.
UF Caliman (Mountains)
UF Calimanului Mountains
BT1 Transylvania
BT1 Romania
BT1 Southern Europe
BT1 Europe
DT2 Eastern Carpathians
BT2 Carpathians
BT2 Europe

Calimanului Mountains
use Calimani Mountains

caliper logging (1978)
Before 1978, also search well-logging AND caliper.
UF logging, caliper
BT well-logging

Calistoga
No longer a valid term for GeoRef. Valid through 1988. Search in combination with state term. After 1988, use specific city-state term.

Calistoga California (1989)
City NE of San Francisco Bay in W central California. Before 1989, search Calistoga AND California.
CO N383700N383700 W1223500W1223500
BT Napa County California
BT California
BT United States

Callander Bay (1978)
W edge of Lake Nipissing near city of North Bay E of Georgian Bay.
BT Ontario
BT Eastern Canada
BT Canada

Callianassa (1978)
Genus.
BT Malacostraca
BT Crustacea
BT Mandibulata
BT Arthropoda
BT Invertebrata

Callipteris (1978)
Genus which paleobotanically determines Permian.
BT Pteridospermae
BT gymnosperms
BT Spermatophyta
BT Plantae
SA Filicopsida
SA pteridophytes

Callisto Satellite (1985)
One of the satellites of Jupiter. Before 1985, also search Callisto AND Jupiter.
SA Galilean satellites
SA icy satellites
SA Jupiter
SA satellites
SA Voyager Program

Callovian (1978)
Europe. Above Bathonian, below Oxfordian.
BT Middle Jurassic
BT Jurassic
BT Mesozoic
SA Dogger

calorimetry (1985)
SA chemical analysis
SA thermochemical properties

Caloris Basin (1989)
Impact feature on Mercury Planet.
BT Mercury Planet

Calpionella (1978)
Genus. This term has multiple hierarchies.
BT1 Tintinnidae
BT1 Protista
BT1 Invertebrata
BT2 Tintinnidae
BT2 Protista
BT2 microfossils

Calpionellidae (1978)
This term has multiple hierarchies.
BT1 Tintinnidae
BT1 Protista
BT1 Invertebrata
BT2 Tintinnidae
BT2 Protista
BT2 microfossils

Calpionellites (1978)
Genus. This term has multiple hierarchies.
BT1 Tintinnidae
BT1 Protista
BT1 Invertebrata
BT2 Tintinnidae
BT2 Protista
BT2 microfossils

Caltanissetta
No longer a valid term for GeoRef. See Caltanissetta Italy.

Caltanissetta Italy (1993)
Province and city in central Sicily.
BT Sicily Italy
BT Italy
BT Southern Europe
BT Europe

Calvados
No longer a valid term for GeoRef. See Calvados France.

Calvados France (1993)
CO N485000N493500 E0004000W0011500
BT France
BT Western Europe
BT Europe
NT Caen France
NT May-sur-Orne France

SA Normandy
SA Seine Estuary

Calvert Bluff Formation (1989)
In Wilcox Group.
BT lower Eocene
BT Eocene
BT Paleogene
BT Tertiary
BT Cenozoic
SA Texas
SA Wilcox Group

Calvert County
Valid through 1988. Search in combination with state term. After 1988, use specific county-state term.

Calvert County Maryland (1989)
S Maryland. Before 1989, also search Calvert County AND Maryland.
CO N381800N384700
   W0762230W0764200
BT Maryland
BT United States

Calvert Formation (1978)
In Chesapeake Group. Delaware, E Maryland, and Virginia.
BT middle Miocene
BT Miocene
BT Neogene
BT Tertiary
BT Cenozoic
SA Chesapeake Group
SA Delaware
SA Maryland
SA Virginia

Calyptogena (1993)
Genus.
BT Venerida
BT Bivalvia
BT Mollusca
BT Invertebrata

Camaguey
No longer a valid term for GeoRef. See Camaguey Cuba.

Camaguey Cuba (1993)
Province including the city in E central Cuba.
BT Cuba
BT Greater Antilles
BT Antilles
BT West Indies
BT Caribbean region

Camarines Norte
No longer a valid term for GeoRef. As of 1993 see Camarines Norte Philippine Islands.

Camarines Norte Philippine Islands (1993)
Province in SE Luzon. Before 1993 also search Camarines Norte AND Philippine Islands.
BT Luzon
BT Philippine Islands
BT Far East
BT Asia

Camaroidea (1978)
IN Index Hemichordata or Invertebrata as applicable.
BT Graptolithina
SA Hemichordata
SA Invertebrata

Camas County Idaho (1993)
S central Idaho. Before 1989, also search Camas County AND Idaho.
CO N431200N435200
   W1142100W1150600
BT Idaho

BT United States
SA Sawtooth Range

Cambay
No longer a valid term for GeoRef. Former Indian state at N end of Gulf of Cambay in central Gujarat, W India. See Cambay India.

Cambay Basin (1978)
Gulf of Cambay area.
SA Gujarat India
SA Gulf of Cambay

Cambay India (1993)
City at N end of Gulf of Cambay in central Gujarat, W India.
BT Gujarat India
BT India
BT Indian Peninsula
BT Asia

Cambisols (1993)
BT soils

Cambodia (1978)
Also search Khmer Republic or Kampuchea. Official name was changed back to State of Cambodia in 1990.
CO N103000N144000
   E1074000E1020000
UF Kampuchea
UF Khmer Republic
BT Far East
BT Asia
NT Battambang Cambodia
SA Indochina

Cambria County
Valid through 1988. Search in combination with state term. After 1988, use specific county-state term.

Cambria County Pennsylvania (1989)
Central Pennsylvania. Before 1989, also search Cambria County AND Pennsylvania.
CO N401500N404300
   W0782000W0790500
BT Pennsylvania
BT United States

Cambrian (1978)
Above Precambrian, below Ordovician. From 1978-1980, Paleozoic was autoposted to this term.
BT Paleozoic
NT Bonanza King Formation
NT Brigham Group
NT Carrara Formation
NT Conasauga Group
NT Glen Mountains Complex
NT Lardeau Group
NT Lower Cambrian
NT Middle Cambrian
NT Mount Read Volcanics
NT Nama System
NT Phuket Group
NT Pioche Shale
NT Semri Series
NT Shadow Lake Formation
NT Tapeats Sandstone
NT Tintic Quartzite
NT Tons Member
NT Upper Cambrian
SA Assyntic Orogeny
SA Baikalian Phase
SA Bhander Group
SA Bowers Supergroup
SA Cadomian Orogeny
SA Collier Shale
SA Cow Head Group
SA Dalradian
SA Elberton Granite
SA Eocambrian
SA Filicopsida

SA Hazens Notch Formation
SA Henderson Gneiss
SA Hoosac Formation
SA Jacobsville Sandstone
SA Kaimur Sandstone
SA Krol Formation
SA Kurnool System
SA New York City Group
SA Penobscot Formation
SA Puncoviscana Formation
SA Robertson Bay Group
SA Sauk Sequence
SA Shoo Fly Complex
SA Talladega Group
SA Vindhyan
SA Yudomian

Cambridge
No longer a valid term for GeoRef. As of 1989, for the city in Massachusetts, use Cambridge Massachusetts. After 1992, also see Cambridge England.

Cambridge Arch (1978)
IN Index states as applicable.
SA Kansas
SA Nebraska
SA United States

Cambridge Argillite (1989)
E Massachusetts. Overlies Roxbury Conglomerate, but the latter is thrust over the Cambridge Argillite in the northern Hingham area.
IN Index ages as applicable.
SA Boston Bay Group
SA Lower Cambrian
SA Massachusetts
SA upper Proterozoic

Cambridge England (1993)
Cities in Cambridgeshire and Gloucestershire, England. Before 1989, search Cambridge in combination with England.
BT England
BT Great Britain
BT United Kingdom
BT Western Europe
BT Europe
SA Cambridgeshire England
SA Gloucestershire England

Cambridge Massachusetts (1989)
City in NE Massachusetts. Before 1989, search Cambridge AND Massachusetts.
CO N422200N422200
   W0710600W0710600
BT Middlesex County Massachusetts
BT Massachusetts
BT United States

Cambridge University (1978)
Located in the city of Cambridge NNE of London.
UF University of Cambridge
SA academic institutions
SA England

Cambridgeshire
No longer a valid term for GeoRef as of 1993.
use Cambridgeshire England

Cambridgeshire England (1993)
County in E England. Before 1989, also search Cambridge in conjunction with England.
UF Cambridgeshire
BT England
BT Great Britain
BT United Kingdom
BT Western Europe
BT Europe

SA Cambridge England

Camden County
Valid through 1988. Search in combination with state term. After 1988, use specific county-state term.

Camden County Georgia (1989)
SE Georgia. Before 1989, also search Camden County AND Georgia.
CO N304000N310700
   W0812600W0815500
BT Georgia
BT United States

Camden County Missouri (1989)
Central Missouri. Before 1989, also search Camden County AND Missouri.
CO N374800N381300
   W0922300W0930500
BT Missouri
BT United States
NT Decaturville Missouri

Camden County New Jersey (1989)
SW New Jersey. Before 1989, also search Camden County AND New Jersey.
CO N393700N400000
   W0744400W0750800
BT New Jersey
BT United States

Camden County North Carolina (1989)
NE North Carolina. Before 1989, also search Camden County AND North Carolina.
CO N361100N363400
   W0755500W0763300
BT North Carolina
BT United States

Camelidae (1978)
Family. Autoposting of Tylopoda, Eutheria and Theria to this term began in 1989.
BT Tylopoda
BT Ruminantia
BT Artiodactyla
BT Eutheria
BT Theria
BT Mammalia
BT Tetrapoda
BT Vertebrata
BT Chordata

Cameron Parish
Valid through 1988. Search in combination with state term. After 1988, use specific county-state term.

Cameron Parish Louisiana (1989)
Extreme SW Louisiana, bordering the Gulf of Mexico. Before 1989, also search Cameron Parish AND Louisiana.
CO N293400N300500
   W0923700W0935400
BT Louisiana
BT United States
SA Calcasieu River
SA Sabine Lake

Cameroon (1978)
Includes former French trust territory and British trust territory of Southern Cameroons. Before 1972, also search Cameroun.
CO N020000N130000
   E0170000E0080000
UF Cameroun
BT West Africa

BT Africa
NT Lake Nyos
NT Mount Cameroon
SA Adamawa
SA Benue Valley
SA Chad Basin
SA Lake Chad
SA Logone River

Cameroon Mountain
   use Mount Cameroon

Cameroun
   Not a valid GeoRef index term after 1971. Was used on level 1 in subfile G.
   use Cameroon

Camp Century (1978)
   On the Greenland Ice Sheet NE of Thule Air Base in NW Greenland.
   BT Greenland
   BT Arctic region

Campania
   No longer a valid term for GeoRef. See Campania Italy.

Campania Italy (1993)
   An autonomous region in S Italy.
   CO N400000N413000
      E0154500E0134500
   BT Italy
   BT Southern Europe
   BT Europe
   NT Bay of Naples
   NT Caserta Italy
   NT Cilento
   NT Monte Somma
   NT Naples Italy
   NT Pozzuoli Italy
   NT Roccamonfina
   NT Salerno Italy
   NT Sorrento Peninsula
   NT Vesuvius
   SA Irpinia earthquake 1980
   SA Ischia
   SA Lucania
   SA Phlegraean Fields

Campania-Basilicata earthquake 1980
   use Irpinia earthquake 1980

Campania-Lucania earthquake 1980
   use Irpinia earthquake 1980

Campanian (1978)
   Europe. Above Santonian, below Maestrichtian. Autoposting of this term began in 1978.
   BT Senonian
   BT Upper Cretaceous
   BT Cretaceous
   BT Mesozoic
   NT upper Campanian
   SA Rosario Formation

Campbell County
   Valid through 1988. Search in combination with state term. After 1988, use specific county-state term.

Campbell County Kentucky (1989)
   N Kentucky. Before 1989, also search Campbell County AND Kentucky.
   CO N384800N390900
      W0841400W0843000
   BT Kentucky
   BT United States

Campbell County South Dakota (1989)
   On North Dakota line in N South Dakota. Before 1989, also search Campbell County AND South Dakota.
   CO N453600N455500
      W0994300W1002500
   BT South Dakota
   BT United States

Campbell County Tennessee (1989)
   NE Tennessee. Before 1989, also search Campbell County AND Tennessee.
   CO N361000N363200
      W0835500W0842200
   BT Tennessee
   BT United States

Campbell County Virginia (1989)
   SW central Virginia. Before 1989, also search Campbell County AND Virginia.
   CO N370200N372500
      W0785000W0792700
   BT Virginia
   BT United States

Campbell County Wyoming (1989)
   NE Wyoming. Before 1989, also search Campbell County AND Wyoming.
   CO N432900N450000
      W1050600W1060000
   BT Wyoming
   BT United States
   SA Hartzog Draw Field

Campbell Island (1981)
   Island in S Pacific Ocean, S of New Zealand. In 1985, broader term changed from Pacific Ocean to West Pacific Ocean Islands.
   IN Index regions as applicable.
   SA West Pacific Ocean Islands

Campeche
   No longer a valid term for GeoRef. As of 1993 see Campeche Mexico.

Campeche Bank (1978)
   N of the Yucatan Peninsula.
   BT Gulf of Mexico
   BT North American Atlantic
   BT North Atlantic
   BT Atlantic Ocean
   SA Campeche Scarp
   SA Yucatan Shelf

Campeche Mexico (1993)
   State on W part of Yucatan Peninsula. Before 1993 also search Campeche AND Mexico.
   BT Mexico
   SA Gulf Coastal Plain
   SA Yucatan Peninsula

Campeche Scarp (1985)
   W Gulf of Mexico.
   BT Gulf of Mexico
   BT North American Atlantic
   BT North Atlantic
   BT Atlantic Ocean
   SA Campeche Bank
   SA DSDP Site 85
   SA DSDP Site 86
   SA DSDP Site 88
   SA DSDP Site 94
   SA Sigsbee Deep

Campi Flegrei
   use Phlegraean Fields

Campine (1978)
   Heathland area.
   IN Index countries as applicable.
   BT Europe

   SA Belgium
   SA Netherlands

Campo del Cielo Meteorite (1981)
   Point of impact N of village of Campo del Cielo in S Argentina. Before 1981, also search Campo del Cielo AND meteorites.
   BT ataxite
   BT iron meteorites
   BT meteorites
   SA Argentina
   SA Chaco
   SA Chaco Argentina

Campobasso
   No longer a valid term for GeoRef. See Campobasso Italy.

Campobasso Italy (1993)
   Province and town in S central Italy.
   BT Molise Italy
   BT Italy
   BT Southern Europe
   BT Europe

Campos Basin (1985)
   South American Atlantic, off Brazil. Autoposting of Atlantic Ocean to this term began in 1990.
   CO S240000S180000
      W0380000W0410000
   BT South American Atlantic
   BT South Atlantic
   BT Atlantic Ocean

camptonite (1978)
   BT lamprophyres
   BT plutonic rocks
   BT igneous rocks

Camptostromatoidea (1978)
   BT Echinozoa
   BT Echinodermata
   BT Invertebrata

Canada (1978)
   Autoposting of this term began in 1978.
   CO N420000N840000
      W0520000W1410000
   NT Carswell Structure
   NT Cassiar Mountains
   NT Clearwater Lake
   NT Cold Lake
   NT Eastern Canada
   NT Elk Point Basin
   NT Grenville Front
   NT Hudson Bay
   NT Hudson Bay Lowlands
   NT Labrador Trough
   NT Liard River
   NT Mackenzie Mountains
   NT North Saskatchewan River
   NT Northumberland Strait
   NT Peel River
   NT Reindeer Lake
   NT Restigouche Estuary
   NT Rice Lake
   NT Richardson Mountains
   NT Selwyn Mountains
   NT Stikinia Terrane
   NT Ungava
   NT Western Canada
   NT Whitehorse Trough
   SA Abee Meteorite
   SA Acadian
   SA Appalachians
   SA Arctic region
   SA Burgess Shale
   SA Christopher Formation
   SA Clearwater River
   SA Cloridorme Formation
   SA Coldbrook Group
   SA Columbia Icefield
   SA Columbian Orogeny
   SA Cumberland Basin
   SA Cypress Hills Formation
   SA Davidsville Group
   SA Disturbed Belt
   SA Eastern Overthrust Belt
   SA Espanola Formation
   SA Flinton Group
   SA Gander Lake Group
   SA Geological Survey of Canada
   SA Gowganda Formation
   SA Great Lakes
   SA Great Lakes region
   SA Great Plains
   SA Harbour Main Group
   SA Hudsonian Orogeny
   SA Huronian
   SA Innisfree Meteorite
   SA Keg River Formation
   SA Kenoran Orogeny
   SA Kootenay Arc
   SA Lardeau Group
   SA Leda Clay
   SA Meguma Group
   SA Melville Island
   SA Michelle Formation
   SA Miette Complex
   SA Miette Group
   SA Milk River Formation
   SA Miramichi earthquake 1982
   SA Missi Group
   SA Mississippi River basin
   SA Missouri River basin
   SA Nahanni Formation
   SA Nicola Group
   SA Nipissing Diabase
   SA North America
   SA North American Cordillera
   SA Onaping Formation
   SA Osler Series
   SA Palliser Formation
   SA Paskapoo Formation
   SA Peel Sound Formation
   SA Prince Albert Group
   SA Ramparts Formation
   SA Ravenscrag Formation
   SA Read Bay Formation
   SA Rice Lake Group
   SA Rocky Mountains
   SA Saint Lawrence River
   SA Shaunavon Formation
   SA Signal Hill Formation
   SA Sokoman Formation
   SA Taconic Allochthon
   SA Taconic Orogeny
   SA Transcontinental Arch
   SA Western Interior
   SA Western Interior Seaway
   SA Windsor Group

Canada Basin (1978)
   N of Alaska and WNW of Canada's Queen Elizabeth Islands.
   UF Canadian Basin
   BT Arctic Ocean

Canadian Basin
   use Canada Basin

Canadian Cordillera (1978)
   This term has multiple hierarchies.
   IN Index provinces as applicable.
   BT1 Western Canada
   BT1 Canada
   BT2 North American Cordillera
   SA Alberta
   SA British Columbia
   SA Coast Mountains
   SA Nahanni earthquake 1985
   SA Rocky Mountains
   SA Western Interior Seaway

Canadian Expedition to Study the Alpha Ridge
   use CESAR

Canadian River (1978)

IN   Index counties or regions as applicable.
UF   South Canadian River
SA   Colorado
SA   New Mexico
SA   Oklahoma
SA   Texas
SA   United States

Canadian Rocky Mountains (1981)
Before 1981, search Canada AND Rocky Mountains. This term began autoposting in 1993. This term has multiple hierarchies.
IN   Index provinces as applicable.
BT1 Western Canada
BT1 Canada
BT2 Rocky Mountains
NT   Selkirk Mountains
SA   Alberta
SA   Athabasca Glacier
SA   British Columbia
SA   Cariboo Mountains
SA   Cassiar Mountains
SA   Columbia Icefield
SA   Gog Group
SA   Mount Erebus
SA   Northern Rocky Mountains
SA   Selwyn Basin
SA   Stikinia Terrane
SA   Whitehorse Trough

Canadian Series (1978)
Provincial series, North America. Above Croixian (Cambrian), below Champlainian. Before 1981, also search Canadian AND Ordovician.
BT   Lower Ordovician
BT   Ordovician
BT   Paleozoic

Canadian Shield (1978)
Also search Canadian AND Shield. Provinces of the Canadian Shield are narrower terms as of 1981.
IN   Index states, Canadian provinces and provinces of the Canadian Shield as applicable.
CO   N430000N790000
     W0560000W1240000
UF   Laurentian Highlands
UF   Laurentian Plateau
UF   Precambrian Shield
BT   North America
NT   Bear Province
NT   Churchill Province
NT   Grenville Province
NT   Nain Province
NT   Slave Province
NT   Southern Province
NT   Superior Province
SA   Adirondack Mountains
SA   Alberta
SA   Eye-Dashwa Lakes Pluton
SA   Hemlo Deposit
SA   Huronian
SA   Laurentia
SA   Manitoba
SA   Michigan
SA   Minnesota
SA   New York
SA   North American Craton
SA   Northwest Territories
SA   Ontario
SA   Quebec
SA   Saskatchewan
SA   Wisconsin

Canal Zone
Not a valid GeoRef index term after 1970. Was used on level 1 in subfile N.
use Panama Canal Zone

canals (1978)
SA   channels
SA   streams
SA   waterways

Cananea
No longer a valid term for GeoRef. As of 1993 see Cananea Mexico.

Cananea Mexico (1993)
City in N Sonora. Before 1993 also search Cananea AND Mexico.
BT   Sonora Mexico
BT   Mexico

Canaries
use Canary Islands

Canary Islands (1978)
Spanish controlled island group off NW Africa. In 1981, broader term changed from Atlantic Ocean to Atlantic Ocean Islands.
CO   N274500N291500
     W0130000W0180000
UF   Canaries
BT   Atlantic Ocean Islands
NT   Fuerteventura
NT   Gomera
NT   Grand Canary
NT   Hierro
NT   Lanzarote
NT   Tenerife
SA   DSDP Site 397
SA   Spain

Canavese Zone (1978)
Mountain region N of Turin.
BT   Italy
BT   Southern Europe
BT   Europe
SA   Piemonte Italy

Canberra
No longer a valid term for GeoRef. See Canberra Australia.

Canberra Australia (1993)
City between Sydney and Melbourne.
BT   Australian Capital Territory
BT   New South Wales Australia
BT   Australia
BT   Australasia

cancrinite (1978)
Autoposting of silicates to this term began in 1989. This term has multiple hierarchies.
BT1 carbonates
BT2 framework silicates
BT2 silicates

Candona (1978)
Genus. As of 1990, microfossils, Crustacea, Mandibulata, and Arthropoda are autoposted to this term. This term has multiple hierarchies.
BT1 Cyprididae
BT1 Cypridocopina
BT1 Podocopida
BT1 Ostracoda
BT1 Crustacea
BT1 Mandibulata
BT1 Arthropoda
BT1 Invertebrata
BT2 Cyprididae
BT2 Cypridocopina
BT2 Podocopida
BT2 Ostracoda
BT2 microfossils

canfieldite (1978)
This term has multiple hierarchies.
BT1 sulfogermanates
BT1 sulfosalts
BT2 sulfostannates
BT2 sulfosalts

Canidae (1978)
Family. Autoposting of Fissipeda, Eutheria and Theria to this term began in 1989.
BT   Fissipeda
BT   Carnivora
BT   Eutheria
BT   Theria
BT   Mammalia
BT   Tetrapoda
BT   Vertebrata
BT   Chordata
NT   Canis

Canis (1978)
Genus. Autoposting of Fissipeda, Eutheria and Theria to this term began in 1989.
BT   Canidae
BT   Fissipeda
BT   Carnivora
BT   Eutheria
BT   Theria
BT   Mammalia
BT   Tetrapoda
BT   Vertebrata
BT   Chordata

Cannanore District
No longer a valid term for GeoRef. As of 1993, see Cannanore India.

Cannanore India (1993)
District, Kerala, SW India. Also search Cannanore District.
BT   Kerala India
BT   India
BT   Indian Peninsula
BT   Asia

Cannikin
No longer a valid term for GeoRef. As of 1993, see Cannikin Alaska.

Cannikin Alaska (1993)
Name of site of nuclear explosion on Amchitka Island in the Aleutian Islands.
BT   Amchitka Island
BT   Aleutian Islands
BT   Southwestern Alaska
BT   Alaska
BT   United States

Canning Basin (1978)
Arid region in northern Western Australia, about 790 miles NNE of Perth. Also search Desert Basin AND Australia.
CO   S240000S180000
     E1280000E1200000
BT   Western Australia
BT   Australia
BT   Australasia

Cannon County
Valid through 1988. Search in combination with state term. After 1988, use specific county-state term.

Cannon County Tennessee (1989)
Central Tennessee. Before 1989, also search Cannon County AND Tennessee.
CO   N354000N355900
     W0855300W0861000
BT   Tennessee
BT   United States
NT   Woodbury Tennessee

Canon City
Valid through 1988. Search in combination with state term. After 1988, use specific city-state term.

Canon City Colorado (1989)
City in central Colorado. Before 1989, search Canon City AND Colorado.
CO   N382700N382700
     W1051400W1051400
BT   Fremont County Colorado
BT   Colorado
BT   United States

canonical analysis (1978)
BT   statistical analysis
SA   analysis

Cantabria Knoll
use Cantabria Seamount

Cantabria Seamount (1978)
W Bay of Biscay. Also search Cantabria.
UF   Cantabria Knoll
BT   Bay of Biscay
BT   European Atlantic
BT   North Atlantic
BT   Atlantic Ocean
SA   DSDP Site 119

Cantabrian Basin (1978)
Region in the Cantabrian Mountains area in N and NW Spain.
IN   Index Asturias Spain, Basque Provinces Spain, and Santander Spain as applicable.
BT   Spain
BT   Iberian Peninsula
BT   Southern Europe
BT   Europe
SA   Asturias Spain
SA   Basque Provinces Spain
SA   Santander Spain

Cantabrian Mountains (1978)
Range extending westward from the Pyrenees along the Bay of Biscay to the Galician Mountains.
CO   N423000N433000
     W0024000W0072000
BT   Spain
BT   Iberian Peninsula
BT   Southern Europe
BT   Europe

Cantabrian region (1981)
BT   Spain
BT   Iberian Peninsula
BT   Southern Europe
BT   Europe

Cantal
No longer a valid term for GeoRef. See Cantal France.

Cantal France (1993)
Department in S central France.
CO   N443000N453000
     E0031500E0020500
BT   France
BT   Western Europe
BT   Europe
NT   Cantal Massif
NT   Massiac France
SA   Central Massif

Cantal Massif (1978)
Part of Central Massif covering central France. This term has multiple hierarchies.
BT1 Cantal France
BT1 France
BT1 Western Europe
BT1 Europe
BT2 Central Massif
BT2 France
BT2 Western Europe
BT2 Europe

Canterbury
No longer a valid term for GeoRef. See Canterbury New Zealand if applicable.

Canterbury New Zealand (1993)
District in E central New Zealand.
IN   Also index South Island.
BT   New Zealand
BT   Australasia
NT   Christchurch New Zealand
SA   South Island

Canton River
use Zhujiang River

Canyon Diablo Meteorites
(1981)
Refers to three meteorites found in Coconino County, Arizona in 1891, 1936, and 1949. (Used for Canon Diablo, Canyon Diablo no. 2, and Canyon Diablo no. 3.) Before 1981, also search Canon Diablo AND meteorites; Canyon Diablo AND meteorites; Canyon Diablo Meteorite. Due to numbering of these meteorites, also truncate when searching.
BT   octahedrite
BT   iron meteorites
BT   meteorites
SA   Arizona
SA   Coconino County Arizona

Canyon Group (1978)
Includes Graford Formation, Winchell Limestone, Brad Formation, and Caddo Creek Formation. Central and N central Texas.
BT   Upper Pennsylvanian
BT   Pennsylvanian
BT   Carboniferous
BT   Paleozoic
SA   Texas
SA   United States

Canyonlands National Park
(1985)
SE Utah. Autoposting of broader terms to this term began in 1989.
IN   Index counties as applicable.
CO   N375500N383000
      W1094000W1101000
BT   Utah
BT   United States
SA   Garfield County Utah
SA   national parks
SA   San Juan County Utah
SA   Wayne County Utah

canyons (1978)
As of 1976, usage restricted to land canyons. Before 1976, also refers to submarine canyons.
UF   land canyons
SA   fluvial features
SA   geomorphology
SA   gorges
SA   solution features
SA   submarine canyons
SA   valleys

Cap Blanc (1978)
IN   Index countries as applicable.
BT   Africa
SA   Mauritania

cap rocks (1978)
Used for salt tectonics: an impervious layer or body of rock over a salt dome.
SA   anhydrite
SA   diapirs
SA   geothermal energy
SA   gypsum
SA   petroleum
SA   reservoir rocks
SA   rocks
SA   salt domes
SA   salt tectonics
SA   sealing

capacity
As of 1981, no longer a valid term for GeoRef. Before 1981, included use as a general term.

Capbreton
No longer a valid term for GeoRef. See Capbreton France.

Capbreton France (1993)
Town in SW France.
BT   Landes France
BT   France
BT   Western Europe
BT   Europe

Cape Ann (1989)
NE Massachusetts.
IN   Index state and county as applicable.
SA   Essex County Massachusetts
SA   Gulf of Maine

Cape Basin (1981)
SE Atlantic Ocean. Autoposting of Atlantic Ocean to this term began in 1990.
CO   S410000S260000
      E0100000E0010000
UF   Walvis Basin
BT   Southeast Atlantic
BT   South Atlantic
BT   Atlantic Ocean
SA   DSDP Site 360
SA   DSDP Site 524
SA   Leg 40

Cape Bojador (1981)
Headland extending into Atlantic Ocean on S coast of Morocco.
CO   N260700N260800
      W0142900W0143000
BT   Morocco
BT   North Africa
BT   Africa
SA   DSDP Site 397

Cape Breton County Nova Scotia
(1993)
Before 1993, also search Cape Breton County AND Nova Scotia.
BT   Nova Scotia
BT   Maritime Provinces
BT   Eastern Canada
BT   Canada
NT   Sydney Nova Scotia

Cape Breton Highlands (1989)
N Cape Breton Island, NE Nova Scotia.
BT   Cape Breton Island
BT   Nova Scotia
BT   Maritime Provinces
BT   Eastern Canada
BT   Canada

Cape Breton Island (1978)
NE Nova Scotia.
CO   N453000N470000
      W0593000W0613000
BT   Nova Scotia
BT   Maritime Provinces
BT   Eastern Canada
BT   Canada
NT   Cape Breton Highlands
SA   Avalon Zone
SA   Loch Lomond
SA   Morien Group
SA   Sydney coal field

Cape Canaveral (1978)
Cape on the Atlantic Ocean in E Florida. Site of John F. Kennedy Space Center. Before 1985, also search Cape Kennedy. Term reintroduced in 1985, after being invalid from 1981-1985. Autoposting of broader terms to this term began in 1989.
CO   N282200N282200

      W0803600W0803600
UF   Cape Kennedy
BT   Brevard County Florida
BT   Florida
BT   United States

Cape Cod (1978)
SE Massachusetts. Coextensive with Barnstable County. Autoposting of broader terms to this term began in 1989.
CO   N413100N420500
      W0695500W0704200
BT   Barnstable County Massachusetts
BT   Massachusetts
BT   United States
NT   Monomoy Island

Cape Dyer (1978)
Cape on Davis Strait on E Baffin Island in Keewatin District. Before 1993, also referred to the village. As of 1993, Cape Dyer Northwest Territories is used for the village.
IN   Index Baffin Island and Keewatin District as applicable.
SA   Baffin Island
SA   Keewatin District Northwest Territories

Cape Elizabeth Formation
(1989)
In Casco Bay Group. SW Maine.
BT   Paleozoic
SA   Casco Bay Group
SA   Maine

Cape Fear Arch (1978)
Extends NW from Cape Fear.
BT   North Carolina
BT   United States

Cape Hatteras (1978)
Promontory on Hatteras Island between Pamlico Sound and the Atlantic Ocean.
BT   Dare County North Carolina
BT   North Carolina
BT   United States

Cape Kennedy
As of 1985, no longer a valid term for GeoRef.
use Cape Canaveral

Cape Lookout (1978)
Headland at S end off Carteret County.
IN   Index counties or regions as applicable.
SA   Carteret County North Carolina
SA   North Carolina

Cape May County
Valid through 1988. Search in combination with state term. After 1988, use specific county-state term.

Cape May County New Jersey
(1989)
S extremity of New Jersey, occupying the Cape May Peninsula between the Atlantic Ocean and Delaware Bay. Before 1989, also search Cape May County AND New Jersey.
CO   N385500N392000
      W0740300W0745900
BT   New Jersey
BT   United States

Cape Mendocino (1978)
On W coast of Humboldt County, extreme W California.
UF   Mendocino (Cape)
BT   Humboldt County California
BT   California

BT   United States

Cape Province
No longer a valid term for GeoRef. As of 1993 see Cape Province South Africa.

Cape Province South Africa
(1993)
Province in S South Africa. Before 1993 also search Cape Province AND South Africa.
CO   S345000S244500
      E0301500E0162500
BT   South Africa
BT   Southern Africa
BT   Africa
NT   Cape Town South Africa
NT   False Bay
NT   Karroo Basin
NT   Kimberley South Africa
NT   Knysna South Africa
NT   Langebaanweg South Africa
NT   Molteno South Africa
NT   Port Elizabeth South Africa
NT   Postmasburg South Africa
NT   Sutherland South Africa
NT   Walvis Bay South Africa
SA   Broken Hill Mine
SA   Ecca Group
SA   Namaqualand
SA   Orange River
SA   Table Mountain
SA   Vaal River
SA   Walvis Bay

Cape Sable (1978)
S point of Cape Sable Island in extreme S Nova Scotia.
IN   Index province or region as applicable.
SA   Nova Scotia

Cape Smith fold belt (1989)
Structural subprovince in N Quebec.
BT   Churchill Province
BT   Canadian Shield
BT   North America
SA   Quebec

Cape Town
No longer a valid term for GeoRef. As of 1993 see Cape Town South Africa.

Cape Town South Africa (1993)
City in extreme SW Cape Province. Before 1993 also search Capetown AND South Africa.
UF   Capetown South Africa
BT   Cape Province South Africa
BT   South Africa
BT   Southern Africa
BT   Africa
SA   Table Mountain

Cape Valse
use False Cape

Cape Verde Atlantic (1981)
Region of the Atlantic Ocean. As of 1990, Atlantic Ocean is autoposted to this term.
BT   North Atlantic
BT   Atlantic Ocean
NT   Cape Verde Basin
NT   Cape Verde Rise
NT   Mazagan Plateau
NT   Sierra Leone Rise
SA   DSDP Site 137
SA   DSDP Site 141
SA   DSDP Site 366
SA   East Atlantic
SA   Equatorial Atlantic
SA   Leg 14
SA   Leg 41
SA   Leg 108
SA   Northeast Atlantic

SA ODP Site 657
SA ODP Site 658
SA ODP Site 659
SA ODP Site 660
SA ODP Site 661
SA ODP Site 664
SA ODP Site 665
SA ODP Site 666
SA ODP Site 667
SA ODP Site 668
SA Orgon III

Cape Verde Basin (1981)
E Atlantic Ocean between the Mid-Atlantic Ridge and West Africa in Cape Verde Atlantic. As of 1990, Atlantic Ocean is autoposted to this term.
CO N080000N350000
W0200000W0420000
BT Cape Verde Atlantic
BT North Atlantic
BT Atlantic Ocean
SA DSDP Site 367
SA DSDP Site 368
SA Leg 41

Cape Verde Islands (1978)
Former Portuguese overseas province W of Dakar, Senegal. This term has multiple hierarchies.
CO N144000N172000
W0223000W0253000
BT1 Atlantic Ocean Islands
BT2 Africa

Cape Verde Rise (1981)
Between Cape Verde Islands and West Africa in Cape Verde Atlantic. As of 1990, Atlantic Ocean is autoposted to this term.
CO N140000N210000
W0180000W0260000
BT Cape Verde Atlantic
BT North Atlantic
BT Atlantic Ocean
SA Leg 41

Cape York Meteorite (1985)
Impact on Cape York, NW Greenland. Before 1985, also search Cape York AND Greenland AND meteorites.
BT iron meteorites
BT meteorites
SA Greenland

Cape York Peninsula (1978)
Between Gulf of Carpentaria and Coral Sea in extreme N Queensland.
IN Index state or region as applicable.
SA Moreton Australia
SA Queensland Australia
SA Weipa Australia

capes (1982)
Autoposting of shore features to this term began in 1989.
BT shore features
SA barrier islands
SA bars
SA beaches
SA coastal environment
SA coastlines
SA geomorphology
SA intertidal environment
SA littoral erosion
SA longshore bars
SA marine environment
SA shorelines
SA spits

Capetown South Africa
use Cape Town South Africa

capillarity (1978)
SA capillary pressure
SA capillary water
SA soils

capillary pressure (1978)
BT pressure
SA capillarity
SA capillary water

capillary water (1981)
Before 1981, also search capillarity.
SA capillarity
SA capillary pressure
SA ground water
SA saturated zone
SA soils
SA unsaturated zone
SA water table

Capistrano Formation (1978)
Upper Miocene and lower Pliocene. Subdivided to include Osa Member. Unconformably underlies Niguel Formation. Southern California between Santa Ana and Oceanside.
IN Index ages as applicable.
BT Neogene
BT Tertiary
BT Cenozoic
SA California
SA lower Pliocene
SA upper Miocene

Capitan Formation (1978)
Massive reef limestone which grades basinward into upper Delaware Mountain Sandstone and lagoonward into Carlsbad Limestone. SE New Mexico and W Texas.
BT Guadalupian
BT Permian
BT Paleozoic
SA Castile Formation
SA New Mexico
SA Texas

Capnic
use Cavnic Romania

Captorhinidae (1993)
BT Captorhinomorpha
BT Cotylosauria
BT Anapsida
BT Reptilia
BT Tetrapoda
BT Vertebrata
BT Chordata

Captorhinomorpha (1981)
Suborder. Autoposting of Cotylosauria to this term began in 1989.
BT Cotylosauria
BT Anapsida
BT Reptilia
BT Tetrapoda
BT Vertebrata
BT Chordata
NT Captorhinidae

capture
use stream capture

Carabobo
No longer a valid term for GeoRef. As of 1993 see Carabobo Venezuela.

Carabobo Venezuela (1993)
State on Caribbean Sea in N Venezuela. Also a village. Before 1993 also search Carabobo AND Venezuela.
BT Venezuela
BT South America
NT Puerto Cabello Venezuela

Caracas
No longer a valid term for GeoRef as of 1993. See Caracas Venezuela.

Caracas Group (1978)
Middle and upper Mesozoic. Includes Las Brisas Formation, Antimano Formation, Tacagua Formation. In Cordillera de la Costa of Venezuela.
BT Mesozoic
SA Caracas Venezuela
SA Federal District Venezuela
SA Venezuela

Caracas Venezuela (1993)
City in N Venezuela.
BT Federal District Venezuela
BT Venezuela
BT South America
SA Caracas Group

Caradoc
use Caradocian

Caradocian (1978)
Europe. Above Llandeilian, below Ashgillian.
UF Caradoc
BT Upper Ordovician
BT Ordovician
BT Paleozoic
SA Bala

carapaces
After 1992, for carapace of an ostracod, use morphology and Ostracoda. See exoskeletons, chitin.

Caravaca
No longer a valid term for GeoRef. As of 1993, see Caravaca Spain.

Caravaca Spain (1993)
City in SE Spain. Before 1993, also search Caravaca and Spain.
BT Murcia Spain
BT Murcia region
BT Spain
BT Iberian Peninsula
BT Southern Europe
BT Europe

carbargilite
use coal

carbides (1978)
Autoposting of this term began in 1981. Before 1993, also search carborundum for silicon carbide if applicable.
BT alloys
NT cohenite
NT moissanite
SA carborundum

carbohydrates (1978)
Autoposting of this term began in 1978.
BT organic materials
NT cellulose
NT sugars

carbon (1978)
UF C
NT C-12
NT C-13
NT C-13/C-12
NT C-14
NT C-14/C-12
SA carbonaceous composition
SA carbonado
SA chaoite
SA charcoal
SA chemical elements
SA diamond
SA diamonds
SA graphite
SA graphitization
SA hydrocarbons
SA isotopes
SA lonsdaleite
SA moissanite
SA organic carbon
SA organic materials
SA stable isotopes

Carbon County
Valid through 1988. Search in combination with state term. After 1988, use specific county-state term.

Carbon County Montana (1989)
S Montana. Before 1989, also search Carbon County AND Montana.
CO N450000N453800
W1081000W1095000
BT Montana
BT United States

Carbon County Pennsylvania (1989)
E Pennsylvania. Before 1989, also search Carbon County AND Pennsylvania.
CO N404300N410800
W0752800W0755900
BT Pennsylvania
BT United States

Carbon County Utah (1989)
Central Utah. Before 1989, also search Carbon County AND Utah.
CO N392800N394800
W1095200W1111800
BT Utah
BT United States
SA Book Cliffs
SA Sunnyside Mine

Carbon County Wyoming (1989)
S Wyoming. Before 1989, also search Carbon County AND Wyoming.
CO N410000N422700
W1060500W1075500
BT Wyoming
BT United States
NT Seminoe Mountains
NT Shirley Basin
SA Great Divide Basin
SA Lost Soldier Field
SA Sullivan Mine

carbon dioxide (1978)
UF CO2
SA carbon monoxide
SA fluid inclusions
SA geochemistry
SA greenhouse effect
SA phase equilibria

carbon monoxide (1978)
UF CO
SA carbon dioxide

carbon-12
use C-12

carbon-13
use C-13

carbon-14
A valid term through 1976 in subfiles E, G, N, T and B.
use C-14

carbonaceous
A valid index term through 1977. After 1977, use carbonaceous chondrites for meteorites, and use carbonaceous composition as a general term referring to carbon and coal.

carbonaceous chondrites (1978)

Before 1978, also search chondrites AND carbonaceous. Autoposting of this term began in 1978.
BT chondrites
BT stony meteorites
BT meteorites
NT Allende Meteorite
NT Efremovka Meteorite
NT Kainsaz Meteorite
NT Leoville Meteorite
NT Murchison Meteorite
NT Murray Meteorite
NT Orgueil Meteorite
NT Vigarano Meteorite
SA ordinary chondrites

carbonaceous composition (1978)
A general term used with carbon and coal. Before 1978, also search carbonaceous.
UF carbonaceous rocks
SA black shale
SA carbon
SA coal
SA composition
SA organic materials

carbonaceous rocks
No longer a valid term for GeoRef. Before 1963, included use in subfiles E and N.
use carbonaceous composition

carbonaceous shale
use black shale

carbonado (1978)
BT native elements
SA carbon
SA diamonds
SA industrial minerals

carbonate
As of 1981, no longer a valid term for GeoRef. See carbonates; carbonate ion; carbonate rocks; carbonate sediments.

carbonate apatite (1978)
UF carbonate-apatites
UF tavistockite
BT phosphates
SA apatite
SA dahllite

carbonate banks (1993)
BT biogenic structures
BT sedimentary structures
SA algal banks
SA banks
SA bioherms
SA mud banks
SA reefs

carbonate compensation
use carbonate compensation depth

carbonate compensation depth (1978)
UF carbonate compensation
UF depth of compensation
SA lysoclines
SA ocean circulation
SA oceanography
SA sea water
SA solution

carbonate composition
As of 1981, no longer a valid term for GeoRef. Now use carbonaceous composition if applicable.

carbonate ion (1981)
SA bicarbonate ion
SA carbonates
SA geochemistry
SA ions

carbonate platforms (1985)
SA carbonate rocks
SA carbonate sediments
SA platform reefs
SA platforms
SA sedimentation
SA shelf environment

carbonate ramps (1989)
SA carbonate rocks
SA carbonate sediments
SA coastal environment
SA ramps
SA shallow-water environment

carbonate rocks (1978)
Autoposting of this term to coquina, caliche and calcrete began in 1981. Autoposting of this term to marl ended in 1981.
BT sedimentary rocks
NT beachrock
NT boundstone
NT calcrete
NT caliche
NT chalk
NT dolostone
NT grainstone
NT limestone
NT packstone
NT travertine
NT wackestone
SA bone beds
SA calcite
SA carbonate platforms
SA carbonate ramps
SA carbonate sediments
SA carbonates
SA carbonatites
SA dolomite
SA floatstone
SA intraclasts
SA marl
SA micritization
SA oolite
SA rocks
SA tufa

carbonate sediments (1978)
BT sediments
SA carbonate platforms
SA carbonate ramps
SA carbonate rocks
SA carbonates
SA coquina
SA dolomite
SA dolostone
SA gravel
SA limestone
SA oolite
SA sand
SA travertine
SA tufa

carbonate-apatites
use carbonate apatite

carbonates (1978)
Before 1981, thomsonite was included as a narrower term. Autoposting of this term began in 1978.
NT alstonite
NT alumohydrocalcite
NT ankerite
NT aragonite
NT artinite
NT azurite
NT bastnaesite
NT calcite
NT cancrinite
NT cerussite
NT dahllite
NT dawsonite
NT dolomite
NT gaylussite
NT huntite

NT hydromagnesite
NT hydrotalcite
NT Iceland spar
NT kutnahorite
NT leadhillite
NT magnesian calcite
NT magnesite
NT malachite
NT meionite
NT monohydrocalcite
NT nahcolite
NT nesquehonite
NT norsethite
NT phosgenite
NT protodolomite
NT pyroaurite
NT rhodochrosite
NT sakhaite
NT shortite
NT siderite
NT sjogrenite
NT smithsonite
NT spurrite
NT strontianite
NT synchisite
NT thaumasite
NT thermonatrite
NT trona
NT vaterite
NT weddellite
NT weloganite
NT whewellite
NT witherite
SA bicarbonate ion
SA calcium carbonate
SA carbonate ion
SA carbonate rocks
SA carbonate sediments
SA minerals
SA sodium carbonate

carbonatite
As of 1989, no longer a valid term for GeoRef. Use carbonatites. From 1981 to 1988, carbonatites was autoposted to this term.

carbonatites (1981)
Before 1981, search lamprophyre and carbonatite family. From 1978-1980, lamprophyre and carbonatite family was autoposted to the rocks that were classified under that term. Before 1989, also search carbonatite. Autoposting of this term to carbonatite ended in 1989.
BT plutonic rocks
BT igneous rocks
NT sovite
SA carbonate rocks

carbonatization
As of 1981, no longer a valid term for GeoRef. See carbonates.

Carbondale
Valid through 1988. Search in combination with state term. After 1988, use specific city-state term.

Carbondale Formation (1978)
In Kewanee Group. Illinois and W Kentucky.
BT Middle Pennsylvanian
BT Pennsylvanian
BT Carboniferous
BT Paleozoic
SA Herrin Coal Member
SA Illinois
SA Kentucky
SA Kewanee Group
SA Springfield Coal Member

Carbondale Illinois (1989)
City in S Illinois. Before 1989, search Carbondale AND Illinois.
CO N374300N374300

W0891200W0891200
BT Jackson County Illinois
BT Illinois
BT United States

Carbonicola (1978)
Genus.
BT Bivalvia
BT Mollusca
BT Invertebrata

Carboniferous (1978)
Above Devonian, below Permian. Autoposting of this term to Mississippian and Pennsylvanian (and their subdivisions) began in 1981. From 1978-1980, Paleozoic was autoposted to this term.
BT Paleozoic
NT Albert Formation
NT Amsden Formation
NT Asturian Orogeny
NT Avonian
NT Bartlesville Sand
NT Benxi Formation
NT Big Snowy Group
NT Burbank Sand
NT Chilliwack Group
NT Culm
NT Diamond Peak Formation
NT Dinantian
NT Ely Limestone
NT Essen Beds
NT Huanglong Formation
NT Jackfork Group
NT Johns Valley Formation
NT Manning Canyon Shale
NT Middle Carboniferous
NT Mispec Group
NT Mississippian
NT Pennsylvanian
NT Schoonover Sequence
NT Silesian
NT Springer Formation
NT Stellarton Group
NT Tesnus Furmation
NT Tubarao Group
NT Upper Carboniferous
SA Baralaba Coal Measures
SA Beacon Supergroup
SA Deer Lake Group
SA Dwyka Formation
SA Geirud Formation
SA Hercynian Orogeny
SA Ishbel Group
SA Itarare Subgroup
SA Karroo Supergroup
SA Krol Formation
SA Lisburne Group
SA lower Gondwana System
SA Ouachita Orogeny
SA Paganzo Group
SA Saint George Batholith
SA Singleton Coal Measures
SA South Mountain Batholith
SA Taiyuan Formation
SA Talchir Formation
SA Waulsortian facies

carbonification
use coalification

carbonitization
use coalification

carbonization
use coalification

carborundum (1982)
Trade name for a synthetic silicon carbide. After 1993, see carbides or moissanite for the mineral.
IN For use as a commodity, index abrasives and industrial minerals or refractory materials as applicable. See List C.
SA abrasives
SA carbides

SA ceramic materials
SA industrial minerals
SA moissanite
SA refractory materials
SA synthetic materials

carboxylic acids (1989)
Organic acids containing one or more carboxyl groups.
BT organic materials
SA acids
SA geochemistry
SA organic acids

Cardiff
No longer a valid term for GeoRef as of 1993. See Cardiff Wales if applicable.

Cardiff Wales (1993)
City on the Bristol Channel in South Glamorgan County, SE Wales. Before 1993, also search Cardiff AND Wales.
BT Glamorgan Wales
BT Wales
BT Great Britain
BT United Kingdom
BT Western Europe
BT Europe

Cardigan Bay (1978)
Large inlet of Saint George's Channel.
BT Wales
BT Great Britain
BT United Kingdom
BT Western Europe
BT Europe
SA Cardiganshire Wales
SA Dyfed Wales
SA Irish Sea
SA Merionethshire Wales

Cardigan County
Not a valid term for GeoRef as of 1993. See Cardiganshire Wales.

Cardiganshire
No longer a valid term for GeoRef as of 1993.
use Cardiganshire Wales

Cardiganshire Wales (1993)
Former county on Cardigan Bay. Before 1993, also search Cardiganshire or Cardigan County.
UF Cardiganshire
BT Dyfed Wales
BT Wales
BT Great Britain
BT United Kingdom
BT Western Europe
BT Europe
NT Aberystwyth Wales
SA Cardigan Bay

Cardiidae (1978)
Autoposting of this term began in 1978.
BT Bivalvia
BT Mollusca
BT Invertebrata
NT Cardium

Carditida (1981)
BT Bivalvia
BT Mollusca
BT Invertebrata

Cardium (1978)
Genus. Autoposting of this term began in 1978.
BT Cardiidae
BT Bivalvia
BT Mollusca
BT Invertebrata
NT Cardium edule

Cardium edule (1978)
BT Cardium
BT Cardiidae
BT Bivalvia
BT Mollusca
BT Invertebrata

Cardium Formation (1989)
Also search Cardium Sandstone NOT Texas. For Miocene formation of offshore Texas, search Cardium Sandstone AND Texas.
BT Upper Cretaceous
BT Cretaceous
BT Mesozoic
SA Alberta
SA British Columbia

Cargo Muchacho Mountains (1989)
SE California.
BT Imperial County California
BT California
BT United States

Cariaco Basin (1978)
W of Cariaco Bay and between Isla la Tortuga and Venezuelan coast.
UF Cariaco Trench
BT Caribbean Sea
BT North American Atlantic
BT North Atlantic
BT Atlantic Ocean

Cariaco Trench
use Cariaco Basin

Caribbean Mountain Range (1978)
Islands comprising the Antilles stretching from Florida to the coast of Venezuela constituting the remainder of several submerged Andean spurs meeting in Puerto Rico.
BT West Indies
BT Caribbean region
SA Andes
SA Antilles
SA Puerto Rico

Caribbean Plate (1978)
The Caribbean Sea including the Caribbean islands and across Central America to the Cocos plate just to the W.
SA Caribbean region
SA plate tectonics
SA plates

Caribbean region (1978)
Also search Caribbean. Before 1981, Netherlands Antilles was a narrower term. As of 1993, West Indies is autoposting Caribbean region as its broader term.
IN Index Antilles, Caribbean Sea, Central America, Colombia, and Venezuela as applicable.
CO N070000N230000
   W0580000W0900000
NT West Indies
SA Caribbean Plate
SA Caribbean Sea
SA Central America
SA Colombia
SA Venezuela
SA Yucatan Peninsula

Caribbean Sea (1978)
Bounded by the West Indies on the N and E, South America on the S, and Central America on the W. Also search Caribbean. Before 1981, Grand Cayman Island was included as a narrower term.
CO N090000N220000
   W0600000W0780000
BT North American Atlantic

BT North Atlantic
BT Atlantic Ocean
NT Aves Island
NT Aves Ridge
NT Beata Ridge
NT Cariaco Basin
NT Cayman Trough
NT Los Testigos
NT Mid-Cayman Rise
NT Nicaragua Rise
NT Venezuelan Basin
NT Yucatan Basin
SA Barracuda Ridge
SA Caribbean region
SA Colombian Basin
SA DSDP Site 146
SA DSDP Site 147
SA DSDP Site 147A
SA DSDP Site 148
SA DSDP Site 149
SA DSDP Site 150
SA DSDP Site 151
SA DSDP Site 152
SA DSDP Site 153
SA DSDP Site 154
SA DSDP Site 502
SA DSDP Site 541
SA DSDP Site 542
SA Grand Cayman Island
SA Gulf of Cariaco
SA Leg 4
SA Leg 15
SA Leg 77
SA Leg 78A

Cariboo Mountains (1978)
Range of the Rocky Mountains in E central British Columbia.
BT British Columbia
BT Western Canada
BT Canada
SA Canadian Rocky Mountains
SA Rocky Mountains

Caribou County
Valid through 1988. Search in combination with state term. After 1988, use specific county-state term.

Caribou County Idaho (1989)
SE Idaho. Before 1989, also search Caribou County AND Idaho.
CO N422300N430200
   W1110300W1121000
BT Idaho
BT United States

Carinthia
No longer a valid term for GeoRef. As of 1993, see Carinthia Austria.

Carinthia Austria (1993)
CO N463000N470000
   E0150000E0123000
BT Austria
BT Central Europe
BT Europe
NT Bleiberg Austria
NT Krappfeld
NT Sau Alps
NT Villach Austria
NT Wolfsberg Austria
SA Carnic Alps
SA Gailtal Alps
SA Karawanken
SA Koralpe Range
SA South Austrian Alps

Carixian (1978)
Europe.
BT Lower Jurassic
BT Jurassic
BT Mesozoic

Carleton County New Brunswick (1993)

Before 1993, also search Carleton County AND New Brunswick.
BT New Brunswick
BT Maritime Provinces
BT Eastern Canada
BT Canada
NT Woodstock New Brunswick

Carlile Shale (1978)
In Colorado Group. E Colorado, Kansas, SE Montana, Nebraska, NE New Mexico, South Dakota, E Wyoming.
BT Upper Cretaceous
BT Cretaceous
BT Mesozoic
SA Benton Formation
SA Codell Sandstone Member
SA Colorado
SA Colorado Group
SA Kansas
SA Montana
SA Nebraska
SA New Mexico
SA South Dakota
SA Wyoming

Carlin
Valid through 1988. Search in combination with state term. After 1988, use specific city-state term.

Carlin Mine (1978)
Gold ores. NE Nevada. Also search Carlin AND Nevada.
BT Nevada
BT United States
SA gold ores
SA mines

carlin model
use carlin-type

Carlin Nevada (1989)
Town in NE Nevada. Before 1989, search Carlin AND Nevada.
CO N404400N404400
   W1160600W1160600
BT Elko County Nevada
BT Nevada
BT United States

carlin type
use carlin-type

carlin-type (1985)
UF carlin model
UF carlin type
UF noseeum
SA disseminated deposits
SA epithermal processes
SA gold ores
SA mineral deposits, genesis

Carlsbad
As of 1989, use Carlsbad New Mexico for the city in New Mexico. For the town in Czechoslovakia, use Karlovy Vary.

Carlsbad Caverns (1978)
In SW Eddy County just N of Texas border.
BT Eddy County New Mexico
BT New Mexico
BT United States

Carlsbad New Mexico (1989)
City in SE New Mexico. Before 1989, search Carlsbad AND New Mexico.
CO N322500N322500
   W1041400W1041400
BT Eddy County New Mexico
BT New Mexico
BT United States

Carlsberg Ridge (1978)

Between Somalia and the Seychelles to the W, and the Laccadive and Maldive islands to the E.
CO S090000N100000
E0700000E0570000
UF Arabian Ridge
UF Arabian-Indian Midoceanic Ridge
UF Arabian-Indian Ridge
BT Indian Ocean
SA mid-ocean ridges

Carlton Rhyolite (1989)
SW Oklahoma.
BT Middle Cambrian
BT Cambrian
BT Paleozoic
SA Oklahoma

Carmarthen County
Not a valid term for GeoRef as of 1993. See Carmarthenshire Wales.

Carmarthenshire
No longer a valid term for GeoRef as of 1993.
use Carmarthenshire Wales

Carmarthenshire Wales (1993)
Former county in S Wales on Bristol Channel. As of April of 1974, considered part of Dyfed. Before 1993, also search Carmarthenshire or Carmarthen County.
UF Carmarthenshire
BT Dyfed Wales
BT Wales
BT Great Britain
BT United Kingdom
BT Western Europe
BT Europe

Carmel
Valid through 1988. Search in combination with state term. After 1988, use specific city-state term.

Carmel Bay (1978)
NW Monterey County.
IN Index counties or regions as applicable.
SA Monterey County California

Carmel California (1989)
City in W California. Before 1989, search Carmel AND California.
CO N363400N363400
W1215600W1215600
UF Carmel-by-the-Sea
BT Monterey County California
BT California
BT United States

Carmel Formation (1985)
Middle and Upper Jurassic. In San Rafael Group. Utah, NE Arizona, W Colorado and NW New Mexico.
BT Jurassic
BT Mesozoic
SA Arizona
SA Colorado
SA New Mexico
SA San Rafael Group
SA Utah

Carmel-by-the-Sea
use Carmel California

Carmona
No longer a valid term for GeoRef. As of 1993, see Carmona Spain.

Carmona Spain (1993)
City in SW Spain. Before 1993, also search Carmona and Spain.
BT Seville Spain
BT Andalusia Spain
BT Spain
BT Iberian Peninsula
BT Southern Europe
BT Europe

carnallite (1978)
BT chlorides
BT halides

Carnarvon Basin (1978)
S of Lake McCleod in Carnarvon region in W Western Australia.
CO S270000S210000
E1160000E1143000
BT Western Australia
BT Australia
BT Australasia

Carnegie Ridge (1978)
Just S of the equator between Ecuador and Galapagos Islands. As of 1990, Pacific Ocean is autoposted to this term.
CO S030000N010000
W0810000W0920000
BT South American Pacific
BT Pacific Ocean
SA Leg 16

carnelian (1985)
BT silica minerals
BT framework silicates
BT silicates
SA chalcedony
SA gems

Carnian (1978)
Europe. Above Ladinian, below Norian.
BT Upper Triassic
BT Triassic
BT Mesozoic
SA Hallstatt Limestone

Carnic Alps (1978)
Range of Eastern Alps.
IN Index countries as applicable.
BT Eastern Alps
BT Alps
BT Europe
SA Carinthia Austria
SA Friuli-Venezia Giulia Italy
SA Karawanken
SA Veneto Italy

Carnivora (1978)
Order. Autoposting of Eutheria and Theria to this term began in 1989.
BT Eutheria
BT Theria
BT Mammalia
BT Tetrapoda
BT Vertebrata
BT Chordata
NT Fissipeda
NT Pinnipedia

Carnmenellis Granite (1989)
SA England
SA Scotland

carnotite (1978)
BT vanadates
SA uranium ores
SA vanadium ores

Carolina Bays (1981)
Depressions on the E coast of the U.S. from Florida to New Jersey that are of controversial origin.
IN Index states as applicable.
BT United States
SA Atlantic Coastal Plain
SA Delaware
SA Florida
SA geomorphology
SA Georgia
SA Maryland
SA New Jersey
SA North Carolina
SA South Carolina
SA Virginia

Carolina slate belt (1978)
W North Carolina and NW South Carolina.
IN Index states as applicable.
BT Appalachians
BT North America
SA Avalon Terrane
SA Charlotte Belt
SA Kings Mountain Belt
SA Kiokee Belt
SA North Carolina
SA Raleigh Belt
SA South Carolina

Caroline Islands (1978)
Central and W island group of U.S. Trust Territory of the Pacific Islands. In 1981, broader term changed from Pacific Ocean to Micronesia. Before 1970, was a level 1 term in subfiles B and E.
CO N050000N100000
E1660000E1300000
UF Carolines
BT Micronesia
BT Oceania
NT Belau
SA Pacific Ocean

Carolines
use Caroline Islands

carotenoids (1978)
BT pigments
BT organic materials

Carpathian Foredeep (1978)
Elongated depressions bordering Carpathians. Also search Carpathian AND foredeep.
IN Index countries as applicable.
BT Europe
SA Carpathians
SA Czechoslovakia
SA Poland
SA Romania
SA Ukraine

Carpathian Foreland (1978)
Stable area marginal to Carpathians. Also search Carpathian AND foreland.
IN Index countries as applicable.
BT Europe
SA Carpathians
SA Czechoslovakia
SA Poland
SA Romania
SA Subcarpathians
SA Ukraine

Carpathian Mountains
use Carpathians

Carpathians (1978)
Mountain system of E and central Europe enclosing the Alfold. Before 1973, also search Carpathian Mountains.
IN Index countries as applicable.
CO N460000N495000
E0280000E0180000
UF Carpathian Mountains
BT Europe
NT Beshchady Mountains
NT Beskid Mountains
NT Bistrita Mountains
NT Bukk Mountains
NT Eastern Carpathians
NT Matra Mountains
NT Polish Carpathians
NT Slovakian Carpathians
NT Subcarpathians
NT Tatra Mountains
NT Ukrainian Carpathians
NT Vrancea
NT Western Carpathians
SA Carpathian Foredeep
SA Carpathian Foreland
SA Czechoslovakia
SA Hungary
SA Low Tatra Mountains
SA Paring Mountains
SA Pieniny Klippen Belt
SA Poland
SA Rodna Mountains
SA Romania
SA Spis-Gemer
SA Tokaj-Eperjes Mountains
SA Transylvanian Alps
SA Ukraine

Carpentaria Basin (1978)
Region around Gulf of Carpentaria.
IN Index Northern Territory and/or Queensland as applicable.
CO S190000S150000
E1430000E1400000
BT Australia
BT Australasia
SA Northern Territory Australia
SA Queensland Australia

Carpentarian (1978)
SE Australia. Autoposting of middle Precambrian to this term ended in 1989. As of 1989, lower Proterozoic and Proterozoic are autoposted to this term. As of 1990, upper Precambrian and Precambrian are autoposted to this term.
BT lower Proterozoic
BT Proterozoic
BT upper Precambrian
BT Precambrian

carpholite (1978)
BT chain silicates
BT silicates
SA aluminosilicates
SA ferrocarpholite

Carrara Formation (1989)
Lower to Middle Cambrian. California and S Nevada. Also search Carrara Marble NOT Italy. For Carrara Marble in Italy, search Carrara Marble AND Italy.
BT Cambrian
BT Paleozoic
SA California
SA Nevada

Carrara Marble (1993)
From marble deposits at Carrara, NW Italy. Material used throughout the world for monuments and sculptures.
SA marble deposits
SA materials
SA ornamental materials

Carriacou (1978)
Largest island of the Grenadines in the Windward Islands. Before 1981, West Indies was the broader term.
IN Index Windward Islands.
BT Lesser Antilles
BT Antilles
BT West Indies
BT Caribbean region
SA Grenada
SA Windward Islands

Carrizo Mountain Formation (1985)
W Texas.
BT Precambrian
SA Texas

Carrizo Sand (1989)

In Claiborne Group.
BT middle Eocene
BT Eocene
BT Paleogene
BT Tertiary
BT Cenozoic
SA Arkansas
SA Claiborne Group
SA Louisiana
SA Texas

Carroll County
Valid through 1988. Search in combination with state term. After 1988, use specific county-state term.

Carroll County Arkansas (1989)
NW Arkansas. Before 1989, also search Carroll County AND Arkansas.
CO N361100N363000 W0931600W0935300
BT Arkansas
BT United States

Carroll County Georgia (1989)
W Georgia. Before 1989, also search Carroll County AND Georgia.
CO N332500N334700 W0844900W0852000
BT Georgia
BT United States

Carroll County Illinois (1989)
NW Illinois. Before 1989, also search Carroll County AND Illinois.
CO N415600N421500 W0894100W0901300
BT Illinois
BT United States

Carroll County Indiana (1989)
NW central Indiana. Before 1989, also search Carroll County AND Indiana.
CO N402500N404300 W0862300W0864500
BT Indiana
BT United States

Carroll County Iowa (1989)
W central Iowa. Before 1989, also search Carroll County AND Iowa.
CO N415300N421300 W0943800W0950600
BT Iowa
BT United States

Carroll County Kentucky (1989)
N Kentucky. Before 1989, also search Carroll County AND Kentucky.
CO N383600N384500 W0845500W0852500
BT Kentucky
BT United States
SA Eagle Station Meteorite

Carroll County Maryland (1989)
N Maryland. Before 1989, also search Carroll County AND Maryland.
CO N392200N394330 W0764800W0771900
BT Maryland
BT United States

Carroll County Mississippi (1989)
Central Mississippi. Before 1989, also search Carroll County AND Mississippi.
CO N331200N334100 W0894000W0901300
BT Mississippi
BT United States

Carroll County Missouri (1989)
NW central Missouri. Before 1989, also search Carroll County AND Missouri.
CO N391200N393700 W0930700W0934800
BT Missouri
BT United States

Carroll County New Hampshire (1989)
On Maine line in E New Hampshire. Before 1989, also search Carroll County AND New Hampshire.
CO N432800N441700 W0705600W0713400
BT New Hampshire
BT United States
NT Ossipee Mountains
SA Mirror Lake
SA Saco River

Carroll County Ohio (1989)
E Ohio. Before 1989, also search Carroll County AND Ohio.
CO N402500N404300 W0805100W0811900
BT Ohio
BT United States

Carroll County Tennessee (1989)
NW Tennessee. Before 1989, also search Carroll County AND Tennessee.
CO N354700N360900 W0881200W0884200
BT Tennessee
BT United States

Carroll County Virginia (1989)
SW Virginia. Before 1989, also search Carroll County AND Virginia.
CO N363300N365600 W0802700W0810300
BT Virginia
BT United States

carrollite (1978)
UF sychnodymite
BT sulfides
SA cobalt ores
SA linnaeite

Carry-le-Rouet (1978)
A section in S Bouches-du-Rhone.
BT Bouches-du-Rhone France
BT France
BT Western Europe
BT Europe

carst
use karst

Carswell Structure (1978)
Lake Athabasca area of NW Saskatchewan, and the Baker Lake area in Keewatin District of the Northwest Territories.
IN Index province and territory as applicable.
BT Canada
SA Northwest Territories
SA Saskatchewan

Cartagena
No longer a valid term for GeoRef. As of 1993, see Cartagena Colombia.

Cartagena Colombia (1993)
City on the Caribbean Sea. Before 1993, search Bolivar AND Colombia.
BT Bolivar Colombia
BT Colombia
BT South America

Carter County
Valid through 1988. Search in combination with state term. After 1988, use specific county-state term.

Carter County Kentucky (1989)
NE Kentucky. Before 1989, also search Carter County AND Kentucky.
CO N381000N383000 W0824700W0832300
BT Kentucky
BT United States

Carter County Missouri (1989)
S Missouri. Before 1989, also search Carter County AND Missouri.
CO N364800N370700 W0903900W0911500
BT Missouri
BT United States

Carter County Montana (1989)
SE Montana. Before 1989, also search Carter County AND Montana.
CO N450000N460800 W1040300W1050200
BT Montana
BT United States
SA Bell Creek Field

Carter County Oklahoma (1989)
S Oklahoma. Before 1989, also search Carter County AND Oklahoma.
CO N340400N343100 W0965600W0973300
BT Oklahoma
BT United States
SA Arbuckle Mountains

Carter County Tennessee (1989)
NE Tennessee. Before 1989, also search Carter County AND Tennessee.
CO N360700N363000 W0815600W0822300
BT Tennessee
BT United States

Carteret County
Valid through 1988. Search in combination with state term. After 1988, use specific county-state term.

Carteret County North Carolina (1989)
E North Carolina, on the Atlantic Ocean. Before 1989, also search Carteret County AND North Carolina.
CO N343500N350500 W0760200W0771000
BT North Carolina
BT United States
NT Beaufort North Carolina
SA Cape Lookout

Carterinacea (1978)
Autoposting of microfossils and Protista to this term began in 1990. This term has multiple hierarchies.
BT1 Rotaliina
BT1 foraminifera
BT1 Protista
BT1 Invertebrata
BT2 Rotaliina
BT2 foraminifera
BT2 Protista
BT2 microfossils

Cartersville
Valid through 1988. Search in combination with state term. After 1988, use specific city-state term.

Cartersville Georgia (1989)
City in NW Georgia. Before 1989, search Cartersville AND Georgia.
CO N340900N340900 W0844900W0844900
BT Bartow County Georgia
BT Georgia
BT United States

cartographic scales (1982)
UF map scales
UF scales, cartographic
SA cartography
SA maps
SA scale factor

cartography (1978)
As of 1993, restricted to the art and science of mapmaking. For methods, instruments, programs. Before 1972, also search geological exploration; before 1976 also search geologic mapping; before 1978 also search mapping. As of 1993, see mapping for the processes of gathering data for the production of a specific map.
UF chartology
NT digital cartography
SA aerial photography
SA altimetry
SA cartographic scales
SA coordinates
SA geodesy
SA geomorphology
SA global maps
SA legends
SA leveling
SA mapping
SA maps
SA photogeologic methods
SA photogeology
SA photogrammetry
SA stereographic projection
SA topographic correction
SA topography
SA triangulation
SA trilateration

Carya (1989)
Genus.
BT Dicotyledoneae
BT angiosperms
BT Spermatophyta
BT Plantae

Caryophyllidae (1981)
BT Dicotyledoneae
BT angiosperms
BT Spermatophyta
BT Plantae

Casa Diablo (1989)
Hot springs in Long Valley Caldera, E California.
IN Index counties or regions as applicable.
UF Casa Diablo hot springs
SA Long Valley Caldera
SA Mono County California

Casa Diablo hot springs
use Casa Diablo

Casablanca
No longer a valid term for GeoRef.

Casablanca Morocco (1993)
City bordering the Atlantic Ocean in W Morocco.
BT Morocco
BT North Africa
BT Africa

Before then, variants of the term may occur in GeoRef.

Casapalca
No longer a valid term for GeoRef. As of 1993 see Casapalca Peru.

Casapalca Peru (1993)
Town NE of city of Lima. Before 1993 also search Casapalca AND Peru.
BT Lima Peru
BT Peru
BT South America

Cascade County
Valid through 1988. Search in combination with state term. After 1988, use specific county-state term.

Cascade County Montana (1989)
W central Montana. Before 1989, also search Cascade County AND Montana.
CO N464900N474100 W1103900W1120300
BT Montana
BT United States
SA Little Belt Mountains

Cascade Mountains
Not a valid term for GeoRef. Before 1971, was used in subfile N.
use Cascade Range

Cascade Range (1978)
Northern continuation of the Sierra Nevada Mountains. Before 1971, also search Cascade Mountains. The Coast Mountains continue the range in British Columbia.
IN Index states as applicable.
CO N400000N510000 W1203000W1230000
UF Cascade Mountains
UF Cascades
BT United States
NT Mount Hood
NT Mount Rainier
NT Mount Saint Helens
NT Mount Shasta
NT Newberry Volcano
SA Brothers fault zone
SA California
SA Coast Mountains
SA Glacier Peak
SA Golden Horn Batholith
SA Mount Adams
SA Mount Baker
SA Mount Jefferson
SA North American Cordillera
SA Oregon
SA Shuksan Thrust
SA Sierra Nevada
SA Straight Creek Fault
SA Washington

cascades (1978)
BT fluvial features
SA geomorphology
SA hydrology
SA rapids
SA streams
SA waterfalls

Cascades
use Cascade Range

Cascadia Basin (1978)
Off the coast of Washington.
BT Pacific Ocean
SA Cascadia subduction zone

Cascadia Channel (1978)
Off coast of Oregon SW of Cascadia Basin.
BT Pacific Ocean

Cascadia subduction zone (1993)
Juan de Fuca Plate is being subducted beneath North American Plate. Extends from Vancouver Island to the N California coast. Before 1993, also search Cascadia and subduction.
SA Cascadia Basin
SA Juan de Fuca Plate
SA North American Pacific
SA North American Plate
SA plate tectonics

Casco Bay Group (1989)
Lower and middle Paleozoic. SW Maine. Includes Cape Elizabeth Formation, Spring Point Greenstone, Diamond Island Slate, Scarboro Phyllite, Spurwink Limestone, Jewell Phyllite and Mackworth Slate.
BT Paleozoic
SA Cape Elizabeth Formation
SA Maine

case histories
As of 1981, no longer a valid term for GeoRef.
use case studies

case studies (1978)
Before 1981, also search case histories.
UF case histories
UF studies, case
SA pollution
SA rock mechanics
SA soil mechanics

Caserta
No longer a valid term for GeoRef. See Caserta Italy.

Caserta Italy (1993)
Province and city NE of Naples.
BT Campania Italy
BT Italy
BT Southern Europe
BT Europe

Caseyville Formation (1978)
In McCormick Group. SE Illinois and W Kentucky.
BT Lower Pennsylvanian
BT Pennsylvanian
BT Carboniferous
BT Paleozoic
SA Illinois
SA Kentucky

cash transactions (1981)
SA economics

Cashel
No longer a valid term for GeoRef. See Cashel Ireland.

Cashel Ireland (1993)
Town at the base of the Rock of Cashel in S central Tipperary, S central Ireland.
BT Tipperary Ireland
BT Ireland
BT Western Europe
BT Europe

Casper
Valid through 1988. Search in combination with state term. After 1988, use specific city-state term.

Casper Formation (1978)
Pennsylvanian and Permian. Includes all beds from top of Madison Group to base of Permian (Opeche Shale) red shales. In SE Wyoming.
IN Index ages as applicable.
BT Paleozoic
SA Pennsylvanian
SA Permian
SA Wyoming

Casper Mountain (1989)
Central Wyoming.
IN Index counties or regions as applicable.
SA Natrona County Wyoming

Casper Wyoming (1989)
City in central Wyoming. Before 1989, search Casper AND Wyoming.
CO N425000N425000 W1062000W1062000
BT Natrona County Wyoming
BT Wyoming
BT United States

Caspian Basin (1978)
Region including the Caspian Sea and surrounding areas. Also search Caspian.
IN Index Asia or Europe; and Caspian Sea, former Soviet republics and/or Iran as applicable.
CO N363000N490000 E0553000E0450000
UF Caspian region
UF Peri-Caspian Depression
SA Azerbaidzhan
SA Caspian Depression
SA Caspian Sea
SA Eurasia
SA Iran
SA Kazakhstan
SA Russian Federation
SA Turkmenia

Caspian Depression (1978)
That part of the Caspian Basin lying N of the Caspian Sea; it is a large lowland area which was submerged by the Caspian Sea at one time. Also search Caspian.
IN Index former Soviet republics as applicable.
CO N450000N520000 E0570000E0423000
UF Caspian Lowland
UF Caspian Syneclise
BT Central Asia
BT Asia
SA Astrakhan Arch
SA Caspian Basin
SA Karachaganak Field
SA Kazakhstan
SA Russian Federation

Caspian Lowland
use Caspian Depression

Caspian region
use Caspian Basin

Caspian Sea (1978)
Largest inland sea in the world. Also search Caspian.
IN Index Iran and/or the former Soviet republics as applicable.
CO N360000N480000 E0530000E0470000
SA Caspian Basin
SA Iran
SA Kara-Bogaz Gulf
SA Neftyanyye Kamni
SA Peschanyy
SA Russian Federation

Caspian Syneclise
use Caspian Depression

Cass County
Valid through 1988. Search in combination with state term. After 1988, use specific county-state term.

Cass County Illinois (1989)
W central Illinois. Before 1989, also search Cass County AND Illinois.
CO N395200N400800 W0900000W0903000
BT Illinois
BT United States

Cass County Indiana (1989)
N central Indiana. Before 1989, also search Cass County AND Indiana.
CO N403400N405400 W0861000W0863500
BT Indiana
BT United States

Cass County Iowa (1989)
SW Iowa. Before 1989, also search Cass County AND Iowa.
CO N411000N413000 W0944200W0950900
BT Iowa
BT United States

Cass County Michigan (1989)
SW Michigan. Before 1989, also search Cass County AND Michigan.
CO N414500N420500 W0854500W0861600
BT Michigan Lower Peninsula
BT Michigan
BT United States

Cass County Minnesota (1989)
N central Minnesota. Before 1989, also search Cass County AND Minnesota.
CO N462000N472800 W0934500W0944800
BT Minnesota
BT United States

Cass County Missouri (1989)
W Missouri Before 1989, also search Cass County AND Missouri.
CO N382700N384900 W0940300W0943700
BT Missouri
BT United States

Cass County Nebraska (1989)
SE Nebraska. Before 1989, also search Cass County AND Nebraska.
CO N404700N410600 W0954700W0962900
BT Nebraska
BT United States

Cass County North Dakota (1989)
E North Dakota. Before 1989, also search Cass County AND North Dakota.
CO N463800N471500 W0964700W0974300
BT North Dakota
BT United States

Cass County Texas (1989)
NE Texas. Before 1989, also search Cass County AND Texas.
CO N325200N331600 W0940200W0944000
BT Texas
BT United States

Cassia County
Valid through 1988. Search in combination with state term. After 1988, use specific county-state term.

Cassia County Idaho (1989)

On Utah and Nevada borders in S central Idaho. Before 1989, also search Cassia County AND Idaho.
  CO N420000N424000
     W1130000W1142000
  BT Idaho
  BT United States

Cassiar
  use Cassiar Mountains

Cassiar Mountains (1985)
  N British Columbia and S Yukon Territory.
  UF Cassiar
  BT Canada
  SA British Columbia
  SA Canadian Rocky Mountains
  SA Yukon Territory

Cassidulina (1985)
  Genus. Autoposting of microfossils and Protista to this term began in 1990. This term has multiple hierarchies.
  BT1 Cassidulinacea
  BT1 Rotaliina
  BT1 foraminifera
  BT1 Protista
  BT1 Invertebrata
  BT2 Cassidulinacea
  BT2 Rotaliina
  BT2 foraminifera
  BT2 Protista
  BT2 microfossils

Cassidulinacea (1978)
  Autoposting of microfossils and Protista to this term began in 1990. This term has multiple hierarchies.
  BT1 Rotaliina
  BT1 foraminifera
  BT1 Protista
  BT1 Invertebrata
  BT2 Rotaliina
  BT2 foraminifera
  BT2 Protista
  BT2 microfossils
  NT Anomalinidae
  NT Cassidulina
  NT Globocassidulina subglobosa

cassiterite (1978)
  BT oxides
  SA heavy minerals
  SA tin ores

Castaic Formation (1985)
  S California.
  BT upper Miocene
  BT Miocene
  BT Neogene
  BT Tertiary
  BT Cenozoic
  SA California

Castellane
  No longer a valid term for GeoRef. See Castellane France.

Castellane France (1993)
  Village in SE France.
  BT Alpes-de-Haute Provence France
  BT France
  BT Western Europe
  BT Europe

Castellon de la Plana
  No longer a valid term for GeoRef. As of 1993, see Castellon de la Plana City Spain.

Castellon de la Plana City Spain (1993)
  Refers to only the city in E Spain. Before 1993, also search Castellon de la Plana and Spain.
  BT Castellon de la Plana Spain

  BT Valencia region
  BT Spain
  BT Iberian Peninsula
  BT Southern Europe
  BT Europe

Castellon de la Plana Province
  No longer a valid term for GeoRef.
  use Castellon de la Plana Spain

Castellon de la Plana Spain (1993)
  Refers only to the province in E Spain. From 1981-1992, also search Castellon de la Plana Province and Spain. Before 1981, also search Castellon de la Plana and Spain. For the city, see Castellon de la Plana City Spain.
  CO N394200N404800
     E0003000W0005300
  UF Castellon de la Plana Province
  BT Valencia region
  BT Spain
  BT Iberian Peninsula
  BT Southern Europe
  BT Europe
  NT Castellon de la Plana City Spain
  SA Maestrazgo Spain

Castile
  No longer a valid term for GeoRef. As of 1993, see Castile Spain.

Castile Formation (1978)
  Formation, as now defined, includes beds in lower part of Ochoa Series that are confined in extent to Delaware Basin and overlap sloping surface of Capitan Formation along its margins. SE New Mexico, and W Texas.
  BT Permian
  BT Paleozoic
  SA Capitan Formation
  SA New Mexico
  SA Texas

Castile Spain (1993)
  Region and ancient kingdom of central and N central part of country consisting of 13 modern provinces. Before 1993, also search Castile and Spain.
  BT Spain
  BT Iberian Peninsula
  BT Southern Europe
  BT Europe
  NT New Castile Spain
  NT Old Castile Spain

Castilla la Nueva
  use New Castile Spain

Castilla la Vieja
  use Old Castile Spain

Castillon Massif (1978)
  In Ariege Pyrenees in S France. This term has multiple hierarchies.
  BT1 Ariege France
  BT1 France
  BT1 Western Europe
  BT1 Europe
  BT2 Pyrenees
  BT2 Europe
  SA French Pyrenees

Castle Hayne
  Valid through 1988. Search in combination with state term. After 1988, use specific city-state term.

Castle Hayne Limestone (1978)
  Middle and upper Eocene. Underlies Yorktown Formation near Pollicksville, North Carolina. Coastal Plain of North Carolina and South Carolina.

  BT Eocene
  BT Paleogene
  BT Tertiary
  BT Cenozoic
  SA North Carolina
  SA South Carolina

Castle Hayne North Carolina (1989)
  City in extreme SE North Carolina. Before 1989, search Castle Hayne AND North Carolina.
  CO N342000N342000
     W0775400W0775400
  BT New Hanover County North Carolina
  BT North Carolina
  BT United States

Castleton
  Valid through 1988. Search in combination with state term. After 1988, use specific city-state term.

Castleton Vermont (1989)
  City in S central Vermont. Before 1989, search Castleton AND Vermont.
  CO N433600N433600
     W0731100W0731100
  BT Rutland County Vermont
  BT Vermont
  BT United States

Castoridae (1978)
  Family. Autoposting of Eutheria and Theria to this term began in 1989. Autoposting of Sciuromorpha ended in 1989.
  BT Rodentia
  BT Eutheria
  BT Theria
  BT Mammalia
  BT Tetrapoda
  BT Vertebrata
  BT Chordata

casts (1982)
  Includes use as the sedimentary structure as well as the paleontologic feature.
  SA bedding plane irregularities
  SA chemical fossils
  SA flute casts
  SA fossilization
  SA fossils
  SA groove casts
  SA ichnofossils
  SA lebensspuren
  SA load casts
  SA paleontology
  SA scour casts
  SA sedimentary structures
  SA soft sediment deformation
  SA turbidity current structures

casts, flute
  use flute casts

casts, groove
  use groove casts

casts, load
  use load casts

cataclasis (1981)
  SA brittle deformation
  SA deformation
  SA metamorphism

cataclasites (1978)
  BT metamorphic rocks

catagenesis (1978)
  UF katagenesis
  SA sedimentary rocks

Catahoula Formation (1985)
  Louisiana, Mississippi, S Alabama and E Texas.
  BT Tertiary

  BT Cenozoic
  SA Alabama
  SA Louisiana
  SA Miocene
  SA Mississippi
  SA Oligocene
  SA Texas

Catalina Schist (1989)
  S California. Mesozoic or older.
  IN Index age as applicable.
  SA California
  SA Mesozoic
  SA Paleozoic

catalog
  No longer a valid term for GeoRef. Before 1972, included use on level 1 in subfile G. Before 1971, also included use on level 1 in subfile N.
  use catalogs

catalogs (1978)
  IN Index language if not English.
  UF catalog
  SA Arabic
  SA associations
  SA atlas
  SA bibliography
  SA book reviews
  SA collections
  SA education
  SA exhibits
  SA French
  SA geology
  SA German
  SA glossaries
  SA lexicons
  SA libraries
  SA maps
  SA mineralogy
  SA monographs
  SA museums
  SA oceanography
  SA paleobotany
  SA paleontology
  SA petrology
  SA publications
  SA Russian
  SA sedimentary petrology
  SA stratigraphy
  SA structural geology
  SA survey organizations
  SA volcanology

Catalonia
  No longer a valid term for GeoRef. As of 1993, see Catalonia Spain.

Catalonia Spain (1993)
  Historical region in NE Spain. Before 1993, also search Catalonia and Spain.
  IN Index provinces as applicable.
  CO N403000N423000
     E0033000E0000000
  BT Spain
  BT Iberian Peninsula
  BT Southern Europe
  BT Europe
  NT Barcelona Spain
  NT Gerona Spain
  NT Lerida Spain
  NT Tarragona Spain
  SA Catalonian Coastal Ranges
  SA Ter River basin

Catalonian Coastal Ranges (1981)
  BT Spain
  BT Iberian Peninsula
  BT Southern Europe
  BT Europe
  SA Catalonia Spain

catalysis (1978)

SA chemical analysis
SA geochemistry

Catamarca
No longer a valid term for GeoRef. See Catamarca Argentina.

Catamarca Argentina (1993)
Province including its capital, NW Argentina.
CO S300000S250000
    W0650000W0690000
BT Argentina
BT South America
SA Pampean Mountains

Catanda
No longer a valid term for GeoRef. After 1992, use Catanda Angola.

Catanda Angola (1993)
Village in E Angola.
BT Angola
BT Central Africa
BT Africa

Catanzaro
No longer a valid term for GeoRef. See Catanzaro Italy.

Catanzaro Italy (1993)
Province and city in central Calabria.
BT Calabria Italy
BT Italy
BT Southern Europe
BT Europe

cataphoresis
use electrophoresis

catastrophes (1978)
SA destruction
SA geologic hazards
SA mass extinctions

catastrophic
A valid term through 1977.

catastrophic advance
use glacier surges

catastrophic waves (1978)
Before 1978, also search catastrophic AND ocean waves.
BT ocean waves
SA tsunamis
SA waves

catastrophism (1978)
SA creationism
SA uniformitarianism

catazonal metamorphism (1982)
UF katazonal metamorphism
BT metamorphism
SA high-grade metamorphism

catchment hydrodynamics (1981)
SA drainage basins
SA ground water
SA hydrodynamics
SA hydrology

catchments
use drainage basins

catenas (1978)
Soil association of a given area developed from common parent material.
SA soils

cathodoluminescence (1978)
SA luminescence

cation exchange
A valid term through 1978. After 1978, use ion exchange or cation exchange capacity.

cation exchange capacity (1978)

Also search cation exchange.
SA clay mineralogy
SA geochemistry
SA ion exchange
SA ions

cations (1978)
Term reintroduced in 1981. Was valid through May 1978. Also search ions.
SA anions
SA geochemistry
SA ions

Catoche Formation (1989)
W Newfoundland.
BT Ordovician
BT Paleozoic
SA Newfoundland Island

Catoctin Formation (1978)
Divided into Warrenton Agglomerate Member at base and basalt flows now altered to metabasalt or greenstone schist. W Maryland, N Virginia, and NE West Virginia.
BT Precambrian
SA Maryland
SA Virginia
SA West Virginia

Catron County
Valid through 1988. Search in combination with state term. After 1988, use specific county-state term.

Catron County New Mexico (1989)
On Arizona border in W New Mexico. Before 1989, also search Catron County AND New Mexico.
CO N331000N343500
    W1074500W1090400
BT New Mexico
BT United States
NT Mogollon Mountains
SA San Francisco Mountains

Catskill Delta (1978)
Devonian age complex.
IN Index states as applicable.
BT United States
SA Maryland
SA New York
SA Pennsylvania
SA West Virginia

Catskill Formation (1978)
Middle Devonian to Lower Mississippian. In Susquehanna Group.
IN Index ages as applicable.
BT Paleozoic
SA Devonian
SA Lower Mississippian
SA Maryland
SA New York
SA Pennsylvania
SA Virginia

Catskill Mountains (1978)
Group of the Appalachians in SE New York. This term has multiple hierarchies.
BT1 Appalachians
BT2 New York
BT2 United States
SA Appalachian Plateau

Cattaraugus County
Valid through 1988. Search in combination with state term. After 1988, use specific county-state term.

Cattaraugus County New York (1989)

SW New York. Before 1989, also search Cattaraugus County AND New York.
CO N420000N423300
    W0781900W0790400
BT New York
BT United States
NT West Valley New York

cattierite (1978)
BT sulfides

Caucasian Foreland
use Caucasus Foreland

Caucasus (1978)
Mountain system between the Black and Caspian seas. Autoposting of this term began in 1981.
IN Index Greater Caucasus, Lesser Caucasus, Northern Caucasus, and former Soviet republics as applicable.
CO N410000N440000
    E0480000E0400000
UF Caucasus Mountains
BT Europe
NT Greater Caucasus
NT Lesser Caucasus
NT Northern Caucasus
SA Armenia
SA Azerbaidzhan
SA Elbrus
SA Georgian Republic
SA Ossetia
SA Russian Federation
SA Transcaucasia

Caucasus Foreland (1978)
The Northern Caucasus or Ciscaucasia area which includes the northern slopes and foothills of the Caucasus plus the Kuban steppes to the north. This term has multiple hierarchies.
UF Caucasian Foreland
BT1 Russian Federation
BT1 Commonwealth of Independent States
BT2 Europe
SA Kuban
SA Northern Caucasus

Caucasus Mountains
use Caucasus

cauldrons (1978)
BT volcanic features
SA calderas
SA crater lakes
SA geomorphology
SA volcanism
SA volcanology

causes (1978)
Appropriate to a large number of general topics.

Causses (1978)
An arid Jurassic limestone plateau of S Central Massif in S and SW France.
UF South Massif
BT France
BT Western Europe
BT Europe
SA Central Massif

caustobiolith (1978)
This term has multiple hierarchies.
UF caustobioliths
BT1 organic residues
BT2 sedimentary rocks
SA organic materials

caustobioliths
use caustobiolith

Cauvery Basin (1978)

River basin in S Karnataka and Tamil Nadu, S India.
IN Index states as applicable.
UF Kaveri Basin
BT India
BT Indian Peninsula
BT Asia
SA Karnataka India
SA Tamil Nadu India

Caux (1978)
Chalky tableland in N France.
BT Seine-Maritime France
BT France
BT Western Europe
BT Europe

cave ecology
use cave environment

cave environment (1985)
UF cave ecology
SA caves
SA depositional environment
SA ecology
SA environment
SA paleoecology
SA sedimentation
SA solution features
SA speleology
SA terrestrial environment

cave maps (1981)
Before 1981, also search caves AND maps.
BT maps
SA caverns
SA caves

Cavellinidae (1978)
As of 1990, microfossils, Crustacea, Mandibulata, and Arthropoda are autoposted to this term. This term has multiple hierarchies.
BT1 Podocopida
BT1 Ostracoda
BT1 Crustacea
BT1 Mandibulata
BT1 Arthropoda
BT1 Invertebrata
BT2 Podocopida
BT2 Ostracoda
BT2 microfossils
SA Healdiidae

caverns (1978)
Autoposting of solution features to this term began in 1989.
BT solution features
SA cave maps
SA caves
SA geomorphology
SA stalactites
SA stalagmites
SA underground cavities
SA underground streams

caves (1978)
SA cave environment
SA cave maps
SA caverns
SA exsurgence
SA geomorphology
SA karst
SA shore features
SA solution
SA solution cavities
SA solution features
SA speleology
SA speleothems
SA stalactites
SA stalagmites
SA underground cavities
SA underground streams
SA volcanic features

cavities
Used as a general term through 1978. After 1978, see specific terms, e.g. underground space or solution cavities.

cavities, solution
use solution cavities

Cavnic
No longer a valid term for GeoRef. As of 1993 see Cavnic Romania.

Cavnic Romania (1993)
Village in Maramures County in N Transylvania. Before 1993 also search Cavnic AND Romania.
UF Capnic
BT Maramures Romania
BT Transylvania
BT Romania
BT Southern Europe
BT Europe

Cayman Islands (1981)
BT West Indies
BT Caribbean region
NT Grand Cayman Island
SA Bluff Formation

Cayman Trench
use Cayman Trough

Cayman Trough (1978)
South of the Cayman Islands and Cuba.
UF Cayman Trench
BT Caribbean Sea
BT North American Atlantic
BT North Atlantic
BT Atlantic Ocean

Caytoniales (1978)
Autoposting of this term began in 1978.
BT gymnosperms
BT Spermatophyta
BT Plantae
NT Pachypteris

Cayuga County
Valid through 1988. Search in combination with state term. After 1988, use specific county-state term.

Cayuga County New York (1989)
Central New York. Before 1989, also search Cayuga County AND New York.
CO N423700N432500
W0761800W0764500
BT New York
BT United States
SA Cayuga Lake

Cayuga Lake (1989)
One of the Finger Lakes in Cayuga, Seneca and Tompkins counties, W central New York.
IN Index counties as applicable.
CO N422600N425500
W0762800W0764800
BT Finger Lakes
BT New York
BT United States
SA Cayuga County New York
SA Seneca County New York
SA Tompkins County New York

Cayugan (1978)
Provincial series, North America. Above Niagaran, below Ulsterian (Devonian).
BT Upper Silurian
BT Silurian
BT Paleozoic
NT Tonoloway Limestone
NT Williamsport Sandstone

SA Keyser Limestone

Cd
use cadmium

CDP method
use common-depth-point method

Ce
use cerium

Ceara
No longer a valid term for GeoRef. See Ceara Brazil.

Ceara Brazil (1993)
State in NE Brazil.
CO S075000S025000
W0371000W0412000
BT Brazil
BT South America
NT Jaguaribe
NT Jaguaribe Brazil

Cebu
No longer a valid term for GeoRef. As of 1993 see Cebu Philippine Islands.

Cebu Philippine Islands (1993)
Province, city and island in Visayan Islands in E central Philippine Islands. Before 1993 also search Cebu AND Philippine Islands.
BT Philippine Islands
BT Far East
BT Asia

Cedar City Formation (1978)
BT Devonian
BT Paleozoic
SA Missouri

Cedar Creek Anticline (1989)
E Montana, W North Dakota and W South Dakota.
IN Index states as applicable.
BT United States
SA Montana
SA North Dakota
SA South Dakota
SA Williston Basin

Cedar Mountain Formation (1978)
Together with Buckhorn Conglomerate comprises Cedar Mountain Group. NW Colorado and central Utah.
BT Lower Cretaceous
BT Cretaceous
BT Mesozoic
SA Colorado
SA Utah

Cedar Valley Formation (1978)
Includes Linwood, Littleton, Coralville, Solon and Rapid members. E Iowa, SW Illinois, and SE Minnesota.
BT Middle Devonian
BT Devonian
BT Paleozoic
SA Illinois
SA Iowa
SA Minnesota

celadonite (1978)
BT mica group
BT sheet silicates
BT silicates
SA glauconite

Celebes (1978)
Island E of Borneo. Sulawesi is the official Indonesian name.
CO S053700N014500
E1250500E1184500
UF Sulawesi
BT Indonesia

BT Far East
BT Asia
SA Celebes Sea

Celebes Sea (1978)
Bounded on N by Philippine Islands, W by Borneo, and S by island of Celebes. Also search Celebes.
BT Indonesian Seas
BT West Pacific
BT Pacific Ocean
SA Celebes
SA Darvel Bay
SA Leg 124
SA Malay Archipelago
SA ODP Site 767
SA ODP Site 770

celestite (1978)
BT sulfates
SA strontium ores

Celje (1978)
Town in NW Yugoslavia.
BT Slovenia
BT Yugoslavia
BT Southern Europe
BT Europe

cell dimensions (1978)
SA bonding
SA crystal structure
SA dimensions
SA lattice
SA lattice parameters
SA minerals

cell, unit
use unit cell

cellulose (1978)
BT carbohydrates
BT organic materials
SA lignin

celsian (1978)
BT barium feldspar
BT feldspar group
BT framework silicates
BT silicates
SA paracelsian

Celtic Sea (1978)
Body of water over the continental shelf between the S coast of Ireland and Brittany. It is arbitrarily divided from the English Channel by a line from Land's End to Ushant Island off Brittany.
BT Atlantic Ocean

cement (1978)
Term restricted to use as the material. Use cementation for the process.
SA cement materials
SA cementation
SA cementation deposits
SA sedimentary rocks
SA sediments

cement materials (1978)
BT construction materials
SA cement
SA concrete
SA limestone deposits
SA materials

cementation (1978)
BT diagenesis
SA cement
SA cementation deposits
SA clastic sediments
SA consolidation
SA deposition
SA lithification
SA sedimentation

cementation deposits (1981)
SA cement

SA cementation
SA mineral deposits, genesis

Cenari Zone
use Ceneri Zone

Ceneri Zone (1978)
In European Southern Alps.
IN Index countries as applicable.
UF Cenari Zone
BT Europe
SA Italy
SA Paleozoic
SA Switzerland

Cenomanian (1978)
Europe. Above Albian, below Turonian. Autoposting of this term began in 1978.
BT Upper Cretaceous
BT Cretaceous
BT Mesozoic
NT lower Cenomanian
NT upper Cenomanian
SA Middle Cretaceous

Cenozoic (1978)
From 1978-1980, Phanerozoic was autoposted to this term. Autoposting of this term to Tertiary and Quaternary ended in 1981.
UF Cainozoic
NT Blancan
NT Bronze Age
NT Chalcolithic
NT Croatan Formation
NT Glenns Ferry Formation
NT Iron Age
NT Kobiwako Group
NT Koobi Fora Formation
NT Leda Clay
NT Loup Fork Group
NT lower Cenozoic
NT Matuyama Epoch
NT middle Cenozoic
NT Omo Group
NT Osaka Group
NT Quaternary
NT San Juan Formation
NT Santa Clara Formation
NT Santa Fe Group
NT Santa Maria Formation
NT Saugus Formation
NT Shungura Formation
NT Siwalik System
NT Stone Age
NT Tertiary
NT Tulare Formation
NT Uonuma Group
NT upper Cenozoic
NT Verde Formation
NT Wildcat Group
NT Yakataga Formation
SA Alpine Orogeny
SA Himalayan Orogeny
SA Phanerozoic

centers, color
use color centers

centers, spreading
use spreading centers

Central Africa (1978)
Before 1978, search Africa AND central. Autoposting of this term began in 1981.
CO S180000N113000
E0313000E0083000
BT Africa
NT Angola
NT Burundi
NT Central African Republic
NT Congo
NT Congo River
NT Equatorial Guinea
NT Gabon
NT Lake Kivu

NT Rwanda
NT Zaire
SA East Africa
SA North Africa
SA Southern Africa
SA West Africa

Central African Republic (1978)
Formerly French territory of Ubangi-Shari. Became independent in 1960. Before 1959, also search Ubangi-Shari; Oubangui-Chari.
CO N030000N120000
E0273000E0140000
UF Oubangui-Chari
UF Ubangi-Shari
BT Central Africa
BT Africa
NT Koto
SA Congo Basin

Central Alps (1978)
Ranges of the Alps extending from the Great Saint Bernard Pass in the W to Innsbruck and the Brenner Pass in the E.
IN Index countries as applicable.
CO N460000N470000
E0113000E0090000
BT Alps
BT Europe
NT Bavarian Alps
NT Bernese Alps
NT Glarus Alps
NT Lepontine Alps
NT Pennine Alps
NT Rhaetian Alps
SA Austria
SA Germany
SA Italy
SA Switzerland

Central America (1978)
Autoposting of this term began in 1978.
IN Index countries as applicable.
CO N071000N183000
W0771000W0921500
NT Belize
NT Costa Rica
NT El Salvador
NT Guatemala
NT Honduras
NT Nicaragua
NT Panama
NT Panama Canal Zone
SA America
SA Caribbean region
SA Latin America

Central Andes (1978)
Before 1978, also search Andes AND central. As of 1990, South America was autoposted to this term.
BT Andes
BT South America
SA Northern Andes
SA Southern Andes

central Apennines

Central Appalachians (1978)
Before 1978, also search Appalachians AND central. Autoposting of North America to this term began in 1990.
BT Appalachians
BT North America
SA Northern Appalachians
SA Reading Prong
SA Southern Appalachians

Central Asia (1978)
Region formerly known as Soviet Central Asia. Autoposting of this term began in 1981. As of 1993, Asia is autoposted to this term.

IN Index former Soviet republics as applicable.
CO N370000N530000
E0860000E0430000
UF Soviet Central Asia
BT Asia
NT Aral region
NT Aral Sea
NT Caspian Depression
NT Fergana Basin
NT Kazakhstan
NT Kyzylkum
NT Pamirs
NT Turgay Basin
NT Ustyurt
NT Zeravshan River
SA Kyrgyzstan
SA Siberia
SA Tadzhikistan
SA Tien Shan
SA Turanian Platform
SA Turkestan
SA Turkmenia
SA Uzbekistan

Central Atlantic Coastal Plain (1981)
Before 1981, also search Atlantic Coastal Plain AND Central.
CO N340000N373000
W0752000W0800000
BT Atlantic Coastal Plain
BT United States

Central Austrian Alps (1981)
This term has multiple hierarchies.
CO N463500N474600
E0163500E0095200
BT1 Alps
BT1 Europe
BT2 Austria
BT2 Central Europe
BT2 Europe

Central Basin (1978)
Central Tennessee. Also in other areas.
IN Index state or country as applicable.
SA Tennessee

Central Basin Platform (1993)
In Permian Basin, West Texas and SE New Mexico.
IN Index counties or regions as applicable.
SA Andrews County Texas
SA Crane County Texas
SA Ector County Texas
SA Gaines County Texas
SA Lea County New Mexico
SA New Mexico
SA Pecos County Texas
SA Texas
SA United States
SA Ward County Texas
SA Winkler County Texas

Central Bohemian Pluton (1978)
Part of the Bohemian Massif in central Bohemia.
CO N484500N494000
E0153200E0144200
BT Bohemia
BT Czech Republic
BT Czechoslovakia
BT Central Europe
BT Europe
SA Bohemian Massif

Central Bulgaria (1981)
CO N414000N431000
E0200000E0224500
BT Bulgaria
BT Southern Europe
BT Europe

Central California (1981)

Autoposting of broader terms to this term began in 1989. Before 1981, also search California AND Central.
BT California
BT United States
SA Shoo Fly Complex

central Chile earthquake 1985
  use Chile earthquake 1985

Central Europe (1978)
Autoposting of this term began in 1981. As of 1993, term is autoposted to Erzgebirge, Molasse Basin, Rhine Graben, Sudeten Mountains, and Vienna Basin.
CO N453000N550000
E0240000E0060000
BT Europe
NT Allgau Alps
NT Austria
NT Bavarian Alps
NT Beskid Mountains
NT Bohemian Massif
NT Czechoslovakia
NT Erzgebirge
NT Germany
NT Hungary
NT Inn Valley
NT Lake Constance
NT Lake Neusiedler
NT Liechtenstein
NT Molasse Basin
NT North Sudetic Basin
NT Poland
NT Pomerania
NT Rhine Graben
NT Salzach River
NT Silesia
NT Spis
NT Sudeten Mountains
NT Sudetic Basin
NT Switzerland
NT Tatra Mountains
NT Vienna Basin

Central Florida (1981)
Autoposting of broader terms began in 1989. Before 1981, also search Florida AND Central.
CO N270000N290000
W0800500W0825000
BT Florida
BT United States

Central Gneiss Belt (1993)
Grenville Province, E Canada.
IN Index provinces as applicable.
BT Grenville Province
BT Canadian Shield
BT North America
SA Central Metasedimentary Belt
SA Eastern Canada
SA Ontario
SA Quebec

Central Graben (1993)
Extends from Viking Graben in N North Sea onshore to Netherlands and NW Germany.
IN Index North Sea and country or countries as applicable.
SA Atlantic Ocean
SA Germany
SA Netherlands
SA North Sea
SA Scotland

central granite
  use batholiths

Central Greece
  use Sterea Ellas

Central Indian Ridge (1981)
CO S400000S050000
E0800000E0630000

BT Mid-Indian Ridge
BT Indian Ocean
SA mid-ocean ridges

Central Kazakhstan (1981)
This term has multiple hierarchies.
CO N450000N550000
E0790000E0650000
BT1 Kazakhstan
BT1 Central Asia
BT1 Asia
BT2 Kazakhstan
BT2 Commonwealth of Independent States

Central Massif (1978)
An old, eroded plateau region in S central France. As of 1993, a restricted term. Autoposting of this term began in 1993.
IN Index departments as applicable.
CO N440000N470000
E0050000E0004500
UF Massif Central
BT France
BT Western Europe
BT Europe
NT Cantal Massif
NT Millevaches Plateau
NT Montagne Noire
NT Morvan
NT Sidobre Massif
SA Aveyron France
SA Cantal France
SA Causses
SA Haute-Loire France
SA Velay

Central Metasedimentary Belt (1993)
SE Ontario and SW Quebec.
IN Index provinces as applicable.
BT Grenville Province
BT Canadian Shield
BT North America
SA Central Gneiss Belt
SA Eastern Canada
SA Ontario
SA Quebec

Central Pacific (1981)
BT Pacific Ocean

Central Polish Glaciation (1978)
The Elster and Saale stages of glaciation both of which covered central Poland. They were respectively the oldest and the middle stage of 3 stages which covered northern Europe.
SA Elsterian
SA Pleistocene
SA Poland
SA Saalian

Central Rand Group (1993)
Precambrian. In South Africa. In Witwatersrand Supergroup.
BT Precambrian
SA South Africa
SA Witwatersrand Supergroup

Central Range
Not a valid term for GeoRef. As of 1989, use Taiwanese Central Range for mountain range in Taiwan.

Central region Scotland (1993)
Administrative region in central Scotland formed in 1975 from Stirlingshire, and parts of West Lothian County and Perthshire.
BT Scotland
BT Great Britain
BT United Kingdom
BT Western Europe

BT Europe
SA Loch Lomond
SA Perthshire Scotland

Central Rocky Mountains (1978)
Before 1978, also search Rocky Mountains AND central. Autoposting of North America to this term began in 1990.
BT Rocky Mountains
BT North America
SA Northern Rocky Mountains
SA Southern Rocky Mountains
SA U. S. Rocky Mountains

Central Siberian Plateau
use Siberian Platform

Central Swiss Alps (1981)
This term has multiple hierarchies.
BT1 Swiss Alps
BT1 Switzerland
BT1 Central Europe
BT1 Europe
BT2 Swiss Alps
BT2 Alps
BT2 Europe

Central Transdanubia (1981)
Autoposting of Central Europe and Europe to this term began in 1990.
CO N464200N474900
    E0191000E0171000
BT Transdanubia
BT Hungary
BT Central Europe
BT Europe

Central Urals (1981)
Autoposting of Russian Republic to this term began in 1989. This term has multiple hierarchies.
CO N530000N600000
    E0623000E0563000
BT1 Russian Federation
BT1 Commonwealth of Independent States
BT2 Urals
BT2 Commonwealth of Independent States
NT Ilmen Mountains
SA Yekaterinburg Russian Federation

Central Valley (1978)
Valley of Sacramento and San Joaquin rivers in California, and a 600 mile long valley between the Andes and the Coastal Range in central Chile. Before 1978, also search Great Valley for the valley in California.
IN Index California and/or Chile as applicable.
SA California
SA Chile
SA Sacramento Valley
SA San Joaquin Valley

centrallasite
use gyrolite

Centre County
Valid through 1988. Search in combination with state term. After 1988, use specific county-state term.

Centre County Pennsylvania (1989)
Central Pennsylvania. Before 1989, also search Centre County AND Pennsylvania.
CO N404000N414700
    W0771000W0782200
BT Pennsylvania
BT United States
SA Nittany Valley

Centre de Recherches Petrographiques et Geochimiques
use CRPG

Centre National de la Recherche Scientifique
use CNRS

centrifuge methods (1993)
Before 1993, also search centrifuge and centrifugal.
SA chemical analysis
SA geochemistry
SA methods
SA physical methods
SA sample preparation

centrum (seismology)
use focus

Cephalodiscida (1978)
IN Index Hemichordata or Invertebrata as applicable.
BT Pterobranchia
SA Hemichordata
SA Invertebrata

Cephalonia (1978)
Island in center of Ionian Islands in Ionian Sea W of Greek mainland. This term has multiple hierarchies.
UF Kefallinia
BT1 Greek Ionian Islands
BT1 Greece
BT1 Southern Europe
BT1 Europe
BT2 Greek Ionian Islands
BT2 Ionian Islands
BT2 Mediterranean region

Cephalopoda (1978)
Class. Autoposting of this term began in 1978.
BT Mollusca
BT Invertebrata
NT Dibranchiata
NT Tetrabranchiata
SA cephalopods

cephalopods (1981)
Common name for Cephalopoda. Autoposting of mollusks to this term began in 1989.
BT mollusks
BT invertebrates
NT ammonoids
NT belemnites
SA biostratigraphy
SA Cephalopoda

ceramic materials (1978)
SA andalusite deposits
SA baddeleyite
SA carborundum
SA clay mineralogy
SA glass materials
SA kyanite deposits
SA materials
SA ornamental materials
SA refractory materials
SA sillimanite deposits
SA spodumene

Ceratites (1978)
Genus. Autoposting of Ceratitida, Ammonoidea and Tetrabranchiata to this term began in 1989.
BT Ceratitida
BT Ammonoidea
BT Tetrabranchiata
BT Cephalopoda
BT Mollusca
BT Invertebrata
SA Ammonites

Ceratitida (1981)
Autoposting of this term to Ceratites began in 1989.

BT Ammonoidea
BT Tetrabranchiata
BT Cephalopoda
BT Mollusca
BT Invertebrata
NT Ceratites

Ceratolithus (1978)
Autoposting of thallophytes and microfossils began in 1990. This term has multiple hierarchies.
BT1 Coccolithophoraceae
BT1 algae
BT1 thallophytes
BT1 Plantae
BT2 Coccolithophoraceae
BT2 algae
BT2 microfossils

Ceratomorpha (1981)
Suborder. Autoposting of Eutheria and Theria to this term began in 1989.
BT Perissodactyla
BT Eutheria
BT Theria
BT Mammalia
BT Tetrapoda
BT Vertebrata
BT Chordata
NT Rhinocerotidae

Ceratopsia (1993)
Suborder. This term has multiple hierarchies.
BT1 Ornithischia
BT1 Archosauria
BT1 Diapsida
BT1 Reptilia
BT1 Tetrapoda
BT1 Vertebrata
BT1 Chordata
BT2 Ornithischia
BT2 dinosaurs
BT2 Tetrapoda
BT2 Vertebrata
BT2 Chordata
NT Ceratopsidae

Ceratopsidae (1993)
Family. This term has multiple hierarchies.
BT1 Ceratopsia
BT1 Ornithischia
BT1 Archosauria
BT1 Diapsida
BT1 Reptilia
BT1 Tetrapoda
BT1 Vertebrata
BT1 Chordata
BT2 Ceratopsia
BT2 Ornithischia
BT2 dinosaurs
BT2 Tetrapoda
BT2 Vertebrata
BT2 Chordata

CERCLA
use Superfund

Cercopithecidae (1993)
Family? Commonly referred to as the Old World monkeys. Before 1993, also search Cynomorpha.
UF Cynomorpha
BT simians
BT Primates
BT Eutheria
BT Theria
BT Mammalia
BT Tetrapoda
BT Vertebrata
BT Chordata

Ceriantipatharia (1978)
BT Anthozoa
BT Coelenterata
BT Invertebrata

BT Ammonoidea
BT Tetrabranchiata
BT Cephalopoda
BT Mollusca
BT Invertebrata
NT Ceratites

Cerithiidae (1978)
Autoposting of Mesogastropoda to this term began in 1989.
BT Mesogastropoda
BT Gastropoda
BT Mollusca
BT Invertebrata
NT Cerithium

Cerithium (1978)
Autoposting of Mesogastropoda to this term began in 1989.
BT Cerithiidae
BT Mesogastropoda
BT Gastropoda
BT Mollusca
BT Invertebrata

cerium (1978)
Chemical element. As of 1982, use cerium ores for cerium as a commodity. Autoposting of rare earths and metals to this term began in 1989.
UF Ce
BT rare earths
BT metals
SA cerium ores
SA lanthanum
SA lithophile elements

cerium ores (1982)
Before 1982, also search (cerium OR Ce) AND (deposit OR deposits OR ore OR ores OR economic) in the basic index. Autoposting of rare earth deposits and metal ores to this term began in 1989.
IN Index for all non-paleontologic studies of fossils.
BT rare earth deposits
BT metal ores
SA cerium
SA monazite deposits

cerolite (1978)
Before 1993, also search kerolite. Before 1993, serpentine group was a broader term. Discredited as a mineral. Considered by some to be a mixture of stevensite and serpentine, by others, a variety of talc.
UF kerolite
BT sheet silicates
BT silicates
SA serpentine
SA stevensite
SA talc

Cerozem
use Sierozems

Cerro de Pasco
No longer a valid term for GeoRef. As of 1993 see Cerro de Pasco Peru.

Cerro de Pasco Peru (1993)
Town in central Peru. Before 1993 also search Cerro de Pasco AND Peru.
BT Junin Peru
BT Peru
BT South America

Cerro del Hoyazo
No longer a valid term for GeoRef. As of 1993, see Cerro del Hoyazo Spain.

Cerro del Hoyazo Spain (1993)
In the Sierra Alhamilla area in SE Spain. Before 1993, also search Cerro del Hoyazo and Spain.
BT Almeria Spain
BT Andalusia Spain
BT Spain
BT Iberian Peninsula

BT Southern Europe
BT Europe
Cerro Prieto (1978)
Geothermal area near Mexicali in N Baja California state.
UF Cerro Prieto Field
BT Baja California Mexico
BT Mexico
SA Baja California
SA Cerro Prieto Fault
SA geothermal fields
Cerro Prieto Fault (1989)
S California and Baja California.
IN Index California and/or Mexico as applicable.
BT North America
SA Baja California
SA California
SA Cerro Prieto
SA Mexico

Cerro Prieto Field
 use Cerro Prieto
cerussite (1978)
BT carbonates
Cervidae (1978)
Family. Autoposting of this term to Rangifer began in 1989. Autoposting of Eutheria and Theria to this term began in 1989.
BT Ruminantia
BT Artiodactyla
BT Eutheria
BT Theria
BT Mammalia
BT Tetrapoda
BT Vertebrata
BT Chordata
NT Cervus
NT Rangifer
Cervus (1978)
Genus. Autoposting of this term to Cervus elaphas began in 1989. Autoposting of Eutheria and Theria to this term began in 1989.
BT Cervidae
BT Ruminantia
BT Artiodactyla
BT Eutheria
BT Theria
BT Mammalia
BT Tetrapoda
BT Vertebrata
BT Chordata
NT Cervus elaphas
Cervus elaphas (1985)
Species. Autoposting of Cervus, Eutheria and Theria to this term began in 1989.
BT Cervus
BT Cervidae
BT Ruminantia
BT Artiodactyla
BT Eutheria
BT Theria
BT Mammalia
BT Tetrapoda
BT Vertebrata
BT Chordata
CESAR (1989)
Acronym.
UF Canadian Expedition to Study the Alpha Ridge
SA Alpha Cordillera
SA Arctic Ocean
SA expeditions
cesium (1978)
Autoposting of alkali metals and metals to this term began in 1989.
UF caesium
UF Cs
BT alkali metals

BT metals
NT Cs-134
NT Cs-137
Cetacea (1978)
Order. Autoposting of Eutheria and Theria to this term began in 1989.
BT Eutheria
BT Theria
BT Mammalia
BT Tetrapoda
BT Vertebrata
BT Chordata

Cette
Not a valid term for GeoRef. See Sete France.
Ceuta (1981)
Seaport and Spanish military installation bordering the E part of the Strait of Gibraltar in N Morocco.
BT North Africa
BT Africa
SA Cadiz Spain
SA Morocco
Cevennes (1978)
Mountain range in S France.
BT France
BT Western Europe
BT Europe
NT Aigoual Massif
SA Ardeche France
SA Gard France
SA Herault France

Ceylon
A valid term through 1973.
 use Sri Lanka

Cf
 use californium

Ch'ang-hsing China
 use Changxing China

Ch'ing Hai China
 use Qinghai China

Ch'ing-hai China
 use Qinghai China
chabazite (1978)
BT zeolite group
BT framework silicates
BT silicates
Chablais (1978)
Limestone massif of Savoy Alps and region in SE part of country.
BT France
BT Western Europe
BT Europe
SA Haute-Savoie France
SA Savoy Alps
Chaco (1978)
Region. As of 1993, use Chaco Argentina for the province in N Argentina and Chaco Paraguay for the province in NW Paraguay.
IN Index countries and provinces as applicable.
UF Gran Chaco
BT South America
SA Argentina
SA Campo del Cielo Meteorite
SA Paraguay
Chaco Argentina (1993)
Province, N Argentina. Before 1993, also search Chaco AND Argentina.
BT Argentina
BT South America
SA Campo del Cielo Meteorite
Chaco Paraguay (1993)

Region in Paraguay. Before 1993 also search Chaco AND Paraguay.
BT Paraguay
BT South America
Chad (1978)
Before 1972, also search Chad Republic.
CO N072000N233000 E0240000E0133000
UF Chad Republic
BT West Africa
BT Africa
NT Kanem Chad
NT Tibesti Massif
SA Chad Basin
SA Lake Chad
SA Logone River
SA Sahara
SA Sahel
Chad Basin (1978)
IN Index countries as applicable.
BT Africa
SA Cameroon
SA Chad
SA Lake Chad
SA Niger
SA Nigeria

Chad Republic
Not a valid GeoRef index term after 1971. Was used on level 1 in subfile G.
 use Chad

Chadak
No longer a valid term for GeoRef. As of 1993, see Chadak Uzbekistan.
Chadak Uzbekistan (1993)
Village in E Uzbekistan. This term has multiple hierarchies.
BT1 Uzbekistan
BT1 Asia
BT2 Uzbekistan
BT2 Commonwealth of Independent States
Chadobets Uplift (1978)
SE Krasnoyarsk Kray in S Siberia. This term has multiple hierarchies.
BT1 Russian Federation
BT1 Commonwealth of Independent States
BT2 Asia
SA Krasnoyarsk Russian Federation
Chadron Arch (1978)
NW Nebraska and SW South Dakota.
IN Index states as applicable.
BT United States
SA Nebraska
SA South Dakota
Chadron Formation (1978)
In White River Group. NE Colorado, W Nebraska, W South Dakota, and E Wyoming.
BT lower Oligocene
BT Oligocene
BT Paleogene
BT Tertiary
BT Cenozoic
SA Chadronian
SA Colorado
SA Nebraska
SA South Dakota
SA White River Group
SA Wyoming
Chadronian (1978)
North America.
BT lower Oligocene
BT Oligocene
BT Paleogene

BT Tertiary
BT Cenozoic
SA Chadron Formation
Chaetetidae (1978)
Family. Autoposting of Zoantharia to this term began in 1989.
BT Tabulata
BT Zoantharia
BT Anthozoa
BT Coelenterata
BT Invertebrata
Chaetognatha (1978)
BT worms
BT Invertebrata
Chaetopoda (1978)
BT Annelida
BT worms
BT Invertebrata
SA Polychaetia

Chaffee County
Valid through 1988. Search in combination with state term. After 1988, use specific county-state term.
Chaffee County Colorado (1989)
Central Colorado. Before 1989, also search Chaffee County AND Colorado.
CO N382200N390400 W1055000W1063800
BT Colorado
BT United States
NT Mount Antero
NT Salida Colorado
Chain Deep (1978)
BT Red Sea
BT Indian Ocean
Chain Lakes Massif (1989)
W Maine.
IN Index state and counties as applicable.
BT Maine
BT United States
SA Appalachians
chain silicates (1978)
Autoposting of this term began in 1978.
UF inosilicates
BT silicates
NT aenigmatite
NT amphibole group
NT babingtonite
NT bustamite
NT carpholite
NT deerite
NT ferrocarpholite
NT ferrosilite
NT howieite
NT jade
NT neptunite
NT pectolite
NT plancheite
NT pyroxene group
NT pyroxferroite
NT pyroxmangite
NT rhodonite
NT rhonite
NT scawtite
NT shattuckite
NT tobermorite
NT wollastonite
NT xonotlite
SA asbestos
Chaine des Puys (1981)
Division of Auvergne Mountains in S central part of country. Before 1981, also search Monts Dome.
BT Puy-de-Dome France
BT France
BT Western Europe
BT Europe

Chainman Shale (1989)
Great Basin region of E California, Nevada and W Utah.
BT Mississippian
BT Carboniferous
BT Paleozoic
SA California
SA Great Basin
SA Nevada
SA Utah

Chainpur Meteorite (1981)
Impact in W Bihar. Before 1981, also search Chainpur AND meteorites.
BT LL chondrites
BT chondrites
BT stony meteorites
BT meteorites
SA Bihar India
SA India

chalcanthite (1978)
BT sulfates

chalcedony (1978)
BT silica minerals
BT framework silicates
BT silicates
SA agate
SA carnelian
SA chert
SA chrysoprase
SA gems
SA jasper
SA jasperoid
SA onyx
SA quartz

chalcocite (1978)
BT sulfides
SA copper
SA copper minerals
SA copper ores
SA copper sulfides
SA digenite

chalcogens
use chalcophile elements

Chalcolithic (1985)
Archeological classification.
UF Copper Age
BT Cenozoic
SA archaeology
SA Holocene
SA Quaternary

chalcophile elements (1985)
Chemical elements which tend to concentrate in sulfide minerals and ores. Also, elements concentrated in sulfide phases of meteorites.
UF chalcogens
SA arsenic
SA cadmium
SA chemical elements
SA copper
SA iron
SA lead
SA lithophile elements
SA polonium
SA selenium
SA siderophile elements
SA silver
SA sulfur
SA tellurium
SA zinc

chalcopyrite (1978)
UF copper pyrites
BT sulfides
SA copper
SA copper minerals
SA copper ores

chalcostibite (1978)
BT sulfantimonites

BT sulfosalts

Chaleur Bay (1978)
Inlet of Gulf of Saint Lawrence between New Brunswick and the Gaspe Peninsula.
IN Index provinces or regions as applicable.
SA Eastern Canada
SA Gulf of Saint Lawrence
SA New Brunswick
SA Quebec
SA Restigouche Estuary

Chalfant Valley earthquake 1986 (1989)
Chalfant Valley, Inyo County, E California.
BT earthquakes
SA California
SA Inyo County California

chalk (1978)
As of 1981, use chalk deposits for chalk as a commodity.
BT carbonate rocks
BT sedimentary rocks
SA calcite
SA chalk deposits
SA limestone

Chalk
As of 1981, no longer a valid term for GeoRef. Use Upper Cretaceous for the age.

chalk deposits (1981)
Before 1981, search chalk AND deposits.
SA chalk
SA limestone deposits

Chalk River (1985)
Tributary of Ottawa River, SE Ontario. For the town, see Chalk River Ontario.
SA Ontario

Chalk River Ontario (1993)
Town near Ottawa River in Renfrew County.
BT Renfrew County Ontario
BT Ontario
BT Eastern Canada
BT Canada

Challenger Knoll (1978)
BT Gulf of Mexico
BT North American Atlantic
BT North Atlantic
BT Atlantic Ocean

Challenger Plateau (1989)
Tasman Sea.
BT Tasman Sea
BT West Pacific
BT Pacific Ocean
SA DSDP Site 593
SA Leg 90

Challis Idaho (1989)
City in central Idaho. Before 1989, search Challis AND Idaho.
CO N443000N443000
 W1141500W1141500
BT Custer County Idaho
BT Idaho
BT United States

Challis Volcanics (1985)
S central Idaho.
BT Tertiary
BT Cenozoic
SA Eocene
SA Idaho
SA Miocene
SA Oligocene

chalmersite
use cubanite

chalybite
use siderite

Chama Basin (1978)
River basin in N New Mexico.
BT New Mexico
BT United States

Chamba
No longer a valid term for GeoRef. See Chamba India.

Chamba India (1993)
Town and district in NW Himachal Pradesh, N India.
BT Himachal Pradesh India
BT India
BT Indian Peninsula
BT Asia

Chambers County
Valid through 1988. Search in combination with state term. After 1988, use specific county-state term.

Chambers County Alabama (1989)
E Alabama. Before 1989, also search Chambers County AND Alabama.
CO N324200N330700
 W0851000W0853700
BT Alabama
BT United States

Chambers County Texas (1989)
SE Texas. Before 1989, also search Chambers County AND Texas.
CO N292700N295300
 W0942300W0950500
BT Texas
BT United States

chambers, magma
use magma chambers

chamosite (1978)
BT chlorite group
BT sheet silicates
BT silicates
SA berthierine

Champaign County Illinois (1989)
E Illinois. Before 1989, also search Champaign County AND Illinois.
CO N395200N402300
 W0875400W0882700
BT Illinois
BT United States
NT Urbana Illinois

Champaign County Ohio (1989)
W central Ohio. Before 1989, also search Champaign County AND Ohio.
CO N400000N401600
 W0833000W0840400
BT Ohio
BT United States
NT Urbana Ohio

Champlain Sea (1985)
Large, postglacial water body which existed in Pleistocene time in the Saint Lawrence Lowlands.
IN Index states and Canadian provinces as applicable.
BT Pleistocene
BT Quaternary
BT Cenozoic
SA Lake Champlain
SA New York
SA North America
SA Ontario
SA Quebec
SA Saint Lawrence Lowlands

SA Vermont

Champlain Valley (1978)
Lake Champlain area between Adirondack Mountains and Green Mountains and extending into S Quebec.
IN Index Quebec and states as applicable.
BT North America
SA Great Appalachian Valley
SA Lake Champlain
SA New York
SA Quebec
SA Vermont

Champlainian (1978)
Provincial series, North America. Above Canadian Series.
BT Middle Ordovician
BT Ordovician
BT Paleozoic
SA Trentonian

Chanares
No longer a valid term for GeoRef. See Chanares Argentina.

Chanares Argentina (1993)
Village in central Argentina.
BT Cordoba Argentina
BT Argentina
BT South America

Chanares Formation (1978)
BT Triassic
BT Mesozoic
SA Argentina
SA La Rioja Argentina

Chandigarh
No longer a valid term for GeoRef. See Chandigarh India.

Chandigarh India (1993)
City in E Punjab State, N India.
BT Punjab India
BT India
BT Indian Peninsula
BT Asia

Chandler wobble (1978)
SA Earth
SA motions

Chang Jiang River
use Yangtze River

Changai Mountains
use Hangay Mountains

Changcheng System (1989)
Includes Changzhougou and Chuanlinggou formations.
BT Precambrian
SA Changzhougou Formation
SA China
SA Chuanlinggou Formation

changes (1978)
SA alteration
SA conversion
SA fluctuations
SA paleoecology
SA transformations
SA transient phenomena
SA variations

changes in level
use changes of level

changes of level (1978)
Used mostly under stratigraphy for sea-level (sometimes lake-level) changes from Miocene through Quaternary. For pre-Miocene, use transgression or regression.
UF changes in level
UF level, changes of
UF sea-level changes
SA continental shelf

SA eustacy
SA geodesy
SA geomorphology
SA glacial geology
SA glaciation
SA greenhouse effect
SA isostasy
SA land bridges
SA neotectonics
SA paleoecology
SA paleogeography
SA reefs
SA regression
SA sedimentation
SA shorelines
SA stratigraphy
SA submergence
SA terraces
SA transgression

Changjiang Estuary
 use Yangtze River

Changjiang River
 use Yangtze River

Changwat Nakhon Phanom
 No longer a valid term for GeoRef. As of 1993 see Changwat Nakhon Phanom Thailand.

Changwat Nakhon Phanom Thailand (1993)
 Province in NE Thailand. Nakhon Phanom is also a city.
 BT Thailand
 BT Far East
 BT Asia

Changxing
 No longer a valid term for GeoRef. See Changxing China.

Changxing China (1993)
 N Zhejiang near Jiangsu border, E China. Before 1993, also search Ch'ang-hsing or Changxing.
 CO N310300N310300
  E1195400E1195400
 UF Ch'ang-hsing China
 BT Zhejiang China
 BT China
 BT Far East
 BT Asia

Changzhougou Formation (1989)
 Proterozoic? In Changcheng System.
 BT Precambrian
 SA Changcheng System
 SA China

channel geometry (1978)
 For channel cross-section only.
 SA channels
 SA fluvial features
 SA geometry
 SA geomorphology
 SA hydrology
 SA rivers
 SA rivers and streams
 SA streams

Channel Islands (1978)
 Used for the Santa Barbara Islands off SW California coast between Santa Barbara and San Diego. Before 1981, term referred to the Channel Islands in the English Channel as well, and was indexed under California or England as applicable. Beginning in 1981, use English Channel Islands for the British islands.
 CO N324000N341000
  W1181500W1203000
 UF Santa Barbara Islands
 BT California
 BT United States
 NT San Clemente Island
 SA English Channel Islands
 SA Los Angeles County California
 SA San Miguel Island
 SA San Nicolas Island
 SA Santa Barbara Basin
 SA Santa Barbara Channel
 SA Santa Barbara County California
 SA Santa Catalina Island
 SA Santa Cruz Island
 SA Ventura County California

channel order
 use stream order

channelization (1989)
 SA agriculture
 SA channels
 SA fluvial features
 SA streams
 SA waterways

channels (1978)
 SA braided streams
 SA buried channels
 SA canals
 SA channel geometry
 SA channelization
 SA fluvial features
 SA furrows
 SA geomorphology
 SA gorges
 SA grooves
 SA gullies
 SA lava channels
 SA levees
 SA planar bedding structures
 SA rills
 SA rivers
 SA sedimentary structures
 SA sedimentation
 SA sinuosity
 SA streamflow
 SA streams
 SA thalwegs
 SA tidal channels
 SA troughs
 SA underground channels
 SA waterways

Chao Phraya River (1989)
 W Thailand. Formed by the combination of the Ping and Wang, and Nam and Yom rivers near Nakhon Sawan. Flows approximately 140 miles, past many cities (including Bangkok) to the Gulf of Siam.
 BT Thailand
 BT Far East
 BT Asia
 SA Gulf of Siam

chaoite (1978)
 BT native elements
 SA carbon
 SA diamond
 SA graphite
 SA lonsdaleite
 SA meteorites

Chappel Limestone (1989)
 Central Texas. Includes King Creek Marl, Ives Conglomerate, Epsey Creek Limestone and Whites Crossing Coquina.
 BT Lower Mississippian
 BT Mississippian
 BT Carboniferous
 BT Paleozoic
 SA Texas

Char (1978)
 Rock belt in the Lake Zaisan and Kalba Range areas of NE Kazakhstan.
 IN Index regions as applicable.
 SA Kazakhstan

Characeae (1978)
 Family. Autoposting of thallophytes and microfossils began in 1990. This term has multiple hierarchies.
 BT1 Charophyta
 BT1 Chlorophyta
 BT1 algae
 BT1 thallophytes
 BT1 Plantae
 BT2 Charophyta
 BT2 Chlorophyta
 BT2 algae
 BT2 microfossils

characterization (1978)
 SA soils
 SA water regimes

charcoal (1978)
 SA absolute age
 SA carbon
 SA coal

Chardzhou Turkmenia (1993)
 Oblast and city in E Turkmenia. Also search Chardzhov. This term has multiple hierarchies.
 BT1 Turkmenia
 BT1 Asia
 BT2 Turkmenia
 BT2 Commonwealth of Independent States
 NT Gaurdak Turkmenia
 NT Kugitang-Tau

Chardzhov
 No longer a valid term for GeoRef. As of 1993, see Chardzhou Turkmenia.

Charente (1978)
 River in W France. Previously used for the department in W France. As of 1993, use Charente France for the department.
 BT France
 BT Western Europe
 BT Europe
 SA Charente France
 SA Charente-Maritime France

Charente France (1993)
 Department in W France.
 CO N451000N461000
  E0005500W0003000
 BT France
 BT Western Europe
 BT Europe
 NT Angouleme France
 NT Saint-Severin France
 SA Charente
 SA Charentes
 SA Saint-Severin Meteorite

Charente-Inferieure France
 use Charente-Maritime France

Charente-Maritime
 As of 1993, no longer a valid term for GeoRef.
 use Charente-Maritime France

Charente-Maritime France (1993)
 Department in W France. Before 1993, also search Charente-Maritime. Before 1978, also search Charente-Inferieure.
 CO N450500N462000
  E0000200W0013500
 UF Charente-Inferieure France
 UF Charente-Maritime
 BT France
 BT Western Europe
 BT Europe
 SA Charente

SA Charentes

Charentes (1978)
 Region in W and W central France.
 BT France
 BT Western Europe
 BT Europe
 SA Charente France
 SA Charente-Maritime France

Charles County
 Valid through 1988. Search in combination with state term. After 1988, use specific county-state term.

Charles County Maryland (1989)
 S Maryland. Before 1989, also search Charles County AND Maryland.
 CO N381500N384200
  W0764000W0771700
 BT Maryland
 BT United States

Charles Darwin
 use Darwin, Charles

Charles Formation (1989)
 In Madison Group.
 BT Mississippian
 BT Carboniferous
 BT Paleozoic
 SA Madison Group
 SA Montana
 SA North Dakota
 SA Saskatchewan
 SA South Dakota

Charles Lyell
 use Lyell, Charles

Charleston
 Valid through 1988. Search in combination with state term. After 1988, use specific city-state term.

Charleston County
 Valid through 1988. Search in combination with state term. After 1988, use specific county-state term.

Charleston County South Carolina (1989)
 Along the Atlantic Ocean in SE South Carolina. Before 1989, also search Charleston County AND South Carolina.
 CO N322800N331300
  W0791500W0802500
 BT South Carolina
 BT United States
 NT Ashley River Fault
 NT Charleston South Carolina

Charleston earthquake 1886 (1985)
 Epicenter near Charleston, South Carolina.
 BT earthquakes
 SA Charleston South Carolina
 SA South Carolina

Charleston South Carolina (1989)
 City in SE South Carolina. Before 1989, search Charleston AND South Carolina.
 CO N324800N324800
  W0795800W0795800
 BT Charleston County South Carolina
 BT South Carolina
 BT United States
 SA Charleston earthquake 1886

Charleston West Virginia (1989)

City in W central West Virginia.
Before 1989, search Charleston
AND West Virginia.
  CO N382300N382300
     W0814000W0814000
  BT Kanawha County West Virginia
  BT West Virginia
  BT United States
Charlevoix (1978)
As of 1989, used only for the cryptoexplosion or impact feature in Quebec. Use Charlevoix Michigan for city on Lake Michigan.
  UF Charlevoix structure
  BT Quebec
  BT Eastern Canada
  BT Canada
  SA cryptoexplosion features
  SA impact features
  SA Michigan
Charlevoix County
Valid through 1988. Search in combination with state term. After 1988, use specific county-state term.
Charlevoix County Michigan (1989)
NW Michigan. Before 1989, also search Charlevoix County AND Michigan.
  CO N450800N455000
     W0844500W0854500
  BT Michigan Lower Peninsula
  BT Michigan
  BT United States
  NT Charlevoix Michigan
Charlevoix County, Quebec
  Use Charlevoix-Est County Quebec or Charlevoix-Ouest County Quebec as applicable.
Charlevoix East
  use Charlevoix-Est County Quebec
Charlevoix Michigan (1989)
City on Lake Michigan in NW Michigan. Before 1989, also search Charlevoix AND Michigan.
  CO N451900N451900
     W0851600W0851600
  BT Charlevoix County Michigan
  BT Michigan Lower Peninsula
  BT Michigan
  BT United States
Charlevoix structure
  use Charlevoix
Charlevoix West
  use Charlevoix-Ouest County Quebec
Charlevoix-Est County Quebec (1993)
E central Quebec. Before 1993, also search Charlevoix-Est County, Charlevoix Est, East Charlevoix County, East Charlevoix, or Charlevoix East in conjunction with Quebec.
  UF East Charlevoix County Quebec
  UF Charlevoix East
  BT Quebec
  BT Eastern Canada
  BT Canada
  SA Charlevoix-Ouest County Quebec
Charlevoix-Ouest County Quebec (1993)
E central Quebec. Before 1993, also search Charlevoix-Ouest County, Charlevoix Ouest, West

Charlevoix County, West Charlevoix, or Charlevoix West in conjunction with Quebec.
  UF Charlevoix West
  BT Quebec
  BT Eastern Canada
  BT Canada
  SA Charlevoix-Est County Quebec
Charlie Gibbs fracture zone
  use Gibbs fracture zone
Charlotte
Valid through 1988. Search in combination with state term. After 1988, use specific city-state term.
Charlotte Belt (1985)
Metamorphic belt, extending from central North Carolina across NW South Carolina to N Georgia.
  IN Index states and regions as applicable.
  SA Appalachians
  SA Carolina slate belt
  SA Georgia
  SA North Carolina
  SA Piedmont
  SA South Carolina
  SA Southern Appalachians
  SA United States
Charlotte County
No longer a valid term for GeoRef. As of 1993, use Charlotte County New Brunswick, Charlotte County Florida, or Charlotte County Virginia.
Charlotte County Florida (1989)
On Gulf of Mexico in S Florida. Before 1989, also search Charlotte County AND Florida.
  CO N264500N270200
     W0813400W0822200
  BT Florida
  BT United States
Charlotte County New Brunswick (1993)
In extreme S New Brunswick. Before 1993, also search Charlotte County and New Brunswick.
  BT New Brunswick
  BT Maritime Provinces
  BT Eastern Canada
  BT Canada
Charlotte County Virginia (1989)
S Virginia. Before 1989, also search Charlotte County AND Virginia.
  CO N364300N371500
     W0782700W0785500
  BT Virginia
  BT United States
Charlotte North Carolina (1989)
City in S North Carolina. Before 1989, search Charlotte AND North Carolina.
  CO N350300N350300
     W0805000W0805000
  BT Mecklenburg County North Carolina
  BT North Carolina
  BT United States
charnockite (1978)
  BT granites
  BT plutonic rocks
  BT igneous rocks
charoite (1989)
  BT sheet silicates
  BT silicates
Charon Satellite (1989)

Satellite of Pluto. Before 1989, also search Charon AND Pluto.
  SA Pluto
  SA satellites
Charophyta (1978)
Before 1981, also search charophytes. As of 1981, charophytes is reserved for use as a common name. Autoposting of this term to Characeae began in 1981. Autoposting of thallophytes and microfossils began in 1990. This term has multiple hierarchies.
  BT1 Chlorophyta
  BT1 algae
  BT1 thallophytes
  BT1 Plantae
  BT2 Chlorophyta
  BT2 algae
  BT2 microfossils
  NT Characeae
  SA charophytes
charophytes (1989)
Common name for Charophyta. Autoposting of microfossils to this term began in 1990. This term has multiple hierarchies.
  BT1 algal flora
  BT1 plants
  BT2 algal flora
  BT2 microfossils
  SA biostratigraphy
  SA Charophyta
Charters Towers
No longer a valid term for GeoRef. See Charters Towers Australia.
Charters Towers Australia (1993)
City just E of the Great Dividing Range in NE central Queensland.
  BT Queensland Australia
  BT Australia
  BT Australasia
chartology
  use cartography
Chase County
Valid through 1988. Search in combination with state term. After 1988, use specific county-state term.
Chase County Kansas (1989)
E central Kansas. Before 1989, also search Chase County AND Kansas.
  CO N380600N383000
     W0962400W0965000
  BT Kansas
  BT United States
Chase County Nebraska (1989)
S Nebraska. Before 1989, also search Chase County AND Nebraska.
  CO N402100N404200
     W1012000W1020400
  BT Nebraska
  BT United States
Chase Group (1978)
Includes Odell Formation, Nolans Formation, Wreford Limestone, Matfield Shale, Doyle Shale, Winfield Limestone, Herrington Limestone, Fort Riley Limestone and Barneston Limestone. E Kansas, N central Oklahoma, and SE Nebraska.
  BT Permian
  BT Paleozoic
  SA Fort Riley Limestone
  SA Kansas

  SA Nebraska
  SA Oklahoma
  SA Wreford Limestone
chassignite (1981)
  BT achondrites
  BT stony meteorites
  BT meteorites
  NT Chassigny Meteorite
  SA SNC Meteorites
Chassigny Meteorite (1985)
Impact at Chassigny in Haute-Marne, France. Before 1985, also search Chassigny AND meteorites.
  BT chassignite
  BT achondrites
  BT stony meteorites
  BT meteorites
  SA France
  SA Haute-Marne France
  SA SNC Meteorites
Chateaulin Basin (1978)
In extreme W Brittany.
  BT Finistere France
  BT France
  BT Western Europe
  BT Europe
Chatham County
Valid through 1988. Search in combination with state term. After 1988, use specific county-state term.
Chatham County Georgia (1989)
E Georgia. Before 1989, also search Chatham County AND Georgia.
  CO N314300N321500
     W0805100W0812500
  BT Georgia
  BT United States
Chatham County North Carolina (1989)
Central North Carolina. Before 1989, also search Chatham County AND North Carolina.
  CO N353100N355300
     W0785400W0793300
  BT North Carolina
  BT United States
Chatham Rise (1978)
500 miles E of New Zealand.
  CO S443000S430000
     W1770000E1743000
  BT Pacific Ocean
  SA DSDP Site 594
Chatkal Range (1978)
Branch of NW Tien Shan. Before 1978, also search Chatkal. This term has multiple hierarchies.
  BT1 Kyrgyzstan
  BT1 Asia
  BT2 Kyrgyzstan
  BT2 Commonwealth of Independent States
  BT3 Tien Shan
  BT3 Asia
Chatsworth Formation (1985)
S California.
  BT Cretaceous
  BT Mesozoic
  SA California
Chattahoochee River (1981)
In Georgia, Alabama, and Florida.
  CO N304000N345000
     W0833000W0851500
  BT United States
  SA Alabama
  SA Florida
  SA Georgia

Chattanooga Shale (1978)
Upper Devonian and Mississippian. Consists of lower Dowelltown Member and upper Gassaway Member. In some areas, includes Hardin Sandstone Member below the Dowelltown. N Alabama, Arkansas, NW Georgia, Illinois, E and W Kentucky, NE Mississippi, Missouri, Oklahoma, and Tennessee.
IN   Index ages as applicable.
BT   Paleozoic
SA   Alabama
SA   Arkansas
SA   Devonian
SA   Georgia
SA   Illinois
SA   Kentucky
SA   Mississippi
SA   Mississippian
SA   Missouri
SA   Oklahoma
SA   Tennessee
SA   Upper Devonian

Chattian (1978)
Europe. Above Rupelian, below Aquitanian (Miocene).
BT   upper Oligocene
BT   Oligocene
BT   Paleogene
BT   Tertiary
BT   Cenozoic

Chaturi Georgian Republic
    use Chiatura Georgian Republic

Chautauqua County
    Valid through 1988. Search in combination with state term. After 1988, use specific county-state term.

Chautauqua County Kansas (1989)
SE Kansas. Before 1989, also search Chautauqua County AND Kansas.
CO   N370000N371800
      W0955700W0963100
BT   Kansas
BT   United States

Chautauqua County New York (1989)
Extreme W New York. Before 1989, also search Chautauqua County AND New York.
CO   N420000N423500
      W0790500W0794700
BT   New York
BT   United States

Chaves County
    Valid through 1988. Search in combination with state term. After 1988, use specific county-state term.

Chaves County New Mexico (1989)
SE New Mexico. Before 1989, also search Chaves County AND New Mexico.
CO   N323000N340500
      W1033000W1052200
BT   New Mexico
BT   United States

Chaves Island (1981)
One of the Galapagos Islands. Not to be confused with Santa Cruz Island which is one of the Channel Islands. Also search Santa Cruz Island AND Galapagos.
UF   Indefatigable Island
BT   Galapagos Islands

BT   East Pacific Ocean Islands
SA   Santa Cruz Island

Chaya Massif (1978)
A gabbro-peridotite massif in northern Baikal region. This term has multiple hierarchies.
BT1   Russian Federation
BT1   Commonwealth of Independent States
BT2   Asia
SA   Baikal region

Chazy Group (1989)
E New York, W Vermont and S Quebec.
BT   Middle Ordovician
BT   Ordovician
BT   Paleozoic
SA   New York
SA   Quebec
SA   Vermont

Chazyan (1978)
North America. Below Mohawkian, above lower Ordovician.
BT   Middle Ordovician
BT   Ordovician
BT   Paleozoic

Chechen-Ingush
    No longer a valid term for GeoRef as of 1993.
    use Chechen-Ingush Russian Federation

Chechen-Ingush Russian Federation (1993)
Former Autonomous Soviet Socialist Republic on N slopes of Caucasus. Before 1993, also search Chechen-Ingush. Before 1978, also search Chechen- This term has multiple hierarchies.
CO   N423000N440000
      E0463000E0443000
UF   Chechen-Ingush
UF   Checheno-Ingush Russian Federation
BT1   Russian Federation
BT1   Commonwealth of Independent States
BT2   Europe
NT   Grozny Russian Federation

Checheno-Ingush Russian Federation
    use Chechen-Ingush Russian Federation

Chedabucto Bay (1978)
Inlet of Atlantic Ocean in NE just S of Cape Breton Island.
BT   Nova Scotia
BT   Maritime Provinces
BT   Eastern Canada
BT   Canada

Cheilostomata (1978)
BT   Bryozoa
BT   Invertebrata

Cheju Island (1978)
In East China Sea S of mainland.
BT   South Korea
BT   Korea
BT   Far East
BT   Asia

Chekalin
    No longer a valid term for GeoRef. As of 1993, see Chekalin Russian Federation.

Chekalin Russian Federation (1993)
Town in W Tula Oblast in SW European Russian Federation. Before 1993, also search Chekalin or Likhvin. This term has multiple hierarchies.

UF   Likhvin Russian Federation
BT1   Tula Russian Federation
BT1   Russian Federation
BT1   Commonwealth of Independent States
BT2   Tula Russian Federation
BT2   Europe

Chekiang China
    use Zhejiang China

Chelan County
    Valid through 1988. Search in combination with state term. After 1988, use specific county-state term.

Chelan County Washington (1989)
Central Washington. Before 1989, also search Chelan County AND Washington.
CO   N471700N483500
      W1195000W1211000
BT   Washington
BT   United States

chelation (1985)
SA   bonding
SA   geochemistry
SA   ligands
SA   metals
SA   organic materials

Cheleken
    No longer a valid term for GeoRef. As of 1993, see Cheleken Turkmenia.

Cheleken Peninsula (1978)
In Krasnovodsk Oblast on Caspian Sea. This term has multiple hierarchies.
BT1   Turkmenia
BT1   Asia
BT2   Turkmenia
BT2   Commonwealth of Independent States

Cheleken Turkmenia (1993)
Town on the Caspian Sea in Ashkhabad Oblast, SW Turkmenia. This term has multiple hierarchies.
BT1   Ashkhabad Turkmenia
BT1   Turkmenia
BT1   Asia
BT2   Ashkhabad Turkmenia
BT2   Turkmenia
BT2   Commonwealth of Independent States

Chelicerata (1978)
Subphylum. Autoposting of this term began in 1978.
BT   Arthropoda
BT   Invertebrata
NT   Arachnida
NT   Merostomata

Chelmsford Formation (1989)
Located in the Sudbury Basin.
BT   Precambrian
SA   Ontario

Chelmsford Granite (1989)
BT   Devonian
BT   Paleozoic
SA   Maine
SA   Massachusetts
SA   New Hampshire

Chelonia (1978)
Order and genus. Autoposting of this term as the order began in 1981.
BT   Anapsida
BT   Reptilia
BT   Tetrapoda
BT   Vertebrata
BT   Chordata
NT   Pelomedusidae

Chelyabinsk
    As of 1993, no longer a valid term for GeoRef.
    use Chelyabinsk Russian Federation

Chelyabinsk Russian Federation (1993)
City and oblast in Southern Urals. Before 1993, also search Chelyabinsk.
UF   Chelyabinsk
BT   Russian Federation
BT   Commonwealth of Independent States
NT   Kochkar Russian Federation
NT   Magnitogorsk Russian Federation
NT   Rudnichny Russian Federation
NT   Yuryuzan Russian Federation
SA   Bakal Russian Federation
SA   Ilmen Mountains
SA   Ufaley Russian Federation
SA   Ural region

chemical
    A valid term through 1977. After 1977, use more specific term, e.g. chemical composition, or chemical properties, or chemical analysis, or chemical weathering.

chemical analyses
    Not a valid GeoRef index term after 1970. Was used on level 1 in subfile N. Now use chemical analysis.

chemical analysis (1978)
Used only for methodology. For data resulting from chemical analysis of a material, use chemical composition AND name of material. Also search chemical analyses; analysis AND chemical. Before 1981, also search chemical methods.
IN   Index specific type of analysis where applicable.
SA   accelerator mass spectroscopy
SA   activation analysis
SA   alpha-ray spectroscopy
SA   analysis
SA   atomic absorption
SA   Auger spectroscopy
SA   automated analysis
SA   calorimetry
SA   catalysis
SA   centrifuge methods
SA   chemical composition
SA   chromatography
SA   color
SA   colorimetry
SA   crystal chemistry
SA   decrepitation
SA   detection
SA   differential thermal analysis
SA   diffusion
SA   electrolytic analysis
SA   electron microscopy
SA   electron paramagnetic resonance
SA   electron probe
SA   electrophoresis
SA   emission spectroscopy
SA   flame photometry
SA   Fraunhofer line discriminators
SA   gamma-ray spectroscopy
SA   geochemical methods
SA   geochemistry
SA   inductively coupled plasma methods
SA   infrared methods
SA   infrared spectra
SA   infrared spectroscopy

SA interlaboratory comparison
SA ion chromatography
SA ion probe
SA isotopes
SA laser methods
SA laser ranging
SA lithogeochemistry
SA major-element analyses
SA mass spectroscopy
SA microwave spectroscopy
SA minor-element analyses
SA Mossbauer spectroscopy
SA neutron activation analysis
SA neutron probe
SA nuclear magnetic resonance
SA optical spectroscopy
SA photometry
SA polarography
SA potentiometry
SA qualitative analysis
SA quantitative analysis
SA radio-frequency spectroscopy
SA radon emanometry
SA Raman spectroscopy
SA Rock-Eval
SA sample preparation
SA scanning electron microscopy
SA secondary ion mass spectroscopy
SA soil gases
SA spectroscopy
SA standard materials
SA synchrotron radiation
SA thermal analysis
SA thermogravimetric analysis
SA titration
SA trace-element analyses
SA transmission electron microscopy
SA ultraviolet spectroscopy
SA vacuum fusion analysis
SA voltammetry
SA wet methods
SA X-ray analysis
SA X-ray diffraction analysis
SA X-ray fluorescence
SA X-ray spectroscopy

chemical buffers
  use buffers

chemical composition (1978)
  Used to discuss data and results of methods. Also search chemical AND composition.
  SA analysis
  SA chemical analysis
  SA composition
  SA gas chromatography
  SA geochemistry
  SA hydrochemistry
  SA ion chromatography
  SA manganese composition
  SA soils

chemical demagnetization (1981)
  BT demagnetization
  SA geophysics
  SA magnetization
  SA paleomagnetism
  SA remanent magnetization

chemical dispersion (1993)
  From 1981-1992, also search dispersion.
  SA dispersivity
  SA fresh water
  SA geochemistry
  SA ground water
  SA hydrochemistry
  SA salt water
  SA salt-water intrusion
  SA sea water

chemical elements (1993)
  Before 1993, also search elements. See list D.
  SA boron
  SA carbon
  SA chalcophile elements
  SA cosmochemistry
  SA cosmogenic elements
  SA geochemical anomalies
  SA geochemistry
  SA halogens
  SA hydrogen
  SA isotopes
  SA lithophile elements
  SA major elements
  SA major-element analyses
  SA metals
  SA migration of elements
  SA minor elements
  SA minor-element analyses
  SA nitrogen
  SA nonmetals
  SA oxygen
  SA pathfinders
  SA phosphorus
  SA precious metals
  SA selenium
  SA siderophile elements
  SA silicon
  SA sulfur
  SA tellurium
  SA trace elements
  SA trace-element analyses
  SA volatile elements

chemical erosion
  use corrosion

chemical evolution
  Not a valid term for GeoRef. As of 1993, see biologic evolution or life origin or biochemistry if applicable. Also see chemical properties, atmosphere or paleoatmosphere, which can be combined with evolution.

chemical explosions (1978)
  BT explosions

chemical fossils (1978)
  SA biomarkers
  SA casts
  SA fossilization
  SA fossils
  SA ichnofossils
  SA paleontology
  SA problematic fossils

chemical fractionation (1985)
  For fractionation of isotopes, use fractionation. For fractionation of magmas, use fractional crystallization. Before 1985, also search fractionation.
  SA crystal fractionation
  SA fractional crystallization
  SA fractionation
  SA geochemistry
  SA partitioning

chemical industry (1981)
  SA economic geology
  SA industry

chemical magnetization
  use chemical remanent magnetization

chemical methods
  As of 1981, no longer a valid term for GeoRef. See geochemical methods; chemical analysis.

chemical properties (1978)
  SA coal
  SA geochemistry
  SA physicochemical properties
  SA properties
  SA thermochemical properties

SA valency

chemical ratios (1993)
  Before 1993, also search ratios and geochemistry.
  SA geochemistry
  SA partition coefficients
  SA Sr/Ca

chemical reactions (1993)
  Before 1993, also search reactions.
  SA coupling
  SA endothermic reactions
  SA geochemistry
  SA reaction rims
  SA sinks
  SA stoichiometry

chemical remanence
  use chemical remanent magnetization

chemical remanent magnetization (1978)
  Autoposting of magnetization to this term began in 1989.
  UF chemical magnetization
  UF chemical remanence
  UF CRM
  UF crystallization magnetization
  UF crystallization remanent magnetization
  BT remanent magnetization
  BT magnetization
  SA paleomagnetism

chemical sedimentation (1981)
  BT sedimentation

chemical weathering (1978)
  BT weathering
  SA corrosion
  SA embayments
  SA karstification
  SA pits
  SA weathering rinds

chemically precipitated rocks (1978)
  As of 1981, terms that were classified under chemically precipitated sediments are included here or under carbonate rocks. Those added here are duricrust, ferricrete, silcrete, siliceous sinter and weathering crust. Autoposting of this term to itabirite and phosphorite ended in 1981. Autoposting of this term to bone beds ended in 1985. Autoposting of this term to potash and salt began in 1981.
  UF chemically precipitated sediments
  BT sedimentary rocks
  NT chert
  NT duricrust
  NT evaporites
  NT ferricrete
  NT flint
  NT iron formations
  NT ironstone
  NT jaspilite
  NT phosphate rocks
  NT silcrete
  NT siliceous sinter
  NT taconite
  NT tufa
  NT weathering crust
  SA anhydrite
  SA dolomite
  SA gypsum
  SA halite
  SA iron-rich composition
  SA itabirite
  SA novaculite
  SA rocks

SA travertine

chemically precipitated sediments
  As of 1981, no longer a valid term for GeoRef. From 1978-1980, was autoposted to the specific sediments that were classified under it: calcrete, caliche, duricrust, ferricrete, silcrete, siliceous sinter and weathering crust.
  use chemically precipitated rocks

chemistry
  As of 1981, no longer a valid term for GeoRef. See geochemistry.

Chemnitz Germany (1993)
  City in Saxony, E Germany. Before 1993, also search Karl-Marx-Stadt or Chemnitz. From 1981 through 1992, Karl-Marx-Stadt referred to the city in SE East Germany. Before 1981, it also included the East German district.
  UF Karl-Marx-Stadt
  BT Saxony Germany
  BT Germany
  BT Central Europe
  BT Europe

chemostratigraphy (1989)
  SA geochemistry
  SA stratigraphy

Chemung County New York (1989)
  S New York. Before 1989, also search Chemung County AND New York.
  CO N420000N423600
     W0763200W0765700
  BT New York
  BT United States
  SA Chemung River

Chemung Formation (1989)
  BT Upper Devonian
  BT Devonian
  BT Paleozoic
  SA Maryland
  SA New York
  SA Pennsylvania
  SA Virginia
  SA West Virginia

Chemung River (1989)
  S New York and N Pennsylvania. Formed at the junction of the Cohocton and Tioga rivers near Painted Post, New York. Flows about 45 miles SE to the Susquehanna River near Athens, Pennsylvania.
  IN Index counties as applicable.
  BT United States
  SA Bradford County Pennsylvania
  SA Chemung County New York
  SA New York
  SA Pennsylvania

Chenango River (1978)
  In S central New York.
  BT New York
  BT United States

chenier plains (1993)
  This term has multiple hierarchies.
  BT1 plains
  BT2 shore features
  SA beaches
  SA cheniers
  SA coastlines
  SA eolian features
  SA mud flats

cheniers (1981)
  Autoposting of shore features to this term began in 1989.
  BT shore features
  SA beach ridges

SA chenier plains
SA geomorphology
SA marine terraces
SA shorelines

Cher (1978)
River in central France. As of 1993, see Cher France for the Department.
BT France
BT Western Europe
BT Europe
SA Cher France

Cher France (1993)
Department in central France.
CO N462500N474000
E0030700E0014500
BT France
BT Western Europe
BT Europe
NT Bourges France
SA Cher

Cheremkhovo Basin
use Irkutsk Basin

Cheremkovo coal basin
use Irkutsk Basin

Cherkasi Ukraine
use Cherkassy Ukraine

Cherkassy
No longer a valid term for GeoRef. As of 1993, see Cherkassy Ukraine.

Cherkassy Ukraine (1993)
City and oblast in central Ukraine. Also search Cherkasi. This term has multiple hierarchies.
UF Cherkasi Ukraine
BT1 Ukraine
BT1 Europe
BT2 Ukraine
BT2 Commonwealth of Independent States
NT Korsun-Shevchenkovski Ukraine

Chernigov
No longer a valid term for GeoRef. As of 1993, see Chernigov Ukraine.

Chernigov Ukraine (1993)
Oblast in N Ukraine. This term has multiple hierarchies.
BT1 Ukraine
BT1 Europe
BT2 Ukraine
BT2 Commonwealth of Independent States

Chernobyl nuclear accident (1993)
Use for the 1986 accident which released radioactive materials into the atmosphere. After 1992, search in combination with Chernobyl for local studies. Before 1993, also search Chernobyl AND (radioactivity or fallout) if applicable.
IN For local studies, also index Chernobyl.
SA fallout
SA radioactive isotopes
SA radioactive tracers
SA radioactivity

Chernobyl Ukraine (1993)
Site of nuclear reactor accident in Kiev oblast, Ukraine. This term has multiple hierarchies.
UF Tschernobyl Ukraine
BT1 Kiev Ukraine
BT1 Ukraine
BT1 Europe
BT2 Kiev Ukraine

BT2 Ukraine
BT2 Commonwealth of Independent States

Chernozems (1978)
UF Black earth
UF Chernozyom
UF Tchornozem
UF Tschernosem
UF Tschernosiom
BT soils
SA Chestnut soils
SA soil group
SA Zonal soils

Chernozyom
use Chernozems

Cherokee County
Valid through 1988. Search in combination with state term. After 1988, use specific county-state term.

Cherokee County Alabama (1989)
NE Alabama. Before 1989, also search Cherokee County AND Alabama.
CO N335600N343000
W0852300W0855300
BT Alabama
BT United States

Cherokee County Georgia (1989)
NW Georgia. Before 1989, also search Cherokee County AND Georgia.
CO N340500N342400
W0841500W0844000
BT Georgia
BT United States

Cherokee County Iowa (1989)
NW Iowa. Before 1989, also search Cherokee County AND Iowa.
CO N423300N425400
W0952300W0955300
BT Iowa
BT United States

Cherokee County Kansas (1989)
Extreme SE Kansas. Before 1989, also search Cherokee County AND Kansas.
CO N370000N372100
W0942400W0950500
BT Kansas
BT United States

Cherokee County North Carolina (1989)
Extreme W North Carolina. Before 1989, also search Cherokee County AND North Carolina.
CO N345900N351800
W0834200W0842100
BT North Carolina
BT United States

Cherokee County Oklahoma (1989)
E Oklahoma. Before 1989, also search Cherokee County AND Oklahoma.
CO N353800N361000
W0944800W0951500
BT Oklahoma
BT United States

Cherokee County South Carolina (1989)
N South Carolina. Before 1989, also search Cherokee County AND South Carolina.
CO N345000N351300
W0812200W0815300
BT South Carolina

BT United States

Cherokee County Texas (1989)
E Texas. Before 1989, also search Cherokee County AND Texas.
CO N312500N320800
W0945000W0952600
BT Texas
BT United States

Cherokee Group (1978)
Divided into Riverton, Warner, Rowe, Dry Wood, Bluejacket, Seville, Weir, Tebo, Scammon, Mineral, Robinson, Branch, Fleming, Croweburg, Verdigris, Bevier, Lagonda, Mulky, and Excello formations. SW Iowa, E Kansas, W Missouri, and SE Nebraska.
BT Desmoinesian
BT Middle Pennsylvanian
BT Pennsylvanian
BT Carboniferous
BT Paleozoic
SA Cabaniss Formation
SA Iowa
SA Kansas
SA Missouri
SA Nebraska

Cherry Canyon Formation (1985)
In Delaware Mountain Group. W Texas and S New Mexico. This term has multiple hierarchies.
BT1 Guadalupian
BT1 Permian
BT1 Paleozoic
BT2 Lower Permian
BT2 Permian
BT2 Paleozoic
SA Delaware Mountain Group
SA New Mexico
SA Ontario
SA Texas

chert (1978)
Autoposting of this term began in 1981.
UF hornstein
UF hornstone
BT chemically precipitated rocks
BT sedimentary rocks
NT jasperoid
SA chalcedony
SA flint
SA jasper
SA metachert

chertification (1978)
BT diagenesis
SA silicification

Chesapeake Bay (1978)
Inlet of Atlantic Ocean in E United States.
IN Index states as applicable.
CO N365500N393700
W0754000W0763300
BT United States
SA Atlantic Coastal Plain
SA Maryland
SA North American Atlantic
SA Virginia

Chesapeake Group (1985)
Middle and upper Miocene. Virginia, North Carolina, Delaware and E Maryland.
BT Miocene
BT Neogene
BT Tertiary
BT Cenozoic
SA Calvert Formation
SA Choptank Formation
SA Chowan River Formation
SA Delaware
SA Eastover Formation

SA Maryland
SA North Carolina
SA Saint Marys Formation
SA Virginia

Cheshire
No longer a valid term for GeoRef as of 1993.
use Cheshire England

Cheshire England (1993)
County in NW England. Before 1993, also search Cheshire.
CO N530000N533000
W0020000W0032000
UF Cheshire
BT England
BT Great Britain
BT United Kingdom
BT Western Europe
BT Europe
SA Mersey River
SA Mersey Valley

Cheshire Formation (1989)
W Connecticut, W Massachusetts, SE New York, SW Vermont and S Quebec.
BT Lower Cambrian
BT Cambrian
BT Paleozoic
SA Connecticut
SA Massachusetts
SA New York
SA Quebec
SA Vermont

Chesil Beach
No longer a valid term for GeoRef. As of 1993, see Chesil Beach England.

Chesil Beach England (1993)
On the English Channel in Dorset. Before 1993, also search Chesil Beach.
BT Dorset England
BT England
BT Great Britain
BT United Kingdom
BT Western Europe
BT Europe

Chester County
Valid through 1988. Search in combination with state term. After 1988, use specific county-state term.

Chester County Pennsylvania (1989)
SE Pennsylvania. Before 1989, also search Chester County AND Pennsylvania.
CO N394300N401400
W0752200W0760700
BT Pennsylvania
BT United States

Chester County South Carolina (1989)
N South Carolina. Before 1989, also search Chester County AND South Carolina.
CO N343300N345000
W0805200W0812800
BT South Carolina
BT United States

Chester County Tennessee (1989)
SW Tennessee. Before 1989, also search Chester County AND Tennessee.
CO N351400N353400
W0882300W0885200
BT Tennessee
BT United States

Chester Series
use Chesterian

Chesterfield County South Carolina (1989)
N South Carolina. Before 1989, search Chesterfield County AND South Carolina.
CO N342300N344900
W0794800W0803400
BT South Carolina
BT United States
NT Pageland South Carolina

Chesterfield County Virginia (1989)
E central Virginia. Before 1989, search Chesterfield County AND Virginia.
CO N371400N373400
W0771500W0775300
BT Virginia
BT United States

Chesterian (1978)
Provincial series, North America. Formations included in standard section: Aux Vases Sandstone, Renault Limestone, Bethel Sandstone, Point Creek Formation. N Alabama, Illinois, S Indiana, Kentucky, E Missouri, and Tennessee.
UF Chester Series
BT Upper Mississippian
BT Mississippian
BT Carboniferous
BT Paleozoic
NT Aux Vases Sandstone
NT Beech Creek Limestone Member
NT Golconda Formation
NT Imo Formation
NT Renault Formation
SA Alabama
SA Illinois
SA Indiana
SA Kentucky
SA Meramecian
SA Missouri
SA Morrowan
SA Springer Formation
SA Tennessee

Chestnut soils (1978)
BT soils
SA Chernozems
SA soil group
SA Zonal soils

Cheviot Hills (1978)
Range of hills along the English-Scottish border. In 1981, Great Britain was added as a broader term.
IN Index England and/or Scotland as applicable.
BT Great Britain
BT United Kingdom
BT Western Europe
BT Europe
SA England
SA Scotland

chevkinite (1978)
Autoposting of sorosilicates began in 1985.
BT sorosilicates
BT orthosilicates
BT silicates

chevron folds (1978)
Before 1978, also search folds AND chevron.
BT folds
SA kink folds

chevron marks (1985)
BT bedding plane irregularities
BT sedimentary structures

Cheyenne County
Valid through 1988. Search in combination with state term. After 1988, use specific county-state term.

Cheyenne County Colorado (1989)
E Colorado. Before 1989, also search Cheyenne County AND Colorado.
CO N383600N390200
W1020200W1030500
BT Colorado
BT United States

Cheyenne County Kansas (1989)
Extreme NW Kansas. Before 1989, also search Cheyenne County AND Kansas.
CO N393300N400000
W1012400W1020300
BT Kansas
BT United States

Cheyenne County Nebraska (1989)
W Nebraska. Before 1989, also search Cheyenne County AND Nebraska.
CO N410000N412500
W1023500W1032300
BT Nebraska
BT United States

Cheyenne Sandstone (1978)
Underlies Kiowa Formation; overlies Morrison Formation. SW Kansas, SE Colorado, and W Oklahoma.
BT Lower Cretaceous
BT Cretaceous
BT Mesozoic
SA Colorado
SA Kansas
SA Oklahoma

Chhindwara
No longer a valid term for GeoRef. See Chhindwara India.

Chhindwara District
No longer a valid term for GeoRef. As of 1993, see Chhindwara India.

Chhindwara India (1993)
District and town in S central Madhya Pradesh, central India. Before 1993, also search Chhindwara District.
BT Madhya Pradesh India
BT India
BT Indian Peninsula
BT Asia

Chiang Mai
No longer a valid term for GeoRef. As of 1993 see Chiang Mai Thailand.

Chiang Mai Thailand (1993)
City and province in NW Thailand. Before 1993 also search Chiang Mai AND Thailand.
BT Thailand
BT Far East
BT Asia

Chiapas
No longer a valid term for GeoRef. As of 1993 see Chiapas Mexico.

Chiapas Mexico (1993)
State in SE Mexico. Before 1993 also search Chiapas AND Mexico.
BT Mexico
NT El Chichon

Chiatura
No longer a valid term for GeoRef. As of 1993, see Chiatura Georgian Republic.

Chiatura Georgian Republic (1993)
City in central Georgian Republic. Before 1993, also search Chaturi or Chiaturi.
UF Chaturi Georgian Republic
UF Chiaturi Georgian Republic
BT Georgian Republic
BT Europe

Chiaturi Georgian Republic
use Chiatura Georgian Republic

Chiayi-Hsinying area (1978)
BT Taiwan
BT Far East
BT Asia

Chiba
No longer a valid term for GeoRef. See Chiba Japan.

Chiba earthquake 1987
use Chibaken-Toho-Oki earthquake 1987

Chiba Japan (1993)
City and prefecture SE of Tokyo in central Honshu.
BT Honshu
BT Japan
BT Far East
BT Asia
NT Choshi Japan
SA Chiba Peninsula
SA Kanto Plain
SA Tokyo Bay
SA Yamanashi Japan

Chiba Peninsula (1978)
Between Tokyo Bay and Sagami Sea.
UF Boso Peninsula
BT Honshu
BT Japan
BT Far East
BT Asia
SA Chiba Japan
SA Chibaken-Toho-Oki earthquake 1987
SA Kazusa Group

Chiba-ken Toho-Oki earthquake 1987
use Chibaken-Toho-Oki earthquake 1987

Chibaken-Toho-Oki earthquake 1987 (1993)
Epicenter off the east coast of Chiba, Japan. Before 1993, also search Chiba earthquake 1987 and Chiba-ken Toho-Oki earthquake 1987.
UF Chiba earthquake 1987
UF Chiba-ken Toho-Oki earthquake 1987
BT earthquakes
SA Chiba Peninsula
SA Honshu
SA Japan
SA Oki Islands
SA West Pacific

Chibougamau
No longer a valid term for GeoRef. As of 1993, see Chibougamau Quebec.

Chibougamau Quebec (1993)
Village SE of Lake Mistassini in central Quebec. Before 1993, also search Chibougamau AND Quebec.
BT Abitibi County Quebec
BT Quebec
BT Eastern Canada
BT Canada
SA Henderson Mine

Chicago
Valid through 1988. Search in combination with state term. After 1988, use specific city-state term.

Chicago Illinois (1989)
City on Lake Michigan in NE Illinois. Before 1989, search Chicago AND Illinois.
CO N415000N415000
W0874500W0874500
BT Cook County Illinois
BT Illinois
BT United States

Chichagof Island (1985)
SE Alaska.
CO N572000N581500
W1344500W1363500
BT Alexander Archipelago
BT Southeastern Alaska
BT Alaska
BT United States

Chichibu
No longer a valid term for GeoRef. See Chichibu Japan.

Chichibu Belt (1978)
Middle segment of Butsuzo Tectonic Line. Kochi Prefecture, central Shikoku.
BT Kochi Japan
BT Shikoku
BT Japan
BT Far East
BT Asia
SA Kurosegawa Zone

Chichibu Japan (1993)
City NW of Tokyo in central Honshu.
BT Saitama Japan
BT Honshu
BT Japan
BT Far East
BT Asia

Chichibu Mine (1978)
Saitama Prefecture, central Honshu.
BT Saitama Japan
BT Honshu
BT Japan
BT Far East
BT Asia
SA metal ores
SA mines

Chickamauga Group (1978)
Includes limestone strata of Middle and Upper Ordovician age. N Alabama, NW Georgia, E Tennessee, and SW Virginia.
BT Ordovician
BT Paleozoic
SA Alabama
SA Georgia
SA Tennessee
SA Virginia

Chickasawhay Formation (1978)
In Limestone Creek Group. In SE Mississippi.
BT upper Oligocene
BT Oligocene
BT Paleogene
BT Tertiary
BT Cenozoic
SA Mississippi

Chickmagalur
No longer a valid term for GeoRef. See Chickmagalur India.

Chickmagalur India (1993)
 Village in SW Karnataka, S India.
 BT Karnataka India
 BT India
 BT Indian Peninsula
 BT Asia

Chicot Aquifer (1989)
 SW Louisiana and SE Texas.
 IN   Index states as applicable.
 BT United States
 SA Louisiana
 SA Texas

Chieti
 No longer a valid term for GeoRef. See Chieti Italy.

Chieti Italy (1993)
 Province and town in E Abruzzi.
 BT Abruzzi Italy
 BT Italy
 BT Southern Europe
 BT Europe

Chihsia Formation (1989)
 Fujian, Hubei, Jiangsu and Zhejiang provinces, E and S central China.
 BT Lower Permian
 BT Permian
 BT Paleozoic
 SA China
 SA Fujian China
 SA Hubei China
 SA Jiangsu China
 SA Zhejiang China

Chihuahua
 No longer a valid term for GeoRef. As of 1993 see Chihuahua Mexico.

Chihuahua Mexico (1993)
 State in northern Mexico. Also a city. Before 1993 also search Chihuahua AND Mexico.
 CO N254500N320000
    W1030000W1080000
 BT Mexico
 NT Sierra Pena Blanca
 SA Allende Meteorite
 SA Chihuahua tectonic belt
 SA Pedregosa Basin
 SA Rio Grande
 SA Rio Grande Rift
 SA Rio Grande Valley
 SA San Antonio Mine
 SA Sierra Madre Occidental

Chihuahua tectonic belt (1978)
 IN   Index Mexican and U. S. states as applicable.
 BT North America
 SA Chihuahua Mexico
 SA Coahuila Mexico
 SA New Mexico
 SA Rio Grande Rift
 SA Texas

Chile (1978)
 CO S560000S174500
    W0670000W0760000
 BT South America
 NT Antofagasta Chile
 NT Arauco Chile
 NT Atacama Chile
 NT Atacama Desert
 NT Copiapo
 NT Coquimbo Chile
 NT Magallanes Chile
 NT Santiago Chile
 NT Tarapaca Chile
 NT Valparaiso Chile
 SA Andes
 SA Central Valley
 SA Chile earthquake 1960
 SA Chile earthquake 1985
 SA Colorado River
 SA Patagonia
 SA Patagonian Andes
 SA Patagonian Batholith
 SA Precordillera
 SA Tierra del Fuego
 SA Western Cordillera

Chile earthquake 1960 (1993)
 Central Andes. To search before 1993, see earthquakes. Before 1993, also search Chilean earthquake 1960.
 BT earthquakes
 SA Chile
 SA Peru-Chile Trench
 SA shallow-focus earthquakes

Chile earthquake 1985 (1989)
 Central Chile.
 UF central Chile earthquake 1985
 UF Chilean earthquake 1985
 BT earthquakes
 SA Chile
 SA Santiago Chile
 SA Valparaiso Chile

Chile Ridge (1978)
 Runs SE-NW off Chilean coast between 80° W-105° W. Autoposting of Pacific Ocean to this term began in 1990.
 CO S453000S350000
    W0800000W1050000
 UF Chile Rise
 BT South American Pacific
 BT Pacific Ocean
 SA mid-ocean ridges

Chile Rise
 use Chile Ridge

Chile Trench
 use Peru-Chile Trench

Chilean earthquake 1985
 use Chile earthquake 1985

Chilhowee Group (1978)
 Includes Loudoun Formation, Weverton Sandstone, Harpers Shale, Antietam Formation, Unicoi Formation, Hampton Shale, Erwin Quartzite, Vann Quartzite, Sandsuck Shale, Cochran Quartzite, Nebo Quartzite, Murry Shale, Hesse Quartzite, Shady Dolomite, Rome Formation, Nichols Shale, Helenmode Formation.
 BT Lower Cambrian
 BT Cambrian
 BT Paleozoic
 SA Antietam Formation
 SA Maryland
 SA North Carolina
 SA Shady Dolomite
 SA Tennessee
 SA Virginia

Chilliwack Group (1985)
 S British Columbia and N Washington.
 BT Carboniferous
 BT Paleozoic
 SA British Columbia
 SA Washington

Chilly Buttes (1989)
 Near Borah Peak in central Idaho.
 IN   Index county as applicable.
 SA Custer County Idaho

Chilton County
 Valid through 1988. Search in combination with state term. After 1988, use specific county-state term.

Chilton County Alabama (1989)
 Central Alabama. Before 1989, also search Chilton County AND Alabama.
 CO N324000N330500
    W0862500W0870200
 BT Alabama
 BT United States

chimneys (1993)
 Use for the landform. As of 1993, use vents for volcanic chimneys and hydrothermal vents for hydrothermal chimneys.
 SA fumaroles
 SA geomorphology
 SA hydrothermal vents
 SA mining geology
 SA ore bodies
 SA pipes
 SA shore features
 SA vents

China (1978)
 Peoples Republic of China. Before 1993, Manchuria, Eastern China, Northern China, Northwestern China, South-Central China, and Southwestern China were narrower terms.
 CO N200000N530000
    E1350000E0740000
 BT Far East
 BT Asia
 NT Anhui China
 NT Dabie Mountains
 NT Dongpu Depression
 NT Fujian China
 NT Gansu China
 NT Guangdong China
 NT Guangxi China
 NT Guizhou China
 NT Hainan China
 NT Han River basin
 NT Hebei China
 NT Heilongjiang China
 NT Henan China
 NT Huang Ho
 NT Hubei China
 NT Hunan China
 NT Inner Mongolia China
 NT Jiangsu China
 NT Jiangxi China
 NT Jilin China
 NT Kunlun Mountains
 NT Liaohe Basin
 NT Liaoning China
 NT Nanling
 NT Ningxia China
 NT North China Platform
 NT Qilian Mountains
 NT Qinghai China
 NT Qinghai-Xizang Plateau
 NT Shaanxi China
 NT Shandong China
 NT Shanghai China
 NT Shanxi China
 NT Sichuan Basin
 NT Sichuan China
 NT Songliao Basin
 NT Taihang Mountains
 NT Tancheng-Lujiang Fault
 NT Tangshan China
 NT Xianshuihe fault zone
 NT Xichang China
 NT Xinjiang China
 NT Xisha Islands
 NT Xizang China
 NT Yangtze Platform
 NT Yangtze River
 NT Yangtze River valley
 NT Yishu Fault
 NT Yunnan China
 NT Zhejiang China
 SA Altai Mountains
 SA Anshan Group
 SA Benxi Formation
 SA Brahmaputra River
 SA Changcheng System
 SA Changzhougou Formation
 SA Chihsia Formation
 SA Chuanlinggou Formation
 SA Dengying Formation
 SA Doushantuo Formation
 SA Haicheng earthquake 1975
 SA Haiyuan earthquake 1920
 SA Huanghua Depression
 SA Huanglong Formation
 SA Hutuo Group
 SA Indosinian Orogeny
 SA Indus-Yarlung Zangbo suture zone
 SA Jilin Meteorite
 SA Kunyang Group
 SA Lancang-Gengma earthquake 1988
 SA Luoquan Formation
 SA Maokou Formation
 SA Mount Everest
 SA Nantuo Formation
 SA Nenjiang Formation
 SA Nihewan Formation
 SA Qingshankou Formation
 SA Qingzhen Meteorite
 SA Qixia Formation
 SA Quantou Formation
 SA Shahejie Formation
 SA Sino-Korean Platform
 SA Taihua Group
 SA Taiyuan Formation
 SA Tancheng-Lujiang fault zone
 SA Tangshan earthquake 1976
 SA Tonghai earthquake 1970
 SA Wufeng Formation
 SA Wumishan Formation
 SA Xingtai earthquake 1966
 SA Yangtze Plate

China Sea (1978)
 Bordering on China and divided by Taiwan into East China Sea to N and South China Sea extending to Malaysia in the S.
 IN   Index seas as applicable.
 BT Pacific Ocean
 SA East China Sea
 SA South China Sea

Chinghai China
 use Qinghai China

Chingis-Tau (1978)
 Mountain range in Semipalatinsk Oblast in E Kazakhstan. This term has multiple hierarchies.
 UF Chingiz Range
 BT1 Semipalatinsk Kazakhstan
 BT1 Kazakhstan
 BT1 Central Asia
 BT1 Asia
 BT2 Semipalatinsk Kazakhstan
 BT2 Kazakhstan
 BT2 Commonwealth of Independent States

Chingiz Range
 use Chingis-Tau

Chinkuashih
 No longer a valid term for GeoRef. As of 1993 see Chinkuashih Taiwan.

Chinkuashih Mine (1978)
 N Taiwan. Also search Chinkuashih.
 UF Kinwashih Mine
 BT Taiwan
 BT Far East
 BT Asia
 SA copper ores
 SA gold ores
 SA mines

Chinkuashih Taiwan (1993)
 Village in N Taiwan.
 UF Kinnashih
 BT Taiwan
 BT Far East

BT Asia

Chinle Formation (1978)
In Dockum Group. N Arizona, SW Colorado, SE Nevada, N New Mexico, S Utah.
BT Upper Triassic
BT Triassic
BT Mesozoic
SA Arizona
SA Colorado
SA Dockum Group
SA Nevada
SA New Mexico
SA Petrified Forest Member
SA Shinarump Member
SA Utah

Chios (1978)
Island department in Aegean Sea off W coast of Turkey. As of 1993, use Chios Greece for the town. Before 1993 also search Chios or Khios AND Greece. This term has multiple hierarchies.
UF Khios
BT1 Greek Aegean Islands
BT1 Greece
BT1 Southern Europe
BT1 Europe
BT2 Greek Aegean Islands
BT2 Aegean Islands
BT2 Mediterranean region

Chipola Formation (1978)
In Alum Bluff Group. SE Alabama and NW Florida.
BT lower Miocene
BT Miocene
BT Neogene
BT Tertiary
BT Cenozoic
SA Alabama
SA Florida

Chippewa County Michigan (1989)
Extreme E Michigan Upper Peninsula, NE Michigan. Before 1989, also search Chippewa County AND Michigan.
CO N455500N464800 W0833000W0851500
BT Michigan Upper Peninsula
BT Michigan
BT United States
NT Sault Sainte Marie Michigan

Chirchik
No longer a valid term for GeoRef. As of 1993, see Chirchik Uzbekistan.

Chirchik River (1978)
Flows into the Syr Darya at Chinaz. Also search Chirchik.
IN Index former Soviet republics as applicable.
BT Asia
SA Kazakhstan
SA Uzbekistan

Chirchik Uzbekistan (1993)
City in Tashkent Oblast in E Uzbekistan. This term has multiple hierarchies.
BT1 Tashkent Uzbekistan
BT1 Uzbekistan
BT1 Asia
BT2 Tashkent Uzbekistan
BT2 Uzbekistan
BT2 Commonwealth of Independent States

Chiroptera (1978)
Order. Autoposting of Eutheria and Theria to this term began in 1989.
BT Eutheria

BT Theria
BT Mammalia
BT Tetrapoda
BT Vertebrata
BT Chordata

Chita
No longer a valid term for GeoRef. As of 1993, see Chita Russian Federation.

Chita Russian Federation (1993)
City and oblast E of Lake Baikal on the Mongolian and Manchurian borders. Before 1993, also search Chita. This term has multiple hierarchies.
BT1 Russian Federation
BT1 Commonwealth of Independent States
BT2 Asia
NT Darasun Russian Federation
NT Itaka Russian Federation
NT Khapcheranga Russian Federation
NT Shilka Valley
NT Unda Russian Federation
SA Baikal-Amur Railroad region
SA Balei Russian Federation
SA Klichka Russian Federation
SA Kodar Range
SA Olekma
SA Olekma-Vitim Highlands
SA Transbaikalia
SA Udokan Series

Chitaldroog India
use Chitradurga India

Chitaldrug
Not a valid term for GeoRef. See Chitradurga India or Chitradurga schist belt.

Chitaldrug India
use Chitradurga India

Chitaldrug schist belt
use Chitradurga schist belt

chitin (1989)
Organic compound which is a common constituent of various invertebrate exoskeletons and tests.
SA cuticles
SA exoskeletons
SA Invertebrata
SA invertebrates
SA organic compounds
SA tests

Chitinozoa (1978)
Before 1985, also search chitinozoans. As of 1985, use chitinozoans as the common name. Autoposting of microfossils to this term began in 1989.
BT palynomorphs
BT microfossils
SA chitinozoans

chitinozoans (1985)
Common name for Chitinozoa. Autoposting of palynomorphs and microfossils to this term began in 1989.
BT palynomorphs
BT microfossils
SA biostratigraphy
SA Chitinozoa

Chitistone Pass (1978)
In the Saint Elias Mountains.
IN Index Alaska and/or Yukon Territory as applicable.
BT North America
SA Alaska
SA Saint Elias Mountains
SA Yukon Territory

Chitradurga
No longer a valid term for GeoRef. See Chitradurga India.

Chitradurga India (1993)
District and town in N Karnataka, SW India. Before 1993, also search Chitradurga or Chitaldroog and Chitaldrug.
UF Chitaldroog India
UF Chitaldrug India
BT Karnataka India
BT India
BT Indian Peninsula
BT Asia
NT Ingaldhal

Chitradurga schist belt (1978)
Karnataka. Crystalline rocks in this area are considered to be Dharwars.
UF Chitaldrug schist belt
SA Dharwar Craton
SA Dharwars
SA Karnataka India

Chitral (1978)
River, North-West Frontier Province, Pakistan. As of 1993, see Chitral Pakistan for the city.
BT North-West Frontier Pakistan
BT Pakistan
BT Indian Peninsula
BT Asia
SA Chitral Pakistan

Chitral Pakistan (1993)
City, NW Pakistan.
BT North-West Frontier Pakistan
BT Pakistan
BT Indian Peninsula
BT Asia
SA Chitral

Chittenden County
Valid through 1988. Search in combination with state term. After 1988, use specific county-state term.

Chittenden County Vermont (1989)
On Lake Champlain in NW Vermont. Before 1989, also search Chittenden County AND Vermont.
CO N440800N444300 W0724800W0735000
BT Vermont
BT United States

Chiuzbaia (1978)
Region in Maramures County in N Transylvania.
BT Maramures Romania
BT Transylvania
BT Romania
BT Southern Europe
BT Europe

Chkalov Russian Federation
use Orenburg Russian Federation

chkalovite (1978)
BT framework silicates
BT silicates

chladnite (1981)
Used for enstatite achondrites. Before 1981, also search aubrite.
UF aubrite
UF enstatite achondrites
BT achondrites
BT stony meteorites
BT meteorites
NT Norton County Meteorite

chlorapatite (1978)
BT phosphates
SA apatite

chloride ion (1978)

Before 1978, search chloride.
SA chlorides
SA ions

chlorides (1978)
Autoposting of this term began in 1981.
BT halides
NT atacamite
NT bideauxite
NT bischofite
NT boracite
NT carnallite
NT eudialyte
NT halite
NT kainite
NT mimetite
NT paratacamite
NT phosgenite
NT pyromorphite
NT rinneite
NT sylvite
NT vanadinite
SA calcium chloride
SA chloride ion
SA sodium chloride

chlorinated hydrocarbons (1989)
NT DDT
NT dibromochloropropane
NT PCBs
NT tetrachloroethylene
NT trichloroethane
NT trichloroethylene
SA hydrocarbons
SA pollutants
SA volatile organic compounds

chlorine (1978)
Autoposting of halogens to this term began in 1989.
UF Cl
BT halogens
NT Cl-36
NT Cl-37/Cl-35
SA salt

chlorite (1978)
BT chlorite group
BT sheet silicates
BT silicates
SA chloritization

chlorite group (1978)
Autoposting of this term began in 1981.
BT sheet silicates
BT silicates
NT chamosite
NT chlorite
NT clinochlore
NT cookeite
NT kammererite
NT sudoite

chlorite schist (1978)
BT schists
BT metamorphic rocks

chloritization (1978)
BT metasomatism
SA chlorite
SA processes

chloritoid (1978)
Autoposting of nesosilicates began in 1985.
BT nesosilicates
BT orthosilicates
BT silicates
SA aluminosilicates

chloropal
use nontronite

chlorophaeite (1978)
BT silicates

chlorophenothane
use DDT

Chlorophyceae (1981)
Autoposting of thallophytes and microfossils began in 1990. This term has multiple hierarchies.
BT1 Chlorophyta
BT1 algae
BT1 thallophytes
BT1 Plantae
BT2 Chlorophyta
BT2 algae
BT2 microfossils
NT Codiaceae
NT Dasycladaceae
NT Receptaculitaceae

chlorophyll (1978)
BT pigments
BT organic materials

Chlorophyta (1978)
Autoposting of this term to Tasmanites, Charophyta, Dasycladaceae and Receptaculitaceae began in 1981. Autoposting of thallophytes and microfossils began in 1990. This term has multiple hierarchies.
UF green algae
BT1 algae
BT1 thallophytes
BT1 Plantae
BT2 algae
BT2 microfossils
NT Botryococcus
NT Charophyta
NT Chlorophyceae
NT Desmidiales
NT Tasmanites

Choc Nappe (1985)
W Carpathians, Czechoslovakia.
CO N480000N490000
E0200000E0170000
BT Czechoslovakia
BT Central Europe
BT Europe
SA Slovakia
SA Slovakian Carpathians
SA Veporides
SA Western Carpathians

Chocolate Mountains (1985)
Mountain range in E Yuma County, Arizona. Also a mountain range in N central Imperial County, California.
IN Index counties or regions and mountains as applicable.
SA Arizona
SA California
SA Imperial County California
SA Yuma County Arizona

Choctaw County
Valid through 1988. Search in combination with state term. After 1988, use specific county-state term.

Choctaw County Alabama (1989)
SW Alabama. Before 1989, also search Choctaw County AND Alabama.
CO N314100N321700
W0875700W0882500
BT Alabama
BT United States

Choctaw County Mississippi (1989)
Central Mississippi. Before 1989, also search Choctaw County AND Mississippi.
CO N330800N333300
W0890600W0892800

BT Mississippi
BT United States

Choctaw County Oklahoma (1989)
SE Oklahoma. Before 1989, also search Choctaw County AND Oklahoma.
CO N335200N340800
W0950900W0955700
BT Oklahoma
BT United States

Choctawhatchee Bay (1978)
Inlet of Gulf of Mexico in the Florida panhandle in NW Florida.
IN Index counties as applicable.
BT Florida
BT United States
SA Walton County Florida

Chokai (1989)
Mountain and volcano between Yamagata and Akita Prefecture. N Honshu, Japan.
IN Index prefecture as applicable.
CO N390800N390800
E1400400E1400400
UF Chokai-san
BT Honshu
BT Japan
BT Far East
BT Asia
SA Akita Japan
SA Yamagata Japan

Chokai-san
use Chokai

Cholame
Valid through 1988. Search in combination with state term. After 1988, use specific city-state term.

Cholame California (1989)
Town on the Pacific in S California. Before 1989, search Cholame AND California.
CO N354400N354400
W1201600W1201600
BT San Luis Obispo County California
BT California
BT United States

Chomolungma
use Mount Everest

Chondrichthyes (1978)
Class. Autoposting of this term began in 1978.
BT Pisces
BT Vertebrata
BT Chordata
NT Elasmobranchii
NT Eubradyodonti
NT Euselachii
NT Holocephali

chondrites (1978)
Search chondrites AND meteorites to distinguish from the ichnofossil. Autoposting of this term began in 1978.
BT stony meteorites
BT meteorites
NT carbonaceous chondrites
NT enstatite chondrites
NT H chondrites
NT HL chondrites
NT Jilin Meteorite
NT L chondrites
NT LL chondrites
NT ordinary chondrites
NT Saint-Severin Meteorite
NT Sharps Meteorite
NT Weston Meteorite
SA achondrites
SA Chondrites ichnofossils

SA chondrules
SA Elephant Moraine Meteorites
SA iron meteorites
SA stony irons

Chondrites ichnofossils (1981)
Used to distinguish the ichnofossil from the meteorite. Before 1981, search Chondrites AND NOT meteorites.
SA chondrites
SA ichnofossils

Chondrostei (1981)
Infraclass.
BT Actinopterygii
BT Osteichthyes
BT Pisces
BT Vertebrata
BT Chordata

chondrules (1978)
SA chondrites
SA meteorites

Chopawamsic Formation (1985)
E central Virginia.
BT lower Paleozoic
BT Paleozoic
SA Virginia

Choptank Formation (1978)
In Chesapeake Group. E Maryland and Virginia.
BT middle Miocene
BT Miocene
BT Neogene
BT Tertiary
BT Cenozoic
SA Chesapeake Group
SA Maryland
SA Virginia

Choptank River (1978)
IN Index states as applicable.
BT United States
SA Delaware
SA Maryland

Chor
No longer a valid term for GeoRef. See Chor Pakistan.

Chor Pakistan (1993)
Village in SE Pakistan.
BT Sind Pakistan
BT Pakistan
BT Indian Peninsula
BT Asia

Chordata (1978)
Term is to be used only when more specific terms do not apply. Has narrower terms as of 1981.
NT Protochordata
NT Vertebrata
SA Hemichordata

chorology
use biogeography

Chorzow
No longer a valid term for GeoRef. As of 1993 see Chorzow Poland.

Chorzow Poland (1993)
City in N Katowice, S Poland. Before 1993 also search Chorzow AND Poland.
BT Katowice Poland
BT Poland
BT Central Europe
BT Europe

Choshi
No longer a valid term for GeoRef. See Choshi Japan.

Choshi Japan (1993)

City in E Chiba Prefecture, SE Honshu, Japan. Before 1993, also search Choshi or Chosi AND Japan.
CO N354300N354300
E1045100E1045100
UF Chosi Japan
BT Chiba Japan
BT Honshu
BT Japan
BT Far East
BT Asia

Chosi Japan
use Choshi Japan

Choukoutien
use Zhoukoudian

Chouteau County
Valid through 1988. Search in combination with state term. After 1988, use specific county-state term.

Chouteau County Montana (1989)
N central Montana. Before 1989, also search Chouteau County AND Montana.
CO N472400N482000
W1093000W1112500
BT Montana
BT United States

Chouteau Limestone (1993)
Kinderhookian or Lower Mississippian. Central and E Missouri, SW Illinois, and Kansas. Underlies Burlington Limestone in Missouri, Sedalia Formation in Illinois; overlies Hannibal Formation in Missouri and Illinois.
IN Index ages as applicable.
BT Lower Mississippian
BT Mississippian
BT Carboniferous
BT Paleozoic
SA Illinois
SA Kansas
SA Kinderhookian
SA Missouri

Chowan River Formation (1989)
In Chesapeake Group.
BT upper Pliocene
BT Pliocene
BT Neogene
BT Tertiary
BT Cenozoic
SA Chesapeake Group
SA North Carolina
SA Virginia

Choybalsan
No longer a valid term for GeoRef. As of 1993 see Choybalsan Mongolia.

Choybalsan Mongolia (1993)
Town on Kerulen River in E Mongolia. Before 1993 also search Choybalsan AND Mongolia.
UF Kerulen
BT Mongolia
BT Far East
BT Asia

Christchurch
No longer a valid term for GeoRef. See Christchurch New Zealand.

Christchurch New Zealand (1993)
City on Pacific Ocean in E Canterbury.
IN Also index South Island.
BT Canterbury New Zealand
BT New Zealand

BT Australasia
SA South Island

Christian County
Valid through 1988. Search in combination with state term. After 1988, use specific county-state term.

Christian County Illinois (1989)
Central Illinois. Before 1989, also search Christian County AND Illinois.
CO N391800N395000
W0890100W0893400
BT Illinois
BT United States

Christian County Kentucky (1989)
SW Kentucky. Before 1989, also search Christian County AND Kentucky.
CO N363700N371000
W0871400W0874200
BT Kentucky
BT United States

Christian County Missouri (1989)
SW Missouri. Before 1989, also search Christian County AND Missouri.
CO N364900N370700
W0925500W0933700
BT Missouri
BT United States

Christiansund
use Kristiansund Norway

Christmas Island (1978)
An external territory of Australia, about 225 miles S of W end of Java. Also one of the Line Islands in the central Pacific Ocean, S of Hawaii. In 1985, broader term changed from Indian Ocean to Indian Ocean Islands and Line Islands and Kiribati. Search with appropriate broader terms.
IN Index Line Islands and Kiribati; or Indian Ocean Islands as applicable.
SA Indian Ocean Islands
SA Kiribati
SA Line Islands

Christmas Mountains (1978)
In Big Bend region in SW Texas.
BT Texas
BT United States
SA Brewster County Texas

Christopher Formation (1978)
Ellef Ringnes Island in Franklin District.
BT Lower Cretaceous
BT Cretaceous
BT Mesozoic
SA Canada
SA Ellef Ringnes Island
SA Northwest Territories

chromates (1978)
Autoposting of this term began in 1978.
NT crocoite
NT hemihedrite
SA minerals

chromatograms (1981)
SA chromatography
SA gas chromatograms

chromatography (1978)
Before 1971, also search paper chromatography.
UF paper chromatography
SA adsorption
SA analysis
SA chemical analysis
SA chromatograms
SA gas chromatography
SA ion chromatography
SA liquid chromatography
SA spectroscopy

chrome diopside (1978)
UF chrome-diopside
BT clinopyroxene
BT pyroxene group
BT chain silicates
BT silicates
SA diopside

chrome spinel (1978)
UF picotite
BT oxides
SA spinel

chrome-diopside
use chrome diopside

chromite (1978)
As of 1981, use chromite ores for chromite as a commodity.
BT oxides
SA chromite ores
SA chromitite
SA chromium

chromite ores (1981)
Before 1981, also search (chromite OR chromium OR Cr) AND (deposit OR deposits OR ore OR ores OR economic) in the basic index. Autoposting of metal ores to this term began in 1985.
IN Commodity. See List C.
UF chromium ores
BT metal ores
SA chromite
SA chromium
SA podiform deposits

chromitite (1978)
BT ultramafics
BT plutonic rocks
BT igneous rocks
SA chromite

chromium (1978)
Chemical element. As of 1981, use chromite ores for the commodity; before 1981, chromite was used for that purpose. Use chromite for the mineral. Autoposting of metals to this term began in 1989.
UF Cr
BT metals
NT Cr-51
NT Cr-53/Cr-52
SA chromite
SA chromite ores

chromium ores
use chromite ores

chronology (1978)
Used as a general term.
SA geochronology

chronostratigraphy (1978)
IN Index age terms as applicable.
SA stratigraphy

Chryse
use Chryse Planitia

Chryse Basin
use Chryse Planitia

Chryse Planitia (1985)
Low plain on Mars.
UF Chryse
UF Chryse Basin
UF Chryse region
BT Mars

Chryse region
use Chryse Planitia

chrysoberyl (1978)
BT oxides
SA alexandrite

chrysocolla (1978)
BT sheet silicates
BT silicates
SA copper minerals
SA copper ores

Chrysophyta (1978)
Including golden-brown algae. Autoposting of thallophytes and microfossils began in 1990. This term has multiple hierarchies.
UF golden-brown algae
BT1 algae
BT1 thallophytes
BT1 Plantae
BT2 algae
BT2 microfossils

chrysoprase (1985)
BT silica minerals
BT framework silicates
BT silicates
SA chalcedony

chrysotile (1978)
UF chrysotile asbestos
BT serpentine group
BT sheet silicates
BT silicates
SA asbestos
SA asbestos deposits

chrysotile asbestos
use chrysotile

Chu (1978)
River in Dzhambul Oblast in S Kazakhstan. After 1992, see Chu Kazakhstan for the town.
IN Index Kazakhstan as applicable.
SA Chu-Sarysu Depression
SA Kazakhstan

Chu Kazakhstan (1993)
Town in Dzhambul Oblast in S Kazakhstan. This term has multiple hierarchies.
BT1 Kazakhstan
BT1 Central Asia
BT1 Asia
BT2 Kazakhstan
BT2 Commonwealth of Independent States

Chu-Ili Mountains (1978)
Branch of the Tien Shan in Dzhambul Oblast in S Kazakhstan. This term has multiple hierarchies.
BT1 Kazakhstan
BT1 Central Asia
BT1 Asia
BT2 Kazakhstan
BT2 Commonwealth of Independent States
BT3 Tien Shan
BT3 Asia

Chu-Sarysu Basin
use Chu-Sarysu Depression

Chu-Sarysu Depression (1989)
This term has multiple hierarchies.
CO N424500N480000
E0780000E0650000
UF Chu-Sarysu Basin
BT1 Kazakhstan
BT1 Central Asia
BT1 Asia
BT2 Kazakhstan
BT2 Commonwealth of Independent States
SA Chu
SA Sarysu

Chuanlinggou Formation (1989)
Proterozoic? In Changcheng System.
BT Precambrian
SA Changcheng System
SA China

Chuar Group (1978)
Grand Canyon Series. Grand Canyon region in N Arizona.
BT Precambrian
SA Arizona
SA Jupiter Member

Chubu Japan (1993)
Region W-central Honshu. Includes Aichi, Fukui, Gifu, Ishikawa, Nagano, Niigata, Shizuoka, Toyama and Yamanashi prefectures and Hokuriku, Tokai, and Tosan districts.
IN Index prefecture or prefectures as applicable.
BT Honshu
BT Japan
BT Far East
BT Asia
SA Aichi Japan
SA Fukui Japan
SA Gifu Japan
SA Hokuriku
SA Ishikawa Japan
SA Japanese Alps
SA Nagano Japan
SA Niigata Japan
SA Shizuoka Japan
SA Tokai Japan
SA Toyama Japan
SA Yamanashi Japan

Chubut
No longer a valid term for GeoRef. See Chubut Argentina.

Chubut Argentina (1993)
Province in S Argentina.
CO S460000S420000
W0640000W0730000
BT Argentina
BT South America
SA Patagonia

Chuckanut Formation (1978)
Overlies schists and igneous rocks. San Juan Islands in NW Washington.
BT Eocene
BT Paleogene
BT Tertiary
BT Cenozoic
SA San Juan Islands
SA Washington

Chudleigh
No longer a valid term for GeoRef as of 1993. Chudleigh England.

Chudleigh England (1993)
Town in Devonshire, SW England. Before 1993, also search Chudleigh England.
BT Devonshire England
BT England
BT Great Britain
BT United Kingdom
BT Western Europe
BT Europe

Chugach Mountains (1978)
Along S coast Alaska.
BT Southern Alaska
BT Alaska
BT United States
SA Coast Ranges
SA Columbia Glacier

SA Martin River Glacier

Chugoku (1978)
Region of mountains in SW Honshu which contains Hiroshima, Okayama, Shimane, Tottori and Yamaguchi prefectures.
IN Index prefectures as applicable.
BT Honshu
BT Japan
BT Far East
BT Asia
SA Hiroshima Japan
SA Okayama Japan
SA Shimane Japan
SA Tottori Japan
SA Yamaguchi Japan

Chugwater Formation (1989)
IN Index ages as applicable.
SA Colorado
SA Montana
SA Permian
SA Red Peak Formation
SA Triassic
SA Wyoming

Chukchi Peninsula (1978)
NE extremity of Siberia in Chukchi National Okrug of Magadan Oblast. Also search Chukchi. As of 1981, term includes Wrangel Island. This term has multiple hierarchies.
CO N640000N720000
   W1700000E1530000
UF Chukot Peninsula
UF Chukotka (Peninsula)
UF Chukotsk Peninsula
UF Chukotski Peninsula
UF Chukotskiy Peninsula
BT1 Russian Federation
BT1 Commonwealth of Independent States
BT2 Asia
SA Anadyr Basin
SA Magadan Russian Federation
SA Okhotsk-Chukchi
SA Wrangel Island

Chukchi Range
  use Anadyr Range

Chukchi Sea (1978)
N of Bering strait. Between Alaska and Siberia. Also search Chukchi.
CO N660000N713000
   W1560000E1770000
BT Arctic Ocean
SA Bering Strait
SA Beringia

Chukot Peninsula
  use Chukchi Peninsula

Chukotka (Peninsula)
  use Chukchi Peninsula

Chukotsk Peninsula
  use Chukchi Peninsula

Chukotski Peninsula
  use Chukchi Peninsula

Chukotskiy Peninsula
  use Chukchi Peninsula

Chuma River
  use Uda River

Chumstick Formation (1989)
BT Eocene
BT Paleogene
BT Tertiary
BT Cenozoic
SA Washington

Chuquicamata
No longer a valid term for GeoRef. See Chiqiocamata Chile.

Chuquicamata Chile (1993)
Mining settlement in N Chile.
BT Antofagasta Chile
BT Chile
BT South America

Church Stretton
No longer a valid term for GeoRef as of 1993. See Church Stretton England.

Church Stretton England (1993)
Urban district in S central Shropshire in W England. Before 1993, also search Church Stretton.
BT Shropshire England
BT England
BT Great Britain
BT United Kingdom
BT Western Europe
BT Europe

Churchill
No longer a valid term for GeoRef. See Churchill Manitoba.

Churchill County
Valid through 1988. Search in combination with state term. After 1988, use specific county-state term.

Churchill County Nevada (1989)
W central Nevada. Before 1989, also search Churchill County AND Nevada.
CO N390400N400000
   W1172500W1191500
BT Nevada
BT United States
SA Dixie Valley

Churchill Falls (1978)
On Churchill River in SW and S Labrador.
UF Grand Falls
BT Labrador
BT Newfoundland
BT Eastern Canada
BT Canada

Churchill Manitoba (1993)
Town on Hudson Bay in NE Manitoba.
BT Manitoba
BT Western Canada
BT Canada

Churchill Province (1978)
Structural province of the Canadian Shield.
IN Index Labrador and provinces as applicable.
CO N533000N790000
   W0640000W1140000
BT Canadian Shield
BT North America
NT Cape Smith fold belt
SA Alberta
SA Athabasca Formation
SA Labrador
SA Manitoba
SA Northwest Territories
SA Ontario
SA Quebec
SA Saskatchewan
SA Wyoming Province

Chuya
No longer a valid term for GeoRef. As of 1993, see Chuya Russian Federation.

Chuya Alps (1978)
Range in Altai Mountains near the Mongolian and Xinjiang China borders. This term has multiple hierarchies.
BT1 Russian Federation
BT1 Commonwealth of Independent States
BT2 Asia
BT3 Altai Mountains
BT3 Asia
SA Altai Russian Federation

Chuya Basin (1978)
River basin in Altai Kray near the Mongolian and Xinjiang China borders. This term has multiple hierarchies.
BT1 Altai Russian Federation
BT1 Russian Federation
BT1 Commonwealth of Independent States
BT2 Altai Russian Federation
BT2 Asia

Chuya Russian Federation (1993)
Village in NE Irkutsk Oblast near the SW Yakutia border. This term has multiple hierarchies.
BT1 Irkutsk Russian Federation
BT1 Russian Federation
BT1 Commonwealth of Independent States
BT2 Irkutsk Russian Federation
BT2 Asia

Chvaletice (1978)
Mine site in E Bohemia.
BT Bohemia
BT Czech Republic
BT Czechoslovakia
BT Central Europe
BT Europe

Cibicides (1978)
Autoposting of microfossils and Protista to this term began in 1990. This term has multiple hierarchies.
BT1 Orbitoidacea
BT1 Rotaliina
BT1 foraminifera
BT1 Protista
BT1 Invertebrata
BT2 Orbitoidacea
BT2 Rotaliina
BT2 foraminifera
BT2 Protista
BT2 microfossils

Cibola County
Valid through 1988. Search in combination with state term. After 1988, use specific county-state term.

Cibola County New Mexico (1989)
On Arizona border in W central New Mexico. Before 1989, also search Cibola County AND New Mexico. Created 1981 from W part of Valencia County.
CO N343500N352000
   W1070500W1090000
BT New Mexico
BT United States
SA Valencia County New Mexico
SA Zuni Mountains

Cicatricosisporites (1978)
Autoposting of microfossils to this term began in 1990.
BT miospores
BT palynomorphs
BT microfossils

Cieszyn
No longer a valid term for GeoRef. As of 1993 see Cieszyn Poland.

Cieszyn Poland (1993)
City in S Poland. Before 1993 also search Cieszyn AND Poland.
BT Bielsko Poland
BT Poland
BT Central Europe
BT Europe
SA Katowice Poland

Cilento (1978)
Area in SW Italy.
BT Campania Italy
BT Italy
BT Southern Europe
BT Europe

Cima d'Asta (1978)
Peak in the Dolomites in N Italy. This term has multiple hierarchies.
BT1 Dolomites
BT1 Eastern Alps
BT1 Alps
BT1 Europe
BT2 Dolomites
BT2 Italy
BT2 Southern Europe
BT2 Europe
BT3 Trentino-Alto Adige Italy
BT3 Italy
BT3 Southern Europe
BT3 Europe

Cima volcanic field (1989)
SE California.
BT San Bernardino County California
BT California
BT United States

Cimarron County
Valid through 1988. Search in combination with state term. After 1988, use specific county-state term.

Cimarron County Oklahoma (1989)
Extreme NW Oklahoma. Before 1989, also search Cimarron County AND Oklahoma.
CO N363000N370000
   W1020200W1030000
BT Oklahoma
BT United States
SA Oklahoma Panhandle

Ciminna Basin (1981)
Structural basin in NW Sicily.
CO N375200N375600
   E0133700E0132700
BT Sicily Italy
BT Italy
BT Southern Europe
BT Europe

Cimmerian (1978)
Black Sea area. Above Meotian, below Kuyalnikian. For Mesozoic orogenic event, use Cimmerian Orogeny.
UF Kimmerian
BT Pliocene
BT Neogene
BT Tertiary
BT Cenozoic

Cimmerian Orogeny (1985)
Two-phase orogenic event, Upper Triassic-Upper Jurassic. Before 1985, also search Cimmerian AND orogeny.
IN Index ages as applicable.
UF Kimmerian Orogeny
BT Mesozoic
SA Jurassic
SA orogeny
SA tectonics
SA Upper Triassic

Cincinnati
Valid through 1988. Search in combination with state term. After 1988, use specific city-state term.

Cincinnati Arch (1978)
IN Index states as applicable.
BT United States
SA Indiana
SA Kentucky
SA Ohio

Cincinnati Ohio (1989)
City on the Ohio River in SW Ohio. Before 1989, search Cincinnati AND Ohio.
CO N391000N391000
    W0843000W0843000
BT Hamilton County Ohio
BT Ohio
BT United States
SA Ohio River

Cincinnatian (1978)
Provincial series, North America. Above Champlainian, below Alexandrian (Silurian). Autoposting of this term began in 1978.
BT Upper Ordovician
BT Ordovician
BT Paleozoic
NT Maysvillian
NT Richmondian

cinder cones (1978)
UF cones, cinder
BT volcanic features
SA cones
SA geomorphology
SA volcanism
SA volcanoes

cinematics
 use kinematics

cinerite (1978)
BT clastic rocks
BT sedimentary rocks
SA terrigenous materials

cinnabar (1978)
BT sulfides
SA mercury
SA mercury ores

circulation (1978)
A valid level 1 term through 1977 used in combination with oceans (i.e. oceans, circulation). After 1977, includes use with meteorology.
SA angular momentum
SA atmosphere
SA climate-induced circulation
SA meteorology
SA movement
SA ocean circulation
SA ocean waves
SA paleocirculation
SA permeability
SA springs
SA thermal circulation
SA thermohaline circulation
SA winds

Circum-Antarctic region (1978)
SA Antarctica

Circum-Pacific Belt
 use Circum-Pacific region

Circum-Pacific region (1978)
The region that borders the Pacific Ocean along the continental margin of Asia and the Americas. Before 1978, also search Circum-Pacific Belt; Circum-Pacific.
UF Circum-Pacific Belt
BT Pacific Ocean
SA Pacific region

cirques (1978)
Autoposting of glacial features to this term began in 1989.
BT glacial features
SA eskers
SA glacial geology
SA glacial lakes
SA kames

Cirripedia (1978)
BT Crustacea
BT Mandibulata
BT Arthropoda
BT Invertebrata
SA barnacles

CIS
Not a valid term for GeoRef. See Commonwealth of Independent States.

Ciscaucasia
 use Northern Caucasus

Cisco Group (1985)
Central and N central Texas.
BT Upper Pennsylvanian
BT Pennsylvanian
BT Carboniferous
BT Paleozoic
SA Texas

cistern rock
 use laccoliths

Citronelle Formation (1978)
Recognized to be equivalent of Willis Formation of Texas and Louisiana. Graham Ferry Formation of Pliocene and Pleistocene age disconformably underlies Citronelle Formation. Gulf Coastal Plain from W Florida and S Georgia to E Texas inclusive.
BT Pliocene
BT Neogene
BT Tertiary
BT Cenozoic
SA Alabama
SA Florida
SA Georgia
SA Louisiana
SA Mississippi
SA Texas

Citrus County Florida (1993)
W Florida. Before 1989, also search Citrus County AND Florida.
CO N283900N290200
    W0821100W0824400
BT Florida
BT United States

Ciudad de Valles
 use Valles Mexico

Ciudad Real
No longer a valid term for GeoRef. As of 1993, see Ciudad Real Spain.

Ciudad Real Spain (1993)
Province in S central New Castile Spain. Before 1993, also search Ciudad Real and Spain.
CO N382100N393700
    W0024000W0050500
BT New Castile Spain
BT Castile Spain
BT Spain
BT Iberian Peninsula
BT Southern Europe
BT Europe
NT Alcudia Valley
NT Almaden Spain
SA Montes de Toledo

Ciudad Trujillo
 use Santo Domingo Dominican Republic

civil engineering (1978)
UF engineering, civil
SA engineering geology

Cl
 use chlorine

Cl-36 (1985)
Autoposting of broader terms began in 1989. This term has multiple hierarchies.
BT1 radioactive isotopes
BT1 isotopes
BT2 chlorine
BT2 halogens

Cl-37/Cl-35 (1989)
Autoposting of broader terms began in 1989. This term has multiple hierarchies.
BT1 stable isotopes
BT1 isotopes
BT2 chlorine
BT2 halogens
SA isotope ratios

Clackamas County
Valid through 1988. Search in combination with state term. After 1988, use specific county-state term.

Clackamas County Oregon (1989)
NW Oregon. Before 1989, also search Clackamas County AND Oregon.
CO N445000N453000
    W1214000W1225000
BT Oregon
BT United States
SA Mount Hood

clades
 use cladistics

cladism
 use cladistics

cladistic analysis
 use cladistics

cladistic taxonomy
 use cladistics

cladistics (1985)
UF clades
UF cladism
UF cladistic analysis
UF cladistic taxonomy
UF cladograms
SA biologic evolution
SA biometry
SA mathematical methods
SA paleobotany
SA paleontology
SA phylogeny
SA statistical analysis
SA taxonomy

Cladocopina (1981)
Suborder. Autoposting of Myodocopida to this term began in 1989. As of 1990, microfossils, Crustacea, Mandibulata, and Arthropoda are autoposted to this term. This term has multiple hierarchies.
BT1 Myodocopida
BT1 Ostracoda
BT1 Crustacea
BT1 Mandibulata
BT1 Arthropoda
BT1 Invertebrata
BT2 Myodocopida
BT2 Ostracoda
BT2 microfossils

cladograms
 use cladistics

Cladophlebis (1978)
Before 1985, miospores and palynomorphs were autoposted to this term.
BT Filicopsida
BT pteridophytes
BT Plantae
SA miospores
SA palynomorphs

Claiborne County
Valid through 1988. Search in combination with state term. After 1988, use specific county-state term.

Claiborne County Mississippi (1989)
SW Mississippi. Before 1989, also search Claiborne County AND Mississippi.
CO N314700N321200
    W0904400W0911300
BT Mississippi
BT United States

Claiborne County Tennessee (1989)
NE Tennessee. Before 1989, also search Claiborne County AND Tennessee.
CO N362300N363600
    W0832300W0840200
BT Tennessee
BT United States

Claiborne Group (1978)
Includes Carrizo Sand, Reklaw Formation, Queen City Sand, Weches Glauconitic Marl, Viesca Glauconitic Marl, Sparta Sand, Stone City Beds, Crockett Formation, Landrum Shale, Spiller Sand, Yegua Formation, Gosport Sand, Tallahatta Formation, Zilpha Clay, Winona Sand, Wautubbee Formation, Cockfield Formation, Avon Park Limestone, Tallahassee Limestone, Lake City Limestone, Lisbon Formation, Neshoba Sand, Cook Mountain Formation, Meridian Formation, Kosciusko Formation, Cane River Formation. Gulf Coastal Plain from Georgia to S Texas.
BT middle Eocene
BT Eocene
BT Paleogene
BT Tertiary
BT Cenozoic
SA Alabama
SA Carrizo Sand
SA Cockfield Formation
SA Cook Mountain Formation
SA Florida
SA Georgia
SA Lisbon Formation
SA Louisiana
SA Mississippi
SA Queen City Formation
SA Sparta Sand
SA Tallahatta Formation
SA Texas
SA Yegua Formation

Claiborne Parish
Valid through 1988. Search in combination with state term. After 1988, use specific county-state term.

Claiborne Parish Louisiana (1989)
NW Louisiana. Before 1989, also search Claiborne Parish AND Louisiana.
CO N323500N330000
    W0924400W0931400
BT Louisiana
BT United States

Clallam County
Valid through 1988. Search in combination with state term. After 1988, use specific county-state term.

Clallam County Washington (1989)
NW Washington. Before 1989, also search Clallam County AND Washington.
CO N475100N482500
W1225500W1244500
BT Washington
BT United States
SA Olympic Mountains

Clara Mine (1993)
Near Oberwolfach, Germany.
IN Index Baden-Wurttemberg Germany or region as applicable.
SA Baden-Wurttemberg Germany
SA barite deposits
SA metal ores
SA mines

clarain (1989)
Coal lithotype.
BT macerals

Clare
No longer a valid term for GeoRef. See Clare Ireland.

Clare Ireland (1993)
County on the Atlantic in W Ireland.
BT Ireland
BT Western Europe
BT Europe

Clarendonian (1985)
North America. Miocene, above Barstovian and below Hemphillian.
BT Miocene
BT Neogene
BT Tertiary
BT Cenozoic
SA middle Miocene
SA upper Miocene

Clarion fracture zone (1985)
BT North Pacific
BT Pacific Ocean

Clark County
Valid through 1988. Search in combination with state term. After 1988, use specific county-state term.

Clark County Arkansas (1989)
S central Arkansas. Before 1989, also search Clark County AND Arkansas.
CO N334300N342000
W0925300W0933100
BT Arkansas
BT United States

Clark County Idaho (1989)
E Idaho. Before 1989, also search Clark County AND Idaho.
CO N435800N443500
W1113800W1130000
BT Idaho
BT United States

Clark County Illinois (1989)
E Illinois. Before 1989, also search Clark County AND Illinois.
CO N391000N393000
W0873000W0880100
BT Illinois
BT United States

Clark County Indiana (1989)
SE Indiana. Before 1989, also search Clark County AND Indiana.
CO N381500N383700
W0852500W0860000
BT Indiana
BT United States

Clark County Kansas (1989)
SW Kansas. Before 1989, also search Clark County AND Kansas.
CO N370000N372800
W0993300W1000700
BT Kansas
BT United States

Clark County Kentucky (1989)
Central Kentucky. Before 1989, also search Clark County AND Kentucky.
CO N374900N380700
W0835700W0842200
BT Kentucky
BT United States

Clark County Missouri (1989)
Extreme NE Missouri. Before 1989, also search Clark County AND Missouri.
CO N403100N403800
W0912600W0915800
BT Missouri
BT United States

Clark County Nevada (1989)
SE Nevada. Before 1989, also search Clark County AND Nevada.
CO N350200N365300
W1140300W1155300
BT Nevada
BT United States
NT Las Vegas Nevada
NT Moapa Valley
SA Lake Mead
SA Las Vegas Valley
SA Lost City
SA River Mountains
SA Spring Mountains

Clark County Ohio (1989)
W central Ohio. Before 1989, also search Clark County AND Ohio.
CO N394700N400300
W0833200W0840300
BT Ohio
BT United States
NT Springfield Ohio

Clark County South Dakota (1989)
E central South Dakota. Before 1989, also search Clark County AND South Dakota.
CO N443300N450900
W0972900W0975900
BT South Dakota
BT United States

Clark County Washington (1989)
SW Washington. Before 1989, also search Clark County AND Washington.
CO N453400N460400
W1221500W1224800
BT Washington
BT United States
NT Vancouver Washington

Clark County Wisconsin (1989)
Central Wisconsin. Before 1989, also search Clark County AND Wisconsin.
CO N442500N450200
W0901900W0905500
BT Wisconsin
BT United States

Clark Fork (1985)
Tributary of the Pend Oreille River, NW Montana and N Idaho.
IN Index states as applicable.
SA Clark's Fork Basin
SA Idaho
SA Montana
SA United States

Clark's Fork Basin (1985)
NW Wyoming and S central Montana.
IN Index states as applicable.
BT United States
SA Clark Fork
SA Montana
SA Wyoming

Clarke County
Valid through 1988. Search in combination with state term. After 1988, use specific county-state term.

Clarke County Alabama (1989)
SW Alabama. Before 1989, also search Clarke County AND Alabama.
CO N311000N315800
W0873200W0880800
BT Alabama
BT United States

Clarke County Georgia (1989)
NE central Georgia. Before 1989, also search Clarke County AND Georgia.
CO N335000N340300
W0831500W0833300
BT Georgia
BT United States
NT Athens Georgia

Clarke County Iowa (1989)
S Iowa. Before 1989, also search Clarke County AND Iowa.
CO N405300N410900
W0933400W0940200
BT Iowa
BT United States

Clarke County Mississippi (1989)
E Mississippi. Before 1989, also search Clarke County AND Mississippi.
CO N314800N321300
W0882600W0885400
BT Mississippi
BT United States

Clarke County Virginia (1989)
N Virginia. Before 1989, also search Clarke County AND Virginia.
CO N385800N391500
W0775000W0780900
BT Virginia
BT United States

Clarno
Valid through 1988. Search in combination with state term. After 1988, use specific city-state term.

Clarno Formation (1978)
Subdivided into four units. N central Oregon.
BT Eocene
BT Paleogene
BT Tertiary
BT Cenozoic
SA Oregon

Clarno Oregon (1989)
Village in N central Oregon. Before 1989, search Clarno AND Oregon.
CO N445600N445600
W1202700W1202700
BT Wheeler County Oregon
BT Oregon
BT United States

classification (1978)
SA definition
SA identification
SA miscellanea
SA nomenclature
SA taxonomy
SA terrain classification

Classopollis (1978)
Autoposting of microfossils to this term began in 1990.
BT miospores
BT palynomorphs
BT microfossils

clastic dikes (1978)
BT soft sediment deformation
BT sedimentary structures
SA dikes
SA sandstone dikes

clastic rocks (1978)
Valid index term for GeoRef since 1976. Before 1976, search clastics AND terrigenous or clastics AND nonterrigenous in combination with sedimentary rocks. From 1976 through 1977, search clastic rocks AND terrigenous or nonterrigenous. Also search clastic. Autoposting of this term to bituminous shale, diatomite and pyroclastics ended in 1981. Autoposting of this term to turbidite ended in 1985.
BT sedimentary rocks
NT arenite
NT argillite
NT arkose
NT bentonite
NT black shale
NT breccia
NT cinerite
NT claystone
NT conglomerate
NT contourite
NT diamictite
NT diatomaceous earth
NT eolianite
NT fanglomerate
NT flysch
NT gaize
NT graywacke
NT marl
NT microbreccia
NT mixtite
NT molasse
NT mudstone
NT novaculite
NT orthoquartzite
NT porcellanite
NT radiolarite
NT red beds
NT sandstone
NT saprolite
NT shale
NT siltstone
NT sparagmite
NT spongolite
NT subarkose
NT subgraywacke
NT tillite
NT tilloid
NT tonstein
SA bauxite
SA bone beds
SA clastic sediments
SA coarse-grained materials
SA diatomite
SA fines
SA fragments
SA greensand
SA pyroclastics
SA siliceous composition
SA siliciclastics
SA tempestite

SA terrigenous materials
SA turbidite
SA volcaniclastics
SA wildflysch

clastic sediments (1978)
Valid index term for GeoRef since 1976. Before 1976, search clastics AND terrigenous or clastics AND nonterrigenous. Before 1978, included use in combination with terrigenous or nonterrigenous (i.e. Also search clastic. Autoposting of this term to kaolin and erratics began in 1981. Autoposting of this term to coquina ended in 1981.
BT sediments
NT alluvium
NT boulder clay
NT boulders
NT clay
NT cobbles
NT colluvium
NT diamicton
NT drift
NT dust
NT eluvium
NT erratics
NT flint clay
NT gravel
NT kaolin
NT loess
NT mud
NT ooze
NT outwash
NT pebbles
NT proluvium
NT quartz sand
NT residual clays
NT residuum
NT sand
NT shingle
NT silt
NT till
SA ash flows
SA bentonite
SA cementation
SA clastic rocks
SA coarse-grained materials
SA coquina
SA fines
SA flysch
SA gaize
SA greensand
SA particulate materials
SA pyroclastic flows
SA pyroclastics
SA siliciclastics
SA tempestite
SA terrigenous materials
SA turbidite
SA volcanic ash
SA volcaniclastics
SA wildflysch

clastic wedges (1989)
SA geosynclines
SA orogenic belts
SA sedimentation

clastics
Used through 1975 in combination with terrigenous or nonterrigenous to indicate sedimentary rocks or sediments (i.e. clastics, terrigenous or clastics, nonterrigenous). After 1975 through 1977, clastic rocks and clastic sediments were used in combination with terrigenous or nonterrigenous (i.e. clastic rocks, terrigenous). After 1977, use clastic rocks or clastic sediments as applicable.

clasts (1978)
SA fragments

SA sedimentary rocks
SA sediments

clathrate compounds
use clathrates

clathrate inclusion compounds
use clathrates

clathrates (1985)
UF clathrate compounds
UF clathrate inclusion compounds
SA clay mineralogy
SA gas hydrates
SA natural gas
SA organic materials
SA petroleum
SA zeolite group

Clatsop County
Valid through 1988. Search in combination with state term. After 1988, use specific county-state term.

Clatsop County Oregon (1989)
Extreme NW Oregon. Before 1989, also search Clatsop County AND Oregon.
CO N454500N461500
W1232000W1240200
BT Oregon
BT United States

clay (1978)
As of 1981, includes use as a material in engineering geology; before 1981, also search clays for engineering applications. When of economic value, use clays.
BT clastic sediments
BT sediments
SA alluvium
SA argillaceous texture
SA armored mud balls
SA bentonite
SA clay mineralogy
SA clays
SA claystone
SA disposal barriers
SA expansive materials
SA fireclay
SA flint clay
SA fuller's earth
SA kaolin
SA loam
SA marl
SA mud
SA mud lumps
SA quick clay
SA residual clays
SA sensitive clays
SA silt
SA soft clays
SA soil mechanics
SA stiff clays
SA terrigenous materials
SA thixotropy
SA underclay

clay (soil)
use Clay soils

Clay County
Valid through 1988. Search in combination with state term. After 1988, use specific county-state term.

Clay County Alabama (1989)
E Alabama. Before 1989, also search Clay County AND Alabama.
CO N330500N332900
W0854100W0860800
BT Alabama
BT United States

Clay County Arkansas (1989)

Extreme NE Arkansas. Before 1989, also search Clay County AND Arkansas.
CO N361100N363200
W0900500W0905000
BT Arkansas
BT United States

Clay County Florida (1989)
NE Florida. Before 1989, also search Clay County AND Florida.
CO N294200N301100
W0813600W0820200
BT Florida
BT United States

Clay County Georgia (1989)
SW Georgia. Before 1989, also search Clay County AND Georgia.
CO N313000N314700
W0844800W0850500
BT Georgia
BT United States

Clay County Illinois (1989)
S central Illinois. Before 1989, also search Clay County AND Illinois.
CO N383600N385400
W0881700W0884200
BT Illinois
BT United States
SA Iola Field

Clay County Iowa (1989)
NW Iowa. Before 1989, also search Clay County AND Iowa.
CO N425400N431500
W0945500W0952300
BT Iowa
BT United States

Clay County Kansas (1989)
N Kansas. Before 1989, also search Clay County AND Kansas.
CO N390800N393300
W0965700W0972300
BT Kansas
BT United States

Clay County Kentucky (1989)
SE Kentucky. Before 1989, also search Clay County AND Kentucky.
CO N365600N372000
W0832900W0835800
BT Kentucky
BT United States

Clay County Minnesota (1989)
W Minnesota. Before 1989, also search Clay County AND Minnesota.
CO N463800N470900
W0961000W0965000
BT Minnesota
BT United States

Clay County Mississippi (1989)
E Mississippi. Before 1989, also search Clay County AND Mississippi.
CO N333000N335300
W0882600W0890300
BT Mississippi
BT United States

Clay County Missouri (1989)
W Missouri. Before 1989, also search Clay County AND Missouri.
CO N390800N392600
W0941200W0943700
BT Missouri
BT United States
SA Kansas City Missouri

Clay County Nebraska (1989)

S Nebraska. Before 1989, also search Clay County AND Nebraska.
CO N402300N404300
W0975100W0981600
BT Nebraska
BT United States

Clay County North Carolina (1989)
W North Carolina. Before 1989, also search Clay County AND North Carolina.
CO N345800N351000
W0833200W0840000
BT North Carolina
BT United States
SA Lake Chatuge

Clay County South Dakota (1989)
SE South Dakota. Before 1989, also search Clay County AND South Dakota.
CO N424300N430600
W0864800W0971000
BT South Dakota
BT United States

Clay County Tennessee (1989)
N Tennessee. Before 1989, also search Clay County AND Tennessee.
CO N362500N363700
W0851600W0854800
BT Tennessee
BT United States

Clay County Texas (1989)
N Texas. Before 1989, also search Clay County AND Texas.
CO N332800N341200
W0975700W0982700
BT Texas
BT United States

Clay County West Virginia (1989)
Central West Virginia. Before 1989, also search Clay County AND West Virginia.
CO N381600N384000
W0804900W0811800
BT West Virginia
BT United States

clay liners
use disposal barriers

clay mineralogy (1978)
SA adsorption
SA areal studies
SA argillaceous texture
SA bentonite
SA cation exchange capacity
SA ceramic materials
SA clathrates
SA clay
SA clay minerals
SA clays
SA crystal chemistry
SA crystal growth
SA crystal structure
SA crystallinity
SA crystallography
SA differential thermal analysis
SA electron microscopy
SA flocculation
SA fuller's earth
SA glauconite
SA ion exchange
SA kaolin
SA mineral data
SA mineralogy
SA minerals
SA ocher
SA paragenesis
SA sedimentary petrology

SA sedimentary rocks
SA sediments
SA sheet silicates
SA silicates
SA soils
SA spectroscopy
SA thermal analysis
SA transformations
SA vermiculite
SA weathering
SA X-ray analysis

clay minerals (1978)
Autoposting of this term began in 1981.
BT sheet silicates
BT silicates
NT allophane
NT beidellite
NT corrensite
NT dickite
NT halloysite
NT hectorite
NT illite
NT imogolite
NT kaolinite
NT metahalloysite
NT montmorillonite
NT nacrite
NT nontronite
NT palygorskite
NT rectorite
NT saponite
NT sepiolite
NT smectite
NT stevensite
NT tosudite
NT vermiculite
SA argillaceous texture
SA argillization
SA bauxite
SA bentonite
SA clay mineralogy
SA differential thermal analysis
SA hisingerite
SA hydromica
SA hydromuscovite
SA illitization
SA kaolin
SA kaolinization
SA metabentonite
SA metakaolin
SA mixed-layer minerals

Clay soils (1978)
UF clay (soil)
BT soils
SA soil group

clay stone
use claystone

clays (1978)
Before 1981, included use when of economic value or of engineering application. As of 1981, use only when of economic value; use clay for material in engineering geology. Before 1981, search clays AND deposits.
IN Commodity. See List C.
SA bentonite deposits
SA clay
SA clay mineralogy
SA engineering geology
SA flint clay
SA fuller's earth
SA insulation materials
SA kaolin
SA kaolin deposits
SA ornamental materials
SA quick clay
SA refractory materials
SA residual clays
SA sensitive clays
SA shale
SA soft clays

SA stiff clays
SA vermiculite deposits

Clays Ferry Formation (1989)
Middle and Upper Devonian.
BT Devonian
BT Paleozoic
SA Kentucky
SA Ohio

claystone (1978)
UF clay stone
BT clastic rocks
BT sedimentary rocks
SA clay
SA mudstone
SA terrigenous materials

Clayton Formation (1978)
In Midway Group. S Alabama, SW Georgia, NE Mississippi, SE Missouri, and S Tennessee.
BT Paleocene
BT Paleogene
BT Tertiary
BT Cenozoic
SA Alabama
SA Georgia
SA Midway Group
SA Mississippi
SA Missouri
SA Tennessee

Clear Creek (1978)
River in N central Colorado.
IN Index state and counties as applicable.
SA Colorado

Clear Creek County
Valid through 1988. Search in combination with state term. After 1988, use specific county-state term.

Clear Creek County Colorado (1989)
N central Colorado. Before 1989, also search Clear Creek County AND Colorado.
CO N393500N395000
    W1052500W1055500
BT Colorado
BT United States
NT Idaho Springs Colorado
NT Idaho Springs mining district
SA Henderson Mine

Clear Fork Group (1985)
Central and N central Texas.
BT Leonardian
BT Lower Permian
BT Permian
BT Paleozoic
SA Texas

Clear Lake (1978)
Lake County in NW central.
IN Index state and counties as applicable.
SA California
SA Lake County California
SA Modoc County California

Clearfield County
Valid through 1988. Search in combination with state term. After 1988, use specific county-state term.

Clearfield County Pennsylvania (1989)
Central Pennsylvania. Before 1989, also search Clearfield County AND Pennsylvania.
CO N404000N411800
    W0780400W0784600
BT Pennsylvania
BT United States

Clearwater County
Valid through 1988. Search in combination with state term. After 1988, use specific county-state term.

Clearwater County Idaho (1989)
N Idaho. Before 1989, also search Clearwater County AND Idaho.
CO N461600N465400
    W1143700W1162600
BT Idaho
BT United States

Clearwater County Minnesota (1989)
NW Minnesota. Before 1989, also search Clearwater County AND Minnesota.
CO N470900N480200
    W0951000W0953500
BT Minnesota
BT United States

Clearwater Formation (1989)
In Mannville Group. Includes Wabiskaw Member.
BT Lower Cretaceous
BT Cretaceous
BT Mesozoic
SA Alberta
SA Mannville Group

Clearwater Lake (1989)
N Quebec and E central British Columbia.
IN Index provinces as applicable.
BT Canada
SA British Columbia
SA Quebec

Clearwater River (1985)
Tributary of the Snake River, N Idaho Panhandle. Also a tributary of the Athabasca River, NW Saskatchewan and NE Alberta, and a tributary of the Red Lake River, NW Minnesota.
IN Index states or Canadian provinces as applicable.
SA Alberta
SA Athabasca River
SA Canada
SA Idaho
SA Minnesota
SA Saskatchewan
SA Snake River
SA United States

cleat spar
use ankerite

cleats (1993)
System of joints found in many coal seams. Usually two systems occur at right angles. Before 1993, also search cleating.
SA coal
SA coal seams
SA coalbed methane
SA cross fractures
SA fractures
SA joints
SA strike

cleavage (1978)
As of 1981, use only for cleavage of rocks produced by deformation or metamorphism. Use mineral cleavage for minerals. Before 1978, term also referred to cleavage folds.
SA axial-plane structures
SA cleavage folds
SA deformation
SA flow cleavage
SA folds
SA foliation

SA fracture cleavage
SA lineation
SA mineral cleavage
SA schistosity
SA slaty cleavage
SA slip cleavage
SA structural analysis

cleavage folds (1978)
Before 1978, also search cleavage AND folds.
UF shear-cleavage folds
BT folds
SA cleavage

Cleburne County
Valid through 1988. Search in combination with state term. After 1988, use specific county-state term.

Cleburne County Alabama (1989)
E Alabama. Before 1989, also search Cleburne County AND Alabama.
CO N332700N335700
    W0851800W0855200
BT Alabama
BT United States

Cleburne County Arkansas (1989)
N central Arkansas. Before 1989, also search Cleburne County AND Arkansas.
CO N352300N354300
    W0914800W0921500
BT Arkansas
BT United States

Clermont
No longer a valid term for GeoRef. See Clermont France.

Clermont France (1993)
City in N France.
BT Oise France
BT France
BT Western Europe
BT Europe

Clermont-Ferrand
No longer a valid term for GeoRef. See Clermont-Ferrand France.

Clermont-Ferrand France (1993)
City in S central France.
BT Puy-de-Dome France
BT France
BT Western Europe
BT Europe

Cleveland
Valid through 1988. Search in combination with state term. After 1988, use specific city-state term. As of 1993, use Cleveland England for the county in England.

Cleveland County
Valid through 1988. Search in combination with state term. After 1988, use specific county-state term.

Cleveland County Arkansas (1989)
S central Arkansas. Before 1989, also search Cleveland County AND Arkansas.
CO N334000N340300
    W0910300W0922900
BT Arkansas
BT United States

Cleveland County North Carolina (1989)

SW North Carolina. Before 1989, also search Cleveland County AND North Carolina.
CO N350900N353600
    W0812000W0814700
BT North Carolina
BT United States

Cleveland County Oklahoma (1989)
Central Oklahoma. Before 1989, also search Cleveland County AND Oklahoma.
CO N345500N352300
    W0970800W0974000
BT Oklahoma
BT United States

Cleveland Member (1993)
In Ohio Shale. Upper Devonian. N Ohio. Overlies Chagrin Member; underlies Bedford Shale.
BT Upper Devonian
BT Devonian
BT Paleozoic
SA Ohio
SA Ohio Shale

Cleveland Ohio (1989)
City on Lake Erie in N Ohio. Before 1989, search Cleveland AND Ohio.
CO N413000N413000
    W0814100W0814100
BT Cuyahoga County Ohio
BT Ohio
BT United States

cliff of displacement
  use fault scarps

cliffs (1978)
SA bluffs
SA coastlines
SA cuestas
SA erosion features
SA fault scarps
SA fluvial features
SA geomorphology
SA gullies
SA scarps
SA shore features
SA slopes
SA talus slopes

climate (1978)
Used for Holocene climate. For pre-Holocene climate, use paleoclimatology.
SA alpine environment
SA arctic environment
SA arid environment
SA atmosphere
SA atmospheric precipitation
SA climate effects
SA climatic controls
SA climatologic maps
SA drought
SA equatorial region
SA factors
SA global warming
SA greenhouse effect
SA humid environment
SA humidity
SA latitude
SA length of day
SA meteorology
SA obliquity of the ecliptic
SA paleoatmosphere
SA paleoclimatology
SA pedogenesis
SA sedimentation
SA semi-arid environment
SA soils
SA storms
SA subtropical environment
SA taiga environment
SA temperate environment

SA temperature
SA tropical environment
SA winds

climate effects (1981)
Not to be used for climates of the past (paleoclimatology).
SA climate
SA effects
SA global warming

climate maps
  use climatologic maps

climate-induced circulation (1978)
Before 1978, also search oceans AND circulation AND climate induced or climate-induced.
SA circulation
SA ocean circulation

climatic controls (1993)
Before 1993, also search climate and controls. Applies to Holocene.
SA climate
SA controls
SA ecology
SA erosion
SA geomorphologic controls
SA global warming
SA sedimentation

climatologic maps (1985)
Before 1985, also search climate AND maps.
UF climate maps
UF climatological maps
BT maps
SA climate
SA paleoclimate maps
SA paleoclimatology

climatological maps
  use climatologic maps

climatology, paleo-
  use paleoclimatology

Climax
Valid through 1988. Search in combination with state term. After 1988, use specific city-state term. Also see Climax Mine.

Climax Colorado (1989)
Village in central Colorado. Before 1989, search Climax AND Colorado.
CO N392200N392200
    W1061100W1061100
BT Lake County Colorado
BT Colorado
BT United States

Climax Mine (1993)
Molybdenum porphyry. Lake County Colorado. Also occurs in other locations.
IN Index location and commodities as applicable.
SA Colorado
SA Lake County Colorado
SA mines
SA molybdenum ores

Climax Porphyry (1985)
UF Climax Stock
BT Tertiary
BT Cenozoic
SA Colorado

Climax Stock
  use Climax Porphyry

Clinch Sandstone (1989)
E Tennessee, SW Virginia and S West Virginia. Lower and Middle Silurian. Includes Hagan Shale Member.
BT Silurian

BT Paleozoic
SA Tennessee
SA Virginia
SA West Virginia

clinoamphibole (1978)
Autoposting of this term began in 1981.
BT amphibole group
BT chain silicates
BT silicates
NT actinolite
NT amosite
NT arfvedsonite
NT crocidolite
NT crossite
NT cummingtonite
NT ferrohastingsite
NT glaucophane
NT grunerite
NT hastingsite
NT hornblende
NT kaersutite
NT magnesioriebeckite
NT nephrite
NT pargasite
NT richterite
NT riebeckite
NT tirodite
NT tremolite
NT tschermakite

clinochlore (1978)
BT chlorite group
BT sheet silicates
BT silicates

clinoenstatite (1978)
BT clinopyroxene
BT pyroxene group
BT chain silicates
BT silicates

clinohumite (1978)
From 1978-1980, halides and orthosilicates were autoposted to this term. Autoposting of fluorides and humite group began in 1981. Autoposting of halides, nesosilicates, orthosilicates and silicates to this term began in 1989. This term has multiple hierarchies.
BT1 fluorides
BT1 halides
BT2 humite group
BT2 nesosilicates
BT2 orthosilicates
BT2 silicates
SA titanoclinohumite

clinohypersthene (1978)
BT clinopyroxene
BT pyroxene group
BT chain silicates
BT silicates

clinoptilolite (1978)
BT zeolite group
BT framework silicates
BT silicates
SA heulandite

clinopyroxene (1978)
Autoposting of this term began in 1981.
BT pyroxene group
BT chain silicates
BT silicates
NT acmite
NT aegirine
NT augite
NT chrome diopside
NT clinoenstatite
NT clinohypersthene
NT diopside
NT fassaite
NT hedenbergite
NT jadeite

NT johannsenite
NT kunzite
NT omphacite
NT pigeonite
NT salite
NT spodumene
NT titanaugite

clinopyroxenite (1978)
BT ultramafics
BT plutonic rocks
BT igneous rocks
SA pyroxenite

clinozoisite (1978)
BT epidote group
BT sorosilicates
BT orthosilicates
BT silicates
SA aluminosilicates
SA zoisite

Clinton County
Valid through 1988. Search in combination with state term. After 1988, use specific county-state term.

Clinton County Illinois (1989)
S Illinois. Before 1989, also search Clinton County AND Illinois.
CO N382500N384500
    W0890800W0894300
BT Illinois
BT United States

Clinton County Indiana (1989)
Central Indiana. Before 1989, also search Clinton County AND Indiana.
CO N401100N402500
    W0861500W0864300
BT Indiana
BT United States

Clinton County Iowa (1989)
E Iowa. Before 1989, also search Clinton County AND Iowa.
CO N414300N420300
    W0901000W0905400
BT Iowa
BT United States

Clinton County Kentucky (1989)
S Kentucky. Before 1989, also search Clinton County AND Kentucky.
CO N363500N365200
    W0845900W0851500
BT Kentucky
BT United States

Clinton County Michigan (1989)
S central Michigan. Before 1989, also search Clinton County AND Michigan.
CO N424300N430800
    W0842200W0845200
BT Michigan Lower Peninsula
BT Michigan
BT United States
SA Lansing Michigan

Clinton County Missouri (1989)
NW Missouri. Before 1989, also search Clinton County AND Missouri.
CO N392700N394300
    W0941200W0943700
BT Missouri
BT United States

Clinton County New York (1989)
Extreme NE New York. Before 1989, also search Clinton County AND New York.
CO N442600N450200
    W0732200W0740200
BT New York

BT United States
Clinton County Ohio (1989)
SW Ohio. Before 1989, also search Clinton County AND Ohio.
CO N391800N393400
W0833500W0840000
BT Ohio
BT United States

Clinton County Pennsylvania
(1989)
N central Pennsylvania. Before 1989, also search Clinton County AND Pennsylvania.
CO N405700N412800
W0770900W0780600
BT Pennsylvania
BT United States
SA Nittany Valley

Clinton Group (1978)
Includes Willowvale Shale, Rose Hill Formation, Keefer Sandstone, Rochester Formation, Osgood Formation, Laurel Formation, Waldron Formation, Reynales Formation, Irondequoit Formation, Decew Formation, Neahga Formation, Thorold Formation, Maplewood Formation, Furnaceville Formation, Sodus Formation, Williamson Formation. New York to NE Tennessee; also Michigan and Ontario. Also search Clinton Formation NOT Australia.
BT Middle Silurian
BT Silurian
BT Paleozoic
SA Keefer Sandstone
SA Maryland
SA Michigan
SA New York
SA Ohio
SA Ontario
SA Pennsylvania
SA Rochester Formation
SA Rose Hill Formation
SA Virginia
SA West Virginia

clintonite (1978)
UF xanthophyllite
BT mica group
BT sheet silicates
BT silicates

Clipperton fracture zone (1985)
S of Clarion fracture zone, N of Galapagos Rift.
BT Equatorial Pacific
BT Pacific Ocean
SA DSDP Site 70
SA DSDP Site 71

Clitheroe
No longer a valid term for GeoRef as of 1993. See Clitheroe England.

Clitheroe England (1993)
City E of Liverpool in E Lancashire, NW England. Before 1993, also search Clitheroe AND England.
BT Lancashire England
BT England
BT Great Britain
BT United Kingdom
BT Western Europe
BT Europe

Cloridorme Formation (1978)
Gaspe Peninsula.
BT Middle Ordovician
BT Ordovician
BT Paleozoic
SA Canada
SA Quebec

closed fractures (1978)
Before 1978, also search fractures AND closed.
BT fractures

closed mines
use abandoned mines

closed systems (1978)
Used as a general term.
SA open systems
SA systems

Closepet Granite (1993)
Archean. Karnataka, India. Intrudes Dharwars, Peninsular Gneiss, Champion Gneiss, and Charnockite Series.
BT Karnataka India
BT India
BT Indian Peninsula
BT Asia
SA Archean
SA Dharwars
SA Peninsular Gneiss

Cloud County
Valid through 1988. Search in combination with state term. After 1988, use specific county-state term.

Cloud County Kansas (1989)
N central Kansas. Before 1989, also search Cloud County AND Kansas.
CO N392000N394000
W0972300W0975500
BT Kansas
BT United States

clouds (1978)
SA droplets
SA meteorology
SA raindrops
SA water

Clough Formation (1989)
N central Massachusetts, W New Hampshire and SE Vermont.
BT Lower Silurian
BT Silurian
BT Paleozoic
SA Massachusetts
SA New Hampshire
SA Vermont

Cloverly Formation (1989)
BT Lower Cretaceous
BT Cretaceous
BT Mesozoic
SA Colorado
SA Montana
SA Utah
SA Wyoming

Clovis (1978)
An informal archaeological subdivision of the Quaternary, based on artifacts. Upper Pleistocene-lower Holocene.
BT Quaternary
BT Cenozoic
SA archaeology
SA artifacts
SA lower Holocene
SA upper Pleistocene

Cluj
No longer a valid term for GeoRef. As of 1993 see Cluj Romania.

Cluj Romania (1993)
County and city in central Transylvania. Before 1993 also search Cluj AND Romania.
BT Transylvania
BT Romania
BT Southern Europe
BT Europe

NT Gilau Romania

cluster analysis (1978)
BT statistical analysis
SA analysis
SA mathematical geology

Clymeniida (1981)
BT Ammonoidea
BT Tetrabranchiata
BT Cephalopoda
BT Mollusca
BT Invertebrata

Cm
use curium

Cnidaria
Not a valid term for GeoRef. In some taxonomic classifications, includes Anthozoa, Hydrozoa and Scyphozoa. Search and index under appropriate group.

CNRS (1985)
Acronym.
UF Centre National de la Recherche Scientifique
BT government agencies
SA CRPG
SA France

Co
use cobalt

CO
use carbon monoxide

Co-60 (1985)
Autoposting of broader terms began in 1989. This term has multiple hierarchies.
BT1 radioactive isotopes
BT1 isotopes
BT2 cobalt
BT2 metals

CO2
use carbon dioxide

Coahuila
No longer a valid term for GeoRef. As of 1993 see Coahuila Mexico.

Coahuila Mexico (1993)
State in NE Mexico. Before 1993 also search Coahuila AND Mexico.
CO N243000N295000
W1000000W1040000
BT Mexico
NT Parras Basin
SA Chihuahua tectonic belt
SA Difunta Group
SA Rio Grande
SA Rio Grande Rift
SA Rio Grande Valley
SA Sierra Madre Oriental
SA Zuloaga Limestone

coal (1978)
Autoposting of this term began in 1981. Autoposting of this term to humic coal began in 1985. This term has multiple hierarchies.
UF carbargilite
BT1 organic residues
BT2 sedimentary rocks
NT anthracite
NT bituminous coal
NT coke coal
NT humic coal
NT lignite
NT sapropelite
NT steam coal
NT subbituminous coal
SA anthraxolite
SA biogenic structures
SA carbonaceous composition
SA charcoal

SA chemical properties
SA cleats
SA coal balls
SA coal deposit maps
SA coal exploration
SA coal fields
SA coal mines
SA coal seams
SA coalbed methane
SA coalification
SA degasification
SA diwa-type
SA energy sources
SA gasification
SA humodetrinite
SA lithification
SA longwall mining
SA macerals
SA National Coal Resources Data System
SA organic materials
SA peat
SA rank
SA reflectance
SA Sakoa Basin
SA strip mining
SA volatiles

coal balls (1978)
UF balls, coal
SA coal
SA coal seams
SA concretions

coal basins
As of 1981, no longer a valid term for GeoRef. See coal and coal mines and coal fields.

coal beds
use coal seams

coal clay
use underclay

Coal County
Valid through 1988. Search in combination with state term. After 1988, use specific county-state term.

Coal County Oklahoma (1989)
S central Oklahoma. Before 1989, also search Coal County AND Oklahoma.
CO N342500N344500
W0960500W0963300
BT Oklahoma
BT United States

coal deposit maps (1982)
Before 1982, search coal AND maps.
BT maps
SA coal
SA economic geology maps

coal exploration (1993)
Before 1993, also search coal AND exploration.
SA coal
SA exploration
SA peat

coal fields (1978)
UF coalfields
UF fields, coal
SA Amasra Basin
SA Bokaro
SA Borsod Basin
SA coal
SA coal mines
SA Irkutsk Basin
SA Jharia coal field
SA Joban coal field
SA Karaganda Basin
SA Karanpura coal field
SA Kizel coal basin
SA Kushiro coal field
SA Kuznetsk Basin

SA Minusinsk Basin
SA Moscow Basin
SA Pechora Basin
SA Pernik coal basin
SA Petrosani Basin
SA Pittsburgh coal basin
SA Ramgarh coal field
SA Rampur coal field
SA Raniganj coal field
SA Saint-Etienne coal basin
SA Silesian coal basin
SA South Wales coal field
SA Southern coal field
SA Suchan Basin
SA Sydney coal field
SA Talchir coal field
SA Timok Basin
SA Tulameen coal area
SA Tunguska Basin
SA Ube coal field
SA Upper Silesian coal basin
SA Warrior coal field

Coal Measures   (1978)
Europe.
BT Upper Carboniferous
BT Carboniferous
BT Paleozoic

coal measures
As of 1981, no longer a valid term for GeoRef. See coal.

coal mines   (1993)
Before 1993, also search coal AND mines or mining geology.
BT mines
SA coal
SA coal fields
SA economic geology
SA longwall mining
SA mining geology
SA strip mining

coal seams   (1978)
Also search coal deposits and coal beds.
UF coal beds
UF seams, coal
SA cleats
SA coal
SA coal balls
SA coalbed methane
SA degasification
SA underclay

coalbed methane   (1993)
Before 1993, also search natural gas and coal.
IN Commodity term. See List C.
BT natural gas
BT petroleum
SA cleats
SA coal
SA coal seams
SA methane

coalfields
use coal fields

coalification   (1978)
UF carbonification
UF carbonitization
UF carbonization
UF incarbonization
UF incoalation
SA coal
SA diagenesis
SA processes

Coalinga
Valid through 1988. Search in combination with state term. After 1988, use specific city-state term.

Coalinga California   (1989)
City in S central California. Before 1989, search Coalinga AND California.

CO N360800N360800 W1202200W1202200
BT Fresno County California
BT California
BT United States
SA Coalinga earthquake 1983

Coalinga earthquake 1983   (1985)
Epicenter near Coalinga, S central California.
BT earthquakes
SA California
SA Coalinga California

Coalsack Bluff   (1978)
Site of paleontological investigations in Beardmore Glacier area around NW end of Queen Maud Range off the Ross Ice Shelf in Ross Dependency.
BT Antarctica
SA Beardmore Glacier
SA Queen Maud Land
SA Ross Dependency
SA Ross Ice Shelf

coarse-grained materials   (1982)
SA clastic rocks
SA clastic sediments
SA fines
SA grain size
SA granular materials
SA materials
SA particulate materials
SA sedimentary rocks
SA sediments
SA soil mechanics
SA soils
SA textures
SA unconsolidated materials

coast
A valid term through 1978. After 1978, see coastal environment, coastal dunes, coastal effects, coastal sedimentation, beaches, coastlines, coastal plains, shore features, or shorelines as applicable.

Coast Mountains   (1978)
Extend from Yukon Territory to the Fraser River serving as border between NW British Columbia and the Alaska Panhandle.
IN Index North America or Alaska, British Columbia, and Yukon Territory as applicable.
SA Alaska
SA Alaska Panhandle
SA British Columbia
SA Canadian Cordillera
SA Cascade Range
SA Coast Ranges
SA North America
SA North American Cordillera
SA Yukon Territory

Coast plutonic complex   (1993)
Cretaceous. Coast Mountains of SE Alaska, W British Columbia, and W Washington.
BT North America
SA Alaska
SA British Columbia
SA Cretaceous
SA Washington

Coast Range Ophiolite   (1993)
Middle-Upper Jurassic. California and Mexico.
IN Index ages as applicable.
BT Jurassic
BT Mesozoic
SA California
SA Mexico
SA Middle Jurassic

SA Upper Jurassic

Coast Ranges   (1978)
Mountains extending along the Pacific Ocean from California to Alaska. For Venezuela, see Cordillera de la Costa.
IN Index North America OR states and/or British Columbia as applicable.
SA Alaska
SA British Columbia
SA California
SA Chugach Mountains
SA Coast Mountains
SA Condrey Mountain Schist
SA Diablo Range
SA Gabilan Range
SA Kenai Peninsula
SA Klamath Mountains
SA Kodiak Island
SA Kula Plate
SA North America
SA North American Cordillera
SA Olympic Mountains
SA Oregon
SA Peninsular Ranges
SA Peninsular Ranges Batholith
SA Queen Charlotte Islands
SA Saint Elias Mountains
SA Salinian Block
SA San Bernardino Mountains
SA San Gabriel Mountains
SA San Gorgonio Mountain
SA San Jacinto Mountains
SA Santa Ana Mountains
SA Santa Cruz Mountains
SA Santa Lucia Range
SA Santa Monica Mountains
SA Santa Ynez Mountains
SA Straight Creek Fault
SA Vancouver Island
SA Washington

coastal
A valid term through 1977. After 1977, see coastal environment, coastal dunes, coastal effects, coastal plains, coastal sedimentation, coastlines, beaches, shore features or shorelines as applicable.

coastal dunes   (1981)
This term has multiple hierarchies.
BT1 shore features
BT2 eolian features
BT3 dunes
SA foredunes
SA geomorphology

coastal effects   (1981)
Use in relation to the Earth's magnetic field.
SA effects
SA magnetic field

coastal environment   (1978)
Before 1978, also search coastal; search coast.
UF shore environment
SA beaches
SA capes
SA carbonate ramps
SA coastal plains
SA coastal sedimentation
SA coastlines
SA depositional environment
SA ecology
SA environment
SA inlets
SA lagoons
SA littoral erosion
SA playas
SA polders
SA sedimentation
SA shorelines

SA spits
SA subtidal environment
SA tidal flats

coastal features
use shore features

coastal plains   (1978)
SA coastal environment
SA fluvial features
SA geomorphology
SA Gilbert-type deltas
SA plains
SA shore features

Coastal Range
Use Taiwanese Coastal Range if applicable. Use Coast Ranges for Pacific Coast of North America.

Coastal Range, Taiwan
use Taiwanese Coastal Range

coastal sedimentation   (1989)
BT sedimentation
SA coastal environment
SA coastlines
SA shorelines

coastlines   (1983)
Use to distinguish from shorelines. Use shorelines for engineering aspects of coastlines. Use coastlines for geomorphology studies. Autoposting of shore features to this term began in 1989.
BT shore features
NT emergent coastlines
SA barrier islands
SA bars
SA bays
SA beaches
SA berms
SA bluffs
SA capes
SA chenier plains
SA cliffs
SA coastal environment
SA coastal sedimentation
SA embayments
SA fjords
SA geomorphology
SA inlets
SA intertidal environment
SA littoral erosion
SA longshore bars
SA marine environment
SA marine terraces
SA ocean waves
SA paleogeography
SA shoals
SA shorelines
SA spits
SA tides
SA tombolos

Cobalt
No longer a valid term for GeoRef. As of 1993, see Cobalt Ontario.

cobalt   (1978)
Chemical element. As of 1981, use cobalt ores for cobalt as a commodity. Autoposting of metals to this term began in 1989.
UF Co
BT metals
NT Co-60
SA cobalt ores
SA cobaltite
SA heavy metals
SA siderophile elements

Cobalt Group   (1989)
BT lower Proterozoic
BT Proterozoic
BT upper Precambrian
BT Precambrian
SA Lorrain Formation

SA Ontario

Cobalt Ontario (1993)
Mining town in SE Ontario. Before 1993, also search Cobalt AND Ontario.
BT Timiskaming District Ontario
BT Ontario
BT Eastern Canada
BT Canada

cobalt ores (1981)
Before 1981, also search (cobalt OR Co) AND (deposit OR deposits OR ore OR ores OR economic) in the basic index. Autoposting of metal ores to this term began in 1985.
IN Commodity. See List C.
BT metal ores
SA Blackbird mining district
SA carrollite
SA cobalt
SA cobaltite

cobalt pyrites
  use linnaeite

cobaltite (1978)
This term has multiple hierarchies.
BT1 arsenides
BT2 sulfides
SA cobalt
SA cobalt ores

Cobb Seamount (1978)
Approximately 550 miles W of Washington.
BT Pacific Ocean

cobbles (1978)
BT clastic sediments
BT sediments
SA boulders
SA gravel
SA pebbles
SA shingle
SA terrigenous materials

Cobequid Bay (1978)
Eastern arm of Minas Basin off Bay of Fundy.
BT Nova Scotia
BT Maritime Provinces
BT Eastern Canada
BT Canada
SA Minas Basin

Cobequid Fault (1989)
UF Cobequid-Chedabucto Fault
BT Nova Scotia
BT Maritime Provinces
BT Eastern Canada
BT Canada

Cobequid Highlands (1989)
N Nova Scotia.
CO N450000N460000
   W0624500W0640000
UF Cobequid Mountains
BT Nova Scotia
BT Maritime Provinces
BT Eastern Canada
BT Canada

Cobequid Mountains
  use Cobequid Highlands

Cobequid-Chedabucto Fault
  use Cobequid Fault

Coburg
  No longer a valid term for GeoRef. See Coburg Germany.

Coburg Germany (1993)
City in N Bavaria.
BT Bavaria Germany
BT Germany
BT Central Europe
BT Europe

Coccolithophoraceae (1978)
Before 1981, also search coccoliths. Autoposting of thallophytes and microfossils to this term began in 1990. This term has multiple hierarchies.
BT1 algae
BT1 thallophytes
BT1 Plantae
BT2 algae
BT2 microfossils
NT Ceratolithus
NT Coccolithus
NT Emiliania huxleyi
NT Gephyrocapsa oceanica
SA coccoliths
SA Discoasteridae
SA discoasters
SA nannofossils

coccoliths (1978)
As of 1981, restricted to use as the common name for Coccolithophoraceae. Autoposting of algal flora to this term began in 1989. Autoposting of microfossils to this term began in 1990. This term has multiple hierarchies.
BT1 algal flora
BT1 plants
BT2 algal flora
BT2 microfossils
SA biostratigraphy
SA Coccolithophoraceae

Coccolithus (1978)
Autoposting of thallophytes and microfossils began in 1990. This term has multiple hierarchies.
BT1 Coccolithophoraceae
BT1 algae
BT1 thallophytes
BT1 Plantae
BT2 Coccolithophoraceae
BT2 algae
BT2 microfossils

Cochabamba
  No longer a valid term for GeoRef. See Cochabamba Bolivia.

Cochabamba Bolivia (1993)
Department including the city in E central Bolivia.
BT Bolivia
BT South America

Cochise County
  Valid through 1988. Search in combination with state term. After 1988, use specific county-state term.

Cochise County Arizona (1989)
SE Arizona. Before 1989, also search Cochise County AND Arizona.
CO N312000N322300
   W1090400W1102800
BT Arizona
BT United States
NT Bisbee Arizona
NT Tombstone Arizona
SA Rincon Mountains
SA Sullivan Mine

Cochrane
  As of 1985, no longer a valid term for GeoRef.
  use Cochrane District Ontario

Cochrane District
  No longer a valid term for GeoRef. As of 1993, see Cochrane District Ontario.

Cochrane District Ontario (1993)

N Ontario. Before 1993, also search Cochrane District, District of Cochrane or Cochrane in conjunction with Ontario.
CO N483000N522000
   W0792000W0864000
UF Cochrane
UF District of Cochrane
BT Ontario
BT Eastern Canada
BT Canada
NT Beatty Ontario
NT Timmins Ontario
SA BRIM

Cockfield Formation (1985)
In Claiborne Group. Mississippi, NW Louisiana and E Texas.
BT middle Eocene
BT Eocene
BT Paleogene
BT Tertiary
BT Cenozoic
SA Claiborne Group
SA Louisiana
SA Mississippi
SA Texas

Coconino County
  Valid through 1988. Search in combination with state term. After 1988, use specific county-state term.

Coconino County Arizona (1989)
N Arizona. Before 1989, also search Coconino County AND Arizona.
CO N341500N370000
   W1104500W1132000
BT Arizona
BT United States
NT Flagstaff Arizona
SA Canyon Diablo Meteorites
SA Lake Powell
SA Marble Canyon
SA O'Leary Peak
SA San Francisco Mountain
SA Walker Lake

Coconino Sandstone (1978)
In Aubrey Group. N Arizona, S Utah, and SE Nevada.
BT Permian
BT Paleozoic
SA Arizona
SA Nevada
SA Utah

COCORP (1985)
Acronym.
UF Consortium for Continental Reflection Profiling
SA continental crust
SA geophysical profiles
SA reflection methods
SA seismic methods
SA seismic surveys
SA United States
SA velocity structure

Cocos Plate (1978)
In the Pacific Ocean W of the Caribbean Plate, E and S of the Pacific Plate, and N of the Nazca Plate which is S of the Galapagos Islands.
BT Pacific Ocean
SA Galapagos Rift
SA plate tectonics
SA plates

Cocos Ridge (1978)
Between Costa Rica and the Galapagos Islands. Autoposting of Pacific Ocean to this term began in 1990.
CO N000000N090000
   W0830000W0923000
BT South American Pacific
BT Pacific Ocean
SA Leg 16

COCRUST (1989)
Acronym.
UF Consortium for Crustal Reconnaissance Using Seismic Techniques
SA elastic waves
SA seismic methods
SA seismic surveys

coda waves (1978)
BT surface waves
BT elastic waves
SA seismograms
SA seismology
SA waves

Codell Sandstone Member (1989)
In Carlile Shale. E Colorado, W Kansas, New Mexico and SE South Dakota.
BT Upper Cretaceous
BT Cretaceous
BT Mesozoic
SA Benton Formation
SA Carlile Shale
SA Colorado
SA Kansas
SA New Mexico
SA South Dakota

Codiaceae (1978)
Autoposting of thallophytes and microfossils began in 1990. This term has multiple hierarchies.
BT1 Chlorophyceae
BT1 Chlorophyta
BT1 algae
BT1 thallophytes
BT1 Plantae
BT2 Chlorophyceae
BT2 Chlorophyta
BT2 algae
BT2 microfossils
NT Halimeda

Cody Shale (1978)
In Colorado Group or Montana Group. Bighorn Basin in N Wyoming.
BT Upper Cretaceous
BT Cretaceous
BT Mesozoic
SA Colorado Group
SA Montana Group
SA Wyoming

coefficient of permeability
  use hydraulic conductivity

coefficient of transmissibility
  use transmissivity

coefficient, correlation
  use correlation coefficient

coefficients, partition
  use partition coefficients

Coelacanthini
  use Actinistia

Coelenterata (1978)
Autoposting of this term began in 1978.
BT Invertebrata
NT Anthozoa
NT Hydrozoa
NT Scyphozoa
NT Stromatoporoidea
SA ahermatypic taxa
SA atolls
SA bioherms
SA biostromes
SA corals
SA hermatypic taxa

SA Receptaculitaceae
SA reef builders
SA reefs

coelenterates
use corals

Coelodonta antiquitatis (1978)
Species. Autoposting of Rhinocerotidae, Ceratomorpha, Eutheria and Theria to this term began in 1989.
BT Rhinocerotidae
BT Ceratomorpha
BT Perissodactyla
BT Eutheria
BT Theria
BT Mammalia
BT Tetrapoda
BT Vertebrata
BT Chordata

coercive force
use coercivity

coercivity (1978)
Before 1978, also search coercive force.
UF coercive force
SA magnetic field
SA paleomagnetism
SA remanent magnetization

coesite (1978)
BT silica minerals
BT framework silicates
BT silicates

Coeur d'Alene
No longer a valid term for GeoRef. See Coeur d'Alene mining district.

Coeur d'Alene District
No longer a valid term for GeoRef.
use Coeur d'Alene mining district

Coeur d'Alene mining district (1993)
N Idaho and W Montana. Before 1985, also search Coeur d'Alene. Before 1993, also search Coeur d'Alene District.
UF Coeur d'Alene District
SA Bunker Hill Mine
SA Idaho
SA Kootenai County Idaho
SA Mineral County Montana
SA mines
SA Montana
SA Sanders County Montana
SA Shoshone County Idaho

Coeur d'Alene River (1978)
Flows into Coeur d'Alene Lake in Kootenai County in N Idaho.
BT Idaho
BT United States

coexisting
A valid term through 1977. After 1977, see coexisting minerals; coexisting materials.

coexisting materials
A valid term index term from 1978-1981.

coexisting minerals (1978)
Before 1978, also search coexisting AND minerals.
SA minerals

Coeymans Formation (1978)
Includes Stormville Sandstone Member, Elbow Ridge Sandstone Member. W Maryland, E New York, E Pennsylvania, W Virginia, N West Virginia, and New Jersey.
BT Lower Devonian
BT Devonian
BT Paleozoic

SA Helderberg Group
SA Maryland
SA New York
SA Pennsylvania
SA Virginia
SA West Virginia

Coffee County
Valid through 1988. Search in combination with state term. After 1988, use specific county-state term.

Coffee County Alabama (1989)
SE Alabama. Before 1989, also search Coffee County AND Alabama.
CO N310800N313500
W0854600W0860900
BT Alabama
BT United States

Coffee County Georgia (1989)
S central Georgia. Before 1989, also search Coffee County AND Georgia.
CO N312200N315000
W0823500W0830800
BT Georgia
BT United States

Coffee County Tennessee (1989)
Central Tennessee. Before 1989, also search Coffee County AND Tennessee.
CO N352000N354300
W0855300W0862100
BT Tennessee
BT United States

coffinite (1978)
Autoposting of nesosilicates began in 1985.
BT nesosilicates
BT orthosilicates
BT silicates
SA uranium ores

Cohansey Formation (1989)
Overlies Kirkwood Formation, underlies Beacon Hill Gravel.
IN Index ages as applicable.
BT Neogene
BT Tertiary
BT Cenozoic
SA Delaware
SA Maryland
SA Miocene
SA New Jersey
SA Pennsylvania
SA Pliocene

cohenite (1978)
BT carbides
BT alloys
SA meteorites

cohesion
use shear strength

cohesionless materials (1981)
SA engineering geology
SA materials

cohesive materials (1978)
Introduced in 1978 as general term mostly used in engineering geology. Before 1978, also search cohesive.
SA engineering geology
SA materials
SA rock mechanics
SA soil mechanics

coiling (1978)
SA fossils
SA morphology

Coimbatore
No longer a valid term for GeoRef. See Coimbatore India.

Coimbatore India (1993)
City in W Tamil Nadu, S India.
BT Tamil Nadu India
BT India
BT Indian Peninsula
BT Asia
NT Sivamalai

Coimbra
No longer a valid term for GeoRef. As of 1993 see Coimbra Portugal.

Coimbra Portugal (1993)
District in W Portugal. Also a city. Before 1993 also search Coimbra AND Portugal.
BT Portugal
BT Iberian Peninsula
BT Southern Europe
BT Europe

Cojedes
No longer a valid term for GeoRef. As of 1993 see Cojedes Venezuela.

Cojedes Venezuela (1993)
State in NW central Venezuela. Before 1993 also search Cojedes AND Venezuela.
BT Venezuela
BT South America
NT Tinaquillo Venezuela

coke coal (1981)
Also search coke. This term has multiple hierarchies.
UF natural coke
BT1 coal
BT1 organic residues
BT2 coal
BT2 sedimentary rocks

Coker Formation (1989)
W Alabama and E Mississippi.
BT Upper Cretaceous
BT Cretaceous
BT Mesozoic
SA Alabama
SA Mississippi

Colchis (1978)
Ancient country on the Black Sea S of the Caucasus.
BT Georgian Republic
BT Europe

Cold Lake (1981)
On Alberta-Saskatchewan border.
CO N542500N544000
W1095000W1101500
BT Canada
SA Alberta
SA Saskatchewan

Coldbrook Group (1989)
BT Proterozoic
BT upper Precambrian
BT Precambrian
SA Canada
SA New Brunswick

Coldwater Meteorites (1981)
Refers to two meteorites found in Comanche County, S Kansas. One is an iron meteorite (octahedrite); the other is a chondrite (H chondrite). Before 1981, also search Coldwater AND meteorites.
IN Index octahedrite and/or H chondrites as applicable.
BT meteorites
SA Comanche County Kansas
SA H chondrites
SA Kansas

SA octahedrite

colemanite (1978)
BT borates

Coleoidea
As of 1989, no longer a valid term for GeoRef. Subclass equivalent to Dibranchiata.
use Dibranchiata

Coleoptera (1978)
BT Coleopteroida
BT Insecta
BT Mandibulata
BT Arthropoda
BT Invertebrata

Coleopteroida (1981)
BT Insecta
BT Mandibulata
BT Arthropoda
BT Invertebrata
NT Coleoptera

Coles Bay (1978)
SA Tasmania Australia

Colfax County
Valid through 1988. Search in combination with state term. After 1988, use specific county-state term.

Colfax County Nebraska (1989)
E Nebraska. Before 1989, also search Colfax County AND Nebraska.
CO N412200N414300
W0965400W0971700
BT Nebraska
BT United States

Colfax County New Mexico (1989)
NE New Mexico. Before 1989, also search Colfax County AND New Mexico.
CO N361300N370000
W1040000W1052400
BT New Mexico
BT United States
SA Palisades Sill

Colima
No longer a valid term for GeoRef. As of 1993 see Colima Mexico.

Colima Mexico (1993)
State in W Mexico. Also a city. Before 1993 also search Colima AND Mexico.
BT Mexico
SA Mexican volcanic belt
SA Sierra Madre Occidental

collagen (1978)
BT proteins
BT organic materials

collapse structures (1978)
SA crater lakes
SA geomorphology
SA land subsidence
SA slump structures

collecting (1978)
SA collections
SA fossil localities
SA mineral localities
SA mineralogy
SA minerals
SA paleontology
SA popular geology

collections (1978)
Usage restricted to collections in museums, etc. Before 1971, also search mineral collections.
SA associations
SA catalogs
SA collecting

SA exhibits
SA geology
SA libraries
SA mineralogy
SA minerals
SA museums
SA paleontology

college level
A valid term through 1977. After 1977, use college-level education.

college-level education (1978)
Before 1978, search college level.
SA education
SA elementary school
SA high school
SA junior high school

colleges
use academic institutions

Collenia (1978)
Genus. Autoposting of thallophytes and microfossils began in 1990. This term has multiple hierarchies.
BT1 stromatolites
BT1 biogenic structures
BT1 sedimentary structures
BT2 algae
BT2 thallophytes
BT2 Plantae
BT3 algae
BT3 microfossils
SA Cyanophyta
SA ichnofossils

Collier County
Valid through 1988. Search in combination with state term. After 1988, use specific county-state term.

Collier County Florida (1989)
On Gulf of Mexico in S Florida. Before 1989, also search Collier County AND Florida.
CO N254500N263000 W0805000W0815000
BT Florida
BT United States

Collier Shale (1989)
SW Arkansas and SE Oklahoma.
IN Index ages as applicable.
BT Paleozoic
SA Arkansas
SA Cambrian
SA Oklahoma
SA Ordovician

collinite (1981)
BT vitrinite
BT macerals

collision
No longer a valid term for GeoRef as of 1981. Use plate collision or impacts if applicable.

colloidal materials (1978)
Before 1978, search colloidal; colloids.
UF colloids
NT gels
SA flocculation
SA materials
SA thixotropy

colloids
use colloidal materials

colloquia
use symposia

colluvium (1978)
BT clastic sediments
BT sediments
SA alluvium
SA erosion features
SA geomorphology

SA soils
SA talus slopes
SA terrigenous materials

Cologne
No longer a valid term for GeoRef. See Cologne Germany.

Cologne Germany (1993)
City on the left bank of Rhine River in W North Rhine-Westphalia.
UF Koln Germany
BT North Rhine-Westphalia Germany
BT Germany
BT Central Europe
BT Europe

Colombia (1978)
CO S040000N121500 W0670000W0790000
BT South America
NT Amazonas Colombia
NT Antioquia Colombia
NT Bogota Colombia
NT Bolivar Colombia
NT Magdalena Colombia
NT Magdalena Delta
NT Magdalena River
NT Magdalena Valley
NT Nevado del Ruiz
NT Sabana de Bogota
NT Santander Colombia
NT Sierra Nevada de Santa Marta
NT Sucre Colombia
SA Amazon Basin
SA Amazon River
SA Andes
SA Caribbean region
SA Eastern Cordillera
SA Guajira Peninsula
SA La Luna Formation
SA Llanos
SA Orinoco River
SA Orinoco River basin
SA Rio Negro
SA San Juan River
SA Sierra de Perija
SA Western Cordillera

Colombia Basin
use Colombian Basin

Colombian Basin (1978)
N of Colombia in S Caribbean.
IN Index Caribbean Sea as applicable.
UF Colombia Basin
UF Colombian Plain
SA Caribbean Sea
SA Leg 15

Colombian Plain
use Colombian Basin

Colon Archipelago
use Galapagos Islands

colonial taxa (1978)
Before 1978, also search colonial; colonies.
UF colonies
UF taxa, colonial
SA colonization
SA ecology
SA fossils
SA paleoecology

colonies
use colonial taxa

colonization (1989)
SA biogeography
SA colonial taxa
SA ecology
SA paleoecology

color (1978)
SA chemical analysis
SA colorimetry

SA crystal field
SA soils

color centers (1978)
UF centers, color
SA crystal structure
SA crystallography
SA minerals
SA optical properties

color imagery (1981)
BT imagery
SA remote sensing

Colorado (1978)
Autoposting of this term began in 1978.
CO N370000N410000 W1020000W1090000
BT United States
NT Adams County Colorado
NT Alamosa County Colorado
NT Arapahoe County Colorado
NT Archuleta County Colorado
NT Boulder County Colorado
NT Cache La Poudre River
NT Chaffee County Colorado
NT Cheyenne County Colorado
NT Clear Creek County Colorado
NT Colorado mineral belt
NT Conejos County Colorado
NT Custer County Colorado
NT Delta County Colorado
NT Denver County Colorado
NT Douglas County Colorado
NT Eagle County Colorado
NT El Paso County Colorado
NT Elbert County Colorado
NT Fremont County Colorado
NT Front Range urban corridor
NT Garfield County Colorado
NT Gilpin County Colorado
NT Grand County Colorado
NT Gunnison County Colorado
NT Hinsdale County Colorado
NT Huerfano County Colorado
NT Jackson County Colorado
NT Jefferson County Colorado
NT Kiowa County Colorado
NT La Plata County Colorado
NT Lake County Colorado
NT Larimer County Colorado
NT Las Animas County Colorado
NT Lincoln County Colorado
NT Logan County Colorado
NT Mahogany Zone
NT Mesa County Colorado
NT Mineral County Colorado
NT Moffat County Colorado
NT Montezuma County Colorado
NT Montrose County Colorado
NT Morgan County Colorado
NT Otero County Colorado
NT Ouray County Colorado
NT Park County Colorado
NT Phillips County Colorado
NT Piceance Creek basin
NT Pikes Peak Batholith
NT Pitkin County Colorado
NT Pueblo County Colorado
NT Rangely Anticline
NT Rio Blanco County Colorado
NT Rio Grande County Colorado
NT Rocky Mountain National Park
NT Routt County Colorado
NT Saguache County Colorado
NT San Juan County Colorado
NT San Juan volcanic field
NT San Miguel County Colorado
NT Sawatch Range
NT Sedgwick County Colorado
NT Spanish Peaks
NT Summit County Colorado
NT Teller County Colorado
NT Vail Pass
NT Washington County Colorado

NT Wattenberg Field
NT Weld County Colorado
NT Wet Mountains
NT Yampa River
NT Yuma County Colorado
SA Albuquerque Basin
SA Arikaree Group
SA Arkansas River
SA Arkansas River valley
SA Bald Mountain
SA Battle Mountain
SA Benton Formation
SA Book Cliffs
SA Bridge Creek Limestone Member
SA Bridger Formation
SA Browns Park Formation
SA Brule Formation
SA Brushy Basin Shale Member
SA Bull Lake Glaciation
SA Burro Canyon Formation
SA Canadian River
SA Carlile Shale
SA Carmel Formation
SA Cedar Mountain Formation
SA Chadron Formation
SA Cheyenne Sandstone
SA Chinle Formation
SA Chugwater Formation
SA Clear Creek
SA Climax Mine
SA Climax Porphyry
SA Cloverly Formation
SA Codell Sandstone Member
SA Colorado Group
SA Colorado Lineament
SA Colorado Plateau
SA Colorado River
SA Colorado River basin
SA Colorado School of Mines
SA Council Grove Group
SA Cutler Formation
SA Dakota Formation
SA Denver Basin
SA Denver Formation
SA Desert Creek Zone
SA Dinosaur National Monument
SA Dockum Group
SA Duchesne River Formation
SA Eagle River
SA Elk Mountains
SA Entrada Sandstone
SA Fish Canyon Tuff
SA Fort Hays Limestone Member
SA Fort Union Formation
SA Fountain Formation
SA Four Corners
SA Fox Hills Formation
SA Front Range
SA Frontier Formation
SA Fruitland Formation
SA Glen Canyon Group
SA Goose Egg Formation
SA Gore Range
SA Graneros Shale
SA Great Plains
SA Green River
SA Green River basin
SA Green River Formation
SA Greenhorn Limestone
SA Henderson Mine
SA Hermosa Formation
SA High Plains Aquifer
SA Homestake Mine
SA Idaho Springs Formation
SA Illinois River
SA Ismay Zone
SA Kayenta Formation
SA Kiowa Formation
SA Kirtland Shale
SA Lake Uinta
SA Lance Formation
SA Laney Shale Member
SA Laramie Formation
SA Laramie Mountains

SA Leadville Formation
SA Lewis Creek
SA Lewis Shale
SA Loup Fork Group
SA Lyons Sandstone
SA Madera Formation
SA Madison Group
SA Mancos Shale
SA Manitou Formation
SA Maroon Formation
SA Medicine Bow Mountains
SA Menefee Formation
SA Mesaverde Group
SA Midcontinent
SA Minturn Formation
SA Missouri River basin
SA Moenkopi Formation
SA Montana Group
SA Morrison Formation
SA Mosquito Range
SA Mullen Creek-Nash Fork shear zone
SA Navajo Sandstone
SA Needle Mountains
SA Niobrara Formation
SA Ogallala Aquifer
SA Ogallala Formation
SA Parachute Creek Member
SA Paradox Basin
SA Paradox Member
SA Park City Formation
SA Pictured Cliffs Sandstone
SA Pierre Shale
SA Pinedale Glaciation
SA Platte River basin
SA Point Lookout Sandstone
SA Price River Formation
SA Rangely Field
SA Raton Basin
SA Raton Formation
SA Red Mountain
SA Rio Grande
SA Rio Grande Rift
SA Rio Grande Valley
SA Salt Wash Sandstone Member
SA San Jose Formation
SA San Juan Basin
SA San Juan Formation
SA San Juan Mountains
SA San Juan River
SA San Luis Valley
SA San Rafael Group
SA Sand Wash Basin
SA Sandia Formation
SA Sangre de Cristo Mountains
SA Santa Fe Group
SA Smoky Hill Chalk Member
SA South Platte River
SA South Platte River valley
SA Steamboat Springs
SA Sundance Formation
SA Sunlight
SA Sunnyside Mine
SA Todilto Formation
SA Twilight Gneiss
SA Twin Lakes
SA U. S. Rocky Mountains
SA Uinta Basin
SA Uinta Formation
SA Uinta Mountain Group
SA Uncompahgre Uplift
SA Wasatch Formation
SA Washakie Basin
SA Weber Sandstone
SA Western Overthrust Belt
SA Westwater Canyon Sandstone Member
SA White Limestone
SA White River Group
SA White River Plateau
SA Williams Fork Formation
SA Wingate Sandstone
SA Yule Marble

Colorado Desert (1989)

SE California. Part of the Great Basin. Includes Coachella Valley, Imperial Valley and Salton Sea.
BT California
BT United States
SA Great Basin
SA Imperial Valley
SA Salton Sea

Colorado Group (1978)
Includes Thermopolis Shale, Mowry Shale, Warm Creek Shale, Graneros Shale, Greenhorn Limestone, Carlile Shale, Niobrara Formation, Belle Fourche Shale, Marias River Formation, Skull Creek Shale, Fall River Sandstone, Newcastle Sandstone, and Telegraph Shale.
BT Cretaceous
BT Mesozoic
SA Benton Formation
SA Blackleaf Formation
SA Carlile Shale
SA Cody Shale
SA Colorado
SA Frontier Formation
SA Graneros Shale
SA Greenhorn Limestone
SA Idaho
SA Iowa
SA Kansas
SA Montana
SA Mowry Shale
SA Nebraska
SA New Mexico
SA Newcastle Sandstone
SA Niobrara Formation
SA North Dakota
SA Skull Creek Shale
SA South Dakota
SA Wyoming

Colorado Lineament (1985)
Proterozoic fault system extending from Nebraska to Arizona.
IN Index states as applicable.
BT United States
SA Arizona
SA Colorado
SA Colorado Plateau
SA Nebraska
SA Rocky Mountains
SA South Dakota
SA Utah
SA Wyoming

Colorado mineral belt (1989)
W Colorado. Produces uranium and molybdenum ores.
BT Colorado
BT United States
SA Leadville mining district
SA molybdenum ores
SA uranium ores

Colorado Plateau (1978)
IN Index states as applicable.
CO N334000N403000
   W1062000W1140000
BT United States
SA Albuquerque Basin
SA Arizona
SA Colorado
SA Colorado Lineament
SA Jemez Lineament
SA New Mexico
SA Price River basin
SA Salt Valley
SA Taos Plateau
SA Utah

Colorado River (1978)
Rises in N Colorado and flows SW, then W through the Grand Canyon, and finally S into the Gulf of California. Also a river in Texas. Also occurs in South America.
IN Index U. S. states or regions as applicable.
SA Argentina
SA Arizona
SA Baja California Mexico
SA Brazil
SA California
SA Chile
SA Colorado
SA Nevada
SA Sonora Mexico
SA Texas
SA Utah

Colorado River basin (1981)
Drainage basin. In Southwestern U. S. and NW Mexico. May occur in other locations.
IN Index North America or U. S. or Mexican states as applicable.
SA Arizona
SA California
SA Colorado
SA Mexico
SA Nevada
SA New Mexico
SA North America
SA Utah
SA Wyoming

Colorado River delta (1978)
At N tip of Gulf of California.
IN Index states as applicable.
BT Mexico
SA Baja California Mexico
SA Sonora Mexico

Colorado School of Mines (1985)
In Golden, Jefferson County, Colorado.
UF CSM
SA academic institutions
SA Colorado

Colorado Springs
Valid through 1988. Search in combination with state term. After 1988, use specific city-state term.

Colorado Springs Colorado (1989)
City at junction of Fountain and Monument creeks near Pikes Peak in central Colorado. Before 1989, search Colorado Springs AND Colorado.
CO N385000N385000
   W1044900W1044900
BT El Paso County Colorado
BT Colorado
BT United States
SA Pikes Peak

coloradoite (1978)
Autoposting of sulfides to this term ended in 1989.
BT tellurides

colorimetric analysis
Not a valid GeoRef index term after 1970. Was used on level 1 in subfile N.
use colorimetry

colorimetry (1978)
Before 1971, also search colorimetric analysis.
UF colorimetric analysis
SA analysis
SA chemical analysis
SA color
SA spectroscopy

Colton Formation (1985)
Central Utah.
BT Eocene
BT Paleogene
BT Tertiary
BT Cenozoic
SA Utah

columbates
use niobates

Columbia
Valid through 1988. Search in combination with state term. After 1988, use specific city-state term.

Columbia Basin
use Columbia River basin

Columbia Connecticut (1989)
Town in NE Connecticut. Before 1989, search Columbia AND Connecticut.
CO N414300N414300
   W0721800W0721800
BT Tolland County Connecticut
BT Connecticut
BT United States

Columbia County
Valid through 1988. Search in combination with state term. After 1988, use specific county-state term.

Columbia County Arkansas (1989)
SW Arkansas. Before 1989, also search Columbia County AND Arkansas.
CO N330100N332300
   W0920100W0932900
BT Arkansas
BT United States
SA Walker Creek Field

Columbia County Florida (1989)
N Florida. Before 1989, also search Columbia County AND Florida.
CO N295000N303300
   W0822500W0824500
BT Florida
BT United States
SA Okefenokee Swamp

Columbia County Georgia (1989)
E Georgia. Before 1989, also search Columbia County AND Georgia.
CO N332300N334000
   W0820200W0822600
BT Georgia
BT United States

Columbia County New York (1989)
SE New York. Before 1989, also search Columbia County AND New York.
CO N415800N423200
   W0732200W0735400
BT New York
BT United States
SA Becraft Mountain

Columbia County Oregon (1989)
NW Oregon. Before 1989, also search Columbia County AND Oregon.
CO N454300N461300
   W1224700W1232200
BT Oregon
BT United States

Columbia County Pennsylvania (1989)
E central Pennsylvania. Before 1989, also search Columbia County AND Pennsylvania.
CO N404600N411800
   W0761200W0763800

BT Pennsylvania
BT United States

Columbia County Washington (1989)
SE Washington. Before 1989, also search Columbia County AND Washington.
CO N460000N463600 W1173700W1181500
BT Washington
BT United States

Columbia County Wisconsin (1989)
S central Wisconsin. Before 1989, also search Columbia County AND Wisconsin.
CO N431700N433800 W0890100W0894800
BT Wisconsin
BT United States

Columbia Glacier (1985)
Chugach Mountains, S coast of Alaska.
IN Index Southern Alaska as applicable.
SA Chugach Mountains
SA Columbia Icefield
SA Southern Alaska

Columbia Icefield (1985)
Surrounding Mount Columbia in Jasper and Banff national parks, Alberta, and extending into British Columbia.
IN Index provinces as applicable.
SA Alberta
SA Banff National Park
SA British Columbia
SA Canada
SA Canadian Rocky Mountains
SA Columbia Glacier
SA Jasper National Park

Columbia Illinois (1989)
City in SW Illinois. Before 1989, search Columbia AND Illinois.
CO N382600N382600 W0901100W0901100
BT Monroe County Illinois
BT Illinois
BT United States

Columbia Kentucky (1989)
City in S Kentucky. Before 1989, search Columbia AND Kentucky.
CO N370500N370500 W0851900W0851900
BT Adair County Kentucky
BT Kentucky
BT United States

Columbia Louisiana (1989)
City in Caldwell Parish, NE central Louisiana. Before 1989, search Columbia AND Louisiana.
CO N320500N320500 W0920400W0920400
BT Caldwell Parish Louisiana
BT Louisiana
BT United States

Columbia Missouri (1989)
City in central Missouri. Before 1989, search Columbia AND Missouri.
CO N385800N385800 W0922000W0922000
BT Boone County Missouri
BT Missouri
BT United States

Columbia Plateau (1978)
Lava basin. W United States.
IN Index states as applicable.
CO N440000N483000 W1150000W1200000

UF Columbia River plateau
BT United States
SA Idaho
SA Oregon
SA Picture Gorge Basalt
SA Washington
SA Yakima fold belt

Columbia River (1978)
Rises in SE British Columbia and flows S and then W into the Pacific Ocean on the Oregon-Washington border. Also search Columbia AND appropriate area.
IN Index British Columbia and states as applicable.
SA British Columbia
SA Cowlitz River
SA Deschutes River
SA Grand Coulee Dam
SA Oregon
SA Snake River
SA Washington
SA Willamette River

Columbia River Basalt
As of 1993, no longer a valid term for GeoRef.
use Columbia River Basalt Group

Columbia River Basalt Group (1993)
Miocene. The Columbia River Basalt was formerly also considered Pliocene (?). N Idaho, Oregon, and Washington. Before 1993, also search Columbia River Basalt. Has been divided into the Yakima Basalt (subgroup) containing the Wanapum Basalt, the Grande Ronde Basalt, and the Saddle Mountains Basalt; and the Imnaha Basalt.
UF Columbia River Basalt
BT Miocene
BT Neogene
BT Tertiary
BT Cenozoic
SA Grande Ronde Basalt
SA Idaho
SA Oregon
SA Picture Gorge Basalt
SA Pliocene
SA Saddle Mountains Basalt
SA Wanapum Basalt
SA Washington
SA Yakima Basalt
SA Yakima fold belt

Columbia River basin (1978)
Drainage basin.
IN Index North America or British Columbia and/or states as applicable.
UF Columbia Basin
SA British Columbia
SA Idaho
SA Montana
SA Nevada
SA North America
SA Oregon
SA Utah
SA Washington
SA Wyoming

Columbia River estuary (1978)
Tidal mouth of the Columbia River.
IN Index counties or regions as applicable.
BT United States
SA Oregon
SA Washington

Columbia River plateau
use Columbia Plateau

Columbia South Carolina (1989)
City in central South Carolina. Before 1989, search Columbia AND South Carolina.
CO N340000N340000 W0810000W0810000
BT Richland County South Carolina
BT South Carolina
BT United States

Columbia Tennessee (1989)
City in central Tennessee. Before 1989, search Columbia AND Tennessee.
CO N353700N353700 W0870200W0870200
BT Maury County Tennessee
BT Tennessee
BT United States

Columbian Orogeny (1989)
Refers to orogeny affecting Western Canada during the Cretaceous.
BT Cretaceous
BT Mesozoic
SA Canada
SA orogeny

columbite (1978)
Autoposting of niobotantalates to this term began in 1989. This term has multiple hierarchies.
BT1 niobotantalates
BT1 niobates
BT1 oxides
BT2 niobotantalates
BT2 tantalates
BT2 oxides
SA niobium ores
SA tantalum ores

columbium
use niobium

Columbus
Valid through 1988. Search in combination with state term. After 1988, use specific city-state term.

Columbus Georgia (1989)
City in W Georgia. Before 1989, search Columbus AND Georgia.
CO N322800N322800 W0845900W0845900
BT Muscogee County Georgia
BT Georgia
BT United States

Columbus Limestone (1978)
Comprises Bellepoint, Eversole (?), Delhi, Marblehead, and Venice members. Named for exposure at Columbus in Franklin County, Ohio.
BT Middle Devonian
BT Devonian
BT Paleozoic
SA Ohio

Columbus Ohio (1989)
City in central Ohio. Before 1989, search Columbus AND Ohio.
CO N395900N395900 W0830300W0830300
BT Franklin County Ohio
BT Ohio
BT United States

columnar basalt (1978)
Before 1978, also search columnar AND basalt.
BT basalts
BT volcanic rocks
BT igneous rocks

columnar jointing
use columnar joints

columnar joints (1978)
Before 1978, also search fractures AND columnar.
UF columnar jointing
BT fractures
SA bedding
SA joints

Colusa County
Valid through 1988. Search in combination with state term. After 1988, use specific county-state term.

Colusa County California (1989)
N central California. Before 1989, also search Colusa County AND California.
CO N385500N392500 W1215000W1224800
BT California
BT United States

Colville Group (1989)
N Alaska. Overlies Nanushuk Group; underlies Sagavanirktok and Gubik formations.
BT Upper Cretaceous
BT Cretaceous
BT Mesozoic
SA Alaska

Colville River (1978)
Flows E along N slope of Brooks Range and then N into Arctic Ocean in N Alaska. Also in Stevens County Washington.
IN Index Stevens County Washington or Northern Alaska as applicable.
SA Alaska
SA Colville River delta
SA Northern Alaska
SA Stevens County Washington
SA Washington

Colville River delta (1978)
On Arctic Ocean in N Alaska.
IN Index Northern Alaska if applicable.
SA Colville River
SA Northern Alaska

comae (1993)
Restricted to cometary atmosphere.
SA comets
SA interplanetary space
SA planetology
SA solar wind

Comal County
Valid through 1988. Search in combination with state term. After 1988, use specific county-state term.

Comal County Texas (1989)
S central Texas. Before 1989, also search Comal County AND Texas.
CO N293500N301000 W0975500W0984500
BT Texas
BT United States
SA Edwards Aquifer

Comanche County
Valid through 1988. Search in combination with state term. After 1988, use specific county-state term.

Comanche County Kansas (1989)
S Kansas. Before 1989, also search Comanche County AND Kansas.
CO N370000N372300 W0990000W0993200
BT Kansas

BT United States
SA Coldwater Meteorites

Comanche County Oklahoma
(1989)
SW Oklahoma. Before 1989, also search Comanche County AND Oklahoma.
CO N342400N345200
W0980500W0985000
BT Oklahoma
BT United States
SA Meers Fault
SA Slick Hills
SA Wichita Mountains

Comanche County Texas (1989)
Central Texas. Before 1989, also search Comanche County AND Texas.
CO N313800N321800
W0980800W0985500
BT Texas
BT United States

Comanche Peak Limestone
(1993)
Fredericksburg Group. Comanchean. E Texas and New Mexico. Underlies Edwards Formation?; overlies Walnut Clay. This term has multiple hierarchies.
BT1 Comanchean
BT1 Cretaceous
BT1 Mesozoic
BT2 Lower Cretaceous
BT2 Cretaceous
BT2 Mesozoic
SA Fredericksburg Group
SA New Mexico
SA Texas

Comanchean (1978)
Provincial series, North America. Below Gulfian. Autoposting of Lower Cretaceous to this term began in 1989 and ended in 1992.
IN Index Lower Cretaceous or Upper Cretaceous as applicable.
UF Comanchian
BT Cretaceous
BT Mesozoic
NT Antlers Sands
NT Bisbee Group
NT Buda Limestone
NT Comanche Peak Limestone
NT Edwards Formation
NT Fredericksburg Group
NT Georgetown Formation
NT Glen Rose Formation
NT Mentor Beds
NT Mural Limestone
NT Paluxy Formation
NT Pearsall Formation
NT Rodessa Formation
NT Sunniland Limestone
NT Travis Peak Formation
NT Trinity Group
NT Washita Group
SA Lower Cretaceous
SA Upper Cretaceous

Comanchian
use Comanchean

combustion (1978)
SA geochemistry
SA processes
SA retorting

comendite (1978)
BT rhyolites
BT volcanic rocks
BT igneous rocks
SA pantellerite

comets (1978)
NT Halley's Comet
SA comae

SA interplanetary space
SA meteorites
SA meteors
SA parent bodies
SA solar system
SA solar wind
SA tektites

Commander Islands
use Komandorski Islands

Commelinidae (1981)
BT Monocotyledoneae
BT angiosperms
BT Spermatophyta
BT Plantae

commencements, sudden
Not a valid term for GeoRef. Usually out-of-scope. Used as sudden commencements with magnetosphere.

commodity exchange (1981)
SA economics

common mica
use muscovite

common salt
use halite

common-depth-point method
(1978)
Sometimes abbreviated as CDP.
UF CDP method
BT reflection methods
SA geophysical methods
SA moveout
SA normal moveout
SA seismic methods
SA velocity analysis

Commonwealth of Independent States (1993)
Grouper for independent republics of the former Soviet Union including Armenia, Azerbaidzhan, Belarus, Kazakhstan, Kyrgyzstan, Moldova, Russian Federation, Tadzhikistan, Turkmenia, Ukraine and Uzbekistan. Term excludes Estonia, Latvia, Lithuania, and the Georgian Republic. Abbreviation: CIS.
IN Index republics as applicable.
NT Aral region
NT Aral Sea
NT Armenia
NT Astrakhan Arch
NT Azerbaidzhan
NT Azov Sea
NT Belarus
NT Dnieper Basin
NT Dnieper River
NT Dnieper-Donets Basin
NT Dniester River
NT Donets Basin
NT Fergana Basin
NT Ishim
NT Kazakhstan
NT Kulunda Steppe
NT Kurama Range
NT Kyrgyzstan
NT Kyzylkum
NT Moldova
NT Moscow-Pechora Syneclise
NT Neman River basin
NT Ob-Irtysh Interfluve
NT Pamirs
NT Polesye
NT Pripet Basin
NT Pskem Range
NT Rudny Altai
NT Russian Federation
NT Scythian Platform
NT Shakhdag Range
NT Steppes region
NT Surkhan Darya

NT Surkhan Darya basin
NT Syr Darya
NT Tadzhikistan
NT Timan Ridge
NT Turan
NT Turanian Platform
NT Turkmenia
NT Ukraine
NT Ural River
NT Urals
NT Ustyurt
NT Uzbekistan
NT Uzen
NT Volga-Ural region
NT Voronezh-Volga Anteclise
NT West Siberia
NT Zangezur
NT Zeravshan River
SA Asia
SA Europe
SA USSR Academy of Sciences

communist countries (1981)
SA industrialized countries
SA Western World

communities (1978)
SA assemblages
SA biocenoses
SA biotopes
SA ecology
SA paleoecology
SA succession
SA trophic analysis

Como (1978)
Lake in NW Lombardy. As of 1993, see Como Italy for the province.
BT Lombardy Italy
BT Italy
BT Southern Europe
BT Europe

Como Italy (1993)
Province and city in NW Lombardy. Before 1993, also search Como.
BT Lombardy Italy
BT Italy
BT Southern Europe
BT Europe

Comoro Islands (1978)
Group of volcanic islands in N Mozambique Channel between Mozambique and Madagascar. Former overseas territory of France. Before 1981, broader term was Indian Ocean. As of 1993, Anjouan Island is a narrower term.
CO S130000S110000
E0450000E0430000
BT Indian Ocean Islands
NT Anjouan Island

compaction (1978)
SA bearing capacity
SA compactness
SA consolidation
SA diagenesis
SA hardground
SA lithification
SA porosity
SA processes
SA soil mechanics

compactness (1982)
SA compaction
SA compressibility
SA compression
SA compressive strength
SA consolidation
SA consolidometer tests
SA density
SA engineering geology
SA land subsidence

SA mechanical properties
SA overconsolidated materials
SA physical properties
SA porosity
SA sediments
SA settlement
SA soil mechanics
SA soils
SA textures
SA unconsolidated materials

companies (1982)
UF corporations
SA company diversification
SA company integration
SA company mergers
SA corporate policy
SA economic geology
SA economics
SA industry

company diversification (1981)
SA companies
SA economics

company integration (1981)
SA companies
SA economics

company mergers (1981)
SA companies
SA economics

comparison
Used as a general term through May 1978. No longer a valid term for GeoRef. See interplanetary comparison or terrestrial comparison or laboratory comparison if applicable.

compensation
No longer a valid term for GeoRef. See isostatic compensation or carbonate compensation depth.

competence
As of 1981, no longer a valid term for GeoRef. See sedimentation, transport, hydrology, hydraulics and competent materials. From 1978-1980, was restricted to water transport of detritus. Prior to 1978, wind transport was included.

competent materials (1978)
SA deformation
SA materials
SA rock mechanics
SA soil mechanics

complexes (1978)
Before 1981, also search structural complexes.
SA igneous rocks
SA metamorphic rocks
SA metamorphism
SA ring complexes

complexing (1978)
SA geochemistry
SA ligands
SA metals
SA organic materials
SA processes

components (1978)
Used as a general term.
SA composition
SA principal components analysis

Compositae (1989)
Genus.
UF Compositales
BT Dicotyledoneae
BT angiosperms
BT Spermatophyta
BT Plantae
SA miospores

Compositales
    use Compositae

composite cone
    use stratovolcanoes

composite volcano
    use stratovolcanoes

composition   (1978)
    UF content
    SA acidic composition
    SA alkalic composition
    SA andesitic composition
    SA arkosic composition
    SA basaltic composition
    SA calc-alkalic composition
    SA calc-silicate composition
    SA calcareous composition
    SA calcic composition
    SA carbonaceous composition
    SA chemical composition
    SA components
    SA dacitic composition
    SA dolomitic composition
    SA ferromanganese composition
    SA ferruginous composition
    SA gabbroic composition
    SA glauconitic composition
    SA granitic composition
    SA iron-rich composition
    SA lithogeochemistry
    SA mafic composition
    SA manganese composition
    SA mineral composition
    SA peralkalic composition
    SA peraluminous composition
    SA phosphate composition
    SA potassic composition
    SA rhyolitic composition
    SA saline composition
    SA siliceous composition
    SA tholeiitic composition
    SA ultramafic composition

compounds   (1978)
    Used as a general term.
    SA acids
    SA hydrogen peroxide
    SA organic compounds
    SA polymers
    SA volatile organic compounds

Comprehensive Environmental Response, Compensation and Liability Act
    use Superfund

compressibility   (1978)
    UF modulus of compression
    SA compactness
    SA compression
    SA elasticity
    SA physical properties
    SA plasticity
    SA soil mechanics

compression   (1978)
    As of 1982, use compression tectonics if applicable. Before 1982, compression was used for compression tectonics as well as for the general concept.
    SA compactness
    SA compressibility
    SA compression tectonics
    SA deformation
    SA flow cleavage
    SA hysteresis
    SA lithification
    SA pressure
    SA tension
    SA yield strength

compression tectonics   (1982)
    Before 1982, search compression AND tectonics.
    BT tectonics
    SA compression

    SA contraction
    SA deformation
    SA extension tectonics
    SA plate tectonics
    SA thrust faults
    SA transpression

compressional waves
    use P-waves

compressive strength   (1978)
    SA compactness
    SA deformation
    SA rock mechanics
    SA soil mechanics
    SA strength

computer languages   (1985)
    UF programming languages
    NT BASIC
    NT Fortran
    NT Fortran 77
    NT Fortran IV
    SA computer programs
    SA data processing
    SA Pascal

computer methods
    A valid term through 1971.
    use data processing

computer networks   (1993)
    SA arrays
    SA computers
    SA data processing
    SA data retrieval
    SA geographic information systems
    SA information systems
    SA microcomputers
    SA minicomputers
    SA networks

computer programs   (1978)
    UF software
    NT PHREEQE
    NT WATEQF
    SA algorithms
    SA BASIC
    SA computer languages
    SA computers
    SA data processing
    SA expert systems
    SA Fortran
    SA Fortran 77
    SA Fortran IV
    SA linear programming
    SA Pascal
    SA programs
    SA systems analysis

Computerized Resources Information Bank
    use CRIB

computers   (1978)
    For documents discussing the machine itself. Autoposting of this term began in 1993.
    NT microcomputers
    NT minicomputers
    NT workstations
    SA computer networks
    SA computer programs
    SA data processing
    SA hand calculators
    SA programs

Conasauga Group   (1989)
    Middle and Upper Cambrian. N Alabama, NE Georgia, North Carolina, E Tennessee and SW Virginia.
    BT Cambrian
    BT Paleozoic
    SA Alabama
    SA Georgia
    SA North Carolina
    SA Tennessee
    SA Virginia

concentration   (1978)
    Used as a general term.
    SA dilution
    SA geochemistry
    SA localization
    SA saturation
    SA solubility
    SA solution
    SA solution transport

concentric folds   (1978)
    Before 1978, also search concentric AND folds; search parallel AND folds.
    UF parallel folds
    BT folds
    SA similar folds

concentric fractures   (1978)
    Before 1978, also search fractures AND concentric.
    BT fractures

concepts   (1978)
    SA dynamos

conceptual models
    use theoretical models

conchiolin   (1978)
    BT proteins
    BT organic materials

Conchostraca
    As of 1981, no longer a valid term for GeoRef.
    use Branchiopoda

concrete   (1978)
    BT construction materials
    SA cement materials
    SA concrete dams
    SA engineering geology

concrete dams   (1982)
    BT dams
    SA concrete

concretions   (1978)
    As of 1981, used for nodules other than those on the ocean floor.
    BT secondary structures
    BT sedimentary structures
    SA coal balls
    SA cone-in-cone
    SA geodes
    SA growth rates
    SA nodules
    SA sedimentary rocks
    SA sediments
    SA septaria

condensates   (1978)
    Liquid hydrocarbon that emanates from a gas well or from the gas-cap of an oil well. Not to be used for other condensed materials.
    UF gas condensates
    SA gas hydrates
    SA heavy oil
    SA natural gas
    SA petroleum

condensation   (1978)
    SA geochemistry
    SA Moon
    SA planetology
    SA planets

conditions, P-T
    use P-T conditions

Condrey Mountain Schist   (1989)
    Klamath Mountains of N California and SW Oregon.
    BT Mesozoic
    SA California
    SA Coast Ranges
    SA Klamath Mountains
    SA Oregon

conduction
    Not a valid term for GeoRef. See heat flow, geothermal gradient, conductivity, electrical conductivity, hydraulic conductivity, thermal conductivity, or resistivity as applicable.

conductive materials   (1981)
    SA engineering geology
    SA geophysical methods
    SA materials

conductivity   (1978)
    A level 2 term under heat flow until 1976. Whenever applicable, use electrical conductivity, thermal conductivity or hydraulic conductivity.
    UF specific conductance
    SA apparent resistivity
    SA electrical conductivity
    SA electrical logging
    SA geothermal gradient
    SA heat flow
    SA hydraulic conductivity
    SA measurement
    SA resistivity
    SA thermal conductivity

Condylarthra   (1978)
    Order. Autoposting of this term to Acreodi and Arctocyonia began in 1989. Autoposting of Eutheria and Theria to this term began in 1989.
    BT Eutheria
    BT Theria
    BT Mammalia
    BT Tetrapoda
    BT Vertebrata
    BT Chordata
    NT Acreodi
    NT Arctocyonia
    SA Ungulata

cone penetration tests   (1985)
    BT penetration tests
    SA cones
    SA engineering geology
    SA soil mechanics

cone-in-cone   (1978)
    BT secondary structures
    BT sedimentary structures
    SA concretions
    SA cones
    SA septaria

Conecuh County
    Valid through 1988. Search in combination with state term. After 1988, use specific county-state term.

Conecuh County Alabama   (1989)
    S Alabama. Before 1989, also search Conecuh County AND Alabama.
    CO N311000N314500
        W0864000W0872800
    BT Alabama
    BT United States

Conejos County
    Valid through 1988. Search in combination with state term. After 1988, use specific county-state term.

Conejos County Colorado   (1989)
    S Colorado. Before 1989, also search Conejos County AND Colorado.
    CO N370000N372200
        W1054000W1064500
    BT Colorado
    BT United States
    SA San Luis Valley

Conemaugh (1978)
 River in SW Pennsylvania.
 IN Index state and counties as applicable.
 SA Pennsylvania

Conemaugh Group (1978)
 Divided into nine parts: Mahoning, Buffalo, Saltsburg, Grafron, Barton, Morgantown, Lonaconing, Connellsville, and one not yet named. Also includes Nadine Limestone, Woods Run Limestone, Carnahan Run Shale, and Brush Creek Limestone. W Maryland, E Ohio, Pennsylvania, N Virginia, and W West Virginia.
 BT Pennsylvanian
 BT Carboniferous
 BT Paleozoic
 SA Ames Limestone
 SA Maryland
 SA Ohio
 SA Pennsylvania
 SA Virginia
 SA West Virginia

cones (1993)
 Use more specific term if applicable.
 SA alluvial fans
 SA cinder cones
 SA cone penetration tests
 SA cone-in-cone
 SA conifers
 SA debris cones
 SA gymnosperm flora
 SA gymnosperms
 SA seeds
 SA shatter cones
 SA spores
 SA stratovolcanoes
 SA submarine fans

cones, cinder
 use cinder cones

cones, shatter
 use shatter cones

conferences
 use symposia

configuration
 As of 1981, no longer a valid term for GeoRef.

confined aquifers (1982)
 BT aquifers
 SA aquitards
 SA artesian waters
 SA confining pressure
 SA ground water

confining pressure (1993)
 SA confined aquifers
 SA pressure
 SA reservoir properties
 SA shear
 SA soil mechanics
 SA triaxial tests

congelifluction
 use gelifluction

congelifraction (1978)
 UF frost bursting
 UF frost shattering
 UF frost splitting
 UF frost weathering
 UF frost wedging
 UF gelifraction
 UF gelivation
 SA cryoturbation
 SA frost action
 SA frost heaving
 SA geomorphology
 SA patterned ground
 SA periglacial features
 SA rock mechanics

 SA soil mechanics

congeliturbation
 use cryoturbation

conglomerate (1978)
 Autoposting of this term began in 1978.
 UF conglomerite
 BT clastic rocks
 BT sedimentary rocks
 NT quartz-pebble conglomerate
 SA agglomerate
 SA breccia
 SA calcrete
 SA ferricrete
 SA metaconglomerate
 SA terrigenous materials

conglomerite
 use conglomerate

Congo (1978)
 This term is now limited to the present Congo (People's Republic of the Congo). Formerly it also included what is now Zaire. Before 1972, also search Congo Republic.
 CO S050000N034000
     E0184000E0110000
 UF Congo Republic
 UF People's Republic of the Congo
 BT Central Africa
 BT Africa
 NT Brazzaville Congo
 SA Congo Basin
 SA Congo River
 SA Zaire

Congo (River)
 use Congo River

Congo Basin (1978)
 River drainage basin.
 IN Index countries as applicable.
 BT Africa
 SA Angola
 SA Central African Republic
 SA Congo
 SA Zaire
 SA Zambia

Congo Republic
 use Congo

Congo River (1978)
 Rises as the Lualaba River in SE Zaire and flows N, then W, and finally SW into the Atlantic Ocean.
 IN Index countries as applicable.
 UF Congo (River)
 UF Kongo River
 UF Zaire River
 BT Central Africa
 BT Africa
 SA Angola
 SA Congo
 SA Lower Zaire
 SA Zaire

congresses
 Not a valid term for GeoRef. See under associations and under symposia.

Coniacian (1978)
 Europe. Above Turonian, below Santonian.
 BT Senonian
 BT Upper Cretaceous
 BT Cretaceous
 BT Mesozoic

conical fractures (1978)
 Before 1978, also search fractures AND conical.
 BT fractures

conichalcite (1978)

 BT arsenates

Coniferae (1978)
 BT gymnosperms
 BT Spermatophyta
 BT Plantae
 SA Coniferales

Coniferales (1978)
 Before 1981, Coniferae was included as a narrower term. Autoposting of this term began in 1978.
 BT gymnosperms
 BT Spermatophyta
 BT Plantae
 NT Buriadia
 NT Cupressaceae
 NT Picea
 NT Pinaceae
 NT Pseudotsuga
 NT Taxodiaceae
 NT Taxodium
 NT Tsuga
 SA Coniferae
 SA conifers
 SA Dadoxylon

conifers (1993)
 Common name for the order Coniferales.
 BT gymnosperm flora
 BT plants
 SA cones
 SA Coniferales

conjugate faults (1993)
 BT faults
 SA joints
 SA structural analysis

conjugate folds (1978)
 Before 1978, search folds AND conjugate.
 BT folds
 SA joints

conjugate fractures (1993)
 BT fractures

conjugate systems
 As of 1993, see conjugate faults or conjugate folds or conjugate fractures.

connate water
 Not a valid GeoRef index term after 1971. Was used on level 1 in subfiles N and B.
 use connate waters

connate waters (1978)
 UF connate water
 UF formation waters
 SA fossil waters
 SA ground water
 SA meteoric water
 SA pore water
 SA water

Connecticut (1978)
 Autoposting of this term began in 1978.
 CO N405900N420300
     W0714800W0734400
 BT United States
 NT Fairfield County Connecticut
 NT Hartford County Connecticut
 NT Litchfield County Connecticut
 NT Middlesex County Connecticut
 NT New Haven County Connecticut
 NT New London County Connecticut
 NT Tolland County Connecticut
 NT Windham County Connecticut
 SA Atlantic Coastal Plain
 SA Avalon Terrane
 SA Cheshire Formation
 SA Connecticut River

 SA Connecticut Valley
 SA East Berlin Formation
 SA Fordham Gneiss
 SA Hampden Basalt
 SA Hartford Basin
 SA Hartland Formation
 SA Holyoke Basalt
 SA Hoosac Formation
 SA Lake Char Fault
 SA Madrid Formation
 SA Manhattan Formation
 SA Narragansett Pier Granite
 SA New York City Group
 SA Newark Supergroup
 SA Portland Formation
 SA Rangeley Formation
 SA Thames River
 SA Westerly Granite
 SA Weston Meteorite

Connecticut River (1978)
 Rises in N New Hampshire and flows S emptying into Long Island Sound.
 IN Index states as applicable.
 BT United States
 SA Connecticut
 SA Massachusetts
 SA New Hampshire
 SA Vermont

Connecticut River valley
 use Connecticut Valley

Connecticut Valley (1978)
 River valley. Also search Connecticut.
 IN Index states as applicable.
 CO N413000N452000
     W0721500W0724500
 UF Connecticut River valley
 BT United States
 SA Connecticut
 SA Massachusetts
 SA New Hampshire
 SA Vermont

Connemara (1978)
 Barren, mountainous coastal region in County Galway on the Atlantic Ocean.
 CO N531500N534000
     W0090000W0102000
 BT Galway Ireland
 BT Ireland
 BT Western Europe
 BT Europe
 SA Galway Granite

Conococheague Formation (1989)
 W Maryland, S central Pennsylvania, Tennessee, NW Virginia and West Virginia.
 BT Upper Cambrian
 BT Cambrian
 BT Paleozoic
 SA Knox Group
 SA Maryland
 SA Pennsylvania
 SA Tennessee
 SA Virginia
 SA West Virginia

Conodonta (1981)
 Before 1981, search conodonts.
 BT microfossils
 NT Adetognathus
 NT Gnathodus
 NT Gondolella
 NT Hindeodella
 NT Icriodus
 NT Idiognathodus
 NT Idiognathoides
 NT Neognathodus
 NT Neogondolella
 NT Ozarkodina
 NT Palmatolepis

Before then, variants of the term may occur in GeoRef.

NT Panderodus
NT Polygnathus
NT Spathognathodus
NT Streptognathodus
SA apparatus
SA conodonts

conodonts (1978)
As of 1981, restricted to use as the common name for Conodonta. Before 1981, included use on levels 1 and 2 as a fossil term.
IN Index for all non-paleontologic studies of fossils.
BT microfossils
SA biostratigraphy
SA Conodonta

Conrad discontinuity (1978)
SA crust
SA discontinuities
SA Earth
SA seismology
SA velocity

conservation (1978)
Used for environmental conservation of natural resources.
SA basin management
SA deforestation
SA depletion
SA energy conservation
SA environmental geology
SA erosion control
SA impact statements
SA Indian reservations
SA land leases
SA land use
SA national parks
SA natural resources
SA pollution
SA protection
SA RARE II regions
SA reclamation
SA Resource Conservation and Recovery Act
SA resources
SA soils
SA urbanization
SA waste disposal
SA water management
SA wetlands
SA wilderness areas

consistency
As of 1981, no longer a valid term for GeoRef. Use consistent materials if applicable.

consistency limits
use Atterberg limits

consistent materials (1981)
SA engineering geology
SA materials
SA stiffness

consolidated materials
As of 1993, use consolidation, compaction, unconsolidated materials, or overconsolidated materials as applicable.

consolidation (1978)
Before 1978, also search solidification if applicable.
SA cementation
SA compaction
SA compactness
SA diagenesis
SA lithification
SA overconsolidated materials
SA soil mechanics
SA unconsolidated materials

consolidometer tests (1981)
SA compactness
SA engineering geology
SA soil mechanics
SA testing

Consortium for Continental Reflection Profiling
use COCORP

Consortium for Crustal Reconnaissance Using Seismic Techniques
use COCRUST

Constance Lake
use Lake Constance

constant, dielectric
use dielectric constant

Constanta
No longer a valid term for GeoRef. As of 1993 see Constanta Romania.

Constanta Romania (1993)
County in SE Romania. Also a city on the Black Sea. Before 1993 also search Constanta AND Romania.
UF Constantsa
BT Romania
BT Southern Europe
BT Europe
NT Mangalia Romania
SA Romanian Dobruja

Constantine
No longer a valid term for GeoRef. Valid 1978-1980. As of 1993, see Constantine Algeria for the department.

Constantine Algeria (1993)
Department. NE Algeria. Before 1993, also search Constantine region or Constantine.
BT Algeria
BT North Africa
BT Africa
SA Hodna Basin

Constantine region
No longer a valid term for GeoRef. Used as large department of French Algeria. As of 1993, see Algeria for departments.

constants, elastic
use elastic constants

Constantsa
use Constanta Romania

constitution
Not a valid term for GeoRef. Was sometimes used for composition; sometimes as the document.

constitutive equations (1985)
SA elasticity
SA equations
SA mathematical models
SA physical properties
SA plasticity
SA properties
SA rheology
SA rock mechanics
SA soil mechanics

construction (1978)
SA backfill
SA construction materials
SA dams
SA dredging
SA foundations
SA highways
SA marine installations
SA railroads
SA reservoirs
SA shorelines
SA tunnels
SA underground installations

construction materials (1978)
Autoposting of this term began in 1978.
IN Commodity. See List C.
UF constructional materials
NT building stone
NT cement materials
NT concrete
NT dimension stone
NT insulation materials
SA aggregate
SA alloys
SA construction
SA granite deposits
SA gravel deposits
SA limestone deposits
SA marble deposits
SA materials
SA ornamental materials
SA perlite
SA refractory materials
SA sand
SA sands
SA sandstone deposits
SA shale
SA trap rocks

constructional materials
No longer a valid term for GeoRef. In 1949, was used in subfile N.
use construction materials

consumption (1982)
SA demand
SA economics
SA supply
SA utilization

contact
As of 1981, no longer a valid term for GeoRef. See boundary; contact metamorphism. Before 1977, used to indicate type of metamorphism.

contact aureole
use aureoles

contact metamorphism (1978)
Not a valid term from 1973 through 1977. Also search metamorphism AND contact. Before 1971, also search contact phenomena.
UF contact phenomena
BT metamorphism
SA aureoles
SA country rocks
SA hornfels facies
SA intrusions
SA pneumatolysis

contact phenomena
No longer a valid GeoRef index term. Before 1971, was used on level 1 in subfiles E and N.
use contact metamorphism

contact zone
use aureoles

containment liners
use disposal barriers

contamination (1978)
SA anatexis
SA assimilation
SA country rocks
SA decontamination
SA emplacement
SA hybridization
SA intrusions
SA magmas
SA palingenesis
SA pollution

contemporaneous faults
use growth faults

content
use composition

conterminous regions (1978)
Used to indicate regions with a common boundary. Used especially for the United States.
SA United States

Conterminous United States Mineral Assessment Program
use CUSMAP

continental
A valid term through 1977. See continental crust; continental type.

continental borderland (1978)
The area of the continental margin between the shoreline and the continental slope which is topographically more complex than the continental shelf.
UF borderland, continental
SA continental margin
SA continental shelf
SA continental slope

continental crust (1981)
Also search crust AND continental type; crust AND continental; continental AND genesis.
UF continental type
BT crust
SA CALCRUST
SA COCORP
SA Illinois Deep Hole Project
SA Lg-waves
SA microcontinents
SA plate tectonics
SA Sn-waves
SA transition zones
SA underplating

continental displacement
use continental drift

continental drift (1978)
For general discussions of "classical" drift concepts as well as plate tectonic reconstructions. Also search continental AND drift.
IN Also index paleogeographic area as applicable [e.g. Gondwana, Iapetus, Laurasia, Laurentia, Mesogaea, Pangaea, Paratethys, Tethys]
UF continental displacement
UF continental migration
UF displacement theory
UF epeirophoresis theory
UF Wegener hypothesis
SA apparent polar wandering
SA biogeography
SA continents
SA geosynclines
SA geotectonic maps
SA Gondwana
SA land bridges
SA Laurasia
SA Laurentia
SA mantle
SA Mohorovicic discontinuity
SA ocean basins
SA paleogeography
SA paleomagnetism
SA paleontology
SA Pangaea
SA Paratethys
SA plate tectonics
SA polar wandering
SA reconstruction
SA sea-floor spreading
SA stratigraphy
SA tectonics
SA tectonophysics
SA Tethys
SA volcanology
SA Wrangellia

continental dunes (1981)
This term has multiple hierarchies.

BT1 dunes
BT2 eolian features
SA foredunes
SA geomorphology
SA shore features

continental environment
use terrestrial environment

continental margin (1978)
Use term to denote both shelf and slope. Also search continental AND margin.
UF margin, continental
SA active margins
SA AMCOR
SA aseismic margins
SA back-arc basins
SA continental borderland
SA continental margin sedimentation
SA Continental Offshore Stratigraphic Test
SA continental rise
SA continental shelf
SA continental slope
SA continents
SA cratons
SA island arcs
SA ocean floors
SA passive margins
SA plate tectonics
SA submarine canyons
SA tectonics
SA tectonophysics

continental margin sedimentation (1989)
SA continental margin
SA continental shelf
SA continental slope
SA sedimentation

continental migration
use continental drift

Continental Offshore Stratigraphic Test (1989)
Also search COST in conjunction with appropriate state or region.
SA Alaska
SA Atlantic Ocean
SA continental margin
SA continental shelf
SA continental slope
SA cores
SA cost
SA marine drilling
SA oceanography
SA offshore
SA stratigraphy

continental platform
use continental shelf

continental rise (1978)
UF rise, continental
SA abyssal plains
SA continental margin
SA continental slope
SA HEBBLE
SA marine geology
SA nepheloid layer
SA ocean floors

Continental Scientific Drilling Program (1989)
UF CSDP
SA drilling
SA programs

continental seas
use epicontinental seas

continental shelf (1978)
For studies of geological processes taking place on the shelf. Restricted to the modern shelf area. Also search continental AND shelf. Autoposting of this term began in 1978.
IN Index both land and sea geographic terms.
UF continental platform
UF continental terrace
UF shelf, continental
NT inner shelf
NT outer shelf
SA AMCOR
SA changes of level
SA continental borderland
SA continental margin
SA continental margin sedimentation
SA Continental Offshore Stratigraphic Test
SA continental slope
SA continents
SA currents
SA ecology
SA epicontinental seas
SA insular shelf
SA marine geology
SA ocean circulation
SA ocean currents
SA ocean floors
SA oceanography
SA offshore
SA Polar Continental Shelf
SA reefs
SA sedimentation
SA shelf environment
SA shelf-slope break
SA slope environment
SA slumping
SA submarine canyons

continental slope (1978)
For modern slope area. Also search continental AND slope. Autoposting of this term began in 1978.
IN Index with geographic terms for both land and sea.
UF slope, continental
NT inner slope
NT outer slope
SA AMCOR
SA continental borderland
SA continental margin
SA continental margin sedimentation
SA Continental Offshore Stratigraphic Test
SA continental rise
SA continental shelf
SA continents
SA ecology
SA HEBBLE
SA hemipelagic environment
SA marginal basins
SA marine geology
SA ocean circulation
SA ocean floors
SA oceanography
SA offshore
SA sedimentation
SA shelf-slope break
SA slope environment
SA slopes
SA slumping
SA submarine canyons

continental terrace
use continental shelf

continental type
use continental crust

continents (1978)
Used for discussions of continents as units (e.g. geometry of a continent).
SA continental drift
SA continental margin
SA continental shelf
SA continental slope
SA cratons
SA Earth
SA epeirogeny
SA epicontinental seas
SA island arcs
SA islands
SA isostasy
SA marginal seas
SA tectonic platforms

continuous filters (1981)
SA analog filters
SA filters

continuous materials (1981)
SA engineering geology
SA materials

contour maps (1978)
BT maps
SA structure contour maps
SA topographic maps

contourite (1978)
BT clastic rocks
BT sedimentary rocks
SA terrigenous materials

Contra Costa County
Valid through 1988. Search in combination with state term. After 1988, use specific county-state term.

Contra Costa County California (1989)
On San Francisco Bay in W California. Before 1989, also search Contra Costa County AND California.
CO N374300N380700 W1213300W1223000
BT California
BT United States
SA Greenville Fault
SA San Francisco Bay region

contraction (1978)
Used as a general term.
SA compression tectonics
SA processes

control, erosion
use erosion control

control, production
use production control

controlled-source audio-magnetotelluric technique
use CSAMT methods

controls (1978)
SA climatic controls
SA erosion control
SA geochemical controls
SA geomorphologic controls
SA hydrogeological controls
SA lithologic controls
SA mechanical controls
SA metallotects
SA mineral deposits, genesis
SA paleogeographic controls
SA sediment supply
SA slope stability
SA stratigraphic controls
SA structural controls
SA tectonic controls
SA waste disposal
SA waterways

Contwoyto Lake (1989)
Central Northwest Territories.
BT Northwest Territories
BT Western Canada
BT Canada

Conularida (1981)
BT Scyphozoa
BT Coelenterata

BT Invertebrata

convection (1978)
SA advection
SA aerosols
SA angular momentum
SA atmosphere
SA boundary conditions
SA convection currents
SA heat flow
SA mantle
SA meteorology
SA oceanography
SA plate tectonics
SA tectonophysics

convection cells
As of 1981, no longer a valid term for GeoRef.
use convection currents

convection currents (1978)
Before 1981, also search convection cells.
UF convection cells
BT currents
SA atmosphere
SA convection
SA density
SA mantle
SA plate tectonics
SA tectonophysics

convergence
As of 1981, no longer a valid term for GeoRef. Use plate convergence if applicable.

Converse County
Valid through 1988. Search in combination with state term. After 1988, use specific county-state term.

Converse County Wyoming (1989)
E Wyoming. Before 1989, also search Converse County AND Wyoming.
CO N421500N433000 W1045000W1061000
BT Wyoming
BT United States

conversion (1978)
Used as a general term.
SA changes

convoluted beds (1978)
UF beds, convoluted
UF convoluted deformation
BT soft sediment deformation
BT sedimentary structures
SA slump structures
SA turbidity current structures

convoluted deformation
use convoluted beds

Coo
Not a valid term for GeoRef. Use Kos for the Greek Island.

Cooch Behar
No longer a valid term for GeoRef. As of 1947, a former state in E India. See Cooch Behar India.

Cooch Behar India (1993)
City just N of Bangladesh in West Bengal, E India.
BT West Bengal India
BT India
BT Indian Peninsula
BT Asia

Cook County
Valid through 1988. Search in combination with state term. After 1988, use specific county-state term.

Cook County Georgia (1989)
S Georgia. Before 1989, also search Cook County AND Georgia.
CO N310200N312000
W0831800W0833400
BT Georgia
BT United States

Cook County Illinois (1989)
NE Illinois. Before 1989, also search Cook County AND Illinois.
CO N412800N421000
W0873100W0881700
BT Illinois
BT United States
NT Chicago Illinois

Cook County Minnesota (1989)
NE Minnesota. Before 1989, also search Cook County AND Minnesota.
CO N472800N481500
W0893500W0910300
BT Minnesota
BT United States
SA Saganaga Lake

Cook Inlet (1978)
Arm of the Pacific Ocean W of Kenai Peninsula in S Alaska.
IN Index Southern Alaska or Southwestern Alaska as applicable.
SA Alaska
SA Anchorage Alaska
SA Augustine
SA Knik Arm
SA Southern Alaska
SA Southwestern Alaska
SA Turnagain Arm

Cook Islands (1978)
Group of 15 islands W of French Polynesia and E of Samoa and Tonga islands. In 1981, broader term changed from Pacific Ocean to Polynesia.
CO S230000S180000
W1560000W1650000
UF Southern Cook Islands
BT Polynesia
BT Oceania
SA Pacific Ocean

Cook Mountain Formation (1978)
In Claiborne Group. NW Louisiana, S and E Texas.
BT middle Eocene
BT Eocene
BT Paleogene
BT Tertiary
BT Cenozoic
SA Claiborne Group
SA Louisiana
SA Texas

cookeite (1978)
BT chlorite group
BT sheet silicates
BT silicates
SA aluminosilicates
SA lepidolite

cooling (1978)
Used as a general term.
SA temperature

Cooma
No longer a valid term for GeoRef. See Cooma Australia.

Cooma Australia (1993)
Town in SE New South Wales.
BT New South Wales Australia
BT Australia
BT Australasia

Cooper Basin (1978)
IN Index states as applicable.
UF Cooper's Creek basin
SA Australia
SA Queensland Australia
SA South Australia

Cooper's Creek basin
use Cooper Basin

cooperation, international
use international cooperation

coordinate systems (1978)
SA systems

coordinates (1978)
For geographic coordinates, see specific area.
SA cartography
SA geodesy
SA geodetic coordinates
SA maps
SA Moon

coordination (1978)
SA crystal chemistry

Coorong Lagoon (1978)
Off Lacepede Bay of Indian Ocean in SE South Australia.
UF The Coorong
BT South Australia
BT Australia
BT Australasia

Coos Bay (1978)
Inlet on coast of Coos County on the Pacific Ocean in SW Oregon.
IN Index state and county as applicable.
SA Coos County Oregon

Coos County
Valid through 1988. Search in combination with state term. After 1988, use specific county-state term.

Coos County New Hampshire (1989)
N New Hampshire. Before 1989, also search Coos County AND New Hampshire.
CO N440500N451900
W0710100W0714600
BT New Hampshire
BT United States
SA Mount Adams
SA Mount Jefferson
SA Rangeley Lakes

Coos County Oregon (1989)
SW Oregon. Before 1989, also search Coos County AND Oregon.
CO N423900N433700
W1234300W1242900
BT Oregon
BT United States
SA Coos Bay

Coosa County
Valid through 1988. Search in combination with state term. After 1988, use specific county-state term.

Coosa County Alabama (1989)
Central Alabama. Before 1989, also search Coosa County AND Alabama.
CO N324500N330600
W0860000W0863300
BT Alabama
BT United States

Copenhagen
No longer a valid term for GeoRef. See Copenhagen Denmark.

Copenhagen Denmark (1993)
County and city on Zealand Island in E Denmark. Before 1993, also search Copenhagen or Kobenhavn.
UF Kobenhavn Denmark
BT Denmark
BT Scandinavia
BT Western Europe
BT Europe
SA Zealand

Copepoda (1978)
BT Crustacea
BT Mandibulata
BT Arthropoda
BT Invertebrata

copiapite (1978)
BT sulfates

Copiapo (1978)
River in N central Chile. Before 1993, also used for the town. See Copiapo Chile for town.
BT Chile
BT South America
SA Atacama Chile

Copiapo Chile (1993)
Town in N central Chile. Also search Selva or San Francisco de la Selva.
UF San Francisco de la Selva
BT Atacama Chile
BT Chile
BT South America

Copley Greenstone (1989)
N California. Conformably underlies Balaklala Rhyolite and interfingers with it in some areas.
BT Middle Devonian
BT Devonian
BT Paleozoic
SA Balaklala Rhyolite
SA California

copper (1978)
Chemical element. As of 1981, use copper ores for copper as a commodity. Autoposting of metals to this term began in 1989.
UF Cu
BT metals
SA bornite
SA chalcocite
SA chalcophile elements
SA chalcopyrite
SA copper minerals
SA copper ores
SA enargite
SA heavy metals
SA native elements
SA porphyry copper

Copper Age
use Chalcolithic

Copper Canyon (1978)
Canyon in Lander County in N central Nevada.
IN Index county or regions as applicable.
SA Lander County Nevada

Copper Harbor Conglomerate (1978)
Upper Keweenawan. Keweenaw County on the Upper Peninsula, NW Michigan.
BT Keweenawan
BT Proterozoic
BT upper Precambrian
BT Precambrian
SA Michigan

copper minerals (1985)
SA azurite
SA bornite
SA chalcocite
SA chalcopyrite
SA chrysocolla
SA copper
SA copper ores
SA covellite
SA cuprite
SA enargite
SA luzonite
SA malachite
SA minerals
SA tennantite
SA tetrahedrite
SA turquoise

Copper Mountain (1978)
Location of copper deposit in British Columbia. Also used for uranium deposit in Wyoming. Also used in other locations.
IN Index province, state and county or country as applicable.
SA British Columbia
SA California
SA Wyoming

copper ores (1981)
Before 1981, also search (copper OR Cu) AND (deposit OR deposits OR ore OR ores OR economic) in the basic index. Autoposting of metal ores to this term began in 1985.
IN Commodity. See List C.
BT metal ores
SA Agrokipia Deposit
SA Ashio Mine
SA azurite
SA base metals
SA besshi-type
SA Blackbird mining district
SA bornite
SA chalcocite
SA chalcopyrite
SA Chinkuashih Mine
SA chrysocolla
SA copper
SA copper minerals
SA Coronation Mine
SA covellite
SA cuprite
SA cyprus-type
SA enargite
SA Equity Mine
SA Hitachi Deposit
SA Horne Mine
SA kupferschiefer-type
SA malachite
SA Mosaboni Mines
SA native elements
SA Norilsk region
SA nuggets
SA Olympic Dam
SA porphyry copper
SA slag
SA Strathcona Mine
SA Sullivan Mine
SA Temagami Mine
SA tennantite
SA tetrahedrite
SA Urup mining district
SA White Pine Mine

copper pyrites
use chalcopyrite

Copper River basin (1989)
S central Alaska.
IN Index Southern Alaska or regions as applicable.
SA Southern Alaska
SA Talkeetna Formation

copper sulfides (1978)
BT sulfides
SA chalcocite
SA covellite
SA gossan

Copperbelt (1978)
　Area in S Zambia.
　IN　Index Zambia as applicable.
　SA　Zambia

Coppermine River (1978)
　Flows into Coronation Gulf in N Mackenzie District, E central Northwest Territories.
　BT　Northwest Territories
　BT　Western Canada
　BT　Canada
　SA　Mackenzie District Northwest Territories

coprolite
　Not a valid GeoRef index term after 1970. Was used on level 1 in subfile N.
　use coprolites

coprolites (1978)
　From 1978-1984, biogenic structures and sedimentary structures were autoposted to this term. Before 1971, also search coprolite.
　UF　coprolite
　SA　biogenic structures
　SA　fecal pellets
　SA　ichnofossils
　SA　lebensspuren
　SA　sedimentary structures
　SA　Vertebrata

coquimbite (1978)
　BT　sulfates

Coquimbo
　No longer a valid term for GeoRef.
　See Coquimbo Chile.

Coquimbo Chile (1993)
　Province and city in central Chile.
　BT　Chile
　BT　South America

coquina (1978)
　BT　limestone
　BT　carbonate rocks
　BT　sedimentary rocks
　SA　carbonate sediments
　SA　clastic sediments
　SA　terrigenous materials

coral pinnacle
　use pinnacle reefs

coral reefs
　A valid term through 1974. Was used in subfiles N, E, G, T and B.
　use reefs

Coral Sea (1978)
　Between Queensland, Australia, on the W and New Hebrides and New Caledonia on the E.
　CO　S300000S040000
　　　E1700000E1423000
　BT　West Pacific
　BT　Pacific Ocean
　NT　Coral Sea Basin
　NT　Great Barrier Reef
　SA　Leg 21
　SA　Leg 133

Coral Sea Basin (1978)
　Between NE Queensland and Solomon Islands.
　BT　Coral Sea
　BT　West Pacific
　BT　Pacific Ocean
　SA　DSDP Site 210

Corallimorpharia (1978)
　Order. Autoposting of Zoantharia to this term began in 1989.
　BT　Zoantharia
　BT　Anthozoa
　BT　Coelenterata
　BT　Invertebrata

Corallinaceae (1978)
　Autoposting of thallophytes and microfossils to this term began in 1990. This term has multiple hierarchies.
　BT1　Rhodophyta
　BT1　algae
　BT1　thallophytes
　BT1　Plantae
　BT2　Rhodophyta
　BT2　algae
　BT2　microfossils
　NT　Archaeolithothamnium
　NT　Lithophyllum
　NT　Lithothamnium

corals (1978)
　As of 1981, restricted to use as the common name for Coelenterata. Autoposting of this term began in 1978.
　IN　Index for all non-paleontologic studies of fossils.
　UF　coelenterates
　BT　invertebrates
　NT　stromatoporoids
　SA　biostratigraphy
　SA　Coelenterata
　SA　encrustations

Corbieres (1978)
　Outliers of the E Pyrenees in S Aude.
　UF　Corbieres Mountains
　BT　Aude France
　BT　France
　BT　Western Europe
　BT　Europe
　SA　French Pyrenees

Corbieres Mountains
　use Corbieres

Cordaitales (1978)
　Autoposting of this term began in 1978.
　BT　gymnosperms
　BT　Spermatophyta
　BT　Plantae
　NT　Cordaites
　SA　Dadoxylon

Cordaites (1978)
　Genus.
　BT　Cordaitales
　BT　gymnosperms
　BT　Spermatophyta
　BT　Plantae

Cordevolian (1985)
　Europe; Triassic.
　BT　Triassic
　BT　Mesozoic
　SA　Lower Triassic
　SA　Middle Triassic

cordierite (1978)
　BT　ring silicates
　BT　silicates
　SA　aluminosilicates

Cordillera
　As of 1981, no longer a valid term for GeoRef. See North American Cordillera; Canadian Cordillera.

Cordillera Betica
　use Betic Cordillera

Cordillera de la Costa (1978)
　Along Caribbean Sea in central N Venezuela.
　BT　Venezuela
　BT　South America

Cordillera Marianica
　use Betic Cordillera

Cordillera Occidental
　use Western Cordillera

Cordillera Oriental
　use Eastern Cordillera

Cordilleran Geosyncline (1978)
　Before 1978, search geosynclines AND Cordilleran.
　SA　Cordilleran Orogeny
　SA　geosynclines

Cordilleran ice sheet (1993)
　North American Cordillera.
　SA　North America
　SA　North American Cordillera

Cordilleran Orogeny (1978)
　Before 1978, search Cordilleran AND orogeny.
　SA　Cordilleran Geosyncline
　SA　Laramide Orogeny
　SA　orogeny
　SA　tectonics

Cordoba
　No longer a valid term for GeoRef. Referred to cities in Spain and in Argentina. Before 1981, also referred to the province in Spain and the department in Argentina. As of 1993, see Cordoba Argentina and Cordoba Spain.

Cordoba Argentina (1993)
　Province including the city in central Argentina. Before 1981, also search Cordoba AND Argentina. Before 1993, also search Cordoba Department or Cordoba AND Argentina.
　CO　S350000S292000
　　　W0613000W0654500
　BT　Argentina
　BT　South America
　NT　Chanares Argentina
　NT　Pampean Mountains

Cordoba Department
　No longer a valid term for GeoRef. See Cordoba Argentina, also occurs in Colombia.

Cordoba Province
　No longer a valid term for GeoRef. As of 1993, see Cordoba Spain.

Cordoba Spain (1993)
　In S central Spain. From 1981-1992, also search Cordoba Province and Spain. Before 1981, also search Cordoba AND Spain.
　CO　N371000N385000
　　　W0035700W0053500
　BT　Andalusia Spain
　BT　Spain
　BT　Iberian Peninsula
　BT　Southern Europe
　BT　Europe
　NT　Los Pedroches Spain
　NT　Valsequillo Spain

core (1978)
　Used for the Earth's core and, when applicable, the cores of the other planets and the Moon. Autoposting of this term began in 1978.
　NT　inner core
　NT　outer core
　SA　core-mantle boundary
　SA　crust
　SA　dynamos
　SA　Earth
　SA　geophysics
　SA　heat flow
　SA　heat sources
　SA　interior
　SA　lithosphere
　SA　magnetic field
　SA　magnetohydrodynamics
　SA　mantle

　SA　Mohorovicic discontinuity
　SA　phase transitions
　SA　seismology
　SA　tectonophysics
　SA　transition zones

core complexes
　use metamorphic core complexes

core-mantle boundary (1989)
　UF　core-mantle couple
　UF　core-mantle interface
　UF　mantle-core boundary
　SA　core
　SA　coupling
　SA　lower mantle
　SA　mantle
　SA　outer core
　SA　seismology
　SA　tectonophysics
　SA　transition zones

core-mantle couple
　use core-mantle boundary

core-mantle interface
　use core-mantle boundary

cores (1978)
　Before 1978, also search borings.
　UF　drill cores
　SA　AMCOR
　SA　boreholes
　SA　Continental Offshore Stratigraphic Test
　SA　cuttings
　SA　Iceland Research Drilling Project
　SA　Illinois Deep Hole Project
　SA　well-logging
　SA　wells

Corinth
　No longer a valid term for GeoRef. As of 1993 see Corinth Greece.

Corinth Greece (1993)
　City on Gulf of Corinth in Corinth Department in NE Peloponnesus. Before 1993 also search Corinth AND Greece.
　BT　Peloponnesus Greece
　BT　Greece
　BT　Southern Europe
　BT　Europe
　SA　Gulf of Corinth

Coriolis force
　As of 1993, no longer a valid term for GeoRef. Usually out-of-scope. Used with ocean circulation.

Cork
　No longer a valid term for GeoRef. See Cork Ireland.

Cork Ireland (1993)
　County including the city on the Atlantic Ocean in S Ireland. Before 1993, also search Cork AND Ireland. Before 1978, also search County Cork.
　CO　N513000N522000
　　　W0081500W0103000
　UF　County Cork Ireland
　BT　Ireland
　BT　Western Europe
　BT　Europe

Cormeilles-en-Parisis
　No longer a valid term for GeoRef.
　use Cormeilles-en-Parisis France

Cormeilles-en-Parisis France (1993)
　Town just N of Paris in N central France.
　UF　Cormeilles-en-Parisis
　BT　Val-d'Oise France
　BT　France
　BT　Western Europe

BT Europe

Cornwall
No longer a valid term for GeoRef as of 1993. See Cornwall England.

Cornwall England (1993)
Former county in extreme SW England. Became part of Cornwall and Isles of Scilly in 1974.
CO N500000N505000
 W0041500W0063000
BT England
BT Great Britain
BT United Kingdom
BT Western Europe
BT Europe
NT Land's End
NT Padstow England
SA Tamar Estuary
SA Tamar Valley

Cornwallis Island (1978)
One of the Parry Islands in central Franklin District, N central Northwest Territories.
CO N743000N753000
 W0933000W0970000
BT Northwest Territories
BT Western Canada
BT Canada
SA Franklin District Northwest Territories
SA Read Bay Formation

Coromandel Coast (1978)
SE coast from Point Calimere in the S to mouths of the Krishna River in Andhra Pradesh. Also search Coromandel.
IN Index states as applicable.
BT India
BT Indian Peninsula
BT Asia
SA Andhra Pradesh India
SA Pondicherry India
SA South Arcot
SA Tamil Nadu India

Coromandel Peninsula (1978)
ENE of Auckland. Also search Coromandel.
IN Also index North Island.
BT New Zealand
BT Australasia
SA North Island

corona
As of 1993, no longer a valid term for GeoRef. Usually out-of-scope. Used with astrophysics and solar physics. For coronas in rocks, use reaction rims.

coronadite (1978)
BT oxides
SA hollandite

coronae (1993)
Ovoid features on other planetary bodies. Before 1993, also search corona and planet name.
SA arcuate structures
SA remote sensing
SA surface features
SA Uranus
SA Venus

Coronation Mine (1978)
Copper and zinc ores. E central Saskatchewan.
IN Index province or region as applicable.
SA copper ores
SA mines
SA Saskatchewan
SA zinc ores

Corophioides (1978)
BT ichnofossils

corporate policy (1982)
SA companies
SA economics
SA management
SA policy

corporations
 use companies

Corpus Christi Bay (1978)
Inlet of Gulf of Mexico between Neuces and San Patricio counties in SE.
IN Index state and counties as applicable.
SA Nueces County Texas
SA San Patricio County Texas
SA Texas

corrections (1978)
Used as a general term.
SA accuracy
SA errors
SA topographic correction

correlation (1978)
Formerly a general term appropriate to a large number of topics. As of 1976, this term is restricted to stratigraphic correlation. Before 1969, also search correlation tables; correlations.
UF correlation tables
UF stratigraphic correlation
SA biochronology
SA stratigraphy

correlation coefficient (1978)
UF coefficient, correlation
BT statistical analysis

correlation tables
 use correlation

correlations
No longer a valid GeoRef index term. Before 1971, was used on level I in subfile N. Now use correlation.

corrensite (1978)
BT clay minerals
BT sheet silicates
BT silicates
SA aluminosilicates

correspondence analysis (1978)
BT statistical analysis
SA analysis

Correze
No longer a valid term for GeoRef. See Correze France.

Correze France (1993)
Department in S central France.
CO N445500N454500
 E0023000E0011000
BT France
BT Western Europe
BT Europe
NT Brive
SA Millevaches Plateau

Corrientes
No longer a valid term for GeoRef. See Corrientes Argentina.

Corrientes Argentina (1993)
Province including the city between the Parana River and the Uruguay River.
CO S300000S270000
 W0560000W0600000
BT Argentina
BT South America

corrosion (1978)
UF chemical erosion
SA chemical weathering

SA destruction
SA embayments
SA erosion
SA geomorphology
SA hydrogen peroxide
SA pits

Corse du Sud France (1993)
Department in the southern part of Corsica.
BT Corsica
BT France
BT Western Europe
BT Europe

Corsica (1978)
Island in the Mediterranean Sea. After 1992, use Haute-Corse France or Corse du Sud France for the departments. Before 1993, also included use for the departments.
CO N411500N430000
 E0093000E0083000
BT France
BT Western Europe
BT Europe
NT Corse du Sud France
NT Haute-Corse France
SA Mediterranean region
SA Mediterranean Sea

Cortez
Valid through 1988. Search in combination with state term. After 1988, use specific city-state term.

Cortez Colorado (1989)
City in extreme SW Colorado. Before 1989, search Cortez AND Colorado.
CO N372200N372200
 W1083600W1083600
BT Montezuma County Colorado
BT Colorado
BT United States

Cortez Mountains (1978)
Eureka County in central Nevada. Also search Cortez.
IN Index county or regions as applicable.
SA Eureka County Nevada
SA Nevada

Cortina
Not a valid term for GeoRef. See Cortina D'Ampezzo Italy.

Cortina D'Ampezzo
No longer a valid term for GeoRef. See Cortina D'Ampezzo Italy.

Cortina D'Ampezzo Italy (1993)
Town in Dolomites in N Veneto, N Italy. Before 1993, also search Cortina D'Ampezzo. Before 1978, also search Cortina.
UF Cortina Italy
BT Veneto Italy
BT Italy
BT Southern Europe
BT Europe

Cortina Italy
 use Cortina D'Ampezzo Italy

Cortlandt Complex (1989)
E New York.
BT Upper Ordovician
BT Ordovician
BT Paleozoic
SA New York

corundum (1978)
As of 1981, use corundum deposits for corundum as a commodity (abrasives).
BT oxides
SA alumina

SA amethyst
SA corundum deposits
SA gems
SA rubies
SA sapphire

corundum deposits (1981)
Before 1981, search corundum AND deposits. For corundum as a gem, use corundum and gems.
IN Commodity. See List C. For corundum as a gem, index the specific type of gem.
SA abrasives
SA corundum
SA industrial minerals

Corylus (1989)
Genus.
BT Dicotyledoneae
BT angiosperms
BT Spermatophyta
BT Plantae
SA palynomorphs

Corynexochida (1978)
BT Trilobita
BT Trilobitomorpha
BT Arthropoda
BT Invertebrata

Cos
Not a valid term for GeoRef. Use Kos for the Greek Island.

cosalite (1978)
BT sulfobismuthites
BT sulfosalts

Cosenza
No longer a valid term for GeoRef. See Cosenza Italy.

Cosenza Italy (1993)
Town and province in W central Calabria.
BT Calabria Italy
BT Italy
BT Southern Europe
BT Europe

Coshocton County
Valid through 1988. Search in combination with state term. After 1988, use specific county-state term.

Coshocton County Ohio (1989)
Central Ohio. Before 1989, also search Coshocton County AND Ohio.
CO N400700N402700
 W0813700W0821300
BT Ohio
BT United States

cosmic dust (1978)
UF zodiacal dust
SA asteroids
SA dust
SA extraterrestrial geology
SA interplanetary dust
SA interplanetary space
SA meteoroids
SA particles
SA planetology

cosmic rays
As of 1993, no longer a valid term for GeoRef. Usually out-of-scope. Used with extraterrestrial geology, aeronomy, magnetosphere, planetology, and interplanetary space.

cosmic-ray methods
Not a valid GeoRef index term after 1971. Was used on level 1 in subfile G. Now use cosmic rays or particle-track dating as applicable.

cosmic-ray tracks
 use particle tracks

cosmochemistry (1978)
For "geochemistry" of extraterrestrial materials.
SA asteroids
SA chemical elements
SA cosmochronology
SA cosmogenic elements
SA extraterrestrial geology
SA geochemistry
SA planetology

cosmochronology (1985)
SA absolute age
SA cosmochemistry
SA expanding universe theory
SA extraterrestrial geology
SA geochronology
SA planetology
SA relative age
SA solar system

cosmogenic elements (1978)
Before 1978, search cosmogenic; cosmogenic AND isotopes.
SA chemical elements
SA cosmochemistry
SA isotopes

cosmogeny
Not a valid GeoRef index term after 1971. Was used on level 1 in subfile G. Now use cosmochronology, cosmogenic elements or cosmochemistry.

cosmolites
use meteorites

Coso Hot Springs
use Coso Hot Springs KGRA

Coso Hot Springs KGRA (1981)
Known Geothermal Resources Area. Before 1981, also search Coso Hot Springs. Autoposting of broader terms began in 1989.
CO N360200N360300
W1174500W1174700
UF Coso Hot Springs
UF KGRA, Coso Hot Springs
BT Inyo County California
BT California
BT United States

Coso Range (1989)
S California.
BT Inyo County California
BT California
BT United States

cost (1981)
SA Continental Offshore Stratigraphic Test
SA economics
SA price

Costa Rica (1978)
CO N080000N111500
W0823000W0860000
BT Central America
NT Arenal
NT Irazu
NT Nicoya Peninsula
NT Poas
SA Aguacate Group
SA Nicoya Complex

Costa Rica Rift (1985)
Equatorial Pacific, off Colombia.
BT Equatorial Pacific
BT Pacific Ocean

Cote d'Or France
use Cote-d'Or France

Cote-d'Or
No longer a valid term for GeoRef.
use Cote-d'Or France

Cote-d'Or France (1993)
Department in E central France. Also search Cote d'Or.
CO N465500N480500
E0053000E0040200
UF Cote d'Or France
UF Cote-d'Or
BT France
BT Western Europe
BT Europe
NT Dijon France
NT Pouilly-en-Auxois France
SA Morvan

Cotentin Peninsula (1978)
Peninsula in Normandy jutting into the English Channel in N Manche, NW France.
BT Manche France
BT France
BT Western Europe
BT Europe

Cotes-du-Nord
No longer a valid term for GeoRef.
use Cotes-du-Nord France

Cotes-du-Nord France (1993)
Department on the English Channel in N Brittany.
CO N480000N490000
W0020000W0034500
UF Cotes-du-Nord
BT France
BT Western Europe
BT Europe
SA Brittany

Cotswold Hills (1978)
Range of hills in Gloucestershire, Avon, and Hereford and Worcester in W central England.
IN Index counties as applicable.
CO N514500N521500
W0011500W0014500
UF Cotswolds
BT England
BT Great Britain
BT United Kingdom
BT Western Europe
BT Europe
SA Avon England
SA Gloucestershire England
SA Worcestershire England

Cotswolds
use Cotswold Hills

Cottageville Field (1985)
W West Virginia. Autoposting of broader terms began in 1989.
CO N384500N385500
W0814500W0815500
BT Jackson County West Virginia
BT West Virginia
BT United States
SA oil and gas fields

Cottbus
No longer a valid term for GeoRef.
See Cottbus Germany.

Cottbus Bezirk
No longer a valid term for GeoRef. District in SE East Germany. Before 1981, search Cottbus. As of 1993, see Saxony Germany and Upper Lusatia.

Cottbus Germany (1993)
City in Brandenburg, E Germany. From 1981-1992, Cottbus was used for this city. Before 1981, Cottbus also included the district.
BT Brandenburg Germany
BT Germany
BT Central Europe
BT Europe

Cottian Alps (1978)
Division of Western Alps. This term has multiple hierarchies.
IN Index countries as applicable.
BT1 Western Alps
BT1 Alps
BT1 Europe
BT2 Western Europe
BT2 Europe
SA Dauphine Alps
SA France
SA Italy
SA Piedmont Alps

Cotton County
Valid through 1988. Search in combination with state term. After 1988, use specific county-state term.

Cotton County Oklahoma (1989)
SW Oklahoma. Before 1989, also search Cotton County AND Oklahoma.
CO N340400N343100
W0981000W0984000
BT Oklahoma
BT United States

Cotton Valley Group (1978)
Includes Schuler Formation, and Bossier Formation. S Arkansas, N Louisiana, W Mississippi, and E Texas.
BT Upper Jurassic
BT Jurassic
BT Mesozoic
SA Arkansas
SA Bossier Formation
SA Louisiana
SA Mississippi
SA Schuler Formation
SA Texas

Cottonwood Limestone (1978)
In Council Grove Group. E Kansas, SE Nebraska, and central northern Oklahoma.
BT Permian
BT Paleozoic
SA Council Grove Group
SA Kansas
SA Nebraska
SA Oklahoma

Cotylosauria (1981)
Order. Autoposting of this term to Captorhinomorpha and Pareiasauria began in 1989.
BT Anapsida
BT Reptilia
BT Tetrapoda
BT Vertebrata
BT Chordata
NT Captorhinomorpha
NT Pareiasauria

Coulomb's modulus
use shear modulus

Council Grove Group (1993)
Lower Permian. E Kansas, SE Nebraska, N Oklahoma, Colorado. In Kansas includes (ascending) Foraker Limestone, Johnson Shale, Red Eagle Limestone, Roca Formation, Grenola Limestone, Eskridge Shale, Beattie Limestone, Stearns Shale, Bader Limestone, Easly Creek Shale, Crouse Limestone, Blue Rapids Shale, Funston Limestone and Speiser Shale. Overlies Admire Group; underlies Chase Group.
BT Permian
BT Paleozoic
SA Colorado
SA Cottonwood Limestone
SA Eskridge Shale
SA Kansas
SA Nebraska
SA Oklahoma
SA Red Eagle Limestone
SA Roca Formation
SA Stearns Shale

country rocks (1978)
SA assimilation
SA aureoles
SA contact metamorphism
SA contamination
SA host rocks
SA hybridization
SA igneous rocks
SA inclusions
SA intrusions
SA magmas
SA rocks
SA wall rocks
SA wall-rock alteration

counts
No longer a valid GeoRef index term. Before 1971, was used on level 2 in subfile N. See granulometry; statistical analysis.

County Cork Ireland
use Cork Ireland

coupled processes
use coupling

coupling (1993)
UF coupled processes
SA chemical reactions
SA core-mantle boundary
SA Earth-Moon couple
SA geochemistry
SA induced polarization
SA inductively coupled plasma methods
SA seismicity

Coupon Bight (1978)
Florida Keys area.
IN Index counties or regions as applicable.
SA Florida
SA Florida Keys

Courland Spit (1978)
Narrow sandspit on the Baltic Sea in both Kaliningrad Oblast and Lithuania.
UF Kurland Spit
BT Baltic region
BT Europe
SA Kaliningrad Russian Federation
SA Lithuania

Couvinian (1978)
Europe. Above Emsian, below Givetian.
BT Middle Devonian
BT Devonian
BT Paleozoic
SA Eifelian

covariance analysis (1993)
BT statistical analysis
SA analysis
SA geostatistics
SA semivariograms
SA variance analysis

Covasna
No longer a valid term for GeoRef. As of 1993 see Covasna Romania.

Covasna Romania (1993)
County in SE Transylvania. Before 1993 also search Covasna AND Romania.
BT Transylvania
BT Romania
BT Southern Europe

BT Europe

covelline
  use covellite

covellite (1978)
  UF covelline
  BT sulfides
  SA copper minerals
  SA copper ores
  SA copper sulfides

cover, sedimentary
  use sedimentary cover

Cow Head Group (1989)
  W Newfoundland.
  IN Index ages as applicable.
  BT Paleozoic
  SA Cambrian
  SA Newfoundland Island
  SA Ordovician

Cowley County
  Valid through 1988. Search in combination with state term. After 1988, use specific county-state term.

Cowley County Kansas (1989)
  S Kansas. Before 1989, also search Cowley County AND Kansas.
  CO N370000N372700
    W0963500W0971000
  BT Kansas
  BT United States
  NT Arkansas City Kansas

Cowlitz County
  Valid through 1988. Search in combination with state term. After 1988, use specific county-state term.

Cowlitz County Washington (1989)
  On the Columbia River in SW Washington. Before 1989, also search Cowlitz County AND Washington.
  CO N455000N462200
    W1221500W1231500
  BT Washington
  BT United States
  SA Cowlitz River
  SA Toutle River

Cowlitz Formation (1985)
  SW Washington and NW Oregon.
  BT upper Eocene
  BT Eocene
  BT Paleogene
  BT Tertiary
  BT Cenozoic
  SA Oregon
  SA Washington

Cowlitz River (1985)
  Tributary of the Columbia River in SW Washington. Autoposting of broader terms began in 1989.
  IN Index counties as applicable.
  BT Washington
  BT United States
  SA Columbia River
  SA Cowlitz County Washington
  SA Lewis County Washington
  SA Toutle River

Coyote Creek Fault (1978)
  W California.
  BT California
  BT United States

Coyote Lake (1985)
  SE California.
  IN Index county or regions as applicable.
  SA Mojave Desert

  SA San Bernardino County California

Coyote Lake earthquake 1979 (1989)
  Epicenter at Coyote Lake, Santa Clara County, Central California.
  BT earthquakes
  SA California
  SA Santa Clara County California

Cr
  use chromium

Cr-51 (1985)
  Autoposting of broader terms began in 1989. This term has multiple hierarchies.
  BT1 radioactive isotopes
  BT1 isotopes
  BT2 chromium
  BT2 metals

Cr-53/Cr-52 (1989)
  Autoposting of broader terms began in 1989. This term has multiple hierarchies.
  BT1 stable isotopes
  BT1 isotopes
  BT2 chromium
  BT2 metals
  SA isotope ratios

Crab Orchard Formation (1989)
  Lower and Middle Silurian. Kentucky and SW Ohio.
  BT Silurian
  BT Paleozoic
  SA Kentucky
  SA Ohio

Crab Orchard Mountain Group
  use Crab Orchard Mountains Group

Crab Orchard Mountains Group (1989)
  NW Georgia and E Tennessee.
  UF Crab Orchard Mountain Group
  BT Lower Pennsylvanian
  BT Pennsylvanian
  BT Carboniferous
  BT Paleozoic
  SA Georgia
  SA Tennessee

cracks (1978)
  NT microcracks
  SA fissures
  SA fractures
  SA open fractures
  SA rock mechanics

Cracow
  No longer a valid term for GeoRef. As of 1993 see Cracow Poland

Cracow Poland (1993)
  Province and city in S Poland. Before 1989, also search Krakow. Before 1993 also search Cracow AND Poland.
  UF Krakow
  BT Poland
  BT Central Europe
  BT Europe
  NT Krzeszowice Poland
  NT Miechow Poland
  NT Pilica Poland
  NT Podhale
  NT Wieliczka Poland
  SA Andrychow Poland
  SA Cracow-Czestochowa Jura
  SA Nowy Targ Poland

Cracow-Czestochowa Jura (1989)

Mountain range in S Poland between Cracow and Czestochowa. Before 1989, also search Krakow-Czestochowa Jura.
  UF Krakow-Czestochowa Jura
  BT Poland
  BT Central Europe
  BT Europe
  SA Cracow Poland
  SA Czestochowa Poland
  SA Katowice Poland

Crai (1978)
  Forest in Apuseni Mountains in WSW.
  IN Index Romania and Transylvania as applicable.
  SA Romania
  SA Transylvania

crandallite (1978)
  BT phosphates
  SA goyazite

Crane County Texas (1993)
  W Central Texas. Before 1989, also search Crane County AND Texas.
  CO N310000N314000
    W1021700W1024600
  BT Texas
  BT United States
  SA Central Basin Platform

Crassostrea virginica (1978)
  Species. Autoposting of Ostreidae and Ostreacea to this term began in 1989.
  BT Ostreidae
  BT Ostreacea
  BT Bivalvia
  BT Mollusca
  BT Invertebrata

Crater Flat Tuff (1989)
  S Nevada. Includes Bullfrog Member and Tram Member.
  BT Miocene
  BT Neogene
  BT Tertiary
  BT Cenozoic
  SA Bullfrog Member
  SA Nevada

Crater Lake (1978)
  In Cascade Range in Klamath County in SW Oregon.
  IN Index county or regions as applicable.
  SA Klamath County Oregon
  SA Mount Mazama
  SA Oregon

crater lakes (1993)
  BT lakes
  BT lacustrine features
  SA calderas
  SA cauldrons
  SA collapse structures
  SA craters
  SA lava lakes
  SA meteor craters
  SA subsidence
  SA volcanic features

cratering (1978)
  SA craters
  SA cryptoexplosion features
  SA earthquakes
  SA geomorphology
  SA impact features

craters (1978)
  SA astroblemes
  SA calderas
  SA crater lakes
  SA cratering
  SA cryptoexplosion features
  SA geomorphology
  SA impact features

  SA impacts
  SA lunar craters
  SA maars
  SA meteor craters
  SA microcraters
  SA Moon
  SA volcanic features
  SA volcanology

Craters of the Moon (1989)
  National Monument. SW Butte and NE Blaine counties, S central Idaho.
  IN Index counties as applicable.
  CO N432000N433000
    W1132500W1134200
  BT Idaho
  BT United States
  SA Blaine County Idaho
  SA Butte County Idaho
  SA national monuments

cratons (1978)
  SA continental margin
  SA continents
  SA crust
  SA forelands
  SA intracratonic basins
  SA shields
  SA tectonic platforms
  SA tectonic units
  SA tectonics

Craven County
  Valid through 1988. Search in combination with state term. After 1988, use specific county-state term.

Craven County North Carolina (1989)
  On both sides of the Neuse River in E North Carolina. Before 1989, also search Craven County AND North Carolina.
  CO N345000N352000
    W0764000W0772500
  BT North Carolina
  BT United States

Crawford County
  Valid through 1988. Search in combination with state term. After 1988, use specific county-state term.

Crawford County Arkansas (1989)
  NW Arkansas. Before 1989, also search Crawford County AND Arkansas.
  CO N352000N354500
    W0935400W0943300
  BT Arkansas
  BT United States

Crawford County Georgia (1989)
  Central Georgia. Before 1989, also search Crawford County AND Georgia.
  CO N323200N325100
    W0834000W0840800
  BT Georgia
  BT United States

Crawford County Illinois (1989)
  SE Illinois. Before 1989, also search Crawford County AND Illinois.
  CO N385200N391000
    W0873000W0875600
  BT Illinois
  BT United States

Crawford County Indiana (1989)
  S Indiana. Before 1989, also search Crawford County AND Indiana.
  CO N380700N382400

    W0861700W0864100
  BT Indiana
  BT United States

Crawford County Iowa (1989)
  W Iowa. Before 1989, also search Crawford County AND Iowa.
  CO N415100N421300
    W0950500W0954000
  BT Iowa
  BT United States

Crawford County Kansas (1989)
  SE Kansas. Before 1989, also search Crawford County AND Kansas.
  CO N372000N374000
    W0942200W0950500
  BT Kansas
  BT United States

Crawford County Michigan (1989)
  N central Michigan. Before 1989, also search Crawford County AND Michigan.
  CO N443000N445200
    W0842200W0845200
  BT Michigan Lower Peninsula
  BT Michigan
  BT United States

Crawford County Missouri (1989)
  E central Missouri. Before 1989, also search Crawford County AND Missouri.
  CO N374200N381300
    W0910600W0913200
  BT Missouri
  BT United States

Crawford County Ohio (1989)
  N central Ohio. Before 1989, also search Crawford County AND Ohio.
  CO N404300N405900
    W0824300W0830700
  BT Ohio
  BT United States
  SA Sandusky River basin

Crawford County Pennsylvania (1989)
  NW Pennsylvania. Before 1989, also search Crawford County AND Pennsylvania.
  CO N412800N415200
    W0793700W0803200
  BT Pennsylvania
  BT United States

Crawford County Wisconsin (1989)
  SW Wisconsin. Before 1989, also search Crawford County AND Wisconsin.
  CO N425900N432500
    W0904000W0911300
  BT Wisconsin
  BT United States

Crawford Thrust (1989)
  E Idaho and W Wyoming.
  IN Index states as applicable.
  BT United States
  SA Idaho
  SA Wyoming

Crazy Mountain Basin
  use Crazy Mountains Basin

Crazy Mountains (1989)
  Range of the Rocky Mountains in central Montana. Also in the region approximately 60 miles NE of Fairbanks, Alaska.
  IN Index U. S. Rocky Mountains, Montana and applicable counties; or Alaska as applicable.
  SA Alaska
  SA Montana
  SA Park County Montana
  SA U. S. Rocky Mountains

Crazy Mountains Basin (1989)
  S central Montana.
  IN Index counties as applicable.
  UF Crazy Mountain Basin
  BT Montana
  BT United States
  SA Meagher County Montana
  SA Park County Montana
  SA Sweet Grass County Montana

creationism (1985)
  UF scientific creationism
  SA biologic evolution
  SA catastrophism
  SA education
  SA geology
  SA life origin
  SA paleontology
  SA philosophy
  SA uniformitarianism

Cree Lake (1981)
  N Saskatchewan.
  CO N572000N574000
    W1060000W1071000
  BT Saskatchewan
  BT Western Canada
  BT Canada

Creede Colorado (1989)
  Town in SW Colorado. Before 1989, search Creede AND Colorado.
  CO N375200N375200
    W1065600W1065600
  BT Mineral County Colorado
  BT Colorado
  BT United States

Creede District
  No longer a valid term for GeoRef.
  use Creede mining district

Creede mining district (1993)
  Mining district in SW Colorado. Before 1993, also search Creede District.
  UF Creede District
  BT Mineral County Colorado
  BT Colorado
  BT United States
  SA mines

creep (1978)
  Autoposting of mass movements to this term began in 1989. Before 1971, also search rock creep.
  UF creeping
  UF rock creep
  BT mass movements
  SA deformation
  SA elastic limit
  SA elastic strain
  SA etching
  SA geomorphology
  SA landslides
  SA mechanical properties
  SA permafrost
  SA slope stability
  SA slumping
  SA soils
  SA solifluction
  SA stress

creeping
  use creep

Creil
  No longer a valid term for GeoRef.
  See Creil France.

Creil France (1993)
  Town N of Paris, N France.
  BT Oise France
  BT France
  BT Western Europe
  BT Europe

crenulation cleavage
  use slip cleavage

Creodonta (1978)
  Order. Autoposting of Eutheria and Theria to this term began in 1989.
  BT Eutheria
  BT Theria
  BT Mammalia
  BT Tetrapoda
  BT Vertebrata
  BT Chordata

creosote (1989)
  SA ground water
  SA organic materials
  SA pollutants
  SA pollution
  SA waste disposal

Crescent Formation (1985)
  NW Washington.
  BT Eocene
  BT Paleogene
  BT Tertiary
  BT Cenozoic
  SA Washington

Crested Butte
  Valid through 1988. Search in combination with state term. After 1988, use specific city-state term.

Crested Butte Colorado (1989)
  Town in W central Colorado. Before 1989, search Crested Butte AND Colorado.
  CO N385200N385200
    W1065800W1065800
  BT Gunnison County Colorado
  BT Colorado
  BT United States

Crestmore (1978)
  Area in Riverside County, S California.
  IN Index county or regions as applicable.
  SA Riverside County California

Cretaceous (1978)
  Above Jurassic, below Tertiary. From 1978-1980, Mesozoic was autoposted to this term.
  BT Mesozoic
  NT Alisitos Formation
  NT Asu River Group
  NT Benton Formation
  NT Blairmore Group
  NT Bulldog Shale
  NT Calera Limestone
  NT Chatsworth Formation
  NT Colorado Group
  NT Columbian Orogeny
  NT Comanchean
  NT Dakota Formation
  NT Eureka Sound Formation
  NT Gosau Formation
  NT Graneros Shale
  NT Hatteras Formation
  NT Judea Group
  NT Kuskokwim Group
  NT Lower Cretaceous
  NT Lower Greensand
  NT Mancos Shale
  NT Middle Cretaceous
  NT Mishash Formation
  NT Nanushuk Group
  NT Nenjiang Formation
  NT Nubian Sandstone
  NT Potomac Group
  NT Qingshankou Formation
  NT Quantou Formation
  NT Robles Formation
  NT San Felipe Formation
  NT Santana Formation
  NT Shiranish Formation
  NT Tananao Schist
  NT Toolebuc Formation
  NT Upper Cretaceous
  NT Valdez Group
  NT Viking Formation
  NT Vraconian
  NT Weald Clay
  NT Whitemud Formation
  NT Yacoraite Formation
  SA Briançonnais Zone
  SA Coast plutonic complex
  SA Franciscan Complex
  SA Great Valley Sequence
  SA Ionian Zone
  SA Laramide Orogeny
  SA Lucas Formation
  SA Maiolica Limestone
  SA Mardin Formation
  SA Mist Mountain Formation
  SA Muro Group
  SA Paskapoo Formation
  SA Peace River Formation
  SA Rincon Formation
  SA Samail Ophiolite
  SA Scaglia Formation
  SA Serra Geral Formation
  SA Sobral Formation
  SA Tal Formation
  SA Tertiary
  SA Tor Formation
  SA upper Gondwana System
  SA Yanshanian

Cretaceous-Tertiary boundary
  use K-T boundary

Crete (1978)
  Administrative region and island in E Mediterranean Sea.
  CO N345500N354500
    E0263000E0231500
  BT Greece
  BT Southern Europe
  BT Europe
  SA Hellenic Arc
  SA Hellenides
  SA Mediterranean Sea

Creuse
  No longer a valid term for GeoRef.
  See Creuse France.

Creuse France (1993)
  Department in central France.
  CO N453500N462800
    E0023500E0012000
  BT France
  BT Western Europe
  BT Europe
  SA Millevaches Plateau

Crevasse Canyon Formation (1985)
  In Mesaverde Group. NW New Mexico.
  BT Upper Cretaceous
  BT Cretaceous
  BT Mesozoic
  SA Mesaverde Group
  SA New Mexico

crevasses
  See fissures and rupture.

CRIB (1985)
  Acronym.
  UF Computerized Resources Information Bank
  SA data bases
  SA data processing
  SA geodetic coordinates
  SA geography
  SA information systems

Cricetidae (1978)
Family. Autoposting of this term to Neotoma began in 1989. Autoposting of Eutheria and Theria to this term began in 1989.
BT Myomorpha
BT Rodentia
BT Eutheria
BT Theria
BT Mammalia
BT Tetrapoda
BT Vertebrata
BT Chordata
NT Microtus
NT Neotoma
NT Pliomys

Crimea
No longer a valid term for GeoRef. See Crimea Ukraine.

Crimea Ukraine (1993)
Oblast and region jutting into the Black Sea. This term has multiple hierarchies.
CO N440000N460000
E0370000E0320000
BT1 Ukraine
BT1 Europe
BT2 Ukraine
BT2 Commonwealth of Independent States
NT Alma River
NT Bakhchisarai Ukraine
NT Crimean Mountains
NT Crimean Plain
NT Kerch Peninsula
NT Kerch Ukraine
NT Simferopol Ukraine
NT Sivash
NT Tarkhankut Peninsula
NT Yalta Ukraine
SA Krymka Meteorite
SA Maikop Series
SA Russian Platform

Crimean Mountains (1978)
Range in S Crimea along Black Sea Coast. This term has multiple hierarchies.
BT1 Crimea Ukraine
BT1 Ukraine
BT1 Europe
BT2 Crimea Ukraine
BT2 Ukraine
BT2 Commonwealth of Independent States

Crimean Plain (1978)
Dry, level steppe covering the northern 80% of Crimean Peninsula. This term has multiple hierarchies.
BT1 Crimea Ukraine
BT1 Ukraine
BT1 Europe
BT2 Crimea Ukraine
BT2 Ukraine
BT2 Commonwealth of Independent States

crinanite (1978)
BT alkali basalts
BT basalts
BT volcanic rocks
BT igneous rocks

Criner Hills (1985)
S central Oklahoma.
CO N334500N340500
W0964500W0973000
BT Oklahoma
BT United States

Crinoidea (1978)
Autoposting of this term began in 1978.
BT Crinozoa
BT Echinodermata
BT Invertebrata
NT Delocrinus
NT Inadunata
SA crinoids

crinoids (1985)
Common name for Crinoidea. Autoposting of echinoderms to this term began in 1989.
BT echinoderms
BT invertebrates
SA biostratigraphy
SA Crinoidea

Crinozoa (1978)
Autoposting of this term to Edrioblastoidea, Lepidocystoidea, Parablastoidea and Paracrinoidea began in 1981. Before 1989, also search Pelmatozoa.
UF Pelmatozoa
BT Echinodermata
BT Invertebrata
NT Blastoidea
NT Crinoidea
NT Cystoidea
NT Eocrinoidea
NT Lepidocystoidea
NT Parablastoidea
NT Paracrinoidea

Cripple Creek
Valid through 1988. Search in combination with state term. After 1988, use specific city-state term.

Cripple Creek Colorado (1989)
City in central Colorado. Before 1989, search Cripple Creek AND Colorado.
CO N384400N384400
W1051100W1051100
BT Teller County Colorado
BT Colorado
BT United States

Crisana-Maramures (1978)
Historical province in NW Transylvania.
BT Transylvania
BT Romania
BT Southern Europe
BT Europe

cristobalite (1978)
BT silica minerals
BT framework silicates
BT silicates
SA opal-CT
SA quartz
SA tridymite

critical flow (1982)
SA hydraulics
SA hydrodynamics
SA hydrology
SA laminar flow
SA turbulence

critical review (1981)
SA review

Crittenden County
Valid through 1988. Search in combination with state term. After 1988, use specific county-state term.

Crittenden County Arkansas (1989)
E Arkansas. Before 1989, also search Crittenden County AND Arkansas.
CO N345100N352600
W0900400W0902800
BT Arkansas
BT United States

Crittenden County Kentucky (1989)
W Kentucky. Before 1989, also search Crittenden County AND Kentucky.
CO N370900N374300
W0874800W0882400
BT Kentucky
BT United States

CRM
use chemical remanent magnetization

Croatan Formation (1989)
E North Carolina.
IN Index ages as applicable.
BT Cenozoic
SA North Carolina
SA Pleistocene
SA Pliocene

Croatia (1978)
Constituent republic in N Yugoslavia.
CO N423000N464500
E0191500E0153000
BT Yugoslavia
BT Southern Europe
BT Europe
NT Limski Channel
NT Northern Limestone Alps
NT Split
NT Velebit Mountains
SA Adriatic region
SA Dinaric Alps
SA Istria
SA Neretva Valley
SA Osogovo Mountains

crocidolite (1978)
Variety of riebeckite. Autoposting of amphibole group to this term ended in 1985 and began again in 1989.
UF krokidolite
BT clinoamphibole
BT amphibole group
BT chain silicates
BT silicates
SA riebeckite

Crocodile River basin
use Limpopo Basin

Crocodilia (1978)
Order. Autoposting of this term began in 1978.
BT Archosauria
BT Diapsida
BT Reptilia
BT Tetrapoda
BT Vertebrata
BT Chordata
NT Teleosauridae

crocoite (1978)
BT chromates

Cromerian (1978)
Glaciation. Europe. Autoposting of Quaternary to this term began in 1989.
BT upper Pleistocene
BT Pleistocene
BT Quaternary
BT Cenozoic
NT South Polish Glaciation

Crook County
Valid through 1988. Search in combination with state term. After 1988, use specific county-state term.

Crook County Oregon (1989)
Central Oregon. Before 1989, also search Crook County AND Oregon.
CO N434100N443400
W1194000W1210700
BT Oregon
BT United States

Crook County Wyoming (1989)
NE Wyoming. Before 1989, also search Crook County AND Wyoming.
CO N441100N450000
W1040500W1050600
BT Wyoming
BT United States

Crooked Creek Formation (1978)
In Meade Group. Southwestern Kansas and northwestern Oklahoma.
BT Pleistocene
BT Quaternary
BT Cenozoic
SA Kansas
SA Oklahoma

cross folds
use superposed folds

cross fractures (1978)
Before 1978, search cross AND fractures. Before 1981, also search cross joints.
UF cross joints
BT fractures
SA cleats
SA joints

cross joints
As of 1981, no longer a valid term for GeoRef.
use cross fractures

Cross Lake (1989)
W central New York and central Manitoba.
IN Index New York or Manitoba as applicable.
SA Finger Lakes
SA Manitoba
SA New York

cross sections (1993)
Before 1993, also search sections and cross-sections. Sections was used for this concept from 1978-1992.
UF cross-sections
SA borehole sections
SA pit sections
SA polished sections
SA stratigraphic columns
SA stratigraphy
SA thin sections
SA type sections

cross-bedding (1978)
Before 1971, also search cross-bedding.
UF crossbedding
BT planar bedding structures
BT sedimentary structures
SA bedding
SA cross-laminations
SA cross-stratification
SA laminations

cross-correlation
use crosscorrelation

cross-hole methods
use crosshole methods

cross-laminations (1978)
UF diagonal lamination
BT planar bedding structures
BT sedimentary structures
SA cross-bedding
SA cross-stratification
SA laminations
SA ripple drift-cross laminations

cross-sections
use cross sections

cross-stratification (1978)

BT planar bedding structures
BT sedimentary structures
SA cross-bedding
SA cross-laminations
SA hummocky cross-stratification
SA stratification

crossbedding
No longer a valid GeoRef index term. Before 1971, was used on level 1 in subfile N.
use cross-bedding

crosscorrelation (1985)
UF cross-correlation
BT statistical analysis
SA autocorrelation
SA geostatistics
SA mathematical geology
SA mathematical methods

crosshole methods (1993)
Before 1993, also search crosshole methods.
UF cross-hole methods
SA boreholes
SA electrical methods
SA electromagnetic methods
SA geophysical methods
SA resistivity
SA seismic methods
SA tomography

crossite (1978)
BT clinoamphibole
BT amphibole group
BT chain silicates
BT silicates
SA glaucophane

Crossopterygii (1978)
Order. Autoposting of this term to Actinistia began in 1989. Autoposting of this term to Rhipidistia ended in 1980 and began again in 1989.
BT Sarcopterygii
BT Osteichthyes
BT Pisces
BT Vertebrata
BT Chordata
NT Actinistia
NT Rhipidistia

Crowsnest Pass (1978)
In SE British Columbia.
BT British Columbia
BT Western Canada
BT Canada

Crozet Islands (1978)
Five small French islands SE of South Africa. In 1985, broader term changed from Indian Ocean to Indian Ocean Islands.
BT Indian Ocean Islands

Crozon Peninsula (1978)
S of Brest in Brittany in W France.
BT Finistere France
BT France
BT Western Europe
BT Europe

CRPG (1985)
Acronym. Located in Vandoeuvre-les-Nancy, France.
UF Centre de Recherches Petrographiques et Geochimiques
BT government agencies
SA CNRS
SA France

crude oil (1989)
This term is used for petroleum after it has left the ground. It occurs infrequently in GeoRef, largely in the context of pollution and engineering. For studies of oil before it leaves the ground, see petroleum. For discussions of oil resources and reserves, see petroleum; reserves; resources.
SA hydrocarbons
SA oil spills
SA organic materials
SA petroleum
SA volatile organic compounds

crust (1978)
Used for the Earth's crust, and the crust of the planets and their satellites. As of 1981, for the crust of the Moon, use lunar crust. Before 1969, also search Earth crust. Also search crusts. Autoposting of this term to oceanic crust and continental crust began in 1985.
NT continental crust
NT lower crust
NT oceanic crust
NT upper crust
SA asthenosphere
SA basaltic layer
SA basement
SA Conrad discontinuity
SA core
SA cratons
SA crustal shortening
SA crustal thinning
SA discontinuities
SA Earth
SA earthquakes
SA epeirogeny
SA geophysics
SA geosynclines
SA geothermal gradient
SA granitic layer
SA heat flow
SA heat sources
SA horizontal discontinuities
SA isostasy
SA isostatic compensation
SA lithosphere
SA low-velocity zones
SA lunar crust
SA magmas
SA mantle
SA mobile belts
SA Mohorovicic discontinuity
SA neotectonics
SA ocean basins
SA paleomagnetism
SA phase transitions
SA plate tectonics
SA plates
SA sea-floor spreading
SA seismic surveys
SA seismology
SA shields
SA tangential discontinuities
SA tectonics
SA tectonophysics
SA thickness
SA thin-skinned tectonics
SA transition zones
SA undation
SA velocity structure
SA vertical discontinuities
SA volcanology

crust, weathering
use weathering crust

Crustacea (1978)
Autoposting of this term began in 1978.
BT Mandibulata
BT Arthropoda
BT Invertebrata
NT Branchiopoda
NT Cirripedia
NT Copepoda
NT Malacostraca
NT Ostracoda
SA crustaceans
SA Trilobita

crustaceans (1981)
Common name for Crustacea. Autoposting of arthropods to this term began in 1989.
BT arthropods
BT invertebrates
NT barnacles
NT ostracods
SA biostratigraphy
SA Crustacea

crustal shortening (1978)
Before 1978, also search shortening AND crustal.
UF shortening, crustal
SA crust
SA gravity sliding
SA plate tectonics
SA tectonics
SA thrust faults
SA transpression

crustal structure
A valid term through 1978. See crust.

crustal studies
A valid term through 1978. After 1978, see crust.

crustal thinning (1993)
SA crust
SA plate tectonics
SA tectonics
SA tectonophysics
SA thickness
SA thin-skinned tectonics

Cruziana (1978)
BT ichnofossils
SA Trilobita

cryogeology
Not a valid GeoRef index term after 1969. Was used on level 1 in subfile B. Now use glacial geology or permafrost as applicable.

cryokarst
use thermokarst

cryolite (1978)
As of 1981, use cryolite deposits for cryolite as a commodity.
BT fluorides
BT halides
SA cryolite deposits

cryolite deposits (1981)
Before 1981, search cryolite AND deposits.
SA cryolite
SA fluorspar

cryopedology (1978)
SA engineering geology
SA frost action
SA glacial geology
SA periglacial features
SA permafrost

Cryosols (1985)
BT soils

cryotectonics
use glaciotectonics

cryoturbation (1978)
UF congeliturbation
UF frost churning
UF frost stirring
UF geliturbation
SA congelifraction
SA frost action
SA frost heaving
SA patterned ground
SA periglacial features
SA rock mechanics
SA soil mechanics

cryptoexplosion features (1978)
Before 1973, also search cryptoexplosion structures. Used for structures formed by the explosive release of energy, that exhibit intense rock deformation, and that show no obvious genetic relation to volcanism, tectonics, or meteorite impact. Autoposting of this term began in 1978.
UF cryptoexplosion structures
UF features, cryptoexplosion
NT shatter cones
SA Charlevoix
SA cratering
SA craters
SA explosions
SA geomorphology
SA impact craters
SA impact features
SA impactite
SA impacts
SA Manicouagan Crater
SA meteor craters
SA shock metamorphism
SA suevite
SA zhamanshinite

cryptoexplosion structures
A valid term through 1972.
use cryptoexplosion features

cryptogam
Not a valid GeoRef index term after 1969. Was used on level 1 in subfile B. Now use thallophytes, bryophytes or pteridophytes as applicable.

Cryptolithus (1978)
Genus.
BT Ptychopariida
BT Trilobita
BT Trilobitomorpha
BT Arthropoda
BT Invertebrata

cryptomelane (1978)
BT oxides
SA manganese minerals

cryptoperthite (1978)
Autoposting of alkali feldspar to this term began in 1989.
BT alkali feldspar
BT feldspar group
BT framework silicates
BT silicates
SA perthite

Cryptostomata (1978)
Autoposting of this term began in 1978.
BT Bryozoa
BT Invertebrata
NT Fenestellidae
NT Rhabdomesidae

crystal chemistry (1978)
Used for the relations among chemical composition, structure and properties of crystals.
UF mineral chemistry
UF stereochemistry
SA bonding
SA chemical analysis
SA clay mineralogy
SA coordination
SA crystal field
SA crystal growth
SA crystal structure
SA crystal systems
SA crystal zoning
SA crystallography
SA electron microscopy
SA exsolution
SA geochemistry
SA grain boundaries

SA ion exchange
SA isomorphism
SA lattice parameters
SA mineralogy
SA minerals
SA order-disorder
SA partition coefficients
SA partitioning
SA phase equilibria
SA piezoelectric properties
SA reaction rims
SA substitution
SA symmetry
SA transformations

crystal defects
Not a valid term for GeoRef. See defects, point defects, crystal dislocations, or surface defects, if applicable.

crystal dislocations (1981)
Before 1981, also search dislocations AND minerals.
UF line defects
SA crystal growth
SA crystal structure
SA crystals
SA defects
SA minerals
SA point defects
SA surface defects

crystal field (1978)
Interactions of individual atoms and their respective electron configurations, which, collectively, influence crystal structure and properties.
UF field, crystal
SA color
SA crystal chemistry
SA crystal structure
SA minerals
SA optical properties
SA symmetry

crystal form (1978)
UF form, crystal
SA crystal growth
SA crystal systems
SA goniometry
SA habit

crystal fractionation (1993)
Refers to differentiation of magmas resulting from the floating or settling, under gravity, of mineral crystals being formed. Also see fractional crystallization.
UF gravitational differentiation
SA chemical fractionation
SA crystallization
SA fractional crystallization
SA fractionation
SA magmas
SA magmatic differentiation
SA segregation

crystal growth (1978)
Used for studies on natural or artificial growth of crystals.
SA clay mineralogy
SA crystal chemistry
SA crystal dislocations
SA crystal form
SA crystal structure
SA crystal zoning
SA crystallography
SA electron microscopy
SA epitaxy
SA fluid inclusions
SA geochemistry
SA grain boundaries
SA growth
SA growth spirals
SA habit
SA inclusions

SA intergrowths
SA lattice parameters
SA mineralogy
SA minerals
SA nucleation
SA overgrowths
SA phase equilibria
SA reaction rims
SA single-crystal method
SA symmetry
SA symplectite
SA synthesis
SA twinning

crystal habit
use habit

crystal lattice
use lattice

Crystal River Formation (1978)
In Ocala Group. N and W Florida.
BT Eocene
BT Paleogene
BT Tertiary
BT Cenozoic
SA Florida
SA Ocala Group

crystal structure (1978)
Used for the internal structure of a crystal. Also search atomic structure.
UF crystalline structure
SA atomic packing
SA bonding
SA cell dimensions
SA clay mineralogy
SA color centers
SA crystal chemistry
SA crystal dislocations
SA crystal field
SA crystal growth
SA crystal systems
SA crystallinity
SA crystallography
SA defects
SA etching
SA exsolution
SA goniometry
SA growth spirals
SA inclusions
SA ion exchange
SA lattice
SA lattice parameters
SA metamict minerals
SA metamictization
SA mineralogy
SA minerals
SA molecular structure
SA perovskite structure
SA point defects
SA polyhedra
SA polymorphism
SA polytypism
SA refinement
SA single-crystal method
SA space groups
SA stacking
SA superstructure
SA surface defects
SA symmetry
SA unit cell
SA X-ray analysis

crystal systems (1983)
NT hexagonal system
NT monoclinic system
NT orthorhombic system
NT triclinic system
SA atomic packing
SA crystal chemistry
SA crystal form
SA crystal structure
SA crystallography
SA crystals
SA lattice

SA minerals
SA symmetry
SA unit cell

crystal zoning (1982)
See zoning. Before 1982, also search zoning; zonation; zones. Before 1975, also search mineral zoning.
SA crystal chemistry
SA crystal growth
SA crystallization
SA crystallography
SA crystals
SA fractional crystallization
SA grain boundaries
SA magmas
SA magmatic differentiation
SA mineralogy
SA minerals
SA petrography
SA phase equilibria
SA reaction rims
SA zoning

crystal, quartz
use quartz crystal

crystalline limestone
use marbles

crystalline rocks (1978)
Used when rocks cannot be distinguished as igneous or metamorphic.
SA igneous rocks
SA metamorphic rocks
SA rocks
SA symplectite

crystalline structure
use crystal structure

crystallinity (1978)
Degree to which a clay mineral, such as illite, is crystalline.
SA clay mineralogy
SA crystal structure
SA diagenesis
SA illite

crystallites (1978)
SA magmas
SA volcanic glass

crystallization (1978)
SA crystal fractionation
SA crystal zoning
SA crystals
SA fractional crystallization
SA genesis
SA granitization
SA intergrowths
SA intrusions
SA lithification
SA magmas
SA magmatic differentiation
SA precipitation
SA reaction rims
SA recrystallization
SA water of crystallization

crystallization magnetization
use chemical remanent magnetization

crystallization remanent magnetization
use chemical remanent magnetization

crystallography (1978)
Used for general studies on the discipline of crystallography as a whole.
SA atomic packing
SA clay mineralogy
SA color centers
SA crystal chemistry
SA crystal growth
SA crystal structure

SA crystal systems
SA crystal zoning
SA crystals
SA defects
SA electron microscopy
SA geochemistry
SA goniometry
SA grain boundaries
SA growth spirals
SA hexagonal system
SA holography
SA intergrowths
SA lattice parameters
SA mineralogy
SA minerals
SA monoclinic system
SA neutron diffraction analysis
SA orthorhombic system
SA paragenesis
SA petrology
SA phase equilibria
SA pleochroism
SA point defects
SA reaction rims
SA reflectance
SA spectroscopy
SA standard materials
SA surface defects
SA symmetry
SA triclinic system
SA X-ray analysis

crystals (1978)
Used as a general term.
SA crystal dislocations
SA crystal systems
SA crystal zoning
SA crystallization
SA crystallography
SA defects
SA grain boundaries
SA intergrowths
SA minerals
SA phenocrysts
SA piezoelectric properties
SA point defects
SA reaction rims
SA surface defects
SA xenocrysts

Cs
use cesium

Cs-134 (1985)
Autoposting of broader terms began in 1989. This term has multiple hierarchies.
BT1 radioactive isotopes
BT1 isotopes
BT2 cesium
BT2 alkali metals
BT2 metals

Cs-137 (1978)
Autoposting of broader terms began in 1989. This term has multiple hierarchies.
BT1 radioactive isotopes
BT1 isotopes
BT2 cesium
BT2 alkali metals
BT2 metals

CSAMT methods (1993)
Acronym.
UF controlled-source audio-magnetotelluric technique
BT audiomagnetotelluric methods
SA Earth-current methods
SA geophysical methods
SA geophysical surveys
SA magnetic methods
SA magnetotelluric methods
SA magnetotelluric surveys

CSDP
use Continental Scientific Drilling Program

CSM
   use Colorado School of Mines

Ctenodontida (1981)
   BT Bivalvia
   BT Mollusca
   BT Invertebrata

Ctenostomata (1978)
   BT Bryozoa
   BT Invertebrata

Cu
   use copper

Cuanza Basin (1978)
   River basin in W central Angola.
   UF Kwanza Basin
   BT Angola
   BT Central Africa
   BT Africa

Cuanza-Sul
   No longer a valid term for GeoRef.
   use Cuanza-Sul Angola

Cuanza-Sul Angola (1993)
   District on the South Atlantic in W Angola.
   UF Cuanza-Sul
   BT Angola
   BT Central Africa
   BT Africa

Cuba (1978)
   CO N195000N231500
      W0740000W0850000
   BT Greater Antilles
   BT Antilles
   BT West Indies
   BT Caribbean region
   NT Camaguey Cuba
   NT La Habana Cuba
   NT Pinar del Rio Cuba
   NT Santiago de Cuba
   NT Villa Clara Cuba
   SA Lucas Formation

cubanite (1978)
   UF chalmersite
   BT sulfides

cube spar
   use anhydrite

Cuddalore Series (1978)
   Upper Miocene to Pliocene.
   IN Index ages as applicable.
   BT Neogene
   BT Tertiary
   BT Cenozoic
   SA India
   SA Neyveli Lignite
   SA Pliocene
   SA Tamil Nadu India
   SA upper Miocene

Cuddapah
   No longer a valid term for GeoRef.
   See Cuddapah India.

Cuddapah Basin (1978)
   S central Andhra Pradesh. Also search Cuddapah.
   BT Andhra Pradesh India
   BT India
   BT Indian Peninsula
   BT Asia

Cuddapah India (1993)
   Town in SW Andhra Pradesh, S India.
   BT Andhra Pradesh India
   BT India
   BT Indian Peninsula
   BT Asia

Cuddapah System (1978)
   Now included in Purana Group. Subdivided into Kistna Series, Nallamalai Series, Cheyair Series and Papaghni Series. Also search Cuddapah.
   BT Precambrian
   SA Andhra Pradesh India
   SA India
   SA Kurnool System

Cue
   No longer a valid term for GeoRef.
   See Cue Australia.

Cue Australia (1993)
   Town in W central Western Australia.
   BT Western Australia
   BT Australia
   BT Australasia

Cuenca
   No longer a valid term for GeoRef. As of 1993, see Cuenca City Spain.

Cuenca City Spain (1993)
   Refers to only the city in E central Spain. Before 1993, also search Cuenca and Spain.
   BT Cuenca Spain
   BT New Castile Spain
   BT Castile Spain
   BT Spain
   BT Iberian Peninsula
   BT Southern Europe
   BT Europe

Cuenca Province
   No longer a valid term for GeoRef. As of 1993, see Cuenca Spain.

Cuenca Spain (1993)
   Refers only to the province in east-central Spain. From 1981-1992, also search Cuenca Province and Spain. Before 1981, also search Cuenca and Spain. For the city, see Cuenca City Spain.
   CO N391300N403800
      W0011000W0031300
   BT New Castile Spain
   BT Castile Spain
   BT Spain
   BT Iberian Peninsula
   BT Southern Europe
   BT Europe
   NT Cuenca City Spain
   SA Serrania de Cuenca

cuestas (1982)
   Autoposting of erosion features to this term began in 1989.
   BT erosion features
   SA cliffs
   SA geomorphology
   SA scarps
   SA slopes

Cuisian (1978)
   Europe. Above Ypresian, below Lutetian.
   BT lower Eocene
   BT Eocene
   BT Paleogene
   BT Tertiary
   BT Cenozoic

Culberson County
   Valid through 1988. Search in combination with state term. After 1988, use specific county-state term.

Culberson County Texas (1989)
   Extreme W Texas. Before 1989, also search Culberson County AND Texas.
   CO N304000N320000
      W1040300W1050000
   BT Texas
   BT United States
   NT Van Horn Texas

SA Marble Canyon

Culebra Dolomite Member (1989)
   Of Rustler Formation. SE New Mexico and W Texas.
   BT Permian
   BT Paleozoic
   SA New Mexico
   SA Rustler Formation
   SA Texas

Culm (1978)
   Provincial series, Europe.
   BT Carboniferous
   BT Paleozoic
   SA Avonian
   SA Dinantian

Culpeper Basin (1985)
   N Virginia. W Maryland.
   IN Index counties and states as applicable.
   CO N381500N391500
      W0771500W0781500
   BT United States
   SA Maryland
   SA Virginia

culture, living
   use living taxa

Cumana
   No longer a valid term for GeoRef. See Cumana Venezuela.

Cumana Venezuela (1993)
   City on the Caribbean Sea in Sucre, N Venezuela.
   BT Sucre Venezuela
   BT Venezuela
   BT South America

Cumberland
   No longer a valid term for GeoRef. As of 1993, see Cumberland England for the former county in England. As of 1989, for locations in the United States, use specific city-state term.

Cumberland Basin (1985)
   Maritime Provinces of Canada and adjacent North American Atlantic.
   IN Index provinces as applicable.
   SA Bay of Fundy
   SA Canada
   SA New Brunswick
   SA North American Atlantic
   SA Nova Scotia
   SA Prince Edward Island

Cumberland County
   Valid through 1988. Search in combination with state term. After 1988, use specific county-state term.

Cumberland County Illinois (1989)
   SE central Illinois. Before 1989, also search Cumberland County AND Illinois.
   CO N391000N392400
      W0880100W0882900
   BT Illinois
   BT United States

Cumberland County Kentucky (1989)
   S Kentucky. Before 1989, also search Cumberland County AND Kentucky.
   CO N363600N365600
      W0851200W0853500
   BT Kentucky
   BT United States

Cumberland County Maine (1989)
   SW Maine. Before 1989, also search Cumberland County AND Maine.
   CO N433400N440900
      W0695200W0705300
   BT Maine
   BT United States
   NT Portland Maine
   SA Saco River

Cumberland County New Jersey (1989)
   Bounded by Delaware Bay in S New Jersey. Before 1989, also search Cumberland County AND New Jersey.
   CO N391100N393400
      W0745200W0752500
   BT New Jersey
   BT United States

Cumberland County North Carolina (1989)
   S central North Carolina. Before 1989, also search Cumberland County AND North Carolina.
   CO N345000N351500
      W0783000W0790700
   BT North Carolina
   BT United States

Cumberland County Pennsylvania (1989)
   S Pennsylvania. Before 1989, also search Cumberland County AND Pennsylvania.
   CO N395700N402000
      W0765200W0773800
   BT Pennsylvania
   BT United States

Cumberland County Tennessee (1989)
   E central Tennessee. Before 1989, also search Cumberland County AND Tennessee.
   CO N354400N361000
      W0844200W0851800
   BT Tennessee
   BT United States

Cumberland County Virginia (1989)
   Central Virginia. Before 1989, also search Cumberland County AND Virginia.
   CO N371700N374500
      W0780500W0782700
   BT Virginia
   BT United States
   NT Cumberland Virginia

Cumberland England (1993)
   Former county in NW England. In 1974, became part of Cumbria. Before 1993, search Cumberland AND England.
   BT Cumbria England
   BT England
   BT Great Britain
   BT United Kingdom
   BT Western Europe
   BT Europe
   NT Bewcastle England
   SA Lake District

Cumberland Group (1989)
   E New Brunswick and N Nova Scotia.
   IN Index provinces as applicable.
   BT Pennsylvanian
   BT Carboniferous
   BT Paleozoic
   SA New Brunswick
   SA Nova Scotia

Cumberland Kentucky (1989)
City in SE Kentucky. Before 1989, search Cumberland AND Kentucky.
CO N365800N365800 W0825900W0825900
BT Harlan County Kentucky
BT Kentucky
BT United States

Cumberland Maryland (1989)
City in W Maryland. Before 1989, search Cumberland AND Maryland.
CO N394000N394000 W0784700W0784700
BT Allegany County Maryland
BT Maryland
BT United States

Cumberland Mountains
use Cumberland Plateau

Cumberland Peninsula (1978)
On easternmost Baffin Island on Davis Strait in Franklin District. Also search Cumberland AND peninsula.
IN Index district and island or region as applicable.
SA Baffin Island
SA Franklin District Northwest Territories
SA Northwest Territories

Cumberland Plateau (1978)
Southwesternmost division of the Appalachians. Also search Cumberland AND plateau.
IN Index states and Appalachians as applicable.
UF Cumberland Mountains
SA Alabama
SA Appalachian Plateau
SA Appalachians
SA Kentucky
SA Pine Mountain Window
SA Tennessee
SA United States
SA Virginia
SA West Virginia

Cumberland Rhode Island (1989)
Town in NE Rhode Island. Before 1989, search Cumberland AND Rhode Island.
CO N415800N415800 W0712600W0712600
BT Providence County Rhode Island
BT Rhode Island
BT United States

Cumberland Virginia (1989)
Village in central Virginia. Before 1989, search Cumberland County AND Virginia.
CO N373100N373100 W0781600W0781600
BT Cumberland County Virginia
BT Virginia
BT United States

Cumbria
No longer a valid term for GeoRef as of 1993. See Cumbria England.

Cumbria England (1993)
County in NW England. Formed in 1974 from Cumberland and Westmorland counties. Before 1993, also search Cumberland or Westmorland or Cumbria AND England as applicable.
CO N540500N551500 W0021000W0034000
BT England
BT Great Britain
BT United Kingdom
BT Western Europe
BT Europe
NT Alston Block
NT Cumberland England
NT Westmorland England
SA Howgill Fells
SA Lake District
SA Morecambe Bay

cummingtonite (1978)
BT clinoamphibole
BT amphibole group
BT chain silicates
BT silicates
SA anthophyllite

cumulates (1978)
SA igneous rocks
SA magmas

Cupido Formation (1989)
N Mexico and S Texas.
BT Upper Cretaceous
BT Cretaceous
BT Mesozoic
SA Mexico
SA Texas

Cupressaceae (1978)
Family.
BT Coniferales
BT gymnosperms
BT Spermatophyta
BT Plantae

cuprite (1978)
BT oxides
SA copper minerals
SA copper ores

Curaca River basin (1978)
Just S of the Sao Francisco River in NE Bahia, E Brazil.
BT Bahia Brazil
BT Brazil
BT South America

Curacao (1978)
Island in the Netherlands Antilles just N of W Venezuela.
BT Netherlands Antilles
BT Lesser Antilles
BT Antilles
BT West Indies
BT Caribbean region

Curie point (1978)
UF Curie temperature
SA heat flow
SA magnetic properties
SA magnetic susceptibility
SA paleomagnetism
SA thermoremanent magnetization

Curie temperature
use Curie point

curium (1978)
Autoposting of actinides and metals to this term began in 1989.
UF Cm
BT actinides
BT metals

current directions (1978)
UF directions, current
SA currents
SA ocean circulation
SA sedimentation

current lineations
use parting lineation

current markings (1978)
UF markings, current
BT bedding plane irregularities
BT sedimentary structures
SA flute casts
SA scour marks
SA tool marks

current partings
use parting lineation

current research (1978)
SA annual report
SA associations
SA geology
SA news
SA progress report
SA report
SA research
SA survey organizations

current transport
No longer a valid GeoRef index term. Before 1967, was used on level 2 in subfile T. See marine transport; stream transport; currents; paleocurrents.

currents (1978)
Now use ocean currents if applicable. As of 1993, electrical currents is no longer a narrower term.
NT bottom currents
NT convection currents
NT density currents
NT fluvial currents
NT ocean currents
NT turbidity currents
SA continental shelf
SA current directions
SA eddy flow
SA flowmeters
SA magnetic field
SA paleocurrents
SA upwelling

curricula (1978)
SA education

Curry County
Valid through 1988. Search in combination with state term. After 1988, use specific county-state term.

Curry County New Mexico (1989)
E New Mexico. Before 1989, also search Curry County AND New Mexico.
CO N341800N345600 W1030300W1034400
BT New Mexico
BT United States

Curry County Oregon (1989)
SW Oregon. Before 1989, also search Curry County AND Oregon.
CO N420000N425700 W1234400W1243500
BT Oregon
BT United States
NT Sixes River

curves, traveltime
use traveltime curves

CUSMAP (1985)
Acronym. U. S. Geological Survey mineral resource exploration and mapping project. Part of a data base.
IN Index states as applicable.
UF Conterminous United States Mineral Assessment Program
SA mineral exploration
SA programs
SA U. S. Geological Survey
SA United States

Custer County
Valid through 1988. Search in combination with state term. After 1988, use specific county-state term.

Custer County Colorado (1989)
S central Colorado. Before 1989, also search Custer County AND Colorado.
CO N375400N381600 W1050200W1054500
BT Colorado
BT United States
SA Wet Mountains

Custer County Idaho (1989)
Central Idaho. Before 1989, also search Custer County AND Idaho.
CO N433500N445400 W1131900W1151800
BT Idaho
BT United States
NT Borah Peak
NT Challis Idaho
NT Mackay Idaho
SA Borah Peak earthquake 1983
SA Chilly Buttes
SA Lost River Fault
SA Lost River Range
SA Pioneer Mountains
SA Sawtooth Range

Custer County Montana (1989)
SE Montana. Before 1989, also search Custer County AND Montana.
CO N454700N465300 W1044500W1061000
BT Montana
BT United States

Custer County Nebraska (1989)
Central Nebraska. Before 1989, also search Custer County AND Nebraska.
CO N410300N414400 W0991200W1001800
BT Nebraska
BT United States

Custer County Oklahoma (1989)
W Oklahoma. Before 1989, also search Custer County AND Oklahoma.
CO N352800N354800 W0983800W0992300
BT Oklahoma
BT United States

Custer County South Dakota (1989)
SW South Dakota. Before 1989, also search Custer County AND South Dakota.
CO N432800N435200 W1024000W1040400
BT South Dakota
BT United States

customs duty (1981)
SA economics

cut and fill (1978)
BT planar bedding structures
BT sedimentary structures

cut and fill mining
use cut-and-fill mining

cut-and-fill mining (1985)
UF cut and fill mining
UF mining, cut-and-fill
BT mining
SA backfill
SA mines
SA mining geology
SA surface mining
SA underground mining

Cutch
No longer a valid term for GeoRef. Former Indian state. See Cutch India.

Cutch India (1993)
District in E Gujarat. Rann of Cutch is to the N and NE. Before 1993, also search Cutch. Before 1978, also search Kutch.
UF Kutch India
BT Gujarat India
BT India
BT Indian Peninsula
BT Asia
NT Bhuj India

cuticles (1993)
SA Arthropoda
SA chitin
SA morphology
SA Plantae

cutinite (1981)
BT exinite
BT macerals

Cutler Formation (1978)
Includes Halgaito Tongue, Cedar Mesa Sandstone Tongue, Organ Rock Tongue, White Rim Sandstone Member, DeChelly Sandstone Member, Hoskinnini Tongue, White River Sandstone, Rico transition facies. NE Arizona, SW Colorado, NW New Mexico, and SE Utah.
BT Permian
BT Paleozoic
SA Arizona
SA Colorado
SA New Mexico
SA Utah

cutoff grade (1982)
SA economic geology
SA economics
SA feasibility studies
SA grade
SA ore grade
SA profitability

cutoff rigidities
As of 1993, no longer a valid term for GeoRef. Usually out-of-scope. Used with interplanetary space and magnetosphere.

Cuttack
No longer a valid term for GeoRef. See Cuttack India.

Cuttack India (1993)
City in E Orissa, E India.
BT Orissa India
BT India
BT Indian Peninsula
BT Asia

cuttings (1978)
UF drill cuttings
SA boreholes
SA cores
SA well-logging
SA wells

Cuvier abyssal plain (1978)
Off Western Australia. Also search Cuvier.
BT Indian Ocean
SA DSDP Site 263

Cuyahoga County
Valid through 1988. Search in combination with state term. After 1988, use specific county-state term.

Cuyahoga County Ohio (1989)
N Ohio. Before 1989, also search Cuyahoga County AND Ohio.
CO N411700N414000 W0812000W0815800
BT Ohio
BT United States
NT Cleveland Ohio

Cuyahoga Formation (1993)
Lower Mississippian. Ohio and W Pennsylvania, New York.
BT Lower Mississippian
BT Mississippian
BT Carboniferous
BT Paleozoic
SA New York
SA Ohio
SA Pennsylvania

Cuyama Basin (1989)
S California.
BT California
BT United States

CYANA (1989)
French submersible.
SA marine geology
SA oceanography
SA research vessels
SA submersibles

cyanides (1982)
SA geochemistry
SA pollutants
SA pollution
SA toxic materials

cyanobacteria
use Cyanophyta

Cyanophyta (1978)
Used for Schizophyta which includes blue and blue-green algae. Autoposting of thallophytes and microfossils to this term began in 1990. This term has multiple hierarchies.
UF blue algae
UF blue-green algae
UF cyanobacteria
UF Schizophyta
BT1 algae
BT1 thallophytes
BT1 Plantae
BT2 algae
BT2 microfossils
NT Renalcis
SA Collenia
SA prokaryotes

Cycadales (1978)
Autoposting of this term began in 1978.
BT gymnosperms
BT Spermatophyta
BT Plantae
NT Ptilophyllum
SA Pachypteris
SA Taeniopteris

Cycadofilicales (1981)
BT gymnosperms
BT Spermatophyta
BT Plantae

Cyclades (1978)
Group of islands in S Aegean Sea. Also a Greek department. This term has multiple hierarchies.
BT1 Greek Aegean Islands
BT1 Greece
BT1 Southern Europe
BT1 Europe
BT2 Greek Aegean Islands
BT2 Aegean Islands
BT2 Mediterranean region
NT Milos
NT Naxos
NT Santorin
NT Thera

cycles (1978)
Use geochemical cycle or hydrologic cycle where applicable.
SA atmosphere
SA Brazilian Cycle
SA cyclic processes
SA cyclothems
SA erosion cycle
SA geochemical cycle
SA hydrologic cycle
SA paleoclimatology
SA punctuated aggradational cycles
SA solar cycles
SA Wilson cycle

cyclic
A valid level 2 term through 1977.

cyclic loading (1978)
BT loading
SA engineering geology
SA load tests

cyclic processes (1978)
Before 1978, also search sedimentation AND cyclic.
SA cycles
SA cyclothems
SA periodicity
SA processes
SA punctuated aggradational cycles
SA sedimentation

Cyclocystoidea (1978)
BT Echinozoa
BT Echinodermata
BT Invertebrata

cyclosilicates
use ring silicates

Cyclostomata (1978)
Autoposting of Bryozoa to this term ended in 1989.
SA Agnatha
SA Bryozoa
SA Cystoporata

cyclothems (1978)
Autoposting of this term began in 1978.
BT planar bedding structures
BT sedimentary structures
NT megacyclothems
SA cycles
SA cyclic processes
SA rhythmic bedding
SA rhythmite
SA sedimentation

cylindrical folds (1978)
Before 1978, search folds AND cylindrical.
UF cylindroidal fold
BT folds

cylindrical structures (1978)
BT sedimentary structures

cylindrite (1978)
BT sulfantimonates
BT sulfosalts

cylindroidal fold
use cylindrical folds

cymrite (1978)
BT sheet silicates
BT silicates
SA aluminosilicates

Cynodontia (1978)
Infraorder.
BT Therapsida
BT Synapsida
BT Reptilia
BT Tetrapoda
BT Vertebrata
BT Chordata

Cynomorpha
As of 1993, no longer a valid term for GeoRef.
use Cercopithecidae

Cyperaceae (1989)
Genus.
BT Monocotyledoneae
BT angiosperms
BT Spermatophyta
BT Plantae
SA miospores

Cypress Hills Formation (1978)
BT Oligocene
BT Paleogene
BT Tertiary
BT Cenozoic
SA Canada
SA Saskatchewan

Cyprididae (1978)
Family. As of 1990, microfossils, Crustacea, Mandibulata and Arthropoda are autoposted to this term. This term has multiple hierarchies.
BT1 Cypridocopina
BT1 Podocopida
BT1 Ostracoda
BT1 Crustacea
BT1 Mandibulata
BT1 Arthropoda
BT1 Invertebrata
BT2 Cypridocopina
BT2 Podocopida
BT2 Ostracoda
BT2 microfossils
NT Candona

Cypridocopina (1981)
Includes the families Cyprididae, Cyclocyprididae, Eucandonidae, Ilyocyprididae, Notodromadidae, Paracyprididae, and Pontocyprididae. Autoposting of this term to narrower terms began in 1985. Autoposting of Podocopida to this term began in 1989. As of 1990, microfossils, Crustacea, Mandibulata and Arthropoda are autoposted to this term. This term has multiple hierarchies.
BT1 Podocopida
BT1 Ostracoda
BT1 Crustacea
BT1 Mandibulata
BT1 Arthropoda
BT1 Invertebrata
BT2 Podocopida
BT2 Ostracoda
BT2 microfossils
NT Cyprididae

Cyprinidae (1978)
Family.
BT Teleostei
BT Actinopterygii
BT Osteichthyes
BT Pisces
BT Vertebrata
BT Chordata

Cyprus (1978)
Island nation in the Mediterranean Sea.
CO N343000N354000 E0343000E0323000
BT Middle East
BT Asia
NT Agrokipia Deposit
NT Troodos Massif
SA Mediterranean region
SA Mediterranean Sea
SA Near East
SA Troodos Ophiolite

cyprus-type (1989)

SA copper ores
SA lead-zinc deposits
SA massive deposits
SA mineral deposits, genesis

Cyrenaica (1978)
Easternmost Libya. Former province under the Italians.
BT Libya
BT North Africa
BT Africa

Cyrtodontida (1981)
Autoposting of this term to Arcina, Ostreacea and Pteriina ended in 1989.
BT Bivalvia
BT Mollusca
BT Invertebrata
SA Arcina

cyrtolite (1978)
Autoposting of nesosilicates began in 1985.
BT nesosilicates
BT orthosilicates
BT silicates
SA zircon

Cyrtospirifer (1978)
BT Spiriferida
BT Articulata
BT Brachiopoda
BT Invertebrata

Cystoidea (1978)
BT Crinozoa
BT Echinodermata
BT Invertebrata

Cystoporata (1978)
BT Bryozoa
BT Invertebrata
SA Cyclostomata

cysts
As of 1993, for cysts of dinoflagellates, use Dinoflagellata and morphology. Use palynomorphs if appropriate.

Cytheracea (1978)
Autoposting of this term to Loxoconcha began in 1989. Autoposting of microfossils, Crustacea, Mandibulata, and Arthropoda to this term began in 1990. This term has multiple hierarchies.
BT1 Cytherocopina
BT1 Podocopida
BT1 Ostracoda
BT1 Crustacea
BT1 Mandibulata
BT1 Arthropoda
BT1 Invertebrata
BT2 Cytherocopina
BT2 Podocopida
BT2 Ostracoda
BT2 microfossils
NT Cytheridae
NT Leptocythere
NT Loxoconcha
NT Trachyleberididae

Cytherella (1978)
Genus. Autoposting of Cytherocopina to this term ended in 1989. Autoposting of Platycopida began in 1989. As of 1990, microfossils, Crustacea, Mandibulata and Arthropoda are autoposted to this term. This term has multiple hierarchies.
BT1 Cytherellidae
BT1 Platycopida
BT1 Podocopida
BT1 Ostracoda
BT1 Crustacea
BT1 Mandibulata

BT1 Arthropoda
BT1 Invertebrata
BT2 Cytherellidae
BT2 Platycopida
BT2 Podocopida
BT2 Ostracoda
BT2 microfossils

Cytherellidae (1978)
Autoposting of Cytherocopina to this term ended in 1989. Autoposting of Platycopida began in 1989. As of 1990, microfossils, Crustacea, Mandibulata and Arthropoda are autoposted to this term. This term has multiple hierarchies.
BT1 Platycopida
BT1 Podocopida
BT1 Ostracoda
BT1 Crustacea
BT1 Mandibulata
BT1 Arthropoda
BT1 Invertebrata
BT2 Platycopida
BT2 Podocopida
BT2 Ostracoda
BT2 microfossils
NT Cytherella
NT Cytherelloidea

Cytherelloidea (1978)
Genus. Autoposting of Cytherocopina to this term ended in 1989. Autoposting of Platycopida began in 1989. As of 1990, microfossils, Crustacea, Mandibulata and Arthropoda are autoposted to this term. This term has multiple hierarchies.
BT1 Cytherellidae
BT1 Platycopida
BT1 Podocopida
BT1 Ostracoda
BT1 Crustacea
BT1 Mandibulata
BT1 Arthropoda
BT1 Invertebrata
BT2 Cytherellidae
BT2 Platycopida
BT2 Podocopida
BT2 Ostracoda
BT2 microfossils

Cytheridae (1978)
Autoposting of microfossils, Crustacea, Mandibulata, and Arthropoda to this term began in 1990. This term has multiple hierarchies.
BT1 Cytheracea
BT1 Cytherocopina
BT1 Podocopida
BT1 Ostracoda
BT1 Crustacea
BT1 Mandibulata
BT1 Arthropoda
BT1 Invertebrata
BT2 Cytheracea
BT2 Cytherocopina
BT2 Podocopida
BT2 Ostracoda
BT2 microfossils

Cytherocopina (1981)
Includes Superfamily Cytheracea. Autoposting of this term to narrower terms began in 1985. Autoposting of Podocopida to this term began in 1989. As of 1990, microfossils, Crustacea, Mandibulata and Arthropoda are autoposted to this term. This term has multiple hierarchies.
BT1 Podocopida
BT1 Ostracoda
BT1 Crustacea

BT1 Mandibulata
BT1 Arthropoda
BT1 Invertebrata
BT2 Podocopida
BT2 Ostracoda
BT2 microfossils
NT Cytheracea

Czech and Slovak Federal Republic
use Czechoslovakia

Czech Bohemian Forest (1981)
Mountain range in W part of country. Indicates that part of the Bohemian Forest that lies in Czechoslovakia. Also search Czechoslovakia AND Bohemian Forest.
CO N483000N500300
   E0150000E0123000
BT Czechoslovakia
BT Central Europe
BT Europe

Czech Erzgebirge (1981)
Autoposting of Czechoslovakia to this term began in 1989. As of 1990, Central Europe and Europe are autoposted to this term. This term has multiple hierarchies.
CO N494500N505000
   E0143500E0121000
BT1 Czechoslovakia
BT1 Central Europe
BT1 Europe
BT2 Erzgebirge
BT2 Central Europe
BT2 Europe

Czech Republic (1993)
Before 1993, also search Bohemia AND Moravia.
IN Index Bohemia and/or Moravia as applicable.
BT Czechoslovakia
BT Central Europe
BT Europe
NT Bohemia
NT Moravia
SA Sudeten Mountains

Czech Sudeten Mountains (1981)
Autoposting of Czechoslovakia to this term began in 1989. As of 1990, Central Europe and Europe are autoposted to this term. This term has multiple hierarchies.
CO N493000N510000
   E0181500E0142000
BT1 Czechoslovakia
BT1 Central Europe
BT1 Europe
BT2 Sudeten Mountains
BT2 Central Europe
BT2 Europe
SA Izera Mountains

Czechoslovakia (1978)
CO N473000N510000
   E0224500E0120000
UF Czech and Slovak Federal Republic
BT Central Europe
BT Europe
NT Barrandian Basin
NT Berounka System
NT Bohemian Basin
NT Choc Nappe
NT Czech Bohemian Forest
NT Czech Erzgebirge
NT Czech Republic
NT Czech Sudeten Mountains
NT Slovakia
NT Slovakian Carpathians
NT Slovakian Pannonian Basin
NT Sudeten
NT Veporides

SA Beshchady Mountains
SA Bohemian Massif
SA Carpathian Foredeep
SA Carpathian Foreland
SA Carpathians
SA Danube River
SA Danube Valley
SA Elbe River
SA Elbe Valley
SA Erzgebirge
SA Izera Mountains
SA Karkonosze Mountains
SA Krizna Nappe
SA Moldanubian
SA Morava River valley
SA North Sudetic Basin
SA Oder Valley
SA Pannonian Basin
SA Pieniny Klippen Belt
SA Silesia
SA Snieznik
SA Spis
SA Subcarpathians
SA Sudeten Mountains
SA Sudetic Basin
SA Tatra Mountains
SA Tieschitz Meteorite
SA Vienna Basin
SA Western Carpathians

Czekanowskiales (1978)
BT Ginkgoales
BT gymnosperms
BT Spermatophyta
BT Plantae

Czestochowa
No longer a valid term for GeoRef. As of 1993 see Czestochowa Poland.

Czestochowa Poland (1993)
Province and city in S Poland. Before 1993 also search Czestochowa AND Poland.
BT Poland
BT Central Europe
BT Europe
SA Cracow-Czestochowa Jura
SA Czestochowa-Zawiercie Basin

Czestochowa-Zawierce Basin
As of 1989, no longer a valid term for GeoRef.
use Czestochowa-Zawiercie Basin

Czestochowa-Zawiercie Basin (1989)
Czestochowa and NE Katowice provinces, S Poland. Before 1989, also search Czestochowa-Zawierce Basin.
IN Index provinces as applicable.
UF Czestochowa-Zawierce Basin
BT Poland
BT Central Europe
BT Europe
SA Czestochowa Poland
SA Katowice Poland
SA Zawiercie Poland

D

D-J Basin
use Denver Basin

D-region
As of 1993, no longer a valid term for GeoRef. Usually out-of-scope. Used with ionosphere.

D/H (1978)

Autoposting of broader terms began in 1989. Deuterium was autoposted to this term 1989-1992. This term has multiple hierarchies.
BT1 stable isotopes
BT1 isotopes
BT2 hydrogen
SA deuterium
SA isotope ratios
SA tracers

Da Hinggan Ling (1993)
Mountainous system in N Inner Mongolia. Before 1993, also search Khingan, Khingan Range, Greater Khingan Range, Da Hingan Ling, and Daxinganling.
UF Daxinganling
UF Great Khingan Range
UF Greater Khingan Mountains
UF Khingan Range
BT Inner Mongolia China
BT China
BT Far East
BT Asia

Da Lat
use Dalat Vietnam

Da-ching Field
use Daqing Field

Dabie Mountains (1993)
South-central China, W Hubei, SE Anhui, SW Henan. Before 1993, also search Dabieshan and Dabie Shan.
CO N283000N320000
 E1172000E1140000
BT China
BT Far East
BT Asia
SA Anhui China
SA Henan China
SA Hubei China

Dachang Deposit (1993)
Nandan County, NW Guangxi.
SA antimony ores
SA gold ores
SA Guangxi China
SA polymetallic ores
SA tin ores

Dacht-e-Nawar
use Dasht-i-Nawar

Dacian (1978)
Europe.
BT Pliocene
BT Neogene
BT Tertiary
BT Cenozoic

Dacian Basin (1978)
Region roughly covering modern Romania.
BT Romania
BT Southern Europe
BT Europe

dacite
As of 1989, no longer a valid term for GeoRef. From 1981 to 1988, dacites and volcanic rocks were autoposted to this term.
use dacites

dacite porphyry (1978)
BT dacites
BT volcanic rocks
BT igneous rocks
SA porphyry

dacites (1981)
Before 1989, also search dacite. Autoposting of this term to dacite ended in 1989.
UF dacite
BT volcanic rocks
BT igneous rocks
NT dacite porphyry
SA dacitic composition
SA metadacite

dacitic composition (1982)
Before 1982, also search dacitic.
SA composition
SA dacites
SA igneous rocks

Dacryoconarida (1978)
Order.
BT problematic fossils
SA Mollusca

Dactylioceratidae (1978)
BT Ammonoidea
BT Tetrabranchiata
BT Cephalopoda
BT Mollusca
BT Invertebrata

Dade County
Valid through 1988. Search in combination with state term. After 1988, use specific county-state term.

Dade County Florida (1989)
S Florida. Before 1989, also search Dade County AND Florida.
CO N251000N260000
 W0800800W0805400
BT Florida
BT United States
NT Biscayne Bay
NT Miami Florida
SA Biscayne Aquifer

Dade County Georgia (1989)
Extreme NW Georgia. Before 1989, also search Dade County AND Georgia.
CO N343800N345900
 W0852500W0853800
BT Georgia
BT United States

Dade County Missouri (1989)
SW Missouri. Before 1989, also search Dade County AND Missouri.
CO N371800N373600
 W0933900W0940500
BT Missouri
BT United States

Dadoxylon (1978)
Genus.
BT gymnosperms
BT Spermatophyta
BT Plantae
SA Coniferales
SA Cordaitales
SA Ginkgoales
SA Glossopteris

Dagestan
No longer a valid term for GeoRef. As of 1993, see Dagestan Russian Federation.

Dagestan Russian Federation (1993)
Former Autonomous Soviet Socialist Republic on W shore of Caspian Sea. Before 1978, also search Daghestan. This term has multiple hierarchies.
CO N410000N450000
 E0470000E0460000
UF Daghestan Russian Federation
BT1 Russian Federation
BT1 Commonwealth of Independent States
BT2 Europe
NT Makhachkala Russian Federation

Daggett County
Valid through 1988. Search in combination with state term. After 1988, use specific county-state term.

Daggett County Utah (1989)
NE Utah. Before 1989, also search Daggett County AND Utah.
CO N404000N410000
 W1090500W1100000
BT Utah
BT United States

Daghestan Russian Federation
use Dagestan Russian Federation

dahllite (1978)
This term has multiple hierarchies.
BT1 carbonates
BT2 phosphates
SA carbonate apatite

Dahomey
A valid index term through 1976.
use Benin

Daiichi-Kashima Seamount
use Kashima Seamount

Daiiti-Kashima Seamount
use Kashima Seamount

daily variation
use diurnal variations

Daito Ridge (1981)
NW Philippine Sea.
CO N215000N250000
 E1360000E1313000
BT Philippine Sea
BT West Pacific
BT Pacific Ocean
SA DSDP Site 445

Dakar
No longer a valid term for GeoRef. As of 1993 see Dakar Senegal.

Dakar Senegal (1993)
City on the Atlantic Ocean. Before 1993 also search Dakar AND Senegal.
BT Senegal
BT West Africa
BT Africa

Dakhla Oasis (1989)
Oasis in Western Desert of central Egypt.
CO N243000N270000
 E0300000E0280000
BT Egypt
BT North Africa
BT Africa
SA Western Desert

Dakhla Shale (1989)
Upper Cretaceous to lower Tertiary.
IN Index ages as applicable.
SA Egypt
SA lower Tertiary
SA Upper Cretaceous

Dakota Formation (1978)
E Colorado, NE New Mexico, North Dakota, Kansas, Minnesota, SE Montana, Nebraska, South Dakota, W Oklahoma, E Wyoming.
BT Cretaceous
BT Mesozoic
SA Colorado
SA Kansas
SA Mesa Rica Sandstone
SA Minnesota
SA Montana
SA Moxa Arch
SA Nebraska
SA New Mexico
SA North Dakota
SA Oklahoma
SA South Dakota
SA Wyoming

Dalarna (1978)
Region in W central Sweden.
UF Dalarne
UF Dalecarlia
BT Sweden
BT Scandinavia
BT Western Europe
BT Europe

Dalarne
use Dalarna

Dalat
No longer a valid term for GeoRef. As of 1993 see Dalat Vietnam.

Dalat Vietnam (1993)
Town NE of Saigon in S Vietnam. Before 1993 also search Dalat AND Vietnam.
UF Da Lat
BT Vietnam
BT Far East
BT Asia

Dalecarlia
use Dalarna

Dalgaranga (1978)
Region. Also mesosiderite meteorite and meteorite crater. As of 1993, use Dalgaranga Australia for the place. Use Dalgaranga Meteorite for the meteorite.
SA mesosiderite
SA Western Australia

Dalhart Basin (1985)
Texas Panhandle.
BT Texas
BT United States

Dallas
Valid through 1988. Search in combination with state term. After 1988, use specific city-state term.

Dallas County
Valid through 1988. Search in combination with state term. After 1988, use specific county-state term.

Dallas County Alabama (1989)
S central Alabama. Before 1989, also search Dallas County AND Alabama.
CO N320400N324000
 W0864900W0873000
BT Alabama
BT United States

Dallas County Arkansas (1989)
S central Arkansas. Before 1989, also search Dallas County AND Arkansas.
CO N334600N340700
 W0922300W0925700
BT Arkansas
BT United States

Dallas County Iowa (1989)
Central Iowa. Before 1989, also search Dallas County AND Iowa.
CO N413000N415200
 W0934900W0941700
BT Iowa
BT United States

Dallas County Missouri (1989)
SW central Missouri. Before 1989, also search Dallas County AND Missouri.

CO N372500N375400
W0925100W0931200
BT Missouri
BT United States

Dallas County Texas (1989)
N Texas. Before 1989, also search Dallas County AND Texas.
CO N323200N325700
W0963000W0970300
BT Texas
BT United States
NT Dallas Texas

Dallas Texas (1989)
City in NE central Texas. Before 1989, search Dallas AND Texas.
CO N324700N324700
W0964800W0964800
BT Dallas County Texas
BT Texas
BT United States

Dalmatia (1978)
Region, including many islands, which extends along the Adriatic Sea from Zadar on the N to near the Albanian border on the S. Sometimes the name is applied to most of the Yugoslav coast.
BT Yugoslavia
BT Southern Europe
BT Europe

Dalradian (1978)
Ireland and Scotland.
IN Index ages as applicable.
UF Dalradian Supergroup
SA Cambrian
SA Devonian
SA Ireland
SA Precambrian
SA Scotland
SA Silurian

Dalradian Supergroup
use Dalradian

dam sites
Not a valid term for GeoRef. Before 1971, was used in subfiles E, G and N.
use dams

damage (1981)
Used for damage resulting from natural phenomena.
SA earthquakes
SA effects
SA engineering geology
SA fires
SA geologic hazards

damage, radiation
use radiation damage

Damara Orogeny (1993)
Upper Proterozoic to lower Paleozoic. Southern Africa.
IN Index age or ages as applicable.
SA lower Paleozoic
SA Namibia
SA orogeny
SA South Africa
SA Southern Africa
SA tectonics
SA upper Proterozoic

Damara System (1978)
Includes Khomas Series, Marble Series, and Quartzite Series, Chuos Tillite, and Otavi Series. In W Damaraland region. NW of Walvis Bay.
BT Proterozoic
BT upper Precambrian
BT Precambrian
SA Namibia

Damascus
No longer a valid term for GeoRef. As of 1993 see Damascus Syria.

Damascus Syria (1993)
City in SW Syria. Before 1993 also search Damascus AND Syria.
BT Syria
BT Middle East
BT Asia

Damodar Valley (1978)
River valley.
IN Index states as applicable.
BT India
BT Indian Peninsula
BT Asia
SA Bihar India
SA West Bengal India

Dampier Ridge (1978)
NE of Sidney in N Tasman Sea.
BT Tasman Sea
BT West Pacific
BT Pacific Ocean

damping
As of 1981, no longer a valid term for GeoRef.
use attenuation

dams (1978)
Before 1965, also search dams and dam sites. Before 1971, also search dam sites.
UF dam sites
UF dams and dam sites
UF damsites
NT arch dams
NT concrete dams
NT earth dams
NT rockfill dams
NT tailings dams
SA construction
SA design
SA embankments
SA engineering geology
SA foundations
SA geomembranes
SA geotextiles
SA ground-water dams
SA grouting
SA land subsidence
SA reinforced materials
SA reservoirs
SA rock mechanics
SA seismic response
SA site exploration
SA slope stability
SA soil mechanics
SA water storage

dams and dam sites
No longer a valid term for GeoRef. Included use from 1949 to 1964 in subfile N.
use dams

damsites
use dams

Damuda Series (1978)
Forms middle part of lower Gondwanas. Subdivided into three stages: Raniganj (series), Ironstone shales, and lower Damudas, afterwards called Barakars.
BT Permian
BT Paleozoic
SA Barakar Stage
SA Bihar India
SA India
SA Karharbari Stage
SA Madhya Pradesh India
SA Panchet Series
SA Raniganj Formation

SA Talchir Series

Dan River basin (1978)
IN Index states as applicable.
BT United States
SA North Carolina
SA Virginia

Danakil
use Afar

Danakil Depression
use Afar Depression

danburite (1978)
BT framework silicates
BT silicates

Dane County
Valid through 1988. Search in combination with state term. After 1988, use specific county-state term.

Dane County Wisconsin (1989)
S Wisconsin. Before 1989, also search Dane County AND Wisconsin.
CO N425000N431700
W0890000W0895100
BT Wisconsin
BT United States
NT Madison Wisconsin
SA Lake Mendota

Dangerous Islands
use Tuamotu Islands

Danian (1978)
Europe. Above Maestrichtian (Cretaceous), below Montian.
BT lower Paleocene
BT Paleocene
BT Paleogene
BT Tertiary
BT Cenozoic
NT Niniyur Group
SA Ekofisk Formation

Daniel's Harbour
No longer a valid term for GeoRef.
use Daniel's Harbour Newfoundland

Daniel's Harbour Newfoundland (1993)
Village on upper Gulf of Saint Lawrence in NW Newfoundland. Before 1993, also search Daniel's Harbour AND Newfoundland.
UF Daniel's Harbour
BT Newfoundland Island
BT Newfoundland
BT Eastern Canada
BT Canada

Dankalia
use Afar

Danube Delta (1978)
On the Black Sea primarily in Romania.
IN Index Romania and/or Ukraine as applicable.
UF Danube River delta
BT Europe
SA Romania
SA Ukraine

Danube Plain (1978)
Covers the Walachian Plain and the N Bulgaria lowlands. Also search Danube.
IN Index countries as applicable.
BT Southern Europe
BT Europe
SA Bulgaria
SA Romania
SA Walachia

Danube River (1978)
Flows E from the Black Forest and turns S just N of Budapest. It then flows S, SE, N, and finally E into the Black Sea. Also search Danube.
IN Index countries as applicable.
UF Donau River
UF Duna River
UF Dunai River
UF Dunarea River
UF Dunau River
BT Europe
SA Austria
SA Bulgaria
SA Czechoslovakia
SA Germany
SA Hungary
SA Romania
SA Ukraine
SA Yugoslavia

Danube River delta
use Danube Delta

Danube Stade (1982)
Replaces Danube Stage. Before 1982, also search Danube Stage. Danube Stage was introduced in 1978.
UF Danube Stage
BT lower Pleistocene
BT Pleistocene
BT Quaternary
BT Cenozoic

Danube Stage
use Danube Stade

Danube Valley (1978)
River valley. Also search Danube.
IN Index countries as applicable.
BT Europe
SA Austria
SA Bulgaria
SA Czechoslovakia
SA Germany
SA Hungary
SA Romania
SA Ukraine
SA Yugoslavia

Danville
Valid through 1988. Search in combination with state term. After 1988, use specific city-state term.

Danville Virginia (1989)
City on the North Carolina border in S Virginia. Before 1989, search Danville AND Virginia.
CO N363400N363400
W0792500W0792500
BT Pittsylvania County Virginia
BT Virginia
BT United States

Danzig
use Gdansk Poland

Daonella (1978)
Autoposting of Pectinacea and Pteriina to this term began in 1989.
BT Pectinacea
BT Pteriina
BT Bivalvia
BT Mollusca
BT Invertebrata

Daqing Field (1985)
Oil field, Heilongjiang, Manchuria.
CO N460000N470000
E1260000E1250000
UF Da-ching Field
BT Heilongjiang China
BT China
BT Far East
BT Asia
SA oil and gas fields

Darasun
No longer a valid term for GeoRef. As of 1993, see Darasun Russian Federation.

Darasun Russian Federation (1993)
Town in S central Chita Oblast. N of Mongolia. Before 1993, also search Darasun. This term has multiple hierarchies.
BT1 Chita Russian Federation
BT1 Russian Federation
BT1 Commonwealth of Independent States
BT2 Chita Russian Federation
BT2 Asia

Darcy's law (1978)
SA aquifers
SA fluid phase
SA gases
SA ground water
SA hydrodynamics
SA permeability
SA petroleum
SA porosity
SA reservoir properties

Dare County
Valid through 1988. Search in combination with state term. After 1988, use specific county-state term.

Dare County North Carolina (1989)
NE North Carolina. Before 1989, also search Dare County AND North Carolina.
CO N351000N361500 W0752500W0760200
BT North Carolina
BT United States
NT Cape Hatteras

Darfur
No longer a valid term for GeoRef. As of 1993 see Darfur Sudan.

Darfur Sudan (1993)
Province in W Sudan. Before 1993 also search Darfur AND Sudan.
BT Sudan
BT East Africa
BT Africa

Darien (1978)
E part of isthmus between Gulf of Darien on E and Gulf of San Miguel on W.
BT Panama
BT Central America

Darjeeling
No longer a valid term for GeoRef. See Darjeeling India.

Darjeeling India (1993)
Town, hill station, and district near Sikkim border in extreme N West Bengal, E India. Before 1993, also search Darjeeling. Before 1978, also search Darjiling.
UF Darjiling India
BT West Bengal India
BT India
BT Indian Peninsula
BT Asia

Darjiling India
use Darjeeling India

Darling Downs (1978)
Tableland in SE Queensland.
BT Queensland Australia
BT Australia
BT Australasia

Darling Range (1978)
In SW Western Australia.
BT Western Australia
BT Australia
BT Australasia

Darmstadt
No longer a valid term for GeoRef. See Darmstadt Germany.

Darmstadt Germany (1993)
City in S Hesse.
BT Hesse Germany
BT Germany
BT Central Europe
BT Europe

Dartmoor (1978)
Tableland in Devonshire, SW England.
BT Devonshire England
BT England
BT Great Britain
BT United Kingdom
BT Western Europe
BT Europe

Darvaz (1978)
Peak in the W Pamirs in central Tadzhikistan. This term has multiple hierarchies.
UF Kaganovich Peak
BT1 Tadzhikistan
BT1 Asia
BT2 Tadzhikistan
BT2 Commonwealth of Independent States
BT3 Pamirs
BT3 Commonwealth of Independent States
BT4 Pamirs
BT4 Central Asia
BT4 Asia

Darvaza
No longer a valid term for GeoRef. As of 1993, see Darvaza Turkmenia.

Darvaza Range (1978)
Branch of Alai Range N of Afghanistan in S central Tadzhikistan. This term has multiple hierarchies.
IN Index mountains as applicable.
BT1 Tadzhikistan
BT1 Asia
BT2 Tadzhikistan
BT2 Commonwealth of Independent States
SA Alai Range
SA Pamirs

Darvaza Turkmenia (1993)
Town in Ashkhabad Oblast in S Turkmenia. This term has multiple hierarchies.
BT1 Ashkhabad Turkmenia
BT1 Turkmenia
BT1 Asia
BT2 Ashkhabad Turkmenia
BT2 Turkmenia
BT2 Commonwealth of Independent States

Darvel Bay (1978)
Inlet of Celebes Sea in Sabah in East Malaysia. This term has multiple hierarchies.
BT1 Sabah Malaysia
BT1 East Malaysia
BT1 Borneo
BT1 Far East
BT1 Asia
BT2 Sabah Malaysia
BT2 East Malaysia
BT2 Borneo
BT2 Malay Archipelago
BT3 Sabah Malaysia

BT3 East Malaysia
BT3 Malaysia
BT3 Far East
BT3 Asia
SA Celebes Sea

Darwin glass (1978)
UF queenstownite
BT tektites
SA glasses
SA impactite

Darwin, Charles (1985)
Also search Darwin AND biography.
UF Charles Darwin
SA biography
SA Darwinism

Darwinism (1989)
Theory on origin of species; evolution is the result of variation and survival of the fittest by natural selection.
SA biologic evolution
SA Darwin, Charles
SA life origin
SA natural selection
SA paleoecology
SA paleontology

Darwinula (1978)
As of 1990, microfossils, Crustacea, Mandibulata and Arthropoda are autoposted to this term. This term has multiple hierarchies.
BT1 Podocopida
BT1 Ostracoda
BT1 Crustacea
BT1 Mandibulata
BT1 Arthropoda
BT1 Invertebrata
BT2 Podocopida
BT2 Ostracoda
BT2 microfossils

Dashkesan
No longer a valid term for GeoRef. As of 1993, see Dashkesan Azerbaidzhan.

Dashkesan Azerbaidzhan (1993)
City in W Azerbaidzhan. This term has multiple hierarchies.
BT1 Azerbaidzhan
BT1 Europe
BT2 Azerbaidzhan
BT2 Commonwealth of Independent States

Dasht-e Bayaz
use Dasht-e-Bayaz

Dasht-e-Bayaz (1978)
Region in NE part of country. Locale of earthquake in 1968.
UF Dasht-e Bayaz
BT Iran
BT Middle East
BT Asia
SA earthquakes

Dasht-e-Lut Basin
use Lut Desert

Dasht-e-Nawar
use Dasht-i-Nawar

Dasht-i-Lut
use Lut Desert

Dasht-i-Nawar (1978)
Volcanic region in central Afghanistan.
UF Dacht-e-Nawar
UF Dasht-e-Nawar
BT Afghanistan
BT Indian Peninsula
BT Asia

Dastakert
No longer a valid term for GeoRef. As of 1993, see Dastakert Armenia.

Dastakert Armenia (1993)
Village in S Armenia. This term has multiple hierarchies.
BT1 Armenia
BT1 Commonwealth of Independent States
BT2 Armenia
BT2 Europe

Dasycladaceae (1978)
Autoposting of thallophytes and microfossils to this term began in 1990. This term has multiple hierarchies.
BT1 Chlorophyceae
BT1 Chlorophyta
BT1 algae
BT1 thallophytes
BT1 Plantae
BT2 Chlorophyceae
BT2 Chlorophyta
BT2 algae
BT2 microfossils

data (1978)
Used as a general term. Before 1975 also search analytical data.
SA data acquisition
SA data bases
SA data processing
SA data retrieval
SA DTA data
SA electron microscopy data
SA electron probe data
SA ion probe data
SA mineral data
SA neutron activation analysis data
SA neutron diffraction data
SA neutron probe data
SA new data
SA SEM data
SA TEM data
SA TGA data
SA X-ray data
SA X-ray diffraction data

data acquisition (1978)
UF acquisition, data
SA data
SA data processing
SA data storage

data analysis
As of 1981, no longer a valid term for GeoRef. From 1981 to 1984, see automatic data processing. After 1984, see data processing.

data bases (1978)
UF data systems
UF databases
SA CRIB
SA data
SA data handling
SA data processing
SA data retrieval
SA data storage
SA geographic information systems
SA GeoRef
SA Geoscan
SA information systems
SA National Coal Resources Data System
SA NAWDEX
SA Pascal

data handling (1978)
UF handling, data
SA data bases
SA data processing
SA data retrieval

SA information systems

data processing (1985)
Replaces automatic data processing. Used for computer applications and machine modelling of geological data. Before 1985, also search automatic data processing. Before 1972, also search computer methods; computer techniques.
UF automatic data processing
UF computer methods
UF electronic data processing
SA algorithms
SA analog simulation
SA artificial intelligence
SA automated analysis
SA BASIC
SA Bayesian analysis
SA computer languages
SA computer networks
SA computer programs
SA computers
SA CRIB
SA data
SA data acquisition
SA data bases
SA data handling
SA data retrieval
SA data storage
SA digital cartography
SA digital line graphs
SA digital simulation
SA digital terrain models
SA expert systems
SA Fortran
SA Fortran 77
SA Fortran IV
SA Fourier analysis
SA geographic information systems
SA GeoRef
SA Geoscan
SA graphic display
SA hand calculators
SA Hilbert transformations
SA holography
SA image enhancement
SA information systems
SA interactive techniques
SA IRIS network
SA kriging
SA linear programming
SA mathematical geology
SA mathematical methods
SA mathematical models
SA microcomputers
SA minicomputers
SA models
SA multichannel methods
SA NAWDEX
SA nomograms
SA Pascal
SA pattern recognition
SA PHREEQE
SA pixels
SA punch cards
SA Seabeam
SA simulation
SA stacking
SA statistical analysis
SA systems analysis
SA transfer functions
SA trend-surface analysis
SA well-logging
SA workstations

data retrieval (1978)
UF retrieval, data
SA computer networks
SA data
SA data bases
SA data handling
SA data processing
SA data storage

SA information systems

data storage (1978)
SA data acquisition
SA data bases
SA data processing
SA data retrieval
SA information systems
SA storage

data systems
use data bases

databases
use data bases

Date Creek basin (1985)
W central Arizona.
IN Index counties as applicable.
BT Arizona
BT United States
SA Yavapai County Arizona
SA Yuma County Arizona

dates (1978)
Used only to indicate specific absolute age dates.
SA absolute age
SA geochronology

Datil-Mogollon volcanic field (1981)
SW New Mexico.
CO N321000N343000
W1063000W1090300
UF Mogollon-Datil volcanic field
BT New Mexico
BT United States
SA Mogollon Mountains

dating, fission-track
use fission-track dating

dating, particle-track
use particle-track dating

datolite (1978)
Autoposting of nesosilicates began in 1985.
BT nesosilicates
BT orthosilicates
BT silicates

Dauphin County Pennsylvania (1989)
S central Pennsylvania. Before 1989, search Dauphin County AND Pennsylvania.
CO N400800N403900
W0763300W0770100
BT Pennsylvania
BT United States
NT Harrisburg Pennsylvania

Dauphine (1978)
Historical region and former province in SE France.
BT France
BT Western Europe
BT Europe
SA Bas-Dauphine
SA Devoluy

Dauphine Alps (1978)
W offshoot of Cottian Alps in SE France. This term has multiple hierarchies.
BT1 Western Alps
BT1 Alps
BT1 Europe
BT2 France
BT2 Western Europe
BT2 Europe
NT Pelvoux Massif
NT Vercors
SA Bas-Dauphine
SA Cottian Alps
SA Devoluy

Davenport
Valid through 1988. Search in combination with state term. After 1988, use specific city-state term.

Davenport Iowa (1989)
City on the Mississippi River in E Iowa. Before 1989 search Davenport AND Iowa.
CO N413200N413200
W0903600W0903600
BT Scott County Iowa
BT Iowa
BT United States

Davidson
Valid through 1988. Search in combination with state term. After 1988, use specific city-state term.

Davidson County
Valid through 1988. Search in combination with state term. After 1988, use specific county-state term.

Davidson County North Carolina (1989)
Central North Carolina. Before 1989, also search Davidson County AND North Carolina.
CO N353000N360200
W0800300W0802800
BT North Carolina
BT United States

Davidson County Tennessee (1989)
N central Tennessee. Before 1989, also search Davidson County AND Tennessee.
CO N355700N362400
W0863200W0870300
BT Tennessee
BT United States
NT Nashville Tennessee

Davidson North Carolina (1989)
Town on the South Carolina border in S North Carolina. Before 1989, search Davidson AND North Carolina.
CO N353000N353000
W0990500W0990500
BT Mecklenburg County North Carolina
BT North Carolina
BT United States

Davidsville Group (1978)
In NE Newfoundland.
BT Ordovician
BT Paleozoic
SA Canada
SA Newfoundland Island

Davis County
Valid through 1988. Search in combination with state term. After 1988, use specific county-state term.

Davis County Iowa (1989)
SE Iowa. Before 1989, also search Davis County AND Iowa.
CO N403500N405400
W0921100W0923800
BT Iowa
BT United States

Davis County Utah (1989)
N Utah. Before 1989, also search Davis County AND Utah.
CO N404700N410900
W1114500W1123300
BT Utah
BT United States
SA Great Salt Lake

Davis Mountains (1978)
Jeff Davis County in W Texas.

IN Index county or regions as applicable.
SA Jeff Davis County Texas

Davis Strait (1978)
Between SW Greenland and Baffin Island.
BT Atlantic Ocean
SA Godthaabsfjord
SA Kangerdlugssuaq
SA Labrador Current

Davos
No longer a valid term for GeoRef. As of 1993, see Davos Switzerland.

Davos Switzerland (1993)
Town consisting of Davos Platz and Davos Dorf in N central Graubunden. Before 1993, also search Davos AND Switzerland.
BT Graubunden Switzerland
BT Switzerland
BT Central Europe
BT Europe

Dawson (1978)
City in W central Yukon Territory.
BT Yukon Territory
BT Western Canada
BT Canada

Dawson County
Valid through 1988. Search in combination with state term. After 1988, use specific county-state term.

Dawson County Georgia (1989)
N Georgia. Before 1989, also search Dawson County AND Georgia.
CO N342000N343700
W0835900W0842200
BT Georgia
BT United States

Dawson County Nebraska (1989)
S central Nebraska. Before 1989, also search Dawson County AND Nebraska.
CO N404200N410300
W0992500W1001400
BT Nebraska
BT United States

Dawson County Texas (1989)
NW Texas. Before 1989, also search Dawson County AND Texas.
CO N323200N325700
W1014200W1021100
BT Texas
BT United States

dawsonite (1978)
BT carbonates

Daxinganling
use Da Hinggan Ling

DDT (1989)
Abbreviation used for the pesticide 1, 1, 1-trichloro-2, 2-bis(p-chlorophenyl) ethane and related compounds. Autoposting of chlorinated hydrocarbons to this term began in 1989.
UF chlorophenothane
UF dichloro-diphenyl-trichloro-ethane
UF dicophane
UF methoxychlor
BT chlorinated hydrocarbons
SA environmental geology
SA ethane
SA pesticides
SA trichloroethane
SA waste disposal

De Baca County
  Valid through 1988. Search in combination with state term. After 1988, use specific county-state term.
De Baca County New Mexico (1989)
  E New Mexico. Before 1989, also search De Baca County AND New Mexico.
  CO N340000N344600
    W1035500W1045300
  BT New Mexico
  BT United States

De Kalb County
  Valid through 1988. Search in combination with state term. After 1988, use specific county-state term.
De Kalb County Alabama (1989)
  NE Alabama. Before 1989, also search De Kalb County AND Alabama.
  CO N341300N345200
    W0853400W0860700
  BT Alabama
  BT United States

De Kalb County Georgia (1989)
  NW central Georgia. Before 1989, also search De Kalb County AND Georgia.
  CO N333700N335700
    W0840200W0842200
  BT Georgia
  BT United States
  SA Stone Mountain

De Kalb County Illinois (1989)
  N Illinois. Before 1989, also search De Kalb County AND Illinois.
  CO N413800N421100
    W0883800W0885700
  BT Illinois
  BT United States

De Kalb County Indiana (1989)
  NE Indiana. Before 1989, also search De Kalb County AND Indiana.
  CO N411500N413200
    W0844800W0851200
  BT Indiana
  BT United States

De Kalb County Missouri (1989)
  NW Missouri. Before 1989, also search De Kalb County AND Missouri.
  CO N394400N400300
    W0941200W0943600
  BT Missouri
  BT United States

De Kalb County Tennessee (1989)
  Central Tennessee. Before 1989, also search De Kalb County AND Tennessee.
  CO N354800N360800
    W0853800W0860400
  BT Tennessee
  BT United States

De Long Mountains (1985)
  NW Alaska. W part of the Brooks Range.
  CO N680000N690000
    W1570000W1650000
  UF DeLong Mountains
  BT Brooks Range
  BT Northern Alaska
  BT Alaska
  BT United States

De Soto County
  Valid through 1988. Search in combination with state term. After 1988, use specific county-state term.
De Soto County Florida (1989)
  S central Florida. Before 1989, also search De Soto County AND Florida.
  CO N270200N273500
    W0813600W0820200
  BT Florida
  BT United States
De Soto County Mississippi (1989)
  Extreme NW Mississippi. Before 1989, also search De Soto County AND Mississippi.
  CO N344300N350000
    W0894300W0901900
  BT Mississippi
  BT United States

De Twente
  use Twente

Dead Sea (1978)
  A salt lake lying in part of the Great Rift Valley and constituting the lowest point on the Earth's surface. Also search Dead Sea Rift.
  IN Index countries as applicable.
  BT Middle East
  BT Asia
  SA Dead Sea Rift
  SA Great Rift Valley
  SA Israel
  SA Jordan

Dead Sea Rift (1978)
  BT Middle East
  BT Asia
  SA Dead Sea
  SA Great Rift Valley
  SA Israel
  SA Jordan
  SA Red Sea Rift

Deadwood Formation (1978)
  Upper Cambrian and Lower Ordovician. In N Black Hills, conformably underlies Aladdin Sandstone. SE Montana, W South Dakota (Black Hills), and E Wyoming.
  IN Index ages as applicable.
  BT Paleozoic
  SA Lower Ordovician
  SA Montana
  SA South Dakota
  SA Upper Cambrian
  SA Wyoming

Deaf Smith County
  Valid through 1988. Search in combination with state term. After 1988, use specific county-state term.
Deaf Smith County Texas (1989)
  Extreme N Texas. Before 1989, also search Deaf Smith County AND Texas.
  CO N344500N352000
    W1021000W1030000
  BT Texas
  BT United States

death assemblages
  use thanatocenoses

Death Valley (1978)
  Valley in San Bernardino and Inyo counties, California. If applicable, also index/search Death Valley National Monument which includes a larger area extending into Nevada.
  IN Index counties as applicable.
  CO N353500N370000
    W1161500W1173000
  BT California
  BT United States
  SA Death Valley Fault
  SA Esmeralda County Nevada
  SA Furnace Creek
  SA Inyo County California
  SA Nevada
  SA Nye County Nevada
  SA San Bernardino County California

Death Valley Fault (1989)
  S California.
  IN Index counties as applicable.
  BT California
  BT United States
  SA Death Valley

Debrecen
  No longer a valid term for GeoRef. See Debrecen Hungary.

Debrecen Hungary (1993)
  City in E Hungary.
  BT Hungary
  BT Central Europe
  BT Europe

debris (1978)
  SA debris avalanches
  SA debris cones
  SA debris flows
  SA detritus
  SA geomorphology
  SA regolith

debris avalanches (1989)
  Autoposting of mass movements to this term began in 1989.
  BT mass movements
  SA avalanches
  SA debris
  SA debris flows
  SA geomorphology
  SA slope stability

debris cones (1982)
  SA ablation
  SA alluvial fans
  SA cones
  SA debris
  SA debris flows
  SA fluvial features
  SA geomorphology
  SA glacial features
  SA glacial geology
  SA glaciers
  SA landslides

debris flows (1978)
  Autoposting of mass movements to this term began in 1989.
  BT mass movements
  SA debris
  SA debris avalanches
  SA debris cones
  SA flows
  SA geomorphology
  SA slope stability

debris slopes
  use talus slopes

Decade of North American Geology
  use DNAG

Decapoda
  As of 1981, no longer a valid term for GeoRef.
  use Malacostraca

Decatur County
  Valid through 1988. Search in combination with state term. After 1988, use specific county-state term.
Decatur County Georgia (1989)
  SW Georgia. Before 1989, also search Decatur County AND Georgia.
  CO N304100N310500
    W0842300W0844800
  BT Georgia
  BT United States

Decatur County Indiana (1989)
  SE central Indiana. Before 1989, also search Decatur County AND Indiana.
  CO N390800N392700
    W0851800W0851200
  BT Indiana
  BT United States

Decatur County Iowa (1989)
  S Iowa. Before 1989, also search Decatur County AND Iowa.
  CO N403400N405400
    W0933400W0940100
  BT Iowa
  BT United States

Decatur County Kansas (1989)
  NW Kansas. Before 1989, also search Decatur County AND Kansas.
  CO N393200N400000
    W1001100W1004500
  BT Kansas
  BT United States
  SA Leoville Meteorite

Decatur County Tennessee (1989)
  W Tennessee. Before 1989, also search Decatur County AND Tennessee.
  CO N352300N355000
    W0875800W0881700
  BT Tennessee
  BT United States

Decaturville
  Valid through 1988. Search in combination with state term. After 1988, use specific city-state term.
Decaturville Missouri (1989)
  Village in central Missouri. Before 1989, search Decaturville AND Missouri.
  CO N375400N375400
    W0924400W0924400
  BT Camden County Missouri
  BT Missouri
  BT United States

decay, radioactive
  use radioactive decay

Decazeville
  No longer a valid term for GeoRef. See Decazeville France.

Decazeville France (1993)
  Town in S central France.
  BT Aveyron France
  BT France
  BT Western Europe
  BT Europe

Deccan Intertrappean
  use Intertrappean Beds

Deccan Intertrappean Beds
  use Intertrappean Beds

Deccan Intertrappean Series
  use Intertrappean Beds

Deccan Plateau (1978)

Triangular tableland between the Eastern and Western Ghats and between the Narmada River on the N and the Krishna River on the S. Also search Deccan.
  IN   Index states as applicable.
  UF   Dekkan Plateau
  BT   India
  BT   Indian Peninsula
  BT   Asia
  SA   Andhra Pradesh India
  SA   Indian Shield
  SA   Karnataka India
  SA   Lonar Crater
  SA   Madhya Pradesh India
  SA   Maharashtra India
  SA   Orissa India

Deccan Traps   (1978)
Upper Cretaceous to lower Eocene. Named after the step-like aspect of the flat topped hills of the Deccan Plateau.
  IN   Index ages as applicable.
  SA   Badami Series
  SA   India
  SA   Intertrappean Beds
  SA   K-T boundary
  SA   lower Eocene
  SA   Upper Cretaceous

Deception Island   (1978)
One of Britain's South Shetland Islands off N Antarctic Peninsula.
  BT   South Shetland Islands
  BT   Scotia Sea Islands
  BT   Antarctica

Dechenellidae   (1978)
  BT   Ptychopariida
  BT   Trilobita
  BT   Trilobitomorpha
  BT   Arthropoda
  BT   Invertebrata

decision-making   (1981)
  SA   legislation
  SA   policy
  SA   regulations

declination
  As of 1981, no longer a valid term for GeoRef. Use magnetic declination.

decollement   (1978)
  UF   detachment
  SA   boundary faults
  SA   deformation
  SA   disharmonic folds
  SA   duplexes
  SA   faults
  SA   flexural-slip
  SA   folds
  SA   mechanics
  SA   tectonics

decomposition
  After 1993, see types of weathering, e.g. chemical weathering, physical weathering, or degradation or alteration.

decontamination   (1982)
  SA   contamination
  SA   desalinization
  SA   dilution
  SA   drinking water
  SA   ground water
  SA   impurities
  SA   polluted water
  SA   pollution
  SA   potability
  SA   purification
  SA   reclamation
  SA   surface water
  SA   waste water
  SA   water
  SA   water management

  SA   water quality
  SA   water resources
  SA   water treatment

deconvolution   (1978)
  SA   distortion
  SA   elastic waves
  SA   filters
  SA   Kalman filters
  SA   seismology

Decorah Shale   (1989)
W Illinois, NE Iowa, S Minnesota, N Missouri and SW Wisconsin.
  BT   Middle Ordovician
  BT   Ordovician
  BT   Paleozoic
  SA   Illinois
  SA   Iowa
  SA   Minnesota
  SA   Missouri
  SA   Wisconsin

decrepitation   (1978)
  SA   chemical analysis
  SA   geologic thermometry
  SA   mineralogy
  SA   sample preparation

Dedham Granodiorite   (1989)
E Massachusetts and SE New Hampshire.
  BT   Proterozoic
  BT   upper Precambrian
  BT   Precambrian
  SA   Massachusetts
  SA   New Hampshire

dedolomitization   (1978)
  SA   calcitization
  SA   diagenesis
  SA   dolomitization
  SA   processes

deep drilling   (1989)
  BT   drilling
  SA   boreholes
  SA   engineering geology
  SA   KTB

deep earthquakes
  use deep-focus earthquakes

deep faults
  use deep-seated structures

deep focus
  A valid term through 1977. After 1977, use deep-focus earthquakes.

deep magnetic sounding   (1981)
  SA   deep sounding
  SA   geophysical methods
  SA   geophysical surveys
  SA   magnetic methods
  SA   magnetic surveys
  SA   sounding

Deep Sea Drilling Project   (1978)
Autoposting of this term began in 1985. Autoposting of specific legs to specific DSDP sites began in 1989. See IPOD or Ocean Drilling Program for listings of additional legs and sites.
  UF   DSDP
  NT   IPOD
  NT   Leg 1
  NT   Leg 2
  NT   Leg 3
  NT   Leg 4
  NT   Leg 5
  NT   Leg 6
  NT   Leg 7
  NT   Leg 8
  NT   Leg 9
  NT   Leg 10
  NT   Leg 11
  NT   Leg 12
  NT   Leg 13
  NT   Leg 14
  NT   Leg 15
  NT   Leg 16
  NT   Leg 17
  NT   Leg 18
  NT   Leg 19
  NT   Leg 20
  NT   Leg 21
  NT   Leg 22
  NT   Leg 23
  NT   Leg 24
  NT   Leg 25
  NT   Leg 26
  NT   Leg 27
  NT   Leg 28
  NT   Leg 29
  NT   Leg 30
  NT   Leg 31
  NT   Leg 32
  NT   Leg 33
  NT   Leg 34
  NT   Leg 35
  NT   Leg 36
  NT   Leg 37
  NT   Leg 38
  NT   Leg 39
  NT   Leg 40
  NT   Leg 41
  NT   Leg 42A
  NT   Leg 42B
  NT   Leg 43
  NT   Leg 44
  SA   Glomar Challenger
  SA   JOIDES
  SA   marine geology
  SA   ocean basins
  SA   Ocean Drilling Program
  SA   ocean floors
  SA   oceanography
  SA   research vessels

deep seismic sounding   (1978)
  UF   DSS
  UF   seismic sounding
  SA   BIRPS
  SA   deep sounding
  SA   explosions
  SA   geophysical methods
  SA   geophysical surveys
  SA   geophysics
  SA   seismic methods
  SA   seismology
  SA   sounding

deep sounding   (1978)
  SA   deep magnetic sounding
  SA   deep seismic sounding
  SA   echo sounding
  SA   electrical sounding
  SA   frequency sounding
  SA   geophysical methods
  SA   geophysical surveys
  SA   sounding

deep structure
  use deep-seated structures

deep-focus earthquakes   (1978)
Before 1978, search earthquakes AND deep focus.
  UF   deep earthquakes
  BT   earthquakes
  SA   focus
  SA   intermediate-focus earthquakes
  SA   shallow-focus earthquakes

deep-sea
  A valid term through 1977. After 1977, use deep-sea environment or (as of 1981) deep-sea sedimentation.

deep-sea environment   (1978)
Before 1978, search deep-sea or deep sea. As of 1981, used for abyssal environment. Before 1981, also search abyssal environment.
  UF   abyssal environment
  SA   abyssal hills
  SA   bottom features
  SA   deep-sea sedimentation
  SA   deep-water environment
  SA   depositional environment
  SA   environment
  SA   hemipelagic environment
  SA   sedimentation

deep-sea fans
  use submarine fans

deep-sea sedimentation   (1981)
  BT   sedimentation
  SA   deep-sea environment

deep-sea sediments
  No longer a valid GeoRef index term. Before 1971, was used on level 2 in subfile N. See marine sediments; deep-sea sedimentation.

deep-seated structures   (1978)
Before 1978, search deep structure; deep-seated.
  UF   deep faults
  UF   deep structure
  SA   tectonics

deep-tow methods   (1978)
Before 1978, search deep-tow.
  SA   acoustical methods
  SA   geophysical methods

deep-water environment   (1993)
Use deep-sea environment if applicable.
  SA   anaerobic environment
  SA   aquatic environment
  SA   brackish-water environment
  SA   deep-sea environment
  SA   deltaic environment
  SA   depositional environment
  SA   environment
  SA   estuarine environment
  SA   fluvial environment
  SA   fresh-water environment
  SA   glaciofluvial environment
  SA   hemipelagic environment
  SA   hypersaline environment
  SA   lacustrine environment
  SA   lagoonal environment
  SA   pelagic environment
  SA   reef environment
  SA   shallow-water environment
  SA   slope environment

deepening
  use dredging

Deer Lake Group   (1989)
Carboniferous formation in W Newfoundland. Also a Precambrian formation in SE Wyoming.
  IN   Index ages as applicable.
  SA   Carboniferous
  SA   Newfoundland Island
  SA   Precambrian
  SA   Wyoming

Deer Lodge County
  Valid through 1988. Search in combination with state term. After 1988, use specific county-state term.

Deer Lodge County Montana   (1989)
SW Montana. Before 1989, also search Deer Lodge County AND Montana.
  CO   N454500N462000 W1123000W1133000
  BT   Montana
  BT   United States

deerite   (1978)

BT chain silicates
BT silicates

defects (1978)
SA crystal dislocations
SA crystal structure
SA crystallography
SA crystals
SA deformation
SA minerals
SA point defects
SA stacking
SA surface defects

definition (1978)
Used as a general term.
SA classification
SA identification
SA nomenclature
SA photogrammetry

Deflandrea (1978)
Autoposting of microfossils to this term began in 1990.
BT Dinoflagellata
BT palynomorphs
BT microfossils

deflection of the vertical
No longer a valid GeoRef index term. Before 1972, was used on level 2 in subfile G.

deforestation (1978)
SA conservation
SA ecology
SA fires
SA forestry
SA forests
SA land use
SA reclamation

deformation (1978)
Used for the process of folding, faulting, shearing, compression, or extension of the rocks as a result of various forces. Only small-scale deformations considered here. See structural analysis for mid-scale. See tectonics for large-scale.
NT brittle deformation
NT ductile deformation
NT plastic deformation
SA anelasticity
SA Atterberg limits
SA axial-plane structures
SA boudinage
SA boundary conditions
SA brittle materials
SA brittleness
SA bulk modulus
SA cataclasis
SA cleavage
SA competent materials
SA compression
SA compression tectonics
SA compressive strength
SA creep
SA decollement
SA defects
SA dilatancy
SA distortion
SA ductility
SA dynamic loading
SA elastic constants
SA elastic limit
SA elastic materials
SA elastic properties
SA elastic strain
SA elastic waves
SA elasticity
SA elastodynamic properties
SA engineering geology
SA enterolithic folds
SA extension
SA extension tectonics
SA extensometers

SA fabric
SA failures
SA faults
SA field studies
SA finite strain analysis
SA flexure
SA flow lines
SA folds
SA foliation
SA fracture strength
SA fractures
SA geodesy
SA geophysics
SA Hooke's law
SA hysteresis
SA isoclinal folds
SA kink-band structures
SA lamellae
SA lineation
SA mechanical properties
SA melange
SA metamorphic rocks
SA metamorphism
SA meteor craters
SA microcracks
SA mobilization
SA mylonitization
SA orogeny
SA petrofabrics
SA plastic flow
SA plasticity
SA plate tectonics
SA point defects
SA pressure
SA recrystallization
SA rheology
SA rock mechanics
SA rupture
SA salt tectonics
SA seismology
SA shatter cones
SA shear
SA shear stress
SA shock waves
SA sintering
SA soil mechanics
SA stiffness
SA strain
SA strainmeters
SA strength
SA stress
SA stress fields
SA stressmeters
SA structural analysis
SA tectonics
SA tectonite
SA tectonophysics
SA tensile strength
SA tension
SA torsion
SA triaxial tests
SA viscoelasticity
SA viscosity
SA wavelength
SA yield strength

degasification (1989)
As of 1989, restricted to process related to the energy sources such as coal seams, petroleum, natural gas, and oil and gas fields. For process involving the mantle, magmas or atmosphere, use degassing.
SA coal
SA coal seams
SA degassing
SA natural gas
SA oil and gas fields
SA petroleum

degassing (1978)
As of 1989, restricted to outgassing; that is, processes related to the mantle, magmas, atmosphere, etc. For processes concerning the

energy sources such as coal seams, petroleum, natural gas, or oil and gas fields, use degasification.
UF outgassing
SA atmosphere
SA degasification
SA mantle

deglaciation (1978)
UF glacial recession
SA glaciation
SA glaciers

degradation (1978)
UF deterioration
SA biodegradation
SA denudation
SA durability
SA erosion
SA geomorphology
SA leaching
SA soils
SA weathering

Dehra Dun
No longer a valid term for GeoRef. See Dehra Dun India.

Dehra Dun India (1993)
City in NW Uttar Pradesh, N India. Before 1993, also search Dehra Dun. Before 1978, also search Dehradun.
UF Dehradun India
BT Uttar Pradesh India
BT India
BT Indian Peninsula
BT Asia

Dehradun India
use Dehra Dun India

dehydration (1978)
UF drying
SA desiccation
SA geochemistry
SA hydration
SA sediments
SA water of dehydration

Deimos
As of 1989, no longer a valid term for GeoRef.
use Deimos Satellite

Deimos Satellite (1989)
One of satellites of Mars. Before 1989, also search Deimos.
UF Deimos
SA Mars
SA satellites

Deinotherioidea (1981)
Suborder. Autoposting of Eutheria and Theria to this term began in 1989.
BT Proboscidea
BT Eutheria
BT Theria
BT Mammalia
BT Tetrapoda
BT Vertebrata
BT Chordata

Dekkan Plateau
use Deccan Plateau

DEKORP (1993)
Acronym.
UF Deutsches Kontinentales Reflexionsseismisches Profil
SA Germany
SA reflection methods
SA seismic methods
SA seismic surveys

Del Monte Beach (1978)
On Monterey Peninsula in Monterey County in W California.

IN Index county or regions as applicable.
BT Monterey County California
BT California
BT United States

Del Norte County
Valid through 1988. Search in combination with state term. After 1988, use specific county-state term.

Del Norte County California (1989)
NW California. Before 1989, also search Del Norte County AND California.
CO N412200N420000
W1233500W1242000
BT California
BT United States

Del Rio
Valid through 1988. Search in combination with state term. After 1988, use specific city-state term.

Del Rio Texas (1989)
City on the Rio Grande in SW Texas. Before 1989, search Del Rio AND Texas.
CO N292300N292300
W1005600W1005600
BT Val Verde County Texas
BT Texas
BT United States

delafossite (1978)
BT oxides
SA iron oxides

Delaware (1978)
Autoposting of this term began in 1978.
CO N382700N395200
W0750300W0754800
BT United States
NT Kent County Delaware
NT New Castle County Delaware
NT Sussex County Delaware
SA Aquia Formation
SA Atlantic Coastal Plain
SA Calvert Formation
SA Carolina Bays
SA Chesapeake Group
SA Choptank River
SA Cohansey Formation
SA Delaware Bay
SA Delaware River
SA Delaware River basin
SA Delmarva Peninsula
SA Indian River
SA Magothy Aquifer
SA Magothy Formation
SA Marshalltown Formation
SA Monmouth Group
SA Navesink Formation
SA Newark Supergroup
SA Piney Point Formation
SA Potomac Group
SA Raritan Formation
SA Saint Marys Formation
SA Severn Formation
SA Wicomico Formation
SA Wilmington Complex
SA Wissahickon Formation

Delaware Basin (1978)
SE New Mexico and SW Texas.
IN Index states as applicable.
BT United States
SA New Mexico
SA Permian Basin
SA Texas

Delaware Bay (1978)
Arm of Atlantic Ocean.
IN Index states as applicable.
BT United States

SA Delaware
SA New Jersey
SA North American Atlantic
Delaware County
Valid through 1988. Search in combination with state term. After 1988, use specific county-state term.
Delaware County Indiana (1989)
E Indiana. Before 1989, also search Delaware County AND Indiana.
CO N400500N402300
W0851300W0853500
BT Indiana
BT United States
Delaware County Iowa (1989)
E Iowa. Before 1989, also search Delaware County AND Iowa.
CO N421800N420900
W0910800W0913700
BT Iowa
BT United States
Delaware County New York (1989)
S New York. Before 1989, also search Delaware County AND New York.
CO N415200N423200
W0742500W0752400
BT New York
BT United States
Delaware County Ohio (1989)
Central Ohio. Before 1989, also search Delaware County AND Ohio.
CO N400700N402700
W0824400W0831500
BT Ohio
BT United States
Delaware County Oklahoma (1989)
NE Oklahoma. Before 1989, also search Delaware County AND Oklahoma.
CO N361000N364000
W0943400W0950100
BT Oklahoma
BT United States
Delaware County Pennsylvania (1989)
SE Pennsylvania. Before 1989, also search Delaware County AND Pennsylvania.
CO N394800N400400
W0751300W0753700
BT Pennsylvania
BT United States
Delaware Limestone (1978)
Overlies Columbus Limestone; underlies Olentangy Shale. Central and N Ohio.
BT Middle Devonian
BT Devonian
BT Paleozoic
SA Ohio
Delaware Mountain Group (1985)
W Texas.
BT Guadalupian
BT Permian
BT Paleozoic
SA Bell Canyon Formation
SA Cherry Canyon Formation
SA Texas
Delaware River (1978)
Flows from S New York SE into Delaware Bay.
IN Index states as applicable.
BT United States

SA Delaware
SA New Jersey
SA New York
SA Pennsylvania
Delaware River basin (1978)
IN Index states as applicable.
BT United States
SA Delaware
SA New Jersey
SA New York
SA Pennsylvania
Delhi
No longer a valid term for GeoRef. See Delhi India.
Delhi India (1993)
Territory in N central India including the city NNE of New Delhi.
BT India
BT Indian Peninsula
BT Asia
NT New Delhi India
Delhi Supergroup (1993)
From the Delhi System which included Ajabgarh Series, Hornstone Breccia, Kushalgarh Limestone, Alwar Series, Raialo limestones and quartzes. Hornstone Breccia and Kushalgarh Limestone later confined to NE Rajputana, and Raialos excluded from Delhis to form separate unit. Named after city of Delhi. Before 1993, also search Delhi System.
UF Delhi System
BT Precambrian
SA Haryana India
SA India
SA Rajasthan India
Delhi System
No longer a valid term for GeoRef.
use Delhi Supergroup
Delinesti
No longer a valid term for GeoRef. see Delinesti Romania.
Delinesti Romania (1993)
Village in Caras-Severin County in W part of country. This term has multiple hierarchies.
BT1 Banat
BT1 Southern Europe
BT1 Europe
BT2 Romania
BT2 Southern Europe
BT2 Europe
dellenite (1978)
BT rhyodacites
BT volcanic rocks
BT igneous rocks
Dellys
No longer a valid term for GeoRef. As of 1993, see Dellys Algeria.
Dellys Algeria (1993)
City on Mediterranean Sea E of Algiers in N Algeria.
BT Algeria
BT North Africa
BT Africa
Delmarva Peninsula (1978)
Between Chesapeake and Delaware bays.
IN Index states as applicable.
BT United States
SA Delaware
SA Maryland
SA Virginia

Delmas
No longer a valid term for GeoRef. As of 1993, see Delmas Saskatchewan.
Delmas Saskatchewan (1993)
Village in W central Saskatchewan. Before 1993, also search Delmas AND Saskatchewan.
BT Saskatchewan
BT Western Canada
BT Canada
Delocrinus (1978)
Genus.
BT Crinoidea
BT Crinozoa
BT Echinodermata
BT Invertebrata
DeLong Mountains
use De Long Mountains
Delta Area (1978)
Netherlands.
IN Index regions as applicable.
SA Netherlands
Delta County
Valid through 1988. Search in combination with state term. After 1988, use specific county-state term.
Delta County Colorado (1989)
W Colorado. Before 1989, also search Delta County AND Colorado.
CO N383900N391000
W1073000W1082200
BT Colorado
BT United States
Delta County Michigan (1989)
N Michigan. Before 1989, also search Delta County AND Michigan.
CO N452700N461000
W0862600W0872200
BT Michigan Upper Peninsula
BT Michigan
BT United States
Delta County Texas (1989)
NE Texas. Before 1989, also search Delta County AND Texas.
CO N331600N332900
W0952000W0955100
BT Texas
BT United States
delta fans
use fan deltas
Delta River (1978)
In E Alaska. Joins the Tanana River at Fairbanks.
IN Index Southern Alaska or East-Central Alaska as applicable.
BT Alaska
BT United States
SA East-Central Alaska
SA Fairbanks Alaska
SA Southern Alaska
SA Tanana River
deltaic
A valid term through 1977. After 1977, use deltas or (as of 1981) deltaic environment or deltaic sedimentation.
deltaic environment (1981)
SA deep-water environment
SA deltaic sedimentation
SA deltas
SA depositional environment
SA ecology
SA environment
SA paleoecology

SA sedimentation
deltaic sedimentation (1981)
BT sedimentation
SA deltaic environment
SA deltas
SA Gilbert-type deltas
deltas (1978)
As of 1981, use deltaic environment for type of environment.
NT fan deltas
NT Gilbert-type deltas
SA deltaic environment
SA deltaic sedimentation
SA ecology
SA fluvial features
SA geomorphology
SA paleoecology
SA sedimentation
SA shore features
demagnetization (1978)
NT adiabatic demagnetization
NT alternating field demagnetization
NT chemical demagnetization
NT thermal demagnetization
SA geophysics
SA magnetic hysteresis
SA magnetization
SA paleomagnetism
SA remagnetization
SA remanent magnetization
demand (1981)
SA consumption
SA economics
SA markets
Demopolis Chalk (1989)
Alabama and E Mississippi. Includes Bluffport Marl Member.
BT Upper Cretaceous
BT Cretaceous
BT Mesozoic
SA Alabama
SA Bluffport Marl Member
SA Mississippi
Demospongea (1978)
BT Porifera
BT Invertebrata
Denali Fault (1978)
Mount McKinley area in S central Alaska. Extends into Yukon Territory.
IN Index Alaska or Yukon Territory as applicable.
BT North America
SA Alaska
SA Yukon Territory
Denbighshire
No longer a valid term for GeoRef as of 1993.
use Denbighshire Wales
Denbighshire Wales (1993)
Former county in N Wales. Before 1993, also search Denbighshire. As of 1975, part of Clwyd.
UF Denbighshire
BT Wales
BT Great Britain
BT United Kingdom
BT Western Europe
BT Europe
dendrochronology
use tree rings
dendrograms (1978)
BT statistical analysis
Dendroidea (1978)
IN Index Hemichordata or Invertebrata as applicable.
BT Graptolithina
SA Hemichordata

SA Invertebrata

Dengizkul (1978)
Lake SW of Bukhara.
IN Index Uzbekistan as applicable.
SA Uzbekistan

Dengying Formation (1989)
SW China.
BT Sinian
BT Proterozoic
BT upper Precambrian
BT Precambrian
SA China

Denison Trough (1985)
SE Queensland, Australia.
IN Index state or region as applicable.
SA Queensland Australia

denitrification (1978)
SA geochemistry

Denmark (1978)
Faeroe Islands is considered a narrower term as of 1981, but Denmark is not autoposted to it.
CO N541500N580000
E0130000E0080000
BT Scandinavia
BT Western Europe
BT Europe
NT Bornholm
NT Copenhagen Denmark
NT Djursland
NT Helsingor Denmark
NT Stevns Klint
NT Vendsyssel
NT Zealand
SA Baltic Glaciation
SA Baltic region
SA Baltic Shield
SA European Platform
SA Fennoscandia
SA Jutland
SA North Sea Coast
SA North Sea region
SA Oresund
SA Zechstein

dense nonaqueous phase liquids (1993)
Before 1993, also search DNAPL.
UF DNAPLs
SA density
SA leaking underground storage tanks
SA liquid phase
SA nonaqueous phase liquids
SA oil spills
SA pollutants

densities
use density

densities and temperatures
As of 1993, no longer a valid term for GeoRef. Usually out-of-scope. Used with aeronomy.

density (1978)
UF densities
SA bulk density
SA buoyancy
SA compactness
SA convection currents
SA dense nonaqueous phase liquids
SA density logging
SA isostasy
SA sea water
SA specific gravity

density currents (1978)
BT currents
SA turbidity currents

density logging (1985)

SA density
SA well-logging

Dentalium (1978)
Genus.
BT Scaphopoda
BT Mollusca
BT Invertebrata

denudation (1978)
SA degradation
SA erosion
SA erosion features
SA geomorphology
SA processes
SA water erosion
SA weathering

Denver
Valid through 1988. Search in combination with state term. After 1988, use specific city-state term.

Denver Basin (1978)
Primarily in central and NE Colorado.
IN Index states as applicable.
UF D-J Basin
UF Denver Julesberg Basin
BT United States
SA Colorado
SA Nebraska
SA Wattenberg Field
SA Wyoming

Denver Colorado (1989)
City in NE central Colorado. Before 1989, search Denver AND Colorado.
CO N394500N394500
W1050000W1050000
UF Denver region
BT Denver County Colorado
BT Colorado
BT United States

Denver County
Valid through 1988. Search in combination with state term. After 1988, use specific county-state term.

Denver County Colorado (1989)
N central Colorado, coextensive with the city of Denver. Before 1989, also search Denver County AND Colorado.
CO N394000N394800
W1045000W1050500
BT Colorado
BT United States
NT Denver Colorado

Denver Formation (1985)
Upper Cretaceous and Paleocene. E Colorado.
IN Index ages as applicable.
SA Colorado
SA K-T boundary
SA Paleocene
SA Upper Cretaceous

Denver Julesberg Basin
use Denver Basin

Denver region
use Denver Colorado

deoxyribonucleic acid
use DNA

Department of Energy, U.S.
use U. S. Department of Energy

depletion (1978)
As of 1981, restricted to resources. Before 1981, used as a general term.
SA conservation
SA mineral resources
SA natural resources

SA petroleum
SA reclamation

deposition (1978)
SA accretion
SA alluvial fans
SA alluvial plains
SA cementation
SA diagenesis
SA glacial geology
SA glaciation
SA ice movement
SA reservoirs
SA reworking
SA sedimentary processes
SA sedimentary structures
SA sedimentation
SA sediments
SA siltation
SA stratification
SA uniformitarianism

depositional environment (1993)
Before 1993, also search environment AND deposition.
UF sedimentary environment
SA aerobic environment
SA alpine environment
SA anaerobic environment
SA aquatic environment
SA arctic environment
SA arid environment
SA boreal environment
SA brackish-water environment
SA cave environment
SA coastal environment
SA deep-sea environment
SA deep-water environment
SA deltaic environment
SA environment
SA estuarine environment
SA fluvial environment
SA fresh-water environment
SA glacial environment
SA glaciofluvial environment
SA glaciolacustrine environment
SA glaciomarine environment
SA hemipelagic environment
SA high-energy environment
SA humid environment
SA hypersaline environment
SA interglacial environment
SA intertidal environment
SA lacustrine environment
SA lagoonal environment
SA low-energy environment
SA marine environment
SA nearshore environment
SA paludal environment
SA pelagic environment
SA periglacial environment
SA postglacial environment
SA reef environment
SA sebkha environment
SA sedimentation
SA semi-arid environment
SA shallow-water environment
SA shelf environment
SA slope environment
SA storm environment
SA subaerial environment
SA subalpine environment
SA subglacial environment
SA submarine environment
SA subtidal environment
SA subtropical environment
SA supratidal environment
SA taiga environment
SA temperate environment
SA terrestrial environment
SA tropical environment
SA urban environment

depositional fault
use growth faults

depositional remanent magnetization (1978)
Autoposting of magnetization to this term began in 1989.
UF detrital remanent magnetization
BT remanent magnetization
BT magnetization
SA paleomagnetism

deposits
As of 1981, no longer a valid term for GeoRef. See list C; mineral deposits, genesis; metal ores; nonmetal deposits. Before 1981, included use for all papers on nonmetal commodities.

depressions (1978)
SA basins
SA folds
SA geomorphology

depth (1978)
SA depth indicators
SA dimensions
SA Mohorovicic discontinuity
SA shallow aquifers
SA shallow depth

depth indicators (1981)
SA burial
SA burial diagenesis
SA depth
SA indicators

depth of compensation
use carbonate compensation depth

Derby
No longer a valid term for GeoRef as of 1993. See Derby England.

Derby England (1993)
City in Derbyshire in central England. Before 1993, also search Derby AND England.
BT Derbyshire England
BT England
BT Great Britain
BT United Kingdom
BT Western Europe
BT Europe

Derbyshire
No longer a valid term for GeoRef as of 1993.
use Derbyshire England

Derbyshire England (1993)
County in central England. Before 1993, also search Derbyshire.
CO N524500N533000
W0011500W0020000
UF Derbyshire
BT England
BT Great Britain
BT United Kingdom
BT Western Europe
BT Europe
NT Derby England
SA Midlands
SA Peak District

Dermapteroida (1981)
BT Insecta
BT Mandibulata
BT Arthropoda
BT Invertebrata

Dermoptera (1981)
Suborder. Autoposting of Insectivora, Eutheria and Theria to this term began in 1989.
BT Insectivora
BT Eutheria
BT Theria
BT Mammalia
BT Tetrapoda

BT Vertebrata
BT Chordata

Des Moines Lobe (1985)
Part of Laurentide ice sheet during Wisconsinan glaciation.
SA Iowa
SA Laurentide ice sheet
SA Minnesota
SA South Dakota

Des Moines Series
use Desmoinesian

desalination
use desalinization

desalinization (1978)
UF desalination
SA brines
SA decontamination
SA dilution
SA drinking water
SA ground water
SA hydrochemistry
SA leaching
SA polluted water
SA potability
SA salinity
SA salt
SA sea water
SA soil treatment
SA soils
SA water management
SA water quality
SA water resources
SA water treatment

Descartes
No longer a valid term for GeoRef. As of 1993, see Descartes Algeria.

Descartes Algeria (1993)
Village in NW Algeria. Before 1993, also search Descartes and Algeria.
BT Oran Algeria
BT Algeria
BT North Africa
BT Africa

Deschutes County
Valid through 1988. Search in combination with state term. After 1988, use specific county-state term.

Deschutes County Oregon (1989)
Central Oregon. Before 1989, also search Deschutes County AND Oregon.
CO N433500N442500
  W1195000W1215500
BT Oregon
BT United States
NT Newberry Volcano

Deschutes Formation (1989)
IN Index ages as applicable.
BT Neogene
BT Tertiary
BT Cenozoic
SA Miocene
SA Oregon
SA Pliocene

Deschutes River (1985)
Tributary of the Columbia River, N central Oregon. Also a river, in NW Washington, which flows into Puget Sound.
IN Index states as applicable.
SA Columbia River
SA Oregon
SA Puget Sound
SA United States
SA Washington

descloizite (1978)

BT vanadates

description (1978)
Used as a general term.
SA landform description
SA morphology

Desert Basin
Not a valid term for GeoRef. Use Canning Basin if applicable.

Desert Creek Zone (1989)
Paradox Basin of Colorado and Utah.
IN Index states as applicable.
SA Colorado
SA Paradox Basin
SA United States
SA Utah

desert pavement (1993)
Reintroduced. Was not a valid GeoRef index term between 1971 and 1992. Before 1971, used on level 1 in subfile N. Before 1993, also search eolian features and deserts.
BT eolian features
SA deserts
SA talus slopes
SA wind erosion

desert pediplains
use pediplains

desert plains
use pediplains

Desert soils (1978)
BT soils
SA soil group
SA Zonal soils

desert varnish (1989)
SA arid environment
SA deserts

desertification (1978)
SA deserts
SA ecology
SA geomorphology
SA processes

deserts (1978)
Autoposting of eolian features to this term began in 1989.
BT eolian features
SA arid environment
SA barchans
SA desert pavement
SA desert varnish
SA desertification
SA ecology
SA ergs
SA geomorphology
SA playas
SA sand seas
SA sand sheets
SA wadis
SA yardangs

desiccation (1978)
SA dehydration
SA ecology
SA moisture
SA pore water
SA soils

design (1978)
SA aseismic design
SA dams
SA development
SA engineering geology
SA foundations
SA instruments
SA marine installations
SA nuclear facilities
SA reservoirs
SA safety
SA shorelines
SA tunnels

SA underground installations
SA waterways

Desmidiales (1981)
Before 1981, search desmids. Autoposting of thallophytes and microfossils to this term began in 1990. This term has multiple hierarchies.
BT1 Chlorophyta
BT1 algae
BT1 thallophytes
BT1 Plantae
BT2 Chlorophyta
BT2 algae
BT2 microfossils
SA desmids

desmids (1978)
As of 1981, restricted to use as the common name for Desmidiales. Autoposting of algal flora to this term began in 1989. Autoposting of microfossils to this term began in 1990. This term has multiple hierarchies.
BT1 algal flora
BT1 plants
BT2 algal flora
BT2 microfossils
SA biostratigraphy
SA Desmidiales

desmine
use stilbite

Desmoceratida (1981)
BT Ammonoidea
BT Tetrabranchiata
BT Cephalopoda
BT Mollusca
BT Invertebrata

Desmoinesian (1978)
Comprises Cherokee Group, Marmaton Group, Cabaniss Group, and Krebs Group. Subdivided into Veteran and Cygnian substages.
UF Des Moines Series
UF Desmoinesian Series
BT Middle Pennsylvanian
BT Pennsylvanian
BT Carboniferous
BT Paleozoic
NT Boggy Shale
NT Cabaniss Formation
NT Cherokee Group
NT Hartshorne Sandstone
NT Krebs Group
NT Labette Shale
NT Marmaton Group
NT Savanna Formation
NT Spiro Sandstone
NT Wewoka Formation
SA Arkansas
SA Bevier Coal
SA Francis Creek Shale
SA Iowa
SA Kansas
SA Manakacha Formation
SA Missouri
SA Nebraska
SA Oklahoma

Desmoinesian Series
use Desmoinesian

Desmostylia (1978)
Order. Autoposting of Eutheria and Theria to this term began in 1989.
BT Eutheria
BT Theria
BT Mammalia
BT Tetrapoda
BT Vertebrata
BT Chordata

Desolation Islands
use Kerguelen Islands

desorption (1978)
SA adsorption
SA geochemistry

desoxyribonucleic acid
use DNA

destruction (1978)
Used as a general term.
SA catastrophes
SA corrosion
SA geologic hazards

detachment
use decollement

detachment faults (1989)
UF sole faults
BT faults

detection (1978)
Appropriate to a large number of general topics.
SA background level
SA chemical analysis
SA earthquakes

detergents (1978)
SA environmental geology
SA liquid waste
SA waste disposal
SA waste water

deterioration
use degradation

determination
Not a valid term for GeoRef. For absolute age determination, use absolute age in conjunction with dates.

detrital deposits (1981)
SA detritus
SA mineral deposits, genesis

detrital fan
use alluvial fans

detrital minerals
use detritus

detrital remanent magnetization
use depositional remanent magnetization

detrital sedimentation (1981)
BT sedimentation

detritus (1978)
Also search detrital.
UF detrital minerals
SA abrasion
SA debris
SA detrital deposits
SA erosion
SA weathering

Detroit River (1993)
Flows from Lake Saint Clair to Lake Erie. International boundary separates SW Ontario and SE Michigan.
IN Index Michigan and/or Ontario as applicable.
BT North America
SA Essex County Ontario
SA Lake Erie
SA Lake Saint Clair
SA Michigan
SA Ontario
SA Wayne County Michigan
SA Windsor Ontario

Detroit River Group (1978)
Includes Sylvania Formation, Amherstburg Formation, Lucas Formation. SE Michigan, N Ohio, and W Ontario.
BT Middle Devonian
BT Devonian

BT Paleozoic
SA Lucas Formation
SA Michigan
SA Ohio
SA Ontario

deuterium (1978)
This term has multiple hierarchies.
UF H-2
BT1 stable isotopes
BT1 isotopes
BT2 hydrogen
SA D/H

Deutsche Demokratische Republik
Not a valid term for GeoRef. As of 1993, see Germany and also Mecklenburg-Western Pomerania Germany, Brandenburg Germany, Saxony-Anhalt Germany, Thuringia Germany, and Saxony Germany.

Deutsches Kontinentales Reflexionsseismisches Profil
use DEKORP

Deux-Sevres
No longer a valid term for GeoRef. See Deux-Sevres France.

Deux-Sevres France (1993)
CO N460000N471000
E0001500W0010000
BT France
BT Western Europe
BT Europe
SA Poitou

developing countries
use Third World

development (1978)
Used as a general term.
SA design
SA economic geology
SA exploration
SA ground water

Devensian (1978)
Europe.
BT upper Pleistocene
BT Pleistocene
BT Quaternary
BT Cenozoic

deviation, standard
use standard deviation

Devils Icebox (1978)
Region in Boone County in central Missouri.
IN Index county as applicable.
SA Boone County Missouri

devitrification (1978)
SA geochemistry
SA glasses
SA lithogeochemistry
SA volcanic glass

Devoluy (1978)
Limestone range of the Dauphine Pre-Alps in SE France.
BT France
BT Western Europe
BT Europe
SA Dauphine
SA Dauphine Alps

Devon
Not a valid term for GeoRef. As of 1993, see Devonshire England.

Devon Island (1978)
N of Baffin Island off Baffin Bay in Franklin District. Also search Devon not England.
CO N743000N763000
W0793000W0970000
UF North Devon Island

BT Franklin District Northwest Territories
BT Northwest Territories
BT Western Canada
BT Canada
NT Radstock Bay

Devonian (1978)
Above Silurian, below Carboniferous. From 1978-1980, Paleozoic was autoposted to this term.
BT Paleozoic
NT Acadian Phase
NT Ackley Granite
NT Barre Granite
NT Cedar City Formation
NT Chelmsford Granite
NT Clays Ferry Formation
NT Genesee Group
NT Gile Mountain Formation
NT Guilmette Formation
NT Heemskirk Granite
NT Hunsruck Shale
NT Keg River Formation
NT Levis Shale
NT Lilydale Limestone
NT Lost Burro Formation
NT Lower Devonian
NT Martin Formation
NT Middle Devonian
NT Miette Complex
NT Millboro Shale
NT Muth Quartzite
NT Old Red Sandstone
NT Ramparts Formation
NT Slave Point Formation
NT Swan Hills Formation
NT Traverse Group
NT Upper Devonian
NT Upper Old Red Sandstone
NT Waterways Formation
NT Zlichovian
SA Antler Orogeny
SA Arkansas Novaculite
SA Beacon Supergroup
SA Bedford Shale
SA Berea Sandstone
SA Broken River Formation
SA Catskill Formation
SA Chattanooga Shale
SA Dalradian
SA Downtonian
SA Earn Group
SA Galway Granite
SA Geirud Formation
SA Henderson Gneiss
SA Hercynian Orogeny
SA Horton Group
SA Hunton Group
SA Kaskaskia Sequence
SA Peel Sound Formation
SA Rangeley Formation
SA Ringerike Sandstone
SA Ross Formation
SA Shoo Fly Complex
SA South Mountain Batholith
SA Table Mountain Group
SA Talladega Group
SA Tor Formation
SA Waits River Formation
SA Woodford Shale

Devonshire
No longer a valid term for GeoRef as of 1993.
use Devonshire England

Devonshire England (1993)
County in SW England. Before 1978, also search Devon AND England. Before 1993, also search Devonshire.
CO N501500N511500
W0030000W0043000
UF Devonshire
BT England

BT Great Britain
BT United Kingdom
BT Western Europe
BT Europe
NT Chudleigh England
NT Dartmoor
NT Exeter England
NT Okehampton England
NT Teign Valley
NT Torquay England
SA Tamar Estuary
SA Tamar Valley

dewatering
As of 1981, no longer a valid term for GeoRef. See drawdown. Also see dehydration.

Dewey County
Valid through 1988. Search in combination with state term. After 1988, use specific county-state term.

Dewey County Oklahoma (1989)
W Oklahoma. Before 1989, also search Dewey County AND Oklahoma.
CO N354800N361000
W0983800W0992200
BT Oklahoma
BT United States

Dewey County South Dakota (1989)
N central South Dakota. Before 1989, also search Dewey County AND South Dakota.
CO N444500N452800
W1001800W1013000
BT South Dakota
BT United States

dextral faults
use right-lateral faults

Dhajala Meteorite (1989)
Impact at Dhajala, Gujarat, W India.
BT H chondrites
BT chondrites
BT stony meteorites
BT meteorites
SA Gujarat India
SA India

Dhanbad
No longer a valid term for GeoRef. See Dhanbad India.

Dhanbad India (1993)
Town in SE Bihar, E India.
BT Bihar India
BT India
BT Indian Peninsula
BT Asia

Dharwar
No longer a valid term for GeoRef. See Dharwar India.

Dharwar Craton (1993)
S Indian Shield.
BT India
BT Indian Peninsula
BT Asia
SA Chitradurga schist belt
SA Kolar schist belt

Dharwar India (1993)
City in NW Karnataka. Now part of the joint municipality of Hubli-Dharwar.
BT Karnataka India
BT India
BT Indian Peninsula
BT Asia

Dharwars (1978)

"Sub metamorphic" crystalline rocks of the Dharwar System, named after city of Dharwar in Karnataka, but occurring throughout the country.
UF Dharwars Supergroup
BT Archean
BT Precambrian
SA Chitradurga schist belt
SA Closepet Granite
SA India

Dharwars Supergroup
use Dharwars

Dhenkanal
No longer a valid term for GeoRef. See Dhenkanal India.

Dhenkanal India (1993)
Town and district in E central Orissa, E India.
BT Orissa India
BT India
BT Indian Peninsula
BT Asia

diabase (1978)
Before 1981, also search dolerite. Autoposting of this term began in 1981.
UF dolerite
BT plutonic rocks
BT igneous rocks
NT diabase porphyry
NT olivine diabase
NT quartz diabase
NT tholeiitic dolerite
SA basalts
SA metadiabase
SA ophite
SA trap rocks

diabase porphyry (1978)
BT diabase
BT plutonic rocks
BT igneous rocks
SA porphyry

Diablo Bolson
use Diablo Platform

Diablo Platform (1978)
Extreme W Texas.
IN Index state and counties as applicable.
UF Diablo Bolson
SA Texas
SA West Texas

Diablo Range (1978)
One of the Coast Ranges in W central California.
IN Index Coast Ranges and counties as applicable.
SA California
SA Coast Ranges
SA Greenville Fault

diachronism (1978)
Before 1993, also search heterochrony, a concept used for the appearance of a flora or fauna in a new region during a time period significantly different from the time when it inhabited the old region.
UF heterochrony
SA biologic evolution
SA biostratigraphy
SA ontogeny
SA paleoecology
SA stratigraphy

diagenesis (1978)
Used for all the chemical, physical and biological changes, modifications, or transformations undergone by a sediment after its initial deposition, and during and after

its lithification, exclusive of weathering and metamorphism. Autoposting of this term began in 1978.
NT burial diagenesis
NT cementation
NT chertification
NT early diagenesis
NT late diagenesis
NT syngenesis
SA alteration
SA authigenesis
SA calcification
SA calcitization
SA coalification
SA compaction
SA consolidation
SA crystallinity
SA dedolomitization
SA deposition
SA dolomitization
SA halmyrolysis
SA indicators
SA lithification
SA materials
SA metamorphism
SA metasomatism
SA phosphatization
SA pressure solution
SA processes
SA pyritization
SA sedimentary petrology
SA sedimentary rocks
SA sedimentary structures
SA sedimentation
SA sediments
SA silicification
SA stylolitization
SA weathering

diagonal lamination
use cross-laminations

diagonal-slip faults (1978)
Before 1978, search faults AND diagonal-slip.
BT faults

diagrams
As of 1981, no longer a valid term for GeoRef. See graphic methods.

diallagite (1978)
BT ultramafics
BT plutonic rocks
BT igneous rocks

dialogite
use rhodochrosite

diamictite (1978)
Also search mixtite.
BT clastic rocks
BT sedimentary rocks
SA mixtite
SA terrigenous materials

diamicton (1978)
UF symmicton
BT clastic sediments
BT sediments
SA till

diamond (1978)
When of economic value, use diamonds.
BT native elements
SA carbon
SA chaoite
SA diamonds
SA gems
SA graphite
SA kimberlite
SA lonsdaleite

diamond anvil cells (1993)
High pressure experimental instrument. As of 1993, use for the instrument or methods. Before 1993, also search diamond-anvil cells.
UF diamond-anvil cells
SA geophysics
SA high pressure
SA instruments
SA methods
SA phase equilibria

Diamond Peak Formation (1989)
E California, N Nevada and N Utah.
BT Carboniferous
BT Paleozoic
SA California
SA Nevada
SA Utah

diamond-anvil cells
use diamond anvil cells

diamonds (1978)
Use only when of economic value.
IN Commodity. See List C.
SA abrasives
SA carbon
SA carbonado
SA diamond
SA gems
SA kimberlite
SA placers

diaphorite
use ultrabasite

diaphthoresis
use retrograde metamorphism

diapiric fold
use diapirs

diapirism (1978)
SA diapirs
SA intrusions
SA salt domes
SA salt tectonics

diapirs (1978)
UF diapiric fold
UF piercement
UF piercing fold
SA anticlines
SA cap rocks
SA diapirism
SA domes
SA folds
SA salt
SA salt domes
SA salt tectonics

Diapsida (1993)
Subclass. Before 1993, also search Euryapsida. Autoposted to its divisions, Placodontia, Ichthyosauria, Sauropterygia, Araeoscelidia, Rhynchosauria, and to the Trilophosauria and Archosauria (Archosauromorpha) as of 1993.
UF Euryapsida
BT Reptilia
BT Tetrapoda
BT Vertebrata
BT Chordata
NT Araeoscelidia
NT Archosauria
NT Eosuchia
NT Ichthyosauria
NT Lepidosauria
NT Placodontia
NT Rhynchosauria
NT Sauropterygia
NT Trilophosauria
SA diapsids

diapsids (1993)
Reptiles with 2 temporal openings (two-arched) in the skull.
BT reptiles
BT tetrapods
BT vertebrates
SA Diapsida
SA dinosaurs

diaspore (1978)
BT oxides

diastrophism
As of 1981, no longer a valid term for GeoRef. See orogeny; epeirogeny.

diatom flora (1981)
Common name for diatoms. Autoposting of algal flora to this term began in 1989. Autoposting of microfossils to this term began in 1990. This term has multiple hierarchies.
BT1 algal flora
BT1 plants
BT2 algal flora
BT2 microfossils
SA biostratigraphy
SA diatoms

diatomaceous earth (1978)
Before 1981, also search diatomite AND sedimentary rocks.
UF earth, diatomaceous
UF tripolite
BT clastic rocks
BT sedimentary rocks
SA diatomite
SA diatoms

diatomite (1978)
As of 1981, to be used only if of economic value; otherwise, use diatomaceous earth. From 1978-1980, clastic rocks was autoposted to this term.
IN See List C for use as a commodity term.
SA abrasives
SA clastic rocks
SA diatomaceous earth

diatoms (1978)
Autoposting of thallophytes and microfossils began in 1990. This term has multiple hierarchies.
BT1 algae
BT1 thallophytes
BT1 Plantae
BT2 algae
BT2 microfossils
NT Melosira
NT Navicula
NT Nitzschia
SA diatom flora
SA diatomaceous earth
SA frustules
SA phytoplankton
SA plankton

diatremes (1978)
BT intrusions
SA breccia pipes
SA pipes
SA volcanic features
SA volcanic necks

Dibranchiata (1981)
Autoposting of this term to Belemnitidae and Belemnopsis began in 1985. Before 1989, also search Coleoidea.
UF Coleoidea
BT Cephalopoda
BT Mollusca
BT Invertebrata
NT Belemnoidea

dibromochloropropane (1989)
BT chlorinated hydrocarbons
SA propane

dichloro-diphenyl-trichloro-ethane
use DDT

Dickenson County
Valid through 1988. Search in combination with state term. After 1988, use specific county-state term.

Dickenson County Virginia (1989)
SW Virginia. Before 1989, also search Dickenson County AND Virginia.
CO N365500N371700 W0821000W0823500
BT Virginia
BT United States

Dickinson County
Valid through 1988. Search in combination with state term. After 1988, use specific county-state term.

Dickinson County Iowa (1989)
NW Iowa. Before 1989, also search Dickinson County AND Iowa.
CO N431600N433000 W0945500W0952300
BT Iowa
BT United States
SA Spirit Lake

Dickinson County Kansas (1989)
Central Kansas. Before 1989, also search Dickinson County AND Kansas.
CO N383500N390900 W0965200W0972300
BT Kansas
BT United States
NT Abilene Kansas

Dickinson County Michigan (1989)
N Michigan. Before 1989, also search Dickinson County AND Michigan.
CO N454200N461500 W0873700W0880700
BT Michigan Upper Peninsula
BT Michigan
BT United States

dickite (1978)
BT clay minerals
BT sheet silicates
BT silicates

dicophane
use DDT

Dicotyledoneae (1978)
Before 1981, also search dicotyledons. Autoposting of this term began in 1978.
BT angiosperms
BT Spermatophyta
BT Plantae
NT Acer
NT Alnus
NT Annonaceae
NT Artemisia
NT Asteridae
NT Betula
NT Carya
NT Caryophyllidae
NT Compositae
NT Corylus
NT Dilleniidae
NT Dryas
NT Ericaceae
NT Euphorbiaceae
NT Fagus
NT Hamamelididae
NT Juglandaceae
NT Juglans
NT Lauraceae
NT Leguminosae

NT Magnoliidae
NT Myrica
NT Nothofagus
NT Platanus
NT Platycarya
NT Pterocarya
NT Quercus
NT Rosidae
NT Salix
NT Tilia
NT Ulmaceae
NT Ulmus
SA Monocotyledoneae

Dicroidium (1978)
BT Filicopsida
BT pteridophytes
BT Plantae
SA gymnosperms
SA Pteridospermae

dictionaries (1978)
SA Arabic
SA French
SA German
SA glossaries
SA lexicons
SA monographs
SA publications
SA Russian

Dictyocha (1978)
This term has multiple hierarchies.
BT1 Silicoflagellata
BT1 Protista
BT1 Invertebrata
BT2 Silicoflagellata
BT2 Protista
BT2 microfossils

Dictyonellidina (1981)
BT Articulata
BT Brachiopoda
BT Invertebrata

Didacna (1978)
Genus.
BT Bivalvia
BT Mollusca
BT Invertebrata

Didymograptina (1978)
Autoposting of this term began in 1981.
IN Index Hemichordata or Invertebrata as applicable.
BT Graptoloidea
BT Graptolithina
NT Didymograptus
NT Isograptus
SA Hemichordata
SA Invertebrata

Didymograptus (1978)
Genus.
IN Index Hemichordata or Invertebrata as applicable.
BT Didymograptina
BT Graptoloidea
BT Graptolithina
SA Hemichordata
SA Invertebrata

dielectric constant (1978)
UF constant, dielectric
SA dielectric properties
SA electrical field
SA electrical methods
SA electrical properties

dielectric properties (1978)
SA dielectric constant
SA electrical field
SA electrical methods
SA electrical properties
SA properties

Dieppe
No longer a valid term for GeoRef. See Dieppe France.

Dieppe France (1993)
City on English Channel in N France.
BT Seine-Maritime France
BT France
BT Western Europe
BT Europe

diet (1985)
SA biochemistry
SA metabolism
SA middens
SA nutrients
SA nutrition
SA paleontology
SA predation
SA trophic analysis

differential thermal analysis (1978)
Was used for methodology through 1977.
UF thermography
SA analysis
SA chemical analysis
SA clay mineralogy
SA clay minerals
SA DTA data
SA electron microscopy
SA sample preparation
SA spectroscopy
SA thermal analysis
SA thermogravimetric analysis
SA X-ray diffraction analysis

differential thermal analysis data
use DTA data

differential weathering (1978)
Before, 1978, also search weathering AND differential.
BT weathering
SA differentiation

differentiation (1978)
As of 1993, see differential weathering, horizon differentiation, magmatic differentiation, and metamorphic differentiation for those processes.
SA differential weathering
SA Earth
SA horizon differentiation
SA magmatic differentiation

diffraction (1978)
SA diffractograms
SA neutron diffraction analysis
SA ocean waves
SA reflection
SA refraction
SA seismic migration
SA seismology

diffractograms (1985)
SA diffraction
SA electron diffraction analysis
SA neutron diffraction analysis
SA X-ray diffraction analysis

diffusion (1978)
IN See also under specific elements.
SA absolute age
SA advection
SA chemical analysis
SA diffusivity
SA geochemistry
SA hydrochemistry
SA ion exchange
SA meteorology
SA ocean circulation
SA oceanography
SA osmosis
SA permeability
SA polarography
SA processes
SA sintering

diffusivity (1985)
Use thermal diffusivity when applicable.
SA diffusion
SA geochemistry
SA hydrochemistry
SA ion exchange
SA migration of elements
SA thermal diffusivity

diffusivity, thermal
use thermal diffusivity

Difunta Group (1978)
Upper Cretaceous to lower Paleocene. Mexico.
IN Index ages as applicable.
SA Coahuila Mexico
SA K-T boundary
SA lower Paleocene
SA Mexico
SA Upper Cretaceous

digenite (1978)
UF alpha chalcocite
UF blue chalcocite
BT sulfides
SA chalcocite

Digha (1978)
Coastal region in West Bengal along the Bay of Bengal.
BT West Bengal India
BT India
BT Indian Peninsula
BT Asia

digital cartography (1993)
Before 1993, search automatic cartography.
UF automatic cartography
BT cartography
SA data processing
SA maps
SA multispectral scanner
SA thematic mapper

digital elevation models
use digital terrain models

digital line graphs (1993)
Before 1993, also search DLG.
SA data processing
SA graphic methods

digital relief models
use digital terrain models

digital simulation (1978)
Before 1978, search digital AND simulation.
SA data processing
SA mathematical geology
SA mathematical models
SA models
SA pixels
SA simulation

digital techniques
As of 1981, no longer a valid term for GeoRef. See data processing.

digital terrain models (1993)
Before 1993, also search digital elevation models.
UF digital elevation models
UF digital relief models
SA data processing
SA geomorphology
SA models
SA relief
SA terrain classification
SA terrains

Digne
No longer a valid term for GeoRef. See Digne France.

Digne France (1993)
City in SE France.
BT Alpes-de-Haute Provence France
BT France
BT Western Europe
BT Europe

Dijon
No longer a valid term for GeoRef. See Dijon France.

Dijon France (1993)
City in E central France.
BT Cote-d'Or France
BT France
BT Western Europe
BT Europe

dike swarms (1978)
SA dikes
SA igneous rocks
SA intrusions
SA ring dikes
SA swarms

dikes (1978)
UF dykes
BT intrusions
NT ring dikes
SA batholiths
SA clastic dikes
SA dike swarms
SA igneous rocks
SA laccoliths
SA ring complexes
SA ring structures
SA sills

dilatancy (1978)
SA deformation
SA dilation
SA dilatometers
SA earthquakes
SA geophysics
SA seismology

dilatation
use dilation

dilatational wave
use P-waves

dilation (1978)
Used as a general term.
UF dilatation
SA dilatancy
SA dilatometers

dilatometers (1985)
SA dilatancy
SA dilation
SA instruments
SA seismology

Dilleniidae (1981)
BT Dicotyledoneae
BT angiosperms
BT Spermatophyta
BT Plantae

Dillon
Valid through 1988. Search in combination with state term. After 1988, use specific city-state term.

Dillon Colorado (1989)
Town in N central Colorado. Before 1989, search Dillon AND Colorado.
CO N393700N393700 W1060600W1060600
BT Summit County Colorado
BT Colorado
BT United States

dilution (1982)
SA concentration
SA decontamination
SA desalinization
SA geochemistry
SA polluted water

Dicroidium (1978)
SA polluMagNdliidae
NT Myrica
NT Nothofagus
NT Platanus
NT Platycarya
NT Pterocarya
NT Quercus
NT Rosidae
NT Salix
NT Tilia
NT Ulmaceae
NT Ulmus
SA Monocotyledoneae

Dicroidium (1978)
BT Filicopsida
BT pteridophytes
BT Plantae
SA gymnosperms
SA Pteridospermae

dictionaries (1978)
SA Arabic
SA French
SA German
SA glossaries
SA lexicons
SA monographs
SA publications
SA Russian

Dictyocha (1978)
This term has multiple hierarchies.
BT1 Silicoflagellata
BT1 Protista
BT1 Invertebrata
BT2 Silicoflagellata
BT2 Protista
BT2 microfossils

Dictyonellidina (1981)
BT Articulata
BT Brachiopoda
BT Invertebrata

Didacna (1978)
Genus.
BT Bivalvia
BT Mollusca
BT Invertebrata

Didymograptina (1978)
Autoposting of this term began in 1981.
IN Index Hemichordata or Invertebrata as applicable.
BT Graptoloidea
BT Graptolithina
NT Didymograptus
NT Isograptus
SA Hemichordata
SA Invertebrata

Didymograptus (1978)
Genus.
IN Index Hemichordata or Invertebrata as applicable.
BT Didymograptina
BT Graptoloidea
BT Graptolithina
SA Hemichordata
SA Invertebrata

dielectric constant (1978)
UF constant, dielectric
SA dielectric properties
SA electrical field
SA electrical methods
SA electrical properties

dielectric properties (1978)
SA dielectric constant
SA electrical field
SA electrical methods
SA electrical properties
SA properties

Dieppe
No longer a valid term for GeoRef. See Dieppe France.

Dieppe France (1993)
City on English Channel in N France.
BT Seine-Maritime France
BT France
BT Western Europe
BT Europe

diet (1985)
SA biochemistry
SA metabolism
SA middens
SA nutrients
SA nutrition
SA paleontology
SA predation
SA trophic analysis

differential thermal analysis (1978)
Was used for methodology through 1977.
UF thermography
SA analysis
SA chemical analysis
SA clay mineralogy
SA clay minerals
SA DTA data
SA electron microscopy
SA sample preparation
SA spectroscopy
SA thermal analysis
SA thermogravimetric analysis
SA X-ray diffraction analysis

differential thermal analysis data
use DTA data

differential weathering (1978)
Before, 1978, also search weathering AND differential.
BT weathering
SA differentiation

differentiation (1978)
As of 1993, see differential weathering, horizon differentiation, magmatic differentiation, and metamorphic differentiation for those processes.
SA differential weathering
SA Earth
SA horizon differentiation
SA magmatic differentiation

diffraction (1978)
SA diffractograms
SA neutron diffraction analysis
SA ocean waves
SA reflection
SA refraction
SA seismic migration
SA seismology

diffractograms (1985)
SA diffraction
SA electron diffraction analysis
SA neutron diffraction analysis
SA X-ray diffraction analysis

diffusion (1978)
IN See also under specific elements.
SA absolute age
SA advection
SA chemical analysis
SA diffusivity
SA geochemistry
SA hydrochemistry
SA ion exchange
SA meteorology
SA ocean circulation
SA oceanography
SA osmosis
SA permeability
SA polarography
SA processes
SA sintering

diffusivity (1985)
Use thermal diffusivity when applicable.
SA diffusion
SA geochemistry
SA hydrochemistry
SA ion exchange
SA migration of elements
SA thermal diffusivity

diffusivity, thermal
use thermal diffusivity

Difunta Group (1978)
Upper Cretaceous to lower Paleocene. Mexico.
IN Index ages as applicable.
SA Coahuila Mexico
SA K-T boundary
SA lower Paleocene
SA Mexico
SA Upper Cretaceous

digenite (1978)
UF alpha chalcocite
UF blue chalcocite
BT sulfides
SA chalcocite

Digha (1978)
Coastal region in West Bengal along the Bay of Bengal.
BT West Bengal India
BT India
BT Indian Peninsula
BT Asia

digital cartography (1993)
Before 1993, search automatic cartography.
UF automatic cartography
BT cartography
SA data processing
SA maps
SA multispectral scanner
SA thematic mapper

digital elevation models
use digital terrain models

digital line graphs (1993)
Before 1993, also search DLG.
SA data processing
SA graphic methods

digital relief models
use digital terrain models

digital simulation (1978)
Before 1978, search digital AND simulation.
SA data processing
SA mathematical geology
SA mathematical models
SA models
SA pixels
SA simulation

digital techniques
As of 1981, no longer a valid term for GeoRef. See data processing.

digital terrain models (1993)
Before 1993, also search digital elevation models.
UF digital elevation models
UF digital relief models
SA data processing
SA geomorphology
SA models
SA relief
SA terrain classification
SA terrains

Digne
No longer a valid term for GeoRef. See Digne France.

Digne France (1993)
City in SE France.
BT Alpes-de-Haute Provence France
BT France
BT Western Europe
BT Europe

Dijon
No longer a valid term for GeoRef. See Dijon France.

Dijon France (1993)
City in E central France.
BT Cote-d'Or France
BT France
BT Western Europe
BT Europe

dike swarms (1978)
SA dikes
SA igneous rocks
SA intrusions
SA ring dikes
SA swarms

dikes (1978)
UF dykes
BT intrusions
NT ring dikes
SA batholiths
SA clastic dikes
SA dike swarms
SA igneous rocks
SA laccoliths
SA ring complexes
SA ring structures
SA sills

dilatancy (1978)
SA deformation
SA dilation
SA dilatometers
SA earthquakes
SA geophysics
SA seismology

dilatation
use dilation

dilatational wave
use P-waves

dilation (1978)
Used as a general term.
UF dilatation
SA dilatancy
SA dilatometers

dilatometers (1985)
SA dilatancy
SA dilation
SA instruments
SA seismology

Dilleniidae (1981)
BT Dicotyledoneae
BT angiosperms
BT Spermatophyta
BT Plantae

Dillon
Valid through 1988. Search in combination with state term. After 1988, use specific city-state term.

Dillon Colorado (1989)
Town in N central Colorado. Before 1989, search Dillon AND Colorado.
CO N393700N393700 W1060600W1060600
BT Summit County Colorado
BT Colorado
BT United States

dilution (1982)
SA concentration
SA decontamination
SA desalinization
SA geochemistry
SA polluted water

SA pollution
SA saturation
SA water treatment

dimension stone (1978)
IN Used as a commodity term.
UF stone, dimension
BT construction materials
SA building stone
SA granite deposits
SA limestone deposits
SA marble deposits
SA sandstone deposits

dimensions (1978)
SA cell dimensions
SA depth
SA grain size
SA grains

dimethyl sulfide (1989)
UF methanethiomethane
UF methyl sulfide
SA geochemistry

Dimitrovo coal basin
use Pernik coal basin

Dimmit County
Valid through 1988. Search in combination with state term. After 1988, use specific county-state term.

Dimmit County Texas (1989)
SW Texas. Before 1989, also search Dimmit County AND Texas.
CO N281000N284000
W0992500W1001000
BT Texas
BT United States

dimorphism
As of 1981, no longer a valid term for GeoRef. See polymorphism or sexual dimorphism.

Dinant
No longer a valid term for GeoRef. See Dinant Belgium.

Dinant Basin (1978)
S Belgium.
BT Namur Belgium
BT Belgium
BT Western Europe
BT Europe

Dinant Belgium (1993)
Town in S Belgium.
BT Namur Belgium
BT Belgium
BT Western Europe
BT Europe

Dinantian (1978)
Europe. Includes Tournaisian and Visean. As of 1981, used for Lower Carboniferous. Before 1981, also search Lower Carboniferous. Autoposting of this term began in 1978.
UF Lower Carboniferous
BT Carboniferous
BT Paleozoic
NT Asbian
NT Tournaisian
NT Visean
SA Avonian
SA Culm
SA Horton Group
SA Middle Carboniferous
SA Mississippian
SA Upper Carboniferous
SA Waulsortian facies

Dinaric Alps (1978)
SE division of Eastern Alps along Adriatic Sea. Also search Dinaric. In 1981, broader term changed from Yugoslavia to Europe. This term has multiple hierarchies.
IN Index countries or regions as applicable.
CO N430000N450000
E0180000E0150000
BT1 Southern Europe
BT1 Europe
BT2 Eastern Alps
BT2 Alps
BT2 Europe
NT Northern Limestone Alps
NT Velebit Mountains
NT Zlatibor Mountains
SA Albania
SA Croatia
SA Slovenia
SA Trieste Italy
SA Yugoslavia

Dingle Peninsula (1978)
Juts into Atlantic in W County Kerry, SW Ireland.
CO N520500N521500
W0094500W0103000
BT Kerry Ireland
BT Ireland
BT Western Europe
BT Europe

Dinocerata (1981)
Suborder. Autoposting of Amblypoda, Eutheria and Theria to this term began in 1989.
BT Amblypoda
BT Eutheria
BT Theria
BT Mammalia
BT Tetrapoda
BT Vertebrata
BT Chordata
SA Ungulata

Dinoflagellata (1978)
Hystrichosphaeridae is included here and under acritarchs. Before 1981, also search dinoflagellates. Autoposting of microfossils to this term began in 1989.
UF Dinophyceae
BT palynomorphs
BT microfossils
NT Deflandrea
NT Hystrichosphaeridae
NT Peridinium
SA acritarchs
SA dinoflagellates
SA plankton
SA Pyrrhophyta

dinoflagellates (1981)
Common name for Dinoflagellata. Autoposting of palynomorphs and microfossils to this term began in 1989.
BT palynomorphs
BT microfossils
SA biostratigraphy
SA Dinoflagellata

Dinophyceae
use Dinoflagellata

Dinosaur National Monument (1985)
NE Utah and NW Colorado.
IN Index states and counties as applicable.
CO N402500N404500
W1083000W1092500
BT United States
SA Colorado
SA Moffat County Colorado
SA national monuments
SA Uintah County Utah
SA Utah

Dinosaur Provincial Park (1993)
E of the town of Brooks in SE Alberta.
CO N504200N505500
W1114500W1112000
BT Alberta
BT Western Canada
BT Canada

Dinosauria
As of 1978, no longer a valid term for GeoRef. Used in subfiles B, N, and E.
use dinosaurs

dinosaurs (1978)
Also search Dinosauria. As of 1993, this term began autoposting to Ornithischia and Saurischia.
IN Index Reptilia or reptiles as applicable.
UF Dinosauria
BT Tetrapoda
BT Vertebrata
BT Chordata
NT Ornithischia
NT Saurischia
SA biostratigraphy
SA diapsids
SA reptiles
SA Reptilia

Dinwiddie County Virginia (1989)
SE central Virginia. Before 1989, search Dinwiddie County AND Virginia.
CO N365300N371700
W0772400W0775400
BT Virginia
BT United States
NT Petersburg Virginia

Dinwoody Formation (1985)
W Wyoming, SE Idaho, SW Montana and NE Utah.
BT Lower Triassic
BT Triassic
BT Mesozoic
SA Beaverhead Formation
SA Idaho
SA Montana
SA Utah
SA Wyoming

diogenite (1981)
BT achondrites
BT stony meteorites
BT meteorites

Diois (1978)
Massif in SE France.
BT Drome France
BT France
BT Western Europe
BT Europe

Dione Satellite (1985)
One of the satellites of Saturn. Before 1985, also search Dione AND Saturn.
SA icy satellites
SA satellites
SA Saturn
SA Voyager Program

diopside (1978)
BT clinopyroxene
BT pyroxene group
BT chain silicates
BT silicates
SA chrome diopside
SA salite

diorite
As of 1989, no longer a valid term for GeoRef. From 1981 to 1988, diorites was autoposted to this term.
use diorites

diorite family
As of 1981, no longer a valid term for GeoRef. Use diorites, alkali diorites, or quartz diorites as applicable. From 1978-1980, was autoposted to the rocks that were classified under it.

diorite porphyry (1978)
From 1978-1980, volcanic rocks was autoposted to this term.
BT diorites
BT plutonic rocks
BT igneous rocks
SA porphyry
SA volcanic rocks

diorites (1981)
Before 1981, search diorite family. From 1978-1980, diorite family was autoposted to the rocks that were classified under that term. Before 1989, also search diorite. Autoposting of this term to diorite ended in 1989.
UF diorite
BT plutonic rocks
BT igneous rocks
NT alkali diorites
NT diorite porphyry
NT ferrodiorite
NT mangerite
NT microdiorite
NT monzodiorite
NT plagiogranite
NT quartz diorites
NT syenodiorite
NT tonalite
NT trondhjemite
SA appinite
SA metadiorite

dip (1978)
Before 1978, included use for dip faults and dip fractures.
UF angle of dip
SA dip-slip faults
SA dipmeter logging
SA orientation
SA strike
SA structural analysis

dip faults (1978)
Before 1978, search faults AND dip.
BT faults

dip fractures (1978)
Before 1978, search dip AND fractures.
BT fractures

dip moveout (1993)
Difference in source-to-geophone arrival times from reflector dip.
SA arrival time
SA moveout
SA reflection
SA reflection methods
SA seismic methods
SA seismic migration
SA seismic surveys
SA traveltime

dip-slip faults (1978)
Before 1978, search faults AND dip-slip.
BT faults
SA dip
SA displacements
SA transfer faults

Diplocraterion (1989)
Genus.
BT ichnofossils

Diplograptina (1978)

IN Index Hemichordata or Invertebrata as applicable.
BT Graptoloidea
BT Graptolithina
SA Hemichordata
SA Invertebrata

dipmeter logging (1978)
Before 1978, also search well-logging AND dipmeter or dip-meter.
UF logging, dipmeter
BT well-logging
SA boreholes
SA dip

Dipnoi (1981)
Order.
BT Sarcopterygii
BT Osteichthyes
BT Pisces
BT Vertebrata
BT Chordata

dipole moment (1978)
SA magnetic field
SA paleomagnetism

Diprotodonta (1981)
Suborder. Autoposting of Marsupialia, Metatheria and Theria to this term began in 1989.
BT Marsupialia
BT Metatheria
BT Theria
BT Mammalia
BT Tetrapoda
BT Vertebrata
BT Chordata
NT Macropodidae

Diptera (1978)
Before 1993, also search Dipteri and Insecta.
BT Insecta
BT Mandibulata
BT Arthropoda
BT Invertebrata
SA Raphidioidea

dipyrite
 use pyrrhotite

direct current methods
 use direct-current methods

direct hydrocarbon indicators
 use hydrocarbon indicators

direct magnetization (1981)
BT magnetization

direct problem (1981)
Before 1993, also search forward problem.
UF forward problem
SA geophysics
SA inverse problem
SA seismology

direct-current methods (1985)
UF direct current methods
SA electrical methods
SA electrical surveys
SA geophysical methods
SA geophysical surveys
SA potentiometry

directions, current
 use current directions

directory (1978)
SA guidebook
SA monographs
SA publications

discharge (1978)
Before 1981, also used for effluents.
SA effluents
SA flowmeters
SA ground water
SA hydrologic cycle
SA hydrology
SA recharge
SA runoff
SA sand boils
SA springs
SA streamflow
SA water balance
SA water recovery

Discoasteridae (1985)
Before 1985. also search discoasters; discoaster. As of 1985, discoasters may be used as the common name for Discoasteridae. Autoposting of thallophytes and microfossils began in 1990. This term has multiple hierarchies.
BT1 nannofossils
BT1 algae
BT1 thallophytes
BT1 Plantae
BT2 nannofossils
BT2 algae
BT2 microfossils
BT3 nannofossils
BT3 algal flora
BT3 plants
BT4 nannofossils
BT4 algal flora
BT4 microfossils
NT Ethmodiscus
SA Coccolithophoraceae
SA discoasters
SA Sphenolithus

discoasters (1978)
As of 1985, to be used as the common name for Discoasteridae. Autoposting of this term to Ethmodiscus ended in 1985. Autoposting of nannofossils and algal flora to this term began in 1989. This term has multiple hierarchies.
BT1 nannofossils
BT1 algae
BT1 thallophytes
BT1 Plantae
BT2 nannofossils
BT2 algae
BT2 microfossils
BT3 nannofossils
BT3 algal flora
BT3 plants
BT4 nannofossils
BT4 algal flora
BT4 microfossils
SA biostratigraphy
SA Coccolithophoraceae
SA Discoasteridae

Discocyclina (1978)
Genus. Autoposting of microfossils and Protista to this term began in 1990. This term has multiple hierarchies.
BT1 Orbitoidacea
BT1 Rotaliina
BT1 foraminifera
BT1 Protista
BT1 Invertebrata
BT2 Orbitoidacea
BT2 Rotaliina
BT2 foraminifera
BT2 Protista
BT2 microfossils

disconformities
 use erosional unconformities

discontinuities (1978)
Use for interface in seismology. For stratigraphic discontinuities use unconformities.
NT horizontal discontinuities
NT tangential discontinuities
NT vertical discontinuities
SA Conrad discontinuity
SA crust
SA curved seismic interface
SA inclined seismic interface
SA mantle
SA Mohorovicic discontinuity
SA seismology

discontinuous materials (1981)
SA engineering geology
SA materials

Discorbacea (1978)
Autoposting of microfossils and Protista to this term began in 1990. This term has multiple hierarchies.
BT1 Rotaliina
BT1 foraminifera
BT1 Protista
BT1 Invertebrata
BT2 Rotaliina
BT2 foraminifera
BT2 Protista
BT2 microfossils

discordant folds (1978)
Before 1978, also search folds AND discordant.
BT folds
SA orientation

discoveries (1978)
As of 1981, restricted to economic geology. Before 1981, used as a general term.
UF discovery
SA mineral deposits, genesis
SA placers

discovery
 use discoveries

Discovery Bay (1978)
Bay in Jefferson County, W Washington. Also an inlet of the Indian Ocean bordering South Australia and Victoria. As of 1993, see Discovery Bay Jamaica for the village.
IN Index regions as applicable.
SA Discovery Bay Jamaica
SA Jefferson County Washington
SA South Australia
SA Victoria Australia
SA Washington

Discovery Bay Jamaica (1993)
Village on central N coast of Jamaica. Before 1993, also search Discovery Bay AND Jamaica.
BT Jamaica
BT Greater Antilles
BT Antilles
BT West Indies
BT Caribbean region
SA Discovery Bay

Discovery Deep (1978)
Also search Discovery AND deeps.
BT Red Sea
BT Indian Ocean

discrete filters (1981)
SA analog filters
SA filters
SA Fourier analysis

discriminant analysis (1978)
UF discriminant function analysis
BT statistical analysis
SA analysis
SA Bayesian analysis

discriminant function analysis
 use discriminant analysis

disequilibrium
 use equilibrium

disharmonic folds (1978)
Before 1978, also search folds AND disharmonic.
BT folds
SA decollement
SA flexural-slip
SA harmonic folds

disjunctive folds (1982)
BT folds

Disko Island (1978)
Davis Strait just off SW Greenland. Also search Disko.
CO N683000N702000 W0520000W0550000
BT Greenland
BT Arctic region

dislocations
As of 1981, no longer a valid term for GeoRef. For dislocations within crystals, use crystal dislocations.

dispersion
No longer a valid term for GeoRef. See chemical dispersion and wave dispersion. From 1981 through 1992 was used for the chemical process of dispersion. Before 1981, included use as a general term for dispersal; also search dispersal.

dispersion patterns (1978)
SA dispersivity
SA geochemical methods
SA haloes
SA leakage anomalies
SA mineral exploration
SA patterns
SA primary dispersion
SA secondary dispersion

dispersion, wave
 use wave dispersion

dispersive filters (1981)
SA filters

dispersivity (1989)
Aquifer rock property.
SA aquifers
SA chemical dispersion
SA dispersion patterns
SA properties
SA rock mechanics
SA waste disposal

displacement theory
 use continental drift

displacements (1978)
SA dip-slip faults
SA drag folds
SA faults
SA nappes
SA overthrust faults
SA reactivation
SA reverse faults
SA seismic moment
SA slip rates
SA transcurrent faults
SA transform faults
SA wrench faults

display, graphic
 use graphic display

disposal barriers (1993)
Before 1993, also search clay liners, soil liners, earthen liners, liners and barriers with waste disposal.
UF clay liners
UF containment liners
UF liners
UF soil liners
SA clay
SA grouting

SA landfills
SA permeability
SA reclamation
SA soils
SA tailings
SA waste disposal

disposal, waste
use waste disposal

disseminated deposits (1978)
Before 1978, search disseminated AND ore deposits; search deposits AND disseminated.
SA carlin-type
SA mineral deposits, genesis

dissertations
use theses

dissipation
As of 1981, no longer a valid term for GeoRef. Before 1981, used as a general term.

dissociation (1978)
SA electrolysis
SA geochemistry
SA processes

dissolution
A valid term through May 1978. After May 1978, use solution or solutions.

dissolved materials (1978)
Before 1978, search dissolved.
SA hydrochemistry
SA materials
SA solubility
SA solutes
SA suspended materials

Distephanus (1978)
This term has multiple hierarchies.
BT1 Silicoflagellata
BT1 Protista
BT1 Invertebrata
BT2 Silicoflagellata
BT2 Protista
BT2 microfossils

distortion (1978)
SA amplitude distortion
SA deconvolution
SA deformation
SA elastic waves
SA filters
SA linear distortion
SA non-linear distortion
SA phase distortion
SA seismology
SA signal distortion
SA strain
SA wave absorption
SA wave dispersion

distribution (1978)
SA asymmetric distribution
SA size distribution
SA spatial distribution
SA statistical distribution
SA temporal distribution

distribution coefficients
use partition coefficients

District of Algoma
use Algoma District Ontario

District of Cochrane
use Cochrane District Ontario

District of Columbia (1978)
CO N384700N390000
W0765400W0771000
BT United States
SA Anacostia River basin
SA Potomac River
SA Potomac River basin
SA Rock Creek Park
SA Smithsonian Institution

SA Wicomico Formation

District of Franklin
No longer a valid term for GeoRef as of 1985.
use Franklin District Northwest Territories

District of Keewatin
No longer a valid term for GeoRef as of 1985.
use Keewatin District Northwest Territories

District of Kenora
No longer a valid term for GeoRef as of 1985.
use Kenora District Ontario

District of Mackenzie
No longer a valid term for GeoRef as of 1985.
use Mackenzie District Northwest Territories

District of Nipissing
use Nipissing District Ontario

District of Sudbury
use Sudbury District Ontario

District of Temiscamingue
use Timiskaming District Ontario

District of Temiskaming
use Timiskaming District Ontario

District of Thunder Bay
use Thunder Bay District Ontario

District of Timiskaming
use Timiskaming District Ontario

Distrito Federal
Not a valid term for GeoRef. Term used where it is larger than the city it contains. As of 1993, see Buenos Aires Argentina, Federal District Mexico, or Federal District Venezuela as applicable.

disturbances
As of 1993, no longer a valid term for GeoRef. Usually out-of-scope. Used with ionosphere and magnetosphere. Term was restricted to extraterrestrial phenomena.

Disturbed Belt (1985)
Structural unit, W Montana and S Alberta.
BT North America
SA Alberta
SA Belt Supergroup
SA Canada
SA Montana
SA Rocky Mountains
SA United States

diterpanes (1989)
BT hydrocarbons
BT organic materials
SA terpanes
SA triterpanes

Ditrau
No longer a valid term for GeoRef. As of 1993 see Ditrau Romania.

Ditrau Romania (1993)
Town in Mures County in Transylvania, E central Romania. Before 1993 also search Ditrau AND Romania.
BT Mures Romania
BT Transylvania
BT Romania
BT Southern Europe
BT Europe

Dittonian (1978)

In Old Red Sandstone, England: upper Gedinnian; above Downtonian.
BT Lower Devonian
BT Devonian
BT Paleozoic
SA Old Red Sandstone

diurnal variations (1978)
Also search diurnal.
UF daily variation
SA magnetic field
SA variations

divergence, plate
use plate divergence

divergent plate boundaries
use accreting plate boundary

diversity (1978)
SA species diversity

diwa-type (1989)
SA coal
SA mineral deposits, genesis
SA petroleum

Dixie Valley (1978)
Churchill County in W Nevada.
IN Index county or regions as applicable.
SA Churchill County Nevada

Djadjerud Valley (1978)
BT Iran
BT Middle East
BT Asia

Djadokhta Formation (1978)
In the Gobi Desert.
BT Upper Cretaceous
BT Cretaceous
BT Mesozoic
SA Mongolia

Djeddah
use Jeddah Saudi Arabia

Djibouti (1978)
Became independent June 27, 1977. Formerly French Territory of the Afars and Issas. Capital city also named Djibouti. Before 1978, also search Afars and Issas Territory; Afars and Issas.
CO N105000N125000
E0433000E0414000
UF Afars and Issas
UF Afars and Issas Territory
UF French Somaliland
BT East Africa
BT Africa
SA Afar
SA Asal Rift
SA Red Sea Basin

Djiddah
use Jeddah Saudi Arabia

djurleite (1978)
BT sulfides

Djursland (1978)
Peninsula of Arhus County jutting into the Kattegat in E Jutland. This term has multiple hierarchies.
BT1 Denmark
BT1 Scandinavia
BT1 Western Europe
BT1 Europe
BT2 Jutland
BT2 Europe

DLG
Not a valid term for GeoRef. After 1993, see digital line graphs.

DNA (1989)
Acronym.
UF deoxyribonucleic acid
UF desoxyribonucleic acid

SA acids
SA biologic evolution
SA life origin

DNAG (1989)
Acronym. Series of projects sponsored by the Geological Society of America.
UF Decade of North American Geology
SA GSA
SA North America

DNAPLs
use dense nonaqueous phase liquids

Dnepropetrovsk
No longer a valid term for GeoRef. As of 1993, see Dnepropetrovsk Ukraine.

Dnepropetrovsk Ukraine (1993)
Oblast in E central Ukraine. Also the city. This term has multiple hierarchies.
BT1 Ukraine
BT1 Europe
BT2 Ukraine
BT2 Commonwealth of Independent States
NT Krivoy Rog Ukraine
NT Nikopol Ukraine
NT Pavlograd Ukraine
NT Saksagan River
NT Verkhovtsevo Ukraine
SA Krivoy Rog Basin

Dnieper Basin (1978)
River drainage basin. Before 1978, also search Dnieper. This term has multiple hierarchies.
IN Index former Soviet republics as applicable.
UF Dnieper region
BT1 Commonwealth of Independent States
BT2 Europe
SA Belarus
SA Dnieper-Donets Basin
SA Russian Federation
SA Ukraine

Dnieper region
use Dnieper Basin

Dnieper River (1978)
Rises in S Valdai Hills and flows S, then SE into a big bend at Dnepropetrovsk, and finally into the Black Sea near Kherson. Also search Dnieper. This term has multiple hierarchies.
IN Index former Soviet republics as applicable.
BT1 Commonwealth of Independent States
BT2 Europe
SA Belarus
SA Russian Federation
SA Ukraine

Dnieper-Donets Basin (1978)
Region comprising the Donets Basin and that part of the Dnieper Basin in the nearby Dnepropetrovsk area. Also search Dnieper-Donets. This term has multiple hierarchies.
IN Index former Soviet republics as applicable.
CO N460000N520000
E0420000E0300000
UF Dnieper-Donets Depression
UF Dnieper-Donets region
BT1 Russian Platform
BT2 Commonwealth of Independent States
BT3 Europe

SA Dnieper Basin
SA Donets Basin
SA Korenevskaya Formation
SA Russian Federation
SA Ukraine

Dnieper-Donets Depression
use Dnieper-Donets Basin

Dnieper-Donets region
use Dnieper-Donets Basin

Dniester River (1978)
Flows into Black Sea SW of Odessa. Also search Dniester, Dniester region, Dniester Valley. This term has multiple hierarchies.
IN Index former Soviet republics as applicable.
BT1 Commonwealth of Independent States
BT2 Europe
SA Moldavia
SA Moldova
SA Ukraine

Dniester-Prut Interfluve (1978)
Area between the two rivers. Includes all of Moldavia in Moldova which formerly was Bessarabia.
IN Index republics as applicable.
BT Europe
SA Moldavia
SA Moldova
SA Prut River
SA Ukraine

Doberlug
No longer a valid term for GeoRef. See Doberlug Germany.

Doberlug Germany (1993)
Town in Brandenburg, E Germany. Before 1993, Doberlug was used with Cottbus Bezirk in East Germany. Before 1993, also search Doberlug or Doberlug-Kirchhain.
BT Brandenburg Germany
BT Germany
BT Central Europe
BT Europe
SA Upper Lusatia

Doboy Sound (1978)
On the Atlantic Ocean in McIntosh County, SE Georgia.
BT McIntosh County Georgia
BT Georgia
BT United States

Dobrogea
use Romanian Dobruja

Dobrudja
use Dobruja Basin

Dobrudzha
use Dobruja Basin

Dobruja Basin (1978)
Black Sea coastal strip S of the Danube River including NE Bulgaria. Also search Dobruja. Before 1993, also search Dobrudja, and Dobrudzha.
IN Index countries as applicable.
UF Dobrudja
UF Dobrudzha
BT Southern Europe
BT Europe
NT Bulgarian Dobruja
NT Romanian Dobruja
SA Black Sea region
SA Bulgaria
SA Romania

Dobrzyn
No longer a valid term for GeoRef. As of 1993 see Dobrzyn Poland.

Dobrzyn Poland (1993)
Town in Wlockawek Province, N central Poland. Before 1993 also search Dobrzyn AND Poland.
BT Poland
BT Central Europe
BT Europe
SA Bydgoszcz Poland

Dockum Group (1978)
Includes Sloan Canyon Formation, Sheep Pen Canyon Formation, Tecovas Shale, Santa Rosa Sandstone, Chinle Formation, Pierce Canyon Redbeds, Baldy Hill Formation, Travesser Formation. Colorado, Kansas, Oklahoma, New Mexico, and W Texas.
BT Upper Triassic
BT Triassic
BT Mesozoic
SA Chinle Formation
SA Colorado
SA Kansas
SA New Mexico
SA Oklahoma
SA Santa Rosa Sandstone
SA Texas

Docodonta (1978)
Order.
BT Mammalia
BT Tetrapoda
BT Vertebrata
BT Chordata

Doda (1978)
Village in SW Jammu and Kashmir, disputed territory between India and Pakistan. As of 1993, India is no longer a broader term.
IN Index Jammu and Kashmir AND India and/or Pakistan as applicable.
SA India
SA Jammu and Kashmir
SA Pakistan

Dodecanese (1978)
Group of islands in SE Aegean Sea. Also a Greek department. This term has multiple hierarchies.
BT1 Greek Aegean Islands
BT1 Greece
BT1 Southern Europe
BT1 Europe
BT2 Greek Aegean Islands
BT2 Aegean Islands
BT2 Mediterranean region
NT Karpathos
NT Rhodes
SA Kos

Dodge County
Valid through 1988. Search in combination with state term. After 1988, use specific county-state term.

Dodge County Georgia (1989)
S central Georgia. Before 1989, also search Dodge County AND Georgia.
CO N312500N325500
W0824700W0832000
BT Georgia
BT United States

Dodge County Minnesota (1989)
SE Minnesota. Before 1989, also search Dodge County AND Minnesota.
CO N435100N441200
W0924200W0930400
BT Minnesota
BT United States

Dodge County Nebraska (1989)
E Nebraska. Before 1989, also search Dodge County AND Nebraska.
CO N412300N414300
W0961700W0965400
BT Nebraska
BT United States

Dodge County Wisconsin (1989)
S central Wisconsin. Before 1989, also search Dodge County AND Wisconsin.
CO N431200N433800
W0881900W0890100
BT Wisconsin
BT United States

DOE
use U. S. Department of Energy

Dogger (1978)
Europe. Above Lias (Lower Jurassic), below Malm.
BT Middle Jurassic
BT Jurassic
BT Mesozoic
SA Aalenian
SA Bajocian
SA Bathonian
SA Callovian

Dogger Bank (1978)
Submerged sand bank.
BT North Sea
BT European Atlantic
BT North Atlantic
BT Atlantic Ocean

Dogo (1978)
Volcano on Hiroshima-Tottori prefectural border, W Honshu. Before 1993, also used for the largest of the Oki Islands off W coast of Honshu in Japan Sea. After 1992, see Dogo Island.
BT Honshu
BT Japan
BT Far East
BT Asia
SA Dogo Island
SA Hiroshima Japan
SA Shimane Japan
SA Tottori Japan

Dogo Island (1993)
Largest of the Oki Islands off W coast of Honshu in Japan Sea. Before 1993, also search Dogo AND Oki Islands.
BT Oki Islands
BT Shimane Japan
BT Honshu
BT Japan
BT Far East
BT Asia
SA Dogo

Dogtooth Mountains (1978)
In SE British Columbia.
BT British Columbia
BT Western Canada
BT Canada

dolerite
As of 1981, no longer a valid term for GeoRef.
use diabase

Dolgan-Nenets Russian Federation
use Taymyr Dolgan-Nenets Russian Federation

Dolina
No longer a valid term for GeoRef. As of 1993, see Dolina Ukraine.

Dolina Ukraine (1993)
City in Lvov Oblast in W Ukraine. This term has multiple hierarchies.
BT1 Ukraine
BT1 Europe
BT2 Ukraine
BT2 Commonwealth of Independent States
SA Lvov Ukraine

dolinen
use dolines

dolines (1978)
UF dolinen
SA hydrogeology
SA karst
SA sinkholes

dolomite (1978)
For sediments, use carbonate sediments; dolomitic composition. To indicate the rock made of dolomite, use dolostone. Before 1976, also search dolomite for the commodity, the rock, and the sediment.
UF bitter spar
UF magnesian spar
BT carbonates
SA ankerite
SA calcite
SA carbonate rocks
SA carbonate sediments
SA chemically precipitated rocks
SA dolomitic composition
SA dolomitic limestone
SA dolomitization
SA dolostone
SA evaporites
SA kutnahorite
SA light minerals
SA marbles

Dolomite Alps
use Dolomites

dolomite marble
A valid metamorphic rock term through 1977. After 1977, use marbles AND dolomite.

Dolomites (1978)
Mountain range of the Eastern Alps between the Adige and Piave river valleys in NE Italy. This term has multiple hierarchies.
IN Index autonomous regions as applicable.
CO N460000N464500
E0121500E0113000
UF Dolomite Alps
BT1 Eastern Alps
BT1 Alps
BT1 Europe
BT2 Italy
BT2 Southern Europe
BT2 Europe
NT Cima d'Asta
NT Latemar Massif
NT Lessini Mountains
SA Marmolada Glacier
SA Trentino-Alto Adige Italy
SA Veneto Italy

dolomitic composition (1978)
Before 1978, also search composition AND dolomitic.
SA composition
SA dolomite
SA dolomitization
SA dolostone
SA dolostone deposits
SA sedimentary rocks
SA sediments

dolomitic limestone (1981)
Before 1981, also search magnesian limestone.
UF magnesian limestone
BT limestone
BT carbonate rocks

BT sedimentary rocks
SA dolomite
SA dolomitization
SA dolostone

dolomitite
use dolostone

dolomitization (1978)
SA calcitization
SA dedolomitization
SA diagenesis
SA dolomite
SA dolomitic composition
SA dolomitic limestone
SA dolostone
SA limestone
SA metasomatism
SA processes
SA sedimentary rocks
SA sedimentation
SA sediments

dolostone (1978)
To be used for rocks made of dolomite (mineral). As of 1981, use dolostone deposits for dolostone as a commodity. For sediments, use carbonate sediments; dolomitic composition. Before 1976, also search dolomite for the rock, sediment and commodity.
UF dolomitite
BT carbonate rocks
BT sedimentary rocks
SA carbonate sediments
SA dolomite
SA dolomitic composition
SA dolomitic limestone
SA dolomitization
SA dolostone deposits

dolostone deposits (1981)
Before 1981, search dolostone or dolomite AND deposits.
IN Commodity. See List C.
SA dolomitic composition
SA dolostone
SA limestone deposits
SA marble deposits

domain structure
As of 1981, no longer a valid term for GeoRef. See magnetic domains or crystal structure.

domains (1978)
Used as a general term.
SA environment
SA facies
SA magnetic domains
SA sedimentary petrology

Domerian (1978)
Europe.
BT middle Liassic
BT Lower Jurassic
BT Jurassic
BT Mesozoic

domes (1978)
BT folds
SA anticlines
SA arches
SA batholiths
SA diapirs
SA geomorphology
SA intrusions
SA laccoliths
SA morphostructures
SA salt domes
SA uplifts
SA volcanoes

Dominica (1981)
Island between Guadeloupe and Martinique in the Lesser Antilles.
BT Lesser Antilles
BT Antilles

BT West Indies
BT Caribbean region

Dominican Republic (1978)
Occupies eastern two thirds of island of Hispaniola in the Greater Antilles.
CO N173000N200000
W0682000W0720000
BT Hispaniola
BT Greater Antilles
BT Antilles
BT West Indies
BT Caribbean region
NT Santo Domingo Dominican Republic

Don Basin (1978)
River basin in central S and S Russian Federation. Also search Don. May also occur in other locations.
IN Index countries as applicable.
SA Don River
SA Russian Federation

Don Juan Pond (1978)
In Wright Valley on W shore of Ross Sea near Ross Ice Sheet in N Victoria Land.
BT Antarctica

Don River (1978)
Rises SE of Tula and flows SE into a big bend near the Volga River then flows SW into the Azov Sea at Rostov. Before 1978, also search Don. Also in other locations. Search in combination with Russian Federation or Russian Republic.
IN Index country as applicable.
SA Don Basin
SA Rostov-na-Donu Russian Federation
SA Russian Federation

Dona Ana County
Valid through 1988. Search in combination with state term. After 1988, use specific county-state term.

Dona Ana County New Mexico (1989)
SW New Mexico. Before 1989, also search Dona Ana County AND New Mexico.
CO N314500N330500
W1062200W1072000
BT New Mexico
BT United States
NT Kilbourne Hole
SA Jornada del Muerto
SA Orogrande Basin
SA White Sands

Donau River
use Danube River

Donbas
use Donets Basin

Donbass
use Donets Basin

Donegal
No longer a valid term for GeoRef. See Donegal Ireland.

Donegal Ireland (1993)
County W of Northern Ireland in N Ireland.
BT Ireland
BT Western Europe
BT Europe
SA Ardara Pluton
SA Rosses Granite

Donets Basin (1978)

Industrial region in plain of Donets River primarily in E Ukraine. Also search Donets. This term has multiple hierarchies.
IN Index former Soviet republics as applicable.
CO N470000N483000
E0400000E0373000
UF Donbas
UF Donbass
BT1 Commonwealth of Independent States
BT2 Europe
SA Dnieper-Donets Basin
SA Gorlovka Basin
SA Novorayskoe Formation
SA Protopivskaya Formation
SA Russian Federation
SA Ukraine

Dong Hai
use East China Sea

Dongpu Depression (1989)
Subbasin of the Bohai Basin located in NE Henan and SW Shandong.
BT China
BT Far East
BT Asia
SA Henan China
SA Shandong China

Doniphan County
Valid through 1988. Search in combination with state term. After 1988, use specific county-state term.

Doniphan County Kansas (1989)
Extreme NE Kansas. Before 1989, also search Doniphan County AND Kansas.
CO N393700N400000
W0944900W0952200
BT Kansas
BT United States

Donjek Glacier (1978)
In SW Yukon Territory.
BT Yukon Territory
BT Western Canada
BT Canada

Donjek River (1978)
In SW Yukon Territory.
BT Yukon Territory
BT Western Canada
BT Canada

Door County
Valid through 1988. Search in combination with state term. After 1988, use specific county-state term.

Door County Wisconsin (1989)
NE Wisconsin, on Door Peninsula. Before 1989, also search Door County AND Wisconsin.
CO N444000N452500
W0864500W0874500
BT Wisconsin
BT United States

Doornik Belgium
use Tournai Belgium

Dorchester County
No longer a valid term for GeoRef. As of 1993, see Dorchester County Quebec, Dorchester County Maryland, or Dorchester County South Carolina.

Dorchester County Maryland (1989)

E Maryland. Before 1989, also search Dorchester County AND Maryland.
CO N380500N424400
W0710200W0715300
BT Maryland
BT United States

Dorchester County Quebec (1993)
S Quebec. Before 1993, also search Dorchester or Dorchester County in conjunction with Quebec.
BT Quebec
BT Eastern Canada
BT Canada

Dorchester County South Carolina (1989)
SE central South Carolina. Before 1989, also search Dorchester County AND South Carolina.
CO N324800N332000
W0800500W0804800
BT South Carolina
BT United States

Dordogne
No longer a valid term for GeoRef. See Dordogne France.

Dordogne France (1993)
CO N443000N454500
E0021500W0010000
BT France
BT Western Europe
BT Europe
NT Perigord
SA Les Eyzies region

Dore Lake Complex (1978)
BT Quebec
BT Eastern Canada
BT Canada
SA Precambrian

Dorog
No longer a valid term for GeoRef. See Dorog Hungary.

Dorog Basin (1978)
Also search Dorog.
BT Hungary
BT Central Europe
BT Europe

Dorog Hungary (1993)
Town NW of Budapest in N Hungary.
BT Hungary
BT Central Europe
BT Europe

Dorpat Estonia
use Tartu Estonia

Dorset
No longer a valid term for GeoRef. As of 1993, see Dorset England.

Dorset England (1993)
County on the English Channel. Before 1993, also search Dorsetshire or Dorset.
CO N503000N511000
W0014500W0030000
UF Dorsetshire England
BT England
BT Great Britain
BT United Kingdom
BT Western Europe
BT Europe
NT Chesil Beach England

Dorsetshire England
use Dorset England

double refraction
use birefringence

double-refracting spar
  use Iceland spar
Doubs
  No longer a valid term for GeoRef. See Doubs France.
Doubs France (1993)
  Department in E France.
  CO N463500N473700
  E0070000E0054000
  BT France
  BT Western Europe
  BT Europe
  NT Besancon France
  SA Franche-Comte
  SA Jura Mountains
Douglas County
  Valid through 1988. Search in combination with state term. After 1988, use specific county-state term.
Douglas County Colorado (1989)
  Central Colorado. Before 1989, also search Douglas County AND Colorado.
  CO N390700N393200
  W1043900W1052000
  BT Colorado
  BT United States
Douglas County Georgia (1989)
  NW central Georgia. Before 1989, also search Douglas County AND Georgia.
  CO N333600N334800
  W0843500W0845300
  BT Georgia
  BT United States
Douglas County Illinois (1989)
  E central Illinois. Before 1989, also search Douglas County AND Illinois.
  CO N393900N395300
  W0875600W0882600
  BT Illinois
  BT United States
Douglas County Kansas (1989)
  E Kansas. Before 1989, also search Douglas County AND Kansas.
  CO N384400N390500
  W0950300W0953000
  BT Kansas
  BT United States
Douglas County Minnesota (1989)
  W Minnesota. Before 1989, also search Douglas County AND Minnesota.
  CO N454300N460800
  W0950800W0954600
  BT Minnesota
  BT United States
Douglas County Missouri (1989)
  S Missouri. Before 1989, also search Douglas County AND Missouri.
  CO N364300N370500
  W0920600W0925500
  BT Missouri
  BT United States
Douglas County Nebraska (1989)
  E Nebraska. Before 1989, also search Douglas County AND Nebraska.
  CO N411200N412500
  W0955000W0963000
  BT Nebraska
  BT United States
Douglas County Nevada (1989)
  W Nevada. Before 1989, also search Douglas County AND Nevada.
  CO N383100N390800
  W1192000W1195700
  BT Nevada
  BT United States
  SA Lake Tahoe
Douglas County Oregon (1989)
  SW Oregon. Before 1989, also search Douglas County AND Oregon.
  CO N424100N435700
  W1215800W1241100
  BT Oregon
  BT United States
  NT Riddle Oregon
Douglas County South Dakota (1989)
  SE South Dakota. Before 1989, also search Douglas County AND South Dakota.
  CO N431200N433000
  W0980700W0984300
  BT South Dakota
  BT United States
Douglas County Washington (1989)
  Central Washington. Before 1989, also search Douglas County AND Washington.
  CO N471300N480900
  W1185700W1202000
  BT Washington
  BT United States
Douglas County Wisconsin (1989)
  Extreme NW Wisconsin. Before 1989, also search Douglas County AND Wisconsin.
  CO N460900N464500
  W0913300W0922000
  BT Wisconsin
  BT United States
Douglas Group (1978)
  Includes Stranger Formation, and Lawrence Formation. SW Iowa, E Kansas, NW Missouri, and SE Nebraska.
  BT Virgilian
  BT Upper Pennsylvanian
  BT Pennsylvanian
  BT Carboniferous
  BT Paleozoic
  SA Iowa
  SA Kansas
  SA Lawrence Formation
  SA Missouri
  SA Nebraska
Douro River
  As of 1985, not a valid term for GeoRef. Term was introduced in 1981.
  use Duero River
Doushantuo Formation (1989)
  Hubei, South-Central China.
  BT Sinian
  BT Proterozoic
  BT upper Precambrian
  BT Precambrian
  SA China
  SA Hubei China
down hole
  use downhole methods
down-hole methods
  use downhole methods
downhole methods (1985)
  UF deepening
  UF down-hole methods
  SA boreholes
  SA geothermal energy
  SA measurement-while-drilling
  SA methods
  SA well-logging
Downtonian (1978)
  In Old Red Sandstone, England. Uppermost Silurian or lowermost Devonian; Schmidt, 1960--lowermost Gedinnian (Lower Devonian).
  BT Paleozoic
  SA Devonian
  SA Lower Devonian
  SA Old Red Sandstone
  SA Silurian
  SA Upper Silurian
Drac Valley (1978)
  River valley in SE France.
  BT France
  BT Western Europe
  BT Europe
drag folds (1978)
  Before 1978, also search drag AND folds.
  BT folds
  SA displacements
  SA faults
  SA orientation
drainage (1978)
  SA acid mine drainage
  SA drainage basins
  SA drainage patterns
  SA dredging
  SA geomembranes
  SA geotextiles
  SA ground water
  SA lakes
  SA lysimeters
  SA mine dewatering
  SA mine drainage
  SA seepage
  SA soils
  SA underground channels
  SA water
  SA water regimes
drainage basins (1978)
  Autoposting of fluvial features to this term began in 1989. Before 1993, also search river basins if applicable.
  UF basins, drainage
  UF catchments
  UF feeding ground
  BT fluvial features
  SA basin management
  SA basins
  SA catchment hydrodynamics
  SA drainage
  SA drainage patterns
  SA geomorphology
  SA poljes
  SA precipitation
  SA representative basins
  SA rivers
  SA rivers and streams
  SA tributaries
  SA valleys
  SA water balance
  SA water yield
  SA watersheds
drainage changes
  A valid term through 1974. After 1974, see landform evolution and fluvial features.
drainage networks
  use drainage patterns
drainage patterns (1978)
  Autoposting of fluvial features to this term began in 1989.
  UF drainage networks
  BT fluvial features
  SA drainage
  SA drainage basins
  SA geomorphology
  SA glacial geology
  SA patterns
  SA rivers
  SA segmentation
  SA stream order
  SA streams
  SA trees
  SA tributaries
Drake Passage (1978)
  Strait between Cape Horn and South Shetland Islands N of Antarctica.
  UF Drake Strait
  BT Antarctic Ocean
  SA Leg 36
Drake Strait
  use Drake Passage
Drammen
  No longer a valid term for GeoRef. As of 1993 see Drammen Norway.
Drammen Norway (1993)
  City on a branch of Oslo Fjord in S Norway. Before 1993 also search Drammen AND Norway.
  BT Norway
  BT Scandinavia
  BT Western Europe
  BT Europe
drape folding
  use drape folds
drape folds (1978)
  UF drape folding
  BT folds
dravite (1978)
  BT ring silicates
  BT silicates
  SA borosilicates
  SA schorl
  SA tourmaline
drawdown (1978)
  SA ground water
  SA hydraulics
  SA levels
  SA mine dewatering
  SA potentiometric surface
  SA pumping
  SA reservoirs
  SA water wells
dredge spoils
  Not a valid term for GeoRef. After 1993, see dredged materials.
dredged materials (1993)
  Before 1993, also search dredge spoils if applicable.
  SA dredged samples
  SA dredging
  SA mining
  SA spoils
  SA tailings
  SA waste disposal
dredged samples (1985)
  SA dredged materials
  SA dredging
  SA samples
  SA sampling
dredging (1978)
  UF deepening
  SA construction
  SA drainage
  SA dredged materials
  SA dredged samples
  SA engineering geology
  SA excavations
  SA mining
  SA mining geology

SA ocean floors
SA siltation
SA spoils
SA tailings
SA waterways

Dresbachian (1985)
North America. Upper Cambrian, above Albertan and below Franconian.
BT Upper Cambrian
BT Cambrian
BT Paleozoic

Dresden
No longer a valid term for GeoRef. See Dresden Germany.

Dresden Bezirk
No longer a valid term for GeoRef. District in SE East Germany. Before 1981, search Dresden. See Saxony Germany and Upper Lusatia.

Dresden Germany (1993)
City in Saxony, E Germany. As of 1981, Dresden refers to only the city. Before 1981, it also included the East German district.
BT Saxony Germany
BT Germany
BT Central Europe
BT Europe

drift (1978)
Before 1971, also search drift deposits.
UF drift deposits
UF glacial drift
BT clastic sediments
BT sediments
SA buried channels
SA buried valleys
SA fluvial features
SA glacial features
SA glacial geology
SA glacial transport
SA glaciofluvial sedimentation
SA loess
SA moraines
SA terrigenous materials
SA till

drift deposits
use drift

drill cores
use cores

drill cuttings
use cuttings

drill holes
A valid term through 1975.
use boreholes

drilling (1978)
Used for the process.
NT deep drilling
NT horizontal drilling
NT marine drilling
SA blowouts
SA borehole breakouts
SA boreholes
SA Continental Scientific Drilling Program
SA engineering geology
SA measurement-while-drilling
SA petroleum
SA petroleum engineering
SA sand equivalent tests
SA sludge
SA well-logging
SA wells

drinking water (1983)
SA bottling
SA decontamination
SA desalinization

SA fresh water
SA ground water
SA impurities
SA potability
SA purification
SA surface water
SA water
SA water quality
SA water resources
SA water supply

Drome
No longer a valid term for GeoRef. See Drome France.

Drome France (1993)
Department in SE France.
CO N441500N452500
    E0055000E0043500
BT France
BT Western Europe
BT Europe
NT Baronnies
NT Diois
NT Valence France
SA Vercors

Dronning Maud Land
use Queen Maud Land

droplets (1978)
SA clouds
SA meteorology
SA raindrops
SA water

dropstone
use argillite

drought (1982)
SA atmospheric precipitation
SA climate
SA environmental geology
SA hydrology
SA rainfall
SA water balance
SA water resources
SA water supply

Drum Mountains (1985)
W central Utah.
IN Index counties and mountains as applicable.
SA Utah

Drumheller
No longer a valid term for GeoRef. See Drumheller Alberta.

Drumheller Alberta (1993)
City in S Alberta.
BT Alberta
BT Western Canada
BT Canada

drumlins (1978)
Autoposting of glacial features to this term began in 1989.
BT glacial features
SA glacial geology
SA moraines
SA till

Dry Branch Formation (1989)
Includes Twiggs Clay and Griffins Landing Member.
BT upper Eocene
BT Eocene
BT Paleogene
BT Tertiary
BT Cenozoic
SA DSDP Site 603
SA Georgia
SA South Carolina
SA Twiggs Clay

dry delta
use alluvial fans

dry hot rocks
use hot dry rocks

Dry Tortugas (1985)
Island group at W end of Florida Keys.
CO N244000N244800
    W0824500W0825500
BT Monroe County Florida
BT Florida
BT United States
SA Gulf of Mexico

Dry Valley Drilling Project (1978)
In different locations such as Lake Vanda, Lake Vida, New Harbor, Ross Island, and McMurdo Sound off SW Ross Ice Sheet in the Victoria Land area of Ross Dependency. Autoposting of Polar regions to this term began in 1990.
UF DVDP
BT Antarctica
SA programs

dry valleys (1989)
Streamless valleys.
SA erosion features
SA fluvial features
SA geomorphology
SA valleys

Dryas (1978)
Genus. Before 1978, term also included use for the intervals of late-glacial time. Before 1985, miospores and palynomorphs were autoposted to this term.
BT Dicotyledoneae
BT angiosperms
BT Spermatophyta
BT Plantae
SA miospores
SA Older Dryas
SA Oldest Dryas
SA palynomorphs
SA Younger Dryas

drying
use dehydration

DSDP
use Deep Sea Drilling Project

DSDP Site 3 (1989)
Sigsbee abyssal plain, Gulf of Mexico.
CO N230100N230100
    W0920124W0920124
BT Leg 1
BT Deep Sea Drilling Project
SA Gulf of Mexico
SA Sigsbee Deep

DSDP Site 4 (1989)
NW Atlantic, in Horizon A area, E of San Salvador, Bahamas.
CO N242840N242841
    W0734731W0734732
BT Leg 1
BT Deep Sea Drilling Project
SA Atlantic Ocean
SA Northwest Atlantic

DSDP Site 19 (1993)
Ridge, S edge of Brazil Basin, South American Atlantic. Before 1982, also search Site 19 AND Deep Sea Drilling Project.
CO S283205S283204
    W0234037W0234038
BT Leg 3
BT Deep Sea Drilling Project
SA Atlantic Ocean
SA Brazil Basin
SA South American Atlantic

DSDP Site 20 (1993)

Valley, S edge of Brazil Basin, South American Atlantic. Before 1981, also search Site 20 AND Deep Sea Drilling Project.
CO S283125S283124
    W0265034W0265035
BT Leg 3
BT Deep Sea Drilling Project
SA Atlantic Ocean
SA Brazil Basin
SA South American Atlantic

DSDP Site 21 (1989)
Brazil Basin, SW Atlantic.
CO S283506S283506
    W0303551W0303551
BT Leg 3
BT Deep Sea Drilling Project
SA Atlantic Ocean
SA Brazil Basin
SA Southwest Atlantic

DSDP Site 32 (1993)
Off California coast, North American Pacific. Before 1981, also search Site 32 AND Deep Sea Drilling Project.
CO N370737N370738
    W1273322W1273323
BT Leg 5
BT Deep Sea Drilling Project
SA North American Pacific
SA Pacific Ocean

DSDP Site 35 (1993)
Gorda Rise, North American Pacific.
CO N404025N404026
    W1272828W1272829
BT Leg 5
BT Deep Sea Drilling Project
SA Gorda Rise
SA North American Pacific
SA Pacific Ocean

DSDP Site 40 (1993)
E North American Pacific, between Mendocino and Molokai fracture zones.
CO N194734N194735
    W1395404W1395405
BT Leg 5
BT Deep Sea Drilling Project
SA North American Pacific
SA Pacific Ocean

DSDP Site 51 (1993)
NW West Pacific, abyssal plain near Shatsky Rise.
CO N332830N332830
    E1532418E1532418
BT Leg 6
BT Deep Sea Drilling Project
SA Pacific Ocean
SA West Pacific

DSDP Site 61 (1985)
Autoposting of Leg 7 to this term began in 1989. W Equatorial Pacific, SE of the Mariana Trench.
CO N120501N120501
    E1470342E1470342
BT Leg 7
BT Deep Sea Drilling Project
SA Equatorial Pacific
SA Mariana Trench
SA Pacific Ocean

DSDP Site 62 (1985)
Autoposting of Leg 7 to this term began in 1989. W Equatorial Pacific, N of New Guinea.
CO N015212N015212
    E1415618E1415618
BT Leg 7
BT Deep Sea Drilling Project
SA Equatorial Pacific
SA Pacific Ocean

DSDP Site 63 (1985)
Autoposting of Leg 7 to this term began in 1989. W Equatorial Pacific, N of New Guinea.
CO N005008N005008
E1475323E1475323
BT Leg 7
BT Deep Sea Drilling Project
SA Equatorial Pacific
SA Pacific Ocean

DSDP Site 64 (1985)
Autoposting of Leg 7 to this term began in 1989. W Equatorial Pacific, N of the Solomon Islands.
CO S014432S014432
E1583635E1583635
BT Leg 7
BT Deep Sea Drilling Project
SA Equatorial Pacific
SA Pacific Ocean

DSDP Site 65 (1985)
Autoposting of Leg 7 to this term began in 1989. Central Equatorial Pacific, N of Kiribati.
CO N042113N042113
E1765909E1765909
BT Leg 7
BT Deep Sea Drilling Project
SA Equatorial Pacific
SA Pacific Ocean

DSDP Site 66 (1985)
Autoposting of Leg 7 to this term began in 1989. Central Equatorial Pacific, near Line Islands.
CO N022337N022337
W1660718W1660718
BT Leg 7
BT Deep Sea Drilling Project
SA Equatorial Pacific
SA Pacific Ocean

DSDP Site 67 (1985)
Autoposting of Leg 7 to this term began in 1989. N central Pacific Ocean, N of Hawaii.
CO N242334N242334
W1573853W1573853
BT Leg 7
BT Deep Sea Drilling Project
SA North Pacific
SA Pacific Ocean

DSDP Site 69 (1985)
Autoposting of Leg 8 to this term began in 1989. Central Equatorial Pacific, between Hawaii and Line Islands.
CO N060000N060000
W1525156W1525156
BT Leg 8
BT Deep Sea Drilling Project
SA Equatorial Pacific
SA Pacific Ocean

DSDP Site 70 (1985)
Autoposting of Leg 8 to this term began in 1989. E Equatorial Pacific, near Clipperton fracture zone.
CO N062005N062005
W1402143W1402143
BT Leg 8
BT Deep Sea Drilling Project
SA Clipperton fracture zone
SA Equatorial Pacific
SA Pacific Ocean

DSDP Site 71 (1985)
Autoposting of Leg 8 to this term began in 1989. E Equatorial Pacific, near Clipperton fracture zone.
CO N042817N042817
W1401855W1401855
BT Leg 8
BT Deep Sea Drilling Project

SA Clipperton fracture zone
SA Equatorial Pacific
SA Pacific Ocean

DSDP Site 72 (1985)
Autoposting of Leg 8 to this term began in 1989. E Equatorial Pacific, N of Marquesas Islands.
CO N002629N002629
W1385201W1385201
BT Leg 8
BT Deep Sea Drilling Project
SA Equatorial Pacific
SA Pacific Ocean

DSDP Site 73 (1985)
Autoposting of Leg 8 to this term began in 1989. E Equatorial Pacific, N of Marquesas Islands.
CO S015535S015435
W1372807W1372807
BT Leg 8
BT Deep Sea Drilling Project
SA Equatorial Pacific
SA Pacific Ocean

DSDP Site 74 (1985)
Autoposting of Leg 8 to this term began in 1989. E Equatorial Pacific, NE of Marquesas Islands.
CO S061412S061412
W1360548W1360548
BT Leg 8
BT Deep Sea Drilling Project
SA Equatorial Pacific
SA Pacific Ocean

DSDP Site 75 (1985)
Autoposting of Leg 8 to this term began in 1989. E Equatorial Pacific, E of Marquesas Islands.
CO S123100S123100
W1341600W1341600
BT Leg 8
BT Deep Sea Drilling Project
SA Equatorial Pacific
SA Pacific Ocean

DSDP Site 77 (1985)
Autoposting of Leg 9 to this term began in 1989. E Equatorial Pacific, NE of Marquesas Islands. Before 1985, also search Site 77 AND Deep Sea Drilling Project.
CO N002854N002854
W1331342W1331342
BT Leg 9
BT Deep Sea Drilling Project
SA Equatorial Pacific
SA Pacific Ocean

DSDP Site 85 (1985)
Autoposting of Leg 10 to this term began in 1989. Campeche Scarp.
CO N225029N225029
W0912522W0912522
BT Leg 10
BT Deep Sea Drilling Project
SA Campeche Scarp
SA Gulf of Mexico

DSDP Site 86 (1985)
Autoposting of Leg 10 to this term began in 1989. Campeche Scarp.
CO N225229N225229
W0905745W0905745
BT Leg 10
BT Deep Sea Drilling Project
SA Campeche Scarp
SA Gulf of Mexico

DSDP Site 88 (1985)
Autoposting of Leg 10 to this term began in 1989. Campeche Scarp.
CO N212256N212256
W0940013W0940013
BT Leg 10
BT Deep Sea Drilling Project
SA Campeche Scarp

SA Gulf of Mexico

DSDP Site 89 (1985)
Autoposting of Leg 10 to this term began in 1989. Bay of Campeche, at foot of Campeche Scarp.
CO N205321N205321
W0950644W0950644
BT Leg 10
BT Deep Sea Drilling Project
SA Gulf of Mexico

DSDP Site 90 (1985)
Autoposting of Leg 10 to this term began in 1989. W Gulf of Mexico.
CO N234748N234748
W0944606W0944606
BT Leg 10
BT Deep Sea Drilling Project
SA Gulf of Mexico

DSDP Site 91 (1985)
Autoposting of Leg 10 to this term began in 1989. Sigsbee Deep.
CO N234624N234624
W0932046W0932046
BT Leg 10
BT Deep Sea Drilling Project
SA Gulf of Mexico
SA Sigsbee Deep

DSDP Site 92 (1985)
Autoposting of Leg 10 to this term began in 1989. Central Gulf of Mexico.
CO N255041N255041
W0914917W0914917
BT Leg 10
BT Deep Sea Drilling Project
SA Gulf of Mexico

DSDP Site 94 (1985)
Autoposting of Leg 10 to this term began in 1989. Campeche Scarp.
CO N243138N243138
W0882809W0882809
BT Leg 10
BT Deep Sea Drilling Project
SA Campeche Scarp
SA Gulf of Mexico

DSDP Site 95 (1985)
Autoposting of Leg 10 to this term began in 1989. E Gulf of Mexico.
CO N240900N240900
W0862351W0862351
BT Leg 10
BT Deep Sea Drilling Project
SA Gulf of Mexico

DSDP Site 96 (1985)
Autoposting of Leg 10 to this term began in 1989. E Gulf of Mexico, NW of Cuba.
CO N234434N234434
W0854548W0854548
BT Leg 10
BT Deep Sea Drilling Project
SA Gulf of Mexico

DSDP Site 97 (1985)
Autoposting of Leg 10 to this term began in 1989. E Gulf of Mexico, NW of Cuba.
CO N235303N235303
W0842645W0842645
BT Leg 10
BT Deep Sea Drilling Project
SA Gulf of Mexico

DSDP Site 100 (1993)
Blake-Bahama Outer Ridge, North American Atlantic. Before 1981, also search Site 100 AND Deep Sea Drilling Project.
CO N244116N244117
W0734157W0734157
BT Leg 11
BT Deep Sea Drilling Project
SA Atlantic Ocean

SA Blake-Bahama Outer Ridge
SA North American Atlantic

DSDP Site 102 (1993)
Blake-Bahama Outer Ridge, North American Atlantic. Before 1981, also search Site 102 AND Deep Sea Drilling Project.
CO N304355N304356
W0742708W0742709
BT Leg 11
BT Deep Sea Drilling Project
SA Atlantic Ocean
SA Blake-Bahama Outer Ridge
SA North American Atlantic

DSDP Site 103 (1993)
Blake-Bahama Outer Ridge, North American Atlantic.
CO N302704N302705
W0743459W0743500
BT Leg 11
BT Deep Sea Drilling Project
SA Atlantic Ocean
SA Blake-Bahama Outer Ridge
SA North American Atlantic

DSDP Site 104 (1993)
Blake-Bahama Outer Ridge, North American Atlantic. Before 1981, also search Site 104 AND Deep Sea Drilling Project.
CO N304939N304939
W0741958W0741959
BT Leg 11
BT Deep Sea Drilling Project
SA Atlantic Ocean
SA Blake-Bahama Outer Ridge
SA North American Atlantic

DSDP Site 105 (1981)
Autoposting of Leg 11 to this term began in 1989. NW Atlantic Ocean. Before 1981, search Site 105 AND Deep Sea Drilling Project.
CO N345343N345343
W0691024W0691024
BT Leg 11
BT Deep Sea Drilling Project
SA Atlantic Ocean

DSDP Site 108 (1993)
Continental slope off New Jersey, North American Atlantic.
CO N384816N384817
W0723912W0723913
BT Leg 11
BT Deep Sea Drilling Project
SA Atlantic Ocean
SA North American Atlantic

DSDP Site 111 (1993)
Orphan Knoll, NE of Newfoundland, North American Atlantic. Before 1982, also search Site 111 AND Deep Sea Drilling Project.
CO N502534N502535
W0462203W0462203
BT Leg 12
BT Deep Sea Drilling Project
SA Atlantic Ocean
SA North American Atlantic

DSDP Site 116 (1989)
Rockall Plateau, NE Atlantic.
CO N572942N572942
W0155530W0155530
BT Leg 12
BT Deep Sea Drilling Project
SA Atlantic Ocean
SA Northeast Atlantic
SA Rockall Plateau

DSDP Site 117 (1993)
W Rockall Bank, near margin of Hatton-Rockall Basin, European Atlantic. Before 1981, also search

Site 117 AND Deep Sea Drilling Project.
　CO N571930N571930
　　W0152300W0152300
　BT Leg 12
　BT Deep Sea Drilling Project
　SA Atlantic Ocean
　SA European Atlantic
　SA Rockall Bank
　SA Rockall Plateau

DSDP Site 119　(1989)
Cantabria Seamount, W Bay of Biscay, off the N coast of Spain.
　CO N450218N450218
　　W0075848W0075848
　BT Leg 12
　BT Deep Sea Drilling Project
　SA Atlantic Ocean
　SA Bay of Biscay
　SA Cantabria Seamount

DSDP Site 120　(1985)
Autoposting of Leg 13 to this term began in 1989. European Atlantic, near Strait of Gibraltar.
　CO N364213N364213
　　W0112557W0112557
　BT Leg 13
　BT Deep Sea Drilling Project
　SA Atlantic Ocean
　SA European Atlantic

DSDP Site 121　(1985)
Autoposting of Leg 13 to this term began in 1989. W Mediterranean Sea.
　CO N360939N360939
　　W0042226W0042226
　BT Leg 13
　BT Deep Sea Drilling Project
　SA Mediterranean Sea
　SA West Mediterranean

DSDP Site 122　(1985)
Autoposting of Leg 13 to this term began in 1989. W Mediterranean Sea.
　CO N402652N402652
　　E0023728E0023728
　BT Leg 13
　BT Deep Sea Drilling Project
　SA Mediterranean Sea
　SA West Mediterranean

DSDP Site 123　(1985)
Autoposting of Leg 13 to this term began in 1989. W Mediterranean Sea.
　CO N403750N403750
　　E0025016E0025016
　BT Leg 13
　BT Deep Sea Drilling Project
　SA Mediterranean Sea
　SA West Mediterranean

DSDP Site 124　(1985)
Autoposting of Leg 13 to this term began in 1989. W Mediterranean Sea.
　CO N385223N385223
　　E0045941E0045941
　BT Leg 13
　BT Deep Sea Drilling Project
　SA Mediterranean Sea
　SA West Mediterranean

DSDP Site 125　(1981)
Autoposting of Leg 13 to this term began in 1989. E Mediterranean Sea. Before 1981, search Site 125 AND Deep Sea Drilling Project.
　CO N343719N343719
　　E0202541E0202541
　BT Leg 13
　BT Deep Sea Drilling Project
　SA East Mediterranean

　SA Mediterranean Sea

DSDP Site 126　(1985)
Autoposting of Leg 13 to this term began in 1989. SE Mediterranean Sea.
　CO N350943N350943
　　E0212538E0212538
　BT Leg 13
　BT Deep Sea Drilling Project
　SA East Mediterranean
　SA Mediterranean Sea

DSDP Site 127　(1985)
Autoposting of Leg 13 to this term began in 1989. E Mediterranean Sea.
　CO N354354N354354
　　E0222949E0222949
　BT Leg 13
　BT Deep Sea Drilling Project
　SA East Mediterranean
　SA Mediterranean Sea

DSDP Site 128　(1985)
Autoposting of Leg 13 to this term began in 1989. E Mediterranean Sea.
　CO N354235N354235
　　E0222406E0222406
　BT Leg 13
　BT Deep Sea Drilling Project
　SA East Mediterranean
　SA Mediterranean Sea

DSDP Site 129　(1985)
Autoposting of Leg 13 to this term began in 1989. E Mediterranean Sea.
　CO N342057N342057
　　E0270455E0270455
　BT Leg 13
　BT Deep Sea Drilling Project
　SA East Mediterranean
　SA Mediterranean Sea
　SA Strabo Trench

DSDP Site 130　(1985)
Autoposting of Leg 13 to this term began in 1989. SE Mediterranean Sea.
　CO N333619N333619
　　E0275159E0275159
　BT Leg 13
　BT Deep Sea Drilling Project
　SA East Mediterranean
　SA Mediterranean Sea

DSDP Site 131　(1989)
Nile cone region, Mediterranean Sea.
　CO N330619N330620
　　E0285242E0285241
　BT Leg 13
　BT Deep Sea Drilling Project
　SA Mediterranean Sea

DSDP Site 132　(1981)
Autoposting of Leg 13 to this term began in 1989. Tyrrhenian Basin, Mediterranean Sea. Before 1981, search Site 132 AND Deep Sea Drilling Project.
　CO N401540N401540
　　E0112628E0112628
　BT Leg 13
　BT Deep Sea Drilling Project
　SA Mediterranean Sea
　SA ODP Site 653
　SA Tyrrhenian Sea

DSDP Site 133　(1985)
Autoposting of Leg 13 to this term began in 1989. W Mediterranean Sea.
　CO N391159N391159
　　E0072008E0072008
　BT Leg 13
　BT Deep Sea Drilling Project

　SA Mediterranean Sea
　SA West Mediterranean

DSDP Site 134　(1985)
Autoposting of Leg 13 to this term began in 1989. W Mediterranean Sea. Before 1981, also search Site 134 AND Deep Sea Drilling Project.
　CO N391142N391142
　　E0071815E0071815
　BT Leg 13
　BT Deep Sea Drilling Project
　SA Mediterranean Sea
　SA West Mediterranean

DSDP Site 137　(1989)
Abyssal hills in Cape Verde Atlantic, near the foot of the continental rise about 1000 km W of Cap Blanc, W Africa.
　CO N255531N255532
　　W0270338W0270339
　BT Leg 14
　BT Deep Sea Drilling Project
　SA Atlantic Ocean
　SA Cape Verde Atlantic

DSDP Site 141　(1989)
Lower continental rise, about 200 km N of Cape Verde Islands in Cape Verde Atlantic.
　CO N192509N192510
　　W0235954W0235955
　BT Leg 14
　BT Deep Sea Drilling Project
　SA Atlantic Ocean
　SA Cape Verde Atlantic

DSDP Site 146　(1985)
Autoposting of Leg 15 to this term began in 1989. E Caribbean Sea.
　CO N150659N150659
　　W0692240W0692240
　BT Leg 15
　BT Deep Sea Drilling Project
　SA Caribbean Sea

DSDP Site 147　(1985)
Autoposting of Leg 15 to this term began in 1989. E Caribbean Sea.
　CO N104229N104229
　　W0651029W0651029
　BT Leg 15
　BT Deep Sea Drilling Project
　SA Caribbean Sea

DSDP Site 147A　(1985)
Autoposting of Leg 15 to this term began in 1989. E Caribbean Sea.
　CO N104241N104241
　　W0651027W0651927
　BT Leg 15
　BT Deep Sea Drilling Project
　SA Caribbean Sea

DSDP Site 148　(1985)
Autoposting of Leg 15 to this term began in 1989. E Caribbean Sea.
　CO N132507N132507
　　W0634315W0634315
　BT Leg 15
　BT Deep Sea Drilling Project
　SA Caribbean Sea

DSDP Site 149　(1985)
Autoposting of Leg 15 to this term began in 1989. E Caribbean Sea. Before 1981, also search Site 149 AND Deep Sea Drilling Project.
　CO N150615N150615
　　W0692151W0692151
　BT Leg 15
　BT Deep Sea Drilling Project
　SA Caribbean Sea

DSDP Site 150　(1985)
Autoposting of Leg 15 to this term began in 1989. E Caribbean Sea. Before 1981, also search Site 150 AND Deep Sea Drilling Project.
　CO N143041N143041
　　W0692121W0692121
　BT Leg 15
　BT Deep Sea Drilling Project
　SA Caribbean Sea

DSDP Site 151　(1985)
Autoposting of Leg 15 to this term began in 1989. Central Caribbean Sea.
　CO N150101N150101
　　W0732435W0732435
　BT Leg 15
　BT Deep Sea Drilling Project
　SA Caribbean Sea

DSDP Site 152　(1985)
Autoposting of Leg 15 to this term began in 1989. Central Caribbean Sea. Before 1981, also search Site 152 AND Deep Sea Drilling Project.
　CO N155243N155243
　　W0743628W0743628
　BT Leg 15
　BT Deep Sea Drilling Project
　SA Caribbean Sea

DSDP Site 153　(1985)
Autoposting of Leg 15 to this term began in 1989. E Caribbean Sea. Before 1981, also search Site 153 AND Deep Sea Drilling Project.
　CO N135820N135820
　　W0722605W0722605
　BT Leg 15
　BT Deep Sea Drilling Project
　SA Caribbean Sea

DSDP Site 154　(1985)
Autoposting of Leg 15 to this term began in 1989. W Caribbean Sea.
　CO N110518N110518
　　W0802245W0802245
　BT Leg 15
　BT Deep Sea Drilling Project
　SA Caribbean Sea

DSDP Site 157　(1981)
Autoposting of Leg 16 to this term began in 1989. E Equatorial Pacific. Before 1981, search Site 157 AND Deep Sea Drilling Project.
　CO S014542S014542
　　W0855412W0855412
　BT Leg 16
　BT Deep Sea Drilling Project
　SA Equatorial Pacific
　SA Pacific Ocean

DSDP Site 158　(1981)
Autoposting of Leg 16 to this term began in 1989. E Equatorial Pacific. Before 1981, search Site 158 AND Deep Sea Drilling Project.
　CO N063724N063724
　　W0851412W0851412
　BT Leg 16
　BT Deep Sea Drilling Project
　SA Equatorial Pacific
　SA Pacific Ocean

DSDP Site 159　(1989)
East Pacific Rise, E Equatorial Pacific.
　CO N121954N121954
　　W1221718W1221718
　BT Leg 16
　BT Deep Sea Drilling Project
　SA East Pacific Rise
　SA Equatorial Pacific
　SA Northeast Pacific
　SA Pacific Ocean

DSDP Site 163　(1981)

Autoposting of Leg 16 to this term began in 1989. E Equatorial Pacific. Before 1981, search Site 163 AND Deep Sea Drilling Project.
CO N111442N111442
  W1501730W1501730
BT Leg 16
BT Deep Sea Drilling Project
SA Equatorial Pacific
SA Pacific Ocean

DSDP Site 167 (1981)
Autoposting of Leg 17 to this term began in 1989. On Magellan Rise in central Equatorial Pacific. Before 1981, search Site 167 AND Deep Sea Drilling Project.
CO N070406N070406
  W1764930W1764930
BT Leg 17
BT Deep Sea Drilling Project
SA Equatorial Pacific
SA Pacific Ocean

DSDP Site 171 (1993)
Horizon Guyot, central West Pacific. Before 1981, also search Site 171 AND Deep Sea Drilling Project.
CO N190754N190754
  W1692736W1692736
BT Leg 17
BT Deep Sea Drilling Project
SA Pacific Ocean
SA West Pacific

DSDP Site 172 (1985)
Autoposting of Leg 18 to this term began in 1989. North American Pacific.
CO N313214N313214
  W1332222W1332222
BT Leg 18
BT Deep Sea Drilling Project
SA North American Pacific
SA Pacific Ocean

DSDP Site 173 (1985)
Autoposting of Leg 18 to this term began in 1989. North American Pacific. Before 1981, also search Site 173 AND Deep Sea Drilling Project.
CO N395742N395742
  W1252707W1252707
BT Leg 18
BT Deep Sea Drilling Project
SA North American Pacific
SA Pacific Ocean

DSDP Site 174 (1985)
Autoposting of Leg 18 to this term began in 1989. North American Pacific.
CO N445330N445330
  W1262048W1262048
BT Leg 18
BT Deep Sea Drilling Project
SA North American Pacific
SA Pacific Ocean

DSDP Site 175 (1985)
Autoposting of Leg 18 to this term began in 1989. North American Pacific.
CO N445012N445012
  W1251430W1251430
BT Leg 18
BT Deep Sea Drilling Project
SA North American Pacific
SA Pacific Ocean

DSDP Site 176 (1985)
Autoposting of Leg 18 to this term began in 1989. North American Pacific.
CO N455600N455600
  W1243700W1243700
BT Leg 18
BT Deep Sea Drilling Project
SA North American Pacific
SA Pacific Ocean

DSDP Site 177 (1985)
Autoposting of Leg 18 to this term began in 1989. North American Pacific.
CO N502811N502811
  W1301218W1301218
BT Leg 18
BT Deep Sea Drilling Project
SA North American Pacific
SA Pacific Ocean

DSDP Site 178 (1985)
Autoposting of Leg 18 to this term began in 1989. Gulf of Alaska. Before 1981, also search Site 178 AND Deep Sea Drilling Project.
CO N565723N565723
  W1470752W1470752
BT Leg 18
BT Deep Sea Drilling Project
SA Gulf of Alaska
SA Pacific Ocean

DSDP Site 179 (1985)
Autoposting of Leg 18 to this term began in 1989. Gulf of Alaska.
CO N562432N562432
  W1455919W1455919
BT Leg 18
BT Deep Sea Drilling Project
SA Gulf of Alaska
SA Pacific Ocean

DSDP Site 180 (1985)
Autoposting of Leg 18 to this term began in 1989. Gulf of Alaska.
CO N572146N572146
  W1475122W1475122
BT Leg 18
BT Deep Sea Drilling Project
SA Gulf of Alaska
SA Pacific Ocean

DSDP Site 181 (1985)
Autoposting of Leg 18 to this term began in 1989. Gulf of Alaska.
CO N572618N572618
  W1482753W1482753
BT Leg 18
BT Deep Sea Drilling Project
SA Gulf of Alaska
SA Pacific Ocean

DSDP Site 182 (1985)
Autoposting of Leg 18 to this term began in 1989. Gulf of Alaska.
CO N575258N575258
  W1484259W1484259
BT Leg 18
BT Deep Sea Drilling Project
SA Gulf of Alaska
SA Pacific Ocean

DSDP Site 183 (1989)
Aleutian abyssal plain, W Pacific Ocean. Drilled to provide information on the geologic history of the Pacific and Kula plates and the formation of the Aleutian Trench.
CO N523417N523419
  E1611220E1611219
BT Leg 19
BT Deep Sea Drilling Project
SA Aleutian Trench
SA Pacific Ocean
SA West Pacific

DSDP Site 186 (1993)
Aleutian Terrace-Aleutian Trench, N North American Pacific. Before 1981, also search Site 186 AND Deep Sea Drilling Project.
CO N510748N510749
  W1740020W1740021
BT Leg 19

BT Deep Sea Drilling Project
SA North American Pacific
SA Pacific Ocean

DSDP Site 192 (1989)
Northernmost seamount in the Emperor Seamounts, W Pacific.
CO N530034N530035
  E1644249E1644248
BT Leg 19
BT Deep Sea Drilling Project
SA Emperor Seamounts
SA Pacific Ocean
SA West Pacific

DSDP Site 194 (1985)
Autoposting of Leg 20 to this term began in 1989. NW Pacific Ocean, off Japan.
CO N335841N335841
  E1484839E1484839
BT Leg 20
BT Deep Sea Drilling Project
SA Northwest Pacific
SA Pacific Ocean

DSDP Site 195 (1985)
Autoposting of Leg 20 to this term began in 1989. NW Pacific Ocean, off Japan.
CO N324624N324624
  E1465844E1465844
BT Leg 20
BT Deep Sea Drilling Project
SA Northwest Pacific
SA Pacific Ocean

DSDP Site 196 (1985)
Autoposting of Leg 20 to this term began in 1989. NW Pacific Ocean.
CO N300658N300658
  E1483429E1483429
BT Leg 20
BT Deep Sea Drilling Project
SA Northwest Pacific
SA Pacific Ocean

DSDP Site 197 (1985)
Autoposting of Leg 20 to this term began in 1989. NW Pacific Ocean.
CO N301726N301726
  E1474027E1474027
BT Leg 20
BT Deep Sea Drilling Project
SA Northwest Pacific
SA Pacific Ocean

DSDP Site 198 (1985)
Autoposting of Leg 20 to this term began in 1989. NW Pacific Ocean.
CO N254933N254933
  E1543503E1543503
BT Leg 20
BT Deep Sea Drilling Project
SA Northwest Pacific
SA Pacific Ocean

DSDP Site 199 (1985)
Autoposting of Leg 20 to this term began in 1989. W Equatorial Pacific.
CO N133047N133047
  E1561020E1561020
BT Leg 20
BT Deep Sea Drilling Project
SA Equatorial Pacific
SA Pacific Ocean

DSDP Site 200 (1985)
Autoposting of Leg 20 to this term began in 1989. W Equatorial Pacific.
CO N125012N125012
  E1564658E1564658
BT Leg 20

BT Deep Sea Drilling Project
SA Equatorial Pacific
SA Pacific Ocean

DSDP Site 202 (1985)
Autoposting of Leg 20 to this term began in 1989. W Equatorial Pacific.
CO N124854N124854
  E1565709E1565709
BT Leg 20
BT Deep Sea Drilling Project
SA Equatorial Pacific
SA Pacific Ocean

DSDP Site 206 (1989)
New Caledonian Basin, W Pacific.
CO S320045S320045
  E1652709E1652709
BT Leg 21
BT Deep Sea Drilling Project
SA Pacific Ocean
SA West Pacific

DSDP Site 207 (1989)
S Lord Howe Rise, Tasman Sea, W Pacific.
CO S365746S365744
  E1652604E1652603
BT Leg 21
BT Deep Sea Drilling Project
SA Lord Howe Rise
SA Pacific Ocean
SA Tasman Sea
SA West Pacific

DSDP Site 208 (1989)
N Lord Howe Rise, SW Pacific.
CO S260637S260636
  E1611317E1611316
BT Leg 21
BT Deep Sea Drilling Project
SA Lord Howe Rise
SA Pacific Ocean
SA Southwest Pacific
SA West Pacific

DSDP Site 209 (1993)
Queensland Plateau, NW South Polynesian Pacific. Before 1981, also search Site 209 AND Deep Sea Drilling Project.
CO S155612S155611
  E1521117E1521116
BT Leg 21
BT Deep Sea Drilling Project
SA Pacific Ocean
SA South Polynesian Pacific

DSDP Site 210 (1989)
Coral Sea Basin, W Pacific.
CO S134600S134559
  E1525347E1525346
BT Leg 21
BT Deep Sea Drilling Project
SA Coral Sea Basin
SA Pacific Ocean
SA West Pacific

DSDP Site 211 (1989)
About 130 nautical miles S of the axis of the Java Trench. NE Indian Ocean.
CO S094632S094631
  E1024158E1024156
BT Leg 22
BT Deep Sea Drilling Project
SA Indian Ocean

DSDP Site 212 (1989)
Wharton Basin, Indian Ocean.
CO S191121S191121
  E0991751E0991750
BT Leg 22
BT Deep Sea Drilling Project
SA Indian Ocean
SA Wharton Basin

DSDP Site 213 (1989)

About 300 nautical miles E of Ninetyeast Ridge, Indian Ocean.
CO S101243S101242
    E0935347E0935346
BT Leg 22
BT Deep Sea Drilling Project
SA Indian Ocean
SA Ninetyeast Ridge

DSDP Site 214 (1981)
Autoposting of Leg 22 to this term began in 1989. On crest of Ninetyeast Ridge, Indian Ocean. Before 1981, search Site 214 AND Deep Sea Drilling Project.
CO S112013S112013
    E0884305E0884305
BT Leg 22
BT Deep Sea Drilling Project
SA Indian Ocean

DSDP Site 215 (1989)
W of Ninetyeast Ridge in E Indian Ocean.
CO S080718S080718
    E0864730E0864730
BT Leg 22
BT Deep Sea Drilling Project
SA Indian Ocean
SA Ninetyeast Ridge

DSDP Site 216 (1989)
Crest of Ninetyeast Ridge near the Equator in E Indian Ocean.
CO N012743N012744
    E0901229E0901228
BT Leg 22
BT Deep Sea Drilling Project
SA Indian Ocean
SA Ninetyeast Ridge

DSDP Site 217 (1989)
E flank of Ninetyeast Ridge, E Indian Ocean.
CO N085534N085535
    E0903220E0903219
BT Leg 22
BT Deep Sea Drilling Project
SA Indian Ocean
SA Ninetyeast Ridge

DSDP Site 218 (1989)
Central Bengal Fan, E Indian Ocean.
CO N080025N080026
    E0861659E0861658
BT Leg 22
BT Deep Sea Drilling Project
SA Indian Ocean

DSDP Site 219 (1989)
In Nine Degree Channel, which lies between Laccadive and Maldive islands in Arabian Sea.
CO N090146N090144
    E0725241E0725240
BT Leg 23
BT Deep Sea Drilling Project
SA Arabian Sea

DSDP Site 220 (1993)
W Chagos-Laccadive Plateau, Arabian Sea. Before 1981, also search Site 220 AND Deep Sea Drilling Project.
CO N063058N063059
    E0705902E0705901
BT Leg 23
BT Deep Sea Drilling Project
SA Arabian Sea
SA Indian Ocean

DSDP Site 225 (1981)
Autoposting of Leg 23 to this term began in 1989. Red Sea. Before 1981, search Site 225 AND Deep Sea Drilling Project.
CO N211835N211835
    E0381507E0381507

BT Leg 23
BT Deep Sea Drilling Project
SA Red Sea

DSDP Site 227 (1981)
Autoposting of Leg 23 to this term began in 1989. E Atlantis II Deep, Red Sea. Before 1981, search Site 227 AND Deep Sea Drilling Project.
CO N211952N211952
    E0380758E0380758
BT Leg 23
BT Deep Sea Drilling Project
SA Atlantis II Deep
SA Red Sea

DSDP Site 228 (1981)
Autoposting of Leg 23 to this term began in 1989. Red Sea. Before 1981, search Site 228 AND Deep Sea Drilling Project.
CO N190510N190510
    E0390012E0390012
BT Leg 23
BT Deep Sea Drilling Project
SA Red Sea

DSDP Site 231 (1993)
S Gulf of Aden, Indian Ocean. Before 1981, also search Site 231 AND Deep Sea Drilling Project.
CO N115304N115305
    E0481443E0481442
BT Leg 24
BT Deep Sea Drilling Project
SA Gulf of Aden
SA Indian Ocean

DSDP Site 232 (1993)
Illaue-Fartak Trench, E Gulf of Aden, Arabian Sea. Before 1982, also search Site 232 AND Deep Sea Drilling Project.
CO N142856N142856
    E0515452E0515452
BT Leg 24
BT Deep Sea Drilling Project
SA Gulf of Aden
SA Indian Ocean

DSDP Site 233 (1993)
Illaue-Fartak Trench, E Gulf of Aden, Arabian Sea. Before 1981, also search Site 233 AND Deep Sea Drilling Project.
CO N141940N141941
    E0520811E0520811
BT Leg 24
BT Deep Sea Drilling Project
SA Arabian Sea
SA Gulf of Aden
SA Indian Ocean

DSDP Site 236 (1993)
Somali Basin between Carlsberg Ridge and Seychelles Bank, W Indian Ocean. Before 1981, also search Site 236 AND Deep Sea Drilling Project.
CO S014038S014037
    E0573851E0573851
BT Leg 24
BT Deep Sea Drilling Project
SA Indian Ocean
SA Somali Basin

DSDP Site 237 (1989)
Mascarene Plateau, W Indian Ocean.
CO S070500S070459
    E0580729E0580728
BT Leg 24
BT Deep Sea Drilling Project
SA Indian Ocean

DSDP Site 238 (1989)
Extreme NE Argo fracture zone, W Indian Ocean.

CO S110913S110912
    E0703134E0703133
BT Leg 24
BT Deep Sea Drilling Project
SA Indian Ocean

DSDP Site 241 (1989)
East African continental rise, W Indian Ocean.
CO S022215S022214
    E0444047E0444046
BT Leg 25
BT Deep Sea Drilling Project
SA Indian Ocean

DSDP Site 245 (1989)
S Madagascar Basin, W Indian Ocean.
CO S313202S313201
    E0521807E0521806
BT Leg 25
BT Deep Sea Drilling Project
SA Indian Ocean
SA Madagascar Basin

DSDP Site 248 (1993)
Mozambique Basin, SW Indian Ocean. Before 1981, also search Site 248 AND Deep Sea Drilling Project.
CO S293147S293146
    E0372829E0372828
BT Leg 25
BT Deep Sea Drilling Project
SA Indian Ocean

DSDP Site 251 (1989)
Indian Ocean Ridge, SW Indian Ocean.
CO S363016S363015
    E0492905E0492904
BT Leg 26
BT Deep Sea Drilling Project
SA Indian Ocean

DSDP Site 253 (1981)
Autoposting of Leg 26 to this term began in 1989. Ninetyeast Ridge, SE Indian Ocean. Before 1981, search Site 253 AND Deep Sea Drilling Project.
CO S245239S245239
    E0872158E0872158
BT Leg 26
BT Deep Sea Drilling Project
SA Indian Ocean
SA Ninetyeast Ridge

DSDP Site 254 (1981)
Autoposting of Leg 26 to this term began in 1989. Ninetyeast Ridge, SE Indian Ocean. Before 1981, search Site 254 AND Deep Sea Drilling Project.
CO S305809S305809
    E0875343E0875343
BT Leg 26
BT Deep Sea Drilling Project
SA Indian Ocean
SA Ninetyeast Ridge

DSDP Site 255 (1993)
S Broken Ridge, SE Indian Ocean. Before 1981, also search Site 255 AND Deep Sea Drilling Project.
CO S310753S310752
    E0934344E0934343
BT Leg 26
BT Deep Sea Drilling Project
SA Indian Ocean

DSDP Site 256 (1989)
S Wharton Basin, E Indian Ocean.
CO S232722S232720
    E1004628E1004627
BT Leg 26
BT Deep Sea Drilling Project
SA Indian Ocean

SA Wharton Basin

DSDP Site 257 (1989)
SE Wharton Basin, E Indian Ocean.
CO S305910S305909
    E1082100E1082059
BT Leg 26
BT Deep Sea Drilling Project
SA Indian Ocean
SA Wharton Basin

DSDP Site 259 (1981)
Autoposting of Leg 27 to this term began in 1989. Perth abyssal plain, E Indian Ocean. Before 1981, search Site 259 AND Deep Sea Drilling Project.
CO S293700S293700
    E1124200E1124200
BT Leg 27
BT Deep Sea Drilling Project
SA Indian Ocean
SA Perth abyssal plain

DSDP Site 260 (1981)
Autoposting of Leg 27 to this term began in 1989. Gascoyne abyssal plain, E Indian Ocean. Before 1981, search Site 260 AND Deep Sea Drilling Project.
CO S160900S160900
    E1101800E1101800
BT Leg 27
BT Deep Sea Drilling Project
SA Gascoyne abyssal plain
SA Indian Ocean

DSDP Site 261 (1981)
Autoposting of Leg 27 to this term began in 1989. NE Argo abyssal plain, E Indian Ocean. Before 1981, search Site 261 AND Deep Sea Drilling Project.
CO S125700S125700
    E1175400E1175400
BT Leg 27
BT Deep Sea Drilling Project
SA Argo abyssal plain
SA Indian Ocean

DSDP Site 262 (1981)
Autoposting of Leg 27 to this term began in 1989. Timor Trough, E Indian Ocean. Before 1981, search Site 262 AND Deep Sea Drilling Project.
CO S105200S105200
    E1235100E1235100
BT Leg 27
BT Deep Sea Drilling Project
SA Indian Ocean
SA Timor Trough

DSDP Site 263 (1981)
Autoposting of Leg 27 to this term began in 1989. E Cuvier abyssal plain, E Indian Ocean. Before 1981, search Site 263 AND Deep Sea Drilling Project.
CO S232000S232000
    E1105800E1105800
BT Leg 27
BT Deep Sea Drilling Project
SA Cuvier abyssal plain
SA Indian Ocean

DSDP Site 264 (1981)
Autoposting of Leg 28 to this term began in 1989. Naturaliste Plateau, E Indian Ocean. Before 1981, search Site 264 AND Deep Sea Drilling Project.
CO S345808S345808
    E1120241E1120241
BT Leg 28
BT Deep Sea Drilling Project
SA Indian Ocean

Before then, variants of the term may occur in GeoRef.

DSDP Site 265 (1993)
Southeast Indian Ridge, Indian Ocean. Before 1983, also search Site 265 AND Deep Sea Drilling Project.
CO S533227S533227
     E1095645E1095644
BT Leg 28
BT Deep Sea Drilling Project
SA Indian Ocean
SA Southeast Indian Ridge

DSDP Site 266 (1989)
Off the coast of Antarctica, about 800 km S of SE Indian Ridge crest in Antarctic Ocean.
CO S562408S562407
     E1100643E1100641
BT Leg 28
BT Deep Sea Drilling Project
SA Antarctic Ocean

DSDP Site 270 (1981)
Autoposting of Leg 28 to this term began in 1989. SE Ross Sea, Antarctic Ocean. Before 1981, search Deep Sea Drilling Project AND Site 270.
CO S772629S772629
     W1783011W1783011
BT Leg 28
BT Deep Sea Drilling Project
SA Antarctic Ocean
SA Ross Sea

DSDP Site 271 (1993)
SE Ross Sea, Antarctic Ocean. Before 1981, also search Site 271 AND Deep Sea Drilling Project.
CO S764317S764316
     W1750251W1750252
BT Leg 28
BT Deep Sea Drilling Project
SA Antarctic Ocean
SA Ross Sea

DSDP Site 272 (1989)
SE Ross Sea, Antarctic Ocean.
CO S770738S770737
     W1764536W1764537
BT Leg 28
BT Deep Sea Drilling Project
SA Antarctic Ocean
SA Ross Sea

DSDP Site 273 (1993)
W central Ross Sea, Antarctic Ocean. Before 1982, also search Site 273 AND Deep Sea Drilling Project.
CO S743218S743217
     E1743735E1743734
BT Leg 28
BT Deep Sea Drilling Project
SA Antarctic Ocean
SA Ross Sea

DSDP Site 274 (1981)
Autoposting of Leg 28 to this term began in 1989. Ross Sea, Antarctic Ocean. Before 1981, search Site 274 AND Deep Sea Drilling Project.
CO S685949S685949
     E1732538E1732538
BT Leg 28
BT Deep Sea Drilling Project
SA Antarctic Ocean
SA Ross Sea

DSDP Site 277 (1989)
Campbell Plateau, extreme SW Polynesian Pacific.
CO S521326S521325
     E1661129E1661128
BT Leg 29
BT Deep Sea Drilling Project
SA Pacific Ocean
SA Polynesian Pacific

DSDP Site 284 (1981)
Autoposting of Leg 29 to this term began in 1989. W of New Zealand in the Tasman Sea. Before 1981, search Site 284 AND Deep Sea Drilling Project.
CO S403029S403029
     E1674049E1674049
BT Leg 29
BT Deep Sea Drilling Project
SA Pacific Ocean
SA Tasman Sea

DSDP Site 285 (1989)
S Fiji Basin, W Pacific.
CO S264910S264909
     E1754815E1754814
BT Leg 30
BT Deep Sea Drilling Project
SA Pacific Ocean
SA West Pacific

DSDP Site 286 (1989)
New Hebrides Basin, W Pacific.
CO S163156S163155
     E1662211E1662210
BT Leg 30
BT Deep Sea Drilling Project
SA Pacific Ocean
SA Vanuatu
SA West Pacific

DSDP Site 288 (1993)
SE margin Ontong Java Plateau, West Pacific. Before 1982, also search Site 288 AND Deep Sea Drilling Project.
CO S055821S055821
     E1614932E1614931
BT Leg 30
BT Deep Sea Drilling Project
SA Ontong Java Plateau
SA Pacific Ocean
SA West Pacific

DSDP Site 289 (1989)
N Ontong Java Plateau, SW Pacific Ocean. Before 1981, also search Site 289 AND Deep Sea Drilling Project.
CO S002956S002955
     E1583042E1583041
BT Leg 30
BT Deep Sea Drilling Project
SA Ontong Java Plateau
SA Pacific Ocean

DSDP Site 291 (1993)
Outer Philippine Trench, S West Philippine Basin, Phillipine Sea.
CO N124825N124827
     E1274959E1274951
BT Leg 31
BT Deep Sea Drilling Project
SA Pacific Ocean
SA Philippine Sea
SA West Philippine Basin

DSDP Site 292 (1981)
Autoposting of Leg 31 to this term began in 1989. Benham Rise, W Philippine Sea. Before 1981, search Site 292 AND Deep Sea Drilling Project.
CO N154906N154907
     E1243904E1243902
BT Leg 31
BT Deep Sea Drilling Project
SA Pacific Ocean
SA Philippine Sea

DSDP Site 293 (1989)
W Philippine Sea, W Pacific.
CO N202115N202115
     E1240539E1240539
BT Leg 31
BT Deep Sea Drilling Project
SA Pacific Ocean
SA Philippine Sea

DSDP Site 294 (1989)
W Philippine Basin, W Pacific.
CO N223444N223445
     E1312308E1312307
BT Leg 31
BT Deep Sea Drilling Project
SA Pacific Ocean
SA Philippine Sea

DSDP Site 296 (1981)
Autoposting of Leg 31 to this term began in 1989. Kyushu-Palau Ridge, N Philippine Sea. Before 1981, search Site 296 AND Deep Sea Drilling Project.
CO N292025N292025
     E1333131E1333131
BT Leg 31
BT Deep Sea Drilling Project
SA Kyushu-Palau Ridge
SA Pacific Ocean
SA Philippine Sea

DSDP Site 297 (1981)
Autoposting of Leg 31 to this term began in 1989. Nankai Trough, N Philippine Sea. Before 1981, search Site 297 AND Deep Sea Drilling Project.
CO N305222N305222
     E1340953E1340953
BT Leg 31
BT Deep Sea Drilling Project
SA Nankai Trough
SA Pacific Ocean
SA Philippine Sea

DSDP Site 298 (1989)
Nankai Trough, NW Pacific.
CO N314255N314256
     E1333614E1333613
BT Leg 31
BT Deep Sea Drilling Project
SA Nankai Trough
SA Northwest Pacific
SA Pacific Ocean

DSDP Site 299 (1989)
Yamato Basin, Japan Sea.
CO N392941N392942
     E1373944E1373943
BT Leg 31
BT Deep Sea Drilling Project
SA Japan Sea
SA Pacific Ocean

DSDP Site 302 (1989)
Yamato Basin, Japan Sea.
CO N402007N402008
     E1365401E1365400
BT Leg 31
BT Deep Sea Drilling Project
SA Japan Sea
SA Pacific Ocean

DSDP Site 305 (1989)
Shatsky Rise, W Pacific. Before 1981, also search Site 305 AND Deep Sea Drilling Project.
CO N320007N320008
     E1575101E1575059
BT Leg 32
BT Deep Sea Drilling Project
SA Pacific Ocean
SA Shatsky Rise
SA West Pacific

DSDP Site 310 (1989)
Hess Rise, N Pacific.
CO N365206N365207
     E1765406E1765405
BT Leg 32
BT Deep Sea Drilling Project
SA Hess Rise
SA North Pacific
SA Pacific Ocean
SA West Pacific

DSDP Site 313 (1989)
NE Mid-Pacific Rise, N Polynesian Pacific. Before 1981, also search Site 313 AND Deep Sea Drilling Project.
CO N201031N201032
     W1705708W1705710
BT Leg 32
BT Deep Sea Drilling Project
SA Mid-Pacific Rise
SA Pacific Ocean
SA Polynesian Pacific

DSDP Site 315 (1981)
Autoposting of Leg 33 to this term began in 1989. Central Equatorial Pacific near the Line Islands. Before 1981, search Site 315 AND Deep Sea Drilling Project.
CO N041016N041016
     W1583132W1583132
BT Leg 33
BT Deep Sea Drilling Project
SA Equatorial Pacific
SA Line Islands
SA Pacific Ocean

DSDP Site 315A (1989)
Near the Line Islands, Polynesian Pacific. Before 1981, also search Site 315A AND Deep Sea Drilling Project.
CO N041015N041016
     W1583132W1583133
BT Leg 33
BT Deep Sea Drilling Project
SA Equatorial Pacific
SA Line Islands
SA Pacific Ocean
SA Polynesian Pacific

DSDP Site 317 (1989)
Manihiki Plateau, Polynesian Pacific. Before 1981, also search Site 317 AND Deep Sea Drilling Project.
CO S110006S110005
     W1621546W1621547
BT Leg 33
BT Deep Sea Drilling Project
SA Manihiki Plateau
SA Pacific Ocean
SA Polynesian Pacific

DSDP Site 319 (1981)
Autoposting of Leg 34 to this term began in 1989. E of the East Pacific Rise, on W flank of Galapagos Rise in SE Pacific Ocean. Before 1981, search Site 319 AND Deep Sea Drilling Project.
CO S130102S130102
     W1013128W1013128
BT Leg 34
BT Deep Sea Drilling Project
SA East Pacific Rise
SA Pacific Ocean

DSDP Site 320 (1981)
Autoposting of Leg 34 to this term began in 1989. Off the coast of Peru, SE Pacific Ocean. Before 1981, search Site 320 AND Deep Sea Drilling Project.
CO S090024S090024
     W0833148W0833148
BT Leg 34
BT Deep Sea Drilling Project
SA Nazca Plate
SA Pacific Ocean

DSDP Site 321 (1981)
Autoposting of Leg 34 to this term began in 1989. Off the coast of Peru, SE Pacific Ocean. Before 1981, search Site 321 AND Deep Sea Drilling Project.
CO S120117S120117
     W0815414W0815414

BT Leg 34
BT Deep Sea Drilling Project
SA Nazca Plate
SA Pacific Ocean

DSDP Site 322 (1981)
Autoposting of Leg 35 to this term began in 1989. Antarctic Ocean between S tip of South America and the Antarctic Peninsula. Before 1981, search Site 322 AND Deep Sea Drilling Project.
CO S600127S600127 W0792529W0792529
BT Leg 35
BT Deep Sea Drilling Project
SA Antarctic Ocean

DSDP Site 323 (1981)
Autoposting of Leg 35 to this term began in 1989. Antarctic Ocean, SW of South America in Bellingshausen abyssal plain. Before 1981, search Site 323 AND Deep Sea Drilling Project.
CO S634050S634050 W0975941W0975941
BT Leg 35
BT Deep Sea Drilling Project
SA Antarctic Ocean

DSDP Site 325 (1981)
Autoposting of Leg 35 to this term began in 1989. Antarctic Ocean, NW of the Antarctic Peninsula. Before 1981, search Site 325 AND Deep Sea Drilling Project.
CO S650247S650247 W0734024W0734024
BT Leg 35
BT Deep Sea Drilling Project
SA Antarctic Ocean

DSDP Site 327 (1981)
Autoposting of Leg 36 to this term began in 1989. E Falkland Plateau in SW Atlantic Ocean, N of the Scotia Sea. Before 1981, search Site 327 AND Deep Sea Drilling Project.
CO S505223S505223 W0464701W0464701
BT Leg 36
BT Deep Sea Drilling Project
SA Atlantic Ocean
SA Falkland Plateau

DSDP Site 328 (1989)
Falkland outer basin, SW Atlantic. Before 1989, also search Site 328 AND Deep Sea Drilling Project.
CO S494841S494840 W0363931W0363932
BT Leg 36
BT Deep Sea Drilling Project
SA Atlantic Ocean
SA Southwest Atlantic

DSDP Site 329 (1981)
Autoposting of Leg 36 to this term began in 1989. E Falkland Plateau in SW Atlantic Ocean, N of the Scotia Sea. Before 1981, search Site 329 AND Deep Sea Drilling Project.
CO S503919S503919 W0460544W0460544
BT Leg 36
BT Deep Sea Drilling Project
SA Atlantic Ocean
SA Falkland Plateau

DSDP Site 330 (1981)
Autoposting of Leg 36 to this term began in 1989. E Falkland Plateau in SW Atlantic Ocean, N of the Scotia Sea. Before 1981, search Site 330 AND Deep Sea Drilling Project.
CO S505511S505511 W0465300W0465300
BT Leg 36
BT Deep Sea Drilling Project
SA Atlantic Ocean
SA Falkland Plateau

DSDP Site 332 (1981)
Autoposting of Leg 37 to this term began in 1989. Near crest of Mid-Atlantic Ridge. Before 1981, search Site 332 AND Deep Sea Drilling Project.
CO N365243N365246 W0333828W0333834
BT Leg 37
BT Deep Sea Drilling Project
SA Atlantic Ocean
SA Mid-Atlantic Ridge

DSDP Site 332B (1989)
North Atlantic Ridge, W of the Azores. Before 1989, also search Site 332B AND Deep Sea Drilling Project.
CO N365245N365246 W0333834W0333835
BT Leg 37
BT Deep Sea Drilling Project
SA Atlantic Ocean
SA Mid-Atlantic Ridge
SA North Atlantic Ridge

DSDP Site 333 (1981)
Autoposting of Leg 37 to this term began in 1989. Near crest of Mid-Atlantic Ridge. Before 1981, search Site 333 AND Deep Sea Drilling Project.
CO N365027N365027 W0334003W0334003
BT Leg 37
BT Deep Sea Drilling Project
SA Atlantic Ocean
SA Mid-Atlantic Ridge

DSDP Site 334 (1981)
Autoposting of Leg 37 to this term began in 1989. W flank of Mid-Atlantic Ridge. Before 1981, search Site 334 AND Deep Sea Drilling Project.
CO N370208N370208 W0342452W0342452
BT Leg 37
BT Deep Sea Drilling Project
SA Atlantic Ocean
SA Mid-Atlantic Ridge

DSDP Site 335 (1981)
Autoposting of Leg 37 to this term began in 1989. W flank of the Mid-Atlantic Ridge. Before 1981, search Site 335 AND Deep Sea Drilling Project.
CO N371744N371744 W0351155W0351155
BT Leg 37
BT Deep Sea Drilling Project
SA Atlantic Ocean
SA Mid-Atlantic Ridge

DSDP Site 336 (1989)
N flank of Iceland-Faeroe Ridge, Norwegian Sea. Before 1989, also search Site 336 AND Deep Sea Drilling Project.
CO N632103N632104 W0074716W0074717
BT Leg 38
BT Deep Sea Drilling Project
SA Arctic Ocean
SA Iceland-Faeroe Ridge
SA Norwegian Sea

DSDP Site 338 (1989)
Outer Voring Plateau, Norwegian Sea. Before 1981, also search Site 338 AND Deep Sea Drilling Project.
CO N674706N674707 E0052316E0052315
BT Leg 38
BT Deep Sea Drilling Project
SA Arctic Ocean
SA Norwegian Sea
SA Voring Plateau

DSDP Site 343 (1989)
At the foot of Voring Plateau in Lofoten Basin, Norwegian Sea. Before 1981, also search Site 343 AND Deep Sea Drilling Project.
CO N684254N684255 E0054544E0054543
BT Leg 38
BT Deep Sea Drilling Project
SA Arctic Ocean
SA Norwegian Sea
SA Voring Plateau

DSDP Site 344 (1989)
Knipovich Ridge, Greenland Sea. Before 1981, also search Site 344 AND Deep Sea Drilling Project.
CO N760858N760859 E0075232E0075231
BT Leg 38
BT Deep Sea Drilling Project
SA Arctic Ocean
SA Greenland Sea
SA Knipovich Ridge

DSDP Site 345 (1989)
Lofoten Basin, Norwegian Sea. Before 1981, also search Site 345 AND Deep Sea Drilling Project.
CO N695013N695014 W0011415W0011416
BT Leg 38
BT Deep Sea Drilling Project
SA Arctic Ocean
SA Norwegian Sea

DSDP Site 346 (1989)
Jan Mayen Ridge, Norwegian Sea. Before 1981, also search Site 346 AND Deep Sea Drilling Project.
CO N695321N695322 W0084107W0084108
BT Leg 38
BT Deep Sea Drilling Project
SA Arctic Ocean
SA Jan Mayen Ridge
SA Norwegian Sea

DSDP Site 348 (1989)
Icelandic Plateau, Arctic Ocean. Before 1981, also search Site 348 AND Deep Sea Drilling Project.
CO N683010N683011 W0122743W0122744
BT Leg 38
BT Deep Sea Drilling Project
SA Arctic Ocean
SA Icelandic Plateau

DSDP Site 354 (1989)
Ceara Rise, Guyanese Atlantic. Before 1981, also search Site 354 AND Deep Sea Drilling Project.
CO N055356N055358 W0441146W0441147
BT Leg 39
BT Deep Sea Drilling Project
SA Atlantic Ocean
SA Guyanese Atlantic

DSDP Site 355 (1989)
Brazil Basin, South American Atlantic. Before 1981, also search Site 355 AND Deep Sea Drilling Project.
CO S154236S154235 W0303601W0303602
BT Leg 39
BT Deep Sea Drilling Project
SA Atlantic Ocean
SA Brazil Basin
SA South American Atlantic

DSDP Site 356 (1981)
Autoposting of Leg 39 to this term began in 1989. Sao Paulo Plateau, SW Atlantic Ocean. Before 1981, search Site 356 AND Deep Sea Drilling Project.
CO S281713S281713 W0410517W0410517
BT Leg 39
BT Deep Sea Drilling Project
SA Atlantic Ocean
SA South Atlantic

DSDP Site 357 (1981)
Autoposting of Leg 39 to this term began in 1989. Rio Grande Rise, SW Atlantic Ocean. Before 1981, search Site 357 AND Deep Sea Drilling Project.
CO S300015S300015 W0353335W0353335
BT Leg 39
BT Deep Sea Drilling Project
SA Atlantic Ocean
SA Rio Grande Rise

DSDP Site 360 (1989)
Cape Basin, SE Atlantic. Before 1981, also search Site 360 AND Deep Sea Drilling Project.
CO S355046S355044 E0180548E0180547
BT Leg 40
BT Deep Sea Drilling Project
SA Atlantic Ocean
SA Cape Basin
SA Southeast Atlantic

DSDP Site 361 (1981)
Autoposting of Leg 40 to this term began in 1989. Off Cape of Good Hope, SE Atlantic Ocean. Before 1981, search Site 361 AND Deep Sea Drilling Project.
CO S350358S350358 E0152655E0152655
BT Leg 40
BT Deep Sea Drilling Project
SA Atlantic Ocean

DSDP Site 362 (1981)
Autoposting of Leg 40 to this term began in 1989. Walvis Ridge, SE Atlantic Ocean. Before 1981, search Site 362 AND Deep Sea Drilling Project.
CO S194527S194527 E0103157E0103157
BT Leg 40
BT Deep Sea Drilling Project
SA Atlantic Ocean
SA Walvis Ridge

DSDP Site 363 (1981)
Autoposting of Leg 40 to this term began in 1989. Walvis Ridge, SE Atlantic Ocean. Before 1981, search Site 363 AND Deep Sea Drilling Project.
CO S193845S193845 E0090248E0090248
BT Leg 40
BT Deep Sea Drilling Project
SA Atlantic Ocean
SA Walvis Ridge

DSDP Site 364 (1981)
Autoposting of Leg 40 to this term began in 1989. Angola Basin, SE Atlantic Ocean. Before 1981,

search Site 364 AND Deep Sea Drilling Project.
CO S113419S113419
E0115818E0115818
BT Leg 40
BT Deep Sea Drilling Project
SA Angola Basin
SA Atlantic Ocean

DSDP Site 365 (1989)
Angola Basin, SE Atlantic. Before 1981, also search Site 365 AND Deep Sea Drilling Project.
CO S113907S113905
E0115344E0115343
BT Leg 40
BT Deep Sea Drilling Project
SA Angola Basin
SA Atlantic Ocean
SA Southeast Atlantic

DSDP Site 366 (1989)
Sierra Leone Rise, Cape Verde Atlantic. Before 1981, search Site 366 AND Deep Sea Drilling Project.
CO N054040N054041
W0195104W0195105
BT Leg 41
BT Deep Sea Drilling Project
SA Atlantic Ocean
SA Cape Verde Atlantic
SA Equatorial Atlantic
SA ODP Site 668
SA Sierra Leone Rise

DSDP Site 367 (1981)
Autoposting of Leg 41 to this term began in 1989. Cape Verde Basin, NE Atlantic Ocean. Before 1981, search Site 367 AND Deep Sea Drilling Project.
CO N122913N122913
W0200250W0200250
BT Leg 41
BT Deep Sea Drilling Project
SA Atlantic Ocean
SA Cape Verde Basin

DSDP Site 368 (1981)
Autoposting of Leg 41 to this term began in 1989. Cape Verde Rise, NE Atlantic Ocean. Before 1981, search Site 368 AND Deep Sea Drilling Project.
CO N173026N173026
W0212114W0212114
BT Leg 41
BT Deep Sea Drilling Project
SA Atlantic Ocean
SA Cape Verde Basin

DSDP Site 369 (1981)
Autoposting of Leg 41 to this term began in 1989. SE of Canary Islands, NE Atlantic Ocean. Before 1981, search Site 369 AND Deep Sea Drilling Project.
CO N263533N263533
W0145955W0145955
BT Leg 41
BT Deep Sea Drilling Project
SA Atlantic Ocean

DSDP Site 370 (1981)
Autoposting of Leg 41 to this term began in 1989. Off Morocco, NE Atlantic Ocean. Before 1981, search Site 370 AND Deep Sea Drilling Project.
CO N325015N325015
W0104634W0104634
BT Leg 41
BT Deep Sea Drilling Project
SA Atlantic Ocean

DSDP Site 372 (1981)
Autoposting of Leg 42A to this term began in 1989. Balearic Basin, Mediterranean Sea. Before 1981, search Site 372 AND Deep Sea Drilling Project.
CO N400152N400154
E0044747E0044747
BT Leg 42A
BT Deep Sea Drilling Project
SA Balearic Basin
SA Mediterranean Sea

DSDP Site 374 (1981)
Autoposting of Leg 42A to this term began in 1989. Ionian Sea. Before 1981, search Site 374 AND Deep Sea Drilling Project.
CO N355052N355052
E0181147E0181147
BT Leg 42A
BT Deep Sea Drilling Project
SA Ionian Sea
SA Mediterranean Sea

DSDP Site 375 (1989)
Florence Rise, E Mediterranean Sea. Before 1981, also search Site 375 AND Deep Sea Drilling Project.
CO N344544N344545
E0304535E0304534
BT Leg 42A
BT Deep Sea Drilling Project
SA Florence Rise
SA Mediterranean Sea

DSDP Site 376 (1989)
Florence Rise, E Mediterranean Sea. Before 1981, also search Site 376 AND Deep Sea Drilling Project.
CO N345219N345220
E0314828E0314826
BT Leg 42A
BT Deep Sea Drilling Project
SA Florence Rise
SA Mediterranean Sea

DSDP Site 378 (1993)
Cretan Basin, East Mediterranean. Before 1981, also search Site 378 AND Deep Sea Drilling Project.
CO N355640N355641
E0250659E0250658
BT Leg 42A
BT Deep Sea Drilling Project
SA East Mediterranean
SA Mediterranean Sea

DSDP Site 379 (1981)
Autoposting of Leg 42B to this term began in 1989. E Black Sea. Before 1981, search Site 379 AND Deep Sea Drilling Project.
CO N430017N430017
E0360041E0360041
BT Leg 42B
BT Deep Sea Drilling Project
SA Black Sea

DSDP Site 380 (1981)
Autoposting of Leg 42B to this term began in 1989. W Black Sea. Before 1981, search Site 380 AND Deep Sea Drilling Project.
CO N420559N420559
E0293654E0293654
BT Leg 42B
BT Deep Sea Drilling Project
SA Black Sea

DSDP Site 381 (1981)
Autoposting of Leg 42B to this term began in 1989. W Black Sea. Before 1981, search Site 381 AND Deep Sea Drilling Project.
CO N414015N414015
E0292458E0292458
BT Leg 42B
BT Deep Sea Drilling Project
SA Black Sea

DSDP Site 382 (1993)
Nashville Seamount, New England Seamounts, North American Atlantic. Before 1981, also search Site 382 AND Deep Sea Drilling Project.
CO N342502N342503
W0563215W0563215
BT Leg 43
BT Deep Sea Drilling Project
SA Atlantic Ocean
SA New England Seamounts
SA North American Atlantic

DSDP Site 384 (1989)
J-anomaly Ridge, Grand Banks, North American Atlantic. Before 1981, also search Site 384 AND Deep Sea Drilling Project.
CO N402139N402139
W0513948W0513948
BT Leg 43
BT Deep Sea Drilling Project
SA Atlantic Ocean
SA Grand Banks
SA North American Atlantic

DSDP Site 385 (1993)
Vogel Seamount, New England Seamounts, North American Atlantic. Before 1981, also search Site 385 AND Deep Sea Drilling Project.
CO N372210N372211
W0600927W0600927
BT Leg 43
BT Deep Sea Drilling Project
SA Atlantic Ocean
SA New England Seamounts
SA North American Atlantic

DSDP Site 386 (1989)
Central Bermuda Rise, Northwest Atlantic. Before 1981, also search Site AND Deep Sea Drilling Project.
CO N311112N311113
W0641456W0641457
BT Leg 43
BT Deep Sea Drilling Project
SA Atlantic Ocean
SA Bermuda Rise
SA Northwest Atlantic

DSDP Site 387 (1989)
W Bermuda Rise, NW Atlantic. Before 1981, also search Site 387 AND Deep Sea Drilling Project.
CO N321912N321912
W0674000W0674000
BT Leg 43
BT Deep Sea Drilling Project
SA Atlantic Ocean
SA Bermuda Rise
SA Northwest Atlantic

DSDP Site 390 (1989)
Blake Nose, NW Atlantic. Before 1981, also search Site 390 AND Deep Sea Drilling Project.
CO N300832N300833
W0760644W0760645
BT Leg 44
BT Deep Sea Drilling Project
SA Atlantic Ocean
SA Blake Nose
SA Northwest Atlantic

DSDP Site 391 (1981)
Autoposting of Leg 44 to this term began in 1989. Blake-Bahama Basin, NW Atlantic Ocean. Before 1981, search Site 391 AND Deep Sea Drilling Project.
CO N281340N281340
W0753653W0753653
BT Leg 44
BT Deep Sea Drilling Project
SA Atlantic Ocean
SA Blake-Bahama Basin

DSDP Site 392 (1993)
Blake Nose, Blake Plateau, North American Atlantic. Before 1981, also search Site 392 AND Deep Sea Drilling Project.
CO N295437N295438
W0761040W0761041
BT Leg 44
BT Deep Sea Drilling Project
SA Atlantic Ocean
SA Blake Nose
SA North American Atlantic

DSDP Site 395 (1981)
N central Atlantic Ocean near the Mid-Atlantic Ridge axis. Holes 395 and 395A were drilled on Leg 45. Hole 395A was reentered and Hole 395B was drilled on Leg 78B. Hole 395A was reentered on Leg 109. Autoposting of Leg 45 to this term began in 1989. Autoposting of Legs 78B and 109 began in 1993. Before 1981, search Site 395 AND Deep Sea Drilling Project. This term has multiple hierarchies.
CO N224521N224521
W0460454W0460454
BT1 Leg 45
BT1 IPOD
BT1 Deep Sea Drilling Project
BT2 Leg 78B
BT2 IPOD
BT2 Deep Sea Drilling Project
BT3 Leg 109
BT3 Ocean Drilling Program
SA Atlantic Ocean
SA Kane fracture zone
SA Mid-Atlantic Ridge
SA North Atlantic Ridge

DSDP Site 396 (1981)
N central Atlantic Ocean near the Mid-Atlantic Ridge axis. Autoposting of Leg 45 to this term began in 1989. Autoposting of Leg 46 began in 1993. Before 1981, search Site 396 AND Deep Sea Drilling Project. This term has multiple hierarchies.
CO N225853N225853
W0433057W0433057
BT1 Leg 45
BT1 IPOD
BT1 Deep Sea Drilling Project
BT2 Leg 46
BT2 IPOD
BT2 Deep Sea Drilling Project
SA Atlantic Ocean
SA Mid-Atlantic Ridge

DSDP Site 397 (1989)
Cape Bojador in NE Atlantic Ocean, near Canary Islands. Before 1981, search Site 397 AND Deep Sea Drilling Project.
CO N265042N265042
W0151048W0151048
BT Leg 47
BT IPOD
BT Deep Sea Drilling Project
SA Atlantic Ocean
SA Canary Islands
SA Cape Bojador
SA Northeast Atlantic

DSDP Site 398 (1989)
Off the coast of N Portugal in NE Atlantic. Before 1981, also search Site 398 AND Deep Sea Drilling Project.

CO N405736N405736
   W0104306W0104306
BT Leg 47
BT IPOD
BT Deep Sea Drilling Project
SA Atlantic Ocean
SA European Atlantic
SA Northeast Atlantic

DSDP Site 400   (1989)
Biscay abyssal plain, NE Atlantic. Before 1981, also search Site 400 AND Deep Sea Drilling Project.
CO N472254N472254
   W0091154W0091154
BT Leg 48
BT IPOD
BT Deep Sea Drilling Project
SA Atlantic Ocean
SA Northeast Atlantic

DSDP Site 401   (1989)
Edge of Meriadzek Terrace, NE Atlantic. Before 1981, also search Site 401 AND Deep Sea Drilling Project.
CO N472538N472540
   W0084837W0084838
BT Leg 48
BT IPOD
BT Deep Sea Drilling Project
SA Atlantic Ocean
SA Northeast Atlantic

DSDP Site 402   (1989)
Continental slope of Bay of Biscay, European Atlantic. Before 1981, also search Site 402 AND Deep Sea Drilling Project.
CO N475228N475229
   W0085026W0085027
BT Leg 48
BT IPOD
BT Deep Sea Drilling Project
SA Atlantic Ocean
SA Bay of Biscay
SA European Atlantic

DSDP Site 403   (1989)
SW margin of Rockall Plateau, NE Atlantic. Before 1981, also search Site 403 AND Deep Sea Drilling Project.
CO N560818N560819
   W0231738W0231739
BT Leg 48
BT IPOD
BT Deep Sea Drilling Project
SA Atlantic Ocean
SA Northeast Atlantic
SA Rockall Plateau

DSDP Site 404   (1989)
SW margin of Rockall Plateau, NE Atlantic. Before 1981, also search Site 404 AND Deep Sea Drilling Project.
CO N560307N560308
   W0231456W0231458
BT Leg 48
BT IPOD
BT Deep Sea Drilling Project
SA Atlantic Ocean
SA Northeast Atlantic
SA Rockall Plateau

DSDP Site 405   (1989)
S Rockall Plateau, NE Atlantic. Before 1981, also search Site 405 AND Deep Sea Drilling Project.
CO N552010N552011
   W0220329W0220330
BT Leg 48
BT IPOD
BT Deep Sea Drilling Project
SA Atlantic Ocean
SA Northeast Atlantic
SA Rockall Plateau

DSDP Site 406   (1989)
S Rockall Plateau, NE Atlantic. Before 1981, also search Site 406 AND Deep Sea Drilling Project.
CO N551529N551531
   W0220524W0220525
BT Leg 48
BT IPOD
BT Deep Sea Drilling Project
SA Atlantic Ocean
SA Northeast Atlantic
SA Rockall Plateau

DSDP Site 407   (1989)
Reykjanes Ridge, N Atlantic. Before 1981, also search Site 407 AND Deep Sea Drilling Project.
CO N635619N635620
   W0303433W0303434
BT Leg 49
BT IPOD
BT Deep Sea Drilling Project
SA Atlantic Ocean
SA Mid-Atlantic Ridge
SA North Atlantic
SA Reykjanes Ridge

DSDP Site 408   (1989)
Reykjanes Ridge, N Atlantic. Before 1981, also search Site 408 AND Deep Sea Drilling Project.
CO N632237N632238
   W0285442W0285443
BT Leg 49
BT IPOD
BT Deep Sea Drilling Project
SA Atlantic Ocean
SA Mid-Atlantic Ridge
SA North Atlantic
SA Reykjanes Ridge

DSDP Site 410   (1989)
Crest of Mid-Atlantic Ridge in N Atlantic. Before 1981, also search Site 410 AND Deep Sea Drilling Project.
CO N453030N453031
   W0292833W0292834
BT Leg 49
BT IPOD
BT Deep Sea Drilling Project
SA Atlantic Ocean
SA Mid-Atlantic Ridge
SA North Atlantic

DSDP Site 416   (1989)
Off coast of Morocco in NE Atlantic Ocean. Before 1981, also search Site 416 AND Deep Sea Drilling Project.
CO N325012N325012
   W0104806W0104806
BT Leg 50
BT IPOD
BT Deep Sea Drilling Project
SA Atlantic Ocean
SA Northeast Atlantic

DSDP Site 417   (1981)
Near S end of Bermuda Rise, NW Atlantic Ocean. Before 1981, search Site 417 AND Deep Sea Drilling Project. Autoposting of legs to this term began in 1993. Site drilled on Leg 51A, Leg 51B, and Leg 52. This term has multiple hierarchies.
CO N250000N250000
   W0680000W0680000
BT1 Leg 51
BT1 IPOD
BT1 Deep Sea Drilling Project
BT2 Leg 52
BT2 IPOD
BT2 Deep Sea Drilling Project
SA Atlantic Ocean
SA Bermuda Rise

DSDP Site 418   (1981)
Near S end of Bermuda Rise, NW Atlantic Ocean. Before 1981, search Site 418 AND Deep Sea Drilling Project. Autoposting of broader terms began in 1993. This term has multiple hierarchies.
CO N250205N250207
   W0680326W0680327
BT1 Leg 52
BT1 IPOD
BT1 Deep Sea Drilling Project
BT2 Leg 53
BT2 IPOD
BT2 Deep Sea Drilling Project
BT3 Leg 102
BT3 Ocean Drilling Program
SA Atlantic Ocean
SA Bermuda Rise

DSDP Site 419   (1993)
East Pacific Rise, North American Pacific. Before 1982, also search Site 419 AND Deep Sea Drilling Project.
CO N035557N085558
   W1054110W1054111
BT Leg 54
BT IPOD
BT Deep Sea Drilling Project
SA East Pacific Rise
SA North American Pacific
SA Pacific Ocean

DSDP Site 420   (1993)
East Pacific Rise, North American Pacific. Before 1982, also search Site 420 AND Deep Sea Drilling Project.
CO N090006N090006
   W1060646W1060647
BT Leg 54
BT IPOD
BT Deep Sea Drilling Project
SA East Pacific Rise
SA North American Pacific
SA Pacific Ocean

DSDP Site 424   (1989)
South American Pacific. Before 1981, also search Site AND Deep Sea Drilling Project.
CO N003537N003538
   W0860749W0860750
BT Leg 54
BT IPOD
BT Deep Sea Drilling Project
SA Pacific Ocean
SA South American Pacific

DSDP Site 430   (1989)
Ojin Seamount, W Pacific. Before 1981, also search Site AND Deep Sea Drilling Project.
CO N375852N375853
   E1703528E1703526
BT Leg 55
BT IPOD
BT Deep Sea Drilling Project
SA Emperor Seamounts
SA Pacific Ocean
SA West Pacific

DSDP Site 431   (1989)
Seamount C, W Pacific. Before 1981, also search Site 431 AND Deep Sea Drilling Project.
CO N422526N422527
   E1703241E1703240
BT Leg 55
BT IPOD
BT Deep Sea Drilling Project
SA Emperor Seamounts
SA Pacific Ocean
SA West Pacific

DSDP Site 432   (1989)
Nintoku Seamount, W Pacific. Before 1981, also search Site 432 AND Deep Sea Drilling Project.
CO N412001N412002
   E1702245E1702244
BT Leg 55
BT IPOD
BT Deep Sea Drilling Project
SA Emperor Seamounts
SA Pacific Ocean
SA West Pacific

DSDP Site 433   (1989)
Suiko Seamount, W Pacific. Before 1981, also search Site 433 AND Deep Sea Drilling Project.
CO N444635N444637
   E1700116E1700115
BT Leg 55
BT IPOD
BT Deep Sea Drilling Project
SA Emperor Seamounts
SA Pacific Ocean
SA West Pacific

DSDP Site 434   (1989)
Japan Trench, NW Pacific. Before 1981, also search Site 434 AND Deep Sea Drilling Project.
CO N394445N394446
   E1440608E1440607
BT Leg 56
BT IPOD
BT Deep Sea Drilling Project
SA Japan Trench
SA Northwest Pacific
SA Pacific Ocean

DSDP Site 435   (1989)
Japan Trench, NW Pacific. Before 1981, also search Site 435 AND Deep Sea Drilling Project.
CO N394405N394406
   E1434732E1434731
BT Leg 56
BT IPOD
BT Deep Sea Drilling Project
SA Japan Trench
SA Northwest Pacific
SA Pacific Ocean

DSDP Site 436   (1989)
NW Pacific. Before 1981, also search Site 436 AND Deep Sea Drilling Project.
CO N395557N395558
   E1453329E1453328
BT Leg 56
BT IPOD
BT Deep Sea Drilling Project
SA Northwest Pacific
SA Pacific Ocean

DSDP Site 438   (1989)
Japan Trench, NW Pacific Ocean.
CO N403748N403748
   E1431448E1431448
BT Leg 57
BT IPOD
BT Deep Sea Drilling Project
SA Japan Trench
SA Pacific Ocean

DSDP Site 439   (1989)
Near Japan Trench in NW Pacific. Before 1981, also search Site 439 AND Deep Sea Drilling Project.
BT Leg 57
BT IPOD
BT Deep Sea Drilling Project
SA Northwest Pacific
SA Pacific Ocean

DSDP Site 440   (1989)
Japan Trench, NW Pacific. Before 1981, also search Site 440 AND Deep Sea Drilling Project.
BT Leg 57

Before then, variants of the term may occur in GeoRef.

BT IPOD
BT Deep Sea Drilling Project
SA Japan Trench
SA Northwest Pacific
SA Pacific Ocean

DSDP Site 442 (1989)
Shikoku Basin, N Philippine Sea. Before 1981, also search Site 442 AND Deep Sea Drilling Project.
CO N285859N285901
   E1360326E1360325
BT Leg 58
BT IPOD
BT Deep Sea Drilling Project
SA Philippine Sea
SA Shikoku Basin

DSDP Site 445 (1993)
Small basin on Daito Ridge, NW Philippine Sea. Before 1982, also search Site 445 AND Deep Sea Drilling Project.
CO N253121N253122
   E1331230E1331229
BT Leg 58
BT IPOD
BT Deep Sea Drilling Project
SA Daito Ridge
SA Pacific Ocean
SA Philippine Sea

DSDP Site 446 (1993)
Daito Basin, Philippine Sea. Before 1982, also search Site 446 AND Deep Sea Drilling Project.
CO N244202N244203
   E1324630E1324629
BT Leg 58
BT IPOD
BT Deep Sea Drilling Project
SA Pacific Ocean
SA Philippine Sea

DSDP Site 450 (1993)
E Parece Vela Basin, Philippine Sea. Before 1982, also search Site 450 AND Deep Sea Drilling Project.
CO N180001N180002
   E1404721E1404720
BT Leg 59
BT IPOD
BT Deep Sea Drilling Project
SA Pacific Ocean
SA Parece Vela Basin
SA Philippine Sea

DSDP Site 453 (1993)
Mariana Trough, Philippine Sea. Before 1981, also search Site 453 AND Deep Sea Drilling Project.
CO N175425N175426
   E1434057E1435057
BT Leg 60
BT IPOD
BT Deep Sea Drilling Project
SA Mariana Trough
SA Pacific Ocean
SA Philippine Sea

DSDP Site 458 (1985)
Autoposting of Leg 60 to this term began in 1989. NW Pacific Ocean.
CO N175151N175151
   E1465603E1465603
BT Leg 60
BT IPOD
BT Deep Sea Drilling Project
SA Northwest Pacific
SA Pacific Ocean

DSDP Site 459 (1985)
Autoposting of Leg 60 to this term began in 1989. NW Pacific Ocean.
CO N175145N175145
   E1471805E1471805

BT Leg 60
BT IPOD
BT Deep Sea Drilling Project
SA Northwest Pacific
SA Pacific Ocean

DSDP Site 462 (1985)
N Nauru Basin, W Equatorial Pacific. Autoposting of Leg 61 to this term began in 1989. Autoposting of Leg 89 began in 1993. Before 1993, search in combination with Deep Sea Drilling Project or Ocean Drilling Program and applicable Leg. This term has multiple hierarchies.
CO N071415N071415
   E1650150E1650150
BT1 Leg 61
BT1 IPOD
BT1 Deep Sea Drilling Project
BT2 Leg 89
BT2 IPOD
BT2 Deep Sea Drilling Project
SA Equatorial Pacific
SA Nauru Basin
SA Pacific Ocean

DSDP Site 463 (1985)
Autoposting of Leg 62 to this term began in 1989. NW Pacific Ocean.
CO N212101N212101
   E1744004E1744004
BT Leg 62
BT IPOD
BT Deep Sea Drilling Project
SA Northwest Pacific
SA Pacific Ocean

DSDP Site 464 (1985)
Autoposting of Leg 62 to this term began in 1989. N Hess Rise, N central Pacific Ocean.
CO N395139N395139
   E1735320E1735320
BT Leg 62
BT IPOD
BT Deep Sea Drilling Project
SA Hess Rise
SA Northwest Pacific
SA Pacific Ocean

DSDP Site 465 (1985)
Autoposting of Leg 62 to this term began in 1989. S Hess Rise, N central Pacific Ocean.
CO N334914N334914
   E1785508E1785508
BT Leg 62
BT IPOD
BT Deep Sea Drilling Project
SA Hess Rise
SA Northwest Pacific
SA Pacific Ocean

DSDP Site 466 (1985)
Autoposting of Leg 62 to this term began in 1989. S Hess Rise, N central Pacific Ocean. Before 1981, also search Site 466 AND Deep Sea Drilling Project.
CO N341128N341128
   E1791521E1791521
BT Leg 62
BT IPOD
BT Deep Sea Drilling Project
SA Hess Rise
SA Northwest Pacific
SA Pacific Ocean

DSDP Site 467 (1985)
Autoposting of Leg 63 to this term began in 1989. North American Pacific, off S California.
CO N335058N335058
   W1204528W1204528
BT Leg 63

BT IPOD
BT Deep Sea Drilling Project
SA North American Pacific
SA Pacific Ocean

DSDP Site 468 (1985)
Autoposting of Leg 63 to this term began in 1989. North American Pacific, off Baja California.
CO N323702N323702
   W1200704W1200704
BT Leg 63
BT IPOD
BT Deep Sea Drilling Project
SA North American Pacific
SA Pacific Ocean

DSDP Site 469 (1985)
Autoposting of Leg 63 to this term began in 1989. North American Pacific, off Baja California.
CO N323700N323700
   W1203254W1203254
BT Leg 63
BT IPOD
BT Deep Sea Drilling Project
SA North American Pacific
SA Pacific Ocean

DSDP Site 470 (1985)
Autoposting of Leg 63 to this term began in 1989. North American Pacific, off Baja California.
CO N285428N285428
   W1173107W1173107
BT Leg 63
BT IPOD
BT Deep Sea Drilling Project
SA North American Pacific
SA Pacific Ocean

DSDP Site 471 (1985)
Autoposting of Leg 63 to this term began in 1989. North American Pacific, off Baja California.
CO N232856N232856
   W1122947W1122947
RT Leg 63
BT IPOD
BT Deep Sea Drilling Project
SA North American Pacific
SA Pacific Ocean

DSDP Site 472 (1985)
Autoposting of Leg 63 to this term began in 1989. North American Pacific, off Baja California.
CO N230021N230021
   W1135943W1135943
BT Leg 63
BT IPOD
BT Deep Sea Drilling Project
SA North American Pacific
SA Pacific Ocean

DSDP Site 473 (1985)
Autoposting of Leg 63 to this term began in 1989. North American Pacific, off Mexico.
CO N205755N205755
   W1070349W1070349
BT Leg 63
BT IPOD
BT Deep Sea Drilling Project
SA North American Pacific
SA Pacific Ocean

DSDP Site 474 (1985)
Autoposting of Leg 64 to this term began in 1989. North American Pacific, off Mexico.
CO N225743N225743
   W1085851W1085851
BT Leg 64
BT IPOD
BT Deep Sea Drilling Project
SA North American Pacific
SA Pacific Ocean

DSDP Site 475 (1985)
Autoposting of Leg 64 to this term began in 1989. North American Pacific, off Mexico.
CO N230302N230302
   W1090312W1090312
BT Leg 64
BT IPOD
BT Deep Sea Drilling Project
SA North American Pacific
SA Pacific Ocean

DSDP Site 476 (1985)
Autoposting of Leg 64 to this term began in 1989. North American Pacific, off Mexico.
CO N230226N230226
   W1091521W1091521
BT Leg 64
BT IPOD
BT Deep Sea Drilling Project
SA North American Pacific
SA Pacific Ocean

DSDP Site 477 (1985)
Autoposting of Leg 64 to this term began in 1989. Gulf of California.
CO N270151N270151
   W1112401W1112401
BT Leg 64
BT IPOD
BT Deep Sea Drilling Project
SA Gulf of California
SA North American Pacific
SA Pacific Ocean

DSDP Site 478 (1985)
Autoposting of Leg 64 to this term began in 1989. Gulf of California.
CO N270548N270548
   W1113027W1113027
BT Leg 64
BT IPOD
BT Deep Sea Drilling Project
SA Gulf of California
SA North American Pacific
SA Pacific Ocean

DSDP Site 479 (1985)
Autoposting of Leg 64 to this term began in 1989. Gulf of California.
CO N275046N275046
   W1113729W1113729
BT Leg 64
BT IPOD
BT Deep Sea Drilling Project
SA Gulf of California
SA North American Pacific
SA Pacific Ocean

DSDP Site 480 (1985)
Autoposting of Leg 64 to this term began in 1989. Gulf of California.
CO N275406N275406
   W1103921W1103921
BT Leg 64
BT IPOD
BT Deep Sea Drilling Project
SA Gulf of California
SA North American Pacific
SA Pacific Ocean

DSDP Site 481 (1985)
Autoposting of Leg 64 to this term began in 1989. Gulf of California.
CO N271511N271511
   W1103028W1103028
BT Leg 64
BT IPOD
BT Deep Sea Drilling Project
SA Gulf of California
SA North American Pacific
SA Pacific Ocean

DSDP Site 482 (1985)
Autoposting of Leg 65 to this term began in 1989. Gulf of California.
CO N224723N224723
   W1075938W1075938

BT Leg 65
BT IPOD
BT Deep Sea Drilling Project
SA Gulf of California
SA North American Pacific
SA Pacific Ocean

DSDP Site 483 (1985)
Autoposting of Leg 65 to this term began in 1989. Gulf of California.
CO N225300N225300
W1084450W1084450
BT Leg 65
BT IPOD
BT Deep Sea Drilling Project
SA Gulf of California
SA North American Pacific
SA Pacific Ocean

DSDP Site 484 (1985)
Autoposting of Leg 65 to this term began in 1989. Gulf of California.
CO N231119N231119
W1082336W1082336
BT Leg 65
BT IPOD
BT Deep Sea Drilling Project
SA Gulf of California
SA North American Pacific
SA Pacific Ocean

DSDP Site 485 (1985)
Autoposting of Leg 65 to this term began in 1989. Gulf of California.
CO N224457N224457
W1075413W1075413
BT Leg 65
BT IPOD
BT Deep Sea Drilling Project
SA Gulf of California
SA North American Pacific
SA Pacific Ocean

DSDP Site 494 (1985)
Autoposting of Leg 67 to this term began in 1989. E Equatorial Pacific, off Central America.
CO N124258N124258
W0905558W0905558
BT Leg 67
BT IPOD
BT Deep Sea Drilling Project
SA Equatorial Pacific
SA Pacific Ocean

DSDP Site 495 (1985)
Autoposting of Leg 67 to this term began in 1989. E Equatorial Pacific, off Central America.
CO N122947N122947
W0910216W0910216
BT Leg 67
BT IPOD
BT Deep Sea Drilling Project
SA Equatorial Pacific
SA Pacific Ocean

DSDP Site 496 (1985)
Autoposting of Leg 67 to this term began in 1989. E Equatorial Pacific, off Central America.
CO N130349N130349
W0904743W0904743
BT Leg 67
BT IPOD
BT Deep Sea Drilling Project
SA Equatorial Pacific
SA Pacific Ocean

DSDP Site 497 (1985)
Autoposting of Leg 67 to this term began in 1989. E Equatorial Pacific, off Central America.
CO N125914N125914
W0904941W0904941
BT Leg 67
BT IPOD
BT Deep Sea Drilling Project

SA Equatorial Pacific
SA Pacific Ocean

DSDP Site 498 (1985)
Autoposting of Leg 67 to this term began in 1989. E Equatorial Pacific, off Central America.
CO N124241N124241
W0905457W0905457
BT Leg 67
BT IPOD
BT Deep Sea Drilling Project
SA Equatorial Pacific
SA Pacific Ocean

DSDP Site 499 (1985)
Autoposting of Leg 67 to this term began in 1989. E Equatorial Pacific, off Central America.
CO N124017N124017
W0905641W0905641
BT Leg 67
BT IPOD
BT Deep Sea Drilling Project
SA Equatorial Pacific
SA Pacific Ocean

DSDP Site 500 (1985)
Autoposting of Leg 67 to this term began in 1989. E Equatorial Pacific, off Central America.
CO N124121N124121
W0905629W0905629
BT Leg 67
BT IPOD
BT Deep Sea Drilling Project
SA Equatorial Pacific
SA Pacific Ocean

DSDP Site 501 (1985)
Autoposting of Leg 68 to this term began in 1989. E Equatorial Pacific, off Colombia.
CO N011338N011338
W0834403W0834403
BT Leg 68
BT IPOD
BT Deep Sea Drilling Project
SA Equatorial Pacific
SA Pacific Ocean

DSDP Site 502 (1985)
Autoposting of Leg 68 to this term began in 1989. W Caribbean Sea, off Central America.
CO N112925N112925
W0792247W0792247
BT Leg 68
BT IPOD
BT Deep Sea Drilling Project
SA Caribbean Sea

DSDP Site 503 (1985)
Autoposting of Leg 68 to this term began in 1989. E Equatorial Pacific, off Central America. Before 1981, also search Site 503 AND Deep Sea Drilling Project.
CO N040303N040303
W0953813W0953813
BT Leg 68
BT IPOD
BT Deep Sea Drilling Project
SA Equatorial Pacific
SA Pacific Ocean

DSDP Site 504 (1985)
Autoposting of Leg 69 to this term began in 1989. E Equatorial Pacific, off Colombia.
CO N011335N011335
W0834356W0834356
BT Leg 69
BT IPOD
BT Deep Sea Drilling Project
SA Equatorial Pacific
SA Pacific Ocean

DSDP Site 504B (1985)

E Equatorial Pacific, off Colombia. Autoposting of broader terms began in 1993. Redrilled on Legs 70, 83, 92, 111, 137, and 140. This term has multiple hierarchies.
CO N011338N011338
W0834349W0834349
BT1 Leg 69
BT1 IPOD
BT1 Deep Sea Drilling Project
BT2 Leg 70
BT2 IPOD
BT2 Deep Sea Drilling Project
BT3 Leg 83
BT3 IPOD
BT3 Deep Sea Drilling Project
BT4 Leg 92
BT4 IPOD
BT4 Deep Sea Drilling Project
BT5 Leg 111
BT5 Ocean Drilling Program
SA Equatorial Pacific
SA Pacific Ocean

DSDP Site 505 (1985)
Autoposting of Leg 69 to this term began in 1989. E Equatorial Pacific, off Colombia.
CO N015458N015458
W0834724W0834724
BT Leg 69
BT IPOD
BT Deep Sea Drilling Project
SA Equatorial Pacific
SA Pacific Ocean

DSDP Site 506 (1985)
Autoposting of Leg 70 to this term began in 1989. E Equatorial Pacific.
CO N003635N003635
W0860559W0860559
BT Leg 70
BT IPOD
BT Deep Sea Drilling Project
SA Equatorial Pacific
SA Pacific Ocean

DSDP Site 507 (1985)
Autoposting of Leg 70 to this term began in 1989. E Equatorial Pacific.
CO N003400N003400
W0860524W0860524
BT Leg 70
BT IPOD
BT Deep Sea Drilling Project
SA Equatorial Pacific
SA Pacific Ocean

DSDP Site 508 (1985)
Autoposting of Leg 70 to this term began in 1989. E Equatorial Pacific.
CO N003000N003000
W0860600W0860600
BT Leg 70
BT IPOD
BT Deep Sea Drilling Project
SA Equatorial Pacific
SA Pacific Ocean

DSDP Site 509 (1985)
Autoposting of Leg 70 to this term began in 1989. E Equatorial Pacific.
CO N003521N003521
W0860753W0860753
BT Leg 70
BT IPOD
BT Deep Sea Drilling Project
SA Equatorial Pacific
SA Pacific Ocean

DSDP Site 510 (1985)
Autoposting of Leg 70 to this term began in 1989. E Equatorial Pacific.

CO N013648N013648
W0862436W0862436
BT Leg 70
BT IPOD
BT Deep Sea Drilling Project
SA Equatorial Pacific
SA Pacific Ocean

DSDP Site 511 (1985)
Autoposting of Leg 71 to this term began in 1989. S Atlantic Ocean.
CO S510017S510017
W0465818W0465818
BT Leg 71
BT IPOD
BT Deep Sea Drilling Project
SA Atlantic Ocean
SA South Atlantic

DSDP Site 512 (1985)
Autoposting of Leg 71 to this term began in 1989. S Atlantic Ocean.
CO S495211S495211
W0405043W0405043
BT Leg 71
BT IPOD
BT Deep Sea Drilling Project
SA Atlantic Ocean
SA South Atlantic

DSDP Site 513 (1985)
Autoposting of Leg 71 to this term began in 1989. S Atlantic Ocean.
CO S473500S473500
W0243824W0243824
BT Leg 71
BT IPOD
BT Deep Sea Drilling Project
SA Atlantic Ocean
SA South Atlantic

DSDP Site 514 (1985)
Autoposting of Leg 71 to this term began in 1989. S Atlantic Ocean.
CO S460246S460246
W0265118W0265118
BT Leg 71
BT IPOD
BT Deep Sea Drilling Project
SA Atlantic Ocean
SA South Atlantic

DSDP Site 515 (1989)
Brazil Basin, S Atlantic.
CO S061418S061418
W0263012W0263012
BT Leg 72
BT IPOD
BT Deep Sea Drilling Project
SA Antarctic bottom water
SA Atlantic Ocean
SA Brazil Basin
SA South American Atlantic

DSDP Site 516 (1989)
Rio Grande Rise, S Atlantic.
CO S301636S301635
W0351706W0351707
BT Leg 72
BT IPOD
BT Deep Sea Drilling Project
SA Atlantic Ocean
SA Rio Grande Rise
SA South Atlantic

DSDP Site 517 (1989)
W flank of Rio Grande Rise, S Atlantic Ocean.
CO S305649S305648
W0380228W0380229
BT Leg 72
BT IPOD
BT Deep Sea Drilling Project
SA Atlantic Ocean
SA Rio Grande Rise
SA South Atlantic

DSDP Site 518 (1989)

Before then, variants of the term may occur in GeoRef.

E flank of Vema Channel in South
Atlantic.
  CO S295826S295825
      W0380807W0380808
  BT Leg 72
  BT IPOD
  BT Deep Sea Drilling Project
  SA Atlantic Ocean
  SA South Atlantic
  SA Vema Channel
DSDP Site 519   (1989)
Mid-Atlantic Ridge in S Atlantic.
  CO S260813S260811
      W0113958W0113959
  BT Leg 73
  BT IPOD
  BT Deep Sea Drilling Project
  SA Atlantic Ocean
  SA Mid-Atlantic Ridge
  SA South Atlantic
DSDP Site 520   (1989)
Mid-Atlantic Ridge in S Atlantic.
  CO S253125S253123
      W0111108W0111109
  BT Leg 73
  BT IPOD
  BT Deep Sea Drilling Project
  SA Atlantic Ocean
  SA Mid-Atlantic Ridge
  SA South Atlantic
DSDP Site 521   (1989)
Mid-Atlantic Ridge in S Atlantic.
  CO S260426S260425
      W0101552W0101553
  BT Leg 73
  BT IPOD
  BT Deep Sea Drilling Project
  SA Atlantic Ocean
  SA Mid-Atlantic Ridge
  SA South Atlantic
DSDP Site 522   (1989)
Angola Basin, S Atlantic.
  CO S260651S260650
      W0050646W0050647
  BT Leg 73
  BT IPOD
  BT Deep Sea Drilling Project
  SA Angola Basin
  SA Atlantic Ocean
  SA South Atlantic
DSDP Site 523   (1989)
Angola Basin, just W of Walvis
Ridge in S Atlantic.
  CO S283308S283307
      W0021504W0021505
  BT Leg 73
  BT IPOD
  BT Deep Sea Drilling Project
  SA Angola Basin
  SA Atlantic Ocean
  SA South Atlantic
  SA Walvis Ridge
DSDP Site 524   (1989)
Cape Basin, E of Walvis Ridge in
S Atlantic.
  CO S292904S292902
      W0023044W0023045
  BT Leg 73
  BT IPOD
  BT Deep Sea Drilling Project
  SA Atlantic Ocean
  SA Cape Basin
  SA South Atlantic
  SA Walvis Ridge
DSDP Site 525   (1989)
Walvis Ridge, S Atlantic.
  CO S290415S290414
      E0025908E0025907
  BT Leg 74
  BT IPOD
  BT Deep Sea Drilling Project
  SA Atlantic Ocean

  SA South Atlantic
  SA Walvis Ridge
DSDP Site 526   (1989)
Walvis Ridge, S Atlantic.
  CO S300724S300724
      E0030818E0030818
  BT Leg 74
  BT IPOD
  BT Deep Sea Drilling Project
  SA Atlantic Ocean
  SA South Atlantic
  SA Walvis Ridge
DSDP Site 527   (1989)
Walvis Ridge, S Atlantic.
  CO S280230S280229
      E0014549E0014547
  BT Leg 74
  BT IPOD
  BT Deep Sea Drilling Project
  SA Atlantic Ocean
  SA South Atlantic
  SA Walvis Ridge
DSDP Site 528   (1989)
W Walvis Ridge, S Atlantic.
  CO S283130S283129
      E0021927E0021926
  BT Leg 74
  BT IPOD
  BT Deep Sea Drilling Project
  SA Atlantic Ocean
  SA South Atlantic
  SA Walvis Ridge
DSDP Site 529   (1989)
Walvis Ridge, S Atlantic.
  CO S285548S285548
      E0024606E0024606
  BT Leg 74
  BT IPOD
  BT Deep Sea Drilling Project
  SA Atlantic Ocean
  SA South Atlantic
  SA Walvis Ridge
DSDP Site 530   (1989)
N Walvis Ridge, S Atlantic.
  CO S191116S191115
      E0092310E0092308
  BT Leg 75
  BT IPOD
  BT Deep Sea Drilling Project
  SA Atlantic Ocean
  SA South Atlantic
  SA Walvis Ridge
DSDP Site 531   (1989)
Crest of Walvis Ridge, near coast
of Africa.
  CO S193827S193826
      E0093519E0093518
  BT Leg 75
  BT IPOD
  BT Deep Sea Drilling Project
  SA Atlantic Ocean
  SA South Atlantic
  SA Walvis Ridge
DSDP Site 532   (1989)
Walvis Ridge, S Atlantic.
  CO S194437S194436
      E0103108E0103107
  BT Leg 75
  BT IPOD
  BT Deep Sea Drilling Project
  SA Atlantic Ocean
  SA South Atlantic
  SA Walvis Ridge
DSDP Site 533   (1985)
Autoposting of Leg 76 to this term
began in 1989. North American
Atlantic, off Florida.
  CO N311536N311536
      W0745211W0745211
  BT Leg 76
  BT IPOD

  BT Deep Sea Drilling Project
  SA Atlantic Ocean
  SA North American Atlantic
DSDP Site 534   (1985)
Autoposting of Leg 76 to this term
began in 1989. North American
Atlantic, off Florida.
  CO N282036N282036
      W0752254W0752254
  BT Leg 76
  BT IPOD
  BT Deep Sea Drilling Project
  SA Atlantic Ocean
  SA North American Atlantic
DSDP Site 535   (1989)
Gulf of Mexico.
  CO N234225N234226
      W0843058W0843059
  BT Leg 77
  BT IPOD
  BT Deep Sea Drilling Project
  SA Gulf of Mexico
DSDP Site 536   (1989)
Gulf of Mexico.
  CO N232917N232918
      W0851234W0851235
  BT Leg 77
  BT IPOD
  BT Deep Sea Drilling Project
  SA Gulf of Mexico
DSDP Site 537   (1989)
Gulf of Mexico.
  CO N235600N235601
      W0852737W0852738
  BT Leg 77
  BT IPOD
  BT Deep Sea Drilling Project
  SA Gulf of Mexico
DSDP Site 538   (1989)
N flank of Catoche Knoll, Gulf of
Mexico.
  CO N235058N235059
      W0852737W0852738
  BT Leg 77
  BT IPOD
  BT Deep Sea Drilling Project
  SA Gulf of Mexico
DSDP Site 540   (1989)
Gulf of Mexico.
  CO N234943N234944
      W0842214W0842216
  BT Leg 77
  BT IPOD
  BT Deep Sea Drilling Project
  SA Gulf of Mexico
DSDP Site 541   (1989)
Barbados Ridge, near border of E
Caribbean Sea.
  CO N153111N153113
      W0584339W0584340
  BT Leg 78A
  BT IPOD
  BT Deep Sea Drilling Project
  SA Atlantic Ocean
  SA Barbados Ridge
  SA Caribbean Sea
DSDP Site 542   (1989)
Barbados Ridge, near border of E
Caribbean Sea.
  CO N153111N153112
      W0584247W0584248
  BT Leg 78A
  BT IPOD
  BT Deep Sea Drilling Project
  SA Atlantic Ocean
  SA Barbados Ridge
  SA Caribbean Sea
DSDP Site 543   (1989)
Located on Tiburon Rise, near
Barbados Ridge.
  CO N154243N154244

      W0583914W0583915
  BT Leg 78A
  BT IPOD
  BT Deep Sea Drilling Project
  SA Atlantic Ocean
  SA Barbados Ridge
DSDP Site 544   (1989)
Mazagan Plateau, off NW coast of
Morocco.
  CO N334607N334608
      W0092417W0092418
  BT Leg 79
  BT IPOD
  BT Deep Sea Drilling Project
  SA Atlantic Ocean
  SA Mazagan Plateau
DSDP Site 545   (1989)
Mazagan Plateau, off NW coast of
Morocco.
  CO N333951N333952
      W0092152W0092153
  BT Leg 79
  BT IPOD
  BT Deep Sea Drilling Project
  SA Atlantic Ocean
  SA Mazagan Plateau
DSDP Site 546   (1989)
Mazagan Plateau, off NW coast of
Morocco.
  CO N334642N334643
      W0093351W0093352
  BT Leg 79
  BT IPOD
  BT Deep Sea Drilling Project
  SA Atlantic Ocean
  SA Mazagan Plateau
DSDP Site 547   (1989)
Located to the west of the
Mazagan Plateau, off the NW
coast of Morocco.
  CO N334650N334651
      W0092058W0092059
  BT Leg 79
  BT IPOD
  BT Deep Sea Drilling Project
  SA Atlantic Ocean
  SA Mazagan Plateau
DSDP Site 548   (1989)
Goban Spur, near Pendragon Es-
carpment in NE Atlantic.
  CO N485456N485459
      W0120950W0120951
  BT Leg 80
  BT IPOD
  BT Deep Sea Drilling Project
  SA Atlantic Ocean
  SA European Atlantic
  SA Goban Spur
DSDP Site 549   (1993)
Off Goban Spur, E European At-
lantic.
  CO N490516N490518
      W0130552W0130554
  BT Leg 80
  BT IPOD
  BT Deep Sea Drilling Project
  SA Atlantic Ocean
  SA European Atlantic
DSDP Site 550   (1989)
Located 10 km SW of Goban
Spur in NE Atlantic.
  CO N483054N483055
      W0132622W0132623
  BT Leg 80
  BT IPOD
  BT Deep Sea Drilling Project
  SA Atlantic Ocean
  SA European Atlantic
  SA Goban Spur
DSDP Site 551   (1989)

Located about 5 km SW of Goban Spur in NE Atlantic.
CO N485438N485439
W0133005W0133006
BT Leg 80
BT IPOD
BT Deep Sea Drilling Project
SA Atlantic Ocean
SA European Atlantic
SA Goban Spur

DSDP Site 552 (1989)
W Rockall Plateau, NE Atlantic.
CO N560233N560234
W0231323W0231324
BT Leg 81
BT IPOD
BT Deep Sea Drilling Project
SA Atlantic Ocean
SA Northeast Atlantic
SA Rockall Plateau

DSDP Site 553 (1989)
W Rockall Plateau, NE Atlantic.
CO N560519N560520
W0232036W0232037
BT Leg 81
BT IPOD
BT Deep Sea Drilling Project
SA Atlantic Ocean
SA Northeast Atlantic
SA Rockall Plateau

DSDP Site 554 (1989)
W Rockall Plateau, NE Atlantic.
CO N561724N561725
W0233141W0233142
BT Leg 81
BT IPOD
BT Deep Sea Drilling Project
SA Atlantic Ocean
SA Northeast Atlantic
SA Rockall Plateau

DSDP Site 555 (1989)
W Rockall Plateau, NE Atlantic.
CO N563341N563343
W0204655W0204656
BT Leg 81
BT IPOD
BT Deep Sea Drilling Project
SA Atlantic Ocean
SA Northeast Atlantic
SA Rockall Plateau

DSDP Site 556 (1989)
W flank of Mid-Atlantic Ridge near the Azores, N Atlantic.
CO N385622N385623
W0344107W0344108
BT Leg 82
BT IPOD
BT Deep Sea Drilling Project
SA Atlantic Ocean
SA Azores
SA Mid-Atlantic Ridge
SA North Atlantic

DSDP Site 558 (1989)
W flank of Mid-Atlantic Ridge about 30 miles S of Pico fracture zone, near the Azores in N Atlantic.
CO N374614N374615
W0372036W0372037
BT Leg 82
BT IPOD
BT Deep Sea Drilling Project
SA Atlantic Ocean
SA Azores
SA Mid-Atlantic Ridge
SA North Atlantic

DSDP Site 561 (1989)
W flank of Mid-Atlantic Ridge, between Hayes and Oceanographer fracture zones. Near the Azores in N Atlantic.
CO N344706N344707
W0390141W0390143
BT Leg 82
BT IPOD
BT Deep Sea Drilling Project
SA Atlantic Ocean
SA Azores
SA Hayes fracture zone
SA Mid-Atlantic Ridge
SA North Atlantic
SA Oceanographer fracture zone

DSDP Site 562 (1989)
W flank of Mid-Atlantic Ridge, about 60 miles S of the Hayes fracture zone. Near the Azores in N Atlantic.
CO N330829N330830
W0414045W0414046
BT Leg 82
BT IPOD
BT Deep Sea Drilling Project
SA Atlantic Ocean
SA Azores
SA Hayes fracture zone
SA Mid-Atlantic Ridge
SA North Atlantic

DSDP Site 563 (1989)
W flank of Mid-Atlantic Ridge, S of Hayes fracture zone, near the Azores in N Atlantic.
CO N333831N333832
W0434602W0434603
BT Leg 82
BT IPOD
BT Deep Sea Drilling Project
SA Atlantic Ocean
SA Azores
SA Hayes fracture zone
SA Mid-Atlantic Ridge
SA North Atlantic

DSDP Site 564 (1989)
W flank of Mid-Atlantic Ridge, S of Hayes fracture zone. Near Azores in N Atlantic.
CO N334421N334422
W0434601W0434602
BT Leg 82
BT IPOD
BT Deep Sea Drilling Project
SA Atlantic Ocean
SA Azores
SA Hayes fracture zone
SA Mid-Atlantic Ridge
SA North Atlantic

DSDP Site 565 (1993)
Middle America Trench, off Nicoya Peninsula, Costa Rica, NE Pacific.
CO N094341N094342
W0860526W0860527
BT Leg 84
BT IPOD
BT Deep Sea Drilling Project
SA Middle America Trench
SA Pacific Ocean

DSDP Site 568 (1993)
Middle America Trench upper slope, off Guatemala, NE Pacific.
CO N130419N130420
W0904800W0904800
BT Leg 84
BT IPOD
BT Deep Sea Drilling Project
SA Middle America Trench
SA Pacific Ocean

DSDP Site 570 (1993)
Middle America Trench upper slope, off Guatemala, NE Pacific.
CO N131707N131708
W0912334W0912335
BT Leg 84
BT IPOD
BT Deep Sea Drilling Project
SA Middle America Trench
SA Pacific Ocean

DSDP Site 572 (1989)
E Equatorial Pacific.
CO N012606N012606
W1135030W1135030
BT Leg 85
BT IPOD
BT Deep Sea Drilling Project
SA Equatorial Pacific
SA Pacific Ocean

DSDP Site 573 (1989)
E Equatorial Pacific.
CO N002954N002955
W1331834W1331835
BT Leg 85
BT IPOD
BT Deep Sea Drilling Project
SA Equatorial Pacific
SA Pacific Ocean

DSDP Site 574 (1989)
E Equatorial Pacific.
CO N041231N041232
W1331948W1331949
BT Leg 85
BT IPOD
BT Deep Sea Drilling Project
SA Equatorial Pacific
SA Pacific Ocean

DSDP Site 575 (1993)
S of Clipperton fracture zone, S North American Pacific.
CO N055100N055100
W1350209W1350210
BT Leg 85
BT IPOD
BT Deep Sea Drilling Project
SA North American Pacific
SA Pacific Ocean

DSDP Site 576 (1989)
Located approximately 300 km E of the Shatsky Rise, NW Pacific.
CO N322121N322122
E1641633E1641632
BT Leg 86
BT IPOD
BT Deep Sea Drilling Project
SA Northwest Pacific
SA Pacific Ocean

DSDP Site 577 (1993)
W flank of Shatsky Rise, central West Pacific. Before 1984, also search Site 577 AND Deep Sea Drilling Project.
CO N322628N322632
E1574324E1574323
BT Leg 86
BT IPOD
BT Deep Sea Drilling Project
SA Pacific Ocean
SA Shatsky Rise
SA West Pacific

DSDP Site 578 (1993)
S Northwest Pacific Basin, West Pacific. Before 1984, also search Site 578 AND Deep Sea Drilling Project.
CO N335533N335534
E1513745E1513744
BT Leg 86
BT IPOD
BT Deep Sea Drilling Project
SA Pacific Ocean
SA West Pacific

DSDP Site 579 (1993)
Central Northwest Pacific Basin. Before 1984, also search Site 579 AND Deep Sea Drilling Project.
CO N383736N383741
E1535017E1535010
BT Leg 86
BT IPOD
BT Deep Sea Drilling Project
SA Pacific Ocean
SA West Pacific

DSDP Site 580 (1993)
N Northwest Pacific Basin. Before 1984, also search Site 580 AND Deep Sea Drilling Project.
CO N413728N413729
E1535835E1535834
BT Leg 86
BT IPOD
BT Deep Sea Drilling Project
SA Pacific Ocean
SA West Pacific

DSDP Site 581 (1989)
Located in Hokkaido fracture zone, NW Pacific. Autoposting of Legs began in 1993. This term has multiple hierarchies.
CO N435537N435538
E1594746E1594745
BT1 Leg 86
BT1 IPOD
BT1 Deep Sea Drilling Project
BT2 Leg 88
BT2 IPOD
BT2 Deep Sea Drilling Project
SA Northwest Pacific
SA Pacific Ocean

DSDP Site 582 (1989)
Off SW coast of Japan in Nankai Trough, NW Pacific Ocean.
CO N314730N314730
E1335448E1335448
BT Leg 87
BT IPOD
BT Deep Sea Drilling Project
SA Nankai Trough
SA Northwest Pacific
SA Pacific Ocean

DSDP Site 583 (1989)
Off SW coast of Japan in Nankai Trough, NW Pacific.
CO N314945N315008
E1335133E1335115
BT Leg 87
BT IPOD
BT Deep Sea Drilling Project
SA Nankai Trough
SA Northwest Pacific
SA Pacific Ocean

DSDP Site 584 (1989)
Off N Honshu in Japan Trench, NW Pacific.
CO N402800N402801
E1435736E1435606
BT Leg 87
BT IPOD
BT Deep Sea Drilling Project
SA Honshu
SA Japan Trench
SA Northwest Pacific
SA Pacific Ocean

DSDP Site 585 (1989)
E Mariana Trough, Pacific Ocean.
CO N132859N132901
E1564855E1564854
BT Leg 89
BT IPOD
BT Deep Sea Drilling Project
SA Mariana Trough
SA Pacific Ocean

DSDP Site 586 (1989)
NE slope of Ontong Java Plateau, Equatorial Pacific.
CO S002951S002950
E1582954E1582953
BT Leg 89
BT IPOD

Before then, variants of the term may occur in GeoRef.

BT Deep Sea Drilling Project
SA Equatorial Pacific
SA Ontong Java Plateau
SA Pacific Ocean

DSDP Site 587   (1989)
S Landsdowne Bank, SW Pacific.
CO S211153S211152
   E1612000E1611959
BT Leg 90
BT IPOD
BT Deep Sea Drilling Project
SA Pacific Ocean
SA Southwest Pacific

DSDP Site 588   (1989)
N Lord Howe Rise, SW Pacific.
CO S260642S260642
   E1611336E1611336
BT Leg 90
BT IPOD
BT Deep Sea Drilling Project
SA Lord Howe Rise
SA Pacific Ocean
SA Southwest Pacific

DSDP Site 589   (1989)
Central Lord Howe Rise, SW Pacific.
CO S304244S304243
   E1633837E1633835
BT Leg 90
BT IPOD
BT Deep Sea Drilling Project
SA Lord Howe Rise
SA Pacific Ocean
SA Southwest Pacific

DSDP Site 590   (1989)
Central Lord Howe Rise, SW Pacific.
CO S311002S311001
   E1632131E1632130
BT Leg 90
BT IPOD
BT Deep Sea Drilling Project
SA Lord Howe Rise
SA Pacific Ocean
SA Southwest Pacific

DSDP Site 591   (1989)
E Lord Howe Rise, SW Pacific.
CO S313504S313503
   E1642656E1642655
BT Leg 90
BT IPOD
BT Deep Sea Drilling Project
SA Lord Howe Rise
SA Pacific Ocean
SA Southwest Pacific

DSDP Site 592   (1989)
S Lord Howe Rise, SW Pacific.
CO S362825S362823
   E1652632E1652631
BT Leg 90
BT IPOD
BT Deep Sea Drilling Project
SA Lord Howe Rise
SA Pacific Ocean
SA Southwest Pacific

DSDP Site 593   (1989)
Located on Challenger Plateau, NW of New Zealand in SW Pacific. Reoccupation of DSDP Site 284 in Leg 29.
CO S403029S403028
   E1674029E1674028
BT Leg 90
BT IPOD
BT Deep Sea Drilling Project
SA Challenger Plateau
SA Pacific Ocean
SA Southwest Pacific

DSDP Site 594   (1989)
S Chatham Rise, E of New Zealand's South Island in SW Pacific.
CO S453129S463128
   E1745653E1745652
BT Leg 90
BT IPOD
BT Deep Sea Drilling Project
SA Chatham Rise
SA Pacific Ocean
SA Southwest Pacific

DSDP Site 595   (1989)
E of Tonga Trench in SW Pacific.
CO S234921S234921
   W1653151W1653151
BT Leg 91
BT IPOD
BT Deep Sea Drilling Project
SA Pacific Ocean
SA Southwest Pacific
SA Tonga Trench

DSDP Site 596   (1989)
E of Tonga Trench in SW Pacific.
CO S235113S235111
   W1653916W1653917
BT Leg 91
BT IPOD
BT Deep Sea Drilling Project
SA Pacific Ocean
SA Southwest Pacific
SA Tonga Trench

DSDP Site 597   (1989)
W flank of East Pacific Rise near Tahiti. S Pacific.
CO S184823S184822
   W1294613W1294614
BT Leg 92
BT IPOD
BT Deep Sea Drilling Project
SA East Pacific Rise
SA Pacific Ocean
SA South Pacific
SA Tahiti

DSDP Site 598   (1989)
East Pacific Rise, E of Tahiti in S Pacific.
CO S190017S190016
   W1244036W1244037
BT Leg 92
BT IPOD
BT Deep Sea Drilling Project
SA East Pacific Rise
SA Pacific Ocean
SA South Pacific
SA Tahiti

DSDP Site 599   (1989)
East Pacific Rise, E of Tahiti in S Pacific.
CO S192706S192705
   W1195252W1195253
BT Leg 92
BT IPOD
BT Deep Sea Drilling Project
SA East Pacific Rise
SA Pacific Ocean
SA South Pacific
SA Tahiti

DSDP Site 600   (1989)
East Pacific Rise, E of Tahiti in S Pacific.
CO S185545S185544
   W1165022W1165023
BT Leg 92
BT IPOD
BT Deep Sea Drilling Project
SA East Pacific Rise
SA Pacific Ocean
SA South Pacific
SA Tahiti

DSDP Site 601   (1989)
East Pacific Rise, E of Tahiti in S Pacific.
CO S185514S185513
   W1165206W1165207
BT Leg 92
BT IPOD
BT Deep Sea Drilling Project
SA East Pacific Rise
SA Pacific Ocean
SA South Pacific
SA Tahiti

DSDP Site 602   (1989)
East Pacific Rise, E of Tahiti in S Pacific.
CO S185425S185424
   W1165440W1165441
BT Leg 92
BT IPOD
BT Deep Sea Drilling Project
SA East Pacific Rise
SA Pacific Ocean
SA South Pacific
SA Tahiti

DSDP Site 603   (1989)
About 270 miles E of Cape Hatteras, North Carolina, on lower continental rise in North American Atlantic. Autoposting of Leg 93 began in 1989. This term has multiple hierarchies.
CO N352939N352940
   W0700142W0700143
BT1 Leg 93
BT1 IPOD
BT1 Deep Sea Drilling Project
BT2 Leg 95
BT2 IPOD
BT2 Deep Sea Drilling Project
SA Atlantic Ocean
SA Blake-Bahama Formation
SA Dry Branch Formation
SA Hatteras Formation
SA North American Atlantic

DSDP Site 604   (1989)
Upper continental rise in Hudson Canyon, North American Atlantic. About 100 miles SE of Atlantic City, New Jersey.
CO N384247N384248
   W0723256W0723258
BT Leg 93
BT IPOD
BT Deep Sea Drilling Project
SA Atlantic Ocean
SA Hatteras Formation
SA Hudson Canyon
SA North American Atlantic

DSDP Site 605   (1989)
Upper continental rise in Hudson Canyon, North American Atlantic. About 100 miles SE of Atlantic City, New Jersey.
CO N384431N384432
   W0723632W0723634
BT Leg 93
BT IPOD
BT Deep Sea Drilling Project
SA Atlantic Ocean
SA Hatteras Formation
SA Hudson Canyon
SA North American Atlantic

DSDP Site 606   (1993)
North Atlantic Ridge, WSW of the Azores, North Atlantic.
CO N372017N372020
   W0352959W0353002
BT Leg 94
BT IPOD
BT Deep Sea Drilling Project
SA Atlantic Ocean
SA Mid-Atlantic Ridge
SA North Atlantic Ridge

DSDP Site 607   (1993)
North Atlantic Ridge area, NW of the Azores.
CO N410004N410005
   W0325726W0325727
BT Leg 94
BT IPOD
BT Deep Sea Drilling Project
SA Atlantic Ocean
SA Mid-Atlantic Ridge
SA North Atlantic Ridge

DSDP Site 608   (1993)
King's Trough complex area, S European Atlantic.
CO N425012N435013
   W0230515W0230515
BT Leg 94
BT IPOD
BT Deep Sea Drilling Project
SA Atlantic Ocean
SA European Atlantic

DSDP Site 609   (1993)
E flank of Mid-Atlantic Ridge, near Gibbs fracture zone, central European Atlantic.
CO N495240N495241
   W0241417W0241418
BT Leg 94
BT IPOD
BT Deep Sea Drilling Project
SA Atlantic Ocean
SA European Atlantic
SA Mid-Atlantic Ridge

DSDP Site 610   (1993)
Feni Ridge, Rockall Trough, central European Atlantic.
CO N531318N531329
   W0185312W0185342
BT Leg 94
BT IPOD
BT Deep Sea Drilling Project
SA Atlantic Ocean
SA European Atlantic
SA Rockall Trough

DSDP Site 611   (1993)
S tip of Gardar Ridge, E of Maury Channel, NW European Atlantic.
CO N525009N525029
   W0301834W0301906
BT Leg 94
BT IPOD
BT Deep Sea Drilling Project
SA Atlantic Ocean
SA European Atlantic

DSDP Site 612   (1993)
Submarine channel, New Jersey mid-continental slope, North American Atlantic.
CO N384912N384913
   W0724625W0724626
BT Leg 95
BT IPOD
BT Deep Sea Drilling Project
SA Atlantic Ocean
SA North American Atlantic

DSDP Site 613   (1993)
Near toe of upper continental rise wedge, New Jersey coast, North American Atlantic. Before 1985, also search Site 613 AND Deep Sea Drilling Project.
CO N384615N384616
   W0723025W0723026
BT Leg 95
BT IPOD
BT Deep Sea Drilling Project
SA Atlantic Ocean
SA North American Atlantic

DSDP Site 614   (1989)
Mississippi Fan, Gulf of Mexico.
CO N250404N250405
   W0860812W0860813
BT Leg 96

BT IPOD
BT Deep Sea Drilling Project
SA Gulf of Mexico
SA Mississippi Fan

DSDP Site 615 (1989)
<u>Mississippi Fan</u>, <u>Gulf of Mexico</u>.
CO N251320N251321
W0855931W0855932
BT Leg 96
BT IPOD
BT Deep Sea Drilling Project
SA Gulf of Mexico
SA Mississippi Fan

DSDP Site 616 (1989)
<u>Mississippi Fan</u>, <u>Gulf of Mexico</u>.
CO N264840N264841
W0865249W0865250
BT Leg 96
BT IPOD
BT Deep Sea Drilling Project
SA Gulf of Mexico
SA Mississippi Fan

DSDP Site 617 (1989)
<u>Mississippi Fan</u>, <u>Gulf of Mexico</u>.
CO N264155N264156
W0883140W0883141
BT Leg 96
BT IPOD
BT Deep Sea Drilling Project
SA Gulf of Mexico
SA Mississippi Fan

DSDP Site 618 (1989)
<u>Orca Basin</u>, <u>Gulf of Mexico</u>.
CO N270040N270041
W0911543W0911544
BT Leg 96
BT IPOD
BT Deep Sea Drilling Project
SA Gulf of Mexico
SA Orca Basin

DSDP Site 619 (1989)
<u>Pigmy Basin</u>, <u>Gulf of Mexico</u>.
CO N271136N271137
W0912432W0912433
BT Leg 96
BT IPOD
BT Deep Sea Drilling Project
SA Gulf of Mexico
SA Pigmy Basin

DSDP Site 620 (1989)
<u>Mississippi Fan</u>, <u>Gulf of Mexico</u>.
CO N265007N265008
W0882214W0882216
BT Leg 96
BT IPOD
BT Deep Sea Drilling Project
SA Gulf of Mexico
SA Mississippi Fan

DSDP Site 621 (1989)
<u>Mississippi Fan</u>, <u>Gulf of Mexico</u>.
CO N264351N264352
W0882945W0882946
BT Leg 96
BT IPOD
BT Deep Sea Drilling Project
SA Gulf of Mexico
SA Mississippi Fan

DSDP Site 622 (1989)
<u>Mississippi Fan</u>, <u>Gulf of Mexico</u>.
CO N264124N264125
W0882849W0882850
BT Leg 96
BT IPOD
BT Deep Sea Drilling Project
SA Gulf of Mexico
SA Mississippi Fan

DSDP Site 623 (1989)
<u>Mississippi Fan</u>, <u>Gulf of Mexico</u>.
CO N254605N254606
W0861350W0861351
BT Leg 96
BT IPOD
BT Deep Sea Drilling Project
SA Gulf of Mexico
SA Mississippi Fan

DSDP Site 624 (1989)
<u>Mississippi Fan</u>, <u>Gulf of Mexico</u>.
CO N254514N254515
W0861637W0861638
BT Leg 96
BT IPOD
BT Deep Sea Drilling Project
SA Gulf of Mexico
SA Mississippi Fan

DSDP Site 625
For sites drilled after <u>DSDP Site 624</u>, see ODP Site.

DSS
use deep seismic sounding

DTA data (1978)
UF differential thermal analysis data
SA data
SA differential thermal analysis
SA endothermic reactions
SA thermal analysis

Dubai (1978)
Emirate. One of federation of 7 states at S end of <u>Persian Gulf</u>.
BT United Arab Emirates
BT Arabian Peninsula
BT Asia

Dublin
No longer a valid term for GeoRef. See <u>Dublin Ireland</u>.

Dublin Ireland (1993)
County and city on <u>Irish Sea</u> in E <u>Ireland</u>.
BT Ireland
BT Western Europe
BT Europe

Dubuque Formation (1989)
NW <u>Illinois</u>, E <u>Iowa</u>, SE <u>Minnesota</u> and SW <u>Wisconsin</u>.
BT Middle Ordovician
BT Ordovician
BT Paleozoic
SA Illinois
SA Iowa
SA Minnesota
SA Wisconsin

Duchesne County
Valid through 1988. Search in combination with state term. After 1988, use specific county-state term.

Duchesne County Utah (1989)
NE <u>Utah</u>. Before 1989, also search Duchesne County AND <u>Utah</u>.
CO N395000N405000
W1095800W1105500
BT Utah
BT United States

Duchesne River Formation (1978)
<u>Eocene</u> or <u>Oligocene</u>. Unconformably underlies Bishop Conglomerate; unconformably overlies Uinta and Uinta (?) Formation. NW <u>Colorado</u> and NE <u>Utah</u>.
IN Index ages as applicable.
BT Paleogene
BT Tertiary
BT Cenozoic
SA Colorado
SA Eocene
SA Oligocene
SA Uinta Formation

SA Utah

Ducktown
Valid through 1988. Search in combination with state term. After 1988, use specific city-state term.

Ducktown Tennessee (1989)
Town in extreme SE <u>Tennessee</u>. Before 1989, search Ducktown AND <u>Tennessee</u>.
CO N350200N350200
W0842200W0842200
BT Polk County Tennessee
BT Tennessee
BT United States

ductile deformation (1989)
BT deformation
SA ductility
SA engineering geology
SA fracture strength
SA mechanical properties
SA rock mechanics

ductility (1978)
SA deformation
SA ductile deformation
SA rock mechanics

Dudley
No longer a valid term for GeoRef as of 1993. See <u>Dudley England</u>.

Dudley England (1993)
Town SW of Birmingham in West <u>Midlands</u>, W central <u>England</u>. Also occurs in other locations.
IN Index county as applicable.
BT England
BT Great Britain
BT United Kingdom
BT Western Europe
BT Europe
SA Midlands
SA Northumberland England
SA Staffordshire England

Duero Basin (1981)
BT Iberian Peninsula
BT Southern Europe
BT Europe
SA Duero River
SA Portugal
SA Spain

Duero River (1985)
River in <u>Spain</u> and <u>Portugal</u>. Rises in central <u>Spain</u>, flows W to NE <u>Portugal</u>, then S along border, and W to <u>Atlantic Ocean</u>. Before 1985, also search Douro River.
UF Douro River
BT Iberian Peninsula
BT Southern Europe
BT Europe
SA Duero Basin
SA Portugal
SA Spain

Dufek Intrusion (1978)
In the Dufek Massif area of Forrestal Range in the <u>Pensacola Mountains</u> of Edith Ronne Land S of the <u>Weddell Sea</u>.
BT Antarctica
SA Pensacola Mountains

dufrenoysite (1978)
BT sulfarsenites
BT sulfosalts

Dugger Formation (1989)
SE <u>Indiana</u>.
BT Middle Pennsylvanian
BT Pennsylvanian
BT Carboniferous
BT Paleozoic
SA Indiana

Dukla
No longer a valid term for GeoRef. As of 1993 see <u>Dukla Poland</u>.

Dukla Poland (1993)
Town in Krosno Province, SE <u>Poland</u>. Before 1993 also search Dukla AND <u>Poland</u>.
BT Poland
BT Central Europe
BT Europe

Duluth
Valid through 1988. Search in combination with state term. After 1988, use specific city-state term.

Duluth Complex (1978)
In NE <u>Minnesota</u>.
IN Index counties as applicable.
BT Minnesota
BT United States
SA Precambrian

Duluth Minnesota (1989)
City at extreme W end of <u>Lake Superior</u> in NE <u>Minnesota</u>. Before 1989, search Duluth AND <u>Minnesota</u>.
CO N464500N464500
W0921000W0921000
BT Saint Louis County Minnesota
BT Minnesota
BT United States

dumortierite (1978)
Autoposting of <u>nesosilicates</u> began in 1985.
BT nesosilicates
BT orthosilicates
BT silicates

Duna River
use Danube River

Dunai River
use Danube River

Dunantul
use Transdanubia

Dunarea River
use Danube River

Dunau River
use Danube River

Dundee
No longer a valid term for GeoRef as of 1993. See <u>Dundee Scotland</u>.

Dundee Limestone (1989)
S <u>Michigan</u>, N <u>Ohio</u> and S <u>Ontario</u>.
BT Middle Devonian
BT Devonian
BT Paleozoic
SA Michigan
SA Ohio
SA Ontario

Dundee Scotland (1993)
City in Angus County, Tayside region, on Firth of Tay on <u>North Sea</u>. Before 1993, also search Dundee AND <u>Scotland</u>.
BT Scotland
BT Great Britain
BT United Kingdom
BT Western Europe
BT Europe

Dunderberg Shale (1989)
In <u>Nopah Formation</u>.
BT Upper Cambrian
BT Cambrian
BT Paleozoic
SA California
SA Nevada
SA Nopah Formation
SA Utah

dune rock
  use eolianite

dune structures (1984)
  Before 1984, search dunes AND sedimentary structures.
  BT bedding plane irregularities
  BT sedimentary structures
  SA dunes

Dunedin
  No longer a valid term for GeoRef. See Dunedin New Zealand.

Dunedin New Zealand (1993)
  City on Pacific Ocean in SE South Island.
  IN Also index South Island.
  BT Otago New Zealand
  BT New Zealand
  BT Australasia
  SA South Island

dunes (1978)
  As of 1984, use dune structures for the sedimentary structure. Autoposting of this term began in 1978.
  UF sand dunes
  NT barchans
  NT coastal dunes
  NT continental dunes
  NT foredunes
  NT submarine dunes
  SA antidunes
  SA bedding plane irregularities
  SA blowouts
  SA dune structures
  SA eluvium
  SA eolian features
  SA ergs
  SA geomorphology
  SA sand ridges
  SA sand seas
  SA sand sheets
  SA sand waves
  SA shore features

Dunham Dolomite (1989)
  S Quebec; NW and W central Vermont.
  BT Lower Cambrian
  BT Cambrian
  BT Paleozoic
  SA Quebec
  SA Vermont

dunite (1978)
  BT peridotites
  BT ultramafics
  BT plutonic rocks
  BT igneous rocks

Dunkard Basin (1978)
  In West Virginia, Ohio, and Pennsylvania. Structural basin.
  IN Index states as applicable.
  BT United States
  SA Ohio
  SA Pennsylvania
  SA West Virginia

Dunkard Group (1978)
  Pennsylvanian and Permian. Includes Washington and Greene formations. W Maryland, E Ohio, SW Pennsylvania, and N West Virginia.
  IN Index ages as applicable.
  BT Paleozoic
  SA Maryland
  SA Ohio
  SA Pennsylvania
  SA Pennsylvanian
  SA Permian
  SA West Virginia

Dunkerque France
  use Dunkirk France

Dunkirk
  No longer a valid term for GeoRef. See Dunkirk France.

Dunkirk France (1993)
  City at S end of North Sea. Before 1978, also search Dunkerque.
  UF Dunkerque France
  BT Nord France
  BT France
  BT Western Europe
  BT Europe

Dunn County
  Valid through 1988. Search in combination with state term. After 1988, use specific county-state term.

Dunn County North Dakota (1989)
  W central North Dakota. Before 1989, also search Dunn County AND North Dakota.
  CO N465800N474800 W1020800W1030700
  BT North Dakota
  BT United States

Dunn County Wisconsin (1989)
  W Wisconsin. Before 1989, also search Dunn County AND Wisconsin.
  CO N444100N451300 W0913800W0921000
  BT Wisconsin
  BT United States

Dunnage Melange (1978)
  SA Eastern Canada

Duperow Formation (1985)
  In Jefferson Group. North Dakota, NE Montana, NW South Dakota, Manitoba and Saskatchewan.
  BT Upper Devonian
  BT Devonian
  BT Paleozoic
  SA Jefferson Group
  SA Manitoba
  SA Montana
  SA North Dakota
  SA Saskatchewan
  SA South Dakota

duplexes (1993)
  Before 1993, also search duplex fault zones and duplex structures if applicable.
  SA bedding faults
  SA decollement
  SA en echelon faults
  SA fault zones
  SA faults
  SA imbricate tectonics
  SA klippen
  SA thrust faults

Duplin Formation (1978)
  Base of Duplin is an unconformity. Florida, E Georgia, E North Carolina, and E South Carolina.
  BT upper Miocene
  BT Miocene
  BT Neogene
  BT Tertiary
  BT Cenozoic
  SA Florida
  SA Georgia
  SA North Carolina
  SA South Carolina

durability (1989)
  Used in engineering studies.
  SA degradation
  SA engineering properties
  SA mechanical properties
  SA rock mechanics
  SA soil mechanics

durain (1978)
  BT macerals
  SA organic residues

Durance
  No longer a valid term for GeoRef. See Durance France.

Durance Basin (1978)
  River basin in SE France. Also search Durance.
  BT France
  BT Western Europe
  BT Europe

Durance France (1993)
  Village in SW France.
  BT Lot-et-Garonne France
  BT France
  BT Western Europe
  BT Europe

Durango
  No longer a valid term for GeoRef. As of 1993 see Durango Mexico.

Durango Mexico (1993)
  State in W central Mexico. Also a city. Before 1993 also search Durango Mexico.
  BT Mexico
  NT Mapimi Mexico
  SA Sierra Madre Occidental

Durban
  No longer a valid term for GeoRef. As of 1993 see Durban South Africa.

Durban South Africa (1993)
  City on the Indian Ocean. Before 1993 also search Durban AND South Africa.
  BT Natal South Africa
  BT South Africa
  BT Southern Africa
  BT Africa

Durham
  No longer a valid term for GeoRef as of 1993. See Durham England.

Durham County
  Valid through 1988. Search in combination with state term. After 1988, use specific county-state term.

Durham County North Carolina (1989)
  N central North Carolina. Before 1989, also search Durham County AND North Carolina.
  CO N355000N361400 W0784100W0790100
  BT North Carolina
  BT United States

Durham England (1993)
  County in NE England. In 1974, parts of Durham went to form Cleveland County and Tyne and Wear County. Before 1993, also search Durham.
  CO N543000N545700 W0011500W0022200
  BT England
  BT Great Britain
  BT United Kingdom
  BT Western Europe
  BT Europe
  SA Ingleton England

duricrust (1978)
  BT chemically precipitated rocks
  BT sedimentary rocks
  SA calcrete
  SA caliche
  SA ferricrete
  SA silcrete
  SA soils
  SA weathering
  SA weathering crust

Dushambe Tadzhikistan
  use Dushanbe Tadzhikistan

Dushanbe
  No longer a valid term for GeoRef. As of 1993, see Dushanbe Tadzhikistan.

Dushanbe Tadzhikistan (1993)
  City in W Tadzhikistan. Before 1978, also search Dushambe or Stalinabad. This term has multiple hierarchies.
  UF Dushambe Tadzhikistan
  UF Stalinabad Tadzhikistan
  BT1 Tadzhikistan
  BT1 Asia
  BT2 Tadzhikistan
  BT2 Commonwealth of Independent States

dust (1978)
  BT clastic sediments
  BT sediments
  SA cosmic dust
  SA dust storms
  SA interplanetary dust
  SA terrigenous materials
  SA volcanic ash

dust clouds
  Use dust storms or planetesimals if applicable.

dust storms (1978)
  Also search dust clouds NOT extraterrestrial geology if applicable.
  SA dust
  SA storms

Dutch East Indies
  No longer a valid GeoRef index term. Before 1960, was used on level 1 in subfile E.
  use Indonesia

Dutch Guiana
  use Surinam

Dutch New Guinea
  use Irian Jaya Indonesia

Dutch West Indies
  No longer a valid GeoRef index term. Before 1969, was used on level 1 in subfile E.
  use Netherlands Antilles

Dutchess County
  Valid through 1988. Search in combination with state term. After 1988, use specific county-state term.

Dutchess County New York (1989)
  SE New York. Before 1989, also search Dutchess County AND New York.
  CO N412500N420500 W0735900W0740000
  BT New York
  BT United States

Duval County
  Valid through 1988. Search in combination with state term. After 1988, use specific county-state term.

Duval County Florida (1989)
  On the Atlantic Ocean in NE Florida. Before 1989, also search Duval County AND Florida.
  CO N300700N303000 W0812100W0820200
  BT Florida
  BT United States

Duval County Texas (1989)
S Texas. Before 1989, also search Duval County AND Texas.
CO N271500N280500
W0981600W0985300
BT Texas
BT United States

Duwi Formation (1989)
BT Upper Cretaceous
BT Cretaceous
BT Mesozoic
SA Egypt

DVDP
use Dry Valley Drilling Project

Dvina River (1978)
The Western Dvina River which flows into the Gulf of Riga and the Northern Dvina River in the White Sea Basin.
IN Index former Soviet republics as applicable.
BT Europe
SA Belarus
SA Latvia
SA Russian Federation

Dwyka Formation (1993)
Of the Karroo Supergroup. Formerly referred to as Dwyka Series which included Upper Shales, Conglomerate (now Tillite), and Lower Shales.
IN Index ages as applicable.
UF Dwyka Series
UF Dwyka Tillite
BT upper Paleozoic
BT Paleozoic
SA Africa
SA Botswana
SA Carboniferous
SA Karroo Supergroup
SA Namibia
SA Permian
SA South Africa
SA Triassic
SA Zimbabwe

Dwyka Series
No longer a valid term for GeoRef as of 1992.
use Dwyka Formation

Dwyka Tillite
use Dwyka Formation

Dy
use dysprosium

dye tracers (1982)
SA fluorescence
SA fluorimetry
SA ground water
SA hydrogeology
SA hydrology
SA surface water
SA tracers

Dyfed Wales (1993)
Welsh county formed in 1974 from Cardiganshire, Pembrokeshire and Carmarthenshire. Before 1978, also search Cardigan County, Pembroke or Pembroke County, and Carmarthen County where applicable. Also search Cardiganshire, Pembrokeshire, and Carmarthenshire.
BT Wales
BT Great Britain
BT United Kingdom
BT Western Europe
BT Europe
NT Cardiganshire Wales
NT Carmarthenshire Wales
NT Llandovery Wales
NT Pembrokeshire Wales

SA Cardigan Bay
SA South Wales

dykes
use dikes

dynamic
A valid term through 1977.

dynamic loading (1985)
BT loading
SA deformation
SA dynamics
SA load pressure
SA rock mechanics
SA soil mechanics

dynamic metamorphism (1978)
Before 1978, also search metamorphism AND dynamic.
BT metamorphism
SA regional metamorphism
SA retrograde metamorphism

dynamic properties (1981)
SA dynamics
SA elastodynamic properties
SA properties

dynamics (1978)
SA dynamic loading
SA dynamic properties
SA geodynamics
SA kinetics
SA shorelines
SA soil dynamics

dynamo theory
use dynamos

dynamos (1989)
Concept used to explain the presence and behavior of magnetic field on planetary bodies such as the Earth.
UF dynamo theory
UF geodynamo
SA concepts
SA core
SA Earth
SA geophysics
SA interior
SA magnetic field
SA patterns

dyscrasite (1978)
Autoposting of sulfides to this term ended in 1989.
BT antimonides

dysprosium (1978)
Autoposting of rare earths and metals to this term began in 1989.
UF Dy
BT rare earths
BT metals

Dzereg
use Altan Teeli Mongolia

Dzhagdy Range (1978)
NW of Komsomolsk in Khabarovsk Kray in Russian Far East. This term has multiple hierarchies.
BT1 Russian Federation
BT1 Commonwealth of Independent States
BT2 Asia
SA Russian Far East

Dzhezkazgan
No longer a valid term for GeoRef. As of 1993, see Dzhezkazgan Kazakhstan.

Dzhezkazgan Kazakhstan (1993)
Town in Karaganda Oblast in central Kazakhstan. This term has multiple hierarchies.
BT1 Karaganda Kazakhstan
BT1 Kazakhstan

BT1 Central Asia
BT1 Asia
BT2 Karaganda Kazakhstan
BT2 Kazakhstan
BT2 Commonwealth of Independent States
SA Ulu-Tau

Dzhida River (1978)
S of Lake Baikal in Buryatia. This term has multiple hierarchies.
BT1 Russian Federation
BT1 Commonwealth of Independent States
BT2 Asia
SA Buryat Russian Federation

Dzhugdzhur region (1981)
Borders the Okhotsk Sea and includes the Dzhugdzhur Range. This term has multiple hierarchies.
CO N540000N620000
E1563000E1330000
BT1 Russian Federation
BT1 Commonwealth of Independent States
BT2 Asia
SA Okhotsk region

Dzhungaria
No longer a valid term for GeoRef.
use Junggar

Dzhungarian Alatau (1978)
Northernmost branch of the Tien Shan.
IN Index Kazakhstan and/or Xinjiang as applicable.
CO N445000N453000
E0823000E0784500
UF Dzungar Alatau
UF Dzungarian Ala-Tau
UF Dzungarian Alatau
BT Asia
SA Kazakhstan
SA Xinjiang China

Dzirula Massif (1978)
Central Georgian Republic.
BT Georgian Republic
BT Europe

Dzungar Alatau
use Dzhungarian Alatau

Dzungaria
use Junggar

Dzungarian Ala-Tau
use Dzhungarian Alatau

Dzungarian Alatau
use Dzhungarian Alatau

E

E-region
As of 1993, no longer a valid term for GeoRef. Usually out-of-scope. Used with ionosphere.

Eagle Bay Formation (1989)
Cambrian-Ordovician? Triassic? S British Columbia.
IN Index ages as applicable.
SA British Columbia

Eagle County
Valid through 1988. Search in combination with state term. After 1988, use specific county-state term.

Eagle County Colorado (1989)

W central Colorado. Before 1989, also search Eagle County AND Colorado.
CO N392000N395500
W1061200W1075200
BT Colorado
BT United States
SA Battle Mountain
SA Vail Pass

Eagle Ford Formation (1989)
Texas, W Louisiana, SE Oklahoma and N Mexico.
UF Eagleford Formation
BT Gulfian
BT Upper Cretaceous
BT Cretaceous
BT Mesozoic
SA Louisiana
SA Mexico
SA Oklahoma
SA Texas

Eagle Lake (1989)
N California, SE Maine, W central Minnesota and NW Ontario.
IN Index states and counties or provinces as applicable.
SA California
SA Maine
SA Minnesota
SA Ontario

Eagle River (1989)
River in Colorado and Yukon Territory.
IN Index state or province as applicable.
SA Colorado
SA Yukon Territory

Eagle Sandstone (1985)
In Montana Group. Montana and N central Wyoming.
BT Upper Cretaceous
BT Cretaceous
BT Mesozoic
SA Montana
SA Montana Group
SA Wyoming

Eagle Station Meteorite (1989)
Found in Carroll County, N Kentucky.
BT pallasite
BT stony irons
BT meteorites
SA Carroll County Kentucky
SA Kentucky

Eagleford Formation
use Eagle Ford Formation

early diagenesis (1978)
BT diagenesis
SA late diagenesis

Earn Group (1993)
Devonian to Mississippian. British Columbia and Yukon Territory.
IN Index ages as applicable.
BT Paleozoic
SA British Columbia
SA Devonian
SA Mississippian
SA Yukon Territory

Earth (1978)
Treated as a whole. For general concepts or studies.
SA angular momentum
SA asthenosphere
SA atmosphere
SA biosphere
SA Chandler wobble
SA Conrad discontinuity
SA continents
SA core
SA crust

SA differentiation
SA dynamos
SA Earth tides
SA Earth-Moon couple
SA eccentricity
SA expansion
SA figure of Earth
SA free oscillations
SA geodesy
SA geodynamics
SA geoid
SA geophysics
SA geothermal gradient
SA gravity field
SA heat flow
SA hydrosphere
SA interior
SA isostasy
SA lithosphere
SA magnetic field
SA mantle
SA mass
SA Milankovitch theory
SA Mohorovicic discontinuity
SA Moon
SA motions
SA nutation
SA obliquity of the ecliptic
SA oscillations
SA paleoatmosphere
SA paleomagnetism
SA phase transitions
SA planetology
SA planets
SA plate tectonics
SA reversals
SA rotation
SA seismology
SA solar system
SA spherical harmonic analysis
SA tectonophysics
SA terrestrial comparison
SA terrestrial planets

Earth crust
No longer a valid GeoRef index term. Before 1969, was used on level 1 in subfile E. Now use crust.

Earth current exploration
Not a valid GeoRef index term after 1970. Was used on level 1 in subfile N. Now use Earth-current methods or Earth-current surveys.

Earth currents
Not a valid GeoRef index term after 1970. Was used on level 1 in subfile N. Now use Earth-current methods or Earth-current surveys.

earth dams (1981)
UF earthdams
BT dams
SA earthworks

earth flows
use earthflows

Earth Observing System (1993)
Before 1993, also search EOS not asteroids.
SA information systems
SA satellite methods

earth pressure (1978)
SA overpressure
SA pressure
SA soil mechanics

Earth sciences
No longer a valid GeoRef index term. Before 1971, was used on levels 1 and 2 in subfile N. See geology; geography. See also list P.

Earth tides (1978)
UF bodily tide
SA Earth
SA geodesy
SA geophysics
SA tectonophysics
SA tides

Earth waves
No longer a valid GeoRef index term. Before 1971, was used on level 1 in subfile N.
use elastic waves

earth, diatomaceous
use diatomaceous earth

Earth-current exploration
Not a valid term for GeoRef after 1971. Was used on level 1 in subfiles G and N. Now use Earth-current methods or Earth-current surveys.

Earth-current methods (1978)
UF telluric methods
BT geophysical methods
SA audiomagnetotelluric methods
SA CSAMT methods
SA Earth-current surveys
SA electrical field
SA magnetic field
SA magnetotelluric methods

Earth-current surveys (1978)
UF telluric surveys
BT geophysical surveys
BT surveys
SA Earth-current methods
SA magnetotelluric surveys

Earth-Moon couple (1985)
Used for discussions of the Earth and Moon together, e.g. motions or genesis.
UF Earth-Moon system
SA coupling
SA Earth
SA extraterrestrial geology
SA Moon
SA planetology
SA solar system

Earth-Moon system
use Earth-Moon couple

earthdams
use earth dams

earthflows (1978)
Autoposting of mass movements to this term began in 1989.
UF earth flows
BT mass movements
SA geomorphology
SA mudslides
SA slope stability

earthquake magnitude
use magnitude

earthquake prediction (1993)
Before 1993, also search earthquakes AND prediction.
SA earthquakes
SA induced earthquakes
SA microzonation
SA Parkfield earthquakes
SA precursors
SA radon emanometry
SA seismic gaps
SA soil gases
SA stress drops

earthquake record
use seismograms

earthquake sea wave
use tsunamis

earthquake swarms
use swarms

earthquakes (1978)
Used for earthquakes on all planetary bodies and their natural satellites, except for the Moon and Mars. For earthquakes on the Moon and Mars, see moonquakes or marsquakes, respectively. Before 1985, search earthquakes AND year earthquake occurred AND location (e.g. earthquakes AND 1966 AND Parkfield).
NT Alaska earthquake 1964
NT Andreanof Islands earthquake 1986
NT Borah Peak earthquake 1983
NT Borrego Mountain earthquake 1968
NT Chalfant Valley earthquake 1986
NT Charleston earthquake 1886
NT Chibaken-Toho-Oki earthquake 1987
NT Chile earthquake 1960
NT Chile earthquake 1985
NT Coalinga earthquake 1983
NT Coyote Lake earthquake 1979
NT deep-focus earthquakes
NT Edgecumbe earthquake 1987
NT El Asnam earthquake 1980
NT El Centro earthquake 1940
NT Friuli earthquake 1976
NT Gazli earthquake 1976
NT Gazli earthquake 1984
NT Goodnow earthquake 1983
NT Grand Banks earthquake 1929
NT Guatemala earthquake 1976
NT Gulf of Corinth earthquake 1981
NT Haicheng earthquake 1975
NT Haiyuan earthquake 1920
NT Hebgen Lake earthquake 1959
NT Imperial Valley earthquake 1979
NT induced earthquakes
NT intermediate-focus earthquakes
NT Irpinia earthquake 1980
NT Izu-Oshima earthquake 1978
NT Japan Sea earthquake 1983
NT Kalamata earthquake 1986
NT Kanto earthquake 1923
NT Kettleman Hills earthquake 1985
NT Lancang-Gengma earthquake 1988
NT Liege earthquake 1983
NT Loma Prieta earthquake 1989
NT Mammoth Lakes earthquakes
NT Mexico City earthquake 1985
NT microearthquakes
NT Miramichi earthquake 1982
NT Miyagi earthquake 1978
NT Montenegro earthquake 1979
NT Morgan Hill earthquake 1984
NT Mudurnu earthquake 1967
NT Nagano earthquake 1984
NT Nahanni earthquake 1985
NT Nankaido earthquake 1946
NT New Madrid earthquakes 1811-1812
NT Niigata earthquake 1964
NT North Palm Springs earthquake 1986
NT Oroville earthquake 1975
NT Parkfield earthquake 1966
NT Parkfield earthquakes
NT Saguenay earthquake 1988
NT San Fernando earthquake 1971
NT San Francisco earthquake 1906
NT San Salvador earthquake 1986
NT shallow-focus earthquakes
NT Spitak earthquake 1988
NT Superstition Hills earthquake 1987
NT Tangshan earthquake 1976
NT Tokachi-Oki earthquake 1968
NT Tonghai earthquake 1970
NT Victoria earthquake 1980
NT volcanic earthquakes
NT Vrancea earthquake 1977
NT Whittier Narrows earthquake 1987
NT Xingtai earthquake 1966
SA accelerograms
SA acoustical emissions
SA aftershocks
SA arrival time
SA b-values
SA cratering
SA crust
SA damage
SA Dasht-e-Bayaz
SA detection
SA dilatancy
SA earthquake prediction
SA effects
SA elastic waves
SA engineering geology
SA epicenters
SA explosions
SA faults
SA fluid injection
SA focal mechanism
SA focus
SA foreshocks
SA foundations
SA geologic hazards
SA GEOS
SA ground motion
SA icequakes
SA intraplate processes
SA isoseismal maps
SA magnitude
SA magnitude-frequency ratio
SA main shocks
SA mantle
SA marsquakes
SA microseisms
SA microzonation
SA modified Mercalli scale
SA Mohorovicic discontinuity
SA moonquakes
SA New Madrid region
SA nuclear facilities
SA paleoseismicity
SA periodicity
SA plate tectonics
SA Q
SA radon emanometry
SA reservoirs
SA Richter Scale
SA rock mechanics
SA seismic energy
SA seismic gaps
SA seismic intensity
SA seismic moment
SA seismic response
SA seismic sources
SA seismic zoning
SA seismograms
SA seismographs
SA seismology
SA slope stability
SA soil mechanics
SA stick-slip
SA stress drops
SA surface waves
SA swarms
SA teleseismic signals
SA tiltmeters
SA tsunamis
SA vibration

earthworks (1978)
SA earth dams
SA foundations

East Africa (1978)

Autoposting of this term began in 1981.
CO S271000N223000
   E0513000E0220000
BT Africa
NT Djibouti
NT Ethiopia
NT Kenya
NT Lake Malawi
NT Lake Natron
NT Lake Turkana
NT Lake Victoria
NT Malawi
NT Mozambique
NT Somali Republic
NT Sudan
NT Tanzania
NT Turkana Basin
NT Uganda
NT Zambia
SA Central Africa
SA North Africa
SA Red Sea Hills
SA Southern Africa
SA West Africa

East African Lakes (1981)
Grouper for all lakes in E Africa.
BT Africa
NT Lake Albert
NT Lake Baringo
NT Lake Edward
NT Lake Kariba
NT Lake Kivu
NT Lake Magadi
NT Lake Malawi
NT Lake Natron
NT Lake Tanganyika
NT Lake Turkana
NT Lake Victoria
SA East African Rift

East African Rift (1978)
Part of Great Rift Valley.
IN Index countries and lakes as applicable.
UF East African Rift system
UF East African Rift valley
UF East African Rift zone
BT Africa
SA Afar Depression
SA Burundi
SA East African Lakes
SA Ethiopia
SA Ethiopian Rift
SA Great Rift Valley
SA Gregory Rift
SA Kenya
SA Lake Albert
SA Lake Edward
SA Lake Kivu
SA Lake Magadi
SA Lake Malawi
SA Lake Natron
SA Lake Tanganyika
SA Lake Turkana
SA Malawi
SA Mozambique
SA Rwanda
SA Tanzania
SA Uganda

East African Rift system
 use East African Rift

East African Rift valley
 use East African Rift

East African Rift zone
 use East African Rift

East Anatolian Fault (1989)
E Turkey.
BT Turkey
BT Middle East
BT Asia

East Anglia (1978)
Region including counties of Norfolk and Suffolk on the North Sea in extreme E England.
BT England
BT Great Britain
BT United Kingdom
BT Western Europe
BT Europe
NT Norfolk England
NT Suffolk England

East Antarctica (1981)
E of Transantarctic Mountains.
CO S900000S650000
   E1710000W0400000
BT Antarctica

East Atlantic (1978)
Before 1978, also search Atlantic Ocean AND east or eastern.
BT Atlantic Ocean
SA Cape Verde Atlantic
SA European Atlantic
SA Northeast Atlantic
SA Northwest Atlantic
SA South Atlantic
SA Southeast Atlantic
SA Southwest Atlantic
SA West Atlantic

East Baton Rouge Parish Louisiana (1989)
S central Louisiana. Before 1989, search East Baton Rouge Parish AND Louisiana.
BT Louisiana
BT United States
NT Baton Rouge Louisiana

East Bay (1981)
In the Gulf of Mexico off the Mississippi Delta.
IN Index county or regions as applicable.
SA Louisiana
SA Mississippi Delta
SA Plaquemines Parish Louisiana

East Berlin Formation (1989)
In Newark Group. Previously thought to be Upper Triassic in age. Central Connecticut and central Massachusetts.
BT Lower Jurassic
BT Jurassic
BT Mesozoic
SA Connecticut
SA Massachusetts
SA Newark Supergroup

East Carpathians
 use Eastern Carpathians

East Charlevoix
No longer a valid term for GeoRef. As of 1993, see Charlevoix-Est County Quebec.

East Charlevoix County Quebec
 use Charlevoix-Est County Quebec

East China Sea (1978)
Enclosed by China, Korea, Japan, and Taiwan.
CO N240000N333000
   E1303000E1200000
UF Dong Hai
BT West Pacific
BT Pacific Ocean
SA China Sea
SA Kuroshio
SA Okinawa Trough

East Coast
Not a valid term for GeoRef. Search in conjunction with appropriate region.

East European Plain
 use Russian Plain

East European Platform
No longer a valid term for GeoRef. As of 1993 see Russian Platform.
 use Russian Platform

East Frisian Islands (1978)
In North Sea off NW coast of West Germany.
BT Lower Saxony Germany
BT Germany
BT Central Europe
BT Europe

East Germany
No longer a valid term for GeoRef. Officially German Democratic Republic or Deutsche Demokratische Republik. In N central Europe, bounded on N by the Baltic Sea, on E by Poland, on S by Czechoslovakia and West Germany, and on W by West Germany. Introduced as a level 1 area term in 1978. As of 1993, see Germany and also Mecklenburg-Western Pomerania Germany, Brandenburg Germany, Saxony-Anhalt Germany, Thuringia Germany, and Saxony Germany.

East Greenland (1978)
Before 1978, search Greenland AND east. Autoposting of Arctic region to this term began in 1990.
BT Greenland
BT Arctic region
SA South Greenland

East Indian Ocean (1978)
Before 1978, also search Indian Ocean AND (east OR eastern).
BT Indian Ocean

East Lothian Scotland (1993)
Formerly Haddingtonshire. Former county on S bank of Firth of Forth. As of 1974, part of Lothian region Scotland. Before 1993, also search East Lothian.
CO N555000N560500
   W0022000W0030000
UF Haddington County
UF Haddingtonshire
BT Lothian region Scotland
BT Scotland
BT Great Britain
BT United Kingdom
BT Western Europe
BT Europe
SA Tyne River

East Malaysia (1978)
That part of the Federation of Malaysia which is composed of the states of Sabah and Sarawak on the island of Borneo. This term has multiple hierarchies.
BT1 Borneo
BT1 Far East
BT1 Asia
BT2 Borneo
BT2 Malay Archipelago
BT3 Malaysia
BT3 Far East
BT3 Asia
NT Sabah Malaysia
NT Sarawak Malaysia
SA Brunei

East Mediterranean (1978)
Before 1978, search Mediterranean Sea AND east. As of 1981, includes the Black Sea. Autoposting of this term began in 1981.
CO N301000N464000
   E0420000E0100000
BT Mediterranean Sea
NT Adriatic Sea
NT Aegean Sea
NT Bannock Basin
NT Black Sea
NT Gulf of Gabes
NT Ionian Sea
NT Pelagian Sea
NT Strabo Trench
SA DSDP Site 125
SA DSDP Site 126
SA DSDP Site 127
SA DSDP Site 128
SA DSDP Site 129
SA DSDP Site 130
SA DSDP Site 378
SA Hellenic Trench
SA West Mediterranean

East Mesa
 use East Mesa KGRA

East Mesa KGRA (1981)
Known Geothermal Resources Area. In SE California. Before 1981, also search East Mesa.
CO N324500N325000
   W1151000W1151500
UF East Mesa
UF KGRA, East Mesa
BT Imperial County California
BT California
BT United States
SA Imperial Valley

East Midlands (1978)
Eastern part of the Midlands of central England.
BT England
BT Great Britain
BT United Kingdom
BT Western Europe
BT Europe
SA Leicestershire England
SA Midlands
SA Northamptonshire England
SA Nottinghamshire England

East Ongul Island (1978)
In E Lutzow-Holm Bay off Prince Olav Coast in Norwegian Sector on Indian Ocean side.
BT Antarctica

East Pacific (1978)
Before 1978, search Pacific Ocean AND east or eastern.
BT Pacific Ocean
SA California Current
SA Equatorial Pacific
SA North Pacific
SA Northeast Pacific
SA Northwest Pacific
SA Orozco fracture zone
SA South Pacific
SA Southeast Pacific
SA Southwest Pacific
SA West Pacific

East Pacific Ocean Islands (1981)
Before 1981, Easter Island and Galapagos Islands were narrower terms of Pacific Ocean. Before 1985, Hawaii, Midway and Guadalupe Island were narrower terms of Pacific Ocean.
NT Easter Island
NT Galapagos Islands
NT Guadalupe Island
NT Hawaii
NT Midway
SA Johnston Island
SA Kiribati
SA Pacific Ocean
SA Seymour Island

East Pacific Rise (1978)

In S Pacific. Indicates the mid-ocean ridge extending from the Pacific-Antarctic Ridge to the Gulf of California.
UF Albatross Cordillera
UF Easter Island Cordillera
BT Pacific Ocean
SA DSDP Site 159
SA DSDP Site 319
SA DSDP Site 419
SA DSDP Site 420
SA DSDP Site 597
SA DSDP Site 598
SA DSDP Site 599
SA DSDP Site 600
SA DSDP Site 601
SA DSDP Site 602
SA Leg 16
SA Leg 92
SA Leg 122
SA mid-ocean ridges
SA Orozco fracture zone
SA Pacific-Antarctic Ridge
SA Rivera Ocean Seismic Experiment

East Pakistan
A valid term through 1972.
use Bangladesh

East Rift Zone (1989)
Of Kilauea.
IN Index county or region as applicable.
SA Hawaii County Hawaii
SA Hawaii Island
SA Kilauea
SA Puu Oo

East Rudolf (1978)
Basin in N Kenya.
BT Kenya
BT East Africa
BT Africa

East Sayan
use Eastern Sayan

East Siberian Sea (1978)
Between New Siberian Islands and Wrangel Island N of Magadan Oblast and NE Yakutia.
CO N690000N770000
E1800000E1390000
BT Arctic Ocean

East Texas (1985)
Autoposting of broader terms to this term began in 1989.
UF eastern Texas
BT Texas
BT United States

East Texas Field (1978)
Oil field in E Texas.
BT Texas
BT United States
SA oil and gas fields

East-Central Alaska (1993)
Artificial region based on U.S. Geological Survey quadrangle designations. Also search specific quadrangles if applicable.
IN Index quadrangles as applicable.
CO N630000N680000
W1410000W1530000
BT Alaska
BT United States
NT Birch Creek Alaska
NT Fairbanks Alaska
NT Fairbanks District
NT Fairbanks mining district
SA Delta River
SA Trans-Alaska Pipeline
SA Yukon-Koyukuk Basin
SA Yukon-Tanana Upland

East-Central Nigeria
No longer a valid term for GeoRef. Valid 1981-1992.

Easter Island (1978)
About 2,000 miles W of the Chilean coast. Before 1981, a narrower term of Pacific Ocean.
CO S280000S270000
W1090000W1100000
BT East Pacific Ocean Islands
SA Pacific Ocean

Easter Island Cordillera
use East Pacific Rise

Eastern Alps (1978)
Ranges of the Alps extending from Innsbruck in the W to near Vienna and including the Dinaric Alps of Yugoslavia.
IN Index countries as applicable.
BT Alps
BT Europe
NT Carnic Alps
NT Dinaric Alps
NT Dolomites
NT Hohe Tauern
NT Julian Alps
NT Karawanken
NT Koralpe Range
NT Otztal Alps
NT Wiener Wald
NT Zillertal Alps
SA Austria
SA Gailtal Alps
SA Gosau Formation
SA Penninic Zone
SA Salzkammergut
SA Tauern Window
SA Wechsel
SA Yugoslavia

Eastern Canada (1981)
Includes Maritime Provinces (New Brunswick, Nova Scotia, and Prince Edward Island), Newfoundland, Ontario, and Quebec.
CO N420000N640000
W0523000W0950000
BT Canada
NT Avalon Zone
NT Gander Zone
NT James Bay
NT James Bay Lowlands
NT Lake Timiskaming
NT Maritime Provinces
NT Newfoundland
NT Ontario
NT Ottawa Valley
NT Quebec
SA Central Gneiss Belt
SA Central Metasedimentary Belt
SA Chaleur Bay
SA Dunnage Melange

Eastern Carpathians (1978)
Ranges of the Carpathians in SW Ukraine and Moldavia.
IN Index Ukraine and/or Romania as applicable.
UF East Carpathians
BT Carpathians
BT Europe
NT Calimani Mountains
SA Haghimas Syncline
SA Romania
SA Tulghes Series
SA Ukraine
SA Ukrainian Carpathians

Eastern China
No longer a valid term for GeoRef. Used from 1985-1992 as a broader term for the following provinces: Anhui, Fujian, Jiangsu, Jiangxi, Shandong, Shanghai, and Zhejiang.

Eastern Cordillera (1978)
Eastern range of the Andes. Restricted term as of 1993.
IN Index countries as applicable.
UF Cordillera Oriental
BT Andes
BT South America
SA Bolivia
SA Colombia
SA Ecuador
SA Peru
SA Sierra de Perija

Eastern Desert (1978)
Between Nile on W and Gulf of Suez and Red Sea on E.
CO N220000N302000
E0365500E0305000
UF Arabian Desert
BT Egypt
BT North Africa
BT Africa

Eastern Europe
Not a valid term for GeoRef. See Central Europe.

Eastern Gas Shales Project (1985)
UF EGSP
SA Appalachian Basin
SA Appalachians
SA Midwest
SA natural gas
SA Ohio
SA shale
SA United States
SA West Virginia

Eastern Ghats (1978)
Low mountain range extending about 500 miles along SE and E coast as far N as Mahanadi River.
BT Ghats
BT India
BT Indian Peninsula
BT Asia

Eastern Goldfields (1978)
SA Western Australia

Eastern Hemisphere (1978)
Used when discussing many large areas too numerous to mention.
SA Northern Hemisphere
SA Southern Hemisphere
SA Western Hemisphere

Eastern Kazakhstan (1981)
This term has multiple hierarchies.
CO N400000N500000
E0860000E0723000
BT1 Kazakhstan
BT1 Central Asia
BT1 Asia
BT2 Kazakhstan
BT2 Commonwealth of Independent States
SA Rudny Altai

Eastern Overthrust Belt (1985)
Thrust belt, primarily in E United States. Has been described as extending from Quebec to Texas.
IN Index countries, states and provinces as applicable.
BT North America
SA Appalachians
SA Canada
SA United States
SA Western Overthrust Belt

Eastern Panhandle, West Virginia
use West Virginia Panhandle

Eastern Sayan (1978)
Before 1978, use Sayan AND (east OR eastern).
CO N523000N560000
E1040000E0920000

UF East Sayan
BT Sayan
BT Siberian fold belt
BT Asia
SA Altai-Sayan region

Eastern Sea (1981)
Before 1981, also search Mare Orientale.
UF Mare Orientale
BT Moon

Eastern Swiss Alps (1981)
This term has multiple hierarchies.
BT1 Swiss Alps
BT1 Switzerland
BT1 Central Europe
BT1 Europe
BT2 Swiss Alps
BT2 Alps
BT2 Europe

eastern Texas
use East Texas

Eastern Townships (1989)
Informally used for the townships S of the Saint Lawrence River in S Quebec.
IN Index the names of specific towns and counties if available.
BT Quebec
BT Eastern Canada
BT Canada

Eastern U.S. (1978)
As of 1993, term is used only in discussions of broad general areas. Also see individual states or regions. Before 1978, also search United States AND east or eastern. From 1981 through 1992, this term was autoposted to New England, Southeastern U.S., Delaware, District of Columbia, Maryland, New Jersey, New York, Pennsylvania, and West Virginia.
BT United States
NT Southeastern U.S.

Eastern Venezuela (1981)
CO N074000N105000
W0595000W0654000
BT Venezuela
BT South America

Eastover Formation (1989)
In Chesapeake Group.
BT upper Miocene
BT Miocene
BT Neogene
BT Tertiary
BT Cenozoic
SA Chesapeake Group
SA Maryland
SA North Carolina
SA Virginia

Eaton
Valid through 1988. Search in combination with state term. After 1988, use specific city-state term.

Eaton Colorado (1989)
Town in NE Colorado. Before 1989, search Eaton AND Colorado.
CO N403200N403200
W1044300W1044300
BT Weld County Colorado
BT Colorado
BT United States

Eaton County Michigan (1989)
S central Michigan. Before 1989, search Eaton County AND Michigan.
CO N422500N424400
W0843700W0850500
BT Michigan Lower Peninsula

BT Michigan
BT United States
SA Lansing Michigan

Eau Claire Formation (1989)
NE Illinois, Indiana, Iowa, W Michigan, SE Minnesota, SW Ohio and W Wisconsin.
BT Upper Cambrian
BT Cambrian
BT Paleozoic
SA Illinois
SA Indiana
SA Iowa
SA Michigan
SA Minnesota
SA Mount Simon Sandstone
SA Ohio
SA Wisconsin

Ebbe Anticlinorium (1985)
E North Rhine-Westphalia.
CO N510000N512000
W008200W0072000
BT North Rhine-Westphalia Germany
BT Germany
BT Central Europe
BT Europe
SA Rhenish Schiefergebirge
SA Sauerland

Ebino
No longer a valid term for GeoRef. See Ebino Japan.

Ebino Japan (1993)
Region around the city in Miyazaki Prefecture, E Kyushu.
BT Miyazaki Japan
BT Kyushu
BT Japan
BT Far East
BT Asia
SA Kirishima
SA Sakura-jima

ebridians (1978)
Autoposting of this term began in 1978. This term has multiple hierarchies.
BT1 Protista
BT1 Invertebrata
BT2 Protista
BT2 microfossils
NT Archaeomonadaceae

Ebro Basin (1978)
River basin in NE Spain extending from Santander Province to Catalonia.
UF Ebro River basin
BT Spain
BT Iberian Peninsula
BT Southern Europe
BT Europe

Ebro River (1978)
Rises in the Cantabrian Mountains in N Spain and flows SE into the Mediterranean Sea SW of Barcelona.
BT Spain
BT Iberian Peninsula
BT Southern Europe
BT Europe

Ebro River basin
use Ebro Basin

Eburnean Orogeny
use Pan-African Orogeny

Ecca Group (1993)
Karroo Basin, South Africa. Often referred to a part of Karroo Supergroup.
UF Ecca Series
BT Permian
BT Paleozoic
SA Botswana
SA Cape Province South Africa
SA Karroo Supergroup
SA Mozambique
SA Vryheid Formation
SA Zimbabwe

Ecca Series
No longer a valid term for GeoRef as of 1993.
use Ecca Group

eccentricity (1978)
Used as a general term.
SA Earth
SA geometry
SA Milankovitch theory
SA motions

Echinodermata (1978)
Autoposting of this term began in 1978.
BT Invertebrata
NT Asterozoa
NT Crinozoa
NT Echinozoa
NT Homalozoa
SA echinoderms
SA Receptaculitaceae
SA reef builders
SA sclerites

echinoderms (1981)
Common name for Echinodermata.
IN Index for all non-paleontologic studies of fossils.
BT invertebrates
NT crinoids
NT echinoids
SA biostratigraphy
SA Echinodermata

Echinoidea (1978)
BT Echinozoa
BT Echinodermata
BT Invertebrata
SA echinoids

echinoids (1985)
Common name for Echinoidea. Autoposting of echinoderms to this term began in 1989.
BT echinoderms
BT invertebrates
SA biostratigraphy
SA Echinoidea

Echinozoa (1978)
Autoposting of this term to Camptostromatoidea, Cyclocystoidea, Helicoplacoidea and Ophiocistioidea began in 1981. Autoposting of this term to Edrioblastoidea began in 1989.
BT Echinodermata
BT Invertebrata
NT Camptostromatoidea
NT Cyclocystoidea
NT Echinoidea
NT Edrioasteroidea
NT Edrioblastoidea
NT Helicoplacoidea
NT Holothuroidea
NT Ophiocistioidea

Echiurida (1981)
BT worms
BT Invertebrata

Echo Bay
No longer a valid term for GeoRef. As of 1993, see Echo Bay Ontario.

Echo Bay Ontario (1993)
Village just E of Sault Sainte Marie in S Ontario. Before 1993, also search Echo Bay AND Ontario.
BT Algoma District Ontario
BT Ontario
BT Eastern Canada
BT Canada

echo sounding (1978)
SA acoustical methods
SA bathymetry
SA deep sounding
SA geophysical methods
SA geophysical surveys
SA oceanography
SA sonar methods
SA sounding

eclogite (1978)
From 1978-1980, granulites was autoposted to this term.
BT metamorphic rocks
SA eclogite facies
SA granulites

eclogite facies (1982)
UF eklogite facies
BT facies
SA eclogite
SA metamorphic rocks

ecology (1978)
Used for the study of Holocene or modern relationships between organisms and their environments, including the study of communities, patterns of life, natural cycles, relationships of organisms to each other, biogeography, and population changes.
IN Index with type of fauna or flora and type of environment where applicable.
SA adaptation
SA alpine environment
SA alpine-type
SA anaerobic taxa
SA aquatic environment
SA arctic environment
SA arid environment
SA behavior
SA benthonic taxa
SA biofacies
SA biogenic processes
SA biogeography
SA biology
SA biomass
SA biota
SA biotopes
SA biotypes
SA boreal environment
SA brackish water
SA brackish-water environment
SA cave environment
SA climatic controls
SA coastal environment
SA colonial taxa
SA colonization
SA communities
SA continental shelf
SA continental slope
SA deforestation
SA deltaic environment
SA deltas
SA desertification
SA deserts
SA desiccation
SA ecosystems
SA endemic taxa
SA environment
SA environmental geology
SA epibiotism
SA equatorial region
SA estuaries
SA estuarine environment
SA eutrophication
SA fires
SA fresh water
SA fresh-water environment
SA geomorphology
SA glacial environment
SA grasslands
SA greenhouse effect
SA growth rates
SA habitat
SA hemipelagic environment
SA herbivorous taxa
SA human ecology
SA humid environment
SA hydrology
SA hypersaline environment
SA impact statements
SA intertidal environment
SA juvenile taxa
SA lacustrine environment
SA lagoonal environment
SA lagoons
SA lakes
SA living taxa
SA mangrove swamps
SA marine environment
SA paleoecology
SA paludal environment
SA peat bogs
SA phenotypes
SA ponds
SA prairies
SA predators
SA provinciality
SA punctuated aggradational cycles
SA reefs
SA revegetation
SA savannas
SA semi-arid environment
SA shallow-water environment
SA soils
SA subalpine environment
SA substrates
SA subtropical environment
SA succession
SA supratidal environment
SA swamps
SA taiga environment
SA temperate environment
SA terrestrial environment
SA trophic analysis
SA tropical environment
SA tundra
SA urban environment
SA vegetation
SA weathering
SA wetlands
SA wilderness areas

economic agreements (1981)
SA aid projects
SA economics

economic geology (1978)
For the discipline as a whole. See also specific commodities (list C).
SA chemical industry
SA coal mines
SA companies
SA cutoff grade
SA development
SA economics
SA energy sources
SA evaluation
SA geology
SA industry
SA Law of the Sea
SA legislation
SA metal ores
SA metallogenic epochs
SA metallogenic provinces
SA metallogeny
SA mineral deposits, genesis
SA mineral economics
SA mineral exploration
SA mineral resources
SA mines
SA mining
SA mining geology
SA nonmetal deposits

SA oil and gas fields
SA ore grade
SA ore minerals
SA paragenesis
SA potential deposits
SA pumping
SA raw materials
SA recycling
SA regulations
SA strategic minerals
SA utilization
SA water resources
SA water wells

economic geology maps (1978)
Before 1978, search maps AND economic geology.
BT maps
SA coal deposit maps
SA industrial minerals maps
SA metallogenic maps
SA petroleum maps

economics (1978)
As of 1985, restricted to certain commodities (list C): all of the energy sources, mineral resources (as a term) and water resources.
SA aid projects
SA antitrust legislation
SA borrowing
SA cash transactions
SA commodity exchange
SA companies
SA company diversification
SA company integration
SA company mergers
SA consumption
SA corporate policy
SA cost
SA customs duty
SA cutoff grade
SA demand
SA economic agreements
SA economic geology
SA European Community
SA evaluation
SA exchange rate
SA Exclusive Economic Zone
SA exploitation
SA export
SA export value
SA financing
SA futures market
SA import
SA import value
SA industrial patents
SA industrialized countries
SA International Monetary Fund
SA investment
SA investment legislation
SA loans
SA Lome Convention
SA London Metal Exchange
SA manpower
SA market stabilization
SA markets
SA mine reactivation
SA mineral economics
SA negotiations
SA ore grade
SA pilot plants
SA possibilities
SA prediction maps
SA price
SA production value
SA productive capacity
SA productivity
SA profitability
SA purchases
SA recycling
SA sales
SA strategic minerals
SA subsidies
SA supply
SA taxes
SA technology transfer
SA Third World
SA UNCTAD
SA United States Exclusive Economic Zone
SA World Bank

ecosystems (1978)
SA ecology
SA environment
SA herbivorous taxa
SA succession
SA trophic analysis

ecoulement
use gravity sliding

Ecrins-Pelvoux Massif
use Pelvoux Massif

ectinite (1978)
BT migmatites
BT metamorphic rocks

Ectoprocta
As of 1989, no longer a valid term for GeoRef. A synonym for Bryozoa.
use Bryozoa

Ector County
Valid through 1988. Search in combination with state term. After 1988, use specific county-state term.

Ector County Texas (1989)
W Texas. Before 1989, also search Ector County AND Texas.
CO N313800N320400
W1021800W1024800
BT Texas
BT United States
NT Odessa Texas
SA Central Basin Platform

Ectotropha (1978)
BT Insecta
BT Mandibulata
BT Arthropoda
BT Invertebrata

Ecuador (1978)
CO S050000N013000
W0750000W0810000
BT South America
SA Amazon Basin
SA Andes
SA Eastern Cordillera
SA Galapagos Islands
SA Western Cordillera

Eddy County
Valid through 1988. Search in combination with state term. After 1988, use specific county-state term.

Eddy County New Mexico (1989)
SE New Mexico. Before 1989, also search Eddy County AND New Mexico.
CO N320000N325700
W1034400W1045200
BT New Mexico
BT United States
NT Carlsbad Caverns
NT Carlsbad New Mexico
NT Waste Isolation Pilot Plant

Eddy County North Dakota (1989)
Central North Dakota. Before 1989, also search Eddy County AND North Dakota.
CO N473600N475100
W0983000W0992000
BT North Dakota
BT United States

eddy flow (1982)
SA currents
SA hydraulics
SA hydrodynamics
SA hydrology

Eden Shale (1978)
Includes Economy Formation, Southgate Formation, McMicken Formation. S Indiana, central Kentucky, and SW Ohio.
BT Edenian
BT Upper Ordovician
BT Ordovician
BT Paleozoic
SA Indiana
SA Kentucky
SA Ohio

Edenian (1978)
North America. Above Mohawkian, below Maysvillian.
BT Upper Ordovician
BT Ordovician
BT Paleozoic
NT Eden Shale

Edentata (1978)
Order. Autoposting of this term to Xenarthra began in 1989. Autoposting of Eutheria and Theria to this term began in 1989.
BT Eutheria
BT Theria
BT Mammalia
BT Tetrapoda
BT Vertebrata
BT Chordata
NT Xenarthra

Edgecumbe earthquake 1987 (1993)
North Island, New Zealand.
BT earthquakes
SA Bay of Plenty
SA New Zealand
SA North Island

Ediacaran (1993)
Late Vendian, above Varangian. Before 1993, also search Ediacara.
BT Vendian
BT upper Proterozoic
BT Proterozoic
BT upper Precambrian
BT Precambrian
SA Eocambrian
SA Infracambrian
SA Riphean

Edinburgh
No longer a valid term for GeoRef as of 1993. See Edinburgh Scotland.

Edinburgh Scotland (1993)
City near the Firth of Forth in Midlothian, Lothian region, in SE Scotland. Before 1993, also search Edinburgh AND Scotland. This term has multiple hierarchies.
BT1 Lothian region Scotland
BT1 Scotland
BT1 Great Britain
BT1 United Kingdom
BT1 Western Europe
BT1 Europe
BT2 Midlothian Scotland
BT2 Scotland
BT2 Great Britain
BT2 United Kingdom
BT2 Western Europe
BT2 Europe

edingtonite (1978)
BT zeolite group
BT framework silicates
BT silicates

Edmondia (1978)
BT Bivalvia
BT Mollusca
BT Invertebrata

Edmonson County
Valid through 1988. Search in combination with state term. After 1988, use specific county-state term.

Edmonson County Kentucky (1989)
Central Kentucky. Before 1989, also search Edmonson County AND Kentucky.
CO N370300N372200
W0860400W0863000
BT Kentucky
BT United States
NT Mammoth Cave

Edmonton
No longer a valid term for GeoRef. See Edmonton Alberta.

Edmonton Alberta (1993)
City in S central Alberta.
BT Alberta
BT Western Canada
BT Canada

Edmonton Formation (1978)
Now considered equivalent of either Bearpaw Shale or Fox Hills Sandstone of Montana.
BT Upper Cretaceous
BT Cretaceous
BT Mesozoic
SA Alberta
SA Bearpaw Formation
SA Fox Hills Formation
SA Northwest Territories
SA Saskatchewan

Edo
Not a valid term for GeoRef. See Tokyo Japan.

Edrioasteroidea (1978)
BT Echinozoa
BT Echinodermata
BT Invertebrata

Edrioblastoidea (1981)
Autoposting of Echinozoa to this term began in 1989. Autoposting of Crinozoa ended in 1989.
BT Echinozoa
BT Echinodermata
BT Invertebrata

education (1978)
Before 1971, also search study and teaching; educational.
IN Index also level of education [e.g. college-level education].
UF study and teaching
SA academic institutions
SA associations
SA catalogs
SA college-level education
SA creationism
SA curricula
SA elementary geology
SA elementary school
SA geology
SA glossaries
SA graduate-level education
SA high school
SA historical geology
SA junior high school
SA libraries
SA materials
SA museums
SA oceanography
SA petrology
SA popular geology
SA programs
SA stratigraphy
SA textbooks

SA theses

educational
No longer a valid GeoRef index term. Before 1971, was used on level 1 in subfiles E and N. Now use education.

Edwards Aquifer (1989)
Central Texas.
IN Index counties as applicable.
BT Texas
BT United States
SA Bell County Texas
SA Bexar County Texas
SA Comal County Texas
SA Hamilton County Texas
SA Hays County Texas
SA Kinney County Texas
SA Medina County Texas
SA Travis County Texas
SA Uvalde County Texas
SA Williamson County Texas

Edwards County
Valid through 1988. Search in combination with state term. After 1988, use specific county-state term.

Edwards County Illinois (1989)
SE Illinois. Before 1989, also search Edwards County AND Illinois.
CO N381500N383500 W0870400W0880900
BT Illinois
BT United States

Edwards County Kansas (1989)
S central Kansas. Before 1989, also search Edwards County AND Kansas.
CO N374100N380700 W0990000W0993300
BT Kansas
BT United States

Edwards County Texas (1989)
SW Texas. Before 1989, also search Edwards County AND Texas.
CO N294000N301600 W0994700W1004400
BT Texas
BT United States

Edwards Formation (1978)
Overlies Walnut Clay and underlies Georgetown Limestone of Washita Group. In S Texas. This term has multiple hierarchies.
BT1 Comanchean
BT1 Cretaceous
BT1 Mesozoic
BT2 Lower Cretaceous
BT2 Cretaceous
BT2 Mesozoic
SA Texas
SA Washita Group

Edwards Plateau (1978)
Highland region of W central Texas.
CO N300000N313000 W0990000W1020000
BT Texas
BT United States

Eel River (1978)
Rises in Mendocino County in NW California and flows NW into the Pacific Ocean.
IN Index counties or regions as applicable.
SA California
SA Humboldt County California
SA Mendocino County California
SA Trinity County California

Eemian (1978)
Europe.
BT upper Pleistocene
BT Pleistocene
BT Quaternary
BT Cenozoic

EETA
use Elephant Moraine Meteorites

EETA 79001 (1989)
One of the Elephant Moraine Meteorites. This term has multiple hierarchies.
UF EETA79001
BT1 Elephant Moraine Meteorites
BT1 meteorites
BT2 achondrites
BT2 stony meteorites
BT2 meteorites
SA Antarctica
SA Elephant Moraine

EETA79001
use EETA 79001

EEZ
use Exclusive Economic Zone

effects (1978)
SA biogenic processes
SA climate effects
SA coastal effects
SA damage
SA earthquakes
SA geomorphologic effects
SA greenhouse effect
SA influence
SA relief effects
SA thermal effects
SA volcanic risk

efficiency (1978)
Used as a general term.
SA accuracy
SA streams

effluents (1981)
Liquids discharged as waste. Before 1981, also search discharge AND pollution; effluent.
SA discharge
SA liquid waste
SA pollutants
SA pollution
SA waste disposal
SA waste water

effusion (1993)
Emission of fluid lava onto the Earth's surface. May also refer to flow of gases.
SA extrusive rocks
SA flows
SA gases
SA lava
SA lava flows
SA magmas
SA viscosity

Efremouka Meteorite
use Efremovka Meteorite

Efremovka Meteorite (1993)
Impact in NE Kazakhstan. Before 1993, also search Yefremovka Meteorite and Efremouka Meteorite.
UF Efremouka Meteorite
UF Yefremovka Meteorite
BT carbonaceous chondrites
BT chondrites
BT stony meteorites
BT meteorites
SA Kazakhstan

Egan Range (1985)
E central Nevada.

IN Index counties and mountains as applicable.
CO N381500N395000 W1145000W1151000
BT Nevada
BT United States
SA Lincoln County Nevada
SA Nye County Nevada
SA White Pine County Nevada

Eger
No longer a valid term for GeoRef. See Eger Hungary.

Eger Hungary (1993)
City NE of Budapest in N central Hungary.
BT Hungary
BT Central Europe
BT Europe

Egerian (1978)
Europe.
BT upper Oligocene
BT Oligocene
BT Paleogene
BT Tertiary
BT Cenozoic

Eggenburgian (1978)
Europe.
BT lower Miocene
BT Miocene
BT Neogene
BT Tertiary
BT Cenozoic

eggs (1978)
SA Aves
SA Insecta
SA Pisces
SA Reptilia

eggstone
use oolite

EGSP
use Eastern Gas Shales Project

Egypt (1978)
CO N220000N320000 E0353000E0250000
BT North Africa
BT Africa
NT Alexandria Egypt
NT Aswan Egypt
NT Bahariya Oasis
NT Cairo Egypt
NT Dakhla Oasis
NT Eastern Desert
NT Fayum Depression
NT Fayum Egypt
NT Gavish Sebkha
NT Kharga Oasis
NT Kosseir Egypt
NT Middle Nile Valley
NT Nile Delta
NT Safaga Egypt
NT Sinai Egypt
NT Solar Lake
NT Suez Canal
NT Western Desert
SA Dakhla Shale
SA Duwi Formation
SA Esna Shale
SA Libyan Desert
SA Mediterranean region
SA Middle East
SA Nakhla Meteorite
SA Near East
SA Nile River
SA Nile Valley
SA Nubia
SA Nubian Sandstone
SA Nubian Shield
SA Red Sea Basin
SA Red Sea Hills
SA Red Sea Rift
SA Sahara

SA Sinai
SA Thebes Formation

Eh (1978)
UF oxidation-reduction potential
UF redox
UF redox potential
SA geochemistry
SA oxidation
SA oxidation zone
SA reduction

Ehime
No longer a valid term for GeoRef. See Ehime Japan.

Ehime Japan (1993)
Prefecture, Shikoku.
BT Shikoku
BT Japan
BT Far East
BT Asia
NT Matsuyama Japan

Eichkogel (1978)
Region in NE Austria.
BT Lower Austria
BT Austria
BT Central Europe
BT Europe

Eichstatt
No longer a valid term for GeoRef. See Eichstatt Germany.

Eichstatt Germany (1993)
Town in W central Bavaria.
BT Bavaria Germany
BT Germany
BT Central Europe
BT Europe

Eifel (1978)
Hilly region in W Germany between Rhine River and Moselle River. Northwestern part of Rhenish Schiefergebirge.
UF Eifel Mountains
BT Rhineland-Palatinate Germany
BT Germany
BT Central Europe
BT Europe
SA Rhenish Schiefergebirge

Eifel Mountains
use Eifel

Eifelian (1978)
Europe. Above Emsian, below Givetian.
BT Middle Devonian
BT Devonian
BT Paleozoic
SA Couvinian

eigenvalues (1978)
SA mathematical methods
SA statistical analysis

Eilat Israel
use Elath Israel

Eilenriede (1978)
Forest area E of Hanover.
BT Lower Saxony Germany
BT Germany
BT Central Europe
BT Europe

einsteinium (1989)
Chemical element.
UF Es
BT actinides
BT metals

Eire
use Ireland

ejecta (1978)
SA lapilli
SA mud volcanoes

SA pyroclastics
SA volcanic ash
SA volcaniclastics
SA volcanology

ekanite (1978)
BT ring silicates
BT silicates

Ekaterinburg Russian Federation
use Yekaterinburg Russian Federation

Ekibastuz
No longer a valid term for GeoRef. As of 1993, see Ekibastuz Kazakhstan.

Ekibastuz Kazakhstan (1993)
Town in Pavlodar Oblast in NE Kazakhstan. This term has multiple hierarchies.
BT1 Pavlodar Kazakhstan
BT1 Kazakhstan
BT1 Central Asia
BT1 Asia
BT2 Pavlodar Kazakhstan
BT2 Kazakhstan
BT2 Commonwealth of Independent States

eklogite facies
use eclogite facies

Ekman spiral
As of 1993, no longer a valid term for GeoRef. Usually out-of-scope. Used with ocean circulation.

Ekofisk Field (1989)
North Sea.
BT North Sea
BT European Atlantic
BT North Atlantic
BT Atlantic Ocean
SA oil and gas fields

Ekofisk Formation (1993)
Lower Paleocene? North Sea and Norway.
IN Index age as applicable.
SA Danian
SA North Sea
SA Norway
SA Paleocene

El Alboran Island (1978)
Spanish islet 135 miles E of Gibraltar in SW Mediterranean Sea. Mediterranean Sea was a broader term until 1993 when Mediterranean region replaced it.
BT Mediterranean region
SA Alboran Sea
SA Mediterranean Sea

El Asnam earthquake 1980 (1985)
Epicenter near El Asnam, N Algeria.
BT earthquakes
SA Algeria

El Centro
Valid through 1988. Search in combination with state term. After 1988, use specific city-state term.

El Centro California (1989)
City in Imperial Valley of S California. Before 1989, search El Centro AND California.
CO N324700N324700
W1153300W1153300
BT Imperial County California
BT California
BT United States
SA El Centro earthquake 1940
SA Imperial Valley

El Centro earthquake 1940 (1993)
Imperial Valley, California. Before 1985, also search (El Centro OR Imperial Valley) AND earthquakes AND 1940.
UF Imperial Valley earthquake 1940
BT earthquakes
SA California
SA El Centro California
SA Imperial Valley

El Chichon (1985)
Volcano, N Chiapas, Mexico.
CO N172000N173000
W0931000W0932000
BT Chiapas Mexico
BT Mexico

El Dorado County
Valid through 1988. Search in combination with state term. After 1988, use specific county-state term.

El Dorado County California (1989)
E and E central California. Before 1989, also search El Dorado County AND California.
CO N383000N390500
W1195300W1211000
BT California
BT United States
SA Lake Tahoe

El Kharga
use Kharga Oasis

El Nino (1989)
Periodic oceanic phenomenon accompanied by heavy storms. In northern Peru, a warm equatorial current with low salinity occurring annually in Peruvian regions north of S06°. Every 6 to 8 years, the current reaches S12°, killing numerous fish and plankton, which in turn affects birds, agriculture, etc. The 1982-1983 event was particularly severe.
UF ENSO
SA Pacific Ocean
SA Peru
SA South American Pacific

El Paso
Valid through 1988. Search in combination with state term. After 1988, use specific city-state term.

El Paso County
Valid through 1988. Search in combination with state term. After 1988, use specific county-state term.

El Paso County Colorado (1989)
E central Colorado. Before 1989, also search El Paso County AND Colorado.
CO N383100N390600
W1040100W1050400
BT Colorado
BT United States
NT Colorado Springs Colorado
SA Pikes Peak

El Paso County Texas (1989)
Westernmost county in Texas. Before 1989, also search El Paso County AND Texas.
CO N312000N320000
W1055700W1063700
BT Texas
BT United States
NT El Paso Texas
SA Orogrande Basin

El Paso Group (1989)
SE Arizona, S New Mexico and W Texas. Also search El Paso Formation NOT Argentina.
BT Lower Ordovician
BT Ordovician
BT Paleozoic
SA Arizona
SA New Mexico
SA Texas

El Paso Mountains (1985)
Inyo and Kern counties, S California.
IN Index counties as applicable.
SA California
SA Inyo County California
SA Kern County California

El Paso Texas (1989)
City in extreme W Texas. Before 1989, search El Paso AND Texas.
CO N314500N314500
W1063000W1063000
BT El Paso County Texas
BT Texas
BT United States

El Salvador (1978)
CO N131000N142800
W0873800W0901000
BT Central America
NT Izalco
NT Izalco El Salvador
NT San Salvador El Salvador
SA San Salvador earthquake 1986

El Salvador earthquake 1986
use San Salvador earthquake 1986

Elasmobranchii (1978)
Subclass. Autoposting of this term began in 1978.
BT Chondrichthyes
BT Pisces
BT Vertebrata
BT Chordata
NT Bradyodonti
NT Selachii

elastic constants (1978)
Autoposting of this term began in 1978.
UF constants, elastic
UF elastic moduli
UF modulus of elasticity
NT bulk modulus
NT Poisson's ratio
NT shear modulus
NT Young's modulus
SA deformation
SA elastic properties
SA elasticity
SA geophysics
SA rock mechanics
SA soil mechanics

elastic limit (1978)
UF limit, elastic
SA creep
SA deformation
SA elastic properties
SA elastic strain
SA elasticity
SA fracture strength
SA Hooke's law
SA Hugoniot analysis
SA mechanical properties
SA strain
SA viscoelasticity
SA yield strength

elastic materials (1978)
Before 1978, also search elastic AND materials.
SA deformation
SA elasticity
SA elastoplastic materials

SA elastoviscoplastic materials
SA explosions
SA geophysics
SA hysteresis
SA materials
SA plastic materials
SA rock mechanics
SA viscoelastic materials
SA viscoelasticity

elastic moduli
use elastic constants

elastic properties (1978)
SA deformation
SA elastic constants
SA elastic limit
SA elastic strain
SA elasticity
SA elastodynamic properties
SA geophysics
SA Hooke's law
SA mechanical properties
SA properties
SA seismology
SA stiffness
SA tensile strength
SA thermoelastic properties

elastic strain (1978)
SA creep
SA deformation
SA elastic limit
SA elastic properties
SA elasticity
SA elastodynamic properties
SA engineering geology
SA Hooke's law
SA strain
SA viscoelasticity

elastic waves (1978)
Before 1976, also search seismic waves. Before 1971, also search Earth waves. As of 1989, also search seismic signals. Autoposting of this term began in 1978.
UF Earth waves
UF seismic waves
NT Airy waves
NT body waves
NT long-period waves
NT short-period waves
NT surface waves
SA acoustical waves
SA amplitude
SA amplitude distortion
SA arrival time
SA asthenosphere
SA attenuation
SA BIRPS
SA COCRUST
SA deconvolution
SA deformation
SA distortion
SA earthquakes
SA explosions
SA filters
SA focus
SA foreshocks
SA frequency domain analysis
SA group velocity
SA Hilbert transformations
SA impedance
SA linear distortion
SA mantle
SA Mohorovicic discontinuity
SA moveout
SA non-linear distortion
SA normal moveout
SA phase distortion
SA phase velocity
SA propagation
SA Q
SA relaxation energy
SA seismic methods
SA seismic sources

SA seismology
SA signal distortion
SA traveltime
SA traveltime curves
SA Vibroseis
SA wave absorption
SA wave dispersion
SA waveforms
SA wavelength
SA waves

elasticity (1978)
SA anelasticity
SA bulk modulus
SA compressibility
SA constitutive equations
SA deformation
SA elastic constants
SA elastic limit
SA elastic materials
SA elastic properties
SA elastic strain
SA elastoplastic materials
SA elastoviscoplastic materials
SA engineering geology
SA mechanical properties
SA plasticity
SA Poisson's ratio
SA rigidity
SA rock mechanics
SA shear modulus
SA soil mechanics
SA strain
SA stress
SA viscoelastic materials
SA viscoelasticity
SA yield strength
SA Young's modulus

elastodynamic properties (1985)
SA deformation
SA dynamic properties
SA elastic properties
SA elastic strain
SA properties

elastoplastic materials (1981)
SA elastic materials
SA elasticity
SA elastoviscoplastic materials
SA engineering geology
SA materials
SA plastic materials
SA plasticity

elastoviscoplastic materials (1982)
SA elastic materials
SA elasticity
SA elastoplastic materials
SA engineering geology
SA materials
SA plastic materials
SA plasticity
SA viscoelastic materials
SA viscoelasticity
SA viscoplastic materials
SA viscosity
SA viscous materials

Elat Israel
  use Elath Israel

Elath
  No longer a valid term for GeoRef.
  See Elath Israel.

Elath Israel (1993)
  City at head of Gulf of Aqaba in S Israel. Before 1993, also search Elath. Before 1978, also search Elat and Eilat.
  UF Eilat Israel
  UF Elat Israel
  BT Israel
  BT Middle East
  BT Asia

Elba (1978)

Island in Mediterranean Sea between Corsica and mainland of Italy.
SA Monte Capanne
SA Tuscany Italy

elbaite (1978)
BT ring silicates
BT silicates
SA schorl
SA tourmaline

Elbe River (1978)
  Rises in N Bohemia and flows NW through Hamburg into the North Sea. Also search Elbe.
  IN Index countries as applicable.
  BT Europe
  SA Czechoslovakia
  SA Germany
  SA Hamburg Germany

Elbe Valley (1978)
  Also search Elbe.
  IN Index countries as applicable.
  BT Europe
  SA Czechoslovakia
  SA Germany

Elbert County
  Valid through 1988. Search in combination with state term. After 1988, use specific county-state term.

Elbert County Colorado (1989)
  E central Colorado. Before 1989, also search Elbert County AND Colorado.
  CO N385200N393300
       W1034400W1043900
  BT Colorado
  BT United States

Elbert County Georgia (1989)
  NE Georgia. Before 1989, also search Elbert County AND Georgia.
  CO N335800N341600
       W0823500W0880700
  BT Georgia
  BT United States

Elberton Granite (1985)
  W Georgia.
  BT lower Paleozoic
  BT Paleozoic
  SA Cambrian
  SA Georgia
  SA Ordovician

Elbingerode
  No longer a valid term for GeoRef.
  See Elbingerode Germany.

Elbingerode Germany (1993)
  Town in central Germany.
  BT Saxony-Anhalt Germany
  BT Germany
  BT Central Europe
  BT Europe

Elbistan
  No longer a valid term for GeoRef.
  As of 1993 see Elbistan Turkey.

Elbistan Turkey (1993)
  Town in Maras Province, S central Turkey. Before 1993 also search Elbistan AND Turkey.
  BT Turkey
  BT Middle East
  BT Asia
  SA Anatolia

Elborus
  use Elbrus

Elbrus (1978)

Peak in a N subsidiary spur of the main range of the Caucasus. Highest peak in Europe.
IN Index Asia and Russian Federation and Caucasus as applicable.
UF Elborus
UF Elbrus Mountain
UF Mount Elbrus
UF Mt. Elbrus
SA Asia
SA Caucasus
SA Russian Federation

Elbrus Mountain
  use Elbrus

Elburz (1978)
  Range in N parallel to S shore of Caspian Sea.
  UF Alborz Mountains
  UF Elburz Mountains
  BT Iran
  BT Middle East
  BT Asia
  SA Geirud Formation
  SA Talysh Mountains

Elburz Mountains
  use Elburz

Eldorado
  No longer a valid term for GeoRef.
  As of 1993, see Eldorado Saskatchewan.

Eldorado Saskatchewan (1993)
  Village in NE Saskatchewan. Before 1993, also search Eldorado AND Saskatchewan.
  BT Saskatchewan
  BT Western Canada
  BT Canada

electric currents
  Not a valid term for GeoRef. Usually out-of-scope. Used as electrical currents with aurora or meteorology.

electric exploration
  Not a valid GeoRef index term after 1970. Was used on level 1 in subfile N. Now use electrical logging, electrical methods or electrical surveys.

electric fields
  A valid term through 1974.
  use electrical field

electrical
  A valid term through 1977.

electrical anomalies (1981)
SA anomalies
SA electrical methods
SA electrical surveys
SA geophysical methods
SA geophysical surveys

electrical conductivity (1978)
SA conductivity
SA electrical field
SA geophysical methods
SA resistivity
SA thermal conductivity

electrical currents
  As of 1993, no longer a valid term for GeoRef. Usually out-of-scope. Used with aurora or meteorology.

electrical exploration
  Not a valid GeoRef index term after 1971. Was used on level 1 in subfiles G and N. Now use electrical logging, electrical methods or electrical surveys.

electrical field (1978)

Before 1975, also search electric fields.
UF electric fields
UF field, electrical
SA alternating field demagnetization
SA dielectric constant
SA dielectric properties
SA Earth-current methods
SA electrical conductivity
SA electromagnetic field
SA electromagnetic induction
SA electromagnetic waves
SA induced polarization
SA magnetic field
SA magnetotelluric methods
SA potential field

electrical logging (1978)
  A valid term since 1978. Not a valid term from 1971 through 1977. Before 1978, also search well-logging AND electrical.
  UF logging, electrical
  BT well-logging
  SA apparent resistivity
  SA boreholes
  SA conductivity
  SA electrical methods
  SA electrical surveys
  SA Laterolog
  SA resistivity

electrical methods (1978)
  BT geophysical methods
  SA apparent resistivity
  SA arrays
  SA crosshole methods
  SA dielectric constant
  SA dielectric properties
  SA direct-current methods
  SA electrical anomalies
  SA electrical logging
  SA electrical properties
  SA electrical sounding
  SA electrical surveys
  SA electromagnetic methods
  SA induced polarization
  SA potentiometry
  SA resistivity
  SA Schlumberger methods
  SA self-potential methods

electrical phenomena
  As of 1993, no longer a valid term for GeoRef. Usually out-of-scope. Used with meteorology.

electrical properties (1978)
SA dielectric constant
SA dielectric properties
SA electrical methods
SA electrical surveys
SA electrochemical properties
SA piezoelectric properties
SA properties
SA surface properties

electrical resistivity
  use resistivity

electrical sounding (1982)
SA deep sounding
SA electrical methods
SA electrical surveys
SA geophysical methods
SA geophysical surveys
SA resistivity
SA sounding

electrical surveys (1978)
BT geophysical surveys
BT surveys
SA apparent resistivity
SA direct-current methods
SA electrical anomalies
SA electrical logging
SA electrical methods

SA electrical properties
SA electrical sounding
SA electromagnetic logging
SA electromagnetic methods
SA induced polarization
SA resistivity
SA self-potential methods

electro-osmosis (1981)
Before 1981, also search electroosmosis.
SA geochemistry
SA ground water
SA hydrochemistry
SA osmosis
SA processes

electrochemical properties
(1978)
SA electrical properties
SA electrolytic analysis
SA geochemistry
SA potentiometry
SA properties
SA voltammetry

electrodes (1978)
Used as a general term.
SA electrolysis
SA electrophoresis

electrojet
As of 1993, no longer a valid term for GeoRef. Usually out-of-scope. Used with ionosphere and aurora.

electrolysis (1978)
SA dissociation
SA electrodes
SA electrolytic analysis
SA geochemistry
SA polarography
SA processes
SA solution

electrolytes (1978)
SA anions
SA electrolytic analysis
SA geochemistry
SA ion exchange
SA ions

electrolytic analysis (1985)
SA analysis
SA chemical analysis
SA electrochemical properties
SA electrolysis
SA electrolytes
SA electrophoresis
SA voltammetry
SA wet methods

electromagnetic
A valid term through 1977.

electromagnetic field (1978)
UF field, electromagnetic
SA electrical field
SA electromagnetic induction
SA geophysical methods
SA geophysical surveys
SA magnetic field
SA magnetotelluric methods
SA potential field
SA synchrotron radiation

electromagnetic induction
(1981)
Also search induction AND electromagnetic.
SA electrical field
SA electromagnetic field
SA electromagnetic logging
SA electromagnetic methods
SA magnetic field

electromagnetic logging (1978)
Before 1978, also search well-logging AND electromagnetic.
UF logging, electromagnetic
BT well-logging

SA electrical surveys
SA electromagnetic induction
SA electromagnetic methods
SA electromagnetic surveys
SA geophysical methods
SA geophysical surveys
SA magnetic logging
SA SQUID
SA transient methods

electromagnetic methods (1978)
UF electromagnetic prospecting
BT geophysical methods
SA broad-band spectra
SA crosshole methods
SA electrical methods
SA electrical surveys
SA electromagnetic induction
SA electromagnetic logging
SA electromagnetic survey maps
SA electromagnetic surveys
SA frequency domain analysis
SA Green function
SA impedance
SA induction
SA laser methods
SA magnetotelluric methods
SA magnetotelluric surveys
SA radar methods
SA SQUID
SA synchrotron radiation
SA synchrotrons
SA transient methods
SA variometers

electromagnetic prospecting
use electromagnetic methods

electromagnetic radiation
(1978)
SA electromagnetic waves
SA gamma rays
SA interplanetary space
SA luminescence
SA magnetic field
SA particles
SA radiation
SA radio wave methods
SA radioactivity
SA solar wind
SA synchrotrons
SA X-rays

electromagnetic seismographs
(1981)
BT seismographs

electromagnetic survey maps
(1985)
BT maps
SA electromagnetic methods
SA electromagnetic surveys
SA geophysical survey maps

electromagnetic surveys (1978)
BT geophysical surveys
BT surveys
SA electromagnetic logging
SA electromagnetic methods
SA electromagnetic survey maps
SA magnetic field
SA magnetotelluric surveys
SA RADAM
SA transient methods

electromagnetic waves (1978)
SA electrical field
SA electromagnetic radiation
SA magnetic field
SA meteorology
SA propagation
SA radio-wave methods
SA waves

electron content
As of 1993, no longer a valid term for GeoRef. Usually out-of-scope. Used with ionosphere.

electron diffraction analysis
(1978)
Before 1971, also search electron-diffraction analysis.
UF electron-diffraction analysis
SA analysis
SA diffractograms
SA electrons
SA neutron diffraction analysis
SA X-ray diffraction analysis

electron microprobe
use electron probe

electron microscopy (1978)
Used for methodology. For data, use electron microscopy data.
IN May be used for methodology, not for data.
UF microscopy, electron
SA chemical analysis
SA clay mineralogy
SA crystal chemistry
SA crystal growth
SA crystallography
SA differential thermal analysis
SA electron microscopy data
SA paleobotany
SA paleontology
SA scanning electron microscopy
SA SEM data
SA spectroscopy
SA structural analysis
SA TEM data
SA thermal analysis
SA thin sections
SA transmission electron microscopy
SA wave dispersion
SA X-ray analysis

electron microscopy data (1978)
Used to distinguish electron microscopy as a method from the data.
SA data
SA electron microscopy
SA SEM data
SA TEM data

electron paramagnetic resonance
(1978)
As of 1978, term is used only for methodology. For data, see EPR spectra.
UF electron spin resonance
BT spectroscopy
SA analysis
SA chemical analysis
SA EPR spectra
SA nuclear magnetic resonance
SA relative age
SA resonance

electron paramagnetic resonance spectra
use EPR spectra

electron probe (1978)
Since 1978, term has been restricted to methodology. For data, use electron probe data. Before 1971, also search electron-probe analysis.
UF electron microprobe
UF electron probe analysis
UF electron-probe analysis
UF microprobe, electron
UF probe, electron
SA analysis
SA chemical analysis
SA electron probe data
SA ion probe
SA neutron probe
SA spectroscopy

electron probe analysis
use electron probe

electron probe data (1978)
Before 1978, search electron probe.
UF electron probe spectra
SA data
SA electron probe
SA spectroscopy

electron probe spectra
use electron probe data

electron spin resonance
use electron paramagnetic resonance

electron-diffraction analysis
Not a valid GeoRef index term after 1970. Was used on level 1 in subfile N.
use electron diffraction analysis

electron-probe analysis
Not a valid GeoRef index term after 1970. Was used on level 1 in subfile N.
use electron probe

electronic data processing
use data processing

electrons (1978)
Used as a general term.
SA electron diffraction analysis
SA neutrons
SA particles
SA protons

electrophoresis (1989)
UF cataphoresis
SA chemical analysis
SA electrodes
SA electrolytic analysis
SA geochemistry

electrum (1993)
BT alloys
SA gold
SA precious metals
SA silver

elementary geology (1978)
Used for simplified, technical, geological discussions written for the lay person. Before 1978, search popular and elementary geology.
SA education
SA geology
SA popular geology

elementary school (1978)
SA college-level education
SA education
SA high school
SA junior high school

elements
No longer a valid term for GeoRef. See chemical elements, native elements, or tectonic elements.

Elephant Moraine (1989)
On the East Antarctic ice sheet.
CO S761735S761735
E1572005E1572005
BT Antarctica
SA Antarctic ice sheet
SA EETA 79001
SA Elephant Moraine Meteorites

Elephant Moraine Meteorite
use Elephant Moraine Meteorites

Elephant Moraine Meteorites
(1989)
Due to abbreviation of meteorite name, (for example, EETA 79004) also search EETA in combination with number.

IN Index types of meteorites as applicable. Also index names of individual meteorites if available.
UF EETA
UF Elephant Moraine Meteorite
BT meteorites
NT EETA 79001
SA achondrites
SA Antarctica
SA chondrites
SA Elephant Moraine

Elephantidae (1978)
Family. Autoposting of Eutheria and Theria to this term began in 1989.
BT Elephantoidea
BT Proboscidea
BT Eutheria
BT Theria
BT Mammalia
BT Tetrapoda
BT Vertebrata
BT Chordata
NT Elephas
NT Mammuthus
NT Palaeoloxodon naumanni
NT Stegodon

Elephantoidea (1981)
Suborder. Autoposting of Eutheria and Theria to this term began in 1989.
UF Euelephantoidea
BT Proboscidea
BT Eutheria
BT Theria
BT Mammalia
BT Tetrapoda
BT Vertebrata
BT Chordata
NT Elephantidae

Elephas (1978)
Genus. Autoposting of Eutheria and Theria to this term began in 1989.
BT Elephantidae
BT Elephantoidea
BT Proboscidea
BT Eutheria
BT Theria
BT Mammalia
BT Tetrapoda
BT Vertebrata
BT Chordata
NT Elephas antiquus

Elephas antiquus (1978)
Species. Autoposting of Eutheria and Theria to this term began in 1989.
BT Elephas
BT Elephantidae
BT Elephantoidea
BT Proboscidea
BT Eutheria
BT Theria
BT Mammalia
BT Tetrapoda
BT Vertebrata
BT Chordata

elevation
use altitude

ELF
As of 1981, no longer a valid term for GeoRef. See electromagnetic waves.

Elisavetgrad
Not a valid term for GeoRef. See Kirovograd Ukraine.

Elk County
Valid through 1988. Search in combination with state term. After 1988, use specific county-state term.

Elk County Kansas (1989)
SE Kansas. Before 1989, also search Elk County AND Kansas.
CO N371700N373000 W0955700W0963300
BT Kansas
BT United States

Elk County Pennsylvania (1989)
N central Pennsylvania. Before 1989, also search Elk County AND Pennsylvania.
CO N411700N413800 W0781500W0790700
BT Pennsylvania
BT United States

Elk Hills Field (1978)
U. S. naval oil reservation in S California. Autoposting of broader terms began in 1989. Also search Elk Hills.
BT Kern County California
BT California
BT United States
SA oil and gas fields

Elk Lake (1989)
IN Index states or provinces as applicable.
SA Alberta
SA Minnesota
SA Ontario
SA Washington

Elk Mountains (1978)
Range of Rocky Mountains in W central Colorado.
IN Index counties and U. S. Rocky Mountains or regions as applicable.
SA Colorado
SA U. S. Rocky Mountains

Elk Point Basin (1978)
IN Index provinces as applicable.
BT Canada
SA Alberta
SA Manitoba
SA Saskatchewan

Elk Point Group (1978)
Includes Ashern Formation, Elm Point Formation, Winnipegosis Formation, and Prairie Evaporite.
BT Middle Devonian
BT Devonian
BT Paleozoic
SA Alberta
SA Manitoba
SA Montana
SA North Dakota
SA Prairie Evaporite
SA Saskatchewan
SA South Dakota
SA Winnipegosis Formation

Elk River (1985)
Tributary of the Verdigris River, SE Kansas. Also a tributary of the Kanawha River, central West Virginia, and several other rivers.
IN Index states or regions as applicable.
SA Kansas
SA United States
SA West Virginia

Elkhorn Mountains Volcanics (1989)
SW Montana.
BT Upper Cretaceous

BT Cretaceous
BT Mesozoic
SA Montana

Elko County
Valid through 1988. Search in combination with state term. After 1988, use specific county-state term.

Elko County Nevada (1989)
NE Nevada. Before 1989, also search Elko County AND Nevada.
CO N400700N420000 W1140500W1170000
BT Nevada
BT United States
NT Carlin Nevada
NT Pequop Mountains
SA Independence Mountains
SA Pilot Range
SA Ruby Mountains

Ellef Ringnes Island (1978)
One of the Sverdrup Islands in Franklin District.
CO N773000N790000 W0980000W1060000
BT Franklin District Northwest Territories
BT Northwest Territories
BT Western Canada
BT Canada
SA Christopher Formation

ellenbergerite (1993)
BT orthosilicates
BT silicates

Ellenburger Group (1989)
W and central Texas, SE New Mexico.
BT Lower Ordovician
BT Ordovician
BT Paleozoic
SA New Mexico
SA Texas

Ellensburg Formation (1985)
Central Washington.
BT lower Pliocene
BT Pliocene
BT Neogene
BT Tertiary
BT Cenozoic
SA Washington

Ellesmere Island (1978)
W of NW Greenland in Franklin District.
CO N760000N840000 W0600000W0920000
BT Franklin District Northwest Territories
BT Northwest Territories
BT Western Canada
BT Canada
NT Tanquary Fiord
SA Nares Strait
SA Otto Fjord

Ellice Islands
use Tuvalu

Elliot Formation (1989)
Upper Triassic to Lower Jurassic in South Africa. Also occurs in Maine, Ontario and Queensland.
IN Index ages as applicable.
SA Australia
SA Lower Jurassic
SA Maine
SA Ontario
SA Queensland Australia
SA South Africa
SA Upper Triassic

Elliot Lake
No longer a valid term for GeoRef. As of 1993, see Elliot Lake Ontario.

Elliot Lake Ontario (1993)
Town in S central Ontario. Before 1993, also search Elliot Lake AND Ontario.
BT Algoma District Ontario
BT Ontario
BT Eastern Canada
BT Canada

Elliott County
Valid through 1988. Search in combination with state term. After 1988, use specific county-state term.

Elliott County Kentucky (1989)
E Kentucky. Before 1989, also search Elliott County AND Kentucky.
CO N375900N381500 W0825400W0831800
BT Kentucky
BT United States

ellipticity (1978)
General property.
SA geodesy
SA geometry

Ellis Bay Formation (1985)
Anticosti Island, Quebec.
BT Paleozoic
SA Anticosti Island
SA Lower Silurian
SA Ordovician
SA Quebec
SA Silurian
SA Upper Ordovician

Ellis County
Valid through 1988. Search in combination with state term. After 1988, use specific county-state term.

Ellis County Kansas (1989)
W central Kansas. Before 1989, also search Ellis County AND Kansas.
CO N384000N390700 W0990400W0993500
BT Kansas
BT United States

Ellis County Oklahoma (1989)
NW Oklahoma. Before 1989, also search Ellis County AND Oklahoma.
CO N355100N363600 W0992200W1000000
BT Oklahoma
BT United States

Ellis County Texas (1989)
N Texas. Before 1989, also search Ellis County AND Texas.
CO N320400N323400 W0962300W0970500
BT Texas
BT United States

Ellsworth County
Valid through 1988. Search in combination with state term. After 1988, use specific county-state term.

Ellsworth County Kansas (1989)
Central Kansas. Before 1989, also search Ellsworth County AND Kansas.
CO N383000N385000 W0975500W0983000
BT Kansas
BT United States

Ellsworth Highland
　use Ellsworth Land

Ellsworth Land　(1978)
　High plateau just S and SW of the Antarctic Peninsula.
　UF　Ellsworth Highland
　UF　James W. Ellsworth Land
　BT　Antarctica
　SA　Jones Mountains

Ellsworth Mountains　(1978)
　Range S of Ellsworth Land and SW of the Antarctic Peninsula.
　BT　Antarctica

ellsworthite
　use betafite

Elmore County
　Valid through 1988. Search in combination with state term. After 1988, use specific county-state term.

Elmore County Alabama　(1989)
　E central Alabama. Before 1989, also search Elmore County AND Alabama.
　CO　N322600N324600
　　　 W0855300W0862700
　BT　Alabama
　BT　United States

Elmore County Idaho　(1989)
　SW Idaho. Before 1989, also search Elmore County AND Idaho.
　CO　N424600N440700
　　　 W1145900W1161800
　BT　Idaho
　BT　United States
　SA　Sawtooth Range

Elmworth Field　(1989)
　Gas field. NW Alberta.
　IN　Index province or region as applicable.
　SA　Alberta
　SA　oil and gas fields

elongate minerals　(1978)
　SA　lineation
　SA　minerals
　SA　structural analysis

Elphidium　(1978)
　Genus. Autoposting of microfossils and Protista to this term began in 1990. This term has multiple hierarchies.
　BT1　Rotaliacea
　BT1　Rotaliina
　BT1　foraminifera
　BT1　Protista
　BT1　Invertebrata
　BT2　Rotaliacea
　BT2　Rotaliina
　BT2　foraminifera
　BT2　Protista
　BT2　microfossils
　NT　Elphidium excavatum

Elphidium excavatum　(1989)
　Species. Autoposting of microfossils and Protista to this term began in 1990. This term has multiple hierarchies.
　BT1　Elphidium
　BT1　Rotaliacea
　BT1　Rotaliina
　BT1　foraminifera
　BT1　Protista
　BT1　Invertebrata
　BT2　Elphidium
　BT2　Rotaliacea
　BT2　Rotaliina
　BT2　foraminifera
　BT2　Protista
　BT2　microfossils

Elsinore Fault　(1978)
　Extending from a point E of Los Angeles to a point E of San Diego.
　BT　California
　BT　United States

Elsterian　(1985)
　Middle Pleistocene glacial stage in NW Europe. Above Menapian and below Saalian. Approximately equivalent to Mindel and Illinoian.
　BT　middle Pleistocene
　BT　Pleistocene
　BT　Quaternary
　BT　Cenozoic
　SA　Central Polish Glaciation
　SA　Illinoian
　SA　Mindel

Eltanin fracture zone　(1989)
　South Pacific.
　CO　S600000S500000
　　　 W1000000W1500000
　BT　Pacific Ocean
　SA　South Pacific

eluvium　(1978)
　BT　clastic sediments
　BT　sediments
　SA　alluvium
　SA　dunes
　SA　soils
　SA　terrigenous materials

Elvas
　No longer a valid term for GeoRef. As of 1993 see Elvas Portugal.

Elvas Portugal　(1993)
　City near the Spanish border in S Portugal. Before 1993 also search Elvas AND Portugal.
　BT　Portugal
　BT　Iberian Peninsula
　BT　Southern Europe
　BT　Europe

Ely
　Valid through 1988. Search in combination with state term. After 1988, use specific city-state term.

Ely Limestone　(1989)
　Upper Mississippian to Middle Pennsylvanian. E Nevada and W Utah.
　IN　Index ages as applicable.
　BT　Carboniferous
　BT　Paleozoic
　SA　Nevada
　SA　Pennsylvanian
　SA　Upper Mississippian
　SA　Utah

Ely Minnesota　(1989)
　City in NE Minnesota. Before 1989, search Ely AND Minnesota.
　CO　N475300N475300
　　　 W0915200W0915200
　BT　Saint Louis County Minnesota
　BT　Minnesota
　BT　United States

Ely Nevada　(1989)
　City in E Nevada. Before 1989, search Ely AND Nevada.
　CO　N391500N391500
　　　 W1145300W1145300
　BT　White Pine County Nevada
　BT　Nevada
　BT　United States

Ely Springs Dolomite　(1989)
　Middle and Upper Ordovician. Great Basin region of SE California, E Nevada and W Utah.
　BT　Ordovician
　BT　Paleozoic

　SA　California
　SA　Nevada
　SA　Utah

Elysium　(1993)
　Volcanic province on Mars which contains Elysium Mons.
　IN　Index specific features if applicable.
　BT　Mars
　SA　Elysium Mons

Elysium Mons　(1989)
　Volcanic feature on Mars.
　BT　Mars
　SA　Elysium

Emba
　No longer a valid term for GeoRef. As of 1993, see Emba Kazakhstan.

Emba Field　(1981)
　Oil and gas field in Aktyubinsk Oblast in NW Kazakhstan. This term has multiple hierarchies.
　CO　N470000N490000
　　　 E0560000E0523000
　BT1　Aktyubinsk Kazakhstan
　BT1　Kazakhstan
　BT1　Central Asia
　BT1　Asia
　BT2　Aktyubinsk Kazakhstan
　BT2　Kazakhstan
　BT2　Commonwealth of Independent States
　SA　oil and gas fields

Emba Kazakhstan　(1993)
　Town in Aktyubinsk Oblast in NW Kazakhstan. This term has multiple hierarchies.
　BT1　Aktyubinsk Kazakhstan
　BT1　Kazakhstan
　BT1　Central Asia
　BT1　Asia
　BT2　Aktyubinsk Kazakhstan
　BT2　Kazakhstan
　BT2　Commonwealth of Independent States

Emba River　(1978)
　Flows into NE Caspian Sea. There are oil fields in its lower course. Before 1978, also search Emba. This term has multiple hierarchies.
　BT1　Kazakhstan
　BT1　Central Asia
　BT1　Asia
　BT2　Kazakhstan
　BT2　Commonwealth of Independent States

embankments　(1978)
　SA　berms
　SA　dams
　SA　geomembranes
　SA　geotextiles
　SA　highways
　SA　levees
　SA　railroads
　SA　reinforced materials
　SA　reservoirs
　SA　retaining walls
　SA　revetments
　SA　slope stability
　SA　slopes

embayments　(1993)
　SA　bays
　SA　chemical weathering
　SA　coastlines
　SA　corrosion
　SA　etching
　SA　fjords
　SA　folds
　SA　inlets
　SA　shore features

　SA　solution
　SA　submergence

embrechite　(1981)
　BT　migmatites
　BT　metamorphic rocks

Embrithopoda　(1978)
　Order. Autoposting of Eutheria and Theria to this term began in 1989.
　BT　Eutheria
　BT　Theria
　BT　Mammalia
　BT　Tetrapoda
　BT　Vertebrata
　BT　Chordata

emerald　(1978)
　Variety of beryl.
　UF　smaragd
　BT　ring silicates
　BT　silicates
　SA　beryl
　SA　gems

emergent coastlines　(1989)
　Autoposting of shore features to this term began in 1990.
　BT　coastlines
　BT　shore features
　SA　landform evolution

Emery
　Valid through 1988. Search in combination with state term. After 1988, use specific city-state term.

Emery County
　Valid through 1988. Search in combination with state term. After 1988, use specific county-state term.

Emery County Utah　(1989)
　Central Utah. Before 1989, also search Emery County AND Utah.
　CO　N383000N394500
　　　 W1100000W1112000
　BT　Utah
　BT　United States
　NT　Emery Utah

Emery Utah　(1989)
　Town in central Utah. Before 1989, search Emery AND Utah.
　CO　N385600N385600
　　　 W1111400W1111400
　BT　Emery County Utah
　BT　Utah
　BT　United States

Emilia-Romagna
　As of 1993, no longer a valid term for GeoRef.
　use Emilia-Romagna Italy

Emilia-Romagna Italy　(1993)
　Autonomous region in N Italy. Before 1993, also search Emilia or Emilia-Romagna.
　CO　N434500N451000
　　　 E0125000E0091000
　UF　Emilia-Romagna
　BT　Italy
　BT　Southern Europe
　BT　Europe
　NT　Bologna Italy
　NT　Modena Italy
　NT　Parma Italy
　NT　Piacenza Italy
　NT　Romagna
　NT　Taro Valley
　SA　Po River
　SA　Po Valley
　SA　San Marino
　SA　Vigarano Meteorite

Emiliana huxleyi
　use Emiliania huxleyi

Emiliania huxleyi (1989)
 Species. Autoposting of thallophytes and microfossils began in 1990. This term has multiple hierarchies.
 UF Emiliana huxleyi
 BT1 Coccolithophoraceae
 BT1 algae
 BT1 thallophytes
 BT1 Plantae
 BT2 Coccolithophoraceae
 BT2 algae
 BT2 microfossils

emission
 A valid term through 1977 used to denote the method. After 1977, use emission spectroscopy.

emission spectra (1989)
 Used for data. For methodology, use emission spectroscopy.
 BT spectra
 SA emission spectroscopy
 SA spectroscopy

emission spectroscopy (1978)
 Before 1978, also search spectroscopy AND emission. As of 1989, used only for methodology. For data, use emission spectra.
 BT spectroscopy
 SA absorption
 SA analysis
 SA chemical analysis
 SA emission spectra
 SA inductively coupled plasma methods
 SA plasma emission spectroscopy

emission, thermal
 use thermal emission

emissions
 As of 1981, no longer a valid term for GeoRef.

emissions, acoustical
 use acoustical emissions

Emperor Seamount Chain
 use Emperor Seamounts

Emperor Seamounts (1978)
 SE of Kamchatka Peninsula. Autoposting of Pacific Ocean to this term began in 1990.
 UF Emperor Seamount Chain
 BT West Pacific
 BT Pacific Ocean
 SA DSDP Site 192
 SA DSDP Site 430
 SA DSDP Site 431
 SA DSDP Site 432
 SA DSDP Site 433

Empire Formation (1985)
 In Missoula Group of Belt Supergroup. W central Montana.
 BT middle Proterozoic
 BT Proterozoic
 BT upper Precambrian
 BT Precambrian
 SA Belt Supergroup
 SA Missoula Group
 SA Montana

emplacement (1978)
 SA assimilation
 SA contamination
 SA hybridization
 SA igneous rocks
 SA intrusions
 SA magmas
 SA plutons
 SA salt tectonics

emplectite (1978)
 BT sulfobismuthites

 BT sulfosalts

Emporia
 Valid through 1988. Search in combination with state term. After 1988, use specific city-state term.

Emporia Kansas (1989)
 City in E central Kansas. Before 1989, search Emporia AND Kansas.
 CO N382400N382400 W0773300W0773300
 BT Lyon County Kansas
 BT Kansas
 BT United States

Ems River (1978)
 Flows into the North Sea. Also search Ems.
 IN Index German states as applicable.
 BT Germany
 BT Central Europe
 BT Europe
 SA Lower Saxony Germany
 SA North Rhine-Westphalia Germany
 SA Weser-Ems

Emsian (1978)
 Europe. Above Siegenian, below Eifelian.
 BT Lower Devonian
 BT Devonian
 BT Paleozoic

Emsland (1978)
 Swampy region between Ems River and Netherlands border in Lower Saxony.
 BT Lower Saxony Germany
 BT Germany
 BT Central Europe
 BT Europe

en echelon faults (1978)
 Before 1978, also search faults AND en echelon.
 BT faults
 SA duplexes
 SA pull-apart basins
 SA systems

en echelon folds (1978)
 Before 1978, search folds AND en echelon.
 BT folds

ENADIMSA (1981)
 BT survey organizations
 BT government agencies
 SA Spain

enargite (1978)
 BT sulfarsenates
 BT sulfosalts
 SA copper
 SA copper minerals
 SA copper ores

Enari
 Not a valid term for GeoRef. See Inari Finland.

Enceladus
 use Enceladus Satellite

Enceladus Satellite (1985)
 One of the satellites of Saturn. Before 1985, also search Enceladus AND Saturn.
 UF Enceladus
 SA icy satellites
 SA satellites
 SA Saturn
 SA Voyager Program

enclaves
 Not a valid term for GeoRef. For petrology, see inclusions for specific type.

Encounter Bay (1978)
 Inlet of Indian Ocean SE of Adelaide in SE South Australia.
 BT South Australia
 BT Australia
 BT Australasia

encroachment (ground water)
 use salt-water intrusion

encrustations (1989)
 UF incrustations
 SA algal flora
 SA bryozoans
 SA corals
 SA indicators
 SA minerals
 SA paleoclimatology
 SA stratigraphy

Encruzilhada
 Not a valid term for GeoRef. See Encruzilhada do Sul Brazil.

Encruzilhada do Sul
 No longer a valid term for GeoRef. See Encruzilhada do Sul Brazil.

Encruzilhada do Sul Brazil (1993)
 City in SE Rio Grande do Sul. Before 1993, also search Encruzilhada or Encruzilhada do Sul.
 BT Rio Grande do Sul Brazil
 BT Brazil
 BT South America

Endako
 No longer a valid term for GeoRef. See Endako British Columbia.

Endako British Columbia (1993)
 Village in central British Columbia. Molybdenum mining center.
 BT British Columbia
 BT Western Canada
 BT Canada

endellionite
 use bournonite

endellite
 use halloysite

endemic taxa (1978)
 UF endemism
 UF taxa, endemic
 SA biogeography
 SA ecology
 SA paleoecology

endemism
 use endemic taxa

enderbite (1978)
 BT granites
 BT plutonic rocks
 BT igneous rocks

Enderby Land (1978)
 Semicircular projection of land on the Indian Ocean side claimed by Australia.
 BT Antarctica
 NT Prince Olav Coast
 SA Napier Complex
 SA Queen Maud Land

Endicott Group (1989)
 N Alaska. Upper Devonian to Mississippian. Includes Kayak Shale and Kekiktuk Conglomerate.
 IN Index ages as applicable.
 BT Paleozoic
 SA Alaska
 SA Kayak Shale

 SA Kekiktuk Conglomerate
 SA Mississippian
 SA North Slope
 SA Upper Devonian

Endicott Mountains (1989)
 N central Alaska. Central part of Brooks Range. Autoposting of Brooks Range began in 1993.
 CO N670000N684500 W1490000W1550000
 BT Brooks Range
 BT Northern Alaska
 BT Alaska
 BT United States

endogene processes (1978)
 Before 1981, also search hypogene processes.
 UF hypogene processes
 SA mineral deposits, genesis
 SA processes

endolithic taxa (1989)
 Organisms that live in minute burrows in corals, shells or reef rock.
 UF endoliths
 SA algae
 SA fungi

endoliths
 use endolithic taxa

endothermic reactions (1978)
 Before 1978, also search endothermic.
 SA chemical reactions
 SA DTA data
 SA stoichiometry
 SA thermal analysis

Endothyra (1978)
 Genus. Autoposting of microfossils and Protista to this term began in 1990. This term has multiple hierarchies.
 BT1 Fusulinina
 BT1 foraminifera
 BT1 Protista
 BT1 Invertebrata
 BT2 Fusulinina
 BT2 foraminifera
 BT2 Protista
 BT2 microfossils

energy (1978)
 Used as a general term.
 SA activation energy
 SA energy sources
 SA fission
 SA free energy
 SA geothermal energy
 SA hydroelectric energy
 SA kinetics
 SA new energy sources
 SA nuclear energy
 SA relaxation energy
 SA seismic energy
 SA solar energy
 SA tidal energy
 SA wind energy

energy conservation (1981)
 SA conservation
 SA environmental geology

energy sources (1978)
 Used as a grouper for energy-related terms such as petroleum, natural gas, coal, geothermal energy, and uranium ores.
 UF sources, energy
 SA coal
 SA economic geology
 SA energy
 SA gas sands
 SA geothermal energy
 SA heat flow
 SA heat storage

SA hydroelectric energy
SA natural gas
SA new energy sources
SA nuclear energy
SA oil sands
SA oil shale
SA petroleum
SA potential deposits
SA power plants
SA resources
SA solar energy
SA tidal energy
SA tight sands
SA water resources
SA wind energy

Enewetak
  use Enewetak Atoll

Enewetak Atoll   (1985)
  Part of U. S. Trust Territory of the Pacific Islands. Before 1985, search Eniwetok Atoll.
  UF Enewetak
  UF Eniwetok
  UF Eniwetok Atoll
  BT Marshall Islands
  BT Micronesia
  BT Oceania

engineering geology   (1978)
  Used for geology as applied to engineering practice, especially mining and civil engineering.
  UF geologic engineering
  SA accelerograms
  SA accelerometers
  SA airfields
  SA anchors
  SA artificial islands
  SA aseismic design
  SA asperities
  SA backfill
  SA berms
  SA blasting
  SA blowouts
  SA breakwaters
  SA bridges
  SA civil engineering
  SA clays
  SA cohesionless materials
  SA cohesive materials
  SA compactness
  SA concrete
  SA conductive materials
  SA cone penetration tests
  SA consistent materials
  SA consolidometer tests
  SA continuous materials
  SA cryopedology
  SA cyclic loading
  SA damage
  SA dams
  SA deep drilling
  SA deformation
  SA design
  SA discontinuous materials
  SA dredging
  SA drilling
  SA ductile deformation
  SA earthquakes
  SA elastic strain
  SA elasticity
  SA elastoplastic materials
  SA elastoviscoplastic materials
  SA engineering geology maps
  SA environmental geology
  SA excavations
  SA explosions
  SA footings
  SA foundations
  SA fractures
  SA frost action
  SA frozen ground
  SA gas storage
  SA geodesy

SA geologic hazards
SA geology
SA geomembranes
SA geomorphology
SA geophysical methods
SA geophysical surveys
SA geophysics
SA geotextiles
SA granular materials
SA ground water
SA grouting
SA highways
SA hydraulics
SA hydrology
SA impact statements
SA induced earthquakes
SA industry
SA injection
SA laboratory studies
SA land subsidence
SA land use
SA landfills
SA landslides
SA layered materials
SA legislation
SA linear materials
SA liquefaction potential
SA liquid waste
SA load pressure
SA load tests
SA maps
SA marine drilling
SA marine installations
SA materials
SA mechanical properties
SA microcracks
SA mining geology
SA non-linear materials
SA nuclear explosions
SA nuclear facilities
SA oil storage
SA overconsolidated materials
SA penetration
SA penetration tests
SA permafrost
SA permeability
SA petroleum engineering
SA photogeology
SA photogrammetry
SA plastic deformation
SA plastic materials
SA platforms
SA Poisson's ratio
SA pore pressure
SA pumping
SA quick clay
SA quicksand
SA radioactive waste
SA radioactivity
SA railroads
SA regulations
SA reinforced materials
SA remote sensing
SA reservoirs
SA revetments
SA road tests
SA rock mechanics
SA safety
SA sand boils
SA sealing
SA sedimentation
SA seepage
SA seismic response
SA seismic risk
SA seismic zoning
SA settlement
SA shear stress
SA shorelines
SA site exploration
SA slope stability
SA soft rocks
SA soil mechanics
SA soil-structure interface
SA soils
SA solid waste

SA solifluction
SA stability
SA strain
SA stressmeters
SA structural analysis
SA structures
SA submarine installations
SA surface waves
SA tensile strength
SA testing
SA thermal properties
SA thermal regime
SA thixotropy
SA tunnel boring machines
SA tunnels
SA underground installations
SA uniaxial tests
SA vane tests
SA viscoelastic materials
SA viscoplastic materials
SA volcanic risk
SA waste disposal
SA water storage
SA water wells
SA waterways
SA weak rocks
SA weathering
SA wells

engineering geology maps   (1982)
  Before 1982, also search engineering geology AND maps.
  BT maps
  NT geotechnical maps
  SA engineering geology
  SA geophysical surveys

engineering properties   (1978)
  SA durability
  SA foundations
  SA highways
  SA lateral loading
  SA permafrost
  SA properties
  SA railroads
  SA sediments
  SA soil mechanics
  SA soils
  SA stiffness

engineering, civil
  use civil engineering

engineering, petroleum
  use petroleum engineering

England   (1978)
  Welsh Borderland is considered a narrower term as of 1981.
  CO N500000N554500
     E001300W0063000
  BT Great Britain
  BT United Kingdom
  BT Western Europe
  BT Europe
  NT Avon England
  NT Bedfordshire England
  NT Berkshire England
  NT Birmingham England
  NT Cambridge England
  NT Cambridgeshire England
  NT Cheshire England
  NT Cornwall England
  NT Cotswold Hills
  NT Cumbria England
  NT Derbyshire England
  NT Devonshire England
  NT Dorset England
  NT Dudley England
  NT Durham England
  NT East Anglia
  NT East Midlands
  NT Essex England
  NT Gloucestershire England
  NT Hampshire England
  NT Hertfordshire England

NT Howgill Fells
NT Humberside England
NT Ingleton England
NT Isle of Wight England
NT Kent England
NT Lancashire England
NT Leicestershire England
NT Lincolnshire England
NT Liverpool Bay
NT London Basin
NT London England
NT Malvern England
NT Malvern Hills
NT Manchester England
NT Morecambe Bay
NT Newcastle England
NT Northamptonshire England
NT Northumberland England
NT Nottinghamshire England
NT Oxfordshire England
NT Pennines
NT Shropshire England
NT Somerset England
NT South-West England
NT Staffordshire England
NT Surrey England
NT Sussex England
NT The Weald
NT Warwickshire England
NT Welsh Borderland
NT Wessex Basin
NT Wiltshire England
NT Wolverhampton England
NT Worcestershire England
NT Yorkshire England
SA Aberystwyth Grits
SA Barton Beds
SA Barwell Meteorite
SA Bembridge Marls
SA Birmingham University
SA Borrowdale Volcanic Series
SA Bracklesham Beds
SA Brandon
SA Bristol Channel
SA British Museum
SA Cadeby Formation
SA Cambridge University
SA Carnmenellis Granite
SA Cheviot Hills
SA Esk Trough
SA Gibraltar Point
SA Great Oolite Series
SA Great Scar Limestone
SA Hampshire Basin
SA Kimmeridge Clay
SA Lincolnshire Limestone
SA London Clay
SA Lower Greensand
SA Mercia Mudstone
SA Mersey River
SA Mersey Valley
SA Midlands
SA New Red Sandstone
SA North Sea Coast
SA North Sea region
SA Oxford Clay
SA Scarborough Formation
SA Severn Estuary
SA Severn Valley
SA Shap Granite
SA Sherwood Sandstone
SA Skiddaw Slates
SA Southern Uplands
SA Speeton Clay
SA Tamar Estuary
SA Tamar Valley
SA Thames Estuary
SA Thames River
SA Trent Valley
SA Tyne River
SA Weald Clay
SA Wenlock Limestone
SA Wye Valley

English
   As if 1993, no longer a valid term for GeoRef. Used to indicate language of document.

English Channel (1978)
   Strait between England and France connecting Atlantic Ocean with the North Sea.
   CO N480000N510000
      E0020000W0060000
   BT European Atlantic
   BT North Atlantic
   BT Atlantic Ocean
   SA Bay of Saint-Michel
   SA Bay of the Seine

English Channel Islands (1981)
   British group of islands in the English Channel off the W coast of Normandy. Before 1981, "Channel Islands" was used for both these and the Santa Barbara Islands off California. Before 1981, search Channel Islands AND England.
   CO N491000N494500
      W0020000W0024500
   BT United Kingdom
   BT Western Europe
   BT Europe
   SA Channel Islands

English Lake District
   use Lake District

English River Belt (1981)
   Subprovince of Superior Province, Canadian Shield. E Manitoba and W Ontario. Before 1981, also search English River Subprovince.
   CO N500000N520000
      W0860000W0970000
   UF English River Subprovince
   BT Superior Province
   BT Canadian Shield
   BT North America
   SA Manitoba
   SA Ontario

English River Subprovince
   use English River Belt

enhanced oil recovery
   use enhanced recovery

enhanced recovery (1985)
   UF enhanced oil recovery
   NT secondary recovery
   NT tertiary recovery
   SA fluid injection
   SA gas injection
   SA hydraulic fracturing
   SA mercury injection
   SA petroleum engineering
   SA recovery
   SA reservoir rocks
   SA steam injection
   SA thermal recovery
   SA waterflooding

enigmatite
   use aenigmatite

Enisei Range
   use Yenisei Ridge

Enisei Ridge
   use Yenisei Ridge

Enisei River
   use Yenisei River

Eniwetok
   use Enewetak Atoll

Eniwetok Atoll
   As of 1985, not a valid term for GeoRef.
   use Enewetak Atoll

Enna
   No longer a valid term for GeoRef. See Enna Italy.

Enna Italy (1993)
   Province and city in central Sicily.
   BT Sicily Italy
   BT Italy
   BT Southern Europe
   BT Europe

Enns Valley (1978)
   River valley. Also search Enns.
   IN Index Austrian states as applicable.
   BT Austria
   BT Central Europe
   BT Europe
   SA Lower Austria
   SA Upper Austria

enrichment (1978)
   SA geochemistry
   SA metasomatism
   SA migration of elements
   SA mineral deposits, genesis
   SA supergene processes
   SA weathering

ENSO
   use El Nino

enstatite (1978)
   BT orthopyroxene
   BT pyroxene group
   BT chain silicates
   BT silicates
   SA bronzite
   SA protoenstatite

enstatite achondrites
   use chladnite

enstatite chondrites (1978)
   Autoposting of this term began in 1978.
   BT chondrites
   BT stony meteorites
   BT meteorites
   NT Abee Meteorite
   NT Indarch Meteorite
   NT Qingzhen Meteorite
   SA ordinary chondrites

Enteletacea (1978)
   Superfamily. Autoposting of this term began in 1978.
   BT Orthida
   BT Articulata
   BT Brachiopoda
   BT Invertebrata
   NT Schizophoria

enterolithic folding
   use enterolithic folds

enterolithic folds (1978)
   UF enterolithic folding
   BT folds
   SA deformation

Enteropneusta (1981)
   IN Index Hemichordata or Invertebrata as applicable.
   SA Hemichordata
   SA Invertebrata

enthalpy (1978)
   UF heat content
   SA entropy
   SA geochemistry
   SA Kirchhoff integral
   SA thermodynamic properties

Entisols (1978)
   BT soils
   SA soil group

Entomotaeniata (1981)
   BT Gastropoda
   BT Mollusca
   BT Invertebrata

Entomozocopina (1981)
   Includes Superfamily Entomozoacea. Autoposting of Myodocopida to this term began in 1989. As of 1990, microfossils, Crustacea, Mandibulata, and Arthropoda are autoposted to this term. This term has multiple hierarchies.
   BT1 Myodocopida
   BT1 Ostracoda
   BT1 Crustacea
   BT1 Mandibulata
   BT1 Arthropoda
   BT1 Invertebrata
   BT2 Myodocopida
   BT2 Ostracoda
   BT2 microfossils

Entotropha (1981)
   BT Insecta
   BT Mandibulata
   BT Arthropoda
   BT Invertebrata

Entrada Sandstone (1985)
   In San Rafael Group. Utah, Colorado, NE Arizona and NW New Mexico.
   BT Upper Jurassic
   BT Jurassic
   BT Mesozoic
   SA Arizona
   SA Colorado
   SA New Mexico
   SA San Rafael Group
   SA Utah

entropy (1978)
   SA enthalpy
   SA phase equilibria
   SA sediments
   SA thermodynamic properties

environment (1978)
   Autoposting of this term to various environments began in 1978 and ended in 1992.
   SA aerobic environment
   SA alpine environment
   SA anaerobic environment
   SA aquatic environment
   SA arctic environment
   SA arid environment
   SA biotopes
   SA boreal environment
   SA brackish-water environment
   SA cave environment
   SA coastal environment
   SA deep-sea environment
   SA deep-water environment
   SA deltaic environment
   SA depositional environment
   SA domains
   SA ecology
   SA ecosystems
   SA environmental analysis
   SA environmental geology
   SA equatorial region
   SA estuarine environment
   SA far-field
   SA fluvial environment
   SA fresh-water environment
   SA geomorphology
   SA glacial environment
   SA glaciofluvial environment
   SA glaciolacustrine environment
   SA glaciomarine environment
   SA habitat
   SA hemipelagic environment
   SA high-energy environment
   SA humid environment
   SA hypersaline environment
   SA interglacial environment
   SA intertidal environment
   SA lacustrine environment
   SA lagoonal environment
   SA latitude
   SA low-energy environment
   SA marine environment
   SA near-field
   SA nearshore environment
   SA paleoecology
   SA paleoenvironment
   SA paludal environment
   SA pelagic environment
   SA periglacial environment
   SA postglacial environment
   SA reclamation
   SA reef environment
   SA sebkha environment
   SA sedimentation
   SA semi-arid environment
   SA shallow-water environment
   SA shelf environment
   SA slope environment
   SA storm environment
   SA subaerial environment
   SA subalpine environment
   SA subglacial environment
   SA submarine environment
   SA subtidal environment
   SA subtropical environment
   SA supratidal environment
   SA taiga environment
   SA temperate environment
   SA terrestrial environment
   SA tropical environment
   SA urban environment

environment of deposition
   Not a valid GeoRef term.

environmental analyses
   use environmental analysis

environmental analysis (1978)
   For papers discussing the paleoenvironmental implications of several rock types and for papers whose main purpose is to determine environment.
   UF environmental analyses
   SA analysis
   SA environment
   SA environmental geology
   SA paleoenvironment
   SA punctuated aggradational cycles
   SA sedimentary rocks
   SA sedimentary structures
   SA sedimentation
   SA sediments

environmental geology (1978)
   Used for studies involving the collection, analysis, and application of geologic data and principles to problems created by human occupancy and use of the physical environment.
   IN General term. Use for studies of the discipline.
   SA acid mine drainage
   SA airfields
   SA atmosphere
   SA conservation
   SA DDT
   SA detergents
   SA drought
   SA ecology
   SA energy conservation
   SA engineering geology
   SA environment
   SA environmental analysis
   SA environmental geology maps
   SA floods
   SA geochemistry
   SA geologic hazards
   SA geology
   SA geomorphology
   SA ground water
   SA hazardous waste
   SA herbicides

SA human ecology
SA hydrology
SA impact statements
SA industrial waste
SA industry
SA injection
SA land use
SA land use maps
SA landfills
SA legislation
SA liquid waste
SA maps
SA marine geology
SA medical geology
SA mine drainage
SA mineral resources
SA mining geology
SA monitoring
SA natural resources
SA pesticides
SA pollution
SA preventive measures
SA protection
SA radioactive waste
SA railroads
SA reclamation
SA recycling
SA regional planning
SA regulations
SA remote sensing
SA Resource Conservation and Recovery Act
SA seismic zoning
SA soils
SA solid waste
SA spoils
SA strip mining
SA surface water
SA toxic materials
SA toxicity
SA trace metals
SA urban environment
SA urban planning
SA urbanization
SA waste disposal
SA waste water
SA water
SA water resources
SA water wells
SA waterways
SA wilderness areas

environmental geology maps (1985)
BT maps
SA environmental geology
SA geologic hazards maps
SA land cover maps
SA land use maps

Environmental Protection Agency, U.S.
use U.S. Environmental Protection Agency

enzymes (1978)
BT proteins
BT organic materials

Eocambrian (1978)
Uppermost Precambrian(?)/lowermost Cambrian(?).
BT Precambrian
SA Cambrian
SA Ediacaran
SA Infracambrian
SA Riphean
SA Sinian
SA Vendian
SA Yudomian

Eocene (1978)
World. Above Paleocene, below Oligocene. Autoposting of this term began in 1978. Autoposting of this term to extinct lakes began in 1993.
BT Paleogene

BT Tertiary
BT Cenozoic
NT Absaroka Supergroup
NT Ameki Formation
NT Annot Sandstone
NT Baca Formation
NT Biarritzian
NT Bridger Formation
NT Bridgerian
NT Castle Hayne Limestone
NT Chuckanut Formation
NT Chumstick Formation
NT Clarno Formation
NT Colton Formation
NT Crescent Formation
NT Crystal River Formation
NT Flournoy Formation
NT Golden Valley Formation
NT Green River Formation
NT Jacksonian
NT Lake Gosiute
NT Lake Uinta
NT Llajas Formation
NT Lookingglass Formation
NT lower Eocene
NT Lutetian
NT Maden Complex
NT middle Eocene
NT Mirador Formation
NT Nanjemoy Formation
NT Narizian
NT Parachute Creek Member
NT Punta Mosquito Formation
NT Pyrenean Orogeny
NT Refugian
NT Rose Canyon Formation
NT Subathu Formation
NT Swauk Formation
NT Uintan
NT Umpqua Formation
NT upper Eocene
NT Vermillion Creek coal bed
NT Wapiti Formation
NT Werillup Formation
NT Wilkins Peak Member
NT Yamhill Formation
SA Beaverhead Formation
SA Brianconnais Zone
SA Challis Volcanics
SA Duchesne River Formation
SA Flagstaff Formation
SA Floridan Aquifer
SA Hanna Formation
SA Huber Formation
SA lower Paleogene
SA Mount Scopus Group
SA Nanaimo Group
SA Orca Group
SA Paskapoo Formation
SA Rundle Group
SA San Lorenzo Formation
SA Scaglia Formation
SA Severn Formation
SA Sheep Pass Formation
SA upper Paleogene
SA Wasatch Formation

Eocrinoidea (1978)
Class.
BT Crinozoa
BT Echinodermata
BT Invertebrata

Eogene
use Paleogene

eolian
A valid term through 1977 for type of environment. See eolian environment and eolian sedimentation (now invalid terms); wind erosion; wind transport; winds.

eolian environment
As of 1982, no longer a valid term for GeoRef. Term was introduced in 1981. See wind transport; wind erosion; winds; environment.

eolian features (1978)
Autoposting of this term began in 1978.
UF features, eolian
NT barchans
NT coastal dunes
NT continental dunes
NT desert pavement
NT deserts
NT ergs
NT sand sheets
SA chenier plains
SA dunes
SA foredunes
SA geomorphology
SA loess
SA sand seas
SA wind erosion
SA wind transport
SA winds
SA yardangs

eolian sedimentation
As of 1982, no longer a valid term for GeoRef. Term was introduced in 1981. See wind transport; wind erosion; winds; sedimentation.

eolian transport
use wind transport

eolianite (1978)
UF aeolianite
UF dune rock
BT clastic rocks
BT sedimentary rocks
SA terrigenous materials

EOS
Not a valid term for GeoRef. As of 1993, see Earth Observing System.

eosphorite (1978)
BT phosphates

Eosuchia (1981)
Order. In Lepidosauromorpha. Before 1993, Lepidosauria was autoposted to this term. Before 1993, also search Eusuchia or Younginiformes.
UF Eusuchia
UF Younginiformes
BT Diapsida
BT Reptilia
BT Tetrapoda
BT Vertebrata
BT Chordata
SA Lepidosauria

Eoulx Basin (1978)
IN Index departments as applicable.
BT France
BT Western Europe
BT Europe
SA Alpes-de-Haute Provence France
SA Hautes-Alpes France

Eozoon
No longer a valid GeoRef index term. Before 1971, was used on level 1 in subfile N. See problematic fossils; foraminifera.

EPA
use U.S. Environmental Protection Agency

epeiric seas
use epicontinental seas

epeirogenesis
A valid term through 1974.
use epeirogeny

epeirogeny (1978)
Used for specific epeirogenic activity. For general treatment, see tectonics. Before 1975, also search epeirogenesis.
UF epeirogenesis
SA continents
SA crust
SA eustacy
SA isostasy
SA neotectonics
SA orogeny
SA tectonics
SA uplifts

epeirophoresis theory
use continental drift

Ephedrales (1981)
BT gymnosperms
BT Spermatophyta
BT Plantae

ephemeral streams (1978)
Autoposting of fluvial features to this term began in 1990.
BT streams
BT fluvial features
SA arroyos
SA wadis

Ephemeropteroida (1981)
BT Insecta
BT Mandibulata
BT Arthropoda
BT Invertebrata

epibiotism (1981)
SA biology
SA ecology
SA parasites
SA symbiosis

epicenters (1978)
SA earthquakes
SA focus
SA seismology

epicontinental seas (1993)
Before 1993, also search epeiric seas.
UF continental seas
UF inland seas
UF epeiric seas
UF seas, epeiric
UF seas, epicontinental
SA continental shelf
SA continents
SA marine geology

epidiorite (1978)
BT schists
BT metamorphic rocks

epidote (1978)
UF arendalite
UF pistacite
BT epidote group
BT sorosilicates
BT orthosilicates
BT silicates
SA aluminosilicates

epidote group (1978)
Autoposting of this term began in 1978.
BT sorosilicates
BT orthosilicates
BT silicates
NT allanite
NT clinozoisite
NT epidote
NT piemontite
NT tanzanite
NT zoisite

epidote-amphibolite facies (1978)
  BT facies
  SA amphibolite facies
  SA metamorphic rocks

epifauna
  Use epibiotism or benthonic taxa if applicable.

epigene processes (1978)
  As of 1993, restricted to economic studies.
  UF epigenic processes
  SA mineral deposits, genesis
  SA processes

epigenic processes
  use epigene processes

epimerization (1989)
  SA geochronology
  SA organic materials
  SA processes
  SA racemization
  SA relative age

Epiphyton (1989)
  Genus. Autoposting of thallophytes and microfossils began in 1990. This term has multiple hierarchies.
  BT1 algae
  BT1 thallophytes
  BT1 Plantae
  BT2 algae
  BT2 microfossils
  SA algal mounds

Epirus
  No longer a valid term for GeoRef. As of 1993 see Epirus Greece.

Epirus Greece (1993)
  Administrative region in NW Greece. Before 1993 also search Epirus AND Greece.
  CO N385500N402200
     E0213000E0204000
  BT Greece
  BT Southern Europe
  BT Europe
  SA Pindus Mountains

epistilbite (1978)
  BT zeolite group
  BT framework silicates
  BT silicates

epitaxy (1978)
  SA crystal growth
  SA twinning

epithermal processes (1978)
  Also search epithermal, or epithermal AND processes.
  SA carlin-type
  SA hydrothermal processes
  SA mineral deposits, genesis
  SA processes

epizoic taxa
  Use epibiotism.

epizonal metamorphism (1982)
  BT metamorphism
  SA low-grade metamorphism
  SA prograde metamorphism

EPR spectra (1978)
  Before 1978, also search electron paramagnetic resonance for data. After 1978, use electron paramagnetic resonance only for methodology.
  UF electron paramagnetic resonance spectra
  BT spectra
  SA electron paramagnetic resonance
  SA spectroscopy

epsomite (1978)
  BT sulfates

Equateur Zaire
  use Equatorial Zaire

equations (1978)
  Used as a general term.
  SA algorithms
  SA Archie's law
  SA constitutive equations
  SA equations of state
  SA formula
  SA functions
  SA Green function
  SA Hilbert transformations
  SA Kirchhoff integral
  SA Langmuir equation
  SA mathematical geology
  SA mathematical methods
  SA mathematical models
  SA Richards equation
  SA statistical analysis
  SA Universal Soil Loss Equation

equations of state (1978)
  SA equations
  SA geochemistry
  SA Hugoniot analysis
  SA phase equilibria

equatorial
  A valid term through 1977. See equatorial region.

Equatorial Africa
  No longer a valid GeoRef index term. Before 1969, was used on level 1 in subfile E. Now index Africa or countries as applicable; e.g. Cameroon, Congo, Gabon.

Equatorial Atlantic (1985)
  BT Atlantic Ocean
  SA Cape Verde Atlantic
  SA DSDP Site 366
  SA North Atlantic
  SA South Atlantic

Equatorial Guinea (1978)
  Formerly Spanish Guinea. Comprises province of Rio Muni on the mainland and the province of Fernando Po consisting of the islands of Fernando Po and Annobon. Before 1976, also search Spanish Guinea.
  CO N010000N022000
     E0113000E0092000
  UF Spanish Guinea
  BT Central Africa
  BT Africa
  NT Fernando Po

Equatorial Pacific (1978)
  Before 1978, search Pacific Ocean AND equatorial.
  BT Pacific Ocean
  NT Clipperton fracture zone
  NT Costa Rica Rift
  NT Siqueiros fracture zone
  SA DSDP Site 61
  SA DSDP Site 62
  SA DSDP Site 63
  SA DSDP Site 64
  SA DSDP Site 65
  SA DSDP Site 66
  SA DSDP Site 69
  SA DSDP Site 70
  SA DSDP Site 71
  SA DSDP Site 72
  SA DSDP Site 73
  SA DSDP Site 74
  SA DSDP Site 75
  SA DSDP Site 77
  SA DSDP Site 157
  SA DSDP Site 158
  SA DSDP Site 159
  SA DSDP Site 163
  SA DSDP Site 167
  SA DSDP Site 199
  SA DSDP Site 200
  SA DSDP Site 202
  SA DSDP Site 315
  SA DSDP Site 315A
  SA DSDP Site 462
  SA DSDP Site 494
  SA DSDP Site 495
  SA DSDP Site 496
  SA DSDP Site 497
  SA DSDP Site 498
  SA DSDP Site 499
  SA DSDP Site 500
  SA DSDP Site 501
  SA DSDP Site 503
  SA DSDP Site 504
  SA DSDP Site 504B
  SA DSDP Site 505
  SA DSDP Site 506
  SA DSDP Site 507
  SA DSDP Site 508
  SA DSDP Site 509
  SA DSDP Site 510
  SA DSDP Site 572
  SA DSDP Site 573
  SA DSDP Site 574
  SA DSDP Site 586
  SA East Pacific
  SA Leg 16
  SA Leg 17
  SA Leg 33
  SA Leg 67
  SA Leg 68
  SA Leg 69
  SA Leg 70
  SA Leg 83
  SA Leg 84
  SA Leg 85
  SA Leg 89
  SA Leg 111
  SA Nauru Basin
  SA North Pacific
  SA Northeast Pacific
  SA Northwest Pacific
  SA ODP Site 677
  SA ODP Site 678
  SA Rivera Ocean Seismic Experiment
  SA South Pacific
  SA Southeast Pacific
  SA Southwest Pacific
  SA West Pacific

equatorial region (1978)
  Before 1978, also search equatorial.
  SA climate
  SA ecology
  SA environment
  SA polar regions
  SA tropical environment

Equatorial Zaire (1981)
  The province of Equateur in NW Zaire.
  CO S023000N051000
     E0243000E0162000
  UF Equateur Zaire
  BT Zaire
  BT Central Africa
  BT Africa

Equidae (1978)
  Family. Autoposting of Hippomorpha, Eutheria and Theria to this term began in 1989.
  BT Hippomorpha
  BT Perissodactyla
  BT Eutheria
  BT Theria
  BT Mammalia
  BT Tetrapoda
  BT Vertebrata
  BT Chordata
  NT Equus
  NT Hipparion

equilibrium (1978)
  Used as a general term.
  UF disequilibrium
  SA phase equilibria
  SA steady-state processes

Equisetales (1978)
  Order. Autoposting of this term began in 1978.
  BT Sphenopsida
  BT pteridophytes
  BT Plantae
  NT Calamites

Equity Mine (1989)
  Silver ores; also some copper and gold ores. W central British Columbia. Also in other locations.
  IN Index British Columbia if applicable.
  SA British Columbia
  SA copper ores
  SA gold ores
  SA mines
  SA silver ores

Equus (1978)
  Genus. Autoposting of Hippomorpha, Eutheria and Theria to this term began in 1989.
  BT Equidae
  BT Hippomorpha
  BT Perissodactyla
  BT Eutheria
  BT Theria
  BT Mammalia
  BT Tetrapoda
  BT Vertebrata
  BT Chordata

Er
  use erbium

Er Rif
  use Rif

Er Riff
  use Rif

Eratosthenian (1978)
  Lunar stratigraphy. Eratosthenian rocks are older than those of the Copernican System but younger than those of the Imbrian System.
  SA Moon

erbium (1978)
  Autoposting of rare earths and metals to this term began in 1989.
  UF Er
  BT rare earths
  BT metals

Erevan Armenia
  use Yerevan Armenia

Erfurt
  No longer a valid term for GeoRef. See Erfurt Germany.

Erfurt Bezirk
  No longer a valid term for GeoRef. See Thuringia Germany. Used for the district in SW East Germany. Before 1981, search Erfurt.

Erfurt Germany (1993)
  City in Thuringia, central Germany. As of 1981, Erfurt refers to only the city in Germany. Before 1981, it also included the East German district.
  BT Thuringia Germany
  BT Germany
  BT Central Europe
  BT Europe

ergs (1993)
  BT eolian features
  SA deserts

SA dunes
SA Sahara
SA sand
SA sand ridges
SA sand seas
SA sand sheets
SA sedimentary structures

Ericaceae (1978)
Family.
BT Dicotyledoneae
BT angiosperms
BT Spermatophyta
BT Plantae

Eridostraca (1981)
Includes Superfamily Leperditellacea. As of 1990, microfossils, Crustacea, Mandibulata, and Arthropoda are autoposted to this term. This term has multiple hierarchies.
BT1 Ostracoda
BT1 Crustacea
BT1 Mandibulata
BT1 Arthropoda
BT1 Invertebrata
BT2 Ostracoda
BT2 microfossils

Erie County
Valid through 1988. Search in combination with state term. After 1988, use specific county-state term.

Erie County New York (1989)
W New York. Before 1989, also search Erie County AND New York.
CO N422600N430700 W0782800W0790900
BT New York
BT United States

Erie County Ohio (1989)
N Ohio. Before 1989, also search Erie County AND Ohio.
CO N411800N412800 W0822000W0825300
BT Ohio
BT United States

Erie County Pennsylvania (1989)
NW Pennsylvania. Before 1989, also search Erie County AND Pennsylvania.
CO N415000N421600 W0793700W0803200
BT Pennsylvania
BT United States
NT Presque Isle

Erimo Seamount (1978)
Off S coast of Hokkaido, Japan.
BT Pacific Ocean

erionite (1978)
BT zeolite group
BT framework silicates
BT silicates

Eritrea
Not a valid GeoRef index term after 1971. Was used on level 1 in subfile G.
use Ethiopia

Erivan Armenia
use Yerevan Armenia

Erlangen
No longer a valid term for GeoRef. See Erlangen Germany.

Erlangen Germany (1993)
City in N central Bavaria.
BT Bavaria Germany
BT Germany
BT Central Europe

BT Europe

erodability
use erodibility

erodibility (1978)
A measurement of the tendency for or degree of a material to erode.
UF erodability
UF erosibility
SA erosion
SA Universal Soil Loss Equation

Eromanga Basin (1978)
In SW Queensland.
BT Queensland Australia
BT Australia
BT Australasia
SA Birkhead Formation

erosibility
use erodibility

erosion (1978)
NT glacial erosion
NT littoral erosion
NT planation
NT scour
NT sheet erosion
NT soil erosion
NT water erosion
NT wind erosion
SA abrasion
SA climatic controls
SA corrosion
SA degradation
SA denudation
SA detritus
SA erodibility
SA erosion control
SA erosion cycle
SA erosion features
SA erosion rates
SA erosion surfaces
SA erosional unconformities
SA etching
SA exfoliation
SA fires
SA geologic hazards
SA geomorphology
SA glaciation
SA grinding
SA landform evolution
SA landslides
SA mass movements
SA mobility
SA mudflows
SA nivation
SA piping
SA processes
SA rates
SA rills
SA sediment yield
SA shorelines
SA slope stability
SA slumping
SA talus slopes
SA Universal Soil Loss Equation
SA valleys
SA waterways
SA weathering

erosion control (1978)
UF control, erosion
SA conservation
SA controls
SA erosion
SA geomembranes
SA geotextiles
SA soil erosion
SA soils
SA water erosion
SA wind erosion

erosion cycle (1982)
SA cycles
SA erosion
SA geomorphology

SA landform evolution
SA peneplains

erosion features (1978)
Before 1969, also search erosion forms. Autoposting of this term began in 1978.
UF erosion forms
UF erosional features
UF features, erosion
NT arroyos
NT badlands
NT cuestas
NT erosion surfaces
NT gullies
NT inselbergs
NT mesas
NT pediments
NT peneplains
NT talus slopes
SA benches
SA berms
SA blowouts
SA bluffs
SA boulders
SA cliffs
SA colluvium
SA denudation
SA dry valleys
SA erosion
SA furrows
SA geomorphology
SA highlands
SA klippen
SA natural curiosities
SA outliers
SA plains
SA potholes
SA rills
SA scarps
SA terraces
SA tors
SA valleys
SA volcanic necks
SA water erosion
SA wave-cut platforms
SA windows
SA yardangs

erosion forms
No longer a valid GeoRef index term. Before 1969, was used on level 1 in subfile E.
use erosion features

erosion rates (1993)
Before 1993, also search erosion AND rates.
SA erosion
SA rates
SA sediment transport
SA sedimentation
SA soil erosion
SA soils

erosion surfaces (1978)
Autoposting of erosion features to this term began in 1989.
UF planation surfaces
UF surfaces, erosion
BT erosion features
NT pediplains
SA erosion
SA geomorphology
SA pediments
SA peneplains
SA planation
SA wave-cut platforms

erosion transport
As of 1993, not a valid term for GeoRef. See sediment transport.

erosional features
use erosion features

erosional unconformities (1982)
UF disconformities
BT unconformities

SA erosion

erratic blocks
use erratics

erratic boulders
use erratics

erratics (1978)
UF erratic blocks
UF erratic boulders
BT clastic sediments
BT sediments
SA boulders
SA glacial features
SA glacial geology
SA glacial transport
SA glaciers

errors (1978)
Used as a general term.
SA accuracy
SA corrections
SA precision
SA reliability

ERTS
A valid term through 1977. After 1977, use Landsat.

ERTS-1
A valid term through 1977. After 1977, use Landsat.

erubescite
use bornite

eruptions (1978)
SA explosive eruptions
SA hawaiian-type eruptions
SA lava
SA nuees ardentes
SA plinian-type eruptions
SA pyroclastics
SA radon emanometry
SA strombolian-type eruptions
SA volcanic earthquakes
SA volcanic risk
SA volcanism
SA volcanoes
SA volcanology
SA vulcanian-type eruptions

eruptive rocks
No longer a valid term for GeoRef. Used in subfiles B, N, and E.
use volcanic rocks

Ervine Creek Limestone (1978)
Of Deer Creek Limestone. SW Iowa, NE Kansas, NW Missouri and SE Nebraska.
BT Virgilian
BT Upper Pennsylvanian
BT Pennsylvanian
BT Carboniferous
BT Paleozoic
SA Iowa
SA Kansas
SA Missouri
SA Nebraska

Erzgebirge (1978)
Mountain range on the border of Saxony, Germany and Bohemia.
IN Index countries as applicable.
CO N501000N510000 E0141000E0120000
UF Krusne Hory
UF Krusnehory Mountains
UF Krusny Hory Mountains
BT Central Europe
BT Europe
NT Czech Erzgebirge
SA Bohemia
SA Czechoslovakia
SA Saxony Germany

Es
use einsteinium

Esan Cape (1978)
At E entrance of Tsugaru Strait in SW Hokkaido. Also search Esan.
UF Ezan Cape
BT Hokkaido
BT Japan
BT Far East
BT Asia
SA Oshima Peninsula

Escambia County
Valid through 1988. Search in combination with state term. After 1988, use specific county-state term.

Escambia County Alabama (1989)
S Alabama. Before 1989, also search Escambia County AND Alabama.
CO N310000N311500
W0864300W0873800
BT Alabama
BT United States

Escambia County Florida (1989)
Extreme NW Florida. Before 1989, also search Escambia County AND Florida.
CO N301500N310000
W0871300W0875500
BT Florida
BT United States
NT Pensacola Florida

Escanaba Trough (1993)
S Gorda Rise area, offshore N California.
BT North American Pacific
BT Pacific Ocean
SA Gorda Rise

escarpments
use scarps

eschinite
use aeschynite

eschynite
use aeschynite

Escondido Formation (1978)
In Navarro Group. In S Texas.
BT Gulfian
BT Upper Cretaceous
BT Cretaceous
BT Mesozoic
SA Navarro Group
SA Texas

Esk Trough (1978)
Also search Esk. In 1985, Great Britain was added as a broader term.
IN Index England and/or Scotland as applicable.
BT Great Britain
BT United Kingdom
BT Western Europe
BT Europe
SA England
SA Scotland

Eskdale
Valid through 1988. Search in combination with state term. After 1988, use specific city-state term.

Eskdale West Virginia (1989)
Village in W West Virginia. Before 1989, search Eskdale AND West Virginia.
CO N380600N380600
W0812700W0812700
BT Kanawha County West Virginia
BT West Virginia
BT United States

eskebornite (1978)

Autoposting of sulfides to this term ended in 1989.
BT selenides

eskers (1978)
Autoposting of glaciofluvial features and glacial features to this term began in 1989. This term has multiple hierarchies.
UF os
BT1 glacial features
BT2 fluvial features
SA cirques
SA glacial geology
SA kames

Eskisehir
No longer a valid term for GeoRef. As of 1993 see Eskisehir Turkey.

Eskisehir Turkey (1993)
Province in NW central Anatolia. Also a city. Before 1993 also search Eskisehir AND Turkey.
BT Turkey
BT Middle East
BT Asia
SA Anatolia

Eskridge Shale (1978)
In Council Grove Group. E Kansas, SE Nebraska, and central N Oklahoma.
BT Permian
BT Paleozoic
SA Council Grove Group
SA Kansas
SA Nebraska
SA Oklahoma

Esmeralda County
Valid through 1988. Search in combination with state term. After 1988, use specific county-state term.

Esmeralda County Nevada (1989)
SW Nevada. Before 1989, also search Esmeralda County AND Nevada.
CO N365600N382800
W1171000W1182800
BT Nevada
BT United States
NT Goldfield Nevada
NT Silver Peak Mountains
SA Death Valley

Esna Shale (1989)
BT Tertiary
BT Cenozoic
SA Egypt
SA Thebes Formation

Espanola Basin (1985)
N New Mexico.
IN Index counties and state as applicable.
SA New Mexico
SA Rio Grande Rift

Espanola Formation (1978)
Of the Quirke Lake Group. Divisible into three members. S central Ontario.
BT Precambrian
SA Canada
SA Ontario

Espirito Santo
No longer a valid term for GeoRef. See Espirito Santo Brazil.

Espirito Santo Brazil (1993)
State on the Atlantic Ocean NE of Rio de Janeiro.
CO S212500S175000
W0393000W0421500
BT Brazil

BT South America
NT Vitoria Brazil
SA Barreiras Formation
SA Sao Francisco Craton

Esplanade Range (1978)
Range of Selkirk Mountains in SE British Columbia. This term has multiple hierarchies.
BT1 Selkirk Mountains
BT1 Canadian Rocky Mountains
BT1 Western Canada
BT1 Canada
BT2 Selkirk Mountains
BT2 Canadian Rocky Mountains
BT2 Rocky Mountains
BT3 Selkirk Mountains
BT3 British Columbia
BT3 Western Canada
BT3 Canada

Esplanade Sandstone Member (1985)
Member of Supai Formation. N Arizona.
BT Permian
BT Paleozoic
SA Arizona
SA Supai Formation

Essei Russian Federation
use Yessey Russian Federation

Essen Beds (1978)
Ruhr District. Also search Essen.
BT Carboniferous
BT Paleozoic
SA Germany
SA North Rhine-Westphalia Germany
SA Westphalian

Essex
No longer a valid term for GeoRef as of 1993. See Essex England.

Essex County
Valid through 1988. Search in combination with state term. After 1988, use specific county-state term. After 1993, use specific county-province term.

Essex County Massachusetts (1989)
NE Massachusetts. Before 1989, also search Essex County AND Massachusetts.
CO N422400N425400
W0703500W0711500
BT Massachusetts
BT United States
SA Cape Ann
SA Merrimack River valley
SA Plum Island

Essex County New Jersey (1989)
NE New Jersey. Before 1989, also search Essex County AND New Jersey.
CO N404200N405400
W0740700W0742300
BT New Jersey
BT United States
NT Newark New Jersey
SA Watchung Mountains

Essex County New York (1989)
NE New York. Before 1989, also search Essex County AND New York.
CO N434400N443300
W0732000W0742000
BT New York
BT United States
NT Ticonderoga New York
SA Lake George
SA Marcy Massif

Essex County Ontario (1993)
S Ontario. Before 1993, also search Essex County AND Ontario.
BT Ontario
BT Eastern Canada
BT Canada
NT Windsor Ontario
SA Detroit River

Essex County Vermont (1989)
NE Vermont. Before 1989, also search Essex County AND Vermont.
CO N442200N450100
W0713300W0715800
BT Vermont
BT United States

Essex County Virginia (1989)
E Virginia. Before 1989, also search Essex County AND Virginia.
CO N374500N381000
W0764200W0770900
BT Virginia
BT United States

Essex England (1993)
County on the North Sea NE of London. Before 1993, also search Essex AND England.
CO N513000N521000
E0011500E0000000
BT England
BT Great Britain
BT United Kingdom
BT Western Europe
BT Europe
SA Thames Estuary
SA Thames River

essexite (1978)
BT alkali gabbros
BT gabbros
BT plutonic rocks
BT igneous rocks

Essonne
No longer a valid term for GeoRef. See Essonne France.

Essonne France (1993)
Department in N France.
CO N481700N485000
E0023000E0015500
BT France
BT Western Europe
BT Europe

Esterel (1978)
Mountainous forested region.
IN Index departments as applicable.
BT France
BT Western Europe
BT Europe
SA Alpes-Maritimes France
SA Var France

Esterhazy
No longer a valid term for GeoRef. As of 1993, see Esterhazy Saskatchewan.

Esterhazy Saskatchewan (1993)
Town near the Manitoba border in SE Saskatchewan. Before 1993, also search Esterhazy AND Saskatchewan.
BT Saskatchewan
BT Western Canada
BT Canada

esters (1989)
Class of organic compounds with coordination complex -COOR.
BT organic materials

Estevan
No longer a valid term for GeoRef. As of 1993, see Estevan Saskatchewan.

Estevan Saskatchewan (1993)
Town near the North Dakota border in SE Saskatchewan. Before 1993, also search Estevan AND Saskatchewan.
BT Saskatchewan
BT Western Canada
BT Canada

estimation
Used as a general term through 1977. After 1977, see evaluation.

Estonia (1978)
Former Estonian Soviet Socialist Republic. Autoposting of this term began in 1978.
CO N580000N593000
   E0270000E0220000
BT Baltic region
BT Europe
NT Saaremaa
NT Tallinn Estonia
NT Tartu Estonia
SA Baltic Glaciation
SA Baltic Plain
SA Baltic Shield
SA Russian Plain

Estrada Nova Formation (1978)
Formation in Brazil. Upper Permian, Permian and Triassic.
IN Index ages as applicable.
SA Brazil
SA Goias Brazil
SA Parana Brazil
SA Permian
SA Santa Catarina Brazil
SA Sao Paulo Brazil
SA Triassic

Estremadura
No longer a valid term for GeoRef. As of 1993, see Estremadura Portugal.

Estremadura Portugal (1993)
Former province in S Portugal. Before 1993, search Estremadura AND Portugal. For region in W central Spain, use Extremadura Spain.
BT Portugal
BT Iberian Peninsula
BT Southern Europe
BT Europe
SA Extremadura Spain
SA Leiria Portugal
SA Setubal Portugal

estuaries (1978)
As of 1981, use estuarine environment for type of environment. This term has multiple hierarchies.
BT1 fluvial features
BT2 shore features
SA ecology
SA estuarine environment
SA estuarine sedimentation
SA fjords
SA geomorphology
SA oceanography
SA paleoecology
SA rivers
SA rivers and streams
SA sedimentation
SA waterways

estuarine
A valid term through 1977. After 1977, use estuaries or (as of 1981) estuarine environment or estuarine sedimentation.

estuarine environment (1981)
SA aquatic environment
SA deep-water environment
SA depositional environment
SA ecology
SA environment
SA estuaries
SA estuarine sedimentation
SA paleoecology
SA sedimentation

estuarine sedimentation (1981)
BT sedimentation
SA estuaries
SA estuarine environment

Etbai
use Red Sea Hills

etching (1978)
SA creep
SA crystal structure
SA embayments
SA erosion
SA geomorphology
SA landform evolution
SA pits
SA weathering

ethane (1978)
Autoposting of alkanes and aliphatic hydrocarbons to this term began in 1989.
BT alkanes
BT aliphatic hydrocarbons
BT hydrocarbons
BT organic materials
SA DDT
SA trichloroethane

ethene
use ethylene

Ethiopia (1978)
Before 1972, also search Eritrea. Before 1945, also search Abyssinia.
CO N032500N180000
   E0480000E0330000
UF Abyssinia
UF Eritrea
BT East Africa
BT Africa
NT Addis Ababa Ethiopia
NT Afar Depression
NT Ethiopian Rift
NT Fantale
NT Hadar
NT Harar Ethiopia
NT Omo Ethiopia
NT Omo River
NT Omo Valley
NT Tigre Ethiopia
SA Afar
SA Awash Valley
SA Blue Nile
SA East African Rift
SA Hadar Formation
SA Koobi Fora Formation
SA Lake Turkana
SA Nile River
SA Nubian Shield
SA Omo Group
SA Red Sea Basin
SA Red Sea Rift
SA Shungura Formation
SA Turkana Basin

Ethiopian Rift (1978)
That part of the East African Rift extending NE-SW from the Red Sea to Lake Turkana.
UF Abyssinian Rift valley
UF Ethiopian Rift system
UF Ethiopian Rift valley
BT Ethiopia
BT East Africa
BT Africa
SA East African Rift
SA Red Sea Rift

Ethiopian Rift system
use Ethiopian Rift

Ethiopian Rift valley
use Ethiopian Rift

Ethmodiscus (1978)
Autoposting of thallophytes and microfossils began in 1990. This term has multiple hierarchies.
BT1 Discoasteridae
BT1 nannofossils
BT1 algae
BT1 thallophytes
BT1 Plantae
BT2 Discoasteridae
BT2 nannofossils
BT2 algae
BT2 microfossils
BT3 Discoasteridae
BT3 nannofossils
BT3 algal flora
BT3 plants
BT4 Discoasteridae
BT4 nannofossils
BT4 algal flora
BT4 microfossils

ethylene (1989)
UF ethene
BT alkenes
BT aliphatic hydrocarbons
BT hydrocarbons
BT organic materials
SA tetrachloroethylene
SA trichloroethylene

Etna
use Mount Etna

Etowah County
Valid through 1988. Search in combination with state term. After 1988, use specific county-state term.

Etowah County Alabama (1989)
NE Alabama. Before 1989, also search Etowah County AND Alabama.
CO N335000N341200
   W0854400W0862200
BT Alabama
BT United States

ettringite (1978)
BT sulfates

Eu
use europium

Eua Island (1978)
One of the Tongapu group in S Tonga. Also search Eua.
BT Tonga
BT Polynesia
BT Oceania

Euboea (1978)
Island in the Aegean Sea.
CO N375500N390000
   E0244000E0225000
UF Evvoia
BT Greece
BT Southern Europe
BT Europe

Eubradyodonti (1981)
BT Chondrichthyes
BT Pisces
BT Vertebrata
BT Chordata

Eucla Basin (1978)
Also search Eucla.
IN Index states as applicable.
SA Australia
SA South Australia
SA Western Australia

euclase (1978)
Autoposting of nesosilicates began in 1985.
BT nesosilicates
BT orthosilicates
BT silicates
SA aluminosilicates

eucrite (1978)
From 1978-1980, gabbro family was autoposted to this term. Before 1993, search eucrite with gabbros for the igneous rocks. As of 1993, term is restricted to meteorites.
BT achondrites
BT stony meteorites
BT meteorites
NT Juvinas Meteorite
NT Moore County Meteorite
NT Pasamonte Meteorite

eucryptite (1978)
Autoposting of nesosilicates began in 1985.
BT nesosilicates
BT orthosilicates
BT silicates
SA aluminosilicates

eudialite
use eudialyte

eudialyte (1978)
From 1978-1980, halides and ring silicates were autoposted to this term. Autoposting of chlorides began in 1981. As of 1989, halides and silicates are autoposted to this term. This term has multiple hierarchies.
UF eudialite
BT1 chlorides
BT1 halides
BT2 ring silicates
BT2 silicates

Euelephantoidea
use Elephantoidea

Euganean Hills (1978)
Range of hills in Padova Province, NE Italy.
BT Veneto Italy
BT Italy
BT Southern Europe
BT Europe

eugeosynclines (1978)
UF pliomagmatic zone
SA geosynclines
SA miogeosynclines
SA tectonics

Eulengebirge
use Sowie Mountains

eulite (1978)
BT orthopyroxene
BT pyroxene group
BT chain silicates
BT silicates

eulysite (1978)
As of 1993, this term is considered obsolete.
BT ultramafics
BT plutonic rocks
BT igneous rocks

Euphorbiaceae (1978)
Family.
BT Dicotyledoneae
BT angiosperms
BT Spermatophyta
BT Plantae

Euphrates River (1978)
Rises in E central Turkey and at its confluence with the Tigris River it forms the Shatt-al-Arab which flows into the Persian Gulf.

IN   Index countries as applicable.
BT   Asia
SA   Iraq
SA   Syria
SA   Turkey

Eurasia   (1978)
Land mass comprising the continents of Europe and Asia. Baltic region was a narrower term from 1981-1992. Before 1981, Black Sea was included as a narrower term. Autoposting of this term began in 1978 and ended in 1992. Used for general treatments.
IN   Index continents as applicable.
SA   Asia
SA   Black Sea
SA   Black Sea region
SA   Caspian Basin
SA   Eurasian Plate
SA   Europe
SA   Kazakhstan
SA   Laurasia
SA   Russian Federation
SA   Urals

Eurasia Basin   (1978)
Between the North Pole and the major islands lying N of the Eurasian land mass.
UF   Eurasian Basin
BT   Arctic Ocean

Eurasian Basin
   use Eurasia Basin

Eurasian Plate   (1978)
That part of the Earth's crust comprising Eurasia, including the British Isles.
UF   European Plate
SA   Eurasia
SA   plate tectonics
SA   plates
SA   Yangtze Platform

Euratom   (1981)
UF   European Atomic Energy Community
SA   associations
SA   Europe
SA   nuclear energy

Eure
No longer a valid term for GeoRef. See Eure France.

Eure France   (1993)
Department in NW France.
CO   N484000N493000
      E0014500E0001500
BT   France
BT   Western Europe
BT   Europe
SA   Eure Valley
SA   Normandy
SA   Seine Estuary

Eure Valley   (1978)
River valley in NW France.
IN   Index departments as applicable.
BT   France
BT   Western Europe
BT   Europe
SA   Eure France
SA   Eure-et-Loir France
SA   Orne France

Eure-et-Loir
No longer a valid term for GeoRef. See Eure-et-Loir France.

Eure-et-Loir France   (1993)
CO   N475700N490500
      E0020000E0004500
BT   France
BT   Western Europe

BT   Europe
SA   Beauce
SA   Eure Valley

Eureka
Valid through 1989. Search in combination with state term. After 1989, use specific city-state term.

Eureka California   (1989)
City on Humboldt Bay in NW California. Before 1989, search Eureka AND California.
CO   N404900N404900
      W1241000W1241000
BT   Humboldt County California
BT   California
BT   United States

Eureka County
Valid through 1988. Search in combination with state term. After 1988, use specific county-state term.

Eureka County Nevada   (1989)
N central Nevada. Before 1989, also search Eureka County AND Nevada.
CO   N391000N410000
      W1154500W1163500
BT   Nevada
BT   United States
NT   Eureka Nevada
NT   Roberts Mountains Allochthon
SA   Cortez Mountains
SA   Roberts Mountains

Eureka Illinois   (1989)
City in N central Illinois. Before 1989, search Eureka AND Illinois.
CO   N404400N404400
      W0891600W0891600
BT   Woodford County Illinois
BT   Illinois
BT   United States

Eureka Kansas   (1989)
City in SE Kansas. Before 1989, search Eureka AND Kansas.
CO   N375100N375100
      W0961700W0961700
BT   Greenwood County Kansas
BT   Kansas
BT   United States

Eureka Nevada   (1989)
Village in central Nevada. Before 1989, search Eureka AND Nevada.
CO   N393100N393100
      W1155800W1155800
BT   Eureka County Nevada
BT   Nevada
BT   United States

Eureka Quartzite   (1985)
Middle to Upper? Ordovician. N Nevada, E California and NW Utah.
BT   Ordovician
BT   Paleozoic
SA   California
SA   Nevada
SA   Utah

Eureka Sound Formation   (1978)
BT   Cretaceous
BT   Mesozoic
SA   Franklin District Northwest Territories
SA   Northwest Territories

Eureka South Dakota   (1989)
City in N South Dakota. Before 1989, search Eureka AND South Dakota.
CO   N454600N454600
      W0993700W0993700

BT   McPherson County South Dakota
BT   South Dakota
BT   United States

Eureka Utah   (1989)
City in W Utah. Before 1989, search Eureka AND Utah.
CO   N395700N395700
      W1120900W1120900
BT   Juab County Utah
BT   Utah
BT   United States

Europa Satellite   (1985)
One of the satellites of Jupiter. Before 1985, also search Europa AND Jupiter.
SA   Galilean satellites
SA   icy satellites
SA   Jupiter
SA   satellites
SA   Voyager Program

Europe   (1978)
As of 1981, Dinaric Alps and Malta are narrower terms. Autoposting of this term began in 1978. As of 1993, this term is autoposted to the following parts of the former USSR: parts of the Russian Federation and to Armenia, Azerbaidzhan, Baltic region, Belarus, Caucasus, Estonia, Georgian Republic, Moldova, Georgian Republic, Latvia, Lithuania, and Ukraine.
CO   N350000N710000
      E0750000W0250000
NT   Adriatic region
NT   Allarechenskiy
NT   Alps
NT   Arkhangelsk Russian Federation
NT   Armenia
NT   Arve Valley
NT   Astrakhan Russian Federation
NT   Azerbaidzhan
NT   Azov region
NT   Azov Sea
NT   Balkan Peninsula
NT   Baltic Plain
NT   Baltic region
NT   Baltic Shield
NT   Bashkiria Russian Federation
NT   Belarus
NT   Belgorod Russian Federation
NT   Black Sea region
NT   Brest Basin
NT   Brianconnais Zone
NT   Bryansk Russian Federation
NT   Bug region
NT   Bug River
NT   Bukovina
NT   Campine
NT   Carpathian Foredeep
NT   Carpathian Foreland
NT   Carpathians
NT   Caucasus
NT   Caucasus Foreland
NT   Ceneri Zone
NT   Central Europe
NT   Chechen-Ingush Russian Federation
NT   Dagestan Russian Federation
NT   Danube Delta
NT   Danube River
NT   Danube Valley
NT   Dnieper Basin
NT   Dnieper River
NT   Dnieper-Donets Basin
NT   Dniester River
NT   Dniester-Prut Interfluve
NT   Donets Basin
NT   Dvina River
NT   Elbe River

NT   Elbe Valley
NT   European Platform
NT   Fennoscandia
NT   Galitsiya
NT   Georgian Republic
NT   Imandra
NT   Isar Valley
NT   Jura Mountains
NT   Jutland
NT   Kabardin-Balkar Russian Federation
NT   Kaliningrad Russian Federation
NT   Kalmyk Russian Federation
NT   Karst region
NT   Khibiny Mountains
NT   Kola
NT   Kola Peninsula
NT   Komi Russian Federation
NT   Kostroma Russian Federation
NT   Kozhim
NT   Krasnodar Russian Federation
NT   Krizna Nappe
NT   Kuban
NT   Kuma Basin
NT   Kursk magnetic anomaly
NT   Kursk Russian Federation
NT   Lago Maggiore
NT   Lake Geneva
NT   Lake Ladoga
NT   Lake Ladoga region
NT   Lake Onega
NT   Lower Rhine Basin
NT   Lusatia
NT   Maritsa River
NT   Mezen River basin
NT   Moldavia
NT   Moldova
NT   Morava River valley
NT   Moscow Basin
NT   Moscow Russian Federation
NT   Moscow Syneclise
NT   Moselle River
NT   Moselle Valley
NT   Murmansk Russian Federation
NT   Neman River basin
NT   Nizhniy Novgorod Russian Federation
NT   North Sea Coast
NT   Northern Dvina River
NT   Novgorod Russian Federation
NT   Oder Valley
NT   Orenburg Russian Federation
NT   Pachelma Russian Federation
NT   Pannonia
NT   Pannonian Basin
NT   Pechora Basin
NT   Pechora River
NT   Peribaltic Syneclise
NT   Perm Russian Federation
NT   Pieniny Klippen Belt
NT   Polesye
NT   Pripet Basin
NT   Prut River
NT   Pskov Russian Federation
NT   Pyrenees
NT   Rhine Basin
NT   Rhine River
NT   Rhine Valley
NT   Rostov Russian Federation
NT   Roztocze
NT   Russian Plain
NT   Ryazan Russian Federation
NT   Saar Basin
NT   Saint Petersburg Russian Federation
NT   Sakmara
NT   Samara Russian Federation
NT   Saratov Russian Federation
NT   Scythian Platform
NT   Simplon region
NT   Siret River
NT   Smolensk Russian Federation
NT   Snieznik

NT Somes Basin
NT Southern Europe
NT Stavelot-Venn Massif
NT Stavropol Russian Federation
NT Tambov Russian Federation
NT Tatar Arch
NT Tatar Russian Federation
NT Tataria
NT Terek River
NT Thrace
NT Ticino
NT Timan Ridge
NT Timan-Pechora region
NT Tisza River
NT Tokaj-Eperjes Mountains
NT Transcaucasia
NT Tula Russian Federation
NT Tver Russian Federation
NT Tyrny-Auz Russian Federation
NT Udmurtia Russian Federation
NT Ukraine
NT Ukrainian Syneclise
NT Ulyanovsk Russian Federation
NT Valdai
NT Vardar River
NT Variscides
NT Vladimir Russian Federation
NT Volga River
NT Volga-Don region
NT Volga-Urals
NT Volgograd Russian Federation
NT Vologda Russian Federation
NT Voronezh Russian Federation
NT Voronezh-Volga Anteclise
NT Vyatka River
NT Vyatka Russian Federation
NT Vyatka-Kama Interfluve
NT Vychegda River
NT Western Europe
NT Yaroslavl Russian Federation
SA Argun River
SA Caledonides
SA Commonwealth of Independent States
SA Eurasia
SA Euratom
SA Lapland
SA Moscow-Pechora Syneclise
SA Oka River
SA Polar Continental Shelf
SA Tethys
SA Volga region

European
A valid term through 1977 used in combination with USSR. After 1977, see European USSR.

European Atlantic (1981)
Region of the Atlantic Ocean. Autoposting of Atlantic Ocean to this term began in 1990.
BT North Atlantic
BT Atlantic Ocean
NT Baltic Sea
NT Bay of Biscay
NT English Channel
NT Goban Spur
NT Gorringe Bank
NT Irish Sea
NT North Sea
SA DSDP Site 117
SA DSDP Site 120
SA DSDP Site 398
SA DSDP Site 402
SA DSDP Site 548
SA DSDP Site 549
SA DSDP Site 550
SA DSDP Site 551
SA DSDP Site 608
SA DSDP Site 609
SA DSDP Site 610
SA DSDP Site 611
SA East Atlantic
SA Leg 13
SA Leg 80
SA Leg 103
SA Northeast Atlantic
SA ODP Site 637
SA ODP Site 638
SA ODP Site 639
SA ODP Site 640
SA ODP Site 641

European Atomic Energy Community
use Euratom

European Community (1981)
SA economics

European Geotraverse (1989)
UF European Geotraverse Project
SA geophysical surveys
SA geotraverses
SA programs

European Geotraverse Project
use European Geotraverse

European Plate
use Eurasian Plate

European Platform (1978)
Ancient platform of Precambrian crystalline rocks overlain by sedimentary deposits of the North European Plain extending from the Low Countries to the Urals.
IN Index countries as applicable.
BT Europe
SA Belarus
SA Belgium
SA Denmark
SA Germany
SA Netherlands
SA Poland
SA Russian Federation
SA Ukraine

European USSR
As of 1993, no longer a valid term for GeoRef. See W Russian Federation (formerly Russian Republic) and individual republics: Armenia, Azerbaidzhan, Belarus (formerly Byelorussia), Estonia, Georgian Republic, Latvia, Lithuania, Moldova, and Ukraine. Also see Moldavia. Before 1978, also search USSR AND European.

europium (1978)
Autoposting of rare earths and metals to this term began in 1989.
UF Eu
BT rare earths
BT metals

Euryapsida
As of 1993, no longer a valid term for GeoRef. Subclass of Reptilia. Considered obsolete.
use Diapsida

Eurypterida (1978)
BT Merostomata
BT Chelicerata
BT Arthropoda
BT Invertebrata

Euselachii (1982)
BT Chondrichthyes
BT Pisces
BT Vertebrata
BT Chordata

eustacy (1978)
Before 1976, also search eustatism.
UF eustasy
SA changes of level
SA epeirogeny
SA isostasy

eustasy
use eustacy

Eusuchia
use Eosuchia

Eutaw Formation (1985)
In Selma Group. Alabama, E and N Mississippi, W Georgia and W Tennessee.
BT Upper Cretaceous
BT Cretaceous
BT Mesozoic
SA Alabama
SA Georgia
SA Mississippi
SA Selma Group
SA Tennessee

Eutheria (1985)
Infraclass. Autoposting of this term began in 1989. Autoposting of Theria and Mammalia to this term began in 1989.
UF Placentalia
BT Theria
BT Mammalia
BT Tetrapoda
BT Vertebrata
BT Chordata
NT Amblypoda
NT Artiodactyla
NT Astrapotheria
NT Carnivora
NT Cetacea
NT Chiroptera
NT Condylarthra
NT Creodonta
NT Desmostylia
NT Edentata
NT Embrithopoda
NT Hyracoidea
NT Insectivora
NT Lagomorpha
NT Litopterna
NT Notoungulata
NT Perissodactyla
NT Pholidota
NT Primates
NT Proboscidea
NT Rodentia
NT Sirenia
NT Taeniodonta
NT Tillodontia
NT Tubulidentata
NT Ungulata

eutrophication (1978)
SA ecology
SA hydrogen sulfide
SA hydrology
SA lakes

euxenite (1978)
Autoposting of niobotantalates to this term began in 1989. This term has multiple hierarchies.
BT1 niobotantalates
BT1 niobates
BT1 oxides
BT2 niobotantalates
BT2 tantalates
BT2 oxides

evaluation (1978)
Before 1985, also search economic evaluation.
SA economic geology
SA economics
SA mining geology
SA ore grade

Evangeline Aquifer (1989)
SW Louisiana and E Texas.
IN Index states as applicable.
SA Louisiana
SA Texas
SA United States

Evangeline Parish
Valid through 1988. Search in combination with state term. After 1988, use specific county-state term.

Evangeline Parish Louisiana (1989)
Parish in S central Louisiana. Before 1989, also search Evangeline Parish AND Louisiana.
CO N302900N310000 W0921000W0923600
BT Louisiana
BT United States

evaporation (1978)
SA evapotranspiration
SA hydrology
SA sublimation
SA water vapor

evaporite deposits (1981)
Before 1981, search evaporites AND deposits.
IN Commodity. See List C.
SA anhydrite deposits
SA borate deposits
SA evaporites
SA gypsum deposits
SA salt
SA sylvinite

evaporites (1978)
As of 1981, use evaporite deposits for evaporites as a commodity. Autoposting of this term began in 1981.
BT chemically precipitated rocks
BT sedimentary rocks
NT potash
NT salt
NT sylvinite
SA anhydrite
SA dolomite
SA evaporite deposits
SA gypsum
SA halite
SA sediments
SA sodium chloride

evapotranspiration (1978)
SA atmospheric precipitation
SA evaporation
SA hydrologic cycle
SA hydrology
SA water balance

Everglades (1978)
Vast swampy region in six counties lying S of Lake Okeechobee in S Florida.
IN Index counties as applicable.
BT Florida
BT United States

Everton Formation (1978)
In Buffalo River Group. Includes Calico Rock Sandstone Member, Kings River Member, Newton Sandstone Member. N Arkansas, and S Missouri.
BT Middle Ordovician
BT Ordovician
BT Paleozoic
SA Arkansas
SA Missouri

evolution (1978)
As of 1983, not to be used for biologic evolution. See biologic evolution.
SA biologic evolution
SA landform evolution

Evora
No longer a valid term for GeoRef. As of 1993 see Evora Portugal.

Evora Portugal (1993)
District in SE central Portugal. Before 1993 also search Evora AND Portugal.
BT Portugal
BT Iberian Peninsula
BT Southern Europe
BT Europe
NT Alandroal Portugal

Evros River
use Maritsa River

Evvoia
use Euboea

Ewekoro (1978)
Region in SW Nigeria.
BT Nigeria
BT West Africa
BT Africa

Ewekoro Formation (1978)
NW, S, SW, and W Nigeria. Upper Paleocene and possibly lower Eocene. Autoposting of Tertiary to this term began in 1990.
BT Paleogene
BT Tertiary
BT Cenozoic
SA lower Eocene
SA Nigeria
SA upper Paleocene

excavations (1978)
SA blasting
SA dredging
SA engineering geology
SA explosions
SA New Austrian Tunnelling Method
SA rock mechanics
SA slope stability
SA tunnels
SA underground installations

Excello Shale (1978)
SE Kansas, W and N Missouri, and NE Oklahoma.
BT Pennsylvanian
BT Carboniferous
BT Paleozoic
SA Kansas
SA Missouri
SA Oklahoma

exchange capacity
As of 1981, no longer a valid term for GeoRef. Use anion exchange capacity, cation exchange capacity, or ion exchange as applicable.

exchange rate (1981)
As of 1993, restricted to economics studies.
SA economics

Exclusive Economic Zone (1993)
Use United States Exclusive Economic Zone if applicable. Before 1993, also search EEZ.
UF EEZ
SA economics
SA exploration
SA Law of the Sea
SA mineral exploration
SA ocean floors
SA offshore
SA petroleum exploration
SA policy
SA strategic minerals
SA United States Exclusive Economic Zone

excursions
No longer a valid GeoRef index term. Before 1971, was used on level 1 in subfiles E and N. See expeditions; field trips; guidebook; road log.

Exeter
No longer a valid term for GeoRef as of 1993. See Exeter England.

Exeter England (1993)
City in Devonshire, SW England.
BT Devonshire England
BT England
BT Great Britain
BT United Kingdom
BT Western Europe
BT Europe

exfoliation (1978)
SA erosion
SA geomorphology
SA weathering
SA weathering rinds

exhalative processes (1978)
As of 1993, restricted to economic studies.
SA mineral deposits, genesis
SA processes

exhibits (1993)
Use as a general term.
SA catalogs
SA collections
SA museums

exine (1978)
SA furrows
SA miospores
SA palynomorphs
SA pollen
SA spores
SA sporopollenin

exinite (1978)
Autoposting of this term began in 1981.
UF liptinite
BT macerals
NT cutinite
NT resinite
NT sporinite

Exmouth Plateau (1985)
Off N Western Australia.
BT Indian Ocean
SA Leg 122
SA ODP Site 759
SA ODP Site 760
SA ODP Site 761
SA ODP Site 762
SA ODP Site 764
SA Western Australia

exobiology (1978)
SA biology
SA extraterrestrial geology
SA life origin
SA planetology

exogene processes (1978)
As of 1993, restricted to economic studies.
SA mineral deposits, genesis
SA processes

exogenous inclusions
use xenoliths

exomorphic zone
use aureoles

exoskeletons (1989)
SA chitin
SA Invertebrata
SA invertebrates
SA shells

exotic terranes (1989)
BT terranes
SA plate tectonics
SA tectonics

expanding universe theory (1981)
SA cosmochronology
SA extraterrestrial geology
SA steady-state processes

expansion (1978)
As of 1981, to be used only for the Earth as a whole. Before 1981, used also as a general term.
SA Earth

expansive materials (1978)
Before 1978, also search expansive soils.
UF expansive soils
UF swelling soils
SA clay
SA materials
SA soil mechanics
SA soils

expansive soils
use expansive materials

expeditions (1978)
Used as a general term. As of 1981, use only when the organization or progress of an expedition is discussed.
SA areal geology
SA associations
SA CESAR
SA FAMOUS
SA oceanography
SA Orgon III
SA research vessels

experimental investigations
No longer a valid GeoRef index term. Before 1971, was used on level 1 in subfiles E and N.
use experimental studies

experimental studies (1978)
Before 1971, also search experimental investigations.
UF experimental investigations
UF studies, experimental
SA alternating field demagnetization
SA b-values
SA laboratory studies
SA synchrotron radiation
SA theoretical studies

expert systems (1989)
UF knowledge-based systems
SA artificial intelligence
SA computer programs
SA data processing
SA geographic information systems
SA PROSPECTOR

explanatory text (1978)
UF text, explanatory
SA areal geology
SA guidebook
SA maps

exploitation (1978)
A valid term through 1978. Reintroduced in 1981. Used in relation to commodity terms (list C).
SA economics
SA production

exploration (1978)
Before 1972, also search geological exploration. Before 1981, also search prospecting. As of 1983, may be used with certain commodities (list C): all of the energy sources and water resources. As of 1993, see coal exploration and petroleum exploration for those commodities. See mineral exploration for all other commodities.

Continues to be used with Moon and the individual planets.
SA coal exploration
SA development
SA Exclusive Economic Zone
SA geochemical methods
SA geodesy
SA geological methods
SA geophysical methods
SA lunar bases
SA mineral exploration
SA mineral resources
SA Moon
SA offshore
SA onshore
SA pathfinders
SA petroleum exploration
SA photogeology
SA potential deposits
SA site exploration
SA speleology
SA surveys
SA United States Exclusive Economic Zone

Explorer Program (1981)
Autoposting of this term to specific Explorer missions began in 1989.
NT Explorer 35
SA extraterrestrial geology
SA Moon
SA planetology
SA remote sensing

Explorer 35 (1978)
Autoposting of Explorer Program to this term began in 1989.
BT Explorer Program
SA Moon
SA remote sensing

explosion phenomena
A valid term through 1972. After 1972, see cryptoexplosion features and impact features under geomorphology. See explosions and nuclear explosions.

explosions (1978)
Used for geotechnical and seismological studies on the effects of explosions. Before 1978, also search explosion phenomena.
NT chemical explosions
NT nuclear explosions
SA blasting
SA blowouts
SA cryptoexplosion features
SA deep seismic sounding
SA earthquakes
SA elastic materials
SA elastic waves
SA engineering geology
SA excavations
SA foundations
SA geologic hazards
SA land subsidence
SA marine installations
SA mining geology
SA rock mechanics
SA seismic sources
SA seismic surveys
SA seismology
SA shatter cones
SA site exploration
SA slope stability
SA soil mechanics

explosive eruptions (1981)
SA eruptions
SA phreatomagmatism
SA plinian-type eruptions
SA volcanism
SA volcanology

export (1981)
SA economics

export value (1981)
SA economics

exposure age (1978)
Before 1975, also search exposure ages.
SA age
SA geochronology
SA relative age

exsolution (1978)
UF unmixing
SA crystal chemistry
SA crystal structure
SA solution
SA solution transport

exsurgence (1981)
SA caves
SA hydrology
SA springs
SA streams
SA underground streams

extension (1978)
Used to indicate the mechanics of deformation. As of 1982, use extension tectonics if applicable. Before 1978, also used to indicate style of structure. Now use extension faults or extension fractures for style of structure.
SA deformation
SA extension faults
SA extension fractures
SA extension tectonics
SA tension
SA torsion

extension faults (1978)
Before 1978, also search extension AND faults.
BT faults
SA extension
SA tension

extension fractures (1978)
Before 1978, search extension AND fractures.
BT fractures
SA extension
SA tension

extension tectonics (1982)
Before 1982, search extension AND tectonics; tension AND tectonics.
BT tectonics
SA basins
SA compression tectonics
SA deformation
SA extension
SA plate tectonics
SA pull-apart basins
SA rifting
SA tension
SA torsion
SA underplating

extensometers (1978)
SA deformation
SA instruments
SA rock mechanics
SA strainmeters
SA stress

extent
As of 1981, no longer a valid term for GeoRef. See distribution. See glacial extent.

extinct
A valid term through 1977 used in combination with lakes (i.e. lakes, extinct). After 1977, use extinct lakes. When referring to taxa, use extinct taxa.

extinct lakes (1978)
Before 1976, also search extinct AND lakes.
BT lakes
BT lacustrine features
SA geomorphology
SA glacial features
SA glacial geology
SA glacial lakes
SA periglacial features

extinct taxa (1978)
Before 1978, also search extinct.
UF taxa, extinct
SA biologic evolution
SA extinction
SA paleontology
SA taxonomy

extinction (1978)
As of 1981, restricted to paleontology. Before 1981, also used for optical extinction.
SA biochronology
SA biologic evolution
SA extinct taxa
SA mass extinctions
SA optical extinction
SA paleontology

extraction
Used for mining geology through May 1978. After May 1978, use production.

extraterrestrial geology (1978)
Broad topic used for studies about materials and processes outside the Earth, and for the discipline as a whole. For solar system or discussions of more than one planet, use planetology.
SA Apollo Program
SA Apollo-Soyuz Program
SA cosmic dust
SA cosmochemistry
SA cosmochronology
SA Earth-Moon couple
SA exobiology
SA expanding universe theory
SA Explorer Program
SA geology
SA Halley's Comet
SA interplanetary comparison
SA interplanetary space
SA life origin
SA Luna Program
SA lunar bases
SA Magellan Program
SA Mariner Program
SA meteorites
SA Moon
SA Pioneer Program
SA planetary rings
SA planetology
SA planets
SA polar caps
SA Skylab
SA solar system
SA Spacelab Program
SA Surveyor Program
SA terrestrial comparison
SA Venera Program
SA Viking Program
SA Voyager Program

Extremadura
No longer a valid term for GeoRef. As of 1993, see Extremadura Spain.

Extremadura Spain (1993)
Region in W central Spain. Before 1993, also search Extremadura and Spain. Before 1981, also search Estremadura AND Spain. For the province in Portugal, use Estremadura Portugal.
CO N375700N403000
    W0044000W0073000
BT Spain
BT Iberian Peninsula
BT Southern Europe
BT Europe
NT Badajoz Spain
NT Caceres Spain
SA Estremadura Portugal

extremities (1981)
SA anatomy
SA morphology
SA muscles
SA Vertebrata

extrusive rocks (1978)
Not a valid index term through 1977. Before 1978, also search extrusive.
SA effusion
SA igneous rocks
SA lava
SA rocks
SA volcanology

Exuma Sound (1978)
Body of water SE of Nassau.
SA Bahamas
SA Leg 101
SA ODP Site 631
SA ODP Site 632
SA ODP Site 633

Eye-Dashwa Lakes Pluton (1989)
Atikokan, Ontario.
UF Eye-Dashwa Pluton
BT Ontario
BT Eastern Canada
BT Canada
SA Canadian Shield
SA Superior Province

Eye-Dashwa Pluton
use Eye-Dashwa Lakes Pluton

Eyre Peninsula (1978)
Between Great Australian Bight and Spencer Gulf.
BT South Australia
BT Australia
BT Australasia

Ezan Cape
use Esan Cape

F

F
use fluorine

F-region
As of 1993, no longer a valid term for GeoRef. Usually out-of-scope. Used with ionosphere.

fabric (1978)
SA deformation
SA igneous rocks
SA lineation
SA metamorphic rocks
SA orientation
SA petrofabrics
SA preferred orientation
SA sedimentary rocks
SA sediments
SA structural analysis
SA tectonite

fabric analysis
A valid term through 1972. After 1972, use structural analysis or fabric.

facies (1978)
NT actinolite facies
NT amphibolite facies
NT blueschist facies
NT eclogite facies
NT epidote-amphibolite facies
NT granulite facies
NT greenschist facies
NT hornfels facies
NT prehnite-pumpellyite facies
NT zeolite facies
SA biofacies
SA domains
SA grade
SA isograds
SA lithofacies
SA metamorphic rocks
SA microfacies

facilities, nuclear
use nuclear facilities

factor analysis (1978)
BT statistical analysis
SA analysis

factors (1978)
SA biota
SA climate
SA parent materials
SA pedogenesis
SA scale factor
SA soils
SA time factor

faecal pellets
use fecal pellets

Faeroe Islands (1978)
Self governing Danish island group between Iceland and the Shetland Islands. Before 1981, broader term was Atlantic Ocean, which was autoposted. As of 1981, Kuno is a narrower term. Before 1972, also search Faroe Islands.
CO N612000N623000
    W0060000W0080000
UF Faeroes
UF Faroe Islands
BT Atlantic Ocean Islands
NT Kuno
SA Atlantic Ocean

Faeroe-Iceland Ridge
use Iceland-Faeroe Ridge

Faeroes
use Faeroe Islands

Fagaras Mountains (1978)
Highest range in Transylvanian Alps.
BT Romania
BT Southern Europe
BT Europe
SA Transylvania
SA Transylvanian Alps
SA Walachia

Fagus (1978)
Genus.
BT Dicotyledoneae
BT angiosperms
BT Spermatophyta
BT Plantae
SA gymnosperms

failure
As of 1981, no longer a valid term for GeoRef.
use failures

failures (1981)
Before 1981, search failure; rock failure.
UF failure
UF rock failure
SA deformation
SA faults
SA fractures
SA rock mechanics
SA rupture

SA slope stability
SA stress
SA tensile strength
SA weak rocks

Fair Isle   (1978)
Most southerly of the Shetland Islands. In 1985, broader term changed from Scotland to Shetland Islands.
SA Scotland
SA Shetland Islands

Fairbanks
Valid through 1988. Search in combination with state term. After 1988, use specific city-state term.

Fairbanks Alaska   (1989)
City in E central Alaska. Before 1989, search Fairbanks AND Alaska.
CO N645000N645000
W1475000W1475000
BT East-Central Alaska
BT Alaska
BT United States
SA Delta River
SA Fairbanks District

Fairbanks District   (1989)
E central Alaska. As of 1993, see Fairbanks mining district if applicable.
BT East-Central Alaska
BT Alaska
BT United States
SA Fairbanks Alaska

Fairbanks mining district   (1993)
Before 1993, also search Fairbanks District.
BT East-Central Alaska
BT Alaska
BT United States
SA metal ores
SA mines

Fairfax County
Valid through 1988. Search in combination with state term. After 1988, use specific county-state term.

Fairfax County Virginia   (1989)
On Potomac River in NE Virginia. Before 1989, also search Fairfax County AND Virginia.
CO N384000N390500
W0770500W0773500
BT Virginia
BT United States

Fairfield
Valid through 1988. Search in combination with state term. After 1988, use specific city-state term.

Fairfield County
Valid through 1988. Search in combination with state term. After 1988, use specific county-state term.

Fairfield County Connecticut   (1989)
Bounded E by Housatonic River on Long Island Sound in SW Connecticut. Before 1989, also search Fairfield County AND Connecticut.
CO N405900N413900
W0730500W0734400
BT Connecticut
BT United States
SA Weston Meteorite

Fairfield County Ohio   (1989)
Central Ohio. Before 1989, also search Fairfield County AND Ohio.

CO N393300N395700
W0822200W0825200
BT Ohio
BT United States

Fairfield County South Carolina   (1989)
N central South Carolina. Before 1989, also search Fairfield County AND South Carolina.
CO N341200N343500
W0804700W0812500
BT South Carolina
BT United States
SA Monticello Reservoir

Fairfield Utah   (1989)
Village in N central Utah. Before 1989, search Fairfield AND Utah.
CO N401500N401500
W1120600W1120600
BT Utah County Utah
BT Utah
BT United States

Fairview Formation   (1978)
In Maysville Group. Includes Mount Hope Member, Fairmount Member. SE Indiana, N central Kentucky, and SW Ohio.
BT Upper Ordovician
BT Ordovician
BT Paleozoic
SA Indiana
SA Kentucky
SA Ohio

Fairview Peak   (1978)
Location of earthquakes and microearthquakes in central Nevada.
IN Index county or regions as applicable.
SA Nevada

Fairweather Fault   (1978)
In the Fairweather Range area near the British Columbia border in SE Alaska.
IN Index Southeastern Alaska as applicable.
SA Southeastern Alaska

Faiyum
Not a valid term for GeoRef. See Fayum Egypt.

Falcon
No longer a valid term for GeoRef. As of 1993 see Falcon Venezuela.

Falcon Venezuela   (1993)
State on the Caribbean Sea in NW Venezuela. Before 1993 also search Falcon AND Venezuela.
BT Venezuela
BT South America
NT Paraguana Peninsula

Falkenau
use Sokolov

Falkland Islands   (1978)
British colony 300 miles E of Straits of Magellan. Islands are claimed by Argentina. Autoposting of South America began in 1981. Autoposting of Atlantic Ocean to this term ended in 1989. Autoposting of Atlantic Ocean Islands began in 1989. This term has multiple hierarchies.
UF Islas Malvinas
BT1 South America
BT2 Atlantic Ocean Islands
SA Antarctica
SA Atlantic Ocean
SA South Georgia
SA South Orkney Islands
SA South Sandwich Islands

Falkland Plateau   (1981)
E and NE of Falkland Islands.
CO S523000S473000
W0400000W0580000
BT Atlantic Ocean
SA DSDP Site 327
SA DSDP Site 329
SA DSDP Site 330

Falknov
use Sokolov

Fall River County
Valid through 1988. Search in combination with state term. After 1988, use specific county-state term.

Fall River County South Dakota   (1989)
Extreme SW South Dakota. Before 1989, also search Fall River County AND South Dakota.
CO N430000N432700
W1030000W1040500
BT South Dakota
BT United States

Fall River Formation   (1978)
In Inyan Kara Group. Includes Keyhole Sandstone Member. W South Dakota and NE Wyoming.
BT Lower Cretaceous
BT Cretaceous
BT Mesozoic
SA Inyan Kara Group
SA South Dakota
SA Wyoming

fallout   (1978)
UF nuclear fallout
UF radioactive fallout
SA ash falls
SA atmosphere
SA Chernobyl nuclear accident
SA isotopes
SA pollution
SA radioactivity

Falls County
Valid through 1988. Search in combination with state term. After 1988, use specific county-state term.

Falls County Texas   (1989)
E central Texas. Before 1989, also search Falls County AND Texas.
CO N305500N313000
W0963800W0972000
BT Texas
BT United States

False Bay   (1978)
E of Cape Good Hope.
BT Cape Province South Africa
BT South Africa
BT Southern Africa
BT Africa

False Cape   (1978)
SW tip of Dolak Island off S coast of Irian Jaya.
IN Index regions as applicable.
UF Cape Valse
UF Kaap Valsch
SA Irian Jaya Indonesia

famatinite   (1978)
BT sulfantimonates
BT sulfosalts
SA luzonite

Famennian   (1978)
Europe. Above Frasnian, below Tournaisian (Carboniferous).
BT Upper Devonian
BT Devonian
BT Paleozoic

FAMOUS   (1978)
Acronym. French-American expedition to the Mid-Atlantic Ridge, using submersibles.
UF French Mid-Ocean Undersea Study
SA Atlantic Ocean
SA expeditions
SA Mid-Atlantic Ridge
SA oceanography
SA research vessels
SA submersibles

fan deltas   (1993)
Before 1993, also search delta fans. This term has multiple hierarchies.
UF delta fans
BT1 alluvial fans
BT1 fluvial features
BT2 deltas
SA lacustrine features
SA shore features

fanglomerate   (1978)
BT clastic rocks
BT sedimentary rocks
SA terrigenous materials

Fanning Island   (1978)
One of the Line Islands S of Hawaii. Was part of British colony of Gilbert and Ellice Islands. Now part of Kiribati. In 1981, broader term was changed to Line Islands and Polynesia. In 1985, Kiribati was added as a broader term. This term has multiple hierarchies.
BT1 Kiribati
BT1 Oceania
BT2 Line Islands
BT2 Polynesia
BT2 Oceania
SA Pacific Ocean

fans
As of 1981, no longer a valid term for GeoRef. See alluvial fans; submarine fans; deltas.

Fantale   (1978)
In Ethiopian Rift valley in W Ethiopia.
UF Fantale Volcano
BT Ethiopia
BT East Africa
BT Africa

Fantale Volcano
use Fantale

Far East   (1978)
Easternmost Asia along the Pacific Ocean. Autoposting of this term began in 1981. As of 1981, Borneo may be indexed under Far East or Malay Archipelago. Brunei and Malay Peninsula are considered narrower terms as of 1981.
IN Index countries and regions as applicable.
BT Asia
NT Borneo
NT Burma
NT Cambodia
NT China
NT Hong Kong
NT Indochina
NT Indonesia
NT Japan
NT Korea
NT Laos
NT Lesser Sunda Islands
NT Malay Peninsula
NT Malaysia
NT Mongolia
NT Philippine Islands
NT Singapore

Before then, variants of the term may occur in GeoRef.

NT Sino-Korean Platform
NT Taiwan
NT Thailand
NT Vietnam
SA Malay Archipelago
SA Pacific region
SA Russian Far East

far-field (1993)
Before 1993, also search farfield.
UF farfield
SA environment
SA ground water
SA landfills
SA near-field
SA near-field spectra
SA pollution
SA seepage
SA seismic methods
SA seismology
SA underground disposal
SA underground installations
SA underground storage
SA waste disposal

Farallon Plate (1985)
Formerly existed between the Pacific and North American plates. Was subducted beneath North America during Cenozoic time.
SA North American Plate
SA Pacific Plate
SA plate tectonics
SA plates

farfield
use far-field

Faristan Iran
use Fars Iran

Farneto Cave (1981)
In province of Bologna, Italy.
CO N442500N443100
E0113000E0112000
BT Bologna Italy
BT Emilia-Romagna Italy
BT Italy
BT Southern Europe
BT Europe

Faroe Islands
Not a valid GeoRef index term after 1971. Was used on level 1 in subfile G.
use Faeroe Islands

Fars
No longer a valid term for GeoRef. See Fars Iran.

Fars Iran (1993)
Province in SW Iran. Before 1993, also search Fars. Before 1978, also search Faristan.
UF Faristan Iran
BT Iran
BT Middle East
BT Asia
NT Band-e-Amir Iran
NT Lar Iran
NT Shiraz Iran

farside (1978)
SA Moon

farsundite (1978)
As of 1993, this term is considered obsolete.
BT granites
BT plutonic rocks
BT igneous rocks

Farther India
use Indochina

fassaite (1978)
BT clinopyroxene
BT pyroxene group
BT chain silicates
BT silicates

SA augite

fatty acids (1978)
Autoposting of this term began in 1978.
BT organic materials
NT lipids
SA acids
SA amino acids
SA ketones

faujasite (1978)
BT zeolite group
BT framework silicates
BT silicates

fault blocks
use block structures

fault escarpment
use fault scarps

fault ledge
use fault scarps

fault planes (1978)
Before 1972, also search fault-plane solutions.
UF fault-plane solutions
UF planes, fault
SA faults
SA focal mechanism

fault scarps (1978)
UF cliff of displacement
UF fault escarpment
UF fault ledge
SA cliffs
SA faults
SA scarps
SA slopes

fault zones (1978)
UF zones, fault
SA duplexes
SA faults
SA lineaments
SA plate tectonics
SA pull-apart basins
SA rift zones
SA transfer faults

fault-plane solutions
use fault planes

faulting
No longer a valid GeoRef index term. Before 1972, was used on levels 1 and 2 in subfiles E and G.
use faults

faults (1978)
To be used for studies primarily stressing individual faults or systems of faults. For relationships with other structures, see tectonics. Autoposting of this term began in 1978. Before 1972, also search faults and faulting; faulting.
IN Index specific type of fault or system where applicable.
UF faulting
UF faults and faulting
NT active faults
NT arcuate faults
NT bedding faults
NT boundary faults
NT conjugate faults
NT detachment faults
NT diagonal-slip faults
NT dip faults
NT dip-slip faults
NT en echelon faults
NT extension faults
NT growth faults
NT high-angle faults
NT hinge faults
NT lateral faults
NT listric faults
NT low-angle faults
NT normal faults

NT oblique-slip faults
NT overthrust faults
NT parallel faults
NT peripheral faults
NT radial faults
NT reverse faults
NT step faults
NT strike faults
NT strike-slip faults
NT tear faults
NT thrust faults
NT transcurrent faults
NT transfer faults
NT transform faults
NT transverse faults
NT underthrust faults
NT wrench faults
SA allochthons
SA asperities
SA basin range structure
SA block structures
SA breccia
SA decollement
SA deformation
SA displacements
SA drag folds
SA duplexes
SA earthquakes
SA failures
SA fault planes
SA fault scarps
SA fault zones
SA folds
SA foliation
SA foot wall
SA fracture zones
SA fractures
SA geologic hazards
SA gouge
SA grabens
SA half grabens
SA hanging wall
SA horsts
SA klippen
SA lineaments
SA mechanics
SA mylonitization
SA nappes
SA neotectonics
SA nuclear facilities
SA oblique orientation
SA orientation
SA orogeny
SA plate tectonics
SA ramps
SA reactivation
SA rupture
SA salt tectonics
SA scarps
SA segmentation
SA seismic gaps
SA seismic moment
SA seismology
SA shear zones
SA slickensides
SA slip rates
SA stick-slip
SA strike
SA structural analysis
SA structural geology
SA systems
SA tectonic elements
SA tectonic wedges
SA tectonics
SA tectonophysics
SA thrust sheets
SA transpression
SA ultramylonite
SA wall rocks

faults and faulting
No longer a valid GeoRef index term. Before 1971, was used on level 1 in subfile N.
use faults

fauna
As of 1981, no longer a valid term for GeoRef. See faunal studies.

faunal assemblages
use assemblages

faunal list (1978)
UF list, faunal
SA floral list
SA fossils

faunal provinces (1978)
UF provinces, faunal
SA assemblages
SA biogeography
SA biotopes
SA floral provinces
SA land bridges
SA paleontology

faunal studies (1978)
IN May be used for many classes, orders, or genera, or when an appropriate systematic term is not available.
UF studies, faunal
SA biota
SA floral studies
SA fossils
SA growth rates
SA living taxa
SA microorganisms

Favosites (1978)
Genus. Autoposting of Zoantharia to this term began in 1989.
BT Favositidae
BT Tabulata
BT Zoantharia
BT Anthozoa
BT Coelenterata
BT Invertebrata
SA Palaeofavosites

Favositidae (1970)
Family. Autoposting of Zoantharia to this term began in 1989.
BT Tabulata
BT Zoantharia
BT Anthozoa
BT Coelenterata
BT Invertebrata
NT Favosites
NT Palaeofavosites

Fawley
No longer a valid term for GeoRef. As of 1993, see Fawley England.

Fawley England (1993)
Town and parish SSE of Southampton in S England. Before 1993, also search Fawley AND England.
BT Hampshire England
BT England
BT Great Britain
BT United Kingdom
BT Western Europe
BT Europe

fayalite (1978)
BT olivine group
BT nesosilicates
BT orthosilicates
BT silicates

Fayette County
Valid through 1988. Search in combination with state term. After 1988, use specific county-state term.

Fayette County Alabama (1989)
W Alabama. Before 1989, also search Fayette County AND Alabama.
CO N333100N335400
W0872300W0875600

BT Alabama
BT United States

Fayette County Georgia (1989)
W central Georgia. Before 1989, also search Fayette County AND Georgia.
CO N331400N333400
W0842200W0843800
BT Georgia
BT United States

Fayette County Illinois (1989)
S central Illinois. Before 1989, also search Fayette County AND Illinois.
CO N384500N391200
W0884200W0891500
BT Illinois
BT United States

Fayette County Indiana (1989)
E Indiana. Before 1989, also search Fayette County AND Indiana.
CO N393200N394100
W0850200W0851800
BT Indiana
BT United States

Fayette County Iowa (1989)
NE Iowa. Before 1989, also search Fayette County AND Iowa.
CO N423800N430500
W0913800W0920600
BT Iowa
BT United States

Fayette County Kentucky (1989)
Central Kentucky. Before 1989, also search Fayette County AND Kentucky.
CO N375200N381300
W0841800W0844000
BT Kentucky
BT United States

Fayette County Ohio (1989)
S central Ohio. Before 1989, also search Fayette County AND Ohio.
CO N392200N394300
W0831500W0834000
BT Ohio
BT United States

Fayette County Pennsylvania (1989)
SW Pennsylvania. Before 1989, also search Fayette County AND Pennsylvania.
CO N394300N400800
W0791700W0800000
BT Pennsylvania
BT United States

Fayette County Tennessee (1989)
SW Tennessee. Before 1989, also search Fayette County AND Tennessee.
CO N345900N352500
W0891200W0894000
BT Tennessee
BT United States

Fayette County Texas (1989)
S central Texas. Before 1989, also search Fayette County AND Texas.
CO N293800N301100
W0963400W0971800
BT Texas
BT United States

Fayette County West Virginia (1989)
S central West Virginia. Before 1989, also search Fayette County AND West Virginia.
CO N374900N381600
W0804600W0812300
BT West Virginia
BT United States

Fayetteville
Valid through 1988. Search in combination with state term. After 1988, use specific city-state term.

Fayetteville Arkansas (1989)
City in NW Arkansas. Before 1989, search Fayetteville AND Arkansas.
CO N360300N360300
W0842800W0842800
BT Washington County Arkansas
BT Arkansas
BT United States

Fayetteville Formation (1978)
Includes Mayes Limestone Member. N Arkansas, S Missouri; and NE, central, and E Oklahoma.
BT Upper Mississippian
BT Mississippian
BT Carboniferous
BT Paleozoic
SA Arkansas
SA Missouri
SA Oklahoma

Fayetteville Meteorite (1989)
Impact in Washington County, NW Arkansas. Olivine-bronzite chondrite.
BT H chondrites
BT chondrites
BT stony meteorites
BT meteorites
SA Arkansas
SA Washington County Arkansas

Fayum
No longer a valid term for GeoRef. See Fayum Egypt.

Fayum Depression (1978)
Bed of ancient Lake Moeris in N Upper Egypt SW of Cairo. Also search Fayum.
BT Egypt
BT North Africa
BT Africa
SA Fayum Egypt

Fayum Egypt (1993)
Province in Upper Egypt, SW of Cairo. Also search Faiyum.
BT Egypt
BT North Africa
BT Africa
SA Fayum Depression

Fayyum
Not a valid term for GeoRef. See Fayum Egypt.

Fe
use iron

Fe-55 (1978)
Autoposting of broader terms began in 1989. This term has multiple hierarchies.
BT1 radioactive isotopes
BT1 isotopes
BT2 iron
BT2 metals

Fe-57 (1978)
Autoposting of broader terms began in 1989. This term has multiple hierarchies.
BT1 stable isotopes
BT1 isotopes
BT2 iron
BT2 metals

Fe-59 (1989)
This term has multiple hierarchies.
BT1 radioactive isotopes
BT1 isotopes
BT2 iron
BT2 metals

Fe2+
use ferrous iron

Fe3+
use ferric iron

feasibility
use feasibility studies

feasibility studies (1978)
UF feasibility
UF studies, feasibility
SA cutoff grade
SA highways
SA marine installations
SA mineral exploration
SA nuclear facilities
SA reservoirs
SA site exploration
SA tunnels
SA underground installations

Feather River (1981)
N central California.
IN Index counties or regions as applicable.
SA California
SA Oroville Dam

feathers (1989)
SA Aves
SA biostratigraphy
SA birds

features, bottom
use bottom features

features, buried
use buried features

features, cryptoexplosion
use cryptoexplosion features

features, eolian
use eolian features

features, erosion
use erosion features

features, fluvial
use fluvial features

features, frost
use frost features

features, glacial
use glacial features

features, impact
use impact features

features, lacustrine
use lacustrine features

features, periglacial
use periglacial features

features, shore
use shore features

features, solution
use solution features

features, surface
use surface features

features, volcanic
use volcanic features

fecal pellets (1993)
UF faecal pellets
SA biogenic structures
SA coprolites
SA marine sediments

Federal Capital Territory
Not a valid term for GeoRef. See Australian Capital Territory Australia.

Federal District
No longer a valid term for GeoRef. As of 1993, term may be used with country where it is larger than the city it contains. See Buenos Aires Argentina, Federal District Mexico, Federal District Venezuela, Rio de Janeiro City Brazil, or Kinshasa Zaire as applicable.

Federal District Argentina
use Buenos Aires City Argentina

Federal District Mexico (1993)
Central Mexico. Surrounded on west, north, and east by Mexico state, but independent of it. Includes Mexico City Mexico. Before 1993, also search Federal District or Distrito Federal AND Mexico.
UF Mexico D.F.
BT Mexico state
BT Mexico
NT Mexico City Mexico
SA Valley of Mexico

Federal District Venezuela (1993)
N central Venezuela. Region not within other states containing the capital city of Caracas. Before 1993, also search Federal District or Distrito Federal AND Venezuela.
BT Venezuela
BT South America
NT Caracas Venezuela
SA Caracas Group

Federal Republic of Germany
Not a valid term for GeoRef. As of 1993, see Germany and also Mecklenburg-Western Pomerania Germany, Brandenburg Germany, Saxony-Anhalt Germany, Thuringia Germany, and Saxony Germany.

Federated Shan States
use Shan State Burma

Federated States of Micronesia
use Micronesia

Fedorov stage
use universal stage

feeding ground
use drainage basins

feldspar
As of 1981, no longer a valid term for GeoRef. See feldspar group. Use feldspar deposits for feldspar as a commodity.

feldspar deposits (1981)
Before 1981, search feldspar AND deposits.
IN Commodity. See List C.
SA feldspar group
SA feldspathization

feldspar group (1978)
Before 1981, maskelynite was included as a narrower term. Through 1977, also search each individual feldspar mineral to retrieve all feldspars. As of 1978, feldspar group is autoposted to individual feldspar minerals.
BT framework silicates
BT silicates
NT alkali feldspar
NT barium feldspar
NT plagioclase
SA argillization
SA feldspar deposits

feldspathization (1978)

BT metasomatism
SA feldspar deposits
SA granitization
SA metamorphism

feldspathoid rocks (1985)
BT igneous rocks
SA alkalic composition
SA nepheline group
SA sodalite group

Felidae (1978)
Family. Autoposting of Fissipeda, Eutheria and Theria to this term began in 1989.
BT Fissipeda
BT Carnivora
BT Eutheria
BT Theria
BT Mammalia
BT Tetrapoda
BT Vertebrata
BT Chordata
SA Smilodon

felsite (1978)
BT granites
BT plutonic rocks
BT igneous rocks

Feltville Formation (1989)
In Newark Group. NE New Jersey.
BT Lower Jurassic
BT Jurassic
BT Mesozoic
SA New Jersey
SA Newark Supergroup

Fenestellidae (1978)
Family.
BT Cryptostomata
BT Bryozoa
BT Invertebrata

fenite (1978)
BT metasomatic rocks
BT metamorphic rocks
SA fenitization
SA metasomatism

fenitization (1978)
BT metasomatism
SA alteration
SA fenite

Fennoscandia (1978)
Geological usage for that part of N Europe consisting of Denmark, Finland and adjacent part of the Russian Federation, Norway, and Sweden.
IN Index countries as applicable.
BT Europe
NT Russian Fennoscandia
SA Baltic Shield
SA Denmark
SA Finland
SA Norway
SA Russian Federation
SA Sweden

Fennoscandian Shield
No longer a valid term for GeoRef.
use Baltic Shield

fensters
use windows

Fenton Hill (1981)
Los Alamos Scientific Laboratory's site for hot-dry-rock geothermal energy experiments. Adjacent to Valles Caldera and Nacimiento Mountains in NW central New Mexico. Autoposting of broader terms began in 1989.
CO N354300N360000
    W1063700W1065700
BT Sandoval County New Mexico

BT New Mexico
BT United States
SA Los Alamos Scientific Laboratory
SA Nacimiento Mountains
SA Valle Grande Mountains
SA Valles Caldera

ferberite (1978)
BT tungstates

Fergana
Not a valid term for GeoRef. Use Fergana Basin.

Fergana Basin (1978)
Mountain enclosed steppe and desert region. Also search Fergana. This term has multiple hierarchies.
IN Index former Soviet republics as applicable.
CO N394500N413000
    E0730000E0693000
UF Fergana Valley
BT1 Commonwealth of Independent States
BT2 Central Asia
BT2 Asia
SA Kyrgyzstan
SA Tadzhikistan
SA Uzbekistan

Fergana Valley
use Fergana Basin

Fergus County
Valid through 1988. Search in combination with state term. After 1988, use specific county-state term.

Fergus County Montana (1989)
Central Montana. Before 1989, also search Fergus County AND Montana.
CO N464200N475000
    W1082000W1101400
BT Montana
BT United States

fergusonite (1978)
Autoposting of niobotantalates to this term began in 1989. This term has multiple hierarchies.
BT1 niobotantalates
BT1 niobates
BT1 oxides
BT2 niobotantalates
BT2 tantalates
BT2 oxides

Fermanagh
No longer a valid term for GeoRef. As of 1993, see Fermanagh Northern Ireland.

Fermanagh Northern Ireland (1993)
Traditional county in SW Northern Ireland. Includes the smaller administrative region of the same name adopted after 1973.
BT Northern Ireland
BT United Kingdom
BT Western Europe
BT Europe

fermium (1989)
Chemical element.
UF Fm
BT actinides
BT metals

Fernando Po (1978)
Province in Bight of Biafra. Also an island.
UF Fernando Poo
BT Equatorial Guinea
BT Central Africa
BT Africa

Fernando Poo
use Fernando Po

Fernie Basin (1989)
SE British Columbia.
BT British Columbia
BT Western Canada
BT Canada

Fernie Formation (1989)
Rocky Mountains of SW Alberta, SE British Columbia and NW Montana.
BT Jurassic
BT Mesozoic
SA Alberta
SA British Columbia
SA Montana

ferns (1978)
As of 1981, restricted to use as the common name for pteridophytes.
IN Index for all non-paleontologic studies of fossils.
BT plants
SA biostratigraphy
SA pteridophytes

Fernvale Formation (1978)
In Richmond Group and Patterson Ranch Group. N Arkansas, SW Illinois, SE Missouri, W Tennessee, NW Alabama, central E and NE Oklahoma.
BT Upper Ordovician
BT Ordovician
BT Paleozoic
SA Alabama
SA Arkansas
SA Illinois
SA Missouri
SA Oklahoma
SA Richmond Group
SA Tennessee

Ferralites (1982)
BT soils
SA soil group

Ferralsols (1985)
BT soils

Ferrar Group (1978)
In Victoria Land W of Ross Sea on Pacific Ocean side.
BT Jurassic
BT Mesozoic
SA Antarctica
SA Victoria Land

ferric iron (1978)
Used for iron with a valence of 3. Before 1978, search iron AND ferric.
UF Fe3+
SA iron

ferricrete (1978)
BT chemically precipitated rocks
BT sedimentary rocks
SA conglomerate
SA duricrust

ferrierite (1978)
BT zeolite group
BT framework silicates
BT silicates

ferrifayalite
use laihunite

ferrihydrite (1985)
BT oxides

ferrimolybdite (1978)
BT molybdates

ferroan dolomite
use ankerite

ferrocarpholite (1978)
BT chain silicates

BT silicates
SA carpholite

ferrodiorite (1978)
UF ferrogabbro
BT diorites
BT plutonic rocks
BT igneous rocks

ferrogabbro
use ferrodiorite

ferrohastingsite (1978)
BT clinoamphibole
BT amphibole group
BT chain silicates
BT silicates
SA hastingsite

ferromanganese
A valid term through 1977.

ferromanganese composition (1978)
Before 1978, also search ferromanganese.
SA composition
SA iron
SA manganese
SA manganese composition
SA nodules

Ferron Sandstone Member (1978)
Of Mancos Shale. Central E Utah.
BT Upper Cretaceous
BT Cretaceous
BT Mesozoic
SA Mancos Shale
SA Utah

ferropseudobrookite (1978)
BT oxides
SA pseudobrookite

ferrosilicon (1978)
BT silicides
BT alloys
SA iron
SA silicon

ferrosilite (1978)
BT chain silicates
BT silicates

ferrospinel
use hercynite

ferrous iron (1978)
Used for the bivalent iron. Before 1978, search iron AND ferrous.
UF Fe2+
SA iron

ferruginous composition (1978)
Before 1978, also search ferruginous. As of 1981, use only for sedimentary formations.
SA composition
SA iron
SA iron formations
SA iron-rich composition
SA red beds
SA sedimentary rocks
SA sediments

ferruginous quartzite (1978)
Autoposting of quartzites to this term began in 1989.
BT quartzites
BT metamorphic rocks
SA orthoquartzite

Ferruginous soils (1982)
BT soils
SA soil group

Ferry County
Valid through 1988. Search in combination with state term. After 1988, use specific county-state term.

Ferry County Washington
(1989)
On Canadian border in NE Washington. Before 1989, also search Ferry County AND Washington.
CO N475000N490000
    W1181000W1185000
BT Washington
BT United States

fertilization (1978)
SA agriculture
SA fertilizers
SA nutrients
SA soil treatment
SA soils

fertilizers (1978)
SA agriculture
SA fertilization
SA nitrogen
SA nutrients
SA phosphorus
SA soil management
SA soils
SA yields

Ferto to
  use Lake Neusiedler

Fezzan (1978)
Desert region in SW Libya.
BT Libya
BT North Africa
BT Africa

fibrolite
  use sillimanite

Fichtelgebirge (1978)
Mountain range in NE Bavaria.
BT Bavaria Germany
BT Germany
BT Central Europe
BT Europe
SA Munchberg Gneiss Massif

field capacity (1981)
SA ground water
SA soils

field crops
As of 1981, no longer a valid term for GeoRef.

field methods
Not a valid term for GeoRef. As of 1989, use field studies AND methods.

field studies (1978)
Before 1971, also search field geology; field tests; field work.
UF studies, field
SA deformation
SA sampling
SA soil sampling
SA soils

field trips (1981)
SA guidebook
SA road log

field, crystal
  use crystal field

field, electrical
  use electrical field

field, electromagnetic
  use electromagnetic field

field, gravity
  use gravity field

field, magnetic
  use magnetic field

fields, coal
  use coal fields

fields, geothermal
  use geothermal fields

fields, giant
  use giant fields

fields, lava
  use lava fields

fields, oil and gas
  use oil and gas fields

Fife
No longer a valid term for GeoRef as of 1993. See Fife region Scotland.

Fife region Scotland (1993)
Administrative region and former county in E Scotland. Before 1993, also search Fife.
UF Fife Scotland
BT Scotland
BT Great Britain
BT United Kingdom
BT Western Europe
BT Europe
SA Tay Estuary

Fife Scotland
  use Fife region Scotland

Fig Tree Series (1978)
In Barberton Mountain Land in SE Transvaal.
BT Precambrian
SA Barberton greenstone belt
SA South Africa
SA Transvaal South Africa

figure of Earth (1978)
Theoretical approximation of the Earth, using an ellipsoid of revolution whose surface is an equipotential surface which is very similar to the mean surface of the ocean.
SA Earth
SA geodesy
SA geoid

Fiji (1978)
Independent state consisting of island group between New Caledonia and Samoa in S Pacific Ocean.
CO S192000S160000
    W1780000E1770000
UF Fiji Islands
BT Melanesia
BT Oceania
NT Vanua Levu
NT Viti Levu
SA Pacific Ocean

Fiji Islands
  use Fiji

Fiji Plateau (1978)
Just E of Fiji in S Pacific Ocean.
BT Pacific Ocean

Filchner Ice Shelf (1989)
Bordered by Weddell Sea, Ronne Ice Shelf and Coats Land, Antarctica.
CO S830000S773000
    W0300000W0600000
BT Antarctica
SA Ronne Ice Shelf
SA Weddell Sea

Filicales
  use Filicopsida

Filicopsida (1978)
Including Filicales. Before 1981, Callipteris was included as a narrower term. Autoposting of this term began in 1978.
UF Filicales
BT pteridophytes
BT Plantae
NT Archaeopteris

NT Azolla
NT Cladophlebis
NT Dicroidium
NT Gleicheniaceae
NT Pecopteris
NT Sphenopteris
NT Taeniopteris
SA Callipteris
SA Cambrian
SA Juglans
SA soil sampling

Filizchay (1978)
Ore bearing region.
IN Index regions as applicable.
SA Azerbaidzhan

Fillmore County
Valid through 1988. Search in combination with state term. After 1988, use specific county-state term.

Fillmore County Minnesota (1989)
SE Minnesota. Before 1989, also search Fillmore County AND Minnesota.
CO N433100N435200
    W0914300W0922800
BT Minnesota
BT United States
SA Mystery Cave
SA Root River

Fillmore County Nebraska (1989)
SE Nebraska. Before 1989, also search Fillmore County AND Nebraska.
CO N402200N403900
    W0972300W0975100
BT Nebraska
BT United States

Fillmore Formation (1978)
Overlies House Limestone; underlies Wahwah Limestone. E Nevada and W central Utah.
BT Lower Ordovician
BT Ordovician
BT Paleozoic
SA Nevada
SA Pogonip Group
SA Utah

filtering
  use filters

filters (1978)
Used for all types including mechanical, optical, signal or wave, and mathematical.
UF filtering
SA adaptive filters
SA analog filters
SA continuous filters
SA deconvolution
SA discrete filters
SA dispersive filters
SA distortion
SA elastic waves
SA Kalman filters
SA multichannel filters
SA multichannel methods
SA noise
SA numerical filters
SA optimal filters
SA passband filters
SA recursive filters
SA seismic methods
SA seismology
SA signals
SA spatial frequency filters
SA wiener filters

filtration (1978)

As of 1978, term is restricted to use with sediments, i.e. for process of removing suspended material from a liquid.
SA processes
SA sediments

financing (1981)
SA economics

fine-grained materials
  use fines

fines (1978)
UF fine-grained materials
SA agglutinates
SA clastic rocks
SA clastic sediments
SA coarse-grained materials
SA grain size
SA granular materials
SA materials
SA mining
SA particles
SA particulate materials
SA sedimentary rocks
SA sediments
SA soils
SA textures
SA unconsolidated materials

Finger Lakes (1978)
Group of long narrow lakes in W central New York.
BT New York
BT United States
NT Cayuga Lake
NT Onondaga Lake
NT Seneca Lake
SA Cross Lake

Finistere
No longer a valid term for GeoRef. See Finistere France.

Finistere France (1993)
Department on W tip of Brittany.
CO N474000N484000
    W0034000W0050000
BT France
BT Western Europe
BT Europe
NT Brest France
NT Chateaulin Basin
NT Crozon Peninsula
NT Morlaix France
SA Brittany

finite difference analysis (1978)
Before 1978, also search finite difference method; finite difference.
UF finite difference method
BT statistical analysis
SA analysis
SA numerical analysis
SA seismic migration

finite difference method
  use finite difference analysis

finite element analysis (1978)
UF finite-element analysis
BT statistical analysis
SA analysis
SA boundary element analysis
SA Galerkin method
SA numerical analysis

finite strain
As of 1981, no longer a valid term for GeoRef. Use finite strain analysis.

finite strain analysis (1981)
Before 1981, search finite strain.
SA deformation
SA rock mechanics
SA soil mechanics
SA strain
SA stress

finite-element analysis
  use finite element analysis
Finland (1978)
  CO N594500N700000
     E0314500E0190000
  BT Scandinavia
  BT Western Europe
  BT Europe
  NT Abo Finland
  NT Inari Finland
  NT Lake Lappajarvi
  NT Outokumpu Finland
  NT Tampere Finland
  NT Ylojarvi Finland
  SA Ancylus Lake
  SA Arctic region
  SA Baltic ice lake
  SA Baltic region
  SA Baltic Shield
  SA Bjurbole Meteorite
  SA Fennoscandia
  SA Gulf of Bothnia
  SA Gulf of Finland
  SA Haveroe Meteorite
  SA Koli Nappe
  SA Ladoga Series
  SA Lake Ladoga region
  SA Lapland
  SA Litorina Sea
  SA Yoldia Sea
Finney County
  Valid through 1988. Search in combination with state term. After 1988, use specific county-state term.
Finney County Kansas (1989)
  SW Kansas. Before 1989, also search Finney County AND Kansas.
  CO N374300N381500
     W1001500W1010700
  BT Kansas
  BT United States
Finnmark
  No longer a valid term for GeoRef. As of 1993 see Finnmark Norway.
Finnmark Norway (1993)
  County in N Norway. Before 1993 also search Finnmark AND Norway.
  BT Norway
  BT Scandinavia
  BT Western Europe
  BT Europe
  NT Alta Norway
  NT Porsang Fjord
  NT Seiland
  NT Tana Fjord
  NT Varanger Fjord
  NT Varanger Peninsula
  SA Porsanger Dolomite Formation
fiord
  use fjords
Fiordland National Park (1978)
  SW South Island. Also search Fiordland.
  IN Also index South Island.
  UF Sounds National Park
  BT Southland New Zealand
  BT New Zealand
  BT Australasia
  SA national parks
  SA South Island
fire clay
  use fireclay
Fire Island (1978)
  Long, narrow sand spit off S Long Island.
  IN Index county or regions as applicable.
  SA Long Island
  SA New York
  SA Suffolk County New York
fire-clay
  use fireclay
fireclay (1978)
  UF fire clay
  UF fire-clay
  UF refractory clay
  SA clay
  SA refractory materials
  SA underclay
Firenze Italy
  use Florence Italy
fires (1993)
  SA damage
  SA deforestation
  SA ecology
  SA erosion
  SA forests
  SA geologic hazards
  SA revegetation
  SA wilderness areas
firn (1978)
  SA glacial features
  SA glacial geology
  SA glaciers
  SA periglacial features
  SA snow
Firth of Clyde (1978)
  An estuary of the Clyde River in SW Scotland. Also search Clyde River.
  BT Scotland
  BT Great Britain
  BT United Kingdom
  BT Western Europe
  BT Europe
  SA Argyllshire Scotland
  SA Ayrshire Scotland
  SA Strathclyde region Scotland
Firth of Tay
  use Tay Estuary
fish (1978)
  As of 1981, restricted to use as the common name for Pisces. Autoposting of this term began in 1978.
  IN Index for all non-paleontologic studies of fossils.
  BT vertebrates
  NT acanthodians
  NT selachians
  SA biostratigraphy
  SA ichthyoliths
  SA Pisces
Fish Canyon Tuff (1989)
  Located in the San Juan volcanic field, SW Colorado.
  BT Oligocene
  BT Paleogene
  BT Tertiary
  BT Cenozoic
  SA Colorado
  SA San Juan volcanic field
Fish Haven Dolomite (1989)
  Located in the Great Basin of Idaho, Nevada and Utah.
  BT Upper Ordovician
  BT Ordovician
  BT Paleozoic
  SA Idaho
  SA Nevada
  SA Utah
Fiskenaesset
  No longer a valid term for GeoRef. As of 1993 see Fiskenaesset Greenland.
Fiskenaesset Greenland (1993)
  Settlement on SW coast of Greenland. Before 1993 also search Fiskenaesset AND Greenland.
  BT Greenland
  BT Arctic region
Fisset Brook Formation (1989)
  Cape Breton Island, N Nova Scotia. Previously thought to be of Carboniferous age.
  BT Middle Devonian
  BT Devonian
  BT Paleozoic
  SA Nova Scotia
fission (1978)
  UF nuclear fission
  SA energy
  SA fission tracks
  SA fission-track dating
  SA fusion
  SA particles
fission tracks (1978)
  Used for the phenomena. For method, use fission-track dating.
  SA fission
  SA fission-track dating
  SA particle tracks
  SA particle-track dating
  SA pits
  SA radiation damage
  SA radioactivity
fission-track
  A valid term through 1977. After 1977, use fission-track dating.
fission-track dating (1978)
  Before 1978, also search fission-track; fission tracks; fission-track method.
  UF dating, fission-track
  UF fission-track method
  UF spontaneous fission-track dating
  SA fission
  SA fission tracks
  SA geochronology
  SA methods
  SA particle tracks
  SA particle-track dating
  SA radiation damage
  SA relative age
fission-track method
  use fission-track dating
Fissipeda (1981)
  Suborder. Autoposting of this term to Canidae, Felidae, Hyaenidae, Mustelidae and Ursidae began in 1989. Autoposting of Eutheria and Theria to this term began in 1989.
  UF Fissipedia
  BT Carnivora
  BT Eutheria
  BT Theria
  BT Mammalia
  BT Tetrapoda
  BT Vertebrata
  BT Chordata
  NT Canidae
  NT Felidae
  NT Hyaenidae
  NT Mustelidae
  NT Ursidae
Fissipedia
  use Fissipeda
fissures (1978)
  SA cracks
  SA fractures
  SA joints
fixation (1978)
  Used in geochemistry, e.g. for fixation of ions.
  SA geochemistry
  SA ions
fjords (1978)
  This term has multiple hierarchies.
  UF fiord
  UF fyord
  BT1 shore features
  BT2 glacial features
  SA bays
  SA coastlines
  SA embayments
  SA estuaries
  SA geomorphology
  SA glacial geology
  SA glaciation
  SA tidewater glaciers
Flack Lake (1978)
  SA Ontario
Flagstaff
  Valid through 1988. Search in combination with state term. After 1988, use specific city-state term.
Flagstaff Arizona (1989)
  City in N central Arizona. Before 1989, search Flagstaff AND Arizona.
  CO N351200N351200
     W1113800W1113800
  BT Coconino County Arizona
  BT Arizona
  BT United States
Flagstaff Formation (1993)
  Eocene and Paleocene. Central eastern Utah.
  IN Index ages as applicable.
  BT Paleogene
  BT Tertiary
  BT Cenozoic
  SA Eocene
  SA Green River Formation
  SA Paleocene
  SA Utah
flame emission spectrometry
  use flame photometry
flame photometric analysis
  Not a valid GeoRef index term after 1970. Was used on level 1 in subfile N.
  use flame photometry
flame photometry (1978)
  Before 1971, also search flame photometric analysis.
  UF flame emission spectrometry
  UF flame photometric analysis
  SA analysis
  SA chemical analysis
  SA photometry
  SA quantitative analysis
  SA spectroscopy
flame structures (1978)
  BT soft sediment deformation
  BT sedimentary structures
  SA antidunes
  SA load casts
Flandrian (1978)
  Europe.
  BT Holocene
  BT Quaternary
  BT Cenozoic
flares
  A valid term through 1978. Now use solar flares.
flaser bedding (1978)
  BT planar bedding structures
  BT sedimentary structures
  SA bedding

Flathead County
    Valid through 1988. Search in combination with state term. After 1988, use specific county-state term.
Flathead County Montana (1989)
    NW Montana. Before 1989, also search Flathead County AND Montana.
    CO N473500N490000
        W1125200W1150100
    BT Montana
    BT United States
    SA Flathead Lake
Flathead Lake (1978)
    NW Montana.
    IN Index counties or regions as applicable.
    SA Flathead County Montana
    SA Lake County Montana
    SA Montana
Flathead Sandstone (1978)
    In Wyoming, unconformably overlies Precambrian granite and underlies Gallatin Limestone and Gros Ventre Formation. Montana and NW Wyoming.
    BT Middle Cambrian
    BT Cambrian
    BT Paleozoic
    SA Montana
    SA Wyoming
Fleming County
    Valid through 1988. Search in combination with state term. After 1988, use specific county-state term.
Fleming County Kentucky (1989)
    NE Kentucky. Before 1989, also search Fleming County AND Kentucky.
    CO N381000N383500
        W0832500W0835500
    BT Kentucky
    BT United States
    NT Maxey Flats
Fleming Formation (1978)
    In Grand Gulf Group. Includes Lena, Carnahan Bayou, Dough Hills, Williamson Creek, Castor Creek, and Blounts Creek members. Also considered a group name including Oakville and Cuero formations. E Texas, and W Louisiana.
    BT Miocene
    BT Neogene
    BT Tertiary
    BT Cenozoic
    SA Louisiana
    SA Texas
Flemish Cap (1978)
    A marine bank NE of Grand Banks off Newfoundland.
    BT Atlantic Ocean
Fleurieu Peninsula (1978)
    S of Adelaide between Gulf Saint Vincent and Indian Ocean.
    BT South Australia
    BT Australia
    BT Australasia
flexural slip
    use flexural-slip
flexural-slip (1978)
    UF bedding-plane slip
    UF flexural slip
    SA decollement
    SA disharmonic folds
    SA folds
    SA mechanics
flexure (1978)
    Used only to indicate the mechanics of deformation. Before 1978, also used to indicate style of structure. Now use flexure folds or growth faults for style of structure.
    SA deformation
    SA flexure folds
    SA growth faults
flexure faults
    use growth faults
flexure folds (1978)
    Before 1978, also search flexure AND folds.
    BT folds
    SA flexure
    SA monoclines
Flin Flon
    No longer a valid term for GeoRef. See Flin Flon Manitoba.
Flin Flon Manitoba (1993)
    Town near Saskatchewan border in W Manitoba.
    BT Manitoba
    BT Western Canada
    BT Canada
Flinders Island (1978)
    Largest island of the Furneaux Islands off NE Tasmania.
    IN Index state or regions as applicable.
    SA Tasmania Australia
Flinders Range (1993)
    Between Lake Frome and Lake Torrens in E central South Australia. Before 1993, also search Flinders Ranges.
    CO S323000S293000
        E1393000E1380000
    UF Flinders Ranges
    BT South Australia
    BT Australia
    BT Australasia
Flinders Ranges
    No longer a valid term for GeoRef.
    use Flinders Range
flint (1978)
    BT chemically precipitated rocks
    BT sedimentary rocks
    SA chert
flint clay (1978)
    BT clastic sediments
    BT sediments
    SA clay
    SA clays
    SA refractory materials
    SA terrigenous materials
Flint Creek Range (1978)
    Range of Rocky Mountains in SW Montana.
    IN Index counties and U. S. Rocky Mountains as applicable.
    SA Montana
    SA U. S. Rocky Mountains
Flint Hills (1978)
    IN Index states as applicable.
    SA Kansas
    SA Oklahoma
    SA United States
Flinton Group (1978)
    E Ontario.
    BT Precambrian
    SA Canada
    SA Ontario
floatstone (1993)
    As of 1993, use opal for the mineral.
    SA carbonate rocks
    SA opal
    SA quartz
flocculation (1978)
    SA clay mineralogy
    SA colloidal materials
    SA geochemistry
    SA processes
    SA suspended materials
flokite
    use mordenite
flood basalts (1993)
    UF plateau basalts
    BT basalts
    BT volcanic rocks
    BT igneous rocks
    SA basalt flows
    SA lava flows
flood plains
    use floodplains
flood tuff
    use ignimbrite
floodplains (1978)
    Autoposting of fluvial features to this term began in 1989.
    UF flood plains
    BT fluvial features
    SA alluvium
    SA floods
    SA geomorphology
    SA levees
    SA meanders
    SA plains
    SA rivers
    SA terraces
floods (1978)
    SA environmental geology
    SA floodplains
    SA geologic hazards
    SA hydrology
    SA jokulhlaups
    SA paleofloods
    SA reservoirs
    SA rivers and streams
    SA sand boils
    SA surface water
    SA watersheds
    SA waterways
flora
    As of 1981, no longer a valid term for GeoRef. See floral studies.
flora studies
    use floral studies
floral list (1978)
    SA faunal list
floral provinces (1978)
    UF provinces, floral
    SA assemblages
    SA biogeography
    SA biotopes
    SA faunal provinces
    SA land bridges
    SA paleobotany
floral studies (1978)
    IN May be used for many classes, orders, or genera, or when an appropriate systematic term is not available.
    UF flora studies
    UF studies, floral
    SA biota
    SA faunal studies
    SA fossils
    SA living taxa
    SA microorganisms
    SA paleobotany
    SA palynomorphs
    SA Plantae
    SA vegetation
Florence
    As of 1993, no longer a valid term for GeoRef. See Florence Italy.
Florence County
    Valid through 1988. Search in combination with state term. After 1988, use specific county-state term.
Florence County South Carolina (1989)
    E central South Carolina. Before 1989, also search Florence County AND South Carolina.
    CO N334700N341900
        W0791900W0800500
    BT South Carolina
    BT United States
Florence County Wisconsin (1989)
    Extreme NE Wisconsin. Before 1989, also search Florence County AND Wisconsin.
    CO N454300N460200
        W0880300W0884200
    BT Wisconsin
    BT United States
Florence Italy (1993)
    Province and city in E central Tuscany. Also search Florence or Firenze AND Italy.
    UF Firenze Italy
    BT Tuscany Italy
    BT Italy
    BT Southern Europe
    BT Europe
Florence Rise (1981)
    Ridge in the Mediterranean Sea. A submarine extension of the Cyprus Arc. Partially separates the Antalya Basin to the N from the Levantine Basin to the S.
    CO N344000N351000
        E0320000E0313000
    BT Mediterranean Sea
    SA DSDP Site 375
    SA DSDP Site 376
florencite (1978)
    BT phosphates
Florida (1978)
    Autoposting of this term began in 1978.
    CO N243000N310000
        W0800000W0873000
    BT United States
    NT Alachua County Florida
    NT Apalachicola River
    NT Baker County Florida
    NT Biscayne Aquifer
    NT Bradford County Florida
    NT Brevard County Florida
    NT Broward County Florida
    NT Calhoun County Florida
    NT Central Florida
    NT Charlotte County Florida
    NT Choctawhatchee Bay
    NT Citrus County Florida
    NT Clay County Florida
    NT Collier County Florida
    NT Columbia County Florida
    NT Dade County Florida
    NT De Soto County Florida
    NT Duval County Florida
    NT Escambia County Florida
    NT Everglades
    NT Florida Panhandle
    NT Floridan Aquifer
    NT Franklin County Florida
    NT Gadsden County Florida
    NT Gulf County Florida

Before then, variants of the term may occur in GeoRef.

NT Hamilton County Florida
NT Hendry County Florida
NT Hillsborough County Florida
NT Jackson County Florida
NT Jefferson County Florida
NT Lafayette County Florida
NT Lake County Florida
NT Lake Okeechobee
NT Lee County Florida
NT Leon County Florida
NT Levy County Florida
NT Liberty County Florida
NT Madison County Florida
NT Manatee County Florida
NT Marion County Florida
NT Martin County Florida
NT Monroe County Florida
NT Nassau County Florida
NT Orange County Florida
NT Palm Beach County Florida
NT Pasco County Florida
NT Pinellas County Florida
NT Polk County Florida
NT Putnam County Florida
NT Saint Johns County Florida
NT Saint Lucie County Florida
NT Sarasota County Florida
NT Seminole County Florida
NT South Florida Basin
NT Southwest Florida Water Management District
NT Sumter County Florida
NT Tampa Bay
NT Taylor County Florida
NT Union County Florida
NT Walton County Florida
NT Washington County Florida
SA Atlantic Coastal Plain
SA Bone Valley Formation
SA Bucatunna Formation
SA Byram Formation
SA Carolina Bays
SA Chattahoochee River
SA Chipola Formation
SA Citronelle Formation
SA Claiborne Group
SA Coupon Bight
SA Crystal River Formation
SA Duplin Formation
SA Glendon Limestone
SA Gulf Coastal Plain
SA Hawthorn Formation
SA Indian River
SA Jackson Group
SA Key Largo Limestone
SA Marianna Limestone
SA Miami Limestone
SA Midway Group
SA Northern Highlands
SA Ocala Group
SA Okefenokee Swamp
SA Peace River Formation
SA Saint George Island
SA Saint Johns River basin
SA Southern Uplands
SA Sunniland Limestone
SA Suwannee Limestone
SA Tamiami Formation
SA Vicksburg Group
SA Wicomico Formation
SA Wilcox Group

Florida Bay (1978)
Body of water between S tip of Florida and Florida Keys.
BT Gulf of Mexico
BT North American Atlantic
BT North Atlantic
BT Atlantic Ocean

Florida Keys (1978)
150 mile chain of coral limestone islands extending SW off S tip of Florida.
CO N243000N252000
  W0815000W0830000

BT Monroe County Florida
BT Florida
BT United States
NT Big Pine Key
NT Key Largo
SA Coupon Bight

Florida Mountains (1978)
Luna County in SW New Mexico.
BT Luna County New Mexico
BT New Mexico
BT United States

Florida Panhandle (1993)
NW Florida. Before 1993, also search Panhandle AND Florida.
IN Index counties as applicable.
BT Florida
BT United States

Florida Strait
 use Straits of Florida

Florida Straits
 use Straits of Florida

Floridan Aquifer (1978)
N and central Florida. Below the confining bed consists from top to bottom, of limestone in the bottom part of the Hawthorn Formation and limestone, dolomite, and dolomitic limestone in formations of Eocene age that include the Ocala Group, the Avon Park Limestone, the Lake City Limestone, and, in part, the Oldsmar Limestone.
BT Florida
BT United States
SA Eocene
SA Hawthorn Formation
SA Ocala Group
SA Southwest Florida Water Management District

Florinites (1978)
Autoposting of microfossils to this term began in 1990.
BT miospores
BT palynomorphs
BT microfossils
SA angiosperms

flotation (1978)
SA beneficiation
SA buoyancy

Flournoy Formation (1985)
SW Oregon.
BT Eocene
BT Paleogene
BT Tertiary
BT Cenozoic
SA Oregon

flow
As of 1981, no longer a valid term for GeoRef. Before 1981, used to indicate the mechanics of deformation. Before 1978, also used to indicate style of structure. Before 1977, also related to movement of ground water and surface water. See flow structures; flow folds; flow lines; flow mechanism; flow cleavage; fluid dynamics.

flow (volcanic)
 use lava flows

flow cleavage (1978)
BT foliation
SA axial-plane structures
SA cleavage
SA compression
SA schistosity
SA slaty cleavage
SA structural analysis

flow folds (1978)

Before 1978, also search flow AND folds.
UF flowage fold
BT folds

flow lines (1978)
UF lines, flow
SA deformation
SA lineation
SA plastic flow
SA rheology

flow mechanism (1978)
SA lava
SA lava flows
SA mechanism
SA rheology
SA viscosity

flow meters
 use flowmeters

flow regime
As of 1981, no longer a valid term for GeoRef.

flow structures (1978)
BT soft sediment deformation
BT sedimentary structures
SA Bouma sequence
SA sedimentation
SA turbidity current structures

flowage fold
 use flow folds

flowers (1989)
SA fossils
SA leaves
SA paleobotany
SA Plantae
SA plants
SA roots
SA vegetation

flowmeters (1985)
UF flow meters
SA currents
SA discharge
SA fluid dynamics
SA hydrodynamics
SA hydrogeology
SA hydrology
SA instruments

flows (1978)
When discussing lava, use lava flows.
SA ash flows
SA debris flows
SA effusion
SA gravity flows
SA hydraulics
SA lava flows
SA mass movements
SA streamflow
SA velocity

flowstone
 use speleothems

Floyd County
Valid through 1988. Search in combination with state term. After 1988, use specific county-state term.

Floyd County Georgia (1989)
NW Georgia. Before 1989, also search Floyd County AND Georgia.
CO N340500N343500
  W0850100W0852800
BT Georgia
BT United States

Floyd County Indiana (1989)
S Indiana. Before 1989, also search Floyd County AND Indiana.
CO N381100N382500
  W0854700W0860400

BT Indiana
BT United States

Floyd County Iowa (1989)
N Iowa. Before 1989, also search Floyd County AND Iowa.
CO N425400N431400
  W0923300W0930200
BT Iowa
BT United States

Floyd County Kentucky (1989)
E Kentucky. Before 1989, also search Floyd County AND Kentucky.
CO N371800N374300
  W0823600W0825500
BT Kentucky
BT United States

Floyd County Texas (1989)
NW Texas. Before 1989, also search Floyd County AND Texas.
CO N335100N342200
  W1010000W1013100
BT Texas
BT United States

Floyd County Virginia (1989)
SW Virginia. Before 1989, also search Floyd County AND Virginia.
CO N364300N370800
  W0800700W0803800
BT Virginia
BT United States

fluctuations (1978)
Used as a general term.
SA changes
SA paleoclimatology
SA variations

fluid dynamics (1981)
SA flowmeters
SA hydraulics
SA hydrodynamics
SA laminar flow
SA steady flow
SA unsteady flow

fluid inclusions (1978)
Used for gaseous or liquid inclusions contained within crystals. Before 1971, also search liquid inclusions. As of 1989, inclusions is autoposted to this term.
IN Index with type of inclusion and name of host rock.
UF liquid inclusions
BT inclusions
SA analysis
SA carbon dioxide
SA crystal growth
SA gases
SA geochemistry
SA geologic barometry
SA geologic thermometry
SA host rocks
SA igneous rocks
SA lithogeochemistry
SA microthermometry
SA mineral deposits, genesis
SA mineral inclusions
SA minerals
SA P-T conditions
SA paleosalinity
SA petrology
SA pressure
SA rocks
SA temperature
SA xenoliths

fluid injection (1978)
UF injection wells
SA artificial recharge
SA earthquakes
SA enhanced recovery
SA gas injection

SA injection
SA petroleum engineering
SA reinjection wells
SA secondary recovery
SA steam injection
SA waste disposal
SA waterflooding

fluid phase (1978)
Includes both gaseous phase and liquid phase. Also search liquid.
UF fluids
SA Darcy's law
SA gaseous phase
SA geochemistry
SA liquid phase
SA solid phase

fluid pressure (1978)
SA pressure

fluidization (1989)
SA ash flows
SA petroleum
SA processes

fluids
 use fluid phase

fluids, ore-forming
 use ore-forming fluids

flume studies (1985)
UF flumes
SA physical models
SA sediment transport
SA sedimentary petrology
SA sedimentation

flumes
 use flume studies

fluoborates (1978)
Autoposting of halides began in 1981.
BT halides

fluoborite (1978)
From 1978-1980, borates and halides were autoposted to this term. Autoposting of fluorides began in 1981. Autoposting of halides began again in 1989. This term has multiple hierarchies.
BT1 borates
BT2 fluorides
BT2 halides

fluor-phlogopite (1978)
BT mica group
BT sheet silicates
BT silicates
SA phlogopite

fluorapatite (1978)
BT phosphates
SA apatite

fluorescence (1978)
SA dye tracers
SA fluorimetry
SA luminescence
SA optical properties
SA thermoluminescence
SA X-ray fluorescence
SA X-ray fluorescence spectra

fluoride ion (1978)
Before 1978, search fluoride.
SA fluorides
SA fluorine
SA ions

fluorides (1978)
Autoposting of this term began in 1981.
BT halides
NT bastnaesite
NT bideauxite
NT clinohumite
NT cryolite
NT fluoborite

NT fluorite
NT humite
NT leifite
NT lepidolite
NT microlite
NT minyulite
NT neighborite
NT norbergite
NT prosopite
NT pyrochlore
NT synchisite
NT thomsenolite
NT topaz
NT triplite
NT villiaumite
NT weberite
NT zinnwaldite
SA fluoride ion

fluorimetric analysis
 use fluorimetry

fluorimetry (1982)
Before 1971, also search fluorometric analysis.
UF fluorimetric analysis
UF fluorometric analysis
UF fluorometry
UF luminescence analysis
SA dye tracers
SA fluorescence
SA ground water
SA hydrogeology
SA hydrology
SA surface water
SA tracers
SA X-ray fluorescence
SA X-ray fluorescence spectra

fluorine (1978)
Autoposting of halogens to this term began in 1989.
UF F
BT halogens
SA fluoride ion

fluorite (1978)
When of economic value, use fluorspar.
BT fluorides
BT halides
SA fluorite structure
SA fluorspar

fluorite structure (1978)
SA fluorite

fluorometric analysis
Not a valid GeoRef index term after 1970. Was used on level 1 in subfile N.
 use fluorimetry

fluorometry
 use fluorimetry

fluorspar (1978)
Use only when of economic value; otherwise use fluorite.
IN Commodity. See List C.
SA cryolite deposits
SA fluorite

fluosilicates (1978)
Autoposting of halides to this term began in 1981.
BT halides

flute casts (1978)
UF casts, flute
BT bedding plane irregularities
BT sedimentary structures
SA casts
SA current markings
SA load casts
SA scour casts
SA scour marks
SA soft sediment deformation
SA sole marks
SA turbidity current structures

fluvial
A valid term through 1977. After 1977, use fluvial features or (as of 1981) fluvial environment or fluvial sedimentation.

fluvial currents (1985)
BT currents
SA hydrology
SA ocean currents
SA rivers
SA rivers and streams
SA stream transport
SA streams
SA thalwegs

fluvial environment (1981)
SA aquatic environment
SA deep-water environment
SA depositional environment
SA environment
SA fluvial features
SA fluvial sedimentation
SA Gilbert-type deltas
SA paleoecology
SA rivers
SA rivers and streams
SA sedimentation
SA stream sediments
SA streams

fluvial features (1978)
Autoposting of this term began in 1978.
UF features, fluvial
NT alluvial fans
NT alluvial plains
NT arroyos
NT buried channels
NT cascades
NT drainage basins
NT drainage patterns
NT eskers
NT estuaries
NT floodplains
NT meanders
NT outwash plains
NT oxbow lakes
NT point bars
NT rapids
NT rivers
NT stream order
NT streams
NT wadis
NT waterfalls
SA bars
SA bedforms
SA benches
SA bluffs
SA bogs
SA canyons
SA channel geometry
SA channelization
SA channels
SA cliffs
SA coastal plains
SA debris cones
SA deltas
SA drift
SA dry valleys
SA fluvial environment
SA geomorphology
SA glacial features
SA glaciofluvial environment
SA gorges
SA inlets
SA kames
SA levees
SA marshes
SA mud banks
SA mud lumps
SA outwash
SA plains
SA rills
SA rivers and streams
SA salt marshes

SA sand boils
SA sedimentary structures
SA shoals
SA sinuosity
SA stream sediments
SA swamps
SA terraces
SA tributaries
SA valleys
SA wetlands

fluvial sedimentation (1981)
BT sedimentation
NT fluviolacustrine sedimentation
NT glaciofluvial sedimentation
SA fluvial environment
SA point bars
SA stream transport

fluvial sediments
For economic studies, use mineral exploration AND stream sediments; otherwise, use fluvial environment AND sediments.

fluvial transport
 use stream transport

fluvioglacial environment
 use glaciofluvial environment

fluviolacustrine sedimentation (1985)
This term has multiple hierarchies.
BT1 fluvial sedimentation
BT1 sedimentation
BT2 lacustrine sedimentation
BT2 sedimentation

flux
As of 1981, no longer a valid term for GeoRef. See meteorite flux; heat flux.

fly ash
 use ash

flysch (1978)
BT clastic rocks
BT sedimentary rocks
SA clastic sediments
SA terrigenous materials
SA wildflysch

Fm
 use fermium

Foaming Sea (1978)
UF Mare Spumans
BT Moon

focal mechanism (1978)
Also search focus AND mechanism. Also search fault-plane solutions.
SA earthquakes
SA fault planes
SA focus
SA mechanism

foci
 use focus

focus (1978)
UF centrum (seismology)
UF foci
UF hypocenters
UF seismic focus
SA aftershocks
SA deep-focus earthquakes
SA earthquakes
SA elastic waves
SA epicenters
SA focal mechanism
SA intermediate-focus earthquakes
SA seismology
SA shallow-focus earthquakes

focused current logging
 use Laterolog

Foggia
No longer a valid term for GeoRef. See Foggia Italy.

Foggia Italy (1993)
City and province in N Apulia.
BT Apulia Italy
BT Italy
BT Southern Europe
BT Europe

fold and thrust belts (1993)
Before 1993, also search fold belts AND thrust faults or thrust sheets.
SA arcuate structures
SA fold belts
SA mobile belts
SA orogenic belts
SA tectonics
SA thrust faults
SA thrust sheets

fold axes (1978)
UF axes, fold
SA folds
SA structural analysis

fold belts (1978)
Also search belts.
UF belts, fold
UF foldbelts
SA fold and thrust belts
SA mobile belts
SA orogenic belts
SA tectonics

foldbelts
use fold belts

folding
As of 1978, no longer a valid term for GeoRef. Used in subfiles B, N, G, and E.
use folds

folds (1978)
Used for studies primarily emphasizing folds and systems of folds. For relationships with other structures, see tectonics. Before 1978, also search folding. Autoposting of this term began in 1978.
IN Index specific type of fold if possible [e.g. cylindrical folds, plane cylindrical folds, plane non-cylindrical folds, decollement, flexural-slip, flexure, kink-band structures, shear, discordant folds, drag folds, horizontal orientation, inclined folds, nappes, normal folds, oblique orientation, overturned folds, plunging folds, recumbent folds, superposed folds, anticlines, antiform folds, asymmetric folds, basins, chevron folds, cleavage folds, concentric folds, conjugate folds, convolute folds, cross folds, diapirs, disharmonic folds, disjunctive folds, domes, drape folds, enterolithic folds, flexure folds, flow folds, harmonic folds, intrafolial folds, isoclinal folds, kink folds, monoclines, polyclinal folds, ptygmatic folds, sheath folds, similar folds, symmetric folds, synclines, synform folds, troughs anticlinoria, en echelon folds, synclinoria.
UF folding
NT anticlines
NT anticlinoria
NT antiform folds
NT asymmetric folds
NT chevron folds
NT cleavage folds
NT concentric folds
NT conjugate folds
NT cylindrical folds
NT discordant folds
NT disharmonic folds
NT disjunctive folds
NT domes
NT drag folds
NT drape folds
NT en echelon folds
NT enterolithic folds
NT flexure folds
NT flow folds
NT harmonic folds
NT inclined folds
NT intrafolial folds
NT isoclinal folds
NT kink folds
NT monoclines
NT normal folds
NT overturned folds
NT plunging folds
NT ptygmatic folds
NT recumbent folds
NT sheath folds
NT similar folds
NT superposed folds
NT symmetric folds
NT synclines
NT synclinoria
NT synform folds
NT upright folds
SA arches
SA axial-plane structures
SA basins
SA cleavage
SA decollement
SA deformation
SA depressions
SA diapirs
SA embayments
SA faults
SA flexural-slip
SA fold axes
SA foliation
SA fractures
SA geanticlines
SA geometry
SA geosynclines
SA horizontal orientation
SA interference patterns
SA kink-band structures
SA lineation
SA mechanics
SA metamorphic rocks
SA metamorphism
SA nappes
SA neotectonics
SA oblique orientation
SA orientation
SA orogeny
SA salt domes
SA salt tectonics
SA strike
SA structural analysis
SA structural geology
SA style
SA systems
SA tectonic elements
SA tectonics
SA tectonophysics
SA vertical orientation
SA wavelength

foliation (1978)
Used for planar arrangements of textural or structural features in any type of rock. Before 1971, also search foliations. Autoposting of this term began in 1978.
UF foliations
NT flow cleavage
NT fracture cleavage
NT schistosity
NT slaty cleavage
NT slip cleavage
SA axial-plane structures
SA cleavage
SA deformation
SA faults
SA folds
SA fractures
SA interference patterns
SA intrusions
SA laminations
SA lineation
SA metamorphic rocks
SA metamorphism
SA neotectonics
SA structural analysis
SA structural geology
SA style
SA tectonics
SA tectonophysics

foliations
Not a valid GeoRef index term after 1970. Was used on level 1 in subfile N.
use foliation

Folkestone
No longer a valid term for GeoRef. As of 1993, see Folkestone England.

Folkestone England (1993)
City on Strait of Dover in SE England. Before 1993, also search Folkestone AND England.
BT Kent England
BT England
BT Great Britain
BT United Kingdom
BT Western Europe
BT Europe

Folldal
No longer a valid term for GeoRef. As of 1993 see Folldal Norway.

Folldal Norway (1993)
Village in SE Norway. Before 1993 also search Folldal Norway.
BT Norway
BT Scandinavia
BT Western Europe
BT Europe

Fontainebleau
No longer a valid term for GeoRef. See Fontainebleau France.

Fontainebleau France (1993)
Town in N central France.
BT Seine-et-Marne France
BT France
BT Western Europe
BT Europe

foot wall (1989)
SA faults
SA hanging wall

footings (1985)
SA engineering geology
SA foundations
SA marine installations

footprints, fossil
No longer a valid GeoRef index term. Before 1971, was used on level 1 in subfile N. See tracks; trails.

foraminifera (1978)
Autoposting of this term began in 1978. This term has multiple hierarchies.
UF Foraminiferida
BT1 Protista
BT1 Invertebrata
BT2 Protista
BT2 microfossils
NT Allogromiina
NT Fusulinina
NT Miliolina
NT Rotaliina
NT Textulariina
SA foraminifers
SA micropaleontology
SA plankton
SA planktonic taxa
SA Receptaculitaceae

Foraminiferida
use foraminifera

foraminifers (1981)
Common name for foraminifera. This term has multiple hierarchies.
IN Index for all non-paleontologic studies of fossils.
BT1 invertebrates
BT2 microfossils
NT fusulinids
SA biostratigraphy
SA foraminifera

Forbes Formation (1989)
N California.
BT Upper Cretaceous
BT Cretaceous
BT Mesozoic
SA California

force, Coriolis
Not a valid term for GeoRef. Usually out-of-scope. Used as Coriolis force with ocean circulation.

Fordham Gneiss (1989)
W Connecticut and SE New York.
BT middle Proterozoic
BT Proterozoic
BT upper Precambrian
BT Precambrian
SA Connecticut
SA New York
SA New York City Group

fore-arc basins (1985)
UF forearc basins
UF foredeeps
BT basins
SA back-arc basins
SA forelands
SA island arcs
SA orogeny
SA plate tectonics
SA subduction zones
SA tectonics
SA trenches

forearc basins
use fore-arc basins

foredeeps
use fore-arc basins

foredunes (1993)
BT dunes
SA beaches
SA coastal dunes
SA continental dunes
SA eolian features
SA intertidal environment
SA shore features

foreland basins (1989)
Basins in foreland regions.
BT basins
SA forelands
SA orogenic belts
SA sedimentary basins
SA sedimentation
SA tectonics
SA thrust faults

forelands (1993)
SA back-arc basins
SA cratons
SA fore-arc basins
SA foreland basins
SA geomorphology
SA orogenic belts
SA shore features

SA tectonics
foreshocks (1978)
SA aftershocks
SA earthquakes
SA elastic waves
SA main shocks
SA shock waves
Forest City Basin (1978)
IN Index states as applicable.
BT United States
SA Iowa
SA Kansas
SA Missouri
SA Nebraska
Forest County
Valid through 1988. Search in combination with state term. After 1988, use specific county-state term.
Forest County Pennsylvania (1989)
NW Pennsylvania. Before 1989, also search Forest County AND Pennsylvania.
CO N412000N413800
W0785700W0793200
BT Pennsylvania
BT United States
Forest County Wisconsin (1989)
NE Wisconsin. Before 1989, also search Forest County AND Wisconsin.
CO N452200N460500
W0882500W0890400
BT Wisconsin
BT United States
Forest of Ardennes
use Ardennes
Forest of Dean (1978)
Gloucestershire in W central.
BT Gloucestershire England
BT England
BT Great Britain
BT United Kingdom
BT Western Europe
BT Europe
forest soils
Use specific soil group name such as Brown forest soils and Gray forest soils if available. For general discussions, use forests AND soils.
forestry (1978)
IN As of 1982, may be used as a supplemental index term.
SA deforestation
SA forests
SA soils
forests (1978)
SA deforestation
SA fires
SA forestry
SA revegetation
SA soils
SA trees
SA wilderness areas
SA yields
Foresudetic Monocline (1981)
BT Poland
BT Central Europe
BT Europe
Forfar
Not a valid term for GeoRef. As of 1993, see Angus Scotland.
Forfarshire
use Angus Scotland
form, crystal
use crystal form

formation pressure
Not a valid GeoRef index term after 1970. Was used on level 1 in subfile N. Now use pressure or geopressure.
formation waters
use connate waters
formations, geologic
See specific names of geologic formations.
formations, iron
use iron formations
Formosa
Not a valid index term for GeoRef since 1976.
use Taiwan
Formosa Strait (1978)
Channel between Fujian Province of China and Taiwan. Also search Taiwan Strait.
UF Taiwan Strait
BT Pacific Ocean
formula (1978)
Term restricted to use for chemical formulas, e.g. for minerals.
UF formulas
SA equations
SA minerals
formulas
use formula
Forrest County Mississippi (1989)
SE Mississippi. Before 1989, search Forrest County AND Mississippi.
CO N305400N372600
W0890900W0892700
BT Mississippi
BT United States
NT Hattiesburg Mississippi
forsterite (1978)
BT olivine group
BT nesosilicates
BT orthosilicates
BT silicates
Fort Bend County Texas (1993)
SE Texas. Before 1989, also search Fort Bend County AND Texas.
CO N291900N294800
W0952500W0960500
BT Texas
BT United States
Fort Churchill (1978)
Destroyed wooden fort at town of Churchill on Hudson Bay in NE Manitoba.
BT Manitoba
BT Western Canada
BT Canada
Fort Good Hope (1978)
Trading station on Mackenzie River in NW Mackenzie District.
UF Good Hope
BT Mackenzie District Northwest Territories
BT Northwest Territories
BT Western Canada
BT Canada
Fort Gouraud
No longer a valid term for GeoRef. See also Fort Gouraud Mauritania.
Fort Gouraud Mauritania (1993)
Village near Western Sahara border in NW central Mauritania.
UF Idjil Mauritania
BT Mauritania

BT West Africa
BT Africa
Fort Hays Limestone Member (1978)
Of Niobrara Formation. E Colorado, W Kansas, NE New Mexico, and SE South Dakota.
BT Upper Cretaceous
BT Cretaceous
BT Mesozoic
SA Colorado
SA Kansas
SA New Mexico
SA Niobrara Formation
SA South Dakota
Fort McMurray
No longer a valid term for GeoRef. See Fort McMurray Alberta.
Fort McMurray Alberta (1993)
Town on Athabasca River in NE central Alberta.
BT Alberta
BT Western Canada
BT Canada
Fort Payne Formation (1978)
Includes Greasy Creek facies, Short Mountain facies. N and E Alabama, NW Georgia, Kentucky, NE Mississippi, and Tennessee.
BT Lower Mississippian
BT Mississippian
BT Carboniferous
BT Paleozoic
SA Alabama
SA Georgia
SA Kentucky
SA Mississippi
SA Tennessee
Fort Riley Limestone (1978)
In Chase Group. E Kansas, SE Nebraska, and central N Oklahoma.
BT Permian
BT Paleozoic
SA Chase Group
SA Kansas
SA Oklahoma
Fort Ross
No longer a valid term for GeoRef. See Fort Ross Northwest Territories.
Fort Ross Northwest Territories (1993)
Trading post on Somerset Island, W of Baffin Island, in Franklin District. Before 1993, also search Fort Ross.
BT Somerset Island
BT Franklin District Northwest Territories
BT Northwest Territories
BT Western Canada
BT Canada
Fort Ternan
No longer a valid term for GeoRef. As of 1993 see Fort Ternan Kenya.
Fort Ternan Kenya (1993)
Village in Nyanza province, SW Kenya. Before 1993 also search Fort Ternan AND Kenya.
BT Kenya
BT East Africa
BT Africa
Fort Union Formation (1978)
Includes Lebo Shale Member, Tongue River Member, Tullock Member, Ludlow Member, Sentinel Butte Member. NW Colorado,

Montana, North Dakota, NW South Dakota, and Wyoming.
IN Index ages as applicable.
SA Bullion Creek Formation
SA Colorado
SA K-T boundary
SA Lebo Member
SA Ludlow Member
SA Montana
SA North Dakota
SA Paleocene
SA Sentinel Butte Formation
SA South Dakota
SA Tongue River Member
SA Tullock Member
SA Upper Cretaceous
SA Wyoming
Fort Wingate (1981)
NW New Mexico.
CO N352500N353500
W1082800W1083500
BT McKinley County New Mexico
BT New Mexico
BT United States
Fort Worth Basin (1978)
N central Texas.
BT Texas
BT United States
Fortescue Group (1989)
Upper Archean and/or lower Proterozoic.
BT Precambrian
SA Australia
SA Western Australia
Forth Valley (1978)
River valley in S central Scotland and in Tasmania.
IN Index regions as applicable.
SA Scotland
SA Tasmania Australia
Forties Field (1978)
BT North Sea
BT European Atlantic
BT North Atlantic
BT Atlantic Ocean
SA oil and gas fields
Fortran (1978)
BT computer languages
SA computer programs
SA data processing
SA Fortran 77
SA Fortran IV
Fortran 77 (1989)
Before 1989, also search Fortran.
UF Fortran-77
BT computer languages
SA computer programs
SA data processing
SA Fortran
SA Fortran IV
Fortran IV (1978)
BT computer languages
SA computer programs
SA data processing
SA Fortran
SA Fortran 77
SA WATEQF
Fortran-77
use Fortran 77
Fortuna
Valid through 1988. Search in combination with state term. After 1988, use specific city-state term.
Fortuna California (1989)
Town in NW California. Before 1989, search Fortuna AND California.
CO N403600N403600
W1240900W1240900

BT Humboldt County California
BT California
BT United States

forward problem
use direct problem

Fosdick Mountains (1978)
N of Edsel Ford Range in W Marie Byrd Land near the Pacific Ocean.
BT Antarctica
SA Marie Byrd Land

Fossa Magna (1978)
Structural trench crossing the mountain ranges of Honshu.
BT Honshu
BT Japan
BT Far East
BT Asia
SA Akaishi Mountains

fossil
A valid index term through 1977 used in combination with man (i.e. man, fossil). After 1977, use fossil man.

fossil assemblages
use assemblages

fossil ice wedges (1978)
UF ice wedges, fossil
UF ice-wedge cast
UF ice-wedge fill
UF ice-wedge pseudomorph
UF wedges, fossil ice
BT bedding plane irregularities
BT sedimentary structures
SA frost action
SA frost features
SA ice wedges
SA periglacial features

fossil localities (1981)
Use for collecting.
SA collecting
SA popular geology

fossil man (1978)
Used mainly for archaeological and anthropological studies, but as of 1981, is restricted to papers about actual fossil remains. If taxonomy is given, use Mammalia. Also search man AND fossil.
SA anthropology
SA archaeological sites
SA archaeology
SA artifacts
SA Mammalia
SA Neanderthal
SA Piltdown man
SA Primates
SA Vertebrata

fossil meteorite craters
use astroblemes

fossil resins
use resins

fossil soils
use Paleosols

fossil waters (1981)
SA connate waters
SA ground water
SA pore water
SA water

fossil wood (1978)
Includes silicified wood and petrified wood. Before 1981, also search silicified wood; petrified wood.
UF petrified wood
UF silicified wood
SA angiosperms
SA gymnosperms
SA Plantae
SA roots
SA Spermatophyta
SA wood

fossiliferous materials (1978)
Before 1978, also search fossiliferous AND specific sediment type or sedimentary rocks.
SA fossilization
SA fossils
SA materials
SA sedimentary rocks
SA sediments

fossilization (1978)
Includes taphonomy.
SA biomineralization
SA calcification
SA casts
SA chemical fossils
SA fossiliferous materials
SA fossils
SA ichnofossils
SA paleobotany
SA paleontology
SA phytoliths
SA problematic fossils
SA silicification
SA taphonomy

fossils (1978)
A valid level 1 term through 1977 used in combination with problematic or problematical (i.e. fossils, problematic or fossils, problematical). Use problematic fossils where applicable. Also search macrofossils and megafossils. See microfossils.
UF macrofossils
UF megafossils
SA allometry
SA anaerobic taxa
SA assemblages
SA benthonic taxa
SA biochemistry
SA biogeography
SA biometry
SA biostratigraphy
SA bones
SA calcification
SA casts
SA chemical fossils
SA coiling
SA colonial taxa
SA faunal list
SA faunal studies
SA floral studies
SA flowers
SA fossiliferous materials
SA fossilization
SA functional morphology
SA growth rates
SA habitat
SA histology
SA holotypes
SA ichnofossils
SA index fossils
SA jaws
SA juvenile taxa
SA living fossils
SA living taxa
SA locomotion
SA microfossils
SA micropaleontology
SA miscellanea
SA nannofossils
SA neotypes
SA ontogeny
SA otoliths
SA paleobiology
SA paleobotany
SA paleoclimatology
SA paleontology
SA paratypes
SA problematic fossils
SA provinciality
SA range
SA revision
SA reworking
SA roots
SA sclerites
SA septa
SA shells
SA skeletons
SA skulls
SA species diversity
SA sutures
SA typomorphism

foundations (1978)
Used for geological studies on engineering foundations.
SA bearing capacity
SA blasting
SA bridges
SA buildings
SA construction
SA dams
SA design
SA earthquakes
SA earthworks
SA engineering geology
SA engineering properties
SA explosions
SA footings
SA geomembranes
SA geotextiles
SA grouting
SA highways
SA land subsidence
SA marine installations
SA nuclear facilities
SA overconsolidated materials
SA piles
SA pipelines
SA raft foundations
SA railroads
SA reinforced materials
SA retaining walls
SA rock mechanics
SA seepage
SA seismic response
SA settlement
SA site exploration
SA slope stability
SA soil mechanics
SA soil-structure interface
SA soils
SA stability
SA structures
SA tunnels
SA underground installations

Fountain Formation (1985)
Pennsylvanian and Permian. E Colorado and SE Wyoming.
IN Index ages as applicable.
BT upper Paleozoic
BT Paleozoic
SA Colorado
SA Pennsylvanian
SA Permian
SA Wyoming

Four Corners (1978)
Location where boundaries of four states meet.
IN Index states as applicable.
SA Arizona
SA Colorado
SA New Mexico
SA Paradox Basin
SA United States
SA Utah

four-dimensional models (1989)
Before 1989, also search four-dimensional AND models.
SA mathematical models
SA models
SA one-dimensional models
SA three-dimensional models
SA two-dimensional models

Fourchu Group (1989)
Upper Hadrynian. Cape Breton Island, N Nova Scotia.
BT Hadrynian
BT upper Proterozoic
BT Proterozoic
BT upper Precambrian
BT Precambrian
SA Nova Scotia

Fourier analysis (1978)
UF Fourier transformations
UF Fourier transforms
UF harmonic analysis
SA analysis
SA data processing
SA discrete filters
SA Green function
SA Hilbert transformations
SA mathematical geology
SA mathematical methods
SA mathematical transformations

Fourier transformations
use Fourier analysis

Fourier transforms
use Fourier analysis

Foveaux Strait (1978)
Channel between S South Island and Stewart Island.
BT New Zealand
BT Australasia

Fox Glacier (1978)
In the Southern Alps of Tasman National Park in E South Island.
SA New Zealand
SA South Island

Fox Hills Formation (1978)
Includes Milliken Sandstone Member, Colgate Member, Stoneville Member, Trail City Sandstone Member, Timber Lake Sandstone Member, Bullhead Member.
BT Upper Cretaceous
BT Cretaceous
BT Mesozoic
SA Colorado
SA Edmonton Formation
SA Montana
SA North Dakota
SA South Dakota
SA Wyoming

foyaite (1978)
BT syenites
BT plutonic rocks
BT igneous rocks
SA alkali gabbros
SA nepheline syenite

Fr
use francium

Fra Mauro (1978)
UF Fra Mauro crater
UF Fra Mauro Formation
BT Moon

Fra Mauro crater
use Fra Mauro

Fra Mauro Formation
use Fra Mauro

fractal analysis
use fractals

fractal dimensions
use fractals

fractal geometry
use fractals

fractals (1989)
UF fractal analysis
UF fractal dimensions

UF fractal geometry
SA mathematical geology
SA mathematical methods

fractional crystallization (1978)
Crystallization during which early-formed crystals are prevented from equilibrating with the liquid from which they grew. Results in series of residual liquids of more extreme compositions than occur with equilibrium crystallization. Also referred to as fractionation of magmas. Compare with crystal fractionation. Before 1978, search crystallization AND fractional.
SA chemical fractionation
SA crystal fractionation
SA crystal zoning
SA crystallization
SA fractionation
SA magmas
SA magmatic differentiation
SA reaction rims
SA segregation

fractionation (1978)
As of 1985, for isotope fractionation only. Use fractional crystallization for fractionation of magmas; use chemical fractionation for other geochemical fractionation.
SA chemical fractionation
SA crystal fractionation
SA fractional crystallization
SA geochemistry
SA isotopes

fracture cleavage (1978)
BT foliation
SA cleavage
SA fractures
SA joints

fracture strength (1978)
UF breaking strength
UF fracture stress
SA deformation
SA ductile deformation
SA elastic limit
SA strength
SA yield strength

fracture stress
 use fracture strength

fracture zones (1978)
UF zones, fracture
SA faults
SA fractures
SA mid-ocean ridges
SA ocean floors
SA plate tectonics
SA sea-floor spreading
SA segmentation
SA transform faults

fractured materials (1985)
SA fractures
SA fracturing
SA hydraulic fracturing
SA materials
SA rock mechanics

fractures (1978)
Used for a break in a rock usually without displacement. Autoposting of this term began in 1978.
NT closed fractures
NT columnar joints
NT concentric fractures
NT conical fractures
NT conjugate fractures
NT cross fractures
NT dip fractures
NT extension fractures
NT joints
NT open fractures

NT polygonal fractures
NT release fractures
SA asperities
SA bedding
SA brittle deformation
SA cleats
SA cracks
SA deformation
SA engineering geology
SA failures
SA faults
SA fissures
SA folds
SA foliation
SA fracture cleavage
SA fracture zones
SA fractured materials
SA lineation
SA microcracks
SA naturally fractured reservoirs
SA neotectonics
SA oblique orientation
SA rupture
SA shear zones
SA strike
SA structural analysis
SA structural geology
SA style
SA systems
SA tectonic elements
SA tectonics
SA tectonophysics
SA tension
SA thermal waters
SA veins

fracturing (1978)
Term is used for natural fracturing, and not for artificially induced fracturing.
SA fractured materials
SA hydraulic fracturing
SA naturally fractured reservoirs

fragipans (1985)
BT soils

fragmentation (1978)
Used as a general term.
SA fragments
SA processes

fragments (1978)
SA clastic rocks
SA clasts
SA fragmentation
SA particles

Fram Strait (1993)
Connects N Greenland Sea with central Arctic Ocean between NE Greenland and W Svalbard.
CO N780000N810000
E0090000W0080000
BT Arctic Ocean
SA Greenland
SA Greenland Sea
SA Svalbard

framboidal pyrite
 use framboidal texture

framboidal texture (1978)
Before 1978, search framboidal.
UF framboidal pyrite
UF framboids
BT textures
SA pyrite

framboids
 use framboidal texture

framework silicates (1978)
Before 1981, maskelynite was included as a narrower term. Autoposting of this term began in 1978.
UF tectosilicates
BT silicates

NT cancrinite
NT chkalovite
NT danburite
NT feldspar group
NT genthelvite
NT helvite
NT karpinskyite
NT leifite
NT leucite
NT nepheline group
NT pollucite
NT pseudoleucite
NT reedmergnerite
NT scapolite group
NT silica minerals
NT sodalite group
NT wenkite
NT zeolite group

France (1978)
Autoposting of this term began in 1978.
CO N423000N510000
E0083000W0050000
BT Western Europe
BT Europe
NT Adour Basin
NT Aigoual Massif
NT Aiguilles Rouges
NT Ain France
NT Aisne France
NT Allier France
NT Alpes-de-Haute Provence France
NT Alpes-Maritimes France
NT Alsace
NT Anjou
NT Aquitaine
NT Aquitaine Basin
NT Ardeche France
NT Ardennes France
NT Ariege France
NT Arize Massif
NT Armorican Massif
NT Artois
NT Aube France
NT Aubrac
NT Aude France
NT Auvergne
NT Aveyron France
NT Bas-Dauphine
NT Bas-Rhin France
NT Bay of Bourgneuf
NT Bay of Saint-Michel
NT Bay of the Seine
NT Bearn
NT Beauce
NT Belfort France
NT Belledonne Massif
NT Bouches-du-Rhone France
NT Brittany
NT Burgundy
NT Calvados France
NT Cantal France
NT Causses
NT Central Massif
NT Cevennes
NT Chablais
NT Charente
NT Charente France
NT Charente-Maritime France
NT Charentes
NT Cher
NT Cher France
NT Correze France
NT Corsica
NT Cote-d'Or France
NT Cotes-du-Nord France
NT Creuse France
NT Dauphine
NT Dauphine Alps
NT Deux-Sevres France
NT Devoluy
NT Dordogne France
NT Doubs France

NT Drac Valley
NT Drome France
NT Durance Basin
NT Eoulx Basin
NT Essonne France
NT Esterel
NT Eure France
NT Eure Valley
NT Eure-et-Loir France
NT Finistere France
NT Franche-Comte
NT French Alps
NT French Coast
NT French Pyrenees
NT Gard France
NT Garonne River
NT Gers France
NT Gironde Estuary
NT Gironde France
NT Haut-Rhin France
NT Haute-Garonne France
NT Haute-Loire France
NT Haute-Marne France
NT Haute-Saone France
NT Haute-Savoie France
NT Haute-Vienne France
NT Hautes-Alpes France
NT Hautes-Pyrenees France
NT Hauts-de-Seine France
NT Herault France
NT Ille-et-Vilaine France
NT Indre France
NT Indre-et-Loire France
NT Isere France
NT Isere Valley
NT Jura France
NT Landes France
NT Languedoc
NT Laval Basin
NT Lherz
NT Limagne
NT Limousin
NT Loir-et-Cher France
NT Loire France
NT Loire River
NT Loire Valley
NT Loire-Atlantique France
NT Loiret France
NT Lot France
NT Lot-et-Garonne France
NT Lozere France
NT Maine-et-Loire France
NT Manche France
NT Marne France
NT Marne Valley
NT Marseilles Basin
NT Mayenne France
NT Meurthe-et-Moselle France
NT Meuse France
NT Monts du Lyonnais
NT Morbihan France
NT Moselle France
NT Navarre
NT Nievre France
NT Nord France
NT Nord-Pas-de-Calais Basin
NT Normandy
NT Oise France
NT Oise River valley
NT Orne France
NT Paris Basin
NT Paris France
NT Pas-de-Calais France
NT Picardy
NT Poitou
NT Provence
NT Provence Alps
NT Puy-de-Dome France
NT Pyrenees-Atlantiques France
NT Pyrenees-Orientales France
NT Quercy
NT Rhone France
NT Rodez Trough
NT Rouergue
NT Roussillon

NT Sainte-Baume Massif
NT Salat Valley
NT Saone Valley
NT Saone-et-Loire France
NT Saone-Rhone Basin
NT Sarthe France
NT Savoie France
NT Savoy
NT Seine
NT Seine Estuary
NT Seine River
NT Seine Valley
NT Seine-et-Marne France
NT Seine-et-Oise
NT Seine-Maritime France
NT Seine-Saint-Denis France
NT Somme France
NT Somme River valley
NT Tarn France
NT Tarn-et-Garonne France
NT Touraine
NT Val-d'Oise France
NT Val-de-Marne France
NT Var France
NT Vaucluse France
NT Vendee France
NT Vienne France
NT Vivarais
NT Vocontian Trough
NT Vosges France
NT Vosges Mountains
NT Yonne France
NT Yonne Valley
NT Yvelines France
SA Alps
SA Annot Sandstone
SA Arve Valley
SA Brianconnais Zone
SA Chassigny Meteorite
SA CNRS
SA Cottian Alps
SA CRPG
SA Juvinas Meteorite
SA Mediterranean region
SA Meuse River
SA Meuse Valley
SA Moselle River
SA Moselle Valley
SA North Pyrenean Fault
SA North Sea Coast
SA North Sea region
SA Orgueil Meteorite
SA Pyrenees
SA Rhine Basin
SA Rhine River
SA Rhine Valley
SA Rhone River
SA Rhone Valley
SA Saint-Severin Meteorite
SA Scheldt River
SA Voltzia Sandstone
SA Western Alps

Franche-Comte (1978)
Historical region on the Swiss border S and SW of Alsace.
BT France
BT Western Europe
BT Europe
SA Doubs France
SA Jura France

Francis Creek Shale (1978)
W and central Illinois.
BT Pennsylvanian
BT Carboniferous
BT Paleozoic
SA Desmoinesian
SA Illinois

Franciscan Complex (1993)
Upper Jurassic to Upper Cretaceous. W California. Before 1993, also search Franciscan Formation which includes Corral Hollow

Shales, Oakridge Sandstone, and Del Puerto Keratophyre.
IN Index age as applicable.
UF Franciscan Formation
BT Mesozoic
SA Calera Limestone
SA California
SA Cretaceous
SA Jurassic

Franciscan Formation
As of 1993, no longer a valid term for GeoRef.
use Franciscan Complex

francium (1978)
Term reintroduced in 1985. Not a valid GeoRef term 1971-1984. Before 1971, was used on level 1 in subfile N. Autoposting of alkali metals and metals to this term began in 1989.
UF Fr
BT alkali metals
BT metals

francolite (1978)
BT phosphates
SA apatite

Franconia (1978)
Former duchy of S central West Germany.
IN Index states as applicable.
UF Franken
BT Germany
BT Central Europe
BT Europe
SA Baden-Wurttemberg Germany
SA Bavaria Germany
SA Hesse Germany
SA Middle Franconia
SA Upper Franconia Germany

Franconia Formation (1978)
Includes Woodhill, Birkmose, Tomah, Reno and Mazomanie members. N Illinois, SE Minnesota, and SW Wisconsin.
IN Index states as applicable.
BT Upper Cambrian
BT Cambrian
BT Paleozoic
SA Illinois
SA Minnesota
SA Wisconsin

Franconian Alb
use Franconian Jura

Franconian Forest (1978)
S outlier of the Thuringian Forest.
BT Bavaria Germany
BT Germany
BT Central Europe
BT Europe
SA Thuringian Forest

Franconian Jura (1978)
Plateau in central Bavaria. A northern continuation of the Swabian Alb. Autoposting of broader terms began in 1993.
CO N484500N500000
  E0120000E0104500
UF Franconian Alb
BT Bavaria Germany
BT Germany
BT Central Europe
BT Europe
SA Middle Franconia

Franken
use Franconia

Frankenberg
No longer a valid term for GeoRef. See Frankenberg Germany.

Frankenberg Germany (1993)

City in Saxony, E Germany also a city in Hesse.
IN Index states as applicable.
BT Germany
BT Central Europe
BT Europe
SA Hesse Germany
SA Saxony Germany

Frankfurt
No longer a valid term for GeoRef. See Frankfurt Germany.

Frankfurt Bezirk
No longer a valid term for GeoRef. District in E East Germany. Before 1981, search Frankfurt AND East Germany. As of 1993, see Brandenburg Germany.

Frankfurt Germany (1993)
City. Frankfurt am Main in Hesse and Frankfurt an der Oder in Brandenburg. Before 1993, Frankfurt am Main was used with West Germany and Frankfurt an der Oder was used with East Germany.
IN Index states as applicable.
BT Germany
BT Central Europe
BT Europe
SA Brandenburg Germany
SA Hesse Germany

Frankfurt Stade (1978)
Europe. Before 1978, search Frankfurt.
BT Weichselian
BT upper Pleistocene
BT Pleistocene
BT Quaternary
BT Cenozoic

Franklin County
Valid through 1988. Search in combination with state term. After 1988, use specific county-state term.

Franklin County Alabama (1989)
NW Alabama. Before 1989, also search Franklin County AND Alabama.
CO N341700N343400
  W0873300W0881300
BT Alabama
BT United States

Franklin County Arkansas (1989)
NW Arkansas. Before 1989, also search Franklin County AND Arkansas.
CO N351500N354600
  W0933700W0940400
BT Arkansas
BT United States

Franklin County Florida (1989)
NW Florida. Before 1989, also search Franklin County AND Florida.
CO N293500N300000
  W0841800W0851400
BT Florida
BT United States
NT Apalachicola Bay
NT Apalachicola Florida

Franklin County Georgia (1989)
NE Georgia. Before 1989, also search Franklin County AND Georgia.
CO N341300N343200
  W0830300W0832400
BT Georgia
BT United States

Franklin County Idaho (1989)

SE Idaho. Before 1989, also search Franklin County AND Idaho.
CO N420000N422500
  W1112900W1120800
BT Idaho
BT United States
NT Oxford Idaho

Franklin County Illinois (1989)
S Illinois. Before 1989, also search Franklin County AND Illinois.
CO N375100N380800
  W0884300W0890900
BT Illinois
BT United States

Franklin County Indiana (1989)
SE Indiana. Before 1989, also search Franklin County AND Indiana.
CO N391600N393100
  W0845100W0851900
BT Indiana
BT United States

Franklin County Iowa (1989)
N central Iowa. Before 1989, also search Franklin County AND Iowa.
CO N423400N422400
  W0930200W0933000
BT Iowa
BT United States

Franklin County Kansas (1989)
E Kansas. Before 1989, also search Franklin County AND Kansas.
CO N382300N384400
  W0950400W0953000
BT Kansas
BT United States

Franklin County Kentucky (1989)
N central Kentucky. Before 1989, also search Franklin County AND Kentucky.
CO N380700N382200
  W0844500W0850100
BT Kentucky
BT United States

Franklin County Maine (1989)
W Maine. Before 1989, also search Franklin County AND Maine.
CO N442500N453700
  W0695300W0704900
BT Maine
BT United States
SA Rangeley Lakes

Franklin County Massachusetts (1989)
NW Massachusetts. Before 1989, also search Franklin County AND Massachusetts.
CO N421900N424400
  W0721300W0730200
BT Massachusetts
BT United States

Franklin County Mississippi (1989)
SW Mississippi. Before 1989, also search Franklin County AND Mississippi.
CO N312100N313800
  W0903800W0911000
BT Mississippi
BT United States

Franklin County Missouri (1989)
E central Missouri. Before 1989, also search Franklin County AND Missouri.
CO N381300N384400

W0904400W0912200
BT Missouri
BT United States

Franklin County Nebraska
(1989)
S Nebraska. Before 1989, also search Franklin County AND Nebraska.
CO N400000N402200
W0984300W0991000
BT Nebraska
BT United States

Franklin County New York
(1989)
NE New York. Before 1989, also search Franklin County AND New York.
CO N440700N450000
W0735500W0744300
BT New York
BT United States
SA Saint Lawrence Lowlands
SA Saint Lawrence River
SA Saint Lawrence Valley

Franklin County North Carolina
(1989)
N central North Carolina. Before 1989, also search Franklin County AND North Carolina.
CO N355000N361500
W0780000W0783400
BT North Carolina
BT United States

Franklin County Ohio (1989)
Central Ohio. Before 1989, also search Franklin County AND Ohio.
CO N394800N400800
W0824600W0831600
BT Ohio
BT United States
NT Columbus Ohio

Franklin County Pennsylvania
(1989)
S Pennsylvania. Before 1989, also search Franklin County AND Pennsylvania.
CO N394300N401700
W0772700W0780600
BT Pennsylvania
BT United States
NT Waynesboro Pennsylvania

Franklin County Tennessee
(1989)
S Tennessee. Before 1989, also search Franklin County AND Tennessee.
CO N350000N352300
W0855300W0862200
BT Tennessee
BT United States

Franklin County Texas (1989)
NE Texas. Before 1989, also search Franklin County AND Texas.
CO N325900N332300
W0950600W0951500
BT Texas
BT United States

Franklin County Vermont (1989)
NW Vermont. Before 1989, also search Franklin County AND Vermont.
CO N443800N450100
W0723200W0731500
BT Vermont
BT United States

Franklin County Virginia (1989)
S Virginia. Before 1989, also search Franklin County AND Virginia.
CO N364800N371300
W0793600W0801800
BT Virginia
BT United States

Franklin County Washington
(1989)
SW Washington. Before 1989, also search Franklin County AND Washington.
CO N461300N464300
W1181300W1192700
BT Washington
BT United States
SA Hanford Reservation

Franklin District
No longer a valid term for GeoRef. As of 1993, see Franklin District Northwest Territories.

Franklin District Northwest Territories (1993)
Northernmost district in Northwest Territories. Also search Franklin District AND Northwest Territories. Before 1985, also search District of Franklin AND Northwest Territories.
CO N630000N850000
W0600000W1250000
UF District of Franklin
BT Northwest Territories
BT Western Canada
BT Canada
NT Arctic Archipelago
NT Axel Heiberg Island
NT Boothia Peninsula
NT Devon Island
NT Ellef Ringnes Island
NT Ellesmere Island
NT Meighen Island
NT Melville Peninsula
NT Queen Elizabeth Islands
NT Somerset Island
NT Sverdrup Basin
SA Baffin Island
SA Banks Island
SA Barnes ice cap
SA Bathurst Island
SA Cornwallis Island
SA Cumberland Peninsula
SA Eureka Sound Formation
SA Lancaster Sound
SA Melville Island
SA Otto Fjord
SA Peel Sound Formation
SA Read Bay Formation
SA Victoria Island

Franklin Mountains (1978)
Mackenzie District in Canada, and N of El Paso in the United States.
IN Index Northwest Territories and/or Texas as applicable.
SA Northwest Territories
SA Rocky Mountains
SA Texas

franklinite (1978)
BT oxides
SA iron minerals
SA iron oxides
SA manganese minerals

Franz Josef Land (1978)
Archipelago in Arctic Ocean. Part of Arkhangelsk Oblast. This term has multiple hierarchies.
CO N790000N830000
E0660000E0450000
BT1 Arkhangelsk Russian Federation
BT1 Russian Federation

BT1 Commonwealth of Independent States
BT2 Arkhangelsk Russian Federation
BT2 Europe
BT3 Russian Platform

Fraser Range (1978)
SE of Kalgoorlie in S central Western Australia.
IN Index state or region as applicable.
SA Western Australia

Fraser River (1978)
S central and SW British Columbia. Flows into Georgia Strait S of Vancouver.
IN Index British Columbia if applicable.
SA British Columbia
SA Fraser River delta

Fraser River delta (1989)
S British Columbia.
IN Index British Columbia if applicable.
SA British Columbia
SA Fraser River

Frasnian (1978)
Europe. Above Givetian, below Famennian.
BT Upper Devonian
BT Devonian
BT Paleozoic
SA Greenland Gap Group

Fraunhofer line discriminators
(1985)
SA chemical analysis
SA instruments
SA luminescence
SA remote sensing

Freda Sandstone (1989)
N Michigan and N Wisconsin. In Oronto Group. Overlies Nonesuch Shale; underlies Orienta Sandstone.
BT middle Proterozoic
BT Proterozoic
BT upper Precambrian
BT Precambrian
SA Michigan
SA Oronto Group
SA Wisconsin

Frederick County
Valid through 1988. Search in combination with state term. After 1988, use specific county-state term.

Frederick County Maryland
(1989)
N Maryland. Before 1989, also search Frederick County AND Maryland.
CO N391300N394330
W0770700W0774100
BT Maryland
BT United States

Frederick County Virginia (1989)
N Virginia. Before 1989, also search Frederick County AND Virginia.
CO N390100N392700
W0780300W0783500
BT Virginia
BT United States

Fredericksburg Group (1978)
Comprises Finlay Limestone, Kiamichi Formation, Edwards Limestone, Paluxy Sand, Walnut Formation, Comanche Peak Limestone, Goodland Formation.

S Oklahoma and Texas. This term has multiple hierarchies.
BT1 Comanchean
BT1 Cretaceous
BT1 Mesozoic
BT2 Lower Cretaceous
BT2 Cretaceous
BT2 Mesozoic
SA Comanche Peak Limestone
SA New Mexico
SA Oklahoma
SA Texas

Fredericton
No longer a valid term for GeoRef. As of 1993, see Fredericton New Brunswick.

Fredericton New Brunswick
(1993)
SW New Brunswick. Before 1993, also search Fredericton and New Brunswick.
BT New Brunswick
BT Maritime Provinces
BT Eastern Canada
BT Canada

Frederikshaab
use Frederikshab Greenland

Frederikshab
No longer a valid term for GeoRef. As of 1993 see Frederikshab Greenland.

Frederikshab Greenland (1993)
Settlement on Davis Strait in extreme SW Greenland. Before 1993 also search Frederikshab AND Greenland.
UF Frederikshaab
BT Greenland
BT Arctic region

free energy (1978)
UF Gibbs free energy
UF Helmholtz free energy
SA energy
SA geochemistry
SA Kirchhoff integral
SA phase equilibria
SA thermodynamic properties

free oscillations (1978)
UF free vibrations
SA Earth
SA oscillations

free vibrations
use free oscillations

free-air anomalies (1978)
Before 1978, search free-air.
SA anomalies
SA Bouguer anomalies
SA gravity methods

Freeport Formation (1985)
In Allegheny Group.
BT Pennsylvanian
BT Carboniferous
BT Paleozoic
SA Allegheny Group
SA Pennsylvania

Freestone County
Valid through 1988. Search in combination with state term. After 1988, use specific county-state term.

Freestone County Texas (1989)
E central Texas. Before 1989, also search Freestone County AND Texas.
CO N312500N320000
W0954000W0962500
BT Texas
BT United States
SA Oakwood Dome

freeze-and-thaw action
use frost action

freeze-thaw action
use frost action

freezing (1981)
SA frost action
SA permafrost
SA thawing

Freiberg
No longer a valid term for GeoRef. See Freiberg Germany.

Freiberg Germany (1993)
City in Saxony, E Germany.
BT Saxony Germany
BT Germany
BT Central Europe
BT Europe

freibergite (1978)
Steel gray variety of tetrahedrite containing silver.
BT sulfantimonites
BT sulfosalts
SA silver ores
SA tetrahedrite

Freiburg
No longer a valid term for GeoRef. As of 1993, see Freiburg Germany.

Freiburg Germany (1993)
City in Baden-Wurttemberg, SW Germany. Before 1978, also search Freiburg im Breisgau.
UF Freiburg im Breisgau
BT Baden-Wurttemberg Germany
BT Germany
BT Central Europe
BT Europe

Freiburg im Breisgau
use Freiburg Germany

freieslebenite (1978)
BT sulfantimonites
BT sulfosalts

Freites Formation (1989)
Located in the states of Anzoategui and Monagas, N Venezuela.
BT Miocene
BT Neogene
BT Tertiary
BT Cenozoic
SA Venezuela

Fremont County
Valid through 1988. Search in combination with state term. After 1988, use specific county-state term.

Fremont County Colorado (1989)
S central Colorado. Before 1989, also search Fremont County AND Colorado.
CO N381700N384000
    W1045600W1060000
BT Colorado
BT United States
NT Canon City Colorado
SA Wet Mountains

Fremont County Idaho (1989)
E Idaho. Before 1989, also search Fremont County AND Idaho.
CO N435300N444500
    W1110200W1121100
BT Idaho
BT United States
SA Yellowstone National Park

Fremont County Iowa (1989)
Extreme SW Iowa. Before 1989, also search Fremont County AND Iowa.
CO N403600N405400
    W0952300W0955300
BT Iowa
BT United States

Fremont County Wyoming (1989)
W central Wyoming. Before 1989, also search Fremont County AND Wyoming.
CO N421700N440100
    W1073000W1100300
BT Wyoming
BT United States
NT Wind River
NT Wind River basin
SA Bighorn River
SA Gas Hills
SA Granite Mountains
SA Great Divide Basin
SA Lost Soldier Field
SA Owl Creek Mountains

Fremouw Formation (1978)
Beardmore Glacier area near the Ross Ice Shelf at N end of Queen Maud Range in Ross Dependency.
BT Triassic
BT Mesozoic
SA Antarctica
SA Beardmore Glacier
SA Queen Maud Range
SA Ross Dependency
SA Ross Ice Shelf

French (1978)
Used to indicate language of a catalog, dictionary, glossary or lexicon.
SA catalogs
SA dictionaries
SA glossaries
SA lexicons

French Alps (1981)
This term has multiple hierarchies.
BT1 France
BT1 Western Europe
BT1 Europe
BT2 Alps
BT2 Europe
SA Aiguilles Rouges
SA Graian Alps

French Coast (1978)
Along British Channel and Bay of Biscay on the Atlantic Ocean side and the Mediterranean Sea on the S.
BT France
BT Western Europe
BT Europe

French colonial possessions
No longer a valid GeoRef index term. Before 1969, was used on level 1 in subfile E. The variant form French territories was also used.

French Equatorial Africa
No longer a valid GeoRef index term. Before 1960, was used on level 1 in subfile E. Now index countries (Central African Republic, Chad, Congo, Gabon) as applicable.

French Guiana (1978)
Republic on NE coast.
CO N015500N055000
    W0513500W0544000
BT South America
SA Guiana Basin
SA Guianas
SA Guyana Shield

French Indochina
No longer a valid GeoRef index term. Before 1960, was used on level 1 in subfile E.
use Indochina

French Mid-Ocean Undersea Study
use FAMOUS

French Polynesia (1981)
French overseas territory in S Pacific Ocean. Includes Marquesas, Society, Gambier, Tubuai, and Tuamotu islands.
CO S280000S073000
    W1320000W1560000
BT Polynesia
BT Oceania
NT Austral Islands
NT Marquesas Islands
NT Society Islands
NT Tuamotu Islands
SA Pacific Ocean

French Pyrenees (1981)
Autoposting of Pyrenees to this term ended in 1989 and began again in 1992. This term has multiple hierarchies.
BT1 Pyrenees
BT1 Europe
BT2 France
BT2 Western Europe
BT2 Europe
SA Castillon Massif
SA Corbieres
SA North Pyrenean Fault
SA Querigut Massif

French Somaliland
use Djibouti

French West Africa
No longer a valid GeoRef index term. Before 1960, was used on level 1 in subfile E. Now index countries (Bonin, Guinea, Ivory Coast, Mali, Mauritania, Niger, Senegal) as applicable.

Frenchman Springs Member (1989)
Of Wanapum Basalt, which is part of the Columbia River Basalt Group. S and central Washington and NW Oregon.
BT Miocene
BT Neogene
BT Tertiary
BT Cenozoic
SA Oregon
SA Wanapum Basalt
SA Washington

frequency (1978)
Used as a general term.
SA broad-band spectra
SA frequency domain analysis
SA frequency sounding
SA periodicity

frequency distribution
Not a valid term for GeoRef. For statistical frequency distribution, use statistical distribution. For seismology or geophysical methods, use frequency.

frequency domain analysis (1985)
SA analysis
SA elastic waves
SA electromagnetic methods
SA frequency
SA geophysical methods
SA seismic methods
SA seismology
SA time domain analysis

frequency sounding (1982)
SA deep sounding
SA frequency
SA geophysical methods
SA geophysical surveys
SA sounding

fresh water (1978)
UF freshwater
SA chemical dispersion
SA drinking water
SA ecology
SA fresh-water environment
SA ground water
SA hydrochemistry
SA paleoecology
SA potability
SA salt water
SA sedimentation
SA water
SA water quality
SA water resources

fresh-water environment (1978)
Before 1978, search fresh-water or fresh water.
SA aquatic environment
SA deep-water environment
SA depositional environment
SA ecology
SA environment
SA fresh water
SA fresh-water sedimentation
SA paleoecology
SA sedimentation

fresh-water sedimentation (1981)
BT sedimentation
SA fresh-water environment

freshwater
use fresh water

Freshwater
No longer a valid term for GeoRef. See fresh water or fresh-water environment if applicable.

Fresno
Valid through 1988. Search in combination with state term. After 1988, use specific city-state term.

Fresno California (1989)
City in central California. Before 1989, search Fresno AND California.
CO N364100N364100
    W1194700W1194700
BT Fresno County California
BT California
BT United States

Fresno County
Valid through 1988. Search in combination with state term. After 1988, use specific county-state term.

Fresno County California (1989)
In the San Joaquin Valley of central California. Before 1989, also search Fresno County AND California.
CO N355500N373000
    W1182000W1205000
BT California
BT United States
NT Coalinga California
NT Fresno California

friction (1978)
Used as a general term.
SA friction angles
SA mechanical properties
SA physical properties

friction angles (1985)
SA friction
SA shear strength

SA slope stability
SA soil mechanics

Friendly Islands
use Tonga

fringing reefs (1978)
UF shore reef
BT reefs
SA barrier reefs
SA platform reefs

Frio County
Valid through 1988. Search in combination with state term. After 1988, use specific county-state term.

Frio County Texas (1989)
SW Texas. Before 1989, also search Frio County AND Texas.
CO N283900N290500
   W0985000W0992500
BT Texas
BT United States

Frio Formation (1978)
Consists of dark to very dark, varicolored shales and silty shales and massive to thin-bedded strata of sand and silty sand. Subsurface.
BT Oligocene
BT Paleogene
BT Tertiary
BT Cenozoic
SA Louisiana
SA Texas

Friuli earthquake 1976 (1985)
Epicenter in Friuli-Venezia Giulia, NE Italy.
BT earthquakes
SA Friuli-Venezia Giulia Italy
SA Italy

Friuli-Venezia Giulia
No longer a valid term for GeoRef. See Friuli-Venezia Giulia Italy.

Friuli-Venezia Giulia Italy (1993)
Autonomous region in NE Italy.
CO N453500N464000
   E0134500E0122200
BT Italy
BT Southern Europe
BT Europe
NT Tarvisio Italy
NT Trieste Italy
SA Carnic Alps
SA Friuli earthquake 1976
SA Tagliamento Valley
SA Venetia

Frobisher Bay (1985)
Off SE Baffin Island, Arctic Canada.
CO N620000N633000
   W0650000W0683000
BT North American Atlantic
BT North Atlantic
BT Atlantic Ocean
SA Baffin Island

Front Range (1978)
A range of Rocky Mountains in N central Colorado.
IN Index counties and Colorado and U. S. Rocky Mountains as applicable.
SA Colorado
SA Pikes Peak
SA Rocky Mountain National Park
SA U. S. Rocky Mountains

Front Range urban corridor (1978)
BT Colorado
BT United States

Frontenac County
No longer a valid term for GeoRef. As of 1993, see Frontenac County Ontario or Frontenac County Quebec.

Frontenac County Ontario (1993)
Before 1993, search Frontenac County AND Ontario.
BT Ontario
BT Eastern Canada
BT Canada

Frontenac County Quebec (1993)
Before 1993, search Frontenac County AND Quebec.
BT Quebec
BT Eastern Canada
BT Canada

Frontier Formation (1978)
In Colorado Group or Mancos Shale. S Montana, and W Wyoming, Colorado, Idaho, Utah.
BT Upper Cretaceous
BT Cretaceous
BT Mesozoic
SA Colorado
SA Colorado Group
SA Idaho
SA Mancos Shale
SA Montana
SA Utah
SA Wall Creek Member
SA Wyoming

froodite (1978)
Autoposting of sulfides to this term ended in 1989.
BT bismuthides

Frosinone
No longer a valid term for GeoRef. See Frosinone Italy.

Frosinone Italy (1993)
Town and province in the Apennines SE of Rome.
BT Latium Italy
BT Italy
BT Southern Europe
BT Europe

frost action (1978)
UF action, frost
UF freeze-and-thaw action
UF freeze-thaw action
SA congelifraction
SA cryopedology
SA cryoturbation
SA engineering geology
SA fossil ice wedges
SA freezing
SA frozen ground
SA gelifluction
SA geomorphology
SA glacial geology
SA ground ice
SA highways
SA ice wedges
SA palsas
SA patterned ground
SA periglacial environment
SA permafrost
SA pingos
SA polygons
SA rock glaciers
SA rock mechanics
SA soil mechanics
SA thawing
SA thermokarst

frost bursting
use congelifraction

frost churning
use cryoturbation

frost features (1978)
UF features, frost
BT bedding plane irregularities
BT sedimentary structures
SA fossil ice wedges

frost heaving (1978)
UF heaving, frost
SA congelifraction
SA cryoturbation
SA palsas
SA patterned ground
SA permafrost

frost shattering
use congelifraction

frost splitting
use congelifraction

frost stirring
use cryoturbation

frost weathering
use congelifraction

frost wedging
use congelifraction

frozen ground (1978)
Includes use in engineering geology.
UF gelisol
UF ground, frozen
UF merzlota
UF taele
UF tele
UF tjaele
SA engineering geology
SA frost action
SA ground ice
SA ice
SA palsas
SA periglacial features
SA permafrost
SA soil mechanics
SA soils
SA taliks

Fruitland Formation (1985)
N New Mexico and SW Colorado.
BT Upper Cretaceous
BT Cretaceous
BT Mesozoic
SA Colorado
SA New Mexico

fruits (1978)
SA soils
SA yields

Frunze
No longer a valid term for GeoRef. As of 1993, see Bishkek Kyrgyzstan.

frustules (1978)
Refers to morphology of diatoms.
SA diatoms

Fuego (1981)
Active volcano in S central Guatemala.

fuel resources
Valid from 1981 to 1988 as a grouper for petroleum, oil shale, oil sands, and natural gas. As of 1989, use energy sources or specific commodity.

fuel resources maps
A valid term for GeoRef from 1982 to 1988.
use petroleum maps

Fuerteventura (1978)
Island in Spain's Las Palmas Province.
BT Canary Islands
BT Atlantic Ocean Islands

fugacity (1978)

SA geochemistry
SA lithogeochemistry
SA partial pressure
SA thermodynamic properties

Fujairah (1978)
Emirate. One of federation of 7 states at S end of Persian Gulf.
BT United Arab Emirates
BT Arabian Peninsula
BT Asia

Fuji
use Fujiyama

Fuji River (1989)
In Yamanashi and Shizuoka prefectures. Flows south from near Kofu, past Fujiyama, to Suruga Bay.
IN Index prefectures as applicable.
BT Honshu
BT Japan
BT Far East
BT Asia
SA Fujiyama
SA Shizuoka Japan
SA Suruga Bay
SA Yamanashi Japan

Fujian
No longer a valid term for GeoRef. See Fujian China.

Fujian China (1993)
Province, SE China. Before 1993, also search Fujian. Before 1978, also search Fukien.
CO N234000N282000
   E1203000E1154000
UF Fukien China
BT China
BT Far East
BT Asia
SA Chihsia Formation

Fujiyama (1978)
Sacred mountain and quiescent volcano W of Yokohama.
UF Fuji
UF Mount Fuji
UF Mt. Fuji
UF Mt. Fuji volcano
BT Honshu
BT Japan
BT Far East
BT Asia
SA Fuji River
SA Saitama Japan
SA Yamanashi Japan

Fukazawa
use Fukazawa Mine

Fukazawa Mine (1989)
Metal ores; also barite deposits. Hokuroku District, Akita Prefecture, Honshu, N Japan.
UF Fukazawa
BT Hokuroku Japan
BT Akita Japan
BT Honshu
BT Japan
BT Far East
BT Asia
SA barite deposits
SA metal ores
SA mines

Fukien China
use Fujian China

Fukui
No longer a valid term for GeoRef. See Fukui.

Fukui Japan (1993)

Fukuoka
City including the prefecture on the Japan Sea in central Honshu. Before 1993, also search Fukui AND Japan.
BT Honshu
BT Japan
BT Far East
BT Asia
SA Chubu Japan

Fukuoka
No longer a valid term for GeoRef. See Fukuoka Japan.

Fukuoka Japan (1993)
City and prefecture in N Kyushu.
BT Kyushu
BT Japan
BT Far East
BT Asia

Fukushima
No longer a valid term for GeoRef. See Fukushima Japan.

Fukushima Japan (1993)
Prefecture including city in central Honshu. Before 1993, also search Fukushima (the city) and/or Fukushima Prefecture AND Japan.
UF Fukushima Prefecture
BT Honshu
BT Japan
BT Far East
BT Asia
SA Abukuma Mountains
SA Aizu Basin
SA Joban coal field
SA Koma-ga-take
SA Nasudake
SA Tohoku

Fukushima Prefecture
No longer a valid term for GeoRef.
use Fukushima Japan

fulgurite (1989)
Formed from fusion of sand or rock by lightning.
BT metamorphic rocks

fuller's earth (1978)
IN May be used as a commodity term.
UF fullers earth
SA clay
SA clay mineralogy
SA clays
SA kaolin
SA kaolin deposits
SA smectite
SA soils

fullers earth
use fuller's earth

Fulton County
Valid through 1988. Search in combination with state term. After 1988, use specific county-state term.

Fulton County Arkansas (1989)
N Arkansas. Before 1989, also search Fulton County AND Arkansas.
CO N361600N363000
W0912700W0921000
BT Arkansas
BT United States

Fulton County Georgia (1989)
NW central Georgia. Before 1989, also search Fulton County AND Georgia.
CO N333100N340800
W0840700W0845000
BT Georgia
BT United States

NT Atlanta Georgia

Fulton County Illinois (1989)
W central Illinois. Before 1989, also search Fulton County AND Illinois.
CO N401000N404200
W0895100W0902800
BT Illinois
BT United States

Fulton County Indiana (1989)
N Indiana. Before 1989, also search Fulton County AND Indiana.
CO N405400N411000
W0855600W0862800
BT Indiana
BT United States

Fulton County Kentucky (1989)
Extreme SW Kentucky. Before 1989, also search Fulton County AND Kentucky.
CO N363000N363800
W0885100W0893500
BT Kentucky
BT United States

Fulton County New York (1989)
E central New York. Before 1989, also search Fulton County AND New York.
CO N425800N431700
W0740600W0744400
BT New York
BT United States

Fulton County Ohio (1989)
NW Ohio. Before 1989, also search Fulton County AND Ohio.
CO N412800N414300
W0835300W0842500
BT Ohio
BT United States

Fulton County Pennsylvania (1989)
S Pennsylvania. Before 1989, also search Fulton County AND Pennsylvania.
CO N394300N401000
W0775200W0782300
BT Pennsylvania
BT United States

fulvic acids (1978)
BT organic materials
SA acids
SA organic acids

fumaroles (1978)
NT solfataras
SA black smokers
SA chimneys
SA gases
SA geothermal energy
SA geysers
SA heat flow
SA springs
SA sublimates
SA thermal waters
SA vents
SA volcanism
SA volcanology

functional morphology (1978)
BT morphology
SA anatomy
SA bipedalism
SA fossils

functions (1978)
Used as a general term in a mathematical sense.
SA equations
SA fuzzy logic
SA Green function
SA mathematical geology
SA mathematical methods

SA transfer functions

Fundy Basin
use Minas Basin

Funeral Mountains (1989)
SE California. Part of Amargosa Range, which forms the eastern wall of Death Valley.
BT Inyo County California
BT California
BT United States

fungi (1978)
Autoposting of this term began in 1978.
BT thallophytes
BT Plantae
NT Myxomycetes
SA algae
SA bacteria
SA endolithic taxa
SA lichens
SA microorganisms
SA parasites

Furnace Creek (1978)
Small creek which sinks into valley floor in Death Valley National Monument near Nevada border in SE California.
IN Index county as applicable.
SA California
SA Death Valley

furrows (1993)
Use more specific term if available. Sometimes used for extraterrestrial remotely-sensed features.
SA bottom features
SA channels
SA erosion features
SA exine
SA geomorphology
SA grooves
SA ocean floors
SA pits
SA rills
SA sedimentary structures
SA soil erosion
SA striations
SA trenches
SA troughs
SA valleys

fusain (1989)
BT macerals

fusinite (1978)
BT inertinite
BT macerals

fusion (1978)
UF nuclear fusion
SA fission
SA melting

Fusselman Dolomite (1989)
S New Mexico and W Texas.
BT Silurian
BT Paleozoic
SA New Mexico
SA Texas

Fusulinidae (1978)
Autoposting of microfossils and Protista to this term began in 1990. This term has multiple hierarchies.
BT1 Fusulinina
BT1 foraminifera
BT1 Protista
BT1 Invertebrata
BT2 Fusulinina
BT2 foraminifera
BT2 Protista
BT2 microfossils
NT Schwagerina
NT Triticites

SA fusulinids

fusulinids (1985)
Common name for Fusulinidae. Autoposting of foraminifers to this term began in 1989. Autoposting of microfossils to this term began in 1990. This term has multiple hierarchies.
BT1 foraminifers
BT1 invertebrates
BT2 foraminifers
BT2 microfossils
SA biostratigraphy
SA Fusulinidae

Fusulinina (1978)
Autoposting of microfossils and Protista to this term began in 1990. This term has multiple hierarchies.
BT1 foraminifera
BT1 Protista
BT1 Invertebrata
BT2 foraminifera
BT2 Protista
BT2 microfossils
NT Archaediscidae
NT Endothyra
NT Fusulinidae

Futaba Group (1978)
Underlain by the Paleozoic and granitic rocks with a distinct unconformity and covered disconformably by the Paleogene. N part of E border of Abukuma Mountains in E central Honshu.
BT Senonian
BT Upper Cretaceous
BT Cretaceous
BT Mesozoic
SA Abukuma Mountains
SA Honshu
SA Japan

future (1978)
Used as a general term.
SA possibilities

futures market (1981)
SA economics

fuzzy analysis
use fuzzy logic

fuzzy logic (1989)
UF fuzzy analysis
UF fuzzy mathematics
UF fuzzy methods
UF fuzzy sets
SA functions
SA mathematical geology
SA mathematical methods

fuzzy mathematics
use fuzzy logic

fuzzy methods
use fuzzy logic

fuzzy sets
use fuzzy logic

fyord
use fjords

G

G.K. Gilbert
use Gilbert, Grove Karl

Ga
use gallium

gabbro
As of 1989, no longer a valid term for GeoRef. From 1981 to 1988, gabbros was autoposted to this term.
use gabbros

gabbro anorthosite
use gabbroic anorthosite

gabbro family
As of 1981, no longer a valid term for GeoRef. Use gabbros or alkali gabbros as applicable. From 1978-1980, was autoposted to the rocks that were classified under it.

gabbroic anorthosite (1978)
 UF gabbro anorthosite
 BT gabbros
 BT plutonic rocks
 BT igneous rocks
 SA anorthosite

gabbroic composition (1978)
Before 1978, search gabbroic; gabbroic rocks.
 SA composition
 SA gabbros
 SA igneous rocks

gabbros (1981)
Before 1981, search gabbro family. From 1978-1980, gabbro family was autoposted to the rocks that were classified under that term. Before 1989, also search gabbro. Autoposting of this term to gabbro ended in 1989. As of 1993, this is no longer a broader term of eucrite.
 UF gabbro
 BT plutonic rocks
 BT igneous rocks
 NT alkali gabbros
 NT anorthosite
 NT anorthositic gabbro
 NT gabbroic anorthosite
 NT labradoritite
 NT microgabbro
 NT norite
 NT olivine gabbro
 NT rodingite
 NT troctolite
 SA gabbroic composition
 SA metagabbro

Gabilan Range (1978)
One of the Coast Ranges in W California.
 IN Index Coast Ranges.
 BT California
 BT United States
 SA Coast Ranges

Gabon (1978)
Formerly part of French Equatorial Africa.
 CO S040000N023000
  E0143000E0083000
 BT Central Africa
 BT Africa
 NT Oklo

Gabonella (1978)
Genus. Autoposting of microfossils and Protista to this term began in 1990. This term has multiple hierarchies.
 BT1 Buliminacea
 BT1 Rotaliina
 BT1 foraminifera
 BT1 Protista
 BT1 Invertebrata
 BT2 Buliminacea
 BT2 Rotaliina
 BT2 foraminifera
 BT2 Protista
 BT2 microfossils

gadolinite (1978)
Autoposting of nesosilicates began in 1985.
 BT nesosilicates
 BT orthosilicates
 BT silicates

gadolinium (1978)
Autoposting of rare earths and metals to this term began in 1989.
 UF Gd
 BT rare earths
 BT metals

Gadsden County
Valid through 1988. Search in combination with state term. After 1988, use specific county-state term.

Gadsden County Florida (1989)
NW Florida. Before 1989, also search Gadsden County AND Florida.
 CO N302000N304500
  W0841500W0845500
 BT Florida
 BT United States

gahnite (1978)
 UF zinc spinel
 BT oxides
 SA spinel group

Gailtal Alps (1978)
S Austria. This term has multiple hierarchies.
 BT1 Alps
 BT1 Europe
 BT2 Austria
 BT2 Central Europe
 BT2 Europe
 SA Carinthia Austria
 SA Eastern Alps

Gaines County
Valid through 1988. Search in combination with state term. After 1988, use specific county-state term.

Gaines County Texas (1989)
NW Texas. Before 1989, also search Gaines County AND Texas.
 CO N323000N330000
  W1021000W1030000
 BT Texas
 BT United States
 SA Ashmore Meteorite
 SA Central Basin Platform

Gairloch
No longer a valid term for GeoRef as of 1993. See Gairloch Scotland.

Gairloch Scotland (1993)
Town in NW Scotland.
 BT Strathclyde region Scotland
 BT Scotland
 BT Great Britain
 BT United Kingdom
 BT Western Europe
 BT Europe

gaize (1978)
 BT clastic rocks
 BT sedimentary rocks
 SA clastic sediments
 SA sandstone

Galaico Massif (1981)
 IN Index Spain if applicable.
 SA Spain

Galapagos Islands (1978)
A territory of Ecuador about 600 miles W of mainland. Before 1981, was included as a narrower term of Pacific Ocean.
 CO S013000N010000
  W0890000W0920000
 UF Colon Archipelago
 BT East Pacific Ocean Islands
 NT Chaves Island
 SA Ecuador
 SA Seymour Island
 SA South America

Galapagos Rift (1981)
Separates the Cocos and Nazca tectonic plates. Forms triple junction with the East Pacific Rise. Before 1981, also search Galapagos spreading center.
 CO N000000N050000
  W0800000W1020000
 UF Galapagos spreading center
 BT Pacific Ocean
 SA Cocos Plate
 SA Leg 83
 SA mid-ocean ridges
 SA Nazca Plate
 SA plate tectonics

Galapagos spreading center
use Galapagos Rift

galena (1978)
 UF blue lead
 UF galenite
 UF lead glance
 BT sulfides
 SA lead ores

Galena Dolomite (1985)
N Illinois, E Iowa, S Minnesota and SW Wisconsin.
 BT Middle Ordovician
 BT Ordovician
 BT Paleozoic
 SA Illinois
 SA Iowa
 SA Minnesota
 SA Wisconsin

galenite
use galena

galenobismutite (1978)
 BT sulfobismuthites
 BT sulfosalts

Galerkin analysis
use Galerkin method

Galerkin method (1989)
 UF Galerkin analysis
 SA finite element analysis
 SA mathematical geology
 SA mathematical methods

Galesville Sandstone (1989)
Wisconsin, Iowa and S Minnesota.
 BT Upper Cambrian
 BT Cambrian
 BT Paleozoic
 SA Iowa
 SA Minnesota
 SA Wisconsin

Galezice (1978)
Syncline in the Swiety Krzyz Mountains between Vistula River and Pilica River in SE central Poland.
 BT Poland
 BT Central Europe
 BT Europe
 SA Kielce Poland
 SA Swiety Krzyz Mountains

Galice Formation (1985)
SW Oregon and N California.
 BT Upper Jurassic
 BT Jurassic
 BT Mesozoic
 SA California
 SA Josephine Ophiolite
 SA Oregon

Galicia
No longer a valid term for GeoRef. As of 1993, see Galicia Spain.

Galicia Spain (1993)
Refers to only the region and ancient kingdom in NW Spain. For the former Austrian crownland now in SE Poland and NW Ukraine, see Galicia and Poland, or Galicia and Ukraine. Before 1993, also search Galicia AND Spain.
 CO N414800N434800
  W0064500W0092000
 BT Spain
 BT Iberian Peninsula
 BT Southern Europe
 BT Europe
 NT Arosa Bay
 NT La Coruna Spain
 NT Lugo Spain
 NT Orense Spain
 NT Pontevedra Spain
 SA Galitsiya

Galilean satellites (1978)
Includes Callisto, Europa, Ganymede, and Io.
 SA Callisto Satellite
 SA Europa Satellite
 SA Ganymede Satellite
 SA icy satellites
 SA Io Satellite
 SA Jupiter
 SA satellites

Galilee (1978)
Hilly region of N Israel.
 BT Israel
 BT Middle East
 BT Asia
 SA Haifa Bay

Galilee Basin (1981)
Structural basin in central Queensland.
 IN Index state or region as applicable.
 SA Queensland Australia

Galitsiya (1981)
Former Austrian crownland now in SE Poland and NW Ukraine. Before 1981, also search Galicia AND Poland; Galicia AND Ukraine.
 BT Europe
 SA Galicia Spain
 SA Poland
 SA Przemysl Poland
 SA Rzeszow Poland
 SA Ukraine

Gallatin County
Valid through 1988. Search in combination with state term. After 1988, use specific county-state term.

Gallatin County Illinois (1989)
SE Illinois. Before 1989, also search Gallatin County AND Illinois.
 CO N373500N375400
  W0880100W0882300
 BT Illinois
 BT United States

Gallatin County Kentucky (1989)
N Kentucky. Before 1989, also search Gallatin County AND Kentucky.
 CO N383900N385100
  W0843800W0850100
 BT Kentucky

BT United States
Gallatin County Montana (1989)
SW Montana. Before 1989, also search Gallatin County AND Montana.
CO N443000N461200
W1104800W1114500
BT Montana
BT United States
NT Hebgen Lake
NT Three Forks Montana
SA Bridger Range
SA Gallatin Range
SA Hebgen Lake earthquake 1959
SA Madison Range

Gallatin Range (1978)
IN Index states and counties as applicable.
BT United States
SA Gallatin County Montana
SA Montana
SA Park County Montana
SA Park County Wyoming
SA Wyoming
SA Yellowstone National Park

gallium (1978)
Autoposting of metals to this term began in 1989.
UF Ga
BT metals

Gallivare
No longer a valid term for GeoRef. As of 1993 see Gallivare Sweden.

Gallivare Sweden (1993)
Village N of the Arctic Circle. Before 1993 also search Gallivare AND Sweden.
UF Gellnare
BT Norrbotten Sweden
BT Sweden
BT Scandinavia
BT Western Europe
BT Europe

Galloway
No longer a valid term for GeoRef as of 1993. See Galloway Scotland.

Galloway Scotland (1993)
District in SW Scotland. Now part of Dumfries and Galloway region. Before 1993, also search Galloway.
BT Scotland
BT Great Britain
BT United Kingdom
BT Western Europe
BT Europe

Gallup Sandstone (1978)
In Mesaverde Group. NW New Mexico.
BT Upper Cretaceous
BT Cretaceous
BT Mesozoic
SA Mesaverde Group
SA New Mexico

Galveston
Valid through 1988. Search in combination with state term. After 1988, use specific city-state term.

Galveston Bay (1978)
Inlet of Gulf of Mexico protected from gulf by Galveston Island and Bolivar Peninsula.
IN Index counties as applicable.
BT Texas
BT United States

Galveston County
Valid through 1988. Search in combination with state term. After 1988, use specific county-state term.

Galveston County Texas (1989)
S Texas. Before 1989, also search Galveston County AND Texas.
CO N290400N293500
W0942200W0951500
BT Texas
BT United States
NT Galveston Island
NT Galveston Texas

Galveston Island (1978)
30 mile long island at the entrance of Galveston Bay.
BT Galveston County Texas
BT Texas
BT United States
SA Galveston Texas

Galveston Texas (1989)
City on Galveston Island in the Gulf of Mexico, S Texas. Before 1989, search Galveston AND Texas.
CO N291700N291700
W0944800W0944800
BT Galveston County Texas
BT Texas
BT United States
SA Galveston Island

Galway
No longer a valid term for GeoRef. See Galway Ireland.

Galway Granite (1978)
Connemara District in W Galway.
BT Galway Ireland
BT Ireland
BT Western Europe
BT Europe
SA Connemara
SA Devonian

Galway Ireland (1993)
County including the city in W Ireland.
CO N525500N534000
W0080000W0101500
BT Ireland
BT Western Europe
BT Europe
NT Connemara
NT Galway Granite
NT Tynagh Ireland

Gambia (1978)
CO N131200N135000
W0135000W0165000
BT West Africa
BT Africa
NT Georgetown Gambia

Gambier Embayment (1978)
IN Index states as applicable.
SA Australia
SA South Australia
SA Victoria Australia

gamma ray
A valid term through 1977. After 1977 use gamma-ray spectroscopy or gamma-ray methods.

gamma rays (1978)
Use when referring to the rays themselves.
UF rays, gamma
SA alpha rays
SA beta rays
SA electromagnetic radiation
SA gamma-gamma methods
SA gamma-ray methods
SA gamma-ray spectra
SA gamma-ray spectroscopy
SA neutron-gamma methods
SA radioactivity
SA scintillations
SA X-rays

gamma-gamma methods (1978)
Before 1978, search gamma-gamma.
SA gamma rays
SA gamma-ray methods
SA geophysical methods
SA neutron-gamma methods
SA radioactivity
SA well-logging

gamma-ray methods (1978)
Before 1978, search gamma ray or gamma-ray.
SA gamma rays
SA gamma-gamma methods
SA gamma-ray spectroscopy
SA geophysical methods
SA neutron-gamma methods
SA radioactivity
SA radioactivity methods
SA well-logging

gamma-ray spectra (1985)
Used for data. For methodology, use gamma-ray spectroscopy.
BT spectra
SA gamma rays
SA gamma-ray spectroscopy
SA spectroscopy

gamma-ray spectrometry
A valid term through 1978. After 1978, see gamma-ray spectroscopy.

gamma-ray spectroscopy (1978)
Before 1978, also search spectroscopy AND gamma ray.
BT spectroscopy
SA alpha-ray spectroscopy
SA analysis
SA chemical analysis
SA gamma rays
SA gamma-ray methods
SA gamma-ray spectra
SA radioactivity
SA well-logging

Gammon Ferruginous Member (1985)
Of Pierre Shale. NE Wyoming, SE Montana and NW South Dakota.
BT Upper Cretaceous
BT Cretaceous
BT Mesozoic
SA Montana
SA Pierre Shale
SA South Dakota
SA Wyoming

Gander Lake Group (1978)
BT Ordovician
BT Paleozoic
SA Canada
SA Newfoundland Island

Gander Zone (1985)
Eastern Canada.
IN Index provinces as applicable.
BT Eastern Canada
BT Canada
SA New Brunswick
SA Newfoundland Island
SA Nova Scotia

Gandja Azerbaidzhan
use Kirovabad Azerbaidzhan

Gandzha Azerbaidzhan
use Kirovabad Azerbaidzhan

Ganga River
use Ganges River

Gangamopteris (1978)
Genus.
BT Glossopteridales
BT gymnosperms
BT Spermatophyta
BT Plantae
SA Glossopteris

Ganges River (1978)
Sacred river which rises in the Himalayas and flows into the Bay of Bengal.
IN Index countries as applicable.
UF Ganga River
BT Asia
SA Bangladesh
SA India

Gangpur Series (1978)
Subdivided into Ghoriajor Stage, Kumarmunda Stage, Birmitrapur Stage, and Laingar Stage.
BT Archean
BT Precambrian
SA India
SA Orissa India

gangue (1978)
SA mineral deposits, genesis

Gansu
No longer a valid term for GeoRef. See Gansu China.

Gansu China (1993)
Province, NW China. Before 1993, also search Gansu or Kansu.
CO N323000N430000
E1090000E0933000
UF Kansu China
BT China
BT Far East
BT Asia
SA Haiyuan earthquake 1920
SA Huang Ho
SA Qilian Mountains

Ganymede Satellite (1985)
One of the satellites of Jupiter. Before 1985, also search Ganymede AND Jupiter.
SA Galilean satellites
SA icy satellites
SA Jupiter
SA satellites
SA Voyager Program

Gard
No longer a valid term for GeoRef. See Gard France.

Gard France (1993)
Department in S France.
CO N433000N443000
E0050000E0031500
BT France
BT Western Europe
BT Europe
NT Ales France
SA Cevennes

Garden County
Valid through 1988. Search in combination with state term. After 1988, use specific county-state term.

Garden County Nebraska (1989)
W Nebraska. Before 1989, also search Garden County AND Nebraska.
CO N411200N420000
W1020500W1024200
BT Nebraska
BT United States
SA Beaver Lake

Garfield County
   Valid through 1988. Search in combination with state term. After 1988, use specific county-state term.

Garfield County Colorado (1989)
   W Colorado. Before 1989, also search Garfield County AND Colorado.
   CO N392100N400500
      W1070000W1090400
   BT Colorado
   BT United States
   NT Anvil Points Mine
   NT Rifle Colorado
   SA Book Clifts

Garfield County Montana (1989)
   E central Montana. Before 1989, also search Garfield County AND Montana.
   CO N465300N475800
      W1060500W1075700
   BT Montana
   BT United States

Garfield County Nebraska (1989)
   Central Nebraska. Before 1989, also search Garfield County AND Nebraska.
   CO N414300N420500
      W0984500W0991300
   BT Nebraska
   BT United States

Garfield County Oklahoma (1989)
   N Oklahoma. Before 1989, also search Garfield County AND Oklahoma.
   CO N361000N363500
      W0972700W0980600
   BT Oklahoma
   BT United States

Garfield County Utah (1989)
   S Utah. Before 1989, also search Garfield County AND Utah.
   CO N373200N381000
      W1095600W1124200
   BT Utah
   BT United States
   SA Canyonlands National Park
   SA Henry Mountains
   SA Lake Powell

Garfield County Washington (1989)
   SE Washington. Before 1989, also search Garfield County AND Washington.
   CO N460000N464000
      W1171200W1175200
   BT Washington
   BT United States

Gargano (1978)
   Mountain promontory extending into Adriatic Sea in NE Apulia.
   UF Monte Gargano
   BT Apulia Italy
   BT Italy
   BT Southern Europe
   BT Europe

Gargasian (1978)
   Substage in Switzerland: upper Aptian; above Bedoulian Substage.
   BT Lower Cretaceous
   BT Cretaceous
   BT Mesozoic
   SA Aptian

Garhwal
   No longer a valid term for GeoRef. See Garhwal India.

Garhwal Himalaya
   use Garhwal Himalayas

Garhwal Himalayas (1985)
   N Uttar Pradesh, India. This term has multiple hierarchies.
   UF Garhwal Himalaya
   BT1 Himalayas
   BT1 Asia
   BT2 Uttar Pradesh India
   BT2 India
   BT2 Indian Peninsula
   BT2 Asia

Garhwal India (1993)
   District in N Uttar Pradesh. Before 1993, also search Garhwal or Gurhwal.
   UF Gurhwal India
   BT Uttar Pradesh India
   BT India
   BT Indian Peninsula
   BT Asia

Garian
   No longer a valid term for GeoRef as of 1993. See Garian Libya.

Garian Libya (1993)
   Town in extreme NW Libya. Also search Gharian.
   UF Gharian Libya
   BT Tripolitania
   BT Libya
   BT North Africa
   BT Africa

Garland County
   Valid through 1988. Search in combination with state term. After 1988, use specific county-state term.

Garland County Arkansas (1989)
   Central Arkansas. Before 1989, also search Garland County AND Arkansas.
   CO N342200N344700
      W0924700W0932200
   BT Arkansas
   BT United States

Garlock Fault (1978)
   Runs E-W in S central California.
   IN Index counties or regions as applicable.
   BT California
   BT United States

Garm
   No longer a valid term for GeoRef. As of 1993, see Garm Tadzhikistan.

Garm Tadzhikistan (1993)
   Town and former oblast in N central Tadzhikistan. This term has multiple hierarchies.
   BT1 Tadzhikistan
   BT1 Asia
   BT2 Tadzhikistan
   BT2 Commonwealth of Independent States

garnet
   As of 1981, no longer a valid term for GeoRef. See garnet group. Use garnet deposits for garnet as a commodity.

garnet deposits (1981)
   Before 1981, search garnet AND deposits.
   IN Index with industrial minerals or gems for use as a commodity.
   SA abrasives

   SA garnet group
   SA heavy mineral deposits
   SA industrial minerals

garnet group (1978)
   Autoposting of this term began in 1978.
   BT nesosilicates
   BT orthosilicates
   BT silicates
   NT almandine
   NT andradite
   NT grossular
   NT hydrogrossular
   NT majorite
   NT melanite
   NT pyrope
   NT schorlomite
   NT spessartine
   NT uvarovite
   NT vanadium garnet
   SA garnet deposits
   SA gems

garnet lherzolite (1978)
   BT peridotites
   BT ultramafics
   BT plutonic rocks
   BT igneous rocks
   SA lherzolite

garnet peridotite (1978)
   BT peridotites
   BT ultramafics
   BT plutonic rocks
   BT igneous rocks

garnet pyroxenite (1978)
   BT ultramafics
   BT plutonic rocks
   BT igneous rocks
   SA pyroxenite

garnetite (1981)
   BT metamorphic rocks

Garnett
   Valid through 1988. Search in combination with state term. After 1988, use specific city-state term.

Garnett Kansas (1989)
   City in E Kansas. Before 1989, search Garnett AND Kansas.
   CO N381600N381600
      W0951500W0951500
   BT Anderson County Kansas
   BT Kansas
   BT United States

garnierite (1978)
   UF nepouite
   BT serpentine group
   BT sheet silicates
   BT silicates
   SA nickel ores

Garo Hills (1978)
   In Meghalaya state, at bend of Brahmaputra River in NE India.
   CO N252000N254500
      E0904500E0900500
   BT Meghalaya India
   BT Northeastern India
   BT India
   BT Indian Peninsula
   BT Asia
   SA Assam India

Garonne River (1978)
   SW France.
   BT France
   BT Western Europe
   BT Europe
   SA Gironde France

garrelsite (1978)
   Autoposting on nesosilicates began in 1985.
   BT nesosilicates

   BT orthosilicates
   BT silicates

Garrett County
   Valid through 1988. Search in combination with state term. After 1988, use specific county-state term.

Garrett County Maryland (1989)
   Extreme W Maryland. Before 1989, also search Garrett County AND Maryland.
   CO N391200N394326
      W0785500W0792900
   BT Maryland
   BT United States
   NT Savage River

garronite (1978)
   BT zeolite group
   BT framework silicates
   BT silicates

Garvin County Oklahoma (1993)
   S Central Oklahoma. Before 1989, also search Garvin County AND Oklahoma.
   CO N3430N3452W096
      5600W0974200
   BT Oklahoma
   BT United States

Garza County
   Valid through 1988. Search in combination with state term. After 1988, use specific county-state term.

Garza County Texas (1989)
   NW Texas. Before 1989, also search Garza County AND Texas.
   CO N330000N332500
      W1010000W1013200
   BT Texas
   BT United States

gas
   A valid index term through 1977 used in combination with natural (i.e. gas, natural). After 1977, use natural gas.

gas chromatograms (1985)
   Use for data. Use gas chromatography for methodology. Before 1971, also search gas chromatographic analyses.
   UF gas chromatographic analyses
   UF gas chromatographs
   UF gas chromatography data
   SA chromatograms
   SA gas chromatography

gas chromatographic analyses
   Not a valid GeoRef index term after 1970. Was used on level 1 in subfile N.
   use gas chromatograms

gas chromatographs
   use gas chromatograms

gas chromatography (1978)
   Not a valid term from 1975 through 1977.
   SA chemical composition
   SA chromatography
   SA gas chromatograms
   SA gases
   SA ion chromatography
   SA radon emanometry

gas chromatography data
   use gas chromatograms

gas condensates
   use condensates

gas fields
No longer a valid term for GeoRef. Before 1971, included use in subfiles G, N and B.
use oil and gas fields

gas fields, oil and
use oil and gas fields

Gas Hills (1978)
Fremont County in central Wyoming.
IN   Index county or regions as applicable.
SA   Fremont County Wyoming
SA   Wyoming

gas hydrates (1985)
UF   hydrates, gas
SA   clathrates
SA   condensates
SA   hydrates
SA   natural gas

gas injection (1985)
SA   enhanced recovery
SA   fluid injection
SA   injection
SA   petroleum engineering
SA   reinjection wells
SA   tertiary recovery

gas phase
use gaseous phase

gas sands (1989)
This term has multiple hierarchies.
BT1 organic residues
BT2 sedimentary rocks
SA   energy sources
SA   natural gas
SA   oil sands
SA   organic materials

gas storage (1978)
SA   engineering geology
SA   natural gas
SA   oil storage
SA   petroleum engineering
SA   storage
SA   underground installations
SA   underground storage

Gascoyne abyssal plain (1978)
Off western Australia.
BT   Indian Ocean
SA   DSDP Site 260
SA   Leg 123
SA   ODP Site 766

gaseous phase (1978)
UF   gas phase
UF   phase, gaseous
SA   fluid phase
SA   gases
SA   gasification
SA   geochemistry
SA   liquid phase
SA   solid phase

gases (1978)
IN   Not used for the commodity natural gas.
SA   Darcy's law
SA   effusion
SA   fluid inclusions
SA   fumaroles
SA   gas chromatography
SA   gaseous phase
SA   gasification
SA   hydrogen sulfide
SA   magmas
SA   mining geology
SA   natural gas
SA   nitrous oxide
SA   noble gases
SA   nuees ardentes
SA   oil-gas interface
SA   phase equilibria
SA   pneumatolysis
SA   soil gases
SA   solfataras
SA   sublimates
SA   volatilization

gasification (1978)
SA   coal
SA   gaseous phase
SA   gases
SA   natural gas

gasoline
use hydrocarbons

Gaspe
No longer a valid term for GeoRef. As of 1993, see Gaspe Quebec.

Gaspe East
use Gaspe-Est County Quebec

Gaspe Est
use Gaspe-Est County Quebec

Gaspe Ouest
use Gaspe-Ouest County Quebec

Gaspe Peninsula (1978)
Extends into Gulf of Saint Lawrence. Also search Gaspe.
CO   N475000N491500 W0641500W0680000
BT   Quebec
BT   Eastern Canada
BT   Canada
SA   Battery Point Formation
SA   Saint John River

Gaspe Quebec (1993)
Village in E Quebec near E extremity of Gaspe Peninsula. Before 1993, also search Gaspe AND Quebec.
BT   Gaspe-Est County Quebec
BT   Quebec
BT   Eastern Canada
BT   Canada

Gaspe West
use Gaspe-Ouest County Quebec

Gaspe-Est County Quebec (1993)
Before 1993, also search Gaspe-Est County, Gaspe Est, East Gaspe County, East Gaspe, or Gaspe East in conjunction with Quebec.
UF   Gaspe Est
UF   Gaspe East
BT   Quebec
BT   Eastern Canada
BT   Canada
NT   Gaspe Quebec
SA   Gaspe-Ouest County Quebec

Gaspe-Ouest County Quebec (1993)
Before 1993, also search Gaspe-Ouest County, Gaspe Ouest, West Gaspe County, West Gaspe, or Gaspe West in conjunction with Quebec.
UF   Gaspe Ouest
UF   Gaspe West
BT   Quebec
BT   Eastern Canada
BT   Canada
SA   Gaspe-Est County Quebec

gastaldite
use glaucophane

gastroliths (1978)
SA   Pisces
SA   Reptilia

Gastropoda (1978)
Class. Autoposting of this term began in 1978.
BT   Mollusca
BT   Invertebrata
NT   Acoela
NT   Archaeogastropoda
NT   Basommatophora
NT   Entomotaeniata
NT   Harpidae
NT   Mesogastropoda
NT   Muricacea
NT   Naticidae
NT   Neogastropoda
NT   Planorbis
NT   Prosobranchia
NT   Pteropoda
NT   Pupillidae
NT   Sacoglossa
NT   Stylommatophora
NT   Turritellidae
NT   Vivipatus
SA   gastropods

gastropods (1981)
Common name for Gastropoda. Autoposting of mollusks to this term began in 1989.
BT   mollusks
BT   invertebrates
SA   biostratigraphy
SA   Gastropoda

Gates Formation (1989)
Coal-bearing strata. NW Alberta and NE British Columbia. Also search Gates Member AND Canada. Also a Gates Member in the Silurian of W New York.
SA   Alberta
SA   British Columbia

Gatineau County Quebec (1993)
Before 1993, also search Gatineau County AND Quebec.
BT   Quebec
BT   Eastern Canada
BT   Canada
NT   Hull Quebec

Gatineau Valley (1978)
River valley in SW Quebec.
BT   Quebec
BT   Eastern Canada
BT   Canada

gauges
use gauging

gauging (1978)
UF   gauges
SA   hydrology
SA   water

Gault (1978)
Middle Europe. Lower Cretaceous clay formation in Great Britain.
UF   Gault Clay
BT   Lower Cretaceous
BT   Cretaceous
BT   Mesozoic
SA   Great Britain

Gault Clay
use Gault

Gaurdak
No longer a valid term for GeoRef. As of 1993, see Gaurdak Turkmenia.

Gaurdak Turkmenia (1993)
Town in Chardzhou Oblast in NE Turkmenia. This term has multiple hierarchies.
BT1 Chardzhou Turkmenia
BT1 Turkmenia
BT1 Asia
BT2 Chardzhou Turkmenia
BT2 Turkmenia
BT2 Commonwealth of Independent States

Gauss Epoch (1978)
Geomagnetic epoch.
UF   Gauss Normal
BT   Pliocene
BT   Neogene
BT   Tertiary
BT   Cenozoic
SA   paleomagnetism
SA   reversals
SA   upper Tertiary

Gauss Normal
use Gauss Epoch

Gavarnie
No longer a valid term for GeoRef. See Gavarnie France.

Gavarnie France (1993)
IN   Village in SW France.
BT   Hautes-Pyrenees France
BT   France
BT   Western Europe
BT   Europe

Gavish Sebkha (1989)
BT   Egypt
BT   North Africa
BT   Africa
SA   Sinai

Gavrovo Zone (1978)
Central Peloponnesus. Also search Gavrovo.
UF   Gavrovo-Tripolis Zone
BT   Peloponnesus Greece
BT   Greece
BT   Southern Europe
BT   Europe

Gavrovo-Tripolis Zone
use Gavrovo Zone

Gavur Mountains
use Amanos Mountains

Gawler Block
use Gawler Craton

Gawler Craton (1989)
S South Australia.
UF   Gawler Block
BT   South Australia
BT   Australia
BT   Australasia

Gay (1978)
Ore fields in the Southern Urals.
IN   Index Russian Federation as applicable.
SA   Russian Federation

Gaya
No longer a valid term for GeoRef. See Gaya India.

Gaya India (1993)
City in central Bihar, E India.
BT   Bihar India
BT   India
BT   Indian Peninsula
BT   Asia
SA   Bihar mica belt

gaylussite (1978)
BT   carbonates

Gazli earthquake 1976 (1989)
Epicenter in Uzbekistan, Central Asia.
BT   earthquakes
SA   Uzbekistan

Gazli earthquake 1984 (1989)
Epicenter in Uzbekistan, Central Asia.
BT   earthquakes
SA   Uzbekistan

Gd
use gadolinium

Gdansk
No longer a valid term for GeoRef. As of 1993 see Gdansk Poland.

Gdansk Poland (1993)
Province in N Poland. Also a city on the Gulf of Danzig. Before 1993 also search Gdansk or Danzig.
UF Danzig
BT Poland
BT Central Europe
BT Europe
NT Puck Bay

Ge
use germanium

geanticlines (1978)
SA anticlines
SA anticlinoria
SA folds
SA geosynclines
SA synclinoria
SA tectonics

Geary County
Valid through 1988. Search in combination with state term. After 1988, use specific county-state term.

Geary County Kansas (1989)
E central Kansas. Before 1989, also search Geary County AND Kansas.
CO N385200N391400
 W0963000W0965700
BT Kansas
BT United States

Geauga County
Valid through 1988. Search in combination with state term. After 1988, use specific county-state term.

Geauga County Ohio (1989)
NE Ohio. Before 1989, also search Geauga County AND Ohio.
CO N412000N414300
 W0810000W0812300
BT Ohio
BT United States

Gedinnian (1978)
Europe. Above Ludlovian (Silurian), below Siegenian.
BT Lower Devonian
BT Devonian
BT Paleozoic

gedrite (1978)
BT orthoamphibole
BT amphibole group
BT chain silicates
BT silicates
SA amosite
SA anthophyllite

gegenschein
Not a valid GeoRef index term after 1971. Was used on level 1 in subfile G.

gehlenite (1978)
BT melilite group
BT sorosilicates
BT orthosilicates
BT silicates

Gehrden
No longer a valid term for GeoRef. See Gehrden Germany.

Gehrden Germany (1993)
Town in SE central Lower Saxony.
BT Lower Saxony Germany
BT Germany
BT Central Europe
BT Europe

geikielite (1978)

BT oxides

Geirud Formation (1978)
Upper Devonian, Carboniferous and possibly Lower Permian. Central Elburz Mountains in N Iran.
IN Index ages as applicable.
BT Paleozoic
SA Carboniferous
SA Devonian
SA Elburz
SA Iran
SA Permian

Geisel River valley
use Geisel Valley

Geisel Valley (1978)
River valley in NE Germany.
UF Geisel River valley
BT Germany
BT Central Europe
BT Europe
SA Saxony-Anhalt Germany
SA Thuringia Germany

gelifluction (1989)
Indicates periglacial origin by soil flow.
UF congelifluction
UF gelifluxion
SA frost action
SA periglacial features
SA soil mechanics

gelifluxion
use gelifluction

gelifraction
use congelifraction

gelisol
use frozen ground

geliturbation
use cryoturbation

gelivation
use congelifraction

Gellnare
use Gallivare Sweden

gels (1978)
BT colloidal materials
SA thixotropy

Gemer (1978)
Village in S central Slovakia.
BT Slovakia
BT Czechoslovakia
BT Central Europe
BT Europe

gems (1978)
See also specific gems. Before 1971, also search gems and gem materials.
IN Commodity. See List C.
UF gems and gem materials
UF gemstones
SA agate
SA amethyst
SA aquamarine
SA beryl
SA carnelian
SA chalcedony
SA corundum
SA diamond
SA diamonds
SA emerald
SA garnet group
SA jade
SA minerals
SA moonstone
SA onyx
SA pearls
SA peridot
SA rubies
SA sapphire

SA topaz
SA tourmaline
SA turquoise
SA zircon

gems and gem materials
No longer a valid GeoRef index term. Before 1971, was used on level 1 in subfile N.
use gems

gemstones
use gems

General Earthquake Observation System
use GEOS

generation
As of 1981, no longer a valid term for GeoRef. See genesis or propagation.

Genesee County
Valid through 1988. Search in combination with state term. After 1988, use specific county-state term.

Genesee County Michigan (1989)
SE central Michigan. Before 1989, also search Genesee County AND Michigan.
CO N424800N431500
 W0832700W0835500
BT Michigan Lower Peninsula
BT Michigan
BT United States

Genesee County New York (1989)
W New York. Before 1989, also search Genesee County AND New York.
CO N425200N430800
 W0775300W0782800
BT New York
BT United States

Genesee Group (1978)
Middle and Upper Devonian. Comprises Geneseo Shale with Genundewa Limestone Lentil, Standish Flagstone, and West River Shale, Penn Yan Shale, Sherburne Flagstone, Renwick Shale, Ithaca Member. Maryland, New York, Pennsylvania, western Virginia and northern West Virginia.
BT Devonian
BT Paleozoic
SA Maryland
SA New York
SA Pennsylvania
SA Virginia
SA West Virginia

Genesee River (1989)
Allegany, Livingston and Monroe counties, W New York; Potter County, N Pennsylvania.
IN Index states and counties as applicable.
BT United States
SA Allegany County New York
SA Livingston County New York
SA Monroe County New York
SA New York
SA Pennsylvania
SA Potter County Pennsylvania

genesis (1978)
SA crystallization
SA mineral deposits, genesis
SA paragenesis
SA pedogenesis
SA podzolization

Geneva
No longer a valid term for GeoRef. As of 1993, see Geneva Switzerland.

Geneva Switzerland (1993)
Canton surrounding W end of Lake Geneva. Also a city. Before 1993 also search Geneva AND Switzerland.
BT Switzerland
BT Central Europe
BT Europe
SA Jura Mountains
SA Lake Geneva

Gengma-Lancang earthquake 1988
use Lancang-Gengma earthquake 1988

Genoa
No longer a valid term for GeoRef. See Genoa Italy.

Genoa Italy (1993)
Province and city on the Ligurian Sea in NW Italy. Also search Genova.
UF Genova Italy
BT Liguria Italy
BT Italy
BT Southern Europe
BT Europe

Genova Italy
use Genoa Italy

genthelvite (1978)
Autoposting of silicates to this term began in 1989. This term has multiple hierarchies.
BT1 framework silicates
BT1 silicates
BT2 sulfides
SA helvite

geoarchaeology
use archaeology

geobarometry
use geologic barometry

geobotanical methods (1978)
Before 1971, also search geobotanical prospecting.
UF geobotanical prospecting
SA biogeochemical methods
SA geochemical methods
SA methods
SA mineral exploration
SA remote sensing

geobotanical prospecting
Not a valid GeoRef index term after 1970. Was used on level 1 in subfile N.
use geobotanical methods

geochemical
A valid term through 1977. See geochemical maps, geochemical methods, and geochemical prospecting.

geochemical anomalies (1993)
SA anomalies
SA chemical elements
SA geochemistry
SA haloes
SA hydrocarbons
SA lithogeochemistry
SA mineral exploration
SA pollution

geochemical background
Not a valid term for GeoRef. See background level.

geochemical controls (1978)
Also search geochemical.
SA controls
SA mineral deposits, genesis

SA sedimentation

geochemical cycle (1982)
Before 1982, search cycles AND geochemistry.
SA cycles
SA geochemistry
SA migration of elements

geochemical exploration
use geochemical methods

geochemical indicators (1981)
SA biomarkers
SA geochemistry
SA indicators

geochemical investigations
No longer a valid GeoRef index term. Before 1971, was used on level 1 in subfile N. See geochemistry; geochemical methods.

geochemical maps (1978)
Before 1978, search maps AND geochemical.
BT maps

geochemical methods (1978)
Also search geochemical exploration before 1976; geochemical prospecting before 1981. Before 1993, also search lithogeochemical methods if applicable.
UF geochemical exploration
UF geochemical prospecting
NT biogeochemical methods
SA automated analysis
SA Bear-Slave Operation
SA chemical analysis
SA dispersion patterns
SA exploration
SA geobotanical methods
SA geochemistry
SA haloes
SA methods
SA mineral exploration
SA pathfinders
SA radon emanometry
SA soil gases

geochemical models
Not a valid term for GeoRef. Use geochemistry in combination with models. Also use specific names of the geochemical models if available.

Geochemical Ocean Sections Program
use GEOSECS

geochemical profiles (1981)
Before 1981, search profiles AND geochemistry.
SA geochemistry

geochemical prospecting
As of 1981, no longer a valid term for GeoRef. Before 1976, also search geochemical exploration. Term was not used for petroleum.
use geochemical methods

geochemical surveys
A valid term through 1978. After 1978, see surveys with geochemistry.

geochemistry (1978)
Used for broad treatments as well as specific experiments.
NT lithogeochemistry
SA absolute age
SA absorption
SA activity
SA adsorption
SA advection
SA alanine
SA alkalinity
SA alumina

SA ammonia compound
SA ammonium
SA ammonium ion
SA anions
SA aqueous solutions
SA atmosphere
SA background level
SA bicarbonate ion
SA biochemistry
SA biogeochemical methods
SA buffers
SA calcium carbonate
SA calcium chloride
SA calcium ion
SA calcium sulfate
SA carbon dioxide
SA carbonate ion
SA carboxylic acids
SA catalysis
SA cation exchange capacity
SA cations
SA centrifuge methods
SA chelation
SA chemical analysis
SA chemical composition
SA chemical dispersion
SA chemical elements
SA chemical fractionation
SA chemical properties
SA chemical ratios
SA chemical reactions
SA chemostratigraphy
SA combustion
SA complexing
SA concentration
SA condensation
SA cosmochemistry
SA coupling
SA crystal chemistry
SA crystal growth
SA crystallography
SA cyanides
SA dehydration
SA denitrification
SA desorption
SA devitrification
SA diffusion
SA diffusivity
SA dilution
SA dimethyl sulfide
SA dissociation
SA Eh
SA electro-osmosis
SA electrochemical properties
SA electrolysis
SA electrolytes
SA electrophoresis
SA enrichment
SA enthalpy
SA environmental geology
SA equations of state
SA fixation
SA flocculation
SA fluid inclusions
SA fluid phase
SA fractionation
SA free energy
SA fugacity
SA gaseous phase
SA geochemical anomalies
SA geochemical cycle
SA geochemical indicators
SA geochemical methods
SA geochemical profiles
SA geochronology
SA geologic barometry
SA geologic thermometry
SA GEOSECS
SA ground water
SA halmyrolysis
SA hydration
SA hydrochemistry
SA hydrochloric acid
SA hydrofluoric acid
SA hydrolysis

SA hydroxyl ion
SA inclusions
SA ion exchange
SA ions
SA iron-rich composition
SA isomerization
SA isotherms
SA isotopes
SA kinetics
SA Kirchhoff integral
SA Langmuir equation
SA leachate
SA leaching
SA leakage anomalies
SA luminescence
SA magnesium ion
SA mass transfer
SA metals
SA metamorphic rocks
SA metamorphism
SA metasomatism
SA meteorites
SA migration of elements
SA mineral deposits, genesis
SA mineral exploration
SA mobilization
SA modal analysis
SA nitrate ion
SA nitrous oxide
SA organic materials
SA organo-metallics
SA osmosis
SA oxidation
SA paleoatmosphere
SA paleomagnetism
SA paleosalinity
SA paleotemperature
SA partitioning
SA pH
SA phase equilibria
SA phosphate ion
SA photochemistry
SA photosynthesis
SA PHREEQE
SA physicochemical properties
SA pollution
SA polymerization
SA potassium ion
SA precious metals
SA precipitation
SA pyrolysis
SA reagents
SA reduction
SA respiration
SA saturation
SA sea water
SA segregation
SA sinks
SA sodium chloride
SA sodium hydroxide
SA sodium ion
SA solid phase
SA solubility
SA solutes
SA solution
SA solutions
SA sorption
SA specific heat
SA spectra
SA spectroscopy
SA stability
SA standard materials
SA stoichiometry
SA sublimation
SA sulfate ion
SA surveys
SA suspension
SA thermochemical properties
SA thermodynamic properties
SA trace metals
SA valency
SA volatiles
SA volatilization
SA WATEQF
SA water

SA water of crystallization
SA water of dehydration
SA water vapor
SA weathering

geochronology (1978)
Used for non-isotopic dating. For isotopic (radiometric or radiogenic) dating, see absolute age.
IN Index for relative age methods. Include specific methods when possible, [e.g. electron paramagnetic resonance, exposure age, fission-track dating, hydration of glass, lichenometry, optical mineralogy, paleomagnetism, particle-track dating (e.g. cosmic-ray tracks), racemization, radiation damage, tephrochronology, thermoluminescence, tree rings, varves]
UF geologic chronology
SA absolute age
SA biochronology
SA chronology
SA cosmochronology
SA dates
SA epimerization
SA exposure age
SA fission-track dating
SA geochemistry
SA hydration of glass
SA isochrons
SA lichenometry
SA metamorphic rocks
SA metamorphism
SA meteor craters
SA optical mineralogy
SA overprinting
SA paleomagnetism
SA paleontology
SA palynology
SA particle-track dating
SA racemization
SA radiation damage
SA relative age
SA stratigraphy
SA tephrochronology
SA thermoluminescence
SA time scales
SA tree rings
SA varves

geocronite (1978)
This term has multiple hierarchies.
BT1 sulfantimonites
BT1 sulfosalts
BT2 sulfarsenites
BT2 sulfosalts

geodes (1978)
Before 1971, also search thunder eggs.
UF thunder eggs
BT secondary structures
BT sedimentary structures
SA concretions
SA vugs

geodesy (1978)
Used infrequently and for geologic applications only.
SA altimetry
SA cartography
SA changes of level
SA coordinates
SA deformation
SA Earth
SA Earth tides
SA ellipticity
SA engineering geology
SA exploration
SA figure of Earth
SA geodetic coordinates
SA geodetic networks
SA geoid
SA geomorphology

SA geophysical methods
SA geophysical surveys
SA geophysics
SA Geosat
SA geotraverses
SA Global Positioning System
SA gravity field
SA harmonics
SA isostasy
SA Laplace transformations
SA laser ranging
SA latitude
SA leveling
SA lunar laser ranging
SA maps
SA measurement
SA neotectonics
SA satellite measurements
SA seismology
SA selenodesy
SA strainmeters
SA surveys
SA tilt
SA topography
SA triangulation
SA trilateration
SA very long baseline interferometry

geodetic coordinates (1978)
SA coordinates
SA CRIB
SA geodesy
SA geometry
SA geotraverses
SA satellite measurements
SA trilateration
SA very long baseline interferometry

geodetic networks (1993)
Before 1993, also search networks and geodesy.
SA geodesy
SA geophysical methods
SA geophysical surveys
SA networks
SA triangulation
SA very long baseline interferometry

geodynamics (1978)
SA dynamics
SA Earth
SA interior
SA plate tectonics
SA tectonics

geodynamo
  use dynamos

geographic distribution
  A valid term through 1978. For fossils, see distribution or biogeography.

geographic information systems (1989)
Sometimes abbreviated as GIS.
BT information systems
SA computer networks
SA data bases
SA data processing
SA expert systems
SA PROSPECTOR
SA systems

geography (1978)
UF physical geography
SA biogeography
SA CRIB
SA geomorphology
SA paleogeography
SA regional

geoid (1978)
An equipotential surface composed of the undisturbed surface of the oceans and its imaginary extension through the continents.
SA Earth
SA figure of Earth
SA geodesy

geologic
  A valid term through 1977. After 1977, use geologic maps.

geologic barometry (1978)
Before 1971, also search geological barometry. Before 1993, also search geobarometry.
UF barometry, geologic
UF geobarometry
UF geological barometry
SA fluid inclusions
SA geochemistry
SA geologic thermometry
SA inclusions
SA lithogeochemistry
SA P-T conditions

geologic chronology
  use geochronology

geologic cryometry
  Not a valid GeoRef index term after 1969. Was used on level 1 in subfiles E and B. Now use glacial geology or permafrost as applicable.

geologic engineering
  use engineering geology

geologic exploration
  Not a valid index term after 1970. Was used on level 1 in subfile N. Now use exploration, geological methods, mineral exploration.

geologic formations, tables
  No longer a valid GeoRef index term. Before 1971, was used on level 1 in subfiles E and N. Also search geologic formation tables; geologic formations; geologic formations (lists, sections, tables).

geologic hazards (1978)
Used for studies on the initiation, controls and effects of various geological phenomena which may be hazardous to human ecology.
IN Index specific hazard as applicable [e.g. avalanches, catastrophes, earthquakes, explosions, faults, floods, land subsidence, landslides, mudflows, rock bursts, storms, sunspots, tsunamis, volcanoes]
UF hazards, geologic
SA avalanches
SA catastrophes
SA damage
SA destruction
SA earthquakes
SA engineering geology
SA environmental geology
SA erosion
SA explosions
SA faults
SA fires
SA floods
SA geologic hazards maps
SA ground motion
SA human ecology
SA hurricanes
SA impact statements
SA land subsidence
SA land use
SA landslides
SA liquefaction
SA liquefaction potential
SA marine installations
SA microzonation
SA mudflows
SA mudslides
SA nuclear facilities
SA paleoseismicity
SA pollution
SA quicksand
SA reservoirs
SA risk assessment
SA rock bursts
SA rock mechanics
SA safety
SA seismic risk
SA sinkholes
SA site exploration
SA slope stability
SA soil mechanics
SA storms
SA sunspots
SA tsunamis
SA tunnels
SA underground installations
SA volcanic earthquakes
SA volcanic risk
SA volcanoes
SA waste disposal

geologic hazards maps (1993)
Before 1993, also search geologic hazards AND environmental geology maps.
BT maps
SA environmental geology maps
SA geologic hazards

geologic history
  No longer a valid GeoRef index term. Before 1971, was used on level 1 in subfiles E and N. See stratigraphy; historical geology; paleogeography.

geologic literature
  No longer a valid GeoRef index term. Before 1971, was used on level 1 in subfile N. See bibliography; publications.

geologic mapping
  A valid term through 1975. To search, see note under cartography. See maps(1).

geologic maps (1978)
Before 1978, also search maps AND geologic.
BT maps

geologic publications
  use publications

geologic reports
  No longer a valid GeoRef index term. Before 1971, was used on level 1 in subfile N. See report; annual report; progress report.

geologic surveys
  use survey organizations

geologic thermometry (1978)
Before 1974, also search paleotemperature; geothermometry.
UF geothermometry
SA C-13/C-12
SA decrepitation
SA fluid inclusions
SA geochemistry
SA geologic barometry
SA inclusions
SA lithogeochemistry
SA microthermometry
SA O-18/O-16
SA P-T conditions
SA paleotemperature
SA S-34/S-32
SA temperature

geologic time
  No longer a valid GeoRef index term. Before 1971, was used on level 1 in subfiles E and N. See absolute age; geochronology.

geological barometry
  Not a valid GeoRef index term after 1971. Was used on level 1 in subfile G.
  use geologic barometry

geological exploration
  A valid term through 1971. After 1971, use exploration under field of study. See maps(1) cartography(2).

Geological Long-Range Inclined ASDIC
  use GLORIA

geological methods (1978)
SA exploration
SA geomorphological methods
SA methods
SA mineral exploration

geological microbiology
  use geomicrobiology

geological oceanography
  use marine geology

Geological Society of America
  use GSA

Geological Survey of Canada (1978)
BT survey organizations
BT government agencies
SA Canada

geologists (1978)
Before 1975, also search geology as a profession.
SA geology
SA practice

geology (1978)
For general treatments stressing the profession and the discipline.
UF geoscience
SA areal geology
SA associations
SA bibliography
SA catalogs
SA collections
SA creationism
SA current research
SA economic geology
SA education
SA elementary geology
SA engineering geology
SA environmental geology
SA extraterrestrial geology
SA geologists
SA geophysics
SA glossaries
SA historical geology
SA libraries
SA marine geology
SA mathematical geology
SA medical geology
SA military geology
SA mining geology
SA museums
SA physical geology
SA popular geology
SA research
SA structural geology
SA surficial geology
SA symposia
SA uniformitarianism

geology as a profession
  A valid term through 1975. After 1975, use practice. See geology and education.

geomagnetic secular variation
  use secular variations

geomagnetism
  A valid term through 1975. After 1975, see magnetic field under Earth (and other planets) and Moon; see paleomagnetism.

geomembranes (1984)
  SA dams
  SA drainage
  SA embankments
  SA engineering geology
  SA erosion control
  SA foundations
  SA geotextiles
  SA highways
  SA materials
  SA reservoirs
  SA rock mechanics
  SA seepage
  SA slope stability
  SA soil mechanics
  SA waste disposal
  SA waterways

geometry (1978)
  SA channel geometry
  SA eccentricity
  SA ellipticity
  SA folds
  SA geodetic coordinates
  SA mathematical geology
  SA plate geometry
  SA quantitative geomorphology
  SA statistical analysis
  SA structural analysis
  SA topology
  SA tortuosity
  SA waterways

geomicrobiological prospecting
  No longer a valid GeoRef index term. Before 1971, was used on level 1 in subfile N. See geobotanical methods; biogeochemical methods.

geomicrobiology (1982)
  UF geological microbiology
  UF microbiological geology
  UF microbiology, geological
  SA biology
  SA microfossils
  SA microorganisms
  SA micropaleontology
  SA paleobiology
  SA paleobotany
  SA paleontology
  SA palynology

geomorphic geology
  use geomorphology

geomorphologic
  A valid term through 1977. After 1977, use geomorphologic maps.

geomorphologic controls (1989)
  SA climatic controls
  SA controls
  SA mineral deposits, genesis
  SA tectonic controls

geomorphologic effects (1978)
  SA effects
  SA isostasy
  SA neotectonics

geomorphologic maps (1978)
  Before 1978, also search maps AND geomorphologic.
  BT maps
  SA geomorphology
  SA glacial geology maps
  SA physiographic maps
  SA surficial geology maps

geomorphological methods (1978)
  SA geological methods
  SA geophysical methods
  SA methods
  SA mineral exploration

geomorphology (1978)
  Before 1976, also search physiography.
  UF geomorphic geology
  SA aerial photography
  SA aggradation
  SA alluvial fans
  SA alluvial plains
  SA arid environment
  SA arroyos
  SA astroblemes
  SA avalanches
  SA badlands
  SA barchans
  SA barrier beaches
  SA barrier islands
  SA bars
  SA bays
  SA beach ridges
  SA beaches
  SA bedforms
  SA benches
  SA berms
  SA blowouts
  SA bluffs
  SA bogs
  SA boulder trains
  SA boulders
  SA braided streams
  SA bridges
  SA buried channels
  SA buried valleys
  SA calderas
  SA canyons
  SA capes
  SA Carolina Bays
  SA cartography
  SA cascades
  SA cauldrons
  SA caverns
  SA caves
  SA changes of level
  SA channel geometry
  SA channels
  SA cheniers
  SA chimneys
  SA cinder cones
  SA cliffs
  SA coastal dunes
  SA coastal plains
  SA coastlines
  SA collapse structures
  SA colluvium
  SA congelifraction
  SA continental dunes
  SA corrosion
  SA cratering
  SA craters
  SA creep
  SA cryptoexplosion features
  SA cuestas
  SA debris
  SA debris avalanches
  SA debris cones
  SA debris flows
  SA degradation
  SA deltas
  SA denudation
  SA depressions
  SA desertification
  SA deserts
  SA digital terrain models
  SA domes
  SA drainage basins
  SA drainage patterns
  SA dry valleys
  SA dunes
  SA earthflows
  SA ecology
  SA engineering geology
  SA environment
  SA environmental geology
  SA eolian features
  SA erosion
  SA erosion cycle
  SA erosion features
  SA erosion surfaces
  SA estuaries
  SA etching
  SA exfoliation
  SA extinct lakes
  SA fjords
  SA floodplains
  SA fluvial features
  SA forelands
  SA frost action
  SA furrows
  SA geodesy
  SA geography
  SA geomorphologic maps
  SA gilgai
  SA glacial erosion
  SA glacial geology
  SA glacial lakes
  SA glaciated terrains
  SA glaciotectonics
  SA glacis
  SA gullies
  SA highlands
  SA hills
  SA hummocks
  SA hydrology
  SA ice-marginal features
  SA impact craters
  SA impact features
  SA impacts
  SA inlets
  SA inselbergs
  SA isostasy
  SA karren
  SA karst
  SA karst filling
  SA karst hydrology
  SA karstification
  SA klippen
  SA lacustrine features
  SA lagoons
  SA lahars
  SA lakes
  SA landform description
  SA landform evolution
  SA landforms
  SA landscapes
  SA landslides
  SA lava channels
  SA lava fields
  SA lava lakes
  SA lava tubes
  SA levees
  SA limnology
  SA liquefaction
  SA littoral erosion
  SA loess
  SA longshore bars
  SA maars
  SA mangrove swamps
  SA maps
  SA marine terraces
  SA marshes
  SA mass movements
  SA meanders
  SA meromictic lakes
  SA mesas
  SA meteor craters
  SA morphometry
  SA morphostructures
  SA mountains
  SA mud banks
  SA mud volcanoes
  SA mudflows
  SA mudslides
  SA muskegs
  SA natural curiosities
  SA neotectonics
  SA oxbow lakes
  SA paleogeography
  SA paleokarst
  SA paleorelief
  SA patterned ground
  SA pediments
  SA pediplains
  SA peneplains
  SA permafrost
  SA physiographic maps
  SA physiographic provinces
  SA piedmonts
  SA plains
  SA planation
  SA plateaus
  SA playas
  SA point bars
  SA poljes
  SA polygons
  SA ponds
  SA potholes
  SA pseudokarst
  SA quantitative geomorphology
  SA rapids
  SA reefs
  SA relief
  SA relief exposure
  SA relief inversion
  SA rivers
  SA rockfalls
  SA rockslides
  SA salt marshes
  SA sand boils
  SA savannas
  SA scarps
  SA sedimentology
  SA semi-arid environment
  SA shatter cones
  SA shoals
  SA shore features
  SA shorelines
  SA sinkholes
  SA slope stability
  SA slopes
  SA slumping
  SA soil erosion
  SA soils
  SA solution cavities
  SA solution features
  SA speleology
  SA speleothems
  SA spits
  SA stalactites
  SA stalagmites
  SA steppes
  SA stream capture
  SA stream order
  SA streams
  SA surface features
  SA surficial geology
  SA surficial geology maps
  SA swamps
  SA taiga environment
  SA talus slopes
  SA tectonic controls
  SA terraces
  SA terrain classification
  SA terrains
  SA thermokarst
  SA tidal channels
  SA tidal flats
  SA tidal inlets
  SA topography
  SA tors
  SA tributaries
  SA tundra
  SA underground cavities
  SA underground streams
  SA uplands
  SA valleys
  SA volcanic features
  SA volcanic necks
  SA volcanoes
  SA wadis
  SA water erosion
  SA waterfalls

SA waterways
SA wave-cut platforms
SA weathering
SA wetlands
SA wind erosion
SA yardangs

geomorphy
No longer a valid GeoRef index term. Before 1971, was used on level 1 in subfile N. See geomorphology; geodesy.

geophones (1985)
SA acoustical methods
SA AVO methods
SA hydrophones
SA instruments
SA seismic methods
SA seismographs

geophysical
A valid term through 1977. After 1977, use geophysical maps, or geophysical methods, or geophysical surveys.

geophysical exploration
Not a valid GeoRef index term after 1971. Was used on level 1 in subfiles E, N, T and B. Now use geophysical methods or geophysical surveys as applicable.

geophysical investigations
No longer a valid GeoRef index term. Before 1971, was used on level 1 in subfiles E and N. See geophysical methods; geophysical surveys.

geophysical logging
use well-logging

geophysical maps (1978)
Before 1978, search maps AND geophysical.
BT maps
SA geophysical survey maps

geophysical methods (1978)
Used for discussions which stress methodology of applied geophysics. Autoposting of this term to many kinds of platforms and kinds of methods (see below) ended in 1985.
NT acoustical methods
NT Earth-current methods
NT electrical methods
NT electromagnetic methods
NT gravity methods
NT infrared methods
NT magnetic methods
NT magnetotelluric methods
NT neutron-gamma methods
NT neutron-neutron methods
NT radioactivity methods
NT seismic methods
SA air guns
SA airborne methods
SA apparent resistivity
SA arrays
SA audiomagnetotelluric methods
SA automated analysis
SA autoradiography
SA AVO methods
SA bathymetry
SA borehole televiewers
SA Bouguer anomalies
SA CALCRUST
SA common-depth-point method
SA conductive materials
SA crosshole methods
SA CSAMT methods
SA deep magnetic sounding
SA deep seismic sounding
SA deep sounding
SA deep-tow methods
SA direct-current methods
SA echo sounding
SA electrical anomalies
SA electrical conductivity
SA electrical sounding
SA electromagnetic field
SA electromagnetic logging
SA engineering geology
SA exploration
SA frequency domain analysis
SA frequency sounding
SA gamma-gamma methods
SA gamma-ray methods
SA geodesy
SA geodetic networks
SA geomorphological methods
SA geophysical profiles
SA geophysical surveys
SA geophysics
SA GEOS
SA Geosat
SA GLORIA
SA gravimeters
SA gravity anomalies
SA Gravsat
SA Green function
SA ground methods
SA heat flow
SA helicopter methods
SA high-resolution methods
SA homogeneity
SA hydrocarbon indicators
SA induced polarization
SA induction
SA inverse problem
SA isostasy
SA Lageos
SA Landsat
SA laser methods
SA laser ranging
SA Lithoprobe
SA lunar laser ranging
SA Magellan Program
SA magnetic anomalies
SA magnetic field
SA Magsat
SA maps
SA marine geology
SA marine methods
SA methods
SA microseismic methods
SA microwave methods
SA mineral exploration
SA mining geology
SA multichannel methods
SA multispectral scanner
SA near-field spectra
SA neutron methods
SA noise
SA paleomagnetism
SA passive methods
SA POGO
SA potential field
SA potentiometry
SA radar methods
SA raypaths
SA reflection methods
SA refraction methods
SA remote sensing
SA research vessels
SA resistivity
SA satellite methods
SA Schlumberger methods
SA Seabeam
SA SeaMarc
SA Seasat
SA seismic networks
SA seismology
SA self-potential methods
SA Shuttle Imaging Radar
SA side-scanning methods
SA signal-to-noise ratio
SA signals
SA sonar methods
SA sonobuoys
SA sounding
SA SPOT
SA stacking
SA thematic mapper
SA three-dimensional methods
SA time domain analysis
SA tomography
SA topographic correction
SA transient methods
SA ultrasonic methods
SA vertical seismic profiles
SA vertical-gradient methods
SA very long baseline interferometry
SA well-logging

geophysical observations
Not a valid GeoRef index term except during 1969. Was used on level 1 in subfile B. Now use geophysical methods, geophysical surveys or geophysics as applicable.

geophysical profiles (1981)
Before 1981, search geophysical AND profiles.
SA COCORP
SA geophysical methods
SA geophysical surveys
SA seismic profiles
SA vertical seismic profiles

geophysical research
Not a valid GeoRef index term after 1971. Was used on level 1 in subfile G. Now use geophysical surveys, geophysical methods or geophysics as applicable.

geophysical survey maps (1985)
BT maps
SA aeromagnetic maps
SA electromagnetic survey maps
SA geophysical maps
SA geophysical surveys
SA gravity survey maps
SA magnetic survey maps
SA radioactivity survey maps
SA seismic survey maps

geophysical surveys (1978)
For geophysical methods applied to specific areas. The methods are generally those of exploration and primarily concerned with the shallow structure of the Earth. Autoposting of this term began in 1978.
BT surveys
NT acoustical surveys
NT Earth-current surveys
NT electrical surveys
NT electromagnetic surveys
NT gravity surveys
NT infrared surveys
NT magnetic surveys
NT magnetotelluric surveys
NT radioactivity surveys
NT seismic surveys
SA aeromagnetic maps
SA airborne methods
SA apparent resistivity
SA arrays
SA audiomagnetotelluric methods
SA automated analysis
SA bathymetry
SA Bouguer anomalies
SA CSAMT methods
SA deep magnetic sounding
SA deep seismic sounding
SA deep sounding
SA direct-current methods
SA echo sounding
SA electrical anomalies
SA electrical sounding
SA electromagnetic field
SA electromagnetic logging
SA engineering geology
SA engineering geology maps
SA European Geotraverse
SA frequency sounding
SA geodesy
SA geodetic networks
SA geophysical methods
SA geophysical profiles
SA geophysical survey maps
SA geophysics
SA GEOS
SA Geosat
SA Global Positioning System
SA GLORIA
SA gravity anomalies
SA gravity field
SA gravity survey maps
SA Gravsat
SA ground methods
SA ground water
SA heat flow
SA helicopter methods
SA high-resolution methods
SA imagery
SA induced polarization
SA inverse problem
SA isostasy
SA Lageos
SA Landsat
SA laser ranging
SA Lithoprobe
SA lunar laser ranging
SA Magellan Program
SA magnetic anomalies
SA magnetic field
SA magnetic survey maps
SA magnetic susceptibility
SA Magsat
SA maps
SA marine methods
SA mineral exploration
SA mining geology
SA monitoring
SA multichannel methods
SA multispectral scanner
SA passive methods
SA POGO
SA radar methods
SA reflection
SA refraction
SA remote sensing
SA research vessels
SA resistivity
SA satellite methods
SA Schlumberger methods
SA Seabeam
SA SeaMarc
SA Seasat
SA seismic networks
SA self-potential methods
SA Shuttle Imaging Radar
SA signal-to-noise ratio
SA sonar methods
SA sonobuoys
SA sounding
SA SPOT
SA TACT
SA thematic mapper
SA three-dimensional methods
SA ultrasonic methods
SA very long baseline interferometry
SA well-logging

geophysics (1978)
For general treatments stressing the profession and the discipline plus experimental studies on minerals and other materials.
SA adiabatic demagnetization
SA alternating field demagnetization
SA atmosphere

SA chemical demagnetization
SA core
SA crust
SA deep seismic sounding
SA deformation
SA demagnetization
SA diamond anvil cells
SA dilatancy
SA direct problem
SA dynamos
SA Earth
SA Earth tides
SA elastic constants
SA elastic materials
SA elastic properties
SA engineering geology
SA geodesy
SA geology
SA geophysical methods
SA geophysical surveys
SA heat flow
SA inverse problem
SA kinetics
SA mathematical geology
SA meteorology
SA observatories
SA paleomagnetism
SA raypaths
SA seismology
SA structural analysis
SA tectonophysics
SA thermal demagnetization

geopressure (1978)
Related to reservoir rocks. Associated with areas of great thickness of sedimentation.
IN   May be used on with inclusions, commodity terms, or materials.
SA blowouts
SA borehole breakouts
SA overpressure
SA petroleum
SA pressure
SA reservoir rocks
SA sedimentation

GeoRef (1985)
SA bibliography
SA data bases
SA data processing
SA information systems

George River Group (1989)
Cape Breton Island, Nova Scotia. Consists of quartzite, marble, schist and gneiss.
BT Hadrynian
BT upper Proterozoic
BT Proterozoic
BT upper Precambrian
BT Precambrian
SA Nova Scotia

Georges Bank (1978)
E of Massachusetts and S of Nova Scotia.
CO N400000N412000
      W0664500W0700000
BT Atlantic Ocean
SA Georges Bank basin
SA Gulf of Maine

Georges Bank basin (1989)
Off the coast of E Massachusetts and S Nova Scotia in North American Atlantic. Autoposting of this term began in 1990.
BT North American Atlantic
BT North Atlantic
BT Atlantic Ocean
SA Georges Bank
SA Gulf of Maine

Georgetown
No longer a valid term for GeoRef. As of 1989, see Georgetown South Carolina if applicable. As of 1993, see Georgetown Gambia or Georgetown Guyana as applicable.

Georgetown County
Valid through 1988. Search in combination with state term. After 1988, use specific county-state term.

Georgetown County South Carolina (1989)
Bordering the Atlantic Ocean in E South Carolina. Before 1989, also search Georgetown County AND South Carolina.
CO N330700N334700
      W0790000W0794200
BT South Carolina
BT United States
NT Georgetown South Carolina

Georgetown Formation (1993)
Comanchean. In Washita Group. Central and S Texas, Arkansas, Louisiana. This term has multiple hierarchies.
BT1 Comanchean
BT1 Cretaceous
BT1 Mesozoic
BT2 Lower Cretaceous
BT2 Cretaceous
BT2 Mesozoic
SA Arkansas
SA Louisiana
SA Texas
SA Washita Group

Georgetown Gambia (1993)
City on the Atlantic Ocean. Before 1993, also search Georgetown AND Gambia.
BT Gambia
BT West Africa
BT Africa

Georgetown Guyana (1993)
City on the Atlantic Ocean. Before 1993, also search Georgetown AND Guyana.
BT Guyana
BT South America

Georgetown South Carolina (1989)
City in E South Carolina. Before 1989, search Georgetown AND South Carolina.
CO N332300N332300
      W0791800W0791800
BT Georgetown County South Carolina
BT South Carolina
BT United States

Georgia (1978)
Used only for the state in the United States. Before 1978, included use for the Georgian Soviet Socialist Republic in Transcaucasia, USSR. After 1978, use Georgian Republic for region in Transcaucasia. Autoposting of this term began in 1978.
CO N302000N350000
      W0804500W0853500
BT United States
NT Altamaha River
NT Baker County Georgia
NT Baldwin County Georgia
NT Bartow County Georgia
NT Berrien County Georgia
NT Bibb County Georgia
NT Bryan County Georgia
NT Burke County Georgia
NT Calhoun County Georgia
NT Camden County Georgia
NT Carroll County Georgia
NT Chatham County Georgia
NT Cherokee County Georgia
NT Clarke County Georgia
NT Clay County Georgia
NT Coffee County Georgia
NT Columbia County Georgia
NT Cook County Georgia
NT Crawford County Georgia
NT Dade County Georgia
NT Dawson County Georgia
NT De Kalb County Georgia
NT Decatur County Georgia
NT Dodge County Georgia
NT Douglas County Georgia
NT Elbert County Georgia
NT Fayette County Georgia
NT Floyd County Georgia
NT Franklin County Georgia
NT Fulton County Georgia
NT Glynn County Georgia
NT Grady County Georgia
NT Greene County Georgia
NT Hall County Georgia
NT Hancock County Georgia
NT Harris County Georgia
NT Henry County Georgia
NT Houston County Georgia
NT Jackson County Georgia
NT Jasper County Georgia
NT Jeff Davis County Georgia
NT Jefferson County Georgia
NT Johnson County Georgia
NT Jones County Georgia
NT Lamar County Georgia
NT Lee County Georgia
NT Liberty County Georgia
NT Lincoln County Georgia
NT Macon County Georgia
NT Madison County Georgia
NT Marion County Georgia
NT McIntosh County Georgia
NT Mitchell County Georgia
NT Monroe County Georgia
NT Montgomery County Georgia
NT Morgan County Georgia
NT Murray County Georgia
NT Muscogee County Georgia
NT Newton County Georgia
NT Oconee County Georgia
NT Pickens County Georgia
NT Pierce County Georgia
NT Pike County Georgia
NT Polk County Georgia
NT Pulaski County Georgia
NT Putnam County Georgia
NT Rabun County Georgia
NT Randolph County Georgia
NT Seminole County Georgia
NT Stephens County Georgia
NT Sumter County Georgia
NT Talbot County Georgia
NT Taylor County Georgia
NT Towns County Georgia
NT Twiggs County Georgia
NT Union County Georgia
NT Walker County Georgia
NT Walton County Georgia
NT Warren County Georgia
NT Washington County Georgia
NT Wayne County Georgia
NT Webster County Georgia
NT Wheeler County Georgia
NT White County Georgia
NT Wilkinson County Georgia
SA Atlantic Coastal Plain
SA Avalon Terrane
SA Bangor Limestone
SA Barnwell Formation
SA Blue Ridge Mountains
SA Blue Ridge Province
SA Blufftown Formation
SA Brevard Zone
SA Carolina Bays
SA Charlotte Belt
SA Chattahoochee River
SA Chattanooga Shale
SA Chickamauga Group
SA Citronelle Formation
SA Claiborne Group
SA Clayton Formation
SA Conasauga Group
SA Crab Orchard Mountains Group
SA Dry Branch Formation
SA Duplin Formation
SA Elberton Granite
SA Eutaw Formation
SA Fort Payne Formation
SA Gizzard Group
SA Golconda Formation
SA Great Smoky Fault
SA Great Smoky Group
SA Hawthorn Formation
SA Hayesville Fault
SA Hillabee Chlorite Schist
SA Holston Formation
SA Huber Formation
SA Jackson Group
SA Kings Mountain Belt
SA Kiokee Belt
SA Knox Group
SA Lake Chatuge
SA Lisbon Formation
SA Middendorf Formation
SA Midway Group
SA Moodys Branch Formation
SA Murphy Marble
SA Nanafalia Formation
SA New River Formation
SA Ocala Group
SA Ocoee Series
SA Okefenokee Swamp
SA Pennington Formation
SA Piedmont
SA Red Mountain Formation
SA Ripley Formation
SA Roan Supergroup
SA Rockwood Formation
SA Rome Formation
SA Rome Trough
SA Saint Louis Limestone
SA Sainte Genevieve Limestone
SA Sand Hills
SA Savannah River
SA Shady Dolomite
SA Suwannee Limestone
SA Talladega Group
SA Tallahatta Formation
SA Tallulah Falls Formation
SA Trenton Group
SA Tuscaloosa Formation
SA Twiggs Clay
SA Valley and Ridge Province
SA Wicomico Formation
SA Wilcox Group

Georgian Bay (1978)
Inlet of E Lake Huron.
IN   Index Ontario if applicable.
SA Ontario

Georgian Republic (1978)
Formerly the Georgian Soviet Socialist Republic in Transcaucasia. Before 1978, also search USSR AND Georgia.
CO N420000N430000
      E0470000E0400000
BT Europe
NT Abkhazia Georgian Republic
NT Adzharistan Georgian Republic
NT Akhaltsikhe Georgian Republic
NT Alazani
NT Aragvi
NT Chiatura Georgian Republic

NT Colchis
NT Dzirula Massif
NT Inguri River
NT Kakhetia
NT Kartlia
NT Khekordzula Deposit
NT Rioni Basin
NT Shiraki Steppe
NT Svanetia
NT Tbilisi Georgian Republic
NT Tkibuli Georgian Republic
NT Trialet Range
SA Argun River
SA Caucasus
SA Greater Caucasus
SA Kuban River
SA Kuban Valley
SA Kura River
SA Lesser Caucasus
SA Ossetia
SA Terek River
SA Transcaucasia

Georgina Basin (1978)
River drainage basin.
IN Index Northern Territory and/or Queensland as applicable.
SA Australia
SA Northern Territory Australia
SA Queensland Australia

GEOS (1989)
Acronym.
UF General Earthquake Observation System
SA earthquakes
SA geophysical methods
SA geophysical surveys
SA observations
SA seismic methods
SA seismic networks
SA seismic surveys

Geosat (1993)
SA altimetry
SA geodesy
SA geophysical methods
SA geophysical surveys
SA high-resolution methods
SA radar methods
SA remote sensing
SA satellite methods

Geoscan (1989)
Canadian national data base for geological information. Also a remote sensing instrument.
IN Index data bases if applicable.
SA bibliography
SA data bases
SA data processing
SA information systems
SA remote sensing

geoscience
use geology

GEOSECS (1978)
Acronym.
UF Geochemical Ocean Sections Program
SA geochemistry
SA oceanography
SA research vessels

geostatistics (1982)
BT statistical analysis
SA autoregression
SA Bayesian analysis
SA covariance analysis
SA crosscorrelation
SA kriging
SA mathematical geology
SA mathematical methods
SA Monte Carlo analysis
SA numerical analysis
SA outliers
SA semivariograms
SA variograms

geostrophic force
Not a valid term for GeoRef. Usually out-of-scope. Used as Coriolis force with ocean circulation.

geosynclines (1978)
Used for large, mobile downwarping of the crust, which subsides as rocks and sediments accumulate to thicknesses of thousands of meters.
IN Index with specific geosyncline where available.
SA Adelaide Geosyncline
SA Alpine Geosyncline
SA Andean Geosyncline
SA anticlinoria
SA burial metamorphism
SA clastic wedges
SA continental drift
SA Cordilleran Geosyncline
SA crust
SA eugeosynclines
SA folds
SA geanticlines
SA Hercynian Geosyncline
SA igneous activity
SA mantle
SA metallogeny
SA miogeosynclines
SA mobile belts
SA ocean basins
SA orogenic belts
SA orogeny
SA paleogeography
SA plate tectonics
SA Propria Geosyncline
SA sea-floor spreading
SA synclinoria
SA Tasman Geosyncline
SA tectonics
SA tectonophysics
SA thermal history

geotechnical maps (1993)
BT engineering geology maps
BT maps
SA rock mechanics
SA soil mechanics

geotechnical studies
Not a valid term for GeoRef. See engineering properties, soil mechanics, rock mechanics, engineering geology, or materials as applicable.

geotectonic maps (1985)
Used for maps illustrating plate tectonics, continental drift, sea-floor spreading and related topics.
BT maps
SA continental drift
SA paleogeographic maps
SA paleogeography
SA plate tectonics
SA sea-floor spreading
SA tectonic maps
SA tectonophysics

geotectonics
use tectonics

geotextiles (1984)
SA dams
SA drainage
SA embankments
SA engineering geology
SA erosion control
SA foundations
SA geomembranes
SA highways
SA materials
SA reservoirs
SA rock mechanics
SA seepage
SA slope stability
SA soil mechanics
SA waste disposal
SA waterways

geothermal doublets (1982)
SA geothermal energy
SA geothermal fields
SA geothermal systems

geothermal energy (1978)
IN Commodity. See List C.
SA brines
SA cap rocks
SA downhole methods
SA energy
SA energy sources
SA fumaroles
SA geothermal doublets
SA geothermal fields
SA geothermal systems
SA geysers
SA ground water
SA heat flow
SA heat pumps
SA heat sources
SA heat storage
SA high-level geothermal energy
SA hot dry rocks
SA hot springs
SA hydraulic fracturing
SA hydroelectric energy
SA low-level geothermal energy
SA reservoir rocks
SA thermal waters

geothermal fields (1978)
Before 1978, search geothermal.
UF fields, geothermal
SA Baca Field
SA Broadlands
SA Cerro Prieto
SA geothermal doublets
SA geothermal energy
SA geothermal systems
SA ground water
SA Hatchobaru Field
SA heat storage
SA hot dry rocks
SA Larderello
SA Los Azufres
SA oil and gas fields
SA Onikobe Field
SA Otake Field
SA Salton Sea geothermal field
SA The Geysers
SA thermal anomalies
SA thermal waters
SA Travale Field
SA Wairakei

geothermal gradient (1978)
Also search geothermal gradients.
SA conductivity
SA crust
SA Earth
SA heat flow
SA mantle
SA measurement
SA regional patterns
SA thermal conductivity
SA thermal waters
SA thermocline

geothermal phenomena
use geothermal systems

geothermal processes
As of 1981, no longer a valid term for GeoRef. See hydrothermal processes. Term was introduced in 1978.

geothermal regimes
use geothermal systems

geothermal surveys
use heat flow

geothermal systems (1978)
Before 1978, also search geothermal; geothermal regions; geothermal phenomena.
UF geothermal phenomena
UF geothermal regimes
UF hydrothermal systems
SA blowouts
SA brines
SA geothermal doublets
SA geothermal energy
SA geothermal fields
SA ground water
SA heat storage
SA hot dry rocks
SA systems
SA thermal regime
SA thermal waters
SA vents

geothermometry
use geologic thermometry

geotomography
use tomography

geotraverses (1985)
UF transects
UF traverses
SA European Geotraverse
SA geodesy
SA geodetic coordinates
SA Lithoprobe
SA surveys
SA TACT
SA triangulation

Gephyrocapsa oceanica (1989)
Species. Autoposting of thallophytes and microfossils to this term began in 1990. This term has multiple hierarchies.
BT1 Coccolithophoraceae
BT1 algae
BT1 thallophytes
BT1 Plantae
BT2 Coccolithophoraceae
BT2 algae
BT2 microfossils

Gera
No longer a valid term for GeoRef. See Gera Germany.

Gera Bezirk
No longer a valid term for GeoRef. District in SW East Germany. Before 1981, search Gera. See Thuringia Germany.

Gera Germany (1993)
City in Thuringia, E Germany. As of 1981, Gera refers to only the city in the former Gera Bezirk, now in Thuringia. Before 1981, also included the East German district.
BT Thuringia Germany
BT Germany
BT Central Europe
BT Europe

German (1978)
Used to indicate language of a catalog, dictionary, glossary or lexicon.
SA catalogs
SA dictionaries
SA glossaries
SA lexicons

German Continental Deep Drilling Program
use KTB

German Democratic Republic
Not a valid term for GeoRef. As of 1993, see Germany and also Mecklenburg-Western Pomerania Germany, Brandenburg Germany,

Saxony-Anhalt Germany, Thuringia Germany, and Saxony Germany.

German Southwest Africa
use Namibia

germanates (1978)
BT oxides

germanite (1978)
BT sulfogermanates
BT sulfosalts

germanium (1978)
Autoposting of metals to this term began in 1989.
UF Ge
BT metals

Germany (1978)
Country in Central Europe. Reunited in 1989 after being divided in 1949 into West Germany and East Germany. Upon reunification, state names for territory in West Germany were maintained, whereas the bezirks or districts in the territories of East Germany mainly reverted to old state names. Autoposting of this term began in 1978.
BT Central Europe
BT Europe
NT Alpenvorland
NT Alster River
NT Baden-Wurttemberg Germany
NT Bavaria Germany
NT Brandenburg Germany
NT Brocken Massif
NT Ems River
NT Franconia
NT Frankenberg Germany
NT Frankfurt Germany
NT Geisel Valley
NT Hamburg Germany
NT Harz Foreland
NT Harz Mountains
NT Harz region
NT Hesse
NT Hesse Basin
NT Hesse Germany
NT Kyffhauser Range
NT Lahn River
NT Lahn River valley
NT Lower Saxony Germany
NT Main River
NT Mainz Basin
NT Mecklenburg
NT Mecklenburg-Western Pomerania Germany
NT Nahe
NT North German Plain
NT North Rhine-Westphalia Germany
NT Northeastern German Plain
NT Northern German Hills
NT Northwestern German Plain
NT Odenwald
NT Palatinate
NT Rhenish Schiefergebirge
NT Rhine Westphalian Basin
NT Rhineland
NT Rhineland-Palatinate Germany
NT Rhon Mountains
NT Saale River
NT Saar-Nahe Basin
NT Saarland Germany
NT Saxonian Massif
NT Saxony Germany
NT Saxony-Anhalt Germany
NT Saxony-Thuringia
NT Schleswig-Holstein Germany
NT Schwarzenberg Germany
NT Southwestern German Hills
NT Southwestern German Massifs
NT Steinheim Basin
NT Steinheim Germany
NT Swabia
NT Teutoburg Forest
NT Thuringia Germany
NT Thuringian Basin
NT Thuringian Forest
NT Thuringian Hills
NT Thuringian Massif
NT Unstrut River
NT Upper Rhine Valley
NT Vogtland
NT Werra River
NT Weser River
NT Westerwald
NT Westphalia
SA Allgau Alps
SA Alps
SA Baltic Glaciation
SA Baltic region
SA Bohemian Massif
SA Central Alps
SA Central Graben
SA Danube River
SA Danube Valley
SA DEKORP
SA Elbe River
SA Elbe Valley
SA Essen Beds
SA European Platform
SA Hunsruck Shale
SA KTB
SA Lower Rhine Basin
SA Lusatia
SA Molasse Basin
SA Moselle River
SA Moselle Valley
SA North Sea Coast
SA North Sea region
SA Posidonia Shale
SA Poznan Clays
SA Rhine Basin
SA Rhine Graben
SA Rhine River
SA Rhine Valley
SA Silesia
SA Solnhofen Limestone
SA Stavelot-Venn Massif
SA University of Bonn
SA Werra Series
SA Zechstein

Gerona
No longer a valid term for GeoRef. As of 1993, see Gerona City Spain.

Gerona City Spain (1993)
Refers only to the city in NE Spain. Before 1993, also search Gerona and Spain.
BT Gerona Spain
BT Catalonia Spain
BT Spain
BT Iberian Peninsula
BT Southern Europe
BT Europe

Gerona Province
No longer a valid term for GeoRef. As of 1993, see Gerona Spain.

Gerona Spain (1993)
Refers only to the province in NE Spain on the Mediterranean Sea. From 1981-1992, also search Gerona Province and Spain. Before 1981, also search Gerona and Spain. For the city, see Gerona City Spain.
CO N414000N423000 E0032000E0014500
BT Catalonia Spain
BT Spain
BT Iberian Peninsula
BT Southern Europe
BT Europe
NT Ampurdan
NT Gerona City Spain
NT Osor Spain
SA Ter River basin

Geronimo volcanic field (1989)
SE Arizona.
BT Graham County Arizona
BT Arizona
BT United States

Gerrei (1978)
Region in SE Sardinia.
BT Sardinia Italy
BT Italy
BT Southern Europe
BT Europe

Gers
No longer a valid term for GeoRef. See Gers France.

Gers France (1993)
Department in SW France.
CO N431500N440500 E0011000W0002000
BT France
BT Western Europe
BT Europe
SA Adour Basin

gersdorffite (1978)
Autoposting of arsenides to this term began in 1989. This term has multiple hierarchies.
UF nickel glance
BT1 arsenides
BT2 sulfides
SA nickel ores

Getchell Mine (1978)
NW Nevada. Autoposting of broader terms began in 1989.
BT Humboldt County Nevada
BT Nevada
BT United States
SA gold ores
SA mines

Gething Formation (1989)
Peace River District. NE British Columbia and NW Alberta.
BT Lower Cretaceous
BT Cretaceous
BT Mesozoic
SA Alberta
SA British Columbia

Getic Basin
use Getic Nappe

Getic Nappe (1978)
NW part of Transylvanian Alps in Hunedoara County in SW Transylvania and E Banat.
IN Index regions as applicable.
UF Getic Basin
BT Romania
BT Southern Europe
BT Europe
SA Banat
SA Transylvania
SA Transylvanian Alps

Gettysburg Basin (1989)
Triassic to Jurassic rift basin in S central Pennsylvania and Maryland.
IN Index states as applicable.
BT United States
SA Maryland
SA Newark Basin
SA Newark-Gettysburg Basin
SA Pennsylvania

Geula Cave
use Geula Caves

Geula Caves (1978)
Mount Carmel.
UF Geula Cave
SA Haifa Israel
SA Mount Carmel

geysers (1978)
UF gusher
UF pulsating spring
SA fumaroles
SA geothermal energy
SA ground water
SA hot springs
SA springs
SA thermal waters

Geysers, The
use The Geysers

Gezira (1978)
Region between Blue Nile and White Nile rivers in central Sudan.
BT Sudan
BT East Africa
BT Africa

Ghana (1978)
Britain's former Gold Coast colony combined with British trust territory of Togo. Before 1969, also search Gold Coast.
CO N043000N120000 E0013000W0030000
UF Gold Coast
BT West Africa
BT Africa
NT Accra Ghana
NT Bosumtwi Crater
SA Togo
SA Volta Basin

Ghareb Formation (1989)
In Mount Scopus Group.
BT Upper Cretaceous
BT Cretaceous
BT Mesozoic
SA Israel
SA Middle East
SA Mount Scopus Group
SA Syria

Gharian Libya
use Garian Libya

Ghats (1978)
Two mountain ranges in S India forming E and W edges of Deccan Plateau.
BT India
BT Indian Peninsula
BT Asia
NT Eastern Ghats
NT Western Ghats

Ghazni
No longer a valid term for GeoRef. See Ghazni Afghanistan.

Ghazni Afghanistan (1993)
Use for the province in E central Afghanistan. Also a city. Before 1993, also search Ghazni and Afghanistan.
BT Afghanistan
BT Indian Peninsula
BT Asia

Ghost Ranch (1978)
Region in N central New Mexico.
IN Index counties or region as applicable.
SA New Mexico

Ghost Rocks Formation (1989)
BT Paleogene
BT Tertiary
BT Cenozoic
SA Alaska

giant fields (1978)
UF fields, giant
BT oil and gas fields
SA Hartzog Draw Field

SA natural gas
SA petroleum
SA Wilmington Field

Giant's Causeway (1978)
Formation of prismatic basaltic columns extending into sea off N coast of County Antrim.
BT Northern Ireland
BT United Kingdom
BT Western Europe
BT Europe
SA Antrim Northern Ireland

Gibbs Fault
use Gibbs fracture zone

Gibbs fault zone
use Gibbs fracture zone

Gibbs Fracture
use Gibbs fracture zone

Gibbs fracture zone (1978)
S of Reykjanes Ridge which is SE of Iceland.
UF Charlie Gibbs fracture zone
UF Gibbs Fault
UF Gibbs fault zone
UF Gibbs Fracture
BT Atlantic Ocean
SA Mid-Atlantic Ridge

Gibbs free energy
use free energy

Gibbs phase rule
use phase rule

gibbsite (1978)
UF hydrargillite
BT oxides
SA aluminum ores
SA bayerite

Gibeon Meteorite (1981)
Impact near village of Gibeon in S central Namibia. Before 1981, also search Gibeon AND meteorites.
BT octahedrite
BT iron meteorites
BT meteorites
SA Namibia

Gibraltar (1978)
Peninsula and British colony at S tip of Spain.
BT Iberian Peninsula
BT Southern Europe
BT Europe
SA Strait of Gibraltar

Gibraltar Point (1978)
Cape in Lincolnshire on N side of entrance to the Wash on the North Sea. Also location in Toronto Ontario on Lake Ontario.
IN Index regions as applicable.
SA England
SA Lincolnshire England
SA Ontario
SA Toronto Ontario

Gibson Dome (1989)
In Paradox Basin, SE Utah.
IN Index county as applicable.
SA Paradox Basin
SA San Juan County Utah

Gifu
No longer a valid term for GeoRef. See Gifu Japan.

Gifu Japan (1993)
City and prefecture in central Honshu, Japan. Before 1993, also search Gifu or Gifu-ken AND Japan.
UF Gifu-ken Japan
BT Honshu
BT Japan

BT Far East
BT Asia
NT Atera Fault
NT Hida Japan
NT Hiyoshi Japan
NT Kamioka Mine
NT Mizunami Japan
NT Toki Japan
SA Chubu Japan
SA Hakusan
SA Hida Mountains
SA Japanese Alps
SA Ontake
SA Yahagi River
SA Yake-Dake

Gifu-ken Japan
use Gifu Japan

Gila County
Valid through 1988. Search in combination with state term. After 1988, use specific county-state term.

Gila County Arizona (1989)
E central Arizona. Before 1989, also search Gila County AND Arizona.
CO N325700N342500
   W1100000W1114500
BT Arizona
BT United States
NT Sierra Ancha
NT Tonto Basin
SA San Carlos Olivine

Gila River (1978)
Rises in SW New Mexico and empties into the Colorado River near Yuma, Arizona.
IN Index states as applicable.
BT United States
SA Arizona
SA New Mexico

Gilau
No longer a valid term for GeoRef. As of 1993 see Gilau Romania.

Gilau Romania (1993)
Village in Cluj County in central Transylvania. Before 1993 also search Gilau AND Romania.
BT Cluj Romania
BT Transylvania
BT Romania
BT Southern Europe
BT Europe

Gilbert Epoch (1978)
Geomagnetic epoch.
UF Gilbert Reversed
BT lower Pliocene
BT Pliocene
BT Neogene
BT Tertiary
BT Cenozoic
SA paleomagnetism
SA reversals
SA upper Tertiary

Gilbert Islands
As of 1985, no longer a valid GeoRef term.
use Kiribati

Gilbert Reversed
use Gilbert Epoch

Gilbert, Grove Karl (1985)
UF G.K. Gilbert
UF Grove Karl Gilbert
SA biography

Gilbert-type deltas (1993)
Used to distinguish and discuss deltaic environments classified based on Gilbert's definition of deltas with topset, foreset and bottomset beds from usage in the literature ascribing the term deltaic to sedimentary successions which involve transition from marine to alluvial environments.
BT deltas
SA coastal plains
SA deltaic sedimentation
SA fluvial environment
SA marine environment
SA paleoecology
SA shore features

Gile Mountain Formation (1989)
E Vermont, W New Hampshire.
BT Devonian
BT Paleozoic
SA New Hampshire
SA Vermont

Giles Complex (1978)
IN Index states as applicable.
SA Australia
SA South Australia
SA Western Australia

Giles County
Valid through 1988. Search in combination with state term. After 1988, use specific county-state term.

Giles County Tennessee (1989)
S Tennessee. Before 1989, also search Giles County AND Tennessee.
CO N350000N352700
   W0864800W0871300
BT Tennessee
BT United States

Giles County Virginia (1989)
SW Virginia. Before 1989, also search Giles County AND Virginia.
CO N371100N372800
   W0802600W0810100
BT Virginia
BT United States

gilgai (1981)
Microrelief of mounds and depressions developed on shrinking and swelling clays.
SA geomorphology
SA mudcracks
SA soils

gillespite (1978)
BT sheet silicates
BT silicates

Gilman Formation (1989)
S Quebec.
BT Archean
BT Precambrian
SA Quebec

Gilpin County
Valid through 1988. Search in combination with state term. After 1988, use specific county-state term.

Gilpin County Colorado (1989)
N central Colorado. Before 1989, also search Gilpin County AND Colorado.
CO N394500N395000
   W1052000W1053500
BT Colorado
BT United States

gilsonite (1989)
Type of solid bitumen.
UF uintahite
UF uintaite
BT bitumens
BT organic materials
SA organic compounds

Ginkgo (1978)
Genus.
BT Ginkgoales
BT gymnosperms
BT Spermatophyta
BT Plantae

Ginkgoales (1978)
Autoposting of this term began in 1978.
BT gymnosperms
BT Spermatophyta
BT Plantae
NT Czekanowskiales
NT Ginkgo
SA Dadoxylon

Gippsland
No longer a valid term for GeoRef. See Gippsland Australia.

Gippsland Australia (1993)
Region in SE Victoria. Includes part of the Gippsland Basin.
BT Victoria Australia
BT Australia
BT Australasia
SA Gippsland Basin

Gippsland Basin (1978)
Similar to area covered by Gippsland, however the Basin also includes an area offshore to S of Victoria.
CO S390000S380000
   E1490000E1470000
BT Victoria Australia
BT Australia
BT Australasia
SA Gippsland Australia
SA Latrobe Group
SA Latrobe Valley

Girnar Hills (1978)
S central Kathiawar Peninsula in W Gujarat. Also search Girnar.
BT Gujarat India
BT India
BT Indian Peninsula
BT Asia
SA Kathiawar

Gironde
No longer a valid term for GeoRef. See Gironde France.

Gironde Estuary (1978)
On Bay of Biscay. Formed by confluence of Garonne River and Dordogne River.
BT France
BT Western Europe
BT Europe
SA Bay of Biscay
SA Gironde France

Gironde France (1993)
Department on Bay of Biscay.
CO N441500N453000
   E0003000W0013000
BT France
BT Western Europe
BT Europe
NT Arcachon Basin
NT Bordeaux France
NT Bordelais
NT Medoc
SA Garonne River
SA Gironde Estuary

Girvan
No longer a valid term for GeoRef as of 1993. See Girvan Scotland.

Girvan Scotland (1993)
Town on Firth of Clyde in Ayrshire, SW Scotland.
BT Ayrshire Scotland
BT Strathclyde region Scotland
BT Scotland

BT Great Britain
BT United Kingdom
BT Western Europe
BT Europe

girvanella (1978)
BT biogenic structures
BT sedimentary structures
SA algal biscuits

GIS
Not a valid term for GeoRef. Sometimes the acronym for geographic information systems; Geoscience Information Society.

Gisborne
No longer a valid term for GeoRef. See Gisborne New Zealand.

Gisborne New Zealand (1993)
Seaport city in E North Island.
IN Also index North Island.
BT New Zealand
BT Australasia
SA North Island

Gissar Range
use Hissar Range

Givet
No longer a valid term for GeoRef. See Givet France.

Givet France (1993)
Town in NE France in the Ardennes.
CO N500800N500800 E0044900E0044900
BT Ardennes France
BT France
BT Western Europe
BT Europe
SA Ardennes

Givetian (1978)
Europe. Above Eifelian, below Frasnian.
BT Middle Devonian
BT Devonian
BT Paleozoic
NT Horn Plateau Formation

Gizzard Group (1989)
N Georgia and E Tennessee.
BT Lower Pennsylvanian
BT Pennsylvanian
BT Carboniferous
BT Paleozoic
SA Georgia
SA Tennessee

glacial
A valid term through 1977. After 1977, see glacial environment, glacial features, glacial geology, glacial sedimentaion.

glacial deposits
Not a valid GeoRef index term after 1970. Was used on level 1 in subfile N.

glacial drift
use drift

glacial environment (1978)
Before 1978, search glacial.
SA depositional environment
SA ecology
SA environment
SA glacial erosion
SA glacial extent
SA glacial geology
SA glacial sedimentation
SA glaciated terrains
SA glaciofluvial environment
SA glaciolacustrine environment
SA glaciomarine environment
SA interglacial environment
SA periglacial environment

SA postglacial environment
SA sedimentation
SA subglacial environment

glacial erosion (1982)
Before 1982, search glacial geology AND erosion.
BT erosion
SA geomorphology
SA glacial environment
SA glacial features
SA glacial geology
SA glacial transport
SA glaciated terrains
SA glaciation
SA glaciers
SA grinding
SA scour

glacial extent (1985)
Before 1985, also search glaciation AND extent.
SA glacial environment
SA glacial geology
SA glaciation
SA glaciers
SA ice caps
SA ice sheets

glacial features (1978)
Before 1969, also search glacial land forms. Autoposting of this term began in 1978.
UF features, glacial
UF glacial land forms
NT boulder trains
NT cirques
NT drumlins
NT eskers
NT fjords
NT glacial lakes
NT kames
NT kettles
NT nunataks
NT outwash plains
SA boulders
SA buried valleys
SA debris cones
SA drift
SA erratics
SA extinct lakes
SA firn
SA fluvial features
SA glacial erosion
SA glacial geology
SA glaciated terrains
SA glaciation
SA glaciers
SA glaciofluvial environment
SA gorges
SA ice
SA ice caps
SA ice fields
SA ice lenses
SA ice streams
SA ice-marginal features
SA icebergs
SA jokulhlaups
SA lakes
SA lodgement till
SA moraines
SA outwash
SA oxides
SA periglacial features
SA sole marks
SA striations
SA taliks
SA till
SA tillite
SA valleys
SA varves

glacial geology (1978)
Before 1971, also search glaciology.

IN Includes Quaternary, as well as ancient ice ages.
UF glaciology
SA ablation
SA ancient ice ages
SA atmosphere
SA boulder clay
SA boulder trains
SA boulders
SA buried channels
SA buried valleys
SA changes of level
SA cirques
SA cryopedology
SA debris cones
SA deposition
SA drainage patterns
SA drift
SA drumlins
SA erratics
SA eskers
SA extinct lakes
SA firn
SA fjords
SA frost action
SA geomorphology
SA glacial environment
SA glacial erosion
SA glacial extent
SA glacial features
SA glacial geology maps
SA glacial lakes
SA glacial rebound
SA glacial transport
SA glaciated terrains
SA glaciation
SA glacier surges
SA glaciers
SA glaciofluvial environment
SA glaciolacustrine environment
SA glaciomarine environment
SA glaciotectonics
SA gravity flows
SA hydrology
SA ice caps
SA ice movement
SA ice sheets
SA ice shelves
SA ice streams
SA ice wedges
SA ice-marginal features
SA interglacial environment
SA isostasy
SA isostatic rebound
SA jokulhlaups
SA kames
SA kettles
SA lodgement till
SA mass balance
SA meltwater
SA Milankovitch theory
SA moraines
SA nunataks
SA outwash
SA outwash plains
SA paleoclimatology
SA palsas
SA patterned ground
SA periglacial environment
SA periglacial features
SA permafrost
SA pingos
SA polygons
SA postglacial environment
SA rock glaciers
SA sedimentary structures
SA sedimentation
SA sediments
SA snow
SA soils
SA sole marks
SA solifluction
SA stratigraphy
SA striations
SA subglacial environment

SA suffosion
SA taliks
SA thawing
SA thermokarst
SA tidewater glaciers
SA till
SA tors
SA tributaries
SA valleys

glacial geology maps (1985)
BT maps
SA geomorphologic maps
SA glacial geology

glacial lakes (1978)
This term has multiple hierarchies.
BT1 glacial features
BT2 lakes
BT2 lacustrine features
SA cirques
SA extinct lakes
SA geomorphology
SA glacial geology
SA glaciolacustrine environment
SA glaciolacustrine sedimentation
SA kettles
SA limnology
SA paleolimnology
SA periglacial features
SA varves

glacial land forms
No longer a valid GeoRef index term. Before 1969, was used on level 1 in subfile E.
use glacial features

glacial outwash
use outwash

glacial rebound (1993)
UF post-glacial rebound
BT isostatic rebound
SA glacial geology
SA glaciation
SA glaciotectonics
SA uplifts

glacial recession
use deglaciation

glacial sedimentation (1981)
BT sedimentation
NT glaciofluvial sedimentation
NT glaciolacustrine sedimentation
NT glaciomarine sedimentation
SA glacial environment
SA glacial transport

glacial sediments
Use glacial environment AND sediments.

glacial transport (1978)
SA drift
SA erratics
SA glacial erosion
SA glacial geology
SA glacial sedimentation
SA sediment transport
SA sedimentation
SA till

glaciated terrains (1983)
SA boulder trains
SA geomorphology
SA glacial environment
SA glacial erosion
SA glacial features
SA glacial geology
SA glaciation
SA mineral deposits, genesis
SA mineral exploration
SA terrains

glaciation (1978)
SA abrasion
SA changes of level
SA deglaciation
SA deposition

SA erosion
SA fjords
SA glacial erosion
SA glacial extent
SA glacial features
SA glacial geology
SA glacial rebound
SA glaciated terrains
SA glaciers
SA ice movement
SA ice sheets
SA ice shelves
SA Milankovitch theory
SA paleoclimatology

Glacier Bay   (1978)
Narrow inlet of Pacific Ocean in N part of SE Alaska. In center of Glacier Bay National Park.
IN   Index Southeastern Alaska as applicable.
SA Glacier Bay National Park
SA Southeastern Alaska

Glacier Bay National Monument
As of 1993, no longer a valid term for GeoRef.
use Glacier Bay National Park

Glacier Bay National Park   (1993)
At S end of Saint Elias Mountains in SE Alaska. Declared as a national monument. Before 1993, also search Glacier Bay National Monument.
CO N581500N592000
      W1351500W1380000
UF Glacier Bay National Monument
BT Southeastern Alaska
BT Alaska
BT United States
SA Glacier Bay
SA national monuments
SA national parks
SA Saint Elias Mountains

glacier bursts
use jokulhlaups

Glacier County
Valid through 1988. Search in combination with state term. After 1988, use specific county-state term.

Glacier County Montana   (1989)
N Montana. Before 1989, also search Glacier County AND Montana.
CO N481900N490000
      W1121100W1140500
BT Montana
BT United States

glacier floods
use jokulhlaups

glacier flows
Not a valid term for GeoRef. See ice movement.

Glacier National Park   (1978)
NW Montana. Also in SE British Columbia, in Selkirk Mountains.
IN   Index state or province as applicable.
BT North America
SA British Columbia
SA Montana
SA national parks
SA Selkirk Mountains
SA Vulture Mountain

glacier outburst floods
use jokulhlaups

Glacier Peak   (1985)
NW central Washington.

IN   Index county and Cascade Range or region as applicable.
SA Cascade Range
SA Snohomish County Washington

glacier surges   (1978)
UF catastrophic advance
SA glacial geology
SA glaciers
SA gravity flows
SA ice movement
SA surges

glaciers   (1978)
Autoposting of this term began in 1978.
IN   Used for present day glaciers.
NT rock glaciers
NT tidewater glaciers
SA ablation
SA debris cones
SA deglaciation
SA erratics
SA firn
SA glacial erosion
SA glacial extent
SA glacial features
SA glacial geology
SA glaciation
SA glacier surges
SA glaciofluvial environment
SA glaciolacustrine environment
SA glaciomarine environment
SA gravity flows
SA hydrology
SA hydrosphere
SA ice
SA ice caps
SA ice fields
SA ice movement
SA ice sheets
SA ice shelves
SA ice streams
SA ice-marginal features
SA icebergs
SA icequakes
SA jokulhlaups
SA kames
SA mass balance
SA meltwater
SA moraines
SA nivation
SA nunataks
SA periglacial environment
SA periglacial features
SA postglacial environment
SA snow
SA subglacial environment
SA tributaries

glaciofluvial environment   (1978)
Before 1978, search glaciofluvial; fluvioglacial; fluvioglacial environment.
UF fluvioglacial environment
SA deep-water environment
SA depositional environment
SA environment
SA fluvial features
SA glacial environment
SA glacial features
SA glacial geology
SA glaciers
SA glaciofluvial sedimentation
SA sedimentation

glaciofluvial features
No longer a valid term for GeoRef. Valid 1981-1992. As of 1993, use glacial features AND fluvial features.

glaciofluvial sedimentation   (1981)
This term has multiple hierarchies.

BT1 glacial sedimentation
BT1 sedimentation
BT2 fluvial sedimentation
BT2 sedimentation
SA drift
SA glaciofluvial environment
SA outwash

glaciolacustrine environment   (1981)
SA depositional environment
SA environment
SA glacial environment
SA glacial geology
SA glacial lakes
SA glaciers
SA glaciolacustrine sedimentation
SA lacustrine features
SA lakes
SA periglacial environment
SA sedimentation

glaciolacustrine sedimentation   (1985)
This term has multiple hierarchies.
BT1 glacial sedimentation
BT1 sedimentation
BT2 lacustrine sedimentation
BT2 sedimentation
SA glacial lakes
SA glaciolacustrine environment
SA lake sediments
SA varves

glaciology
use glacial geology

glaciomarine environment   (1981)
SA depositional environment
SA environment
SA glacial environment
SA glacial geology
SA glaciers
SA glaciomarine sedimentation
SA marine environment
SA sedimentation

glaciomarine sedimentation   (1981)
This term has multiple hierarchies.
BT1 glacial sedimentation
BT1 sedimentation
BT2 marine sedimentation
BT2 sedimentation
SA glaciomarine environment

glaciotectonics   (1981)
UF cryotectonics
SA geomorphology
SA glacial geology
SA glacial rebound
SA icequakes
SA tectonics

glacis   (1978)
Autoposting of slopes to this term began in 1989.
BT slopes
SA geomorphology

Glamorgan
No longer a valid term for GeoRef as of 1993. As of 1993, use Glamorgan Wales if applicable.

Glamorgan Wales   (1993)
Former county on Bristol Channel in S Wales which was split into three counties: West Glamorgan, South Glamorgan and Mid Galamorgan in April 1974. Before 1993, search Glamorgan or Glamorganshire AND Wales.
UF Glamorganshire Wales
BT Wales
BT Great Britain
BT United Kingdom
BT Western Europe

BT Europe
NT Cardiff Wales
NT Gower Peninsula
NT Swansea Bay
NT Swansea Wales
SA South Wales

Glamorganshire Wales
use Glamorgan Wales

Glarus
No longer a valid term for GeoRef. As of 1993, see Glarus Switzerland.

Glarus Alps   (1978)
N division of Central Alps. Chiefly in Glarus Canton. Also search Glarus. This term has multiple hierarchies.
BT1 Switzerland
BT1 Central Europe
BT1 Europe
BT2 Central Alps
BT2 Alps
BT2 Europe
SA Glarus Switzerland
SA Swiss Alps

Glarus Switzerland   (1993)
Canton in E central Switzerland. Before 1993 also search Glarus AND Switzerland.
BT Switzerland
BT Central Europe
BT Europe
SA Glarus Alps

Glasford Formation   (1985)
Wisconsin and N central Illinois.
BT Pleistocene
BT Quaternary
BT Cenozoic
SA Illinoian
SA Illinois
SA upper Pleistocene
SA Wisconsin

Glasgow
No longer a valid term for GeoRef as of 1993. See Glasgow Scotland.

Glasgow Scotland   (1993)
City on Clyde River in W central Scotland.
BT Strathclyde region Scotland
BT Scotland
BT Great Britain
BT United Kingdom
BT Western Europe
BT Europe

glass
use glasses

glass materials   (1982)
Before 1971, also search glass sand.
IN   May be used as a commodity term with sands.
UF glass sand
SA ceramic materials
SA insulation materials
SA materials
SA sands
SA silica

Glass Mountains   (1978)
Brewster and Pecos counties in W Texas.
IN   Index counties or regions as applicable.
SA Brewster County Texas
SA Pecos County Texas
SA Texas

glass sand
use glass materials

glass, hydration of
  use hydration of glass
glass, volcanic
  use volcanic glass
glasses (1978)
  Before 1981, search pyroclastics and glasses. From 1978-1980, pyroclastics and glasses was autoposted to the rocks that were classified under that term. Autoposting of this term began in 1981.
  UF glass
  BT volcanic rocks
  BT igneous rocks
  NT obsidian
  NT palagonite
  NT perlite
  NT pitchstone
  NT volcanic glass
  NT zirkelite
  SA Darwin glass
  SA devitrification
  SA maskelynite
  SA palagonitization
  SA pyroclastics
  SA tektites
glassy feldspar
  use sanidine
glauberite (1978)
  BT sulfates
glauconite (1978)
  As of 1981, use glauconite deposits for glauconite as a commodity.
  BT mica group
  BT sheet silicates
  BT silicates
  SA celadonite
  SA clay mineralogy
  SA glauconite deposits
  SA glauconitic composition
  SA glauconitization
  SA greensand
glauconite deposits (1981)
  Before 1981, search glauconite AND deposits.
  IN Commodity. See List C.
  SA glauconite
  SA glauconitization
  SA mica deposits
glauconitic composition (1978)
  Before 1978, also search glauconitic AND specific sediment type or sedimentary rock.
  SA composition
  SA glauconite
  SA greensand
  SA sedimentary rocks
  SA sediments
glauconitic sand
  A valid term through 1974.
  use greensand
glauconitic sandstone
  use greensand
glauconitization (1978)
  BT metamorphism
  SA glauconite
  SA glauconite deposits
glaucophane (1978)
  UF gastaldite
  BT clinoamphibole
  BT amphibole group
  BT chain silicates
  BT silicates
  SA aluminosilicates
  SA crossite
glaucophane schist (1978)
  BT schists
  BT metamorphic rocks
  SA blueschist
Glavatsi (1978)
  Region in NW Bulgaria.
  BT Bulgaria
  BT Southern Europe
  BT Europe
Gleicheniaceae (1978)
  Family.
  BT Filicopsida
  BT pteridophytes
  BT Plantae
Glen Canyon Group (1989)
  N Arizona, W Colorado, W New Mexico and S Utah. Includes Kayenta Formation and Wingate Sandstone.
  IN Index ages as applicable.
  BT Mesozoic
  SA Arizona
  SA Colorado
  SA Jurassic
  SA Kayenta Formation
  SA Moenave Formation
  SA Navajo Sandstone
  SA New Mexico
  SA Triassic
  SA Utah
  SA Wingate Sandstone
Glen Coe (1978)
  Glen in Argyllshire, S Highland region in W Scotland. Also search Glencoe AND Scotland.
  IN Index Argyllshire Scotland as applicable.
  SA Argyllshire Scotland
Glen Mountains Complex (1989)
  S Oklahoma.
  UF Glen Mountains layered complex
  BT Cambrian
  BT Paleozoic
  SA Oklahoma
Glen Mountains layered complex
  use Glen Mountains Complex
Glen Rose Formation (1978)
  In Trinity Group. Includes Lower Glen Rose Formation, Glen Rose Anhydrite, upper Glen Rose Formation. This term has multiple hierarchies.
  BT1 Comanchean
  BT1 Cretaceous
  BT1 Mesozoic
  BT2 Lower Cretaceous
  BT2 Cretaceous
  BT2 Mesozoic
  SA Texas
  SA Trinity Group
Glenarm Series (1978)
  Includes Wakefield Marble, Silver Run Limestone, Ijamsville Phyllite, Urbana Phyllite, Marburg Schist, Setters Quartzite, Cockeysville Marble, Wissahickon Formation, Peters Creek Schist, Cardiff Conglomerate, Peach Bottom Slate. Lower Paleozoic (?).
  BT lower Paleozoic
  BT Paleozoic
  SA Maryland
  SA New Jersey
  SA Pennsylvania
  SA Virginia
  SA Wissahickon Formation
Glencoe
  Not a valid term for GeoRef. See Glen Coe.
Glendon Limestone (1989)
  Member of Byram Formation in Vicksburg Group. Alabama, Florida and Mississippi.
  BT Oligocene
  BT Paleogene
  BT Tertiary
  BT Cenozoic
  SA Alabama
  SA Byram Formation
  SA Florida
  SA Mississippi
  SA Vicksburg Group
Glennie
  Valid through 1988. Search in combination with state term. After 1988, use specific city-state term.
Glennie Michigan (1989)
  Village in E Michigan. Before 1989, search Glennie AND Michigan.
  CO N443400N443400 W0834300W0834300
  BT Alcona County Michigan
  BT Michigan Lower Peninsula
  BT Michigan
  BT United States
Glenns Ferry Formation (1989)
  IN Index ages as applicable.
  BT Cenozoic
  SA Idaho
  SA Oregon
  SA Pleistocene
  SA Pliocene
  SA Wyoming
Glenwood Shale (1989)
  In Platteville Group. NE Iowa, W Illinois, Michigan, S Minnesota, Ohio and S Wisconsin.
  BT Middle Ordovician
  BT Ordovician
  BT Paleozoic
  SA Illinois
  SA Iowa
  SA Michigan
  SA Minnesota
  SA Ohio
  SA Platteville Formation
  SA Wisconsin
Gley soils
  use Gleys
Gleys (1978)
  UF Gley soils
  BT soils
  SA Pseudogleys
  SA soil group
gliding (tectonics)
  use gravity sliding
GLIMPCE (1993)
  Acronym.
  UF Great Lakes International Multidisciplinary Program on Crustal Evolution
  SA Great Lakes region
  SA programs
Glinsk
  No longer a valid term for GeoRef. As of 1993, see Glinsk Ukraine.
Glinsk Ukraine (1993)
  Village in SW Suny Oblast, N Ukraine. This term has multiple hierarchies.
  BT1 Ukraine
  BT1 Europe
  BT2 Ukraine
  BT2 Commonwealth of Independent States
Gliridae (1978)
  Family. Autoposting of Eutheria and Theria to this term began in 1989.
  BT Myomorpha
  BT Rodentia
  BT Eutheria
  BT Theria
  BT Mammalia
  BT Tetrapoda
  BT Vertebrata
  BT Chordata
global (1978)
  General term used in a geographic sense. Before 1972, also search worldwide.
  UF world
  UF worldwide
  SA world ocean
global maps (1993)
  Before 1993, also search global with map type or maps.
  BT maps
  SA cartography
  SA mapping
Global Positioning System (1989)
  UF GPS
  UF GPS Standard Positioning System
  UF GPS tracking
  UF NAVSTAR GPS
  SA geodesy
  SA geophysical surveys
  SA remote sensing
  SA satellite methods
global warming (1993)
  SA climate
  SA climate effects
  SA climatic controls
  SA greenhouse effect
  SA temperature
Globigerina (1978)
  Genus. Autoposting of microfossils and Protista to this term began in 1990. This term has multiple hierarchies.
  BT1 Globigerinidae
  BT1 Globigerinacea
  BT1 Rotaliina
  BT1 foraminifera
  BT1 Protista
  BT1 Invertebrata
  BT2 Globigerinidae
  BT2 Globigerinacea
  BT2 Rotaliina
  BT2 foraminifera
  BT2 Protista
  BT2 microfossils
  NT Globigerina bulloides
  NT Globigerina pachyderma
Globigerina bulloides (1978)
  Autoposting of microfossils and Protista to this term began in 1990. This term has multiple hierarchies.
  BT1 Globigerina
  BT1 Globigerinidae
  BT1 Globigerinacea
  BT1 Rotaliina
  BT1 foraminifera
  BT1 Protista
  BT1 Invertebrata
  BT2 Globigerina
  BT2 Globigerinidae
  BT2 Globigerinacea
  BT2 Rotaliina
  BT2 foraminifera
  BT2 Protista
  BT2 microfossils
Globigerina pachyderma (1978)

Autoposting of microfossils and Protista to this term began in 1990. This term has multiple hierarchies.
BT1 Globigerina
BT1 Globigerinidae
BT1 Globigerinacea
BT1 Rotaliina
BT1 foraminifera
BT1 Protista
BT1 Invertebrata
BT2 Globigerina
BT2 Globigerinidae
BT2 Globigerinacea
BT2 Rotaliina
BT2 foraminifera
BT2 Protista
BT2 microfossils

Globigerinacea (1978)
Autoposting of microfossils and Protista to this term began in 1990. This term has multiple hierarchies.
BT1 Rotaliina
BT1 foraminifera
BT1 Protista
BT1 Invertebrata
BT2 Rotaliina
BT2 foraminifera
BT2 Protista
BT2 microfossils
NT Globigerinidae
NT Globorotaliidae
NT Globotruncanidae
NT Hantkenina
NT Hedbergella
NT Heterohelicidae
NT Neogloboquadrina pachyderma
NT Rotalipora

Globigerinidae (1978)
Family. Autoposting of microfossils and Protista to this term began in 1990. This term has multiple hierarchies.
BT1 Globigerinacea
BT1 Rotaliina
BT1 foraminifera
BT1 Protista
BT1 Invertebrata
BT2 Globigerinacea
BT2 Rotaliina
BT2 foraminifera
BT2 Protista
BT2 microfossils
NT Globigerina
NT Globigerinoides
NT Orbulina
NT Sphaeroidinella dehiscens

Globigerinoides (1978)
Genus. Autoposting of microfossils and Protista to this term began in 1990. This term has multiple hierarchies.
BT1 Globigerinidae
BT1 Globigerinacea
BT1 Rotaliina
BT1 foraminifera
BT1 Protista
BT1 Invertebrata
BT2 Globigerinidae
BT2 Globigerinacea
BT2 Rotaliina
BT2 foraminifera
BT2 Protista
BT2 microfossils
NT Globigerinoides ruber
NT Globigerinoides sacculifer
NT Globigerinoides trilobus

Globigerinoides ruber (1978)
Autoposting of microfossils and Protista to this term began in 1990. This term has multiple hierarchies.
BT1 Globigerinoides
BT1 Globigerinidae
BT1 Globigerinacea
BT1 Rotaliina
BT1 foraminifera
BT1 Protista
BT1 Invertebrata
BT2 Globigerinoides
BT2 Globigerinidae
BT2 Globigerinacea
BT2 Rotaliina
BT2 foraminifera
BT2 Protista
BT2 microfossils

Globigerinoides sacculifer (1989)
Species. Autoposting of microfossils and Protista to this term began in 1990. This term has multiple hierarchies.
BT1 Globigerinoides
BT1 Globigerinidae
BT1 Globigerinacea
BT1 Rotaliina
BT1 foraminifera
BT1 Protista
BT1 Invertebrata
BT2 Globigerinoides
BT2 Globigerinidae
BT2 Globigerinacea
BT2 Rotaliina
BT2 foraminifera
BT2 Protista
BT2 microfossils

Globigerinoides trilobus (1978)
Autoposting of microfossils and Protista to this term began in 1990. This term has multiple hierarchies.
BT1 Globigerinoides
BT1 Globigerinidae
BT1 Globigerinacea
BT1 Rotaliina
BT1 foraminifera
BT1 Protista
BT1 Invertebrata
BT2 Globigerinoides
BT2 Globigerinidae
BT2 Globigerinacea
BT2 Rotaliina
BT2 foraminifera
BT2 Protista
BT2 microfossils

Globocassidulina subglobosa (1989)
Species. Autoposting of microfossils and Protista to this term began in 1990. This term has multiple hierarchies.
BT1 Cassidulinacea
BT1 Rotaliina
BT1 foraminifera
BT1 Protista
BT1 Invertebrata
BT2 Cassidulinacea
BT2 Rotaliina
BT2 foraminifera
BT2 Protista
BT2 microfossils

Globorotalia (1978)
Genus. Autoposting of microfossils and Protista to this term began in 1990. This term has multiple hierarchies.
BT1 Globorotaliidae
BT1 Globigerinacea
BT1 Rotaliina
BT1 foraminifera
BT1 Protista
BT1 Invertebrata
BT2 Globorotaliidae
BT2 Globigerinacea
BT2 Rotaliina
BT2 foraminifera
BT2 Protista
BT2 microfossils
NT Globorotalia inflata
NT Globorotalia menardii
NT Globorotalia pachyderma
NT Globorotalia truncatulinoides

Globorotalia inflata (1989)
Species. Autoposting of microfossils and Protista to this term began in 1990. This term has multiple hierarchies.
BT1 Globorotalia
BT1 Globorotaliidae
BT1 Globigerinacea
BT1 Rotaliina
BT1 foraminifera
BT1 Protista
BT1 Invertebrata
BT2 Globorotalia
BT2 Globorotaliidae
BT2 Globigerinacea
BT2 Rotaliina
BT2 foraminifera
BT2 Protista
BT2 microfossils

Globorotalia menardii (1978)
Autoposting of microfossils and Protista to this term began in 1990. This term has multiple hierarchies.
BT1 Globorotalia
BT1 Globorotaliidae
BT1 Globigerinacea
BT1 Rotaliina
BT1 foraminifera
BT1 Protista
BT1 Invertebrata
BT2 Globorotalia
BT2 Globorotaliidae
BT2 Globigerinacea
BT2 Rotaliina
BT2 foraminifera
BT2 Protista
BT2 microfossils

Globorotalia pachyderma (1978)
Autoposting of microfossils and Protista to this term began in 1990. This term has multiple hierarchies.
BT1 Globorotalia
BT1 Globorotaliidae
BT1 Globigerinacea
BT1 Rotaliina
BT1 foraminifera
BT1 Protista
BT1 Invertebrata
BT2 Globorotalia
BT2 Globorotaliidae
BT2 Globigerinacea
BT2 Rotaliina
BT2 foraminifera
BT2 Protista
BT2 microfossils

Globorotalia truncatulinoides (1978)
Autoposting of microfossils and Protista to this term began in 1990. This term has multiple hierarchies.
BT1 Globorotalia
BT1 Globorotaliidae
BT1 Globigerinacea
BT1 Rotaliina
BT1 foraminifera
BT1 Protista
BT1 Invertebrata
BT2 Globorotalia
BT2 Globorotaliidae

BT2 Globigerinacea
BT2 Rotaliina
BT2 foraminifera
BT2 Protista
BT2 microfossils

Globorotaliidae (1978)
Family. Autoposting of microfossils and Protista to this term began in 1990. This term has multiple hierarchies.
BT1 Globigerinacea
BT1 Rotaliina
BT1 foraminifera
BT1 Protista
BT1 Invertebrata
BT2 Globigerinacea
BT2 Rotaliina
BT2 foraminifera
BT2 Protista
BT2 microfossils
NT Globorotalia

Globotruncana (1978)
Genus. Autoposting of microfossils and Protista to this term began in 1990. This term has multiple hierarchies.
BT1 Globotruncanidae
BT1 Globigerinacea
BT1 Rotaliina
BT1 foraminifera
BT1 Protista
BT1 Invertebrata
BT2 Globotruncanidae
BT2 Globigerinacea
BT2 Rotaliina
BT2 foraminifera
BT2 Protista
BT2 microfossils

Globotruncanidae (1978)
Family. Autoposting of microfossils and Protista to this term began in 1990. This term has multiple hierarchies.
BT1 Globigerinacea
BT1 Rotaliina
BT1 foraminifera
BT1 Protista
BT1 Invertebrata
BT2 Globigerinacea
BT2 Rotaliina
BT2 foraminifera
BT2 Protista
BT2 microfossils
NT Globotruncana

Glomar Challenger (1985)
SA Deep Sea Drilling Project
SA marine geology
SA oceanography
SA research vessels

GLORIA (1993)
Acronym.
UF Geological Long-Range Inclined ASDIC
SA acoustical methods
SA acoustical surveys
SA geophysical methods
SA geophysical surveys
SA instruments
SA marine methods
SA side-scanning methods
SA sonar methods

Glorieta Sandstone (1985)
N central New Mexico.
BT Permian
BT Paleozoic
SA New Mexico

glossaries (1978)
IN Use for glossaries and atlases. Index with language of text.
SA Arabic
SA atlas

SA bibliography
SA catalogs
SA dictionaries
SA education
SA French
SA geology
SA German
SA lexicons
SA monographs
SA paleontology
SA publications
SA Russian
SA sedimentary petrology
SA stratigraphy
SA structural geology
SA textbooks

Glossograptina (1978)
IN Index Hemichordata or Invertebrata as applicable.
BT Graptoloidea
BT Graptolithina
SA Hemichordata
SA Invertebrata

Glossopteridales (1981)
BT gymnosperms
BT Spermatophyta
BT Plantae
NT Gangamopteris
NT Glossopteris

Glossopteris (1978)
Genus. Gangamopteris was included as a narrower term before 1981. Autoposting of this term began in 1978.
BT Glossopteridales
BT gymnosperms
BT Spermatophyta
BT Plantae
NT Glossopteris flora
SA Dadoxylon
SA Gangamopteris
SA Pecopteris
SA Taeniopteris

Glossopteris flora (1978)
BT Glossopteris
BT Glossopteridales
BT gymnosperms
BT Spermatophyta
BT Plantae

Gloucester County
Valid through 1988. Search in combination with state term. After 1988, use specific county-state term.

Gloucester County New Brunswick (1993)
Before 1993, also search Gloucester County AND New Brunswick.
BT New Brunswick
BT Maritime Provinces
BT Eastern Canada
BT Canada
NT Bathurst mining district
NT Bathurst New Brunswick

Gloucester County New Jersey (1989)
SW New Jersey. Before 1989, also search Gloucester County AND New Jersey.
CO N393200N395300
W0745300W0752700
BT New Jersey
BT United States
NT Woodbury New Jersey

Gloucester County Virginia (1989)
E Virginia. Before 1989, also search Gloucester County AND Virginia.
CO N371600N373700
W0762300W0764300
BT Virginia
BT United States

Gloucestershire
No longer a valid term for GeoRef as of 1993.
use Gloucestershire England

Gloucestershire England (1993)
County in W England. Before 1978, also search Gloucester. In 1975, part of Gloucestershire was included in Avon, a newly formed county.
CO N513000N521000
W0013000W0024000
UF Gloucestershire
BT England
BT Great Britain
BT United Kingdom
BT Western Europe
BT Europe
NT Forest of Dean
SA Avon England
SA Bristol England
SA Cambridge England
SA Cotswold Hills

glowing avalanche
use ash flows

glycine (1989)
Principal amino acid in sugar cane.
BT amino acids
BT organic materials
SA organic acids

Glycymeris (1978)
BT Bivalvia
BT Mollusca
BT Invertebrata

Glynn County
Valid through 1988. Search in combination with state term. After 1988, use specific county-state term.

Glynn County Georgia (1989)
On the Atlantic Ocean in SE. Before 1989, also search Glynn County AND Georgia.
CO N310000N312600
W0811800W0814800
BT Georgia
BT United States
NT Brunswick Georgia

gmelinite (1978)
BT zeolite group
BT framework silicates
BT silicates

Gnathodus (1989)
Genus. Autoposting of microfossils to this term began in 1990.
BT Conodonta
BT microfossils

gneiss
As of 1981, no longer a valid term for GeoRef. From 1978-1980, gneisses was autoposted to this term.
use gneisses

gneisses (1978)
Before 1981, also search gneiss. Autoposting of this term to gneiss ended in 1981.
UF gneiss
BT metamorphic rocks
NT augen gneiss
NT banded gneiss
NT biotite gneiss
NT granite gneiss
NT orthogneiss
NT paragneiss
NT sillimanite gneiss
NT tonalite gneiss

gneissic texture (1978)
Before 1978, search gneissic.
BT textures
SA metamorphic rocks

Gnetales (1978)
BT gymnosperms
BT Spermatophyta
BT Plantae

Goa
No longer a valid term for GeoRef. Former Portuguese possession on the Arabian Sea annexed by India in 1962. See Goa India.

Goa India (1993)
As of 1987, state in W India. Previously part of the Territory of Goa, Daman, and Diu.
BT India
BT Indian Peninsula
BT Asia

Goban Spur (1989)
Plateau region on the continental margin located about 250 km SW of mainland Ireland in NE Atlantic. As of 1990, Atlantic Ocean is autoposted to this term.
BT European Atlantic
BT North Atlantic
BT Atlantic Ocean
SA DSDP Site 548
SA DSDP Site 550
SA DSDP Site 551
SA Leg 80

Gobi Desert (1978)
Also search Gobi.
IN Index Mongolia and/or Inner Mongolia China as applicable.
CO N420000N460000
E1170000E1020000
BT Asia
SA Bayn Dzak
SA Inner Mongolia China
SA Mongolia

Goczalkowice (1978)
Borehole region in Upper Silesian coal basin in S part of country.
BT Katowice Poland
BT Poland
BT Central Europe
BT Europe

Godavari Basin
use Krishna-Godavari Basin

Godavari River (1978)
Rises in NW Maharashtra and flows SE into the Bay of Bengal. Also search Godavari.
IN Index states as applicable.
BT Indian Peninsula
BT Asia
SA Andhra Pradesh India
SA Bay of Bengal
SA Godavari Valley
SA Maharashtra India

Godavari Valley (1978)
River valley in central India. Also search Godavari.
IN Index states as applicable.
BT India
BT Indian Peninsula
BT Asia
SA Andhra Pradesh India
SA Godavari River
SA Krishna-Godavari Basin
SA Maharashtra India
SA Pranhita-Godavari Valley

Godthaab
No longer a valid term for GeoRef. As of 1993 see Godthaab Greenland.

Godthaab Greenland (1993)
Town on the SW coast of Greenland. Before 1993 also search Godthaab or Godthab.
UF Godthab Greenland
BT Greenland
BT Arctic region
SA Godthaabsfjord

Godthaabsfjord (1989)
SW Greenland. Inlet of Davis Strait.
UF Godthabsfjord
BT Greenland
BT Arctic region
SA Davis Strait
SA Godthaab Greenland

Godthab Greenland
use Godthaab Greenland

Godthabsfjord
use Godthaabsfjord

goethite (1978)
UF gothite
UF xanthosiderite
BT oxides
SA iron minerals
SA iron ores
SA iron oxides

Gog Group (1993)
Lower Cambrian. Alberta and British Columbia.
BT Lower Cambrian
BT Cambrian
BT Paleozoic
SA Alberta
SA British Columbia
SA Canadian Rocky Mountains

Gogebic County
Valid through 1988. Search in combination with state term. After 1988, use specific county-state term.

Gogebic County Michigan (1989)
On Lake Superior at W end of Michigan Upper Peninsula, NW Michigan. Before 1989, also search Gogebic County AND Michigan.
CO N461000N464000
W0890000W0902500
BT Michigan Upper Peninsula
BT Michigan
BT United States
SA Porcupine Mountains

Goias
No longer a valid term for GeoRef. See Goias Brazil.

Goias Brazil (1993)
State in central Brazil.
CO S193000S050500
W0460000W0532000
BT Brazil
BT South America
SA Brazilian Shield
SA Estrada Nova Formation
SA Irati Formation
SA Maranhao Basin
SA Tocantins River region

Golan Heights (1978)
Hilly region in SW Syria.
BT Syria
BT Middle East
BT Asia

Golconda Formation (1985)

S Illinois, NW Alabama, NW Georgia, S Indiana, W Kentucky, SE Missouri and W Tennessee.
BT Chesterian
BT Upper Mississippian
BT Mississippian
BT Carboniferous
BT Paleozoic
SA Alabama
SA Beech Creek Limestone Member
SA Georgia
SA Illinois
SA Indiana
SA Kentucky
SA Missouri
SA Tennessee

gold (1978)
Chemical element. As of 1981, use gold ores for gold as a commodity. Autoposting of metals to this term began in 1989.
UF Au
BT metals
SA electrum
SA gold ores
SA heavy metals
SA native elements
SA siderophile elements

Gold Coast
No longer a valid GeoRef index term. Before 1969, was used on level 1 in subfile E.
use Ghana

gold ores (1981)
Before 1981, also search gold AND (deposit OR deposits OR ore OR ores OR economic) in the basic index. Autoposting of metal ores to this term began in 1985.
IN Commodity. See List C.
BT metal ores
SA Alligator Ridge Mine
SA calaverite
SA Carlin Mine
SA carlin-type
SA Chinkuashih Mine
SA Dachang Deposit
SA Equity Mine
SA Getchell Mine
SA gold
SA Hemlo Deposit
SA Homestake Mine
SA Horne Mine
SA Jabiluka Australia
SA Klerksdorp Field
SA Kolar Gold Fields India
SA La Ronge Domain
SA Larder Lake District Ontario
SA Leadville mining district
SA Lupin Mine
SA native elements
SA Norseman-Wiluna Belt
SA nuggets
SA Olympic Dam
SA Pilbara gold field
SA placers
SA Salsigne Mine
SA San Antonio Mine
SA sandstone-type
SA Sigma Mine
SA Sunnyside Mine
SA Temagami Mine
SA Yilgarn

Golden
Valid through 1988. Search in combination with state term. After 1988, use specific city-state term.

Golden Colorado (1989)
City in N central Colorado. Before 1989, search Golden AND Colorado.
CO N394500N394500
W1165800W1165800
BT Jefferson County Colorado
BT Colorado
BT United States

Golden Horn Batholith (1978)
In the Cascade Range of Washington.
BT Washington
BT United States
SA Cascade Range

Golden Trout Wilderness (1985)
E Tulare County and W Inyo County, California.
IN Index counties as applicable.
CO N360500N363000
W1180000W1184500
BT California
BT United States
SA Inyo County California
SA Tulare County California
SA wilderness areas

Golden Valley County
Valid through 1988. Search in combination with state term. After 1988, use specific county-state term.

Golden Valley County Montana (1989)
Central Montana. Before 1989, also search Golden Valley County AND Montana.
CO N460300N464500
W1084600W1094000
BT Montana
BT United States

Golden Valley County North Dakota (1989)
W North Dakota. Before 1989, also search Golden Valley County AND North Dakota.
CO N463200N471800
W1033700W1040300
BT North Dakota
BT United States

Golden Valley Formation (1978)
Includes White Earth, South Ross, Lakeside, and East Tioga Clay Beds. SW North Dakota.
BT Eocene
BT Paleogene
BT Tertiary
BT Cenozoic
SA North Dakota

golden-brown algae
use Chrysophyta

Goldenville Formation (1989)
Consists of metasedimentary rocks. Basal unit of the Meguma Group. Probably Upper Cambrian in age.
BT Upper Cambrian
BT Cambrian
BT Paleozoic
SA Meguma Group
SA Nova Scotia

Goldfield
Valid through 1988. Search in combination with state term. After 1988, use specific city-state term.

Goldfield Nevada (1989)
Village in SW Nevada. Before 1989, search Goldfield AND Nevada.
CO N374200N374200
W1171500W1171500
BT Esmeralda County Nevada
BT Nevada
BT United States

Goliad Sand (1985)
S Texas.
BT Pliocene
BT Neogene
BT Tertiary
BT Cenozoic
SA Texas

Golodnaya Step
use Bet-Pak-Dala

Golovanevsk
No longer a valid term for GeoRef. As of 1993, see Golovanevsk Russian Federation.

Golovanevsk Russian Federation (1993)
Village in Voronezh Oblast in S central European Russian Federation. This term has multiple hierarchies.
BT1 Voronezh Russian Federation
BT1 Russian Federation
BT1 Commonwealth of Independent States
BT2 Voronezh Russian Federation
BT2 Europe

Gomera (1978)
Island belonging to Spain.
BT Canary Islands
BT Atlantic Ocean Islands

gondite (1978)
BT metamorphic rocks

Gondolella (1985)
Genus. Autoposting of microfossils to this term began in 1990.
BT Conodonta
BT microfossils

Gondwana (1978)
Theoretical ancient continent including India, Australia, Antarctica; and parts of S Africa and South America. It is supposed to have fragmented and drifted apart in Post-Carboniferous time. Before 1974, also search Gondwanaland. Also search lower Gondwana.
IN Index countries and continents as applicable.
UF Gondwanaland
SA Africa
SA Antarctica
SA Australia
SA continental drift
SA India
SA Laurasia
SA lower Gondwana System
SA Pangaea
SA South America
SA Southern Hemisphere
SA upper Gondwana System

Gondwana System (1978)
Upper Carboniferous to Lower Cretaceous. Peninsular India. Autoposting of this term began in 1978.
IN Index ages as applicable.
NT lower Gondwana System
NT upper Gondwana System
SA India
SA Lower Jurassic
SA Permian
SA Triassic
SA Upper Carboniferous

Gondwanaland
A valid term through 1973.
use Gondwana

Goniatites (1978)
BT Goniatitidae

BT Goniatitida
BT Ammonoidea
BT Tetrabranchiata
BT Cephalopoda
BT Mollusca
BT Invertebrata

Goniatitida (1978)
Autoposting of this term began in 1978.
BT Ammonoidea
BT Tetrabranchiata
BT Cephalopoda
BT Mollusca
BT Invertebrata
NT Goniatitidae

Goniatitidae (1978)
Autoposting of this term began in 1978.
BT Goniatitida
BT Ammonoidea
BT Tetrabranchiata
BT Cephalopoda
BT Mollusca
BT Invertebrata
NT Goniatites

goniometry (1982)
SA crystal form
SA crystal structure
SA crystallography
SA microscope methods
SA mineralogy
SA optical mineralogy
SA petrography
SA X-ray analysis

Gonzales County
Valid through 1988. Search in combination with state term. After 1988, use specific county-state term.

Gonzales County Texas (1989)
S central Texas. Before 1989, also search Gonzales County AND Texas.
CO N290500N294500
W0970500W0975000
BT Texas
BT United States

Good Hope
use Fort Good Hope

Goodnow earthquake 1983 (1993)
In Adirondack Mountains, NE New York.
BT earthquakes
SA Adirondack Mountains
SA New York

Goose Egg Formation (1989)
Permian and Lower Triassic.
IN Index ages as applicable.
SA Colorado
SA Lower Triassic
SA Montana
SA Permian
SA Utah
SA Wyoming

Gorda Ridge
use Gorda Rise

Gorda Rise (1978)
Off N California E of the Mendocino Escarpment.
UF Gorda Ridge
BT North American Pacific
BT Pacific Ocean
SA DSDP Site 35
SA Escanaba Trough

Gore Range (1978)
Part of Park Range in N central Colorado.

IN Index counties and U.S. Rocky Mountains or regions as applicable.
SA Colorado
SA U. S. Rocky Mountains
SA Vail Pass

gorges (1993)
BT valleys
SA canyons
SA channels
SA fluvial features
SA glacial features
SA poljes
SA rilles
SA solution features
SA troughs
SA waterways

Gorki
No longer a valid term for GeoRef. As of 1993, see Nizhniy Novgorod Russian Federation.

Gorki Russian Federation
use Nizhniy Novgorod Russian Federation

Gorkiy Russian Federation
use Nizhniy Novgorod Russian Federation

Gorky Russian Federation
use Nizhniy Novgorod Russian Federation

Gorleben (1993)
Former salt mine used for radioactive waste disposal.
CO N530300N530300
  E0112100E0112100
BT Lower Saxony Germany
BT Germany
BT Central Europe
BT Europe
SA radioactive waste
SA waste disposal
SA waste disposal sites

Gorlovka Basin (1978)
Industrial basin in the Donets Basin in E Ukraine. This term has multiple hierarchies.
BT1 Ukraine
BT1 Europe
BT2 Ukraine
BT2 Commonwealth of Independent States
SA Donets Basin

Gornaya Shoriya (1978)
Mountainous region in Kemerovo Oblast just S of Kuznetsk Basin. This term has multiple hierarchies.
BT1 Russian Federation
BT1 Commonwealth of Independent States
BT2 Asia
SA Kemerovo Russian Federation

Gorny Altai (1978)
Mountain range in Altai Mountains of Gorno-Altai Autonomous Oblast. This term has multiple hierarchies.
CO N484500N523000
  E0900000E0840000
BT1 Altai Russian Federation
BT1 Russian Federation
BT1 Commonwealth of Independent States
BT2 Altai Russian Federation
BT2 Asia
BT3 Altai Mountains
BT3 Asia

Gorringe Bank (1985)
European Atlantic, off SW Portugal. As of 1990, Atlantic Ocean is autoposted to this term.

BT European Atlantic
BT North Atlantic
BT Atlantic Ocean

Goryachiy Plyazh (1978)
Beach on Kunashir Island of the Kuril Islands NE of Japan. As of 1993, Kunashir Island is a broader term. This term has multiple hierarchies.
BT1 Kunashir Island
BT1 Kuril Islands
BT1 Sakhalin Russian Federation
BT1 Russian Federation
BT1 Commonwealth of Independent States
BT2 Kunashir Island
BT2 Kuril Islands
BT2 Sakhalin Russian Federation
BT2 Asia
BT3 Kunashir Island
BT3 Kuril Islands
BT3 Russian Pacific region
BT3 Russian Federation
BT3 Commonwealth of Independent States
BT4 Kunashir Island
BT4 Kuril Islands
BT4 Russian Pacific region
BT4 Asia

Gosau
No longer a valid term for GeoRef. See Gosau Austria.

Gosau Austria (1993)
Village in S Upper Austria.
BT Upper Austria
BT Austria
BT Central Europe
BT Europe

Gosau Formation (1981)
Before 1981, also search Gosau Beds.
BT Cretaceous
BT Mesozoic
SA Austria
SA Eastern Alps

Gosford
No longer a valid term for GeoRef. See Gosford Australia.

Gosford Australia (1993)
Town in SE New South Wales.
BT New South Wales Australia
BT Australia
BT Australasia

Goshen County
Valid through 1988. Search in combination with state term. After 1988, use specific county-state term.

Goshen County Wyoming (1989)
SE Wyoming. Before 1989, also search Goshen County AND Wyoming.
CO N413300N423700
  W1040500W1044300
BT Wyoming
BT United States

gossan (1982)
Also search gossans.
SA copper sulfides
SA indicators
SA iron
SA iron oxides
SA iron sulfides
SA leaching
SA metal ores
SA mineral exploration
SA ore guides
SA sulfides
SA weathering
SA zinc sulfides

Gosses Bluff (1978)
Meteor crater.
UF Gosses Bluff Meteor Crater
BT Northern Territory Australia
BT Australia
BT Australasia

Gosses Bluff Meteor Crater
use Gosses Bluff

Gota Valley (1978)
River valley between Lake Vanern and Kattegat in SW Sweden.
BT Sweden
BT Scandinavia
BT Western Europe
BT Europe

Goteborg
No longer a valid term for GeoRef. As of 1993 see Goteborg Sweden.

Goteborg Sweden (1993)
City on the Kattegat in SW Sweden. Before 1993 also search Goteborg AND Sweden.
UF Gothenburg
BT Sweden
BT Scandinavia
BT Western Europe
BT Europe
SA Vastergotland

Gothard
use Gotthard Massif

Gothenburg
use Goteborg Sweden

gothite
use goethite

Gothland
use Gotland Sweden

Gothlandian
use Silurian

Gotland
No longer a valid term for GeoRef. As of 1993 see Gotland Sweden.

Gotland Sweden (1993)
Island and county in Baltic Sea off SE coast of Sweden. Before 1993 also search Gotland AND Sweden.
CO N570000N580000
  E0193000E0180000
UF Gothland
UF Gottland
BT Sweden
BT Scandinavia
BT Western Europe
BT Europe

Gotlandian
use Silurian

Goto Islands (1978)
Island chain extending about 100 miles SW from NW Kyushu.
BT Nagasaki Japan
BT Kyushu
BT Japan
BT Far East
BT Asia

Gotthard Massif (1978)
Mountain range of the Lepontine Alps in SE central Switzerland. This term has multiple hierarchies.
UF Gothard
UF Saint Gotthard
UF St. Gotthard
BT1 Lepontine Alps
BT1 Central Alps
BT1 Alps
BT1 Europe
BT2 Switzerland

BT2 Central Europe
BT2 Europe

Gottingen
No longer a valid term for GeoRef. See Gottingen Germany.

Gottingen Germany (1993)
City in SE Lower Saxony.
BT Lower Saxony Germany
BT Germany
BT Central Europe
BT Europe

Gottland
use Gotland Sweden

gouge (1978)
SA breccia
SA faults

Gough Island (1989)
One of the five islands in the Tristan da Cunha group. S Atlantic. A British claim, made a dependency of Saint Helena Island in 1938.
CO S402000S402000
  W0100000W0100000
BT Tristan da Cunha
BT Atlantic Ocean Islands
SA Saint Helena

Gour Oumelalen (1978)
Region in NE Ahaggar, SE Algeria.
BT Algeria
BT North Africa
BT Africa

Gouverneur
Valid through 1988. Search in combination with state term. After 1988, use specific city-state term.

Gouverneur New York (1989)
Village in N New York. Before 1989, search Gouverneur AND New York.
CO N442100N442100
  W0752900W0752900
BT Saint Lawrence County New York
BT New York
BT United States

Gove County
Valid through 1988. Search in combination with state term. After 1988, use specific county-state term.

Gove County Kansas (1989)
NW central. Before 1989, also search Gove County AND Kansas.
CO N384000N391000
  W1001000W1005000
BT Kansas
BT United States

government agencies (1989)
Used for government agencies, whether national, state, provincial, municipal, etc. For geological survey organizations use survey organizations.
NT CNRS
NT CRPG
NT NASA
NT NOAA
NT NSF
NT Oak Ridge National Laboratory
NT survey organizations
NT U. S. Bureau of Mines
NT U. S. Department of Energy
NT U. S. Environmental Protection Agency
NT U. S. Nuclear Regulatory Commission

NT USSR Academy of Sciences
SA associations
SA organization
SA regulations

Gower Peninsula (1978)
Extends S into Bristol Channel from West Glamorgan County, S Wales.
BT Glamorgan Wales
BT Wales
BT Great Britain
BT United Kingdom
BT Western Europe
BT Europe

Gowganda
No longer a valid term for GeoRef. As of 1993, see Gowganda Ontario.

Gowganda Formation (1978)
Consists of paraconglomerate, argillite, siltstone, subarkose, and greywacke. W Ontario.
BT Huronian
BT Proterozoic
BT upper Precambrian
BT Precambrian
SA Canada
SA Ontario

Gowganda Ontario (1993)
Village N of Georgian Bay. Before 1993, also search Gowganda AND Ontario.
BT Ontario
BT Eastern Canada
BT Canada

goyazite (1993)
Before 1993, also search hamlinite.
UF hamlinite
BT phosphates
SA crandallite

GPR
Not a valid term for GeoRef. See ground-penetrating radar if applicable.

GPS
use Global Positioning System

GPS Standard Positioning System
use Global Positioning System

GPS tracking
use Global Positioning System

grabens (1978)
SA aulacogens
SA faults
SA half grabens
SA horsts
SA systems

grade (1978)
UF metamorphic grade
SA cutoff grade
SA facies
SA high-grade metamorphism
SA isograds
SA low-grade metamorphism
SA metamorphism
SA ore grade
SA P-T conditions

graded bedding (1978)
BT turbidity current structures
BT sedimentary structures
SA bedding
SA Bouma sequence
SA planar bedding structures
SA varves

gradient
A valid term through 1978. After 1978, see geothermal gradient, or stream gradient for streams.

gradualism (1985)
UF phyletic gradualism
SA biologic evolution
SA paleontology
SA phylogeny
SA punctuated equilibria

graduate-level education (1978)
Before 1978, also search graduate level.
SA education

Grady County
Valid through 1988. Search in combination with state term. After 1988, use specific county-state term.

Grady County Georgia (1989)
SW Georgia. Before 1989, also search Grady County AND Georgia.
CO N304000N310500
  W0840500W0842300
BT Georgia
BT United States

Grady County Oklahoma (1989)
Central Oklahoma. Before 1989, also search Grady County AND Oklahoma.
CO N344100N352200
  W0974000W0980500
BT Oklahoma
BT United States

Grafton County
Valid through 1988. Search in combination with state term. After 1988, use specific county-state term.

Grafton County New Hampshire (1989)
W New Hampshire. Before 1989, also search Grafton County AND New Hampshire.
CO N433200N442300
  W0712200W0722000
BT New Hampshire
BT United States

graftonite (1978)
BT phosphates

Graham Coast
use Graham Land

Graham County
Valid through 1988. Search in combination with state term. After 1988, use specific county-state term.

Graham County Arizona (1989)
SE Arkansas. Before 1989, also search Graham County AND Arizona.
CO N322300N333900
  W1090600W1102800
BT Arizona
BT United States
NT Geronimo volcanic field

Graham County Kansas (1989)
NW Kansas. Before 1989, also search Graham County AND Kansas.
CO N390700N393300
  W0993700W1001100
BT Kansas
BT United States

Graham County North Carolina (1989)
Extreme W North Carolina. Before 1989, also search Graham County AND North Carolina.
CO N351300N352800
  W0833600W0840300
BT North Carolina

BT United States

Graham Island (1993)
Queen Charlotte Islands, W of British Columbia. Also an island in Norwegian Bay in the Queen Elizabeth Islands, Northwest Territories.
IN Index island group and British Columbia or Northwest Territories as applicable.
SA British Columbia
SA Northwest Territories
SA Queen Charlotte Islands
SA Queen Elizabeth Islands

Graham Land (1978)
Part of N Antarctic Peninsula S of Cape Horn.
UF Graham Coast
BT Antarctica

grahamite (1985)
Used for the solid bitumen. Use mesosiderite for the meteorite. Autoposting of broader terms to this term began in 1989.
BT bitumens
BT organic materials
SA organic compounds

Graian Alps (1985)
Division of Western Alps. On French-Italian border.
IN Index countries as applicable.
BT Western Alps
BT Alps
BT Europe
SA French Alps
SA Piedmont Alps
SA Piemonte Italy
SA Savoie France
SA Savoy Alps
SA Valle d'Aosta Italy

grain boundaries (1983)
SA crystal chemistry
SA crystal growth
SA crystal zoning
SA crystallography
SA crystals
SA grain size
SA grains
SA minerals
SA phase equilibria
SA shape analysis
SA textures

grain size (1978)
Before 1981, also search size.
SA coarse-grained materials
SA dimensions
SA fines
SA grain boundaries
SA granulometry
SA micritization
SA sediments
SA size distribution
SA soils

grain-size analysis
use granulometry

Grainger County
Valid through 1988. Search in combination with state term. After 1988, use specific county-state term.

Grainger County Tennessee (1989)
E Tennessee. Before 1989, also search Grainger County AND Tennessee.
CO N360500N362500
  W0831500W0834500
BT Tennessee
BT United States

grains (1978)

SA dimensions
SA grain boundaries
SA granulometry
SA particles
SA recrystallization
SA sand
SA sedimentary rocks
SA sediments
SA shape analysis
SA sorting

grainstone (1978)
BT carbonate rocks
BT sedimentary rocks

Graisivaudan
use Gresivaudan

gramenite
use nontronite

Gramineae (1978)
Family.
BT Monocotyledoneae
BT angiosperms
BT Spermatophyta
BT Plantae
NT Spartina alterniflora

Grampian Highlands (1981)
CO N551500N574500
  W0014500W0055000
BT Scotland
BT Great Britain
BT United Kingdom
BT Western Europe
BT Europe
SA Cairngorm Mountains
SA Grampian region Scotland
SA Moray Firth
SA Scottish Northern Highlands

Grampian region Scotland (1993)
Administrative region in NE Scotland since 1974. It replaced the former counties Banffshire, Morayshire, Aberdeenshire, and Kincardineshire.
BT Scotland
BT Great Britain
BT United Kingdom
BT Western Europe
BT Europe
NT Aberdeenshire Scotland
NT Banffshire Scotland
NT Huntly Scotland
SA Cairngorm Mountains
SA Grampian Highlands
SA Moray Firth

Gran Canaria
use Grand Canary

Gran Chaco
use Chaco

Granada
No longer a valid term for GeoRef. As of 1993, see Granada City Spain.

Granada City Spain (1993)
Refers only to the city. In S Spain. Before 1993, also search Granada and Spain.
BT Granada Spain
BT Andalusia Spain
BT Spain
BT Iberian Peninsula
BT Southern Europe
BT Europe

Granada Depression (1978)
S part of country.
BT Granada Spain
BT Andalusia Spain
BT Spain
BT Iberian Peninsula
BT Southern Europe

BT Europe

Granada Province
No longer a valid term for GeoRef. As of 1993, see Granada Spain.

Granada Spain (1993)
Refers only to the province. From 1981-1992, also search Granada Province and Spain. Before 1981, also search Granada and Spain. In S Spain. For the city, see Granada City Spain.
CO N364000N380500
 W0021100W0041800
BT Andalusia Spain
BT Spain
BT Iberian Peninsula
BT Southern Europe
BT Europe
NT Granada City Spain
NT Granada Depression

Grand Atlas Mountains
use High Atlas

Grand Banks (1978)
Shoal or banks E and S of Newfoundland.
IN Index North American Atlantic as applicable.
SA DSDP Site 384
SA Grand Banks earthquake 1929
SA Hibernia Field
SA Jeanne d'Arc Basin
SA North American Atlantic

Grand Banks earthquake 1929 (1993)
Continental slope off Newfoundland. Before 1985, also search Grand Banks AND earthquakes AND 1929.
BT earthquakes
SA Grand Banks
SA Newfoundland Island
SA North American Atlantic

Grand Canary (1978)
One of Spain's Canary Islands NW of Western Sahara.
UF Gran Canaria
BT Canary Islands
BT Atlantic Ocean Islands

Grand Canyon (1978)
Gorge in Colorado River in NW Arizona.
IN Index counties and state as applicable.
UF Grand Canyon region
SA Arizona
SA Marble Canyon
SA national parks
SA Snake River canyon
SA Stanton's Cave

Grand Canyon region
use Grand Canyon

Grand Cayman Island (1978)
Largest of the Cayman Islands. About 200 miles NW of Jamaica. Also search Grand Cayman. Before 1981, Caribbean Sea was the broader term.
BT Cayman Islands
BT West Indies
BT Caribbean region
SA Caribbean Sea

Grand Coulee Dam (1985)
On Columbia River, NE Washington.
IN Index counties as applicable.
CO N475800N480000
 W1185700W1185900
BT Washington
BT United States

SA Columbia River

Grand County
Valid through 1988. Search in combination with state term. After 1988, use specific county-state term.

Grand County Colorado (1989)
N Colorado. Before 1989, also search Grand County AND Colorado.
CO N393900N402800
 W1053800W1063900
BT Colorado
BT United States
SA Book Cliffs
SA Rocky Mountain National Park

Grand County Utah (1989)
E Utah. Before 1989, also search Grand County AND Utah.
CO N383000N393100
 W1090300W1101300
BT Utah
BT United States
NT Moab Utah
SA Book Cliffs
SA Salt Valley

Grand Falls
use Churchill Falls

Grand Forks
Valid through 1988. Search in combination with state term. After 1988, use specific city-state term.

Grand Forks County North Dakota (1989)
E North Dakota. Before 1989, search Grand Forks County AND North Dakota.
CO N474100N481200
 W0965400W0975500
BT North Dakota
BT United States
NT Grand Forks North Dakota

Grand Forks North Dakota (1989)
City on the Red River of the North in E North Dakota. Before 1989, search Grand Forks AND North Dakota.
CO N475700N475700
 W1183000W1183000
BT Grand Forks County North Dakota
BT North Dakota
BT United States

Grand Isle (1978)
Island in Jefferson Parish off SE coast of Louisiana.
IN Index parish or regions as applicable.
SA Jefferson Parish Louisiana

Grand Junction
Valid through 1988. Search in combination with state term. After 1988, use specific city-state term.

Grand Junction Colorado (1989)
City in W Colorado. Before 1989, search Grand Junction AND Colorado.
CO N390400N390400
 W1083300W1083300
BT Mesa County Colorado
BT Colorado
BT United States

Grand Manitoulin
use Manitoulin Island

Grand Rapids
Valid through 1988. Search in combination with state term. After 1988, use specific city-state term.

Grand Rapids Michigan (1989)
City in SW Michigan. Before 1989, search Grand Rapids AND Michigan.
CO N425700N425700
 W0864000W0864000
BT Kent County Michigan
BT Michigan Lower Peninsula
BT Michigan
BT United States

Grand River (1978)
River in numerous locations.
IN Index Ontario and states as applicable.
SA Louisiana
SA Michigan
SA Missouri
SA Neosho River valley
SA Oklahoma
SA Ontario
SA South Dakota

Grand Teton National Park (1985)
NW Wyoming.
CO N433000N441000
 W1102500W1105500
BT Teton County Wyoming
BT Wyoming
BT United States
SA national parks
SA Teton National Forest

Grand Traverse County
Valid through 1988. Search in combination with state term. After 1988, use specific county-state term.

Grand Traverse County Michigan (1989)
NW Michigan. Before 1989, also search Grand Traverse County AND Michigan.
CO N443000N445900
 W0852500W0855000
BT Michigan Lower Peninsula
BT Michigan
BT United States

Grande River
use Rio Grande

Grande Ronde Basalt (1985)
SE Washington, Oregon and W Idaho. Lower to middle Miocene. Has been considered part of the Yakima Basalt of the Columbia River Basalt Group.
IN Index ages as applicable.
BT Miocene
BT Neogene
BT Tertiary
BT Cenozoic
SA Columbia River Basalt Group
SA Hanford Reservation
SA Idaho
SA lower Miocene
SA middle Miocene
SA Oregon
SA Picture Gorge Basalt
SA Washington
SA Yakima Basalt

Grandfather Mountain (1978)
In the Blue Ridge Mountains in NW North Carolina.
IN Index county and Blue Ridge Mountains as applicable.
SA Blue Ridge Mountains
SA North Carolina

grandidierite (1978)
Autoposting of nesosilicates began in 1985.
BT nesosilicates
BT orthosilicates
BT silicates

Graneros Shale (1978)
In Colorado Group. E Colorado, Kansas, SE Montana, Nebraska, NE New Mexico, South Dakota, and E Wyoming.
BT Cretaceous
BT Mesozoic
SA Benton Formation
SA Colorado
SA Colorado Group
SA Kansas
SA Montana
SA Nebraska
SA New Mexico
SA South Dakota
SA Wyoming

granite
As of 1989, no longer a valid term for GeoRef. As of 1981, use granite deposits for granite as a commodity. From 1981 to 1988, granites was autoposted to this term.
use granites

Granite County
Valid through 1988. Search in combination with state term. After 1988, use specific county-state term.

Granite County Montana (1989)
W Montana. Before 1989, also search Granite County AND Montana.
CO N455500N465000
 W1130300W1135000
BT Montana
BT United States

granite deposits (1981)
Before 1981, search granite AND deposits.
IN Commodity. See List C.
SA building stone
SA construction materials
SA dimension stone
SA granites
SA granitization
SA ornamental materials

granite gneiss (1978)
UF granitic gneiss
BT gneisses
BT metamorphic rocks

Granite Mountains (1978)
Central Wyoming.
IN Index counties as applicable.
SA Fremont County Wyoming
SA Natrona County Wyoming
SA Wyoming

granite porphyry (1978)
BT granites
BT plutonic rocks
BT igneous rocks
SA porphyry

granite-granodiorite family
As of 1981, no longer a valid term for GeoRef. Use granites, alkali granites, or granodiorites as applicable. From 1978-1980, was autoposted to the rocks that were classified under it.

granites (1981)
Before 1981, search granite-granodiorite family. From 1978-1980, granite-granodiorite family was autoposted to the rocks that were classified under that term. Before 1989, also search granite. Autoposting of this term to granite ended in 1989.
UF granite
UF granitic rocks
BT plutonic rocks

BT igneous rocks
NT A-type granites
NT adamellite
NT alaskite
NT alkali granites
NT aplite
NT apogranite
NT biotite granite
NT charnockite
NT enderbite
NT farsundite
NT felsite
NT granite porphyry
NT granodiorites
NT granosyenite
NT graphic granite
NT I-type granites
NT leucogranite
NT microgranite
NT micropegmatite
NT monzogranite
NT muscovite granite
NT pegmatite
NT quartz monzonite
NT rapakivi
NT S-type granites
NT two-mica granite
SA albite
SA beresite
SA granite deposits
SA granitic composition
SA granitization
SA metagranite

granitic
A valid term through 1977. After 1977, use granitic composition or granitic layer.

granitic composition (1978)
Before 1978, also search granitic; granitic rocks.
SA composition
SA granites
SA granitic layer
SA igneous rocks

granitic gneiss
 use granite gneiss

granitic layer (1978)
Before 1978, also search granitic; sial.
UF layer, granitic
UF sial
SA basaltic layer
SA crust
SA granitic composition
SA lower crust

granitic rocks
 use granites

granitification
 use granitization

granitization (1978)
UF granitification
BT metamorphism
SA crystallization
SA feldspathization
SA granite deposits
SA granites
SA metasomatism
SA processes

granitoid
A valid term through 1978. After 1978, use granite or granitic composition.

granodiorite
As of 1989, no longer a valid term for GeoRef. From 1981 to 1988, granodiorites was autoposted to this term.
 use granodiorites

granodiorite porphyry (1978)
BT granodiorites

BT granites
BT plutonic rocks
BT igneous rocks
SA porphyry

granodiorites (1981)
Before 1981, search granite-granodiorite family. From 1978-1980, granite-granodiorite family was autoposted to the rocks that were classified under that term. Before 1989, also search granodiorite. Autoposting of this term to granodiorite ended in 1989.
UF granodiorite
BT granites
BT plutonic rocks
BT igneous rocks
NT granodiorite porphyry
SA beresite

granophyre (1978)
BT volcanic rocks
BT igneous rocks

granosyenite (1978)
This term has multiple hierarchies.
BT1 granites
BT1 plutonic rocks
BT1 igneous rocks
BT2 syenites
BT2 plutonic rocks
BT2 igneous rocks

Grant County
Valid through 1988. Search in combination with state term. After 1988, use specific county-state term.

Grant County Arkansas (1989)
Central Arkansas. Before 1989, also search Grant County AND Arkansas.
CO N340400N342800
 W0921400W0924100
BT Arkansas
BT United States

Grant County Indiana (1989)
E central Indiana. Before 1989, also search Grant County AND Indiana.
CO N402300N403900
 W0852700W0855200
BT Indiana
BT United States

Grant County Kansas (1989)
SW Kansas. Before 1989, also search Grant County AND Kansas.
CO N372200N374100
 W1010600W1013200
BT Kansas
BT United States

Grant County Kentucky (1989)
N Kentucky. Before 1989, also search Grant County AND Kentucky.
CO N382800N384700
 W0842800W0844700
BT Kentucky
BT United States

Grant County Minnesota (1989)
W Minnesota. Before 1989, also search Grant County AND Minnesota.
CO N454400N460800
 W0954700W0961700
BT Minnesota
BT United States

Grant County Nebraska (1989)
W central Nebraska. Before 1989, also search Grant County AND Nebraska.
CO N414200N420700

 W1012700W1020600
BT Nebraska
BT United States

Grant County New Mexico
 (1989)
SW New Mexico. Before 1989, also search Grant County AND New Mexico.
CO N314700N331200
 W1073700W1090300
BT New Mexico
BT United States
NT Santa Rita New Mexico
NT Silver City New Mexico
SA Black Range

Grant County North Dakota (1989)
S North Dakota. Before 1989, also search Grant County AND North Dakota.
CO N460200N464300
 W1010300W1020600
BT North Dakota
BT United States

Grant County Oklahoma (1989)
N Oklahoma. Before 1989, also search Grant County AND Oklahoma.
CO N363500N370000
 W0972800W0980600
BT Oklahoma
BT United States

Grant County Oregon (1989)
E central Oregon. Before 1989, also search Grant County AND Oregon.
CO N435700N445900
 W1181400W1194100
BT Oregon
BT United States

Grant County South Dakota (1989)
NE South Dakota. Before 1989, also search Grant County AND South Dakota.
CO N445800N452000
 W0962700W0971500
BT South Dakota
BT United States

Grant County Washington (1989)
E central Washington. Before 1989, also search Grant County AND Washington.
CO N463700N475800
 W1185800W1200200
BT Washington
BT United States
SA Hanford Reservation

Grant County West Virginia (1989)
E West Virginia. Before 1989, also search Grant County AND West Virginia.
CO N384800N391800
 W0785800W0792800
BT West Virginia
BT United States

Grant County Wisconsin (1989)
Extreme SW Wisconsin. Before 1989, also search Grant County AND Wisconsin.
CO N423000N431300
 W0902500W0911000
BT Wisconsin
BT United States

Grant Range (1978)
IN NE Nye County, central Nevada.

IN Index county and range as applicable.
SA Nevada
SA Nye County Nevada

Grants
Valid through 1988. Search in combination with state term. After 1988, use specific city-state term.

Grants District
No longer a valid term for GeoRef.
 use Grants mining district

Grants mineral belt (1981)
Mineralized region in NW New Mexico.
BT New Mexico
BT United States
SA Ambrosia Lake mining district
SA Grants mining district

Grants mining district (1993)
Mining district in NW New Mexico. Before 1993, also search Grants District.
UF Grants District
BT McKinley County New Mexico
BT New Mexico
BT United States
SA Grants mineral belt
SA mines
SA uranium ores

Grants New Mexico (1989)
Town in W New Mexico. Before 1989, search Grants AND New Mexico.
CO N351000N351000
 W1075100W1075100
BT Valencia County New Mexico
BT New Mexico
BT United States

granular materials (1978)
Also search granular.
SA coarse-grained materials
SA engineering geology
SA fines
SA granulometry
SA materials
SA particulate materials
SA soil mechanics

granulite
As of 1981, no longer a valid term for GeoRef. From 1978-1980, granulites was autoposted to this term.
 use granulites

granulite facies (1978)
BT facies
SA metamorphic rocks

granulites (1978)
Autoposting of this term to eclogite and granulite ended in 1981. Before 1981, also search granulite.
UF granulite
BT metamorphic rocks
NT kinzigite
NT leptite
NT pyroxene granulite
SA eclogite

granulometry (1978)
UF grain-size analysis
SA grain size
SA grains
SA granular materials
SA particles
SA sedimentation
SA sediments
SA shape analysis
SA size distribution

Granville County North Carolina (1989)

N North Carolina. Before 1989, also search Granville AND North Carolina.
CO N360300N363500 W0782500W0785000
BT North Carolina
BT United States
NT Oxford North Carolina

grapestone (1978)
Used to indicate a texture.
BT sedimentary rocks
SA textures

graphic display (1978)
Before 1978, also search graphic AND display.
UF display, graphic
SA data processing

graphic granite (1978)
BT granites
BT plutonic rocks
BT igneous rocks
SA micropegmatite
SA pegmatite

graphic methods (1978)
Before 1978, also search graphic.
SA digital line graphs
SA histograms
SA methods

graphic texture (1978)
Before 1978, search graphic.
BT textures
SA igneous rocks

graphite (1978)
As of 1981, use graphite deposits for graphite as a commodity.
UF black lead
BT native elements
SA carbon
SA chaoite
SA diamond
SA graphite deposits
SA graphitization
SA lonsdaleite

graphite deposits (1981)
Before 1981, search graphite AND deposits.
IN Commodity. See List C.
SA graphite
SA industrial minerals
SA refractory materials

graphitization (1989)
SA annealing
SA carbon
SA graphite
SA metal ores
SA processes

graptolites (1981)
Common name for Graptolithina.
IN Index for all non-paleontologic studies of fossils.
BT invertebrates
SA biostratigraphy
SA Graptolithina

Graptolithina (1978)
IN Index Hemichordata or Invertebrata as applicable.
NT Camaroidea
NT Dendroidea
NT Graptoloidea
NT Stolonoidea
NT Tuboidea
SA graptolites
SA Hemichordata
SA Invertebrata

Graptoloidea (1978)
Autoposting of this term to Didymograptina began in 1981.
IN Index Hemichordata or Invertebrata as applicable.
BT Graptolithina
NT Didymograptina
NT Diplograptina
NT Glossograptina
NT Monograptina
SA Hemichordata
SA Invertebrata

grasslands (1978)
SA ecology
SA herbivorous taxa
SA prairies
SA revegetation
SA savannas
SA steppes
SA wetlands

Gratiot County
Valid through 1988. Search in combination with state term. After 1988, use specific county-state term.

Gratiot County Michigan (1989)
Central Michigan. Before 1989, also search Gratiot County AND Michigan.
CO N430600N432900 W0842200W0845100
BT Michigan Lower Peninsula
BT Michigan
BT United States

gratonite (1978)
BT sulfarsenites
BT sulfosalts

Gratz Austria
use Graz Austria

Graubunden
As of 1993, no longer a valid term for GeoRef.
use Graubunden Switzerland

Graubunden Switzerland (1993)
Canton in E Switzerland. Before 1993, also search Graubunden or Grisons AND Switzerland.
UF Graubunden
UF Grisons Switzerland
BT Switzerland
BT Central Europe
BT Europe
NT Davos Switzerland
NT Oberhalbstein
NT Tavetsch
SA Inn Valley
SA Lepontine Alps
SA Oberalp Pass
SA Rhaetian Alps

grauwacke
use graywacke

Gravberg Well (1993)
Deep borehole, Siljan Ring, central Sweden.
UF Gravberg-1 Well
BT Sweden
BT Scandinavia
BT Western Europe
BT Europe
SA Siljan Ring

Gravberg-1 Well
use Gravberg Well

gravel (1978)
As of 1981, use gravel deposits for gravel as a commodity.
BT clastic sediments
BT sediments
SA alluvium
SA boulders
SA breccia
SA carbonate sediments
SA cobbles
SA gravel deposits
SA pebbles
SA sand
SA shingle
SA terrigenous materials

gravel deposits (1981)
Includes sand when sand is used as a construction material. Before 1981, search gravel AND deposits.
IN Commodity. See List C.
SA aggregate
SA construction materials
SA gravel
SA sands

Gravelly Range (1985)
SW Montana. Autoposting of broader terms to this term began in 1989. This term has multiple hierarchies.
IN Index counties as applicable.
BT1 U. S. Rocky Mountains
BT1 Rocky Mountains
BT2 U. S. Rocky Mountains
BT2 United States
BT3 Montana
BT3 United States
SA Beaverhead County Montana
SA Madison County Montana

gravimeters (1978)
UF gravity meters
NT Askania gravimeters
NT LaCoste-Romberg gravimeters
NT marine gravimeters
NT pendulum gravimeters
NT quartz-fiber gravimeters
NT spring gravimeters
NT vibrating-string gravimeters
NT Western gravimeters
NT Worden gravimeters
SA geophysical methods
SA gravity methods
SA instruments

gravitation
use gravity field

gravitational differentiation
use crystal fractionation

gravitational gliding
use gravity sliding

gravitational sliding
use gravity sliding

gravity
A valid term through mid-1978. Use specific term, e.g. gravity field or gravity platforms. For gravity faults, use the term normal faults.

gravity anomalies (1978)
Also search gravity surveys AND anomalies; gravity methods AND anomalies. Autoposting of this term began in 1978.
NT Bouguer anomalies
SA anomalies
SA geophysical methods
SA geophysical surveys
SA gravity methods
SA gravity surveys
SA magnetic anomalies
SA residual anomalies

gravity exploration
Not a valid index term after 1971. Was used on level 1 in subfiles G and N. Now use gravity methods or gravity surveys.

gravity faults
use normal faults

gravity field (1978)
Before 1971, also search gravitation.
UF field, gravity
UF gravitation
SA Earth
SA geodesy
SA geophysical surveys
SA Moon
SA splines

gravity field, Earth
Not a valid GeoRef index term after 1971. Was used on level 1 in subfiles E, G, N, T and B.

gravity field, Moon
Not a valid GeoRef index term after 1971. Was used on level 1 in subfiles G and B.

gravity flows (1993)
SA bottom features
SA flows
SA glacial geology
SA glacier surges
SA glaciers
SA gravity sliding
SA ice movement
SA landslides
SA mass movements
SA rock glaciers
SA sedimentation
SA slumping

gravity gliding
use gravity sliding

gravity logging (1984)
BT well-logging
SA gravity methods
SA gravity surveys

gravity meters
use gravimeters

gravity methods (1978)
BT geophysical methods
SA free-air anomalies
SA gravimeters
SA gravity anomalies
SA gravity logging
SA gravity survey maps
SA gravity surveys
SA Gravsat
SA variometers

gravity platforms (1978)
Used in engineering geology. Before 1978, search gravity AND platforms.
BT marine installations
SA marine platforms
SA platforms

gravity sliding (1978)
Also search sliding.
UF ecoulement
UF gliding (tectonics)
UF gravitational gliding
UF gravitational sliding
UF gravity gliding
UF sliding, gravity
SA crustal shortening
SA gravity flows
SA nappes
SA tectonics

gravity survey maps (1978)
Also search maps AND gravity surveys.
BT maps
SA geophysical survey maps
SA geophysical surveys
SA gravity methods
SA gravity surveys

gravity surveys (1978)
BT geophysical surveys
BT surveys
SA gravity anomalies
SA gravity logging
SA gravity survey maps
SA Gravsat

gravity tectonics
Not a valid GeoRef index term after 1970. Was used on level 1 in subfile N.

Gravsat (1985)
SA geophysical methods
SA geophysical surveys
SA gravity methods
SA gravity surveys
SA remote sensing
SA satellite methods

gray antimony
use jamesonite

Gray desert soil
use Sierozems

Gray earth
use Sierozems

Gray forest soils (1978)
UF Grey forest soils
BT soils
SA soil group

gray hematite
use specularite

Grayburg Formation (1989)
Guadalupian. SE New Mexico, W Texas.
BT Guadalupian
BT Permian
BT Paleozoic
SA New Mexico
SA Texas

Grays Harbor (1978)
Inlet of Pacific Ocean on SW coast of Grays Harbor County.
IN Index county as applicable.
SA Grays Harbor County Washington
SA Washington

Grays Harbor County
Valid through 1988. Search in combination with state term. After 1988, use specific county-state term.

Grays Harbor County Washington (1989)
On Pacific Ocean in W central Washington. Before 1989, also search Grays Harbor County AND Washington.
CO N464500N473000
W1231000W1242500
BT Washington
BT United States
SA Grays Harbor

graywacke (1978)
UF grauwacke
UF greywacke
BT clastic rocks
BT sedimentary rocks
SA metagraywacke
SA sandstone
SA subgraywacke
SA terrigenous materials

Graz
No longer a valid term for GeoRef. See Graz Austria.

Graz Austria (1993)
City in Styrian Alps in SE Styria. Also search Gratz.
UF Gratz Austria
BT Styria Austria
BT Austria
BT Central Europe
BT Europe

Great Alfold
use Alfold

Great Appalachian Valley (1978)
Before 1977, also search Great Valley. A longitudinal chain of lowlands of the Appalachians extending from Canada on NE to Alabama on SW. In 1985, Appalachians was added as a broader term.
BT Appalachians
BT North America
SA Champlain Valley
SA Hudson Valley
SA Saint Lawrence Lowlands
SA Saint Lawrence Valley
SA Shenandoah Valley
SA Tennessee Valley
SA Valley and Ridge Province

Great Artesian Basin (1978)
IN Index Northern Territory and states as applicable.
CO S200000S180000
E1490000E1370000
BT Australia
BT Australasia
SA New South Wales Australia
SA Northern Territory Australia
SA Queensland Australia
SA South Australia

Great Australian Bight (1978)
Bay on S coast of Australia.
BT Indian Ocean

Great Bahama Bank (1978)
Large shoal in the Bahamas SE of Miami and extending between Cuba and Andros Island.
BT Atlantic Ocean
SA Little Bahama Bank
SA Northwest Providence Channel

Great Barrier Reef (1978)
Largest coral reef in the world extending 1250 miles off NE Queensland, Australia.
CO S230000S100000
E1530000E1430000
BT Coral Sea
BT West Pacific
BT Pacific Ocean

Great Basin (1978)
Interior region between Sierra Nevada Mountains and S Cascade Range on W, and Wasatch Range and W face of Colorado Plateau on E.
IN Index states as applicable.
CO N350000N430000
W1110000W1200000
BT Basin and Range Province
BT United States
SA Battle Mountain High
SA California
SA Chainman Shale
SA Colorado Desert
SA Idaho
SA Nevada
SA Oregon
SA Utah

Great Bay (1978)
Inland tidal bay in SE New Hampshire and inlet of the Atlantic Ocean in SE New Jersey.
IN Index states and counties as applicable.
SA Atlantic County New Jersey
SA New Hampshire
SA New Jersey
SA Ocean County New Jersey
SA Rockingham County New Hampshire
SA United States
SA York County Maine

Great Bear Lake (1978)
NW central Mackenzie District.
BT Mackenzie District Northwest Territories
BT Northwest Territories
BT Western Canada
BT Canada
SA Bear-Slave Operation

Great Britain (1978)
Autoposting of this term began in 1981. Cheviot Hills is considered a narrower term as of 1981. Bristol Channel, Esk Trough, Severn Valley and Wye Valley are considered narrower terms as of 1985.
IN Index political divisions as applicable.
CO N500000N580000
E0013000W0080000
BT United Kingdom
BT Western Europe
BT Europe
NT Bristol Channel
NT Cheviot Hills
NT England
NT Esk Trough
NT Scotland
NT Tyne River
NT Wales
NT Wye Valley
SA Ballantrae Complex
SA Cadeby Formation
SA Caledonides
SA Gault
SA Upper Old Red Sandstone

Great Divide Basin (1985)
S central Wyoming. Autoposting of broader terms to this term began in 1989.
IN Index counties as applicable.
CO N414000N422000
W1070000W1090000
BT Wyoming
BT United States
SA Carbon County Wyoming
SA Fremont County Wyoming
SA Red Desert
SA Sweetwater County Wyoming
SA U. S. Rocky Mountains

Great Dyke (1978)
Part of great South African plateau.
BT Zimbabwe
BT Southern Africa
BT Africa

Great Glen Fault (1978)
Depression extending along ancient fault NE-SW across Scotland from Moray Firth to Loch Linnhe. Part of Caledonian Canal.
BT Scotland
BT Great Britain
BT United Kingdom
BT Western Europe
BT Europe
SA Highland region Scotland
SA Moray Firth

Great Hungarian Plain
use Alfold

Great Kabylia (1978)
W part of Kabylia which is a mountainous coastal region.
BT Algeria
BT North Africa
BT Africa
SA Kabylia

Great Khingan Range
use Da Hinggan Ling

Great Konya Basin
use Konya Basin

Great Lake (1978)
Largest lake on the island in central Tasmania.
SA Tasmania Australia

Great Lakes (1978)
Largest group of fresh-water lakes in the world. Autoposting of this term began in 1978. See East African Lakes if applicable.
IN Index lakes as applicable.
CO N414000N490000
W0760000W0922000
BT North America
NT Lake Erie
NT Lake Huron
NT Lake Michigan
NT Lake Ontario
NT Lake Superior
SA Canada
SA Great Lakes region
SA United States

Great Lakes International Multidisciplinary Program on Crustal Evolution
use GLIMPCE

Great Lakes region (1978)
Also search Great Lakes.
IN Index Great Lakes, states, and provinces as applicable.
CO N410000N500000
W0750000W0930000
BT North America
SA Canada
SA GLIMPCE
SA Great Lakes
SA Illinois
SA Indiana
SA Lake Chicago
SA Lake Maumee
SA Lake Saint Clair
SA Lake Superior region
SA Michigan
SA Minnesota
SA New York
SA Niagara Escarpment
SA Ohio
SA Ontario
SA Pennsylvania
SA United States
SA Wisconsin

Great Meteor Seamount (1978)
About 800 miles S of Azores.
BT Atlantic Ocean

Great Northern Peninsula (1985)
N Newfoundland.
CO N493500N513500
W0552500W0580000
BT Newfoundland Island
BT Newfoundland
BT Eastern Canada
BT Canada

Great Oolite Series (1978)
Consists of upper Estuarine Series, Great Oolite Limestone, Great Oolite Clays and Cornbrash. Oxfordshire and Gloucestershire in W central England.
BT Bathonian
BT Middle Jurassic
BT Jurassic
BT Mesozoic
SA England

Great Plains (1978)
Sloping plateau extending from Rocky Mountains in W to the margin of the Central Plains in the U. S. and to the margin of the Laurentian Highlands (Canadian Shield) in Canada. Until 1978, High Plains was used for the Great Plains from Nebraska

southward. Now use Great Plains.
Autoposting of this term began in 1978.
  IN   Index states and provinces as applicable.
  CO N290000N560000
       W0970000W1150000
  UF High Plains
  BT North America
  NT Northern Great Plains
  NT Southern Great Plains
  SA Alberta
  SA Canada
  SA Colorado
  SA High Plains Aquifer
  SA Kansas
  SA Madison Aquifer
  SA Marfa Basin
  SA Missouri Plateau
  SA Montana
  SA Nebraska
  SA New Mexico
  SA North Dakota
  SA Oklahoma
  SA Palo Duro Basin
  SA Platte River basin
  SA Saskatchewan
  SA South Dakota
  SA Texas
  SA Transcontinental Arch
  SA United States
  SA Wyoming
  SA Yellowstone River

Great Rift Valley   (1978)
  A great fault system extending from the Sea of Galilee in SW Asia to Mozambique.
  IN   Index East African Rift, and SW Asian countries, lakes, gulfs and seas.
  UF Rift Valley
  SA Afar Depression
  SA Dead Sea
  SA Dead Sea Rift
  SA East African Rift
  SA Gulf of Aden
  SA Gulf of Aqaba
  SA Israel
  SA Jordan
  SA Jordan Valley
  SA Kenya Rift valley
  SA Lake Baringo
  SA Red Sea
  SA Red Sea Rift
  SA Sea of Galilee
  SA Syria
  SA Wadi Araba

Great Salt Lake   (1978)
  In NW near Salt Lake City.
  CO N404000N414000
       W1121000W1125500
  BT Utah
  BT United States
  SA Box Elder County Utah
  SA Davis County Utah
  SA Lake Bonneville
  SA Salt Lake County Utah
  SA Tooele County Utah
  SA Weber County Utah

Great Scar Limestone   (1978)
  Overlain by the Yoredale Series. Yorkshire in N England.
  BT Visean
  BT Dinantian
  BT Carboniferous
  BT Paleozoic
  SA England

Great Slave Lake   (1978)
  S Mackenzie District.
  CO N610000N630000
       W1100000W1170000
  BT Mackenzie District Northwest Territories
  BT Northwest Territories
  BT Western Canada
  BT Canada
  SA Bear-Slave Operation
  SA Mackenzie River

Great Smokies
  use Great Smoky Mountains

Great Smoky Fault   (1989)
  N Georgia and SE Tennessee.
  IN   Index states and counties as applicable.
  BT United States
  SA Georgia
  SA Tennessee

Great Smoky Group   (1985)
  N central Georgia, E Tennessee, W North Carolina.
  BT Precambrian
  SA Georgia
  SA North Carolina
  SA Ocoee Series
  SA Tennessee

Great Smoky Mountains   (1978)
  Range of the Appalachians.
  IN   Index counties and Appalachians as applicable.
  UF Great Smokies
  UF Smokies
  BT United States
  SA Appalachians
  SA national parks
  SA North Carolina
  SA Tennessee

Great South Bay   (1978)
  Long narrow inlet of Atlantic Ocean between Fire Island and S shore of Long Island.
  IN   Index county as applicable.
  SA New York
  SA Suffolk County New York

Great Valley
  A valid index term through 1977. After 1977, see entry for Central Valley in California, and see entry for Great Appalachian Valley.

Great Valley Sequence   (1978)
  In California.
  BT Mesozoic
  SA California
  SA Cretaceous
  SA Jurassic
  SA Upper Cretaceous

Greater Antilles   (1978)
  Includes the major islands of the Antilles. As of 1981, the following terms (and their respective narrower terms) are considered narrower terms: Cuba, Hispaniola, Jamaica and Puerto Rico. Autoposting of this term began in 1978.
  IN   Index islands as applicable.
  CO N173000N273000
       W0650000W0850000
  BT Antilles
  BT West Indies
  BT Caribbean region
  NT Cuba
  NT Hispaniola
  NT Jamaica
  NT Puerto Rico

Greater Caucasus   (1978)
  The main range of the Caucasus separating the Northern Caucasus from the Colchis and Kura Lowland.
  IN   Index former Soviet republics as applicable.
  CO N390000N443000
       E0503000E0370000
  BT Caucasus
  BT Europe
  SA Argun River
  SA Azerbaidzhan
  SA Georgian Republic
  SA Russian Federation
  SA Sunzha
  SA Svanetia
  SA Transcaucasia
  SA Urup mining district

Greater Khingan Mountains
  use Da Hinggan Ling

Greater London England
  use London England

Greece   (1978)
  Before 1981, Aegean Islands and Ionian Islands were narrower terms.
  CO N345500N414500
       E0284500E0193000
  BT Southern Europe
  BT Europe
  NT Athens Greece
  NT Crete
  NT Epirus Greece
  NT Euboea
  NT Greek Aegean Islands
  NT Greek Ionian Islands
  NT Greek Macedonia
  NT Greek Thrace
  NT Hellenic Arc
  NT Hellenides
  NT Othrys
  NT Peloponnesus Greece
  NT Pindus Mountains
  NT Sterea Ellas
  NT Thessaly Greece
  SA Balkan Peninsula
  SA Gulf of Corinth
  SA Gulf of Corinth earthquake 1981
  SA Ionian Zone
  SA Kalamata earthquake 1986
  SA Macedonia
  SA Mediterranean region
  SA Near East
  SA Struma River valley
  SA Thrace
  SA Vardar River
  SA Vardar Zone

Greek Aegean Islands   (1981)
  Greek islands in the Aegean Sea. This term has multiple hierarchies.
  CO N352000N405000
       E0283000E0230000
  BT1 Greece
  BT1 Southern Europe
  BT1 Europe
  BT2 Aegean Islands
  BT2 Mediterranean region
  NT Chios
  NT Cyclades
  NT Dodecanese
  NT Lesbos
  NT Samos
  SA Aegean Sea
  SA Hellenic Arc
  SA Mediterranean Sea

Greek Ionian Islands   (1981)
  Islands off the W coast of Greece, including Kithira (Cerigo). They constitute a Greek region. This term has multiple hierarchies.
  CO N360500N395500
       E0231000E0192000
  BT1 Greece
  BT1 Southern Europe
  BT1 Europe
  BT2 Ionian Islands
  BT2 Mediterranean region
  NT Cephalonia
  NT Ithaca
  SA Ionian Sea
  SA Mediterranean Sea

Greek Macedonia   (1981)
  Autoposting of Greece to this term began in 1989. Before 1981, search Macedonia AND Greece. As of 1990, Southern Europe and Europe are autoposted to this term. This term has multiple hierarchies.
  CO N395000N413000
       E0244000E0204500
  BT1 Greece
  BT1 Southern Europe
  BT1 Europe
  BT2 Macedonia
  BT2 Southern Europe
  BT2 Europe
  NT Salonika Greece
  NT Vourinos
  SA Rhodope Mountains

Greek Thrace   (1981)
  Autoposting of Greece to this term began in 1989. As of 1990, Southern Europe and Europe are autoposted to this term. This term has multiple hierarchies.
  CO N404000N414500
       E0264000E0243000
  BT1 Greece
  BT1 Southern Europe
  BT1 Europe
  BT2 Thrace
  BT2 Europe
  NT Rhodope Greece
  SA Maritsa River
  SA Rhodope Mountains

Greeley
  Valid through 1988. Search in combination with state term. After 1988, use specific city-state term.

Greeley Colorado   (1989)
  City in N Colorado. Before 1989, search Greeley AND Colorado.
  CO N402600N402600
       W1044300W1044300
  BT Weld County Colorado
  BT Colorado
  BT United States

green algae
  use Chlorophyta

Green Bay   (1978)
  Inlet of NW Lake Michigan in NE Wisconsin and NW Michigan.
  IN   Index Michigan and/or Wisconsin as applicable.
  SA Lake Michigan
  SA Michigan
  SA Wisconsin

Green Bay Wisconsin   (1989)
  City in E Wisconsin. Before 1989, search Green Bay AND Wisconsin.
  CO N443200N443200
       W0880000W0880000
  BT Brown County Wisconsin
  BT Wisconsin
  BT United States

Green County
  Valid through 1988. Search in combination with state term. After 1988, use specific county-state term.

Green County Kentucky   (1989)
  Central Kentucky. Before 1989, also search Green County AND Kentucky.
  CO N370700N372800
       W0852100W0854000
  BT Kentucky
  BT United States

Green County Wisconsin   (1989)

S Wisconsin. Before 1989, also search Green County AND Wisconsin.
CO N423000N425100 W0892300W0895200
BT Wisconsin
BT United States
NT Monticello Wisconsin

Green function (1985)
UF Green's functions
SA acoustical methods
SA electromagnetic methods
SA equations
SA Fourier analysis
SA functions
SA geophysical methods
SA inverse problem
SA mathematical models
SA seismic methods
SA seismology

Green Gully (1978)
Near Keilor just NE of Melbourne, in Victoria Australia.
IN Index state or region as applicable.
SA Victoria Australia

Green Lake (1978)
Green Lake County in S central Wisconsin.
IN Index county or region as applicable.
SA Wisconsin

Green Mountains (1978)
Range of the Appalachians.
IN Index Appalachians and Quebec and/or U.S. states as applicable.
SA Appalachians
SA Massachusetts
SA Quebec
SA Vermont

Green River (1978)
Rises in the Wind River Range in W central Wyoming and flows S into the Colorado River N of Lake Powell. Also rivers in Illinois, Kentucky, and Washington.
IN Index counties, states or regions as applicable.
SA Colorado
SA Utah
SA Wyoming
SA Yampa River

Green River basin (1978)
IN Index states or regions as applicable.
SA Colorado
SA Moxa Arch
SA Pinedale Anticline
SA Sand Wash Basin
SA United States
SA Utah
SA Wyoming

Green River Formation (1978)
Lower and middle Eocene. Subdivided into Tipton Tongue, Laney Shale Member, and Morrow Creek Member. NW and central W Colorado, E Utah, and SW Wyoming.
BT Eocene
BT Paleogene
BT Tertiary
BT Cenozoic
SA Colorado
SA Flagstaff Formation
SA Lake Gosiute
SA Laney Shale Member
SA Parachute Creek Member
SA Utah
SA Wilkins Peak Member

SA Wyoming

green tuff (1978)
BT pyroclastics
BT volcanic rocks
BT igneous rocks
SA tuff

Green Tuff Formation (1978)
Underlies Daijima Formation.
BT Miocene
BT Neogene
BT Tertiary
BT Cenozoic
SA Hokkaido
SA Honshu
SA Japan
SA Kyushu

Green tuff region (1978)
Neogene volcanic region which covers large portions of Hokkaido and Honshu.
BT Japan
BT Far East
BT Asia
SA Hokkaido
SA Honshu

Green's functions
  use Green function

Greenbrier County
  Valid through 1988. Search in combination with state term. After 1988, use specific county-state term.

Greenbrier County West Virginia (1989)
SE West Virginia. Before 1989, also search Greenbrier County AND West Virginia.
CO N374000N381600 W0795700W0805400
BT West Virginia
BT United States

Greenbrier Limestone (1985)
Virginia, N West Virginia, E Kentucky, W Maryland and S Pennsylvania.
BT Upper Mississippian
BT Mississippian
BT Carboniferous
BT Paleozoic
SA Kentucky
SA Maryland
SA Pennsylvania
SA Virginia
SA West Virginia

Greene County
  Valid through 1988. Search in combination with state term. After 1988, use specific county-state term.

Greene County Alabama (1989)
W Alabama. Before 1989, also search Greene County AND Alabama.
CO N323100N330600 W0874500W0880900
BT Alabama
BT United States

Greene County Arkansas (1989)
NE Arkansas. Before 1989, also search Greene County AND Arkansas.
CO N355600N361600 W0901200W0905200
BT Arkansas
BT United States

Greene County Georgia (1989)
NE central Georgia. Before 1989, also search Greene County AND Georgia.

CO N332300N334400 W0825500W0832400
BT Georgia
BT United States

Greene County Illinois (1989)
SW central Illinois. Before 1989, also search Greene County AND Illinois.
CO N390800N393000 W0900900W0903700
BT Illinois
BT United States

Greene County Indiana (1989)
SW Indiana. Before 1989, also search Greene County AND Indiana.
CO N385500N391000 W0864200W0871500
BT Indiana
BT United States

Greene County Iowa (1989)
Central Iowa. Before 1989, also search Greene County AND Iowa.
CO N415200N421300 W0940900W0943800
BT Iowa
BT United States

Greene County Mississippi (1989)
SE Mississippi. Before 1989, also search Greene County AND Mississippi.
CO N310000N312500 W0882700W0885200
BT Mississippi
BT United States

Greene County Missouri (1989)
SW Missouri. Before 1989, also search Greene County AND Missouri.
CO N370700N372500 W0930500W0933700
BT Missouri
BT United States
NT Springfield Missouri

Greene County New York (1989)
SE New York. Before 1989, also search Greene County AND New York.
CO N420600N422800 W0734700W0743200
BT New York
BT United States

Greene County North Carolina (1989)
E central North Carolina. Before 1989, also search Greene County AND North Carolina.
CO N352100N353900 W0772800W0775000
BT North Carolina
BT United States

Greene County Ohio (1989)
SW central Ohio. Before 1989, also search Greene County AND Ohio.
CO N393300N395100 W0833900W0840700
BT Ohio
BT United States

Greene County Pennsylvania (1989)
SW Pennsylvania. Before 1989, also search Greene County AND Pennsylvania.
CO N394300N400200 W0795400W0803200
BT Pennsylvania
BT United States

Greene County Tennessee (1989)
NE Tennessee. Before 1989, also search Greene County AND Tennessee.
CO N355500N362600 W0823600W0831100
BT Tennessee
BT United States

Greene County Virginia (1989)
N central Virginia. Before 1989, also search Greene County AND Virginia.
CO N381200N382800 W0781800W0784000
BT Virginia
BT United States

Greenhorn Limestone (1978)
Of Colorado Group. Includes Lincoln Limestone, Jetmore Chalk, Hartland Shale, Pfeifer Shale, Bridge Creek Limestone Member, Orman Lake Limestone Member.
BT Upper Cretaceous
BT Cretaceous
BT Mesozoic
SA Benton Formation
SA Bridge Creek Limestone Member
SA Colorado
SA Colorado Group
SA Kansas
SA Montana
SA Nebraska
SA New Mexico
SA South Dakota
SA Wyoming

greenhouse effect (1985)
SA atmosphere
SA carbon dioxide
SA changes of level
SA climate
SA ecology
SA effects
SA global warming
SA temperature

Greenland (1978)
Largest island in the world and a Danish province. As of 1990, Polar regions is autoposted to this term.
CO N600000N840000 W0200000W0700000
BT Arctic region
NT Agto Greenland
NT Camp Century
NT Disko Island
NT East Greenland
NT Fiskenaesset Greenland
NT Frederikshab Greenland
NT Godthaab Greenland
NT Godthaabsfjord
NT Greenland ice sheet
NT Igaliko Greenland
NT Ivigtut Greenland
NT Jameson Land
NT Kangerdlugssuaq
NT Kuhn Island
NT Milne Land
NT Northern Greenland
NT Nugssuaq
NT Peary Land
NT Scoresby Land
NT Scoresby Sound
NT Skaergaard Intrusion
NT Sondre Strom Fjord
NT South Greenland
NT Thule Greenland
NT Umanak Greenland
NT West Greenland
SA Amitsoq Gneiss
SA Arctic Coastal Plain
SA Cape York Meteorite

SA Fram Strait
SA Ilimaussaq
SA Laurentia
SA Nares Strait
SA Nuk Gneiss
SA Qorqut Granite

Greenland Gap Group (1978)
BT Upper Devonian
BT Devonian
BT Paleozoic
SA Frasnian
SA Maryland
SA Virginia
SA West Virginia

Greenland ice sheet (1978)
Located in basin surrounded by coastal mountains covering roughly 85% of the island. Also search Greenland.
BT Greenland
BT Arctic region

Greenland Sea (1978)
Section of Arctic Ocean between NE coast of Greenland and Spitsbergen.
BT Arctic Ocean
NT Icelandic Plateau
SA DSDP Site 344
SA Fram Strait
SA Jan Mayen

Greenlee County
Valid through 1988. Search in combination with state term. After 1988, use specific county-state term.

Greenlee County Arizona (1989)
SE Arizona. Before 1989, also search Greenlee County AND Arizona.
CO N322500N334800
W1090300W1093000
BT Arizona
BT United States

greenockite (1978)
UF cadmium blende
UF cadmium ocher
UF xanthochroite
BT sulfides

greensand (1978)
Before 1975, also search glauconitic sand.
UF glauconitic sand
UF glauconitic sandstone
SA clastic rocks
SA clastic sediments
SA glauconite
SA glauconitic composition
SA sand
SA sandstone
SA sedimentary rocks

greenschist (1978)
UF prasinite
BT schists
BT metamorphic rocks
SA greenschist facies

greenschist facies (1978)
BT facies
SA blueschist facies
SA greenschist
SA metamorphic rocks
SA metamorphism

greenstone (1978)
Also used as mineral.
BT schists
BT metamorphic rocks
SA greenstone belts
SA nephrite

greenstone belts (1978)
UF belts, greenstone
SA arcuate structures

SA greenstone
SA supracrustals
SA terranes

Greenville Fault (1985)
Central California. Autoposting of broader terms to this term began in 1989.
IN Index counties as applicable.
BT California
BT United States
SA Alameda County California
SA Contra Costa County California
SA Diablo Range
SA San Francisco Bay

Greenwood County
Valid through 1988. Search in combination with state term. After 1988, use specific county-state term.

Greenwood County Kansas (1989)
SE Kansas. Before 1989, also search Greenwood County AND Kansas.
CO N373700N381000
W0955700W0963100
BT Kansas
BT United States
NT Eureka Kansas
NT Hamilton Quarry

Greenwood County South Carolina (1989)
W South Carolina. Before 1989, also search Greenwood County AND South Carolina.
CO N335700N342300
W0815300W0822100
BT South Carolina
BT United States

Greer County
Valid through 1988. Search in combination with state term. After 1988, use specific county-state term.

Greer County Oklahoma (1989)
SW Oklahoma. Before 1989, also search Greer County AND Oklahoma.
CO N344300N350800
W0991500W0995300
BT Oklahoma
BT United States

Gregory Rift (1978)
That part of the East African Rift extending SW from Lake Turkana into NW central Tanzania.
IN Index countries as applicable.
UF Gregory Rift valley
BT Africa
SA East African Rift
SA Kenya
SA Tanzania

Gregory Rift valley
use Gregory Rift

greigite (1978)
UF melnikovite
BT sulfides

greisen (1978)
Compositional term. Autoposted to beresite as of 1981.
BT metasomatic rocks
BT metamorphic rocks
NT beresite
SA greisenization

greisenization (1978)
BT metasomatism
SA greisen
SA hydrothermal alteration

Grenada (1978)
Island, southernmost of Windward Islands in British West Indies. Before 1981, West Indies was the broader term.
IN Index Windward Islands.
BT Lesser Antilles
BT Antilles
BT West Indies
BT Caribbean region
NT Grenville Grenada
SA Carriacou

Grenoble
No longer a valid term for GeoRef. See Grenoble France.

Grenoble France (1993)
City in SE France.
BT Isere France
BT France
BT Western Europe
BT Europe

Grenville
No longer a valid term for GeoRef. As of 1993, see Grenville County Ontario or Grenville Grenada. Until 1977, used to indicate provincial series. Now use Grenvillian Orogeny for Grenvillian. Use Grenville Province for the province of the Canadian Shield.

Grenville County Ontario (1993)
County in SE Ontario. Before 1993, search Grenville AND Ontario.
BT Ontario
BT Eastern Canada
BT Canada
SA Grenville Province

Grenville Front (1978)
IN Index provinces as applicable.
BT Canada
SA Ontario
SA Quebec

Grenville Grenada (1993)
Town on E coast of Grenada in West Indies. Before 1993, search Grenville AND Grenada.
IN Index Windward Islands.
BT Grenada
BT Lesser Antilles
BT Antilles
BT West Indies
BT Caribbean region
SA Windward Islands

Grenville Orogeny
use Grenvillian Orogeny

Grenville Province (1978)
Structural province of the Canadian Shield. As of 1993, see Grenville County Ontario for the county.
IN Index New York, Labrador, Newfoundland, and provinces as applicable.
CO N430000N550000
W0560000W0820000
BT Canadian Shield
BT North America
NT Central Gneiss Belt
NT Central Metasedimentary Belt
SA Grenville County Ontario
SA Grenvillian Orogeny
SA Labrador
SA New York
SA Newfoundland Island
SA Ontario
SA Quebec

Grenvillian
use Grenvillian Orogeny

Grenvillian Orogeny (1978)
Before 1978, also search Grenvillian; Grenville AND orogeny.
UF Grenville Orogeny
UF Grenvillian
BT Precambrian
SA Grenville Province
SA orogeny

Gresivaudan (1978)
Alpine glacial trough in SE France.
UF Graisivaudan
BT Isere France
BT France
BT Western Europe
BT Europe

Grey forest soils
use Gray forest soils

greywacke
use graywacke

Grimmertingen (1978)
Sands.
BT Oligocene
BT Paleogene
BT Tertiary
BT Cenozoic
SA Belgium

Grimsby Sandstone (1989)
In Medina Formation.
BT Lower Silurian
BT Silurian
BT Paleozoic
SA Medina Formation
SA New York
SA Ontario
SA Pennsylvania

Grimsvotn (1989)
Mountain in Vatnajokull, Iceland.
UF Grimsvotn Volcano
BT Iceland
BT Western Europe
BT Europe
SA Vatnajokull

Grimsvotn Volcano
use Grimsvotn

Grindelwald (1981)
Valley in the Bernese Alps of central Switzerland. After 1992, use Grindelwald Switzerland for the town of Grindelwald.
CO N463700N464000
E0080500E0080000
BT Switzerland
BT Central Europe
BT Europe
SA Bern Switzerland
SA Bernese Alps

grinding (1978)
SA abrasion
SA erosion
SA glacial erosion

griquaite (1978)
BT ultramafics
BT plutonic rocks
BT igneous rocks

Grisons Switzerland
use Graubunden Switzerland

groins (1978)
BT marine installations
SA jetties

Groix (1978)
Groix Island was used through 1978. In Bay of Biscay off Brittany coast. Before 1978, also search Groix Island.
UF Groix Island
SA Bay of Biscay
SA Morbihan France

Groix Island
    use Groix

Groningen
    No longer a valid term for GeoRef. As of 1993 see Groningen Netherlands.

Groningen Netherlands (1993)
    Province and city in NE Netherlands. Before 1993 also search Gronongen AND Netherlands.
    BT Netherlands
    BT Western Europe
    BT Europe

groove casts (1978)
    UF casts, groove
    BT bedding plane irregularities
    BT sedimentary structures
    SA casts
    SA grooves

grooves (1978)
    BT bedding plane irregularities
    BT sedimentary structures
    SA channels
    SA furrows
    SA groove casts
    SA striations

Grosmont Formation (1989)
    N Alberta.
    BT Upper Devonian
    BT Devonian
    BT Paleozoic
    SA Alberta

Grosseto
    No longer a valid term for GeoRef. See Grosseto Italy.

Grosseto Italy (1993)
    City and province in S Tuscany.
    BT Tuscany Italy
    BT Italy
    BT Southern Europe
    BT Europe

grossular (1978)
    UF grossularite
    BT garnet group
    BT nesosilicates
    BT orthosilicates
    BT silicates
    SA hydrogrossular
    SA vanadium garnet

grossularite
    use grossular

ground
    A valid term through 1977. After 1977, use ground methods.

ground ice (1978)
    SA frost action
    SA frozen ground
    SA ice
    SA ice lenses
    SA ice wedges
    SA permafrost
    SA suffosion
    SA taliks
    SA thermokarst

ground mass
    use matrix

ground methods (1978)
    Before 1978, also search ground AND methods.
    SA airborne methods
    SA geophysical methods
    SA geophysical surveys
    SA ground-penetrating radar
    SA methods

ground motion (1978)
    UF motion, ground
    SA earthquakes
    SA geologic hazards
    SA microzonation
    SA seismology
    SA strong motion
    SA surface waves

ground penetrating radar
    use ground-penetrating radar

ground probing radar
    use ground-penetrating radar

ground truth (1978)
    SA imagery
    SA remote sensing
    SA satellite methods

ground water (1978)
    For economically oriented discussions of ground water, use water resources. This term is used for subsurface waters, especially for studies on specific areas. It does not include the unsaturated zone. For 1965 and 1966, also search groundwater. Before 1945, also search underground water.
    UF groundwater
    UF underground water
    SA age
    SA alluvium aquifers
    SA anchimetamorphism
    SA aquifers
    SA aquitards
    SA artesian waters
    SA artificial recharge
    SA background level
    SA basin management
    SA blowouts
    SA boundary conditions
    SA capillary water
    SA catchment hydrodynamics
    SA chemical dispersion
    SA confined aquifers
    SA connate waters
    SA creosote
    SA Darcy's law
    SA decontamination
    SA desalinization
    SA development
    SA discharge
    SA drainage
    SA drawdown
    SA drinking water
    SA dye tracers
    SA electro-osmosis
    SA engineering geology
    SA environmental geology
    SA far-field
    SA field capacity
    SA fluorimetry
    SA fossil waters
    SA fresh water
    SA geochemistry
    SA geophysical surveys
    SA geothermal energy
    SA geothermal fields
    SA geothermal systems
    SA geysers
    SA ground-water dams
    SA ground-water provinces
    SA heat storage
    SA hot springs
    SA hydraulics
    SA hydrochemistry
    SA hydrodynamics
    SA hydrogeology
    SA hydrologic cycle
    SA hydrology
    SA hydrosphere
    SA hydrothermal alteration
    SA impurities
    SA infiltration
    SA infiltration galleries
    SA intermittent springs
    SA isotopes
    SA juvenile water
    SA Kalman filters
    SA laminar flow
    SA leaking underground storage tanks
    SA leaky aquifers
    SA levels
    SA liquid waste
    SA lysimeters
    SA meteoric water
    SA mineral waters
    SA movement
    SA multiple aquifers
    SA natural recharge
    SA NAWDEX
    SA near-field
    SA nonpoint sources
    SA observation wells
    SA perched aquifers
    SA percolation
    SA permeability
    SA polluted water
    SA pollution
    SA pore water
    SA porosity
    SA potability
    SA potentiometric surface
    SA pump tests
    SA pumping
    SA purification
    SA radioactive tracers
    SA RASA
    SA recharge
    SA representative basins
    SA Resource Conservation and Recovery Act
    SA Richards equation
    SA saline composition
    SA salt water
    SA salt-water intrusion
    SA saturated zone
    SA seepage
    SA shallow aquifers
    SA shallow depth
    SA springs
    SA storage
    SA storage coefficient
    SA submarine springs
    SA suffosion
    SA surface water
    SA surveys
    SA systems analogs
    SA thalwegs
    SA thermal waters
    SA tracers
    SA transmissivity
    SA unconfined aquifers
    SA underground channels
    SA underground streams
    SA unsaturated zone
    SA waste disposal
    SA WATEQF
    SA water
    SA water hardness
    SA water management
    SA water quality
    SA water recovery
    SA water regimes
    SA water resources
    SA water rights
    SA water table
    SA water treatment
    SA water wells
    SA well screens

ground water models
    Not a valid term for GeoRef. Use ground water in combination with models. Also use names of the specific models if available.

ground, frozen
    use frozen ground

ground, patterned
    use patterned ground

ground-penetrating radar (1993)
    Time-domain impulse radar which transmits broad bandwidth pulses into geologic media and acts as a sounding device. Before 1993, also search radar methods and ground methods, GPR, ground probing radar and ground penetrating radar.
    UF ground penetrating radar
    UF ground probing radar
    BT radar methods
    SA ground methods
    SA high-resolution methods
    SA microwave methods
    SA radio-wave methods
    SA reflection methods

ground-water dams (1982)
    UF subsurface dams
    SA dams
    SA ground water

ground-water increment
    use recharge

ground-water movement
    A valid term through 1972. After 1972, see movement with ground water.

ground-water provinces (1982)
    UF provinces, ground-water
    SA ground water
    SA hydrogeology
    SA water resources

ground-water recharge
    use recharge

ground-water replenishment
    use recharge

ground-water resources
    No longer a valid GeoRef index term. Before 1967, was used on level 1 in subfile T. See ground water; water resources.

ground-water tracers
    No longer a valid GeoRef index term. Before 1971, was used on level 2 in subfile N. See ground water; tracers; hydrochemistry.

groundmass
    use matrix

groundwater
    Not a valid GeoRef index term except during 1965 and 1966. Was used on level 1 in subfile T.
    use ground water

group velocity (1981)
    SA elastic waves
    SA seismology

grout
    use grouting

grouting (1978)
    UF grout
    SA dams
    SA disposal barriers
    SA engineering geology
    SA foundations

groutite (1978)
    BT oxides

Grove Karl Gilbert
    use Gilbert, Grove Karl

growth (1978)
    As of 1978, term is restricted to use for biological growth. For crystals, use crystal growth.
    SA crystal growth

growth faults (1978)
    Before 1978, also search flexure AND faults; contemporaneous faults; slump faults.

UF contemporaneous faults
UF depositional fault
UF flexure faults
UF sedimentary fault
BT faults
SA flexure
SA listric faults

growth rates (1989)
Used as a general term applicable to many topics.
SA concretions
SA ecology
SA faunal studies
SA fossils
SA ontogeny
SA paleoecology

growth spirals (1978)
UF spirals, growth
SA crystal growth
SA crystal structure
SA crystallography

Grozny
No longer a valid term for GeoRef. As of 1993, see Grozny Russian Federation.

Grozny Russian Federation (1993)
City in former Chechen-Ingush A.S.S.R. in Northern Caucasus. This term has multiple hierarchies.
BT1 Chechen-Ingush Russian Federation
BT1 Russian Federation
BT1 Commonwealth of Independent States
BT2 Chechen-Ingush Russian Federation
BT2 Europe

Grudziadz
No longer a valid term for GeoRef. As of 1993 see Grudziadz Poland.

Grudziadz Poland (1993)
City in N central Poland. Before 1993 also search Grudziadz AND Poland.
BT Torun Poland
BT Poland
BT Central Europe
BT Europe
SA Bydgoszcz Poland

Grundy County
Valid through 1988. Search in combination with state term. After 1988, use specific county-state term.

Grundy County Illinois (1989)
NE Illinois. Before 1989, also search Grundy County AND Illinois.
CO N410700N412700
   W0881500W0883500
BT Illinois
BT United States

Grundy County Iowa (1989)
Central Iowa. Before 1989, also search Grundy County AND Iowa.
CO N421300N423400
   W0923200W0930200
BT Iowa
BT United States

Grundy County Missouri (1989)
N Missouri. Before 1989, also search Grundy County AND Missouri.
CO N395700N401700
   W0932200W0934800
BT Missouri
BT United States

Grundy County Tennessee (1989)
SE central Tennessee. Before 1989, also search Grundy County AND Tennessee.
CO N351400N353300
   W0853200W0855400
BT Tennessee
BT United States

Gruneisen parameters (1981)
SA seismology

grunerite (1978)
BT clinoamphibole
BT amphibole group
BT chain silicates
BT silicates

Gryphaea (1978)
Genus.
BT Bivalvia
BT Mollusca
BT Invertebrata

Grzybow (1978)
Sulfur mining region in S Poland.
BT Nowy Sacz Poland
BT Poland
BT Central Europe
BT Europe
SA sulfur deposits

GSA (1985)
Acronym.
UF Geological Society of America
SA associations
SA DNAG

Guadalajara
No longer a valid term for GeoRef. As of 1993, see Guadalajara City Spain, or see Guadalajara Mexico.

Guadalajara City
No longer a valid term for GeoRef. As of 1993 see Guadalajara Mexico.

Guadalajara City Spain (1993)
Refers to only the city in central Spain. Before 1993, also search Guadalajara and Spain.
BT Guadalajara Spain
BT New Castile Spain
BT Castile Spain
BT Spain
BT Iberian Peninsula
BT Southern Europe
BT Europe

Guadalajara Mexico (1993)
City in Mexico. Before 1981, also search Guadalajara AND Mexico. Also search Guadalajara City.
BT Jalisco Mexico
BT Mexico

Guadalajara Province
No longer a valid term for GeoRef. As of 1993, see Guadalajara Spain.

Guadalajara Spain (1993)
Refers only to the province in central Spain. From 1981-1992, also search Guadalajara Province and Spain. Before 1981, also search Guadalajara and Spain. For the city, see Guadalajara City Spain.
CO N401000N412000
   W0013200W0033500
BT New Castile Spain
BT Castile Spain
BT Spain
BT Iberian Peninsula
BT Southern Europe
BT Europe
NT Guadalajara City Spain
SA Serrania de Cuenca

Guadalquivir (1978)
River in S Spain.
UF Guadalquivir River
BT Spain
BT Iberian Peninsula
BT Southern Europe
BT Europe
SA Guadalquivir Basin

Guadalquivir Basin (1978)
River basin in S and SW Spain. Also search Guadalquivir River.
UF Guadalquivir River basin
BT Spain
BT Iberian Peninsula
BT Southern Europe
BT Europe
SA Guadalquivir

Guadalquivir River
use Guadalquivir

Guadalquivir River basin
use Guadalquivir Basin

Guadalupe
Not a valid index term after 1969. Was used on level 1 in subfiles E and B. Now use Guadalupe Island, Guadalupian, Guadalupe River or Guadeloupe.

Guadalupe County
Valid through 1988. Search in combination with state term. After 1988, use specific county-state term.

Guadalupe County New Mexico (1989)
E central New Mexico. Before 1989, also search Guadalupe County AND New Mexico.
CO N342100N351300
   W1040800W1052100
BT New Mexico
BT United States

Guadalupe County Texas (1989)
S central Texas. Before 1989, also search Guadalupe County AND Texas.
CO N292100N295100
   W0973700W0982000
BT Texas
BT United States

Guadalupe Island (1978)
Mexican island 180 miles off W central Baja California. In 1985, broader term changed from Pacific Ocean to East Pacific Ocean Islands.
BT East Pacific Ocean Islands

Guadalupe Mountains (1978)
IN Index counties or states as applicable.
SA national parks
SA New Mexico
SA Sacramento Mountains
SA Texas
SA United States

Guadalupe River (1978)
SE Texas.
IN Index counties as applicable.
SA Texas

Guadalupian (1978)
Provincial series, North America. Above Leonardian, below Ochoan.
BT Permian
BT Paleozoic
NT Bell Canyon Formation
NT Capitan Formation
NT Cherry Canyon Formation
NT Delaware Mountain Group
NT Grayburg Formation
NT Queen Formation
NT Seven Rivers Formation
NT Tansill Formation
SA Lower Permian
SA San Andres Formation
SA Upper Permian

Guadeloupe (1978)
Combined islands of Basse-Terre and Grande-Terre which constitute an overseas department of France. Before 1981, West Indies was the broader term. After 1992, Caribbean region is autoposted.
IN Index Leeward Islands.
BT Lesser Antilles
BT Antilles
BT West Indies
BT Caribbean region
NT La Grande Soufriere
SA Leeward Islands

Guajira Peninsula (1978)
N Colombia and NW Venezuela.
BT South America
SA Colombia
SA Venezuela
SA Zulia Venezuela

Guam (1978)
Island and unincorporated U. S. territory.
BT Mariana Islands
BT Micronesia
BT Oceania

Guanabara
No longer a valid term for GeoRef. See Rio de Janeiro Brazil. Former state in SE Brazil. Combined with Rio de Janeiro State in 1975.

Guanabara Bay (1985)
Inlet of Atlantic Ocean on SE Coast of Brazil.
CO S225500S224000
   W0430000W0431800
BT Rio de Janeiro Brazil
BT Brazil
BT South America
SA South American Atlantic

Guanajuato
No longer a valid term for GeoRef. As of 1993 see Guanajuato Mexico.

Guanajuato Mexico (1993)
State in central Mexico.
BT Mexico
NT Leon Mexico
NT Sierra Gorda
SA Michoacan-Guanajuato volcanic field

Guandong China
use Guangdong China

Guangdong
No longer a valid term for GeoRef. See Guangdong China.

Guangdong China (1993)
Province, South-Central China. As of 1992, Hainan had become a separate province. Hence, Guangdong no longer includes Hainan as a narrower term. Before 1993, also search Guandong, Kwangtung, and Guangdong.
CO N201000N253000
   E1171000E1100000
UF Guandong China
UF Kwangtung China
BT China
BT Far East
BT Asia
NT Zhujiang River
SA Hainan China
SA Nanling

Guangxi
No longer a valid term for GeoRef.
Valid 1992. See Guangxi China.

Guangxi China (1993)
Autonomous region, S central China. Before 1993, also search Guangxi, Guangxi Zhuangsu, and Kwangsi Chuang.
CO N213000N262000
E1120000E1043000
UF Guangxi Zhuangsu
UF Kwangsi Chuang
BT China
BT Far East
BT Asia
NT Guilin
SA Dachang Deposit
SA Nanling
SA Xichang China

Guangxi Zhuangsu
No longer a valid term for GeoRef.
use Guangxi China

guano (1978)
This term has multiple hierarchies.
BT1 organic residues
BT2 sediments

Guapore Brazil
use Rondonia Brazil

guard log
use Laterolog

Guarico
No longer a valid term for GeoRef.
As of 1993 see Guarico Venezuela.

Guarico Venezuela (1993)
State in N central Venezuela. Before 1993 also search Guarico AND Venezuela.
BT Venezuela
BT South America
SA Orinoco River

Guatemala (1978)
CO N135500N174500
W0881500W0921500
BT Central America
NT Guatemala City Guatemala
NT Huehuetenango Guatemala
NT Motagua Fault
NT Pacaya
NT Santiaguito
SA Guatemala earthquake 1976

Guatemala Basin (1981)
Off the coast of Central America in the Pacific Ocean.
CO N100000N140000
W0920000W0990000
BT Pacific Ocean

Guatemala City
No longer a valid term for GeoRef.
See Guatemala City Guatemala.

Guatemala City Guatemala (1993)
City in S central Guatemala.
BT Guatemala
BT Central America

Guatemala earthquake 1976 (1993)
Before 1985, also search Guatemala AND earthquakes AND 1976.
BT earthquakes
SA Guatemala

Guaymas Basin (1985)
Central Gulf of California.
BT Gulf of California
BT North American Pacific
BT Pacific Ocean

Gubbio
No longer a valid term for GeoRef.
See Gubbio Italy.

Gubbio Italy (1993)
Commune in Perugia province, Umbria, central Italy.
CO N431500N432500
E0134200E0133500
BT Perugia Italy
BT Umbria Italy
BT Italy
BT Southern Europe
BT Europe

Gubik Formation (1985)
N Alaska.
BT Pleistocene
BT Quaternary
BT Cenozoic
SA Alaska

Gudbrandsdal
use Gudbrandsdalen

Gudbrandsdalen (1978)
Valley in S central Norway.
UF Gudbrandsdal
BT Norway
BT Scandinavia
BT Western Europe
BT Europe

gudmundite (1978)
BT sulfides

Guelma
No longer a valid term for GeoRef.
As of 1993, see Guelma Algeria.

Guelma Algeria (1993)
Department and town in Constantine Mountains in NE Algeria. Before 1993, search Guelma AND Algeria.
BT Algeria
BT North Africa
BT Africa

Guelph
No longer a valid term for GeoRef.
As of 1993, see Guelph Ontario.

Guelph Formation (1989)
Michigan and S Ontario.
BT Middle Silurian
BT Silurian
BT Paleozoic
SA Michigan
SA Ontario

Guelph Ontario (1993)
City in Wellington County, SE Ontario. Before 1993, also search Guelph AND Ontario.
BT Wellington County Ontario
BT Ontario
BT Eastern Canada
BT Canada

Guerrero
No longer a valid term for GeoRef.
As of 1993 see Guerrero Mexico.

Guerrero Mexico (1993)
State in SW Mexico. Before 1993 also search Guerrero AND Mexico.
BT Mexico
SA Mexican volcanic belt
SA Sierra Madre del Sur

guest element
use trace elements

Guiana (Shield)
use Guyana Shield

Guiana Basin (1978)
Region between the Orinoco River, the Rio Negro, and the Amazon River and the Atlantic Ocean. Also search Guiana.
IN Index countries as applicable.
BT South America
SA Brazil
SA French Guiana
SA Guyana
SA Surinam

Guiana Highland
use Guyana Shield

Guiana Massif
use Guyana Shield

Guianas (1978)
IN Index countries as applicable.
BT South America
SA French Guiana
SA Guyana
SA Surinam

Guichon Batholith
use Guichon Creek Batholith

Guichon Creek Batholith (1978)
S central British Columbia.
UF Guichon Batholith
BT British Columbia
BT Western Canada
BT Canada

guide fossils
use index fossils

guidebook (1978)
Before 1970, also search guidebooks.
UF guidebooks
SA areal geology
SA directory
SA explanatory text
SA field trips
SA manuals
SA monographs
SA publications
SA road log

guidebooks
No longer a valid term for GeoRef.
use guidebook

guides, ore
use ore guides

Guilin (1993)
Karst region surrounding city in NE Guangxi. After 1992, use Guilin China for the city.
CO N244000N254000
E1100900E1104200
UF Kweilin
BT Guangxi China
BT China
BT Far East
BT Asia

Guilmette Formation (1989)
W Utah. Middle and Upper Devonian.
BT Devonian
BT Paleozoic
SA Utah

Guinea (1978)
Formerly French Guinea.
CO N071500N124000
W0074000W0150000
BT West Africa
BT Africa
NT Los Islands
SA Niger Basin
SA Niger Valley
SA Nimba Mountains
SA Senegal Basin
SA Senegal River

Guinea-Bissau (1978)
Formerly Portuguese Guinea. Before 1975, also search Portuguese Guinea.
CO N110000N130000
W0130000W0170000
UF Portuguese Guinea
BT West Africa
BT Africa

Guipuzcoa
No longer a valid term for GeoRef.
As of 1993, see Guipuzcoa Spain.

Guipuzcoa Spain (1993)
One of the Basque Provinces in N Spain. Before 1993, also search Guipuzcoa and Spain.
CO N425600N432500
W0014500W0023200
BT Basque Provinces Spain
BT Spain
BT Iberian Peninsula
BT Southern Europe
BT Europe
NT San Sebastian Spain

Guizhou
No longer a valid term for GeoRef.
See Guizhou China.

Guizhou China (1993)
Province, SW China. Before 1993, also search Guizhou or Kweichow.
CO N244000N292000
E1093000E1030000
UF Kweichow China
BT China
BT Far East
BT Asia
SA Qingzhen Meteorite
SA Wufeng Formation
SA Yangtze Platform

Gujarat
No longer a valid term for GeoRef.
See Gujarat India.

Gujarat India (1993)
State in W India.
CO N200000N245000
E0743000E0680000
BT India
BT Indian Peninsula
BT Asia
NT Amba Dongar
NT Baroda India
NT Broach India
NT Cambay India
NT Cutch India
NT Girnar Hills
NT Kachchh India
NT Mount Girnar
NT Panch Mahals India
NT Saurashtra
NT Wadia
SA Aravalli System
SA Bagh Beds
SA Bhuj Series
SA Cambay Basin
SA Dhajala Meteorite
SA Kathiawar
SA Nari Series
SA Narmada River
SA Narmada Valley
SA Narmada-Son Lineament

Gulbarga
No longer a valid term for GeoRef.
See Gulbarga India.

Gulbarga India (1993)
Town in NE Karnataka, S India.
BT Karnataka India
BT India
BT Indian Peninsula
BT Asia

Before then, variants of the term may occur in GeoRef.

Guldsmedshyttan
  use Stripa region

Gulf Basin
  Not a valid term for GeoRef. For basin S of Mississippi, Louisiana, and Texas, use Gulf of Mexico or Gulf Coastal Plain. For other gulf basins, search or index specific gulf or region.

Gulf Coast
  use Gulf Coastal Plain

Gulf Coastal Plain   (1978)
  IN   Index North America or Mexican and U.S. states as applicable.
  UF   Gulf Coast
  SA   Alabama
  SA   Atchafalaya Bay
  SA   Avery Island
  SA   Barataria Bay
  SA   Campeche Mexico
  SA   Florida
  SA   Louisiana
  SA   Mississippi
  SA   Mississippi Embayment
  SA   Mississippi Sound
  SA   Mobile Bay
  SA   North America
  SA   Oakwood Dome
  SA   Pleasant Bayou
  SA   Quintana Roo Mexico
  SA   Richton Dome
  SA   Sabine Uplift
  SA   Sanibel Island
  SA   South Florida Basin
  SA   Southwest Florida Water Management District
  SA   Tabasco Mexico
  SA   Tamaulipas Mexico
  SA   Texas
  SA   Tombigbee River
  SA   United States
  SA   Vacherie Dome
  SA   Veracruz Mexico
  SA   Wiggins Arch
  SA   Yucatan Mexico

Gulf County
  Valid through 1988. Search in combination with state term. After 1988, use specific county-state term.

Gulf County Florida   (1989)
  NW Florida. Before 1989, also search Gulf County AND Florida.
  CO   N294000N301300
         W0850000W0852400
  BT   Florida
  BT   United States

Gulf of Aden   (1978)
  Between S coast of Arabian Peninsula and Somali Republic, E Africa. Gulf of Tadjoura is included as a narrower term as of 1985.
  CO   N103000N150000
         E0520000E0430000
  BT   Arabian Sea
  BT   Indian Ocean
  NT   Gulf of Tadjoura
  SA   DSDP Site 231
  SA   DSDP Site 232
  SA   DSDP Site 233
  SA   Great Rift Valley
  SA   Leg 24
  SA   Red Sea Rift

Gulf of Alaska   (1978)
  Between Alaska Peninsula and Alaskan panhandle. Autoposting of Pacific Ocean to this term began in 1990.
  CO   N540000N611500
         W1350000W1630000
  BT   North American Pacific
  BT   Pacific Ocean
  SA   DSDP Site 178
  SA   DSDP Site 179
  SA   DSDP Site 180
  SA   DSDP Site 181
  SA   DSDP Site 182
  SA   Icy Bay
  SA   Middleton Island
  SA   Shelikof Strait
  SA   Sitka Sound
  SA   Yakutat Bay

Gulf of Aqaba   (1978)
  Between NW Saudi Arabia and the Sinai Peninsula.
  CO   N280000N293000
         E0350000E0342000
  UF   Gulf of Eilat
  UF   Gulf of Elat
  BT   Red Sea
  BT   Indian Ocean
  SA   Great Rift Valley

Gulf of Bothnia   (1978)
  Between Finland and Sweden.
  CO   N600000N660000
         E0253000E0170000
  UF   Bothnian Sea
  BT   Baltic Sea
  BT   European Atlantic
  BT   North Atlantic
  BT   Atlantic Ocean
  SA   Alno
  SA   Finland

Gulf of California   (1978)
  Between peninsula of Baja California and the Mexican states of Sonora and Sinaloa.
  CO   N224500N320000
         W1060000W1150000
  BT   North American Pacific
  BT   Pacific Ocean
  NT   Guaymas Basin
  SA   DSDP Site 477
  SA   DSDP Site 478
  SA   DSDP Site 479
  SA   DSDP Site 480
  SA   DSDP Site 481
  SA   DSDP Site 482
  SA   DSDP Site 483
  SA   DSDP Site 484
  SA   DSDP Site 485
  SA   Leg 64
  SA   Leg 65
  SA   Mexico

Gulf of Cambay   (1978)
  Inlet on W coast of India SE of Kathiawar Peninsula of Gujarat State.
  BT   Arabian Sea
  BT   Indian Ocean
  SA   Cambay Basin

Gulf of Cariaco   (1978)
  Inlet on NE coast of Venezuela S of Araya Peninsula in Sucre State. Venezuela is a broader term as of 1993.
  BT   Sucre Venezuela
  BT   Venezuela
  BT   South America
  SA   Caribbean Sea

Gulf of Chihli
  use Bohai Bay

Gulf of Corinth   (1989)
  Inlet of Ionian Sea located between Peloponnesus and central Greece.
  CO   N375800N382800
         E0231200E0214000
  BT   Ionian Sea
  BT   East Mediterranean
  BT   Mediterranean Sea
  SA   Corinth Greece
  SA   Greece
  SA   Gulf of Corinth earthquake 1981
  SA   Peloponnesus Greece

Gulf of Corinth earthquake 1981   (1989)
  Epicenter in the Gulf of Corinth, central Greece.
  BT   earthquakes
  SA   Greece
  SA   Gulf of Corinth

Gulf of Eilat
  use Gulf of Aqaba

Gulf of Elat
  use Gulf of Aqaba

Gulf of Finland   (1978)
  Between Finland and Estonia.
  CO   N590000N603000
         E0302000E0223000
  BT   Baltic Sea
  BT   European Atlantic
  BT   North Atlantic
  BT   Atlantic Ocean
  SA   Finland

Gulf of Gabes   (1985)
  Inlet of Mediterranean Sea, on E coast of Tunisia.
  CO   N333000N345000
         E0110000E0100000
  BT   East Mediterranean
  BT   Mediterranean Sea
  SA   Pelagian Sea
  SA   Tunisia

Gulf of Gascony
  No longer a valid term for GeoRef as of 1981.
  use Bay of Biscay

Gulf of Guinea   (1978)
  Wide inlet on W coast of Africa just south of the continent's great bulge. As of 1990, Atlantic Ocean is autoposted to this term.
  BT   North Atlantic
  BT   Atlantic Ocean

Gulf of Lion   (1978)
  Wide bay extending from French-Spanish border to Toulon.
  CO   N420000N433500
         E0061000E0030500
  UF   Gulf of Lions
  BT   West Mediterranean
  BT   Mediterranean Sea
  SA   Rhone Delta

Gulf of Lions
  use Gulf of Lion

Gulf of Maine   (1978)
  Off New England and S of Nova Scotia in Georges Bank area.
  BT   Atlantic Ocean
  SA   Cape Ann
  SA   Georges Bank
  SA   Georges Bank basin

Gulf of Mexico   (1978)
  Relatively shallow oceanic-type basin encircled by Cuba, Mexico, and the United States. Also search Gulf Basin.
  CO   N180000N300400
         W0803000W0980000
  BT   North American Atlantic
  BT   North Atlantic
  BT   Atlantic Ocean
  NT   Alacran Reef
  NT   Alaminos Canyon
  NT   Buccaneer Field
  NT   Campeche Bank
  NT   Campeche Scarp
  NT   Challenger Knoll
  NT   Florida Bay
  NT   Mississippi Fan
  NT   Orca Basin
  NT   Pigmy Basin
  NT   Sigsbee Deep
  NT   Yucatan Shelf
  SA   Apalachicola Bay
  SA   Atchafalaya Bay
  SA   Dry Tortugas
  SA   DSDP Site 3
  SA   DSDP Site 85
  SA   DSDP Site 86
  SA   DSDP Site 88
  SA   DSDP Site 89
  SA   DSDP Site 90
  SA   DSDP Site 91
  SA   DSDP Site 92
  SA   DSDP Site 94
  SA   DSDP Site 95
  SA   DSDP Site 96
  SA   DSDP Site 97
  SA   DSDP Site 535
  SA   DSDP Site 536
  SA   DSDP Site 537
  SA   DSDP Site 538
  SA   DSDP Site 540
  SA   DSDP Site 614
  SA   DSDP Site 615
  SA   DSDP Site 616
  SA   DSDP Site 617
  SA   DSDP Site 618
  SA   DSDP Site 619
  SA   DSDP Site 620
  SA   DSDP Site 621
  SA   DSDP Site 622
  SA   DSDP Site 623
  SA   DSDP Site 624
  SA   Gulf Stream
  SA   Laguna Madre
  SA   Leg 1
  SA   Leg 10
  SA   Leg 96
  SA   Leg 100
  SA   Matagorda Bay
  SA   Mississippi Sound
  SA   Mobile Bay
  SA   Saint George Island
  SA   Sanibel Island
  SA   United States Exclusive Economic Zone
  SA   Wiggins Arch

Gulf of Oman   (1978)
  Extends between N Oman on the Arabian Peninsula and SE Iran.
  CO   N223000N260000
         E0620000E0560000
  BT   Arabian Sea
  BT   Indian Ocean

Gulf of Panama   (1978)
  On S coast of Panama.
  BT   Pacific Ocean

Gulf of Pozzuoli   (1978)
  NW inlet of Bay of Naples.
  BT   Mediterranean Sea

Gulf of Riga   (1978)
  Between Estonia and Latvia.
  CO   N570000N583000
         E0243000E0220000
  BT   Baltic Sea
  BT   European Atlantic
  BT   North Atlantic
  BT   Atlantic Ocean
  SA   Baltic region

Gulf of Saint Lawrence   (1978)
  Off E coast of Canada encircled by Newfoundland, New Brunswick, Nova Scotia, and Quebec. Before 1972, also search Gulf of St. Lawrence. As of 1990, Atlantic Ocean is autoposted to this term.
  CO   N450000N513000
         W0570000W0650000
  UF   Gulf of St. Lawrence
  BT   North American Atlantic
  BT   North Atlantic
  BT   Atlantic Ocean

SA Bay of Islands
SA Chaleur Bay
SA Magdalen Islands
SA Miramichi Bay

Gulf of Siam (1978)
Between S extension of Thailand on the Malay Peninsula on the W and Cambodia and Vietnam on the E.
CO N060000N133000
   E1044000E0991000
UF Gulf of Thailand
BT South China Sea
BT West Pacific
BT Pacific Ocean
SA Chao Phraya River

Gulf of St. Lawrence
Not a valid GeoRef index term after 1971. Was used on level 1 in subfiles G and N.
use Gulf of Saint Lawrence

Gulf of Suez (1978)
Between the Sinai Peninsula and the Arabian Desert of Egypt. Joined to the Mediterranean Sea by the Suez Canal.
CO N274000N300000
   E0341000E0322000
BT Red Sea
BT Indian Ocean
SA Red Sea Rift

Gulf of Tadjoura (1978)
Inlet of the Gulf of Aden on coast of Djibouti, E Africa. In 1985, broader term changed from Arabian Sea to Gulf of Aden.
UF Gulf of Tadjura
BT Gulf of Aden
BT Arabian Sea
BT Indian Ocean
SA Asal Rift

Gulf of Tadjura
use Gulf of Tadjoura

Gulf of Thailand
use Gulf of Siam

Gulf of Tonkin (1981)
Arm of the South China Sea, E of North Vietnam.
CO N170000N220000
   E1100000E1054000
UF Tonkin Gulf
BT South China Sea
BT West Pacific
BT Pacific Ocean
SA Hainan China

Gulf Stream (1978)
Warm ocean current flowing out of Gulf of Mexico along E U.S. to mid-Atlantic Ocean where it merges with North Atlantic drift current and influences climate of Europe as far N as Norway.
BT Atlantic Ocean
SA Gulf of Mexico

Gulfian (1978)
North America, provincial series.
BT Upper Cretaceous
BT Cretaceous
BT Mesozoic
NT Aguja Formation
NT Austin Chalk
NT Austin Group
NT Eagle Ford Formation
NT Escondido Formation
NT Navarro Group
NT Olmos Formation
NT San Miguel Formation
NT Taylor Marl
NT Woodbine Formation

Gull River Formation (1989)
S Ontario.
BT Ordovician
BT Paleozoic
SA Ontario

gullies (1978)
Autoposting of erosion features to this term began in 1989.
BT erosion features
SA arroyos
SA channels
SA cliffs
SA geomorphology
SA rills

Gumma
No longer a valid term for GeoRef. See Gumma Japan.

Gumma Japan (1993)
Prefecture in central Honshu. Also search Gunma.
UF Gunma Japan
BT Honshu
BT Japan
BT Far East
BT Asia
NT Haruna
SA Asama
SA Ashio Japan
SA Ashio Mine
SA Kanto Plain

Gunflint Iron Formation (1978)
In Animikie Group. Gunflint Lake region and Vermilion District in NE Minnesota.
UF Gunflint Iron-Formation
BT Precambrian
SA Animikie Group
SA Minnesota

Gunflint Iron-Formation
use Gunflint Iron Formation

Gunma Japan
use Gumma Japan

Gunnedah Basin (1989)
NE New South Wales.
BT New South Wales Australia
BT Australia
BT Australasia
SA Sydney Basin

Gunnison County
Valid through 1988. Search in combination with state term. After 1988, use specific county-state term.

Gunnison County Colorado (1989)
W central Colorado. Before 1989, also search Gunnison County AND Colorado.
CO N380900N391600
   W1060900W1073000
BT Colorado
BT United States
NT Crested Butte Colorado
SA Homestake Mine

Guntur
No longer a valid term for GeoRef. See Guntur India.

Guntur India (1993)
City and district in E central Andhra Pradesh, S India.
BT Andhra Pradesh India
BT India
BT Indian Peninsula
BT Asia

Gunz (1978)
Europe. Above Astian (Pliocene), below Mindel.
BT Pleistocene
BT Quaternary
BT Cenozoic

Gurev Kazakhstan
use Guryev Kazakhstan

Gurghiu
No longer a valid term for GeoRef. As of 1993 see Gurghiu Romania.

Gurghiu Mountains (1978)
In SE Transylvania.
UF Gurghiului Mountains
BT Transylvania
BT Romania
BT Southern Europe
BT Europe

Gurghiu Romania (1993)
Village in Mures County in E central Transylvania. Before 1993 also search Gurghiu AND Romania.
BT Mures Romania
BT Transylvania
BT Romania
BT Southern Europe
BT Europe

Gurghiului Mountains
use Gurghiu Mountains

Gurhwal
Not a valid term for GeoRef. See Garhwal India.

Gurhwal India
use Garhwal India

Guryev
No longer a valid term for GeoRef. As of 1993, see Guryev Kazakhstan.

Guryev Kazakhstan (1993)
Oblast and city at mouth of Ural River at N end of Caspian Sea. Before 1978, also search Gurev. This term has multiple hierarchies.
UF Gurev Kazakhstan
BT1 Kazakhstan
BT1 Central Asia
BT1 Asia
BT2 Kazakhstan
BT2 Commonwealth of Independent States

gusher
use geysers

gustavite (1978)
Autoposting of sulfobismuthites began in 1989.
BT sulfobismuthites
BT sulfosalts

Gutii Mountains (1978)
This term has multiple hierarchies.
BT1 Moldavia
BT1 Europe
BT2 Romania
BT2 Southern Europe
BT2 Europe

Guyana (1978)
Formerly British Guiana. Gained independence in 1966.
CO N011000N083000
   W0562500W0612500
UF British Guiana
BT South America
NT Georgetown Guyana
SA Guiana Basin
SA Guianas
SA Guyana Shield
SA Roraima Formation

Guyana Shield (1978)
Highland area extending from E Venezuela across N Brazil and the Guianas. Also search Guiana.
IN Index countries as applicable.
UF Guiana (Shield)
UF Guiana Highland
UF Guiana Massif
BT South America
SA Brazil
SA French Guiana
SA Guyana
SA Surinam

Guyanese Atlantic (1981)
As of 1990, Atlantic Ocean is autoposted to this term.
BT North Atlantic
BT Atlantic Ocean
SA DSDP Site 354
SA Leg 14
SA Leg 39

guyots
use seamounts

Gymnocodiaceae (1981)
Autoposting of thallophytes and microfossils to this term began in 1990. This term has multiple hierarchies.
BT1 Rhodophyta
BT1 algae
BT1 thallophytes
BT1 Plantae
BT2 Rhodophyta
BT2 algae
BT2 microfossils

gymnosperm flora (1982)
Common name for gymnosperms.
IN Index for all non-paleontologic studies of fossils.
BT plants
NT conifers
SA biostratigraphy
SA cones
SA gymnosperms

Gymnospermae
use gymnosperms

gymnosperms (1978)
Callipteris is a narrower term as of 1981. Autoposting of this term began in 1978. Before 1993, also search Gymnospermae.
UF Gymnospermae
BT Spermatophyta
BT Plantae
NT Bennettitales
NT Caytoniales
NT Coniferae
NT Coniferales
NT Cordaitales
NT Cycadales
NT Cycadofilicales
NT Dadoxylon
NT Ephedrales
NT Ginkgoales
NT Glossopteridales
NT Gnetales
NT Nilssoniales
NT Pentoxylales
NT Pteridospermae
NT Welwitschiales
SA angiosperms
SA cones
SA Dicroidium
SA Fagus
SA fossil wood
SA gymnosperm flora
SA Juglans
SA Pecopteris
SA Pteropsida
SA seeds
SA sporangia
SA Taeniopteris

gypsite
use gypsum

gypsum (1978)
As of 1981, use gypsum deposits for gypsum as a commodity.
UF gypsite

UF plaster of paris
UF plaster stone
BT sulfates
SA alabaster
SA anhydrite
SA calcium sulfate
SA cap rocks
SA chemically precipitated rocks
SA evaporites
SA gypsum deposits
SA halite
SA sedimentary rocks
SA sediments
SA selenite

gypsum deposits (1981)
Before 1981, search gypsum AND deposits.
IN Commodity. See List C.
SA anhydrite deposits
SA evaporite deposits
SA gypsum

gyrolite (1978)
UF centrallasite
BT sheet silicates
BT silicates

gyttja (1978)
This term has multiple hierarchies.
BT1 organic residues
BT2 sediments
SA sapropel
SA soils

Gzhelian (1981)
BT Upper Carboniferous
BT Carboniferous
BT Paleozoic

H

H
  use hydrogen

H chondrites (1981)
UF H-group chondrites
BT chondrites
BT stony meteorites
BT meteorites
NT Ashmore Meteorite
NT Dhajala Meteorite
NT Fayetteville Meteorite
NT Haviland Meteorite
NT Tieschitz Meteorite
NT Ybbsitz Meteorite
SA Coldwater Meteorites
SA Odessa Meteorite

H-2
  use deuterium

H-3
  use tritium

H-group chondrites
  use H chondrites

H2SO4
  use sulfuric acid

Haast River (1978)
W South Island.
IN Also index South Island.
BT New Zealand
BT Australasia
SA South Island

habit (1978)
UF crystal habit
SA crystal form
SA crystal growth
SA lattice
SA minerals

habitat (1978)
SA biotopes
SA ecology
SA environment
SA fossils
SA wilderness areas

Hachijo-jima (1978)
Second largest island in Izu-shichito group about 180 miles S of Tokyo.
BT Izu-shichito
BT Honshu
BT Japan
BT Far East
BT Asia

Hachinohe
No longer a valid term for GeoRef. See Hachinohe Japan.

Hachinohe Japan (1993)
City in N Honshu.
BT Aomori Japan
BT Honshu
BT Japan
BT Far East
BT Asia

Hackberry Formation (1989)
Louisiana and SE Texas.
BT Oligocene
BT Paleogene
BT Tertiary
BT Cenozoic
SA Louisiana
SA Texas

Hadar (1989)
Archaeological site for fossil hominids in central Ethiopia.
BT Ethiopia
BT East Africa
BT Africa
SA Afar

Hadar Formation (1978)
In the Afar region of Ethiopia.
BT Pliocene
BT Neogene
BT Tertiary
BT Cenozoic
SA Afar
SA Ethiopia

Haddington County
  use East Lothian Scotland

Haddingtonshire
  use East Lothian Scotland

Hadley Rill
  use Hadley Rille

Hadley Rille (1978)
UF Hadley Rill
BT Moon

Hadrosauridae (1978)
Family. This term has multiple hierarchies.
BT1 Ornithischia
BT1 Archosauria
BT1 Diapsida
BT1 Reptilia
BT1 Tetrapoda
BT1 Vertebrata
BT1 Chordata
BT2 Ornithischia
BT2 dinosaurs
BT2 Tetrapoda
BT2 Vertebrata
BT2 Chordata

Hadrynian (1978)
North America. Upper Proterozoic, above Helikian. Autoposting of upper Precambrian and Precambrian to this term ended in 1989. As of 1989, upper Proterozoic and Proterozoic are autoposted to this term. As of 1990, upper Precambrian and Precambrian are autoposted to this term.
BT upper Proterozoic
BT Proterozoic
BT upper Precambrian
BT Precambrian
NT Fourchu Group
NT George River Group

hafnium (1978)
Autoposting of metals to this term began in 1989.
UF Hf
BT metals
NT Hf-177/Hf-176

Hagendorf (1978)
Region in E Bavaria.
BT Bavaria Germany
BT Germany
BT Central Europe
BT Europe

Haghimas Syncline (1978)
BT Romania
BT Southern Europe
BT Europe
SA Eastern Carpathians

Hai-ch'eng China
  use Haicheng China

Hai-nan Island
  use Hainan China

Haicheng
No longer a valid term for GeoRef. See Haicheng China.

Haicheng China (1993)
Town and county in province of Liaoning, S Manchuria. Before 1993, also search Haicheng or Hai-ch'eng.
UF Hai-ch'eng China
BT Liaoning China
BT China
BT Far East
BT Asia
SA Haicheng earthquake 1975

Haicheng earthquake 1975 (1989)
Epicenter in Liaoning.
BT earthquakes
SA China
SA Haicheng China
SA Liaoning China

haidingerite (1978)
BT arsenates

Haifa
No longer a valid term for GeoRef. See Haifa Israel.

Haifa Bay (1989)
Inlet of the Mediterranean Sea off the coast of Galilee, N Israel.
CO N324800N325800
   E0350500E0345500
UF Bay of Haifa
BT Israel
BT Middle East
BT Asia
SA Galilee
SA Mediterranean Sea

Haifa Israel (1993)
District in NE Israel. Including the city on the Mediterranean Sea.
BT Israel
BT Middle East
BT Asia
SA Geula Caves

Hail
No longer a valid term for GeoRef. As of 1993 see Hail Saudi Arabia.

Hail Saudi Arabia (1993)
Town and oasis. Before 1993 also search Hail AND Saudia Arabia.
UF Hayel
BT Saudi Arabia
BT Arabian Peninsula
BT Asia

Hainan
No longer a valid term for GeoRef. See Hainan China.

Hainan China (1993)
Province. In South China Sea off the S coast of Guangdong China and E of Gulf of Tonkin. Until the late 1980's, was part of Guangdong province. Before 1993, also search Hainan, Hainan Island, and Hai-nan Island.
CO N180000N202000
   E1110000E1083000
UF Hai-nan Island
UF Hainan Island
BT China
BT Far East
BT Asia
SA Guangdong China
SA Gulf of Tonkin
SA South China Sea

Hainan Island
As of 1993, no longer a valid term for GeoRef.
  use Hainan China

Hainaut
No longer a valid term for GeoRef. See Hainaut Belgium.

Hainaut Belgium (1993)
Province in SW Belgium.
BT Belgium
BT Western Europe
BT Europe
NT Tournai Belgium
SA Mons Basin

Haiti (1978)
Occupies western third of island of Hispaniola.
CO N180000N202000
   W0714000W0743000
BT Hispaniola
BT Greater Antilles
BT Antilles
BT West Indies
BT Caribbean region

Haiyuan earthquake 1920 (1993)
Before 1985, also search Haiyuan AND earthquakes.
BT earthquakes
SA China
SA Gansu China
SA Ningxia China
SA Qinghai-Xizang Plateau

Hakodate
No longer a valid term for GeoRef. See Hakodate Japan.

Hakodate Japan (1993)
City on Tsugaru Strait in Oshima sub-prefecture, SW Hokkaido.
BT Hokkaido
BT Japan
BT Far East
BT Asia
SA Oshima Peninsula

Hakone (1978)
Hot springs and mountain pass in Kanagawa Prefecture, SE Honshu. Before 1993, also used for nearby village. As of 1993, use Mount Hakone for the volcano and Hakone Japan for the village.
SA Honshu
SA Kanagawa Japan

SA Mount Hakone

Hakone Volcano
 use Mount Hakone

Hakusan (1978)
 Extinct volcano on border of Ishikawa and Gifu prefectures, W Honshu.
 IN Index prefectures as applicable.
 UF Hakusan Volcano
 BT Honshu
 BT Japan
 BT Far East
 BT Asia
 SA Gifu Japan
 SA Ishikawa Japan

Hakusan Volcano
 use Hakusan

Halberstadt
 No longer a valid term for GeoRef.
 See Halberstadt Germany.

Halberstadt Germany (1993)
 City in Saxony-Anhalt, central Germany.
 BT Saxony-Anhalt Germany
 BT Germany
 BT Central Europe
 BT Europe

Haleakala (1989)
 Volcano on Maui Island, Hawaii. Also a national park. This term has multiple hierarchies.
 UF Haleakala Volcano
 BT1 Maui County Hawaii
 BT1 Hawaii
 BT1 United States
 BT2 Maui County Hawaii
 BT2 Hawaii
 BT2 Polynesia
 BT2 Oceania
 BT3 Maui County Hawaii
 BT3 Hawaii
 BT3 East Pacific Ocean Islands
 SA Maui
 SA national parks
 SA Red Hill

Haleakala Volcano
 use Haleakala

Halecostomi (1981)
 Autoposting of this term to Amiidae ended in 1989.
 UF Pholidophoriformes
 BT Holostei
 BT Actinopterygii
 BT Osteichthyes
 BT Pisces
 BT Vertebrata
 BT Chordata

half grabens (1993)
 UF half-grabens
 SA faults
 SA grabens
 SA horsts
 SA listric faults
 SA systems
 SA tectonics

half space
 use half-space

half-grabens
 use half grabens

half-space (1978)
 UF half space
 SA impedance
 SA seismology
 SA velocity

Haliburton County
 No longer a valid term for GeoRef. As of 1993, see Haliburton County Ontario.

Haliburton County Ontario (1993)
 Before 1993, also search Haliburton County or Haliburton in conjunction with Ontario.
 CO N444500N453500
     W0775600W0790100
 BT Ontario
 BT Eastern Canada
 BT Canada

halides (1978)
 Before 1981, sarcopside was included as a narrower term. Autoposting of this term to fluoborates and fluosilicates began in 1981.
 UF halogenide
 NT bromides
 NT chlorides
 NT fluoborates
 NT fluorides
 NT fluosilicates
 NT iodides
 NT zunyite
 SA calcium chloride
 SA minerals
 SA sodium chloride

Halifax
 No longer a valid term for GeoRef. As of 1993, see Halifax Nova Scotia.

Halifax County
 No longer a valid term for GeoRef. As of 1993, see Halifax County Nova Scotia.

Halifax County Nova Scotia (1993)
 Central Nova Scotia on the Atlantic Ocean. Before 1993, also search Halifax County AND Nova Scotia.
 CO N442500N451600
     W0620900W0641500
 BT Nova Scotia
 BT Maritime Provinces
 BT Eastern Canada
 BT Canada
 NT Halifax Nova Scotia

Halifax Formation (1989)
 In Meguma Group. Overlies Goldenville Formation.
 BT Tremadocian
 BT Lower Ordovician
 BT Ordovician
 BT Paleozoic
 SA Meguma Group
 SA Nova Scotia

Halifax Nova Scotia (1993)
 City on the Atlantic Ocean. Before 1993, also search Halifax AND Nova Scotia.
 BT Halifax County Nova Scotia
 BT Nova Scotia
 BT Maritime Provinces
 BT Eastern Canada
 BT Canada

Halimeda (1978)
 Autoposting of thallophytes and microfossils began in 1990. This term has multiple hierarchies.
 BT1 Codiaceae
 BT1 Chlorophyceae
 BT1 Chlorophyta
 BT1 algae
 BT1 thallophytes
 BT1 Plantae
 BT2 Codiaceae
 BT2 Chlorophyceae
 BT2 Chlorophyta
 BT2 algae
 BT2 microfossils

halite (1978)
 Mineral.
 UF common salt
 UF rock salt
 BT chlorides
 BT halides
 SA chemically precipitated rocks
 SA evaporites
 SA gypsum
 SA salt
 SA sediments
 SA sodium chloride
 SA sylvinite

Hall County
 Valid through 1988. Search in combination with state term. After 1988, use specific county-state term.

Hall County Georgia (1989)
 NE Georgia. Before 1989, also search Hall County AND Georgia.
 CO N340600N343100
     W0833800W0840300
 BT Georgia
 BT United States

Hall County Nebraska (1989)
 S central Nebraska. Before 1989, also search Hall County AND Nebraska.
 CO N404000N410300
     W0981700W0984300
 BT Nebraska
 BT United States

Hall County Texas (1989)
 NW Texas. Before 1989, also search Hall County AND Texas.
 CO N342100N344900
     W1002700W1005800
 BT Texas
 BT United States

Halland
 No longer a valid term for GeoRef. As of 1993 see Halland Sweden.

Halland Sweden (1993)
 County in SW Sweden. Before 1993 also search Halland AND Sweden.
 BT Sweden
 BT Scandinavia
 BT Western Europe
 BT Europe

Halle
 No longer a valid term for GeoRef. See Halle Germany.

Halle Bezirk
 No longer a valid term for GeoRef. District in SW central East Germany. Before 1981, search Halle. See Saxony-Anhalt Germany and Thuringia Germany.

Halle Germany (1993)
 City in Saxony-Anhalt, E central Germany. As of 1981, Halle refers to only the city. Before 1981, it also included the East German district.
 BT Saxony-Anhalt Germany
 BT Germany
 BT Central Europe
 BT Europe

Halley's Comet (1989)
 UF P/Halley
 BT comets
 SA extraterrestrial geology
 SA interplanetary space

 SA solar system

halloysite (1978)
 Before 1982, also search metahalloysite.
 UF endellite
 BT clay minerals
 BT sheet silicates
 BT silicates
 SA metahalloysite

Hallstatt Limestone (1978)
 BT Triassic
 BT Mesozoic
 SA Austria
 SA Carnian
 SA Ladinian
 SA Upper Triassic

halmyrolysis (1978)
 UF halmyrosis
 UF submarine weathering
 SA diagenesis
 SA geochemistry
 SA rock-water interface
 SA sea water

halmyrosis
 use halmyrolysis

Halobia (1978)
 Autoposting of Pectinacea and Pteriina to this term began in 1989.
 BT Pectinacea
 BT Pteriina
 BT Bivalvia
 BT Mollusca
 BT Invertebrata

haloes (1978)
 For intrusions and metamorphism, use aureoles.
 UF halos
 SA aureoles
 SA dispersion patterns
 SA geochemical anomalies
 SA geochemical methods
 SA mineral deposits, genesis
 SA mineral exploration
 SA zoning

Halog
 No longer a valid term for GeoRef. See Halog India.

Halog India (1993)
 Settlement in S Himachal Pradesh, N India.
 BT Himachal Pradesh India
 BT India
 BT Indian Peninsula
 BT Asia

halogenide
 use halides

halogens (1978)
 Term reintroduced in 1982. Not a valid term from 1972 through 1981. Five chemical elements from group VIIA in the periodic table. Autoposting of this term began in 1982.
 NT astatine
 NT bromine
 NT chlorine
 NT fluorine
 NT iodine
 SA chemical elements

halokinesis
 use salt tectonics

Halomorphic soils (1982)
 BT soils
 SA soil group

halos
 use haloes

halotrichite (1978)

UF iron alum
BT sulfates

Halsingborg
No longer a valid term for GeoRef. As of 1993 see Halsingborg Sweden.

Halsingborg Sweden (1993)
City on Oresund in SW Sweden. Before 1993 also search Halsingborg AND Sweden.
UF Helsingborg
BT Malmohus Sweden
BT Sweden
BT Scandinavia
BT Western Europe
BT Europe

Hamamelididae (1981)
BT Dicotyledoneae
BT angiosperms
BT Spermatophyta
BT Plantae

Hamburg
No longer a valid term for GeoRef. See Hamburg Germany.

Hamburg Germany (1993)
City state on both sides of the Elbe River in N Germany on the Schleswig-Holstein-Lower Saxony border.
BT Germany
BT Central Europe
BT Europe
SA Alster River
SA Elbe River
SA Lower Saxony Germany
SA North German Plain
SA Schleswig-Holstein Germany

Hamersley Basin (1978)
BT Western Australia
BT Australia
BT Australasia

Hamersley Group (1978)
BT Precambrian
SA Australia
SA Western Australia

Hamersley Range (1978)
In NW Western Australia.
BT Western Australia
BT Australia
BT Australasia

Hamilton
See Hamilton Ontario if applicable. Before 1989, for U.S. cities, search in combination with state, e.g. Hamilton AND New York. After 1989, search combined with state term e.g. Hamilton New York. After 1992, search in combination with country term for countries outside the United States and Canada.

Hamilton County
Valid through 1988. Search in combination with state term. After 1988, use specific county-state term.

Hamilton County Florida (1989)
N Florida. Before 1989, also search Hamilton County AND Florida.
CO N302000N303600
W0823800W0831500
BT Florida
BT United States

Hamilton County Illinois (1989)
SE Illinois. Before 1989, also search Hamilton County AND Illinois.
CO N375400N381500
W0882300W0884300
BT Illinois
BT United States

Hamilton County Indiana (1989)
Central Indiana. Before 1989, also search Hamilton County AND Indiana.
CO N395600N401300
W0855200W0861500
BT Indiana
BT United States

Hamilton County Iowa (1989)
Central Iowa. Before 1989, also search Hamilton County AND Iowa.
CO N421300N423400
W0932700W0935800
BT Iowa
BT United States

Hamilton County Kansas (1989)
SW Kansas. Before 1989, also search Hamilton County AND Kansas.
CO N374200N381400
W1013100W1020200
BT Kansas
BT United States

Hamilton County Nebraska (1989)
SE central Nebraska. Before 1989, also search Hamilton County AND Nebraska.
CO N404200N411000
W0975000W0981600
BT Nebraska
BT United States

Hamilton County New York (1989)
NE central New York. Before 1989, also search Hamilton County AND New York.
CO N431200N440700
W0740500W0744700
RT New York
BT United States

Hamilton County Ohio (1989)
Extreme SW Ohio. Before 1989, also search Hamilton County AND Ohio.
CO N390200N391800
W0841700W0845000
BT Ohio
BT United States
NT Cincinnati Ohio

Hamilton County Tennessee (1989)
SE Tennessee. Before 1989, also search Hamilton County AND Tennessee.
CO N345800N352800
W0845700W0852900
BT Tennessee
BT United States

Hamilton County Texas (1989)
Central Texas. Before 1989, also search Hamilton County AND Texas.
CO N312300N320000
W0974500W0982700
BT Texas
BT United States
SA Edwards Aquifer

Hamilton Group (1978)
Divisions include Onondaga Limestone, Marcellus Shale, Mahantango Formation, Speeds Formation, Deputy Formation, Silver Creek Formation, Swanville Formation, Beachwood Formation, Backoven (Bakoven) Shale, Mount Marion Formation, Ashokan Formation, Kiskatom Formation, Skaneateles Shale, Ludlowville Formation, Moscow Formation, Montebello Formation, Sherman Ridge Formation.
BT Middle Devonian
BT Devonian
BT Paleozoic
SA Ludlowville Formation
SA Mahantango Formation
SA Marcellus Shale
SA Maryland
SA Moscow Formation
SA New York
SA Onondaga Limestone
SA Pennsylvania
SA West Virginia

Hamilton Ontario (1993)
Area around and including city of Hamilton. At W end of Lake Ontario, S Ontario, Canada. Before 1993, also search Hamilton or Hamilton region AND Ontario.
BT Ontario
BT Eastern Canada
BT Canada

Hamilton Quarry (1993)
Upper Pennsylvanian fossil locality in SE Kansas.
BT Greenwood County Kansas
BT Kansas
BT United States

Hamilton region
No longer a valid term for GeoRef. See Hamilton Ontario if applicable.

hamlinite
use goyazite

hammarite (1978)
BT sulfobismuthites
BT sulfosalts

Hampden Basalt (1989)
In Newark Group. Central Connecticut and central Massachusetts. Overlies East Berlin Formation. Previously thought to be Upper Triassic in age.
BT Lower Jurassic
BT Jurassic
BT Mesozoic
SA Connecticut
SA Massachusetts
SA Newark Supergroup

Hampden County
Valid through 1988. Search in combination with state term. After 1988, use specific county-state term.

Hampden County Massachusetts (1989)
SW Massachusetts. Before 1989, also search Hampden County AND Massachusetts.
CO N420000N422200
W0720800W0730500
BT Massachusetts
BT United States
NT Springfield Massachusetts

Hampshire
No longer a valid term for GeoRef. As of 1993, see Hampshire England or Hampshire Basin.

Hampshire Basin (1978)
IN Index England if applicable.
SA England

Hampshire County
Valid through 1988. Search in combination with state term. After 1988, use specific county-state term.

Hampshire County Massachusetts (1989)
W central Massachusetts. Before 1989, also search Hampshire County AND Massachusetts.
CO N421100N423400
W0721300W0730400
BT Massachusetts
BT United States

Hampshire County West Virginia (1989)
NE West Virginia. Before 1989, also search Hampshire County AND West Virginia.
CO N390600N393300
W0782200W0785800
BT West Virginia
BT United States

Hampshire England (1993)
County in S England. In 1974, the Isle of Wight became a separate county. Before 1993, also search Hampshire AND England.
CO N504200N512300
W0004500W0015500
BT England
BT Great Britain
BT United Kingdom
BT Western Europe
BT Europe
NT Fawley England
SA Isle of Wight England

Hampshire Formation (1978)
W Maryland, Pennsylvania, W Virginia, E West Virginia.
BT Upper Devonian
BT Devonian
BT Paleozoic
SA Maryland
SA Pennsylvania
SA Virginia
SA West Virginia

Han River basin (1978)
Also search Han River; Han Kiang.
IN Index provinces as applicable.
BT China
BT Far East
BT Asia
SA Hubei China
SA Shaanxi China

Hanaoka Mine (1993)
In Hokuroku District.
BT Akita Japan
BT Honshu
BT Japan
BT Far East
BT Asia
SA Hokuroku Japan
SA metal ores
SA mines

Hanawa Mine (1978)
Metal ores. Akita Prefecture in N Honshu.
BT Akita Japan
BT Honshu
BT Japan
BT Far East
BT Asia
SA metal ores
SA mines

Hancock County
Valid through 1988. Search in combination with state term. After 1988, use specific county-state term.

Hancock County Georgia (1989)
E central Georgia. Before 1989, also search Hancock County AND Georgia.
CO N330300N332800
W0824500W0831700
BT Georgia
BT United States

Hancock County Illinois (1989)
W Illinois. Before 1989, also search Hancock County AND Illinois.
CO N401000N403700
W0905500W0913100
BT Illinois
BT United States

Hancock County Indiana (1989)
Central Indiana. Before 1989, also search Hancock County AND Indiana.
CO N394200N395600
W0853500W0855700
BT Indiana
BT United States

Hancock County Iowa (1989)
N Iowa. Before 1989, also search Hancock County AND Iowa.
CO N425400N431600
W0933000W0935800
BT Iowa
BT United States

Hancock County Kentucky (1989)
NW Kentucky. Before 1989, also search Hancock County AND Kentucky.
CO N373900N375800
W0863800W0865800
BT Kentucky
BT United States

Hancock County Maine (1989)
S and SE Maine. Before 1989, also search Hancock County AND Maine.
CO N440600N451400
W0675800W0685100
BT Maine
BT United States

Hancock County Mississippi (1989)
SE Mississippi. Before 1989, also search Hancock County AND Mississippi.
CO N301000N304100
W0892200W0894200
BT Mississippi
BT United States
SA Mississippi Sound

Hancock County Ohio (1989)
NW Ohio. Before 1989, also search Hancock County AND Ohio.
CO N404800N411000
W0832500W0835300
BT Ohio
BT United States

Hancock County Tennessee (1989)
NE Tennessee. Before 1989, also search Hancock County AND Tennessee.
CO N362500N363600
W0825200W0832700
BT Tennessee
BT United States

Hancock County West Virginia (1989)
Northernmost county in West Virginia. Before 1989, also search Hancock County AND West Virginia.
CO N402300N403800
W0803200W0804000
BT West Virginia
BT United States

hand calculators (1981)
UF pocket calculators
SA computers
SA data processing
SA microcomputers
SA minicomputers

handbooks
use manuals

handling, data
use data handling

Hanford Reservation (1978)
Atomic energy facility in Benton, Franklin, and Grant counties in SE Washington.
IN Index counties as applicable.
BT Washington
BT United States
SA Benton County Washington
SA Franklin County Washington
SA Grande Ronde Basalt
SA Grant County Washington

Hangay
use Hangay Mountains

Hangay Mountains (1978)
W central Mongolia.
CO N460000N481500
E1023000E0970000
UF Changai Mountains
UF Hangay
UF Hangay Range
UF Hangayn Mountains
UF Khangai
UF Khangai Mountains
BT Mongolia
BT Far East
BT Asia
SA Bayan-Nuurin-khotnor Basin

Hangay Range
use Hangay Mountains

Hangayn Mountains
use Hangay Mountains

hanging wall (1989)
SA faults
SA foot wall

Hanna Basin (1978)
S central Wyoming. Also search Hanna.
IN Index counties or regions as applicable.
SA Hanna Formation
SA Wyoming

Hanna Formation (1989)
S central Wyoming.
IN Index ages as applicable.
BT Paleogene
BT Tertiary
BT Cenozoic
SA Eocene
SA Hanna Basin
SA Paleocene
SA Wyoming

Hannover Germany
use Hanover Germany

Hanover
No longer a valid term for GeoRef.
See Hanover Germany.

Hanover County
Valid through 1988. Search in combination with state term. After 1988, use specific county-state term.

Hanover County Virginia (1989)
E central Virginia. Before 1989, also search Hanover County AND Virginia.
CO N373300N380000
W0770700W0774800
BT Virginia
BT United States

Hanover Germany (1993)
City in Lower Saxony. Former state of NW Germany. Before 1978, also search Hannover.
UF Hannover Germany
BT Lower Saxony Germany
BT Germany
BT Central Europe
BT Europe

Hanson Creek Formation (1989)
Central Nevada. Middle Ordovician to Silurian.
IN Index ages as applicable.
BT Paleozoic
SA Nevada
SA Ordovician
SA Silurian

Hanson Lake (1978)
Near Manitoba border in E Saskatchewan.
IN Index Saskatchewan if applicable.
SA Saskatchewan

Hantkenina (1978)
Genus. Autoposting of microfossils and Protista to this term began in 1990. This term has multiple hierarchies.
BT1 Globigerinacea
BT1 Rotaliina
BT1 foraminifera
BT1 Protista
BT1 Invertebrata
BT2 Globigerinacea
BT2 Rotaliina
BT2 foraminifera
BT2 Protista
BT2 microfossils

Haplophragmoides (1978)
Genus. Autoposting of microfossils and Protista to this term began in 1990. This term has multiple hierarchies.
BT1 Lituolidae
BT1 Lituolacea
BT1 Textulariina
BT1 foraminifera
BT1 Protista
BT1 Invertebrata
BT2 Lituolidae
BT2 Lituolacea
BT2 Textulariina
BT2 foraminifera
BT2 Protista
BT2 microfossils

Harar
No longer a valid term for GeoRef.
See Harar Ethiopia.

Harar Ethiopia (1993)
Province and city in E Ethiopia. Before 1978, also search Harer or Harrar.
UF Harer Ethiopia
UF Harrar Ethiopia
BT Ethiopia
BT East Africa
BT Africa

Harare Zimbabwe (1993)
City in NE Zimbabwe. Formerly Salisbury. Before 1993 also search (Harare OR Salisbury) AND (Zimbabwe OR Rhodesia).
BT Zimbabwe
BT Southern Africa
BT Africa

harbors (1978)
SA shorelines
SA waterways

Harbour Main Group (1978)
Overlain with angular unconformity by Lower Cambrian strata. The Conception Group conformably overlies the Harbour Main Group except near the Holyrood Granite where it is unconformable. E Newfoundland.
BT Precambrian
SA Canada
SA Newfoundland Island

hard ground
use hardground

hard-ground
use hardground

Hardanger Plateau
use Hardangervidda

Hardangervidda (1978)
Plateau in Hordaland, SW Norway.
UF Hardanger Plateau
UF Vidda
BT Norway
BT Scandinavia
BT Western Europe
BT Europe

Hardeman Basin (1989)
N Texas and SW Oklahoma.
IN Index counties or regions as applicable.
SA Oklahoma
SA Texas
SA United States

Hardeman County
Valid through 1988. Search in combination with state term. After 1988, use specific county-state term.

Hardeman County Tennessee (1989)
SW Tennessee. Before 1989, also search Hardeman County AND Tennessee.
CO N345900N352600
W0884800W0891300
BT Tennessee
BT United States

Hardeman County Texas (1989)
N Texas. Before 1989, also search Hardeman County AND Texas.
CO N340600N344000
W0993000W1000000
BT Texas
BT United States

hardground (1982)
UF hard ground
UF hard-ground
SA biogenic structures
SA compaction
SA lithofacies
SA lithostratigraphy
SA marine environment
SA ocean floors
SA paleoecology
SA sedimentary rocks
SA sedimentary structures
SA sediments

SA stratigraphic gaps
SA unconformities
Hardin County
Valid through 1988. Search in combination with state term. After 1988, use specific county-state term.
Hardin County Illinois (1989)
SE Illinois. Before 1989, also search Hardin County AND Illinois.
CO N372400N373500
W0880500W0882500
BT Illinois
BT United States
Hardin County Iowa (1989)
Central Iowa. Before 1989, also search Hardin County AND Iowa.
CO N421300N423400
W0930000W0933000
BT Iowa
BT United States
Hardin County Kentucky (1989)
Central and N Kentucky. Before 1989, also search Hardin County AND Kentucky.
CO N372400N380000
W0853900W0861400
BT Kentucky
BT United States
Hardin County Ohio (1989)
W central Ohio. Before 1989, also search Hardin County AND Ohio.
CO N403100N404800
W0832500W0835300
BT Ohio
BT United States
Hardin County Tennessee (1989)
SW Tennessee. Before 1989, also search Hardin County AND Tennessee.
CO N350000N352500
W0875800W0882300
BT Tennessee
BT United States
Hardin County Texas (1989)
E Texas. Before 1989, also search Hardin County AND Texas.
CO N300800N303600
W0940400W0944400
BT Texas
BT United States
Harding County
Valid through 1988. Search in combination with state term. After 1988, use specific county-state term.
Harding County New Mexico (1989)
NE New Mexico. Before 1989, also search Harding County AND New Mexico.
CO N352200N361400
W1032200W1042700
BT New Mexico
BT United States
Harding County South Dakota (1989)
NW South Dakota. Before 1989, also search Harding County AND South Dakota.
CO N451300N455500
W1025500W1040200
BT South Dakota
BT United States
hardness (1978)
SA mechanical properties
SA microhardness

SA minerals
SA physical properties
SA water hardness
Hardy County
Valid through 1988. Search in combination with state term. After 1988, use specific county-state term.
Hardy County West Virginia (1989)
E West Virginia. Before 1989, also search Hardy County AND West Virginia.
CO N384500N391500
W0783000W0791500
BT West Virginia
BT United States
Hare Bay (1978)
Inlet near N tip of Newfoundland.
SA Atlantic Ocean
SA Newfoundland Island
Harebell Formation (1978)
Unconformable contacts with overlying Pinyon Conglomerate and underlying Bacon Ridge Sandstone. Teton County in NW Wyoming.
BT Upper Cretaceous
BT Cretaceous
BT Mesozoic
SA Wyoming
Harer Ethiopia
use Harar Ethiopia
Harford County
Valid through 1988. Search in combination with state term. After 1988, use specific county-state term.
Harford County Maryland (1989)
NE Maryland. Before 1989, also search Harford County AND Maryland.
CO N391100N394326
W0760400W0763500
BT Maryland
BT United States
Harghita
No longer a valid term for GeoRef. As of 1993 see Harghita Romania.
Harghita Mountains (1978)
E Transylvania.
UF Harghitei Mountains
BT Transylvania
BT Romania
BT Southern Europe
BT Europe
Harghita Romania (1993)
County in E Transylvania. Before 1993 also search Harghita AND Romania.
BT Transylvania
BT Romania
BT Southern Europe
BT Europe
Harghitei Mountains
use Harghita Mountains
Hariana India
use Haryana India
harkerite (1978)
Autoposting of nesosilicates began in 1985.
BT nesosilicates
BT orthosilicates
BT silicates

Harlan County
Valid through 1988. Search in combination with state term. After 1988, use specific county-state term.
Harlan County Kentucky (1989)
SE Kentucky. Before 1989, also search Harlan County AND Kentucky.
CO N364000N370200
W0825200W0832900
BT Kentucky
BT United States
NT Benham Kentucky
NT Cumberland Kentucky
Harlan County Nebraska (1989)
S Nebraska. Before 1989, also search Harlan County AND Nebraska.
CO N400000N402200
W0991000W0993800
BT Nebraska
BT United States
Harlech
No longer a valid term for GeoRef as of 1993. See Harlech Wales or Harlech Dome.
Harlech Dome (1993)
N Wales. Before 1993, also search Harlech.
BT Wales
BT Great Britain
BT United Kingdom
BT Western Europe
BT Europe
Harlech Stage (1978)
Before 1978, search Harlech.
BT Lower Cambrian
BT Cambrian
BT Paleozoic
Harlech Wales (1993)
Village in Merionethshire (Gwynedd), W Wales. Before 1993, term sometimes used to refer to Harlech Dome. Before 1993, also search Harlech.
BT Merionethshire Wales
BT Wales
BT Great Britain
BT United Kingdom
BT Western Europe
BT Europe
harmonic analysis
use Fourier analysis
harmonic folds (1978)
Before 1978, search folds AND harmonic.
BT folds
SA disharmonic folds
harmonics (1978)
SA geodesy
harmotome (1978)
BT zeolite group
BT framework silicates
BT silicates
Harney County
Valid through 1988. Search in combination with state term. After 1988, use specific county-state term.
Harney County Oregon (1989)
SE central Oregon. Before 1989, also search Harney County AND Oregon.
CO N420000N440500
W1181000W1195500
BT Oregon
BT United States
NT Steens Mountain

SA Borax Lake
Harney Peak Granite (1989)
SW South Dakota.
BT South Dakota
BT United States
SA lower Proterozoic
SA Proterozoic
Harper County
Valid through 1988. Search in combination with state term. After 1988, use specific county-state term.
Harper County Kansas (1989)
S Kansas. Before 1989, also search Harper County AND Kansas.
CO N370000N372400
W0974800W0982200
BT Kansas
BT United States
Harper County Oklahoma (1989)
NW Oklahoma. Before 1989, also search Harper County AND Oklahoma.
CO N363600N370000
W0991600W1000000
BT Oklahoma
BT United States
Harpidae (1978)
BT Gastropoda
BT Mollusca
BT Invertebrata
Harrar Ethiopia
use Harar Ethiopia
Harrington Sound (1978)
BT Bermuda
BT Atlantic Ocean Islands
Harris (1978)
Southern section of the island of Lewis in the Outer Hebrides off NW Scotland. In Western Isles region.
IN Index Outer Hebrides as applicable.
SA Outer Hebrides
Harris County
Valid through 1988. Search in combination with state term. After 1988, use specific county-state term.
Harris County Georgia (1989)
W Georgia. Before 1989, also search Harris County AND Georgia.
CO N323300N325000
W0844000W0850900
BT Georgia
BT United States
Harris County Texas (1989)
S Texas. Before 1989, also search Harris County AND Texas.
CO N293600N301000
W0945400W0960000
BT Texas
BT United States
NT Houston Texas
Harrisburg
Valid through 1988. Search in combination with state term. After 1988, use specific city-state term.
Harrisburg Pennsylvania (1989)
City in SE central Pennsylvania. Before 1989, search Harrisburg AND Pennsylvania.
CO N401800N401800
W0764900W0764900
BT Dauphin County Pennsylvania
BT Pennsylvania
BT United States

Harrison County
Valid through 1988. Search in combination with state term. After 1988, use specific county-state term.

Harrison County Indiana (1989)
S Indiana. Before 1989, also search Harrison County AND Indiana.
CO N375700N382500
W0855400W0862000
BT Indiana
BT United States

Harrison County Iowa (1989)
W Iowa. Before 1989, also search Harrison County AND Iowa.
CO N413000N415100
W0952900W0960700
BT Iowa
BT United States

Harrison County Kentucky (1989)
N Kentucky. Before 1989, also search Harrison County AND Kentucky.
CO N381800N383600
W0840800W0843500
BT Kentucky
BT United States

Harrison County Mississippi (1989)
SE Mississippi. Before 1989, also search Harrison County AND Mississippi.
CO N302000N304200
W0885000W0892100
BT Mississippi
BT United States
SA Mississippi Sound

Harrison County Missouri (1989)
NW Missouri. Before 1989, also search Harrison County AND Missouri.
CO N400800N403500
W0935000W0941200
BT Missouri
BT United States

Harrison County Ohio (1989)
E Ohio. Before 1989, also search Harrison County AND Ohio.
CO N401000N402500
W0805100W0812000
BT Ohio
BT United States

Harrison County Texas (1989)
E Texas. Before 1989, also search Harrison County AND Texas.
CO N322000N324600
W0940300W0944000
BT Texas
BT United States

Harrison County West Virginia (1989)
N West Virginia. Before 1989, also search Harrison County AND West Virginia.
CO N390700N392800
W0801200W0803700
BT West Virginia
BT United States

Harrison Lake (1978)
In S British Columbia.
IN Index British Columbia if applicable.
SA British Columbia

Harrodsburg Limestone (1989)
Indiana and N Kentucky.
BT Mississippian

BT Carboniferous
BT Paleozoic
SA Indiana
SA Kentucky

Hartberg
No longer a valid term for GeoRef. See Hartberg Austria.

Hartberg Austria (1993)
Town in Styria, SE Austria.
BT Styria Austria
BT Austria
BT Central Europe
BT Europe

Hartford Basin (1989)
Triassic to Jurassic rift basin in central Connecticut and W Massachusetts.
IN Index states as applicable.
SA Connecticut
SA Massachusetts
SA United States

Hartford County
Valid through 1988. Search in combination with state term. After 1988, use specific county-state term.

Hartford County Connecticut (1989)
N central Connecticut. Before 1989, also search Hartford County AND Connecticut.
CO N413300N420300
W0722400W0730200
BT Connecticut
BT United States
NT Suffield Connecticut
SA Roaring Brook Valley

Hartland Formation (1989)
W Connecticut.
BT Paleozoic
SA Connecticut

Harts Range (1989)
SE central Northern Territory, Australia.
CO S231500S230000
E1350000E1340000
BT Northern Territory Australia
BT Australia
BT Australasia

Hartselle Sandstone (1985)
N, central and E Alabama.
BT Upper Mississippian
BT Mississippian
BT Carboniferous
BT Paleozoic
SA Alabama

Hartshorne Sandstone (1985)
In Krebs Group. E Oklahoma and W Arkansas.
BT Desmoinesian
BT Middle Pennsylvanian
BT Pennsylvanian
BT Carboniferous
BT Paleozoic
SA Arkansas
SA Krebs Group
SA Oklahoma

Hartzog Draw Field (1989)
Campbell and Johnson counties, Wyoming. Also search Hartzog Draw.
IN Index counties as applicable.
CO N430000N450000
W1050000W1070000
BT Wyoming
BT United States
SA Campbell County Wyoming
SA giant fields
SA Johnson County Wyoming
SA oil and gas fields

SA Powder River basin

Haruna (1978)
Volcano in Gumma Prefecture in N central Honshu. Also a lake.
UF Haruna Volcano
BT Gumma Japan
BT Honshu
BT Japan
BT Far East
BT Asia

Haruna Volcano
use Haruna

Harvey County
Valid through 1988. Search in combination with state term. After 1988, use specific county-state term.

Harvey County Kansas (1989)
S central Kansas. Before 1989, also search Harvey County AND Kansas.
CO N375500N381000
W0971000W0974300
BT Kansas
BT United States

Haryana
No longer a valid term for GeoRef. See Haryana India.

Haryana India (1993)
State in N central India. Before 1993, also search Hariana.
CO N274000N310000
E0773000E0743000
UF Hariana India
BT India
BT Indian Peninsula
BT Asia
NT Hissar India
NT Narnaul India
SA Delhi Supergroup

Harz Foreland (1978)
Central Germany. Also search Harz.
IN Index states as applicable.
BT Germany
BT Central Europe
BT Europe
SA Harz region
SA Lower Saxony Germany
SA Saxony Germany

Harz Mountains (1978)
Mountain group between Elbe River and Weser River in SE Lower Saxony, and S Saxony-Anhalt, central Germany. Also search Harz.
IN Index states as applicable.
CO N512500N515700
E0114000E0101000
BT Germany
BT Central Europe
BT Europe
NT Rammelsberg
SA Brocken Massif
SA Harz region
SA Lower Saxony Germany
SA Saxony-Anhalt Germany

Harz region (1978)
Harz Foreland and Harz Mountains area of central Germany. Also search Harz.
IN Index states as applicable.
BT Germany
BT Central Europe
BT Europe
SA Harz Foreland
SA Harz Mountains
SA Lower Saxony Germany
SA Saxony Germany
SA Saxony-Anhalt Germany

harzburgite (1978)
BT peridotites
BT ultramafics
BT plutonic rocks
BT igneous rocks

Haskell County
Valid through 1988. Search in combination with state term. After 1988, use specific county-state term.

Haskell County Kansas (1989)
SW Kansas. Before 1989, also search Haskell County AND Kansas.
CO N372300N374300
W1003900W1010500
BT Kansas
BT United States

Haskell County Oklahoma (1989)
E Oklahoma. Before 1989, also search Haskell County AND Oklahoma.
CO N350300N352500
W0945000W0952700
BT Oklahoma
BT United States

Haskell County Texas (1989)
NW central Texas. Before 1989, also search Haskell County AND Texas.
CO N325800N332500
W0992800W0995900
BT Texas
BT United States

Hassan
No longer a valid term for GeoRef. See Hassan India.

Hassan India (1993)
Town in S Karnataka, S India.
BT Karnataka India
BT India
BT Indian Peninsula
BT Asia

Hassi Messaoud Field (1978)
Oil field near Great Eastern Erg in NE Algeria.
UF Hassi-Messaoud
BT Algeria
BT North Africa
BT Africa
SA oil and gas fields

Hassi-Messaoud
use Hassi Messaoud Field

Hastings County
No longer a valid term for GeoRef. As of 1993, see Hastings County Ontario.

Hastings County Ontario (1993)
SE Ontario. Before 1993, also search Hastings County AND Ontario.
CO N440700N452800
W0770000W0781400
BT Ontario
BT Eastern Canada
BT Canada
NT Bancroft Ontario
NT Madoc Ontario

hastingsite (1978)
BT clinoamphibole
BT amphibole group
BT chain silicates
BT silicates
SA ferrohastingsite

Hat Creek Basalt (1978)
Recent. Hat Creek Valley 10 miles NNE of Lassen Peak. Also search Hat Creek.

BT Holocene
BT Quaternary
BT Cenozoic
SA California

Hatay
No longer a valid term for GeoRef. As of 1993 see Hatay Turkey.

Hatay Turkey (1993)
Province in S Anatolia on the NW Syrian border. Before 1993 also search Hatay AND Turkey.
BT Turkey
BT Middle East
BT Asia
SA Anatolia

Hatchetigbee Formation (1985)
In Wilcox Group. S Alabama and E Mississippi.
BT lower Eocene
BT Eocene
BT Paleogene
BT Tertiary
BT Cenozoic
SA Alabama
SA Mississippi
SA Wilcox Group

hatchettolite
 use betafite

Hatchobaru Field (1989)
Central Kyushu, Japan.
CO N330000N330000
  E1310000E1310000
BT Kyushu
BT Japan
BT Far East
BT Asia
SA geothermal fields
SA Otake Field

Hateg
No longer a valid term for GeoRef. As of 1993 see Hateg Romania.

Hateg Romania (1993)
Town in Hunedoara county in SW Transylvania. Before 1993 also search Hateg AND Romania.
BT Transylvania
BT Romania
BT Southern Europe
BT Europe

Hatteras abyssal plain (1978)
Off SE coast of U. S.
BT Atlantic Ocean

Hatteras Formation (1989)
N American Atlantic. Located in the Outer Hatteras Rise off Cape Hatteras.
BT Cretaceous
BT Mesozoic
SA Atlantic Ocean
SA DSDP Site 603
SA DSDP Site 604
SA DSDP Site 605
SA Leg 93
SA North American Atlantic

Hattiesburg
Valid through 1988. Search in combination with state term. After 1988, use specific city-state term.

Hattiesburg Mississippi (1989)
City in S Mississippi. Before 1989, search Hattiesburg AND Mississippi.
CO N312000N312000
  W0891900W0891900
BT Forrest County Mississippi
BT Mississippi
BT United States

hausmannite (1978)
BT oxides
SA manganese minerals

Haut-Rhin
No longer a valid term for GeoRef. See Haut-Rhin France.

Haut-Rhin France (1993)
Department in NE France.
CO N473000N481500
  E0074500E0070000
BT France
BT Western Europe
BT Europe
NT Sainte-Marie-aux-Mines France
SA Alsace
SA Vosges Mountains

Haut-Zaire
 use Upper Zaire

Haute-Corse France (1993)
Department in the northern part of Corsica.
BT Corsica
BT France
BT Western Europe
BT Europe

Haute-Garonne
No longer a valid term for GeoRef.
 use Haute-Garonne France

Haute-Garonne France (1993)
Department in S France.
CO N424000N435300
  E0020000E0002500
UF Haute-Garonne
BT France
BT Western Europe
BT Europe
NT Toulouse France
SA Rodez Trough
SA Salat Valley

Haute-Loire
No longer a valid term for GeoRef.
 use Haute-Loire France

Haute-Loire France (1993)
Department in S central France.
CO N445500N451500
  E0043000E0030000
UF Haute-Loire
BT France
BT Western Europe
BT Europe
NT Velay
SA Central Massif

Haute-Marne
No longer a valid term for GeoRef.
 use Haute-Marne France

Haute-Marne France (1993)
CO N473500N484000
  E0055200E0043500
UF Haute-Marne
BT France
BT Western Europe
BT Europe
SA Chassigny Meteorite

Haute-Saone
No longer a valid term for GeoRef.
 use Haute-Saone France

Haute-Saone France (1993)
CO N471500N480200
  E0065000E0052500
UF Haute-Saone
BT France
BT Western Europe
BT Europe

Haute-Savoie
No longer a valid term for GeoRef.
 use Haute-Savoie France

Haute-Savoie France (1993)
Department in E France.
CO N454500N463000
  E0070000E0054500
UF Haute-Savoie
BT France
BT Western Europe
BT Europe
NT Annecy France
NT Argentiere Glacier
NT Rumilly France
NT Savoy Alps
SA Chablais
SA Lake Geneva
SA Savoy

Haute-Vienne
No longer a valid term for GeoRef.
 use Haute-Vienne France

Haute-Vienne France (1993)
Department of W central France.
CO N460300N471000
  E0011000W0000800
UF Haute-Vienne
BT France
BT Western Europe
BT Europe
NT Rochechouart France
NT Saint-Sylvestre Massif

Hauterivian (1978)
Europe. Above Valanginian, below Barremian.
BT Lower Cretaceous
BT Cretaceous
BT Mesozoic
SA Neocomian
SA Zubair Formation

Hautes-Alpes
No longer a valid term for GeoRef.
 use Hautes-Alpes France

Hautes-Alpes France (1993)
Department in SE France.
CO N441000N451000
  E0070200E0052500
UF Hautes-Alpes
BT France
BT Western Europe
BT Europe
NT Briancon France
SA Eoulx Basin
SA Pelvoux Massif

Hautes-Pyrenees
No longer a valid term for GeoRef.
 use Hautes-Pyrenees France

Hautes-Pyrenees France (1993)
Department in SW France.
CO N423500N434000
  E0005000W0001500
UF Hautes-Pyrenees
BT France
BT Western Europe
BT Europe
NT Gavarnie France
NT Lourdes France

Hauts-de-Seine
No longer a valid term for GeoRef.
 use Hauts-de-Seine France

Hauts-de-Seine France (1993)
Department in N France.
CO N484200N490000
  E0022000E0020800
UF Hauts-de-Seine
BT France

BT Western Europe
BT Europe

hauyne (1978)
From 1978-1980, framework silicates and sulfates were autoposted to this term. Autoposting of sodalite group and silicates began in 1981.
UF hauynite
BT sodalite group
BT framework silicates
BT silicates
SA sulfates

hauynite
 use hauyne

Havana
No longer a valid term for GeoRef. See Havana Cuba.

Havana Cuba (1993)
City on Straits of Florida in NW Cuba.
BT La Habana Cuba
BT Cuba
BT Greater Antilles
BT Antilles
BT West Indies
BT Caribbean region

Haveroe Meteorite (1981)
Impact in Turku area in SW Finland. Before 1981, also search Haveroe AND meteorites.
BT ureilite
BT achondrites
BT stony meteorites
BT meteorites
SA Finland

Haviland
Valid through 1988. Search in combination with state term. After 1988, use specific city-state term.

Haviland Kansas (1989)
Village in S central Kansas. Before 1989, search Haviland AND Kansas.
CO N373700N373700
  W0990600W0990600
BT Kiowa County Kansas
BT Kansas
BT United States

Haviland Meteorite (1981)
Impact in Kiowa County, Kansas. Before 1981, also search Haviland AND meteorites.
BT H chondrites
BT chondrites
BT stony meteorites
BT meteorites
SA Kansas
SA Kiowa County Kansas

Hawaii (1978)
State including all of the Hawaiian Islands. Before 1971, also search Hawaiian Islands. In 1985, Pacific Ocean was removed as an autoposted broader term, and East Pacific Ocean Islands was added. Before 1989, was also used for southernmost and largest of the islands. Hawaii began autoposting in 1989. From 1989 through 1992, Hawaii County Hawaii was used for the island. As of 1993, see Hawaii Island for the island. This term has multiple hierarchies.
CO N190000N283000
  W1550000W1790000
UF Hawaiian Islands
BT1 United States
BT2 Polynesia
BT2 Oceania
BT3 East Pacific Ocean Islands

NT Hawaii County Hawaii
NT Honolulu County Hawaii
NT Kauai
NT Maui County Hawaii
NT Mauna Loa
NT Molokai
SA Pacific Ocean
SA Red Hill

Hawaii County
Valid through 1988. Search in combination with state term. After 1988, use specific county-state term.

Hawaii County Hawaii (1989)
Entire island of Hawaii. Before 1989, also search Hawaii County AND Hawaii. As of 1993, Hawaii Island is a narrower term. This term has multiple hierarchies.
CO N185500N201600
 W1545000W1560500
BT1 Hawaii
BT1 United States
BT2 Hawaii
BT2 Polynesia
BT2 Oceania
BT3 Hawaii
BT3 East Pacific Ocean Islands
NT Hawaii Island
SA East Rift Zone
SA Mauna Loa

Hawaii Island (1993)
Southernmost and largest of islands of Hawaii. Coterminous with Hawaii County Hawaii. Before 1993, also search Hawaii County Hawaii. This term has multiple hierarchies.
CO N185500N201600
 W1545000W1560500
BT1 Hawaii County Hawaii
BT1 Hawaii
BT1 United States
BT2 Hawaii County Hawaii
BT2 Hawaii
BT2 Polynesia
BT2 Oceania
BT3 Hawaii County Hawaii
BT3 Hawaii
BT3 East Pacific Ocean Islands
NT Alae Crater
NT Hilo Hawaii
NT Hualalai
NT Kilauea
NT Kona
NT Mauna Kea
NT Puu Oo
SA East Rift Zone
SA Loihi Seamount
SA Mauna Loa

Hawaii Ridge
 use Hawaiian Ridge

Hawaiian Arch
 use Hawaiian Ridge

Hawaiian Islands
No longer a valid GeoRef index term. Before 1971, was used on level 1 in subfile N.
 use Hawaii

Hawaiian Ridge (1978)
NW of Hawaiian Islands.
UF Hawaii Ridge
UF Hawaiian Arch
BT Pacific Ocean

hawaiian-type eruptions (1981)
SA eruptions
SA volcanism
SA volcanology

hawaiite (1978)
BT alkali basalts
BT basalts
BT volcanic rocks
BT igneous rocks

Hawke's Bay
No longer a valid term for GeoRef. See Hawke's Bay New Zealand.

Hawke's Bay New Zealand (1993)
Provincial district on E North Island.
IN Also index North Island.
BT New Zealand
BT Australasia
SA North Island

Hawkesbury Sandstone (1978)
BT Triassic
BT Mesozoic
SA Australia
SA New South Wales Australia

Hawthorn Formation (1978)
In Alum Bluff Group. Lower and middle Miocene. Central northern, N and S Florida; S and SE Georgia; South Carolina.
UF Hawthorne Formation
BT Miocene
BT Neogene
BT Tertiary
BT Cenozoic
SA Florida
SA Floridan Aquifer
SA Georgia
SA Peace River Formation
SA South Carolina

Hawthorne Formation
 use Hawthorn Formation

Hayel
 use Hail Saudi Arabia

Hayes fracture zone (1989)
Mid-Atlantic Ridge; North Atlantic.
BT Mid-Atlantic Ridge
BT Atlantic Ocean
SA DSDP Site 561
SA DSDP Site 562
SA DSDP Site 563
SA DSDP Site 564

Hayes River Group (1989)
E Manitoba.
BT Archean
BT Precambrian
SA Manitoba

Hayesville Fault (1989)
IN Index states as applicable.
SA Georgia
SA North Carolina
SA Tennessee

Hayfork Terrane (1985)
Tectonostratigraphic unit, N California and S Oregon.
CO N410000N420000
 W1220000W1240000
BT United States
SA California
SA Klamath Mountains
SA Oregon
SA Pacific Coast

Haymond Formation (1978)
Overlies Dimple Limestone; underlies Gaptank Formation. Marathon region in Brewster County in SW Texas.
BT Lower Pennsylvanian
BT Pennsylvanian
BT Carboniferous
BT Paleozoic
SA Texas

Haynesville Formation (1985)
BT Upper Jurassic
BT Jurassic
BT Mesozoic
SA Alabama
SA Arkansas
SA Louisiana
SA Mississippi
SA Texas

Hays County
Valid through 1988. Search in combination with state term. After 1988, use specific county-state term.

Hays County Texas (1989)
S central Texas. Before 1989, also search Hays County AND Texas.
CO N294700N302000
 W0974000W0982000
BT Texas
BT United States
SA Barton Springs
SA Edwards Aquifer

Hayward
Valid through 1988. Search in combination with state term. After 1988, use specific city-state term.

Hayward California (1989)
City in W California. Before 1989, search Hayward AND California.
CO N374000N374000
 W1220700W1220700
BT Alameda County California
BT California
BT United States

Hayward Fault (1978)
NE, E, and SE of San Francisco Bay.
IN Index counties or regions as applicable.
SA California

Hazara
No longer a valid term for GeoRef. See Hazara Pakistan.

Hazara Pakistan (1993)
District.
BT North-West Frontier Pakistan
BT Pakistan
BT Indian Peninsula
BT Asia

hazardous waste (1989)
SA environmental geology
SA pollution
SA radioactive waste
SA Superfund
SA toxic materials
SA waste disposal
SA waste disposal sites

hazards, geologic
 use geologic hazards

Hazaribagh
No longer a valid term for GeoRef. See Hazaribagh India.

Hazaribagh India (1993)
Town and district in central Bihar, E India.
BT Bihar India
BT India
BT Indian Peninsula
BT Asia
SA Bihar mica belt

Hazel Formation (1985)
W Texas.
BT Precambrian
SA Texas

Hazelton Group (1985)
Jurassic? SE Alaska and British Columbia.
SA Alaska
SA British Columbia

SA Jurassic

Hazen's Notch Formation
 use Hazens Notch Formation

Hazens Notch Formation (1989)
Cambrian?.
UF Hazen's Notch Formation
SA Cambrian
SA Vermont

HDR
Not a valid term for GeoRef. See hot dry rocks.

He
 use helium

He-3 (1978)
Autoposting of broader terms began in 1989. This term has multiple hierarchies.
BT1 stable isotopes
BT1 isotopes
BT2 helium
BT2 noble gases
SA He-4/He-3

He-3/He-4
 use He-4/He-3

He-4 (1978)
Autoposting of broader terms began in 1989. This term has multiple hierarchies.
BT1 stable isotopes
BT1 isotopes
BT2 helium
BT2 noble gases
SA He-4/He-3

He-4/He-3 (1978)
Isotopic ratio. As of 1989, use He/He for age determination method. Autoposting of broader terms began in 1989. This term has multiple hierarchies.
UF He-3/He-4
BT1 stable isotopes
BT1 isotopes
BT2 helium
BT2 noble gases
SA absolute age
SA He-3
SA He-4
SA isotope ratios

He/He (1989)
Isotopic ratio used in age determination. Before 1989, also search He-4/He-3.
SA absolute age
SA helium
SA isotope ratios

Healdiidae (1978)
As of 1990, microfossils, Crustacea, Mandibulata, and Arthropoda are autoposted to this term. This term has multiple hierarchies.
BT1 Podocopida
BT1 Ostracoda
BT1 Crustacea
BT1 Mandibulata
BT1 Arthropoda
BT1 Invertebrata
BT2 Podocopida
BT2 Ostracoda
BT2 microfossils
SA Cavellinidae

health
 use medical geology

Heard Plateau
 use Kerguelen Plateau

Heart Mountain Fault (1978)
NW Wyoming.
BT Wyoming

BT United States

heat capacity (1978)
SA Kirchhoff integral
SA specific heat
SA thermal properties

heat conduction
No longer a valid GeoRef index term. Was used on level 1 in subfile N. See heat flow; heat transfer; thermal conductivity.

heat content
use enthalpy

heat flow (1978)
Used for the product of the thermal conductivity of a substance and the thermal gradient in the direction of the flow of heat. Before 1972, also search geothermal surveys; temperature methods; temperature surveys; thermal methods; thermal surveys.
UF geothermal surveys
UF temperature methods
UF temperature surveys
UF thermal surveys
SA Battle Mountain High
SA boreholes
SA boundary conditions
SA conductivity
SA convection
SA core
SA crust
SA Curie point
SA Earth
SA energy sources
SA fumaroles
SA geophysical methods
SA geophysical surveys
SA geophysics
SA geothermal energy
SA geothermal gradient
SA heat flux
SA heat pumps
SA heat sources
SA heat transfer
SA hot springs
SA Illinois Deep Hole Project
SA intrusions
SA mantle
SA measurement
SA plate tectonics
SA radioactivity
SA regional patterns
SA sea-floor spreading
SA tectonophysics
SA temperature
SA temperature logging
SA thermal anomalies
SA thermal conductivity
SA thermal diffusivity
SA thermal inertia
SA thermal regime
SA thermal waters
SA well-logging

heat flux (1978)
Heat flow per unit of time.
SA heat flow

heat pumps (1985)
SA geothermal energy
SA heat flow
SA heat transfer

heat sources (1978)
UF sources, heat
SA core
SA crust
SA geothermal energy
SA heat flow
SA mantle
SA regional patterns
SA thermal waters
SA volcanism

heat storage (1982)
UF thermal energy storage
SA aquifers
SA energy sources
SA geothermal energy
SA geothermal fields
SA geothermal systems
SA ground water
SA new energy sources
SA storage

heat transfer (1978)
UF transfer, heat
SA heat flow
SA heat pumps

heat treatment
use thermal recovery

heat, specific
use specific heat

Heath Formation (1989)
In Big Snowy Group. N central Montana and W North Dakota.
BT Upper Mississippian
BT Mississippian
BT Carboniferous
BT Paleozoic
SA Big Snowy Group
SA Montana
SA North Dakota

heaving, frost
use frost heaving

heavy metals (1978)
Includes use for environmental studies.
IN Index individual metals if possible.
SA bismuth
SA cadmium
SA cobalt
SA copper
SA gold
SA iron
SA lead
SA manganese
SA nickel
SA platinum
SA pollution
SA precious metals
SA silver
SA tantalum
SA tellurium
SA zinc

heavy mineral deposits (1981)
Before 1981, search heavy minerals AND deposits.
IN Commodity. See List C.
SA garnet deposits
SA heavy minerals
SA kyanite deposits
SA monazite deposits
SA placers
SA Prakasam India
SA stream placers
SA zircon deposits

heavy minerals (1978)
See also individual heavy minerals. As of 1981, use heavy mineral deposits for heavy minerals as a commodity.
SA accessory minerals
SA apatite
SA biotite
SA cassiterite
SA heavy mineral deposits
SA ilmenite
SA kyanite
SA light minerals
SA magnetite
SA minerals
SA monazite
SA nonmagnetic minerals
SA placers

SA rutile
SA sedimentation
SA sediments
SA titanite
SA tourmaline
SA zircon

heavy oil (1978)
SA condensates
SA petroleum

heazlewoodite (1978)
BT sulfides

HEBBLE (1985)
Acronym.
UF High Energy Benthic Boundary Layer Experiment
SA Atlantic Ocean
SA boundary layer
SA continental rise
SA continental slope
SA high-energy environment
SA North American Atlantic
SA ocean circulation
SA oceanography
SA sedimentation
SA turbidity currents
SA turbulence

Hebei
No longer a valid term for GeoRef. See Hebei China.

Hebei China (1993)
Province in NE China. Also search Hopei.
CO N360000N424000 E1195000E1131000
UF Hopei China
BT China
BT Far East
BT Asia
NT Beijing China
NT Jixian China
NT Tianjin China
NT Yanshan Range
SA Bohai Bay
SA Huang Ho
SA Hutuo Group
SA Nihewan Formation
SA Taihang Mountains
SA Tangshan China
SA Tangshan earthquake 1976
SA Wumishan Formation
SA Xingtai earthquake 1966

Hebgen Lake (1978)
W of Yellowstone National Park in Gallatin County.
BT Gallatin County Montana
BT Montana
BT United States
SA Hebgen Lake earthquake 1959

Hebgen Lake earthquake 1959 (1989)
Hebgen Lake, Gallatin County, SW Montana. Also search Hebgen Lake AND earthquakes.
BT earthquakes
SA Gallatin County Montana
SA Hebgen Lake
SA Montana

Hebrides (1978)
Islands in Atlantic Ocean W of Scotland.
CO N553000N583000 W0053000W0084000
UF Western Islands
BT Scotland
BT Great Britain
BT United Kingdom
BT Western Europe
BT Europe
NT Inner Hebrides
NT Outer Hebrides

hebronite
use amblygonite

Hecla
use Hekla

Hecla Hoek Formation (1978)
Spitsbergen Island.
BT Precambrian
SA Arctic region
SA Spitsbergen
SA Spitsbergen Island

Hector Formation (1978)
Precambrian formation in Alberta and Tertiary formation in the Mojave Desert area, S California.
IN Index ages as applicable.
SA Alberta
SA California
SA Miocene
SA Mojave Desert
SA Oligocene
SA Precambrian
SA Tertiary

hectorite (1978)
BT clay minerals
BT sheet silicates
BT silicates

Hedbergella (1978)
Genus. Autoposting of microfossils and Protista to this term began in 1990. This term has multiple hierarchies.
BT1 Globigerinacea
BT1 Rotaliina
BT1 foraminifera
BT1 Protista
BT1 Invertebrata
BT2 Globigerinacea
BT2 Rotaliina
BT2 foraminifera
BT2 Protista
BT2 microfossils

hedenbergite (1978)
BT clinopyroxene
BT pyroxene group
BT chain silicates
BT silicates

Hedjaz Meteorite (1989)
Impact at Et-Tlahi, Saudi Arabia.
BT L chondrites
BT chondrites
BT stony meteorites
BT meteorites
SA Hejaz Saudi Arabia

hedleyite (1978)
From 1978-1980, sulfides was autoposted to this term. Autoposting of tellurides and alloys began in 1981. This term has multiple hierarchies.
BT1 tellurides
BT2 alloys

Heemskirk Granite (1978)
W Tasmania.
BT Devonian
BT Paleozoic
SA Australia
SA Tasmania Australia

Hegau (1978)
Region in SW Germany.
BT Baden-Wurttemberg Germany
BT Germany
BT Central Europe
BT Europe

Heidelberg
No longer a valid term for GeoRef. See Heidelberg Germany.

Heidelberg Germany (1993)

City on Neckar River in NW Baden-Wurttemberg.
BT Baden-Wurttemberg Germany
BT Germany
BT Central Europe
BT Europe

Heilongjiang
No longer a valid term for GeoRef. See Heilongjiang China.

Heilongjiang China (1993)
Province in NE China. Before 1993, also search Heilongjiang or Heilungkiang or Heilungjiang.
CO N433000N533000 E1350000E1213000
UF Heilungjiang China
BT China
BT Far East
BT Asia
NT Daqing Field
NT Xiao Hinggan Ling
SA Amur River
SA Khanka Lake
SA Tancheng-Lujiang fault zone

Heilungjiang China
use Heilongjiang China

Heilungkiang
No longer a valid term for GeoRef. See Heilongjiang China.

Heimaey (1978)
Island a few miles off S Iceland. Largest of Vestmannaeyjar group.
BT Vestmannaeyjar
BT Iceland
BT Western Europe
BT Europe

heintzite
use kaliborite

Hejaz
No longer a valid term for GeoRef. As of 1993 see Hejaz Saudi Arabia.

Hejaz Saudi Arabia (1993)
Province in W Saudi Arabia. Before 1993 also search Hejaz AND Saudi Arabia.
BT Saudi Arabia
BT Arabian Peninsula
BT Asia
NT Jeddah Saudi Arabia
SA Hedjaz Meteorite

Hekla (1978)
Volcano in SW Iceland.
UF Hecla
BT Iceland
BT Western Europe
BT Europe

Helderberg Group (1978)
Includes Coeymans Formation, New Scotland Limestone, Mandata Shale, Becraft Shale, Port Ewen Shale, Keyser Limestone, Port Jervis Limestone, Licking Creek Limestone, Alsen, Kalkberg, Manlius. New York, E Pennsylvania, W Maryland, Western Virginia, and N West Virginia.
BT Lower Devonian
BT Devonian
BT Paleozoic
SA Coeymans Formation
SA Keyser Limestone
SA Manlius Formation
SA Maryland
SA New York
SA Pennsylvania
SA Virginia
SA West Virginia

Helderbergian (1978)
North America. Above Upper Silurian, below Deerparkian.
BT Lower Devonian
BT Devonian
BT Paleozoic

Helena
Valid through 1988. Search in combination with state term. After 1988, use specific city-state term.

Helena Formation (1989)
In Belt Supergroup. W central Montana.
UF Siyeh Formation
BT middle Proterozoic
BT Proterozoic
BT upper Precambrian
BT Precambrian
SA Belt Supergroup
SA Montana

Helena Montana (1989)
City in W Montana. Before 1989, search Helena AND Montana.
CO N463500N463500 W1120000W1120000
BT Lewis and Clark County Montana
BT Montana
BT United States

Helgoland (1978)
Island in North Sea 28 miles off mainland.
BT Schleswig-Holstein Germany
BT Germany
BT Central Europe
BT Europe
SA North Sea

Helicoplacoidea (1978)
BT Echinozoa
BT Echinodermata
BT Invertebrata

Helicopontosphaera (1978)
Autoposting of thallophytes and microfossils began in 1990. This term has multiple hierarchies.
BT1 nannofossils
BT1 algae
BT1 thallophytes
BT1 Plantae
BT2 nannofossils
BT2 algae
BT2 microfossils
BT3 nannofossils
BT3 algal flora
BT3 plants
BT4 nannofossils
BT4 algal flora
BT4 microfossils

helicopter methods (1981)
SA geophysical methods
SA geophysical surveys
SA methods
SA remote sensing

Helikian (1978)
North America. Autoposting of Precambrian to this term ended in 1989. As of 1989, middle Proterozoic and Proterozoic are autoposted to this term. As of 1990, upper Precambrian and Precambrian are autoposted to this term.
BT middle Proterozoic
BT Proterozoic
BT upper Precambrian
BT Precambrian
NT Neohelikian

Heliolites (1978)
Genus. Autoposting of Zoantharia to this term began in 1989.
BT Heliolitidae
BT Tabulata
BT Zoantharia
BT Anthozoa
BT Coelenterata
BT Invertebrata

Heliolitidae (1978)
Family. Autoposting of Zoantharia to this term began in 1989.
BT Tabulata
BT Zoantharia
BT Anthozoa
BT Coelenterata
BT Invertebrata
NT Heliolites

helium (1978)
Chemical element. As of 1981, use helium gas for helium as a commodity. Autoposting of noble gases to this term began in 1989.
UF He
BT noble gases
NT He-3
NT He-4
NT He-4/He-3
SA He/He
SA helium gas
SA isotopes
SA U/He

helium gas (1981)
Before 1981, search helium or He AND deposits.
IN Commodity. See List C.
SA helium

Hell Creek Formation (1978)
Includes Bull Creek Sand, Isabel-Firesteal Coal Member. E, N, and central S Montana; SW North Dakota; and NW and N South Dakota.
BT Upper Cretaceous
BT Cretaceous
BT Mesozoic
SA Montana
SA North Dakota
SA South Dakota

Hell's Canyon
use Snake River canyon

Hellenic Arc (1985)
Island arc from Peloponnesus to Rhodes. As of 1993, Greece is a broader term.
CO N345000N382000 E0281000E0212000
BT Greece
BT Southern Europe
BT Europe
SA Crete
SA Greek Aegean Islands
SA Hellenic Trench
SA Mediterranean Sea
SA Peloponnesus Greece
SA Rhodes

Hellenic Trench (1985)
S of Hellenic Arc.
BT Mediterranean Sea
SA East Mediterranean
SA Hellenic Arc

Hellenides (1978)
Deep rooted mountain chain of Greece which is part of the Alpine orogenic belt. Bordered by Rhodope Mountains in N Greece.
BT Greece
BT Southern Europe
BT Europe
SA Alps
SA Crete
SA Peloponnesus Greece
SA Pindus Mountains

Helmholtz free energy
use free energy

Helmstedt
No longer a valid term for GeoRef. See Helmstedt Germany.

Helmstedt Germany (1993)
City near Saxony-Anhalt border in E Lower Saxony.
BT Lower Saxony Germany
BT Germany
BT Central Europe
BT Europe

Helsingborg
use Halsingborg Sweden

Helsingor
No longer a valid term for GeoRef. See Helsingor Denmark.

Helsingor Denmark (1993)
City on the Oresund N of Copenhagen.
BT Denmark
BT Scandinavia
BT Western Europe
BT Europe

Helvetian (1978)
Europe.
BT middle Miocene
BT Miocene
BT Neogene
BT Tertiary
BT Cenozoic
SA Tortonian

helvine
use helvite

helvite (1978)
Autoposting of framework silicates to this term began in 1989. This term has multiple hierarchies.
UF helvine
BT1 framework silicates
BT1 silicates
BT2 sulfides
SA genthelvite

hematite (1978)
Before 1981, included use as a commodity term under iron.
IN As of 1981, may be used as a commodity term with iron ores.
BT oxides
SA iron
SA iron minerals
SA iron ores
SA iron oxides
SA martite
SA ocher
SA specularite

Hemichordata (1978)
Term is to be used only when more specific terms do not apply. Has narrower terms as of 1981.
UF Stomachorda
SA Camaroidea
SA Cephalodiscida
SA Chordata
SA Dendroidea
SA Didymograptina
SA Didymograptus
SA Diplograptina
SA Enteropneusta
SA Glossograptina
SA Graptolithina
SA Graptoloidea
SA Isograptus
SA Monograptina
SA Monograptus
SA Monograptus uniformis
SA Pterobranchia
SA Rhabdopleurida
SA Stolonoidea
SA Tuboidea

hemihedrite (1978)
　BT chromates

hemimorphite (1978)
　Autoposting of sorosilicates began in 1985.
　BT sorosilicates
　BT orthosilicates
　BT silicates
　SA zinc ores

Hemingfordian (1978)
　BT lower Miocene
　BT Miocene
　BT Neogene
　BT Tertiary
　BT Cenozoic
　SA Kirkwood Formation

hemipelagic environment (1985)
　SA continental slope
　SA deep-sea environment
　SA deep-water environment
　SA depositional environment
　SA ecology
　SA environment
　SA oceanography
　SA paleoecology
　SA pelagic environment
　SA sedimentation
　SA slope environment
　SA terrigenous materials

Hemipteroida (1981)
　BT Insecta
　BT Mandibulata
　BT Arthropoda
　BT Invertebrata

Hemlo
　As of 1993, no longer a valid term for GeoRef. See Hemlo Deposit.

Hemlo Deposit (1993)
　Gold ores. Canadian Shield of Ontario. Before 1993, also search Hemlo and Hemlo region.
　BT Ontario
　BT Eastern Canada
　BT Canada
　SA Canadian Shield
　SA gold ores

Hemlock Conglomerate (1989)
　In Kenai Group. S Alaska. Before 1989, also search Hemlock Formation AND Alaska.
　BT Oligocene
　BT Paleogene
　BT Tertiary
　BT Cenozoic
　SA Alaska
　SA Kenai Group

Hemlock Formation (1989)
　In Baraga Group. Lower Proterozoic. N Michigan and N Wisconsin. Before 1989, was sometimes used for Hemlock Conglomerate in Alaska.
　BT lower Proterozoic
　BT Proterozoic
　BT upper Precambrian
　BT Precambrian
　SA Baraga Group
　SA Michigan
　SA Wisconsin

Hemphillian (1978)
　Upper Miocene to lower Pliocene. Above Clarendonian, below Blancan.
　IN Index ages as applicable.
　BT Neogene
　BT Tertiary
　BT Cenozoic
　SA lower Pliocene
　SA upper Miocene

Henan
　No longer a valid term for GeoRef. See Henan China.

Henan China (1993)
　Province, South-Central China. Before 1985, also search Henan Sheng or Honan.
　CO N313000N362000
　　E1163000E1101000
　UF Henan Sheng China
　UF Honan China
　BT China
　BT Far East
　BT Asia
　SA Benxi Formation
　SA Dabie Mountains
　SA Dongpu Depression
　SA Huang Ho
　SA Luoquan Formation
　SA Taihua Group
　SA Taiyuan Formation

Henan Sheng China
　use Henan China

Henbury Meteorite (1981)
　Impact in Northern Territory. Before 1981, also search Henbury AND meteorites.
　BT octahedrite
　BT iron meteorites
　BT meteorites
　SA Australia
　SA Northern Territory Australia

Henderson County
　Valid through 1988. Search in combination with state term. After 1988, use specific county-state term.

Henderson County Illinois (1989)
　W Illinois. Before 1989, also search Henderson County AND Illinois.
　CO N403600N410500
　　W0904800W0911300
　BT Illinois
　BT United States

Henderson County Kentucky (1989)
　W Kentucky. Before 1989, also search Henderson County AND Kentucky.
　CO N373800N375800
　　W0871700W0875400
　BT Kentucky
　BT United States

Henderson County North Carolina (1989)
　W North Carolina. Before 1989, also search Henderson County AND North Carolina.
　CO N350900N353000
　　W0821700W0824500
　BT North Carolina
　BT United States

Henderson County Tennessee (1989)
　W Tennessee. Before 1989, also search Henderson County AND Tennessee.
　CO N352500N354800
　　W0881200W0883700
　BT Tennessee
　BT United States

Henderson County Texas (1989)
　E Texas. Before 1989, also search Henderson County AND Texas.
　CO N320000N322200
　　W0952500W0962400
　BT Texas
　BT United States

Henderson Gneiss (1989)
　W North Carolina and NW South Carolina.
　IN Index ages as applicable.
　BT lower Paleozoic
　BT Paleozoic
　SA Cambrian
　SA Devonian
　SA North Carolina
　SA Ordovician
　SA Silurian
　SA South Carolina

Henderson Mine (1981)
　Molybdenum ores in Clear Creek County, Colorado. Polymetallic ores in region near Chibougamau, Quebec. Talc deposits near Madoc, Ontario.
　IN Index Colorado and Clear Creek County Colorado; or Quebec or Ontario as applicable.
　SA Chibougamau Quebec
　SA Clear Creek County Colorado
　SA Colorado
　SA mines
　SA molybdenum ores
　SA Ontario
　SA polymetallic ores
　SA Quebec
　SA talc deposits

Hendry County
　Valid through 1988. Search in combination with state term. After 1988, use specific county-state term.

Hendry County Florida (1989)
　SW of Lake Okeechobee in S Florida. Before 1989, also search Hendry County AND Florida.
　CO N261500N264500
　　W0805000W0813500
　BT Florida
　BT United States

Hengchun Peninsula (1989)
　Extreme S Taiwan.
　BT Taiwan
　BT Far East
　BT Asia

Hennepin County
　Valid through 1988. Search in combination with state term. After 1988, use specific county-state term.

Hennepin County Minnesota (1989)
　E Minnesota. Before 1989, also search Hennepin County AND Minnesota.
　CO N444500N451500
　　W0931000W0935000
　BT Minnesota
　BT United States
　NT Minneapolis Minnesota

Hennessey Formation (1978)
　Overlies Wichita Group and underlies Duncan Sandstone, or Flowerpot Shale, where the Duncan is absent. Central and SW Oklahoma.
　BT Permian
　BT Paleozoic
　SA Oklahoma

Henrico County Virginia (1989)
　E central Virginia. Before 1989, search Henrico County AND Virginia.
　CO N372200N374300
　　W0771100W0773900
　BT Virginia
　BT United States
　NT Richmond Virginia

Henry County
　Valid through 1988. Search in combination with state term. After 1988, use specific county-state term.

Henry County Alabama (1989)
　SE Alabama. Before 1989, also search Henry County AND Alabama.
　CO N311700N314500
　　W0850300W0852400
　BT Alabama
　BT United States

Henry County Georgia (1989)
　N central Georgia. Before 1989, also search Henry County AND Georgia.
　CO N331700N333800
　　W0835500W0842000
　BT Georgia
　BT United States

Henry County Illinois (1989)
　NW Illinois. Before 1989, also search Henry County AND Illinois.
　CO N410900N413600
　　W0895100W0902800
　BT Illinois
　BT United States

Henry County Indiana (1989)
　E Indiana. Before 1989, also search Henry County AND Indiana.
　CO N394800N400500
　　W0851300W0853700
　BT Indiana
　BT United States

Henry County Iowa (1989)
　SE Iowa. Before 1989, also search Henry County AND Iowa.
　CO N404900N411000
　　W0912300W0914100
　BT Iowa
　BT United States

Henry County Kentucky (1989)
　N Kentucky. Before 1989, also search Henry County AND Kentucky.
　CO N382100N383700
　　W0845400W0852300
　BT Kentucky
　BT United States

Henry County Missouri (1989)
　W central Missouri. Before 1989, also search Henry County AND Missouri.
　CO N381300N383500
　　W0933000W0940400
　BT Missouri
　BT United States

Henry County Ohio (1989)
　NW Ohio. Before 1989, also search Henry County AND Ohio.
　CO N411000N412800
　　W0835300W0842100
　BT Ohio
　BT United States

Henry County Tennessee (1989)
　NW Tennessee. Before 1989, also search Henry County AND Tennessee.
　CO N360800N363000
　　W0880100W0883300
　BT Tennessee
　BT United States

Henry County Virginia (1989)
　S Virginia. Before 1989, also search Henry County AND Virginia.
　CO N363300N365100
　　W0793800W0800800

BT Virginia
BT United States

Henry Mountains (1978)
Garfield County in S central Utah.
IN Index county and mountain ranges or regions as applicable.
SA Garfield County Utah
SA Utah

Henryhouse Formation (1989)
S central Oklahoma.
BT Silurian
BT Paleozoic
SA Hunton Group
SA Oklahoma

Hepaticae (1978)
BT bryophytes
BT Plantae

Herault
No longer a valid term for GeoRef.
See Herault France.

Herault France (1993)
Department in S France.
CO N431500N440000
 E0041500E0023000
BT France
BT Western Europe
BT Europe
NT Bedarieux France
NT Cabrieres France
NT Lodeve Basin
NT Lodeve France
NT Montpellier France
NT Saint-Chinian France
NT Sete France
SA Cevennes

herbicides (1993)
SA atrazine
SA environmental geology
SA pesticides
SA pollutants
SA waste disposal

herbivores
use herbivorous taxa

herbivorous taxa (1993)
Before 1993, also search herbivores.
UF herbivores
SA ecology
SA ecosystems
SA grasslands
SA paleoecology

Hercegovina
use Herzegovina

Hercinico Centro (1981)
BT Spain
BT Iberian Peninsula
BT Southern Europe
BT Europe

Hercinico Sur (1981)
BT Spain
BT Iberian Peninsula
BT Southern Europe
BT Europe

Hercoglossa danica (1978)
BT Nautiloidea
BT Tetrabranchiata
BT Cephalopoda
BT Mollusca
BT Invertebrata

Hercynian
A valid index term through 1977. After 1977, use Hercynian Orogeny.

Hercynian Geosyncline (1978)
Before 1978, search geosynclines AND Hercynian.
SA geosynclines
SA Hercynian Orogeny

Hercynian Orogeny (1978)
Devonian-Lower Triassic. Also search orogeny AND Hercynian; Variscan; Abean.
IN Index ages as applicable.
UF Abean Orogeny
UF Variscan Orogeny
BT Paleozoic
SA Carboniferous
SA Devonian
SA Hercynian Geosyncline
SA Lower Triassic
SA orogeny
SA Permian
SA tectonics
SA Variscides

Hercynides
use Variscides

hercynite (1978)
UF ferrospinel
UF iron spinel
BT oxides
SA spinel group

herderite (1978)
BT phosphates

Herja (1978)
Mining region in Maramures County in N Transylvania.
BT Maramures Romania
BT Transylvania
BT Romania
BT Southern Europe
BT Europe

Herkimer County
Valid through 1988. Search in combination with state term. After 1988, use specific county-state term.

Herkimer County New York (1989)
Central and N central New York. Before 1989, also search Herkimer County AND New York.
CO N425000N440500
 W0744000W0751500
BT New York
BT United States

hermatypic taxa (1978)
Before 1978, also search hermatypic.
UF taxa, hermatypic
SA ahermatypic taxa
SA Coelenterata
SA reef builders
SA reefs

Hermosa Formation (1985)
Middle Pennsylvanian of Arizona, Colorado, New Mexico and Utah.
BT Middle Pennsylvanian
BT Pennsylvanian
BT Carboniferous
BT Paleozoic
SA Arizona
SA Colorado
SA Honaker Trail Formation
SA New Mexico
SA Paradox Member
SA Utah

Heron Island (1978)
Largest of Capricorn Islands in Coral Sea 40 miles off SE coast.
IN Index state or region as applicable.
SA Queensland Australia

Herrin Coal Member (1985)
Of Carbondale Formation.
UF Herrin No. 6 Coal
BT Pennsylvanian
BT Carboniferous
BT Paleozoic

SA Carbondale Formation
SA Illinois

Herrin No. 6 Coal
use Herrin Coal Member

Hertfordshire
No longer a valid term for GeoRef as of 1993.
use Hertfordshire England

Hertfordshire England (1993)
SE England. Before 1993, also search Hertfordshire.
CO N513000N521000
 E0001000W0004500
UF Hertfordshire
BT England
BT Great Britain
BT United Kingdom
BT Western Europe
BT Europe

hervidero
use mud volcanoes

Herzegovina (1978)
Region in N Bosnia and Herzegovina.
UF Hercegovina
BT Bosnia-Herzegovina
BT Yugoslavia
BT Southern Europe
BT Europe

herzenbergite (1978)
UF kolbeckine
BT sulfides

Hesperian (1993)
Intermediate system of three in the Martian stratigraphic column.
SA Mars

Hess Mountains (1978)
E central Yukon Territory.
BT Yukon Territory
BT Western Canada
BT Canada

Hess Rise (1985)
N central Pacific Ocean. Autoposting of Pacific Ocean to this term began in 1990.
BT North Pacific
BT Pacific Ocean
SA DSDP Site 310
SA DSDP Site 464
SA DSDP Site 465
SA DSDP Site 466

Hesse (1978)
Region in central Germany, comprising the state of Hesse and the former Prussian province of Hesse-Nassau. As of 1993, use Hesse Germany for the state.
BT Germany
BT Central Europe
BT Europe
SA Hesse Germany

Hesse Basin (1981)
Between Kassel and Wetterau.
CO N501500N514000
 E0103500E0083700
BT Germany
BT Central Europe
BT Europe

Hesse Germany (1993)
State. Before 1978, also search Hessen.
CO N492300N514000
 E0101500E0074500
BT Germany
BT Central Europe
BT Europe
NT Bergstrasse
NT Bundenbach
NT Darmstadt Germany

NT Kassel Germany
NT Marburg Germany
NT Messel Germany
NT Rheingau
NT Taunus
NT Vogelsberg
NT Waldeck Germany
NT Wetterau
NT Wiesbaden Germany
SA Franconia
SA Frankenberg Germany
SA Frankfurt Germany
SA Hesse
SA Lahn River
SA Lahn River valley
SA Main River
SA Mainz Basin
SA Odenwald
SA Palatinate
SA Rhenish Schiefergebirge
SA Rhon Mountains
SA Southwestern German Massifs
SA Steinheim Germany
SA Swabia
SA Upper Rhine Valley
SA Werra River
SA Weser River
SA Westerwald
SA Westphalia

Hessen
Not a valid term for GeoRef. See Hesse Germany or Hesse.

hessite (1978)
Autoposting of sulfides to this term ended in 1989.
BT tellurides

heterochrony
No longer a valid term for GeoRef. Valid 1989-1992.
use diachronism

Heterocorallia (1978)
Order. Autoposting of Zoantharia to this term began in 1989.
BT Zoantharia
BT Anthozoa
BT Coelenterata
BT Invertebrata

heterogeneity (1981)
UF inhomogeneity
SA homogeneity
SA seismology

heterogeneous materials (1978)
Before 1978, search heterogeneous.
SA homogeneous materials
SA materials

heterogenite (1978)
UF stainierite
BT oxides

Heterohelicidae (1978)
Family. Autoposting of microfossils and Protista to this term began in 1990. This term has multiple hierarchies.
BT1 Globigerinacea
BT1 Rotaliina
BT1 foraminifera
BT1 Protista
BT1 Invertebrata
BT2 Globigerinacea
BT2 Rotaliina
BT2 foraminifera
BT2 Protista
BT2 microfossils

heteromorphic taxa (1978)
Before 1978, search heteromorphs.
UF heteromorphs
UF taxa, heteromorphic

SA taxonomy

heteromorphite (1978)
BT sulfantimonites
BT sulfosalts

heteromorphs
use heteromorphic taxa

Heteroptera (1978)
BT Insecta
BT Mandibulata
BT Arthropoda
BT Invertebrata

Heterostegina (1978)
Autoposting of microfossils and Protista to this term began in 1990. This term has multiple hierarchies.
BT1 Rotaliacea
BT1 Rotaliina
BT1 foraminifera
BT1 Protista
BT1 Invertebrata
BT2 Rotaliacea
BT2 Rotaliina
BT2 foraminifera
BT2 Protista
BT2 microfossils
NT Heterostegina depressa
SA Nummulitidae

Heterostegina depressa (1978)
Autoposting of microfossils and Protista to this term began in 1990. This term has multiple hierarchies.
BT1 Heterostegina
BT1 Rotaliacea
BT1 Rotaliina
BT1 foraminifera
BT1 Protista
BT1 Invertebrata
BT2 Heterostegina
BT2 Rotaliacea
BT2 Rotaliina
BT2 foraminifera
BT2 Protista
BT2 microfossils

Heterostraci (1978)
Order.
BT Agnatha
BT Vertebrata
BT Chordata
SA Pisces

Hettangian (1978)
Europe. Above Rhaetian (Triassic), below Sinemurian.
BT lower Liassic
BT Lower Jurassic
BT Jurassic
BT Mesozoic

heulandite (1978)
BT zeolite group
BT framework silicates
BT silicates
SA clinoptilolite

Hexacorallia
use Scleractinia

Hexactiniaria (1981)
Autoposting of Zoantharia to this term began in 1989.
BT Zoantharia
BT Anthozoa
BT Coelenterata
BT Invertebrata

hexagonal system (1985)
Before 1985, also search hexagonal.
BT crystal systems
SA crystallography

hexahedrite (1981)

Before 1981, also search hexahedrites.
UF hexahedrites
BT iron meteorites
BT meteorites

hexahedrites
No longer a valid term for GeoRef as of 1981.
use hexahedrite

hexahydrite (1978)
BT sulfates

Hf
use hafnium

HF
As of 1981, no longer a valid term for GeoRef. See electromagnetic waves. For the acid with this chemical formula, use hydrofluoric acid.

Hf-177/Hf-176 (1989)
Autoposting of broader terms began in 1989. This term has multiple hierarchies.
BT1 stable isotopes
BT1 isotopes
BT2 hafnium
BT2 metals
SA isotope ratios

Hg
use mercury

Hiatella arctica (1989)
Species belonging to the order Myoida.
BT Bivalvia
BT Mollusca
BT Invertebrata

Hibernia Field (1989)
Oil field. Offshore of Newfoundland in Jeanne d'Arc Basin, Grand Banks, N Atlantic Ocean.
IN Index North American Atlantic and Grand Banks as applicable.
SA Grand Banks
SA Jeanne d'Arc Basin
SA North American Atlantic
SA offshore
SA oil and gas fields

hibonite (1985)
BT oxides

hibschite
use hydrogrossular

Hida
No longer a valid term for GeoRef. See Hida Japan or Hida Mountains.

Hida Japan (1993)
Former province in central Honshu. Now part of Gifu Prefecture. Before 1993, also search Hida AND Japan.
BT Gifu Japan
BT Honshu
BT Japan
BT Far East
BT Asia
SA Hida Mountains

Hida metamorphic belt (1978)
Central Honshu. From NE Toyama to Oki Islands in Shimane Prefecture. Exposed mainly in Hida Mountains.
BT Honshu
BT Japan
BT Far East
BT Asia
SA Hida Mountains

Hida Mountains (1978)

Central Honshu.
BT Honshu
BT Japan
BT Far East
BT Asia
SA Atotsugawa Fault
SA Gifu Japan
SA Hida Japan
SA Hida metamorphic belt
SA Japanese Alps
SA Toyama Japan

Hidaka
No longer a valid term for GeoRef. See Hidaka Japan, Hidaka metamorphic belt or Hidaka Mountains.

Hidaka Japan (1993)
Town in Hyogo Prefecture, S Honshu. Also a subprefecture in Hokkaido. Search in combination with Honshu or Hokkaido as applicable.
BT Japan
BT Far East
BT Asia
SA Hidaka Mountains
SA Hokkaido
SA Honshu
SA Hyogo Japan

Hidaka metamorphic belt (1978)
In S Hokkaido.
BT Hokkaido
BT Japan
BT Far East
BT Asia

Hidaka Mountains (1978)
In S Hokkaido. Also search Hidaka.
BT Hokkaido
BT Japan
BT Far East
BT Asia
SA Hidaka Japan
SA Tokachi

Hidalgo
No longer a valid term for GeoRef. As of 1993 see Hidalgo Mexico.

Hidalgo County
Valid through 1988. Search in combination with state term. After 1988, use specific county-state term.

Hidalgo County New Mexico (1989)
Extreme SW New Mexico. Before 1989, also search Hidalgo County AND New Mexico.
CO N312000N324500 W1081300W1090400
BT New Mexico
BT United States
SA Animas Mountains

Hidalgo County Texas (1989)
Extreme S Texas. Before 1989, also search Hidalgo County AND Texas.
CO N260300N264600 W0975300W0984200
BT Texas
BT United States
NT McAllen Ranch Field

Hidalgo Mexico (1993)
State in E central Mexico. Before 1993 also search Hidalgo AND Mexico.
BT Mexico
NT Pachuca Mexico
SA Sierra Madre Oriental

Hidas
No longer a valid term for GeoRef. See Hidas Hungary.

Hidas Hungary (1993)
Town in S Hungary. In Baranya.
BT Hungary
BT Central Europe
BT Europe

Hidden Valley Dolomite (1978)
Silurian and Lower Devonian. Underlies Lost Burro Formation; overlies Ely Springs Dolomite. S California. Also search Hidden Valley.
IN Index ages as applicable.
BT Paleozoic
SA California
SA Lower Devonian
SA Silurian

Hierro (1978)
Spanish island off Western Sahara. Westernmost of the Canary Islands.
BT Canary Islands
BT Atlantic Ocean Islands

High Atlas (1978)
A range of the Atlas Mountains containing the highest peaks in the entire mountain system. In W and S Morocco. This term has multiple hierarchies.
UF Grand Atlas Mountains
UF High Atlas Mountains
BT1 Moroccan Atlas Mountains
BT1 Atlas Mountains
BT1 North Africa
BT1 Africa
BT2 Moroccan Atlas Mountains
BT2 Morocco
BT2 North Africa
BT2 Africa

High Atlas Mountains
use High Atlas

High Bridge Group (1985)
Central Kentucky.
BT Middle Ordovician
BT Ordovician
BT Paleozoic
SA Kentucky

High Energy Benthic Boundary Layer Experiment
use HEBBLE

High Plains
A valid term through mid-1978 used to indicate the Great Plains from Nebraska southward.
use Great Plains

High Plains Aquifer (1989)
IN Index states as applicable.
BT United States
SA Colorado
SA Great Plains
SA Kansas
SA New Mexico
SA Oklahoma
SA South Dakota
SA Texas
SA Wyoming

high pressure (1978)
Before 1971, also search high-pressure research.
UF high-pressure research
BT pressure
SA diamond anvil cells
SA low pressure
SA metamorphism
SA phase equilibria
SA pore pressure

high school (1978)
SA college-level education

SA education
SA elementary school
SA junior high school

High Tatra
use Tatra Mountains

High Tatra Mountains
use Tatra Mountains

High Tatras
use Tatra Mountains

high temperature (1978)
UF high-temperature
BT temperature
SA metamorphism
SA phase equilibria

high-angle faults (1989)
BT faults
SA low-angle faults

high-energy environment (1978)
Before 1978, also search high-energy or high energy.
SA depositional environment
SA environment
SA HEBBLE
SA low-energy environment
SA sedimentation

high-grade metamorphism (1978)
Before 1978, also search high-grade AND metamorphism; high grade AND metamorphism.
BT metamorphism
SA catazonal metamorphism
SA grade
SA mesozonal metamorphism

high-level geothermal energy (1985)
SA geothermal energy

high-level mode (1981)
SA oscillations
SA seismographs
SA seismology

high-level waste (1985)
SA low-level waste
SA radioactive waste
SA underground storage
SA waste disposal

high-pressure research
use high pressure

high-resolution methods (1981)
SA geophysical methods
SA geophysical surveys
SA Geosat
SA ground-penetrating radar
SA methods
SA paleomagnetism
SA remote sensing
SA seismology
SA total-field methods
SA vertical-gradient methods

high-temperature
use high temperature

Highland Boundary Fault (1989)
IN Index regions as applicable.
SA Highland region Scotland
SA Ireland
SA North Sea
SA Scotland
SA Shetland Islands

Highland County
Valid through 1988. Search in combination with state term. After 1988, use specific county-state term.

Highland County Ohio (1989)
SW Ohio. Before 1989, also search Highland County AND Ohio.
CO N390200N392200
   W0832200W0835300
BT Ohio
BT United States

Highland County Virginia (1989)
NW Virginia. Before 1989, also search Highland County AND Virginia.
CO N381200N383600
   W0791900W0794800
BT Virginia
BT United States

Highland region Scotland (1993)
Administrative region in NW Scotland created in 1975 from Caithness, Sutherland, Ross and Cromarty excluding Lewis Island in the Outer Hebrides and Nairn counties, and Inverness-shire. Before 1993, also search Highland.
BT Scotland
BT Great Britain
BT United Kingdom
BT Western Europe
BT Europe
NT Ardnamurchan
NT Caithness Scotland
NT Inverness-shire Scotland
NT Ross-shire Scotland
NT Sutherland Scotland
SA Argyllshire Scotland
SA Cairngorm Mountains
SA Great Glen Fault
SA Highland Boundary Fault
SA Inner Hebrides

Highland Rim (1978)
IN Index counties or regions as applicable.
SA Alabama

Highland Series (1978)
BT Precambrian
SA Sri Lanka

highlands (1978)
Before 1981, search highlands AND NOT Scotland. See Highlands. Before 1993, may also include lunar highlands. Autoposting of erosion features to this term began in 1989 and ended in 1992.
SA erosion features
SA geomorphology
SA lunar highlands
SA maria
SA mesas
SA Moon
SA mountains
SA plateaus
SA uplands
SA wrinkle ridges

Highlands
As of 1981, no longer a valid term for GeoRef. Use Grampian Highlands and/or Scottish Northern Highlands. Was used for that part of Scotland lying NW of a line from Loch Lomond to just S of Aberdeen.

highways (1978)
Used for geological studies on highways and materials.
SA bridges
SA construction
SA embankments
SA engineering geology
SA engineering properties
SA feasibility studies
SA foundations
SA frost action
SA geomembranes
SA geotextiles
SA land subsidence
SA planning
SA railroads
SA reinforced materials
SA road tests
SA rock mechanics
SA seismic response
SA site exploration
SA slope stability
SA soil mechanics
SA soils
SA tunnels

Hikurangi Trench
use Hikurangi Trough

Hikurangi Trough (1989)
UF Hikurangi Trench
BT New Zealand
BT Australasia

Hilbert transformations (1989)
Relate wave phase and amplitude, Fourier analysis and time domain analysis.
UF Hilbert transforms
SA amplitude
SA analysis
SA data processing
SA elastic waves
SA equations
SA Fourier analysis
SA mathematical geology
SA mathematical methods
SA mathematical transformations
SA seismology
SA time domain analysis
SA waves

Hilbert transforms
use Hilbert transformations

Hildesheim
No longer a valid term for GeoRef. See Hildesheim Germany.

Hildesheim Germany (1993)
City in SE Lower Saxony.
BT Lower Saxony Germany
BT Germany
BT Central Europe
BT Europe

Hildoceratacea (1978)
Autoposting of this term to Hildoceratidae began in 1989.
BT Ammonoidea
BT Tetrabranchiata
BT Cephalopoda
BT Mollusca
BT Invertebrata
NT Hildoceratidae

Hildoceratidae (1978)
Autoposting of Hildoceratacea to this term began in 1989.
BT Hildoceratacea
BT Ammonoidea
BT Tetrabranchiata
BT Cephalopoda
BT Mollusca
BT Invertebrata

Hill County
Valid through 1988. Search in combination with state term. After 1988, use specific county-state term.

Hill County Montana (1989)
N Montana. Before 1989, also search Hill County AND Montana.
CO N480900N490000
   W1092800W1104600
BT Montana
BT United States

Hill County Texas (1989)
N central Texas. Before 1989, also search Hill County AND Texas.
CO N314200N321600
   W0964300W0972800
BT Texas
BT United States

Hillabee Chlorite Schist (1985)
E Alabama and Georgia. Sometimes may be considered a member of the Talladega Group.
BT middle Paleozoic
BT Paleozoic
SA Alabama
SA Georgia
SA Talladega Front
SA Talladega Group

Hilliard Shale (1989)
NE Utah and SW Wyoming. Overlies Frontier Formation; underlies Adaville Formation, Blair Formation. Sometimes replaces Blair Formation and/or intertongues with overlying Rock Springs Formation or undifferentiated Mesaverde Group.
BT Upper Cretaceous
BT Cretaceous
BT Mesozoic
SA Mesaverde Group
SA Rock Springs Formation
SA Utah
SA Wyoming

hills (1978)
SA geomorphology
SA landforms
SA mesas
SA mountains
SA topography
SA tors

Hillsborough County
Valid through 1988. Search in combination with state term. After 1988, use specific county-state term.

Hillsborough County Florida (1989)
On Gulf of Mexico in W Florida. Before 1989, also search Hillsborough County AND Florida.
CO N273700N280900
   W0820200W0823700
BT Florida
BT United States
SA Tampa Bay

Hillsborough County New Hampshire (1989)
S New Hampshire. Before 1989, also search Hillsborough County AND New Hampshire.
CO N424200N431800
   W0712300W0720500
BT New Hampshire
BT United States
NT Manchester New Hampshire
SA Merrimack River valley

Hilo
Valid through 1988. Search in combination with state term. After 1988, use specific city-state term.

Hilo Hawaii (1989)
City on E coast of island of Hawaii. Before 1989, search Hilo AND Hawaii. This term has multiple hierarchies.
CO N194200N194200
   W1550400W1550400
BT1 Hawaii Island
BT1 Hawaii County Hawaii
BT1 Hawaii
BT1 United States

BT2 Hawaii Island
BT2 Hawaii County Hawaii
BT2 Hawaii
BT2 Polynesia
BT2 Oceania
BT3 Hawaii Island
BT3 Hawaii County Hawaii
BT3 Hawaii
BT3 East Pacific Ocean Islands

Hilton Creek Fault (1989)
E California.
BT California
BT United States

Himachal Pradesh
As of 1993, no longer a valid term for GeoRef.
use Himachal Pradesh India

Himachal Pradesh India (1993)
State in the Himalayas of N India.
CO N302000N332000
E0790000E0754000
UF Himachal Pradesh
BT India
BT Indian Peninsula
BT Asia
NT Chamba India
NT Halog India
NT Kangra India
NT Kinnaur India
NT Mahasu India
NT Mandi India
NT Simla Hills
NT Simla India
SA Beas River
SA Bilaspur India
SA Jumna River
SA Kasauli Series
SA Kumaun Himalayas
SA Subathu Formation

Himalaya
use Himalayas

Himalaya Mountains
use Himalayas

Himalayan Orogeny (1981)
IN Index ages as applicable.
SA Cenozoic
SA Himalayas
SA Mesozoic
SA orogeny
SA tectonics

Himalayas (1978)
Mountain system extending from Jammu and Kashmir in the W to Assam in the E.
IN Index Xizang China and/or countries as applicable.
CO N270000N370000
E0970000E0720000
UF Himalaya
UF Himalaya Mountains
UF The Himalaya
BT Asia
NT Garhwal Himalayas
NT Kumaun Himalayas
NT Lesser Himalayas
NT Mount Everest
SA Arunachal Pradesh India
SA Bhutan
SA Himalayan Orogeny
SA India
SA Indo-Australian Plate
SA Krol Formation
SA Main Boundary Fault
SA Main Central Thrust
SA Mount Namjagbarwa
SA Mussoorie Syncline
SA Nepal
SA Sikkim India
SA Siwalik Range

Hindenburg
use Zabrze Poland

Hindenburg in Oberschlesien
use Zabrze Poland

Hindeodella (1978)
Genus. Autoposting of microfossils to this term began in 1990.
BT Conodonta
BT microfossils

Hinds County
Valid through 1988. Search in combination with state term. After 1988, use specific county-state term.

Hinds County Mississippi (1989)
W central Mississippi. Before 1989, also search Hinds County AND Mississippi.
CO N320500N323500
W0901000W0904500
BT Mississippi
BT United States

Hindu Kush (1978)
Mountain range in central Asia.
IN Index countries as applicable.
BT Asia
SA Afghanistan
SA Kohistan
SA Pakistan
SA Tadzhikistan

hinge faults (1978)
Before 1978, search faults AND hinge.
BT faults
SA listric faults

Hinsdale County
Valid through 1988. Search in combination with state term. After 1988, use specific county-state term.

Hinsdale County Colorado (1989)
SW Colorado. Before 1989, also search Hinsdale County AND Colorado.
CO N372500N381000
W1070000W1073500
BT Colorado
BT United States

hinsdalite (1978)
This term has multiple hierarchies.
BT1 phosphates
BT2 sulfates

hintzeite
use kaliborite

Hipparion (1978)
Genus. Autoposting of Eutheria and Theria to this term began in 1989.
BT Equidae
BT Hippomorpha
BT Perissodactyla
BT Eutheria
BT Theria
BT Mammalia
BT Tetrapoda
BT Vertebrata
BT Chordata

Hippomorpha (1981)
Suborder. Autoposting of Eutheria and Theria to this term began in 1989.
BT Perissodactyla
BT Eutheria
BT Theria
BT Mammalia
BT Tetrapoda
BT Vertebrata
BT Chordata
NT Equidae

Hippopotamus (1978)
Genus. Autoposting of Eutheria and Theria to this term began in 1989.
BT Artiodactyla
BT Eutheria
BT Theria
BT Mammalia
BT Tetrapoda
BT Vertebrata
BT Chordata

Hippuritacea (1978)
BT Bivalvia
BT Mollusca
BT Invertebrata

Hirnantian (1993)
Europe. Upper stage of Ashgillian. Below Llandoverian.
BT Ashgillian
BT Upper Ordovician
BT Ordovician
BT Paleozoic
SA Bala

Hiroshima
No longer a valid term for GeoRef. See Hiroshima Japan.

Hiroshima Japan (1993)
Prefecture and city on the Inland Sea in W Honshu.
BT Honshu
BT Japan
BT Far East
BT Asia
NT Otake Japan
SA Chugoku
SA Dogo

hisingerite (1978)
BT sheet silicates
BT silicates
SA clay minerals

Hispaniola (1978)
Island of the Greater Antilles between Cuba and Puerto Rico. Autoposting of this term began in 1978.
IN Index countries as applicable.
BT Greater Antilles
BT Antilles
BT West Indies
BT Caribbean region
NT Dominican Republic
NT Haiti

Hissar
No longer a valid term for GeoRef. See Hissar India.

Hissar India (1993)
City WNW of Delhi in W Haryana, N India.
BT Haryana India
BT India
BT Indian Peninsula
BT Asia

Hissar Range (1978)
A branch of the Alai Range in NW Tadzhikistan. Also search Hissar. This term has multiple hierarchies.
CO N380000N381000
E0700000E0673000
UF Gissar Range
BT1 Tadzhikistan
BT1 Asia
BT2 Tadzhikistan
BT2 Commonwealth of Independent States
BT3 Alai Range
BT3 Tien Shan
BT3 Asia
SA Zeravshan-Hissar

histograms (1978)
BT statistical analysis

SA graphic methods
SA kurtosis

histology (1978)
SA fossils

historical (stratigraphic) geology
No longer a valid GeoRef index term. Before 1971, was used on level 1 in subfiles E and N. See stratigraphy; historical geology.

historical geology (1978)
IN See list E (age terms).
SA biostratigraphy
SA education
SA geology
SA paleoenvironment
SA paleontology
SA physical geology
SA stratigraphy

history (1978)
Use term under various disciplines.
SA survey organizations
SA uniformitarianism

Histosols (1978)
BT soils
SA Bog soils
SA Hydromorphic soils
SA peat
SA soil group

Hitachi
No longer a valid term for GeoRef. See Hitachi Japan.

Hitachi Deposit (1993)
Polymetallic deposits including copper and iron deposits, Ibaraki, E central Honshu. Also search Hitachi Mine.
IN Index commodity as applicable.
BT Ibaraki Japan
BT Honshu
BT Japan
BT Far East
BT Asia
SA copper ores
SA Hitachi Japan
SA iron ores
SA polymetallic ores

Hitachi Japan (1993)
City on the Pacific Ocean in E Honshu, E Japan. Hitachi has also been a copper mining area.
BT Ibaraki Japan
BT Honshu
BT Japan
BT Far East
BT Asia
SA Hitachi Deposit

Hitchcock County
Valid through 1988. Search in combination with state term. After 1988, use specific county-state term.

Hitchcock County Nebraska (1989)
S Nebraska. Before 1989, also search Hitchcock County AND Nebraska.
CO N400000N402000
W1004500W1012000
BT Nebraska
BT United States

Hiyoshi
No longer a valid term for GeoRef. See Hiyoshi Japan.

Hiyoshi Japan (1993)
An area of Mizunami city in Gifu Prefecture in central Honshu, central Japan. Mining locality. Before

1993, also search Hiyoshi AND Japan.
BT Gifu Japan
BT Honshu
BT Japan
BT Far East
BT Asia
SA Mizunami Japan

HL chondrites (1981)
UF HL-group chondrites
BT chondrites
BT stony meteorites
BT meteorites

HL-group chondrites
use HL chondrites

Ho
use holmium

Hobart
No longer a valid term for GeoRef. See Hobart Australia.

Hobart Australia (1993)
Coastal city in SE Tasmania.
BT Tasmania Australia
BT Australia
BT Australasia

Hobdo
use Kobdo Mongolia

Hobsogol
use Khubsugul Mongolia

Hochschwab (1985)
Mountain group in N Styria, Austria. This term has multiple hierarchies.
CO N473000N474000 E0152000E0145000
BT1 Alps
BT1 Europe
BT2 Austria
BT2 Central Europe
BT2 Europe
SA North Austrian Alps

Hockley County
Valid through 1988. Search in combination with state term. After 1988, use specific county-state term.

Hockley County Texas (1989)
NW Texas. Before 1989, also search Hockley County AND Texas.
CO N332000N335200 W1020500W1023800
BT Texas
BT United States

Hodgeman County
Valid through 1988. Search in combination with state term. After 1988, use specific county-state term.

Hodgeman County Kansas (1989)
SW central Kansas. Before 1989, also search Hodgeman County AND Kansas.
CO N375500N381500 W0993700W1001500
BT Kansas
BT United States

Hodna Basin (1978)
Interior drainage basin in NE Algeria.
BT Algeria
BT North Africa
BT Africa
SA Constantine Algeria

hodographs
use traveltime curves

Hodrusa (1978)
Ore bearing region.
BT Slovakia
BT Czechoslovakia
BT Central Europe
BT Europe

hoegbomite (1993)
Before 1993, also search hogbomite.
UF hogbomite
BT oxides
SA boehmite
SA iron oxides

hogbomite
use hoegbomite

Hoggar Mountains
use Ahaggar

Hoghiz
No longer a valid term for GeoRef. As of 1993 see Hoghiz Romania.

Hoghiz Romania (1993)
Village in Brasov County in SE Transylvania. Before 1993 also search Hoghiz AND Romania
BT Brasov Romania
BT Transylvania
BT Romania
BT Southern Europe
BT Europe

Hohe Tauern (1978)
Range of Eastern Alps in SW Austria. This term has multiple hierarchies.
CO N470000N471500 E0131500E0120000
BT1 Eastern Alps
BT1 Alps
BT1 Europe
BT2 Austria
BT2 Central Europe
BT2 Europe
SA Penninic Zone
SA Salzburg State Austria
SA Tauern Tunnel
SA Tauern Window

Hohe Venn (1978)
Range of low mountains in Liege, E Belgium.
BT Liege Belgium
BT Belgium
BT Western Europe
BT Europe
SA Stavelot-Venn Massif

Hokkaido (1978)
Northernmost of the four main islands.
CO N412500N453000 E1455500E1394000
BT Japan
BT Far East
BT Asia
NT Abashiri Japan
NT Esan Cape
NT Hakodate Japan
NT Hidaka metamorphic belt
NT Hidaka Mountains
NT Ikushumbetsu
NT Ishikari Bay
NT Ishikari Plain
NT Kamikawa Japan
NT Kamuikotan Belt
NT Kitami Basin
NT Kushiro coal field
NT Nemuro Peninsula
NT Okushiri Island
NT Oshima Peninsula
NT Shikotsu
NT Sorachi
NT Tarumae
NT Tokachi
NT Tokachi Plain
NT Toyoha Mine
NT Usu
SA Green Tuff Formation
SA Green tuff region
SA Hidaka Japan
SA Honshu Arc
SA Koma-ga-take
SA Mitsuishi Japan
SA Tokachi-Oki earthquake 1968
SA Tsugaru Strait

Hoko Gunto
use Penghu Islands

Hoko Shoto
use Penghu Islands

Hokuriku (1978)
Coastal district of the Chubu region, Noto Peninsula, central Honshu, Japan. Region of volcanic activity in Miocene.
BT Honshu
BT Japan
BT Far East
BT Asia
SA Atotsugawa Fault
SA Chubu Japan
SA Noto Peninsula

Hokuroku
No longer a valid term for GeoRef. See Hokuroku Japan.

Hokuroku Japan (1993)
District in Akita Prefecture, N Honshu, Japan.
BT Akita Japan
BT Honshu
BT Japan
BT Far East
BT Asia
NT Fukazawa Mine
SA Hanaoka Mine
SA Shakanai Mine

Holbrook
Valid through 1988. Search in combination with state term. After 1988, use specific city-state term.

Holbrook Arizona (1989)
Town in E central Arizona. Before 1989, search Holbrook AND Arizona.
CO N345400N345400 W1101100W1101100
BT Navajo County Arizona
BT Arizona
BT United States

Holderness (1978)
Peninsula in Humberside County in NE England. Before 1974, it was in Yorkshire County.
IN Index Humberside England or regions as applicable.
SA Humberside England
SA Yorkshire England

Holland
use Netherlands

hollandite (1978)
BT oxides
SA coronadite
SA manganese minerals

Hollister
Valid through 1988. Search in combination with state term. After 1988, use specific city-state term.

Hollister California (1989)
City in W central California. Before 1989, search Hollister AND California.
CO N364900N364900 W1212500W1212500
BT San Benito County California
BT California
BT United States

holmium (1978)
Autoposting of rare earths and metals to this term began in 1989.
UF Ho
BT rare earths
BT metals

holmquistite (1978)
BT orthoamphibole
BT amphibole group
BT chain silicates
BT silicates

Holocene (1978)
Recent. Late Quaternary, from Pleistocene to present time. Autoposting of this term began in 1978.
UF Post-glacial
UF Postglacial
UF Recent
BT Quaternary
BT Cenozoic
NT Atlantic
NT Boreal
NT Flandrian
NT Hat Creek Basalt
NT lower Holocene
NT Mesolithic
NT middle Holocene
NT Neoglacial
NT Neolithic
NT Preboreal
NT Subboreal
NT upper Holocene
NT Versilian
SA Bronze Age
SA Chalcolithic
SA Iron Age
SA Lake Bonneville
SA modern
SA Yakataga Formation

Holocephali (1981)
Subclass.
BT Chondrichthyes
BT Pisces
BT Vertebrata
BT Chordata

holography (1978)
Used as a general term in crystallography and geophysics.
SA crystallography
SA data processing

Holostei (1978)
Infraclass. Autoposting of this term began in 1981.
BT Actinopterygii
BT Osteichthyes
BT Pisces
BT Vertebrata
BT Chordata
NT Amiidae
NT Halecostomi

Holothuroidea (1978)
BT Echinozoa
BT Echinodermata
BT Invertebrata
SA sclerites

holotypes (1981)
SA fossils
SA paleobotany
SA paleontology
SA paratypes
SA taxonomy
SA type specimens

Holstein (1978)
Region in N Germany.
BT Schleswig-Holstein Germany
BT Germany
BT Central Europe
BT Europe

Holsteinian (1978)
  Europe.
  BT upper Pleistocene
  BT Pleistocene
  BT Quaternary
  BT Cenozoic

Holston Formation (1985)
  E Tennessee, NW Georgia, W North Carolina and W Virginia.
  BT Middle Ordovician
  BT Ordovician
  BT Paleozoic
  SA Georgia
  SA North Carolina
  SA Tennessee
  SA Virginia

Holy Cross Mountains
  use Swiety Krzyz Mountains

Holyoke Basalt (1989)
  In Newark Group. Central Connecticut and central Massachusetts. Previously thought to be Upper Triassic in age.
  BT Lower Jurassic
  BT Jurassic
  BT Mesozoic
  SA Connecticut
  SA Massachusetts
  SA Newark Supergroup

Holz Shale (1989)
  In Ladd Formation. S California.
  BT Upper Cretaceous
  BT Cretaceous
  BT Mesozoic
  SA California
  SA Ladd Formation

Holzmaden region (1985)
  E Baden-Wurttemberg, West Germany.
  BT Baden-Wurttemberg Germany
  BT Germany
  BT Central Europe
  BT Europe
  SA Southwestern German Hills
  SA Swabian Alb

Homalozoa (1978)
  Autoposting of this term to Homoiostelea, Homostelea, Machaeridia and Stylophora began in 1981.
  BT Echinodermata
  BT Invertebrata
  NT Homoiostelea
  NT Homostelea
  NT Machaeridia
  NT Stylophora

Homestake Mine (1978)
  Largest gold mine in U. S., located in Lead, Lawrence County, W South Dakota. Also gold ores in Napa County, California; Gunnison County, Colorado, and San Juan County, Utah.
  IN Index states and counties as applicable.
  SA California
  SA Colorado
  SA gold ores
  SA Gunnison County Colorado
  SA Lawrence County South Dakota
  SA mines
  SA Napa County California
  SA San Juan County Utah
  SA South Dakota
  SA United States
  SA Utah

Hominidae (1978)
  Family. Autoposting of this term to Australopithecinae began in 1989. Autoposting of Eutheria and Theria to this term began in 1989. Considered a narrower term of Anthropoidea as of 1993.
  BT Primates
  BT Eutheria
  BT Theria
  BT Mammalia
  BT Tetrapoda
  BT Vertebrata
  BT Chordata
  NT Australopithecinae
  NT Homo
  NT Paranthropus
  NT Pithecanthropus
  NT pre-Neanderthal
  NT Ramapithecus

Homo (1978)
  Genus. Autoposting of Eutheria and Theria to this term began in 1989. As of 1993, Neanderthal is a narrower term.
  BT Hominidae
  BT Primates
  BT Eutheria
  BT Theria
  BT Mammalia
  BT Tetrapoda
  BT Vertebrata
  BT Chordata
  NT Homo erectus
  NT Homo habilis
  NT Homo sapiens
  NT Neanderthal

Homo erectus (1978)
  Species. Autoposting of Eutheria and Theria to this term began in 1989.
  BT Homo
  BT Hominidae
  BT Primates
  BT Eutheria
  BT Theria
  BT Mammalia
  BT Tetrapoda
  BT Vertebrata
  BT Chordata

Homo habilis (1978)
  Species. Autoposting of Eutheria and Theria to this term began in 1989.
  BT Homo
  BT Hominidae
  BT Primates
  BT Eutheria
  BT Theria
  BT Mammalia
  BT Tetrapoda
  BT Vertebrata
  BT Chordata

Homo neanderthalensis
  use Neanderthal

Homo sapiens (1978)
  Species. Autoposting of Eutheria and Theria to this term began in 1989.
  BT Homo
  BT Hominidae
  BT Primates
  BT Eutheria
  BT Theria
  BT Mammalia
  BT Tetrapoda
  BT Vertebrata
  BT Chordata

homogeneity (1978)
  SA geophysical methods
  SA heterogeneity
  SA homogeneous materials
  SA seismology

homogeneous materials (1981)
  SA heterogeneous materials
  SA homogeneity

homogenization (1978)
  Used as a general term.
  SA petrology

Homoiostelea (1978)
  BT Homalozoa
  BT Echinodermata
  BT Invertebrata

homonymy (1978)
  SA nomenclature
  SA taxonomy

Homostelea (1978)
  BT Homalozoa
  BT Echinodermata
  BT Invertebrata

Honaker Trail Formation (1978)
  Pennsylvanian and Permian. In Hermosa Formation. SE Utah.
  IN Index ages as applicable.
  BT Paleozoic
  SA Hermosa Formation
  SA Pennsylvanian
  SA Permian
  SA Utah

Honan China
  use Henan China

Honduras (1978)
  CO N130000N160000
    W0831500W0891500
  BT Central America

Honey Hill Fault (1989)
  E Connecticut.
  BT New London County Connecticut
  BT Connecticut
  BT United States

Hong Kong (1978)
  British crown colony which includes the island of Hong Kong; and Kowloon Peninsula and New Territories on mainland.
  CO N220500N223500
    E1143000E1135000
  BT Far East
  BT Asia

Honghe Fault
  use Red River Fault

Honolulu
  Valid through 1988. Search in combination with state term. After 1988, use specific city-state term.

Honolulu County
  Valid through 1988. Search in combination with state term. After 1988, use specific county-state term.

Honolulu County Hawaii (1989)
  Coterminous with island of Oahu. Before 1989, also search Honolulu County AND Hawaii. This term has multiple hierarchies.
  CO N211500N214300
    W1573900W1581800
  BT1 Hawaii
  BT1 United States
  BT2 Hawaii
  BT2 Polynesia
  BT2 Oceania
  BT3 Hawaii
  BT3 East Pacific Ocean Islands
  NT Oahu

Honolulu Hawaii (1989)
  City on island of Oahu. Before 1989, search Honolulu AND Hawaii. This term has multiple hierarchies.
  CO N211900N211900
    W1575000W1575000
  BT1 Oahu
  BT1 Honolulu County Hawaii
  BT1 Hawaii
  BT1 United States
  BT2 Oahu
  BT2 Honolulu County Hawaii
  BT2 Hawaii
  BT2 Polynesia
  BT2 Oceania
  BT3 Oahu
  BT3 Honolulu County Hawaii
  BT3 Hawaii
  BT3 East Pacific Ocean Islands

Honshu (1978)
  Largest of the Japanese islands, with adjacent small islands.
  CO N332000N413000
    E1420000E1305000
  UF Honshu Island
  BT Japan
  BT Far East
  BT Asia
  NT Abukuma Mountains
  NT Aichi Japan
  NT Akita Japan
  NT Akiyoshi
  NT Aomori Japan
  NT Asama
  NT Ashio Japan
  NT Ashio Mine
  NT Atotsugawa Fault
  NT Awashima
  NT Chiba Japan
  NT Chiba Peninsula
  NT Chokai
  NT Chubu Japan
  NT Chugoku
  NT Dogo
  NT Fossa Magna
  NT Fuji River
  NT Fujiyama
  NT Fukui Japan
  NT Fukushima Japan
  NT Gifu Japan
  NT Gumma Japan
  NT Hakusan
  NT Hida metamorphic belt
  NT Hida Mountains
  NT Hiroshima Japan
  NT Hokuriku
  NT Hyogo Japan
  NT Ibaraki Japan
  NT Ishikawa Japan
  NT Itoigawa-Shizuoka tectonic line
  NT Iwate
  NT Iwate Japan
  NT Izu Peninsula
  NT Izu-shichito
  NT Japanese Alps
  NT Joban coal field
  NT Kanagawa Japan
  NT Kanto Mountains
  NT Kanto Plain
  NT Kii Peninsula
  NT Kinki Japan
  NT Kiso Mountains
  NT Kyoto Japan
  NT Maizuru Belt
  NT Matsushiro Japan
  NT Mie Japan
  NT Miho Bay
  NT Mino Belt
  NT Miura Peninsula
  NT Miyagi Japan
  NT Mount Hakone
  NT Nagano City Japan
  NT Nagano Japan
  NT Naka-no-umi
  NT Nara Japan
  NT Narugo Japan
  NT Nasudake
  NT Niigata Japan
  NT Nohi Rhyolite
  NT Noto Peninsula
  NT Okayama Japan

NT Ontake
NT Osaka Japan
NT Ryoke Belt
NT Sado Island
NT Sagami Bay
NT Saitama Japan
NT Sangun Zone
NT Shiga Japan
NT Shimane Japan
NT Shirasu River
NT Shizuoka Japan
NT Suruga Bay
NT Takanuki Japan
NT Tamagawa
NT Tamba Plateau
NT Tanzawa Mountains
NT Tochigi Japan
NT Tohoku
NT Tokyo Bay
NT Tokyo Japan
NT Tono
NT Tottori Japan
NT Towada
NT Toyama Japan
NT Toyoma Japan
NT Wakayama Japan
NT Yahagi River
NT Yamagata Japan
NT Yamaguchi Japan
NT Yamanashi Japan
NT Yamasaki Fault
NT Yatsugatake
NT Zao
SA Aizu Basin
SA Akaishi Mountains
SA Akiyoshi Limestone
SA Chibaken-Toho-Oki earthquake 1987
SA DSDP Site 584
SA Futaba Group
SA Green Tuff Formation
SA Green tuff region
SA Hakone
SA Hidaka Japan
SA Honshu Arc
SA Inland Sea
SA Ise Bay
SA Izu-Oshima earthquake 1978
SA Izumi Group
SA Japan Sea earthquake 1983
SA Kanto earthquake 1923
SA Kazusa Group
SA Kobiwako Group
SA Koma-ga-take
SA Median Tectonic Line
SA Mikabu System
SA Mitsuishi Japan
SA Miyagi earthquake 1978
SA Mizunami Group
SA Motojuku Formation
SA Nagano earthquake 1984
SA Narita Formation
SA Niigata earthquake 1964
SA Osaka Bay
SA Osaka Group
SA Rokko Mountains
SA Ryujima
SA Sambagawa Belt
SA Shimanto Belt
SA Tsugaru Strait
SA Uonuma Group
SA Usuginu Conglomerate
SA Yake-Dake

Honshu Arc (1993)
 Central Japan.
 BT Japan
 BT Far East
 BT Asia
 SA Hokkaido
 SA Honshu

Honshu Island
 use Honshu

Hood River County
 Valid through 1988. Search in combination with state term. After 1988, use specific county-state term.

Hood River County Oregon
 (1989)
 On Columbia River in NE Oregon. Before 1989, also search Hood River County AND Oregon.
 CO N451500N454000 W1212500W1215500
 BT Oregon
 BT United States
 SA Mount Hood

Hooke's law (1978)
 SA deformation
 SA elastic limit
 SA elastic properties
 SA elastic strain
 SA strain
 SA stress

Hoosac Formation (1993)
 Lower Cambrian and upper Proterozoic. W Massachusetts, W Connecticut and S Vermont. Schist.
 IN Index ages as applicable.
 UF Hoosac Schist
 SA Cambrian
 SA Connecticut
 SA Massachusetts
 SA Proterozoic
 SA Vermont

Hoosac Schist
 use Hoosac Formation

hopanes (1985)
 From 1985 to 1989, alkanes and aliphatic hydrocarbons were autoposted to this term.
 BT hydrocarbons
 BT organic materials
 SA saturated hydrocarbons

Hope
 No longer a valid term for GeoRef. See Hope British Columbia.

Hope British Columbia (1993)
 Town in SW British Columbia.
 BT British Columbia
 BT Western Canada
 BT Canada

Hopei China
 use Hebei China

Hopi Buttes Field (1978)
 Volcanic field in Hopi and Navajo Indian reservations in Navajo County.
 BT Navajo County Arizona
 BT Arizona
 BT United States

horizon differentiation (1978)
 SA differentiation
 SA horizons
 SA morphology
 SA pedogenesis
 SA soils

horizons (1978)
 UF soil horizon
 UF soil zone
 SA horizon differentiation
 SA morphology
 SA parent materials
 SA podzolization
 SA soils

horizontal
 A valid term through 1977. After 1977, use horizontal orientation.

horizontal discontinuities (1981)
 Used for a type of interface in seismology.
 BT discontinuities
 SA crust
 SA mantle
 SA seismology

horizontal drilling (1993)
 Before 1993, also search horizontal wells if applicable.
 BT drilling
 SA boreholes
 SA mineral exploration
 SA mining geology
 SA petroleum engineering
 SA pipelines
 SA well-logging
 SA wells

horizontal movements (1981)
 SA neotectonics
 SA tectonics
 SA vertical movements

horizontal orientation (1978)
 Reference to folds defined on the basis of orientation in relation to the geographic horizontal plane. Before 1978, also search folds AND horizontal.
 SA folds
 SA orientation

horizontal wells
 As of 1993, use horizontal drilling.

horizontal-component seismographs (1981)
 BT seismographs

Horlick Mountains (1978)
 SE of S tip of Ross Ice Shelf.
 BT Antarctica

Horn Plateau Formation (1978)
 S Mackenzie District.
 BT Givetian
 BT Middle Devonian
 BT Devonian
 BT Paleozoic
 SA Mackenzie District Northwest Territories
 SA Northwest Territories

hornblende (1978)
 BT clinoamphibole
 BT amphibole group
 BT chain silicates
 BT silicates
 SA kaersutite

hornblende gneiss
 A valid metamorphic rock term through 1977. Use gneisses and hornblende.

hornblende schist (1978)
 Term reintroduced in 1982. Was valid through 1977. Before 1982, also search hornblende AND schist; hornblende AND schists.
 BT schists
 BT metamorphic rocks

hornblendite (1978)
 BT ultramafics
 BT plutonic rocks
 BT igneous rocks

Hornbrook Formation (1989)
 N California and SW Oregon.
 BT Upper Cretaceous
 BT Cretaceous
 BT Mesozoic
 SA California
 SA Oregon

Horne Mine (1978)
 Copper and gold ores. Noranda area in SW Quebec.
 BT Quebec
 BT Eastern Canada
 BT Canada
 SA copper ores
 SA gold ores
 SA mines
 SA Noranda Quebec

hornfels (1978)
 BT metamorphic rocks
 SA hornfels facies

hornfels facies (1982)
 BT facies
 SA contact metamorphism
 SA hornfels
 SA metamorphic rocks

hornstein
 use chert

hornstone
 use chert

Hornsund (1978)
 Inlet on SW coast of Spitsbergen Island.
 BT Spitsbergen Island
 BT Spitsbergen
 BT Svalbard
 BT Arctic region

Horry County
 Valid through 1988. Search in combination with state term. After 1988, use specific county-state term.

Horry County South Carolina (1989)
 Extreme E South Carolina. Before 1989, also search Horry County AND South Carolina.
 CO N333500N342000 W0783500W0792300
 BT South Carolina
 BT United States

horseflesh ore
 use bornite

Horseshoe Canyon Formation (1989)
 BT Upper Cretaceous
 BT Cretaceous
 BT Mesozoic
 SA Alberta

Horsethief Creek Group (1989)
 E British Columbia.
 BT upper Proterozoic
 BT Proterozoic
 BT upper Precambrian
 BT Precambrian
 SA British Columbia

Horsham
 No longer a valid term for GeoRef. As of 1993, see Horsham Australia or Horsham England.

Horsham Australia (1993)
 Town in W Victoria. Before 1993, also search Horsham AND Australia.
 BT Victoria Australia
 BT Australia
 BT Australasia

Horsham England (1993)
 City in W Sussex, S England. As of 1993, also search Horsham AND England.
 BT Sussex England
 BT England
 BT Great Britain
 BT United Kingdom
 BT Western Europe
 BT Europe

horsts (1978)
 SA block structures
 SA faults

SA grabens
SA half grabens
SA systems

Horton Group (1989)
Devonian to Dinantian.
IN Index ages as applicable.
BT Paleozoic
SA Devonian
SA Dinantian
SA New Brunswick
SA Nova Scotia

Hosgri Fault (1989)
Santa Maria Basin, SW California. Component of San Andreas fault system.
IN Index counties as applicable.
BT California
BT United States
SA San Andreas Fault
SA San Gregorio Fault
SA San Luis Obispo County California
SA Santa Barbara County California
SA Santa Maria Basin

Hosokura Mine (1993)
Lead-zinc deposits, Miyagi, N Honshu.
BT Miyagi Japan
BT Honshu
BT Japan
BT Far East
BT Asia
SA lead-zinc deposits
SA mines

Hosston Formation (1985)
Louisiana, Arkansas, and E and S Texas.
BT Lower Cretaceous
BT Cretaceous
BT Mesozoic
SA Arkansas
SA Louisiana
SA Texas

host materials
As of 1981, no longer a valid term for GeoRef. Use host rocks or country rocks.

host rocks (1978)
Before 1981, also search host materials. Includes use for rock serving as a host for mineral deposits.
SA country rocks
SA fluid inclusions
SA inclusions
SA mineral deposits, genesis
SA mineral inclusions
SA rocks
SA veins
SA wall rocks
SA xenoliths

hot dry rocks (1981)
Before 1981, also search dry hot rocks. Before 1993, also search HDR.
UF dry hot rocks
SA geothermal energy
SA geothermal fields
SA geothermal systems

hot spots (1978)
SA magmas
SA mantle
SA plate tectonics
SA plumes
SA thermal anomalies
SA volcanoes

Hot Spring County
Valid through 1988. Search in combination with state term. After 1988, use specific county-state term.

Hot Spring County Arkansas (1989)
Central Arkansas. Before 1989, also search Hot Spring County AND Arkansas.
CO N340800N343000 W0924000W0932500
BT Arkansas
BT United States

hot springs (1978)
Before 1972, also search thermal springs.
UF thermal springs
BT springs
SA geothermal energy
SA geysers
SA ground water
SA heat flow
SA hydrothermal processes
SA thermal anomalies
SA thermal waters

Hot Springs County
Valid through 1988. Search in combination with state term. After 1988, use specific county-state term.

Hot Springs County Wyoming (1989)
N central Wyoming. Before 1989, also search Hot Springs County AND Wyoming.
CO N432500N440800 W1073500W1092000
BT Wyoming
BT United States
SA Bighorn River
SA Owl Creek Mountains

Houghton County
Valid through 1988. Search in combination with state term. After 1988, use specific county-state term.

Houghton County Michigan (1989)
NW Michigan Upper Peninsula. NW Michigan. Before 1989, also search Houghton County AND Michigan.
CO N462300N472000 W0881500W0890000
BT Michigan Upper Peninsula
BT Michigan
BT United States

House Range (1978)
Millard County in W Utah.
BT Millard County Utah
BT Utah
BT United States

Houston
Valid through 1988. Search in combination with state term. After 1988, use specific city-state term.

Houston County
Valid through 1988. Search in combination with state term. After 1988, use specific county-state term.

Houston County Alabama (1989)
Extreme SE Alabama. Before 1989, also search Houston County AND Alabama.
CO N310000N311700 W0850000W0854300
BT Alabama

BT United States

Houston County Georgia (1989)
Central Georgia. Before 1989, also search Houston County AND Georgia.
CO N321700N324000 W0832700W0835000
BT Georgia
BT United States

Houston County Minnesota (1989)
Extreme SE Minnesota. Before 1989, also search Houston County AND Minnesota.
CO N433200N435000 W0911400W0914300
BT Minnesota
BT United States
SA Root River

Houston County Tennessee (1989)
NW Tennessee. Before 1989, also search Houston County AND Tennessee.
CO N361100N362200 W0873200W0875700
BT Tennessee
BT United States

Houston County Texas (1989)
E Texas. Before 1989, also search Houston County AND Texas.
CO N305600N313300 W0945600W0954600
BT Texas
BT United States

Houston Texas (1989)
City in E Texas. Before 1989, search Houston AND Texas.
CO N294500N294500 W0952500W0952500
BT Harris County Texas
BT Texas
BT United States

Howard County
Valid through 1988. Search in combination with state term. After 1988, use specific county-state term.

Howard County Arkansas (1989)
SW Arkansas. Before 1989, also search Howard County AND Arkansas.
CO N334500N342400 W0935100W0941700
BT Arkansas
BT United States

Howard County Indiana (1989)
Central Indiana. Before 1989, also search Howard County AND Indiana.
CO N402200N403400 W0855200W0862300
BT Indiana
BT United States

Howard County Iowa (1989)
NE Iowa. Before 1989, also search Howard County AND Iowa.
CO N431400N433000 W0920600W0923400
BT Iowa
BT United States

Howard County Maryland (1989)
Central Maryland. Before 1989, also search Howard County AND Maryland.
CO N390700N392200 W0764200W0771200

BT Maryland
BT United States

Howard County Missouri (1989)
Central Missouri. Before 1989, also search Howard County AND Missouri.
CO N385800N392200 W0922500W0925500
BT Missouri
BT United States

Howard County Nebraska (1989)
E central Nebraska. Before 1989, also search Howard County AND Nebraska.
CO N410300N412300 W0981700W0984400
BT Nebraska
BT United States

Howard County Texas (1989)
NW Texas. Before 1989, also search Howard County AND Texas.
CO N320500N323200 W1011000W1014100
BT Texas
BT United States

howardite (1981)
Before 1981, also search howardites.
UF howardites
BT achondrites
BT stony meteorites
BT meteorites
NT Kapoeta Meteorite
NT Malvern Meteorite

howardites
As of 1981, no longer a valid term for GeoRef.
use howardite

Howgill Fells (1978)
Fossil locality. N England.
IN Index counties as applicable.
BT England
BT Great Britain
BT United Kingdom
BT Western Europe
BT Europe
SA Cumbria England
SA North Yorkshire England

howieite (1978)
BT chain silicates
BT silicates

Hoxnian (1978)
Europe.
BT upper Pleistocene
BT Pleistocene
BT Quaternary
BT Cenozoic

Hsinchu (1978)
Region in N Taiwan.
BT Taiwan
BT Far East
BT Asia

Hsingtai earthquake 1966
use Xingtai earthquake 1966

Hualalai (1989)
Volcano on island of Hawaii. This term has multiple hierarchies.
BT1 Hawaii Island
BT1 Hawaii County Hawaii
BT1 Hawaii
BT1 United States
BT2 Hawaii Island
BT2 Hawaii County Hawaii
BT2 Hawaii
BT2 Polynesia
BT2 Oceania
BT3 Hawaii Island
BT3 Hawaii County Hawaii

BT3 Hawaii
BT3 East Pacific Ocean Islands

Huang Hai
use Yellow Sea

Huang He
use Huang Ho

Huang Ho (1985)
River. Source in E Qinghai. Flows into Bohai Bay on E coast of China. Also search Yellow River AND China.
UF Huang He
UF Huanghe River
UF Hwang Ho
UF Yellow River, China
BT China
BT Far East
BT Asia
SA Gansu China
SA Hebei China
SA Henan China
SA Shaanxi China
SA Shandong China
SA Shanxi China

Huang-hua Depression
use Huanghua Depression

Huanghai Sea
use Yellow Sea

Huanghe River
use Huang Ho

Huanghua Depression (1989)
E China, in the Bohai Bay region.
UF Huang-hua Depression
SA Bohai Bay
SA China

Huanglong Formation (1989)
E and SW China.
BT Carboniferous
BT Paleozoic
SA China

Hubei
No longer a valid term for GeoRef. See Hubei China.

Hubei China (1993)
Province, South-Central China. Also search Hupei or Hupeh.
CO N290000N332000
   E1160000E1083000
UF Hupeh China
UF Hupei China
BT China
BT Far East
BT Asia
SA Chihsia Formation
SA Dabie Mountains
SA Doushantuo Formation
SA Han River basin
SA Jianghan Basin
SA Tancheng-Lujiang fault zone
SA Wufeng Formation
SA Yangtze Platform

Huber Formation (1989)
Paleocene to Eocene.
IN Index ages as applicable.
BT Paleogene
BT Tertiary
BT Cenozoic
SA Eocene
SA Georgia
SA Paleocene
SA South Carolina

Hubusugul
use Khubsugul Mongolia

Hudson Bay (1978)
Administratively a part of Keewatin District. Atlantic Ocean is considered a broader term as of 1981. This term has multiple hierarchies.
CO N550000N650000
   W0760000W0950000
BT1 North American Atlantic
BT1 North Atlantic
BT1 Atlantic Ocean
BT2 Canada
SA Belcher Islands
SA Hudson Bay Lowlands
SA James Bay
SA Keewatin District Northwest Territories
SA Northwest Territories

Hudson Bay Lowlands (1978)
IN Index Northwest Territories and/or provinces as applicable.
BT Canada
SA Hudson Bay
SA Manitoba
SA Northwest Territories
SA Ontario
SA Quebec

Hudson Canyon (1989)
Submarine canyon off the coast of New Jersey.
IN Index North American Atlantic as applicable.
SA DSDP Site 604
SA DSDP Site 605
SA Leg 93
SA North American Atlantic

Hudson County
Valid through 1988. Search in combination with state term. After 1988, use specific county-state term.

Hudson County New Jersey (1989)
NE New Jersey. Before 1989, also search Hudson County AND New Jersey.
CO N403800N404800
   W0735900W0741000
BT New Jersey
BT United States

Hudson Highlands (1978)
Part of Appalachians lying along both banks of the Hudson River S of Newburgh.
IN Index counties and Appalachians as applicable.
SA Appalachians
SA New York

Hudson River (1978)
In New York and New Jersey.
IN Index counties or states as applicable.
SA Hudson Valley
SA New Jersey
SA New York
SA United States

Hudson Valley (1978)
River valley primarily in New York.
IN Index states as applicable.
SA Great Appalachian Valley
SA Hudson River
SA New Jersey
SA New York
SA United States

Hudsonian
use Hudsonian Orogeny

Hudsonian Orogeny (1978)
Name proposed for a time of plutonism, metamorphism, and deformation during the Precambrian in the Canadian Shield.
UF Hudsonian
UF Trans-Hudsonian Orogeny
BT Precambrian
SA Canada
SA orogeny
SA Wyoming Province

Hudspeth County
Valid through 1988. Search in combination with state term. After 1988, use specific county-state term.

Hudspeth County Texas (1989)
W Texas. Before 1989, also search Hudspeth County AND Texas.
CO N303700N320000
   W1045500W1060000
BT Texas
BT United States
NT Quitman Mountains
SA Hueco Mountains
SA Orogrande Basin

Hueco Limestone (1989)
Wolfcampian. S New Mexico and W Texas.
BT Wolfcampian
BT Lower Permian
BT Permian
BT Paleozoic
SA New Mexico
SA Texas

Hueco Mountains (1989)
Hudspeth County, W Texas; Otero County, S New Mexico.
IN Index states and counties as applicable.
SA Hudspeth County Texas
SA New Mexico
SA Otero County New Mexico
SA Texas
SA United States

Huehuetenango
No longer a valid term for GeoRef. See Huehuetenango Guatemala.

Huehuetenango Guatemala (1993)
Department including the mining town in Guatemala.
BT Guatemala
BT Central America

Huelva
No longer a valid term for GeoRef. As of 1993, see Huelva City Spain.

Huelva City Spain (1993)
Refers to only the city in SW Spain on the Gulf of Cadiz. Before 1993, also search Huelva and Spain. For the province, see Huelva Spain.
BT Huelva Spain
BT Andalusia Spain
BT Spain
BT Iberian Peninsula
BT Southern Europe
BT Europe

Huelva Province
No longer a valid term for GeoRef. As of 1993, see Huelva Spain.

Huelva Spain (1993)
Refers only to the province. From 1981-1992, also search Huelva Province and Spain. Before 1981, also search Huelva and Spain. In SW Spain. For the city, see Huelva City Spain.
CO N364800N381000
   W0060800W0073200
BT Andalusia Spain
BT Spain
BT Iberian Peninsula
BT Southern Europe
BT Europe
NT Aracena Spain
NT Huelva City Spain
NT Puebla de Guzman Spain
NT Rio Tinto Spain

Huerfano County
Valid through 1988. Search in combination with state term. After 1988, use specific county-state term.

Huerfano County Colorado (1989)
S central Colorado. Before 1989, also search Huerfano County AND Colorado.
CO N372000N380500
   W1042000W1053000
BT Colorado
BT United States
SA Spanish Peaks
SA Wet Mountains

Huesca
No longer a valid term for GeoRef. As of 1993, see Huesca City Spain.

Huesca City Spain (1993)
Refers to only the city in NE Spain. Before 1993, also search Huesca and Spain.
BT Huesca Spain
BT Aragon Spain
BT Spain
BT Iberian Peninsula
BT Southern Europe
BT Europe

Huesca Province
No longer a valid term for GeoRef. As of 1993, see Huesca Spain.

Huesca Spain (1993)
Refers only to the province. From 1981-1992, also search Huesca Province and Spain. Before 1981, also search Huesca and Spain. In NE Spain. For the city, see Huesca City Spain.
CO N411700N425700
   E0004500W0005700
BT Aragon Spain
BT Spain
BT Iberian Peninsula
BT Southern Europe
BT Europe
NT Huesca City Spain

Hughes County
Valid through 1988. Search in combination with state term. After 1988, use specific county-state term.

Hughes County Oklahoma (1989)
Central Oklahoma. Before 1989, also search Hughes County AND Oklahoma.
CO N344600N351700
   W0955800W0962900
BT Oklahoma
BT United States

Hughes County South Dakota (1989)
Central South Dakota. Before 1989, also search Hughes County AND South Dakota.
CO N440600N443300
   W0993700W1003400
BT South Dakota
BT United States
NT Pierre South Dakota

Hughes Creek Shale (1978)
Of Foraker Limestone. NE Kansas and SE Nebraska.
BT Permian

BT Paleozoic
  SA Kansas
  SA Nebraska

Hugoniot analysis (1985)
  UF Hugoniot curve
  UF Hugoniot data
  SA analysis
  SA elastic limit
  SA equations of state
  SA mathematical geology
  SA mathematical models
  SA phase equilibria
  SA shock metamorphism

Hugoniot curve
  use Hugoniot analysis

Hugoniot data
  use Hugoniot analysis

Hugoton
  Valid through 1988. Search in combination with state term. After 1988, use specific city-state term.

Hugoton Embayment (1978)
  SW Kansas and Oklahoma panhandle. Also search Hugoton.
  IN Index states as applicable.
  BT United States
  SA Kansas
  SA Oklahoma

Hugoton Field (1978)
  SW Kansas. Also search Hugoton.
  UF Hugoton-Panhandle Field
  BT Kansas
  BT United States
  SA oil and gas fields

Hugoton Kansas (1989)
  City in SW Kansas. Before 1989, search Hugoton AND Kansas.
  CO N371100N371100
     W1012200W1012200
  BT Stevens County Kansas
  BT Kansas
  BT United States

Hugoton-Panhandle Field
  use Hugoton Field

Hull
  No longer a valid term for GeoRef. As of 1993, see Hull Quebec.

Hull Quebec (1993)
  City across Ottawa River from Ottawa, Ontario, in S Quebec. Before 1993, also search Hull AND Quebec.
  BT Gatineau County Quebec
  BT Quebec
  BT Eastern Canada
  BT Canada

hulsite (1978)
  BT borates

human activity (1978)
  Used for the effects of humans on the environment.
  UF anthropogenic (activity)
  UF human impact
  UF human influence
  UF human interference
  SA human ecology
  SA Indian reservations

human ecology (1978)
  Used for the effect of the environment on humans.
  SA ecology
  SA environmental geology
  SA geologic hazards
  SA human activity
  SA human waste
  SA land use
  SA medical geology
  SA pollution
  SA urban environment

human impact
  use human activity

human influence
  use human activity

human interference
  use human activity

human waste (1978)
  UF waste, human
  SA human ecology
  SA waste disposal
  SA waste disposal sites

humates (1978)
  BT organic materials
  SA humic acids

Humber Arm Allochthon (1989)
  W Newfoundland.
  BT Newfoundland Island
  BT Newfoundland
  BT Eastern Canada
  BT Canada

Humber Estuary (1978)
  On E coast opening on North Sea.
  IN Index Humberside England or regions as applicable.
  SA Humberside England

Humberside England (1993)
  County in E England. Formed in 1974 from parts of Yorkshire and Lincolnshire. Before 1993, also search Humberside.
  BT England
  BT Great Britain
  BT United Kingdom
  BT Western Europe
  BT Europe
  SA Holderness
  SA Humber Estuary
  SA Lincolnshire England
  SA Yorkshire England

Humboldt County
  Valid through 1988. Search in combination with state term. After 1988, use specific county-state term.

Humboldt County California (1989)
  On Pacific Ocean in NW California. Before 1989, also search Humboldt County AND California.
  CO N400000N412700
     W1232800W1243000
  BT California
  BT United States
  NT Cape Mendocino
  NT Eureka California
  NT Fortuna California
  SA Eel River
  SA Redwood Creek
  SA Trinity River

Humboldt County Iowa (1989)
  N central Iowa. Before 1989, also search Humboldt County AND Iowa.
  CO N423800N425400
     W0935800W0942700
  BT Iowa
  BT United States

Humboldt County Nevada (1989)
  NW Nevada. Before 1989, also search Humboldt County AND Nevada.
  CO N403100N420000
     W1170000W1192100
  BT Nevada
  BT United States
  NT Getchell Mine
  NT Osgood Mountains
  NT Santa Rosa Range
  NT Winnemucca Nevada
  SA Black Rock Desert

Humboldt Range (1989)
  W Nevada.
  IN Index county or region as applicable.
  SA Pershing County Nevada

Humboldt River valley (1978)
  N central Nevada.
  BT Nevada
  BT United States

humic acids (1978)
  BT organic materials
  SA acids
  SA humates
  SA humus
  SA organic acids

humic coal (1978)
  Also search humite AND coal. As of 1990, organic residues is autoposted to this term. This term has multiple hierarchies.
  UF humite (coal)
  BT1 coal
  BT1 organic residues
  BT2 coal
  BT2 sedimentary rocks

humid environment (1978)
  Before 1978, search humid.
  SA climate
  SA depositional environment
  SA ecology
  SA environment
  SA humidity
  SA paleoclimatology

humidity (1978)
  SA atmosphere
  SA climate
  SA humid environment
  SA meteorology
  SA moisture
  SA raindrops
  SA water
  SA water vapor

huminite (1989)
  As of 1993, humodetrinite is a narrower term.
  BT macerals
  NT humodetrinite
  SA lignite

humite (1978)
  From 1978-1980, halides and orthosilicates were autoposted to this term. Autoposting of fluorides and humite group began in 1981. Autoposting of halides, nesosilicates, orthosilicates and silicates began in 1989. This term has multiple hierarchies.
  BT1 fluorides
  BT1 halides
  BT2 humite group
  BT2 nesosilicates
  BT2 orthosilicates
  BT2 silicates

humite (coal)
  use humic coal

humite group (1978)
  Autoposting of this term to clinohumite, humite and norbergite began in 1981.
  BT nesosilicates
  BT orthosilicates
  BT silicates
  NT clinohumite
  NT humite
  NT norbergite
  NT titanoclinohumite

hummocks (1978)
  SA geomorphology
  SA periglacial features
  SA permafrost
  SA taliks

hummocky cross-stratification (1985)
  BT planar bedding structures
  BT sedimentary structures
  SA cross-stratification

humodetrinite (1989)
  BT huminite
  BT macerals
  SA coal
  SA lignite

Humphreys Peak
  use San Francisco Mountain

humus (1978)
  UF soil ulmin
  BT organic materials
  SA humic acids
  SA leaves
  SA Mor
  SA Mull
  SA peat
  SA pedons
  SA soils

Hunan
  No longer a valid term for GeoRef. See Hunan China.

Hunan China (1993)
  Province, South-Central China.
  CO N244000N301000
     E1141500E1090000
  BT China
  BT Far East
  BT Asia
  SA Nanling
  SA Wufeng Formation
  SA Yangtze Platform

Hungarian Basin
  use Alfold

Hungarian Great Plain
  use Alfold

Hungarian Plain
  use Alfold

Hungary (1978)
  CO N454500N483000
     E0230000E0161000
  BT Central Europe
  BT Europe
  NT Alfold
  NT Bakony Mountains
  NT Balaton region
  NT Borsod Basin
  NT Borzsony Mountains
  NT Budapest Hungary
  NT Bukk Mountains
  NT Debrecen Hungary
  NT Dorog Basin
  NT Dorog Hungary
  NT Eger Hungary
  NT Hidas Hungary
  NT Matra Mountains
  NT Mecsek Mountains
  NT Northeastern Hungarian Hills
  NT Szolnok Hungary
  NT Tokaj Hungary
  NT Tokaj Mountains
  NT Transdanubia
  NT Vertes Hungary
  NT Vertes Mountains
  NT Veszprem Hungary
  NT Villany Mountains
  SA Banat
  SA Buda Marl
  SA Carpathians
  SA Danube River
  SA Danube Valley

SA Lake Neusiedler
SA Pannonia
SA Pannonian Basin
SA Somes Basin
SA Subcarpathians
SA Tisza River
SA Tokaj-Eperjes Mountains

Hunger Steppe
 use Bet-Pak-Dala

Hunsruck (1978)
 Mountain region S of the Moselle River between the Rhine River and the Saar River.
 BT Rhineland-Palatinate Germany
 BT Germany
 BT Central Europe
 BT Europe

Hunsruck Shale (1978)
 UF Hunsrueck Shale
 BT Devonian
 BT Paleozoic
 SA Germany
 SA Rhineland-Palatinate Germany

Hunsrueck Shale
 use Hunsruck Shale

Hunter Valley (1978)
 River valley in E central New South Wales. Also in other locations.
 IN Index county or state or region as applicable.
 SA New South Wales Australia

Hunterdon County
 Valid through 1988. Search in combination with state term. After 1988, use specific county-state term.

Hunterdon County New Jersey (1989)
 W New Jersey. Before 1989, also search Hunterdon County AND New Jersey.
 CO N402000N404700
    W0744300W0751200
 BT New Jersey
 BT United States
 SA New Jersey Highlands

huntite (1978)
 BT carbonates

Huntly (1978)
 Coal mines in Waikato Basin, New Zealand, and bauxite deposits in Western Australia. As of 1993, use Huntly Scotland for the town in Grampian region, Scotland.
 IN Index regions as applicable.
 SA New Zealand
 SA Waikato Basin
 SA Western Australia

Huntly Scotland (1993)
 Town in Grampian region. Before 1993, also search Huntly AND Scotland.
 BT Grampian region Scotland
 BT Scotland
 BT Great Britain
 BT United Kingdom
 BT Western Europe
 BT Europe

Hunton Group (1978)
 Silurian and Devonian. Includes Chimneyhill Limestone, Henryhouse Formation, Haragan Shale, Bois d'Arc Limestone, Frisco Limestone, Kite Group with Haragan and Cravatt formations.
 SE Oklahoma.
 IN Index ages as applicable.
 BT Paleozoic
 SA Devonian
 SA Henryhouse Formation
 SA Oklahoma
 SA Silurian

Huon Peninsula (1978)
 On E coast between Astrolabe Bay and Huon Gulf.
 BT Papua New Guinea
 BT Australasia

Hupeh China
 use Hubei China

Hupei China
 use Hubei China

hureaulite (1978)
 BT phosphates

Huron Member (1985)
 Of Ohio Shale.
 BT Upper Devonian
 BT Devonian
 BT Paleozoic
 SA Ohio
 SA Ohio Shale

Huronian (1978)
 Great Lakes region. Division of the Proterozoic in the Canadian Shield. Autoposting of Precambrian to this term ended in 1989. As of 1989, Proterozoic is autoposted to this term. As of 1990, upper Precambrian and Precambrian are autoposted to this term.
 UF Huronian Supergroup
 BT Proterozoic
 BT upper Precambrian
 BT Precambrian
 NT Gowganda Formation
 NT Onaping Formation
 NT Signal Hill Formation
 SA Canada
 SA Canadian Shield
 SA United States

Huronian Supergroup
 use Huronian

Hurricane Ridge Syncline (1978)
 Southwest Virginia and SE West Virginia.
 IN Index states as applicable.
 BT United States
 SA Virginia
 SA West Virginia

hurricanes (1978)
 SA Agnes
 SA geologic hazards
 SA meteorology
 SA monsoons
 SA storms
 SA winds

Hurwitz Group (1978)
 Keewatin District.
 BT Aphebian
 BT lower Proterozoic
 BT Proterozoic
 BT upper Precambrian
 BT Precambrian
 SA Keewatin District Northwest Territories
 SA Northwest Territories

Hut Point Peninsula (1978)
 On Ross Island which is just off NW edge of Ross Ice Shelf in Ross Dependency.
 BT Antarctica
 SA Ross Island

Hutchinson
 Valid through 1988. Search in combination with state term. After 1988, use specific city-state term.

Hutchinson Kansas (1989)
 City in S central Kansas. Before 1989, search Hutchinson AND Kansas.
 CO N380300N380300
    W0975600W0975600
 BT Reno County Kansas
 BT Kansas
 BT United States

Hutchinson Salt Member (1978)
 Of Wellington Formation. Salt beds largely within Hutchinson city limits in Reno County in S central Kansas.
 BT Permian
 BT Paleozoic
 SA Kansas
 SA Wellington Formation

Hutton, James (1985)
 UF James Hutton
 SA biography

Hutuo Group (1989)
 Hebei and Shanxi provinces, N China.
 BT Proterozoic
 BT upper Precambrian
 BT Precambrian
 SA China
 SA Hebei China
 SA Shanxi China

Hwang Hae
 use Yellow Sea

Hwang Hai
 use Yellow Sea

Hwang Ho
 use Huang Ho

Hyaenidae (1978)
 Family. Autoposting of Fissipeda, Eutheria and Theria to this term began in 1989.
 BT Fissipeda
 BT Carnivora
 BT Eutheria
 BT Theria
 BT Mammalia
 BT Tetrapoda
 BT Vertebrata
 BT Chordata

hyalite (1978)
 UF water opal
 BT silica minerals
 BT framework silicates
 BT silicates
 SA opal

hyaloclastite (1985)
 BT pyroclastics
 BT volcanic rocks
 BT igneous rocks

hyalophane (1978)
 Autoposting of feldspar group to this term ended in 1985 and began again in 1989.
 BT barium feldspar
 BT feldspar group
 BT framework silicates
 BT silicates

Hyalospongea (1978)
 BT Porifera
 BT Invertebrata

hybridization (1978)
 Used in relation to magmas.
 SA assimilation
 SA contamination
 SA country rocks
 SA emplacement
 SA intrusions
 SA magmas
 SA magmatic differentiation

Hyderabad
 No longer a valid term for GeoRef. Princely state prior to Indian independence. See Hyderabad India or Hyderabad Pakistan.

Hyderabad India (1993)
 City in Andhra Pradesh State in India. Before 1993, also search Hyderabad AND India.
 BT Andhra Pradesh India
 BT India
 BT Indian Peninsula
 BT Asia

Hyderabad Pakistan (1993)
 City in Sind Province, Pakistan. Before 1993, also search Hyderabad AND Pakistan.
 BT Sind Pakistan
 BT Pakistan
 BT Indian Peninsula
 BT Asia

hydrargillite
 use gibbsite

hydrates (1985)
 SA gas hydrates
 SA minerals

hydrates, gas
 use gas hydrates

hydration (1978)
 SA dehydration
 SA geochemistry
 SA processes
 SA water

hydration of glass (1978)
 Before 1972, also search obsidian hydration.
 UF glass, hydration of
 UF obsidian hydration
 SA geochronology
 SA relative age

hydraulic conductivity (1978)
 UF coefficient of permeability
 UF permeability coefficient
 SA conductivity
 SA hydraulics
 SA permeability
 SA transmissivity

hydraulic fracturing (1978)
 Term is to be used for artificial fracturing induced by hydraulic pressure.
 SA enhanced recovery
 SA fractured materials
 SA fracturing
 SA geothermal energy
 SA massive hydraulic fracturing
 SA mining geology
 SA petroleum engineering
 SA rock mechanics
 SA tertiary recovery

hydraulic gradients
 use hydraulics

hydraulic maps (1978)
 Before 1978, search maps AND hydraulic.
 BT maps

hydraulic pressure (1978)
 Before 1978, search hydraulic AND pressure.
 BT pressure

hydraulic properties
 use hydraulics

hydraulics (1978)

hydrobiotite • hydrology

Used when discussing channels, or engineering and practical applications. For mathematical aspects of movement, use hydrodynamics.
- UF hydraulic gradients
- UF hydraulic properties
- SA critical flow
- SA drawdown
- SA eddy flow
- SA engineering geology
- SA flows
- SA fluid dynamics
- SA ground water
- SA hydraulic conductivity
- SA hydrodynamics
- SA hydrographs
- SA hydrostatic pressure
- SA laminar flow
- SA multiphase flow
- SA roughness
- SA shorelines
- SA soil mechanics
- SA steady flow
- SA unsteady flow
- SA waterways

hydrobiotite (1978)
- BT sheet silicates
- BT silicates
- SA aluminosilicates
- SA mica group

hydroboracite (1978)
- BT borates

hydrocarbon indicators (1993)
Seismic method. Before 1993, also search bright spots.
- UF direct hydrocarbon indicators
- SA AVO methods
- SA geophysical methods
- SA indicators
- SA petroleum exploration
- SA seismic methods
- SA seismic stratigraphy

hydrocarbons (1978)
Used for small amounts of hydrocarbons in rocks or sediments. Also used for pollution studies of petroleum products. For pollution studies, before 1993, also search petroleum and pollution. For economic deposits, see petroleum; natural gas; bitumens; asphalt; oil sands; oil shale. Autoposting of this term began in 1978. Before 1993, also search gasoline if applicable.
- UF gasoline
- UF petroleum products
- BT organic materials
- NT aliphatic hydrocarbons
- NT aromatic hydrocarbons
- NT diterpanes
- NT hopanes
- NT saturated hydrocarbons
- NT steranes
- NT terpanes
- NT triterpanes
- SA aromatization
- SA carbon
- SA chlorinated hydrocarbons
- SA crude oil
- SA geochemical anomalies
- SA isoprenoids
- SA leaking underground storage tanks
- SA natural gas
- SA petroleum
- SA pollutants
- SA pollution
- SA Rock-Eval
- SA sapropel
- SA volatile organic compounds

hydrochemistry (1978)
Before 1971, also search hydrogeochemistry.
- UF hydrogeochemistry
- SA aqueous solutions
- SA brackish water
- SA brines
- SA chemical composition
- SA chemical dispersion
- SA desalinization
- SA diffusion
- SA diffusivity
- SA dissolved materials
- SA electro-osmosis
- SA fresh water
- SA geochemistry
- SA ground water
- SA hydrogeology
- SA hydrology
- SA ion exchange
- SA ions
- SA lysimeters
- SA mass transfer
- SA osmosis
- SA polluted water
- SA potability
- SA salinity
- SA salt water
- SA saturation
- SA springs
- SA surface water
- SA suspended materials
- SA thermal waters
- SA WATEQF
- SA water
- SA water hardness
- SA water quality

hydrochloric acid (1985)
- UF muriatic acid
- SA acids
- SA geochemistry
- SA inorganic acids

hydrodynamics (1978)
Used when discussing mathematical aspects of ground water movement. As of 1993, restricted to natural processes. For discussion of channels, engineering and practical applications, see hydraulics.
- SA catchment hydrodynamics
- SA critical flow
- SA Darcy's law
- SA eddy flow
- SA flowmeters
- SA fluid dynamics
- SA ground water
- SA hydraulics
- SA Kalman filters
- SA laminar flow
- SA multiphase flow
- SA Richards equation
- SA steady flow
- SA storage coefficient
- SA unsteady flow

hydroelectric energy (1993)
Before 1993, also search hydroelectric power, hydropower and water power.
- UF hydropower
- UF water power
- SA energy
- SA energy sources
- SA geothermal energy
- SA power plants
- SA water harnessing
- SA water resources
- SA waterways

hydrofluoric acid (1989)
Chemical formula is HF.
- UF hydrogen fluoride
- SA acids
- SA geochemistry
- SA inorganic acids

hydrogen (1978)
- UF H
- NT D/H
- NT deuterium
- NT tritium
- SA chemical elements
- SA isotopes
- SA stable isotopes

hydrogen dioxide
use hydrogen peroxide

hydrogen fluoride
use hydrofluoric acid

hydrogen peroxide (1989)
An explosive, corrosive compound.
- UF hydrogen dioxide
- UF peroxide
- SA compounds
- SA corrosion

hydrogen sulfide (1978)
- SA eutrophication
- SA gases
- SA lakes
- SA natural gas
- SA sulfides

hydrogeochemical prospecting
No longer a valid GeoRef index term. Before 1971, was used on level 1 in subfile N. See geochemical methods; hydrological methods; hydrochemistry.

hydrogeochemistry
No longer a valid GeoRef index term. Before 1971, was used on level 1 in subfile N.
use hydrochemistry

hydrogeologic
A valid term through 1977. After 1977, use hydrogeologic maps.

hydrogeologic maps (1978)
Before 1978, search hydrogeologic AND maps.
- BT maps
- SA hydrogeology
- SA potentiometric surface

hydrogeological controls (1978)
IN As of 1993, restricted to economic studies.
- SA controls
- SA hydrogeology
- SA mineral deposits, genesis

hydrogeology (1978)
Used for general studies. For the science that deals with subsurface waters and related geologic aspects of surface waters. To search for the hydrology of a particular river, search for the name of the river in combination with hydrology or hydrogeology. To search for a particular aquifer, search for the name of the aquifer OR area in combination with ground water.
- SA aquifers
- SA artesian waters
- SA dolines
- SA dye tracers
- SA flowmeters
- SA fluorimetry
- SA ground water
- SA ground-water provinces
- SA hydrochemistry
- SA hydrogeologic maps
- SA hydrogeological controls
- SA hydrologic cycle
- SA hydrological methods
- SA hydrology
- SA hydrosphere
- SA lysimeters
- SA monitoring
- SA pumping
- SA radioactive tracers
- SA regulations
- SA representative basins
- SA reservoirs
- SA rivers
- SA rivers and streams
- SA shallow aquifers
- SA shallow depth
- SA springs
- SA streams
- SA surface water
- SA thermal regime
- SA thermal waters
- SA tracers
- SA transmissivity
- SA underground streams
- SA unsaturated zone
- SA water resources
- SA water table
- SA water wells

hydrogoethite (1978)
- BT oxides
- SA iron ores
- SA limonite

hydrographic maps (1985)
- BT maps
- SA waterways

hydrographs (1978)
- SA hydraulics
- SA hydrology

hydrogrossular (1978)
- UF hibschite
- BT garnet group
- BT nesosilicates
- BT orthosilicates
- BT silicates
- SA grossular

hydrologic budget
use water balance

hydrologic cycle (1978)
Before 1983, also search cycles AND hydrology.
- UF water cycle
- SA atmospheric precipitation
- SA cycles
- SA discharge
- SA evapotranspiration
- SA ground water
- SA hydrogeology
- SA hydrology
- SA hydrosphere
- SA lysimeters
- SA meteorology
- SA recharge
- SA surface water
- SA water
- SA water balance

hydrologic maps (1978)
Before 1978, search maps AND hydrologic.
- BT maps
- SA hydrology

hydrological methods (1978)
- SA hydrogeology
- SA hydrology
- SA methods
- SA mineral exploration
- SA water

hydrology (1978)
Used for continental surface water. In cases where the interchange between ground and surface water is intimate, this term may be used with hydrologic cycle. (Before 1983, cycles was used this way.) To search for the hydrology of a particular river, search for the name of the river in

combination with hydrology or hydrogeology.
SA atmosphere
SA atmospheric precipitation
SA basin management
SA bedforms
SA cascades
SA catchment hydrodynamics
SA channel geometry
SA critical flow
SA discharge
SA drought
SA dye tracers
SA ecology
SA eddy flow
SA engineering geology
SA environmental geology
SA eutrophication
SA evaporation
SA evapotranspiration
SA exsurgence
SA floods
SA flowmeters
SA fluorimetry
SA fluvial currents
SA gauging
SA geomorphology
SA glacial geology
SA glaciers
SA ground water
SA hydrochemistry
SA hydrogeology
SA hydrographs
SA hydrologic cycle
SA hydrologic maps
SA hydrological methods
SA hydrosphere
SA hydrostatic pressure
SA ice
SA infiltration
SA intermittent springs
SA karst hydrology
SA lakes
SA laminar flow
SA limnology
SA low water
SA lysimeters
SA meltwater
SA meteorology
SA multiphase flow
SA NAWDEX
SA paleohydrology
SA paleokarst
SA percolation
SA pollution
SA precipitation
SA pump tests
SA purification
SA radioactive tracers
SA rainfall
SA rapids
SA representative basins
SA reservoirs
SA retention
SA Richards equation
SA rills
SA rivers
SA rivers and streams
SA runoff
SA sediment supply
SA sediment traps
SA sedimentation
SA sediments
SA snow
SA springs
SA steady flow
SA surface water
SA surveys
SA suspension
SA tidewater glaciers
SA tracers
SA transmissivity
SA underground streams
SA unsteady flow
SA wadis

SA water
SA water balance
SA water quality
SA water resources
SA water supply
SA water table
SA waterlogging
SA watersheds
SA waterways

hydrolysis (1978)
SA geochemistry
SA oxidation
SA processes
SA weathering

hydromagnesite (1978)
BT carbonates

hydrometallurgy (1981)
SA metallurgy

hydromica (1978)
BT sheet silicates
BT silicates
SA clay minerals
SA illite
SA mica group

hydromica schist
Not a valid term for GeoRef. Use schists and hydromica.

Hydromorphic soils (1978)
BT soils
SA Bog soils
SA Histosols
SA Intrazonal soils
SA soil group

hydromuscovite (1978)
BT sheet silicates
BT silicates
SA clay minerals
SA mica group
SA muscovite

hydrophones (1985)
SA acoustical methods
SA geophones
SA instruments
SA marine methods
SA ocean bottom seismographs
SA pressuremeters
SA seismic methods

hydropower
use hydroelectric energy

hydrosodalite (1978)
Artificial mineral.
BT sodalite group
BT framework silicates
BT silicates
SA aluminosilicates

hydrosphere (1978)
SA atmosphere
SA biosphere
SA Earth
SA glaciers
SA ground water
SA hydrogeology
SA hydrologic cycle
SA hydrology
SA ice
SA lakes
SA rivers
SA rivers and streams
SA snow
SA streams
SA surface water
SA tectonosphere
SA water

hydrostatic pressure (1978)
BT pressure
SA hydraulics
SA hydrology
SA overpressure

hydrotalcite (1993)
BT carbonates
SA pyroaurite

hydrothermal
A valid term through 1977. See hydrothermal conditions; ore-forming fluids; hydrothermal alteration; hydrothermal processes.

hydrothermal alteration (1978)
As of 1981, use hydrothermal processes if related to the genesis of mineral deposits.
BT metasomatism
SA alteration
SA autometamorphism
SA greisenization
SA ground water
SA hydrothermal conditions
SA hydrothermal processes
SA leakage anomalies
SA ore-forming fluids
SA propylitization
SA pyritization
SA pyrometasomatism
SA sericitization
SA serpentinization
SA thermal alteration
SA thermal waters
SA tourmalinization
SA uralitization
SA wall-rock alteration
SA zeolitization

hydrothermal chimneys
use hydrothermal vents

hydrothermal conditions (1981)
As of 1981, use hydrothermal processes if related to the genesis of mineral deposits.
SA black smokers
SA hydrothermal alteration
SA hydrothermal processes
SA hydrothermal vents
SA metasomatism
SA phase equilibria

hydrothermal fluids
Not a valid term for GeoRef. As of 1981, use ore-forming fluids if applicable. Before 1981, hydrothermal solutions was used for this term.

hydrothermal processes (1978)
As of 1981, restricted to economic geology. If paper is not economically oriented, use hydrothermal conditions or hydrothermal alteration.
SA alteration
SA epithermal processes
SA hot springs
SA hydrothermal alteration
SA hydrothermal conditions
SA igneous processes
SA mineral deposits, genesis
SA ore-forming fluids
SA processes
SA water

hydrothermal solutions
As of 1981, no longer a valid term for GeoRef. Use ore-forming fluids or thermal waters if applicable.

hydrothermal systems
use geothermal systems

hydrothermal vents (1993)
Before 1993, also search vents or chimneys in conjunction with hydrothermal conditions.
UF hydrothermal chimneys
NT black smokers
SA bottom features
SA chimneys

SA hydrothermal conditions
SA vents

hydroxides (1978)
BT oxides

hydroxyapatite
use hydroxylapatite

hydroxyl ion (1985)
SA geochemistry
SA ions
SA sodium hydroxide

hydroxylapatite (1978)
UF hydroxyapatite
BT phosphates
SA apatite

Hydrozoa (1978)
BT Coelenterata
BT Invertebrata

Hymenoptera (1978)
BT Hymenopteroida
BT Insecta
BT Mandibulata
BT Arthropoda
BT Invertebrata
SA Raphidiodea

Hymenopteroida (1981)
BT Insecta
BT Mandibulata
BT Arthropoda
BT Invertebrata
NT Hymenoptera

Hyogo
No longer a valid term for GeoRef. See Hyogo Japan.

Hyogo Japan (1993)
Before 1993, also search Hyogo or Hyogo Prefecture AND Japan.
UF Hyogo Prefecture
BT Honshu
BT Japan
BT Far East
BT Asia
NT Akenobe Mine
NT Ikuno Mine
SA Hidaka Japan
SA Kinki Japan
SA Osaka Bay
SA Rokko Mountains
SA Yamasaki Fault

Hyogo Prefecture
use Hyogo Japan

Hyolithes (1981)
Used as the control term for terms such as Hyolothida, Hyolithidae, Hyolithimorpha and Hyolitha; these terms may be used in addition to Hyolithes. Before 1981, may have been classified under problematic fossils.
BT Mollusca
BT Invertebrata
SA problematic fossils

hypabyssal
A valid term through 1977.

hypabyssal processes (1981)
SA hypabyssal rocks
SA mineral deposits, genesis
SA processes

hypabyssal rocks (1978)
Before 1978, also search hypabyssal AND igneous rocks.
BT igneous rocks
SA beresite
SA hypabyssal processes
SA intrusions
SA plutonic rocks
SA rocks

Before then, variants of the term may occur in GeoRef.

hypergene processes
use supergene processes

Hyperion Satellite (1989)
One of the satellites of Saturn. Before 1989, also search Hyperion AND Saturn.
SA icy satellites
SA satellites
SA Saturn
SA Voyager Program

hypersaline environment (1985)
SA deep-water environment
SA depositional environment
SA ecology
SA environment
SA lacustrine environment
SA lagoonal environment
SA paleoecology
SA salinity
SA sebkha environment
SA sedimentation

hypersthene (1978)
BT orthopyroxene
BT pyroxene group
BT chain silicates
BT silicates

hypocenters
use focus

hypogene processes
As of 1981, no longer a valid term for GeoRef.
use endogene processes

hypothermal processes (1981)
UF katathermal processes
SA mineral deposits, genesis
SA processes

Hyracoidea (1981)
Order. Autoposting of Eutheria and Theria to this term began in 1989.
BT Eutheria
BT Theria
BT Mammalia
BT Tetrapoda
BT Vertebrata
BT Chordata

hysteresis (1989)
For the magnetic property of a rock, use magnetic hysteresis.
SA compression
SA deformation
SA elastic materials
SA magnetic hysteresis
SA mechanical properties
SA rock mechanics
SA strain
SA stress

Hystrichosphaeridae (1978)
Also search hystrichospheres; hystrichosphaerids. Autoposting of microfossils to this term began in 1990. This term has multiple hierarchies.
BT1 acritarchs
BT1 palynomorphs
BT1 microfossils
BT2 Dinoflagellata
BT2 palynomorphs
BT2 microfossils

Hystricomorpha (1981)
Suborder. Autoposting of Eutheria and Theria to this term began in 1989.
BT Rodentia
BT Eutheria
BT Theria
BT Mammalia
BT Tetrapoda
BT Vertebrata
BT Chordata

I

I
use iodine

I-129 (1978)
Autoposting of broader terms began in 1989. This term has multiple hierarchies.
BT1 radioactive isotopes
BT1 isotopes
BT2 iodine
BT2 halogens

I-131 (1978)
Autoposting of broader terms began in 1989. This term has multiple hierarchies.
BT1 radioactive isotopes
BT1 isotopes
BT2 iodine
BT2 halogens

I-type granites (1989)
Derived mainly from igneous source material.
BT granites
BT plutonic rocks
BT igneous rocks
SA A-type granites
SA calc-alkalic composition
SA S-type granites

I/Xe (1989)
Isotopic ratio used in age determination.
SA absolute age
SA isotope ratios
SA isotopes
SA xenon

Iacobeni
No longer a valid term for GeoRef. As of 1993 see Iacobeni Romania.

Iacobeni Romania (1993)
Village in Suceava County in N Romania. Before 1993 also search Iacobeni AND Romania. This term has multiple hierarchies.
BT1 Moldavia
BT1 Europe
BT2 Romania
BT2 Southern Europe
BT2 Europe

Iapetus (1978)
Proto-Atlantic Ocean. Search Iapetus NOT Saturn. For satellite of Saturn, use Iapetus Satellite.
UF Iapetus Ocean
UF Iapetus Suture
UF Proto-Atlantic Ocean
SA Atlantic Ocean
SA Iapetus Satellite

Iapetus Ocean
use Iapetus

Iapetus Satellite (1978)
One of the satellites of Saturn. Used to distinguish from Iapetus, the proto-Atlantic Ocean. Also search Iapetus AND Saturn.
SA Iapetus
SA icy satellites
SA satellites
SA Saturn
SA Voyager Program

Iapetus Suture
use Iapetus

Ibadan
No longer a valid term for GeoRef. As of 1993 see Ibadan Nigeria.

Ibadan Nigeria (1993)
City in SW Nigeria. In Western Provinces. Before 1993 also search Ibadan AND Nigeria.
BT Nigeria
BT West Africa
BT Africa

Ibaragi Complex (1978)
Granitic complex in Osaka Prefecture.
BT Osaka Japan
BT Honshu
BT Japan
BT Far East
BT Asia

Ibaraki
No longer a valid term for GeoRef. See Ibaraki Japan.

Ibaraki Japan (1993)
Prefecture in N central Honshu.
BT Honshu
BT Japan
BT Far East
BT Asia
NT Hitachi Deposit
NT Hitachi Japan
NT Kakioka Japan
NT Tokai Japan
SA Abukuma Mountains
SA Kanto Plain

Ibbenbueren
Not a valid term for GeoRef. See Ibbenburen Germany.

Ibbenburen
No longer a valid term for GeoRef. See Ibbenburen Germany.

Ibbenburen Germany (1993)
Town in N North Rhine-Westphalia. Before 1993, also search Ibbenburen or Ibbenbueren.
BT North Rhine-Westphalia Germany
BT Germany
BT Central Europe
BT Europe

Iberia
use Iberian Peninsula

Iberia Parish
Valid through 1988. Search in combination with state term. After 1988, use specific county-state term.

Iberia Parish Louisiana (1989)
On Vermilion Bay in S central Louisiana. Before 1989, also search Iberia Parish AND Louisiana.
CO N293000N301500 W0911500W0920000
BT Louisiana
BT United States

Iberian Cordillera
use Iberian Mountains

Iberian Mountains (1978)
Mountain system on E edge of great central plateau.
UF Iberian Cordillera
BT Spain
BT Iberian Peninsula
BT Southern Europe
BT Europe

Iberian Peninsula (1978)
Extreme SW Europe S of the Pyrenees and the Bay of Biscay. Autoposting of this term began in 1981. As of 1990, Europe is autoposted to this term.
IN Index countries as applicable.
CO N360000N435000 E0032000W0093000
UF Iberia
BT Southern Europe
BT Europe
NT Duero Basin
NT Duero River
NT Gibraltar
NT Portugal
NT Spain
NT Tagus Basin
NT Tagus River

Iberica (1981)
BT Spain
BT Iberian Peninsula
BT Southern Europe
BT Europe

Ibiza (1978)
Third largest of Spain's Balearic Islands off E coast of Spanish mainland. This term has multiple hierarchies.
UF Iviza
BT1 Balearic Islands
BT1 Mediterranean region
BT2 Balearic Islands
BT2 Spain
BT2 Iberian Peninsula
BT2 Southern Europe
BT2 Europe

ice (1978)
Before 1972, was sometimes used on level 1 in combination with nonglacial or non-glacial (i.e. ice, nonglacial or ice, non-glacial).
IN For the mineral, also index oxides.
SA frozen ground
SA glacial features
SA glaciers
SA ground ice
SA hydrology
SA hydrosphere
SA ice fields
SA ice lenses
SA ice rafting
SA meltwater
SA oxides
SA periglacial features
SA permafrost
SA rock glaciers
SA scour
SA sea ice
SA snow
SA thawing
SA water

ice ages
A valid index term through 1977 used in combination with ancient (i.e. After 1978, use ancient ice ages.

ice caps (1978)
SA glacial extent
SA glacial features
SA glacial geology
SA glaciers
SA ice sheets
SA ice streams

ice fields (1993)
UF icefields
SA glacial features
SA glaciers
SA ice
SA ice sheets
SA ice shelves
SA ice-marginal features

SA periglacial features
SA sea ice
SA snow

ice flows
As of 1993, use ice movement.

ice islands
Not a valid GeoRef index term after 1970. Was used on level 1 in subfile N. Now see artificial islands, sea ice, ice shelves, ice rafting or icebergs.

ice lenses (1993)
SA glacial features
SA ground ice
SA ice
SA ice wedges
SA lenses
SA palsas
SA periglacial features
SA permafrost

ice mantle
use ice sheets

ice movement (1978)
Before 1976, also search ice movements. Before 1993, also search gravity flows AND glaciers if applicable.
SA deposition
SA glacial geology
SA glaciation
SA glacier surges
SA glaciers
SA gravity flows
SA ice streams
SA movement

ice rafting (1978)
UF ice-rafting
UF rafting, ice
SA ice
SA ice streams
SA raft foundations
SA sediment transport
SA sedimentation

ice scouring
Use glacial erosion AND scour if applicable.

ice sheets (1978)
UF ice mantle
SA glacial extent
SA glacial geology
SA glaciation
SA glaciers
SA ice caps
SA ice fields
SA ice streams

ice shelves (1978)
UF shelves, ice
SA glacial geology
SA glaciation
SA glaciers
SA ice fields
SA marine environment

ice streams (1993)
SA glacial features
SA glacial geology
SA glaciers
SA ice caps
SA ice movement
SA ice rafting
SA ice sheets
SA rivers and streams
SA streams

ice veins
use ice wedges

ice wedges (1978)
Autoposting of periglacial features to this term began in 1989.
UF ice veins
UF wedges, ice

BT periglacial features
SA fossil ice wedges
SA frost action
SA glacial geology
SA ground ice
SA ice lenses
SA permafrost
SA taliks

ice wedges, fossil
use fossil ice wedges

ice-marginal features (1985)
SA geomorphology
SA glacial features
SA glacial geology
SA glaciers
SA ice fields
SA periglacial features

ice-rafting
use ice rafting

ice-wedge cast
use fossil ice wedges

ice-wedge fill
use fossil ice wedges

ice-wedge pseudomorph
use fossil ice wedges

icebergs (1978)
SA glacial features
SA glaciers
SA sea ice
SA tidewater glaciers

icefields
use ice fields

Iceland (1978)
Independent country and island 155 miles SE of Greenland. As of 1981, Atlantic Ocean is no longer autoposted.
CO N634000N663000
W0133000W0244500
BT Western Europe
BT Europe
NT Breidamerkurjokull
NT Grimsvotn
NT Hekla
NT Kolbeinsey Island
NT Krafla
NT Reykjanes Peninsula
NT Skeidararjokull
NT Snaefellsnes Peninsula
NT Surtsey
NT Vatnajokull
NT Vestmannaeyjar
SA Arctic region
SA Atlantic Ocean
SA Atlantic Ocean Islands
SA Iceland Research Drilling Project

Iceland crystal
use Iceland spar

Iceland Research Drilling Project (1985)
Borehole at Reydarfjordur, Iceland.
SA Atlantic Ocean
SA boreholes
SA cores
SA Iceland
SA Mid-Atlantic Ridge
SA North Atlantic
SA oceanic crust
SA sea-floor spreading
SA well-logging

Iceland spar (1978)
Variety of calcite.
UF double-refracting spar
UF Iceland crystal
BT carbonates
SA calcite

Iceland-Faeroe Ridge (1978)
Between Iceland and Faeroe Islands.
UF Faeroe-Iceland Ridge
UF Iceland-Faeroe Rise
UF Iceland-Faeroes Ridge
BT Atlantic Ocean
SA DSDP Site 336

Iceland-Faeroe Rise
use Iceland-Faeroe Ridge

Iceland-Faeroes Ridge
use Iceland-Faeroe Ridge

Icelandic Plateau (1981)
W of Jan Mayen Ridge, N of Iceland.
CO N670000N700000
W0093000W0160000
BT Greenland Sea
BT Arctic Ocean
SA DSDP Site 348

icequakes (1981)
SA earthquakes
SA glaciers
SA glaciotectonics

ichnofossils (1978)
As of 1981, is not to be used under sedimentary structures, and refers only to studies in which the emphasis is paleontology, not sedimentary petrology. For ichnofossils as components of sediments or sedimentary rocks, use lebensspuren. From 1978-1980, biogenic structures and sedimentary structures were autoposted to this term. Before 1973, also search trace fossils.
NT Arenicolites
NT Corophioides
NT Cruziana
NT Diplocraterion
NT Nereites
NT Palaeophycus
NT Planolites
NT Rhizocorallium
NT Rusophycus
NT Skolithos
NT Teichichnus
NT Thalassinoides
NT Trypanites
NT Zoophycos
SA biogenic structures
SA borings
SA burrows
SA casts
SA chemical fossils
SA Chondrites ichnofossils
SA Collenia
SA coprolites
SA fossilization
SA fossils
SA Invertebrata
SA lebensspuren
SA morphology
SA Ophiomorpha
SA Ophiomorpha nodosa
SA tracks
SA trails

ichthyoliths (1989)
Fish skeletal debris.
SA biostratigraphy
SA fish
SA microfossils
SA Pisces
SA skeletons

Ichthyopterygia
As of 1993, no longer a valid term for GeoRef.
use Ichthyosauria

Ichthyosauria (1981)

Order. Before 1993, also search Ichthyopterygia. Before 1993, Ichthyopterygia was autoposted to this term.
UF Ichthyopterygia
BT Diapsida
BT Reptilia
BT Tetrapoda
BT Vertebrata
BT Chordata
NT Ichthyosaurus

Ichthyosaurus (1978)
Genus.
BT Ichthyosauria
BT Diapsida
BT Reptilia
BT Tetrapoda
BT Vertebrata
BT Chordata

Icriodus (1978)
Genus. Autoposting of microfossils to this term began in 1990.
BT Conodonta
BT microfossils

Icy Bay (1981)
Inlet of Gulf of Alaska, SE Alaska, W of Yakutat Bay.
IN Index Southern Alaska as applicable.
SA Gulf of Alaska
SA Southern Alaska

icy moons
use icy satellites

icy satellites (1993)
Natural satellites. Before 1993, also search icy moons.
UF icy moons
BT satellites
SA Amalthea Satellite
SA Ariel Satellite
SA Callisto Satellite
SA Dione Satellite
SA Enceladus Satellite
SA Europa Satellite
SA Galilean satellites
SA Ganymede Satellite
SA Hyperion Satellite
SA Iapetus Satellite
SA Mimas Satellite
SA Miranda Satellite
SA Oberon Satellite
SA planetology
SA Rhea Satellite
SA Tethys Satellite
SA Titan Satellite
SA Titania Satellite
SA Triton Satellite

Idaho (1978)
Autoposting of this term began in 1978.
CO N420000N490000
W1110500W1171500
BT United States
NT Ada County Idaho
NT Adams County Idaho
NT Albion Range
NT Bannock County Idaho
NT Bannock Range
NT Bingham County Idaho
NT Blaine County Idaho
NT Boise County Idaho
NT Bonner County Idaho
NT Bonneville County Idaho
NT Boundary County Idaho
NT Butte County Idaho
NT Camas County Idaho
NT Caribou County Idaho
NT Cassia County Idaho
NT Clark County Idaho
NT Clearwater County Idaho
NT Coeur d'Alene River
NT Craters of the Moon

NT Custer County Idaho
NT Elmore County Idaho
NT Franklin County Idaho
NT Fremont County Idaho
NT Idaho County Idaho
NT Idaho Panhandle
NT Jefferson County Idaho
NT Jerome County Idaho
NT Kootenai County Idaho
NT Latah County Idaho
NT Lemhi County Idaho
NT Lemhi Range
NT Lewis County Idaho
NT Lincoln County Idaho
NT Lost River Fault
NT Lost River Range
NT Madison County Idaho
NT Oneida County Idaho
NT Owyhee County Idaho
NT Pioneer Mountains
NT Power County Idaho
NT Salmon River breaks
NT Seven Devils Mountains
NT Shoshone County Idaho
NT Snake River plain
NT Teton County Idaho
NT Valley County Idaho
NT Washington County Idaho
SA Absaroka Fault
SA Aldridge Formation
SA Bald Mountain
SA Basin and Range Province
SA Battle Mountain High
SA Bear Lake
SA Bear River basin
SA Bear River Formation
SA Bear River Range
SA Beaverhead Formation
SA Beaverhead Mountains
SA Belt Basin
SA Belt Supergroup
SA Big Snowy Group
SA Bitterroot Range
SA Blue Mountain
SA Blue Mountains
SA Borah Peak earthquake 1983
SA Brigham Group
SA Bull Lake Glaciation
SA Cache Valley
SA Challis Volcanics
SA Clark Fork
SA Clearwater River
SA Coeur d'Alene mining district
SA Colorado Group
SA Columbia Plateau
SA Columbia River Basalt Group
SA Columbia River basin
SA Crawford Thrust
SA Dinwoody Formation
SA Fish Haven Dolomite
SA Frontier Formation
SA Glenns Ferry Formation
SA Grande Ronde Basalt
SA Great Basin
SA Idaho Batholith
SA Jefferson Group
SA Karmutsen Group
SA Lake Bonneville
SA Lake Missoula
SA Lewis and Clark Lineament
SA Madison Group
SA Meade Peak Member
SA Mission Canyon Limestone
SA Missoula Group
SA Mount Shields Formation
SA Newland Limestone
SA Nugget Sandstone
SA Oquirrh Formation
SA Owyhee Mountains
SA Park City Formation
SA Phosphoria Formation
SA Pocatello Formation
SA Prichard Formation
SA Purcell Mountains
SA Purcell System
SA Raft River
SA Raft River basin
SA Ravalli Group
SA Retort Phosphatic Shale Member
SA Revett Quartzite
SA Reynolds Creek
SA Riggins Group
SA Saddle Mountains Basalt
SA Saint Regis Formation
SA Salmon River
SA Salt Lake Formation
SA Sawtooth Range
SA Selway-Bitterroot Wilderness
SA Sevier orogenic belt
SA Snake River
SA Snake River basin
SA Snake River canyon
SA Stump Formation
SA Sunnyside Mine
SA Swan Peak Formation
SA Thaynes Formation
SA Twin Creek Limestone
SA U. S. Rocky Mountains
SA Wallace Formation
SA Wanapum Basalt
SA Wasatch fault zone
SA Wasatch Front
SA Wasatch Range
SA Wells Formation
SA Western Interior
SA Western Interior Seaway
SA Western Overthrust Belt
SA Windermere System
SA Wood River Formation
SA Wyoming Province
SA Yakima Basalt
SA Yellowjacket Formation
SA Yellowstone National Park

Idaho Batholith (1978)
In Bitterroot Range.
IN Index states as applicable.
BT United States
SA Bitterroot Range
SA Idaho
SA Montana

Idaho County
Valid through 1988. Search in combination with state term. After 1988, use specific county-state term.

Idaho County Idaho (1989)
Central Idaho. Before 1989, also search Idaho County AND Idaho.
CO N450800N464300
  W1142300W1165000
BT Idaho
BT United States
SA Selway-Bitterroot Wilderness
SA Seven Devils Mountains

Idaho Panhandle (1993)
N Idaho between Washington and Montana. Before 1993, also search Idaho AND Panhandle.
IN Index counties as applicable.
BT Idaho
BT United States
SA Bonner County Idaho
SA Boundary County Idaho
SA Kootenai County Idaho
SA Shoshone County Idaho

Idaho Springs
No longer a valid term for GeoRef. See Idaho Springs mining district if applicable.

Idaho Springs Colorado (1989)
Town in N central Colorado. Before 1989, also search Idaho Springs AND Colorado.
CO N394400N394400
  W1053100W1053100
BT Clear Creek County Colorado
BT Colorado
BT United States

Idaho Springs Formation (1985)
In Gunnison River Series. N central Colorado.
BT Precambrian
SA Colorado

Idaho Springs mining district (1993)
N central Colorado. Before 1993, also search Idaho Springs or Idaho Springs District.
BT Clear Creek County Colorado
BT Colorado
BT United States
SA metal ores
SA mines

idaite (1978)
BT sulfides

ideal waves (1978)
Before 1978, also search ocean waves AND ideal.
BT ocean waves
SA waves

identification (1978)
SA classification
SA definition
SA nomenclature
SA pattern recognition

Idiognathodus (1989)
Genus. Autoposting of microfossils to this term began in 1990.
BT Conodonta
BT microfossils

Idiognathoides (1989)
Genus. Autoposting of microfossils to this term began in 1990.
BT Conodonta
BT microfossils

Idjil Mauritania
use Fort Gouraud Mauritania

idocrase
use vesuvianite

Igaliko
No longer a valid term for GeoRef. As of 1993 see Igaliko Greenland.

Igaliko Greenland (1993)
Settlement 25 miles NE of Julianenhaab near S tip of island. Before 1993 also search Igaliko AND Greenland.
BT Greenland
BT Arctic region

Igarka
No longer a valid term for GeoRef. As of 1993, see Igarka Russian Federation.

Igarka Russian Federation (1993)
City on the Yenisei River north of Arctic Circle in Krasnoyarsk Kray. Before 1993, also search Igarka. This term has multiple hierarchies.
BT1 Krasnoyarsk Russian Federation
BT1 Russian Federation
BT1 Commonwealth of Independent States
BT2 Krasnoyarsk Russian Federation
BT2 Asia

IGBP (1993)
Acronym.
UF International Geosphere-Biosphere Program
SA biosphere
SA programs

IGC
use International Geological Congress

IGCP (1985)
Acronym.
IN To be used as an index term for all papers associated with the International Geological Correlation Programme.
UF International Geological Correlation Programme
SA associations
SA IUGS

Iglesiente (1978)
Region in SW Sardinia.
BT Sardinia Italy
BT Italy
BT Southern Europe
BT Europe

igneous activity (1978)
UF activity, igneous
SA geosynclines
SA igneous processes
SA magmas
SA nuees ardentes
SA orogeny
SA phreatomagmatism
SA volcanology

igneous and metamorphic rocks
No longer a valid GeoRef index term. Before 1972, was used on level 2 in subfile G. See igneous rocks; metamorphic rocks; crystalline rocks.

igneous and volcanic rocks
No longer a valid GeoRef index term. Before 1971, was used on level 1 in subfiles E and N. See igneous rocks; volcanic rocks.

igneous intrusions
Not a valid index term for GeoRef. Use intrusions and igneous rocks.

Igneous petrology
Not a valid GeoRef index term after 1970. Was used on level 1 in subfile N. Now use petrology or igneous rocks as applicable.

igneous processes (1978)
As of 1993, restricted to economic studies.
SA hydrothermal processes
SA igneous activity
SA magmas
SA mineral deposits, genesis
SA plutonic rocks
SA processes
SA volcanic processes

igneous rocks (1978)
Used to describe crystalline rocks of primary origin (list H). Before 1981, beresite was included as a narrower term. Autoposting of this term began in 1978.
IN See List H.
NT feldspathoid rocks
NT hypabyssal rocks
NT picrite
NT picrite porphyry
NT plutonic rocks
NT porphyry
NT pyrolite
NT vitrophyre
NT volcanic rocks
SA acidic composition
SA agglomerate
SA alkalic composition
SA alteration
SA amygdules
SA andesitic composition
SA aphanitic texture
SA ash flows

SA autometamorphism
SA banded structures
SA basaltic composition
SA basement
SA batholiths
SA calc-alkalic composition
SA calcic composition
SA complexes
SA country rocks
SA crystalline rocks
SA cumulates
SA dacitic composition
SA dike swarms
SA dikes
SA emplacement
SA extrusive rocks
SA fabric
SA fluid inclusions
SA gabbroic composition
SA granitic composition
SA graphic texture
SA inclusions
SA intrusions
SA labradorite
SA lava
SA layered intrusions
SA lineation
SA lithogeochemistry
SA lopoliths
SA mafic composition
SA magmas
SA magmatic differentiation
SA matrix
SA megacrysts
SA metaigneous rocks
SA metamorphism
SA metasomatism
SA nuees ardentes
SA ophiolite
SA ophite
SA paragenesis
SA peralkalic composition
SA peraluminous composition
SA petrography
SA petrology
SA phase equilibria
SA phenocrysts
SA pipes
SA plutons
SA porphyritic texture
SA protoliths
SA pyroclastic flows
SA rhyolitic composition
SA schlieren
SA spinifex texture
SA standard materials
SA stocks
SA symplectite
SA tholeiitic composition
SA ultramafic composition
SA vesicular texture
SA volcaniclastics
SA volcanology
SA wehrlite
SA xenocrysts

ignimbrite (1978)
UF flood tuff
BT pyroclastics
BT volcanic rocks
BT igneous rocks
SA ash flows
SA ash-flow tuff
SA pyroclastic flows
SA shirasu
SA tuff
SA volcanic ash
SA volcanic glass
SA welded tuff

ijolite (1978)
BT alkali gabbros
BT gabbros
BT plutonic rocks
BT igneous rocks
SA melteigite

SA urtite

Ikh-Khayrkhan (1978)
Ore bearing region in central Mongolia.
BT Mongolia
BT Far East
BT Asia

Iki Island (1978)
In Tsushima Strait off NW coast of Kyushu.
UF Iki-shima
BT Kyushu
BT Japan
BT Far East
BT Asia
SA Nagasaki Japan

Iki-shima
use Iki Island

Ikuno Mine (1978)
Metal ores. Hyogo Prefecture in central Honshu.
BT Hyogo Japan
BT Honshu
BT Japan
BT Far East
BT Asia
SA metal ores
SA mines

Ikushumbetsu (1978)
Coal mine and fossil locality within town area of Mikasa in Ishikari coal field, W central Hokkaido. After 1993, Ikushumbetsu Mine is used for the mine.
BT Hokkaido
BT Japan
BT Far East
BT Asia

Il'Intsa (1978)
Paleovolcano in Vinnitsa area of the Ukrainian Shield in W central Ukraine. This term has multiple hierarchies.
BT1 Ukraine
BT1 Europe
BT2 Ukraine
BT2 Commonwealth of Independent States

Ilerdian (1978)
Europe.
BT Paleogene
BT Tertiary
BT Cenozoic

Iles de Loos
use Los Islands

Iles Madeleine
use Magdalen Islands

Ili
No longer a valid term for GeoRef. As of 1993, see Ili Kazakhstan.

Ili Basin (1978)
River basin.
IN Index Kazakhstan and/or Xinjiang China as applicable.
BT Asia
SA Kazakhstan
SA Xinjiang China

Ili Kazakhstan (1993)
Town on Ili River in E Kazakhstan. This term has multiple hierarchies.
BT1 Kazakhstan
BT1 Central Asia
BT1 Asia
BT2 Kazakhstan
BT2 Commonwealth of Independent States

Ili Maussaq
use Ilimaussaq

Ilimaussaq (1978)
Alkaline intrusion in S Greenland, and Lovozero area of the Kola Peninsula in NW European Russia. Index Greenland and/or Russian Federation.
UF Ili Maussaq
UF Ilimaussaq Intrusion
UF Ilmaussaq
SA Greenland
SA Kola Peninsula
SA Lovozero Russian Federation
SA Russian Federation

Ilimaussaq Intrusion
use Ilimaussaq

Ilion
use Troy

Ilium
use Troy

Illawarra Coal Measures (1978)
Top of the measures is marked by Bulli Seam, which is conformably overlain by rocks of the Triassic Narrabeen Group. Illawarra District in SE New South Wales. Also search Illawarra.
BT Permian
BT Paleozoic
SA Australia
SA Bulli Seam
SA Narrabeen Group
SA New South Wales Australia

Ille-et-Vilaine
No longer a valid term for GeoRef. See Ille-et-Vilane France.

Ille-et-Vilaine France (1993)
Department in Brittany.
CO N473500N484500
W0010300W0022000
BT France
BT Western Europe
BT Europe
NT Redon France
NT Rennes France

Illinoian (1978)
Pertaining to third glacial stage of Pleistocene Epoch in North America.
UF Illinoisan
BT upper Pleistocene
BT Pleistocene
BT Quaternary
BT Cenozoic
SA Elsterian
SA Glasford Formation

Illinois (1978)
Autoposting of this term began in 1978.
CO N370000N423000
W0873000W0913000
BT United States
NT Adams County Illinois
NT Boone County Illinois
NT Brown County Illinois
NT Bureau County Illinois
NT Calhoun County Illinois
NT Carroll County Illinois
NT Cass County Illinois
NT Champaign County Illinois
NT Christian County Illinois
NT Clark County Illinois
NT Clay County Illinois
NT Clinton County Illinois
NT Cook County Illinois
NT Crawford County Illinois
NT Cumberland County Illinois
NT De Kalb County Illinois
NT Douglas County Illinois
NT Edwards County Illinois
NT Fayette County Illinois
NT Franklin County Illinois
NT Fulton County Illinois
NT Gallatin County Illinois
NT Greene County Illinois
NT Grundy County Illinois
NT Hamilton County Illinois
NT Hancock County Illinois
NT Hardin County Illinois
NT Henderson County Illinois
NT Henry County Illinois
NT Iroquois County Illinois
NT Jackson County Illinois
NT Jasper County Illinois
NT Jefferson County Illinois
NT Johnson County Illinois
NT Kane County Illinois
NT Knox County Illinois
NT La Salle County Illinois
NT Lake County Illinois
NT Lawrence County Illinois
NT Lee County Illinois
NT Livingston County Illinois
NT Logan County Illinois
NT Macon County Illinois
NT Madison County Illinois
NT Marion County Illinois
NT Marshall County Illinois
NT Mason County Illinois
NT McDonough County Illinois
NT McHenry County Illinois
NT McLean County Illinois
NT Mercer County Illinois
NT Monroe County Illinois
NT Montgomery County Illinois
NT Morgan County Illinois
NT Perry County Illinois
NT Pike County Illinois
NT Pope County Illinois
NT Pulaski County Illinois
NT Putnam County Illinois
NT Randolph County Illinois
NT Richland County Illinois
NT Rock Island County Illinois
NT Saint Clair County Illinois
NT Saline County Illinois
NT Sangamon
NT Sangamon County Illinois
NT Scott County Illinois
NT Shelby County Illinois
NT Stark County Illinois
NT Stephenson County Illinois
NT Tazewell County Illinois
NT Union County Illinois
NT Warren County Illinois
NT Washington County Illinois
NT Wayne County Illinois
NT White County Illinois
NT Will County Illinois
NT Williamson County Illinois
NT Winnebago County Illinois
NT Woodford County Illinois
SA Aux Vases Sandstone
SA Bainbridge Formation
SA Brassfield Formation
SA Burlington Limestone
SA Cache Valley
SA Carbondale Formation
SA Caseyville Formation
SA Cedar Valley Formation
SA Chattanooga Shale
SA Chesterian
SA Chouteau Limestone
SA Decorah Shale
SA Dubuque Formation
SA Eau Claire Formation
SA Fernvale Formation
SA Francis Creek Shale
SA Franconia Formation
SA Galena Dolomite
SA Glasford Formation
SA Glenwood Shale
SA Golconda Formation
SA Great Lakes region
SA Herrin Coal Member
SA Illinois Basin
SA Illinois Deep Hole Project

Before then, variants of the term may occur in GeoRef.

SA Illinois River
SA Illinois State Geological Survey
SA Imo Formation
SA Keokuk Limestone
SA Kewanee Group
SA La Salle Limestone
SA Lake Chicago
SA Lake Michigan
SA Lewis Creek
SA Maquoketa Formation
SA Mattoon Formation
SA Mazon Creek
SA Michigan Basin
SA Midcontinent
SA Midway Group
SA Mississippi Embayment
SA Mississippi River
SA Mississippi Valley
SA Mount Simon Sandstone
SA New Madrid region
SA Niagara Escarpment
SA Ohio River
SA Ohio River basin
SA Ohio River valley
SA Peoria Loess
SA Platteville Formation
SA Porters Creek Formation
SA Pottsville Group
SA Prairie du Chien Group
SA Racine Dolomite
SA Renault Formation
SA Richmond Group
SA Ripley Formation
SA Roxana Silt
SA Saint Laurent Limestone
SA Saint Louis Limestone
SA Saint Peter Sandstone
SA Sainte Genevieve Limestone
SA Salem Limestone
SA Shakopee Formation
SA Spergen Formation
SA Spoon Formation
SA Springfield Coal Member
SA Upper Mississippi Valley
SA Warsaw Formation
SA Wedron Formation
SA Wilcox Group
SA Woodfordian

Illinois Basin (1978)
Structural basin.
IN Index states or regions as applicable.
SA Aux Vases Sandstone
SA Illinois
SA Indiana
SA Kentucky
SA United States

Illinois Deep Hole Project (1985)
Borehole in Stephenson County, NW Illinois.
SA boreholes
SA continental crust
SA cores
SA heat flow
SA Illinois

Illinois River (1989)
River in many locations. In N and W Illinois, formed by the confluence of Des Plaines and Kankakee rivers and flows W and SW to the Mississippi River. Also in Jackson County Colorado, N Colorado. Also in Benton County Arkansas, NW Arkansas; and Adair County, Sequoyah and Cherokee County Oklahoma, NE Oklahoma. Also in Josephine County Oregon, SW Oregon.
IN Index states as applicable.
SA Arkansas
SA Colorado
SA Illinois

SA Oklahoma
SA Oregon
SA United States

Illinois State Geological Survey (1978)
BT survey organizations
BT government agencies
SA Illinois

Illinoisan
use Illinoian

illite (1978)
BT clay minerals
BT sheet silicates
BT silicates
SA crystallinity
SA hydromica
SA illitization
SA muscovite

illitization (1989)
SA alteration
SA clay minerals
SA illite
SA processes

Ilmaussaq
use Ilimaussaq

Ilmen Mountains (1978)
Near Chelyabinsk in the Central Urals. This term has multiple hierarchies.
BT1 Central Urals
BT1 Russian Federation
BT1 Commonwealth of Independent States
BT2 Central Urals
BT2 Urals
BT2 Commonwealth of Independent States
SA Chelyabinsk Russian Federation

ilmenite (1978)
UF menaccanite
BT oxides
SA heavy minerals
SA iron minerals
SA magnetic minerals
SA titanium ores

Iloilo
No longer a valid term for GeoRef. As of 1993 see Iloilo Philippine Islands.

Iloilo Basin (1978)
SW Panay.
BT Philippine Islands
BT Far East
BT Asia
SA Panay Island

Iloilo Philippine Islands (1993)
Province, SE Panay Island and Guimaras Island, Philippine Islands. Also a city. Before 1993 also search Iloilo AND Philippine Islands.
CO N102500N113500 E1231000E1220000
BT Philippine Islands
BT Far East
BT Asia
SA Panay Island

ilvaite (1978)
Autoposting of sorosilicates began in 1985.
BT sorosilicates
BT orthosilicates
BT silicates

image enhancement (1989)
Used in relation to remote sensing. Also search enhancement AND (imagery OR remote sensing).

SA data processing
SA imagery
SA Landsat
SA Magellan Program
SA radar methods
SA remote sensing
SA satellite methods
SA SeaMarc
SA Shuttle Imaging Radar

imagery (1978)
NT color imagery
SA geophysical surveys
SA ground truth
SA image enhancement
SA Landsat
SA Magellan Program
SA mosaics
SA pattern recognition
SA pixels
SA remote sensing
SA SeaMarc
SA seismic migration
SA Shuttle Imaging Radar

Imandra (1978)
Lake on W Kola Peninsula, Murmansk Oblast. This term has multiple hierarchies.
IN Index Murmansk Russian Federation as applicable.
BT1 Russian Federation
BT1 Commonwealth of Independent States
BT2 Europe
SA Kola Peninsula
SA Murmansk Russian Federation

Imataca Complex (1978)
In Serrania de Imataca S of Orinoco Delta.
BT Bolivar Venezuela
BT Venezuela
BT South America

imbricate tectonics (1982)
Before 1982, also search imbrication AND tectonics.
UF schuppen texture
UF tectonic imbrication
BT tectonics
SA duplexes
SA imbrication
SA reverse faults
SA tectonic wedges
SA thrust faults

imbrication (1978)
As of 1982, use imbricate tectonics for tectonic imbrication.
BT planar bedding structures
BT sedimentary structures
SA bedding plane irregularities
SA imbricate tectonics

Imbrium Basin
use Sea of Rains

IMF
Not a valid term for GeoRef. Acronym for many concepts including intense magnetic field, internal magnetic focus, International Monetary Fund, interplanetary magnetic field and inventory master file.

immiscibility (1978)
UF miscibility
SA magmas
SA mixing
SA phase equilibria
SA solubility

Imo Formation (1989)
Before 1989, also search Imo Shale AND United States.
BT Chesterian
BT Upper Mississippian

BT Mississippian
BT Carboniferous
BT Paleozoic
SA Arkansas
SA Illinois
SA Oklahoma

Imo Shale (1989)
Paleocene to lower Eocene. Before 1989, also search Imo Formation AND Africa.
IN Index ages as applicable.
BT Paleogene
BT Tertiary
BT Cenozoic
SA Benin
SA lower Eocene
SA Nigeria
SA Paleocene

imogolite (1978)
BT clay minerals
BT sheet silicates
BT silicates

impact craters (1993)
BT impact features
NT astroblemes
NT meteor craters
SA cryptoexplosion features
SA geomorphology

impact features (1978)
Autoposting of this term began in 1978.
UF features, impact
NT impact craters
SA Charlevoix
SA cratering
SA craters
SA cryptoexplosion features
SA geomorphology
SA impactite
SA impacts
SA meteorites
SA shatter cones
SA suevite

impact phenomena
A valid term through 1972. From 1972 through 1978 impact was used. After 1978, use impacts.

impact statements (1978)
For official documents detailing the impact of contemplated works or land use on the environment.
SA conservation
SA ecology
SA engineering geology
SA environmental geology
SA geologic hazards
SA land use
SA nuclear facilities
SA pollution
SA reclamation
SA risk assessment
SA waste disposal

impactite (1978)
Before 1972, also search impactites.
UF impactites
SA cryptoexplosion features
SA Darwin glass
SA impact features
SA metamorphic rocks
SA meteorites
SA suevite
SA zhamanshinite

impactites
Not a valid GeoRef index term after 1971. Was used on level 1 in subfiles G and N.
use impactite

impacts (1978)
Before 1973, also search impact phenomena; impact.

SA craters
SA cryptoexplosion features
SA geomorphology
SA impact features
SA meteor craters
SA meteorites
SA planetology

impedance (1985)
SA acoustical logging
SA acoustical methods
SA acoustical waves
SA elastic waves
SA electromagnetic methods
SA half-space
SA inverse problem
SA seismic logging
SA seismic methods
SA seismology

Imperial County
Valid through 1988. Search in combination with state term. After 1988, use specific county-state term.

Imperial County California (1989)
S California. Before 1989, also search Imperial County AND California.
CO N323500N332500
W1142500W1161000
BT California
BT United States
NT Brawley Fault
NT Cargo Muchacho Mountains
NT East Mesa KGRA
NT El Centro California
NT Imperial Fault
NT Salton Sea geothermal field
SA Chocolate Mountains
SA Mexicali Valley
SA Salton Sea
SA San Jacinto Fault

Imperial Fault (1989)
Imperial Valley, California.
BT Imperial County California
BT California
BT United States
SA Imperial Valley

Imperial Formation (1985)
Pliocene and Miocene in S California. Formerly considered Pleistocene. Also occurs in other locations.
IN Index ages as applicable.
SA California
SA Miocene
SA Pliocene

Imperial Valley (1978)
Imperial County in extreme SE California and Baja California, Mexico.
IN Index counties or regions as applicable.
SA Baja California
SA Borrego Mountain earthquake 1968
SA California
SA Colorado Desert
SA East Mesa KGRA
SA El Centro California
SA El Centro earthquake 1940
SA Imperial Fault
SA Imperial Valley earthquake 1979
SA Mexicali Valley

Imperial Valley earthquake 1940
use El Centro earthquake 1940

Imperial Valley earthquake 1979 (1985)
Epicenter in Imperial Valley, S California.

BT earthquakes
SA California
SA Imperial Valley

import (1981)
UF imports
SA economics

import value (1981)
SA economics

imports
use import

impregnated deposits (1981)
SA mineral deposits, genesis

impurities (1978)
A general term used mostly in reference to water.
SA decontamination
SA drinking water
SA ground water
SA polluted water
SA pollution
SA potability
SA purification
SA water
SA water quality
SA water treatment

In
use indium

in situ (1978)
Applicable to many fields, especially rocks, soils, or fossils when in the situation in which they were originally formed or deposited.
UF in-situ
SA retorting

in-situ
use in situ

Inadunata (1978)
Subclass.
BT Crinoidea
BT Crinozoa
BT Echinodermata
BT Invertebrata

Inangahua
No longer a valid term for GeoRef. See Inangahua New Zealand.

Inangahua New Zealand (1993)
Village in NW South Island.
IN Also index South Island.
BT Nelson New Zealand
BT New Zealand
BT Australasia
SA South Island

Inari
No longer a valid term for GeoRef. See Inari Finland.

Inari Finland (1993)
Town on Lake Inari in N Finland. Before 1993, also search Inari. Before 1978, also search Enari.
BT Finland
BT Scandinavia
BT Western Europe
BT Europe

Inarticulata (1978)
Autoposting of this term began in 1978.
BT Brachiopoda
BT Invertebrata
NT Lingula
SA Articulata

incarbonization
use coalification

Inceptisols (1978)
Autoposting of this term began in 1978.
BT soils
NT Andepts

SA soil group

incertae sedis
use miscellanea

inclination, magnetic
use magnetic inclination

inclined folds (1978)
Before 1978, also search folds AND inclined.
BT folds
SA orientation

inclined seismic interface (1981)
SA curved seismic interface
SA discontinuities
SA seismology

inclinometers (1985)
SA instruments
SA magnetic inclination

inclusions (1978)
Used for studies on enclosures or enclaves in rocks or minerals. For liquid or gaseous inclusions contained within crystals, use fluid inclusions. Autoposting of this term began in 1989. Before 1993, also search enclaves.
IN Also index specific type of inclusion if applicable [e.g. mineral inclusions, xenoliths].
NT calcium-aluminum inclusions
NT fluid inclusions
NT mineral inclusions
NT xenoliths
SA assimilation
SA country rocks
SA crystal growth
SA crystal structure
SA geochemistry
SA geologic barometry
SA geologic thermometry
SA host rocks
SA igneous rocks
SA isotopes
SA lava
SA magmas
SA mineral deposits, genesis
SA minerals
SA P-T conditions
SA petrology
SA phase equilibria
SA rocks
SA schlieren
SA xenocrysts

incoalation
use coalification

incompressibility modulus
use bulk modulus

Incorporated Research Institutions for Seismology network
use IRIS network

increment
use recharge

incrustations
use encrustations

Indarch Meteorite (1989)
Impact at Shusta, Elisavetpol, Azerbaidzhan, Transcaucasia.
BT enstatite chondrites
BT chondrites
BT stony meteorites
BT meteorites
SA Azerbaidzhan

Indefatigable Island
use Chaves Island

Independence County Arkansas (1989)
NE central Arkansas. Before 1989, search Independence County AND Arkansas.

CO N353200N355700
W0911200W0915300
BT Arkansas
BT United States
NT Batesville Arkansas

Independence Mountains (1989)
NE Nevada.
IN Index county or regions as applicable.
SA Elko County Nevada

index beds
use key beds

index fauna
use index fossils

index flora
use index fossils

index fossils (1978)
Before 1993, also search index flora and index fauna.
UF guide fossils
UF index fauna
UF index flora
SA assemblages
SA biostratigraphy
SA fossils
SA microfossils
SA paleontology
SA stratigraphy

index maps (1978)
BT maps
SA site location maps

index of refraction
use refractive index

India (1978)
CO N070000N370000
E0970000E0680000
BT Indian Peninsula
BT Asia
NT Ajay River
NT Andhra Pradesh India
NT Beas River
NT Bengal Islands
NT Bihar India
NT Bilaspur India
NT Bundelkhand
NT Cauvery Basin
NT Coromandel Coast
NT Damodar Valley
NT Deccan Plateau
NT Delhi India
NT Dharwar Craton
NT Ghats
NT Goa India
NT Godavari Valley
NT Gujarat India
NT Haryana India
NT Himachal Pradesh India
NT Jumna River
NT Karnataka India
NT Kerala India
NT Kolar schist belt
NT Krishna
NT Laccadive Islands
NT Madhya Pradesh India
NT Mahanadi Valley
NT Maharashtra India
NT Narmada River
NT Narmada Valley
NT Narmada-Son Lineament
NT Northeastern India
NT Orissa India
NT Pondicherry India
NT Pranhita-Godavari Valley
NT Punjab India
NT Rajasthan India
NT Satpura Range
NT Sikkim India
NT Singhbhum shear zone
NT Son Valley
NT Tamil Nadu India
NT Trans-Aravalli Vindhyan Basin

Before then, variants of the term may occur in GeoRef.

NT Uttar Pradesh India
NT West Bengal India
SA Anantnag
SA Aravalli System
SA Ariyalur Stage
SA Athgarh Sandstone
SA Badami Series
SA Bagh Beds
SA Banded Gneissic Complex
SA Barail Group
SA Barakar Stage
SA Barren Measures
SA Bengal
SA Bhander Group
SA Bhuj Series
SA Bijawar System
SA Brahmaputra River
SA Chainpur Meteorite
SA Cuddalore Series
SA Cuddapah System
SA Damuda Series
SA Deccan Traps
SA Delhi Supergroup
SA Dhajala Meteorite
SA Dharwars
SA Doda
SA Ganges River
SA Gangpur Series
SA Gondwana
SA Gondwana System
SA Himalayas
SA Indian Plate
SA Indian Shield
SA Indo-Australian Plate
SA Indus-Yarlung Zangbo suture zone
SA Intertrappean Beds
SA Iron Ore Group
SA Jabalpur Series
SA Jammu
SA Jammu and Kashmir
SA Kaimur Sandstone
SA Kaladgi System
SA Kamthi Formation
SA Karakoram
SA Karewa Group
SA Karharbari Stage
SA Kasauli Series
SA Kashmir
SA Kashmir Valley
SA Kishtwar
SA Kohistan
SA Krol Formation
SA Kunavaram Series
SA Kurnool System
SA Ladakh
SA Lathi Formation
SA Lesser Himalayas
SA Main Boundary Fault
SA Main Central Thrust
SA Muth Quartzite
SA Naga Hills
SA Nari Series
SA Naubug
SA Neyveli Lignite
SA Niniyur Group
SA Panchet Series
SA Peninsular Gneiss
SA Raghavapuram Shales
SA Rajmahal Series
SA Raniganj Formation
SA Riasi
SA Sausar Series
SA Semarkona Meteorite
SA Semri Series
SA Shergotty Meteorite
SA Singhbhum Granite
SA Siwalik Range
SA Siwalik System
SA Srinagar
SA Subathu Formation
SA Surma Group
SA Tal Formation
SA Talchir Formation
SA Talchir Series

SA Tawi Valley
SA Tons Member
SA Vindhyan
SA Vindhyan Basin
SA Warkalli Formation

Indian Ocean (1978)
Many islands, formerly narrower terms of Indian Ocean, are now narrower terms of Indian Ocean Islands. For some, the change was made in 1981; for others, in 1985. Red Sea became a narrower term in 1981. Autoposting of this term began in 1978.
CO S700000N250000
  E1200000E0200000
NT Agulhas Bank
NT Andaman Sea
NT Arabian Sea
NT Argo abyssal plain
NT Bay of Bengal
NT Bengal Fan
NT Carlsberg Ridge
NT Cuvier abyssal plain
NT East Indian Ocean
NT Exmouth Plateau
NT Gascoyne abyssal plain
NT Great Australian Bight
NT Kerguelen Plateau
NT Madagascar Basin
NT Mascarene Basin
NT Mid-Indian Ridge
NT Mozambique Channel
NT Ninetyeast Ridge
NT Owen fracture zone
NT Perth abyssal plain
NT Red Sea
NT Somali Basin
NT Southwest Indian Ridge
NT West Indian Ocean
NT Wharton Basin
SA Antarctic Ocean
SA Browse Basin
SA DSDP Site 211
SA DSDP Site 212
SA DSDP Site 213
SA DSDP Site 214
SA DSDP Site 215
SA DSDP Site 216
SA DSDP Site 217
SA DSDP Site 218
SA DSDP Site 220
SA DSDP Site 231
SA DSDP Site 232
SA DSDP Site 233
SA DSDP Site 236
SA DSDP Site 237
SA DSDP Site 238
SA DSDP Site 241
SA DSDP Site 245
SA DSDP Site 248
SA DSDP Site 251
SA DSDP Site 253
SA DSDP Site 254
SA DSDP Site 255
SA DSDP Site 256
SA DSDP Site 257
SA DSDP Site 259
SA DSDP Site 260
SA DSDP Site 261
SA DSDP Site 262
SA DSDP Site 263
SA DSDP Site 264
SA DSDP Site 265
SA Indian Ocean Islands
SA Indian Plate
SA Indo-Australian Plate
SA Kerguelen Islands
SA Leg 22
SA Leg 23
SA Leg 24
SA Leg 25
SA Leg 26
SA Leg 27

SA Leg 28
SA Leg 115
SA Leg 116
SA Leg 117
SA Leg 118
SA Leg 119
SA Leg 120
SA Leg 121
SA Leg 122
SA Leg 123
SA Mahe Island
SA ODP Site 705
SA ODP Site 706
SA ODP Site 707
SA ODP Site 708
SA ODP Site 709
SA ODP Site 710
SA ODP Site 711
SA ODP Site 712
SA ODP Site 713
SA ODP Site 714
SA ODP Site 715
SA ODP Site 716
SA ODP Site 717
SA ODP Site 718
SA ODP Site 719
SA ODP Site 720
SA ODP Site 721
SA ODP Site 722
SA ODP Site 723
SA ODP Site 724
SA ODP Site 725
SA ODP Site 726
SA ODP Site 727
SA ODP Site 728
SA ODP Site 729
SA ODP Site 730
SA ODP Site 731
SA ODP Site 732
SA ODP Site 733
SA ODP Site 734
SA ODP Site 735
SA ODP Site 736
SA ODP Site 737
SA ODP Site 747
SA ODP Site 752
SA ODP Site 753
SA ODP Site 754
SA ODP Site 755
SA ODP Site 756
SA ODP Site 757
SA ODP Site 758
SA ODP Site 759
SA ODP Site 760
SA ODP Site 761
SA ODP Site 762
SA ODP Site 764
SA ODP Site 765
SA ODP Site 766
SA Seychelles
SA Shark Bay
SA Spencer Gulf
SA Table Mountain Group
SA Timor Sea

Indian Ocean Islands (1981)
Before 1981, many of these islands were considered narrower terms of Indian Ocean; many others were so considered until 1985.
NT Amsterdam Island
NT Comoro Islands
NT Crozet Islands
NT Madagascar
NT Maldive Islands
NT Marion Island
NT Mascarene Islands
NT Mauritius
NT Prince Edward Island Group
NT Seychelles
SA Bengal Islands
SA Christmas Island
SA Indian Ocean
SA Laccadive Islands

SA Saint Paul Island

Indian Ocean Ridge
  use Mid-Indian Ridge

Indian Peninsula (1981)
BT Asia
NT Afghanistan
NT Bangladesh
NT Bhutan
NT Godavari River
NT India
NT Jammu and Kashmir
NT Kohistan
NT Nepal
NT Pakistan
NT Sri Lanka
SA Brahmaputra River
SA Indian Shield
SA Indo-Australian Plate
SA Indus-Yarlung Zangbo suture zone
SA Kosi Basin
SA Main Boundary Fault
SA Main Central Thrust

Indian Plate (1978)
Redefined. As of 1993, ancient plate pre-dating modern Indo-Australian Plate. Before 1993, also used for the modern tectonic plate extending from the Himalayas to the Tonga Trench in the Pacific Ocean which included Australia.
SA Bangladesh
SA India
SA Indian Ocean
SA Pacific Ocean
SA Pakistan
SA plate tectonics
SA plates
SA Xizang China

Indian reservations (1985)
IN Index specific reservations and states as applicable.
SA conservation
SA human activity
SA land use
SA legislation
SA Navajo Indian Reservation
SA policy
SA San Carlos Indian Reservation
SA United States
SA Yakima Indian Reservation

Indian Ridge
  use Mid-Indian Ridge

Indian River (1989)
In many locations. Search in combination with appropriate state or province.
IN Index states and/or provinces as applicable.
SA British Columbia
SA Delaware
SA Florida
SA Ontario
SA Yukon Territory

Indian Shield (1978)
Plateau region in S and central India plus Sri Lanka.
IN Index countries as applicable.
UF Peninsular Shield
SA Deccan Plateau
SA India
SA Indian Peninsula
SA Sri Lanka

Indian-Antarctic Ridge
  use Southeast Indian Ridge

Indiana (1978)
Autoposting of this term began in 1978.
CO N374500N414500
  W0844500W0881000

BT United States
NT Adams County Indiana
NT Benton County Indiana
NT Blackford County Indiana
NT Boone County Indiana
NT Brown County Indiana
NT Carroll County Indiana
NT Cass County Indiana
NT Clark County Indiana
NT Clinton County Indiana
NT Crawford County Indiana
NT De Kalb County Indiana
NT Decatur County Indiana
NT Delaware County Indiana
NT Fayette County Indiana
NT Floyd County Indiana
NT Franklin County Indiana
NT Fulton County Indiana
NT Grant County Indiana
NT Greene County Indiana
NT Hamilton County Indiana
NT Hancock County Indiana
NT Harrison County Indiana
NT Henry County Indiana
NT Howard County Indiana
NT Jackson County Indiana
NT Jasper County Indiana
NT Jefferson County Indiana
NT Johnson County Indiana
NT Lake County Indiana
NT Lawrence County Indiana
NT Madison County Indiana
NT Marion County Indiana
NT Marshall County Indiana
NT Martin County Indiana
NT Monroe County Indiana
NT Montgomery County Indiana
NT Morgan County Indiana
NT Newton County Indiana
NT Noble County Indiana
NT Orange County Indiana
NT Perry County Indiana
NT Pike County Indiana
NT Porter County Indiana
NT Posey County Indiana
NT Pulaski County Indiana
NT Putnam County Indiana
NT Randolph County Indiana
NT Ripley County Indiana
NT Rush County Indiana
NT Scott County Indiana
NT Shelby County Indiana
NT Sullivan County Indiana
NT Tippecanoe County Indiana
NT Union County Indiana
NT Vigo County Indiana
NT Warren County Indiana
NT Washington County Indiana
NT Wayne County Indiana
NT White County Indiana
NT Whitley County Indiana
SA Aux Vases Sandstone
SA Beech Creek Limestone Member
SA Borden Group
SA Brassfield Formation
SA Chesterian
SA Cincinnati Arch
SA Dugger Formation
SA Eau Claire Formation
SA Eden Shale
SA Fairview Formation
SA Golconda Formation
SA Great Lakes region
SA Harrodsburg Limestone
SA Illinois Basin
SA Jeffersonville Limestone
SA Kittanning Formation
SA Kope Formation
SA Lake Chicago
SA Lake Maumee
SA Lake Michigan
SA Mansfield Formation
SA Maumee River valley
SA Michigan Basin

SA Midcontinent
SA Mount Simon Sandstone
SA New Albany Shale
SA New Madrid region
SA Ohio River
SA Ohio River basin
SA Ohio River valley
SA Petersburg Formation
SA Ramp Creek Formation
SA Renault Formation
SA Richmond Group
SA Saint Louis Limestone
SA Saint Peter Sandstone
SA Sainte Genevieve Limestone
SA Salem Limestone
SA Spergen Formation
SA Staunton Formation
SA Waldron Shale
SA Warsaw Formation
SA Wedron Formation
SA White River
SA Whitewater River valley

Indiana County
Valid through 1988. Search in combination with state term. After 1988, use specific county-state term.

Indiana County Pennsylvania (1989)
W central Pennsylvania. Before 1989, also search Indiana County AND Pennsylvania.
CO N402200N405500
W0784800W0792700
BT Pennsylvania
BT United States

indicators (1978)
SA depth indicators
SA diagenesis
SA encrustations
SA geochemical indicators
SA gossan
SA hydrocarbon indicators
SA paleoclimatology
SA paleoecology
SA plate tectonics

Indigirka River (1978)
NE Yakutia. Flows into East Siberian Sea. Also search Indigirka. This term has multiple hierarchies.
BT1 Yakutia Russian Federation
BT1 Russian Federation
BT1 Commonwealth of Independent States
BT2 Yakutia Russian Federation
BT2 Asia

indium (1978)
Autoposting of metals to this term began in 1989.
UF In
BT metals

Indo-Antarctic Ridge
use Southeast Indian Ridge

Indo-Australian Plate (1993)
Modern plate. Before 1993, also search Indian Plate and Australian Plate.
SA Australia
SA Australian Plate
SA Bangladesh
SA Himalayas
SA India
SA Indian Ocean
SA Indian Peninsula
SA New Guinea
SA New Zealand
SA Pacific Ocean
SA Pakistan
SA plate tectonics
SA plates
SA Xizang China

Indo-China (French)
No longer a valid GeoRef index term. Before 1945, was used on level 1 in subfile E.
use Indochina

Indo-Sinian Orogeny
use Indosinian Orogeny

Indochina (1978)
The SE peninsula of Asia including the Malay Peninsula. (French Indochina consisted of current Cambodia, Laos, and Vietnam). Before 1960, also search French Indochina; Indo-China (French).
IN Index countries as applicable.
UF Farther India
UF French Indochina
UF Indo-China (French)
BT Far East
BT Asia
SA Burma
SA Cambodia
SA Indosinian Orogeny
SA Laos
SA Malay Peninsula
SA Malaysia
SA Thailand
SA Vietnam
SA West Malaysia

indochinite (1978)
BT tektites

Indonesia (1978)
Before 1960, also search Dutch East Indies. Before 1945, also search Netherland India.
CO S113000N063000
E1413000E0950000
UF Dutch East Indies
UF Netherland India
BT Far East
BT Asia
NT Bali
NT Bangka
NT Billiton
NT Celebes
NT Irian Jaya Indonesia
NT Java
NT Kalimantan Indonesia
NT Krakatoa
NT Moluccas
NT Pantar
NT Sumatra
NT Sunda Arc
SA Arafura Sea
SA ASEAN
SA Borneo
SA Indonesian Seas
SA Kabuh Formation
SA Kalibeng Formation
SA Lesser Sunda Islands
SA Malay Archipelago
SA Pucangan Formation
SA Timor

Indonesian Seas (1981)
Includes small seas of the Pacific Ocean within Indonesia. Autoposting of Pacific Ocean to this term began in 1990.
BT West Pacific
BT Pacific Ocean
NT Banda Sea
NT Celebes Sea
NT Java Sea
SA Banda Arc
SA Indonesia

Indosinian Orogeny (1993)
Triassic and Lower Jurassic. China, E and central Xizang and Indochina. Before 1993, also search Indo-Sinian Orogeny, Indosinian and Indosinian Movement.

UF Indo-Sinian Orogeny
BT Mesozoic
SA China
SA Indochina
SA Lower Jurassic
SA orogeny
SA tectonics
SA Triassic

Indre
No longer a valid term for GeoRef. See Indre France.

Indre France (1993)
Department in W central France.
CO N462000N471700
E0021000E0005000
BT France
BT Western Europe
BT Europe
SA Touraine

Indre-et-Loire
No longer a valid term for GeoRef.
use Indre-et-Loire France

Indre-et-Loire France (1993)
Department in N central France.
CO N464500N474500
E0012000E0000000
UF Indre-et-Loire
BT France
BT Western Europe
BT Europe
NT Tours France
SA Touraine

induced
No longer a valid GeoRef index term. Before 1972, was used on level 2 in subfile G. See induced earthquakes; induced magnetization; induced polarization; induction.

induced earthquakes (1981)
BT earthquakes
SA earthquake prediction
SA engineering geology
SA Parkfield earthquakes
SA seismology

induced magnetization (1981)
BT magnetization

induced polarization (1978)
SA coupling
SA electrical field
SA electrical methods
SA electrical surveys
SA geophysical methods
SA geophysical surveys
SA polarization
SA well-logging

induction (1978)
Used for geophysical properties.
SA electromagnetic methods
SA geophysical methods
SA magnetic field
SA paleomagnetism
SA well-logging

inductively coupled plasma
use inductively coupled plasma methods

inductively coupled plasma methods (1985)
UF inductively coupled plasma
UF inductively coupled plasma spectroscopy
BT spectroscopy
SA chemical analysis
SA coupling
SA emission spectroscopy
SA mass spectroscopy
SA plasma emission spectroscopy

inductively coupled plasma spectroscopy
  use inductively coupled plasma methods

Indus River (1978)
  Rises in SW Tibet and flows into Arabian Sea.
  IN Index Jammu and Kashmir, Pakistan, and Xinjiang China as applicable.
  BT Asia
  SA Jammu and Kashmir
  SA Pakistan
  SA Xizang China

Indus suture zone
  use Indus-Yarlung Zangbo suture zone

Indus-Tsangpo suture zone
  use Indus-Yarlung Zangbo suture zone

Indus-Yarlung Zangbo suture zone (1993)
  Extends from Kohistan through Ladakh and Xizang. Before 1993, also search Indus Zone, Indus suture zone, Indus-Tsangpo suture zone for sections primarily outside Xizang, and Yarlung Zangbo suture zone for sections in Xizang.
  IN Index countries and regions as applicable.
  UF Indus suture zone
  UF Indus-Tsangpo suture zone
  UF Yarlung Zangbo suture zone
  BT Asia
  SA China
  SA India
  SA Indian Peninsula
  SA Jammu and Kashmir
  SA Kohistan
  SA Ladakh
  SA Pakistan
  SA Qinghai-Xizang Plateau
  SA Xizang China

industrial ash
  use ash

industrial materials
  Not a valid GeoRef index term after 1968. Was used on level 1 in subfile B.
  use industrial minerals

industrial minerals (1978)
  Before 1969, also search industrial materials.
  UF industrial materials
  SA abrasives
  SA absorbent materials
  SA carbonado
  SA carborundum
  SA corundum deposits
  SA garnet deposits
  SA graphite deposits
  SA industrial minerals maps
  SA minerals
  SA nonmetal deposits
  SA refractory materials
  SA zeolite deposits

industrial minerals maps (1982)
  Before 1982, also search industrial minerals AND maps.
  BT maps
  SA economic geology maps
  SA industrial minerals

industrial patents (1981)
  SA economics
  SA industry

industrial waste (1978)
  UF waste, industrial
  SA environmental geology
  SA liquid waste
  SA PCBs
  SA pollution
  SA radioactive waste
  SA solid waste
  SA waste disposal
  SA waste disposal sites
  SA waste water

industrialized countries (1985)
  SA communist countries
  SA economics
  SA international cooperation
  SA policy
  SA Third World
  SA Western World

industry (1978)
  Used as a general term.
  SA annual report
  SA chemical industry
  SA companies
  SA economic geology
  SA engineering geology
  SA environmental geology
  SA industrial patents

inert gases
  use noble gases

inertial seismographs (1981)
  UF inertial seismometers
  BT seismographs

inertial seismometers
  use inertial seismographs

inertinite (1978)
  Autoposting of this term began in 1981.
  BT macerals
  NT fusinite
  NT micrinite
  NT sclerotinite

infiltration (1978)
  SA ground water
  SA hydrology
  SA percolation
  SA recharge
  SA Richards equation
  SA soils
  SA unsaturated zone
  SA water regimes
  SA waterlogging

infiltration galleries (1981)
  SA ground water
  SA underground channels

infinite models (1981)
  SA mathematical models
  SA models
  SA semi-infinite models

influence (1981)
  SA effects

information systems (1978)
  NT geographic information systems
  SA computer networks
  SA CRIB
  SA data bases
  SA data handling
  SA data processing
  SA data retrieval
  SA data storage
  SA Earth Observing System
  SA GeoRef
  SA Geoscan
  SA Pascal
  SA systems
  SA systems analysis

Infracambrian (1978)
  Autoposting of upper Precambrian and Precambrian to this term ended in 1989. As of 1989, upper Proterozoic and Proterozoic are autoposted to this term. Autoposting of upper Precambrian and Precambrian to this term began in 1990.
  BT upper Proterozoic
  BT Proterozoic
  BT upper Precambrian
  BT Precambrian
  SA Ediacaran
  SA Eocambrian
  SA Vendian

infrared
  A valid term through 1977. After 1977, see infrared methods, infrared spectra, infrared spectroscopy or infrared surveys as applicable.

infrared exploration
  Not a valid index term after 1971. Was used on level 1 in subfiles G and N. Now use infrared methods or infrared surveys.

infrared methods (1978)
  BT geophysical methods
  SA chemical analysis
  SA infrared spectroscopy
  SA infrared surveys
  SA remote sensing
  SA thermal emission

infrared spectra (1978)
  Before 1978, also search infrared.
  UF IR spectra
  BT spectra
  SA chemical analysis
  SA infrared spectroscopy
  SA spectroscopy

infrared spectroscopy (1978)
  Before 1978, also search spectroscopy AND infrared.
  BT spectroscopy
  SA analysis
  SA chemical analysis
  SA infrared methods
  SA infrared spectra

infrared surveys (1978)
  BT geophysical surveys
  BT surveys
  SA infrared methods
  SA remote sensing

Ingaladhal
  use Ingaldhal

Ingaldhal (1978)
  Ore bearing region in Chitradurga District in central Karnataka.
  UF Ingaladhal
  BT Chitradurga India
  BT Karnataka India
  BT India
  BT Indian Peninsula
  BT Asia

Ingham County
  Valid through 1988. Search in combination with state term. After 1988, use specific county-state term.

Ingham County Michigan (1989)
  S central Michigan. Before 1989, also search Ingham County AND Michigan.
  CO N422300N424700 W0841000W0844000
  BT Michigan Lower Peninsula
  BT Michigan
  BT United States
  SA Lansing Michigan

Ingleborough (1978)
  Mountain in W North Yorkshire in N England.
  CO N541000N541000 W0022300W0022300
  BT North Yorkshire England
  BT Yorkshire England
  BT England
  BT Great Britain
  BT United Kingdom
  BT Western Europe
  BT Europe

Ingleton
  No longer a valid term for GeoRef as of 1993. See Ingleton England.

Ingleton England (1993)
  Towns in North Yorkshire and Durham in N England. Before 1993, also search Ingleton.
  IN Index counties as applicable.
  BT England
  BT Great Britain
  BT United Kingdom
  BT Western Europe
  BT Europe
  SA Durham England
  SA North Yorkshire England

Inglewood
  Valid through 1988. Search in combination with state term. After 1988, use specific city-state term.

Inglewood California (1989)
  City in S California. Before 1989, search Inglewood AND California.
  CO N335800N335800 W1182200W1182200
  BT Los Angeles County California
  BT California
  BT United States

Inglewood Fault
  use Newport-Inglewood Fault

Ingul River (1978)
  S Ukraine. Also search Ingul. This term has multiple hierarchies.
  BT1 Ukraine
  BT1 Europe
  BT2 Ukraine
  BT2 Commonwealth of Independent States

Ingulets River (1978)
  S Ukraine. Also search Ingulets. This term has multiple hierarchies.
  BT1 Ukraine
  BT1 Europe
  BT2 Ukraine
  BT2 Commonwealth of Independent States
  SA Krivoy Rog Basin

Ingur (River)
  use Inguri River

Inguri River (1978)
  Flows into Black Sea. Also search Inguri.
  UF Ingur (River)
  BT Georgian Republic
  BT Europe

inhomogeneity
  use heterogeneity

injection (1978)
  SA artificial recharge
  SA engineering geology
  SA environmental geology
  SA fluid injection
  SA gas injection
  SA mercury injection
  SA reinjection wells
  SA waste disposal

injection wells
  use fluid injection

Inland Sea (1978)
  Extends E-W with W Honshu on the N, Shikoku on the S, and Kyushu on the W. Connected with Pacific Ocean by 4 channels.
  UF Seto Inland Sea
  UF Seto Sea
  UF Seto-chi-umi

UF Seto-Naikai
UF Setouchi
BT Japan
BT Far East
BT Asia
SA Honshu
SA Kyushu
SA Osaka Bay
SA Shikoku

inland seas
use epicontinental seas

inlets (1978)
BT shore features
SA bays
SA coastal environment
SA coastlines
SA embayments
SA fluvial features
SA geomorphology
SA lacustrine features
SA shorelines
SA tidal inlets

Inn Valley (1978)
River valley.
BT Central Europe
BT Europe
SA Baumkirchen
SA Bavaria Germany
SA Graubunden Switzerland
SA Tyrol Austria

innelite (1978)
Autoposting of sorosilicates began in 1985.
BT sorosilicates
BT orthosilicates
BT silicates

inner core (1978)
BT core
SA outer core

Inner Hebrides (1978)
Islands immediately off W coast of Scotland. Separated from Outer Hebrides by the Straits of The Minch, the Little Minch, and the Sea of Hebrides.
BT Hebrides
BT Scotland
BT Great Britain
BT United Kingdom
BT Western Europe
BT Europe
NT Islay
NT Isle of Skye
NT Mull Island
NT Raasay
NT Rhum
SA Highland region Scotland
SA Outer Hebrides
SA Strathclyde region Scotland

Inner Mongolia
No longer a valid term for GeoRef.
use Inner Mongolia China

Inner Mongolia China (1993)
Autonomous region N China. Before 1993, also search Inner Mongolia. Before 1978, also search Nei Mongol and Nei Monggol.
CO N374000N533000
 E1260000E0980000
UF Inner Mongolia
UF Nei Monggol
UF Nei Mongol
BT China
BT Far East
BT Asia
NT Da Hinggan Ling
NT Ordos Basin
SA Amur Basin
SA Amur River
SA Gobi Desert
SA Kerulen River

inner shelf (1978)
BT continental shelf
SA outer shelf

inner slope (1978)
BT continental slope
SA outer slope
SA shelf-slope break

Inner Slovakian Carpathians (1993)
Central Slovakia. This term has multiple hierarchies.
CO N480400N491700
 E0211900E0165600
BT1 Slovakian Carpathians
BT1 Carpathians
BT1 Europe
BT2 Slovakian Carpathians
BT2 Czechoslovakia
BT2 Central Europe
BT2 Europe
BT3 Slovakia
BT3 Czechoslovakia
BT3 Central Europe
BT3 Europe

inner transition elements
use rare earths

Innisfree Meteorite (1985)
Impact at Innisfree, E central Alberta, Canada. Before 1985, also search Innisfree AND Alberta AND meteorites.
BT LL chondrites
BT chondrites
BT stony meteorites
BT meteorites
SA Alberta
SA Canada

Innsbruck
No longer a valid term for GeoRef.
See Innsbruck Austria.

Innsbruck Austria (1993)
City in central Tyrol, W Austria.
BT Tyrol Austria
BT Austria
BT Central Europe
BT Europe

Inocerami (1981)
Includes the family Inoceramidae. Autoposting of this term to Inoceramidae began in 1989. Autoposting of Cyrtodontida to this term ended in 1989.
BT Pteriina
BT Bivalvia
BT Mollusca
BT Invertebrata
NT Inoceramidae

Inoceramidae (1978)
Family. Autoposting of this term to Inoceramus began in 1989. Autoposting of Inocerami and Pteriina to this term began in 1989.
BT Inocerami
BT Pteriina
BT Bivalvia
BT Mollusca
BT Invertebrata
NT Inoceramus

Inoceramus (1978)
Genus. Autoposting of Inoceramidae, Inocerami and Pteriina to this term began in 1989.
BT Inoceramidae
BT Inocerami
BT Pteriina
BT Bivalvia
BT Mollusca
BT Invertebrata

inorganic
A valid term through 1977. After 1977, use inorganic materials.

inorganic acids (1981)
SA acids
SA hydrochloric acid
SA hydrofluoric acid
SA organic acids
SA sulfuric acid

inorganic materials (1978)
Before 1978, also search inorganic AND materials.
SA ligands
SA materials
SA organic materials

inosilicates
use chain silicates

Inowroclaw
No longer a valid term for GeoRef. As of 1993 see Inowroclaw Poland.

Inowroclaw Poland (1993)
City in N central Poland. Before 1993 also search Inowroclaw AND Poland.
BT Bydgoszcz Poland
BT Poland
BT Central Europe
BT Europe

INQUA (1978)
Acronym.
UF International Association for Quaternary Research
SA associations

Insecta (1978)
Class. Autoposting of this term began in 1978.
BT Mandibulata
BT Arthropoda
BT Invertebrata
NT Archostemata
NT Blattopteroida
NT Coleopteroida
NT Dermapteroida
NT Diptera
NT Ectotropha
NT Entotropha
NT Ephemeropteroida
NT Hemipteroida
NT Heteroptera
NT Hymenopteroida
NT Lepidopteroida
NT Mecopteroida
NT Neuropteroida
NT Odonatopteroida
NT Orthopteroida
NT Palaeodictyopteroida
NT Plecoptera
NT Plectoptera
NT Protorthoptera
NT Psocopteroida
NT Raphidiodea
NT Siphonapteroida
NT Thysanopteroida
SA eggs
SA insects

Insectivora (1978)
Order. Autoposting of this term to Dermoptera, Macroscelida and Proteutheria began in 1989. Autoposting of Eutheria and Theria to this term began in 1989.
BT Eutheria
BT Theria
BT Mammalia
BT Tetrapoda
BT Vertebrata
BT Chordata
NT Dermoptera
NT Macroscelida
NT Proteutheria

insects (1985)
Common name for Insecta.
IN Index for all non-paleontologic studies of fossils.
BT arthropods
BT invertebrates
NT termites
SA biostratigraphy
SA Insecta

inselbergs (1978)
Autoposting of erosion features to this term began in 1989.
BT erosion features
SA geomorphology

insolation (1978)
SA mechanical weathering
SA Milankovitch theory
SA weathering

insoluble residues (1978)
Sometimes abbreviated as IR.
UF residues, insoluble
SA sedimentary rocks
SA siliceous composition
SA solubility

instabilities, plasma
IN Not a valid term for GeoRef. Usually out-of-scope. Used as plasma instabilities with magnetosphere.

installations, marine
use marine installations

installations, submarine
use submarine installations

installations, underground
use underground installations

institutions (1981)
SA academic institutions
SA associations

instruments (1978)
SA accelerator mass spectroscopy
SA accelerometers
SA air guns
SA borehole televiewers
SA calibration
SA design
SA diamond anvil cells
SA dilatometers
SA extensometers
SA flowmeters
SA Fraunhofer line discriminators
SA geophones
SA GLORIA
SA gravimeters
SA hydrophones
SA inclinometers
SA magnetometers
SA multispectral scanner
SA penetrometers
SA precision
SA pressuremeter tests
SA pressuremeters
SA Rock-Eval
SA samplers
SA Seabeam
SA sediment traps
SA Seislog
SA seismographs
SA SQUID
SA stressmeters
SA synchrotrons
SA techniques
SA tensiometers
SA thematic mapper
SA tiltmeters
SA tunnel boring machines
SA variometers

insular shelf (1985)
UF island shelf
SA continental shelf

SA islands
SA ocean floors
SA shelf environment

insulation materials (1982)
IN May be used as a commodity term.
BT construction materials
SA asbestos deposits
SA clays
SA glass materials
SA materials
SA mica deposits

intensities
use intensity

intensity (1978)
General term as of 1981. For intensity of earthquakes, use seismic intensity.
UF intensities
SA interplanetary space
SA magnetic intensity
SA magnitude
SA particles
SA seismic intensity
SA solar cycles

interactions, boundary
use boundary interactions

interactive analysis
use interactive techniques

interactive methods
use interactive techniques

interactive techniques (1985)
UF interactive analysis
UF interactive methods
SA data processing

interface, air-sea
use air-sea interface

interface, oil-gas
use oil-gas interface

interface, oil-water
use oil-water interface

interface, sediment-water
use sediment-water interface

interfaces (1981)
SA air-sea interface
SA oil-gas interface
SA oil-water interface
SA rock-water interface
SA sediment-water interface
SA soil-structure interface

interference patterns (1985)
SA folds
SA foliation
SA patterns
SA petrofabrics
SA structural analysis

interferometry (1978)
General term appropriate for many topics.
SA very long baseline interferometry

interglacial
A valid term through 1977. After 1977, see interglacial environment or specific interglacial stage.

interglacial environment (1978)
Before 1978, search interglacial.
SA depositional environment
SA environment
SA glacial environment
SA glacial geology
SA paleoclimatology
SA sedimentation

intergrowths (1978)
SA crystal growth
SA crystallization
SA crystallography

SA crystals

interior (1978)
As of 1981, restricted to the Earth.
SA core
SA dynamos
SA Earth
SA geodynamics
SA lunar interior
SA Moon
SA planetary interiors
SA seismology

interlaboratory comparison (1989)
Used for comparison of standard materials between laboratories.
SA chemical analysis
SA materials
SA minerals
SA petrology
SA soil mechanics
SA spectroscopy
SA standard materials
SA standard rocks
SA standardization
SA thermal analysis

Interlake Formation (1989)
Located in the Williston Basin.
BT Llandoverian
BT Lower Silurian
BT Silurian
BT Paleozoic
SA Manitoba
SA Montana
SA North Dakota
SA Saskatchewan

intermediate earthquake
use intermediate-focus earthquakes

intermediate-focus earthquakes (1978)
Before 1978, search intermediate-focus.
UF intermediate earthquake
BT earthquakes
SA deep-focus earthquakes
SA focus
SA shallow-focus earthquakes

intermittent springs (1982)
BT springs
SA ground water
SA hydrology

intermontane basins (1978)
BT basins

internal constitution
No longer a valid GeoRef index term. Before 1972, was used on level 2 in subfile G. See interior; planetary interiors.

internal waves (1978)
BT ocean waves
SA waves

International Association for Quaternary Research
use INQUA

international cooperation (1978)
UF cooperation, international
SA aid projects
SA ASEAN
SA associations
SA industrialized countries
SA International Monetary Fund
SA Law of the Sea
SA negotiations
SA policy
SA strategic minerals
SA technical cooperation
SA UNCTAD
SA World Bank

International Decade of Ocean Exploration (1978)
SA oceanography
SA research vessels

International Geological Congress (1978)
UF IGC
SA associations

International Geological Correlation Programme
use IGCP

International Geosphere-Biosphere Program
use IGBP

International Monetary Fund (1981)
Sometimes abbreviated as IMF.
SA associations
SA economics
SA international cooperation
SA United Nations
SA World Bank

International Phase of Ocean Drilling
use IPOD

International Union of Geological Sciences
use IUGS

interplanetary comparison (1981)
SA extraterrestrial geology
SA planetology
SA planets
SA terrestrial comparison

interplanetary dust (1978)
SA cosmic dust
SA dust
SA interplanetary space
SA meteorites
SA particles

interplanetary space (1978)
Usually out-of-scope for GeoRef.
UF outer space
UF space
SA albedo
SA asteroids
SA comae
SA comets
SA cosmic dust
SA electromagnetic radiation
SA extraterrestrial geology
SA Halley's Comet
SA intensity
SA interplanetary dust
SA meteoroids
SA meteors
SA particles
SA planetology
SA shock waves
SA solar cycles
SA solar system
SA solar wind

interpretation (1978)
Appropriate to a large number of general topics.

interstitial water
use pore water

intertidal environment (1978)
As of 1981, used for littoral environment. Before 1981, also search littoral environment; tidal environment. Before 1978, search intertidal; littoral.
UF littoral environment
SA capes
SA coastlines
SA depositional environment
SA ecology
SA environment

SA foredunes
SA intertidal sedimentation
SA littoral erosion
SA sedimentation
SA wave-cut platforms

intertidal sedimentation (1981)
BT sedimentation
SA intertidal environment

Intertrappean Beds (1978)
Upper Cretaceous to lower Eocene. Sedimentary deposits of fossiliferous cherts, freshwater limestone or bole intercalated between the flows of Deccan Trap Basalt. Also search Intertrappean.
IN Index ages as applicable.
UF Deccan Intertrappean
UF Deccan Intertrappean Beds
UF Deccan Intertrappean Series
SA Deccan Traps
SA India
SA lower Eocene
SA Paleocene
SA Upper Cretaceous

intraclastic texture
use intraclasts

intraclasts (1978)
UF intraclastic texture
SA carbonate rocks
SA textures

intracontinental belts (1981)
SA orogenic belts
SA orogeny

intracratonic basins (1981)
BT basins
SA cratons
SA tectonics

intrafolial folds (1978)
Before 1978, search folds AND intrafolial.
BT folds

intraformational structures (1982)
SA sedimentary structures

intramagmatic deposits (1981)
SA mineral deposits, genesis

intraplate
No longer a valid term for GeoRef.
use intraplate processes

intraplate processes (1993)
Before 1993, also search intraplate, midplate or mid-plate. Before 1993, search intraplate tectonics if applicable. May be searched in combination with earthquakes, seismicity, volcanism, tectonics, etc.
UF intraplate
SA earthquakes
SA plate tectonics
SA seismicity
SA tectonics
SA volcanism
SA volcanoes

intraplate tectonics
No longer a valid term for GeoRef. See intraplate processes.

Intrazonal soils (1978)
BT soils
SA Bog soils
SA Brown forest soils
SA Hydromorphic soils
SA Meadow soils
SA Planosols
SA soil group
SA Solonchak soils
SA Solonetz soils

intrusion (ground water)
 use salt-water intrusion

intrusions (1978)
 Used for the igneous rocks mass formed by emplacement of magma in pre-existing rock. It is also used for the process of emplacement. Autoposting of this term began in 1978.
 UF invasion (intrusion)
 UF irruption (intrusion)
 NT batholiths
 NT diatremes
 NT dikes
 NT laccoliths
 NT layered intrusions
 NT lopoliths
 NT pipes
 NT plugs
 NT plutons
 NT ring complexes
 NT sills
 NT stocks
 SA assimilation
 SA aureoles
 SA contact metamorphism
 SA contamination
 SA country rocks
 SA crystallization
 SA diapirism
 SA dike swarms
 SA domes
 SA emplacement
 SA foliation
 SA heat flow
 SA hybridization
 SA hypabyssal rocks
 SA igneous rocks
 SA lineation
 SA magma chambers
 SA magmas
 SA magmatic differentiation
 SA metamorphic rocks
 SA metamorphism
 SA metasomatism
 SA petrology
 SA ring structures
 SA swarms
 SA veins
 SA volcanic necks
 SA volcanology
 SA wall rocks
 SA zoning

intrusive
 A valid term through 1977. Included use in subfiles E and N. After 1977, use intrusive rocks.

intrusive mountain
 use batholiths

intrusive rocks
 As of 1989, no longer a valid term for GeoRef. Before 1978, the term intrusive was used in conjuction with igneous rocks. Now use igneous rocks, intrusions, hypabyssal rocks and/or plutonic rocks as applicable.

intrusives
 No longer a valid term for GeoRef. Before 1971, included use in subfiles E, G, N, T and B. Now use igneous rocks and/or intrusions as applicable.

Inuvik
 As of 1993, no longer a valid term for GeoRef. See Inuvik Northwest Territories.

Inuvik Northwest Territories (1993)
 Village in NW Mackenzie District, W Northwest Territories. Before 1993, also search Inuvik.
 BT Mackenzie District Northwest Territories
 BT Northwest Territories
 BT Western Canada
 BT Canada

invasion (intrusion)
 use intrusions

inventory
 No longer a valid term for GeoRef. As of 1993, see mineral inventory.

Inverness
 No longer a valid term for GeoRef as of 1993. See Inverness Scotland.

Inverness Scotland (1993)
 City in Inverness-shire, Highland region, N Scotland. Before 1993, also search Inverness AND Scotland.
 CO N564000N580000
  W0033000W0073000
 BT Inverness-shire Scotland
 BT Highland region Scotland
 BT Scotland
 BT Great Britain
 BT United Kingdom
 BT Western Europe
 BT Europe

Inverness-shire
 No longer a valid term for GeoRef as of 1993.
 use Inverness-shire Scotland

Inverness-shire Scotland (1993)
 Former county in NW Scotland.
 UF Inverness-shire
 BT Highland region Scotland
 BT Scotland
 BT Great Britain
 BT United Kingdom
 BT Western Europe
 BT Europe
 NT Inverness Scotland
 NT Isle of Skye
 NT Raasay
 NT Rhum

inverse magnetization (1981)
 BT magnetization

inverse problem (1978)
 UF inversion (seismology)
 SA direct problem
 SA geophysical methods
 SA geophysical surveys
 SA geophysics
 SA Green function
 SA impedance
 SA numerical filters
 SA seismic migration
 SA seismology

inverse thermoremanent magnetization (1981)
 Autoposting of magnetization to this term began in 1989.
 BT remanent magnetization
 BT magnetization
 SA paleomagnetism
 SA partial thermoremanent magnetization
 SA thermoremanent magnetization

inversion (seismology)
 use inverse problem

Invertebrata (1978)
 Protista and worms are considered narrower terms as of 1981.

IN Term is to be used only when more specific terms do not apply or are too numerous to be recorded.
 NT Archaeocyatha
 NT Arthropoda
 NT Brachiopoda
 NT Bryozoa
 NT Coelenterata
 NT Echinodermata
 NT Mollusca
 NT Porifera
 NT Protista
 NT worms
 SA affinities
 SA Camaroidea
 SA Cephalodiscida
 SA chitin
 SA Dendroidea
 SA Didymograptina
 SA Didymograptus
 SA Diplograptina
 SA Enteropneusta
 SA exoskeletons
 SA Glossograptina
 SA Graptolithina
 SA Graptoloidea
 SA ichnofossils
 SA invertebrates
 SA Isograptus
 SA Metazoa
 SA Monograptina
 SA Monograptus
 SA Monograptus uniformis
 SA Pterobranchia
 SA Rhabdopleurida
 SA shells
 SA Stolonoidea
 SA tests
 SA Tuboidea

invertebrates (1981)
 Common name for Invertebrata.
 IN Index for all non-paleontologic studies of fossils.
 NT arthropods
 NT brachiopods
 NT bryozoans
 NT corals
 NT echinoderms
 NT foraminifers
 NT graptolites
 NT mollusks
 NT radiolarians
 NT silicoflagellates
 NT sponges
 SA biostratigraphy
 SA chitin
 SA exoskeletons
 SA Invertebrata

investment (1981)
 SA economics
 SA investment legislation

investment legislation (1981)
 SA economics
 SA investment
 SA legislation

Inyan Kara Group (1985)
 NE Wyoming, North Dakota and W South Dakota.
 BT Lower Cretaceous
 BT Cretaceous
 BT Mesozoic
 SA Fall River Formation
 SA Lakota Formation
 SA North Dakota
 SA South Dakota
 SA Wyoming

Inyo County
 Valid through 1988. Search in combination with state term. After 1988, use specific county-state term.

Inyo County California (1989)
 E California. Before 1989, also search Inyo County AND California.
 CO N354500N373000
  W1154000W1184500
 BT California
 BT United States
 NT Coso Hot Springs KGRA
 NT Coso Range
 NT Funeral Mountains
 NT Inyo Mountains
 NT Panamint Range
 SA Chalfant Valley earthquake 1986
 SA Death Valley
 SA El Paso Mountains
 SA Golden Trout Wilderness
 SA Long Valley
 SA Nopah Range
 SA Owens Valley
 SA Searles Lake

Inyo Domes (1989)
 In Long Valley Caldera, E California.
 BT Mono County California
 BT California
 BT United States
 SA Long Valley Caldera
 SA Obsidian Dome
 SA Owens Valley

Inyo Mountains (1978)
 Range in E California.
 UF White-Inyo Mountains
 UF White-Inyo Range
 BT Inyo County California
 BT California
 BT United States

Io
 Not a valid term for GeoRef. As of 1985, for isotope ionium, use Th-230. For satellite of Jupiter, use Io Satellite.

Io Satellite (1985)
 One of the satellites of Jupiter. Before 1985, also search Io AND Jupiter. Use Th-230 for isotope ionium (Io).
 SA Galilean satellites
 SA Jupiter
 SA satellites
 SA Voyager Program

Io/Th (1978)
 Isotopic ratio used in age determination.
 SA absolute age
 SA isotope ratios
 SA Th-230
 SA thorium

Io/U (1978)
 Isotopic ratio used in age determination.
 SA absolute age
 SA isotope ratios
 SA Th-230
 SA thorium
 SA uranium

iodates (1978)
 SA minerals

iodides (1981)
 BT halides

iodine (1978)
 Chemical element. As of 1981, use iodine deposits for iodine as a commodity. Autoposting of halogens to this term began in 1989.
 UF I
 BT halogens
 NT I-129
 NT I-131

SA brines
SA I/Xe
SA iodine deposits

iodine deposits (1981)
Before 1981, search iodine AND deposits.
IN Commodity. See List C.
SA brines
SA iodine
SA salt

Iola Field (1978)
Gas field in Clay County in S central Illinois. Before 1978, search Iola; Iola gas field.
IN Index county or regions as applicable.
UF Iola gas field
SA Clay County Illinois
SA oil and gas fields

Iola gas field
use Iola Field

ion chromatography (1989)
SA analysis
SA aqueous solutions
SA chemical analysis
SA chemical composition
SA chromatography
SA gas chromatography
SA liquid chromatography

ion densities
Not a valid term for GeoRef. Usually out-of-scope. Used as ion densities and temperatures with aeronomy.

ion densities and temperatures
As of 1993, no longer a valid term for GeoRef. Usually out-of-scope. Used with ionosphere.

ion exchange (1978)
Also search cation exchange.
UF base exchange
SA anions
SA cation exchange capacity
SA clay mineralogy
SA crystal chemistry
SA crystal structure
SA diffusion
SA diffusivity
SA electrolytes
SA geochemistry
SA hydrochemistry
SA ions
SA sea water
SA soils

ion probe (1978)
Term is used for methodology, not data.
UF probe, ion
SA analysis
SA chemical analysis
SA electron probe
SA ion probe data
SA ions
SA neutron probe
SA spectroscopy

ion probe data (1978)
Before 1978, search ion probe AND data.
SA data
SA ion probe
SA spectroscopy

Ionian Islands (1978)
In 1981, broader term changed from Greece to Mediterranean region.
BT Mediterranean region
NT Greek Ionian Islands

Ionian Sea (1978)
Between SE coast of Italy and W Greece.

CO N362000N403000
E0222000E0150000
BT East Mediterranean
BT Mediterranean Sea
NT Gulf of Corinth
SA DSDP Site 374
SA Greek Ionian Islands

Ionian Zone (1978)
S, SW, and W Albania and W Greece.
IN Index countries or regions as applicable.
BT Southern Europe
BT Europe
SA Albania
SA Cretaceous
SA Greece
SA Jurassic
SA Mesozoic
SA Tertiary

ionium
As of 1985, not a valid term for GeoRef.
use Th-230

ionization
As of 1993, no longer a valid term for GeoRef. Usually out-of-scope. Used with aeronomy and interplanetary space.

ionosphere
As of 1993, no longer a valid term for GeoRef. Usually out-of-scope. Sometimes used on level 1 for special bibliographies before 1993.

ions (1978)
Also search cations; anions.
SA ammonium ion
SA anions
SA bicarbonate ion
SA bromide ion
SA calcium ion
SA carbonate ion
SA cation exchange capacity
SA cations
SA chloride ion
SA electrolytes
SA fixation
SA fluoride ion
SA geochemistry
SA hydrochemistry
SA hydroxyl ion
SA ion exchange
SA ion probe
SA ligands
SA magnesium ion
SA meteorology
SA nitrate ion
SA particles
SA phosphate ion
SA potassium ion
SA secondary ion mass spectroscopy
SA sodium ion
SA sulfate ion

Iowa (1978)
Autoposting of this term began in 1978.
CO N402500N433000
W0901000W0963500
BT United States
NT Adams County Iowa
NT Benton County Iowa
NT Boone County Iowa
NT Buchanan County Iowa
NT Butler County Iowa
NT Calhoun County Iowa
NT Carroll County Iowa
NT Cass County Iowa
NT Cherokee County Iowa
NT Clarke County Iowa
NT Clay County Iowa

NT Clinton County Iowa
NT Crawford County Iowa
NT Dallas County Iowa
NT Davis County Iowa
NT Decatur County Iowa
NT Delaware County Iowa
NT Dickinson County Iowa
NT Fayette County Iowa
NT Floyd County Iowa
NT Franklin County Iowa
NT Fremont County Iowa
NT Greene County Iowa
NT Grundy County Iowa
NT Hamilton County Iowa
NT Hancock County Iowa
NT Hardin County Iowa
NT Harrison County Iowa
NT Henry County Iowa
NT Howard County Iowa
NT Humboldt County Iowa
NT Iowa County Iowa
NT Jackson County Iowa
NT Jasper County Iowa
NT Jefferson County Iowa
NT Johnson County Iowa
NT Jones County Iowa
NT Lee County Iowa
NT Linn County Iowa
NT Lucas County Iowa
NT Lyon County Iowa
NT Madison County Iowa
NT Mahaska County Iowa
NT Marion County Iowa
NT Marshall County Iowa
NT Mills County Iowa
NT Mitchell County Iowa
NT Monroe County Iowa
NT Montgomery County Iowa
NT Page County Iowa
NT Plymouth County Iowa
NT Pocahontas County Iowa
NT Polk County Iowa
NT Scott County Iowa
NT Shelby County Iowa
NT Sioux County Iowa
NT Taylor County Iowa
NT Union County Iowa
NT Warren County Iowa
NT Washington County Iowa
NT Wayne County Iowa
NT Webster County Iowa
NT Winnebago County Iowa
SA Beil Limestone Member
SA Burlington Limestone
SA Cedar Valley Formation
SA Cherokee Group
SA Colorado Group
SA Decorah Shale
SA Des Moines Lobe
SA Desmoinesian
SA Douglas Group
SA Dubuque Formation
SA Eau Claire Formation
SA Ervine Creek Limestone
SA Forest City Basin
SA Galena Dolomite
SA Galesville Sandstone
SA Glenwood Shale
SA Kansas City Group
SA Keokuk Limestone
SA Labette Shale
SA Lawrence Formation
SA Lecompton Limestone
SA Lime Creek Formation
SA Maquoketa Formation
SA Marmaton Group
SA Midcontinent
SA Midcontinent geophysical anomaly
SA Mississippi River
SA Mississippi Valley
SA Missouri River
SA Missouri River basin
SA Missouri River valley
SA Mount Simon Sandstone

SA Neda Formation
SA Niagara Escarpment
SA Oread Limestone
SA Peoria Loess
SA Platte River
SA Platteville Formation
SA Plattsburg Limestone
SA Plattsmouth Limestone Member
SA Pleasanton Group
SA Prairie du Chien Group
SA Saint Louis Limestone
SA Saint Peter Sandstone
SA Sainte Genevieve Limestone
SA Salem Limestone
SA Shakopee Formation
SA Sioux Quartzite
SA Spergen Formation
SA Spirit Lake
SA Stanton Formation
SA Upper Mississippi Valley
SA Wabaunsee Group
SA Warsaw Formation
SA Wonewoc Formation
SA Wyandotte Limestone

Iowa County
Valid through 1988. Search in combination with state term. After 1988, use specific county-state term.

Iowa County Iowa (1989)
E central Iowa. Before 1989, also search Iowa County AND Iowa.
CO N413100N415200
W0915000W0921800
BT Iowa
BT United States

Iowa County Wisconsin (1989)
S Wisconsin. Before 1989, also search Iowa County AND Wisconsin.
CO N424800N431200
W0895200W0902500
BT Wisconsin
BT United States

iozite
use wuestite

IPOD (1981)
Phase of Deep Sea Drilling Project extending from Leg 45 to Leg 96. Before 1981, also search International Phase of Ocean Drilling. Autoposting of this term began in 1985. Autoposting of specific legs to specific DSDP sites began in 1989.
UF International Phase of Ocean Drilling
BT Deep Sea Drilling Project
NT Leg 45
NT Leg 46
NT Leg 47
NT Leg 48
NT Leg 49
NT Leg 50
NT Leg 51
NT Leg 52
NT Leg 53
NT Leg 54
NT Leg 55
NT Leg 56
NT Leg 57
NT Leg 58
NT Leg 59
NT Leg 60
NT Leg 61
NT Leg 62
NT Leg 63
NT Leg 64
NT Leg 65
NT Leg 66
NT Leg 67
NT Leg 68

NT Leg 69
NT Leg 70
NT Leg 71
NT Leg 72
NT Leg 73
NT Leg 74
NT Leg 75
NT Leg 76
NT Leg 77
NT Leg 78A
NT Leg 78B
NT Leg 79
NT Leg 80
NT Leg 81
NT Leg 82
NT Leg 83
NT Leg 84
NT Leg 85
NT Leg 86
NT Leg 87
NT Leg 88
NT Leg 89
NT Leg 90
NT Leg 91
NT Leg 92
NT Leg 93
NT Leg 94
NT Leg 95
NT Leg 96
SA JOIDES
SA Ocean Drilling Program
SA research vessels

Ipswich
No longer a valid term for GeoRef. See Ipswich Australia if applicable.

Ipswich Australia (1993)
City in SE Queensland.
BT Queensland Australia
BT Australia
BT Australasia

Ipswich Coal Measures (1978)
SE Queensland.
BT Triassic
BT Mesozoic
SA Australia
SA Queensland Australia

Ipswichian (1978)
Europe.
BT upper Pleistocene
BT Pleistocene
BT Quaternary
BT Cenozoic

Ir
use iridium

IR spectra
use infrared spectra

Iran (1978)
CO N250000N393000
E0632000E0440000
UF Persia
BT Middle East
BT Asia
NT Anarak Iran
NT Azerbaijan
NT Dasht-e-Bayaz
NT Djadjerud Valley
NT Elburz
NT Fars Iran
NT Kerman Iran
NT Lar
NT Lut Desert
NT Talysh Mountains
NT Teheran Iran
SA Bakhtiari Formation
SA Caspian Basin
SA Caspian Sea
SA Geirud Formation
SA Kopet-Dag Range
SA Kura Lowland
SA Zagros

Iraq (1978)
CO N290000N371500
E0483000E0390000
BT Middle East
BT Asia
NT Kirkuk Iraq
SA Bakhtiari Formation
SA Euphrates River
SA Mesopotamia
SA Near East
SA Shiranish Formation
SA Sinjar Formation
SA Tigris River
SA Zagros
SA Zubair Formation

Irati Formation (1978)
BT Permian
BT Paleozoic
SA Brazil
SA Goias Brazil
SA Mato Grosso Brazil
SA Minas Gerais Brazil
SA Parana Brazil
SA Passa Dois Group
SA Sao Paulo Brazil

Irazu (1989)
Volcano. central Costa Rica.
CO N095900N095900
W0835200W0835200
UF Irazu Volcano
BT Costa Rica
BT Central America

Irazu Volcano
use Irazu

Ireland (1978)
The Republic of Ireland, also called Eire, which occupies the 26 counties in the S, central, and NW of the island of Ireland.
CO N513000N552000
W0063000W0103000
UF Eire
BT Western Europe
BT Europe
NT Clare Ireland
NT Cork Ireland
NT Donegal Ireland
NT Dublin Ireland
NT Galway Ireland
NT Kerry Ireland
NT Kildare Ireland
NT Leinster
NT Limerick Ireland
NT Mayo Ireland
NT Meath Ireland
NT Sligo Ireland
NT Tipperary Ireland
NT Wexford Ireland
NT Wicklow Ireland
NT Wicklow Mountains
SA Caledonides
SA Dalradian
SA Highland Boundary Fault
SA Leinster Granite
SA Northern Ireland
SA Rosses Granite

irghizite (1993)
Tektite group associated with Zhamanshin Crater in Kazakhstan. Before 1993, also search irgizite.
BT tektites
SA Aktyubinsk Kazakhstan
SA Kazakhstan
SA Zhamanshin Crater

Irgiz
No longer a valid term for GeoRef. As of 1993, see Irgiz Kazakhstan.

Irgiz Kazakhstan (1993)
Village on Irgiz River in Aktyubinsk Oblast in E Kazakhstan. This term has multiple hierarchies.
BT1 Aktyubinsk Kazakhstan
BT1 Kazakhstan
BT1 Central Asia
BT1 Asia
BT2 Aktyubinsk Kazakhstan
BT2 Kazakhstan
BT2 Commonwealth of Independent States

Irian Barat
use Irian Jaya Indonesia

Irian Jaya
No longer a valid term for GeoRef.
use Irian Jaya Indonesia

Irian Jaya Indonesia (1993)
W half of island of New Guinea. This term has multiple hierarchies.
CO S091500S001000
E1410000E1305000
UF Dutch New Guinea
UF Irian Barat
UF Irian Jaya
UF Netherlands New Guinea
UF West Irian
UF West New Guinea
BT1 Indonesia
BT1 Far East
BT1 Asia
BT2 New Guinea
BT2 Malay Archipelago
SA False Cape

iridium (1978)
Chemical element. As of 1985, use iridium ores for iridium as a commodity. Autoposting of platinum group and metals to this term began in 1989.
UF Ir
BT platinum group
BT metals
SA iridium ores

iridium ores (1985)
Before 1985, also search iridium AND (deposit OR deposits OR ore OR ores OR economic) in the basic index. Autoposting of metal ores to this term began in 1985.
IN Commodity. See List C.
BT metal ores
SA iridium
SA platinum ores

IRIS network (1993)
Acronym. Before 1993, also search IRIS AND seismology or geophysical surveys. Consortium founded in 1984, collects and disseminates seismic data from around the world. Three major programs are supported by the consortium, IRIS: GSN (Global Seismic Network), PASSCAL (Program for Array Seismic Studies of the Continental Lithosphere) and DMS (Data Management System).
UF Incorporated Research Institutions for Seismology network
SA data processing
SA PASSCAL
SA programs
SA seismic networks
SA seismology

Irish Sea (1978)
Between England and Ireland.
CO N530000N552000
W0030000W0063000
BT European Atlantic
BT North Atlantic
BT Atlantic Ocean
SA Cardigan Bay
SA Morecambe Bay

Irkutsk
No longer a valid term for GeoRef. As of 1993, see Irkutsk Russian Federation.

Irkutsk Amphitheater
use Irkutsk Basin

Irkutsk Basin (1978)
Coal Basin extending 150 miles NW along Trans-Siberian RR from Lake Baikal. This term has multiple hierarchies.
UF Cheremkhovo Basin
UF Cheremkovo coal basin
UF Irkutsk Amphitheater
BT1 Russian Federation
BT1 Commonwealth of Independent States
BT2 Asia
SA coal fields

Irkutsk Russian Federation (1993)
Oblast W and N of Lake Baikal. Before 1993, also search Irkutsk. This term has multiple hierarchies.
CO N510000N640000
E1190000E0960000
BT1 Russian Federation
BT1 Commonwealth of Independent States
BT2 Asia
NT Chuya Russian Federation
NT Mama Russian Federation
NT Slyudyanka Russian Federation
NT Ulkan Russian Federation
NT Ust-Kut Russian Federation
SA Angara-Lena Basin
SA Baikal Mountains
SA Baikal region
SA Baikal rift zone
SA Biryusa
SA Bodaibo Russian Federation
SA Iya River
SA Lake Baikal
SA Oka River
SA Patom Plateau
SA Uda River

iron (1978)
Chemical element. As of 1981, use iron ores for iron as a commodity. Autoposting of metals to this term began in 1989.
UF Fe
BT metals
NT Fe-55
NT Fe-57
NT Fe-59
SA awaruite
SA chalcophile elements
SA ferric iron
SA ferromanganese composition
SA ferrosilicon
SA ferrous iron
SA ferruginous composition
SA gossan
SA heavy metals
SA hematite
SA iron meteorites
SA iron minerals
SA iron ores
SA iron-rich composition
SA josephinite
SA kamacite
SA native elements
SA siderophile elements

Iron Age (1985)
Archaeological classification.
BT Cenozoic
SA archaeology

SA Holocene
SA Quaternary

iron alum
use halotrichite

Iron County
Valid through 1988. Search in combination with state term. After 1988, use specific county-state term.

Iron County Michigan (1989)
SW Michigan Upper Peninsula. NW Michigan. Before 1989, also search Iron County AND Michigan.
CO N455500N462400
  W0880700W0890000
BT Michigan Upper Peninsula
BT Michigan
BT United States
SA Perch Lake

Iron County Missouri (1989)
SE central Missouri. Before 1989, also search Iron County AND Missouri.
CO N371700N374500
  W0903300W0911000
BT Missouri
BT United States

Iron County Utah (1989)
SW Utah. Before 1989, also search Iron County AND Utah.
CO N372800N381000
  W1123000W1140300
BT Utah
BT United States
SA Wah Wah Mountains

Iron County Wisconsin (1989)
N Wisconsin. Before 1989, also search Iron County AND Wisconsin.
CO N455800N463600
  W0895500W0903300
BT Wisconsin
BT United States

iron formations (1978)
Before 1989, also search banded iron formations if applicable.
UF banded iron formations
UF formations, iron
UF iron-formations
BT chemically precipitated rocks
BT sedimentary rocks
SA algoma-type
SA ferruginous composition
SA iron-rich composition
SA itabirite
SA jaspilite
SA taconite

iron meteorites (1978)
Before 1974, also search octahedrites. Autoposting of this term began in 1978.
UF metallic meteorites
UF meteoric iron
UF siderite (meteorite)
BT meteorites
NT ataxite
NT Cape York Meteorite
NT hexahedrite
NT octahedrite
NT Santa Clara Meteorite
SA achondrites
SA chondrites
SA iron
SA kamacite
SA stony irons

iron minerals (1985)
SA franklinite
SA goethite
SA hematite
SA ilmenite
SA iron
SA iron ores
SA jacobsite
SA limonite
SA magnetite
SA marcasite
SA minerals
SA pyrite
SA pyrrhotite
SA siderite

Iron Ore Group (1989)
E India.
BT Archean
BT Precambrian
SA India

iron ores (1981)
Before 1981, also search (iron OR Fe) AND (deposit OR deposits OR ore OR ores OR economic) in the basic index. Autoposting of metal ores to this term began in 1985.
BT metal ores
SA Agrokipia Deposit
SA algoma-type
SA Benson Mines
SA besshi-type
SA goethite
SA hematite
SA Hitachi Deposit
SA hydrogoethite
SA iron
SA iron minerals
SA Konrad Mine
SA limonite
SA magnetite
SA martite
SA native elements
SA ocher
SA pyrite
SA pyrite ores
SA pyrrhotite
SA siderite
SA slag
SA stilpnomelane
SA titanomagnetite

iron oxides (1978)
BT oxides
SA akaganeite
SA delafossite
SA franklinite
SA goethite
SA gossan
SA hematite
SA hoegbomite
SA jacobsite
SA lepidocrocite
SA limonite
SA maghemite
SA magnesioferrite
SA magnetite
SA martite
SA ocher
SA priderite
SA trevorite
SA ulvospinel
SA wuestite

iron spinel
use hercynite

iron sulfides (1978)
BT sulfides
SA gossan
SA tochilinite

iron-formations
use iron formations

iron-rich composition (1978)
For sedimentary rocks, use ferruginous composition. Before 1978, search iron-rich rocks; iron-rich sediments.
SA chemically precipitated rocks
SA composition
SA ferruginous composition
SA geochemistry
SA iron
SA iron formations
SA lithogeochemistry
SA sedimentary rocks
SA sediments

iron-rich rocks
A valid term through mid-1978. Use iron-rich composition or iron formations.

iron-rich sediments
A valid term through mid-1978. Use ferruginous composition under sediments.

iron-stony meteorite
use stony irons

ironstone (1978)
BT chemically precipitated rocks
BT sedimentary rocks

Iroquois County
Valid through 1988. Search in combination with state term. After 1988, use specific county-state term.

Iroquois County Illinois (1989)
E Illinois. Before 1989, also search Iroquois County AND Illinois.
CO N402800N410100
  W0873000W0881000
BT Illinois
BT United States

Irpinia earthquake 1980 (1993)
Before 1985, also search (Irpinia OR Lucania OR Campania OR southern Italy) AND earthquakes AND 1980. Before 1993, also search Campania-Lucania earthquake 1980, Campania-Basilicata earthquake 1980 Irpinia-Basilicata earthquake 1980, and southern Italy earthquake 1980.
UF Campania-Basilicata earthquake 1980
UF Campania-Lucania earthquake 1980
UF southern Italy earthquake 1980
BT earthquakes
SA Basilicata Italy
SA Campania Italy
SA Italy
SA Lucania

irradiation (1981)
SA techniques
SA thermoluminescence
SA X-ray analysis

irrigation (1978)
SA ponds
SA reinjection wells
SA soil treatment
SA soils
SA surface water
SA underground channels
SA water
SA water quality
SA water regimes
SA water supply
SA waterways

irrotational wave
use P-waves

irruption (intrusion)
use intrusions

Irtish River
use Irtysh River

Irtysh River (1978)
Largest tributary of the Ob River. Also search Irtysh.
IN Index Xinjiang China and former Soviet republics as applicable.
UF Irtish River
BT Asia
SA Kazakhstan
SA Ob River
SA Ob-Irtysh Interfluve
SA Russian Federation
SA Xinjiang China

Irvingtonian (1989)
North American continental stage. Above Blancan and below Rancholabrean.
BT Pleistocene
BT Quaternary
BT Cenozoic

Isachsen Formation (1978)
Conformably overlies Deer Bay Formation on Axel Heiberg and Ellef Ringnes Islands. On Isachsen Peninsula of Ellef Rignes Island in Franklin District.
BT Lower Cretaceous
BT Cretaceous
BT Mesozoic
SA Northwest Territories

Isar Valley (1978)
River valley. Also search Isar River.
IN Index Austrian and German states as applicable.
BT Europe
SA Bavaria Germany
SA Tyrol Austria

Ischia (1978)
Island in Tyrrhenian Sea outside Bay of Naples.
SA Campania Italy

Ise Bay (1978)
Inlet of Pacific Ocean on S coast of Honshu.
SA Aichi Japan
SA Honshu
SA Mie Japan

Isera Mountains
use Izera Mountains

Isere
No longer a valid term for GeoRef. See Isere France.

Isere France (1993)
Department in SE France.
CO N444500N455000
  E0063000E0044500
BT France
BT Western Europe
BT Europe
NT Grenoble France
NT Gresivaudan
SA Pelvoux Massif
SA Vercors

Isere Valley (1978)
River valley in SE France. Also search Isere.
BT France
BT Western Europe
BT Europe
SA Tarentaise

Isergebirge
As of 1985, no longer a valid term for GeoRef.
use Izera Mountains

Ishbel Group (1989)
Includes Ranger Canyon and Mowitch formations. Carboniferous to Permian. SW Alberta and SE British Columbia.
IN Index ages as applicable.
BT Paleozoic
SA Alberta

SA British Columbia
SA Carboniferous
SA Permian
SA Ranger Canyon Formation

Ishikari Bay (1978)
Inlet of Sea of Japan on W Hokkaido. Also search Ishikari.
BT Hokkaido
BT Japan
BT Far East
BT Asia
SA Ishikari Plain

Ishikari Plain (1978)
SE of Ishikari Bay in SW Hokkaido. Also search Ishikari.
BT Hokkaido
BT Japan
BT Far East
BT Asia
SA Ishikari Bay

Ishikawa
No longer a valid term for GeoRef. See Ishikawa Japan.

Ishikawa Japan (1993)
Prefecture on the Sea of Japan N of Nagoya in central Honshu. Before 1993, also search Ishikawa AND Japan.
BT Honshu
BT Japan
BT Far East
BT Asia
NT Kanazawa Japan
NT Nanao Japan
SA Chubu Japan
SA Hakusan
SA Noto Peninsula

Ishim (1978)
River which flows into the Irtysh River. This term has multiple hierarchies.
IN Index former Soviet republics as applicable.
BT1 Commonwealth of Independent States
BT2 Asia
SA Kazakhstan
SA Russian Federation

Ishtar
use Ishtar Terra

Ishtar Terra (1989)
Physiographic province containing Maxwell Montes on Venus.
UF Ishtar
BT Venus
SA Maxwell Montes
SA Pioneer Program

Iskar River
use Isker River

Isker River (1978)
NW central Bulgaria. Also search Isker.
UF Iskar River
UF Iskur River
BT Bulgaria
BT Southern Europe
BT Europe

Iskur River
use Isker River

island arcs (1978)
UF arcs, island
UF island-arc areas
UF volcanic arcs
SA arcuate structures
SA back-arc basins
SA continental margin
SA continents
SA fore-arc basins
SA islands

SA plate tectonics
SA tectonophysics
SA trenches

island shelf
use insular shelf

island-arc areas
use island arcs

islands (1978)
SA atolls
SA barrier islands
SA continents
SA insular shelf
SA island arcs

Islas Malvinas
use Falkland Islands

Islay (1978)
Southernmost island of Inner Hebrides off W coast of Scotland. This term has multiple hierarchies.
CO N553500N555500
    W0060000W0063000
BT1 Inner Hebrides
BT1 Hebrides
BT1 Scotland
BT1 Great Britain
BT1 United Kingdom
BT1 Western Europe
BT1 Europe
BT2 Argyllshire Scotland
BT2 Scotland
BT2 Great Britain
BT2 United Kingdom
BT2 Western Europe
BT2 Europe
BT3 Strathclyde region Scotland
BT3 Scotland
BT3 Great Britain
BT3 United Kingdom
BT3 Western Europe
BT3 Europe

Isle of Man (1978)
Island in Irish Sea.
CO N541000N543000
    W0041500W0044500
BT United Kingdom
BT Western Europe
BT Europe

Isle of Pines (1978)
Island in Caribbean S of W Cuba.
BT La Habana Cuba
BT Cuba
BT Greater Antilles
BT Antilles
BT West Indies
BT Caribbean region

Isle of Sheppey (1989)
Island in N Kent, located at the mouth of the Thames River in Thames Estuary, SE England.
CO N512200N512800
    E005500E004200
BT Kent England
BT England
BT Great Britain
BT United Kingdom
BT Western Europe
BT Europe
SA Thames Estuary
SA Thames River

Isle of Skye (1978)
Island of Inner Hebrides off NW coast. This term has multiple hierarchies.
CO N570000N574000
    W0054000W0064500
UF Skye
BT1 Inner Hebrides
BT1 Hebrides
BT1 Scotland
BT1 Great Britain

BT1 United Kingdom
BT1 Western Europe
BT1 Europe
BT2 Inverness-shire Scotland
BT2 Highland region Scotland
BT2 Scotland
BT2 Great Britain
BT2 United Kingdom
BT2 Western Europe
BT2 Europe

Isle of Wight
No longer a valid term for GeoRef as of 1993.
use Isle of Wight England

Isle of Wight England (1993)
Island in English Channel just S of Southampton. In 1974, it became a county separate from Hampshire. Before 1993, also search Isle of Wight.
CO N504000N504500
    W0010000W0013000
UF Isle of Wight
BT England
BT Great Britain
BT United Kingdom
BT Western Europe
BT Europe
SA Bembridge Marls
SA Hampshire England

Isle Royale National Park (1978)
In NW Lake Superior. Includes Isle Royale and neighboring islands.
BT Keweenaw County Michigan
BT Michigan Upper Peninsula
BT Michigan
BT United States
SA national parks

Isles Dernieres (1993)
SE Louisiana.
CO N290400N291100
    W0903500W0910000
BT Terrebonne Parish Louisiana
BT Louisiana
BT United States

Ismay Zone (1989)
Paradox Basin, Utah and Colorado.
IN Index states as applicable.
BT United States
SA Colorado
SA Paradox Basin
SA Utah

isobath maps (1985)
BT maps
SA bathymetric maps
SA bathymetry
SA ocean floors

isochrons (1978)
SA absolute age
SA geochronology
SA relative age
SA seismology

isoclinal
A valid term through 1977. After 1977, use isoclinal folds.

isoclinal folds (1978)
Before 1978, also search folds AND isoclinal; isoclines.
UF isoclines
BT folds
SA deformation

isoclines
use isoclinal folds

isograd maps (1978)
Before 1978, search maps AND isograds.
BT maps

SA isograds

isograds (1978)
SA facies
SA grade
SA isograd maps
SA maps
SA metamorphism

Isograptus (1978)
Genus.
IN Index Hemichordata or Invertebrata as applicable.
BT Didymograptina
BT Graptoloidea
BT Graptolithina
SA Hemichordata
SA Invertebrata

isolation (1981)
SA radiation

isoleucine (1985)
BT amino acids
BT organic materials
SA organic acids

isomerization (1989)
SA geochemistry
SA processes

isomorphism (1978)
Used as a general term.
SA crystal chemistry
SA paleontology

isopach maps (1978)
Before 1978, also search maps AND isopach.
BT maps
SA isopleth maps
SA stratigraphy

isopleth maps (1978)
BT maps
SA isopach maps

isoprenoids (1978)
Compounds based on the isoprene structure. Classified according to composition and thus may be various types of materials such as hydrocarbons and alcohols.
BT organic materials
SA hydrocarbons
SA phytane

isoseismic maps (1978)
BT maps
SA earthquakes
SA seismology

isostacy
use isostasy

isostasy (1978)
To be used for the condition of equilibrium of the crust above the mantle, and related features.
UF isostacy
SA asthenosphere
SA changes of level
SA continents
SA crust
SA density
SA Earth
SA epeirogeny
SA eustacy
SA geodesy
SA geomorphologic effects
SA geomorphology
SA geophysical methods
SA geophysical surveys
SA glacial geology
SA isostatic compensation
SA isostatic rebound
SA mantle
SA Mohorovicic discontinuity
SA neotectonics
SA tectonics
SA tectonophysics

SA vertical movements
isostatic compensation (1993)
Before 1993, also search isostasy AND compensation.
SA crust
SA isostasy
SA isostatic rebound
isostatic rebound (1993)
Before 1993, also search rebound and isostasy.
NT glacial rebound
SA glacial geology
SA isostasy
SA isostatic compensation
SA neotectonics
SA tectonics
SA uplifts
isothermal magnetization (1981)
BT magnetization
isothermal remanent magnetization (1978)
Autoposting of magnetization to this term began in 1989.
BT remanent magnetization
BT magnetization
isotherms (1989)
Curves indicating relationships between two variables at constant temperature.
SA geochemistry
SA P-T conditions
SA temperature
SA thermodynamic properties
isotope ratios (1993)
A valid term through 1972. From 1972 through 1992, also search ratios AND isotopes. Also see isotopes, absolute age and name of chemical element.
SA Al-27/Al-26
SA Ar-38/Ar-36
SA Ar-40/Ar-36
SA Ar-40/Ar-39
SA Ar/Ar
SA Be-10/Be-9
SA C-13/C-12
SA C-14/C-12
SA Cl-37/Cl-35
SA Cr-53/Cr-52
SA D/H
SA He-4/He-3
SA He/He
SA Hf-177/Hf-176
SA I/Xe
SA Io/Th
SA Io/U
SA isotopes
SA K/Ar
SA Kr/Kr
SA Mg-25/Mg-24
SA Mg-26/Mg-24
SA N-15/N-14
SA Nd-144/Nd-143
SA Ne-22/Ne-20
SA Ne-22/Ne-21
SA O-17/O-16
SA O-18/O-16
SA Os-187/Os-186
SA Pb-206/Pb-204
SA Pb-207/Pb-204
SA Pb-207/Pb-206
SA Pb-208/Pb-204
SA Pb-208/Pb-206
SA Pb/Pb
SA Pb/Th
SA Pu-240/Pu-239
SA Ra-228/Ra-226
SA Rb-87/Sr-86
SA Rb/Sr
SA Re/Os
SA S-34/S-32
SA Sm-147/Nd-144
SA Sm/Nd
SA Sr-87/Sr-86
SA Sr/Sr
SA Th-232/Th-230
SA Th/Th
SA Th/U
SA U-234/Th-230
SA U-238/Pb-206
SA U-238/Th-230
SA U-238/U-234
SA U-238/U-235
SA U/He
SA U/Pb
SA U/Th/Pb
SA Xe-130/Xe-129
SA Xe-132/Xe-129
SA Xe-136/Xe-130
SA Xe-136/Xe-132
isotopes (1978)
Used for the application of the study of radioactive and stable isotopes to geology, excluding absolute age dating. Also search isotopic. Autoposting of this term began in 1978.
IN Index with specific isotope and material where available.
NT radioactive isotopes
NT stable isotopes
SA absolute age
SA activation analysis
SA analysis
SA argon
SA background level
SA carbon
SA chemical analysis
SA chemical elements
SA cosmogenic elements
SA fallout
SA fractionation
SA geochemistry
SA ground water
SA helium
SA hydrogen
SA I/Xe
SA Inclusions
SA isotope ratios
SA K/Rb
SA Kr/Kr
SA krypton
SA lead
SA metals
SA meteorites
SA mineral deposits, genesis
SA neodymium
SA neon
SA noble gases
SA organic materials
SA overprinting
SA oxygen
SA paleoclimatology
SA potassium
SA radioactive decay
SA radioactive tracers
SA radioactivity
SA radium
SA radon
SA rubidium
SA sea water
SA spallation
SA standard materials
SA strontium
SA sulfur
SA tektites
SA thorium
SA tracers
SA uranium
SA uranium disequilibrium
SA xenon
isotropic materials (1978)
Before 1978, search isotropic.
SA materials
isotropy (1981)
Use with minerals and crystals.
SA optical properties
Israel (1978)
CO N293000N333000 E0353000E0342000
BT Middle East
BT Asia
NT Elath Israel
NT Galilee
NT Haifa Bay
NT Haifa Israel
NT Jerusalem Israel
NT Makhtesh Ramon
NT Mount Carmel
NT Negev
NT Sea of Galilee
SA Dead Sea
SA Dead Sea Rift
SA Ghareb Formation
SA Great Rift Valley
SA Jordan River
SA Jordan Valley
SA Judea Group
SA Mediterranean region
SA Menuha Formation
SA Mishash Formation
SA Mount Scopus Group
SA Palestine
SA Red Sea Basin
SA Samaria
SA Sinai
Issyk-kul Lake (1978)
NE Kyrgyzstan. Before 1978, also search Issyk-Kul. This term has multiple hierarchies.
BT1 Kyrgyzstan
BT1 Asia
BT2 Kyrgyzstan
BT2 Commonwealth of Independent States
Istanbul
No longer a valid term for GeoRef. As of 1993 see Istanbul Turkey.
Istanbul Turkey (1993)
Province and city on both sides of Bosporus in NW Turkey. Before 1993 also search Istanbul AND Turkey.
BT Turkey
BT Middle East
BT Asia
SA Bosporus
Istranca Mountains (1978)
Range along Black Sea coast in SE Bulgaria, and Turkey in Europe.
IN Index countries as applicable.
UF Strandzha Mountains
SA Bulgaria
SA Turkey
Istria (1978)
Peninsula in NW Yugoslavia jutting into Adriatic Sea.
IN Index republics as applicable.
BT Yugoslavia
BT Southern Europe
BT Europe
SA Croatia
SA Slovenia
Isua Belt (1985)
SW Greenland. Also search Isua.
CO N600000N640000 W0500000W0600000
BT West Greenland
BT Greenland
BT Arctic region
Itabira
No longer a valid term for GeoRef. See Itabira Brazil.
Itabira Brazil (1993)
City in E central Minas Gerais. Before 1993, also search Itabira and Brazil. Before 1978, also search Presidente Vargas.
BT Minas Gerais Brazil
BT Brazil
BT South America
itabirite (1978)
From 1978-1980, chemically precipitated rocks was autoposted to this term.
BT metamorphic rocks
SA chemically precipitated rocks
SA iron formations
Itaborai
No longer a valid term for GeoRef. See Itaborai Brazil.
Itaborai Brazil (1993)
City in S central Rio de Janeiro, SE Brazil.
BT Rio de Janeiro Brazil
BT Brazil
BT South America
Itaka
No longer a valid term for GeoRef. As of 1993, see Itaka Russian Federation.
Itaka Russian Federation (1993)
Town in central Chita Oblast in Transbaikalia. Before 1993, also search Itaka. This term has multiple hierarchies.
BT1 Chita Russian Federation
BT1 Russian Federation
BT1 Commonwealth of Independent States
BT2 Chita Russian Federation
BT2 Asia
Italy (1978)
Apennines is a narrower term as of 1981. Autoposting of this term began in 1978.
CO N363000N473000 E0190000E0063000
BT Southern Europe
BT Europe
NT Abruzzi Italy
NT Apennines
NT Apulia Italy
NT Argentera Italy
NT Basilicata Italy
NT Calabria Italy
NT Campania Italy
NT Canavese Zone
NT Dolomites
NT Emilia-Romagna Italy
NT Friuli-Venezia Giulia Italy
NT Ivrea-Verbano Zone
NT Langhe
NT Latium Italy
NT Liguria Italy
NT Lombardy Italy
NT Lucania
NT Marches Italy
NT Marmolada Glacier
NT Molise Italy
NT Piemonte Italy
NT Po River
NT Po Valley
NT Sardinia Italy
NT Sesia Valley
NT Sesia-Lanzo Zone
NT Sicily Italy
NT Tagliamento Valley
NT Tiber Valley
NT Trentino-Alto Adige Italy
NT Tuscany Italy
NT Umbria Italy
NT Valle d'Aosta Italy
NT Venetia
NT Veneto Italy
SA Adriatic region

SA Alps
SA Apennine Front
SA Brianconnais Zone
SA Ceneri Zone
SA Central Alps
SA Cottian Alps
SA Friuli earthquake 1976
SA Irpinia earthquake 1980
SA Julian Alps
SA Karst region
SA Maiolica Limestone
SA Mediterranean region
SA Prealps
SA San Marino
SA Scaglia Formation
SA Simplon region
SA Ticino
SA University of Rome
SA Val Gardena Sandstone
SA Vigarano Meteorite
SA Voltri Group
SA Western Alps

Itarare Subgroup   (1985)
S Brazil.
BT Paleozoic
SA Brazil
SA Carboniferous
SA Lower Permian
SA Parana Basin
SA Permian
SA Rio Grande do Sul Brazil
SA Sao Paulo Brazil
SA Upper Carboniferous

Itasca County
Valid through 1988. Search in combination with state term. After 1988, use specific county-state term.

Itasca County Minnesota   (1989)
N Minnesota. Before 1989, also search Itasca County AND Minnesota.
CO N470200N475300
W0930400W0942500
BT Minnesota
BT United States
NT Keewatin Minnesota
SA Mesabi Range

iterative methods   (1981)
Also search iteration; iterative.
SA mathematical geology
SA mathematical methods
SA mathematical models
SA statistical analysis

Ithaca   (1978)
One of Greek Ionian Islands. Search in conjunction with Greece. Before 1989, was also used for the city in New York. As of 1989, use Ithaca New York for the city in New York. This term has multiple hierarchies.
BT1 Greek Ionian Islands
BT1 Greece
BT1 Southern Europe
BT1 Europe
BT2 Greek Ionian Islands
BT2 Ionian Islands
BT2 Mediterranean region
SA Ithaca New York

Ithaca New York   (1989)
City in W central New York. Before 1989, search Ithaca AND New York.
CO N422600N422600
W0763000W0763000
BT Tompkins County New York
BT New York
BT United States
SA Ithaca

Ito
No longer a valid term for GeoRef. See Ito Japan.

Ito Japan   (1993)
City on E Izu Peninsula on Sagami Sea in central Honshu. Before 1993, also search Ito AND Japan.
BT Shizuoka Japan
BT Honshu
BT Japan
BT Far East
BT Asia
SA Izu Peninsula

Itoigawa-Shizuoka tectonic line   (1989)
Central Honshu.
UF Shizuoka-Itoigawa tectonic line
BT Honshu
BT Japan
BT Far East
BT Asia
SA Shizuoka Japan

Iturup Island   (1978)
Largest of the Kuril Islands, SW of Kamchatka. Also search Iturup. This term has multiple hierarchies.
BT1 Kuril Islands
BT1 Sakhalin Russian Federation
BT1 Russian Federation
BT1 Commonwealth of Independent States
BT2 Kuril Islands
BT2 Sakhalin Russian Federation
BT2 Asia
BT3 Kuril Islands
BT3 Russian Pacific region
BT3 Russian Federation
BT3 Commonwealth of Independent States
BT4 Kuril Islands
BT4 Russian Pacific region
BT4 Asia

IUGS   (1978)
Acronym.
UF International Union of Geological Sciences
SA associations
SA IGCP

Ivano-Frankovsk
No longer a valid term for GeoRef. As of 1993, see Ivano-Frankovsk Ukraine.

Ivano-Frankovsk Ukraine   (1993)
City and oblast in W Ukraine. Formerly part of Poland. Before 1978, also search Stanislav or Stanislawow. This term has multiple hierarchies.
UF Stanislav Ukraine
UF Stanislawow Ukraine
BT1 Ukraine
BT1 Europe
BT2 Ukraine
BT2 Commonwealth of Independent States
SA Kosov Ukraine

Ivigtut
No longer a valid term for GeoRef. As of 1993 see Ivigtut Greenland.

Ivigtut Greenland   (1993)
Settlement on SW coast of Greenland. Before 1993 also search Ivigtut AND Greenland.
BT Greenland
BT Arctic region

Ivishak Formation   (1989)
In Sadlerochit Group. Permian to Lower Triassic. N Alaska.

IN Index ages as applicable.
SA Alaska
SA Lower Triassic
SA North Slope
SA Permian
SA Prudhoe Bay Field
SA Sadlerochit Group

Iviza
use Ibiza

Ivory Coast   (1978)
CO N042000N104000
W0023000W0084500
BT West Africa
BT Africa
NT Bandama River
NT Toumodi Ivory Coast
SA Nimba Mountains
SA Trou Sans Fond
SA Volta Basin

Ivrea
No longer a valid term for GeoRef. See Ivrea Italy.

Ivrea Italy   (1993)
City NE of Turin in central Piemonte, NW Italy.
BT Piemonte Italy
BT Italy
BT Southern Europe
BT Europe

Ivrea Zone
As of 1983, no longer a valid term for GeoRef. Used for the region in central Italy.
use Ivrea-Verbano Zone

Ivrea-Verbano Zone   (1978)
Central and NW Italy. Before 1983, also search Ivrea Zone.
UF Ivrea Zone
BT Italy
BT Southern Europe
BT Europe
SA Piemonte Italy

Iwaki   (1978)
Mountain peak in Aomori Prefecture in N Honshu. Before 1993, also used for river and city.
SA Aomori Japan

Iwate   (1978)
Dormant volcano. Before 1993, also included use for Iwate Prefecture.
BT Honshu
BT Japan
BT Far East
BT Asia
SA Nohi Rhyolite

Iwate Japan   (1993)
Prefecture in NW Honshu. Before 1993, also search Iwate AND Honshu.
BT Honshu
BT Japan
BT Far East
BT Asia
NT Kamaishi Mine
NT Kitakami Mountains
NT Mizusawa Japan
NT Omine Mine
NT Onikobe Field
NT Shizukuishi Japan
SA Koma-ga-take
SA Nohi Rhyolite
SA Tohoku

Iwo-Jima   (1978)
Center island of volcano group located 660 nautical miles S of Tokyo. Returned to Japan in 1968.
BT Pacific Ocean
SA Izu-Bonin Arc

ixiolite   (1978)
Autoposting of niobotantalates to this term began in 1989. This term has multiple hierarchies.
BT1 niobotantalates
BT1 niobates
BT1 oxides
BT2 niobotantalates
BT2 tantalates
BT2 oxides

Iya River   (1978)
SW Irkutsk Oblast. Also search Iya. This term has multiple hierarchies.
BT1 Russian Federation
BT1 Commonwealth of Independent States
BT2 Asia
SA Irkutsk Russian Federation

Izalco   (1989)
Volcano in SW El Salvador. As of 1993, use Izalco El Salvador for the town.
CO N134300N134300
W0894000W0894000
UF Izalco Volcano
BT El Salvador
BT Central America

Izalco El Salvador   (1993)
Town in El Salvador, use Izalco for the volcano.
CO N134300N134300
W0894000W0894000
BT El Salvador
BT Central America

Izalco Volcano
use Izalco

Izera Mountains   (1978)
Range of the Sudeten Mountains in N Bohemia and Lower Silesia. Before 1985, also search Isergebirge.
IN Index countries, provinces, and mountains as applicable.
UF Isera Mountains
UF Isergebirge
UF Jizerske Hory
BT Sudeten Mountains
BT Central Europe
BT Europe
SA Bohemia
SA Czech Sudeten Mountains
SA Czechoslovakia
SA Lower Silesia
SA Poland
SA Polish Sudeten Mountains

Izu Islands
use Izu-shichito

Izu Oshima
use Izu-Oshima

Izu Peninsula   (1978)
Extends into Pacific Ocean between Suruga Bay and Sagami Bay in S central Honshu. Also search Izu.
BT Honshu
BT Japan
BT Far East
BT Asia
SA Ito Japan
SA Kanagawa Japan
SA Sagami Bay
SA Shizuoka Japan
SA Suruga Bay
SA Tokai Japan

Izu-Bonin Arc   (1993)
Island arc bordered by Izu-Bonin Trench and Shikoku Basin in the West Pacific.
UF Izu-Ogasawara Arc

BT West Pacific
BT Pacific Ocean
SA Bonin Islands
SA Iwo-Jima
SA Izu-shichito
SA Leg 125
SA Leg 126
SA Mariana Islands
SA Shikoku Basin

Izu-Ogasawara Arc
use Izu-Bonin Arc

Izu-Oshima (1993)
Largest island in Izu-shichito group off Izu Peninsula of Honshu. Location of Mount Mihara volcanic cone. Before 1993 also search O-shima and Oshima.
UF Izu Oshima
UF Vries Island
BT Izu-shichito
BT Honshu
BT Japan
BT Far East
BT Asia
NT Mount Mihara
SA O-shima
SA Oshima

Izu-Oshima earthquake 1978 (1989)
Epicenter near the Izu-shichito, Honshu, Japan.
BT earthquakes
SA Honshu
SA Izu-shichito
SA Japan
SA O-shima
SA Pacific Ocean

Izu-shichito (1978)
Group of volcanic islands in Pacific Ocean off Izu Peninsula.
UF Izu Islands
BT Honshu
BT Japan
DT Far East
BT Asia
NT Hachijo-jima
NT Izu-Oshima
NT Miyake-Jima
SA Izu-Bonin Arc
SA Izu-Oshima earthquake 1978
SA O-shima

Izumi
No longer a valid term for GeoRef. See Izumi Japan.

Izumi Group (1978)
Kii Peninsula in S central Honshu, and Shikoku.
BT Senonian
BT Upper Cretaceous
BT Cretaceous
BT Mesozoic
SA Honshu
SA Japan
SA Shikoku

Izumi Japan (1993)
City in Osaka Prefecture.
BT Osaka Japan
BT Honshu
BT Japan
BT Far East
BT Asia

J

J-M Reef (1989)
In Stillwater Complex. S central Montana.
BT Archean
BT Precambrian
SA Montana
SA Stillwater Complex

Jabal Idsas (1985)
Mountain, central Saudi Arabia.
CO N232000N233000
 E0452000E0450000
BT Saudi Arabia
BT Arabian Peninsula
BT Asia

Jabalpur
No longer a valid term for GeoRef. See Jabalpur India.

Jabalpur India (1993)
City and district in central Madhya Pradesh, central India. Before 1993, also search Jabalpur. Before 1978, also search Jubbulpore.
UF Jubbulpore India
BT Madhya Pradesh India
BT India
BT Indian Peninsula
BT Asia

Jabalpur Series (1978)
Consists of clays, shales, earthy sandstones, and thin coal seams. Near city of Jabalpur.
BT Upper Jurassic
BT Jurassic
BT Mesozoic
SA India
SA Madhya Pradesh India

Jabiluka Australia (1993)
Uranium and gold ores. Alligator Rivers region, N Northern Territory, Australia. Before 1993, also search Jabiluka region.
UF Jabiluka region
BT Northern Territory Australia
BT Australia
BT Australasia
SA Alligator Rivers Field
SA gold ores
SA uranium ores

Jabiluka region
No longer a valid term for GeoRef.
use Jabiluka Australia

Jachymov (1978)
Town in Erzgebirge in Bohemia, NW Czechoslovakia.
BT Bohemia
BT Czech Republic
BT Czechoslovakia
BT Central Europe
BT Europe

Jackfork Group (1978)
Divided into Wildhorse Mountain Formation, Prairie Mountain Formation, Markham Mill Formation, Wesley Formation, Game Refuge Formation. SE and central S Oklahoma; and SW Arkansas.
BT Carboniferous
BT Paleozoic
SA Arkansas
SA Oklahoma

Jackson County
Valid through 1988. Search in combination with state term. After 1988, use specific county-state term.

Jackson County Alabama (1989)
NE Alabama. Before 1989, also search Jackson County AND Alabama.
CO N342900N345900
 W0853500W0861900
BT Alabama
BT United States
SA Russell Cave

Jackson County Arkansas (1989)
NE Arkansas. Before 1989, also search Jackson County AND Arkansas.
CO N352200N355300
 W0910300W0913500
BT Arkansas
BT United States

Jackson County Colorado (1989)
N Colorado. Before 1989, also search Jackson County AND Colorado.
CO N402100N410000
 W1055300W1065200
BT Colorado
BT United States

Jackson County Florida (1989)
NW Florida. Before 1989, also search Jackson County AND Florida.
CO N303400N310000
 W0850000W0853200
BT Florida
BT United States

Jackson County Georgia (1989)
NE central Georgia. Before 1989, also search Jackson County AND Georgia.
CO N335700N341900
 W0832200W0834900
BT Georgia
BT United States

Jackson County Illinois (1989)
SW Illinois. Before 1989, also search Jackson County AND Illinois.
CO N373300N375600
 W0890800W0894000
BT Illinois
BT United States
NT Carbondale Illinois

Jackson County Indiana (1989)
S Indiana. Before 1989, also search Jackson County AND Indiana.
CO N384300N390300
 W0854800W0862000
BT Indiana
BT United States

Jackson County Iowa (1989)
E Iowa. Before 1989, also search Jackson County AND Iowa.
CO N420300N422300
 W0901000W0905400
BT Iowa
BT United States

Jackson County Kansas (1989)
NE Kansas. Before 1989, also search Jackson County AND Kansas.
CO N391300N394000
 W0953300W0960300
BT Kansas
BT United States

Jackson County Kentucky (1989)
SE central Kentucky. Before 1989, also search Jackson County AND Kentucky.
CO N371600N373400
 W0834900W0841300
BT Kentucky
BT United States

Jackson County Michigan (1989)
S Michigan. Before 1989, also search Jackson County AND Michigan.
CO N420400N422500
 W0840800W0844300
BT Michigan Lower Peninsula
BT Michigan
BT United States

Jackson County Minnesota (1989)
SW Minnesota. Before 1989, also search Jackson County AND Minnesota.
CO N433000N435100
 W0945200W0952800
BT Minnesota
BT United States

Jackson County Mississippi (1989)
Extreme SE Mississippi. Before 1989, also search Jackson County AND Mississippi.
CO N301800N304300
 W0882400W0885300
BT Mississippi
BT United States
SA Mississippi Sound

Jackson County Missouri (1989)
W Missouri. Before 1989, also search Jackson County AND Missouri.
CO N385000N391500
 W0940700W0943700
BT Missouri
BT United States
SA Kansas City Missouri

Jackson County North Carolina (1989)
W North Carolina. Before 1989, also search Jackson County AND North Carolina.
CO N350000N353200
 W0825400W0832200
BT North Carolina
BT United States

Jackson County Ohio (1989)
S Ohio. Before 1989, also search Jackson County AND Ohio.
CO N384900N391300
 W0822500W0824900
BT Ohio
BT United States

Jackson County Oklahoma (1989)
SW Oklahoma. Before 1989, also search Jackson County AND Oklahoma.
CO N342100N345000
 W0990300W0994800
BT Oklahoma
BT United States

Jackson County Oregon (1989)
SW Oregon. Before 1989, also search Jackson County AND Oregon.
CO N420000N430000
 W1221600W1231500
BT Oregon
BT United States

Jackson County South Dakota (1989)
SW central South Dakota. Before 1989, also search Jackson County AND South Dakota.
CO N432400N440000
 W1010500W1020900
BT South Dakota

BT United States
Jackson County Tennessee
(1989)
N central Tennessee. Before 1989, also search Jackson County AND Tennessee.
CO N361300N363000 W0853100W0855100
BT Tennessee
BT United States

Jackson County Texas (1989)
S Texas. Before 1989, also search Jackson County AND Texas.
CO N283900N291800 W0962000W0965700
BT Texas
BT United States

Jackson County West Virginia (1989)
W West Virginia. Before 1989, also search Jackson County AND West Virginia.
CO N383400N390600 W0813000W0815500
BT West Virginia
BT United States
NT Cottageville Field

Jackson County Wisconsin (1989)
W central Wisconsin. Before 1989, also search Jackson County AND Wisconsin.
CO N440500N443600 W0901900W0911000
BT Wisconsin
BT United States

Jackson Group (1978)
Includes Caddell (Moodys Marl) Formation, Wellborn Formation, Manning Formation, and Whitsett Formation, Gosport Sand, Yazoo Clay, Ocata Limestone, Barnwell Formation, McElroy Formation, Mosley Hill Formation. Gulf Coastal Plain from South Carolina to S Texas.
BT upper Eocene
BT Eocene
BT Paleogene
BT Tertiary
BT Cenozoic
SA Alabama
SA Barnwell Formation
SA Florida
SA Georgia
SA Louisiana
SA Mississippi
SA Moodys Branch Formation
SA South Carolina
SA Texas
SA Whitsett Formation
SA Yazoo Clay

Jackson Hole (1978)
Valley in Teton County in NW Wyoming.
BT Teton County Wyoming
BT Wyoming
BT United States

Jackson Parish Louisiana (1989)
N Louisiana. Before 1989, also search Jackson Parish; Jackson County AND Louisiana.
CO N321000N323000 W0922200W0924500
BT Louisiana
BT United States

Jacksonian (1989)
North American stage. Above Claibornian, below Vicksburgian.

BT Eocene
BT Paleogene
BT Tertiary
BT Cenozoic

jacobsite (1978)
BT oxides
SA iron minerals
SA iron oxides
SA manganese minerals
SA spinel group

Jacobsville Sandstone (1978)
Precambrian or Cambrian. Underlies Munising Formation. N Michigan.
SA Cambrian
SA Michigan
SA Precambrian

jade (1978)
Gemstone composed of jadeite or nephrite. From 1978-1980, in addition to chain silicates and silicates, pyroxene group was autoposted to this term.
BT chain silicates
BT silicates
SA amphibole group
SA gems
SA jadeite
SA nephrite
SA pyroxene group

Jade Bay (1978)
Inlet of North Sea on N coast of Oldenburg region.
SA Lower Saxony Germany

jadeite (1978)
BT clinopyroxene
BT pyroxene group
BT chain silicates
BT silicates
SA jade
SA ornamental materials

Jaen
No longer a valid term for GeoRef. As of 1993, see Jaen City Spain.

Jaen City Spain (1993)
Refers to only the city in S central Spain. Before 1993, also search Jaen and Spain.
BT Jaen Spain
BT Andalusia Spain
BT Spain
BT Iberian Peninsula
BT Southern Europe
BT Europe

Jaen Province
No longer a valid term for GeoRef. As of 1993, see Jaen Spain.

Jaen Spain (1993)
Refers only to the province. From 1981-1992, also search Jaen Province and Spain. Before 1981, also search Jaen and Spain. In S central Spain. For the city, see Jaen City Spain.
CO N372300N383200 W0022800W0041800
BT Andalusia Spain
BT Spain
BT Iberian Peninsula
BT Southern Europe
BT Europe
NT Jaen City Spain

Jaguaribe (1978)
River in Ceara, NE Brazil. After 1992, see Jaguaribe Brazil for the town.
BT Ceara Brazil
BT Brazil
BT South America

Jaguaribe Brazil (1993)
Town in Ceara, NE Brazil.
BT Ceara Brazil
BT Brazil
BT South America

Jaipur
No longer a valid term for GeoRef. Former princely state in E central Rajasthan, NW India. See Jaipur India.

Jaipur India (1993)
City in E central Rajasthan, NW India.
BT Rajasthan India
BT India
BT Indian Peninsula
BT Asia

Jaisalmer
No longer a valid term for GeoRef. Former princely state in W Rajasthan, NW India. See Jaisalmer India.

Jaisalmer India (1993)
Town in W Rajasthan, NW India.
BT Rajasthan India
BT India
BT Indian Peninsula
BT Asia

Jalisco
No longer a valid term for GeoRef. As of 1993 see Jalisco Mexico.

Jalisco Mexico (1993)
State in W central Mexico. Before 1993 also search Jalisco AND Mexico.
BT Mexico
NT Guadalajara Mexico
SA Mexican volcanic belt
SA Sierra Madre Occidental

Jamaica (1978)
CO N174000N183000 W0761000W0782500
BT Greater Antilles
BT Antilles
BT West Indies
BT Caribbean region
NT Discovery Bay Jamaica
NT Maroon Town Jamaica
SA Blue Mountains

James Bay (1978)
Southern extension of Hudson Bay between NE Ontario and W Quebec. Islands within James Bay lie within the Northwest Territories.
IN Index provinces as applicable.
CO N510000N550000 W0783000W0823000
BT Eastern Canada
BT Canada
SA Hudson Bay
SA James Bay Lowlands
SA North American Atlantic
SA Northwest Territories
SA Ontario
SA Quebec

James Bay basin
use James Bay Lowlands

James Bay Lowlands (1985)
Area surrounding James Bay.
IN Index provinces as applicable.
UF James Bay basin
UF James Bay region
BT Eastern Canada
BT Canada
SA James Bay
SA Ontario
SA Quebec

James Bay region
use James Bay Lowlands

James Hutton
use Hutton, James

James River (1978)
E central Virginia. Also in North Dakota and South Dakota.
IN Index counties or regions as applicable.
SA North Dakota
SA South Dakota
SA Virginia

James Ross Island (1989)
In Weddell Sea, off the coast of the Antarctic Peninsula.
CO S643000S634000 W0570000W0590000
BT Antarctica
SA Lopez de Bertodano Formation
SA Sobral Formation
SA Weddell Sea

James W. Ellsworth Land
use Ellsworth Land

Jameson Land (1978)
Central E coast of Greenland.
BT Greenland
BT Arctic region

jamesonite (1978)
UF gray antimony
BT sulfantimonites
BT sulfosalts

Jamkhandi
No longer a valid term for GeoRef. Former princely state in W central India. See Jamkhandi India.

Jamkhandi India (1993)
Town in W central India.
BT Maharashtra India
BT India
BT Indian Peninsula
BT Asia

Jammu (1978)
City and district in S Jammu and Kashmir, disputed territory between India and Pakistan. Before 1989, India was a broader term.
IN Index Jammu and Kashmir AND India and/or Pakistan as applicable.
UF Jummoo
SA India
SA Jammu and Kashmir
SA Pakistan

Jammu and Kashmir (1978)
Term redefined in 1989. Disputed region divided into two parts. State in N India. Also part of NE Pakistan.
IN Index India and/or Pakistan as applicable.
CO N321000N370000 E0802300E0722000
UF Kashmir and Jammu
BT Indian Peninsula
BT Asia
SA Anantnag
SA Doda
SA India
SA Indus River
SA Indus-Yarlung Zangbo suture zone
SA Jammu
SA Karakoram
SA Karewa Group
SA Kashmir
SA Kashmir Valley
SA Kishtwar
SA Ladakh
SA Muth Quartzite

SA Naubug
SA Pakistan
SA Riasi
SA Srinagar
SA Subathu Formation
SA Tawi Valley

Jamtland
No longer a valid term for GeoRef. As of 1993 see Jamtland Sweden.

Jamtland Sweden (1993)
County in W Sweden. Before 1993 also search Jamtland AND Sweden.
BT Sweden
BT Scandinavia
BT Western Europe
BT Europe

Jan Mayen (1978)
Norwegian volcanic island between Greenland and Norway. In 1981, broader terms changed from Arctic Ocean and Atlantic Ocean to Arctic region. As of 1981, Beerenberg is a narrower term. As of 1990, Polar regions is autoposted to this term.
CO N700000N720000
W0073000W0090000
BT Arctic region
NT Beerenberg
SA Arctic Coastal Plain
SA Arctic Ocean
SA Atlantic Ocean
SA Greenland Sea
SA Norwegian Sea

Jan Mayen Ridge (1981)
Off the coast of Jan Mayen, to the S.
CO N673000N713000
W0070000W0100000
BT Norwegian Sea
BT Arctic Ocean
SA DSDP Site 346

Janggun Mine (1993)
Bonghwa area, South Korea.
BT South Korea
BT Korea
BT Far East
BT Asia
SA lead-zinc deposits
SA mines

Janus Satellite (1989)
One of the satellites of Saturn. Sometimes referred to as 1980S1. Before 1989, also search Janus AND Saturn.
SA satellites
SA Saturn
SA Voyager Program

Japan (1978)
CO N300000N450000
E1470000E1290000
BT Far East
BT Asia
NT Green tuff region
NT Hidaka Japan
NT Hokkaido
NT Honshu
NT Honshu Arc
NT Inland Sea
NT Koma-ga-take
NT Kyushu
NT Median Tectonic Line
NT Mitsuishi Japan
NT O-shima
NT Ryukyu Islands
NT Shikoku
NT Shimanto Belt
NT Tsugaru Strait
SA Akiyoshi Limestone
SA Ata Caldera
SA Bonin Islands
SA Chibaken-Toho-Oki earthquake 1987
SA Futaba Group
SA Green Tuff Formation
SA Izu-Oshima earthquake 1978
SA Izumi Group
SA Japan Sea earthquake 1983
SA Kanto earthquake 1923
SA Kazusa Group
SA Kobiwako Group
SA Kuroshio
SA Mikabu System
SA Miyagi earthquake 1978
SA Mizunami Group
SA Motojuku Formation
SA Muro Group
SA Nagano earthquake 1984
SA Nankaido earthquake 1946
SA Narita Formation
SA Niigata earthquake 1964
SA Okinawa Trough
SA Osaka Group
SA Ryukyu Group
SA Semata Formation
SA Shimanto Group
SA Shimosa Group
SA Taishu Group
SA Tokachi-Oki earthquake 1968
SA Toyoura Sand
SA University of Tokyo
SA Uonuma Group
SA Usuginu Conglomerate
SA Yangtze Plate

Japan Current
use Kuroshio

Japan Sea (1978)
Between Japan in the E; and Korea and Primorye Kray in the Russian Far East on the W. Before 1974, also search Sea of Japan.
CO N340000N450000
E1420000E1271500
UF Sea of Japan
BT West Pacific
BT Pacific Ocean
SA DSDP Site 299
SA DSDP Site 302
SA Japan Sea earthquake 1983
SA Kuroshio
SA Leg 31
SA Leg 127
SA Leg 128
SA Miho Bay
SA ODP Site 794
SA ODP Site 796
SA ODP Site 797
SA ODP Site 798
SA ODP Site 799
SA Tokachi-Oki earthquake 1968

Japan Sea earthquake 1983 (1989)
Epicenter in Honshu, Japan. Before 1993, also search Nihonkai-Chubu earthquake 1983 and Nihonkai-Chuba earthquake 1983.
UF Nihonkai-Chuba earthquake 1983
UF Nihonkai-Chubu earthquake 1983
BT earthquakes
SA Honshu
SA Japan
SA Japan Sea

Japan Trench (1978)
Submarine depression extending from Kuril Islands along E coast of Hokkaido and E coast of N and central Honshu.
BT West Pacific
BT Pacific Ocean
SA DSDP Site 434
SA DSDP Site 435
SA DSDP Site 438
SA DSDP Site 440
SA DSDP Site 584
SA Kashima Seamount
SA Leg 87

Japanese Alps (1981)
Before 1981, also search Japan AND Alps. Ranges extending N-S through central Honshu. The Hida Mountains are the Northern Alps of Japan, the Kiso Mountains are the Central Alps, and the Akaishi Mountains are the Southern Alps. Chubu-Sangaku (Japanese Alps) national park is in the Hida mountain range.
UF Alps, Japanese
UF Nihon'arupusu
BT Honshu
BT Japan
BT Far East
BT Asia
SA Akaishi Mountains
SA Chubu Japan
SA Gifu Japan
SA Hida Mountains
SA Kiso Mountains
SA Nagano Japan
SA Niigata Japan
SA Toyama Japan

Jaramillo Event (1978)
Geomagnetic event.
BT lower Pleistocene
BT Pleistocene
BT Quaternary
BT Cenozoic
SA paleomagnetism
SA reversals

Jarlsberg
use Vestfold Norway

jarosite (1978)
UF utahite
BT sulfates
SA alunite

jasper (1978)
BT silica minerals
BT framework silicates
BT silicates
SA chalcedony
SA chert
SA jasperoid

Jasper County
Valid through 1988. Search in combination with state term. After 1988, use specific county-state term.

Jasper County Georgia (1989)
Central Georgia. Before 1989, also search Jasper County AND Georgia.
CO N330800N333300
W0833200W0835100
BT Georgia
BT United States

Jasper County Illinois (1989)
SE central Illinois. Before 1989, also search Jasper County AND Illinois.
CO N385200N391100
W0875700W0882300
BT Illinois
BT United States

Jasper County Indiana (1989)
NW Indiana. Before 1989, also search Jasper County AND Indiana.
CO N404300N411700
W0865500W0871700
BT Indiana
BT United States

Jasper County Iowa (1989)
Central Iowa. Before 1989, also search Jasper County AND Iowa.
CO N413100N415300
W0924500W0935200
BT Iowa
BT United States

Jasper County Mississippi (1989)
E central Mississippi. Before 1989, also search Jasper County AND Mississippi.
CO N314700N321300
W0885300W0892000
BT Mississippi
BT United States

Jasper County Missouri (1989)
SW Missouri. Before 1989, also search Jasper County AND Missouri.
CO N370400N372200
W0940400W0943700
BT Missouri
BT United States
SA Joplin Missouri

Jasper County South Carolina (1989)
Extreme S South Carolina. Before 1989, also search Jasper County AND South Carolina.
CO N320500N324500
W0805000W0811800
BT South Carolina
BT United States

Jasper County Texas (1989)
E Texas. Before 1989, also search Jasper County AND Texas.
CO N301800N311000
W0935000W0942700
BT Texas
BT United States

Jasper National Park (1978)
On British Columbia border in W Alberta. Also search Jasper.
BT Alberta
BT Western Canada
BT Canada
SA Columbia Icefield
SA Mount Erebus
SA national parks

jasperoid (1978)
BT chert
BT chemically precipitated rocks
BT sedimentary rocks
SA chalcedony
SA jasper
SA limestone
SA siliceous composition

jaspilite (1978)
UF jaspillite
BT chemically precipitated rocks
BT sedimentary rocks
SA iron formations

jaspillite
use jaspilite

Java (1978)
Island in the Greater Sunda group. Most populous and economically most important.
CO S090000S054000
E1144000E1050500
BT Indonesia
BT Far East
BT Asia
NT Merapi
SA Kabuh Formation
SA Kalibeng Formation
SA Pucangan Formation

SA Sunda Arc

Java Sea (1978)
N of Java and S of Borneo. Autoposting of Pacific Ocean to this term began in 1990.
BT Indonesian Seas
BT West Pacific
BT Pacific Ocean
NT Sunda Shelf

jaws (1978)
SA anatomy
SA bones
SA fossils
SA muscles
SA paleontology
SA paleopathology
SA skulls
SA teeth
SA Vertebrata

Jaxartes River
use Syr Darya

Jeanne d'Arc Basin (1993)
In Grand Banks area, North American Atlantic.
BT North American Atlantic
BT North Atlantic
BT Atlantic Ocean
SA Grand Banks
SA Hibernia Field

Jeddah
No longer a valid term for GeoRef. As of 1993 see Jeddah Saudi Arabia.

Jeddah Saudi Arabia (1993)
W Saudi Arabia. Port city on the Red Sea in central Hejaz. Before 1993 also search Jeddah AND Saudi Arabia.
CO N212900N212900
    E0391100E0391100
UF Djeddah
UF Djiddah
UF Jidda
UF Jiddah
BT Hejaz Saudi Arabia
BT Saudi Arabia
BT Arabian Peninsula
BT Asia
SA Red Sea

Jeff Davis County
Valid through 1988. Search in combination with state term. After 1988, use specific county-state term.

Jeff Davis County Georgia (1989)
SE central Georgia. Before 1989, also search Jeff Davis County AND Georgia.
CO N314100N315800
    W0822500W0825000
BT Georgia
BT United States

Jeff Davis County Texas (1989)
Extreme W Texas. Before 1989, also search Jeff Davis County AND Texas.
CO N302500N310800
    W1032300W1050200
BT Texas
BT United States
SA Davis Mountains

Jefferson County
Valid through 1988. Search in combination with state term. After 1988, use specific county-state term.

Jefferson County Alabama (1989)
N central Alabama. Before 1989, also search Jefferson County AND Alabama.
CO N331500N334700
    W0863000W0871800
BT Alabama
BT United States
NT Birmingham Alabama
SA Red Mountain

Jefferson County Arkansas (1989)
Central Arkansas. Before 1989, also search Jefferson County AND Arkansas.
CO N340400N342800
    W0912600W0921700
BT Arkansas
BT United States

Jefferson County Colorado (1989)
Central Colorado. Before 1989, also search Jefferson County AND Colorado.
CO N391200N395400
    W1050200W1052400
BT Colorado
BT United States
NT Golden Colorado
NT Schwartzwalder Mine

Jefferson County Florida (1989)
NW Florida. Before 1989, also search Jefferson County AND Florida.
CO N300700N304000
    W0833100W0840500
BT Florida
BT United States

Jefferson County Georgia (1989)
E Georgia. Before 1989, also search Jefferson County AND Georgia.
CO N324500N331900
    W0821500W0824000
BT Georgia
BT United States

Jefferson County Idaho (1989)
E Idaho. Before 1989, also search Jefferson County AND Idaho.
CO N433700N440500
    W1113000W1124500
BT Idaho
BT United States

Jefferson County Illinois (1989)
S Illinois. Before 1989, also search Jefferson County AND Illinois.
CO N380800N383000
    W0884300W0890800
BT Illinois
BT United States

Jefferson County Indiana (1989)
SE Indiana. Before 1989, also search Jefferson County AND Indiana.
CO N383600N385500
    W0851300W0854300
BT Indiana
BT United States

Jefferson County Iowa (1989)
SE Iowa. Before 1989, also search Jefferson County AND Iowa.
CO N405300N410900
    W0914000W0921000
BT Iowa
BT United States

Jefferson County Kansas (1989)
NE Kansas. Before 1989, also search Jefferson County AND Kansas.
CO N390300N392600
    W0950800W0953300
BT Kansas
BT United States

Jefferson County Kentucky (1989)
N Kentucky. Before 1989, also search Jefferson County AND Kentucky.
CO N380100N382300
    W0852600W0855700
BT Kentucky
BT United States
NT Louisville Kentucky

Jefferson County Mississippi (1989)
SW Mississippi. Before 1989, also search Jefferson County AND Mississippi.
CO N313700N315300
    W0904500W0912200
BT Mississippi
BT United States

Jefferson County Missouri (1989)
E Missouri. Before 1989, also search Jefferson County AND Missouri.
CO N380100N383000
    W0901500W0904700
BT Missouri
BT United States

Jefferson County Montana (1989)
SW central Montana. Before 1989, also search Jefferson County AND Montana.
CO N454500N463500
    W1114000W1123700
BT Montana
BT United States

Jefferson County Nebraska (1989)
SE Nebraska. Before 1989, also search Jefferson County AND Nebraska.
CO N400000N402200
    W0965500W0972300
BT Nebraska
BT United States

Jefferson County New York (1989)
N New York. Before 1989, also search Jefferson County AND New York.
CO N434000N442300
    W0752700W0762400
BT New York
BT United States
SA Saint Lawrence Lowlands
SA Saint Lawrence River
SA Saint Lawrence Valley

Jefferson County Ohio (1989)
E Ohio. Before 1989, also search Jefferson County AND Ohio.
CO N400900N403700
    W0803600W0805700
BT Ohio
BT United States
NT Steubenville Ohio

Jefferson County Oklahoma (1989)
S Oklahoma. Before 1989, also search Jefferson County AND Oklahoma.
CO N335100N341800
    W0973200W0980800
BT Oklahoma

BT United States

Jefferson County Oregon (1989)
Central Oregon. Before 1989, also search Jefferson County AND Oregon.
CO N442300N444800
    W1202200W1214700
BT Oregon
BT United States
SA Mount Jefferson

Jefferson County Pennsylvania (1989)
W central Pennsylvania. Before 1989, also search Jefferson County AND Pennsylvania.
CO N405500N412200
    W0784200W0791300
BT Pennsylvania
BT United States

Jefferson County Tennessee (1989)
E Tennessee. Before 1989, also search Jefferson County AND Tennessee.
CO N355400N361200
    W0831500W0834300
BT Tennessee
BT United States

Jefferson County Texas (1989)
SE Texas. Before 1989, also search Jefferson County AND Texas.
CO N293400N301200
    W0935200W0942600
BT Texas
BT United States
NT Beaumont Texas
SA Sabine Lake

Jefferson County Washington (1989)
W Washington. Before 1989, also search Jefferson County AND Washington.
CO N473100N480900
    W1223800W1243700
BT Washington
BT United States
SA Discovery Bay
SA Olympic Mountains

Jefferson County West Virginia (1989)
Northeasternmost county in West Virginia. Before 1989, also search Jefferson County AND West Virginia.
CO N390800N393000
    W0774000W0780200
BT West Virginia
BT United States

Jefferson County Wisconsin (1989)
S Wisconsin. Before 1989, also search Jefferson County AND Wisconsin.
CO N425100N431300
    W0883300W0890100
BT Wisconsin
BT United States

Jefferson Group (1985)
Montana, Idaho, N Utah and W Wyoming.
BT Upper Devonian
BT Devonian
BT Paleozoic
SA Birdbear Formation
SA Duperow Formation
SA Idaho
SA Montana
SA Utah
SA Wyoming

Jefferson Parish
Valid through 1988. Search in combination with state term. After 1988, use specific county-state term.

Jefferson Parish Louisiana (1989)
W and S of New Orleans. Before 1989, also search Jefferson Parish; Jefferson County AND Louisiana.
CO N291200N300300
  W0895500W0901600
BT Louisiana
BT United States
SA Calcasieu River
SA Grand Isle

Jeffersonville Limestone (1985)
Indiana and N central Kentucky.
BT Middle Devonian
BT Devonian
BT Paleozoic
SA Indiana
SA Kentucky

Jemez Lineament (1985)
N New Mexico.
BT New Mexico
BT United States
SA Colorado Plateau

Jemez Mountains
  use Valle Grande Mountains

Jerome
Valid through 1988. Search in combination with state term. After 1988, use specific city-state term.

Jerome Arizona (1989)
Town in central Arizona. Before 1989, search Jerome AND Arizona.
CO N344600N344600
  W1120700W1120700
BT Yavapai County Arizona
BT Arizona
BT United States

Jerome County Idaho (1989)
S Idaho. Before 1989, search Jerome County AND Idaho.
CO N423100N425200
  W1135600W1143600
BT Idaho
BT United States
NT Jerome Idaho
SA Wilson Lake

Jerome Idaho (1989)
City in S Idaho. Before 1989, search Jerome AND Idaho.
CO N424400N424400
  W1143100W1143100
BT Jerome County Idaho
BT Idaho
BT United States

Jerusalem
No longer a valid term for GeoRef. See Jerusalem Israel.

Jerusalem Israel (1993)
District in E central Israel and capital city of Israel. Before 1967, part of the city belonged to Jordan.
BT Israel
BT Middle East
BT Asia
SA Jordan

jetties (1978)
BT marine installations
SA groins

Jhabua
No longer a valid term for GeoRef. Former princely state in W Madhya Pradesh, central India. See Jhabua India.

Jhabua India (1993)
District and village in W Madhya Pradesh, central India.
BT Madhya Pradesh India
BT India
BT Indian Peninsula
BT Asia

Jhansi
No longer a valid term for GeoRef. See Jhansi India.

Jhansi India (1993)
City and division in S Uttar Pradesh, N central India.
BT Uttar Pradesh India
BT India
BT Indian Peninsula
BT Asia

Jharia
No longer a valid term for GeoRef. See Jharia India.

Jharia coal field (1978)
In SE Bihar. Also search Jharia.
UF Jharia Coalfield
SA Bihar India
SA coal fields
SA Jharia India

Jharia Coalfield
  use Jharia coal field

Jharia India (1993)
Town in SE Bihar, E India. Before 1993, also search Jharia. Before 1978, also search Jherria.
BT Bihar India
BT India
BT Indian Peninsula
BT Asia
SA Jharia coal field

Jherria
Not a valid term for GeoRef. See Jharia India.

Jhunjhunu
No longer a valid term for GeoRef. See Jhunjhunu India.

Jhunjhunu India (1993)
Town and district in NE Rajasthan, NW India.
BT Rajasthan India
BT India
BT Indian Peninsula
BT Asia

Jianghan Basin (1989)
Petroliferous region in Hubei.
SA Hubei China

Jiangsu
No longer a valid term for GeoRef. See Jiangsu China.

Jiangsu China (1993)
Province. E China. Before 1993, also search Jiangsu. Before 1986, also search Kiangsu.
CO N304500N350000
  E1220000E1163000
BT China
BT Far East
BT Asia
NT Nanjing China
SA Chihsia Formation
SA Shanghai China
SA Tancheng-Lujiang Fault
SA Tancheng-Lujiang fault zone
SA Wufeng Formation
SA Yangtze Platform

Jiangxi
No longer a valid term for GeoRef. See Jiangxi China.

Jiangxi China (1993)
Province. SE China.
CO N243000N300000
  E1183000E1134000
BT China
BT Far East
BT Asia
SA Nanling
SA Yangtze Platform

Jiaodong Peninsula
  use Shandong Peninsula

Jidda
  use Jeddah Saudi Arabia

Jiddah
  use Jeddah Saudi Arabia

Jilin
No longer a valid term for GeoRef. See Jilin China.

Jilin China (1993)
Province. NE China.
CO N410000N464000
  E1310000E1220000
UF Kirin, China
BT China
BT Far East
BT Asia
SA Anshan Group
SA Jilin Meteorite
SA Tancheng-Lujiang fault zone
SA Yishu Fault

Jilin Meteorite (1985)
Impact in Jilin, Manchuria, NW China. Before 1985, also search (Jilin OR Kirin) AND meteorites.
UF Kirin Meteorite
BT chondrites
BT stony meteorites
BT meteorites
SA China
SA Jilin China

Jinning
No longer a valid term for GeoRef. See Jinning China.

Jinning China (1993)
Location in Yunnan.
BT Yunnan China
BT China
BT Far East
BT Asia

Jiu River valley (1978)
Also search Jiu; Jiu River.
IN Index regions as applicable.
BT Romania
BT Southern Europe
BT Europe
SA Transylvania
SA Walachia

Jixian
No longer a valid term for GeoRef. See Jixian China.

Jixian China (1993)
County in province of Hebei, N China.
BT Hebei China
BT China
BT Far East
BT Asia

Jiyang Depression (1989)
N Shandong, E China.
BT Shandong China
BT China
BT Far East
BT Asia

Jizerske Hory
  use Izera Mountains

Joana Limestone (1989)
In White Pine Shale. E Nevada and W Utah. Above Pilot Shale; below Chainman Shale.
BT Lower Mississippian
BT Mississippian
BT Carboniferous
BT Paleozoic
SA Nevada
SA Utah

joaquinite (1978)
Autoposting of sorosilicates began in 1985.
BT sorosilicates
BT orthosilicates
BT silicates

Joban coal field (1978)
A major coal field on Pacific Ocean in central Honshu. Also search Joban.
UF Joban Coalfield
BT Honshu
BT Japan
BT Far East
BT Asia
SA coal fields
SA Fukushima Japan

Joban Coalfield
  use Joban coal field

Jodhpur
No longer a valid term for GeoRef. Former princely state, NW India. See Jodhpur India.

Jodhpur India (1993)
City and region in central Rajasthan, NW India.
BT Rajasthan India
BT India
BT Indian Peninsula
BT Asia

Johanna Island
  use Anjouan Island

Johannesburg
No longer a valid term for GeoRef. As of 1993 see Johannesburg South Africa.

Johannesburg South Africa (1993)
City in S central Transvaal, NE South Africa. Before 1993 also search Johannesburg AND South Africa.
BT Transvaal South Africa
BT South Africa
BT Southern Africa
BT Africa

johannsenite (1978)
BT clinopyroxene
BT pyroxene group
BT chain silicates
BT silicates

John Day Formation (1978)
Upper Oligocene and lower Miocene. Tentatively divided into three unnamed members. N central Oregon.
IN Index ages as applicable.
BT Tertiary
BT Cenozoic
SA lower Miocene
SA Oregon
SA upper Oligocene

Johnnie Formation (1978)
Underlies Stirling Quartzite and overlies Noonday Dolomite. E California and SE Nevada.
BT Precambrian
SA California
SA Nevada

Johns Valley Formation (1978)
Scott County, Arkansas, and Ouachita region of SE Oklahoma.
BT Carboniferous
BT Paleozoic
SA Arkansas
SA Mississippian
SA Oklahoma
SA Pennsylvanian

Johnson County
Valid through 1988. Search in combination with state term. After 1988, use specific county-state term.

Johnson County Arkansas (1989)
NW Arkansas. Before 1989, also search Johnson County AND Arkansas.
CO N352200N354600
  W0931200W0934300
BT Arkansas
BT United States

Johnson County Georgia (1989)
E central Georgia. Before 1989, also search Johnson County AND Georgia.
CO N323100N325000
  W0822500W0825500
BT Georgia
BT United States

Johnson County Illinois (1989)
S Illinois. Before 1989, also search Johnson County AND Illinois.
CO N371700N373500
  W0884200W0890300
BT Illinois
BT United States

Johnson County Indiana (1989)
Central Indiana. Before 1989, also search Johnson County AND Indiana.
CO N392100N393800
  W0855700W0861600
BT Indiana
BT United States

Johnson County Iowa (1989)
E Iowa. Before 1989, also search Johnson County AND Iowa.
CO N412500N415200
  W0912300W0915000
BT Iowa
BT United States

Johnson County Kansas (1989)
E Kansas. Before 1989, also search Johnson County AND Kansas.
CO N384500N390500
  W0944000W0950300
BT Kansas
BT United States

Johnson County Kentucky (1989)
E Kentucky. Before 1989, also search Johnson County AND Kentucky.
CO N374300N375900
  W0823800W0830000
BT Kentucky
BT United States

Johnson County Missouri (1989)
W central Missouri. Before 1989, also search Johnson County AND Missouri.
CO N383400N385700
  W0932900W0940800
BT Missouri
BT United States

Johnson County Nebraska (1989)
SE Nebraska. Before 1989, also search Johnson County AND Nebraska.
CO N401800N403000
  W0960400W0962800
BT Nebraska
BT United States

Johnson County Tennessee (1989)
Extreme NE Tennessee. Before 1989, also search Johnson County AND Tennessee.
CO N362200N363700
  W0813800W0820300
BT Tennessee
BT United States

Johnson County Texas (1989)
N central Texas. Before 1989, also search Johnson County AND Texas.
CO N320900N323500
  W0970600W0973700
BT Texas
BT United States

Johnson County Wyoming (1989)
N central Wyoming. Before 1989, also search Johnson County AND Wyoming.
CO N432900N443400
  W1060000W1072300
BT Wyoming
BT United States
SA Hartzog Draw Field

Johnston County
Valid through 1988. Search in combination with state term. After 1988, use specific county-state term.

Johnston County North Carolina (1989)
Central North Carolina. Before 1989, also search Johnston County AND North Carolina.
CO N351500N354800
  W0780500W0784200
BT North Carolina
BT United States

Johnston County Oklahoma (1989)
S Oklahoma. Before 1989, also search Johnston County AND Oklahoma.
CO N340700N343100
  W0962400W0965500
BT Oklahoma
BT United States

Johnston Island (1985)
Term reintroduced in 1985. Not a valid term 1971-1984. NE Pacific Ocean, S of Hawaii. United States territory.
CO N153000N173000
  W1680000W1700000
SA East Pacific Ocean Islands
SA Northeast Pacific

Johore
No longer a valid term for GeoRef. As of 1993, see Johore Malaysia.

Johore Malaysia (1993)
State just N of Singapore on the Malay Peninsula. This term has multiple hierarchies.
BT1 West Malaysia
BT1 Malay Peninsula
BT1 Far East
BT1 Asia
BT2 West Malaysia
BT2 Malaysia
BT2 Far East
BT2 Asia

JOIDES (1989)
Acronym.
UF Joint Oceanographic Institutions for Deep Earth Sampling
SA associations
SA Deep Sea Drilling Project
SA IPOD
SA Ocean Drilling Program

Joint Oceanographic Institutions for Deep Earth Sampling
use JOIDES

jointing
No longer a valid GeoRef index term. Before 1971, was used on level 1 in subfiles E and N.
use joints

joints (1978)
Before 1971, also search jointing.
UF jointing
BT fractures
SA bedding joints
SA cleats
SA columnar joints
SA conjugate faults
SA conjugate folds
SA cross fractures
SA fissures
SA fracture cleavage
SA oblique orientation
SA release fractures
SA strike
SA structural analysis

jokulhlaups (1989)
Icelandic for glacier outburst floods.
UF glacier bursts
UF glacier floods
UF glacier outburst floods
SA floods
SA glacial features
SA glacial geology
SA glaciers
SA meltwater

Joliba River
use Niger River

Joliba Valley
use Niger Valley

Jones County
Valid through 1988. Search in combination with state term. After 1988, use specific county-state term.

Jones County Georgia (1989)
Central Georgia. Before 1989, also search Jones County AND Georgia.
CO N325000N331100
  W0832000W0834700
BT Georgia
BT United States

Jones County Iowa (1989)
E Iowa. Before 1989, also search Jones County AND Iowa.
CO N415700N421800
  W0905400W0912300
BT Iowa
BT United States
NT Monticello Iowa

Jones County Mississippi (1989)
SE Mississippi. Before 1989, also search Jones County AND Mississippi.
CO N312600N314900
  W0885500W0892400
BT Mississippi
BT United States

Jones County North Carolina (1989)
E North Carolina. Before 1989, also search Jones County AND North Carolina.
CO N344700N351400
  W0770700W0774300
BT North Carolina
BT United States

Jones County South Dakota (1989)
S central South Dakota. Before 1989, also search Jones County AND South Dakota.
CO N434300N441100
  W1002200W1010500
BT South Dakota
BT United States

Jones County Texas (1989)
W central Texas. Before 1989, also search Jones County AND Texas.
CO N323300N325900
  W0993700W1000900
BT Texas
BT United States

Jones Mountains (1978)
On Eights Coast in Ellsworth Land on Pacific Ocean side.
SA Antarctica
SA Ellsworth Land

Joplin
Valid through 1988. Search in combination with state term. After 1988, use specific city-state term.

Joplin Missouri (1989)
City in Jasper and Newton counties, SW Missouri. Before 1989, search Joplin AND Missouri.
IN Index counties as applicable.
CO N370400N370400
  W0943100W0943100
BT Missouri
BT United States
SA Jasper County Missouri
SA Newton County Missouri

Jordan (1978)
Before 1969, also search Trans-Jordan.
CO N290000N333000
  E0390000E0343000
UF Trans-Jordan
BT Middle East
BT Asia
SA Dead Sea
SA Dead Sea Rift
SA Great Rift Valley
SA Jerusalem Israel
SA Jordan River
SA Jordan Valley
SA Judea Group
SA Menuha Formation
SA Mount Scopus Group
SA Near East
SA Palestine
SA Red Sea Basin
SA Samaria
SA Wadi Araba

Jordan Rift valley
use Jordan Valley

Jordan River (1978)
Rises in Syria and flows S through the Sea of Galilee to N end of the Dead Sea.
IN Index countries as applicable.
BT Middle East
BT Asia
SA Israel

SA Jordan
SA Syria

Jordan River valley
use Jordan Valley

Jordan Valley (1978)
River valley between Sea of Galilee and Dead Sea which is part of the Great Rift Valley.
IN Index countries as applicable.
UF Jordan Rift valley
UF Jordan River valley
UF Tiberias-Dead Sea Rift valley
BT Middle East
BT Asia
SA Great Rift Valley
SA Israel
SA Jordan
SA Syria

jordanite (1978)
BT sulfarsenites
BT sulfosalts

Jordanow
No longer a valid term for GeoRef. As of 1993 see Jordanow Poland.

Jordanow Poland (1993)
Town in S Poland. Before 1993 also search Jordanow AND Poland.
BT Nowy Sacz Poland
BT Poland
BT Central Europe
BT Europe

jordisite (1978)
BT sulfides

Jornada del Muerto (1989)
Arid region in Dona Ana, Sierra and Socorro counties, S New Mexico.
IN Index counties as applicable.
BT New Mexico
BT United States
SA Dona Ana County New Mexico
SA Sierra County New Mexico
SA Socorro County New Mexico

Jos Plateau (1978)
Central Nigeria.
UF Bauchi Plateau
BT Nigeria
BT West Africa
BT Africa

joseite (1978)
This term has multiple hierarchies.
BT1 tellurides
BT2 sulfides

Joseph Bonaparte Gulf (1978)
Inlet of Timor Sea.
IN Index Northern Territory Australia and/or Western Australia as applicable.
BT Australia
BT Australasia
SA Bonaparte Gulf basin
SA Northern Territory Australia
SA Timor Sea
SA Western Australia

Josephine County
Valid through 1988. Search in combination with state term. After 1988, use specific county-state term.

Josephine County Oregon (1989)
SW Oregon. Before 1989, also search Josephine County AND Oregon.
CO N420000N424700
W1231300W1240500
BT Oregon

BT United States

Josephine Ophiolite (1989)
N California and SW Oregon. Conformably overlain by the Galice Formation. Includes Josephine Peridotite.
BT Upper Jurassic
BT Jurassic
BT Mesozoic
SA California
SA Galice Formation
SA Josephine Peridotite
SA Northern California
SA Oregon

Josephine Peridotite (1989)
N California and SW Oregon. Basal unit of the Josephine Ophiolite.
BT Upper Jurassic
BT Jurassic
BT Mesozoic
SA California
SA Josephine Ophiolite
SA Northern California
SA Oregon

josephinite (1978)
BT alloys
SA iron
SA nickel

Jotnian (1978)
Europe. Autoposting of upper Precambrian and Precambrian to this term ended in 1989. As of 1989, upper Proterozoic and Proterozoic are autoposted to this term. Autoposting of upper Precambrian and Precambrian to this term began in 1990.
BT upper Proterozoic
BT Proterozoic
BT upper Precambrian
BT Precambrian

Jotunheim Mountains (1978)
Range in S central Norway. Also search Jotunheimen.
UF Jotunheimen Mountains
BT Norway
BT Scandinavia
BT Western Europe
BT Europe

Jotunheimen Mountains
use Jotunheim Mountains

jouravskite (1978)
BT sulfates

Juab County
Valid through 1988. Search in combination with state term. After 1988, use specific county-state term.

Juab County Utah (1989)
W Utah. Before 1989, also search Juab County AND Utah.
CO N392000N400200
W1114000W1140500
BT Utah
BT United States
NT Eureka Utah
SA Sevier Desert
SA Thomas Range
SA Tintic mining district

Juan de Fuca Plate (1985)
Small plate in NE Pacific Ocean, E of the Pacific Plate and W of the North American Plate.
BT Pacific Ocean
SA Cascadia subduction zone
SA Juan de Fuca Ridge
SA plate tectonics
SA plates

Juan de Fuca Ridge (1978)

Off coast of Washington State.
UF Juan de Fuca Rise
BT Pacific Ocean
SA Axial Seamount
SA Juan de Fuca Plate
SA mid-ocean ridges

Juan de Fuca Rise
use Juan de Fuca Ridge

Juan de Fuca Strait (1978)
Between Vancouver Island on the N and mainland of Washington on the S.
IN Index British Columbia and/or Washington as applicable.
UF Strait of Juan de Fuca
BT North America
SA British Columbia
SA Washington

Jubbulpore India
use Jabalpur India

Judea Group (1989)
BT Cretaceous
BT Mesozoic
SA Israel
SA Jordan
SA Lebanon
SA Syria

Judith Basin County
Valid through 1988. Search in combination with state term. After 1988, use specific county-state term.

Judith Basin County Montana (1989)
Central Montana. Before 1989, also search Judith Basin County AND Montana.
CO N464500N472300
W1094500W1104500
BT Montana
BT United States
SA Little Belt Mountains

Judith River Formation (1978)
In Montana Group. Comprises two members: Parkman Sandstone and an upper unnamed member. N, central, S and SE Montana; NW Wyoming; Alberta.
BT Upper Cretaceous
BT Cretaceous
BT Mesozoic
SA Alberta
SA Montana
SA Montana Group
SA Parkman Sandstone
SA Wyoming

Juglandaceae (1978)
BT Dicotyledoneae
BT angiosperms
BT Spermatophyta
BT Plantae
SA Juglans

Juglans (1978)
Genus.
BT Dicotyledoneae
BT angiosperms
BT Spermatophyta
BT Plantae
SA Filicopsida
SA gymnosperms
SA Juglandaceae
SA pteridophytes

Jujuy
No longer a valid term for GeoRef. See Jujuy Argentina.

Jujuy Argentina (1993)
Province including city in extreme NW Argentina.
CO S250000S220000
W0640000W0673000

BT Argentina
BT South America
SA Yacoraite Formation

julgoldite (1978)
Autoposting of sorosilicates began in 1985.
BT sorosilicates
BT orthosilicates
BT silicates

Julian Alps (1978)
Division of Eastern Alps.
IN Index countries as applicable.
BT Eastern Alps
BT Alps
BT Europe
SA Italy
SA Slovenia
SA Yugoslavia

Jumilla
No longer a valid term for GeoRef. As of 1993, see Jumilla Spain.

Jumilla Spain (1993)
City in Murcia Province, SE Spain. Before 1993, also search Jumilla and Spain.
BT Murcia Spain
BT Murcia region
BT Spain
BT Iberian Peninsula
BT Southern Europe
BT Europe

Jummoo
use Jammu

Jumna River (1978)
N central India.
IN Index states as applicable.
UF Yamuna River
BT India
BT Indian Peninsula
BT Asia
SA Himachal Pradesh India
SA Uttar Pradesh India

Juneau
Valid through 1988. Search in combination with state term. After 1988, use specific city-state term.

Juneau Alaska (1989)
City in panhandle near border of Yukon Territory. Before 1989, search Juneau AND Alaska.
CO N582000N582000
W1342000W1342000
BT Southeastern Alaska
BT Alaska
BT United States

Juneau Glacier
use Juneau ice field

Juneau ice field (1978)
N of Juneau in panhandle.
UF Juneau Glacier
UF Juneau Icefield
BT Southeastern Alaska
BT Alaska
BT United States

Juneau Icefield
use Juneau ice field

Junggar (1993)
Region in N Xinjiang. Before 1993, also search Junggar, Dzhungaria, Dzungaria, or Zungaria.
UF Dzhungaria
UF Dzungaria
UF Zungaria
BT Xinjiang China
BT China
BT Far East
BT Asia

Junggar Basin (1989)
NW Xinjiang, NW China.
BT Xinjiang China
BT China
BT Far East
BT Asia

Juniata Formation (1978)
Includes East Waterford Red Sandstone Member, Plummer Hollow Red Mudstone, Run Gap Red Sandstone Member. S central and E Pennsylvania, W Maryland, E Tennessee, W Virginia, and E West Virginia.
BT Upper Ordovician
BT Ordovician
BT Paleozoic
SA Maryland
SA Pennsylvania
SA Richmond Group
SA Tennessee
SA Virginia
SA West Virginia

Junin
No longer a valid term for GeoRef. As of 1993 see Junin Peru.

Junin Peru (1993)
Department in W central Peru. Before 1993 also search Junin AND Peru.
BT Peru
BT South America
NT Cerro de Pasco Peru
SA Pucara Group

junior high school (1978)
Before 1978, search junior high.
SA college-level education
SA education
SA elementary school
SA high school

Jupiter (1978)
SA Adrastea Satellite
SA Amalthea Satellite
SA Callisto Satellite
SA Europa Satellite
SA Galilean satellites
SA Ganymede Satellite
SA Io Satellite
SA outer planets
SA Pioneer 10
SA Pioneer 11
SA Pioneer Program
SA planetology
SA planets
SA satellites
SA solar system
SA Voyager 1
SA Voyager 2
SA Voyager Program

Jupiter Member (1985)
Of Galeros Formation. N Arizona.
BT Precambrian
SA Arizona
SA Chuar Group

Jura
No longer a valid term for GeoRef. See Jura France.

Jura Canton
No longer a valid term for GeoRef.
use Jura Switzerland

Jura France (1993)
Department on the Swiss border.
CO N461500N472000
 E0061500E0051500
BT France
BT Western Europe
BT Europe
NT Morez France
SA Franche-Comte
SA Jura Mountains

SA Jura Switzerland

Jura Mountains (1978)
Also search Jura.
IN Index countries as applicable.
CO N455000N473500
 E0081500E0044500
BT Europe
NT Swiss Jura Mountains
SA Ain France
SA Doubs France
SA Geneva Switzerland
SA Jura France
SA Neuchatel Switzerland
SA Vaud Switzerland

Jura Switzerland (1993)
Formerly part of Bern Canton. Became a separate canton in 1979. Use Jura France for the department in France. Before 1993 also search Jura Canton AND Switzerland.
CO N470800N473100
 E0073500E0065000
UF Jura Canton
BT Switzerland
BT Central Europe
BT Europe
SA Jura France
SA Swiss Jura Mountains

Jurassic (1978)
Above Triassic, below Cretaceous. From 1978-1980, Mesozoic was autoposted to this term.
BT Mesozoic
NT Arapien Shale
NT Athgarh Sandstone
NT Aztec Sandstone
NT Bazhenov Formation
NT Birkhead Formation
NT Bonanza Group
NT Bowser Formation
NT Carmel Formation
NT Coast Range Ophiolite
NT Fernie Formation
NT Ferrar Group
NT Kingak Shale
NT Kirkpatrick Basalt
NT La Quinta Formation
NT Ladner Group
NT Lathi Formation
NT Lower Jurassic
NT Mansalay Formation
NT Middle Jurassic
NT Norphlet Formation
NT Posidonia Shale
NT Rajmahal Series
NT San Rafael Group
NT Shaunavon Formation
NT Smartville Complex
NT Solnhofen Limestone
NT Twin Creek Limestone
NT Upper Jurassic
NT Younger Granites
SA Beacon Supergroup
SA Briançonnais Zone
SA Cimmerian Orogeny
SA Franciscan Complex
SA Glen Canyon Group
SA Great Valley Sequence
SA Hazelton Group
SA Ionian Zone
SA Karroo Supergroup
SA Khorat Group
SA Lincolnshire Limestone
SA Louann Salt
SA Maiolica Limestone
SA Mist Mountain Formation
SA Navajo Sandstone
SA Nevadan Orogeny
SA Passaic Formation
SA Pucara Group
SA Serra Geral Formation

SA Tal Formation
SA upper Gondwana System
SA Yanshanian

Jutland (1978)
Peninsula including Schleswig in N Germany and the Danish mainland. Politically it applies only to the mainland of Denmark.
IN Index Denmark and/or Schleswig-Holstein Germany as applicable.
UF Jylland
BT Europe
NT Djursland
SA Denmark
SA Schleswig-Holstein Germany
SA Vendsyssel

juvenile taxa (1989)
SA ecology
SA fossils
SA ontogeny
SA paleoecology

juvenile water (1981)
SA ground water
SA magmas
SA water

Juvinas Meteorite (1985)
Impact at Juvinas in Ardeche, France. Before 1985, also search Juvinas AND meteorites.
BT eucrite
BT achondrites
BT stony meteorites
BT meteorites
SA Ardeche France
SA France

Jylland
use Jutland

K

K
use potassium

K'un-ming China
use Kunming China

K-40 (1978)
Autoposting of broader terms began in 1989. This term has multiple hierarchies.
BT1 radioactive isotopes
BT1 isotopes
BT2 potassium
BT2 alkali metals
BT2 metals

K-feldspar (1978)
UF potash feldspar
UF potassium feldspar
BT alkali feldspar
BT feldspar group
BT framework silicates
BT silicates
SA adularia
SA microcline
SA orthoclase
SA sanidine

K-T boundary (1993)
Before 1993, also search Cretaceous-Tertiary; Upper Cretaceous AND lower Paleocene AND boundary; K/T boundary; and Cretaceous AND Tertiary AND boundary. This term has multiple hierarchies.
UF Cretaceous-Tertiary boundary
UF K/T boundary

BT1 lower Paleocene
BT1 Paleocene
BT1 Paleogene
BT1 Tertiary
BT1 Cenozoic
BT2 Upper Cretaceous
BT2 Cretaceous
BT2 Mesozoic
SA Bauru Formation
SA Beaverhead Formation
SA biostratigraphy
SA Deccan Traps
SA Denver Formation
SA Difunta Group
SA Fort Union Formation
SA Mardin Formation
SA Mount Scopus Group
SA Muro Group
SA Nanaimo Group
SA North Horn Formation
SA Paskapoo Formation
SA Phanerozoic
SA Raton Formation
SA Scaglia Formation
SA Severn Formation
SA stratigraphic boundary

K/Ar (1978)
Isotopic ratio used in age determination. Before 1972, also search potassium-argon.
UF potassium-argon
SA absolute age
SA argon
SA isotope ratios
SA potassium

K/Rb (1989)
SA isotopes
SA potassium
SA rubidium

K/T boundary
use K-T boundary

Kaap Valsch
use False Cape

Kaapvaal Craton (1985)
Archean, S Africa.
IN Index countries as applicable.
CO S270000S180000
 E0300000E0180000
BT Southern Africa
BT Africa
SA Archean
SA Barberton greenstone belt
SA Botswana
SA Namibia
SA South Africa
SA Swaziland

Kabardia (1978)
N part of Kabardin-Balkar Autonomous Republic which is on N slopes of Caucasus. This term has multiple hierarchies.
BT1 Kabardin-Balkar Russian Federation
BT1 Russian Federation
BT1 Commonwealth of Independent States
BT2 Kabardin-Balkar Russian Federation
BT2 Europe

Kabardin-Balkar
No longer a valid term for GeoRef as of 1993.
use Kabardin-Balkar Russian Federation

Kabardin-Balkar Russian Federation (1993)
Former Autonomous Soviet Socialist Republic. Before 1993, also search Kabardin-Balkar, Kabardinia-Balkar, or Kabarnino-

Balkar. This term has multiple hierarchies.
 UF Kabardin-Balkar
 UF Kabardinia-Balkar Russian Republic
 UF Kabarnino-Balkar Russian Republic
 BT Russian Federation
 BT Commonwealth of Independent States
 BT2 Europe
 NT Kabardia
 SA Peredovoy Range
 SA Transcaucasia

Kabardinia-Balkar Russian Republic
 use Kabardin-Balkar Russian Federation

Kabarnino-Balkar Russian Republic
 use Kabardin-Balkar Russian Federation

Kabinda Angola
 use Cabinda Angola

Kabuh Formation   (1989)
 Central Java.
 BT Quaternary
 BT Cenozoic
 SA Indonesia
 SA Java

Kabul
 No longer a valid term for GeoRef. See Kabul Afghanistan.

Kabul Afghanistan   (1993)
 Use for the province in NE Afghanistan. Also a city. Before 1993, also search Kabul and Afghanistan.
 BT Afghanistan
 BT Indian Peninsula
 BT Asia

Kabylia   (1978)
 Mountainous coastal region in N Algeria.
 BT Algeria
 BT North Africa
 BT Africa
 SA Great Kabylia

Kachchh
 No longer a valid term for GeoRef. See Kachchh India.

Kachchh India   (1993)
 District in Gujarat, W India.
 BT Gujarat India
 BT India
 BT Indian Peninsula
 BT Asia

Kadah
 No longer a valid term for GeoRef. As of 1993, see Kedah Malaysia.

Kadzharan
 No longer a valid term for GeoRef. As of 1993, see Kadzharan Armenia.

Kadzharan Armenia   (1993)
 Town in S Armenia. This term has multiple hierarchies.
 BT1 Armenia
 BT1 Commonwealth of Independent States
 BT2 Armenia
 BT2 Europe

kaersutite   (1978)
 BT clinoamphibole
 BT amphibole group
 BT chain silicates
 BT silicates
 SA hornblende

Kafan
 No longer a valid term for GeoRef. As of 1993, see Kafan Armenia.

Kafan Armenia   (1993)
 City in SE Armenia. This term has multiple hierarchies.
 BT1 Armenia
 BT1 Commonwealth of Independent States
 BT2 Armenia
 BT2 Europe

Kafue   (1978)
 Township and river in N Zambia.
 BT Zambia
 BT East Africa
 BT Africa

Kaganovich Peak
 use Darvaz

Kagawa
 No longer a valid term for GeoRef. See Kagawa Japan.

Kagawa Japan   (1993)
 Prefecture on the Inland Sea in NE Shikoku, S Japan.
 BT Shikoku
 BT Japan
 BT Far East
 BT Asia

Kagoshima
 No longer a valid term for GeoRef. See Kagoshima Japan.

Kagoshima Japan   (1993)
 City and prefecture in S Kyushu, S Japan.
 BT Kyushu
 BT Japan
 BT Far East
 BT Asia
 NT Aira Caldera
 NT Oshima
 NT Osumi Peninsula
 NT Sakura-jima
 NT Satsuma Peninsula
 NT Tanega-shima
 SA Kirishima
 SA Ontake
 SA Ryukyu Islands

Kaibab Formation   (1978)
 Divided vertically into three members. N Arizona, SE California, SE Nevada and W Utah.
 BT Permian
 BT Paleozoic
 SA Arizona
 SA California
 SA Nevada
 SA Utah

Kaikoura   (1978)
 Mountain range in NE South Island. Also a peninsula.
 IN Also index South Island.
 BT Marlborough New Zealand
 BT New Zealand
 BT Australasia
 SA South Island

Kaimur Sandstone   (1978)
 Constitutes part of the Semri Series in the Gwalior and Mewar areas. Upper Kaimur Sandstone and lower Kaimur Sandstone are included in Kaimur Series.
 IN Index ages as applicable.
 SA Cambrian
 SA India
 SA Madhya Pradesh India
 SA Precambrian
 SA Rajasthan India
 SA Semri Series
 SA Uttar Pradesh India

SA Vindhyan

kainite   (1978)
 From 1978-1980, halides and sulfates were autoposted to this term. Autoposting of chlorides began in 1981. Autoposting of halides began again in 1989. This term has multiple hierarchies.
 BT1 chlorides
 BT1 halides
 BT2 sulfates

Kainsaz Meteorite   (1989)
 Impact at Muslyumov, Tatar. Black olivine-pigeonite chondrite.
 BT carbonaceous chondrites
 BT chondrites
 BT stony meteorites
 BT meteorites
 SA Russian Federation
 SA Tatar Russian Federation

Kaipara Harbor   (1978)
 Inlet of Pacific Ocean on W coast of N extension of North Island. Also search Kaipara.
 IN Also index North Island.
 BT New Zealand
 BT Australasia
 SA North Island

Kaiparowits Plateau   (1978)
 W of Colorado River between Escalante and Paria rivers in S central Utah.
 BT Utah
 BT United States

Kaiserstuhl   (1978)
 Mountain group in SW Baden-Wurttemberg, SW West Germany.
 BT Baden-Wurttemberg Germany
 BT Germany
 BT Central Europe
 BT Europe

Kakagi Lake   (1978)
 East of Lake of the Woods in W Ontario.
 BT Ontario
 BT Eastern Canada
 BT Canada

Kakanui   (1978)
 Mountains in SE New Zealand, north of Dunedin.
 IN Also Index South Island.
 BT Otago New Zealand
 BT New Zealand
 BT Australasia
 SA South Island

Kakhetia   (1978)
 Region in E Georgian Republic.
 BT Georgian Republic
 BT Europe

Kakioka
 No longer a valid term for GeoRef. See Kakioka Japan.

Kakioka Japan   (1993)
 Town in Ibaraki prefecture, central Honshu. Before 1993, also search Kakioka AND Japan.
 CO N361400N361400
  E1401100E1401100
 BT Ibaraki Japan
 BT Honshu
 BT Japan
 BT Far East
 BT Asia

kakortokite   (1978)
 BT syenites
 BT plutonic rocks
 BT igneous rocks
 SA nepheline syenite

Kakuto
 No longer a valid term for GeoRef. As of 1993 see Kakuto Uganda.

Kakuto Uganda   (1993)
 Village in SE Uganda on Lake Victoria. Before 1993 also search Kakuto AND Uganda.
 BT Uganda
 BT East Africa
 BT Africa

Kaladgi
 No longer a valid term for GeoRef. See Kaladgi India.

Kaladgi India   (1993)
 Village in W central Karnataka, SW India.
 BT Karnataka India
 BT India
 BT Indian Peninsula
 BT Asia

Kaladgi System   (1978)
 Precambrian? Divided into upper Kaladgi and lower Kaladgi series.
 BT Precambrian
 SA India
 SA Karnataka India
 SA Maharashtra India

Kalahari Desert   (1978)
 Plateau and desert region.
 IN Index countries as applicable.
 BT Africa
 SA Botswana
 SA Namibia
 SA South Africa

Kalamata earthquake 1986   (1993)
 In Peloponnesus, Greece.
 BT earthquakes
 SA Aegean Sea
 SA Greece
 SA Peloponnesus Greece

Kalamazoo County
 Valid through 1988. Search in combination with state term. After 1988, use specific county-state term.

Kalamazoo County Michigan   (1989)
 SW Michigan. Before 1989, also search Kalamazoo County AND Michigan.
 CO N420500N422500
  W0852000W0854700
 BT Michigan Lower Peninsula
 BT Michigan
 BT United States

Kalba Range   (1978)
 W branch of Altai Mountains in E Kazakhstan. Also search Kalba. This term has multiple hierarchies.
 CO N490000N492000
  E0830000E0823000
 BT1 Kazakhstan
 BT1 Central Asia
 BT1 Asia
 BT2 Kazakhstan
 BT2 Commonwealth of Independent States
 BT3 Altai Mountains
 BT3 Asia

Kalgoorlie
 No longer a valid term for GeoRef. See Kalgoorlie Australia.

Kalgoorlie Australia   (1993)
 Municipality in S Western Australia.
 CO S293000S293000
  E1220000E1220000
 BT Western Australia

BT Australia
BT Australasia

Kalgoorlie System (1978)
Includes "Older Greenstones", Black Flag Group, Yindarlgooda Group, Kundana Group, and "Younger Greenstones". S Western Australia.
BT Archean
BT Precambrian
SA Western Australia

Kali Gandaki Valley (1978)
River valley in central Nepal.
BT Nepal
BT Indian Peninsula
BT Asia

Kalibeng Formation (1989)
Central Java, in the Sambungmacan region.
BT Quaternary
BT Cenozoic
SA Indonesia
SA Java

kaliborite (1978)
UF heintzite
UF hintzeite
UF paternoite
BT borates

Kalimantan
No longer a valid term for GeoRef. As of 1993 see Kalimantan Indonesia.

Kalimantan Indonesia (1993)
All of island of Borneo excepting Sarawak, Sabah and Brunei. Before 1993 also search Kalimantan AND Indonesia. This term has multiple hierarchies.
CO S042000N043000
   E1190000E1085500
BT1 Borneo
BT1 Far East
BT1 Asia
BT2 Borneo
BT2 Malay Archipelago
BT3 Indonesia
BT3 Far East
BT3 Asia
NT Mahakam Delta

Kalinin
No longer a valid term for GeoRef. As of 1993, see Tver Russian Federation.

Kalinin Russian Federation
use Tver Russian Federation

Kaliningrad
As of 1993, no longer a valid term for GeoRef.
use Kaliningrad Russian Federation

Kaliningrad Russian Federation (1993)
City and oblast, an exclave of the Russian Federation. Before 1993, also search Kaliningrad. This term has multiple hierarchies.
UF Kaliningrad
BT1 Russian Federation
BT1 Commonwealth of Independent States
BT2 Europe
SA Baltic region
SA Courland Spit

kaliophilite (1978)
BT nepheline group
BT framework silicates
BT silicates
SA aluminosilicates

kalium
use potassium

Kalman filters (1985)
SA algorithms
SA deconvolution
SA filters
SA ground water
SA hydrodynamics
SA mathematical models
SA recursive filters
SA seismic methods
SA seismology

Kalmar
No longer a valid term for GeoRef. As of 1993 see Kalmar Sweden.

Kalmar Sweden (1993)
County in S Sweden. Before 1993 also search Kalmar AND Sweden.
BT Sweden
BT Scandinavia
BT Western Europe
BT Europe
NT Oland
NT Vastervik Sweden
SA Smaland

Kalmyk
No longer a valid term for GeoRef. As of 1993, see Kalmyk Russian Federation.

Kalmyk Russian Federation (1993)
Former Autonomous Soviet Socialist Republic on NW shore of Caspian Sea. This term has multiple hierarchies.
BT1 Russian Federation
BT1 Commonwealth of Independent States
BT2 Europe

kalsilite (1978)
BT nepheline group
BT framework silicates
BT silicates
SA aluminosilicates

Kaltag Fault (1989)
W central Alaska and Bering Sea.
IN Index land or ocean regions as applicable.
CO N620000N631500
   W1650000W1693000
SA Alaska
SA Bering Sea
SA Yukon-Koyukuk Basin

Kama
No longer a valid term for GeoRef. As of 1993, see Kama Zaire or Kama Burma. See Kama River for the river.

Kama Burma (1993)
Village in Thayetmyo district, Upper Burma. Before 1993, also search Kama in combination with Burma.
BT Burma
BT Far East
BT Asia

Kama River (1978)
Rises in central Urals and flows into Volga River. Before 1993, also search Kama AND USSR.
CO N550000N610000
   E0560000E0520000
BT Russian Federation
BT Commonwealth of Independent States
SA Vyatka-Kama Interfluve

Kama Zaire (1993)
Village in Kivu, Zaire. Before 1993, search Kama AND Zaire.

BT Kivu Zaire
BT Zaire
BT Central Africa
BT Africa

kamacite (1978)
Meteoric mineral.
BT alloys
SA iron
SA iron meteorites
SA meteorites
SA nickel
SA plessite
SA taenite

Kamaishi Mine (1993)
BT Iwate Japan
BT Honshu
BT Japan
BT Far East
BT Asia
SA metal ores
SA mines

Kambalda
No longer a valid term for GeoRef. See Kambalda Australia.

Kambalda Australia (1993)
Settlement S of Kalgoorlie in S Western Australia.
BT Western Australia
BT Australia
BT Australasia

Kamchatka
No longer a valid term for GeoRef. As of 1993, see Kamchatka Russian Federation for the oblast.

Kamchatka Peninsula (1978)
Between Okhotsk Sea and Bering Sea. As of 1981, also includes the Komandorski Islands and Karaginskiy Island. This term has multiple hierarchies.
CO N510000N600000
   E1680000E1550000
BT1 Russian Pacific region
BT1 Russian Federation
BT1 Commonwealth of Independent States
BT2 Russian Pacific region
BT2 Asia
NT Karaginskiy Island
NT Komandorski Islands
NT Tolbachik
SA Avacha
SA Kamchatka Russian Federation
SA Karymskaya Sopka
SA Klyuchevskaya Sopka
SA Okhotsk-Chukchi volcanic belt
SA Russian Far East
SA Sheveluch
SA Uzon

Kamchatka Russian Federation (1993)
Large oblast in NE Russian Federation. Includes Kamchatka Peninsula. This term has multiple hierarchies.
CO N510000N650000
   E1750000E1550000
BT1 Russian Federation
BT1 Commonwealth of Independent States
BT2 Asia
NT Avacha
NT Klyuchevskaya Sopka
NT Paratunka Russian Federation
NT Sheveluch
SA Kamchatka Peninsula
SA Okhotsk region
SA Okhotsk-Chukchi volcanic belt
SA Penzhina Bay
SA Uzon

Kamen
No longer a valid term for GeoRef. See Kamen Germany.

Kamen Germany (1993)
City in the Ruhr in W North Rhine-Westphalia, W Germany.
BT North Rhine-Westphalia Germany
BT Germany
BT Central Europe
BT Europe

Kamenka River (1978)
Flows into the Angara River in S Krasnoyarsk Kray. This term has multiple hierarchies.
BT1 Russian Federation
BT1 Commonwealth of Independent States
BT2 Asia
SA Krasnoyarsk Russian Federation

kames (1978)
Autoposting of glaciofluvial features and glacial features to this term began in 1989.
IN Index fluvial features and/or lacustrine features as applicable.
BT glacial features
SA cirques
SA eskers
SA fluvial features
SA glacial geology
SA glaciers
SA lacustrine features

Kamikawa
No longer a valid term for GeoRef. See Kamikawa Japan.

Kamikawa Japan (1993)
Sub-prefecture in central Hokkaido, N Japan.
BT Hokkaido
BT Japan
BT Far East
BT Asia

Kaminak Lake (1978)
SW Keewatin District, SE Northwest Territories.
BT Keewatin District Northwest Territories
BT Northwest Territories
BT Western Canada
BT Canada

Kamioka Mine (1993)
BT Gifu Japan
BT Honshu
BT Japan
BT Far East
BT Asia
SA lead-zinc deposits
SA mines
SA silver ores

Kamloops
No longer a valid term for GeoRef. See Kamloops British Columbia.

Kamloops British Columbia (1993)
City in S British Columbia.
BT British Columbia
BT Western Canada
BT Canada

kammererite (1978)
BT chlorite group
BT sheet silicates
BT silicates

Kamoto (1978)
Mine five miles W of village of Kolwezi in S Zaire.
BT Shaba Zaire
BT Zaire

BT Central Africa
BT Africa
SA mines
Kamp Valley (1978)
River valley in NE Austria.
BT Lower Austria
BT Austria
BT Central Europe
BT Europe
Kampinos Forest (1978)
E central Poland.
BT Poland
BT Central Europe
BT Europe
Kampuchea
use Cambodia
Kamthi Formation (1989)
BT Permian
BT Paleozoic
SA India
SA Maharashtra India
SA Orissa India
Kamuikotan
use Kamuikotan Belt
Kamuikotan Belt (1989)
Tectonic zone in Ishikari sub-prefecture of W Hokkaido.
UF Kamuikotan
UF Kamuikotan Zone
BT Hokkaido
BT Japan
BT Far East
BT Asia
Kamuikotan Zone
use Kamuikotan Belt
Kamyshin
No longer a valid term for GeoRef. As of 1993, see Kamyshin Russian Federation.
Kamyshin Russian Federation (1993)
City on the Volga River in Volgograd Oblast. Before 1993, also search Kamyshin. This term has multiple hierarchies.
BT1 Volgograd Russian Federation
BT1 Russian Federation
BT1 Commonwealth of Independent States
BT2 Volgograd Russian Federation
BT2 Europe
Kanagawa
No longer a valid term for GeoRef. See Kanagawa Japan.
Kanagawa Japan (1993)
Prefecture in SE Japan.
BT Honshu
BT Japan
BT Far East
BT Asia
NT Yokohama Japan
SA Hakone
SA Izu Peninsula
SA Kanto Plain
SA Miura Peninsula
SA Sagami Bay
SA Tanzawa Mountains
SA Tokyo Bay
Kanara (1978)
Region. Former district of North Kanara in W Karnataka, SW India.
BT Karnataka India
BT India
BT Indian Peninsula
BT Asia

Kanawha County
Valid through 1988. Search in combination with state term. After 1988, use specific county-state term.
Kanawha County West Virginia (1989)
W West Virginia. Before 1989, also search Kanawha County AND West Virginia.
CO N375800N383700 W0811300W0815500
BT West Virginia
BT United States
NT Charleston West Virginia
NT Eskdale West Virginia
Kanawha Formation (1989)
BT Middle Pennsylvanian
BT Pennsylvanian
BT Carboniferous
BT Paleozoic
SA Kentucky
SA Virginia
SA West Virginia
Kanayut Conglomerate (1985)
N Alaska.
BT Upper Devonian
BT Devonian
BT Paleozoic
SA Alaska
SA Lisburne Group
Kanazawa
No longer a valid term for GeoRef. For the city see Kanazawa Japan.
Kanazawa Japan (1993)
City in W Honshu, Japan.
BT Ishikawa Japan
BT Honshu
BT Japan
BT Far East
BT Asia
Kane County
Valid through 1988. Search in combination with state term. After 1988, use specific county-state term.
Kane County Illinois (1989)
NE Illinois. Before 1989, also search Kane County AND Illinois.
CO N414200N421000 W0881700W0883800
BT Illinois
BT United States
Kane County Utah (1989)
S Utah. Before 1989, also search Kane County AND Utah.
CO N370000N373200 W1104000W1125500
BT Utah
BT United States
SA Lake Powell
Kane fracture zone (1989)
Part of the Mid-Atlantic Ridge in N Atlantic.
CO N220000N250000 W0420000W0470000
BT Mid-Atlantic Ridge
BT Atlantic Ocean
SA DSDP Site 395
SA Leg 78B
SA Leg 109
SA North Atlantic
SA North Atlantic Ridge
SA ODP Site 669
Kanem
No longer a valid term for GeoRef. See Kanem Chad.
Kanem Chad (1993)

Prefecture NE of Lake Chad in W Chad.
BT Chad
BT West Africa
BT Africa
Kaneohe Bay (1978)
Wide inlet on E coast of Oahu Island. Autoposting of broader terms to this term began in 1989. This term has multiple hierarchies.
BT1 Oahu
BT1 Honolulu County Hawaii
BT1 Hawaii
BT1 United States
BT2 Oahu
BT2 Honolulu County Hawaii
BT2 Hawaii
BT2 Polynesia
BT2 Oceania
BT3 Oahu
BT3 Honolulu County Hawaii
BT3 Hawaii
BT3 East Pacific Ocean Islands
Kanev
No longer a valid term for GeoRef. As of 1993, see Kanev Ukraine.
Kanev Ukraine (1993)
City in Kiev Oblast in N central Ukraine. This term has multiple hierarchies.
BT1 Kiev Ukraine
BT1 Ukraine
BT1 Europe
BT2 Kiev Ukraine
BT2 Ukraine
BT2 Commonwealth of Independent States
Kangaroo Island (1978)
In Indian Ocean S of Yorke Peninsula.
BT South Australia
BT Australia
BT Australasia
Kangerdlugssuak
use Kangerdlugssuaq
Kangerdlugssuaq (1989)
SE Greenland. Inlet of Davis Strait.
CO N682000N682000 W0321500W0321500
UF Kangerdlugssuak
BT Greenland
BT Arctic region
SA Davis Strait
Kangra
No longer a valid term for GeoRef. See Kangra India.
Kangra India (1993)
Town in W Himachal Pradesh, N India.
BT Himachal Pradesh India
BT India
BT Indian Peninsula
BT Asia
Kangwon
No longer a valid term for GeoRef. As of 1993 see Kangwon South Korea or Kangwon North Korea.
Kangwon North Korea (1993)
Province in SE North Korea. Before 1993, also search Kangwon AND North Korea.
BT North Korea
BT Korea
BT Far East
BT Asia
SA Kangwon South Korea
Kangwon South Korea (1993)

Province in NE South Korea. Before 1993 also search Kangwon AND South Korea.
BT South Korea
BT Korea
BT Far East
BT Asia
SA Kangwon North Korea
Kanhan Valley (1978)
River valley in central Madhya Pradesh. Also search Kanhan.
BT Madhya Pradesh India
BT India
BT Indian Peninsula
BT Asia
Kanin Peninsula (1978)
Projects into Barents Sea on N coast of Nenets National Okrug in Arkhangelsk Oblast. This term has multiple hierarchies.
BT1 Nenets Russian Federation
BT1 Arkhangelsk Russian Federation
BT1 Russian Federation
BT1 Commonwealth of Independent States
BT2 Nenets Russian Federation
BT2 Arkhangelsk Russian Federation
BT2 Europe
Kansan (1978)
North America. Pertains to second glacial stage of Pleistocene Epoch.
BT lower Pleistocene
BT Pleistocene
BT Quaternary
BT Cenozoic
Kansas (1978)
Autoposting of this term began in 1978.
CO N370000N400000 W0943500W1020000
RT United States
NT Allen County Kansas
NT Anderson County Kansas
NT Atchison County Kansas
NT Barber County Kansas
NT Barton County Kansas
NT Brown County Kansas
NT Butler County Kansas
NT Chase County Kansas
NT Chautauqua County Kansas
NT Cherokee County Kansas
NT Cheyenne County Kansas
NT Clark County Kansas
NT Clay County Kansas
NT Cloud County Kansas
NT Comanche County Kansas
NT Cowley County Kansas
NT Crawford County Kansas
NT Decatur County Kansas
NT Dickinson County Kansas
NT Doniphan County Kansas
NT Douglas County Kansas
NT Edwards County Kansas
NT Elk County Kansas
NT Ellis County Kansas
NT Ellsworth County Kansas
NT Finney County Kansas
NT Franklin County Kansas
NT Geary County Kansas
NT Gove County Kansas
NT Graham County Kansas
NT Grant County Kansas
NT Greenwood County Kansas
NT Hamilton County Kansas
NT Harper County Kansas
NT Harvey County Kansas
NT Haskell County Kansas
NT Hodgeman County Kansas
NT Hugoton Field
NT Jackson County Kansas

NT Jefferson County Kansas
NT Johnson County Kansas
NT Kansas River
NT Kansas River valley
NT Kingman County Kansas
NT Kiowa County Kansas
NT Labette County Kansas
NT Lane County Kansas
NT Leavenworth County Kansas
NT Lincoln County Kansas
NT Linn County Kansas
NT Logan County Kansas
NT Lyon County Kansas
NT Marion County Kansas
NT Marshall County Kansas
NT McPherson County Kansas
NT Meade County Kansas
NT Mitchell County Kansas
NT Montgomery County Kansas
NT Morris County Kansas
NT Morton County Kansas
NT Nemaha County Kansas
NT Ness County Kansas
NT Norton County Kansas
NT Osage County Kansas
NT Osborne County Kansas
NT Ottawa County Kansas
NT Pawnee County Kansas
NT Phillips County Kansas
NT Pottawatomie County Kansas
NT Pratt County Kansas
NT Rawlins County Kansas
NT Reno County Kansas
NT Republic County Kansas
NT Rice County Kansas
NT Riley County Kansas
NT Rooks County Kansas
NT Rush County Kansas
NT Russell County Kansas
NT Saline County Kansas
NT Scott County Kansas
NT Sedgwick County Kansas
NT Seward County Kansas
NT Shawnee County Kansas
NT Sheridan County Kansas
NT Sherman County Kansas
NT Smith County Kansas
NT Smoky Hill River basin
NT Stafford County Kansas
NT Stanton County Kansas
NT Stevens County Kansas
NT Sumner County Kansas
NT Trego County Kansas
NT Wabaunsee County Kansas
NT Wallace County Kansas
NT Washington County Kansas
NT Wichita County Kansas
NT Wilson County Kansas
NT Woodson County Kansas
NT Wyandotte County Kansas
SA Americus Limestone Member
SA Anadarko Basin
SA Arkansas River
SA Arkansas River valley
SA Ash Hollow Formation
SA Bartlesville Sand
SA Beil Limestone Member
SA Benton Formation
SA Bevier Coal
SA Bridge Creek Limestone Member
SA Burbank Sand
SA Cabaniss Formation
SA Cambridge Arch
SA Carlile Shale
SA Chase Group
SA Cherokee Group
SA Cheyenne Sandstone
SA Chouteau Limestone
SA Codell Sandstone Member
SA Coldwater Meteorites
SA Colorado Group
SA Cottonwood Limestone
SA Council Grove Group
SA Crooked Creek Formation
SA Dakota Formation
SA Desmoinesian
SA Dockum Group
SA Douglas Group
SA Elk River
SA Ervine Creek Limestone
SA Eskridge Shale
SA Excello Shale
SA Flint Hills
SA Forest City Basin
SA Fort Hays Limestone Member
SA Fort Riley Limestone
SA Graneros Shale
SA Great Plains
SA Greenhorn Limestone
SA Haviland Meteorite
SA High Plains Aquifer
SA Hughes Creek Shale
SA Hugoton Embayment
SA Hutchinson Salt Member
SA Kansas City Group
SA Kingsdown Formation
SA Kiowa Formation
SA Krebs Group
SA Labette Shale
SA Lansing Group
SA Laramie Formation
SA Lawrence Formation
SA Lecompton Limestone
SA Leoville Meteorite
SA Loup Fork Group
SA Marmaton Group
SA McPherson Formation
SA Meade Basin
SA Mentor Beds
SA Midcontinent
SA Midcontinent geophysical anomaly
SA Missouri River
SA Missouri River basin
SA Missouri River valley
SA Montana Group
SA Morrison Formation
SA Nemaha Ridge
SA Neosho River valley
SA Niobrara Formation
SA Norton County Meteorite
SA Ogallala Aquifer
SA Ogallala Formation
SA Oread Limestone
SA Peoria Loess
SA Plattsburg Limestone
SA Plattsmouth Limestone Member
SA Pleasanton Group
SA Red Eagle Limestone
SA Rexroad Formation
SA Roca Formation
SA Rock Lake Shale Member
SA Saint Peter Sandstone
SA Sedgwick Basin
SA Shawnee Group
SA Smoky Hill Chalk Member
SA Stanton Formation
SA Stearns Shale
SA Stone Corral Formation
SA Tonganoxie Sandstone
SA Valentine Formation
SA Verdigris River valley
SA Wabaunsee Group
SA Wellington Formation
SA Wreford Limestone
SA Wyandotte Limestone

Kansas City
  Valid through 1988. Search in combination with state term. After 1988, use specific city-state term.

Kansas City Group (1978)
  Subdivided into Bronson, Linn, and Zarah subgroups. Includes Hertha, Ladore, Swope, Galesburg, Dennis, Fontana, Sarpy, Drum, Chanute, Iola, Lane, Wyandotte and Bonner Springs formations.
  BT Missourian
  BT Upper Pennsylvanian
  BT Pennsylvanian
  BT Carboniferous
  BT Paleozoic
  SA Iowa
  SA Kansas
  SA Missouri
  SA Nebraska
  SA Wyandotte Limestone

Kansas City Kansas (1989)
  City in NE Kansas. Before 1989, search Kansas City AND Kansas.
  CO N390500N390500 W0943700W0943700
  BT Wyandotte County Kansas
  BT Kansas
  BT United States

Kansas City Missouri (1989)
  City in Clay and Jackson counties, W Missouri. Before 1989, search Kansas City AND Missouri.
  IN Index counties as applicable.
  CO N390200N390200 W0943300W0943300
  BT Missouri
  BT United States
  SA Clay County Missouri
  SA Jackson County Missouri

Kansas River (1978)
  Flows into Missouri River at Kansas City.
  BT Kansas
  BT United States

Kansas River valley (1978)
  NE central Kansas.
  BT Kansas
  BT United States

kansite
  use mackinawite

Kansk
  No longer a valid term for GeoRef. As of 1993, see Kansk Russian Federation.

Kansk Russian Federation (1993)
  City E of Krasnoyarsk on the Trans-Siberian R.R. in S Krasnoyarsk Kray. Before 1993, also search Kansk. This term has multiple hierarchies.
  BT1 Krasnoyarsk Russian Federation
  BT1 Russian Federation
  BT1 Commonwealth of Independent States
  BT2 Krasnoyarsk Russian Federation
  BT2 Asia

Kansk-Achinsk Basin (1978)
  S and SW Krasnoyarsk Kray along Trans-Siberian R.R. This term has multiple hierarchies.
  BT1 Russian Federation
  BT1 Commonwealth of Independent States
  BT2 Asia
  SA Krasnoyarsk Russian Federation

Kansu China
  use Gansu China

Kanto
  use Kanto Plain

Kanto earthquake 1923 (1993)
  Honshu, Japan. Before 1985, also search Kanto or Kwanto AND earthquakes.
  UF Kwanto earthquake 1923
  BT earthquakes
  SA Honshu
  SA Japan
  SA Kanto Plain

Kanto Mountains (1989)
  Honshu, Japan.
  UF Kwanto Mountains
  BT Honshu
  BT Japan
  BT Far East
  BT Asia
  SA Nagano Japan
  SA Yamanashi Japan

Kanto Plain (1989)
  Largest and most densely populated plain in country. In central part of Honshu. Includes Tokyo-Yokohama industrial region. As of 1993, use also for the region or district including Chiba, Gumma, Ibaraki, Kanagawa, Saitama, and Tochigi prefecture. Before 1989, search Kwanto Plain.
  UF Kanto
  UF Kwanto
  UF Kwanto Plain
  BT Honshu
  BT Japan
  BT Far East
  BT Asia
  SA Chiba Japan
  SA Gumma Japan
  SA Ibaraki Japan
  SA Kanagawa Japan
  SA Kanto earthquake 1923
  SA Kazusa Group
  SA Saitama Japan
  SA Tochigi Japan
  SA Tokyo Japan

kaolin (1978)
  As of 1981, use kaolin deposits for kaolin as a commodity.
  BT clastic sediments
  BT sediments
  SA clay
  SA clay mineralogy
  SA clay minerals
  SA clays
  SA fuller's earth
  SA kaolin deposits
  SA kaolinite
  SA kaolinization
  SA metakaolin

kaolin deposits (1981)
  Before 1981, search kaolin AND deposits.
  IN Commodity. See List C.
  SA clays
  SA fuller's earth
  SA kaolin
  SA kaolinization
  SA refractory materials

kaolinisation
  use kaolinization

kaolinite (1978)
  BT clay minerals
  BT sheet silicates
  BT silicates
  SA kaolin

kaolinitization
  use kaolinization

kaolinization (1978)
  UF kaolinisation
  UF kaolinitization
  UF kaolisation
  BT metasomatism
  SA alteration
  SA clay minerals
  SA kaolin
  SA kaolin deposits

SA processes
SA weathering

kaolisation
use kaolinization

Kapoeta Meteorite (1981)
Impact near Kapoeta in S Sudan. Before 1981, also search Kapoeta AND meteorites.
BT howardite
BT achondrites
BT stony meteorites
BT meteorites
SA Sudan

Kapuskasing structural zone
use Kapuskasing Zone

Kapuskasing Zone (1989)
Structural region in N Ontario.
UF Kapuskasing structural zone
BT Superior Province
BT Canadian Shield
BT North America
SA Ontario

Kara Sea (1978)
Between Novaya Zemlya Island on W; and Yamal and Tamyr peninsulas on the E.
CO N680000N820000 E1050000E0553000
BT Arctic Ocean

Kara-Bogaz Gulf (1978)
Large shallow gulf of the Caspian Sea in NW Krasnovodsk Kray. This term has multiple hierarchies.
UF Kara-Bogaz-Gol Gulf
UF Kara-Bogaz-Gol region
BT1 Turkmenia
BT1 Asia
BT2 Turkmenia
BT2 Commonwealth of Independent States
SA Caspian Sea

Kara-Bogaz-Gol
No longer a valid term for GeoRef. As of 1993, see Kara-Bogaz-Gol Turkmenia.

Kara-Bogaz-Gol Gulf
use Kara-Bogaz Gulf

Kara-Bogaz-Gol region
use Kara-Bogaz Gulf

Kara-Bogaz-Gol Turkmenia (1993)
Town on the Caspian Sea at entrance to Kara-Bogaz Gulf. This term has multiple hierarchies.
BT1 Turkmenia
BT1 Asia
BT2 Turkmenia
BT2 Commonwealth of Independent States

Kara-Kum
use Karakum

Kara-Mazar
No longer a valid term for GeoRef. As of 1993, see Kara-Mazar Tadzhikistan.

Kara-Mazar Tadzhikistan (1993)
Village in Leninabad Oblast in N Tadzhikistan. This term has multiple hierarchies.
BT1 Leninabad Tadzhikistan
BT1 Tadzhikistan
BT1 Asia
BT2 Leninabad Tadzhikistan
BT2 Tadzhikistan
BT2 Commonwealth of Independent States

Kara-Tau (Range)
use Karatau Range

Karabakh
No longer a valid term for GeoRef. As of 1993, see Nagorno-Karabakh Azerbaidzhan for the autonomous oblast in S Azerbaidzhan.

Karachaganak Field (1993)
Astrakhan Arch area, Caspian Depression. Includes gas condensate deposits. This term has multiple hierarchies.
BT1 Kazakhstan
BT1 Central Asia
BT1 Asia
BT2 Kazakhstan
BT2 Commonwealth of Independent States
SA Caspian Depression
SA natural gas
SA oil and gas fields

Karafuto
use Sakhalin

Karaganda
No longer a valid term for GeoRef. As of 1993, see Karaganda Kazakhstan.

Karaganda Basin (1978)
Coal basin in Karaganda Oblast in NE central Kazakhstan. Also search Karaganda. This term has multiple hierarchies.
BT1 Karaganda Kazakhstan
BT1 Kazakhstan
BT1 Central Asia
BT1 Asia
BT2 Karaganda Kazakhstan
BT2 Kazakhstan
BT2 Commonwealth of Independent States
SA coal fields

Karaganda Kazakhstan (1993)
Oblast and city in NE central Kazakhstan. This term has multiple hierarchies.
BT1 Kazakhstan
BT1 Central Asia
BT1 Asia
BT2 Kazakhstan
BT2 Commonwealth of Independent States
NT Dzhezkazgan Kazakhstan
NT Karaganda Basin
NT Uspenskiy Kazakhstan
SA Akchatau Kazakhstan

Karaginski Island
use Karaginskiy Island

Karaginskiy Island (1978)
In Bering Sea just off coast of N Kamchatka Peninsula. This term has multiple hierarchies.
UF Karaginski Island
BT1 Kamchatka Peninsula
BT1 Russian Pacific region
BT1 Russian Federation
BT1 Commonwealth of Independent States
BT2 Kamchatka Peninsula
BT2 Russian Pacific region
BT2 Asia
SA Bering Sea

Karakoram (1978)
Mountain range in N Jammu and Kashmir.
IN Index Jammu and Kashmir and countries as applicable.
UF Karakorum
SA India
SA Jammu and Kashmir
SA Pakistan

Karakorum
use Karakoram

Karakum (1978)
Desert area S of the Aral Sea stretching from the Caspian Sea to the Amu Darya River. Autoposting of Turkmenia to this term began in 1989. This term has multiple hierarchies.
CO N350000N433000 E0660000E0540000
UF Kara-Kum
UF Karakumy
UF Qaraqum
BT1 Turkmenia
BT1 Asia
BT2 Turkmenia
BT2 Commonwealth of Independent States

Karakumy
use Karakum

Karamazar
No longer a valid term for GeoRef. As of 1993, see Karamazar Tadzhikistan.

Karamazar Tadzhikistan (1993)
Village in Leninabad Oblast in N Tadzhikistan. This term has multiple hierarchies.
BT1 Leninabad Tadzhikistan
BT1 Tadzhikistan
BT1 Asia
BT2 Leninabad Tadzhikistan
BT2 Tadzhikistan
BT2 Commonwealth of Independent States

Karanpura
No longer a valid term for GeoRef. See Karanpura India.

Karanpura Basin
use Karanpura coal field

Karanpura coal field (1978)
Central Bihar. Also search Karanpura.
UF Karanpura Basin
UF Karanpura Coalfield
SA Bihar India
SA coal fields

Karanpura Coalfield
use Karanpura coal field

Karanpura India (1993)
Village in central Bihar, E India.
BT Bihar India
BT India
BT Indian Peninsula
BT Asia

Karatau
use Karatau Range

Karatau Range (1978)
W branch of the Tien Shan in SE Kazakhstan. This term has multiple hierarchies.
CO N420000N443000 E0710000E0670000
UF Kara-Tau (Range)
UF Karatau
BT1 Kazakhstan
BT1 Central Asia
BT1 Asia
BT2 Kazakhstan
BT2 Commonwealth of Independent States
BT3 Tien Shan
BT3 Asia

Karategin Range (1978)
SW Tadzhikistan. This term has multiple hierarchies.
BT1 Tadzhikistan
BT1 Asia
BT2 Tadzhikistan
BT2 Commonwealth of Independent States

Karawanken (1978)
Range of Eastern Alps. A continuation of Carnic Alps.
IN Index countries as applicable.
UF Karawanken Alps
UF Karawanken Mountains
BT Eastern Alps
BT Alps
BT Europe
SA Carinthia Austria
SA Carnic Alps
SA Slovenia

Karawanken Alps
use Karawanken

Karawanken Mountains
use Karawanken

Karelia
As of 1993, no longer a valid term for GeoRef.
use Karelia Russian Federation

Karelia Russian Federation (1993)
This term has multiple hierarchies.
CO N600000N680000 E0400000E0300000
UF Karelia
BT1 Russian Federation
BT1 Commonwealth of Independent States
BT2 Asia
NT Belomorsk Russian Federation
NT Segozero
NT Vuoriyarvi Russian Federation
SA Ladoga Series
SA Lake Ladoga region
SA Lake Onega
SA Plotina Deposit

Karelian (1978)
Autoposting of middle Precambrian and Precambrian to this term ended in 1989. As of 1989, lower Proterozoic and Proterozoic are autoposted to this term. Autoposting of upper Precambrian and Precambrian to this term began in 1990.
BT lower Proterozoic
BT Proterozoic
BT upper Precambrian
BT Precambrian
SA Karelian Orogeny

Karelian Orogeny (1978)
Before 1978, also search Karelian AND orogeny.
BT Proterozoic
BT upper Precambrian
BT Precambrian
SA Karelian
SA orogeny

Karewa Group (1989)
Kashmir, N India.
BT Pleistocene
BT Quaternary
BT Cenozoic
SA India
SA Jammu and Kashmir
SA Kashmir

Karharbari
No longer a valid term for GeoRef. See Karharbari India.

Karharbari India (1993)
Town in E central Bihar, E India.
BT Bihar India
BT India
BT Indian Peninsula
BT Asia

Karharbari Stage (1978)

Basal member of the Damuda Series. Named after Karharbari in E central Bihar.
BT Upper Carboniferous
BT Carboniferous
BT Paleozoic
SA Bihar India
SA Damuda Series
SA India

Kariba
No longer a valid term for GeoRef. As of 1993 see Kariba Zimbabwe.

Kariba Lake
use Lake Kariba

Kariba Zimbabwe (1993)
Town on Lake Kariba in NW Zimbabwe. Before 1993 also search Kariba AND Zimbabwe.
BT Zimbabwe
BT Southern Africa
BT Africa

Karibib
No longer a valid term for GeoRef. As of 1993 see Karibib Namibia.

Karibib Namibia (1993)
Town in W central Namibia. Before 1993 also search Karibib AND Namibia.
BT Namibia
BT Southern Africa
BT Africa

Karimskaia Sopka
use Karymskaya Sopka

Karkonosze
use Karkonosze Mountains

Karkonosze Mountains (1978)
Highest range of the Sudeten Mountains.
IN Index Bohemia and/or Poland as applicable.
CO N493000N495800
    E0160000E0152400
UF Karkonosze
UF Krkonose Mountains
UF Riesengebirge
BT Sudeten Mountains
BT Central Europe
BT Europe
SA Bohemia
SA Czechoslovakia
SA Poland

Karl-Marx-Stadt
No longer a valid term for GeoRef. use Chemnitz Germany

Karl-Marx-Stadt Bezirk
No longer a valid term for GeoRef. District in SE East Germany. Before 1981, search Karl-Marx-Stadt. See Germany.

Karlovy Vary (1978)
Town and health resort in W Bohemia. Used for Carlsbad when referring to Czechoslovakia. Before 1977, also search Carlsbad AND Czechoslovakia.
BT Bohemia
BT Czech Republic
BT Czechoslovakia
BT Central Europe
BT Europe

Karlsruhe
No longer a valid term for GeoRef. See Karlsruhe Germany.

Karlsruhe Germany (1993)
City in NW central Baden-Wurttemberg, SW Germany.
BT Baden-Wurttemberg Germany
BT Germany

BT Central Europe
BT Europe

Karmutsen Group (1989)
Vancouver Island and Queen Charlotte Islands, British Columbia; W Idaho.
BT Upper Triassic
BT Triassic
BT Mesozoic
SA British Columbia
SA Idaho
SA Queen Charlotte Islands
SA Vancouver Island

Karnataka
No longer a valid term for GeoRef. See Karnataka India.

Karnataka India (1993)
State in SW India since 1973. Before 1973, also search Mysore.
CO N113000N183000
    E0784500E0741000
BT India
BT Indian Peninsula
BT Asia
NT Bagalkot India
NT Bangalore India
NT Belgaum India
NT Bellary India
NT Bijapur India
NT Chickmagalur India
NT Chitradurga India
NT Closepet Granite
NT Dharwar India
NT Gulbarga India
NT Hassan India
NT Kaladgi India
NT Kanara
NT Kolar Gold Fields India
NT Kolar India
NT Mangalore India
NT Mysore India
NT Sandur India
SA Badami Series
SA Cauvery Basin
SA Chitradurga schist belt
SA Deccan Plateau
SA Kaladgi System
SA Kolar schist belt
SA Krishna

Karnes County
Valid through 1988. Search in combination with state term. After 1988, use specific county-state term.

Karnes County Texas (1989)
S Texas. Before 1989, also search Karnes County AND Texas.
CO N284000N291600
    W0973500W0981200
BT Texas
BT United States

Karnul India
use Kurnool India

Karoo
use Karroo Basin

Karpathos (1981)
Island in the Dodecanese. Before 1981, also search Karpathos Island. This term has multiple hierarchies.
BT1 Dodecanese
BT1 Greek Aegean Islands
BT1 Greece
BT1 Southern Europe
BT1 Europe
BT2 Dodecanese
BT2 Greek Aegean Islands
BT2 Aegean Islands
BT2 Mediterranean region

karpinskyite (1978)
BT framework silicates
BT silicates

karren (1978)
Autoposting of solution features to this term began in 1989.
BT solution features
SA geomorphology
SA karst

Karroo Basin (1978)
An arid tableland region consisting of North Karroo along the Orange River in N, Great Karroo or Central Karroo in S central, and Southern Karroo or Little Karoo along S coast.
UF Karoo
BT Cape Province South Africa
BT South Africa
BT Southern Africa
BT Africa

Karroo Supergroup (1993)
Formerly the Karroo System: Carboniferous to Jurassic of Namibia and South Africa. Divided into Dwyka Formation, Ecca Group, Beaufort Group, and Stormberg Series. Before 1993, also search Karroo System. Before 1978, also search Karroo.
IN Index ages as applicable.
UF Karroo System
SA Africa
SA Beaufort Group
SA Botswana
SA Carboniferous
SA Dwyka Formation
SA Ecca Group
SA Jurassic
SA Namibia
SA Permian
SA South Africa
SA Stormberg Series
SA Triassic

Karroo System
No longer a valid term for GeoRef as of 1993.
use Karroo Supergroup

Karshi
No longer a valid term for GeoRef. As of 1993, see Karshi Uzbekistan.

Karshi Steppe (1978)
In Kashkadarya Oblast in S Uzbekistan. Before 1978, also search Karshi. This term has multiple hierarchies.
BT1 Uzbekistan
BT1 Asia
BT2 Uzbekistan
BT2 Commonwealth of Independent States

Karshi Uzbekistan (1993)
Town in Kashkadarya Oblast in S Uzbekistan. Before 1978, also search Bek-Budi. This term has multiple hierarchies.
UF Bek-Budi Uzbekistan
BT1 Uzbekistan
BT1 Asia
BT2 Uzbekistan
BT2 Commonwealth of Independent States

karst (1978)
Autoposting of solution features to this term began in 1989. Before 1977, Karst was infrequently used to indicate the region in Europe; now use Karst region for that.
UF carst
UF karst topography

BT solution features
SA caves
SA dolines
SA geomorphology
SA karren
SA karst filling
SA karst hydrology
SA karstification
SA paleokarst
SA poljes
SA pseudokarst
SA sinkholes
SA solution cavities
SA thermokarst
SA topography
SA underground streams

karst filling (1982)
SA geomorphology
SA karst
SA solution features

karst hydrology (1982)
UF karst waters
SA geomorphology
SA hydrology
SA karst
SA paleokarst
SA solution features

Karst region (1978)
Limestone plateau N of Trieste and E of Isonzo River. Also search Karst regions. Before 1977, Karst was used to indicate the region.
IN Index countries as applicable.
BT Europe
SA Italy
SA Yugoslavia

karst topography
use karst

karst waters
use karst hydrology

karstification (1985)
SA chemical weathering
SA geomorphology
SA karst
SA landform evolution
SA processes
SA solution
SA solution transport
SA weathering

Kartalinia
use Kartlia

Kartlia (1978)
Hilly central region.
UF Kartalinia
BT Georgian Republic
BT Europe

Karvina (1978)
City in NE Moravia.
UF Karvinna
BT Moravia
BT Czech Republic
BT Czechoslovakia
BT Central Europe
BT Europe

Karvinna
use Karvina

Karymskaya Sopka (1985)
Volcano, S Kamchatka Peninsula, Russian Federation. This term has multiple hierarchies.
CO N540400N540600
    E1592300E1592100
UF Karimskaia Sopka
UF Karymskiy
BT1 Russian Federation
BT1 Commonwealth of Independent States
BT2 Asia

SA Kamchatka Peninsula
SA Russian Far East

Karymskiy
use Karymskaya Sopka

Kasai (1981)
Territory in Zaire comprising two provinces, Kasai-Occidental, and Kasai-Oriental.
CO S081000S014500
E0261500E0194000
BT Zaire
BT Central Africa
BT Africa

Kasai River (1978)
Flows into the Congo. Also search Kasai.
IN Index countries as applicable.
BT Africa
SA Angola
SA Zaire

Kasauli Series (1978)
BT lower Miocene
BT Miocene
BT Neogene
BT Tertiary
BT Cenozoic
SA Himachal Pradesh India
SA India

Kasei Valles
use Kasei Vallis

Kasei Vallis (1989)
E-W trending valley on Mars located on the N border of Lunae Planum.
UF Kasei Valles
BT Mars
SA Lunae Planum

Kashima Seamount (1993)
NW Pacific near junction of Japan Trench and Izu-Bonin Trench.
CO N354400N355400
E1422500E1422430
UF Daiichi-Kashima Seamount
UF Daiiti-Kashima Seamount
UF Kasima Seamount
BT West Pacific
BT Pacific Ocean
SA Japan Trench

Kashiwazaki
No longer a valid term for GeoRef. See Kashiwazaki Japan.

Kashiwazaki Japan (1993)
City on Japan Sea in central Honshu, Japan. Before 1993, also search Kashiwazaki AND Japan.
BT Niigata Japan
BT Honshu
BT Japan
BT Far East
BT Asia

Kashmir (1978)
Part of Jammu and Kashmir.
IN Index Jammu and Kashmir AND India and/or Pakistan as applicable.
SA India
SA Jammu and Kashmir
SA Karewa Group
SA Pakistan

Kashmir and Jammu
use Jammu and Kashmir

Kashmir Valley (1978)
Intermontane valley in the Himalayas of W Jammu and Kashmir. As of 1993, India is no longer a broader term.

IN Index Jammu and Kashmir AND India and/or Pakistan as applicable.
UF Vale of Kashmir
SA India
SA Jammu and Kashmir
SA Pakistan

Kasima Seamount
use Kashima Seamount

Kasimovian (1981)
BT Upper Carboniferous
BT Carboniferous
BT Paleozoic

Kaskaskia Sequence (1985)
Devonian and Mississippian. Central and western U.S.
IN Index ages as applicable.
BT upper Paleozoic
BT Paleozoic
SA Devonian
SA Mississippian

Kaskawulsh Glacier (1978)
In the Saint Elias Mountains in SW Yukon Territory.
BT Yukon Territory
BT Western Canada
BT Canada

kasolite (1978)
Autoposting of nesosilicates began in 1985.
BT nesosilicates
BT orthosilicates
BT silicates

Kassel
No longer a valid term for GeoRef. See Kassel Germany.

Kassel Germany (1993)
City in N Hesse, central Germany.
BT Hesse Germany
BT Germany
BT Central Europe
BT Europe

katagenesis
use catagenesis

Katanga Basin, USSR
use Khatanga Basin

Katanga River, USSR
use Khatanga River

Katanga Russian Federation
use Khatanga Russian Federation

Katanga, Zaire
use Shaba Zaire

Katangan Orogeny (1978)
Before 1978, search Katangan AND orogeny.
BT Precambrian
SA Africa
SA orogeny
SA Zaire

katathermal processes
use hypothermal processes

katazonal metamorphism
use catazonal metamorphism

Katherine
No longer a valid term for GeoRef. See Katherine Australia. As of 1993, use Katherine River for the river.

Katherine Australia (1993)
Settlement in N central Northern Territory.
BT Northern Territory Australia
BT Australia
BT Australasia

Kathiawar (1978)
Peninsula jutting into Arabian Sea.
SA Girnar Hills
SA Gujarat India
SA Mount Girnar
SA Wadia

Katmai (1989)
Volcano in Katmai National Monument, S Aleutian Range, NE Alaska Peninsula, SW Alaska.
CO N582000N582000
W1545900W1545900
BT Southwestern Alaska
BT Alaska
BT United States
SA Alaska Peninsula
SA Katmai National Monument

Katmai National Monument (1978)
At N end of Alaska Peninsula in S Alaska.
BT Southwestern Alaska
BT Alaska
BT United States
SA Katmai
SA national monuments
SA Valley of Ten Thousand Smokes

Katowice
No longer a valid term for GeoRef. As of 1993 see Katowice Poland.

Katowice Poland (1993)
Province and city in S central Poland. Before 1993 also search Katowice AND Poland.
BT Poland
BT Central Europe
BT Europe
NT Bytom Poland
NT Chorzow Poland
NT Goczalkowice
NT Olkusz Poland
NT Raciborz Poland
NT Ruda Poland
NT Rybnik Poland
NT Sosnowiec Poland
NT Upper Silesian coal basin
NT Zabrze Poland
NT Zawiercie Poland
SA Bielsko Poland
SA Bielsko-Biala Poland
SA Cieszyn Poland
SA Cracow-Czestochowa Jura
SA Czestochowa-Zawiercie Basin
SA Silesian coal basin

Kattegat (1978)
Strait between Sweden and Jutland, Denmark.
UF Kattegatt
BT North Sea
BT European Atlantic
BT North Atlantic
BT Atlantic Ocean
SA Oresund

Kattegatt
use Kattegat

Katzenbuckel (1978)
Mountain. Highest point in the Odenwald in N Baden-Wurttemberg, SW West Germany.
BT Baden-Wurttemberg Germany
BT Germany
BT Central Europe
BT Europe
SA Odenwald

Kauai (1978)
One of the major Hawaiian Islands. Part of Kauai County, NW Hawaii. Autoposting of broader terms to this term began in 1989. This term has multiple hierarchies.
BT1 Hawaii
BT1 United States
BT2 Hawaii
BT2 Polynesia
BT2 Oceania
BT3 Hawaii
BT3 East Pacific Ocean Islands

Kavalerovo
No longer a valid term for GeoRef. As of 1993, see Kavalerovo Russian Federation.

Kavalerovo Russian Federation (1993)
Village in SE Primorye Kray in Russian Far East. This term has multiple hierarchies.
BT1 Primorye Russian Federation
BT1 Russian Federation
BT1 Commonwealth of Independent States
BT2 Primorye Russian Federation
BT2 Asia

Kaveri Basin
use Cauvery Basin

Kayak Shale (1989)
In Endicott Group. N Alaska and N Yukon Territory.
BT Lower Mississippian
BT Mississippian
BT Carboniferous
BT Paleozoic
SA Alaska
SA Endicott Group
SA Lisburne Group
SA Yukon Territory

Kayenta Formation (1978)
In Glen Canyon Group. Upper Triassic(?). NE Arizona, SW Colorado, and S and SE Utah.
BT Upper Triassic
BT Triassic
BT Mesozoic
SA Arizona
SA Colorado
SA Glen Canyon Group
SA Utah

Kazakhstan (1978)
Formerly the Kazakh Soviet Socialist Republic. In Soviet Central Asia. Autoposting of this term began in 1978. This term has multiple hierarchies.
CO N410000N550000
E0860000E0470000
BT1 Central Asia
BT1 Asia
BT2 Commonwealth of Independent States
NT Akbastau Kazakhstan
NT Akchatau Kazakhstan
NT Aksu Kazakhstan
NT Aktyubinsk Kazakhstan
NT Alma-Ata Kazakhstan
NT Atasu Kazakhstan
NT Balkhash Kazakhstan
NT Balkhash region
NT Bet-Pak-Dala
NT Biikzhal
NT Buzachi Peninsula
NT Central Kazakhstan
NT Chu Kazakhstan
NT Chu-Ili Mountains
NT Chu-Sarysu Depression
NT Eastern Kazakhstan
NT Emba River
NT Guryev Kazakhstan
NT Ili Kazakhstan
NT Kalba Range
NT Karachaganak Field
NT Karaganda Kazakhstan

NT Karatau Range
  NT Kokchetav Kazakhstan
  NT Kustanay Kazakhstan
  NT Lake Balkhash
  NT Leninogorsk Kazakhstan
  NT Mangyshlak Peninsula
  NT Mirgalimsay Kazakhstan
  NT Mugodzhar Hills
  NT Muyunkum
  NT Pavlodar Kazakhstan
  NT Sayak Kazakhstan
  NT Semipalatinsk Kazakhstan
  NT Tekeli Kazakhstan
  NT Tengiz
  NT Tokrau Synclinorium
  NT Turgay Kazakhstan
  NT Ulu-Tau
  NT Zaisan
  NT Zaisan Basin
  NT Zaisan Kazakhstan
  NT Zhetybay Kazakhstan
  NT Zyryanovsk Kazakhstan
  SA Alakol
  SA Altai Mountains
  SA Aral region
  SA Astrakhan Arch
  SA Barsakelmes
  SA Caspian Basin
  SA Caspian Depression
  SA Char
  SA Chirchik River
  SA Chu
  SA Dzhungarian Alatau
  SA Efremovka Meteorite
  SA Eurasia
  SA Ili Basin
  SA irghizite
  SA Irtysh River
  SA Ishim
  SA Kulunda Steppe
  SA Kyzylkum
  SA Ob-Irtysh Interfluve
  SA Pskem Range
  SA Rudny Altai
  SA Sarysu
  SA Siberia
  SA Siberian Lowland
  SA Southern Urals
  SA Steppes region
  SA Syr Darya
  SA Tarbagatay Range
  SA Turan
  SA Turanian Platform
  SA Turgay Basin
  SA Turkestan
  SA Ural River
  SA Urals
  SA Ustyurt
  SA Uzen
  SA Volga-Ural region
  SA West Siberia

Kazan
  No longer a valid term for GeoRef. As of 1993, Kazan Russian Federation.

Kazan Russian Federation (1993)
  City on left bank of Volga in Tatar Autonomous Republic. Before 1993, also search Kazan. This term has multiple hierarchies.
  BT1 Tatar Russian Federation
  BT1 Russian Federation
  BT1 Commonwealth of Independent States
  BT2 Tatar Russian Federation
  BT2 Europe

Kazanian (1978)
  Europe. Above Kungurian, below Tatarian.
  BT Upper Permian
  BT Permian
  BT Paleozoic

Kazusa Group (1989)
  Chiba Peninsula, Kanto Plain and Miura Peninsula, Honshu.
  BT Quaternary
  BT Cenozoic
  SA Chiba Peninsula
  SA Honshu
  SA Japan
  SA Kanto Plain
  SA Miura Peninsula

keatite (1978)
  BT silica minerals
  BT framework silicates
  BT silicates

Keban Mine (1978)
  In E central Turkey. Also search Keban.
  BT Turkey
  BT Middle East
  BT Asia
  SA lead-zinc deposits
  SA mines

Kebnekaise (1978)
  Peak in Kjolen Mountains in NW Norrbotten. Highest in country.
  BT Norrbotten Sweden
  BT Sweden
  BT Scandinavia
  BT Western Europe
  BT Europe

Kedabek
  No longer a valid term for GeoRef. As of 1993, see Kedabek Azerbaidzhan.

Kedabek Azerbaidzhan (1993)
  Town in W Azerbaidzhan. This term has multiple hierarchies.
  BT1 Azerbaidzhan
  BT1 Europe
  BT2 Azerbaidzhan
  BT2 Commonwealth of Independent States

Kedah Malaysia (1993)
  Northernmost state in West Malaysia on the Malay Peninsula. Before 1993, also search Kadah. This term has multiple hierarchies.
  BT1 West Malaysia
  BT1 Malay Peninsula
  BT1 Far East
  BT1 Asia
  BT2 West Malaysia
  BT2 Malaysia
  BT2 Far East
  BT2 Asia
  SA Langkawi Islands

Keefer Sandstone (1978)
  In Clinton Group. W Maryland, and from central Pennsylvania to NE West Virginia.
  BT Middle Silurian
  BT Silurian
  BT Paleozoic
  SA Clinton Group
  SA Maryland
  SA Pennsylvania
  SA Rochester Formation
  SA West Virginia

Keel
  No longer a valid term for GeoRef. See Keel Ireland.

Keel Ireland (1993)
  Village on Achill Island in Atlantic Ocean.
  BT Mayo Ireland
  BT Ireland
  BT Western Europe
  BT Europe

Keewatin
  No longer a valid term for GeoRef. As of 1993, see Keewatin Ontario. As of 1989, see Keewatin Minnesota. Before 1978, also used for Keewatin District Northwest Territories.

Keewatin District
  No longer a valid term for GeoRef. As of 1993, see Keewatin District Northwest Territories.

Keewatin District Northwest Territories (1993)
  District in SE Northwest Territories. From 1985-1992, also search Keewatin District AND Northwest Territories. Before 1985, also search District of Keewatin AND Northwest Territories. Before 1978, also search Keewatin AND Northwest Territories.
  CO N600000N690000
      W0830000W1020000
  UF District of Keewatin
  BT Northwest Territories
  BT Western Canada
  BT Canada
  NT Kaminak Lake
  NT Narpaing Fjord
  NT Wager
  SA Baffin Island
  SA Baker Lake
  SA Cape Dyer
  SA Hudson Bay
  SA Hurwitz Group

Keewatin Minnesota (1993)
  Village in NE Minnesota. Before 1989, search Keewatin AND Minnesota.
  BT Itasca County Minnesota
  BT Minnesota
  BT United States

Keewatin Ontario (1993)
  Town in W. Ontario. Before 1993, search Keewatin AND Ontario.
  BT Kenora District Ontario
  BT Ontario
  BT Eastern Canada
  BT Canada

Kefallinia
  use Cephalonia

Keg River Formation (1978)
  Subsurface. Includes the Rainbow (reef) Member. Eastward in Saskatchewan, the Keg River Formation is termed the Winnipegosis Formation. NW Alberta.
  BT Devonian
  BT Paleozoic
  SA Alberta
  SA Canada
  SA Winnipegosis Formation

Kekiktuk Conglomerate (1989)
  In Endicott Group. N Alaska.
  BT Lower Mississippian
  BT Mississippian
  BT Carboniferous
  BT Paleozoic
  SA Alaska
  SA Endicott Group

Kelantan
  No longer a valid term for GeoRef. As of 1993, see Kelantan Malaysia.

Kelantan Malaysia (1993)
  State on Thailand border on the Malay Peninsula in West Malaysia. This term has multiple hierarchies.
  BT1 West Malaysia
  BT1 Malay Peninsula
  BT1 Far East
  BT1 Asia
  BT2 West Malaysia
  BT2 Malaysia
  BT2 Far East
  BT2 Asia

Kelheim
  No longer a valid term for GeoRef. See Kelheim Germany.

Kelheim Germany (1993)
  Town in central Bavaria, SE Germany.
  BT Bavaria Germany
  BT Germany
  BT Central Europe
  BT Europe

kelyphytic rims
  use reaction rims

Kemerovo
  No longer a valid term for GeoRef. As of 1993, see Kemerovo Russian Federation.

Kemerovo Russian Federation (1993)
  Oblast in S central Siberia. Before 1993, also search Kemerovo. This term has multiple hierarchies.
  BT1 Russian Federation
  BT1 Commonwealth of Independent States
  BT2 Asia
  NT Prokopyevsk Russian Federation
  NT Salair Russian Federation
  NT Taiga Russian Federation
  SA Gornaya Shoriya
  SA Kiya River
  SA Kuznetsk Basin
  SA Mariinsk Russian Federation
  SA Salair Ridge

Kenai Group (1978)
  Eocene(?) and Oligocene, Central S Alaska.
  BT Paleogene
  BT Tertiary
  BT Cenozoic
  SA Alaska
  SA Hemlock Conglomerate

Kenai Peninsula (1978)
  Between Cook Inlet on W and Prince William Sound on E in S Alaska.
  BT Southern Alaska
  BT Alaska
  BT United States
  SA Coast Ranges

Kenedy County
  Valid through 1988. Search in combination with state term. After 1988, use specific county-state term.

Kenedy County Texas (1989)
  On Gulf of Mexico in S Texas. Before 1989, also search Kenedy County AND Texas.
  CO N263500N271000
      W0971000W0980000
  BT Texas
  BT United States
  SA Baffin Bay
  SA Laguna Madre

Kennebec County
  Valid through 1988. Search in combination with state term. After 1988, use specific county-state term.

Kennebec County Maine (1989)

Keno Hill
S Maine. Before 1989, also search Kennebec County AND Maine.
CO N441000N444000
W0692500W0701000
BT Maine
BT United States
NT Augusta Maine

Keno Hill
No longer a valid term for GeoRef. As of 1993, see Keno Hill Yukon Territory.

Keno Hill Yukon Territory (1993)
Village in central Yukon Territory.
BT Yukon Territory
BT Western Canada
BT Canada

Kenora
No longer a valid term for GeoRef. As of 1993, see Kenora Ontario.

Kenora District
No longer a valid term for GeoRef. As of 1993, see Kenora District Ontario.

Kenora District Ontario (1993)
Extreme N, NW and W Ontario. Before 1993, also search Kenora District, District of Kenora, or Kenora in conjunction with Ontario.
UF District of Kenora
BT Ontario
BT Eastern Canada
BT Canada
NT Keewatin Ontario
NT Kenora Ontario
NT Uchi Lake

Kenora Ontario (1993)
City in W Ontario. Before 1993, also search Kenora AND Ontario.
BT Kenora District Ontario
BT Ontario
BT Eastern Canada
BT Canada

Kenoran Orogeny (1978)
A time of plutonism, metamorphism, and deformation during the Precambrian of the Canadian Shield. Before 1978, also search Kenoran AND orogeny.
BT Precambrian
SA Canada
SA orogeny

Kent
No longer a valid term for GeoRef. As of 1993, see Kent England.

Kent County
Valid through 1988. For U.S. counties, search in combination with state term. After 1988, use specific county-state term.

Kent County Delaware (1989)
Central Delaware. Before 1989, also search Kent County AND Delaware.
CO N385000N391700
W0752000W0754500
BT Delaware
BT United States

Kent County Maryland (1989)
E Maryland. Before 1989, also search Kent County AND Maryland.
CO N385900N392300
W0754500W0761800
BT Maryland
BT United States

Kent County Michigan (1989)
SW Michigan. Before 1989, also search Kent County AND Michigan.
CO N424300N431700
W0851800W0854500
BT Michigan Lower Peninsula
BT Michigan
BT United States
NT Grand Rapids Michigan

Kent County Rhode Island (1989)
On Narragansett Bay in W and central Rhode Island. Before 1989, also search Kent County AND Rhode Island.
CO N413600N414600
W0712200W0714800
BT Rhode Island
BT United States

Kent County Texas (1989)
NW Texas. Before 1989, also search Kent County AND Texas.
CO N325800N332400
W1003100W1010200
BT Texas
BT United States

Kent England (1993)
County in SE England. Before 1993, also search Kent AND England.
CO N505400N513100
E0013000E0010000
BT England
BT Great Britain
BT United Kingdom
BT Western Europe
BT Europe
NT Folkestone England
NT Isle of Sheppey
SA Thames Estuary
SA Thames River
SA The Weald

Kentland
Valid through 1988. Search in combination with state term. After 1988, use specific city-state term.

Kentland Indiana (1989)
Town in NW Indiana. Before 1989, search Kentland AND Indiana.
CO N404600N404600
W0872600W0872600
BT Newton County Indiana
BT Indiana
BT United States

Kentucky (1978)
Autoposting of this term began in 1978.
CO N363000N391000
W0820000W0894000
BT United States
NT Adair County Kentucky
NT Allen County Kentucky
NT Anderson County Kentucky
NT Bath County Kentucky
NT Bell County Kentucky
NT Boone County Kentucky
NT Breathitt County Kentucky
NT Butler County Kentucky
NT Caldwell County Kentucky
NT Campbell County Kentucky
NT Carroll County Kentucky
NT Carter County Kentucky
NT Christian County Kentucky
NT Clark County Kentucky
NT Clay County Kentucky
NT Clinton County Kentucky
NT Crittenden County Kentucky
NT Cumberland County Kentucky
NT Edmonson County Kentucky
NT Elliott County Kentucky
NT Fayette County Kentucky
NT Fleming County Kentucky
NT Floyd County Kentucky
NT Franklin County Kentucky
NT Fulton County Kentucky
NT Gallatin County Kentucky
NT Grant County Kentucky
NT Green County Kentucky
NT Hancock County Kentucky
NT Hardin County Kentucky
NT Harlan County Kentucky
NT Harrison County Kentucky
NT Henderson County Kentucky
NT Henry County Kentucky
NT Jackson County Kentucky
NT Jefferson County Kentucky
NT Johnson County Kentucky
NT Kentucky River
NT Kentucky River basin
NT Kentucky River Fault
NT Kentucky River valley
NT Knott County Kentucky
NT Lawrence County Kentucky
NT Lee County Kentucky
NT Leslie County Kentucky
NT Letcher County Kentucky
NT Lewis County Kentucky
NT Lincoln County Kentucky
NT Livingston County Kentucky
NT Logan County Kentucky
NT Lyon County Kentucky
NT Madison County Kentucky
NT Marion County Kentucky
NT Marshall County Kentucky
NT Martin County Kentucky
NT Mason County Kentucky
NT McCreary County Kentucky
NT McLean County Kentucky
NT Meade County Kentucky
NT Mercer County Kentucky
NT Monroe County Kentucky
NT Montgomery County Kentucky
NT Moorman Syncline
NT Morgan County Kentucky
NT Nelson County Kentucky
NT Owsley County Kentucky
NT Pendleton County Kentucky
NT Perry County Kentucky
NT Pike County Kentucky
NT Powell County Kentucky
NT Pulaski County Kentucky
NT Rough Creek fault zone
NT Rowan County Kentucky
NT Russell County Kentucky
NT Scott County Kentucky
NT Shelby County Kentucky
NT Taylor County Kentucky
NT Union County Kentucky
NT Warren County Kentucky
NT Washington County Kentucky
NT Wayne County Kentucky
NT Webster County Kentucky
NT Whitley County Kentucky
NT Woodford County Kentucky
SA Allegheny Group
SA Allegheny Plateau
SA Ames Limestone
SA Appalachian Basin
SA Appalachian Plateau
SA Bass Islands Dolomite
SA Bedford Shale
SA Beech Creek Limestone Member
SA Berea Sandstone
SA Borden Group
SA Boyle Dolomite
SA Brassfield Formation
SA Breathitt Formation
SA Burlington Limestone
SA Carbondale Formation
SA Caseyville Formation
SA Chattanooga Shale
SA Chesterian
SA Cincinnati Arch
SA Clays Ferry Formation
SA Crab Orchard Formation
SA Cumberland Plateau
SA Eagle Station Meteorite
SA Eden Shale
SA Fairview Formation
SA Fort Payne Formation
SA Golconda Formation
SA Greenbrier Limestone
SA Harrodsburg Limestone
SA High Bridge Group
SA Illinois Basin
SA Jeffersonville Limestone
SA Kanawha Formation
SA Keokuk Limestone
SA Kittanning Formation
SA Kope Formation
SA Lee Formation
SA Lewis Creek
SA Lexington Limestone
SA Lockport Formation
SA Logan Formation
SA Midcontinent
SA Midway Group
SA Mississippi Embayment
SA Mississippi River
SA Mississippi Valley
SA Murray Meteorite
SA Nashville Dome
SA New Albany Shale
SA New Madrid region
SA Newman Limestone
SA Ohio River
SA Ohio River basin
SA Ohio River valley
SA Ohio Shale
SA Pennington Formation
SA Pine Mountain Window
SA Porters Creek Formation
SA Pottsville Group
SA Ramp Creek Formation
SA Renault Formation
SA Richmond Group
SA Ripley Formation
SA Rome Trough
SA Roxana Silt
SA Saint Louis Limestone
SA Saint Peter Sandstone
SA Sainte Genevieve Limestone
SA Salem Limestone
SA Salt River valley
SA Spergen Formation
SA Sturgis Formation
SA Sunbury Shale
SA Tennessee River
SA Tennessee Valley
SA Tradewater Formation
SA Waldron Shale
SA Warsaw Formation
SA Wilcox Group

Kentucky River (1978)
Formed by junction of North and Middle forks 4 miles ENE of Beattyville.
BT Kentucky
BT United States
SA Kentucky River valley

Kentucky River basin (1981)
Drainage basin.
CO N365000N384200
W0823500W0853000
BT Kentucky
BT United States
SA Kentucky River valley

Kentucky River Fault (1989)
Central Kentucky.
BT Kentucky
BT United States

Kentucky River valley (1978)
NE central Kentucky.
BT Kentucky
BT United States
SA Kentucky River
SA Kentucky River basin

Kenya (1978)
   CO S043000N043000
      E0420000E0340000
   BT East Africa
   BT Africa
   NT East Rudolf
   NT Fort Ternan Kenya
   NT Kenya Rift valley
   NT Lake Magadi
   NT Mount Kenya
   NT Nairobi Kenya
   NT Nakuru Basin
   NT Turkana District
   SA East African Rift
   SA Gregory Rift
   SA Koobi Fora Formation
   SA Lake Baringo
   SA Lake Natron
   SA Lake Turkana
   SA Lake Victoria
   SA Ngorora Formation
   SA Omo Group
   SA Shungura Formation
   SA Turkana Basin
   SA Umba

Kenya Rift valley (1978)
   That segment of the Great Rift Valley running N and S through W central Kenya.
   BT Kenya
   BT East Africa
   BT Africa
   SA Great Rift Valley
   SA Lake Baringo

kenyaite (1978)
   BT scapolite group
   BT framework silicates
   BT silicates

Keokuk Limestone (1978)
   Uppermost formation of Osagian Series. Iowa, Illinois, W Kentucky, and E Missouri.
   BT Osagian
   BT Lower Mississippian
   BT Mississippian
   BT Carboniferous
   BT Paleozoic
   SA Illinois
   SA Iowa
   SA Kentucky
   SA Missouri

Keonjhar
   No longer a valid term for GeoRef.
   See Keonjhar India.

Keonjhar India (1993)
   District in N Orissa, E India.
   BT Orissa India
   BT India
   BT Indian Peninsula
   BT Asia

kerabitumen
   use kerogen

Kerala
   No longer a valid term for GeoRef.
   See Kerala India.

Kerala India (1993)
   State on Arabian Sea in SW India.
   CO N080000N130000
      E0773000E0744000
   BT India
   BT Indian Peninsula
   BT Asia
   NT Cannanore India
   NT Trivandrum India
   SA Warkalli Formation

keratophyre (1978)
   BT trachytes
   BT volcanic rocks
   BT igneous rocks
   SA albitophyre

   SA quartz keratophyre

Kerch
   No longer a valid term for GeoRef.
   As of 1993, see Kerch Ukraine.

Kerch Peninsula (1978)
   E section of the Crimea between the Azov Sea on the N and the Black Sea on the S. Also search Kerch. This term has multiple hierarchies.
   BT1 Crimea Ukraine
   BT1 Ukraine
   BT1 Europe
   BT2 Crimea Ukraine
   BT2 Ukraine
   BT2 Commonwealth of Independent States

Kerch Ukraine (1993)
   City on the Kerch Peninsula in E Crimea Oblast. This term has multiple hierarchies.
   BT1 Crimea Ukraine
   BT1 Ukraine
   BT1 Europe
   BT2 Crimea Ukraine
   BT2 Ukraine
   BT2 Commonwealth of Independent States

Kerguelen Islands (1978)
   French island group, 1,400 miles off Antarctic mainland. In 1981, broader term changed from Indian Ocean to Antarctica.
   CO S500000S480000
      E0710000E0680000
   UF Desolation Islands
   BT Antarctica
   SA Indian Ocean

Kerguelen Plateau (1993)
   Submarine feature, S Indian Ocean. Kerguelen Islands and Heard Island area.
   CO S550000S450000
      E0800000E0620000
   UF Heard Plateau
   BT Indian Ocean
   SA Leg 119
   SA Leg 120
   SA ODP Site 736
   SA ODP Site 744
   SA ODP Site 745
   SA ODP Site 746
   SA ODP Site 747
   SA ODP Site 748
   SA ODP Site 749
   SA ODP Site 750
   SA ODP Site 751

Kermadec Islands (1978)
   Volcanic island group 600 miles NNW of New Zealand. Annexed to New Zealand in 1887. Also search Kermadec. In 1985, broader term changed from Pacific Ocean to West Pacific Ocean Islands. Raoul Island is included as a narrower term as or 1985.
   BT West Pacific Ocean Islands
   NT Raoul Island
   SA Kermadec Trench

Kermadec Trench (1989)
   In S Pacific Ocean, located NE of New Zealand.
   BT Pacific Ocean
   SA Kermadec Islands

Kerman
   No longer a valid term for GeoRef.
   See Kerman Iran.

Kerman Iran (1993)
   Province including the city in SE central Iran.

   BT Iran
   BT Middle East
   BT Asia
   NT Songhor Iran
   SA Zarand Basin

Kermine
   No longer a valid term for GeoRef.
   As of 1993, see Kermine Uzbekistan.

Kermine Uzbekistan (1993)
   Town in Bukhara Oblast in central Uzbekistan. This term has multiple hierarchies.
   BT1 Bukhara Uzbekistan
   BT1 Uzbekistan
   BT1 Asia
   BT2 Bukhara Uzbekistan
   BT2 Uzbekistan
   BT2 Commonwealth of Independent States

Kern County
   Valid through 1988. Search in combination with state term. After 1988, use specific county-state term.

Kern County California (1989)
   S central California. Before 1989, also search Kern County AND California.
   CO N344800N354800
      W1174000W1201500
   BT California
   BT United States
   NT Bakersfield California
   NT Elk Hills Field
   NT White Wolf Fault
   SA El Paso Mountains
   SA Mojave Desert
   SA San Emigdio Mountains
   SA Tehachapi Mountains
   SA Temblor Range

kernite (1978)
   UF rasorite
   BT borates

kerogen (1978)
   UF kerabitumen
   UF petrologen
   BT organic materials
   SA bitumens
   SA oil shale
   SA petroleum
   SA Rock-Eval
   SA sedimentary rocks

kerolite
   use cerolite

Kerry
   No longer a valid term for GeoRef.
   See Kerry Ireland.

Kerry Ireland (1993)
   County in SW Ireland.
   BT Ireland
   BT Western Europe
   BT Europe
   NT Dingle Peninsula

kersantite (1978)
   BT lamprophyres
   BT plutonic rocks
   BT igneous rocks

Kerulen
   use Choybalsan Mongolia

Kerulen River (1978)
   A headstream of the Amur River. Also search Kerulen.
   IN Index Inner Mongolia and/or Mongolia as applicable.
   BT Asia
   SA Inner Mongolia China
   SA Mongolia

kesterite (1978)
   BT sulfides

ketones (1985)
   BT organic materials
   SA fatty acids
   SA organic acids

Kettleman Hills (1978)
   Along W side of San Joaquin Valley.
   BT California
   BT United States

Kettleman Hills earthquake 1985 (1993)
   Central California, Parkfield area.
   BT earthquakes
   SA California
   SA Parkfield California
   SA shallow-focus earthquakes

kettles (1985)
   Autoposting of glacial features to this term began in 1989.
   BT glacial features
   SA glacial geology
   SA glacial lakes
   SA potholes

Keuper (1978)
   Europe. Above Muschelkalk, below Jurassic.
   BT Upper Triassic
   BT Triassic
   BT Mesozoic

Kewanee Group (1985)
   Illinois.
   BT Middle Pennsylvanian
   BT Pennsylvanian
   BT Carboniferous
   BT Paleozoic
   SA Carbondale Formation
   SA Illinois
   SA Spoon Formation

Keweenaw County
   Valid through 1988. Search in combination with state term. After 1988, use specific county-state term.

Keweenaw County Michigan (1989)
   NW Michigan Upper Peninsula. NW Michigan. Before 1989, also search Keweenaw County AND Michigan.
   CO N471300N482000
      W0873500W0892000
   BT Michigan Upper Peninsula
   BT Michigan
   BT United States
   NT Isle Royale National Park

Keweenaw Peninsula (1978)
   Juts into Lake Superior in NW Upper Peninsula.
   BT Michigan Upper Peninsula
   BT Michigan
   BT United States
   SA Lake Superior

Keweenawan (1978)
   Provincial series. Michigan, Minnesota, Wisconsin and Ontario. Autoposting of Precambrian to this term ended in 1989. As of 1989, Proterozoic is autoposted to this term. Autoposting of upper Precambrian and Precambrian to this term began in 1990.
   BT Proterozoic
   BT upper Precambrian
   BT Precambrian
   NT Copper Harbor Conglomerate
   NT Portage Lake Lava Series
   SA Michigan
   SA Minnesota

Before then, variants of the term may occur in GeoRef.

SA North Shore Volcanics
SA Ontario
SA Wisconsin

Keweenawan Rift   (1989)
Rift zone during the Proterozoic.
UF Mid-continent Rift
UF Midcontinent Rift
UF Midcontinent Rift System
BT North America

key beds   (1978)
Use for stratigraphy. Before 1978, search marker beds.
UF beds, key
UF index beds
SA marker beds
SA stratigraphy

Key Lake
No longer a valid term for GeoRef. As of 1993, see Key Lake Deposits.

Key Lake Deposits   (1993)
Location of uranium deposit in NE Saskatchewan. Before 1993, also search Key Lake.
CO N570000N570000
W1040000W1040000
BT Saskatchewan
BT Western Canada
BT Canada
SA uranium ores

Key Largo   (1978)
Island off extreme SE coast. Largest of the Florida Keys and the first island link of the Overseas Highway.
BT Florida Keys
BT Monroe County Florida
BT Florida
BT United States

Key Largo Limestone   (1978)
Pleistocene deposit older than Pamlico Sand. S Florida.
BT Pleistocene
BT Quaternary
BT Cenozoic
SA Florida

Keyser Limestone   (1985)
Upper Silurian to Lower Devonian? Pennsylvania, N West Virginia, W Maryland and W Virginia.
BT Paleozoic
SA Cayugan
SA Helderberg Group
SA Lower Devonian
SA Maryland
SA Pennsylvania
SA Upper Silurian
SA Virginia
SA West Virginia

KGRA, Coso Hot Springs
use Coso Hot Springs KGRA

KGRA, East Mesa
use East Mesa KGRA

KGRA, Roosevelt Hot Springs
use Roosevelt Hot Springs KGRA

Khabarovsk region
No longer a valid term for GeoRef. As of 1993, see Khabarovsk Russian Federation.

Khabarovsk Russian Federation   (1993)
Territory or kray in SE Russian Federation. Before 1993, also search Khabarovsk region. Before 1981, also search Khabarovsk. This term has multiple hierarchies.
BT1 Russian Federation
BT1 Commonwealth of Independent States
BT2 Asia
NT Okhotsk Russian Federation
NT Ozernoye Russian Federation
SA Amur region
SA Komsomolsk Russian Federation
SA Maya River basin
SA Omolon
SA Russian Far East
SA Sette-Daban Range
SA Sikhote-Alin Range
SA Stanovoy Range
SA Uchur River basin
SA Yudoma

Khamar-Daban Range   (1978)
S of Lake Baikal in Buryatia. Also search Khamar-Daban. This term has multiple hierarchies.
BT1 Russian Federation
BT1 Commonwealth of Independent States
BT2 Asia
SA Buryat Russian Federation

Khammam
No longer a valid term for GeoRef. See Khammam India.

Khammam India   (1993)
Town and district in E central Andhra Pradesh, central India.
BT Andhra Pradesh India
BT India
BT Indian Peninsula
BT Asia

Khangai
use Hangay Mountains

Khangai Mountains
use Hangay Mountains

Khanka Lake   (1978)
N of Vladivostok. Also search Khanka.
IN Index Heilongjiang China and/or Russian Federation as applicable.
BT Asia
SA Heilongjiang China
SA Russian Federation

Khapcheranga
No longer a valid term for GeoRef. As of 1993, see Khapcheranga Russian Federation.

Khapcheranga Russian Federation   (1993)
Town in SW Chita Oblast in Transbaikalia. Before 1993, also search Khapcheranga. This term has multiple hierarchies.
BT1 Chita Russian Federation
BT1 Russian Federation
BT1 Commonwealth of Independent States
BT2 Chita Russian Federation
BT2 Asia

Kharga
use Kharga Oasis

Kharga Oasis   (1978)
Valley and oasis in S central Egypt.
UF El Kharga
UF Kharga
UF The Great Oasis
BT Egypt
BT North Africa
BT Africa

Kharkov
No longer a valid term for GeoRef. As of 1993, see Kharkov Ukraine.

Kharkov Ukraine   (1993)
Oblast and city in NE Ukraine. This term has multiple hierarchies.
BT1 Ukraine
BT1 Europe
BT2 Ukraine
BT2 Commonwealth of Independent States

Khaskovo
No longer a valid term for GeoRef. See Khaskovo Bulgaria.

Khaskovo Bulgaria   (1993)
Province including the city in S Bulgaria.
CO N411000N421000
E0264000E0244000
BT Bulgaria
BT Southern Europe
BT Europe

Khatanga
No longer a valid term for GeoRef. As of 1993, see Khatanga Russian Federation or Shaba Zaire.

Khatanga Basin   (1978)
River basin in E Taymyr National Okrug in N central Siberia. Also search Khatanga. Also search Katanga AND USSR. This term has multiple hierarchies.
CO N713000N740000
E1080000E1013000
UF Katanga Basin, USSR
BT1 Russian Federation
BT1 Commonwealth of Independent States
BT2 Asia
SA Kotui
SA Siberia
SA Taymyr Dolgan-Nenets Russian Federation
SA Yenisei-Khatanga basin

Khatanga River   (1978)
Formed by the union of the Kheta River and the Kotui in N central Siberia. Flows into Khatanga Gulf, an inlet of the Laptev Sea. Also search Khatanga. Also search Katanga AND USSR.
IN Index Taymyr Dolgan-Nenets Russian Federation if applicable.
UF Katanga River, USSR
SA Kheta River
SA Kotui
SA Taymyr Dolgan-Nenets Russian Federation

Khatanga Russian Federation   (1993)
Town in Taymyr National Okrug about 150 miles from mouth of Khatanga River in N central Siberia. Before 1993, also search Khatanga. Before 1978, also search Katanga. This term has multiple hierarchies.
UF Katanga Russian Federation
BT1 Taymyr Dolgan-Nenets Russian Federation
BT1 Krasnoyarsk Russian Federation
BT1 Russian Federation
BT1 Commonwealth of Independent States
BT2 Taymyr Dolgan-Nenets Russian Federation
BT2 Krasnoyarsk Russian Federation
BT2 Asia

Khawr al Bazam
use Khor al Bazam

Khekordzula Deposit   (1989)
Zeolite deposits.
BT Georgian Republic
BT Europe
SA zeolite deposits

Kheta River   (1978)
Chief tributary of the Khatanga River in Taymyr National Okrug in N central Siberia. This term has multiple hierarchies.
BT1 Krasnoyarsk Russian Federation
BT1 Russian Federation
BT1 Commonwealth of Independent States
BT2 Krasnoyarsk Russian Federation
BT2 Asia
SA Khatanga River
SA Taymyr Dolgan-Nenets Russian Federation

Khetri
No longer a valid term for GeoRef. See Khetri India.

Khetri copper belt   (1978)
In N Aravalli Range in central Rajasthan, India. Also search Khetri.
SA Aravalli Range
SA Rajasthan India

Khetri India   (1993)
Town in E Rajasthan, NW India.
BT Rajasthan India
BT India
BT Indian Peninsula
BT Asia

Khewra
No longer a valid term for GeoRef. See Khewra Pakistan.

Khewra Pakistan   (1993)
Village in N Punjab Province, E central Pakistan.
BT Punjab Pakistan
BT Pakistan
BT Indian Peninsula
BT Asia

Khibiny Massif
use Khibiny Mountains

Khibiny Mountains   (1978)
Central Kola Peninsula in extreme NW European Russian Federation. Also search Khibiny. This term has multiple hierarchies.
CO N670000N680000
E0340000E0330000
UF Khibiny Massif
BT1 Russian Federation
BT1 Commonwealth of Independent States
BT2 Europe

Khingan Range
No longer a valid term for GeoRef.
use Da Hinggan Ling

Khios
use Chios

Khiva
No longer a valid term for GeoRef. As of 1993, see Khiva Uzbekistan for the town. The former Khanate of Khiva is coextensive with current Khorezm Oblast.

Khiva Uzbekistan   (1993)
Town in oasis region on left bank of the lower Amu Darya in Khorezm Oblast in NW Uzbekistan. The former Khanate of Khiva is coextensive with current Khorezm Oblast. This term has multiple hierarchies.

BT1 Uzbekistan
BT1 Asia
BT2 Uzbekistan
BT2 Commonwealth of Independent States

Khmer Republic
use Cambodia

Khobdo
use Kobdo Mongolia

khondalite (1978)
BT metasedimentary rocks
BT metamorphic rocks

Khor al Bazam (1978)
A strait off coast of central United Arab Emirates W of Abu Dhabi.
UF Khawr al Bazam
BT Persian Gulf
BT Arabian Sea
BT Indian Ocean

Khorat Group (1978)
Includes the Kamawkala Limestone. May also include beds of Permian as well as Cretaceous age or younger. Widespread throughout the country.
IN Index ages as applicable.
BT Mesozoic
SA Jurassic
SA Thailand
SA Triassic

Khorat Plateau (1978)
Tableland between the Mekong River on N and E and Cambodia on the S.
UF Korat Plateau
BT Thailand
BT Far East
BT Asia

Khovu-Aksy
No longer a valid term for GeoRef. As of 1993, see Khovu-Aksy Russian Federation.

Khovu-Aksy Russian Federation (1993)
Village in Tuva Autonomous Republic just N of W Mongolia. Before 1993, also search Khovu-Aksy. This term has multiple hierarchies.
BT1 Tuva Russian Federation
BT1 Russian Federation
BT1 Commonwealth of Independent States
BT2 Tuva Russian Federation
BT2 Siberian fold belt
BT2 Asia

Khubsugul
No longer a valid term for GeoRef. As of 1993 see Khubsugul Mongolia.

Khubsugul Mongolia (1993)
Province in N Mongolia. Before 1993 also search Khubusugul AND Mongolia.
UF Hobsogol
UF Hubusugul
BT Mongolia
BT Far East
BT Asia

Kiamba Philippine Islands (1993)
City on S Mindanao, South Cotabato Province, Philippine Islands. Before 1993, also search Kiamba.
BT Mindanao
BT Philippine Islands
BT Far East
BT Asia

Kiangsi
Not a valid term for GeoRef. See Jiangxi China.

Kiangsu
Not a valid term for GeoRef. See Jiangsu China.

Kicking Horse River valley (1978)
In Banff National Park area. Also search Kicking Horse River.
IN Index provinces as applicable.
BT Western Canada
BT Canada
SA Alberta
SA Banff National Park
SA British Columbia

Kiel
No longer a valid term for GeoRef. See Kiel Germany.

Kiel Bay (1978)
Off NE coast of Schleswig-Holstein.
BT Baltic Sea
BT European Atlantic
BT North Atlantic
BT Atlantic Ocean

Kiel Germany (1993)
City on the Baltic Sea in N Germany.
BT Schleswig-Holstein Germany
BT Germany
BT Central Europe
BT Europe

Kielce
No longer a valid term for GeoRef. As of 1993 see Kielce Poland.

Kielce Poland (1993)
BT Poland
BT Central Europe
BT Europe
NT Nida Basin
NT Ostrowiec Swietokrzyski Poland
SA Galezice
SA Swiety Krzyz Mountains

kieserite (1978)
BT sulfates

Kiev
No longer a valid term for GeoRef. See Kiev Ukraine.

Kiev Member (1978)
BT lower Tertiary
BT Tertiary
BT Cenozoic
SA Ukraine

Kiev Ukraine (1993)
City and oblast in N central Ukraine. This term has multiple hierarchies.
BT1 Ukraine
BT1 Europe
BT2 Ukraine
BT2 Commonwealth of Independent States
NT Chernobyl Ukraine
NT Kanev Ukraine

Kiglapait (1978)
Cape in N Labrador.
BT Labrador
BT Newfoundland
BT Eastern Canada
BT Canada
SA Kiglapait Intrusion

Kiglapait Intrusion (1989)
Iron-rich layered intrusion located in NE Labrador.
CO N563000N573000
W0612000W0614000
BT Labrador
BT Newfoundland
BT Eastern Canada
BT Canada
SA Kiglapait

Kii Peninsula (1978)
Between Kii Channel on W and Kumano Sea on E in S Honshu. Also search Kii.
BT Honshu
BT Japan
BT Far East
BT Asia
SA Mie Japan
SA Nara Japan
SA Wakayama Japan

Kilauea (1978)
Active crater on E side of Mauna Loa in Volcanoes National Park, S central island of Hawaii. Autoposting of broader terms to this term began in 1989. This term has multiple hierarchies.
CO N191500N192500
    W1550500W1552000
UF Kilauea Iki
UF Kilauea Volcano
BT1 Hawaii Island
BT1 Hawaii County Hawaii
BT1 Hawaii
BT1 United States
BT2 Hawaii Island
BT2 Hawaii County Hawaii
BT2 Hawaii
BT2 Polynesia
BT2 Oceania
BT3 Hawaii Island
BT3 Hawaii County Hawaii
BT3 Hawaii
BT3 East Pacific Ocean Islands
SA Alae Crater
SA East Rift Zone
SA Mauna Loa
SA Puu Oo

Kilauea Iki
use Kilauea

Kilauea Volcano
use Kilauea

Kilbourne Hole (1981)
Exploratory drill hole in S New Mexico. Autoposting of broader terms to this term began in 1989.
CO N315700N315900
    W1065700W1065900
BT Dona Ana County New Mexico
BT New Mexico
BT United States

kilchoanite (1978)
Autoposting of sorosilicates began in 1985.
BT sorosilicates
BT orthosilicates
BT silicates

Kildare
No longer a valid term for GeoRef. See Kildare Ireland.

Kildare Ireland (1993)
County including the town, E central Ireland.
BT Ireland
BT Western Europe
BT Europe

Kilimanjaro (1978)
Highest mountain in Africa. Near Kenya border in NE Tanzania.
BT Tanzania
BT East Africa
BT Africa

Kilo Zaire
use Bambu Zaire

Kilo-Mines Zaire
use Bambu Zaire

Kimberley
No longer a valid term for GeoRef. As of 1993, see Kimberley South Africa or Kimberley British Columbia.

Kimberley British Columbia (1993)
Town in SE British Columbia with silver, lead and zinc mines nearby. Before 1993, search Kimberley AND British Columbia.
BT British Columbia
BT Western Canada
BT Canada

Kimberley South Africa (1993)
City near Orange Free State border in Cape Province. World's diamond center with mines nearby. Before 1993 also search Kimberley AND South Africa.
BT Cape Province South Africa
BT South Africa
BT Southern Africa
BT Africa

kimberlite (1978)
BT ultramafics
BT plutonic rocks
BT igneous rocks
SA diamond
SA diamonds

kimberlite pipes
Not a valid term for GeoRef. Use kimberlite in conjunction with pipes.

Kimmerian
use Cimmerian

Kimmerian Orogeny
use Cimmerian Orogeny

Kimmeridge Clay (1989)
BT Upper Jurassic
BT Jurassic
BT Mesozoic
SA England
SA North Sea
SA Scotland

Kimmeridgian (1978)
Europe. Above Oxfordian, below Portlandian.
BT Upper Jurassic
BT Jurassic
BT Mesozoic
SA Malm

Kinderhookian (1978)
Provincial series, North America. Above Chautauquan (Devonian), below Osagian.
BT Lower Mississippian
BT Mississippian
BT Carboniferous
BT Paleozoic
NT Banff Formation
SA Chouteau Limestone

kinematics (1978)
Branch of dynamics that deals with aspects of motion apart from considerations of mass and force.
UF cinematics
SA kinetics
SA velocity

kinetics (1978)
SA dynamics
SA energy
SA geochemistry
SA geophysics

SA kinematics
SA velocity

King County
Valid through 1988. Search in combination with state term. After 1988, use specific county-state term.

King County Texas (1989)
NW Texas. Before 1989, also search King County AND Texas.
CO N332500N335100
W0995800W1003000
BT Texas
BT United States

King County Washington (1989)
W central Washington. Before 1989, also search King County AND Washington.
CO N470600N474600
W1210500W1223300
BT Washington
BT United States
NT Seattle Washington
SA Lake Washington

King George Island (1978)
Largest of the South Shetland Islands in British Antarctic Territory N of the Antarctic Peninsula.
UF Waterloo Island
BT South Shetland Islands
BT Scotia Sea Islands
BT Antarctica

Kingak Shale (1989)
NE Alaska.
BT Jurassic
BT Mesozoic
SA Alaska

Kingfisher County
Valid through 1988. Search in combination with state term. After 1988, use specific county-state term.

Kingfisher County Oklahoma (1989)
Central Oklahoma. Before 1989, also search Kingfisher County AND Oklahoma.
CO N354300N361000
W0974000W0981300
BT Oklahoma
BT United States

Kingman County
Valid through 1988. Search in combination with state term. After 1988, use specific county-state term.

Kingman County Kansas (1989)
S Kansas. Before 1989, also search Kingman County AND Kansas.
CO N372500N374500
W0974700W0982900
BT Kansas
BT United States

Kings Bay (1978)
Inlet on NW coast of Spitsbergen Island.
IN Index island or region as applicable.
SA Spitsbergen Island

Kings County
Valid through 1988. Search in combination with state term. After 1988, use specific county-state term.

Kings County California (1989)
S central California. Before 1989, also search Kings County AND California.
CO N354800N363000
W1193000W1202200
BT California
BT United States

Kings County New York (1989)
Coextensive with the borough of Brooklyn in New York City on Long Island in SE New York. Before 1989, also search Kings County AND New York.
IN Index Long Island.
CO N403600N404400
W0735200W0740400
BT New York City New York
BT New York
BT United States
SA Long Island

Kings Mountain (1978)
Isolated ridge.
IN Index counties or regions as applicable.
SA Kings Mountain Belt
SA North Carolina
SA South Carolina
SA United States

Kings Mountain Belt (1985)
Metamorphic belt, SW North Carolina, NW South Carolina and NE Georgia.
IN Index states as applicable.
BT United States
SA Appalachians
SA Carolina slate belt
SA Georgia
SA Kings Mountain
SA North Carolina
SA Piedmont
SA South Carolina
SA Southern Appalachians

Kingsdown Formation (1978)
In Sanborn Group. Overlies Crooked Creek Formation; underlies Vanhem Formation. SW Kansas.
BT Pleistocene
BT Quaternary
BT Cenozoic
SA Kansas

Kingsport Formation (1978)
In Knox Group. E Tennessee and SW Virginia.
BT Lower Ordovician
BT Ordovician
BT Paleozoic
SA Knox Group
SA Tennessee
SA Virginia

Kingston Peak Formation (1985)
S California.
BT Precambrian
SA California

Kingston Range (1989)
S California.
IN Index county or regions as applicable.
SA San Bernardino County California

kink folds (1978)
Before 1978, also search folds AND kink.
BT folds
SA chevron folds
SA kink-band structures

kink-band structures (1978)
Also search kink-band.
UF kink-bands
SA deformation
SA folds
SA kink folds

kink-bands
use kink-band structures

Kinki District
No longer a valid term for GeoRef.
use Kinki Japan

Kinki Japan (1993)
The Osaka, Kobe and Kyoto region in S central Honshu. Includes Hyogo, Kyoto, Mie, Nara, Osaka, Shiga, and Wakayama prefectures. Before 1993, also search Kinki District and Kinki.
CO N332000N353000
E1363000E1350000
UF Kinki District
BT Honshu
BT Japan
BT Far East
BT Asia
SA Hyogo Japan
SA Kyoto Japan
SA Mie Japan
SA Nara Japan
SA Osaka Japan
SA Wakayama Japan

Kinnashih
use Chinkuashih Taiwan

Kinnaur
No longer a valid term for GeoRef. See Kinnaur India.

Kinnaur India (1993)
District in Himachal Pradesh.
CO N310500N320000
E0790000E0774500
BT Himachal Pradesh India
BT India
BT Indian Peninsula
BT Asia

Kinney County
Valid through 1988. Search in combination with state term. After 1988, use specific county-state term.

Kinney County Texas (1989)
SW Texas. Before 1989, also search Kinney County AND Texas.
CO N290100N294000
W1000700W1005200
BT Texas
BT United States
SA Edwards Aquifer

Kinshasa
No longer a valid term for GeoRef. As of 1993 see Kinshasa Zaire.

Kinshasa Zaire (1993)
City and Federal District on left bank of the Congo River in W Zaire. Before 1993 also search Kinshasa AND Zaire.
UF Leopoldville
BT Zaire
BT Central Africa
BT Africa

Kinta Valley (1989)
River valley in NW West Malaysia. This term has multiple hierarchies.
BT1 Perak Malaysia
BT1 West Malaysia
BT1 Malay Peninsula
BT1 Far East
BT1 Asia
BT2 Perak Malaysia
BT2 West Malaysia
BT2 Malaysia
BT2 Far East
BT2 Asia

Kintyre (1981)
Peninsula in Argyllshire, SW Scotland. This term has multiple hierarchies.
CO N551800N555200
W0051900W0054700
BT1 Argyllshire Scotland
BT1 Scotland
BT1 Great Britain
BT1 United Kingdom
BT1 Western Europe
BT1 Europe
BT2 Strathclyde region Scotland
BT2 Scotland
BT2 Great Britain
BT2 United Kingdom
BT2 Western Europe
BT2 Europe

Kinwashih Mine
use Chinkuashih Mine

Kinzers Formation (1978)
Overlies Vintage Limestone; underlies Ledger Dolomite. Maryland, SE Pennsylvania, and Virginia.
BT Lower Cambrian
BT Cambrian
BT Paleozoic
SA Maryland
SA Pennsylvania
SA Virginia

kinzigite (1978)
BT granulites
BT metamorphic rocks

Kiokee Belt (1985)
Metamorphic belt, central North Carolina to E central Georgia.
IN Index states as applicable.
BT United States
SA Appalachians
SA Carolina slate belt
SA Georgia
SA North Carolina
SA Piedmont
SA South Carolina

Kiowa County
Valid through 1988. Search in combination with state term. After 1988, use specific county-state term.

Kiowa County Colorado (1989)
E Colorado. Before 1989, also search Kiowa County AND Colorado.
CO N381400N383600
W1020200W1032900
BT Colorado
BT United States

Kiowa County Kansas (1989)
S Kansas. Before 1989, also search Kiowa County AND Kansas.
CO N372200N374300
W0990000W0993500
BT Kansas
BT United States
NT Haviland Kansas
SA Haviland Meteorite

Kiowa County Oklahoma (1989)
SW Oklahoma. Before 1989, also search Kiowa County AND Oklahoma.
CO N343500N350700
W0983700W0992300
BT Oklahoma
BT United States
SA Meers Fault
SA Slick Hills
SA Wichita Mountains

Kiowa Formation (1978)

Overlies Cheyenne Sandstone; underlies Dakota Formation. Central S Kansas, E Colorado, W Oklahoma.
BT Lower Cretaceous
BT Cretaceous
BT Mesozoic
SA Colorado
SA Kansas
SA Oklahoma

Kipawa Lake (1978)
SW Quebec. Also search Kipawa.
BT Temiscamingue County Quebec
BT Quebec
BT Eastern Canada
BT Canada

Kirchhoff integral (1989)
Shows relationships between enthalpy, heat capacity and temperature for a reaction.
SA enthalpy
SA equations
SA free energy
SA geochemistry
SA heat capacity
SA mathematical models
SA seismic migration
SA temperature
SA thermodynamic properties

Kirghiz Soviet Socialist Republic
use Kyrgyzstan

Kirghizia
As of 1993, no longer a valid term for GeoRef.
use Kyrgyzstan

Kirgizia
use Kyrgyzstan

Kiribati (1985)
Formerly Gilbert Islands, under British administration. Now independent. Also includes Phoenix Islands and many formerly British Line Islands. Before 1985, also search Gilbert Islands.
UF Gilbert Islands
BT Oceania
NT Fanning Island
SA Christmas Island
SA East Pacific Ocean Islands
SA Line Islands
SA Micronesia
SA Polynesia
SA Vostok
SA West Pacific Ocean Islands

Kirin Meteorite
use Jilin Meteorite

Kirin, China
use Jilin China

Kirishima (1978)
Mountain and geothermal area in Kirishima Range in S Kyushu.
UF Kirishima-Yama
BT Kyushu
BT Japan
BT Far East
BT Asia
SA Ebino Japan
SA Kagoshima Japan
SA Miyazaki Japan

Kirishima-Yama
use Kirishima

Kirkbyocopina (1981)
Includes the superfamily Kirkbyacea. As of 1990, microfossils, Crustacea, Mandibulata, and Arthropoda are autoposted to this term. This term has multiple hierarchies.
BT1 Ostracoda
BT1 Crustacea
BT1 Mandibulata
BT1 Arthropoda
BT1 Invertebrata
BT2 Ostracoda
BT2 microfossils

Kirkcudbrightshire
No longer a valid term for GeoRef as of 1993.
use Kirkcudbrightshire Scotland

Kirkcudbrightshire Scotland (1993)
Former county in S Scotland. Now part of Dumfries and Galloway region.
CO N544500N551500 W0033000W0043000
UF Kirkcudbrightshire
BT Scotland
BT Great Britain
BT United Kingdom
BT Western Europe
BT Europe

Kirkland Lake
No longer a valid term for GeoRef. As of 1993, see Kirkland Lake Ontario.

Kirkland Lake Ontario (1993)
Town in E Ontario near Quebec border. Before 1993, also search Kirkland Lake AND Ontario.
BT Timiskaming District Ontario
BT Ontario
BT Eastern Canada
BT Canada

Kirkpatrick Basalt (1989)
Queen Alexandra Range and Victoria Land, Antarctica.
BT Jurassic
BT Mesozoic
SA Antarctica
SA Queen Alexandra Range
SA Victoria Land

Kirkuk
No longer a valid term for GeoRef. See Kirkuk Iraq.

Kirkuk Iraq (1993)
Town near oil fields in NE Iraq.
BT Iraq
BT Middle East
BT Asia

Kirkwood Formation (1978)
Arikareean and Hemingfordian. E New Jersey.
BT middle Miocene
BT Miocene
BT Neogene
BT Tertiary
BT Cenozoic
SA Arikareean
SA Hemingfordian
SA New Jersey

Kirov
No longer a valid term for GeoRef. As of 1993, see Vyatka Russian Federation.

Kirov Russian Federation
use Vyatka Russian Federation

Kirovabad
No longer a valid term for GeoRef. As of 1993, see Kirovabad Azerbaidzhan.

Kirovabad Azerbaidzhan (1993)
City in W Azerbaidzhan. Before 1993, also search Gandja or Gandzha. This term has multiple hierarchies.
UF Gandja Azerbaidzhan
UF Gandzha Azerbaidzhan
BT1 Azerbaidzhan
BT1 Europe
BT2 Azerbaidzhan
BT2 Commonwealth of Independent States

Kirovograd
No longer a valid term for GeoRef. As of 1993, see Kirovograd Ukraine.

Kirovograd Ukraine (1993)
City and oblast in central S Ukraine. Historical names include Zinovievsk, Elisavetgrad, and Yelizavetgrad. This term has multiple hierarchies.
BT1 Ukraine
BT1 Europe
BT2 Ukraine
BT2 Commonwealth of Independent States

Kirtland Shale (1985)
NW New Mexico and SW Colorado.
BT Upper Cretaceous
BT Cretaceous
BT Mesozoic
SA Colorado
SA New Mexico

Kiruna
No longer a valid term for GeoRef. As of 1993 see Kiruna Sweden.

Kiruna Sweden (1993)
City in N central Norrbotten, N Sweden. Before 1993 also search Kiruna AND Sweden.
BT Norrbotten Sweden
BT Sweden
BT Scandinavia
BT Western Europe
BT Europe

Kirwin
Valid through 1988. Search in combination with state term. After 1988, use specific city-state term.

Kirwin Kansas (1989)
Town in N Kansas. Before 1989, search Kirwin AND Kansas.
CO N394100N394100 W0990600W0990600
BT Phillips County Kansas
BT Kansas
BT United States

Kishtwar (1978)
Town in SW Jammu and Kashmir, disputed territory between India and Pakistan. As of 1993, India is no longer a broader term.
IN Index Jammu and Kashmir AND India and/or Pakistan as applicable.
SA India
SA Jammu and Kashmir
SA Pakistan

Kiso Mountains (1978)
In central Honshu, Japan. Also search Kiso.
BT Honshu
BT Japan
BT Far East
BT Asia
SA Aichi Japan
SA Japanese Alps
SA Nagano Japan

Kisseynew Complex (1989)
Archean? Subdivided into Nokomis and Sherridon groups.
BT Precambrian
SA Manitoba
SA Nokomis Group

SA Saskatchewan
SA Sherridon Group

Kistna River
use Krishna

Kitakami Massif
use Kitakami Mountains

Kitakami Mountains (1978)
N Honshu. Also search Kitakami.
UF Kitakami Massif
BT Iwate Japan
BT Honshu
BT Japan
BT Far East
BT Asia

Kitami
No longer a valid term for GeoRef. See Kitami Japan or Kitami Basin.

Kitami Basin (1993)
NE Hokkaido, N Japan. Before 1993, also search Kitami.
BT Hokkaido
BT Japan
BT Far East
BT Asia
SA Abashiri Japan

Kitami Japan (1993)
City in Abashiri sub-prefecture, Hokkaido. Before 1993, also search Kitami AND Japan.
BT Abashiri Japan
BT Hokkaido
BT Japan
BT Far East
BT Asia

Kitsap County
Valid through 1988. Search in combination with state term. After 1988, use specific county-state term.

Kitsap County Washington (1989)
On Puget Sound in NW Washington. Before 1989, also search Kitsap County AND Washington.
CO N472500N475500 W1223000W1230000
BT Washington
BT United States
NT Kitsap Peninsula

Kitsap Peninsula (1978)
Bounded on W by Hood Canal and on E by Puget Sound. W of Seattle in NW Washington.
BT Kitsap County Washington
BT Washington
BT United States

Kittanning Formation (1989)
In Allegheny Group.
BT Pennsylvanian
BT Carboniferous
BT Paleozoic
SA Allegheny Group
SA Indiana
SA Kentucky
SA Maryland
SA Ohio
SA Pennsylvania
SA West Virginia

Kittery Formation (1989)
Ordovician and Silurian. In Merrimack Group. SW Maine and SE New Hampshire.
IN Index ages as applicable.
BT Paleozoic
SA Maine
SA Merrimack Group
SA New Hampshire
SA Ordovician
SA Silurian

Kittitas County
Valid through 1988. Search in combination with state term. After 1988, use specific county-state term.

Kittitas County Washington (1989)
Central Washington. Before 1989, also search Kittitas County AND Washington.
CO N464300N473600
W1195600W1213000
BT Washington
BT United States

Kivu
No longer a valid term for GeoRef. As of 1993 see Kivu Zaire.

Kivu Zaire (1993)
Province in E Zaire. Before 1993 also search Kivu AND Zaire.
CO S050000N010000
E0300000E0244000
BT Zaire
BT Central Africa
BT Africa
NT Kama Zaire
NT Nyiragongo
SA Lake Kivu

Kiya River (1978)
In Kemerovo and Tomsk oblasts E and NE of Novosibirsk. This term has multiple hierarchies.
BT1 Russian Federation
BT1 Commonwealth of Independent States
BT2 Asia
SA Kemerovo Russian Federation
SA Tomsk Russian Federation

Kizel coal basin (1978)
In Central Urals of European Russian Federation in Perm Oblast. Also search Kizel. This term has multiple hierarchies.
BT1 Perm Russian Federation
BT1 Russian Federation
BT1 Commonwealth of Independent States
BT2 Perm Russian Federation
BT2 Europe
SA coal fields

Kizil Kum
use Kyzylkum

Kizil-Dere
use Kizyl-Dere

Kizildere
use Kizyl-Dere

Kizyl-Dere (1978)
Gorge near Iranian border in S Turkmenia. This term has multiple hierarchies.
UF Kizil-Dere
UF Kizildere
BT1 Turkmenia
BT1 Asia
BT2 Turkmenia
BT2 Commonwealth of Independent States

Kladno (1978)
City NW of Prague in W Czechoslovakia.
BT Bohemia
BT Czech Republic
BT Czechoslovakia
BT Central Europe
BT Europe

Klamath County
Valid through 1988. Search in combination with state term. After 1988, use specific county-state term.

Klamath County Oregon (1989)
S Oregon. Before 1989, also search Klamath County AND Oregon.
CO N420000N433500
W1205000W1221500
BT Oregon
BT United States
NT Klamath Falls
NT Klamath Falls Oregon
NT Mount Mazama
SA Crater Lake

Klamath Falls (1989)
Waterfall in S Oregon. For city, use Klamath Falls Oregon.
BT Klamath County Oregon
BT Oregon
BT United States
SA Klamath Falls Oregon

Klamath Falls Oregon (1989)
City in S Oregon. Before 1989, also search Klamath Falls AND Oregon.
CO N421400N421400
W1214700W1214700
BT Klamath County Oregon
BT Oregon
BT United States
SA Klamath Falls

Klamath Mountains (1978)
Part of the Coast Ranges.
IN Index Coast Ranges and states as applicable.
CO N410000N430000
W1230000W1243000
BT United States
SA California
SA Coast Ranges
SA Condrey Mountain Schist
SA Hayfork Terrane
SA Marble Mountains
SA Oregon
SA Preston Peak
SA Rattlesnake Creek Terrane
SA Trinity Complex
SA Yolla Bolly Terrane

Kleberg County
Valid through 1988. Search in combination with state term. After 1988, use specific county-state term.

Kleberg County Texas (1989)
S Texas. Before 1989, also search Kleberg County AND Texas.
CO N271000N274000
W0974000W0980500
BT Texas
BT United States
SA Baffin Bay

Klerksdorp Field (1993)
Gold field in Witwatersrand. Area also contains uranium ores.
BT South Africa
BT Southern Africa
BT Africa
SA gold ores
SA mines
SA Orange Free State South Africa
SA Transvaal South Africa
SA uranium ores
SA Witwatersrand

Kletno
No longer a valid term for GeoRef. See Kletno Belarus.

Kletno Belarus (1993)
Village in central Belarus. This term has multiple hierarchies.
BT1 Belarus
BT1 Europe
BT2 Belarus
BT2 Commonwealth of Independent States

Klichka
No longer a valid term for GeoRef. As of 1993, see Klichka Russian Federation.

Klichka Russian Federation (1993)
Village in S Chita Oblast near Inner Mongolian border. Before 1993, also search Klichka.
IN Index regions as applicable.
BT Russian Federation
BT Commonwealth of Independent States
SA Chita Russian Federation

Klickitat County
Valid through 1988. Search in combination with state term. After 1988, use specific county-state term.

Klickitat County Washington (1989)
On Columbia River in S central Washington. Before 1989, also search Klickitat County AND Washington.
CO N453500N460500
W1195000W1213500
BT Washington
BT United States
SA Yakima Indian Reservation

klippe
use klippen

klippen (1978)
UF klippe
SA allochthons
SA duplexes
SA erosion features
SA faults
SA geomorphology
SA nappes
SA overthrust faults
SA tectonics

Klippen Belt
use Pieniny Klippen Belt

Klodawa
No longer a valid term for GeoRef. As of 1993 see Klodawa Poland.

Klodawa Poland (1993)
Town in central Poland. Before 1993 also search Klodawa AND Poland.
BT Poznan Poland
BT Poland
BT Central Europe
BT Europe

Klodzko Unit (1985)
Metasedimentary sequence, S Poland.
IN Index ages as applicable.
SA Paleozoic
SA Poland
SA Polish Sudeten Mountains
SA Proterozoic
SA Sowie Mountains

Klondike (1978)
Region in Yukon River basin in SW central Yukon Territory.
BT Yukon Territory
BT Western Canada
BT Canada

Kluane Lake (1978)
Along N slope of Saint Elias Mountains in SW Yukon Territory.
BT Yukon Territory
BT Western Canada
BT Canada

Klyuchevskaya Sopka (1978)
Active volcano in Eastern Range of E central Kamchatka Peninsula. Highest mountain in Siberia. Also search Klyuchevskaya. This term has multiple hierarchies.
UF Klyuchevskoy
BT1 Kamchatka Russian Federation
BT1 Russian Federation
BT1 Commonwealth of Independent States
BT2 Kamchatka Russian Federation
BT2 Asia
SA Kamchatka Peninsula

Klyuchevskoy
use Klyuchevskaya Sopka

Knee Lake (1978)
NE Manitoba.
BT Manitoba
BT Western Canada
BT Canada

Knife Lake Group (1978)
Comprises rocks formerly called Ogishke Conglomerate, Agawa Iron-formation, and Knife Lake Slates. Includes many members which cannot be separated into the three former divisions.
BT Precambrian
SA Minnesota

Knik Arm (1978)
Extension of Cook Inlet N of Anchorage.
UF Knik River
BT Southern Alaska
BT Alaska
BT United States
SA Cook Inlet

Knik River
use Knik Arm

Knipovich Ridge (1989)
In Norwegian Sea, Arctic Ocean.
CO N703000N714000
W0030000W0090000
BT Norwegian Sea
BT Arctic Ocean
SA DSDP Site 344

Knott County
Valid through 1988. Search in combination with state term. After 1988, use specific county-state term.

Knott County Kentucky (1989)
E Kentucky. Before 1989, also search Knott County AND Kentucky.
CO N371000N373200
W0824300W0830900
BT Kentucky
BT United States

knowledge-based systems
use expert systems

Known Geothermal Resources Area
Not a valid term for GeoRef. See specific area.

Knox County
Valid through 1988. Search in combination with state term. After 1988, use specific county-state term.

Knox County Illinois (1989)

NW central Illinois. Before 1989, also search Knox County AND Illinois.
CO N404400N411000
 W0895900W0902800
BT Illinois
BT United States

Knox County Tennessee (1993)
E Tennessee. Before 1989, also search Knox County AND Tennessee.
CO N354900N361200
 W0833900W0841600
BT Tennessee
BT United States

Knox Group (1978)
Upper Cambrian and Lower Ordovician. Includes Copper Ridge Dolomite, Chepultepec Dolomite or Limestone, Longview or Nittany Dolomite, Kingsport Formation, Mascot Dolomite, Newala Formation, Conococheague Formation, Jonesboro Limestone. NW Georgia, W North Carolina, E Tennessee, and SW Virginia.
IN Index ages as applicable.
BT Paleozoic
SA Conococheague Formation
SA Georgia
SA Kingsport Formation
SA Lower Ordovician
SA Mascot Dolomite
SA North Carolina
SA Tennessee
SA Upper Cambrian
SA Virginia

Knysna
No longer a valid term for GeoRef. As of 1993 see Knysna South Africa.

Knysna South Africa (1993)
Town in S Cape Province, S South Africa. Before 1993 also search Knysna AND South Africa.
BT Cape Province South Africa
BT South Africa
BT Southern Africa
BT Africa

Kobdo (1978)
River in W Mongolia. Use Kobdo Mongolia for the town.
BT Mongolia
BT Far East
BT Asia

Kobdo Mongolia (1993)
Town in W Mongolia. Before 1993 also search Kobdo AND Mongolia.
UF Hobdo
UF Khobdo
BT Mongolia
BT Far East
BT Asia

kobellite (1978)
BT sulfantimonites
BT sulfosalts

Kobenhavn Denmark
use Copenhagen Denmark

Kobiwako Group (1989)
Includes Iga Formation (Pliocene) and Katata Formation (Pleistocene).
IN Index ages as applicable.
BT Cenozoic
SA Honshu
SA Japan
SA Pleistocene
SA Pliocene

Kobystan (1978)

Region near Baku in E Azerbaidzhan. This term has multiple hierarchies.
BT1 Azerbaidzhan
BT1 Europe
BT2 Azerbaidzhan
BT2 Commonwealth of Independent States

Kocaeli
No longer a valid term for GeoRef. As of 1993 see Kocaeli Turkey.

Kocaeli Turkey (1993)
Province in NW Anatolia. Before 1993 also search Kocaeli AND Turkey.
BT Turkey
BT Middle East
BT Asia
SA Anatolia

Kochi
No longer a valid term for GeoRef. See Kochi Japan.

Kochi Japan (1993)
City and prefecture in S Shikoku, S Japan.
BT Shikoku
BT Japan
BT Far East
BT Asia
NT Chichibu Belt
SA Tosa Bay
SA Yoshino River

Kochkar
No longer a valid term for GeoRef. As of 1993, see Kockhar Russian Federation.

Kochkar Russian Federation (1993)
Village in central Chelyabinsk Oblast in W Siberia.
BT Chelyabinsk Russian Federation
BT Russian Federation
BT Commonwealth of Independent States

Kodaikanal
No longer a valid term for GeoRef. See Kodaikanal India.

Kodaikanal India (1993)
City in NE Tamil Nadu, SE India.
BT Tamil Nadu India
BT India
BT Indian Peninsula
BT Asia

Kodar Range (1978)
NE of Lake Baikal in N Chita Oblast. Also search Kodar. This term has multiple hierarchies.
BT1 Russian Federation
BT1 Commonwealth of Independent States
BT2 Asia
SA Chita Russian Federation

Kodiak Formation (1989)
Kodiak Island, Alaska.
BT Upper Cretaceous
BT Cretaceous
BT Mesozoic
SA Alaska
SA Kodiak Island

Kodiak Island (1978)
In Gulf of Alaska SE of Alaska Peninsula.
BT Southwestern Alaska
BT Alaska
BT United States
SA Coast Ranges
SA Kodiak Formation

Kodor River basin (1978)

In Abkhazia in W Georgian Republic.
BT Abkhazia Georgian Republic
BT Georgian Republic
BT Europe

Koenigsberger ratio (1981)
SA magnetic field
SA magnetic susceptibility
SA paleomagnetism
SA remanent magnetization

Kohistan (1985)
Region N of Kabul, NE Afghanistan. May also occur in neighboring Pakistan and India.
IN Index country as applicable.
BT Indian Peninsula
BT Asia
SA Afghanistan
SA Hindu Kush
SA India
SA Indus-Yarlung Zangbo suture zone
SA Pakistan

Kokchetav
No longer a valid term for GeoRef. As of 1993, see Kokshatav Kazakhstan.

Kokchetav Kazakhstan (1993)
Oblast including city in N Kazakhstan. This term has multiple hierarchies.
BT1 Kazakhstan
BT1 Central Asia
BT1 Asia
BT2 Kazakhstan
BT2 Commonwealth of Independent States
NT Stepnyak Kazakhstan
NT Zerenda Kazakhstan

Kola (1978)
River in Murmansk Oblast on the Kola Peninsula in extreme NW European Russian Federation. After 1993, see Kola Russian Federation for the town. This term has multiple hierarchies.
BT1 Russian Federation
BT1 Commonwealth of Independent States
BT2 Europe
SA Kola Peninsula
SA Murmansk Russian Federation

Kola Peninsula (1978)
Between White Sea and Barents Sea in extreme NW European Russian Federation. This term has multiple hierarchies.
CO N660000N700000
 E0420000E0320000
BT1 Russian Federation
BT1 Commonwealth of Independent States
BT2 Europe
SA Allarechenskiy
SA Baltic Glaciation
SA Baltic ice lake
SA Ilimaussaq
SA Imandra
SA Kola
SA Lapland
SA Lovozero Massif
SA Yoldia Sea

Kola Russian Federation (1993)
Town in Murmansk Oblast on the Kola Peninsula in extreme NW European Russian Federation. Before 1993, also search Kola. This term has multiple hierarchies.
BT1 Murmansk Russian Federation

BT1 Russian Federation
BT1 Commonwealth of Independent States
BT2 Murmansk Russian Federation
BT2 Europe

Kolar
No longer a valid term for GeoRef. See Kolar India.

Kolar Gold Fields
No longer a valid term for GeoRef. use Kolar Gold Fields India

Kolar Gold Fields India (1993)
City in E Karnataka. Center of gold-mining industry. Before 1993, also search Kolar Gold Fields. Before 1978, also search Kolar or Kolar Goldfields.
UF Kolar Gold Fields
UF Kolar Goldfields
BT Karnataka India
BT India
BT Indian Peninsula
BT Asia
SA gold ores

Kolar Goldfields
use Kolar Gold Fields India

Kolar India (1993)
Town and district in E Karnataka, SW India.
BT Karnataka India
BT India
BT Indian Peninsula
BT Asia

Kolar schist belt (1993)
Archaean. S India.
IN Index states as applicable.
BT India
BT Indian Peninsula
BT Asia
SA Dharwar Craton
SA Karnataka India

kolbeckine
use herzenbergite

Kolbeinsey Island (1978)
Off N coast in Norwegian Sea.
BT Iceland
BT Western Europe
BT Europe
SA Norwegian Sea

Koli Nappe (1989)
Located in the Scandinavian Caledonides.
IN Index countries as applicable.
BT Scandinavia
BT Western Europe
BT Europe
SA Caledonides
SA Finland
SA Norway
SA Sweden

Kolima Mountains
use Kolyma Uplift

Kolima River
use Kolyma River

Kolima River basin
use Kolyma River basin

Koln Germany
use Cologne Germany

Kolyma Massif
use Kolyma Uplift

Kolyma River (1978)
Rises in SE Cherskiy Range and flows generally N into the East Siberian Sea. Also search Kolyma. This term has multiple hierarchies.
UF Kolima River

BT1 Russian Federation
BT1 Commonwealth of Independent States
BT2 Asia
SA Magadan Russian Federation
SA Yakutia Russian Federation

Kolyma River basin (1978)
N Khabarovsk Kray and E Yakutia. Also search Kolyma. This term has multiple hierarchies.
UF Kolima River basin
BT1 Russian Federation
BT1 Commonwealth of Independent States
BT2 Asia
SA Yakutia Russian Federation

Kolyma Uplift (1978)
Mountain range N of the Okhotsk Sea in Magadan Oblast and NE Yakutia. This term has multiple hierarchies.
CO N610000N673000 E1650000E1420000
UF Kolima Mountains
UF Kolyma Massif
BT1 Russian Federation
BT1 Commonwealth of Independent States
BT2 Asia
SA Magadan Russian Federation
SA Yakutia Russian Federation

Kom Ombo
No longer a valid term for GeoRef. See Kom Ombo Egypt.

Kom Ombo Egypt (1993)
Village in S Egypt. Before 1993, also search Kom Ombo. Before 1978, also search Kum Umbu.
UF Kum Umbu Egypt
BT Aswan Egypt
BT Egypt
BT North Africa
BT Africa

Koma-ga-take (1993)
Term reintroduced in 1993. Valid to 1980. From 1981 through 1992, also search Koma-ga-take 1 for the volcano on Oshima Peninsula, Hokkaido or Koma-ga-take 2 for the various volcanoes and mountain peaks on Honshu: on the Iwate/Akita or Nagano/Yamanashi prefectural borders, or within Akita, Nagano or Fukushima prefectures. Search in combination with Honshu or Hokkaido as applicable.
IN Index island and prefectures as applicable.
UF Koma-ga-take 1
UF Koma-ga-take 2
BT Japan
BT Far East
BT Asia
SA Akita Japan
SA Fukushima Japan
SA Hokkaido
SA Honshu
SA Iwate Japan
SA Nagano City Japan
SA Oshima Peninsula
SA Yamanashi Japan

Koma-ga-take 1
No longer a valid term for GeoRef.
use Koma-ga-take

Koma-ga-take 2
No longer a valid term for GeoRef.
use Koma-ga-take

Komandor Islands
use Komandorski Islands

Komandorski Islands (1985)
Island group east of Kamchatka Peninsula in SW Bering Sea. Before 1985, also search Commander Islands; Komandor Islands; Komandorskie Islands; Komandorskiy Islands; Komandorskiye Islands. This term has multiple hierarchies.
UF Commander Islands
UF Komandor Islands
UF Komandorskie Islands
UF Komandorskiy Islands
UF Komandorskiye Islands
BT1 Kamchatka Peninsula
BT1 Russian Pacific region
BT1 Russian Federation
BT1 Commonwealth of Independent States
BT2 Kamchatka Peninsula
BT2 Russian Pacific region
BT2 Asia
SA Bering Sea

Komandorskie Islands
use Komandorski Islands

Komandorskiy Islands
use Komandorski Islands

Komandorskiye Islands
use Komandorski Islands

komatiite (1978)
BT ultramafics
BT plutonic rocks
BT igneous rocks
SA metakomatiite

Kombolgie Formation (1989)
BT Proterozoic
BT upper Precambrian
BT Precambrian
SA Australia
SA Northern Territory Australia

Komi
No longer a valid term for GeoRef. As of 1993, see Komi Russian Federation.

Komi Russian Federation (1993)
Former Autonomous Soviet Socialist Republic in NE Russian Federation, west of the Northern Urals. Before 1993, also search Komi. This term has multiple hierarchies.
BT1 Russian Federation
BT1 Commonwealth of Independent States
BT2 Europe
NT Pechora Russian Federation
NT Vorkuta Russian Federation
SA Kozhim
SA Mezen River basin
SA Northern Dvina River
SA Pechora Basin
SA Timan-Pechora region

Kommunar
No longer a valid term for GeoRef. As of 1993, see Kommunar Russian Federation.

Kommunar Russian Federation (1993)
Town in Khakass Autonomous Oblast in SW Krasnoyarsk Kray. Before 1993, also search Kommunar.
IN Index regions as applicable.
BT Russian Federation
BT Commonwealth of Independent States
SA Krasnoyarsk Russian Federation

Komsomolsk
No longer a valid term for GeoRef. As of 1993, see Komsomolsk Russian Federation.

Komsomolsk Russian Federation (1993)
City in S Khabarovsk Kray in Russian Far East. Before 1993, also search Komsomolsk-na-Amure or Komsomolsk-on-Amure.
IN Index regions as applicable.
UF Komsomolsk-na-Amure Russian Federation
BT Russian Federation
BT Commonwealth of Independent States
SA Khabarovsk Russian Federation
SA Russian Far East

Komsomolsk-na-Amure Russian Federation
use Komsomolsk Russian Federation

Kona (1978)
Includes divisions of Hawaii County, that is, North Kona and South Kona, on W and SW side of island of Hawaii. Autoposting of broader terms to this term began in 1989. This term has multiple hierarchies.
UF North Kona
UF South Kona
BT1 Hawaii Island
BT1 Hawaii County Hawaii
BT1 Hawaii
BT1 United States
BT2 Hawaii Island
BT2 Hawaii County Hawaii
BT2 Hawaii
BT2 Polynesia
BT2 Oceania
BT3 Hawaii Island
BT3 Hawaii County Hawaii
BT3 Hawaii
BT3 East Pacific Ocean Islands

Kondapalli
No longer a valid term for GeoRef. See Kondapalli India.

Kondapalli India (1993)
Town in NE Tamil Nadu, SE India.
BT Tamil Nadu India
BT India
BT Indian Peninsula
BT Asia

Kongo River
use Congo River

Kongsberg
No longer a valid term for GeoRef. As of 1993 see Kongsberg Norway.

Kongsberg Norway (1993)
Town in SE Norway. Before 1993 also search Kongsberg AND Norway.
BT Norway
BT Scandinavia
BT Western Europe
BT Europe

Konia
use Konya Turkey

Konieh
use Konya Turkey

Konin
No longer a valid term for GeoRef. As of 1993 see Konin Poland.

Konin Poland (1993)
Town in central Poland. Before 1993 also search Konin AND Poland.
BT Poznan Poland
BT Poland
BT Central Europe
BT Europe

Konrad Mine (1993)
Iron mine near Salzgitter-Lebenstedt. Radioactive waste disposal site.
IN Index region as applicable.
BT Lower Saxony Germany
BT Germany
BT Central Europe
BT Europe
SA iron ores
SA radioactive waste
SA waste disposal
SA waste disposal sites

Konstantinovka
No longer a valid term for GeoRef. As of 1993, see Konstantinovka Ukraine.

Konstantinovka Ukraine (1993)
City in Donetsk Oblast in SE Ukraine. This term has multiple hierarchies.
BT1 Ukraine
BT1 Europe
BT2 Ukraine
BT2 Commonwealth of Independent States

Kontinentales Tiefbohrprogramm der Bundesrepublik Deutschland
use KTB

Konya
No longer a valid term for GeoRef. As of 1993 see Konya Turkey.

Konya Basin (1985)
Central Turkey.
CO N370000N392000 E0350000E0320000
UF Great Konya Basin
BT Turkey
BT Middle East
BT Asia
SA Anatolia
SA Konya Turkey

Konya Turkey (1993)
Province in SW central Anatolia. Also a city. Before 1993 also search Konya AND Turkey.
UF Konia
UF Konieh
BT Turkey
BT Middle East
BT Asia
SA Anatolia
SA Konya Basin

Koobi Fora Formation (1989)
Pliocene to Pleistocene.
IN Index ages as applicable.
BT Cenozoic
SA Ethiopia
SA Kenya
SA Pleistocene
SA Pliocene
SA Tanzania

Koochiching County Minnesota (1993)
NE Minnesota. Before 1989, also search Koochiching County AND Minnesota.
CO N475100N484100 W0930300W0942400
BT Minnesota
BT United States
SA Rainy Lake

Koolau Range (1978)
Extends along E side of Oahu Island, Hawaii. Autoposting of broader terms to this term began in 1989. This term has multiple hierarchies.
BT1 Oahu
BT1 Honolulu County Hawaii
BT1 Hawaii
BT1 United States
BT2 Oahu
BT2 Honolulu County Hawaii
BT2 Hawaii
BT2 Polynesia
BT2 Oceania
BT3 Oahu
BT3 Honolulu County Hawaii
BT3 Hawaii
BT3 East Pacific Ocean Islands

Kootenai County
Valid through 1988. Search in combination with state term. After 1988, use specific county-state term.

Kootenai County Idaho (1989)
N Idaho. Before 1989, also search Kootenai County AND Idaho.
CO N472200N480000
W1162000W1170500
BT Idaho
BT United States
SA Coeur d'Alene mining district
SA Idaho Panhandle

Kootenai Formation
use Kootenay Formation

Kootenay Arc (1985)
Folded and metamorphosed area. SE British Columbia, NE Washington and NW Montana.
IN Index states and British Columbia as applicable.
CO N480000N520000
W1140000W1170000
BT North America
SA British Columbia
SA Canada
SA Montana
SA North American Cordillera
SA United States
SA Washington

Kootenay Formation (1978)
SW Alberta, SE British Columbia and Montana.
UF Kootenai Formation
BT Lower Cretaceous
BT Cretaceous
BT Mesozoic
SA Alberta
SA British Columbia
SA Mist Mountain Formation
SA Montana

Kootenay Lake (1978)
SE British Columbia.
UF Kutenai Lake
BT British Columbia
BT Western Canada
BT Canada

Kopaonik (1978)
Central Serbia.
UF Kopaonik Mountains
BT Serbia
BT Yugoslavia
BT Southern Europe
BT Europe

Kopaonik Mountains
use Kopaonik

Kope Formation (1989)
BT Upper Ordovician
BT Ordovician
BT Paleozoic

SA Indiana
SA Kentucky
SA Ohio

Kopet Dag (Range)
use Kopet-Dag Range

Kopet Dagh (Range)
use Kopet-Dag Range

Kopet-Dag Range (1978)
In SW Turkmenia. Autoposting of Central Asia to this term began in 1989. Before 1981, term included extension in Iran. This term has multiple hierarchies.
CO N350000N393000
E0630000E0523000
UF Kopet Dag (Range)
UF Kopet Dagh (Range)
BT1 Turkmenia
BT1 Asia
BT2 Turkmenia
BT2 Commonwealth of Independent States
SA Iran

Koralpe Range (1978)
Small range of Noric Alps in S Austria. Also search Koralpe. This term has multiple hierarchies.
BT1 Eastern Alps
BT1 Alps
BT1 Europe
BT2 Austria
BT2 Central Europe
BT2 Europe
SA Carinthia Austria
SA Styria Austria

Koraput
No longer a valid term for GeoRef. See Koraput India.

Koraput India (1993)
Village and district in SW Orissa, E India.
BT Orissa India
BT India
BT Indian Peninsula
BT Asia

Korat Plateau
use Khorat Plateau

Kordofan
No longer a valid term for GeoRef. As of 1993 see Kordofan Sudan.

Kordofan Sudan (1993)
Province in central Sudan. Before 1993 also search Kordofan AND Sudan.
BT Sudan
BT East Africa
BT Africa

Korea (1978)
A peninsula on E coast of Asia, since 1948 partitioned into two republics.
IN Use North Korea and/or South Korea.
CO N330000N430000
E1300000E1241500
BT Far East
BT Asia
NT North Korea
NT South Korea
SA Sino-Korean Platform

Korenevskaya Formation (1978)
Dnieper-Donets region.
IN Index age as applicable.
SA Dnieper-Donets Basin
SA Permian
SA Russian Federation
SA Triassic
SA Ukraine

kornerupine (1978)

Autoposting of nesosilicates began in 1985.
UF prismatine
BT nesosilicates
BT orthosilicates
BT silicates

Korosten
No longer a valid term for GeoRef. As of 1993, see Korosten Ukraine.

Korosten Ukraine (1993)
City in Zhitomir Oblast in NW central Ukraine. This term has multiple hierarchies.
BT1 Zhitomir Ukraine
BT1 Ukraine
BT1 Europe
BT2 Zhitomir Ukraine
BT2 Ukraine
BT2 Commonwealth of Independent States

Korsun
No longer a valid term for GeoRef. As of 1993, see Korsun-Shevchenkovski Ukraine.

Korsun-Shevchenkovski Ukraine (1993)
City in Cherkassy Oblast in central Ukraine. Before 1993, also search Korsun or Korsun-Shevchenkovski. This term has multiple hierarchies.
BT1 Cherkassy Ukraine
BT1 Ukraine
BT1 Europe
BT2 Cherkassy Ukraine
BT2 Ukraine
BT2 Commonwealth of Independent States

Koryak Range (1978)
Extends from neck of Kamchatka Peninsula in Khabarovsk Kray into the Chukchi National Okrug. Also search Koryak. This term has multiple hierarchies.
CO N600000N650000
E1800000E1650000
BT1 Russian Pacific region
BT1 Russian Federation
BT1 Commonwealth of Independent States
BT2 Russian Pacific region
BT2 Asia

Korytnica (1978)
Health resort on E slope of the Tatra Mountains.
BT Slovakia
BT Czechoslovakia
BT Central Europe
BT Europe

Kos (1978)
One of the islands of the Dodecanese off SW Turkey. Also search Coo or Cos.
IN Index Dodecanese as applicable.
SA Dodecanese

Kosaka Mine (1978)
Metal ores. Near town of Kosaka in Akita Prefecture, N Honshu.
BT Akita Japan
BT Honshu
BT Japan
BT Far East
BT Asia
SA metal ores
SA mines

Kosi Basin (1978)
River basin. Also search Kosi River.

IN Index Bihar India and/or Nepal as applicable.
SA Bihar India
SA Indian Peninsula
SA Nepal

Kosov
No longer a valid term for GeoRef. As of 1993, see Kosov Ukraine.

Kosov Ukraine (1993)
Town in Ivano-Frankovsk Oblast in SE Ukraine. This term has multiple hierarchies.
IN Index regions as applicable.
BT1 Ukraine
BT1 Europe
BT2 Ukraine
BT2 Commonwealth of Independent States
SA Ivano-Frankovsk Ukraine

Kosovo-Metohija (1981)
Autonomous region in S part of country.
BT Serbia
BT Yugoslavia
BT Southern Europe
BT Europe
SA Metohija

Kosseir
No longer a valid term for GeoRef. See Kosseir Egypt.

Kosseir Egypt (1993)
E Egypt, port on Red Sea. Before 1993, also search Kossier. Before 1978, also search Al-Qusayr, Al-Quseir, or Quseir.
UF Al-Qusayr Egypt
UF Al-Quseir Egypt
UF Quseir Egypt
BT Egypt
BT North Africa
BT Africa

Kostolac (1978)
Village and arm of the Danube River in E Serbia.
UF Kostolats
BT Serbia
BT Yugoslavia
BT Southern Europe
BT Europe

Kostolats
use Kostolac

Kostroma
No longer a valid term for GeoRef. As of 1993, see Kostroma Russian Federation.

Kostroma Russian Federation (1993)
Oblast including the city on left Bank of the Volga River. Before 1993, also search Kostroma. This term has multiple hierarchies.
BT1 Russian Federation
BT1 Commonwealth of Independent States
BT2 Europe

Koszalin
No longer a valid term for GeoRef. As of 1993 see Koszalin Poland.

Koszalin Poland (1993)
Province and city in NW Poland. Before 1993 also search Koszalin AND Poland.
BT Poland
BT Central Europe
BT Europe

Kota
No longer a valid term for GeoRef. See Kota India.

Kota India (1993)
Town in E central Madhya Pradesh, central India.
BT Madhya Pradesh India
BT India
BT Indian Peninsula
BT Asia

Koto (1978)
River in E and S Central African Republic which flows into the Ubangi River.
UF Kotto
BT Central African Republic
BT Central Africa
BT Africa

kotoite (1978)
BT borates

Kotto
use Koto

Kotui (1978)
River in Evenk and Taymyr National okrugs in N central Siberia. Along with the Kheta River, it forms Khatanga River. This term has multiple hierarchies.
UF Kotuy
UF Kotuy River
BT1 Krasnoyarsk Russian Federation
BT1 Russian Federation
BT1 Commonwealth of Independent States
BT2 Krasnoyarsk Russian Federation
BT2 Asia
SA Khatanga Basin
SA Khatanga River
SA Taymyr Dolgan-Nenets Russian Federation

Kotuy
use Kotui

Kotuy River
use Kotui

Kouchibouguac Bay (1978)
On N North Umberland Strait in E New Brunswick.
BT New Brunswick
BT Maritime Provinces
BT Eastern Canada
BT Canada

Koyna (1978)
River in W Maharashtra, W central India.
SA Maharashtra India

Koyukuk Basin
use Yukon-Koyukuk Basin

Kozhim (1978)
River W of the Northern Urals in former Komi A.S.S.R. Before 1993, also used for village. After 1993, use Kozhim Russian Federation for the village. This term has multiple hierarchies.
IN Index Komi Russian Federation as applicable.
BT1 Russian Federation
BT1 Commonwealth of Independent States
BT2 Europe
SA Komi Russian Federation

Kr
use krypton

Kr-81 (1989)
Autoposting of broader terms began in 1989. This term has multiple hierarchies.
BT1 radioactive isotopes
BT1 isotopes
BT2 krypton

BT2 noble gases

Kr-84 (1989)
Autoposting of broader terms began in 1989. This term has multiple hierarchies.
BT1 stable isotopes
BT1 isotopes
BT2 krypton
BT2 noble gases

Kr-85 (1989)
Autoposting of broader terms began in 1989. This term has multiple hierarchies.
BT1 radioactive isotopes
BT1 isotopes
BT2 krypton
BT2 noble gases

Kr/Kr (1989)
Isotopic ratio used in age determination.
SA absolute age
SA isotope ratios
SA isotopes
SA krypton

Krafla (1985)
Mountain, NE Iceland.
CO N654200N654400 W0164300W0164500
BT Iceland
BT Western Europe
BT Europe

Kraishte
use Krajiste

Krajiste (1978)
Highland.
IN Index countries as applicable.
UF Kraishte
UF Krayishte
BT Southern Europe
BT Europe
SA Bulgaria
SA Yugoslavia

Krakatao
use Krakatoa

Krakatau
use Krakatoa

Krakatoa (1978)
Island volcano in center of Sunda Strait between Sumatra and Java.
UF Krakatao
UF Krakatau
BT Indonesia
BT Far East
BT Asia

Krakow
No longer a valid term for GeoRef as of 1989.
use Cracow Poland

Krakow-Czestochowa Jura
As of 1989, no longer a valid term for GeoRef.
use Cracow-Czestochowa Jura

Krappfeld (1978)
Region in S Austria.
BT Carinthia Austria
BT Austria
BT Central Europe
BT Europe

Krasnodar
No longer a valid term for GeoRef. As of 1993, see Krasnodar City Russian Federation for the city or Krasnodar Russian Federation for the territory.

Krasnodar City Russian Federation (1993)
City in Krasnodar Kray in the Northern Caucasus. Before 1993, also search Krasnodar. This term has multiple hierarchies.
BT1 Krasnodar Russian Federation
BT1 Russian Federation
BT1 Commonwealth of Independent States
BT2 Krasnodar Russian Federation
BT2 Europe

Krasnodar Russian Federation (1993)
Kray in the Northern Caucasus. Before 1993, also search Krasnodar. This term has multiple hierarchies.
BT1 Russian Federation
BT1 Commonwealth of Independent States
BT2 Europe
NT Anapa Russian Federation
NT Krasnodar City Russian Federation
NT Laba Basin
NT Maikop Russian Federation
NT Sochi Russian Federation
NT Taman Peninsula
NT Tuapse Russian Federation
SA Stony Tunguska River

Krasnoyarsk
No longer a valid term for GeoRef. As of 1993, see Krasnoyarsk Russian Federation.

Krasnoyarsk City Russian Federation (1993)
City in Krasnoyarsk Kray in S central Siberia. This term has multiple hierarchies.
BT1 Krasnoyarsk Russian Federation
BT1 Russian Federation
BT1 Commonwealth of Independent States
BT2 Krasnoyarsk Russian Federation
BT2 Asia

Krasnoyarsk Russian Federation (1993)
Kray in S central Siberia. Before 1993, also search Krasnoyarsk. This term has multiple hierarchies.
BT1 Russian Federation
BT1 Commonwealth of Independent States
BT2 Asia
NT Igarka Russian Federation
NT Kansk Russian Federation
NT Kheta River
NT Kotui
NT Krasnoyarsk City Russian Federation
NT Kureyka River region
NT Minusinsk Russian Federation
NT Taymyr Dolgan-Nenets Russian Federation
NT Turukhansk Russian Federation
NT Yessey Russian Federation
SA Chadobets Uplift
SA Kamenka River
SA Kansk-Achinsk Basin
SA Kommunar Russian Federation
SA Minusinsk Basin
SA Turukhan

Krayishte
use Krajiste

Krebs Group (1985)

Central and NE Oklahoma, SE Kansas and W Missouri.
BT Desmoinesian
BT Middle Pennsylvanian
BT Pennsylvanian
BT Carboniferous
BT Paleozoic
SA Boggy Shale
SA Hartshorne Sandstone
SA Kansas
SA Missouri
SA Oklahoma
SA Savanna Formation

KREEP (1978)
Acronym for a basaltic lunar rock type first found in Apollo 12 fines and breccias and characterized by unusually high contents of potassium (K), rare-earth elements (REE), phosphorus (P), and other trace elements in comparison to other lunar rock types.
UF nonmare basalt
SA Moon
SA phosphorus
SA potassium
SA rare earths

Krefeld
No longer a valid term for GeoRef. See Krefeld Germany.

Krefeld Germany (1993)
City in W North Rhine-Westphalia, W Germany.
BT North Rhine-Westphalia Germany
BT Germany
BT Central Europe
BT Europe

Kremasta
No longer a valid term for GeoRef. As of 1993, see Kremasta Greece.

Kremasta Greece (1993)
Village in NW Greece. Before 1993 also search Kremasta AND Greece.
BT Sterea Ellas
BT Greece
BT Southern Europe
BT Europe

Kremenchug
No longer a valid term for GeoRef. As of 1993, see Kremenchug Ukraine.

Kremenchug Ukraine (1993)
City in Poltava Oblast in NE central Ukraine. This term has multiple hierarchies.
BT1 Poltava Ukraine
BT1 Ukraine
BT1 Europe
BT2 Poltava Ukraine
BT2 Ukraine
BT2 Commonwealth of Independent States

Kremnica (1978)
Town in W central Slovakia.
BT Slovakia
BT Czechoslovakia
BT Central Europe
BT Europe

Kremnica Mountains (1978)
In W central Slovakia.
CO N483100N484800 E0191000E0184700
BT Slovakia
BT Czechoslovakia
BT Central Europe
BT Europe

Krems
  No longer a valid term for GeoRef. See Krems Austria.

Krems an der Donau
  use Krems Austria

Krems Austria (1993)
  City on the Danube River at the mouth of the Krems River.
  UF Krems an der Donau
  BT Lower Austria
  BT Austria
  BT Central Europe
  BT Europe

Kreuth
  No longer a valid term for GeoRef. See Kreuth Germany.

Kreuth Germany (1993)
  Resort in Bavarian Alps in S Bavaria, SE Germany.
  BT Upper Bavaria Germany
  BT Bavaria Germany
  BT Germany
  BT Central Europe
  BT Europe

kriging (1985)
  BT statistical analysis
  SA data processing
  SA geostatistics
  SA mathematical geology
  SA mathematical methods
  SA mathematical models
  SA stochastic processes
  SA variance analysis

Krishna (1978)
  River which flows into the Bay of Bengal.
  IN Index states as applicable.
  UF Kistna River
  BT India
  BT Indian Peninsula
  BT Asia
  SA Andhra Pradesh India
  SA Karnataka India
  SA Krishna-Godavari Basin
  SA Maharashtra India

Krishna-Godavari Basin (1993)
  Basin in Andhra Pradesh, partially offshore in Bay of Bengal as well as the basins of the Krishna and Godavari rivers in E India. Before 1993, also search Godavari Basin.
  IN Index land or ocean regions as applicable.
  CO N150000N173000
      E0840000E0783000
  UF Godavari Basin
  SA Andhra Pradesh India
  SA Bay of Bengal
  SA Godavari Valley
  SA Krishna

Kristianstad
  No longer a valid term for GeoRef. As of 1993 see Kristianstad Sweden.

Kristianstad Sweden (1993)
  County and city in S Sweden. Before 1993 also search Kristianstad Sweden.
  BT Sweden
  BT Scandinavia
  BT Western Europe
  BT Europe
  SA Skane

Kristiansund
  No longer a valid term for GeoRef. As of 1993 see Kristiansund Norway.

Kristiansund Norway (1993)
  Seaport city, W Norway. Before 1993 also search Kristiansund AND Norway.
  UF Christiansund
  BT Norway
  BT Scandinavia
  BT Western Europe
  BT Europe

Krivoi Rog Basin
  use Krivoy Rog Basin

Krivoi Rog region
  use Krivoy Rog Basin

Krivoi Rog Series
  use Krivoy Rog Series

Krivoi Rog Ukraine
  use Krivoy Rog Ukraine

Krivoy Rog
  No longer a valid term for GeoRef. As of 1993, see Krivoy Rog Ukraine.

Krivoy Rog Basin (1978)
  Basin of the Ingulets River in the Krivoy Rog region. An industrial basin in W Dnepropetrovsk Oblast in S central Ukraine. This term has multiple hierarchies.
  CO N473000N480000
      E0353000E0323000
  UF Krivoi Rog Basin
  UF Krivoi Rog region
  BT1 Ukraine
  BT1 Europe
  BT2 Ukraine
  BT2 Commonwealth of Independent States
  SA Dnepropetrovsk Ukraine
  SA Ingulets River

Krivoy Rog Series (1978)
  UF Krivoi Rog Series
  BT Archean
  BT Precambrian
  SA Ukraine

Krivoy Rog Ukraine (1993)
  City in W Dnepropetrovsk Oblast in S central Ukraine. This term has multiple hierarchies.
  UF Krivoi Rog Ukraine
  BT1 Dnepropetrovsk Ukraine
  BT1 Ukraine
  BT1 Europe
  BT2 Dnepropetrovsk Ukraine
  BT2 Ukraine
  BT2 Commonwealth of Independent States

Krizna Nappe (1985)
  W Czechoslovakia and SW Poland.
  BT Europe
  SA Czechoslovakia
  SA Poland
  SA Slovakia
  SA Slovakian Carpathians
  SA Tatra Mountains
  SA Western Carpathians

Krkonose Mountains
  use Karkonosze Mountains

krokidolite
  use crocidolite

Krol Formation (1993)
  Has been variously considered as Permian and Upper Carboniferous; more recent literature has assigned it to the Proterozoic? and Cambrian. Himalayas of India.
  IN Index ages as applicable.
  SA Cambrian
  SA Carboniferous
  SA Himalayas
  SA India

  SA Permian
  SA Proterozoic
  SA Uttar Pradesh India

Krosno Beds (1978)
  Lattorfian or Rupelian of some authors. In the Polish Carpathians.
  BT Oligocene
  BT Paleogene
  BT Tertiary
  BT Cenozoic
  SA Lattorfian
  SA Poland
  SA Rupelian
  SA Tongrian

Krusne Hory
  use Erzgebirge

Krusnehory Mountains
  use Erzgebirge

Krusny Hory Mountains
  use Erzgebirge

Krymka Meteorite (1989)
  Impact in the Nicholayev region, Crimea, Ukraine.
  BT LL chondrites
  BT chondrites
  BT stony meteorites
  BT meteorites
  SA Crimea Ukraine

krypton (1978)
  Autoposting of noble gases to this term began in 1989.
  UF Kr
  BT noble gases
  NT Kr-81
  NT Kr-84
  NT Kr-85
  SA isotopes
  SA Kr/Kr

Krzeszowice
  No longer a valid term for GeoRef. As of 1993 see Krzeszowice Poland.

Krzeszowice Poland (1993)
  Town in S Poland. Before 1993 also search Krzeszowice AND Poland.
  BT Cracow Poland
  BT Poland
  BT Central Europe
  BT Europe

Ksiaz
  No longer a valid term for GeoRef. As of 1993 see Ksiaz Poland.

Ksiaz Poland (1993)
  Town in W central Poland. Before 1993 also search Ksiaz AND Poland.
  UF Ksiaz Wielkopolski
  BT Poznan Poland
  BT Poland
  BT Central Europe
  BT Europe

Ksiaz Wielkopolski
  use Ksiaz Poland

KTB (1993)
  Acronym. Hauptbohrung and Vorbohrung boreholes are located near Windischeschenbach in Bavarian Upper Palatinate at edge of Bohemian Massif. Before 1993, also search German Continental Deep Drilling Program, Kontinentales Tiefbohrprogramm der Bundesrepublik Deutschland, or Continental Deep Drilling Program AND Germany.
  UF German Continental Deep Drilling Program

  UF Kontinentales Tiefbohrprogramm der Bundesrepublik Deutschland
  SA Bavaria Germany
  SA Bohemian Massif
  SA boreholes
  SA deep drilling
  SA Germany
  SA programs

Kuala Lumpur
  No longer a valid term for GeoRef. As of 1993, see Kuala Lumpur Malaysia.

Kuala Lumpur Malaysia (1993)
  Capital city and federal territory in SE West Malaysia on the Malay Peninsula. This term has multiple hierarchies.
  BT1 Selangor Malaysia
  BT1 West Malaysia
  BT1 Malay Peninsula
  BT1 Far East
  BT1 Asia
  BT2 Selangor Malaysia
  BT2 West Malaysia
  BT2 Malaysia
  BT2 Far East
  BT2 Asia

Kuban (1978)
  Steppe region in the W Northern Caucasus. Before the Revolution, also an oblast. This term has multiple hierarchies.
  BT1 Russian Federation
  BT1 Commonwealth of Independent States
  BT2 Europe
  SA Caucasus Foreland

Kuban River (1978)
  Rises in Georgian Republic in the Caucasus and flows N and NW into the Azov Sea. Also search Kuban.
  IN Index former Soviet republics as applicable.
  BT Asia
  SA Georgian Republic
  SA Russian Federation

Kuban Valley (1978)
  River valley. Also search Kuban.
  IN Index former Soviet republics as applicable.
  BT Asia
  SA Georgian Republic
  SA Russian Federation

Kuchi-no-erabu
  use Kuchinoerabu-Jima

Kuchinoerabu-Jima (1978)
  Small island in Osumi Islands off S coast of Kyushu.
  UF Kuchi-no-erabu
  BT Kyushu
  BT Japan
  BT Far East
  BT Asia

Kufara
  use Kufra Basin

Kufra Basin (1978)
  Group of 5 oases in central Libyan Desert.
  UF Al-Kufrah
  UF Kufara
  BT Libya
  BT North Africa
  BT Africa

Kugitang (-Tau)
  use Kugitang-Tau

Kugitang-Tau (1978)

Mountain range in E Chardzou Oblast in E Turkmenia. S spur of Baisun-Tau. This term has multiple hierarchies.
UF Kugitang (-Tau)
BT1 Chardzhou Turkmenia
BT1 Turkmenia
BT1 Asia
BT2 Chardzhou Turkmenia
BT2 Turkmenia
BT2 Commonwealth of Independent States

Kuhn Island (1978)
Off NE coast.
BT Greenland
BT Arctic region

Kuibyshev
No longer a valid term for GeoRef. As of 1993, see Samara Russian Federation.

Kuibyshev Russian Federation
use Samara Russian Federation

Kujawy (1978)
Region in N central Poland.
BT Poland
BT Central Europe
BT Europe

Kuju (1978)
Mountain in Oita Prefecture in central Kyushu, S Japan.
BT Oita Japan
BT Kyushu
BT Japan
BT Far East
BT Asia

kukersite (1993)
Found in Ordovician of Estonia. This term has multiple hierarchies.
BT1 oil shale
BT1 organic residues
BT2 oil shale
BT2 sedimentary rocks

Kul'dzhuktau (1978)
Mountains in central Uzbekistan. This term has multiple hierarchies.
UF Kuldzhuktau
BT1 Uzbekistan
BT1 Asia
BT2 Uzbekistan
BT2 Commonwealth of Independent States

Kula Plate (1989)
Small plate in N Pacific involved in the development of the Coast Ranges.
BT Pacific Ocean
SA Coast Ranges
SA plate tectonics
SA plates

Kular Range (1978)
N central Yakutia. Also search Kular. This term has multiple hierarchies.
BT1 Yakutia Russian Federation
BT1 Russian Federation
BT1 Commonwealth of Independent States
BT2 Yakutia Russian Federation
BT2 Asia

Kuldzhuktau
use Kul'dzhuktau

Kulunda Steppe (1978)
Between Ob River in Novosibirsk Oblast and Altai Kray on the E, and E Pavlodar Oblast in Kazakhstan on the W. This term has multiple hierarchies.

IN Index former Soviet republics as applicable.
BT1 Commonwealth of Independent States
BT2 Asia
SA Kazakhstan
SA Russian Federation

Kulyab
No longer a valid term for GeoRef. As of 1993, see Kulyab Tadzhikistan.

Kulyab Tadzhikistan (1993)
Town in SW Tadzhikistan. This term has multiple hierarchies.
BT1 Tadzhikistan
BT1 Asia
BT2 Tadzhikistan
BT2 Commonwealth of Independent States

Kum Umbu Egypt
use Kom Ombo Egypt

Kuma Basin (1978)
River basin in Northern Caucasus. Also search Kuma; Kuma River. This term has multiple hierarchies.
BT1 Russian Federation
BT1 Commonwealth of Independent States
BT2 Europe

Kumamoto
No longer a valid term for GeoRef. See Kumamoto Japan.

Kumamoto Japan (1993)
City and prefecture in W central Kyushu, S Japan.
BT Kyushu
BT Japan
BT Far East
BT Asia
NT Aso
NT Aso Caldera
SA Oguni Japan

Kumaon Himalaya
use Kumaun Himalayas

Kumaon Himalayas
use Kumaun Himalayas

Kumaun Himalayas (1978)
W central subdivision of the Himalayas.
IN Index Nepal, Xizang China, and/or Indian states as applicable.
UF Kumaon Himalaya
UF Kumaon Himalayas
UF Kumuan Himalaya
BT Himalayas
BT Asia
NT Simla Hills
SA Himachal Pradesh India
SA Nepal
SA Uttar Pradesh India
SA Xizang China

Kumuan Himalaya
use Kumaun Himalayas

Kunashir Island (1978)
Second largest of the Kuril Islands which are administratively part of Sakhalin Oblast. Also search Kunashir. This term has multiple hierarchies.
BT1 Kuril Islands
BT1 Sakhalin Russian Federation
BT1 Russian Federation
BT1 Commonwealth of Independent States
BT2 Kuril Islands
BT2 Sakhalin Russian Federation
BT2 Asia
BT3 Kuril Islands
BT3 Russian Pacific region

BT3 Russian Federation
BT3 Commonwealth of Independent States
BT4 Kuril Islands
BT4 Russian Pacific region
BT4 Asia
NT Goryachiy Plyazh
NT Mendeleyev Volcano

Kunavaram Series (1978)
SA India

Kundelungu Plateau (1978)
SE Zaire.
BT Zaire
BT Central Africa
BT Africa

Kunga Group (1993)
Upper Triassic and Lower Jurassic. Queen Charlotte Islands, British Columbia.
IN Index ages as applicable.
BT Mesozoic
SA British Columbia
SA Lower Jurassic
SA Queen Charlotte Islands
SA Upper Triassic

Kungurian (1978)
Europe. Above Artinskian, below Kazanian.
BT Permian
BT Paleozoic
SA Lower Permian
SA Middle Permian

Kunlun Mountains (1989)
Xizang and Qinghai, SW China. One of the great mountain systems of central Asia, located between the Himalayas and Tien Shan.
IN Index regions as applicable.
BT China
BT Far East
BT Asia
SA Qinghai China
SA Qinghai-Xizang Plateau
SA Xizang China

Kunming
No longer a valid term for GeoRef. See Kunming China.

Kunming China (1993)
City in Yunnan, SW China. Also search K'un-ming.
CO N250400N250400 E1024100E1024100
UF K'un-ming China
BT Yunnan China
BT China
BT Far East
BT Asia

Kuno (1978)
Danish island in the Faeroe Islands. In 1981, broader term changed from Atlantic Ocean to Faeroe Islands.
BT Faeroe Islands
BT Atlantic Ocean Islands

Kunyang Group (1989)
SW China. Includes Etouchang Formation, Luoxue Formation.
BT Proterozoic
BT upper Precambrian
BT Precambrian
SA China
SA Yunnan China

kunzite (1978)
Transparent gem variety of spodumene.
BT clinopyroxene
BT pyroxene group
BT chain silicates
BT silicates

SA spodumene

Kupferschiefer (1981)
Lithofacies in Europe. A copper-bearing shale/schist.
SA Permian
SA Zechstein

kupferschiefer type
use kupferschiefer-type

kupferschiefer-type (1989)
UF kupferschiefer type
SA copper ores
SA mineral deposits, genesis

Kura Depression
use Kura Lowland

Kura Lowland (1978)
Extensive plain along SW shores of Caspian Sea. Also search Kura.
IN Index Azerbaidzhan and/or Iran as applicable.
UF Kura Depression
SA Azerbaidzhan
SA Iran
SA Mugan Steppe

Kura River (1978)
Flows into Caspian Sea S of Baku. Also search Kura.
IN Index Turkey and former Soviet republics as applicable.
BT Asia
SA Azerbaidzhan
SA Georgian Republic
SA Turkey

Kurama Range (1978)
Branch of the Tien Shan. Also search Kurama. This term has multiple hierarchies.
IN Index former Soviet republics as applicable.
UF Kuraminskiy Range
BT1 Commonwealth of Independent States
BT2 Asia
SA Tadzhikistan
SA Tien Shan
SA Uzbekistan

Kuraminskiy Range
use Kurama Range

kurchatovite (1978)
BT borates

Kureika River (region)
use Kureyka River region

Kureyka River region (1978)
N Krasnoyarsk Kray in NW central Siberia. Also search Kureyka. This term has multiple hierarchies.
UF Kureika River (region)
BT1 Krasnoyarsk Russian Federation
BT1 Russian Federation
BT1 Commonwealth of Independent States
BT2 Krasnoyarsk Russian Federation
BT2 Asia

Kurgan
No longer a valid term for GeoRef. As of 1993, see Kurgan Russian Federation.

Kurgan Russian Federation (1993)
Oblast and city in Russian Federation E of the Southern Urals. Before 1993, also search Kurgan.
SA Ural region

Kuril Islands (1978)

Group of 56 islands extending from Kamchatka Peninsula to Hokkaido. Administratively part of Sakhalin Oblast. This term has multiple hierarchies.
CO N433000N510000
E1570000E1450000
UF Kurile Islands
UF Kurils
BT1 Sakhalin Russian Federation
BT1 Russian Federation
BT1 Commonwealth of Independent States
BT2 Sakhalin Russian Federation
BT2 Asia
BT3 Russian Pacific region
BT3 Russian Federation
BT3 Commonwealth of Independent States
BT4 Russian Pacific region
BT4 Asia
NT Iturup Island
NT Kunashir Island
NT Paramushir
SA Kuril Trench
SA Okhotsk region
SA Russian Far East
SA Sakhalin
SA Urup

Kuril Trench (1978)
E of the Kuril Islands and along E coast of SE Kamchatka Peninsula. Also search Kuril-Kamchatka.
UF Kuril-Kamchatka Trench
BT Pacific Ocean
SA Kuril Islands

Kuril-Kamchatka Trench
use Kuril Trench

Kurile Islands
use Kuril Islands

Kurils
use Kuril Islands

Kurland Spit
use Courland Spit

kurnakovite (1978)
BT borates

Kurnool
No longer a valid term for GeoRef. See Kurnool India.

Kurnool India (1993)
Town in W Andhra Pradesh, central India. Before 1993, also search Kurnool. Before 1978, also search Karnul.
UF Karnul India
BT Andhra Pradesh India
BT India
BT Indian Peninsula
BT Asia

Kurnool System (1978)
Included in the Purana Group. Subdivided into Kundair Stage, Paniam Stage, Jammalamadugu Stage, Banganapalli Stage. Named after town of Kurnool. Unconformably overlies the Cuddapah System.
IN Index ages as applicable.
UF Kurnul System
SA Andhra Pradesh India
SA Cambrian
SA Cuddapah System
SA India
SA Precambrian

Kurnul System
use Kurnool System

Kuro Siwo Current
use Kuroshio

Kuroko
No longer a valid term for GeoRef. Used for town and mine in Hokuroku District, Honshu and for various mines with kuroko-type ores throughout the Green Tuff region. As of 1993, use Kuroko Japan for town and Kuroko Mine for mine of that name. Also index specific mine name and prefecture if applicable. After 1976, usage restricted to geographic place name. Before 1976, term was also used for kuroko-type ore deposits.

kuroko-type (1978)
Type of massive, base metal sulfide ore. Before 1977, also search kuroko.
SA massive deposits
SA metal ores
SA mineral deposits, genesis
SA stratabound deposits

Kurosegawa Zone (1989)
Structural zone in Shikoku, S Japan.
CO N323500N342500
E1344500E1320000
BT Shikoku
BT Japan
BT Far East
BT Asia
SA Chichibu Belt

Kuroshio (1978)
Ocean current flowing from the eastern coast of N Philippines through East China Sea to E Japan. Connected to North Pacific Current (Kuroshio Extension). Autoposting of West Pacific to this term began in 1989. Autoposting of Pacific Ocean to this term began in 1990.
UF Japan Current
UF Kuro Siwo Current
UF Kuroshio Current
BT West Pacific
BT Pacific Ocean
SA East China Sea
SA Japan
SA Japan Sea
SA Leg 86
SA North Pacific
SA Philippine Islands
SA Ryukyu Islands

Kuroshio Current
use Kuroshio

Kursk
No longer a valid term for GeoRef. As of 1993, see Kursk Russian Federation.

Kursk magnetic anomaly (1978)
SW European Russian Federation. Extensive iron-ore region, 90 miles long and 10 miles wide, in central SE Kursk Oblast. Also search Kursk. This term has multiple hierarchies.
BT1 Russian Federation
BT1 Commonwealth of Independent States
BT2 Europe
SA Kursk Russian Federation

Kursk Russian Federation (1993)
Oblast including the city in SW European Russian Federation. This term has multiple hierarchies.
CO N510000N521500
E0380000E0341500
BT1 Russian Federation

BT1 Commonwealth of Independent States
BT2 Europe
SA Kursk magnetic anomaly

Kursk Series (1978)
Named after city in Kursk Oblast in SW Russian Federation.
BT Proterozoic
BT upper Precambrian
BT Precambrian
SA Russian Federation

Kurt
No longer a valid term for GeoRef. As of 1993, see Kurt Turkmenia.

Kurt Turkmenia (1993)
Village S of Aral Sea. This term has multiple hierarchies.
BT1 Turkmenia
BT1 Asia
BT2 Turkmenia
BT2 Commonwealth of Independent States

kurtosis (1978)
BT statistical analysis
SA histograms
SA skewness

Kushiro coal field (1978)
Kushiro sub-prefecture, SE Hokkaido. Also search Kushiro.
BT Hokkaido
BT Japan
BT Far East
BT Asia
SA coal fields

Kushva
No longer a valid term for GeoRef. As of 1993, Kushva Russian Federation.

Kushva Russian Republic (1993)
City in Ural Mountains in W Yekaterinburg Oblast in West Siberia. Before 1993, also search Kushva.
BT Yekaterinburg Russian Federation
BT Russian Federation
BT Commonwealth of Independent States

Kuskokwim Group (1989)
W Alaska.
BT Cretaceous
BT Mesozoic
SA Alaska

Kustanai Kazakhstan
use Kustanay Kazakhstan

Kustanay
No longer a valid term for GeoRef. As of 1993, see Kustanay Kazakhstan.

Kustanay Kazakhstan (1993)
Oblast and city in NW Kazakhstan. Before 1978, also search Kustanai. This term has multiple hierarchies.
UF Kustanai Kazakhstan
BT1 Kazakhstan
BT1 Central Asia
BT1 Asia
BT2 Kazakhstan
BT2 Commonwealth of Independent States

Kutch India
use Cutch India

Kutenai Lake
use Kootenay Lake

Kutna Hora (1978)
Town in central Bohemia, W Czechoslovakia.
BT Bohemia
BT Czech Republic
BT Czechoslovakia
BT Central Europe
BT Europe

kutnahorite (1989)
Isomorphous with dolomite.
UF kutnohorite
BT carbonates
SA dolomite

kutnohorite
use kutnahorite

Kuwait (1978)
CO N283000N301500
E0484500E0463000
BT Arabian Peninsula
BT Asia
SA Middle East
SA Near East

Kuybyshev Russian Federation
use Samara Russian Federation

Kuzbas
use Kuznetsk Basin

Kuznetsk Alatau (1978)
Mountain system along borders of Khakass Autonomous Oblast of Krasnoyarsk Kray, and Kemerovo Oblast in SSW Siberia. An outlier of the Altai Mountains. Autoposting of Russian Republic to this term began in 1989. This term has multiple hierarchies.
CO N520000N561500
E0890000E0823000
BT1 Russian Federation
BT1 Commonwealth of Independent States
BT2 West Siberia
BT2 Commonwealth of Independent States
BT3 West Siberia
BT3 Asia
BT4 Altai Mountains
BT4 Asia
NT Batenev Ridge

Kuznetsk Basin (1978)
Coal basin, 150 miles long and 65 miles wide, in Kemerovo Oblast extending from Tomsk to Novokuznetsk in SSW Siberia. This term has multiple hierarchies.
CO N520000N561500
E0890000E0840000
UF Kuzbas
BT1 Russian Federation
BT1 Commonwealth of Independent States
BT2 Asia
SA coal fields
SA Kemerovo Russian Federation
SA Tomsk Russian Federation

Kwangsi Chuang
use Guangxi China

Kwangtung China
use Guangdong China

Kwanto
use Kanto Plain

Kwanto earthquake 1923
use Kanto earthquake 1923

Kwanto Mountains
use Kanto Mountains

Kwanto Plain
As of 1989, no longer a valid term for GeoRef.
use Kanto Plain

Kwanza Basin
  use Cuanza Basin

Kweichow China
  use Guizhou China

Kweilin
  use Guilin

Kworra River
  use Niger River

Kworra Valley
  use Niger Valley

kyanite (1978)
  Autoposting of nesosilicates began in 1985. As of 1981, use kyanite deposits for kyanite as a commodity.
  BT nesosilicates
  BT orthosilicates
  BT silicates
  SA andalusite
  SA heavy minerals
  SA kyanite deposits
  SA sillimanite

kyanite deposits (1981)
  Before 1981, search kyanite AND deposits.
  SA ceramic materials
  SA heavy mineral deposits
  SA kyanite

Kyffhauser Range (1978)
  In central Germany, Along the border of Thuringia and Saxony-Anhalt. Also search Kyffhauser.
  BT Germany
  BT Central Europe
  BT Europe
  SA Saxony-Anhalt Germany
  SA Thuringia Germany

Kyongsang Basin (1978)
  In North Kyongsang and South Kyongsang provinces in SE South Korea.
  BT South Korea
  DT Korea
  BT Far East
  BT Asia

Kyoto
  No longer a valid term for GeoRef. See Kyoto Japan.

Kyoto Japan (1993)
  Prefecture including the city in central Honshu. Before 1993, also search Kyoto AND Japan.
  BT Honshu
  BT Japan
  BT Far East
  BT Asia
  NT Amagase Japan
  NT Maizuru Japan
  SA Kinki Japan
  SA Maizuru Belt
  SA Tamba Plateau

Kyrgyzstan (1993)
  Former Kirghiz Soviet Socialist Republic in Soviet Central Asia. Before 1993, also search Kirghizia. Before 1978, also search Kirgizia. This term has multiple hierarchies.
  CO N390000N420000
      E0800000E0690000
  UF Kirghiz Soviet Socialist Republic
  UF Kirghizia
  UF Kirgizia
  BT1 Asia
  BT2 Commonwealth of Independent States
  NT Bishkek Kyrgyzstan
  NT Chatkal Range
  NT Issyk-kul Lake
  NT Naryn Kyrgyzstan
  NT Suzak Kyrgyzstan
  NT Talas Range
  SA Alai Range
  SA Central Asia
  SA Fergana Basin
  SA Pskem Range
  SA Syr Darya
  SA Tien Shan

Kyushu (1978)
  Southernmost of the four main islands.
  CO N305500N340000
      E1320500E1292500
  BT Japan
  BT Far East
  BT Asia
  NT Ariake Bay
  NT Fukuoka Japan
  NT Hatchobaru Field
  NT Iki Island
  NT Kagoshima Japan
  NT Kirishima
  NT Kuchinoerabu-Jima
  NT Kumamoto Japan
  NT Miyazaki Japan
  NT Nagasaki Japan
  NT Oita Japan
  NT Saga Japan
  NT Tsushima
  NT Usuki Japan
  SA Ata Caldera
  SA Green Tuff Formation
  SA Inland Sea
  SA Median Tectonic Line
  SA Ontake
  SA Otake Japan
  SA Sambagawa Belt
  SA Shimanto Belt
  SA Shimanto Group
  SA shirasu
  SA Taishu Group

Kyushu-Palau Ridge (1981)
  NW Philippine Sea.
  CO N110000N200000
      E1370000E1340000
  UF Palau-Kyushu Ridge
  BT Philippine Sea
  BT West Pacific
  BT Pacific Ocean
  SA DSDP Site 296
  SA West Philippine Basin

Kyzyl Kum
  use Kyzylkum

Kyzyl Kum Desert
  use Kyzylkum

Kyzyl-Kum
  use Kyzylkum

Kyzylkum (1978)
  Desert SE of Aral Sea between the Amu Darya and the Syr Darya. This term has multiple hierarchies.
  IN Index former Soviet republics as applicable.
  CO N400000N480000
      E0733000E0580000
  UF Kizil Kum
  UF Kyzyl Kum
  UF Kyzyl Kum Desert
  UF Kyzyl-Kum
  UF Qizil Qum
  BT1 Commonwealth of Independent States
  BT2 Central Asia
  BT2 Asia
  NT Tamdytau
  SA Kazakhstan
  SA Uzbekistan

L

L chondrites (1981)
  UF L-group chondrites
  BT chondrites
  BT stony meteorites
  BT meteorites
  NT Barwell Meteorite
  NT Bjurbole Meteorite
  NT Bruderheim Meteorite
  NT Hedjaz Meteorite
  NT Rio Negro Meteorite
  NT Tsarev Meteorite

L waves
  use surface waves

L'vov Ukraine
  use Lvov Ukraine

L'vov Volyn Basin
  use Lvov-Volyn Basin

L-group chondrites
  use L chondrites

La
  use lanthanum

La Brea tar pits
  use Rancho La Brea

La Caridad
  No longer a valid term for GeoRef. As of 1993, see La Caridad Spain.

La Caridad Spain (1993)
  Settlement in NW central Spain. Before 1993, also search La Caridad and Spain.
  BT Asturias Spain
  BT Spain
  BT Iberian Peninsula
  BT Southern Europe
  BT Europe

La Coruna
  No longer a valid term for GeoRef. As of 1993, see La Coruna City Spain.

La Coruna City Spain (1993)
  Refers to only the city in extreme NW Spain on the Atlantic Ocean. Before 1993, also search La Coruna and Spain.
  BT La Coruna Spain
  BT Galicia Spain
  BT Spain
  BT Iberian Peninsula
  BT Southern Europe
  BT Europe

La Coruna Province
  No longer a valid term for GeoRef. As of 1993, see La Coruna Spain.

La Coruna Spain (1993)
  Refers only to the province in extreme NW Spain. From 1981-1992, also search La Coruna Province and Spain. Before 1981, also search La Coruna and Spain. For the city, see La Coruna City Spain.
  CO N423200N434800
      W0073800W0092000
  BT Galicia Spain
  BT Spain
  BT Iberian Peninsula
  BT Southern Europe
  BT Europe
  NT Cabo Ortegal
  NT La Coruna City Spain
  NT Santiago de Compostela Spain
  SA Arosa Bay

La Grande Soufriere (1981)
  Volcano. Before 1981, also search Soufriere AND Guadeloupe.
  IN Index Leeward Islands.
  BT Guadeloupe
  BT Lesser Antilles
  BT Antilles
  BT West Indies
  BT Caribbean region
  SA Leeward Islands
  SA Soufriere

La Habana
  No longer a valid term for GeoRef. See La Habana Cuba.

La Habana Cuba (1993)
  Province in W Cuba. Before 1981, also search Havana.
  BT Cuba
  BT Greater Antilles
  BT Antilles
  BT West Indies
  BT Caribbean region
  NT Havana Cuba
  NT Isle of Pines

La Jolla
  No longer a valid term for GeoRef. As of 1993, see La Jolla California.

La Jolla California (1993)
  NW section of San Diego in San Diego County. Before 1993, also search La Jolla AND California.
  CO N324800N325200
      W1171500W1171800
  BT San Diego County California
  BT California
  BT United States
  SA San Diego California

La Ligua (1978)
  River in central Chile. As of 1993, use only for the river. See La Ligua Chile for the town.
  BT Valparaiso Chile
  BT Chile
  BT South America

La Ligua Chile (1993)
  Town in central Chile. In former Aconcagua Province.
  BT Valparaiso Chile
  BT Chile
  BT South America

La Luna Formation (1989)
  N Colombia, W Venezuela. Divided into Salada, Pujamana and Galembo members.
  BT Upper Cretaceous
  BT Cretaceous
  BT Mesozoic
  SA Colombia
  SA Venezuela

La Meseta Formation (1989)
  Seymour Island and Vicecomodoro Marambio Island, Antarctica.
  BT upper Eocene
  BT Eocene
  BT Paleogene
  BT Tertiary
  BT Cenozoic
  SA Antarctica
  SA Seymour Island

La Pampa
  No longer a valid term for GeoRef. See La Pampa Argentina.

La Pampa Argentina (1993)

Province in central Argentina.
BT Argentina
BT South America
SA Salado Basin

La Paz
No longer a valid term for GeoRef. As of 1993, see La Paz County Arizona. After 1992, see La Paz Bolivia. Also occurs as towns in Argentina and Mexico and as department of El Salvador. For places outside the United States, search in combination with country.

La Paz Bolivia (1993)
Department and city in W Bolivia.
CO S180000S115000
    W0670000W0693000
BT Bolivia
BT South America
SA Lake Titicaca

La Paz County
Valid through 1988. Search in combination with state term. After 1988, use specific county-state term.

La Paz County Arizona (1989)
W central Arizona. Officially became a county in 1983. Before 1989, also search La Paz County AND Arizona.
CO N332500N341800
    W1132000W1144000
BT Arizona
BT United States

La Plata County
Valid through 1988. Search in combination with state term. After 1988, use specific county-state term.

La Plata County Colorado (1989)
SW Colorado. Before 1989, also search La Plata County AND Colorado.
CO N370000N374000
    W1073000W1082500
BT Colorado
BT United States
SA Needle Mountains

La Quinta Formation (1989)
Venezuela and S Mexico.
BT Jurassic
BT Mesozoic
SA Mexico
SA Venezuela

La Rioja
No longer a valid term for GeoRef. See La Rioja Argentina or Logrono Spain.

La Rioja Argentina (1993)
Province in NW Argentina. Before 1993, also search La Rioja AND Argentina.
CO S330000S270000
    W0650000W0700000
UF La Rioja Province
BT Argentina
BT South America
NT Paganzo Argentina
SA Chanares Formation
SA Paganzo Basin
SA Pampean Mountains
SA Salado Basin

La Rioja Province
use La Rioja Argentina

La Ronge
No longer a valid term for GeoRef. As of 1993, see La Ronge Saskatchewan.

La Ronge Belt
use La Ronge Domain

La Ronge Domain (1989)
N central Saskatchewan. Gold ores.
UF La Ronge Belt
BT Saskatchewan
BT Western Canada
BT Canada
SA gold ores

La Ronge Saskatchewan (1993)
Village and lake in N central Saskatchewan. Before 1993, also search La Ronge AND Saskatchewan.
BT Saskatchewan
BT Western Canada
BT Canada
SA Lac La Ronge

La Salle County
Valid through 1988. Search in combination with state term. After 1988, use specific county-state term.

La Salle County Illinois (1989)
N Illinois. Before 1989, also search La Salle County AND Illinois.
CO N405500N413800
    W0883500W0891000
BT Illinois
BT United States

La Salle County Texas (1989)
S Texas. Before 1989, also search La Salle County AND Texas.
CO N280300N283900
    W0985300W0992900
BT Texas
BT United States

La Salle Limestone (1978)
Middle and Upper Pennsylvanian. In McLeansboro Group. NE Illinois.
BT Pennsylvanian
BT Carboniferous
BT Paleozoic
SA Illinois

La Ventana Sandstone (1978)
Of Mesaverde Group. NW New Mexico.
BT Upper Cretaceous
BT Cretaceous
BT Mesozoic
SA Mesaverde Group
SA New Mexico

Laacher See (1989)
Rhineland-Palatinate, W Germany.
CO N502400N502400
    E0071800E0071800
UF Lake of Laach
BT Rhineland-Palatinate Germany
BT Germany
BT Central Europe
BT Europe

Laba Basin (1978)
River basin in Krasnodar Kray of the Northern Caucasus. Also search Laba. This term has multiple hierarchies.
BT1 Krasnodar Russian Federation
BT1 Russian Federation
BT1 Commonwealth of Independent States
BT2 Krasnodar Russian Federation
BT2 Europe

Labette County
Valid through 1988. Search in combination with state term. After 1988, use specific county-state term.

Labette County Kansas (1989)
SE Kansas. Before 1989, also search Labette County AND Kansas.
CO N370000N372200
    W0950500W0953500
BT Kansas
BT United States

Labette Shale (1978)
In Marmaton Group.
BT Desmoinesian
BT Middle Pennsylvanian
BT Pennsylvanian
BT Carboniferous
BT Paleozoic
SA Iowa
SA Kansas
SA Missouri
SA Marmaton Group
SA Oklahoma

laboratory studies (1978)
As of 1981, for engineering application only. Prior to 1981, applicable to a large number of topics.
UF studies, laboratory
SA engineering geology
SA experimental studies
SA theoretical studies

Laborcita Formation (1989)
In Magdalena Group. S New Mexico.
IN   Index ages as applicable.
BT Paleozoic
SA Lower Permian
SA Magdalena Group
SA New Mexico
SA Upper Pennsylvanian

Labrador (1978)
As of 1993, Newfoundland is a broader term.
CO N513000N610000
    W0553000W0673000
BT Newfoundland
BT Eastern Canada
BT Canada
NT Churchill Falls
NT Kiglapait
NT Kiglapait Intrusion
NT Mistastin Lake
NT Nain Massif
NT Nain Newfoundland
NT Torngat Mountains
NT Wabush
SA Churchill Province
SA Grenville Province
SA Labrador Current
SA Labrador Trough
SA Nain Province
SA Sokoman Formation
SA Superior Province
SA Ungava

Labrador Basin (1978)
E of Labrador and S of Greenland.
BT Atlantic Ocean

Labrador Current (1989)
Formed by West Greenland Current and Baffin Island Current at Davis Strait, flows SE along coast of Labrador and Newfoundland, and meets North Atlantic Current

at about N42˚ W50˚. As of 1990, Atlantic Ocean is autoposted to this term.
BT North American Atlantic
BT North Atlantic
BT Atlantic Ocean
SA Davis Strait
SA Labrador
SA Labrador Sea
SA Newfoundland Island

Labrador Sea (1978)
Between Labrador and SW Greenland. As of 1990, Atlantic Ocean is autoposted to this term.
CO N480000N600000
    W0440000W0640000
BT North American Atlantic
BT North Atlantic
BT Atlantic Ocean
SA Labrador Current
SA Leg 105
SA ODP Site 646
SA ODP Site 647

Labrador Trough (1981)
In Quebec and Labrador.
BT Canada
SA Labrador
SA Quebec

labradorite (1978)
BT plagioclase
BT feldspar group
BT framework silicates
BT silicates
SA igneous rocks

labradoritite (1978)
BT gabbros
BT plutonic rocks
BT igneous rocks

labuntsovite (1978)
Autoposting of sorosilicates began in 1985.
BT sorosilicates
BT orthosilicates
BT silicates

Labyrinthodontia (1978)
Subclass.
BT Amphibia
BT Tetrapoda
BT Vertebrata
BT Chordata
NT Temnospondyli

Lac du Bonnet (1989)
Lake in SE Manitoba. As of 1993, use Lac du Bonnet Manitoba for the town.
CO N501000N503000
    W0954500W0961500
BT Manitoba
BT Western Canada
BT Canada
SA Lac du Bonnet Batholith
SA Lac du Bonnet Manitoba

Lac du Bonnet Batholith (1989)
Intrusion; site for nuclear waste disposal management in SE Manitoba.
BT Manitoba
BT Western Canada
BT Canada
SA Lac du Bonnet

Lac du Bonnet Manitoba (1993)
Town in SE Manitoba. Before 1993, also search Lac du Bonnet AND Manitoba.
BT Manitoba
BT Western Canada
BT Canada
SA Lac du Bonnet

Lac La Ronge (1978)
Lake in N central Saskatchewan.

UF Lake La Ronge
BT Saskatchewan
BT Western Canada
BT Canada
SA La Ronge Saskatchewan

Laccadive Islands (1981)
Off SW coast of India in Lakshadweep Territory.
CO N100000N122000
   E0740000E0720000
BT India
BT Indian Peninsula
BT Asia
SA Indian Ocean Islands

laccoliths (1978)
UF cistern rock
BT intrusions
SA batholiths
SA dikes
SA domes
SA lopoliths

Lacertilia (1978)
Suborder.
BT Squamata
BT Lepidosauria
BT Diapsida
BT Reptilia
BT Tetrapoda
BT Vertebrata
BT Chordata

Lachlan fold belt (1981)
E Australia, from E central Queensland to NE Tasmania.
CO S420000S300000
   E1501500E1430000
BT Australia
BT Australasia
SA New South Wales Australia
SA Queensland Australia
SA Tasmania Australia
SA Victoria Australia

LaCoste-Romberg gravimeters (1981)
BT gravimeters

lacustrine
A valid term through 1977. After 1977, use lacustrine features or (as of 1981) lacustrine environment.

lacustrine environment (1981)
SA aquatic environment
SA deep-water environment
SA depositional environment
SA ecology
SA environment
SA hypersaline environment
SA lacustrine features
SA lacustrine sedimentation
SA lakes
SA meromictic lakes
SA paleoecology
SA sedimentation

lacustrine features (1978)
Also search lacustrine. Autoposting of this term began in 1978.
UF features, lacustrine
NT lakes
NT playas
SA beaches
SA bedforms
SA bogs
SA fan deltas
SA geomorphology
SA glaciolacustrine environment
SA inlets
SA kames
SA lacustrine environment
SA limnology
SA marshes
SA salt marshes
SA shoals
SA swamps
SA terraces
SA wetlands

lacustrine sedimentation (1981)
BT sedimentation
NT fluviolacustrine sedimentation
NT glaciolacustrine sedimentation
SA lacustrine environment

lacustrine sediments
As of 1993, for economic discussions, use lake sediments. For other discussions, use sediments AND lacustrine environment.

Ladak
use Ladakh

Ladakh (1978)
Region in Jammu and Kashmir. S Ladakh divided between India and Pakistan.
IN Index Jammu and Kashmir and countries as applicable.
UF Ladak
SA India
SA Indus-Yarlung Zangbo suture zone
SA Jammu and Kashmir
SA Pakistan

Ladd Formation (1989)
S California. Includes Holz Shale and Baker Conglomerate Member.
BT Upper Cretaceous
BT Cretaceous
BT Mesozoic
SA California
SA Holz Shale

Ladinian (1978)
Europe: upper Middle Triassic. Above Anisian, below Carnian.
BT Middle Triassic
BT Triassic
BT Mesozoic
SA Hallstatt Limestone

Ladner Group (1989)
S British Columbia and N Washington.
BT Jurassic
BT Mesozoic
SA British Columbia
SA Washington

Ladoga Series (1981)
Used for the Precambrian metamorphic unit in the Lake Ladoga region of the Russian Federation and Finland. Not to be confused with Quaternary deposits or Cambrian-Silurian sedimentary rocks of the same name in the same region. (These units will be assigned unique names as needed according to frequency.) Before 1981, also search Ladoga Formation; Ladoga. Age may be upper Archean or lower Proterozoic.
BT Precambrian
SA Baltic Shield
SA Finland
SA Karelia Russian Federation
SA Russian Federation

Laetoli (1989)
Pliocene of Tanzania.
BT Tanzania
BT East Africa
BT Africa

Laetoli Beds (1993)
Pleistocene. Tanzania.
UF Laetolil Beds
BT Pleistocene
BT Quaternary
BT Cenozoic
SA Tanzania

Laetolil Beds
use Laetoli Beds

Lafayette County
Valid through 1988. Search in combination with state term. After 1988, use specific county-state term.

Lafayette County Arkansas (1989)
SW Arkansas. Before 1989, also search Lafayette County AND Arkansas.
CO N330200N332700
   W0932200W0935200
BT Arkansas
BT United States
SA Walker Creek Field

Lafayette County Florida (1989)
Bounded by Suwannee River in N Florida. Before 1989, also search Lafayette County AND Florida.
CO N295000N301500
   W0825100W0832200
BT Florida
BT United States

Lafayette County Mississippi (1989)
N Mississippi. Before 1989, also search Lafayette County AND Mississippi.
CO N341000N343300
   W0891500W0894300
BT Mississippi
BT United States
NT Oxford Mississippi

Lafayette County Missouri (1989)
W central Missouri. Before 1989, also search Lafayette County AND Missouri.
CO N385400N391700
   W0932800W0940700
BT Missouri
BT United States

Lafayette County Wisconsin (1989)
S Wisconsin. Before 1989, also search Lafayette County AND Wisconsin.
CO N423000N424800
   W0895200W0902500
BT Wisconsin
BT United States

Lafayette Parish Louisiana (1989)
S central Louisiana. Before 1989, also search Lafayette Parish; Lafayette County AND Louisiana.
CO N300300N302600
   W0915300W0922000
BT Louisiana
BT United States

Lafourche Parish Louisiana (1989)
Extreme SE Louisiana, bordering on the Gulf of Mexico. Before 1989, search Lafourche Parish AND Louisiana.
CO N290300N295500
   W0900100W0910100
BT Louisiana
BT United States
SA Barataria Bay

Lagenidae (1978)
Autoposting of microfossils and Protista to this term began in 1990. This term has multiple hierarchies.
BT1 Rotaliina
BT1 foraminifera
BT1 Protista
BT1 Invertebrata
BT2 Rotaliina
BT2 foraminifera
BT2 Protista
BT2 microfossils

Lageos (1985)
UF Laser Geodynamics Satellite
SA geophysical methods
SA geophysical surveys
SA laser ranging
SA remote sensing
SA satellite methods

Laghouat (1978)
Oasis in N central Algeria. After 1993, use Laghouat Algeria for the town.
BT Algeria
BT North Africa
BT Africa

Lago Maggiore (1978)
Lake in N Italy and S Switzerland.
IN Index countries and canton or provinces as applicable.
BT Europe
SA Lombardy Italy
SA Piemonte Italy
SA Ticino Switzerland

Lagomorpha (1978)
Autoposting of Eutheria and Theria to this term began in 1989.
BT Eutheria
BT Theria
BT Mammalia
BT Tetrapoda
BT Vertebrata
BT Chordata

lagoonal
A valid term through 1977. After 1977, use lagoons or (as of 1981) lagoonal environment or lagoonal sedimentation.

lagoonal environment (1981)
SA deep-water environment
SA depositional environment
SA ecology
SA environment
SA hypersaline environment
SA lagoonal sedimentation
SA lagoons
SA paleoecology
SA sedimentation

lagoonal sedimentation (1981)
BT sedimentation
SA lagoonal environment

lagoons (1978)
As of 1981, use lagoonal environment for type of environment. Autoposting of shore features to this term began in 1989.
BT shore features
SA atolls
SA barrier reefs
SA beaches
SA coastal environment
SA ecology
SA geomorphology
SA lagoonal environment
SA lakes
SA paleoecology
SA reefs
SA sedimentation
SA segmentation

Lagos Lagoon (1978)
Between Lagos Island just off the Gulf of Guinea and the mainland. City of Lagos at W end of lagoon. Also search Lagos.
BT Nigeria
BT West Africa

BT Africa

Laguna Madre (1989)
Inlet of the Gulf of Mexico located between Padre Island and mainland S Texas, S of Corpus Christi. Also an inlet of the Gulf of Mexico off the coast of Tamaulipas, E Mexico.
IN Index regions as applicable.
SA Gulf of Mexico
SA Kenedy County Texas
SA Mexico
SA Padre Island
SA Tamaulipas Mexico
SA Texas

lahars (1978)
Autoposting of mass movements to this term began in 1989.
BT mass movements
SA geomorphology
SA lava flows
SA mudflows
SA volcanism

Lahn River (1978)
Flows S and SW into the Rhine River SE of Koblenz. Also search Lahn.
IN Index states as applicable.
BT Germany
BT Central Europe
BT Europe
SA Hesse Germany
SA Rhineland-Palatinate Germany

Lahn River valley (1978)
Also search Lahn.
IN Index states as applicable.
BT Germany
BT Central Europe
BT Europe
SA Hesse Germany
SA Rhineland-Palatinate Germany

laihunite (1989)
UF ferrifayalite
BT nesosilicates
BT orthosilicates
BT silicates

Laisvall
No longer a valid term for GeoRef. As of 1993 see Laisvall Sweden.

Laisvall Sweden (1993)
Village in N Sweden. Before 1993 also search Laisvall AND Sweden.
BT Norrbotten Sweden
BT Sweden
BT Scandinavia
BT Western Europe
BT Europe

Lake Agassiz (1978)
Glacial lake which existed in Pleistocene Epoch. Covered parts of N central U.S. and S central Canada.
IN Index Manitoba and U.S. states as applicable.
BT Pleistocene
BT Quaternary
BT Cenozoic
SA Manitoba
SA Minnesota
SA North America
SA North Dakota

Lake Albert (1978)
Before 1993, also search Lake Mobutu.
IN Index Uganda and/or Zaire as applicable.
UF Lake Mobutu
BT East African Lakes

BT Africa
SA East African Rift
SA Uganda
SA Upper Zaire
SA Zaire

Lake Algonquin (1985)
Pleistocene glacial lake. Occupied sites of modern lakes Superior, Michigan and Huron, and adjacent land areas.
IN Index Ontario and U.S. states as applicable.
BT Pleistocene
BT Quaternary
BT Cenozoic
SA Lake Superior region
SA Michigan
SA Minnesota
SA North America
SA Ontario
SA Wisconsin

Lake Athabasca (1985)
NW Saskatchewan and NE Alberta.
IN Index provinces as applicable.
CO N583000N592000
    W1071000W1111500
BT Western Canada
BT Canada
SA Alberta
SA Athabasca District
SA Saskatchewan

Lake Baikal (1978)
Largest freshwater lake in Eurasia. Bounded on N, E, and S by the Buryatia, and on the W by Irkutsk Oblast. Also search Baikal AND lake or lakes. This term has multiple hierarchies.
CO N520000N560000
    E1100000E1040000
UF Baikal (Lake)
BT1 Russian Federation
BT1 Commonwealth of Independent States
BT2 Asia
SA Baikal region
SA Baikal rift zone
SA Buryat Russian Federation
SA Irkutsk Russian Federation

Lake Balaton (1978)
W Hungary.
UF Balaton (Lake)
BT Veszprem Hungary
BT Hungary
BT Central Europe
BT Europe
SA Balaton region

Lake Balkhash (1978)
E Kazakhstan. This term has multiple hierarchies.
CO N450000N470000
    E0790000E0730000
UF Balkhash (Lake)
BT1 Kazakhstan
BT1 Central Asia
BT1 Asia
BT2 Kazakhstan
BT2 Commonwealth of Independent States
SA Balkhash region

Lake Baringo (1981)
W central Kenya, in Great Rift Valley.
CO N003000N004500
    E0360700E0360000
UF Baringo Lake
BT East African Lakes
BT Africa
SA Great Rift Valley
SA Kenya

SA Kenya Rift valley

lake biscuits
use algal biscuits

Lake Biwa (1978)
W central Honshu. Also search Biwa.
UF Biwa Lake
UF Omi (Lake)
BT Shiga Japan
BT Honshu
BT Japan
BT Far East
BT Asia

Lake Bonneville (1978)
Large prehistoric body of water in present Great Salt Lake area.
IN Index states and age or ages as applicable.
SA Great Salt Lake
SA Holocene
SA Idaho
SA Nevada
SA United States
SA upper Pleistocene
SA Utah

Lake Bonney (1978)
Lake in SE South Australia.
BT South Australia
BT Australia
BT Australasia

Lake Chad (1978)
IN Index countries as applicable.
CO N123000N143000
    E0160000E0123000
BT West Africa
BT Africa
SA Cameroon
SA Chad
SA Chad Basin
SA Niger
SA Nigeria

Lake Champlain (1978)
Between New York and Vermont extending into Quebec.
IN Index Quebec and states as applicable.
BT North America
SA Champlain Sea
SA Champlain Valley
SA New York
SA Quebec
SA Vermont

Lake Char Fault (1985)
SE New England.
IN Index states as applicable.
BT United States
SA Connecticut
SA Massachusetts
SA Rhode Island

Lake Chatuge (1978)
IN Index counties as applicable.
BT United States
SA Clay County North Carolina
SA Georgia
SA North Carolina
SA Towns County Georgia

Lake Chicago (1989)
Pleistocene glacial lake in the Great Lakes region.
IN Index states as applicable.
BT Pleistocene
BT Quaternary
BT Cenozoic
SA Great Lakes region
SA Illinois
SA Indiana
SA Michigan
SA United States
SA Wisconsin

Lake Constance (1978)

IN Index countries as applicable.
UF Bodensee
UF Constance Lake
BT Central Europe
BT Europe
SA Baden-Wurttemberg Germany
SA Bavaria Germany
SA Rhine Basin
SA Rhine River
SA Switzerland
SA Vorarlberg Austria

Lake County
Valid through 1988. Search in combination with state term. After 1988, use specific county-state term.

Lake County California (1989)
NW California. Before 1989, also search Lake County AND California.
CO N384000N393400
    W1222400W1230400
BT California
BT United States
SA Clear Lake

Lake County Colorado (1989)
Central Colorado. Before 1989, also search Lake County AND Colorado.
CO N390400N392300
    W1060800W1063700
BT Colorado
BT United States
NT Climax Colorado
NT Leadville Colorado
NT Leadville mining district
SA Climax Mine

Lake County Florida (1989)
Bounded by Saint Johns River in central Florida. Before 1989, also search Lake County AND Florida.
CO N282000N291500
    W0812300W0815800
BT Florida
BT United States

Lake County Illinois (1989)
Bounded by Lake Michigan in extreme NE Illinois. Before 1989, also search Lake County AND Illinois.
CO N421000N423000
    W0874400W0881400
BT Illinois
BT United States

Lake County Indiana (1989)
Extreme NW Indiana. Before 1989, also search Lake County AND Indiana.
CO N411000N414500
    W0871400W0873200
BT Indiana
BT United States

Lake County Michigan (1989)
W Michigan. Before 1989, also search Lake County AND Michigan.
CO N434800N441200
    W0853300W0860300
BT Michigan Lower Peninsula
BT Michigan
BT United States

Lake County Minnesota (1989)
NE Minnesota. Before 1989, also search Lake County AND Minnesota.
CO N465700N481300
    W0910000W0914800
BT Minnesota
BT United States
NT Two Harbors Minnesota
SA Vermilion Range

Lake County Montana (1989)
 NW Montana. Before 1989, also search Lake County AND Montana.
 CO N470800N480400
  W1133200W1143300
 BT Montana
 BT United States
 SA Flathead Lake

Lake County Ohio (1989)
 NE Ohio. Before 1989, also search Lake County AND Ohio.
 CO N413500N415200
  W0810000W0813000
 BT Ohio
 BT United States

Lake County Oregon (1989)
 S Oregon. Before 1989, also search Lake County AND Oregon.
 CO N420000N433700
  W1192200W1212300
 BT Oregon
 BT United States

Lake County South Dakota (1989)
 E South Dakota. Before 1989, also search Lake County AND South Dakota.
 CO N435200N441300
  W0965500W0972300
 BT South Dakota
 BT United States

Lake County Tennessee (1989)
 Extreme NW Tennessee. Before 1989, also search Lake County AND Tennessee.
 CO N361100N363000
  W0892300W0894300
 BT Tennessee
 BT United States
 SA Reelfoot Lake

Lake District (1978)
 Mountain and lake region in Cumbria, NW England. Until 1975, parts were included in Lancashire, Cumberland County and Westmorland County.
 IN Index England and counties as applicable.
 UF English Lake District
 SA Cumberland England
 SA Cumbria England
 SA Westmorland England
 SA Worcestershire England

Lake Edward (1978)
 S of Lake Albert.
 IN Index countries as applicable.
 BT East African Lakes
 BT Africa
 SA East African Rift
 SA Uganda
 SA Zaire

Lake Erie (1978)
 As of 1990, North America is autoposted to this term.
 IN Index Ontario and states as applicable.
 CO N412230N425500
  W0785100W0832800
 BT Great Lakes
 BT North America
 SA Detroit River
 SA Michigan
 SA New York
 SA Ohio
 SA Ontario
 SA Pennsylvania

Lake Eyre (1978)
 NE South Australia.
 BT South Australia
 BT Australia

 BT Australasia

Lake Frome (1989)
 E South Australia.
 CO S311500S300000
  E1401000E1392000
 BT South Australia
 BT Australia
 BT Australasia

Lake Geneva (1978)
 IN Index countries as applicable.
 UF Lake Leman
 UF Leman Lake
 BT Europe
 SA Geneva Switzerland
 SA Haute-Savoie France
 SA Rhone River
 SA Rhone Valley
 SA Valais Switzerland
 SA Vaud Switzerland

Lake George (1978)
 NE New York.
 IN Index counties or regions as applicable.
 SA Essex County New York
 SA New York
 SA Warren County New York
 SA Washington County New York

Lake Gosiute (1981)
 Eocene lake in SW Wyoming.
 CO N403000N424500
  W1080000W1104000
 BT Eocene
 BT Paleogene
 BT Tertiary
 BT Cenozoic
 SA Green River Formation
 SA Sweetwater County Wyoming

Lake Huron (1978)
 As of 1990, North America is autoposted to this term.
 IN Index Michigan and/or Ontario.
 BT Great Lakes
 BT North America
 SA Michigan
 SA Ontario

Lake Jocassee (1981)
 Reservoir on Keowee River, NW South Carolina, near Seneca. Autoposting of broader terms began in 1989.
 IN Index counties as applicable.
 CO N345710N350230
  W0825330W0825945
 BT South Carolina
 BT United States
 SA Oconee County South Carolina
 SA Pickens County South Carolina

Lake Kariba (1978)
 Formed by dam in Kariba Gorge of Zambezi River. Also search Kariba.
 IN Index countries as applicable.
 UF Kariba Lake
 BT East African Lakes
 BT Africa
 SA Zambia
 SA Zimbabwe

Lake Kinneret
 use Sea of Galilee

Lake Kivu (1978)
 This term has multiple hierarchies.
 IN Index countries as applicable.
 BT1 East African Lakes
 BT1 Africa
 BT2 Central Africa
 BT2 Africa
 SA East African Rift
 SA Kivu Zaire

 SA Rwanda
 SA Zaire

Lake La Ronge
 use Lac La Ronge

Lake Ladoga (1978)
 E and NE of Saint Petersburg. This term has multiple hierarchies.
 CO N595000N615000
  E0330000E0295000
 BT1 Russian Federation
 BT1 Commonwealth of Independent States
 BT2 Europe

Lake Ladoga region (1981)
 Region surrounding Lake Ladoga, extending into Finland.
 IN Index countries as applicable.
 CO N595000N620000
  E0333000E0283000
 BT Europe
 SA Baltic Shield
 SA Finland
 SA Karelia Russian Federation
 SA Russian Federation

Lake Lahontan (1985)
 Extinct lake. Occupied parts of NW Nevada, NE California and SE Oregon during Pleistocene time.
 IN Index states as applicable.
 BT Pleistocene
 BT Quaternary
 BT Cenozoic
 SA California
 SA Nevada
 SA Oregon
 SA United States

Lake Lappajarvi (1978)
 In Vaasa, SW central Finland.
 BT Finland
 BT Scandinavia
 BT Western Europe
 BT Europe

Lake Leman
 use Lake Geneva

Lake Louise (1978)
 Village and small lake in Banff National Park in SW Alberta. As of 1993, use Lake Louise Alberta for the village. Also in other locations.
 IN Index province or regions as applicable.
 SA Alberta
 SA Banff National Park

Lake Magadi (1978)
 Soda lake in S central Kenya. This term has multiple hierarchies.
 BT1 East African Lakes
 BT1 Africa
 BT2 Kenya
 BT2 East Africa
 BT2 Africa
 SA East African Rift

Lake Malawi (1978)
 This term has multiple hierarchies.
 IN Index countries as applicable.
 UF Lake Nyasa
 BT1 East African Lakes
 BT1 Africa
 BT2 East Africa
 BT2 Africa
 SA East African Rift
 SA Malawi
 SA Mozambique
 SA Tanzania

Lake Maracaibo (1978)
 Lake in NW Venezuela.
 BT Venezuela
 BT South America
 SA Maracaibo Basin

 SA Trujillo Venezuela
 SA Zulia Venezuela

Lake Matagami
 use Matagami

Lake Maumee (1989)
 Pleistocene glacial lake in Ontario, Ohio, Michigan and Indiana.
 IN Index states or provinces as applicable.
 BT Pleistocene
 BT Quaternary
 BT Cenozoic
 SA Great Lakes region
 SA Indiana
 SA Michigan
 SA North America
 SA Ohio
 SA Ontario

Lake Mead (1978)
 Within Lake Mead National Recreation Area.
 IN Index states and counties as applicable.
 SA Arizona
 SA Clark County Nevada
 SA Lake Mead Fault
 SA Lost City
 SA Mohave County Arizona
 SA Nevada
 SA United States

Lake Mead Fault (1989)
 S Nevada.
 BT Nevada
 BT United States
 SA Lake Mead

Lake Mendota (1978)
 In Dane County, S central Wisconsin.
 IN Index county or regions as applicable.
 SA Dane County Wisconsin

Lake Michigan (1978)
 As of 1990, North America is autoposted to this term.
 IN Index states as applicable.
 CO N414500N460000
  W0844500W0875000
 BT Great Lakes
 BT North America
 SA Green Bay
 SA Illinois
 SA Indiana
 SA Michigan
 SA United States
 SA Wedron Formation
 SA Wisconsin

Lake Michigan Basin
 use Michigan Basin

Lake Missoula (1985)
 Pleistocene glacial lake in NW United States.
 IN Index states as applicable.
 BT Pleistocene
 BT Quaternary
 BT Cenozoic
 SA Idaho
 SA Montana
 SA Oregon
 SA United States
 SA Washington

Lake Mobutu
 use Lake Albert

Lake Natron (1978)
 This term has multiple hierarchies.
 IN Index countries as applicable.
 BT1 East African Lakes
 BT1 Africa
 BT2 East Africa
 BT2 Africa
 SA East African Rift

SA Kenya
SA Tanzania

Lake Neusiedler (1985)
E Austria and NW Hungary.
IN Index countries as applicable.
CO N474000N475500
E0165000E0164000
UF Ferto to
UF Neusiedler Lake
UF Neusiedlersee
BT Central Europe
BT Europe
SA Austria
SA Burgenland Austria
SA Hungary

Lake Nipissing (1989)
Nipissing District, SE Ontario.
BT Nipissing District Ontario
BT Ontario
BT Eastern Canada
BT Canada

Lake Nyasa
 use Lake Malawi

Lake Nyos (1993)
Volcanic crater lake in NW Cameroon. Site of carbon dioxide outgassing natural disasters.
UF Nyos
BT Cameroon
BT West Africa
BT Africa

Lake of Bienne
 use Biel Lake

Lake of Laach
 use Laacher See

Lake of the Woods (1989)
NW Ontario, SE Manitoba and N Minnesota.
IN Index provinces and/or Minnesota and counties as applicable.
BT North America
SA Lake of the Woods region
SA Manitoba
SA Minnesota
SA Ontario

Lake of the Woods region (1985)
N Minnesota, SW Ontario and SE Manitoba.
IN Index Minnesota and counties and/or Canadian provinces as applicable.
CO N484500N494500
W0940000W0952000
BT North America
SA Lake of the Woods
SA Manitoba
SA Minnesota
SA Ontario

Lake of Zurich (1978)
N Switzerland.
BT Switzerland
BT Central Europe
BT Europe
SA Zurich Switzerland

Lake Okeechobee (1978)
S central Florida.
BT Florida
BT United States

Lake Onega (1978)
In Karelia Autonomous Republic. Also search Onega. This term has multiple hierarchies.
BT1 Russian Federation
BT1 Commonwealth of Independent States
BT2 Europe
SA Karelia Russian Federation

SA Onega Russian Federation

Lake Ontario (1978)
As of 1990, North America is autoposted to this term.
IN Index New York and/or Ontario as applicable.
BT Great Lakes
BT North America
SA New York
SA Ontario

Lake Pontchartrain (1978)
S Louisiana.
IN Index parishes as applicable.
BT Louisiana
BT United States

Lake Powell (1978)
Artificial lake in S Utah and N Arizona, in Glen Canyon National Recreation Area.
IN Index states and counties as applicable.
BT United States
SA Arizona
SA Coconino County Arizona
SA Garfield County Utah
SA Kane County Utah
SA San Juan County Utah
SA Utah

Lake Rudolf
 use Lake Turkana

Lake Rudolph
 use Lake Turkana

Lake Saint Clair (1989)
SE Michigan and SW Ontario.
IN Index North America or Michigan and U.S. counties or and/or Ontario as applicable.
SA Detroit River
SA Great Lakes region
SA Macomb County Michigan
SA Michigan
SA Michigan Lower Peninsula
SA North America
SA Ontario
SA Saint Clair County Michigan
SA Thames River
SA Wayne County Michigan

lake sediments (1981)
Restricted to mineral exploration only. Before 1981, also search sediments AND lakes.
SA glaciolacustrine sedimentation
SA lakes
SA mineral exploration
SA sediments

Lake Sevan (1978)
N Armenia. This term has multiple hierarchies.
BT1 Armenia
BT1 Commonwealth of Independent States
BT2 Armenia
BT2 Europe
SA Sevan Armenia

Lake Superior (1978)
As of 1990, North America is autoposted to this term.
IN Index Ontario and states as applicable.
CO N462500N490000
W0842000W0921500
BT Great Lakes
BT North America
SA Keweenaw Peninsula
SA Lake Superior region
SA Michigan
SA Minnesota
SA Ontario
SA Thunder Bay
SA Wisconsin

Lake Superior Basin
 use Lake Superior region

Lake Superior region (1978)
Term reintroduced in 1985. Also search Lake Superior. Not a valid term from 1972 through 1984.
IN Index Lake Superior, Ontario and states as applicable.
UF Lake Superior Basin
BT North America
SA Great Lakes region
SA Lake Algonquin
SA Lake Superior
SA Michigan
SA Minnesota
SA Ontario
SA Wisconsin

Lake Suwa (1978)
W central Honshu.
BT Nagano Prefecture Japan
BT Honshu
BT Japan
BT Far East
BT Asia

Lake Tahoe (1978)
E California and W Nevada.
IN Index states and counties as applicable.
BT United States
SA California
SA Douglas County Nevada
SA El Dorado County California
SA Nevada
SA Placer County California
SA Washoe County Nevada

Lake Tanganyika (1978)
IN Index countries as applicable.
BT East African Lakes
BT Africa
SA Burundi
SA East African Rift
SA Tanzania
SA Zaire
SA Zambia

Lake Taupo (1989)
Central North Island, New Zealand.
IN Also index North Island.
CO S390500S384000
E1761000E1753000
BT New Zealand
BT Australasia
SA North Island

Lake Temiscamingue
 use Lake Timiskaming

Lake Temiskaming
 use Lake Timiskaming

Lake Tiberias
 use Sea of Galilee

Lake Timiskaming (1981)
In SW Quebec and E central Ontario. Before 1985, also search Timiskaming.
IN Index provinces as applicable.
UF Lake Temiscamingue
UF Lake Temiskaming
UF Temiscamingue Lake
UF Temiskaming Lake
UF Timiskaming Lake
BT Eastern Canada
BT Canada
SA Ontario
SA Quebec
SA Temiscamingue County Quebec
SA Timiskaming District Ontario

Lake Titicaca (1978)
IN Index countries as applicable.
BT South America

SA Altiplano
SA La Paz Bolivia
SA Puno Peru

Lake Torrens (1978)
E central South Australia.
BT South Australia
BT Australia
BT Australasia

Lake Turkana (1978)
New name for Lake Rudolf. Primarily in Kenya. This term has multiple hierarchies.
IN Index countries as applicable.
UF Lake Rudolf
UF Lake Rudolph
UF Rudolf (Lake)
BT1 East African Lakes
BT1 Africa
BT2 East Africa
BT2 Africa
SA East African Rift
SA Ethiopia
SA Kenya
SA Sudan
SA Turkana Basin

Lake Uinta (1985)
Extinct lake. Occupied parts of W Colorado, SW Wyoming and E Utah during Eocene time.
IN Index states as applicable.
BT Eocene
BT Paleogene
BT Tertiary
BT Cenozoic
SA Colorado
SA United States
SA Utah
SA Wyoming

Lake Valley Formation (1978)
Includes Alamogordo, Arcente, Dona Ana, and Nunn members.
BT Lower Mississippian
BT Mississippian
BT Carboniferous
BT Paleozoic
SA New Mexico

Lake Vanda (1981)
Saline lake in dry-valleys region of S Victoria Land, near McMurdo Sound.
CO S773500S773500
E1613900E1613900
BT Victoria Land
BT Antarctica

Lake Victoria (1978)
This term has multiple hierarchies.
IN Index countries as applicable.
BT1 East African Lakes
BT1 Africa
BT2 East Africa
BT2 Africa
SA Kenya
SA Tanzania
SA Uganda

Lake Washington (1989)
NW Washington, near Seattle; Meeker County, S central Minnesota, and elsewhere.
IN Index states and counties as applicable.
SA King County Washington
SA Minnesota
SA Washington

Lake Winnipeg (1978)
S central Manitoba.
BT Manitoba
BT Western Canada
BT Canada

lakes (1978)

Used for the geomorphologic feature. As of 1981, use lacustrine environment for type of environment. Before 1972, was sometimes used in combination with extinct (i.e. lakes, extinct). Now use extinct lakes. Autoposting of lacustrine features to this term began in 1989.
BT lacustrine features
NT crater lakes
NT extinct lakes
NT glacial lakes
NT meromictic lakes
SA drainage
SA ecology
SA eutrophication
SA geomorphology
SA glacial features
SA glaciolacustrine environment
SA hydrogen sulfide
SA hydrology
SA hydrosphere
SA lacustrine environment
SA lagoons
SA lake sediments
SA lava lakes
SA limnology
SA paleoecology
SA periglacial features
SA ponds
SA reservoirs
SA salt lakes
SA sedimentation
SA seiches
SA surface water
SA thermocline
SA water balance
SA water resources
SA water supply
SA watersheds

Lakota Formation (1985)
In Inyan Kara Group. Montana, W South Dakota, NE Nebraska and E Wyoming.
BT Lower Cretaceous
BT Cretaceous
BT Mesozoic
SA Inyan Kara Group
SA Montana
SA Nebraska
SA South Dakota
SA Wyoming

Lakshmi Planum (1993)
High plateau region. Before 1993, also search Lakshmi Plateau AND Venus.
BT Venus

Lamar County
Valid through 1988. Search in combination with state term. After 1988, use specific county-state term.

Lamar County Alabama (1989)
W Alabama. Before 1989, also search Lamar County AND Alabama.
CO N330300N340400 W0875600W0881700
BT Alabama
BT United States

Lamar County Georgia (1989)
W central Georgia. Before 1989, also search Lamar County AND Georgia.
CO N325500N331200 W0840200W0841400
BT Georgia
BT United States

Lamar County Mississippi (1989)
SE Mississippi. Before 1989, also search Lamar County AND Mississippi.
CO N305900N312600 W0892200W0894000
BT Mississippi
BT United States

Lamar County Texas (1989)
NE Texas. Before 1989, also search Lamar County AND Texas.
CO N332300N335500 W0952200W0955300
BT Texas
BT United States

Lambton County
No longer a valid term for GeoRef. As of 1993, see Lambton County Ontario.

Lambton County Ontario (1993)
S Ontario. Before 1993, also search Lambton County AND Ontario.
CO N423000N432000 W0814500W0824000
BT Ontario
BT Eastern Canada
BT Canada

lamellae (1978)
Restricted to use with minerals. Not used in biological sense.
SA deformation
SA minerals

laminar flow (1982)
SA critical flow
SA fluid dynamics
SA ground water
SA hydraulics
SA hydrodynamics
SA hydrology
SA sediment transport
SA sedimentation

laminations (1978)
Before 1976, also search lamination.
BT planar bedding structures
BT sedimentary structures
SA bedding
SA Bouma sequence
SA cross-bedding
SA cross-laminations
SA foliation
SA ripple drift-cross laminations
SA stratification

Lamont-Doherty Geological Observatory (1985)
In Palisades, SE New York. Part of Columbia University.
SA academic institutions
SA New York

Lamott Sandstone
use Lamotte Sandstone

Lamotte Sandstone (1985)
E and central Missouri.
UF Lamott Sandstone
BT Upper Cambrian
BT Cambrian
BT Paleozoic
SA Missouri

Lampang
No longer a valid term for GeoRef. As of 1993 see Lampang Thailand.

Lampang Thailand (1993)
City and province in NW Thailand. Before 1993 also search Lampang AND Thailand.
BT Thailand
BT Far East
BT Asia

lamproite (1978)
BT lamprophyres
BT plutonic rocks
BT igneous rocks

lamproite pipes
Not a valid term for GeoRef. Search lamproite AND (pipes OR diatremes).

lamprophyre
As of 1989, no longer a valid term for GeoRef. From 1981 to 1988, lamprophyres was autoposted to this term.
use lamprophyres

lamprophyre and carbonatite family
As of 1981, no longer a valid term for GeoRef. Use lamprophyres or carbonatites as applicable. From 1978-1980, was autoposted to the rocks that were classified under it.

lamprophyres (1981)
Before 1981, search lamprophyre and carbonatite family. From 1978-1980, lamprophyre and carbonatite family was autoposted to the rocks that were classified under that term. Before 1989, also search lamprophyre. Autoposting of this term to lamprophyre ended in 1989.
UF lamprophyre
BT plutonic rocks
BT igneous rocks
NT alnoite
NT camptonite
NT kersantite
NT lamproite
NT minette
NT monchiquite
NT spessartite
NT vogesite

Lanark County Ontario (1993)
Before 1993, also search Lanark County AND Ontario.
BT Ontario
BT Eastern Canada
BT Canada
NT Perth Ontario

Lancang-Gengma earthquake 1988 (1993)
E Yunnan, China.
UF Gengma-Lancang earthquake 1988
BT earthquakes
SA China
SA Yunnan China

Lancara Formation (1978)
Lower Acadian. Overlies Herreira Sandstone. NW Spain.
BT Acadian
BT Middle Cambrian
BT Cambrian
BT Paleozoic
SA Spain

Lancashire
No longer a valid term for GeoRef as of 1993.
use Lancashire England

Lancashire England (1993)
County in NW England. In 1974, parts were taken to form two new counties, Merseyside and Greater Manchester. Before 1993, also search Lancashire AND England.
CO N532000N543000 W0020000W0031500
UF Lancashire
BT England
BT Great Britain
BT United Kingdom
BT Western Europe
BT Europe
NT Clitheroe England
SA Liverpool Bay
SA Manchester England
SA Morecambe Bay

Lancaster County
Valid through 1988. Search in combination with state term. After 1988, use specific county-state term.

Lancaster County Nebraska (1989)
SE Nebraska. Before 1989, also search Lancaster County AND Nebraska.
CO N403200N410300 W0962800W0965500
BT Nebraska
BT United States

Lancaster County Pennsylvania (1989)
SE Pennsylvania. Before 1989, also search Lancaster County AND Pennsylvania.
CO N394300N401800 W0755300W0764300
BT Pennsylvania
BT United States

Lancaster County South Carolina (1989)
N South Carolina. Before 1989, also search Lancaster County AND South Carolina.
CO N342800N350500 W0802400W0805400
BT South Carolina
BT United States

Lancaster County Virginia (1989)
E Virginia. Before 1989, also search Lancaster County AND Virginia.
CO N373700N375100 W0761700W0763800
BT Virginia
BT United States

Lancaster Sound (1978)
Channel between Devon Island and N Baffin Island in E Franklin District.
BT Northwest Territories
BT Western Canada
BT Canada
SA Franklin District Northwest Territories

Lance Formation (1978)
Includes Colgate Sandstone Member, Cannonball Marine Member, Ludlow Lignitic Member, Tullock Member, Hell Creek Member, Torrington Member, and Ilo Ridge Member. Montana, W North Dakota, W South Dakota, and Wyoming.
BT Upper Cretaceous
BT Cretaceous
BT Mesozoic
SA Colorado
SA Montana
SA North Dakota
SA South Dakota
SA Wyoming

land bridges (1985)
SA biogeography
SA changes of level
SA continental drift
SA faunal provinces
SA floral provinces
SA migration
SA paleogeography

land canyons
  use canyons

land cover maps (1993)
  BT land use maps
  BT maps
  SA environmental geology maps
  SA land use
  SA natural resources

land forms
  use landforms

land leases (1978)
  SA conservation
  SA reclamation

land reclamation
  use reclamation

land subsidence (1978)
  Used for geotechnical studies on land subsidence.
  SA abandoned mines
  SA collapse structures
  SA compactness
  SA dams
  SA engineering geology
  SA explosions
  SA foundations
  SA geologic hazards
  SA highways
  SA mechanics
  SA mines
  SA oil and gas fields
  SA railroads
  SA rock mechanics
  SA settlement
  SA site exploration
  SA slope stability
  SA soil mechanics
  SA solution features
  SA stability
  SA subsidence
  SA tunnels
  SA underground installations

land use (1978)
  SA agriculture
  SA airfields
  SA arid environment
  SA basin management
  SA conservation
  SA deforestation
  SA engineering geology
  SA environmental geology
  SA geologic hazards
  SA human ecology
  SA impact statements
  SA Indian reservations
  SA land cover maps
  SA land use maps
  SA legislation
  SA management
  SA mine reactivation
  SA mines
  SA national parks
  SA natural resources
  SA planning
  SA polders
  SA pollution
  SA preservation
  SA railroads
  SA RARE II regions
  SA reclamation
  SA recreation
  SA regional planning
  SA regulations
  SA remote sensing
  SA soils
  SA strip mining
  SA surface mining
  SA underground space
  SA urban planning
  SA urbanization
  SA waste disposal
  SA wilderness areas

land use maps (1983)
  Before 1983, also search land use AND maps.
  BT maps
  NT land cover maps
  SA environmental geology
  SA environmental geology maps
  SA land use

Land's End (1978)
  Cape. SW end of Cornwall. Use for Lands End.
  UF Lands End
  BT Cornwall England
  BT England
  BT Great Britain
  BT United Kingdom
  BT Western Europe
  BT Europe

Landenian (1978)
  Belgium.
  BT upper Paleocene
  BT Paleocene
  BT Paleogene
  BT Tertiary
  BT Cenozoic

Lander County
  Valid through 1988. Search in combination with state term. After 1988, use specific county-state term.

Lander County Nevada (1989)
  N central Nevada. Before 1989, also search Lander County AND Nevada.
  CO N391000N410000
    W1163500W1174500
  BT Nevada
  BT United States
  SA Battle Mountain
  SA Copper Canyon
  SA Shoshone Mountains
  SA Toiyabe Range
  SA Toquima Range

Landes
  No longer a valid term for GeoRef. See Landes France.

Landes France (1993)
  Department in SW France.
  CO N433000N443000
    E0000800W0013000
  BT France
  BT Western Europe
  BT Europe
  NT Capbreton France
  SA Adour Basin

landfills (1978)
  NT sanitary landfills
  SA disposal barriers
  SA engineering geology
  SA environmental geology
  SA far-field
  SA near-field
  SA underground disposal
  SA waste disposal
  SA waste disposal sites

landform description (1978)
  Before 1975, also search surficial geology.
  SA aerial photography
  SA description
  SA geomorphology
  SA landform evolution
  SA landforms
  SA landscapes
  SA morphometry
  SA mountains
  SA relief
  SA surficial geology

landform evolution (1978)
  UF landscape evolution
  SA beaches
  SA emergent coastlines
  SA erosion
  SA erosion cycle
  SA etching
  SA evolution
  SA geomorphology
  SA karstification
  SA landform description
  SA landforms
  SA paleorelief
  SA relief
  SA relief inversion
  SA soil erosion

landforms (1978)
  UF land forms
  UF relief features
  SA geomorphology
  SA hills
  SA landform description
  SA landform evolution
  SA landscapes
  SA mountains
  SA plains
  SA plateaus
  SA relief exposure
  SA steppes
  SA tundra
  SA valleys

landing sites (1978)
  UF sites, landing
  SA Apollo Program
  SA lunar bases
  SA Mars
  SA Moon
  SA Venus
  SA Viking Program

Lands End
  use Land's End

Landsat (1978)
  Before 1978, also search ERTS; ERTS-1.
  SA geophysical methods
  SA geophysical surveys
  SA image enhancement
  SA imagery
  SA multispectral scanner
  SA remote sensing
  SA satellite methods
  SA telemetry
  SA thematic mapper

landscape evolution
  use landform evolution

landscapes (1978)
  SA geomorphology
  SA landform description
  SA landforms
  SA tundra

landslides (1978)
  Autoposting of mass movements to this term began in 1989.
  BT mass movements
  SA avalanches
  SA creep
  SA debris cones
  SA engineering geology
  SA erosion
  SA geologic hazards
  SA geomorphology
  SA gravity flows
  SA mudflows
  SA mudslides
  SA retaining walls
  SA rockslides
  SA slope stability
  SA slopes
  SA slumping
  SA soil erosion
  SA soils
  SA talus slopes

Lane County
  Valid through 1988. Search in combination with state term. After 1988, use specific county-state term.

Lane County Kansas (1989)
  W central Kansas. Before 1989, also search Lane County AND Kansas.
  CO N381600N384200
    W1001500W1004300
  BT Kansas
  BT United States

Lane County Oregon (1989)
  On Pacific Ocean in W Oregon. Before 1989, also search Lane County AND Oregon.
  CO N432600N441700
    W1214700W1240800
  BT Oregon
  BT United States

Lanersbach
  No longer a valid term for GeoRef. See Lanersbach Austria.

Lanersbach Austria (1993)
  Village SE of Innsbruck in central Tyrol, W Austria.
  BT Tyrol Austria
  BT Austria
  BT Central Europe
  BT Europe

Laneuville-devant-Nancy Boring (1978)
  Boring at or near outer SE suburb of Nancy.
  SA Meurthe-et-Moselle France

Laney Shale Member (1978)
  Of Green River Formation. SW Wyoming.
  BT middle Eocene
  BT Eocene
  BT Paleogene
  BT Tertiary
  BT Cenozoic
  SA Colorado
  SA Green River Formation
  SA Wyoming

langbeinite (1978)
  BT sulfates

Langebaanweg
  No longer a valid term for GeoRef. As of 1993 see Langebaanweg South Africa.

Langebaanweg South Africa (1993)
  Village in SW South Africa. Before 1993 also search Langebaanweg AND South Africa.
  BT Cape Province South Africa
  BT South Africa
  BT Southern Africa
  BT Africa

Langesund Fjord (1978)
  Inlet of the Skaggerak.
  BT Telemark Norway
  BT Norway
  BT Scandinavia
  BT Western Europe
  BT Europe

Langhe (1978)
  Region NW of Genoa.
  BT Italy
  BT Southern Europe
  BT Europe
  SA Piemonte Italy

Langhian (1978)
  Europe.
  BT Miocene
  BT Neogene

BT Tertiary
BT Cenozoic

Langkawi Islands (1978)
Group in Andaman Sea near entrance to Strait of Malacca. Administered by Kedah state in West Malaysia. Also search Langkawi. This term has multiple hierarchies.
BT1 West Malaysia
BT1 Malay Peninsula
BT1 Far East
BT1 Asia
BT2 West Malaysia
BT2 Malaysia
BT2 Far East
BT2 Asia
SA Kedah Malaysia

Langmuir equation (1985)
SA adsorption
SA equations
SA geochemistry
SA soils

Languedoc (1978)
Historical region of S France.
BT France
BT Western Europe
BT Europe

Lansing
Valid through 1988. Search in combination with state term. After 1988, use specific city-state term.

Lansing Group (1978)
Includes Lane Shale, Plattsburg Limestone, Vilas Shale, Stanton Limestone.
BT Missourian
BT Upper Pennsylvanian
BT Pennsylvanian
BT Carboniferous
BT Paleozoic
SA Kansas
SA Plattsburg Limestone
SA Stanton Formation

Lansing Michigan (1989)
City in Clinton, Eaton, and Ingham counties, S Michigan. Before 1989, search Lansing AND Michigan.
IN Index counties as applicable.
CO N424400N424400
 W0853400W0853400
BT Michigan Lower Peninsula
BT Michigan
BT United States
SA Clinton County Michigan
SA Eaton County Michigan
SA Ingham County Michigan

lanthanide series
use rare earths

lanthanides
use rare earths

lanthanoans
use rare earths

lanthanum (1978)
Autoposting of rare earths and metals to this term began in 1989.
UF La
BT rare earths
BT metals
SA cerium
SA lithophile elements

Lanzarote (1978)
Easternmost of the Spanish controlled Canary Islands.
BT Canary Islands
BT Atlantic Ocean Islands

Lanzo Massif (1978)
NW of Turin in Piemonte.

BT Piemonte Italy
BT Italy
BT Southern Europe
BT Europe

Laos (1978)
CO N140000N224000
 E1074000E1000000
BT Far East
BT Asia
SA Indochina

lapilli (1978)
SA ejecta
SA pyroclastics
SA volcaniclastics

Laplace azimuths
use Laplace transformations

Laplace points
use Laplace transformations

Laplace tidal equation
use Laplace transformations

Laplace transformations (1978)
UF Laplace azimuths
UF Laplace points
UF Laplace tidal equation
UF Laplace transforms
UF transformations, Laplace
SA geodesy

Laplace transforms
use Laplace transformations

Lapland (1978)
Region N of the Arctic Circle including the Kola Peninsula of the Russian Federation.
IN Index countries as applicable.
SA Europe
SA Finland
SA Kola Peninsula
SA Norway
SA Russian Federation
SA Sweden

Laptev Sea (1978)
Along N coast of Siberia between Taymyr Peninsula and New Siberian Islands.
CO N710000N813000
 E1420000E0963000
BT Arctic Ocean

Lar (1978)
River valley in N Iran. As of 1993, use Lar Iran for the city.
BT Iran
BT Middle East
BT Asia
SA Lar Iran

Lar Iran (1993)
City in Laristan region, Fars in S Iran.
BT Fars Iran
BT Iran
BT Middle East
BT Asia
SA Lar

Lara
No longer a valid term for GeoRef. As of 1993 see Lara Venezuela.

Lara Venezuela (1993)
State in NW Venezuela. Before 1993 also search Lara AND Venezuela.
BT Venezuela
BT South America

Laramian Orogenic Phase
use Laramide Orogeny

Laramian Orogeny
use Laramide Orogeny

Laramic Orogeny
use Laramide Orogeny

Laramide Orogeny (1978)
A time of deformation typically developed in the eastern Rocky Mountains of the United States, whose several phases extended from Upper Cretaceous until the end of the Paleocene. Before 1978, search orogeny AND Laramide; Laramian.
IN Index ages as applicable.
UF Laramian Orogenic Phase
UF Laramian Orogeny
UF Laramic Orogeny
UF Laramide Revolution
SA Cordilleran Orogeny
SA Cretaceous
SA orogeny
SA Paleocene
SA tectonics
SA Tertiary
SA Upper Cretaceous

Laramide Revolution
use Laramide Orogeny

Laramie anorthosite complex (1989)
BT Precambrian
SA Wyoming

Laramie Basin (1978)
High plateau in SE between Laramie Mountains on E and Medicine Bow Mountains on the W.
BT Wyoming
BT United States

Laramie County
Valid through 1988. Search in combination with state term. After 1988, use specific county-state term.

Laramie County Wyoming (1989)
SE Wyoming. Before 1989, also search Laramie County AND Wyoming.
CO N410000N414000
 W1040500W1052000
BT Wyoming
BT United States

Laramie Formation (1978)
Overlies Fox Hills Formation; unconformably underlies Denver Formation. Denver Basin region of E Colorado. Also in Arizona, N Colorado, Kansas, Montana, South Dakota and Wyoming.
BT Upper Cretaceous
BT Cretaceous
BT Mesozoic
SA Arizona
SA Colorado
SA Kansas
SA Montana
SA South Dakota
SA Wyoming

Laramie Mountains (1978)
A range of the Rocky Mountains. This term has multiple hierarchies.
IN Index states as applicable.
BT1 U. S. Rocky Mountains
BT1 Rocky Mountains
BT2 U. S. Rocky Mountains
BT2 United States
SA Colorado
SA Wyoming

Lardeau Group (1978)
Includes Triune and Sharon Creek Formations.
BT Cambrian
BT Paleozoic
SA British Columbia

SA Canada
SA Precambrian

Larder Lake District Ontario (1993)
E Ontario. Before 1993, also search Larder Lake region, Larder Lake District, or Larder Lake in conjunction with Ontario.
CO N480300N481000
 W0793000W0794500
UF Larder Lake region
BT Ontario
BT Eastern Canada
BT Canada
SA gold ores
SA Timiskaming District Ontario

Larder Lake region
No longer a valid term for GeoRef.
use Larder Lake District Ontario

Larderello (1978)
Geothermal field and geothermal region in central Italy.
UF Larderello Field
UF Larderello-Travale geothermal field
BT Tuscany Italy
BT Italy
BT Southern Europe
BT Europe
SA geothermal fields

Larderello Field
use Larderello

Larderello-Travale geothermal field
use Larderello

Large Aperture Seismic Array
use LASA

Largentiere
No longer a valid term for GeoRef. See Largentiere France.

Largentiere France (1993)
Village in SE France.
BT Ardeche France
BT France
BT Western Europe
BT Europe

Larimer County
Valid through 1988. Search in combination with state term. After 1988, use specific county-state term.

Larimer County Colorado (1989)
N Colorado. Before 1989, also search Larimer County AND Colorado.
CO N401500N410000
 W1045000W1062000
BT Colorado
BT United States
NT Loveland Colorado
SA Rocky Mountain National Park

Larix (1989)
Genus.
BT Pinaceae
BT Coniferales
BT gymnosperms
BT Spermatophyta
BT Plantae

larnite (1978)
Autoposting of nesosilicates began in 1985.
BT nesosilicates
BT orthosilicates
BT silicates

larvae (1985)
SA ontogeny
SA paleontology
SA reproduction

Larvik
No longer a valid term for GeoRef. As of 1993 see Larvik Norway.

Larvik Norway (1993)
Town at head of Larvik Fjord near entrance to Oslofjord. Before 1993 also search Larvik AND Norway.
BT Vestfold Norway
BT Norway
BT Scandinavia
BT Western Europe
BT Europe

Las Animas County
Valid through 1988. Search in combination with state term. After 1988, use specific county-state term.

Las Animas County Colorado (1989)
SE Colorado. Before 1989, also search Las Animas County AND Colorado.
CO N370000N375000 W1030500W1051500
BT Colorado
BT United States
SA Spanish Peaks

Las Vegas
Valid through 1988. Search in combination with state term. After 1988, use specific city-state term.

Las Vegas Nevada (1989)
City in extreme S Nevada. Before 1989, search Las Vegas AND Nevada.
CO N361000N361000 W1151000W1151000
BT Clark County Nevada
BT Nevada
BT United States

Las Vegas Valley (1981)
S Nevada.
IN Index county or regions as applicable.
SA Clark County Nevada

Las Villas
No longer a valid term for GeoRef. See Villa Clara Cuba.

LASA (1978)
Array of seismographs.
UF Large Aperture Seismic Array
SA arrays
SA Montana
SA seismic networks
SA seismographs

Laser Geodynamics Satellite
use Lageos

laser methods (1978)
Term reintroduced in 1982. Before 1982, also search laser probe. Before 1971, also search laser surveys. Laser methods was not a valid term from 1972 through 1981.
UF laser probe
UF laser surveys
UF probe, laser
SA analysis
SA chemical analysis
SA electromagnetic methods
SA geophysical methods
SA laser ranging
SA lunar laser ranging
SA methods
SA remote sensing
SA spectroscopy

laser probe
use laser methods

laser ranging (1989)
As of 1993, use in combination with satellite methods and/or space shuttle methods. Also see lunar laser ranging.
SA analysis
SA chemical analysis
SA geodesy
SA geophysical methods
SA geophysical surveys
SA Lageos
SA laser methods
SA lunar laser ranging
SA methods
SA remote sensing
SA satellite methods
SA spectroscopy

laser surveys
use laser methods

Lassen County
Valid through 1988. Search in combination with state term. After 1988, use specific county-state term.

Lassen County California (1989)
NE California. Before 1989, also search Lassen County AND California.
CO N394000N411000 W1200000W1212500
BT California
BT United States
SA Lassen Volcanic National Park

Lassen Peak (1978)
Active volcano in Lassen Volcanic National Park in NE California.
UF Mount Lassen
BT Shasta County California
BT California
BT United States

Lassen Volcanic National Park (1978)
NE California.
IN Index counties as applicable.
BT California
BT United States
SA Lassen County California
SA national parks
SA Plumas County California
SA Shasta County California
SA Tehama County California

Latah County
Valid through 1988. Search in combination with state term. After 1988, use specific county-state term.

Latah County Idaho (1989)
N Idaho. Before 1989, also search Latah County AND Idaho.
CO N463000N471000 W1162200W1170500
BT Idaho
BT United States

Latdorfian
use Lattorfian

late diagenesis (1978)
BT diagenesis
SA burial diagenesis
SA early diagenesis

Latemar Massif (1978)
In Dolomites. This term has multiple hierarchies.
BT1 Dolomites
BT1 Eastern Alps
BT1 Alps
BT1 Europe
BT2 Dolomites
BT2 Italy
BT2 Southern Europe
BT2 Europe
BT3 Trentino-Alto Adige Italy
BT3 Italy
BT3 Southern Europe
BT3 Europe

lateral faults (1989)
BT faults
NT left-lateral faults
NT right-lateral faults

lateral heterogeneity (1989)
SA mantle
SA upper mantle
SA velocity structure

lateral loading (1989)
BT loading
SA engineering properties
SA load tests
SA soil mechanics
SA testing

lateral logging
use Laterolog

laterite
Not a valid GeoRef term after 1970. Was used on level 1 in subfile N.
use laterites

Laterite soils
use laterites

laterites (1978)
Before 1971, also search laterite. Before 1978, also search lateritic soils.
UF laterite
UF Laterite soils
UF lateritic soils
BT soils
SA bauxite
SA bauxitization
SA biorhexistasy
SA laterization
SA Latosols
SA ocher
SA soil group

lateritic soils
No longer a valid term for GeoRef. Used in subfiles B, N, and E.
use laterites

laterization (1978)
SA bauxitization
SA laterites
SA mineral deposits, genesis
SA soils

Laterolog (1984)
UF focused current logging
UF guard log
UF lateral logging
SA electrical logging
SA resistivity
SA well-logging

Lathi Formation (1978)
BT Jurassic
BT Mesozoic
SA India
SA Rajasthan India

Latimer County Oklahoma (1993)
SE Oklahoma. Before 1989, also search Latimer County AND Oklahoma.
CO N344100N350300 W0945500W0953100
BT Oklahoma
BT United States
SA Ouachita Mountains

Latin America (1981)
Region comprising South America except the Guianas; Central America except Belize; Mexico, Cuba, Puerto Rico, Dominican Republic and some of the islands in the West Indies.
IN Used only when a more specific area term is not given.
SA America
SA Central America
SA South America

latite (1978)
BT trachyandesites
BT volcanic rocks
BT igneous rocks
SA quartz latite

latitude (1993)
Includes latitude effects.
SA altitude
SA climate
SA environment
SA geodesy
SA paleolatitude

latitude, paleo-
use paleolatitude

Latium
No longer a valid term for GeoRef. See Latium Italy.

Latium Italy (1993)
Autonomous region in W central Italy.
CO N411400N425000 E0140000E0113000
BT Italy
BT Southern Europe
BT Europe
NT Alban Hills
NT Frosinone Italy
NT Rome Italy
NT Sabatini Mountains
NT Tolfa Hills
NT Viterbo Italy
SA Tiber Valley

Latosols (1978)
BT soils
SA laterites
SA soil group

Latrobe Group (1989)
Gippsland Basin of Victoria.
BT Paleocene
BT Paleogene
BT Tertiary
BT Cenozoic
SA Australia
SA Gippsland Basin
SA Victoria Australia

Latrobe Valley (1989)
Coal mining region in S Victoria.
SA Gippsland Basin
SA Victoria Australia

lattice (1978)
UF Bravais lattice
UF crystal lattice
UF space lattice
UF translation lattice
SA cell dimensions
SA crystal structure
SA crystal systems
SA habit
SA lattice parameters
SA substitution
SA twinning
SA unit cell

lattice constants
use lattice parameters

lattice parameters (1978)
Before 1979, also search lattice AND parameters; parameters AND crystal structure.
UF lattice constants
SA cell dimensions
SA crystal chemistry
SA crystal growth

SA crystal structure
SA crystallography
SA lattice
SA unit cell

Lattorfian (1978)
Europe.
UF Latdorfian
BT upper Eocene
BT Eocene
BT Paleogene
BT Tertiary
BT Cenozoic
SA Krosno Beds
SA Sannoisian
SA Tongrian

Latvia (1978)
Formerly the Latvian Soviet Socialist Republic.
CO N560000N580000
    E0280000E0210000
BT Baltic region
BT Europe
NT Riga Latvia
SA Baltic Glaciation
SA Baltic Plain
SA Baltic Shield
SA Dvina River
SA Russian Plain
SA Russian Platform

Lau Basin (1978)
West of the Tonga Islands and S of the Lau group of the Fiji Islands.
BT Pacific Ocean
SA Leg 135

Lauderdale County
Valid through 1988. Search in combination with state term. After 1988, use specific county-state term.

Lauderdale County Alabama (1989)
Extreme NW Alabama. Before 1989, also search Lauderdale County AND Alabama.
CO N344500N350000
    W0871300W0881300
BT Alabama
BT United States

Lauderdale County Mississippi (1989)
E Mississippi. Before 1989, also search Lauderdale County AND Mississippi.
CO N321300N323600
    W0882500W0885300
BT Mississippi
BT United States

Lauderdale County Tennessee (1989)
W Tennessee. Before 1989, also search Lauderdale County AND Tennessee.
CO N353300N355700
    W0892200W0895500
BT Tennessee
BT United States

laughing gas
  use nitrous oxide

laumonite
  use laumontite

laumontite (1978)
UF laumonite
UF lomonite
UF lomontite
BT zeolite group
BT framework silicates
BT silicates
SA leonhardite

Lauraceae (1978)

Family.
BT Dicotyledoneae
BT angiosperms
BT Spermatophyta
BT Plantae

Laurasia (1978)
The protocontinent of the Northern Hemisphere which is a combination of Laurentia and Eurasia. It included most of North America, Greenland, and much of Eurasia excluding India.
SA continental drift
SA Eurasia
SA Gondwana
SA Laurentia
SA Northern Hemisphere
SA Pangaea

Laurentia (1993)
Ancient continent which formed in the early Proterozoic. It preceeded Laurasia. Included much of the Canadian Shield and Greenland and NW Scotland.
IN Index countries and continents as applicable.
SA Atlantic region
SA Canadian Shield
SA continental drift
SA Greenland
SA Laurasia
SA North America
SA plate tectonics
SA Scotland

Laurentian Highlands
  use Canadian Shield

Laurentian Plateau
  use Canadian Shield

Laurentide ice sheet (1978)
SA Des Moines Lobe
SA Saint Lawrence Lowlands
SA Saint Lawrence Valley

Laurion
No longer a valid term for GeoRef. As of 1993 see Laurion Greece.

Laurion Greece (1993)
City SE of Athens. Before 1993 also search Laurion AND Greece.
UF Laurium
UF Lavrion
BT Sterea Ellas
BT Greece
BT Southern Europe
BT Europe

Laurium
  use Laurion Greece

Lausitz
  use Lusatia

lava (1978)
This set is used for all molten extrusives and the rocks that solidify from them.
NT aa lava
NT pahoehoe
NT pillow lava
SA basalts
SA effusion
SA eruptions
SA extrusive rocks
SA flow mechanism
SA igneous rocks
SA inclusions
SA lava channels
SA lava fields
SA lava flows
SA lava lakes
SA lava tubes
SA magmas
SA mud volcanoes

SA paleomagnetism
SA perlite
SA petrology
SA phreatomagmatism
SA pillow structure
SA plugs
SA vesicular texture
SA viscosity
SA volcanic features
SA volcanic rocks
SA volcanism
SA volcanoes
SA volcanology

lava channels (1978)
BT volcanic features
SA channels
SA geomorphology
SA lava

lava domes
  use shield volcanoes

lava fields (1978)
UF fields, lava
BT volcanic features
SA geomorphology
SA lava
SA volcanic fields

lava flows (1978)
Also search specific types, such as basalt flows or rhyolite flows as applicable.
UF flow (volcanic)
UF nappe (volcanic)
SA basalt flows
SA effusion
SA flood basalts
SA flow mechanism
SA flows
SA lahars
SA lava
SA lava tubes
SA volcanic features
SA volcanology

lava lakes (1978)
BT volcanic features
SA crater lakes
SA geomorphology
SA lakes
SA lava

lava tubes (1978)
Before 1981, also search lava tunnels.
UF lava tunnels
UF tubes, lava
BT volcanic features
SA geomorphology
SA lava
SA lava flows

lava tunnels
As of 1981, no longer a valid term for GeoRef.
  use lava tubes

Lavaca County Texas (1993)
SE Texas. Before 1989, also search Lavaca County AND Texas.
CO N290500N293800
    W0963000W0971300
BT Texas
BT United States

Laval Basin (1978)
BT France
BT Western Europe
BT Europe

Lavery Till (1985)
NW Pennsylvania.
BT Wisconsinan
BT upper Pleistocene
BT Pleistocene
BT Quaternary
BT Cenozoic

SA Pennsylvania

Lavrion
  use Laurion Greece

Law of the Sea (1982)
Before 1993, also search UNCLOS.
UF UNCLOS
UF United Nations Convention on the Law of the Sea
SA economic geology
SA Exclusive Economic Zone
SA international cooperation
SA legislation
SA mineral resources
SA mining legislation
SA nodules
SA ocean floors
SA oceanography
SA United States Exclusive Economic Zone
SA world ocean

Lawrence County
Valid through 1988. Search in combination with state term. After 1988, use specific county-state term.

Lawrence County Alabama (1989)
NW Alabama. Before 1989, also search Lawrence County AND Alabama.
CO N341700N344500
    W0870600W0873000
BT Alabama
BT United States

Lawrence County Arkansas (1989)
NE Arkansas. Before 1989, also search Lawrence County AND Arkansas.
CO N355300N361300
    W0904600W0912300
BT Arkansas
BT United States

Lawrence County Illinois (1989)
Bounded by Wabash River in SE Illinois. Before 1989, also search Lawrence County AND Illinois.
CO N383500N385200
    W0873000W0875500
BT Illinois
BT United States

Lawrence County Indiana (1989)
S Indiana. Before 1989, also search Lawrence County AND Indiana.
CO N384200N385900
    W0861800W0864200
BT Indiana
BT United States

Lawrence County Kentucky (1989)
NE Kentucky. Before 1989, also search Lawrence County AND Kentucky.
CO N375200N381700
    W0823200W0830200
BT Kentucky
BT United States

Lawrence County Mississippi (1989)
S central Mississippi. Before 1989, also search Lawrence County AND Mississippi.
CO N312000N314600
    W0895800W0901600
BT Mississippi
BT United States

Lawrence County Missouri (1989)

SW Missouri. Before 1989, also search Lawrence County AND Missouri.
CO N365500N371800
   W0933700W0940400
BT Missouri
BT United States

Lawrence County Ohio (1989)
S Ohio. Before 1989, also search Lawrence County AND Ohio.
CO N382300N385100
   W0821800W0824800
BT Ohio
BT United States

Lawrence County Pennsylvania (1989)
W Pennsylvania. Before 1989, also search Lawrence County AND Pennsylvania.
CO N405200N410800
   W0800700W0803200
BT Pennsylvania
BT United States

Lawrence County South Dakota (1989)
W South Dakota. Before 1989, also search Lawrence County AND South Dakota.
CO N440900N443600
   W1032700W1040400
BT South Dakota
BT United States
SA Homestake Mine
SA Whitewood Creek

Lawrence County Tennessee (1989)
S Tennessee. Before 1989, also search Lawrence County AND Tennessee.
CO N350000N352800
   W0871300W0873500
BT Tennessee
BT United States

Lawrence Formation (1978)
Includes two unnamed shale members separated by Amazonia Limestone Member, Ireland Limestone Member. SW Iowa, E Kansas, NW Missouri, and SE Nebraska.
BT Virgilian
BT Upper Pennsylvanian
BT Pennsylvanian
BT Carboniferous
BT Paleozoic
SA Douglas Group
SA Iowa
SA Kansas
SA Missouri
SA Nebraska

lawrencium (1989)
Chemical element.
UF Lr
BT actinides
BT metals

laws
Not a valid term for GeoRef. As of 1993, use legislation or principles as applicable.

lawsonite (1978)
Autoposting of sorosilicates began in 1985.
BT sorosilicates
BT orthosilicates
BT silicates
SA aluminosilicates

Laxfordian (1978)
Europe. Autoposting of upper Precambrian and Precambrian to this term ended in 1988. As of 1989, middle Proterozoic and Proterozoic are autoposted to this term. Autoposting of upper Precambrian and Precambrian to this term began in 1990.
BT middle Proterozoic
BT Proterozoic
BT upper Precambrian
BT Precambrian

layer, active
   use active layer

layer, basaltic
   use basaltic layer

layer, boundary
   use boundary layer

layer, granitic
   use granitic layer

layer, nepheloid
   use nepheloid layer

layered
A valid index term through 1977. After 1977, use layered intrusions or layered materials.

layered intrusions (1978)
Not a valid term from 1975 through 1977. Before 1978, also search intrusions AND layered.
BT intrusions
SA igneous rocks

layered materials (1981)
Before 1981, also search layered media; layered; layers. After 1992, use multilayered systems for aquifers.
UF layered media
SA engineering geology
SA materials
SA multiple aquifers

layered media
As of 1981, no longer a valid term for GeoRef.
use layered materials

layers
A valid term through 1978. See layered materials.

lazulite (1985)
BT phosphates

lazurite (1978)
From 1978-1980, framework silicates, sulfates and sulfides were autoposted to this term. Autoposting of sodalite group began in 1981. Autoposting of framework silicates and silicates began in 1989. This term has multiple hierarchies.
BT1 sodalite group
BT1 framework silicates
BT1 silicates
BT2 sulfates
BT3 sulfides

Le Flore County
Valid through 1988. Search in combination with state term. After 1988, use specific county-state term.

Le Flore County Oklahoma (1989)
SE Oklahoma. Before 1989, also search Le Flore County AND Oklahoma.
CO N343000N352300
   W0942500W0950500
BT Oklahoma
BT United States

Le Havre
No longer a valid term for GeoRef. See Le Havre France.

Le Havre France (1993)
City on the English Channel in N France.
BT Seine-Maritime France
BT France
BT Western Europe
BT Europe

Le Mans
No longer a valid term for GeoRef. See Le Mans France.

Le Mans France (1993)
City in NE central France.
BT Sarthe France
BT France
BT Western Europe
BT Europe

Le Mont-Dore France
   use Mont-Dore France

Le Trou Sans Fond
   use Trou Sans Fond

Lea County
Valid through 1988. Search in combination with state term. After 1988, use specific county-state term.

Lea County New Mexico (1989)
Extreme SE New Mexico. Before 1989, also search Lea County AND New Mexico.
CO N320000N333500
   W1030300W1035000
BT New Mexico
BT United States
SA Central Basin Platform

leachate (1993)
Use leaching or solution for the process.
SA geochemistry
SA leaching
SA mining geology
SA pollutants
SA pollution
SA seepage
SA soils
SA solutions
SA waste disposal

leaching (1978)
UF lixiviation
SA alteration
SA degradation
SA desalinization
SA geochemistry
SA gossan
SA leachate
SA mining geology
SA pedogenesis
SA percolation
SA soils
SA solution
SA solution mining
SA solution transport
SA weathering

lead (1978)
Chemical element. As of 1981, use lead ores for lead as a commodity. Autoposting of metals to this term began in 1989.
UF Pb
BT metals
NT Pb-204
NT Pb-206
NT Pb-206/Pb-204
NT Pb-207
NT Pb-207/Pb-204
NT Pb-207/Pb-206
NT Pb-208
NT Pb-208/Pb-204
NT Pb-208/Pb-206
NT Pb-210
NT U-238/Pb-204
NT U-238/Pb-206
SA chalcophile elements
SA heavy metals
SA isotopes
SA lead ores
SA lead-zinc deposits
SA Pb/Pb
SA Pb/Th
SA U/Pb
SA U/Th/Pb

lead glance
   use galena

lead ores (1981)
Before 1981, also search (lead OR Pb) AND (deposit OR deposits OR ore OR ores OR economic) in the basic index. Autoposting of metal ores to this term began in 1985.
IN Commodity. See List C.
BT metal ores
SA base metals
SA Bunker Hill Mine
SA galena
SA lead
SA lead-zinc deposits
SA mississippi valley-type

lead-lead
Not a valid term for GeoRef. Before 1971, was used in subfiles E, G, N and B.
use Pb/Pb

lead-zinc deposits (1978)
Before 1978, also search lead-zinc AND (deposit OR deposits OR ore OR ores OR economic) in the basic index. Autoposting of metal ores to this term began in 1985.
IN Commodity. See List C.
UF lead-zinc ores
BT metal ores
SA Aberfeldy
SA Bluebell Mine
SA Broken Hill Mine
SA Bunker Hill Mine
SA cyprus-type
SA Hosokura Mine
SA Janggun Mine
SA Kamioka Mine
SA Keban Mine
SA lead
SA lead ores
SA Leadville mining district
SA mississippi valley-type
SA Pine Point mining district
SA Sullivan Mine
SA Toyoha Mine
SA Trepca Mine
SA zinc
SA zinc ores

lead-zinc ores
   use lead-zinc deposits

leadhillite (1989)
This term has multiple hierarchies.
BT1 carbonates
BT2 sulfates

Leadville
Valid through 1988. Search in combination with state term. After 1988, use specific city-state term.

Leadville Colorado (1989)
City in central Colorado. Before 1989, search Leadville AND Colorado.
CO N391400N391400
   W1061800W1061800
BT Lake County Colorado

BT Colorado
BT United States
Leadville Formation (1978)
Includes Gilman Sandstone Member.
BT Mississippian
BT Carboniferous
BT Paleozoic
SA Colorado
SA Yule Marble
Leadville mining district (1993)
In the central Colorado mineral belt, Leadville area, Lake County Colorado. Lead-zinc, silver, and gold deposits. Before 1993, also search Leadville District or Leadville and commodity.
BT Lake County Colorado
BT Colorado
BT United States
SA Colorado mineral belt
SA gold ores
SA lead-zinc deposits
SA mines
SA polymetallic ores
SA silver ores
leaf
 use leaves
leakage
 use seepage
leakage anomalies (1981)
SA anomalies
SA dispersion patterns
SA geochemistry
SA hydrothermal alteration
SA mineral deposits, genesis
leaking underground storage tanks (1993)
Before 1993, also search LUST or leaking underground storage tanks.
UF LUST
SA dense nonaqueous phase liquids
SA ground water
SA hydrocarbons
SA nonaqueous phase liquids
SA oil storage
SA polluted water
SA pollution
SA reclamation
SA remediation
SA seepage
SA underground installations
SA underground storage
SA waste disposal
SA water supply
leaky aquifers (1985)
SA aquifers
SA aquitards
SA ground water
SA multiple aquifers
SA seepage
least squares analysis
 use least-squares analysis
least-squares analysis (1978)
UF least squares analysis
BT statistical analysis
SA analysis
SA regression analysis
Leavenworth County
Valid through 1988. Search in combination with state term. After 1988, use specific county-state term.
Leavenworth County Kansas (1989)
NE Kansas. Before 1989, also search Leavenworth County AND Kansas.

CO N385500N392500
W0944500W0951200
BT Kansas
BT United States
leaves (1978)
UF leaf
SA flowers
SA humus
SA paleobotany
SA Plantae
SA vegetation
Leba (1978)
River in N Poland. Use Leba Poland for the town.
BT Poland
BT Central Europe
BT Europe
SA Leba Poland
Leba Poland (1993)
Town on the Baltic in Slupsk province. Use Leba for the river. Before 1993 also search Leba AND Poland.
BT Poland
BT Central Europe
BT Europe
SA Leba
Lebanon (1978)
CO N330000N344500
E0364500E0350000
BT Middle East
BT Asia
NT Beirut Lebanon
SA Judea Group
SA Mediterranean region
Lebanon County
Valid through 1988. Search in combination with state term. After 1988, use specific county-state term.
Lebanon County Pennsylvania (1989)
SE central Pennsylvania. Before 1989, also search Lebanon County AND Pennsylvania.
CO N401000N403500
W0761000W0764200
BT Pennsylvania
BT United States
Lebedin
No longer a valid term for GeoRef. As of 1993, see Lebedin Ukraine.
Lebedin Ukraine (1993)
City in Sumy Oblast in NE central Ukraine. This term has multiple hierarchies.
BT1 Ukraine
BT1 Europe
BT2 Ukraine
BT2 Commonwealth of Independent States
lebensspuren (1978)
Before 1981, also search ichnofossils. Before 1973, also search trace fossils. As of 1981, used when the emphasis is sedimentary petrology, not paleontology. Use ichnofossils for paleontologic studies.
BT biogenic structures
BT sedimentary structures
SA borings
SA burrows
SA casts
SA coprolites
SA ichnofossils
SA tracks
SA trails
Lebo Member (1993)

Paleocene. Considered Eocene earlier. Montana. In Fort Union Formation.
BT Paleocene
BT Paleogene
BT Tertiary
BT Cenozoic
SA Fort Union Formation
SA Montana
Lebombo Mountains (1978)
Also search Lebombo.
IN Index Mozambique and/or Transvaal as applicable.
BT Africa
SA Mozambique
SA South Africa
SA Transvaal South Africa
Lecce
No longer a valid term for GeoRef. See Lecce Italy.
Lecce Italy (1993)
City and province in heel of Italian boot, SE Italy.
BT Apulia Italy
BT Italy
BT Southern Europe
BT Europe
lechatelierite (1978)
BT silica minerals
BT framework silicates
BT silicates
Lechtal Alps (1978)
W Tyrol. This term has multiple hierarchies.
BT1 Tyrol Austria
BT1 Austria
BT1 Central Europe
BT1 Europe
BT2 Alps
BT2 Europe
Lecompton Limestone (1978)
In Shawnee Group. Comprises Spring Branch Limestone, Doniphan Shale, Big Springs Limestone, Queen Hill Shale, Beil Limestone, King Hill Shale, Avoca Limestone members. E Kansas, NW Missouri, SE Nebraska, SW Iowa, central N Oklahoma.
BT Virgilian
BT Upper Pennsylvanian
BT Pennsylvanian
BT Carboniferous
BT Paleozoic
SA Beil Limestone Member
SA Iowa
SA Kansas
SA Missouri
SA Nebraska
SA Oklahoma
SA Shawnee Group
Leczyca
No longer a valid term for GeoRef. As of 1993 see Leczyca Poland.
Leczyca Poland (1993)
Town in Plock Province, central Poland. Before 1993 also search Leczyca AND Poland
BT Plock Poland
BT Poland
BT Central Europe
BT Europe
Leda Clay (1978)
BT Cenozoic
SA Canada
SA Ontario
SA Ottawa Ontario
Ledbetter Slate (1985)
NE Washington.

BT Ordovician
BT Paleozoic
SA Washington
Leduc
No longer a valid term for GeoRef. See Leduc Alberta.
Leduc Alberta (1993)
Town S of Edmonton in central Alberta.
BT Alberta
BT Western Canada
BT Canada
Lee County
Valid through 1988. Search in combination with state term. After 1988, use specific county-state term.
Lee County Alabama (1989)
E Albama. Before 1989, also search Lee County AND Alabama.
CO N322500N324400
W0850000W0854200
BT Alabama
BT United States
Lee County Arkansas (1989)
E Arkansas. Before 1989, also search Lee County AND Arkansas.
CO N343800N345400
W0902500W0910600
BT Arkansas
BT United States
Lee County Florida (1989)
On Gulf of Mexico in SW Florida. Before 1989, also search Lee County AND Florida.
CO N261900N264400
W0813500W0821900
BT Florida
BT United States
NT Sanibel Island
Lee County Georgia (1989)
SW central Georgia. Before 1989, also search Lee County AND Georgia.
CO N313800N315500
W0835400W0842000
BT Georgia
BT United States
Lee County Illinois (1989)
N Illinois. Before 1989, also search Lee County AND Illinois.
CO N413500N415400
W0885600W0893800
BT Illinois
BT United States
Lee County Iowa (1989)
Extreme SE Iowa. Before 1989, also search Lee County AND Iowa.
CO N402300N404800
W0910700W0914200
BT Iowa
BT United States
Lee County Kentucky (1989)
E central Kentucky. Before 1989, also search Lee County AND Kentucky.
CO N372800N374300
W0833000W0835300
BT Kentucky
BT United States
Lee County Mississippi (1989)
NE Mississippi. Before 1989, also search Lee County AND Mississippi.
CO N340500N343000
W0883200W0884800
BT Mississippi

BT United States

Lee County North Carolina
(1989)
Central North Carolina. Before 1989, also search Lee County AND North Carolina.
CO N351700N353800
W0785800W0792200
BT North Carolina
BT United States

Lee County South Carolina
(1989)
NE central South Carolina. Before 1989, also search Lee County AND South Carolina.
CO N335700N342300
W0800000W0802800
BT South Carolina
BT United States

Lee County Texas (1989)
S central Texas. Before 1989, also search Lee County AND Texas.
CO N300300N303500
W0963800W0972000
BT Texas
BT United States

Lee County Virginia (1989)
Extreme SW Virginia. Before 1989, also search Lee County AND Virginia.
CO N363600N365300
W0824700W0834200
BT Virginia
BT United States

Lee Creek Mine (1989)
E North Carolina.
BT Beaufort County North Carolina
BT North Carolina
BT United States
SA mines
SA phosphate deposits

Lee Formation (1978)
Includes Rockcastle Sandstone. Defined as all of lower Pennsylvanian rocks from Mississippian-Pennsylvanian contact to top of Zachariah Coal Bed. E Kentucky, E Tennessee, and SW Virginia.
BT Lower Pennsylvanian
BT Pennsylvanian
BT Carboniferous
BT Paleozoic
SA Kentucky
SA Tennessee
SA Virginia

Leech River Fault (1989)
Vancouver Island, British Columbia; NW Washington.
IN Index British Columbia and/or Washington as applicable.
BT North America
SA British Columbia
SA Vancouver Island
SA Washington

Leeward Islands (1978)
The northern chain of islands in the Lesser Antilles extending from the Virgin Islands in N to Dominica in S. Also occurs in Society Islands. From 1981 through 1992, the following terms (and their respective narrower terms) were narrower terms: Guadeloupe, Montserrat and Virgin Islands. This term was autoposted to Guadalupe and Virgin Islands frrom 1978 through 1992. Search in combination with Lesser Antilles or Society Islands.
IN Index Lesser Antilles or Society Islands as applicable.
SA Guadeloupe
SA La Grande Soufriere
SA Lesser Antilles
SA Montserrat
SA Netherlands Antilles
SA Saint John Island
SA Society Islands

left-lateral faults (1985)
As of 1989, lateral faults is autoposted to this term.
BT lateral faults
BT faults
SA right-lateral faults
SA strike-slip faults

Leg 1 (1981)
Gulf of Mexico and NW Atlantic Ocean, off Florida and SW of Bermuda. Sites 1-7. Autoposting of this term to individual sites began in 1989.
CO N230000N310000
W0670000W0930000
BT Deep Sea Drilling Project
NT DSDP Site 3
NT DSDP Site 4
SA Atlantic Ocean
SA Gulf of Mexico

Leg 2 (1981)
N Atlantic Ocean. Sites 8-12, from NW of Bermuda to NW of Cape Verde Islands.
CO N190000N360000
W0253000W0680000
BT Deep Sea Drilling Project
SA Atlantic Ocean
SA North Atlantic

Leg 3 (1981)
S Atlantic Ocean. Sites 13-22, primarily spanning the Mid-Atlantic Ridge at approximately S30°, W to the Rio Grande Rise. Autoposting of this term to individual sites began in 1989.
CO S310000N100000
W0055000W0360000
BT Deep Sea Drilling Project
NT DSDP Site 19
NT DSDP Site 20
NT DSDP Site 21
SA Atlantic Ocean

Leg 4 (1981)
W Atlantic Ocean, NE of Brazil to the Puerto Rico Trench area and the Caribbean Sea. Sites 23-31.
CO S070000N210000
W0300000W0730000
BT Deep Sea Drilling Project
SA Atlantic Ocean
SA Caribbean Sea

Leg 5 (1981)
NE Pacific Ocean. Sites 32-43, off California and across the Mendocino, Murray, and Clarion fracture zones at approximately W140°. Autoposting of this term to individual sites began in 1993.
CO N130000N410000
W1260000W1520000
BT Deep Sea Drilling Project
NT DSDP Site 32
NT DSDP Site 35
NT DSDP Site 40
SA Pacific Ocean

Leg 6 (1981)
NW Pacific Ocean and Philippine Sea. Sites 44-60 from N of the Mid-Pacific Mountains to the Shatsky Rise, the E Philippine Sea, and the Mariana Trench. Autoposting of this term to individual sites began in 1993.
CO N080000N340000
W1680000E1400000
BT Deep Sea Drilling Project
NT DSDP Site 51
SA Pacific Ocean
SA Philippine Sea

Leg 7 (1981)
W equatorial and N central Pacific Ocean. Sites 61-67. Autoposting of this term to individual sites began in 1989.
CO S020000N250000
W1570000E1410000
BT Deep Sea Drilling Project
NT DSDP Site 61
NT DSDP Site 62
NT DSDP Site 63
NT DSDP Site 64
NT DSDP Site 65
NT DSDP Site 66
NT DSDP Site 67
SA Pacific Ocean

Leg 8 (1981)
E equatorial Pacific Ocean. Sites 68-75. Autoposting of this term to individual sites began in 1989.
CO S130000N170000
W1340000W1650000
BT Deep Sea Drilling Project
NT DSDP Site 69
NT DSDP Site 70
NT DSDP Site 71
NT DSDP Site 72
NT DSDP Site 73
NT DSDP Site 74
NT DSDP Site 75
SA Pacific Ocean

Leg 9 (1981)
E equatorial Pacific Ocean. Sites 76-84. Autoposting of this term to individual sites began in 1989.
CO S150000N080000
W0820000W1460000
BT Deep Sea Drilling Project
NT DSDP Site 77
SA Pacific Ocean

Leg 10 (1981)
Gulf of Mexico. Sites 85-97. Autoposting of this term to individual sites began in 1989.
CO N200000N260000
W0840000W0960000
BT Deep Sea Drilling Project
NT DSDP Site 85
NT DSDP Site 86
NT DSDP Site 88
NT DSDP Site 89
NT DSDP Site 90
NT DSDP Site 91
NT DSDP Site 92
NT DSDP Site 94
NT DSDP Site 95
NT DSDP Site 96
NT DSDP Site 97
SA Gulf of Mexico

Leg 11 (1981)
NW Atlantic Ocean. Sites 98-108, from N of the Bahamas to NW of Bermuda near the continental margin of the United States. Autoposting of this term to individual sites began in 1989.
CO N230000N404500
W0690000W0801000
BT Deep Sea Drilling Project
NT DSDP Site 100
NT DSDP Site 102
NT DSDP Site 103
NT DSDP Site 104
NT DSDP Site 105
NT DSDP Site 108
SA Atlantic Ocean
SA North American Atlantic

Leg 12 (1981)
N Atlantic Ocean. Sites 111-119, from N of the Grand Banks to the Labrador Sea, to SW of Iceland, to W of Scotland, to N of Spain. Autoposting of this term to individual sites began in 1989.
CO N450000N600000
W0070000W0490000
BT Deep Sea Drilling Project
NT DSDP Site 111
NT DSDP Site 116
NT DSDP Site 117
NT DSDP Site 119
SA Atlantic Ocean
SA Northeast Atlantic
SA Northwest Atlantic

Leg 13 (1981)
Mediterranean Sea. Sites 120-134. (Site 120 is nearby in the Atlantic Ocean.) Autoposting of this term to individual sites began in 1989.
CO N320000N410000
E0290000W0120000
BT Deep Sea Drilling Project
NT DSDP Site 120
NT DSDP Site 121
NT DSDP Site 122
NT DSDP Site 123
NT DSDP Site 124
NT DSDP Site 125
NT DSDP Site 126
NT DSDP Site 127
NT DSDP Site 128
NT DSDP Site 129
NT DSDP Site 130
NT DSDP Site 131
NT DSDP Site 132
NT DSDP Site 133
NT DSDP Site 134
SA Atlantic Ocean
SA Bannock Basin
SA European Atlantic
SA Mediterranean Sea
SA Strabo Trench

Leg 14 (1981)
N Atlantic Ocean off the coasts of Africa and South America. Sites 135-144. Autoposting of this term to individual sites began in 1989.
CO N030000N360000
W0100000W0550000
BT Deep Sea Drilling Project
NT DSDP Site 137
NT DSDP Site 141
SA Atlantic Ocean
SA Cape Verde Atlantic
SA Guyanese Atlantic

Leg 15 (1981)
Caribbean Sea. Sites 145-154. Autoposting of this term to individual sites began in 1989.
CO N100000N170000
W0630000W0810000
BT Deep Sea Drilling Project
NT DSDP Site 146
NT DSDP Site 147
NT DSDP Site 147A
NT DSDP Site 148
NT DSDP Site 149
NT DSDP Site 150
NT DSDP Site 151
NT DSDP Site 152
NT DSDP Site 153
NT DSDP Site 154
SA Aves Ridge
SA Beata Ridge

Before then, variants of the term may occur in GeoRef.

SA Caribbean Sea
SA Colombian Basin
SA Nicaragua Rise
SA Venezuelan Basin

Leg 16 (1981)
E Equatorial Pacific. Sites 155-163. Autoposting of this term to individual sites began in 1989.
CO S020000N150000
W0800000W1510000
BT Deep Sea Drilling Project
NT DSDP Site 157
NT DSDP Site 158
NT DSDP Site 159
NT DSDP Site 163
SA Carnegie Ridge
SA Cocos Ridge
SA East Pacific Rise
SA Equatorial Pacific
SA Pacific Ocean

Leg 17 (1981)
N central and Equatorial Pacific. Sites 164-171. Autoposting of this term to individual sites began in 1989.
CO N020000N200000
W1610000E1730000
BT Deep Sea Drilling Project
NT DSDP Site 167
NT DSDP Site 171
SA Equatorial Pacific
SA Pacific Ocean
SA Polynesian Pacific

Leg 18 (1978)
NE Pacific Ocean. Sites 172-182, from off California and Oregon to the Gulf of Alaska. Autoposting of this term to individual sites began in 1989.
CO N290000N590000
W1220000W1350000
BT Deep Sea Drilling Project
NT DSDP Site 172
NT DSDP Site 173
NT DSDP Site 174
NT DSDP Site 175
NT DSDP Site 176
NT DSDP Site 177
NT DSDP Site 178
NT DSDP Site 179
NT DSDP Site 180
NT DSDP Site 181
NT DSDP Site 182
SA North American Pacific
SA Pacific Ocean

Leg 19 (1981)
Bering Sea and N Pacific Ocean. Sites 183-193, roughly following Aleutian Trench to E of the Kuril Trench. Autoposting of this term to individual sites began in 1989.
CO N450000N580000
W1610000E1550000
BT Deep Sea Drilling Project
NT DSDP Site 183
NT DSDP Site 186
NT DSDP Site 192
SA Aleutian Trench
SA Bering Sea
SA North Pacific
SA Pacific Ocean

Leg 20 (1981)
NW Pacific Ocean. Sites 194-202, from E of the Japan and Bonin trenches, to E of the Mariana Trench, and W of the Marshall Islands. Autoposting of this term to individual sites began in 1989.
CO N120000N340000
E1570000E1460000
BT Deep Sea Drilling Project
NT DSDP Site 194
NT DSDP Site 195

NT DSDP Site 196
NT DSDP Site 197
NT DSDP Site 198
NT DSDP Site 199
NT DSDP Site 200
NT DSDP Site 202
SA Pacific Ocean

Leg 21 (1981)
SW Pacific Ocean. Sites 203-210, primarily W of the Tonga and Kermadec trenches and New Zealand, and in the Coral Sea. Autoposting of this term to individual sites began in 1989.
CO S370000S130000
W1740000E1520000
BT Deep Sea Drilling Project
NT DSDP Site 206
NT DSDP Site 207
NT DSDP Site 208
NT DSDP Site 209
NT DSDP Site 210
SA Coral Sea
SA Pacific Ocean

Leg 22 (1981)
E Indian Ocean. Sites 211-218. Autoposting of this term to individual sites began in 1989.
CO S200000N090000
E1030000E0860000
BT Deep Sea Drilling Project
NT DSDP Site 211
NT DSDP Site 212
NT DSDP Site 213
NT DSDP Site 214
NT DSDP Site 215
NT DSDP Site 216
NT DSDP Site 217
NT DSDP Site 218
SA Indian Ocean

Leg 23 (1981)
Arabian Sea and Red Sea. Sites 219-230. Autoposting of this term to individual sites began in 1989.
CO N060000N220000
E0730000E0380000
BT Deep Sea Drilling Project
NT DSDP Site 219
NT DSDP Site 220
NT DSDP Site 225
NT DSDP Site 227
NT DSDP Site 228
SA Arabian Sea
SA Indian Ocean
SA Red Sea

Leg 24 (1981)
W Indian Ocean and Gulf of Aden. Sites 231-238. Autoposting of this term to individual sites began in 1989.
CO S120000N150000
E0710000E0480000
BT Deep Sea Drilling Project
NT DSDP Site 231
NT DSDP Site 232
NT DSDP Site 233
NT DSDP Site 236
NT DSDP Site 237
NT DSDP Site 238
SA Arabian Sea
SA Gulf of Aden
SA Indian Ocean

Leg 25 (1978)
W Indian Ocean. Sites 239-249. Autoposting of this term to individual sites began in 1989.
CO S340000S020000
E0530000E0360000
BT Deep Sea Drilling Project
NT DSDP Site 241
NT DSDP Site 245
NT DSDP Site 248
SA Indian Ocean

Leg 26 (1981)
S Indian Ocean. Sites 250-258. Autoposting of this term to individual sites began in 1989.
CO S380000S230000
E1130000E0390000
BT Deep Sea Drilling Project
NT DSDP Site 251
NT DSDP Site 253
NT DSDP Site 254
NT DSDP Site 255
NT DSDP Site 256
NT DSDP Site 257
SA Indian Ocean

Leg 27 (1978)
E Indian Ocean. Sites 259-263. Autoposting of this term to individual sites began in 1989.
CO S300000S100000
E1240000E1100000
BT Deep Sea Drilling Project
NT DSDP Site 259
NT DSDP Site 260
NT DSDP Site 261
NT DSDP Site 262
NT DSDP Site 263
SA Indian Ocean

Leg 28 (1981)
E Indian Ocean, and Antarctic Ocean S of Australia and New Zealand. Sites 264-274. Autoposting of this term to individual sites began in 1989.
CO S780000S340000
E1790000E1040000
BT Deep Sea Drilling Project
NT DSDP Site 264
NT DSDP Site 265
NT DSDP Site 266
NT DSDP Site 270
NT DSDP Site 271
NT DSDP Site 272
NT DSDP Site 273
NT DSDP Site 274
SA Antarctic Ocean
SA Indian Ocean

Leg 29 (1981)
Pacific and Antarctic oceans, S of Australia and New Zealand. Sites 275-284. Autoposting of this term to individual sites began in 1989.
CO S570000S400000
E1770000E1430000
BT Deep Sea Drilling Project
NT DSDP Site 277
NT DSDP Site 284
SA Antarctic Ocean

Leg 30 (1981)
SW Pacific Ocean. Sites 285-289 in the South Fiji Basin, New Hebrides, Coral Sea, and Ontong Java Plateau. Autoposting of this term to individual sites began in 1989.
CO S270000S000000
E1760000E1530000
BT Deep Sea Drilling Project
NT DSDP Site 285
NT DSDP Site 286
NT DSDP Site 288
NT DSDP Site 289
SA Pacific Ocean

Leg 31 (1981)
Japan Sea and W Philippine Sea. Sites 290-302. Autoposting of this term to individual sites began in 1989.
CO N120000N420000
E1380000E1240000
BT Deep Sea Drilling Project
NT DSDP Site 291
NT DSDP Site 292
NT DSDP Site 293
NT DSDP Site 294

NT DSDP Site 296
NT DSDP Site 297
NT DSDP Site 298
NT DSDP Site 299
NT DSDP Site 302
SA Japan Sea
SA Pacific Ocean
SA Philippine Sea
SA West Philippine Basin

Leg 32 (1981)
NW Pacific Ocean. Sites 303-313, from E of the Kuril and Bonin trenches to the Emperor Seamounts and the Necker Ridge. Autoposting of this term to individual sites began in 1989.
CO N200000N410000
E1800000E1540000
BT Deep Sea Drilling Project
NT DSDP Site 305
NT DSDP Site 310
NT DSDP Site 313
SA Pacific Ocean

Leg 33 (1981)
Central equatorial Pacific Ocean. Sites 314-318. Autoposting of this term to individual sites began in 1989.
CO S150000N160000
W1460000W1690000
BT Deep Sea Drilling Project
NT DSDP Site 315
NT DSDP Site 315A
NT DSDP Site 317
SA Equatorial Pacific
SA Pacific Ocean

Leg 34 (1981)
SE Pacific Ocean. Sites 319-321. Autoposting of this term to individual sites began in 1989.
CO S140000S090000
W0810000W1020000
BT Deep Sea Drilling Project
NT DSDP Site 319
NT DSDP Site 320
NT DSDP Site 321
SA Pacific Ocean
SA South American Pacific

Leg 35 (1981)
Antarctic Ocean, S of the SE Pacific Ocean. Sites 322-325. Autoposting of this term to individual sites began in 1989.
CO S700000S600000
W0730000W0990000
BT Deep Sea Drilling Project
NT DSDP Site 322
NT DSDP Site 323
NT DSDP Site 325
SA Antarctic Ocean

Leg 36 (1981)
Drake Passage, Scotia Sea, and SW Atlantic Ocean. Sites 326-331. Autoposting of this term to individual sites began in 1989.
CO S570000S370000
W0360000W0660000
BT Deep Sea Drilling Project
NT DSDP Site 327
NT DSDP Site 328
NT DSDP Site 329
NT DSDP Site 330
SA Antarctic Ocean
SA Atlantic Ocean
SA Drake Passage
SA Scotia Sea
SA South American Atlantic

Leg 37 (1981)
In vicinity of the Mid-Atlantic Ridge, W of the Azores, N Atlantic Ocean. Sites 332-335. Autopost-

ing of this term to individual sites began in 1989.
CO N360000N380000
   W0330000W0360000
BT Deep Sea Drilling Project
NT DSDP Site 332
NT DSDP Site 332B
NT DSDP Site 333
NT DSDP Site 334
NT DSDP Site 335
SA Atlantic Ocean
SA Mid-Atlantic Ridge
SA North Atlantic
SA North Atlantic Ridge

Leg 38 (1981)
Norwegian and Greenland seas. Sites 336-352. Autoposting of this term to individual sites began in 1989.
CO N630000N770000
   E0080000W0130000
BT Deep Sea Drilling Project
NT DSDP Site 336
NT DSDP Site 338
NT DSDP Site 343
NT DSDP Site 344
NT DSDP Site 345
NT DSDP Site 346
NT DSDP Site 348
SA Arctic Ocean
SA Atlantic Ocean

Leg 39 (1981)
W and SE Atlantic Ocean. Sites 353-359. Autoposting of this term to individual sites began in 1989.
CO S380000N110000
   W0040000W0450000
BT Deep Sea Drilling Project
NT DSDP Site 354
NT DSDP Site 355
NT DSDP Site 356
NT DSDP Site 357
SA Argentine Basin
SA Atlantic Ocean
SA Brazil Basin
SA Guyanese Atlantic
SA Mid-Atlantic Ridge
SA North Atlantic Ridge
SA Rio Grande Rise
SA South American Atlantic
SA Southeast Atlantic
SA Vema fracture zone
SA Walvis Ridge

Leg 40 (1981)
SE Atlantic Ocean. Sites 360-365. Autoposting of this term to individual sites began in 1989.
CO S360000S110000
   E0190000E0090000
BT Deep Sea Drilling Project
NT DSDP Site 360
NT DSDP Site 361
NT DSDP Site 362
NT DSDP Site 363
NT DSDP Site 364
NT DSDP Site 365
SA Angola Basin
SA Atlantic Ocean
SA Cape Basin
SA Southeast Atlantic
SA Walvis Ridge

Leg 41 (1981)
Cape Verde Atlantic, off the coast of West Africa. Sites 366-370. Autoposting of this term to individual sites began in 1989.
CO N050000N330000
   W0100000W0220000
BT Deep Sea Drilling Project
NT DSDP Site 366
NT DSDP Site 367
NT DSDP Site 368
NT DSDP Site 369
NT DSDP Site 370
SA Atlantic Ocean
SA Cape Verde Atlantic
SA Cape Verde Basin
SA Cape Verde Rise
SA Sierra Leone Rise

Leg 42A (1981)
Mediterranean Sea. Sites 371-378. Autoposting of this term to individual sites began in 1989.
CO N340000N410000
   E0320000E0040000
BT Deep Sea Drilling Project
NT DSDP Site 372
NT DSDP Site 374
NT DSDP Site 375
NT DSDP Site 376
NT DSDP Site 378
SA Bannock Basin
SA Mediterranean Sea

Leg 42B (1981)
Black Sea. Sites 379-381. Autoposting of this term to individual sites began in 1989.
CO N230000N440000
   E0370000E0290000
BT Deep Sea Drilling Project
NT DSDP Site 379
NT DSDP Site 380
NT DSDP Site 381
SA Black Sea

Leg 43 (1981)
North American Atlantic. Sites 382-387. Autoposting of this term to individual sites began in 1989.
CO N310000N410000
   W0510000W0680000
BT Deep Sea Drilling Project
NT DSDP Site 382
NT DSDP Site 384
NT DSDP Site 385
NT DSDP Site 386
NT DSDP Site 387
SA Atlantic Ocean
SA Bermuda Rise
SA North American Atlantic

Leg 44 (1981)
Central NW Atlantic Ocean. Sites 388-392. Autoposting of this term to individual sites began in 1989.
CO N280000N360000
   W0690000W0770000
BT Deep Sea Drilling Project
NT DSDP Site 390
NT DSDP Site 391
NT DSDP Site 392
SA Atlantic Ocean
SA Blake Nose
SA Blake-Bahama Basin
SA Blake-Bahama Outer Ridge
SA North American Atlantic

Leg 45 (1978)
First cruise of the International Phase of Ocean Drilling (IPOD) of the Deep Sea Drilling Project. N central Atlantic Ocean. Sites 395-396. Autoposting of this term to individual sites began in 1989.
CO N220000N230000
   W0430000W0470000
BT IPOD
BT Deep Sea Drilling Project
NT DSDP Site 395
NT DSDP Site 396
SA Atlantic Ocean
SA North Atlantic
SA North Atlantic Ridge

Leg 46 (1978)
N central Atlantic Ocean. Site 396B, near the Mid-Atlantic Ridge axis.
CO N225200N230300
   W0442700W0443700
BT IPOD
BT Deep Sea Drilling Project
NT DSDP Site 396
SA Atlantic Ocean

Leg 47 (1981)
NE Atlantic Ocean. Sites 397-398, off NW Africa near the Canary Islands and off SW Europe near Portugal. Autoposting of this term to individual sites began in 1989.
CO N260000N410000
   W0100000W0160000
BT IPOD
BT Deep Sea Drilling Project
NT DSDP Site 397
NT DSDP Site 398
SA Atlantic Ocean

Leg 48 (1978)
Bay of Biscay and Rockall Plateau, NE Atlantic Ocean. Sites 399-406. Autoposting of this term to individual sites began in 1989.
CO N470000N570000
   W0080000W0240000
BT IPOD
BT Deep Sea Drilling Project
NT DSDP Site 400
NT DSDP Site 401
NT DSDP Site 402
NT DSDP Site 403
NT DSDP Site 404
NT DSDP Site 405
NT DSDP Site 406
SA Atlantic Ocean

Leg 49 (1981)
N Atlantic Ocean near Iceland and along the Mid-Atlantic Ridge. Sites 407-414. Autoposting of this term to individual sites began in 1989.
CO N320000N640000
   W0250000W0340000
BT IPOD
BT Deep Sea Drilling Project
NT DSDP Site 407
NT DSDP Site 408
NT DSDP Site 410
SA Atlantic Ocean

Leg 50 (1981)
NE Atlantic Ocean, off Morocco. Sites 415-416. Autoposting of this term to individual sites began in 1989.
CO N310000N330000
   W0100000W0120000
BT IPOD
BT Deep Sea Drilling Project
NT DSDP Site 416
SA Atlantic Ocean

Leg 51 (1981)
NW Atlantic Ocean, E of Florida, N of Puerto Rico. Site 417.
CO N240000N260000
   W0670000W0690000
BT IPOD
BT Deep Sea Drilling Project
NT DSDP Site 417
SA Atlantic Ocean
SA Bermuda Rise
SA North American Atlantic

Leg 52 (1981)
NW Atlantic Ocean, E of Florida, N of Puerto Rico. Sites 417-418.
CO N240000N260000
   W0670000W0690000
BT IPOD
BT Deep Sea Drilling Project
NT DSDP Site 418
SA Atlantic Ocean
SA Bermuda Rise
SA North American Atlantic

Leg 53 (1981)
NW Atlantic Ocean. Site 418.
CO N240000N260000
   W0670000W0690000
BT IPOD
BT Deep Sea Drilling Project
NT DSDP Site 418
SA Atlantic Ocean
SA Bermuda Rise
SA North American Atlantic

Leg 54 (1981)
E Equatorial Pacific. Sites 419-429, on the East Pacific Rise and Galapagos Rift. Autoposting of this term to individual sites began in 1989.
CO N000000N100000
   W0860000W1070000
BT IPOD
BT Deep Sea Drilling Project
NT DSDP Site 419
NT DSDP Site 420
NT DSDP Site 424
SA Pacific Ocean

Leg 55 (1981)
NW Pacific Ocean. Sites 430-433, in the Emperor Seamounts. Autoposting of this term to individual sites began in 1989.
CO N370000N450000
   E1710000E1700000
BT IPOD
BT Deep Sea Drilling Project
NT DSDP Site 430
NT DSDP Site 431
NT DSDP Site 432
NT DSDP Site 433
SA Pacific Ocean

Leg 56 (1981)
NW Pacific Ocean. Sites 434-437, near the Japan Trench. Autoposting of this term to individual sites began in 1989.
CO N390000N400000
   E1460000E1430000
BT IPOD
BT Deep Sea Drilling Project
NT DSDP Site 434
NT DSDP Site 435
NT DSDP Site 436
SA Pacific Ocean

Leg 57 (1981)
NW Pacific Ocean. Sites 438-441, near the Japan Trench. Autoposting of this term to individual sites began in 1989.
CO N390000N410000
   E1450000E1430000
BT IPOD
BT Deep Sea Drilling Project
NT DSDP Site 438
NT DSDP Site 439
NT DSDP Site 440
SA Pacific Ocean

Leg 58 (1981)
N Philippine Sea. Sites 442-446. Autoposting of this term to individual sites began in 1989.
CO N240000N300000
   E1380000E1320000
BT IPOD
BT Deep Sea Drilling Project
NT DSDP Site 442
NT DSDP Site 445
NT DSDP Site 446
SA Pacific Ocean
SA Philippine Sea

Leg 59 (1981)

Philippine Sea. Sites 447-451. Autoposting of this term to individual sites began in 1993.
CO N150000N190000
    E1450000E1320000
BT IPOD
BT Deep Sea Drilling Project
NT DSDP Site 450
SA Pacific Ocean
SA Philippine Sea
SA West Philippine Basin

Leg 60 (1981)
SE Philippine Sea. Sites 452-461 from the Mariana Trough to E of the Mariana Trench. Autoposting of this term to individual sites began in 1989.
CO N170000N190000
    E1490000E1430000
BT IPOD
BT Deep Sea Drilling Project
NT DSDP Site 453
NT DSDP Site 458
NT DSDP Site 459
SA Pacific Ocean
SA Philippine Sea

Leg 61 (1981)
W central Pacific Ocean. Site 462 in the Nauru Basin. Autoposting of this term to individual sites began in 1989.
CO N070000N080000
    E1660000E1650000
BT IPOD
BT Deep Sea Drilling Project
NT DSDP Site 462
SA Pacific Ocean

Leg 62 (1981)
N central Pacific Ocean. Sites 463-466 in the Mid-Pacific Mountains and on Hess Rise. Autoposting of this term to individual sites began in 1989.
CO N200000N400000
    E1792000E1735000
DT IPOD
BT Deep Sea Drilling Project
NT DSDP Site 463
NT DSDP Site 464
NT DSDP Site 465
NT DSDP Site 466
SA Pacific Ocean

Leg 63 (1981)
NE Pacific Ocean. Sites 467-473 off S California and Baja California. Autoposting of this term to individual sites began in 1989.
BT IPOD
BT Deep Sea Drilling Project
NT DSDP Site 467
NT DSDP Site 468
NT DSDP Site 469
NT DSDP Site 470
NT DSDP Site 471
NT DSDP Site 472
NT DSDP Site 473
SA Pacific Ocean

Leg 64 (1981)
Gulf of California. Sites 474-481. Autoposting of this term to individual sites began in 1989.
BT IPOD
BT Deep Sea Drilling Project
NT DSDP Site 474
NT DSDP Site 475
NT DSDP Site 476
NT DSDP Site 477
NT DSDP Site 478
NT DSDP Site 479
NT DSDP Site 480
NT DSDP Site 481
SA Gulf of California
SA Pacific Ocean

Leg 65 (1981)
At mouth of the Gulf of California. Sites 482-485. Autoposting of this term to individual sites began in 1989.
BT IPOD
BT Deep Sea Drilling Project
NT DSDP Site 482
NT DSDP Site 483
NT DSDP Site 484
NT DSDP Site 485
SA Gulf of California
SA Pacific Ocean

Leg 66 (1981)
NE Pacific Ocean off S Mexico. Sites 486-493.
BT IPOD
BT Deep Sea Drilling Project
SA Pacific Ocean

Leg 67 (1985)
E Equatorial Pacific, off Central America. Sites 494-500. Autoposting of this term to individual sites began in 1989.
CO N124000N131000
    W0905000W0911000
BT IPOD
BT Deep Sea Drilling Project
NT DSDP Site 494
NT DSDP Site 495
NT DSDP Site 496
NT DSDP Site 497
NT DSDP Site 498
NT DSDP Site 499
NT DSDP Site 500
SA Equatorial Pacific
SA Nicoya Complex
SA Pacific Ocean

Leg 68 (1985)
E Equatorial Pacific. Sites 501-503. Autoposting of this term to individual sites began in 1989.
CO N011000N113000
    W0792000W0954500
BI IPOD
BT Deep Sea Drilling Project
NT DSDP Site 501
NT DSDP Site 502
NT DSDP Site 503
SA Equatorial Pacific
SA Pacific Ocean

Leg 69 (1985)
E Equatorial Pacific, off Ecuador and Colombia. Sites 504-505. Autoposting of this term to individual sites began in 1989.
CO N011000N020000
    W0834000W0835000
BT IPOD
BT Deep Sea Drilling Project
NT DSDP Site 504
NT DSDP Site 504B
NT DSDP Site 505
SA Equatorial Pacific
SA Pacific Ocean

Leg 70 (1985)
E Equatorial Pacific, off Ecuador. Sites 506-510. Autoposting of this term to individual sites began in 1989.
CO N003000N014000
    W0860000W0863000
BT IPOD
BT Deep Sea Drilling Project
NT DSDP Site 504B
NT DSDP Site 506
NT DSDP Site 507
NT DSDP Site 508
NT DSDP Site 509
NT DSDP Site 510
SA Equatorial Pacific
SA Pacific Ocean

Leg 71 (1985)
S Atlantic Ocean. Sites 511-514. Autoposting of this term to individual sites began in 1989.
CO S511000S460000
    W0243000W0470000
BT IPOD
BT Deep Sea Drilling Project
NT DSDP Site 511
NT DSDP Site 512
NT DSDP Site 513
NT DSDP Site 514
SA Atlantic Ocean
SA South Atlantic

Leg 72 (1985)
S Atlantic Ocean, off Brazil. Sites 515-518. Autoposting of this term to individual sites began in 1989.
CO S302000S261000
    W0351500W0381000
BT IPOD
BT Deep Sea Drilling Project
NT DSDP Site 515
NT DSDP Site 516
NT DSDP Site 517
NT DSDP Site 518
SA Atlantic Ocean
SA Brazil Basin
SA Rio Grande Rise
SA South American Atlantic
SA Vema Channel

Leg 73 (1985)
E flank of Mid-Atlantic Ridge, S Atlantic Ocean. Sites 519-524. Autoposting of this term to individual sites began in 1989.
CO S293000S253000
    E0034000W0114000
BT IPOD
BT Deep Sea Drilling Project
NT DSDP Site 519
NT DSDP Site 520
NT DSDP Site 521
NT DSDP Site 522
NT DSDP Site 523
NT DSDP Site 524
SA Atlantic Ocean
SA South Atlantic

Leg 74 (1985)
SE Atlantic Ocean. Sites 525-529. Autoposting of this term to individual sites began in 1989.
CO S301000S280000
    E0031000E0014500
BT IPOD
BT Deep Sea Drilling Project
NT DSDP Site 525
NT DSDP Site 526
NT DSDP Site 527
NT DSDP Site 528
NT DSDP Site 529
SA Atlantic Ocean
SA Southeast Atlantic

Leg 75 (1985)
SE Atlantic Ocean, off Namibia. Sites 530-532. Autoposting of this term to individual sites began in 1989.
CO S194500S191000
    E0103500E0092000
BT IPOD
BT Deep Sea Drilling Project
NT DSDP Site 530
NT DSDP Site 531
NT DSDP Site 532
SA Atlantic Ocean
SA Southeast Atlantic

Leg 76 (1985)
North American Atlantic, off Florida. Sites 533-534. Autoposting of this term to individual sites began in 1989.
CO N282000N312000
    W0745000W0752500
BT IPOD
BT Deep Sea Drilling Project
NT DSDP Site 533
NT DSDP Site 534
SA Atlantic Ocean
SA North American Atlantic

Leg 77 (1985)
W Caribbean Sea. Sites 535-540. Autoposting of this term to individual sites began in 1989.
CO N232000N240000
    W0842000W0853000
BT IPOD
BT Deep Sea Drilling Project
NT DSDP Site 535
NT DSDP Site 536
NT DSDP Site 537
NT DSDP Site 538
NT DSDP Site 540
SA Caribbean Sea

Leg 78A (1989)
In Atlantic Ocean, E of the Caribbean Sea, including sites in Barbados Ridge. Sites 541-543.
CO N153111N154245
    W0583912W0584340
BT IPOD
BT Deep Sea Drilling Project
NT DSDP Site 541
NT DSDP Site 542
NT DSDP Site 543
SA Atlantic Ocean
SA Barbados Ridge
SA Caribbean Sea
SA Leg 110

Leg 78B (1989)
Near axis of Mid-Atlantic Ridge, just S of Kane fracture zone in N central Atlantic. Site 395A, which was previously drilled during Leg 45.
BT IPOD
BT Deep Sea Drilling Project
NT DSDP Site 395
SA Atlantic Ocean
SA Kane fracture zone
SA Mid-Atlantic Ridge

Leg 79 (1989)
Mazagan Plateau, off NW coast of Morocco. Sites 544-547.
CO N333951N334651
    W0092058W0093352
BT IPOD
BT Deep Sea Drilling Project
NT DSDP Site 544
NT DSDP Site 545
NT DSDP Site 546
NT DSDP Site 547
SA Atlantic Ocean
SA Mazagan Plateau

Leg 80 (1989)
Goban Spur, N Biscay region in NE Atlantic. Sites 548-551. Autoposting of this term to individual sites began in 1989.
CO N483054N485457
    W0120950W0133006
BT IPOD
BT Deep Sea Drilling Project
NT DSDP Site 548
NT DSDP Site 549
NT DSDP Site 550
NT DSDP Site 551
SA Atlantic Ocean
SA European Atlantic
SA Goban Spur

Leg 81 (1989)
Rockall Plateau, NE Atlantic. Sites 552-555.
CO N560233N563342
    W0204655W0233142

BT IPOD
BT Deep Sea Drilling Project
NT DSDP Site 552
NT DSDP Site 553
NT DSDP Site 554
NT DSDP Site 555
SA Atlantic Ocean
SA Northeast Atlantic
SA Rockall Plateau

Leg 82 (1989)
Mid-Atlantic Ridge near the Azores. Sites 556-564.
CO N330829N385623
    W0323334W0434603
BT IPOD
BT Deep Sea Drilling Project
NT DSDP Site 556
NT DSDP Site 558
NT DSDP Site 561
NT DSDP Site 562
NT DSDP Site 563
NT DSDP Site 564
SA Atlantic Ocean
SA Azores
SA Mid-Atlantic Ridge
SA North Atlantic
SA Oceanographer fracture zone

Leg 83 (1989)
Located about 201 km S of Costa Rica Rift in E Equatorial Pacific. Site 504B.
CO N011337N011338
    W0834348W0834349
BT IPOD
BT Deep Sea Drilling Project
NT DSDP Site 504B
SA Equatorial Pacific
SA Galapagos Rift
SA Pacific Ocean

Leg 84 (1989)
Middle America Trench off coast of Guatemala. Sites 565-570. Autoposting of this term to individual sites began in 1993.
CO N094341N131708
    W0860526W0912335
BT IPOD
BT Deep Sea Drilling Project
NT DSDP Site 565
NT DSDP Site 568
NT DSDP Site 570
SA Equatorial Pacific
SA Middle America Trench
SA Pacific Ocean

Leg 85 (1989)
E Equatorial Pacific. Sites 571-575.
CO N002954N055100
    W1135030W1332010
BT IPOD
BT Deep Sea Drilling Project
NT DSDP Site 572
NT DSDP Site 573
NT DSDP Site 574
NT DSDP Site 575
SA Equatorial Pacific
SA Pacific Ocean

Leg 86 (1989)
NW Pacific. Sites 576-581.
CO N322121N435538
    E1641633E1513743
BT IPOD
BT Deep Sea Drilling Project
NT DSDP Site 576
NT DSDP Site 577
NT DSDP Site 578
NT DSDP Site 579
NT DSDP Site 580
NT DSDP Site 581
SA Kuroshio
SA Northwest Pacific
SA Pacific Ocean

Leg 87 (1989)
Nankai Trough and Japan Trench. Sites 582-584.
CO N314730N402800
    E1435736E1335115
BT IPOD
BT Deep Sea Drilling Project
NT DSDP Site 582
NT DSDP Site 583
NT DSDP Site 584
SA Japan Trench
SA Nankai Trough
SA Northwest Pacific
SA Pacific Ocean

Leg 88 (1989)
NW Pacific Basin, S of Hokkaido fracture zone. Site 581.
CO N435537N435538
    E1594746E1594745
BT IPOD
BT Deep Sea Drilling Project
NT DSDP Site 581
SA Northwest Pacific
SA Pacific Basin
SA Pacific Ocean

Leg 89 (1989)
Mariana Trough, Ontong Java Plateau and Nauru Basin, Equatorial Pacific. Sites 585, 586 and 462A.
CO S002951N132900
    E1650154E1564855
BT IPOD
BT Deep Sea Drilling Project
NT DSDP Site 462
NT DSDP Site 585
NT DSDP Site 586
SA Equatorial Pacific
SA Mariana Trough
SA Nauru Basin
SA Ontong Java Plateau
SA Pacific Ocean

Leg 90 (1989)
SW Pacific, off E coast of Australia and NW of New Zealand. Sites 587-594B.
CO S453129S211153
    E1745653E1611336
BT IPOD
BT Deep Sea Drilling Project
NT DSDP Site 587
NT DSDP Site 588
NT DSDP Site 589
NT DSDP Site 590
NT DSDP Site 591
NT DSDP Site 592
NT DSDP Site 593
NT DSDP Site 594
SA Challenger Plateau
SA Lord Howe Rise
SA Pacific Ocean
SA Southwest Pacific

Leg 91 (1989)
E of Tonga Trench in SW Pacific. Sites 595-596B.
CO S235112S234920
    W1653137W1653917
BT IPOD
BT Deep Sea Drilling Project
NT DSDP Site 595
NT DSDP Site 596
SA Pacific Ocean
SA Southwest Pacific
SA Tonga Trench

Leg 92 (1989)
East Pacific Rise. Sites 597-602B. Also includes Site 504B near coast of Colombia.
CO S192706S184822
    W0834349W1294614
BT IPOD
BT Deep Sea Drilling Project

NT DSDP Site 504B
NT DSDP Site 597
NT DSDP Site 598
NT DSDP Site 599
NT DSDP Site 600
NT DSDP Site 601
NT DSDP Site 602
SA East Pacific Rise
SA Pacific Ocean
SA South Pacific

Leg 93 (1989)
Continental rise off E coast of United States. Sites 603-605.
CO N352939N384432
    W0700142W0723633
BT IPOD
BT Deep Sea Drilling Project
NT DSDP Site 603
NT DSDP Site 604
NT DSDP Site 605
SA Atlantic Ocean
SA Blake-Bahama Formation
SA Hatteras Formation
SA Hudson Canyon
SA North American Atlantic
SA North Atlantic

Leg 94 (1989)
Mid-Atlantic Ridge, Feni Drift and Gardar Drift, NE Atlantic. Sites 606-611. Autoposting of this term to individual sites began in 1993.
CO N371724N534600
    W0180000W0351154
BT IPOD
BT Deep Sea Drilling Project
NT DSDP Site 606
NT DSDP Site 607
NT DSDP Site 608
NT DSDP Site 609
NT DSDP Site 610
NT DSDP Site 611
SA Atlantic Ocean
SA Mid-Atlantic Ridge
SA Northeast Atlantic

Leg 95 (1989)
Part of the New Jersey Transect in Baltimore Canyon, North American Atlantic. Sites 612-613. Autoposting of this term to individual sites began in 1993.
CO N384600N384930
    W0723000W0724700
BT IPOD
BT Deep Sea Drilling Project
NT DSDP Site 603
NT DSDP Site 612
NT DSDP Site 613
SA Atlantic Ocean
SA Baltimore Canyon
SA North American Atlantic
SA North Atlantic

Leg 96 (1989)
Mostly in the Mississippi Fan, a region in the Gulf of Mexico which extends from the Mississippi Delta to the Sigsbee Deep and Florida abyssal plain. Sites 614-624A.
CO N250404N271137
    W0855931W0912433
BT IPOD
BT Deep Sea Drilling Project
NT DSDP Site 614
NT DSDP Site 615
NT DSDP Site 616
NT DSDP Site 617
NT DSDP Site 618
NT DSDP Site 619
NT DSDP Site 620
NT DSDP Site 621
NT DSDP Site 622
NT DSDP Site 623
NT DSDP Site 624
SA Gulf of Mexico

SA Mississippi Fan
SA Pigmy Basin

Leg 100 (1989)
First cruise of the Ocean Drilling Program (ODP), the continuation of IPOD and DSDP. W of the Florida shelf in the Gulf of Mexico. Site 625.
BT Ocean Drilling Program
SA Gulf of Mexico

Leg 101 (1989)
Bahamas carbonate platform, including Blake Plateau, Straits of Florida, Little Bahama Bank, Exuma Sound, and Northwest Providence Channel. Sites 626-636.
CO N233512N273806
    W0752606W0793247
BT Ocean Drilling Program
NT ODP Site 626
NT ODP Site 627
NT ODP Site 628
NT ODP Site 629
NT ODP Site 630
NT ODP Site 631
NT ODP Site 632
NT ODP Site 633
NT ODP Site 634
NT ODP Site 635
SA Atlantic Ocean
SA Bahamas
SA Blake Plateau
SA Exuma Sound
SA Little Bahama Bank
SA North Atlantic
SA Northwest Providence Channel
SA Straits of Florida

Leg 102 (1989)
NW Atlantic. Re-entry of DSDP Site 418.
CO N250205N250207
    W0680326W0680327
BT Ocean Drilling Program
NT DSDP Site 418
SA Atlantic Ocean
SA Northwest Atlantic

Leg 103 (1989)
Galicia Bank margin, NW of Iberian Peninsula, European Atlantic. Sites 637-641. Autoposting of this term to individual sites began in 1993.
CO N420042N420918
    W0121054W0125148
BT Ocean Drilling Program
NT ODP Site 637
NT ODP Site 638
NT ODP Site 639
NT ODP Site 640
NT ODP Site 641
SA Atlantic Ocean
SA European Atlantic
SA Northeast Atlantic

Leg 104 (1989)
Voring Plateau, Norwegian Sea. Sites 642-644. Autoposting of this term to individual sites began in 1993.
CO N664042N674254
    E0043436E0010200
BT Ocean Drilling Program
NT ODP Site 642
NT ODP Site 643
NT ODP Site 644
SA Arctic Ocean
SA Norwegian Sea
SA Voring Plateau

Leg 105 (1989)

Baffin Bay and Labrador Sea.
Sites 645-647. Autoposting of this
term to individual sites began in
1993.
CO N531954N702730
 W0451542W0643924
BT Ocean Drilling Program
NT ODP Site 645
NT ODP Site 646
NT ODP Site 647
SA Arctic Ocean
SA Baffin Bay
SA Labrador Sea
SA North American Atlantic

Leg 106 (1993)
North Atlantic Ridge, S of Kane
fracture zone. Sites 648-649.
CO N225518N232210
 W0445649W0445704
BT Ocean Drilling Program
NT ODP Site 648
NT ODP Site 649
SA Atlantic Ocean
SA Mid-Atlantic Ridge
SA North Atlantic Ridge

Leg 107 (1993)
Vavilov and Marsili basins,
Tyrrhenian Sea, West Mediterranean. Sites 650-656.
CO N393000N404000
 E0141000E0103500
BT Ocean Drilling Program
NT ODP Site 652
NT ODP Site 653
NT ODP Site 654
NT ODP Site 655
NT ODP Site 656
SA Mediterranean Sea
SA Tyrrhenian Basin
SA Tyrrhenian Sea

Leg 108 (1993)
E Equatorial Atlantic and NW
margin of Africa, Cape Verde Atlantic and Southeast Atlantic.
ODP Sites 657-668.
CO S012325N211954
 W0114421W0231630
BT Ocean Drilling Program
NT ODP Site 657
NT ODP Site 658
NT ODP Site 659
NT ODP Site 660
NT ODP Site 661
NT ODP Site 662
NT ODP Site 663
NT ODP Site 664
NT ODP Site 665
NT ODP Site 666
NT ODP Site 667
NT ODP Site 668
SA Atlantic Ocean
SA Cape Verde Atlantic
SA Southeast Atlantic

Leg 109 (1989)
Mid-Atlantic Ridge, S of Kane
fracture zone in N Atlantic. Sites
669 and 670A; also re-entry of
Sites 648 and 395A.
CO N224521N233102
 W0445651W0460454
BT Ocean Drilling Program
NT DSDP Site 395
NT ODP Site 648
NT ODP Site 669
NT ODP Site 670
SA Atlantic Ocean
SA Kane fracture zone
SA Mid-Atlantic Ridge
SA North Atlantic

Leg 110 (1989)
Near Tiburon Rise, E of Barbados
in Barbados Forearc. Sites 671-
676. Autoposting of this term to
individual sites began in 1993.
CO N153133N153224
 W0584207W0585106
BT Ocean Drilling Program
NT ODP Site 671
NT ODP Site 672
NT ODP Site 673
NT ODP Site 674
NT ODP Site 675
NT ODP Site 676
SA Atlantic Ocean
SA Leg 78A
SA Northwest Atlantic

Leg 111 (1989)
E Equatorial Pacific. Sites 677-
678, plus DSDP Site 504B. Autoposting of this term to individual
sites began in 1993.
BT Ocean Drilling Program
NT DSDP Site 504B
NT ODP Site 677
NT ODP Site 678
SA Equatorial Pacific
SA Pacific Ocean

Leg 112 (1989)
Peru continental margin. Sites
679-688. Autoposting of this term
to individual sites began in 1993.
CO S132849S085929
 W0765329W0803501
BT Ocean Drilling Program
NT ODP Site 679
NT ODP Site 680
NT ODP Site 681
NT ODP Site 682
NT ODP Site 683
NT ODP Site 684
NT ODP Site 685
NT ODP Site 686
NT ODP Site 687
NT ODP Site 688
SA Pacific Ocean
SA South American Pacific

Leg 113 (1989)
Weddell Sea, near Antarctica.
Sites 689-697. Autoposting of this
term to individual sites began in
1993.
CO S704954S614837
 E0030559W0432706
BT Ocean Drilling Program
NT ODP Site 689
NT ODP Site 690
NT ODP Site 691
NT ODP Site 692
NT ODP Site 693
NT ODP Site 694
NT ODP Site 695
NT ODP Site 696
NT ODP Site 697
SA Antarctic Ocean
SA Weddell Sea

Leg 114 (1989)
Islas Orcadas Rise, Meteor Rise,
Northeast Georgia Rise. Sites
698-704. Autoposting of this term
to individual sites began in 1993.
CO S515905S465245
 E0075341W0330558
BT Ocean Drilling Program
NT ODP Site 698
NT ODP Site 699
NT ODP Site 700
NT ODP Site 701
NT ODP Site 702
NT ODP Site 703
NT ODP Site 704
SA Antarctic Ocean
SA Atlantic Ocean
SA South Atlantic

Leg 115 (1989)
Chagos Bank, Madingley Rise,
Maldives Ridge and Mascarene
Plateau in central Indian Ocean.
Sites 705-716. Autoposting of this
term to individual sites began in
1993.
CO S131002N050454
 E0734953E0590100
BT Ocean Drilling Program
NT ODP Site 705
NT ODP Site 706
NT ODP Site 707
NT ODP Site 708
NT ODP Site 709
NT ODP Site 710
NT ODP Site 711
NT ODP Site 712
NT ODP Site 713
NT ODP Site 714
NT ODP Site 715
NT ODP Site 716
SA Indian Ocean

Leg 116 (1989)
Distal end of Bengal Fan in central Indian Ocean. Sites 717-719.
Autoposting of this term to individual sites began in 1993.
CO S010116S005547
 E0812404E0812324
BT Ocean Drilling Program
NT ODP Site 717
NT ODP Site 718
NT ODP Site 719
SA Bengal Fan
SA Indian Ocean

Leg 117 (1993)
Indus Fan, Owen Ridge, Oman
continental margin, central Arabian Sea. Sites 720-731. Autoposting of this term to individual
sites began in 1993.
CO N160748N182912
 E0604438E0572212
BT Ocean Drilling Program
NT ODP Site 720
NT ODP Site 721
NT ODP Site 722
NT ODP Site 723
NT ODP Site 724
NT ODP Site 725
NT ODP Site 726
NT ODP Site 727
NT ODP Site 728
NT ODP Site 729
NT ODP Site 730
NT ODP Site 731
SA Arabian Sea
SA Indian Ocean

Leg 118 (1993)
Atlantis II fracture zone, Southwest Indian Ridge, Indian Ocean.
Sites 732-735.
CO S330518S320655
 E0571618E0565839
BT Ocean Drilling Program
NT ODP Site 732
NT ODP Site 733
NT ODP Site 734
NT ODP Site 735
SA Indian Ocean
SA Southwest Indian Ridge

Leg 119 (1989)
Kerguelen Plateau and Prydz
Bay, S Indian Ocean and Antarctic Ocean. Sites 736-746. Autoposting of this term to individual
sites began in 1993.
CO S684114S492407
 E0855147E0713936
BT Ocean Drilling Program
NT ODP Site 736
NT ODP Site 737
NT ODP Site 738
NT ODP Site 739
NT ODP Site 741
NT ODP Site 743
NT ODP Site 744
NT ODP Site 745
NT ODP Site 746
SA Antarctic Ocean
SA Indian Ocean
SA Kerguelen Plateau
SA Prydz Bay

Leg 120 (1993)
Kerguelen Plateau, S Indian
Ocean-Antarctic Ocean. Sites
747-751.
CO S584302S544838
 E0811426E0762427
BT Ocean Drilling Program
NT ODP Site 747
NT ODP Site 748
NT ODP Site 749
NT ODP Site 750
NT ODP Site 751
SA Antarctic Ocean
SA Indian Ocean
SA Kerguelen Plateau

Leg 121 (1993)
Broken Ridge and Ninetyeast
Ridge, Indian Ocean. Sites 752-
758.
CO S310148N052304
 E0933525E0873548
BT Ocean Drilling Program
NT ODP Site 752
NT ODP Site 753
NT ODP Site 754
NT ODP Site 755
NT ODP Site 756
NT ODP Site 757
NT ODP Site 758
SA Indian Ocean
SA Ninetyeast Ridge

Leg 122 (1989)
Exmouth Plateau and Wombat
Plateau, East Pacific Rise. Sites
759-764. Autoposting of this term
to individual sites began in 1993.
CO S203512S163357
 E1153337E1121231
BT Ocean Drilling Program
NT ODP Site 759
NT ODP Site 760
NT ODP Site 761
NT ODP Site 762
NT ODP Site 764
SA East Pacific Rise
SA Exmouth Plateau
SA Indian Ocean
SA Pacific Ocean

Leg 123 (1993)
Argo abyssal plain and Gascoyne
abyssal plain, E Indian Ocean.
Sites 765-766.
CO S195557S155831
 E1173430E1102714
BT Ocean Drilling Program
NT ODP Site 765
NT ODP Site 766
SA Argo abyssal plain
SA Gascoyne abyssal plain
SA Indian Ocean

Leg 124 (1993)
Celebes Sea and Sulu Sea, West
Pacific. Sites 767-771.
CO N044727N084709
 E1234015E1204046
BT Ocean Drilling Program
NT ODP Site 767
NT ODP Site 768
NT ODP Site 769
NT ODP Site 770
NT ODP Site 771
SA Celebes Sea

SA Pacific Ocean
SA Sulu Sea
SA West Pacific

Leg 124E (1993)
Philippine Sea, South China Sea, and Mariana Trench-Trough area, West Pacific. Sites 772-777.
CO N163900N203554
   E1484148E1194200
BT Ocean Drilling Program
NT ODP Site 776
SA Mariana Trench
SA Mariana Trough
SA Pacific Ocean
SA Philippine Sea
SA South China Sea
SA West Pacific

Leg 125 (1993)
Mariana and Izu-Bonin forearc regions, West Pacific. Sites 778-786.
CO N192955N304929
   E1464146E1405509
BT Ocean Drilling Program
NT ODP Site 778
NT ODP Site 779
NT ODP Site 780
NT ODP Site 781
NT ODP Site 782
NT ODP Site 783
NT ODP Site 785
SA Izu-Bonin Arc
SA Pacific Ocean
SA West Pacific

Leg 126 (1993)
Sumisu Rift and Izu-Bonin forearc basin, Izu-Bonin arc-trench system, NW Philippine Sea. Sites 787-793.
CO N305457N322359
   E1405317E1395039
BT Ocean Drilling Program
NT ODP Site 787
NT ODP Site 788
NT ODP Site 789
NT ODP Site 790
NT ODP Site 791
NT ODP Site 792
NT ODP Site 793
SA Izu-Bonin Arc
SA Pacific Ocean
SA Philippine Sea

Leg 127 (1993)
Japan Basin and Yamato Basin, E Japan Sea. Sites 794-797.
CO N383655N435912
   E1392451E1343209
BT Ocean Drilling Program
NT ODP Site 794
NT ODP Site 796
NT ODP Site 797
SA Japan Sea
SA Pacific Ocean

Leg 128 (1993)
S Japan Sea. Sites 794, 798-799. Reentry of Hole 794C.
CO N370218N401130
   E1381356E1335200
BT Ocean Drilling Program
NT ODP Site 794
NT ODP Site 798
NT ODP Site 799
SA Japan Sea
SA Pacific Ocean

Leg 129 (1993)
Pigafetta and East Mariana basins, central West Pacific. Sites 800-802.
CO N120546N215523
   E1562136E1521919
BT Ocean Drilling Program
NT ODP Site 800

NT ODP Site 801
NT ODP Site 802
SA Mariana Trough
SA Pacific Ocean
SA West Pacific

Leg 130 (1993)
Ontong Java Plateau, West Pacific. Sites 803-807.
CO N001906N033626
   E1613538E1563728
BT Ocean Drilling Program
NT ODP Site 803
NT ODP Site 804
NT ODP Site 805
NT ODP Site 806
NT ODP Site 807
SA Ontong Java Plateau
SA Pacific Ocean
SA West Pacific

Leg 131 (1993)
Nankai accretionary prism, SE of Shikoku, Japan, West Pacific. Site 808.
CO N322105N322111
   E1345646E1345634
BT Ocean Drilling Program
NT ODP Site 808
SA Pacific Ocean
SA West Pacific

Leg 132 (1993)
NW Philippine Sea and Shatsky Rise, N West Pacific. Sites 808-810.
CO N310326N322524
   E1575045E1345636
BT Ocean Drilling Program
NT ODP Site 808
NT ODP Site 809
SA Pacific Ocean
SA Philippine Sea
SA Shatsky Rise
SA West Pacific

Leg 133 (1993)
W Coral Sea, S West Pacific. Sites 811-826.
CO S191332S162522
   E1500232E1461721
BT Ocean Drilling Program
SA Coral Sea
SA Pacific Ocean

Leg 134 (1993)
D'Entrecasteaux Zone and New Hebrides island arc, Vanuatu region, S West Pacific. Sites 827-833.
CO S160034S144746
   E1675247E1661657
BT Ocean Drilling Program
SA Pacific Ocean
SA West Pacific

Leg 135 (1993)
Lau Basin and Tonga Ridge, SW Pacific. Sites 834-841.
CO S232045S183003
   W1751752W1775145
BT Ocean Drilling Program
SA Lau Basin
SA Pacific Ocean
SA West Pacific

Leg 136 (1993)
About 350 km S of the Hawaiian Islands, N South Polynesian Pacific. Sites 842-843.
CO N192010N192042
   W1590519W1590544
BT Ocean Drilling Program
SA Pacific Ocean
SA South Polynesian Pacific

Leg 138 (1993)

Transect in Galapagos Islands-East Pacific Rise area and transect 650-1000 km west of the East Pacific Rise, E Equatorial Pacific. Sites 844-854.
CO S030548N111326
   W0902851W1103419
BT Ocean Drilling Program
SA Pacific Ocean

legal
See legislation or practice if applicable.

legends (1981)
Before 1981, search legend.
SA cartography
SA maps

Leghorn Italy
 use Livorno Italy

legislation (1978)
SA antitrust legislation
SA decision-making
SA economic geology
SA engineering geology
SA environmental geology
SA Indian reservations
SA investment legislation
SA land use
SA Law of the Sea
SA mining legislation
SA national parks
SA policy
SA regulations
SA strategic minerals
SA water rights

legrandite (1978)
BT arsenates

Leguminosae (1978)
Family.
BT Dicotyledoneae
BT angiosperms
BT Spermatophyta
BT Plantae

Lehigh County
Valid through 1988. Search in combination with state term. After 1988, use specific county-state term.

Lehigh County Pennsylvania (1989)
E Pennsylvania. Before 1989, also search Lehigh County AND Pennsylvania.
CO N402500N404700
   W0752000W0755400
BT Pennsylvania
BT United States

lehm
 use loess

Leicestershire
No longer a valid term for GeoRef as of 1993.
 use Leicestershire England

Leicestershire England (1993)
Central England. After 1974, includes former counties of Leicester and Rutland.
CO N523000N530000
   W0004000W0014000
UF Leicestershire
BT England
BT Great Britain
BT United Kingdom
BT Western Europe
BT Europe
SA Barwell Meteorite
SA East Midlands
SA Midlands

leifite (1978)

From 1978-1980, framework silicates and halides were autoposted to this term. Autoposting of fluorides began in 1981. Autoposting of halides (again) and silicates began in 1989. This term has multiple hierarchies.
BT1 fluorides
BT1 halides
BT2 framework silicates
BT2 silicates
SA aluminosilicates

Leine Valley (1978)
River valley in SE and central Lower Saxony. Also search Leine.
BT Lower Saxony Germany
BT Germany
BT Central Europe
BT Europe

Leinster (1978)
Traditional province in E comprising 12 counties.
BT Ireland
BT Western Europe
BT Europe

Leinster Granite (1993)
Middle Paleozoic. SE Ireland.
IN Index ages as applicable.
BT Paleozoic
SA Ireland

Leipzig
No longer a valid term for GeoRef. See Leipzig Germany.

Leipzig Bezirk
No longer a valid term for GeoRef. District in S central East Germany. Before 1981, search Leipzig. See Germany.

Leipzig Germany (1993)
In Saxony, E Germany. As of 1981, Leipzig refers to only the city in S central East Germany. Before 1981, it also included the East German district.
BT Saxony Germany
BT Germany
BT Central Europe
BT Europe

Leiria
No longer a valid term for GeoRef. As of 1993 see Leiria Portugal.

Leiria Portugal (1993)
District and city in W Portugal. Before 1993 also search Leiria AND Portugal.
BT Portugal
BT Iberian Peninsula
BT Southern Europe
BT Europe
SA Estremadura Portugal

Leman Lake
 use Lake Geneva

Lemhi County
Valid through 1988. Search in combination with state term. After 1988, use specific county-state term.

Lemhi County Idaho (1989)
E Idaho. Before 1989, also search Lemhi County AND Idaho.
CO N441500N454500
   W1125000W1145000
BT Idaho
BT United States
NT Blackbird mining district
NT Salmon Idaho
SA Lemhi Range

Lemhi Range (1978)

E central Idaho. Primarily in Lemhi and Butte counties.
BT Idaho
BT United States
SA Butte County Idaho
SA Lemhi County Idaho

Lena Basin (1978)
River basin. E central and N Siberia including areas in Irkutsk Oblast, and Yakutia. Also search Lena. This term has multiple hierarchies.
UF Lena region
UF Lena River basin
BT1 Russian Federation
BT1 Commonwealth of Independent States
BT2 Asia
SA Angara-Lena Basin

Lena region
use Lena Basin

Lena River (1978)
Rises in the Baikal Mountains W of Lake Baikal in Irkutsk Oblast and flows NE into a great bend in Yakutia, and then NW through a large delta into the Laptev Sea. Also search Lena. This term has multiple hierarchies.
BT1 Russian Federation
BT1 Commonwealth of Independent States
BT2 Asia

Lena River basin
use Lena Basin

length of day (1981)
SA climate
SA paleoclimatology
SA rotation

Leninabad
No longer a valid term for GeoRef. As of 1993, see Leninabad Tadzhikistan.

Leninabad Tadzhikistan (1993)
Oblast and city in NW Tadzhikistan. This term has multiple hierarchies.
BT1 Tadzhikistan
BT1 Asia
BT2 Tadzhikistan
BT2 Commonwealth of Independent States
NT Kara-Mazar Tadzhikistan
NT Karamazar Tadzhikistan
NT Tary-Ekam Tadzhikistan
SA Shugnan Tadzhikistan

Leningrad
No longer a valid term for GeoRef. As of 1993 see Saint Petersburg Russian Federation
use Saint Petersburg Russian Federation

Leningrad Mining Institute (1978)
In Leningrad.
SA academic institutions
SA Russian Federation

Leninogorsk
No longer a valid term for GeoRef. As of 1993, see Leninogorsk Kazakhstan.

Leninogorsk Kazakhstan (1993)
City in Vostochno (East) Kazakhstan Oblast in extreme E Kazakhstan. Before 1978, also search Ridder. This term has multiple hierarchies.
UF Ridder Kazakhstan
BT1 Kazakhstan
BT1 Central Asia
BT1 Asia
BT2 Kazakhstan
BT2 Commonwealth of Independent States

Lenkoran
No longer a valid term for GeoRef. As of 1993, see Lenkoran Azerbaidzhan.

Lenkoran Azerbaidzhan (1993)
City on Caspian Sea near Iranian border. This term has multiple hierarchies.
BT1 Azerbaidzhan
BT1 Europe
BT2 Azerbaidzhan
BT2 Commonwealth of Independent States

Lenoir Limestone (1985)
In Stones River Group. E Tennessee, N Alabama and W Virginia.
BT Middle Ordovician
BT Ordovician
BT Paleozoic
SA Alabama
SA Stones River Group
SA Tennessee
SA Virginia

lenses (1978)
Any lense-shaped deposit.
SA ice lenses
SA mineral deposits, genesis
SA sedimentary rocks

Lenticulina (1978)
Genus. Autoposting of microfossils and Protista to this term began in 1990. This term has multiple hierarchies.
BT1 Nodosariidae
BT1 Nodosariacea
BT1 Rotaliina
BT1 foraminifera
BT1 Protista
BT1 Invertebrata
BT2 Nodosariidae
BT2 Nodosariacea
BT2 Rotaliina
BT2 foraminifera
BT2 Protista
BT2 microfossils

Leon
No longer a valid term for GeoRef. As of 1993, see Leon City Spain for the city. For the department in Nicaragua, see Leon Nicaragua. For the city in Mexico, see Leon Mexico.

Leon City
No longer a valid term for GeoRef. As of 1993 see Leon Mexico.

Leon City Spain (1993)
Refers to only the city in Spain. Before 1993, also search Leon and Spain.
BT Leon Spain
BT Leon region
BT Spain
BT Iberian Peninsula
BT Southern Europe
BT Europe
SA Mexico
SA Nicaragua

Leon County
Valid through 1988. Search in combination with state term. After 1988, use specific county-state term.

Leon County Florida (1989)
Bounded by Ochlockonee River and Lake Talquin in NW Florida. Before 1989, also search Leon County AND Florida.
CO N301600N304200 W0835900W0844100
BT Florida
BT United States

Leon County Texas (1989)
E central Texas. Before 1989, also search Leon County AND Texas.
CO N305700N313800 W0944900W0952600
BT Texas
BT United States
SA Oakwood Dome

Leon Department
No longer a valid term for GeoRef. As of 1993 see Leon Nicaragua.

Leon Mexico (1993)
City in Guanajuato, Mexico. Before 1993 also search Leon City, and Leon AND Mexico.
BT Guanajuato Mexico
BT Mexico

Leon Nicaragua (1993)
Department including the city in W Nicaragua. Before 1981, search Leon AND Nicaragua. Before 1993, also search Leon Department and Nicaragua.
BT Nicaragua
BT Central America

Leon Province
No longer a valid term for GeoRef. As of 1993, see Leon Spain.

Leon region (1981)
Before 1981, search Leon AND Spain. In NW Spain.
CO N401500N431500 W0044500W0070000
BT Spain
BT Iberian Peninsula
BT Southern Europe
BT Europe
NT Leon Spain
NT Salamanca Spain
NT Zamora Spain

Leon Spain (1993)
Refers only to the province in NW Spain. From 1981-1992, also search Leon Province and Spain. Before 1981, also search Leon and Spain. For the city, see Leon City Spain.
CO N420000N431500 W0044500W0070500
BT Leon region
BT Spain
BT Iberian Peninsula
BT Southern Europe
BT Europe
NT Leon City Spain

Leonardian (1978)
Provincial series, North America. Above Wolfcampian, below Guadalupian.
BT Lower Permian
BT Permian
BT Paleozoic
NT Bone Spring Limestone
NT Clear Fork Group
SA Abo Formation
SA San Andres Formation

leonardite
use leonhardite

leonhardite (1978)
UF leonardite
BT zeolite group
BT framework silicates
BT silicates
SA laumontite

leopoldite
use sylvite

Leopoldville
use Kinshasa Zaire

Leoville
No longer a valid term for GeoRef. As of 1993, see Leoville Saskatchewan.

Leoville Meteorite (1989)
Impact in Decatur County, NW Kansas.
BT carbonaceous chondrites
BT chondrites
BT stony meteorites
BT meteorites
SA Decatur County Kansas
SA Kansas

Leoville Saskatchewan (1993)
Village in W central Saskatchewan. Before 1993, also search Leoville AND Saskatchewan.
BT Saskatchewan
BT Western Canada
BT Canada

Leperditicopida (1978)
Order. As of 1990, microfossils, Crustacea, Mandibulata, and Arthropoda are autoposted to this term. This term has multiple hierarchies.
BT1 Ostracoda
BT1 Crustacea
BT1 Mandibulata
BT1 Arthropoda
BT1 Invertebrata
BT2 Ostracoda
BT2 microfossils

lepidocrocite (1978)
BT oxides
SA iron oxides

Lepidocyclina (1978)
Autoposting of Orbitoididae to this term ended in 1989. Autoposting of microfossils and Protista to this term began in 1990. This term has multiple hierarchies.
BT1 Orbitoidacea
BT1 Rotaliina
BT1 foraminifera
BT1 Protista
BT1 Invertebrata
BT2 Orbitoidacea
BT2 Rotaliina
BT2 foraminifera
BT2 Protista
BT2 microfossils

Lepidocystoidea (1978)
BT Crinozoa
BT Echinodermata
BT Invertebrata

Lepidodendron (1989)
Genus.
BT Lycopsida
BT pteridophytes
BT Plantae

lepidolite (1978)
From 1978-1980, halides and sheet silicates were autoposted to this term. Autoposting of fluorides and mica group began in 1981. Autoposting of halides, sheet silicates and silicates began in 1989. This term has multiple hierarchies.
BT1 fluorides
BT1 halides

BT2 mica group
BT2 sheet silicates
BT2 silicates
SA cookeite

lepidomelane (1978)
BT mica group
BT sheet silicates
BT silicates

Lepidoptera (1978)
BT Lepidopteroida
BT Insecta
BT Mandibulata
BT Arthropoda
BT Invertebrata
SA Raphidioidea

Lepidopteroida (1981)
BT Insecta
BT Mandibulata
BT Arthropoda
BT Invertebrata
NT Lepidoptera

Lepidosauria (1978)
In Lepidosauromorpha. Autoposting of this term began in 1978. Before 1993, considered a narrower term of Eosuchia. Before 1993, Rhynchocephalia was a narrower term.
BT Diapsida
BT Reptilia
BT Tetrapoda
BT Vertebrata
BT Chordata
NT Squamata
SA Eosuchia
SA Rhynchosauria

Lepontine Alps (1978)
Division of Central Alps.
IN Index countries as applicable.
CO N460000N464500
 E0091500E0081500
BT Central Alps
BT Alps
BT Europe
NT Gotthard Massif
SA Graubunden Switzerland
SA Piedmont Alps
SA Piemonte Italy
SA San Giorgio Mountain
SA Ticino Switzerland

Lepospondyli (1978)
Subclass.
BT Amphibia
BT Tetrapoda
BT Vertebrata
BT Chordata

leptite (1978)
Before 1978, also search leptynite.
UF leptynite
BT granulites
BT metamorphic rocks

Leptocythere (1978)
Genus. As of 1990, microfossils, Crustacea, Mandibulata, and Arthropoda are autoposted to this term. This term has multiple hierarchies.
BT1 Cytheracea
BT1 Cytherocopina
BT1 Podocopida
BT1 Ostracoda
BT1 Crustacea
BT1 Mandibulata
BT1 Arthropoda
BT1 Invertebrata
BT2 Cytheracea
BT2 Cytherocopina
BT2 Podocopida
BT2 Ostracoda
BT2 microfossils

leptynite
 use leptite

Lerida
 No longer a valid term for GeoRef. As of 1993, see Lerida Spain.

Lerida Spain (1993)
Province in Catalonia in NE Spain. Before 1993, also search Lerida and Spain.
CO N411500N425200
 E0015100E0002000
BT Catalonia Spain
BT Spain
BT Iberian Peninsula
BT Southern Europe
BT Europe
NT Montsech
NT Pobla de Segur Spain
NT Tremp Spain
SA Segre Valley

Les Eyzies region (1985)
S Dordogne, France.
UF Les Eyzies-de-Tayac
SA Dordogne France

Les Eyzies-de-Tayac
 use Les Eyzies region

Lesbos (1978)
Island in E Aegean Sea off NW Anatolia. Also a Greek department. This term has multiple hierarchies.
BT1 Greek Aegean Islands
BT1 Greece
BT1 Southern Europe
BT1 Europe
BT2 Greek Aegean Islands
BT2 Aegean Islands
BT2 Mediterranean region

Leslie County
 Valid through 1988. Search in combination with state term. After 1988, use specific county-state term.

Leslie County Kentucky (1989)
SE Kentucky. Before 1989, also search Leslie County AND Kentucky.
CO N365200N372000
 W0831000W0833500
BT Kentucky
BT United States
SA Lewis Creek

Lesna
 No longer a valid term for GeoRef. As of 1993 see Lesna Poland.

Lesna Poland (1993)
Town in Lower Silesia in SW Poland. Before 1993 also search Lesna AND Poland.
BT Wroclaw Poland
BT Poland
BT Central Europe
BT Europe

Lesotho (1978)
Former British colony of Basutoland.
CO S304000S283500
 E0293000E0270000
UF Basutoland
BT Southern Africa
BT Africa
SA Orange River

Lesser Antilles (1978)
The smaller islands of the Antilles extending in an arc from Puerto Rico to the islands N of Venezuela. As of 1981, the following terms (and their respective narrower terms) are considered narrower terms: Barbados, Dominica, Leeward Islands, Netherlands Antilles, Trinidad and Tobago, and Windward Islands. Martinique has been a narrower term since before 1981. Autoposting of this term began in 1978. Leeward Islands and Windward Islands do not autopost after 1992.
IN Index Barbados, Trinidad and island groups as applicable.
CO N100000N190000
 W0590000W0700200
BT Antilles
BT West Indies
BT Caribbean region
NT Barbados
NT Carriacou
NT Dominica
NT Grenada
NT Guadeloupe
NT Martinique
NT Montserrat
NT Netherlands Antilles
NT Saint Lucia
NT Saint Vincent
NT Trinidad and Tobago
NT Virgin Islands
SA Leeward Islands
SA Saint John Island
SA Windward Islands

Lesser Caucasus (1978)
Mountain system formed by the N frontal ranges of the Armenian Highland and separated from the Greater Caucasus by the Colchis and Kura lowlands.
IN Index former Soviet republics as applicable.
CO N380000N420000
 E0490000E0413000
BT Caucasus
BT Europe
NT Shakhdag Range
NT Trialet Range
NT Zangezur
SA Armenia
SA Azerbaidzhan
SA Georgian Republic
SA Nagorno-Karabakh Azerbaidzhan
SA Transcaucasia

Lesser Himalayas (1978)
The central range of the Himalayas running parallel to the Greater Himalayas to the N.
IN Index countries as applicable.
BT Himalayas
BT Asia
SA Bhutan
SA India
SA Nepal
SA Sikkim India

Lesser Khingan Mountains
 No longer a valid term for GeoRef.
 use Xiao Hinggan Ling

Lesser Sunda Islands (1978)
Chain of islands E from Bali to and including Alor and Timor, but not Wetar. Bali and Timor are narrower terms as of 1981.
IN Index Indonesia as applicable.
CO S110000S080000
 E1273000E1154000
BT Far East
BT Asia
NT Bali
NT Pantar
NT Timor
SA Indonesia
SA Sunda Arc

Lesser Walachia
 use Oltenia

Lessini Mountains (1978)
In SW Dolomites in NE Italy. Before 1982, also search Monti Lessini. This term has multiple hierarchies.
BT1 Dolomites
BT1 Eastern Alps
BT1 Alps
BT1 Europe
BT2 Dolomites
BT2 Italy
BT2 Southern Europe
BT2 Europe
SA Veneto Italy

Letcher County
 Valid through 1988. Search in combination with state term. After 1988, use specific county-state term.

Letcher County Kentucky (1989)
SE Kentucky. Before 1989, also search Letcher County AND Kentucky.
CO N365700N372100
 W0823400W0831000
BT Kentucky
BT United States

letter
 A valid term through mid-1978; used in title annotation when document is correspondence. No longer a valid index term.

leucite (1978)
UF Vesuvian garnet
BT framework silicates
BT silicates

leucitite (1978)
From 1978-1980, alkali basalt family was autoposted to this term.
BT alkali gabbros
BT gabbros
BT plutonic rocks
BT igneous rocks
SA alkali basalts
SA ultramafics

leucogranite (1978)
BT granites
BT plutonic rocks
BT igneous rocks

leucosomes (1989)
Light-colored part of a migmatite.
BT migmatites
BT metamorphic rocks

leucosphenite (1978)
BT sheet silicates
BT silicates

leucoxene (1985)
BT oxides
SA titanium ores

Leuven Belgium
 use Louvain Belgium

Levalloisian (1985)
Archaeological classification. Europe, Middle East and the former USSR.
BT Paleolithic
BT Stone Age
BT Cenozoic
SA archaeology
SA Pleistocene
SA Quaternary

Levant
 As of 1981, no longer a valid term for GeoRef. Use Middle East.

Levantinian (1978)

Austria.
BT upper Pliocene
BT Pliocene
BT Neogene
BT Tertiary
BT Cenozoic

levees (1978)
SA channels
SA embankments
SA floodplains
SA fluvial features
SA geomorphology
SA marine geology
SA retaining walls
SA sand boils
SA streams
SA submarine canyons
SA waterways

level, changes of
use changes of level

leveling (1978)
UF levelling
SA cartography
SA geodesy
SA trilateration

leveling networks
Not a valid GeoRef index term after 1971. Was used on level 1 in subfile G. Now use leveling or networks.

levelling
use leveling

levels (1978)
SA background level
SA background radiation
SA drawdown
SA ground water
SA potentiometric surface
SA reservoirs
SA shallow aquifers
SA shallow depth
SA water table
SA water wells

Levis Shale (1978)
BT Devonian
BT Paleozoic
SA Quebec

Levy County
Valid through 1988. Search in combination with state term. After 1988, use specific county-state term.

Levy County Florida (1989)
Bounded by Gulf of Mexico in N Florida. Before 1989, also search Levy County AND Florida.
CO N290000N293500
W0822500W0831300
BT Florida
BT United States

Lewis and Clark County
Valid through 1988. Search in combination with state term. After 1988, use specific county-state term.

Lewis and Clark County Montana (1989)
W central Montana. Before 1989, also search Lewis and Clark County AND Montana.
CO N462200N475500
W1113500W1131000
BT Montana
BT United States
NT Helena Montana
SA Sawtooth Range

Lewis and Clark Line
use Lewis and Clark Lineament

Lewis and Clark Lineament (1989)
N Idaho to central Montana.
IN Index states as applicable.
UF Lewis and Clark Line
UF Montana Lineament
BT United States
SA Idaho
SA Montana

Lewis Cliff Meteorites (1993)
Meteorites found in or near ice tongue at Lewis Cliff, Beardmore Glacier region near the edge of the Ross Ice Shelf, Ross Dependency. Due to abbreviations for these meteorites, such as LEW 86010 and LEW86010, also search LEW in combination with number. Before 1993, also search Lewis Cliff AND meteorites and Lewis Cliffs Meteorites.
UF Lewis Cliffs Meteorites
BT meteorites
SA Antarctica
SA Beardmore Glacier
SA Ross Dependency
SA Ross Ice Shelf

Lewis Cliffs Meteorites
use Lewis Cliff Meteorites

Lewis County
Valid through 1988. Search in combination with state term. After 1988, use specific county-state term.

Lewis County Idaho (1989)
W Idaho. Before 1989, also search Lewis County AND Idaho.
CO N460000N462900
W1160200W1164600
BT Idaho
BT United States

Lewis County Kentucky (1989)
NE Kentucky. Before 1989, also search Lewis County AND Kentucky.
CO N381400N384200
W0830300W0833700
BT Kentucky
BT United States

Lewis County Missouri (1989)
NE Missouri. Before 1989, also search Lewis County AND Missouri.
CO N395600N401700
W0912600W0915700
BT Missouri
BT United States

Lewis County New York (1989)
N central New York. Before 1989, also search Lewis County AND New York.
CO N432500N441300
W0750700W0755200
BT New York
BT United States

Lewis County Tennessee (1989)
Central Tennessee. Before 1989, also search Lewis County AND Tennessee.
CO N352400N353900
W0871600W0874200
BT Tennessee
BT United States

Lewis County Washington (1989)
SW Washington. Before 1989, also search Lewis County AND Washington.
CO N462400N464700
W1212300W1232200
BT Washington
BT United States
SA Cowlitz River
SA Mount Rainier National Park
SA Yakima Indian Reservation

Lewis County West Virginia (1989)
Central W Virginia. Before 1989, also search Lewis County AND West Virginia.
CO N384300N391000
W0801900W0804400
BT West Virginia
BT United States

Lewis Creek (1989)
Paleobotanical locality in Leslie County, E Kentucky. Also a Lewis Creek in Colorado and Illinois. Search in combination with appropriate state.
IN Index states as applicable.
SA Colorado
SA Illinois
SA Kentucky
SA Leslie County Kentucky

Lewis Shale (1978)
Lewis and Meeteetse formations throughout N central Wyoming considered stratigraphic equivalents with Lewis being applied generally in areas where marine strata are present and Meeteetse in areas where rocks are nonmarine in character. W Colorado, NW New Mexico, and S and central Wyoming.
BT Upper Cretaceous
BT Cretaceous
BT Mesozoic
SA Colorado
SA New Mexico
SA Wyoming

Lewis thrust fault (1989)
NW Montana and SW Alberta.
IN Index Montana and/or Alberta or region as applicable.
SA Alberta
SA Montana
SA North America

Lewisian (1978)
Europe. Autoposting of Precambrian to this term ended in 1989. As of 1989, Proterozoic is autoposted to this term. Autoposting of upper Precambrian and Precambrian to this term began in 1990.
BT Proterozoic
BT upper Precambrian
BT Precambrian

Lewisian Complex (1989)
NW Scotland. Includes Loch Maree Group.
IN Index ages as applicable.
BT Precambrian
SA Archean
SA Proterozoic
SA Scotland

lexicon
Not a valid GeoRef index term after 1970. Was used on level 1 in subfile N.
use lexicons

lexicons (1978)
Used for the systematic arrangement of words in one or more languages, and their definitions. Before 1971, also search lexicon.
IN Index language if not English.
UF lexicon
SA Arabic
SA bibliography
SA catalogs
SA dictionaries
SA French
SA German
SA glossaries
SA mineralogy
SA monographs
SA paleontology
SA petrology
SA publications
SA Russian
SA sedimentary petrology
SA stratigraphy
SA structural geology

Lexington Limestone (1978)
Comprises Curdsville Limestone, Logana Formation, Jessamine Limestone, Benson Limestone, Brannon Limestone with Woodburn Phosphatic Member. Central Kentucky.
BT Middle Ordovician
BT Ordovician
BT Paleozoic
SA Kentucky

Lg-waves (1989)
Higher-mode short-period Love and Rayleigh waves which only travel through continental crust.
BT surface waves
BT elastic waves
SA continental crust
SA Love waves
SA Rayleigh waves
SA seismology
SA short-period waves
SA waves

Lhasa
No longer a valid term for GeoRef. See Lhasa China.

Lhasa Block (1989)
Qinghai-Xizang Plateau, SW China.
BT Xizang China
BT China
BT Far East
BT Asia
SA Lhasa China
SA Qinghai-Xizang Plateau

Lhasa China (1993)
City in SE Xizang.
CO N294100N294100
E0911000E0911000
BT Xizang China
BT China
BT Far East
BT Asia
SA Lhasa Block

Lherz (1978)
Lake in the Pyrenees in S France.
BT France
BT Western Europe
BT Europe
SA Ariege France

lherzolite (1978)
BT peridotites
BT ultramafics
BT plutonic rocks
BT igneous rocks
SA garnet lherzolite
SA spinel lherzolite

Li
use lithium

Li-6 (1978)
Autoposting of broader terms began in 1989. This term has multiple hierarchies.
BT1 stable isotopes
BT1 isotopes
BT2 lithium
BT2 alkali metals
BT2 metals

Li-7 (1978)
  Autoposting of broader terms began in 1989. This term has multiple hierarchies.
  BT1 stable isotopes
  BT1 isotopes
  BT2 lithium
  BT2 alkali metals
  BT2 metals

Liaodong Peninsula (1989)
  S Liaoning, NE China. Extends into Bohai Bay.
  CO N383000N410000
      E1250000E1220000
  UF Liaotung Peninsula
  BT Liaoning China
  BT China
  BT Far East
  BT Asia
  SA Bohai Bay

Liaohe Basin (1989)
  NE China.
  BT China
  BT Far East
  BT Asia

Liaoning
  No longer a valid term for GeoRef. See Liaoning China.

Liaoning China (1993)
  Province in NE China.
  CO N390000N433000
      E1254000E1190000
  BT China
  BT Far East
  BT Asia
  NT Anshan China
  NT Haicheng China
  NT Liaodong Peninsula
  SA Anshan Group
  SA Bohai Bay
  SA Haicheng earthquake 1975
  SA Tancheng-Lujiang fault zone

Liaotung Peninsula
  use Liaodong Peninsula

Liard River (1978)
  IN   Index British Columbia and territories as applicable.
  BT Canada
  SA British Columbia
  SA Northwest Territories
  SA Yukon Territory

Lias
  use Lower Jurassic

Liassic
  As of 1981, no longer a valid term for GeoRef.
  use Lower Jurassic

Liberia (1978)
  CO N040000N080000
      W0070000W0120000
  BT West Africa
  BT Africa
  NT Bong Range
  NT Liberian Shield
  SA Nimba Mountains
  SA Saint John River

Liberian Shield (1978)
  Inland plateau.
  BT Liberia
  BT West Africa
  BT Africa

Liberty County
  Valid through 1988. Search in combination with state term. After 1988, use specific county-state term.

Liberty County Florida (1989)
  Bounded by Ochlockonee River and Apalachicola River in NW Florida. Before 1989, also search Liberty County AND Florida.
  CO N295800N303600
      W0843100W0851000
  BT Florida
  BT United States

Liberty County Georgia (1989)
  SE Georgia. Before 1989, also search Liberty County AND Georgia.
  CO N313400N320600
      W0811000W0815000
  BT Georgia
  BT United States

Liberty County Montana (1989)
  E Montana. Before 1989, also search Liberty County AND Montana.
  CO N481100N490000
      W1104500W1112500
  BT Montana
  BT United States

Liberty County Texas (1989)
  E Texas. Before 1989, also search Liberty County AND Texas.
  CO N295400N303300
      W0942600W0951000
  BT Texas
  BT United States

Libia
  No longer a valid GeoRef index term. Before 1969, was used on level 1 in subfile E.
  use Libya

libraries (1985)
  SA bibliography
  SA catalogs
  SA collections
  SA education
  SA geology

Libya (1978)
  Before 1969, also search Libia.
  CO N193000N330000
      E0250000E0093000
  UF Libia
  BT North Africa
  BT Africa
  NT Cyrenaica
  NT Fezzan
  NT Kufra Basin
  NT Murzuk Basin
  NT Sirte Basin
  NT Tibesti Libya
  NT Tripoli Libya
  NT Tripolitania
  NT Zelten Libya
  SA Libyan Desert
  SA Mediterranean region
  SA Near East
  SA Sahara
  SA Tibesti Massif

Libyan Desert (1978)
  Desert area of the E Sahara, W of the Nile.
  IN   Index countries as applicable.
  BT Africa
  SA Baharlya Oasis
  SA Egypt
  SA Libya
  SA Sudan
  SA Western Desert

lichenometry (1978)
  SA geochronology
  SA lichens
  SA relative age

lichens (1978)
  BT thallophytes
  BT Plantae
  SA algae
  SA fungi
  SA lichenometry

Lichida (1978)
  BT Trilobita
  BT Trilobitomorpha
  BT Arthropoda
  BT Invertebrata

Liechtenstein (1978)
  Independent principality between NE Switzerland and W Austria. As of 1990, Europe is autoposted to this term.
  BT Central Europe
  BT Europe
  SA Rhine Basin
  SA Rhine River

Liege
  No longer a valid term for GeoRef. See Liege Belgium.

Liege Belgium (1993)
  Province and city in E Belgium.
  CO N501400N504800
      E0063000E0045800
  BT Belgium
  BT Western Europe
  BT Europe
  NT Hohe Venn
  NT Stavelot Belgium
  NT Vesdre Valley
  SA Liege earthquake 1983
  SA Limburg Belgium

Liege earthquake 1983 (1989)
  Epicenter in Liege, E Belgium.
  BT earthquakes
  SA Belgium
  SA Liege Belgium

life
  A valid index term through 1977 used in combination with origin.

life assemblage
  use biocenoses

life origin (1978)
  Before 1978, also search origin AND life.
  SA biologic evolution
  SA creationism
  SA Darwinism
  SA DNA
  SA exobiology
  SA extraterrestrial geology
  SA paleontology

ligands (1993)
  In a chelate or coordination compound, the molecule, ion or group bound to the central atom.
  SA bonding
  SA chelation
  SA complexing
  SA inorganic materials
  SA ions
  SA metals
  SA organic materials

light microscopy
  As of 1993, see optical spectra, optical spectroscopy and microscope methods.

light minerals (1978)
  SA calcite
  SA dolomite
  SA heavy minerals
  SA minerals
  SA muscovite
  SA opaque minerals
  SA quartz

lightning (1978)
  SA meteorology

lightweight aggregate
  use aggregate

lignin (1993)
  BT organic materials
  SA cellulose
  SA seeds
  SA wood

lignite (1978)
  Before 1981, also search brown coal. This term has multiple hierarchies.
  IN   See list C for use as a commodity term.
  BT1 coal
  BT1 organic residues
  BT2 coal
  BT2 sedimentary rocks
  SA Adamow Mine
  SA anthracite
  SA Belchatow Mine
  SA huminite
  SA humodetrinite
  SA organic materials
  SA peat

Liguria
  No longer a valid term for GeoRef. See Liguria Italy.

Liguria Italy (1993)
  Autonomous region in NW Italy.
  CO N434500N444000
      E0100500E0074500
  BT Italy
  BT Southern Europe
  BT Europe
  NT Genoa Italy
  NT Ligurian Alps
  NT Ligurian Apennines
  NT Savona Italy

Ligurian Alps (1978)
  E extension of Maritime Alps along coast of Ligurian Sea. This term has multiple hierarchies.
  BT1 Liguria Italy
  BT1 Italy
  BT1 Southern Europe
  BT1 Europe
  BT2 Western Alps
  BT2 Alps
  BT2 Europe
  SA Maritime Alps

Ligurian Apennines (1978)
  Extend from near Savona SE to just N of La Spezia along coast of Ligurian Sea. This term has multiple hierarchies.
  BT1 Liguria Italy
  BT1 Italy
  BT1 Southern Europe
  BT1 Europe
  BT2 Apennines
  BT2 Italy
  BT2 Southern Europe
  BT2 Europe
  NT Monte Antola

Ligurian Sea (1978)
  Enclosed by Italian autonomous regions of Liguria and Tuscany on the N and E, and by Corsica on the S.
  BT Mediterranean Sea

Likhvin Russian Federation
  use Chekalin Russian Federation

Liliidae (1981)
  BT Monocotyledoneae
  BT angiosperms
  BT Spermatophyta
  BT Plantae

Lille
  No longer a valid term for GeoRef. See Lille France.

Lille France (1993)
City near Belgian frontier in N France.
BT Nord France
BT France
BT Western Europe
BT Europe

lillianite (1978)
BT sulfobismuthites
BT sulfosalts

Lilydale Limestone (1978)
Upper Yarra District in S central Victoria.
BT Devonian
BT Paleozoic
SA Australia
SA Victoria Australia

Lima
No longer a valid term for GeoRef. As of 1993 see Lima Peru.

Lima Peru (1993)
Department and city in W Peru. Before 1993 also search Lima Peru.
BT Peru
BT South America
NT Casapalca Peru
NT Santa Eulalia Peru
SA Pucara Group

Limagne (1978)
Fertile lowland in central France.
IN Index departments as applicable.
BT France
BT Western Europe
BT Europe
SA Allier France
SA Puy-de-Dome France

Limbourg
As of 1993, see Limburg, Limburg Belgium, and Limburg Netherlands.

Limburg (1978)
As of 1993, use only for the region in W Europe. Before then was used also as the name of town and province in Belgium and province in SE Netherlands. Also search Limbourg.
IN Index countries and provinces as applicable.
SA Limburg Belgium
SA Limburg Netherlands

Limburg Belgium (1993)
Name of province in E Belgium. As of 1993, use Limbourg Belgium for town in Liege Belgium.
BT Belgium
BT Western Europe
BT Europe
SA Liege Belgium
SA Limburg
SA Limburg Netherlands

Limburg Netherlands (1993)
Province in SE Netherlands.
BT Netherlands
BT Western Europe
BT Europe
NT Maastricht Netherlands
SA Limburg
SA Limburg Belgium

limburgite (1978)
BT ultramafics
BT plutonic rocks
BT igneous rocks

lime (1978)
IN Used as a commodity term.
SA calcium carbonate
SA limestone deposits

Lime Creek Formation (1978)
Includes Juniper Hill, Cerro Gordo, and Owen members. Central N Iowa.
BT Upper Devonian
BT Devonian
BT Paleozoic
SA Iowa

Limerick
No longer a valid term for GeoRef. See Limerick Ireland.

Limerick Ireland (1993)
County in W central Ireland.
BT Ireland
BT Western Europe
BT Europe

limestone (1978)
As of 1981, use limestone deposits for limestone as a commodity. Autoposting of this term to calcarenite, biocalcarenite, biomicrite, biosparite, calcilutite, coquina and micrite began in 1985.
BT carbonate rocks
BT sedimentary rocks
NT algal limestone
NT biocalcarenite
NT biomicrite
NT biosparite
NT calcarenite
NT calcilutite
NT coquina
NT dolomitic limestone
NT micrite
NT microcrystalline limestone
NT oolitic limestone
SA algal mounds
SA calcite
SA calcium carbonate
SA carbonate sediments
SA chalk
SA dolomitization
SA jasperoid
SA limestone deposits
SA metalimestone
SA oolite
SA oolitic texture
SA ophicalcite
SA pisolites
SA travertine

Limestone Alps (1978)
This term has multiple hierarchies.
BT1 Alps
BT1 Europe
BT2 Austria
BT2 Central Europe
BT2 Europe

Limestone County
Valid through 1988. Search in combination with state term. After 1988, use specific county-state term.

Limestone County Alabama (1989)
N Alabama. Before 1989, also search Limestone County AND Alabama.
CO N343300N350000
W0864800W0871500
BT Alabama
BT United States

Limestone County Texas (1989)
E central Texas. Before 1989, also search Limestone County AND Texas.
CO N311300N314700
W0861200W0965500
BT Texas
BT United States

limestone deposits (1981)
Before 1981, search limestone AND deposits.
IN Commodity. See List C.
SA building stone
SA cement materials
SA chalk deposits
SA construction materials
SA dimension stone
SA dolostone deposits
SA lime
SA limestone
SA marble deposits
SA ornamental materials

limit, elastic
use elastic limit

limits, Atterberg
use Atterberg limits

Limnocardium (1978)
Genus.
BT Bivalvia
BT Mollusca
BT Invertebrata

limnology (1978)
SA geomorphology
SA glacial lakes
SA hydrology
SA lacustrine features
SA lakes
SA paleolimnology
SA ponds
SA seiches

limonite (1978)
Before 1981, included use as a commodity term under iron.
IN As of 1981, may be used as a commodity term.
BT oxides
SA hydrogoethite
SA iron minerals
SA iron ores
SA iron oxides
SA ocher

Limousin (1978)
Historical region in S central France.
CO N450000N461000
E0030000E0003000
BT France
BT Western Europe
BT Europe

Limpopo Basin (1978)
River basin. Also search Limpopo.
IN Index Transvaal and countries as applicable.
UF Crocodile River basin
BT Africa
SA Botswana
SA Mozambique
SA South Africa
SA Transvaal South Africa
SA Zimbabwe

Limpopo Belt (1981)
Orogenic belt at the N margin of the Kaapvaal Craton in S Africa. Consists of Precambrian metamorphic rocks (the Messina Formation in South Africa).
IN Index countries as applicable.
BT Africa
SA Botswana
SA Mozambique
SA Precambrian
SA South Africa
SA Zimbabwe

Limski Channel (1978)
On Istrian Peninsula SW of Rijeka.
BT Croatia
BT Yugoslavia
BT Southern Europe
BT Europe

Lincoln County
Valid through 1988. Search in combination with state term. After 1988, use specific county-state term.

Lincoln County Arkansas (1989)
SE Arkansas. Before 1989, also search Lincoln County AND Arkansas.
CO N334500N340900
W0912600W0915900
BT Arkansas
BT United States

Lincoln County Colorado (1989)
E central Colorado. Before 1989, also search Lincoln County AND Colorado.
CO N383000N393300
W1030700W1040100
BT Colorado
BT United States

Lincoln County Georgia (1989)
NE Georgia. Before 1989, also search Lincoln County AND Georgia.
CO N334300N335800
W0821500W0823800
BT Georgia
BT United States

Lincoln County Idaho (1989)
S Idaho. Before 1989, also search Lincoln County AND Idaho.
CO N424400N431200
W1134400W1143600
BT Idaho
BT United States

Lincoln County Kansas (1989)
Central Kansas. Before 1989, also search Lincoln County AND Kansas.
CO N385200N391300
W0975500W0982900
BT Kansas
BT United States

Lincoln County Kentucky (1989)
Central Kentucky. Before 1989, also search Lincoln County AND Kentucky.
CO N371200N373600
W0842800W0845200
BT Kentucky
BT United States

Lincoln County Maine (1989)
S Maine. Before 1989, also search Lincoln County AND Maine.
CO N434500N442000
W0691900W0694800
BT Maine
BT United States

Lincoln County Minnesota (1989)
SW Minnesota. Before 1989, also search Lincoln County AND Minnesota.
CO N441300N443800
W0960700W0962800
BT Minnesota
BT United States

Lincoln County Mississippi (1989)
SW Mississippi. Before 1989, also search Lincoln County AND Mississippi.
CO N312200N314300
W0901600W0904500
BT Mississippi
BT United States

Lincoln County Missouri (1989)

E Missouri. Before 1989, also search Lincoln County AND Missouri.
CO N385200N391300
W0903900W0911700
BT Missouri
BT United States

Lincoln County Montana (1989)
Extreme NW Montana. Before 1989, also search Lincoln County AND Montana.
CO N475300N490000
W1143700W1160300
BT Montana
BT United States
SA Purcell Mountains

Lincoln County Nebraska (1989)
SW central Nebraska. Before 1989, also search Lincoln County AND Nebraska.
CO N404200N412300
W1001400W1011500
BT Nebraska
BT United States

Lincoln County Nevada (1989)
SE Nevada. Before 1989, also search Lincoln County AND Nevada.
CO N365300N384000
W1140300W1155300
BT Nevada
BT United States
SA Egan Range

Lincoln County New Mexico (1989)
S central New Mexico. Before 1989, also search Lincoln County AND New Mexico.
CO N330800N342100
W1045400W1062200
BT New Mexico
BT United States
SA Orogrande Basin
SA Sierra Blanca

Lincoln County North Carolina (1989)
W central North Carolina. Before 1989, also search Lincoln County AND North Carolina.
CO N352400N353400
W0805700W0813200
BT North Carolina
BT United States

Lincoln County Oklahoma (1989)
Central Oklahoma. Before 1989, also search Lincoln County AND Oklahoma.
CO N352700N355600
W0963700W0970800
BT Oklahoma
BT United States

Lincoln County Oregon (1989)
W Oregon. Before 1989, also search Lincoln County AND Oregon.
CO N441600N450300
W1233700W1240600
BT Oregon
BT United States
NT Newport Oregon
NT Yaquina Bay

Lincoln County South Dakota (1989)
SE South Dakota. Before 1989, also search Lincoln County AND South Dakota.
CO N430600N433000
W0962600W0965600
BT South Dakota
BT United States

Lincoln County Tennessee (1989)
S Tennessee. Before 1989, also search Lincoln County AND Tennessee.
CO N345900N352300
W0862200W0865000
BT Tennessee
BT United States

Lincoln County Washington (1989)
E Washington. Before 1989, also search Lincoln County AND Washington.
CO N471500N475600
W1175000W1185800
BT Washington
BT United States
NT Odessa Washington

Lincoln County West Virginia (1989)
W West Virginia. Before 1989, also search Lincoln County AND West Virginia.
CO N375700N382200
W0814700W0821700
BT West Virginia
BT United States

Lincoln County Wisconsin (1989)
N central Wisconsin. Before 1989, also search Lincoln County AND Wisconsin.
CO N450700N453300
W0892500W0900300
BT Wisconsin
BT United States

Lincoln County Wyoming (1989)
W Wyoming. Before 1989, also search Lincoln County AND Wyoming.
CO N413500N431900
W1100300W1110300
BT Wyoming
BT United States
SA Moxa Arch

Lincoln Creek Formation (1978)
Vaguely defined. SW Washington.
UF Lincoln Formation
BT Oligocene
BT Paleogene
BT Tertiary
BT Cenozoic
SA Washington

Lincoln Formation
use Lincoln Creek Formation

Lincoln Parish Louisiana (1989)
N Louisiana. Before 1989, also search Lincoln Parish; Lincoln County AND Louisiana.
CO N322600N324500
W0922600W0925100
BT Louisiana
BT United States

Lincolnshire
No longer a valid term for GeoRef. As of 1993, see Lincolnshire England.

Lincolnshire England (1993)
County on the North Sea in E England. In 1974, Humberside County was formed from part. Before 1993, also search Lincolnshire.
CO N524000N534500
E0002000W0010000
BT England
BT Great Britain
BT United Kingdom
BT Western Europe

BT Europe
SA Gibraltar Point
SA Humberside England

Lincolnshire Limestone (1978)
Ordovician formation in SW Virginia and NW Tennessee, and a Jurassic formation in England. Also search Lincolnshire Formation for the one in the United States; also search Lincolnshire Limestone Formation for the one in England.
IN Index ages as applicable.
SA England
SA Jurassic
SA Ordovician
SA Tennessee
SA Virginia

Lindlar
No longer a valid term for GeoRef. See Lindlar Germany.

Lindlar Germany (1993)
Village in North Rhine-Westphalia, W Germany.
BT North Rhine-Westphalia Germany
BT Germany
BT Central Europe
BT Europe

line defects
use crystal dislocations

Line Islands (1978)
Group S of Hawaiian Islands, N and S of Equator. Some belong to U.S., some to Kiribati. In 1981, broader term changed from Pacific Ocean to Polynesia. Fanning Island was added as a narrower term in 1981.
BT Polynesia
BT Oceania
NT Fanning Island
SA Christmas Island
SA DSDP Site 315
SA DSDP Site 315A
SA Kiribati
SA Pacific Ocean
SA Vostok

line of strike
use strike

lineaments (1978)
UF tectonic lines
SA fault zones
SA faults
SA lineation
SA remote sensing
SA tectonics

linear distortion (1981)
SA distortion
SA elastic waves
SA seismology

linear materials (1981)
SA engineering geology
SA materials
SA stiffness

linear orientation (1978)
Before 1978, search linear.
SA orientation

linear programming (1985)
SA computer programs
SA data processing

lineation (1978)
Small-scale linear features. For large-scale features, see lineaments.
UF lineations
SA bedding
SA boudinage
SA cleavage

SA deformation
SA elongate minerals
SA fabric
SA flow lines
SA folds
SA foliation
SA fractures
SA igneous rocks
SA intrusions
SA lineaments
SA metamorphic rocks
SA metamorphism
SA mullions
SA orientation
SA parting lineation
SA preferred orientation
SA slickensides
SA structural analysis
SA structural geology
SA style
SA tectonics

lineations
use lineation

liners
use disposal barriers

lines, flow
use flow lines

Lingula (1978)
Genus.
BT Inarticulata
BT Brachiopoda
BT Invertebrata

Linn County
Valid through 1988. Search in combination with state term. After 1988, use specific county-state term.

Linn County Iowa (1989)
E Iowa. Before 1989, also search Linn County AND Iowa.
CO N415200N421800
W0912300W0915000
BT Iowa
BT United States

Linn County Kansas (1989)
E Kansas. Before 1989, also search Linn County AND Kansas.
CO N380200N382300
W0943800W0950500
BT Kansas
BT United States

Linn County Missouri (1989)
N central Missouri. Before 1989, also search Linn County AND Missouri.
CO N394300N400300
W0925200W0932200
BT Missouri
BT United States

Linn County Oregon (1989)
W Oregon. Before 1989, also search Linn County AND Oregon.
CO N441200N444700
W1214700W1231400
BT Oregon
BT United States
SA Mount Jefferson

linnaeite (1978)
UF cobalt pyrites
UF linneite
BT sulfides
SA carrollite
SA polydymite
SA siegenite

linneite
use linnaeite

Linz
No longer a valid term in GeoRef. See Linz Austria.

Before then, variants of the term may occur in GeoRef.

Linz Austria (1993)
City in NW Upper Austria.
BT Upper Austria
BT Austria
BT Central Europe
BT Europe

Lipari Island (1978)
BT Lipari Islands
BT Sicily Italy
BT Italy
BT Southern Europe
BT Europe

Lipari Islands (1978)
Group of small volcanic islands in SE Tyrrhenian Sea off N coast of Sicily. Also search Lipari.
CO N382000N385000
E0152000E0142000
UF Aeolian Islands
BT Sicily Italy
BT Italy
BT Southern Europe
BT Europe
NT Lipari Island
NT Stromboli
NT Vulcano

liparite (1978)
BT rhyolites
BT volcanic rocks
BT igneous rocks

liparite porphyry (1978)
BT rhyolites
BT volcanic rocks
BT igneous rocks
SA porphyry

lipids (1978)
BT fatty acids
BT organic materials

Lippe (1978)
Former German state (now part of North Rhine-Westphalia) in NW West Germany. Also a river.
BT North Rhine-Westphalia Germany
BT Germany
BT Central Europe
BT Europe

liptinite
use exinite

liquefaction (1978)
Not to be used under phase equilibria.
SA geologic hazards
SA geomorphology
SA liquefaction potential
SA mass movements
SA sediments
SA slope stability
SA slumping
SA soil mechanics

liquefaction potential (1981)
SA engineering geology
SA geologic hazards
SA liquefaction
SA slope stability

liquid
use liquid phase

liquid chromatography (1985)
SA chromatography
SA ion chromatography

liquid inclusions
No longer a valid GeoRef index term. Before 1971, was used on level 1 in subfile N.
use fluid inclusions

liquid phase (1978)
Before 1978, also search liquid.
UF liquid
SA dense nonaqueous phase liquids
SA fluid phase
SA gaseous phase
SA melts
SA nonaqueous phase liquids
SA solid phase

liquid waste (1978)
UF waste, liquid
SA detergents
SA effluents
SA engineering geology
SA environmental geology
SA ground water
SA industrial waste
SA radioactive waste
SA Resource Conservation and Recovery Act
SA sludge
SA solid waste
SA waste disposal
SA waste disposal sites
SA waste water

Lisbon Formation (1985)
In Claiborne Group. Georgia, Mississippi and SW Alabama.
BT middle Eocene
BT Eocene
BT Paleogene
BT Tertiary
BT Cenozoic
SA Alabama
SA Claiborne Group
SA Georgia
SA Mississippi

Lisburne Group (1978)
Mississippian to Permian. Includes Wachsmuth Limestone, Alapah Limestone, Kayak Shale, Kanayut Conglomerate. N Alaska.
IN Index ages as applicable.
BT Paleozoic
SA Alaska
SA Carboniferous
SA Kanayut Conglomerate
SA Kayak Shale
SA Permian

Lissamphibia (1978)
Subclass. Autoposting of this term began in 1978.
BT Amphibia
BT Tetrapoda
BT Vertebrata
BT Chordata
NT Anura

list, faunal
use faunal list

listric faults (1985)
BT faults
SA growth faults
SA half grabens
SA hinge faults
SA normal faults
SA slumping

listvenite (1978)
BT mica group
BT sheet silicates
BT silicates

listwanite (1978)
Compositional term.
BT schists
BT metamorphic rocks

Litchfield County
Valid through 1988. Search in combination with state term. After 1988, use specific county-state term.

Litchfield County Connecticut (1989)
NW Connecticut. Before 1989, also search Litchfield County AND Connecticut.
CO N412800N420400
W0725300W0733200
BT Connecticut
BT United States
NT Woodbury Connecticut

lithic texture
As of 1981, no longer a valid term for GeoRef. See textures.

lithification (1978)
SA cementation
SA coal
SA compaction
SA compression
SA consolidation
SA crystallization
SA diagenesis
SA recrystallization
SA sedimentation
SA sediments

lithiophilite (1978)
BT phosphates

lithiophorite (1978)
BT oxides

lithium (1978)
Chemical element. As of 1981, use lithium ores for lithium as a commodity. Autoposting of alkali metals and metals to this term began in 1989.
UF Li
BT alkali metals
BT metals
NT Li-6
NT Li-7
SA amblygonite
SA lithium ores

lithium ores (1981)
Before 1981, also search (lithium OR Li) AND (deposit OR deposits OR ore OR ores OR economic) in the basic index. Autoposting of metal ores to this term began in 1989.
BT metal ores
SA lithium
SA petalite
SA spodumene

lithofacies (1978)
SA facies
SA hardground
SA reefs
SA sedimentary rocks
SA sediments
SA seismic stratigraphy
SA stratigraphic wedges
SA Waulsortian facies

lithofacies maps
use lithologic maps

lithogeochemical methods
Not a valid term for GeoRef. See geochemical methods or lithogeochemistry.

lithogeochemistry (1993)
before 1993, also search geochemistry and rock type.
BT geochemistry
SA chemical analysis
SA composition
SA devitrification
SA fluid inclusions
SA fugacity
SA geochemical anomalies
SA geologic barometry
SA geologic thermometry
SA igneous rocks
SA iron-rich composition
SA metamorphic rocks
SA metamorphism
SA metasomatism
SA migration of elements
SA mineral deposits, genesis
SA mineral exploration
SA paleotemperature
SA partitioning
SA sedimentary rocks
SA water of crystallization
SA water of dehydration
SA weathering

lithologic controls (1978)
Also search lithologic.
SA controls
SA mineral deposits, genesis
SA stratigraphic controls

lithologic maps (1978)
Before 1978, search maps AND lithologic.
UF lithofacies maps
BT maps

lithologic traps
use stratigraphic traps

lithology
As of 1981, no longer a valid term for GeoRef.

lithophile elements (1978)
Chemical elements which tend to concentrate in silicate phases of meteorites. Also, an element possessing greater free energy of oxidation per gram of oxygen than iron. Before 1978, also search lithophile.
UF oxyphile elements
SA aluminum
SA boron
SA calcium
SA cerium
SA chalcophile elements
SA chemical elements
SA lanthanum
SA manganese
SA potassium
SA rubidium
SA selenium
SA siderophile elements
SA sodium
SA uranium

Lithophyllum (1978)
Autoposting of thallophytes and microfossils to this term began in 1990. This term has multiple hierarchies.
BT1 Corallinaceae
BT1 Rhodophyta
BT1 algae
BT1 thallophytes
BT1 Plantae
BT2 Corallinaceae
BT2 Rhodophyta
BT2 algae
BT2 microfossils

Lithoprobe (1989)
Program for seismic reflection profiling of the lithosphere.
SA geophysical methods
SA geophysical surveys
SA geotraverses
SA lithosphere
SA reflection methods
SA seismic methods
SA seismic profiles
SA seismic surveys

lithosiderite
use stony irons

lithosphere (1978)
SA asthenosphere
SA Benioff zone
SA biosphere
SA core

SA crust
SA Earth
SA Lithoprobe
SA lower crust
SA mantle
SA Mohorovicic discontinuity
SA ophiolite complexes
SA PASSCAL
SA plate tectonics
SA tectonophysics
SA tectonosphere
SA underplating

lithostratigraphy (1978)
UF petrostratigraphy
UF rock-stratigraphy
SA hardground
SA metamorphic rocks
SA outliers
SA sedimentary rocks
SA sediments
SA seismic stratigraphy
SA stratigraphic gaps
SA stratigraphic units
SA stratigraphic wedges
SA stratigraphy
SA tectonostratigraphic units
SA unconformities

Lithothamnion
use Lithothamnium

Lithothamnium (1978)
Genus. Autoposting of thallophytes and microfossils began in 1990. This term has multiple hierarchies.
UF Lithothamnion
BT1 Corallinaceae
BT1 Rhodophyta
BT1 algae
BT1 thallophytes
BT1 Plantae
BT2 Corallinaceae
BT2 Rhodophyta
BT2 algae
BT2 microfossils

Lithuania (1978)
Former Lithuanian Soviet Socialist Republic.
CO N530000N560000
E0270000E0210000
BT Baltic region
BT Europe
NT Sventoji River
NT Vilna Lithuania
SA Baltic Glaciation
SA Courland Spit
SA Neman River basin
SA Russian Plain

Litopterna (1978)
Order. Autoposting of Eutheria and Theria to this term began in 1989.
BT Eutheria
BT Theria
BT Mammalia
BT Tetrapoda
BT Vertebrata
BT Chordata
SA Ungulata

Litorina Sea (1993)
Lower Holocene. Stage of Baltic Glaciation. Baltic region.
IN Index countries as applicable.
BT lower Holocene
BT Holocene
BT Quaternary
BT Cenozoic
SA Baltic Glaciation
SA Baltic region
SA Finland
SA Norway
SA Sweden

Little Bahama Bank (1985)
Shoal in N Bahamas, between Grand Bahama and Great Abaco islands. As of 1990, Atlantic Ocean is autoposted to this term.
CO N260000N264000
W0771000W0781000
BT North Atlantic
BT Atlantic Ocean
SA Bahamas
SA Great Bahama Bank
SA Leg 101
SA Northwest Providence Channel
SA ODP Site 628
SA ODP Site 629
SA ODP Site 630

Little Belt Mountains (1978)
Range of the Rocky Mountains in central Montana. This term has multiple hierarchies.
BT1 Montana
BT1 United States
BT2 U. S. Rocky Mountains
BT2 Rocky Mountains
BT3 U. S. Rocky Mountains
BT3 United States
SA Cascade County Montana
SA Judith Basin County Montana

Little Falls Formation (1978)
Represents an offshore, more carbonate phase of Potsdam Sandstone, Galway Formation, and perhaps even part of Hoyt Limestone. E central and E New York.
BT Upper Cambrian
BT Cambrian
BT Paleozoic
SA New York
SA Potsdam Sandstone

Little Ice Age
use Neoglacial

Little Khingan Mountains
use Xiao Hinggan Ling

Little Maria Mountains (1985)
SE California. Autoposting of broader terms to this term began in 1989.
CO N334500N340000
W1144500W1150000
BT Riverside County California
BT California
BT United States
SA Big Maria Mountains

Little Missouri River basin (1978)
Also search Little Missouri River.
IN Index states as applicable.
BT United States
SA Montana
SA North Dakota
SA South Dakota
SA Wyoming

Little Walachia
use Oltenia

Littleton Formation (1978)
Includes Gove Member, Hubbard Hill Member, May Pond Member, Dakin Hill Member, Pittsfield Member, Jenness Pond Member, Durgin Brook Member. Central and S New Hampshire, N central Massachusetts, and SE Vermont.
BT Lower Devonian
BT Devonian
BT Paleozoic
SA Massachusetts
SA New Hampshire
SA Vermont

littoral
A valid term through 1977. After 1977, use littoral environment; as of 1981, use intertidal environment.

littoral drift (1978)
UF longshore drift
UF shore drift
SA longshore currents
SA progradation
SA sediments

littoral environment
As of 1981, no longer a valid term for GeoRef. Term was introduced in 1978.
use intertidal environment

littoral erosion (1982)
Before 1982, also search beach AND erosion; beaches AND erosion; beach erosion.
UF beach erosion
BT erosion
SA barrier islands
SA bars
SA beaches
SA benches
SA capes
SA coastal environment
SA coastlines
SA geomorphology
SA intertidal environment
SA marine environment
SA marine terraces
SA nearshore environment
SA ocean waves
SA shorelines
SA spits
SA tides
SA wave-cut platforms

Lituolacea (1978)
Autoposting of microfossils and Protista to this term began in 1990. This term has multiple hierarchies.
BT1 Textulariina
BT1 foraminifera
BT1 Protista
BT1 Invertebrata
BT2 Textulariina
BT2 foraminifera
BT2 Protista
BT2 microfossils
NT Ammobaculites
NT Ataxophragmiidae
NT Lituolidae
NT Orbitolinidae
NT Trochammina

Lituolidae (1978)
Family. Autoposting of microfossils and Protista to this term began in 1990. This term has multiple hierarchies.
BT1 Lituolacea
BT1 Textulariina
BT1 foraminifera
BT1 Protista
BT1 Invertebrata
BT2 Lituolacea
BT2 Textulariina
BT2 foraminifera
BT2 Protista
BT2 microfossils
NT Haplophragmoides

Live Oak County
Valid through 1988. Search in combination with state term. After 1988, use specific county-state term.

Live Oak County Texas (1989)
S Texas. Before 1989, also search Live Oak County AND Texas.
CO N280500N285000
W0975000W0982500
BT Texas
BT United States

Liverpool Bay (1978)
Inlet of Irish Sea off Liverpool. Also search Liverpool.
BT England
BT Great Britain
BT United Kingdom
BT Western Europe
BT Europe
SA Lancashire England

living culture
As of 1981, no longer a valid term for GeoRef. See living taxa.

living fossils (1981)
SA fossils
SA living taxa

living materials
As of 1981, no longer a valid term for GeoRef. Term was introduced in 1978 to distinguish from fossiliferous materials. Before 1978, search living.
use living taxa

living taxa (1981)
Used when organisms are studied while they are still living. Before 1981, also search living materials; living culture.
UF culture, living
UF living materials
SA biota
SA ecology
SA faunal studies
SA floral studies
SA fossils
SA living fossils

Livingston County
Valid through 1988. Search in combination with state term. After 1988, use specific county-state term.

Livingston County Illinois (1989)
E central Illinois. Before 1989, also search Livingston County AND Illinois.
CO N403500N410700
W0881500W0885500
BT Illinois
BT United States

Livingston County Kentucky (1989)
SW Kentucky. Before 1989, also search Livingston County AND Kentucky.
CO N370000N372700
W0881300W0883500
BT Kentucky
BT United States

Livingston County Michigan (1989)
SE Michigan. Before 1989, also search Livingston County AND Michigan.
CO N422400N424100
W0833800W0840700
BT Michigan Lower Peninsula
BT Michigan
BT United States

Livingston County Missouri (1989)
N central Missouri. Before 1989, also search Livingston County AND Missouri.

CO N393700N395700
   W0931600W0934800
BT Missouri
BT United States

Livingston County New York (1989)
W central New York. Before 1989, also search Livingston County AND New York.
CO N422800N425900
   W0772800W0780400
BT New York
BT United States
NT Springwater New York
SA Genesee River

Livingston Island (1978)
One of South Shetland Islands in British Antarctic Territory off the Antarctic Peninsula.
BT South Shetland Islands
BT Scotia Sea Islands
BT Antarctica

Livingston Parish Louisiana (1989)
SE Louisiana. Before 1989, also search Livingston Parish; Livingston County AND Louisiana.
CO N301100N304100
   W0902800W0905900
BT Louisiana
BT United States

Livorno
No longer a valid term for GeoRef. See Livorno Italy.

Livorno Italy (1993)
Province in NW Tuscany. Also a city on the Ligurian Sea now known as Leghorn.
UF Leghorn Italy
BT Tuscany Italy
BT Italy
BT Southern Europe
BT Europe

lixiviation
 use leaching

lizardite (1978)
BT serpentine group
BT sheet silicates
BT silicates

Ljubljana (1978)
City in central Slovenia, NW Yugoslavia.
UF Lyublyana
BT Slovenia
BT Yugoslavia
BT Southern Europe
BT Europe

LL chondrites (1981)
UF LL-group chondrites
BT chondrites
BT stony meteorites
BT meteorites
NT Chainpur Meteorite
NT Innisfree Meteorite
NT Krymka Meteorite
NT Semarkona Meteorite

LL-group chondrites
 use LL chondrites

Llajas Formation (1989)
Lower and middle Eocene. S California.
BT Eocene
BT Paleogene
BT Tertiary
BT Cenozoic
SA California

Llandeilian (1978)
Europe. Above Llanvirnian, below lower Caradocian.
BT Middle Ordovician
BT Ordovician
BT Paleozoic
SA Borrowdale Volcanic Series

Llandoverian (1978)
Valentian. Europe. Above Ashgillian (Ordovician), below Wenlockian. Before 1978, also search Llandovery.
UF Valentian
BT Lower Silurian
BT Silurian
BT Paleozoic
NT Interlake Formation
SA Llandovery Wales

Llandovery
No longer a valid term for GeoRef as of 1993. See Llandovery Wales. Before 1978, this term was often used for the age term, Llandoverian.

Llandovery Wales (1993)
Town in S Wales. Before 1993, search in combination with Wales. Before 1978, Llandovery term was often used for the age term.
BT Dyfed Wales
BT Wales
BT Great Britain
BT United Kingdom
BT Western Europe
BT Europe
SA Llandoverian

Llano
Valid through 1988. Search in combination with state term. After 1988, use specific city-state term.

Llano County
Valid through 1988. Search in combination with state term. After 1988, use specific county-state term.

Llano County Texas (1989)
Central Texas. Before 1989, also search Llano County AND Texas.
CO N303000N305500
   W0982000W0990000
BT Texas
BT United States
NT Llano Texas
SA Llano Uplift

Llano Estacado (1978)
Vast plateau.
IN Index states as applicable.
UF Staked Plain
BT United States
SA New Mexico
SA Oklahoma
SA Southern Great Plains
SA Texas

Llano Texas (1989)
Town in central Texas. Before 1989, search Llano AND Texas.
CO N304900N304900
   W0984200W0984200
BT Llano County Texas
BT Texas
BT United States

Llano Uplift (1978)
BT Texas
BT United States
SA Llano County Texas

Llanos (1978)
Vast plains drained by the Orinoco River.
IN Index countries as applicable.
UF Llanos Basin
SA Colombia
SA Orinoco River basin
SA South America
SA Venezuela

Llanos Basin
 use Llanos

Llanvirnian (1978)
Europe. Above Arenigian, below Llandeilian.
BT Middle Ordovician
BT Ordovician
BT Paleozoic
SA Borrowdale Volcanic Series
SA Buchans Group

Llewn Promontory
 use Lleyn Peninsula

Lleyn Peninsula (1978)
Headland extending SW into Saint George's Channel from NW Wales.
UF Llewn Promontory
BT Caernarvonshire Wales
BT Wales
BT Great Britain
BT United Kingdom
BT Western Europe
BT Europe

Llobregat River basin (1978)
In Catalonia, NE Spain. Also search Llobregat River.
IN Index Spain if applicable.
SA Barcelona Spain

load casts (1978)
UF casts, load
BT turbidity current structures
BT sedimentary structures
SA bedding plane irregularities
SA casts
SA flame structures
SA flute casts
SA soft sediment deformation
SA sole marks

load pressure (1978)
SA dynamic loading
SA engineering geology
SA load tests
SA pressure

load tests (1982)
SA cyclic loading
SA engineering geology
SA lateral loading
SA load pressure
SA loading
SA rock mechanics
SA soil mechanics
SA testing

loading (1978)
NT cyclic loading
NT dynamic loading
NT lateral loading
SA load tests
SA soil mechanics

loam (1978)
BT soils
SA clay
SA loess
SA sand
SA silt

Loanda Angola
 use Luanda Angola

loans (1981)
SA economics

localities
Not a valid term for GeoRef. As of 1993, see fossil localities, mineral localities, or type localities.

localization (1978)
Used for concentration of minerals.
SA concentration
SA metasomatism
SA mineral deposits, genesis
SA mineralization
SA minerals
SA sedimentation

Loch Lomond (1978)
S central Scotland. Also a lake in Nova Scotia. Before 1993 also used for various town names.
IN Index regions as applicable.
SA Cape Breton Island
SA Central region Scotland
SA Nova Scotia
SA Strathclyde region Scotland

Loch Lomond Stade (1989)
Europe.
UF Loch Lomond Stadial
BT Weichselian
BT upper Pleistocene
BT Pleistocene
BT Quaternary
BT Cenozoic

Loch Lomond Stadial
 use Loch Lomond Stade

Lochkovian (1978)
Europe.
BT Lower Devonian
BT Devonian
BT Paleozoic

Lockatong Formation (1989)
In Newark Supergroup. New Jersey, S New York and SE Pennsylvania.
BT Upper Triassic
BT Triassic
BT Mesozoic
SA New Jersey
SA New York
SA Newark Supergroup
SA Pennsylvania

Lockport Formation (1978)
Includes Gasport Dolomite, Suspension Bridge Dolomite and Eramosa Dolomite members at Niagara Gorge. In New York, includes DeCew Waterlime, Gasport Limestone, Goat Island, Oak Orchard members, Devils Hole Dolomite, and Oakfield Limestone.
BT Niagaran
BT Middle Silurian
BT Silurian
BT Paleozoic
SA Kentucky
SA Michigan
SA New York
SA Ohio
SA Ontario
SA Pennsylvania
SA West Virginia

locomotion (1978)
SA bipedalism
SA fossils

Lodeve
No longer a valid term for GeoRef. See Lodeve France.

Lodeve Basin (1978)
S France. Also search Lodeve.
BT Herault France
BT France
BT Western Europe
BT Europe

Lodeve France (1993)
Town in N Herault, S France.
BT Herault France
BT France
BT Western Europe
BT Europe

lodgement till (1989)
UF lodgment till

BT till
BT clastic sediments
BT sediments
SA glacial features
SA glacial geology

Lodgepole Formation (1978)
In Madison Group. Includes Paine member, Woodhurst Member, Little Chief Canyon Member. SW Montana, NE Utah, and E Wyoming.
BT Lower Mississippian
BT Mississippian
BT Carboniferous
BT Paleozoic
SA Madison Group
SA Montana
SA Utah
SA Wyoming

lodgment till
use lodgement till

Lodz
No longer a valid term for GeoRef. As of 1993 see Lodz Poland.

Lodz Poland (1993)
Province and city in central Poland. Before 1993 also search Lodz AND Poland.
BT Poland
BT Central Europe
BT Europe
SA Belchatow Mine
SA Belchatow Poland
SA Widawka Basin

loellingite
use lollingite

loess (1978)
UF lehm
BT clastic sediments
BT sediments
SA drift
SA eolian features
SA geomorphology
SA loam
SA marl
SA silt
SA soil mechanics
SA terrigenous materials

Lofoten Islands (1978)
Island group in Norwegian Sea off NW mainland. Also search Lofoten.
BT Nordland Norway
BT Norway
BT Scandinavia
BT Western Europe
BT Europe

log, road
use road log

Logan County
Valid through 1988. Search in combination with state term. After 1988, use specific county-state term.

Logan County Arkansas (1989)
W Arkansas. Before 1989, also search Logan County AND Arkansas.
CO N350200N352400
    W0931500W0940900
BT Arkansas
BT United States

Logan County Colorado (1989)
NE Colorado. Before 1989, also search Logan County AND Colorado.
CO N402400N410000
    W1023800W1033300
BT Colorado

BT United States

Logan County Illinois (1989)
Central Illinois. Before 1989, also search Logan County AND Illinois.
CO N395500N402000
    W0890800W0893500
BT Illinois
BT United States

Logan County Kansas (1989)
W Kansas. Before 1989, also search Logan County AND Kansas.
CO N384200N390800
    W1004900W1013000
BT Kansas
BT United States

Logan County Kentucky (1989)
S Kentucky. Before 1989, also search Logan County AND Kentucky.
CO N363800N370500
    W0863800W0870300
BT Kentucky
BT United States

Logan County Nebraska (1989)
Central Nebraska. Before 1989, also search Logan County AND Nebraska.
CO N412300N414300
    W1001700W1004400
BT Nebraska
BT United States

Logan County North Dakota (1989)
S North Dakota. Before 1989, also search Logan County AND North Dakota.
CO N461700N463800
    W0990300W0995500
BT North Dakota
BT United States

Logan County Ohio (1989)
W central Ohio. Before 1989, also search Logan County AND Ohio.
CO N401300N403300
    W0833200W0840200
BT Ohio
BT United States

Logan County Oklahoma (1989)
Central Oklahoma. Before 1989, also search Logan County AND Oklahoma.
CO N354300N361000
    W0970900W0974200
BT Oklahoma
BT United States

Logan County West Virginia (1989)
SW West Virginia. Before 1989, also search Logan County AND West Virginia.
CO N373800N380200
    W0813700W0821300
BT West Virginia
BT United States

Logan Formation (1978)
Comprises Beyer, Allensville, and Vinton members. NE Kentucky, and Ohio.
BT Mississippian
BT Carboniferous
BT Paleozoic
SA Kentucky
SA Ohio

Logar
No longer a valid term for GeoRef. See Logar Afghanistan.

Logar Afghanistan (1993)

Use for the province in E Afghanistan. Before 1993, also search Logar and Afghanistan.
BT Afghanistan
BT Indian Peninsula
BT Asia

logging, acoustical
use acoustical logging

logging, caliper
use caliper logging

logging, dipmeter
use dipmeter logging

logging, electrical
use electrical logging

logging, electromagnetic
use electromagnetic logging

Logone River (1978)
Flows into the Shari River at Fort Lamy in Chad S of Lake Chad. Also search Logone.
IN Index countries as applicable.
BT West Africa
BT Africa
SA Cameroon
SA Chad

Logrono
No longer a valid term for GeoRef. As of 1993, see Logrono City Spain.

Logrono City Spain (1993)
Refers to only the city in N central Spain. Before 1993, also search Logrono and Spain.
BT Logrono Spain
BT Old Castile Spain
BT Castile Spain
BT Spain
BT Iberian Peninsula
BT Southern Europe
BT Europe

Logrono Province
No longer a valid term for GeoRef. As of 1993, see Logrono Spain.

Logrono Spain (1993)
Refers only to the province in N central Spain. From 1981-1992, also search Logrono Province and Spain. Before 1981, also search Logrono and Spain. For the city, see Logrono City Spain.
CO N415600N424000
    W0014100W0030800
BT Old Castile Spain
BT Castile Spain
BT Spain
BT Iberian Peninsula
BT Southern Europe
BT Europe
NT Logrono City Spain

Logudoro (1978)
Region in NW Italy.
BT Sardinia Italy
BT Italy
BT Southern Europe
BT Europe

Loihi
use Loihi Seamount

Loihi Seamount (1985)
N central Pacific Ocean, S of Hawaii. Autoposting of Pacific Ocean to this term began in 1990.
CO N185400N185600
    W1551400W1551600
UF Loihi
BT North Pacific
BT Pacific Ocean
SA Hawaii Island

Loir-et-Cher
No longer a valid term for GeoRef. use Loir-et-Cher France

Loir-et-Cher France (1993)
Department in W central France.
CO N471200N481000
    E0021500E0003200
UF Loir-et-Cher
BT France
BT Western Europe
BT Europe
SA Beauce

Loire
No longer a valid term for GeoRef. See Loire France.

Loire France (1993)
Department in S central France.
CO N452000N462000
    E0050000E0034000
BT France
BT Western Europe
BT Europe
NT Saint-Etienne coal basin
SA Monts du Lyonnais

Loire River (1978)
Longest river in France. Rises in Ardeche Department in SE and flows N and NE into Loiret Department where it turns E emptying into the Bay of Biscay at Saint-Nazaire.
BT France
BT Western Europe
BT Europe

Loire Valley (1978)
River valley in S central, central, and E central France. Also search Loire.
BT France
BT Western Europe
BT Europe

Loire-Atlantique
No longer a valid term for GeoRef. use Loire-Atlantique France

Loire-Atlantique France (1993)
Department in W France. Before 1976, also search Loire-Inferieure.
CO N470000N475000
    W0005000W0023000
UF Loire-Atlantique
UF Loire-Inferieure
BT France
BT Western Europe
BT Europe
NT Ancenis France
NT Nantes France
SA Bay of Bourgneuf

Loire-Inferieure
A valid term through 1976.
use Loire-Atlantique France

Loiret
No longer a valid term for GeoRef. See Loiret France.

Loiret France (1993)
Department in N central France.
CO N473000N481500
    E0031000E0013000
BT France
BT Western Europe
BT Europe
NT Orleans France

lollingite (1978)
Autoposting of sulfides to this term ended in 1989.
UF loellingite
BT arsenides
SA arsenopyrite

Lom Depression (1978)

In vicinity of Lom River and city of Lom in NW Bulgaria.
BT Vidin Bulgaria
BT Bulgaria
BT Southern Europe
BT Europe

Loma Prieta earthquake 1989
(1993)
Santa Cruz County, W California.
BT earthquakes
SA California
SA San Francisco Bay region
SA Santa Cruz County California

Lombardy
No longer a valid term for GeoRef. See Lombardy Italy.

Lombardy Italy (1993)
Autonomous region in N Italy.
CO N444000N464000
E0112500E0083000
BT Italy
BT Southern Europe
BT Europe
NT Adamello Massif
NT Bergamo Italy
NT Brescia Italy
NT Como
NT Como Italy
NT Milan Italy
NT Pavia Italy
NT Sondrio Italy
NT Trompia Valley
NT Valtellina
NT Varese Italy
SA Lago Maggiore
SA Maiolica Limestone
SA Po River
SA Po Valley
SA Rhaetian Alps
SA Sesia-Lanzo Zone

Lome Convention (1981)
SA economics

lomonite
use laumontite

Lomonosov Range
use Lomonosov Ridge

Lomonosov Ridge (1978)
N of New Siberian Islands across top of world to N of Greenland.
UF Lomonosov Range
BT Arctic Ocean
SA Makarov Basin

lomontite
use laumontite

Lompoc
Valid through 1988. Search in combination with state term. After 1988, use specific city-state term.

Lompoc California (1989)
City in SW California. Before 1989, search Lompoc AND California.
CO N343800N343800
W1202700W1202700
BT Santa Barbara County California
BT California
BT United States

Lonar Crater (1981)
Impact crater in the basalt lava flows of the Deccan Plateau, Maharashtra.
CO N195845N195845
E0763400E0763400
BT Maharashtra India
BT India
BT Indian Peninsula
BT Asia
SA Deccan Plateau

London
No longer a valid term for GeoRef. As of 1993, see London England if applicable.

London Basin (1978)
SE England.
CO N503500N544000
E0014500W0031000
BT England
BT Great Britain
BT United Kingdom
BT Western Europe
BT Europe

London Clay (1978)
Rests on Reading, Woolwich, Blackheath or Oldhaven Beds except in Norfolk to the west of Great Yarmouth. Overlain by the Claygate or Bagshot Beds, except in East Anglia. Occupies greater part of a large triangular area in London Basin bounded roughly by line from Herne Bay in Kent to Croyden, Farnham, and Newbury and from there NE to Aldeburgh in Suffolk.
BT Ypresian
BT lower Eocene
BT Eocene
BT Paleogene
BT Tertiary
BT Cenozoic
SA England
SA United Kingdom

London England (1993)
City on the Thames River in SE England. As of 1993, also used for the urban county of Greater London. Before 1993, also search London or Greater London in combination with England.
UF Greater London England
BT England
BT Great Britain
BT United Kingdom
BT Western Europe
BT Europe
SA Thames Estuary
SA Thames River

London Metal Exchange (1981)
SA associations
SA economics

Londonderry
No longer a valid term for GeoRef. After 1993, see Londonderry Northern Ireland.

Londonderry Northern Ireland (1993)
Traditional county in NW Northern Ireland. Includes the smaller administrative region of the same name adopted after 1973. Also includes the city.
BT Northern Ireland
BT United Kingdom
BT Western Europe
BT Europe

Long Beach
Valid through 1988. Search in combination with state term. After 1988, use specific city-state term.

Long Beach California (1989)
City in S California. Before 1989, search Long Beach AND California.
CO N334700N334700
W1181500W1181500
BT Los Angeles County California
BT California
BT United States

Long Island (1978)
Lying between Long Island Sound on N and Atlantic Ocean on S in S New York. Also in various other locations.
IN Index counties or regions as applicable.
SA Bay Park New York
SA Fire Island
SA Kings County New York
SA New York
SA Outer Hebrides
SA Queens County New York

Long Island Sound (1978)
Body of water between S shore of Connecticut and N shore of Long Island, New York.
BT Atlantic Ocean

long period
use long-period waves

Long Valley (1993)
E California, Mono Lake and Owens Valley area. Also in N central New Jersey. Also in White Pine County, E Nevada.
IN Index counties as applicable.
SA California
SA Inyo County California
SA Mono County California
SA Morris County New Jersey
SA Nevada
SA New Jersey
SA New Jersey Highlands
SA White Pine County Nevada

Long Valley Caldera (1985)
S Mono County, E California.
CO N374100N374600
W1184300W1185200
BT Mono County California
BT California
BT United States
SA Casa Diablo
SA Inyo Domes
SA Mammoth Lakes earthquakes
SA Obsidian Dome
SA Sierra Nevada

long-period seismographs (1981)
BT seismographs

long-period waves (1978)
Before 1978, search long period or long-period AND waves.
UF long period
BT elastic waves
SA seismology
SA traveltime
SA waves

long-wall mining
use longwall mining

longitudinal
A valid term through 1977. See longitudinal orientation.

longitudinal faults
use strike faults

longitudinal orientation
As of 1981, no longer a valid term for GeoRef. Before 1978, also search longitudinal.

longitudinal wave
use P-waves

longshore bars (1978)
Autoposting of bars to this term began in 1989.
IN Index shore features as applicable.
UF barrier bars
BT bars
SA barrier beaches
SA barrier islands
SA capes
SA coastlines
SA geomorphology
SA sand ridges
SA shore features

longshore currents (1978)
BT ocean currents
BT currents
SA littoral drift
SA ocean circulation
SA ocean waves
SA tidal currents

longshore drift
use littoral drift

longwall mining (1985)
UF long-wall mining
SA coal
SA coal mines
SA mines
SA mining geology
SA underground mining

lonsdaleite (1978)
Meteorite mineral.
BT native elements
SA carbon
SA chaoite
SA diamond
SA graphite
SA meteorites

Lookingglass Formation (1985)
SW Oregon.
UF Lookinglass Formation
BT Eocene
BT Paleogene
BT Tertiary
BT Cenozoic
SA Oregon

Lookinglass Formation
use Lookingglass Formation

loparite (1978)
BT niobates
BT oxides

Lopez de Bertodano Formation (1989)
Vicecomodoro Marambio Island, James Ross Island, Seymour Island and Cockburn Island, Antarctica.
BT Upper Cretaceous
BT Cretaceous
BT Mesozoic
SA Antarctica
SA James Ross Island
SA Seymour Island

lopoliths (1978)
BT intrusions
SA batholiths
SA igneous rocks
SA laccoliths

Lord Howe Island (1978)
Volcanic island 435 miles NE of Sydney. Dependency of New South Wales. In 1985, broader term changed from Pacific Ocean to West Pacific Ocean Islands.
BT West Pacific Ocean Islands

Lord Howe Rise (1978)
Extends from SW of New Caledonia to W of New Zealand.
UF Lord Howe-Chesterfield Ridge
UF Lord Howe-New Zealand Ridge
UF Lord Howe-New Zealand Rise
BT Pacific Ocean
SA DSDP Site 207
SA DSDP Site 208
SA DSDP Site 588
SA DSDP Site 589
SA DSDP Site 590
SA DSDP Site 591
SA DSDP Site 592

SA Leg 90

Lord Howe-Chesterfield Ridge
 use Lord Howe Rise

Lord Howe-New Zealand Ridge
 use Lord Howe Rise

Lord Howe-New Zealand Rise
 use Lord Howe Rise

Lorrain Formation (1978)
 Huronian Supergroup of the Cobalt Group. Overlies the Gowganda Formation conformably and gradationally.
 BT Proterozoic
 BT upper Precambrian
 BT Precambrian
 SA Cobalt Group
 SA Ontario
 SA Quebec

Lorraine (1978)
 Region and former province on Belgium and Luxembourg borders in NE France.
 SA Meurthe-et-Moselle France
 SA Moselle France
 SA Vosges France

Los Alamos County
 Valid through 1988. Search in combination with state term. After 1988, use specific county-state term.

Los Alamos County New Mexico (1989)
 N central New Mexico. Before 1989, also search Los Alamos County AND New Mexico.
 CO N354500N355700
   W1061500W1062600
 BT New Mexico
 BT United States
 NT Los Alamos Scientific Laboratory

Los Alamos Scientific Laboratory (1978)
 Los Alamos County, 35 miles NW of Santa Fe, in N central New Mexico.
 BT Los Alamos County New Mexico
 BT New Mexico
 BT United States
 SA Fenton Hill
 SA University of California

Los Angeles
 Valid through 1988. Search in combination with state term. After 1988, use specific city-state term.

Los Angeles Basin (1978)
 Greater Los Angeles area.
 BT California
 BT United States
 SA Cajon Pass
 SA Whittier Narrows earthquake 1987
 SA Wilmington Field

Los Angeles California (1989)
 City in S California. Before 1989, search Los Angeles AND California.
 CO N340000N340000
   W1181500W1181500
 BT Los Angeles County California
 BT California
 BT United States

Los Angeles County
 Valid through 1988. Search in combination with state term. After 1988, use specific county-state term.

Los Angeles County California (1989)
 S California. Before 1989, also search Los Angeles County AND California.
 CO N334500N344500
   W1174000W1185000
 BT California
 BT United States
 NT Inglewood California
 NT Long Beach California
 NT Los Angeles California
 NT Pacoima Dam
 NT Palmdale California
 NT Palos Verdes Hills
 NT Palos Verdes Peninsula
 NT Pasadena California
 NT Puente Hills
 NT Rancho La Brea
 NT San Clemente Island
 NT San Fernando California
 NT Wilmington California
 SA Channel Islands
 SA Mojave Desert
 SA San Gabriel Mountains
 SA San Pedro
 SA Santa Catalina Island
 SA Santa Clara Valley
 SA Sylmar Fault
 SA Wilmington Field

Los Azufres (1989)
 Geothermal field. Michoacan, SW Mexico.
 BT Michoacan Mexico
 BT Mexico
 SA geothermal fields

Los Humeros (1989)
 Puebla, S central Mexico.
 CO N193000N195000
   W0971500W0974000
 BT Puebla Mexico
 BT Mexico

Los Islands (1978)
 Group of small islands in Atlantic Ocean off Conakry.
 UF Iles de Loos
 BT Guinea
 BT West Africa
 BT Africa

Los Pedroches
 No longer a valid term for GeoRef. As of 1993, see Los Pedroches Spain.

Los Pedroches Spain (1993)
 Region N of city of Cordoba in SW central Spain. Before 1993, also search Los Pedroches and Spain.
 BT Cordoba Spain
 BT Andalusia Spain
 BT Spain
 BT Iberian Peninsula
 BT Southern Europe
 BT Europe

Los Testigos (1978)
 Small group of islands SW of Grenada and N of Sucre, Venezuela.
 BT Caribbean Sea
 BT North American Atlantic
 BT North Atlantic
 BT Atlantic Ocean
 SA Venezuela

Lost Burro Formation (1978)
 Middle and Upper Devonian. Includes Lippincott Member, Quartz Spring Sandstone Member. S California.
 BT Devonian
 BT Paleozoic
 SA California

Lost City (1978)
 Ancient Indian city now covered by Lake Mead. Relics now housed in Overton in Clark County 5 miles N of original location in S Nevada.
 IN Index county or regions as applicable.
 SA Clark County Nevada
 SA Lake Mead

Lost River Fault (1989)
 E central Idaho.
 IN Index counties as applicable.
 CO N434000N450000
   W1123000W1141000
 BT Idaho
 BT United States
 SA Butte County Idaho
 SA Custer County Idaho

Lost River Range (1989)
 E central Idaho, in Butte and Custer counties.
 IN Index counties as applicable.
 BT Idaho
 BT United States
 NT Borah Peak
 SA Butte County Idaho
 SA Custer County Idaho

Lost Soldier Field (1989)
 IN Index counties as applicable.
 BT Wyoming
 BT United States
 SA Carbon County Wyoming
 SA Fremont County Wyoming
 SA oil and gas fields
 SA Sweetwater County Wyoming

Lot
 No longer a valid term for GeoRef. See Lot France.

Lot France (1993)
 Department in S central France.
 CO N441000N450500
   E0021100E0005900
 BT France
 BT Western Europe
 BT Europe
 SA Quercy

Lot-et-Garonne
 No longer a valid term for GeoRef.
 use Lot-et-Garonne France

Lot-et-Garonne France (1993)
 Department in SW France.
 CO N435700N444500
   E0010000W0001000
 UF Lot-et-Garonne
 BT France
 BT Western Europe
 BT Europe
 NT Durance France

Lotharingian (1978)
 Europe. Above Sinemurian, below Pliensbachian.
 BT Lower Jurassic
 BT Jurassic
 BT Mesozoic

Lothian region Scotland (1993)
 Administrative region in SE Scotland created from East Lothian (formerly Haddingtonshire), and parts of West Lothian and Midlothian in 1975.
 BT Scotland
 BT Great Britain
 BT United Kingdom
 BT Western Europe
 BT Europe
 NT East Lothian Scotland
 NT Edinburgh Scotland
 SA Midlothian Scotland
 SA Tyne River

Louann Salt (1985)
 UF Louann Tongue
 SA Arkansas
 SA Jurassic
 SA Louisiana
 SA Permian
 SA salt tectonics
 SA Texas
 SA Triassic

Louann Tongue
 use Louann Salt

Louisiade Archipelago (1978)
 Island group in Solomon Sea SE of E tip of New Guinea.
 BT Papua New Guinea
 BT Australasia

Louisiana (1978)
 Autoposting of this term began in 1978.
 CO N290000N330000
   W0890000W0940500
 BT United States
 NT Atchafalaya Bay
 NT Atchafalaya River
 NT Barataria Bay
 NT Calcasieu Parish Louisiana
 NT Calcasieu River
 NT Caldwell Parish Louisiana
 NT Cameron Parish Louisiana
 NT Claiborne Parish Louisiana
 NT East Baton Rouge Parish Louisiana
 NT Evangeline Parish Louisiana
 NT Iberia Parish Louisiana
 NT Jackson Parish Louisiana
 NT Jefferson Parish Louisiana
 NT Lafayette Parish Louisiana
 NT Lafourche Parish Louisiana
 NT Lake Pontchartrain
 NT Lincoln Parish Louisiana
 NT Livingston Parish Louisiana
 NT Mississippi Delta
 NT Orleans Parish Louisiana
 NT Ouachita Parish Louisiana
 NT Plaquemines Parish Louisiana
 NT Rapides Parish Louisiana
 NT Richland Parish Louisiana
 NT Saint Bernard Parish Louisiana
 NT Saint Mary Parish Louisiana
 NT Terrebonne Parish Louisiana
 NT Vacherie Dome
 NT Vermilion Parish Louisiana
 NT Winnfield salt dome
 SA Amite River
 SA Anahuac Formation
 SA Austin Chalk
 SA Avery Island
 SA Beaumont Clay
 SA Bossier Formation
 SA Buckner Formation
 SA Byram Formation
 SA Carrizo Sand
 SA Catahoula Formation
 SA Chicot Aquifer
 SA Citronelle Formation
 SA Claiborne Group
 SA Cockfield Formation
 SA Cook Mountain Formation
 SA Cotton Valley Group
 SA Eagle Ford Formation
 SA East Bay
 SA Evangeline Aquifer
 SA Fleming Formation
 SA Frio Formation
 SA Georgetown Formation
 SA Grand River
 SA Gulf Coastal Plain
 SA Hackberry Formation
 SA Haynesville Formation
 SA Hosston Formation
 SA Jackson Group
 SA Louann Salt
 SA Mississippi River
 SA Mississippi Valley

SA Moodys Branch Formation
SA Norphlet Formation
SA Pearl River
SA Pearsall Formation
SA Queen City Formation
SA Red River
SA Red River valley
SA Rodessa Formation
SA Sabine Lake
SA Sabine Uplift
SA Schuler Formation
SA Sligo Formation
SA Sparta Sand
SA Taylor Marl
SA Trinity Group
SA Tuscaloosa Formation
SA Vicksburg Group
SA Washita Group
SA Wilcox Group
SA Woodbine Formation
SA Yazoo Clay
SA Yegua Formation

Louisville
Valid through 1988. Search in combination with state term. After 1988, use specific city-state term.

Louisville Kentucky (1989)
City on the Ohio River in N Kentucky. Before 1989, search Louisville AND Kentucky.
CO N381300N381300
W0854800W0854800
BT Jefferson County Kentucky
BT Kentucky
BT United States

Louisville Ridge (1989)
S Pacific. Autoposting of Pacific Ocean to this term began in 1990.
BT South Pacific
BT Pacific Ocean

Loup Fork Group (1978)
Miocene, Pliocene, and Pleistocene(?).
IN Index ages as applicable.
BT Cenozoic
SA Colorado
SA Kansas
SA Miocene
SA Nebraska
SA New Mexico
SA Pleistocene
SA Pliocene
SA South Dakota
SA Texas
SA Wyoming

Lourdes
No longer a valid term for GeoRef. See Lourdes France.

Lourdes France (1993)
Town in SW France.
BT Hautes-Pyrenees France
BT France
BT Western Europe
BT Europe

Louvain
No longer a valid term for GeoRef. See Louvain Belgium.

Louvain Belgium (1993)
City E of Brussels in central Belgium. Before 1993, also search Louvain and Belgium. Before 1978, also search Leuven AND Belgium.
UF Leuven Belgium
BT Brabant Belgium
BT Belgium
BT Western Europe
BT Europe

Love waves (1978)
UF Q waves
BT surface waves
BT elastic waves
SA Lg-waves
SA Q
SA seismology
SA waves

Loveland
Valid through 1988. Search in combination with state term. After 1988, use specific city-state term.

Loveland Colorado (1989)
City in N Colorado. Before 1989, search Loveland AND Colorado.
CO N402400N402400
W1050600W1050600
BT Larimer County Colorado
BT Colorado
BT United States

Lovelock
Valid through 1988. Search in combination with state term. After 1988, use specific city-state term.

Lovelock Nevada (1989)
City in NW Nevada. Before 1989, search Lovelock AND Nevada.
CO N401200N401200
W1182800W1182800
BT Pershing County Nevada
BT Nevada
BT United States

Lovozero
No longer a valid term for GeoRef. As of 1993, see Lovozero Russian Federation.

Lovozero Massif (1978)
Murmansk Oblast on the Kola Peninsula in NW European Russian Federation. This term has multiple hierarchies.
BT1 Murmansk Russian Federation
BT1 Russian Federation
BT1 Commonwealth of Independent States
BT2 Murmansk Russian Federation
BT2 Europe
SA Kola Peninsula

Lovozero Russian Federation (1993)
Village in Murmansk Oblast on Kola Peninsula in NW Russian Federation. Before 1993, also search Lovozero. This term has multiple hierarchies.
CO N670000N680000
E0350000E0340000
BT1 Murmansk Russian Federation
BT1 Russian Federation
BT1 Commonwealth of Independent States
BT2 Murmansk Russian Federation
BT2 Europe
SA Ilimaussaq

Low Archipelago
use Tuamotu Islands

Low Jesenik Mountains
use Nizky Jezenik Mountains

low pressure (1978)
BT pressure
SA high pressure
SA metamorphism
SA phase equilibria

Low Tatra
use Low Tatra Mountains

Low Tatra Mountains (1978)
Section of the Carpathians parallel to and S of the Tatra Mountains in N central Slovakia.
CO N484400N500500
E0201200E0190800
UF Low Tatra
UF Low Tatras
UF Lower Tatra
UF Nizke Tatra
BT Slovakia
BT Czechoslovakia
BT Central Europe
BT Europe
SA Carpathians
SA Tatra Mountains

Low Tatras
use Low Tatra Mountains

low temperature (1978)
Before 1971, also search low-temperature analysis.
UF low-temperature analysis
BT temperature
SA metamorphism
SA phase equilibria

low velocity zones
use low-velocity zones

low water (1981)
Before 1993, also search low-water levels.
UF low-water levels
SA hydrology
SA rivers and streams
SA tides
SA water

low-angle faults (1978)
Before 1978, search faults AND low-angle.
BT faults
SA high-angle faults

low-energy environment (1978)
Before 1978, search low-energy.
SA depositional environment
SA environment
SA high-energy environment
SA sedimentation

low-grade
A valid term through 1977. After 1977, use low-grade metamorphism.

low-grade metamorphism (1978)
Before 1978, search low-grade AND metamorphism.
BT metamorphism
SA epizonal metamorphism
SA grade

low-level geothermal energy (1985)
SA geothermal energy

low-level waste (1985)
SA high-level waste
SA radioactive waste
SA underground storage
SA waste disposal

low-temperature analysis
use low temperature

low-velocity layer
As of 1981, no longer a valid term for GeoRef. Use low-velocity zones.

low-velocity layers
use low-velocity zones

low-velocity zones (1978)
General term, i.e. applicable to crust, mantle, etc.
UF low velocity zones
UF low-velocity layers
UF zones, low-velocity
SA crust
SA mantle
SA seismology

low-water levels
use low water

lower Archean (1985)
From 1985 to 1989, Precambrian was autoposted to this term. Autoposting of Precambrian to this term began in 1990.
BT Archean
BT Precambrian

Lower Austria (1978)
State in NE Austria.
CO N472500N490200
E0170500E0142500
BT Austria
BT Central Europe
BT Europe
NT Bad Deutsch Altenburg Austria
NT Eichkogel
NT Kamp Valley
NT Krems Austria
NT Schwechat Valley
NT Spitz Austria
NT Vienna Austria
NT Wiener Wald
SA Enns Valley
SA North Austrian Crystallines
SA Semmering
SA Wechsel
SA Ybbsitz Meteorite

lower boundary
A valid general term through 1976. For stratigraphic meaning use boundary.

Lower Cambrian (1978)
Autoposting of this term began in 1978.
BT Cambrian
BT Paleozoic
NT Aldanian
NT Antietam Formation
NT Atdabanian
NT Cheshire Formation
NT Chilhowee Group
NT Dunham Dolomite
NT Gog Group
NT Harlech Stage
NT Kinzers Formation
NT Murphy Marble
NT Pinney Hollow Formation
NT Rome Formation
NT Shady Dolomite
NT Tommotian
NT Usa Series
NT Yudoma Series
NT Zabriskie Quartzite
SA Boston Bay Group
SA Cambridge Argillite
SA Middle Cambrian
SA Porsanger Dolomite Formation
SA Puncoviscana Formation
SA Roxbury Conglomerate
SA Underhill Formation
SA Upper Cambrian
SA Wood Canyon Formation
SA Yudomian

Lower Carboniferous
As of 1981, no longer a valid term for GeoRef.
use Dinantian

lower Cenomanian (1985)
BT Cenomanian
BT Upper Cretaceous
BT Cretaceous
BT Mesozoic

lower Cenozoic (1978)
BT Cenozoic

Lower Cretaceous (1978)

Autoposting of this term began in 1978.
BT Cretaceous
BT Mesozoic
NT Albian
NT Antlers Sands
NT Aptian
NT Areado Formation
NT Baquero Formation
NT Barremian
NT Bear River Formation
NT Berriasian
NT Bisbee Group
NT Blackleaf Formation
NT Blake-Bahama Formation
NT Bluesky Formation
NT Burro Canyon Formation
NT Cadomin Formation
NT Cedar Mountain Formation
NT Cheyenne Sandstone
NT Christopher Formation
NT Clearwater Formation
NT Cloverly Formation
NT Comanche Peak Limestone
NT Edwards Formation
NT Fall River Formation
NT Fredericksburg Group
NT Gargasian
NT Gault
NT Georgetown Formation
NT Gething Formation
NT Glen Rose Formation
NT Hauterivian
NT Hosston Formation
NT Inyan Kara Group
NT Isachsen Formation
NT Kiowa Formation
NT Kootenay Formation
NT Lakota Formation
NT Mannville Group
NT McMurray Formation
NT Mentor Beds
NT Mesa Rica Sandstone
NT Moosebar Formation
NT Muddy Sandstone
NT Mural Limestone
NT Neocomian
NT Newcastle Sandstone
NT Paluxy Formation
NT Pasayten Group
NT Pearsall Formation
NT Pebble Shale
NT Purbeckian
NT Rodessa Formation
NT Skull Creek Shale
NT Sligo Formation
NT Speeton Clay
NT Spirit River Formation
NT Sunniland Limestone
NT Thamama Group
NT Torok Formation
NT Travis Peak Formation
NT Trinity Group
NT Urgonian
NT Valanginian
NT Wealden
NT Zubair Formation
SA Comanchean
SA Middle Cretaceous
SA Nevadan Orogeny
SA Nicoya Complex
SA Upper Cretaceous
SA upper Gondwana System
SA Vraconian
SA Washita Group

lower crust (1978)
BT crust
SA asthenosphere
SA basaltic layer
SA granitic layer
SA lithosphere
SA upper crust
SA upper mantle

Lower Devonian (1978)
Autoposting of this term began in 1978.
BT Devonian
BT Paleozoic
NT Battery Point Formation
NT Coeymans Formation
NT Dittonian
NT Emsian
NT Gedinnian
NT Helderberg Group
NT Helderbergian
NT Littleton Formation
NT Lochkovian
NT Manlius Formation
NT Matagamon Sandstone
NT Oriskany Sandstone
NT Pragian
NT Seboomook Formation
NT Siegenian
NT York River Formation
SA Downtonian
SA Hidden Valley Dolomite
SA Keyser Limestone
SA Middle Devonian
SA Read Bay Formation
SA Road River Formation
SA Rondout Formation
SA Ross Formation
SA Upper Devonian
SA Zlichovian

lower Eocene (1978)
Autoposting of this term began in 1978.
BT Eocene
BT Paleogene
BT Tertiary
BT Cenozoic
NT Ager Formation
NT Aquia Formation
NT Calvert Bluff Formation
NT Cuisian
NT Hatchetigbee Formation
NT Misoa Formation
NT Nanafalia Formation
NT San Jose Formation
NT Sparnacian
NT Thebes Formation
NT Wasatchian
NT Wilcox Group
NT Willwood Formation
NT Wind River Formation
NT Ypresian
SA Deccan Traps
SA Ewekoro Formation
SA Imo Shale
SA Intertrappean Beds
SA middle Eocene
SA Santa Susana Formation
SA Sinjar Formation
SA upper Eocene

lower Gondwana
  use lower Gondwana System

lower Gondwana System (1985)
Peninsular India.
IN Index age as applicable.
UF lower Gondwana
BT Gondwana System
SA Carboniferous
SA Gondwana
SA Mesozoic
SA Paleozoic
SA Permian
SA Triassic
SA Upper Carboniferous
SA upper Gondwana System

Lower Greensand (1978)
Aptian and lower Albian.
BT Cretaceous
BT Mesozoic
SA England
SA United Kingdom

lower Holocene (1978)
BT Holocene
BT Quaternary
BT Cenozoic
NT Ancylus Lake
NT Litorina Sea
SA Baltic Glaciation
SA Clovis

Lower Jurassic (1978)
Autoposting of this term began in 1978. Before 1981, also search Liassic.
UF Lias
UF Liassic
BT Jurassic
BT Mesozoic
NT Carixian
NT East Berlin Formation
NT Feltville Formation
NT Hampden Basalt
NT Holyoke Basalt
NT Lotharingian
NT lower Liassic
NT Luxembourg Sandstone
NT middle Liassic
NT Novorayskoe Formation
NT Nugget Sandstone
NT Portland Formation
NT Sunrise Formation
NT Talkeetna Formation
NT upper Liassic
SA Aalenian
SA Brunswick Formation
SA Elliot Formation
SA Gondwana System
SA Indosinian Orogeny
SA Kunga Group
SA Middle Jurassic
SA Moenave Formation
SA Passaic Formation
SA Upper Jurassic

lower Liassic (1981)
BT Lower Jurassic
BT Jurassic
BT Mesozoic
NT Hettangian
NT Sinemurian

Lower Lusatia (1978)
That part of Lusatia between the Neisse and the Bober rivers in SW Poland.
BT Poland
BT Central Europe
BT Europe
SA Lusatia

lower mantle (1978)
BT mantle
SA core-mantle boundary
SA upper mantle

lower Mesozoic (1978)
BT Mesozoic
SA Monashee Complex
SA Shuswap Complex

lower Miocene (1978)
Autoposting of this term began in 1978.
BT Miocene
BT Neogene
BT Tertiary
BT Cenozoic
NT Altonian
NT Aquitanian
NT Awamoan
NT Burdigalian
NT Chipola Formation
NT Eggenburgian
NT Hemingfordian
NT Kasauli Series
NT Ottnangian
NT Saucesian
NT Waitemata Group
SA Anahuac Formation
SA Arikareean
SA Badenian
SA Grande Ronde Basalt
SA John Day Formation
SA middle Miocene
SA Relizian
SA upper Miocene
SA Vaqueros Formation

Lower Mississippi Valley (1978)
Autoposting of United States and North America to this term began in 1990.
BT Mississippi Valley
BT United States

Lower Mississippian (1978)
Autoposting of this term began in 1978.
BT Mississippian
BT Carboniferous
BT Paleozoic
NT Chappel Limestone
NT Chouteau Limestone
NT Cuyahoga Formation
NT Fort Payne Formation
NT Joana Limestone
NT Kayak Shale
NT Kekiktuk Conglomerate
NT Kinderhookian
NT Lake Valley Formation
NT Lodgepole Formation
NT Osagian
NT Pocono Formation
SA Antrim Shale
SA Bakken Formation
SA Catskill Formation
SA Pilot Shale
SA Rampart Group
SA Rundle Group
SA Upper Mississippian
SA Valmeyeran
SA Waulsortian facies

lower Neogene
  use Miocene

lower Oligocene (1978)
Autoposting of this term began in 1978.
BT Oligocene
BT Paleogene
BT Tertiary
BT Cenozoic
NT Chadron Formation
NT Chadronian
NT Rupelian
NT Sannoisian
NT Stampian
SA middle Oligocene
SA upper Oligocene

Lower Ordovician (1978)
Autoposting of this term began in 1978.
BT Ordovician
BT Paleozoic
NT Arenigian
NT Beekmantown Group
NT Canadian Series
NT El Paso Group
NT Ellenburger Group
NT Fillmore Formation
NT Kingsport Formation
NT Manitou Formation
NT Mascot Dolomite
NT Prairie du Chien Group
NT Saint George Group
NT Shakopee Formation
NT Smithville Formation
NT Tremadocian
NT White Limestone
SA Arbuckle Group
SA Deadwood Formation
SA Knox Group
SA Middle Ordovician
SA Upper Ordovician

lower Paleocene (1978)

Autoposting of this term began in 1978.
- BT Paleocene
- BT Paleogene
- BT Tertiary
- BT Cenozoic
- NT Danian
- NT K-T boundary
- NT Puercan
- NT Tiffanian
- NT Torrejonian
- SA Difunta Group
- SA upper Paleocene

lower Paleogene (1985)
- BT Paleogene
- BT Tertiary
- BT Cenozoic
- SA Eocene
- SA Paleocene

lower Paleozoic (1978)
- BT Paleozoic
- NT Ashe Formation
- NT Bay of Islands Ophiolite
- NT Berwick Formation
- NT Chopawamsic Formation
- NT Elberton Granite
- NT Glenarm Series
- NT Henderson Gneiss
- NT Penobscot Formation
- NT Wilmington Complex
- SA Damara Orogeny
- SA Sparagmite Group
- SA Tallulah Falls Formation
- SA Wissahickon Formation

Lower Peninsula, Michigan
  use Michigan Lower Peninsula

Lower Pennsylvanian (1978)
Autoposting of this term began in 1978.
- BT Pennsylvanian
- BT Carboniferous
- BT Paleozoic
- NT Caseyville Formation
- NT Crab Orchard Mountains Group
- NT Gizzard Group
- NT Haymond Formation
- NT Lee Formation
- NT Morrowan
- NT New River Formation
- NT Pocahontas Formation
- SA Manning Canyon Shale
- SA Middle Pennsylvanian
- SA Upper Pennsylvanian

Lower Permian (1978)
Autoposting of this term began in 1978.
- BT Permian
- BT Paleozoic
- NT Abo Formation
- NT Artinskian
- NT Autunian
- NT Barakar Stage
- NT Cherry Canyon Formation
- NT Chihsia Formation
- NT Leonardian
- NT Opeche Shale
- NT Qixia Formation
- NT Sakmarian
- NT Wichita Group
- NT Wolfcampian
- SA Bird Spring Formation
- SA Guadalupian
- SA Itarare Subgroup
- SA Kungurian
- SA Laborcita Formation
- SA Middle Permian
- SA Oquirrh Formation
- SA Pictou Group
- SA Rotliegendes
- SA Stockton Formation
- SA Tensleep Sandstone

- SA Upper Permian

lower Pleistocene (1978)
Autoposting of this term began in 1978.
- BT Pleistocene
- BT Quaternary
- BT Cenozoic
- NT Aftonian
- NT Calabrian
- NT Danube Stade
- NT Jaramillo Event
- NT Kansan
- NT Nebraskan
- NT Olduvai Event
- NT Villafranchian
- NT Waccamaw Formation
- SA Matuyama Epoch
- SA Uonuma Group
- SA upper Pleistocene
- SA Wildcat Group

lower Pliocene (1978)
Autoposting of this term began in 1978.
- BT Pliocene
- BT Neogene
- BT Tertiary
- BT Cenozoic
- NT Ellensburg Formation
- NT Gilbert Epoch
- NT Plaisancian
- NT Tabianian
- SA Capistrano Formation
- SA Hemphillian
- SA Meotian
- SA middle Pliocene
- SA Ridge Route Formation
- SA upper Pliocene
- SA Yakima Basalt

lower Precambrian
  As of 1981, no longer a valid term for GeoRef.
  use Archean

lower Proterozoic (1978)
Autoposting of Precambrian to this term ended in 1989. Autoposting of upper Precambrian and Precambrian to this term began in 1990.
- BT Proterozoic
- BT upper Precambrian
- BT Precambrian
- NT Aphebian
- NT Birrimian
- NT Carpentarian
- NT Cobalt Group
- NT Hemlock Formation
- NT Karelian
- NT Marquette Range Supergroup
- NT Michigamme Formation
- NT Svecofennian
- NT Thomson Formation
- NT Virginia Formation
- NT Willyama Complex
- NT Wollaston Group
- SA Harney Peak Granite
- SA Rocknest Formation
- SA Tyler Formation
- SA Ventersdorp Supergroup
- SA Wopmay Orogeny

lower Quaternary (1978)
- BT Quaternary
- BT Cenozoic

Lower Rhine Basin (1978)
That section of the Rhine Basin between Bonn and the North Sea; in the Netherlands it includes the basins of the Lower Rhine, Lek, and Waal rivers. Also search Lower Rhine.
  IN Index countries as applicable.
- BT Europe
- SA Germany

- SA Netherlands
- SA Rhine Basin
- SA Rhine Valley

Lower Rhine Graben (1978)
- BT Rhine Graben
- BT Central Europe
- BT Europe

Lower Saxony
  No longer a valid term for GeoRef. See Lower Saxony Germany.

Lower Saxony Germany (1993)
State in NW Germany.
  CO N511500N535000
     E0113000E0070000
- BT Germany
- BT Central Europe
- BT Europe
- NT Asse Mine
- NT Bad Pyrmont Germany
- NT Bramsche Germany
- NT Bremen Germany
- NT Brunswick Germany
- NT East Frisian Islands
- NT Eilenriede
- NT Emsland
- NT Gehrden Germany
- NT Gorleben
- NT Gottingen Germany
- NT Hanover Germany
- NT Helmstedt Germany
- NT Hildesheim Germany
- NT Konrad Mine
- NT Leine Valley
- NT Osnabruck Germany
- NT Rammelsberg
- NT Weser-Ems
- SA Alster River
- SA Ems River
- SA Hamburg Germany
- SA Harz Foreland
- SA Harz Mountains
- SA Harz region
- SA Jade Bay
- SA North German Plain
- SA Teutoburg Forest
- SA Weser River
- SA Westphalia

Lower Silesia (1978)
Most of former German Silesia which became part of Poland following World War II. It lies on both sides of the upper Oder River centering on Wroclaw with Czechoslovakia and the Neisse River constituting the SE and E borders respectively.
- BT Poland
- BT Central Europe
- BT Europe
- SA Izera Mountains
- SA Polkowice Poland
- SA Silesia
- SA Silesian coal basin
- SA Snieznik
- SA Strzegom-Sobotka Massif
- SA Upper Silesia

Lower Silurian (1978)
Autoposting of this term began in 1978.
- BT Silurian
- BT Paleozoic
- NT Alexandrian
- NT Brassfield Formation
- NT Clough Formation
- NT Grimsby Sandstone
- NT Llandoverian
- NT Medina Formation
- NT Tuscarora Formation
- NT Wenlockian
- NT Whirlpool Sandstone
- SA Ellis Bay Formation

- SA Matapedia Group
- SA Middle Silurian
- SA Upper Silurian

Lower Tatra
  use Low Tatra Mountains

lower Tertiary (1978)
- BT Tertiary
- BT Cenozoic
- NT Kiev Member
- NT Taishu Group
- SA Bauru Formation
- SA Dakhla Shale
- SA middle Tertiary
- SA Paleogene
- SA upper Tertiary

Lower Triassic (1978)
Autoposting of this term began in 1978.
- BT Triassic
- BT Mesozoic
- NT Bunter
- NT Dinwoody Formation
- NT Panchet Series
- NT Scythian
- NT Serebryanka Formation
- NT Smithian
- NT Spathian
- NT Thaynes Formation
- NT Werfenian
- SA Cordevolian
- SA Goose Egg Formation
- SA Hercynian Orogeny
- SA Ivishak Formation
- SA Middle Triassic
- SA Sadlerochit Group
- SA Upper Triassic

Lower Tunguska River (1978)
Rises in N central Irkutsk Oblast and flows N crossing into Evenk National Okrug then W into the Yenisei River at Turukhansk. This term has multiple hierarchies.
- BT1 Russian Federation
- BT1 Commonwealth of Independent States
- BT2 Asia
- SA Tunguska
- SA Tunguska River

lower Turonian (1989)
- BT Turonian
- BT Upper Cretaceous
- BT Cretaceous
- BT Mesozoic

Lower Zaire (1981)
The province of Bas-Zaire.
  CO S060000S040000
     E0163000E0121000
- UF Bas-Zaire
- BT Zaire
- BT Central Africa
- BT Africa
- NT Matadi Zaire
- SA Congo River

Lowicz
  No longer a valid term for GeoRef. As of 1993 see Lowicz Poland.

Lowicz Poland (1993)
Town in Skierniewice Province, central Poland. Before 1993 also search Lowicz AND Poland.
- BT Poland
- BT Central Europe
- BT Europe

lowlands
  use plains

Lowndes County Alabama (1993)

S Central Alabama. Before 1989, also search Lowndes County AND Alabama.
CO N315700N322400
W0862300W0865400
BT Alabama
BT United States

Lowndes County Mississippi (1993)
NE Mississippi. Before 1989, also search Lowndes County AND Mississippi.
CO N331700N334400
W0881400W0884100
BT Mississippi
BT United States

Loxoconcha (1978)
Autoposting of Cytheracea and Cytherocopina to this term began in 1989. As of 1990, microfossils, Crustacea, Mandibulata, and Arthropoda are autoposted to this term. This term has multiple hierarchies.
BT1 Cytheracea
BT1 Cytherocopina
BT1 Podocopida
BT1 Ostracoda
BT1 Crustacea
BT1 Mandibulata
BT1 Arthropoda
BT1 Invertebrata
BT2 Cytheracea
BT2 Cytherocopina
BT2 Podocopida
BT2 Ostracoda
BT2 microfossils

Loyalty Islands (1978)
Island group 60 miles E of New Caledonia belonging to France. In 1985, broader term changed from Pacific Ocean to West Pacific Ocean Islands.
BT West Pacific Ocean Islands

Lozere
No longer a valid term for GeoRef.
See Lozere France.

Lozere France (1993)
Department in S France.
CO N440800N450000
E0040000E0030000
BT France
BT Western Europe
BT Europe

Lr
use lawrencium

Lu
use lutetium

Luanda
No longer a valid term for GeoRef.
use Luanda Angola

Luanda Angola (1993)
District including the city in NW Angola. Before 1993, also search Luanda. Before 1978, also search Loanda.
UF Loanda Angola
UF Luanda
BT Angola
BT Central Africa
BT Africa

Lubbock
Valid through 1988. Search in combination with state term. After 1988, use specific city-state term.

Lubbock County
Valid through 1988. Search in combination with state term. After 1988, use specific county-state term.

Lubbock County Texas (1989)
NW Texas. Before 1989, also search Lubbock County AND Texas.
CO N332000N335200
W1013000W1020400
BT Texas
BT United States
NT Lubbock Lake
NT Lubbock Texas

Lubbock Lake (1989)
Archaeological site in W Texas.
BT Lubbock County Texas
BT Texas
BT United States
SA archaeology

Lubbock Texas (1989)
City in NW Texas. Before 1989, search Lubbock AND Texas.
CO N333500N333500
W1015300W1015300
BT Lubbock County Texas
BT Texas
BT United States

Luben Poland
use Lubin Poland

Lubin
No longer a valid term for GeoRef.
As of 1993 see Lubin Poland.

Lubin Legnicki (1978)
Region in and around cities of Lubin and Legnica in SW Poland.
BT Poland
BT Central Europe
BT Europe

Lubin Poland (1993)
City in central Legnica, SW Poland. Before 1993 also search Lubin or Luben.
UF Luben Poland
BT Poland
BT Central Europe
BT Europe

Lublin
No longer a valid term for GeoRef.
As of 1993 see Lublin Poland.

Lublin Poland (1993)
Province and city in E Poland. Before 1993 also search Lublin AND Poland.
BT Poland
BT Central Europe
BT Europe
NT Lublin Upland
NT Lukow Poland

Lublin Upland (1978)
SW Lublin. Also search Lublin.
BT Lublin Poland
BT Poland
BT Central Europe
BT Europe

Lucania (1978)
Ancient district in S including Basilicata autonomous region and part of Salerno Province of Campania.
IN Index autonomous regions as applicable.
BT Italy
BT Southern Europe
BT Europe
SA Basilicata Italy
SA Campania Italy
SA Irpinia earthquake 1980

SA Salerno Italy

Lucas County
Valid through 1988. Search in combination with state term. After 1988, use specific county-state term.

Lucas County Iowa (1989)
S Iowa. Before 1989, also search Lucas County AND Iowa.
CO N405400N410900
W0930600W0933300
BT Iowa
BT United States

Lucas County Ohio (1989)
NW Ohio. Before 1989, also search Lucas County AND Ohio.
CO N412400N414300
W0831100W0835300
BT Ohio
BT United States
NT Toledo Ohio
SA Ottawa River

Lucas Formation (1989)
In Detroit River Group. Ohio, SE Michigan and Ontario. Includes Richfield Member. Also a Cretaceous formation in Cuba.
IN Index ages as applicable.
SA Cretaceous
SA Cuba
SA Detroit River Group
SA Michigan
SA Middle Devonian
SA Ohio
SA Ontario

Lucca
No longer a valid term for GeoRef.
See Lucca Italy.

Lucca Italy (1993)
City and province in central Italy.
BT Tuscany Italy
BT Italy
BT Southern Europe
BT Europe

Lucern Switzerland
use Lucerne Switzerland

Lucerne
No longer a valid term for GeoRef.
As of 1993, see Lucern Switzerland.

Lucerne Switzerland (1993)
Canton in N central Switzerland. Also a city. Before 1993 also search Lucern.
UF Lucern Switzerland
BT Switzerland
BT Central Europe
BT Europe

Lucinidae (1978)
Family.
BT Bivalvia
BT Mollusca
BT Invertebrata

Luderitz
No longer a valid term for GeoRef.
As of 1993 see Luderitz Namibia.

Luderitz Namibia (1993)
Town in SW Namibia. Formerly Angra Pequena. Before 1993 also search Luderitz AND Namibia.
BT Namibia
BT Southern Africa
BT Africa

Ludian (1978)
Europe. Above Bartonian, below Tongrian (Oligocene).
BT upper Eocene
BT Eocene

BT Paleogene
BT Tertiary
BT Cenozoic
SA Priabonian

Ludlovian (1978)
Europe. Above Wenlockian, below Gedinnian (Devonian).
BT Upper Silurian
BT Silurian
BT Paleozoic

Ludlow Member (1989)
In Fort Union Formation. NW and N South Dakota, NE Montana and SW North Dakota.
BT Paleocene
BT Paleogene
BT Tertiary
BT Cenozoic
SA Fort Union Formation
SA Montana
SA North Dakota
SA South Dakota

Ludlowville Formation (1989)
In Hamilton Group. W and central New York. Includes Ledyard Shale Member; Wanakah Member.
BT Middle Devonian
BT Devonian
BT Paleozoic
SA Hamilton Group
SA New York

ludwigite (1978)
BT borates

Lufeng
No longer a valid term for GeoRef.
See Lufeng China.

Lufeng China (1993)
Town in central Yunnan, SW China.
CO N250700N250700
E1021000E1021000
BT Yunnan China
BT China
BT Far East
BT Asia

Lugo
No longer a valid term for GeoRef.
As of 1993, see Lugo Spain.

Lugo Spain (1993)
Province in NW Spain. Before 1993, also search Lugo and Spain.
CO N421800N434600
W0065000W0075800
BT Galicia Spain
BT Spain
BT Iberian Peninsula
BT Southern Europe
BT Europe

Luisian (1985)
North America. Middle Miocene, above Relizian and below Mohnian.
BT middle Miocene
BT Miocene
BT Neogene
BT Tertiary
BT Cenozoic

lujavrite (1978)
BT syenites
BT plutonic rocks
BT igneous rocks
SA nepheline syenite

Lukov Poland
use Lukow Poland

Lukow
No longer a valid term for GeoRef.
As of 1993 see Lukow Poland.

Lukow Poland (1993)
Town in E Poland. Before 1993 also search Lukow or Lukov.
UF Lukov Poland
BT Lublin Poland
BT Poland
BT Central Europe
BT Europe

Lukuga River valley (1978)
E Zaire.
BT Zaire
BT Central Africa
BT Africa

luminescence (1978)
Before 1971, also search phosphorescence.
UF phosphorescence
SA cathodoluminescence
SA electromagnetic radiation
SA fluorescence
SA Fraunhofer line discriminators
SA geochemistry
SA Moon
SA thermoluminescence
SA X-ray fluorescence

luminescence analysis
use fluorimetry

Luna County
Valid through 1988. Search in combination with state term. After 1988, use specific county-state term.

Luna County New Mexico (1989)
SW New Mexico. Before 1989, also search Luna County AND New Mexico.
CO N314500N323500
W1072000W1081500
BT New Mexico
BT United States
NT Florida Mountains

Luna Program (1981)
Autoposting of this term to specific Luna missions began in 1989.
NT Luna 9
NT Luna 16
NT Luna 20
NT Luna 24
SA extraterrestrial geology
SA Moon
SA planetology
SA remote sensing

Luna 9 (1989)
BT Luna Program
SA Moon
SA remote sensing

Luna 16 (1978)
Autoposting of Luna Program to this term began in 1989.
BT Luna Program
SA Moon
SA remote sensing

Luna 20 (1978)
Autoposting of Luna Program to this term began in 1989.
BT Luna Program
SA Moon
SA remote sensing

Luna 24 (1981)
Autoposting of Luna Program to this term began in 1989.
BT Luna Program
SA Moon
SA remote sensing

Lunae Planum (1985)
Large plateau on Mars.
BT Mars
SA Kasei Vallis

lunar bases (1989)
SA exploration
SA extraterrestrial geology
SA landing sites
SA Moon
SA observations

lunar breccia (1981)
Before 1981, also search Moon AND breccia.
SA breccia
SA lunar samples
SA Moon

lunar craters (1981)
Before 1981, search Moon AND craters.
SA craters
SA Moon
SA wrinkle ridges

lunar crust (1981)
Also search Moon AND crust.
SA crust
SA lunar samples
SA Moon

lunar highlands (1993)
Before 1993, also search highlands AND Moon.
SA highlands
SA Moon
SA wrinkle ridges

lunar interior (1981)
Before 1981, also search Moon AND interior.
SA interior
SA Moon
SA planetary interiors

lunar laser ranging (1993)
Before 1993, also search laser ranging.
SA geodesy
SA geophysical methods
SA geophysical surveys
SA laser methods
SA laser ranging
SA Moon
SA remote sensing

lunar mantle (1981)
Also search Moon AND mantle.
SA mantle
SA Moon

lunar materials
As of 1981, no longer a valid term for GeoRef. Use Moon and type of material. Use lunar samples for actual samples.

lunar meteorites
After 1992, use meteorites AND Moon.

lunar rocks
use lunar samples

lunar samples (1978)
Before 1981, also search lunar materials.
UF lunar rocks
SA lunar breccia
SA lunar crust
SA lunar soils
SA Moon
SA samples

lunar soils (1978)
Term reintroduced in 1981. A valid term through mid-1978. Before 1981, also search Moon AND soils.
SA lunar samples
SA Moon
SA soils

Lund
No longer a valid term for GeoRef. As of 1993 see Lund Sweden.

Lund Sweden (1993)
City in extreme S Sweden. Before 1993 also search Lund AND Sweden.
BT Malmohus Sweden
BT Sweden
BT Scandinavia
BT Western Europe
BT Europe

Luning Formation (1985)
SW Nevada.
BT Upper Triassic
BT Triassic
BT Mesozoic
SA Nevada

Luoquan Formation (1989)
Located in Henan, Shaanxi, Sichuan and Yunnan provinces.
BT Precambrian
SA China
SA Henan China
SA Shaanxi China
SA Sichuan China
SA Yunnan China

Lupin Mine (1989)
Gold ores. E Mackenzie District, Northwest Territories.
BT Mackenzie District Northwest Territories
BT Northwest Territories
BT Western Canada
BT Canada
SA gold ores
SA mines

Lusaka
No longer a valid term for GeoRef. As of 1993 see Lusaka Zambia.

Lusaka Zambia (1993)
City in central Zambia. Before 1993 also search Lusaka AND Zambia.
BT Zambia
DT East Africa
BT Africa

Lusatia (1978)
Region in E Germany, and in SW Poland. It comprises both Lower Lusatia and Upper Lusatia.
IN Index countries as applicable.
UF Lausitz
BT Europe
SA Germany
SA Lower Lusatia
SA Poland
SA Upper Lusatia

Lush's Bight Group
use Lushs Bight Group

Lushs Bight Group (1978)
UF Lush's Bight Group
BT Ordovician
BT Paleozoic
SA Newfoundland Island

Lusitanian (1978)
Europe. Above Oxfordian, below Kimmeridgian.
BT Upper Jurassic
BT Jurassic
BT Mesozoic
SA Malm
SA Rauracian

LUST
Acronym.
use leaking underground storage tanks

Lut Desert (1978)
Great sandy and stony desert in E central Iran.
UF Dasht-e-Lut Basin
UF Dasht-i-Lut
BT Iran
BT Middle East
BT Asia

Lutetian (1978)
Europe. Above Cuisian, below Auversian.
BT Eocene
BT Paleogene
BT Tertiary
BT Cenozoic

lutetium (1978)
Autoposting of rare earths and metals to this term began in 1989.
UF Lu
BT rare earths
BT metals

Lutzow-Holm Bay (1978)
Between Prince Olav Coast and Prince Harald Coast in Queen Maud Land of the Norwegian Sector.
UF Lutzow-Holmbukta
BT Antarctica
SA Prince Olav Coast
SA Queen Maud Land
SA Syowa Station

Lutzow-Holmbukta
use Lutzow-Holm Bay

Luvisols (1978)
BT soils
SA soil group

Luxembourg (1978)
CO N493000N501500
E0063000E0054500
BT Western Europe
BT Europe
SA Ardennes
SA Luxembourg Sandstone
SA Moselle River
SA Moselle Valley

Luxembourg Belgium (1993)
Province in S Belgium. Before 1993, also search Luxembourg Province. Before 1978, also search Luxembourg and Belgium.
CO N493000N502800
E0060200E0045100
UF Luxembourg Province
BT Belgium
BT Western Europe
BT Europe
NT Bastogne Belgium
SA Ardennes

Luxembourg Province
No longer a valid term for GeoRef.
use Luxembourg Belgium

Luxembourg Sandstone (1978)
Hettangian and Sinemurian. Also called Hettange Sandstone or Orval Sandstone.
BT Lower Jurassic
BT Jurassic
BT Mesozoic
SA Belgium
SA Luxembourg

Luzerne County
Valid through 1988. Search in combination with state term. After 1988, use specific county-state term.

Luzerne County Pennsylvania (1989)
E central Pennsylvania. Before 1989, also search Luzerne County AND Pennsylvania.
CO N405300N412500
W0753800W0762000
BT Pennsylvania
BT United States

Luzon (1978)
Northernmost and most important island.
BT Philippine Islands
BT Far East
BT Asia
NT Bataan Philippine Islands
NT Bondoc Peninsula
NT Camarines Norte Philippine Islands
NT Mayon
NT Taal
SA Sierra Madre
SA Zambales Ophiolite

luzonite (1978)
Autoposting of sulfantimonites and sulfarsenites began in 1989. This term has multiple hierarchies.
BT1 sulfantimonites
BT1 sulfosalts
BT2 sulfarsenites
BT2 sulfosalts
SA copper minerals
SA famatinite

Lvov
No longer a valid term for GeoRef. See Lvov Ukraine.

Lvov Basin (1978)
In Lvov Oblast on the Polish border. Also search Lvov. This term has multiple hierarchies.
BT1 Lvov Ukraine
BT1 Ukraine
BT1 Europe
BT2 Lvov Ukraine
BT2 Ukraine
BT2 Commonwealth of Independent States

Lvov Ukraine (1993)
Oblast including the city in W Ukraine. Before 1993, also search Lvov and Ukraine. Before 1978, also search L'vov or Lwow in combination with Ukraine. This term has multiple hierarchies.
UF L'vov Ukraine
UF Lwow Ukraine
BT1 Ukraine
BT1 Europe
BT2 Ukraine
BT2 Commonwealth of Independent States
NT Borislav Ukraine
NT Lvov Basin
NT Stebnik Ukraine
SA Dolina Ukraine
SA Lvov-Volyn Basin
SA Rozdol Ukraine
SA Roztocze
SA Rudki Ukraine

Lvov-Volyn Basin (1978)
Includes parts of Lvov Oblast and adjoining Volyn Oblast along the Polish border. This term has multiple hierarchies.
UF L'vov Volyn Basin
BT1 Ukraine
BT1 Europe
BT2 Ukraine
BT2 Commonwealth of Independent States
SA Lvov Ukraine
SA Volyn Ukraine

Lwow Ukraine
use Lvov Ukraine

Lyangar
No longer a valid term for GeoRef. As of 1993, see Lyangar Uzbekistan.

Lyangar Uzbekistan (1993)
Town in Samarkand Oblast in S central Uzbekistan. This term has multiple hierarchies.
BT1 Samarkand Uzbekistan
BT1 Uzbekistan
BT1 Asia
BT2 Samarkand Uzbekistan
BT2 Uzbekistan
BT2 Commonwealth of Independent States

Lycian Taurus (1978)
Mountains in ancient district of SW Anatolia called Lycia.
BT Turkey
BT Middle East
BT Asia

Lycopoda
use Lycopsida

Lycopodiales
use Lycopsida

Lycopodium (1989)
Genus.
BT Lycopsida
BT pteridophytes
BT Plantae

Lycopodiumsporites (1989)
Genus. Belongs to the family Lycopodiaceae. Autoposting of microfossils to this term began in 1990.
BT miospores
BT palynomorphs
BT microfossils

lycopods
use Lycopsida

Lycopsida (1978)
Before 1981, also search lycopods; Lycopoda. Autoposting of this term began in 1978.
UF Lycopoda
UF Lycopodiales
UF lycopods
BT pteridophytes
BT Plantae
NT Lepidodendron
NT Lycopodium
NT Sigillaria
NT Stigmaria
SA vascular taxa

Lydonia Canyon (1985)
Submarine canyon in N Atlantic continental slope off New England.
CO N401000N404000 W0673000W0675000
BT North American Atlantic
BT North Atlantic
BT Atlantic Ocean
SA Atlantic Coastal Plain
SA New England
SA submarine canyons

Lyell, Charles (1985)
UF Charles Lyell
SA biography

Lynchburg Formation (1989)
N North Carolina and Virginia.
BT upper Proterozoic
BT Proterozoic
BT upper Precambrian
BT Precambrian
SA North Carolina
SA Virginia

Lyngen Peninsula (1978)
Along Lyngen Fjord 30 miles E of Tromso in N Norway.
BT Troms Norway
BT Norway
BT Scandinavia
BT Western Europe
BT Europe

Lynn Lake
No longer a valid term for GeoRef. See Lynn Lake Manitoba.

Lynn Lake Manitoba (1993)
Town in NW Manitoba.
BT Manitoba
BT Western Canada
BT Canada

Lyon
Not a valid term for GeoRef. See Lyons France.

Lyon County
Valid through 1988. Search in combination with state term. After 1988, use specific county-state term.

Lyon County Iowa (1989)
Extreme NW Iowa. Before 1989, also search Lyon County AND Iowa.
CO N431600N433000 W0955100W0963700
BT Iowa
BT United States

Lyon County Kansas (1989)
E central Kansas. Before 1989, also search Lyon County AND Kansas.
CO N381100N384400 W0955700W0962200
BT Kansas
BT United States
NT Admire Kansas
NT Emporia Kansas

Lyon County Kentucky (1989)
W Kentucky. Before 1989, also search Lyon County AND Kentucky.
CO N365300N371200 W0875100W0881500
BT Kentucky
BT United States

Lyon County Minnesota (1989)
SW Minnesota. Before 1989, also search Lyon County AND Minnesota.
CO N441300N443800 W0953600W0960700
BT Minnesota
BT United States

Lyon County Nevada (1989)
W Nevada. Before 1989, also search Lyon County AND Nevada.
CO N382300N394500 W1184100W1194200
BT Nevada
BT United States
NT Silver City Nevada
NT Yerington Nevada

Lyons
No longer a valid term for GeoRef. See Lyons France.

Lyons France (1993)
City in E central France. Before 1978, also search Lyon.
BT Rhone France
BT France
BT Western Europe
BT Europe

Lyons Sandstone (1985)
In Cassa Group. N central Colorado.
BT Permian
BT Paleozoic
SA Colorado

lysimeters (1982)
SA drainage
SA ground water
SA hydrochemistry
SA hydrogeology
SA hydrologic cycle
SA hydrology
SA soils
SA water balance
SA water regimes

lysoclines (1978)
SA carbonate compensation depth
SA solution

Lystrosaurus (1978)
Genus.
BT Therapsida
BT Synapsida
BT Reptilia
BT Tetrapoda
BT Vertebrata
BT Chordata

Lytoceratida (1981)
BT Ammonoidea
BT Tetrabranchiata
BT Cephalopoda
BT Mollusca
BT Invertebrata

Lyublyana
use Ljubljana

M

M-discontinuity
use Mohorovicic discontinuity

maars (1978)
BT volcanic features
SA craters
SA geomorphology
SA phreatomagmatism
SA volcanism

Maas River
use Meuse River

Maas Valley
use Meuse Valley

Maastricht
No longer a valid term for GeoRef. See Maastricht Netherlands.

Maastricht Netherlands (1993)
City near the German border in extreme SE Netherlands. Before 1993, also search Maastricht. Before 1978, also search Maestricht.
BT Limburg Netherlands
BT Netherlands
BT Western Europe
BT Europe

Maastrichtian
use Maestrichtian

Macacu River (1978)
Central Rio de Janeiro.
SA Rio de Janeiro Brazil

macaluba
use mud volcanoes

Macanao Peninsula (1978)
On Margarita Island 15 miles off NE coast of Venezuela.
BT Nueva Esparta Venezuela
BT Venezuela
BT South America
SA Margarita Island

Macao
 Not a valid GeoRef index term after 1971. Was used on level 1 in subfile G. See China; Guangdong China.

Macedonia (1978)
 Region in Bulgaria, Greece, and Yugoslavia.
 BT Southern Europe
 BT Europe
 NT Greek Macedonia
 NT Yugoslav Macedonia
 SA Bulgaria
 SA Greece
 SA Yugoslavia

macerals (1978)
 Autoposting of this term to vitrain and durain began in 1981.
 UF micropetrological unit
 NT alginite
 NT clarain
 NT durain
 NT exinite
 NT fusain
 NT huminite
 NT inertinite
 NT vitrain
 NT vitrinite
 SA coal
 SA petrography

Machaeridia (1978)
 BT Homalozoa
 BT Echinodermata
 BT Invertebrata

Mackay
 Valid through 1988. Search in combination with state term. After 1988, use specific city-state term.

Mackay Idaho (1989)
 Town in S central Idaho. Before 1989, search Mackay AND Idaho.
 CO N435600N435600
  W1133700W1133700
 BT Custer County Idaho
 BT Idaho
 BT United States

Mackenzie
 Not a valid term for GeoRef. Use Mackenzie Mountains, Mackenzie Delta, Mackenzie District Northwest Teritories, Mackenzie River or Mackenzie River valley as applicable.

Mackenzie Delta (1978)
 On Beaufort Sea in Mackenzie District. Also search Mackenzie.
 UF Mackenzie River delta
 BT Mackenzie District Northwest Territories
 BT Northwest Territories
 BT Western Canada
 BT Canada
 SA Mackenzie River

Mackenzie District
 No longer a valid term for GeoRef. As of 1993, see Mackenzie District Northwest Territories.

Mackenzie District Northwest Territories (1993)
 Central and W Northwest Territories. From 1985-1992, also search Mackenzie District AND Northwest Territories. Before 1985, also search District of Mackenzie or Mackenzie AND Northwest Territories.
 CO N600000N710000
  W1020000W1370000
 UF District of Mackenzie
 BT Northwest Territories
 BT Western Canada

 BT Canada
 NT Fort Good Hope
 NT Great Bear Lake
 NT Great Slave Lake
 NT Inuvik Northwest Territories
 NT Lupin Mine
 NT Mackenzie Delta
 NT Mackenzie River
 NT Mackenzie River valley
 NT Muskox Intrusion
 NT Norman Wells Northwest Territories
 NT Pine Point mining district
 NT Richards Island
 NT South Nahanni River
 NT Tuktoyaktuk Peninsula
 NT Yellowknife Northwest Territories
 SA Bear-Slave Operation
 SA Coppermine River
 SA Horn Plateau Formation
 SA Root River

Mackenzie Mountains (1978)
 Also search Mackenzie.
 IN Index territories as applicable.
 BT Canada
 SA Northwest Territories
 SA Rocky Mountains
 SA Yukon Territory

Mackenzie River (1985)
 W Mackenzie District. Flows from Great Slave Lake N to Beaufort Sea.
 BT Mackenzie District Northwest Territories
 BT Northwest Territories
 BT Western Canada
 BT Canada
 SA Great Slave Lake
 SA Mackenzie Delta
 SA Mackenzie River valley

Mackenzie River delta
 use Mackenzie Delta

Mackenzie River valley (1978)
 W Mackenzie District. Also search Mackenzie; Mackenzie River.
 UF Mackenzie Valley
 BT Mackenzie District Northwest Territories
 BT Northwest Territories
 BT Western Canada
 BT Canada
 SA Mackenzie River

Mackenzie Valley
 use Mackenzie River valley

mackinawite (1978)
 UF kansite
 BT sulfides

Macleay River (1978)
 E New South Wales.
 BT New South Wales Australia
 BT Australia
 BT Australasia

Macomb County
 Valid through 1988. Search in combination with state term. After 1988, use specific county-state term.

Macomb County Michigan (1989)
 SE Michigan. Before 1989, also search Macomb County AND Michigan.
 CO N422500N425200
  W0824500W0830800
 BT Michigan Lower Peninsula
 BT Michigan
 BT United States
 SA Lake Saint Clair

Macon
 Valid through 1988. Search in combination with state term. After 1988, use specific city-state term.

Macon County
 Valid through 1988. Search in combination with state term. After 1988, use specific county-state term.

Macon County Alabama (1989)
 E Alabama. Before 1989, also search Macon County AND Alabama.
 CO N321400N323600
  W0852500W0860200
 BT Alabama
 BT United States

Macon County Georgia (1989)
 Central Georgia. Before 1989, also search Macon County AND Georgia.
 CO N321000N323000
  W0835000W0841400
 BT Georgia
 BT United States

Macon County Illinois (1989)
 Central Illinois. Before 1989, also search Macon County AND Illinois.
 CO N393600N400400
  W0884500W0891200
 BT Illinois
 BT United States

Macon County Missouri (1989)
 N central Missouri. Before 1989, also search Macon County AND Missouri.
 CO N393700N400300
  W0921800W0925200
 BT Missouri
 BT United States

Macon County North Carolina (1989)
 W North Carolina. Before 1989, also search Macon County AND North Carolina.
 CO N345900N352000
  W0830700W0834400
 BT North Carolina
 BT United States

Macon County Tennessee (1989)
 N Tennessee. Before 1989, also search Macon County AND Tennessee.
 CO N362400N363800
  W0854700W0861400
 BT Tennessee
 BT United States

Macon Georgia (1989)
 City in central Georgia. Before 1989, search Macon AND Georgia.
 CO N324900N324900
  W0833700W0833700
 BT Bibb County Georgia
 BT Georgia
 BT United States

Macquarie Island (1978)
 850 miles SE of Tasmania by whom it is administered. In 1985, broader term changed from Pacific Ocean to West Pacific Ocean Islands.
 BT West Pacific Ocean Islands

Macquarie Ridge (1978)
 Extends between Macquarie Island on S and South Island of New Zealand on N.
 BT Pacific Ocean

macrofossils
 use fossils

Macropodidae (1978)
 Family. Autoposting of Diprotodonta, Metatheria and Theria to this term began in 1989.
 BT Diprotodonta
 BT Marsupialia
 BT Metatheria
 BT Theria
 BT Mammalia
 BT Tetrapoda
 BT Vertebrata
 BT Chordata

Macroscelida (1981)
 Autoposting of Insectivora, Eutheria and Theria to this term began in 1989.
 UF Macroscelidea
 BT Insectivora
 BT Eutheria
 BT Theria
 BT Mammalia
 BT Tetrapoda
 BT Vertebrata
 BT Chordata

Macroscelidea
 use Macroscelida

Mactra (1978)
 Genus.
 BT Bivalvia
 BT Mollusca
 BT Invertebrata

Madagascar (1993)
 A valid term through 1974. Became Malagasy Republic in 1958. Before 1993, also search Malagasy Republic. This term has multiple hierarchies.
 CO S254000S115200
  E0503000E0430000
 UF Malagasy Republic
 BT1 Indian Ocean Islands
 BT2 Africa
 NT Ambre Mountain
 NT Andriamena Madagascar
 NT Antananarivo Madagascar
 NT Betsiboka Basin
 NT Sakoa Basin
 NT Tulear Basin

Madagascar Basin (1978)
 SE of Madagascar. Also search Madagascar.
 BT Indian Ocean
 SA DSDP Site 245

Madan
 No longer a valid term for GeoRef. See Madan Bulgaria.

Madan Bulgaria (1993)
 Village in SE Rhodope Mountains in S Bulgaria.
 BT Plovdiv Bulgaria
 BT Bulgaria
 BT Southern Europe
 BT Europe
 SA Bulgarian Rhodope Mountains

Madeira (1978)
 Island group belonging to Portugal 600 miles SW of Lisbon and 400 miles W of Morocco. In 1981, broader term changed from Atlantic Ocean to Atlantic Ocean Islands.
 CO N323000N330700
  W0161300W0173000
 UF Madeira Archipelago
 UF Madeira Island
 UF Madeira Islands
 UF Madeiras
 BT Atlantic Ocean Islands
 NT Porto Santo Island

Madeira abyssal plain (1989)
E North Atlantic. As of 1990, Atlantic Ocean is autoposted to this term.
BT   North Atlantic
BT   Atlantic Ocean

Madeira Archipelago
use Madeira

Madeira Island
use Madeira

Madeira Islands
use Madeira

Madeiras
use Madeira

Maden Complex (1989)
BT   Eocene
BT   Paleogene
BT   Tertiary
BT   Cenozoic
SA   Turkey

Madera County
Valid through 1988. Search in combination with state term. After 1988, use specific county-state term.

Madera County California (1989)
Central California. Before 1989, also search Madera County AND California.
CO   N364500N374500
       W1190100W1203800
BT   California
BT   United States
SA   Yosemite National Park

Madera Formation (1989)
In Magdalena Group. N and central New Mexico and S Colorado.
BT   Pennsylvanian
BT   Carboniferous
BT   Paleozoic
SA   Colorado
SA   Magdalena Group
SA   New Mexico

Madhya Pradesh
No longer a valid term for GeoRef as of 1993.
use Madhya Pradesh India

Madhya Pradesh India (1993)
State in central India.
CO   N180000N270000
       E0840000E0740000
UF   Madhya Pradesh
BT   India
BT   Indian Peninsula
BT   Asia
NT   Balaghat India
NT   Bastar India
NT   Chhindwara India
NT   Jabalpur India
NT   Jhabua India
NT   Kanhan Valley
NT   Kota India
NT   Pauni India
NT   Pench Valley
NT   Raipur India
NT   Rewa India
NT   Sagar India
NT   Satna India
NT   Sidhi India
NT   Umaria India
NT   Umrer India
SA   Bagh Beds
SA   Bhander Group
SA   Bijawar System
SA   Bilaspur India
SA   Bundelkhand
SA   Damuda Series
SA   Deccan Plateau
SA   Jabalpur Series
SA   Kaimur Sandstone
SA   Mahanadi Valley
SA   Narmada River
SA   Narmada Valley
SA   Narmada-Son Lineament
SA   Satpura Range
SA   Semarkona Meteorite
SA   Semri Series
SA   Son Valley
SA   Trans-Aravalli Vindhyan Basin

Madison
Valid through 1988. Search in combination with state term. After 1988, use specific city-state term.

Madison Aquifer (1989)
IN   Index states as applicable.
SA   Bighorn Basin
SA   Great Plains
SA   Madison Group
SA   Montana
SA   North Dakota
SA   South Dakota
SA   United States
SA   Wyoming

Madison County
Valid through 1988. Search in combination with state term. After 1988, use specific county-state term.

Madison County Alabama (1989)
N Alabama. Before 1989, also search Madison County AND Alabama.
CO   N343000N350000
       W0861600W0864800
BT   Alabama
BT   United States

Madison County Arkansas (1989)
NW Arkansas. Before 1989, also search Madison County AND Arkansas.
CO   N354500N361800
       W0932800W0935700
BT   Arkansas
BT   United States

Madison County Florida (1989)
N Florida. Before 1989, also search Madison County AND Florida.
CO   N301500N303700
       W0831000W0834500
BT   Florida
BT   United States

Madison County Georgia (1989)
NE Georgia. Before 1989, also search Madison County AND Georgia.
CO   N340000N341500
       W0825800W0832500
BT   Georgia
BT   United States

Madison County Idaho (1989)
E Idaho. Before 1989, also search Madison County AND Idaho.
CO   N433700N435600
       W1112400W1120000
BT   Idaho
BT   United States

Madison County Illinois (1989)
Bounded by Mississippi River in SW Illinois. Before 1989, also search Madison County AND Illinois.
CO   N384000N390000
       W0893500W0901700
BT   Illinois
BT   United States
NT   Wood River Illinois

Madison County Indiana (1989)
E central Indiana. Before 1989, also search Madison County AND Indiana.
CO   N395600N402300
       W0853500W0855200
BT   Indiana
BT   United States

Madison County Iowa (1989)
S central Iowa. Before 1989, also search Madison County AND Iowa.
CO   N410900N413100
       W0934700W0941500
BT   Iowa
BT   United States

Madison County Kentucky (1989)
Central Kentucky. Before 1989, also search Madison County AND Kentucky.
CO   N373100N375400
       W0840500W0843800
BT   Kentucky
BT   United States

Madison County Mississippi (1989)
Central Mississippi. Before 1989, also search Madison County AND Mississippi.
CO   N322200N325200
       W0894300W0902800
BT   Mississippi
BT   United States

Madison County Missouri (1989)
SE Missouri. Before 1989, also search Madison County AND Missouri.
CO   N371900N373900
       W0900800W0903300
BT   Missouri
BT   United States

Madison County Montana (1989)
SW Montana. Before 1989, also search Madison County AND Montana.
CO   N444200N455200
       W1112200W1124400
BT   Montana
BT   United States
NT   Tobacco Root Mountains
SA   Gravelly Range
SA   Madison Range

Madison County Nebraska (1989)
NE central Nebraska. Before 1989, also search Madison County AND Nebraska.
CO   N414500N420700
       W0972200W0975000
BT   Nebraska
BT   United States

Madison County New York (1989)
Central New York. Before 1989, also search Madison County AND New York.
CO   N424300N431200
       W0751200W0755900
BT   New York
BT   United States
SA   Oneida Lake

Madison County North Carolina (1989)
W North Carolina. Before 1989, also search Madison County AND North Carolina.
CO   N354200N360500
       W0822400W0822800
BT   North Carolina
BT   United States

Madison County Tennessee (1989)
W Tennessee. Before 1989, also search Madison County AND Tennessee.
CO   N352600N354700
       W0883700W0890700
BT   Tennessee
BT   United States

Madison County Texas (1989)
E central Texas. Before 1989, also search Madison County AND Texas.
CO   N305000N310700
       W0953700W0961400
BT   Texas
BT   United States

Madison County Virginia (1989)
N Virginia. Before 1989, also search Madison County AND Virginia.
CO   N381400N383800
       W0780600W0782700
BT   Virginia
BT   United States

Madison Group (1978)
Lower and Upper Mississippian. Consists of Lodgepole Limestone, Mission Canyon Limestone, and Charles Formation.
BT   Mississippian
BT   Carboniferous
BT   Paleozoic
SA   Beaverhead Formation
SA   Charles Formation
SA   Colorado
SA   Idaho
SA   Lodgepole Formation
SA   Madison Aquifer
SA   Mission Canyon Limestone
SA   Montana
SA   North Dakota
SA   South Dakota
SA   Utah
SA   Wyoming

Madison Range (1985)
SW Montana.
IN   Index counties and U. S. Rocky Mountains or regions as applicable.
SA   Gallatin County Montana
SA   Madison County Montana
SA   Montana
SA   Rocky Mountains
SA   U. S. Rocky Mountains

Madison Wisconsin (1989)
City in S central Wisconsin. Before 1989, search Madison AND Wisconsin.
CO   N430400N430400
       W0892200W0892200
BT   Dane County Wisconsin
BT   Wisconsin
BT   United States

Madoc
No longer a valid term for GeoRef. As of 1993, see Madoc Ontario.

Madoc Ontario (1993)
Village in SE Ontario. Before 1993, also search Madoc AND Ontario.
BT   Hastings County Ontario
BT   Ontario
BT   Eastern Canada
BT   Canada

Madonie Mountains (1978)
NW central Sicily.
BT   Sicily Italy
BT   Italy

BT Southern Europe
BT Europe

Madras
No longer a valid term for GeoRef. State in India Before 1956. See Madras India for the city.

Madras India (1993)
City on the Bay of Bengal in NE Tamil Nadu, SE India.
BT Tamil Nadu India
BT India
BT Indian Peninsula
BT Asia

Madreporaria
use Scleractinia

madrepores
use Scleractinia

Madrid
No longer a valid term for GeoRef. As of 1993, see Madrid City Spain.

Madrid City Spain (1993)
Refers to only the city in central Spain. Before 1993, also search Madrid and Spain.
BT Madrid Spain
BT New Castile Spain
BT Castile Spain
BT Spain
BT Iberian Peninsula
BT Southern Europe
BT Europe

Madrid Formation (1989)
Silurian of Connecticut, Maine, Massachusetts and New Hampshire. Also a formation in Mexico.
IN Index age as applicable.
SA Connecticut
SA Maine
SA Massachusetts
SA Mexico
SA New Hampshire
SA Silurian

Madrid Province
No longer a valid term for GeoRef. As of 1993, see Madrid Spain.

Madrid Spain (1993)
Refers only to the province in central Spain. From 1981-1992, also search Madrid Province and Spain. Before 1981, also search Madrid and Spain. For the city, see Madrid City Spain.
CO N395700N411000
    W0030500W0043600
BT New Castile Spain
BT Castile Spain
BT Spain
BT Iberian Peninsula
BT Southern Europe
BT Europe
NT Madrid City Spain

Maestrazgo
No longer a valid term for GeoRef. As of 1993, see Maestrazgo Spain.

Maestrazgo Spain (1993)
Mountainous district in E Spain. Before 1993, also search Maestrazgo and Spain.
IN Index provinces as applicable.
BT Spain
BT Iberian Peninsula
BT Southern Europe
BT Europe
SA Castellon de la Plana Spain
SA Teruel Spain

Maestricht
Not a valid term for GeoRef. See Maestricht Netherlands.

Maestrichtian (1978)
Europe. Above Campanian, below Danian (Tertiary). Autoposting of this term began in 1978.
UF Maastrichtian
BT Senonian
BT Upper Cretaceous
BT Cretaceous
BT Mesozoic
NT upper Maestrichtian
SA Rosario Formation

mafic
A valid term through 1977.

mafic and ultramafic rocks
No longer a valid GeoRef index term. Before 1972, was used on level 2 in subfile G. See ultramafics; ultramafic composition; mafic composition.

mafic composition (1978)
Before 1978, also search igneous rocks AND mafic. Also search basic rocks; mafic rocks.
SA composition
SA igneous rocks
SA mafic magmas
SA ophiolite
SA ultramafic composition

mafic magmas (1981)
Autoposting of magmas to this term began in 1989.
BT magmas
SA mafic composition

mafic rocks
No longer a valid GeoRef index term. Before 1971, was used on level 2 in subfile N. Now use mafic composition.

Magadan
No longer a valid term for GeoRef. As of 1993, see Magadan Russian Federation.

Magadan Russian Federation (1993)
Oblast including the city on N shore of Okhotsk Sea. Before 1993, also search Magadan. This term has multiple hierarchies.
BT1 Russian Federation
BT1 Commonwealth of Independent States
BT2 Asia
NT   Okhotsk Massif
NT   Taygonos Peninsula
SA   Chukchi Peninsula
SA   Kolyma River
SA   Kolyma Uplift
SA   Okhotsk region
SA   Okhotsk-Chukchi volcanic belt
SA   Omulevka
SA   Penzhina Bay
SA   Verkhoyansk region

magadiite (1978)
BT scapolite group
BT framework silicates
BT silicates

Magallanes
No longer a valid term for GeoRef. See Magallanes Chile.

Magallanes Chile (1993)
Southernmost province.
BT Chile
BT South America
SA Patagonia
SA Staten Island
SA Tierra del Fuego

Magdalen Islands (1989)
Group of 13 islands in S central Gulf of Saint Lawrence in E Quebec.
CO N471300N474900
    W0612300W0620100
UF Iles Madeleine
BT Quebec
BT Eastern Canada
BT Canada
SA Gulf of Saint Lawrence

Magdalena
No longer a valid term for GeoRef. As of 1989, use in combination with state name for the village in Socorro County New Mexico. As of 1993, see Magdalena Colombia for the department in N Colombia; use in combination with country name for the towns in Buenos Aires Argentina, Sonora Mexico, Bolivia and Peru.

Magdalena Colombia (1993)
Department in N Colombia.
BT Colombia
BT South America
NT Santa Marta Colombia
SA Magdalena River

Magdalena Delta (1978)
At the mouth of the Magdalena River on the Caribbean Sea near Barranquilla. Also search Magdalena.
BT Colombia
BT South America
SA Bolivar Colombia

Magdalena Group (1989)
New Mexico and W Texas. Includes Laborcita Formation, Madera Formation, Sandia Formation.
IN Index ages as applicable.
BT Paleozoic
SA Laborcita Formation
SA Madera Formation
SA New Mexico
SA Pennsylvanian
SA Permian
SA Sandia Formation
SA Texas

Magdalena Mountains (1978)
In Socorro County in W central New Mexico.
IN Index county or regions as applicable.
SA Socorro County New Mexico

Magdalena New Mexico (1993)
Village and mining center in W central New Mexico. Before 1989, also search Magdalena AND New Mexico.
BT Socorro County New Mexico
BT New Mexico
BT United States

Magdalena River (1978)
Rises on E slope of Andes in S Colombia and flows N into the Caribbean Sea near Barranquilla. Also search Magdalena AND appropriate area.
BT Colombia
BT South America
SA Magdalena Colombia
SA Magdalena Valley

Magdalena Valley (1978)
River valley extending from NW to SW Colombia.
BT Colombia
BT South America
SA Magdalena River

Magdalenian (1978)
Archaeological classification. Europe, North Africa. Upper Paleolithic.
BT Paleolithic
BT Stone Age
BT Cenozoic
SA archaeology
SA Pleistocene
SA Quaternary

Magdeburg
No longer a valid term for GeoRef. See Magdeburg Germany.

Magdeburg Bezirk
No longer a valid term for GeoRef. District in W central East Germany. Before 1981, search Magdeburg. See Germany.

Magdeburg Germany (1993)
City in Saxony-Anhalt, central Germany. As of 1981, Magdeburg refers to only the city in W East Germany. Before 1981, it also included the East German district.
BT Saxony-Anhalt Germany
BT Germany
BT Central Europe
BT Europe

Magellan Program (1993)
Before 1993, also search Venus Radar Mapper.
UF Venus Radar Mapper
SA extraterrestrial geology
SA geophysical methods
SA geophysical surveys
SA image enhancement
SA imagery
SA photogrammetry
SA planetology
SA radar methods
SA remote sensing
SA Venus

maghemite (1978)
UF oxymagnite
BT oxides
SA iron oxides
SA spinel group
SA titanomaghemite

magma
Not a valid GeoRef index term after 1971. Was used on level 1 in subfile G.
use magmas

magma chambers (1978)
UF chambers, magma
UF magma reservoir
SA intrusions
SA magmas

magma reservoir
use magma chambers

magmas (1978)
Used for naturally occurring, mobile rock material, generated within the Earth and capable of intrusion and extrusion. Before 1972, also search magma. Autoposting of this term began in 1978.
UF magma
NT acidic magmas
NT mafic magmas
SA anatexis
SA assimilation
SA asthenosphere
SA contamination
SA country rocks
SA crust
SA crystal fractionation
SA crystal zoning
SA crystallites
SA crystallization

SA cumulates
SA effusion
SA emplacement
SA fractional crystallization
SA gases
SA hot spots
SA hybridization
SA igneous activity
SA igneous processes
SA igneous rocks
SA immiscibility
SA inclusions
SA intrusions
SA juvenile water
SA lava
SA magma chambers
SA magmatic associations
SA magmatic differentiation
SA melts
SA palingenesis
SA partial melting
SA petrology
SA phase equilibria
SA phreatomagmatism
SA pneumatolysis
SA pyrolite
SA reaction rims
SA segregation
SA viscosity
SA volcanology

magmas and magmatic differentiation
Not a valid GeoRef index term after 1970. Was used on level 1 in subfile N. See magmatic differentiation.

magmatic associations (1981)
SA magmas
SA magmatic differentiation

magmatic differentiation (1993)
Before 1993, also search magmas AND differentiation. Before 1970, also search magmas and magmatic differentiation.
SA crystal fractionation
SA crystal zoning
SA crystallization
SA differentiation
SA fractional crystallization
SA hybridization
SA igneous rocks
SA intrusions
SA magmas
SA magmatic associations
SA melts
SA partial melting
SA reaction rims

magmatism
A valid term through 1974. Use igneous activity or magmas.

magnesian calcite (1978)
UF magnesium calcite
BT carbonates
SA calcite

magnesian limestone
As of 1981, no longer a valid term for GeoRef.
use dolomitic limestone

magnesian spar
use dolomite

magnesioferrite (1978)
UF magnoferrite
BT oxides
SA iron oxides
SA spinel group

magnesiorieebeckite (1978)
BT clinoamphibole
BT amphibole group
BT chain silicates
BT silicates

magnesite (1978)
As of 1981, use magnesite deposits for magnesite as a commodity.
BT carbonates
SA magnesite deposits

magnesite deposits (1981)
Before 1981, search magnesite AND deposits.
IN See List C for use as a commodity term.
SA magnesite
SA refractory materials

magnesium (1978)
Chemical element. As of 1981, use magnesium ores for magnesium as a commodity. Autoposting of alkaline earth metals and metals to this term began in 1989.
UF Mg
BT alkaline earth metals
BT metals
NT Mg-25/Mg-24
NT Mg-26/Mg-24
SA magnesium ion
SA magnesium ores

magnesium calcite
use magnesian calcite

magnesium ion (1985)
SA geochemistry
SA ions
SA magnesium

magnesium ores (1981)
Before 1981, also search (magnesium OR Mg) AND (deposit OR deposits OR ore OR ores OR economic) in the basic index. Autoposting of metal ores to this term began in 1985.
BT metal ores
SA magnesium

magnetic anomalies (1978)
SA anomalies
SA geophysical methods
SA geophysical surveys
SA gravity anomalies
SA magnetic field
SA magnetic methods
SA magnetic surveys
SA nonmagnetic minerals
SA paleomagnetism
SA POGO
SA sea-floor spreading

magnetic declination (1981)
Also search declination AND magnetic field; declination AND paleomagnetism.
SA magnetic field
SA paleomagnetism

magnetic domains (1978)
NT multidomains
NT pseudo-single domains
NT single domains
SA domains
SA magnetic properties
SA paleomagnetism

magnetic exploration
Not a valid index term after 1971. Was used on level 1 in subfiles G and N. Now use magnetic methods, magnetic surveys, or magnetic logging.

magnetic field (1978)
Before 1974, also search magnetic fields. Before 1972, was sometimes used in combination with Earth, Moon, Sun, names of planets or other extraterrestrial terms (i.e. magnetic field, Earth; magnetic field, Mars; magnetic field, asteroids; etc.).
UF field, magnetic
SA alternating field demagnetization
SA annual variations
SA coastal effects
SA coercivity
SA core
SA currents
SA dipole moment
SA diurnal variations
SA dynamos
SA Earth
SA Earth-current methods
SA electrical field
SA electromagnetic field
SA electromagnetic induction
SA electromagnetic radiation
SA electromagnetic surveys
SA electromagnetic waves
SA geophysical methods
SA geophysical surveys
SA induction
SA Koenigsberger ratio
SA magnetic anomalies
SA magnetic declination
SA magnetic inclination
SA magnetic intensity
SA magnetic methods
SA magnetic surveys
SA magnetic susceptibility
SA magnetization
SA magnetohydrodynamics
SA magnetometers
SA magnetotelluric methods
SA magnetotelluric surveys
SA Moon
SA natural remanent magnetization
SA paleomagnetism
SA pole positions
SA potential field
SA relief effects
SA remanent magnetization
SA reversals
SA seasonal variations
SA secular variations
SA solar wind
SA spherical harmonic analysis
SA Thellier Method
SA thermoremanent magnetization
SA variations

magnetic field of the Earth
Not a valid GeoRef index term after 1970. Was used on level 1 in subfile N.

magnetic hysteresis (1981)
Before 1981, also search hysteresis AND paleomagnetism; hysteresis AND magnetization; hysteresis AND demagnetization; hysteresis AND magnetic.
SA demagnetization
SA hysteresis
SA magnetization
SA nonmagnetic minerals
SA paleomagnetism
SA properties

magnetic inclination (1981)
Also search inclination AND magnetic field; inclination AND paleomagnetism.
UF inclination, magnetic
SA inclinometers
SA magnetic field
SA paleomagnetism

magnetic intensity (1981)
SA intensity
SA magnetic field
SA paleomagnetism

magnetic iron ore
use magnetite

magnetic logging (1984)
BT well-logging
SA electromagnetic logging
SA magnetic methods
SA magnetic surveys

magnetic methods (1978)
BT geophysical methods
SA airborne methods
SA audiomagnetotelluric methods
SA CSAMT methods
SA deep magnetic sounding
SA magnetic anomalies
SA magnetic field
SA magnetic logging
SA magnetic properties
SA magnetic surveys
SA magnetometers
SA Magsat
SA nonmagnetic minerals
SA POGO
SA SQUID
SA total-field methods
SA variometers
SA vertical-gradient methods

magnetic minerals (1978)
SA ilmenite
SA magnetic properties
SA magnetite
SA minerals
SA titanomagnetite

magnetic polarization
use magnetization

magnetic properties (1978)
SA Curie point
SA magnetic domains
SA magnetic methods
SA magnetic minerals
SA magnetic susceptibility
SA nonmagnetic minerals
SA properties
SA remanent magnetization

magnetic storms
As of 1993, no longer a valid term for GeoRef. Usually out-of-scope. Used with magnetosphere.

magnetic survey maps (1978)
Before 1978, also search maps AND magnetic surveys.
BT maps
SA aeromagnetic maps
SA geophysical survey maps
SA geophysical surveys
SA magnetic surveys

magnetic surveys (1978)
Before 1974, also search aeromagnetic surveys; aeromagnetic.
BT geophysical surveys
BT surveys
SA aeromagnetic maps
SA airborne methods
SA deep magnetic sounding
SA magnetic anomalies
SA magnetic field
SA magnetic logging
SA magnetic methods
SA magnetic survey maps
SA magnetometers
SA Magsat
SA nonmagnetic minerals
SA POGO
SA SQUID
SA total-field methods
SA vertical-gradient methods

magnetic susceptibility (1978)
UF magnetic susceptibility anisotropy
UF volume susceptibility (magnetic)
SA anisotropy
SA Curie point
SA geophysical surveys

SA Koenigsberger ratio
SA magnetic field
SA magnetic properties
SA nonmagnetic minerals
SA paleomagnetism

magnetic susceptibility anisotropy
use magnetic susceptibility

magnetic tail
As of 1993, no longer a valid term for GeoRef. Usually out-of-scope. Used with magnetosphere.

magnetism, paleo-
use paleomagnetism

magnetite (1978)
IN May be used as a mineral. As of 1981, may be used as a commodity term.
UF magnetic iron ore
UF octahedral iron ore
BT oxides
SA heavy minerals
SA iron minerals
SA iron ores
SA iron oxides
SA magnetic minerals
SA spinel group
SA titanomagnetite

magnetization (1978)
The magnetic moment per unit volume. Autoposting of this term began in 1978.
UF magnetic polarization
NT direct magnetization
NT induced magnetization
NT inverse magnetization
NT isothermal magnetization
NT primary magnetization
NT remanent magnetization
NT saturation magnetization
NT secondary magnetization
NT spontaneous magnetization
SA adiabatic demagnetization
SA chemical demagnetization
SA demagnetization
SA magnetic field
SA magnetic hysteresis
SA paleomagnetism
SA remagnetization
SA thermal demagnetization

magnetohydrodynamics (1978)
SA core
SA magnetic field

magnetometers (1978)
SA instruments
SA magnetic field
SA magnetic methods
SA magnetic surveys
SA SQUID

magnetopause
As of 1993, no longer a valid term for GeoRef. Usually out-of-scope. Used with magnetosphere.

magnetosheath
As of 1993, no longer a valid term for GeoRef. Usually out-of-scope. Used with magnetosphere.

magnetosphere
As of 1993, no longer a valid term for GeoRef. Usually out-of-scope. Sometimes used on level 1 before 1993.

magnetostratigraphy (1978)
SA paleomagnetism
SA stratigraphy

magnetotelluric exploration
Not a valid GeoRef index term after 1971. Was used on level 1 in subfiles G and N. Now use mag-

netotelluric methods or magnetotelluric surveys.

magnetotelluric methods (1978)
BT geophysical methods
SA audiomagnetotelluric methods
SA CSAMT methods
SA Earth-current methods
SA electrical field
SA electromagnetic field
SA electromagnetic methods
SA magnetic field
SA magnetotelluric surveys
SA SQUID
SA variometers

magnetotelluric surveys (1978)
BT geophysical surveys
BT surveys
SA audiomagnetotelluric methods
SA CSAMT methods
SA Earth-current surveys
SA electromagnetic methods
SA electromagnetic surveys
SA magnetic field
SA magnetotelluric methods

Magnitogorsk
No longer a valid term for GeoRef. As of 1993, see Magnitogorsk Russian Federation.

Magnitogorsk Russian Federation (1993)
City just E of the South Urals in Chelyabinsk Oblast. Before 1993, also search Magnitogorsk.
BT Chelyabinsk Russian Federation
BT Russian Federation
BT Commonwealth of Independent States

magnitude (1978)
As of 1981, restricted to magnitude of earthquakes.
UF earthquake magnitude
SA earthquakes
SA intensity
SA Richter Scale
SA scale factor
SA seismic intensity

magnitude-frequency ratio (1981)
SA earthquakes
SA seismology

magnoferrite
use magnesioferrite

Magnoliidae (1981)
BT Dicotyledoneae
BT angiosperms
BT Spermatophyta
BT Plantae

Magothy Aquifer (1989)
IN Index states as applicable.
BT United States
SA Delaware
SA Magothy Formation
SA Maryland
SA New Jersey
SA New York

Magothy Formation (1978)
Consists essentially of light-gray crossbedded coarse sand containing small amounts of glauconite and pyrite; particles of carbonaceous matter or lignite common throughout.
BT Upper Cretaceous
BT Cretaceous
BT Mesozoic
SA Delaware
SA Magothy Aquifer
SA Maryland
SA New Jersey

SA New York

Magsat (1985)
SA geophysical methods
SA geophysical surveys
SA magnetic methods
SA magnetic surveys
SA remote sensing
SA satellite methods

Magura
No longer a valid term for GeoRef. See Magura Bangladesh.

Magura Bangladesh (1993)
Village W of Dacca in Bangladesh.
BT Bangladesh
BT Indian Peninsula
BT Asia

Mahakam Delta (1989)
E Kalimantan, Indonesia. This term has multiple hierarchies.
CO S010000S002000 E1174000E1170000
BT1 Kalimantan Indonesia
BT1 Borneo
BT1 Far East
BT1 Asia
BT2 Kalimantan Indonesia
BT2 Borneo
BT2 Malay Archipelago
BT3 Kalimantan Indonesia
BT3 Indonesia
BT3 Far East
BT3 Asia

Mahanadi Valley (1978)
River valley. Also search Mahanadi River.
IN Index states as applicable.
BT India
BT Indian Peninsula
BT Asia
SA Madhya Pradesh India
SA Orissa India

Mahantango Formation (1978)
Central Pennsylvania. In Hamilton Group. Consists principally of greenish-gray, thin- to medium-bedded slightly carbonaceous shale. Sporadic sandstone zones appear throughout area and thicken eastward. In central Pennsylvania.
BT Middle Devonian
BT Devonian
BT Paleozoic
SA Hamilton Group
SA Pennsylvania

Maharashtra
No longer a valid term for GeoRef. See Maharashtra India.

Maharashtra India (1993)
State in W central India.
CO N153000N220000 E0804000E0723000
BT India
BT Indian Peninsula
BT Asia
NT Bhandara India
NT Bombay India
NT Jamkhandi India
NT Lonar Crater
NT Nagpur India
NT Poona India
NT Wardha River valley
SA Deccan Plateau
SA Godavari River
SA Godavari Valley
SA Kaladgi System
SA Kamthi Formation
SA Koyna
SA Krishna

SA Pranhita-Godavari Valley
SA Satpura Range

Mahaska County
Valid through 1988. Search in combination with state term. After 1988, use specific county-state term.

Mahaska County Iowa (1989)
S central Iowa. Before 1989, also search Mahaska County AND Iowa.
CO N411000N413100 W0922500W0925300
BT Iowa
BT United States

Mahasu
No longer a valid term for GeoRef. See Mahasu India.

Mahasu India (1993)
District in N India.
BT Himachal Pradesh India
BT India
BT Indian Peninsula
BT Asia

Mahe Island (1978)
Most important of Seychelles, a British Colony, 1100 miles E of Kenya and 700 miles N of the Malagasy Republic. Also search Mahe. In 1981, broader term changed from Indian Ocean to Seychelles.
BT Seychelles
BT Indian Ocean Islands
SA Indian Ocean

Mahogany Zone (1985)
NW Colorado.
IN Index counties or regions as applicable.
BT Colorado
BT United States
SA U. S. Rocky Mountains

Mahoning County
Valid through 1988. Search in combination with state term. After 1988, use specific county-state term.

Mahoning County Ohio (1989)
E Ohio. Before 1989, also search Mahoning County AND Ohio.
CO N405300N410730 W0803200W0810500
BT Ohio
BT United States

Maikop
No longer a valid term for GeoRef. As of 1993, Maikop Russian Federation.

Maikop Russian Federation (1993)
City in Adygey Autonomous Oblast in the Northern Caucusus. Before 1993, also search Maikop or Maykop. This term has multiple hierarchies.
UF Maykop Russian Federation
BT1 Krasnodar Russian Federation
BT1 Russian Federation
BT1 Commonwealth of Independent States
BT2 Krasnodar Russian Federation
BT2 Europe

Maikop Series (1978)
Oligocene-Miocene. The following horizons have been established: Zuramakent, Riki, Mutsidakal, Miatly, and Khadum. Northern Caucusus and Crimea.

BT Tertiary
BT Cenozoic
SA Crimea Ukraine
SA Russian Federation
SA Ukraine

Maimecha
use Maymecha

Maimecha-Kotui
use Maymecha-Kotuy

Main Boundary Fault (1989)
IN Index countries as applicable.
BT Asia
SA Himalayas
SA India
SA Indian Peninsula
SA Nepal
SA Pakistan

Main Central Thrust (1989)
IN Index countries as applicable.
BT Asia
SA Himalayas
SA India
SA Indian Peninsula
SA Nepal
SA Pakistan

main phase
As of 1993, no longer a valid term for GeoRef. Usually out-of-scope. Used with magnetic storms.

Main River (1978)
Also search Main.
IN Index states as applicable.
BT Germany
BT Central Europe
BT Europe
SA Baden-Wurttemberg Germany
SA Bavaria Germany
SA Hesse Germany

main shocks (1985)
SA aftershocks
SA earthquakes
SA foreshocks
SA seismology

Maine (1978)
Autoposting of this term began in 1978.
CO N430000N473000
W0670000W0710500
BT United States
NT Androscoggin County Maine
NT Aroostook County Maine
NT Chain Lakes Massif
NT Cumberland County Maine
NT Franklin County Maine
NT Hancock County Maine
NT Kennebec County Maine
NT Lincoln County Maine
NT Norumbega fault zone
NT Oxford County Maine
NT Penobscot Bay
NT Penobscot County Maine
NT Piscataquis County Maine
NT Sagadahoc County Maine
NT Somerset County Maine
NT Waldo County Maine
NT Washington County Maine
NT York County Maine
SA Avalon Terrane
SA Berwick Formation
SA Bucksport Formation
SA Cape Elizabeth Formation
SA Casco Bay Group
SA Chelmsford Granite
SA Eagle Lake
SA Elliot Formation
SA Kittery Formation
SA Madrid Formation
SA Matagamon Sandstone
SA Matapedia Group
SA Merrimack Group
SA Merrimack Synclinorium
SA Penobscot Formation
SA Perry Mountain Formation
SA Presumpscot Formation
SA Rangeley Formation
SA Rangeley Lakes
SA Rye Formation
SA Saco River
SA Saint John River
SA Sangerville Formation
SA Seboomook Formation
SA Vassalboro Formation
SA Waterville Formation

Maine-et-Loire
No longer a valid term for GeoRef.
use Maine-et-Loire France

Maine-et-Loire France (1993)
Department in NW central France.
CO N470000N474500
E0001500W0012000
UF Maine-et-Loire
BT France
BT Western Europe
BT Europe

maintenance (1978)
SA reservoirs
SA siltation

Mainz
No longer a valid term for GeoRef.
See Mainz Germany.

Mainz Basin (1978)
Area around confluence of the Main and Rhine Rivers.
IN Index states as applicable.
BT Germany
BT Central Europe
BT Europe
SA Hesse Germany
SA Mainz Germany
SA Rhineland-Palatinate Germany

Mainz Germany (1993)
City on the Rhine River in E Rhineland-Palatinate, W West Germany.
BT Rhineland-Palatinate Germany
BT Germany
BT Central Europe
BT Europe
SA Mainz Basin

Maiolica Limestone (1978)
Jurassic-Cretaceous. White, marly limestone. In Lombardy, Italy.
IN Index ages as applicable.
BT Mesozoic
SA Cretaceous
SA Italy
SA Jurassic
SA Lombardy Italy

Maizuru
No longer a valid term for GeoRef.
See Maizuru Japan.

Maizuru Belt (1989)
S Honshu, Japan.
BT Honshu
BT Japan
BT Far East
BT Asia
SA Kyoto Japan

Maizuru Japan (1993)
City in Kyoto Prefecture in central Honshu.
BT Kyoto Japan
BT Honshu
BT Japan
BT Far East
BT Asia

Majdan Pek
use Majdanpek

Majdanpek (1978)
City in E Serbia.
UF Majdan Pek
BT Serbia
BT Yugoslavia
BT Southern Europe
BT Europe

Major County
Valid through 1988. Search in combination with state term. After 1988, use specific county-state term.

Major County Oklahoma (1989)
NW Oklahoma. Before 1989, also search Major County AND Oklahoma.
CO N361000N363000
W0980800W0985600
BT Oklahoma
BT United States

major elements (1978)
SA chemical elements
SA major-element analyses
SA minor elements
SA nutrients
SA trace elements

major-element analyses (1978)
For analytical methods. For data, see material name.
SA analysis
SA chemical analysis
SA chemical elements
SA major elements
SA minor-element analyses
SA spectroscopy
SA trace-element analyses

Majorca (1978)
Largest of Spain's Balearic Islands off E coast of Spanish mainland. This term has multiple hierarchies.
CO N391000N400000
E0033000E0021500
UF Mallorca
BT1 Balearic Islands
BT1 Mediterranean region
BT2 Balearic Islands
BT2 Spain
BT2 Iberian Peninsula
BT2 Southern Europe
BT2 Europe
NT Palma Spain

majorite (1993)
Meteorite mineral.
BT garnet group
BT nesosilicates
BT orthosilicates
BT silicates
SA meteorites

Majunga Basin
use Betsiboka Basin

Makarov Basin (1989)
BT Arctic Ocean
SA Lomonosov Ridge
SA North Pole
SA polar regions

Makhachkala
No longer a valid term for GeoRef. As of 1993, see Makhachkala Russian Federation.

Makhachkala Russian Federation (1993)
City in former Dagestan Autonomous Soviet Socialist Republic on Caspian Sea. Before 1978, also search Petrovsk. This term has multiple hierarchies.
UF Petrovsk Russian Federation
BT1 Dagestan Russian Federation
BT1 Russian Federation
BT1 Commonwealth of Independent States
BT2 Dagestan Russian Federation
BT2 Europe

Makhtesh Ramon (1978)
Canyon in the Negev, Southern Israel.
BT Israel
BT Middle East
BT Asia
SA Negev

Makushin (1989)
Volcano. Unalaska Island, Fox Islands, Aleutian Islands, SW Alaska.
CO N535200N535200
W1665500W1665500
UF Makushin Volcano
BT Unalaska Island
BT Aleutian Islands
BT Southwestern Alaska
BT Alaska
BT United States

Makushin Volcano
use Makushin

Malacca Strait
use Strait of Malacca

Malacca Straits
use Strait of Malacca

malachite (1978)
BT carbonates
SA copper minerals
SA copper ores

Malacostraca (1978)
As of 1981, used for Decapoda. Before 1981, also search Decapoda; decapods. Autoposting of this term began in 1978.
UF Decapoda
BT Crustacea
BT Mandibulata
BT Arthropoda
BT Invertebrata
NT Brachyura
NT Callianassa
NT Ophiomorpha

Malaga
No longer a valid term for GeoRef. As of 1993, see Malaga City Spain.

Malaga City Spain (1993)
Refers to only the city in S Spain on the Mediterranean Sea. Before 1993, also search Malaga and Spain.
BT Malaga Spain
BT Andalusia Spain
BT Spain
BT Iberian Peninsula
BT Southern Europe
BT Europe

Malaga Province
No longer a valid term for GeoRef. As of 1993, see Malaga Spain.

Malaga Spain (1993)
Refers only to the province. From 1981-1992, also search Malaga Province and Spain. Before 1981, also search Malaga and Spain. In S central Spain. For the city, see Malaga City Spain.
CO N361800N371800
W0034500W0053000
BT Andalusia Spain
BT Spain
BT Iberian Peninsula
BT Southern Europe

BT Europe
NT Malaga City Spain
SA Serrania de Ronda

Malagasy Republic
No longer a valid term for GeoRef.
use Madagascar

Malaita (1978)
Long, narrow island in SE Solomon Islands. In 1981, broader term changed from Pacific Ocean to Solomon Islands.
BT Solomon Islands
BT Melanesia
BT Oceania
SA Pacific Ocean

Malakal
No longer a valid term for GeoRef. As of 1993 see Malakal Sudan.

Malakal Sudan (1993)
Town in S Sudan. Before 1993 also search Malakal AND Sudan.
BT Sudan
BT East Africa
BT Africa

Malatya
No longer a valid term for GeoRef. As of 1993 see Malatya Turkey.

Malatya Turkey (1993)
Province in E central Anatolia. Also a city. Before 1993 also search Malatya AND Turkey.
BT Turkey
BT Middle East
BT Asia

Malawi (1978)
Formerly Nyasaland. At one time it was part of British Central African Protectorate and later part of the Federation of Rhodesia and Nyasaland.
CO S173000S090000
   E0363000E0330000
UF Nyasaland
BT East Africa
BT Africa
SA East African Rift
SA Lake Malawi

Malay Archipelago (1978)
Largest island group in world off SE coast of Asia between Pacific and Indian oceans. Before 1981, a narrower term of Asia. Includes New Guinea as of 1981. As of 1981, Borneo may be indexed under Far East or Malay Archipelago. Before 1981, Borneo and Brunei were indexed under this term. Timor considered a narrower term as of 1981. Autoposting of this term began in 1978.
IN Index countries as applicable.
NT Borneo
NT New Guinea
NT Timor
SA Asia
SA Celebes Sea
SA Far East
SA Indonesia
SA Malaysia
SA Oceania
SA Papua New Guinea
SA Sumatra

Malay Peninsula (1978)
Comprises West Malaysia and SW part of Thailand. Before 1981, broader term was Asia.
IN Index countries as applicable.
BT Far East

BT Asia
NT West Malaysia
SA Indochina
SA Malaysia
SA Thailand

Malaya
A valid index term through 1976. Former Federation of Malaya which was a federation of 9 Malay states of the Malay Peninsula, plus 2 of the Straits settlements, Malacca and Penang. It constituted what is now West Malaysia of the present Federation of Malaysia.
use Malaysia

malayaite (1978)
BT ring silicates
BT silicates

Malaysia (1978)
Or officially known as Federation of Malaysia. Independent federation, SE Asia, consisting of eleven states (West Malaysia) on the Malay Peninsula and two states (East Malaysia) on the island of Borneo.
CO N013000N073000
   E1190000E1000000
UF Malaya
BT Far East
BT Asia
NT East Malaysia
NT West Malaysia
SA ASEAN
SA Borneo
SA Indochina
SA Malay Archipelago
SA Malay Peninsula

Maldive Archipelago
No longer a valid GeoRef index term. Before 1969, was used on level 1 in subfile E.
use Maldive Islands

Maldive Islands (1978)
Group of 19 atolls in the Indian Ocean 300 miles SW of southern tip of India. In 1981, broader term changed from Indian Ocean to Indian Ocean Islands. Before 1969, also search Maldive Archipelago.
CO S010000N072000
   E0740000E0720000
UF Maldive Archipelago
BT Indian Ocean Islands

Malekula (1978)
An island of Vanuatu which lies E of Queensland, Australia, between the Solomon Islands and New Caledonia. In 1981, broader term changed from Pacific Ocean to New Hebrides. In 1985, broader term changed to Vanuatu.
BT Vanuatu
BT Melanesia
BT Oceania
SA Pacific Ocean

Malgobek
No longer a valid term for GeoRef. As of 1993, see Malgobek Russian Federation.

Malgobek Russian Federation (1993)
City in former North Ossetian Autonomous Soviet Socialist Republic in the Northern Caucasus. A petroleum producing center.
IN Index regions as applicable.
BT Russian Federation

BT Commonwealth of Independent States
SA North Ossetia Russian Federation

Malheur County
Valid through 1988. Search in combination with state term. After 1988, use specific county-state term.

Malheur County Oregon (1989)
SE Oregon. Before 1989, also search Malheur County AND Oregon.
CO N420000N442500
   W1170500W1181500
BT Oregon
BT United States
SA Owyhee Mountains

Mali (1978)
CO N100500N250000
   E0040000W0130000
BT West Africa
BT Africa
NT Taoudenni
SA Adrar des Iforas
SA Mali-Niger Syneclise
SA Mauritanides
SA Niger River
SA Niger Valley
SA Sahara
SA Sahel
SA Senegal Basin
SA Senegal River
SA Tanezrouft
SA West African Shield

Mali-Niger Syneclise (1978)
IN Index countries as applicable.
UF Mali-Nigeria Syneclise
BT West Africa
BT Africa
SA Benin
SA Mali
SA Niger
SA Nigeria

Mali-Nigeria Syneclise
use Mali-Niger Syneclise

Malines
No longer a valid term for GeoRef. See Malines Belgium.

Malines Belgium (1993)
City N of Brussels. Before 1993, also search Malines. Before 1978, also search Mechelen or Mechlin AND Belgium.
BT Antwerp Belgium
BT Belgium
BT Western Europe
BT Europe

Mallorca
use Majorca

Malm (1978)
Middle Europe. Above Dogger, below Neocomian (Cretaceous).
BT Upper Jurassic
BT Jurassic
BT Mesozoic
SA Kimmeridgian
SA Lusitanian
SA Oxfordian
SA Portlandian
SA Volgian

Malmani Subgroup (1993)
Proterozoic. South Africa. In Transvaal Supergroup.
BT Proterozoic
BT upper Precambrian
BT Precambrian
SA South Africa
SA Transvaal Supergroup

Malmohus
No longer a valid term for GeoRef. As of 1993 see Malmohus Sweden.

Malmohus Sweden (1993)
County in S Sweden. Before 1993 also search Malmohus AND Sweden.
BT Sweden
BT Scandinavia
BT Western Europe
BT Europe
NT Halsingborg Sweden
NT Lund Sweden
SA Skane

Malta (1978)
Comprises 3 islands. In 1981, broader term changed from Mediterranean Sea to Europe.
CO N354500N361000
   E0144000E0141000
BT Southern Europe
BT Europe
SA Mediterranean region
SA Mediterranean Sea

Malvern
No longer a valid term for GeoRef. As of 1993, see Malvern England.

Malvern England (1993)
City in Hereford and Worcester County in W central England. Before 1993, also search Malvern AND England.
BT England
BT Great Britain
BT United Kingdom
BT Western Europe
BT Europe
SA Worcestershire England

Malvern Hills (1978)
In Hereford and Worcester County in W England.
BT England
BI Great Britain
BT United Kingdom
BT Western Europe
BT Europe
SA Worcestershire England

Malvern Meteorite (1993)
Ca-rich howardite. Orange Free State, South Africa. Before 1993, also search Malvern AND (meteorites OR achondrites OR howardites).
BT howardite
BT achondrites
BT stony meteorites
BT meteorites
SA Orange Free State South Africa
SA South Africa

Mama
No longer a valid term for GeoRef. As of 1993, see Mama Russian Federation.

Mama River (1978)
Rises NE of Lake Baikal and flows NE into Vitim River at Mama. Also search Mama. This term has multiple hierarchies.
BT1 Russian Federation
BT1 Commonwealth of Independent States
BT2 Asia
SA Vitim

Mama Russian Federation (1993)
Town on the Vitim River at the mouth of the Mama River in NE Irkutsk Oblast. Before 1993, also

search Mama. This term has multiple hierarchies.
BT1 Irkutsk Russian Federation
BT1 Russian Federation
BT1 Commonwealth of Independent States
BT2 Irkutsk Russian Federation
BT2 Asia

Mamainse Point (1978)
On Lake Superior N of Sault Sainte Marie.
BT Algoma District Ontario
BT Ontario
BT Eastern Canada
BT Canada

Mammalia (1978)
Autoposting of this term to Eutheria began in 1989.
BT Tetrapoda
BT Vertebrata
BT Chordata
NT Docodonta
NT Monotremata
NT Multituberculata
NT Paucituberculata
NT Theria
NT Triconodonta
SA fossil man
SA mammals

mammals (1981)
Common name for Mammalia.
IN Index for all non-paleontologic studies of fossils.
BT tetrapods
BT vertebrates
NT rodents
SA biostratigraphy
SA Mammalia

Mammoth Cave (1978)
Cave and national park in Edmonson County in SW central Kentucky.
BT Edmonson County Kentucky
BT Kentucky
BT United States
SA national parks

Mammoth Lakes earthquakes (1985)
Series of earthquakes, with epicenters near town of Mammoth Lakes, California.
BT earthquakes
SA California
SA Long Valley Caldera
SA Mono County California

Mammut
use Mastodon

Mammuthus (1978)
Genus. Autoposting of Eutheria and Theria to this term began in 1989. Before 1981, also search mammoths. As of 1981, search mammoths only for biostratigraphy.
BT Elephantidae
BT Elephantoidea
BT Proboscidea
BT Eutheria
BT Theria
BT Mammalia
BT Tetrapoda
BT Vertebrata
BT Chordata
NT Mammuthus primigenii

Mammuthus primigenius (1978)
Species. Autoposting of Eutheria and Theria to this term began in 1989.
BT Mammuthus
BT Elephantidae
BT Elephantoidea

BT Proboscidea
BT Eutheria
BT Theria
BT Mammalia
BT Tetrapoda
BT Vertebrata
BT Chordata

Mammutidae
use Mastodontidae

man
As of 1981, no longer a valid term for GeoRef. See fossil man; Mammalia. Before 1978, included use in combination with fossil (i.e. man, fossil).

man-made islands
use artificial islands

management (1978)
Used as a general term.
SA basin management
SA corporate policy
SA land use
SA planning
SA regulations
SA shorelines
SA soil management
SA water management

Managua
No longer a valid term for GeoRef. As of 1993 see Managua Nicaragua.

Managua Nicaragua (1993)
Department and city in W Nicaragua. Before 1993 also search Managua AND Nicaragua.
BT Nicaragua
BT Central America

Manakacha Formation (1985)
NW Arizona.
BT Middle Pennsylvanian
BT Pennsylvanian
BT Carboniferous
BT Paleozoic
SA Arizona
SA Atokan
SA Desmoinesian

Manatee County
Valid through 1988. Search in combination with state term. After 1988, use specific county-state term.

Manatee County Florida (1989)
On Gulf of Mexico and Tampa Bay in SW Florida. Before 1989, also search Manatee County AND Florida.
CO N271500N273500 W0820500W0824500
BT Florida
BT United States

Manawatu River valley (1978)
SW North Island. Also search Manawatu.
IN Also index North Island.
BT New Zealand
BT Australasia
SA North Island

Manche
No longer a valid term for GeoRef. See Manche France.

Manche France (1993)
Department on the English Channel in NW France.
CO N483000N494500 W0004500W0020000
BT France
BT Western Europe
BT Europe
NT Cotentin Peninsula

SA Bay of Saint-Michel
SA Normandy

Manchester
No longer a valid term for GeoRef. As of 1993, see Manchester England. As of 1989, for the city in New Hampshire, use Manchester New Hampshire.

Manchester England (1993)
City in Greater Manchester in NW England. Before 1993, also search Manchester in conjunction with England.
BT England
BT Great Britain
BT United Kingdom
BT Western Europe
BT Europe
SA Lancashire England

Manchester New Hampshire (1989)
City in S New Hampshire. Before 1989, search Manchester AND New Hampshire.
CO N425900N425900 W0712800W0712800
BT Hillsborough County New Hampshire
BT New Hampshire
BT United States

Manchukuo
No longer a valid GeoRef index term. Before 1960, was used on level 1 in subfile E. See Manchuria.

Manchuria
As of 1993, no longer a valid term for GeoRef. See individual provinces: Heilongjiang China, Jilin China, Liaoning China. Autoposting of this term began in 1985 and ended in 1992. Before 1960, also search Manchukuo.

Mancos Shale (1978)
Lower and Upper Cretaceous. In Colorado, it is restricted to thick succession of shale, sandy shale, and thin-bedded sandstone overlying Niobrara Formation and underlying Mesaverde Group. NE Arizona, W Colorado, NW New Mexico, E Utah, and S and central Wyoming.
BT Cretaceous
BT Mesozoic
SA Arizona
SA Colorado
SA Ferron Sandstone Member
SA Frontier Formation
SA Mesaverde Group
SA New Mexico
SA Niobrara Formation
SA Utah
SA Wyoming

Mandi
No longer a valid term for GeoRef. See Mandi India.

Mandi India (1993)
City and district in N India.
BT Himachal Pradesh India
BT India
BT Indian Peninsula
BT Asia

Mandibulata (1981)
BT Arthropoda
BT Invertebrata
NT Crustacea
NT Insecta
NT Myriapoda

Mangala Valles (1989)

Valley on Mars.
UF Mangala Vallis
BT Mars

Mangala Vallis
use Mangala Valles

Mangalia
No longer a valid term for GeoRef. As of 1993 see Mangalia Romania.

Mangalia Romania (1993)
Town on the Black Sea in Constanta County. Before 1993 also search Mangalia AND Romania.
BT Constanta Romania
BT Romania
BT Southern Europe
BT Europe
SA Romanian Dobruja

Mangalore
No longer a valid term for GeoRef. See Mangalore India.

Mangalore India (1993)
City in SW Karnataka, SW India.
BT Karnataka India
BT India
BT Indian Peninsula
BT Asia

manganblende
use alabandite

manganese (1978)
Chemical element. As of 1981, use manganese ores for manganese as a commodity. Autoposting of metals to this term began in 1989.
UF Mn
BT metals
NT Mn-53
NT Mn-54
SA ferromanganese composition
SA heavy metals
SA lithophile elements
SA manganese composition
SA manganese minerals
SA manganese ores
SA manganese oxides

manganese composition (1985)
SA chemical composition
SA composition
SA ferromanganese composition
SA manganese
SA nodules

manganese minerals (1985)
SA alabandite
SA cryptomelane
SA franklinite
SA hausmannite
SA hollandite
SA jacobsite
SA manganese
SA manganese ores
SA manganite
SA manganosite
SA minerals
SA piemontite
SA psilomelane
SA pyrolusite
SA rhodochrosite
SA rhodonite
SA tephroite
SA todorokite
SA vernadite

Manganese Nodule Project
use MANOP

manganese nodules
No longer a valid GeoRef index term. Before 1972, was used on level 2 in subfile G.

manganese ores (1981)

Before 1981, also search (manganese OR Mn) AND (deposit OR deposits OR ore OR ores OR economic) in the basic index. Autoposting of metal ores to this term began in 1985.
IN Commodity. See List C.
BT metal ores
SA braunite
SA manganese
SA manganese minerals
SA MANOP
SA rhodochrosite
SA rhodonite

manganese oxides (1978)
BT oxides
SA asbolite
SA buserite
SA manganese
SA psilomelane

manganite (1978)
BT oxides
SA manganese minerals

manganosite (1978)
BT oxides
SA manganese minerals

mangerite (1978)
BT diorites
BT plutonic rocks
BT igneous rocks

Mangishlak Peninsula
use Mangyshlak Peninsula

Mangla Dam (1978)
SE of Rawalpindi in N Pakistan. Also search Mangla.
SA Punjab Pakistan

mangrove swamps (1978)
Autoposting of swamps and shore features to this term began in 1989. This term has multiple hierarchies.
BT1 swamps
BT2 shore features
SA ecology
SA geomorphology

Mangyshlak Peninsula (1978)
On E coast of N Caspian Sea. Also search Mangyshlak. This term has multiple hierarchies.
CO N423000N450000
   E0543000E0500000
UF Mangishlak Peninsula
BT1 Kazakhstan
BT1 Central Asia
BT1 Asia
BT2 Kazakhstan
BT2 Commonwealth of Independent States

Manhattan (1978)
Island and borough of New York City at N end of New York Bay. Coextensive with New York County, SE New York. Search in combination with state.
IN Index New York City New York and New York County New York as applicable.
BT New York City New York
BT New York
BT United States

Manhattan Formation (1978)
Dominantly a garnetiferous quartz-biotite-plagioclase gneiss characterized by sillimanite and locally much muscovite. SE New York, and W Connecticut.
BT Paleozoic
SA Connecticut
SA New York
SA New York City Group

Manhattan Kansas (1989)
City in NE central Kansas. Before 1989, search Manhattan AND Kansas.
CO N391100N391100
   W0963500W0963500
BT Riley County Kansas
BT Kansas
BT United States

Manicouagan Crater (1981)
Shock-metamorphic structure in E central Quebec. Before 1981, also search Manicouagan cryptoexplosion structure.
BT Quebec
BT Eastern Canada
BT Canada
SA cryptoexplosion features
SA Manicouagan Lake

Manicouagan Lake (1981)
Lake and reservoir on the Manicouagan River.
BT Quebec
BT Eastern Canada
BT Canada
SA Manicouagan Crater

Manicouagan River (1981)
E central Quebec.
BT Quebec
BT Eastern Canada
BT Canada

Manihiki Plateau (1978)
Undersea feature in the N Cook Islands area E of the Samoa Islands.
BT Pacific Ocean
SA DSDP Site 317

Manila Trench (1978)
Just W of Luzon.
BT South China Sea
BT West Pacific
BT Pacific Ocean

Manildra
No longer a valid term for GeoRef. See Manildra Australia.

Manildra Australia (1993)
Village in E central New South Wales.
BT New South Wales Australia
BT Australia
BT Australasia

Manipur India (1993)
State in Northeastern India.
CO N234500N260000
   E0953000E0925500
BT Northeastern India
BT India
BT Indian Peninsula
BT Asia

Manistee County
Valid through 1988. Search in combination with state term. After 1988, use specific county-state term.

Manistee County Michigan (1989)
NW Michigan. Before 1989, also search Manistee County AND Michigan.
CO N441000N443000
   W0855000W0862500
BT Michigan Lower Peninsula
BT Michigan
BT United States

Manitoba (1978)
CO N490000N600000
   W0893000W1020000
BT Western Canada
BT Canada
NT Beresford Lake
NT Bernic Lake
NT Bird River Manitoba
NT Brandon Manitoba
NT Churchill Manitoba
NT Flin Flon Manitoba
NT Fort Churchill
NT Knee Lake
NT Lac du Bonnet
NT Lac du Bonnet Batholith
NT Lac du Bonnet Manitoba
NT Lake Winnipeg
NT Lynn Lake Manitoba
NT Nelson River
NT Nelson River basin
NT Riding Mountain National Park
NT Setting Lake
NT Snow Lake Manitoba
NT Steinbach Manitoba
NT Tanco Pegmatite
NT Thompson nickel belt
NT Winnipeg Manitoba
SA Amisk Group
SA Assiniboine River
SA Assiniboine River valley
SA Bakken Formation
SA Canadian Shield
SA Churchill Province
SA Cross Lake
SA Duperow Formation
SA Elk Point Basin
SA Elk Point Group
SA English River Belt
SA Hayes River Group
SA Hudson Bay Lowlands
SA Interlake Formation
SA Kisseynew Complex
SA Lake Agassiz
SA Lake of the Woods
SA Lake of the Woods region
SA Missi Group
SA Nokomis Group
SA Prairie Evaporite
SA Red River
SA Red River Formation
SA Red River valley
SA Reindeer Lake
SA Rice Lake
SA Rice Lake Group
SA San Antonio Mine
SA Saskatchewan River
SA Sherridon Group
SA Sullivan Mine
SA Superior Province
SA Wabigoon Belt
SA Wasekwan Group
SA Williston Basin
SA Winnipeg Formation
SA Winnipegosis Formation

Manitou Formation (1978)
Consists of finely to coarsely crystalline limestone and minor amounts of dolomite. E Colorado.
BT Lower Ordovician
BT Ordovician
BT Paleozoic
SA Colorado

Manitoulin District Ontario (1993)
Before 1993, also search Manitoulin District AND Ontario.
BT Ontario
BT Eastern Canada
BT Canada
NT Manitoulin Island

Manitoulin Island (1978)
Largest lake island in world. In northern Lake Huron at NW end of Georgian Bay. Also search Manitoulin.
UF Grand Manitoulin
BT Manitoulin District Ontario
BT Ontario
BT Eastern Canada
BT Canada
SA Niagara Escarpment

Manitouwadge
No longer a valid term for GeoRef. As of 1993, see Manitouwadge Ontario.

Manitouwadge Ontario (1993)
Town N of Lake Superior in central Ontario. Before 1993, also search Manitouwadge AND Ontario.
BT Thunder Bay District Ontario
BT Ontario
BT Eastern Canada
BT Canada

Manlius Formation (1978)
According to the latest published Lexique, the age of Manlius Limestone (in Helderberg Group) is Lower Devonian.
BT Lower Devonian
BT Devonian
BT Paleozoic
SA Helderberg Group
SA New York

Manning Canyon Shale (1985)
Upper Mississippian and Lower Pennsylvanian. N Utah.
IN Index ages as applicable.
BT Carboniferous
BT Paleozoic
SA Lower Pennsylvanian
SA Upper Mississippian
SA Utah

Manning Park (1978)
Provincial park in Cascade Mountains E of Vancouver.
BT British Columbia
BT Western Canada
BT Canada

Mannville Formation
As of 1989, no longer a valid term for GeoRef.
use Mannville Group

Mannville Group (1989)
Before 1989, also search Mannville Formation. Includes Clearwater Formation.
UF Mannville Formation
UF Manville Formation
BT Lower Cretaceous
BT Cretaceous
BT Mesozoic
SA Alberta
SA Clearwater Formation
SA McMurray Formation
SA Montana
SA Saskatchewan

MANOP (1985)
Acronym.
UF Manganese Nodule Project
SA manganese ores
SA marine geology
SA mineral exploration
SA nodules
SA ocean floors
SA oceanography

manpower (1981)
SA economics

Mansalay Formation (1978)
Composed of hard calcareous mudstones, and calcareous and siliceous carbonaceous sandstone. S Mindoro Island.
BT Jurassic
BT Mesozoic
SA Mindoro
SA Philippine Islands

Mansehra
No longer a valid term for GeoRef.
See Mansehra Pakistan.

Mansehra Pakistan (1993)
Village in N Pakistan.
BT Punjab Pakistan
BT Pakistan
BT Indian Peninsula
BT Asia

Mansfeld Syncline (1978)
Also search Mansfeld.
BT Saxony-Anhalt Germany
BT Germany
BT Central Europe
BT Europe

Mansfield Formation (1978)
Lower and Middle Pennsylvanian. Contains large amounts of shale, thin beds of coal under clay, and limestone, and is only locally predominantly sandstone.
BT Pennsylvanian
BT Carboniferous
BT Paleozoic
SA Indiana

mantle (1978)
Used for the zone of Earth below the crust and above the core (to a depth of 3480 km). Autoposting of this term began in 1978.
NT lower mantle
NT upper mantle
SA asthenosphere
SA continental drift
SA convection
SA convection currents
SA core
SA core-mantle boundary
SA crust
SA degassing
SA discontinuities
SA Earth
SA earthquakes
SA elastic waves
SA geosynclines
SA geothermal gradient
SA heat flow
SA heat sources
SA horizontal discontinuities
SA hot spots
SA isostasy
SA lateral heterogeneity
SA lithosphere
SA low-velocity zones
SA lunar mantle
SA Mohorovicic discontinuity
SA partial melting
SA phase transitions
SA plate tectonics
SA plates
SA sea-floor spreading
SA seismology
SA tangential discontinuities
SA tectonophysics
SA transition zones
SA velocity structure
SA vertical discontinuities

mantle-core boundary
use core-mantle boundary

mantos (1989)
SA mineral deposits, genesis
SA stratabound deposits

manuals (1978)
UF handbooks
SA guidebook
SA monographs
SA publications
SA textbooks

Manville Formation
use Mannville Group

Manzano Mountains (1985)
Central New Mexico. Autoposting of broader terms to this term began in 1989.
IN Index counties as applicable.
CO N343000N352000
W1061500W1063500
BT New Mexico
BT United States
SA Basin and Range Province
SA Bernalillo County New Mexico
SA Torrance County New Mexico
SA Valencia County New Mexico

Maokou Formation (1989)
BT Permian
BT Paleozoic
SA China

map scales
use cartographic scales

Mapimi
No longer a valid term for GeoRef.
As of 1993 see Mapimi Mexico.

Mapimi Mexico (1993)
Town in NE Durango, W central Mexico. Before 1993 also search Mapimi AND Mexico.
BT Durango Mexico
BT Mexico

mapping (1993)
As of 1993, use for the process of gathering data for the production of a specific map. Before 1993, also search maps or specific map type and the geographic area of interest. A valid term through 1977.
SA cartography
SA global maps
SA maps

maps (1978)
In general, used only for maps larger than a standard manuscript page. Before 1993, also used for the process of making maps, and for global maps. Autoposting of this term began in 1978.
NT aeromagnetic maps
NT base maps
NT bathymetric maps
NT biogeographic maps
NT cave maps
NT climatologic maps
NT coal deposit maps
NT contour maps
NT economic geology maps
NT electromagnetic survey maps
NT engineering geology maps
NT environmental geology maps
NT geochemical maps
NT geologic hazards maps
NT geologic maps
NT geomorphologic maps
NT geophysical maps
NT geophysical survey maps
NT geotectonic maps
NT glacial geology maps
NT global maps
NT gravity survey maps
NT hydraulic maps
NT hydrogeologic maps
NT hydrographic maps
NT hydrologic maps
NT index maps
NT industrial minerals maps
NT isobath maps
NT isograd maps
NT isopach maps
NT isopleth maps
NT isoseismic maps
NT land use maps
NT lithologic maps
NT magnetic survey maps
NT marine geology maps
NT metallogenic maps
NT paleoclimate maps
NT paleocurrent maps
NT paleogeographic maps
NT petroleum maps
NT photogeologic maps
NT physiographic maps
NT prediction maps
NT radioactivity survey maps
NT seismic survey maps
NT seismicity maps
NT seismotectonic maps
NT shaded relief maps
NT site location maps
NT soils maps
NT stratigraphic maps
NT structural maps
NT structure contour maps
NT surficial geology maps
NT tectonic maps
NT topographic maps
SA aerial photography
SA areal geology
SA atlas
SA cartographic scales
SA cartography
SA catalogs
SA coordinates
SA digital cartography
SA engineering geology
SA environmental geology
SA explanatory text
SA geodesy
SA geomorphology
SA geophysical methods
SA geophysical surveys
SA isograds
SA legends
SA mapping
SA mineral exploration
SA Moon
SA photogeologic methods
SA photogeology
SA publications
SA remote sensing
SA road log
SA stereographic projection
SA structural geology
SA surveys

Maputo
No longer a valid term for GeoRef.
As of 1993 see Maputo Mozambique.

Maputo Mozambique (1993)
New name for city of Lourenco Marques located on Delagoa Bay in S Mozambique. Before 1993 also search Maputo AND Mozambique.
BT Mozambique
BT East Africa
BT Africa

Maquoketa Formation (1978)
In Missouri, it is typically thin laminated shale interbedded with shaly limestone members. W Illinois, E Iowa, S Minnesota, E Missouri, and SW Wisconsin.
BT Upper Ordovician
BT Ordovician
BT Paleozoic
SA Illinois
SA Iowa
SA Minnesota
SA Missouri
SA Wisconsin

Maracaibo Basin (1978)
NW part of country.
IN Index Lake Maracaibo and states as applicable.
BT Venezuela
BT South America
SA Lake Maracaibo
SA Merida Venezuela
SA Trujillo Venezuela
SA Zulia Venezuela

Maracaibo Falcon Plain (1981)
CO N081000N121500
W0693000W0730000
BT Venezuela
BT South America

Marajo (1978)
Largest island in Amazon Delta.
BT Para Brazil
BT Brazil
BT South America
SA Amazon River

Marambio Island
use Seymour Island

Maramures
No longer a valid term for GeoRef.
As of 1993 see Maramures Romania.

Maramures Romania (1993)
County in N Romania. Before 1993 also search Maramures AND Romania.
BT Transylvania
BT Romania
BT Southern Europe
BT Europe
NT Baia Mare Romania
NT Baita Romania
NT Cavnic Romania
NT Chiuzbaia
NT Herja

Maranhao
No longer a valid term for GeoRef.
See Maranhao Brazil.

Maranhao Basin (1978)
River basin in central Brazil.
SA Goias Brazil

Maranhao Brazil (1993)
State in NE Brazil.
CO S103000S010000
W0415000W0490000
BT Brazil
BT South America
NT Barreirinhas Basin
SA Brazilian Shield
SA Parnaiba Basin
SA Tocantins River region

Marathon Basin (1978)
North of Big Bend National Park in W Texas.
IN Index counties or regions as applicable.
SA Texas

Marathon County
Valid through 1988. Search in combination with state term. After 1988, use specific county-state term.

Marathon County Wisconsin (1989)
Central Wisconsin. Before 1989, also search Marathon County AND Wisconsin.
CO N444000N451000
W0891300W0902000
BT Wisconsin
BT United States

Marathon Geosyncline (1981)
W Texas.
CO N294000N320000
W1020000W1043000
BT Texas
BT United States
SA Tesnus Formation
SA Trans-Pecos

Maravillas Formation (1989)
Upper Ordovician of W Texas. Also, an Upper Cretaceous formation in Puerto Rico.
IN Index ages as applicable.
SA Puerto Rico
SA Texas
SA Upper Cretaceous
SA Upper Ordovician

marble
As of 1981, no longer a valid term for GeoRef. Use marbles or marble deposits as applicable. From 1978-1980, marbles was autoposted to this term.

Marble Canyon (1978)
Gorge along the Colorado River. Often considered the upper part of Grand Canyon. Established as National Monument in 1969. Abolished as a national monument and added to Grand Canyon national park in 1975. Also a canyon in Culberson County, Texas.
IN Index states and counties as applicable.
SA Arizona
SA Coconino County Arizona
SA Culberson County Texas
SA Grand Canyon
SA national monuments
SA Texas

marble deposits (1981)
Before 1981, search marble AND deposits.
SA building stone
SA calcite deposits
SA Carrara Marble
SA construction materials
SA dimension stone
SA dolostone deposits
SA limestone deposits
SA marbles
SA ornamental materials

Marble Falls Group (1978)
Lower and middle Pennsylvanian. Comprises Sloan Formation below and Big Saline Formation above. Central Texas.
BT Pennsylvanian
BT Carboniferous
BT Paleozoic
SA Texas

Marble Mountains (1989)
Klamath Mountains, N California.
IN Index Siskiyou County California and Klamath Mountains as applicable.
SA Klamath Mountains
SA Siskiyou County California

marbles (1978)
As of 1981, use marble deposits for marble as a commodity. Autoposting of this term to marble ended in 1981.
UF crystalline limestone
BT metamorphic rocks
NT brucite marble
NT calciphyre
NT ophicalcite
SA dolomite
SA marble deposits
SA ophite

Marburg
No longer a valid term for GeoRef. See Marburg Germany.

Marburg an der Lahn
use Marburg Germany

Marburg Germany (1993)

City on Lahn River N of Frankfurt. Before 1978, also search Marburg an der Lahn.
UF Marburg an der Lahn
BT Hesse Germany
BT Germany
BT Central Europe
BT Europe

Marca Shale Member (1978)
Of Moreno Formation. Underlies Dos Palos Shale Member; overlies Tierra Loma Shale Member. S California.
BT Upper Cretaceous
BT Cretaceous
BT Mesozoic
SA California
SA Moreno Formation

marcasite (1978)
BT sulfides
SA iron minerals

Marcellus Shale (1985)
In Hamilton Group.
BT Middle Devonian
BT Devonian
BT Paleozoic
SA Hamilton Group
SA Maryland
SA New Jersey
SA New York
SA Ohio
SA Pennsylvania
SA Virginia
SA West Virginia

Marche
As of 1981, no longer a valid term for GeoRef. Referred to the historical region in central France. Also referred to region in central Italy; now use Marches Italy for this.

Marches
No longer a valid term for GeoRef. See Marches Italy.

Marches Italy (1993)
Autonomous region in E central Italy. Before 1993, also search Marches AND Italy. Before 1978, also search The Marches.
CO N424200N435800
    E0135700E0120800
BT Italy
BT Southern Europe
BT Europe
NT Ancona Italy
NT Ascoli Piceno Italy
NT Pesaro Italy
SA San Marino

Marcy Massif (1978)
Mount Marcy. Highest peak in the Adirondack Mountains.
IN Index counties and Adirondack Mountains as applicable.
SA Adirondack Mountains
SA Essex County New York
SA New York

Mardin Formation (1978)
Cretaceous to Paleocene. Composed of clayey schist and marly sandstone. In Mardin in SE Anatolia.
IN Index ages as applicable.
SA Cretaceous
SA K-T boundary
SA Paleocene
SA Turkey

mare
use maria

Mare Crisium
use Sea of Crises

Mare Fecunditatis
use Sea of Fertility

Mare Imbrium
use Sea of Rains

Mare Nectaris
use Sea of Nectar

Mare Orientale
use Eastern Sea

mare ridges
use wrinkle ridges

Mare Serenitatis
use Sea of Serenity

Mare Spumans
use Foaming Sea

Mare Tranquillitatis
use Sea of Tranquillity

Mare Undarum
use Sea of Waves

Marfa Basin (1985)
SW Texas.
BT Texas
BT United States
SA Great Plains
SA Permian Basin

Margarita Island (1978)
In the Caribbean Sea 15 miles off NE coast.
BT Nueva Esparta Venezuela
BT Venezuela
BT South America
SA Macanao Peninsula

margarite (1978)
BT mica group
BT sheet silicates
BT silicates

margin, continental
use continental margin

marginal basins (1978)
BT basins
SA bottom features
SA continental slope
SA marginal seas
SA ocean basins
SA ocean floors
SA plate tectonics
SA subduction

marginal faults
use boundary faults

marginal seas (1978)
UF seas, marginal
SA continents
SA marginal basins
SA plate tectonics
SA subduction

marginal trench
use trenches

margins
No longer a valid term for GeoRef as of 1981. Use plate boundaries.

Marguerite Bay (1978)
Inlet on W coast of the Antarctic Peninsula between Adelaide and Alexander I Islands.
BT Antarctica

maria (1978)
Plural of mare.
UF mare
SA highlands
SA mascons
SA Moon
SA wrinkle ridges

Mariana Basin
use Mariana Trough

Mariana Islands (1978)

Group of islands 1500 miles E of Philippine Islands. The islands, not including Guam, formerly were part of the U.S. Trust Territory of the Pacific Islands. The islands excluding Guam achieved commonwealth status within the U.S. as the Northern Mariana Islands in 1986.
CO N120000N210000
    E1470000E1440000
UF Marianas
UF Marianas Islands
BT Micronesia
BT Oceania
NT Guam
NT Saipan
SA Izu-Bonin Arc
SA Pacific Ocean

Mariana Trench (1981)
E of the Mariana Islands.
CO N100000N223000
    E1490000E1440000
BT Pacific Ocean
SA DSDP Site 61
SA Leg 124E

Mariana Trough (1981)
Basin W of the Mariana Islands.
CO N120000N223000
    E1453000E1430000
UF Mariana Basin
BT Pacific Ocean
SA DSDP Site 453
SA DSDP Site 585
SA Leg 89
SA Leg 124E
SA Leg 129
SA ODP Site 776
SA ODP Site 802

Marianas
use Mariana Islands

Marianas Islands
use Mariana Islands

Marianna Limestone (1978)
In Vicksburg Group. Consists of soft chalky limestone, locally called chimney rock; some local, hard limestone; tough to hard ledges in the chimney rock; sandy glauconitic limestone; marl; calcareous sand; and lignitic clay. S Alabama, W Florida, and S Mississippi.
BT middle Oligocene
BT Oligocene
BT Paleogene
BT Tertiary
BT Cenozoic
SA Alabama
SA Florida
SA Mississippi
SA Vicksburg Group

Mariano Lake (1985)
NW New Mexico. Autoposting of broader terms to this term began in 1989.
BT McKinley County New Mexico
BT New Mexico
BT United States

Maricopa County
Valid through 1988. Search in combination with state term. After 1988, use specific county-state term.

Maricopa County Arizona (1989)
SW central Arizona. Before 1989, also search Maricopa County AND Arizona.
CO N323000N340400
    W1110400W1132000
BT Arizona

BT United States
NT Phoenix Arizona
SA Salt River valley

Marie Byrd Land (1978)
Large section E of Ross Ice Shelf and Ross Sea and extending E to Ellsworth Land. Claimed for U.S. by Richard E. Byrd in 1924.
BT Antarctica
SA Fosdick Mountains

Mariinsk
No longer a valid term for GeoRef. As of 1993, see Mariinsk Russian Federation.

Mariinsk Russian Federation (1993)
City in Kemerovo Oblast on Trans-Siberian R.R. Before 1993, also search Mariinsk.
IN Index regions as applicable.
BT Russian Federation
BT Commonwealth of Independent States
SA Kemerovo Russian Federation

Marilia
No longer a valid term for GeoRef. See Marilia Brazil.

Marilia Brazil (1993)
City in W central Sao Paulo, SE Brazil.
BT Sao Paulo Brazil
BT Brazil
BT South America

Marin County
Valid through 1988. Search in combination with state term. After 1988, use specific county-state term.

Marin County California (1989)
W California. Before 1989, also search Marin County AND California.
CO N374800N382000
 W1223000W1230000
BT California
BT United States
NT Bolinas Lagoon
NT Point Reyes
NT Tomales Bay
SA San Francisco Bay region

Marinduque
No longer a valid term for GeoRef. As of 1993 see Marinduque Philippine Islands.

Marinduque Philippine Islands (1993)
Island and province just S of Luzon. Before 1993 also search Marinduque AND Philippine Islands.
BT Philippine Islands
BT Far East
BT Asia

marine
A valid index term through 1977. After 1977, use marine environment for type of environment or marine methods for geophysical method, or use marine platforms for type of platform.

marine drilling (1985)
BT drilling
SA AMCOR
SA boreholes
SA Continental Offshore Stratigraphic Test
SA engineering geology
SA marine geology
SA marine installations

SA marine methods
SA oceanography
SA offshore

marine environment (1978)
Before 1978, also search marine.
SA aquatic environment
SA capes
SA coastlines
SA depositional environment
SA ecology
SA environment
SA Gilbert-type deltas
SA glaciomarine environment
SA hardground
SA ice shelves
SA littoral erosion
SA marine sedimentation
SA paleoecology
SA reef environment
SA sedimentation
SA spits
SA submarine environment
SA submarine springs
SA tidal flats

marine features
A valid term through 1975. Use shore features or ocean floors, or use the specific feature.

marine geology (1978)
Used for the interdisciplinary study of the ocean floor and the ocean-continent border. Before 1971, also search submarine geology. Compare with oceanography, ocean floors, and ocean basins.
UF geological oceanography
UF submarine geology
SA bathymetry
SA benthonic taxa
SA bottom features
SA continental rise
SA continental shelf
SA continental slope
SA CYANA
SA Deep Sea Drilling Project
SA environmental geology
SA epicontinental seas
SA geology
SA geophysical methods
SA Glomar Challenger
SA levees
SA MANOP
SA marine drilling
SA marine geology maps
SA nodules
SA ocean basins
SA ocean circulation
SA Ocean Drilling Program
SA ocean floors
SA ocean waves
SA oceanography
SA Orgon III
SA reefs
SA research vessels
SA Rivera Ocean Seismic Experiment
SA rock-water interface
SA sea water
SA sea-floor spreading
SA Seasat
SA sedimentary petrology
SA sedimentation
SA sedimentology
SA sediments
SA submarine canyons
SA submersibles

marine geology maps (1985)
BT maps
SA marine geology
SA oceanography
SA offshore

marine gravimeters (1981)
BT gravimeters

marine installations (1976)
UF installations, marine
NT breakwaters
NT gravity platforms
NT groins
NT jetties
NT marine platforms
NT piers
NT submarine installations
SA artificial islands
SA construction
SA design
SA engineering geology
SA explosions
SA feasibility studies
SA footings
SA foundations
SA geologic hazards
SA marine drilling
SA nuclear facilities
SA pipelines
SA rock mechanics
SA seismic response
SA shorelines
SA site exploration
SA soil mechanics
SA waste disposal
SA waterways

marine methods (1978)
Before 1978, also search marine.
SA geophysical methods
SA geophysical surveys
SA GLORIA
SA hydrophones
SA marine drilling
SA methods
SA research vessels
SA Rivera Ocean Seismic Experiment
SA Seabeam
SA sonobuoys

marine platforms (1978)
Before 1978, search platforms AND marine.
BT marine installations
SA gravity platforms
SA platforms

marine sedimentation (1981)
BT sedimentation
NT glaciomarine sedimentation
SA marine environment
SA marine sediments
SA marine transport

marine sediments (1981)
BT sediments
SA AMCOR
SA beach placers
SA fecal pellets
SA marine sedimentation
SA Orgon III

marine terraces (1982)
Before 1982, also search terraces AND shore features; terraces AND marine. Autoposting of shore features to this term began in 1989.
BT shore features
SA beach ridges
SA benches
SA cheniers
SA coastlines
SA geomorphology
SA littoral erosion
SA mud flats
SA shoals
SA shorelines
SA terraces
SA wave-cut platforms

marine transport (1978)

SA marine sedimentation
SA sediment transport
SA sedimentation
SA solution transport
SA tidal currents

Mariner IV
use Mariner 4

Mariner Program (1981)
Autoposting of this term to specific Mariner missions began in 1989.
NT Mariner 4
NT Mariner 5
NT Mariner 6
NT Mariner 7
NT Mariner 9
NT Mariner 10
NT Mariner 11
SA extraterrestrial geology
SA Mars
SA Mercury Planet
SA planetology
SA remote sensing
SA Venus

Mariner V observations
use Mariner 5

Mariner 4 (1989)
UF Mariner IV
BT Mariner Program
SA Mars
SA remote sensing

Mariner 5 (1989)
UF Mariner V observations
BT Mariner Program
SA remote sensing
SA Venus

Mariner 6 (1978)
Autoposting of Mariner Program to this term began in 1989.
BT Mariner Program
SA Mars
SA remote sensing

Mariner 7 (1978)
Autoposting of Mariner Program to this term began in 1989.
BT Mariner Program
SA Mars
SA remote sensing

Mariner 9 (1978)
Autoposting of Mariner Program to this term began in 1989.
BT Mariner Program
SA Mars
SA remote sensing

Mariner 10 (1978)
Autoposting of Mariner Program to this term began in 1989.
BT Mariner Program
SA Mars
SA Mercury Planet
SA remote sensing
SA Venus

Mariner 11 (1982)
Autoposting of Mariner Program to this term began in 1989.
BT Mariner Program
SA Mars
SA remote sensing
SA satellite methods

Marion County
Valid through 1988. Search in combination with state term. After 1988, use specific county-state term.

Marion County Alabama (1989)
NW Alabama. Before 1989, also search Marion County AND Alabama.
CO N335400N341800
 W0873700W0881300

BT Alabama
BT United States

Marion County Arkansas (1989)
N Arkansas. Before 1989, also search Marion County AND Arkansas.
CO N360300N362500
W0922500W0925400
BT Arkansas
BT United States

Marion County Florida (1989)
N central Florida. Before 1989, also search Marion County AND Florida.
CO N285800N292700
W0814200W0823200
BT Florida
BT United States

Marion County Georgia (1989)
W Georgia. Before 1989, also search Marion County AND Georgia.
CO N320700N323300
W0842100W0843800
BT Georgia
BT United States

Marion County Illinois (1989)
S central Illinois. Before 1989, also search Marion County AND Illinois.
CO N383000N385000
W0884100W0890800
BT Illinois
BT United States

Marion County Indiana (1989)
Central Indiana. Before 1989, also search Marion County AND Indiana.
CO N393800N395500
W0855600W0862000
BT Indiana
BT United States

Marion County Iowa (1989)
S central Iowa. Before 1989, also search Marion County AND Iowa.
CO N411000N413100
W0925300W0932000
BT Iowa
BT United States

Marion County Kansas (1989)
E central Kansas. Before 1989, also search Marion County AND Kansas.
CO N380500N383600
W0965000W0972400
BT Kansas
BT United States

Marion County Kentucky (1989)
Central Kentucky. Before 1989, also search Marion County AND Kentucky.
CO N372400N374300
W0850200W0853200
BT Kentucky
BT United States

Marion County Mississippi (1989)
S Mississippi. Before 1989, also search Marion County AND Mississippi.
CO N310000N312300
W0894000W0900300
BT Mississippi
BT United States

Marion County Missouri (1989)
NE Missouri. Before 1989, also search Marion County AND Missouri.
CO N393800N395700
W0911800W0915000

BT Missouri
BT United States

Marion County Ohio (1989)
Central Ohio. Before 1989, also search Marion County AND Ohio.
CO N402500N404300
W0825200W0832500
BT Ohio
BT United States

Marion County Oregon (1989)
NW Oregon. Before 1989, also search Marion County AND Oregon.
CO N444000N451700
W1214500W1231100
BT Oregon
BT United States
SA Mount Jefferson

Marion County South Carolina (1989)
E South Carolina. Before 1989, also search Marion County AND South Carolina.
CO N334200N341800
W0790800W0793500
BT South Carolina
BT United States

Marion County Tennessee (1989)
SE Tennessee. Before 1989, also search Marion County AND Tennessee.
CO N345800N352000
W0852400W0855300
BT Tennessee
BT United States

Marion County Texas (1989)
E Texas. Before 1989, also search Marion County AND Texas.
CO N323900N325200
W0940200W0944200
BT Texas
BT United States

Marion County West Virginia (1989)
N West Virginia. Before 1989, also search Marion County AND West Virginia.
CO N392300N393800
W0795700W0803000
BT West Virginia
BT United States

Marion Island (1978)
Subantarctic island 1200 miles SE of Capetown and just SW of Prince Edward Islands. Annexed by South Africa. In 1985, broader term changed from Indian Ocean to Indian Ocean Islands.
BT Indian Ocean Islands
SA Pacific Ocean

Mariposa County
Valid through 1988. Search in combination with state term. After 1988, use specific county-state term.

Mariposa County California (1989)
Central California. Before 1989, also search Mariposa County AND California.
CO N371000N375000
W1192000W1202500
BT California
BT United States
SA Yosemite National Park

Maritime Alps (1978)
S division of Western Alps. This term has multiple hierarchies.

IN Index countries as applicable.
BT1 Western Alps
BT1 Alps
BT1 Europe
BT2 Western Europe
BT2 Europe
SA Alpes-Maritimes France
SA Ligurian Alps
SA Piedmont Alps
SA Piemonte Italy
SA Provence Alps

Maritime Kray
use Primorye Russian Federation

Maritime Provinces (1978)
Often called the Maritimes. Introduced as level 1 area term in 1978. Includes New Brunswick, Nova Scotia, and Prince Edward Island. Autoposting of this term began in 1981.
IN Index provinces as applicable.
UF Maritimes
BT Eastern Canada
BT Canada
NT New Brunswick
NT Nova Scotia
NT Prince Edward Island

Maritimes
use Maritime Provinces

Maritsa River (1978)
Rises in Bulgaria and flows into Aegean Sea. Also search Maritsa.
IN Index countries as applicable.
UF Evros River
UF Meric River
BT Europe
SA Bulgaria
SA Greek Thrace
SA Turkey

Mariupol Ukraine
use Zhdanov Ukraine

Marjum Formation (1989)
W Utah.
BT Middle Cambrian
BT Cambrian
BT Paleozoic
SA Utah

marker beds (1978)
Before 1978, term also included use for key beds (stratigraphy).
IN As of 1978, term is restricted for use in seismology.
UF beds, marker
SA key beds
SA seismology

market stabilization (1981)
SA economics
SA markets

markets (1982)
SA demand
SA economics
SA market stabilization
SA supply

markings on rocks
No longer a valid GeoRef index term. Before 1945, was used on level 1 in subfiles E and N. See sedimentary structures; textures; petrofabrics.

markings, current
use current markings

Markov chain analysis (1981)
BT statistical analysis

marl (1978)
From 1978-1980, carbonate rocks and clastic rocks were autoposted to this term.
UF calcareous clay

UF marlstone
BT clastic rocks
BT sedimentary rocks
SA argillaceous texture
SA carbonate rocks
SA clay
SA loess
SA terrigenous materials

Marlborough
No longer a valid term for GeoRef. See Marlborough New Zealand if applicable.

Marlborough New Zealand (1993)
Provincial district, NE South Island, New Zealand. Search in conjunction with New Zealand.
IN Also index South Island.
BT New Zealand
BT Australasia
NT Kaikoura
SA South Island

marlstone
use marl

Marmaton Group (1978)
Includes (ascending) Fort Scott, Labette, Pawnee, Bandera, Altamont, Nowata, and Lenapan formations. Also includes Wewoka Formation. SW Iowa, E Kansas, W Missouri, and NE Oklahoma.
BT Desmoinesian
BT Middle Pennsylvanian
BT Pennsylvanian
BT Carboniferous
BT Paleozoic
SA Iowa
SA Kansas
SA Labette Shale
SA Missouri
SA Oklahoma
SA Wewoka Formation

Marmolada Glacier (1978)
At Marmolada Peak, the highest in the Dolomites.
IN Index autonomous regions as applicable.
BT Italy
BT Southern Europe
BT Europe
SA Dolomites
SA Trentino-Alto Adige Italy
SA Veneto Italy

Marne
No longer a valid term for GeoRef. See Marne France.

Marne France (1993)
Department in NE France.
CO N483000N493000
E0051000E0031500
BT France
BT Western Europe
BT Europe

Marne Valley (1978)
River valley in NE central France. Also search Marne River.
BT France
BT Western Europe
BT Europe
SA Val-de-Marne France

Maroon Formation (1989)
W central Colorado.
IN Index ages as applicable.
BT Paleozoic
SA Colorado
SA Pennsylvanian
SA Permian

Maroon Town
  No longer a valid term for GeoRef.
  See Maroon Town Jamaica.
Maroon Town Jamaica (1993)
  Town in NW Jamaica.
  BT Jamaica
  BT Greater Antilles
  BT Antilles
  BT West Indies
  BT Caribbean region
Marquesas Islands (1981)
  Group of ten islands in French Polynesia.
  CO S110000S080000
    W1380000W1420000
  BT French Polynesia
  BT Polynesia
  BT Oceania
Marquette
  Valid through 1988. Search in combination with state term. After 1988, use specific city-state term.
Marquette County
  Valid through 1988. Search in combination with state term. After 1988, use specific county-state term.
Marquette County Michigan (1989)
  N Michigan Upper Peninsula, NW Michigan. Before 1989, also search Marquette County AND Michigan.
  CO N455800N465400
    W0870700W0880700
  BT Michigan Upper Peninsula
  BT Michigan
  BT United States
  NT Marquette Michigan
  NT Marquette Range
Marquette County Wisconsin (1989)
  S central Wisconsin. Before 1989, also search Marquette County AND Wisconsin.
  CO N433800N435800
    W0891000W0893700
  BT Wisconsin
  BT United States
Marquette Iron Range
  use Marquette Range
Marquette Michigan (1989)
  City on Lake Superior in NW Michigan Upper Peninsula, N Michigan. Before 1989, search Marquette AND Michigan.
  CO N463300N463300
    W0872300W0872300
  BT Marquette County Michigan
  BT Michigan Upper Peninsula
  BT Michigan
  BT United States
Marquette Range (1978)
  Iron range in Marquette County in Upper Peninsula.
  UF Marquette Iron Range
  BT Marquette County Michigan
  BT Michigan Upper Peninsula
  BT Michigan
  BT United States
Marquette Range Supergroup (1989)
  BT lower Proterozoic
  BT Proterozoic
  BT upper Precambrian
  BT Precambrian
  SA Michigan
  SA Minnesota
  SA Ontario
  SA Wisconsin

Mars (1978)
  NT Alba Patera
  NT Arsia Mons
  NT Ascraeus Mons
  NT Chryse Planitia
  NT Elysium
  NT Elysium Mons
  NT Kasei Vallis
  NT Lunae Planum
  NT Mangala Valles
  NT Olympus Mons
  NT Tharsis
  NT Tharsis Montes
  NT Valles Marineris
  SA Amazonian
  SA Deimos Satellite
  SA Hesperian
  SA landing sites
  SA Mariner 4
  SA Mariner 6
  SA Mariner 7
  SA Mariner 9
  SA Mariner 10
  SA Mariner 11
  SA Mariner Program
  SA marsquakes
  SA Moon
  SA Noachian
  SA Northern Highlands
  SA Phobos Satellite
  SA planetology
  SA planets
  SA polar caps
  SA polar regions
  SA satellites
  SA solar system
  SA terrestrial planets
  SA Viking Program
Marseilles
  No longer a valid term for GeoRef.
  See Marseilles France.
Marseilles Basin (1981)
  Structural basin in SE France.
  CO N431000N432500
    E0054000E0050000
  BT France
  BT Western Europe
  BT Europe
Marseilles France (1993)
  City on NE shore of Gulf of Lion in SE France.
  BT Bouches-du-Rhone France
  BT France
  BT Western Europe
  BT Europe
Marshall County
  Valid through 1988. Search in combination with state term. After 1988, use specific county-state term.
Marshall County Alabama (1989)
  NE Alabama. Before 1989, also search Marshall County AND Alabama.
  CO N340800N343500
    W0860500W0863400
  BT Alabama
  BT United States
Marshall County Illinois (1989)
  N central Illinois. Before 1989, also search Marshall County AND Illinois.
  CO N405600N411000
    W0890300W0893800
  BT Illinois
  BT United States
Marshall County Indiana (1989)
  N Indiana. Before 1989, also search Marshall County AND Indiana.
  CO N410900N412800
    W0860500W0862700
  BT Indiana
  BT United States
Marshall County Iowa (1989)
  Central Iowa. Before 1989, also search Marshall County AND Iowa.
  CO N415200N421300
    W0924600W0931400
  BT Iowa
  BT United States
Marshall County Kansas (1989)
  NE Kansas. Before 1989, also search Marshall County AND Kansas.
  CO N393300N400000
    W0961400W0964800
  BT Kansas
  BT United States
Marshall County Kentucky (1989)
  SW Kentucky. Before 1989, also search Marshall County AND Kentucky.
  CO N364500N370500
    W0880900W0882900
  BT Kentucky
  BT United States
Marshall County Minnesota (1989)
  NW Minnesota. Before 1989, also search Marshall County AND Minnesota.
  CO N481200N483200
    W0953400W0971000
  BT Minnesota
  BT United States
Marshall County Mississippi (1989)
  N Mississippi. Before 1989, also search Marshall County AND Mississippi.
  CO N343000N345900
    W0891400W0894400
  BT Mississippi
  BT United States
Marshall County Oklahoma (1989)
  S Oklahoma. Before 1989, also search Marshall County AND Oklahoma.
  CO N335000N341000
    W0963500W0965800
  BT Oklahoma
  BT United States
Marshall County South Dakota (1989)
  NE South Dakota. Before 1989, also search Marshall County AND South Dakota.
  CO N453400N455600
    W0971500W0975800
  BT South Dakota
  BT United States
Marshall County Tennessee (1989)
  Central Tennessee. Before 1989, also search Marshall County AND Tennessee.
  CO N351700N354200
    W0863400W0865700
  BT Tennessee
  BT United States
Marshall County West Virginia (1989)
  N West Virginia. Before 1989, also search Marshall County AND West Virginia.
  CO N394300N400300
    W0803200W0805300
  BT West Virginia
  BT United States
Marshall Islands (1978)
  Group of 34 atolls and coral islands, including Enewetak Atoll and Kwajalein, SW of Hawaii and E of Guam. Part of the U.S. Trust Territory of the Pacific Islands.
  CO N043000N150000
    E1730000E1610000
  BT Micronesia
  BT Oceania
  NT Enewetak Atoll
  SA Pacific Ocean
Marshall Lake (1985)
  N Ontario.
  CO N502500N503000
    W0872500W0873500
  BT Thunder Bay District Ontario
  BT Ontario
  BT Eastern Canada
  BT Canada
Marshalltown Formation (1978)
  In Matawan Group. Overlies Englishtown Formation; underlies Wenonah Sand.
  BT Upper Cretaceous
  BT Cretaceous
  BT Mesozoic
  SA Delaware
  SA New Jersey
marshes (1978)
  As of 1981, use paludal environment for type of environment.
  BT shore features
  NT salt marshes
  SA bogs
  SA fluvial features
  SA geomorphology
  SA lacustrine features
  SA paludal environment
  SA swamps
  SA wetlands
Marsica (1978)
  Mountain range in E central Italy.
  BT Abruzzi Italy
  BT Italy
  BT Southern Europe
  BT Europe
marsquakes (1981)
  SA earthquakes
  SA Mars
Marsupialia (1978)
  Order. Autoposting of this term to Diprotodontia and Polyprotodontia began in 1989. Autoposting of Metatheria and Theria to this term began in 1989.
  BT Metatheria
  BT Theria
  BT Mammalia
  BT Tetrapoda
  BT Vertebrata
  BT Chordata
  NT Diprotodonta
  NT Polyprotodontia
Martha's Vineyard (1978)
  Island in Atlantic Ocean off SW coast of Cape Cod. In Dukes County.
  BT Massachusetts
  BT United States
Martin County
  Valid through 1988. Search in combination with state term. After 1988, use specific county-state term.
Martin County Florida (1989)

SE Florida. Before 1989, also search Martin County AND Florida.
CO N265800N271400
W0800300W0804000
BT Florida
BT United States

Martin County Indiana (1989)
SW Indiana. Before 1989, also search Martin County AND Indiana.
CO N383000N385400
W0864200W0865500
BT Indiana
BT United States

Martin County Kentucky (1989)
E Kentucky. Before 1989, also search Martin County AND Kentucky.
CO N374100N375700
W0822100W0824100
BT Kentucky
BT United States

Martin County Minnesota (1989)
S Minnesota. Before 1989, also search Martin County AND Minnesota.
CO N433100N435100
W0941700W0945200
BT Minnesota
BT United States

Martin County North Carolina (1989)
E North Carolina. Before 1989, also search Martin County AND North Carolina.
CO N354000N360500
W0764700W0772400
BT North Carolina
BT United States

Martin County Texas (1989)
W Texas. Before 1989, also search Martin County AND Texas.
CO N320500N323300
W1014200W1021100
BT Texas
BT United States

Martin Formation (1978)
Composed of three members (ascending): conglomerate sandstone and dolomite limestone member; sandstone and limestone member; and calcareous sandstone, sandy limestone and shale member. Central and E Arizona.
BT Devonian
BT Paleozoic
SA Arizona
SA Middle Devonian
SA Upper Devonian

Martin Lake (1978)
Formed by dam across Tallapoosa River in E central Alabama.
IN Index counties or regions as applicable.
SA Alabama

Martin River Glacier (1978)
NE of Cordova in S Alaska.
BT Southern Alaska
BT Alaska
BT United States
SA Chugach Mountains

Martinique (1978)
Island and department of France in Windward Islands. Autoposting of this term began in 1978.
BT Lesser Antilles
BT Antilles
BT West Indies

BT Caribbean region
NT Mount Pelee
SA Windward Islands

Martinsburg Formation (1978)
According to the latest published Lexique, the age of Martinsburg Formation is Middle and Upper Ordovician. In Pennsylvania, comprises (ascending) Cocalio Shale, unnamed and undifferentiated dark shales, Jonestown Beds (new), and Fairview Shochary Sandstones. Maryland, New Jersey, SE Pennsylvania, Tennessee, W Virginia, and West Virginia.
BT Ordovician
BT Paleozoic
SA Maryland
SA New Jersey
SA Pennsylvania
SA Virginia
SA West Virginia

martite (1978)
BT oxides
SA hematite
SA iron ores
SA iron oxides

Mary Lee Coal (1993)
Pennsylvanian. Black Warrior Basin, Alabama. In Pottsville Group.
BT Pennsylvanian
BT Carboniferous
BT Paleozoic
SA Alabama
SA Black Warrior Basin
SA Pottsville Group

Maryborough Basin (1978)
SE Queensland. Also search Maryborough.
IN Index state or region as applicable.
SA Queensland Australia

Maryland (1978)
Autoposting of this term began in 1978.
CO N375500N394300
W0750500W0793000
BT United States
NT Allegany County Maryland
NT Anne Arundel County Maryland
NT Baltimore County Maryland
NT Calvert County Maryland
NT Carroll County Maryland
NT Charles County Maryland
NT Dorchester County Maryland
NT Frederick County Maryland
NT Garrett County Maryland
NT Harford County Maryland
NT Howard County Maryland
NT Kent County Maryland
NT Montgomery County Maryland
NT Patuxent River
NT Prince Georges County Maryland
NT Saint Marys County Maryland
NT Somerset County Maryland
NT Talbot County Maryland
NT Washington County Maryland
NT Worcester County Maryland
SA Allegheny Front
SA Allegheny Group
SA Allegheny Mountains
SA Ames Limestone
SA Anacostia River basin
SA Antietam Formation
SA Appalachian Basin
SA Appalachian Plateau
SA Aquia Formation
SA Atlantic Coastal Plain
SA Baltimore Gneiss

SA Beekmantown Group
SA Bloomsburg Formation
SA Blue Ridge Province
SA Calvert Formation
SA Carolina Bays
SA Catoctin Formation
SA Catskill Delta
SA Catskill Formation
SA Chemung Formation
SA Chesapeake Bay
SA Chesapeake Group
SA Chilhowee Group
SA Choptank Formation
SA Choptank River
SA Clinton Group
SA Coeymans Formation
SA Cohansey Formation
SA Conemaugh Group
SA Conococheague Formation
SA Culpeper Basin
SA Delmarva Peninsula
SA Dunkard Group
SA Eastover Formation
SA Genesee Group
SA Gettysburg Basin
SA Glenarm Series
SA Greenbrier Limestone
SA Greenland Gap Group
SA Hamilton Group
SA Hampshire Formation
SA Helderberg Group
SA Juniata Formation
SA Keefer Sandstone
SA Keyser Limestone
SA Kinzers Formation
SA Kittanning Formation
SA Magothy Aquifer
SA Magothy Formation
SA Marcellus Shale
SA Martinsburg Formation
SA Mauch Chunk Formation
SA McKenzie Formation
SA Millboro Shale
SA Monmouth Group
SA Monongahela Group
SA Nanjemoy Formation
SA Newark Supergroup
SA Newark-Gettysburg Basin
SA Onondaga Limestone
SA Oriskany Sandstone
SA Piedmont
SA Piney Point Formation
SA Pocono Formation
SA Potomac Group
SA Potomac River
SA Potomac River basin
SA Pottsville Group
SA Raritan Formation
SA Rochester Formation
SA Rock Creek Park
SA Rose Hill Formation
SA Saint Marys Formation
SA Salisbury Embayment
SA Severn Estuary
SA Severn Formation
SA Sharon Conglomerate
SA South Mountain
SA Stones River Group
SA Susquehanna River
SA Susquehanna River basin
SA Tonoloway Limestone
SA Tuscarora Formation
SA Valley and Ridge Province
SA Wicomico Formation
SA Williamsport Sandstone
SA Wilmington Complex
SA Wissahickon Formation
SA Yorktown Formation

Marysvale
Valid through 1988. Search in combination with state term. After 1988, use specific city-state term.

Marysvale Utah (1989)

City in SE Utah. Before 1989, also search Marysvale AND Utah.
CO N382800N382800
W1121400W1121400
BT Piute County Utah
BT Utah
BT United States

Masaya (1989)
Volcano. S Nicaragua.
UF Masaya Volcano
BT Nicaragua
BT Central America

Masaya Volcano
use Masaya

Mascarene Basin (1978)
E of N Madagascar.
BT Indian Ocean

Mascarene Islands (1978)
Group between 400 and 500 miles E of the Madagascar. In 1985, broader term changed from Indian Ocean to Indian Ocean Islands. Before 1981, Piton de la Fournaise, Piton des Neiges and Saint-Louis were narrower terms. As of 1981, Rodriguez Island is a narrower term. After 1992, this term is autoposted to Reunion.
IN Index islands as applicable.
BT Indian Ocean Islands
NT Reunion
NT Rodriguez Island
SA Mauritius

mascons (1978)
SA maria
SA Moon

Mascot Dolomite (1978)
In Knox Group. Consists of light- and dark-gray dolomite and limestone; moderately cherty; base marked by chert matrix sandstone. E Tennessee and Virginia.
IN Index states as applicable.
BT Lower Ordovician
BT Ordovician
BT Paleozoic
SA Knox Group
SA Tennessee
SA Virginia

maskelynite (1978)
Meteorite mineral. From 1978-1980, feldspar group, framework silicates and silicates were autoposted to this term. As of 1989, aluminosilicates and silicates are autoposted to this term.
BT aluminosilicates
BT silicates
SA glasses
SA meteorites
SA plagioclase

Mason County
Valid through 1988. Search in combination with state term. After 1988, use specific county-state term.

Mason County Illinois (1989)
Central Illinois. Before 1989, also search Mason County AND Illinois.
CO N400400N402800
W0893700W0902200
BT Illinois
BT United States

Mason County Kentucky (1989)
NE Kentucky. Before 1989, also search Mason County AND Kentucky.
CO N382800N384500
W0833700W0835900

BT Kentucky
BT United States

Mason County Michigan (1989)
W Michigan. Before 1989, also search Mason County AND Michigan.
CO N434800N441100
W0860300W0863000
BT Michigan Lower Peninsula
BT Michigan
BT United States

Mason County Texas (1989)
Central Texas. Before 1989, also search Mason County AND Texas.
CO N303000N305400
W0985900W0994500
BT Texas
BT United States

Mason County Washington (1989)
W Washington. Before 1989, also search Mason County AND Washington.
CO N470500N473600
W1224700W1233000
BT Washington
BT United States
SA Olympic Mountains

Mason County West Virginia (1989)
W West Virginia. Before 1989, also search Mason County AND West Virginia.
CO N382800N390300
W0814700W0821300
BT West Virginia
BT United States

mass (1978)
SA Earth
SA mass balance
SA mass spectra
SA mass spectroscopy
SA specific surface

mass balance (1978)
UF balance, mass
UF mass budget
SA ablation
SA glacial geology
SA glaciers
SA mass
SA water balance

mass budget
use mass balance

mass extinctions (1985)
SA biochronology
SA catastrophes
SA extinction
SA paleontology

mass movements (1978)
Before 1981, also search mass wasting. Before 1971, also search mass wastage. Autoposting of this term began in 1978.
UF mass wastage
UF mass wasting
UF movements, mass
NT avalanches
NT creep
NT debris avalanches
NT debris flows
NT earthflows
NT lahars
NT landslides
NT mudflows
NT rockfalls
NT rockslides
SA erosion
SA flows
SA geomorphology

SA gravity flows
SA liquefaction
SA slope stability
SA slumping
SA talus slopes

mass spectra (1981)
Used for data. For methodology, use mass spectroscopy.
BT spectra
SA mass
SA mass spectroscopy
SA spectroscopy

mass spectrometry
A valid term through 1973.
use mass spectroscopy

mass spectroscopy (1978)
Before 1978, also search mass spectrometry; spectroscopy AND mass.
UF mass spectrometry
BT spectroscopy
NT accelerator mass spectroscopy
SA chemical analysis
SA inductively coupled plasma methods
SA mass
SA mass spectra
SA secondary ion mass spectroscopy

mass transfer (1978)
As of 1981, use only in the chemical sense. Before 1981, included use as a general term.
UF transfer, mass
SA geochemistry
SA hydrochemistry

mass transport
Not a valid term for GeoRef. As of 1989, see mass movements; sedimentation; transport.

mass wastage
Not a valid GeoRef index term after 1970. Was used on level 1 in subfile N.
use mass movements

mass wasting
As of 1981, no longer a valid term for GeoRef.
use mass movements

Massachusetts (1978)
Autoposting of this term began in 1978.
CO N411500N425500
W0695500W0733000
BT United States
NT Barnstable County Massachusetts
NT Berkshire County Massachusetts
NT Buzzards Bay
NT Essex County Massachusetts
NT Franklin County Massachusetts
NT Hampden County Massachusetts
NT Hampshire County Massachusetts
NT Martha's Vineyard
NT Middlesex County Massachusetts
NT Nantucket Island
NT Norfolk County Massachusetts
NT Plymouth County Massachusetts
NT Suffolk County Massachusetts
NT Wachusett-Marlborough Tunnel
NT Worcester County Massachusetts
SA Ammonoosuc Volcanics

SA Atlantic Coastal Plain
SA Avalon Terrane
SA Blackstone Series
SA Bloody Bluff Fault
SA Boston Bay Group
SA Bronson Hill Anticlinorium
SA Cambridge Argillite
SA Chelmsford Granite
SA Cheshire Formation
SA Clough Formation
SA Connecticut River
SA Connecticut Valley
SA Dedham Granodiorite
SA East Berlin Formation
SA Green Mountains
SA Hampden Basalt
SA Hartford Basin
SA Holyoke Basalt
SA Hoosac Formation
SA Lake Char Fault
SA Littleton Formation
SA Madrid Formation
SA Merrimack Group
SA Merrimack River valley
SA Merrimack Synclinorium
SA Narragansett Basin
SA Normanskill Formation
SA Partridge Formation
SA Plum Island
SA Portland Formation
SA Rangeley Formation
SA Roxbury Conglomerate
SA Waits River Formation
SA Woods Hole Oceanographic Institution

Massiac
No longer a valid term for GeoRef. See Massiac France.

Massiac France (1993)
Village in S central France.
BT Cantal France
BT France
BT Western Europe
BT Europe

Massif Central
use Central Massif

massifs
As of 1981, no longer a valid term for GeoRef. Use complexes.

massive
A valid index term through 1977. After 1977, use massive bedding for the sedimentary structure, and use massive deposits for ore deposits.

massive bedding (1978)
Before 1978, also search massive AND bedding.
BT planar bedding structures
BT sedimentary structures
SA bedding

massive deposits (1978)
Before 1978, also search massive AND ore deposits; massive AND deposits.
SA cyprus-type
SA kuroko-type
SA mineral deposits, genesis

massive hydraulic fracturing (1981)
SA hydraulic fracturing

Mastodon (1978)
Genus. Autoposting of Eutheria and Theria to this term began in 1989.
UF Mammut
BT Mastodontidae
BT Mastodontoidea
BT Proboscidea
BT Eutheria

BT Theria
BT Mammalia
BT Tetrapoda
BT Vertebrata
BT Chordata

Mastodontidae (1978)
Family. Autoposting of Eutheria and Theria to this term began in 1989.
UF Mammutidae
BT Mastodontoidea
BT Proboscidea
BT Eutheria
BT Theria
BT Mammalia
BT Tetrapoda
BT Vertebrata
BT Chordata
NT Mastodon

Mastodontoidea (1981)
Autoposting of Eutheria and Theria to this term began in 1989.
BT Proboscidea
BT Eutheria
BT Theria
BT Mammalia
BT Tetrapoda
BT Vertebrata
BT Chordata
NT Mastodontidae

Matadi
No longer a valid term for GeoRef. As of 1993 see Matadi Zaire.

Matadi Zaire (1993)
Port city on Congo River in W Zaire. Bas-Zaire.
BT Lower Zaire
BT Zaire
BT Central Africa
BT Africa

Matagami (1978)
Lake in Abitibi Territory SE of James Bay.
UF Mattagami
UF Lake Matagami
BT Quebec
BT Eastern Canada
BT Canada

Matagamon Sandstone (1978)
BT Lower Devonian
BT Devonian
BT Paleozoic
SA Maine

Matagorda Bay (1978)
Inlet of the Gulf of Mexico in SE Texas.
IN Index counties as applicable.
BT Texas
BT United States
SA Calhoun County Texas
SA Gulf of Mexico

Matanuska Valley (1978)
River valley NE of Anchorage in S Alaska. Also search Matanuska River.
BT Southern Alaska
BT Alaska
BT United States

Matapedia Group (1989)
E Maine, N New Brunswick and Quebec.
IN Index ages as applicable.
BT Paleozoic
SA Maine
SA New Brunswick
SA Quebec
SA Upper Ordovician

Matera
No longer a valid term for GeoRef. See Matera Italy.

Matera Italy (1993)
City and province in S Italy.
BT Basilicata Italy
BT Italy
BT Southern Europe
BT Europe

materials (1978)
Before 1981, also search media.
UF media
SA absorbent materials
SA amorphous materials
SA anelastic materials
SA anisotropic materials
SA backfill
SA banded materials
SA brittle materials
SA Carrara Marble
SA cement materials
SA ceramic materials
SA coarse-grained materials
SA cohesionless materials
SA cohesive materials
SA colloidal materials
SA competent materials
SA conductive materials
SA consistent materials
SA construction materials
SA continuous materials
SA diagenesis
SA discontinuous materials
SA dissolved materials
SA education
SA elastic materials
SA elastoplastic materials
SA elastoviscoplastic materials
SA engineering geology
SA expansive materials
SA fines
SA fossiliferous materials
SA fractured materials
SA geomembranes
SA geotextiles
SA glass materials
SA granular materials
SA heterogeneous materials
SA inorganic materials
SA insulation materials
SA interlaboratory comparison
SA isotropic materials
SA layered materials
SA linear materials
SA natural materials
SA non-linear materials
SA organic materials
SA ornamental materials
SA Ottawa Sand
SA overconsolidated materials
SA parent materials
SA particulate materials
SA plastic materials
SA polycrystalline materials
SA porous materials
SA properties
SA raw materials
SA refractory materials
SA reinforced materials
SA relict materials
SA rock mechanics
SA San Carlos Olivine
SA saturated materials
SA soil mechanics
SA standard materials
SA suspended materials
SA synthetic materials
SA terrigenous materials
SA toxic materials
SA unconsolidated materials
SA viscoelastic materials
SA viscoplastic materials
SA viscous materials
SA weathered materials

mathematical geology (1978)
Used only for studies specifically discussing the intricacies of the mathematics itself, rather than equations applied to a particular subject.
SA algorithms
SA analog simulation
SA autocorrelation
SA autoregression
SA Bayesian analysis
SA biometry
SA cluster analysis
SA crosscorrelation
SA data processing
SA digital simulation
SA equations
SA Fourier analysis
SA fractals
SA functions
SA fuzzy logic
SA Galerkin method
SA geology
SA geometry
SA geophysics
SA geostatistics
SA Hilbert transformations
SA Hugoniot analysis
SA iterative methods
SA kriging
SA mathematical methods
SA mathematical models
SA mathematical transformations
SA Monte Carlo analysis
SA pattern recognition
SA probability
SA semivariograms
SA simulation
SA statistical analysis
SA time series analysis
SA topology
SA transfer functions
SA variograms

mathematical methods (1978)
Used as a general term. Before 1981, also search quantitative methods. Before 1978, also search quantitative; quantitative AND methods.
SA algorithms
SA Archie's law
SA Bayesian analysis
SA cladistics
SA crosscorrelation
SA data processing
SA eigenvalues
SA equations
SA Fourier analysis
SA fractals
SA functions
SA fuzzy logic
SA Galerkin method
SA geostatistics
SA Hilbert transformations
SA iterative methods
SA kriging
SA mathematical geology
SA mathematical models
SA mathematical transformations
SA methods
SA numerical analysis
SA probability
SA semivariograms
SA sensitivity analysis
SA splines
SA statistical analysis
SA transfer functions
SA Universal Soil Loss Equation
SA variograms

mathematical models (1978)
SA Bayesian analysis
SA boundary element analysis
SA constitutive equations
SA data processing
SA digital simulation
SA equations
SA four-dimensional models
SA Green function
SA Hugoniot analysis
SA infinite models
SA iterative methods
SA Kalman filters
SA Kirchhoff integral
SA kriging
SA mathematical geology
SA mathematical methods
SA models
SA numerical models
SA one-dimensional models
SA PHREEQE
SA Richards equation
SA semi-infinite models
SA sensitivity analysis
SA simulation
SA spherical models
SA three-dimensional models
SA transfer functions
SA two-dimensional models

mathematical transformations (1981)
SA Fourier analysis
SA Hilbert transformations
SA mathematical geology
SA mathematical methods
SA Radon transforms
SA seismology

matildite (1978)
Autoposting of sulfobismuthites to this term began in 1981.
UF plenargyrite
UF schapbachite
UF schapbacite
BT sulfobismuthites
BT sulfosalts

Mato Grosso
No longer a valid term for GeoRef. See Mato Grosso Brazil.

Mato Grosso Brazil (1993)
State in SW central Brazil.
CO S240000S072000
   W0501000W0613000
BT Brazil
BT South America
SA Botucatu Formation
SA Brazilian Shield
SA Irati Formation

Matra Mountains (1978)
S spur of the Carpathians. Also search Matra. This term has multiple hierarchies.
BT1 Hungary
BT1 Central Europe
BT1 Europe
BT2 Carpathians
BT2 Europe

matrix (1978)
The fine-grained interstitial material of an igneous rock.
UF ground mass
UF groundmass
SA igneous rocks
SA megacrysts
SA phenocrysts

Matsukawa
No longer a valid term for GeoRef. See Matsukawa Japan.

Matsukawa Japan (1993)
Town in geothermal region, N central Honshu. Before 1993, Also search Matsukawa.
BT Nagano Prefecture Japan
BT Honshu
BT Japan
BT Far East
BT Asia

Matsushiro
No longer a valid term for GeoRef. See Matsushiro Japan.

Matsushiro Japan (1993)
Town and observatory in central Honshu.
BT Honshu
BT Japan
BT Far East
BT Asia

Matsuyama
No longer a valid term for GeoRef. See Matsuyama Japan.

Matsuyama Japan (1993)
City in Ehime, W Shikoku, S Japan. Near Inland Sea.
BT Ehime Japan
BT Shikoku
BT Japan
BT Far East
BT Asia

Mattagami
use Matagami

Matto Grosso
Not a valid term for GeoRef. See Mato Grosso Brazil.

Mattoon Formation (1978)
McLeansboro Group. Predominantly shale and sandstone.
BT Pennsylvanian
BT Carboniferous
BT Paleozoic
SA Illinois

maturity (1978)
Used as a general term.
SA age
SA ontogeny
SA thermal maturity

Matuyama Epoch (1978)
Geomagnetic epoch.
UF Matuyama Reversed
BT Cenozoic
SA lower Pleistocene
SA Neogene
SA paleomagnetism
SA Quaternary
SA reversals
SA upper Pliocene
SA upper Tertiary

Matuyama Reversed
use Matuyama Epoch

Mauch Chunk Formation (1985)
Pennsylvania, W Maryland and N West Virginia.
BT Upper Mississippian
BT Mississippian
BT Carboniferous
BT Paleozoic
SA Maryland
SA Pennsylvania
SA West Virginia

maucherite (1978)
Autoposting of sulfides to this term ended in 1989.
BT arsenides

Maui (1978)
Island in the Hawaiian Islands NW of the island of Hawaii. Autoposting of broader terms to this term began in 1989. This term has multiple hierarchies.
CO N203500N210300
   W1555700W1564300
BT1 Maui County Hawaii
BT1 Hawaii
BT1 United States
BT2 Maui County Hawaii
BT2 Hawaii
BT2 Polynesia

BT2 Oceania
BT3 Maui County Hawaii
BT3 Hawaii
BT3 East Pacific Ocean Islands
SA Haleakala
SA Red Hill

Maui County Hawaii (1989)
Includes islands of Kahoolawe, Lanai, Maui and Molokai. Before 1989, search Maui County AND Hawaii. This term has multiple hierarchies.
CO N203000N211300
W1555830W1572000
BT1 Hawaii
BT1 United States
BT2 Hawaii
BT2 Polynesia
BT2 Oceania
BT3 Hawaii
BT3 East Pacific Ocean Islands
NT Haleakala
NT Maui
SA Mauna Loa
SA Molokai

Maumee River valley (1978)
NE Indiana and NW Ohio. Also search Maumee River.
IN Index states and counties as applicable.
BT United States
SA Indiana
SA Ohio

Mauna Kea (1978)
Extinct volcano on N central island of Hawaii. Autoposting of broader terms to this term began in 1989. This term has multiple hierarchies.
CO N193000N194000
W1552000W1553500
UF Mauna Kea Volcano
BT1 Hawaii Island
BT1 Hawaii County Hawaii
BT1 Hawaii
BT1 United States
BT2 Hawaii Island
BT2 Hawaii County Hawaii
BT2 Hawaii
BT2 Polynesia
BT2 Oceania
BT3 Hawaii Island
BT3 Hawaii County Hawaii
BT3 Hawaii
BT3 East Pacific Ocean Islands

Mauna Kea Volcano
use Mauna Kea

Mauna Loa (1978)
Volcano on S central island of Hawaii. Autoposting of broader terms to this term began in 1989. May also be used for mountain on Molokai. This term has multiple hierarchies.
IN Also index county as applicable.
CO N192000N193500
W1552000W1554000
UF Mauna Loa Volcano
BT1 Hawaii
BT1 United States
BT2 Hawaii
BT2 Polynesia
BT2 Oceania
BT3 Hawaii
BT3 East Pacific Ocean Islands
SA Hawaii County Hawaii
SA Hawaii Island
SA Kilauea
SA Maui County Hawaii
SA Molokai

Mauna Loa Volcano
use Mauna Loa

Maures Massif (1978)
On the Mediterranean coast at W end of the Riviera in Var. Also search Maures.
UF Monts des Maures
BT Var France
BT France
BT Western Europe
BT Europe

Maurice Ewing Bank (1985)
S Atlantic Ocean, N of South Georgia. As of 1990, Atlantic Ocean is autoposted to this term.
BT South Atlantic
BT Atlantic Ocean

Mauritania (1978)
CO N145500N273000
W0045500W0173000
BT West Africa
BT Africa
NT Adrar
NT Akjoujt Mauritania
NT Aouelloul
NT Fort Gouraud Mauritania
NT Nouakchott Mauritania
NT Richat Mountain
SA Cap Blanc
SA Mauritanides
SA Reguibat Ridge
SA Sahara
SA Sahel
SA Senegal Basin
SA Senegal River
SA Western Sahara

Mauritanide Orogen
use Mauritanides

Mauritanides (1993)
Orogen in West Africa which has undergone polyphase evolution.
IN Index countries as applicable.
UF Mauritanide Orogen
BT West Africa
BT Africa
SA Mali
SA Mauritania
SA Pan-African Orogeny
SA Senegal

Mauritius (1978)
An independent state in the Mascarene Islands about 450 miles E of the Madagascar. In 1981, broader term changed from Indian Ocean to Indian Ocean Islands.
CO S204000S195000
E0580000E0570000
BT Indian Ocean Islands
SA Mascarene Islands

Maury Channel (1978)
Sea channel E of central Reykjanes Ridge and S of Iceland.
BT Atlantic Ocean

Maury County
Valid through 1988. Search in combination with state term. After 1988, use specific county-state term.

Maury County Tennessee (1989)
Central Tennessee. Before 1989, also search Maury County AND Tennessee.
CO N352300N355000
W0864700W0872300
BT Tennessee
BT United States
NT Columbia Tennessee

Maverick Basin (1981)
Structural basin in SW Texas and NE Mexico.
IN Index North America or states as applicable.
SA Mexico
SA North America
SA Texas

Maverick County
Valid through 1988. Search in combination with state term. After 1988, use specific county-state term.

Maverick County Texas (1989)
SW Texas. Before 1989, also search Maverick County AND Texas.
CO N281000N290500
W1001000W1004500
BT Texas
BT United States

mawsonite (1978)
BT sulfides

Maxey Flats (1985)
Radioactive waste disposal site in NE Kentucky. Autoposting of broader terms to this term began in 1989.
BT Fleming County Kentucky
BT Kentucky
BT United States

maximum entropy analysis (1981)
SA seismology

Maxwell Montes (1989)
Volcanic feature on Venus.
BT Venus
SA Ishtar Terra
SA Venera Program

May-sur-Orne
No longer a valid term for GeoRef. See May-sur-Orne France.

May-sur-Orne France (1993)
Village on the Orne River in NW France.
BT Calvados France
BT France
BT Western Europe
BT Europe

Maya Mountains (1978)
Range in S Belize. Also search Maya AND appropriate area term.
BT Belize
BT Central America

Maya River basin (1978)
Chiefly in central Khabarovsk Kray but also in Yakutia. Also search Maya River. This term has multiple hierarchies.
BT1 Russian Federation
BT1 Commonwealth of Independent States
BT2 Asia
SA Khabarovsk Russian Federation
SA Yakutia Russian Federation

Mayaguez Bay (1978)
Off city of Mayaguez in W Puerto Rico.
BT Puerto Rico
BT Greater Antilles
BT Antilles
BT West Indies
BT Caribbean region

Mayenne
No longer a valid term for GeoRef. See Mayenne France.

Mayenne France (1993)
Department in NW France.
CO N474500N483500
W0000200W0011000
BT France
BT Western Europe
BT Europe

Maykop Russian Federation
use Maikop Russian Federation

Maymecha (1978)
River. Rises in Evenki National Okrug and flows N into the Kheta River in the Taymyr National Okrug. This term has multiple hierarchies.
UF Maimecha
BT1 Russian Federation
BT1 Commonwealth of Independent States
BT2 Asia

Maymecha Kotuy
use Maymecha-Kotuy

Maymecha-Kotui
use Maymecha-Kotuy

Maymecha-Kotuy (1978)
Region in valleys of the Maymecha, Kotui, and Kheta River in the Taymyr and Evenki national okrugs. This term has multiple hierarchies.
CO N700000N711500
E1033000E0993000
UF Maimecha-Kotui
UF Maymecha Kotuy
UF Maymecha-Kotui
BT1 Russian Federation
BT1 Commonwealth of Independent States
BT2 Asia

Mayo
No longer a valid term for GeoRef. See Mayo Ireland.

Mayo Ireland (1993)
County in NW Ireland.
CO N533000N542000
W0084000W0101500
BT Ireland
BT Western Europe
BT Europe
NT Keel Ireland

Mayon (1993)
Stratovolcano. Albay Province, SE Luzon.
CO N130000N133000
E1234000E1233000
BT Luzon
BT Philippine Islands
BT Far East
BT Asia

Maysvillian (1985)
North America. Upper Ordovician, below Richmondian and above Edenian.
BT Cincinnatian
BT Upper Ordovician
BT Ordovician
BT Paleozoic

Mayurbhanj
No longer a valid term for GeoRef. See Mayurbhanj India.

Mayurbhanj India (1993)
District in E India.
BT Orissa India
BT India
BT Indian Peninsula
BT Asia

Mazagan Plateau (1989)
Off the coast of Morocco. As of 1990, Atlantic Ocean is autoposted to this term.
BT Cape Verde Atlantic

BT North Atlantic
BT Atlantic Ocean
SA DSDP Site 544
SA DSDP Site 545
SA DSDP Site 546
SA DSDP Site 547
SA Leg 79

Mazama Ash (1978)
Refers to any or all of several Quaternary volcanic ash beds, commonly used as geochronologic markers in Oregon, Washington, Montana, Alberta and Saskatchewan. Erupted from prehistoric Mount Mazama.
BT Quaternary
BT Cenozoic
SA Alberta
SA Montana
SA Mount Mazama
SA Oregon
SA Saskatchewan
SA Washington

Mazatlan
No longer a valid term for GeoRef as of 1993 see Mazatlan Mexico

Mazatlan Mexico (1993)
City on the Pacific Ocean. Before 1993 also search Mazatlan AND Mexico.
BT Sinaloa Mexico
BT Mexico

Mazon Creek (1978)
NE Illinois.
IN Index counties or state as applicable.
SA Illinois

McAllen Ranch Field (1985)
S Texas. Autoposting of broader terms to this term began in 1989.
CO N261000N264000
W0980000W0984000
BT Hidalgo County Texas
BT Texas
BT United States
SA oil and gas fields

McArthur Basin (1985)
W Queensland, E Northern Territory and S Gulf of Carpentaria, Australia.
IN Index regions as applicable.
SA Arafura Sea
SA Australia
SA Northern Territory Australia
SA Queensland Australia

McCloud Limestone (1989)
N California. In some cases, age is given as Mississippian.
BT Permian
BT Paleozoic
SA California

McCone County
Valid through 1988. Search in combination with state term. After 1988, use specific county-state term.

McCone County Montana (1989)
NE Montana. Before 1989, also search McCone County AND Montana.
CO N470800N481000
W1051300W1063500
BT Montana
BT United States

McCoy Creek Group (1989)
Great Basin of NE Nevada and N Utah.
BT upper Proterozoic
BT Proterozoic
BT upper Precambrian
BT Precambrian
SA Nevada
SA Utah

McCreary County
Valid through 1988. Search in combination with state term. After 1988, use specific county-state term.

McCreary County Kentucky (1989)
S Kentucky. Before 1989, also search McCreary County AND Kentucky.
CO N363800N365500
W0842000W0844700
BT Kentucky
BT United States

McDonough County
Valid through 1988. Search in combination with state term. After 1988, use specific county-state term.

McDonough County Illinois (1989)
W Illinois. Before 1989, also search McDonough County AND Illinois.
CO N401500N403800
W0902500W0905500
BT Illinois
BT United States

McDowell County
Valid through 1988. Search in combination with state term. After 1988, use specific county-state term.

McDowell County North Carolina (1989)
W central North Carolina. Before 1989, also search McDowell County AND North Carolina.
CO N353200N355900
W0815000W0821700
BT North Carolina
BT United States

McDowell County West Virginia (1989)
S West Virginia. Before 1989, also search McDowell County AND West Virginia.
CO N371200N373300
W0812000W0820000
BT West Virginia
BT United States

McHenry County
Valid through 1988. Search in combination with state term. After 1988, use specific county-state term.

McHenry County Illinois (1989)
NE Illinois. Before 1989, also search McHenry County AND Illinois.
CO N421000N423000
W0881300W0884500
BT Illinois
BT United States

McHenry County North Dakota (1989)
N central North Dakota. Before 1989, also search McHenry County AND North Dakota.
CO N475100N483700
W1001200W1010400
BT North Dakota
BT United States

McHugh Complex (1989)
S Alaska.
BT Mesozoic
SA Alaska

McIntosh County
Valid through 1988. Search in combination with state term. After 1988, use specific county-state term.

McIntosh County Georgia (1989)
SE Georgia. Before 1989, also search McIntosh County AND Georgia.
CO N311800N314300
W0811200W0814000
BT Georgia
BT United States
NT Doboy Sound
NT Sapelo Island

McIntosh County North Dakota (1989)
S North Dakota. Before 1989, also search McIntosh County AND North Dakota.
CO N455700N461700
W0990000W0995300
BT North Dakota
BT United States

McIntosh County Oklahoma (1989)
E Oklahoma. Before 1989, also search McIntosh County AND Oklahoma.
CO N350800N353300
W0952100W0955800
BT Oklahoma
BT United States

McKenzie County
Valid through 1988. Search in combination with state term. After 1988, use specific county-state term.

McKenzie County North Dakota (1989)
W North Dakota. Before 1989, also search McKenzie County AND North Dakota.
CO N471900N480800
W1023400W1040300
BT North Dakota
BT United States

McKenzie Formation (1978)
In Maryland, it includes Rabble Run Sandstone Member. Underlies Bloomsburg Formation and overlies Rochester Formation. W Maryland, central Pennsylvania, N Virginia, and NE West Virginia.
BT Middle Silurian
BT Silurian
BT Paleozoic
SA Maryland
SA Pennsylvania
SA Virginia
SA West Virginia

McKinley County
Valid through 1988. Search in combination with state term. After 1988, use specific county-state term.

McKinley County New Mexico (1989)
NW New Mexico. Before 1989, also search McKinley County AND New Mexico.
CO N345700N360000
W1072000W1090400
BT New Mexico
BT United States
NT Ambrosia Lake mining district
NT Fort Wingate
NT Grants mining district
NT Mariano Lake
SA Rio Puerco
SA Zuni Mountains

McLean County
Valid through 1988. Search in combination with state term. After 1988, use specific county-state term.

McLean County Illinois (1989)
Central Illinois. Before 1989, also search McLean County AND Illinois.
CO N401800N404600
W0882500W0891700
BT Illinois
BT United States

McLean County Kentucky (1989)
W Kentucky. Before 1989, also search McLean County AND Kentucky.
CO N372100N374100
W0870200W0872900
BT Kentucky
BT United States

McLean County North Dakota (1989)
Central North Dakota. Before 1989, also search McLean County AND North Dakota.
CO N471000N475100
W1003700W1021600
BT North Dakota
BT United States

McLennan County Texas (1989)
E central Texas. Before 1989, search McLennan County AND Texas.
CO N311500N314900
W0965000W0973500
BT Texas
BT United States
NT Waco Texas

McMurdo Ice Shelf (1978)
Part of Ross Ice Shelf in McMurdo Sound area in Ross Dependency on the Pacific Ocean side.
BT Antarctica
SA Ross Ice Shelf

McMurdo Sound (1978)
Inlet of SW Ross Sea at the edge of the Ross Ice Shelf beween Ross Island and coast of Victoria Land. Site of major U.S. research and exploration base in Ross Dependency on the Pacific Ocean side.
BT Antarctic Ocean
SA Taylor Valley

McMurray Formation (1985)
In Mannville Group. E Alberta.
BT Lower Cretaceous
BT Cretaceous
BT Mesozoic
SA Alberta
SA Athabasca Oil Sands
SA Mannville Group

McPherson County
Valid through 1988. Search in combination with state term. After 1988, use specific county-state term.

McPherson County Kansas (1989)
Central Kansas. Before 1989, also search McPherson County AND Kansas.
CO N380900N383600
W0972300W0975500
BT Kansas
BT United States

McPherson County Nebraska (1989)
W central Nebraska. Before 1989, also search McPherson County AND Nebraska.
CO N412300N414400
W1004300W1012300
BT Nebraska
BT United States

McPherson County South Dakota (1989)
N South Dakota. Before 1989, also search McPherson County AND South Dakota.
CO N453600N455600
W0984300W0994300
BT South Dakota
BT United States
NT Eureka South Dakota

McPherson Formation (1978)
Consists of early Pleistocene stream deposits, later and coarser Pleistocene stream channel deposits, and still later Pleistocene silt, clay, and fine sand. Central Kansas.
BT Pleistocene
BT Quaternary
BT Cenozoic
SA Kansas

Md
use mendelevium

Meade Basin (1978)
Located in Meade County, SW Kansas. Also a river basin S of Point Barrow in N Alaska, with coal mining in area.
IN Index states and counties as applicable.
SA Alaska
SA Kansas
SA Meade County Kansas
SA Northern Alaska

Meade County
Valid through 1988. Search in combination with state term. After 1988, use specific county-state term.

Meade County Kansas (1989)
SW Kansas. Before 1989, also search Meade County AND Kansas.
CO N370000N372300
W1000500W1003800
BT Kansas
BT United States
SA Meade Basin

Meade County Kentucky (1989)
NW Kentucky. Before 1989, also search Meade County AND Kentucky.
CO N374900N381300
W0855800W0863000
BT Kentucky
BT United States

Meade County South Dakota (1989)
W central South Dakota. Before 1989, also search Meade County AND South Dakota.
CO N440800N450300
W1020000W1033500
BT South Dakota
BT United States
SA Whitewood Creek

Meade Peak Member (1989)
Of Phosphoria Formation. E Idaho, SW Montana, E Utah and W Wyoming.
BT Permian
BT Paleozoic
SA Idaho
SA Montana
SA Phosphoria Formation
SA Utah
SA Wyoming

Meadow soils (1978)
BT soils
SA Intrazonal soils
SA soil group

Meagher County
Valid through 1988. Search in combination with state term. After 1988, use specific county-state term.

Meagher County Montana (1989)
Central Montana. Before 1989, also search Meagher County AND Montana.
CO N461000N470700
W1102000W1114500
BT Montana
BT United States
SA Bridger Range
SA Crazy Mountains Basin

meandering
use meanders

meanders (1978)
Autoposting of fluvial features to this term began in 1989.
UF meandering
BT fluvial features
SA floodplains
SA geomorphology
SA oxbow lakes
SA point bars
SA rivers
SA rivers and streams
SA sinuosity
SA streams

measurement (1978)
UF measurements
SA conductivity
SA geodesy
SA geothermal gradient
SA heat flow
SA measurement-while-drilling
SA satellite measurements
SA telemetry
SA thermal conductivity

measurement while drilling
use measurement-while-drilling

measurement-while-drilling (1993)
Before 1993, also search MWD and measurement while drilling.
UF measurement while drilling
SA boreholes
SA downhole methods
SA drilling
SA measurement
SA methods
SA sounding
SA techniques
SA well-logging

measurements
use measurement

Meath
No longer a valid term for GeoRef. See Meath Ireland.

Meath Ireland (1993)
County on the Irish Sea in NE Ireland.
BT Ireland
BT Western Europe
BT Europe

mechanical analysis
No longer a valid GeoRef index term. Before 1971, was used on level 1 in subfile N. See mechanism; mechanics.

mechanical controls (1978)
As of 1993, restricted to economic studies.
SA controls
SA mineral deposits, genesis

mechanical erosion
use abrasion

mechanical properties (1978)
SA anisotropy
SA brittleness
SA compactness
SA creep
SA deformation
SA ductile deformation
SA durability
SA elastic limit
SA elastic properties
SA elasticity
SA engineering geology
SA friction
SA hardness
SA hysteresis
SA physical properties
SA plasticity
SA Poisson's ratio
SA properties
SA rigidity
SA shear
SA shear strength
SA soil mechanics
SA strain
SA strength
SA tensile strength
SA thermoelastic properties
SA thermomechanical properties
SA thixotropy
SA weathering
SA yield strength

mechanical weathering (1978)
Before 1978, also search weathering AND mechanical.
BT weathering
SA insolation
SA physical weathering

mechanics (1978)
SA decollement
SA faults
SA flexural-slip
SA folds
SA land subsidence
SA plate tectonics
SA rock mechanics
SA shear
SA soil mechanics
SA stick-slip

mechanism (1978)
UF mechanisms
SA flow mechanism
SA focal mechanism

mechanisms
use mechanism

Mechelen
Not a valid term for GeoRef. See Malines Belgium.

Mechlin
Not a valid term for GeoRef. See Malines Belgium.

Mecklenburg (1978)
Region on the Baltic Sea in Mecklenburg-Western Pomerania, NE Germany. A former German state.
BT Germany
BT Central Europe
BT Europe
SA Barth Germany
SA Mecklenburg-Western Pomerania Germany

Mecklenburg County
Valid through 1988. Search in combination with state term. After 1988, use specific county-state term.

Mecklenburg County North Carolina (1989)
S North Carolina. Before 1989, also search Mecklenburg County AND North Carolina.
CO N350000N353100
W0803300W0810400
BT North Carolina
BT United States
NT Charlotte North Carolina
NT Davidson North Carolina

Mecklenburg County Virginia (1989)
S Virginia. Before 1989, also search Mecklenburg County AND Virginia.
CO N363300N365300
W0780200W0784300
BT Virginia
BT United States

Mecklenburg-Vorpommern Germany
use Mecklenburg-Western Pomerania Germany

Mecklenburg-Western Pomerania Germany (1993)
State in NE Germany. Before 1993, also search Mecklenburg-West Pomerania or Mecklenburg-Vorpommern.
UF Mecklenburg-Vorpommern Germany
BT Germany
BT Central Europe
BT Europe
NT Barth Germany
NT Rugen Island
NT Schwerin Germany
SA Mecklenburg
SA North German Plain
SA Oder Valley
SA Pomerania

Mecopteroida (1981)
BT Insecta
BT Mandibulata
BT Arthropoda
BT Invertebrata

Mecsek Mountains (1978)
Near Croatian border in S Hungary. In Baranya. Also search Mecsek.
UF Baranya Mountains
BT Hungary
BT Central Europe
BT Europe

media
As of 1981, no longer a valid term for GeoRef.
use materials

median masses
As of 1981, no longer a valid term for GeoRef. See tectonics; orogenic belts.

Median Tectonic Line (1978)
Southwest Japan.
BT Japan
BT Far East
BT Asia
SA Honshu
SA Kyushu
SA plate tectonics
SA Shikoku

SA tectonics

median valley (1978)
UF valley, median
SA Mid-Atlantic Ridge
SA mid-ocean ridges
SA rift zones
SA rifting

medical geology (1978)
UF health
SA environmental geology
SA geology
SA human ecology
SA pollution

medicinal waters (1981)
SA mineral waters
SA thermal waters
SA water
SA water quality
SA water resources

Medicine Bow Mountains (1978)
Range of the Rocky Mountains. This term has multiple hierarchies.
IN Index states as applicable.
BT1 U. S. Rocky Mountains
BT1 Rocky Mountains
BT2 U. S. Rocky Mountains
BT2 United States
SA Colorado
SA Mullen Creek-Nash Fork shear zone
SA Wyoming

Medicine Hat
No longer a valid term for GeoRef. See Medicine Hat Alberta.

Medicine Hat Alberta (1993)
City in SE Alberta.
BT Alberta
BT Western Canada
BT Canada

Medicine Lake (1978)
Lake and volcano in N California. Also a lake in Sheridan County, NE Montana and in Codington County, NE South Dakota.
IN Index states and counties as applicable.
UF Medicine Lake Volcano
SA California
SA Medicine Lake Highland
SA Montana
SA Sheridan County Montana
SA Siskiyou County California
SA South Dakota

Medicine Lake Highland (1985)
Plateau surrounding Medicine Lake, N California. Autoposting of broader terms to this term began in 1989.
CO N413000N414500
W1212500W1215000
BT Siskiyou County California
BT California
BT United States
SA Medicine Lake

Medicine Lake Volcano
use Medicine Lake

Medina County
Valid through 1988. Search in combination with state term. After 1988, use specific county-state term.

Medina County Ohio (1989)
N Ohio. Before 1989, also search Medina County AND Ohio.
CO N405800N411800
W0814200W0821200
BT Ohio
BT United States

Medina County Texas (1989)
SW Texas. Before 1989, also search Medina County AND Texas.
CO N290700N294300
W0985000W0992900
BT Texas
BT United States
SA Edwards Aquifer

Medina Formation (1978)
BT Lower Silurian
BT Silurian
BT Paleozoic
SA Grimsby Sandstone
SA Michigan
SA New York
SA Ontario
SA Pennsylvania

Mediterranean (region)
use Mediterranean region

Mediterranean region (1978)
The Mediterranean Sea and its islands, and parts of those countries along its shores in S Europe, N Africa, and the Middle East. Also search Mediterranean. In 1981, Aegean Islands and Ionian Islands became narrower terms.
IN Index Mediterranean Sea, countries and islands as applicable.
CO N300000N473000
E0380000W0050000
UF Mediterranean (region)
NT Aegean Islands
NT Balearic Islands
NT El Alboran Island
NT Ionian Islands
SA Albania
SA Algeria
SA Corsica
SA Cyprus
SA Egypt
SA France
SA Greece
SA Israel
SA Italy
SA Lebanon
SA Libya
SA Malta
SA Mediterranean Sea
SA Mesogaea
SA Morocco
SA Sardinia Italy
SA Sicily Italy
SA Spain
SA Syria
SA Tunisia
SA Turkey
SA Yugoslavia

Mediterranean Ridge (1978)
Between Crete and the coast of E Libya.
BT Mediterranean Sea

Mediterranean Sea (1978)
Enclosed by Europe on the W and N, Asia on the E, and Africa on the S. As of 1981, includes the Black Sea. Before 1981, Balearic Islands and Malta were included as narrower terms. As of 1981, Black Sea is a narrower term.
CO N301000N464000
E0420000W0053000
NT Balearic Basin
NT East Mediterranean
NT Florence Rise
NT Gulf of Pozzuoli
NT Hellenic Trench
NT Ligurian Sea
NT Mediterranean Ridge
NT Minorca Rise
NT Strait of Sicily
NT Tyrrhenian Basin
NT West Mediterranean
SA Balearic Islands
SA Corsica
SA Crete
SA Cyprus
SA DSDP Site 121
SA DSDP Site 122
SA DSDP Site 123
SA DSDP Site 124
SA DSDP Site 125
SA DSDP Site 126
SA DSDP Site 127
SA DSDP Site 128
SA DSDP Site 129
SA DSDP Site 130
SA DSDP Site 131
SA DSDP Site 132
SA DSDP Site 133
SA DSDP Site 134
SA DSDP Site 372
SA DSDP Site 374
SA DSDP Site 375
SA DSDP Site 376
SA DSDP Site 378
SA El Alboran Island
SA Greek Aegean Islands
SA Greek Ionian Islands
SA Haifa Bay
SA Hellenic Arc
SA Leg 13
SA Leg 42A
SA Leg 107
SA Malta
SA Mediterranean region
SA ODP Site 652
SA ODP Site 653
SA ODP Site 654
SA ODP Site 655
SA ODP Site 656
SA Sardinia Italy
SA Sicily Italy

Mediterranean soils (1982)
BT soils
SA soil group

Medo
Not a valid term for GeoRef. See Tokyo Japan.

Medoc (1978)
District between the Gironde River and the Bay of Biscay in NW Gironde, SW France.
BT Gironde France
BT France
BT Western Europe
BT Europe

Meerfeld crater lake (1989)
W Germany.
CO N500300N500700
E0065000E0064500
BT Rhineland-Palatinate Germany
BT Germany
BT Central Europe
BT Europe

Meers Fault (1989)
SW Oklahoma.
IN Index counties as applicable.
SA Caddo County Oklahoma
SA Comanche County Oklahoma
SA Kiowa County Oklahoma
SA Oklahoma

meerschaum
use sepiolite

Meerut
No longer a valid term for GeoRef. See Meerut India.

Meerut India (1993)
City, district, and division in N India.
BT Uttar Pradesh India

BT India
BT Indian Peninsula
BT Asia

meetings
use symposia

megacrysts (1978)
SA igneous rocks
SA matrix
SA metamorphic rocks
SA textures

megacyclothems (1978)
BT cyclothems
BT planar bedding structures
BT sedimentary structures

megafossils
use fossils

Megalopolis Basin (1978)
In Arcadia Department in central Peloponnesus. Also search Megalopolis.
BT Peloponnesus Greece
BT Greece
BT Southern Europe
BT Europe

megaripples (1978)
BT bedding plane irregularities
BT sedimentary structures
SA sand ridges
SA sand waves

megaspores (1978)
Autoposting of microfossils to this term began in 1989.
BT palynomorphs
BT microfossils
SA biostratigraphy
SA miospores
SA spores

Meghalaya
No longer a valid term for GeoRef. See Meghalaya India.

Meghalaya India (1993)
State in NE India. Became a state in 1972. Formerly in Assam.
CO N250000N280000
E0961500E0893000
BT Northeastern India
BT India
BT Indian Peninsula
BT Asia
NT Garo Hills
NT Shillong India
SA Assam India
SA Shillong Plateau

Megion Field (1978)
Oil field in SE Khanty-Mansi National Okrug in West Siberian Plain. Also search Megion. This term has multiple hierarchies.
BT1 Russian Federation
BT1 Commonwealth of Independent States
BT2 Asia
SA oil and gas fields

Megri
No longer a valid term for GeoRef. As of 1993, see Megri Armenia.

Megri Armenia (1993)
Town in S Armenia near Iranian border. This term has multiple hierarchies.
BT1 Armenia
BT1 Commonwealth of Independent States
BT2 Armenia
BT2 Europe

Megrinskiy Pluton (1978)

In Megrinskiy Mountains in S Armenia. This term has multiple hierarchies.
BT1 Armenia
BT1 Commonwealth of Independent States
BT2 Armenia
BT2 Europe

Meguma Group (1978)
Composed of Goldenville and Halifax formations. In Nova Scotia.
BT Ordovician
BT Paleozoic
SA Canada
SA Goldenville Formation
SA Halifax Formation
SA Nova Scotia

Mehadia
No longer a valid term for GeoRef. See Mehadia Romania.

Mehadia Romania (1993)
Village at W edge of the Transylvanian Alps. This term has multiple hierarchies.
BT1 Banat
BT1 Southern Europe
BT1 Europe
BT2 Romania
BT2 Southern Europe
BT2 Europe

Mehedinti Plateau (1978)
SW Romania.
IN Index regions as applicable.
BT Romania
BT Southern Europe
BT Europe
SA Banat
SA Walachia

Meighen Island (1978)
W of Axel Heiberg Island in Franklin District.
BT Franklin District Northwest Territories
BT Northwest Territories
BT Western Canada
BT Canada

meimechite (1978)
Before 1975, also search meymechite.
UF meymechite
BT ultramafics
BT plutonic rocks
BT igneous rocks

meionite (1978)
From 1978-1980, carbonates and framework silicates were autoposted to this term. Autoposting of scapolite group began in 1981. As of 1989, framework silicates and silicates are autoposted to this term. This term has multiple hierarchies.
BT1 carbonates
BT2 scapolite group
BT2 framework silicates
BT2 silicates

Mekong Delta (1978)
On the South China Sea S of Saigon.
BT Vietnam
BT Far East
BT Asia

Melanesia (1978)
Collective name for islands NE of Australia. In 1981, broader term changed from Pacific Ocean to Oceania. New Caledonia is a narrower term as of 1981. Autoposting of this term began in 1978.

IN Index island groups as applicable.
CO S230000S050000
 W1710000E1550000
BT Oceania
NT Fiji
NT New Caledonia
NT Solomon Islands
NT Vanuatu
SA Bismarck Archipelago
SA Pacific Ocean

melange (1978)
UF block clay
SA boudinage
SA deformation
SA plate tectonics
SA structural analysis
SA tectonostratigraphic units

melanite (1978)
BT garnet group
BT nesosilicates
BT orthosilicates
BT silicates

melanocerite (1978)
Autoposting of nesosilicates began in 1985.
BT nesosilicates
BT orthosilicates
BT silicates

melanophlogite (1978)
BT silica minerals
BT framework silicates
BT silicates

melanterite (1978)
BT sulfates

melaphyre (1978)
BT basalts
BT volcanic rocks
BT igneous rocks
SA porphyry

Melbourne
No longer a valid term for GeoRef. See Melbourne Australia if applicable.

Melbourne Australia (1993)
City on N Port Phillip Bay, SE Australia at mouth of Yarra River. Before 1993, also search Melbourne or Melbourne region AND Australia.
CO S374500S374500
 E1445800E1445800
BT Victoria Australia
BT Australia
BT Australasia

Melbourne region
No longer a valid term for GeoRef. See Melbourne Australia if applicable.

Meldon Aplite (1978)
Northumberland in extreme NE England.
BT Northumberland England
BT England
BT Great Britain
BT United Kingdom
BT Western Europe
BT Europe

melilite (1978)
UF mellilite
BT melilite group
BT sorosilicates
BT orthosilicates
BT silicates

melilite group (1978)
Autoposting of this term began in 1978.
BT sorosilicates
BT orthosilicates

BT silicates
NT akermanite
NT gehlenite
NT melilite

melilitite (1978)
From 1978-1980, alkali basalt family was autoposted to this term.
BT volcanic rocks
BT igneous rocks
SA alkali basalts
SA ultramafics

Melilla (1981)
Spanish presidio on north coast of Morocco.
BT North Africa
BT Africa
SA Morocco

mellilite
use melilite

melnikovite
use greigite

Melones Fault (1981)
Major structural feature in the Sierra Nevada of California.
CO N372500N400700
 W1195000W1211500
BT California
BT United States
SA Sierra Nevada

Melos
Use Milos for the Greek island.

Melosira (1978)
Autoposting of thallophytes and microfossils began in 1990. This term has multiple hierarchies.
BT1 diatoms
BT1 algae
BT1 thallophytes
BT1 Plantae
BT2 diatoms
BT2 algae
BT2 microfossils

melt water
use meltwater

melteigite (1978)
BT alkali gabbros
BT gabbros
BT plutonic rocks
BT igneous rocks
SA ijolite
SA urtite

melting (1978)
SA fusion
SA partial melting
SA phase equilibria

melting relations
A valid term through mid-1978. Use melting or phase equilibria.

melts (1978)
SA liquid phase
SA magmas
SA magmatic differentiation
SA partial melting
SA petrology
SA phase equilibria

meltwater (1978)
UF melt water
SA glacial geology
SA glaciers
SA hydrology
SA ice
SA jokulhlaups
SA snow
SA water

Melville Island (1989)

One of the Queen Elizabeth Islands in W Franklin District, Northwest Territories. Also an island off the NW coast of Northern Territory, Australia.
IN Index regions as applicable.
SA Australia
SA Canada
SA Franklin District Northwest Territories
SA Northern Territory Australia
SA Northwest Territories
SA Queen Elizabeth Islands

Melville Peninsula (1978)
N projection of Canadian mainland in Franklin District between Committee Bay on the W and Foxe Basin on the E.
BT Franklin District Northwest Territories
BT Northwest Territories
BT Western Canada
BT Canada

Memel River basin
use Neman River basin

Memphis Tennessee (1989)
City in SW Tennessee. Before 1989, search Memphis AND Tennessee.
CO N350800N351000
 W0900000W0900500
BT Shelby County Tennessee
BT Tennessee
BT United States

menaccanite
use ilmenite

mendelevium (1989)
Chemical element.
UF Md
BT actinides
BT metals

Mendeleyev Ridge (1978)
N of Alaska and E of Siberia.
BT Arctic Ocean

Mendeleyev Volcano (1978)
On Kunashir Island, southernmost of the Kuril Islands. This term has multiple hierarchies.
UF Mendeleyeva Volcano
BT1 Kunashir Island
BT1 Kuril Islands
BT1 Sakhalin Russian Federation
BT1 Russian Federation
BT1 Commonwealth of Independent States
BT2 Kunashir Island
BT2 Kuril Islands
BT2 Sakhalin Russian Federation
BT2 Asia
BT3 Kunashir Island
BT3 Kuril Islands
BT3 Russian Pacific region
BT3 Russian Federation
BT3 Commonwealth of Independent States
BT4 Kunashir Island
BT4 Kuril Islands
BT4 Russian Pacific region
BT4 Asia

Mendeleyeva Volcano
use Mendeleyev Volcano

Mendip Hills (1978)
SW England. Also search Mendip.
BT Somerset England
BT England
BT Great Britain
BT United Kingdom
BT Western Europe
BT Europe

Mendocino (Cape)
use Cape Mendocino

Mendocino County
Valid through 1988. Search in combination with state term. After 1988, use specific county-state term.

Mendocino County California (1989)
On the Pacific Ocean in NW California. Before 1989, also search Mendocino County AND California.
CO N384500N400000 W1224700W1240000
BT California
BT United States
NT Point Arena California
SA Eel River
SA Point Arena

Mendocino fracture zone (1978)
Running in an E-W direction approximately 1250 miles W of N California.
UF Mendocino triple junction
BT Pacific Ocean

Mendocino triple junction
use Mendocino fracture zone

Mendoza
No longer a valid term for GeoRef. See Mendoza Argentina.

Mendoza Argentina (1993)
Province including city in W Argentina. Before 1993, also search Mendoza or Mendoza Province AND Argentina.
CO S380000S330000 W0660000W0710000
UF Mendoza Province
BT Argentina
BT South America
NT Payun Matru
SA Salado Basin

Mendoza Province
use Mendoza Argentina

Menefee Formation (1978)
In Mesaverde Group. Includes Cleary Coal Member and above that a bed formerly called Allison Barren Member. SW Colorado, and NW New Mexico.
BT Upper Cretaceous
BT Cretaceous
BT Mesozoic
SA Colorado
SA Mesaverde Group
SA New Mexico

meneghinite (1978)
BT sulfantimonites
BT sulfosalts

Menorca
use Minorca

Menorca Rise
use Minorca Rise

Mentor Beds (1978)
According to the latest published Lexique, the age of Mentor Sandstone Member (or Belvedere Formation) is Lower Cretaceous. In central Kansas. This term has multiple hierarchies.
BT1 Comanchean
BT1 Cretaceous
BT1 Mesozoic
BT2 Lower Cretaceous
BT2 Cretaceous
BT2 Mesozoic
SA Kansas

Menuha Formation (1989)
BT Upper Cretaceous
BT Cretaceous
BT Mesozoic
SA Israel
SA Jordan
SA Middle East
SA Syria

Meotian (1978)
European stage, Black Sea area. Above Sarmatian, below Cimmerian.
BT upper Miocene
BT Miocene
BT Neogene
BT Tertiary
BT Cenozoic
SA lower Pliocene
SA Pliocene

Meramecian (1978)
Provincial series, North America. Above Osagian, below Chesterian.
BT Upper Mississippian
BT Mississippian
BT Carboniferous
BT Paleozoic
NT Saint Louis Limestone
NT Sainte Genevieve Limestone
NT Salem Limestone
NT Spergen Formation
NT Warsaw Formation
SA Chesterian
SA Valmeyeran

Merano
No longer a valid term for GeoRef. See Merano Italy.

Merano Italy (1993)
City in Bolzano Province in NE Italy.
BT Bolzano Italy
BT Trentino-Alto Adige Italy
BT Italy
BT Southern Europe
BT Europe

Merapi (1989)
Volcano in Java, Indonesia.
UF Merapi Volcano
BT Java
BT Indonesia
BT Far East
BT Asia

Merapi Volcano
use Merapi

Mercalli scale, modified
use modified Mercalli scale

Merced County
Valid through 1988. Search in combination with state term. After 1988, use specific county-state term.

Merced County California (1989)
Central California. Before 1989, also search Merced County AND California.
CO N364500N374000 W1200600W1211500
BT California
BT United States

Mercenaria (1978)
Genus.
BT Bivalvia
BT Mollusca
BT Invertebrata

Mercer County
Valid through 1988. Search in combination with state term. After 1988, use specific county-state term.

Mercer County Illinois (1989)
NW Illinois. Before 1989, also search Mercer County AND Illinois.
CO N410500N412000 W0902800W0910900
BT Illinois
BT United States

Mercer County Kentucky (1989)
Central Kentucky. Before 1989, also search Mercer County AND Kentucky.
CO N374000N375800 W0844200W0850200
BT Kentucky
BT United States

Mercer County Missouri (1989)
N Missouri. Before 1989, also search Mercer County AND Missouri.
CO N401700N403500 W0933200W0934000
BT Missouri
BT United States

Mercer County New Jersey (1989)
W New Jersey. Before 1989, also search Mercer County AND New Jersey.
CO N400800N402500 W0742800W0745600
BT New Jersey
BT United States
NT Princeton New Jersey
NT Trenton New Jersey

Mercer County North Dakota (1989)
Central North Dakota. Before 1989, also search Mercer County AND North Dakota.
CO N465800N473400 W1011500W1021300
BT North Dakota
BT United States

Mercer County Ohio (1989)
W Ohio. Before 1989, also search Mercer County AND Ohio.
CO N402200N404300 W0842700W0845100
BT Ohio
BT United States

Mercer County Pennsylvania (1989)
NW Pennsylvania. Before 1989, also search Mercer County AND Pennsylvania.
CO N410500N412900 W0800000W0803200
BT Pennsylvania
BT United States

Mercer County West Virginia (1989)
S West Virginia. Before 1989, also search Mercer County AND West Virginia.
CO N371700N373600 W0805200W0812200
BT West Virginia
BT United States

Mercia Mudstone (1989)
BT Upper Triassic
BT Triassic
BT Mesozoic
SA England
SA Scotland

mercury (1978)
Chemical element. As of 1981, use mercury ores for mercury as a commodity. Use Mercury Planet for the planet. Autoposting of metals to this term began in 1989.
UF Hg
UF quicksilver
BT metals
SA cinnabar
SA mercury ores
SA Mercury Planet
SA native elements

mercury injection (1985)
UF mercury injection method
SA enhanced recovery
SA injection
SA petroleum engineering
SA tertiary recovery

mercury injection method
use mercury injection

mercury ores (1981)
Before 1981, also search (mercury OR quicksilver OR Hg) AND (deposit OR deposits OR ore OR ores OR economic) in the basic index. Autoposting of metal ores to this term began in 1985.
IN Commodity. See List C.
BT metal ores
SA Borkut Deposit
SA cinnabar
SA mercury
SA Mercury Planet
SA native elements

Mercury Planet (1978)
Before 1972, also search Mercury in combination with the category (extraterrestrial geology).
NT Caloris Basin
SA Mariner 10
SA Mariner Program
SA mercury
SA mercury ores
SA planetology
SA planets
SA solar system
SA terrestrial planets

Merensky Reef (1985)
Ore body, Transvaal, South Africa.
BT Transvaal South Africa
BT South Africa
BT Southern Africa
BT Africa
SA Bushveld Complex

Meric River
use Maritsa River

Merida
No longer a valid term for GeoRef. As of 1993 see Merida Mexico or Merida Venezuela.

Merida Mexico (1993)
City in Yucatan. Before 1993 also search Merida AND Mexico.
BT Yucatan Mexico
BT Mexico

Merida Venezuela (1993)
State in W Venezuela. Before 1993 also search Merida AND Venezuela.
BT Venezuela
BT South America
SA Maracaibo Basin

Merionethshire
No longer a valid term for GeoRef as of 1993.
use Merionethshire Wales

Merionethshire Wales (1993)
Former county in W Wales. As of April 1974, part of Gwynedd. Before 1993, also search Merionethshire or Merioneth County.
CO N523000N530000 W0031500W0041000
UF Merionethshire

BT Wales
BT Great Britain
BT United Kingdom
BT Western Europe
BT Europe
NT Arenig
NT Harlech Wales
SA Bala
SA Cardigan Bay
SA Tremadoc Bay

meromictic lakes (1989)
As of 1990, lacustrine features is autoposted to this term.
BT lakes
BT lacustrine features
SA geomorphology
SA lacustrine environment

Merostomata (1978)
Autoposting of this term began in 1978.
BT Chelicerata
BT Arthropoda
BT Invertebrata
NT Eurypterida
NT Xiphosura

merrillite
use whitlockite

Merrimack Group (1989)
SW Maine, NE Massachusetts and SE New Hampshire. Includes Berwick Formation, Kittery Formation.
IN Index ages as applicable.
BT Paleozoic
SA Berwick Formation
SA Kittery Formation
SA Maine
SA Massachusetts
SA New Hampshire
SA Ordovician
SA Silurian

Merrimack River valley (1978)
IN Index states and counties as applicable.
BT United States
SA Essex County Massachusetts
SA Hillsborough County New Hampshire
SA Massachusetts
SA Middlesex County Massachusetts
SA New Hampshire

Merrimack Synclinorium (1978)
IN Index states as applicable.
BT United States
SA Maine
SA Massachusetts
SA New Hampshire

Merseburg
No longer a valid term for GeoRef.
See Merseburg Germany.

Merseburg Germany (1993)
City in E central Germany.
BT Saxony-Anhalt Germany
BT Germany
BT Central Europe
BT Europe

Mersey River (1978)
Name of rivers in N Tasmania and in NW England.
IN Index England or Tasmania Australia as applicable.
SA Cheshire England
SA England
SA Mersey Valley
SA Tasmania Australia

Mersey Valley (1978)
River valley in NW England and in Tasmania.

IN Index regions as applicable.
SA Australia
SA Cheshire England
SA England
SA Mersey River
SA Tasmania Australia

merwinite (1978)
Autoposting of nesosilicates began in 1985.
BT nesosilicates
BT orthosilicates
BT silicates

merzlota
use frozen ground

Mesa County
Valid through 1988. Search in combination with state term. After 1988, use specific county-state term.

Mesa County Colorado (1989)
W central Colorado. Before 1989, also search Mesa County AND Colorado.
CO N383000N392000 W1072000W1090500
BT Colorado
BT United States
NT Grand Junction Colorado
SA Book Cliffs

Mesa Rica Sandstone (1993)
Lower Cretaceous. May be in Purgatoire Formation or Dakota Formation. NE New Mexico and Texas.
BT Lower Cretaceous
BT Cretaceous
BT Mesozoic
SA Dakota Formation
SA New Mexico
SA Texas

Mesabi Range (1978)
Iron-ore bearing range of low hills in NE Minnesota.
BT Minnesota
BT United States
SA Itasca County Minnesota
SA Saint Louis County Minnesota

mesas (1993)
BT erosion features
SA benches
SA geomorphology
SA highlands
SA hills
SA mountains
SA plateaus
SA uplands

Mesaverde Group (1978)
Throughout San Juan Basin of SW Colorado and NW New Mexico, composed of Menefee Formation, Point Lookout Sandstone, Cliff House Sandstone, Gallup Sandstone, and Crevasse Canyon Formation. Arizona, W Colorado, S central Montana, W New Mexico, E Utah, and Wyoming.
BT Upper Cretaceous
BT Cretaceous
BT Mesozoic
SA Almond Formation
SA Arizona
SA Blackhawk Formation
SA Blair Formation
SA Colorado
SA Crevasse Canyon Formation
SA Gallup Sandstone
SA Hilliard Shale
SA La Ventana Sandstone
SA Mancos Shale
SA Menefee Formation
SA Montana

SA New Mexico
SA Point Lookout Sandstone
SA Price River Formation
SA Rock Springs Formation
SA Star Point Sandstone
SA Utah
SA Williams Fork Formation
SA Wyoming

Meseta (1978)
Geographic term for the entire interior of Spain covering almost 3/4 of country and consisting of an immense plateau with Madrid at its center.
IN Index Spain if applicable.
SA Spain

Mesogaea (1978)
SA Mediterranean region
SA paleogeography
SA Tethys

Mesogastropoda (1981)
Autoposting of this term to Cerithiidae began in 1989.
BT Gastropoda
BT Mollusca
BT Invertebrata
NT Cerithiidae

mesolite (1978)
BT zeolite group
BT framework silicates
BT silicates

Mesolithic (1978)
Archaeological classification. This term has multiple hierarchies.
UF Middle Stone Age
BT1 Stone Age
BT1 Cenozoic
BT2 Holocene
BT2 Quaternary
BT2 Cenozoic
SA archaeology

Mesopotamia (1978)
Region in SW Asia between Tigris River and Euphrates River. Primarily in Iraq but partly in NE Syria.
IN Index countries as applicable.
BT Middle East
BT Asia
SA Iraq
SA Syria

Mesosauria (1981)
Order. Autoposting of Anapsida to this term began in 1989. Autoposting of Synapsida to this term ended in 1989. Autoposting of this term to Proganosauria ended in 1989.
BT Anapsida
BT Reptilia
BT Tetrapoda
BT Vertebrata
BT Chordata

mesosiderite (1984)
Before 1984, also search mesosiderites. Use grahamite for the bitumen.
UF mesosiderites
BT stony irons
BT meteorites
SA Dalgaranga

mesosiderites
As of 1984, no longer a valid term for GeoRef.
use mesosiderite

mesothermal processes (1981)
Also search mesothermal.
SA mineral deposits, genesis
SA processes

Mesozoic (1978)
From the end of the Paleozoic to the beginning of the Cenozoic. From 1978-1980, Phanerozoic was autoposted to this term. Autoposting of this term to Triassic, Jurassic and Cretaceous ended in 1981 and resumed in 1993.
NT Antalya Complex
NT Brunswick Formation
NT Caracas Group
NT Cimmerian Orogeny
NT Condrey Mountain Schist
NT Cretaceous
NT Franciscan Complex
NT Glen Canyon Group
NT Great Valley Sequence
NT Indosinian Orogeny
NT Jurassic
NT Khorat Group
NT Kunga Group
NT lower Mesozoic
NT Maiolica Limestone
NT McHugh Complex
NT middle Mesozoic
NT Mist Mountain Formation
NT Moenave Formation
NT Navajo Sandstone
NT Nevadan Orogeny
NT Newark Supergroup
NT Nicoya Complex
NT Orocopia Schist
NT Passaic Formation
NT Pictou Group
NT Pucara Group
NT Serra Geral Formation
NT Shimanto Group
NT Statfjord Formation
NT Triassic
NT upper Mesozoic
NT Yanshanian
SA Alpine Orogeny
SA Brianconnais Zone
SA Catalina Schist
SA Himalayan Orogeny
SA Ionian Zone
SA lower Gondwana System
SA New England Orogeny
SA Phanerozoic
SA Riggins Group
SA upper Gondwana System
SA Voltri Group

mesozonal metamorphism (1982)
BT metamorphism
SA high-grade metamorphism

Messel
No longer a valid term for GeoRef.
See Messel Germany.

Messel Germany (1993)
Fossil locality and village near Darmstadt.
BT Hesse Germany
BT Germany
BT Central Europe
BT Europe

Messina
No longer a valid term for GeoRef.
See Messina Italy.

Messina Italy (1993)
City and province in NE Sicily.
BT Sicily Italy
BT Italy
BT Southern Europe
BT Europe

Messinian (1978)
BT upper Miocene
BT Miocene
BT Neogene
BT Tertiary
BT Cenozoic

meta-andesite (1978)
UF metaandesite
BT metaigneous rocks
BT metamorphic rocks

meta-anorthosite (1989)
UF metaanorthosite
BT metaigneous rocks
BT metamorphic rocks
SA anorthosite

meta-arkose (1978)
UF metaarkose
BT metasedimentary rocks
BT metamorphic rocks
SA arkose

meta-limestone
  use metalimestone

meta-ophiolite
  use ophiolite

metaandesite
  use meta-andesite

metaanorthosite
  use meta-anorthosite

metaarkose
  use meta-arkose

metabasalt (1978)
UF metabasalts
BT metaigneous rocks
BT metamorphic rocks
SA basalts

metabasalts
  use metabasalt

metabasite (1978)
BT metaigneous rocks
BT metamorphic rocks

metabentonite (1978)
From 1978-1980, metasedimentary rocks was autoposted to this term.
IN May be used as a rock and a mineral.
SA bentonite
SA clay minerals
SA metamorphic rocks
SA metasedimentary rocks
SA minerals

metabolism (1978)
SA diet
SA nutrition

metachert (1985)
BT metasedimentary rocks
BT metamorphic rocks
SA chert

metacinnabar (1978)
UF metacinnabarite
BT sulfides
SA onofrite

metacinnabarite
  use metacinnabar

metaconglomerate (1978)
BT metasedimentary rocks
BT metamorphic rocks
SA conglomerate

metadacite (1985)
BT metaigneous rocks
BT metamorphic rocks
SA dacites
SA metavolcanic rocks

metadiabase (1978)
Before 1981, also search metadolerite.
UF metadolerite
BT metaigneous rocks
BT metamorphic rocks
SA diabase

metadiorite (1978)
BT metaigneous rocks
BT metamorphic rocks
SA diorites

metadolerite
As of 1981, no longer a valid term for GeoRef. From 1978-1980, metaigneous rocks was autoposted to this term.
  use metadiabase

metagabbro (1978)
BT metaigneous rocks
BT metamorphic rocks
SA gabbros

metagranite (1978)
BT metaigneous rocks
BT metamorphic rocks
SA granites

metagraywacke (1978)
UF metagreywacke
BT metasedimentary rocks
BT metamorphic rocks
SA graywacke

metagreywacke
  use metagraywacke

metahalloysite (1982)
Before 1982, also search halloysite; endellite.
BT clay minerals
BT sheet silicates
BT silicates
SA halloysite

metaigneous
A valid index term through 1977. After 1977, use metaigneous rocks.

metaigneous rocks (1978)
Before 1978, also search metaigneous. Autoposting of this term to metadolerite ended in 1981.
BT metamorphic rocks
NT meta-andesite
NT meta-anorthosite
NT metabasalt
NT metabasite
NT metadacite
NT metadiabase
NT metadiorite
NT metagabbro
NT metagranite
NT metakomatiite
NT metaperidotite
NT metapyroxenite
NT metarhyolite
NT metatuff
NT serpentinite
SA igneous rocks
SA rocks

metakaolin (1978)
UF metakaolinite
SA clay minerals
SA kaolin

metakaolinite
  use metakaolin

metakomatiite (1993)
BT metaigneous rocks
BT metamorphic rocks
SA komatiite

metal
Not a valid GeoRef index term after 1969. Was used on level 1 in subfiles E and B. Now use metals or metal ores as applicable.

metal ores (1981)
Before 1981, search metals AND (deposit OR deposits OR ore OR ores OR economic) in the basic index. Autoposting of this term to specific metal ores began in 1985. Autoposting of this term to rare earth deposits began in 1989. Before 1971, also search metallic minerals.
IN Commodity. See List C.
UF metallic minerals
NT aluminum ores
NT antimony ores
NT arsenic ores
NT base metals
NT beryllium ores
NT bismuth ores
NT cadmium ores
NT chromite ores
NT cobalt ores
NT copper ores
NT gold ores
NT iridium ores
NT iron ores
NT lead ores
NT lead-zinc deposits
NT lithium ores
NT magnesium ores
NT manganese ores
NT mercury ores
NT molybdenum ores
NT nickel ores
NT niobium ores
NT osmium ores
NT palladium ores
NT platinum ores
NT polymetallic ores
NT pyrite ores
NT rare earth deposits
NT rhodium ores
NT ruthenium ores
NT silver ores
NT strontium ores
NT tantalum ores
NT thorium ores
NT tin ores
NT titanium ores
NT tungsten ores
NT uranium ores
NT vanadium ores
NT zinc ores
SA Alma mining district
SA annealing
SA Bathurst mining district
SA Boquira Mine
SA Broken Hill
SA Broken Hill Mine
SA Chichibu Mine
SA Clara Mine
SA economic geology
SA Fairbanks mining district
SA Fukazawa Mine
SA gossan
SA graphitization
SA Hanaoka Mine
SA Hanawa Mine
SA Idaho Springs mining district
SA Ikuno Mine
SA Kamaishi Mine
SA Kosaka Mine
SA kuroko-type
SA metallogenic epochs
SA metallogenic maps
SA metallogenic provinces
SA metallogeny
SA metallurgy
SA metals
SA native elements
SA nodules
SA nonmetal deposits
SA Omine Mine
SA ore bodies
SA ore grade
SA ore guides
SA ore microscopy
SA ore minerals
SA ore reflectivity
SA ore transport
SA placers
SA precious metals
SA Shakanai Mine
SA slag
SA Sunnyside Mine
SA Tintic mining district
SA West Shasta mining district

metal oxides
As of 1993, for mineralogical discussions use oxides AND metals. See also heavy metals and heavy minerals.

metalimestone (1978)
UF meta-limestone
BT metasedimentary rocks
BT metamorphic rocks
SA limestone

Metaline Falls
Valid through 1988. Search in combination with state term. After 1988, use specific city-state term.

Metaline Falls Washington (1989)
Town in NE Washington. Before 1989, search Metaline Falls AND Washington.
CO N485200N485200 W1172000W1172000
BT Pend Oreille County Washington
BT Washington
BT United States

Metaline Limestone (1985)
NE Washington.
BT Middle Cambrian
BT Cambrian
BT Paleozoic
SA Washington

metallic meteorites
  use iron meteorites

metallic minerals
  use metal ores

metallogenesis
  use metallogeny

metallogenic epochs (1982)
SA economic geology
SA metal ores
SA metallogenic provinces
SA metallogeny
SA mineral deposits, genesis
SA mineralization

metallogenic maps (1993)
BT maps
SA economic geology maps
SA metal ores
SA metallogenic provinces
SA metallogeny
SA mineral deposits, genesis

metallogenic provinces (1978)
UF provinces, metallogenic
SA economic geology
SA metal ores
SA metallogenic epochs
SA metallogenic maps
SA mineral deposits, genesis
SA mineralization

metallogeny (1978)
UF metallogenesis
SA economic geology
SA geosynclines
SA metal ores
SA metallogenic epochs
SA metallogenic maps
SA metallotects
SA mineral deposits, genesis
SA mineralization

metallography
Not a valid index term for GeoRef. See mineralogy and petrology.

metallotects (1989)

Any geologic feature which has influenced concentration of elements, resulting in formation of mineral deposits.
- SA controls
- SA metallogeny
- SA mineral deposits, genesis

metallurgy (1978)
- SA alloys
- SA annealing
- SA hydrometallurgy
- SA metal ores
- SA metals

metals (1978)
As of 1981, use metal ores for metals as a commodity. Autoposting of this term began in 1978.
- NT actinides
- NT alkali metals
- NT alkaline earth metals
- NT aluminum
- NT antimony
- NT arsenic
- NT bismuth
- NT cadmium
- NT chromium
- NT cobalt
- NT copper
- NT gallium
- NT germanium
- NT gold
- NT hafnium
- NT indium
- NT iron
- NT lead
- NT manganese
- NT mercury
- NT molybdenum
- NT nickel
- NT niobium
- NT platinum group
- NT polonium
- NT precious metals
- NT rare earths
- NT rhenium
- NT silver
- NT tantalum
- NT technetium
- NT thallium
- NT tin
- NT titanium
- NT tungsten
- NT vanadium
- NT zinc
- NT zirconium
- SA alloys
- SA base metals
- SA chelation
- SA chemical elements
- SA complexing
- SA geochemistry
- SA isotopes
- SA ligands
- SA metal ores
- SA metallurgy
- SA native elements
- SA nonmetals
- SA ore minerals
- SA pollution
- SA trace metals

metamict
No longer a valid term for GeoRef as of 1981. Use metamict minerals.

metamict minerals (1981)
Before 1981, search metamict.
- SA crystal structure
- SA metamictization
- SA minerals
- SA radiation damage

metamictization (1978)
- SA crystal structure
- SA metamict minerals
- SA minerals
- SA radiation damage

metamorphic aureoles
use aureoles

metamorphic belts (1978)
- UF belts, metamorphic
- SA metamorphic core complexes
- SA plate tectonics
- SA terranes

metamorphic core complexes (1985)
- UF core complexes
- SA metamorphic belts
- SA metamorphic rocks
- SA plate tectonics
- SA shear zones
- SA tectonics
- SA terranes

metamorphic differentiation
Not a valid index term for GeoRef. See migration of elements or zoning.

metamorphic grade
use grade

metamorphic petrology
Not a valid GeoRef index term after 1970. Was used on level 1 in subfile N. Now use petrology or metamorphic rocks as applicable.

metamorphic processes (1984)
Before 1984, also search metamorphism AND mineral deposits, genesis. As of 1993, restricted to economic studies.
- SA metamorphism
- SA mineral deposits, genesis
- SA processes

metamorphic rocks (1978)
Autoposting of this term to quartzite and anatexite ended in 1981. Itabirite is a narrower term as of 1981. Autoposting of this term to metasomatic rocks began in 1993.
- NT amphibolites
- NT cataclasites
- NT eclogite
- NT fulgurite
- NT garnetite
- NT gneisses
- NT gondite
- NT granulites
- NT hornfels
- NT itabirite
- NT marbles
- NT metaigneous rocks
- NT metaplutonic rocks
- NT metasedimentary rocks
- NT metasomatic rocks
- NT metavolcanic rocks
- NT migmatites
- NT mylonites
- NT phyllites
- NT phyllonites
- NT quartzites
- NT schists
- NT slates
- NT supracrustals
- SA actinolite facies
- SA amphibolite facies
- SA banded structures
- SA basement
- SA basement tectonics
- SA blueschist facies
- SA calc-silicate composition
- SA complexes
- SA crystalline rocks
- SA deformation
- SA eclogite facies
- SA epidote-amphibolite facies
- SA fabric
- SA facies
- SA folds
- SA foliation
- SA geochemistry
- SA geochronology
- SA gneissic texture
- SA granulite facies
- SA greenschist facies
- SA hornfels facies
- SA impactite
- SA intrusions
- SA lineation
- SA lithogeochemistry
- SA lithostratigraphy
- SA megacrysts
- SA metabentonite
- SA metamorphic core complexes
- SA metamorphism
- SA metasomatism
- SA mineral assemblages
- SA mullions
- SA occurrence
- SA ophiolite
- SA ophite
- SA paragenesis
- SA pelitic texture
- SA petrography
- SA petrology
- SA phase equilibria
- SA porphyroblastic texture
- SA porphyroclastic texture
- SA prehnite-pumpellyite facies
- SA protoliths
- SA relict materials
- SA rocks
- SA schistosity
- SA suevite
- SA symplectite
- SA zeolite facies
- SA zhamanshinite

metamorphic zone
use aureoles

metamorphism (1978)
Autoposting of this term began in 1978.
IN Index with specific type of metamorphism [e.g. anchimetamorphism, autometamorphism, burial metamorphism, contact metamorphism, dynamic metamorphism, grade, polymetamorphism, prograde metamorphism, regional metamorphism, retrograde metamorphism, rheomorphism, shock metamorphism, thermal metamorphism, ultrametamorphism].
- NT anchimetamorphism
- NT autometamorphism
- NT burial metamorphism
- NT catazonal metamorphism
- NT contact metamorphism
- NT dynamic metamorphism
- NT epizonal metamorphism
- NT glauconitization
- NT granitization
- NT high-grade metamorphism
- NT low-grade metamorphism
- NT mesozonal metamorphism
- NT migmatization
- NT phyllonitization
- NT polymetamorphism
- NT prograde metamorphism
- NT regional metamorphism
- NT retrograde metamorphism
- NT shock metamorphism
- NT thermal metamorphism
- NT ultrametamorphism
- SA anatexis
- SA aureoles
- SA cataclasis
- SA complexes
- SA deformation
- SA diagenesis
- SA feldspathization
- SA folds
- SA foliation
- SA geochemistry
- SA geochronology
- SA grade
- SA greenschist facies
- SA high pressure
- SA high temperature
- SA igneous rocks
- SA intrusions
- SA isograds
- SA lineation
- SA lithogeochemistry
- SA low pressure
- SA low temperature
- SA metamorphic processes
- SA metamorphic rocks
- SA metaplutonic rocks
- SA metasomatism
- SA meteor craters
- SA migration of elements
- SA orogeny
- SA overprinting
- SA P-T conditions
- SA paragenesis
- SA petrology
- SA phase equilibria
- SA protoliths
- SA recrystallization
- SA rheomorphism
- SA shatter cones
- SA shock waves
- SA slaty cleavage
- SA supracrustals
- SA uralitization
- SA zoning

metapelite (1978)
- BT metasedimentary rocks
- BT metamorphic rocks

metaperidotite (1978)
- BT metaigneous rocks
- BT metamorphic rocks
- SA peridotites

metaplutonic rocks (1978)
Before 1978, also search metamorphic rocks AND metaplutonic.
- BT metamorphic rocks
- SA metamorphism
- SA plutonic rocks
- SA rocks

metapyroxenite (1978)
- BT metaigneous rocks
- BT metamorphic rocks
- SA pyroxenite

metarhyolite (1978)
- BT metaigneous rocks
- BT metamorphic rocks
- SA rhyolites

metasandstone (1978)
- BT metasedimentary rocks
- BT metamorphic rocks
- SA sandstone

metasedimentary
A valid index term through 1977. After 1977, use metasedimentary rocks.

metasedimentary rocks (1978)
Before 1978, also search metasedimentary. Autoposting of this term to metabentonite ended in 1981.
- BT metamorphic rocks
- NT khondalite
- NT meta-arkose
- NT metachert
- NT metaconglomerate
- NT metagraywacke
- NT metalimestone
- NT metapelite
- NT metasandstone

NT metasiltstone
NT paragneiss
SA metabentonite
SA rocks
SA sedimentary rocks
SA supracrustals

metasediments
A valid term through 1974. Use metasedimentary rocks.

metasiltstone (1978)
BT metasedimentary rocks
BT metamorphic rocks
SA siltstone

metasomatic rocks (1978)
As of 1978, term is used for any rock produced by replacement processes at constant volume with little disturbance of textural or structural features. Beresite is considered a narrower term as of 1981. Autoposting of this term began in 1978. As of 1993, autoposting of metamorphic rocks was begun.
UF metasomatite
UF metasomatites
BT metamorphic rocks
NT fenite
NT greisen
NT propylite
NT serpentinite
NT skarn
NT steatite
NT tactite
NT talc rock
SA alteration
SA metasomatism
SA petrology
SA rocks

metasomatism (1978)
For documents before 1976, also search replacement. Autoposting of this term began in 1978. As of 1993, use hydrothermal processes for economic studies.
IN Index material and specific process where available.
NT albitization
NT alunitization
NT amphibolitization
NT argillization
NT chloritization
NT feldspathization
NT fenitization
NT greisenization
NT hydrothermal alteration
NT kaolinization
NT microclinization
NT muscovitization
NT palagonitization
NT propylitization
NT pyritization
NT pyrometasomatism
NT scapolitization
NT sericitization
NT serpentinization
NT spilitization
NT tourmalinization
NT uralitization
NT wall-rock alteration
NT zeolitization
SA alteration
SA diagenesis
SA dolomitization
SA enrichment
SA fenite
SA geochemistry
SA granitization
SA hydrothermal conditions
SA igneous rocks
SA intrusions
SA lithogeochemistry
SA localization

SA metamorphic rocks
SA metamorphism
SA metasomatic rocks
SA mineral deposits, genesis
SA P-T conditions
SA palingenesis
SA paragenesis
SA petrology
SA phase equilibria
SA plutons
SA pneumatolysis
SA silicification
SA zoning

metasomatite
use metasomatic rocks

metasomatites
use metasomatic rocks

metastibnite (1978)
BT sulfides
SA stibnite

Metatheria (1985)
Autoposting of this term began in 1989. Autoposting of Theria and Mammalia to this term began in 1989.
BT Theria
BT Mammalia
BT Tetrapoda
BT Vertebrata
BT Chordata
NT Marsupialia

metatuff (1978)
BT metaigneous rocks
BT metamorphic rocks
SA tuff

metavolcanic
A valid term through 1977.

metavolcanic rocks (1978)
Before 1978, also search metamorphic rocks AND metavolcanic.
BT metamorphic rocks
SA metadacite
SA rocks
SA supracrustals
SA volcanic rocks

Metazoa (1978)
Used as a general fossil term. Also search individual phyla and groups.
SA Invertebrata
SA Protista

meteor craters (1978)
Used for impact craters formed by the falling of a large meteorite onto a surface. Autoposting of impact features to this term began in 1989. Autoposting of impact craters to this term began in 1993.
UF meteorite crater
BT impact craters
BT impact features
SA astroblemes
SA breccia
SA crater lakes
SA craters
SA cryptoexplosion features
SA deformation
SA geochronology
SA geomorphology
SA impacts
SA metamorphism
SA meteorites
SA relative age
SA shatter cones
SA shock metamorphism
SA tectonics

meteoric iron
use iron meteorites

meteoric water (1989)
UF meteoric waters

SA connate waters
SA ground water
SA oil-water interface
SA pore water
SA reservoir rocks
SA water

meteoric waters
use meteoric water

meteorite crater
use meteor craters

meteorite flux (1981)
Before 1981, search meteorites AND flux.
SA meteorites

meteorites (1978)
Used for any meteoroid that has fallen to the Earth's surface in one piece or in fragments without being completely vaporized.
IN Index specific meteorite or group of meteorites if applicable.
UF cosmolites
UF skystones
NT Allan Hills Meteorites
NT Coldwater Meteorites
NT Elephant Moraine Meteorites
NT iron meteorites
NT Lewis Cliff Meteorites
NT micrometeorites
NT Odessa Meteorite
NT stony irons
NT stony meteorites
NT Yamato Meteorites
SA asteroids
SA australite
SA calcium-aluminum inclusions
SA chaoite
SA chondrules
SA cohenite
SA comets
SA extraterrestrial geology
SA geochemistry
SA impact features
SA Impactite
SA impacts
SA interplanetary dust
SA isotopes
SA kamacite
SA lonsdaleite
SA majorite
SA maskelynite
SA meteor craters
SA meteorite flux
SA meteoroids
SA meteors
SA moissanite
SA Moon
SA niningerite
SA oldhamite
SA parent bodies
SA petrology
SA planetology
SA plessite
SA relict materials
SA roedderite
SA schreibersite
SA shatter cones
SA sintering
SA solar system
SA suevite
SA taenite
SA tektites
SA tetrataenite
SA troilite

meteoroids (1978)
SA asteroids
SA cosmic dust
SA interplanetary space
SA meteorites
SA meteors
SA parent bodies
SA planets

meteorology (1978)
Primarily used for special indexes.
SA aerosols
SA air-sea interface
SA atmosphere
SA atmospheric precipitation
SA boundary interactions
SA circulation
SA climate
SA clouds
SA convection
SA diffusion
SA droplets
SA electromagnetic waves
SA geophysics
SA humidity
SA hurricanes
SA hydrologic cycle
SA hydrology
SA ions
SA lightning
SA particles
SA raindrops
SA storms
SA turbidity
SA turbulence
SA water
SA waves
SA winds

meteors (1978)
UF shooting star
SA comets
SA interplanetary space
SA meteorites
SA meteoroids
SA planetology

methane (1978)
Autoposting of alkanes and aliphatic hydrocarbons to this term began in 1989.
BT alkanes
BT aliphatic hydrocarbons
BT hydrocarbons
BT organic materials
SA coalbed methane
SA natural gas

methanethiomethane
use dimethyl sulfide

methodology
use methods

methods (1978)
This term applies to theoretical and experimental studies, while the term techniques is used for discussion of samples.
UF methodology
SA applications
SA beneficiation
SA biogeochemical methods
SA centrifuge methods
SA diamond anvil cells
SA downhole methods
SA fission-track dating
SA geobotanical methods
SA geochemical methods
SA geological methods
SA geomorphological methods
SA geophysical methods
SA graphic methods
SA ground methods
SA helicopter methods
SA high-resolution methods
SA hydrological methods
SA laser methods
SA laser ranging
SA marine methods
SA mathematical methods
SA measurement-while-drilling
SA microscope methods
SA microwave methods
SA multichannel methods
SA neutron methods

SA New Austrian Tunnelling Method
SA new methods
SA optimization
SA passive methods
SA photogeologic methods
SA physical methods
SA pillars
SA powder method
SA precision
SA quantitative analysis
SA radio-wave methods
SA satellite methods
SA side-scanning methods
SA single-crystal method
SA sonar methods
SA statistical analysis
SA techniques
SA Thellier Method
SA total-field methods
SA ultrasonic methods
SA vertical-gradient methods
SA wet methods

Methow Basin (1989)
Sedimentary basin extending from N Washington to SW British Columbia.
IN Index Washington and/or British Columbia as applicable.
BT North America
SA British Columbia
SA Methow River valley
SA Washington

Methow River valley (1978)
Okanogan County on British Columbia border. Also search Methow River.
BT Okanogan County Washington
BT Washington
BT United States
SA Methow Basin

methoxychlor
use DDT

methyl sulfide
use dimethyl sulfide

Metohija (1978)
District in SW Serbia, SE Yugoslavia forming part of autonomous province of Kosovo-Metohija.
BT Serbia
BT Yugoslavia
BT Southern Europe
BT Europe
SA Kosovo-Metohija

Meurthe-et-Moselle
No longer a valid term for GeoRef. See Meurthe-et-Moselle France.

Meurthe-et-Moselle France (1993)
Department in NE France.
CO N483000N493000
      E0071500E0053500
BT France
BT Western Europe
BT Europe
SA Laneuville-devant-Nancy Boring
SA Lorraine

Meuse
No longer a valid term for GeoRef. See Meuse France.

Meuse France (1993)
Department in NE France.
CO N482500N493500
      E0055200E0045000
BT France
BT Western Europe
BT Europe
NT Verdun France

Meuse River (1978)
Rises in NE France, flows N across E Belgium and, as the Maas River, flows through the Netherlands entering the North Sea through its estuary, the Hollandsch Diep. Also search Meuse.
IN Index countries as applicable.
UF Maas River
BT Western Europe
BT Europe
SA Belgium
SA France
SA Netherlands

Meuse Valley (1978)
River valley. Known as the Maas Valley in the Netherlands.
IN Index countries as applicable.
UF Maas Valley
BT Western Europe
BT Europe
SA Belgium
SA France
SA Netherlands

Mexicali
No longer a valid term for GeoRef. As of 1993 see Mexicali Mexico.

Mexicali Mexico (1993)
BT Baja California Mexico
BT Mexico

Mexicali Valley (1989)
Baja California, NW Mexico and extends into Imperial Valley irrigation area, SE California.
IN Index regions as applicable.
BT North America
SA Baja California
SA California
SA Imperial County California
SA Imperial Valley
SA Mexico

Mexican neovolcanic belt
use Mexican volcanic belt

Mexican volcanic belt (1989)
SW Mexico.
IN Index states as applicable.
CO N160000N220000
      W0940000W1050000
UF Mexican neovolcanic belt
BT Mexico
SA Colima Mexico
SA Guerrero Mexico
SA Jalisco Mexico
SA Michoacan Mexico
SA Nayarit Mexico
SA Oaxaca Mexico
SA Puebla Mexico
SA Sierra Madre del Sur
SA Sierra Madre Occidental

Mexico (1978)
CO N143000N324300
      W0864500W1170000
NT Baja California
NT Baja California Mexico
NT Baja California Sur Mexico
NT Campeche Mexico
NT Chiapas Mexico
NT Chihuahua Mexico
NT Coahuila Mexico
NT Colima Mexico
NT Colorado River delta
NT Durango Mexico
NT Guanajuato Mexico
NT Guerrero Mexico
NT Hidalgo Mexico
NT Jalisco Mexico
NT Mexican volcanic belt
NT Mexico state
NT Michoacan Mexico
NT Michoacan-Guanajuato volcanic field
NT Nayarit Mexico
NT Nuevo Leon Mexico
NT Oaxaca Mexico
NT Popocatepetl
NT Puebla Mexico
NT Quintana Roo Mexico
NT Sabinas Basin
NT San Luis Potosi Mexico
NT Sierra Madre del Sur
NT Sierra Madre Occidental
NT Sierra Madre Oriental
NT Sinaloa Mexico
NT Sonora Mexico
NT Tabasco Mexico
NT Tamaulipas Mexico
NT Valley of Mexico
NT Veracruz Mexico
NT Yucatan Mexico
NT Yucatan Peninsula
NT Zacatecas Mexico
NT Zaragoza Mexico
SA Agua Blanca Fault
SA Aguacate Group
SA Alisitos Formation
SA Altiplano
SA Buda Limestone
SA California Current
SA Cerro Prieto Fault
SA Coast Range Ophiolite
SA Colorado River basin
SA Cupido Formation
SA Difunta Group
SA Eagle Ford Formation
SA Gulf of California
SA La Quinta Formation
SA Laguna Madre
SA Leon City Spain
SA Madrid Formation
SA Maverick Basin
SA Mexicali Valley
SA Mexico City earthquake 1985
SA North America
SA North American Cordillera
SA Peninsular Ranges Batholith
SA Rio Grande Valley
SA Rosario Formation
SA San Antonio Mine
SA San Felipe Formation
SA San Jacinto Fault
SA San Juan River
SA Santa Clara Meteorite
SA Santa Susana Formation
SA Sierra Madre
SA Sonoran Desert
SA Toluca Meteorite
SA Transcontinental Arch
SA Victoria earthquake 1980
SA Western Overthrust Belt

Mexico Basin
Not a valid term for GeoRef. See Sigsbee Deep or Valley of Mexico.

Mexico City
No longer a valid term for GeoRef as of 1993.
use Mexico City Mexico

Mexico City earthquake 1985 (1993)
Michoacan, Mexico. Before 1993, also search Mexico earthquake 1985 and Michoacan earthquake 1985.
UF Michoacan earthquake 1985
BT earthquakes
SA Mexico
SA Michoacan Mexico

Mexico City Mexico (1993)
Central Mexico. In Federal District Mexico. Before 1993, also search Mexico City, and Mexico D.F.
UF Mexico City
BT Federal District Mexico
BT Mexico state
BT Mexico
SA Valley of Mexico

Mexico D.F.
use Federal District Mexico

Mexico state (1978)
State N, E and W of Federal District.
BT Mexico
NT Federal District Mexico
SA Popocatepetl
SA Valley of Mexico

meymechite
use meimechite

Mezen
No longer a valid term for GeoRef. As of 1993, see Mezen Russian Federation.

Mezen River basin (1978)
N central Arkhangelsk Oblast in NW European Russian Federation. Also search Mezen River. This term has multiple hierarchies.
BT1 Russian Federation
BT1 Commonwealth of Independent States
BT2 Europe
SA Arkhangelsk Russian Federation
SA Komi Russian Federation

Mezen Russian Federation (1993)
Town on the Mezen River in Arkhangelsk Oblast in NW European Russian Federation. Before 1993, also search Mezen. This term has multiple hierarchies.
BT1 Arkhangelsk Russian Federation
BT1 Russian Federation
BT1 Commonwealth of Independent States
BT2 Arkhangelsk Russian Federation
BT2 Europe

Mezhrechye
No longer a valid term for GeoRef. As of 1993, see Mezhrechye Belarus.

Mezhrechye Belarus (1993)
Village in W Belarus. This term has multiple hierarchies.
BT1 Belarus
BT1 Europe
BT2 Belarus
BT2 Commonwealth of Independent States

Mg
use magnesium

Mg-24/Mg-25
use Mg-25/Mg-24

Mg-24/Mg-26
use Mg-26/Mg-24

Mg-25/Mg-24 (1985)
Autoposting of broader terms began in 1989. This term has multiple hierarchies.
UF Mg-24/Mg-25
BT1 stable isotopes
BT1 isotopes
BT2 magnesium
BT2 alkaline earth metals
BT2 metals
SA isotope ratios

Mg-26/Mg-24 (1985)

Autoposting of broader terms began in 1989. This term has multiple hierarchies.
UF Mg-24/Mg-26
BT1 stable isotopes
BT1 isotopes
BT2 magnesium
BT2 alkaline earth metals
BT2 metals
SA isotope ratios

MHD
Not a valid term for GeoRef. See magnetohydrodynamics if applicable.

Miami
Valid through 1988. Search in combination with state term. After 1988, use specific city-state term.

Miami Florida (1989)
City in SE Florida. Before 1989, search Miami AND Florida.
CO N254500N254500 W0801500W0801500
BT Dade County Florida
BT Florida
BT United States

Miami Limestone (1978)
According to the latest published Lexique, the age of Miami Oolite is Pleistocene. It is a soft white oolitic limestone, containing streaks of thin irregular layers of calcite separating less crystalline streaks. In S Florida.
BT Pleistocene
BT Quaternary
BT Cenozoic
SA Florida

Mianwali
No longer a valid term for GeoRef. See Mianwali Pakistan.

Mianwali Pakistan (1993)
Town and district in NW Punjab Province, NE Pakistan.
BT Punjab Pakistan
BT Pakistan
BT Indian Peninsula
BT Asia

Miaoli
No longer a valid term for GeoRef. As of 1993 see Miaoli Taiwan.

Miaoli Taiwan (1993)
Town in NW Taiwan.
BT Taiwan
BT Far East
BT Asia

miargyrite (1978)
BT sulfantimonites
BT sulfosalts

miaskite (1978)
BT syenites
BT plutonic rocks
BT igneous rocks

mica
As of 1981, no longer a valid term for GeoRef. See mica group. Use mica deposits for mica as a commodity.

mica deposits (1981)
Before 1981, search mica AND deposits.
SA glauconite deposits
SA insulation materials
SA mica group
SA Plotina Deposit
SA vermiculite deposits

mica group (1978)

Autoposting of this term began in 1981.
BT sheet silicates
BT silicates
NT biotite
NT celadonite
NT clintonite
NT fluor-phlogopite
NT glauconite
NT lepidolite
NT lepidomelane
NT listvenite
NT margarite
NT muscovite
NT paragonite
NT phengite
NT phlogopite
NT polylithionite
NT sericite
NT zinnwaldite
SA hydrobiotite
SA hydromica
SA hydromuscovite
SA mica deposits
SA vermiculite

mica schist
As of 1981, no longer a valid term for GeoRef. Use terms that indicate specific mica (i.e. biotite schist or muscovite schist). From 1978-1980, schists was autoposted to this term.

Michelle Formation (1978)
Composed of black calcareous shale and aphanitic silty limestone. In W Northwest Territories and Yukon Territory.
BT Middle Devonian
BT Devonian
BT Paleozoic
SA Canada
SA Northwest Territories
SA Yukon Territory

michenerite (1978)
Autoposting of bismuthides to this term began in 1981. Autoposting of sulfides to this term ended in 1989.
BT bismuthides

Michigamme Formation (1993)
Lower Proterozoic. In Baraga Group. Ontario, NW Michigan, NE Wisconsin.
BT lower Proterozoic
BT Proterozoic
BT upper Precambrian
BT Precambrian
SA Baraga Group
SA Michigan
SA Ontario
SA Wisconsin

Michigan (1978)
Autoposting of this term began in 1978.
CO N414500N473000 W0823000W0901500
BT United States
NT Michigan Lower Peninsula
NT Michigan Upper Peninsula
SA Animikie Group
SA Antrim Shale
SA Baraga Group
SA Bass Islands Dolomite
SA Berea Sandstone
SA Canadian Shield
SA Charlevoix
SA Clinton Group
SA Copper Harbor Conglomerate
SA Detroit River
SA Detroit River Group
SA Dundee Limestone
SA Eau Claire Formation
SA Freda Sandstone
SA Glenwood Shale
SA Grand River
SA Great Lakes region
SA Green Bay
SA Guelph Formation
SA Hemlock Formation
SA Jacobsville Sandstone
SA Keweenawan
SA Lake Algonquin
SA Lake Chicago
SA Lake Erie
SA Lake Huron
SA Lake Maumee
SA Lake Michigan
SA Lake Saint Clair
SA Lake Superior
SA Lake Superior region
SA Lockport Formation
SA Lucas Formation
SA Marquette Range Supergroup
SA Medina Formation
SA Michigamme Formation
SA Michigan Basin
SA Midcontinent
SA Mount Simon Sandstone
SA Negaunee Iron Formation
SA Niagara Escarpment
SA Nonesuch Shale
SA Oronto Group
SA Penokean Orogeny
SA Perch Lake
SA Portage Lake Lava Series
SA Racine Dolomite
SA Saginaw Formation
SA Saint Clair River
SA Saint Clair River delta
SA Saint Peter Sandstone
SA Salina Group
SA Southern Province
SA Sylvania Formation
SA Traverse Group
SA Trenton Group
SA Tyler Formation
SA University of Michigan
SA Utica Shale

Michigan Basin (1978)
Term revised in 1989. Structural basin with its center in the Michigan Lower Peninsula. Located in NE Illinois, N Indiana, Michigan, NW Ohio, E Wisconsin and W Ontario.
IN Index states and province as applicable.
UF Lake Michigan Basin
BT North America
SA Bass Islands Dolomite
SA Illinois
SA Indiana
SA Michigan
SA Michigan Lower Peninsula
SA Michigan Upper Peninsula
SA Ohio
SA Ontario
SA Wisconsin

Michigan Lower Peninsula (1981)
S part of state S of Straits of Mackinack. Autoposting of Michigan and United States began in 1989. Autoposting of this term to counties and cities began in 1989. Before 1981, search Michigan AND Lower Peninsula; Southern Peninsula.
UF Lower Peninsula, Michigan
UF Southern Peninsula, Michigan
BT Michigan
BT United States
NT Alcona County Michigan
NT Alpena County Michigan
NT Berrien County Michigan
NT Calhoun County Michigan
NT Cass County Michigan
NT Charlevoix County Michigan
NT Clinton County Michigan
NT Crawford County Michigan
NT Eaton County Michigan
NT Genesee County Michigan
NT Grand Traverse County Michigan
NT Gratiot County Michigan
NT Ingham County Michigan
NT Jackson County Michigan
NT Kalamazoo County Michigan
NT Kent County Michigan
NT Lake County Michigan
NT Lansing Michigan
NT Livingston County Michigan
NT Macomb County Michigan
NT Manistee County Michigan
NT Mason County Michigan
NT Midland County Michigan
NT Monroe County Michigan
NT Otsego County Michigan
NT Ottawa County Michigan
NT Presque Isle County Michigan
NT Roscommon County Michigan
NT Saint Clair County Michigan
NT Washtenaw County Michigan
NT Wayne County Michigan
SA Lake Saint Clair
SA Michigan Basin
SA Michigan Upper Peninsula

Michigan Upper Peninsula (1981)
N part of state between Lake Michigan and Lake Superior. Autoposting of Michigan and United States began in 1989. Autoposting of this term to counties and cities began in 1989. Before 1981, search Michigan AND Upper Peninsula.
UF Northern Peninsula, Michigan
UF Upper Peninsula, Michigan
BT Michigan
BT United States
NT Baraga County Michigan
NT Chippewa County Michigan
NT Delta County Michigan
NT Dickinson County Michigan
NT Gogebic County Michigan
NT Houghton County Michigan
NT Iron County Michigan
NT Keweenaw County Michigan
NT Keweenaw Peninsula
NT Marquette County Michigan
NT Ontonagon County Michigan
NT Porcupine Mountains
SA Michigan Basin
SA Michigan Lower Peninsula
SA Perch Lake

Michipicoten Belt (1993)
Greenstone belt in Wawa subprovince of the Superior Province, Wawa and Gamitagama area of S Ontario.
BT Superior Province
BT Canadian Shield
BT North America
SA Ontario
SA Wawa Belt

Michipicoten Island (1978)
NE Lake Superior.
BT Ontario
BT Eastern Canada
BT Canada
SA Algoma District Ontario
SA Thunder Bay District Ontario

Michoacan
No longer a valid term for GeoRef. As of 1993 see Michoacan Mexico.

Michoacan earthquake 1985
use Mexico City earthquake 1985

Michoacan Mexico (1993)
State in SW Mexico, bordering on the Pacific Ocean. Before 1993 also search Michoacan AND Mexico.
CO N175800N202500 W1001500W1034700
BT Mexico
NT Los Azufres
NT Paricutin
SA Mexican volcanic belt
SA Mexico City earthquake 1985
SA Michoacan-Guanajuato volcanic field

Michoacan-Guanajuato volcanic field (1989)
Michoacan and Guanajuato, S central Mexico.
IN Index states as applicable.
BT Mexico
SA Guanajuato Mexico
SA Michoacan Mexico

micrinite (1978)
BT inertinite
BT macerals

micrite (1978)
BT limestone
BT carbonate rocks
BT sedimentary rocks
SA biomicrite
SA micritization
SA microcrystalline limestone

micritization (1989)
SA carbonate rocks
SA grain size
SA micrite

microbiological geology
use geomicrobiology

microbiology, geological
use geomicrobiology

microbreccia (1978)
BT clastic rocks
BT sedimentary rocks
SA breccia
SA sandstone
SA terrigenous materials

microcline (1978)
BT alkali feldspar
BT feldspar group
BT framework silicates
BT silicates
SA amazonite
SA K-feldspar
SA microclinization

microclinization (1989)
BT metasomatism
SA alteration
SA microcline
SA processes

Microcodium (1978)
Autoposting of thallophytes and microfossils began in 1990. This term has multiple hierarchies.
BT1 algae
BT1 thallophytes
BT1 Plantae
BT2 algae
BT2 microfossils

microcomputers (1978)
Before 1981, also search microprocessors. Before 1993, also search personal computers or PCs if applicable.
UF microprocessors
BT computers
SA computer networks
SA data processing
SA hand calculators
SA minicomputers
SA workstations

microcontinents (1993)
SA aseismic ridges
SA bottom features
SA continental crust
SA ocean floors
SA plate tectonics
SA tectonics

microcracks (1978)
Before 1981, also search microfractures.
UF microfractures
BT cracks
SA deformation
SA engineering geology
SA fractures
SA rock mechanics

microcraters (1978)
SA craters
SA Moon

microcrystalline limestone (1982)
BT limestone
BT carbonate rocks
BT sedimentary rocks
SA micrite

microdiorite (1978)
BT diorites
BT plutonic rocks
BT igneous rocks

microearthquakes (1978)
BT earthquakes
SA seismology

microelement
use trace elements

microfacies (1978)
SA facies
SA micropaleontology
SA petrography
SA thin sections

microfauna
As of 1981, no longer a valid term for GeoRef.
use microfossils

microflora
use microfossils

microfossils (1978)
Before 1981, also search microflora; microfauna. Autoposting of this term began in 1978. As of 1993, this term is autoposted to Protista.
UF microfauna
UF microflora
NT algae
NT algal flora
NT Conodonta
NT conodonts
NT foraminifers
NT Ostracoda
NT ostracods
NT palynomorphs
NT problematic microfossils
NT Protista
NT radiolarians
NT scolecodonts
SA fossils
SA geomicrobiology
SA ichthyoliths
SA index fossils
SA microorganisms
SA micropaleontology
SA otoliths
SA pollen

microfractures
As of 1981, no longer a valid term for GeoRef.
use microcracks

microgabbro (1978)
BT gabbros
BT plutonic rocks
BT igneous rocks

microgranite (1978)
BT granites
BT plutonic rocks
BT igneous rocks

microhardness (1978)
SA hardness
SA minerals
SA physical properties

microlite (1978)
From 1978-1980, halides and oxides were autoposted to this term. Autoposting of fluorides, niobates and tantalates began in 1981. As of 1989, niobotantalates, halides and oxides are autoposted to this term. This term has multiple hierarchies.
BT1 fluorides
BT1 halides
BT2 niobotantalates
BT2 niobates
BT2 oxides
BT3 niobotantalates
BT3 tantalates
BT3 oxides
SA pyrochlore

micrometeorites (1978)
Very small meteorite or meteoritic particle with diameter generally less than a millimeter.
BT meteorites

micromorphology (1978)
SA morphology
SA soils

Micronesia (1978)
Collective name for islands E of the Philippine Islands and S of Japan. In 1981, broader term changed from Pacific Ocean to Oceania. As of 1981, Caroline Islands is considered a narrower term. Gilbert Islands was a narrower term from 1981 to 1985, when Gilbert Islands became Kiribati. Kiribati is a narrower term as of 1993. Autoposting of this term began in 1978.
IN Index island groups as applicable.
CO S040000N210000 E1770000E1440000
UF Federated States of Micronesia
BT Oceania
NT Caroline Islands
NT Mariana Islands
NT Marshall Islands
NT Nauru Island
SA Kiribati
SA Pacific Ocean

microorganisms (1978)
For several types.
SA algae
SA bacteria
SA biota
SA faunal studies
SA floral studies
SA fungi
SA geomicrobiology
SA microfossils
SA pollen
SA Protista
SA soils

micropaleontology (1978)
Used for the discipline as a whole.
SA biologic evolution
SA foraminifera
SA fossils
SA geomicrobiology
SA microfacies
SA microfossils
SA paleobotany
SA paleontology
SA palynology
SA sample preparation
SA stratigraphy

micropegmatite (1978)
BT granites
BT plutonic rocks
BT igneous rocks
SA graphic granite
SA pegmatite

microperthite (1978)
Autoposting of alkali feldspar to this term began in 1989.
BT alkali feldspar
BT feldspar group
BT framework silicates
BT silicates
SA perthite

micropetrological unit
use macerals

microplankton
use plankton

microplates (1978)
SA plate tectonics
SA plates

microprobe, electron
use electron probe

microprocessors
As of 1981, no longer a valid term for GeoRef.
use microcomputers

micropulsations
As of 1993, no longer a valid term for GeoRef. Usually out-of-scope. Used with aurora.

microscope methods (1978)
Before 1972, also search microscope techniques; microscopic methods.
UF microscope techniques
UF microscopic methods
SA goniometry
SA methods
SA mineralogy
SA optical mineralogy
SA ore microscopy
SA petrography
SA petrology
SA polished sections
SA reflectance
SA sedimentary petrology
SA thin sections
SA undulatory extinction
SA universal stage

microscope techniques
Not a valid GeoRef index term after 1970. Was used on level 1 in subfile N.
use microscope methods

microscopic methods
Not a valid GeoRef index term after 1971. Was used on level 1 in subfiles E, N and B.
use microscope methods

microscopy, electron
use electron microscopy

microscopy, ore
use ore microscopy

microseismic methods (1985)
SA geophysical methods
SA microseisms

SA seismic methods
SA seismic surveys

microseismicity
use seismicity

microseisms (1978)
SA earthquakes
SA microseismic methods
SA noise
SA seismology

microstructure
A valid term through 1978. After 1978, use ultrastructure.

microstylolites (1978)
BT secondary structures
BT sedimentary structures
SA stylolites
SA stylolitization

microsyenite (1978)
BT syenites
BT plutonic rocks
BT igneous rocks

microtectonics
A valid term through 1975. After 1975, use structural analysis.

microtektites (1978)
BT tektites

microthermometry (1985)
SA fluid inclusions
SA geologic thermometry
SA phase equilibria

Microtus (1993)
Genus.
BT Cricetidae
BT Myomorpha
BT Rodentia
BT Eutheria
BT Theria
BT Mammalia
BT Tetrapoda
BT Vertebrata
BT Chordata

microwave exploration
use microwave methods

microwave methods (1978)
Before 1978, also search microwave. From 1981-1984, geophysical methods was autoposted to this term. Before 1971, also search microwave exploration.
UF microwave exploration
UF microwave surveys
SA geophysical methods
SA ground-penetrating radar
SA methods
SA radar methods
SA remote sensing
SA telemetry
SA well-logging

microwave spectroscopy (1978)
Before 1978, also search spectroscopy AND microwave.
BT spectroscopy
SA analysis
SA chemical analysis

microwave surveys
use microwave methods

microzonation (1985)
SA earthquake prediction
SA earthquakes
SA geologic hazards
SA ground motion
SA seismic response
SA seismic risk
SA seismic zoning

Mid-Antarctic Ridge (1981)
BT Antarctic Ocean

Mid-Arctic Ocean Ridge (1981)
BT Arctic Ocean
SA mid-ocean ridges

Mid-Atlantic Ridge (1978)
Extends parallel to the continental margins in mid ocean in both the North and South Atlantic Ocean. Rises 6000 ft. above the ocean floor and surfaces as the Azores, Ascension, Saint Helena, and Tristan de Cunha islands.
CO S570000N600000
  E0030000W0620000
BT Atlantic Ocean
NT Hayes fracture zone
NT Kane fracture zone
NT North Atlantic Ridge
NT South Atlantic Ridge
SA Atlantis fracture zone
SA DSDP Site 332
SA DSDP Site 332B
SA DSDP Site 333
SA DSDP Site 334
SA DSDP Site 335
SA DSDP Site 395
SA DSDP Site 396
SA DSDP Site 407
SA DSDP Site 408
SA DSDP Site 410
SA DSDP Site 519
SA DSDP Site 520
SA DSDP Site 521
SA DSDP Site 556
SA DSDP Site 558
SA DSDP Site 561
SA DSDP Site 562
SA DSDP Site 563
SA DSDP Site 564
SA DSDP Site 606
SA DSDP Site 607
SA DSDP Site 609
SA FAMOUS
SA Gibbs fracture zone
SA Iceland Research Drilling Project
SA Leg 37
SA Leg 39
SA Leg 78B
SA Leg 82
SA Leg 94
SA Leg 106
SA Leg 109
SA median valley
SA mid-ocean ridges
SA Oceanographer fracture zone
SA ODP Site 648
SA ODP Site 649
SA ODP Site 662
SA ODP Site 663
SA ODP Site 664
SA ODP Site 669
SA ODP Site 701
SA TAG hydrothermal field
SA Vema fracture zone

Mid-Cayman Rise (1981)
NW Caribbean Sea, S of Cuba.
CO N190000N200000
  W0740000W0820000
BT Caribbean Sea
BT North American Atlantic
BT North Atlantic
BT Atlantic Ocean

Mid-Continent
use Midcontinent

Mid-continent Rift
use Keweenawan Rift

Mid-Indian Ocean Ridge
use Mid-Indian Ridge

Mid-Indian Ridge (1978)
Extends from S of the Maldives, S and SE into Antarctic waters. Autoposting of this term began in 1978.
UF Indian Ocean Ridge
UF Indian Ridge
UF Mid-Indian Ocean Ridge
BT Indian Ocean
NT Central Indian Ridge
NT Southeast Indian Ridge
SA mid-ocean ridges

mid-ocean ridge basalts (1989)
UF MORB
BT basalts
BT volcanic rocks
BT igneous rocks
SA mid-ocean ridges
SA sea-floor spreading
SA spreading centers

mid-ocean ridges (1978)
UF mid-oceanic ridges
UF midoceanic ridges
UF ocean ridges
SA aseismic ridges
SA black smokers
SA bottom features
SA Carlsberg Ridge
SA Central Indian Ridge
SA Chile Ridge
SA East Pacific Rise
SA fracture zones
SA Galapagos Rift
SA Juan de Fuca Ridge
SA median valley
SA Mid-Arctic Ocean Ridge
SA Mid-Atlantic Ridge
SA Mid-Indian Ridge
SA mid-ocean ridge basalts
SA ocean floors
SA Pacific-Antarctic Ridge
SA sea-floor spreading
SA Southeast Indian Ridge
SA Southwest Indian Ridge
SA TAG hydrothermal field
SA transform faults

mid-oceanic ridges
use mid-ocean ridges

Mid-Pacific Rise (1978)
SA DSDP Site 313
SA Pacific Ocean

Midcontinent (1978)
Mid-America between the Appalachians and the Rocky Mountains.
IN Index states as applicable.
UF Mid-Continent
BT United States
SA Arkansas
SA Colorado
SA Illinois
SA Indiana
SA Iowa
SA Kansas
SA Kentucky
SA Michigan
SA Midcontinent geophysical anomaly
SA Minnesota
SA Missouri
SA Montana
SA Nebraska
SA New Mexico
SA North American Craton
SA North Dakota
SA Ohio
SA Oklahoma
SA Reelfoot Rift
SA South Dakota
SA Tennessee
SA Texas
SA Transcontinental Arch
SA Wisconsin
SA Wyoming

Midcontinent geophysical anomaly (1985)
Central and W Iowa, NE Kansas and SE Nebraska.
IN Index states as applicable.
UF Midcontinent high
BT United States
SA Iowa
SA Kansas
SA Midcontinent
SA Nebraska

Midcontinent high
use Midcontinent geophysical anomaly

Midcontinent Rift
use Keweenawan Rift

Midcontinent Rift System
use Keweenawan Rift

Middendorf Formation (1985)
E Georgia and E South Carolina.
BT Upper Cretaceous
BT Cretaceous
BT Mesozoic
SA Atlantic Coastal Plain
SA Georgia
SA South Carolina

middens (1985)
SA archaeological sites
SA archaeology
SA diet
SA paleoecology

Middle America Trench (1978)
Off W coast of Mexico and Central America.
BT Pacific Ocean
SA DSDP Site 565
SA DSDP Site 568
SA DSDP Site 570
SA Leg 84

Middle Atlantic Bight (1985)
Off E coast of United States, from Cape Hatteras to Cape Cod.
CO N351500N420500
  W0695500W0760000
BT United States
SA Atlantic Coastal Plain
SA North American Atlantic

Middle Atlas (1978)
Range of the Atlas Mountains N and E of the High Atlas Mountains in N central Morocco. This term has multiple hierarchies.
BT1 Moroccan Atlas Mountains
BT1 Atlas Mountains
BT1 North Africa
BT1 Africa
BT2 Moroccan Atlas Mountains
BT2 Morocco
BT2 North Africa
BT2 Africa

Middle Cambrian (1978)
Autoposting of this term began in 1978.
BT Cambrian
BT Paleozoic
NT Acadian
NT Barrandian
NT Bright Angel Shale
NT Burgess Shale
NT Carlton Rhyolite
NT Flathead Sandstone
NT Marjum Formation
NT Metaline Limestone
NT Pagoda Formation
NT Wheeler Formation
SA Lower Cambrian
SA Upper Cambrian

Middle Carboniferous (1978)
BT Carboniferous
BT Paleozoic
SA Dinantian
SA Upper Carboniferous

middle Cenozoic (1989)
BT Cenozoic

Middle Cretaceous (1978)
BT Cretaceous
BT Mesozoic
NT Bhuj Series
NT Winton Formation
SA Cenomanian
SA Lower Cretaceous
SA Turonian
SA Upper Cretaceous

Middle Devonian (1978)
Autoposting of this term began in 1978.
BT Devonian
BT Paleozoic
NT Balaklala Rhyolite
NT Boyle Dolomite
NT Cedar Valley Formation
NT Columbus Limestone
NT Copley Greenstone
NT Couvinian
NT Delaware Limestone
NT Detroit River Group
NT Dundee Limestone
NT Eifelian
NT Elk Point Group
NT Fisset Brook Formation
NT Givetian
NT Hamilton Group
NT Jeffersonville Limestone
NT Ludlowville Formation
NT Mahantango Formation
NT Marcellus Shale
NT Michelle Formation
NT Moscow Formation
NT Nahanni Formation
NT Onondaga Limestone
NT Prairie Evaporite
NT Saint Laurent Limestone
NT Sylvania Formation
NT Tioga Bentonite
NT Tully Limestone
NT Winnipegosis Formation
SA Lower Devonian
SA Lucas Formation
SA Martin Formation
SA Ross Formation
SA Swan Hills Formation
SA Upper Devonian
SA Zlichovian

Middle East (1978)
An indefinite and unofficial term comprising a region including Cyprus and the countries of SW Asia. Autoposting of this term to Iran began in 1981.
IN Index countries as applicable.
CO N250000N420000
    E0632000E0253000
BT Asia
NT Cyprus
NT Dead Sea
NT Dead Sea Rift
NT Iran
NT Iraq
NT Israel
NT Jordan
NT Jordan River
NT Jordan Valley
NT Lebanon
NT Mesopotamia
NT Palestine
NT Samaria
NT Syria
NT Tigris River
NT Turkey
NT Wadi Araba
NT Zagros
SA Arabian Peninsula
SA Egypt
SA Ghareb Formation
SA Kuwait
SA Menuha Formation
SA Oman
SA Qatar

SA Red Sea Basin
SA Sinjar Formation
SA Troodos Ophiolite
SA United Arab Emirates
SA Yemen
SA Zubair Formation

middle Eocene (1978)
Autoposting of this term began in 1978.
BT Eocene
BT Paleogene
BT Tertiary
BT Cenozoic
NT Aycross Formation
NT Carrizo Sand
NT Claiborne Group
NT Cockfield Formation
NT Cook Mountain Formation
NT Laney Shale Member
NT Lisbon Formation
NT Queen City Formation
NT Santee Limestone
NT Sparta Sand
NT Tallahatta Formation
NT Tyee Formation
NT Ulatisian
NT Yegua Formation
SA lower Eocene
SA Narizian
SA upper Eocene

Middle Franconia (1978)
Part of old historical region of Franconia. A hilly region in Franconian Jura in N central Bavaria, S Germany.
BT Bavaria Germany
BT Germany
BT Central Europe
BT Europe
SA Franconia
SA Franconian Jura

middle Holocene (1978)
BT Holocene
BT Quaternary
BT Cenozoic

Middle Jurassic (1978)
Autoposting of this term began in 1978.
BT Jurassic
BT Mesozoic
NT Aalenian
NT Bajocian
NT Bathonian
NT Callovian
NT Dogger
SA Coast Range Ophiolite
SA Lower Jurassic
SA Scarborough Formation
SA Upper Jurassic

middle Liassic (1981)
BT Lower Jurassic
BT Jurassic
BT Mesozoic
NT Domerian
NT Pliensbachian

middle Mesozoic (1978)
BT Mesozoic
SA upper Mesozoic

middle Miocene (1978)
Autoposting of this term began in 1978.
BT Miocene
BT Neogene
BT Tertiary
BT Cenozoic
NT Calvert Formation
NT Choptank Formation
NT Helvetian
NT Kirkwood Formation
NT Luisian
NT San Onofre Breccia

NT Serravallian
SA Badenian
SA Clarendonian
SA Grande Ronde Basalt
SA lower Miocene
SA Mohnian
SA Picture Gorge Basalt
SA Relizian
SA upper Miocene
SA Wanapum Basalt

Middle Mississippian (1978)
BT Mississippian
BT Carboniferous
BT Paleozoic

Middle Nile Valley (1981)
Used for the Nile Valley in Egypt.
CO N220000N301000
    E0330000E0303000
BT Egypt
BT North Africa
BT Africa
SA Nile Valley

middle Oligocene (1978)
BT Oligocene
BT Paleogene
BT Tertiary
BT Cenozoic
NT Bucatunna Formation
NT Byram Formation
NT Marianna Limestone
NT Vicksburg Group
SA lower Oligocene
SA upper Oligocene

Middle Ordovician (1978)
Autoposting of this term began in 1978.
BT Ordovician
BT Paleozoic
NT Ammonoosuc Volcanics
NT Bigby-Cannon Limestone
NT Black River Group
NT Blackriverian
NT Bromide Formation
NT Champlainian
NT Chazy Group
NT Chazyan
NT Cloridorme Formation
NT Decorah Shale
NT Dubuque Formation
NT Everton Formation
NT Galena Dolomite
NT Glenwood Shale
NT High Bridge Group
NT Holston Formation
NT Lenoir Limestone
NT Lexington Limestone
NT Llandeilian
NT Llanvirnian
NT Normanskill Formation
NT Partridge Formation
NT Platteville Formation
NT Saint Peter Sandstone
NT Schenectady Formation
NT Simpson Group
NT Stones River Group
NT Swan Peak Formation
NT Table Head Group
NT Trentonian
NT Winnipeg Formation
NT Woods Hollow Shale
SA Bala
SA Lower Ordovician
SA Upper Ordovician

middle Paleocene (1978)
BT Paleocene
BT Paleogene
BT Tertiary
BT Cenozoic

middle Paleozoic (1978)
BT Paleozoic
NT Hillabee Chlorite Schist

Middle Pennsylvanian (1978)
Autoposting of this term began in 1978.
BT Pennsylvanian
BT Carboniferous
BT Paleozoic
NT Allegheny Group
NT Atokan
NT Breathitt Formation
NT Carbondale Formation
NT Desmoinesian
NT Dugger Formation
NT Hermosa Formation
NT Kanawha Formation
NT Kewanee Group
NT Manakacha Formation
NT Paradox Member
NT Petersburg Formation
NT Spoon Formation
NT Staunton Formation
NT Tradewater Formation
SA Lower Pennsylvanian
SA Upper Pennsylvanian

Middle Permian (1978)
BT Permian
BT Paleozoic
SA Barren Measures
SA Kungurian
SA Lower Permian
SA Rotliegendes
SA Upper Permian

middle Pleistocene (1978)
Autoposting of this term began in 1978.
BT Pleistocene
BT Quaternary
BT Cenozoic
NT Elsterian

middle Pliocene (1978)
BT Pliocene
BT Neogene
BT Tertiary
BT Cenozoic
NT Bone Valley Formation
NT Peace Valley Beds
SA lower Pliocene
SA upper Pliocene

middle Precambrian (1978)
BT Precambrian
SA Aphebian

middle Proterozoic (1978)
Autoposting of Precambrian to this term ended in 1989. Autoposting of upper Precambrian and Precambrian to this term began in 1990.
BT Proterozoic
BT upper Precambrian
BT Precambrian
NT Aldridge Formation
NT Altyn Limestone
NT Apache Group
NT Belt Supergroup
NT Bonner Formation
NT Empire Formation
NT Fordham Gneiss
NT Freda Sandstone
NT Helena Formation
NT Helikian
NT Laxfordian
NT Missoula Group
NT Mount Shields Formation
NT Newland Limestone
NT Ravalli Group
NT Revett Quartzite
NT Saint Regis Formation
NT Shepard Formation
NT Snowslip Formation
NT Spokane Formation
NT Wallace Formation

middle Quaternary (1978)
BT Quaternary

BT Cenozoic
Middle Silurian (1978)
Autoposting of this term began in 1978.
BT Silurian
BT Paleozoic
NT Clinton Group
NT Guelph Formation
NT Keefer Sandstone
NT McKenzie Formation
NT Niagaran
NT Roberts Mountains Formation
NT Rochester Formation
NT Rose Hill Formation
NT Waldron Shale
SA Lower Silurian
SA Upper Silurian

Middle Stone Age
use Mesolithic

middle Tertiary (1978)
BT Tertiary
BT Cenozoic
SA lower Tertiary
SA upper Tertiary

Middle Triassic (1978)
Autoposting of this term began in 1978.
BT Triassic
BT Mesozoic
NT Anisian
NT Ladinian
NT Muschelkalk
SA Cordevolian
SA Lower Triassic
SA Serebryanka Formation
SA Upper Triassic

Middle Tunguska River
use Stony Tunguska River

Middle West (United States)
use Midwest

Middlesex County
Valid through 1988. Search in combination with state term. After 1988, use specific county-state term.

Middlesex County Connecticut (1989)
S Connecticut. Before 1989, also search Middlesex County AND Connecticut.
CO N411600N413800
W0721800W0724500
BT Connecticut
BT United States

Middlesex County Massachusetts (1989)
NE Massachusetts. Before 1989, also search Middlesex County AND Massachusetts.
CO N420800N424400
W0710200W0715300
BT Massachusetts
BT United States
NT Cambridge Massachusetts
SA Merrimack River valley

Middlesex County New Jersey (1989)
E New Jersey. Before 1989, also search Middlesex County AND New Jersey.
CO N401500N403700
W0741300W0743700
BT New Jersey
BT United States
SA Raritan Bay

Middlesex County Virginia (1989)
E Virginia. Before 1989, also search Middlesex County AND Virginia.

CO N373200N374700
W0761800W0764500
BT Virginia
BT United States

Middleton Island (1989)
S Alaska, in Gulf of Alaska.
IN Index Southern Alaska as applicable.
SA Gulf of Alaska
SA Southern Alaska
SA Yakataga Formation

Midland
Valid through 1988. Search in combination with state term. After 1988, use specific city-state term.

Midland Basin (1978)
W Texas. Also search Midland AND Texas.
IN Index counties or regions as applicable.
SA Permian Basin
SA Texas

Midland County
Valid through 1988. Search in combination with state term. After 1988, use specific county-state term.

Midland County Michigan (1989)
E central Michigan. Before 1989, also search Midland County AND Michigan.
CO N432400N434900
W0841000W0843700
BT Michigan Lower Peninsula
BT Michigan
BT United States

Midland County Texas (1989)
W Texas. Before 1989, also search Midland County AND Texas.
CO N314000N320600
W1014500W1021700
BT Texas
BT United States
NT Midland Texas
SA Azalea Field

Midland Texas (1989)
City in W Texas. Before 1989, search Midland AND Texas.
CO N320000N320000
W1020900W1020900
BT Midland County Texas
BT Texas
BT United States

Midland Valley (1981)
S Scotland. Also in New Brunswick.
IN Index regions as applicable.
SA New Brunswick
SA Scotland

Midlands (1978)
English region including the highly industrialized central counties. Refers especially to Derbyshire, Leicestershire, Northamptonshire, Nottinghamshire, Rutlandshire, Staffordshire, and Warwickshire. May occur in other locations.
IN Index England or regions as applicable.
SA Derbyshire England
SA Dudley England
SA East Midlands
SA England
SA Leicestershire England
SA Northamptonshire England
SA Nottinghamshire England
SA Pennines
SA Staffordshire England
SA Warwickshire England
SA Worcestershire England

Midlothian
No longer a valid term for GeoRef as of 1993. See Midlothian Scotland.

Midlothian Scotland (1993)
Formerly Edinburghshire. Former county in SE Scotland on S shore of Firth of Forth. Split between Borders region and Lothian region in 1974.
BT Scotland
BT Great Britain
BT United Kingdom
BT Western Europe
BT Europe
NT Edinburgh Scotland
SA Lothian region Scotland

Midnite Mine (1985)
Uranium mine in NE Washington.
CO N474500N480000
W1180000W1181000
BT Washington
BT United States
SA mines
SA Spokane County Washington
SA Stevens County Washington
SA uranium ores

midoceanic ridges
use mid-ocean ridges

Midway (1978)
Two small islands (Eastern and Sand), parts of a low coral atoll under U.S. Navy control 1300 miles WNW of Honolulu. In 1985, broader term changed from Pacific Ocean to East Pacific Ocean Islands. Before 1985, search Midway AND Pacific Ocean. Before 1971, also search Midway Islands.
UF Midway Atoll
UF Midway Islands
BT East Pacific Ocean Islands

Midway Atoll
use Midway

Midway Group (1981)
In SE United States from W Georgia to S Texas, Arkansas, SW Illinois, Kentucky, SE Missouri, and W Tennessee.
BT Paleocene
BT Paleogene
BT Tertiary
BT Cenozoic
SA Alabama
SA Arkansas
SA Clayton Formation
SA Florida
SA Georgia
SA Illinois
SA Kentucky
SA Mississippi
SA Missouri
SA Naheola Formation
SA Porters Creek Formation
SA Tennessee
SA Texas

Midway Islands
Not a valid GeoRef index term after 1970. Was used on level 1 in subfile N.
use Midway

Midwest (1978)
As of 1993, term is used only for broad general discussions. N part of central U.S. Region comprising states N of the Ohio and Missouri Rivers plus the E edge of the Great Plains. From 1981 through 1992, this term was autoposted to Illinois, Indiana, Iowa, Kansas, Michigan, Minnesota, Missouri, Nebraska, North Dakota, Ohio, and South Dakota. Before 1978, also search United States AND middle west or north-central or north central. As of 1993, for complete search, also search states.
CO N360000N490000
W0803000W1040000
UF Middle West (United States)
UF North Central (United States)
UF North-Central (United States)
BT United States
SA Eastern Gas Shales Project
SA Mississippi Valley
SA Sauk Sequence

Midwest Lake (1985)
N Saskatchewan.
IN Index province or region as applicable.
SA Athabasca District
SA Saskatchewan

Midwest Lake Deposits (1993)
N Saskatchewan. Before 1993, also search Midwest Lake.
BT Saskatchewan
BT Western Canada
BT Canada
SA Athabasca District
SA uranium ores

Mie
No longer a valid term for GeoRef. See Mie Japan.

Mie Japan (1993)
Prefecture in S central Honshu. Before 1993, also search Mie or Miye AND Japan.
UF Miye Japan
BT Honshu
BT Japan
BT Far East
BT Asia
SA Ise Bay
SA Kii Peninsula
SA Kinki Japan

Miechow
No longer a valid term for GeoRef. As of 1993 see Miechow Poland.

Miechow Poland (1993)
Town in S Poland. Before 1993 also search Miechow AND Poland.
BT Cracow Poland
BT Poland
BT Central Europe
BT Europe

Miette Complex (1978)
Devonian reef complex. Distinguish from Miette Group.
BT Devonian
BT Paleozoic
SA Canada
SA Miette Group

Miette Group (1978)
Composed of a basal unit of argillite and argillaceous sandstone, succeeded by 2, 000 feet of sandstone, grit, conglomerate, and argillite, which is overlain by 3, 000 feet of argillite, and at the top by a carbonate. Distinguish from the Devonian Miette Complex.
BT Proterozoic
BT upper Precambrian
BT Precambrian
SA Alberta
SA British Columbia
SA Canada
SA Miette Complex

migmatite
As of 1981, no longer a valid term for GeoRef. From 1978-1980, migmatites was autoposted to this term.
use migmatites

migmatites (1978)
Before 1981, also search migmatite. Autoposting of this term to anatexite began in 1981. Autoposting of this term to migmatite ended in 1981.
UF migmatite
BT metamorphic rocks
NT agmatite
NT anatexite
NT ectinite
NT embrechite
NT leucosomes
SA migmatization
SA ptygmatic folds

migmatitization
use migmatization

migmatization (1978)
UF migmatitization
BT metamorphism
SA migmatites

migration (1978)
As of 1993, see seismic migration if applicable.
SA land bridges
SA petroleum
SA petroleum accumulation
SA petroleum exploration
SA primary migration
SA secondary migration
SA seismic migration

migration (seismic)
use seismic migration

migration of elements (1978)
SA chemical elements
SA diffusivity
SA enrichment
SA geochemical cycle
SA geochemistry
SA lithogeochemistry
SA metamorphism
SA mobilization
SA thermal waters
SA weathering

Mihara (Mount)
use Mount Mihara

Mihara Volcano
use Mount Mihara

Mihara-Yama
use Mount Mihara

Miho Bay (1978)
Inlet of Japan Sea in S Honshu.
BT Honshu
BT Japan
BT Far East
BT Asia
SA Japan Sea

Mikabu System (1978)
According to the Japan Lexique, the age of Mikabu Series is Paleozoic. Consists of amphibole, pyroxene, semi-schists, and phyllites with limestones and quartzites.
BT Paleozoic
SA Honshu
SA Japan

Mikawa (1978)
Former province in central Honshu. Now part of Aichi Prefecture.
SA Aichi Japan

Mikhailovka
No longer a valid term for GeoRef. As of 1993, see Mikhailovka Russian Federation.

Mikhailovka Russian Federation (1993)
City in Volgograd Oblast 110 miles NW of Volgograd in European Russian Federation. Before 1993, also search Mikhailovka. This term has multiple hierarchies.
BT1 Volgograd Russian Federation
BT1 Russian Federation
BT1 Commonwealth of Independent States
BT2 Volgograd Russian Federation
BT2 Europe

Milam County
Valid through 1988. Search in combination with state term. After 1988, use specific county-state term.

Milam County Texas (1989)
Central Texas. Before 1989, also search Milam County AND Texas.
CO N302500N311000
    W0964000W0972000
BT Texas
BT United States

Milan
No longer a valid term for GeoRef. See Milan Italy.

Milan Italy (1993)
Province including the city in W Lombardy, N Italy. Before 1993, also search Milan and Italy. Before 1978, also search Milano.
UF Milano Italy
BT Lombardy Italy
BT Italy
BT Southern Europe
BT Europe

Milankovitch theory (1985)
SA atmosphere
SA Earth
SA eccentricity
SA glacial geology
SA glaciation
SA insolation
SA motions
SA obliquity of the ecliptic
SA paleoclimatology
SA periodicity

Milano Italy
use Milan Italy

milarite (1978)
BT ring silicates
BT silicates
SA aluminosilicates

Mildura
No longer a valid term for GeoRef. See Mildura Australia

Mildura Australia (1993)
Town on the Murray River 300 miles NW of Melbourne in NW Victoria.
BT Victoria Australia
BT Australia
BT Australasia

Miliolacea (1978)
Autoposting of this term to Alveolinellidae began in 1989. Autoposting of microfossils and Protista to this term began in 1990. This term has multiple hierarchies.
BT1 Miliolina

BT1 foraminifera
BT1 Protista
BT1 Invertebrata
BT2 Miliolina
BT2 foraminifera
BT2 Protista
BT2 microfossils
NT Alveolinellidae
NT Miliolidae
NT Ophthalmidium
NT Quinqueloculina

Miliolidae (1978)
Family. Autoposting of microfossils and Protista to this term began in 1990. This term has multiple hierarchies.
BT1 Miliolacea
BT1 Miliolina
BT1 foraminifera
BT1 Protista
BT1 Invertebrata
BT2 Miliolacea
BT2 Miliolina
BT2 foraminifera
BT2 Protista
BT2 microfossils

Miliolina (1978)
Autoposting of microfossils and Protista to this term began in 1990. This term has multiple hierarchies.
BT1 foraminifera
BT1 Protista
BT1 Invertebrata
BT2 foraminifera
BT2 Protista
BT2 microfossils
NT Miliolacea

military geology (1981)
SA geology

Milk River Formation (1978)
Composed of well-sorted sandstones. In S Alberta.
BT Upper Cretaceous
BT Cretaceous
BT Mesozoic
SA Alberta
SA Canada

Mill Creek (1978)
Rises in Lassen Volcanic National Park and flows into Sacramento River.
IN Index counties or regions as applicable.
SA California

Millard County
Valid through 1988. Search in combination with state term. After 1988, use specific county-state term.

Millard County Utah (1989)
W central Utah. Before 1989, also search Millard County AND Utah.
CO N383500N393500
    W1120300W1140500
BT Utah
BT United States
NT House Range
SA Sevier Desert
SA Wah Wah Mountains

Millboro Shale (1989)
Middle and Upper Devonian. Maryland, S Pennsylvania, Virginia and E West Virginia.
BT Devonian
BT Paleozoic
SA Maryland
SA Pennsylvania
SA Virginia
SA West Virginia

millerite (1978)
UF nickel pyrites
BT sulfides
SA nickel ores

Millevaches Plateau (1981)
Tableland of the Central Massif, in Correze and Creuse departments, central France.
CO N451500N460000
    E0023000E0014500
BT Central Massif
BT France
BT Western Europe
BT Europe
SA Correze France
SA Creuse France

Mills County
Valid through 1988. Search in combination with state term. After 1988, use specific county-state term.

Mills County Iowa (1989)
SW Iowa. Before 1989, also search Mills County AND Iowa.
CO N405300N410900
    W0952300W0955300
BT Iowa
BT United States

Mills County Texas (1989)
Central Texas. Before 1989, also search Mills County AND Texas.
CO N311300N314000
    W0981700W0985800
BT Texas
BT United States

Millstone Grit (1978)
Provincial series, Europe.
BT Upper Carboniferous
BT Carboniferous
BT Paleozoic

Milne Land (1978)
An island in Scoresby Sound which is a deep inlet of Greenland Sea on E coast.
BT Greenland
BT Arctic region

Milos (1993)
SE Aegean Islands. Before 1993, also search Melos and Greece. This term has multiple hierarchies.
CO N363500N364500
    E0244000E0241500
BT1 Cyclades
BT1 Greek Aegean Islands
BT1 Greece
BT1 Southern Europe
BT1 Europe
BT2 Cyclades
BT2 Greek Aegean Islands
BT2 Aegean Islands
BT2 Mediterranean region

Milwaukee
Valid through 1988. Search in combination with state term. After 1988, use specific city-state term.

Milwaukee County
Valid through 1988. Search in combination with state term. After 1988, use specific county-state term.

Milwaukee County Wisconsin (1989)
SE Wisconsin. Before 1989, also search Milwaukee County AND Wisconsin.
CO N425000N431300
    W0874800W0880500
BT Wisconsin
BT United States
NT Milwaukee Wisconsin

Milwaukee Wisconsin (1989)
City on Lake Michigan in SE Wisconsin. Before 1989, search Milwaukee AND Wisconsin.
CO N430300N430300
W0875600W0875600
BT Milwaukee County Wisconsin
BT Wisconsin
BT United States

Mimas
use Mimas Satellite

Mimas Satellite (1985)
One of satellites of Saturn. Before 1985, also search Mimas AND Saturn.
UF Mimas
SA icy satellites
SA satellites
SA Saturn
SA Voyager Program

mimetite (1978)
From 1978-1980, arsenates and halides were autoposted to this term. Autoposting of chlorides began in 1981. As of 1989, halides is autoposted to this term. This term has multiple hierarchies.
BT1 arsenates
BT2 chlorides
BT2 halides

Minas Basin (1978)
NE extension of Bay of Fundy in central Nova Scotia. Also search Minas AND appropriate area term.
UF Fundy Basin
BT Nova Scotia
BT Maritime Provinces
BT Eastern Canada
BT Canada
SA Bay of Fundy
SA Cobequid Bay

Minas Gerais
No longer a valid term for GeoRef. See Minas Gerais Brazil.

Minas Gerais Brazil (1993)
State in E central Brazil.
CO S233000S140000
W0394500W0510000
BT Brazil
BT South America
NT Araxa Brazil
NT Itabira Brazil
NT Ouro Preto Brazil
NT Pocos de Caldas Brazil
SA Areado Formation
SA Bambui Group
SA Bauru Formation
SA Brazilian Shield
SA Irati Formation
SA Sao Francisco Basin
SA Sao Francisco Craton
SA Serra do Espinhaco

Minas Series (1978)
BT Precambrian
SA Brazil

Mindanao (1978)
Second largest and southernmost major island of the Philippine Islands.
BT Philippine Islands
BT Far East
BT Asia
NT Kiamba Philippine Islands
SA Surigao del Norte Philippine Islands

Mindel (1978)
Europe. Above Gunz, below Riss.
BT Pleistocene
BT Quaternary
BT Cenozoic

SA Elsterian

Mindel-Riss
use Mindel/Riss Interglacial

Mindel-Riss Interglacial
use Mindel/Riss Interglacial

Mindel/Riss Interglacial (1978)
Europe. Term applied in the Alps to the second interglacial stage of the Pleistocene Epoch, after the Mindel glacial stage and before the Riss.
UF Mindel-Riss
UF Mindel-Riss Interglacial
BT Pleistocene
BT Quaternary
BT Cenozoic

Mindoro (1978)
Island in central Philippines SW of Luzon.
BT Philippine Islands
BT Far East
BT Asia
SA Mansalay Formation

mine dewatering (1993)
Restricted to artificial methods.
SA drainage
SA drawdown

mine drainage (1981)
As of 1993, use for natural drainage.
SA acid mine drainage
SA drainage
SA environmental geology
SA mines

mine reactivation (1981)
SA economics
SA land use
SA mines

mineragraphy
A valid term through 1976.
use ore microscopy

mineral assemblages (1978)
SA autometamorphism
SA metamorphic rocks
SA paragenesis

mineral associations
A valid term through 1977. Use mineral assemblages.

mineral chemistry
use crystal chemistry

mineral cleavage (1981)
Before 1981, search cleavage AND minerals.
SA cleavage
SA minerals

mineral collecting
As of 1981, no longer a valid term for GeoRef. Use collecting.

mineral collections
Not a valid GeoRef index term after 1970. Was used on level 1 in subfile N. Now use collections.

mineral composition (1978)
SA composition
SA modal analysis

Mineral County
Valid through 1988. Search in combination with state term. After 1988, use specific county-state term.

Mineral County Colorado (1989)
SW Colorado. Before 1989, also search Mineral County AND Colorado.
CO N372200N375900
W1064500W1070700
BT Colorado
BT United States
NT Creede Colorado
NT Creede mining district

Mineral County Montana (1989)
W Montana. Before 1989, also search Mineral County AND Montana.
CO N464000N473200
W1142500W1154500
BT Montana
BT United States
SA Coeur d'Alene mining district

Mineral County Nevada (1989)
W Nevada. Before 1989, also search Mineral County AND Nevada.
CO N375300N390500
W1174000W1190900
BT Nevada
BT United States
SA Walker Lake

Mineral County West Virginia (1989)
E West Virginia. Before 1989, also search Mineral County AND West Virginia.
CO N391400N393800
W0783900W0791600
BT West Virginia
BT United States

mineral data (1978)
Before 1971, also search mineral descriptions.
UF mineral descriptions
SA areal studies
SA clay mineralogy
SA data
SA mineralogy
SA minerals

mineral deposits, genesis (1978)
Used for substantial discussions of the genesis of ore deposits. Does not include water resources or energy sources. Before 1971, was sometimes used with origin (i.e. mineral deposits, origin).
IN Index specific type of process or control where applicable.
SA algoma-type
SA alpine-type
SA bauxitization
SA beneficiation
SA besshi-type
SA blind deposits
SA carlin-type
SA cementation deposits
SA controls
SA cyprus-type
SA detrital deposits
SA discoveries
SA disseminated deposits
SA diwa-type
SA economic geology
SA endogene processes
SA enrichment
SA epigene processes
SA epithermal processes
SA exhalative processes
SA exogene processes
SA fluid inclusions
SA gangue
SA genesis
SA geochemical controls
SA geochemistry
SA geomorphologic controls
SA glaciated terrains
SA haloes
SA host rocks
SA hydrogeological controls
SA hydrothermal processes
SA hypabyssal processes
SA hypothermal processes
SA igneous processes
SA impregnated deposits
SA inclusions
SA intramagmatic deposits
SA isotopes
SA kupferschiefer-type
SA kuroko-type
SA laterization
SA leakage anomalies
SA lenses
SA lithogeochemistry
SA lithologic controls
SA localization
SA mantos
SA massive deposits
SA mechanical controls
SA mesothermal processes
SA metallogenic epochs
SA metallogenic maps
SA metallogenic provinces
SA metallogeny
SA metallotects
SA metamorphic processes
SA metasomatism
SA mineral exploration
SA mineral resources
SA mineralization
SA mining
SA mining geology
SA mississippi valley-type
SA nuggets
SA ore bodies
SA ore reflectivity
SA ore transport
SA ore-forming fluids
SA oxidation zone
SA paleogeographic controls
SA paragenesis
SA phase equilibria
SA placers
SA pneumatolysis
SA podiform deposits
SA porphyry copper
SA porphyry molybdenum
SA possibilities
SA potential deposits
SA quartz veins
SA roll-type
SA sandstone-type
SA sedimentary processes
SA stockwork deposits
SA stratabound deposits
SA stratiform deposits
SA stratigraphic controls
SA structural controls
SA sublimation
SA supergene processes
SA syngenesis
SA telethermal processes
SA unconformity-type
SA veins
SA volcanic processes
SA wall-rock alteration
SA weathering

mineral descriptions
use mineral data

mineral economics (1978)
As of 1985, used for economic discussions of commodities (list C), excluding the energy sources, mineral resources (as a term), and water resources. For economics of the energy sources, mineral resources (as a term), and water resources, use economics.
SA economic geology
SA economics
SA mineral resources
SA ore grade
SA possibilities

mineral exploration (1978)

Used primarily for the application of methods in relation to the commodity terms (list C), excluding the energy sources and water resources.
IN   May be used for substantial discussions of exploration for the commodities, excluding the energy sources and water resources. Index specific method if possible [i.e. biogeochemical methods, geobotanical methods, geochemical methods, geological methods, geomorphological methods, geophysical methods, geophysical surveys, hydrological methods, photogeologic methods, photogeology, remote sensing]
UF   ore exploration
SA   Bear-Slave Operation
SA   biogeochemical methods
SA   blind deposits
SA   CUSMAP
SA   dispersion patterns
SA   economic geology
SA   Exclusive Economic Zone
SA   exploration
SA   feasibility studies
SA   geobotanical methods
SA   geochemical anomalies
SA   geochemical methods
SA   geochemistry
SA   geological methods
SA   geomorphological methods
SA   geophysical methods
SA   geophysical surveys
SA   glaciated terrains
SA   gossan
SA   haloes
SA   horizontal drilling
SA   hydrological methods
SA   lake sediments
SA   lithogeochemistry
SA   MANOP
SA   maps
SA   mineral deposits, genesis
SA   mineral resources
SA   minerals
SA   objectives
SA   ore grade
SA   ore guides
SA   panning
SA   pathfinders
SA   photogeologic methods
SA   photogeology
SA   pit sections
SA   placers
SA   potential deposits
SA   primary dispersion
SA   programs
SA   PROSPECTOR
SA   radon emanometry
SA   remote sensing
SA   secondary dispersion
SA   soil sampling
SA   stream sediments
SA   type localities
SA   United States Exclusive Economic Zone
SA   well-logging

mineral facies
A valid term through 1976. Use facies.

mineral inclusions   (1978)
Used for a solid-phase mineral within a solid-phase mineral. As of 1989, inclusions is autoposted to this term.
BT   inclusions
SA   fluid inclusions
SA   host rocks
SA   San Carlos Olivine
SA   xenocrysts
SA   xenoliths

mineral inventory   (1993)
Used for inventories of minerals. Before 1993, also search miscellaneous minerals or sundry minerals (before 1971) as applicable. Includes use for the case of several mineral groups when the topic is not emphasized or for a mineral of unknown affinity.
UF   mineral list
UF   miscellaneous minerals
UF   sundry minerals
SA   mineral resources
SA   mineralogy
SA   minerals

mineral list
use mineral inventory

mineral localities   (1981)
Use for collecting.
SA   collecting
SA   popular geology

Mineral Mountains   (1985)
SW   Utah.
CO   N381500N383500
        W1124500W1130000
BT   Beaver County Utah
BT   Utah
BT   United States
SA   Basin and Range Province

mineral resins
use resins

mineral resources   (1978)
For very general treatments of both metal and nonmetal deposits.
IN   Commodity. See List C.
SA   affinities
SA   depletion
SA   economic geology
SA   environmental geology
SA   exploration
SA   Law of the Sea
SA   mineral deposits, genesis
SA   mineral economics
SA   mineral exploration
SA   mineral inventory
SA   minerals
SA   ocean floors
SA   potential deposits
SA   raw materials
SA   resources
SA   strategic minerals
SA   stream placers

mineral sequence
use paragenesis

mineral soap
use bentonite

mineral synthesis
A valid term through 1970. After 1970, see synthesis under crystal growth and under phase equilibria.

mineral waters   (1978)
SA   brines
SA   ground water
SA   medicinal waters
SA   springs
SA   thermal waters
SA   water

mineral zoning
A valid term through 1975. See zoning; crystal zoning.

mineralization   (1978)
Restricted to ore deposits. Not used for paleontology.
SA   localization
SA   metallogenic epochs
SA   metallogenic provinces
SA   metallogeny
SA   mineral deposits, genesis
SA   mineralized mud
SA   ore-forming fluids

mineralized mud   (1985)
SA   mineralization
SA   mud

mineralogy   (1978)
Used for the discipline as a whole.
SA   atomic packing
SA   bibliography
SA   catalogs
SA   clay mineralogy
SA   collecting
SA   collections
SA   crystal chemistry
SA   crystal growth
SA   crystal structure
SA   crystal zoning
SA   crystallography
SA   decrepitation
SA   goniometry
SA   lexicons
SA   microscope methods
SA   mineral data
SA   mineral inventory
SA   minerals
SA   optical mineralogy
SA   optical properties
SA   ore microscopy
SA   petrology
SA   phase equilibria
SA   reaction rims
SA   reflectance
SA   type localities

minerals   (1978)
Used for descriptions of mineral occurrence or of the minerals themselves. For searching techniques, see list L.
IN   General term. Index only for general studies. Specific mineral names or groups should be used when applicable.
SA   accessory minerals
SA   alloys
SA   antimonates
SA   antimonides
SA   antimonites
SA   arsenates
SA   arsenides
SA   arsenites
SA   authigenesis
SA   authigenic minerals
SA   bismuthides
SA   borates
SA   carbonates
SA   cell dimensions
SA   chromates
SA   clay mineralogy
SA   coexisting minerals
SA   collecting
SA   collections
SA   color centers
SA   copper minerals
SA   crystal chemistry
SA   crystal dislocations
SA   crystal field
SA   crystal growth
SA   crystal structure
SA   crystal systems
SA   crystal zoning
SA   crystallography
SA   crystals
SA   defects
SA   elongate minerals
SA   encrustations
SA   fluid inclusions
SA   formula
SA   gems
SA   grain boundaries
SA   habit
SA   halides
SA   hardness
SA   heavy minerals
SA   hydrates
SA   inclusions
SA   industrial minerals
SA   interlaboratory comparison
SA   iodates
SA   iron minerals
SA   lamellae
SA   light minerals
SA   localization
SA   magnetic minerals
SA   manganese minerals
SA   metabentonite
SA   metamict minerals
SA   metamictization
SA   microhardness
SA   mineral cleavage
SA   mineral data
SA   mineral exploration
SA   mineral inventory
SA   mineral resources
SA   mineralogy
SA   mixed-layer minerals
SA   molybdates
SA   native elements
SA   new minerals
SA   nitrates
SA   nonmagnetic minerals
SA   opaque minerals
SA   optical properties
SA   ore minerals
SA   organic compounds
SA   overgrowths
SA   oxalates
SA   oxides
SA   paragenesis
SA   phenocrysts
SA   phosphates
SA   platinum minerals
SA   polymorphism
SA   polytypism
SA   pseudomorphism
SA   reaction rims
SA   secondary minerals
SA   selenates
SA   selenides
SA   selenites
SA   silicates
SA   single-crystal method
SA   spectra
SA   standard materials
SA   sulfates
SA   sulfides
SA   sulfosalts
SA   surface defects
SA   symplectite
SA   synthesis
SA   synthetic materials
SA   tellurates
SA   tellurides
SA   tellurites
SA   transformations
SA   tungstates
SA   typomorphism
SA   uranium minerals
SA   vanadates
SA   varieties
SA   water of crystallization
SA   water of dehydration
SA   wehrlite

minerals described
No longer a valid GeoRef index term. Before 1969, was used on level 1 in subfile E. See minerals; mineral data.

minerals in meteorites
Not a valid index term. Place minerals in appropriate group.

mines   (1978)
This term is limited to papers on mine engineering. For specific mines, search: term for specific

mine OR (term for region around the mine AND term for commodity mined). See list C for commodity terms.
IN   Restricted to engineering use only.
NT abandoned mines
NT coal mines
NT small mines
SA acid mine drainage
SA Adamow Mine
SA Akenobe Mine
SA Alligator Ridge Mine
SA Alligator Rivers Field
SA Alma mining district
SA Ambrosia Lake mining district
SA Anvil Points Mine
SA Ashio Mine
SA Balmat-Edwards mining district
SA Bathurst mining district
SA Belchatow Mine
SA Benson Mines
SA Blackbird mining district
SA Bluebell Mine
SA Boquira Mine
SA Broken Hill Mine
SA Bunker Hill Mine
SA Butte mining district
SA Carlin Mine
SA Chichibu Mine
SA Chinkuashih Mine
SA Clara Mine
SA Climax Mine
SA Coeur d'Alene mining district
SA Coronation Mine
SA Creede mining district
SA cut-and-fill mining
SA economic geology
SA Equity Mine
SA Fairbanks mining district
SA Fukazawa Mine
SA Getchell Mine
SA Grants mining district
SA Hanaoka Mine
SA Hanawa Mine
SA Henderson Mine
SA Homestake Mine
SA Horne Mine
SA Hosokura Mine
SA Idaho Springs mining district
SA Ikuno Mine
SA Janggun Mine
SA Kamaishi Mine
SA Kamioka Mine
SA Kamoto
SA Keban Mine
SA Klerksdorp Field
SA Kosaka Mine
SA land subsidence
SA land use
SA Leadville mining district
SA Lee Creek Mine
SA longwall mining
SA Lupin Mine
SA Midnite Mine
SA mine drainage
SA mine reactivation
SA mining
SA mining geology
SA Mosaboni Mines
SA Nasliden Mine
SA Omine Mine
SA open-pit mining
SA Pilbara gold field
SA Pine Point mining district
SA quarries
SA Questa Mine
SA Ranger Mine
SA roof control
SA Salsigne Mine
SA San Antonio Mine
SA Schwartzwalder Mine
SA Shakanai Mine
SA Sigma Mine
SA Skellefte mining district
SA Slocan mining camp
SA solution mining
SA Strathcona Mine
SA strip mining
SA Sullivan Mine
SA Sunnyside Mine
SA surface mining
SA Temagami Mine
SA Thetford Mines
SA Toyoha Mine
SA Trepca Mine
SA underground installations
SA underground mining
SA White Pine Mine

minette (1978)
BT lamprophyres
BT plutonic rocks
BT igneous rocks

minicomputers (1978)
BT computers
SA computer networks
SA data processing
SA hand calculators
SA microcomputers
SA workstations

mining (1978)
Autoposting of this term began in 1993.
NT cut-and-fill mining
NT solution mining
NT surface mining
NT underground mining
SA dredged materials
SA dredging
SA economic geology
SA fines
SA mineral deposits, genesis
SA mines
SA mining geology
SA mining legislation
SA New Austrian Tunnelling Method
SA pits
SA quarries
SA quartz diabase
SA recovery
SA spoils
SA tailings
SA tailings ponds
SA tunnels

mining districts
Not a valid GeoRef term. Use names of individual districts as applicable.

mining engineering
Not a valid term for GeoRef. See engineering geology; mining geology.

mining geology (1978)
Geology applied to mining operations.
SA abandoned mines
SA backfill
SA chimneys
SA coal mines
SA cut-and-fill mining
SA dredging
SA economic geology
SA engineering geology
SA environmental geology
SA evaluation
SA explosions
SA gases
SA geology
SA geophysical methods
SA geophysical surveys
SA horizontal drilling
SA hydraulic fracturing
SA leachate
SA leaching
SA longwall mining
SA mineral deposits, genesis
SA mines
SA mining
SA open-pit mining
SA overburden
SA pillars
SA production control
SA quarries
SA reclamation
SA recovery
SA rock bursts
SA roof control
SA small mines
SA soft rocks
SA solution mining
SA strip mining
SA surface mining
SA tailings dams
SA tailings ponds
SA technology
SA underground mining

mining legislation (1981)
SA Law of the Sea
SA legislation
SA mining

mining, cut-and-fill
use cut-and-fill mining

mining, open-pit
use open-pit mining

Minneapolis
Valid through 1988. Search in combination with state term. After 1988, use specific city-state term.

Minneapolis Minnesota (1989)
City in SE central Minnesota. Before 1989, search Minneapolis AND Minnesota.
CO N450000N450000
      W0931500W0931500
BT Hennepin County Minnesota
BT Minnesota
BT United States

Minnehaha County
Valid through 1988. Search in combination with state term. After 1988, use specific county-state term.

Minnehaha County South Dakota (1989)
E South Dakota. Before 1989, also search Minnehaha County AND South Dakota.
CO N433000N435000
      W0962500W0971000
BT South Dakota
BT United States

Minnelusa Formation (1978)
Pennsylvanian and Permian. Composed of sandstones and evaporites. W South Dakota, and NE Wyoming.
IN   Index ages as applicable.
BT Paleozoic
SA Pennsylvanian
SA Permian
SA South Dakota
SA Wyoming

Minnesota (1978)
Autoposting of this term began in 1978.
CO N433000N490000
      W0894500W0971000
BT United States
NT Anoka County Minnesota
NT Beltrami County Minnesota
NT Benton County Minnesota
NT Brown County Minnesota
NT Cass County Minnesota
NT Clay County Minnesota
NT Clearwater County Minnesota
NT Cook County Minnesota
NT Dodge County Minnesota
NT Douglas County Minnesota
NT Duluth Complex
NT Fillmore County Minnesota
NT Grant County Minnesota
NT Hennepin County Minnesota
NT Houston County Minnesota
NT Itasca County Minnesota
NT Jackson County Minnesota
NT Koochiching County Minnesota
NT Lake County Minnesota
NT Lincoln County Minnesota
NT Lyon County Minnesota
NT Marshall County Minnesota
NT Martin County Minnesota
NT Mesabi Range
NT Minnesota River valley
NT Murray County Minnesota
NT Pennington County Minnesota
NT Polk County Minnesota
NT Pope County Minnesota
NT Ramsey County Minnesota
NT Renville County Minnesota
NT Rice County Minnesota
NT Saint Louis County Minnesota
NT Scott County Minnesota
NT Sherburne County Minnesota
NT Stearns County Minnesota
NT Stevens County Minnesota
NT Vermilion Range
NT Washington County Minnesota
NT Winona County Minnesota
NT Wright County Minnesota
SA Animikie Group
SA Benton Formation
SA Biwabik Iron Formation
SA Canadian Shield
SA Cedar Valley Formation
SA Clearwater River
SA Dakota Formation
SA Decorah Shale
SA Des Moines Lobe
SA Dubuque Formation
SA Eagle Lake
SA Eau Claire Formation
SA Elk Lake
SA Franconia Formation
SA Galena Dolomite
SA Galesville Sandstone
SA Glenwood Shale
SA Great Lakes region
SA Gunflint Iron Formation
SA Keweenawan
SA Knife Lake Group
SA Lake Agassiz
SA Lake Algonquin
SA Lake of the Woods
SA Lake of the Woods region
SA Lake Superior
SA Lake Superior region
SA Lake Washington
SA Maquoketa Formation
SA Marquette Range Supergroup
SA Midcontinent
SA Mississippi River
SA Mississippi Valley
SA Mount Simon Sandstone
SA Mystery Cave
SA Niobrara Formation
SA North Shore Volcanics
SA Oronto Group
SA Penokean Orogeny
SA Pierre Shale
SA Platteville Formation
SA Prairie du Chien Group
SA Rainy Lake
SA Rainy River
SA Red River
SA Red River valley
SA Root River
SA Saganaga Lake
SA Saint Peter Sandstone
SA Shakopee Formation

SA Sioux Quartzite
SA Southern Province
SA Sturgeon Lake
SA Superior Province
SA Thomson Formation
SA Upper Mississippi Valley
SA Vermilion granitic complex
SA Virginia Formation

Minnesota River valley (1978)
S Minnesota.
BT Minnesota
BT United States

Mino Belt (1989)
Orogenic belt in central Honshu, central Japan.
CO N350000N360000
 E1373000E1350000
UF Mino Terrane
BT Honshu
BT Japan
BT Far East
BT Asia

Mino Terrane
 use Mino Belt

minor elements (1978)
SA chemical elements
SA major elements
SA nutrients
SA trace elements

minor planets
 use asteroids

minor-element analyses (1978)
Before 1974, also search minor element analyses.
SA analysis
SA chemical analysis
SA chemical elements
SA major-element analyses
SA trace-element analyses

Minorca (1978)
Second largest of Spain's Balearic Islands off E coast of Spanish mainland. This term has multiple hierarchies.
UF Menorca
BT1 Balearic Islands
BT1 Mediterranean region
BT2 Balearic Islands
BT2 Spain
BT2 Iberian Peninsula
BT2 Southern Europe
BT2 Europe

Minorca Rise (1981)
Off Minorca in the Mediterranean Sea. Before 1981, also search Menorca Rise.
CO N393500N403000
 E0050000E0034500
UF Menorca Rise
BT Mediterranean Sea

Minsk
No longer a valid term for GeoRef. As of 1993, see Minsk Belarus.

Minsk Belarus (1993)
Oblast and city in central Belarus. This term has multiple hierarchies.
BT1 Belarus
BT1 Europe
BT2 Belarus
BT2 Commonwealth of Independent States

Minturn Formation (1978)
Consists chiefly of grayish to greenish sandstones, conglomerates, and shales or siltstones. Central Colorado.
BT Pennsylvanian
BT Carboniferous
BT Paleozoic

SA Colorado

Minusinsk
No longer a valid term for GeoRef. As of 1993, see Minusinsk Russian Federation.

Minusinsk Basin (1978)
Coal basin in extreme S Krasnoyarsk Kray in SSW Siberia. A good agricultural region. Also search Minusinsk. This term has multiple hierarchies.
CO N520000N553000
 E0950000E0870000
BT1 West Siberia
BT1 Commonwealth of Independent States
BT2 West Siberia
BT2 Asia
BT3 Russian Federation
BT3 Commonwealth of Independent States
SA coal fields
SA Krasnoyarsk Russian Federation

Minusinsk Russian Federation (1993)
Town in SW Krasnoyarsk Kray on the Yenisei River in SSW Siberia. Before 1993, also search Minusinsk. This term has multiple hierarchies.
BT1 Krasnoyarsk Russian Federation
BT1 Russian Federation
BT1 Commonwealth of Independent States
BT2 Krasnoyarsk Russian Federation
BT2 Asia

minyulite (1978)
From 1978-1980, halides and phosphates were autoposted to this term. Autoposting of fluorides began in 1981. As of 1989, halides is autoposted to this term. This term has multiple hierarchies.
BT1 fluorides
BT1 halides
BT2 phosphates

Miocene (1978)
World. Above Oligocene, below Pliocene. Autoposting of this term began in 1978.
UF lower Neogene
BT Neogene
BT Tertiary
BT Cenozoic
NT Arikaree Group
NT Astoria Formation
NT Badenian
NT Barstovian
NT Barstow Formation
NT Bullfrog Member
NT Chesapeake Group
NT Clarendonian
NT Columbia River Basalt Group
NT Crater Flat Tuff
NT Fleming Formation
NT Freites Formation
NT Frenchman Springs Member
NT Grande Ronde Basalt
NT Green Tuff Formation
NT Hawthorn Formation
NT Langhian
NT lower Miocene
NT middle Miocene
NT Mizunami Group
NT Mohnian
NT Monterey Formation
NT Motojuku Formation
NT Ngorora Formation
NT Oakville Sandstone

NT Oficina Formation
NT Oktemberyan Series
NT Paintbrush Tuff
NT Peach Springs Tuff
NT Picture Gorge Basalt
NT Pirabas Formation
NT Pungo River Formation
NT Relizian
NT Saddle Mountains Basalt
NT Saint Marys Formation
NT Surma Group
NT Temblor Formation
NT Topopah Spring Member
NT upper Miocene
NT Vallesian
NT Vindobonian
NT Wanapum Basalt
NT Wood Mountain Formation
NT Yakima Basalt
SA Aguacate Group
SA Bluff Formation
SA Browns Park Formation
SA Catahoula Formation
SA Challis Volcanics
SA Cohansey Formation
SA Deschutes Formation
SA Hector Formation
SA Imperial Formation
SA Loup Fork Group
SA Muddy Creek Formation
SA Neyveli Lignite
SA Poltava Series
SA Purisima Formation
SA Rincon Formation
SA Salt Lake Formation
SA Santa Fe Group
SA Siwalik System
SA Snoqualmie Batholith
SA Twin River Formation
SA Yakataga Formation

miogeoclines
 use miogeosynclines

miogeosynclines (1978)
Before 1993, also search miogeoclines.
UF miogeoclines
UF miomagmatic zone
SA eugeosynclines
SA geosynclines
SA tectonics

Miogypsina (1978)
Genus. Autoposting of microfossils and Protista to this term began in 1990. This term has multiple hierarchies.
BT1 Miogypsinidae
BT1 Rotaliacea
BT1 Rotaliina
BT1 foraminifera
BT1 Protista
BT1 Invertebrata
BT2 Miogypsinidae
BT2 Rotaliacea
BT2 Rotaliina
BT2 foraminifera
BT2 Protista
BT2 microfossils

Miogypsinidae (1978)
Family. Autoposting of microfossils and Protista to this term began in 1990. This term has multiple hierarchies.
BT1 Rotaliacea
BT1 Rotaliina
BT1 foraminifera
BT1 Protista
BT1 Invertebrata
BT2 Rotaliacea
BT2 Rotaliina
BT2 foraminifera
BT2 Protista
BT2 microfossils
NT Miogypsina

miomagmatic zone
 use miogeosynclines

miospores (1978)
Autoposting of this term to Cladophlebis and Dryas ended in 1985. Autoposting of microfossils to this term began in 1989.
BT palynomorphs
BT microfossils
NT Cicatricosisporites
NT Classopollis
NT Florinites
NT Lycopodiumsporites
NT Momipites
NT Normapolles
SA Artemisia
SA biostratigraphy
SA Cladophlebis
SA Compositae
SA Cyperaceae
SA Dryas
SA exine
SA megaspores
SA monolete taxa
SA spores
SA sporopollenin

Mir Pipe (1989)
Kimberlite pipe. This term has multiple hierarchies.
BT1 Yakutia Russian Federation
BT1 Russian Federation
BT1 Commonwealth of Independent States
BT2 Yakutia Russian Federation
BT2 Asia
SA Siberian Platform
SA Vilyuy Syneclise

mirabilite (1978)
BT sulfates

Mirador Formation (1978)
BT Eocene
BT Paleogene
BT Tertiary
BT Cenozoic
SA Venezuela

Miramichi Bay (1978)
Inlet of Gulf of Saint Lawrence. Also search Miramichi.
BT New Brunswick
BT Maritime Provinces
BT Eastern Canada
BT Canada
SA Gulf of Saint Lawrence

Miramichi earthquake 1982 (1989)
Epicenter at Miramichi, New Brunswick, Canada.
UF New Brunswick earthquake 1982
BT earthquakes
SA Canada
SA New Brunswick

Miranda
No longer a valid term for GeoRef. As of 1993 see Miranda Venezuela.

Miranda Satellite (1989)
One of the satellites of Uranus. Before 1989, also search Miranda and Uranus.
SA icy satellites
SA satellites
SA Uranus
SA Voyager Program

Miranda Venezuela (1993)
State on the Caribbean Sea. Before 1993 also search Miranda AND Venezuela.
BT Venezuela
BT South America

mires
: Not a valid GeoRef index term. See bogs; marshes; swamps.

Mirgalimsay
: No longer a valid term for GeoRef. As of 1993, see Mirgalimsay Kazakhstan.

Mirgalimsay Kazakhstan (1993)
: Village in S near Syr Darya River. This term has multiple hierarchies.
  BT1 Kazakhstan
  BT1 Central Asia
  BT1 Asia
  BT2 Kazakhstan
  BT2 Commonwealth of Independent States

Mirny
: use Mirnyy Station

Mirnyy
: No longer a valid term for GeoRef. As of 1993, see Mirnyy Russian Federation.

Mirnyy Russian Federation (1993)
: City in Yakutia. Before 1993, also search Mirnyy. This term has multiple hierarchies.
  BT1 Yakutia Russian Federation
  BT1 Russian Federation
  BT1 Commonwealth of Independent States
  BT2 Yakutia Russian Federation
  BT2 Asia

Mirnyy Station (1981)
: USSR IGY station on Davis Sea near Shackleton Ice Shelf on the Indian Ocean side. Before 1981, also search Mirnyy AND Antarctica; Mirny AND Antarctica.
  UF Mirny
  BT Antarctica

Mirror Lake (1989)
: E central New Hampshire.
  IN Index county or regions as applicable.
  SA Carroll County New Hampshire

mirrorstone
: use muscovite

Mirsk
: No longer a valid term for GeoRef. As of 1993 see Mirsk Poland.

Mirsk Poland (1993)
: Town near Czechoslovak border in Jelenia Gora Province. Before 1993 also search Mirsk AND Poland.
  BT Poland
  BT Central Europe
  BT Europe
  SA Polish Sudeten Mountains

Mirzapur
: No longer a valid term for GeoRef. See Mirzapur India.

Mirzapur India (1993)
: City in SE Uttar Pradesh, N central India.
  BT Uttar Pradesh India
  BT India
  BT Indian Peninsula
  BT Asia

miscellanea (1978)
: IN May be used if fossil is not yet classified.
  UF incertae sedis
  SA classification
  SA fossils
  SA paleobotany
  SA paleontology
  SA taxonomy

miscellaneous minerals
: No longer a valid term for GeoRef. Included use for the case of several mineral groups when the topic is not emphasized or for a mineral of unknown affinity.
  use mineral inventory

miscibility
: use immiscibility

miscibility gap (1978)
: SA phase equilibria

Mishash Formation (1978)
: BT Cretaceous
  BT Mesozoic
  SA Israel

Misoa Formation (1989)
: NW Venezuela.
  BT lower Eocene
  BT Eocene
  BT Paleogene
  BT Tertiary
  BT Cenozoic
  SA Venezuela

Mispec Group (1989)
: S New Brunswick. Overlain by Lancaster Formation.
  BT Carboniferous
  BT Paleozoic
  SA New Brunswick

Missi Group (1978)
: Its basal unit is normally composed of a conglomerate consisting of angular to subrounded boulders or pebbles along with minor amounts of chert, jasper and quartz. The unit grades upwards to graywacke and subgraywacke with thin beds of conglomerate. In the Flin Flon area of W Manitoba and E Saskatchewan.
  BT Precambrian
  SA Amisk Group
  SA Canada
  SA Manitoba
  SA Saskatchewan

Mission Canyon Limestone (1985)
: In Madison Group.
  BT Mississippian
  BT Carboniferous
  BT Paleozoic
  SA Idaho
  SA Madison Group
  SA Montana
  SA North Dakota
  SA Utah

Mission Creek Fault (1989)
: S California.
  BT Riverside County California
  BT California
  BT United States

Mission Valley Formation (1985)
: S California.
  BT upper Eocene
  BT Eocene
  BT Paleogene
  BT Tertiary
  BT Cenozoic
  SA California

Mississippi (1978)
: Autoposting of this term began in 1978.
  CO N301500N350000 W0880500W0914000
  BT United States
  NT Adams County Mississippi
  NT Benton County Mississippi
  NT Calhoun County Mississippi
  NT Carroll County Mississippi
  NT Choctaw County Mississippi
  NT Claiborne County Mississippi
  NT Clarke County Mississippi
  NT Clay County Mississippi
  NT De Soto County Mississippi
  NT Forrest County Mississippi
  NT Franklin County Mississippi
  NT Greene County Mississippi
  NT Hancock County Mississippi
  NT Harrison County Mississippi
  NT Hinds County Mississippi
  NT Jackson County Mississippi
  NT Jasper County Mississippi
  NT Jefferson County Mississippi
  NT Jones County Mississippi
  NT Lafayette County Mississippi
  NT Lamar County Mississippi
  NT Lauderdale County Mississippi
  NT Lawrence County Mississippi
  NT Lee County Mississippi
  NT Lincoln County Mississippi
  NT Lowndes County Mississippi
  NT Madison County Mississippi
  NT Marion County Mississippi
  NT Marshall County Mississippi
  NT Monroe County Mississippi
  NT Montgomery County Mississippi
  NT Newton County Mississippi
  NT Panola County Mississippi
  NT Pascagoula River basin
  NT Perry County Mississippi
  NT Pike County Mississippi
  NT Pontotoc County Mississippi
  NT Rankin County Mississippi
  NT Scott County Mississippi
  NT Smith County Mississippi
  NT Tishomingo County Mississippi
  NT Union County Mississippi
  NT Warren County Mississippi
  NT Washington County Mississippi
  NT Wayne County Mississippi
  NT Webster County Mississippi
  NT Wilkinson County Mississippi
  SA Amite River
  SA Arkansas River
  SA Black Warrior Basin
  SA Bluffport Marl Member
  SA Bucatunna Formation
  SA Buckner Formation
  SA Byram Formation
  SA Catahoula Formation
  SA Chattanooga Shale
  SA Chickasawhay Formation
  SA Citronelle Formation
  SA Claiborne Group
  SA Clayton Formation
  SA Cockfield Formation
  SA Coker Formation
  SA Cotton Valley Group
  SA Demopolis Chalk
  SA Eutaw Formation
  SA Fort Payne Formation
  SA Glendon Limestone
  SA Gulf Coastal Plain
  SA Hatchetigbee Formation
  SA Haynesville Formation
  SA Jackson Group
  SA Lisbon Formation
  SA Marianna Limestone
  SA Midway Group
  SA Mississippi Embayment
  SA Mississippi River
  SA Mississippi Sound
  SA Mississippi Valley
  SA Moodys Branch Formation
  SA Mooreville Chalk
  SA Naheola Formation
  SA Nanafalia Formation
  SA New Madrid region
  SA Pearl River
  SA Porters Creek Formation
  SA Pottsville Group
  SA Reelfoot Rift
  SA Ripley Formation
  SA Rodessa Formation
  SA Ross Formation
  SA Selma Group
  SA Shubuta Member
  SA Smackover Formation
  SA Sparta Sand
  SA Stones River Group
  SA Talladega Group
  SA Tallahatta Formation
  SA Tombigbee River
  SA Tuscaloosa Formation
  SA Vicksburg Group
  SA Warsaw Formation
  SA Wiggins Arch
  SA Wilcox Group
  SA Yazoo Clay
  SA Yegua Formation

Mississippi Delta (1978)
: On Gulf of Mexico 100 miles SE of New Orleans.
  UF Mississippi River delta
  BT Louisiana
  BT United States
  SA Avery Island
  SA East Bay
  SA Mississippi Fan
  SA Mississippi River

Mississippi Embayment (1978)
: A geosynclinal area including a section of the U S Gulf Coast and extending northward up the Mississippi Valley.
  IN Index states as applicable.
  BT United States
  SA Alabama
  SA Arkansas
  SA Gulf Coastal Plain
  SA Illinois
  SA Kentucky
  SA Mississippi
  SA Mississippi Valley
  SA Missouri
  SA New Madrid earthquakes 1811-1812
  SA New Madrid region
  SA Tennessee

Mississippi Fan (1989)
: In Gulf of Mexico, extending from the Mississippi Delta to Sigsbee Deep and Florida abyssal plain.
  CO N234700N284500 W0843000W0911500
  BT Gulf of Mexico
  BT North American Atlantic
  BT North Atlantic
  BT Atlantic Ocean
  SA DSDP Site 614
  SA DSDP Site 615
  SA DSDP Site 616
  SA DSDP Site 617
  SA DSDP Site 620
  SA DSDP Site 621
  SA DSDP Site 622
  SA DSDP Site 623
  SA DSDP Site 624
  SA Leg 96
  SA Mississippi Delta
  SA Mississippi River
  SA Mississippi River basin
  SA Sigsbee Deep

Mississippi River (1978)
: Rises in Lake Itasca in NW Minnesota and flows SE and S into the Gulf of Mexico.
  IN Index states as applicable.
  BT United States
  SA Arkansas
  SA Illinois
  SA Iowa

SA Kentucky
SA Louisiana
SA Minnesota
SA Mississippi
SA Mississippi Delta
SA Mississippi Fan
SA Mississippi River basin
SA Mississippi Valley
SA Missouri
SA Tennessee
SA Wisconsin

Mississippi River basin (1978)
Drainage basin including the Ohio and Missouri river systems comprises 2/5 of the total U.S. area covering all or parts of 31 states plus S Alberta and Saskatchewan. It drains much of the interior lowlands, Great Plains, central Gulf Coastal Plain, a portion of the Appalachians, and the Rocky Mountains W to the Continental Divide.
IN Index countries as applicable.
BT North America
SA Canada
SA Mississippi Fan
SA Mississippi River
SA Ohio River basin
SA United States

Mississippi River delta
  use Mississippi Delta

Mississippi River valley
  use Mississippi Valley

Mississippi Sound (1985)
Arm of Gulf of Mexico, off Mississippi and Alabama.
IN Index states as applicable.
CO N301500N303000
   W0874500W0893000
BT United States
SA Alabama
SA Gulf Coastal Plain
SA Gulf of Mexico
SA Hancock County Mississippi
SA Harrison County Mississippi
SA Jackson County Mississippi
SA Mississippi
SA Mobile Bay
SA Mobile County Alabama

Mississippi Valley (1978)
River valley. Autoposting of this term began in 1978.
IN Index states as applicable.
UF Mississippi River valley
BT United States
NT Lower Mississippi Valley
NT Upper Mississippi Valley
SA Arkansas
SA Illinois
SA Iowa
SA Kentucky
SA Louisiana
SA Midwest
SA Minnesota
SA Mississippi
SA Mississippi Embayment
SA Mississippi River
SA Missouri
SA Reelfoot Rift
SA Southern U.S.
SA Tennessee
SA Wisconsin

mississippi valley type
  use mississippi valley-type

mississippi valley-type (1978)
UF mississippi valley type
UF MVT
SA lead ores
SA lead-zinc deposits
SA mineral deposits, genesis
SA zinc ores

Mississippian (1978)
After the Devonian and before the Pennsylvanian. Approximate equivalent of Lower Carboniferous of European usage. Autoposting of Carboniferous to this term began in 1981. From 1978-1980, Paleozoic was autoposted to this term.
BT Carboniferous
BT Paleozoic
NT Barnett Formation
NT Bear Gulch Limestone Member
NT Boone Formation
NT Borden Group
NT Chainman Shale
NT Charles Formation
NT Harrodsburg Limestone
NT Leadville Formation
NT Logan Formation
NT Lower Mississippian
NT Madison Group
NT Middle Mississippian
NT Mission Canyon Limestone
NT Monte Cristo Limestone
NT Newman Limestone
NT Price Formation
NT Ramp Creek Formation
NT Rampart Group
NT Redwall Limestone
NT Stanley Group
NT Sunbury Shale
NT Upper Mississippian
NT Valmeyeran
NT Windsor Group
NT Yule Marble
SA Amsden Formation
SA Antler Orogeny
SA Arkansas Novaculite
SA Bartlesville Sand
SA Bedford Shale
SA Berea Sandstone
SA Chattanooga Shale
SA Dinantian
SA Earn Group
SA Endicott Group
SA Johns Valley Formation
SA Kaskaskia Sequence
SA Pennsylvanian
SA Rundle Group
SA Saint George Batholith
SA Springer Formation
SA Waulsortian facies
SA Woodford Shale

Missoula County
Valid through 1988. Search in combination with state term. After 1988, use specific county-state term.

Missoula County Montana (1989)
W Montana. Before 1989, also search Missoula County AND Montana.
CO N463500N473500
   W1131500W1144500
BT Montana
BT United States
SA Selway-Bitterroot Wilderness

Missoula Group (1978)
In Belt Supergroup. Includes numerous named formations: among these are Marsh Shale in Helena region; Striped Peak and Libby formations in NW Montana, five formations near Missoula, and others in and S of Glacier National Park. In N Idaho and NW Montana.
BT middle Proterozoic
BT Proterozoic
BT upper Precambrian

BT Precambrian
SA Belt Supergroup
SA Bonner Formation
SA Empire Formation
SA Idaho
SA Montana
SA Mount Shields Formation
SA Shepard Formation
SA Snowslip Formation
SA Spokane Formation

Missouri (1978)
Autoposting of this term began in 1978.
CO N360000N403500
   W0890500W0954500
BT United States
NT Atchison County Missouri
NT Barton County Missouri
NT Benton County Missouri
NT Boone County Missouri
NT Buchanan County Missouri
NT Butler County Missouri
NT Caldwell County Missouri
NT Camden County Missouri
NT Carroll County Missouri
NT Carter County Missouri
NT Cass County Missouri
NT Christian County Missouri
NT Clark County Missouri
NT Clay County Missouri
NT Clinton County Missouri
NT Crawford County Missouri
NT Dade County Missouri
NT Dallas County Missouri
NT De Kalb County Missouri
NT Douglas County Missouri
NT Franklin County Missouri
NT Greene County Missouri
NT Grundy County Missouri
NT Harrison County Missouri
NT Henry County Missouri
NT Howard County Missouri
NT Iron County Missouri
NT Jackson County Missouri
NT Jasper County Missouri
NT Jefferson County Missouri
NT Johnson County Missouri
NT Joplin Missouri
NT Kansas City Missouri
NT Lafayette County Missouri
NT Lawrence County Missouri
NT Lewis County Missouri
NT Lincoln County Missouri
NT Linn County Missouri
NT Livingston County Missouri
NT Macon County Missouri
NT Madison County Missouri
NT Marion County Missouri
NT Mercer County Missouri
NT Monroe County Missouri
NT Montgomery County Missouri
NT Morgan County Missouri
NT New Madrid County Missouri
NT Newton County Missouri
NT Osage County Missouri
NT Perry County Missouri
NT Phelps County Missouri
NT Pike County Missouri
NT Platte County Missouri
NT Polk County Missouri
NT Pulaski County Missouri
NT Putnam County Missouri
NT Randolph County Missouri
NT Reynolds County Missouri
NT Saint Charles County Missouri
NT Saint Clair County Missouri
NT Saint Francois County Missouri
NT Saint Francois Mountains
NT Saint Louis County Missouri
NT Saline County Missouri
NT Scott County Missouri
NT Shannon County Missouri
NT Shelby County Missouri

NT Sullivan County Missouri
NT Texas County Missouri
NT Viburnum Trend
NT Warren County Missouri
NT Washington County Missouri
NT Wayne County Missouri
NT Webster County Missouri
SA Aux Vases Sandstone
SA Bainbridge Formation
SA Bonneterre Formation
SA Boone Formation
SA Burlington Limestone
SA Cabaniss Formation
SA Cedar City Formation
SA Chattanooga Shale
SA Cherokee Group
SA Chesterian
SA Chouteau Limestone
SA Clayton Formation
SA Decorah Shale
SA Desmoinesian
SA Douglas Group
SA Ervine Creek Limestone
SA Everton Formation
SA Excello Shale
SA Fayetteville Formation
SA Fernvale Formation
SA Forest City Basin
SA Golconda Formation
SA Grand River
SA Kansas City Group
SA Keokuk Limestone
SA Krebs Group
SA Labette Shale
SA Lamotte Sandstone
SA Lawrence Formation
SA Lecompton Limestone
SA Maquoketa Formation
SA Marmaton Group
SA Midcontinent
SA Midway Group
SA Mississippi Embayment
SA Mississippi River
SA Mississippi Valley
SA Missouri River
SA Missouri River basin
SA Missouri River valley
SA New Madrid region
SA Oread Limestone
SA Ozark Mountains
SA Platte River
SA Plattsburg Limestone
SA Plattsmouth Limestone Member
SA Pleasanton Group
SA Porters Creek Formation
SA Red Eagle Limestone
SA Reelfoot Rift
SA Renault Formation
SA Richmond Group
SA Ripley Formation
SA Rock Lake Shale Member
SA Saint Laurent Limestone
SA Saint Louis Limestone
SA Saint Peter Sandstone
SA Sainte Genevieve Limestone
SA Salem Limestone
SA Smithville Formation
SA Spergen Formation
SA Stanton Formation
SA Tonganoxie Sandstone
SA Upper Mississippi Valley
SA Wabaunsee Group
SA Warsaw Formation
SA White River
SA Wilcox Group
SA Wyandotte Limestone

Missouri Plateau (1985)
N Great Plains.
IN Index states as applicable.
BT United States
SA Great Plains
SA Montana
SA North Dakota

SA South Dakota

Missouri River (1978)
Rises in S Montana and flows E and SE to join the Mississippi River N of Saint Louis.
IN Index states as applicable.
BT United States
SA Iowa
SA Kansas
SA Missouri
SA Missouri River basin
SA Missouri River valley
SA Montana
SA Nebraska
SA North Dakota
SA South Dakota

Missouri River basin (1978)
Drainage basin.
IN Index states and provinces as applicable.
BT North America
SA Alberta
SA Canada
SA Colorado
SA Iowa
SA Kansas
SA Missouri
SA Missouri River
SA Montana
SA Nebraska
SA North Dakota
SA Saskatchewan
SA South Dakota
SA United States
SA Wyoming

Missouri River valley (1978)
IN Index states as applicable.
BT United States
SA Iowa
SA Kansas
SA Missouri
SA Missouri River
SA Montana
SA Nebraska
SA North Dakota
SA South Dakota

Missourian (1978)
Provincial series, North America: lower Upper Pennsylvanian, above Desmoinesian, below Virgilian.
UF Missourian Series
BT Upper Pennsylvanian
BT Pennsylvanian
BT Carboniferous
BT Paleozoic
NT Kansas City Group
NT Lansing Group
NT Plattsburg Limestone
NT Pleasanton Group
NT Rock Lake Shale Member
NT Seminole Formation
NT Stanton Formation
NT Wann Formation
NT Wyandotte Limestone

Missourian Series
use Missourian

Mist Mountain Formation (1989)
In Kootenay Formation.
IN Index ages as applicable.
BT Mesozoic
SA Alberta
SA British Columbia
SA Cretaceous
SA Jurassic
SA Kootenay Formation

Mistastin Lake (1978)
N Labrador.
BT Labrador
BT Newfoundland
BT Eastern Canada
BT Canada

Mitchell County
Valid through 1988. Search in combination with state term. After 1988, use specific county-state term.

Mitchell County Georgia (1989)
SW Georgia. Before 1989, also search Mitchell County AND Georgia.
CO N310500N312700
W0835900W0842900
BT Georgia
BT United States

Mitchell County Iowa (1989)
N Iowa. Before 1989, also search Mitchell County AND Iowa.
CO N431300N433000
W0923300W0930200
BT Iowa
BT United States

Mitchell County Kansas (1989)
N Kansas. Before 1989, also search Mitchell County AND Kansas.
CO N391300N393300
W0975500W0983000
BT Kansas
BT United States

Mitchell County North Carolina (1989)
W North Carolina. Before 1989, also search Mitchell County AND North Carolina.
CO N355000N360900
W0815800W0822400
BT North Carolina
BT United States

Mitchell County Texas (1989)
W Texas. Before 1989, also search Mitchell County AND Texas.
CO N320500N323500
W1004000W1011000
BT Texas
BT United States

Mitsuishi
No longer a valid term for GeoRef. See Mitsuishi Japan.

Mitsuishi Japan (1993)
Town in Okayama Prefecture, SW Honshu. Also town in S Hokkaido. Before 1993, also search Mitsuishi in combination with Japan. Also Before 1978, also search Mituisi.
UF Mituisi Japan
BT Japan
BT Far East
BT Asia
SA Hokkaido
SA Honshu
SA Okayama Japan

Mitterberg
No longer a valid term for GeoRef. See Mitterberg Austria.

Mitterberg Austria (1993)
Village in Salzburg State, W central Austria.
BT Salzburg State Austria
BT Austria
BT Central Europe
BT Europe

Mituisi Japan
use Mitsuishi Japan

Miura Peninsula (1978)
SE Honshu. Extends into Sagami Sea S of Yokohama and Tokyo Bay.
BT Honshu
BT Japan

BT Far East
BT Asia
SA Kanagawa Japan
SA Kazusa Group

mixed crystals
use solid solution

mixed-layer minerals (1978)
Before 1978, search clay minerals AND mixed-layer.
SA clay minerals
SA minerals

mixing (1978)
Used as a general term.
SA immiscibility
SA processes
SA separation

mixite (1978)
BT arsenites
SA mixtite

mixtite (1978)
BT clastic rocks
BT sedimentary rocks
SA diamictite
SA mixite

Miyagi
No longer a valid term for GeoRef. See Miyagi Japan.

Miyagi earthquake 1978 (1989)
Epicenter in Miyagi, NE Honshu.
UF Miyagi-ken-oki earthquake, 1978
UF Miyagiken-oki earthquake, 1978
BT earthquakes
SA Honshu
SA Japan
SA Miyagi Japan

Miyagi Japan (1993)
Prefecture in N Honshu.
BT Honshu
BT Japan
BT Far East
BT Asia
NT Hosokura Mine
NT Oshika Peninsula
NT Sendai Japan
SA Abukuma Mountains
SA Miyagi earthquake 1978
SA Tohoku
SA Zao

Miyagi-ken-oki earthquake, 1978
use Miyagi earthquake 1978

Miyagiken-oki earthquake, 1978
use Miyagi earthquake 1978

Miyake-Jima (1978)
Island of Izu-shichito group in Greater Tokyo, S of O-shima Island and Sagami Sea. Active volcano on island.
UF Miyaki Island
UF Miyaki-Jima Island
BT Izu-shichito
BT Honshu
BT Japan
BT Far East
BT Asia

Miyaki Island
use Miyake-Jima

Miyaki-Jima Island
use Miyake-Jima

Miyazaki
No longer a valid term for GeoRef. See Miyazaki Japan.

Miyazaki Japan (1993)
City and Prefecture in SE Honshu.
BT Kyushu
BT Japan

BT Far East
BT Asia
NT Ebino Japan
SA Kirishima

Miye Japan
use Mie Japan

Mizunami
No longer a valid term for GeoRef. See Mizunami Japan.

Mizunami Group (1985)
Central Honshu, Japan.
BT Miocene
BT Neogene
BT Tertiary
BT Cenozoic
SA Honshu
SA Japan
SA Mizunami Japan

Mizunami Japan (1993)
Town in central Honshu. Before 1993, also search Mizunami AND Japan.
BT Gifu Japan
BT Honshu
BT Japan
BT Far East
BT Asia
SA Hiyoshi Japan
SA Mizunami Group

Mizusawa
No longer a valid term for GeoRef.
use Mizusawa Japan

Mizusawa Japan (1993)
Town in N Honshu. Before 1993, also search Mizusawa AND Japan.
UF Mizusawa
BT Iwate Japan
BT Honshu
BT Japan
BT Far East
BT Asia

MM scale
use modified Mercalli scale

Mn
use manganese

Mn-53 (1978)
Autoposting of broader terms began in 1989. This term has multiple hierarchies.
BT1 radioactive isotopes
BT1 isotopes
BT2 manganese
BT2 metals

Mn-54 (1978)
Autoposting of broader terms began in 1989. This term has multiple hierarchies.
BT1 radioactive isotopes
BT1 isotopes
BT2 manganese
BT2 metals

Mo
use molybdenum

Moab
Valid through 1988. Search in combination with state term. After 1988, use specific city-state term.

Moab Utah (1989)
City in E Utah. Before 1989, search Moab AND Utah.
CO N383500N383500
W1093400W1093400
BT Grand County Utah
BT Utah
BT United States

Moapa Valley (1978)

NE of Las Vegas in Clark County in S Nevada.
BT Clark County Nevada
BT Nevada
BT United States

Mobile
Valid through 1988. Search in combination with state term. After 1988, use specific city-state term.

Mobile Alabama (1989)
City on Mobile Bay in S Alabama. Before 1989, search Mobile AND Alabama.
CO N304000N304000
  W0880500W0880500
BT Mobile County Alabama
BT Alabama
BT United States

Mobile Bay (1985)
Inlet of Gulf of Mexico.
CO N301400N304500
  W0875000W0881000
BT Alabama
BT United States
SA Gulf Coastal Plain
SA Gulf of Mexico
SA Mississippi Sound
SA Tombigbee River

mobile belts (1978)
UF belts, mobile
SA crust
SA fold and thrust belts
SA fold belts
SA geosynclines
SA orogenic belts
SA tectonics

Mobile County
Valid through 1988. Search in combination with state term. After 1988, use specific county-state term.

Mobile County Alabama (1989)
Extreme SW Alabama. Before 1989, also search Mobile County AND Alabama.
CO N301500N311000
  W0875500W0883200
BT Alabama
BT United States
NT Mobile Alabama
SA Mississippi Sound

mobility (1978)
Used as a general term.
SA erosion
SA tectonics

mobilization (1978)
Restricted to geochemical meaning.
SA deformation
SA geochemistry
SA migration of elements

Mocamedes
Not a valid term for GeoRef. River. Before 1993, also used for the province and city which were renamed. See Namibe Angola for the province and city.
use Mossamedes

Moctezuma
No longer a valid term for GeoRef. As of 1993 see Moctezuma Mexico.

Moctezuma Mexico (1993)
Town in E central Sonora, NW Mexico. Before 1993 also search Moctezuma AND Mexico.
BT Sonora Mexico
BT Mexico

modal analysis (1978)

SA analysis
SA geochemistry
SA mineral composition
SA petrology

model studies
 use models

modeling
 use models

modelling
 use models

Modelo Formation (1989)
S California.
BT upper Miocene
BT Miocene
BT Neogene
BT Tertiary
BT Cenozoic
SA California

models (1978)
Before 1971, also search model studies. From 1978 to 1992, this term was autoposted to its narrower terms.
UF model studies
UF modeling
UF modelling
SA analog simulation
SA data processing
SA digital simulation
SA digital terrain models
SA four-dimensional models
SA infinite models
SA mathematical models
SA numerical models
SA one-dimensional models
SA physical models
SA representative basins
SA scale models
SA semi-infinite models
SA simulation
SA spherical models
SA theoretical models
SA three-dimensional models
SA two-dimensional models
SA two-layer models
SA two-phase models

Modena
No longer a valid term for GeoRef. See Modena Italy.

Modena Italy (1993)
City and province in central Emilia-Romagna, N Italy.
BT Emilia-Romagna Italy
BT Italy
BT Southern Europe
BT Europe

modern (1978)
Used for present-day studies in conjunction with Holocene.
SA Holocene

modified Mercalli scale (1978)
UF Mercalli scale, modified
UF MM scale
SA earthquakes
SA seismic intensity
SA seismology

Modiomorphoida
 use Actinodontida

Modoc County
Valid through 1988. Search in combination with state term. After 1988, use specific county-state term.

Modoc County California (1989)
NE California. Before 1989, also search Modoc County AND California.
CO N411200N420000
  W1200000W1212500

BT California
BT United States
SA Clear Lake

Modoc Plateau (1978)
High, semiarid, and volcanic plateau in NE California. Also search Modoc.
BT California
BT United States

modulus of compression
 use compressibility

modulus of elasticity
 use elastic constants

modulus of incompressibility
 use bulk modulus

modulus of rigidity
 use shear modulus

modulus, bulk
 use bulk modulus

modulus, shear
 use shear modulus

Moenave Formation (1993)
Lower Jurassic and Upper Triassic. In NE Arizona, S Utah, and Nevada. Member of Glen Canyon Group.
IN Index ages as applicable.
BT Mesozoic
SA Arizona
SA Glen Canyon Group
SA Lower Jurassic
SA Nevada
SA Upper Triassic
SA Utah

Moenkopi Formation (1978)
Lower and middle (?) Triassic or Triassic (?). In SW Utah, subdivided into (ascending) Timpoweap, Lower Red, Virgin Limestone, Middle Red, Shnabkaib, and Upper Red members. Arizona, California, Colorado, Nevada, and S Utah.
BT Triassic
BT Mesozoic
SA Arizona
SA California
SA Colorado
SA Nevada
SA Utah

Moeritherioidea (1981)
Suborder. Autoposting of Eutheria and Theria to this term began in 1989.
BT Proboscidea
BT Eutheria
BT Theria
BT Mammalia
BT Tetrapoda
BT Vertebrata
BT Chordata

Moesia (1978)
Ancient region S of the lower Danube. Includes Bulgaria, SE Romania, and Serbia.
IN Index Serbia and countries as applicable.
BT Southern Europe
BT Europe
SA Bulgaria
SA Romania
SA Serbia

Moesian Platform (1978)
IN Index countries as applicable.
BT Southern Europe
BT Europe
SA Bulgaria
SA Romania

Moffat County
Valid through 1988. Search in combination with state term. After 1988, use specific county-state term.

Moffat County Colorado (1989)
Extreme NW Colorado. Before 1989, also search Moffat County AND Colorado.
CO N401500N410000
  W1072000W1090500
BT Colorado
BT United States
SA Dinosaur National Monument
SA Yampa River

Mogilev
No longer a valid term for GeoRef. As of 1993, see Mogilev Belarus.

Mogilev Belarus (1993)
Oblast and city on the Dnieper River. This term has multiple hierarchies.
UF Mogilev on the Dnieper
BT1 Belarus
BT1 Europe
BT2 Belarus
BT2 Commonwealth of Independent States

Mogilev on the Dnieper
 use Mogilev Belarus

Mogilno
No longer a valid term for GeoRef. As of 1993 see Mogilno Poland.

Mogilno Poland (1993)
Town in Bydgoszcz, NW central Poland. Before 1993 also search Mogilno AND Poland.
BT Bydgoszcz Poland
BT Poland
BT Central Europe
BT Europe

Mogollon Mountains (1981)
SW New Mexico. Autoposting of broader terms to this term began in 1989.
CO N330000N334000
  W1083000W1085000
BT Catron County New Mexico
BT New Mexico
BT United States
SA Datil-Mogollon volcanic field
SA Mogollon Plateau
SA Mogollon Rim

Mogollon Plateau (1978)
Tableland S of Winslow in E central Arizona.
BT Arizona
BT United States
SA Mogollon Mountains
SA Mogollon Rim

Mogollon Rim (1981)
S escarpment of Mogollon Plateau.
CO N335000N344000
  W1090000W1113500
BT Arizona
BT United States
SA Mogollon Mountains
SA Mogollon Plateau

Mogollon-Datil volcanic field
 use Datil-Mogollon volcanic field

Mohave County
Valid through 1988. Search in combination with state term. After 1988, use specific county-state term.

Mohave County Arizona (1989)

NW Arizona. Before 1989, also search Mohave County AND Arizona.
CO N341000N370000
W1123500W1144500
BT Arizona
BT United States
SA Lake Mead
SA Mohave Mountains

Mohave Desert
use Mojave Desert

Mohave Mountains (1985)
W Arizona and SE California. Autoposting of broader terms to this term began in 1989.
IN Index states and counties as applicable.
CO N342500N345000
W1140000W1144000
UF Mojave Mountains
BT United States
SA Arizona
SA California
SA Mohave County Arizona
SA San Bernardino County California

Mohawk River valley
use Mohawk Valley

Mohawk Valley (1978)
River valley in E central New York.
UF Mohawk River valley
BT New York
BT United States
SA Appalachian Plateau

Mohnian (1985)
North America. Miocene, above Luisian and below Delmontian.
BT Miocene
BT Neogene
BT Tertiary
BT Cenozoic
SA middle Miocene
SA upper Miocene

Mohns Ridge (1981)
NE of Jan Mayen in the Arctic Ocean.
CO N700000N740000
E0100000W0080000
BT Arctic Ocean

Moho
use Mohorovicic discontinuity

Mohorovicic discontinuity (1978)
UF M-discontinuity
UF Moho
SA asthenosphere
SA basement
SA continental drift
SA core
SA crust
SA depth
SA discontinuities
SA Earth
SA earthquakes
SA elastic waves
SA isostasy
SA lithosphere
SA mantle
SA plate tectonics
SA sea-floor spreading
SA seismology
SA Sn-waves
SA tectonophysics

Moine thrust zone (1989)
BT Scotland
BT Great Britain
BT United Kingdom
BT Western Europe
BT Europe

Moinian (1978)
Europe. Autoposting of upper Precambrian and Precambrian to this term ended in 1989. As of 1989, upper Proterozoic and Proterozoic are autoposted to this term. Autoposting of upper Precambrian and Precambrian to this term began in 1990.
BT upper Proterozoic
BT Proterozoic
BT upper Precambrian
BT Precambrian

moissanite (1978)
Meteorite mineral.
BT carbides
BT alloys
SA carbon
SA carborundum
SA meteorites
SA silicides
SA silicon

moisture (1978)
UF moisture content
UF water content
SA atmospheric precipitation
SA desiccation
SA humidity
SA soils
SA water
SA water regimes
SA water vapor
SA wettability

moisture content
use moisture

Mojave Desert (1978)
Arid basin in S California. Also search Mojave.
UF Mohave Desert
BT California
BT United States
SA Antelope Valley
SA Aztec Sandstone
SA Cady Mountains
SA Cajon Pass
SA Calico Mountains
SA Coyote Lake
SA Hector Formation
SA Kern County California
SA Los Angeles County California
SA Old Woman Mountains
SA Rand Mountains
SA San Bernardino County California

Mojave Mountains
use Mohave Mountains

Mokoia
No longer a valid term for GeoRef. See Mokoia New Zealand.

Mokoia New Zealand (1993)
Village in SW North Island.
IN Also index North Island.
BT Taranaki New Zealand
BT New Zealand
BT Australasia
SA North Island

molasse (1978)
BT clastic rocks
BT sedimentary rocks
SA terrigenous materials

Molasse Basin (1978)
A geologic-geographic term which refers to the feature in Europe. Also search molasse in combination with country name.
IN Index countries as applicable.
BT Central Europe
BT Europe
NT Swiss Molasse Basin
SA Austria
SA Germany
SA Switzerland

Moldanubian (1978)
UF Moldanubicum
UF Moldanubicum Complex
UF Moldanubikum
BT Precambrian
SA Archean
SA Czechoslovakia

Moldanubicum
use Moldanubian

Moldanubicum Complex
use Moldanubian

Moldanubikum
use Moldanubian

Moldava Valley (1978)
River valley. Also search Moldava River. This term has multiple hierarchies.
UF Moldova Valley
BT1 Moldavia
BT1 Europe
BT2 Romania
BT2 Southern Europe
BT2 Europe

Moldavia (1978)
Region E of Transylvania and N of E Walachia in Romania, and the republic of Moldova, the former Moldavian Soviet Socialist Republic.
IN Index countries as applicable.
BT Europe
NT Bacau Romania
NT Bicaz Valley
NT Bistrita Valley
NT Gutii Mountains
NT Iacobeni Romania
NT Moldava Valley
NT Moldavian Platform
NT Piatra-Neamt Romania
NT Suceava Romania
NT Teleajen Valley
NT Trotus Valley
SA Dniester River
SA Dniester-Prut Interfluve
SA Moldova
SA Prut River
SA Rarau Massif
SA Romania
SA Romanian Plain
SA Russian Plain
SA Russian Platform
SA Siret River
SA Tiraspol Moldova

Moldavian Plateau
use Moldavian Platform

Moldavian Platform (1978)
This term has multiple hierarchies.
UF Moldavian Plateau
BT1 Moldavia
BT1 Europe
BT2 Romania
BT2 Southern Europe
BT2 Europe
SA Siret River

moldavite (1978)
BT tektites

Moldova (1993)
Republic. Former Moldavian Soviet Socialist Republic. Before 1993, also search Bessarabian or Bessarabia if applicable; also search Moldavia AND USSR. This term has multiple hierarchies.
BT1 Europe
BT2 Commonwealth of Independent States
NT Tiraspol Moldova
SA Black Sea region
SA Dniester River
SA Dniester-Prut Interfluve
SA Moldavia
SA Prut River
SA Russian Plain

Moldova Noua
No longer a valid term for GeoRef. See Moldova Noua Romania.

Moldova Noua Romania (1993)
Town in SW near the Serbian border in Caras-Severin County. Before 1993, also search Moldova Noua. Before 1978, also search Moldova-Noua. This term has multiple hierarchies.
BT1 Banat
BT1 Southern Europe
BT1 Europe
BT2 Romania
BT2 Southern Europe
BT2 Europe

Moldova Valley
use Moldava Valley

Moldova-Noua
Not a valid term for GeoRef. See Moldova Noua Romania.

molecular fossils
use biomarkers

molecular structure (1978)
SA crystal structure

Molise
No longer a valid term for GeoRef. See Molise Italy.

Molise Italy (1993)
Autonomous region on the Adriatic Sea.
CO N412300N420500
E0151000E0140000
BT Italy
BT Southern Europe
BT Europe
NT Campobasso Italy

mollisol
use active layer

Mollisols (1978)
BT soils
SA soil group

Mollusca (1978)
Before 1981, also search mollusks. Autoposting of this term to Tentaculites, Tentaculitidae and Tentaculitida began in 1981.
BT Invertebrata
NT Aplacophora
NT Bivalvia
NT Cephalopoda
NT Gastropoda
NT Hyolithes
NT Monoplacophora
NT Polyplacophora
NT Rostroconchia
NT Scaphopoda
NT Tentaculitida
SA Dacryoconarida
SA mollusks
SA pearls
SA reef builders

mollusks (1978)
As of 1981, restricted to use as the common name for Mollusca. Autoposting of this term began in 1978.
IN Index for all non-paleontologic studies of fossils.
BT invertebrates
NT bivalves
NT cephalopods
NT gastropods

SA biostratigraphy
SA Mollusca

Molodezhnaya Station (1978)
USSR station on Alasheyev Bight in Enderby Land between Lutzow-Holm Bay and Cape Ann on the Indian Ocean side. Also search Molodezhnaya.
BT Antarctica

Molokai (1978)
Island in Hawaii between Oahu and Maui. Autoposting of broader terms to this term began in 1989. This term has multiple hierarchies.
IN Index Maui County Hawaii or Kalawao County Hawaii as applicable.
CO N210400N211500
 W1564200W1572200
BT1 Hawaii
BT1 United States
BT2 Hawaii
BT2 Polynesia
BT2 Oceania
BT3 Hawaii
BT3 East Pacific Ocean Islands
SA Maui County Hawaii
SA Mauna Loa

Molotov
No longer a valid term for GeoRef. As of 1993, see Molotov Russian Federation.

Molotov Russian Federation (1993)
City W of the Urals in Perm Oblast. Before 1993, also search Molotov. This term has multiple hierarchies.
BT1 Perm Russian Federation
BT1 Russian Federation
BT1 Commonwealth of Independent States
BT2 Perm Russian Federation
BT2 Europe

Molteno
No longer a valid term for GeoRef. As of 1993 see Molteno South Africa.

Molteno South Africa (1993)
Town in E Cape Province, S South Africa. Before 1993 also search Molteno AND South Africa.
BT Cape Province South Africa
BT South Africa
BT Southern Africa
BT Africa

Moluccas (1978)
Group of islands between Celebes and New Guinea.
CO S071000N030000
 E1350000E1240000
UF Spice Islands
BT Indonesia
BT Far East
BT Asia

Moluya
use Moulouya River

molybdates (1978)
Autoposting of this term began in 1978.
NT ferrimolybdite
NT powellite
NT umohoite
NT wulfenite
SA minerals
SA tungstates

molybdenite (1978)
BT sulfides
SA molybdenum
SA molybdenum ores

molybdenite porphyry
use porphyry molybdenum

molybdenum (1978)
Chemical element. As of 1981, use molybdenum ores for molybdenum as a commodity. Autoposting of metals to this term began in 1989.
UF Mo
BT metals
SA molybdenite
SA molybdenum ores
SA porphyry molybdenum

molybdenum ores (1981)
Before 1981, also search (molybdenum OR Mo) AND (deposit OR deposits OR ore OR ores OR economic) in the basic index. Autoposting of metal ores to this term began in 1985.
IN Commodity. See List C.
BT metal ores
SA Climax Mine
SA Colorado mineral belt
SA Henderson Mine
SA molybdenite
SA molybdenum
SA porphyry molybdenum
SA Questa Mine

moment tensors
Not a valid term for GeoRef. See seismic moment.

Momipites (1989)
Autoposting of microfossils to this term began in 1990.
BT miospores
BT palynomorphs
BT microfossils

Monaco (1978)
On the Mediterranean Sea near the French-Italian border.
BT Western Europe
BT Europe

Monagas
No longer a valid term for GeoRef. As of 1993 see Monagas Venezuela.

Monagas Venezuela (1993)
State in NE Venezuela. Before 1993 also search Monagas AND Venezuela.
CO N082500N102000
 W0620000W0641000
BT Venezuela
BT South America
SA Orinoco River
SA San Juan River

Monashee Complex (1989)
In Shuswap Complex of SE British Columbia.
IN Index ages as applicable.
SA British Columbia
SA lower Mesozoic
SA Shuswap Complex
SA upper Paleozoic

monazite (1978)
As of 1981, use monazite deposits for monazite as a commodity.
BT phosphates
SA heavy minerals
SA monazite deposits

monazite deposits (1981)
Before 1981, search monazite AND deposits.
IN Commodity. See List C.
SA cerium ores
SA heavy mineral deposits
SA monazite
SA placers
SA rare earth deposits

SA thorium ores

Monchegorsk
No longer a valid term for GeoRef. As of 1993, see Monchegorsk Russian Federation.

Monchegorsk Russian Federation (1993)
City in Murmansk Oblast about 70 miles S of Murmansk in extreme NW European Russian Federation. Before 1993, also search Monchegorsk. This term has multiple hierarchies.
BT1 Murmansk Russian Federation
BT1 Russian Federation
BT1 Commonwealth of Independent States
BT2 Murmansk Russian Federation
BT2 Europe

monchiquite (1978)
BT lamprophyres
BT plutonic rocks
BT igneous rocks

Moncton Basin (1989)
Carboniferous structural basin in New Brunswick.
UF Moncton Subbasin
BT New Brunswick
BT Maritime Provinces
BT Eastern Canada
BT Canada

Moncton Subbasin
use Moncton Basin

Moneron Island (1978)
Soviet island in Japan Sea 30 miles off SW Sakhalin. This term has multiple hierarchies.
BT1 Russian Federation
BT1 Commonwealth of Independent States
BT2 Asia
SA Sakhalin

monetite (1978)
BT phosphates

Monghyr
No longer a valid term for GeoRef. See Monghyr India.

Monghyr India (1993)
City in NE central Bihar, E India.
BT Bihar India
BT India
BT Indian Peninsula
BT Asia

Mongolia (1978)
CO N420000N520000
 E1200000E0870000
UF Outer Mongolia
BT Far East
BT Asia
NT Altan Teeli Mongolia
NT Bayan-Nuurin-khotnor Basin
NT Bayn Dzak
NT Choybalsan Mongolia
NT Hangay Mountains
NT Ikh-Khayrkhan
NT Khubsugul Mongolia
NT Kobdo
NT Kobdo Mongolia
NT Mongolian Altai
NT Selenga Mongolia
NT Ulan Bator Mongolia
SA Altai Mountains
SA Altai-Sayan region
SA Amur Basin
SA Djadokhta Formation
SA Gobi Desert
SA Kerulen River
SA Selenga River valley

SA Tannu-Ola Range
SA Yenisei Basin
SA Yenisei-Khatanga basin

Mongolian Altai (1978)
Southeastern extension of the Altai Mountains. This term has multiple hierarchies.
BT1 Mongolia
BT1 Far East
BT1 Asia
BT2 Altai Mountains
BT2 Asia

monitoring (1978)
SA environmental geology
SA geophysical surveys
SA hydrogeology
SA networks
SA observation wells
SA pollution
SA water wells

monitoring wells
Not a valid term for GeoRef. See observation wells or monitoring.

Monmouth County
Valid through 1988. Search in combination with state term. After 1988, use specific county-state term.

Monmouth County New Jersey (1989)
On the Atlantic Ocean in E New Jersey. Before 1989, also search Monmouth County AND New Jersey.
CO N400400N402900
 W0735700W0743800
BT New Jersey
BT United States
SA Raritan Bay
SA Sandy Hook

Monmouth Group (1978)
Includes (ascending) Mount Laurel, Navesink, Red Bank, and Tinton formations. Delaware, NE Maryland, and New Jersey.
BT Upper Cretaceous
BT Cretaceous
BT Mesozoic
SA Delaware
SA Maryland
SA Navesink Formation
SA New Jersey

Monmouthshire
No longer a valid term for GeoRef as of 1993.
use Monmouthshire Wales

Monmouthshire Wales (1993)
Former county in SE Wales. As of April 1974, part of Gwent. Before 1993, also search Monmouthshire.
CO N513000N520000
 W0024000W0032000
UF Monmouthshire
BT Wales
BT Great Britain
BT United Kingdom
BT Western Europe
BT Europe

Mono Basin (1978)
Mono County in E California.
IN Index county or region as applicable.
SA California
SA Mono County California

Mono County
Valid through 1988. Search in combination with state term. After 1988, use specific county-state term.

Mono County California (1989)
  E California. Before 1989, also search Mono County AND California.
  CO N373000N384000
       W1175000W1193000
  BT California
  BT United States
  NT Inyo Domes
  NT Long Valley Caldera
  NT Mono Craters
  NT Obsidian Dome
  SA Casa Diablo
  SA Long Valley
  SA Mammoth Lakes earthquakes
  SA Mono Basin
  SA Mono Lake

Mono Craters (1978)
  Range of about 20 geologically recent volcanic cones just S of Mono Lake, Mono County in E California.
  BT Mono County California
  BT California
  BT United States

Mono Lake (1978)
  In central Mono County in E California.
  IN Index county or regions as applicable.
  SA California
  SA Mono County California

monoclines (1978)
  Not valid term from 1975 through 1977. Before 1978, also search folds AND monoclinal.
  BT folds
  SA flexure folds

monoclinic system (1978)
  Before 1978, search monoclinic.
  BT crystal systems
  SA crystallography

Monocotyledoneae (1978)
  Autoposting of this term began in 1978.
  BT angiosperms
  BT Spermatophyta
  BT Plantae
  NT Alismidae
  NT Arecidae
  NT Commelinidae
  NT Cyperaceae
  NT Gramineae
  NT Liliidae
  NT Palmae
  SA Dicotyledoneae

monographs (1982)
  UF books
  SA bibliography
  SA book reviews
  SA catalogs
  SA dictionaries
  SA directory
  SA glossaries
  SA guidebook
  SA lexicons
  SA manuals
  SA publications
  SA textbooks
  SA theses

Monograptina (1978)
  IN Index Hemichordata or Invertebrata as applicable.
  BT Graptoloidea
  BT Graptolithina
  NT Monograptus
  SA Hemichordata
  SA Invertebrata

Monograptus (1978)
  Genus.
  IN Index Hemichordata or Invertebrata as applicable.
  BT Monograptina
  BT Graptoloidea
  BT Graptolithina
  NT Monograptus uniformis
  SA Hemichordata
  SA Invertebrata

Monograptus uniformis (1978)
  IN Index Hemichordata or Invertebrata as applicable.
  BT Monograptus
  BT Monograptina
  BT Graptoloidea
  BT Graptolithina
  SA Hemichordata
  SA Invertebrata

monohydrocalcite (1978)
  BT carbonates

monolete taxa (1978)
  Before 1978, also search monoletes.
  UF monoletes
  SA miospores
  SA palynomorphs

monoletes
  use monolete taxa

Monomoy Island (1978)
  A 10 mile sandspit S of Chantham on Cape Cod. Sometimes it is connected to land and is not an island.
  BT Cape Cod
  BT Barnstable County Massachusetts
  BT Massachusetts
  BT United States

Monongahela Group (1978)
  In Pennsylvania, contains Pittsburgh Coal Seam at base; top of Waynesburg Coal marks upper boundary. W Maryland, E Ohio, W Pennsylvania, W Virginia, and West Virginia.
  BT Pennsylvanian
  BT Carboniferous
  BT Paleozoic
  SA Maryland
  SA Ohio
  SA Pennsylvania
  SA Pittsburgh Coal
  SA Virginia
  SA West Virginia

Monongalia County
  Valid through 1988. Search in combination with state term. After 1988, use specific county-state term.

Monongalia County West Virginia (1989)
  N West Virginia. Before 1989, also search Monongalia County AND West Virginia.
  CO N392500N394300
       W0794700W0802500
  BT West Virginia
  BT United States

Monoplacophora (1978)
  BT Mollusca
  BT Invertebrata

Monotis (1978)
  Genus. Autoposting of Pectinacea and Pteriina to this term began in 1989.
  BT Pectinacea
  BT Pteriina
  BT Bivalvia
  BT Mollusca
  BT Invertebrata

Monotremata (1978)
  Order.
  BT Mammalia
  BT Tetrapoda
  BT Vertebrata
  BT Chordata

Monroe County
  Valid through 1988. Search in combination with state term. After 1988, use specific county-state term.

Monroe County Alabama (1989)
  SW Alabama. Before 1989, also search Monroe County AND Alabama.
  CO N311300N314800
       W0865400W0874800
  BT Alabama
  BT United States

Monroe County Arkansas (1989)
  E central Arkansas. Before 1989, also search Monroe County AND Arkansas.
  CO N342000N350000
       W0910000W0912700
  BT Arkansas
  BT United States

Monroe County Florida (1989)
  S Florida. Before 1989, also search Monroe County AND Florida.
  CO N243500N254700
       W0801500W0821100
  BT Florida
  BT United States
  NT Dry Tortugas
  NT Florida Keys

Monroe County Georgia (1989)
  Central Georgia. Before 1989, also search Monroe County AND Georgia.
  CO N325200N331200
       W0834100W0840500
  BT Georgia
  BT United States

Monroe County Illinois (1989)
  SW Illinois. Before 1989, also search Monroe County AND Illinois.
  CO N380500N383000
       W0895400W0902200
  BT Illinois
  BT United States
  NT Columbia Illinois

Monroe County Indiana (1989)
  S central Indiana. Before 1989, also search Monroe County AND Indiana.
  CO N385900N392100
       W0862000W0864200
  BT Indiana
  BT United States

Monroe County Iowa (1989)
  S Iowa. Before 1989, also search Monroe County AND Iowa.
  CO N405400N411000
       W0923800W0930600
  BT Iowa
  BT United States

Monroe County Kentucky (1989)
  S Kentucky. Before 1989, also search Monroe County AND Kentucky.
  CO N363600N364800
       W0852500W0855800
  BT Kentucky
  BT United States

Monroe County Michigan (1989)
  Extreme SE Michigan. Before 1989, also search Monroe County AND Michigan.
  CO N414100N420700
       W0831200W0834500
  BT Michigan Lower Peninsula
  BT Michigan
  BT United States

Monroe County Mississippi (1989)
  E Mississippi. Before 1989, also search Monroe County AND Mississippi.
  CO N333800N340500
       W0881200W0884100
  BT Mississippi
  BT United States

Monroe County Missouri (1989)
  NE central Missouri. Before 1989, also search Monroe County AND Missouri.
  CO N391900N393900
       W0914300W0921800
  BT Missouri
  BT United States

Monroe County New York (1989)
  W New York. Before 1989, also search Monroe County AND New York.
  CO N425600N432200
       W0772300W0775900
  BT New York
  BT United States
  NT Rochester New York
  SA Genesee River

Monroe County Ohio (1989)
  E Ohio. Before 1989, also search Monroe County AND Ohio.
  CO N393300N395200
       W0804700W0811800
  BT Ohio
  BT United States

Monroe County Pennsylvania (1989)
  E Pennsylvania. Before 1989, also search Monroe County AND Pennsylvania.
  CO N404800N411500
       W0745800W0753900
  BT Pennsylvania
  BT United States

Monroe County Tennessee (1989)
  SE Tennessee. Before 1989, also search Monroe County AND Tennessee.
  CO N351300N354200
       W0835500W0843200
  BT Tennessee
  BT United States

Monroe County West Virginia (1989)
  SE West Virginia. Before 1989, also search Monroe County AND West Virginia.
  CO N372300N374400
       W0801300W0805200
  BT West Virginia
  BT United States

Monroe County Wisconsin (1989)
  W central Wisconsin. Before 1989, also search Monroe County AND Wisconsin.
  CO N434300N441000
       W0901900W0905800
  BT Wisconsin
  BT United States

Mons Basin (1978)
  Coal mining area near the French border. Also search Mons.
  SA Hainaut Belgium

monsoons (1978)
SA hurricanes
SA storms

Mont Blanc (1978)
In Savoy Alps near Italian border. Highest mountain in Alps. This term has multiple hierarchies.
BT1 Savoy Alps
BT1 Haute-Savoie France
BT1 France
BT1 Western Europe
BT1 Europe
BT2 Savoy Alps
BT2 Western Alps
BT2 Alps
BT2 Europe
SA Argentiere Glacier
SA Valle d'Aosta Italy

Mont Dore France
use Mont-Dore France

Mont Pelee
use Mount Pelee

Mont-Dore
No longer a valid term for GeoRef. See Mont-Dore France.

Mont-Dore France (1993)
Town and thermal station in SE Puy-de-Dome, S central France. Before 1993, also search Mont-Dore. Before 1978, also search Le Mont-Dore, Mont Dore, and Mont-Dore-les-Bains.
UF Le Mont-Dore France
UF Mont Dore France
UF Mont-Dore-les-Bains France
BT Puy-de-Dome France
BT France
BT Western Europe
BT Europe

Mont-Dore-les-Bains France
use Mont-Dore France

Mont-Saint-Michel Bay
use Bay of Saint-Michel

Montagne Noire (1978)
Southernmost range of Central Massif.
IN Index departments as applicable.
BT Central Massif
BT France
BT Western Europe
BT Europe
SA Aude France
SA Tarn France

Montague Island (1978)
At entrance of Prince William Sound, SE of Anchorage in S Alaska. Also occurs in Gulf of Mexico and off New South Wales, Australia.
IN Index Southern Alaska or other regions as applicable.
SA Alaska
SA Prince William Sound
SA Southern Alaska

Montana (1978)
Autoposting of this term began in 1978.
CO N443000N490000 W1040200W1160200
BT United States
NT Beaverhead County Montana
NT Big Horn County Montana
NT Blaine County Montana
NT Boulder Batholith
NT Broadwater County Montana
NT Butte mining district
NT Carbon County Montana
NT Carter County Montana
NT Cascade County Montana
NT Chouteau County Montana
NT Crazy Mountains Basin
NT Custer County Montana
NT Deer Lodge County Montana
NT Fergus County Montana
NT Flathead County Montana
NT Gallatin County Montana
NT Garfield County Montana
NT Glacier County Montana
NT Golden Valley County Montana
NT Granite County Montana
NT Gravelly Range
NT Hill County Montana
NT Jefferson County Montana
NT Judith Basin County Montana
NT Lake County Montana
NT Lewis and Clark County Montana
NT Liberty County Montana
NT Lincoln County Montana
NT Little Belt Mountains
NT Madison County Montana
NT McCone County Montana
NT Meagher County Montana
NT Mineral County Montana
NT Missoula County Montana
NT Park County Montana
NT Phillips County Montana
NT Powder River County Montana
NT Powell County Montana
NT Prairie County Montana
NT Ravalli County Montana
NT Richland County Montana
NT Roosevelt County Montana
NT Rosebud County Montana
NT Sanders County Montana
NT Sheridan County Montana
NT Shonkin Sag Laccolith
NT Silver Bow County Montana
NT Stillwater County Montana
NT Sweet Grass County Montana
NT Teton County Montana
NT Toole County Montana
NT Valley County Montana
NT Yellowstone County Montana
SA Absaroka Range
SA Absaroka Supergroup
SA Aldridge Formation
SA Altyn Limestone
SA Amsden Formation
SA Arikaree Group
SA Bakken Formation
SA Battle Mountain High
SA Bear Gulch Limestone Member
SA Bearpaw Formation
SA Beartooth Mountains
SA Beaverhead Formation
SA Beaverhead Mountains
SA Bell Creek Field
SA Belt Basin
SA Belt Supergroup
SA Benton Formation
SA Big Belt Mountains
SA Big Snowy Group
SA Big Snowy Mountains
SA Bighorn Basin
SA Bighorn Mountains
SA Bighorn River
SA Birdbear Formation
SA Bitterroot Range
SA Blackleaf Formation
SA Bonner Formation
SA Bridger Range
SA Bull Lake Glaciation
SA Carlile Shale
SA Cedar Creek Anticline
SA Charles Formation
SA Chugwater Formation
SA Clark Fork
SA Clark's Fork Basin
SA Cloverly Formation
SA Coeur d'Alene mining district
SA Colorado Group
SA Columbia River basin
SA Crazy Mountains
SA Dakota Formation
SA Deadwood Formation
SA Dinwoody Formation
SA Disturbed Belt
SA Duperow Formation
SA Eagle Sandstone
SA Elk Point Group
SA Elkhorn Mountains Volcanics
SA Empire Formation
SA Fernie Formation
SA Flathead Lake
SA Flathead Sandstone
SA Flint Creek Range
SA Fort Union Formation
SA Fox Hills Formation
SA Frontier Formation
SA Gallatin Range
SA Gammon Ferruginous Member
SA Glacier National Park
SA Goose Egg Formation
SA Graneros Shale
SA Great Plains
SA Greenhorn Limestone
SA Heath Formation
SA Hebgen Lake earthquake 1959
SA Helena Formation
SA Hell Creek Formation
SA Idaho Batholith
SA Interlake Formation
SA J-M Reef
SA Jefferson Group
SA Judith River Formation
SA Kootenay Arc
SA Kootenay Formation
SA Lake Missoula
SA Lakota Formation
SA Lance Formation
SA Laramie Formation
SA LASA
SA Lebo Member
SA Lewis and Clark Lineament
SA Lewis thrust fault
SA Little Missouri River basin
SA Lodgepole Formation
SA Ludlow Member
SA Madison Aquifer
SA Madison Group
SA Madison Range
SA Mannville Group
SA Mazama Ash
SA Meade Peak Member
SA Medicine Lake
SA Mesaverde Group
SA Midcontinent
SA Mission Canyon Limestone
SA Missoula Group
SA Missouri Plateau
SA Missouri River
SA Missouri River basin
SA Missouri River valley
SA Montana Group
SA Morrison Formation
SA Mount Morgan
SA Mount Shields Formation
SA Mowry Shale
SA Muddy Sandstone
SA Newcastle Sandstone
SA Newland Limestone
SA Niobrara Formation
SA Opeche Shale
SA Pagoda Formation
SA Park City Formation
SA Parkman Sandstone
SA Phosphoria Formation
SA Pierre Shale
SA Pilgrim Formation
SA Pinedale Glaciation
SA Powder River basin
SA Prairie Evaporite
SA Prichard Formation
SA Purcell Mountains
SA Purcell System
SA Ravalli Group
SA Red River Formation
SA Retort Phosphatic Shale Member
SA Revett Quartzite
SA Rocky Mountain Trench
SA Ruby Range
SA Rundle Group
SA Saint Regis Formation
SA Sawtooth Range
SA Selway-Bitterroot Wilderness
SA Sentinel Butte Formation
SA Shannon Sandstone Member
SA Shepard Formation
SA Skull Creek Shale
SA Snowslip Formation
SA Spokane Formation
SA Stillwater Complex
SA Sundance Formation
SA Sweetgrass Arch
SA Swift Formation
SA Tensleep Sandstone
SA Thaynes Formation
SA Tongue River
SA Tongue River Member
SA Tullock Member
SA Two Medicine Formation
SA Tyler Formation
SA U. S. Rocky Mountains
SA Vulture Mountain
SA Wallace Formation
SA Wapiti Formation
SA Wasatch Formation
SA Western Interior
SA Western Interior Seaway
SA Western Overthrust Belt
SA White River Group
SA Williston Basin
SA Windermere System
SA Winnipeg Formation
SA Winnipegosis Formation
SA Wyoming Province
SA Yellowstone National Park
SA Yellowstone River

Montana Group (1978)
Includes Bearpaw Shale, Claggett Shale, Cody Shale (upper part), Eagle Sandstone, Fox Hills Sandstone, Horsethief Sandstone, Judith River Formation, Lennep Sandstone, Parkman Sandstone, Pierre Shale, Trinidad Sandstone, Two Medicine Formation, and Virgelle Sandstone.
BT Upper Cretaceous
BT Cretaceous
BT Mesozoic
SA Bearpaw Formation
SA Cody Shale
SA Colorado
SA Eagle Sandstone
SA Judith River Formation
SA Kansas
SA Montana
SA New Mexico
SA North Dakota
SA Parkman Sandstone
SA Pierre Shale
SA South Dakota
SA Two Medicine Formation
SA Utah
SA Wyoming

Montana Lineament
use Lewis and Clark Lineament

Montastrea (1978)
Genus. Autoposting of Zoantharia to this term began in 1989.
BT Scleractinia
BT Zoantharia
BT Anthozoa
BT Coelenterata
BT Invertebrata

NT Montastrea annularis

Montastrea annularis (1989)
Species.
BT Montastrea
BT Scleractinia
BT Zoantharia
BT Anthozoa
BT Coelenterata
BT Invertebrata

Montauban Meteorite
use Orgueil Meteorite

Monte Amiata (1978)
Extinct volcano in the Apennines. This term has multiple hierarchies.
BT1 Tuscany Italy
BT1 Italy
BT1 Southern Europe
BT1 Europe
BT2 Apennines
BT2 Italy
BT2 Southern Europe
BT2 Europe

Monte Antola (1978)
Peak in Ligurian Apennines in NW Italy. This term has multiple hierarchies.
BT1 Ligurian Apennines
BT1 Liguria Italy
BT1 Italy
BT1 Southern Europe
BT1 Europe
BT2 Ligurian Apennines
BT2 Apennines
BT2 Italy
BT2 Southern Europe
BT2 Europe

Monte Baldo (1978)
Mountain range between Lake Garda and Adige River in NE Italy.
BT Veneto Italy
BT Italy
BT Southern Europe
BT Europe

Monte Capanne (1978)
W end of island of Elba in Mediterranean Sea off Tuscany.
BT Tuscany Italy
BT Italy
BT Southern Europe
BT Europe
SA Elba

Monte Carlo analysis (1985)
Statistical method. Before 1985, also search Monte Carlo (truncated) AND statistical analysis.
UF Monte Carlo methods
BT statistical analysis
SA geostatistics
SA mathematical geology

Monte Carlo methods
use Monte Carlo analysis

Monte Cristo Limestone (1978)
Lower and Upper Mississippian. Consists of (ascending) Dawn Limestone, Anchor Limestone, Bullion Limestone, and Yellowpine Limestone members. SE California and SE Nevada.
IN Index states as applicable.
BT Mississippian
BT Carboniferous
BT Paleozoic
SA California
SA Nevada

Monte Gargano
use Gargano

Monte Pisano (1978)
Mountain group near Pisa in NW.
BT Tuscany Italy

BT Italy
BT Southern Europe
BT Europe

Monte Rosa (1978)
Highest mountain group in the Pennine Alps.
IN Index countries and Pennine Alps as applicable.
SA Pennine Alps
SA Piemonte Italy
SA Valais Switzerland

Monte Somma (1978)
Semicircular ridge on N and E sides of Vesuvius in E Italy.
BT Campania Italy
BT Italy
BT Southern Europe
BT Europe
SA Vesuvius

Monteagle Limestone (1978)
BT Upper Mississippian
BT Mississippian
BT Carboniferous
BT Paleozoic
SA Alabama
SA Tennessee

montebrasite (1978)
BT phosphates
SA amblygonite

Montenegro (1978)
Constituent republic in SW Yugoslavia.
CO N415000N433800
    E0202000E0183000
BT Yugoslavia
BT Southern Europe
BT Europe
NT Niksic
SA Montenegro earthquake 1979

Montenegro earthquake 1979 (1993)
Montenegro, Yugoslavia-Albania region.
BT earthquakes
SA Albania
SA Montenegro
SA Yugoslavia

Monteregian Hills (1989)
Located in the Saint Lawrence Lowlands of Quebec. Composed of alkalic igneous intrusive rocks of Upper Jurassic to Lower Cretaceous age.
BT Quebec
BT Eastern Canada
BT Canada
SA Saint Lawrence Lowlands

Monterey
Valid through 1988. Search in combination with state term. After 1988, use specific city-state term.

Monterey Bay (1978)
Inlet of Pacific Ocean in Santa Cruz and Monterey counties.
BT California
BT United States
SA Monterey County California
SA Santa Cruz County California

Monterey California (1989)
City on the Pacific Ocean in W California. Before 1989, search Monterey AND California.
CO N363500N363500
    W1215500W1215500
BT Monterey County California
BT California
BT United States

Monterey Canyon (1978)
Submarine canyon just off coast of central California.

UF Monterey Deep-Sea Channel
UF Monterey Gorge
UF Monterey Seavalley
UF Monterey Submarine Canyon
UF Monterey Trough
BT Pacific Ocean
SA submarine canyons

Monterey County
Valid through 1988. Search in combination with state term. After 1988, use specific county-state term.

Monterey County California (1989)
On the Pacific Ocean and Monterey Bay in W California. Before 1989, also search Monterey County AND California.
CO N354800N365500
    W1201200W1215800
BT California
BT United States
NT Carmel California
NT Del Monte Beach
NT Monterey California
NT Parkfield California
NT Salinas California
SA Carmel Bay
SA Monterey Bay
SA Salinas Valley
SA Santa Lucia Range

Monterey Deep-Sea Channel
use Monterey Canyon

Monterey Deep-Sea Fan
use Monterey Fan

Monterey Fan (1978)
Off coast of central California just to the W of the Monterey Canyon.
UF Monterey Deep-Sea Fan
BT Pacific Ocean

Monterey Formation (1978)
Middle and upper Miocene. Comprises (ascending) Gould Shale, Devilwater Silt, McDonald Shale, Antelope Shale, and Chico-Martinez Chert members. W California.
BT Miocene
BT Neogene
BT Tertiary
BT Cenozoic
SA California

Monterey Gorge
use Monterey Canyon

Monterey Seavalley
use Monterey Canyon

Monterey Submarine Canyon
use Monterey Canyon

Monterey Trough
use Monterey Canyon

Montes de Toledo (1978)
Mountain range in S central Spain.
IN Index provinces as applicable.
BT Spain
BT Iberian Peninsula
BT Southern Europe
BT Europe
SA Ciudad Real Spain
SA Toledo Spain

Montesano Formation (1978)
Largely massive coarse-grained light-brown sandstones, with many intercalated lenses of conglomerate and grit; shales subordinate in lower part but common in upper part. SW and NW Washington.

BT upper Miocene
BT Miocene
BT Neogene
BT Tertiary
BT Cenozoic
SA Washington

Montevideo
No longer a valid term for GeoRef. As of 1993 see Montevideo Uruguay.

Montevideo Uruguay (1993)
Department in S Uruguay. Also a city on the Rio de la Plata. Before 1993 also search Montevideo AND Uruguay.
BT Uruguay
BT South America

Montezuma County
Valid through 1988. Search in combination with state term. After 1988, use specific county-state term.

Montezuma County Colorado (1989)
SW Colorado. Before 1989, also search Montezuma County AND Colorado.
CO N370000N373800
    W1075800W1090230
BT Colorado
BT United States
NT Cortez Colorado

Montgomery County
Valid through 1988. Search in combination with state term. After 1988, use specific county-state term.

Montgomery County Alabama (1989)
SE central Alabama. Before 1989, also search Montgomery County AND Alabama.
CO N315700N322700
    W0855600W0863000
BT Alabama
BT United States

Montgomery County Arkansas (1989)
W Arkansas. Before 1989, also search Montgomery County AND Arkansas.
CO N342000N344100
    W0932300W0935600
BT Arkansas
BT United States

Montgomery County Georgia (1989)
E central Georgia. Before 1989, also search Montgomery County AND Georgia.
CO N315600N322100
    W0822400W0823900
BT Georgia
BT United States

Montgomery County Illinois (1989)
S central Illinois. Before 1989, also search Montgomery County AND Illinois.
CO N390000N393500
    W0890800W0894300
BT Illinois
BT United States

Montgomery County Indiana (1989)
W central Indiana. Before 1989, also search Montgomery County AND Indiana.
CO N395200N401300
    W0864000W0870500

BT Indiana
BT United States

Montgomery County Iowa (1989)
SW Iowa. Before 1989, also search Montgomery County AND Iowa.
CO N405300N410900
  W0945500W0952300
BT Iowa
BT United States

Montgomery County Kansas (1989)
SE Kansas. Before 1989, also search Montgomery County AND Kansas.
CO N370000N372200
  W0953000W0955700
BT Kansas
BT United States

Montgomery County Kentucky (1989)
NE central Kentucky. Before 1989, also search Montgomery County AND Kentucky.
CO N375200N381200
  W0834600W0840600
BT Kentucky
BT United States

Montgomery County Maryland (1989)
Central Maryland. Before 1989, also search Montgomery County AND Maryland.
CO N385600N392100
  W0765300W0773200
BT Maryland
BT United States

Montgomery County Mississippi (1989)
Central Mississippi. Before 1989, also search Montgomery County AND Mississippi.
CO N331700N333900
  W0892300W0894500
BT Mississippi
BT United States

Montgomery County Missouri (1989)
E central Missouri. Before 1989, also search Montgomery County AND Missouri.
CO N384200N390800
  W0911700W0913700
BT Missouri
BT United States

Montgomery County New York (1989)
E central New York. Before 1989, also search Montgomery County AND New York.
CO N424700N430300
  W0740600W0744300
BT New York
BT United States

Montgomery County North Carolina (1989)
Central North Carolina. Before 1989, also search Montgomery County AND North Carolina.
CO N350800N353100
  W0793700W0801000
BT North Carolina
BT United States

Montgomery County Ohio (1989)
W Ohio. Before 1989, also search Montgomery County AND Ohio.
CO N393500N395500
  W0840300W0842800

BT Ohio
BT United States

Montgomery County Pennsylvania (1989)
SE Pennsylvania. Before 1989, also search Montgomery County AND Pennsylvania.
CO N395800N402700
  W0750100W0754300
BT Pennsylvania
BT United States

Montgomery County Tennessee (1989)
N Tennessee. Before 1989, also search Montgomery County AND Tennessee.
CO N362000N363800
  W0870700W0873900
BT Tennessee
BT United States

Montgomery County Texas (1989)
E Texas. Before 1989, also search Montgomery County AND Texas.
CO N300200N303900
  W0950500W0954900
BT Texas
BT United States

Montgomery County Virginia (1989)
SW Virginia. Before 1989, also search Montgomery County AND Virginia.
CO N370000N372100
  W0800800W0803700
BT Virginia
BT United States
NT Blacksburg Virginia

Monti Berici (1978)
Range of volcanic hills SW of Vicenza in NE Italy.
BT Veneto Italy
BT Italy
BT Southern Europe
BT Europe

Monti Caronie
  use Nebrodi Mountains

Monti Lessini
  As of 1982, no longer a valid term for GeoRef. Use Lessini Mountains.

Montian (1978)
Europe. Above Danian, below Thanetian.
BT Paleocene
BT Paleogene
BT Tertiary
BT Cenozoic

monticellite (1978)
Autoposting of olivine group and nesosilicates to this term began in 1981.
BT olivine group
BT nesosilicates
BT orthosilicates
BT silicates

Monticello
  Valid through 1988. Search in combination with state term. After 1988, use specific city-state term.

Monticello Iowa (1989)
City on Maquoketa River in E Iowa. Before 1989, search Monticello AND Iowa.
CO N421400N421400
  W0911200W0911200
BT Jones County Iowa
BT Iowa

BT United States

Monticello Kentucky (1989)
Town in S Kentucky. Before 1989, search Monticello AND Kentucky.
CO N365000N365000
  W0845000W0845000
BT Wayne County Kentucky
BT Kentucky
BT United States

Monticello Minnesota (1989)
Village on Mississippi River in S central Minnesota. Before 1989, search Monticello AND Minnesota.
CO N451700N451700
  W0934400W0934400
BT Wright County Minnesota
BT Minnesota
BT United States

Monticello New York (1989)
Resort village in SE New York. Before 1989, search Monticello AND New York.
CO N413900N413900
  W0744100W0744100
BT Sullivan County New York
BT New York
BT United States

Monticello Reservoir (1989)
Located in Fairfield County, N central South Carolina. Also in San Juan County, SE Utah. Search in conjunction with respective state.
IN Index states and counties as applicable.
BT United States
SA Fairfield County South Carolina
SA San Juan County Utah
SA South Carolina
SA Utah

Monticello Utah (1989)
Town in SE Utah. Before 1989, search Monticello AND Utah.
CO N375300N375300
  W1092200W1092200
BT San Juan County Utah
BT Utah
BT United States

Monticello Wisconsin (1989)
Village in S Wisconsin. Before 1989, search Monticello AND Wisconsin.
CO N424300N424300
  W0893600W0893600
BT Green County Wisconsin
BT Wisconsin
BT United States

montmorillonite (1978)
UF Na-montmorillonite
BT clay minerals
BT sheet silicates
BT silicates
SA beidellite
SA nontronite
SA smectite

Montoya Group (1989)
Middle and Upper Ordovician. Arizona, S New Mexico and W Texas. Includes Cable Canyon Sandstone, Upham Dolomite, Aleman Formation and Cutter Formation.
BT Ordovician
BT Paleozoic
SA Arizona
SA New Mexico
SA Texas

Montpellier
  No longer a valid term for GeoRef. See Montpellier France.

Montpellier France (1993)
City in E Herault, S France.
BT Herault France
BT France
BT Western Europe
BT Europe

Montreal
  No longer a valid term for GeoRef. As of 1993, see Montreal Quebec.

Montreal and Jesus Islands County Quebec (1993)
Before 1993, also search Montreal and Jesus Islands County AND Quebec.
BT Quebec
BT Eastern Canada
BT Canada
NT Montreal Quebec

Montreal Quebec (1993)
City on Montreal Island in the Saint Lawrence River in S Quebec. Before 1993, also search Montreal AND Quebec.
BT Montreal and Jesus Islands County Quebec
BT Quebec
BT Eastern Canada
BT Canada

Montrose County
  Valid through 1988. Search in combination with state term. After 1988, use specific county-state term.

Montrose County Colorado (1989)
W Colorado. Before 1989, also search Montrose County AND Colorado.
CO N381000N383500
  W1073000W1090500
BT Colorado
BT United States

Monts des Maures
  use Maures Massif

Monts Dome
  As of 1981, use Chaine des Puys.

Monts du Lyonnais (1978)
IN Index departments as applicable.
BT France
BT Western Europe
BT Europe
SA Loire France
SA Rhone France

Montsech (1978)
Foothills of the Pyrenees.
BT Lerida Spain
BT Catalonia Spain
BT Spain
BT Iberian Peninsula
BT Southern Europe
BT Europe

Montserrat (1978)
Volcanic island in the Leeward Islands belonging to the United Kingdom. Before 1981, West Indies was the broader term.
IN Index Leeward Islands.
BT Lesser Antilles
BT Antilles
BT West Indies
BT Caribbean region
SA Leeward Islands

monzodiorite (1978)
BT diorites
BT plutonic rocks

BT igneous rocks
monzogranite (1985)
BT granites
BT plutonic rocks
BT igneous rocks
SA monzonites

monzonite
As of 1989, no longer a valid term for GeoRef. From 1981 to 1988, monzonites and syenites were autoposted to this term.
use monzonites

monzonites (1981)
Before 1989, search monzonite. Autoposting of this term to monzonite ended in 1989.
UF monzonite
BT syenites
BT plutonic rocks
BT igneous rocks
SA monzogranite

Moodys Branch Formation (1985)
In Jackson Group.
UF Moodys Marl
BT upper Eocene
BT Eocene
BT Paleogene
BT Tertiary
BT Cenozoic
SA Alabama
SA Georgia
SA Jackson Group
SA Louisiana
SA Mississippi

Moodys Marl
use Moodys Branch Formation

Moon (1978)
NT Aristarchus
NT Eastern Sea
NT Foaming Sea
NT Fra Mauro
NT Hadley Rille
NT North Ray Crater
NT Ocean of Storms
NT Sculptured Hills
NT Sea of Crises
NT Sea of Fertility
NT Sea of Nectar
NT Sea of Rains
NT Sea of Serenity
NT Sea of Tranquillity
NT Sea of Waves
NT Shorty Crater
NT South Ray Crater
NT Taurus-Littrow
SA accretion
SA agglutinates
SA albedo
SA Apennine Front
SA Apollo 8
SA Apollo 9
SA Apollo 11
SA Apollo 12
SA Apollo 14
SA Apollo 15
SA Apollo 16
SA Apollo 17
SA Apollo Program
SA asteroids
SA atmosphere
SA condensation
SA coordinates
SA craters
SA Earth
SA Earth-Moon couple
SA Eratosthenian
SA exploration
SA Explorer 35
SA Explorer Program
SA extraterrestrial geology
SA farside

SA gravity field
SA highlands
SA interior
SA KREEP
SA landing sites
SA luminescence
SA Luna 9
SA Luna 16
SA Luna 20
SA Luna 24
SA Luna Program
SA lunar bases
SA lunar breccia
SA lunar craters
SA lunar crust
SA lunar highlands
SA lunar interior
SA lunar laser ranging
SA lunar mantle
SA lunar samples
SA lunar soils
SA magnetic field
SA maps
SA maria
SA Mars
SA mascons
SA meteorites
SA microcraters
SA moonquakes
SA motions
SA obliquity of the ecliptic
SA orange material
SA particles
SA planetology
SA planets
SA polar caps
SA rilles
SA satellites
SA selenodesy
SA shaded relief maps
SA solar system
SA solar wind
SA Southern Highlands
SA Sun
SA surface features
SA surface properties
SA Surveyor 3
SA Surveyor 5
SA Surveyor 7
SA Surveyor Program
SA terrestrial comparison
SA wrinkle ridges

moon rocks
A valid term through mid-1978. Use lunar samples.

moonquakes (1978)
SA earthquakes
SA Moon

moonstone (1978)
BT alkali feldspar
BT feldspar group
BT framework silicates
BT silicates
SA gems
SA peristerite

Moore County
Valid through 1988. Search in combination with state term. After 1988, use specific county-state term.

Moore County Meteorite (1981)
Before 1981, also search Moore County AND meteorites.
BT eucrite
BT achondrites
BT stony meteorites
BT meteorites
SA Moore County North Carolina
SA North Carolina

Moore County North Carolina (1989)

Central North Carolina. Before 1989, also search Moore County AND North Carolina.
CO N350300N353200
 W0790600W0794700
BT North Carolina
BT United States
SA Moore County Meteorite

Moore County Tennessee (1989)
S Tennessee. Before 1989, also search Moore County AND Tennessee.
CO N350800N352500
 W0861800W0863500
BT Tennessee
BT United States

Moore County Texas (1989)
Extreme N Texas. Before 1989, also search Moore County AND Texas.
CO N353700N360600
 W1014000W1020900
BT Texas
BT United States

Mooreville Chalk (1989)
In Selma Group. W Alabama and NE Mississippi.
BT Upper Cretaceous
BT Cretaceous
BT Mesozoic
SA Alabama
SA Mississippi
SA Selma Group

Moorman Syncline (1985)
Central and W Kentucky.
BT Kentucky
BT United States

Moose River basin (1978)
SW of lower James Bay.
BT Ontario
BT Eastern Canada
RT Canada

Moosebar Formation (1989)
Alberta and NE British Columbia.
BT Lower Cretaceous
BT Cretaceous
BT Mesozoic
SA Alberta
SA British Columbia

Mor (1978)
UF raw humus
BT soils
SA humus
SA Mull

Mora County
Valid through 1988. Search in combination with state term. After 1988, use specific county-state term.

Mora County New Mexico (1989)
NE New Mexico. Before 1989, also search Mora County AND New Mexico.
CO N354500N361600
 W1042000W1054500
BT New Mexico
BT United States

morainal plains
use outwash plains

moraines (1978)
SA drift
SA drumlins
SA glacial features
SA glacial geology
SA glaciers
SA outwash plains
SA periglacial features
SA till

Morava
use Moravia

Morava River valley (1978)
River valley in NE Austria and central Czechoslovakia. Also a river valley in NE central Serbia. Also search Morava; Morava River.
IN Index Austria and/or Czechoslovakian regions or Serbia as applicable.
CO N481000N500000
 E0174800E0165000
BT Europe
SA Austria
SA Czechoslovakia
SA Moravia
SA Serbia
SA Slovakia
SA Yugoslavia

Moravia (1978)
Region between Bohemia and Slovakia.
CO N483500N502800
 E0185000E0151500
UF Morava
BT Czech Republic
BT Czechoslovakia
BT Central Europe
BT Europe
NT Brno
NT Karvina
NT Moravian Karst
NT Nizky Jezenik Mountains
NT Ostrava
NT Ostrava-Karvina
NT Stramberk
NT Svratka
SA Morava River valley
SA Oder Valley
SA Silesia
SA Sudeten
SA Sudeten Mountains
SA Sudetic Basin
SA Tieschitz Meteorite

Moravian Karst (1978)
CO N491500N493300
 E0165200E0163200
BT Moravia
BT Czech Republic
BT Czechoslovakia
BT Central Europe
BT Europe

Moravska Ostrava
use Ostrava

Moray Firth (1989)
Inlet of the North Sea bordered by Grampian and Highland regions in NE Scotland.
BT Scotland
BT Great Britain
BT United Kingdom
BT Western Europe
BT Europe
SA Grampian Highlands
SA Grampian region Scotland
SA Great Glen Fault
SA North Sea

MORB
use mid-ocean ridge basalts

Morbihan
No longer a valid term for GeoRef. See Morbihan France.

Morbihan France (1993)
Department in NW France.
CO N471500N481500
 W0020300W0035000
BT France
BT Western Europe
BT Europe
SA Brittany

SA Groix
SA Vilaine Bay

mordenite (1978)
UF arduinite
UF ashtonite
UF flokite
UF ptilolite
BT zeolite group
BT framework silicates
BT silicates
SA aluminosilicates

Morecambe Bay (1978)
Inlet of the Irish Sea in NW England.
BT England
BT Great Britain
BT United Kingdom
BT Western Europe
BT Europe
SA Cumbria England
SA Irish Sea
SA Lancashire England
SA Westmorland England

morencite
 use nontronite

Moreno Formation (1978)
In Chico Group. Upper Cretaceous and Paleocene (?). Includes four members (ascending): Dosados Sand and Shale, Tierra Loma Shale (including Mercy Sandstone lens), Marca Shale Member, and Dos Palos Shale (includes Cima Sandstone lens). S California.
BT Upper Cretaceous
BT Cretaceous
BT Mesozoic
SA California
SA Marca Shale Member
SA Paleocene

Moreton
 No longer a valid term for GeoRef. See Moreton Australia.

Moreton Australia (1993)
Village on Cape York Peninsula in N Queensland.
BT Queensland Australia
BT Australia
BT Australasia
SA Cape York Peninsula

Moreton Bay (1978)
In SE Queensland just N of Brisbane.
IN Index state or regions as applicable.
SA Queensland Australia

Morez
 No longer a valid term for GeoRef. See Morez France.

Morez France (1993)
Town in in SE Jura, E France.
BT Jura France
BT France
BT Western Europe
BT Europe

Morgan County
 Valid through 1988. Search in combination with state term. After 1988, use specific county-state term.

Morgan County Alabama (1989)
N Alabama. Before 1989, also search Morgan County AND Alabama.
CO N341800N343800
  W0863400W0870500
BT Alabama
BT United States

Morgan County Colorado (1989)
NE Colorado. Before 1989, also search Morgan County AND Colorado.
CO N400000N403100
  W1034100W1040800
BT Colorado
BT United States

Morgan County Georgia (1989)
N central Georgia. Before 1989, also search Morgan County AND Georgia.
CO N332600N334900
  W0831600W0834100
BT Georgia
BT United States

Morgan County Illinois (1989)
W central Illinois. Before 1989, also search Morgan County AND Illinois.
CO N393100N395200
  W0895500W0903700
BT Illinois
BT United States

Morgan County Indiana (1989)
Central Indiana. Before 1989, also search Morgan County AND Indiana.
CO N392100N393800
  W0861600W0864200
BT Indiana
BT United States

Morgan County Kentucky (1989)
E Kentucky. Before 1989, also search Morgan County AND Kentucky.
CO N374300N380700
  W0825700W0833700
BT Kentucky
BT United States

Morgan County Missouri (1989)
Central Missouri. Before 1989, also search Morgan County AND Missouri.
CO N381100N384200
  W0923700W0930500
BT Missouri
BT United States
NT Versailles Missouri

Morgan County Ohio (1989)
E central Ohio. Before 1989, also search Morgan County AND Ohio.
CO N392700N394700
  W0813600W0820500
BT Ohio
BT United States

Morgan County Tennessee (1989)
NE central Tennessee. Before 1989, also search Morgan County AND Tennessee.
CO N355400N362200
  W0842200W0845500
BT Tennessee
BT United States

Morgan County Utah (1989)
N Utah. Before 1989, also search Morgan County AND Utah.
CO N404800N412100
  W1111500W1115300
BT Utah
BT United States

Morgan County West Virginia (1989)
E West Virginia. Before 1989, also search Morgan County AND West Virginia.
CO N392300N394200
  W0780200W0782800
BT West Virginia

BT United States

Morgan Hill earthquake 1984 (1989)
Epicenter at Morgan Hill, Santa Clara County, Central California.
BT earthquakes
SA California
SA Santa Clara County California

Morien Group (1989)
Cape Breton Island.
BT Pennsylvanian
BT Carboniferous
BT Paleozoic
SA Cape Breton Island
SA Nova Scotia

Morin Complex (1978)
CO N454500N463000
  W0733000W0743000
BT Quebec
BT Eastern Canada
BT Canada

Morlaix
 No longer a valid term for GeoRef. See Morlaix France.

Morlaix France (1993)
Town on inlet of English Channel.
BT Finistere France
BT France
BT Western Europe
BT Europe

Mormoiron
 No longer a valid term for GeoRef. See Mormoiron France.

Mormoiron France (1993)
Village in SE France.
BT Vaucluse France
BT France
BT Western Europe
BT Europe

Mormon Mountains (1978)
SE Nevada.
BT Nevada
BT United States

Moroccan Atlas Mountains (1981)
Atlas Mountains in Morocco. Autoposting of Morocco to this term began in 1989. As of 1990, North Africa and Africa are autoposted to this term. As of 1993, Anti-Atlas, High Atlas, and Middle Atlas are narrower terms. This term has multiple hierarchies.
CO N303000N351100
  W0010000W0100000
BT1 Morocco
BT1 North Africa
BT1 Africa
BT2 Atlas Mountains
BT2 North Africa
BT2 Africa
NT Anti-Atlas
NT High Atlas
NT Middle Atlas

Morocco (1978)
Before 1960, also search Spanish Morocco.
CO N274000N360000
  W0010000W0131500
UF Spanish Morocco
BT North Africa
BT Africa
NT Bou Azzer
NT Cape Bojador
NT Casablanca Morocco
NT Moroccan Atlas Mountains
NT Moulouya River
NT Rabat Morocco
NT Rehamna
NT Rif

NT Tafilalt
NT Tangier Morocco
NT Taourirt Morocco
NT Tarfaya Morocco
NT Taza Morocco
SA Atlas Mountains
SA Ceuta
SA Mediterranean region
SA Melilla
SA Sahara
SA Western Sahara

Moroto Mountain (1978)
Just E of town of Moroto near Kenya border.
BT Uganda
BT East Africa
BT Africa

morphology (1978)
NT functional morphology
SA anatomy
SA coiling
SA cuticles
SA description
SA extremities
SA horizon differentiation
SA horizons
SA ichnofossils
SA micromorphology
SA muscles
SA nervous system
SA paleopathology
SA physiology
SA septa
SA soils
SA spinal column
SA sutures
SA valves
SA vascular taxa

morphometry (1978)
Used as a general term.
SA geomorphology
SA landform description
SA quantitative geomorphology

morphostructures (1978)
SA basins
SA domes
SA geomorphology
SA trenches

Morris County
 Valid through 1988. Search in combination with state term. After 1988, use specific county-state term.

Morris County Kansas (1989)
E central Kansas. Before 1989, also search Morris County AND Kansas.
CO N383000N385300
  W0962200W0965500
BT Kansas
BT United States

Morris County New Jersey (1989)
N New Jersey. Before 1989, also search Morris County AND New Jersey.
CO N403800N410500
  W0741700W0745300
BT New Jersey
BT United States
SA Long Valley
SA New Jersey Highlands

Morris County Texas (1989)
NE Texas. Before 1989, also search Morris County AND Texas.
CO N325300N332100
  W0943800W0944800
BT Texas
BT United States

Morrison Formation (1978)

In Plata Mountains of Colorado, comprises (ascending) Pony Express Limestone, Bilk Creek Sandstone, Wanakah Marl (restricted), and Junction Creek Sandstone.
BT Upper Jurassic
BT Jurassic
BT Mesozoic
SA Arizona
SA Brushy Basin Shale Member
SA Colorado
SA Kansas
SA Montana
SA New Mexico
SA Oklahoma
SA Salt Wash Sandstone Member
SA South Dakota
SA Utah
SA Westwater Canyon Sandstone Member
SA Wyoming

Morrow County
Valid through 1988. Search in combination with state term. After 1988, use specific county-state term.

Morrow County Ohio (1989)
Central Ohio. Before 1989, also search Morrow County AND Ohio.
CO N402000N404300
    W0823800W0830200
BT Ohio
BT United States

Morrow County Oregon (1989)
N Oregon. Before 1989, also search Morrow County AND Oregon.
CO N445900N455300
    W1190700W1200100
BT Oregon
BT United States

Morrow Formation (1978)
Composed of shales with some thin limestones and occasional sandstones. SE Arkansas, and central E and NE Oklahoma.
BT Pennsylvanian
BT Carboniferous
BT Paleozoic
SA Arkansas
SA Morrowan
SA Oklahoma

Morrow Series
use Morrowan

Morrowan (1978)
Provincial series, North America. Above Chesterian (Mississippian), below Atokan.
UF Morrow Series
UF Morrowan Series
BT Lower Pennsylvanian
BT Pennsylvanian
BT Carboniferous
BT Paleozoic
NT Bloyd Formation
SA Chesterian
SA Morrow Formation
SA Springer Formation
SA Watahomigi Formation

Morrowan Series
use Morrowan

Mortagne
No longer a valid term for GeoRef. See Mortagne France.

Mortagne France (1993)
Town in SE Orne, NW France. Before 1989, also search Mortagne-au-Perchey.
UF Mortagne-au-Perchey
BT Orne France
BT France
BT Western Europe
BT Europe

Mortagne-au-Perchey
use Mortagne France

Morton County
Valid through 1988. Search in combination with state term. After 1988, use specific county-state term.

Morton County Kansas (1989)
Extreme SW Kansas. Before 1989, also search Morton County AND Kansas.
CO N370000N372200
    W1013400W1020400
BT Kansas
BT United States

Morton County North Dakota (1989)
Central North Dakota. Before 1989, also search Morton County AND North Dakota.
CO N461700N465800
    W1003500W1020600
BT North Dakota
BT United States

Morvan (1978)
Northernmost spur of the Central Massif.
IN Index departments as applicable.
UF Morvan Massif
UF Morvan Mountains
BT Central Massif
BT France
BT Western Europe
BT Europe
SA Cote-d'Or France
SA Nievre France
SA Saone-et-Loire France
SA Yonne France

Morvan Massif
use Morvan

Morvan Mountains
use Morvan

Mosabani Mine
use Mosaboni Mines

Mosaboni copper mines
As of 1989, no longer a valid term for GeoRef.
use Mosaboni Mines

Mosaboni Mines (1978)
SE Bihar, India. This term was invalid 1979-1988. Before 1989, also search Mosaboni copper mines. Also search Mosaboni.
UF Mosabani Mine
UF Mosaboni copper mines
UF Musabani copper mines
UF Mushaboni copper mines
SA Bihar India
SA copper ores
SA mines

mosaics (1982)
UF photomosaics
SA aerial photography
SA imagery
SA multispectral analysis
SA multispectral scanner
SA photogeologic maps
SA photogeologic methods
SA photogeology
SA photogrammetry
SA photography
SA remote sensing
SA SAR
SA space photography
SA thematic mapper

Mosasauridae (1978)
Family. Before 1981, also search mosasaurs. Also search Mosasaurus.
BT Squamata
BT Lepidosauria
BT Diapsida
BT Reptilia
BT Tetrapoda
BT Vertebrata
BT Chordata

Mosbas
use Moscow Basin

Moscovian (1978)
Russia.
BT Upper Carboniferous
BT Carboniferous
BT Paleozoic

Moscow
No longer a valid term for GeoRef. As of 1993, see Moscow Russian Federation.

Moscow Basin (1978)
Lignite basin extending over 600 miles in an arc from Borovichi in Saint Petersburg Oblast to Skopin, SE of Moscow, in Ryazan Oblast. Before 1978, also search Moscow. This term has multiple hierarchies.
UF Mosbas
UF Moscow coal basin
BT1 Russian Federation
BT1 Commonwealth of Independent States
BT2 Europe
SA coal fields

Moscow coal basin
use Moscow Basin

Moscow Formation (1989)
In Hamilton Group. W and central New York and Pennsylvania.
BT Middle Devonian
BT Devonian
BT Paleozoic
SA Hamilton Group
SA New York
SA Pennsylvania

Moscow River
use Moskva River

Moscow Russian Federation (1993)
Oblast and city in W central Russian Federation. Before 1993, also search Moscow. Before 1978, also search Moskva. This term has multiple hierarchies.
UF Moskva Russian Federation
BT1 Russian Federation
BT1 Commonwealth of Independent States
BT2 Europe

Moscow State University
use Moscow University

Moscow Syneclise (1978)
SE European Russian Federation. Before 1978, also search Moscow. This term has multiple hierarchies.
CO N540000N573000
    E0400000E0350000
BT1 Russian Federation
BT1 Commonwealth of Independent States
BT2 Europe
SA Moscow-Pechora Syneclise

Moscow University (1978)

In Moscow.
UF Moscow State University
SA academic institutions
SA Russian Federation

Moscow-Pechora Syneclise (1981)
This term has multiple hierarchies.
IN Index countries and Europe if applicable.
CO N503000N693000
    E0620000E0300000
BT1 Commonwealth of Independent States
BT2 Russian Platform
SA Europe
SA Moscow Syneclise
SA Pechora Basin
SA Pechora River
SA Russian Federation

Mosel Valley
use Moselle Valley

Moselle
No longer a valid term for GeoRef. See Moselle France.

Moselle France (1993)
Department in NE France.
CO N483000N493200
    E0074000E0055300
BT France
BT Western Europe
BT Europe
SA Lorraine
SA Saar Basin

Moselle River (1978)
Rises in NE France and flows N and NE entering the Rhine River at Koblenz. Also search Moselle.
IN Index countries as applicable.
BT Europe
SA France
SA Germany
SA Luxembourg

Moselle Valley (1978)
River valley.
IN Index countries as applicable.
UF Mosel Valley
BT Europe
SA France
SA Germany
SA Luxembourg

Moskva River (1978)
Flows through Moscow and then SE into the Oka River. This term has multiple hierarchies.
UF Moscow River
BT1 Russian Federation
BT1 Commonwealth of Independent States
BT2 Asia

Moskva Russian Federation
use Moscow Russian Federation

Mosquito Range (1978)
S part of Park Range near Leadville.
IN Index U. S. Rocky Mountains and counties or regions as applicable.
SA Colorado
SA U. S. Rocky Mountains

Mossamedes (1978)
Desert and basin in extreme SW part of Angola. Before 1993, also used for the province and city. As of 1993, see Namibe Angola for the province and city.
UF Mocamedes
BT Angola
BT Central Africa
BT Africa
SA Namibe Angola

Mossbauer
A valid index term through 1977. After 1977, use Mossbauer spectroscopy.

Mossbauer analysis
Not a valid GeoRef index term after 1970. Was used on level 1 in subfile N.
use Mossbauer spectroscopy

Mossbauer effect
Not a valid GeoRef index term after 1971. Was used on level 1 in subfile G. Now use Mossbauer spectra or Mossbauer spectroscopy.

Mossbauer spectra (1978)
Before 1978, also search Mossbauer.
BT spectra
SA Mossbauer spectroscopy
SA spectroscopy

Mossbauer spectroscopy (1978)
Before 1978, also search Mossbauer AND spectroscopy. Before 1971, also search Mossbauer analysis.
UF Mossbauer analysis
BT spectroscopy
SA analysis
SA chemical analysis
SA Mossbauer spectra

mosses
use bryophytes

Motagua Fault (1981)
SE Guatemala.
CO N144500N153000
W0884000W0905000
BT Guatemala
BT Central America

motion, ground
use ground motion

motion, plasma
Not a valid term for GeoRef. Usually out-of-scope. Used as plasma motion with magnetosphere.

motion, strong
use strong motion

motions (1978)
SA acceleration
SA angular momentum
SA Chandler wobble
SA Earth
SA eccentricity
SA Milankovitch theory
SA Moon
SA nutation
SA obliquity of the ecliptic
SA oscillations
SA rotation

Motojuku Formation (1978)
BT Miocene
BT Neogene
BT Tertiary
BT Cenozoic
SA Honshu
SA Japan

Moty Formation (1989)
Siberia.
SA Russian Federation
SA Siberia

Moulouya River (1978)
Rises in central part of country and flows NE into Mediterranean Sea. Also search Mouloua.
UF Moluya
UF Mulwiya
BT Morocco
BT North Africa

BT Africa

mounds (1978)
BT bedding plane irregularities
BT sedimentary structures
SA mud mounds

mounds, algal
use algal mounds

Mount Adams (1985)
Mountain in Presidential Range of White Mountains, Coos County, N New Hampshire. Also a mountain in Cascade Range, Yakima County, S Washington. Autoposting of United States to this term began in 1989.
IN Index states and counties and mountain ranges as applicable.
BT United States
SA Cascade Range
SA Coos County New Hampshire
SA New Hampshire
SA Washington
SA White Mountains
SA Yakima County Washington

Mount Agoeng
use Mount Agung

Mount Agung (1978)
Highest volcanic peak on Bali. This term has multiple hierarchies.
UF Mount Agoeng
BT1 Bali
BT1 Lesser Sunda Islands
BT1 Far East
BT1 Asia
BT2 Bali
BT2 Indonesia
BT2 Far East
BT2 Asia

Mount Amiata (1989)
Tuscany, western Italy.
CO N422000N442000
E0122500E0094000
BT Tuscany Italy
BT Italy
BT Southern Europe
BT Europe

Mount Antero (1985)
Peak in Sawatch Range, central Colorado. Autoposting of broader terms to this term began in 1989. This term has multiple hierarchies.
CO N384000N384200
W1061500W1061700
BT1 Chaffee County Colorado
BT1 Colorado
BT1 United States
BT2 Sawatch Range
BT2 Colorado
BT2 United States
BT3 Sawatch Range
BT3 U. S. Rocky Mountains
BT3 Rocky Mountains
BT4 Sawatch Range
BT4 U. S. Rocky Mountains
BT4 United States

Mount Asama
No longer a valid term for GeoRef.
use Asama

Mount Baker (1985)
Peak in Cascade Range, NW Washington. Also a peak in the Ruwenzori mountain group in W Uganda.
IN Index Washington, Whatcom County Washington, and Cascade Range; or Uganda and Ruwenzori as applicable.
SA Cascade Range
SA Uganda

SA Washington
SA Whatcom County Washington

Mount Borah
use Borah Peak

Mount Borah earthquake, 1983
use Borah Peak earthquake 1983

Mount Cameroon (1989)
Mountain. SW Cameroon.
UF Cameroon Mountain
BT Cameroon
BT West Africa
BT Africa

Mount Carmel (1978)
Mountain in N near Mediterranean coast.
BT Israel
BT Middle East
BT Asia
SA Geula Caves

Mount Elbrus
use Elbrus

Mount Erebus (1985)
Mountain in SW Alberta. Also a volcano on Ross Island, Antarctica.
IN Index Alberta and Canadian Rocky Mountains; or Antarctica and Ross Island as applicable.
SA Alberta
SA Antarctica
SA Canadian Rocky Mountains
SA Jasper National Park
SA Ross Island

Mount Etna (1978)
Active volcano in NE Sicily.
CO N374500N374500
E0150100E0150100
UF Etna
UF Mt. Etna
BT Sicily Italy
BT Italy
BT Southern Europe
BT Europe

Mount Everest (1985)
On China-Nepal border.
IN Index Nepal and/or Xizang China as applicable.
CO N275900N280000
E0865800E0865600
UF Chomolungma
UF Qomolangma Feng
BT Himalayas
BT Asia
SA China
SA Nepal
SA Xizang China

Mount Fuji
use Fujiyama

Mount Girnar (1978)
S central Kathiawar Peninsula.
BT Gujarat India
BT India
BT Indian Peninsula
BT Asia
SA Kathiawar

Mount Hakone (1978)
Extinct volcano in central Honshu. Also search Hakone.
UF Hakone Volcano
BT Honshu
BT Japan
BT Far East
BT Asia
SA Hakone

Mount Hood (1978)
Peak in Cascade Range in Clackamas and Hood River counties. This term has multiple hierarchies.

CO N452500N452500
W1214000W1214000
UF Mt. Hood
BT1 Oregon
BT1 United States
BT2 Cascade Range
BT2 United States
SA Clackamas County Oregon
SA Hood River County Oregon

Mount Isa
No longer a valid term for GeoRef. See Mount Isa Australia.

Mount Isa Australia (1993)
Town and mining center in W Queensland, Australia.
CO S204500S204500
E1393000E1393000
BT Queensland Australia
BT Australia
BT Australasia

Mount Jefferson (1989)
On the border of Jefferson, Linn and Marion counties, Cascade Range, N central Oregon; also in Toquima Range, Nye County, central Nevada; and in Presidential Range of White Mountains, Coos County, New Hampshire.
IN Index states, counties, and mountain ranges as applicable.
SA Cascade Range
SA Coos County New Hampshire
SA Jefferson County Oregon
SA Linn County Oregon
SA Marion County Oregon
SA Nevada
SA New Hampshire
SA Nye County Nevada
SA Oregon
SA Toquima Range
SA United States
SA White Mountains

Mount Katahdin (1989)
N terminus of the Appalachian Trail in N central Maine.
IN Index mountain range as applicable.
CO N455500N455500
W0685700W0685700
BT Piscataquis County Maine
BT Maine
BT United States

Mount Kenya (1978)
Extinct volcano in central Kenya.
BT Kenya
BT East Africa
BT Africa

Mount Lassen
use Lassen Peak

Mount Lofty Ranges (1978)
SE South Australia.
UF Mt. Lofty Ranges
BT South Australia
BT Australia
BT Australasia

Mount Lyell (1978)
Peak in Sierra Nevada near junction of Madera, Mariposa, and Tuolumne counties in E central California.
IN Index Sierra Nevada as applicable.
UF Mt. Lyell
BT California
BT United States
SA Sierra Nevada

Mount Mazama (1985)

Prehistoric volcano in S Oregon. Caldera now occupied by Crater Lake. Autoposting of broader terms to this term began in 1989.
BT Klamath County Oregon
BT Oregon
BT United States
SA Crater Lake
SA Mazama Ash

Mount Mihara (1978)
Use for the volcano on Izu-Oshima of Izu-shichito island group S of Sagami Bay. Also search Mihara. Before 1993, O-shima was a broader term. Izu-Oshima was also used to refer to the volcano before 1993.
CO N344300N344400
  E1392300E1392200
UF Mihara (Mount)
UF Mihara Volcano
UF Mihara-Yama
UF Mt. Mihara
BT Izu-Oshima
BT Izu-shichito
BT Honshu
BT Japan
BT Far East
BT Asia
SA O-shima

Mount Morgan (1978)
Peak in Glacier National Park in NW Montana.
IN Index county or region and mountain range as applicable.
SA Montana

Mount Namjagbarwa (1989)
Himalayas of Xizang, SW China.
SA Himalayas
SA Xizang China

Mount Ontake
  use Ontake

Mount Pelee (1989)
Volcano. N Martinique.
CO N144800N144800
  W0611000W0611000
UF Mont Pelee
BT Martinique
BT Lesser Antilles
BT Antilles
BT West Indies
BT Caribbean region

Mount Rainier (1978)
Volcanic peak of Cascade Range in Mount Rainier National Park in W central Washington. This term has multiple hierarchies.
UF Mt. Rainier
BT1 Pierce County Washington
BT1 Washington
BT1 United States
BT2 Cascade Range
BT2 United States
SA Nisqually Glacier

Mount Rainier National Park (1978)
W central part of state. Mt. Rainier occupies one fourth of park area in W central Washington.
UF Mt. Rainier National Park
BT Washington
BT United States
SA Lewis County Washington
SA national parks
SA Pierce County Washington

Mount Read Volcanics (1993)
Cambrian. W Tasmania.
BT Cambrian
BT Paleozoic
SA Tasmania Australia

Mount Saint Helens (1978)
Peak in NW Skamania County in S Washington. This term has multiple hierarchies.
UF Mount St. Helens
UF Mt. St. Helens
BT1 Skamania County Washington
BT1 Washington
BT1 United States
BT2 Cascade Range
BT2 United States

Mount Scopus Group (1989)
Senonian to Eocene.
IN Index ages as applicable.
SA Eocene
SA Ghareb Formation
SA Israel
SA Jordan
SA K-T boundary
SA Paleocene
SA Syria
SA Upper Cretaceous

Mount Shasta (1985)
Cascade Range, N California. Autoposting of broader terms began in 1989. This term has multiple hierarchies.
CO N412400N412600
  W1221100W1221300
BT1 Siskiyou County California
BT1 California
BT1 United States
BT2 Cascade Range
BT2 United States

Mount Shields Formation (1989)
In Missoula Group of Belt Supergroup.
BT middle Proterozoic
BT Proterozoic
BT upper Precambrian
BT Precambrian
SA Belt Supergroup
SA Idaho
SA Missoula Group
SA Montana

Mount Simon Sandstone (1989)
Unconformably overlies Precambrian basement; underlies Eau Claire Formation.
BT Upper Cambrian
BT Cambrian
BT Paleozoic
SA Eau Claire Formation
SA Illinois
SA Indiana
SA Iowa
SA Michigan
SA Minnesota
SA Ohio
SA Wisconsin

Mount St. Helens
  use Mount Saint Helens

Mount Wood (1978)
Peak of Saint Elias Mountains in SW Yukon Territory.
BT Yukon Territory
BT Western Canada
BT Canada

mountain building
A valid term through 1973. After 1973, use orogeny.

mountain crystal
  use quartz crystal

mountains (1978)
SA alpine environment
SA geomorphology
SA highlands
SA hills
SA landform description
SA landforms
SA mesas

Mousterian (1978)
Archaeological classification.
BT Paleolithic
BT Stone Age
BT Cenozoic
SA archaeology
SA Pleistocene
SA Quaternary

Mouthoumet Massif (1978)
S France. Also search Mouthoumet.
BT Aude France
BT France
BT Western Europe
BT Europe

Mouydir (1978)
Mountains in S central Algeria.
BT Algeria
BT North Africa
BT Africa

movement (1978)
UF movements
SA circulation
SA ground water
SA ice movement
SA plate tectonics
SA soils
SA water regimes

movements
  use movement

movements, mass
  use mass movements

movements, vertical
  use vertical movements

moveout (1989)
UF moveouts
SA arrival time
SA common-depth-point method
SA dip moveout
SA elastic waves
SA normal moveout
SA reflection
SA reflection methods
SA seismic methods
SA seismic surveys
SA traveltime

moveouts
  use moveout

Mowry Shale (1978)
In Colorado Group. Consists of hard lighter-gray shales and thin bedded sandstones that weather light gray and form ridges. Montana, W South Dakota, and Wyoming.
IN Index states as applicable.
BT Upper Cretaceous
BT Cretaceous
BT Mesozoic
SA Colorado Group
SA Montana
SA South Dakota
SA Wyoming

Moxa Arch (1993)
NS trending anticlinorium in Green River basin of SW Wyoming and NE Utah. Known for petroleum production.
IN Index counties as applicable.
BT United States
SA Box Elder County Utah
SA Dakota Formation
SA Green River basin
SA Lincoln County Wyoming
SA Sublette County Wyoming
SA Sweetwater County Wyoming
SA Uinta County Wyoming
SA Utah
SA Wyoming

Mozambique (1978)
Before 1969, also search Portuguese East Africa.
CO S270000S100000
  E0410000E0300000
UF Portuguese East Africa
BT East Africa
BT Africa
NT Maputo Mozambique
NT Mozambique Belt
NT Tete Mozambique
NT Zambezia Mozambique
SA East African Rift
SA Ecca Group
SA Lake Malawi
SA Lebombo Mountains
SA Limpopo Basin
SA Limpopo Belt
SA Zambezi Valley

Mozambique Belt (1978)
Orogenic belt. Mountains in N extend to the rim of the Great Rift Valley.
BT Mozambique
BT East Africa
BT Africa

Mozambique Channel (1978)
Strait between the Malagasy Republic and Mozambique.
CO S261500S103000
  E0491000E0323000
BT Indian Ocean

mroseite (1978)
From 1978-1980, tellurates and tellurites was autoposted to this term. Autoposting of tellurites began in 1981.
BT tellurites

MSS
Not a valid term for GeoRef. Acronym for various entities including multispectral scanner; Minnesota Speleological Survey.

Mt. Elbrus
  use Elbrus

Mt. Etna
  use Mount Etna

Mt. Fuji
  use Fujiyama

Mt. Fuji volcano
  use Fujiyama

Mt. Hood
  use Mount Hood

Mt. Lofty Ranges
  use Mount Lofty Ranges

Mt. Lyell
  use Mount Lyell

Mt. Mihara
  use Mount Mihara

Mt. Rainier
  use Mount Rainier

Mt. Rainier National Park
  use Mount Rainier National Park

Mt. St. Helens
  use Mount Saint Helens

Mubarek
No longer a valid term for GeoRef. As of 1993, see Mubarek Uzbekistan.

Mubarek Uzbekistan (1993)
Village SW of Samarkand. This term has multiple hierarchies.
BT1 Uzbekistan
BT1 Asia
BT2 Asia
BT2 Commonwealth of Independent States

mud (1978)
BT clastic sediments
BT sediments
SA clay
SA mineralized mud
SA mud banks
SA mud lumps
SA mud mounds
SA mud volcanoes
SA ooze
SA silt
SA sludge
SA terrigenous materials

mud balls, armored
use armored mud balls

mud banks (1978)
UF mudbanks
SA banks
SA carbonate banks
SA fluvial features
SA geomorphology
SA mud
SA shore features

mud cracks
use mudcracks

mud flats (1978)
UF mudflats
SA chenier plains
SA marine terraces
SA shore features
SA tidal flats

mud flows
use mudflows

mud lumps (1978)
UF mudlumps
BT bedding plane irregularities
BT sedimentary structures
SA clay
SA fluvial features
SA mud
SA mud mounds
SA shore features

mud mounds (1993)
UF mudmounds
BT bioherms
BT biogenic structures
BT sedimentary structures
SA bryophytes
SA mounds
SA mud
SA mud lumps

mud volcanoes (1978)
Used for an accumulation, usually conical, of mud and rock ejected by volcanic gases; also for a similar accumulation formed by escaping petroliferous gases. Before 1970, also search sand volcanoes.
UF hervidero
UF macaluba
UF sand volcanoes
SA ejecta
SA geomorphology
SA lava
SA mud
SA volcanoes
SA volcanology

mud waves
After 1992, use mud AND bottom features. See also ripple marks and submarine dunes.

mudbanks
use mud banks

mudcracks (1978)
Before 1981, also search shrinkage cracks; mud cracks.
UF mud cracks
UF shrinkage cracks

BT bedding plane irregularities
BT sedimentary structures
SA gilgai

Muddy Creek Formation (1985)
Miocene or Pliocene. SE Nevada, NW Arizona and SW Utah.
BT Tertiary
BT Cenozoic
SA Arizona
SA Miocene
SA Nevada
SA Pliocene
SA Utah

Muddy Sandstone (1978)
Of Thermopolis Shale. Overlies a lower black shale member and underlies a black shale member. Surface and subsurface in central N Wyoming and subsurface in central S Montana.
BT Lower Cretaceous
BT Cretaceous
BT Mesozoic
SA Montana
SA Wyoming

mudflats
use mud flats

mudflows (1978)
Autoposting of mass movements to this term began in 1989.
UF mud flows
BT mass movements
NT mudslides
SA erosion
SA geologic hazards
SA geomorphology
SA lahars
SA landslides
SA slope stability
SA soil erosion

mudlumps
use mud lumps

mudmounds
use mud mounds

mudrocks
No longer a valid GeoRef index term. Before 1971, was used on level 1 in subfile N.
use mudstone

mudslides (1978)
Autoposting of mudflows and mass movements to this term began in 1989.
BT mudflows
BT mass movements
SA earthflows
SA geologic hazards
SA geomorphology
SA landslides
SA slope stability

mudstone (1978)
Before 1971, also search mudrocks.
UF mudrocks
BT clastic rocks
BT sedimentary rocks
SA claystone
SA siltstone
SA terrigenous materials

Mudurnu
No longer a valid term for GeoRef. See Mudurnu earthquake 1967.

Mudurnu earthquake 1967 (1993)
In Bolu province, NW Turkey. Before 1993, also search Mudurnu and Mudurnu Valley.
BT earthquakes
SA Anatolia

SA Turkey

Muenster
Not a valid term for GeoRef. See Munster Germany.

Muenster in Westfalen
use Munster Germany

Muensterland
use Munsterland

Mugan Steppe (1978)
Part of Kura Lowland S of Aras River and Kura River in SE Azerbaidzhan. Also search Mugan. This term has multiple hierarchies.
BT1 Azerbaidzhan
BT1 Europe
BT2 Azerbaidzhan
BT2 Commonwealth of Independent States
SA Kura Lowland

mugearite (1978)
BT alkali basalts
BT basalts
BT volcanic rocks
BT igneous rocks

Mugodzhar Hills (1978)
Southernmost extension of the Urals in Aktyubinsk Oblast. Also search Mugodzhar. As of 1990, Central Asia is autoposted to this term. This term has multiple hierarchies.
CO N480000N520000
 E0610000E0550000
UF Mugodzhar
UF Mugodzhary
BT1 Kazakhstan
BT1 Central Asia
BT1 Asia
BT2 Kazakhstan
BT2 Commonwealth of Independent States
BT3 Urals
BT3 Commonwealth of Independent States
SA Aktyubinsk Kazakhstan
SA Southern Urals

Mugodzhars
use Mugodzhar Hills

Mugodzhary
use Mugodzhar Hills

Mukhor-Tala
No longer a valid term for GeoRef. As of 1993, see Mukhor-Tala Russian Federation.

Mukhor-Tala Russian Federation (1993)
Village E of Lake Baikal in former Buryat A.S.S.R., SE Russian Federation. Before 1993, also search Mukhor-Tala. This term has multiple hierarchies.
BT1 Buryat Russian Federation
BT1 Russian Federation
BT1 Commonwealth of Independent States
BT2 Buryat Russian Federation
BT2 Asia

Mule Ear Diatreme (1978)
Mountain in Mule Ear Peaks, SW Texas. Also search Mule Ear.
BT Brewster County Texas
BT Texas
BT United States

Mull (1978)
BT soils
SA humus
SA Mor
SA Mull Island

Mull Island (1978)
Island of the Inner Hebrides, Argyllshire, Scotland. Before 1978, also search Mull. After 1978, Mull is restricted for use as the soil. This term has multiple hierarchies.
CO N561500N564000
 W0054000W0063000
BT1 Inner Hebrides
BT1 Hebrides
BT1 Scotland
BT1 Great Britain
BT1 United Kingdom
BT1 Western Europe
BT1 Europe
BT2 Argyllshire Scotland
BT2 Scotland
BT2 Great Britain
BT2 United Kingdom
BT2 Western Europe
BT2 Europe
BT3 Strathclyde region Scotland
BT3 Scotland
BT3 Great Britain
BT3 United Kingdom
BT3 Western Europe
BT3 Europe
SA Mull

Mullen Creek-Nash Fork shear zone (1985)
SE Wyoming and NE Colorado.
IN Index states as applicable.
UF Nash Fork-Mullen Creek shear zone
BT United States
SA Colorado
SA Medicine Bow Mountains
SA Sierra Madre Range
SA U. S. Rocky Mountains
SA Wyoming

mullions (1978)
SA lineation
SA metamorphic rocks
SA structural analysis

mullite (1978)
Autoposting of nesosilicates began in 1985.
BT nesosilicates
BT orthosilicates
BT silicates

multichannel filters (1981)
SA filters

multichannel methods (1985)
SA data processing
SA filters
SA geophysical methods
SA geophysical surveys
SA methods
SA seismic methods
SA seismic surveys
SA seismology

multidomains (1981)
Before 1993, also search magnetic multidomains.
BT magnetic domains

multilayer hydrologic systems
use multiple aquifers

multilayered systems
Not a valid term for GeoRef. As of 1993, see multiple aquifers.

multiphase flow (1981)
SA aquifers
SA hydraulics
SA hydrodynamics
SA hydrology
SA multiple aquifers

multiple aquifers (1993)

Before 1993, also search multilayer hydrologic systems, multilayered systems, or multilayer aquifers.
UF multilayer hydrologic systems
SA aquifers
SA ground water
SA layered materials
SA leaky aquifers
SA multiphase flow

multispectral analysis (1978)
Before 1978, search multispectral AND analysis.
UF multispectral methods
SA analysis
SA mosaics
SA multispectral scanner
SA remote sensing
SA space photography
SA SPOT
SA telemetry
SA thematic mapper

multispectral methods
use multispectral analysis

multispectral scanner (1989)
Sometimes abbreviated as MSS.
SA airborne methods
SA digital cartography
SA geophysical methods
SA geophysical surveys
SA instruments
SA Landsat
SA mosaics
SA multispectral analysis
SA radar methods
SA remote sensing
SA SAR
SA satellite methods
SA SLAR
SA thematic mapper

Multituberculata (1978)
Order.
BT Mammalia
BT Tetrapoda
BT Vertebrata
BT Chordata

multivariate analysis (1978)
BT statistical analysis
SA analysis
SA univariate analysis

Multnomah County
Valid through 1988. Search in combination with state term. After 1988, use specific county-state term.

Multnomah County Oregon (1989)
NW Oregon. Before 1989, also search Multnomah County AND Oregon.
CO N452700N454500
  W1214900W1225500
BT Oregon
BT United States
NT Portland Oregon

Mulwiya
use Moulouya River

Munchberg Gneiss Massif (1978)
In Fichtelgebirge area in NE Bavaria. Also search Munchberg Gneiss.
BT Bavaria Germany
BT Germany
BT Central Europe
BT Europe
SA Fichtelgebirge

Munchen Germany
use Munich Germany

Mundrabilla
No longer a valid term for GeoRef. See Mundrabilla Australia.

Mundrabilla Australia (1993)
Village in SE Western Australia.
BT Western Australia
BT Australia
BT Australasia
SA Nullarbor Plain

Munich
No longer a valid term for GeoRef. See Munich Germany.

Munich Germany (1993)
City in S Bavaria. Before 1993, also search Munich and Germany. Before 1978, also search Munchen.
UF Munchen Germany
BT Upper Bavaria Germany
BT Bavaria Germany
BT Germany
BT Central Europe
BT Europe

Munster
No longer a valid term for GeoRef. See Munster Germany.

Munster Germany (1993)
City in N North Rhine-Westphalia. Before 1993, also search Muenster and Muenster in Westfalen. Search in combination with Germany.
UF Muenster in Westfalen
BT North Rhine-Westphalia Germany
BT Germany
BT Central Europe
BT Europe

Munsterland (1978)
Area around Munster in N North Rhine-Westphalia.
UF Muensterland
BT North Rhine-Westphalia Germany
BT Germany
BT Central Europe
BT Europe

Muntenia (1978)
Region and former province in E Walachia, S Romania.
BT Walachia
BT Romania
BT Southern Europe
BT Europe

Muntii Metalici (1978)
Metal Mountains. S part of Apuseni Mountains in W central Romania.
BT Transylvania
BT Romania
BT Southern Europe
BT Europe
SA Apuseni Mountains

muong nong type (1978)
Also search muong nong.
UF muong nong-type
BT tektites

muong nong-type
use muong nong type

Mural Limestone (1989)
In Bisbee Group. Overlies Morita Formation; underlies Cintura Formation. SE Arizona. This term has multiple hierarchies.
BT1 Comanchean
BT1 Cretaceous
BT1 Mesozoic
BT2 Lower Cretaceous
BT2 Cretaceous
BT2 Mesozoic
SA Arizona
SA Bisbee Group

Murchison Meteorite (1981)
Impact at Murchison, Victoria. Before 1981, also search Murchison AND meteorites.
BT carbonaceous chondrites
BT chondrites
BT stony meteorites
BT meteorites
SA Australia
SA Victoria Australia

Murcia
As of 1981, no longer a valid term for GeoRef. Use Murcia Province or Murcia region as applicable. Before 1981, used for the province.

Murcia Province
No longer a valid term for GeoRef. As of 1993, see Murcia Spain.

Murcia region (1981)
In SE Spain.
CO N372200N392800
  W0004000W0025500
BT Spain
BT Iberian Peninsula
BT Southern Europe
BT Europe
NT Albacete Spain
NT Murcia Spain

Murcia Spain (1993)
Refers only to the province in SE Spain. From 1981-1993, also search Murcia Province and Spain. Before 1981, also search Murcia and Spain.
CO N372200N384500
  W0004000W0022000
BT Murcia region
BT Spain
BT Iberian Peninsula
BT Southern Europe
BT Europe
NT Caravaca Spain
NT Jumilla Spain

Mures
No longer a valid term for GeoRef. As of 1993 see Mures Romania.

Mures Province
use Mures Romania

Mures Romania (1993)
County in N central Romania. Before 1993 also search Mures AND Romania.
UF Mures Province
BT Transylvania
BT Romania
BT Southern Europe
BT Europe
NT Ditrau Romania
NT Gurghiu Romania

Murgab Basin (1978)
River basin. Also search Murgab.
IN Index Afghanistan and/or Turkmenia as applicable.
UF Murghab (Basin)
BT Asia
SA Afghanistan
SA Turkmenia

Murge (1978)
Plateau region SW of Bari near the heel of the Italian boot.
BT Apulia Italy
BT Italy
BT Southern Europe
BT Europe

Murghab (Basin)
use Murgab Basin

muriatic acid
use hydrochloric acid

Muricacea (1978)
BT Gastropoda
BT Mollusca
BT Invertebrata

Muridae (1978)
Family. Autoposting of Eutheria and Theria to this term began in 1989.
BT Myomorpha
BT Rodentia
BT Eutheria
BT Theria
BT Mammalia
BT Tetrapoda
BT Vertebrata
BT Chordata

Murmansk
No longer a valid term for GeoRef. As of 1993, see Murmansk Russian Federation.

Murmansk Russian Federation (1993)
Oblast including the city on the Arctic Ocean in extreme NW European Russian Federation. Before 1993, also search Murmansk. This term has multiple hierarchies.
BT1 Russian Federation
BT1 Commonwealth of Independent States
BT2 Europe
NT Kola Russian Federation
NT Lovozero Massif
NT Lovozero Russian Federation
NT Monchegorsk Russian Federation
NT Pechenga
SA Imandra
SA Kola

Muro Group (1978)
IN Index age as applicable.
SA Cretaceous
SA Japan
SA K-T boundary
SA Tertiary

Murphy Marble (1978)
Sequence (ascending) is Valleytown Formation, Murphy Marble, Andrews Schist, and Nottely Quartzite. N Georgia, W North Carolina and E Tennessee.
BT Lower Cambrian
BT Cambrian
BT Paleozoic
SA Georgia
SA North Carolina
SA Tennessee

Murray Basin (1978)
River basin. Also search Murray AND basin.
IN Index states as applicable.
SA Australia
SA New South Wales Australia
SA South Australia
SA Victoria Australia

Murray County
Valid through 1988. Search in combination with state term. After 1988, use specific county-state term.

Murray County Georgia (1989)
NW Georgia. Before 1989, also search Murray County AND Georgia.
CO N343700N345900
  W0843600W0845600

BT Georgia
BT United States

Murray County Minnesota
(1989)
SW Minnesota. Before 1989, also search Murray County AND Minnesota.
CO N435100N441300
W0952800W0960500
BT Minnesota
BT United States

Murray County Oklahoma (1989)
S Oklahoma. Before 1989, also search Murray County AND Oklahoma.
CO N341900N343800
W0965000W0972100
BT Oklahoma
BT United States
SA Arbuckle Mountains

Murray fracture zone (1978)
NE of Hawaii. Also search Murray AND fracture zone.
UF Murray Seascarp
BT Pacific Ocean

Murray Meteorite (1981)
Impact in Calloway County, Kentucky. Before 1981, also search Murray AND meteorites.
BT carbonaceous chondrites
BT chondrites
BT stony meteorites
BT meteorites
SA Kentucky

Murray River (1978)
The major river of Australia. It rises in E Victoria and flows NW and then S into Encounter Bay on the Indian Ocean. Also search Murray.
IN Index states as applicable.
SA Australia
SA New South Wales Australia
SA South Australia
SA Victoria Australia

Murray Seascarp
use Murray fracture zone

Murrumbidgee River (1978)
Flows W from Great Dividing Range near Canberra to join Murray River in S.
BT New South Wales Australia
BT Australia
BT Australasia

Murrumbidgee Valley (1978)
River valley in S New South Wales.
BT New South Wales Australia
BT Australia
BT Australasia

Mururoa Atoll (1978)
Tuamotu Islands considered a broader term as of 1981.
BT Tuamotu Islands
BT French Polynesia
BT Polynesia
BT Oceania

Murzuk Basin (1978)
Dune region in SW Libya. Also search Murzuk.
UF Murzuq
BT Libya
BT North Africa
BT Africa

Murzuq
use Murzuk Basin

Musabani copper mines
use Mosaboni Mines

Musashino
No longer a valid term for GeoRef. See Musashino Japan.

Musashino Japan (1993)
City in Tokyo Prefecture, central Honshu.
BT Tokyo Japan
BT Honshu
BT Japan
BT Far East
BT Asia

Muschelkalk (1978)
Europe.
BT Middle Triassic
BT Triassic
BT Mesozoic

Musci (1978)
BT bryophytes
BT Plantae
NT Sphagnum

muscles (1993)
SA anatomy
SA extremities
SA jaws
SA morphology
SA physiology
SA skeletons

Muscogee County Georgia (1989)
W Georgia. Before 1989, search Muscogee County AND Georgia.
CO N322200N323600
W0843800W0850100
BT Georgia
BT United States
NT Columbus Georgia

muscovite (1978)
UF common mica
UF mirrorstone
UF moscovite
UF potash mica
UF white mica
BT mica group
BT sheet silicates
BT silicates
SA hydromuscovite
SA illite
SA light minerals
SA muscovitization
SA phengite
SA Plotina Deposit
SA sericite

muscovite granite (1978)
BT granites
BT plutonic rocks
BT igneous rocks

muscovite schist (1981)
BT schists
BT metamorphic rocks

muscovitization (1989)
BT metasomatism
SA alteration
SA muscovite
SA processes

museums (1978)
Used for papers discussing the functioning of a geology museum and/or its history.
IN Index with name of museum in original language.
SA academic institutions
SA associations
SA British Museum
SA catalogs
SA collections
SA education
SA exhibits
SA geology
SA organization
SA Smithsonian Institution

SA survey organizations

Musgrave Ranges (1978)
Also search Musgrave.
IN Index Northern Territory and/or South Australia as applicable.
BT Australia
BT Australasia
SA Northern Territory Australia
SA South Australia

Mushabani copper mines
use Mosaboni Mines

muskegs (1985)
Autoposting of bogs to this term began in 1989.
BT bogs
SA geomorphology
SA paludal environment
SA peat bogs

Muskogee County
Valid through 1988. Search in combination with state term. After 1988, use specific county-state term.

Muskogee County Oklahoma (1989)
E Oklahoma. Before 1989, also search Muskogee County AND Oklahoma.
CO N351500N355000
W0950500W0954600
BT Oklahoma
BT United States

Muskox Intrusion (1978)
In Mackenzie District.
BT Mackenzie District Northwest Territories
BT Northwest Territories
BT Western Canada
BT Canada

Mussoorie
No longer a valid term for GeoRef. See Mussoorie India.

Mussoorie India (1993)
Hill station and sanitarium in N Uttar Pradesh, N central India.
BT Uttar Pradesh India
BT India
BT Indian Peninsula
BT Asia

Mussoorie Syncline (1989)
N Uttar Pradesh, N central India.
SA Himalayas
SA Uttar Pradesh India

Mustang Island (1978)
In Neuces County between Corpus Christi Bay and the Gulf of Mexico.
IN Index county or region as applicable.
SA Nueces County Texas

Mustelidae (1978)
Family. Autoposting of Fissipeda, Eutheria and Theria to this term began in 1989. Before 1981, also search mustelids. As of 1981, search mustelids only for biostratigraphy.
BT Fissipeda
BT Carnivora
BT Eutheria
BT Theria
BT Mammalia
BT Tetrapoda
BT Vertebrata
BT Chordata

Muth Quartzite (1978)

Middle to Upper Devonian. Includes a thick succession of snow-white to greenish quartzites.
BT Devonian
BT Paleozoic
SA India
SA Jammu and Kashmir
SA Uttar Pradesh India

Muyun-Kum
use Muyunkum

Muyunkum (1978)
Sandy desert in Dzhambul Oblast, S of the Chu River near Kyrgyzstan border. This term has multiple hierarchies.
UF Muyun-Kum
BT1 Kazakhstan
BT1 Central Asia
BT1 Asia
BT2 Kazakhstan
BT2 Commonwealth of Independent States

MVT
use mississippi valley-type

MWD
Not a valid term for GeoRef. See measurement-while-drilling if applicable.

Myanmar
use Burma

mylonite
As of 1981, no longer a valid term for GeoRef. From 1981-1980, mylonites was autoposted to this term.
use mylonites

mylonites (1978)
Autoposting of this term to mylonite ended in 1981. Before 1981, also search mylonite.
UF mylonite
BT metamorphic rocks
NT blastomylonite
NT pseudotachylite
NT ultramylonite
SA shear zones

mylonitization (1978)
SA deformation
SA faults
SA phyllonitization

Myodocopida (1978)
Order. Autoposting of this term to Cladocopina, Entomozocopina and Myodocopina began in 1989. Autoposting of Myodocopina to this term ended in 1989. As of 1990, microfossils, Crustacea, Mandibulata, and Arthropoda are autoposted to this term. This term has multiple hierarchies.
BT1 Ostracoda
BT1 Crustacea
BT1 Mandibulata
BT1 Arthropoda
BT1 Invertebrata
BT2 Ostracoda
BT2 microfossils
NT Cladocopina
NT Entomozocopina
NT Myodocopina

Myodocopina (1981)
Suborder. Autoposting of this term began in 1985. Autoposting of Myodocopida to this term began in 1989. As of 1990, microfossils, Crustacea, Mandibulata, and Arthropoda are autoposted to this term. This term has multiple hierarchies.
BT1 Myodocopida

BT1 Ostracoda
BT1 Crustacea
BT1 Mandibulata
BT1 Arthropoda
BT1 Invertebrata
BT2 Myodocopida
BT2 Ostracoda
BT2 microfossils

Myoko Mountain (1978)
W central Honshu.
UF Myoko Volcano
UF Myoko-san
UF Myoko-zan
BT Niigata Japan
BT Honshu
BT Japan
BT Far East
BT Asia

Myoko Volcano
use Myoko Mountain

Myoko-san
use Myoko Mountain

Myoko-zan
use Myoko Mountain

Myomorpha (1981)
Suborder. Autoposting of Eutheria and Theria to this term began in 1989.
BT Rodentia
BT Eutheria
BT Theria
BT Mammalia
BT Tetrapoda
BT Vertebrata
BT Chordata
NT Arvicolidae
NT Cricetidae
NT Gliridae
NT Muridae

Myophoria (1978)
Genus.
BT Bivalvia
BT Mollusca
BT Invertebrata

Myriapoda (1978)
Includes Archipolypoda, Chilopoda, Diploda, Pauropoda, and Symphyla.
BT Mandibulata
BT Arthropoda
BT Invertebrata

Myrica (1989)
Genus.
BT Dicotyledoneae
BT angiosperms
BT Spermatophyta
BT Plantae

myrmekite (1978)
An intergrowth of vermicular quartz in plagioclase feldspar.
BT plagioclase
BT feldspar group
BT framework silicates
BT silicates
SA quartz

Mysore
No longer a valid term for GeoRef. Before 1989, used for the former state on the Arabian Sea in SW India. As of 1989, use Karnataka for the state. See Mysore India.

Mysore India (1993)
City in Karnataka. Before 1989, Mysore was also used for the former state on the Arabian Sea in SW India. As of 1989, use Karnataka for the state.
CO N121800N121800
E0763700E0763700

BT Karnataka India
BT India
BT Indian Peninsula
BT Asia

Mystery Cave (1989)
Cave in Fillmore County, SE Minnesota and also in Sullivan County, SE New York.
IN Index states and counties as applicable.
SA Fillmore County Minnesota
SA Minnesota
SA New York
SA Sullivan County New York

Mytilus (1978)
Genus. Autoposting of this term began in 1978.
BT Bivalvia
BT Mollusca
BT Invertebrata
NT Mytilus edulis

Mytilus edulis (1978)
BT Mytilus
BT Bivalvia
BT Mollusca
BT Invertebrata

Myxomycetes (1981)
BT fungi
BT thallophytes
BT Plantae

Myzostomia (1978)
BT worms
BT Invertebrata

N

N
use nitrogen

N-15 (1985)
Autoposting of broader terms began in 1989. This term has multiple hierarchies.
BT1 stable isotopes
BT1 isotopes
BT2 nitrogen
SA N-15/N-14

N-15/N-14 (1978)
Autoposting of broader terms began in 1989. This term has multiple hierarchies.
BT1 stable isotopes
BT1 isotopes
BT2 nitrogen
SA isotope ratios
SA N-15

n-alkanes (1985)
BT alkanes
BT aliphatic hydrocarbons
BT hydrocarbons
BT organic materials

Na
use sodium

Na-22 (1978)
Autoposting of broader terms began in 1989. This term has multiple hierarchies.
BT1 radioactive isotopes
BT1 isotopes
BT2 sodium
BT2 alkali metals
BT2 metals

Na-24 (1978)
Autoposting of broader terms began in 1989. This term has multiple hierarchies.
BT1 radioactive isotopes
BT1 isotopes
BT2 sodium
BT2 alkali metals
BT2 metals

Na-montmorillonite
After 1992, use montmorillonite and sodium ion if applicable.
use montmorillonite

Nabburg
No longer a valid term for GeoRef. See Nabburg Germany.

Nabburg Germany (1993)
Town in E Bavaria, SE West Germany.
BT Bavaria Germany
BT Germany
BT Central Europe
BT Europe

Nacimiento Formation (1985)
NW New Mexico.
BT Paleocene
BT Paleogene
BT Tertiary
BT Cenozoic
SA New Mexico

Nacimiento Mountains (1978)
Range in NW New Mexico.
BT Rio Arriba County New Mexico
BT New Mexico
BT United States
SA Fenton Hill

Nacozari de Garcia
No longer a valid term for GeoRef. As of 1993 see Nacozari de Garcia Mexico.

Nacozari de Garcia Mexico (1993)
Town in NE Sonora. Before 1993 also search Nacozari de Garcia AND Mexico. Before 1978, also search Nacozari and Mexico.
BT Sonora Mexico
BT Mexico

nacrite (1978)
BT clay minerals
BT sheet silicates
BT silicates

Naga Hills (1993)
Ranges on India-Burma border, N of Manipur Hills.
IN Index country as applicable.
BT Asia
SA Burma
SA India
SA Nagaland India

Nagaland India (1993)
State NE of Bangladesh in NE India. Established in 1961.
CO N253000N271500
E0953000E0940000
BT Northeastern India
BT India
BT Indian Peninsula
BT Asia
SA Naga Hills

Nagano
No longer a valid term for GeoRef. See Nagano Japan and Nagano Prefecture Japan.

Nagano City Japan (1993)
City in Nagano Prefecture, central Honshu. Also a town in Akita Prefecture and in Osaka Prefecture.
IN Index prefectures as applicable.
BT Honshu
BT Japan
BT Far East
BT Asia
SA Akita Japan
SA Koma-ga-take
SA Nagano earthquake 1984
SA Nagano Japan
SA Osaka Japan

Nagano earthquake 1984 (1989)
Epicenter at Nagano, central Honshu. Before 1993, also search western Nagano earthquake 1984.
UF Nagano Prefecture earthquake 1984
UF Naganoken-seibu earthquake 1984
UF western Nagano earthquake 1984
BT earthquakes
SA Honshu
SA Japan
SA Nagano City Japan
SA Ontake

Nagano Japan (1993)
Central Honshu. Before 1993, also search Nagano or Nagano Prefecture.
BT Honshu
BT Japan
BT Far East
BT Asia
SA Akaishi Mountains
SA Asama
SA Chubu Japan
SA Japanese Alps
SA Kanto Mountains
SA Kiso Mountains
SA Nagano City Japan
SA Ontake
SA Ryujima
SA Yake-Dake
SA Yatsugatake

Nagano Prefecture
No longer a valid term for GeoRef.
use Nagano Prefecture Japan

Nagano Prefecture earthquake 1984
use Nagano earthquake 1984

Nagano Prefecture Japan (1993)
Central Honshu. Before 1993, also search Nagano or Nagano Prefecture AND Japan.
UF Nagano Prefecture
NT Lake Suwa
NT Matsukawa Japan
NT Nojiri Lake

Naganoken-seibu earthquake 1984
use Nagano earthquake 1984

Nagasaki
No longer a valid term for GeoRef. See Nagasaki Japan.

Nagasaki Japan (1993)
City and prefecture in W Kyushu, S Japan.
BT Kyushu
BT Japan
BT Far East
BT Asia
NT Goto Islands
NT Sasebo Japan
NT Unzen
SA Iki Island

Nagorno-Karabakh Azerbaidzhan (1993)

Autonomous Oblast. SW Azerbaidzhan. Formerly Karabakh Mountain Area. Before 1993, also search Karabakh. This term has multiple hierarchies.
BT1 Azerbaidzhan
BT1 Europe
BT2 Azerbaidzhan
BT2 Commonwealth of Independent States
SA Lesser Caucasus

Nagpur
No longer a valid term for GeoRef. See Nagpur India.

Nagpur India (1993)
City, district, and division in NE Maharashtra, W central India.
BT Maharashtra India
BT India
BT Indian Peninsula
BT Asia

Naha test well (1978)
In Navajo Indian Reservation in NE Arizona.
BT Arizona
BT United States
SA Navajo Indian Reservation

Nahanni earthquake 1985 (1993)
SW Northwest Territories, Canada.
BT earthquakes
SA Canadian Cordillera
SA Northwest Territories

Nahanni Formation (1978)
British Columbia, Mackenzie District of Northwest Territories, and Yukon Territory.
BT Middle Devonian
BT Devonian
BT Paleozoic
SA British Columbia
SA Canada
SA Northwest Territories
SA Yukon Territory

nahcolite (1978)
BT carbonates

Nahe (1978)
River which flows into Rhine at Bingen.
IN Index states as applicable.
BT Germany
BT Central Europe
BT Europe
SA Rhineland-Palatinate Germany
SA Saar-Nahe Basin
SA Saarland Germany

Naheola Formation (1985)
In Midway Group. S Alabama and E Mississippi.
BT Paleocene
BT Paleogene
BT Tertiary
BT Cenozoic
SA Alabama
SA Midway Group
SA Mississippi
SA Oak Hill Member

Naiba River basin
use Nayba River basin

Nain
No longer a valid term for GeoRef. Use Nain Newfoundland if applicable.

Nain Anorthosite Massif
use Nain Massif

Nain Complex
use Nain Massif

Nain Massif (1978)
Anorthosite massif.
UF Nain Anorthosite Massif
UF Nain Complex
BT Labrador
BT Newfoundland
BT Eastern Canada
BT Canada

Nain Newfoundland (1993)
Village on central coast of Labrador. Before 1993, also search Nain and Labrador.
BT Labrador
BT Newfoundland
BT Eastern Canada
BT Canada
SA Nain Province

Nain Province (1989)
Structural province of the Canadian Shield in N Labrador and E Quebec.
IN Index Canadian Shield, and Labrador and/or Quebec as applicable.
BT Canadian Shield
BT North America
SA Labrador
SA Nain Newfoundland
SA Quebec

Naini Tal
No longer a valid term for GeoRef. See Naini Tal India.

Naini Tal India (1993)
Town, hill station and district in NE Uttar Pradesh. Before 1993, also search Naini Tal. Before 1978, also search Naini-Tal or Nainital.
CO N292000N292500 E0793000E0792200
UF Naini-Tal India
UF Nainital India
BT Uttar Pradesh India
BT India
BT Indian Peninsula
BT Asia

Naini-Tal India
use Naini Tal India

Nainital India
use Naini Tal India

Nairobi
No longer a valid term for GeoRef. See Nairobi Kenya.

Nairobi Kenya (1993)
City in S Kenya. Before 1993 also search Nairobi AND Kenya.
BT Kenya
BT East Africa
BT Africa

Naka-no-umi (1978)
Inlet of Sea of Japan in SW Honshu.
UF Nakano umi
UF Nakanoumi Lake
BT Honshu
BT Japan
BT Far East
BT Asia
SA Shimane Japan
SA Tottori Japan

Nakano umi
use Naka-no-umi

Nakanoumi Lake
use Naka-no-umi

Nakhichevan
No longer a valid term for GeoRef. As of 1993, see Nakhichevan Azerbaidzhan.

Nakhichevan Azerbaidzhan (1993)
Former Autonomous Soviet Socialist Republic and town on the Iranian border separated from Azerbaidzhan proper by a narrow strip of Armenia. This term has multiple hierarchies.
BT1 Azerbaidzhan
BT1 Europe
BT2 Azerbaidzhan
BT2 Commonwealth of Independent States
SA Ordubad Azerbaidzhan

Nakhla
No longer a valid GeoRef term. See Nakhla Qatar.

Nakhla Meteorite (1989)
Impact at Abu Hommos, Alexandria, Egypt. Diopside-olivine achondrite.
BT nakhlite
BT achondrites
BT stony meteorites
BT meteorites
SA Egypt
SA SNC Meteorites

Nakhla Qatar (1993)
Village in Qatar.
BT Qatar
BT Arabian Peninsula
BT Asia

nakhlite (1981)
BT achondrites
BT stony meteorites
BT meteorites
NT Nakhla Meteorite
SA SNC Meteorites

Naknek Formation (1989)
S Alaska.
BT Upper Jurassic
BT Jurassic
BT Mesozoic
SA Alaska

Nakuru Basin (1978)
In Great Rift Valley in Lake Nakuru area of W central Rift Valley Province. Also search Nakuru.
BT Kenya
BT East Africa
BT Africa

Nama System (1978)
In SW and S Africa.
BT Cambrian
BT Paleozoic
SA Namibia
SA South Africa

Namaland
use Namaqualand

Namaqualand (1978)
Coastal region of sandy plains and bare hills.
IN Index Cape Province and/or Namibia as applicable.
UF Namaland
BT Southern Africa
BT Africa
SA Cape Province South Africa
SA Namibia

Namaqualand metamorphic complex (1993)
Proterozoic. Namibia and Cape Province, South Africa.
IN Index country as applicable.
BT Southern Africa
BT Africa
SA Namibia
SA South Africa

Namib Desert (1978)
Arid region extending along entire coast.
BT Namibia
BT Southern Africa
BT Africa

Namibe Angola (1993)
Province, formerly Mocamedes, including the city in SW Angola. Also search Mossamedes or Mocamedes.
BT Angola
BT Central Africa
BT Africa
SA Mossamedes

Namibia (1981)
Replaces South-West Africa. Before 1972, also search South West Africa. Before 1981, also search South-West Africa.
CO S290000S164000 E0251500E0113000
UF German Southwest Africa
UF South West Africa
UF South-West Africa
BT Southern Africa
BT Africa
NT Karibib Namibia
NT Luderitz Namibia
NT Namib Desert
NT Otavi Namibia
NT Tsumeb Namibia
NT Windhoek Namibia
SA Beaufort Group
SA Damara Orogeny
SA Damara System
SA Dwyka Formation
SA Gibeon Meteorite
SA Kaapvaal Craton
SA Kalahari Desert
SA Karroo Supergroup
SA Nama System
SA Namaqualand
SA Namaqualand metamorphic complex
SA Orange River
SA Pongola Supergroup
SA Table Mountain Group
SA Walvis Bay
SA Walvis Bay South Africa
SA Zambezi Valley

Namur
No longer a valid term for GeoRef. See Namur Belgium.

Namur Belgium (1993)
Province in S Belgium.
BT Belgium
BT Western Europe
BT Europe
NT Dinant Basin
NT Dinant Belgium

Namurian (1978)
Europe.
BT Upper Carboniferous
BT Carboniferous
BT Paleozoic

Nan Ling
use Nanling

Nan-king
Not a valid term for GeoRef. See Nanjing China.

Nanafalia Formation (1985)
In Wilcox Group. S Alabama, SW Georgia and E Mississippi.
BT lower Eocene
BT Eocene
BT Paleogene
BT Tertiary
BT Cenozoic
SA Alabama
SA Georgia
SA Mississippi

SA Wilcox Group

Nanaimo Group (1985)
Upper Cretaceous to Eocene. NW Washington and SW British Columbia.
IN Index ages as applicable.
SA British Columbia
SA Eocene
SA K-T boundary
SA Paleocene
SA Upper Cretaceous
SA Washington

Nanao
No longer a valid term for GeoRef. See Nanao Japan.

Nanao Japan (1993)
City on E side of Noto Peninsula, W coast of Honshu. Before 1993, also search Nanao AND Japan.
BT Ishikawa Japan
BT Honshu
BT Japan
BT Far East
BT Asia
SA Noto Peninsula

Nandewar Mountains (1978)
Range in NE New South Wales.
BT New South Wales Australia
BT Australia
BT Australasia

Nanjemoy Formation (1978)
Lower and middle Eocene. In Pamunkey Group. Includes Potapaco Clay Member, Woodstock Greensand, Marlboro Clay Member. E Maryland and E Virginia.
BT Eocene
BT Paleogene
BT Tertiary
BT Cenozoic
SA Maryland
SA Virginia

Nanjing
No longer a valid term for GeoRef. See Nanjing China.

Nanjing China (1993)
City in Jiangsu, E China.
CO N320300N320300
E1184700E1184700
BT Jiangsu China
BT China
BT Far East
BT Asia

Nankai Trough (1985)
Off SE Japan.
BT Northwest Pacific
BT Pacific Ocean
SA DSDP Site 297
SA DSDP Site 298
SA DSDP Site 582
SA DSDP Site 583
SA Leg 87
SA Nankaido earthquake 1946

Nankaido
No longer a valid term for GeoRef. Former division including Awaji Islands, S central part of S coast of Honshu, and all of Shikoku. See Nankaido earthquake 1946.

Nankaido earthquake 1946 (1993)
Nankai Trough, NW Pacific. Before 1993, also search Nankai earthquake 1946. Before 1986, also search Nankaido AND earthquakes.
BT earthquakes
SA Japan

SA Nankai Trough
SA Northwest Pacific
SA Pacific Ocean

Nanling (1989)
Mountain system in S central and E China. Separates Guangdong and Guangxi from Hunan and Jiangxi provinces.
IN Index provinces as applicable.
CO N243000N263000
E1140000E1110000
UF Nan Ling
BT China
BT Far East
BT Asia
SA Guangdong China
SA Guangxi China
SA Hunan China
SA Jiangxi China

nannoconids (1978)
As of 1981, restricted to use as the common name for Nannoconus. Autoposting of broader terms to this term began in 1989. Autoposting of microfossils to this term began in 1990. This term has multiple hierarchies.
BT1 nannofossils
BT1 algae
BT1 thallophytes
BT1 Plantae
BT2 nannofossils
BT2 algae
BT2 microfossils
BT3 nannofossils
BT3 algal flora
BT3 plants
BT4 nannofossils
BT4 algal flora
BT4 microfossils
SA biostratigraphy
SA Nannoconus

Nannoconus (1978)
Genus. Before 1981, also search nannoconids. Autoposting of thallophytes and microfossils began in 1990. This term has multiple hierarchies.
BT1 nannofossils
BT1 algae
BT1 thallophytes
BT1 Plantae
BT2 nannofossils
BT2 algae
BT2 microfossils
BT3 nannofossils
BT3 algal flora
BT3 plants
BT4 nannofossils
BT4 algal flora
BT4 microfossils
SA nannoconids

nannofossils (1978)
Treated as both a systematic name and a common name. Autoposting of thallophytes and microfossils began in 1990. As of 1993, algal flora is autoposted to this term. This term has multiple hierarchies.
UF calcareous nannofossils
BT1 algae
BT1 thallophytes
BT1 Plantae
BT2 algae
BT2 microfossils
BT3 algal flora
BT3 plants
BT4 algal flora
BT4 microfossils
NT Braarudosphaeridae
NT Discoasteridae

NT discoasters
NT Helicopontosphaera
NT nannoconids
NT Nannoconus
NT Sphenolithus
SA biostratigraphy
SA Coccolithophoraceae
SA fossils
SA problematic fossils

nannoplankton (1989)
BT plankton

Nansei-shoto
use Ryukyu Islands

Nantes
No longer a valid term for GeoRef. See Nantes France.

Nantes France (1993)
City on the Loire River in central Loire-Atlantique, W France.
BT Loire-Atlantique France
BT France
BT Western Europe
BT Europe

Nantucket (Island)
use Nantucket Island

Nantucket Island (1978)
In Atlantic Ocean, S of Cape Cod. Also search Nantucket.
UF Nantucket (Island)
BT Massachusetts
BT United States

Nantuo Formation (1989)
BT Sinian
BT Proterozoic
BT upper Precambrian
BT Precambrian
SA China

Nanushuk Group (1985)
N Alaska.
BT Cretaceous
BT Mesozoic
SA Alaska

NaOH
use sodium hydroxide

Napa County
Valid through 1988. Search in combination with state term. After 1988, use specific county-state term.

Napa County California (1989)
W California. Before 1989, also search Napa County AND California.
CO N381000N385000
W1220300W1224000
BT California
BT United States
NT Calistoga California
SA Homestake Mine
SA San Francisco Bay region

naphthalene (1989)
BT polycyclic aromatic hydrocarbons
BT aromatic hydrocarbons
BT hydrocarbons
BT organic materials

Napier Complex (1989)
Enderby Land and Tonagh Island, Antarctica.
BT Archean
BT Precambrian
SA Antarctica
SA Enderby Land

Naples
No longer a valid term for GeoRef. See Naples Italy.

Naples Italy (1993)

Province and city on the Tyrrhenian Sea in W Campania, S Italy. Before 1993, also search Naples and Italy. Before 1978, also search Napoli.
UF Napoli Italy
BT Campania Italy
BT Italy
BT Southern Europe
BT Europe

NAPLs
use nonaqueous phase liquids

Napoli Italy
use Naples Italy

nappe (volcanic)
use lava flows

nappes (1978)
SA allochthons
SA displacements
SA faults
SA folds
SA gravity sliding
SA klippen
SA orientation
SA overthrust faults
SA recumbent folds
SA tectonics

Nara
No longer a valid term for GeoRef. See Nara Japan.

Nara Japan (1993)
City and prefecture E of Osaka in W central Honshu.
BT Honshu
BT Japan
BT Far East
BT Asia
SA Kii Peninsula
SA Kinki Japan

Narbada River
use Narmada River

Narbada Valley
use Narmada Valley

Narbonne
No longer a valid term for GeoRef. See Narbonne France.

Narbonne France (1993)
City near the Mediterranean Sea in E Aude, S France.
BT Aude France
BT France
BT Western Europe
BT Europe

Nares abyssal plain (1989)
W North American Atlantic. As of 1990, Atlantic Ocean is autoposted to this term.
CO N220000N270000
W0600000W0700000
BT North American Atlantic
BT North Atlantic
BT Atlantic Ocean

Nares Strait (1989)
Part of the Arctic Ocean, connecting Lincoln Sea to Baffin Bay between E Ellesmere Island, NW Greenland and NW Baffin Bay.
CO N773000N830000
W0600000W0790000
BT Arctic Ocean
SA Baffin Bay
SA Ellesmere Island
SA Greenland
SA Northwest Territories

Nari Series (1978)
Underlies the Gaj and overlies the Khirthar Series.
BT Oligocene
BT Paleogene

BT Tertiary
BT Cenozoic
SA Gujarat India
SA India
SA Pakistan
SA Sind Pakistan

Narita Formation (1978)
Tokyo Bay area.
BT Pliocene
BT Neogene
BT Tertiary
BT Cenozoic
SA Honshu
SA Japan

Narizian (1978)
North America. Above Ulatisian, below Fresnian.
BT Eocene
BT Paleogene
BT Tertiary
BT Cenozoic
SA middle Eocene
SA upper Eocene

Narmada River (1978)
Rises in the Maikala Range in Madhya Pradesh and flows W into Gulf of Cambay. Also search Narmada.
IN Index states as applicable.
UF Narbada River
UF Nerbudda River
BT India
BT Indian Peninsula
BT Asia
SA Gujarat India
SA Madhya Pradesh India
SA Narmada-Son Lineament

Narmada Valley (1978)
River valley.
IN Index states as applicable.
UF Narbada Valley
UF Nerbuda Valley
BT India
BT Indian Peninsula
BT Asia
SA Gujarat India
SA Madhya Pradesh India
SA Narmada-Son Lineament

Narmada-Son Lineament (1993)
W and central India. Follows line of Narmada and Son rivers.
IN Index states as applicable.
BT India
BT Indian Peninsula
BT Asia
SA Gujarat India
SA Madhya Pradesh India
SA Narmada River
SA Narmada Valley
SA Son Valley

Narnaul
No longer a valid term for GeoRef. See Narnaul India.

Narnaul India (1993)
Town 80 miles WSW of Delhi in S Haryana, N central India.
BT Haryana India
BT India
BT Indian Peninsula
BT Asia

Narpaign Fjord
use Narpaing Fjord

Narpaing Fiord
use Narpaing Fjord

Narpaing Fjord (1978)
Narrow inlet of Davis Strait on Baffin Island, Keewatin District.
IN Index Baffin Island.
UF Narpaign Fjord
UF Narpaing Fiord

BT Keewatin District Northwest Territories
BT Northwest Territories
BT Western Canada
BT Canada
SA Baffin Island

Narrabeen Group (1978)
Includes the Gosford Formation and Clifton Sub-Group. N of Sydney.
BT Triassic
BT Mesozoic
SA Australia
SA Illawarra Coal Measures
SA New South Wales Australia

Narragansett Basin (1985)
S New England.
IN Index states as applicable.
BT United States
SA Massachusetts
SA New England
SA Rhode Island

Narragansett Bay (1978)
Inlet of Atlantic Ocean in SE Rhode Island.
BT Rhode Island
BT United States

Narragansett Pier Granite (1989)
E Connecticut, NE New York and S Rhode Island.
IN Index ages as applicable.
BT upper Paleozoic
BT Paleozoic
SA Connecticut
SA New York
SA Pennsylvanian
SA Permian
SA Rhode Island

Narrows (1978)
Strait between W end of Long Island and Staten Island separating Upper New York Bay from Lower New York Bay. Search in combination with New York.
IN Index regions as applicable.
SA New York

Narugo
No longer a valid term for GeoRef. As of 1993, see Narugo Japan.

Narugo Japan (1993)
Town and geothermal area in N Honshu.
BT Honshu
BT Japan
BT Far East
BT Asia

Narvik
No longer a valid term for GeoRef. As of 1993 see Narvik Norway.

Narvik Norway (1993)
City on Ofot Fjord in N Nordland, N Norway. Before 1993 also search Narvik AND Norway.
BT Nordland Norway
BT Norway
BT Scandinavia
BT Western Europe
BT Europe

Narym
No longer a valid term for GeoRef. As of 1993, Narym Russian Federation.

Narym Russian Federation (1993)
Village on Ob River in central Tomsk Oblast in W central Siberia. Before 1993, also search Narym. This term has multiple hierarchies.

BT1 Tomsk Russian Federation
BT1 Russian Federation
BT1 Commonwealth of Independent States
BT2 Tomsk Russian Federation
BT2 Asia

Naryn
No longer a valid term for GeoRef. As of 1993, see Naryn Kyrgyzstan.

Naryn Kyrgyzstan (1993)
Oblast and town in SE Kyrgyzstan. This term has multiple hierarchies.
BT1 Kyrgyzstan
BT1 Asia
BT2 Kyrgyzstan
BT2 Commonwealth of Independent States

NASA (1985)
Acronym.
UF National Aeronautics and Space Administration
BT government agencies
SA United States

Nasca Plate
use Nazca Plate

Nash Fork-Mullen Creek shear zone
use Mullen Creek-Nash Fork shear zone

Nashville
Valid through 1988. Search in combination with state term. After 1988, use specific city-state term.

Nashville Dome (1989)
IN Index states as applicable.
BT United States
SA Alabama
SA Kentucky
SA Tennessee

Nashville Tennessee (1989)
City on Cumberland River in central Tennessee. Before 1989, search Nashville AND Tennessee.
CO N361000N361000
 W0865000W0865000
BT Davidson County Tennessee
BT Tennessee
BT United States

Nasliden Mine (1985)
Base metals. NE Vasterbotten, Sweden.
BT Vasterbotten Sweden
BT Sweden
BT Scandinavia
BT Western Europe
BT Europe
SA base metals
SA Boliden Sweden
SA mines

Nassau County
Valid through 1988. Search in combination with state term. After 1988, use specific county-state term.

Nassau County Florida (1989)
On the Atlantic Ocean in extreme NE Florida. Before 1989, also search Nassau County AND Florida.
CO N301400N304600
 W0812500W0820300
BT Florida
BT United States

Nassau County New York (1989)
SE New York. Before 1989, also search Nassau County AND New York.
CO N403400N405400

 W0732400W0734700
BT New York
BT United States
NT Bay Park New York

Nassellaria
use Nassellina

Nassellina (1978)
Suborder. Autoposting of microfossils and Protista to this term began in 1990. This term has multiple hierarchies.
UF Nassellaria
BT1 Osculosida
BT1 Radiolaria
BT1 Protista
BT1 Invertebrata
BT2 Osculosida
BT2 Radiolaria
BT2 Protista
BT2 microfossils

nasturan
use pitchblende

Nasu (dake)
use Nasudake

Nasu Mountain
use Nasudake

Nasudake (1978)
Volcanic peak NE of Nikko in N central Honshu.
IN Index prefecture or prefectures as applicable.
UF Nasu (dake)
UF Nasu Mountain
BT Honshu
BT Japan
BT Far East
BT Asia
SA Fukushima Japan
SA Tochigi Japan

Natal
No longer a valid term for GeoRef. As of 1993 see Natal South Africa.

Natal South Africa (1993)
South African province on the Indian Ocean. Before 1993 also search Natal AND South Africa.
CO S311000S264500
 E0325500E0285000
BT South Africa
BT Southern Africa
BT Africa
NT Durban South Africa
NT Tugela Basin
NT Zululand

Naticidae (1989)
Family.
BT Gastropoda
BT Mollusca
BT Invertebrata

National Aeronautics and Space Administration
use NASA

National Coal Resources Data System (1985)
UF NCRDS
SA coal
SA data bases
SA United States

national monuments (1989)
SA Badlands National Monument
SA Craters of the Moon
SA Dinosaur National Monument
SA Glacier Bay National Park
SA Katmai National Monument
SA Marble Canyon
SA national parks
SA Pinnacles National Monument
SA White Sands

SA wilderness areas

National Oceanic and Atmospheric Administration
use NOAA

national parks (1985)
SA Arches National Park
SA Bialowieza
SA Canyonlands National Park
SA conservation
SA Fiordland National Park
SA Glacier Bay National Park
SA Glacier National Park
SA Grand Canyon
SA Grand Teton National Park
SA Great Smoky Mountains
SA Guadalupe Mountains
SA Haleakala
SA Isle Royale National Park
SA Jasper National Park
SA land use
SA Lassen Volcanic National Park
SA legislation
SA Mammoth Cave
SA Mount Rainier National Park
SA national monuments
SA Petrified Forest National Park
SA policy
SA Rocky Mountain National Park
SA wilderness areas
SA Yellowstone National Park
SA Yosemite National Park

National Petroleum Reserve Alaska (1989)
North Slope, N Alaska. Also search National Petroleum Reserve in conjunction with Alaska.
CO N680000N710000
    W1440000W1620000
UF National Petroleum Reserve in Alaska
UF NPRA
BT Northern Alaska
BT Alaska
BT United States
SA North Slope

National Petroleum Reserve in Alaska
use National Petroleum Reserve Alaska

National Science Foundation
use NSF

National Uranium Resource Evaluation Program
use NURE

National Water Data Exchange
use NAWDEX

native elements (1981)
Before 1981, search native elements and alloys. As of 1981, minerals that were classified under native elements and alloys are included here or under alloys.
NT carbonado
NT chaoite
NT diamond
NT graphite
NT lonsdaleite
SA alloys
SA antimony
SA antimony ores
SA arsenic
SA arsenic ores
SA bismuth
SA bismuth ores
SA copper
SA copper ores
SA gold
SA gold ores
SA iron
SA iron ores
SA mercury

SA mercury ores
SA metal ores
SA metals
SA minerals
SA nonmetal deposits
SA nonmetals
SA silver
SA silver ores
SA sulfur
SA sulfur deposits

native elements and alloys
As of 1981, no longer a valid term for GeoRef. Use native elements or alloys as applicable. From 1978-1980, was autoposted to the minerals that were classified under it.

natrium
use sodium

natroalunite (1993)
UF almeriite
BT sulfates
SA alunitization

natroborocalcite
use ulexite

natrolite (1978)
BT zeolite group
BT framework silicates
BT silicates

Natrona County
Valid through 1988. Search in combination with state term. After 1988, use specific county-state term.

Natrona County Wyoming (1989)
Central Wyoming. Before 1989, also search Natrona County AND Wyoming.
CO N422500N433000
    W1060500W1073700
BT Wyoming
BT United States
NT Casper Wyoming
SA Casper Mountain
SA Granite Mountains
SA Rattlesnake Hills
SA Teapot Dome

natural
A valid index term through 1977 used in combination with gas (i.e. gas, natural). After 1977 use natural gas.

natural bridges
Not a valid GeoRef index term after 1970. Was used on level 1 in subfile N.
use natural curiosities

natural coke
use coke coal

natural curiosities (1981)
Before 1970, also search natural bridges.
UF natural bridges
SA bridges
SA erosion features
SA geomorphology
SA popular geology
SA solution features

natural gas (1978)
Not a valid term through 1977. Before 1978, also search gas AND natural; petroleum-gas; petroleum-natural gas.
IN Commodity. See List C.
BT petroleum
NT coalbed methane
SA clathrates
SA condensates

SA degasification
SA Eastern Gas Shales Project
SA energy sources
SA gas hydrates
SA gas sands
SA gas storage
SA gases
SA gasification
SA giant fields
SA hydrocarbons
SA hydrogen sulfide
SA Karachaganak Field
SA methane
SA offshore
SA oil and gas fields
SA oil-gas interface
SA onshore
SA petroleum engineering
SA petroleum maps
SA pipelines
SA reservoir properties
SA reservoir rocks
SA secondary migration
SA source rocks
SA stratigraphic traps
SA stratigraphic wedges
SA structural traps
SA tight sands
SA traps

natural materials (1968)
To distinguish from synthetic materials.
SA materials
SA synthetic materials

natural recharge (1982)
SA artificial recharge
SA ground water
SA recharge

natural remanence
use natural remanent magnetization

natural remanent magnetism
use natural remanent magnetization

natural remanent magnetization (1978)
Autoposting of magnetization to this term began in 1989.
UF natural remanence
UF natural remanent magnetism
UF NRM
BT remanent magnetization
BT magnetization
SA magnetic field
SA paleomagnetism
SA remagnetization

natural resources (1978)
SA conservation
SA depletion
SA environmental geology
SA land cover maps
SA land use
SA reclamation
SA Resource Conservation and Recovery Act
SA resources
SA wilderness areas

natural selection (1978)
Process by which "weaker" or less well-adapted organisms tend to be eliminated from a given population.
UF selection, natural
SA biologic evolution
SA Darwinism

naturally fractured reservoirs (1989)
Sometimes difficult to determine. Also search (reservoir rocks OR subsurface reservoirs OR reservoirs OR reservoir properties)

AND (petroleum OR natural gas OR geothermal energy, as applicable) AND fractures.
SA fractures
SA fracturing
SA petroleum
SA petroleum engineering
SA reservoir properties
SA reservoir rocks
SA reservoirs

Naubug (1978)
Village SE of Srinager in S Jammu and Kashmir, disputed territory between India and Pakistan. As of 1993, India is no longer a broader term.
IN Index Jammu and Kashmir AND India and/or Pakistan as applicable.
SA India
SA Jammu and Kashmir
SA Pakistan

naujaite (1978)
BT syenites
BT plutonic rocks
BT igneous rocks

Nauru Basin (1985)
W Equatorial Pacific Ocean, N and W of Nauru Island.
BT Pacific Ocean
SA DSDP Site 462
SA Equatorial Pacific
SA Leg 89
SA Nauru Island
SA West Pacific

Nauru Island (1981)
UF Pleasant Island
BT Micronesia
BT Oceania
SA Nauru Basin

Nautiloidea (1978)
Autoposting of this term began in 1978.
BT Tetrabranchiata
BT Cephalopoda
BT Mollusca
BT Invertebrata
NT Aturia
NT Hercoglossa danica
NT Nautilus
SA Bactritida

Nautilus (1978)
Genus.
BT Nautiloidea
BT Tetrabranchiata
BT Cephalopoda
BT Mollusca
BT Invertebrata

Navajo County
Valid through 1988. Search in combination with state term. After 1988, use specific county-state term.

Navajo County Arizona (1989)
E Arizona. Before 1989, also search Navajo County AND Arizona.
CO N333200N370000
    W1095000W1104500
BT Arizona
BT United States
NT Holbrook Arizona
NT Hopi Buttes Field
SA Black Mesa
SA Petrified Forest National Park
SA Rio Puerco

Navajo Indian Reservation (1978)
NE Arizona, NW New Mexico, and SE Utah.

IN Index states as applicable.
BT United States
SA Arizona
SA Indian reservations
SA Naha test well
SA New Mexico
SA Utah

Navajo Sandstone (1978)
In Glen Canyon Group. Upper Triassic(?)and Jurassic. N Arizona, W Colorado, NW New Mexico, and SE Utah.
BT Mesozoic
SA Arizona
SA Colorado
SA Glen Canyon Group
SA Jurassic
SA New Mexico
SA Nugget Sandstone
SA Upper Triassic
SA Utah

Navarin Basin (1985)
N Bering Sea.
CO N560000N630000
    W1700000W1800000
BT Bering Sea
BT West Pacific
BT Pacific Ocean

Navarra
No longer a valid term for GeoRef. As of 1993, see Navarra Spain.

Navarra Spain (1993)
Former kingdom and province in NE Spain. Before 1993, also search Navarra and Spain. Before 1981, also search Navarre AND Spain.
CO N415500N431800
    W0004000W0023000
BT Spain
BT Iberian Peninsula
BT Southern Europe
BT Europe
NT Asturreta
NT Pamplona Spain

Navarre (1978)
As of 1981, refers to only the region in SW France. Before 1981, also referred to the former kingdom and province of N Spain. See Navarra Spain. Before 1981, search Navarre AND France.
BT France
BT Western Europe
BT Europe
SA Pyrenees-Atlantiques France

Navarro Group (1985)
E Texas.
BT Gulfian
BT Upper Cretaceous
BT Cretaceous
BT Mesozoic
SA Escondido Formation
SA Olmos Formation
SA Texas

Navesink Formation (1978)
In Monmouth Group.
BT Upper Cretaceous
BT Cretaceous
BT Mesozoic
SA Delaware
SA Monmouth Group
SA New Jersey

Navicula (1978)
Autoposting of thallophytes and microfossils began in 1990. This term has multiple hierarchies.
BT1 diatoms
BT1 algae
BT1 thallophytes
BT1 Plantae
BT2 diatoms
BT2 algae
BT2 microfossils

NAVSTAR GPS
use Global Positioning System

NAWDEX (1985)
Acronym.
UF National Water Data Exchange
SA data bases
SA data processing
SA ground water
SA hydrology
SA United States
SA water resources

Naxos (1978)
Largest island of the Cyclades in the Aegean Sea. This term has multiple hierarchies.
BT1 Cyclades
BT1 Greek Aegean Islands
BT1 Greece
BT1 Southern Europe
BT1 Europe
BT2 Cyclades
BT2 Greek Aegean Islands
BT2 Aegean Islands
BT2 Mediterranean region

Nayarit
No longer a valid term for GeoRef. As of 1993 see Nayarit Mexico.

Nayarit Mexico (1993)
State in W Mexico. Before 1993 also search Nayarit AND Mexico.
CO N203500N230000
    W1041000W1060000
BT Mexico
NT Sanguangey
SA Mexican volcanic belt
SA Sierra Madre Occidental

Nayba River basin (1978)
S Sakhalin Island. Also search Nayba. This term has multiple hierarchies.
UF Naiba River basin
BT1 Russian Federation
BT1 Commonwealth of Independent States
BT2 Asia
SA Sakhalin

Nazca Plate (1978)
S of the Galapagos Islands.
UF Nasca Plate
BT Pacific Ocean
SA DSDP Site 320
SA DSDP Site 321
SA Galapagos Rift
SA plate tectonics
SA plates

Nazca Ridge (1981)
SW of Peru. Autoposting of Pacific Ocean to this term began in 1990.
CO S253000S140000
    W0760000W0873000
BT South American Pacific
BT Pacific Ocean

Nb
use niobium

Nchanga
No longer a valid term for GeoRef. As of 1993 see Nchanga Zambia.

Nchanga Zambia (1993)
Mining township in N near Zaire border. Before 1993 also search Nchanga AND Zambia.
BT Zambia
BT East Africa
BT Africa

NCRDS
use National Coal Resources Data System

Nd
use neodymium

Nd-143/Nd-144
use Nd-144/Nd-143

Nd-144/Nd-143 (1982)
Autoposting of broader terms began in 1989. This term has multiple hierarchies.
UF Nd-143/Nd-144
BT1 stable isotopes
BT1 isotopes
BT2 neodymium
BT2 rare earths
BT2 metals
SA isotope ratios

Nd/Sm
use Sm/Nd

Ne
use neon

Ne-20 (1978)
Autoposting of broader terms began in 1989. This term has multiple hierarchies.
BT1 stable isotopes
BT1 isotopes
BT2 neon
BT2 noble gases
SA Ne-22/Ne-20

Ne-20/Ne-22
use Ne-22/Ne-20

Ne-21 (1978)
Autoposting of broader terms began in 1989. This term has multiple hierarchies.
BT1 stable isotopes
BT1 isotopes
BT2 neon
BT2 noble gases
SA Ne-22/Ne-21

Ne-21/Ne-22
use Ne-22/Ne-21

Ne-22 (1978)
Autoposting of broader terms began in 1989. This term has multiple hierarchies.
BT1 stable isotopes
BT1 isotopes
BT2 neon
BT2 noble gases
SA Ne-22/Ne-20
SA Ne-22/Ne-21

Ne-22/Ne-20 (1985)
Autoposting of broader terms began in 1989. This term has multiple hierarchies.
UF Ne-20/Ne-22
BT1 stable isotopes
BT1 isotopes
BT2 neon
BT2 noble gases
SA isotope ratios
SA Ne-20
SA Ne-22

Ne-22/Ne-21 (1985)
Autoposting of broader terms began in 1989. This term has multiple hierarchies.
UF Ne-21/Ne-22
BT1 stable isotopes
BT1 isotopes
BT2 neon
BT2 noble gases
SA isotope ratios
SA Ne-21
SA Ne-22

NCRDS
use National Coal Resources Data System

Neanderthal (1978)
Autoposting of Eutheria and Theria to this term began in 1989. As of 1993, Homo is autoposted as a broader term.
UF Homo neanderthalensis
UF Neanderthal man
BT Homo
BT Hominidae
BT Primates
BT Eutheria
BT Theria
BT Mammalia
BT Tetrapoda
BT Vertebrata
BT Chordata
SA fossil man

Neanderthal man
use Neanderthal

Near East (1978)
An indefinite and unofficial term including the countries of the Middle East plus Libya, Sudan, and occasionally Greece.
IN Index countries as applicable.
SA Cyprus
SA Egypt
SA Greece
SA Iraq
SA Jordan
SA Kuwait
SA Libya
SA Oman
SA Qatar
SA Sudan
SA Syria
SA Turkey
SA United Arab Emirates
SA Yemen

near-field (1993)
Includes use in relation to radioactive waste and waste disposal and seismology. From 1989 through 1992, near-field environment was used with radioactive waste and waste disposal; nearfield processes was used with seismology.
UF near-field environment
UF nearfield processes
SA environment
SA far-field
SA ground water
SA landfills
SA near-field spectra
SA pollution
SA seepage
SA seismic methods
SA seismology
SA underground disposal
SA underground installations
SA underground storage
SA waste disposal

near-field environment
No longer a valid term for GeoRef.
use near-field

near-field spectra (1985)
SA attenuation
SA far-field
SA geophysical methods
SA near-field
SA seismic methods
SA seismology
SA spectral analysis

nearfield processes
No longer a valid term for GeoRef.
use near-field

nearshore
A valid term through 1977. After 1977, use nearshore environment or (as of 1981) nearshore sedimentation.

nearshore environment (1978)
Before 1978, search nearshore.
SA depositional environment
SA environment
SA littoral erosion
SA nearshore sedimentation
SA sedimentation
SA tidal currents

nearshore sedimentation (1981)
BT sedimentation
SA nearshore environment

Nebraska (1978)
Autoposting of this term began in 1978.
CO N400000N430000
       W0952000W1040500
BT United States
NT Adams County Nebraska
NT Blaine County Nebraska
NT Boone County Nebraska
NT Butler County Nebraska
NT Cass County Nebraska
NT Chase County Nebraska
NT Cheyenne County Nebraska
NT Clay County Nebraska
NT Colfax County Nebraska
NT Custer County Nebraska
NT Dawson County Nebraska
NT Dodge County Nebraska
NT Douglas County Nebraska
NT Fillmore County Nebraska
NT Franklin County Nebraska
NT Garden County Nebraska
NT Garfield County Nebraska
NT Grant County Nebraska
NT Hall County Nebraska
NT Hamilton County Nebraska
NT Harlan County Nebraska
NT Hitchcock County Nebraska
NT Howard County Nebraska
NT Jefferson County Nebraska
NT Johnson County Nebraska
NT Lancaster County Nebraska
NT Lincoln County Nebraska
NT Logan County Nebraska
NT Madison County Nebraska
NT McPherson County Nebraska
NT Nemaha County Nebraska
NT Pawnee County Nebraska
NT Phelps County Nebraska
NT Pierce County Nebraska
NT Platte County Nebraska
NT Polk County Nebraska
NT Saline County Nebraska
NT Seward County Nebraska
NT Sheridan County Nebraska
NT Sherman County Nebraska
NT Sioux County Nebraska
NT Stanton County Nebraska
NT Thurston County Nebraska
NT Valley County Nebraska
NT Washington County Nebraska
NT Wayne County Nebraska
NT Webster County Nebraska
NT Wheeler County Nebraska
NT York County Nebraska
SA Americus Limestone Member
SA Ash Hollow Formation
SA Beil Limestone Member
SA Benton Formation
SA Brule Formation
SA Cambridge Arch
SA Carlile Shale
SA Chadron Arch
SA Chadron Formation
SA Chase Group
SA Cherokee Group
SA Colorado Group
SA Colorado Lineament
SA Cottonwood Limestone
SA Council Grove Group
SA Dakota Formation
SA Denver Basin
SA Desmoinesian
SA Douglas Group
SA Ervine Creek Limestone
SA Eskridge Shale
SA Forest City Basin
SA Graneros Shale
SA Great Plains
SA Greenhorn Limestone
SA Hughes Creek Shale
SA Kansas City Group
SA Lakota Formation
SA Lawrence Formation
SA Lecompton Limestone
SA Loup Fork Group
SA Midcontinent
SA Midcontinent geophysical anomaly
SA Missouri River
SA Missouri River basin
SA Missouri River valley
SA Nemaha Ridge
SA Newcastle Sandstone
SA Niobrara Formation
SA Ogallala Aquifer
SA Ogallala Formation
SA Oread Limestone
SA Peoria Loess
SA Pierre Shale
SA Platte River
SA Platte River basin
SA Plattsburg Limestone
SA Plattsmouth Limestone Member
SA Red Eagle Limestone
SA Roca Formation
SA Rock Lake Shale Member
SA Sand Hills
SA Shawnee Group
SA Simpson Field
SA Sioux Quartzite
SA South Platte River
SA South Platte River valley
SA Spearfish Formation
SA Stanton Formation
SA Stearns Shale
SA Sundance Formation
SA Valentine Formation
SA Wabaunsee Group
SA White River Group
SA Wreford Limestone
SA Wyandotte Limestone

Nebraskan (1978)
Pertaining to first glacial stage of Pleistocene Epoch in North America.
BT lower Pleistocene
BT Pleistocene
BT Quaternary
BT Cenozoic

Nebrodi Mountains (1978)
N Sicily.
UF Monti Caronie
BT Sicily Italy
BT Italy
BT Southern Europe
BT Europe

nebula, solar
    use solar nebula

Neckar River (1978)
Rises in the Black Forest and flows N past Heidelberg into the Rhine River. Also search Neckar.
BT Baden-Wurttemberg Germany
BT Germany
BT Central Europe
BT Europe

Neda Formation (1985)
SE Wisconsin and NE Iowa.
BT Upper Ordovician
BT Ordovician
BT Paleozoic
SA Iowa
SA Wisconsin

Needle Mountains (1989)
San Juan and La Plata counties, SW Colorado.
IN Index counties as applicable.
SA Colorado
SA La Plata County Colorado
SA San Juan County Colorado

Needles
Valid through 1988. Search in combination with state term. After 1988, use specific city-state term.

Needles California (1989)
Town on the Colorado River in SE California. Before 1989, search Needles AND California.
CO N345100N345100
       W1143600W1143600
BT San Bernardino County California
BT California
BT United States

Neftechala
No longer a valid term for GeoRef. As of 1993, see Neftechala Azerbaidzhan.

Neftechala Azerbaidzhan (1993)
Town in SE Azerbaidzhan. This term has multiple hierarchies.
BT1 Azerbaidzhan
BT1 Europe
BT2 Azerbaidzhan
BT2 Commonwealth of Independent States

Neftyanyye Kamni (1978)
Oil well in Caspian Sea off Apsheron Peninsula. This term has multiple hierarchies.
BT1 Azerbaidzhan
BT1 Europe
BT2 Azerbaidzhan
BT2 Commonwealth of Independent States
SA Caspian Sea

Negaunee Iron Formation (1978)
Upper Peninsula, N Michigan. In Menominee Group.
BT Proterozoic
BT upper Precambrian
BT Precambrian
SA Michigan

Negev (1978)
Desert region in S Israel.
UF Negev Desert
BT Israel
BT Middle East
BT Asia
SA Makhtesh Ramon

Negev Desert
    use Negev

negotiations (1981)
SA economics
SA international cooperation

Negro River
    use Rio Negro

Nei Monggol
Not a valid term for GeoRef.
    use Inner Mongolia China

Nei Mongol
Not a valid term for GeoRef.
    use Inner Mongolia China

neighborite (1978)
BT fluorides
BT halides

Neill Island (1978)
Small island E of South Andaman Island in the Andaman Islands in E Bay of Bengal. Also search Neill AND appropriate area term.
IN Index Andaman Islands as applicable.
SA Andaman Islands

Neiveli India
    use Neyveli India

Nellore
No longer a valid term for GeoRef. See Nellore India.

Nellore India (1993)
City in SE Andhra Pradesh, central India.
BT Andhra Pradesh India
BT India
BT Indian Peninsula
BT Asia

Nellore mica belt (1978)
In Nellore area. Also search Nellore.
SA Andhra Pradesh India

Nelson County
Valid through 1988. Search in combination with state term. After 1988, use specific county-state term.

Nelson County Kentucky (1989)
Central Kentucky. Before 1989, also search Nelson County AND Kentucky.
CO N373000N375900
       W0850800W0854500
BT Kentucky
BT United States

Nelson County North Dakota (1989)
E central North Dakota. Before 1989, also search Nelson County AND North Dakota.
CO N474100N481200
       W0975300W0983300
BT North Dakota
BT United States

Nelson County Virginia (1989)
Central Virginia. Before 1989, also search Nelson County AND Virginia.
CO N373200N380300
       W0783800W0791000
BT Virginia
BT United States

Nelson New Zealand (1993)
Provincial district in NW South Island.
IN Also index South Island.
BT New Zealand
BT Australasia
NT Inangahua New Zealand
SA South Island

Nelson River (1978)
Flows NE out of N Lake Winnipeg into Hudson Bay at Fort Nelson.
BT Manitoba
BT Western Canada
BT Canada

Nelson River basin (1978)
N central Manitoba.
BT Manitoba
BT Western Canada
BT Canada

Nemaha County
Valid through 1988. Search in combination with state term. After 1988, use specific county-state term.

Nemaha County Kansas (1989)
NE Kansas. Before 1989, also search Nemaha County AND Kansas.
CO N393500N400000
       W0954800W0961400

BT Kansas
BT United States
Nemaha County Nebraska (1989)
SE Nebraska. Before 1989, also search Nemaha County AND Nebraska.
CO N401500N403100 W0952900W0960500
BT Nebraska
BT United States
Nemaha Ridge (1978)
IN Index states as applicable.
BT United States
SA Kansas
SA Nebraska
SA Oklahoma
Neman River basin (1978)
Also search Neman AND appropriate area term. This term has multiple hierarchies.
IN Index countries as applicable.
UF Memel River basin
UF Niemen River basin
UF Nyeman River basin
BT1 Commonwealth of Independent States
BT2 Europe
SA Belarus
SA Lithuania
Nematoida (1978)
BT worms
BT Invertebrata
Nematomorpha (1978)
BT worms
BT Invertebrata
Nemerta (1978)
BT worms
BT Invertebrata
Nemuro Peninsula (1978)
in Nemuro sub-prefecture, E Hokkaido. Also search Nemuro.
BT Hokkaido
BT Japan
BT Far East
BT Asia
nenadkevichite (1978)
From 1978-1980, orthosilicates and oxides were autoposted to this term. Autoposting of niobates began in 1981. In 1985, autoposting of orthosilicates ended, and autoposting of sorosilicates began. As of 1989, oxides, orthosilicates and silicates are autoposted to this term. This term has multiple hierarchies.
BT1 niobates
BT1 oxides
BT2 sorosilicates
BT2 orthosilicates
BT2 silicates
Nenets
No longer a valid term for GeoRef. As of 1993, see Nenets Russian Federation.
Nenets Russian Federation (1993)
National okrug of NE Arkhangelsk Oblast in N European Russian Federation. Before 1993, also search Nenets. This term has multiple hierarchies.
BT1 Arkhangelsk Russian Federation
BT1 Russian Federation
BT1 Commonwealth of Independent States
BT2 Arkhangelsk Russian Federation

BT2 Europe
NT Bolshezemelskaya Tundra
NT Kanin Peninsula
NT Vaygach Island
SA Timan-Pechora region
Nenjiang Formation (1989)
In Songliao Basin of NE China.
BT Cretaceous
BT Mesozoic
SA China
SA Songliao Basin
Neocomian (1978)
Europe.
BT Lower Cretaceous
BT Cretaceous
BT Mesozoic
SA Berriasian
SA Hauterivian
SA Valanginian
neodymium (1978)
Autoposting of rare earths and metals to this term began in 1989.
UF Nd
BT rare earths
BT metals
NT Nd-144/Nd-143
NT Sm-147/Nd-144
SA isotopes
SA Sm/Nd
Neogastropoda (1981)
BT Gastropoda
BT Mollusca
BT Invertebrata
Neogene (1978)
An interval of geologic time incorporating the Miocene and Pliocene of the Tertiary period; the upper Tertiary. Autoposting of this term began in 1978.
BT Tertiary
BT Cenozoic
NT Aguacate Group
NT Browns Park Formation
NT Capistrano Formation
NT Cohansey Formation
NT Cuddalore Series
NT Deschutes Formation
NT Hemphillian
NT Miocene
NT Pliocene
NT Purisima Formation
NT Ridge Route Formation
NT Salt Lake Formation
NT Sisquoc Formation
NT upper Neogene
SA Matuyama Epoch
SA Peace River Formation
SA upper Tertiary
Neoglacial (1985)
A brief episode of glacial advance during the Holocene.
UF Little Ice Age
BT Holocene
BT Quaternary
BT Cenozoic
SA Bull Lake Glaciation
Neogloboquadrina pachyderma (1989)
Species. Autoposting of microfossils and Protista to this term began in 1990. This term has multiple hierarchies.
BT1 Globigerinacea
BT1 Rotaliina
BT1 foraminifera
BT1 Protista
BT1 Invertebrata
BT2 Globigerinacea
BT2 Rotaliina
BT2 foraminifera
BT2 Protista

BT2 microfossils
Neognathodus (1989)
Genus. Autoposting of microfossils to this term began in 1990.
BT Conodonta
BT microfossils
Neogondolella (1985)
Autoposting of microfossils to this term began in 1990.
BT Conodonta
BT microfossils
Neohelikian (1985)
Autoposting of Precambrian to this term ended in 1989. As of 1989, Proterozoic is autoposted to this term. Autoposting of middle Proterozoic, upper Precambrian and Precambrian to this term began in 1990.
BT Helikian
BT middle Proterozoic
BT Proterozoic
BT upper Precambrian
BT Precambrian
Neolithic (1978)
Archaeological classification. This term has multiple hierarchies.
BT1 Stone Age
BT1 Cenozoic
BT2 Holocene
BT2 Quaternary
BT2 Cenozoic
SA archaeology
neon (1978)
Autoposting of noble gases to this term began in 1989.
UF Ne
BT noble gases
NT Ne-20
NT Ne-21
NT Ne-22
NT Ne-22/Ne-20
NT Ne-22/Ne-21
SA isotopes
Neornithes (1978)
Subclass. Autoposting of this term began in 1978.
BT Aves
BT Tetrapoda
BT Vertebrata
BT Chordata
NT Odontornithes
SA Archaeornithes
Neosho River valley (1978)
The lower course of the Neosho is called Grand River in Oklahoma. Also search Neosho River.
IN Index states as applicable.
BT United States
SA Grand River
SA Kansas
SA Oklahoma
neotectonics (1978)
Used for the study of post-Miocene structures and structural history of the Earth's crust. Before 1976, also search paleotectonics.
UF active tectonics
BT tectonics
SA changes of level
SA crust
SA epeirogeny
SA faults
SA folds
SA foliation
SA fractures
SA geodesy
SA geomorphologic effects
SA geomorphology
SA horizontal movements
SA isostasy

SA isostatic rebound
SA reactivation
SA seismology
SA subsidence
SA tectonophysics
SA thrust sheets
SA triangulation
SA uplifts
SA vertical movements
Neotoma (1985)
Genus. Autoposting of Cricetidae, Eutheria and Theria to this term began in 1989. As of 1990, Myomorpha is autoposted to this term.
BT Cricetidae
BT Myomorpha
BT Rodentia
BT Eutheria
BT Theria
BT Mammalia
BT Tetrapoda
BT Vertebrata
BT Chordata
neotypes (1978)
SA fossils
SA type specimens
Nepal (1978)
CO N270000N301500 E0880000E0800000
BT Indian Peninsula
BT Asia
NT Kali Gandaki Valley
SA Himalayas
SA Kosi Basin
SA Kumaun Himalayas
SA Lesser Himalayas
SA Main Boundary Fault
SA Main Central Thrust
SA Mount Everest
SA Siwalik Range
nepheline (1978)
BT nepheline group
BT framework silicates
BT silicates
nepheline basalt (1978)
BT alkali basalts
BT basalts
BT volcanic rocks
BT igneous rocks
SA ankaratrite
SA olivine nephelinite
nepheline group (1978)
Autoposting of this term began in 1978.
BT framework silicates
BT silicates
NT kaliophilite
NT kalsilite
NT nepheline
NT petalite
SA feldspathoid rocks
nepheline syenite (1978)
BT syenites
BT plutonic rocks
BT igneous rocks
SA foyaite
SA kakortokite
SA lujavrite
nephelinite (1978)
From 1978-1980, alkali basalt family was autoposted to this term.
BT alkali gabbros
BT gabbros
BT plutonic rocks
BT igneous rocks
SA alkali basalts
SA ocean-island basalts
SA olivine nephelinite
SA ultramafics

nepheloid layer (1978)
  Also search nepheloid.
  UF layer, nepheloid
  UF nepheloid zone
  SA continental rise
  SA sea water

nepheloid zone
  use nepheloid layer

nephrite (1978)
  BT clinoamphibole
  BT amphibole group
  BT chain silicates
  BT silicates
  SA greenstone
  SA jade

Nephrolepidina (1978)
  Autoposting of microfossils and Protista to this term began in 1990. This term has multiple hierarchies.
  BT1 Orbitoidacea
  BT1 Rotaliina
  BT1 foraminifera
  BT1 Protista
  BT1 Invertebrata
  BT2 Orbitoidacea
  BT2 Rotaliina
  BT2 foraminifera
  BT2 Protista
  BT2 microfossils

nepouite
  use garnierite

Neptune (1978)
  SA Nereid Satellite
  SA outer planets
  SA planetology
  SA planets
  SA satellites
  SA solar system
  SA Triton Satellite
  SA Voyager 2
  SA Voyager Program

neptunite (1978)
  BT chain silicates
  BT silicates

neptunium (1978)
  Autoposting of actinides and metals to this term began in 1989.
  UF Np
  BT actinides
  BT metals

Nerbuda Valley
  use Narmada Valley

Nerbudda River
  use Narmada River

Nereid Satellite (1989)
  One of satellites of Neptune. Before 1989, also search Nereid AND Neptune.
  SA Neptune
  SA satellites
  SA Triton Satellite
  SA Voyager Program

Nereites (1978)
  Genus.
  BT ichnofossils

Neretva Valley (1978)
  River valley in Bosnia and Herzegovina.
  BT Yugoslavia
  BT Southern Europe
  BT Europe
  SA Bosnia-Herzegovina
  SA Croatia

neritic environment
  Not a valid term for GeoRef. Before 1981, was used in subfile B.
  use subtidal environment

nervous system (1981)
  SA anatomy
  SA morphology
  SA Vertebrata

nesosilicates (1981)
  Autoposting of this term to nesosilicate minerals that are not found in the garnet, humite or olivine groups began in 1985.
  BT orthosilicates
  BT silicates
  NT andalusite
  NT braunite
  NT chloritoid
  NT coffinite
  NT cyrtolite
  NT datolite
  NT dumortierite
  NT euclase
  NT eucryptite
  NT gadolinite
  NT garnet group
  NT garrelsite
  NT grandidierite
  NT harkerite
  NT humite group
  NT kasolite
  NT kornerupine
  NT kyanite
  NT laihunite
  NT larnite
  NT melanocerite
  NT merwinite
  NT mullite
  NT olivine group
  NT phenakite
  NT sapphirine
  NT sillimanite
  NT spurrite
  NT staurolite
  NT stillwellite
  NT thaumasite
  NT thorite
  NT titanite
  NT topaz
  NT uranophane
  NT viridine
  NT willemite
  NT zircon

nesquehonite (1978)
  BT carbonates

Ness County
  Valid through 1988. Search in combination with state term. After 1988, use specific county-state term.

Ness County Kansas (1989)
  W central Kansas. Before 1989, also search Ness County AND Kansas.
  CO N381500N384200
      W0993700W1001500
  BT Kansas
  BT United States

Netherland India
  No longer a valid term for GeoRef. Before 1945, was used on level 1 in subfile E.
  use Indonesia

Netherlands (1978)
  CO N504500N533000
      E0071500E0031500
  UF Holland
  BT Western Europe
  BT Europe
  NT Amsterdam Netherlands
  NT Groningen Netherlands
  NT Limburg Netherlands
  NT Overijssel Netherlands
  NT Utrecht Netherlands
  NT Wadden Zee
  NT Zeeland Netherlands
  SA Campine
  SA Central Graben
  SA Delta Area
  SA European Platform
  SA Lower Rhine Basin
  SA Meuse River
  SA Meuse Valley
  SA North Sea Coast
  SA North Sea region
  SA Rhine Basin
  SA Rhine River
  SA Rhine Valley
  SA Scheldt River

Netherlands Antilles (1978)
  Formerly known as Curacao territory. Islands of Aruba, Bonaire, and Curacao in the Caribbean Sea off the coast of Venezuela, plus the Dutch section of Saint Martin at N end of Leeward Islands. Before 1981, broader term was Caribbean region. Before 1969, also search Dutch West Indies. In 1986, Aruba was separated from the group.
  UF Dutch West Indies
  BT Lesser Antilles
  BT Antilles
  BT West Indies
  BT Caribbean region
  NT Bonaire
  NT Curacao
  SA Leeward Islands
  SA Rincon Formation

Netherlands Guiana
  use Surinam

Netherlands New Guinea
  use Irian Jaya Indonesia

network deposits
  use stockwork deposits

networks (1978)
  As of 1993, see computer networks, geodetic networks, and seismic networks if applicable. Before 1993, also search arrays.
  SA arrays
  SA computer networks
  SA geodetic networks
  SA monitoring
  SA seismic networks
  SA trees

Neubrandenburg Bezirk
  No longer a valid term for GeoRef. District in NE East Germany. See Germany.

Neuburg
  No longer a valid term for GeoRef. See Neuburg Germany.

Neuburg an der Donau
  use Neuburg Germany

Neuburg Germany (1993)
  City in W central Bavaria. Before 1993, also search Neuburg. Before 1978, also search Neuburg an der Donau.
  UF Neuburg an der Donau
  BT Bavaria Germany
  BT Germany
  BT Central Europe
  BT Europe

Neuchatel
  No longer a valid term for GeoRef. As of 1993, see Neuchatel Switzerland.

Neuchatel Switzerland (1993)
  Canton in W Switzerland. Also a city on Lake Neuchatel. Before 1993 also search Neuchatel AND Switzerland.
  BT Switzerland
  BT Central Europe
  BT Europe
  SA Jura Mountains

Neuquen
  No longer a valid term for GeoRef. See Neuquen Argentina.

Neuquen Argentina (1993)
  Province in W central Argentina.
  CO S410000S360000
      W0670000W0730000
  UF Neuquen Province
  BT Argentina
  BT South America
  SA Neuquen Basin

Neuquen Basin (1978)
  River basin in W central Argentina.
  IN Index provinces as applicable.
  BT Argentina
  BT South America
  SA Neuquen Argentina
  SA Rio Negro Argentina

Neuquen Province
  use Neuquen Argentina

Neuropteris (1978)
  BT Pteridospermae
  BT gymnosperms
  BT Spermatophyta
  BT Plantae
  SA pteridophytes

Neuropteroida (1981)
  BT Insecta
  BT Mandibulata
  BT Arthropoda
  BT Invertebrata

Neuse River (1978)
  Rises in the Piedmont and flows SE into Pamlico Sound, E central North Carolina.
  SA North Carolina

Neusiedler Lake
  use Lake Neusiedler

Neusiedlersee
  use Lake Neusiedler

neutral stress
  use pore pressure

neutron
  A valid term through mid-1978 used as a modifier for activation analysis. Use neutron activation analysis.

neutron activation analysis (1978)
  Before 1978, also search activation analysis AND neutron.
  SA activation analysis
  SA analysis
  SA chemical analysis
  SA neutron activation analysis data
  SA neutron probe
  SA quantitative analysis
  SA spectroscopy

neutron activation analysis data (1985)
  Use for data. Use neutron activation analysis for methodology.
  SA data
  SA neutron activation analysis

neutron diffraction analysis (1978)
  SA analysis
  SA crystallography
  SA diffraction
  SA diffractograms
  SA electron diffraction analysis
  SA neutron diffraction data

SA neutron probe
SA X-ray diffraction analysis

neutron diffraction analysis data
  use neutron diffraction data

neutron diffraction data (1989)
  Used only for data. For methodology, use neutron diffraction analysis.
  UF neutron diffraction analysis data
  SA data
  SA neutron diffraction analysis

neutron logging
  use neutron methods

neutron methods (1985)
  UF neutron logging
  SA geophysical methods
  SA methods
  SA neutron-gamma methods
  SA neutron-neutron methods
  SA radioactivity
  SA radioactivity methods
  SA well-logging

neutron probe (1982)
  Used for methodology. For data, use neutron probe data. Before 1985, also search neutron spectroscopy.
  UF neutron spectroscopy
  UF probe, neutron
  SA analysis
  SA chemical analysis
  SA electron probe
  SA ion probe
  SA neutron activation analysis
  SA neutron diffraction analysis
  SA neutron probe data
  SA spectroscopy

neutron probe data (1982)
  Used for data. For methodology, use neutron probe.
  SA data
  SA neutron probe
  SA spectroscopy

neutron spectroscopy
  use neutron probe

neutron-gamma methods (1984)
  BT geophysical methods
  SA gamma rays
  SA gamma-gamma methods
  SA gamma-ray methods
  SA neutron methods
  SA neutron-neutron methods
  SA radioactivity
  SA well-logging

neutron-neutron methods (1984)
  BT geophysical methods
  SA neutron methods
  SA neutron-gamma methods
  SA radioactivity
  SA well-logging

neutrons (1978)
  Used as a general term.
  SA electrons
  SA particles

Nevada (1978)
  Autoposting of this term began in 1978.
  CO N350000N420000
     W1140500W1200000
  BT United States
  NT Arrow Canyon Range
  NT Carlin Mine
  NT Churchill County Nevada
  NT Clark County Nevada
  NT Douglas County Nevada
  NT Egan Range
  NT Elko County Nevada
  NT Esmeralda County Nevada
  NT Eureka County Nevada
  NT Humboldt County Nevada
  NT Humboldt River valley
  NT Lake Mead Fault
  NT Lander County Nevada
  NT Lincoln County Nevada
  NT Lyon County Nevada
  NT Mineral County Nevada
  NT Mormon Mountains
  NT Nevada Test Site
  NT Nye County Nevada
  NT Pershing County Nevada
  NT Shoshone Mountains
  NT Storey County Nevada
  NT Toiyabe Range
  NT Toquima Range
  NT Walker River
  NT Washoe County Nevada
  NT White Pine County Nevada
  NT Yucca Flat
  SA Amargosa Desert
  SA Antelope Valley Limestone
  SA Antler Orogeny
  SA Aztec Sandstone
  SA Basin and Range Province
  SA Battle Mountain
  SA Battle Mountain High
  SA Bird Spring Formation
  SA Black Rock Desert
  SA Bonanza King Formation
  SA Bright Angel Shale
  SA Bullfrog Member
  SA Carrara Formation
  SA Chainman Shale
  SA Chinle Formation
  SA Coconino Sandstone
  SA Colorado River
  SA Colorado River basin
  SA Columbia River basin
  SA Cortez Mountains
  SA Crater Flat Tuff
  SA Death Valley
  SA Diamond Peak Formation
  SA Dunderberg Shale
  SA Ely Limestone
  SA Ely Springs Dolomite
  SA Eureka Quartzite
  SA Fairview Peak
  SA Fillmore Formation
  SA Fish Haven Dolomite
  SA Grant Range
  SA Great Basin
  SA Hanson Creek Formation
  SA Joana Limestone
  SA Johnnie Formation
  SA Kaibab Formation
  SA Lake Bonneville
  SA Lake Lahontan
  SA Lake Mead
  SA Lake Tahoe
  SA Long Valley
  SA Luning Formation
  SA McCoy Creek Group
  SA Moenave Formation
  SA Moenkopi Formation
  SA Monte Cristo Limestone
  SA Mount Jefferson
  SA Muddy Creek Formation
  SA Nopah Formation
  SA Opeche Shale
  SA Paintbrush Tuff
  SA Petrified Forest Member
  SA Pilot Range
  SA Pilot Shale
  SA Pinon Range
  SA Pioche Shale
  SA Pogonip Group
  SA Retort Phosphatic Shale Member
  SA Roberts Mountains Formation
  SA Ruby Mountains
  SA Salt Lake Formation
  SA Schoonover Sequence
  SA Shadow Mountains
  SA Sheep Pass Formation
  SA Shinarump Member
  SA Snake River basin
  SA Steamboat Springs
  SA Stirling Quartzite
  SA Sunrise Formation
  SA Supai Formation
  SA Tapeats Sandstone
  SA Topopah Spring Member
  SA Tor Formation
  SA Toroweap Formation
  SA Truckee River
  SA Vinini Formation
  SA Virgin River valley
  SA White Mountains
  SA Wildcat Group
  SA Wood Canyon Formation
  SA Zabriskie Quartzite

Nevada County
  Valid through 1988. Search in combination with state term. After 1988, use specific county-state term.

Nevada County Arkansas (1989)
  SW Arkansas. Before 1989, also search Nevada County AND Arkansas.
  CO N332300N335500
     W0930700W0932600
  BT Arkansas
  BT United States

Nevada County California (1989)
  E California. Before 1989, also search Nevada County AND California.
  CO N390100N392900
     W1200000W1211600
  BT California
  BT United States

Nevada Test Site (1978)
  NW of Las Vegas in S Nevada.
  IN Index counties as applicable.
  CO N363500N375200
     W1152000W1170500
  BT Nevada
  BT United States
  SA Rainier Mesa
  SA Yucca Flat
  SA Yucca Mountain

Nevadan Orogeny (1985)
  Jurassic to Lower Cretaceous. Before 1985, also search orogeny AND Nevadan.
  IN Index ages as applicable.
  BT Mesozoic
  SA Jurassic
  SA Lower Cretaceous
  SA orogeny
  SA tectonics

Nevado del Ruiz (1989)
  Volcano. Tolima Department, W central Colombia.
  CO N045300N045300
     W0752200W0752200
  BT Colombia
  BT South America

New Albany Shale (1978)
  Includes Devonian Blocher and Blackiston formations, and Mississippian Sanderson, Underwood, and Henryville formations. Indiana and N central Kentucky.
  BT Upper Devonian
  BT Devonian
  BT Paleozoic
  SA Indiana
  SA Kentucky

New Austrian Tunnelling Method (1985)
  SA excavations
  SA methods
  SA mining
  SA tunnels

New Britain (1978)
  Largest island in the Bismarck Archipelago NE of New Guinea. Before 1972, also search New Britain Island.
  UF New Britain Island
  BT Papua New Guinea
  BT Australasia
  SA Bismarck Archipelago
  SA Rabaul Caldera

New Britain Island
  Not a valid GeoRef index term after 1971. Was used on level 1 in subfile G.
  use New Britain

New Brunswick (1978)
  CO N450000N480500
     W0634500W0690000
  BT Maritime Provinces
  BT Eastern Canada
  BT Canada
  NT Carleton County New Brunswick
  NT Charlotte County New Brunswick
  NT Fredericton New Brunswick
  NT Gloucester County New Brunswick
  NT Kouchibouguac Bay
  NT Miramichi Bay
  NT Moncton Basin
  NT Northumberland County New Brunswick
  NT Saint John County New Brunswick
  NT Victoria County New Brunswick
  SA Albert Formation
  SA Avalon Terrane
  SA Avalon Zone
  SA Chaleur Bay
  SA Coldbrook Group
  SA Cumberland Basin
  SA Cumberland Group
  SA Gander Zone
  SA Horton Group
  SA Matapedia Group
  SA Midland Valley
  SA Miramichi earthquake 1982
  SA Mispec Group
  SA Pictou Group
  SA Restigouche Estuary
  SA Saint George Batholith
  SA Saint John River
  SA Tetagouche Group
  SA Windsor Group

New Brunswick earthquake 1982
  use Miramichi earthquake 1982

New Caledonia (1978)
  French overseas territory E of Queensland, Australia. Also its main island. In 1981, broader term changed from Pacific Ocean to Melanesia.
  CO S223000S200000
     E1670000E1640000
  BT Melanesia
  BT Oceania
  NT Noumea New Caledonia
  NT Ouegoa New Caledonia
  SA Pacific Ocean

New Caledonia Basin (1978)
  Between central Lord Howe Rise and Norfolk Island in SW Pacific Ocean.
  BT Pacific Ocean

New Castile
No longer a valid term for GeoRef. As of 1993, see New Castile Spain.

New Castile Spain (1993)
Region in central Spain. Before 1993, also search New Castile and Spain.
CO N382100N412000
    E0010800W0052500
UF Castilla la Nueva
BT Castile Spain
BT Spain
BT Iberian Peninsula
BT Southern Europe
BT Europe
NT Ciudad Real Spain
NT Cuenca Spain
NT Guadalajara Spain
NT Madrid Spain
NT Toledo Spain

New Castle County
Valid through 1988. Search in combination with state term. After 1988, use specific county-state term.

New Castle County Delaware (1989)
N Delaware. Before 1989, also search New Castle County AND Delaware.
CO N391700N395200
    W0752500W0754800
BT Delaware
BT United States
NT Wilmington Delaware

new data (1978)
Used as a general term.
SA data

New Delhi
No longer a valid term for GeoRef. See New Delhi India.

New Delhi India (1993)
Capital city SSW of Delhi in Delhi, N central India.
BT Delhi India
BT India
BT Indian Peninsula
BT Asia

new energy sources (1981)
SA biomass
SA energy
SA energy sources
SA heat storage
SA solar energy
SA tidal energy
SA wind energy

New England (1978)
As of 1993, term is used only for broad general discussions. From 1981 through 1992, this term was autoposted to its narrower terms: Connecticut, Maine, Massachusetts, New Hampshire, Rhode Island, and Vermont.
CO N420000N473000
    W0670000W0733000
BT United States
SA Lydonia Canyon
SA Narragansett Basin

New England Batholith (1978)
SA Australia
SA New England Range
SA New South Wales Australia
SA Queensland Australia

New England fold belt
use New England Orogeny

New England Orogen
use New England Orogeny

New England Orogeny (1989)
Devonian to Triassic orogeny affecting New South Wales and Queensland, Australia. Before 1993, also search New England fold belt.
IN Index ages as applicable.
UF New England fold belt
UF New England Orogen
SA Australia
SA Mesozoic
SA New England Range
SA New South Wales Australia
SA orogeny
SA Paleozoic
SA Queensland Australia

New England Plateau
use New England Range

New England Range (1981)
NE New South Wales. Part of the Great Dividing Range. Before 1981, also search New England AND Australia.
IN Index New South Wales Australia or region as applicable.
UF New England Plateau
SA New England Batholith
SA New England Orogeny
SA New South Wales Australia

New England Seamounts (1989)
W North American Atlantic. As of 1990, Atlantic Ocean is autoposted to this term.
CO N340000N400000
    W0560000W0680000
BT North American Atlantic
BT North Atlantic
BT Atlantic Ocean
SA DSDP Site 382
SA DSDP Site 385

new genera
A valid term through 1973.
use new taxa

new genus
A valid term through 1973.
use new taxa

New Guinea (1978)
Refers to whole island. In 1981, broader term changed from Australasia to Malay Archipelago.
IN Index Irian Jaya and/or Papua New Guinea.
CO S101500S023000
    E1510000E1410000
BT Malay Archipelago
NT Irian Jaya Indonesia
SA Australasia
SA Australian Plate
SA Indo-Australian Plate
SA Pacific mobile belt
SA Papua New Guinea

New Hampshire (1978)
Autoposting of this term began in 1978.
CO N424500N452000
    W0704500W0723500
BT United States
NT Carroll County New Hampshire
NT Coos County New Hampshire
NT Grafton County New Hampshire
NT Hillsborough County New Hampshire
NT Rockingham County New Hampshire
NT Sullivan County New Hampshire
SA Ammonoosuc Volcanics
SA Avalon Terrane
SA Berwick Formation
SA Bronson Hill Anticlinorium
SA Chelmsford Granite
SA Clough Formation
SA Connecticut River
SA Connecticut Valley
SA Dedham Granodiorite
SA Gile Mountain Formation
SA Great Bay
SA Kittery Formation
SA Littleton Formation
SA Madrid Formation
SA Merrimack Group
SA Merrimack River valley
SA Merrimack Synclinorium
SA Mount Adams
SA Mount Jefferson
SA Partridge Formation
SA Perry Mountain Formation
SA Presumpscot Formation
SA Rangeley Formation
SA Rangeley Lakes
SA Rye Formation
SA Saco River
SA Waits River Formation
SA White Mountains

New Hanover County
Valid through 1988. Search in combination with state term. After 1988, use specific county-state term.

New Hanover County North Carolina (1989)
On the Atlantic Ocean in SE North Carolina. Before 1989, also search New Hanover County AND North Carolina.
CO N335500N342500
    W0774500W0780500
BT North Carolina
BT United States
NT Castle Hayne North Carolina
NT Wilmington North Carolina

New Haven County
Valid through 1988. Search in combination with state term. After 1988, use specific county-state term.

New Haven County Connecticut (1989)
S Connecticut. Before 1989, also search New Haven County AND Connecticut.
CO N411000N413800
    W0723200W0732100
BT Connecticut
BT United States

New Hebrides
As of 1985, no longer a valid term for GeoRef.
use Vanuatu

New Ireland (1978)
Island in the Bismarck Archipelago NNE of New Britain in SW Pacific.
BT Papua New Guinea
BT Australasia
SA Bismarck Archipelago

New Jersey (1978)
Autoposting of this term began in 1978.
CO N385500N412100
    W0735300W0753500
BT United States
NT Atlantic County New Jersey
NT Bergen County New Jersey
NT Burlington County New Jersey
NT Camden County New Jersey
NT Cape May County New Jersey
NT Cumberland County New Jersey
NT Essex County New Jersey
NT Gloucester County New Jersey
NT Hudson County New Jersey
NT Hunterdon County New Jersey
NT Mercer County New Jersey
NT Middlesex County New Jersey
NT Monmouth County New Jersey
NT Morris County New Jersey
NT New Jersey Highlands
NT Ocean County New Jersey
NT Passaic County New Jersey
NT Passaic River
NT Passaic River basin
NT Pine Barrens
NT Raritan River
NT Salem County New Jersey
NT Somerset County New Jersey
NT Sussex County New Jersey
NT Union County New Jersey
NT Warren County New Jersey
NT Watchung Mountains
SA Atlantic Coastal Plain
SA Brunswick Formation
SA Carolina Bays
SA Cohansey Formation
SA Delaware Bay
SA Delaware River
SA Delaware River basin
SA Feltville Formation
SA Glenarm Series
SA Great Bay
SA Hudson River
SA Hudson Valley
SA Kirkwood Formation
SA Lockatong Formation
SA Long Valley
SA Magothy Aquifer
SA Magothy Formation
SA Marcellus Shale
SA Marshalltown Formation
SA Martinsburg Formation
SA Monmouth Group
SA Navesink Formation
SA Newark Basin
SA Newark Supergroup
SA Newark-Gettysburg Basin
SA Palisades Sill
SA Passaic Formation
SA Piedmont
SA Piney Point Formation
SA Ramapo Fault
SA Raritan Bay
SA Raritan Formation
SA Reading Prong
SA Shawangunk Formation
SA Stockton Formation
SA Valley and Ridge Province
SA Woodbury Clay

New Jersey Highlands (1989)
N New Jersey, in Hunterdon, Morris, Sussex and Warren counties. NE-SW trending ridges including Musconetcong Mountain, Schooleys Mountain and the Ramapo Mountains.
IN Index counties as applicable.
CO N404000N411000
    W0742500W0751000
BT New Jersey
BT United States
SA Hunterdon County New Jersey
SA Long Valley
SA Morris County New Jersey
SA Sussex County New Jersey
SA Warren County New Jersey

New Liskeard
No longer a valid term for GeoRef. As of 1993, see New Liskeard Ontario.

New Liskeard Ontario (1993)
Town in SE Ontario. Before 1993, also search New Liskeard AND Ontario.
BT Timiskaming District Ontario
BT Ontario

BT Eastern Canada
BT Canada

New London County
Valid through 1988. Search in combination with state term. After 1988, use specific county-state term.

New London County Connecticut (1989)
SE Connecticut. Before 1989, also search New London County AND Connecticut.
CO N411700N414300
W0714800W0722800
BT Connecticut
BT United States
NT Honey Hill Fault
SA Thames River

New Madrid
Valid through 1988. Search in combination with state term. After 1988, use specific city-state term.

New Madrid County
Valid through 1988. Search in combination with state term. After 1988, use specific county-state term.

New Madrid County Missouri (1989)
SE Missouri. Before 1989, also search New Madrid County AND Missouri.
CO N362200N365300
W0891900W0895800
BT Missouri
BT United States
NT New Madrid Missouri

New Madrid earthquakes 1811-1812 (1993)
Epicenter near New Madrid, Missouri. Before 1993, search New Madrid region and earthquakes.
BT earthquakes
SA Mississippi Embayment
SA New Madrid region

New Madrid Missouri (1989)
Town on Mississippi River in extreme SE Missouri. Before 1989, search New Madrid AND Missouri.
CO N363400N363400
W0893200W0893200
BT New Madrid County Missouri
BT Missouri
BT United States

New Madrid region (1981)
Region of earthquakes in the Mississippi Embayment. Before 1993, also search New Madrid Seismic Zone and variations of New Madrid earthquake. After 1992, see New Madrid earthquakes 1811-1812, for discussion of the 1811-1812 earthquakes.
IN Index states as applicable.
UF New Madrid Seismic Zone
BT United States
SA Arkansas
SA earthquakes
SA Illinois
SA Indiana
SA Kentucky
SA Mississippi
SA Mississippi Embayment
SA Missouri
SA New Madrid earthquakes 1811-1812
SA Tennessee

New Madrid Seismic Zone
use New Madrid region

new methods (1978)
SA absolute age
SA methods

New Mexico (1978)
Autoposting of this term began in 1978.
CO N313000N370000
W1030000W1090500
BT United States
NT Ambrosia Lake
NT Animas Mountains
NT Bernalillo County New Mexico
NT Catron County New Mexico
NT Chama Basin
NT Chaves County New Mexico
NT Cibola County New Mexico
NT Colfax County New Mexico
NT Curry County New Mexico
NT Datil-Mogollon volcanic field
NT De Baca County New Mexico
NT Dona Ana County New Mexico
NT Eddy County New Mexico
NT Grant County New Mexico
NT Grants mineral belt
NT Guadalupe County New Mexico
NT Harding County New Mexico
NT Hidalgo County New Mexico
NT Jemez Lineament
NT Jornada del Muerto
NT Lea County New Mexico
NT Lincoln County New Mexico
NT Los Alamos County New Mexico
NT Luna County New Mexico
NT Manzano Mountains
NT McKinley County New Mexico
NT Mora County New Mexico
NT Otero County New Mexico
NT Picuris Range
NT Quay County New Mexico
NT Rio Arriba County New Mexico
NT Roosevelt County New Mexico
NT San Juan County New Mexico
NT San Miguel County New Mexico
NT Sandia Mountains
NT Sandoval County New Mexico
NT Santa Fe County New Mexico
NT Sierra County New Mexico
NT Socorro County New Mexico
NT Taos County New Mexico
NT Taos Plateau
NT Tatum Basin
NT Torrance County New Mexico
NT Tucumcari Basin
NT Tularosa Basin
NT Tusas Mountains
NT Union County New Mexico
NT Valencia County New Mexico
NT Valle Grande Mountains
NT Valles Caldera
NT Zuni Mountains
SA Abo Formation
SA Albuquerque Basin
SA Baca Formation
SA Bandelier Tuff
SA Basin and Range Province
SA Bear Mountain
SA Bell Canyon Formation
SA Benton Formation
SA Bidahochi Formation
SA Bisbee Group
SA Black Range
SA Blue Mountain
SA Bone Spring Limestone
SA Brushy Basin Shale Member
SA Burro Canyon Formation
SA Caballo Mountains
SA Canadian River
SA Capitan Formation
SA Carlile Shale
SA Carmel Formation
SA Castile Formation
SA Central Basin Platform
SA Cherry Canyon Formation
SA Chihuahua tectonic belt
SA Chinle Formation
SA Codell Sandstone Member
SA Colorado Group
SA Colorado Plateau
SA Colorado River basin
SA Comanche Peak Limestone
SA Crevasse Canyon Formation
SA Culebra Dolomite Member
SA Cutler Formation
SA Dakota Formation
SA Delaware Basin
SA Dockum Group
SA El Paso Group
SA Ellenburger Group
SA Entrada Sandstone
SA Espanola Basin
SA Fort Hays Limestone Member
SA Four Corners
SA Fredericksburg Group
SA Fruitland Formation
SA Fusselman Dolomite
SA Gallup Sandstone
SA Ghost Ranch
SA Gila River
SA Glen Canyon Group
SA Glorieta Sandstone
SA Graneros Shale
SA Grayburg Formation
SA Great Plains
SA Greenhorn Limestone
SA Guadalupe Mountains
SA Hermosa Formation
SA High Plains Aquifer
SA Hueco Limestone
SA Hueco Mountains
SA Kirtland Shale
SA La Ventana Sandstone
SA Laborcita Formation
SA Lake Valley Formation
SA Lewis Shale
SA Llano Estacado
SA Loup Fork Group
SA Madera Formation
SA Magdalena Group
SA Mancos Shale
SA Menefee Formation
SA Mesa Rica Sandstone
SA Mesaverde Group
SA Midcontinent
SA Montana Group
SA Montoya Group
SA Morrison Formation
SA Nacimiento Formation
SA Navajo Indian Reservation
SA Navajo Sandstone
SA Niobrara Formation
SA Ogallala Aquifer
SA Ogallala Formation
SA Ojo Alamo Sandstone
SA Orogrande Basin
SA Ortega Group
SA Palisades Sill
SA Palo Duro Basin
SA Paradox Basin
SA Pasamonte Meteorite
SA Pecos River
SA Pecos River valley
SA Pedregosa Basin
SA Permian Basin
SA Petrified Forest Member
SA Pictured Cliffs Sandstone
SA Point Lookout Sandstone
SA Queen Formation
SA Raton Basin
SA Raton Formation
SA Red Hill
SA Red River
SA Red River valley
SA Redonda Formation
SA Rio Grande
SA Rio Grande Rift
SA Rio Grande Valley
SA Rio Puerco
SA Rustler Formation
SA Salado Formation
SA Salt Wash Sandstone Member
SA San Andres Formation
SA San Francisco Mountains
SA San Jose Formation
SA San Juan Basin
SA San Juan Mountains
SA San Juan River
SA San Luis Valley
SA San Mateo Mountains
SA San Rafael Group
SA Sandia Formation
SA Sandia Granite
SA Sangre de Cristo Mountains
SA Santa Fe Group
SA Santa Rosa Sandstone
SA Seven Rivers Formation
SA Shinarump Member
SA Sierra Blanca
SA Smoky Hill Chalk Member
SA Strawn Series
SA Supai Formation
SA Tansill Formation
SA Todilto Formation
SA Trans-Pecos
SA U. S. Rocky Mountains
SA Uncompahgre Uplift
SA Ute Creek
SA Vadito Group
SA Wasatch Formation
SA Western Overthrust Belt
SA Westwater Canyon Sandstone Member
SA White Sands
SA Wingate Sandstone
SA Wolfcampian
SA Yates Formation
SA Yeso Formation

new minerals (1978)
Used to indicate new mineral names.
SA minerals

new names (1978)
Used as a general term.
SA nomenclature
SA taxonomy

New Orleans
Valid through 1988. Search in combination with state term. After 1988, use specific city-state term.

New Orleans Louisiana (1989)
City on the Mississippi River in Orleans Parish, SE Louisiana. Before 1989, search New Orleans AND Louisiana.
CO N300000N300000
W0900300W0900300
BT Orleans Parish Louisiana
BT Louisiana
BT United States

New Quebec
use Ungava

New Red Sandstone (1978)
England. Permian and Triassic.
SA England
SA Permian
SA Triassic

New River (1978)
Rises in NW North Carolina and joins with Gauley River in West Virginia to form the Kanawha River. Also in other locations.
IN Index counties or regions as applicable.
SA North Carolina
SA United States
SA Virginia
SA West Virginia

New River Formation (1989)

In Pottsville Group.
BT Lower Pennsylvanian
BT Pennsylvanian
BT Carboniferous
BT Paleozoic
SA Alabama
SA Georgia
SA Pennsylvania
SA Pottsville Group
SA Virginia
SA West Virginia

New Siberian Islands (1981)
Island group in Arctic Ocean between Laptev Sea and East Siberian Sea. Part of Yakutia. This term has multiple hierarchies.
CO N730000N770000
    E1510000E1350000
BT1 Yakutia Russian Federation
BT1 Russian Federation
BT1 Commonwealth of Independent States
BT2 Yakutia Russian Federation
BT2 Asia
BT3 Russian Arctic
BT3 Russian Federation
BT3 Commonwealth of Independent States
BT4 Russian Arctic
BT4 Arctic region

New South Wales
No longer a valid term for GeoRef. See New South Wales Australia.

New South Wales Australia (1993)
Before 1993, also search New South Wales.
CO S373000S281500
    E1533000E1410000
BT Australia
BT Australasia
NT Armidale Australia
NT Australian Capital Territory
NT Bathurst Australia
NT Bungonia Caves
NT Cooma Australia
NT Gosford Australia
NT Gunnedah Basin
NT Macleay River
NT Manildra Australia
NT Murrumbidgee River
NT Murrumbidgee Valley
NT Nandewar Mountains
NT Newcastle Australia
NT Southern coal field
NT Sydney Australia
NT Tamworth Australia
NT Wagga Wagga Australia
NT Yass
NT Yeoval Australia
SA Broken Bay
SA Broken Hill
SA Broken Hill Block
SA Broken Hill Mine
SA Bulldog Shale
SA Bulli Seam
SA Great Artesian Basin
SA Hawkesbury Sandstone
SA Hunter Valley
SA Illawarra Coal Measures
SA Lachlan fold belt
SA Murray Basin
SA Murray River
SA Narrabeen Group
SA New England Batholith
SA New England Orogeny
SA New England Range
SA Newcastle Coal Measures
SA Round Mountain
SA Singleton Coal Measures
SA Snowy Mountains
SA Southern Highlands

SA Sydney Basin
SA Tasman Geosyncline
SA Tasman orogenic zone
SA Toolebuc Formation
SA Willyama Complex

new species
A valid term through 1973.
use new taxa

new taxa (1978)
Before 1974, also search new genera; new genus; new species as applicable.
IN Should be used whenever a new taxonomic group is discussed.
UF new genera
UF new genus
UF new species
UF taxa, new
SA revision
SA taxonomy

New Ulm
Valid through 1988. Search in combination with state term. After 1988, use specific city-state term.

New Ulm Minnesota (1989)
City in S central Minnesota. Before 1989, search New Ulm AND Minnesota.
CO N441900N441900
    W0942800W0942800
BT Brown County Minnesota
BT Minnesota
BT United States

New World Island (1978)
In Notre Dame Bay off NE Newfoundland.
BT Newfoundland Island
BT Newfoundland
BT Eastern Canada
BT Canada

New York (1978)
Autoposting of this term began in 1978.
CO N403000N450000
    W0715500W0794500
BT United States
NT Adirondack Mountains
NT Albany County New York
NT Allegany County New York
NT Broome County New York
NT Catskill Mountains
NT Cattaraugus County New York
NT Cayuga County New York
NT Chautauqua County New York
NT Chemung County New York
NT Chenango River
NT Clinton County New York
NT Columbia County New York
NT Delaware County New York
NT Dutchess County New York
NT Erie County New York
NT Essex County New York
NT Finger Lakes
NT Franklin County New York
NT Fulton County New York
NT Genesee County New York
NT Greene County New York
NT Hamilton County New York
NT Herkimer County New York
NT Jefferson County New York
NT Lewis County New York
NT Livingston County New York
NT Madison County New York
NT Mohawk Valley
NT Monroe County New York
NT Montgomery County New York
NT Nassau County New York
NT New York Bight
NT New York City New York
NT Niagara County New York
NT Oneida County New York

NT Oneida Lake
NT Onondaga County New York
NT Orange County New York
NT Oswego County New York
NT Otsego County New York
NT Putnam County New York
NT Rensselaer County New York
NT Rockland County New York
NT Saint Lawrence County New York
NT Saratoga County New York
NT Schoharie County New York
NT Seneca County New York
NT Suffolk County New York
NT Sullivan County New York
NT Tioga County New York
NT Tompkins County New York
NT Ulster County New York
NT Warren County New York
NT Washington County New York
NT Wayne County New York
NT Westchester County New York
NT Wyoming County New York
NT Yates County New York
SA Adirondack Anorthosite
SA Allegheny Plateau
SA Appalachian Basin
SA Appalachian Plateau
SA Atlantic Coastal Plain
SA Bass Islands Dolomite
SA Bear Mountain
SA Beekmantown Group
SA Black River Group
SA Blue Mountain Lake
SA Canadian Shield
SA Catskill Delta
SA Catskill Formation
SA Champlain Sea
SA Champlain Valley
SA Chazy Group
SA Chemung Formation
SA Chemung River
SA Cheshire Formation
SA Clinton Group
SA Coeymans Formation
SA Cortlandt Complex
SA Cross Lake
SA Cuyahoga Formation
SA Delaware River
SA Delaware River basin
SA Fire Island
SA Fordham Gneiss
SA Genesee Group
SA Genesee River
SA Goodnow earthquake 1983
SA Great Lakes region
SA Great South Bay
SA Grenville Province
SA Grimsby Sandstone
SA Hamilton Group
SA Helderberg Group
SA Hudson Highlands
SA Hudson River
SA Hudson Valley
SA Lake Champlain
SA Lake Erie
SA Lake George
SA Lake Ontario
SA Lamont-Doherty Geological Observatory
SA Little Falls Formation
SA Lockatong Formation
SA Lockport Formation
SA Long Island
SA Ludlowville Formation
SA Magothy Aquifer
SA Magothy Formation
SA Manhattan Formation
SA Manlius Formation
SA Marcellus Shale
SA Marcy Massif
SA Medina Formation
SA Moscow Formation
SA Mystery Cave
SA Narragansett Pier Granite

SA Narrows
SA New York City Group
SA Newark Basin
SA Newark Supergroup
SA Newark-Gettysburg Basin
SA Niagara Escarpment
SA Niagara Falls
SA Niagara River
SA Normanskill Formation
SA Onondaga Limestone
SA Oriskany Sandstone
SA Ottawa Sand
SA Palisades Sill
SA Passaic Formation
SA Piedmont
SA Plum Island
SA Potsdam Sandstone
SA Queenston Shale
SA Ramapo Fault
SA Raritan Bay
SA Reading Prong
SA Rochester Formation
SA Rondout Formation
SA Saint Lawrence Lowlands
SA Saint Lawrence River
SA Saint Lawrence River basin
SA Saint Lawrence Valley
SA Schenectady Formation
SA Shawangunk Formation
SA Staten Island
SA Stockton Formation
SA Susquehanna River
SA Susquehanna River basin
SA Tioga Bentonite
SA Trenton Group
SA Tully Limestone
SA Utica Shale
SA Valley and Ridge Province
SA West Falls Formation
SA Whirlpool Sandstone

New York Bight (1978)
CO N390000N410000
    W0720000W0750000
BT New York
BT United States
SA Atlantic Coastal Plain

New York City
Valid through 1988. Search in combination with state term. After 1988, use specific city-state term.

New York City Group (1978)
Precambrian to Paleozoic (pre-Upper Devonian). Includes Fordham Gneiss, Inwood Marble, Manhattan Formation. W Connecticut and SE New York.
IN Index ages as applicable.
SA Cambrian
SA Connecticut
SA Fordham Gneiss
SA Manhattan Formation
SA New York
SA Ordovician
SA Proterozoic
SA Silurian

New York City New York (1989)
City on W tip of Long Island, Manhattan Island, Staten Island, and S tip of mainland, SE New York. Composed of five boroughs (Bronx, Brooklyn, Manhattan, Queens and Staten Island) which are coextensive with five counties (Bronx, Kings, New York, Queens and Richmond). Before 1989, search New York City AND New York.
CO N403000N405300
    W0734100W0741700
BT New York
BT United States
NT Kings County New York
NT Manhattan

NT Queens County New York
SA Staten Island

New Zealand (1978)
CO S473000S343000
  E1783000E1663000
BT Australasia
NT Auckland New Zealand
NT Canterbury New Zealand
NT Coromandel Peninsula
NT Foveaux Strait
NT Gisborne New Zealand
NT Haast River
NT Hawke's Bay New Zealand
NT Hikurangi Trough
NT Kaipara Harbor
NT Lake Taupo
NT Manawatu River valley
NT Marlborough New Zealand
NT Nelson New Zealand
NT Northland New Zealand
NT Otago New Zealand
NT Oxford New Zealand
NT Rangitata River
NT Raukumara Peninsula
NT Reefton New Zealand
NT Ruapehu
NT Southland New Zealand
NT Taranaki New Zealand
NT Tarawera volcanic complex
NT Taupo New Zealand
NT Taupo volcanic zone
NT Te Aroha New Zealand
NT Tongariro
NT Waikato Basin
NT Wairakei
NT Wairarapa
NT Wanganui Valley
NT Wellington New Zealand
NT Westland New Zealand
SA Alpine Fault
SA Australian Plate
SA Bay of Plenty
SA Bluff Formation
SA Broadlands
SA Edgecumbe earthquake 1987
SA Fox Glacier
SA Huntly
SA Indo-Australian Plate
SA North Island
SA Oceania
SA Pacific mobile belt
SA Ross Formation
SA South Island
SA Southern Alps
SA Taranaki Basin
SA Tasman orogenic zone
SA Torlesse Supergroup
SA Waitemata Group
SA White Island

Newark
Valid through 1988. Search in combination with state term. After 1988, use specific city-state term.

Newark Basin (1978)
Triassic to Jurassic rift basin.
IN Index states as applicable.
BT United States
SA Gettysburg Basin
SA New Jersey
SA New York
SA Newark-Gettysburg Basin
SA Passaic Formation
SA Pennsylvania
SA Ramapo Fault
SA Watchung Mountains

Newark Group
As of 1993, no longer valid for GeoRef.
use Newark Supergroup

Newark New Jersey (1989)
City on Newark Bay just W of Jersey City and New York City in NE New Jersey. Before 1989, search Newark AND New Jersey.
CO N404400N404400
  W0741200W0741200
BT Essex County New Jersey
BT New Jersey
BT United States

Newark Supergroup (1993)
Lower Jurassic or Upper Triassic. Includes New Oxford Formation, Gettysburg Shale, Stockton Formation, Lockatong Formation, Brunswick Formation, New Haven Arkose, Meriden Formation, Portland Arkose, Pekin Formation, Cumnock Formation, Sanford Formation, Talcott Basalt, Shuttle Meadow Formation, Holyoke Basalt, East Berlin Formation, Hampden Basalt, Otterdale Sandstone. Before 1993, also search Newark Group.
IN Index ages as applicable.
UF Newark Group
BT Mesozoic
SA Brunswick Formation
SA Connecticut
SA Delaware
SA East Berlin Formation
SA Feltville Formation
SA Hampden Basalt
SA Holyoke Basalt
SA Lockatong Formation
SA Maryland
SA New Jersey
SA New York
SA North Carolina
SA Passaic Formation
SA Portland Formation
SA Stockton Formation
SA Virginia

Newark-Gettysburg Basin (1978)
Newark Basin plus southwestward extension into SE Pennsylvania and N central Maryland.
IN Index states as applicable.
BT United States
SA Gettysburg Basin
SA Maryland
SA New Jersey
SA New York
SA Newark Basin
SA Pennsylvania

Newberry Volcano (1985)
Central Oregon. Autoposting of broader terms to this term began in 1989. Also search Newberry AND Oregon. This term has multiple hierarchies.
CO N433500N435000
  W1211000W1212000
BT1 Deschutes County Oregon
BT1 Oregon
BT1 United States
BT2 Cascade Range
BT2 United States

newberyite (1978)
BT phosphates

Newcastle
No longer a valid term for GeoRef. As of 1993, See Newcastle Australia, Newcastle England, or Newcastle New Brunswick.

Newcastle Australia (1993)
City on E coast of New South Wales. Before 1993, also search Newcastle AND New South Wales.
BT New South Wales Australia
BT Australia
BT Australasia

Newcastle Coal Measures (1989)
Permian in the Sydney Basin of New South Wales.
BT Permian
BT Paleozoic
SA Australia
SA New South Wales Australia
SA Sydney Basin

Newcastle England (1993)
City on the Tyne River in Tyne and Wear County in N England. Before 1993, also search Newcastle upon Tyne or Newcastle AND England.
UF Newcastle upon Tyne England
BT England
BT Great Britain
BT United Kingdom
BT Western Europe
BT Europe
SA Northumberland England

Newcastle New Brunswick (1993)
Town, Northumberland County. Lead and zinc mining region. Before 1993, also search Newcastle AND New Brunswick.
BT Northumberland County New Brunswick
BT New Brunswick
BT Maritime Provinces
BT Eastern Canada
BT Canada

Newcastle Sandstone (1989)
SE Montana, W Nebraska, NE North Dakota, W South Dakota and NE Wyoming.
BT Lower Cretaceous
BT Cretaceous
BT Mesozoic
SA Colorado Group
SA Montana
SA Nebraska
SA North Dakota
SA South Dakota
SA Wyoming

Newcastle upon Tyne England
use Newcastle England

Newfoundland (1978)
From 1981 through 1992, referred to the island of Newfoundland. Before 1981 and after 1992, refers to the province, which includes the island of Newfoundland plus Labrador.
CO N463000N610000
  W0523000W0673000
BT Eastern Canada
BT Canada
NT Labrador
NT Newfoundland Island
SA Belle Isle
SA White Bay

Newfoundland Island (1993)
Part of Newfoundland province, E Canada.
CO N463000N514000
  W0523000W0592000
BT Newfoundland
BT Eastern Canada
BT Canada
NT Avalon Peninsula
NT Bay of Islands
NT Betts Cove
NT Buchans Newfoundland
NT Burin Peninsula
NT Burlington Peninsula
NT Daniel's Harbour Newfoundland
NT Great Northern Peninsula
NT Humber Arm Allochthon
NT New World Island
NT Notre Dame Bay
NT Port au Port Peninsula
NT Saint John's Newfoundland
SA Ackley Granite
SA Avalon Terrane
SA Avalon Zone
SA Bay of Islands Ophiolite
SA Bell Island
SA Betts Cove Ophiolite
SA Buchans Group
SA Catoche Formation
SA Cow Head Group
SA Davidsville Group
SA Deer Lake Group
SA Gander Lake Group
SA Gander Zone
SA Grand Banks earthquake 1929
SA Grenville Province
SA Harbour Main Group
SA Hare Bay
SA Labrador Current
SA Lushs Bight Group
SA Saint George Group
SA Saint John Island
SA Signal Hill Formation
SA Table Head Group
SA White Bay

Newland Limestone (1989)
In Piegan Group of Belt Supergroup. Idaho and W central Montana. In central and western Montana, Newland and Wallace formations have been treated as essentially synonymous terms by some authors.
BT middle Proterozoic
BT Proterozoic
BT upper Precambrian
BT Precambrian
SA Belt Supergroup
SA Idaho
SA Montana
SA Wallace Formation

Newman Limestone (1978)
Overlies Grainger Formation; underlies Pennington Formation. E Kentucky, E Tennessee, and SW Virginia.
BT Mississippian
BT Carboniferous
BT Paleozoic
SA Kentucky
SA Tennessee
SA Virginia

Newmarket
use Nowy Targ Poland

Newport
Valid through 1988. Search in combination with state term. After 1988, use specific city-state term.

Newport Bay (1978)
Dredged harbor in Orange County in S California.
IN Index county as applicable.
SA Orange County California

Newport County Rhode Island (1989)
On the Atlantic Ocean in SE Rhode Island. Before 1989, search Newport County AND Rhode Island.
CO N412700N413800
  W0711300W0712300
BT Rhode Island
BT United States
NT Newport Rhode Island

Newport Oregon (1989)
City on the Pacific Ocean in W Oregon. Before 1989, search Newport AND Oregon.
CO N443900N443900
W1240400W1240400
BT Lincoln County Oregon
BT Oregon
BT United States

Newport Rhode Island (1989)
City in SE Rhode Island. Before 1989, search Newport AND Rhode Island.
CO N413000N413000
W0711900W0711900
BT Newport County Rhode Island
BT Rhode Island
BT United States

Newport-Inglewood Fault (1978)
In Los Angeles area in S California. Before 1993, also search Inglewood Fault.
UF Inglewood Fault
BT California
BT United States

news (1985)
IN Used for indexing of news items, usually discussing mineral exploration.
SA current research
SA popular geology

Newton County
Valid through 1988. Search in combination with state term. After 1988, use specific county-state term.

Newton County Arkansas (1989)
NW Arkansas. Before 1989, also search Newton County AND Arkansas.
CO N354500N360700
W0925700W0933100
BT Arkansas
BT United States

Newton County Georgia (1989)
N central Georgia. Before 1989, also search Newton County AND Georgia.
CO N332700N333900
W0834100W0840300
BT Georgia
BT United States

Newton County Indiana (1989)
NW Indiana. Before 1989, also search Newton County AND Indiana.
CO N404400N411200
W0871600W0873200
BT Indiana
BT United States
NT Kentland Indiana

Newton County Mississippi (1989)
E central Mississippi. Before 1989, also search Newton County AND Mississippi.
CO N321300N323600
W0885400W0892100
BT Mississippi
BT United States

Newton County Missouri (1989)
SW Missouri. Before 1989, also search Newton County AND Missouri.
CO N364500N370300
W0940400W0943700
BT Missouri
BT United States
SA Joplin Missouri

Newton County Texas (1989)
E Texas. Before 1989, also search Newton County AND Texas.
CO N301800N311300
W0932800W0935400
BT Texas
BT United States

Neyveli
No longer a valid term for GeoRef. See Neyveli India.

Neyveli India (1993)
Town in E Tamil Nadu, SE India. Before 1993, also search Neyveli. Before 1978, also search Neiveli.
UF Neiveli India
BT Tamil Nadu India
BT India
BT Indian Peninsula
BT Asia

Neyveli Lignite (1978)
Associated with sands, conglomerates and clays of Cuddalore Series. E and NE of Tamil Nadu. Also search Neyveli.
BT Tertiary
BT Cenozoic
SA Cuddalore Series
SA India
SA Miocene
SA Pliocene
SA Tamil Nadu India
SA upper Miocene

Ngalia Basin (1989)
SW Northern Territory.
CO S230000S210000
E1330000E1300000
BT Northern Territory Australia
BT Australia
BT Australasia

Ngorora Formation (1978)
BT Miocene
BT Neogene
BT Tertiary
RT Cenozoic
SA Kenya

Ni
use nickel

Niagara County
Valid through 1988. Search in combination with state term. After 1988, use specific county-state term.

Niagara County New York (1989)
W New York. Before 1989, also search Niagara County AND New York.
CO N430200N432300
W0782700W0790500
BT New York
BT United States

Niagara Escarpment (1978)
Cuesta of Paleozoic rocks sloping away from the Canadian Shield, extending from central New York to N of Lake Huron and Lake Michigan, continuing southwestward into Wisconsin, Illinois, and Iowa.
IN Index states and province as applicable.
BT North America
SA Bruce Peninsula
SA Great Lakes region
SA Illinois
SA Iowa
SA Manitoulin Island
SA Michigan
SA New York
SA Niagara Falls
SA Ontario
SA Wisconsin

Niagara Falls (1978)
Great falls of the Niagara River. Also search Niagara.
IN Index state and/or province as applicable.
CO N430600N430600
W0790300W0790300
BT North America
SA New York
SA Niagara Escarpment
SA Ontario

Niagara River (1978)
Flows between Lake Erie and Lake Ontario going over the Niagara Escarpment. Also search Niagara.
IN Index province and/or state as applicable.
BT North America
SA New York
SA Ontario

Niagaran (1978)
Provincial series, North America. Above Alexandrian, below Cayugan.
BT Middle Silurian
BT Silurian
BT Paleozoic
NT Bainbridge Formation
NT Lockport Formation
NT Racine Dolomite

Nicaragua (1978)
CO N104000N150000
W0830500W0874000
BT Central America
NT Leon Nicaragua
NT Managua Nicaragua
NT Masaya
SA Leon City Spain
SA San Juan River

Nicaragua Rise (1978)
Between Jamaica and Nicaragua.
UF Nicaraguan Rise
BT Caribbean Sea
BT North American Atlantic
BT North Atlantic
BT Atlantic Ocean
SA Leg 15

Nicaraguan Rise
use Nicaragua Rise

niccolite (1978)
Autoposting of sulfides to this term ended in 1989.
UF nickeline
UF nicolite
BT arsenides
SA nickel
SA nickel ores

Nice
No longer a valid term for GeoRef. See Nice France.

Nice France (1993)
City on the Mediterranean Sea on the Riviera in S Alpes-Maritimes, SE France.
BT Alpes-Maritimes France
BT France
BT Western Europe
BT Europe

nickel (1978)
Chemical element. As of 1981, use nickel ores for nickel as a commodity. Autoposting of metals to this term began in 1989.
UF Ni
BT metals
SA awaruite
SA heavy metals
SA josephinite
SA kamacite
SA niccolite
SA nickel ores
SA pentlandite
SA siderophile elements

nickel glance
use gersdorffite

nickel ores (1981)
Before 1981, also search (nickel OR Ni) AND (deposit OR deposits OR ore OR ores OR economic) in the basic index. Autoposting of metal ores to this term began in 1985.
BT metal ores
SA garnierite
SA gersdorffite
SA millerite
SA niccolite
SA nickel
SA Norilsk region
SA pentlandite
SA rammelsbergite
SA Strathcona Mine

nickel pyrites
use millerite

nickel-antimony glance
use ullmannite

nickeline
use niccolite

Nicobar Islands (1978)
Island group in Bay of Bengal NW of Sumatra, forming S part of Andaman and Nicobar Islands.
UF Nicobars
BT Bengal Islands
BT India
BT Indian Peninsula
BT Asia

Nicobars
use Nicobar Islands

Nicola Group (1978)
S British Columbia.
BT Triassic
BT Mesozoic
SA British Columbia
SA Canada

Nicolaus
Valid through 1988. Search in combination with state term. After 1988, use specific city-state term.

Nicolaus California (1989)
Village N of Sacramento in N central California. Before 1989, search Nicolaus AND California.
CO N385500N385500
W1213400W1213400
BT Sutter County California
BT California
BT United States

nicolite
use niccolite

nicopyrite
use pentlandite

Nicoya Complex (1989)
NW Costa Rica.
IN Index ages as applicable.
BT Mesozoic
SA Costa Rica
SA Leg 67
SA Lower Cretaceous
SA Upper Jurassic

Nicoya Peninsula (1989)
Peninsula in W Costa Rica, extending between the Gulf of Nicoya and Pacific Ocean.
CO N093000N101500
W0845200W0854500

BT Costa Rica
BT Central America

Nida Basin (1978)
River basin in S Poland. Also search Nida.
UF Nida River basin
BT Kielce Poland
BT Poland
BT Central Europe
BT Europe

Nida River basin
  use Nida Basin

Niemen River basin
  use Neman River basin

Nievre
No longer a valid term for GeoRef. See Nievre France.

Nievre France (1993)
Department in central France.
CO N463000N473000
   E0041000E0024500
BT France
BT Western Europe
BT Europe
NT Nivernais Plateau
SA Morvan
SA Yonne Valley

Niger (1978)
Before 1972, also search Niger Republic.
CO N120000N233000
   E0160000E0001500
UF Niger Republic
BT West Africa
BT Africa
NT Agades Niger
SA Chad Basin
SA Lake Chad
SA Mali-Niger Syneclise
SA Niger River
SA Niger Valley
SA Sahara
SA Sahel
SA Tibesti Massif
SA West African Shield
SA Younger Granites

Niger Delta (1978)
On the Gulf of Guinea.
BT Nigeria
BT West Africa
BT Africa

Niger Republic
Not a valid GeoRef index term since 1971. Was used on level 1 in subfile G.
  use Niger

Niger River (1978)
Rises in Guinea and flows NE into a great curve and then SE into the Gulf of Guinea.
IN Index countries as applicable.
UF Joliba River
UF Kworra River
BT West Africa
BT Africa
SA Benin
SA Guinea
SA Mali
SA Niger
SA Niger Valley
SA Nigeria

Niger Valley (1978)
River valley. Also search Niger.
IN Index countries as applicable.
UF Joliba Valley
UF Kworra Valley
BT West Africa
BT Africa
SA Benin
SA Guinea

SA Mali
SA Niger
SA Niger River
SA Nigeria

Nigeria (1978)
CO N040000N140000
   E0143000E0023000
BT West Africa
BT Africa
NT Bida Nigeria
NT Ewekoro
NT Ibadan Nigeria
NT Jos Plateau
NT Lagos Lagoon
NT Niger Delta
NT Sokoto Basin
NT Zaria Nigeria
SA Adamawa
SA Ameki Formation
SA Asu River Group
SA Benue Valley
SA Chad Basin
SA Ewekoro Formation
SA Imo Shale
SA Lake Chad
SA Mali-Niger Syneclise
SA Niger River
SA Niger Valley
SA Victoria Island
SA Younger Granites

nigerite (1978)
BT oxides

niggliite (1978)
From 1978-1980, native elements and alloys; sulfides were autoposted to this term. Autoposting of alloys began in 1981.
BT alloys
SA platinum
SA platinum minerals
SA tin

Nihewan Formation (1989)
Hebei Province of N China.
BT Pleistocene
BT Quaternary
BT Cenozoic
SA China
SA Hebei China

Nihon'arupusu
  use Japanese Alps

Nihonkai-Chuba earthquake 1983
  use Japan Sea earthquake 1983

Nihonkai-Chubu earthquake 1983
  use Japan Sea earthquake 1983

Niigata
No longer a valid term for GeoRef. See Niigata Japan.

Niigata earthquake 1964 (1985)
Epicenter near Niigata, NW Honshu, Japan.
BT earthquakes
SA Honshu
SA Japan
SA Niigata Japan

Niigata Japan (1993)
Prefecture including the city in NW Honshu.
BT Honshu
BT Japan
BT Far East
BT Asia
NT Kashiwazaki Japan
NT Myoko Mountain
NT Shinano River
SA Awashima
SA Chubu Japan
SA Japanese Alps
SA Niigata earthquake 1964
SA Sado Island

Nikitovka
No longer a valid term for GeoRef. As of 1993, see Nikitovka Ukraine.

Nikitovka Ukraine (1993)
City in central Donetsk (formerly Stalino) Oblast in Donets Basin. This term has multiple hierarchies.
BT1 Ukraine
BT1 Europe
BT2 Ukraine
BT2 Commonwealth of Independent States

Nikolai Greenstone (1985)
E Alaska.
IN Index ages as applicable.
SA Alaska
SA Permian
SA Triassic

Nikopol
No longer a valid term for GeoRef. As of 1993, see Nikopol Ukraine.

Nikopol Ukraine (1993)
City in Dnepropetrovsk Oblast in E central Ukraine. This term has multiple hierarchies.
BT1 Dnepropetrovsk Ukraine
BT1 Ukraine
BT1 Europe
BT2 Dnepropetrovsk Ukraine
BT2 Ukraine
BT2 Commonwealth of Independent States

Niksic (1978)
Town in W Montenegro, SW Yugoslavia.
BT Montenegro
BT Yugoslavia
BT Southern Europe
BT Europe

Nile Delta (1978)
On Mediterranean Sea E of Alexandria.
CO N301000N314000
   E0322000E0295000
UF Nile River delta
BT Egypt
BT North Africa
BT Africa
SA Nile River

Nile River (1978)
The longest river in the world. Its remotest headstream is the Luvironza River in Burundi. The flow is through Lake Victoria and Lake Albert, and then, as the White Nile, into Sudan where it unites with the Blue Nile from Ethiopia as the Nile proper which flows into the Mediterranean Sea.
IN Index countries as applicable.
CO N220000N310000
   E0370000E0310000
BT Africa
SA Aswan
SA Blue Nile
SA Burundi
SA Egypt
SA Ethiopia
SA Nile Delta
SA Sudan
SA Uganda

Nile River delta
  use Nile Delta

Nile Valley (1978)
River valley of the Nile River proper.
IN Index countries as applicable.
BT Africa
SA Egypt

SA Middle Nile Valley
SA Nubia
SA Sudan

Nilgiri
No longer a valid term for GeoRef. Former Indian state near coast of NW Bay of Bengal. Became part of Orissa state, established 1950.

Nilssoniales (1978)
BT gymnosperms
BT Spermatophyta
BT Plantae

Nimba Mountains (1978)
West African bulge. Also search Nimba.
IN Index countries as applicable.
BT West Africa
BT Africa
SA Guinea
SA Ivory Coast
SA Liberia

Ninetyeast Ridge (1978)
Undersea ridge extending N and S along line of 90° E longitude from just S of Nicobar Islands to approximately 30° S latitude.
CO S350000N080000
   E0920000E0850000
BT Indian Ocean
SA DSDP Site 213
SA DSDP Site 215
SA DSDP Site 216
SA DSDP Site 217
SA DSDP Site 253
SA DSDP Site 254
SA Leg 121
SA ODP Site 756
SA ODP Site 757
SA ODP Site 758

Ning-hsia
Not a valid term for GeoRef. See Ningxia China.

Ning-hsia China
  use Ningxia China

Ninghsia China
  use Ningxia China

Ningsia China
  use Ningxia China

Ningxia
No longer a valid term for GeoRef. See Ningxia China.

Ningxia China (1993)
Autonomous region, NW China. Before 1993, also search Ningxia. Before 1978, also search Ning-hsia, Ninghsia, and Ningsia.
CO N353000N403000
   E1074000E1040000
UF Ning-hsia China
UF Ninghsia China
UF Ningsia China
BT China
BT Far East
BT Asia
SA Haiyuan earthquake 1920

niningerite (1978)
Meteorite mineral.
BT sulfides
SA meteorites

Niniyur Group (1978)
S India.
BT Danian
BT lower Paleocene
BT Paleocene
BT Paleogene
BT Tertiary
BT Cenozoic
SA India
SA Tamil Nadu India

SA Upper Cretaceous
niobates (1978)
Autoposting of this term began in 1981.
UF columbates
BT oxides
NT aeschynite
NT loparite
NT nenadkevichite
NT niobotantalates
NT zirconolite

niobium (1978)
Chemical element. As of 1981, use niobium ores for niobium as a commodity. Autoposting of metals to this term began in 1989.
UF columbium
UF Nb
BT metals
SA niobium ores
SA tantalum

niobium ores (1981)
Before 1981, also search (niobium OR Nb) AND (deposit OR deposits OR ore OR ores OR economic) in the basic index. Autoposting of metal ores to this term began in 1985.
BT metal ores
SA columbite
SA niobium
SA pyrochlore
SA tantalum ores

niobotantalates (1981)
Autoposting of niobates and tantalates to this term began in 1989. This term has multiple hierarchies.
BT1 niobates
BT1 oxides
BT2 tantalates
BT2 oxides
NT betafite
NT columbite
NT euxenite
NT fergusonite
NT ixiolite
NT microlite
NT pyrochlore
NT samarskite
NT stibiotantalite
NT tantalite
NT tapiolite
NT wodginite

Niobrara County
Valid through 1988. Search in combination with state term. After 1988, use specific county-state term.

Niobrara County Wyoming (1989)
E Wyoming. Before 1989, also search Niobrara County AND Wyoming.
CO N423500N432500
    W1040500W1045500
BT Wyoming
BT United States
SA Walker Creek Field

Niobrara Formation (1978)
In Colorado Group. Includes Fort Hays Limestone Member, Smoky Hill Chalk Member.
BT Upper Cretaceous
BT Cretaceous
BT Mesozoic
SA Colorado
SA Colorado Group
SA Fort Hays Limestone Member
SA Kansas
SA Mancos Shale
SA Minnesota
SA Montana
SA Nebraska
SA New Mexico
SA North Dakota
SA Smoky Hill Chalk Member
SA South Dakota
SA Wyoming

Nipissing
As of 1985, no longer a valid term for GeoRef.
use Nipissing District Ontario

Nipissing Diabase (1978)
BT Precambrian
SA Canada
SA Ontario
SA Quebec

Nipissing District
No longer a valid term for GeoRef. As of 1993, see Nipissing District Ontario.

Nipissing District Ontario (1993)
SE central Ontario. Before 1993, also search Nipissing District, District of Nipissing, or Nipissing in conjunction with Ontario.
CO N452000N472000
    W0773000W0802500
UF District of Nipissing
UF Nipissing
BT Ontario
BT Eastern Canada
BT Canada
NT Lake Nipissing
NT North Bay Ontario
SA Brent Crater

Nisku Formation (1989)
In Winterburn Group. Term restricted to region in central Alberta. For Upper Devonian strata in Montana, North Dakota and Saskatchewan, see Birdbear Formation.
BT Upper Devonian
BT Devonian
BT Paleozoic
SA Alberta
SA Birdbear Formation

Nisqually Glacier (1978)
On S slope of Mt Rainier in Mt. Rainier National Park in W central Washington.
BT Pierce County Washington
BT Washington
BT United States
SA Mount Rainier

nitrate deposits (1981)
Before 1981, search nitrates AND deposits.
IN Commodity. See List C.
SA nitrates

nitrate ion (1985)
SA geochemistry
SA ions
SA nitrates

nitrates (1978)
As of 1981, use nitrate deposits for nitrate as a commodity.
SA minerals
SA nitrate deposits
SA nitrate ion

nitrides (1978)
BT alloys

nitrogen (1978)
UF N
NT N-15
NT N-15/N-14
SA chemical elements
SA fertilizers
SA soils

nitrous oxide (1985)
UF laughing gas
SA gases
SA geochemistry
SA oxides

Nittany Valley (1978)
In Centre and Clinton counties in central Pennsylvania.
IN Index counties as applicable.
BT Pennsylvania
BT United States
SA Centre County Pennsylvania
SA Clinton County Pennsylvania

Nitzschia (1978)
Genus. Autoposting of thallophytes and microfossils began in 1990. This term has multiple hierarchies.
BT1 diatoms
BT1 algae
BT1 thallophytes
BT1 Plantae
BT2 diatoms
BT2 algae
BT2 microfossils

nivation (1978)
UF snow-patch erosion
SA erosion
SA glaciers
SA sheet erosion
SA water erosion

Nivernais Hills
use Nivernais Plateau

Nivernais Plateau (1978)
In central France.
UF Nivernais Hills
BT Nievre France
BT France
BT Western Europe
BT Europe

Nix Olympia
use Olympus Mons

Nixon Fork Terrane (1989)
In Medfra Quadrangle, central Alaska.
CO N610000N630000
    W1540000W1570000
BT West-Central Alaska
BT Alaska
BT United States

Nizhni Novgorod Russian Federation
use Nizhniy Novgorod Russian Federation

Nizhniy Novgorod Russian Federation (1993)
Oblast and city, SE Russian Federation. Before 1993, also search Gorki. Before 1978, also search Gorkiy, Gorky, or Nizhni Novgorod. This term has multiple hierarchies.
UF Gorki Russian Federation
UF Gorkiy Russian Federation
UF Gorky Russian Federation
UF Nizhni Novgorod Russian Federation
BT1 Russian Federation
BT1 Commonwealth of Independent States
BT2 Europe

Nizhniy Ufaley Russian Federation
use Ufaley Russian Federation

Nizke Jesenik Mountains
use Nizky Jezenik Mountains

Nizke Tatra
use Low Tatra Mountains

Nizky Jezenik Mountains (1978)
In NW Czechoslovakia near the Polish border. Also search Nizky Jezenik.
CO N494400N501400
    E0174800E0170400
UF Low Jesenik Mountains
UF Nizke Jesenik Mountains
BT Moravia
BT Czech Republic
BT Czechoslovakia
BT Central Europe
BT Europe

NMO
use normal moveout

NMR data
use NMR spectra

NMR spectra (1982)
Used for data. For methodology, use nuclear magnetic resonance.
UF NMR data
UF nuclear magnetic resonance spectra
BT spectra
SA nuclear magnetic resonance
SA spectroscopy

No (element)
use nobelium

NOAA (1989)
Acronym.
UF National Oceanic and Atmospheric Administration
BT government agencies
SA United States

Noachian (1993)
Oldest system of three in the Martian stratigraphic column.
SA Mars

nobelium (1989)
Chemical element.
UF No (element)
BT actinides
BT metals

Noble County
Valid through 1988. Search in combination with state term. After 1988, use specific county-state term.

Noble County Indiana (1989)
NE Indiana. Before 1989, also search Noble County AND Indiana.
CO N411500N413200
    W0851200W0851000
BT Indiana
BT United States

Noble County Ohio (1989)
E Ohio. Before 1989, also search Noble County AND Ohio.
CO N393500N395700
    W0811300W0814300
BT Ohio
BT United States

Noble County Oklahoma (1989)
N Oklahoma. Before 1989, also search Noble County AND Oklahoma.
CO N361000N363500
    W0965300W0972800
BT Oklahoma
BT United States

noble gases (1978)
Autoposting of this term began in 1985.
UF inert gases
UF rare gases
NT argon
NT helium
NT krypton
NT neon

NT radon
NT xenon
SA gases
SA isotopes

Nodosaria (1978)
Genus. Autoposting of microfossils and Protista to this term began in 1990. This term has multiple hierarchies.
BT1 Nodosariidae
BT1 Nodosariacea
BT1 Rotaliina
BT1 foraminifera
BT1 Protista
BT1 Invertebrata
BT2 Nodosariidae
BT2 Nodosariacea
BT2 Rotaliina
BT2 foraminifera
BT2 Protista
BT2 microfossils

Nodosariacea (1978)
Autoposting of microfossils and Protista to this term began in 1990. This term has multiple hierarchies.
BT1 Rotaliina
BT1 foraminifera
BT1 Protista
BT1 Invertebrata
BT2 Rotaliina
BT2 foraminifera
BT2 Protista
BT2 microfossils
NT Nodosariidae

Nodosariidae (1978)
Family. Autoposting of microfossils and Protista to this term began in 1990. This term has multiple hierarchies.
BT1 Nodosariacea
BT1 Rotaliina
BT1 foraminifera
BT1 Protista
BT1 Invertebrata
BT2 Nodosariacea
BT2 Rotaliina
BT2 foraminifera
BT2 Protista
BT2 microfossils
NT Lenticulina
NT Nodosaria

nodules (1978)
As of 1981, restricted to nodules on the ocean floor, nodules within sediments in the ocean, and nodules in lakes. For nodules in non-sedimentary formations, use xenoliths. For nodules within sedimentary formations, such as chert nodules, use concretions. From 1978-1980, secondary structures and sedimentary structures were autoposted to this term.
SA concretions
SA ferromanganese composition
SA Law of the Sea
SA manganese composition
SA MANOP
SA marine geology
SA metal ores
SA ocean floors
SA oceanography
SA secondary structures
SA sedimentary structures
SA sedimentation

Noeggerathiales (1978)
BT pteridophytes
BT Plantae

Nohi Rhyolite (1978)
Complex in central Japan.
BT Honshu
BT Japan
BT Far East
BT Asia
SA Iwate
SA Iwate Japan

noise (1978)
Before 1978, also search seismic noise.
SA filters
SA geophysical methods
SA microseisms
SA seismology
SA signal-to-noise ratio
SA signals

Nojiri Lake (1978)
Central Honshu, near Nagano-Niigata prefectural border.
BT Nagano Prefecture Japan
BT Honshu
BT Japan
BT Far East
BT Asia

Nokomis Group (1989)
Hornblende gneiss in the Kisseynew Complex. Conformably overlain by the Sherridon Group.
BT Precambrian
SA Kisseynew Complex
SA Manitoba
SA Saskatchewan

Nome
Valid through 1988. Search in combination with state term. After 1988, use specific city-state term.

Nome Alaska (1989)
City on S side of Seward Peninsula, W Alaska. Before 1989, search Nome AND Alaska.
CO N643000N643000
 W1653000W1653000
BT West-Central Alaska
BT Alaska
BT United States
SA Seward Peninsula

nomenclature (1978)
Before 1979, also search terminology.
UF terminology
SA classification
SA definition
SA homonymy
SA identification
SA new names

nomograms (1978)
UF nomographs
SA data processing

nomographs
 use nomograms

non-glacial
A valid index term through 1976. See non-glacial ice. Before 1976 also search ice AND non-glacial.

non-glacial ice
As of 1981, no longer a valid term for GeoRef. Use ice.

non-linear distortion (1981)
SA distortion
SA elastic waves
SA seismology

non-linear materials (1981)
SA engineering geology
SA materials

Nonacho Basin (1989)
Lower Proterozoic basin in Northwest Territories.
BT Northwest Territories
BT Western Canada
BT Canada

nonaqueous phase liquids (1993)
Before 1993, also search NAPL.
UF NAPLs
SA dense nonaqueous phase liquids
SA leaking underground storage tanks
SA liquid phase
SA oil spills
SA pollutants

Noncalcarea
 use Silicispongiae

Nonesuch Shale (1978)
In Oronto Group. N Michigan and NE Wisconsin.
BT Precambrian
SA Michigan
SA Oronto Group
SA Wisconsin

Nong-Son
No longer a valid term for GeoRef. As of 1993 see Nong-Son Vietnam.

Nong-Son Vietnam (1993)
Village in Quang Nam Province in N South Vietnam. Before 1993 also search Nong-Son AND Vietnam.
UF Nongson
BT Vietnam
BT Far East
BT Asia

Nongson
 use Nong-Son Vietnam

nonmagnetic minerals (1993)
SA heavy minerals
SA magnetic anomalies
SA magnetic hysteresis
SA magnetic methods
SA magnetic properties
SA magnetic surveys
SA magnetic susceptibility
SA minerals
SA properties

nonmare basalt
 use KREEP

nonmetal deposits (1981)
Before 1981, search nonmetals AND deposits. Before 1971, also search nonmetallic minerals.
UF nonmetallic minerals
SA economic geology
SA industrial minerals
SA metal ores
SA native elements
SA nonmetals

nonmetallic minerals
 use nonmetal deposits

nonmetals (1978)
As of 1981, use nonmetal deposits for nonmetals as a commodity.
SA chemical elements
SA metals
SA native elements
SA nonmetal deposits

nonpoint sources (1985)
SA ground water
SA point sources
SA pollution
SA surface water
SA water quality

nontectonics
 use atectonic processes

nonterrigenous
A valid term through 1977. Included use in combination with clastic rocks or clastic sediments.

nonterrigenous materials
As of 1981, no longer a valid term for GeoRef. Term was introduced in 1978. Before 1978, also search nonterrigenous AND clastic rocks or clastic sediments.

nontronite (1978)
UF chloropal
UF gramenite
UF morencite
UF pinquite
BT clay minerals
BT sheet silicates
BT silicates
SA montmorillonite

Noonday Dolomite (1978)
Underlies Johnnie Formation; overlies Kingston Peak Formation with disconformity or slight angular unconformity.
BT Precambrian
SA California

Nopah Formation (1989)
Occurs in Death Valley of SE California and W Nevada. Includes Dunderberg Shale and Halfpint Member.
BT Upper Cambrian
BT Cambrian
BT Paleozoic
SA California
SA Dunderberg Shale
SA Nevada

Nopah Range (1985)
SE California. Autoposting of broader terms to this term began in 1989.
IN Index counties and mountain range as applicable.
CO N354500N361500
 W1155500W1161500
BT California
BT United States
SA Basin and Range Province
SA Inyo County California
SA San Bernardino County California

Noranda
No longer a valid term for GeoRef. As of 1993, see Noranda Quebec.

Noranda Quebec (1993)
Mining city in SW Quebec. Before 1993, also search Noranda AND Quebec.
BT Quebec
BT Eastern Canada
BT Canada
SA Horne Mine

norbergite (1978)
From 1978-1980, halides and orthosilicates were autoposted to this term. Autoposting of fluorides and humite group began in 1981. Autoposting of halides, nesosilicates, orthosilicates and silicates began in 1989. This term has multiple hierarchies.
BT1 fluorides
BT1 halides
BT2 humite group
BT2 nesosilicates
BT2 orthosilicates
BT2 silicates

Nord Department
No longer a valid term for GeoRef. See Nord Department France.

Nord Fjord (1978)
Inlet of Norwegian Sea.
BT Norway
BT Scandinavia
BT Western Europe
BT Europe

Nord France (1993)
Department in extreme N France. Before 1981, Nord was used for the department. Also search Nord Department.
CO N495800N510500
E0041500E0020800
BT France
BT Western Europe
BT Europe
NT Dunkirk France
NT Lille France
NT Valenciennes France
SA Nord-Pas-de-Calais Basin

Nord-Pas-de-Calais Basin (1978)
In extreme N France.
IN Index departments as applicable.
BT France
BT Western Europe
BT Europe
SA Nord France
SA Pas-de-Calais France

Nord-Trondelag
No longer a valid term for GeoRef. As of 1993 see Nord-Trondelag Norway.

Nord-Trondelag Norway (1993)
County N of Trondheim Fjord in central Norway. Before 1993 also search Nord-Trondelag AND Norway.
BT Norway
BT Scandinavia
BT Western Europe
BT Europe
SA Trondelag

Nordaustlandet (1978)
Norwegian for North East Land. One of the islands of the Spitsbergen group in Barents Sea NE of island of Spitsbergen.
BT Spitsbergen
BT Svalbard
BT Arctic region

Nordland
No longer a valid term for GeoRef. As of 1993 see Nordland Norway.

Nordland Norway (1993)
County in W Norway. Before 1993 also search Nordland AND Norway.
BT Norway
BT Scandinavia
BT Western Europe
BT Europe
NT Bleikvassli
NT Lofoten Islands
NT Narvik Norway
NT Ofoten
NT Rana Fjord
SA Vesteralen

Nordlingen
use Ries Crater

Nordlinger Ries
use Ries Crater

Nordlinger Ries Crater
use Ries Crater

nordstrandite (1978)
BT oxides

NORESS (1993)
Acronym. Small-aperture seismic array.
UF Norwegian Regional Seismic Array
SA arrays
SA Norway
SA Scandinavia
SA seismic networks
SA seismology

Norfolk
No longer a valid term for GeoRef. As of 1993, see Norfolk England for the English county. As of 1989, see Norfolk Virginia if applicable.

Norfolk City
As of 1989, no longer a valid term for GeoRef.
use Norfolk Virginia

Norfolk County
Valid through 1988. Search in combination with state term. After 1988, use specific county-state term.

Norfolk County Massachusetts (1989)
E Massachusetts. Before 1989, also search Norfolk County AND Massachusetts.
CO N415800N422100
W0704800W0713000
BT Massachusetts
BT United States

Norfolk County Virginia (1989)
SE Virginia. Before 1989, also search Norfolk County AND Virginia.
CO N365100N365700
W0761100W0762100
BT Virginia
BT United States
NT Norfolk Virginia

Norfolk England (1993)
County in SE England. Before 1993, also search Norfolk AND England.
CO N522500N530000
E0014500E0001000
BT East Anglia
BT England
BT Great Britain
BT United Kingdom
BT Western Europe
BT Europe
NT Norwich England

Norfolk Island (1978)
Midway between New Caledonia and N New Zealand. An external territory of Australia. In 1985, broader term changed from Pacific Ocean to West Pacific Ocean Islands.
BT West Pacific Ocean Islands

Norfolk Ridge (1978)
Between New Caledonia and Norfolk Island in SW Pacific Ocean.
BT Pacific Ocean

Norfolk Virginia (1989)
City in but independent of Norfolk County in SE Virgina. From 1981 to 1989, search Norfolk City AND Virginia. Before 1981, search Norfolk AND Virginia.
CO N365400N365400
W0761800W0761800
UF Norfolk City
BT Norfolk County Virginia
BT Virginia
BT United States

Norian (1978)

Europe. Above Carnian, below Rhaetian.
BT Upper Triassic
BT Triassic
BT Mesozoic

Noril'sk Russian Federation
use Norilsk Russian Federation

Norilsk
No longer a valid term for GeoRef. As of 1993, see Norisk Russian Federation.

Norilsk region (1978)
A mining area in the Taymyr National Okrug connected to the town of Dudinka on the Yenisei River by a 60 mile narrow gage rail line. Also search Norilsk. This term has multiple hierarchies.
CO N690000N694500
E0890000E0870000
BT1 Taymyr Dolgan-Nenets Russian Federation
BT1 Krasnoyarsk Russian Federation
BT1 Russian Federation
BT1 Commonwealth of Independent States
BT2 Taymyr Dolgan-Nenets Russian Federation
BT2 Krasnoyarsk Russian Federation
BT2 Asia
SA copper ores
SA nickel ores

Norilsk Russian Federation (1993)
Town in Taymyr National Okrug in N Siberia. Before 1993, also search Norilsk. Before 1978, also search Noril'sk. This term has multiple hierarchies.
CO N692000N692200
E0880300E0880100
UF Noril'sk Russian Federation
BT1 Taymyr Dolgan-Nenets Russian Federation
BT1 Krasnoyarsk Russian Federation
BT1 Russian Federation
BT1 Commonwealth of Independent States
BT2 Taymyr Dolgan-Nenets Russian Federation
BT2 Krasnoyarsk Russian Federation
BT2 Asia

norite (1978)
BT gabbros
BT plutonic rocks
BT igneous rocks

normal
A valid term through 1977. After 1977, use normal faults or normal folds.

normal earthquake
use shallow-focus earthquakes

normal faults (1978)
Before 1978, also search faults AND normal; slump faults.
UF gravity faults
UF normal slip faults
BT faults
SA block structures
SA listric faults

normal folds (1978)
Before 1978 search folds AND normal.
BT folds
SA symmetric folds

normal moveout (1989)

Sometimes abbreviated NMO.
UF NMO
UF normal moveouts
SA arrival time
SA common-depth-point method
SA elastic waves
SA moveout
SA reflection
SA reflection methods
SA seismic methods
SA seismic surveys
SA stacking
SA traveltime
SA velocity analysis

normal moveouts
use normal moveout

normal slip faults
use normal faults

Norman Wells
As of 1993, no longer a valid term for GeoRef. See Norman Wells Northwest Territories.

Norman Wells Northwest Territories (1993)
Village on the Mackenzie River in W Mackenzie District, W Northwest Territories. Before 1993, also search Norman Wells.
BT Mackenzie District Northwest Territories
BT Northwest Territories
BT Western Canada
BT Canada

Normandie
use Normandy

Normandy (1978)
Historical region of NW France.
IN Index departments as applicable.
UF Normandie
BT France
BT Western Europe
BT Europe
SA Calvados France
SA Eure France
SA Manche France
SA Orne France
SA Seine-Maritime France

Normanskill Formation (1978)
Comprises Mount Merino Chert and Shale, Austin Glen Grit and Shale. NW Massachusetts, E New York, and SW Vermont.
BT Middle Ordovician
BT Ordovician
BT Paleozoic
SA Massachusetts
SA New York
SA Vermont

Normapolles (1985)
Autoposting of microfossils to this term began in 1990.
BT miospores
BT palynomorphs
BT microfossils

Norphlet Formation (1978)
Underlies Smackover Formation; overlies Louann Salt. Subsurface.
BT Jurassic
BT Mesozoic
SA Arkansas
SA Louisiana
SA Texas

Norrbotten
No longer a valid term for GeoRef. As of 1993 see Norrbotten Sweden.

Norrbotten Sweden (1993)

County in N Sweden. Before 1993 also search Norrbotten AND Sweden.
BT Sweden
BT Scandinavia
BT Western Europe
BT Europe
NT Gallivare Sweden
NT Kebnekaise
NT Kiruna Sweden
NT Laisvall Sweden
SA Skellefte

NORSAR (1989)
Acronym.
UF Norwegian Large Aperture Seismic Array
SA arrays
SA Norway
SA Scandinavia
SA seismic networks
SA seismology

Norseman
No longer a valid term for GeoRef. See Norseman Australia.

Norseman Australia (1993)
Town in S central Western Australia.
BT Western Australia
BT Australia
BT Australasia
SA Norseman-Wiluna Belt

Norseman-Wiluna Belt (1993)
W Australia. Mainly gold ores.
BT Western Australia
BT Australia
BT Australasia
SA gold ores
SA Norseman Australia
SA Wiluna Australia

norsethite (1978)
BT carbonates

North Africa (1978)
A region. Autoposting of this term began in 1981. Autoposting to Atlas Mountains began in 1993.
IN Index countries as applicable.
CO N190000N373000
   E0360000W0153000
BT Africa
NT Algeria
NT Atlas Mountains
NT Ceuta
NT Egypt
NT Libya
NT Melilla
NT Morocco
NT Tunisia
NT Western Sahara
SA Central Africa
SA East Africa
SA Southern Africa
SA West Africa

North America (1978)
To retrieve all documents, individual countries and physiographic regions should also be searched (see list O) before 1993. After 1992, only North America and individual countries, Canada, United States, and Mexico, need to be searched. In 1978, autoposting of this term to Appalachians, Great Lakes, Great Lakes region, Great Plains, Gulf Coastal Plain, Rocky Mountains, and Western Interior began. Autoposting of North America to the following terms began in 1981: Basin and Range Province, Colorado Plateau, Columbia Plateau, Great Basin and Mississippi Valley. Autoposting of this term was extended to additional narrower physiographic terms in 1992.
CO N080000N840000
   W0100000W1730000
NT Appalachian Basin
NT Appalachians
NT Avalon Terrane
NT Belt Basin
NT Canadian Shield
NT Cerro Prieto Fault
NT Champlain Valley
NT Chihuahua tectonic belt
NT Chitistone Pass
NT Coast plutonic complex
NT Denali Fault
NT Detroit River
NT Disturbed Belt
NT Eastern Overthrust Belt
NT Glacier National Park
NT Great Lakes
NT Great Lakes region
NT Great Plains
NT Juan de Fuca Strait
NT Keweenawan Rift
NT Kootenay Arc
NT Lake Champlain
NT Lake of the Woods
NT Lake of the Woods region
NT Lake Superior region
NT Leech River Fault
NT Methow Basin
NT Mexicali Valley
NT Michigan Basin
NT Mississippi River basin
NT Missouri River basin
NT Niagara Escarpment
NT Niagara Falls
NT Niagara River
NT North American Cordillera
NT North American Craton
NT Okanagan Valley
NT Okanogan Range
NT Pedregosa Basin
NT Purcell Mountains
NT Rainy River
NT Rio Grande Depression
NT Rio Grande Rift
NT Rocky Mountain Trench
NT Rocky Mountains
NT Rocky Mountains foreland
NT Saganaga Lake
NT Saint Clair River
NT Saint Clair River delta
NT Saint Elias Mountains
NT Saint Lawrence Lowlands
NT Saint Lawrence River
NT Saint Lawrence River basin
NT Saint Lawrence Valley
NT Saint Pierre and Miquelon
NT Shuksan Thrust
NT Skagit Valley
NT Sonoran Desert
NT Straight Creek Fault
NT Strait of Georgia
NT Sweetgrass Arch
NT Tanana River
NT Tintina Fault
NT Transcontinental Arch
NT Western Canada Basin
NT Western Interior
NT Western Overthrust Belt
NT Williston Basin
NT Yukon River
NT Yukon-Tanana Terrane
SA Agua Blanca Fault
SA America
SA Beringia
SA Canada
SA Champlain Sea
SA Coast Mountains
SA Coast Ranges
SA Colorado River basin
SA Columbia River basin
SA Cordilleran ice sheet
SA DNAG
SA Gulf Coastal Plain
SA Lake Agassiz
SA Lake Algonquin
SA Lake Maumee
SA Lake Saint Clair
SA Laurentia
SA Lewis thrust fault
SA Maverick Basin
SA Mexico
SA North American Plate
SA Pacific mobile belt
SA Peninsular Ranges Batholith
SA Polar Continental Shelf
SA Rainy Lake
SA Red River
SA Rio Grande Valley
SA San Jacinto Fault
SA United States
SA White River

North American Atlantic (1981)
Region of the Atlantic Ocean. As of 1990, Atlantic Ocean is autoposted to this term.
BT North Atlantic
BT Atlantic Ocean
NT Baltimore Canyon
NT Baltimore Canyon Trough
NT Caribbean Sea
NT Frobisher Bay
NT Georges Bank basin
NT Gulf of Mexico
NT Gulf of Saint Lawrence
NT Hudson Bay
NT Jeanne d'Arc Basin
NT Labrador Current
NT Labrador Sea
NT Lydonia Canyon
NT Nares abyssal plain
NT New England Seamounts
NT Sohm abyssal plain
SA AMCOR
SA Blake-Bahama Formation
SA Chesapeake Bay
SA Cumberland Basin
SA Delaware Bay
SA DSDP Site 100
SA DSDP Site 102
SA DSDP Site 103
SA DSDP Site 104
SA DSDP Site 108
SA DSDP Site 111
SA DSDP Site 382
SA DSDP Site 384
SA DSDP Site 385
SA DSDP Site 392
SA DSDP Site 533
SA DSDP Site 534
SA DSDP Site 603
SA DSDP Site 604
SA DSDP Site 605
SA DSDP Site 612
SA DSDP Site 613
SA Grand Banks
SA Grand Banks earthquake 1929
SA Hatteras Formation
SA HEBBLE
SA Hibernia Field
SA Hudson Canyon
SA James Bay
SA Leg 11
SA Leg 43
SA Leg 44
SA Leg 51
SA Leg 52
SA Leg 53
SA Leg 76
SA Leg 93
SA Leg 95
SA Leg 105
SA Middle Atlantic Bight
SA ODP Site 628
SA ODP Site 629
SA ODP Site 630
SA ODP Site 631
SA ODP Site 632
SA ODP Site 633
SA ODP Site 634
SA ODP Site 635
SA ODP Site 671
SA ODP Site 672
SA ODP Site 673
SA ODP Site 674
SA ODP Site 675
SA ODP Site 676
SA United States Exclusive Economic Zone

North American Cordillera (1981)
Before 1981, search Cordillera AND North America.
BT North America
NT Canadian Cordillera
SA Alaska Range
SA Canada
SA Cascade Range
SA Coast Mountains
SA Coast Ranges
SA Cordilleran ice sheet
SA Kootenay Arc
SA Mexico
SA Rocky Mountains
SA Sierra Madre
SA Sierra Nevada
SA U. S. Rocky Mountains
SA United States
SA Western Interior Seaway

North American Craton (1993)
Area of North American continent where basement is buried by Precambrian and Phanerozoic sedimentary cover.
BT North America
SA Canadian Shield
SA Midcontinent

North American Pacific (1981)
Region of the Pacific Ocean.
BT Pacific Ocean
NT Axial Seamount
NT Escanaba Trough
NT Gorda Rise
NT Gulf of Alaska
NT Gulf of California
NT Panama Basin
NT Queen Charlotte Basin
NT Quinault Canyon
NT Rivera fracture zone
NT Tamayo fracture zone
SA Cascadia subduction zone
SA DSDP Site 32
SA DSDP Site 35
SA DSDP Site 40
SA DSDP Site 172
SA DSDP Site 173
SA DSDP Site 174
SA DSDP Site 175
SA DSDP Site 176
SA DSDP Site 177
SA DSDP Site 186
SA DSDP Site 419
SA DSDP Site 420
SA DSDP Site 467
SA DSDP Site 468
SA DSDP Site 469
SA DSDP Site 470
SA DSDP Site 471
SA DSDP Site 472
SA DSDP Site 473
SA DSDP Site 474
SA DSDP Site 475
SA DSDP Site 476
SA DSDP Site 477
SA DSDP Site 478
SA DSDP Site 479
SA DSDP Site 480
SA DSDP Site 481
SA DSDP Site 482

SA DSDP Site 483
SA DSDP Site 484
SA DSDP Site 485
SA DSDP Site 575
SA Leg 18
SA Northeast Pacific
SA Queen Charlotte Fault
SA San Pedro Basin
SA United States Exclusive Economic Zone
SA Willapa Bay

North American Plate (1978)
Includes North America N of central America as well as all the North Atlantic Ocean W of the Mid-Ocean Ridge.
SA Cascadia subduction zone
SA Farallon Plate
SA North America
SA plate tectonics
SA plates

North Anatolian Fault (1989)
W Turkey.
BT Turkey
BT Middle East
BT Asia

North Arcot
No longer a valid term for GeoRef. See North Arcot India.

North Arcot India (1993)
District in E central Tamil Nadu, S India.
BT Tamil Nadu India
BT India
BT Indian Peninsula
BT Asia

North Atlantic (1978)
Before 1978, search Atlantic Ocean AND north. Autoposting of this term began in 1981.
BT Atlantic Ocean
NT Azores-Gibraltar Ridge
NT Cape Verde Atlantic
NT European Atlantic
NT Gulf of Guinea
NT Guyanese Atlantic
NT Little Bahama Bank
NT Madeira abyssal plain
NT North American Atlantic
NT North Atlantic Deep Water
NT North Atlantic Ridge
NT Oceanographer fracture zone
NT Vema fracture zone
SA DSDP Site 407
SA DSDP Site 408
SA DSDP Site 410
SA DSDP Site 556
SA DSDP Site 558
SA DSDP Site 561
SA DSDP Site 562
SA DSDP Site 563
SA DSDP Site 564
SA Equatorial Atlantic
SA Iceland Research Drilling Project
SA Kane fracture zone
SA Leg 2
SA Leg 37
SA Leg 45
SA Leg 82
SA Leg 93
SA Leg 95
SA Leg 101
SA Leg 109
SA Northeast Atlantic
SA Northwest Atlantic
SA South Atlantic
SA Southeast Atlantic
SA Southwest Atlantic
SA West Atlantic

North Atlantic Deep Water (1989)

As of 1990, Atlantic Ocean is autoposted to this term.
BT North Atlantic
BT Atlantic Ocean
SA bottom water
SA ocean circulation

North Atlantic Ridge (1981)
As of 1990, Atlantic Ocean is autoposted to this term. This term has multiple hierarchies.
BT1 Mid-Atlantic Ridge
BT1 Atlantic Ocean
BT2 North Atlantic
BT2 Atlantic Ocean
SA DSDP Site 332B
SA DSDP Site 395
SA DSDP Site 606
SA DSDP Site 607
SA Kane fracture zone
SA Leg 37
SA Leg 39
SA Leg 45
SA Leg 106
SA ODP Site 648
SA ODP Site 649
SA ODP Site 664
SA ODP Site 669
SA ODP Site 670
SA Vema fracture zone

North Australian Seas (1981)
Includes seas of the Pacific Ocean just N of Australia. Autoposting of Pacific Ocean to this term began in 1990.
BT West Pacific
BT Pacific Ocean
NT Arafura Sea
NT Timor Sea

North Austrian Alps (1981)
This term has multiple hierarchies.
CO N470000N482200
 E0162000E0093300
BT1 Austria
BT1 Central Europe
BT1 Europe
BT2 Alps
BT2 Europe
SA Hochschwab

North Austrian Crystallines (1981)
CO N480700N490200
 E0160800E0132500
BT Austria
BT Central Europe
BT Europe
SA Lower Austria
SA Upper Austria

North Austrian Molasse (1981)
CO N475500N482700
 E0161700E0124500
BT Austria
BT Central Europe
BT Europe

North Bay
No longer a valid term for GeoRef. As of 1993, see North Bay Ontario.

North Bay Ontario (1993)
City on Lake Nipissing in SE Ontario. Before 1993, also search North Bay AND Ontario.
BT Nipissing District Ontario
BT Ontario
BT Eastern Canada
BT Canada

North Borneo
use Sabah Malaysia

North Bulgarian Hills (1981)
CO N431200N441300
 E0274500E0223000
BT Bulgaria

BT Southern Europe
BT Europe

North Carolina (1978)
Autoposting of this term began in 1978.
CO N335000N363500
 W0753000W0841500
BT United States
NT Alleghany County North Carolina
NT Beaufort County North Carolina
NT Brunswick County North Carolina
NT Buncombe County North Carolina
NT Burke County North Carolina
NT Cabarrus County North Carolina
NT Caldwell County North Carolina
NT Camden County North Carolina
NT Cape Fear Arch
NT Carteret County North Carolina
NT Chatham County North Carolina
NT Cherokee County North Carolina
NT Clay County North Carolina
NT Cleveland County North Carolina
NT Craven County North Carolina
NT Cumberland County North Carolina
NT Dare County North Carolina
NT Davidson County North Carolina
NT Durham County North Carolina
NT Franklin County North Carolina
NT Graham County North Carolina
NT Granville County North Carolina
NT Greene County North Carolina
NT Henderson County North Carolina
NT Jackson County North Carolina
NT Johnston County North Carolina
NT Jones County North Carolina
NT Lee County North Carolina
NT Lincoln County North Carolina
NT Macon County North Carolina
NT Madison County North Carolina
NT Martin County North Carolina
NT McDowell County North Carolina
NT Mecklenburg County North Carolina
NT Mitchell County North Carolina
NT Montgomery County North Carolina
NT Moore County North Carolina
NT New Hanover County North Carolina
NT Northampton County North Carolina
NT Onslow Bay
NT Onslow County North Carolina
NT Orange County North Carolina
NT Pamlico River
NT Pamlico Sound
NT Polk County North Carolina
NT Pungo River
NT Randolph County North Carolina
NT Rockingham County North Carolina
NT Rowan County North Carolina

NT Stanly County North Carolina
NT Stokes County North Carolina
NT Surry County North Carolina
NT Union County North Carolina
NT Wake County North Carolina
NT Warren County North Carolina
NT Washington County North Carolina
NT Wayne County North Carolina
NT Wilson County North Carolina
SA Appalachian Basin
SA Ashe Formation
SA Atlantic Coastal Plain
SA Avalon Terrane
SA Beaufort Formation
SA Black Creek Formation
SA Blue Ridge Mountains
SA Blue Ridge Province
SA Brevard Zone
SA Cape Lookout
SA Carolina Bays
SA Carolina slate belt
SA Castle Hayne Limestone
SA Charlotte Belt
SA Chesapeake Group
SA Chilhowee Group
SA Chowan River Formation
SA Conasauga Group
SA Croatan Formation
SA Dan River basin
SA Duplin Formation
SA Eastover Formation
SA Grandfather Mountain
SA Great Smoky Group
SA Great Smoky Mountains
SA Hayesville Fault
SA Henderson Gneiss
SA Holston Formation
SA Kings Mountain
SA Kings Mountain Belt
SA Kiokee Belt
SA Knox Group
SA Lake Chatuge
SA Lynchburg Formation
SA Moore County Meteorite
SA Murphy Marble
SA Neuse River
SA New River
SA Newark Supergroup
SA Ocoee Series
SA Outer Banks
SA Peedee Formation
SA Piedmont
SA Pungo River Formation
SA Raleigh Belt
SA Roan Supergroup
SA Rome Formation
SA Salisbury Embayment
SA Sand Hills
SA Santee Limestone
SA Shady Dolomite
SA Talladega Group
SA Tallulah Falls Formation
SA Trent Valley
SA Tuscaloosa Formation
SA Waccamaw Formation
SA Wicomico Formation
SA Yorktown Formation

North Caucasus
use Northern Caucasus

North Central (United States)
use Midwest

North China Platform (1993)
E China.
IN Index Chinese provinces as applicable.
BT China
BT Far East
BT Asia

North Dakota (1978)
Autoposting of this term began in 1978.
CO N455500N490000

W0963500W1040500
BT United States
NT Adams County North Dakota
NT Billings County North Dakota
NT Burke County North Dakota
NT Cass County North Dakota
NT Dunn County North Dakota
NT Eddy County North Dakota
NT Golden Valley County North Dakota
NT Grand Forks County North Dakota
NT Grant County North Dakota
NT Logan County North Dakota
NT McHenry County North Dakota
NT McIntosh County North Dakota
NT McKenzie County North Dakota
NT McLean County North Dakota
NT Mercer County North Dakota
NT Morton County North Dakota
NT Nelson County North Dakota
NT Oliver County North Dakota
NT Pierce County North Dakota
NT Ramsey County North Dakota
NT Renville County North Dakota
NT Richland County North Dakota
NT Sheridan County North Dakota
NT Sioux County North Dakota
NT Slope County North Dakota
NT Stark County North Dakota
NT Stutsman County North Dakota
NT Walsh County North Dakota
NT Ward County North Dakota
SA Amsden Formation
SA Bakken Formation
SA Birdbear Formation
SA Bullion Creek Formation
SA Cedar Creek Anticline
SA Charles Formation
SA Colorado Group
SA Dakota Formation
SA Duperow Formation
SA Elk Point Group
SA Fort Union Formation
SA Fox Hills Formation
SA Golden Valley Formation
SA Great Plains
SA Heath Formation
SA Hell Creek Formation
SA Interlake Formation
SA Inyan Kara Group
SA James River
SA Lake Agassiz
SA Lance Formation
SA Little Missouri River basin
SA Ludlow Member
SA Madison Aquifer
SA Madison Group
SA Midcontinent
SA Mission Canyon Limestone
SA Missouri Plateau
SA Missouri River
SA Missouri River basin
SA Missouri River valley
SA Montana Group
SA Newcastle Sandstone
SA Niobrara Formation
SA Pierre Shale
SA Prairie Evaporite
SA Red River
SA Red River Formation
SA Red River valley
SA Sentinel Butte Formation
SA Skull Creek Shale
SA Souris River basin
SA Swift Formation
SA Tongue River
SA Tongue River Member
SA Tullock Member
SA Tyler Formation
SA Wasatch Formation
SA White River Group
SA Williston Basin

SA Winnipeg Formation
SA Winnipegosis Formation
SA Wyoming Province
SA Yellowstone River

North Devon Island
use Devon Island

North German Plain (1978)
Roughly the northern one-third of Germany.
IN Index states as applicable.
BT Germany
BT Central Europe
BT Europe
SA Berlin Germany
SA Brandenburg Germany
SA Hamburg Germany
SA Lower Saxony Germany
SA Mecklenburg-Western Pomerania Germany
SA North Rhine-Westphalia Germany
SA Northeastern German Plain
SA Saxony-Anhalt Germany
SA Schleswig-Holstein Germany

North Greenland
use Northern Greenland

North Horn Formation (1985)
Upper Cretaceous and Paleocene. Central Utah.
IN Index ages as applicable.
SA K-T boundary
SA Paleocene
SA Upper Cretaceous
SA Utah

North Island (1978)
The northernmost of 3 main islands of New Zealand. Also in other locations.
IN Also index county or country as applicable.
SA Auckland New Zealand
SA Bay of Plenty
SA Broadlands
SA Coromandel Peninsula
SA Edgecumbe earthquake 1987
SA Gisborne New Zealand
SA Hawke's Bay New Zealand
SA Kaipara Harbor
SA Lake Taupo
SA Manawatu River valley
SA Mokoia New Zealand
SA New Zealand
SA Northland New Zealand
SA Raukumara Peninsula
SA Ruapehu
SA Taranaki Basin
SA Taranaki New Zealand
SA Tarawera volcanic complex
SA Taupo New Zealand
SA Taupo volcanic zone
SA Te Aroha New Zealand
SA Tongariro
SA Waikato Basin
SA Wairakei
SA Wairarapa
SA Waitemata Group
SA Wanganui New Zealand
SA Wanganui Valley
SA Wellington New Zealand
SA White Island

North Kona
use Kona

North Korea (1978)
Officially Democratic People's Republic of Korea. Republic, on E coast of Asia, bounded on N by China, on NE by Russian Federation, on E by Japan Sea, on S by South Korea, and on W by the Yellow Sea and Korea Bay.
CO N374500N430000

E1303000E1241500
BT Korea
BT Far East
BT Asia
NT Kangwon North Korea
SA South Korea

North Ossetia
No longer a valid term for GeoRef. As of 1993, see North Ossetia Russian Federation.

North Ossetia Russian Federation (1993)
Former North Ossetian Autonomous Soviet Socialist Republic. On N slopes of central Caucasus Mountains. Before 1993, also search North Ossetia. This term has multiple hierarchies.
BT1 Russian Federation
BT1 Commonwealth of Independent States
BT2 Ossetia
BT2 Asia
SA Malgobek Russian Federation
SA Sadon Russian Federation

North Pacific (1978)
Before 1978, search Pacific Ocean AND north.
BT Pacific Ocean
NT Clarion fracture zone
NT Hess Rise
NT Loihi Seamount
SA California Current
SA DSDP Site 67
SA DSDP Site 310
SA East Pacific
SA Equatorial Pacific
SA Kuroshio
SA Leg 19
SA Northeast Pacific
SA Northwest Pacific
SA Pacific Basin
SA South Pacific
SA Southeast Pacific
SA Southwest Pacific
SA West Pacific

North Palm Springs earthquake 1986 (1989)
Riverside County, S California.
BT earthquakes
SA California
SA Riverside County California

North Polar Sea
use Arctic Ocean

North Pole (1978)
The N extremity of the Earth's axis at 90° N latitude and the point from which all directions are S.
BT Arctic region
SA Makarov Basin
SA polar caps
SA polar regions

North Pole Deposit (1981)
Barite deposit in Western Australia. Also search North Pole AND Australia.
BT Western Australia
BT Australia
BT Australasia
SA barite deposits

North Pyrenean Fault (1989)
IN Index countries as applicable.
CO N420000N430000
E0031500W0020000
BT Western Europe
BT Europe
SA France
SA French Pyrenees
SA Pyrenees
SA Spain
SA Spanish Pyrenees

North Ray Crater (1978)
BT Moon

North Rhine (1978)
North portion of former Prussian Rhine Province which included Aachen, Cologne and Dusseldorf.
SA North Rhine-Westphalia Germany

North Rhine-Westphalia
No longer a valid term for GeoRef.
use North Rhine-Westphalia Germany

North Rhine-Westphalia Germany (1993)
CO N502000N523000
E0093000E0060000
UF North Rhine-Westphalia
BT Germany
BT Central Europe
BT Europe
NT Aachen Germany
NT Bergisch Gladbach Germany
NT Bergisches Land
NT Bochum Germany
NT Cologne Germany
NT Ebbe Anticlinorium
NT Ibbenburen Germany
NT Kamen Germany
NT Krefeld Germany
NT Lindlar Germany
NT Lippe
NT Munster Germany
NT Munsterland
NT Paderborn Germany
NT Plettenberg Germany
NT Ruhr
NT Sauerland
NT Siebengebirge
NT Wiehen Mountains
SA Ems River
SA Essen Beds
SA North German Plain
SA North Rhine
SA Rhenish Schiefergebirge
SA Rhineland
SA Steinheim Germany
SA Teutoburg Forest
SA Weser River
SA Westphalia

North Saskatchewan River (1978)
Rises in Columbia Icefield at foot of Mt. Saskatchewan in Alberta.
IN Index provinces as applicable.
BT Canada
SA Alberta
SA Saskatchewan
SA Saskatchewan River
SA South Saskatchewan River

North Sea (1978)
Between the European continent on the S and E, and Great Britain on the W.
CO N510000N611000
E0110000W0040000
UF North Sea Basin
BT European Atlantic
BT North Atlantic
BT Atlantic Ocean
NT Dogger Bank
NT Ekofisk Field
NT Forties Field
NT Kattegat
NT Norwegian Channel
NT Orcadian Basin
NT Skagerrak
NT Viking Graben
SA Brent Group
SA Central Graben
SA Ekofisk Formation
SA Helgoland

SA　Highland Boundary Fault
　　SA　Kimmeridge Clay
　　SA　Moray Firth
　　SA　North Sea region
　　SA　Statfjord Formation
　　SA　Tor Formation

North Sea Basin
　　use North Sea

North Sea Coast　(1978)
　　IN　Index England, Scotland and European countries as applicable.
　　BT　Europe
　　SA　Belgium
　　SA　Denmark
　　SA　England
　　SA　France
　　SA　Germany
　　SA　Netherlands
　　SA　Norway
　　SA　Scotland
　　SA　Sweden

North Sea region　(1978)
　　IN　Index countries as applicable.
　　SA　Belgium
　　SA　Denmark
　　SA　England
　　SA　France
　　SA　Germany
　　SA　Netherlands
　　SA　North Sea
　　SA　Norway
　　SA　Scotland
　　SA　Sweden

North Shore Volcanics　(1978)
　　In Keweenawan Group. NE Minnesota.
　　BT　Precambrian
　　SA　Keweenawan
　　SA　Minnesota

North Siberian Plain　(1981)
　　This term has multiple hierarchies.
　　CO　N700000N750000
　　　　E1300000E0810000
　　BT1　Siberian Platform
　　BT1　Russian Federation
　　BT1　Commonwealth of Independent States
　　BT2　Siberian Platform
　　BT2　Asia

North Slope　(1978)
　　Arctic plains N of Brooks Range to Arctic Ocean. Autoposting of this term began in 1978.
　　IN　Index Northern Alaska as applicable.
　　SA　Arctic Coastal Plain
　　SA　Barrow Field
　　SA　Endicott Group
　　SA　Ivishak Formation
　　SA　National Petroleum Reserve Alaska
　　SA　Northern Alaska
　　SA　Prudhoe Bay
　　SA　Prudhoe Bay Field
　　SA　Simpson Field
　　SA　Umiat Field

North Sudetic Basin　(1978)
　　North part of the Sudetic Basin in the Sudeten Mountains.
　　IN　Index countries as applicable.
　　CO　N502800N510000
　　　　E0164500E0155000
　　BT　Central Europe
　　BT　Europe
　　SA　Czechoslovakia
　　SA　Poland
　　SA　Sudeten Mountains
　　SA　Sudetic Basin

North Victoria Land
　　use Victoria Land

North Vietnam
　　Not a valid GeoRef index term after 1971. Was used on level 1 in subfile G.
　　use Vietnam

North Yorkshire England　(1993)
　　County in NE England. Formed in 1974 from North Riding in Yorkshire County.
　　BT　Yorkshire England
　　BT　England
　　BT　Great Britain
　　BT　United Kingdom
　　BT　Western Europe
　　BT　Europe
　　NT　Ingleborough
　　NT　York England
　　SA　Howgill Fells
　　SA　Ingleton England

North-Central (United States)
　　use Midwest

North-West Frontier Pakistan (1993)
　　W Pakistan. Before 1993, also search North-West Frontier Province.
　　BT　Pakistan
　　BT　Indian Peninsula
　　BT　Asia
　　NT　Chitral
　　NT　Chitral Pakistan
　　NT　Hazara Pakistan
　　NT　Peshawar Pakistan
　　NT　Swat Pakistan
　　SA　Sulaiman Range

North-West Frontier Province
　　No longer a valid term for GeoRef. See North-West Frontier Pakistan.

Northampton County
　　Valid through 1988. Search in combination with state term. After 1988, use specific county-state term.

Northampton County North Carolina　(1989)
　　NE North Carolina. Before 1989, also search Northampton County AND North Carolina.
　　CO　N361200N363300
　　　　W0770500W0775300
　　BT　North Carolina
　　BT　United States

Northampton County Pennsylvania　(1989)
　　E Pennsylvania. Before 1989, also search Northampton County AND Pennsylvania.
　　CO　N403300N405700
　　　　W0750300W0753800
　　BT　Pennsylvania
　　BT　United States

Northampton County Virginia (1989)
　　E Virginia. Before 1989, also search Northampton County AND Virginia.
　　CO　N370700N373400
　　　　W0754000W0760200
　　BT　Virginia
　　BT　United States

Northamptonshire
　　No longer a valid term for GeoRef as of 1993.
　　use Northamptonshire England

Northamptonshire England (1993)
　　County in central England. Before 1993, also search Northamptonshire.
　　CO　N520000N524000
　　　　W0002000W0012000
　　UF　Northamptonshire
　　BT　England
　　BT　Great Britain
　　BT　United Kingdom
　　BT　Western Europe
　　BT　Europe
　　SA　East Midlands
　　SA　Midlands

Northeast Atlantic　(1978)
　　Before 1978, search Atlantic Ocean AND northeast. Also see European Atlantic or Cape Verde Atlantic as applicable.
　　BT　Atlantic Ocean
　　SA　Cape Verde Atlantic
　　SA　DSDP Site 116
　　SA　DSDP Site 397
　　SA　DSDP Site 398
　　SA　DSDP Site 400
　　SA　DSDP Site 401
　　SA　DSDP Site 403
　　SA　DSDP Site 404
　　SA　DSDP Site 405
　　SA　DSDP Site 406
　　SA　DSDP Site 416
　　SA　DSDP Site 552
　　SA　DSDP Site 553
　　SA　DSDP Site 554
　　SA　DSDP Site 555
　　SA　East Atlantic
　　SA　European Atlantic
　　SA　Leg 12
　　SA　Leg 81
　　SA　Leg 94
　　SA　Leg 103
　　SA　North Atlantic
　　SA　Northwest Atlantic
　　SA　South Atlantic
　　SA　Southeast Atlantic
　　SA　Southwest Atlantic
　　SA　West Atlantic

Northeast Japan
　　use Tohoku

Northeast Pacific　(1978)
　　Before 1978, search Pacific Ocean AND northeast.
　　BT　Pacific Ocean
　　NT　California Current
　　SA　DSDP Site 159
　　SA　East Pacific
　　SA　Equatorial Pacific
　　SA　Johnston Island
　　SA　North American Pacific
　　SA　North Pacific
　　SA　Northwest Pacific
　　SA　South Pacific
　　SA　Southeast Pacific
　　SA　Southwest Pacific
　　SA　West Pacific

Northeast Providence Channel (1978)
　　Strait NE of Nassau between Great Abaco and Eleuthera islands.
　　SA　Bahamas
　　SA　Northwest Providence Channel

Northeastern German Plain (1981)
　　CO　N514000N544000
　　　　E0144500E0104000
　　BT　Germany
　　BT　Central Europe
　　BT　Europe
　　SA　North German Plain

Northeastern Hungarian Hills (1981)
　　CO　N474500N483500
　　　　E0214200E0184500
　　BT　Hungary
　　BT　Central Europe
　　BT　Europe

Northeastern India　(1981)
　　CO　N214000N293500
　　　　E0972000E0880000
　　BT　India
　　BT　Indian Peninsula
　　BT　Asia
　　NT　Arunachal Pradesh India
　　NT　Assam India
　　NT　Manipur India
　　NT　Meghalaya India
　　NT　Nagaland India
　　NT　Tripura India

Northeastern Polish Plain (1981)
　　CO　N505500N542700
　　　　E0240500E0181000
　　BT　Poland
　　BT　Central Europe
　　BT　Europe

Northeastern USSR
　　No longer a valid term for GeoRef. Valid 1978-1992. As of 1993, see Russian Federation. Before 1978, also search USSR AND northeast. After 1980, term referred to the region W of the Soviet Pacific region. Autoposting of this term began in 1981.

Northern Alaska　(1993)
　　Artificial region based on U.S. Geological Survey quadrangle designations. Also search specific quadrangles if applicable.
　　IN　Index quadrangles as applicable.
　　CO　N670000N713000
　　　　W1410000W1670000
　　BT　Alaska
　　BT　United States
　　NT　Arctic National Wildlife Refuge
　　NT　Brooks Range
　　NT　National Petroleum Reserve Alaska
　　NT　Prudhoe Bay
　　NT　Prudhoe Bay Field
　　NT　Umiat Field
　　SA　Arctic Coastal Plain
　　SA　Barrow Field
　　SA　Colville River
　　SA　Colville River delta
　　SA　Meade Basin
　　SA　North Slope
　　SA　Point Barrow
　　SA　Trans-Alaska Pipeline
　　SA　Yukon-Koyukuk Basin

Northern Andes　(1978)
　　Before 1978, search Andes AND north or northern. As of 1990, South America is autoposted to this term.
　　BT　Andes
　　BT　South America
　　SA　Central Andes
　　SA　Southern Andes

Northern Apennines　(1981)
　　Before 1981, also search Apennines AND North. Extend SE to Scheggia Pass.
　　CO　N432000N450000
　　　　E0124500E0073000
　　BT　Apennines
　　BT　Italy
　　BT　Southern Europe
　　BT　Europe

Northern Appalachians　(1978)
　　Before 1978, search Appalachians AND north or northern. As of 1990, North America is autoposted to this term.
　　BT　Appalachians

BT North America
SA Central Appalachians
SA Southern Appalachians

Northern California (1981)
Autoposting of broader terms to this term began in 1989. Before 1981, also search California AND North.
BT California
BT United States
SA Josephine Ophiolite
SA Josephine Peridotite
SA Sacramento River
SA Shoo Fly Complex

Northern Caucasia
use Northern Caucasus

Northern Caucasus (1978)
N part of the Caucasus comprising former Chechen-Ingush ASSR, S half of Krasnodar Kray, and the Daghestan, Kabardin-Balkar, and North Ossetian former ASSRs. Autoposting of Russian Republic to this term began in 1989. This term has multiple hierarchies.
CO N440000N460000
E0450000E0380000
UF Ciscaucasia
UF North Caucasus
UF Northern Caucasia
BT1 Caucasus
BT1 Europe
BT2 Russian Federation
BT2 Commonwealth of Independent States
NT Peredovoy Range
NT Sunzha
SA Caucasus Foreland
SA Stavropol Russian Federation
SA Urup mining district

Northern China
No longer a valid term for GeoRef. See provinces: Hebei China, Inner Mongolia China, Shanxi China, and Beijing China and Tianjin China.

Northern Dvina River (1978)
Chief river of the White Sea Basin in NW European Russian Federation. This term has multiple hierarchies.
BT1 Russian Federation
BT1 Commonwealth of Independent States
BT2 Europe
SA Arkhangelsk Russian Federation
SA Komi Russian Federation

Northern Europe
Not a valid term for GeoRef. Use Western Europe.

Northern German Hills (1981)
CO N511700N522800
E0120700E0073500
BT Germany
BT Central Europe
BT Europe

Northern Great Plains (1978)
Before 1978, search Great Plains AND north or northern. As of 1990, North America is autoposted to this term.
BT Great Plains
BT North America
SA Southern Great Plains

Northern Greenland (1985)
UF North Greenland
BT Greenland
BT Arctic region

Northern Hemisphere (1978)
Used when discussing many large areas too numerous to mention.
SA Eastern Hemisphere
SA Laurasia
SA Pangaea
SA Southern Hemisphere
SA Western Hemisphere

Northern Highlands (1981)
After 1993, see Scottish Northern Highlands if applicable.
IN Index Mars or Florida if applicable.
SA Florida
SA Mars

Northern Ireland (1978)
Divided into 26 administrative regions comprising the NE part of island of Ireland. Before 1974, was divided into the traditional counties of Antrim, Armagh, Down, Fermangh, Londonderry and Tyrone. Before 1978, also search Ulster.
CO N540000N553000
W0053000W0080000
BT United Kingdom
BT Western Europe
BT Europe
NT Antrim Northern Ireland
NT Belfast Northern Ireland
NT Fermanagh Northern Ireland
NT Giant's Causeway
NT Londonderry Northern Ireland
NT Tyrone Northern Ireland
SA Ireland
SA Sherwood Sandstone

Northern Light Lake (1978)
In SW Ontario just N of Minnesota border.
BT Thunder Bay District Ontario
BT Ontario
BT Eastern Canada
BT Canada

Northern Limestone Alps (1978)
The northern part of the Dinaric Alps which are a great belt of limestone ranges and plateaus along the Dalmatian coast of the Adriatic Sea. This term has multiple hierarchies.
BT1 Dinaric Alps
BT1 Southern Europe
BT1 Europe
BT2 Dinaric Alps
BT2 Eastern Alps
BT2 Alps
BT2 Europe
BT3 Croatia
BT3 Yugoslavia
BT3 Southern Europe
BT3 Europe

Northern Nigeria
No longer a valid term for GeoRef. Valid 1981-1992.

Northern Norway (1981)
S boundary is Trondheim.
CO N633000N711000
E0310500E0093000
BT Norway
BT Scandinavia
BT Western Europe
BT Europe

Northern Peninsula, Michigan
use Michigan Upper Peninsula

Northern Range (1978)
N Trinidad.
IN Index Trinidad as applicable.
SA Trinidad
SA Trinidad and Tobago

Northern Rhodesia
use Zambia

Northern Rocky Mountains (1978)
Before 1978, also search Rocky Mountains AND north or northern. As of 1990, North America is autoposted to this term.
BT Rocky Mountains
BT North America
SA Canadian Rocky Mountains
SA Central Rocky Mountains
SA Rocky Mountain Trench
SA Southern Rocky Mountains
SA U. S. Rocky Mountains

Northern Swiss Alps (1981)
This term has multiple hierarchies.
CO N460700N472200
E0093800E0064600
BT1 Swiss Alps
BT1 Switzerland
BT1 Central Europe
BT1 Europe
BT2 Swiss Alps
BT2 Alps
BT2 Europe

Northern Territory
No longer a valid term for GeoRef. See Northern Territory Australia.

Northern Territory Australia (1993)
CO S260000S110000
E1380000E1290000
BT Australia
BT Australasia
NT Alice Springs Australia
NT Alligator Rivers Field
NT Arunta Block
NT Gosses Bluff
NT Harts Range
NT Jabiluka Australia
NT Katherine Australia
NT Ngalia Basin
NT Pine Creek Geosyncline
NT Tennant Creek Australia
SA Amadeus Basin
SA Arunta Complex
SA Bathurst Island
SA Bitter Springs Formation
SA Cahill Formation
SA Carpentaria Basin
SA Georgina Basin
SA Great Artesian Basin
SA Henbury Meteorite
SA Joseph Bonaparte Gulf
SA Kombolgie Formation
SA McArthur Basin
SA Melville Island
SA Musgrave Ranges
SA Ranger Mine
SA Simpson Desert
SA Toolebuc Formation
SA Victoria Valley

Northern Urals (1978)
Autoposting of Russian Republic to this term began in 1989. Before 1978, search Urals AND north or northern. This term has multiple hierarchies.
CO N600000N650000
E0620000E0560000
BT1 Russian Federation
BT1 Commonwealth of Independent States
BT2 Urals
BT2 Commonwealth of Independent States
SA Polar Urals

Northland New Zealand (1993)
Provincial district, N North Island.
IN Also index North Island.
CO S363000S342000

E1743000E1724000
BT New Zealand
BT Australasia
SA North Island

Northumberland
No longer a valid term for GeoRef. As of 1993, see Northumberland England.

Northumberland County
No longer a valid term for GeoRef. As of 1993, see Northumberland County Ontario or Northumberland County New Brunswick. As of 1989, for counties in Pennsylvania and Virginia, use specific county-state term.

Northumberland County New Brunswick (1993)
Bordering Gulf of Saint Lawrence in N central New Brunswick. Before 1993, also search Northumberland or Northumberland County in combination with New Brunswick.
BT New Brunswick
BT Maritime Provinces
BT Eastern Canada
BT Canada
NT Newcastle New Brunswick

Northumberland County Ontario (1993)
Bordering Lake Ontario in S Ontario. Before 1993, also search Northumberland or Northumberland County AND Ontario.
BT Ontario
BT Eastern Canada
BT Canada

Northumberland County Pennsylvania (1989)
E central Pennsylvania. Before 1989, also search Northumberland County AND Pennsylvania.
CO N403700N411100
W0762200W0765600
BT Pennsylvania
BT United States

Northumberland County Virginia (1989)
E Virginia. Before 1989, also search Northumberland County AND Virginia.
CO N374200N380200
W0761500W0763800
BT Virginia
BT United States

Northumberland England (1993)
County in N England on border of Scotland. In 1974, S part of county was taken to form Tyne and Wear. Before 1993, also search Northumberland AND England.
CO N544500N554500
W0013000W0023000
BT England
BT Great Britain
BT United Kingdom
BT Western Europe
BT Europe
NT Meldon Aplite
SA Dudley England
SA Newcastle England

Northumberland Strait (1978)
Channel of the Gulf of Saint Lawrence between Prince Edward Island on the N, and New Brunswick and Nova Scotia on the SE and E respectively.
BT Canada

Northwest Atlantic (1978)
Before 1978, search Atlantic Ocean AND northwest.
BT Atlantic Ocean
NT Bermuda Rise
NT Blake Plateau
SA DSDP Site 4
SA DSDP Site 386
SA DSDP Site 387
SA DSDP Site 390
SA East Atlantic
SA Leg 12
SA Leg 102
SA Leg 110
SA North Atlantic
SA Northeast Atlantic
SA South Atlantic
SA South Sandwich Islands
SA Southeast Atlantic
SA Southwest Atlantic
SA West Atlantic

Northwest Pacific (1978)
Before 1978, search Pacific Ocean AND northwest.
BT Pacific Ocean
NT Nankai Trough
NT Okinawa Trough
SA DSDP Site 194
SA DSDP Site 195
SA DSDP Site 196
SA DSDP Site 197
SA DSDP Site 198
SA DSDP Site 298
SA DSDP Site 434
SA DSDP Site 435
SA DSDP Site 436
SA DSDP Site 439
SA DSDP Site 440
SA DSDP Site 458
SA DSDP Site 459
SA DSDP Site 463
SA DSDP Site 464
SA DSDP Site 465
SA DSDP Site 466
SA DSDP Site 576
SA DSDP Site 581
SA DSDP Site 582
SA DSDP Site 583
SA DSDP Site 584
SA East Pacific
SA Equatorial Pacific
SA Leg 86
SA Leg 87
SA Leg 88
SA Nankaido earthquake 1946
SA North Pacific
SA Northeast Pacific
SA South Pacific
SA Southeast Pacific
SA Southwest Pacific
SA West Pacific

Northwest Providence Channel (1989)
Channel between Grand Bahama Island and Bimini, linking Straits of Florida to Northeast Providence Channel.
SA Bahamas
SA Bimini
SA Great Bahama Bank
SA Leg 101
SA Little Bahama Bank
SA Northeast Providence Channel
SA ODP Site 634
SA ODP Site 635
SA Straits of Florida

Northwest Shelf
Not a valid GeoRef term. See Western Australia; Northern Territory; continental shelf.

Northwest Territories (1978)
As of 1993, Western Canada is autoposted to this term.
IN Index districts as applicable.
CO N600000N840000
    W0600000W1360000
BT Western Canada
BT Canada
NT Agricola Lake
NT Amund Ringnes Island
NT Belcher Islands
NT Contwoyto Lake
NT Coppermine River
NT Cornwallis Island
NT Franklin District Northwest Territories
NT Keewatin District Northwest Territories
NT Lancaster Sound
NT Mackenzie District Northwest Territories
NT Nonacho Basin
SA Arctic Coastal Plain
SA Baffin Bay
SA Baffin Island
SA Banks Island
SA Bathurst Island
SA Bear Province
SA Bear-Slave Operation
SA Cache Creek
SA Canadian Shield
SA Carswell Structure
SA Christopher Formation
SA Churchill Province
SA Cumberland Peninsula
SA Edmonton Formation
SA Eureka Sound Formation
SA Franklin Mountains
SA Graham Island
SA Horn Plateau Formation
SA Hudson Bay
SA Hudson Bay Lowlands
SA Hurwitz Group
SA Isachsen Formation
SA James Bay
SA Liard River
SA Mackenzie Mountains
SA Melville Island
SA Michelle Formation
SA Nahanni earthquake 1985
SA Nahanni Formation
SA Nares Strait
SA Peel River
SA Peel Sound Formation
SA Prince of Wales Island
SA Ramparts Formation
SA Read Bay Formation
SA Richardson Mountains
SA Road River Formation
SA Rocknest Formation
SA Root River
SA Selwyn Mountains
SA Slave Point Formation
SA Slave Province
SA Swan Hills Formation
SA Victoria Island
SA Yellowknife Group

Northwestern China
No longer a valid term for GeoRef. See provinces: Gansu China, Ningxia China, Qinghai China, Shaanxi China, and Xinjiang China.

Northwestern German Plain (1981)
CO N521700N550500
    E0113500E0064000
BT Germany
BT Central Europe
BT Europe

Northwestern Polish Plain (1981)
CO N495500N545500
    E0220000E0141000
BT Poland

BT Central Europe
BT Europe
SA Polish Lowland

Northwestern Transdanubia (1981)
Autoposting of Hungary to this term began in 1989. As of 1990, Central Europe and Europe are autoposted to this term.
CO N465500N480200
    E0182000E0161500
BT Transdanubia
BT Hungary
BT Central Europe
BT Europe

Northwestern USSR
No longer a valid term for GeoRef. As of 1993, see Russian Federation or Siberia. Before 1978, search USSR AND northwest.

Norton Basin (1985)
Off W Coast of Alaska.
IN Index Alaska and Alaska regions as applicable.
SA Alaska
SA Bering Sea
SA Norton Sound

Norton County
Valid through 1988. Search in combination with state term. After 1988, use specific county-state term.

Norton County Kansas (1989)
NW Kansas. Before 1989, also search Norton County AND Kansas.
CO N393500N400000
    W0994000W1001300
BT Kansas
BT United States
SA Norton County Meteorite

Norton County Mcteorite (1978)
Used to distinguish from place name, Norton County in Kansas. Before 1978, also search Norton County for the meteorite.
BT chladnite
BT achondrites
BT stony meteorites
BT meteorites
SA Kansas
SA Norton County Kansas

Norton Sound (1981)
Arm of the Bering Sea, S of Seward Peninsula, W Alaska.
CO N631600N645500
    W1604500W1660000
BT Bering Sea
BT West Pacific
BT Pacific Ocean
SA Norton Basin

Norumbega fault zone (1989)
E Maine.
CO N430000N473000
    W0670000W0710500
BT Maine
BT United States
SA Appalachians

Norway (1978)
CO N580000N710000
    E0310000E0040000
BT Scandinavia
BT Western Europe
BT Europe
NT Arendal Norway
NT Bergen Norway
NT Bjerkrem-Sogndal Massif
NT Drammen Norway
NT Finnmark Norway
NT Folldal Norway

NT Gudbrandsdalen
NT Hardangervidda
NT Jotunheim Mountains
NT Kongsberg Norway
NT Kristiansund Norway
NT Nord Fjord
NT Nord-Trondelag Norway
NT Nordland Norway
NT Northern Norway
NT Oslo Graben
NT Oslo Norway
NT Ringerike
NT Rogaland Norway
NT Sogn
NT Solund Islands
NT Sor-Trondelag Norway
NT Soroy
NT Southern Norway
NT Telemark Norway
NT Troms Norway
NT Trondelag
NT Tunsbergdalsbreen
NT Valdres
NT Vest-Agder Norway
NT Vesteralen
NT Vestfold Norway
SA Arctic region
SA Baltic Glaciation
SA Baltic ice lake
SA Baltic Shield
SA Ekofisk Formation
SA Fennoscandia
SA Koli Nappe
SA Lapland
SA Litorina Sea
SA NORESS
SA NORSAR
SA North Sea Coast
SA North Sea region
SA Porsanger Dolomite Formation
SA Ringerike Sandstone
SA Sparagmite Group
SA Sulitjelma
SA Yoldia Sea

Norwegian Channel (1978)
Undersea feature just off SW Norway.
UF Norwegian Deep
UF Norwegian Trough
BT North Sea
BT European Atlantic
BT North Atlantic
BT Atlantic Ocean

Norwegian Deep
use Norwegian Channel

Norwegian Large Aperture Seismic Array
use NORSAR

Norwegian Regional Seismic Array
use NORESS

Norwegian Sea (1978)
Off coast of Norway, opening N on Greenland Sea, NE on Barents Sea, S on North Sea, and SW on the open Atlantic. In 1981, broader term changed from Atlantic Ocean to Arctic Ocean.
BT Arctic Ocean
NT Jan Mayen Ridge
NT Knipovich Ridge
NT Voring Plateau
SA Atlantic Ocean
SA DSDP Site 336
SA DSDP Site 338
SA DSDP Site 343
SA DSDP Site 345
SA DSDP Site 346
SA Jan Mayen
SA Kolbeinsey Island
SA Leg 104
SA ODP Site 642

SA ODP Site 643
SA ODP Site 644

Norwegian Trough
  use Norwegian Channel

Norwich
  No longer a valid term for GeoRef as of 1993. See Norwich England.

Norwich England (1993)
  City in E England. Before 1993, also search Norwich and England.
  BT Norfolk England
  BT East Anglia
  BT England
  BT Great Britain
  BT United Kingdom
  BT Western Europe
  BT Europe

nosean (1978)
  UF noselite
  BT sodalite group
  BT framework silicates
  BT silicates

noseeum
  use carlin-type

noselite
  use nosean

Nothofagus (1985)
  Genus.
  BT Dicotyledoneae
  BT angiosperms
  BT Spermatophyta
  BT Plantae

Nothosauria (1981)
  Suborder.
  BT Sauropterygia
  BT Diapsida
  BT Reptilia
  BT Tetrapoda
  BT Vertebrata
  BT Chordata

Noto Peninsula (1978)
  Large headland projecting N into Japan Sea in central Honshu.
  IN Index prefecture as applicable.
  BT Honshu
  BT Japan
  BT Far East
  BT Asia
  SA Hokuriku
  SA Ishikawa Japan
  SA Nanao Japan
  SA Toyama Japan

Notoungulata (1978)
  Order. Autoposting of Eutheria and Theria to this term began in 1989.
  UF Notungulata
  BT Eutheria
  BT Theria
  BT Mammalia
  BT Tetrapoda
  BT Vertebrata
  BT Chordata

Notre Dame Bay (1978)
  Inlet of Atlantic Ocean on N coast of island of Newfoundland.
  BT Newfoundland Island
  BT Newfoundland
  BT Eastern Canada
  BT Canada

Nottingham
  No longer a valid term for GeoRef as of 1993. See Nottingham England.

Nottingham England (1993)
  City in N central England. Before 1993, also search Nottingham. Before 1978, Nottingham may have been used for the county.
  BT Nottinghamshire England
  BT England
  BT Great Britain
  BT United Kingdom
  BT Western Europe
  BT Europe

Nottinghamshire
  No longer a valid term for GeoRef as of 1993.
  use Nottinghamshire England

Nottinghamshire England (1993)
  County in N central England. Before 1978, also search Nottingham. Before 1993, also search Nottinghamshire.
  CO N524500N533000 W0004000W0013000
  UF Nottinghamshire
  BT England
  BT Great Britain
  BT United Kingdom
  BT Western Europe
  BT Europe
  NT Nottingham England
  SA East Midlands
  SA Midlands

Notungulata
  use Notoungulata

Nouakchott
  No longer a valid term for GeoRef. See Nouakchott Mauritania.

Nouakchott Mauritania (1993)
  City in SW Mauritania near coast. Before 1993 also search Nouakchott AND Mauritania.
  BT Mauritania
  BT West Africa
  BT Africa

Noumea
  No longer a valid term for GeoRef. As of 1993 see Noumea New Caledonia.

Noumea New Caledonia (1993)
  Town on SW coast of French island of New Caledonia E of Queensland, Australia. Before 1993 also search Noumea AND New Caledonia.
  BT New Caledonia
  BT Melanesia
  BT Oceania

Nova Scotia (1978)
  CO N433000N470000 W0594500W0661500
  BT Maritime Provinces
  BT Eastern Canada
  BT Canada
  NT Antigonish County Nova Scotia
  NT Cape Breton County Nova Scotia
  NT Cape Breton Island
  NT Chedabucto Bay
  NT Cobequid Bay
  NT Cobequid Fault
  NT Cobequid Highlands
  NT Halifax County Nova Scotia
  NT Minas Basin
  NT Pictou County Nova Scotia
  NT Sable Island
  NT Victoria County Nova Scotia
  NT Yarmouth County Nova Scotia
  SA Annapolis Valley
  SA Avalon Terrane
  SA Avalon Zone
  SA Blue Mountain
  SA Cape Sable
  SA Cumberland Basin
  SA Cumberland Group
  SA Fisset Brook Formation
  SA Fourchu Group
  SA Gander Zone
  SA George River Group
  SA Goldenville Formation
  SA Halifax Formation
  SA Horton Group
  SA Loch Lomond
  SA Meguma Group
  SA Morien Group
  SA Pictou Group
  SA South Mountain Batholith
  SA Southern Uplands
  SA Stellarton Group
  SA Sydney coal field
  SA Windsor Group

Nova Scotia Shelf
  use Scotian Shelf

Nova Scotian Shelf
  use Scotian Shelf

novaculite (1978)
  UF razor stone
  BT clastic rocks
  BT sedimentary rocks
  SA chemically precipitated rocks

Novaky (1978)
  Village in W central Slovakia, E Czechoslovakia.
  BT Slovakia
  BT Czechoslovakia
  BT Central Europe
  BT Europe

Novara
  No longer a valid term for GeoRef. See Novara Italy.

Novara Italy (1993)
  City and province in NW Italy.
  BT Piemonte Italy
  BT Italy
  BT Southern Europe
  BT Europe

Novaya Zemlya (1978)
  Two large islands of Arkhangelsk Oblast between Barents and Kara seas. Autoposting of Russian Republic to this term began in 1989. This term has multiple hierarchies.
  CO N700000N770000 E0700000E0500000
  BT1 Arkhangelsk Russian Federation
  BT1 Russian Federation
  BT1 Commonwealth of Independent States
  BT2 Arkhangelsk Russian Federation
  BT2 Europe
  BT3 Urals
  BT3 Commonwealth of Independent States

Novgorod
  No longer a valid term for GeoRef. As of 1993, see Novgorod Russian Federation.

Novgorod Russian Federation (1993)
  Oblast and city S of Saint Petersburg in NW Russian Republic. Before 1993, also search Novgorod. This term has multiple hierarchies.
  BT1 Russian Federation
  BT1 Commonwealth of Independent States
  BT2 Europe

Novoraskoe Series
  use Novorayskoe Formation

Novorayskoe Formation (1978)
  In Donets Basin in E Ukraine.
  UF Novoraskoe Series
  UF Novorayskoye Formation
  BT Lower Jurassic
  BT Jurassic
  BT Mesozoic
  SA Donets Basin
  SA Ukraine

Novorayskoye Formation
  use Novorayskoe Formation

Novosibirsk
  No longer a valid term for GeoRef. As of 1993, see Novosibirsk Russian Federation.

Novosibirsk Russian Federation (1993)
  Oblast and city N of NE Kazakhstan in W Siberia. Before 1993, also search Novosibirsk. This term has multiple hierarchies.
  BT1 Russian Federation
  BT1 Commonwealth of Independent States
  BT2 Asia
  SA Salym
  SA Vodino Russian Federation

Nowa Ruda
  No longer a valid term for GeoRef. As of 1993 see Nowa Ruda Poland.

Nowa Ruda Poland (1993)
  Town near the border of Czechoslovakia in W Walbrzych, SW Poland. Before 1993 also search Nowa Ruda AND Poland.
  BT Walbrzych Poland
  BT Poland
  BT Central Europe
  BT Europe

Nowa Sol
  No longer a valid term for GeoRef. As of 1993 see Nowa Sol Poland.

Nowa Sol Poland (1993)
  City on the Odor River in W central Zielona Gora, W Poland. Before 1993 also search Nowa Sol AND Poland.
  BT Zielona Gora Poland
  BT Poland
  BT Central Europe
  BT Europe

Nowy Sacz
  No longer a valid term for GeoRef. As of 1993 see Nowy Sacz Poland.

Nowy Sacz Poland (1993)
  City and province in N foothills of Carpathians in S Poland. Before 1993 also search Nowy Sacz AND Poland.
  BT Poland
  BT Central Europe
  BT Europe
  NT Grzybow
  NT Jordanow Poland
  NT Nowy Targ Poland
  NT Zakopane Poland
  SA Beskid Mountains
  SA Polish Carpathians

Nowy Targ
  No longer a valid term for GeoRef. As of 1993 see Nowy Targ Poland.

Nowy Targ Poland (1993)
  City at foot of Tatra Mountains in W central Nowy Sacz, S Poland. Before 1993 also search Nowy Targ AND Poland.

UF Newmarkt
BT Nowy Sacz Poland
BT Poland
BT Central Europe
BT Europe
SA Cracow Poland

Np
use neptunium

NPRA
use National Petroleum Reserve Alaska

NRC
use U. S. Nuclear Regulatory Commission

NRM
use natural remanent magnetization

NSF (1985)
Acronym.
UF National Science Foundation
BT government agencies
SA policy
SA United States

nsutite (1978)
BT oxides
SA vernadite

Nubia (1978)
Region in Nile Valley extending from about 16° N to include Aswan and First Cataract.
IN Index countries as applicable.
BT Africa
SA Egypt
SA Nile Valley
SA Sudan

Nubia Group
use Nubian Sandstone

Nubian Sandstone (1978)
Cretaceous?. Also search Nubia Group.
UF Nubia Group
BT Cretaceous
BT Mesozoic
SA Africa
SA Egypt
SA Sudan

Nubian Shield (1981)
E Egypt, E Sudan, and N Ethiopia.
CO N140000N273000
E0410000E0300000
BT Africa
SA Egypt
SA Ethiopia
SA Sudan

nuclear energy (1978)
Also search nuclear AND energy. Before 1971, also search atomic energy.
UF atomic energy
SA energy
SA energy sources
SA Euratom
SA nuclear facilities

nuclear explosions (1978)
BT explosions
SA engineering geology
SA seismology
SA shatter cones

nuclear facilities (1978)
Also search nuclear AND facilities.
UF facilities, nuclear
SA design
SA earthquakes
SA engineering geology
SA faults
SA feasibility studies
SA foundations
SA geologic hazards
SA impact statements
SA marine installations
SA nuclear energy
SA power plants
SA rock mechanics
SA seepage
SA seismic response
SA seismic risk
SA site exploration
SA soil mechanics

nuclear fallout
use fallout

nuclear fission
use fission

nuclear fusion
use fusion

nuclear geology
No longer a valid GeoRef index term. Before 1971, was used on level 1 in subfile N. See nuclear energy; nuclear explosions; nuclear facilities; isotopes.

nuclear magnetic resonance (1978)
As of 1982, for methodology only; use NMR spectra for data.
BT spectroscopy
SA analysis
SA chemical analysis
SA electron paramagnetic resonance
SA NMR spectra
SA resonance

nuclear magnetic resonance spectra
use NMR spectra

nuclear science
Not a valid GeoRef index term after 1970. Was used on level 1 in subfile N. Now use isotopes, nuclear facilities, nuclear energy, or nuclear explosions.

nuclear waste
use radioactive waste

nuclear waste repositories
Not a valid term for GeoRef. After 1993, see storage or underground storage and radioactive waste or waste disposal.

nucleation (1978)
Restricted to use under crystal growth.
SA crystal growth

nucleosynthesis
Not a valid GeoRef index term after 1971. Was used on level 1 in subfile G. Now use cosmochemistry or radioactivity as applicable.

Nuculanidae (1978)
BT Bivalvia
BT Mollusca
BT Invertebrata
SA Nuculidae

Nuculidae (1978)
BT Bivalvia
BT Mollusca
BT Invertebrata
SA Nuculanidae

Nueces County
Valid through 1988. Search in combination with state term. After 1988, use specific county-state term.

Nueces County Texas (1989)
On Gulf of Mexico in S Texas. Before 1989, also search Nueces County AND Texas.
CO N273000N275500
W0965000W0975500
BT Texas
BT United States
SA Corpus Christi Bay
SA Mustang Island

Nueces River (1978)
Flows into Nueces Bay at head of Corpus Christi Bay in S Texas.
BT Texas
BT United States

nuee ardentes
use nuees ardentes

nuees ardentes (1993)
UF nuee ardentes
SA ash flows
SA base surges
SA eruptions
SA gases
SA igneous activity
SA igneous rocks
SA pyroclastic flows
SA volcanic ash
SA volcanism
SA volcanoes

Nueva Esparta
No longer a valid term for GeoRef. As of 1993 see Nueva Esparta Venezuela.

Nueva Esparta Venezuela (1993)
State comprising an island group in the Caribbean Sea off N coast. Before 1993 also search Nueva Esparta AND Venezuela.
BT Venezuela
BT South America
NT Macanao Peninsula
NT Margarita Island

Nuevo Leon
No longer a valid term for GeoRef. As of 1993 see Nuevo Leon Mexico.

Nuevo Leon Mexico (1993)
State in W Mexico. Before 1993 also search Nuevo Leon AND Mexico.
BT Mexico
SA Rio Grande
SA Rio Grande Valley
SA San Juan River
SA Sierra Madre Oriental

Nugget Sandstone (1985)
SW Wyoming, SE Idaho and NE Utah.
BT Lower Jurassic
BT Jurassic
BT Mesozoic
SA Idaho
SA Navajo Sandstone
SA Utah
SA Wyoming

nuggets (1989)
SA copper ores
SA gold ores
SA mineral deposits, genesis
SA placers
SA platinum ores
SA silver ores

Nugssuak
use Nugssuaq

Nugssuaq (1978)
Peninsula in W central coast of Greenland.
UF Nugssuak
BT Greenland
BT Arctic region

Nuk Gneiss (1989)
S West Greenland.
BT Archean
BT Precambrian
SA Greenland
SA West Greenland

Nullarbor Plain (1978)
Extends inland along entire Great Australian Bight in S part of country.
IN Index states as applicable.
SA Australia
SA Mundrabilla Australia
SA South Australia
SA Western Australia

numerical analysis (1978)
Also search numerical methods.
SA analysis
SA finite difference analysis
SA finite element analysis
SA geostatistics
SA mathematical methods
SA numerical models
SA sensitivity analysis
SA statistical analysis

numerical filters (1981)
SA filters
SA inverse problem

numerical methods
As of 1993, not a valid term for GeoRef. See numerical analysis or mathematical methods.

numerical models (1985)
SA mathematical models
SA models
SA numerical analysis
SA PHREEQE
SA Richards equation

numerical seismograms (1981)
BT seismograms
SA synthetic seismograms

Nummulites (1978)
Genus. Autoposting of microfossils and Protista to this term began in 1990. This term has multiple hierarchies.
BT1 Nummulitidae
BT1 Rotaliacea
BT1 Rotaliina
BT1 foraminifera
BT1 Protista
BT1 Invertebrata
BT2 Nummulitidae
BT2 Rotaliacea
BT2 Rotaliina
BT2 foraminifera
BT2 Protista
BT2 microfossils
NT Operculina

Nummulitidae (1978)
Autoposting of microfossils and Protista to this term began in 1990. This term has multiple hierarchies.
BT1 Rotaliacea
BT1 Rotaliina
BT1 foraminifera
BT1 Protista
BT1 Invertebrata
BT2 Rotaliacea
BT2 Rotaliina
BT2 foraminifera
BT2 Protista
BT2 microfossils
NT Nummulites
SA Heterostegina

nunataks (1985)
Autoposting of glacial features to this term began in 1989.
BT glacial features
SA glacial geology
SA glaciers

Nunivak Island (1978)

Second largest island in Bering Sea. Off W coast of Alaska.
BT Southwestern Alaska
BT Alaska
BT United States
SA Bering Sea

Nura-Tau (1978)
Range in N Samarkand Oblast. This term has multiple hierarchies.
CO N380000N400000
    E0660000E0640000
UF Nuratau
BT1 Uzbekistan
BT1 Asia
BT2 Uzbekistan
BT2 Commonwealth of Independent States
SA Samarkand Uzbekistan

Nuratau
use Nura-Tau

NURE (1985)
Acronym.
UF National Uranium Resource Evaluation Program
SA policy
SA programs
SA strategic minerals
SA United States
SA uranium ores

Nurek
No longer a valid term for GeoRef. As of 1993, see Nurek Tadzhikistan.

Nurek Tadzhikistan (1993)
Village in NE Dushanbe region in W Tadzhikistan. This term has multiple hierarchies.
BT1 Tadzhikistan
BT1 Asia
BT2 Tadzhikistan
BT2 Commonwealth of Independent States

Nuremberg
No longer a valid term for GeoRef. See Nuremberg Germany.

Nuremberg Germany (1993)
City in N central Bavaria, SE Germany. Before 1978, also search Nurnberg.
BT Bavaria Germany
BT Germany
BT Central Europe
BT Europe

Nurnberg
Not a valid term for GeoRef. See Nuremberg Germany.

nutation (1978)
Term reintroduced in 1985. Not a valid term from 1972 through 1984.
SA Earth
SA motions

nutrients (1978)
SA diet
SA fertilization
SA fertilizers
SA major elements
SA minor elements
SA nutrition
SA soils
SA trace elements

nutrition (1978)
SA diet
SA metabolism
SA nutrients
SA paleontology
SA predation
SA trophic analysis

Nuwuk
See Barrow Alaska for the Eskimo village in Point Barrow region.

Ny Friesland (1978)
Region on NE Spitsbergen Island.
BT Spitsbergen Island
BT Spitsbergen
BT Svalbard
BT Arctic region

Nyasaland
use Malawi

Nye County
Valid through 1988. Search in combination with state term. After 1988, use specific county-state term.

Nye County Nevada (1989)
S and central Nevada. Before 1989, also search Nye County AND Nevada.
CO N360000N390800
    W1150000W1181500
BT Nevada
BT United States
NT Beatty Nevada
NT Pahute Mesa
NT Rainier Mesa
NT Tonopah Nevada
NT Trap Spring Field
NT Yucca Mountain
SA Death Valley
SA Egan Range
SA Grant Range
SA Mount Jefferson
SA Railroad Valley
SA Shoshone Mountains
SA Toiyabe Range
SA Toquima Range
SA Yucca Flat

Nyeman River basin
use Neman River basin

Nyiragongo (1978)
Volcano at N end of Lake Kivu in E Zaire.
UF Nyiragongo Volcano
BT Kivu Zaire
BT Zaire
BT Central Africa
BT Africa

Nyiragongo Volcano
use Nyiragongo

Nykoeping
use Nykoping Sweden

Nykoping
No longer a valid term for GeoRef. As of 1993 see Nykoping Sweden.

Nykoping Sweden (1993)
City on Baltic Sea in SE Sweden. Before 1993 also search Nykoping AND Sweden.
UF Nykoeping
BT Sweden
BT Scandinavia
BT Western Europe
BT Europe

Nyos
use Lake Nyos

O

O
use oxygen

O'Leary Peak (1978)
Peak in Coconino County, N central Arizona.
IN Index county or region as applicable.
SA Arizona
SA Coconino County Arizona

O'Sullivan Lake (1985)
Lake in N Ontario. Also a lake in E Quebec.
IN Index provinces or regions as applicable.
SA Ontario
SA Quebec

O-16 (1978)
Autoposting of broader terms began in 1989. This term has multiple hierarchies.
BT1 stable isotopes
BT1 isotopes
BT2 oxygen
SA O-17/O-16
SA O-18/O-16

O-16/O-18
use O-18/O-16

O-17/O-16 (1989)
Autoposting of broader terms began in 1989. This term has multiple hierarchies.
BT1 stable isotopes
BT1 isotopes
BT2 oxygen
SA isotope ratios
SA O-16

O-18 (1978)
Autoposting of broader terms began in 1989. This term has multiple hierarchies.
BT1 stable isotopes
BT1 isotopes
BT2 oxygen
SA O-18/O-16

O-18/O-16 (1978)
Autoposting of broader terms began in 1989. This term has multiple hierarchies.
UF O-16/O-18
BT1 stable isotopes
BT1 isotopes
BT2 oxygen
SA geologic thermometry
SA isotope ratios
SA O-16
SA O-18
SA paleoclimatology

O-shima (1978)
Refers to various islands in Japan. After 1992, see Izu-Oshima for the largest island in the Izu-shichito group which is the location of Mount Mihara volcano, Oshima for the Ryukyu Islands group and Oshima Peninsula for the sub-prefecture on Hokkaido. Before 1993, also referred to the largest island of Izu-shichito island group.
IN Index as applicable, including names of appropriate regions or prefectures.
BT Japan
BT Far East
BT Asia
SA Amami-O-shima
SA Izu-Oshima
SA Izu-Oshima earthquake 1978
SA Izu-shichito
SA Mount Mihara
SA Oshima
SA Oshima Peninsula

Oahu (1978)
Third largest of the Hawaiian islands. Island on which Honolulu is located. Principal part of Honolulu County. Autoposting of broader terms to this term began in 1989. This term has multiple hierarchies.
CO N211500N214300
    W1573900W1581800
BT1 Honolulu County Hawaii
BT1 Hawaii
BT1 United States
BT2 Honolulu County Hawaii
BT2 Hawaii
BT2 Polynesia
BT2 Oceania
BT3 Honolulu County Hawaii
BT3 Hawaii
BT3 East Pacific Ocean Islands
NT Honolulu Hawaii
NT Kaneohe Bay
NT Koolau Range
NT Pearl Harbor
NT Waianae
NT Waimanalo Hawaii

Oak Hill Member (1989)
Of Naheola Formation. W Alabama.
BT Paleocene
BT Paleogene
BT Tertiary
BT Cenozoic
SA Alabama
SA Naheola Formation

Oak Ridge National Laboratory (1985)
Near Oak Ridge in Anderson County, Tennessee.
UF ORNL
BT government agencies
SA Tennessee

Oakville Sandstone (1985)
SW Texas.
BT Miocene
BT Neogene
BT Tertiary
BT Cenozoic
SA Texas

Oakwood Dome (1985)
Salt dome, Leon and Freestone counties, E Texas. Autoposting of broader terms to this term began in 1989.
IN Index counties as applicable.
BT Texas
BT United States
SA Freestone County Texas
SA Gulf Coastal Plain
SA Leon County Texas
SA salt domes

Oamaru
No longer a valid term for GeoRef. See Oamaru New Zealand.

Oamaru New Zealand (1993)
Town on Pacific Ocean in SE South Island.
IN Also index South Island.
BT Otago New Zealand
BT New Zealand
BT Australasia
SA South Island

Oasis region
No longer a valid term for GeoRef. Used as large department of French Algeria. As of 1993, see Algeria for departments.

Oaxaca
No longer a valid term for GeoRef. As of 1993 see Oaxaca Mexico.

Oaxaca Mexico (1993)

State and city in S Mexico. Before 1993 also search Oaxaca AND Mexico.
  BT Mexico
  SA Mexican volcanic belt
  SA Sierra Madre del Sur

Ob River (1978)
Rises in Altai Mountains and flows NE and N through West Siberian Plain into Gulf of Ob on Kara Sea. Also search Ob. This term has multiple hierarchies.
  BT1 Russian Federation
  BT1 Commonwealth of Independent States
  BT2 Asia
  SA Irtysh River
  SA Ob-Irtysh Interfluve
  SA Siberia

Ob-Irtysh Interfluve (1978)
Extends from source of both rivers in Altai Mountains to their conjunction in the Siberian Lowland. This term has multiple hierarchies.
  IN Index former Soviet republics as applicable.
  BT1 Commonwealth of Independent States
  BT2 Asia
  SA Irtysh River
  SA Kazakhstan
  SA Ob River
  SA Russian Federation
  SA Siberian Lowland

obduction (1978)
  SA plate tectonics
  SA subduction

Oberalp Pass (1981)
In the Alps, on border of Uri and Graubunden cantons, S central Switzerland.
  CO N463800N464100 E0083800E0084200
  BT Switzerland
  BT Central Europe
  BT Europe
  SA Graubunden Switzerland
  SA Swiss Alps
  SA Uri Switzerland

Oberhalbstein (1978)
Valley NW of Saint Moritz in E Switzerland.
  BT Graubunden Switzerland
  BT Switzerland
  BT Central Europe
  BT Europe

Oberon Satellite (1989)
One of the satellites of Uranus. Before 1989, also search Oberon AND Uranus.
  SA icy satellites
  SA satellites
  SA Uranus
  SA Voyager Program

Oberpfalz
  use Upper Palatinate

Obion County Tennessee (1989)
Extreme NW Tennessee. Before 1989, search Obion County AND Tennessee.
  CO N361200N363000 W0884900W0893200
  BT Tennessee
  BT United States
  SA Reelfoot Lake

objectives (1978)
  SA mineral exploration

oblique orientation (1978)
Includes use as orientation of folds relative to spatially associated macroscopic structures such as large folds, fold systems and orogenic zones. Before 1978, also search folds or faults AND oblique.
  SA faults
  SA folds
  SA fractures
  SA joints
  SA orientation

oblique-slip faults (1989)
  BT faults
  SA strike-slip faults

obliquity
As of 1981, no longer a valid term for GeoRef. See oblique orientation; obliquity of the ecliptic. Included use as a general term and for planet obliquity.

obliquity of the ecliptic (1981)
Before 1981, also search planet name AND tilt.
  SA climate
  SA Earth
  SA Milankovitch theory
  SA Moon
  SA motions
  SA paleoclimatology
  SA tilt

observation wells (1993)
Used especially in ground water and petroleum engineering studies. Before 1993, also search monitoring wells if applicable.
  BT wells
  SA ground water
  SA monitoring
  SA petroleum engineering
  SA water wells

observations (1978)
  SA GEOS
  SA lunar bases
  SA orbital observations

observatories (1978)
  SA geophysics
  SA seismology

obsidian (1978)
  BT glasses
  BT volcanic rocks
  BT igneous rocks
  SA volcanic glass

Obsidian Dome (1989)
One of the Inyo Domes in Owens Valley, Long Valley, E California.
  BT Mono County California
  BT California
  BT United States
  SA Inyo Domes
  SA Long Valley Caldera

obsidian hydration
No longer a valid GeoRef index term. Before 1972, was used on level 2 in subfile G.
  use hydration of glass

Ocala Group (1978)
Includes Tivola Tongue, Inglis Formation, Williston Formation, Crystal River Formation.
  BT upper Eocene
  BT Eocene
  BT Paleogene
  BT Tertiary
  BT Cenozoic
  SA Alabama
  SA Crystal River Formation
  SA Florida
  SA Floridan Aquifer
  SA Georgia

occurrence (1978)
  SA metamorphic rocks

ocean basins (1978)
Used for studies on the major ocean basins, their origin, evolution and present configuration. For basins found within the oceans and for sedimentation studies, see ocean floors.
  SA abyssal plains
  SA age
  SA basins
  SA continental drift
  SA crust
  SA Deep Sea Drilling Project
  SA geosynclines
  SA marginal basins
  SA marine geology
  SA Ocean Drilling Program
  SA ocean floors
  SA oceanography
  SA paleo-oceanography
  SA paleogeography
  SA plate tectonics
  SA sea-floor spreading
  SA submarine volcanoes
  SA tectonophysics

ocean bottom seismographs (1978)
  UF ocean-bottom seismographs
  UF ocean-bottom seismometers
  BT seismographs
  SA hydrophones
  SA seismic methods

ocean circulation (1978)
Primarily used for special indexes. Before 1978, also search circulation AND oceans.
  SA bottom currents
  SA boundary layer
  SA carbonate compensation depth
  SA circulation
  SA climate-induced circulation
  SA continental shelf
  SA continental slope
  SA current directions
  SA diffusion
  SA HEBBLE
  SA longshore currents
  SA marine geology
  SA North Atlantic Deep Water
  SA ocean currents
  SA ocean floors
  SA ocean waves
  SA oceanography
  SA paleocirculation
  SA sedimentation
  SA shorelines
  SA thermal circulation
  SA thermocline
  SA thermohaline circulation
  SA tides
  SA turbulence
  SA upwelling
  SA waterways

Ocean County
Valid through 1988. Search in combination with state term. After 1988, use specific county-state term.

Ocean County New Jersey (1989)
On the Atlantic Ocean in E New Jersey. Before 1989, also search Ocean County AND New Jersey.
  CO N393000N401100 W0740200W0743400
  BT New Jersey
  BT United States
  SA Great Bay

ocean crust
  use oceanic crust

ocean currents (1985)
Before 1985, also search currents AND ocean circulation.
  BT currents
  NT longshore currents
  SA continental shelf
  SA fluvial currents
  SA ocean circulation
  SA paleocurrents
  SA tidal currents
  SA upwelling

Ocean Drilling Program (1989)
A continuation of DSDP and IPOD, beginning with Leg 100.
  UF ODP
  NT Leg 100
  NT Leg 101
  NT Leg 102
  NT Leg 103
  NT Leg 104
  NT Leg 105
  NT Leg 106
  NT Leg 107
  NT Leg 108
  NT Leg 109
  NT Leg 110
  NT Leg 111
  NT Leg 112
  NT Leg 113
  NT Leg 114
  NT Leg 115
  NT Leg 116
  NT Leg 117
  NT Leg 118
  NT Leg 119
  NT Leg 120
  NT Leg 121
  NT Leg 122
  NT Leg 123
  NT Leg 124
  NT Leg 124E
  NT Leg 125
  NT Leg 126
  NT Leg 127
  NT Leg 128
  NT Leg 129
  NT Leg 130
  NT Leg 131
  NT Leg 132
  NT Leg 133
  NT Leg 134
  NT Leg 135
  NT Leg 136
  NT Leg 138
  SA Deep Sea Drilling Project
  SA IPOD
  SA JOIDES
  SA marine geology
  SA ocean basins
  SA ocean floors
  SA oceanography
  SA research vessels

ocean floors (1978)
Used for discussions of features of ocean floors, as well as processes taking place there. For very large scale, perhaps hemispheric, oceanic features and activity, see ocean basins. Before 1993, also search sea floors.
  UF sea floors
  SA abyssal plains
  SA bathymetric maps
  SA bathymetry
  SA black smokers
  SA bottom features
  SA continental margin
  SA continental rise
  SA continental shelf
  SA continental slope
  SA Deep Sea Drilling Project

SA dredging
SA Exclusive Economic Zone
SA fracture zones
SA furrows
SA hardground
SA insular shelf
SA isobath maps
SA Law of the Sea
SA MANOP
SA marginal basins
SA marine geology
SA microcontinents
SA mid-ocean ridges
SA mineral resources
SA nodules
SA ocean basins
SA ocean circulation
SA Ocean Drilling Program
SA oceanography
SA paleo-oceanography
SA paleobathymetry
SA pockmarks
SA relief
SA sea-floor spreading
SA seamounts
SA sedimentation
SA sediments
SA submarine canyons
SA submarine dunes
SA submarine fans
SA topography
SA trenches
SA troughs
SA United States Exclusive Economic Zone

Ocean of Storms (1978)
UF Oceanus Procellarum
BT Moon
SA Aristarchus

ocean ridges
use mid-ocean ridges

ocean waves (1978)
Used for special indexes or for papers relating physical oceanography to marine geology.
NT breaking waves
NT catastrophic waves
NT ideal waves
NT internal waves
SA circulation
SA coastlines
SA diffraction
SA littoral erosion
SA longshore currents
SA marine geology
SA ocean circulation
SA oceanography
SA propagation
SA seiches
SA shoaling
SA shorelines
SA surges
SA transformations
SA tsunamis
SA waves

ocean, world
use world ocean

ocean-bottom seismographs
use ocean bottom seismographs

ocean-bottom seismometers
use ocean bottom seismographs

ocean-floor spreading
use sea-floor spreading

ocean-island basalts (1993)
Tholeiites, alkali basalts and nephelinites found on submarine volcanoes which form ocean islands away from ocean ridges. Chemically distinct from mid-ocean ridge basalts.
UF OIB

BT basalts
BT volcanic rocks
BT igneous rocks
SA alkali basalts
SA nephelinite
SA tholeiite

Oceania (1978)
Collective name for the islands and island groups of the central and south Pacific Ocean. Before 1981, Pacific Ocean was a broader term. Micronesia, Melanesia and Polynesia are narrower terms as of 1981.
IN Index island divisions as applicable: Melanesia, Micronesia, and Polynesia.
UF Oceanica
NT Kiribati
NT Melanesia
NT Micronesia
NT Polynesia
SA Australasia
SA Australia
SA Malay Archipelago
SA New Zealand
SA Pacific Ocean

oceanic
A valid term through 1977. See oceanic type; oceanic crust.

oceanic crust (1981)
Also search oceanic type AND crust; oceanic AND crust; oceanic AND genesis.
UF ocean crust
UF oceanic genesis
BT crust
SA Iceland Research Drilling Project
SA ophiolite complexes
SA plate tectonics
SA Rivera Ocean Seismic Experiment
SA transition zones
SA underplating

oceanic genesis
use oceanic crust

oceanic ridges
A valid term through 1971. After 1971, use mid-ocean ridges.

oceanic trench
use trenches

oceanic type
As of 1981, no longer a valid term for GeoRef. Use oceanic crust.

Oceanica
use Oceania

oceanite (1978)
BT basalts
BT volcanic rocks
BT igneous rocks

Oceanographer fracture zone (1989)
N Atlantic, across the Mid-Atlantic Ridge. As of 1990, Atlantic Ocean is autoposted to this term.
CO N330000N380000 W0280000W0430000
BT North Atlantic
BT Atlantic Ocean
SA DSDP Site 561
SA Leg 82
SA Mid-Atlantic Ridge

oceanography (1978)
Used for special indexes and the discipline as a whole.
UF oceanology
SA abyssal plains
SA air-sea interface

SA AMCOR
SA bathymetry
SA bibliography
SA bottom features
SA bottom water
SA boundary layer
SA carbonate compensation depth
SA catalogs
SA Continental Offshore Stratigraphic Test
SA continental shelf
SA continental slope
SA convection
SA CYANA
SA Deep Sea Drilling Project
SA diffusion
SA echo sounding
SA education
SA estuaries
SA expeditions
SA FAMOUS
SA GEOSECS
SA Glomar Challenger
SA HEBBLE
SA hemipelagic environment
SA International Decade of Ocean Exploration
SA Law of the Sea
SA MANOP
SA marine drilling
SA marine geology
SA marine geology maps
SA nodules
SA ocean basins
SA ocean circulation
SA Ocean Drilling Program
SA ocean floors
SA ocean waves
SA Orgon III
SA paleo-oceanography
SA pelagic environment
SA reefs
SA research vessels
SA Rivera Ocean Seismic Experiment
SA sea ice
SA sea water
SA Seasat
SA sediment traps
SA sedimentary petrology
SA sedimentation
SA shelf-slope break
SA shoals
SA submersibles
SA tidal currents
SA turbulence
SA upwelling

oceanology
use oceanography

oceans
A valid index term through 1977 used in combination with circulation (i.e. oceans, circulation). After 1977 use ocean circulation.

Oceanus Procellarum
use Ocean of Storms

ocher (1985)
UF ochre
SA archaeology
SA clay mineralogy
SA hematite
SA iron ores
SA iron oxides
SA laterites
SA limonite
SA oxides

ochre
use ocher

Ocoee Series (1978)

Provincial series. Virginia, Tennessee, North Carolina, and Georgia.
BT Precambrian
SA Georgia
SA Great Smoky Group
SA North Carolina
SA Tennessee
SA Virginia

Oconee County
Valid through 1988. Search in combination with state term. After 1988, use specific county-state term.

Oconee County Georgia (1989)
NE central Georgia. Before 1989, also search Oconee County AND Georgia.
CO N334100N335700 W0831700W0833800
BT Georgia
BT United States

Oconee County South Carolina (1989)
Extreme NW South Carolina. Before 1989, also search Oconee County AND South Carolina.
CO N343000N350300 W0825300W0832200
BT South Carolina
BT United States
SA Lake Jocassee

octahedral iron ore
use magnetite

octahedrite (1981)
BT iron meteorites
BT meteorites
NT Canyon Diablo Meteorites
NT Gibeon Meteorite
NT Henbury Meteorite
NT Toluca Meteorite
NT Weekeroo Station Meteorite
NT Wolf Creek Meteorite
SA Coldwater Meteorites
SA Odessa Meteorite

octahedrite (mineral)
use anatase

octahedrites
A valid term through 1975. After 1975, use iron meteorites. See octahedrite.

Octocorallia (1978)
BT Anthozoa
BT Coelenterata
BT Invertebrata

Odate
No longer a valid term for GeoRef. See Odate Japan.

Odate Japan (1993)
City in N Honshu. Before 1993, also search Odate AND Japan.
BT Akita Japan
BT Honshu
BT Japan
BT Far East
BT Asia

Odenwald (1978)
Mountainous region in SW Germany.
IN Index states as applicable.
BT Germany
BT Central Europe
BT Europe
SA Baden-Wurttemberg Germany
SA Bavaria Germany
SA Bergstrasse
SA Hesse Germany
SA Katzenbuckel

Oder Valley (1978)

River valley.
IN Index countries as applicable.
UF Odra River valley
BT Europe
SA Brandenburg Germany
SA Czechoslovakia
SA Mecklenburg-Western Pomerania Germany
SA Moravia
SA Poland
SA Warta

Odessa
No longer a valid term for GeoRef. As of 1993, see Odessa City Ukraine or Odessa Texas or Odessa Washington or Odessa Meteorite.

Odessa City Ukraine (1993)
City in S Ukraine. Before 1993, search in conjunction with USSR. Before 1993, also search Odessa. This term has multiple hierarchies.
CO N463000N463000
    E0304600E0304600
BT1 Odessa Ukraine
BT1 Ukraine
BT1 Europe
BT2 Odessa Ukraine
BT2 Ukraine
BT2 Commonwealth of Independent States
SA Odessa Meteorite

Odessa Meteorite (1981)
Refers to both the meteorite found in Texas (octahedrite) and to the meteorite found in the Ukraine (H chondrite). Before 1981, also search Odessa AND meteorites.
IN Index octahedrite or H chondrites as applicable.
BT meteorites
SA H chondrites
SA octahedrite
SA Odessa City Ukraine
SA Texas
SA Ukraine

Odessa Oblast
As of 1993, no longer a valid term for GeoRef.
use Odessa Ukraine

Odessa Texas (1989)
City in W Texas. Before 1989, search Odessa AND Texas.
CO N315000N315000
    W1022300W1022300
BT Ector County Texas
BT Texas
BT United States

Odessa Ukraine (1993)
Oblast on the Black Sea. Before 1981, also search Odessa AND USSR. After 1980, also search Odessa Oblast. This term has multiple hierarchies.
UF Odessa Oblast
BT1 Ukraine
BT1 Europe
BT2 Ukraine
BT2 Commonwealth of Independent States
NT Odessa City Ukraine

Odessa Washington (1989)
Town in E Washington. Before 1989, search Odessa AND Washington.
CO N471900N471900
    W1184000W1184000
BT Lincoln County Washington
BT Washington
BT United States

Odonata (1978)
BT Odonatopteroida
BT Insecta
BT Mandibulata
BT Arthropoda
BT Invertebrata

Odonatopteroida (1981)
BT Insecta
BT Mandibulata
BT Arthropoda
BT Invertebrata
NT Odonata

Odontopleurida (1978)
BT Trilobita
BT Trilobitomorpha
BT Arthropoda
BT Invertebrata

Odontornithes (1978)
Subclass.
BT Neornithes
BT Aves
BT Tetrapoda
BT Vertebrata
BT Chordata

ODP
use Ocean Drilling Program

ODP Site 626 (1989)
Straits of Florida. Leg 101.
CO N253556N253558
    W0793246W0793247
BT Leg 101
BT Ocean Drilling Program
SA Atlantic Ocean
SA Straits of Florida

ODP Site 627 (1989)
S Blake Plateau. Leg 101.
CO N273806N273806
    W0781742W0781742
BT Leg 101
BT Ocean Drilling Program
SA Atlantic Ocean
SA Blake Plateau

ODP Site 628 (1993)
Little Bahama Bank, S. North American Atlantic.
CO N273139N273139
    W0781857W0781857
BT Leg 101
BT Ocean Drilling Program
SA Atlantic Ocean
SA Little Bahama Bank
SA North American Atlantic

ODP Site 629 (1993)
Little Bahama Bank, S. North American Atlantic.
CO N272424N272424
    W0782206W0782206
BT Leg 101
BT Ocean Drilling Program
SA Atlantic Ocean
SA Little Bahama Bank
SA North American Atlantic

ODP Site 630 (1993)
Little Bahama Bank, S. North American Atlantic.
CO N272654N272654
    W0782024W0782024
BT Leg 101
BT Ocean Drilling Program
SA Atlantic Ocean
SA Little Bahama Bank
SA North American Atlantic

ODP Site 631 (1993)
Exuma Sound, Bahamas, S. North American Atlantic.
CO N233512N233512
    W0754436W0754436
BT Leg 101
BT Ocean Drilling Program
SA Atlantic Ocean
SA Bahamas
SA Exuma Sound
SA North American Atlantic

ODP Site 632 (1993)
Exuma Sound, Bahamas, S. North American Atlantic.
CO N235024N235024
    W0752606W0752606
BT Leg 101
BT Ocean Drilling Program
SA Atlantic Ocean
SA Bahamas
SA Exuma Sound
SA North American Atlantic

ODP Site 633 (1993)
Exuma Sound, Bahamas, S. North American Atlantic.
CO N234118N234118
    W0753736W0753736
BT Leg 101
BT Ocean Drilling Program
SA Atlantic Ocean
SA Bahamas
SA Exuma Sound
SA North American Atlantic

ODP Site 634 (1993)
Northwest Providence Channel, Bahamas, S. North American Atlantic.
CO N252300N252301
    W0771852W0771853
BT Leg 101
BT Ocean Drilling Program
SA Atlantic Ocean
SA Bahamas
SA North American Atlantic
SA Northwest Providence Channel

ODP Site 635 (1993)
Northwest Providence Channel, Bahamas, S. North American Atlantic.
CO N252506N252512
    W0771818W0771954
BT Leg 101
BT Ocean Drilling Program
SA Atlantic Ocean
SA Bahamas
SA North American Atlantic
SA Northwest Providence Channel

ODP Site 637 (1993)
W Galicia Bank margin, NW of Iberian Peninsula, SE European Atlantic.
CO N420518N420518
    W0125148W0125148
BT Leg 103
BT Ocean Drilling Program
SA Atlantic Ocean
SA European Atlantic

ODP Site 638 (1993)
W Galicia Bank margin, NW of Iberian Peninsula, SE European Atlantic.
CO N420912N420912
    W0121148W0121148
BT Leg 103
BT Ocean Drilling Program
SA Atlantic Ocean
SA European Atlantic

ODP Site 639 (1993)
W Galicia Bank margin, NW of Iberian Peninsula, SE European Atlantic.
CO N420836N420836
    W0121454W0121524
BT Leg 103
BT Ocean Drilling Program
SA Atlantic Ocean
SA European Atlantic

ODP Site 640 (1993)
Galicia Bank margin, NW of Iberian Peninsula, SE European Atlantic.
CO N420042N420042
    W0122748W0122748
BT Leg 103
BT Ocean Drilling Program
SA Atlantic Ocean
SA European Atlantic

ODP Site 641 (1993)
Galicia Bank margin, NW of Iberian Peninsula, SE European Atlantic.
CO N420918N420918
    W0121054W0121054
BT Leg 103
BT Ocean Drilling Program
SA Atlantic Ocean
SA European Atlantic

ODP Site 642 (1993)
Outer Voring Plateau, Norwegian Sea.
CO N671312N671330
    E0025548E0025542
BT Leg 104
BT Ocean Drilling Program
SA Arctic Ocean
SA Norwegian Sea
SA Voring Plateau

ODP Site 643 (1993)
NW slope of Voring Plateau, Norwegian Sea.
CO N674254N674254
    E0010200E0010200
BT Leg 104
BT Ocean Drilling Program
SA Arctic Ocean
SA Norwegian Sea
SA Voring Plateau

ODP Site 644 (1993)
Inner Voring Plateau, Norwegian Sea.
CO N664042N664042
    E0043436E0043436
BT Leg 104
BT Ocean Drilling Program
SA Arctic Ocean
SA Norwegian Sea
SA Voring Plateau

ODP Site 645 (1993)
NW Baffin Bay, Arctic Ocean. Before 1988, also search Site 645 AND Ocean Drilling Program.
CO N702730N702730
    W0643924W0643924
BT Leg 105
BT Ocean Drilling Program
SA Arctic Ocean
SA Baffin Bay

ODP Site 646 (1993)
SE Labrador Sea, Arctic Ocean. Before 1988, also search Site 646 AND Ocean Drilling Program.
CO N581236N581236
    W0482206W0482206
BT Leg 105
BT Ocean Drilling Program
SA Arctic Ocean
SA Labrador Sea

ODP Site 647 (1993)
N Labrador Sea, Arctic Ocean. Before 1988, also search Site 647 AND Ocean Drilling Program.
CO N531954N531954
    W0451542W0451542
BT Leg 105
BT Ocean Drilling Program
SA Arctic Ocean
SA Labrador Sea

ODP Site 648 (1993)

Serocki volcano, rift valley near Mid-Atlantic Ridge, North Atlantic. This term has multiple hierarchies.
CO N225518N225520
    W0445649W0445650
BT1 Leg 106
BT1 Ocean Drilling Program
BT2 Leg 109
BT2 Ocean Drilling Program
SA Atlantic Ocean
SA Mid-Atlantic Ridge
SA North Atlantic Ridge

ODP Site 649 (1993)
Snake Pit hydrothermal vent area, 25 km S of Kane fracture zone, North Atlantic Ridge.
CO N232208N232210
    W0445702W0445705
BT Leg 106
BT Ocean Drilling Program
SA Atlantic Ocean
SA Mid-Atlantic Ridge
SA North Atlantic Ridge

ODP Site 652 (1993)
Lower Sardinian margin, Tyrrhenian Sea, Mediterranean Sea.
CO N402117N402119
    E0120834E0120836
BT Leg 107
BT Ocean Drilling Program
SA Mediterranean Sea
SA Tyrrhenian Sea

ODP Site 653 (1993)
E rim Cornaglia Basin, W Tyrrhenian Sea, Mediterranean Sea. Near DSDP Site 132.
CO N401551N401552
    E0112700E0112657
BT Leg 107
BT Ocean Drilling Program
SA DSDP Site 132
SA Mediterranean Sea
SA Tyrrhenian Sea

ODP Site 654 (1993)
Upper Sardinia margin, Tyrrhenian Sea, Mediterranean Sea.
CO N403445N403446
    E0104148E0104148
BT Leg 107
BT Ocean Drilling Program
SA Mediterranean Sea
SA Tyrrhenian Sea

ODP Site 655 (1993)
Ridge, W rim Vavilov Basin, central Tyrrhenian Sea, Mediterranean Sea.
CO N401019N401020
    E0122756E0122755
BT Leg 107
BT Ocean Drilling Program
SA Mediterranean Sea
SA Tyrrhenian Sea

ODP Site 656 (1993)
De Marchi Seamount, lower Sardinian margin, central Tyrrhenian Sea, Mediterranean Sea.
CO N401103N401104
    E0121102E0121101
BT Leg 107
BT Ocean Drilling Program
SA Mediterranean Sea
SA Tyrrhenian Sea

ODP Site 657 (1993)
Continental rise, NW Africa, Cape Verde Atlantic.
CO N211953N211954
    W0205655W0205656
BT Leg 108
BT Ocean Drilling Program
SA Atlantic Ocean
SA Cape Verde Atlantic

ODP Site 658 (1993)
Continental slope NW Africa, Cape Verde Atlantic.
CO N204457N204457
    W0183451W0183451
BT Leg 108
BT Ocean Drilling Program
SA Atlantic Ocean
SA Cape Verde Atlantic

ODP Site 659 (1993)
Cape Verde Plateau, Cape Verde Atlantic.
CO N180437N183438
    W0210134W0210135
BT Leg 108
BT Ocean Drilling Program
SA Atlantic Ocean
SA Cape Verde Atlantic

ODP Site 660 (1993)
NE of Kane Gap, Cape Verde Atlantic.
CO N100048N100049
    W0191444W0191445
BT Leg 108
BT Ocean Drilling Program
SA Atlantic Ocean
SA Cape Verde Atlantic

ODP Site 661 (1993)
East of Kane Gap, Cape Verde Atlantic.
CO N092648N092649
    W0192310W0192311
BT Leg 108
BT Ocean Drilling Program
SA Atlantic Ocean
SA Cape Verde Atlantic

ODP Site 662 (1993)
E Mid-Atlantic Ridge, S of Romanche fracture zone, N Southeast Atlantic region..
CO S012325S012324
    W0114421W0112421
BT Leg 108
BT Ocean Drilling Program
SA Atlantic Ocean
SA Mid-Atlantic Ridge
SA South Atlantic Ridge
SA Southeast Atlantic

ODP Site 663 (1993)
E flank of Mid-Atlantic Ridge, S of Romanche fracture zone, N Southeast Atlantic area.
CO S011153S011152
    W0115242W0115243
BT Leg 108
BT Ocean Drilling Program
SA Atlantic Ocean
SA Mid-Atlantic Ridge
SA South Atlantic Ridge
SA Southeast Atlantic

ODP Site 664 (1993)
E flank of the Mid-Atlantic Ridge, N of Romanche fracture zone, Cape Verde Atlantic region.
CO N000626N000627
    W0231339W0231630
BT Leg 108
BT Ocean Drilling Program
SA Atlantic Ocean
SA Cape Verde Atlantic
SA Mid-Atlantic Ridge
SA North Atlantic Ridge

ODP Site 665 (1993)
Near SE margin Sierra Leone Rise, Cape Verde Atlantic.
CO N025704N025705
    W0194004W0194005
BT Leg 108
BT Ocean Drilling Program
SA Atlantic Ocean
SA Cape Verde Atlantic

SA Sierra Leone Rise

ODP Site 666 (1993)
SE margin of Sierra Leone Rise, Cape Verde Atlantic.
CO N032950N032951
    W0201001W0201002
BT Leg 108
BT Ocean Drilling Program
SA Atlantic Ocean
SA Cape Verde Atlantic
SA Sierra Leone Rise

ODP Site 667 (1993)
Sierra Leone Rise, Cape Verde Atlantic.
CO N043408N043409
    W0215440W0215441
BT Leg 108
BT Ocean Drilling Program
SA Atlantic Ocean
SA Cape Verde Atlantic
SA Sierra Leone Rise

ODP Site 668 (1993)
Sierra Leone Rise, Cape Verde Atlantic.
CO N044607N044608
    W0205537W0205538
BT Leg 108
BT Ocean Drilling Program
SA Atlantic Ocean
SA Cape Verde Atlantic
SA DSDP Site 366
SA Sierra Leone Rise

ODP Site 669 (1993)
S Kane fracture zone, N Mid-Atlantic Ridge.
CO N233101N233102
    W0450245W0450245
BT Leg 109
BT Ocean Drilling Program
SA Atlantic Ocean
SA Kane fracture zone
SA Mid-Atlantic Ridge
SA North Atlantic Ridge

ODP Site 670 (1993)
Median valley, W flank of rift valley, North Atlantic Ridge.
CO N230959N231000
    W0450155W0450156
BT Leg 109
BT Ocean Drilling Program
SA Atlantic Ocean
SA North Atlantic Ridge

ODP Site 671 (1993)
Near Tiburon Rise, E of Barbados in Barbados accretionary ridge complex, North American Atlantic.
CO N153133N153133
    W0584357W0584357
BT Leg 110
BT Ocean Drilling Program
SA Atlantic Ocean
SA North American Atlantic

ODP Site 672 (1993)
Atlantic abyssal plain at NW Tiburon Rise, E of Barbados, W North Atlantic.
CO N153224N153224
    W0583827W0583828
BT Leg 110
BT Ocean Drilling Program
SA Atlantic Ocean
SA North American Atlantic

ODP Site 673 (1993)
Near Tiburon Rise, E of Barbados in Barbados Forearc, North American Atlantic.
CO N153155N153156
    W0584823W0584824
BT Leg 110
BT Ocean Drilling Program
SA Atlantic Ocean

SA North American Atlantic

ODP Site 674 (1993)
Near Tiburon Rise, E of Barbados in Barbados Forearc, North American Atlantic.
CO N153217N153218
    W0585105W0585106
BT Leg 110
BT Ocean Drilling Program
SA Atlantic Ocean
SA North American Atlantic

ODP Site 675 (1993)
Near Tiburon Rise, E of Barbados in Barbados Forearc. Near DSDP Site 542.
CO N153146N153147
    W0584300W0584301
BT Leg 110
BT Ocean Drilling Program
SA Atlantic Ocean
SA North American Atlantic

ODP Site 676 (1993)
Near Tiburon Rise, E of Barbados in Barbados Forearc, North American Atlantic.
CO N153150N153151
    W0584211W0584212
BT Leg 110
BT Ocean Drilling Program
SA Atlantic Ocean
SA North American Atlantic

ODP Site 677 (1993)
E Equatorial Pacific, off Colombia. Near DSDP Site 504.
CO N011203N011209
    W0834413W0834414
BT Leg 111
BT Ocean Drilling Program
SA Equatorial Pacific
SA Pacific Ocean

ODP Site 678 (1993)
E Equatorial Pacific, off Colombia. Near DSDP Site 504.
CO N011300N011301
    W0834323W0834323
BT Leg 111
BT Ocean Drilling Program
SA Equatorial Pacific
SA Pacific Ocean

ODP Site 679 (1993)
Peru outer shelf, Lima Basin, South American Pacific.
CO S110348S110348
    W0781618W0781618
BT Leg 112
BT Ocean Drilling Program
SA Pacific Ocean
SA South American Pacific

ODP Site 680 (1993)
Peru continental margin, Lima Basin, South American Pacific.
CO S110354S110354
    W0780436W0780436
BT Leg 112
BT Ocean Drilling Program
SA Pacific Ocean
SA South American Pacific

ODP Site 681 (1993)
Peru continental slope, South American Pacific.
CO S105836S105836
    W0775730W0775730
BT Leg 112
BT Ocean Drilling Program
SA Pacific Ocean
SA South American Pacific

ODP Site 682 (1993)
Peru-Chile Trench, South American Pacific.
CO S105836S105836

W0775730W0775730
  BT Leg 112
  BT Ocean Drilling Program
  SA Pacific Ocean
  SA Peru-Chile Trench
  SA South American Pacific

ODP Site 683   (1993)
  Peru continental margin, Yaquina Basin area, South American Pacific.
  CO S090142S090141
   W0802424W0802424
  BT Leg 112
  BT Ocean Drilling Program
  SA Pacific Ocean
  SA South American Pacific

ODP Site 684   (1993)
  Peru continental slope, South American Pacific.
  CO S085930S085929
   W0795721W0795721
  BT Leg 112
  BT Ocean Drilling Program
  SA Pacific Ocean
  SA South American Pacific

ODP Site 685   (1993)
  Peru continental margin, near Peru-Chile Trench, South American Pacific.
  CO S090647S090646
   W0803500W0803501
  BT Leg 112
  BT Ocean Drilling Program
  SA Pacific Ocean
  SA South American Pacific

ODP Site 686   (1993)
  West Pisco Basin, off Peru, South American Pacific.
  CO S132849S132848
   W0765329W0765329
  BT Leg 112
  BT Ocean Drilling Program
  SA Pacific Ocean
  SA South American Pacific

ODP Site 687   (1993)
  Lima Platform-Lima Basin, South American Pacific.
  CO S125147S125146
   W0765925W0765926
  BT Leg 112
  BT Ocean Drilling Program
  SA Pacific Ocean
  SA South American Pacific

ODP Site 688   (1993)
  Peru continental margin, South American Pacific.
  CO S113216S113215
   W0785634W0785635
  BT Leg 112
  BT Ocean Drilling Program
  SA Pacific Ocean
  SA South American Pacific

ODP Site 689   (1993)
  Maud Rise, Antarctic Ocean.
  CO S643101S643100
   E0030600E0030559
  BT Leg 113
  BT Ocean Drilling Program
  SA Antarctic Ocean

ODP Site 690   (1993)
  SW Maud Rise, Weddell Sea.
  CO S650938S650937
   E0011218E0011218
  BT Leg 113
  BT Ocean Drilling Program
  SA Antarctic Ocean
  SA Weddell Sea

ODP Site 691   (1993)
  E Antarctica continental slope, Weddell Sea margin.
  CO S704439S704438
   W0134840W0134841
  BT Leg 113
  BT Ocean Drilling Program
  SA Antarctic Ocean
  SA Weddell Sea

ODP Site 692   (1993)
  SW of Wegener Canyon, Weddell Sea margin.
  CO S704326S704325
   W0134912W0134912
  BT Leg 113
  BT Ocean Drilling Program
  SA Antarctic Ocean
  SA Weddell Sea

ODP Site 693   (1993)
  SW of Wegener Canyon, SW Weddell Sea margin.
  CO S704954S704953
   W0143424W0143425
  BT Leg 113
  BT Ocean Drilling Program
  SA Antarctic Ocean
  SA Weddell Sea

ODP Site 694   (1993)
  N abyssal plain, Weddell Basin, Antarctic Ocean.
  CO S665050S665049
   W0332645W0332646
  BT Leg 113
  BT Ocean Drilling Program
  SA Antarctic Ocean
  SA Weddell Sea

ODP Site 695   (1993)
  SE South Orkney microcontinent, Antarctic Ocean N of Weddell Sea.
  CO S622329S622328
   W0432706W0432706
  BT Leg 113
  BT Ocean Drilling Program
  SA Antarctic Ocean

ODP Site 696   (1993)
  SE margin South Orkney microcontinent, South Scotia Ridge, Antarctic Ocean, N of Weddell Sea.
  CO S615057S615057
   W0425558W0425559
  BT Leg 113
  BT Ocean Drilling Program
  SA Antarctic Ocean
  SA Scotia Ridge

ODP Site 697   (1993)
  Jane Basin, Antarctic Ocean N of Weddell Sea.
  CO S614838S614837
   W0401743W0401744
  BT Leg 113
  BT Ocean Drilling Program
  SA Antarctic Ocean

ODP Site 698   (1993)
  E Northeast Georgia Rise, South American Atlantic.
  CO S512731S512730
   W0330557W0330558
  BT Leg 114
  BT Ocean Drilling Program
  SA Atlantic Ocean
  SA South American Atlantic

ODP Site 699   (1993)
  NE Northeast Georgia Rise, South American Atlantic.
  CO S513233S513232
   W0304037W0304038
  BT Leg 114
  BT Ocean Drilling Program
  SA Atlantic Ocean
  SA South American Atlantic

ODP Site 700   (1993)
  W East Georgia Basin, on NE slope of Northeast Georgia Rise, South American Atlantic.
  CO S513200S513159
   W0301641W0301642
  BT Leg 114
  BT Ocean Drilling Program
  SA Atlantic Ocean
  SA South American Atlantic

ODP Site 701   (1993)
  W margin South Atlantic Ridge, E of Islas Orcadas Rise.
  CO S515905S515904
   W0231243W0231244
  BT Leg 114
  BT Ocean Drilling Program
  SA Atlantic Ocean
  SA Mid-Atlantic Ridge
  SA South Atlantic Ridge

ODP Site 702   (1993)
  Central Islas Orcadas Rise, South American Atlantic.
  CO S505648S505647
   W0262207W0262207
  BT Leg 114
  BT Ocean Drilling Program
  SA Atlantic Ocean
  SA South American Atlantic

ODP Site 703   (1993)
  Meteor Rise, Southeast Atlantic Ocean.
  CO S470303S470302
   E0075341E0075340
  BT Leg 114
  BT Ocean Drilling Program
  SA Atlantic Ocean
  SA Southeast Atlantic

ODP Site 704   (1993)
  S Meteor Rise, Southeast Atlantic Ocean.
  CO S465246S465245
   E0072515E0072515
  BT Leg 114
  BT Ocean Drilling Program
  SA Atlantic Ocean
  SA Southeast Atlantic

ODP Site 705   (1993)
  E Mascarene Plateau, NE margin Nazareth Bank, W Indian Ocean.
  CO S131002S131001
   E0612302E0612301
  BT Leg 115
  BT Ocean Drilling Program
  SA Indian Ocean

ODP Site 706   (1993)
  E Mascarene Plateau, W Indian Ocean.
  CO S130652S130650
   E0612216E0612215
  BT Leg 115
  BT Ocean Drilling Program
  SA Indian Ocean

ODP Site 707   (1993)
  NW Mascarene Plateau, W Indian Ocean. Near DSDP Site 237.
  CO S073244S073243
   E0590101E0590100
  BT Leg 115
  BT Ocean Drilling Program
  SA Indian Ocean

ODP Site 708   (1993)
  Abyssal plain, W Equatorial Indian Ocean.
  CO S052721S052721
   E0595638E0595637
  BT Leg 115
  BT Ocean Drilling Program
  SA Indian Ocean

ODP Site 709   (1993)
  N central Madingley Rise, W Indian Ocean.
  CO S035454S035454
   E0603306E0603306
  BT Leg 115
  BT Ocean Drilling Program
  SA Indian Ocean

ODP Site 710   (1993)
  Central Madingley Rise, W equatorial Indian Ocean.
  CO S041842S041842
   E0605848E0605848
  BT Leg 115
  BT Ocean Drilling Program
  SA Indian Ocean

ODP Site 711   (1993)
  N Madingley Rise, W equatorial Indian Ocean.
  CO S024434S024433
   E0610947E0610946
  BT Leg 115
  BT Ocean Drilling Program
  SA Indian Ocean

ODP Site 712   (1993)
  N margin Chagos Bank, central equatorial Indian Ocean.
  CO S041300S041259
   E0732423E0732422
  BT Leg 115
  BT Ocean Drilling Program
  SA Indian Ocean

ODP Site 713   (1993)
  N margin Chagos Bank, central equatorial Indian Ocean.
  CO S041135S041134
   E0732339E0732339
  BT Leg 115
  BT Ocean Drilling Program
  SA Indian Ocean

ODP Site 714   (1993)
  E Maldives Ridge, N equatorial Indian Ocean.
  CO N050336N050336
   E0734712E0734712
  BT Leg 115
  BT Ocean Drilling Program
  SA Indian Ocean

ODP Site 715   (1993)
  E Maldives Ridge, N equatorial Indian Ocean.
  CO N050454N050453
   E0734953E0734952
  BT Leg 115
  BT Ocean Drilling Program
  SA Indian Ocean

ODP Site 716   (1993)
  Central Maldives Ridge, N equatorial Indian Ocean.
  CO N045600N045600
   E0731700E0731700
  BT Leg 115
  BT Ocean Drilling Program
  SA Indian Ocean

ODP Site 717   (1993)
  Distal end of Bengal Fan, central Indian Ocean.
  CO S005548S005547
   E0812325E0812324
  BT Leg 116
  BT Ocean Drilling Program
  SA Bengal Fan
  SA Indian Ocean

ODP Site 718   (1993)
  Distal end of Bengal Fan, central Indian Ocean.
  CO S010132S010131
   E0812404E0812403
  BT Leg 116
  BT Ocean Drilling Program
  SA Bengal Fan
  SA Indian Ocean

ODP Site 719   (1993)

Distal end of Bengal Fan, central Indian Ocean.
CO S005739S005738
E0812358E0812358
BT Leg 116
BT Ocean Drilling Program
SA Bengal Fan
SA Indian Ocean

ODP Site 720   (1993)
Mid Indus Fan, Arabian Sea.
CO N160748N160748
E0604438E0604437
BT Leg 117
BT Ocean Drilling Program
SA Arabian Sea
SA Indian Ocean

ODP Site 721   (1993)
Owen Ridge, W Arabian Sea.
CO N164038N164039
E0595147E0595146
BT Leg 117
BT Ocean Drilling Program
SA Arabian Sea
SA Indian Ocean

ODP Site 722   (1993)
Owen Ridge, W Arabian Sea.
CO N163718N163719
E0594746E0594745
BT Leg 117
BT Ocean Drilling Program
SA Arabian Sea
SA Indian Ocean

ODP Site 723   (1993)
Continental margin near Oman, Arabian Sea.
CO N180304N180305
E0573634E0573633
BT Leg 117
BT Ocean Drilling Program
SA Arabian Sea
SA Indian Ocean

ODP Site 724   (1993)
Continental margin near Oman, Arabian Sea.
CO N182742N182743
E0574709E0574708
BT Leg 117
BT Ocean Drilling Program
SA Arabian Sea
SA Indian Ocean

ODP Site 725   (1993)
Continental margin near Oman, Arabian Sea.
CO N182912N182912
E0574202E0574201
BT Leg 117
BT Ocean Drilling Program
SA Arabian Sea
SA Indian Ocean

ODP Site 726   (1993)
Continental margin near Oman, Arabian Sea.
CO N174856N174857
E0572212E0572212
BT Leg 117
BT Ocean Drilling Program
SA Arabian Sea
SA Indian Ocean

ODP Site 727   (1993)
Continental margin near Oman, Arabian Sea.
CO N174605N174606
E0573513E0573512
BT Leg 117
BT Ocean Drilling Program
SA Arabian Sea
SA Indian Ocean

ODP Site 728   (1993)
Continental margin near Oman, Arabian Sea.
CO N174042N174042
E0594934E0594933
BT Leg 117
BT Ocean Drilling Program
SA Arabian Sea
SA Indian Ocean

ODP Site 729   (1993)
Continental margin near Oman, Arabian Sea.
CO N173842N173843
E0575714E0575713
BT Leg 117
BT Ocean Drilling Program
SA Arabian Sea
SA Indian Ocean

ODP Site 730   (1993)
Continental margin near Oman, Arabian Sea.
CO N174353N174354
E0574132E0574131
BT Leg 117
BT Ocean Drilling Program
SA Arabian Sea
SA Indian Ocean

ODP Site 731   (1993)
Owen Ridge, W Arabian Sea.
CO N162813N162814
E0594209E0594208
BT Leg 117
BT Ocean Drilling Program
SA Arabian Sea
SA Indian Ocean

ODP Site 732   (1993)
Ridge at Atlantis II Transform, Southwest Indian Ridge, Indian Ocean.
CO S323251S323248
E0570318E0570242
BT Leg 118
BT Ocean Drilling Program
SA Indian Ocean
SA Southwest Indian Ridge

ODP Site 733   (1993)
Atlantis II Transform, Southwest Indian Ridge, Indian Ocean.
CO S330518S330418
E0565919E0565839
BT Leg 118
BT Ocean Drilling Program
SA Indian Ocean
SA Southwest Indian Ridge

ODP Site 734   (1993)
Atlantis II Transform, Southwest Indian Ridge, Indian Ocean.
CO S320656S320649
E0570808E0570747
BT Leg 118
BT Ocean Drilling Program
SA Indian Ocean
SA Southwest Indian Ridge

ODP Site 735   (1993)
Atlantis II Transform, Southwest Indian Ridge, Indian Ocean.
CO S324327S324318
E0571618E0571557
BT Leg 118
BT Ocean Drilling Program
SA Indian Ocean
SA Southwest Indian Ridge

ODP Site 736   (1993)
N Kerguelen-Heard Plateau, S Indian Ocean.
CO S492408S492407
E0713937E0713936
BT Leg 119
BT Ocean Drilling Program
SA Indian Ocean
SA Kerguelen Plateau

ODP Site 737   (1993)
N Kerguelen-Heard Plateau, S Indian Ocean.
CO S501340S501339
E0730157E0730157
BT Leg 119
BT Ocean Drilling Program
SA Indian Ocean

ODP Site 738   (1993)
S Kerguelen Plateau, N Antarctic Ocean.
CO S624233S624232
E0824715E0824714
BT Leg 119
BT Ocean Drilling Program
SA Antarctic Ocean

ODP Site 739   (1993)
Antarctic shelf, Prydz Bay, Antarctic Ocean.
CO S671635S671634
E0750455E0750454
BT Leg 119
BT Ocean Drilling Program
SA Antarctic Ocean
SA Prydz Bay

ODP Site 741   (1993)
Antarctic shelf, Prydz Bay, Antarctic Ocean.
CO S682310S682309
E0762302E0762301
BT Leg 119
BT Ocean Drilling Program
SA Antarctic Ocean
SA Prydz Bay

ODP Site 743   (1993)
Antarctic slope, Prydz Bay, Antarctic Ocean.
CO S665500S665459
E0744126E0744125
BT Leg 119
BT Ocean Drilling Program
SA Antarctic Ocean
SA Prydz Bay

ODP Site 744   (1993)
S Kerguelen Plateau, Antarctic Ocean.
CO S613440S613439
E0803528E0803527
BT Leg 119
BT Ocean Drilling Program
SA Antarctic Ocean
SA Kerguelen Plateau

ODP Site 745   (1993)
SE slope, Kerguelen Plateau, Antarctic Ocean.
CO S593543S593542
E0855136E0855136
BT Leg 119
BT Ocean Drilling Program
SA Antarctic Ocean
SA Kerguelen Plateau

ODP Site 746   (1993)
SE slope Kerguelen Plateau, Antarctic Ocean.
CO S593250S593249
E0855147E0855146
BT Leg 119
BT Ocean Drilling Program
SA Antarctic Ocean
SA Kerguelen Plateau

ODP Site 747   (1993)
Central Kerguelen Plateau, S Indian Ocean.
CO S544841S544840
E0764739E0764738
BT Leg 120
BT Ocean Drilling Program
SA Indian Ocean
SA Kerguelen Plateau

ODP Site 748   (1993)
W Raggatt Basin, S Kerguelen Plateau, Antarctic Ocean.
CO S582627S582627
E0785854E0785853
BT Leg 120
BT Ocean Drilling Program
SA Antarctic Ocean
SA Kerguelen Plateau

ODP Site 749   (1993)
W flank of Banzare Bank, S Kerguelen Plateau, Antarctic Ocean.
CO S584302S584301
E0762427E0762427
BT Leg 120
BT Ocean Drilling Program
SA Antarctic Ocean
SA Kerguelen Plateau

ODP Site 750   (1993)
E Raggatt Basin, S Kerguelen Plateau, Antarctic Ocean.
CO S573533S573531
E0811426E0811422
BT Leg 120
BT Ocean Drilling Program
SA Antarctic Ocean
SA Kerguelen Plateau

ODP Site 751   (1993)
Central Raggatt Basin, S Kerguelen Plateau, Antarctic Ocean.
CO S574334S574333
E0794854E0794853
BT Leg 120
BT Ocean Drilling Program
SA Antarctic Ocean
SA Kerguelen Plateau

ODP Site 752   (1993)
Broken Ridge, central Indian Ocean.
CO S305329S305328
E0933440E0933439
BT Leg 121
BT Ocean Drilling Program
SA Indian Ocean

ODP Site 753   (1993)
Broken Ridge, central Indian Ocean.
CO S305021S305018
E0933524E0933523
BT Leg 121
BT Ocean Drilling Program
SA Indian Ocean

ODP Site 754   (1993)
Broken Ridge, central Indian Ocean.
CO S305627S305627
E0933400E0933357
BT Leg 121
BT Ocean Drilling Program
SA Indian Ocean

ODP Site 755   (1993)
Broken Ridge, central Indian Ocean.
CO S310148S310147
E0933249E0933248
BT Leg 121
BT Ocean Drilling Program
SA Indian Ocean

ODP Site 756   (1993)
Ninetyeast Ridge, E Indian Ocean.
CO S272120S272115
E0873554E0873548
BT Leg 121
BT Ocean Drilling Program
SA Indian Ocean
SA Ninetyeast Ridge

ODP Site 757   (1993)

Ninetyeast Ridge, E Indian Ocean.
CO S170128S170123
E0881054E0881048
BT Leg 121
BT Ocean Drilling Program
SA Indian Ocean
SA Ninetyeast Ridge

ODP Site 758 (1993)
Ninetyeast Ridge, E Indian Ocean.
CO N052302N052303
E0902141E0902140
BT Leg 121
BT Ocean Drilling Program
SA Indian Ocean
SA Ninetyeast Ridge

ODP Site 759 (1993)
Wombat Plateau, NE Exmouth Plateau, E Indian Ocean.
CO S165712S165712
E1153337E1153336
BT Leg 122
BT Ocean Drilling Program
SA Exmouth Plateau
SA Indian Ocean

ODP Site 760 (1993)
SE Wombat Plateau, NE Exmouth Plateau, E Indian Ocean.
CO S165520S165519
E1153229E1153228
BT Leg 122
BT Ocean Drilling Program
SA Exmouth Plateau
SA Indian Ocean

ODP Site 761 (1993)
Central Wombat Plateau, NE Exmouth Plateau, E Indian Ocean.
CO S164414S164413
E1153206E1153206
BT Leg 122
BT Ocean Drilling Program
SA Exmouth Plateau
SA Indian Ocean

ODP Site 762 (1993)
W central Exmouth Plateau, E Indian Ocean.
CO S195315S195314
E1121515E1121514
BT Leg 122
BT Ocean Drilling Program
SA Exmouth Plateau
SA Indian Ocean

ODP Site 764 (1993)
NE Wombat Plateau, NE Exmouth Plateau, E Indian Ocean.
CO S163358S163357
E1152726E1152725
BT Leg 122
BT Ocean Drilling Program
SA Exmouth Plateau
SA Indian Ocean

ODP Site 765 (1993)
Argo abyssal plain, E Indian Ocean.
CO S155833S155832
E1173430E1173429
BT Leg 123
BT Ocean Drilling Program
SA Argo abyssal plain
SA Indian Ocean

ODP Site 766 (1993)
Gascoyne abyssal plain, E Indian Ocean.
CO S195556S195555
E1102715E1102714
BT Leg 123
BT Ocean Drilling Program
SA Gascoyne abyssal plain
SA Indian Ocean

ODP Site 767 (1993)
Celebes Sea, West Pacific.
CO N044727N044731
E1233014E1233011
BT Leg 124
BT Ocean Drilling Program
SA Celebes Sea
SA Pacific Ocean

ODP Site 768 (1993)
Cagayan Ridge, Sulu Sea, West Pacific.
CO N080001N080004
E1211312E1211309
BT Leg 124
BT Ocean Drilling Program
SA Pacific Ocean
SA Sulu Sea

ODP Site 769 (1993)
Cagayan Ridge, Sulu Sea, West Pacific.
CO N084706N084709
E1211742E1211309
BT Leg 124
BT Ocean Drilling Program
SA Pacific Ocean
SA Sulu Sea

ODP Site 770 (1993)
Celebes Sea, West Pacific.
CO N050840N050843
E1234015E1234005
BT Leg 124
BT Ocean Drilling Program
SA Celebes Sea
SA Pacific Ocean

ODP Site 771 (1993)
Cagayan Ridge, Sulu Sea, West Pacific.
CO N084040N084042
E1204048E1204046
BT Leg 124
BT Ocean Drilling Program
SA Pacific Ocean
SA Sulu Sea

ODP Site 776 (1993)
W Mariana Trough, West Pacific.
CO N175424N175424
E1434057E1434057
BT Leg 124E
BT Ocean Drilling Program
SA Mariana Trough
SA Pacific Ocean
SA West Pacific

ODP Site 778 (1993)
Mariana forearc, West Pacific.
CO N192956N192955
E1463956E1463955
BT Leg 125
BT Ocean Drilling Program
SA Pacific Ocean
SA West Pacific

ODP Site 779 (1993)
Mariana forearc, West Pacific.
CO N193045N193045
E1464145E1464145
BT Leg 125
BT Ocean Drilling Program
SA Pacific Ocean
SA West Pacific

ODP Site 780 (1993)
Mariana forearc, West Pacific.
CO N193228N193233
E1463916E1463912
BT Leg 125
BT Ocean Drilling Program
SA Pacific Ocean
SA West Pacific

ODP Site 781 (1993)
Mariana forearc, West Pacific.
CO N193755N193754
E1463234E1463233
BT Leg 125
BT Ocean Drilling Program

SA Pacific Ocean
SA West Pacific

ODP Site 782 (1993)
Izu-Bonin forearc, West Pacific.
CO N305136N305140
E1411851E1411850
BT Leg 125
BT Ocean Drilling Program
SA Pacific Ocean
SA West Pacific

ODP Site 783 (1993)
Izu-Bonin forearc, West Pacific.
CO N305752N305751
E1414717E1414716
BT Leg 125
BT Ocean Drilling Program
SA Pacific Ocean
SA West Pacific

ODP Site 785 (1993)
Izu-Bonin forearc, West Pacific.
CO N304929N304928
E1405511E1405510
BT Leg 125
BT Ocean Drilling Program
SA Pacific Ocean
SA West Pacific

ODP Site 787 (1993)
NW of Izu-Bonin Trench, N Philippine Sea.
CO N322230N322231
E1404439E1404438
BT Leg 126
BT Ocean Drilling Program
SA Pacific Ocean
SA Philippine Sea

ODP Site 788 (1993)
E margin of Sumisu Rift, NW Philippine Sea.
CO N305521N305523
E1400014E1400010
BT Leg 126
BT Ocean Drilling Program
SA Pacific Ocean
SA Philippine Sea

ODP Site 789 (1993)
E margin of Sumisu Rift, NW Philippine Sea.
CO N305514N305515
E1395951E1395950
BT Leg 126
BT Ocean Drilling Program
SA Pacific Ocean
SA Philippine Sea

ODP Site 790 (1993)
Sumisu Rift, NW Philippine Sea.
CO N305457N305458
E1395042E1395039
BT Leg 126
BT Ocean Drilling Program
SA Pacific Ocean
SA Philippine Sea

ODP Site 791 (1993)
Sumisu Rift, NW Philippine Sea.
CO N305457N305459
E1395212E1395211
BT Leg 126
BT Ocean Drilling Program
SA Pacific Ocean
SA Philippine Sea

ODP Site 792 (1993)
W Izu-Bonin forearc basin, W of Izu-Bonin Trench, NW Philippine Sea.
CO N322355N322359
E1402249E1402246
BT Leg 126
BT Ocean Drilling Program
SA Pacific Ocean
SA Philippine Sea

ODP Site 793 (1993)

W of Izu-Bonin Trench, NW Philippine Sea.
CO N310619N310621
E1405217E1405215
BT Leg 126
BT Ocean Drilling Program
SA Pacific Ocean
SA Philippine Sea

ODP Site 794 (1993)
NE Yamato Basin, E Japan Sea. Drilled on Legs 127 and 128. Hole 794 C was reentered on Leg 128. This term has multiple hierarchies.
CO N401124N401125
E1381356E1385651
BT1 Leg 127
BT1 Ocean Drilling Program
BT2 Leg 128
BT2 Ocean Drilling Program
SA Japan Sea
SA Pacific Ocean

ODP Site 796 (1993)
Okushiri Ridge, E Japan Sea.
CO N425055N425056
E1392451E1392440
BT Leg 127
BT Ocean Drilling Program
SA Japan Sea
SA Pacific Ocean

ODP Site 797 (1993)
SW Yamato Basin, S central Japan Sea.
CO N383655N383657
E1343211E1343209
BT Leg 127
BT Ocean Drilling Program
SA Japan Sea
SA Pacific Ocean

ODP Site 798 (1993)
Oki Ridge, SE Japan Sea.
CO N370218N370218
E1344759E1344758
BT Leg 128
BT Ocean Drilling Program
SA Japan Sea
SA Pacific Ocean

ODP Site 799 (1993)
Kita-Yamato Trough, Yamato Rise, S central Japan Sea.
CO N391311N391314
E1335201E1335200
BT Leg 128
BT Ocean Drilling Program
SA Japan Sea
SA Pacific Ocean

ODP Site 800 (1993)
N Pigafetta Basin, central West Pacific.
CO N215522N215523
E1521920E1521919
BT Leg 129
BT Ocean Drilling Program
SA Pacific Ocean
SA West Pacific

ODP Site 801 (1993)
Central Pigafetta Basin, central West Pacific.
CO N183831N183835
E1562136E1562134
BT Leg 129
BT Ocean Drilling Program
SA Pacific Ocean
SA West Pacific

ODP Site 802 (1993)
Central East Mariana Trough, central West Pacific.
CO N120546N120547
E1531238E1531237
BT Leg 129
BT Ocean Drilling Program
SA Mariana Trough

SA Pacific Ocean
SA West Pacific

ODP Site 803 (1993)
Ontong Java Plateau, West Pacific.
CO N022559N022601
E1603229E1603227
BT Leg 130
BT Ocean Drilling Program
SA Ontong Java Plateau
SA Pacific Ocean
SA West Pacific

ODP Site 804 (1993)
Ontong Java Plateau, West Pacific.
CO N010016N010018
E1613538E1613537
BT Leg 130
BT Ocean Drilling Program
SA Ontong Java Plateau
SA Pacific Ocean
SA West Pacific

ODP Site 805 (1993)
Ontong Java Plateau, West Pacific.
CO N011340N011342
E1603147E1603145
BT Leg 130
BT Ocean Drilling Program
SA Ontong Java Plateau
SA Pacific Ocean
SA West Pacific

ODP Site 806 (1993)
Ontong Java Plateau, West Pacific.
CO N001906N001907
E1592142E1592140
BT Leg 130
BT Ocean Drilling Program
SA Ontong Java Plateau
SA Pacific Ocean
SA West Pacific

ODP Site 807 (1993)
Ontong Java Plateau, West Pacific.
CO N033622N033626
E1563730E1563728
BT Leg 130
BT Ocean Drilling Program
SA Ontong Java Plateau
SA Pacific Ocean
SA West Pacific

ODP Site 808 (1993)
Nankai accretionary prism, SE of Shikoku, Japan, West Pacific. Revisited on Leg 132, Site 808E was redrilled. This term has multiple hierarchies.
CO N322105N322111
E1345646E1345634
BT1 Leg 131
BT1 Ocean Drilling Program
BT2 Leg 132
BT2 Ocean Drilling Program
SA Pacific Ocean
SA West Pacific

ODP Site 809 (1993)
Sumisu Rift, NW Philippine Sea.
CO N310326N310331
E1395244E1395237
BT Leg 132
BT Ocean Drilling Program
SA Pacific Ocean
SA Philippine Sea

Odra River valley
 use Oder Valley

Oerebro
 use Orebro Sweden

Oesel
 use Saaremaa

Oetztal Alps
 use Otztal Alps

Oetztal Massif
 use Otztal Alps

Oetztaler
 use Otztal Alps

off-lying
 use offshore

Officer Basin (1978)
Sedimentary basin.
IN Index states as applicable.
SA Australia
SA South Australia
SA Western Australia

offretite (1978)
BT zeolite group
BT framework silicates
BT silicates
SA aluminosilicates

offsets
 Use AVO if applicable.

offshore (1978)
As of 1985, term is to be used for petroleum and natural gas exploration and production.
UF off-lying
SA AMCOR
SA Continental Offshore Stratigraphic Test
SA continental shelf
SA continental slope
SA Exclusive Economic Zone
SA exploration
SA Hibernia Field
SA marine drilling
SA marine geology maps
SA natural gas
SA onshore
SA petroleum
SA production
SA shorelines
SA United States Exclusive Economic Zone

offshore Alaska
 Not a valid term for GeoRef. As of 1989, search Alaska AND offshore. For indexing purposes, use Alaska and offshore as two separate terms.

Oficina (1978)
Oil field near town of El Tigre in NE central Venezuela.
BT Anzoategui Venezuela
BT Venezuela
BT South America
SA oil and gas fields

Oficina Formation (1989)
E Venezuela.
BT Miocene
BT Neogene
BT Tertiary
BT Cenozoic
SA Venezuela

Ofot Fjord
 use Ofoten

Ofoten (1978)
Fjord on west coast of Norway. NE extension of Vest Fjord on which Narvik is located.
UF Ofot Fjord
BT Nordland Norway
BT Norway
BT Scandinavia
BT Western Europe
BT Europe

Oga Peninsula (1978)
Extends into Japan Sea in N Honshu.
BT Akita Japan
BT Honshu
BT Japan
BT Far East
BT Asia

Ogallala Aquifer (1989)
Used for hydrogeology. For actual formation, use Ogallala Formation.
IN Index states as applicable.
BT United States
SA Colorado
SA Kansas
SA Nebraska
SA New Mexico
SA Ogallala Formation
SA Oklahoma
SA Texas
SA Wyoming

Ogallala Formation (1978)
Includes Burge Sands, Valentine Beds, Ash Hollow Member, Kimball Member. NE Colorado, W and central Kansas, W Nebraska, E New Mexico, W Oklahoma, NW Texas and SE Wyoming. As of 1989, use Ogallala Aquifer for hydrogeological discussions.
BT Pliocene
BT Neogene
BT Tertiary
BT Cenozoic
SA Ash Hollow Formation
SA Colorado
SA Kansas
SA Nebraska
SA New Mexico
SA Ogallala Aquifer
SA Oklahoma
SA Texas
SA Valentine Formation
SA Wyoming

Ogasawara
 As of 1993, no longer a valid term for GeoRef. See Bonin Islands.

Ogasawara Islands
 use Bonin Islands

Ogden Utah (1989)
City in N Utah. Before 1989, search Ogden AND Utah.
CO N411400N411400
W1115900W1115900
BT Weber County Utah
BT Utah
BT United States

Ogdensburg
 Valid through 1988. Search in combination with state term. After 1988, use specific city-state term.

Ogdensburg New York (1989)
City on the Saint Lawrence River in N New York. Before 1989, search Ogdensburg AND New York.
CO N444200N444200
W0753100W0753100
BT Saint Lawrence County New York
BT New York
BT United States

Ogilvie Mountains (1978)
Range in central Yukon Territory.
BT Yukon Territory
BT Western Canada
BT Canada

OGS
 use Ontario Geological Survey

Oguni
 No longer a valid term for GeoRef. See Oguni Japan.

Oguni Japan (1993)
Town on Honshu in Yamagata and geothermal area in Kumamoto Prefectures.
IN Index prefecture as applicable.
SA Kumamoto Japan
SA Yamagata Japan

Ohio (1978)
Autoposting of this term began in 1978.
CO N382500N420000
W0803200W0845000
BT United States
NT Adams County Ohio
NT Allen County Ohio
NT Ashland County Ohio
NT Ashtabula County Ohio
NT Athens County Ohio
NT Butler County Ohio
NT Carroll County Ohio
NT Champaign County Ohio
NT Clark County Ohio
NT Clinton County Ohio
NT Coshocton County Ohio
NT Crawford County Ohio
NT Cuyahoga County Ohio
NT Delaware County Ohio
NT Erie County Ohio
NT Fairfield County Ohio
NT Fayette County Ohio
NT Franklin County Ohio
NT Fulton County Ohio
NT Geauga County Ohio
NT Greene County Ohio
NT Hamilton County Ohio
NT Hancock County Ohio
NT Hardin County Ohio
NT Harrison County Ohio
NT Henry County Ohio
NT Highland County Ohio
NT Jackson County Ohio
NT Jefferson County Ohio
NT Lake County Ohio
NT Lawrence County Ohio
NT Logan County Ohio
NT Lucas County Ohio
NT Mahoning County Ohio
NT Marion County Ohio
NT Medina County Ohio
NT Mercer County Ohio
NT Monroe County Ohio
NT Montgomery County Ohio
NT Morgan County Ohio
NT Morrow County Ohio
NT Noble County Ohio
NT Ottawa County Ohio
NT Perry County Ohio
NT Pike County Ohio
NT Portage County Ohio
NT Putnam County Ohio
NT Richland County Ohio
NT Sandusky County Ohio
NT Sandusky River basin
NT Scioto River basin
NT Seneca County Ohio
NT Shelby County Ohio
NT Stark County Ohio
NT Summit County Ohio
NT Trumbull County Ohio
NT Tuscarawas County Ohio
NT Union County Ohio
NT Warren County Ohio
NT Washington County Ohio
NT Wayne County Ohio
NT Wood County Ohio
NT Wyandot County Ohio
SA Allegheny Group
SA Allegheny Plateau
SA Appalachian Basin
SA Appalachian Plateau
SA Bass Islands Dolomite
SA Bedford Shale
SA Berea Sandstone

SA Brassfield Formation
SA Cincinnati Arch
SA Clays Ferry Formation
SA Cleveland Member
SA Clinton Group
SA Columbus Limestone
SA Conemaugh Group
SA Crab Orchard Formation
SA Cuyahoga Formation
SA Delaware Limestone
SA Detroit River Group
SA Dundee Limestone
SA Dunkard Basin
SA Dunkard Group
SA Eastern Gas Shales Project
SA Eau Claire Formation
SA Eden Shale
SA Fairview Formation
SA Glenwood Shale
SA Great Lakes region
SA Huron Member
SA Kittanning Formation
SA Kope Formation
SA Lake Erie
SA Lake Maumee
SA Lockport Formation
SA Logan Formation
SA Lucas Formation
SA Marcellus Shale
SA Maumee River valley
SA Michigan Basin
SA Midcontinent
SA Monongahela Group
SA Mount Simon Sandstone
SA Ohio River
SA Ohio River basin
SA Ohio River valley
SA Ohio Shale
SA Ohio State University
SA Olentangy Shale
SA Ottawa River
SA Pittsburgh Coal
SA Pottsville Group
SA Queenston Shale
SA Richmond Group
SA Saint Peter Sandstone
SA Salina Group
SA Sharon Conglomerate
SA Sunbury Shale
SA Sylvania Formation
SA Tioga Bentonite
SA Trenton Group
SA Whitewater River valley

Ohio County West Virginia (1989)
N West Virginia. Before 1989, also search Ohio County AND West Virginia.
CO N400100N401100
   W0803100W0804500
BT West Virginia
BT United States
NT Wheeling West Virginia

Ohio Range (1978)
N of the Horlick Mountains and S of Marie Byrd Land.
BT Antarctica

Ohio River (1978)
Formed by confluence of the Allegheny and Monongahela rivers at Pittsburgh. It flows W and SW into the Mississippi River at Cairo, Illinois.
IN Index states as applicable.
BT United States
SA Cincinnati Ohio
SA Illinois
SA Indiana
SA Kentucky
SA Ohio
SA Pennsylvania
SA West Virginia

Ohio River basin (1993)
E United States. Formed by confluence of Allegheny and Monongahela rivers. Chief tributaries are Beaver River, Muskingum River, Hocking River, Scioto River, Miami River, Wabash River, Kanawha River, Guyandot River, Big Sandy River, Licking River, Kentucky River, Salt River, Green River, Cumberland River and Tennessee River.
IN Index states as applicable.
BT United States
SA Illinois
SA Indiana
SA Kentucky
SA Mississippi River basin
SA Ohio
SA Pennsylvania
SA West Virginia

Ohio River valley (1978)
IN Index states as applicable.
UF Ohio Valley
BT United States
SA Illinois
SA Indiana
SA Kentucky
SA Ohio
SA Pennsylvania
SA West Virginia

Ohio Shale (1978)
Includes Cleveland Member, Huron Member, Chagrin Member. Ohio, N central Kentucky, and West Virginia.
BT Upper Devonian
BT Devonian
BT Paleozoic
SA Cleveland Member
SA Huron Member
SA Kentucky
SA Ohio
SA West Virginia

Ohio State University (1978)
In Columbus, Franklin County.
SA academic institutions
SA Ohio

Ohio Valley
   use Ohio River valley

OIB
   use ocean-island basalts

oil (petroleum)
   use petroleum

oil and gas
   use petroleum

oil and gas fields (1978)
For detailed descriptions of individual fields or for discussions of the origin of several fields. Before 1971, also search oil fields; gas fields.
IN Commodity. See List C.
UF fields, oil and gas
UF gas fields
UF gas fields, oil and
UF oil fields
NT giant fields
SA Aberfeldy Field
SA Azalea Field
SA Barrow Field
SA Bell Creek Field
SA blowouts
SA Buccaneer Field
SA Cottageville Field
SA Daqing Field
SA degasification
SA East Texas Field
SA economic geology
SA Ekofisk Field
SA Elk Hills Field
SA Elmworth Field
SA Emba Field
SA Forties Field
SA geothermal fields
SA Hartzog Draw Field
SA Hassi Messaoud Field
SA Hibernia Field
SA Hugoton Field
SA Iola Field
SA Karachaganak Field
SA land subsidence
SA Lost Soldier Field
SA McAllen Ranch Field
SA Megion Field
SA natural gas
SA Oficina
SA petroleum
SA petroleum engineering
SA petroleum exploration
SA Prudhoe Bay Field
SA Rangely Field
SA reservoir properties
SA reservoir rocks
SA Romashkino Field
SA Salym Field
SA Simpson Field
SA stratigraphic traps
SA stratigraphic wedges
SA Surakhany Azerbaidzhan
SA tight sands
SA Trap Spring Field
SA traps
SA Umiat Field
SA Walker Creek Field
SA Wattenberg Field
SA Wilmington Field

oil fields
   No longer a valid term for GeoRef. Before 1971, included use in subfiles E, G, N, T and B.
   use oil and gas fields

oil reserves
   Not a valid term for GeoRef. See petroleum; reserves.

oil resources
   Not a valid term for GeoRef. See petroleum; resources.

oil sand
   Not a valid GeoRef index term after 1971. Was used on level 1 in subfile B.
   use oil sands

oil sands (1978)
Before 1972, also search bituminous sands; oil sand. This term has multiple hierarchies.
IN Commodity. See List C.
UF bituminous sands
UF oil sand
UF tar sands
BT1 organic residues
BT2 sedimentary rocks
SA energy sources
SA gas sands
SA oil shale
SA organic materials
SA petroleum
SA Sunnyside Mine
SA tight sands

oil seeps (1978)
UF petroleum seepage
UF seeps, oil
SA bitumens
SA petroleum
SA seepage

oil shale (1978)
This term has multiple hierarchies.
IN See List C for use as a commodity term.
UF shale oil
BT1 organic residues
BT2 sedimentary rocks
NT kukersite
SA anthraxolite
SA Anvil Points Mine
SA energy sources
SA kerogen
SA oil sands
SA organic materials
SA petroleum
SA retorting
SA shale
SA torbanite

oil spills (1978)
UF spills, oil
SA crude oil
SA dense nonaqueous phase liquids
SA nonaqueous phase liquids
SA petroleum
SA pollution

oil storage (1983)
SA engineering geology
SA gas storage
SA leaking underground storage tanks
SA petroleum
SA petroleum engineering
SA storage
SA underground installations
SA underground storage

oil-gas interface (1978)
Before 1978, search interface AND oil-gas.
UF interface, oil-gas
SA gases
SA interfaces
SA natural gas
SA oil-water interface
SA petroleum

oil-water contact
   use oil-water interface

oil-water interface (1978)
Before 1978, search oil-water AND interface; oil-water contact
UF interface, oil-water
UF oil-water contact
SA interfaces
SA meteoric water
SA oil-gas interface
SA petroleum
SA water

Oise
   No longer a valid term for GeoRef. See Oise France.

Oise France (1993)
Department in N France.
CO N490000N494500
   E0031000E0014000
BT France
BT Western Europe
BT Europe
NT Baron France
NT Clermont France
NT Creil France
SA Oise River valley

Oise River valley (1978)
IN Index departments as applicable.
UF Oise Valley
BT France
BT Western Europe
BT Europe
SA Aisne France
SA Oise France
SA Val-d'Oise France

Oise Valley
   use Oise River valley

Oita
   No longer a valid term for GeoRef. See Oita Japan.

Oita Japan (1993)

City and prefecture in NE Kyushu, S Japan.
BT Kyushu
BT Japan
BT Far East
BT Asia
NT Beppu Japan
NT Kuju
NT Otake Field

Ojika Peninsula
No longer a valid term for GeoRef.
use Oshika Peninsula

Ojo Alamo Sandstone (1978)
NW New Mexico.
BT Upper Cretaceous
BT Cretaceous
BT Mesozoic
SA New Mexico

Ojo de Liebre Lagoon (1978)
Inlet of Sebastian Vizcaino Bay on NW coast.
BT Baja California Mexico
BT Mexico
SA Baja California

Oka Complex (1981)
Carbonatite complex mined for pyrochlore (niobium ores). In S Quebec.
BT Quebec
BT Eastern Canada
BT Canada

Oka River (1978)
River W of Lake Baikal in central Irkutsk Oblast which flows into the Angara River. Also a river in central European Russian Federation which flows into the Volga at Gorki. Before 1978, also search Oka AND USSR.
IN Index Europe or Asia and regions as applicable.
BT Russian Federation
BT Commonwealth of Independent States
SA Asia
SA Europe
SA Irkutsk Russian Federation

Okanagan Valley (1978)
River valley. Called Okanogan Valley in the U.S. Also search Okanogan Valley.
IN Index British Columbia and/or Washington as applicable.
UF Okanagen Valley
UF Okanogan Valley
BT North America
SA British Columbia
SA Washington

Okanagen Valley
use Okanagan Valley

Okanogan County
Valid through 1988. Search in combination with state term. After 1988, use specific county-state term.

Okanogan County Washington (1989)
N Washington. Before 1989, also search Okanogan County AND Washington.
CO N475700N490000
    W1184700W1205300
BT Washington
BT United States
NT Methow River valley

Okanogan Range (1985)
S British Columbia and N Washington.
CO N481500N491500
    W1194000W1201000
BT North America
SA British Columbia
SA Washington

Okanogan Valley
use Okanagan Valley

Okayama
No longer a valid term for GeoRef. See Okayama Japan.

Okayama Japan (1993)
City and prefecture on N side of Inland Sea W of Kobe in W Honshu.
BT Honshu
BT Japan
BT Far East
BT Asia
SA Chugoku
SA Mitsuishi Japan

Okefenokee Swamp (1978)
Mostly in SE Georgia (Charlton, Ware, Clinch, Brantley counties). Partly in NE Florida (Baker and Columbia counties).
IN Index states and counties as applicable.
BT United States
SA Baker County Florida
SA Columbia County Florida
SA Florida
SA Georgia

Okehampton
No longer a valid term for GeoRef as of 1993. See Okehampton England.

Okehampton England (1993)
Town in Devonshire in SW England.
BT Devonshire England
BT England
BT Great Britain
BT United Kingdom
BT Western Europe
BT Europe

Okhotsk
No longer a valid term for GeoRef. As of 1993, see Okhotsk Russian Federation.

Okhotsk Massif (1978)
Just N of Magadan in S Magadan Oblast in E Siberia. This term has multiple hierarchies.
UF Okhotskoye Ploskogor'ye
BT1 Magadan Russian Federation
BT1 Russian Federation
BT1 Commonwealth of Independent States
BT2 Magadan Russian Federation
BT2 Asia

Okhotsk region (1981)
The Okhotsk Sea and the land areas surrounding it. This term has multiple hierarchies.
CO N433000N630000
    E1660000E1340000
BT1 Russian Federation
BT1 Commonwealth of Independent States
BT2 Asia
SA Dzhugdzhur region
SA Kamchatka Russian Federation
SA Kuril Islands
SA Magadan Russian Federation
SA Sakhalin

Okhotsk Russian Federation (1993)
Town on NW coast of Okhotsk Sea in Khabarovsk Kray in E Siberia. Before 1993, also search Okhotsk. This term has multiple hierarchies.
BT1 Khabarovsk Russian Federation
BT1 Russian Federation
BT1 Commonwealth of Independent States
BT2 Khabarovsk Russian Federation
BT2 Asia

Okhotsk Sea (1978)
West of Kamchatka Peninsula and the Kuril Islands. Before 1974, also search Sea of Okhotsk.
CO N440000N630000
    E1650000E1350000
UF Sea of Okhotsk
BT West Pacific
BT Pacific Ocean
SA Tancheng-Lujiang fault zone

Okhotsk-Chukchi (1978)
Large region extending NE of Okhotsk Sea, including area of Chukchi Peninsula. This term has multiple hierarchies.
BT1 Russian Federation
BT1 Commonwealth of Independent States
BT2 Asia
SA Chukchi Peninsula
SA Russian Far East

Okhotsk-Chukchi volcanic belt (1978)
Includes numerous ranges extending S from the Chukchi Peninsula around both sides of Okhotsk Sea. Many volcanoes are still active in Kamchatka Peninsula. Also search Okhotsk-Chukchi. This term has multiple hierarchies.
CO N580000N700000
    W1700000E1400000
UF Okhotsk-Chukot volcanic belt
BT1 Russian Federation
BT1 Commonwealth of Independent States
BT2 Asia
SA Kamchatka Peninsula
SA Kamchatka Russian Federation
SA Magadan Russian Federation

Okhotsk-Chukot volcanic belt
use Okhotsk-Chukchi volcanic belt

Okhotskoye Ploskogor'ye
use Okhotsk Massif

Oki Archipelago
use Oki Islands

Oki Islands (1978)
In SE Japan Sea 44 miles off W coast of Honshu.
CO N360000N370000
    E1340000E1323000
UF Oki Archipelago
UF Oki Retto
BT Shimane Japan
BT Honshu
BT Japan
BT Far East
BT Asia
NT Dogo Island
SA Chibaken-Toho-Oki earthquake 1987
SA Tokachi-Oki earthquake 1968

Oki Retto
use Oki Islands

Okinawa (1978)
Major island in Okinawa Island group midway between the main islands of Japan and Taiwan.
BT Ryukyu Islands
BT Japan
BT Far East
BT Asia

Okinawa Trench
use Okinawa Trough

Okinawa Trough (1989)
CO N233000N330000
    E1320000E1220000
UF Okinawa Trench
BT Northwest Pacific
BT Pacific Ocean
SA East China Sea
SA Japan

Oklahoma (1978)
Autoposting of this term began in 1978.
CO N333500N370000
    W0942500W1030000
BT United States
NT Arbuckle Anticline
NT Arbuckle Mountains
NT Atoka County Oklahoma
NT Beaver County Oklahoma
NT Blaine County Oklahoma
NT Bryan County Oklahoma
NT Caddo County Oklahoma
NT Carter County Oklahoma
NT Cherokee County Oklahoma
NT Choctaw County Oklahoma
NT Cimarron County Oklahoma
NT Cleveland County Oklahoma
NT Coal County Oklahoma
NT Comanche County Oklahoma
NT Cotton County Oklahoma
NT Criner Hills
NT Custer County Oklahoma
NT Delaware County Oklahoma
NT Dewey County Oklahoma
NT Ellis County Oklahoma
NT Garfield County Oklahoma
NT Garvin County Oklahoma
NT Grady County Oklahoma
NT Grant County Oklahoma
NT Greer County Oklahoma
NT Harper County Oklahoma
NT Haskell County Oklahoma
NT Hughes County Oklahoma
NT Jackson County Oklahoma
NT Jefferson County Oklahoma
NT Johnston County Oklahoma
NT Kingfisher County Oklahoma
NT Kiowa County Oklahoma
NT Latimer County Oklahoma
NT Le Flore County Oklahoma
NT Lincoln County Oklahoma
NT Logan County Oklahoma
NT Major County Oklahoma
NT Marshall County Oklahoma
NT McIntosh County Oklahoma
NT Murray County Oklahoma
NT Muskogee County Oklahoma
NT Noble County Oklahoma
NT Oklahoma Panhandle
NT Osage County Oklahoma
NT Ottawa County Oklahoma
NT Pawnee County Oklahoma
NT Payne County Oklahoma
NT Pittsburg County Oklahoma
NT Pontotoc County Oklahoma
NT Pottawatomie County Oklahoma
NT Seminole County Oklahoma
NT Stephens County Oklahoma
NT Texas County Oklahoma
NT Tulsa County Oklahoma
NT Washington County Oklahoma
NT Wichita Mountains
NT Woods County Oklahoma
NT Woodward County Oklahoma

SA Americus Limestone Member
SA Anadarko Basin
SA Antlers Sands
SA Arbuckle Group
SA Ardmore Basin
SA Arkansas Novaculite
SA Arkansas River
SA Arkansas River valley
SA Arkoma Basin
SA Atoka Formation
SA Bartlesville Sand
SA Benton Uplift
SA Blaine Formation
SA Bloyd Formation
SA Boggy Shale
SA Boone Formation
SA Bromide Formation
SA Burbank Sand
SA Cabaniss Formation
SA Cache Creek
SA Canadian River
SA Carlton Rhyolite
SA Chase Group
SA Chattanooga Shale
SA Cheyenne Sandstone
SA Collier Shale
SA Cottonwood Limestone
SA Council Grove Group
SA Crooked Creek Formation
SA Dakota Formation
SA Desmoinesian
SA Dockum Group
SA Eagle Ford Formation
SA Eskridge Shale
SA Excello Shale
SA Fayetteville Formation
SA Fernvale Formation
SA Flint Hills
SA Fort Riley Limestone
SA Fredericksburg Group
SA Glen Mountains Complex
SA Grand River
SA Great Plains
SA Hardeman Basin
SA Hartshorne Sandstone
SA Hennessey Formation
SA Henryhouse Formation
SA High Plains Aquifer
SA Hugoton Embayment
SA Hunton Group
SA Illinois River
SA Imo Formation
SA Jackfork Group
SA Johns Valley Formation
SA Kiowa Formation
SA Krebs Group
SA Labette Shale
SA Lecompton Limestone
SA Llano Estacado
SA Marmaton Group
SA Meers Fault
SA Midcontinent
SA Morrison Formation
SA Morrow Formation
SA Nemaha Ridge
SA Neosho River valley
SA Ogallala Aquifer
SA Ogallala Formation
SA Oread Limestone
SA Ouachita Belt
SA Ouachita Mountains
SA Ozark Mountains
SA Palo Duro Basin
SA Paluxy Formation
SA Pitkin Limestone
SA Pleasanton Group
SA Reagan Sandstone
SA Red Eagle Limestone
SA Red Fork Sandstone
SA Red River
SA Red River valley
SA Rexroad Formation
SA Roca Formation
SA Saint Louis Limestone
SA Saint Peter Sandstone
SA Savanna Formation
SA Seminole Formation
SA Simpson Group
SA Slick Hills
SA Spiro Sandstone
SA Springer Formation
SA Stanley Group
SA Stanton Formation
SA Timbered Hills Group
SA Trinity Group
SA Verdigris River valley
SA Viola Limestone
SA Wabaunsee Group
SA Wann Formation
SA Wapanucka Limestone
SA Washita Group
SA Washita River valley
SA Wellington Formation
SA Wewoka Formation
SA Wichita Group
SA Womble Shale
SA Woodbine Formation
SA Woodford Shale
SA Wreford Limestone

Oklahoma Panhandle (1993)
NW Oklahoma. Before 1993, also search Panhandle AND Oklahoma.
IN Index counties as applicable.
BT Oklahoma
BT United States
SA Beaver County Oklahoma
SA Cimarron County Oklahoma
SA Texas County Oklahoma

Oklo (1978)
Gabon. Site of natural reactor in uranium deposit.
BT Gabon
BT Central Africa
BT Africa
SA uranium ores

Oktemberyan Series (1978)
BT Miocene
BT Neogene
BT Tertiary
BT Cenozoic
SA Armenia

Oktyabr
No longer a valid term for GeoRef. As of 1993, see Oktyabr Uzbekistan.

Oktyabr Uzbekistan (1993)
Town in SE Uzbekistan. This term has multiple hierarchies.
BT1 Uzbekistan
BT1 Asia
BT2 Uzbekistan
BT2 Commonwealth of Independent States

Okujiri
use Okushiri Island

Okushiri Island (1978)
In Japan Sea off SW coast of Hokkaido.
UF Okujiri
UF Okushiri-shima
BT Hokkaido
BT Japan
BT Far East
BT Asia

Okushiri-shima
use Okushiri Island

Ol Doinyo Lengai
use Oldoinyo Lengai

Oland (1978)
Island in Baltic Sea off SE coast of Sweden. Before 1978, also search Oland Island.
UF Oland Island
BT Kalmar Sweden
BT Sweden
BT Scandinavia
BT Western Europe
BT Europe

Oland Island
use Oland

Old Castile
No longer a valid term for GeoRef. As of 1993, see Old Castile Spain.

Old Castile Spain (1993)
Region in N central Spain. Before 1993, also search Old Castile and Spain.
CO N400500N433000
    W0015000W0054500
UF Castilla la Vieja
BT Castile Spain
BT Spain
BT Iberian Peninsula
BT Southern Europe
BT Europe
NT Avila Spain
NT Burgos Spain
NT Logrono Spain
NT Palencia Spain
NT Santander Spain
NT Segovia Spain
NT Soria Spain
NT Valladolid Spain

Old Crow (1978)
Village in N Yukon Territory.
BT Yukon Territory
BT Western Canada
BT Canada

Old Crow Tephra (1993)
Upper Pleistocene. Alaska and Yukon Territory.
BT upper Pleistocene
BT Pleistocene
BT Quaternary
BT Cenozoic
SA Alaska
SA Yukon Territory

Old Faithful Geyser (1978)
In Yellowstone National Park, NW Wyoming. Autoposting of broader terms to this term began in 1989.
BT Teton County Wyoming
BT Wyoming
BT United States
SA Yellowstone National Park

Old Red Sandstone (1978)
Great Britain.
BT Devonian
BT Paleozoic
SA Dittonian
SA Downtonian

Old Stone Age
use Paleolithic

Old Woman Mountains (1993)
BT San Bernardino County California
BT California
BT United States
SA Mojave Desert

Older Dryas (1978)
Term used primarily in Europe for an interval of late glacial time following the Bolling and preceding the Allerod. Before 1978, search Dryas.
BT Weichselian
BT upper Pleistocene
BT Pleistocene
BT Quaternary
BT Cenozoic
SA Dryas

Oldest Dryas (1978)
Used primarily in Europe for an interval of late-glacial time preceding the Bolling. Before 1978, search Dryas.
BT Weichselian
BT upper Pleistocene
BT Pleistocene
BT Quaternary
BT Cenozoic
SA Dryas

oldhamite (1985)
Meteorite mineral.
BT sulfides
SA meteorites

Oldman Formation (1978)
Of the Belly River Group.
BT Upper Cretaceous
BT Cretaceous
BT Mesozoic
SA Alberta

Oldoinyo Lengai (1978)
Volcano in N Tanzania.
UF Ol Doinyo Lengai
BT Tanzania
BT East Africa
BT Africa

Olduvai Event (1981)
Geomagnetic event.
BT lower Pleistocene
BT Pleistocene
BT Quaternary
BT Cenozoic
SA paleomagnetism
SA reversals

Olduvai Gorge (1978)
Site of rich fossil beds in N Tanzania. Also search Olduvai.
CO S031500S031500
    E0353000E0353000
BT Tanzania
BT East Africa
BT Africa

Olekma (1978)
River in N Chita Oblast and S Yakutia in SE central Siberia. This term has multiple hierarchies.
BT1 Russian Federation
BT1 Commonwealth of Independent States
BT2 Asia
SA Chita Russian Federation
SA Yakutia Russian Federation

Olekma-Vitim Highlands (1978)
In NE Buryatia and NW Chita Oblast in N Transbaikalia. This term has multiple hierarchies.
UF Olekma-Vitim mountain country
UF Olekma-Vitim Mountains
BT1 Russian Federation
BT1 Commonwealth of Independent States
BT2 Asia
SA Buryat Russian Federation
SA Chita Russian Federation

Olekma-Vitim mountain country
use Olekma-Vitim Highlands

Olekma-Vitim Mountains
use Olekma-Vitim Highlands

Olenek River (1978)
Rises in Central Siberian Plateau and flows NE through Yakutia into Laptev Sea. Also search Olenek. This term has multiple hierarchies.
BT1 Yakutia Russian Federation
BT1 Russian Federation
BT1 Commonwealth of Independent States
BT2 Yakutia Russian Federation
BT2 Asia

Olenidae (1978)
  Family.
  BT Ptychopariida
  BT Trilobita
  BT Trilobitomorpha
  BT Arthropoda
  BT Invertebrata

Olentangy Shale (1978)
  Plum Brook Shale has been correlated with Olentangy, and so-called Prout Limestone has been regarded as member of Olentangy Shale. Central Ohio.
  BT Upper Devonian
  BT Devonian
  BT Paleozoic
  SA Ohio

Oligocene (1978)
  Worldwide. Above Eocene, below Miocene. Autoposting of this term began in 1978.
  BT Paleogene
  BT Tertiary
  BT Cenozoic
  NT Barail Group
  NT Bembridge Marls
  NT Boom Clay
  NT Brule Formation
  NT Buda Marl
  NT Cypress Hills Formation
  NT Fish Canyon Tuff
  NT Frio Formation
  NT Glendon Limestone
  NT Grimmertingen
  NT Hackberry Formation
  NT Hemlock Conglomerate
  NT Krosno Beds
  NT Lincoln Creek Formation
  NT lower Oligocene
  NT middle Oligocene
  NT Nari Series
  NT upper Oligocene
  NT White River Group
  NT Wiggins Formation
  SA Bluff Formation
  SA Catahoula Formation
  SA Challis Volcanics
  SA Duchesne River Formation
  SA Hector Formation
  SA Poltava Series
  SA Refugian
  SA Rincon Formation
  SA San Lorenzo Formation
  SA Saucesian
  SA Sespe Formation
  SA Twin River Formation
  SA upper Paleogene
  SA Vaqueros Formation

Oligochaetia (1978)
  BT worms
  BT Invertebrata

oligoclase (1978)
  BT plagioclase
  BT feldspar group
  BT framework silicates
  BT silicates

olistoliths (1978)
  BT soft sediment deformation
  BT sedimentary structures
  SA olistostromes
  SA turbidity current structures

olistostromes (1978)
  BT soft sediment deformation
  BT sedimentary structures
  SA olistoliths
  SA turbidity current structures

Oliver County
  Valid through 1988. Search in combination with state term. After 1988, use specific county-state term.

Oliver County North Dakota (1989)
  W central North Dakota. Before 1989, also search Oliver County AND North Dakota.
  CO N465800N471800 W1005300W1014700
  BT North Dakota
  BT United States

olivine (1978)
  A mineral in the olivine group. See also olivine group.
  BT olivine group
  BT nesosilicates
  BT orthosilicates
  BT silicates

olivine basalt (1978)
  BT basalts
  BT volcanic rocks
  BT igneous rocks
  SA alkali olivine basalt

olivine diabase (1978)
  Before 1981, also search olivine dolerite.
  UF olivine dolerite
  BT diabase
  BT plutonic rocks
  BT igneous rocks

olivine dolerite
  As of 1981, no longer a valid term for GeoRef.
  use olivine diabase

olivine gabbro (1978)
  BT gabbros
  BT plutonic rocks
  BT igneous rocks

olivine group (1978)
  Monticellite is a narrower term as of 1981. Autoposting of this term began in 1978.
  BT nesosilicates
  BT orthosilicates
  BT silicates
  NT fayalite
  NT forsterite
  NT monticellite
  NT olivine
  NT peridot
  NT tephroite

olivine nephelinite (1978)
  BT alkali gabbros
  BT gabbros
  BT plutonic rocks
  BT igneous rocks
  SA ankaratrite
  SA nepheline basalt
  SA nephelinite

olivine tholeiite (1978)
  BT basalts
  BT volcanic rocks
  BT igneous rocks
  SA tholeiite

olivinite (1978)
  BT ultramafics
  BT plutonic rocks
  BT igneous rocks

Olkusz
  No longer a valid term for GeoRef. As of 1993 see Olkusz Poland.

Olkusz Poland (1993)
  Town in E Katowice, S Poland. Before 1993 also search Olkusz AND Poland.
  BT Katowice Poland
  BT Poland
  BT Central Europe
  BT Europe

Olmos Formation (1985)
  In Navarro Group. SW Texas.
  BT Gulfian
  BT Upper Cretaceous
  BT Cretaceous
  BT Mesozoic
  SA Navarro Group
  SA Texas

Olsztyn
  No longer a valid term for GeoRef. As of 1993 see Olsztyn Poland.

Olsztyn Poland (1993)
  Province in N Poland. Also city SE of Gdansk. Before 1993 also search Olsztyn AND Poland.
  BT Poland
  BT Central Europe
  BT Europe

Olt River (1978)
  Central and S Romania, in Transylvania and Walachia. Also search Olt and Olt River valley.
  UF Oltul
  BT Romania
  BT Southern Europe
  BT Europe
  SA Olt River valley
  SA Transylvania
  SA Walachia

Olt River valley (1978)
  Also search Olt.
  IN Index regions as applicable.
  UF Olt Valley
  UF Oltul Valley
  BT Romania
  BT Southern Europe
  BT Europe
  SA Olt River
  SA Transylvania
  SA Walachia

Olt Valley
  use Olt River valley

Oltenia (1978)
  A sub-region. W part of Walachia.
  UF Lesser Walachia
  UF Little Walachia
  BT Walachia
  BT Romania
  BT Southern Europe
  BT Europe

Oltrepo Pavese Italy
  use Pavia Italy

Oltul
  use Olt River

Oltul Valley
  use Olt River valley

Olympic Dam (1989)
  Site of copper, uranium and gold ores in S Australia.
  BT South Australia
  BT Australia
  BT Australasia
  SA copper ores
  SA gold ores
  SA uranium ores

Olympic Mountains (1978)
  Part of the Coast Ranges in Olympic National Park on the Olympic Peninsula in NW Washington.
  IN Index Coast Ranges AND counties as applicable.
  BT Washington
  BT United States
  SA Blue Glacier
  SA Clallam County Washington
  SA Coast Ranges
  SA Jefferson County Washington
  SA Mason County Washington

Olympic Peninsula (1978)
  Between the Pacific Ocean and Puget Sound in NW Washington.
  BT Washington
  BT United States

Olympus (1978)
  Highest mountain in Greece in NE Greece.
  BT Thessaly Greece
  BT Greece
  BT Southern Europe
  BT Europe

Olympus Mons (1985)
  Volcano on Mars.
  UF Nix Olympia
  BT Mars

Oman (1978)
  Formerly Muscat and Oman.
  CO N170000N270000 E0600000E0523000
  BT Arabian Peninsula
  BT Asia
  NT Oman Mountains
  SA Middle East
  SA Near East
  SA Samail Ophiolite
  SA Semail Ophiolite
  SA Thamama Group

Oman Mountains (1978)
  BT Oman
  BT Arabian Peninsula
  BT Asia

Omi (Lake)
  use Lake Biwa

Omine Mine (1978)
  Metal ores. Near town of Omine in SW Honshu.
  BT Iwate Japan
  BT Honshu
  BT Japan
  BT Far East
  BT Asia
  SA metal ores
  SA mines

Omineca Belt (1989)
  S British Columbia.
  BT British Columbia
  BT Western Canada
  BT Canada
  SA Valhalla Complex

Omineca Mountains (1978)
  NW British Columbia.
  BT British Columbia
  BT Western Canada
  BT Canada

Omo
  No longer a valid term for GeoRef. See Omo Ethiopia.

Omo Ethiopia (1993)
  Village in SW Ethiopia.
  BT Ethiopia
  BT East Africa
  BT Africa

Omo Group (1989)
  IN Index ages as applicable.
  BT Cenozoic
  SA Ethiopia
  SA Kenya
  SA Pleistocene
  SA Pliocene
  SA Shungura Formation

Omo River (1978)
  Flows through SW Ethiopia into Lake Turkana (Lake Rudolph). Also search Omo.
  BT Ethiopia
  BT East Africa
  BT Africa

Omo Valley (1978)
  River valley in SW Ethiopia. Also search Omo.

BT Ethiopia
BT East Africa
BT Africa

Omolon (1978)
River in N Khabarovsk Kray, W Chukchi National Okrug, and NE Yakutia which flows into the Kolyma River in NE Siberia. This term has multiple hierarchies.
BT1 Russian Federation
BT1 Commonwealth of Independent States
BT2 Asia
SA Khabarovsk Russian Federation
SA Yakutia Russian Federation

Omolon Block (1978)
NE Siberia in the Omolon River basin. Before 1978, also search Omolon. This term has multiple hierarchies.
BT1 Russian Federation
BT1 Commonwealth of Independent States
BT2 Asia
SA Siberia

omphacite (1978)
BT clinopyroxene
BT pyroxene group
BT chain silicates
BT silicates
SA augite

Omsk
No longer a valid term for GeoRef. As of 1993, see Omsk Russian Federation.

Omsk Russian Federation (1993)
Oblast and city in SW Siberia. Before 1993, also search Omsk. This term has multiple hierarchies.
BT1 Russian Federation
BT1 Commonwealth of Independent States
BT2 Asia

Omulevka (1978)
River in N Magadan Oblast in NE Siberia.
IN Index regions as applicable.
SA Magadan Russian Federation
SA Russian Federation

On-take
use Ontake

Onaping Formation (1978)
Of the Whitewater Group.
BT Huronian
BT Proterozoic
BT upper Precambrian
BT Precambrian
SA Canada
SA Ontario

oncolites (1978)
BT biogenic structures
BT sedimentary structures

one-dimensional models (1978)
Before 1978, search one dimensional; one-dimensional AND models.
SA four-dimensional models
SA mathematical models
SA models
SA three-dimensional models
SA two-dimensional models

Onega
No longer a valid term for GeoRef. As of 1993, see Onega Russian Federation.

Onega Russian Federation (1993)
City on Onega Bay at mouth of Onega River in NW Arkhangelsk Oblast in NW European Russian Federation. Before 1993, also search Onega. This term has multiple hierarchies.
BT1 Arkhangelsk Russian Federation
BT1 Russian Federation
BT1 Commonwealth of Independent States
BT2 Arkhangelsk Russian Federation
BT2 Europe
SA Lake Onega

Oneida County
Valid through 1988. Search in combination with state term. After 1988, use specific county-state term.

Oneida County Idaho (1989)
SE Idaho. Before 1989, also search Oneida County AND Idaho.
CO N420000N422500 W1120500W1130000
BT Idaho
BT United States
SA Bannock Range

Oneida County New York (1989)
Central New York. Before 1989, also search Oneida County AND New York.
CO N425200N433700 W0750500W0755300
BT New York
BT United States
SA Oneida Lake

Oneida County Wisconsin (1989)
N Wisconsin. Before 1989, also search Oneida County AND Wisconsin.
CO N452700N455400 W0890400W0900300
BT Wisconsin
BT United States

Oneida Lake (1978)
SE of Lake Ontario in central New York, bounded by Oswego, Oneida, Madison and Onondaga counties.
IN Index counties as applicable.
CO N430900N431600 W0754300W0761000
BT New York
BT United States
SA Madison County New York
SA Oneida County New York
SA Onondaga County New York
SA Oswego County New York

Onikobe
No longer a valid term for GeoRef. See Onikobe Field.

Onikobe Field (1993)
Geothermal field. N Honshu. Before 1993, also search Onikobe.
BT Iwate Japan
BT Honshu
BT Japan
BT Far East
BT Asia
SA geothermal fields

onofrite (1978)
This term has multiple hierarchies.
BT1 selenites
BT2 sulfides
SA metacinnabar
SA selenium

Onondaga County
Valid through 1988. Search in combination with state term. After 1988, use specific county-state term.

Onondaga County New York (1989)
Central New York. Before 1989, also search Onondaga County AND New York.
CO N424500N431500 W0755400W0763000
BT New York
BT United States
NT Onondaga Lake
NT Syracuse New York
SA Oneida Lake

Onondaga Lake (1989)
One of the Finger Lakes in W central New York. This term has multiple hierarchies.
CO N430500N431000 W0761000W0761800
BT1 Onondaga County New York
BT1 New York
BT1 United States
BT2 Finger Lakes
BT2 New York
BT2 United States

Onondaga Limestone (1978)
Includes Springfield Center Member, Babcock Hill Member, Edgecliff Member, Nedrow Member, Moorehouse Member, Seneca Member, Needmore Shale, Selinsgrove Limestone. W Maryland, New York, Pennsylvania, Western Virginia, and N West Virginia.
BT Middle Devonian
BT Devonian
BT Paleozoic
SA Hamilton Group
SA Maryland
SA New York
SA Pennsylvania
SA Virginia
SA West Virginia

onshore (1978)
Term reintroduced in 1985. Not a valid GeoRef term from 1981-1984. To be used for petroleum and natural gas.
SA exploration
SA natural gas
SA offshore
SA petroleum
SA production

Onslow Bay (1978)
Between Cape Lookout and Cape Fear off SE coast of North Carolina.
BT North Carolina
BT United States

Onslow County
Valid through 1988. Search in combination with state term. After 1988, use specific county-state term.

Onslow County North Carolina (1989)
E North Carolina. Before 1989, also search Onslow County AND North Carolina.
CO N342500N345800 W0770800W0774300
BT North Carolina
BT United States

Ontake (1978)
A volcano on the border between Gifu and Nagano prefectures in central Honshu. Before 1993 was also used for a volcano (also called Sakura-jima) on the peninsula of Sakura-jima in Kagoshima Bay, S Kyushu. See Sakura-jima for the volcano on Kyushu. Before 1993, search in combination with Honshu.
IN Index prefectures as applicable.
UF Mount Ontake
UF On-take
UF Ontake Volcano
BT Honshu
BT Japan
BT Far East
BT Asia
SA Aira Caldera
SA Gifu Japan
SA Kagoshima Japan
SA Kyushu
SA Nagano earthquake 1984
SA Nagano Japan
SA Sakura-jima

Ontake Volcano
use Ontake

Ontario (1978)
CO N420000N570000 W0740000W0950000
BT Eastern Canada
BT Canada
NT Algoma District Ontario
NT Batchawana Bay
NT Bruce Peninsula
NT Callander Bay
NT Cochrane District Ontario
NT Essex County Ontario
NT Eye-Dashwa Lakes Pluton
NT Frontenac County Ontario
NT Gowganda Ontario
NT Grenville County Ontario
NT Haliburton County Ontario
NT Hamilton Ontario
NT Hastings County Ontario
NT Hemlo Deposit
NT Kakagi Lake
NT Kenora District Ontario
NT Lambton County Ontario
NT Lanark County Ontario
NT Larder Lake District Ontario
NT Manitoulin District Ontario
NT Michipicoten Island
NT Moose River basin
NT Nipissing District Ontario
NT Northumberland County Ontario
NT Ottawa Ontario
NT Oxford County Ontario
NT Parry Sound District Ontario
NT Rainy River District Ontario
NT Renfrew County Ontario
NT Saint Catharines Ontario
NT Strathcona Mine
NT Sudbury Basin
NT Sudbury District Ontario
NT Sudbury Irruptive
NT Temagami Mine
NT Thunder Bay
NT Thunder Bay District Ontario
NT Timiskaming District Ontario
NT Toronto Ontario
NT Victoria County Ontario
NT Welland Canal
NT Wellington County Ontario
NT Whetstone Lake
SA Abitibi Belt
SA Animikie Group
SA Appalachian Basin
SA Bass Islands Dolomite
SA Beekmantown Group
SA Blake River Group
SA Blue Mountain

SA Brent Crater
SA BRIM
SA Canadian Shield
SA Central Gneiss Belt
SA Central Metasedimentary Belt
SA Chalk River
SA Champlain Sea
SA Chelmsford Formation
SA Cherry Canyon Formation
SA Churchill Province
SA Clinton Group
SA Cobalt Group
SA Detroit River
SA Detroit River Group
SA Dundee Limestone
SA Eagle Lake
SA Elk Lake
SA Elliot Formation
SA English River Belt
SA Espanola Formation
SA Flack Lake
SA Flinton Group
SA Georgian Bay
SA Gibraltar Point
SA Gowganda Formation
SA Grand River
SA Great Lakes region
SA Grenville Front
SA Grenville Province
SA Grimsby Sandstone
SA Guelph Formation
SA Gull River Formation
SA Henderson Mine
SA Hudson Bay Lowlands
SA Indian River
SA James Bay
SA James Bay Lowlands
SA Kapuskasing Zone
SA Keweenawan
SA Lake Algonquin
SA Lake Erie
SA Lake Huron
SA Lake Maumee
SA Lake of the Woods
SA Lake of the Woods region
SA Lake Ontario
SA Lake Saint Clair
SA Lake Superior
SA Lake Superior region
SA Lake Timiskaming
SA Leda Clay
SA Lockport Formation
SA Lorrain Formation
SA Lucas Formation
SA Marquette Range Supergroup
SA Medina Formation
SA Michigamme Formation
SA Michigan Basin
SA Michipicoten Belt
SA Niagara Escarpment
SA Niagara Falls
SA Niagara River
SA Nipissing Diabase
SA O'Sullivan Lake
SA Onaping Formation
SA Ontario Geological Survey
SA Osler Series
SA Ottawa River
SA Ottawa Sand
SA Ottawa Valley
SA Perch Lake
SA Potsdam Sandstone
SA Queenston Shale
SA Quetico Belt
SA Rainy Lake
SA Rainy River
SA Rice Lake
SA Rochester Formation
SA Saganaga Lake
SA Saint Clair River
SA Saint Clair River delta
SA Saint Lawrence Lowlands
SA Saint Lawrence River
SA Saint Lawrence River basin
SA Saint Lawrence Valley
SA Scarborough Formation
SA Shadow Lake Formation
SA Southern Province
SA Sturgeon Lake
SA Superior Province
SA Sylvania Formation
SA Thames River
SA Timiskaming Group
SA Trent Valley
SA Utica Shale
SA Vermilion granitic complex
SA Wabigoon Belt
SA Wawa Belt
SA Whirlpool Sandstone

Ontario Geological Survey (1985)
Formerly Ontario Department of Mines.
UF OGS
BT survey organizations
BT government agencies
SA Ontario

ontogeny (1978)
SA diachronism
SA fossils
SA growth rates
SA juvenile taxa
SA larvae
SA maturity
SA paleoindian
SA paleontology
SA phenotypes
SA phylogeny

Ontonagon County
Valid through 1988. Search in combination with state term. After 1988, use specific county-state term.

Ontonagon County Michigan (1989)
NW Michigan Upper Peninsula, NW Michigan. Before 1989, also search Ontonagon County AND Michigan.
CO N462000N470300 W0885300W0895300
BT Michigan Upper Peninsula
BT Michigan
BT United States
NT White Pine Michigan
NT White Pine Mine
SA Porcupine Mountains

Ontong Java Plateau (1978)
Between Solomon Islands on the W and Lord Howe Islands (Ontong Java Islands) on the E.
UF Antong Java Rise
UF Ontong Java Rise
BT West Pacific
BT Pacific Ocean
SA DSDP Site 288
SA DSDP Site 289
SA DSDP Site 586
SA Leg 89
SA Leg 130
SA ODP Site 803
SA ODP Site 804
SA ODP Site 805
SA ODP Site 806
SA ODP Site 807

Ontong Java Rise
use Ontong Java Plateau

onvarovite
use uvarovite

Onverwacht Group (1978)
BT Precambrian
SA Barberton greenstone belt
SA South Africa

onyx (1985)
BT silica minerals
BT framework silicates
BT silicates
SA chalcedony
SA gems

onyx marble
use alabaster

ooids
use oolite

oolite (1978)
Before 1977, also search oolites.
UF eggstone
UF ooids
UF oolites
UF ooliths
UF roestone
SA carbonate rocks
SA carbonate sediments
SA limestone
SA oolitic limestone
SA oolitic texture
SA pellets
SA sedimentary rocks

oolites
A valid term through 1976.
use oolite

ooliths
use oolite

oolitic limestone (1978)
BT limestone
BT carbonate rocks
BT sedimentary rocks
SA oolite

oolitic texture (1978)
Before 1978, also search oolitic.
BT textures
SA limestone
SA oolite
SA sedimentary rocks
SA sediments

ooze (1978)
BT clastic sediments
BT sediments
SA calcareous composition
SA mud
SA sapropel

opal (1978)
Autoposting of this term began in 1978.
UF opaline
BT silica minerals
BT framework silicates
BT silicates
NT opal-A
NT opal-CT
SA floatstone
SA hyalite

opal A
use opal-A

opal-A (1993)
Amorphous, biogenic opal.
UF opal A
BT opal
BT silica minerals
BT framework silicates
BT silicates
SA amorphous materials
SA opal-CT

opal-CT (1989)
BT opal
BT silica minerals
BT framework silicates
BT silicates
SA cristobalite
SA opal-A
SA tridymite

opaline
use opal

opaque minerals (1978)
SA light minerals
SA minerals

Opeche Shale (1989)
SE Montana, NW Nevada, W South Dakota and Wyoming.
BT Lower Permian
BT Permian
BT Paleozoic
SA Montana
SA Nevada
SA South Dakota
SA Wyoming

Opemisca Group (1989)
BT Archean
BT Precambrian
SA Quebec

open cast mining
use open-pit mining

open cut mining
use open-pit mining

open fractures (1978)
Before 1978, also search fractures AND open.
BT fractures
SA cracks

open pit mines
use open-pit mining

open pit mining
use open-pit mining

open systems (1978)
Used as a general term.
SA closed systems
SA systems

open-file reports
No longer a valid GeoRef index term. Before 1971, was used on level 1 in subfile N. See report; annual report; USGS.

open-pit mining (1978)
UF mining, open-pit
UF open cast mining
UF open cut mining
UF open pit mines
UF open pit mining
UF opencast mining
UF opencut mining
BT surface mining
BT mining
SA mines
SA mining geology
SA pits
SA quarries
SA strip mining

opencast mining
use open-pit mining

opencut mining
use open-pit mining

Operculina (1978)
Autoposting of microfossils and Protista to this term began in 1990. This term has multiple hierarchies.
BT1 Nummulites
BT1 Nummulitidae
BT1 Rotaliacea
BT1 Rotaliina
BT1 foraminifera
BT1 Protista
BT1 Invertebrata
BT2 Nummulites
BT2 Nummulitidae
BT2 Rotaliacea
BT2 Rotaliina
BT2 foraminifera
BT2 Protista
BT2 microfossils

ophicalcite (1978)
BT marbles

BT metamorphic rocks
SA limestone

Ophiocistioidea (1978)
BT Echinozoa
BT Echinodermata
BT Invertebrata

ophiolite (1978)
Group of mafic and ultramafic igneous rocks with a broad compositional range, including rocks rich in serpentinite, chlorite, epidote and albite derived from them later by metamorphism. Before 1993, ultramafics was a broader term. As of 1993, this term is restricted to petrologic studies. Before 1993, also search meta-ophiolite if applicable. After 1993, see ophiolite complexes for tectonic studies.
IN Index metamorphic or igneous rocks and specific rock types as applicable.
UF meta-ophiolite
SA igneous rocks
SA mafic composition
SA metamorphic rocks
SA ultramafics

ophiolite complexes (1993)
Sequence of rock types consisting of deep-sea sediments lying above basaltic pillow lavas, dikes, gabbro, and ultramafic peridotite. They are considered to be fragments of oceanic lithosphere that were tectonically transported to continental margins and island arcs. This term is restricted to use with tectonics and plate tectonics. See ophiolite for petrologic studies.
SA lithosphere
SA oceanic crust
SA plate tectonics
SA tectonics

Ophiomorpha (1978)
Autoposting of this term began in 1978.
BT Malacostraca
BT Crustacea
BT Mandibulata
BT Arthropoda
BT Invertebrata
NT Ophiomorpha nodosa
SA ichnofossils

Ophiomorpha nodosa (1978)
BT Ophiomorpha
BT Malacostraca
BT Crustacea
BT Mandibulata
BT Arthropoda
BT Invertebrata
SA ichnofossils

ophite (1978)
Before 1993, marbles was autoposted to this term. Term may be used with marbles and also with diabase.
IN Index marbles or diabase as applicable.
SA diabase
SA igneous rocks
SA marbles
SA metamorphic rocks

Ophiuroidea (1978)
BT Stelleroidea
BT Asterozoa
BT Echinodermata
BT Invertebrata

Ophthalmidium (1978)

Genus. Autoposting of microfossils and Protista to this term began in 1990. This term has multiple hierarchies.
BT1 Miliolacea
BT1 Miliolina
BT1 foraminifera
BT1 Protista
BT1 Invertebrata
BT2 Miliolacea
BT2 Miliolina
BT2 foraminifera
BT2 Protista
BT2 microfossils

Opole
No longer a valid term for GeoRef. As of 1993 see Opole Poland.

Opole Poland (1993)
Province and city in S Poland. Before 1993 also search Opole AND Poland.
BT Poland
BT Central Europe
BT Europe
SA Polish Sudeten Mountains

optical
A valid term through 1977. After 1977, use optical spectroscopy, optical spectra, optical properties, optical mineralogy or (as of 1981) optical dispersion or optical extinction as applicable.

optical data processing
Not a valid GeoRef index term after 1970. Was used on level 1 in subfile N.

optical dispersion (1981)
SA optical properties
SA refractive index

optical extinction (1981)
Before 1981, also search extinction AND minerals.
SA extinction
SA optical properties
SA undulatory extinction

optical methods
As of 1981, no longer a valid term for GeoRef. See microscope methods.

optical mineralogy (1978)
SA geochronology
SA goniometry
SA microscope methods
SA mineralogy
SA ore microscopy
SA relative age

optical properties (1978)
SA birefringence
SA color centers
SA crystal field
SA fluorescence
SA isotropy
SA mineralogy
SA minerals
SA optical dispersion
SA optical extinction
SA pleochroism
SA polarization
SA properties
SA refractive index
SA wave dispersion

optical spectra (1978)
Used for data. For methodology, use optical spectroscopy. Before 1978, search optical AND spectroscopy.
BT spectra
SA optical spectroscopy
SA spectroscopy

optical spectroscopy (1978)
Before 1978, also search optical AND spectroscopy.
BT spectroscopy
SA analysis
SA chemical analysis
SA optical spectra

optimal filters (1981)
SA filters

optimization (1978)
Used as a general term.
SA beneficiation
SA methods
SA production

Oquirrh Formation (1985)
SE Idaho and N central Utah.
IN Index ages as applicable.
BT Paleozoic
SA Idaho
SA Lower Permian
SA Pennsylvanian
SA Utah

Oquirrh Mountains (1978)
S of Great Salt Lake.
BT Utah
BT United States

Oran
No longer a valid term for GeoRef. Valid 1978-1981. As of 1982, see Oran region for the department of French Algeria. As of 1993, see Oran Algeria for the department.

Oran Algeria (1993)
Department. NW coast. Before 1993, also search Oran or Oran region.
BT Algeria
BT North Africa
BT Africa
NT Descartes Algeria

Oran region
No longer a valid term for GeoRef. Used as large department of French Algeria. As of 1993, see Algeria for departments. Before 1982, also search Oran.

Orange County
Valid through 1988. Search in combination with state term. After 1988, use specific county-state term.

Orange County California (1989)
S California. Before 1989, also search Orange County AND California.
CO N332500N335400
 W1172600W1180800
BT California
BT United States
NT Santa Ana California
SA Newport Bay
SA San Jacinto Fault
SA Santa Ana Mountains
SA Santa Ana River

Orange County Florida (1989)
Central Florida. Before 1989, also search Orange County AND Florida.
CO N282100N284700
 W0805300W0815000
BT Florida
BT United States

Orange County Indiana (1989)
S Indiana. Before 1989, also search Orange County AND Indiana.
CO N382300N384200
 W0862000W0864200

BT Indiana
BT United States

Orange County New York (1989)
SE New York. Before 1989, also search Orange County AND New York.
CO N410800N413800
 W0735800W0744300
BT New York
BT United States
NT Port Jervis New York

Orange County North Carolina (1989)
N central North Carolina. Before 1989, also search Orange County AND North Carolina.
CO N355200N361400
 W0785700W0791500
BT North Carolina
BT United States

Orange County Texas (1989)
SE Texas. Before 1989, also search Orange County AND Texas.
CO N295800N301800
 W0934300W0940700
BT Texas
BT United States
SA Sabine Lake

Orange County Vermont (1989)
Central and E Vermont. Before 1989, also search Orange County AND Vermont.
CO N434700N441300
 W0720200W0724700
BT Vermont
BT United States

Orange County Virginia (1989)
N central Virginia. Before 1989, also search Orange County AND Virginia.
CO N380800N382300
 W0774200W0782200
BT Virginia
BT United States

Orange Free State
No longer a valid term for GeoRef.
use Orange Free State South Africa

Orange Free State South Africa (1993)
Province in E central South Africa. Before 1993 also search Orange Free State AND South Africa.
CO S304500S264000
 E0295000E0241500
UF Orange Free State
BT South Africa
BT Southern Africa
BT Africa
NT Swartkrans
NT Vredefort Dome
SA Klerksdorp Field
SA Malvern Meteorite
SA Orange River
SA Vaal River

orange material (1978)
SA Moon

Orange River (1978)
Rises in the Drakensberg Range in Lesotho and flows W into the Pacific Ocean.
IN Index Namibia, Lesotho and provinces of South Africa.
BT Southern Africa
BT Africa
SA Cape Province South Africa
SA Lesotho
SA Namibia

SA Orange Free State South Africa

Orava Valley (1978)
River valley in N Slovakia. Also search Orava.
BT Slovakia
BT Czechoslovakia
BT Central Europe
BT Europe

orbicular texture (1978)
Before 1978, search orbicular AND texture.
BT textures
SA spherulites

orbital observations (1978)
Before 1978, search observations AND orbital.
SA observations
SA remote sensing
SA satellite methods

Orbitoidacea (1978)
Autoposting of microfossils and Protista to this term began in 1990. This term has multiple hierarchies.
BT1 Rotaliina
BT1 foraminifera
BT1 Protista
BT1 Invertebrata
BT2 Rotaliina
BT2 foraminifera
BT2 Protista
BT2 microfossils
NT Amphistegina
NT Asterocyclina
NT Cibicides
NT Discocyclina
NT Lepidocyclina
NT Nephrolepidina
NT Orbitoididae

Orbitoidae
use Orbitoididae

Orbitoididae (1978)
Autoposting of microfossils and Protista to this term began in 1990. This term has multiple hierarchies.
UF Orbitoidae
BT1 Orbitoidacea
BT1 Rotaliina
BT1 foraminifera
BT1 Protista
BT1 Invertebrata
BT2 Orbitoidacea
BT2 Rotaliina
BT2 foraminifera
BT2 Protista
BT2 microfossils

Orbitolina (1978)
Genus. Autoposting of microfossils and Protista to this term began in 1990. This term has multiple hierarchies.
BT1 Orbitolinidae
BT1 Lituolacea
BT1 Textulariina
BT1 foraminifera
BT1 Protista
BT1 Invertebrata
BT2 Orbitolinidae
BT2 Lituolacea
BT2 Textulariina
BT2 foraminifera
BT2 Protista
BT2 microfossils

Orbitolinidae (1978)
Family. Autoposting of microfossils and Protista to this term began in 1990. This term has multiple hierarchies.
BT1 Lituolacea
BT1 Textulariina
BT1 foraminifera
BT1 Protista
BT1 Invertebrata
BT2 Lituolacea
BT2 Textulariina
BT2 foraminifera
BT2 Protista
BT2 microfossils
NT Orbitolina

Orbulina (1978)
Genus. Autoposting of microfossils and Protista to this term began in 1990. This term has multiple hierarchies.
BT1 Globigerinidae
BT1 Globigerinacea
BT1 Rotaliina
BT1 foraminifera
BT1 Protista
BT1 Invertebrata
BT2 Globigerinidae
BT2 Globigerinacea
BT2 Rotaliina
BT2 foraminifera
BT2 Protista
BT2 microfossils
NT Orbulina universa

Orbulina universa (1978)
Species. Autoposting of microfossils and Protista to this term began in 1990. This term has multiple hierarchies.
BT1 Orbulina
BT1 Globigerinidae
BT1 Globigerinacea
BT1 Rotaliina
BT1 foraminifera
BT1 Protista
BT1 Invertebrata
BT2 Orbulina
BT2 Globigerinidae
BT2 Globigerinacea
BT2 Rotaliina
BT2 foraminifera
BT2 Protista
BT2 microfossils

Orca Basin (1981)
Depression on the continental slope of the N central Gulf of Mexico, containing brines.
CO N265200N270300
W0911300W0912700
BT Gulf of Mexico
BT North American Atlantic
BT North Atlantic
BT Atlantic Ocean
SA DSDP Site 618

Orca Group (1989)
S central Alaska.
IN Index ages as applicable.
BT Paleogene
BT Tertiary
BT Cenozoic
SA Alaska
SA Eocene
SA Paleocene

Orcadian Basin (1989)
Offshore NE Scotland, in the North Sea.
BT North Sea
BT European Atlantic
BT North Atlantic
BT Atlantic Ocean
SA Orkney Islands
SA Scotland

order-disorder (1978)
SA bonding
SA crystal chemistry
SA substitution

ordinary chondrites (1985)
BT chondrites
BT stony meteorites
BT meteorites
SA carbonaceous chondrites
SA enstatite chondrites

Ordos Basin (1989)
Central Inner Mongolia, N China.
BT Inner Mongolia China
BT China
BT Far East
BT Asia

Ordovician (1978)
After the Cambrian, before the Silurian. From 1978-1980, Paleozoic was autoposted to this term.
BT Paleozoic
NT Antelope Valley Limestone
NT Bala
NT Betts Cove Ophiolite
NT Borrowdale Volcanic Series
NT Buchans Group
NT Catoche Formation
NT Chickamauga Group
NT Davidsville Group
NT Ely Springs Dolomite
NT Eureka Quartzite
NT Gander Lake Group
NT Gull River Formation
NT Ledbetter Slate
NT Lower Ordovician
NT Lushs Bight Group
NT Martinsburg Formation
NT Meguma Group
NT Middle Ordovician
NT Montoya Group
NT Pogonip Group
NT Shap Granite
NT Skiddaw Slates
NT Tetagouche Group
NT Upper Ordovician
NT Vinini Formation
NT Viola Limestone
NT Viruan
NT Womble Shale
SA Berwick Formation
SA Bowers Supergroup
SA Collier Shale
SA Cow Head Group
SA Elberton Granite
SA Ellis Bay Formation
SA Hanson Creek Formation
SA Henderson Gneiss
SA Kittery Formation
SA Lincolnshire Limestone
SA Merrimack Group
SA New York City Group
SA Penobscot Formation
SA Road River Formation
SA Robertson Bay Group
SA Shawangunk Formation
SA Shoo Fly Complex
SA Table Mountain Group
SA Taconic Orogeny
SA Talladega Group

Ordubad
No longer a valid term for GeoRef. As of 1993, see Ordubad Azerbaidzhan.

Ordubad Azerbaidzhan (1993)
City in Nakhichevan Autonomous Soviet Socialist Republic (W enclave of Azerbaidzhan). This term has multiple hierarchies.
BT1 Azerbaidzhan
BT1 Europe
BT2 Azerbaidzhan
BT2 Commonwealth of Independent States
SA Nakhichevan Azerbaidzhan

ore bodies (1978)
UF bodies, ore
SA chimneys
SA metal ores
SA mineral deposits, genesis

ore deposits
As of 1981, no longer a valid term for GeoRef. Before 1981, included use for all papers on metal commodities. Was sometimes used in combination with origin (i.e. ore deposits, origin). Now use metal ores; mineral deposits, genesis; mineral resources or specific commodity terms as applicable. See list C for commodity terms.

ore exploration
use mineral exploration

ore finding
Not a valid index term for GeoRef. See mineral exploration.

ore grade (1985)
SA cutoff grade
SA economic geology
SA economics
SA evaluation
SA grade
SA metal ores
SA mineral economics
SA mineral exploration
SA production
SA reserves

ore guides (1978)
UF guides, ore
SA gossan
SA metal ores
SA mineral exploration

ore microscopy (1978)
Before 1976, also search mineragraphy.
UF microscopy, ore
UF mineragraphy
SA metal ores
SA microscope methods
SA mineralogy
SA optical mineralogy
SA universal stage

ore minerals (1978)
SA economic geology
SA metal ores
SA metals
SA minerals
SA ore-forming fluids
SA platinum minerals
SA uranium minerals

ore of sedimentation
use placers

ore reflectivity (1982)
Before 1982, reflectivity was used for this term.
SA metal ores
SA mineral deposits, genesis
SA reflectance

ore rolls
use roll-type

ore sources
As of 1981, no longer a valid term for GeoRef. See mineral deposits, genesis.

ore transport (1978)
Before 1978, search ore deposits AND transport.
SA metal ores
SA mineral deposits, genesis
SA sediment transport
SA solution transport

ore-forming fluids (1978)
Before 1981, also search mineralizers; hydrothermal solutions; hydrothermal fluids.
UF fluids, ore-forming

SA hydrothermal alteration
SA hydrothermal processes
SA mineral deposits, genesis
SA mineralization
SA ore minerals

Oread Limestone (1978)
In Shawnee Group. Includes Toronto Limestone, Snyderville Shale, Leavenworth Limestone, Heebner Shale, Plattsmouth Limestone, Heumader Shale, Kereford Limestone members. SW Iowa, E Kansas, NW Missouri, SE Nebraska, and N Oklahoma.
BT Virgilian
BT Upper Pennsylvanian
BT Pennsylvanian
BT Carboniferous
BT Paleozoic
SA Iowa
SA Kansas
SA Missouri
SA Nebraska
SA Oklahoma
SA Plattsmouth Limestone Member
SA Shawnee Group

Orebro
No longer a valid term for GeoRef. As of 1993 see Orebro Sweden.

Orebro Sweden (1993)
County and city in S central Sweden. Before 1993 also search Orebro AND Sweden.
UF Oerebro
BT Sweden
BT Scandinavia
BT Western Europe
BT Europe
NT Stripa region

Oregon (1978)
Autoposting of this term began in 1978.
CO N420000N462000
  W1163500W1243500
BT United States
NT Baker County Oregon
NT Benton County Oregon
NT Brothers fault zone
NT Clackamas County Oregon
NT Clatsop County Oregon
NT Columbia County Oregon
NT Coos County Oregon
NT Crook County Oregon
NT Curry County Oregon
NT Deschutes County Oregon
NT Douglas County Oregon
NT Grant County Oregon
NT Harney County Oregon
NT Hood River County Oregon
NT Jackson County Oregon
NT Jefferson County Oregon
NT Josephine County Oregon
NT Klamath County Oregon
NT Lake County Oregon
NT Lane County Oregon
NT Lincoln County Oregon
NT Linn County Oregon
NT Malheur County Oregon
NT Marion County Oregon
NT Morrow County Oregon
NT Mount Hood
NT Multnomah County Oregon
NT Polk County Oregon
NT Rogue River
NT Sherman County Oregon
NT Tillamook County Oregon
NT Union County Oregon
NT Wallowa County Oregon
NT Wallowa Mountains
NT Washington County Oregon
NT Wheeler County Oregon
NT Willamette River
NT Willamette Valley
SA Astoria Canyon
SA Astoria Formation
SA Bald Mountain
SA Basin and Range Province
SA Battle Mountain High
SA Blue Mountain
SA Blue Mountains
SA Borax Lake
SA Calera Limestone
SA Cascade Range
SA Clarno Formation
SA Coast Ranges
SA Columbia Plateau
SA Columbia River
SA Columbia River Basalt Group
SA Columbia River basin
SA Columbia River estuary
SA Condrey Mountain Schist
SA Cowlitz Formation
SA Crater Lake
SA Deschutes Formation
SA Deschutes River
SA Flournoy Formation
SA Frenchman Springs Member
SA Galice Formation
SA Glenns Ferry Formation
SA Grande Ronde Basalt
SA Great Basin
SA Hayfork Terrane
SA Hornbrook Formation
SA Illinois River
SA John Day Formation
SA Josephine Ophiolite
SA Josephine Peridotite
SA Klamath Mountains
SA Lake Lahontan
SA Lake Missoula
SA Lookingglass Formation
SA Mazama Ash
SA Mount Jefferson
SA Owyhee Mountains
SA Picture Gorge Basalt
SA Powder River basin
SA Riggins Group
SA Ringold Formation
SA Saddle Mountains Basalt
SA Snake River
SA Snake River basin
SA Snake River canyon
SA Tyee Formation
SA Umpqua Formation
SA Wanapum Basalt
SA Willow Creek
SA Wrangellia
SA Yakima Basalt
SA Yakima fold belt
SA Yamhill Formation

Orenburg
No longer a valid term for GeoRef. As of 1993, see Orenberg Russian Federation.

Orenburg Russian Federation (1993)
Oblast and city in the Southern Urals in European Russian Federation. Before 1993, also search Orenburg. Before 1978, also search Chkalov. This term has multiple hierarchies.
UF Chkalov Russian Federation
BT1 Russian Federation
BT1 Commonwealth of Independent States
BT2 Europe
NT Orsk Russian Federation
NT Sakmara Russian Federation
SA Romashkino Field
SA Ural region

Orense
No longer a valid term for GeoRef. As of 1993, see Orense City Spain.

Orense City Spain (1993)
Refers to only the city in NW Spain. Before 1993, also search Orense and Spain.
BT Orense Spain
BT Galicia Spain
BT Spain
BT Iberian Peninsula
BT Southern Europe
BT Europe

Orense Province
No longer a valid term for GeoRef. As of 1993, see Orense Spain.

Orense Spain (1993)
Refers only to the province in NW Spain. From 1981-1992, also search Orense Province and Spain. Before 1981, also search Orense and Spain. For the city, see Orense City Spain.
CO N414700N423500
  W0064500W0082000
BT Galicia Spain
BT Spain
BT Iberian Peninsula
BT Southern Europe
BT Europe
NT Orense City Spain

ores
Not a valid term for GeoRef. Use specific commodity terms (see list C) or mineral resources or mineral deposits, genesis.

ores in sedimentary rocks
A valid term through 1973. Use sedimentary rocks or sedimentation in conjunction with the individual ore (list C).

ores, polymetallic
  use polymetallic ores

Oresund (1978)
Strait connecting the Kattegat with the Baltic Sea.
IN Index countries as applicable.
SA Baltic Sea
SA Denmark
SA Kattegat
SA Sweden

organic
A valid term through 1977. After 1977, use organic materials.

organic acids (1981)
SA acids
SA alanine
SA amino acids
SA aspartic acid
SA carboxylic acids
SA fulvic acids
SA glycine
SA humic acids
SA inorganic acids
SA isoleucine
SA ketones

organic carbon (1978)
BT organic materials
SA carbon

organic compounds (1978)
Autoposting of this term began in 1978.
NT amber
NT ozocerite
SA chitin
SA compounds
SA gilsonite
SA grahamite
SA minerals
SA porphyrins

organic materials (1978)
Used for discussions of mostly very small concentrations of organic materials in rocks. For economic deposits, use petroleum, coal or other appropriate commodity terms in list C. Also search organic matter. Autoposting of this term began in 1978.
IN Index specific material as applicable [e.g. amino acids, bitumens, carbohydrates, fatty acids, humates, humic acids, hydrocarbons, kerogen, phenols, pigments, proteins]
NT acetate
NT alcohols
NT amino acids
NT bitumens
NT carbohydrates
NT carboxylic acids
NT esters
NT fatty acids
NT fulvic acids
NT humates
NT humic acids
NT humus
NT hydrocarbons
NT isoprenoids
NT kerogen
NT ketones
NT lignin
NT organic carbon
NT organic sulfur
NT peptides
NT phenols
NT pigments
NT proteins
NT resins
NT sapropel
NT sporopollenin
NT steroids
NT volatile organic compounds
SA anthracite
SA biochemistry
SA carbon
SA carbonaceous composition
SA caustobiolith
SA chelation
SA clathrates
SA coal
SA complexing
SA creosote
SA crude oil
SA epimerization
SA gas sands
SA geochemistry
SA inorganic materials
SA isotopes
SA ligands
SA lignite
SA materials
SA oil sands
SA oil shale
SA organic residues
SA organo-metallics
SA Orgon III
SA PCBs
SA peat
SA petroleum
SA polymerization
SA polymers
SA reflectance
SA sedimentary rocks
SA sedimentation
SA sediments
SA soils
SA tight sands

organic mound
  use bioherms

organic residues (1978)

Before 1981, also search organic sediments. As of 1981, terms that were classified under organic sediments are included here: guano, gyttja and peat. Autoposting of this term to macerals, oil shale, oil sands, vitrinite, micrinite, fusinite, inertinite, sporinite, resinite and exinite began in 1981. Autoposting of this term to brown coal ended in 1981.
  IN See lists I and N.
  UF residues, organic
  NT caustobiolith
  NT coal
  NT gas sands
  NT guano
  NT gyttja
  NT oil sands
  NT oil shale
  NT peat
  NT torbanite
  SA alginite
  SA durain
  SA organic materials
  SA sedimentary rocks
  SA sediments

organic sediments
  As of 1981, no longer a valid term for GeoRef. Use organic residues. Term was introduced in 1976. From 1978-1980, was autoposted to the specific sediments that were classified under it: guano, gyttja and peat.

organic sulfur (1993)
  BT organic materials
  SA sulfur

organisms
  use biota

organization (1978)
  SA academic institutions
  SA associations
  SA government agencies
  SA museums
  SA survey organizations
  SA symposia

organo-metallic complexes
  use organo-metallics

organo-metallic compounds
  use organo-metallics

organo-metallics (1981)
  UF organo-metallic complexes
  UF organo-metallic compounds
  SA geochemistry
  SA organic materials
  SA porphyrins

Orgon III (1985)
  Oceanographic expedition.
  SA Atlantic Ocean
  SA Cape Verde Atlantic
  SA expeditions
  SA marine geology
  SA marine sediments
  SA oceanography
  SA organic materials

Orgueil Meteorite (1981)
  Before 1981, also search Montauban AND meteorites; Orgueil AND meteorites.
  UF Montauban Meteorite
  BT carbonaceous chondrites
  BT chondrites
  BT stony meteorites
  BT meteorites
  SA France
  SA Tarn-et-Garonne France

orientation (1978)
  Includes use as attitude of fold elements with respect to external coordinates.
  SA dip
  SA discordant folds
  SA drag folds
  SA fabric
  SA faults
  SA folds
  SA horizontal orientation
  SA inclined folds
  SA linear orientation
  SA lineation
  SA nappes
  SA oblique orientation
  SA overturned folds
  SA petrofabrics
  SA plunging folds
  SA preferred orientation
  SA recumbent folds
  SA schistosity
  SA strike
  SA structural analysis
  SA style
  SA superposed folds
  SA transverse faults
  SA vertical orientation

Oriente
  No longer a valid term for GeoRef. See Santiago de Cuba.

Oriente Cuba
  use Santiago de Cuba

origin
  A valid index term through 1977 used in combination with life. Before 1971, was used sometimes used alone or in combination with ore deposits or mineral deposits (i.e. ore deposits, origin; mineral deposits, origin).

Orinoco Basin
  use Orinoco River basin

Orinoco Belt (1989)
  Petroliferous region in Venezuela.
  BT Venezuela
  BT South America

Orinoco Delta (1989)
  On Atlantic Ocean in Delta Amacuro Territory, NE Venezuela.
  UF Orinoco River delta
  BT Venezuela
  BT South America
  SA Orinoco River

Orinoco River (1978)
  Rises in the Serra Parima of S Venezuela, flows W across Amazonas, then flows N along the Colombia/Venezuela border, turns E in central Venezuela and empties through a wide delta into the Atlantic Ocean. Also search Orinoco.
  IN Index countries as applicable.
  BT South America
  SA Amazonas Brazil
  SA Amazonas Colombia
  SA Amazonas Venezuela
  SA Anzoategui Venezuela
  SA Bolivar Venezuela
  SA Colombia
  SA Guarico Venezuela
  SA Monagas Venezuela
  SA Orinoco Delta
  SA Orinoco River basin
  SA Venezuela

Orinoco River basin (1989)
  Drainage basin of the Orinoco River.
  IN Index countries as applicable.
  UF Orinoco Basin
  BT South America
  SA Colombia
  SA Llanos
  SA Orinoco River
  SA Venezuela

Orinoco River delta
  use Orinoco Delta

Oriskany Sandstone (1978)
  Comprises Shriver Chert, and Ridgeley Sandstone Member. W Maryland, New York, Pennsylvania, Western Virginia and E West Virginia.
  BT Lower Devonian
  BT Devonian
  BT Paleozoic
  SA Maryland
  SA New York
  SA Pennsylvania
  SA Virginia
  SA West Virginia

Orissa
  No longer a valid term for GeoRef. See Orissa India.

Orissa India (1993)
  State on the Bay of Bengal in E India.
  CO N174000N223000
      E0873000E0812000
  BT India
  BT Indian Peninsula
  BT Asia
  NT Bolangir India
  NT Cuttack India
  NT Dhenkanal India
  NT Keonjhar India
  NT Koraput India
  NT Mayurbhanj India
  NT Sambalpur India
  NT Talchir
  SA Deccan Plateau
  SA Gangpur Series
  SA Kamthi Formation
  SA Mahanadi Valley
  SA Rampur coal field
  SA Singhbhum Granite
  SA Singhbhum shear zone
  SA Talchir coal field
  SA Talchir Series

Orkney Islands (1978)
  Archipelago off NE coast of Scotland.
  CO N584500N593000
      W0061000W0080000
  UF Orkney region Scotland
  BT Scotland
  BT Great Britain
  BT United Kingdom
  BT Western Europe
  BT Europe
  SA Orcadian Basin
  SA South Orkney Islands

Orkney region Scotland
  use Orkney Islands

Orleans
  No longer a valid term for GeoRef. See Orleans France or Orleans Parish Louisiana.

Orleans France (1993)
  City in N central France.
  BT Loiret France
  BT France
  BT Western Europe
  BT Europe

Orleans Parish Louisiana (1989)
  SE Louisiana, coextensive with New Orleans. Before 1989, search Orleans Parish AND Louisiana.
  BT Louisiana
  BT United States
  NT New Orleans Louisiana

Orlov
  No longer a valid term for GeoRef. As of 1993, see Orlov Russian Federation.

Orlov Russian Federation (1993)
  Old name for city of Khalturin just W of Vyatka in Vyatka Oblast of European Russian Federation.
  IN Index oblast as applicable.
  BT Russian Federation
  BT Commonwealth of Independent States
  SA Vyatka Russian Federation

ornamental materials (1982)
  IN May be used as a commodity term with terms such as granite deposits, limestone deposits, marble deposits and construction materials.
  SA Carrara Marble
  SA ceramic materials
  SA clays
  SA construction materials
  SA granite deposits
  SA jadeite
  SA limestone deposits
  SA marble deposits
  SA materials
  SA rhodonite

ornamentation (1978)
  SA paleontology
  SA shells

Orne
  No longer a valid term for GeoRef. See Orne France.

Orne France (1993)
  Department in NW France.
  CO N481000N490000
      E0010000W0005000
  BT France
  BT Western Europe
  BT Europe
  NT Mortagne France
  SA Eure Valley
  SA Normandy

Ornithischia (1978)
  Order. Autoposting of this term began in 1978. This term has multiple hierarchies.
  BT1 Archosauria
  BT1 Diapsida
  BT1 Reptilia
  BT1 Tetrapoda
  BT1 Vertebrata
  BT1 Chordata
  BT2 dinosaurs
  BT2 Tetrapoda
  BT2 Vertebrata
  BT2 Chordata
  NT Ceratopsia
  NT Hadrosauridae

ORNL
  use Oak Ridge National Laboratory

oroclines (1989)
  Orogenic belts that curve sharply, possibly due to horizontal bending of the crust.
  BT orogenic belts
  SA plate tectonics
  SA tectonics

Orocopia Schist (1989)
  S California and SW Arizona.
  BT Mesozoic
  SA Arizona
  SA California

orogenesis
Not valid for GeoRef. A valid term through 1974.
use orogeny

orogenic belts (1981)
NT oroclines
SA clastic wedges
SA fold and thrust belts
SA fold belts
SA foreland basins
SA forelands
SA geosynclines
SA intracontinental belts
SA mobile belts
SA orogeny
SA tectonics

orogeny (1978)
Used for discussions of either individual orogenies or detailed general treatments on several orogenies. Before 1974, also search mountain building; before 1975, also search orogenesis. Before 1993, also search tectogenesis.
UF orogenesis
UF tectogenesis
SA absolute age
SA Acadian Phase
SA Alleghany Orogeny
SA Alpine Orogeny
SA Andean Orogeny
SA Antler Orogeny
SA Appalachian Phase
SA Assyntic Orogeny
SA Asturian Orogeny
SA Avalonian Orogeny
SA back-arc basins
SA Baikalian Phase
SA Cadomian Orogeny
SA Caledonian Orogeny
SA Cimmerian Orogeny
SA Columbian Orogeny
SA Cordilleran Orogeny
SA Damara Orogeny
SA deformation
SA epeirogeny
SA faults
SA folds
SA fore-arc basins
SA geosynclines
SA Grenvillian Orogeny
SA Hercynian Orogeny
SA Himalayan Orogeny
SA Hudsonian Orogeny
SA igneous activity
SA Indosinian Orogeny
SA intracontinental belts
SA Karelian Orogeny
SA Katangan Orogeny
SA Kenoran Orogeny
SA Laramide Orogeny
SA metamorphism
SA Nevadan Orogeny
SA New England Orogeny
SA orogenic belts
SA Ouachita Orogeny
SA overprinting
SA Pan-African Orogeny
SA Penokean Orogeny
SA periodicity
SA plate tectonics
SA Pyrenean Orogeny
SA Saalian Phase
SA structural geology
SA Taconic Orogeny
SA taphrogeny
SA tectonics
SA transpression
SA volcanology
SA Wopmay Orogeny

Orogrande Basin (1989)
Petroleum-producing region in S New Mexico and W Texas.
IN Index states and counties as applicable.
BT United States
SA Dona Ana County New Mexico
SA El Paso County Texas
SA Hudspeth County Texas
SA Lincoln County New Mexico
SA New Mexico
SA Otero County New Mexico
SA Sierra County New Mexico
SA Socorro County New Mexico
SA Texas

Oronto Group (1989)
Keweenawan. Includes Freda Sandstone.
BT Proterozoic
BT upper Precambrian
BT Precambrian
SA Freda Sandstone
SA Michigan
SA Minnesota
SA Nonesuch Shale
SA Wisconsin

Orosei (1978)
Gulf of Tyrrhenian Sea. As of 1993, see Orosei Italy for the village.
BT Tyrrhenian Sea
BT West Mediterranean
BT Mediterranean Sea
SA Sardinia Italy

Orosei Italy (1993)
Village in Sardinia.
BT Sardinia Italy
BT Italy
BT Southern Europe
BT Europe

Oroville
Valid through 1988. Search in combination with state term. After 1988, use specific city-state term.

Oroville California (1981)
City in N California. Site of August, 1975 earthquake.
CO N393100N393100 W1213400W1213400
BT Butte County California
BT California
BT United States

Oroville Dam (1978)
Earth fill dam on Feather River which forms Oroville Reservoir in Butte County of N central California. Also search Oroville Reservoir; reservoirs AND Oroville; surface reservoirs AND Oroville.
BT Butte County California
BT California
BT United States
SA Feather River

Oroville earthquake 1975 (1989)
Butte County, N California.
BT earthquakes
SA Butte County California
SA California

Orozco fracture zone (1989)
East Pacific Rise, E Pacific.
CO N050000N220000 W1000000W1200000
BT Pacific Ocean
SA East Pacific
SA East Pacific Rise

orpiment (1978)
UF yellow arsenic
BT sulfides

Orr Formation (1985)
W Utah.
BT Upper Cambrian
BT Cambrian
BT Paleozoic

SA Utah

Orsha
No longer a valid term for GeoRef. As of 1993, see Orsha Belarus.

Orsha Belarus (1993)
City in Vitebsk Oblast in N Belarus. This term has multiple hierarchies.
BT1 Belarus
BT1 Europe
BT2 Belarus
BT2 Commonwealth of Independent States

Orsk
No longer a valid term for GeoRef. As of 1993, see Orsk Russian Federation.

Orsk Russian Federation (1993)
City in Orenburg Oblast in Southern Urals near the Kazakh border. Before 1993, also search Orsk. This term has multiple hierarchies.
BT1 Orenburg Russian Federation
BT1 Russian Federation
BT1 Commonwealth of Independent States
BT2 Orenburg Russian Federation
BT2 Europe

Ortega Group (1989)
N New Mexico.
BT Proterozoic
BT upper Precambrian
BT Precambrian
SA New Mexico

Orthida (1978)
Autoposting of this term began in 1978.
BT Articulata
BT Brachiopoda
BT Invertebrata
NT Enteletacea

orthite
use allanite

ortho-amphibolite
use orthoamphibolite

orthoamphibole (1981)
BT amphibole group
BT chain silicates
BT silicates
NT anthophyllite
NT gedrite
NT holmquistite

orthoamphibolite (1978)
UF ortho-amphibolite
BT amphibolites
BT metamorphic rocks

orthoclase (1978)
BT alkali feldspar
BT feldspar group
BT framework silicates
BT silicates
SA adularia
SA anorthoclase
SA K-feldspar
SA sanidine

orthoenstatite (1978)
BT orthopyroxene
BT pyroxene group
BT chain silicates
BT silicates

orthogneiss (1981)
BT gneisses
BT metamorphic rocks

Orthopteroida (1981)
BT Insecta

BT Mandibulata
BT Arthropoda
BT Invertebrata

orthopyroxene (1978)
Autoposting of this term began in 1981.
BT pyroxene group
BT chain silicates
BT silicates
NT bronzite
NT enstatite
NT eulite
NT hypersthene
NT orthoenstatite
NT protoenstatite

orthoquartzite (1978)
UF orthoquartzitic sandstone
UF sedimentary quartzite
BT clastic rocks
BT sedimentary rocks
SA ferruginous quartzite
SA quartzites
SA sandstone
SA terrigenous materials

orthoquartzitic sandstone
use orthoquartzite

orthorhombic
use orthorhombic system

orthorhombic system (1985)
Before 1985, also search orthorhombic.
UF orthorhombic
BT crystal systems
SA crystallography

orthosilicates (1978)
Sorosilicates and nesosilicates. Autoposting of this term began in 1978.
BT silicates
NT britholite
NT ellenbergerite
NT nesosilicates
NT sorosilicates

Orulgan Mountains (1978)
Extend N and S, E of Lena River in N central Russian Federation. Before 1978, also search Orulgan. This term has multiple hierarchies.
BT1 Yakutia Russian Federation
BT1 Russian Federation
BT1 Commonwealth of Independent States
BT2 Yakutia Russian Federation
BT2 Asia

Oruro
No longer a valid term for GeoRef. See Oruro Bolivia.

Oruro Bolivia (1993)
Department and city in W Bolivia on the Chilean border.
BT Bolivia
BT South America

Os
use osmium

os
use eskers

Os-187/Os-186 (1989)
Autoposting of broader terms began in 1989. This term has multiple hierarchies.
BT1 radioactive isotopes
BT1 isotopes
BT2 stable isotopes
BT2 isotopes
BT3 osmium
BT3 platinum group
BT3 metals
SA isotope ratios

Os/Re
  use Re/Os
Osage County
  Valid through 1988. Search in combination with state term. After 1988, use specific county-state term.
Osage County Kansas (1989)
  E Kansas. Before 1989, also search Osage County AND Kansas.
  CO N382500N385300
      W0953000W0955700
  BT Kansas
  BT United States
Osage County Missouri (1989)
  Central Missouri. Before 1989, also search Osage County AND Missouri.
  CO N381700N384300
      W0913900W0921000
  BT Missouri
  BT United States
Osage County Oklahoma (1989)
  Coextensive with Osage Indian Reservation in N Oklahoma. Before 1989, also search Osage County AND Oklahoma.
  CO N361000N370000
      W0960000W0970500
  BT Oklahoma
  BT United States
  NT Pawhuska Rock Plain
Osagean
  use Osagian
Osagian (1978)
  Provincial series, North America. Above Kinderhookian, below Meramecian. Before 1976, also search Osagean.
  UF Osagean
  BT Lower Mississippian
  BT Mississippian
  BT Carboniferous
  BT Paleozoic
  NT Burlington Limestone
  NT Keokuk Limestone
  SA Boone Formation
  SA Valmeyeran
  SA Waulsortian facies
Osaka
  No longer a valid term for GeoRef. See Osaka Japan.
Osaka Bay (1993)
  Inlet of Pacific Ocean on S coast of Honshu and E of Awaji Island. Kobe and Osaka are major ports.
  CO N341500N344000
      E1352500E1345100
  BT West Pacific
  BT Pacific Ocean
  SA Honshu
  SA Hyogo Japan
  SA Inland Sea
  SA Osaka Japan
Osaka Group (1978)
  BT Cenozoic
  SA Honshu
  SA Japan
  SA Pleistocene
  SA Pliocene
Osaka Japan (1993)
  City and prefecture in S Honshu.
  BT Honshu
  BT Japan
  BT Far East
  BT Asia
  NT Ibaragi Complex
  NT Izumi Japan
  SA Kinki Japan
  SA Nagano City Japan
  SA Osaka Bay
Osborne County
  Valid through 1988. Search in combination with state term. After 1988, use specific county-state term.
Osborne County Kansas (1989)
  N central Kansas. Before 1989, also search Osborne County AND Kansas.
  CO N391000N393500
      W0983000W0990500
  BT Kansas
  BT United States
oscillations (1978)
  As of 1978 term is restricted to refer to motions under Earth.
  SA Earth
  SA free oscillations
  SA high-level mode
  SA motions
  SA resonance
  SA spheroidal mode
  SA toroidal mode
  SA vibration
Osculosida (1978)
  Order. Autoposting of this term to Phaeodarina began in 1989. Autoposting of microfossils and Protista to this term began in 1990. This term has multiple hierarchies.
  BT1 Radiolaria
  BT1 Protista
  BT1 Invertebrata
  BT2 Radiolaria
  BT2 Protista
  BT2 microfossils
  NT Nassellina
  NT Phaeodarina
Osetia
  use Ossetia
Osgood Mountains (1989)
  N Nevada.
  BT Humboldt County Nevada
  BT Nevada
  BT United States
Oshika Peninsula (1993)
  Between Ishinomaki Bay and the Pacific Ocean in NE Honshu. Before 1993, also search Ojika Peninsula.
  UF Ojika Peninsula
  BT Miyagi Japan
  BT Honshu
  BT Japan
  BT Far East
  BT Asia
Oshima (1978)
  Island group, N Ryukyu Islands, Japan. See O-shima and Izu-Oshima for other islands and for the volcano respectively. After 1992, use Oshima Peninsula if applicable to refer to Hokkaido. Oshima also refers to a sub-prefecture of Hokkaido. This term has multiple hierarchies.
  UF Amami
  UF Amami Gunto
  BT1 Ryukyu Islands
  BT1 Japan
  BT1 Far East
  BT1 Asia
  BT2 Kagoshima Japan
  BT2 Kyushu
  BT2 Japan
  BT2 Far East
  BT2 Asia
  NT Amami-O-shima
  SA Izu-Oshima
  SA O-shima
  SA Oshima Peninsula
Oshima Peninsula (1978)
  N of Hakodate in Oshima and Hiyama sub-prefectures in SW Hokkaido, N Japan.
  BT Hokkaido
  BT Japan
  BT Far East
  BT Asia
  SA Amami-O-shima
  SA Esan Cape
  SA Hakodate Japan
  SA Koma-ga-take
  SA O-shima
  SA Oshima
  SA Tsugaru Strait
Osler Series (1978)
  SW Ontario.
  BT Precambrian
  SA Canada
  SA Ontario
Oslo
  No longer a valid term for GeoRef. As of 1993 see Oslo Norway.
Oslo Graben (1993)
  SE Norway.
  BT Norway
  BT Scandinavia
  BT Western Europe
  BT Europe
Oslo Norway (1993)
  County in SE Norway. Also a city on Oslo Fjord, an inlet of the Skagerrak. Before 1993 also search Oslo AND Norway.
  BT Norway
  BT Scandinavia
  BT Western Europe
  BT Europe
osmium (1978)
  Chemical element. As of 1985, use osmium ores for osmium as a commodity. Autoposting of platinum group and metals to this term began in 1989.
  UF Os
  BT platinum group
  BT metals
  NT Os-187/Os-186
  SA osmium ores
  SA Re/Os
osmium ores (1985)
  Before 1985, also search osmium AND (deposit OR deposits OR ore OR ores OR economic) in the basic index. Autoposting of metal ores to this term began in 1985.
  IN Commodity. See List C.
  BT metal ores
  SA osmium
  SA platinum ores
osmosis (1982)
  SA diffusion
  SA electro-osmosis
  SA geochemistry
  SA hydrochemistry
  SA processes
Osnabruck
  No longer a valid term for GeoRef. See Osnabruck Germany.
Osnabruck Germany (1993)
  City in SE Lower Saxony, central Germany. Before 1993, also search Osnabruck. Before 1978, also search Osnabrueck.
  UF Osnabrueck Germany
  BT Lower Saxony Germany
  BT Germany
  BT Central Europe
  BT Europe
Osnabrueck Germany
  use Osnabruck Germany
Osogovo Mountains (1978)
  Also search Osogov; Osogovo.
  IN Index countries as applicable.
  BT Southern Europe
  BT Europe
  SA Bulgaria
  SA Croatia
  SA Serbia
  SA Yugoslavia
Osor
  No longer a valid term for GeoRef. As of 1993, see Osor Spain.
Osor Spain (1993)
  Village in NE Spain. Before 1993, also search Osor and Spain.
  BT Gerona Spain
  BT Catalonia Spain
  BT Spain
  BT Iberian Peninsula
  BT Southern Europe
  BT Europe
Ossetia (1978)
  Region of the central Caucasus. Divided into the former North Ossetian ASSR in the Russian Federation and the South Ossetian Autonomous Oblast of the Georgian Republic.
  IN Index former Soviet republics as applicable.
  UF Osetia
  BT Asia
  NT North Ossetia Russian Federation
  SA Caucasus
  SA Georgian Republic
  SA Russian Federation
Ossipee Mountains (1978)
  In Carroll County, E New Hampshire.
  BT Carroll County New Hampshire
  BT New Hampshire
  BT United States
Ostashkovichi
  No longer a valid term for GeoRef. As of 1993, see Ostashkovichi Belarus.
Ostashkovichi Belarus (1993)
  Village in S Belarus. This term has multiple hierarchies.
  BT1 Belarus
  BT1 Europe
  BT2 Belarus
  BT2 Commonwealth of Independent States
Osteichthyes (1978)
  Class. Acanthodii was a narrower term from 1981 through 1992. Autoposting of this term began in 1978.
  BT Pisces
  BT Vertebrata
  BT Chordata
  NT Actinopterygii
  NT Brachiopterygii
  NT Sarcopterygii
  SA Acanthodii
osteology (1978)
  SA bones
  SA paleontology
Ostracoda (1978)
  Autoposting of this term began in 1978. This term has multiple hierarchies.
  BT1 Crustacea

BT1 Mandibulata
BT1 Arthropoda
BT1 Invertebrata
BT2 microfossils
NT Archeocopida
NT Beyrichicopina
NT Eridostraca
NT Kirkbyocopina
NT Leperditicopida
NT Myodocopida
NT Paleocopida
NT Podocopida
SA ostracods
SA plankton

Ostracodermi (1981)
Before 1981, also search ostracoderms. As of 1981, ostracoderms may be used as the common name for Ostracodermi.
BT Agnatha
BT Vertebrata
BT Chordata
SA ostracoderms

ostracoderms (1978)
As of 1981, restricted to use as the common name for Ostracodermi.
BT agnathans
BT vertebrates
SA biostratigraphy
SA Ostracodermi

ostracods (1981)
Common name for Ostracoda. Autoposting of crustaceans and arthropods to this term began in 1989. This term has multiple hierarchies.
IN Index for all non-paleontologic studies of fossils.
BT1 crustaceans
BT1 arthropods
BT1 invertebrates
BT2 microfossils
SA biostratigraphy
SA Ostracoda

Ostrava (1978)
City in N Moravia.
UF Moravska Ostrava
BT Moravia
BT Czech Republic
BT Czechoslovakia
BT Central Europe
BT Europe

Ostrava-Karvina (1978)
Coal mining and steel producing region near the Polish border in N Moravia.
CO N493800N500000
E0184500E0180200
BT Moravia
BT Czech Republic
BT Czechoslovakia
BT Central Europe
BT Europe

Ostrea (1978)
Genus. Autoposting of Ostreacea to this term began in 1989.
BT Ostreidae
BT Ostreacea
BT Bivalvia
BT Mollusca
BT Invertebrata

Ostreacea (1981)
Superfamily. Autoposting of this term to Ostreidae began in 1989. Autoposting of Cyrtodontida to this term ended in 1989.
BT Bivalvia
BT Mollusca
BT Invertebrata
NT Ostreidae

Ostreidae (1978)
Family. Autoposting of this term to Crassostrea virginica began in 1989. Autoposting of Ostreacea to this term began in 1989.
BT Ostreacea
BT Bivalvia
BT Mollusca
BT Invertebrata
NT Crassostrea virginica
NT Ostrea

Ostrowiec
 use Ostrowiec Swietokrzyski Poland

Ostrowiec Swietokrzyski
No longer a valid term for GeoRef. As of 1993 see Ostrowiec Swietokrzyski Poland.

Ostrowiec Swietokrzyski Poland (1993)
City in NW Kielce, SE central Poland. Before 1993 also search Ostrowiec Swietokrzyski AND Poland.
UF Ostrowiec
BT Kielce Poland
BT Poland
BT Central Europe
BT Europe

Osumi Peninsula (1978)
Between Kagoshima Bay and Osumi Strait in S Kyushu, S Japan.
BT Kagoshima Japan
BT Kyushu
BT Japan
BT Far East
BT Asia

osumilite (1978)
BT ring silicates
BT silicates
SA aluminosilicates

Oswego County
Valid through 1988. Search in combination with state term. After 1988, use specific county-state term.

Oswego County New York (1989)
N central New York. Before 1989, also search Oswego County AND New York.
CO N431400N444400
W0754500W0763700
BT New York
BT United States
SA Oneida Lake

Otago
No longer a valid term for GeoRef. See Otago New Zealand.

Otago New Zealand (1993)
Provincial district in S South Island.
IN Also index South Island.
BT New Zealand
BT Australasia
NT Dunedin New Zealand
NT Kakanui
NT Oamaru New Zealand
NT Otago Peninsula
SA South Island

Otago Peninsula (1978)
On E side of Otago Harbor in Dunedin area on SE South Island.
IN Also index South Island.
BT Otago New Zealand
BT New Zealand
BT Australasia
SA South Island

Otake
No longer a valid term for GeoRef. See Otake Japan and Otake Field.

Otake Field (1993)
Geothermal area in Oita Prefecture, Kyushu. Before 1993, also search Otake AND Kyushu.
BT Oita Japan
BT Kyushu
BT Japan
BT Far East
BT Asia
SA geothermal fields
SA Hatchobaru Field

Otake Japan (1993)
Town on Hiroshima Bay in Hiroshima Prefecture, SW Honshu. Before 1993, also search Otake and Honshu. May also occur in other locations.
BT Hiroshima Japan
BT Honshu
BT Japan
BT Far East
BT Asia
SA Kyushu

Otavi
No longer a valid term for GeoRef. As of 1993 see Otavi Namibia.

Otavi Namibia (1993)
Town in N Namibia. Before 1993 also search Otavi AND Namibia.
BT Namibia
BT Southern Africa
BT Africa

Otero County
Valid through 1988. Search in combination with state term. After 1988, use specific county-state term.

Otero County Colorado (1989)
SE Colorado. Before 1989, also search Otero County AND Colorado.
CO N373800N381500
W1032300W1040300
BT Colorado
BT United States

Otero County New Mexico (1989)
S New Mexico. Before 1989, also search Otero County AND New Mexico.
CO N320000N332300
W1045200W1062300
BT New Mexico
BT United States
SA Hueco Mountains
SA Orogrande Basin
SA Sacramento Mountains
SA Sierra Blanca
SA White Sands

Othris
 use Othrys

Othrys (1978)
Mountain range in central Greece.
IN Index administrative regions as applicable.
UF Othris
UF Othrys Massif
BT Greece
BT Southern Europe
BT Europe
SA Sterea Ellas
SA Thessaly Greece

Othrys Massif
 use Othrys

otoliths (1978)
SA fossils
SA microfossils
SA Pisces

Otranto
No longer a valid term for GeoRef. See Otranto Italy.

Otranto Italy (1993)
Town on Strait of Otranto in heel of Italian boot.
BT Apulia Italy
BT Italy
BT Southern Europe
BT Europe

Otsego County
Valid through 1988. Search in combination with state term. After 1988, use specific county-state term.

Otsego County Michigan (1989)
N Michigan. Before 1989, also search Otsego County AND Michigan.
CO N445200N451500
W0842200W0845200
BT Michigan Lower Peninsula
BT Michigan
BT United States

Otsego County New York (1989)
Central New York. Before 1989, also search Otsego County AND New York.
CO N421800N425300
W0743700W0752400
BT New York
BT United States

Ottawa
No longer a valid term for GeoRef. As of 1993, see Ottawa Ontario.

Ottawa County
Valid through 1988. Search in combination with state term. After 1988, use specific county-state term.

Ottawa County Kansas (1989)
N central Kansas. Before 1989, also search Ottawa County AND Kansas.
CO N385700N391800
W0972300W0975500
BT Kansas
BT United States

Ottawa County Michigan (1989)
SW Michigan. Before 1989, also search Ottawa County AND Michigan.
CO N424500N431200
W0854600W0861600
BT Michigan Lower Peninsula
BT Michigan
BT United States

Ottawa County Ohio (1989)
N Ohio. Before 1989, also search Ottawa County AND Ohio.
CO N412700N413800
W0824300W0832500
BT Ohio
BT United States

Ottawa County Oklahoma (1989)
Extreme NE Oklahoma. Before 1989, also search Ottawa County AND Oklahoma.
CO N364200N370000
W0943800W0950000
BT Oklahoma
BT United States
NT Picher Oklahoma

Ottawa Ontario (1993)

City in Ottawa-Carleton regional municipality, SE Ontario. Before 1993, also search Ottawa AND Ontario.
BT Ontario
BT Eastern Canada
BT Canada
SA Leda Clay

Ottawa River (1989)
River. Forms the border between Ontario and Quebec. Also in Lucas and Wood counties, N Ohio.
IN Index provinces or Ohio and counties as applicable.
SA Lucas County Ohio
SA Ohio
SA Ontario
SA Quebec
SA Saint Lawrence River basin
SA Wood County Ohio

Ottawa Sand (1993)
Silica sand material used for engineering studies. From New York and Ontario.
IN Index age if applicable.
SA materials
SA New York
SA Ontario
SA sand
SA silica

Ottawa Valley (1978)
River valley.
IN Index provinces as applicable.
BT Eastern Canada
BT Canada
SA Ontario
SA Quebec

Ottnangian (1978)
Europe.
BT lower Miocene
BT Miocene
BT Neogene
BT Tertiary
BT Cenozoic

Otto Fiord
  use Otto Fjord

Otto Fjord (1978)
NW Ellesmere Island, Franklin District. Also search Otto.
IN Index Ellesmere Island and district or region as applicable.
UF Otto Fiord
SA Ellesmere Island
SA Franklin District Northwest Territories

Otway Basin (1978)
Primarily in W and SW Victoria.
IN Index states as applicable.
BT Australia
BT Australasia
SA South Australia
SA Victoria Australia

Otztal Alps (1978)
Mountain range of the Eastern Alps.
IN Index countries as applicable.
UF Oetztal Alps
UF Oetztal Massif
UF Oetztaler
UF Otztaler Alps
BT Eastern Alps
BT Alps
BT Europe
NT Stubai Alps
SA Trentino-Alto Adige Italy
SA Tyrol Austria
SA Vernagt Glacier

Otztaler Alps
  use Otztal Alps

Ouachita Belt (1989)
Tectonic zone in S United States.
IN Index states as applicable.
BT United States
SA Arkansas
SA Oklahoma
SA Texas

Ouachita Mountains (1978)
IN Index states as applicable.
CO N340000N344000 W0943000W0960000
BT United States
SA Arkansas
SA Benton Uplift
SA Latimer County Oklahoma
SA Oklahoma

Ouachita Orogeny (1985)
Carboniferous to Permian. Before 1985, also search orogeny AND Ouachita.
BT Paleozoic
SA Benton Uplift
SA Carboniferous
SA orogeny
SA Pennsylvanian
SA Permian

Ouachita Parish Louisiana (1989)
N Louisiana. Before 1989, search Ouachita Parish AND Louisiana.
CO N321500N324500 W0915300W0922500
BT Louisiana
BT United States

Ouachita-Balcones Trend
  use Balcones fault zone

Oubangui-Chari
  As of 1959, no longer a valid term for GeoRef. Used in subfile E.
  use Central African Republic

oueds
  use wadis

Ouegoa
  No longer a valid term for GeoRef. As of 1993 see Ouegoa New Caledonia.

Ouegoa New Caledonia (1993)
Village in NE New Caledonia. Before 1993 also search Ouegoa AND New Caledonia
BT New Caledonia
BT Melanesia
BT Oceania

Ougarta
  No longer a valid term for GeoRef. As of 1993, see Ougarta Algeria.

Ougarta Algeria (1993)
Village W of the Great Western Erg in W central.
BT Algeria
BT North Africa
BT Africa

Ouray County
  Valid through 1988. Search in combination with state term. After 1988, use specific county-state term.

Ouray County Colorado (1989)
SW central Colorado. Before 1989, also search Ouray County AND Colorado.
CO N375300N382000 W1073500W1081000
BT Colorado
BT United States

Ouro Preto
  No longer a valid term for GeoRef. See Ouro Preto Brazil.

Ouro Preto Brazil (1993)
City in S Minas Gerais, E Brazil.
BT Minas Gerais Brazil
BT Brazil
BT South America

outbursts
  Not a valid term for GeoRef. As of 1993, see blowouts, borehole breakouts, jokulhlaups, and rock bursts.

outcropping
  use outcrops

outcrops (1978)
Used as a general term.
UF outcropping
SA bedrock

Outer Banks (1978)
Chain of sandy barrier islands stretching length of coast.
IN Index counties or regions as applicable.
UF The Banks
SA Atlantic Coastal Plain
SA North Carolina

outer continental shelf
  use outer shelf

outer core (1978)
BT core
SA core-mantle boundary
SA inner core

Outer Hebrides (1978)
Outer group of the Hebrides, W of Little Minch. Sometimes called Long Island.
CO N564500N583000 W0061000W0080000
BT Hebrides
BT Scotland
BT Great Britain
BT United Kingdom
BT Western Europe
BT Europe
SA Harris
SA Inner Hebrides
SA Long Island
SA Saint Kilda

outer mantle
  use upper mantle

Outer Mongolia
  use Mongolia

outer planets (1978)
The outer five planets in the solar system: Jupiter, Saturn, Uranus, Neptune and Pluto.
SA Jupiter
SA Neptune
SA planetology
SA planets
SA Pluto
SA Saturn
SA solar system
SA terrestrial planets
SA Uranus
SA Voyager Program

outer shelf (1978)
UF outer continental shelf
BT continental shelf
SA inner shelf
SA shelf-slope break

outer slope (1978)
UF slope, outer
BT continental slope
SA inner slope

outer space
  use interplanetary space

outgassing
  use degassing

outliers (1993)
Use more specific term if applicable.
SA erosion features
SA geostatistics
SA lithostratigraphy
SA shore features
SA statistical analysis
SA stratigraphy

Outokumpu
  No longer a valid term for GeoRef. See Outokumpu Finland.

Outokumpu Finland (1993)
Village in SE Finland.
BT Finland
BT Scandinavia
BT Western Europe
BT Europe

outwash (1978)
UF glacial outwash
UF outwash drift
BT clastic sediments
BT sediments
SA fluvial features
SA glacial features
SA glacial geology
SA glaciofluvial sedimentation
SA periglacial features
SA terrigenous materials

outwash aprons
  use outwash plains

outwash drift
  use outwash

outwash plains (1978)
Autoposting of glaciofluvial features and glacial features to this term began in 1989. This term has multiple hierarchies.
UF morainal plains
UF outwash aprons
UF overwash plain
UF sandurs
BT1 glacial features
BT2 fluvial features
SA glacial geology
SA moraines
SA plains

over printing
  use overprinting

overburden (1978)
SA mining geology
SA regolith
SA sedimentary cover

overconsolidated materials (1978)
Before 1978, search overconsolidated AND materials.
SA compactness
SA consolidation
SA engineering geology
SA foundations
SA materials
SA soil mechanics

overgrowths (1978)
SA crystal growth
SA minerals

Overijssel
  No longer a valid term for GeoRef. As of 1993 see Overijssel Netherlands.

Overijssel Netherlands (1993)
Province in E Netherlands. Before 1993 also search Overijssel AND Netherlands.
BT Netherlands
BT Western Europe

BT Europe
NT Twente

overland flow
As of 1993, use runoff if appropriate.

overpressure (1993)
Abnormally high pore pressure or hydrostatic pressure.
BT pressure
SA earth pressure
SA geopressure
SA hydrostatic pressure
SA pore pressure
SA reservoir properties
SA reservoir rocks

overprinting (1985)
UF over printing
SA absolute age
SA geochronology
SA isotopes
SA metamorphism
SA orogeny
SA relative age

overthrust
A valid term through 1977. After 1977, use overthrust faults.

Overthrust Belt
As of 1981, no longer a valid term for GeoRef. See Western Overthrust Belt; Eastern Overthrust Belt.

overthrust faults (1978)
Before 1978, also search faults AND overthrust.
BT faults
SA allochthons
SA autochthons
SA displacements
SA klippen
SA nappes
SA thrust faults
SA windows

Overton County
Valid through 1988. Search in combination with state term. After 1988, use specific county-state term.

Overton County Tennessee (1989)
N Tennessee. Before 1989, also search Overton County AND Tennessee.
CO N361500N370500
W0850500W0853000
BT Tennessee
BT United States

overturned folds (1978)
Includes use as folds defined on the basis of orientation in relation to the geographic horizontal plane. Before 1978, also search folds AND overturned.
BT folds
SA orientation

overwash
A valid index term from 1977-1980.

overwash plain
use outwash plains

Oviedo Spain
use Asturias Spain

Ovruch Series (1978)
Divided into two subseries.
BT Proterozoic
BT upper Precambrian
BT Precambrian
SA Ukraine

Owen fracture zone (1989)
NW Indian Ocean.
BT Indian Ocean

Owens Valley (1978)
River valley between Sierra Nevada on the W, and White and Inyo Mountains on E.
IN Index counties or regions as applicable.
SA California
SA Inyo County California
SA Inyo Domes

Owl Creek Mountains (1985)
NW Wyoming. Autoposting of broader terms to this term began in 1989. This term has multiple hierarchies.
IN Index counties as applicable.
CO N432500N434000
W1081500W1090000
BT1 U. S. Rocky Mountains
BT1 Rocky Mountains
BT2 U. S. Rocky Mountains
BT2 United States
BT3 Wyoming
BT3 United States
SA Fremont County Wyoming
SA Hot Springs County Wyoming

Owsley County
Valid through 1988. Search in combination with state term. After 1988, use specific county-state term.

Owsley County Kentucky (1989)
E central Kentucky. Before 1989, also search Owsley County AND Kentucky.
CO N371500N373200
W0833000W0835300
BT Kentucky
BT United States

Owyhee County
Valid through 1988. Search in combination with state term. After 1988, use specific county-state term.

Owyhee County Idaho (1989)
SW Idaho. Before 1989, also search Owyhee County AND Idaho.
CO N420000N434000
W1150400W1170200
BT Idaho
BT United States
SA Owyhee Mountains

Owyhee Mountains (1989)
Owyhee County, SW Idaho; Malheur County, SE Oregon.
IN Index states and counties as applicable.
UF Owyhee Upland
BT United States
SA Idaho
SA Malheur County Oregon
SA Oregon
SA Owyhee County Idaho

Owyhee Upland
use Owyhee Mountains

oxalates (1981)
SA minerals

oxbow lakes (1989)
Also search oxbows.
BT fluvial features
SA geomorphology
SA meanders
SA rivers
SA rivers and streams
SA streams

Oxford
No longer a valid term for GeoRef. As of 1993, see Oxford England or Oxford New Zealand. As of 1989, for cities in the U.S., see specific city-state terms.

Oxford Clay (1978)
BT Oxfordian
BT Upper Jurassic
BT Jurassic
BT Mesozoic
SA England
SA United Kingdom

Oxford County
No longer a valid term for GeoRef. As of 1993, see Oxford County Ontario or Oxford County Maine.

Oxford County Maine (1989)
W Maine. Before 1989, also search Oxford County AND Maine.
CO N434700N452100
W0701600W0710600
BT Maine
BT United States
SA Rangeley Lakes
SA Saco River

Oxford County Ohio (1989)
SW Ohio. Before 1989, search Oxford AND Ohio.
CO N393000N393000
W0844500W0844500
BT Butler County Ohio
BT Ohio
BT United States

Oxford County Ontario (1993)
S Ontario. Before 1993, also search Oxford County AND Ontario. For county in Maine, see Oxford County Maine.
BT Ontario
BT Eastern Canada
BT Canada

Oxford England (1993)
City 52 miles WNW of London in central England. Before 1993, also search Oxford AND England.
BT Oxfordshire England
BT England
BT Great Britain
BT United Kingdom
BT Western Europe
BT Europe

Oxford Idaho (1989)
SE Idaho. Before 1989, search Oxford AND Idaho.
CO N421600N421600
W1120100W1120100
BT Franklin County Idaho
BT Idaho
BT United States

Oxford Mississippi (1989)
N Mississippi. Before 1989, search Oxford AND Mississippi.
CO N342100N342100
W0893000W0893000
BT Lafayette County Mississippi
BT Mississippi
BT United States

Oxford New Zealand (1993)
Township on E South Island. Before 1993, search Oxford AND New Zealand.
IN Index South Island.
BT New Zealand
BT Australasia
SA South Island

Oxford North Carolina (1989)
N North Carolina. Before 1989, search Oxford AND North Carolina.
CO N362200N362200
W0783700W0783700
BT Granville County North Carolina
BT North Carolina
BT United States

Oxford Ohio (1989)
SW Ohio. Before 1989, search Oxford AND Ohio.
CO N393000N393000
W0844500W0844500
BT Butler County Ohio
BT Ohio
BT United States

Oxfordian (1978)
Europe. Above Callovian, below Kimmeridgian.
BT Upper Jurassic
BT Jurassic
BT Mesozoic
NT Oxford Clay
SA Malm

Oxfordshire
No longer a valid term for GeoRef as of 1993.
use Oxfordshire England

Oxfordshire England (1993)
County in central England. As of 1975, includes land which was part of Berkshire. Before 1993, also search Oxfordshire.
CO N513000N521000
W0005000W0014500
UF Oxfordshire
BT England
BT Great Britain
BT United Kingdom
BT Western Europe
BT Europe
NT Oxford England
SA Berkshire England

oxidation (1978)
SA Eh
SA geochemistry
SA hydrolysis
SA oxidation zone
SA oxygen
SA processes
SA reduction

oxidation zone (1978)
UF zone, oxidation
SA Eh
SA mineral deposits, genesis
SA oxidation

oxidation-reduction potential
use Eh

oxides (1978)
Before 1981, zirkelite was included as a narrower term. Autoposting of this term began in 1978.
NT akaganeite
NT alexandrite
NT anatase
NT armalcolite
NT asbolite
NT baddeleyite
NT bayerite
NT birnessite
NT bixbyite
NT boehmite
NT brannerite
NT bromellite
NT brookite
NT brucite
NT buserite
NT cassiterite
NT chrome spinel

NT chromite
NT chrysoberyl
NT coronadite
NT corundum
NT cryptomelane
NT cuprite
NT delafossite
NT diaspore
NT ferrihydrite
NT ferropseudobrookite
NT franklinite
NT gahnite
NT geikielite
NT germanates
NT gibbsite
NT goethite
NT groutite
NT hausmannite
NT hematite
NT hercynite
NT heterogenite
NT hibonite
NT hoegbomite
NT hollandite
NT hydrogoethite
NT hydroxides
NT ilmenite
NT iron oxides
NT jacobsite
NT lepidocrocite
NT leucoxene
NT limonite
NT lithiophorite
NT maghemite
NT magnesioferrite
NT magnetite
NT manganese oxides
NT manganite
NT manganosite
NT martite
NT nigerite
NT niobates
NT nordstrandite
NT nsutite
NT periclase
NT perovskite
NT pitchblende
NT priderite
NT pseudobrookite
NT psilomelane
NT pyrolusite
NT pyrophanite
NT ramsdellite
NT rancieite
NT rutile
NT sapphire
NT senarmontite
NT specularite
NT spinel
NT spinel group
NT taaffeite
NT tantalates
NT thorianite
NT titanomaghemite
NT titanomagnetite
NT todorokite
NT trevorite
NT ulvospinel
NT uraninite
NT valentinite
NT vernadite
NT wuestite
NT zincite
SA alumina
SA glacial features
SA ice
SA minerals
SA nitrous oxide
SA ocher

Oxisols (1978)
  BT soils
  SA soil group

Oxnard
  Valid through 1988. Search in combination with state term. After 1988, use specific city-state term.
Oxnard California (1989)
  City in S California. Before 1989, search Oxnard AND California.
  CO N341100N341100
    W1191000W1191000
  BT Ventura County California
  BT California
  BT United States

Oxus
  use Amu Darya

oxygen (1978)
  UF O
  NT O-16
  NT O-17/O-16
  NT O-18
  NT O-18/O-16
  SA chemical elements
  SA isotopes
  SA oxidation
  SA ozone
  SA stable isotopes

oxymagnite
  use maghemite

oxyphile elements
  use lithophile elements

oxysulfides (1978)
  Autoposting of sulfides to this term began in 1981.
  BT sulfides

Ozark Highlands
  use Ozark Mountains

Ozark Mountains (1978)
  Eroded tableland.
  IN Index states as applicable.
  UF Ozark Highlands
  UF Ozark Uplift
  UF Ozarks
  BT United States
  SA Arkansas
  SA Missouri
  SA Oklahoma
  SA Saint Francois Mountains

Ozark Uplift
  use Ozark Mountains

Ozarkodina (1978)
  Genus. Autoposting of microfossils to this term began in 1990.
  BT Conodonta
  BT microfossils

Ozarks
  use Ozark Mountains

Ozaukee County
  Valid through 1988. Search in combination with state term. After 1988, use specific county-state term.

Ozaukee County Wisconsin (1989)
  E Wisconsin. Before 1989, also search Ozaukee County AND Wisconsin.
  CO N431300N433300
    W0874700W0880500
  BT Wisconsin
  BT United States

Ozernoye
  No longer a valid term for GeoRef. As of 1993, see Ozernoye Russian Federation.

Ozernoye Russian Federation (1993)
  Village in SE Kamchatka on Okhotsk Sea just N of Ozernovskiy in Khabarovsk Kray. Before 1993, also search Ozernoye. This term has multiple hierarchies.
  BT1 Khabarovsk Russian Federation
  BT1 Russian Federation
  BT1 Commonwealth of Independent States
  BT2 Khabarovsk Russian Federation
  BT2 Asia

ozocerite (1978)
  BT organic compounds

ozone (1978)
  SA atmosphere
  SA oxygen

Ozun
  use Uzon

P

P
  use phosphorus

P-32 (1985)
  Autoposting of broader terms began in 1989. This term has multiple hierarchies.
  BT1 radioactive isotopes
  BT1 isotopes
  BT2 phosphorus

P-T conditions (1978)
  UF conditions, P-T
  SA fluid inclusions
  SA geologic barometry
  SA geologic thermometry
  SA grade
  SA inclusions
  SA isotherms
  SA metamorphism
  SA metasomatism
  SA paleotemperature
  SA phase equilibria
  SA pressure
  SA temperature

P-waves (1978)
  UF compressional waves
  UF dilatational wave
  UF irrotational wave
  UF longitudinal wave
  UF pressure wave
  UF primary wave
  UF push wave
  UF push-pull wave
  BT body waves
  BT elastic waves
  SA PcP-waves
  SA PKiKP-waves
  SA PKP-waves
  SA PKS-waves
  SA Pn-waves
  SA PP-waves
  SA PPP-waves
  SA PPS-waves
  SA PS-waves
  SA S-waves
  SA seismology
  SA SKP-waves
  SA SKS-waves
  SA waves

P/Halley
  use Halley's Comet

Pa
  use protactinium

Pa-231 (1978)
  Autoposting of broader terms began in 1989. This term has multiple hierarchies.
  BT1 radioactive isotopes
  BT1 isotopes
  BT2 protactinium
  BT2 actinides
  BT2 metals

Pacaya (1978)
  Volcano in S central Guatemala.
  UF Pacaya Volcano
  BT Guatemala
  BT Central America

Pacaya Volcano
  use Pacaya

Pachelma
  No longer a valid term for GeoRef. As of 1993, see Pachelma Russian Federation.

Pachelma Russian Federation (1993)
  Town in Penza Oblast W of Kuibyshev in S central European Russian Federation. Before 1993, also search Pachelma. This term has multiple hierarchies.
  BT1 Russian Federation
  BT1 Commonwealth of Independent States
  BT2 Europe

Pachuca
  No longer a valid term for GeoRef. As of 1993 see Pachuca Mexico.

Pachuca de Soto
  use Pachuca Mexico

Pachuca Mexico (1993)
  City NE of Mexico City in central Mexico. Before 1993 also search Pachuca AND Mexico.
  UF Pachuca de Soto
  BT Hidalgo Mexico
  BT Mexico

Pachypteris (1978)
  BT Caytoniales
  BT gymnosperms
  BT Spermatophyta
  BT Plantae
  SA Cycadales

Pacific Basin (1978)
  Generally the entire ocean floor of both the North and South Pacific up to the continental shelf.
  BT Pacific Ocean
  SA Leg 88
  SA North Pacific
  SA Pacific Plate
  SA South Pacific

Pacific Coast (1978)
  As of 1993, term is used only for broad general discussions. Before 1993, term used for the region comprising those U.S. states fronting on the Pacific Ocean. From 1981 through 1992, this term was autoposted to Washington, Oregon, and California. Before 1978, also search West Coast. As of 1993, for complete search, also search states.
  UF West Coast
  BT Western U.S.
  BT United States
  SA California Current
  SA Hayfork Terrane
  SA Pacific region

Pacific County
Valid through 1988. Search in combination with state term. After 1988, use specific county-state term.

Pacific County Washington (1989)
On Pacific Ocean in SW Washington. Before 1989, also search Pacific County AND Washington.
CO N461500N464500
   W1232000W1240500
BT Washington
BT United States
NT Willapa Bay

Pacific Islands
Not a valid GeoRef index term after 1971. Was used on level 1 in subfiles G and N. Now use West Pacific Ocean Islands or East Pacific Ocean Islands as applicable.

Pacific mobile belt (1978)
Crustal regions of tectonic activity along the coastal fringes of continents and some entire islands of the Pacific.
IN Index continents and islands as applicable.
BT Pacific region
SA Asia
SA New Guinea
SA New Zealand
SA North America
SA South America

Pacific Ocean (1978)
Many islands, formerly included as narrower terms of Pacific Ocean, are now included under East Pacific Ocean Islands or West Pacific Ocean Islands. Autoposting of this term began in 1978.
NT Aleutian Ridge
NT Aleutian Trench
NT Astoria Canyon
NT Bauer Deep
NT Bay of Plenty
NT Blanco fracture zone
NT Bowie Seamount
NT Cascadia Basin
NT Cascadia Channel
NT Central Pacific
NT Chatham Rise
NT China Sea
NT Circum-Pacific region
NT Cobb Seamount
NT Cocos Plate
NT East Pacific
NT East Pacific Rise
NT Eltanin fracture zone
NT Equatorial Pacific
NT Erimo Seamount
NT Fiji Plateau
NT Formosa Strait
NT Galapagos Rift
NT Guatemala Basin
NT Gulf of Panama
NT Hawaiian Ridge
NT Iwo-Jima
NT Juan de Fuca Plate
NT Juan de Fuca Ridge
NT Kermadec Trench
NT Kula Plate
NT Kuril Trench
NT Lau Basin
NT Lord Howe Rise
NT Macquarie Ridge
NT Manihiki Plateau
NT Mariana Trench
NT Mariana Trough
NT Mendocino fracture zone
NT Middle America Trench
NT Monterey Canyon
NT Monterey Fan
NT Murray fracture zone
NT Nauru Basin
NT Nazca Plate
NT New Caledonia Basin
NT Norfolk Ridge
NT North American Pacific
NT North Pacific
NT Northeast Pacific
NT Northwest Pacific
NT Orozco fracture zone
NT Pacific Basin
NT Peru-Chile Trench
NT Polynesian Pacific
NT Redondo Canyon
NT San Diego Trough
NT Santa Barbara Basin
NT Shatsky Rise
NT Shikoku Basin
NT Solomon Sea
NT South American Pacific
NT South Pacific
NT South Polynesian Pacific
NT Southeast Pacific
NT Southwest Pacific
NT Tasman Basin
NT Tonga Trench
NT Torres Strait
NT Wake
NT West Pacific
NT Woodlark Basin
SA Antarctic Ocean
SA Austral Islands
SA Bering Strait
SA Caroline Islands
SA Cook Islands
SA DSDP Site 32
SA DSDP Site 35
SA DSDP Site 40
SA DSDP Site 51
SA DSDP Site 61
SA DSDP Site 62
SA DSDP Site 63
SA DSDP Site 64
SA DSDP Site 65
SA DSDP Site 66
SA DSDP Site 67
SA DSDP Site 69
SA DSDP Site 70
SA DSDP Site 71
SA DSDP Site 72
SA DSDP Site 73
SA DSDP Site 74
SA DSDP Site 75
SA DSDP Site 77
SA DSDP Site 157
SA DSDP Site 158
SA DSDP Site 159
SA DSDP Site 163
SA DSDP Site 167
SA DSDP Site 171
SA DSDP Site 172
SA DSDP Site 173
SA DSDP Site 174
SA DSDP Site 175
SA DSDP Site 176
SA DSDP Site 177
SA DSDP Site 178
SA DSDP Site 179
SA DSDP Site 180
SA DSDP Site 181
SA DSDP Site 182
SA DSDP Site 183
SA DSDP Site 186
SA DSDP Site 192
SA DSDP Site 194
SA DSDP Site 195
SA DSDP Site 196
SA DSDP Site 197
SA DSDP Site 198
SA DSDP Site 199
SA DSDP Site 200
SA DSDP Site 202
SA DSDP Site 206
SA DSDP Site 207
SA DSDP Site 208
SA DSDP Site 209
SA DSDP Site 210
SA DSDP Site 277
SA DSDP Site 284
SA DSDP Site 285
SA DSDP Site 286
SA DSDP Site 288
SA DSDP Site 289
SA DSDP Site 291
SA DSDP Site 292
SA DSDP Site 293
SA DSDP Site 294
SA DSDP Site 296
SA DSDP Site 297
SA DSDP Site 298
SA DSDP Site 299
SA DSDP Site 302
SA DSDP Site 305
SA DSDP Site 310
SA DSDP Site 313
SA DSDP Site 315
SA DSDP Site 315A
SA DSDP Site 317
SA DSDP Site 319
SA DSDP Site 320
SA DSDP Site 321
SA DSDP Site 419
SA DSDP Site 420
SA DSDP Site 424
SA DSDP Site 430
SA DSDP Site 431
SA DSDP Site 432
SA DSDP Site 433
SA DSDP Site 434
SA DSDP Site 435
SA DSDP Site 436
SA DSDP Site 438
SA DSDP Site 439
SA DSDP Site 440
SA DSDP Site 445
SA DSDP Site 446
SA DSDP Site 450
SA DSDP Site 453
SA DSDP Site 458
SA DSDP Site 459
SA DSDP Site 462
SA DSDP Site 463
SA DSDP Site 464
SA DSDP Site 465
SA DSDP Site 466
SA DSDP Site 467
SA DSDP Site 468
SA DSDP Site 469
SA DSDP Site 470
SA DSDP Site 471
SA DSDP Site 472
SA DSDP Site 473
SA DSDP Site 474
SA DSDP Site 475
SA DSDP Site 476
SA DSDP Site 477
SA DSDP Site 478
SA DSDP Site 479
SA DSDP Site 480
SA DSDP Site 481
SA DSDP Site 482
SA DSDP Site 483
SA DSDP Site 484
SA DSDP Site 485
SA DSDP Site 494
SA DSDP Site 495
SA DSDP Site 496
SA DSDP Site 497
SA DSDP Site 498
SA DSDP Site 499
SA DSDP Site 500
SA DSDP Site 501
SA DSDP Site 503
SA DSDP Site 504
SA DSDP Site 504B
SA DSDP Site 505
SA DSDP Site 506
SA DSDP Site 507
SA DSDP Site 508
SA DSDP Site 509
SA DSDP Site 510
SA DSDP Site 565
SA DSDP Site 568
SA DSDP Site 570
SA DSDP Site 572
SA DSDP Site 573
SA DSDP Site 574
SA DSDP Site 575
SA DSDP Site 576
SA DSDP Site 577
SA DSDP Site 578
SA DSDP Site 579
SA DSDP Site 580
SA DSDP Site 581
SA DSDP Site 582
SA DSDP Site 583
SA DSDP Site 584
SA DSDP Site 585
SA DSDP Site 586
SA DSDP Site 587
SA DSDP Site 588
SA DSDP Site 589
SA DSDP Site 590
SA DSDP Site 591
SA DSDP Site 592
SA DSDP Site 593
SA DSDP Site 594
SA DSDP Site 595
SA DSDP Site 596
SA DSDP Site 597
SA DSDP Site 598
SA DSDP Site 599
SA DSDP Site 600
SA DSDP Site 601
SA DSDP Site 602
SA East Pacific Ocean Islands
SA Easter Island
SA El Nino
SA Fanning Island
SA Fiji
SA French Polynesia
SA Hawaii
SA Indian Plate
SA Indo-Australian Plate
SA Izu-Oshima earthquake 1978
SA Leg 5
SA Leg 6
SA Leg 7
SA Leg 8
SA Leg 9
SA Leg 16
SA Leg 17
SA Leg 18
SA Leg 19
SA Leg 20
SA Leg 21
SA Leg 30
SA Leg 31
SA Leg 32
SA Leg 33
SA Leg 34
SA Leg 54
SA Leg 55
SA Leg 56
SA Leg 57
SA Leg 58
SA Leg 59
SA Leg 60
SA Leg 61
SA Leg 62
SA Leg 63
SA Leg 64
SA Leg 65
SA Leg 66
SA Leg 67
SA Leg 68
SA Leg 69
SA Leg 70
SA Leg 83
SA Leg 84
SA Leg 85
SA Leg 86
SA Leg 87
SA Leg 88

SA Leg 89
SA Leg 90
SA Leg 91
SA Leg 92
SA Leg 111
SA Leg 112
SA Leg 122
SA Leg 124
SA Leg 124E
SA Leg 125
SA Leg 126
SA Leg 127
SA Leg 128
SA Leg 129
SA Leg 130
SA Leg 131
SA Leg 132
SA Leg 133
SA Leg 134
SA Leg 135
SA Leg 136
SA Leg 138
SA Line Islands
SA Malaita
SA Malekula
SA Mariana Islands
SA Marion Island
SA Marshall Islands
SA Melanesia
SA Micronesia
SA Mid-Pacific Rise
SA Nankaido earthquake 1946
SA New Caledonia
SA Oceania
SA ODP Site 677
SA ODP Site 678
SA ODP Site 679
SA ODP Site 680
SA ODP Site 681
SA ODP Site 682
SA ODP Site 683
SA ODP Site 684
SA ODP Site 685
SA ODP Site 686
SA ODP Site 687
SA ODP Site 688
SA ODP Site 767
SA ODP Site 768
SA ODP Site 769
SA ODP Site 770
SA ODP Site 771
SA ODP Site 776
SA ODP Site 778
SA ODP Site 779
SA ODP Site 780
SA ODP Site 781
SA ODP Site 782
SA ODP Site 783
SA ODP Site 785
SA ODP Site 787
SA ODP Site 788
SA ODP Site 789
SA ODP Site 790
SA ODP Site 791
SA ODP Site 792
SA ODP Site 793
SA ODP Site 794
SA ODP Site 796
SA ODP Site 797
SA ODP Site 798
SA ODP Site 799
SA ODP Site 800
SA ODP Site 801
SA ODP Site 802
SA ODP Site 803
SA ODP Site 804
SA ODP Site 805
SA ODP Site 806
SA ODP Site 807
SA ODP Site 808
SA ODP Site 809
SA Pacific Plate
SA Pacific region
SA Pacific-Antarctic Ridge
SA Polynesia

SA Rivera Ocean Seismic Experiment
SA Ross Sea
SA Solomon Islands
SA Tonga
SA Tuvalu
SA Vanuatu

Pacific Ocean Islands
Not a valid term for GeoRef. See East Pacific Ocean Islands (containing Easter Island and Galapagos Islands), West Pacific Ocean Islands, Melanesia, Micronesia, Polynesia, Malay Archipelago, Indonesia, Papua New Guinea, Philippine Islands, Japan, New Zealand, Hawaii, Taiwan and terms such as Kodiak Island and Kuril Islands which are indexed under appropriate land terms.

Pacific Plate (1978)
Includes most of the Pacific Basin with the Eurasian Plate on the W and N; the American, Cocos, and Nazca Plates on the E; the Indian Plate on the SW; and the Antarctic Plate on the S.
SA Farallon Plate
SA Pacific Basin
SA Pacific Ocean
SA plate tectonics
SA plates

Pacific region (1978)
NT Pacific mobile belt
SA Circum-Pacific region
SA Far East
SA Pacific Coast
SA Pacific Ocean

Pacific-Antarctic Ridge (1978)
Mid-ocean ridge extending from the Balleny fracture zone S of Macquarie Island to the East Pacific Rise.
SA Antarctic Ocean
SA East Pacific Rise
SA mid-ocean ridges
SA Pacific Ocean

packing (1978)
Before 1978, also used for atomic packing under crystal structure.
SA atomic packing
SA sedimentary rocks
SA sedimentation
SA textures

packstone (1978)
BT carbonate rocks
BT sedimentary rocks

Pacoima Dam (1978)
On Pacoima River in Los Angeles County NW of Los Angeles. Also search Pacoima.
BT Los Angeles County California
BT California
BT United States

Paderborn
No longer a valid term for GeoRef. See Paderborn Germany.

Paderborn Germany (1993)
City in E North Rhine-Westphalia, W central Germany.
BT North Rhine-Westphalia Germany
BT Germany
BT Central Europe
BT Europe

Padre Island (1978)
Narrow barrier island along Gulf of Mexico from SE of Corpus Christi to near the Mexican border.

IN Index counties or regions as applicable.
SA Laguna Madre
SA Texas

Padstow
No longer a valid term for GeoRef as of 1993. See Padstow England.

Padstow England (1993)
Town in SW England near N coast of Cornwall. Before 1993, also search Padstow AND England.
BT Cornwall England
BT England
BT Great Britain
BT United Kingdom
BT Western Europe
BT Europe

Padurea Craiului Mountains (1978)
Range in SW Transylvania. Also search Padurea Craiului.
BT Transylvania
BT Romania
BT Southern Europe
BT Europe

Paganzo
No longer a valid term for GeoRef. See Paganzo Argentina and Paganzo Basin.

Paganzo Argentina (1993)
Village in central La Rioja, NW Argentina.
BT La Rioja Argentina
BT Argentina
BT South America
SA Paganzo Basin

Paganzo Basin (1985)
NW Argentina.
SA La Rioja Argentina
SA Paganzo Argentina
SA San Juan Argentina

Paganzo Group (1978)
BT Paleozoic
SA Argentina
SA Carboniferous
SA Permian

Page County
Valid through 1988. Search in combination with state term. After 1988, use specific county-state term.

Page County Iowa (1989)
SW Iowa. Before 1989, also search Page County AND Iowa.
CO N403500N405400
W0945400W0952300
BT Iowa
BT United States

Page County Virginia (1989)
N Virginia. Before 1989, also search Page County AND Virginia.
CO N382500N384900
W0781800W0784200
BT Virginia
BT United States

Pageland
Valid through 1988. Search in combination with state term. After 1988, use specific city-state term.

Pageland South Carolina (1989)
Town in N South Carolina. Before 1989, search Pageland AND South Carolina.
CO N344600N344600
W0802500W0802500

BT Chesterfield County South Carolina
BT South Carolina
BT United States

Pagoda Formation (1978)
Underlies Pentagon Shale; overlies Dearborn Limestone. NW Montana.
BT Middle Cambrian
BT Cambrian
BT Paleozoic
SA Montana

Pahang
No longer a valid term for GeoRef. As of 1993, see Pahang Malaysia.

Pahang Malaysia (1993)
State on the Malay Peninsula in West Malaysia. This term has multiple hierarchies.
BT1 West Malaysia
BT1 Malay Peninsula
BT1 Far East
BT1 Asia
BT2 West Malaysia
BT2 Malaysia
BT2 Far East
BT2 Asia

pahoehoe (1978)
BT lava
SA aa lava
SA volcanism

Pahrump Series (1985)
S California.
BT Precambrian
SA California

Pahute Mesa (1978)
Tableland in Nye County in S Nevada.
UF Paiute Mesa
BT Nye County Nevada
BT Nevada
BT United States

Pai Khoi
use Pai-Khoi

Pai-Khoi (1978)
Mountain range on Yugor Peninsula between Barents and Kara seas. N extension of Northern Urals. This term has multiple hierarchies.
CO N683000N691500
E0650000E0604500
UF Pai Khoi
UF Pay-Khoy
BT1 Polar Urals
BT1 Russian Federation
BT1 Commonwealth of Independent States
BT2 Polar Urals
BT2 Urals
BT2 Commonwealth of Independent States

Paintbrush Tuff (1989)
S Nevada. Includes Pah Canyon Member, Yucca Mountain Member, Tiva Canyon Member and Topopah Spring Member.
BT Miocene
BT Neogene
BT Tertiary
BT Cenozoic
SA Nevada
SA Topopah Spring Member

Paiute Mesa
use Pahute Mesa

Pajaro Valley (1978)
River valley S of San Jose in W California. Also search Pajaro River.

BT California
BT United States

Pakistan (1978)
Formerly consisted of East Pakistan and West Pakistan which were separated by about 1,000 miles of Indian territory. After East Pakistan became the independent state of Bangladesh in 1971, West Pakistan and Pakistan became coextensive. Before 1973, search West Pakistan or Pakistan AND west.
CO N233500N373000
    E0751500E0601500
UF West Pakistan
BT Indian Peninsula
BT Asia
NT Baluchistan Pakistan
NT North-West Frontier Pakistan
NT Punjab Pakistan
NT Sind Pakistan
NT Sulaiman Range
SA Anantnag
SA Doda
SA Hindu Kush
SA Indian Plate
SA Indo-Australian Plate
SA Indus River
SA Indus-Yarlung Zangbo suture zone
SA Jammu
SA Jammu and Kashmir
SA Karakoram
SA Kashmir
SA Kashmir Valley
SA Kishtwar
SA Kohistan
SA Ladakh
SA Main Boundary Fault
SA Main Central Thrust
SA Nari Series
SA Naubug
SA Riasi
SA Siwalik Range
SA Siwalik System
SA Srinagar
SA Tawi Valley

Paktia
No longer a valid term for GeoRef. See Paktia Afghanistan.

Paktia Afghanistan (1993)
Use for the province in E Afghanistan. Before 1993, also search Paktia and Afghanistan.
BT Afghanistan
BT Indian Peninsula
BT Asia

Palaeodictyopteroida (1981)
BT Insecta
BT Mandibulata
BT Arthropoda
BT Invertebrata

Palaeofavosites (1978)
Genus. Autoposting of Zoantharia to this term began in 1989.
BT Favositidae
BT Tabulata
BT Zoantharia
BT Anthozoa
BT Coelenterata
BT Invertebrata
SA Favosites

Palaeoloxodon naumanni (1978)
Species. Autoposting of Eutheria and Theria to this term began in 1989.
BT Elephantidae
BT Elephantoidea
BT Proboscidea
BT Eutheria
BT Theria

BT Mammalia
BT Tetrapoda
BT Vertebrata
BT Chordata

Palaeophycus (1989)
BT ichnofossils

palagonite (1978)
BT glasses
BT volcanic rocks
BT igneous rocks
SA palagonitization

palagonitization (1989)
BT metasomatism
SA alteration
SA glasses
SA palagonite
SA processes

Palamau
No longer a valid term for GeoRef. See Palamau India.

Palamau India (1993)
District in coal mining area in W Bihar, E India.
BT Bihar India
BT India
BT Indian Peninsula
BT Asia

Palatinate (1978)
Historical region in two parts. Lower or Rhine Palatinate on both sides of Rhine River S of the Main River; and the Upper Palatinate in E Bavaria.
IN Index states as applicable.
BT Germany
BT Central Europe
BT Europe
SA Baden-Wurttemberg Germany
SA Bavaria Germany
SA Hesse Germany
SA Rhineland-Palatinate Germany
SA Upper Palatinate

Palau
use Belau

Palau-Kyushu Ridge
use Kyushu-Palau Ridge

Palawan (1978)
Long, narrow island between N Borneo and W Philippines.
UF Paragua
BT Philippine Islands
BT Far East
BT Asia

Palencia
No longer a valid term for GeoRef. As of 1993, see Palencia City Spain.

Palencia City Spain (1993)
Refers to only the city in S Palencia Province, N Spain. Before 1993, also search Palencia and Spain.
BT Palencia Spain
BT Old Castile Spain
BT Castile Spain
BT Spain
BT Iberian Peninsula
BT Southern Europe
BT Europe

Palencia Province
No longer a valid term for GeoRef. As of 1993, see Palencia Spain.

Palencia Spain (1993)
Refers only to the province in N Spain. From 1981-1992, also search Palencia Province and Spain. Before 1981, also search Palencia and Spain. For the city, see Palencia City Spain.
CO N414600N430300
    W0035200W0050000
BT Old Castile Spain
BT Castile Spain
BT Spain
BT Iberian Peninsula
BT Southern Europe
BT Europe
NT Palencia City Spain

paleo-indian
use paleoindian

paleo-oceanography (1978)
SA bathymetry
SA ocean basins
SA ocean floors
SA oceanography
SA paleocirculation
SA paleohydrology
SA sea-floor spreading

paleoatmosphere (1982)
SA atmosphere
SA climate
SA Earth
SA geochemistry
SA paleoclimatology
SA paleoecology
SA paleoenvironment
SA paleotemperature

paleobathymetry (1978)
SA bathymetry
SA ocean floors
SA paleocirculation
SA paleorelief
SA sea-floor spreading

paleobiochemistry
No longer a valid GeoRef index term. Before 1971, was used on level 1 in subfile N.
use biochemistry

paleobiogeography
Not a valid GeoRef index term after 1970. Was used on level 1 in subfile N.
use biogeography

paleobiology (1978)
SA biochemistry
SA biology
SA fossils
SA geomicrobiology
SA paleontology

paleobotany (1978)
Used for general discussions of fossil plants and for the discipline as a whole.
IN See names of major floral groups (list F).
UF botany, paleo-
SA bibliography
SA biogeography
SA biologic evolution
SA catalogs
SA cladistics
SA electron microscopy
SA floral provinces
SA floral studies
SA flowers
SA fossilization
SA fossils
SA geomicrobiology
SA holotypes
SA leaves
SA micropaleontology
SA miscellanea
SA paleoecology
SA paleontology
SA palynology
SA paratypes
SA photosynthesis
SA phytoliths

SA Plantae
SA roots
SA seeds
SA sporangia
SA staining
SA vegetation
SA vicariance
SA wood

Paleocene (1978)
World. Lower Tertiary, above Gulfian (Cretaceous), below Eocene. Autoposting of this term began in 1978.
BT Paleogene
BT Tertiary
BT Cenozoic
NT Beaufort Formation
NT Bullion Creek Formation
NT Clayton Formation
NT Latrobe Group
NT Lebo Member
NT lower Paleocene
NT Ludlow Member
NT middle Paleocene
NT Midway Group
NT Montian
NT Nacimiento Formation
NT Naheola Formation
NT Oak Hill Member
NT Pinyon Conglomerate
NT Porters Creek Formation
NT Ravenscrag Formation
NT Sentinel Butte Formation
NT Tongue River Member
NT Tullock Member
NT upper Paleocene
SA Beaverhead Formation
SA Brianconnais Zone
SA Denver Formation
SA Ekofisk Formation
SA Flagstaff Formation
SA Fort Union Formation
SA Hanna Formation
SA Huber Formation
SA Imo Shale
SA Intertrappean Beds
SA Laramide Orogeny
SA lower Paleogene
SA Mardin Formation
SA Moreno Formation
SA Mount Scopus Group
SA Nanaimo Group
SA North Horn Formation
SA Orca Group
SA Paskapoo Formation
SA Raton Formation
SA Santa Susana Formation
SA Scaglia Formation
SA Sheep Pass Formation
SA Sinjar Formation
SA Sobral Formation
SA Sparnacian
SA Wasatch Formation

paleochannels
Not a valid index term for GeoRef. See channels under sedimentary structures planar bedding structures (list K). See also buried channels.

paleocirculation (1978)
SA circulation
SA ocean circulation
SA paleo-oceanography
SA paleobathymetry

paleoclimate maps (1985)
Before 1985, also search paleoclimatology AND maps.
UF paleoclimatologic maps
BT maps
SA climatologic maps
SA paleoclimatology
SA photogeology

paleoclimatic controls
  Not a valid term for GeoRef. See paleoclimatology.

paleoclimatic effects
  Not a valid term for GeoRef. See paleoclimatology.

paleoclimatologic maps
  use paleoclimate maps

paleoclimatology (1978)
  Used for treatments of the climate of a given period of time in the geologic past (pre-Holocene).
  IN Index also age and area when possible.
  UF climatology, paleo-
  SA arctic environment
  SA arid environment
  SA atmosphere
  SA Boreal
  SA boreal environment
  SA C-13/C-12
  SA climate
  SA climatologic maps
  SA cycles
  SA encrustations
  SA fluctuations
  SA fossils
  SA glacial geology
  SA glaciation
  SA humid environment
  SA indicators
  SA interglacial environment
  SA isotopes
  SA length of day
  SA Milankovitch theory
  SA O-18/O-16
  SA obliquity of the ecliptic
  SA paleoatmosphere
  SA paleoclimate maps
  SA paleoecology
  SA paleogeography
  SA paleotemperature
  SA S-34/S-32
  SA subtropical environment
  SA taiga environment
  SA temperate environment
  SA tropical environment

Paleocopida (1978)
  As of 1990, microfossils, Crustacea, Mandibulata, and Arthropoda are autoposted to this term. This term has multiple hierarchies.
  BT1 Ostracoda
  BT1 Crustacea
  BT1 Mandibulata
  BT1 Arthropoda
  BT1 Invertebrata
  BT2 Ostracoda
  BT2 microfossils

paleocurrent maps (1985)
  BT maps
  SA paleocurrents
  SA paleogeography
  SA sedimentation

paleocurrents (1978)
  SA currents
  SA ocean currents
  SA paleocurrent maps
  SA paleogeography
  SA paleohydrology
  SA provenance
  SA sedimentation

paleoearthquakes
  use paleoseismicity

paleoecology (1978)
  Used for the study of the relationships between organisms and their environments, the death of organisms, and their burial and postburial history in the geologic past (pre-Holocene), based on fossil fauna and flora and their stratigraphic position. For post-Holocene studies, use ecology.
  IN Index with type of fossil fauna or flora, age, and type of environment if applicable.
  SA adaptation
  SA aerobic environment
  SA anaerobic environment
  SA anaerobic taxa
  SA aquatic environment
  SA archaeology
  SA arctic environment
  SA beaches
  SA behavior
  SA benthonic taxa
  SA biogeography
  SA biologic evolution
  SA biostratigraphy
  SA biotopes
  SA biotypes
  SA boreal environment
  SA cave environment
  SA changes
  SA changes of level
  SA colonial taxa
  SA colonization
  SA communities
  SA Darwinism
  SA deltaic environment
  SA deltas
  SA diachronism
  SA ecology
  SA endemic taxa
  SA environment
  SA estuaries
  SA estuarine environment
  SA fluvial environment
  SA fresh water
  SA fresh-water environment
  SA Gilbert-type deltas
  SA growth rates
  SA hardground
  SA hemipelagic environment
  SA herbivorous taxa
  SA hypersaline environment
  SA indicators
  SA juvenile taxa
  SA lacustrine environment
  SA lagoonal environment
  SA lagoons
  SA lakes
  SA marine environment
  SA middens
  SA paleoatmosphere
  SA paleobotany
  SA paleoclimatology
  SA paleoenvironment
  SA paleofloods
  SA paleogeography
  SA paleontology
  SA paleotemperature
  SA paludal environment
  SA peat bogs
  SA phenotypes
  SA ponds
  SA predators
  SA provinciality
  SA punctuated aggradational cycles
  SA reef environment
  SA reefs
  SA salinity
  SA salt marshes
  SA sedimentation
  SA shelf environment
  SA substrates
  SA succession
  SA taphonomy
  SA terrestrial environment
  SA trophic analysis
  SA vicariance

paleoenvironment (1978)
  Distinguish from paleoecology.
  SA biotopes
  SA environment
  SA environmental analysis
  SA historical geology
  SA paleoatmosphere
  SA paleoecology
  SA paleogeography
  SA paleotemperature
  SA sedimentation

paleofloods (1993)
  Not to be used for modern floods. Before 1993, also search paleohydrology AND floods.
  SA floods
  SA paleoecology
  SA paleohydrology

Paleogene (1978)
  An interval of geologic time incorporating the Oligocene, Eocene, and Paleocene of the Tertiary; the lower Tertiary. Autoposting of this term began in 1978.
  UF Eogene
  BT Tertiary
  BT Cenozoic
  NT Bracklesham Beds
  NT Duchesne River Formation
  NT Eocene
  NT Ewekoro Formation
  NT Flagstaff Formation
  NT Ghost Rocks Formation
  NT Hanna Formation
  NT Huber Formation
  NT Ilerdian
  NT Imo Shale
  NT Kenai Group
  NT lower Paleogene
  NT Oligocene
  NT Orca Group
  NT Paleocene
  NT San Lorenzo Formation
  NT Santa Susana Formation
  NT Sespe Formation
  NT Sinjar Formation
  NT Tyonek Formation
  NT upper Paleogene
  NT Wasatch Formation
  SA lower Tertiary

paleogeographic controls (1978)
  As of 1993, restricted to economic studies.
  SA controls
  SA mineral deposits, genesis
  SA paleogeography

paleogeographic maps (1978)
  Before 1978, also search maps AND paleogeographic.
  BT maps
  SA geotectonic maps
  SA paleogeography

paleogeography (1978)
  Used for the study and description of the physical geography of the geologic past, before the Holocene. For the geographic distribution of fossil and modern organisms, use biogeography.
  SA alluvial plains
  SA basins
  SA biogeography
  SA buried channels
  SA buried valleys
  SA changes of level
  SA coastlines
  SA continental drift
  SA geography
  SA geomorphology
  SA geosynclines
  SA geotectonic maps
  SA land bridges
  SA Mesogaea
  SA ocean basins
  SA paleoclimatology
  SA paleocurrent maps
  SA paleocurrents
  SA paleoecology
  SA paleoenvironment
  SA paleogeographic controls
  SA paleogeographic maps
  SA paleohydrology
  SA plate tectonics
  SA reconstruction
  SA sebkha environment
  SA sedimentary basins
  SA sedimentary structures
  SA sedimentation
  SA shorelines
  SA stratigraphy
  SA transgression

paleohydrology (1985)
  SA hydrology
  SA paleo-oceanography
  SA paleocurrents
  SA paleofloods
  SA paleogeography
  SA paleolimnology

paleoindian (1989)
  UF paleo-indian
  SA anthropology
  SA biogeography
  SA biologic evolution
  SA ontogeny

paleokarst (1982)
  Autoposting of solution features to this term began in 1989.
  BT solution features
  SA buried features
  SA geomorphology
  SA hydrology
  SA karst
  SA karst hydrology

paleolatitude (1978)
  UF latitude, paleo-
  SA apparent polar wandering
  SA latitude
  SA paleomagnetism
  SA polar wandering

paleolimnology (1978)
  SA glacial lakes
  SA limnology
  SA paleohydrology

Paleolithic (1978)
  Archaeological classification. Autoposting of this term began in 1978.
  UF Old Stone Age
  BT Stone Age
  BT Cenozoic
  NT Acheulian
  NT Aurignacian
  NT Levalloisian
  NT Magdalenian
  NT Mousterian
  NT upper Paleolithic
  SA archaeology
  SA Pleistocene
  SA Quaternary
  SA Tertiary

paleomagnetism (1978)
  Used for the study of natural remanent magnetization.
  IN Always include rock type and age as index terms, if possible.
  UF magnetism, paleo-
  SA adiabatic demagnetization
  SA alternating field demagnetization
  SA anhysteretic remanent magnetization
  SA apparent polar wandering
  SA Brunhes Epoch

SA chemical demagnetization
SA chemical remanent magnetization
SA coercivity
SA continental drift
SA crust
SA Curie point
SA demagnetization
SA depositional remanent magnetization
SA dipole moment
SA Earth
SA Gauss Epoch
SA geochemistry
SA geochronology
SA geophysical methods
SA geophysics
SA Gilbert Epoch
SA high-resolution methods
SA induction
SA inverse thermoremanent magnetization
SA Jaramillo Event
SA Koenigsberger ratio
SA lava
SA magnetic anomalies
SA magnetic declination
SA magnetic domains
SA magnetic field
SA magnetic hysteresis
SA magnetic inclination
SA magnetic intensity
SA magnetic susceptibility
SA magnetization
SA magnetostratigraphy
SA Matuyama Epoch
SA natural remanent magnetization
SA Olduvai Event
SA paleolatitude
SA partial thermoremanent magnetization
SA physical remanent magnetization
SA plate tectonics
SA polar wandering
SA pole positions
SA relative age
SA remagnetization
SA remanent magnetization
SA reversals
SA sea-floor spreading
SA secular variations
SA stability
SA stratigraphy
SA tectonophysics
SA Thellier Method
SA thermal demagnetization
SA thermochemical remanent magnetization
SA thermoremanent magnetization
SA viscous remanent magnetization

paleontology (1978)
Used for discipline as a whole.
IN Index as a general term for studies of general paleontological principles.
SA adaptation
SA anatomy
SA bibliography
SA biocenoses
SA biogeographic maps
SA biogeography
SA biologic evolution
SA biometry
SA biostratigraphy
SA bipedalism
SA bones
SA casts
SA catalogs
SA chemical fossils
SA cladistics
SA collecting

SA collections
SA continental drift
SA creationism
SA Darwinism
SA diet
SA electron microscopy
SA extinct taxa
SA extinction
SA faunal provinces
SA fossilization
SA fossils
SA geochronology
SA geomicrobiology
SA glossaries
SA gradualism
SA historical geology
SA holotypes
SA index fossils
SA isomorphism
SA jaws
SA larvae
SA lexicons
SA life origin
SA mass extinctions
SA micropaleontology
SA miscellanea
SA nutrition
SA ontogeny
SA ornamentation
SA osteology
SA paleobiology
SA paleobotany
SA paleoecology
SA paleopathology
SA palynology
SA paratypes
SA phylogeny
SA Piltdown man
SA predation
SA problematic fossils
SA punctuated equilibria
SA radiation
SA reproduction
SA sample preparation
SA shells
SA skeletons
SA skulls
SA speciation
SA species diversity
SA staining
SA stratigraphy
SA synonymy
SA taxonomy
SA thanatocenoses
SA trophic analysis
SA type localities
SA ultrastructure
SA vicariance
SA wood

paleopathology (1985)
SA archaeology
SA bones
SA jaws
SA morphology
SA paleontology
SA shells
SA skeletons
SA skulls
SA taphonomy
SA teeth

paleorelief (1978)
UF paleotopography
SA buried channels
SA buried features
SA buried valleys
SA geomorphology
SA landform evolution
SA paleobathymetry
SA relief

paleosalinity (1978)
SA brackish water
SA fluid inclusions
SA geochemistry

SA salinity
SA salt
SA sea water

paleoseismicity (1993)
Study of past, pre-instrumental earthquakes traceable by structural, geomorphologial or sedimentological features or by geophysical properties in rocks and sediments.
UF paleoearthquakes
UF paleoseismology
SA archaeology
SA earthquakes
SA geologic hazards
SA seismicity
SA seismology
SA seismotectonics

paleoseismology
 use paleoseismicity

Paleosols (1978)
UF fossil soils
BT soils
SA buried features
SA soil group

paleotectonics
A valid term through 1975. After 1975, see tectonics and paleogeography. See also neotectonics.

paleotemperature (1978)
Term reintroduced in 1982. Before 1982, also search temperature; paleotemperatures. Term was valid through 1973. From 1974-1981, temperature was used under paleoclimatology. See geologic thermometry.
SA geochemistry
SA geologic thermometry
SA lithogeochemistry
SA P-T conditions
SA paleoatmosphere
SA paleoclimatology
SA paleoecology
SA paleoenvironment
SA temperature
SA thermal history

paleotemperatures
Not a valid GeoRef index term after 1970. Was used on level 1 in subfile N. Now use temperature or geologic thermometry, paleotemperature, or paleoclimatology.

paleotopography
 use paleorelief

Paleozoic (1978)
From 1978-1980, Phanerozoic was autoposted to this term. Autoposting of this term to its main subdivisions (e.g. Cambrian, Mississippian, Carboniferous) ended in 1981. Autoposting of this term to its main subdivisions resumed in 1993.
NT Antler Orogeny
NT Arbuckle Group
NT Baralaba Coal Measures
NT Bedford Shale
NT Berea Sandstone
NT Bird Spring Formation
NT Bowers Supergroup
NT Broken River Formation
NT Bucksport Formation
NT Caledonian Orogeny
NT Cambrian
NT Cape Elizabeth Formation
NT Carboniferous
NT Casco Bay Group
NT Casper Formation
NT Catskill Formation

NT Chattanooga Shale
NT Collier Shale
NT Cow Head Group
NT Deadwood Formation
NT Devonian
NT Downtonian
NT Dunkard Group
NT Earn Group
NT Ellis Bay Formation
NT Endicott Group
NT Geirud Formation
NT Hanson Creek Formation
NT Hartland Formation
NT Hercynian Orogeny
NT Hidden Valley Dolomite
NT Honaker Trail Formation
NT Horton Group
NT Hunton Group
NT Ishbel Group
NT Itarare Subgroup
NT Keyser Limestone
NT Kittery Formation
NT Knox Group
NT Laborcita Formation
NT Leinster Granite
NT Lisburne Group
NT lower Paleozoic
NT Magdalena Group
NT Manhattan Formation
NT Maroon Formation
NT Matapedia Group
NT Merrimack Group
NT middle Paleozoic
NT Mikabu System
NT Minnelusa Formation
NT Oquirrh Formation
NT Ordovician
NT Ouachita Orogeny
NT Paganzo Group
NT Peel Sound Formation
NT Permian
NT Petersburg Granite
NT Pilot Shale
NT Rangeley Formation
NT Read Bay Formation
NT Ringerike Sandstone
NT Road River Formation
NT Rondout Formation
NT Sambagawa Belt
NT Shawangunk Formation
NT Shoo Fly Complex
NT Silurian
NT Silvretta Group
NT Supai Formation
NT Taconic Orogeny
NT Taiyuan Formation
NT Talchir Formation
NT Talladega Group
NT Tensleep Sandstone
NT Tulghes Series
NT upper Paleozoic
NT Waits River Formation
NT Weber Sandstone
NT Wells Formation
NT Wissahickon Formation
NT Woodford Shale
SA Catalina Schist
SA Ceneri Zone
SA Klodzko Unit
SA lower Gondwana System
SA New England Orogeny
SA Phanerozoic
SA Riggins Group
SA Shuswap Complex

Palermo
No longer a valid term for GeoRef. See Palermo Italy.

Palermo Italy (1993)
City on Bay of Palermo in NW Sicily.
BT Sicily Italy
BT Italy
BT Southern Europe
BT Europe

Palestine (1978)
Approximately coextensive with Israel and the W bank of the Jordan River.
IN   Index countries as applicable.
BT   Middle East
BT   Asia
SA   Israel
SA   Jordan

palingenesis (1978)
As of 1978, term is restricted to meaning in petrology. Term is not used in reference to paleontology.
SA   anatexis
SA   assimilation
SA   contamination
SA   magmas
SA   metasomatism
SA   petrology

Palisades Sill (1989)
Located along the Hudson River in Paterson County, NE New Jersey and Westchester County, SE New York. Also located in Colfax County, N New Mexico.
IN   Index states and counties as applicable.
SA   Colfax County New Mexico
SA   New Jersey
SA   New Mexico
SA   New York
SA   Westchester County New York

palladium (1978)
Chemical element. As of 1985, use palladium ores for palladium as a commodity. Autoposting of platinum group and metals to this term began in 1989.
UF   Pd
BT   platinum group
BT   metals
SA   palladium ores

palladium ores (1985)
Before 1985, also search (palladium OR Pd) AND (deposit OR deposits OR ore OR ores OR economic) in the basic index. Autoposting of metal ores to this term began in 1985.
IN   Commodity. See List C.
BT   metal ores
SA   palladium
SA   platinum ores

pallasite (1984)
Used as an igneous rock before 1978. As of 1978, used for the meteorite. Before 1984, also search pallasites.
UF   pallasites
BT   stony irons
BT   meteorites
NT   Eagle Station Meteorite

pallasites
As of 1984, no longer a valid term for GeoRef.
use pallasite

Palliser Formation (1978)
BT   Upper Devonian
BT   Devonian
BT   Paleozoic
SA   Alberta
SA   Canada

Palm Beach County
Valid through 1988. Search in combination with state term. After 1988, use specific county-state term.

Palm Beach County Florida (1989)
On the Atlantic Ocean in SE Florida. Before 1989, also search Palm Beach County AND Florida.
CO   N262000N265700
      W0800000W0805300
BT   Florida
BT   United States
SA   Biscayne Aquifer

Palm Springs
Valid through 1988. Search in combination with state term. After 1988, use specific city-state term.

Palm Springs California (1989)
City in S California. Before 1989, search Palm Springs AND California.
CO   N334900N334900
      W1163400W1163400
BT   Riverside County California
BT   California
BT   United States

Palma
No longer a valid term for GeoRef. As of 1993, see Palma Spain.

Palma de Majorca
use Palma Spain

Palma Spain (1993)
City on Majorca. Before 1993, also search Palma and Spain, or Palma and Mediterranean region. This term has multiple hierarchies.
UF   Palma de Majorca
BT1   Majorca
BT1   Balearic Islands
BT1   Mediterranean region
BT2   Majorca
BT2   Balearic Islands
BT2   Spain
BT2   Iberian Peninsula
BT2   Southern Europe
BT2   Europe

Palmae (1978)
Family coextensive with order Palmales.
UF   Palmales
BT   Monocotyledoneae
BT   angiosperms
BT   Spermatophyta
BT   Plantae

Palmales
use Palmae

Palmatolepis (1978)
Genus. Autoposting of microfossils to this term began in 1990.
BT   Conodonta
BT   microfossils

Palmdale Bulge
No longer a valid term for GeoRef.
use Palmdale California

Palmdale California (1993)
N Los Angeles County, California. Before 1993 also search Palmdale or Palmdale region or Palmdale Bulge AND California.
UF   Palmdale Bulge
UF   Palmdale region
BT   Los Angeles County California
BT   California
BT   United States

Palmdale region
No longer a valid term for GeoRef.
use Palmdale California

Palmer Peninsula
use Antarctic Peninsula

Palo Alto
Valid through 1988. Search in combination with state term. After 1988, use specific city-state term.

Palo Alto California (1989)
City SSE of San Francisco in W California. Before 1989, search Palo Alto AND California.
CO   N372600N372600
      W1221000W1221000
BT   Santa Clara County California
BT   California
BT   United States

Palo Duro Basin (1985)
E central New Mexico, Texas panhandle and SW Oklahoma.
IN   Index states as applicable.
BT   United States
BT   Great Plains
SA   New Mexico
SA   Oklahoma
SA   Texas

Palos Verdes Hills (1978)
Occupy peninsula between Santa Monica Bay and San Pedro Bay S of Los Angeles.
UF   San Pedro Hills
BT   Los Angeles County California
BT   California
BT   United States
SA   Palos Verdes Peninsula

Palos Verdes Peninsula (1989)
Located between Santa Monica Bay and San Pedro Bay in S California.
CO   N334000N335000
      W1181500W1183000
BT   Los Angeles County California
BT   California
BT   United States
SA   Palos Verdes Hills
SA   San Pedro

palsas (1985)
Autoposting of periglacial features to this term began in 1989.
BT   periglacial features
SA   frost action
SA   frost heaving
SA   frozen ground
SA   glacial geology
SA   ice lenses
SA   peat
SA   taliks
SA   tundra

paludal environment (1981)
SA   depositional environment
SA   ecology
SA   environment
SA   marshes
SA   muskegs
SA   paleoecology
SA   paludal sedimentation
SA   sedimentation
SA   swamps

paludal sedimentation (1981)
BT   sedimentation
SA   paludal environment

Paluxy Formation (1978)
In Trinity Group. SW Arkansas, SE Oklahoma, and E Texas. This term has multiple hierarchies.
BT1   Comanchean
BT1   Cretaceous
BT1   Mesozoic
BT2   Lower Cretaceous
BT2   Cretaceous
BT2   Mesozoic
SA   Arkansas
SA   Oklahoma
SA   Texas
SA   Trinity Group

palygorskite (1978)
UF   attapulgite
BT   clay minerals
BT   sheet silicates
BT   silicates

palynology (1978)
Used for the discipline which is concerned with the study of pollen of seed plants and spores of other embryophytic plants whether living or fossil, including their dispersal and applications.
IN   Index only for general studies of the discipline.
SA   biogeography
SA   geochronology
SA   geomicrobiology
SA   micropaleontology
SA   paleobotany
SA   paleontology
SA   palynomorphs
SA   pollen
SA   pollen analysis
SA   pollen diagrams
SA   relative age
SA   sample preparation
SA   stratigraphy

palynomorphs (1978)
Used for pollen of seed plants and spores of other embryophytic plants, whether living or fossil, including their dispersal and applications. As of 1981, considered both a common name and a systematic name. Autoposting of this term to Cladophlebis and Dryas ended in 1985.
BT   microfossils
NT   acritarch flora
NT   acritarchs
NT   Chitinozoa
NT   chitinozoans
NT   Dinoflagellata
NT   dinoflagellates
NT   megaspores
NT   miospores
NT   problematic palynomorphs
SA   Artemisia
SA   biostratigraphy
SA   Cladophlebis
SA   Corylus
SA   Dryas
SA   exine
SA   floral studies
SA   monolete taxa
SA   palynology
SA   plankton
SA   Plantae
SA   pollen
SA   pollen analysis
SA   pollen diagrams
SA   Pyrrhophyta
SA   spores
SA   Tasmanites

Pambak
No longer a valid term for GeoRef. As of 1993, see Pambak Armenia.

Pambak Armenia (1993)
NE section of city of Kirovakan in N central Armenia. Before 1993, also search Bambak. This term has multiple hierarchies.
UF   Bambak Armenia
BT1   Armenia
BT1   Commonwealth of Independent States
BT2   Armenia
BT2   Europe

Pamir (Range)
use Pamirs

Pamir Range
use Pamirs

Pamirs (1978)

High altitude region, mostly in Tadzhikistan. Before 1981, also included small parts of China, India, and Afghanistan. Also search Pamir. This term has multiple hierarchies.
IN Index former Soviet republics as applicable.
CO N370000N393000
    E0750000E0650000
UF Pamir (Range)
UF Pamir Range
BT1 Commonwealth of Independent States
BT2 Central Asia
BT2 Asia
NT Darvaz
SA Darvaza Range
SA Tadzhikistan
SA Uzbekistan
SA Zeravshan Range

Pamlico River (1978)
Bisects Beaufort County and flows into Pamlico Sound. Estuary of the Tar River.
BT North Carolina
BT United States
SA Beaufort County North Carolina
SA Pungo River

Pamlico Sound (1978)
Between E North Carolina mainland and narrow islands off the coast.
IN Index counties as applicable.
BT North Carolina
BT United States

Pampa
use Pampas

Pampas (1978)
Vast, treeless, fertile plain extending N and S approximately 1000 miles from lower Parana River to Patagonia, and from the Atlantic to the Andean piedmont. The humid Pampas are in the E and the dry Pampas are in the W.
CO S395000S215500
    W0533000W0694000
UF Pampa
BT Argentina
BT South America

Pampean Mountains (1978)
In Cordoba Province, Argentina.
UF Sierras Pampeanas
BT Cordoba Argentina
BT Argentina
BT South America
SA Catamarca Argentina
SA La Rioja Argentina
SA Salta Argentina
SA San Juan Argentina
SA San Luis Argentina
SA Tucuman Argentina

Pamplona
No longer a valid term for GeoRef. As of 1993, see Pamplona Spain.

Pamplona Spain (1993)
City in central Navarra, N Spain. Before 1993, also search Pamplona and Spain.
BT Navarra Spain
BT Spain
BT Iberian Peninsula
BT Southern Europe
BT Europe

Pamunkey River (1989)
E Virginia, about 15 miles NE of Richmond.
BT Virginia

BT United States

Pan-African Orogeny (1978)
North Africa. Before 1978, also search orogeny AND Pan-African or Pan African.
UF Eburnean Orogeny
BT Proterozoic
BT upper Precambrian
BT Precambrian
SA Mauritanides
SA orogeny
SA tectonics

Panagyurishte
No longer a valid term for GeoRef. See Panagyurishte Bulgaria.

Panagyurishte Bulgaria (1993)
Plovdiv district, W Central Bulgaria in central Sredna Gora on Luda Yana River.
BT Plovdiv Bulgaria
BT Bulgaria
BT Southern Europe
BT Europe
SA Sredna Gora

Panama (1978)
CO N071000N094000
    W0771000W0830200
BT Central America
NT Darien
NT Sansan Panama

Panama Basin (1978)
SW of Panama and W of Colombia. Autoposting of Pacific Ocean to this term began in 1990.
BT North American Pacific
BT Pacific Ocean

Panama Canal Zone (1978)
10 mile strip of territory across Panama in which canal is located. Before 1971, also search Canal Zone.
UF Canal Zone
BT Central America

Panamint Range (1978)
W of Death Valley in Inyo County in E California.
BT Inyo County California
BT California
BT United States

Panasqueira
No longer a valid term for GeoRef. As of 1993 see Panasquiera Portugal.

Panasqueira Portugal (1993)
Village in S Portugal. Before 1993 also search Panasqueira AND Portugal.
BT Portugal
BT Iberian Peninsula
BT Southern Europe
BT Europe

Panay Island (1978)
One of the Visayan Islands midway between Luzon and Mindanao. Also search Panay.
BT Philippine Islands
BT Far East
BT Asia
SA Iloilo Basin
SA Iloilo Philippine Islands

Panch Mahals
No longer a valid term for GeoRef. See Panch Mahals India.

Panch Mahals India (1993)
District in Gujarat, W India. Before 1993, also search Panch Mahals. Before 1978, also search Panchmahal or Panchmahals.

UF Panchmahal India
BT Gujarat India
BT India
BT Indian Peninsula
BT Asia

Panchet Series (1978)
Series of rocks overlying the Damuda Series in the Raniganj coal field.
BT Lower Triassic
BT Triassic
BT Mesozoic
SA Bihar India
SA Damuda Series
SA India

Panchmahal India
use Panch Mahals India

Panderodus (1978)
Autoposting of microfossils to this term began in 1990.
BT Conodonta
BT microfossils

Pangaea (1978)
Name proposed for the supercontinent comprising all the landmasses or earth which existed about 300 million years ago prior to continental drift.
UF Pangea
SA continental drift
SA Gondwana
SA Laurasia
SA Northern Hemisphere

Pangea
use Pangaea

Panhandle
No longer a valid term for GeoRef. An area or projection of land like the handle of a pan which occurs in a number of states. As of 1993, see Alaska Panhandle, Florida Panhandle, Idaho Panhandle, Oklahoma Panhandle, Texas Panhandle, or West Virginia Panhandle.

panning (1985)
SA mineral exploration
SA placers
SA sampling

Pannonia (1978)
Ancient Roman province S and W of the Danube.
IN Index countries as applicable.
BT Europe
SA Austria
SA Hungary
SA Pannonian Basin
SA Yugoslavia

Pannonian (1978)
Europe. Before 1993 may have been used for lower Pliocene.
BT upper Miocene
BT Miocene
BT Neogene
BT Tertiary
BT Cenozoic
SA Pliocene

Pannonian Basin (1978)
W Romania, SE Austria, Hungary W of the Danube, N central Yugoslavia, and S central Czechoslovakia. Also search Pannonia.
IN Index countries as applicable.
BT Europe
NT Romanian Pannonian Basin
NT Slovakian Pannonian Basin
SA Austria
SA Czechoslovakia
SA Hungary
SA Pannonia

SA Romania
SA Yugoslavia

Panola County
Valid through 1988. Search in combination with state term. After 1988, use specific county-state term.

Panola County Mississippi (1989)
NW Mississippi. Before 1989, also search Panola County AND Mississippi.
CO N341000N343400
    W0894200W0901200
BT Mississippi
BT United States

Panola County Texas (1989)
E Texas. Before 1989, also search Panola County AND Texas.
CO N315800N322200
    W0940300W0943500
BT Texas
BT United States

Pantar (1978)
Island of the Alor group 60 miles NW of Timor. This term has multiple hierarchies.
BT1 Lesser Sunda Islands
BT1 Far East
BT1 Asia
BT2 Indonesia
BT2 Far East
BT2 Asia

Pantellaria
use Pantelleria

Pantelleria (1978)
Volcanic Italian island in the Mediterranean Sea between Sicily and Tunisia.
UF Pantellaria
BT Sicily Italy
BT Italy
BT Southern Europe
BT Europe

pantellerite (1978)
BT rhyolites
BT volcanic rocks
BT igneous rocks
SA comendite

Pantodonta (1981)
Suborder. Autoposting of Amblypoda, Eutheria and Theria to this term began in 1989.
BT Amblypoda
BT Eutheria
BT Theria
BT Mammalia
BT Tetrapoda
BT Vertebrata
BT Chordata
SA Ungulata

Pantotheria (1978)
Order. Autoposting of Theria to this term began in 1989.
BT Theria
BT Mammalia
BT Tetrapoda
BT Vertebrata
BT Chordata

Panxi Rift (1989)
Sichuan, SW China.
CO N270000N300000
    E1050000E1000000
BT Sichuan China
BT China
BT Far East
BT Asia

Papaloapan River basin (1981)
In state of Veracruz, SE Mexico.

CO N180500N184500
    W0953000W0961000
BT Veracruz Mexico
BT Mexico

paper chromatography
  use chromatography

Papua (1978)
Former Australian territory. Denotes the southern half of new state of Papua New Guinea on the island of New Guinea. Included use as a first order area term until 1976.
BT Papua New Guinea
BT Australasia

Papua New Guinea (1978)
Eastern half of the island of New Guinea. Comprises Papua and the former Australian U.N. trusteeship of The Territory of New Guinea, plus the Bismarck Archipelago, and Bougainville, Buka and Green islands of the W Solomon Islands. Became an independent state on September 16, 1976.
CO S110000S003000
    E1563000E1410000
UF Papua-New Guinea
BT Australasia
NT Bismarck Archipelago
NT Bougainville
NT Huon Peninsula
NT Louisiade Archipelago
NT New Britain
NT New Ireland
NT Papua
NT Port Moresby Papua New Guinea
NT Southern Highlands Papua New Guinea
NT Talasea New Guinea
NT Vanimo Papua New Guinea
SA Malay Archipelago
SA New Guinea
SA Rat Island
SA Solomon Islands
SA Woodlark Basin

Papua-New Guinea
  use Papua New Guinea

Para
No longer a valid term for GeoRef. See Para Brazil.

Para Brazil (1993)
State in N Brazil.
CO S095000N024000
    W0460000W0592000
BT Brazil
BT South America
NT Marajo
SA Barreiras Formation
SA Brazilian Shield
SA Tocantins River region

para-amphibolite (1978)
BT amphibolites
BT metamorphic rocks

Parablastoidea (1978)
BT Crinozoa
BT Echinodermata
BT Invertebrata

Paracel Islands
  use Xisha Islands

paracelsian (1978)
BT barium feldspar
BT feldspar group
BT framework silicates
BT silicates
SA celsian

Parachute Creek Member (1985)
Of Green River Formation. NW Colorado and NE Utah.
BT Eocene
BT Paleogene
BT Tertiary
BT Cenozoic
SA Colorado
SA Green River Formation
SA Utah

Paracrinoidea (1978)
BT Crinozoa
BT Echinodermata
BT Invertebrata

Paradox Basin (1978)
In the Four Corners area of the SW U.S. where the boundaries of 4 states intersect.
IN Index states as applicable.
SA Arizona
SA Bluff Formation
SA Colorado
SA Desert Creek Zone
SA Four Corners
SA Gibson Dome
SA Ismay Zone
SA New Mexico
SA Uncompahgre Uplift
SA United States
SA Utah

Paradox Member (1978)
Of Hermosa Formation. W Colorado and SE Utah.
BT Middle Pennsylvanian
BT Pennsylvanian
BT Carboniferous
BT Paleozoic
SA Colorado
SA Hermosa Formation
SA Utah

paraffins (1985)
Autoposting of alkanes to this term began in 1989.
BT alkanes
BT aliphatic hydrocarbons
BT hydrocarbons
BT organic materials

paragenesis (1978)
For detailed treatment of mineral sequences in metamorphosed or altered rocks and mineral deposits.
IN Index with type of material.
UF mineral sequence
UF paragenetic sequence
SA clay mineralogy
SA crystallography
SA economic geology
SA genesis
SA igneous rocks
SA metamorphic rocks
SA metamorphism
SA metasomatism
SA mineral assemblages
SA mineral deposits, genesis
SA minerals
SA phase equilibria
SA rocks
SA sedimentary rocks

paragenetic sequence
  use paragenesis

paragneiss (1978)
This term has multiple hierarchies.
BT1 gneisses
BT1 metamorphic rocks
BT2 metasedimentary rocks
BT2 metamorphic rocks

paragonite (1978)
BT mica group
BT sheet silicates
BT silicates

Paragua
  use Palawan

Paraguana Peninsula (1978)
Between the Caribbean Sea and the Gulf of Venezuela in N Falcon, NW Venezuela.
BT Falcon Venezuela
BT Venezuela
BT South America

Paraguay (1978)
CO S273000S191000
    W0543000W0624500
BT South America
NT Chaco Paraguay
SA Botucatu Formation
SA Chaco
SA Parana Basin
SA Parana River
SA Serra Geral Formation

Parahiba Brazil
  use Paraiba Brazil

Parahyba Brazil
  use Paraiba Brazil

Paraiba
No longer a valid term for GeoRef. See Paraiba Brazil.

Paraiba Brazil (1993)
State in NE Brazil. Before 1993, also search Paraiba. Before 1978, also search Parahiba and Parahyba.
CO S082000S060000
    W0343000W0384500
UF Parahiba Brazil
UF Parahyba Brazil
BT Brazil
BT South America
SA Borborema
SA Recife-Joao Pessoa

parallel faults (1978)
Before 1978, also search faults AND parallel.
BT faults

parallel folds
  use concentric folds

parameters
A valid general term through 1978. See factors. Also included use for lattice parameters (crystal structure); see lattice parameters.

Paramushir (1978)
Large island in the N Kuril Islands near S tip of Kamchatka Peninsula. This term has multiple hierarchies.
BT1 Kuril Islands
BT1 Sakhalin Russian Federation
BT1 Russian Federation
BT1 Commonwealth of Independent States
BT2 Kuril Islands
BT2 Sakhalin Russian Federation
BT2 Asia
BT3 Kuril Islands
BT3 Russian Pacific region
BT3 Russian Federation
BT3 Commonwealth of Independent States
BT4 Kuril Islands
BT4 Russian Pacific region
BT4 Asia

Parana
No longer a valid term for GeoRef. See Parana Brazil.

Parana Basin (1978)
River basin.
IN Index countries as applicable.
UF Parana River basin
BT South America
SA Argentina
SA Brazil
SA Itarare Subgroup
SA Paraguay

Parana Brazil (1993)
State in S Brazil.
CO S265000S222500
    W0481000W0543000
BT Brazil
BT South America
SA Acungui Group
SA Brazilian Shield
SA Estrada Nova Formation
SA Irati Formation
SA Serra do Mar

Parana River (1978)
Formed in S central Brazil by the confluence of the Rio Grande and the Paranaiba rivers. It flows SW and S into the Rio de la Plata. Also search Parana.
IN Index countries as applicable.
BT South America
SA Argentina
SA Brazil
SA Paraguay

Parana River basin
  use Parana Basin

Paranthropus (1978)
Genus. Autoposting of Eutheria and Theria to this term began in 1989.
BT Hominidae
BT Primates
BT Eutheria
BT Theria
BT Mammalia
BT Tetrapoda
BT Vertebrata
BT Chordata

pararammelsbergite (1978)
Autoposting of sulfides to this term ended in 1989.
BT arsenides
SA rammelsbergite

parasites (1978)
SA bacteria
SA epibiotism
SA fungi

paratacamite (1978)
BT chlorides
BT halides
SA atacamite

Paratethys (1978)
SA continental drift
SA Tethys

Paratunka
No longer a valid term for GeoRef. As of 1993, see Paratunka Russian Federation.

Paratunka Russian Federation (1993)
Village just W of Petropavlosk Kamchatskiy in S Kamchatka Oblast on Kamchatka Peninsula. Before 1993, also search Paratunka. This term has multiple hierarchies.
BT1 Kamchatka Russian Federation
BT1 Russian Federation
BT1 Commonwealth of Independent States
BT2 Kamchatka Russian Federation
BT2 Asia

paratypes (1989)
SA fossils
SA holotypes

SA paleobotany
SA paleontology
SA taxonomy
SA type specimens

Parece Vela Basin (1985)
Central Philippine Sea.
BT Philippine Sea
BT West Pacific
BT Pacific Ocean
SA DSDP Site 450

Pareiasauria (1981)
Autoposting of Cotylosauria to this term began in 1989.
BT Cotylosauria
BT Anapsida
BT Reptilia
BT Tetrapoda
BT Vertebrata
BT Chordata

parent bodies (1993)
As of 1993, use only for meteorites and extraterrestrial bodies.
SA asteroids
SA comets
SA meteorites
SA meteoroids
SA planetology
SA protoliths
SA relict materials

parent materials (1978)
SA factors
SA horizons
SA materials
SA pedogenesis
SA soils
SA weathering

parent rocks
use protoliths

pargasite (1978)
BT clinoamphibole
BT amphibole group
BT chain silicates
BT silicates

Paria Peninsula (1978)
NE Venezuela. Along with island of Trinidad nearly encloses Gulf of Paria.
BT Sucre Venezuela
BT Venezuela
BT South America

Paricutin (1978)
Volcano on site of former village of Paricutin in SW Mexico.
BT Michoacan Mexico
BT Mexico

Paring Mountains (1978)
In the Transylvanian Alps.
CO N452000N452000
    E0234000E0234000
BT Romania
BT Southern Europe
BT Europe
SA Carpathians
SA Transylvanian Alps

Paris
No longer a valid term for GeoRef. See Paris France.

Paris Basin (1978)
Chief depression of N and N central France. Bounded by English Channel on NW, American Massif on W, Massif Central on S, and plateaus of Langres and Lorraine on E. Also search Paris AND basin.
CO N480000N500000
    E0050000E0010000
BT France
BT Western Europe
BT Europe

SA Paris France

Paris France (1993)
Department in N central France. Includes the city of Paris. Before 1993, also search Paris or Paris region AND France.
CO N485000N490000
    E0023000E0021000
BT France
BT Western Europe
BT Europe
SA Paris Basin
SA Seine
SA Seine-et-Oise

Park City
Valid through 1988. Search in combination with state term. After 1988, use specific city-state term.

Park City Formation (1985)
NE Utah, NW Colorado, E Idaho, SW Montana, and central and E Wyoming.
BT Permian
BT Paleozoic
SA Colorado
SA Idaho
SA Montana
SA Utah
SA Wyoming

Park City Utah (1989)
City in N central Utah. Before 1989, search Park City AND Utah.
CO N403900N403900
    W1113000W1113000
BT Summit County Utah
BT Utah
BT United States

Park County
Valid through 1988. Search in combination with state term. After 1988, use specific county-state term.

Park County Colorado (1989)
Central Colorado. Before 1989, also search Park County AND Colorado.
CO N384000N393200
    W1052000W1061000
BT Colorado
BT United States
NT Alma mining district

Park County Montana (1989)
S Montana. Before 1989, also search Park County AND Montana.
CO N450000N461100
    W1095100W1110200
BT Montana
BT United States
SA Bridger Range
SA Crazy Mountains
SA Crazy Mountains Basin
SA Gallatin Range

Park County Wyoming (1989)
NW Wyoming. Before 1989, also search Park County AND Wyoming.
CO N434800N450000
    W1083500W1110300
BT Wyoming
BT United States
NT Powell Wyoming
SA Gallatin Range
SA Sunlight
SA Yellowstone National Park

Parker County
Valid through 1988. Search in combination with state term. After 1988, use specific county-state term.

Parker County Texas (1989)
N Texas. Before 1989, also search Parker County AND Texas.
CO N323500N330000
    W0974000W0980500
BT Texas
BT United States

Parkfield
Valid through 1988. Search in combination with state term. After 1988, use specific city-state term.

Parkfield California (1989)
Village near Fresno County line in W California. Location of earthquakes. Also site of Parkfield Earthquake Prediction Experiment. Before 1989, search Parkfield AND California.
CO N355500N355500
    W1202500W1202500
BT Monterey County California
BT California
BT United States
SA Kettleman Hills earthquake 1985
SA Parkfield earthquake 1966
SA Parkfield earthquakes

Parkfield earthquake 1966 (1989)
Epicenter near Parkfield, California.
BT earthquakes
SA California
SA Parkfield California

Parkfield earthquakes (1993)
Series of earthquakes with epicenters near Parkfield, Central California. Associated with earthquake prediction experiment.
BT earthquakes
SA California
SA oarthquake prediction
SA induced earthquakes
SA Parkfield California
SA San Andreas Fault

Parkman Sandstone (1978)
In Montana Group. N Wyoming and S Montana.
BT Upper Cretaceous
BT Cretaceous
BT Mesozoic
SA Judith River Formation
SA Montana
SA Montana Group
SA Wyoming

Parkwood Formation (1985)
N Alabama.
BT Upper Mississippian
BT Mississippian
BT Carboniferous
BT Paleozoic
SA Alabama

Parma
No longer a valid term for GeoRef. See Parma Italy.

Parma Italy (1993)
City in NW Emilia-Romagna, N central Italy.
BT Emilia-Romagna Italy
BT Italy
BT Southern Europe
BT Europe

Parnaiba Basin (1978)
River basin.
IN Index states as applicable.
SA Maranhao Brazil
SA Piaui Brazil

Parnalee
No longer a valid term for GeoRef. Use Parnallee Meteorite for the LL chondrite which landed in India.

Parnassos
use Parnassus

Parnassus (1978)
One of highest massifs in central Greece. N of Gulf of Corinth.
UF Parnassos
BT Sterea Ellas
BT Greece
BT Southern Europe
BT Europe

Parras Basin (1978)
S Coahuila in NE central Mexico. Also search Parras.
UF Parras de la Fuente
BT Coahuila Mexico
BT Mexico

Parras de la Fuente
use Parras Basin

Parry Sound
No longer a valid term for GeoRef. As of 1993, see Parry Sound Ontario.

Parry Sound District Ontario (1993)
SE Ontario. Before 1993, also search Parry Sound District AND Ontario.
BT Ontario
BT Eastern Canada
BT Canada
NT Parry Sound Ontario

Parry Sound Ontario (1993)
Town in SE Ontario on E shore of Georgian Bay. Before 1993, also search Parry Sound AND Ontario.
BT Parry Sound District Ontario
BT Ontario
BT Eastern Canada
BT Canada

partial melting (1978)
SA magmas
SA magmatic differentiation
SA mantle
SA melting
SA melts
SA phase equilibria
SA plate tectonics

partial pressure (1978)
BT pressure
SA fugacity

partial thermoremanent magnetization (1981)
Autoposting of magnetization to this term began in 1989.
BT remanent magnetization
BT magnetization
SA inverse thermoremanent magnetization
SA paleomagnetism
SA thermoremanent magnetization

particle precipitation
As of 1993, no longer a valid term for GeoRef. Usually out-of-scope. Used with ionosphere.

particle radiation
As of 1993, no longer a valid term for GeoRef. Usually out-of-scope. Used with astrophysics and solar physics.

particle tracks (1982)
Used for the phenomenon. For dating method, use particle-track dating.

UF cosmic-ray tracks
SA fission tracks
SA fission-track dating
SA particle-track dating
SA particles
SA radiation damage
SA radioactivity

particle-track dating (1978)
Cosmic-ray tracks.
UF dating, particle-track
SA fission tracks
SA fission-track dating
SA geochronology
SA particle tracks
SA radiation damage
SA relative age

particles (1978)
General term.
SA activation energy
SA aerosols
SA agglutinates
SA cosmic dust
SA electromagnetic radiation
SA electrons
SA fines
SA fission
SA fragments
SA grains
SA granulometry
SA intensity
SA interplanetary dust
SA interplanetary space
SA ions
SA meteorology
SA Moon
SA neutrons
SA particle tracks
SA precipitation
SA rounding
SA roundness
SA scintillations
SA soils
SA solar wind
SA sorting
SA sphericity
SA synchrotrons

particulate materials (1981)
SA clastic sediments
SA coarse-grained materials
SA fines
SA granular materials
SA materials
SA unconsolidated materials

parting lineation (1978)
Before 1978, also search current lineations.
UF current lineations
UF current partings
BT bedding plane irregularities
BT sedimentary structures
SA lineation

partition coefficients (1978)
Before 1993, also search distribution coefficients.
UF coefficients, partition
UF distribution coefficients
SA chemical ratios
SA crystal chemistry
SA partitioning

partitioning (1978)
SA chemical fractionation
SA crystal chemistry
SA geochemistry
SA lithogeochemistry
SA partition coefficients
SA phase equilibria

Partridge Formation (1989)
BT Middle Ordovician
BT Ordovician
BT Paleozoic
SA Massachusetts

SA New Hampshire
SA Vermont

partridgeite
use bixbyite

Pas-de-Calais
No longer a valid term for GeoRef. See Pas-de-Calais France.

Pas-de-Calais France (1993)
Department in extreme N France.
CO N500000N511500
  E0031000E0014500
BT France
BT Western Europe
BT Europe
NT Boulogne France
NT Boulonnais
SA Nord-Pas-de-Calais Basin

Pasadena
Valid through 1988. Search in combination with state term. After 1988, use specific city-state term.

Pasadena California (1989)
City in S California. Before 1989, search Pasadena AND California.
CO N341000N341000
  W1180900W1180900
BT Los Angeles County California
BT California
BT United States

Pasamonte Meteorite (1981)
Impact near village of Pasamonte in Union County, New Mexico. Before 1981, also search Pasamonte AND meteorites.
BT eucrite
BT achondrites
BT stony meteorites
BT meteorites
SA New Mexico
SA Union County New Mexico

Pasayten Group (1978)
Stratigraphically above Dewdney Creek Formation. S British Columbia and Central N Washington.
BT Lower Cretaceous
BT Cretaceous
BT Mesozoic
SA British Columbia
SA Washington

Pascagoula River basin (1978)
SE Mississippi.
BT Mississippi
BT United States

Pascal (1989)
Computer language. Also PASCAL, a French geological data base.
IN Index computer languages or data bases as applicable.
SA computer languages
SA computer programs
SA data bases
SA data processing
SA information systems

Pasco Basin (1978)
S central Washington.
IN Index counties as applicable.
SA Washington

Pasco County
Valid through 1988. Search in combination with state term. After 1988, use specific county-state term.

Pasco County Florida (1989)
On Gulf of Mexico in W central Florida. Before 1989, also search Pasco County AND Florida.
CO N281000N283000
  W0820500W0824500

BT Florida
BT United States

Paskapoo Formation (1978)
IN Index ages as applicable.
SA Alberta
SA Canada
SA Cretaceous
SA Eocene
SA K-T boundary
SA Paleocene
SA Saskatchewan
SA Upper Cretaceous

Passa Dois Group (1978)
BT Permian
BT Paleozoic
SA Brazil
SA Irati Formation

Passaic County
Valid through 1988. Search in combination with state term. After 1988, use specific county-state term.

Passaic County New Jersey (1989)
N New Jersey. Before 1989, also search Passaic County AND New Jersey.
CO N404800N411300
  W0740800W0743100
BT New Jersey
BT United States

Passaic Formation (1993)
Lower Jurassic and Upper Triassic. SE Pennsylvania, New Jersey, and New York. In Newark Supergroup. Conformably overlies Lockatong Formation; underlies Orange Mountain Basalt. Formerly lower part of Brunswick Formation.
IN Index ages as applicable.
BT Mesozoic
SA Brunswick Formation
SA Jurassic
SA Lower Jurassic
SA New Jersey
SA New York
SA Newark Basin
SA Newark Supergroup
SA Pennsylvania
SA Triassic
SA Upper Triassic

Passaic River (1989)
NE New Jersey.
BT New Jersey
BT United States
SA Passaic River basin

Passaic River basin (1989)
River basin in NE New Jersey.
BT New Jersey
BT United States
SA Passaic River

Passau
No longer a valid term for GeoRef. See Passau Germany.

Passau Germany (1993)
City on the Danube River at the Austrian border in E Bavaria, SE Germany.
BT Bavaria Germany
BT Germany
BT Central Europe
BT Europe

passband filters (1981)
SA filters

PASSCAL (1993)
Acronym.

UF Program for Array Seismic Studies of the Continental Lithosphere
SA arrays
SA IRIS network
SA lithosphere
SA programs
SA seismic methods
SA seismic networks

passive margins (1981)
UF Atlantic-type margins
SA accreting plate boundary
SA aseismic margins
SA continental margin
SA plate boundaries
SA plate divergence
SA plate tectonics
SA rift zones
SA rifting

passive methods (1985)
SA geophysical methods
SA geophysical surveys
SA methods
SA remote sensing
SA seismic methods
SA seismic surveys

Patagonia (1978)
Region. A barren tableland S of the Pampas, between the Atlantic Ocean and the Andes.
IN Index provinces as applicable.
CO S550000S370000
  W0621000W0733000
BT South America
NT Patagonian Andes
SA Argentina
SA Chile
SA Chubut Argentina
SA Magallanes Chile
SA Patagonian Batholith
SA Rio Negro Argentina
SA Santa Cruz Argentina

Patagonia Cordillera
use Patagonian Andes

Patagonian Andes (1978)
Mountain range extending the length of Patagonia. This term has multiple hierarchies.
IN Index countries as applicable.
UF Patagonia Cordillera
BT1 Patagonia
BT1 South America
BT2 Andes
BT2 South America
SA Argentina
SA Chile
SA Patagonian Batholith

Patagonian Batholith (1989)
SA Chile
SA Patagonia
SA Patagonian Andes

patch reefs (1978)
BT reefs
SA platform reefs

paternoite
use kaliborite

pathfinder elements
use pathfinders

pathfinders (1989)
Used in geochemical exploration.
UF pathfinder elements
SA chemical elements
SA exploration
SA geochemical methods
SA mineral exploration

Patom Plateau (1978)

Between Lena River and Vitim in NE Irkutsk Oblast N of Lake Baikal. Also search Patom. This term has multiple hierarchies.
CO N540000N590000
E1180000E1060000
BT1 Russian Federation
BT1 Commonwealth of Independent States
BT2 Asia
SA Irkutsk Russian Federation

pattern recognition (1993)
SA analysis
SA data processing
SA identification
SA imagery
SA mathematical geology
SA patterns
SA statistical analysis

patterned ground (1978)
UF ground, patterned
SA congelifraction
SA cryoturbation
SA frost action
SA frost heaving
SA geomorphology
SA glacial geology
SA periglacial features
SA polygons

patterns (1978)
SA dispersion patterns
SA drainage patterns
SA dynamos
SA interference patterns
SA pattern recognition
SA regional patterns

Patuxent River (1978)
Rises in central Maryland and flows S and SE into Chesapeake Bay.
BT Maryland
BT United States

Pau
No longer a valid term for GeoRef. See Pau France.

Pau France (1993)
City in extreme SW part of country.
BT Pyrenees-Atlantiques France
BT France
BT Western Europe
BT Europe

Paucituberculata (1981)
BT Mammalia
BT Tetrapoda
BT Vertebrata
BT Chordata

Pauni
No longer a valid term for GeoRef. See Pauni India.

Pauni India (1993)
Town in central Madhya Pradesh.
BT Madhya Pradesh India
BT India
BT Indian Peninsula
BT Asia

pavements
This term used in the engineering sense is usually out-of-scope for GeoRef. As of 1993, see construction materials, highways, road materials, or desert pavement if applicable.

Pavia
No longer a valid term for GeoRef. See Pavia Italy.

Pavia Italy (1993)

City and province S of Milan. Before 1993, also search Pavia and Italy. Before 1978, also search Oltrepo Pavese.
UF Oltrepo Pavese Italy
BT Lombardy Italy
BT Italy
BT Southern Europe
BT Europe

Pavlodar
No longer a valid term for GeoRef. As of 1993, see Pavlodar Kazakhstan.

Pavlodar Kazakhstan (1993)
Oblast and town in NE Kazakhstan. This term has multiple hierarchies.
BT1 Kazakhstan
BT1 Central Asia
BT1 Asia
BT2 Kazakhstan
BT2 Commonwealth of Independent States
NT Boshchekul Kazakhstan
NT Ekibastuz Kazakhstan

Pavlof (1989)
Volcano. Alaska Peninsula, SW Alaska.
IN Index Southwestern Alaska as applicable.
UF Pavlof Volcano
SA Alaska Peninsula
SA Southwestern Alaska

Pavlof Volcano
use Pavlof

Pavlograd
No longer a valid term for GeoRef. As of 1993, see Pavlograd Ukraine.

Pavlograd Ukraine (1993)
City in Dnepropetrovsk Oblast in E central Ukraine. This term has multiple hierarchies.
BT1 Dnepropetrovsk Ukraine
BT1 Ukraine
BT1 Europe
BT2 Dnepropetrovsk Ukraine
BT2 Ukraine
BT2 Commonwealth of Independent States

Pawhuska Rock Plain (1978)
Osage County in N Oklahoma.
BT Osage County Oklahoma
BT Oklahoma
BT United States

Pawnee County
Valid through 1988. Search in combination with state term. After 1988, use specific county-state term.

Pawnee County Kansas (1989)
SW central Kansas. Before 1989, also search Pawnee County AND Kansas.
CO N380000N382000
W0985500W0993600
BT Kansas
BT United States

Pawnee County Nebraska (1989)
SE Nebraska. Before 1989, also search Pawnee County AND Nebraska.
CO N400000N401500
W0960100W0962800
BT Nebraska
BT United States

Pawnee County Oklahoma (1989)

N Oklahoma. Before 1989, also search Pawnee County AND Oklahoma.
CO N360900N363300
W0961600W0970300
BT Oklahoma
BT United States

Pay-Khoy
use Pai-Khoi

Payne County
Valid through 1988. Search in combination with state term. After 1988, use specific county-state term.

Payne County Oklahoma (1989)
N central Oklahoma. Before 1989, also search Payne County AND Oklahoma.
CO N355500N361500
W0963700W0972200
BT Oklahoma
BT United States
NT Stillwater Oklahoma

Payson
Valid through 1988. Search in combination with state term. After 1988, use specific city-state term.

Payson Utah (1989)
City in central Utah. Before 1989, search Payson AND Utah.
CO N400400N400400
W1114500W1114500
BT Utah County Utah
BT Utah
BT United States

Payun Matru (1978)
Volcano on Payun Plateau in SW Mendoza.
UF Payun Matru Volcano
UF Payun-Matru
UF Payun-Matru Volcano
BT Mendoza Argentina
RT Argentina
BT South America

Payun Matru Volcano
use Payun Matru

Payun-Matru
use Payun Matru

Payun-Matru Volcano
use Payun Matru

Pb
use lead

Pb-204 (1978)
Autoposting of broader terms began in 1989. This term has multiple hierarchies.
BT1 radioactive isotopes
BT1 isotopes
BT2 lead
BT2 metals
SA Pb-206/Pb-204
SA Pb-207/Pb-204
SA Pb-208/Pb-204
SA U-238/Pb-204

Pb-204/Pb-206
use Pb-206/Pb-204

Pb-204/Pb-208
use Pb-208/Pb-204

Pb-206 (1978)
Autoposting of broader terms began in 1989. This term has multiple hierarchies.
BT1 stable isotopes
BT1 isotopes
BT2 lead
BT2 metals
SA Pb-206/Pb-204
SA Pb-207/Pb-206

SA Pb-208/Pb-206

Pb-206 Pb-207
use Pb-207/Pb-206

Pb-206/Pb-204 (1978)
Autoposting of broader terms began in 1989. This term has multiple hierarchies.
UF Pb-204/Pb-206
BT1 radioactive isotopes
BT2 stable isotopes
BT2 isotopes
BT3 lead
BT3 metals
SA isotope ratios
SA Pb-204
SA Pb-206

Pb-206/Pb-207
use Pb-207/Pb-206

Pb-207 (1978)
Autoposting of broader terms began in 1989. This term has multiple hierarchies.
BT1 stable isotopes
BT1 isotopes
BT2 lead
BT2 metals
SA Pb-207/Pb-204
SA Pb-207/Pb-206

Pb-207/Pb-204 (1978)
Autoposting of broader terms began in 1989. This term has multiple hierarchies.
BT1 radioactive isotopes
BT1 isotopes
BT2 stable isotopes
BT2 isotopes
BT3 lead
BT3 metals
SA isotope ratios
SA Pb-204
SA Pb-207

Pb-207/Pb-206 (1978)
Autoposting of broader terms began in 1989. This term has multiple hierarchies.
UF Pb-206 Pb-207
UF Pb-206/Pb-207
BT1 stable isotopes
BT1 isotopes
BT2 lead
BT2 metals
SA isotope ratios
SA Pb-206
SA Pb-207

Pb-208 (1978)
Autoposting of broader terms began in 1989. This term has multiple hierarchies.
BT1 stable isotopes
BT1 isotopes
BT2 lead
BT2 metals
SA Pb-208/Pb-204
SA Pb-208/Pb-206

Pb-208/Pb-204 (1978)
Autoposting of broader terms began in 1989. This term has multiple hierarchies.
UF Pb-204/Pb-208
BT1 radioactive isotopes
BT1 isotopes
BT2 stable isotopes
BT2 isotopes
BT3 lead
BT3 metals
SA isotope ratios
SA Pb-204
SA Pb-208

Pb-208/Pb-206 (1989)

Autoposting of broader terms began in 1989. This term has multiple hierarchies.
BT1 stable isotopes
BT1 isotopes
BT2 lead
BT2 metals
SA isotope ratios
SA Pb-206
SA Pb-208

Pb-210 (1978)
Isotope used in age determination. Autoposting of broader terms began in 1989. This term has multiple hierarchies.
BT1 radioactive isotopes
BT1 isotopes
BT2 lead
BT2 metals
SA absolute age
SA Pb/Pb
SA Pb/Th
SA U/Pb

Pb/Pb (1978)
Isotopic ratio used in age determination. Before 1971, also search lead-lead.
UF lead-lead
SA absolute age
SA isotope ratios
SA lead
SA Pb-210
SA Pb/Th
SA U/Pb

Pb/Th (1978)
Isotopic ratio used in age determination. Also search Pb AND Th.
UF thorium-lead
SA absolute age
SA isotope ratios
SA lead
SA Pb-210
SA Pb/Pb
SA thorium
SA U/Pb

PCB
  use PCBs

PCB's
  use PCBs

PCBs (1985)
Autoposting of chlorinated hydrocarbons to this term began in 1989.
UF PCB
UF PCB's
UF polychlorinated biphenyls
UF polychlorobiphenyls
BT chlorinated hydrocarbons
SA industrial waste
SA organic materials
SA pollutants
SA pollution
SA toxic materials

PcP waves
  use PcP-waves

PcP-waves (1981)
UF PcP waves
BT body waves
BT elastic waves
SA P-waves
SA seismology
SA waves

PCs
  As of 1993, use microcomputers if applicable. Also see minicomputers.

Pd
  use palladium

Peace River (1978)

Flows into Snake River just N of Lake Athabasca.
IN Index provinces as applicable.
BT Western Canada
BT Canada
SA Alberta
SA British Columbia

Peace River Arch (1989)
W Alberta.
BT Alberta
BT Western Canada
BT Canada

Peace River Formation (1993)
Albian. Alberta and British Columbia. Also Neogene in Florida; in Hawthorn Formation? For Western Canada, also search Peace River oil sands.
IN Index ages and locations as applicable.
SA Alberta
SA Albian
SA British Columbia
SA Cretaceous
SA Florida
SA Hawthorn Formation
SA Neogene

Peace Valley Beds (1985)
S California.
BT middle Pliocene
BT Pliocene
BT Neogene
BT Tertiary
BT Cenozoic
SA California

Peach Springs Tuff (1989)
BT Miocene
BT Neogene
BT Tertiary
BT Cenozoic
SA Arizona
SA California

Peak District (1978)
Plateau region in N central England at S end of Pennine Chain.
IN Index Derbyshire England as applicable.
SA Derbyshire England
SA Pennines

pearceite (1978)
BT sulfarsenites
BT sulfosalts

Pearl Harbor (1978)
Inlet on S coast of Oahu 6 miles W of Honolulu. Autoposting of broader terms to this term began in 1989. This term has multiple hierarchies.
CO N212200N212200
      W1580000W1580000
BT1 Oahu
BT1 Honolulu County Hawaii
BT1 Hawaii
BT1 United States
BT2 Oahu
BT2 Honolulu County Hawaii
BT2 Hawaii
BT2 Polynesia
BT2 Oceania
BT3 Oahu
BT3 Honolulu County Hawaii
BT3 Hawaii
BT3 East Pacific Ocean Islands

Pearl River (1989)
Central and S Mississippi and SE Louisiana. For Pearl River in China, use Zhujiang River.
IN Index states or regions as applicable.
SA Louisiana

SA Mississippi
SA United States
SA Zhujiang River

Pearl River, China
  use Zhujiang River

Pearlette Volcanic Ash (1978)
SW Kansas.
BT Pleistocene
BT Quaternary
BT Cenozoic

pearlite
  use perlite

pearls (1978)
SA aragonite
SA gems
SA Mollusca

Pearsall Formation (1989)
In Trinity Group. S Texas and Louisiana. Includes Cow Creek Member. This term has multiple hierarchies.
BT1 Comanchean
BT1 Cretaceous
BT1 Mesozoic
BT2 Lower Cretaceous
BT2 Cretaceous
BT2 Mesozoic
SA Louisiana
SA Texas
SA Trinity Group

Peary Land (1978)
Region in extreme N Greenland on Arctic Ocean.
BT Greenland
BT Arctic region

peat (1978)
This term has multiple hierarchies.
IN See List C for use as a commodity term.
BT1 organic residues
BT2 sediments
SA biogenic structures
SA Bog soils
SA coal
SA coal exploration
SA Histosols
SA humus
SA lignite
SA organic materials
SA palsas
SA peat bogs

peat bogs (1978)
BT bogs
SA ecology
SA muskegs
SA paleoecology
SA peat
SA Sphagnum
SA swamps

Pebble Shale (1989)
BT Lower Cretaceous
BT Cretaceous
BT Mesozoic
SA Alaska

pebbles (1978)
BT clastic sediments
BT sediments
SA amygdules
SA cobbles
SA gravel
SA rounding
SA shingle
SA terrigenous materials
SA ventifacts

Pechenga (1978)
Territory, formerly in N Finland, which was called Petsamo. Now it is part of Murmansk Oblast. This term has multiple hierarchies.

UF Petsamo
BT1 Murmansk Russian Federation
BT1 Russian Federation
BT1 Commonwealth of Independent States
BT2 Murmansk Russian Federation
BT2 Europe

Pechora
  No longer a valid term for GeoRef. As of 1993, see Pechora Russian Federation.

Pechora Basin (1978)
Coal basin mainly in basin of Usa River in NE Komi Autonomous Republic in NE European Russian Federation. Before 1978, also search Pechora. This term has multiple hierarchies.
CO N650000N680000
      E0600000E0500000
BT1 Russian Federation
BT1 Commonwealth of Independent States
BT2 Europe
SA coal fields
SA Komi Russian Federation
SA Moscow-Pechora Syneclise

Pechora River (1978)
Rises in the Central Urals of N Perm Oblast and flows N through the Komi Autonomous Republic, and the Nenets National Okrug into Pechora Bay. Also search Pechora. This term has multiple hierarchies.
CO N620000N670000
      E0593000E0520000
BT1 Russian Federation
BT1 Commonwealth of Independent States
BT2 Europe
SA Moscow-Pechora Syneclise

Pechora Russian Federation
  (1993)
City in Komi Autonomous Soviet Socialist Republic on left bank of Pechora River in NE European Russian Federation. This term has multiple hierarchies.
BT1 Komi Russian Federation
BT1 Russian Federation
BT1 Commonwealth of Independent States
BT2 Komi Russian Federation
BT2 Europe

Pecopteris (1978)
Genus.
BT Filicopsida
BT pteridophytes
BT Plantae
SA Glossopteris
SA gymnosperms
SA Pteridospermae

Pecos County
  Valid through 1988. Search in combination with state term. After 1988, use specific county-state term.

Pecos County Texas (1989)
Extreme W Texas. Before 1989, also search Pecos County AND Texas.
CO N300400N312500
      W1014500W1034000
BT Texas
BT United States
SA Central Basin Platform
SA Glass Mountains

Pecos River (1985)

E New Mexico and SW Texas.
Joins Rio Grande at Armistad
Reservoir.
IN Index states as applicable.
BT United States
SA New Mexico
SA Texas

Pecos River valley (1978)
Also search Pecos River.
IN Index states as applicable.
BT United States
SA New Mexico
SA Texas

Pectinacea (1978)
Superfamily. Autoposting of this term to Daonella, Halobia, Monotis and Pectinidae began in 1989. Autoposting of Pteriina to this term began in 1989.
BT Pteriina
BT Bivalvia
BT Mollusca
BT Invertebrata
NT Daonella
NT Halobia
NT Monotis
NT Pectinidae

Pectinidae (1978)
Family. Autoposting of Pectinacea and Pteriina to this term began in 1989.
BT Pectinacea
BT Pteriina
BT Bivalvia
BT Mollusca
BT Invertebrata

pectolite (1978)
BT chain silicates
BT silicates

pediments (1978)
Autoposting of erosion features to this term began in 1989.
BT erosion features
SA erosion surfaces
SA geomorphology

pediplains (1978)
Autoposting of erosion surfaces and erosion features to this term began in 1989.
UF desert pediplains
UF desert plains
UF pediplanes
BT erosion surfaces
BT erosion features
SA geomorphology
SA peneplains
SA plains

pediplanes
  use pediplains

pedogenesis (1982)
Before 1982, search soils AND genesis.
SA climate
SA factors
SA genesis
SA horizon differentiation
SA leaching
SA parent materials
SA soils
SA time factor

pedons (1985)
SA humus
SA soil profiles
SA soil sampling
SA soils

Pedregosa Basin (1978)
IN Index Mexican and U.S. states as applicable.
BT North America
SA Arizona

SA Chihuahua Mexico
SA New Mexico
SA Sonora Mexico

Peedee Formation (1978)
Variable gray to green argillaceous sands and impure limestones. E North Carolina and E South Carolina.
BT Upper Cretaceous
BT Cretaceous
BT Mesozoic
SA North Carolina
SA South Carolina

Peekskill
Valid through 1988. Search in combination with state term. After 1988, use specific city-state term.

Peekskill New York (1989)
City in SE New York. Before 1989, search Peekskill AND New York.
CO N411800N411800
    W0735600W0735600
BT Westchester County New York
BT New York
BT United States

Peel River (1978)
Flows into Mackenzie River.
IN Index territories as applicable.
BT Canada
SA Northwest Territories
SA Yukon Territory

Peel Sound Formation (1978)
Franklin District.
BT Paleozoic
SA Canada
SA Devonian
SA Franklin District Northwest Territories
SA Northwest Territories
SA Silurian

pegmatite (1978)
IN See List C for use as a commodity term.
UF pegmatites
BT granites
BT plutonic rocks
BT igneous rocks
SA graphic granite
SA micropegmatite
SA Plotina Deposit

pegmatites
  use pegmatite

Peichang
  use Peikang Taiwan

Peikang
No longer a valid term for GeoRef. As of 1993 see Peikang Taiwan.

Peikang Taiwan (1993)
Town in W central Taiwan.
UF Peichang
BT Taiwan
BT Far East
BT Asia

Peking
Not a valid term for GeoRef. See Beijing China.

Pelagian Sea (1985)
Mediterranean Sea, W of Malta, E of Tunisia and S of Sicily.
CO N303000N364000
    E0143000E0100000
BT East Mediterranean
BT Mediterranean Sea
SA Gulf of Gabes

pelagic environment (1978)
Before 1978, search pelagic.

SA deep-water environment
SA depositional environment
SA environment
SA hemipelagic environment
SA oceanography
SA pelagic sedimentation
SA sedimentation

pelagic sedimentation (1981)
BT sedimentation
SA pelagic environment

Pelecypoda
A valid index term through 1975.
  use Bivalvia

pelecypods
  use bivalves

Pelew
  use Belau

pelite
  use shale

pelitic gneiss
A valid metamorphic rock term through 1977. Use gneisses and pelitic texture.

pelitic rocks
  use pelitic texture

pelitic schist
As of 1982, no longer a valid term for GeoRef.
  use slates

pelitic texture (1978)
Before 1978, search pelitic AND texture.
UF pelitic rocks
BT textures
SA metamorphic rocks
SA sedimentary rocks

pelletization (1981)
SA production

pellets (1978)
SA oolite
SA pisoliths

Pelmatozoa
As of 1989, no longer a valid term for GeoRef. Equivalent to Crinozoa.
  use Crinozoa

Pelomedusidae (1978)
Family.
BT Chelonia
BT Anapsida
BT Reptilia
BT Tetrapoda
BT Vertebrata
BT Chordata

Pelona Schist (1978)
Precambrian(?). S California.
BT Precambrian
SA California

Peloponnese
  use Peloponnesus Greece

Peloponnesos
  use Peloponnesus Greece

Peloponnesus
No longer a valid term for GeoRef. As of 1993 see Peloponnesus Greece.

Peloponnesus Greece (1993)
Administrative region and peninsula forming S Greece. Before 1993 also search Peloponnesus AND Greece.
CO N361500N382000
    E0233000E0210000
UF Peloponnese
UF Peloponnesos
BT Greece

BT Southern Europe
BT Europe
NT Argolis Greece
NT Corinth Greece
NT Gavrovo Zone
NT Megalopolis Basin
NT Tripolis Greece
SA Gulf of Corinth
SA Hellenic Arc
SA Hellenides
SA Kalamata earthquake 1986

Peloritani Mountains (1978)
NE Sicily.
BT Sicily Italy
BT Italy
BT Southern Europe
BT Europe

Pelotas Basin (1978)
River basin.
IN Index states as applicable.
SA Rio Grande do Sul Brazil
SA Santa Catarina Brazil

Pelvoux Massif (1978)
Mountain group which contains Barre das Ecrins, the highest peak in the Dauphine Alps. Also search Pelvoux. This term has multiple hierarchies.
IN Index departments as applicable.
UF Ecrins-Pelvoux Massif
BT1 Dauphine Alps
BT1 Western Alps
BT1 Alps
BT1 Europe
BT2 Dauphine Alps
BT2 France
BT2 Western Europe
BT2 Europe
SA Hautes-Alpes France
SA Isere France

Pelycosauria (1981)
Order.
BT Synapsida
BT Reptilia
BT Tetrapoda
BT Vertebrata
BT Chordata

Pembroke
No longer a valid term for GeoRef as of 1993. See Pembroke Wales if applicable. Before 1978, also used as the Welsh County. Also refers to places in Ontario, Nova Scotia, and Massachusetts.

Pembroke County
Not a valid term for GeoRef as of 1993. See Pembrokeshire Wales.

Pembroke Wales (1993)
Town in Dyfed in SW Wales. Before 1978, Pembroke was also used as the Welsh County. Before 1993, also search Pembroke AND Wales.
BT Pembrokeshire Wales
BT Dyfed Wales
BT Wales
BT Great Britain
BT United Kingdom
BT Western Europe
BT Europe

Pembrokeshire
No longer a valid term for GeoRef as of 1993.
  use Pembrokeshire Wales

Pembrokeshire Wales (1993)
Former county in SW Wales. Part of Dyfed. Before 1993, also search Pembrokeshire or Pembroke County AND Wales. Before

1978, also search Pembroke and Wales.
CO N513500N520500
  W0043000W0051500
UF Pembrokeshire
BT Dyfed Wales
BT Wales
BT Great Britain
BT United Kingdom
BT Western Europe
BT Europe
NT Pembroke Wales
NT Skomer Island

Pench Valley (1978)
River valley in central Madhya Pradesh. Also search Pench.
BT Madhya Pradesh India
BT India
BT Indian Peninsula
BT Asia

Pend Oreille County
Valid through 1988. Search in combination with state term. After 1988, use specific county-state term.

Pend Oreille County Washington (1989)
Extreme NE Washington. Before 1989, also search Pend Oreille County AND Washington.
CO N480300N490000
  W1170400W1173800
BT Washington
BT United States
NT Metaline Falls Washington

Pendleton County
Valid through 1988. Search in combination with state term. After 1988, use specific county-state term.

Pendleton County Kentucky (1989)
N Kentucky. Before 1989, also search Pendleton County AND Kentucky.
CO N383500N385200
  W0841500W0843300
BT Kentucky
BT United States

Pendleton County West Virginia (1989)
E West Virginia. Before 1989, also search Pendleton County AND West Virginia.
CO N382400N385700
  W0790300W0793800
BT West Virginia
BT United States

pendulum gravimeters (1981)
BT gravimeters

peneplains (1978)
Autoposting of erosion features to this term began in 1989. Before 1971, also search peneplanes.
UF base-level peneplain
UF peneplanes
BT erosion features
SA erosion cycle
SA erosion surfaces
SA geomorphology
SA pediplains
SA plains

peneplanes
Not a valid GeoRef term after 1970. Was used on level 1 in subfile N.
use peneplains

penetration (1978)
Not to be used for crystal growth.
SA engineering geology
SA rock mechanics
SA soil mechanics

penetration tests (1981)
NT cone penetration tests
SA California bearing ratio
SA engineering geology
SA penetrometers
SA soil mechanics
SA testing

penetrometers (1985)
SA instruments
SA penetration tests
SA soil mechanics

Penghu Islands (1978)
Group of about 48 islands in Formosa Strait between Taiwan and the mainland of China.
UF Hoko Gunto
UF Hoko Shoto
UF Pescadores
BT Taiwan
BT Far East
BT Asia

Peninsular Gneiss (1989)
S India.
BT Archean
BT Precambrian
SA Closepet Granite
SA India

Peninsular Ranges (1978)
Part of the Coast Ranges S of Los Angeles.
IN Index Coast Ranges and county or regions as applicable.
SA California
SA Coast Ranges
SA Peninsular Ranges Batholith

Peninsular Ranges Batholith (1989)
S California and N Baja California, Mexico.
IN Index North America or states or regions as applicable.
SA Baja California
SA Baja California Mexico
SA California
SA Coast Ranges
SA Mexico
SA North America
SA Peninsular Ranges

Peninsular Shield
use Indian Shield

Pennine Alps (1978)
SW division of Central Alps extending from Great Saint Bernard Pass to Simplon Pass. Also search Pennine.
IN Index countries as applicable.
CO N455000N461000
  E0081500E0070000
BT Central Alps
BT Alps
BT Europe
SA Argentiere Glacier
SA Monte Rosa
SA Piedmont Alps
SA Piemonte Italy
SA Valais Switzerland

Pennine Chain
use Pennines

Pennine Hills
use Pennines

Pennine Range
use Pennines

Pennines (1978)
Long hill range extending from the Cheviot Hills in the N of England to the S Midlands.
CO N520800N554000
  W0004000W0034000
UF Pennine Chain
UF Pennine Hills
UF Pennine Range
BT England
BT Great Britain
BT United Kingdom
BT Western Europe
BT Europe
SA Midlands
SA Peak District

Pennington County
Valid through 1988. Search in combination with state term. After 1988, use specific county-state term.

Pennington County Minnesota (1989)
NW Minnesota. Before 1989, also search Pennington County AND Minnesota.
CO N475400N481200
  W0953300W0961500
BT Minnesota
BT United States

Pennington County South Dakota (1989)
SW South Dakota. Before 1989, also search Pennington County AND South Dakota.
CO N434200N443100
  W1020000W1040400
BT South Dakota
BT United States
NT Rapid City South Dakota

Pennington Formation (1978)
Comprises Stony Gap Sandstone, Avis Limestone, Falls Mills Sandstone members. N Alabama, NW Georgia, E Kentucky, E Tennessee, and SW Virginia.
BT Upper Mississippian
BT Mississippian
BT Carboniferous
BT Paleozoic
SA Alabama
SA Georgia
SA Kentucky
SA Tennessee
SA Virginia

Penninic Zone (1985)
S Austria.
BT Austria
BT Central Europe
BT Europe
SA Eastern Alps
SA Hohe Tauern

penninite (1978)
BT sheet silicates
BT silicates

Pennsylvania (1978)
Autoposting of this term began in 1978.
CO N394300N421800
  W0744000W0803200
BT United States
NT Adams County Pennsylvania
NT Allegheny County Pennsylvania
NT Armstrong County Pennsylvania
NT Beaver County Pennsylvania
NT Bedford County Pennsylvania
NT Berks County Pennsylvania
NT Blair County Pennsylvania
NT Bradford County Pennsylvania
NT Bucks County Pennsylvania
NT Butler County Pennsylvania
NT Cambria County Pennsylvania
NT Carbon County Pennsylvania
NT Centre County Pennsylvania
NT Chester County Pennsylvania
NT Clearfield County Pennsylvania
NT Clinton County Pennsylvania
NT Columbia County Pennsylvania
NT Crawford County Pennsylvania
NT Cumberland County Pennsylvania
NT Dauphin County Pennsylvania
NT Delaware County Pennsylvania
NT Elk County Pennsylvania
NT Erie County Pennsylvania
NT Fayette County Pennsylvania
NT Forest County Pennsylvania
NT Franklin County Pennsylvania
NT Fulton County Pennsylvania
NT Greene County Pennsylvania
NT Indiana County Pennsylvania
NT Jefferson County Pennsylvania
NT Lancaster County Pennsylvania
NT Lawrence County Pennsylvania
NT Lebanon County Pennsylvania
NT Lehigh County Pennsylvania
NT Luzerne County Pennsylvania
NT Mercer County Pennsylvania
NT Monroe County Pennsylvania
NT Montgomery County Pennsylvania
NT Nittany Valley
NT Northampton County Pennsylvania
NT Northumberland County Pennsylvania
NT Perry County Pennsylvania
NT Philadelphia County Pennsylvania
NT Pike County Pennsylvania
NT Pittsburgh coal basin
NT Potter County Pennsylvania
NT Schuylkill County Pennsylvania
NT Somerset County Pennsylvania
NT Sullivan County Pennsylvania
NT Tioga County Pennsylvania
NT Union County Pennsylvania
NT Warren County Pennsylvania
NT Washington County Pennsylvania
NT Wayne County Pennsylvania
NT Westmoreland County Pennsylvania
NT Wyoming County Pennsylvania
NT York County Pennsylvania
SA Allegheny Front
SA Allegheny Group
SA Allegheny Mountains
SA Allegheny Plateau
SA Ames Limestone
SA Antietam Formation
SA Appalachian Basin
SA Appalachian Plateau
SA Baltimore Gneiss
SA Bass Islands Dolomite
SA Bedford Shale
SA Beekmantown Group
SA Berea Sandstone
SA Black River Group
SA Bloomsburg Formation
SA Blue Mountain
SA Blue Ridge Province
SA Brallier Shale
SA Brunswick Formation
SA Catskill Delta
SA Catskill Formation
SA Chemung Formation
SA Chemung River
SA Clinton Group

SA Coeymans Formation
SA Cohansey Formation
SA Conemaugh
SA Conemaugh Group
SA Conococheague Formation
SA Cuyahoga Formation
SA Delaware River
SA Delaware River basin
SA Dunkard Basin
SA Dunkard Group
SA Freeport Formation
SA Genesee Group
SA Genesee River
SA Gettysburg Basin
SA Glenarm Series
SA Great Lakes region
SA Greenbrier Limestone
SA Grimsby Sandstone
SA Hamilton Group
SA Hampshire Formation
SA Helderberg Group
SA Juniata Formation
SA Keefer Sandstone
SA Keyser Limestone
SA Kinzers Formation
SA Kittanning Formation
SA Lake Erie
SA Lavery Till
SA Lockatong Formation
SA Lockport Formation
SA Mahantango Formation
SA Marcellus Shale
SA Martinsburg Formation
SA Mauch Chunk Formation
SA McKenzie Formation
SA Medina Formation
SA Millboro Shale
SA Monongahela Group
SA Moscow Formation
SA New River Formation
SA Newark Basin
SA Newark-Gettysburg Basin
SA Ohio River
SA Ohio River basin
SA Ohio River valley
SA Onondaga Limestone
SA Oriskany Sandstone
SA Passaic Formation
SA Pennsylvania State University
SA Perry Formation
SA Piedmont
SA Pittsburgh Coal
SA Pocono Formation
SA Pottsville Group
SA Queenston Shale
SA Reading Prong
SA Reedsville Formation
SA Rochester Formation
SA Rondout Formation
SA Rose Hill Formation
SA Sharon Conglomerate
SA Shawangunk Formation
SA South Mountain
SA Stockton Formation
SA Stones River Group
SA Susquehanna River
SA Susquehanna River basin
SA Tioga Bentonite
SA Tonoloway Limestone
SA Trenton Group
SA Tully Limestone
SA Tuscarora Formation
SA University of Pennsylvania
SA Valley and Ridge Province
SA Wilmington Complex
SA Wissahickon Formation

Pennsylvania State University (1978)
University Park, Centre County.
SA academic institutions
SA Pennsylvania

Pennsylvanian (1978)
Autoposting of Carboniferous to this term began in 1981. From 1978-1980, Paleozoic was autoposted to this term.
BT Carboniferous
BT Paleozoic
NT Bevier Coal
NT Conemaugh Group
NT Cumberland Group
NT Excello Shale
NT Francis Creek Shale
NT Freeport Formation
NT Herrin Coal Member
NT Kittanning Formation
NT La Salle Limestone
NT Lower Pennsylvanian
NT Madera Formation
NT Mansfield Formation
NT Marble Falls Group
NT Mattoon Formation
NT Middle Pennsylvanian
NT Minturn Formation
NT Monongahela Group
NT Morien Group
NT Morrow Formation
NT Pittsburgh Coal
NT Pottsville Group
NT Red Fork Sandstone
NT Saginaw Formation
NT Sharon Conglomerate
NT Smithwick Shale
NT Springfield Coal Member
NT Strawn Series
NT Tennessee Sandstone
NT Upper Pennsylvanian
NT Wapanucka Limestone
NT Watahomigi Formation
NT Westerly Granite
SA Alleghany Orogeny
SA Amsden Formation
SA Bartlesville Sand
SA Bird Spring Formation
SA Cache Creek Group
SA Casper Formation
SA Dunkard Group
SA Ely Limestone
SA Fountain Formation
SA Honaker Trail Formation
SA Johns Valley Formation
SA Magdalena Group
SA Maroon Formation
SA Minnelusa Formation
SA Mississippian
SA Narragansett Pier Granite
SA Oquirrh Formation
SA Ouachita Orogeny
SA Pictou Group
SA Sandia Formation
SA Sicker Group
SA Springer Formation
SA Supai Formation
SA Talladega Group
SA Tensleep Sandstone
SA Tyler Formation
SA Upper Carboniferous
SA Weber Sandstone
SA Wells Formation
SA Wood River Formation

Penobscot Bay (1978)
Large inlet of Atlantic Ocean in S Maine.
BT Maine
BT United States

Penobscot County
Valid through 1988. Search in combination with state term. After 1988, use specific county-state term.

Penobscot County Maine (1989)
S and central Maine. Before 1989, also search Penobscot County AND Maine.
CO N443500N462500
W0675500W0692500
BT Maine
BT United States

Penobscot Formation (1989)
S central Maine. Cambrian? Later was considered Ordovician. Above Battio Quartzite; below Rockland Formation or Bucksport Formation.
IN Index age as applicable.
BT lower Paleozoic
BT Paleozoic
SA Cambrian
SA Maine
SA Ordovician

Penokean Orogeny (1978)
A time of deformation and granite emplacement during the Precambrian. Before 1978, also search Penokean AND orogeny.
IN Index states as applicable.
BT Precambrian
SA Michigan
SA Minnesota
SA orogeny

Penrhyn Slate (1978)
BT Precambrian
SA United Kingdom
SA Wales

Pensacola
Valid through 1988. Search in combination with state term. After 1988, use specific city-state term.

Pensacola Florida (1989)
City on Pensacola Bay in extreme NW Florida. Before 1989, search Pensacola AND Florida.
CO N302600N302600
W0871200W0871200
BT Escambia County Florida
BT Florida
BT United States

Pensacola Mountains (1978)
In the Horlick Mountains area SE of Ross Ice Shelf on the Pacific Ocean side.
BT Antarctica
SA Dufek Intrusion

Pentameracea (1978)
Superfamily.
BT Pentamerida
BT Articulata
BT Brachiopoda
BT Invertebrata

Pentamerida (1978)
Autoposting of this term began in 1978.
BT Articulata
BT Brachiopoda
BT Invertebrata
NT Pentameracea

pentane (1989)
BT alkanes
BT aliphatic hydrocarbons
BT hydrocarbons
BT organic materials

pentlandite (1978)
UF nicopyrite
BT sulfides
SA nickel
SA nickel ores

Pentoxylales (1981)
BT gymnosperms
BT Spermatophyta
BT Plantae

Penzhina Bay (1978)
NE extension of the Shelikov Gulf of the Okhotsk Sea in NE Siberia. Also search Penzhina. This term has multiple hierarchies.
UF Penzhinskaya Bay
BT1 Russian Federation
BT1 Commonwealth of Independent States
BT2 Asia
SA Kamchatka Russian Federation
SA Magadan Russian Federation

Penzhinskaya Bay
use Penzhina Bay

People's Republic of the Congo
use Congo

Peoples Democratic Republic of Yemen
use Yemen

Peoria Loess (1978)
Underlies Bignell Formation.
UF Peorian Loess
BT Pleistocene
BT Quaternary
BT Cenozoic
SA Illinois
SA Iowa
SA Kansas
SA Nebraska

Peorian Loess
use Peoria Loess

peptides (1985)
BT organic materials
SA amino acids
SA proteins

Pequop Mountains (1978)
In Elko County in NE Nevada.
BT Elko County Nevada
BT Nevada
BT United States

per-aluminous composition
use peraluminous composition

Perak
No longer a valid term for GeoRef. As of 1993, see Perak Malaysia.

Perak Malaysia (1993)
State on W coast of Malay Peninsula in West Malaysia. This term has multiple hierarchies.
BT1 West Malaysia
BT1 Malay Peninsula
BT1 Far East
BT1 Asia
BT2 West Malaysia
BT2 Malaysia
BT2 Far East
BT2 Asia
NT Kinta Valley

peralkalic composition (1989)
UF peralkaline composition
SA composition
SA igneous rocks
SA peraluminous composition

peralkaline composition
use peralkalic composition

peraluminous composition (1989)
UF per-aluminous composition
SA composition
SA igneous rocks
SA peralkalic composition
SA S-type granites

Perch Lake (1985)
Lake near Chalk River in Renfrew County, SE Ontario. Also a lake in Michigan Upper Peninsula, N Michigan.

IN Index Ontario; or Michigan and Iron County Michigan as applicable.
SA Iron County Michigan
SA Michigan
SA Michigan Upper Peninsula
SA Ontario
SA Renfrew County Ontario

perched aquifers (1982)
BT aquifers
SA ground water

perchloroethylene
use tetrachloroethylene

percolation (1978)
SA ground water
SA hydrology
SA infiltration
SA leaching
SA seepage
SA water

Peredovoy Range (1978)
In former Kabardin-Balkar ASSR in Northern Caucasus. Before 1978, also search Peredovoy. This term has multiple hierarchies.
BT1 Northern Caucasus
BT1 Caucasus
BT1 Europe
BT2 Northern Caucasus
BT2 Russian Federation
BT2 Commonwealth of Independent States
SA Kabardin-Balkar Russian Federation

Peremyshl
use Przemysl Poland

Peri-Caspian Depression
use Caspian Basin

Peribaltic Syneclise (1978)
A depressed structure of the continental platform in the Baltic Sea area produced by slow crustal downwarp.
UF Baltic Syneclise
BT Europe
SA Baltic region

periclase (1978)
BT oxides

Peridinium (1978)
Autoposting of microfossils to this term began in 1990.
BT Dinoflagellata
BT palynomorphs
BT microfossils

peridot (1978)
UF peridote
UF peridote (olivine)
BT olivine group
BT nesosilicates
BT orthosilicates
BT silicates
SA gems
SA tourmaline

peridote
use peridot

peridote (olivine)
use peridot

peridotite
As of 1989, no longer a valid term for GeoRef. From 1981 to 1988, peridotites and ultramafics were autoposted to this term.
use peridotites

peridotites (1981)
Before 1989, also search peridotite. Autoposting of this term to peridotite ended in 1989.
UF peridotite

BT ultramafics
BT plutonic rocks
BT igneous rocks
NT dunite
NT garnet lherzolite
NT garnet peridotite
NT harzburgite
NT lherzolite
NT spinel lherzolite
NT spinel peridotite
SA metaperidotite
SA pyrolite
SA wehrlite

periglacial environment (1978)
Before 1978, search periglacial.
SA depositional environment
SA environment
SA frost action
SA glacial environment
SA glacial geology
SA glaciers
SA glaciolacustrine environment
SA sedimentation

periglacial features (1978)
Autoposting of this term began in 1978. Before 1977, also search periglacial phenomena.
UF features, periglacial
UF periglacial phenomena
NT ice wedges
NT palsas
NT pingos
SA congelifraction
SA cryopedology
SA cryoturbation
SA extinct lakes
SA firn
SA fossil ice wedges
SA frozen ground
SA gelifluction
SA glacial features
SA glacial geology
SA glacial lakes
SA glaciers
SA hummocks
SA ice
SA ice fields
SA ice lenses
SA ice-marginal features
SA lakes
SA moraines
SA outwash
SA patterned ground
SA permafrost
SA polygons
SA sand seas
SA solifluction
SA taliks
SA thermokarst
SA tors

periglacial phenomena
A valid term through 1976.
use periglacial features

Perigord (1985)
Region, central Dordogne, France.
BT Dordogne France
BT France
BT Western Europe
BT Europe

periodicity (1978)
SA cyclic processes
SA earthquakes
SA frequency
SA Milankovitch theory
SA orogeny
SA sedimentation
SA seismology
SA volcanology

peripheral faults (1978)

Before 1978, also search faults AND peripheral. Before 1993, also search border faults.
UF border faults
BT faults
SA arcuate faults
SA boundary faults

Perisphinctida (1981)
Before 1989, also search Perisphinctidae.
UF Perisphinctidae
BT Ammonoidea
BT Tetrabranchiata
BT Cephalopoda
BT Mollusca
BT Invertebrata

Perisphinctidae
As of 1989, no longer a valid term for GeoRef. Equivalent to Perisphinctida.
use Perisphinctida

Perissodactyla (1978)
Order. Autoposting of Eutheria and Theria to this term began in 1989.
BT Eutheria
BT Theria
BT Mammalia
BT Tetrapoda
BT Vertebrata
BT Chordata
NT Ceratomorpha
NT Hippomorpha
SA Ungulata

peristerite (1978)
Gem variety of albite. Autoposting of feldspar group began in 1989.
BT plagioclase
BT feldspar group
BT framework silicates
BT silicates
SA albite
SA moonstone

perlite (1978)
IN May be used as an igneous rock with pyroclastics. May also may be used as a commodity term with construction materials.
UF pearlite
BT glasses
BT volcanic rocks
BT igneous rocks
SA construction materials
SA lava
SA volcanic glass

Perm
No longer a valid term for GeoRef. As of 1993, see Perm Russian Federation.

Perm Russian Federation (1993)
Oblast and city W the Urals in W Russian Federation. Before 1993, also search Perm. Oblast was also called Molotov in the 1940s. This term has multiple hierarchies.
BT1 Russian Federation
BT1 Commonwealth of Independent States
BT2 Europe
NT Kizel coal basin
NT Molotov Russian Federation
NT Solikamsk Russian Federation
SA Ural region

permafrost (1978)
Used for geotechnical studies of permafrost as well as geomorphological and glacial studies.
SA active layer
SA creep
SA cryopedology
SA engineering geology

SA engineering properties
SA freezing
SA frost action
SA frost heaving
SA frozen ground
SA geomorphology
SA glacial geology
SA ground ice
SA hummocks
SA ice
SA ice lenses
SA ice wedges
SA periglacial features
SA pingos
SA relict materials
SA rock glaciers
SA site exploration
SA soils
SA solifluction
SA suffosion
SA taliks
SA thawing
SA thermal properties
SA thermokarst
SA tundra
SA Tundra soils
SA underground installations

permeability (1978)
SA aquitards
SA circulation
SA Darcy's law
SA diffusion
SA disposal barriers
SA engineering geology
SA ground water
SA hydraulic conductivity
SA porosity
SA rock mechanics
SA roughness
SA secondary porosity
SA sediments
SA thermal waters
SA tight sands
SA transmissivity
SA wettability

permeability coefficient
use hydraulic conductivity

Permian (1978)
Above Carboniferous, below Triassic (Mesozoic). From 1978-1980, Paleozoic was autoposted to this term.
BT Paleozoic
NT Akiyoshi Limestone
NT Americus Limestone Member
NT Appalachian Phase
NT Asselian
NT Blaine Formation
NT Bulli Seam
NT Castile Formation
NT Chase Group
NT Coconino Sandstone
NT Cottonwood Limestone
NT Council Grove Group
NT Culebra Dolomite Member
NT Cutler Formation
NT Damuda Series
NT Ecca Group
NT Eskridge Shale
NT Esplanade Sandstone Member
NT Fort Riley Limestone
NT Glorieta Sandstone
NT Guadalupian
NT Hennessey Formation
NT Hughes Creek Shale
NT Hutchinson Salt Member
NT Illawarra Coal Measures
NT Irati Formation
NT Kaibab Formation
NT Kamthi Formation
NT Kungurian
NT Lower Permian

NT Lyons Sandstone
NT Maokou Formation
NT McCloud Limestone
NT Meade Peak Member
NT Middle Permian
NT Newcastle Coal Measures
NT Park City Formation
NT Passa Dois Group
NT Phosphoria Formation
NT Ranger Canyon Formation
NT Red Eagle Limestone
NT Retort Phosphatic Shale Member
NT Rio Bonito Formation
NT Rotliegendes
NT Rustler Formation
NT Saalian Phase
NT San Andres Formation
NT Saxonian
NT Stearns Shale
NT Stone Corral Formation
NT Thuringian
NT Toroweap Formation
NT Upper Permian
NT Val Gardena Sandstone
NT Vryheid Formation
NT Wellington Formation
NT Werra Series
NT Wreford Limestone
NT Yates Formation
NT Yeso Formation
SA Alleghany Orogeny
SA Baralaba Coal Measures
SA Baskunchak Series
SA Beacon Supergroup
SA Beaufort Group
SA Botucatu Formation
SA Brianconnais Zone
SA Cache Creek Group
SA Calaveras Formation
SA Casper Formation
SA Chugwater Formation
SA Dunkard Group
SA Dwyka Formation
SA Estrada Nova Formation
SA Fountain Formation
SA Geirud Formation
SA Gondwana System
SA Goose Egg Formation
SA Hercynian Orogeny
SA Honaker Trail Formation
SA Ishbel Group
SA Itarare Subgroup
SA Ivishak Formation
SA Karroo Supergroup
SA Korenevskaya Formation
SA Krol Formation
SA Kupferschiefer
SA Lisburne Group
SA Louann Salt
SA lower Gondwana System
SA Magdalena Group
SA Maroon Formation
SA Minnelusa Formation
SA Narragansett Pier Granite
SA New Red Sandstone
SA Nikolai Greenstone
SA Ouachita Orogeny
SA Paganzo Group
SA Sadlerochit Group
SA Sicker Group
SA Singleton Coal Measures
SA Slovakian Karst
SA Spearfish Formation
SA Supai Formation
SA Taiyuan Formation
SA Talchir Formation
SA Weber Sandstone
SA Wells Formation
SA Wood River Formation

Permian Basin (1981)
Structural basin in W Texas and SE New Mexico. Includes Midland, Delaware, and Marfa basins.
IN Index states or regions as applicable.
CO N283000N333000
W1003000W1062000
BT United States
SA Delaware Basin
SA Marfa Basin
SA Midland Basin
SA New Mexico
SA Texas

Pernambuco
No longer a valid term for GeoRef. See Pernambuco Brazil.

Pernambuco Brazil (1993)
State in NE Brazil.
CO S092000S071500
W0344500W0412500
BT Brazil
BT South America
SA Barreiras Formation
SA Recife-Joao Pessoa
SA Sao Francisco Basin

Pernik coal basin (1978)
15 miles SW of Sofia. Also search Pernik.
UF Dimitrovo coal basin
BT Bulgaria
BT Southern Europe
BT Europe
SA coal fields

perofskite
use perovskite

perovskite (1978)
UF perofskite
BT oxides
SA perovskite structure

perovskite structure (1978)
SA crystal structure
SA perovskite

peroxide
use hydrogen peroxide

perrierite (1978)
Autoposting of sorosilicates began in 1985.
BT sorosilicates
BT orthosilicates
BT silicates

Perris
Valid through 1988. Search in combination with state term. After 1988, use specific city-state term.

Perris California (1989)
City in S California. Before 1989, search Perris AND California.
CO N334800N334800
W1171200W1171200
BT Riverside County California
BT California
BT United States

Perry County
Valid through 1988. Search in combination with state term. After 1988, use specific county-state term.

Perry County Alabama (1989)
W central Alabama. Before 1989, also search Perry County AND Alabama.
CO N321800N324900
W0870100W0873200
BT Alabama
BT United States

Perry County Arkansas (1989)
Central Arkansas. Before 1989, also search Perry County AND Arkansas.
CO N344600N350600
W0923300W0931800
BT Arkansas
BT United States

Perry County Illinois (1989)
SW Illinois. Before 1989, also search Perry County AND Illinois.
CO N375500N381400
W0890800W0893500
BT Illinois
BT United States

Perry County Indiana (1989)
S Indiana. Before 1989, also search Perry County AND Indiana.
CO N374800N381700
W0862600W0865000
BT Indiana
BT United States

Perry County Kentucky (1989)
SE Kentucky. Before 1989, also search Perry County AND Kentucky.
CO N370000N372700
W0830000W0833500
BT Kentucky
BT United States

Perry County Mississippi (1989)
SE Mississippi. Before 1989, also search Perry County AND Mississippi.
CO N305400N312600
W0885200W0890800
BT Mississippi
BT United States
SA Richton Dome

Perry County Missouri (1989)
E Missouri. Before 1989, also search Perry County AND Missouri.
CO N373500N375300
W0892800W0900800
BT Missouri
BT United States

Perry County Ohio (1989)
Central Ohio. Before 1989, also search Perry County AND Ohio.
CO N393300N395500
W0820200W0822800
BT Ohio
BT United States

Perry County Pennsylvania (1989)
S central Pennsylvania. Before 1989, also search Perry County AND Pennsylvania.
CO N401100N403800
W0765500W0774100
BT Pennsylvania
BT United States

Perry County Tennessee (1989)
W central Tennessee. Before 1989, also search Perry County AND Tennessee.
CO N352500N354800
W0873800W0880200
BT Tennessee
BT United States

Perry Formation (1978)
Consists of thick sequence of coarse clastic sediments containing interbedded basalt flows. Central Pennsylvania.
BT Upper Devonian
BT Devonian
BT Paleozoic
SA Pennsylvania

Perry Mountain Formation (1989)
W Maine and SE New Hampshire. Underlies Parmachenee Formation; overlies Johns Pond Formation.
BT Silurian
BT Paleozoic
SA Maine
SA New Hampshire

Persani Mountains (1978)
SE Transylvania.
BT Transylvania
BT Romania
BT Southern Europe
BT Europe

Pershing County
Valid through 1988. Search in combination with state term. After 1988, use specific county-state term.

Pershing County Nevada (1989)
NW central Nevada. Before 1989, also search Pershing County AND Nevada.
CO N400000N405700
W1172000W1192000
BT Nevada
BT United States
NT Lovelock Nevada
SA Black Rock Desert
SA Humboldt Range

Persia
use Iran

Persian Gulf (1978)
Between Arabian Peninsula on the W and S, and Iran on E. Arabian Sea is considered a broader term as of 1981.
CO N240000N310000
E0570000E0480000
UF Arabian Gulf
BT Arabian Sea
BT Indian Ocean
NT Khor al Bazam

personal computers
As of 1993, use microcomputers if applicable. Also see minicomputers.

Perth
No longer a valid term for GeoRef. In various locations. As of 1993, see Perth Australia, Perth Ontario, and Perth Scotland.

Perth abyssal plain (1978)
Off SW Australia.
BT Indian Ocean
SA DSDP Site 259

Perth Australia (1993)
City in Western Australia. Also a town in NE central Tasmania. Before 1993, search Perth in combination with state.
IN Index Western Australia or Tasmania Australia as applicable.
BT Australia
BT Australasia
SA Perth Basin
SA Tasmania Australia
SA Western Australia

Perth Basin (1978)
Along the Indian Ocean in Western Australia from the Perth area on the S to the Carnarvon Basin in the N.
SA Perth Australia

Perth Ontario (1993)
Town in Lanark County Ontario. Before 1993, search Perth in combination with Ontario.
BT Lanark County Ontario
BT Ontario
BT Eastern Canada
BT Canada

Perth Scotland (1993)
Town in Perthshire, Tayside region Scotland. Before 1993, search Perth AND Scotland.
BT Perthshire Scotland
BT Scotland
BT Great Britain
BT United Kingdom
BT Western Europe
BT Europe

Perth Stade (1978)
Europe. Before 1978, also search Perth.
BT Weichselian
BT upper Pleistocene
BT Pleistocene
BT Quaternary
BT Cenozoic

perthite (1978)
Autoposting of alkali feldspar to this term began in 1989.
BT alkali feldspar
BT feldspar group
BT framework silicates
BT silicates
SA antiperthite
SA cryptoperthite
SA microperthite

Perthshire
No longer a valid term for GeoRef as of 1993.
use Perthshire Scotland

Perthshire Scotland (1993)
Former county in central Scotland. As of 1975, became part of the Tayside and Central administrative regions. Before 1993, also search Perthshire.
UF Perthshire
BT Scotland
BT Great Britain
BT United Kingdom
BT Western Europe
BT Europe
NT Aberfeldy
NT Perth Scotland
SA Central region Scotland

perturbations
use variations

Peru (1978)
CO S181500N000000
W0700000W0811000
BT South America
NT Amazonas Peru
NT Arequipa Peru
NT Ayacucho Peru
NT Junin Peru
NT Lima Peru
NT Puno Peru
SA Altiplano
SA Amazon Basin
SA Amazon River
SA Andes
SA Eastern Cordillera
SA El Nino
SA Pucara Group
SA Sandia Formation

Peru Trench
use Peru-Chile Trench

Peru-Chile Trench (1978)
Just off Peruvian and Chilean coasts.
UF Chile Trench
UF Peru Trench
BT Pacific Ocean
SA Chile earthquake 1960
SA ODP Site 682

Perugia
No longer a valid term for GeoRef.
See Perugia Italy.

Perugia Italy (1993)
City and province in central Umbria, central Italy.
BT Umbria Italy
BT Italy
BT Southern Europe
BT Europe
NT Gubbio Italy
NT Spoleto Italy

Pervomaisk
No longer a valid term for GeoRef. As of 1993, see Pervomaisk Ukraine.

Pervomaisk Ukraine (1993)
City in Nikolayev Oblast in S Ukraine. This term has multiple hierarchies.
BT1 Ukraine
BT1 Europe
BT2 Ukraine
BT2 Commonwealth of Independent States

Pesaro
No longer a valid term for GeoRef. See Pesaro Italy.

Pesaro Italy (1993)
City on Adriatic Sea in N Marches, E central Italy.
BT Marches Italy
BT Italy
BT Southern Europe
BT Europe

Pescadores
use Penghu Islands

Peschany Island
use Peschanyy

Peschanyy (1978)
Island in Caspian Sea off S shore of Apsheron Peninsula.
IN Index Azerbaidzhan or regions as applicable.
UF Peschany Island
SA Apsheron Peninsula
SA Azerbaidzhan
SA Caspian Sea

Peshawar
No longer a valid term for GeoRef. See Peshawar Pakistan.

Peshawar Pakistan (1993)
City, district and division near the Khyber Pass.
BT North-West Frontier Pakistan
BT Pakistan
BT Indian Peninsula
BT Asia

pesticides (1978)
SA DDT
SA environmental geology
SA herbicides
SA pollutants
SA waste disposal

petalite (1978)
BT nepheline group
BT framework silicates
BT silicates
SA aluminosilicates
SA lithium ores

Petersburg
Valid through 1988. Search in combination with state term. After 1988, use specific city-state term.

Petersburg Formation (1985)
SW Indiana.
BT Middle Pennsylvanian
BT Pennsylvanian
BT Carboniferous
BT Paleozoic
SA Indiana

Petersburg Granite (1978)
Late Paleozoic. Underlies Triassic sediments. E Virginia.
BT Paleozoic
SA Virginia

Petersburg Virginia (1989)
City S of Richmond. In but independent of Dinwiddie County, SE Virginia. Before 1989, search Petersburg AND Virginia.
CO N371400N371400
W0772400W0772400
BT Dinwiddie County Virginia
BT Virginia
BT United States

Petrified Forest Member (1993)
Upper Triassic. In Chinle Formation. N Arizona, SE Nevada, west-central New Mexico, S Utah.
BT Upper Triassic
BT Triassic
BT Mesozoic
SA Arizona
SA Chinle Formation
SA Nevada
SA New Mexico
SA Utah

Petrified Forest National Park (1978)
In Painted Desert area in E Arizona.
BT Arizona
BT United States
SA Apache County Arizona
SA national parks
SA Navajo County Arizona

petrified moss
use tufa

petrified wood
use fossil wood

petrofabrics (1978)
SA deformation
SA fabric
SA interference patterns
SA orientation
SA preferred orientation
SA structural analysis

petrogenesis
A valid term through 1973. Use igneous rocks and genesis; metamorphic rocks and genesis.

petrogeometry
use structural analysis

petrography (1978)
SA crystal zoning
SA goniometry
SA igneous rocks
SA macerals
SA metamorphic rocks
SA microfacies
SA microscope methods
SA petrology
SA reaction rims
SA sedimentary rocks
SA sediments
SA thin sections
SA ultrastructure

petroleum (1978)
Also search petroleum-gas; petroleum-natural gas. As of 1993, use hydrocarbons or crude oil for pollution studies.
UF oil (petroleum)
UF oil and gas
NT natural gas
SA aromatization
SA benzene
SA bitumens
SA blowouts
SA cap rocks
SA clathrates
SA condensates
SA crude oil
SA Darcy's law
SA degasification
SA depletion
SA diwa-type
SA drilling
SA energy sources
SA fluidization
SA geopressure
SA giant fields
SA heavy oil
SA hydrocarbons
SA kerogen
SA migration
SA naturally fractured reservoirs
SA offshore
SA oil and gas fields
SA oil sands
SA oil seeps
SA oil shale
SA oil spills
SA oil storage
SA oil-gas interface
SA oil-water interface
SA onshore
SA organic materials
SA petroleum accumulation
SA petroleum engineering
SA petroleum exploration
SA petroleum maps
SA pipelines
SA potential deposits
SA primary migration
SA reservoir properties
SA reservoir rocks
SA Rock-Eval
SA secondary migration
SA secondary recovery
SA seepage
SA source rocks
SA stratigraphic traps
SA stratigraphic wedges
SA structural traps
SA tertiary recovery
SA thermal maturity
SA tight sands
SA traps

petroleum accumulation (1982)
UF accumulation, petroleum
SA migration
SA petroleum
SA stratigraphic traps
SA stratigraphic wedges
SA structural traps
SA traps

petroleum engineering (1978)
Used for petroleum and natural gas.
UF engineering, petroleum
SA acidification
SA Archie's law
SA blowouts
SA borehole breakouts
SA drilling
SA engineering geology
SA enhanced recovery
SA fluid injection
SA gas injection
SA gas storage
SA horizontal drilling
SA hydraulic fracturing
SA mercury injection
SA natural gas
SA naturally fractured reservoirs
SA observation wells
SA oil and gas fields
SA oil storage
SA petroleum
SA pumping
SA reservoir rocks
SA retorting
SA sand equivalent tests

SA secondary recovery
SA steam injection
SA tertiary recovery
SA waterflooding
SA well stimulation
SA wettability

petroleum exploration (1993)
Before 1971, was used on level 1 in subfile N. Reintroduced as a valid term in 1993. From 1971 to 1993, also search petroleum AND exploration.
SA Exclusive Economic Zone
SA exploration
SA hydrocarbon indicators
SA migration
SA oil and gas fields
SA petroleum
SA source rocks
SA stratigraphic traps
SA stratigraphic wedges
SA structural traps

petroleum geology
No longer a valid GeoRef index term. Before 1971, was used on level 1 in subfile N. See petroleum; petroleum engineering.

petroleum maps (1989)
Used for maps indicating location of natural gas and oil. Before 1989, search fuel resources maps.
UF fuel resources maps
BT maps
SA economic geology maps
SA natural gas
SA petroleum

petroleum products
use hydrocarbons

petroleum reserves
Not a valid term for GeoRef. See petroleum; reserves.

petroleum resources
Not a valid term for GeoRef. See petroleum; resources.

petroleum seepage
use oil seeps

petrologen
use kerogen

petrology (1978)
Used for the discipline as a whole. Primarily for studies on igneous, metamorphic, or metasomatic rocks.
SA accessory minerals
SA banded structures
SA bibliography
SA catalogs
SA crystallography
SA education
SA fluid inclusions
SA homogenization
SA igneous rocks
SA inclusions
SA interlaboratory comparison
SA intrusions
SA lava
SA lexicons
SA magmas
SA melts
SA metamorphic rocks
SA metamorphism
SA metasomatic rocks
SA metasomatism
SA meteorites
SA microscope methods
SA mineralogy
SA modal analysis
SA palingenesis
SA petrography
SA sedimentary petrology

SA sedimentary rocks
SA sediments
SA silicate rocks
SA staining
SA standard materials
SA tektites
SA volcanology

petromorphology
use structural analysis

petrophysics
As of 1989, no longer a valid term for GeoRef. Use physical properties in combination with type of rock.

Petrosani Basin (1978)
Coal basin in Hunedoara County in the Transylvanian Alps. Also search Petrosani.
UF Petroseni Basin
BT Transylvania
BT Romania
BT Southern Europe
BT Europe
SA coal fields

Petroseni Basin
use Petrosani Basin

petrostratigraphy
use lithostratigraphy

Petrovsk Russian Federation
use Makhachkala Russian Federation

Petsamo
use Pechenga

PGE
use platinum group

pH (1978)
UF acidity
SA acid rain
SA acidic composition
SA acids
SA alkalinity
SA buffers
SA geochemistry
SA physical properties
SA sea water

Phacopida (1978)
Order. Autoposting of this term began in 1978.
BT Trilobita
BT Trilobitomorpha
BT Arthropoda
BT Invertebrata
NT Phacopina

Phacopina (1978)
Suborder or genus. Autoposting of this term to Phacops began in 1989.
BT Phacopida
BT Trilobita
BT Trilobitomorpha
BT Arthropoda
BT Invertebrata
NT Phacops

Phacops (1978)
Genus. Autoposting of Phacopina to this term began in 1989.
BT Phacopina
BT Phacopida
BT Trilobita
BT Trilobitomorpha
BT Arthropoda
BT Invertebrata

Phaeodarina (1978)
Autoposting of Osculosida to this term began in 1989. Autoposting of Protista and microfossils to this term began in 1990. This term has multiple hierarchies.
BT1 Osculosida

BT1 Radiolaria
BT1 Protista
BT1 Invertebrata
BT2 Osculosida
BT2 Radiolaria
BT2 Protista
BT2 microfossils

Phaeophyta (1978)
Including brown algae and seaweed. Autoposting of thallophytes and microfossils began in 1990. This term has multiple hierarchies.
UF brown algae
UF seaweed
BT1 algae
BT1 thallophytes
BT1 Plantae
BT2 algae
BT2 microfossils

Phanerozoic (1978)
Autoposting of this term to Paleozoic, Mesozoic and Cenozoic ended in 1981.
SA Cenozoic
SA K-T boundary
SA Mesozoic
SA Paleozoic

pharmacolite (1978)
BT arsenates

pharmacosiderite (1978)
BT arsenates

phase diagrams
As of 1981, no longer a valid term for GeoRef. See phase equilibria.

phase distortion (1981)
SA distortion
SA elastic waves
SA seismology

phase equilibria (1978)
Primarily used for laboratory studies in petrology and mineralogy. May be used for theoretical or field studies of specific rocks, minerals or localities. For the Earth as a whole, crust, mantle or core, use phase transitions.
IN Include system where available.
SA carbon dioxide
SA crystal chemistry
SA crystal growth
SA crystal zoning
SA crystallography
SA diamond anvil cells
SA entropy
SA equations of state
SA equilibrium
SA free energy
SA gases
SA geochemistry
SA grain boundaries
SA high pressure
SA high temperature
SA Hugoniot analysis
SA hydrothermal conditions
SA igneous rocks
SA immiscibility
SA inclusions
SA low pressure
SA low temperature
SA magmas
SA melting
SA melts
SA metamorphic rocks
SA metamorphism
SA metasomatism
SA microthermometry
SA mineral deposits, genesis
SA mineralogy
SA miscibility gap
SA P-T conditions
SA paragenesis

SA partial melting
SA partitioning
SA phase rule
SA phase transitions
SA pressure
SA reaction rims
SA segregation
SA sintering
SA solid solution
SA temperature
SA transformations
SA WATEQF

phase rule (1978)
UF Gibbs phase rule
SA phase equilibria

phase transitions (1981)
For phase changes within the Earth. For laboratory studies, use phase equilibria.
SA core
SA crust
SA Earth
SA mantle
SA phase equilibria
SA seismology
SA transformations
SA transition zones

phase velocity (1981)
SA elastic waves
SA seismology

phase, gaseous
use gaseous phase

phase, solid
use solid phase

phases
A valid general term through 1978. After 1978, see specific terms, e.g. liquid phase, solid phase, gaseous phase, etc.

Phelps County
Valid through 1988. Search in combination with state term. After 1988, use specific county-state term.

Phelps County Missouri (1989)
Central Missouri. Before 1989, also search Phelps County AND Missouri.
CO N373700N380900
W0913100W0920200
BT Missouri
BT United States

Phelps County Nebraska (1989)
S Nebraska. Before 1989, also search Phelps County AND Nebraska.
CO N402200N403900
W0991000W0993900
BT Nebraska
BT United States

phenacite
use phenakite

phenakite (1978)
Autoposting of nesosilicates began in 1985.
UF phenacite
BT nesosilicates
BT orthosilicates
BT silicates

phenanthrene (1989)
Crystalline tricyclic aromatic hydrocarbon.
BT polycyclic aromatic hydrocarbons
BT aromatic hydrocarbons
BT hydrocarbons
BT organic materials

phengite (1978)
BT mica group

BT sheet silicates
BT silicates
SA muscovite

phenocrysts (1978)
SA crystals
SA igneous rocks
SA matrix
SA minerals
SA porphyritic texture

phenols (1978)
BT organic materials
SA alcohols

phenomena, electrical
Not a valid term for GeoRef. Usually out-of-scope. Used as electrical phenomena with meteorology.

phenomena, surface
Not a valid term for GeoRef. Usually out-of-scope. Used as surface phenomena with astrophysics and solar physics.

phenomena, transient
use transient phenomena

phenotypes (1989)
SA ecology
SA ontogeny
SA paleoecology
SA taxonomy
SA type specimens

phi grade scale
use phi scale

phi scale (1978)
UF phi grade scale
BT statistical analysis
SA sedimentary rocks
SA sediments

Philadelphia
Valid through 1988. Search in combination with state term. After 1988, use specific city-state term.

Philadelphia County
Valid through 1988. Search in combination with state term. After 1988, use specific county-state term.

Philadelphia County Pennsylvania (1989)
SE Pennsylvania. Before 1989, also search Philadelphia County AND Pennsylvania.
CO N395000N400800
 W0745500W0752000
BT Pennsylvania
BT United States
NT Philadelphia Pennsylvania

Philadelphia Pennsylvania (1989)
City on the Delaware River. Coextensive with Philadelphia County, SE Pennsylvania. Before 1989, search Philadelphia AND Pennsylvania.
CO N400000N400000
 W0751000W0751000
BT Philadelphia County Pennsylvania
BT Pennsylvania
BT United States

Philip Island (1978)
In Indian Ocean E of Mornington Peninsula S of Melbourne.
SA Victoria Australia

Philippine Islands (1978)
CO N050000N190000
 E1263000E1170000
UF Philippines
BT Far East
BT Asia

NT Cebu Philippine Islands
NT Iloilo Basin
NT Iloilo Philippine Islands
NT Luzon
NT Marinduque Philippine Islands
NT Mindanao
NT Mindoro
NT Palawan
NT Panay Island
NT Samar
NT Surigao del Norte Philippine Islands
SA ASEAN
SA Kuroshio
SA Mansalay Formation
SA Zambales Ophiolite

Philippine Sea (1978)
That part of the W Pacific Ocean with the Philippines Islands on the W, Taiwan and the Ryukyus on the NW, and the U.S. Trust Territory of the Pacific Islands on the E and SE.
CO N000000N350000
 E1500000E1200000
BT West Pacific
BT Pacific Ocean
NT Daito Ridge
NT Kyushu-Palau Ridge
NT Parece Vela Basin
NT Ryukyu Trench
NT West Philippine Basin
SA DSDP Site 291
SA DSDP Site 292
SA DSDP Site 293
SA DSDP Site 294
SA DSDP Site 296
SA DSDP Site 297
SA DSDP Site 442
SA DSDP Site 445
SA DSDP Site 446
SA DSDP Site 450
SA DSDP Site 453
SA Leg 6
SA Leg 31
SA Leg 58
SA Leg 59
SA Leg 60
SA Leg 124E
SA Leg 126
SA Leg 132
SA ODP Site 787
SA ODP Site 788
SA ODP Site 789
SA ODP Site 790
SA ODP Site 791
SA ODP Site 792
SA ODP Site 793
SA ODP Site 809
SA Philippine Sea Plate

Philippine Sea Plate (1978)
A tectonic plate roughly covering the same area as the Philippine Sea and virtually surrounded by the SE edge of the Eurasian Plate and the Pacific Plate.
SA Philippine Sea
SA plate tectonics
SA plates

Philippines
use Philippine Islands

philippinite (1978)
BT tektites

Phillips County
Valid through 1988. Search in combination with state term. After 1988, use specific county-state term.

Phillips County Arkansas (1989)
E Arkansas. Before 1989, also search Phillips County AND Arkansas.

CO N340800N344000
 W0902700W0910600
BT Arkansas
BT United States

Phillips County Colorado (1989)
NE Colorado. Before 1989, also search Phillips County AND Colorado.
CO N402700N404600
 W1020300W1023700
BT Colorado
BT United States

Phillips County Kansas (1989)
N Kansas. Before 1989, also search Phillips County AND Kansas.
CO N393300N400000
 W0990400W0993900
BT Kansas
BT United States
NT Kirwin Kansas

Phillips County Montana (1989)
N Montana. Before 1989, also search Phillips County AND Montana.
CO N473300N490000
 W1071000W1082800
BT Montana
BT United States

phillipsite (1978)
BT zeolite group
BT framework silicates
BT silicates

philosophy (1978)
Use term with discipline.
SA creationism

Phlegraean Fields (1978)
Volcanic region W of Naples in S Italy.
UF Campi Flegrei
SA Campania Italy

phlogopite (1978)
UF amber mica
UF brown mica
BT mica group
BT sheet silicates
BT silicates
SA fluor-phlogopite

Phobos
As of 1989, no longer a valid term for GeoRef.
use Phobos Satellite

Phobos Satellite (1989)
One of satellites of Mars. Before 1989, also search Phobos.
UF Phobos
SA Mars
SA satellites

Phoebe Satellite (1989)
One of the satellites of Saturn. Before 1989, also search Phoebe AND Saturn.
SA satellites
SA Saturn

Phoenix
Valid through 1988. Search in combination with state term. After 1988, use specific city-state term.

Phoenix Arizona (1989)
City in S central Arizona. Before 1989, search Phoenix AND Arizona.
CO N333000N333000
 W1120300W1120300
BT Maricopa County Arizona
BT Arizona
BT United States

Pholadomyida (1981)

BT Bivalvia
BT Mollusca
BT Invertebrata

Pholidophoriformes
use Halecostomi

Pholidota (1978)
Order. Autoposting of Eutheria and Theria to this term began in 1989.
BT Eutheria
BT Theria
BT Mammalia
BT Tetrapoda
BT Vertebrata
BT Chordata

phonolite
As of 1989, no longer a valid term for GeoRef. From 1981 to 1988, phonolites and volcanic rocks were autoposted to this term.
use phonolites

phonolites (1981)
Before 1981, search trachyte-phonolite family. From 1978-1980, trachyte-phonolite family was autoposted to the rocks that were classified under that term. Before 1989, also search phonolite. Autoposting of this term to phonolite ended in 1989.
UF phonolite
BT volcanic rocks
BT igneous rocks
NT tinguaite

Phoronida (1981)
BT worms
BT Invertebrata

phosgenite (1978)
From 1978-1980, carbonates and halides were autoposted to this term. Autoposting of chlorides began in 1981. As of 1989, halides is autoposted to this term. This term has multiple hierarchies.
BT1 carbonates
BT2 chlorides
BT2 halides

phosphate
As of 1981, no longer a valid term for GeoRef. Use phosphate deposits for the commodity, phosphates for the mineral group, or phosphate rocks. See also phosphate composition.

phosphate composition (1978)
Before 1978, also search nodules AND phosphate.
SA composition

phosphate deposits (1981)
Before 1981, search phosphate AND deposits.
IN Commodity. See List C.
UF apatite ores
SA apatite
SA Lee Creek Mine

phosphate ion (1985)
SA geochemistry
SA ions
SA phosphates

phosphate rocks (1978)
Refers to any rock containing phosphate. Before 1981, also search phosphorite. Autoposting of this term to bone beds ended in 1985.
UF phosphorite
BT chemically precipitated rocks
BT sedimentary rocks
SA bone beds

phosphates (1978)
Sarcopside is considered a narrower term as of 1981. Autoposting of this term began in 1978.
NT amblygonite
NT apatite
NT autunite
NT barbosalite
NT beraunite
NT britholite
NT brushite
NT cacoxenite
NT carbonate apatite
NT chlorapatite
NT crandallite
NT dahllite
NT eosphorite
NT florencite
NT fluorapatite
NT francolite
NT goyazite
NT graftonite
NT herderite
NT hinsdalite
NT hureaulite
NT hydroxylapatite
NT lazulite
NT lithiophilite
NT minyulite
NT monazite
NT monetite
NT montebrasite
NT newberyite
NT plumbogummite
NT pyromorphite
NT rhabdophane
NT rockbridgeite
NT sarcopside
NT scholzite
NT strengite
NT struvite
NT torbernite
NT triphylite
NT triplite
NT turquoise
NT variscite
NT vivianite
NT wavellite
NT whitlockite
NT xenotime
SA minerals
SA phosphate ion

phosphatization (1978)
SA diagenesis

phosphides (1978)
Autoposting of this term to schreibersite began in 1981. Before 1981, was autoposted to sarcopside.
BT alloys
NT schreibersite

phosphorescence
Not a valid GeoRef index term after 1970. Was used on level 1 in subfile N.
use luminescence

Phosphoria Formation (1978)
Includes Sybille tongue (new), Forelle tongue, and Ervay tongue (new).
BT Permian
BT Paleozoic
SA Beaverhead Formation
SA Idaho
SA Meade Peak Member
SA Montana
SA Retort Phosphatic Shale Member
SA Utah
SA Wyoming

phosphorite
As of 1981, no longer a valid term for GeoRef. From 1978-1980, chemically precipitated rocks was autoposted to this term.
use phosphate rocks

phosphorous
Not a valid GeoRef index term after 1970. Was used on level 1 in subfile N.
use phosphorus

phosphorus (1978)
Before 1971, also search phosphorous.
UF P
UF phosphorous
NT P-32
SA chemical elements
SA fertilizers
SA KREEP
SA siderophile elements
SA soils

photochemistry (1985)
SA geochemistry
SA photosynthesis

photogeologic maps (1978)
Before 1978, also search maps AND photogeologic.
BT maps
SA mosaics
SA remote sensing

photogeologic methods (1978)
As of 1985, for methods only. Use photogeology for the application of methods to specific regions. Before 1985, also search photogeology.
IN Also index remote sensing.
SA cartography
SA maps
SA methods
SA mineral exploration
SA mosaics
SA photogeology
SA photography
SA remote sensing
SA video methods

photogeology (1978)
Used for application of photogeologic methods to specific regions. For methods, use photogeologic methods. Before 1985, also search photogeologic methods.
IN Also index remote sensing.
SA aerial photography
SA cartography
SA engineering geology
SA exploration
SA maps
SA mineral exploration
SA mosaics
SA paleoclimate maps
SA photogeologic methods
SA photogrammetry
SA photography
SA remote sensing
SA space photography

photogrammetry (1978)
Before 1975, also search photogrammetric studies.
SA aerial photography
SA cartography
SA definition
SA engineering geology
SA Magellan Program
SA mosaics
SA photogeology
SA photography
SA remote sensing
SA space photography
SA stereographic projection

photographs
A valid term through 1978.
use photography

photography (1978)
Before 1979, also search photographs. Before 1971, also search photomicroscopy.
UF photographs
UF photomicroscopy
SA aerial photography
SA mosaics
SA photogeologic methods
SA photogeology
SA photogrammetry
SA remote sensing
SA space photography
SA video methods

photometric
A valid term through 1977. After 1977, see photometry.

photometry (1978)
SA chemical analysis
SA flame photometry
SA spectroscopy

photomicroscopy
use photography

photomosaics
use mosaics

photosynthesis (1978)
SA geochemistry
SA paleobotany
SA photochemistry
SA processes
SA respiration

phreatic zone
use saturated zone

phreatomagmatic eruptions
use phreatomagmatism

phreatomagmatism (1985)
UF phreatomagmatic eruptions
SA explosive eruptions
SA igneous activity
SA lava
SA maars
SA magmas
SA thermal waters
SA volcanism
SA volcanology

PHREEQE (1989)
Geochemical simulation program applicable to hydrogeology and engineering geology.
BT computer programs
SA data processing
SA geochemistry
SA mathematical models
SA numerical models

Phuket Group (1978)
May be the equivalent of the Megui Series, at least in part, of Burma. W side of peninsular Thailand, including island of Phuket, and lower Burma.
BT Cambrian
BT Paleozoic
SA Asia
SA Burma
SA Thailand

phyletic gradualism
use gradualism

phyllite
As of 1981, no longer a valid term for GeoRef. From 1978-1980, phyllites was autoposted to this term.
use phyllites

phyllites (1978)
Before 1981, also search phyllite. Autoposting of this term to phyllite ended in 1981.
UF phyllite
BT metamorphic rocks

Phylloceratida (1981)
BT Ammonoidea
BT Tetrabranchiata
BT Cephalopoda
BT Mollusca
BT Invertebrata

phyllonite
As of 1981, no longer a valid term for GeoRef. From 1978-1980, phyllonites was autoposted to this term.
use phyllonites

phyllonites (1978)
Before 1981, also search phyllonite. Autoposting of this term to phyllonite ended in 1981.
UF phyllonite
BT metamorphic rocks
SA phyllonitization

phyllonitization (1978)
BT metamorphism
SA mylonitization
SA phyllonites
SA processes
SA recrystallization

phyllosilicates
use sheet silicates

phylogeny (1978)
SA biologic evolution
SA cladistics
SA gradualism
SA ontogeny
SA paleontology
SA punctuated equilibria
SA speciation

physical
A valid term through 1977. After 1977, use physical properties.

physical geography
use geography

physical geology (1978)
SA geology
SA historical geology

physical methods (1978)
SA centrifuge methods
SA methods
SA soils

physical models (1978)
SA flume studies
SA models

physical processes
As of 1993, use geophysical methods, geophysics, physical models, or physical properties as applicable.

physical properties (1978)
See also appropriate entries under material name.
SA acoustical properties
SA b-values
SA bearing capacity
SA compactness
SA compressibility
SA constitutive equations
SA friction
SA hardness
SA mechanical properties
SA microhardness
SA pH
SA physicochemical properties
SA porosity
SA properties
SA secondary porosity
SA soils

SA specific heat
SA thermal properties
SA thermoelastic properties
SA thermomechanical properties

physical remanent magnetization (1981)
Autoposting of magnetization to this term began in 1989.
BT remanent magnetization
BT magnetization
SA paleomagnetism

physical weathering (1978)
Before 1978, search weathering AND physical.
BT weathering
SA mechanical weathering

physico-chemical properties
use physicochemical properties

physicochemical properties (1982)
UF physico-chemical properties
UF physiochemical properties
SA chemical properties
SA geochemistry
SA physical properties
SA properties

physiochemical properties
use physicochemical properties

physiographic geology
No longer a valid GeoRef index term. Before 1971, was used on levels 1 and 2 in subfiles E and N. See geomorphology; sedimentation; hydrology.

physiographic maps (1985)
BT maps
SA geomorphologic maps
SA geomorphology

physiographic provinces (1978)
UF provinces, physiographic
SA geomorphology
SA topography

physiography
A valid term through 1975. After 1975, see geomorphology.

physiology (1978)
IN Index with appropriate fossil term (list F).
SA anatomy
SA morphology
SA muscles

phytane (1985)
2, 6, 10, 14-tetramethylhexadecane.
BT alkanes
BT aliphatic hydrocarbons
BT hydrocarbons
BT organic materials
SA isoprenoids

phytogeography
use biogeography

phytoliths (1978)
Term reintroduced in 1985. Not a valid term from 1969 through 1984.
SA fossilization
SA paleobotany
SA Plantae

phytoplankton (1978)
BT plankton
SA algae
SA diatoms

Piacenza
No longer a valid term for GeoRef. See Piacenza Italy.

Piacenza Italy (1993)
City and province in NW Emilia-Romagna, N Italy.
BT Emilia-Romagna Italy
BT Italy
BT Southern Europe
BT Europe

Piatra-Neamt
No longer a valid term for GeoRef. As of 1993 see Piatra-Neamt Romania.

Piatra-Neamt Romania (1993)
City in Neamt County in NE Romania. This term has multiple hierarchies.
BT1 Moldavia
BT1 Europe
BT2 Romania
BT2 Southern Europe
BT2 Europe

Piaui
No longer a valid term for GeoRef. See Piaui Brazil.

Piaui Brazil (1993)
CO S105000S024500
   W0402500W0460000
BT Brazil
BT South America
SA Parnaiba Basin

Piave Valley (1978)
River valley in N and NE Veneto. Also search Piave River.
BT Veneto Italy
BT Italy
BT Southern Europe
BT Europe

Picacho
Valid through 1988. Search in combination with state term. After 1988, use specific city-state term.

Picacho Arizona (1989)
Town in S central Arizona. Before 1989, search Picacho AND Arizona.
CO N324200N324200
   W1113000W1113000
BT Pinal County Arizona
BT Arizona
BT United States

Picardy (1978)
Historical region in extreme N France. Abuts on English Channel between Calais in N and Treport on S, and extends E along Somme River and upper Oise River to Belgian border.
BT France
BT Western Europe
BT Europe

Picea (1981)
Genus.
BT Coniferales
BT gymnosperms
BT Spermatophyta
BT Plantae
NT Picea glauca

Picea glauca (1989)
Species.
BT Picea
BT Coniferales
BT gymnosperms
BT Spermatophyta
BT Plantae

Piceance Basin
use Piceance Creek basin

Piceance Creek basin (1978)
In NW Colorado.
UF Piceance Basin
BT Colorado
BT United States

Picher
Valid through 1988. Search in combination with state term. After 1988, use specific city-state term.

Picher Oklahoma (1989)
City in Ottawa County, NE Oklahoma. Before 1989, search Picher AND Oklahoma.
CO N365900N365900
   W0945200W0945200
BT Ottawa County Oklahoma
BT Oklahoma
BT United States

Pickens County
Valid through 1988. Search in combination with state term. After 1988, use specific county-state term.

Pickens County Alabama (1989)
W Alabama. Before 1989, also search Pickens County AND Alabama.
CO N325900N333100
   W0875300W0881900
BT Alabama
BT United States

Pickens County Georgia (1989)
N Georgia. Before 1989, also search Pickens County AND Georgia.
CO N342200N343500
   W0841600W0844100
BT Georgia
BT United States

Pickens County South Carolina (1989)
NW South Carolina. Before 1989, also search Pickens County AND South Carolina.
CO N344000N350600
   W0822800W0825700
BT South Carolina
BT United States
SA Lake Jocassee

pickeringite (1978)
BT sulfates

Pico Formation (1978)
Consists of three members: lower Pico marine, middle Pliocene; upper Pico marine, upper Pliocene; and continental upper Pliocene Sunshine Ranch member.
BT Pliocene
BT Neogene
BT Tertiary
BT Cenozoic
SA California

picotite
use chrome spinel

picrite (1978)
From 1978-1980, basalt family was autoposted to this term.
BT igneous rocks
SA basalts
SA ultramafics

picrite porphyry (1978)
BT igneous rocks
SA basalts
SA porphyry
SA ultramafics

Pictou County
No longer a valid term for GeoRef. As of 1993, see Pictou County Nova Scotia.

Pictou County Nova Scotia (1993)
N Nova Scotia on Northumberland Strait. Before 1993, also search Pictou County AND Nova Scotia.
CO N451700N455000
   W0620500W0631000
BT Nova Scotia
BT Maritime Provinces
BT Eastern Canada
BT Canada

Pictou Group (1989)
IN Index ages as applicable.
BT Mesozoic
SA Lower Permian
SA New Brunswick
SA Nova Scotia
SA Pennsylvanian
SA Prince Edward Island

picture elements
use pixels

Picture Gorge Basalt (1993)
Miocene. NE Oregon. In Columbia River Basalt Group. May be considered part of Grande Ronde Basalt.
BT Miocene
BT Neogene
BT Tertiary
BT Cenozoic
SA Columbia Plateau
SA Columbia River Basalt Group
SA Grande Ronde Basalt
SA middle Miocene
SA Oregon

Pictured Cliffs Sandstone (1985)
NW New Mexico and SW Colorado.
BT Upper Cretaceous
BT Cretaceous
BT Mesozoic
SA Colorado
SA New Mexico

Picuris Range (1978)
Mountains in N New Mexico.
BT New Mexico
BT United States

Piedmont (1978)
An upland belt in the U. S. lying E of the Blue Ridge and Appalachian mountains and W of the coastal plain. For region in Italy, use Piemonte. Before 1981, search Piedmont AND United States. In 1985, broader term changed from United States to Appalachians and North America. Before 1993, also search Piedmont Province.
UF Piedmont Province
BT Appalachians
BT North America
SA Alabama
SA Blue Ridge Province
SA Charlotte Belt
SA Georgia
SA Kings Mountain Belt
SA Kiokee Belt
SA Maryland
SA New Jersey
SA New York
SA North Carolina
SA Pennsylvania
SA Raleigh Belt
SA South Carolina
SA Virginia

Piedmont Alps (1978)
Those alpine ranges such as the Maritime Alps, Cottian Alps, Graian Alps, Pennine Alps, and Lepontine Alps, which lie within or on the borders of Piemonte.
BT Alps

BT Europe
SA Cottian Alps
SA Graian Alps
SA Lepontine Alps
SA Maritime Alps
SA Pennine Alps
SA Piemonte Italy

Piedmont Province
use Piedmont

piedmontite
use piemontite

piedmonts (1981)
Generic term.
SA geomorphology

Piemonte
No longer a valid term for GeoRef. See Piemonte Italy.

Piemonte Italy (1993)
Region in NW Italy. Before 1981, search Piedmont AND Italy.
CO N440500N462900
E0091100E0063800
BT Italy
BT Southern Europe
BT Europe
NT Acceglio Italy
NT Acqui Italy
NT Alessandria Italy
NT Argentera Massif
NT Asti Italy
NT Ivrea Italy
NT Lanzo Massif
NT Novara Italy
NT Sesia Zone
NT Turin Italy
NT Vercelli Italy
NT Villafranca d'Asti Italy
SA Argentera Italy
SA Canavese Zone
SA Graian Alps
SA Ivrea-Verbano Zone
SA Lago Maggiore
SA Langhe
SA Lepontine Alps
SA Maritime Alps
SA Monte Rosa
SA Pennine Alps
SA Piedmont Alps
SA Po River
SA Po Valley
SA Sesia Valley
SA Sesia-Lanzo Zone
SA Simplon region

piemontite (1978)
UF piedmontite
BT epidote group
BT sorosilicates
BT orthosilicates
BT silicates
SA manganese minerals

Pieniny Klippen Belt (1978)
Central and E Europe. Also search Pieniny.
IN Index countries as applicable.
UF Klippen Belt
BT Europe
SA Carpathians
SA Czechoslovakia
SA Pieniny Mountains
SA Poland
SA Ukraine

Pieniny Mountains (1978)
Range of the Beskids in S Poland. Also search Pieniny. This term has multiple hierarchies.
BT1 Beskid Mountains
BT1 Carpathians
BT1 Europe
BT2 Beskid Mountains
BT2 Central Europe
BT2 Europe

BT3 Poland
BT3 Central Europe
BT3 Europe
SA Pieniny Klippen Belt

Pierce County
Valid through 1988. Search in combination with state term. After 1988, use specific county-state term.

Pierce County Georgia (1989)
SE Georgia. Before 1989, also search Pierce County AND Georgia.
CO N311300N313200
W0815900W0822400
BT Georgia
BT United States

Pierce County Nebraska (1989)
NE Nebraska. Before 1989, also search Pierce County AND Nebraska.
CO N420700N422700
W0972200W0975100
BT Nebraska
BT United States

Pierce County North Dakota (1989)
N central North Dakota. Before 1989, also search Pierce County AND North Dakota.
CO N475100N483300
W0993000W1001700
BT North Dakota
BT United States

Pierce County Washington (1989)
W central Washington. Before 1989, also search Pierce County AND Washington.
CO N464300N471900
W1212300W1224300
BT Washington
BT United States
NT Mount Rainier
NT Nisqually Glacier
SA Mount Rainier National Park

Pierce County Wisconsin (1989)
W Wisconsin. Before 1989, also search Pierce County AND Wisconsin.
CO N443200N445200
W0920800W0925000
BT Wisconsin
BT United States

piercement
use diapirs

piercing fold
use diapirs

Pierre
Valid through 1988. Search in combination with state term. After 1988, use specific city-state term.

Pierre Shale (1978)
In Montana Group. E Colorado, W Minnesota, E Montana, Nebraska, North Dakota, South Dakota, and E Wyoming.
BT Upper Cretaceous
BT Cretaceous
BT Mesozoic
SA Colorado
SA Gammon Ferruginous Member
SA Minnesota
SA Montana
SA Montana Group
SA Nebraska
SA North Dakota
SA South Dakota
SA Wyoming

Pierre South Dakota (1989)
City on the Missouri River in central South Dakota. Before 1989, search Pierre AND South Dakota.
CO N442300N442300
W1002000W1002000
BT Hughes County South Dakota
BT South Dakota
BT United States

piers (1978)
BT marine installations

piezoelectric properties (1982)
UF piezoelectricity
SA crystal chemistry
SA crystals
SA electrical properties
SA properties

piezoelectricity
use piezoelectric properties

piezometers
use pressuremeters

piezometric surface
use potentiometric surface

piezoremanent magnetization (1981)
Autoposting of remanent magnetization to this term began in 1989.
BT remanent magnetization
BT magnetization

pigeonite (1978)
BT clinopyroxene
BT pyroxene group
BT chain silicates
BT silicates

pigments (1978)
Autoposting of this term began in 1978.
BT organic materials
NT carotenoids
NT chlorophyll
NT porphyrins

Pigmy Basin (1989)
Off the coast of Louisiana in Gulf of Mexico.
BT Gulf of Mexico
BT North American Atlantic
BT North Atlantic
BT Atlantic Ocean
SA DSDP Site 619
SA Leg 96

Pike County
Valid through 1988. Search in combination with state term. After 1988, use specific county-state term.

Pike County Alabama (1989)
SE Alabama. Before 1989, also search Pike County AND Alabama.
CO N313500N320500
W0854500W0861000
BT Alabama
BT United States

Pike County Arkansas (1989)
SW Arkansas. Before 1989, also search Pike County AND Arkansas.
CO N330300N342000
W0932200W0935700
BT Arkansas
BT United States

Pike County Georgia (1989)
W central Georgia. Before 1989, also search Pike County AND Georgia.
CO N325800N331200
W0841400W0843100
BT Georgia

BT United States

Pike County Illinois (1989)
W Illinois. Before 1989, also search Pike County AND Illinois.
CO N392300N395000
W0903600W0912200
BT Illinois
BT United States

Pike County Indiana (1989)
SW Indiana. Before 1989, also search Pike County AND Indiana.
CO N381300N383300
W0870500W0872800
BT Indiana
BT United States

Pike County Kentucky (1989)
E Kentucky. Before 1989, also search Pike County AND Kentucky.
CO N371300N374500
W0815700W0824500
BT Kentucky
BT United States

Pike County Mississippi (1989)
SW Mississippi. Before 1989, also search Pike County AND Mississippi.
CO N310000N312200
W0901700W0903300
BT Mississippi
BT United States

Pike County Missouri (1989)
E Missouri. Before 1989, also search Pike County AND Missouri.
CO N390800N393700
W0904300W0912800
BT Missouri
BT United States

Pike County Ohio (1989)
S Ohio. Before 1989, also search Pike County AND Ohio.
CO N385700N391200
W0824800W0832300
BT Ohio
BT United States

Pike County Pennsylvania (1989)
NE Pennsylvania. Before 1989, also search Pike County AND Pennsylvania.
CO N410600N413600
W0744100W0752200
BT Pennsylvania
BT United States

Pikes Peak (1978)
Mountain W of Colorado Springs in El Paso County in central Colorado. Before 1978, also included use for Pikes Peak Batholith.
IN Index county and Front Range and U. S. Rocky Mountains as applicable.
SA Colorado Springs Colorado
SA El Paso County Colorado
SA Front Range
SA Pikes Peak Batholith
SA U. S. Rocky Mountains

Pikes Peak Batholith (1978)
Before 1978, also search Pikes Peak for the batholith.
BT Colorado
BT United States
SA Pikes Peak

Pilbara Block (1985)
Structural unit, N Western Australia. Before 1993, also search Pilbara Craton.
CO S223000S202000
E1213000E1170000

UF Pilbara Craton
BT Western Australia
BT Australia
BT Australasia
SA Warrawoona Group

Pilbara Craton
use Pilbara Block

Pilbara gold field   (1978)
NW Western Australia. Also search Pilbara.
UF Pilbara Goldfield
BT Western Australia
BT Australia
BT Australasia
SA gold ores
SA mines

Pilbara Goldfield
use Pilbara gold field

piles   (1978)
SA foundations

Pilgrim Formation   (1978)
Threefold subdivision. Underlies Maywood or Red Lion Formation.
BT Upper Cambrian
BT Cambrian
BT Paleozoic
SA Montana

Pilica
No longer a valid term for GeoRef. As of 1993 see Pilica Poland.

Pilica Poland   (1993)
Village in N Cracow, S Poland. Before 1993 also search Pilica AND Poland.
BT Cracow Poland
BT Poland
BT Central Europe
BT Europe

Pilica River basin   (1981)
E central Poland.
CO N502000N515200
   E0212000E0193500
BT Poland
BT Central Europe
BT Europe

pillars   (1978)
Includes use as a general term.
SA methods
SA mining geology

pillow lava   (1978)
BT lava
SA basalts
SA pillow structure
SA spilite

pillow structure   (1978)
Included use under sedimentary structures soft sediment deformation until 1976; now use ball-and-pillow.
SA ball-and-pillow
SA lava
SA pillow lava

pilot plants   (1981)
SA economics

Pilot Range   (1985)
NE Nevada and NW Utah.
IN Index states and counties as applicable.
CO N403000N413000
   W1140000W1160000
BT United States
SA Basin and Range Province
SA Box Elder County Utah
SA Elko County Nevada
SA Nevada
SA Utah

Pilot Shale   (1989)

E Nevada and Utah. Underlies Joana Limestone; overlies Guilmette Formation.
IN Index ages as applicable.
BT Paleozoic
SA Lower Mississippian
SA Nevada
SA Upper Devonian
SA Utah

Pilsen Basin   (1978)
Valley in W Czechoslovakia in which city of Pilsen is located. Also search Pilsen.
UF Plzen Basin
BT Bohemia
BT Czech Republic
BT Czechoslovakia
BT Central Europe
BT Europe

Piltdown
As of 1993, no longer a valid term for GeoRef. See Piltdown man or Piltdown England as applicable.

Piltdown England   (1993)
Locality in East Sussex in SE England.
BT Sussex England
BT England
BT Great Britain
BT United Kingdom
BT Western Europe
BT Europe
SA Piltdown man

Piltdown man   (1993)
Before 1993, also search Piltdown and fossil man.
SA fossil man
SA paleontology
SA Piltdown England

Pima County
Valid through 1988. Search in combination with state term. After 1988, use specific county-state term.

Pima County Arizona   (1989)
S Arizona. Before 1989, also search Pima County AND Arizona.
CO N312400N323000
   W1102800W1132200
BT Arizona
BT United States
NT Tucson Arizona
NT Tucson Mountains
SA Rincon Mountains
SA Santa Catalina Mountains
SA Santa Cruz River

Pinaceae   (1978)
Family. Autoposting of this term began in 1978.
BT Coniferales
BT gymnosperms
BT Spermatophyta
BT Plantae
NT Abies
NT Larix
NT Pinus
SA Taxodiaceae

Pinal County
Valid through 1988. Search in combination with state term. After 1988, use specific county-state term.

Pinal County Arizona   (1989)
S central Arizona. Before 1989, also search Pinal County AND Arizona.
CO N323000N331300
   W1102800W1122600
BT Arizona

BT United States
NT Picacho Arizona
NT San Manuel Arizona
SA Santa Cruz River
SA Superstition Mountains

Pinar del Rio
No longer a valid term for GeoRef. See Pinar del Rio Cuba.

Pinar del Rio Cuba   (1993)
Province and city of W Cuba.
BT Cuba
BT Greater Antilles
BT Antilles
BT West Indies
BT Caribbean region
SA Sierra de los Organos

pinch-outs
Preferred term pinchouts.

Pinchi Lake   (1978)
Central British Columbia.
BT British Columbia
BT Western Canada
BT Canada

pinchouts
use stratigraphic wedges

Pindus Mountains   (1978)
W central and NW Greece. Also search Pindus.
IN Index administrative regions as applicable.
BT Greece
BT Southern Europe
BT Europe
SA Epirus Greece
SA Hellenides
SA Sterea Ellas
SA Thessaly Greece

Pine Barrens   (1978)
Region of coastal plain of S and SE New Jersey.
IN Index counties or regions as applicable.
BT New Jersey
BT United States

Pine Creek Geosyncline   (1985)
N and central Northern Territory, Australia.
BT Northern Territory Australia
BT Australia
BT Australasia
SA Alligator Rivers Field

Pine Mountain Window   (1985)
W Virginia, SE Kentucky and NE Tennessee.
IN Index states as applicable.
BT United States
SA Appalachian Plateau
SA Cumberland Plateau
SA Kentucky
SA Tennessee
SA Virginia

Pine Point District
As of 1993, no longer a valid term for GeoRef.
use Pine Point mining district

Pine Point mining district   (1993)
S central Mackenzie District, Northwest Territories. Mississippi valley-type lead-zinc deposits. Before 1993, also search Pine Point.
UF Pine Point District
BT Mackenzie District Northwest Territories
BT Northwest Territories
BT Western Canada
BT Canada
SA lead-zinc deposits
SA mines

Pine Valley Mountains   (1978)

Washington County in SW Utah.
BT Washington County Utah
BT Utah
BT United States

Pinedale Anticline   (1989)
In Green River basin, W Wyoming.
CO N421500N430000
   W1090400W1104000
BT Sublette County Wyoming
BT Wyoming
BT United States
SA Green River basin

Pinedale Glaciation   (1981)
Before 1981, also search Pinedale.
BT upper Quaternary
BT Quaternary
BT Cenozoic
SA Bull Lake Glaciation
SA Colorado
SA Montana
SA Wyoming

Pinellas County
Valid through 1988. Search in combination with state term. After 1988, use specific county-state term.

Pinellas County Florida   (1989)
W Florida. Before 1989, also search Pinellas County AND Florida.
CO N273500N281000
   W0823500W0825300
BT Florida
BT United States
SA Tampa Bay

Piney Point Formation   (1989)
Above Nanjemoy Formation and below Calvert Formation.
BT upper Eocene
BT Eocene
BT Paleogene
BT Tertiary
BT Cenozoic
SA Delaware
SA Maryland
SA New Jersey
SA Virginia

pingos   (1978)
Autoposting of periglacial features to this term began in 1989.
BT periglacial features
SA frost action
SA glacial geology
SA permafrost
SA taliks

Pinnacle Formation   (1989)
In Camels Hump Group. S Quebec and NW Vermont.
BT Precambrian
SA Quebec
SA Vermont

pinnacle reefs   (1978)
UF coral pinnacle
UF reef pinnacle
BT reefs
SA platform reefs

Pinnacles National Monument   (1978)
W California. Autoposting of broader terms to this term began in 1989.
BT San Benito County California
BT California
BT United States
SA national monuments

Pinney Hollow Formation   (1989)

In Camels Hump Group. Lower Cambrian and/or Proterozoic. Includes Hancock Member.
BT Lower Cambrian
BT Cambrian
BT Paleozoic
SA Vermont

Pinnipedia   (1981)
Suborder. Autoposting of Eutheria and Theria to this term began in 1989.
BT Carnivora
BT Eutheria
BT Theria
BT Mammalia
BT Tetrapoda
BT Vertebrata
BT Chordata

Pinon Range   (1978)
Also search Pinon.
IN   Index county or region as applicable.
SA Nevada

pinquite
  use nontronite

Pinus   (1978)
Genus.
BT Pinaceae
BT Coniferales
BT gymnosperms
BT Spermatophyta
BT Plantae

Pinyon Conglomerate   (1978)
Unconformably overlies Harebell Formation and unconformably underlies Colter Formation. Yellowstone National Park in NW Wyoming.
BT Paleocene
BT Paleogene
BT Tertiary
BT Cenozoic
SA Wyoming

Pinzgau   (1981)
Valley of the upper Salzach River in state of Salzburg. W central Austria.
CO N471500N472000
   E0130400E0121000
BT Salzburg State Austria
BT Austria
BT Central Europe
BT Europe
SA Salzach River

Pioche Shale   (1989)
Lower and Middle Cambrian. NW Arizona, E California, Nevada and W Utah.
BT Cambrian
BT Paleozoic
SA Arizona
SA California
SA Nevada
SA Utah

Pioneer Mountains   (1978)
Custer and Blaine counties in S central Idaho.
IN   Index counties as applicable.
BT Idaho
BT United States
SA Blaine County Idaho
SA Custer County Idaho

Pioneer Program   (1981)
Autoposting of this term to specific Pioneer missions began in 1989.
NT Pioneer 10
NT Pioneer 11
SA Aphrodite Terra
SA extraterrestrial geology
SA Ishtar Terra
SA Jupiter

SA planetology
SA remote sensing
SA Saturn
SA Venus

Pioneer 10   (1978)
Autoposting of Pioneer Program to this term began in 1989.
BT Pioneer Program
SA Jupiter
SA remote sensing

Pioneer 11   (1989)
BT Pioneer Program
SA Jupiter
SA remote sensing
SA Saturn

pipelines   (1978)
SA foundations
SA horizontal drilling
SA marine installations
SA natural gas
SA petroleum
SA rock mechanics
SA soil mechanics
SA transportation
SA underground installations

pipes   (1978)
BT intrusions
NT breccia pipes
SA chimneys
SA diatremes
SA igneous rocks
SA plugs
SA volcanic necks
SA volcanoes

piping   (1989)
Used for the erosion process.
UF tunnel erosion
SA erosion
SA soil erosion
SA water erosion

Pirabas Formation   (1978)
BT Miocene
BT Neogene
BT Tertiary
BT Cenozoic
SA Brazil

piracy
  use stream capture

Pisa
  No longer a valid term for GeoRef. See Pisa Italy.

Pisa Italy   (1993)
City and province in NW Tuscany, W Italy.
BT Tuscany Italy
BT Italy
BT Southern Europe
BT Europe

Piscataquis County
  Valid through 1988. Search in combination with state term. After 1988, use specific county-state term.

Piscataquis County Maine   (1989)
Central Maine. Before 1989, also search Piscataquis County AND Maine.
CO N450000N463500
   W0685000W0694500
BT Maine
BT United States
NT Mount Katahdin

Pisces   (1978)
Before 1981, Agnatha and Heterostraci were included as narrower terms. Autoposting of this term began in 1978.
BT Vertebrata

BT Chordata
NT Acanthodii
NT Chondrichthyes
NT Osteichthyes
NT Placodermi
SA Agnatha
SA eggs
SA fish
SA gastroliths
SA Heterostraci
SA ichthyoliths
SA otoliths

Pisgah Crater   (1978)
IN   Index counties or regions as applicable.
SA California

Pishpek Kyrgyzstan
  use Bishkek Kyrgyzstan

Pismo Basin   (1989)
SW California.
BT San Luis Obispo County California
BT California
BT United States

pisolites   (1978)
SA limestone
SA pisoliths
SA pisolitic texture
SA sedimentary rocks

pisoliths   (1978)
SA pellets
SA pisolites
SA pisolitic texture
SA sedimentary rocks
SA sediments

pisolitic texture   (1978)
Before 1978, also search pisolitic AND specific sediment type or sedimentary rocks.
BT textures
SA pisolites
SA pisoliths
SA sedimentary rocks
SA sediments

pistacite
  use epidote

pit sections   (1981)
SA cross sections
SA mineral exploration
SA soils

pitchblende   (1978)
UF nasturan
BT oxides
SA uraninite
SA uranium ores

pitching folds
  use plunging folds

pitchstone   (1978)
BT glasses
BT volcanic rocks
BT igneous rocks
SA volcanic glass

Pitesti
  No longer a valid term for GeoRef. As of 1993 see Pitesti Romania.

Pitesti Romania   (1993)
City in Arges County in S central Romania. Before 1993 also search Pitesti AND Romania.
BT Walachia
BT Romania
BT Southern Europe
BT Europe

Pithecanthropus   (1978)
Genus. Autoposting of Eutheria and Theria to this term began in 1989.

BT Hominidae
BT Primates
BT Eutheria
BT Theria
BT Mammalia
BT Tetrapoda
BT Vertebrata
BT Chordata

Pitkin County
  Valid through 1988. Search in combination with state term. After 1988, use specific county-state term.

Pitkin County Colorado   (1989)
W central Colorado. Before 1989, also search Pitkin County AND Colorado.
CO N385800N392000
   W1062500W1073000
BT Colorado
BT United States

Pitkin Limestone   (1978)
In Arkansas, underlies Cane Hill Member of Hale Formation. N Arkansas, and E Oklahoma.
BT Upper Mississippian
BT Mississippian
BT Carboniferous
BT Paleozoic
SA Arkansas
SA Oklahoma

Piton de la Fournaise   (1978)
Volcanic peak on Reunion Island. In 1981, broader term changed from Mascarene Islands and Indian Ocean to Reunion.
BT Reunion
BT Mascarene Islands
BT Indian Ocean Islands

Piton des Neiges   (1978)
Highest peak on Reunion Island. In 1981, broader term changed from Mascarene Islands and Indian Ocean to Reunion.
BT Reunion
BT Mascarene Islands
BT Indian Ocean Islands

pits   (1993)
SA chemical weathering
SA corrosion
SA etching
SA fission tracks
SA furrows
SA mining
SA open-pit mining
SA quarries
SA solution
SA solution features
SA waste disposal

Pittsburg County
  Valid through 1988. Search in combination with state term. After 1988, use specific county-state term.

Pittsburg County Oklahoma   (1989)
SE Oklahoma. Before 1989, also search Pittsburg County AND Oklahoma.
CO N343500N351800
   W0952000W0960500
BT Oklahoma
BT United States

Pittsburgh
  Valid through 1988. Search in combination with state term. After 1988, use specific city-state term.

Pittsburgh Coal   (1989)
BT Pennsylvanian
BT Carboniferous

BT Paleozoic
SA Monongahela Group
SA Ohio
SA Pennsylvania
SA West Virginia

Pittsburgh coal basin (1978)
Bituminous coal and steel producing region in the Pittsburgh area.
BT Pennsylvania
BT United States
SA coal fields

Pittsburgh Pennsylvania (1989)
City in SW Pennsylvania. Before 1989, search Pittsburgh AND Pennsylvania.
CO N402600N402600
    W0800000W0800000
BT Allegheny County Pennsylvania
BT Pennsylvania
BT United States

Pittsylvania County Virginia (1989)
S Virginia. Before 1989, search Pittsylvania County AND Virginia.
CO N363300N370800
    W0790600W0794300
BT Virginia
BT United States
NT Danville Virginia

Piute County
Valid through 1988. Search in combination with state term. After 1988, use specific county-state term.

Piute County Utah (1989)
SW central Utah. Before 1989, also search Piute County AND Utah.
CO N380800N383300
    W1115000W1123500
BT Utah
BT United States
NT Marysvale Utah
SA Tushar Mountains

pixels (1993)
Smallest part of an electronically coded picture element.
UF picture elements
SA aerial photography
SA data processing
SA digital simulation
SA imagery
SA remote sensing

PKiKP-waves (1981)
BT body waves
BT elastic waves
SA P-waves
SA seismology
SA waves

PKP-waves (1981)
BT body waves
BT elastic waves
SA P-waves
SA seismology
SA waves

PKS-waves (1981)
BT body waves
BT elastic waves
SA P-waves
SA S-waves
SA seismology
SA waves

Placentalia
    use Eutheria

Placer County
Valid through 1988. Search in combination with state term. After 1988, use specific county-state term.

Placer County California (1989)
Central and E California. Before 1989, also search Placer County AND California.
CO N384200N392000
    W1200000W1213000
BT California
BT United States
SA Lake Tahoe

placers (1978)
Used for a surficial mineral deposit formed by mechanical concentration of mineral particles from weathered debris.
IN Commodity. See List C. Also index type of placer [e.g. diamonds, gold ores, heavy mineral deposits, platinum ores, tin ores].
UF ore of sedimentation
NT beach placers
NT stream placers
SA diamonds
SA discoveries
SA gold ores
SA heavy mineral deposits
SA heavy minerals
SA metal ores
SA mineral deposits, genesis
SA mineral exploration
SA monazite deposits
SA nuggets
SA panning
SA platinum ores
SA sampling
SA sediments
SA tin ores
SA weathering

Placodermi (1978)
Class. Before 1981, Acanthodii was included as a narrower term. Autoposting of this term began in 1978.
BT Pisces
BT Vertebrata
BT Chordata
NT Arthrodira
SA Acanthodii

Placodontia (1981)
Order. Autoposted to Euryapsida 1989-1992. Autoposting of Synapsida to this term ended in 1989.
BT Diapsida
BT Reptilia
BT Tetrapoda
BT Vertebrata
BT Chordata

plagioclase (1978)
Autoposting of this term began in 1981.
BT feldspar group
BT framework silicates
BT silicates
NT albite
NT analbite
NT andesine
NT anorthite
NT bytownite
NT labradorite
NT myrmekite
NT oligoclase
NT peristerite
SA maskelynite

plagioclasite
    use anorthosite

plagiogranite (1978)
Before 1981, plutonic rocks was autoposted to this term. As of 1982, diorites is autoposted to this term.
BT diorites
BT plutonic rocks
BT igneous rocks
SA quartz diorites
SA trondhjemite

plagionite (1978)
BT sulfantimonites
BT sulfosalts

plains (1978)
Before 1993, also search lowlands.
UF lowlands
NT chenier plains
SA abyssal plains
SA alluvial plains
SA coastal plains
SA erosion features
SA floodplains
SA fluvial features
SA geomorphology
SA landforms
SA outwash plains
SA pediplains
SA peneplains
SA plateaus
SA playas
SA prairies
SA savannas
SA shore features
SA steppes
SA topography
SA tundra

Plaisancian (1978)
Europe. Above Pontian (Miocene), below Astian.
UF Plaisanzian
BT lower Pliocene
BT Pliocene
BT Neogene
BT Tertiary
BT Cenozoic

Plaisanzian
    use Plaisancian

Plan de la Tour
No longer a valid term for GeoRef. See Plan de la Tour France.

Plan de la Tour France (1993)
Village in SE France.
BT Var France
BT France
BT Western Europe
BT Europe

Plana (1978)
Town in W Bohemia, W Czechoslovakia.
SA Bohemia

planar bedding structures (1978)
Autoposting of this term to rhythmite began in 1985.
UF bedding structures, planar
BT sedimentary structures
NT bedding
NT cross-bedding
NT cross-laminations
NT cross-stratification
NT cut and fill
NT cyclothems
NT flaser bedding
NT hummocky cross-stratification
NT imbrication
NT laminations
NT massive bedding
NT rhythmic bedding
NT rhythmite
NT ripple drift-cross laminations
NT sand bodies
NT stratification
NT varves
SA bars
SA buried channels
SA channels
SA graded bedding

planation (1978)
BT erosion
SA abrasion
SA erosion surfaces
SA geomorphology

planation surfaces
    use erosion surfaces

plancheite (1978)
BT chain silicates
BT silicates

planes, fault
    use fault planes

planetary interiors (1981)
Before 1981, search planet name AND interior.
SA interior
SA lunar interior
SA planetology
SA planets

planetary rings (1985)
Before 1985, also search rings AND name of planet or planetology.
UF rings, planetary
SA extraterrestrial geology
SA planetology
SA Saturn
SA Uranus

planetesimals (1993)
UF protoplanets
SA accretion
SA planetology
SA planets

planetoids
As of 1993, use planetesimals or asteroids if applicable.

planetology (1978)
Used for studies of more than one planet, the relationship between planets, or the solar system as a whole. See also names of planets.
SA accretion
SA Apollo Program
SA Apollo-Soyuz Program
SA asteroids
SA atmosphere
SA comae
SA condensation
SA cosmic dust
SA cosmochemistry
SA cosmochronology
SA Earth
SA Earth-Moon couple
SA exobiology
SA Explorer Program
SA extraterrestrial geology
SA icy satellites
SA impacts
SA interplanetary comparison
SA interplanetary space
SA Jupiter
SA Luna Program
SA Magellan Program
SA Mariner Program
SA Mars
SA Mercury Planet
SA meteorites
SA meteors
SA Moon
SA Neptune
SA outer planets
SA parent bodies
SA Pioneer Program
SA planetary interiors

SA planetary rings
SA planetesimals
SA planets
SA Pluto
SA polar regions
SA Saturn
SA Skylab
SA solar system
SA Spacelab Program
SA Sun
SA surface features
SA Surveyor Program
SA tektites
SA terrestrial comparison
SA terrestrial planets
SA Uranus
SA Venera Program
SA Venus
SA Viking Program
SA Voyager Program

planets (1978)
SA asteroids
SA condensation
SA Earth
SA extraterrestrial geology
SA interplanetary comparison
SA Jupiter
SA Mars
SA Mercury Planet
SA meteoroids
SA Moon
SA Neptune
SA outer planets
SA planetary interiors
SA planetesimals
SA planetology
SA Pluto
SA Saturn
SA solar system
SA Sun
SA surface features
SA terrestrial comparison
SA terrestrial planets
SA Uranus
SA Venus

plankton (1978)
UF microplankton
NT nannoplankton
NT phytoplankton
NT zooplankton
SA algae
SA diatoms
SA Dinoflagellata
SA foraminifera
SA Ostracoda
SA palynomorphs
SA planktonic taxa
SA Protista
SA Radiolaria

planktonic
A valid term through 1977. After 1977, use planktonic taxa.

planktonic taxa (1978)
Before 1978, also search planktonic.
UF taxa, planktonic
SA foraminifera
SA plankton

planning (1978)
SA associations
SA highways
SA land use
SA management
SA programs
SA railroads
SA regional planning
SA urban planning

Planolites (1978)
Genus.
BT ichnofossils

Planorbis (1978)
BT Gastropoda

BT Mollusca
BT Invertebrata

Planosols (1978)
BT soils
SA Intrazonal soils
SA soil group

Plantae (1978)
IN In indexing, may be used in a general sense when specific plants are unknown.
NT bryophytes
NT pteridophytes
NT Pteropsida
NT Spermatophyta
NT thallophytes
SA affinities
SA cuticles
SA floral studies
SA flowers
SA fossil wood
SA leaves
SA paleobotany
SA palynomorphs
SA phytoliths
SA plants
SA roots
SA vegetation

plants (1985)
Common name for Plantae.
IN Index for all non-paleontologic studies of fossils.
NT algal flora
NT angiosperm flora
NT ferns
NT gymnosperm flora
SA biostratigraphy
SA flowers
SA Plantae
SA roots
SA seeds

Plaquemines Parish Louisiana (1989)
Extreme SE Louisiana, bordering the Gulf of Mexico. Before 1989, search Plaquemines Parish AND Louisiana.
CO N285400N295300 W0890000W0900600
BT Louisiana
BT United States
SA Barataria Bay
SA East Bay
SA South Pass

plasma
As of 1993, no longer a valid term for GeoRef. Usually out-of-scope. Was not to be used for soil plasma. Used with interplanetary space and plasma emission spectroscopy.

plasma emission spectroscopy (1985)
Before 1985, also search emission spectroscopy AND plasma.
SA emission spectroscopy
SA inductively coupled plasma methods
SA spectroscopy

plasma instabilities
As of 1993, no longer a valid term for GeoRef. Usually out-of-scope. Used with magnetosphere.

plasma motion
As of 1993, no longer a valid term for GeoRef. Usually out-of-scope. Used with magnetosphere.

plasmapause
As of 1993, no longer a valid term for GeoRef. Usually out-of-scope. Used with magnetosphere.

plasmasphere
As of 1993, no longer a valid term for GeoRef. Usually out-of-scope. Used with magnetosphere.

plaster of paris
use gypsum

plaster stone
use gypsum

plastic deformation (1989)
BT deformation
SA engineering geology
SA plastic flow
SA plasticity
SA rheology
SA rock mechanics
SA soil mechanics
SA viscoplastic materials

plastic flow (1978)
Also search plastic.
UF plastic strain
SA deformation
SA flow lines
SA plastic deformation
SA plasticity

plastic materials (1978)
Before 1978, also search plastic AND materials.
SA elastic materials
SA elastoplastic materials
SA elastoviscoplastic materials
SA engineering geology
SA materials
SA plasticity
SA rock mechanics
SA soil mechanics
SA viscoplastic materials

plastic strain
use plastic flow

plasticity (1978)
SA Atterberg limits
SA compressibility
SA constitutive equations
SA deformation
SA elasticity
SA elastoplastic materials
SA elastoviscoplastic materials
SA mechanical properties
SA plastic deformation
SA plastic flow
SA plastic materials
SA rock mechanics
SA viscoplastic materials
SA viscosity

Platanus (1989)
Genus.
BT Dicotyledoneae
BT angiosperms
BT Spermatophyta
BT Plantae

plate boundaries (1981)
Before 1981, also search margins; plate margins; plate boundary.
UF plate margins
SA accreting plate boundary
SA active margins
SA passive margins
SA plate convergence
SA plate divergence
SA plate tectonics
SA plates
SA rifting

plate collision (1981)
Before 1981, also search collision; collisions.
SA plate tectonics
SA plates

plate convergence (1981)
Before 1981, search convergence.

SA active margins
SA plate boundaries
SA plate tectonics

plate divergence (1993)
Before 1993, also search divergence.
UF divergence, plate
SA accreting plate boundary
SA accretion
SA passive margins
SA plate boundaries
SA plate tectonics
SA rift zones
SA rifting
SA sea-floor spreading

plate geometry (1981)
Before 1981, also search geometry AND plate tectonics.
SA geometry
SA plate tectonics
SA plates
SA triple junctions

plate margins
use plate boundaries

plate motion
Not a valid term for GeoRef. Use movement in conjunction with plate tectonics.

plate movement
Not a valid term for GeoRef. Use movement in conjunction with plate tectonics.

plate rotation (1982)
Before 1982, search rotation AND plate tectonics.
SA plate tectonics
SA plates
SA rotation

plate tectonics (1978)
Used for global tectonics based on an Earth model characterized by a small number of large, broad, thick plates. Before 1976, articles discussing plate tectonics were included under tectonophysics.
IN Used for extremely large-scale events, usually affecting oceanic regions.
SA accreting plate boundary
SA accretionary wedges
SA active margins
SA African Plate
SA Antarctic Plate
SA apparent polar wandering
SA Arabian Plate
SA aseismic margins
SA asthenosphere
SA Australian Plate
SA back-arc basins
SA Benioff zone
SA Caribbean Plate
SA Cascadia subduction zone
SA Cocos Plate
SA compression tectonics
SA continental crust
SA continental drift
SA continental margin
SA convection
SA convection currents
SA crust
SA crustal shortening
SA crustal thinning
SA deformation
SA Earth
SA earthquakes
SA Eurasian Plate
SA exotic terranes
SA extension tectonics
SA Farallon Plate
SA fault zones
SA faults
SA fore-arc basins

SA fracture zones
SA Galapagos Rift
SA geodynamics
SA geosynclines
SA geotectonic maps
SA heat flow
SA hot spots
SA Indian Plate
SA indicators
SA Indo-Australian Plate
SA intraplate processes
SA island arcs
SA Juan de Fuca Plate
SA Kula Plate
SA Laurentia
SA lithosphere
SA mantle
SA marginal basins
SA marginal seas
SA mechanics
SA Median Tectonic Line
SA melange
SA metamorphic belts
SA metamorphic core complexes
SA microcontinents
SA microplates
SA Mohorovicic discontinuity
SA movement
SA Nazca Plate
SA North American Plate
SA obduction
SA ocean basins
SA oceanic crust
SA ophiolite complexes
SA oroclines
SA orogeny
SA Pacific Plate
SA paleogeography
SA paleomagnetism
SA partial melting
SA passive margins
SA Philippine Sea Plate
SA plate boundaries
SA plate collision
SA plate convergence
SA plate divergence
SA plate geometry
SA plate rotation
SA plates
SA plumes
SA polar wandering
SA rift zones
SA rifting
SA sea-floor spreading
SA seismology
SA seismotectonics
SA slabs
SA South American Plate
SA spherical harmonic analysis
SA spreading centers
SA subduction
SA subduction zones
SA suture zones
SA taphrogeny
SA tectonics
SA tectonophysics
SA tectonosphere
SA terranes
SA tesserae
SA transform faults
SA transition zones
SA transpression
SA trenches
SA triple junctions
SA underplating
SA volcanic belts
SA volcanology
SA West Siberian Plate
SA Wilson cycle
SA Wrangellia
SA Yangtze Plate

plate-bearing tests (1981)
SA soil mechanics
SA testing

plateau basalts
   use flood basalts

plateaus (1978)
SA geomorphology
SA highlands
SA landforms
SA mesas
SA plains
SA topography
SA uplands

plates (1978)
SA active margins
SA African Plate
SA Antarctic Plate
SA Arabian Plate
SA Australian Plate
SA Benioff zone
SA Caribbean Plate
SA Cocos Plate
SA crust
SA Eurasian Plate
SA Farallon Plate
SA Indian Plate
SA Indo-Australian Plate
SA Juan de Fuca Plate
SA Kula Plate
SA mantle
SA microplates
SA Nazca Plate
SA North American Plate
SA Pacific Plate
SA Philippine Sea Plate
SA plate boundaries
SA plate collision
SA plate geometry
SA plate rotation
SA plate tectonics
SA South American Plate
SA subduction zones
SA West Siberian Plate
SA Yangtze Plate

platform reefs (1993)
BT reefs
SA carbonate platforms
SA fringing reefs
SA patch reefs
SA pinnacle reefs

platforms (1978)
SA carbonate platforms
SA engineering geology
SA gravity platforms
SA marine platforms
SA tectonic platforms

platinum (1978)
Chemical element. As of 1981, use platinum ores for platinum as a commodity. Autoposting of platinum group and metals to this term began in 1989.
UF Pt
BT platinum group
BT metals
SA heavy metals
SA niggliite
SA platinum minerals
SA platinum ores
SA siderophile elements

platinum group (1985)
Six elements from group VIII of the periodic table. Autoposting of this term to specific elements in the platinum group began in 1989. Autoposting of metals to this term began in 1989.
UF PGE
UF platinum group elements
UF platinum metals
BT metals
NT iridium
NT osmium
NT palladium
NT platinum

NT rhodium
NT ruthenium
SA platinum minerals

platinum group elements
   use platinum group

platinum metals
   use platinum group

platinum minerals (1982)
IN May be used indexed with commodities (list C).
SA minerals
SA niggliite
SA ore minerals
SA platinum
SA platinum group
SA platinum ores
SA sperrylite

platinum ores (1981)
Before 1981, also search (platinum OR Pt) AND (deposit OR deposits OR ore OR ores OR economic) in the basic index. Autoposting of metal ores to this term began in 1985.
IN Commodity. See List C.
BT metal ores
SA iridium ores
SA nuggets
SA osmium ores
SA palladium ores
SA placers
SA platinum
SA platinum minerals
SA rhodium ores
SA ruthenium ores

Platte County
Valid through 1988. Search in combination with state term. After 1988, use specific county-state term.

Platte County Missouri (1989)
W Missouri. Before 1989, also search Platte County AND Missouri.
CO N391000N393300 W0943700W0950700
BT Missouri
BT United States

Platte County Nebraska (1989)
E central Nebraska. Before 1989, also search Platte County AND Nebraska.
CO N412000N414300 W0971500W0975000
BT Nebraska
BT United States

Platte County Wyoming (1989)
SE Wyoming. Before 1989, also search Platte County AND Wyoming.
CO N413800N423700 W1043800W1051800
BT Wyoming
BT United States

Platte River (1978)
Formed by confluence of the North Platte River and South Platte River in SW central Nebraska. Flows E into the Missouri River below Omaha.
IN Index counties or regions applicable.
SA Iowa
SA Missouri
SA Nebraska
SA Platte River basin
SA South Platte River

Platte River basin (1985)
IN Index states as applicable.
SA Colorado

SA Great Plains
SA Nebraska
SA Platte River
SA United States
SA Wyoming

Platteville Formation (1978)
Includes Mifflin Limestone Member, Quimbys Mill Member, Pecatonica Member, McGregor Member, and Carimona Member, Hidden Falls Member, Magnolia Member, Glenwood Shale. E Iowa, NW Illinois, S Minnesota, and SW Wisconsin.
BT Middle Ordovician
BT Ordovician
BT Paleozoic
SA Glenwood Shale
SA Illinois
SA Iowa
SA Minnesota
SA Saint Peter Sandstone
SA Wisconsin

Plattsburg Limestone (1978)
In Lansing Group. Includes Merriam Limestone, Hickory Creek Shale and Spring Hill Limestone members. SW Iowa, E Kansas, NW Missouri, and SE Nebraska.
BT Missourian
BT Upper Pennsylvanian
BT Pennsylvanian
BT Carboniferous
BT Paleozoic
SA Iowa
SA Kansas
SA Lansing Group
SA Missouri
SA Nebraska

Plattsmouth Limestone Member (1978)
Of Oread Limestone. SW Iowa, E Kansas, NW Missouri, and SE Nebraska.
BT Virgilian
BT Upper Pennsylvanian
BT Pennsylvanian
BT Carboniferous
BT Paleozoic
SA Iowa
SA Kansas
SA Missouri
SA Nebraska
SA Oread Limestone

Platycarya (1989)
Genus.
BT Dicotyledoneae
BT angiosperms
BT Spermatophyta
BT Plantae

Platycopida (1981)
Autoposting of Podocopida to this term began in 1989. As of 1990, microfossils, Crustacea, Mandibulata, and Arthropoda are autoposted to this term. This term has multiple hierarchies.
UF Platycopina
BT1 Podocopida
BT1 Ostracoda
BT1 Crustacea
BT1 Mandibulata
BT1 Arthropoda
BT1 Invertebrata
BT2 Podocopida
BT2 Ostracoda
BT2 microfossils
NT Cytherellidae

Platycopina
   use Platycopida

Platyrrhina (1981)
Infraorder? Commonly referred to as the New World monkeys. Autoposting of Eutheria and Theria to this term began in 1989.
BT simians
BT Primates
BT Eutheria
BT Theria
BT Mammalia
BT Tetrapoda
BT Vertebrata
BT Chordata

playas (1978)
BT lacustrine features
SA basins
SA coastal environment
SA deserts
SA geomorphology
SA plains
SA sebkha environment

Pleasant Bayou (1985)
SE Texas. Autoposting of broader terms to this term began in 1989.
BT Brazoria County Texas
BT Texas
BT United States
SA Gulf Coastal Plain

Pleasant Island
 use Nauru Island

Pleasanton Group (1978)
Comprises Warrensburg Channel Sandstone, Sni Mills Limestone, Dawson Coal Horizon, Wayside Sandstone, Exline Limestone, Knobtown Sandstone, Hepler Sandstone, Checkerboard Limestone, and Ovid Coal. SW Iowa, E Kansas, NW Missouri, and NE Oklahoma.
BT Missourian
BT Upper Pennsylvanian
BT Pennsylvanian
BT Carboniferous
BT Paleozoic
SA Iowa
SA Kansas
SA Missouri
SA Oklahoma

Plecoptera (1978)
BT Insecta
BT Mandibulata
BT Arthropoda
BT Invertebrata

Plectoptera (1978)
BT Insecta
BT Mandibulata
BT Arthropoda
BT Invertebrata

Pleistocene (1978)
Glacial epoch. World. Above Pliocene (Tertiary), below Holocene. Autoposting of this term began in 1978. Autoposting to extinct lakes began in 1993.
BT Quaternary
BT Cenozoic
NT Bandelier Tuff
NT Beaumont Clay
NT Bishop Tuff
NT Bootlegger Cove Clay
NT Champlain Sea
NT Crooked Creek Formation
NT Glasford Formation
NT Gubik Formation
NT Gunz
NT Irvingtonian
NT Karewa Group
NT Key Largo Limestone
NT Kingsdown Formation
NT Laetoli Beds
NT Lake Agassiz
NT Lake Algonquin
NT Lake Chicago
NT Lake Lahontan
NT Lake Maumee
NT Lake Missoula
NT lower Pleistocene
NT McPherson Formation
NT Miami Limestone
NT middle Pleistocene
NT Mindel
NT Mindel/Riss Interglacial
NT Nihewan Formation
NT Pearlette Volcanic Ash
NT Peoria Loess
NT Presumpscot Formation
NT Riss
NT Riss/Wurm Interglacial
NT Roxana Silt
NT Ryukyu Group
NT Santa Barbara Formation
NT Semata Formation
NT Shimosa Group
NT upper Pleistocene
NT Wedron Formation
NT Wicomico Formation
SA Aurignacian
SA Blancan
SA Central Polish Glaciation
SA Croatan Formation
SA Glenns Ferry Formation
SA Kobiwako Group
SA Koobi Fora Formation
SA Levalloisian
SA Loup Fork Group
SA Magdalenian
SA Mousterian
SA Omo Group
SA Osaka Group
SA Paleolithic
SA Ringold Formation
SA San Juan Formation
SA Santa Clara Formation
SA Santa Fe Group
SA Santa Maria Formation
SA Saugus Formation
SA Shungura Formation
SA Siwallk System
SA Tulare Formation
SA upper Paleolithic
SA Verde Formation
SA Yakataga Formation

plenargyrite
 use matildite

pleochroism (1978)
UF polychroism
SA crystallography
SA optical properties

Plesiosauria (1978)
Suborder.
BT Sauropterygia
BT Diapsida
BT Reptilia
BT Tetrapoda
BT Vertebrata
BT Chordata

plessite (1978)
BT alloys
SA kamacite
SA meteorites
SA taenite

Plettenberg
 No longer a valid term for GeoRef. See Plettenberg Germany.

Plettenberg Germany (1993)
City NE of Cologne in W central Germany.
BT North Rhine-Westphalia Germany
BT Germany
BT Central Europe
BT Europe

Pleven
 No longer a valid term for GeoRef. See Pleven Bulgaria.

Pleven Bulgaria (1993)
Province and city in N Bulgaria.
BT Bulgaria
BT Southern Europe
BT Europe
NT Teteven Bulgaria

Pliensbachian (1978)
Europe.
BT middle Liassic
BT Lower Jurassic
BT Jurassic
BT Mesozoic

plinian-type eruptions (1982)
SA eruptions
SA explosive eruptions
SA volcanism
SA volcanology

Pliocene (1978)
World. Above Miocene, below Pleistocene (Quaternary). Autoposting of this term began in 1978.
BT Neogene
BT Tertiary
BT Cenozoic
NT Ash Hollow Formation
NT Bakhtiari Formation
NT Bidahochi Formation
NT Cimmerian
NT Citronelle Formation
NT Dacian
NT Gauss Epoch
NT Goliad Sand
NT Hadar Formation
NT lower Pliocene
NT middle Pliocene
NT Narita Formation
NT Ogallala Formation
NT Pico Formation
NT Ringold Formation
NT San Diego Formation
NT upper Pliocene
NT Uvalde Gravel
NT Valentine Formation
SA Aguacate Group
SA Blancan
SA Browns Park Formation
SA Cohansey Formation
SA Columbia River Basalt Group
SA Croatan Formation
SA Cuddalore Series
SA Deschutes Formation
SA Glenns Ferry Formation
SA Imperial Formation
SA Kobiwako Group
SA Koobi Fora Formation
SA Loup Fork Group
SA Meotian
SA Muddy Creek Formation
SA Neyveli Lignite
SA Omo Group
SA Osaka Group
SA Pannonian
SA Purisima Formation
SA Salt Lake Formation
SA Santa Clara Formation
SA Santa Fe Group
SA Santa Maria Formation
SA Saugus Formation
SA Shungura Formation
SA Sisquoc Formation
SA Siwalik System
SA Snoqualmie Batholith
SA Tulare Formation
SA upper Neogene
SA Verde Formation
SA Warkalli Formation
SA Wildcat Group
SA Yakataga Formation

pliomagmatic zone
 use eugeosynclines

Pliomys (1978)
Genus. Autoposting of Eutheria and Theria to this term began in 1989.
BT Cricetidae
BT Myomorpha
BT Rodentia
BT Eutheria
BT Theria
BT Mammalia
BT Tetrapoda
BT Vertebrata
BT Chordata

Plock
 No longer a valid term for GeoRef. As of 1993 see Plock Poland.

Plock Poland (1993)
Province and city in central Poland. Before 1993 also search Plock AND Poland.
UF Plokz
BT Poland
BT Central Europe
BT Europe
NT Leczyca Poland

Ploesti
 No longer a valid term for GeoRef. As of 1993 see Ploesti Romania.

Ploesti Romania (1993)
City N of Bucharest in Prahova County. Before 1993 also search Ploesti AND Romania.
UF Ploiesti
BT Prahova Romania
BT Walachia
BT Romania
BT Southern Europe
BT Europe

Ploiesti
 use Ploesti Romania

Plokz
 use Plock Poland

Plotina Deposit (1993)
Loukhi Lake region.
IN Index Karelia Russian Federation as applicable.
SA Karelia Russian Federation
SA mica deposits
SA muscovite
SA pegmatite

Plovdiv
 No longer a valid term for GeoRef. See Plovdiv Bulgaria.

Plovdiv Bulgaria (1993)
Province including the city in S central Bulgaria.
BT Bulgaria
BT Southern Europe
BT Europe
NT Madan Bulgaria
NT Panagyurishte Bulgaria

plugs (1978)
BT intrusions
SA breccia pipes
SA lava
SA pipes
SA volcanic necks
SA volcanoes

Plum Island (1989)
E Essex County, NE Massachusetts, bordering the Atlantic Ocean. Also in Suffolk County, SE New York.

IN Index states and counties as applicable.
SA Essex County Massachusetts
SA Massachusetts
SA New York
SA Suffolk County New York

Plumas County
Valid through 1988. Search in combination with state term. After 1988, use specific county-state term.

Plumas County California (1989)
NE California. Before 1989, also search Plumas County AND California.
CO N393500N402500
W1201000W1213000
BT California
BT United States
SA Lassen Volcanic National Park

plumbogummite (1978)
BT phosphates

plumes (1978)
SA hot spots
SA plate tectonics

plunging folds (1978)
Folds defined on the basis of orientation in relation to the geographic horizontal plane. Before 1978, also search folds AND plunging.
UF pitching folds
BT folds
SA orientation

Pluto (1978)
SA Charon Satellite
SA outer planets
SA planetology
SA planets
SA satellites
SA solar system

plutonic
A valid term through 1977.

plutonic rocks (1978)
Before 1978, also search igneous rocks AND plutonic. From 1978-1980, was autoposted to appinite and plagiogranite. As of 1981, autoposted to quartzolite. As of 1993, autoposted to appinite, carbonatites, diabase, diorites, gabbros, granites, lamprophyres, syenites, and ultramafics.
BT igneous rocks
NT appinite
NT carbonatites
NT diabase
NT diorites
NT gabbros
NT granites
NT lamprophyres
NT quartzolite
NT syenites
NT ultramafics
SA hypabyssal rocks
SA igneous processes
SA metaplutonic rocks
SA rocks
SA schlieren

plutonium (1978)
Autoposting of actinides and metals to this term began in 1989.
UF Pu
BT actinides
BT metals
NT Pu-238
NT Pu-239
NT Pu-240
NT Pu-240/Pu-239

NT Pu-244

plutons (1978)
BT intrusions
SA batholiths
SA emplacement
SA igneous rocks
SA metasomatism

Plymouth County
Valid through 1988. Search in combination with state term. After 1988, use specific county-state term.

Plymouth County Iowa (1989)
NW Iowa. Before 1989, also search Plymouth County AND Iowa.
CO N423400N425400
W0955300W0963800
BT Iowa
BT United States

Plymouth County Massachusetts (1989)
SE Massachusetts. Before 1989, also search Plymouth County AND Massachusetts.
CO N413700N421800
W0703200W0710500
BT Massachusetts
BT United States
SA Buzzards Bay

Plzen Basin
use Pilsen Basin

Pm
use promethium

Pn-waves (1985)
BT body waves
BT elastic waves
SA P-waves
SA seismology
SA Sn-waves
SA waves

pneumatolysis (1981)
Also search pneumatolytic.
UF pneumatolytic alteration
SA contact metamorphism
SA gases
SA magmas
SA metasomatism
SA mineral deposits, genesis

pneumatolytic alteration
use pneumatolysis

Po (element)
use polonium

Po Delta (1978)
On Adriatic Sea S of Venice.
UF Po River delta
BT Veneto Italy
BT Italy
BT Southern Europe
BT Europe

Po Hai
use Bohai Bay

Po Hai Wan
use Bohai Bay

Po River (1978)
Rises in the Cottian Alps of W Piemonte and flows E into the Adriatic Sea.
IN Index autonomous regions as applicable.
BT Italy
BT Southern Europe
BT Europe
SA Emilia-Romagna Italy
SA Lombardy Italy
SA Piemonte Italy
SA Po Valley
SA Veneto Italy

Po River delta
use Po Delta

Po Valley (1978)
River valley.
IN Index autonomous regions as applicable.
BT Italy
BT Southern Europe
BT Europe
SA Emilia-Romagna Italy
SA Lombardy Italy
SA Piemonte Italy
SA Po River
SA Veneto Italy

Po-210 (1978)
Autoposting of broader terms began in 1989. This term has multiple hierarchies.
BT1 radioactive isotopes
BT1 isotopes
BT2 polonium
BT2 metals

Poas (1989)
Volcano. Central Costa Rica.
UF Poas Volcano
BT Costa Rica
BT Central America

Poas Volcano
use Poas

Pobla de Segur
No longer a valid term for GeoRef. As of 1993, see Pobla de Segur Spain.

Pobla de Segur Spain (1993)
Town in W Lerida, NE Spain. Before 1993, also search Pobla de Segur and Spain.
BT Lerida Spain
BT Catalonia Spain
BT Spain
BT Iberian Peninsula
BT Southern Europe
BT Europe

Pocahontas County
Valid through 1988. Search in combination with state term. After 1988, use specific county-state term.

Pocahontas County Iowa (1989)
N central Iowa. Before 1989, also search Pocahontas County AND Iowa.
CO N423400N425500
W0942500W0945500
BT Iowa
BT United States

Pocahontas County West Virginia (1989)
E West Virginia. Before 1989, also search Pocahontas County AND West Virginia.
CO N380300N384300
W0793800W0802200
BT West Virginia
BT United States

Pocahontas Formation (1978)
In Pottsville Group. SW Virginia and S West Virginia.
BT Lower Pennsylvanian
BT Pennsylvanian
BT Carboniferous
BT Paleozoic
SA Pottsville Group
SA Virginia
SA West Virginia

Pocatello Formation (1978)
Divided into two series: lower tillite series and upper varved slate series. SE Idaho.

BT Proterozoic
BT upper Precambrian
BT Precambrian
SA Idaho

pocket calculators
use hand calculators

pockmarks (1989)
SA bottom features
SA ocean floors

Pocono Formation (1978)
Includes Benezette Limestone, Patton Shale Member, Peters Mountain Sandstone, Second Mountain Member, Cove Mountain Member, Burgoon Sandstone Member. W Maryland, E Ohio, Pennsylvania, W Virginia, N West Virginia.
BT Lower Mississippian
BT Mississippian
BT Carboniferous
BT Paleozoic
SA Maryland
SA Pennsylvania
SA Virginia
SA West Virginia

Pocos de Caldas
No longer a valid term for GeoRef. See Pocos de Caldas Brazil.

Pocos de Caldas Brazil (1993)
Resort city in SW Minas Gerais, E central Brazil.
BT Minas Gerais Brazil
BT Brazil
BT South America

Podhale (1978)
Highland basin in the Carpathians.
BT Cracow Poland
BT Poland
BT Central Europe
BT Europe
SA Polish Carpathians

podiform deposits (1985)
SA chromite ores
SA mineral deposits, genesis

Podkamennaya Tunguska
use Stony Tunguska River

Podocopida (1978)
Autoposting of this term to Bairdiomorpha began in 1985. Autoposting of this term to Cytherocopina began in 1989. As of 1990, microfossils, Crustacea, Mandibulata, and Arthropoda are autoposted to this term. This term has multiple hierarchies.
BT1 Ostracoda
BT1 Crustacea
BT1 Mandibulata
BT1 Arthropoda
BT1 Invertebrata
BT2 Ostracoda
BT2 microfossils
NT Bairdiomorpha
NT Cavellinidae
NT Cypridocopina
NT Cytherocopina
NT Darwinula
NT Healdiidae
NT Platycopida

Podolia (1978)
Region between the Dniester River and the Southern Bug River nearly coextensive with Khmelnitskiy Oblast in W Ukraine. This term has multiple hierarchies.
BT1 Ukraine
BT1 Europe
BT2 Ukraine

BT2 Commonwealth of Independent States
SA Volyn-Podolia

Podsols
use Podzols

podzolization (1989)
SA genesis
SA horizons
SA Podzols
SA soils
SA Zonal soils

Podzols (1978)
UF Podsols
BT soils
SA podzolization
SA Sod-podzolic soils
SA soil group
SA Zonal soils

POGO (1985)
Acronym.
UF Polar Orbiting Geophysical Observatory
SA geophysical methods
SA geophysical surveys
SA magnetic anomalies
SA magnetic methods
SA magnetic surveys
SA remote sensing
SA satellite methods

Pogonip Group (1978)
Lower and Middle Ordovician. Comprises House Limestone, Fillmore Formation, Wahwah Limestone, Juab Limestone, Kanosh Shale, Lehman Formation, Mazourka Formation, Goodwin Limestone, Ninemile Formation, Antelope Valley Limestone. SE California, E and S Nevada, and W Utah.
BT Ordovician
BT Paleozoic
SA Antelope Valley Limestone
SA California
SA Fillmore Formation
SA Nevada
SA Utah

Pogonophora (1981)
BT worms
BT Invertebrata

Pohang
No longer a valid term for GeoRef. As of 1993 see Pohang South Korea.

Pohang South Korea (1993)
City NNE of Pusan in SE South Korea.
BT South Korea
BT Korea
BT Far East
BT Asia

Poiana Rusca (Mountains)
use Poiana-Rusca Mountains

Poiana Rusca Massif
use Poiana-Rusca Mountains

Poiana-Rusca Massif
use Poiana-Rusca Mountains

Poiana-Rusca Mountains (1978)
Hunedoara County in SW Transylvania. Also search Poiana Rusca.
UF Poiana Rusca (Mountains)
UF Poiana Rusca Massif
UF Poiana-Rusca Massif
BT Transylvania
BT Romania
BT Southern Europe
BT Europe

Point Arena (1978)
A promontory on the Pacific Ocean in NW California.
IN Index county as applicable.
SA Mendocino County California

Point Arena California (1989)
City in NW California. Before 1989, search Point Arena AND California.
CO N385400N385400 W1234000W1234000
BT Mendocino County California
BT California
BT United States

Point Barrow (1993)
Point of land, N Alaska. Also used for Eskimo village (Nuwuk) before 1993. Before 1993, also search Point Barrow Alaska.
IN Index Northern Alaska as applicable.
SA Barrow Alaska
SA Northern Alaska

Point Barrow Alaska
No longer a valid term for GeoRef.
use Barrow Alaska

point bars (1985)
Autoposting of fluvial features to this term began in 1989. This term has multiple hierarchies.
BT1 bars
BT2 fluvial features
SA fluvial sedimentation
SA geomorphology
SA meanders

Point Conception (1985)
S coast of California. Autoposting of broader terms to this term began in 1989.
CO N342700N342900 W1202500W1202700
BT Santa Barbara County California
DT California
BT United States

point defects (1989)
SA crystal dislocations
SA crystal structure
SA crystallography
SA crystals
SA defects
SA deformation

Point Loma (1978)
Peninsula sheltering San Diego Bay.
CO N324000N324400 W1171200W1171600
BT San Diego County California
BT California
BT United States

Point Loma Formation (1985)
In Rosario Group. S California.
BT Upper Cretaceous
BT Cretaceous
BT Mesozoic
SA California

Point Lookout Sandstone (1989)
In Mesaverde Group. SW Colorado and NW New Mexico.
BT Upper Cretaceous
BT Cretaceous
BT Mesozoic
SA Colorado
SA Mesaverde Group
SA New Mexico

Point Mugu (1978)
On the Pacific Ocean in Ventura County W of Los Angeles. Also site of Point Mugu Naval Missile center.
BT Ventura County California
BT California
BT United States

Point Reyes (1978)
On the Pacific Ocean N of San Francisco in Marin County. Part of Point Reyes National Seashore.
BT Marin County California
BT California
BT United States

Point Sal (1981)
On central California coast. Autoposting of broader terms to this term began in 1989.
CO N345200N345800 W1203700W1204400
BT Santa Barbara County California
BT California
BT United States

point sources (1989)
Used only in relation to pollution.
SA nonpoint sources
SA pollution

Poisson's ratio (1978)
BT elastic constants
SA elasticity
SA engineering geology
SA mechanical properties
SA strain
SA stress
SA tunnels

Poitiers
No longer a valid term for GeoRef. See Poitiers France.

Poitiers France (1993)
City in central Vienne, W central France.
BT Vienne France
BT France
BT Western Europe
BT Europe

Poitou (1978)
Historical region of W central France.
IN Index departments as applicable.
BT France
BT Western Europe
BT Europe
SA Deux-Sevres France
SA Vendee France
SA Vienne France

Pokutye (1978)
Upland region between E Beskids on the S and the Dniester River on the N in SW. This term has multiple hierarchies.
BT1 Ukraine
BT1 Europe
BT2 Ukraine
BT2 Commonwealth of Independent States

Poland (1978)
CO N490000N544500 E0241500E0141500
BT Central Europe
BT Europe
NT Andrychow
NT Belchatow Mine
NT Belchatow Poland
NT Bialowieza
NT Bialystok Poland
NT Bielsko Poland
NT Boleslawiec Poland
NT Borzeta
NT Boza Wola
NT Bydgoszcz Poland
NT Cracow Poland
NT Cracow-Czestochowa Jura
NT Czestochowa Poland
NT Czestochowa-Zawiercie Basin
NT Dobrzyn Poland
NT Dukla Poland
NT Foresudetic Monocline
NT Galezice
NT Gdansk Poland
NT Kampinos Forest
NT Katowice Poland
NT Kielce Poland
NT Koszalin Poland
NT Kujawy
NT Leba
NT Leba Poland
NT Lodz Poland
NT Lower Lusatia
NT Lower Silesia
NT Lowicz Poland
NT Lubin Legnicki
NT Lubin Poland
NT Lublin Poland
NT Mirsk Poland
NT Northeastern Polish Plain
NT Northwestern Polish Plain
NT Nowy Sacz Poland
NT Olsztyn Poland
NT Opole Poland
NT Pieniny Mountains
NT Pilica River basin
NT Plock Poland
NT Polish Carpathians
NT Polish Lowland
NT Polish Sudeten Mountains
NT Polkowice Poland
NT Poznan Poland
NT Przemysl Poland
NT Pultusk Poland
NT Rzeszow Poland
NT Siedlce Poland
NT Silesian coal basin
NT Southeastern Polish Hills
NT Sowie Mountains
NT Suwalki Poland
NT Swiety Krzyz Mountains
NT Szczecin Poland
NT Tarnobrzeg Poland
NT Tarnow Poland
NT Tomaszow Lubelski Poland
NT Tomaszow Mazowiecki Poland
NT Torun Poland
NT Upper Silesia
NT Vistula River
NT Vistula River valley
NT Walbrzych Poland
NT Warsaw Poland
NT Warta
NT Widawka Basin
NT Wielun Poland
NT Wroclaw Poland
NT Zielona Gora Poland
SA Baltic Glaciation
SA Beschady Mountains
SA Bohemian Massif
SA Brest Basin
SA Bug region
SA Bug River
SA Carpathian Foredeep
SA Carpathian Foreland
SA Carpathians
SA Central Polish Glaciation
SA European Platform
SA Galitsiya
SA Izera Mountains
SA Karkonosze Mountains
SA Klodzko Unit
SA Krizna Nappe
SA Krosno Beds
SA Lusatia
SA North Sudetic Basin
SA Oder Valley
SA Pieniny Klippen Belt
SA Pomerania
SA Poznan Clays
SA Roztocze
SA Silesia

SA Sniezink
SA South Polish Glaciation
SA Spis
SA Subcarpathians
SA Sudeten Mountains
SA Sudetic Basin
SA Tatra Mountains
SA Werra Series
SA Western Carpathians

polar caps (1989)
As of 1993, term may be used for caps on Earth or on other bodies. From 1989 to 1992, it was restricted to polar caps of extraterrestrial bodies; North Pole, South Pole, Polar regions or Antarctic Polar Cap was used for the Earth. For Mars, search Polar Cap AND Mars. Before 1981, Polar Cap included use for the ice cap centered at the Earth's South Pole and was indexed with Antarctica. Now use Antarctic Polar Cap for this.
SA Antarctic Polar Cap
SA extraterrestrial geology
SA Mars
SA Moon
SA North Pole
SA polar regions
SA South Pole

Polar Continental Shelf (1978)
The continental shelf surrounding the Antarctic Continent. Might apply also to continental shelves of Europe, Asia, and North America facing the Arctic Ocean and lying within the Arctic Circle.
SA Antarctica
SA Asia
SA continental shelf
SA Europe
SA North America
SA polar regions

polar migration
  use polar wandering

polar motion
  use polar wandering

Polar Orbiting Geophysical Observatory
  use POGO

polar regions (1978)
Used for the region surrounding both geographic poles. Before 1993, used with Earth. May now be used with Earth or other planets. Polar regions was autoposted to Antarctica and Arctic region from 1981 through 1992. After 1992, search in combination with planet or continent or region. Before 1978, also search Polar.
IN Index planet or Antarctica or Arctic region as applicable.
SA Antarctica
SA Arctic region
SA equatorial region
SA Makarov Basin
SA Mars
SA North Pole
SA planetology
SA polar caps
SA Polar Continental Shelf
SA South Pole

Polar Urals (1978)
Section of the Urals extending from approximately N65° to the Arctic Ocean, including Vaygach Island. Autoposting of Russian Republic to this term began in 1989. This term has multiple hierarchies.
CO N650000N710000
    E0680000E0573000
BT1 Russian Federation
BT1 Commonwealth of Independent States
BT2 Urals
BT2 Commonwealth of Independent States
NT Pai-Khoi
SA Northern Urals
SA Vaygach Island

polar wandering (1978)
UF polar migration
UF polar motion
UF pole shifting
UF wandering, polar
SA apparent polar wandering
SA continental drift
SA paleolatitude
SA paleomagnetism
SA plate tectonics
SA pole positions

polarization (1978)
A general term used mostly for optical properties. Not used for magnetization.
SA birefringence
SA induced polarization
SA optical properties

polarographic analysis
  use polarography

polarography (1978)
For methods, see under chemical analysis. For data, see appropriate material.
UF polarographic analysis
SA analysis
SA chemical analysis
SA diffusion
SA electrolysis
SA quantitative analysis

polders (1985)
SA coastal environment
SA land use
SA reclamation
SA shorelines

pole positions (1978)
UF positions, pole
SA apparent polar wandering
SA magnetic field
SA paleomagnetism
SA polar wandering
SA reversals

pole shifting
  use polar wandering

Polesie
  use Polesye

Polesye (1978)
Lowland area, formerly in Poland, comprising the Pripet Marshes of the Pripet Basin. This term has multiple hierarchies.
IN Index former Soviet republics as applicable.
CO N513000N530000
    E0320000E0250000
UF Polesie
BT1 Commonwealth of Independent States
BT2 Europe
SA Belarus
SA Pripet Basin
SA Ukraine

polianite
  use pyrolusite

policies
  use policy

policy (1978)
Used as a general term.
UF policies
SA corporate policy
SA decision-making
SA Exclusive Economic Zone
SA Indian reservations
SA industrialized countries
SA international cooperation
SA legislation
SA national parks
SA NSF
SA NURE
SA programs
SA regulations
SA strategic minerals
SA United States Exclusive Economic Zone
SA wilderness areas

Polish Carpathians (1981)
Autoposting of Poland to this term began in 1989. As of 1990, Central Europe and Europe are autoposted to this term. This term has multiple hierarchies.
CO N490000N502200
    E0230200E0183200
BT1 Poland
BT1 Central Europe
BT1 Europe
BT2 Carpathians
BT2 Europe
SA Beskid Mountains
SA Nowy Sacz Poland
SA Podhale

Polish Lowland (1978)
Eastern extension of the North German Plain in NW and N Poland.
CO N513000N544500
    E0200000E0141500
BT Poland
BT Central Europe
BT Europe
SA Northwestern Polish Plain

Polish Sudeten Mountains (1981)
Autoposting of Poland to this term began in 1989. As of 1990, Central Europe and Europe are autoposted to this term. This term has multiple hierarchies.
CO N500500N511800
    E0171000E0145000
BT1 Poland
BT1 Central Europe
BT1 Europe
BT2 Sudeten Mountains
BT2 Central Europe
BT2 Europe
SA Izera Mountains
SA Klodzko Unit
SA Mirsk Poland
SA Opole Poland
SA Sowie Mountains
SA Walbrzych Poland

polished sections (1978)
SA cross sections
SA microscope methods
SA thin sections

polished surface
  use slickensides

poljes (1985)
Autoposting of solution features to this term began in 1989.
BT solution features
SA drainage basins
SA geomorphology
SA gorges
SA karst
SA valleys

Polk County
  Valid through 1988. Search in combination with state term. After 1988, use specific county-state term.

Polk County Arkansas (1989)
W Arkansas. Before 1989, also search Polk County AND Arkansas.
CO N341100N344400
    W0935700W0943100
BT Arkansas
BT United States

Polk County Florida (1989)
Central Florida. Before 1989, also search Polk County AND Florida.
CO N273500N282000
    W0811000W0820600
BT Florida
BT United States

Polk County Georgia (1989)
NW Georgia. Before 1989, also search Polk County AND Georgia.
CO N335300N340600
    W0845500W0852500
BT Georgia
BT United States

Polk County Iowa (1989)
Central Iowa. Before 1989, also search Polk County AND Iowa.
CO N412900N415200
    W0932000W0934800
BT Iowa
BT United States

Polk County Minnesota (1989)
NW Minnesota. Before 1989, also search Polk County AND Minnesota.
CO N473000N481200
    W0953200W0971000
BT Minnesota
BT United States

Polk County Missouri (1989)
SW central Missouri. Before 1989, also search Polk County AND Missouri.
CO N372500N374900
    W0931000W0934100
BT Missouri
BT United States

Polk County Nebraska (1989)
E central Nebraska. Before 1989, also search Polk County AND Nebraska.
CO N410300N412300
    W0972300W0975000
BT Nebraska
BT United States

Polk County North Carolina (1989)
SW North Carolina. Before 1989, also search Polk County AND North Carolina.
CO N351100N352300
    W0815700W0822100
BT North Carolina
BT United States

Polk County Oregon (1989)
NW Oregon. Before 1989, also search Polk County AND Oregon.
CO N444300N450500
    W1230300W1234500
BT Oregon
BT United States

Polk County Tennessee (1989)
SE Tennessee. Before 1989, also search Polk County AND Tennessee.
CO N345800N351800
    W0841900W0844800

BT Tennessee
BT United States
NT Ducktown Tennessee

Polk County Texas (1989)
E Texas. Before 1989, also search Polk County AND Texas.
CO N303100N313700 W0943300W0950900
BT Texas
BT United States

Polk County Wisconsin (1989)
NW Wisconsin. Before 1989, also search Polk County AND Wisconsin.
CO N451300N454300 W0920900W0925400
BT Wisconsin
BT United States

Polkowice
No longer a valid term for GeoRef. As of 1993 see Polkowice Poland.

Polkowice Poland (1993)
Town in Lower Silesia in central Legnica, SW Poland. Before 1993 also search Polkowice AND Poland.
BT Poland
BT Central Europe
BT Europe
SA Lower Silesia

pollen (1978)
SA exine
SA microfossils
SA microorganisms
SA palynology
SA palynomorphs
SA pollen analysis
SA pollen diagrams
SA spores
SA sporopollenin

pollen analysis (1978)
SA analysis
SA palynology
SA palynomorphs
SA pollen
SA pollen diagrams

pollen diagrams (1978)
SA palynology
SA palynomorphs
SA pollen
SA pollen analysis

pollucite (1978)
BT framework silicates
BT silicates

pollutants (1978)
SA atrazine
SA chlorinated hydrocarbons
SA creosote
SA cyanides
SA dense nonaqueous phase liquids
SA effluents
SA herbicides
SA hydrocarbons
SA leachate
SA nonaqueous phase liquids
SA PCBs
SA pesticides
SA pollution
SA volatile organic compounds
SA waste disposal

polluted water (1981)
SA bioremediation
SA decontamination
SA desalinization
SA dilution
SA ground water
SA hydrochemistry
SA impurities

SA leaking underground storage tanks
SA pollution
SA purification
SA reclamation
SA surface water
SA waste water
SA water
SA water management
SA water quality
SA water treatment

pollution (1978)
Used for geological studies on pollution of the environment.
SA acid mine drainage
SA acid rain
SA air
SA atmosphere
SA background level
SA bioremediation
SA case studies
SA conservation
SA contamination
SA creosote
SA cyanides
SA decontamination
SA dilution
SA effluents
SA environmental geology
SA fallout
SA far-field
SA geochemical anomalies
SA geochemistry
SA geologic hazards
SA ground water
SA hazardous waste
SA heavy metals
SA human ecology
SA hydrocarbons
SA hydrology
SA impact statements
SA impurities
SA industrial waste
SA land use
SA leachate
SA leaking underground storage tanks
SA medical geology
SA metals
SA monitoring
SA near-field
SA nonpoint sources
SA oil spills
SA PCBs
SA point sources
SA pollutants
SA polluted water
SA potability
SA protection
SA purification
SA radioactive waste
SA radioactivity
SA reclamation
SA regulations
SA remediation
SA Resource Conservation and Recovery Act
SA sewage
SA sewage sludge
SA sludge
SA soil gases
SA sulfur dioxide
SA sulfuric acid
SA surface water
SA thermal pollution
SA toxic materials
SA trace metals
SA urban environment
SA waste disposal
SA waste disposal sites
SA water
SA water management
SA water supply
SA water treatment

SA water wells

polonium (1978)
Autoposting of metals to this term began in 1989.
UF Po (element)
BT metals
NT Po-210
SA chalcophile elements

Polousnyy (1978)
A ridge in N Yakutia W of the Indigirka River.
IN Index regions as applicable.
SA Yakutia Russian Federation

Poltava
No longer a valid term for GeoRef. As of 1993, see Poltava Ukraine.

Poltava Series (1978)
In Donets Ridge, and W and NW Ukraine.
BT Tertiary
BT Cenozoic
SA Miocene
SA Oligocene
SA Russian Federation
SA Ukraine

Poltava Ukraine (1993)
Oblast and city in NE central Ukraine. Before 1993, also search Pultova or Pultowa. This term has multiple hierarchies.
UF Pultova Ukraine
UF Pultowa Ukraine
BT1 Ukraine
BT1 Europe
BT2 Ukraine
BT2 Commonwealth of Independent States
NT Kremenchug Ukraine

polybasite (1978)
BT sulfantimonites
BT sulfosalts

Polychaetia (1978)
Autoposting of this term began in 1978.
BT worms
BT Invertebrata
NT Serpulidae
SA Chaetopoda

polychlorinated biphenyls
use PCBs

polychlorobiphenyls
use PCBs

polychroism
use pleochroism

polycrystalline materials (1985)
SA materials
SA rock mechanics

polycyclic aromatic hydrocarbons (1989)
BT aromatic hydrocarbons
BT hydrocarbons
BT organic materials
NT naphthalene
NT phenanthrene

polydymite (1978)
BT sulfides
SA linnaeite

Polygnathus (1978)
Genus. Autoposting of microfossils to this term began in 1990.
BT Conodonta
BT microfossils

polygonal fractures (1978)
Before 1978, also search fractures AND polygonal.
BT fractures

polygons (1978)
SA frost action
SA geomorphology
SA glacial geology
SA patterned ground
SA periglacial features

polyhalite (1978)
BT sulfates

polyhedra (1978)
NT tetrahedra
SA crystal structure

polylithionite (1978)
BT mica group
BT sheet silicates
BT silicates

polymerization (1978)
SA geochemistry
SA organic materials
SA polymers

polymers (1989)
Natural or synthetic chemical compounds formed by polymerization.
SA compounds
SA organic materials
SA polymerization

polymetallic ores (1978)
Autoposting of metal ores to this term began in 1985.
IN Commodity. See List C.
UF ores, polymetallic
BT metal ores
SA Akenobe Mine
SA Dachang Deposit
SA Henderson Mine
SA Hitachi Deposit
SA Leadville mining district
SA San Antonio Mine
SA Skellefte mining district
SA Toyoha Mine

polymetamorphism (1978)
UF superimposed metamorphism
BT metamorphism
SA prograde metamorphism
SA retrograde metamorphism

polymorphism (1978)
Before 1981, also search dimorphism.
UF polymorphs
SA crystal structure
SA minerals
SA polytypism

polymorphs
use polymorphism

Polynesia (1978)
Collective name for islands of the central and SE Pacific Ocean. In 1981, broader term changed from Pacific Ocean to Oceania. As of 1981, Hawaii, Cook Islands and Line Islands are considered narrower terms. Autoposting of this term began in 1978.
IN Index islands and island groups as applicable.
CO S280000N290000 W1310000W1790000
BT Oceania
NT Cook Islands
NT French Polynesia
NT Hawaii
NT Line Islands
NT Samoa
NT Tokelau
NT Tonga
NT Tuvalu
SA Ata Caldera
SA Kiribati
SA Pacific Ocean
SA Polynesian Pacific

Polynesian Pacific (1985)
   Region of the Pacific Ocean.
   BT Pacific Ocean
   SA DSDP Site 277
   SA DSDP Site 313
   SA DSDP Site 315A
   SA DSDP Site 317
   SA Leg 17
   SA Polynesia

polyphase processes (1978)
   Also search polyphase. As of 1985, restricted to structural geology. Use polymetamorphism under metamorphism.
   SA processes

Polyplacophora (1978)
   BT Mollusca
   BT Invertebrata

Polyprotodontia (1981)
   Suborder. Autoposting of Marsupialia, Metatheria and Theria to this term began in 1989.
   BT Marsupialia
   BT Metatheria
   BT Theria
   BT Mammalia
   BT Tetrapoda
   BT Vertebrata
   BT Chordata

polytypes
   use polytypism

polytypism (1978)
   UF polytypes
   SA crystal structure
   SA minerals
   SA polymorphism

Polyzoa
   use Bryozoa

Pomerania (1978)
   Historical region on Baltic Sea extending from Stralsund, W of the Oder River, to the Vistula River on the E. Now primarily in Poland.
   IN Index countries as applicable.
   BT Central Europe
   BT Europe
   SA Mecklenburg-Western Pomerania Germany
   SA Poland

Ponce
   No longer a valid term for GeoRef. As of 1993 see Ponce Puerto Rico.

Ponce Puerto Rico (1993)
   City in S Puerto Rico on Caribbean Sea. Before 1993 also search Ponce AND Puerto Rico.
   BT Puerto Rico
   BT Greater Antilles
   BT Antilles
   BT West Indies
   BT Caribbean region

Pondicherry
   No longer a valid term for GeoRef. See Pondicherry India.

Pondicherry India (1993)
   A centrally administered territory known as an Union Territory composed of the four scattered former French settlements of Karikal, Pondicherry, and Yanaon on the Bay of Bengal and Mahe on the Indian Ocean. The city of Pondicherry is the capital.
   BT India
   BT Indian Peninsula
   BT Asia
   SA Coromandel Coast

ponds (1978)
   SA ecology
   SA geomorphology
   SA irrigation
   SA lakes
   SA limnology
   SA paleoecology
   SA reservoirs
   SA surface water
   SA water resources
   SA water supply

Pongidae (1978)
   Family. Autoposting of this term to Sivapithecus began in 1989. Autoposting of Eutheria and Theria to this term began in 1989.
   BT simians
   BT Primates
   BT Eutheria
   BT Theria
   BT Mammalia
   BT Tetrapoda
   BT Vertebrata
   BT Chordata
   NT Sivapithecus
   SA Ramapithecus

Pongola Supergroup (1989)
   BT Precambrian
   SA Namibia
   SA South Africa
   SA Swaziland

Pontevedra
   No longer a valid term for GeoRef. As of 1993, see Pontevedra City Spain.

Pontevedra City Spain (1993)
   Refers to only the city in NW Spain Before 1993, also search Pontevedra and Spain.
   BT Pontevedra Spain
   BT Galicia Spain
   BT Spain
   BT Iberian Peninsula
   BT Southern Europe
   BT Europe

Pontevedra Province
   No longer a valid term for GeoRef. As of 1993, see Pontevedra Spain.

Pontevedra Spain (1993)
   Refers only to the province in NW Spain. From 1981-1992, also search Pontevedra Province and Spain. Before 1981, also search Pontevedra and Spain. For the city, see Pontevedra City Spain.
   CO N415100N425300
       W0075300W0085500
   BT Galicia Spain
   BT Spain
   BT Iberian Peninsula
   BT Southern Europe
   BT Europe
   NT Pontevedra City Spain
   SA Arosa Bay

Pontian (1978)
   Europe. Above Sarmatian, below Plaisancian (Pliocene).
   BT upper Miocene
   BT Miocene
   BT Neogene
   BT Tertiary
   BT Cenozoic

Pontic Mountains (1978)
   In N Turkey.
   CO N395000N420500
       E0423500E0300000
   BT Turkey
   BT Middle East
   BT Asia

Pontnewydd Cave (1989)
   Pleistocene hominid site in Clwyd, N Wales.
   CO N531100N531100
       W0032500W0032500
   BT Wales
   BT Great Britain
   BT United Kingdom
   BT Western Europe
   BT Europe

Pontotoc County
   Valid through 1988. Search in combination with state term. After 1988, use specific county-state term.

Pontotoc County Mississippi (1989)
   N Mississippi. Before 1989, also search Pontotoc County AND Mississippi.
   CO N340500N342300
       W0884800W0891500
   BT Mississippi
   BT United States

Pontotoc County Oklahoma (1989)
   S central Oklahoma. Before 1989, also search Pontotoc County AND Oklahoma.
   CO N343000N345700
       W0962300W0965600
   BT Oklahoma
   BT United States

Poona
   No longer a valid term for GeoRef. See Poona India.

Poona India (1993)
   City, district, and division in W Maharashtra, W central India. Before 1993, also search Poona. Before 1978, also search Pune.
   BT Maharashtra India
   BT India
   BT Indian Peninsula
   BT Asia

Pope County
   Valid through 1988. Search in combination with state term. After 1988, use specific county-state term.

Pope County Arkansas (1989)
   N central Arkansas. Before 1989, also search Pope County AND Arkansas.
   CO N351000N354500
       W0924900W0932200
   BT Arkansas
   BT United States

Pope County Illinois (1989)
   Extreme SE Illinois. Before 1989, also search Pope County AND Illinois.
   CO N370500N373600
       W0882300W0884500
   BT Illinois
   BT United States

Pope County Minnesota (1989)
   W Minnesota. Before 1989, also search Pope County AND Minnesota.
   CO N452400N454500
       W0950800W0954600
   BT Minnesota
   BT United States

Popigay
   No longer a valid term for GeoRef. As of 1993, see Popigay Russian Federation.

Popigay Russian Federation (1993)
   Village on Popigay River of E Taymyr National Okrug in NW Siberia. Before 1993, also search Popigay. This term has multiple hierarchies.
   BT1 Taymyr Dolgan-Nenets Russian Federation
   BT1 Krasnoyarsk Russian Federation
   BT1 Russian Federation
   BT1 Commonwealth of Independent States
   BT2 Taymyr Dolgan-Nenets Russian Federation
   BT2 Krasnoyarsk Russian Federation
   BT2 Asia

Popocatepetl (1989)
   Volcano on the Puebla-Mexico state border, S Mexico.
   IN Index states as applicable.
   CO N190200N190300
       W0983500W0983800
   BT Mexico
   SA Mexico state
   SA Puebla Mexico

popular and elementary geology
   A valid term through 1978. After 1978, use terms separately, i.e. elementary geology.

popular geology (1978)
   Used for discussions written for the hobbyist or collector. Before 1978, search popular and elementary geology.
   SA Atlantis
   SA collecting
   SA education
   SA elementary geology
   SA fossil localities
   SA geology
   SA mineral localities
   SA natural curiosities
   SA news

populations (1978)
   As of 1981, restricted to statistics. Before 1981, used as a general term.
   BT statistical analysis

porcelanite
   use porcellanite

porcellanite (1978)
   UF porcelanite
   BT clastic rocks
   BT sedimentary rocks

Porcupine Mountains (1978)
   Range in Gogebic and Ontonagon counties in NW extremity of Michigan Upper Peninsula, NW Michigan.
   IN Index counties as applicable.
   BT Michigan Upper Peninsula
   BT Michigan
   BT United States
   SA Gogebic County Michigan
   SA Ontonagon County Michigan

pore pressure (1978)
   UF neutral stress
   UF pore-water pressure
   BT pressure
   SA engineering geology
   SA high pressure
   SA overpressure
   SA porous materials
   SA soil mechanics
   SA stress

pore space
   Not a valid term for GeoRef. Use porosity for the open spaces in a rock or soil.

pore spaces
 use porosity

pore structure
 use porosity

pore water (1978)
 UF interstitial water
 SA connate waters
 SA desiccation
 SA fossil waters
 SA ground water
 SA meteoric water
 SA porous materials
 SA sediments
 SA water

pore-water pressure
 use pore pressure

Porifera (1978)
 Autoposting of this term began in 1978.
 UF Spongiae
 BT Invertebrata
 NT Calcispongea
 NT Demospongea
 NT Hyalospongea
 NT Silicispongia
 SA Receptaculitaceae
 SA reef builders
 SA spicules
 SA sponges
 SA Stromatoporoidea

Porites (1985)
 Genus. Autoposting of Zoantharia to this term began in 1989.
 BT Scleractinia
 BT Zoantharia
 BT Anthozoa
 BT Coelenterata
 BT Invertebrata

porosity (1978)
 UF pore spaces
 UF pore structure
 NT secondary porosity
 SA Archie's law
 SA compaction
 SA compactness
 SA Darcy's law
 SA ground water
 SA permeability
 SA physical properties
 SA porous materials
 SA sedimentary rocks
 SA sediments
 SA wettability

porosity traps
 use stratigraphic traps

porous materials (1981)
 Before 1981, also search porous media.
 UF porous media
 SA materials
 SA pore pressure
 SA pore water
 SA porosity
 SA secondary porosity

porous media
 As of 1981, no longer a valid term for GeoRef.
 use porous materials

porphyrins (1978)
 BT pigments
 BT organic materials
 SA organic compounds
 SA organo-metallics

porphyrite
 use porphyry

porphyritic texture (1978)
 Before 1978, search porphyritic AND texture.
 BT textures
 SA igneous rocks
 SA phenocrysts
 SA porphyry
 SA vitrophyre

porphyroblastic texture (1978)
 Before 1978, search porphyroblasts; porphyroblastic AND texture.
 UF porphyroblasts
 BT textures
 SA metamorphic rocks

porphyroblasts
 use porphyroblastic texture

porphyroclastic texture (1989)
 Before 1989, also search porphyroclastic AND texture.
 UF porphyroclasts
 BT textures
 SA metamorphic rocks

porphyroclasts
 use porphyroclastic texture

porphyry (1978)
 UF porphyrite
 BT igneous rocks
 SA andesite porphyry
 SA dacite porphyry
 SA diabase porphyry
 SA diorite porphyry
 SA granite porphyry
 SA granodiorite porphyry
 SA liparite porphyry
 SA melaphyre
 SA picrite porphyry
 SA porphyritic texture
 SA quartz porphyry
 SA rhyolite porphyry
 SA syenite porphyry

porphyry copper (1978)
 SA copper
 SA copper ores
 SA mineral deposits, genesis

porphyry molybdenum (1985)
 UF molybdenite porphyry
 SA mineral deposits, genesis
 SA molybdenum
 SA molybdenum ores

Porsang Fjord (1978)
 Inlet of Arctic Ocean on N coast.
 UF Porsangen Fjord
 UF Porsangerfjord
 BT Finnmark Norway
 BT Norway
 BT Scandinavia
 BT Western Europe
 BT Europe

Porsangen Fjord
 use Porsang Fjord

Porsanger Dolomite Formation (1978)
 Uppermost Precambrian or lowermost Cambrian. Porsanger Dolomite found in upper part of Porsanger Series.
 IN Index age as applicable.
 SA Finnmark Norway
 SA Lower Cambrian
 SA Norway
 SA upper Precambrian

Porsangerfjord
 use Porsang Fjord

Port au Port Peninsula (1978)
 On the Gulf of Saint Lawrence between Port au Port Bay and Saint George Bay.
 BT Newfoundland Island
 BT Newfoundland
 BT Eastern Canada
 BT Canada

Port Campbell
 No longer a valid term for GeoRef. See Port Campbell Australia.

Port Campbell Australia (1993)
 Town on Indian Ocean in SW Victoria.
 BT Victoria Australia
 BT Australia
 BT Australasia

Port Elizabeth
 No longer a valid term for GeoRef. As of 1993 see Port Elizabeth South Africa.

Port Elizabeth South Africa (1993)
 City on Indian Ocean in SE Cape Province. Before 1993 also search Port Elizabeth AND South Africa.
 BT Cape Province South Africa
 BT South Africa
 BT Southern Africa
 BT Africa

Port Jervis
 Valid through 1988. Search in combination with state term. After 1988, use specific city-state term.

Port Jervis New York (1989)
 Resort city on the Delaware River at meeting of borders with New Jersey and Pennsylvania in SE New York. Before 1989, search Port Jervis AND New York.
 CO N412200N412200
   W0744000W0744000
 BT Orange County New York
 BT New York
 BT United States

Port Moresby
 No longer a valid term for GeoRef. As of 1993 see Port Moresby Papua New Guinea.

Port Moresby Papua New Guinea (1993)
 City on Coral Sea in SE Papua New Guinea. Before 1993 also search port Moresby AND New Guinea.
 BT Papua New Guinea
 BT Australasia

Port Phillip Bay (1978)
 Harbor of Melbourne.
 SA Victoria Australia

Port Royal Sound (1978)
 Inlet of Atlantic Ocean between islands of Saint Helena and Hilton Head in Beaufort County.
 IN Index county or region as applicable.
 SA Beaufort County South Carolina

Port Valdez
 use Valdez Alaska

Portage County
 Valid through 1988. Search in combination with state term. After 1988, use specific county-state term.

Portage County Ohio (1989)
 NE Ohio. Before 1989, also search Portage County AND Ohio.
 CO N405800N412100
   W0810000W0812300
 BT Ohio
 BT United States

Portage County Wisconsin (1989)
 Central Wisconsin. Before 1989, also search Portage County AND Wisconsin.
 CO N441400N444200
   W0891300W0895000
 BT Wisconsin
 BT United States

Portage Lake Lava Series (1978)
 Tops of conglomerate beds used for stratigraphic reference include: Saint Louis Conglomerate, The Old Colony and Wolverine sandstones, and the Kingston, Calumet and Hecla, Houghton, Allouez, Pewabic West, and Hancock conglomerates. Upper Peninsula of Michigan. Also search Portage Lake.
 BT Keweenawan
 BT Proterozoic
 BT upper Precambrian
 BT Precambrian
 SA Michigan

Portel
 No longer a valid term for GeoRef. See Portel France.

Portel France (1993)
 Village in S France.
 BT Aude France
 BT France
 BT Western Europe
 BT Europe

Porter County Indiana (1993)
 NW Indiana. Before 1989, also search Porter County AND Indiana.
 CO N411300N414200
   W0865600W0871400
 BT Indiana
 BT United States

Porters Creek Formation (1989)
 In Midway Group. SW Alabama, SW Illinois, W Kentucky, E Mississippi, SE Missouri and W Tennessee.
 BT Paleocene
 BT Paleogene
 BT Tertiary
 BT Cenozoic
 SA Alabama
 SA Illinois
 SA Kentucky
 SA Midway Group
 SA Mississippi
 SA Missouri
 SA Tennessee

Portland
 Valid through 1988. Search in combination with state term. After 1988, use specific city-state term.

Portland Formation (1989)
 In Newark Group. Previously thought to be Upper Triassic in age.
 BT Lower Jurassic
 BT Jurassic
 BT Mesozoic
 SA Connecticut
 SA Massachusetts
 SA Newark Supergroup

Portland Maine (1989)
 City on peninsula near S end of Casco Bay in S Maine. Before 1989, search Portland AND Maine.
 CO N434100N434100
   W0701800W0701800
 BT Cumberland County Maine
 BT Maine
 BT United States

Portland Oregon (1989)

City on the Willamette River in NW Oregon. Before 1989, search Portland AND Oregon.
CO N453200N453200
    W1224000W1224000
BT Multnomah County Oregon
BT Oregon
BT United States

Portlandian (1978)
Europe. Above Kimmeridgian, below Berriasian (Cretaceous). Autoposting of this term began in 1978.
BT Upper Jurassic
BT Jurassic
BT Mesozoic
NT Tithonian
SA Malm

Porto Santo Island (1978)
One of two inhabited islands of Portuguese island group. Also search Porto Santo.
BT Madeira
BT Atlantic Ocean Islands

Portola Valley
Valid through 1988. Search in combination with state term. After 1988, use specific city-state term.

Portola Valley California (1989)
City S of San Francisco in W California. Before 1989, search Portola Valley AND California.
BT San Mateo County California
BT California
BT United States

Portugal (1978)
CO N370000N421000
    W0061000W0093000
BT Iberian Peninsula
BT Southern Europe
BT Europe
NT Alentejo
NT Algarve
NT Alto Alentejo
NT Baixo Alentejo
NT Beira Baixa
NT Braganca Portugal
NT Coimbra Portugal
NT Elvas Portugal
NT Estremadura Portugal
NT Evora Portugal
NT Leiria Portugal
NT Panasqueira Portugal
NT Setubal Portugal
NT Tomar Portugal
NT Tras-os-Montes
NT Vila Real Portugal
NT Viseu Portugal
SA Duero Basin
SA Duero River
SA Tagus Basin
SA Tagus River
SA Timor

Portuguese East Africa
No longer a valid GeoRef index term. Before 1969, was used on level 1 in subfile E.
use Mozambique

Portuguese Guinea
Not a valid index term for GeoRef since 1974.
use Guinea-Bissau

Portuguese Timor
A valid index term through mid-1978. Included use on level 1 as an area term until 1977.
use Timor

Portuguese West Africa
use Angola

Porulosida (1978)
Order. Autoposting of microfossils and Protista to this term began in 1990. This term has multiple hierarchies.
BT1 Radiolaria
BT1 Protista
BT1 Invertebrata
BT2 Radiolaria
BT2 Protista
BT2 microfossils
NT Acantharina

Posen
use Poznan Poland

Posey County Indiana (1993)
SW Indiana. Before 1989, also search Posey County AND Indiana.
CO N374600N381400
    W0872800W0880700
BT Indiana
BT United States

Posidonia Shale (1978)
BT Jurassic
BT Mesozoic
SA Germany

positions, pole
use pole positions

posnjakite (1978)
BT sulfates

Possagno
No longer a valid term for GeoRef. See Possagno Italy.

Possagno Italy (1993)
Village in central Veneto, NE Italy.
BT Veneto Italy
BT Italy
BT Southern Europe
BT Europe

possibilities (1978)
As of 1981, restricted to economic geology, especially referring to the economic potential of a mineral deposit. Before 1981, included use as a general term.
SA economics
SA future
SA mineral deposits, genesis
SA mineral economics
SA prediction maps

Post-glacial
use Holocene

post-glacial rebound
use glacial rebound

Postglacial
use Holocene

postglacial environment (1978)
Before 1978, search postglacial.
SA depositional environment
SA environment
SA glacial environment
SA glacial geology
SA glaciers
SA sedimentation

Postmasburg
No longer a valid term for GeoRef. As of 1993 see Postmasburg South Africa.

Postmasburg South Africa (1993)
Village in N central Cape Province. Before 1993 also search Postmasburg AND South Africa.
BT Cape Province South Africa
BT South Africa
BT Southern Africa
BT Africa

potability (1982)
SA bottling
SA decontamination
SA desalinization
SA drinking water
SA fresh water
SA ground water
SA hydrochemistry
SA impurities
SA pollution
SA purification
SA surface water
SA water
SA water management
SA water quality
SA water resources
SA water supply
SA water treatment

potash (1978)
IN See List C for use as a commodity term.
BT evaporites
BT chemically precipitated rocks
BT sedimentary rocks
SA alunite
SA Searles Lake

potash alum
use alum

potash feldspar
use K-feldspar

potash mica
use muscovite

potassic composition (1978)
Before 1978, search potassic.
SA composition
SA potassium

potassium (1978)
Autoposting of alkali metals and metals to this term began in 1989.
UF K
UF kalium
BT alkali metals
BT metals
NT K-40
SA isotopes
SA K/Ar
SA K/Rb
SA KREEP
SA lithophile elements
SA potassic composition
SA potassium ion

potassium alum
use alum

potassium feldspar
use K-feldspar

potassium ion (1985)
SA geochemistry
SA ions
SA potassium

potassium-argon
No longer a valid term for GeoRef. Before 1972, was used in subfiles E, G, N, T and B.
use K/Ar

potential deposits (1989)
SA economic geology
SA energy sources
SA exploration
SA mineral deposits, genesis
SA mineral exploration
SA mineral resources
SA petroleum

potential field (1989)
SA electrical field
SA electromagnetic field
SA geophysical methods
SA magnetic field

potentiometric surface (1981)
Also search piezometry AND ground water; piezometers AND ground water.
UF piezometric surface
SA drawdown
SA ground water
SA hydrogeologic maps
SA levels
SA water table
SA water wells

potentiometry (1985)
SA chemical analysis
SA direct-current methods
SA electrical methods
SA electrochemical properties
SA geophysical methods

Potenza
No longer a valid term for GeoRef. See Potenza Italy.

Potenza Italy (1993)
City and province in W central Basilicata, S Italy.
BT Basilicata Italy
BT Italy
BT Southern Europe
BT Europe

potholes (1978)
General term.
SA erosion features
SA geomorphology
SA kettles
SA sinkholes

Potiguar Basin (1989)
Basin in Rio Grande do Norte, NE Brazil.
BT Rio Grande do Norte Brazil
BT Brazil
BT South America

Potomac Group (1978)
Consists of Patuxent, Arundel, and Patapsco formations.
BT Cretaceous
BT Mesozoic
SA Delaware
SA Maryland
SA Virginia

Potomac River (1978)
Formed by the confluence of the North and South branches of the Potomac River SE of Cumberland, Maryland. It flows SE into Chesapeake Bay.
IN Index District of Columbia and states as applicable.
BT United States
SA District of Columbia
SA Maryland
SA Virginia
SA West Virginia

Potomac River basin (1978)
IN Index states and District of Columbia as applicable.
BT United States
SA District of Columbia
SA Maryland
SA Virginia
SA West Virginia

Potsdam
No longer a valid term for GeoRef. See Potsdam Germany.

Potsdam Bezirk
No longer a valid term for GeoRef. District in central East Germany. Before 1981, search Potsdam AND East Germany.

Potsdam Germany (1993)

City NE Germany. As of 1981, Potsdam refers to only the city. Before 1981, it also included the East German district.
BT Brandenburg Germany
BT Germany
BT Central Europe
BT Europe

Potsdam Sandstone (1978)
Underlies Ticonderoga Formation. Central and E New York, Vermont, E Ontario and S Quebec.
BT Upper Cambrian
BT Cambrian
BT Paleozoic
SA Little Falls Formation
SA New York
SA Ontario
SA Quebec
SA Vermont

Pottawatomie County
Valid through 1988. Search in combination with state term. After 1988, use specific county-state term.

Pottawatomie County Kansas (1989)
NE Kansas. Before 1989, also search Pottawatomie County AND Kansas.
CO N390700N393500
    W0960200W0964100
BT Kansas
BT United States

Pottawatomie County Oklahoma (1989)
Central Oklahoma. Before 1989, also search Pottawatomie County AND Oklahoma.
CO N345400N352700
    W0963700W0970800
BT Oklahoma
BT United States

Potter County
Valid through 1988. Search in combination with state term. After 1988, use specific county-state term.

Potter County Pennsylvania (1989)
N Pennsylvania. Before 1989, also search Potter County AND Pennsylvania.
CO N412800N420000
    W0773700W0781300
BT Pennsylvania
BT United States
SA Genesee River

Potter County South Dakota (1989)
N central South Dakota. Before 1989, also search Potter County AND South Dakota.
CO N445300N451500
    W0993500W1002400
BT South Dakota
BT United States

Potter County Texas (1989)
Extreme N Texas. Before 1989, also search Potter County AND Texas.
CO N351300N353800
    W1014100W1021200
BT Texas
BT United States

pottery
   use artifacts

Pottsville Group (1978)
Includes Lee Formation, Norton Formation, Gladeville Sandstone, Wise Formation, Harlan Sandstone, Lookout Sandstone, Walden Sandstone, Tumbling Run Formation, Schuylkill Formation, Sharp Mountain Formation. Alabama, S Ohio, S Illinois, S Indiana, Kentucky, Maryland, NE Mississippi, Pennsylvania, Tennessee, Virginia, West Virginia.
BT Pennsylvanian
BT Carboniferous
BT Paleozoic
SA Alabama
SA Breathitt Formation
SA Illinois
SA Kentucky
SA Mary Lee Coal
SA Maryland
SA Mississippi
SA New River Formation
SA Ohio
SA Pennsylvania
SA Pocahontas Formation
SA Sharon Conglomerate
SA Tennessee
SA Virginia
SA West Virginia

Potwar Plateau (1978)
Lies between Indus River and Jhelum River in N Punjab Province, Pakistan. Also search Potwar.
SA Punjab Pakistan

Pouilly-en-Auxois
No longer a valid term for GeoRef. See Pouilly-en-Auxois France.

Pouilly-en-Auxois France (1993)
Village in E central France.
BT Cote-d'Or France
BT France
BT Western Europe
BT Europe

Poway Conglomerate (1978)
Nearly horizontal, and in most places lies unconformably on Cretaceous Peninsular Range Batholith and accompanying metamorphic rocks. S California.
BT upper Eocene
BT Eocene
BT Paleogene
BT Tertiary
BT Cenozoic
SA California

powder method (1985)
SA methods
SA single-crystal method
SA X-ray analysis
SA X-ray diffraction analysis

Powder River basin (1978)
CO N430000N452000
    W1050000W1071500
BT United States
SA Hartzog Draw Field
SA Montana
SA Oregon
SA Walker Creek Field
SA Wyoming

Powder River County
Valid through 1988. Search in combination with state term. After 1988, use specific county-state term.

Powder River County Montana (1989)
SE Montana. Before 1989, also search Powder River County AND Montana.
CO N450000N454800
    W1045700W1062000
BT Montana
BT United States
SA Bell Creek Field

Powell
Valid through 1988. Search in combination with state term. After 1988, use specific city-state term.

Powell County
Valid through 1988. Search in combination with state term. After 1988, use specific county-state term.

Powell County Kentucky (1989)
E central Kentucky. Before 1989, also search Powell County AND Kentucky.
CO N374200N375500
    W0833900W0840100
BT Kentucky
BT United States

Powell County Montana (1989)
W central Montana. Before 1989, also search Powell County AND Montana.
CO N461700N473600
    W1121700W1132900
BT Montana
BT United States

Powell Wyoming (1989)
Town in NW Wyoming. Before 1989, search Powell AND Wyoming.
CO N444500N444500
    W1084600W1084600
BT Park County Wyoming
BT Wyoming
BT United States

powellite (1978)
BT molybdates

Power County
Valid through 1988. Search in combination with state term. After 1988, use specific county-state term.

Power County Idaho (1989)
SE Idaho. Before 1989, also search Power County AND Idaho.
CO N422000N430800
    W1122500W1131500
BT Idaho
BT United States

power plants (1978)
SA energy sources
SA hydroelectric energy
SA nuclear facilities

Powys Wales (1993)
County. Central Wales. Formed from Brecknockshire, Montgomeryshire and Radnorshire in April 1974. Before 1993, also search Powys.
BT Wales
BT Great Britain
BT United Kingdom
BT Western Europe
BT Europe
NT Brecknockshire Wales
NT Radnorshire Wales

Poza Rica
No longer a valid term for GeoRef as of 1993 see Poza Rica Mexico.

Poza Rica de Hidalgo
   use Poza Rica Mexico

Poza Rica Mexico (1993)
City in N Veracruz. Before 1993 also search Poza Rica AND Mexico.
UF Poza Rica de Hidalgo
BT Veracruz Mexico
BT Mexico

Poznan
No longer a valid term for GeoRef. As of 1993 see Poznan Poland.

Poznan Clays (1978)
E Germany and Poland.
BT upper Tertiary
BT Tertiary
BT Cenozoic
SA Germany
SA Poland

Poznan Poland (1993)
Province and city in W central Poland. Before 1993 also search Poznan AND Poland.
UF Posen
BT Poland
BT Central Europe
BT Europe
NT Adamow Mine
NT Klodawa Poland
NT Konin Poland
NT Ksiaz Poland

Pozzuoli
No longer a valid term for GeoRef. See Pozzuoli Italy.

Pozzuoli Italy (1993)
City on inlet of Gulf of Naples.
BT Campania Italy
BT Italy
BT Southern Europe
BT Europe

PP-waves (1989)
BT body waves
BT elastic waves
SA P-waves
SA seismology
SA waves

PPP-waves (1981)
BT body waves
BT elastic waves
SA P-waves
SA seismology
SA waves

PPS-waves (1981)
BT body waves
BT elastic waves
SA P-waves
SA S-waves
SA seismology
SA waves

Pr
   use praseodymium

practice (1978)
After 1975, to be used for geology as a profession.
SA geologists

Praecardiida (1981)
BT Bivalvia
BT Mollusca
BT Invertebrata

Pragian (1985)
Europe. Lower Devonian, above Lochkovian and below Zlichovian.
BT Lower Devonian
BT Devonian
BT Paleozoic

Prague (1978)
City in central Bohemia.
UF Praha
BT Bohemia
BT Czech Republic
BT Czechoslovakia
BT Central Europe
BT Europe

Praha
  use Prague

Prahova
  No longer a valid term for GeoRef. As of 1993 see Prahova Romania.

Prahova Romania (1993)
  County in NE Walachia, S central Romania. Before 1993 also search Prahova AND Romania.
  BT Walachia
  BT Romania
  BT Southern Europe
  BT Europe
  NT Ploesti Romania
  NT Sinaia Romania
  NT Slanic Romania
  SA Teleajen Valley

Prairie County
  Valid through 1988. Search in combination with state term. After 1988, use specific county-state term.

Prairie County Arkansas (1989)
  E central Arkansas. Before 1989, also search Prairie County AND Arkansas.
  CO N343000N350800
      W0912300W0914700
  BT Arkansas
  BT United States

Prairie County Montana (1989)
  E Montana. Before 1989, also search Prairie County AND Montana.
  CO N463300N471200
      W1043700W1060500
  BT Montana
  BT United States

Prairie du Chien Group (1985)
  SW Wisconsin, Illinois, Iowa and S Minnesota.
  BT Lower Ordovician
  BT Ordovician
  BT Paleozoic
  SA Illinois
  SA Iowa
  SA Minnesota
  SA Shakopee Formation
  SA Wisconsin

Prairie Evaporite (1978)
  Salt and anhydrite beds that form upper unit of Elk Point Group. Subsurface. Also search Prairie Formation NOT Louisiana. For Pleistocene formation in Louisiana, search Prairie Formation AND Louisiana.
  BT Middle Devonian
  BT Devonian
  BT Paleozoic
  SA Elk Point Group
  SA Manitoba
  SA Montana
  SA North Dakota
  SA Saskatchewan

prairies (1993)
  SA ecology
  SA grasslands
  SA plains
  SA revegetation
  SA steppes

Prakasam District
  No longer a valid term for GeoRef. As of 1993, see Prakasam India.

Prakasam India (1993)
  Andhra Pradesh, India. Heavy mineral deposits. Also search Prakasam or Prakasam District.
  BT Andhra Pradesh India
  BT India

  BT Indian Peninsula
  BT Asia
  SA heavy mineral deposits

Pranhita-Godavari Valley (1978)
  Combined valleys of the Pranhita and Godavari river systems.
  IN Index states as applicable.
  BT India
  BT Indian Peninsula
  BT Asia
  SA Andhra Pradesh India
  SA Godavari Valley
  SA Maharashtra India

praseodymium (1978)
  Autoposting of rare earths and metals to this term began in 1989.
  UF Pr
  BT rare earths
  BT metals

prasinite
  use greenschist

Pratt County
  Valid through 1988. Search in combination with state term. After 1988, use specific county-state term.

Pratt County Kansas (1989)
  S Kansas. Before 1989, also search Pratt County AND Kansas.
  CO N373000N375000
      W0983000W0990100
  BT Kansas
  BT United States

Prbram
  use Pribram

Pre-Cambrian
  No longer a valid GeoRef index term. Before 1960, was used on level 1 in subfiles E and N.
  use Precambrian

pre-Neanderthal (1981)
  Autoposting of Eutheria and Theria to this term began in 1989.
  BT Hominidae
  BT Primates
  BT Eutheria
  BT Theria
  BT Mammalia
  BT Tetrapoda
  BT Vertebrata
  BT Chordata

Prealps (1978)
  Mostly in Italy.
  UF Alpine Foreland
  BT Alps
  BT Europe
  SA Italy

Prebetic Zone (1978)
  A geographic term with stratigraphic-tectonic connotations.
  BT Spain
  BT Iberian Peninsula
  BT Southern Europe
  BT Europe
  SA Betic Cordillera
  SA Subbetic Zone

Preboreal (1978)
  Europe. An interval of postglacial time following the Younger Dryas of the late-glacial Arctic interval and preceding the Boreal.
  UF Subarctic
  BT Holocene
  BT Quaternary
  BT Cenozoic
  SA Boreal

Precambrian (1978)

As of 1981, Archean is used for lower Precambrian. Before 1960, also search Pre-Cambrian. Autoposting of this term began in 1978.
  UF Pre-Cambrian
  NT Acungui Group
  NT Adirondack Anorthosite
  NT Animikie Group
  NT Anshan Group
  NT Archean
  NT Assyntic Orogeny
  NT Avalonian Orogeny
  NT Baltimore Gneiss
  NT Bijawar System
  NT Biwabik Iron Formation
  NT Brockman Iron Formation
  NT Carrizo Mountain Formation
  NT Catoctin Formation
  NT Central Rand Group
  NT Changcheng System
  NT Changzhougou Formation
  NT Chelmsford Formation
  NT Chuanlinggou Formation
  NT Chuar Group
  NT Cuddapah System
  NT Delhi Supergroup
  NT Eocambrian
  NT Espanola Formation
  NT Fig Tree Series
  NT Flinton Group
  NT Fortescue Group
  NT Great Smoky Group
  NT Grenvillian Orogeny
  NT Gunflint Iron Formation
  NT Hamersley Group
  NT Harbour Main Group
  NT Hazel Formation
  NT Hecla Hoek Formation
  NT Highland Series
  NT Hudsonian Orogeny
  NT Idaho Springs Formation
  NT Johnnie Formation
  NT Jupiter Member
  NT Kaladgi System
  NT Katangan Orogeny
  NT Kenoran Orogeny
  NT Kingston Peak Formation
  NT Kisseynew Complex
  NT Knife Lake Group
  NT Ladoga Series
  NT Laramie anorthosite complex
  NT Lewisian Complex
  NT Luoquan Formation
  NT middle Precambrian
  NT Minas Series
  NT Missi Group
  NT Moldanubian
  NT Nipissing Diabase
  NT Nokomis Group
  NT Nonesuch Shale
  NT Noonday Dolomite
  NT North Shore Volcanics
  NT Ocoee Series
  NT Onverwacht Group
  NT Osler Series
  NT Pahrump Series
  NT Pelona Schist
  NT Penokean Orogeny
  NT Penrhyn Slate
  NT Pinnacle Formation
  NT Pongola Supergroup
  NT Prichard Formation
  NT Prince Albert Group
  NT Purcell System
  NT Roan Supergroup
  NT Sandia Granite
  NT Sherridon Group
  NT Sioux Quartzite
  NT Sokoman Formation
  NT Stillwater Complex
  NT Stirling Quartzite
  NT Swaziland Sequence
  NT Swaziland System
  NT Transvaal Supergroup

  NT Twilight Gneiss
  NT Uinta Mountain Group
  NT Unkar Group
  NT upper Precambrian
  NT Vadito Group
  NT Ventersdorp Supergroup
  NT Wasekwan Group
  NT Waterberg System
  NT Wilcox Formation
  NT Witwatersrand Supergroup
  NT Wyman Formation
  NT Yellowjacket Formation
  SA Baikalian Phase
  SA Bhander Group
  SA Cadomian Orogeny
  SA Dalradian
  SA Deer Lake Group
  SA Dore Lake Complex
  SA Duluth Complex
  SA Hector Formation
  SA Jacobsville Sandstone
  SA Kaimur Sandstone
  SA Kurnool System
  SA Lardeau Group
  SA Limpopo Belt
  SA Puncoviscana Formation
  SA Tal Formation
  SA Talladega Group
  SA Vindhyan
  SA Wood Canyon Formation
  SA Yudomian

Precambrian Shield
  use Canadian Shield

precession (1978)
  Used as a general term.
  SA processes

precious metals (1989)
  Includes gold, silver, platinum group.
  BT metals
  SA alkali metals
  SA base metals
  SA chemical elements
  SA electrum
  SA geochemistry
  SA heavy metals
  SA metal ores

precipitation (1978)
  As of 1978, term refers to geochemistry. For meteorology, use atmospheric precipitation.
  SA atmospheric precipitation
  SA crystallization
  SA drainage basins
  SA geochemistry
  SA hydrology
  SA particles
  SA rivers and streams
  SA runoff
  SA sedimentation
  SA storms
  SA water

precision (1993)
  Use restricted to occurrences where it is a major concept. Also see accuracy.
  SA accuracy
  SA errors
  SA instruments
  SA methods
  SA reliability
  SA techniques

Precordillera (1978)
  IN Index countries and regions as applicable.
  BT South America
  SA Argentina
  SA Chile

precursors (1978)
  Before 1981, also search premonitory phenomena.

SA acoustical emissions
SA earthquake prediction
SA radon emanometry
SA seismology
SA stress drops

predation (1985)
SA diet
SA nutrition
SA paleontology
SA trophic analysis

predators (1989)
SA ecology
SA paleoecology

Predazzo
No longer a valid term for GeoRef. See Predazzo Italy.

Predazzo Italy (1993)
City and resort in Dolomites in S Trento-Alto Adige, NE Italy.
BT Trentino-Alto Adige Italy
BT Italy
BT Southern Europe
BT Europe

prediction (1978)
See earthquake prediction if applicable.

prediction maps (1981)
BT maps
SA economics
SA possibilities
SA reserves

preferred orientation (1978)
SA fabric
SA lineation
SA orientation
SA petrofabrics
SA structural analysis
SA tectonite

prehistory
No longer a valid GeoRef index term. Before 1971, was used on level 1 in subfiles E and N. See archaeology; anthropology.

prehnite (1978)
BT sheet silicates
BT silicates
SA aluminosilicates

prehnite-pumpellyite facies (1978)
BT facies
SA metamorphic rocks

premonitory phenomena
As of 1981, no longer a valid term for GeoRef. Use precursors.

preobrazhenskite (1978)
BT borates

preparation (1978)
Use sample preparation where applicable.
SA sample preparation

preservation (1978)
SA land use
SA regulations
SA taphonomy

Presidente Vargas
Not a valid term for GeoRef. See Itabira Brazil.

Presidio County
Valid through 1988. Search in combination with state term. After 1988, use specific county-state term.

Presidio County Texas (1989)
Extreme W Texas. Before 1989, also search Presidio County AND Texas.
CO N292000N303500

W1034500W1050000
BT Texas
BT United States

Presque Isle (1978)
As of 1989, used only for the small peninsula in Lake Erie in NW Pennsylvania. For the city in Aroostook County, Maine, use Presque Isle Maine. For the county in NE Michigan, use Presque Isle County Michigan.
BT Erie County Pennsylvania
BT Pennsylvania
BT United States

Presque Isle County Michigan (1989)
NE Michigan. Before 1989, also search Presque Isle AND Michigan.
CO N451300N454000
W0832000W0841500
BT Michigan Lower Peninsula
BT Michigan
BT United States

Presque Isle Maine (1989)
City in N Maine. Before 1989, also search Presque Isle AND Maine.
CO N464200N464200
W0680100W0680100
BT Aroostook County Maine
BT Maine
BT United States

pressure (1978)
NT atmospheric pressure
NT capillary pressure
NT high pressure
NT hydraulic pressure
NT hydrostatic pressure
NT low pressure
NT overpressure
NT partial pressure
NT pore pressure
NT water pressure
SA compression
SA confining pressure
SA deformation
SA earth pressure
SA fluid inclusions
SA fluid pressure
SA geopressure
SA load pressure
SA P-T conditions
SA phase equilibria
SA pressuremeters
SA retaining walls
SA stress
SA tension

pressure meters
use pressuremeters

pressure solution (1978)
SA diagenesis
SA solution

pressure wave
use P-waves

pressuremeter tests (1993)
SA instruments
SA pressuremeters

pressuremeters (1985)
UF piezometers
UF pressure meters
SA hydrophones
SA instruments
SA pressure
SA pressuremeter tests

Preston County
Valid through 1988. Search in combination with state term. After 1988, use specific county-state term.

Preston County West Virginia (1989)
N West Virginia. Before 1989, also search Preston County AND West Virginia.
CO N391100N394200
W0792800W0795500
BT West Virginia
BT United States

Preston Peak (1978)
In Siskiyou Mountains in Siskiyou County in extreme N California.
IN Index county and Klamath Mountains as applicable.
SA Klamath Mountains
SA Siskiyou County California

Presumpscot Formation (1989)
Maine and SE New Hampshire.
BT Pleistocene
BT Quaternary
BT Cenozoic
SA Maine
SA New Hampshire

Pretoria
No longer a valid term for GeoRef. As of 1993 see Pretoria South Africa.

Pretoria Group (1993)
Proterozoic. South Africa and Botswana. In Transvaal Supergroup.
BT Proterozoic
BT upper Precambrian
BT Precambrian
SA Botswana
SA South Africa
SA Transvaal Supergroup

Pretoria South Africa (1993)
City in S Transvaal. Before 1993 also search Pretoria AND South Africa.
BT Transvaal South Africa
BT South Africa
BT Southern Africa
BT Africa

prevention
A valid general term through mid-1978.

preventive measures (1981)
SA environmental geology

Priabonian (1978)
Europe.
BT upper Eocene
BT Eocene
BT Paleogene
BT Tertiary
BT Cenozoic
SA Auversian
SA Bartonian
SA Ludian

Priapulida (1981)
BT worms
BT Invertebrata

Pribram (1978)
City SW of Prague in central Bohemia.
UF Prbram
BT Bohemia
BT Czech Republic
BT Czechoslovakia
BT Central Europe
BT Europe

price (1981)
SA cost
SA economics

Price Formation (1978)
Includes Cloyd Conglomerate Member. SW Virginia.
BT Mississippian

BT Carboniferous
BT Paleozoic
SA Virginia

Price River basin (1985)
Central Utah.
IN Index counties as applicable.
SA Colorado Plateau
SA Utah

Price River Formation (1985)
In Mesaverde Group. Central and E central Utah and W central Wyoming.
BT Upper Cretaceous
BT Cretaceous
BT Mesozoic
SA Colorado
SA Mesaverde Group
SA Utah

Prichard Formation (1978)
Belt Supergroup. NE Idaho and NW Montana.
BT Precambrian
SA Belt Supergroup
SA Idaho
SA Montana

priderite (1989)
BT oxides
SA iron oxides

Pridolian (1978)
Europe.
BT Upper Silurian
BT Silurian
BT Paleozoic

Prikumsk
No longer a valid term for GeoRef. As of 1993, see Prikumsk Russian Federation.

Prikumsk Russian Federation (1993)
City in Stavropol Kray in the Northern Caucasus. Before 1993, also search Prikumsk. This term has multiple hierarchies.
UF Budennovsk Russian Federation
BT1 Stavropol Russian Federation
BT1 Russian Federation
BT1 Commonwealth of Independent States
BT2 Stavropol Russian Federation
BT2 Europe

primary dispersion (1978)
Before 1978, search primary AND dispersion.
SA dispersion patterns
SA mineral exploration
SA secondary dispersion

primary magnetization (1981)
BT magnetization

primary migration (1985)
Not for migration of elastic waves.
SA migration
SA petroleum

primary structures (1978)
BT sedimentary structures
SA ball-and-pillow
SA bedding
SA ripple marks
SA secondary structures

primary wave
use P-waves

Primates (1978)
Order. Autoposting of Eutheria and Theria to this term began in 1989.
BT Eutheria
BT Theria

BT Mammalia
BT Tetrapoda
BT Vertebrata
BT Chordata
NT Hominidae
NT Prosimii
NT simians
SA fossil man

Primorski Krai
use Primorye Russian Federation

Primorye
No longer a valid term for GeoRef. As of 1993, see Primorye Russian Federation.

Primorye Russian Federation (1993)
Primorye Kray (territory). New name for Primorski Krai which is the SE section of the Russian Far East between NE China and the Japan Sea. Before 1978, also search Maritime Kray or Maritime Territory. This term has multiple hierarchies.
CO N420000N500000
   E1400000E1300000
UF Maritime Kray
UF Primorski Krai
BT1 Russian Federation
BT1 Commonwealth of Independent States
BT2 Asia
NT Kavalerovo Russian Federation
NT Tetyukhe Russian Federation
NT Vladivostok Russian Federation
SA Barabash Suite
SA Russian Far East
SA Sikhote-Alin Range
SA Suchan Basin

Prince Albert Group (1978)
BT Precambrian
SA Canada
SA Saskatchewan

Prince Charles Mountains (1978)
In Mac-Robertson Land near the Amery Ice Shelf on Indian Ocean side.
BT Antarctica

Prince Edward Island (1978)
As of 1993, restricted to the island in the Gulf of Saint Lawrence constituting a Canadian province. For the island in the Indian Ocean, see Prince Edward Island Group.
CO N455500N470500
   W0620000W0643000
BT Maritime Provinces
BT Eastern Canada
BT Canada
SA Cumberland Basin
SA Pictou Group

Prince Edward Island Group (1993)
SE Indian Ocean. Territory of South Africa. Before 1993, also search Prince Edward Islands or Prince Edward Island with Indian Ocean Islands.
UF Prince Edward Islands
BT Indian Ocean Islands

Prince Edward Islands
use Prince Edward Island Group

Prince Georges County
Valid through 1988. Search in combination with state term. After 1988, use specific county-state term.

Prince Georges County Maryland (1989)
Central Maryland. Before 1989, also search Prince Georges County AND Maryland.
CO N383200N390800
   W0764000W0770500
BT Maryland
BT United States

Prince of Wales Island (1978)
Largest island of Alexander Archipelago in SE Alaska, W of Ketchikan; and an island between Victoria and Somerset Islands in Franklin District, Northwest Territories.
IN Index Alaska and/or Northwest Territories as applicable.
SA Alaska
SA Northwest Territories

Prince Olaf Coast
use Prince Olav Coast

Prince Olav Coast (1989)
Coast of Enderby Land, Antarctica, bordering the Indian Ocean.
CO S700000S680000
   E0450000E0390000
UF Prince Olaf Coast
BT Enderby Land
BT Antarctica
SA Lutzow-Holm Bay
SA Queen Maud Land
SA Syowa Station

Prince Rupert
No longer a valid term for GeoRef. See Prince Rupert British Columbia.

Prince Rupert British Columbia (1993)
City in W British Columbia on Kaien Island in Chatham Sound near extreme SE Alaska.
CO N542000N542000
   W1301800W1301800
BT British Columbia
BT Western Canada
BT Canada

Prince William County
Valid through 1988. Search in combination with state term. After 1988, use specific county-state term.

Prince William County Virginia (1989)
N Virginia. Before 1989, also search Prince William County AND Virginia.
CO N383000N385700
   W0771500W0774500
BT Virginia
BT United States

Prince William Sound (1978)
Inlet of Gulf of Alaska E of Kenai Peninsula.
BT Southern Alaska
BT Alaska
BT United States
SA Montague Island

Prince William Terrane (1989)
S Alaska.
BT Alaska
BT United States

Princess Anne County Virginia (1989)
SE Virginia.
CO N363400N365500
   W0755200W0761500
BT Virginia
BT United States
NT Virginia Beach Virginia

Princeton
No longer a valid term for GeoRef. As of 1993, use Princeton British Columbia for the village in S British Columbia, in the Cascade Range. As of 1989, use Princeton New Jersey for the borough in New Jersey.

Princeton British Columbia (1993)
Village in S British Columbia, in the Cascade Range. Search in conjunction with British Columbia.
CO N492500N492500
   W1203500W1203500
BT British Columbia
BT Western Canada
BT Canada

Princeton New Jersey (1989)
Borough in W central New Jersey. Before 1989, search Princeton AND New Jersey.
CO N402100N402100
   W0744000W0744000
BT Mercer County New Jersey
BT New Jersey
BT United States

principal components analysis (1981)
BT statistical analysis
SA components

Principe (1985)
Island in Gulf of Guinea. Formerly Portuguese territory, now part of independent nation Sao Tome e Principe.
BT Sao Tome e Principe
BT Atlantic Ocean Islands
SA Sao Tome

principles (1978)
Appropriate to a large number of general topics.

Pripet Basin (1978)
River basin most of which comprises the Pripet or Pinsk Marshes which is the largest tract of swamp in Europe. Largely coextensive with Polesye Lowland. Also search Pripet. This term has multiple hierarchies.
IN Index former Soviet republics as applicable.
CO N510000N540000
   E0330000E0240000
UF Pripet Depression
UF Pripyat Basin
BT1 Commonwealth of Independent States
BT2 Europe
SA Belarus
SA Polesye
SA Ukraine

Pripet Depression
use Pripet Basin

Pripyat Basin
use Pripet Basin

prismatine
use kornerupine

pristane (1985)
BT alkanes
BT aliphatic hydrocarbons
BT hydrocarbons
BT organic materials

Privas
No longer a valid term for GeoRef. See Privas France.

Privas France (1993)
Town in E Ardeche, SE central France.
BT Ardeche France
BT France
BT Western Europe
BT Europe

probability (1978)
BT statistical analysis
SA mathematical geology
SA mathematical methods
SA reliability
SA sampling
SA stochastic processes

probe, electron
use electron probe

probe, ion
use ion probe

probe, laser
use laser methods

probe, neutron
use neutron probe

problematic
A valid index term through 1977 used in combination with fossils (i.e. fossils, problematic). After 1977, use problematic fossils.

problematic fossils (1978)
Before 1978, also search fossils AND problematic; fossils AND problematical; problematic organisms. Before 1981, Hyolithidae was included as a narrower term; now use Hyolithes under Mollusca. Also before 1981, Receptaculitaceae was included as a narrower term.
UF problematical fossils
NT Calcisphaerulidae
NT Dacryoconarida
NT problematic microfossils
SA chemical fossils
SA fossilization
SA fossils
SA Hyolithes
SA nannofossils
SA paleontology
SA Receptaculitaceae
SA Stromatoporoidea
SA Tentaculites
SA Tentaculitida
SA Tentaculitidae

problematic microfossils (1981)
This term has multiple hierarchies.
BT1 microfossils
BT2 problematic fossils
NT Bolboforma
NT Tubiphytes

problematic palynomorphs (1978)
Autoposting of microfossils to this term began in 1990.
BT palynomorphs
BT microfossils

problematical fossils
Not a valid GeoRef index term after 1970. Was used on level 1 in subfile N.
use problematic fossils

problems
A valid general term through mid-1978.

Proboscidea (1978)
Order. Autoposting of Eutheria and Theria to this term began in 1989.
BT Eutheria
BT Theria
BT Mammalia
BT Tetrapoda
BT Vertebrata
BT Chordata

NT Barytherioidea
   NT Deinotherioidea
   NT Elephantoidea
   NT Mastodontoidea
   NT Moeritherioidea
procaryotes
   use prokaryotes

processes   (1978)
   SA adsorption
   SA advection
   SA aggradation
   SA albitization
   SA alteration
   SA alunitization
   SA amphibolitization
   SA argillization
   SA aromatization
   SA atectonic processes
   SA autometamorphism
   SA bauxitization
   SA biomineralization
   SA brecciation
   SA burial
   SA calcification
   SA calcitization
   SA chloritization
   SA coalification
   SA combustion
   SA compaction
   SA complexing
   SA contraction
   SA cyclic processes
   SA dedolomitization
   SA denudation
   SA desertification
   SA diagenesis
   SA diffusion
   SA dissociation
   SA dolomitization
   SA electro-osmosis
   SA electrolysis
   SA endogene processes
   SA epigene processes
   SA epimerization
   SA epithermal processes
   SA erosion
   SA exhalative processes
   SA exogene processes
   SA filtration
   SA flocculation
   SA fluidization
   SA fragmentation
   SA granitization
   SA graphitization
   SA hydration
   SA hydrolysis
   SA hydrothermal processes
   SA hypabyssal processes
   SA hypothermal processes
   SA igneous processes
   SA illitization
   SA isomerization
   SA kaolinization
   SA karstification
   SA mesothermal processes
   SA metamorphic processes
   SA microclinization
   SA mixing
   SA muscovitization
   SA osmosis
   SA oxidation
   SA palagonitization
   SA photosynthesis
   SA phyllonitization
   SA polyphase processes
   SA post-tectonic processes
   SA precession
   SA propylitization
   SA reduction
   SA respiration
   SA scapolitization
   SA sedimentary processes
   SA separation
   SA serpentinization
   SA silicification
   SA steady-state processes
   SA stochastic processes
   SA supergene processes
   SA synsedimentary processes
   SA syntectonic processes
   SA telethermal processes
   SA volatilization
   SA volcanic processes

production   (1978)
   When referring to mining geology before May 1978, also search extraction.
   IN   May be used with commodity terms, e.g. petroleum.
   SA aromatization
   SA beneficiation
   SA exploitation
   SA offshore
   SA onshore
   SA optimization
   SA ore grade
   SA pelletization
   SA production control
   SA production value
   SA pumping
   SA water wells

production control   (1978)
   UF control, production
   SA mining geology
   SA production
   SA technology

production value   (1981)
   SA economics
   SA production

productive capacity   (1981)
   SA economics

productivity   (1978)
   Used as a general term.
   SA economics

products
   A valid general term through mid-1978.

Proetidae   (1978)
   Family.
   BT Ptychopariida
   BT Trilobita
   BT Trilobitomorpha
   BT Arthropoda
   BT Invertebrata

profiles
   As of 1981, no longer a valid term for GeoRef. See sections; geophysical profiles; soil profiles; geochemical profiles.

profitability   (1981)
   SA cutoff grade
   SA economics

Proganosauria
   As of 1993, no longer a valid term for GeoRef. Autoposting of Anapsida to this term began in 1989. Autoposting of Mesosauria and Synapsida to this term ended in 1989.

progradation   (1978)
   SA littoral drift
   SA sedimentation
   SA shore features

prograde metamorphism   (1978)
   Before 1978, also search metamorphism AND prograde.
   BT metamorphism
   SA epizonal metamorphism
   SA polymetamorphism
   SA retrograde metamorphism

Program for Array Seismic Studies of the Continental Lithosphere
   use PASSCAL

Program Voyager
   use Voyager Program

programming languages
   use computer languages

programs   (1978)
   UF projects
   SA associations
   SA BRIM
   SA computer programs
   SA computers
   SA Continental Scientific Drilling Program
   SA CUSMAP
   SA Dry Valley Drilling Project
   SA education
   SA European Geotraverse
   SA GLIMPCE
   SA IGBP
   SA IRIS network
   SA KTB
   SA mineral exploration
   SA NURE
   SA PASSCAL
   SA planning
   SA policy
   SA RARE II regions
   SA RASA
   SA reclamation
   SA research
   SA Superfund
   SA TACT

progress report   (1978)
   BT report
   SA annual report
   SA associations
   SA current research
   SA research
   SA survey organizations
   SA symposia

Project Voyager
   use Voyager Program

projects
   use programs

prokaryotes   (1993)
   Member of group of organisms which have no vesicular nucleus or membrane-bounded organelles. Groups included are blue-green algae and bacteria.
   UF procaryotes
   BT thallophytes
   BT Plantae
   SA bacteria
   SA biologic evolution
   SA Cyanophyta

Prokopyevsk
   No longer a valid term for GeoRef. As of 1993, see Prokopyevsk Russian Federation.

Prokopyevsk Russian Federation   (1993)
   City at S end of Kuznetsk Basin in Kemerovo Oblast in S Siberia. Before 1993, also search Prokopyevsk. This term has multiple hierarchies.
   BT1 Kemerovo Russian Federation
   BT1 Russian Federation
   BT1 Commonwealth of Independent States
   BT2 Kemerovo Russian Federation
   BT2 Asia

Prolecanitida   (1981)
   BT Ammonoidea
   BT Tetrabranchiata
   BT Cephalopoda
   BT Mollusca
   BT Invertebrata

proluvium   (1978)
   BT clastic sediments
   BT sediments

prometheum
   use promethium

promethium   (1978)
   Autoposting of rare earths and metals to this term began in 1989.
   UF Pm
   UF prometheum
   BT rare earths
   BT metals

propagation   (1978)
   Before 1981, also search wave propagation. As of 1993, see emplacement for use with intrusions.
   SA elastic waves
   SA electromagnetic waves
   SA ocean waves
   SA seismology
   SA shock waves
   SA waves

propane   (1985)
   BT alkanes
   BT aliphatic hydrocarbons
   BT hydrocarbons
   BT organic materials
   SA dibromochloropropane

properties   (1978)
   SA acoustical properties
   SA behavior
   SA bulk density
   SA chemical properties
   SA constitutive equations
   SA dielectric properties
   SA dispersivity
   SA dynamic properties
   SA elastic properties
   SA elastodynamic properties
   SA electrical properties
   SA electrochemical properties
   SA engineering properties
   SA magnetic hysteresis
   SA magnetic properties
   SA materials
   SA mechanical properties
   SA nonmagnetic minerals
   SA optical properties
   SA physical properties
   SA physicochemical properties
   SA piezoelectric properties
   SA reservoir properties
   SA solubility
   SA specific surface
   SA stiffness
   SA surface properties
   SA thermal properties
   SA thermochemical properties
   SA thermodynamic properties
   SA thermoelastic properties
   SA thixotropy
   SA tortuosity
   SA wettability

Propria Geosyncline   (1978)
   In Propria region of Sergipe State, on the lower Sao Francisco River near the Atlantic Ocean.
   BT Sergipe Brazil
   BT Brazil
   BT South America
   SA geosynclines

propylite   (1978)
   BT metasomatic rocks
   BT metamorphic rocks
   SA andesites
   SA propylitization

propylitization   (1978)
   BT metasomatism
   SA hydrothermal alteration
   SA processes
   SA propylite

Prosimii (1981)
Autoposting of Eutheria and Theria to this term began in 1989.
BT Primates
BT Eutheria
BT Theria
BT Mammalia
BT Tetrapoda
BT Vertebrata
BT Chordata

Prosobranchia (1978)
BT Gastropoda
BT Mollusca
BT Invertebrata

prosopite (1978)
BT fluorides
BT halides

Prospect Mountain Quartzite
use Stirling Quartzite

prospecting
As of 1981, no longer a valid term for GeoRef. Use exploration or mineral exploration as applicable. After 1992, also see coal exploration or petroleum exploration.

PROSPECTOR (1989)
Computer-based consultation system for mineral exploration. Search in conjunction with data AND processing.
SA expert systems
SA geographic information systems
SA mineral exploration

protactinium (1978)
Autoposting of actinides and metals to this term began in 1989.
UF Pa
BT actinides
BT metals
NT Pa-231

protection (1985)
SA conservation
SA environmental geology
SA pollution
SA reclamation
SA regulations
SA Resource Conservation and Recovery Act
SA water quality

proteins (1978)
Autoposting of this term began in 1978.
BT organic materials
NT collagen
NT conchiolin
NT enzymes
NT purines
SA amino acids
SA peptides

Proterozoic (1978)
Autoposting of upper Precambrian to this term began in 1989.
UF Agnotozoic
BT upper Precambrian
BT Precambrian
NT Adelaidean
NT Algonkian
NT Athabasca Formation
NT Badami Series
NT Bambui Group
NT Banded Gneissic Complex
NT Baraga Group
NT Bitter Springs Formation
NT Cahill Formation
NT Coldbrook Group
NT Damara System
NT Dedham Granodiorite
NT Huronian
NT Hutuo Group
NT Karelian Orogeny
NT Keweenawan
NT Kombolgie Formation
NT Kunyang Group
NT Kursk Series
NT Lewisian
NT Lorrain Formation
NT lower Proterozoic
NT Malmani Subgroup
NT middle Proterozoic
NT Miette Group
NT Negaunee Iron Formation
NT Oronto Group
NT Ortega Group
NT Ovruch Series
NT Pan-African Orogeny
NT Pocatello Formation
NT Pretoria Group
NT Sinian
NT Tapley Hill Formation
NT Udokan Series
NT Umberatana Group
NT upper Proterozoic
NT Windermere System
NT Wolkberg Group
NT Wopmay Orogeny
SA Harney Peak Granite
SA Hoosac Formation
SA Klodzko Unit
SA Krol Formation
SA Lewisian Complex
SA New York City Group
SA Shuswap Complex
SA Tallulah Falls Formation
SA Yudomian

Proteutheria (1981)
Suborder. Autoposting of this term to Anagalida and Scandentia began in 1989. Autoposting of Insectivora, Eutheria and Theria to this term began in 1989.
BT Insectivora
BT Eutheria
BT Theria
BT Mammalia
BT Tetrapoda
BT Vertebrata
BT Chordata
NT Anagalida
NT Scandentia

Protista (1978)
Before 1976, also search Protozoa. Foraminifera and Radiolaria added as narrower terms in 1981. Autoposting of this term began in 1978. As of 1993, microfossils is a broader term. This term has multiple hierarchies.
BT1 Invertebrata
BT2 microfossils
NT ebridians
NT foraminifera
NT Radiolaria
NT Silicoflagellata
NT silicoflagellates
NT Thecamoeba
NT Tintinnidae
SA algae
SA bacteria
SA Metazoa
SA microorganisms
SA plankton
SA zooxanthellae

Proto-Atlantic Ocean
use Iapetus

Protochordata (1981)
BT Chordata

protodolomite (1978)
BT carbonates

protoenstatite (1978)
Artificial mineral. Autoposting of pyroxene group to this term ended in 1985 and began again in 1989.
BT orthopyroxene
BT pyroxene group
BT chain silicates
BT silicates
SA enstatite

protoliths (1978)
UF parent rocks
SA igneous rocks
SA metamorphic rocks
SA metamorphism
SA parent bodies

protons (1978)
Used as a general term.
SA electrons

Protopivskaya Formation (1978)
Central Ukraine and NW Donets Basin.
BT Upper Triassic
BT Triassic
BT Mesozoic
SA Donets Basin
SA Russian Federation
SA Ukraine

protoplanets
use planetesimals

Protorthoptera (1978)
BT Insecta
BT Mandibulata
BT Arthropoda
BT Invertebrata

Protozoa
A valid index term through 1976. After 1976, use Protista.

proustite (1978)
BT sulfarsenites
BT sulfosalts

provenance (1978)
UF source areas
UF source regions
UF source terrains
UF source terranes
UF sourceland
SA paleocurrents
SA relict materials
SA sedimentary rocks
SA sedimentation
SA sediments

Provence (1978)
Historical region in SE France.
IN Index departments as applicable.
CO N430000N444500 E0074500E0043000
BT France
BT Western Europe
BT Europe
SA Alpes-de-Haute Provence France
SA Alpes-Maritimes France
SA Bouches-du-Rhone France
SA Provence Alps
SA Sainte-Baume Massif
SA Var France
SA Vaucluse France

Provence Alps (1978)
In SE Provence. This term has multiple hierarchies.
IN Index departments as applicable.
BT1 France
BT1 Western Europe
BT1 Europe
BT2 Western Alps
BT2 Alps
BT2 Europe
SA Alpes-de-Haute Provence France
SA Alpes-Maritimes France
SA Maritime Alps
SA Provence
SA Sainte-Baume Massif
SA Var France

Providence County
Valid through 1988. Search in combination with state term. After 1988, use specific county-state term.

Providence County Rhode Island (1989)
N Rhode Island. Before 1989, also search Providence County AND Rhode Island.
CO N414300N420200 W0712000W0714800
BT Rhode Island
BT United States
NT Cumberland Rhode Island

provinces
A valid general term through 1976. See specific term, e.g. metallogenic provinces, physiographic provinces, etc.

provinces, faunal
use faunal provinces

provinces, floral
use floral provinces

provinces, ground-water
use ground-water provinces

provinces, metallogenic
use metallogenic provinces

provinces, physiographic
use physiographic provinces

provincialism
use provinciality

provinciality (1978)
Used mostly for fossils.
UF provincialism
SA ecology
SA fossils
SA paleoecology

Provo
Valid through 1988. Search in combination with state term. After 1988, use specific city-state term.

Provo Utah (1989)
City on Provo River and SSE of Salt Lake City in N central Utah. Before 1989, search Provo AND Utah.
CO N401500N401500 W1114000W1114000
BT Utah County Utah
BT Utah
BT United States

Prudhoe Bay (1978)
Inlet of the Arctic Ocean on the North Slope.
BT Northern Alaska
BT Alaska
BT United States
SA North Slope
SA Prudhoe Bay Field

Prudhoe Bay Field (1985)
North Slope of Alaska.
CO N700000N702500 W1474000W1490500
BT Northern Alaska
BT Alaska
BT United States
SA Ivishak Formation
SA North Slope
SA oil and gas fields
SA Prudhoe Bay

Prut River (1978)

Rises in W central Carpathians and flows into Danube River below Galati. Also search Prut.
IN Index Romania and Soviet republics as applicable.
UF Pruth River
BT Europe
SA Dniester-Prut Interfluve
SA Moldavia
SA Moldova
SA Romania
SA Ukraine

Pruth River
use Prut River

Prydz Bay (1993)
Off Amery ice shelf, Antarctic Ocean.
CO S690000S660000 E0810000E0730000
BT Antarctic Ocean
SA Leg 119
SA ODP Site 739
SA ODP Site 741
SA ODP Site 743

Przemsyl
As of 1989, no longer a valid term for GeoRef.
use Przemysl Poland

Przemysl
No longer a valid term for GeoRef. As of 1993 see Przemysl Poland.

Przemysl Poland (1993)
Province and city near Ukrainian border in SE Poland. Before 1989, also search Przemsyl.
UF Peremyshl
UF Przemysl
BT Poland
BT Central Europe
BT Europe
SA Galitsiya

PS waves
use PS-waves

PS-waves (1989)
UF PS waves
BT body waves
BT elastic waves
SA P-waves
SA S-waves
SA seismology
SA waves

psammite
use sandstone

psammitic texture
use arenaceous texture

pseudo-karst
use pseudokarst

pseudo-single domains (1981)
BT magnetic domains

pseudobrookite (1978)
BT oxides
SA armalcolite
SA ferropseudobrookite

pseudogalena
use sphalerite

Pseudogleys (1978)
BT soils
SA Gleys
SA soil group

pseudokarst (1985)
UF pseudo-karst
SA geomorphology
SA karst
SA thermokarst
SA topography

pseudoleucite (1978)
BT framework silicates
BT silicates

pseudomorphism (1978)
Before 1976, also search replacement. Before 1971, also search pseudomorphs.
UF pseudomorphs
SA minerals

pseudomorphs
Not a valid GeoRef index term after 1970. Was used on level 1 in subfile N.
use pseudomorphism

pseudotachylite (1978)
BT mylonites
BT metamorphic rocks

Pseudotsuga (1985)
Genus.
BT Coniferales
BT gymnosperms
BT Spermatophyta
BT Plantae

Psiloceratida (1981)
BT Ammonoidea
BT Tetrabranchiata
BT Cephalopoda
BT Mollusca
BT Invertebrata

psilomelane (1978)
BT oxides
SA manganese minerals
SA manganese oxides

Psilophytales
use Psilopsida

Psilopsida (1978)
Including Psilophytales.
UF Psilophytales
BT pteridophytes
BT Plantae
SA vascular taxa

Pskem Range (1978)
Outlier of Kirghiz Range. This term has multiple hierarchies.
IN Index former Soviet republics as applicable.
BT1 Commonwealth of Independent States
BT2 Asia
SA Kazakhstan
SA Kyrgyzstan

Pskov
No longer a valid term for GeoRef. As of 1993, see Pskov Russian Federation.

Pskov Russian Federation (1993)
Oblast including city 155 miles SW of Saint Petersburg in W Russian Federation. This term has multiple hierarchies.
BT1 Russian Federation
BT1 Commonwealth of Independent States
BT2 Europe

Psocopteroida (1981)
BT Insecta
BT Mandibulata
BT Arthropoda
BT Invertebrata

Pt
use platinum

Pteranodon (1978)
Genus.
BT Pterosauria
BT Archosauria
BT Diapsida
BT Reptilia
BT Tetrapoda
BT Vertebrata
BT Chordata

Pteridophyllen (1978)
BT pteridophytes
BT Plantae

Pteridophyta
use pteridophytes

pteridophytes (1978)
Before 1981, Callipteris was included as a narrower term. Autoposting of this term began in 1978. Before 1993, also search Pteridophyta.
UF Pteridophyta
BT Plantae
NT Filicopsida
NT Lycopsida
NT Noeggerathiales
NT Psilopsida
NT Pteridophyllen
NT Sphenopsida
SA Callipteris
SA ferns
SA Juglans
SA Neuropteris
SA Pteropsida
SA sporangia
SA spores
SA vascular taxa

Pteridospermae (1978)
Callipteris is a narrower term as of 1981. Autoposting of this term began in 1978.
BT gymnosperms
BT Spermatophyta
BT Plantae
NT Callipteris
NT Neuropteris
SA Dicroidium
SA Pecopteris
SA Ptilophyllum
SA Sphenopteris
SA Taeniopteris

Pteriina (1981)
Autoposting of this term to Anthraconaia, Anthraconauta and Pectinacea began in 1989. Autoposting of Cyrtodontida to this term ended in 1989.
BT Bivalvia
BT Mollusca
BT Invertebrata
NT Anthraconaia
NT Anthraconauta
NT Inocerami
NT Pectinacea

Pterobranchia (1978)
IN Index Hemichordata or Invertebrata as applicable.
NT Cephalodiscida
NT Rhabdopleurida
SA Hemichordata
SA Invertebrata

Pterocarya (1989)
Genus.
BT Dicotyledoneae
BT angiosperms
BT Spermatophyta
BT Plantae

Pteropoda (1978)
BT Gastropoda
BT Mollusca
BT Invertebrata

Pteropsida (1978)
Used as a general term.
BT Plantae
SA angiosperms
SA gymnosperms
SA pteridophytes
SA thallophytes
SA vascular taxa

Pterosauria (1978)
Order. Autoposting of this term to Pteranodon began in 1981.
BT Archosauria
BT Diapsida
BT Reptilia
BT Tetrapoda
BT Vertebrata
BT Chordata
NT Pteranodon

ptilolite
use mordenite

Ptilophyllum (1978)
BT Cycadales
BT gymnosperms
BT Spermatophyta
BT Plantae
SA Bennettitales
SA Pteridospermae

Ptychopariida (1978)
Autoposting of this term to Proetidae and Dechenellidae began in 1985.
BT Trilobita
BT Trilobitomorpha
BT Arthropoda
BT Invertebrata
NT Asaphidae
NT Cryptolithus
NT Dechenellidae
NT Olenidae
NT Proetidae
NT Scutelluidae

ptygma
use ptygmatic folds

ptygmatic folds (1978)
Before 1978, also search folds AND ptygmatic.
UF ptygma
BT folds
SA migmatites

Pu
use plutonium

Pu'u O'o
use Puu Oo

Pu-238 (1985)
Autoposting of broader terms began in 1989. This term has multiple hierarchies.
BT1 radioactive isotopes
BT1 isotopes
BT2 plutonium
BT2 actinides
BT2 metals

Pu-239 (1978)
Autoposting of broader terms began in 1989. This term has multiple hierarchies.
BT1 radioactive isotopes
BT1 isotopes
BT2 plutonium
BT2 actinides
BT2 metals
SA Pu-240/Pu-239

Pu-240 (1985)
Autoposting of broader terms began in 1989. This term has multiple hierarchies.
BT1 radioactive isotopes
BT1 isotopes
BT2 plutonium
BT2 actinides
BT2 metals
SA Pu-240/Pu-239

Pu-240/Pu-239 (1989)
Autoposting of broader terms began in 1989. This term has multiple hierarchies.
BT1 radioactive isotopes
BT1 isotopes

BT2 plutonium
BT2 actinides
BT2 metals
SA isotope ratios
SA Pu-239
SA Pu-240

Pu-244   (1978)
Autoposting of broader terms began in 1989. This term has multiple hierarchies.
BT1 radioactive isotopes
BT1 isotopes
BT2 plutonium
BT2 actinides
BT2 metals

publication lists
use publications

publications   (1978)
Used as a general term. Before 1971, also search geologic publications; publication lists.
UF geologic publications
UF publication lists
SA associations
SA atlas
SA bibliography
SA book reviews
SA catalogs
SA dictionaries
SA directory
SA glossaries
SA guidebook
SA lexicons
SA manuals
SA maps
SA monographs
SA survey organizations
SA textbooks
SA theses

Pucangan Formation   (1989)
Central Java.
BT Quaternary
BT Cenozoic
SA Indonesia
SA Java

Pucara Group   (1978)
Includes Paria Formation. W Peru. Upper Triassic and Jurassic.
IN Index ages as applicable.
BT Mesozoic
SA Junin Peru
SA Jurassic
SA Lima Peru
SA Peru
SA Upper Triassic

Puck Bay   (1978)
Inlet of the Gulf of Danzig on the Baltic Sea. Also search Puck.
BT Gdansk Poland
BT Poland
BT Central Europe
BT Europe

Puebla
No longer a valid term for GeoRef. As of 1993 see Puebla Mexico.

Puebla de Guzman
No longer a valid term for GeoRef. As of 1993, see Puebla de Guzman Spain.

Puebla de Guzman Spain   (1993)
Town in SW part of country. Before 1993, also search Puebla de Guzman and Spain.
BT Huelva Spain
BT Andalusia Spain
BT Spain
BT Iberian Peninsula
BT Southern Europe
BT Europe

Puebla Mexico   (1993)
State and city SE of Mexico City in SE Mexico. Before 1993 also search Puebla AND Mexico.
BT Mexico
NT Los Humeros
SA Mexican volcanic belt
SA Popocatepetl
SA Sierra Madre Oriental

Pueblo Colorado   (1989)
City in SE central Colorado. Before 1989, search Pueblo AND Colorado.
CO N381700N381700
W1043800W1043800
BT Pueblo County Colorado
BT Colorado
BT United States

Pueblo County
Valid through 1988. Search in combination with state term. After 1988, use specific county-state term.

Pueblo County Colorado   (1989)
S central Colorado. Before 1989, also search Pueblo County AND Colorado.
CO N374500N383000
W1040500W1050500
BT Colorado
BT United States
NT Pueblo Colorado
SA Wet Mountains

Puente Formation   (1978)
Includes Papel Blanco Shale, Blanco Sandstone, Cubierto Shale, Hunter Sandstone and Conglomerate, Peculiar Shale, Mahala Sandstone and Conglomerate, Sycamore Canyon Member, La Vida Member, Soquel Member, Yorba Member. S California.
BT upper Miocene
BT Miocene
BT Neogene
BT Tertiary
BT Cenozoic
SA California

Puente Hills   (1978)
S part of Los Angeles County.
BT Los Angeles County California
BT California
BT United States

Puercan   (1989)
North American continental stage. Above Cretaceous and below Dragonian.
BT lower Paleocene
BT Paleocene
BT Paleogene
BT Tertiary
BT Cenozoic

Puerco River
use Rio Puerco

Puerto Cabello
No longer a valid term for GeoRef. As of 1993 see Puerto Cabello Venezuela.

Puerto Cabello Venezuela   (1993)
City W of Caracas on the Caribbean Sea in N Venezuela. Before 1993 also search Puerto Cabello AND Venezuela.
BT Carabobo Venezuela
BT Venezuela
BT South America

Puerto Deseado   (1978)
Town in Santa Cruz province, Argentina and Bay in S Argentina. As of 1993, use Puerto Deseado Argentina for the town.
BT Santa Cruz Argentina
BT Argentina
BT South America

Puerto Rico   (1978)
Before 1932, also search Porto Rico.
CO N175000N183000
W0654000W0671500
BT Greater Antilles
BT Antilles
BT West Indies
BT Caribbean region
NT Anasco Bay
NT Arecibo Puerto Rico
NT Cabo Rojo Puerto Rico
NT Mayaguez Bay
NT Ponce Puerto Rico
NT San Juan Puerto Rico
SA Caribbean Mountain Range
SA Maravillas Formation
SA Robles Formation
SA San Juan Formation

Puerto Rico Trench   (1978)
NNE of Puerto Rico.
UF Puerto Rico Trough
BT Atlantic Ocean

Puerto Rico Trough
use Puerto Rico Trench

Puget Lowland   (1978)
Area surrounding Puget Sound and the broad trough extending S.
BT Washington
BT United States

Puget Sound   (1978)
Arm of Pacific Ocean extending S from E end of Juan de Fuca Strait.
CO N471000N483000
W1221000W1231500
BT Washington
BT United States
SA Deschutes River

Puglia Italy
use Apulia Italy

Pulaski County
Valid through 1988. Search in combination with state term. After 1988, use specific county-state term.

Pulaski County Arkansas   (1989)
Central Arkansas. Before 1989, also search Pulaski County AND Arkansas.
CO N343000N350200
W0920400W0924500
BT Arkansas
BT United States

Pulaski County Georgia   (1989)
S central Georgia. Before 1989, also search Pulaski County AND Georgia.
CO N320800N322300
W0831700W0833700
BT Georgia
BT United States

Pulaski County Illinois   (1989)
Extreme S Illinois. Before 1989, also search Pulaski County AND Illinois.
CO N370700N371900
W0885500W0891500
BT Illinois
BT United States

Pulaski County Indiana   (1989)
NW Indiana. Before 1989, also search Pulaski County AND Indiana.
CO N405400N411000
W0862800W0865500
BT Indiana
BT United States

Pulaski County Kentucky   (1989)
S Kentucky. Before 1989, also search Pulaski County AND Kentucky.
CO N365000N371800
W0841800W0845400
BT Kentucky
BT United States

Pulaski County Missouri   (1989)
Central Missouri. Before 1989, also search Pulaski County AND Missouri.
CO N373700N380200
W0920200W0922300
BT Missouri
BT United States

Pulaski County Virginia   (1989)
SW Virginia. Before 1989, also search Pulaski County AND Virginia.
CO N365200N371500
W0803100W0805500
BT Virginia
BT United States

Pulaski Fault   (1978)
SW Virginia.
IN Index counties or region as applicable.
SA Virginia

Pulaski thrust sheet   (1989)
SW Virginia and NE Tennessee.
IN Index states as applicable.
BT United States
SA Appalachians
SA Tennessee
SA Virginia

pulaskite   (1978)
BT syenites
BT plutonic rocks
BT igneous rocks

pull apart structures
use boudinage

pull-apart basins   (1989)
Basins associated with strike-slip fault zones.
BT basins
SA en echelon faults
SA extension tectonics
SA fault zones
SA strike-slip faults
SA tectonics

pulsating spring
use geysers

pulsations
As of 1993, no longer a valid term for GeoRef. Usually out-of-scope. Used with magnetosphere and magnetic field.

Pultova Ukraine
use Poltava Ukraine

Pultowa Ukraine
use Poltava Ukraine

Pultusk
No longer a valid term for GeoRef. As of 1993 see Pultusk Poland.

Pultusk Poland   (1993)
City in SE Ciechanow, E central Poland. Before 1993 also search Pultusk AND Poland.
BT Poland
BT Central Europe

BT Europe
pumice (1978)
As of 1981, use pumice deposits for pumice as a commodity.
BT pyroclastics
BT volcanic rocks
BT igneous rocks
SA pumice deposits
SA scoria
SA shirasu
pumice deposits (1981)
Before 1981, search pumice AND deposits.
IN Commodity. See List C.
SA abrasives
SA aggregate
SA pumice
pump tests (1978)
Before 1978, also search pumping; pumping tests.
UF pumping tests
SA ground water
SA hydrology
SA pumping
SA water wells
pumpellyite (1978)
Autoposting of sorosilicates began in 1985.
BT sorosilicates
BT orthosilicates
BT silicates
SA aluminosilicates
pumping (1982)
SA drawdown
SA economic geology
SA engineering geology
SA ground water
SA hydrogeology
SA petroleum engineering
SA production
SA pump tests
SA water recovery
SA water resources
SA water wells
SA well screens
SA wells
pumping tests
  use pump tests
punch cards (1978)
UF punched cards
SA data processing
Punchbowl Formation (1978)
Described in Valyermo Quadrangle as consisting of two facies. S California.
BT upper Miocene
BT Miocene
BT Neogene
BT Tertiary
BT Cenozoic
SA California
punched cards
  use punch cards
Puncoviscana Formation (1993)
Upper Precambrian and Lower Cambrian. NW Argentina.
IN Index ages as applicable.
SA Argentina
SA Cambrian
SA Lower Cambrian
SA Precambrian
SA upper Precambrian
punctuated aggradational cycles (1989)
SA cycles
SA cyclic processes
SA ecology
SA environmental analysis
SA paleoecology
SA sedimentation

SA stratigraphy
punctuated equilibria (1985)
SA biologic evolution
SA gradualism
SA paleontology
SA phylogeny
Pune
  Not a valid term for GeoRef. See Poona India.
Pungo River (1978)
Tidal estuary near the Pamlico River off Pamlico Sound.
BT North Carolina
BT United States
SA Pamlico River
Pungo River Formation (1985)
Middle Miocene. E North Carolina and Virginia.
BT Miocene
BT Neogene
BT Tertiary
BT Cenozoic
SA North Carolina
SA Virginia
Punjab
  No longer a valid term for GeoRef as of 1981. Use Punjab India or Punjab Pakistan as applicable. Before 1981, referred to both the state in India and the province in Pakistan.
Punjab India (1993)
State. NW India. Before 1993, also search Punjab State. Before 1981, also search Punjab AND India.
CO N295000N323000
    E0770000E0735000
BT India
BT Indian Peninsula
BT Asia
NT Chandigarh India
NT Spiti
SA Beas River
Punjab Pakistan (1993)
Province. E central Pakistan. Before 1993, also search Punjab Province. Before 1981, also search Punjab AND Pakistan.
BT Pakistan
BT Indian Peninsula
BT Asia
NT Attock Pakistan
NT Khewra Pakistan
NT Mansehra Pakistan
NT Mianwali Pakistan
NT Ramnagar Pakistan
NT Salt Range
SA Mangla Dam
SA Potwar Plateau
Punjab Province
  No longer a valid term for GeoRef. See Punjab Pakistan.
Punjab State
  No longer a valid term for GeoRef. See Punjab India.
Puno
  No longer a valid term for GeoRef. As of 1993 see Puno Peru.
Puno Peru (1993)
Department in SE Peru. Before 1993 also search Puno AND Peru.
CO S172000S130000
    W0685000W0711400
BT Peru
BT South America
SA Lake Titicaca

Punta Mosquito Formation (1978)
BT Eocene
BT Paleogene
BT Tertiary
BT Cenozoic
SA Venezuela
Pupillidae (1978)
BT Gastropoda
BT Mollusca
BT Invertebrata
Purbeckian (1978)
Europe.
BT Lower Cretaceous
BT Cretaceous
BT Mesozoic
Purcell Mountains (1989)
Mountain range in SE British Columbia; Boundary County, N Idaho and Lincoln County, NW Montana.
IN Index British Columbia and/or states and counties as applicable.
BT North America
SA Boundary County Idaho
SA British Columbia
SA Idaho
SA Lincoln County Montana
SA Montana
Purcell System (1978)
BT Precambrian
SA Alberta
SA British Columbia
SA Idaho
SA Montana
SA Windermere System
purchases (1981)
SA economics
pure coal
  use vitrain
purification (1978)
SA beneficiation
SA decontamination
SA drinking water
SA ground water
SA hydrology
SA impurities
SA polluted water
SA pollution
SA potability
SA water
SA water quality
SA water resources
SA water treatment
purines (1978)
BT proteins
BT organic materials
Purisima Formation (1989)
Miocene-Pliocene. W California.
IN Index age as applicable.
BT Neogene
BT Tertiary
BT Cenozoic
SA California
SA Miocene
SA Pliocene
purple copper ore
  use bornite
Purulia
  No longer a valid term for GeoRef. See Purulia India.
Purulia India (1993)
City in W West Bengal, E India.
BT West Bengal India
BT India
BT Indian Peninsula
BT Asia

push wave
  use P-waves
push-pull wave
  use P-waves
Putnam County
  Valid through 1988. Search in combination with state term. After 1988, use specific county-state term.
Putnam County Florida (1989)
N Florida. Before 1989, also search Putnam County AND Florida.
CO N292000N294700
    W0812900W0820300
BT Florida
BT United States
Putnam County Georgia (1989)
Central Georgia. Before 1989, also search Putnam County AND Georgia.
CO N331000N332900
    W0831000W0833200
BT Georgia
BT United States
Putnam County Illinois (1989)
N central Illinois. Before 1989, also search Putnam County AND Illinois.
CO N410700N412100
    W0891000W0892800
BT Illinois
BT United States
Putnam County Indiana (1989)
W central Indiana. Before 1989, also search Putnam County AND Indiana.
CO N392800N395200
    W0864000W0870100
BT Indiana
BT United States
Putnam County Missouri (1989)
N Missouri. Before 1989, also search Putnam County AND Missouri.
CO N402000N403700
    W0924000W0932300
BT Missouri
BT United States
Putnam County New York (1989)
SE New York. Before 1989, also search Putnam County AND New York.
CO N411900N413300
    W0733200W0735800
BT New York
BT United States
Putnam County Ohio (1989)
NW Ohio. Before 1989, also search Putnam County AND Ohio.
CO N405100N411000
    W0835300W0842400
BT Ohio
BT United States
Putnam County Tennessee (1989)
Central Tennessee. Before 1989, also search Putnam County AND Tennessee.
CO N355800N361800
    W0850800W0854900
BT Tennessee
BT United States
Putnam County West Virginia (1989)
W West Virginia. Before 1989, also search Putnam County AND West Virginia.

CO N381600N384100
    W0814300W0820400
BT West Virginia
BT United States

Putrid Sea
use Sivash

Puu Oo (1993)
Extensively studied eruptive activity site on East Rift Zone of Kilauea volcano. Site is just inside Hawaii Volcanoes national park. Activity began in 1983 and continued for several years. This term has multiple hierarchies.
CO N191500N192500
    W1552000W1550500
UF Pu'u O'o
BT1 Hawaii Island
BT1 Hawaii County Hawaii
BT1 Hawaii
BT1 United States
BT2 Hawaii Island
BT2 Hawaii County Hawaii
BT2 Hawaii
BT2 Polynesia
BT2 Oceania
BT3 Hawaii Island
BT3 Hawaii County Hawaii
BT3 Hawaii
BT3 East Pacific Ocean Islands
SA East Rift Zone
SA Kilauea

Puy-de-Dome
No longer a valid term for GeoRef.
See Puy-de-Dome France.

Puy-de-Dome France (1993)
Department in S central France.
CO N451500N461500
    E0040000E0023000
BT France
BT Western Europe
BT Europe
NT Chaine des Puys
NT Clermont-Ferrand France
NT Mont-Dore France
SA Limagne

Pyramid Lake (1978)
Lake in Washoe County, NW Nevada.
IN Index county or regions as applicable.
SA Washoe County Nevada

pyrargyrite (1978)
BT sulfantimonites
BT sulfosalts

Pyrenean Orogenic Phase
use Pyrenean Orogeny

Pyrenean Orogeny (1978)
One of the 30 or more short-lived orogenies during Phanerozoic time identified by Stille, in this case during the late Eocene, between the Bartonian and Ludian stages. Before 1978, search Pyrenean AND orogeny.
UF Pyrenean Orogenic Phase
BT Eocene
BT Paleogene
BT Tertiary
BT Cenozoic
SA orogeny

Pyrenees (1978)
Mountain range extending from the Bay of Biscay to the SW coast of the Gulf of Lion.
IN Index countries as applicable.
CO N420000N430000
    E0031500W0020000
BT Europe
NT Castillon Massif
NT French Pyrenees
NT Querigut Massif
NT Spanish Pyrenees
SA Andorra
SA France
SA North Pyrenean Fault
SA Spain

Pyrenees-Atlantiques
No longer a valid term for GeoRef.
use Pyrenees-Atlantiques France

Pyrenees-Atlantiques France (1993)
Department in SW France. Before 1978, also search Basses-Pyrenees.
CO N424500N433500
    W0000000W0014500
UF Basses-Pyrenees France
UF Pyrenees-Atlantiques
BT France
BT Western Europe
BT Europe
NT Biarritz France
NT Pau France
SA Navarre

Pyrenees-Orientales
No longer a valid term for GeoRef.
use Pyrenees-Orientales France

Pyrenees-Orientales France (1993)
Department in E Pyrenees.
CO N421500N430000
    E0031500E0013000
UF Pyrenees-Orientales
BT France
BT Western Europe
BT Europe
NT Agly Massif
SA Roussillon

pyrite (1978)
As of 1981, use pyrite ores for pyrite as a commodity.
BT sulfides
SA bravoite
SA framboidal texture
SA iron minerals
SA iron ores
SA pyrite ores

pyrite ores (1981)
Before 1981, also search pyrite AND (deposit OR deposits OR ore OR ores OR economic) in the basic index. Autoposting of metal ores to this term began in 1985.
IN Commodity. See List C.
BT metal ores
SA iron ores
SA pyrite
SA sulfur deposits

pyritization (1978)
BT metasomatism
SA diagenesis
SA hydrothermal alteration

pyroaurite (1978)
BT carbonates
SA hydrotalcite

pyrochlore (1978)
From 1978-1980, halides and oxides were autoposted to this term. Autoposting of fluorides, niobates and tantalates began in 1981. Autoposting of halides, niobotantalates and oxides began in 1989. This term has multiple hierarchies.
BT1 fluorides
BT1 halides
BT2 niobotantalates
BT2 niobates
BT2 oxides
BT3 niobotantalates
BT3 tantalates
BT3 oxides
SA betafite
SA microlite
SA niobium ores

pyroclastic flows (1985)
SA ash flows
SA ash-flow tuff
SA base surges
SA clastic sediments
SA igneous rocks
SA ignimbrite
SA nuees ardentes
SA pyroclastics
SA sediments
SA shirasu
SA tuff
SA volcanic ash
SA volcanic rocks
SA volcaniclastics
SA volcanism

pyroclastics (1978)
As of 1984, not to be used as sediments or sedimentary rocks. Use volcaniclastics when sediments or sedimentary rocks are made up partially of material originating from a volcano. Before 1981, also search pyroclastics and glasses. From 1978-1980, pyroclastics and glasses was autoposted to the specific rocks that were classified under that term. From 1978-1980, clastic rocks was autoposted to this term. Autoposting of this term began in 1981. Autoposting to volcanic ash and ash flows ended in 1985.
UF tephra
BT volcanic rocks
BT igneous rocks
NT andesite tuff
NT ash-flow tuff
NT green tuff
NT hyaloclastite
NT ignimbrite
NT pumice
NT rhyolite tuff
NT scoria
NT tuff
NT tuffite
NT welded tuff
SA agglomerate
SA ash falls
SA ash flows
SA bentonite
SA clastic rocks
SA clastic sediments
SA ejecta
SA eruptions
SA glasses
SA lapilli
SA pyroclastic flows
SA sedimentary rocks
SA sediments
SA shirasu
SA volcanic ash
SA volcanic breccia
SA volcaniclastics

pyroclastics and glasses
As of 1981, no longer a valid term for GeoRef. Use pyroclastics or glasses as applicable. See also volcaniclastics. From 1978-1980, was autoposted to the rocks that were classified under it.

pyrolite (1989)
According to the model proposed by Ringwood, a 3:1 mixture of peridotite to basalt which represents upper mantle composition.
BT igneous rocks
SA basalts
SA magmas
SA peridotites
SA upper mantle

pyrolusite (1978)
UF polianite
BT oxides
SA manganese minerals
SA vernadite

pyrolysis (1978)
SA geochemistry
SA Rock-Eval

pyrometasomatism (1978)
BT metasomatism
SA hydrothermal alteration

pyromorphite (1978)
From 1978-1980, halides and phosphates were autoposted to this term. Autoposting of chlorides began in 1981. As of 1989, halides is autoposted to this term. This term has multiple hierarchies.
BT1 chlorides
BT1 halides
BT2 phosphates
SA apatite

pyrope (1978)
BT garnet group
BT nesosilicates
BT orthosilicates
BT silicates

pyrophanite (1989)
BT oxides

pyrophyllite (1978)
BT sheet silicates
BT silicates

Pyrotheria (1981)
Suborder. Autoposting of Amblypoda, Eutheria and Theria to this term began in 1989.
BT Amblypoda
BT Eutheria
BT Theria
BT Mammalia
BT Tetrapoda
BT Vertebrata
BT Chordata
SA Ungulata

pyroxene
As of 1981, no longer a valid term for GeoRef. Was used in subfiles E, G, N and B.
use pyroxene group

pyroxene andesite (1978)
BT andesites
BT volcanic rocks
BT igneous rocks

pyroxene granulite (1978)
BT granulites
BT metamorphic rocks

pyroxene group (1978)
Before 1981, jade was included as a narrower term. Autoposting of this term began in 1978. Before 1981, also search pyroxene.
UF pyroxene
UF pyroxenes
BT chain silicates
BT silicates
NT alkalic pyroxene
NT clinopyroxene
NT orthopyroxene
SA jade

pyroxenes
use pyroxene group

pyroxenite (1978)
BT ultramafics
BT plutonic rocks

BT igneous rocks
   SA bronzitite
   SA clinopyroxenite
   SA garnet pyroxenite
   SA metapyroxenite
   SA websterite
pyroxferroite (1978)
   BT chain silicates
   BT silicates
pyroxmangite (1978)
   BT chain silicates
   BT silicates
Pyrrhophyta (1978)
   Autoposting of thallophytes and microfossils began in 1990. This term has multiple hierarchies.
   BT1 algae
   BT1 thallophytes
   BT1 Plantae
   BT2 algae
   BT2 microfossils
   SA Dinoflagellata
   SA palynomorphs
pyrrhotine
   use pyrrhotite
pyrrhotite (1978)
   UF dipyrite
   UF pyrrhotine
   BT sulfides
   SA iron minerals
   SA iron ores
   SA troilite

Q

Q (1978)
   Ratio of the peak seismic energy in a cycle to the energy dissipated.
   SA earthquakes
   SA elastic waves
   SA Love waves
   SA seismology
Q waves
   use Love waves
Qaidam Basin (1989)
   Qinghai, NW China.
   CO N360000N383000
      E0980000E0900000
   UF Tsaidam Basin
   BT Qinghai China
   BT China
   BT Far East
   BT Asia
Qaraqum
   use Karakum
Qatar (1978)
   CO N243000N262000
      E0514000E0504000
   BT Arabian Peninsula
   BT Asia
   NT Nakhla Qatar
   SA Middle East
   SA Near East
   SA Thamama Group
Qianshan
   No longer a valid term for GeoRef. See Qianshan China.
Qianshan China (1993)
   County and town in S Anhui, China. Before 1993, also search Qianshan. Before 1985, also search Qianshan Xian.
   UF Qianshan Xian China

   BT Anhui China
   BT China
   BT Far East
   BT Asia
Qianshan Xian China
   use Qianshan China
Qilian Mountains (1989)
   Mountain range in Gansu and Qinghai, NW China.
   IN Index provinces as applicable.
   UF Qilian Shan
   UF Qilianshan
   BT China
   BT Far East
   BT Asia
   SA Gansu China
   SA Qinghai China
Qilian Shan
   use Qilian Mountains
Qilianshan
   use Qilian Mountains
Qing Zang Gaoyuan
   use Qinghai-Xizang Plateau
Qinghai
   No longer a valid term for GeoRef. See Qinghai China.
Qinghai China (1993)
   Province, NW China. Before 1993, also search Qinghai. Before 1985, also search Ch'ing Hai, Ch'ing-hai, Chinghai, and Tsinghai.
   CO N313000N390000
      E1030000E0894000
   UF Ch'ing Hai China
   UF Ch'ing-hai China
   UF Chinghai China
   UF Tsinghai China
   BT China
   BT Far East
   BT Asia
   NT Qaidam Basin
   SA Kunlun Mountains
   SA Qilian Mountains
   SA Qinghai-Xizang Plateau
   SA Xianshuihe fault zone
Qinghai-Tibet Plateau
   use Qinghai-Xizang Plateau
Qinghai-Xizang Plateau (1993)
   W China, between the Himalayas and the Kunlun Mountains. Covers Xizang, SW Qinghai, and W Sichuan. Before 1993, also search Tibetan Plateau, Qing-Zang, Qinghai-Tibet Plateau, or Xizang Plateau.
   IN Index provinces as applicable.
   UF Qing Zang Gaoyuan
   UF Qinghai-Tibet Plateau
   UF Tibetan Plateau
   UF Xizang Plateau
   BT China
   BT Far East
   BT Asia
   SA Haiyuan earthquake 1920
   SA Indus-Yarlung Zangbo suture zone
   SA Kunlun Mountains
   SA Lhasa Block
   SA Qinghai China
   SA Sichuan China
   SA Xizang China
Qingshankou Formation (1989)
   Songliao Basin of NE China.
   BT Cretaceous
   BT Mesozoic
   SA China
   SA Songliao Basin

Qingzhen Meteorite (1989)
   Impact at Yongle, Qingzhen, Guizhou.
   BT enstatite chondrites
   BT chondrites
   BT stony meteorites
   BT meteorites
   SA China
   SA Guizhou China
Qinling Mountains (1989)
   Mountain range in Shaanxi, NW China.
   BT Shaanxi China
   BT China
   BT Far East
   BT Asia
Qixia Formation (1989)
   BT Lower Permian
   BT Permian
   BT Paleozoic
   SA China
Qizil Qum
   use Kyzylkum
Qomolangma Feng
   use Mount Everest
Qorqut Granite (1989)
   S West Greenland.
   BT upper Archean
   BT Archean
   BT Precambrian
   SA Greenland
qualitative analysis (1978)
   SA analysis
   SA chemical analysis
   SA quantitative analysis
quality
   A valid general term through 1978. After 1978 use specific term, e.g. water quality.
quantitative analysis (1978)
   SA analysis
   SA chemical analysis
   SA flame photometry
   SA methods
   SA neutron activation analysis
   SA polarography
   SA qualitative analysis
quantitative geomorphology (1978)
   SA geometry
   SA geomorphology
   SA morphometry
quantitative methods
   As of 1981, no longer a valid term for GeoRef. See mathematical methods. Term was introduced in 1978.
Quantou Formation (1989)
   Songliao Basin of Manchuria.
   BT Cretaceous
   BT Mesozoic
   SA China
   SA Songliao Basin
quark
   Not a valid GeoRef index term after 1971. Was used on level 1 in subfile G.
quarries (1978)
   SA mines
   SA mining
   SA mining geology
   SA open-pit mining
   SA pits
   SA strip mining
   SA surface mining
quartz (1978)

   Used for the mineral. When of economic value, use quartz crystal.
   BT silica minerals
   BT framework silicates
   BT silicates
   SA agate
   SA amethyst
   SA chalcedony
   SA cristobalite
   SA floatstone
   SA light minerals
   SA myrmekite
   SA quartz crystal
   SA smoky quartz
quartz arenite (1989)
   UF quartzarenite
   BT arenite
   BT clastic rocks
   BT sedimentary rocks
quartz crystal (1978)
   IN Commodity. See List C.
   UF berg crystal
   UF crystal, quartz
   UF mountain crystal
   UF rock crystal
   SA quartz
quartz diabase (1978)
   Before 1981, also search quartz dolerite. From 1978-1980, basalt family was autoposted to this term.
   UF quartz dolerite
   BT diabase
   BT plutonic rocks
   BT igneous rocks
   SA basalts
   SA mining
quartz diorite
   As of 1989, no longer a valid term for GeoRef. From 1981 to 1988, quartz diorites was autoposted to this term.
   use quartz diorites
quartz diorites (1981)
   Before 1989, also search quartz diorite. Autoposting of this term to quartz diorite ended in 1989.
   UF quartz diorite
   BT diorites
   BT plutonic rocks
   BT igneous rocks
   SA plagiogranite
   SA tonalite
quartz dolerite
   As of 1981, no longer a valid term for GeoRef.
   use quartz diabase
quartz keratophyre (1978)
   BT trachytes
   BT volcanic rocks
   BT igneous rocks
   SA keratophyre
quartz latite (1978)
   BT rhyodacites
   BT volcanic rocks
   BT igneous rocks
   SA latite
quartz monzonite (1978)
   BT granites
   BT plutonic rocks
   BT igneous rocks
quartz porphyry (1978)
   BT rhyolites
   BT volcanic rocks
   BT igneous rocks
   SA beresite
   SA porphyry
quartz sand (1978)
   BT clastic sediments

BT sediments
SA sand

quartz syenite (1978)
BT syenites
BT plutonic rocks
BT igneous rocks

quartz veins (1985)
BT veins
SA mineral deposits, genesis

quartz-fiber gravimeters (1981)
BT gravimeters

quartz-pebble conglomerate (1989)
BT conglomerate
BT clastic rocks
BT sedimentary rocks

quartzarenite
  use quartz arenite

quartzite
  As of 1981, no longer a valid term for GeoRef. Use quartzites for metamorphic rocks and orthoquartzite for sedimentary rocks. From 1978-1980, metamorphic rocks was autoposted to this term.

quartzites (1981)
  Before 1981, search quartzite AND metamorphic rocks.
  BT metamorphic rocks
  NT ferruginous quartzite
  SA orthoquartzite

quartzolite (1981)
  Before 1981, search silexite.
  BT plutonic rocks
  BT igneous rocks

Quaternary (1978)
  Consists of Pleistocene and Holocene. From 1978-1980, Cenozoic was autoposted to this term. Autoposting of Cenozoic to this term began in 1993.
  BT Cenozoic
  NT Anglian
  NT Brunhes Epoch
  NT Clovis
  NT Holocene
  NT Kabuh Formation
  NT Kalibeng Formation
  NT Kazusa Group
  NT lower Quaternary
  NT Mazama Ash
  NT middle Quaternary
  NT Pleistocene
  NT Pucangan Formation
  NT upper Quaternary
  NT Zhoukoudian
  SA Aurignacian
  SA Bronze Age
  SA Chalcolithic
  SA Iron Age
  SA Levalloisian
  SA Magdalenian
  SA Matuyama Epoch
  SA Mousterian
  SA Paleolithic
  SA Stone Age
  SA Tertiary
  SA upper Cenozoic
  SA upper Paleolithic

Quay County
  Valid through 1988. Search in combination with state term. After 1988, use specific county-state term.

Quay County New Mexico (1989)

E New Mexico. Before 1989, also search Quay County AND New Mexico.
  CO N343600N354500
      W1030300W1040800
  BT New Mexico
  BT United States
  NT Tucumcari New Mexico

Quebec (1978)
  CO N450000N630000
      W0570000W0790000
  BT Eastern Canada
  BT Canada
  NT Abitibi County Quebec
  NT Anticosti Island
  NT Charlevoix
  NT Charlevoix-Est County Quebec
  NT Charlevoix-Ouest County Quebec
  NT Dorchester County Quebec
  NT Dore Lake Complex
  NT Eastern Townships
  NT Frontenac County Quebec
  NT Gaspe Peninsula
  NT Gaspe-Est County Quebec
  NT Gaspe-Ouest County Quebec
  NT Gatineau County Quebec
  NT Gatineau Valley
  NT Horne Mine
  NT Magdalen Islands
  NT Manicouagan Crater
  NT Manicouagan Lake
  NT Manicouagan River
  NT Matagami
  NT Monteregian Hills
  NT Montreal and Jesus Islands County Quebec
  NT Morin Complex
  NT Noranda Quebec
  NT Oka Complex
  NT Quebec City Quebec
  NT Saguenay County Quebec
  NT Saguenay Valley
  NT Saint Lawrence Estuary
  NT Sigma Mine
  NT Temiscamingue County Quebec
  NT Thetford Mines
  SA Abitibi Belt
  SA Appalachian Basin
  SA Battery Point Formation
  SA Beekmantown Group
  SA Blake River Group
  SA Blondeau Formation
  SA Canadian Shield
  SA Cape Smith fold belt
  SA Central Gneiss Belt
  SA Central Metasedimentary Belt
  SA Chaleur Bay
  SA Champlain Sea
  SA Champlain Valley
  SA Chazy Group
  SA Cheshire Formation
  SA Churchill Province
  SA Clearwater Lake
  SA Cloridorme Formation
  SA Dunham Dolomite
  SA Ellis Bay Formation
  SA Gilman Formation
  SA Green Mountains
  SA Grenville Front
  SA Grenville Province
  SA Henderson Mine
  SA Hudson Bay Lowlands
  SA James Bay
  SA James Bay Lowlands
  SA Labrador Trough
  SA Lake Champlain
  SA Lake Timiskaming
  SA Levis Shale
  SA Lorrain Formation
  SA Matapedia Group
  SA Nain Province
  SA Nipissing Diabase
  SA O'Sullivan Lake
  SA Opemisca Group
  SA Ottawa River
  SA Ottawa Valley
  SA Pinnacle Formation
  SA Potsdam Sandstone
  SA Restigouche Estuary
  SA Roy Group
  SA Saguenay earthquake 1988
  SA Saint John River
  SA Saint Lawrence Lowlands
  SA Saint Lawrence River
  SA Saint Lawrence River basin
  SA Saint Lawrence Valley
  SA Seboomook Formation
  SA Sokoman Formation
  SA Superior Province
  SA Timiskaming Group
  SA Ungava
  SA York River Formation

Quebec City
  No longer a valid term for GeoRef. As of 1993, see Quebec City Quebec.

Quebec City Quebec (1993)
  City in S Quebec on the Saint Lawrence River near where the river widens into its estuary. Before 1993, also search Quebec City AND Quebec.
  BT Quebec
  BT Eastern Canada
  BT Canada

Queen Alexandra Range (1978)
  W of the S Ross Ice Shelf and NW of Queen Maud Range on the Pacific Ocean side.
  BT Antarctica
  SA Kirkpatrick Basalt

Queen Charlotte Basin (1993)
  Marine basin, offshore W British Columbia. Queen Charlotte Sound and Queen Charlotte Islands area.
  CO N500000N543000
      W1270000W1340000
  BT North American Pacific
  BT Pacific Ocean
  SA British Columbia
  SA Queen Charlotte Islands
  SA Queen Charlotte Sound

Queen Charlotte Fault (1985)
  Coastal British Columbia and adjacent Pacific Ocean.
  IN Index regions as applicable.
  SA British Columbia
  SA North American Pacific
  SA Queen Charlotte Islands

Queen Charlotte Islands (1978)
  Group of islands in Pacific Ocean separated from mainland by Hecate Strait.
  BT British Columbia
  BT Western Canada
  BT Canada
  SA Coast Ranges
  SA Graham Island
  SA Karmutsen Group
  SA Kunga Group
  SA Queen Charlotte Basin
  SA Queen Charlotte Fault
  SA Queen Charlotte Sound

Queen Charlotte Sound (1989)
  Off the coast of W British Columbia, between N Vancouver Island and S Queen Charlotte Islands.
  BT British Columbia
  BT Western Canada
  BT Canada
  SA Queen Charlotte Basin

SA Queen Charlotte Islands
SA Vancouver Island

Queen City Formation (1978)
  In Claiborne Group. Includes Arp Member, Omen Glauconitic Sandstone Member, and unnamed upper sand member. NW Louisiana and E Texas.
  BT middle Eocene
  BT Eocene
  BT Paleogene
  BT Tertiary
  BT Cenozoic
  SA Claiborne Group
  SA Louisiana
  SA Texas

Queen Elizabeth Islands (1978)
  Large group in Arctic Ocean N of Parry Channel in Franklin District.
  BT Franklin District Northwest Territories
  BT Northwest Territories
  BT Western Canada
  BT Canada
  SA Amund Ringnes Island
  SA Graham Island
  SA Melville Island

Queen Formation (1989)
  SE New Mexico and W Texas.
  BT Guadalupian
  BT Permian
  BT Paleozoic
  SA New Mexico
  SA Texas

Queen Maud Land (1989)
  Large region between Coats Land and Enderby Land in Norwegian Sector on the Atlantic Ocean side of Antarctica. From 1978 to 1989, Queen Maude Land was a valid term.
  UF Dronning Maud Land
  UF Queen Maude Land
  BT Antarctica
  SA Coalsack Bluff
  SA Enderby Land
  SA Lutzow-Holm Bay
  SA Prince Olav Coast
  SA Queen Maud Range
  SA Sor-Rondane Mountains

Queen Maud Mountains
  use Queen Maud Range

Queen Maud Range (1989)
  SSW of S tip of Ross Ice Shelf on the Pacific Ocean side of Antarctica. From 1978 to 1989, Queen Maude Range was a valid term for GeoRef.
  UF Queen Maud Mountains
  UF Queen Maude Range
  BT Antarctica
  SA Beacon Supergroup
  SA Beardmore Glacier
  SA Fremouw Formation
  SA Queen Maud Land
  SA Ross Ice Shelf
  SA Shackleton Glacier

Queen Maude Land
  From 1978 to 1988, a valid term for GeoRef.
  use Queen Maud Land

Queen Maude Range
  From 1978 to 1988, a valid term for GeoRef.
  use Queen Maud Range

Queens County
  Valid through 1988. Search in combination with state term. After 1988, use specific county-state term.

Queens County New York (1989)
Coextensive with the borough of Queens in New York City on Long Island, SE New York. Before 1989, also search Queens County AND New York.
IN Index Long Island.
CO N403300N404800
W0734100W0735900
BT New York City New York
BT New York
BT United States
SA Long Island

Queensland
No longer a valid term for GeoRef.
use Queensland Australia

Queensland Australia (1993)
CO S290000S100000
E1530000E1380000
UF Queensland
BT Australia
BT Australasia
NT Adavale Basin
NT Bowen Australia
NT Brisbane Australia
NT Burdekin Delta
NT Burdekin River
NT Charters Towers Australia
NT Darling Downs
NT Eromanga Basin
NT Ipswich Australia
NT Moreton Australia
NT Mount Isa Australia
NT Rockhampton Australia
NT Selwyn Range
NT Springsure Australia
NT Surat Basin
NT Townsville Australia
NT Vale of Glamorgan
NT Weipa Australia
NT Yarrol Basin
SA Baralaba Coal Measures
SA Birkhead Formation
SA Bowen Basin
SA Broken River Formation
SA Bulldog Shale
SA Cape York Peninsula
SA Carpentaria Basin
SA Cooper Basin
SA Denison Trough
SA Elliot Formation
SA Galilee Basin
SA Georgina Basin
SA Great Artesian Basin
SA Heron Island
SA Ipswich Coal Measures
SA Lachlan fold belt
SA Maryborough Basin
SA McArthur Basin
SA Moreton Bay
SA New England Batholith
SA New England Orogeny
SA Rundle Group
SA Simpson Desert
SA Tasman Geosyncline
SA Tasman orogenic zone
SA Toolebuc Formation
SA Willyama Complex
SA Winton Formation

Queenston Shale (1989)
In Richmond Group. W New York, NE Ohio, NW Pennsylvania and S Ontario.
BT Upper Ordovician
BT Ordovician
BT Paleozoic
SA New York
SA Ohio
SA Ontario
SA Pennsylvania
SA Richmond Group

queenstownite
use Darwin glass

Quercus (1978)
Genus.
BT Dicotyledoneae
BT angiosperms
BT Spermatophyta
BT Plantae

Quercy (1978)
Region in SW France.
IN Index departments as applicable.
BT France
BT Western Europe
BT Europe
SA Lot France
SA Tarn-et-Garonne France

Querigut Massif (1978)
In the eastern Pyrenees. Also search Querigut. This term has multiple hierarchies.
UF Querigut Mountains
BT1 Ariege France
BT1 France
BT1 Western Europe
BT1 Europe
BT2 Pyrenees
BT2 Europe
SA French Pyrenees

Querigut Mountains
use Querigut Massif

Quesnel Lake (1978)
SE British Columbia.
SA British Columbia

Quesnellia
use Quesnellia Terrane

Quesnellia Terrane (1989)
UF Quesnellia
SA British Columbia

Questa Caldera (1985)
N New Mexico. Autoposting of broader terms to this term began in 1989.
BT Taos County New Mexico
BT New Mexico
BT United States
SA Questa Mine
SA Sangre de Cristo Mountains

Questa Mine (1978)
N New Mexico. Autoposting of broader terms began in 1989. Also search Questa AND New Mexico.
BT Taos County New Mexico
BT New Mexico
BT United States
SA mines
SA molybdenum ores
SA Questa Caldera

Quetico Belt (1985)
Subprovince of the Superior Province, Canadian Shield.
CO N484500N513000
W0820000W0890000
UF Quetico Subprovince
BT Superior Province
BT Canadian Shield
BT North America
SA Ontario

Quetico Subprovince
use Quetico Belt

quick clay (1978)
UF quickclays
SA clay
SA clays
SA engineering geology
SA quicksand
SA soil mechanics

quickclays
use quick clay

quicksand (1982)
SA engineering geology
SA geologic hazards
SA quick clay
SA sand
SA soil mechanics

quicksilver
use mercury

Quillan
No longer a valid term for GeoRef. See Quillan France.

Quillan France (1993)
Town in Aude, S France.
BT Aude France
BT France
BT Western Europe
BT Europe

Quinault Canyon (1989)
Submarine canyon off the coast of Washington in the North American Pacific.
CO N472500N472500
W1244500W1244500
BT North American Pacific
BT Pacific Ocean

Quinqueloculina (1978)
Genus. Autoposting of microfossils and Protista to this term began in 1990. This term has multiple hierarchies.
BT1 Miliolacea
BT1 Miliolina
BT1 foraminifera
BT1 Protista
BT1 Invertebrata
BT2 Miliolacea
BT2 Miliolina
BT2 foraminifera
BT2 Protista
BT2 microfossils

Quintana Roo
No longer a valid term for GeoRef. As of 1993 see Quintana Roo Mexico.

Quintana Roo Mexico (1993)
State on the Caribbean Sea on E Yucatan Peninsula.
BT Mexico
SA Gulf Coastal Plain
SA Yucatan Peninsula

Quirke Lake (1978)
N of Lake Huron.
BT Algoma District Ontario
BT Ontario
BT Eastern Canada
BT Canada

Quitman Mountains (1985)
W Texas. Autoposting of broader terms to this term began in 1989.
CO N304000N312000
W1051500W1054000
BT Hudspeth County Texas
BT Texas
BT United States
SA Basin and Range Province

Quseir Egypt
use Kosseir Egypt

R

R waves
use Rayleigh waves

Ra
use radium

Ra-226 (1978)
Autoposting of broader terms began in 1989. This term has multiple hierarchies.
BT1 radioactive isotopes
BT1 isotopes
BT2 radium
BT2 alkaline earth metals
BT2 metals
SA Ra-228/Ra-226

Ra-228 (1985)
Autoposting of broader terms began in 1989. This term has multiple hierarchies.
BT1 radioactive isotopes
BT1 isotopes
BT2 radium
BT2 alkaline earth metals
BT2 metals
SA Ra-228/Ra-226

Ra-228/Ra-226 (1989)
Autoposting of broader terms began in 1989. This term has multiple hierarchies.
BT1 radioactive isotopes
BT1 isotopes
BT2 radium
BT2 alkaline earth metals
BT2 metals
SA isotope ratios
SA Ra-226
SA Ra-228

Raasay (1978)
Island of the Inner Hebrides NE of Skye Island off NW Scotland. This term has multiple hierarchies.
BT1 Inner Hebrides
BT1 Hebrides
BT1 Scotland
BT1 Great Britain
BT1 United Kingdom
BT1 Western Europe
BT1 Europe
BT2 Inverness-shire Scotland
BT2 Highland region Scotland
BT2 Scotland
BT2 Great Britain
BT2 United Kingdom
BT2 Western Europe
BT2 Europe

Rabat
No longer a valid term for GeoRef.

Rabat Morocco (1993)
City on the Atlantic Ocean in NW Morocco.
BT Morocco
BT North Africa
BT Africa

Rabaul Caldera (1978)
On New Britain Island.
BT Bismarck Archipelago
BT Papua New Guinea
BT Australasia
SA New Britain

Rabbit Lake
As of 1993, no longer a valid term for GeoRef. See Rabbit Lake Saskatchewan or Rabbit Lake Deposit.

Rabbit Lake Deposit (1993)
Uranium deposit in central Saskatchewan. Before 1993, also search Rabbit Lake.
BT Saskatchewan
BT Western Canada
BT Canada
SA Rabbit Lake Saskatchewan
SA uranium ores

Rabbit Lake Saskatchewan
(1993)
Town. Location of uranium deposit in central Saskatchewan.
CO N531000N531000
 W1074600W1074600
BT Saskatchewan
BT Western Canada
BT Canada
SA Rabbit Lake Deposit

Rabun County
Valid through 1988. Search in combination with state term. After 1988, use specific county-state term.

Rabun County Georgia (1989)
Extreme NE Georgia. Before 1989, also search Rabun County AND Georgia.
CO N344300N345800
 W0830800W0834000
BT Georgia
BT United States

Raca
 use Racha

racemization (1978)
SA amino acids
SA epimerization
SA geochronology
SA relative age

Racha (1978)
Village in central Serbia.
UF Raca
BT Serbia
BT Yugoslavia
BT Southern Europe
BT Europe

Raciborz
No longer a valid term for GeoRef. As of 1993 see Raciborz Poland.

Raciborz Poland (1993)
City on Oder River in SW Katowice, S Poland. Before 1993 also search Raciborz AND Poland.
UF Ratibor
BT Katowice Poland
BT Poland
BT Central Europe
BT Europe

Racine County Wisconsin
(1989)
Extreme SE Wisconsin. Before 1989, search Racine County AND Wisconsin.
CO N423600N425200
 W0874500W0882000
BT Wisconsin
BT United States
SA Root River

Racine Dolomite (1989)
NE Illinois, N Michigan and SE Wisconsin.
BT Niagaran
BT Middle Silurian
BT Silurian
BT Paleozoic
SA Illinois
SA Michigan
SA Wisconsin

RADAM (1985)
Remote sensing survey of Brazil.
SA Brazil
SA electromagnetic surveys
SA radar methods
SA remote sensing

radar
A valid term through 1977. After 1977, use radar methods.

radar exploration
 use radar methods

radar methods (1978)
Before 1978, search radar.
UF radar exploration
UF radar surveys
NT ground-penetrating radar
SA electromagnetic methods
SA geophysical methods
SA geophysical surveys
SA Geosat
SA image enhancement
SA Magellan Program
SA microwave methods
SA multispectral scanner
SA RADAM
SA radio-wave methods
SA remote sensing
SA SAR
SA Shuttle Imaging Radar
SA side-scanning methods
SA SLAR
SA sonar methods

radar surveys
 use radar methods

radial faults (1978)
Before 1978, also search faults AND radial.
BT faults

radiation (1978)
As of 1978, term is restricted to use for biologic evolution. Before 1978, term could also have been used for radioactivity.
SA background radiation
SA behavior
SA biologic evolution
SA electromagnetic radiation
SA isolation
SA paleontology
SA radioactivity

radiation damage (1978)
UF damage, radiation
SA fission tracks
SA fission-track dating
SA geochronology
SA metamict minerals
SA metamictization
SA particle tracks
SA particle-track dating
SA radioactivity
SA relative age

radio-frequency spectroscopy
(1978)
Before 1978, also search spectroscopy AND radio frequency.
BT spectroscopy
SA analysis
SA chemical analysis

radio-wave methods (1978)
Before 1978, search radio waves. Before 1972, also search radiowave methods; radiowave surveys.
UF radiowave methods
UF radiowave surveys
SA electromagnetic radiation
SA electromagnetic waves
SA ground-penetrating radar
SA methods
SA radar methods

radioactivation analysis
 use activation analysis

radioactive decay (1978)
Before 1978, search decay AND radioactive.
UF decay, radioactive
SA absolute age
SA isotopes

radioactive elements
 use radioactive isotopes

radioactive fallout
 use fallout

radioactive isotopes (1978)
Also search radionuclides.
UF radioactive elements
BT isotopes
NT Al-26
NT Al-27/Al-26
NT Am-241
NT Ar-37
NT Ar-38
NT Ar-38/Ar-36
NT Ar-39
NT Ar-40/Ar-39
NT Be-10
NT Be-10/Be-9
NT Be-7
NT C-14
NT C-14/C-12
NT Cl-36
NT Co-60
NT Cr-51
NT Cs-134
NT Cs-137
NT Fe-55
NT Fe-59
NT I-129
NT I-131
NT K-40
NT Kr-81
NT Kr-85
NT Mn-53
NT Mn-54
NT Na-22
NT Na-24
NT Os-187/Os-186
NT P-32
NT Pa-231
NT Pb-204
NT Pb-206/Pb-204
NT Pb-207/Pb-204
NT Pb-208/Pb-204
NT Pb-210
NT Po-210
NT Pu-238
NT Pu-239
NT Pu-240
NT Pu-240/Pu-239
NT Pu-244
NT Ra-226
NT Ra-228
NT Ra-228/Ra-226
NT Rb-87/Sr-86
NT Rn-222
NT Ru-106
NT S-35
NT Si-32
NT Sm-147/Nd-144
NT Sr-85
NT Sr-89
NT Sr-90
NT Tc-99
NT Th-228
NT Th-230
NT Th-232
NT Th-232/Th-230
NT Th-234
NT tritium
NT U-234
NT U-234/Th-230
NT U-235
NT U-238
NT U-238/Pb-204
NT U-238/Pb-206
NT U-238/Th-230
NT U-238/U-234
NT U-238/U-235
NT Zn-65
NT Zr-95
SA background radiation
SA Chernobyl nuclear accident
SA radioactivity

radioactive materials
No longer a valid GeoRef index term. Before 1969, was used on level 1 in subfile E. See isotopes; radioactivity; radioactive tracers; radioactive waste; radioactive isotopes.

radioactive minerals
No longer a valid GeoRef index term. Before 1971, was used on level 1 in subfiles E and N. See radioactivity; minerals.

radioactive tracers (1978)
SA aerosols
SA Chernobyl nuclear accident
SA ground water
SA hydrogeology
SA hydrology
SA isotopes
SA radioactivity
SA surface water
SA tracers

radioactive waste (1978)
UF nuclear waste
UF waste, radioactive
SA Asse Mine
SA background radiation
SA engineering geology
SA environmental geology
SA Gorleben
SA hazardous waste
SA high-level waste
SA industrial waste
SA Konrad Mine
SA liquid waste
SA low-level waste
SA pollution
SA radioactivity
SA Resource Conservation and Recovery Act
SA solid waste
SA Stripa region
SA underground disposal
SA waste disposal
SA waste disposal sites
SA Waste Isolation Pilot Plant

radioactive waste repositories
Not a valid term for GeoRef. After 1993, see storage or underground storage and radioactive waste or waste disposal.

radioactive-waste disposal
Not a valid GeoRef index term after 1970. Was used on level 1 in subfile N.

radioactivity (1978)
For natural or induced radioactivity. Before 1978, also search radiation.
SA alpha rays
SA alpha-ray spectroscopy
SA autoradiography
SA beta rays
SA Chernobyl nuclear accident
SA electromagnetic radiation
SA engineering geology
SA fallout
SA fission tracks
SA gamma rays
SA gamma-gamma methods
SA gamma-ray methods
SA gamma-ray spectroscopy
SA heat flow
SA isotopes
SA neutron methods
SA neutron-gamma methods
SA neutron-neutron methods
SA particle tracks
SA pollution
SA radiation
SA radiation damage
SA radioactive isotopes

SA radioactive tracers
SA radioactive waste
SA radioactivity methods
SA radioactivity surveys
SA scintillations
SA synchrotron radiation
SA synchrotrons
SA tracers
SA waste disposal
SA well-logging
SA X-rays

radioactivity exploration
Not a valid GeoRef index term after 1971. Was used on level 1 in subfiles G and N. Now use radioactivity, radioactivity methods or radioactivity surveys.

radioactivity logging
Not a valid GeoRef term after 1970. Was used on level 1 in subfile N.

radioactivity methods (1978)
BT geophysical methods
SA gamma-ray methods
SA neutron methods
SA radioactivity
SA radioactivity survey maps
SA radioactivity surveys
SA radon emanometry

radioactivity survey maps (1985)
BT maps
SA geophysical survey maps
SA radioactivity methods
SA radioactivity surveys

radioactivity surveys (1978)
BT geophysical surveys
BT surveys
SA radioactivity
SA radioactivity methods
SA radioactivity survey maps
SA radon emanometry

radiocarbon dating
No longer a valid GeoRef index term. Before 1971, was used on level 1 in subfile N.
use C-14

radiochemical analysis
Not a valid GeoRef index term after 1970. Was used on level 1 in subfile N. Now use chemical analysis or isotopes as applicable.

radiography
use autoradiography

radiography, X-ray
use X-ray radiography

Radiolaria (1978)
Autoposting of this term began in 1978. This term has multiple hierarchies.
BT1 Protista
BT1 Invertebrata
BT2 Protista
BT2 microfossils
NT Osculosida
NT Porulosida
NT Spumellina
SA plankton
SA radiolarians

radiolarians (1981)
Common name for Radiolaria. This term has multiple hierarchies.
IN Index for all non-paleontologic studies of fossils.
BT1 invertebrates
BT2 microfossils
SA biostratigraphy
SA Radiolaria

radiolarite (1978)
BT clastic rocks
BT sedimentary rocks

radiometric properties
As of 1981, no longer a valid term for GeoRef. Term was introduced in 1978. Before 1978, also search radiometric; radiometry. See radioactivity.

radionuclides
A valid term through 1978. After 1978 use radioactive isotopes.

radiowave methods
Not a valid GeoRef index term after 1970. Was used on level 1 in subfile N.
use radio-wave methods

radiowave surveys
use radio-wave methods

radium (1978)
Autoposting of alkaline earth metals and metals to this term began in 1989.
UF Ra
BT alkaline earth metals
BT metals
NT Ra-226
NT Ra-228
NT Ra-228/Ra-226
SA isotopes
SA uranium

Radnorshire
No longer a valid term for GeoRef as of 1993.
use Radnorshire Wales

Radnorshire Wales (1993)
Former county E Wales. As of April 1974, part of Powys. Before 1993, also search Radnorshire.
UF Radnorshire
BT Powys Wales
BT Wales
BT Great Britain
BT United Kingdom
BT Western Europe
BT Europe

radon (1978)
Autoposting of noble gases to this term began in 1989.
UF Rn
BT noble gases
NT Rn-222
SA isotopes
SA radon emanometry

radon emanometry (1985)
SA chemical analysis
SA earthquake prediction
SA earthquakes
SA eruptions
SA gas chromatography
SA geochemical methods
SA mineral exploration
SA precursors
SA radioactivity methods
SA radioactivity surveys
SA radon
SA soil gases

Radon transforms (1993)
Integral of physical property of an object along a given line or energy transit path. Before 1993, also search Radon transformations or slant stacks.
UF slant stacks
SA mathematical transformations
SA seismic methods
SA tomography

radon-222
use Rn-222

Radstock Bay (1978)
SW coast of Devon Island on Lancaster Sound in Franklin District.
BT Devon Island
BT Franklin District Northwest Territories
BT Northwest Territories
BT Western Canada
BT Canada

raft foundations (1985)
SA buildings
SA foundations
SA ice rafting
SA settlement

Raft River (1978)
In N Utah and S Idaho.
IN Index states as applicable.
BT United States
SA Idaho
SA Raft River basin
SA Utah

Raft River basin (1978)
Also search Raft River.
IN Index states as applicable.
SA Idaho
SA Raft River
SA United States
SA Utah

rafting, ice
use ice rafting

Ragavapuram (Shales)
use Raghavapuram Shales

Raghavapuram Shales (1978)
Also search Raghavapuram.
UF Ragavapuram (Shales)
BT Upper Jurassic
BT Jurassic
BT Mesozoic
SA Andhra Pradesh India
SA India

Railroad Valley (1981)
SE central Nevada.
IN Index county or region as applicable.
SA Nye County Nevada

railroads (1989)
Used for geological or geotechnical studies pertaining in some way to railroad tracks and materials.
UF railways
SA construction
SA embankments
SA engineering geology
SA engineering properties
SA environmental geology
SA foundations
SA highways
SA land subsidence
SA land use
SA planning
SA seismic response
SA site exploration
SA slope stability
SA soil mechanics
SA tunnels

railways
use railroads

rain
use rainfall

Rainbow Mountain (1978)
Peak in Salmon River Mountains in Valley County, W central Idaho.
IN Index county and mountain range as applicable.
SA Valley County Idaho

raindrops (1978)
SA atmospheric precipitation
SA clouds
SA droplets
SA humidity
SA meteorology
SA rainfall
SA water

rainfall (1978)
UF rain
SA acid rain
SA atmospheric precipitation
SA drought
SA hydrology
SA raindrops
SA watersheds

Rainier Mesa (1981)
Outlier of the Belted Range, in NW portion of the Nevada Test Site. Autoposting of broader terms to this term began in 1989.
CO N371000N371245 W1161215W1161430
BT Nye County Nevada
BT Nevada
BT United States
SA Nevada Test Site

Rainy Lake (1978)
IN Index North America or Minnesota and/or Ontario as applicable.
SA Koochiching County Minnesota
SA Minnesota
SA North America
SA Ontario
SA Saint Louis County Minnesota

Rainy River (1978)
Flows from Rainy Lake to Lake of the Woods.
IN Index Minnesota and/or Ontario as applicable.
BT North America
SA Minnesota
SA Ontario

Rainy River District
No longer a valid term for GeoRef. As of 1993, see Rainy River District Ontario.

Rainy River District Ontario (1993)
NW Ontario, on the border with Minnesota. Before 1993, also search Rainy River District AND Ontario.
CO N480500N490000 W0910000W0944500
BT Ontario
BT Eastern Canada
BT Canada
NT Atikokan Ontario

Raipur
No longer a valid term for GeoRef. See Raipur India.

Raipur India (1993)
City in SE Madhya Pradesh, central India.
BT Madhya Pradesh India
BT India
BT Indian Peninsula
BT Asia

Rajahmundry
No longer a valid term for GeoRef. See Rajahmundry India.

Rajahmundry India (1993)
City in NE Andhra Pradesh, central India.
BT Andhra Pradesh India
BT India
BT Indian Peninsula
BT Asia

Rajasthan
No longer a valid term for GeoRef. See Rajasthan India.

Rajasthan India (1993)
   State in NW India.
   CO N230000N301000
      E0775000E0693000
   BT India
   BT Indian Peninsula
   BT Asia
   NT Ajmer India
   NT Alwar India
   NT Barmer India
   NT Bhilwara India
   NT Bikaner India
   NT Jaipur India
   NT Jaisalmer India
   NT Jhunjhunu India
   NT Jodhpur India
   NT Khetri India
   NT Sirohi India
   NT Udaipur India
   SA Aravalli Range
   SA Aravalli System
   SA Delhi Supergroup
   SA Kaimur Sandstone
   SA Khetri copper belt
   SA Lathi Formation
   SA Trans-Aravalli Vindhyan Basin

Rajgir
   No longer a valid term for GeoRef. See Rajgir India.

Rajgir India (1993)
   Village in N central Bihar, E India.
   BT Bihar India
   BT India
   BT Indian Peninsula
   BT Asia

Rajmahal
   No longer a valid term for GeoRef. See Rajmahal India.

Rajmahal Hills (1978)
   Low range of hills S and W of the Ganges River in E Bihar, E India.
   SA Amarjola
   SA Bihar India

Rajmahal India (1993)
   Town in E Bihar, E India.
   BT Bihar India
   BT India
   BT Indian Peninsula
   BT Asia

Rajmahal Series (1993)
   Bangladesh and NE India. Includes grits and shales intercalated in trap flows. Also search Rajmahal Traps if applicable.
   BT Jurassic
   BT Mesozoic
   SA Bangladesh
   SA India

Rakhov
   No longer a valid term for GeoRef. As of 1993, see Rakhov Ukraine or Rakhov Massif.

Rakhov Massif (1978)
   In central Carpathian Mountains of Transcarpathian Oblast. Also search Rakhov. This term has multiple hierarchies.
   BT1 Ukrainian Carpathians
   BT1 Ukraine
   BT1 Europe
   BT2 Ukrainian Carpathians
   BT2 Ukraine
   BT2 Commonwealth of Independent States
   BT3 Ukrainian Carpathians
   BT3 Carpathians
   BT3 Europe
   SA Transcarpathia Ukraine

Rakhov Ukraine (1993)
   Town in Transcarpathian Oblast in W Ukraine. This term has multiple hierarchies.
   BT1 Transcarpathia Ukraine
   BT1 Ukraine
   BT1 Europe
   BT2 Transcarpathia Ukraine
   BT2 Ukraine
   BT2 Commonwealth of Independent States

Raleigh Belt (1985)
   Metamorphic belt in central North Carolina.
   IN Index counties or region as applicable.
   SA Appalachians
   SA Carolina slate belt
   SA North Carolina
   SA Piedmont

Raleigh County
   Valid through 1988. Search in combination with state term. After 1988, use specific county-state term.

Raleigh County West Virginia (1989)
   S West Virginia. Before 1989, also search Raleigh County AND West Virginia.
   CO N373100N375900
      W0805200W0813500
   BT West Virginia
   BT United States

Raman
   A valid term through 1977. After 1977, see Raman spectra or Raman spectroscopy.

Raman spectra (1982)
   Used for data. For methodology, use Raman spectroscopy.
   BT spectra
   SA Raman spectroscopy
   SA spectroscopy

Raman spectroscopy (1978)
   Before 1978, also search spectroscopy AND Raman. As of 1982, for methodology only; use Raman spectra for data.
   BT spectroscopy
   SA analysis
   SA chemical analysis
   SA Raman spectra

Ramapithecus (1978)
   Genus. Autoposting of Eutheria and Theria to this term began in 1989. Sometimes classed under the Pongidae.
   IN Index Pongidae if applicable.
   BT Hominidae
   BT Primates
   BT Eutheria
   BT Theria
   BT Mammalia
   BT Tetrapoda
   BT Vertebrata
   BT Chordata
   SA Pongidae

Ramapo Fault (1989)
   N New Jersey and SE New York.
   IN Index states as applicable.
   BT United States
   SA New Jersey
   SA New York
   SA Newark Basin

Ramgarh coal field (1978)
   S Bihar, India. Also search Ramgarh.
   UF Ramgarh Coalfield
   SA Bihar India
   SA coal fields

Ramgarh Coalfield
   use Ramgarh coal field

Rammelsberg (1978)
   Mountain in the Harz Mountains. This term has multiple hierarchies.
   BT1 Lower Saxony Germany
   BT1 Germany
   BT1 Central Europe
   BT1 Europe
   BT2 Harz Mountains
   BT2 Germany
   BT2 Central Europe
   BT2 Europe

rammelsbergite (1978)
   Autoposting of sulfides to this term ended in 1989.
   BT arsenides
   SA nickel ores
   SA pararammelsbergite

Ramnagar
   No longer a valid term for GeoRef. See Ramnagar India and Ramnagar Pakistan.

Ramnagar India (1993)
   Town in SE Uttar Pradesh, India. Before 1993, also search Ramnagar AND India.
   BT Uttar Pradesh India
   BT India
   BT Indian Peninsula
   BT Asia

Ramnagar Pakistan (1993)
   Village NW of Lahore in Punjab, Pakistan. Before 1993, search Ramnagar AND Pakistan.
   BT Punjab Pakistan
   BT Pakistan
   BT Indian Peninsula
   BT Asia

Ramp Creek Formation (1989)
   S Indiana and N Kentucky.
   BT Mississippian
   BT Carboniferous
   BT Paleozoic
   SA Indiana
   SA Kentucky

Rampart Group (1978)
   Probably Lower Mississippian. Yukon-Tanana region in E central and E Alaska.
   BT Mississippian
   BT Carboniferous
   BT Paleozoic
   SA Alaska
   SA Lower Mississippian

Ramparts Formation (1978)
   Mackenzie District. Also search Ramparts.
   BT Devonian
   BT Paleozoic
   SA Canada
   SA Northwest Territories

ramps (1989)
   Used as a general term. Use carbonate ramps or thrust faults where applicable.
   SA carbonate ramps
   SA faults
   SA thrust faults

Rampur coal field (1978)
   NW Orissa. Also search Rampur.
   SA coal fields
   SA Orissa India

ramsdellite (1978)
   BT oxides
   SA vernadite

Ramsey County
   Valid through 1988. Search in combination with state term. After 1988, use specific county-state term.

Ramsey County Minnesota (1989)
   E Minnesota. Before 1989, also search Ramsey County AND Minnesota.
   CO N450000N452400
      W0930100W0933100
   BT Minnesota
   BT United States
   NT Saint Paul Minnesota

Ramsey County North Dakota (1989)
   NE central North Dakota. Before 1989, also search Ramsey County AND North Dakota.
   CO N475600N483200
      W0981800W0991300
   BT North Dakota
   BT United States

Ran
   use Rana Fjord

Rana Fjord (1978)
   On the Norwegian Sea in central part of country. Also search Rana.
   UF Ran
   UF Ranen Fjord
   BT Nordland Norway
   BT Norway
   BT Scandinavia
   BT Western Europe
   BT Europe

Ranchi
   No longer a valid term for GeoRef. See Ranchi India.

Ranchi India (1993)
   City and District in S Bihar, E India.
   BT Bihar India
   BT India
   BT Indian Peninsula
   BT Asia

Rancho La Brea (1985)
   Pleistocene fossil locality in SW California. Autoposting of broader terms to this term began in 1989.
   UF La Brea tar pits
   BT Los Angeles County California
   BT California
   BT United States

Rancholabrean (1989)
   North American continental stage. Above Irvingtonian, below Holocene.
   BT upper Pleistocene
   BT Pleistocene
   BT Quaternary
   BT Cenozoic

rancieite (1978)
   BT oxides

Rand
   use Witwatersrand

Rand Mountains (1978)
   Range in the Mojave Desert. Also search Rand.
   IN Index county or region as applicable.
   SA California
   SA Mojave Desert

Randolph County
   Valid through 1988. Search in combination with state term. After 1988, use specific county-state term.

Randolph County Alabama (1989)
E Alabama. Before 1989, also search Randolph County AND Alabama.
CO N330600N333000
 W0851500W0854000
BT Alabama
BT United States

Randolph County Arkansas (1989)
NE Arkansas. Before 1989, also search Randolph County AND Arkansas.
CO N361400N362900
 W0904600W0912800
BT Arkansas
BT United States

Randolph County Georgia (1989)
SW Georgia. Before 1989, also search Randolph County AND Georgia.
CO N313700N315500
 W0843100W0845500
BT Georgia
BT United States

Randolph County Illinois (1989)
SW Illinois. Before 1989, also search Randolph County AND Illinois.
CO N374700N381300
 W0893500W0901500
BT Illinois
BT United States

Randolph County Indiana (1989)
E Indiana. Before 1989, also search Randolph County AND Indiana.
CO N400000N401800
 W0845000W0851400
BT Indiana
BT United States

Randolph County Missouri (1989)
N central Missouri. Before 1989, also search Randolph County AND Missouri.
CO N391500N393700
 W0921700W0924200
BT Missouri
BT United States

Randolph County North Carolina (1989)
Central North Carolina. Before 1989, also search Randolph County AND North Carolina.
CO N353100N355400
 W0793300W0803300
BT North Carolina
BT United States

Randolph County West Virginia (1989)
E West Virginia. Before 1989, also search Randolph County AND West Virginia.
CO N382400N390700
 W0792200W0801700
BT West Virginia
BT United States

random processes
 use stochastic processes

Ranen Fjord
 use Rana Fjord

range (1978)
Usage is restricted to stratigraphy.
UF stratigraphic range
SA biogeography
SA biostratigraphy
SA fossils
SA stratigraphy
SA taxonomy

range, basin
 use basin range structure

Rangeley Formation (1989)
N Connecticut, W Maine, Massachusetts and S New Hampshire. Silurian and/or Devonian.
BT Paleozoic
SA Connecticut
SA Devonian
SA Maine
SA Massachusetts
SA New Hampshire
SA Silurian

Rangeley Lakes (1978)
Also search Rangeley.
IN Index states and counties as applicable.
BT United States
SA Coos County New Hampshire
SA Franklin County Maine
SA Maine
SA New Hampshire
SA Oxford County Maine

Rangely
Valid through 1988. Search in combination with state term. After 1988, use specific city-state term.

Rangely Anticline (1978)
In NW near Utah border. Also search Rangely.
BT Colorado
BT United States

Rangely Colorado (1989)
Town on White River in NW Colorado. Before 1989, search Rangely AND Colorado.
CO N400400N400400
 W1084800W1084800
BT Rio Blanco County Colorado
BT Colorado
BT United States

Rangely Field (1978)
In NW Colorado near Utah border. Before 1978, also search Rangely; Rangely oil field.
IN Index county or region as applicable.
UF Rangely oil field
SA Colorado
SA oil and gas fields

Rangely oil field
 use Rangely Field

Ranger
No longer a valid term for GeoRef. As of 1993, see Ranger Mine if applicable.

Ranger Canyon Formation (1989)
In Ishbel Group. Rocky Mountains of British Columbia and Alberta.
BT Permian
BT Paleozoic
SA Alberta
SA British Columbia
SA Ishbel Group

Ranger Mine (1993)
Uranium ore deposit and mine in Alligator Rivers region, Northern Territory. Before 1993, also search Ranger AND Australia.
IN Index state or region as applicable.
SA Alligator Rivers Field
SA mines
SA Northern Territory Australia
SA uranium ores

Rangifer (1985)
Genus. Autoposting of Cervidae, Eutheria and Theria to this term began in 1989.
BT Cervidae
BT Ruminantia
BT Artiodactyla
BT Eutheria
BT Theria
BT Mammalia
BT Tetrapoda
BT Vertebrata
BT Chordata

Rangitata River (1978)
E central South Island. Also search Rangitata.
IN Also index South Island.
BT New Zealand
BT Australasia
SA South Island

Raniganj
No longer a valid term for GeoRef. See Raniganj India.

Raniganj (Stage)
 use Raniganj Formation

Raniganj coal field (1978)
West Bengal state. Also search Raniganj.
CO N233500N233900
 E0870300E0870900
UF Raniganj Coalfield
BT West Bengal India
BT India
BT Indian Peninsula
BT Asia
SA coal fields

Raniganj Coalfield
 use Raniganj coal field

Raniganj Formation (1993)
The highest division of the Damuda Series. Before 1993, also search Raniganj Stage. Before 1978, also search Raniganj.
UF Raniganj (Stage)
UF Raniganj Stage
BT Upper Permian
BT Permian
BT Paleozoic
SA Damuda Series
SA India
SA West Bengal India

Raniganj India (1993)
City on N bank of Damodar River.
BT West Bengal India
BT India
BT Indian Peninsula
BT Asia

Raniganj Stage
No longer a valid term for GeoRef.
 use Raniganj Formation

rank (1978)
To be used only for classification of coal.
SA coal

Rankin County
Valid through 1988. Search in combination with state term. After 1988, use specific county-state term.

Rankin County Mississippi (1989)
Central Mississippi. Before 1989, also search Rankin County AND Mississippi.
CO N320300N323500
 W0894500W0901500
BT Mississippi
BT United States

Raoul Island (1978)
Largest of the volcanic Kermadec Islands which are a dependency of New Zealand 500 miles NE of Auckland. In 1985, broader term changed from Pacific Ocean to Kermadec Islands and West Pacific Ocean Islands.
BT Kermadec Islands
BT West Pacific Ocean Islands

rapakivi (1978)
UF wiborgite
BT granites
BT plutonic rocks
BT igneous rocks

Raphidiodea (1978)
BT Insecta
BT Mandibulata
BT Arthropoda
BT Invertebrata
SA Diptera
SA Hymenoptera
SA Lepidoptera

Rapid City
Valid through 1988. Search in combination with state term. After 1988, use specific city-state term.

Rapid City South Dakota (1989)
City in W South Dakota. Before 1989, search Rapid City AND South Dakota.
CO N440600N440600
 W1031400W1031400
BT Pennington County South Dakota
BT South Dakota
BT United States

rapid variations
No longer a valid GeoRef index term. Before 1972, was used on level 2 in subfiles T, G and N. See variations; secular variations; fluctuations.

Rapides Parish
Valid through 1988. Search in combination with state term. After 1988, use specific county-state term.

Rapides Parish Louisiana (1989)
Central Louisiana. Before 1989, also search Rapides Parish AND Louisiana.
CO N305200N313300
 W0920500W0925600
BT Louisiana
BT United States
SA Calcasieu River

rapids (1993)
BT fluvial features
SA cascades
SA geomorphology
SA hydrology
SA rivers
SA streams
SA waterfalls
SA waterways

Rappahannock River (1978)
Rises in the Blue Ridge Mountains and flows into Chesapeake Bay.
BT Virginia
BT United States

Rarau Massif (1978)
Suceava County in N Romania. Also search Rarau.
UF Rarau Mountains
BT Romania
BT Southern Europe
BT Europe
SA Moldavia

SA Suceava Romania

Rarau Mountains
 use Rarau Massif

rare earth deposits (1981)
Before 1981, also search (rare earths OR lanthanides) AND (deposit OR deposits OR ore OR ores OR economic) in the basic index. Autoposting of metal ores to this term began in 1989.
IN Commodity. See List C.
BT metal ores
NT cerium ores
SA bastnaesite
SA monazite deposits
SA rare earths

rare earths (1978)
Series of metallic elements with atomic numbers 57 to 71 (the lanthanides). Also, scandium and yttrium are generally included in this group due to chemical similarities. As of 1981, use rare earth deposits for rare earths as a commodity. Autoposting of this term to specific rare earths began in 1989. Autoposting of metals to this term began in 1989.
UF inner transition elements
UF lanthanide series
UF lanthanides
UF lanthanoans
BT metals
NT cerium
NT dysprosium
NT erbium
NT europium
NT gadolinium
NT holmium
NT lanthanum
NT lutetium
NT neodymium
NT praseodymium
NT promethium
NT samarium
NT scandium
NT terbium
NT thulium
NT ytterbium
NT yttrium
SA KREEP
SA rare earth deposits
SA trace elements

rare gases
 use noble gases

RARE II
 use RARE II regions

RARE II regions (1981)
Tracts of land belonging to the U. S. Forest Service that are being evaluated for their potential as wilderness areas.
UF RARE II
UF Roadless Area Review & Evaluation, Phase II
SA conservation
SA land use
SA programs
SA United States
SA wilderness areas

rare metals
As of 1989, no longer valid for GeoRef. Use metals or rare earths if applicable.

Raritan Bay (1989)
NE New Jersey and SE New York. Inlet of Atlantic Ocean bordered by Staten Island, New York, and Middlesex and Monmouth counties, New Jersey.

IN Index states and counties as applicable.
CO N402300N403300 W0735800W0742000
BT United States
SA Middlesex County New Jersey
SA Monmouth County New Jersey
SA New Jersey
SA New York
SA Staten Island

Raritan Formation (1978)
Comprises Amboy Stonewane Clay, Old Bridge Sand Member, South Amboy Fire Clay, Sayreville Sand Member, Woodbridge Clay, Farrington Sand Member, and Raritan Fire Clay.
BT Upper Cretaceous
BT Cretaceous
BT Mesozoic
SA Delaware
SA Maryland
SA New Jersey

Raritan River (1985)
N New Jersey. Flows into New York Bay S of Staten Island.
BT New Jersey
BT United States

Ras al-Khaimah (1978)
Sheikdom.
BT United Arab Emirates
BT Arabian Peninsula
BT Asia

RASA (1993)
Acronym. U. S. Geological Survey program begun in 1978.
UF Regional Aquifer-System Analysis Program
SA aquifers
SA ground water
SA programs
SA U. S. Geological Survey
SA United States

rasorite
 use kernite

Rat Island (1978)
Small island in center of Rat Islands in W Aleutian Islands. Also an island in the Bismarck Archipelago, Papua New Guinea.
IN Index Rat Islands or Papua New Guinea as applicable.
SA Alaska
SA Aleutian Islands
SA Bismarck Archipelago
SA Papua New Guinea
SA Rat Islands

Rat Islands (1978)
Group in W Aleutian Islands.
BT Aleutian Islands
BT Southwestern Alaska
BT Alaska
BT United States
SA Rat Island

rate of sedimentation
 use sedimentation rates

rates (1978)
As of 1982, use sedimentation rates or erosion rates.
SA erosion
SA erosion rates
SA sedimentation rates

Ratibor
 use Raciborz Poland

ratios
No longer a valid term for GeoRef. After 1992, see chemical ratios or isotope ratios.

Raton Basin (1978)

SE Colorado and NE New Mexico. Also search Raton.
IN Index states or regions as applicable.
SA Colorado
SA New Mexico
SA United States

Raton Formation (1985)
Upper Cretaceous and Paleocene. NE New Mexico and SE Colorado.
IN Index ages as applicable.
SA Colorado
SA K-T boundary
SA New Mexico
SA Paleocene
SA Upper Cretaceous

Rattlesnake Creek Terrane (1989)
In Klamath Mountains of California.
BT California
BT United States
SA Klamath Mountains

Rattlesnake Hills (1985)
Area in SW Natrona County, Wyoming. Also an area in Benton and Yakima counties, Washington. Before 1985, also search Rattlesnake Mountain AND Wyoming for area in Wyoming.
IN Index states and counties as applicable.
SA Benton County Washington
SA Natrona County Wyoming
SA Rattlesnake Mountain
SA United States
SA Washington
SA Wyoming
SA Yakima County Washington

Rattlesnake Mountain (1978)
W Texas. Before 1985, term was sometimes used for Rattlesnake Hills, Wyoming.
IN Index county and mountains as applicable.
SA Big Bend National Park
SA Brewster County Texas
SA Rattlesnake Hills

Raukumara Peninsula (1978)
NE North Island.
IN Also index North Island.
BT New Zealand
BT Australasia
SA North Island

Rauracian (1978)
Europe. Above Argovian, below Sequanian.
BT Upper Jurassic
BT Jurassic
BT Mesozoic
SA Lusitanian

Ravalli County
Valid through 1988. Search in combination with state term. After 1988, use specific county-state term.

Ravalli County Montana (1989)
W Montana. Before 1989, also search Ravalli County AND Montana.
CO N452800N463800 W1133200W1143300
BT Montana
BT United States
SA Selway-Bitterroot Wilderness

Ravalli Group (1978)

Of Belt Supergroup. Includes Altyn Formation, Appekunny Formation, Grinnell Formation. NE Idaho, and NW Montana.
BT middle Proterozoic
BT Proterozoic
BT upper Precambrian
BT Precambrian
SA Altyn Limestone
SA Belt Supergroup
SA Idaho
SA Montana
SA Revett Quartzite
SA Saint Regis Formation

Ravenscrag Formation (1978)
BT Paleocene
BT Paleogene
BT Tertiary
BT Cenozoic
SA Canada
SA Saskatchewan

raw humus
 use Mor

raw materials (1982)
SA economic geology
SA materials
SA mineral resources

Rawlins County
Valid through 1988. Search in combination with state term. After 1988, use specific county-state term.

Rawlins County Kansas (1989)
NW Kansas. Before 1989, also search Rawlins County AND Kansas.
CO N393500N400000 W1004500W1012500
BT Kansas
BT United States

ray paths
 use raypaths

ray tracing (1993)
SA arrival time
SA raypaths
SA seismic methods
SA seismic surveys
SA velocity

Rayleigh waves (1978)
UF R waves
BT surface waves
BT elastic waves
SA Lg-waves
SA seismology
SA Stoneley waves
SA waves

raypaths (1981)
UF ray paths
SA arrival time
SA geophysical methods
SA geophysics
SA ray tracing
SA seismic migration
SA seismology

rays, alpha
 use alpha rays

rays, beta
 use beta rays

rays, cosmic
Not a valid term for GeoRef. Usually out-of-scope. Used as cosmic rays.

rays, gamma
 use gamma rays

Razdan (1978)
River which serves as an outlet of Lake Sevan. This term has multiple hierarchies.

IN Index regions as applicable.
UF Zanga
BT1 Armenia
BT1 Commonwealth of Independent States
BT2 Armenia
BT2 Europe

razor stone
  use novaculite

Rb
  use rubidium

Rb-87/Sr-86 (1985)
  Autoposting of broader terms began in 1989. This term has multiple hierarchies.
  UF Sr-86/Rb-87
  BT1 radioactive isotopes
  BT1 isotopes
  BT2 stable isotopes
  BT2 isotopes
  BT3 rubidium
  BT3 alkali metals
  BT3 metals
  BT4 strontium
  BT4 alkaline earth metals
  BT4 metals
  SA isotope ratios
  SA Sr-86

Rb-Sr
  Not a valid term for GeoRef.
  use Rb/Sr

Rb/Sr (1989)
  Isotopic ratio used in age determination. Before 1990, also search Sr/Rb.
  UF Rb-Sr
  UF rubidium-strontium
  UF Sr/Rb
  SA absolute age
  SA isotope ratios
  SA rubidium
  SA strontium

RCRA
  use Resource Conservation and Recovery Act

Re
  use rhenium

Re/Os (1989)
  Isotopic ratio used in age determination.
  UF Os/Re
  SA absolute age
  SA isotope ratios
  SA osmium
  SA rhenium

reaction rims (1982)
  Before 1993, also search coronas and petrology.
  UF kelyphytic rims
  SA chemical reactions
  SA crystal chemistry
  SA crystal growth
  SA crystal zoning
  SA crystallization
  SA crystallography
  SA crystals
  SA fractional crystallization
  SA magmas
  SA magmatic differentiation
  SA mineralogy
  SA minerals
  SA petrography
  SA phase equilibria

reactions
  No longer a valid term for GeoRef as of 1993. See chemical reactions.

reactivation (1978)
  SA displacements
  SA faults

  SA neotectonics

Read Bay Formation (1978)
  Middle Silurian to Lower Devonian. Cornwallis Island in Franklin District. Also search Read; Read Bay.
  IN Index ages as applicable.
  BT Paleozoic
  SA Canada
  SA Cornwallis Island
  SA Franklin District Northwest Territories
  SA Lower Devonian
  SA Northwest Territories
  SA Silurian

Reading
  Valid through 1988. Search in combination with state term. After 1988, use specific city-state term.

Reading Pennsylvania (1989)
  City in SE central Pennsylvania. Before 1989, search Reading AND Pennsylvania.
  CO N402000N402000 W0755500W0755500
  BT Berks County Pennsylvania
  BT Pennsylvania
  BT United States

Reading Prong (1985)
  Precambrian metamorphic terrane, Central Appalachians.
  IN Index states as applicable.
  BT United States
  SA Appalachians
  SA Central Appalachians
  SA New Jersey
  SA New York
  SA Pennsylvania

Reagan Sandstone (1978)
  In Timbered Hills Group. Central S Oklahoma.
  BT Upper Cambrian
  BT Cambrian
  BT Paleozoic
  SA Oklahoma
  SA Timbered Hills Group

reagents (1978)
  Use this term, and not names of specific reagents, when chemical reagents are discussed.
  SA buffers
  SA geochemistry
  SA solutions

realgar (1978)
  UF red arsenic
  UF red orpiment
  UF sandarac
  BT sulfides

Reasi
  use Riasi

rebound
  As of 1993, see isostatic rebound, glacial rebound, uplifts or elastic properties if applicable.

Recent
  use Holocene

recent deformation
  No longer a valid GeoRef index term. Before 1972, was used on level 2 in subfiles G and N. See deformation; neotectonics.

Receptaculitaceae (1978)
  Autoposting of algae, Chlorophyta and Chlorophyceae began in 1981. From 1978-1980, problematic fossils was autoposted to this term. Autoposting of thallophytes and microfossils began in 1990. This term has multiple hierarchies.

  BT1 Chlorophyceae
  BT1 Chlorophyta
  BT1 algae
  BT1 thallophytes
  BT1 Plantae
  BT2 Chlorophyceae
  BT2 Chlorophyta
  BT2 algae
  BT2 microfossils
  SA Coelenterata
  SA Echinodermata
  SA foraminifera
  SA Porifera
  SA problematic fossils

recharge (1978)
  UF ground-water increment
  UF ground-water recharge
  UF ground-water replenishment
  UF increment
  SA aquifers
  SA artificial recharge
  SA discharge
  SA ground water
  SA hydrologic cycle
  SA infiltration
  SA natural recharge
  SA water balance

Rechitsa
  No longer a valid term for GeoRef. As of 1993, see Rechitsa Belarus.

Rechitsa Belarus (1993)
  City in Gomel Oblast in S Belarus. This term has multiple hierarchies.
  BT1 Belarus
  BT1 Europe
  BT2 Belarus
  BT2 Commonwealth of Independent States

Recife-Joao Pessoa (1978)
  Two city region in neighboring states in NE Brazil. Recife is in Pernambuco and Joao Pessoa is in Paraiba.
  BT Brazil
  BT South America
  SA Paraiba Brazil
  SA Pernambuco Brazil

Recita Romania
  use Resita Romania

reclamation (1978)
  Used for geological studies on the reclamation of the natural environment.
  UF land reclamation
  SA abandoned mines
  SA artificial islands
  SA basin management
  SA bioremediation
  SA conservation
  SA decontamination
  SA deforestation
  SA depletion
  SA disposal barriers
  SA environment
  SA environmental geology
  SA impact statements
  SA land leases
  SA land use
  SA leaking underground storage tanks
  SA mining geology
  SA natural resources
  SA polders
  SA polluted water
  SA pollution
  SA programs
  SA protection
  SA recovery
  SA recycling
  SA regulations
  SA remediation

  SA soil treatment
  SA soils
  SA strip mining
  SA Superfund
  SA tailings
  SA waste disposal
  SA waste disposal sites
  SA waste water
  SA water management
  SA water treatment

reclined
  A valid term through 1976. After 1976, use recumbent folds.

reclined folds
  use recumbent folds

Reconcavo Basin (1978)
  Fertile coastal lowland surrounding Todos os Santos Bay W of Salvador. Also search Reconcavo.
  BT Bahia Brazil
  BT Brazil
  BT South America

reconstruction (1978)
  SA continental drift
  SA paleogeography

recording
  A valid general term through 1976.

recovery (1978)
  General term. Use secondary or tertiary recovery for petroleum if applicable.
  SA enhanced recovery
  SA mining
  SA mining geology
  SA reclamation
  SA retorting
  SA secondary recovery
  SA tertiary recovery
  SA thermal recovery
  SA water recovery
  SA well stimulation

recreation (1978)
  SA land use
  SA soils

recrystallization (1978)
  SA authigenesis
  SA crystallization
  SA deformation
  SA grains
  SA lithification
  SA metamorphism
  SA phyllonitization

rectorite (1978)
  UF allevardite
  BT clay minerals
  BT sheet silicates
  BT silicates

recumbent
  A valid term through 1977. After 1977, use recumbent folds.

recumbent folds (1978)
  Before 1978, search folds AND recumbent; folds AND reclined.
  UF reclined folds
  BT folds
  SA nappes
  SA orientation

recursive filters (1981)
  SA filters
  SA Kalman filters

recycling (1981)
  SA economic geology
  SA economics
  SA environmental geology
  SA reclamation

red algae
    use Rhodophyta

red arsenic
    use realgar

red beds (1978)
    UF red rock
    UF redbeds
    BT clastic rocks
    BT sedimentary rocks
    SA ferruginous composition
    SA sandstone
    SA terrigenous materials

Red Deer River (1978)
    Rises in Banff National Park and flows into South Saskatchewan River near the Saskatchewan border.
    BT Alberta
    BT Western Canada
    BT Canada
    SA Red Deer River valley

Red Deer River valley (1989)
    S Alberta.
    UF Red Deer Valley
    BT Alberta
    BT Western Canada
    BT Canada
    SA Red Deer River

Red Deer Valley
    use Red Deer River valley

Red Desert (1985)
    SW Wyoming.
    IN Index counties or regions as applicable.
    SA Great Divide Basin
    SA Sweetwater County Wyoming
    SA U. S. Rocky Mountains

Red Eagle Limestone (1978)
    In Council Grove Group. Includes Glenrock Limestone Member, Bennett Shale Member, Howe Limestone Member.
    BT Permian
    BT Paleozoic
    SA Council Grove Group
    SA Kansas
    SA Missouri
    SA Nebraska
    SA Oklahoma
    SA Roca Formation

Red Fork Sandstone (1978)
    Said to lie higher than Glenn Sand, lower than Skinner Sand. NE Oklahoma.
    BT Pennsylvanian
    BT Carboniferous
    BT Paleozoic
    SA Oklahoma

Red Hill (1978)
    Mountain in Haleakala National Park on Maui Island. Also hills or mountains in SW Utah, Scotland, and other locations. As of 1993, use Red Hill Mine for the turquoise mine in the White Signal District of New Mexico and mines in California and Western Australia.
    IN Index states and countries as applicable.
    SA Haleakala
    SA Hawaii
    SA Maui
    SA New Mexico
    SA Scotland
    SA Utah
    SA Western Australia
    SA Wyoming

Red Lake (1978)
    In Beltrami County, N Minnesota. Includes both Upper Red Lake and Lower Red Lake.
    IN Index county or region as applicable.
    SA Beltrami County Minnesota

Red Mountain (1978)
    Ridge primarily in Jefferson County, N central Alabama. Also large alunite deposit near Lake City, Colorado. Also in other locations.
    IN Index county or region and mountains as applicable.
    SA Alabama
    SA alunite
    SA Colorado
    SA Jefferson County Alabama

Red Mountain Formation (1978)
    Overlies Chickamauga Group; underlies Frog Mountain Sandstone. N Alabama, and NW Georgia.
    BT Silurian
    BT Paleozoic
    SA Alabama
    SA Georgia

red orpiment
    use realgar

Red Peak Formation (1978)
    In Chugwater Formation. NW Wyoming.
    BT Triassic
    BT Mesozoic
    SA Chugwater Formation
    SA Wyoming

Red River (1978)
    A southern tributary of the Mississippi River, rising in E New Mexico. Also a river rising in Manitoba and flowing into the Minnesota River. Also several other North American rivers. Search in combination with appropriate area term.
    IN Index North America or states and Canadian provinces as applicable.
    SA Arkansas
    SA Louisiana
    SA Manitoba
    SA Minnesota
    SA New Mexico
    SA North America
    SA North Dakota
    SA Oklahoma
    SA Red River valley
    SA South Dakota
    SA Texas

Red River Fault (1989)
    Yunnan, SW China. Before 1993, also search Honghe Fault.
    UF Honghe Fault
    BT Yunnan China
    BT China
    BT Far East
    BT Asia
    SA Vietnam

Red River Formation (1978)
    In Bighorn Group.
    BT Upper Ordovician
    BT Ordovician
    BT Paleozoic
    SA Manitoba
    SA Montana
    SA North Dakota
    SA South Dakota
    SA Wyoming

Red River valley (1978)
    As of 1985, can be used for valley of any of several Red rivers in the United States and Canada.
    IN Index states or provinces as applicable.
    SA Arkansas
    SA Louisiana
    SA Manitoba
    SA Minnesota
    SA New Mexico
    SA North Dakota
    SA Oklahoma
    SA Red River
    SA South Dakota
    SA Texas

red rock
    use red beds

Red Sea (1978)
    Between NE Africa and the Arabian Peninsula connecting the Mediterranean Sea with the Indian Ocean. Considered a narrower term of Indian Ocean as of 1981.
    CO N100000N300000 E0430000E0340000
    BT Indian Ocean
    NT Atlantis II Deep
    NT Chain Deep
    NT Discovery Deep
    NT Gulf of Aqaba
    NT Gulf of Suez
    NT Red Sea Rift
    SA DSDP Site 225
    SA DSDP Site 227
    SA DSDP Site 228
    SA Great Rift Valley
    SA Jeddah Saudi Arabia
    SA Leg 23
    SA Red Sea Basin
    SA Red Sea region

Red Sea Basin (1978)
    IN Index Red Sea, Djibouti and countries as applicable.
    SA Djibouti
    SA Egypt
    SA Ethiopia
    SA Israel
    SA Jordan
    SA Middle East
    SA Red Sea
    SA Red Sea region
    SA Saudi Arabia
    SA Sudan
    SA Yemen

Red Sea Hills (1985)
    Range of hills along Red Sea in E Egypt and NE Sudan.
    IN Index countries as applicable.
    UF Atbay
    UF Etbai
    BT Africa
    SA East Africa
    SA Egypt
    SA Sudan

Red Sea region (1978)
    This is an artificial term used to indicate the coastal region immediately adjacent to the Red Sea and the immediate littoral zone.
    SA Red Sea
    SA Red Sea Basin
    SA Red Sea Rift

Red Sea Rift (1993)
    Part of the Great Rift Valley system which connects the Dead Sea Rift to the East African Rift.
    IN Index countries as applicable.
    BT Red Sea
    BT Indian Ocean
    SA Arabian Peninsula
    SA Dead Sea Rift
    SA Egypt
    SA Ethiopia
    SA Ethiopian Rift
    SA Great Rift Valley
    SA Gulf of Aden
    SA Gulf of Suez
    SA Red Sea region
    SA Saudi Arabia
    SA Sudan
    SA Yemen

Red soils (1978)
    BT soils
    SA soil group
    SA Zonal soils

red zinc ore
    use zincite

redbeds
    use red beds

Redlichiida (1978)
    BT Trilobita
    BT Trilobitomorpha
    BT Arthropoda
    BT Invertebrata

Redon
    No longer a valid term for GeoRef. See Redon France.

Redon France (1993)
    Town in E Brittany, NW France.
    BT Ille-et-Vilaine France
    BT France
    BT Western Europe
    BT Europe

Redonda Formation (1978)
    Unconformably overlies Chinle Formation; unconformably underlies Entrada Sandstone. NE New Mexico.
    BT Upper Triassic
    BT Triassic
    BT Mesozoic
    SA New Mexico

Redondo Canyon (1978)
    Off S California.
    UF Redondo submerged valley
    BT Pacific Ocean
    SA submarine canyons

Redondo submerged valley
    use Redondo Canyon

Redonian (1978)
    North America.
    BT upper Pliocene
    BT Pliocene
    BT Neogene
    BT Tertiary
    BT Cenozoic

redox
    use Eh

redox potential
    use Eh

reduction (1978)
    SA Eh
    SA geochemistry
    SA oxidation
    SA processes

Redwall Limestone (1978)
    Comprises four unnamed members. N Arizona.
    BT Mississippian
    BT Carboniferous
    BT Paleozoic
    SA Arizona

Redwood Creek (1989)
    NW California.
    IN Index county or regions as applicable.
    SA Humboldt County California

reedmergnerite (1978)
    BT framework silicates
    BT silicates

Reedsville Formation (1989)
Central Pennsylvania, E Tennessee, SW Virginia and E West Virginia.
BT Upper Ordovician
BT Ordovician
BT Paleozoic
SA Pennsylvania
SA Tennessee
SA Virginia
SA West Virginia

reef builders (1981)
SA algae
SA Brachiopoda
SA Bryozoa
SA Coelenterata
SA Echinodermata
SA hermatypic taxa
SA Mollusca
SA Porifera
SA reefs
SA stromatolites

reef environment (1993)
For paleoecology. Before 1993, also search reefs and environment.
SA deep-water environment
SA depositional environment
SA environment
SA marine environment
SA paleoecology
SA reefs
SA sedimentation
SA shallow-water environment
SA submarine environment

reef formations
No longer a valid GeoRef index term. Before 1969, was used on level 1 in subfile E.
use reefs

reef pinnacle
use pinnacle reefs

reefs (1978)
As of 1984, not to be for sedimentary structures; use bioherms there. Before 1974, also search coral reefs. Before 1969, also search reef formations. As of 1993, use reef environment for paleoecology.
UF coral reefs
UF reef formations
NT atolls
NT barrier reefs
NT fringing reefs
NT patch reefs
NT pinnacle reefs
NT platform reefs
SA biogenic structures
SA bioherms
SA biostromes
SA carbonate banks
SA changes of level
SA Coelenterata
SA continental shelf
SA ecology
SA geomorphology
SA hermatypic taxa
SA lagoons
SA lithofacies
SA marine geology
SA oceanography
SA paleoecology
SA reef builders
SA reef environment
SA sedimentary rocks
SA sedimentary structures
SA sedimentation
SA shoals
SA spits

Reefton
No longer a valid term for GeoRef. See Reefton New Zealand.

Reefton New Zealand (1993)
Village in NW South Island.
IN Also index South Island.
BT New Zealand
BT Australasia
SA South Island

Reelfoot Lake (1989)
Lake and Obion counties, NW Tennessee.
IN Index counties as applicable.
CO N362000N362800
W0892000W0893000
BT Tennessee
BT United States
SA Lake County Tennessee
SA Obion County Tennessee

Reelfoot Rift (1989)
NW Tennessee, SE Missouri, NE Arkansas and NW Mississippi.
IN Index states as applicable.
BT United States
SA Arkansas
SA Midcontinent
SA Mississippi
SA Mississippi Valley
SA Missouri
SA Tennessee

Reeves County
Valid through 1988. Search in combination with state term. After 1988, use specific county-state term.

Reeves County Texas (1989)
Extreme W Texas. Before 1989, also search Reeves County AND Texas.
CO N304500N320000
W1030000W1041000
BT Texas
BT United States

refinement (1978)
SA crystal structure

reflectance (1978)
Applies to specific samples of coal, organic materials, and sometimes minerals. As of 1982, use ore reflectivity for ore deposits. When discussing methods, also use microscope methods. Before 1993, also search reflectivity. As of 1982, use ore reflectivity for reflectivity of ore deposits. Does not include reflectivity of a rock bed to seismic waves.
UF reflectance spectra
UF reflectivity
SA albedo
SA coal
SA crystallography
SA microscope methods
SA mineralogy
SA ore reflectivity
SA organic materials

reflectance spectra
use reflectance

reflection (1978)
As of 1981, restricted to seismic meaning only. For methodology use reflection methods.
SA diffraction
SA dip moveout
SA geophysical surveys
SA moveout
SA normal moveout
SA reflectograms
SA refraction
SA seismic migration
SA seismic surveys
SA seismology

reflection methods (1978)
Before 1978, search reflection AND seismic methods or geophysical methods.
NT common-depth-point method
SA AVO methods
SA BIRPS
SA CALCRUST
SA COCORP
SA DEKORP
SA dip moveout
SA geophysical methods
SA ground-penetrating radar
SA Lithoprobe
SA moveout
SA normal moveout
SA reflectograms
SA seismic methods

reflection waves
As of 1981, no longer a valid index term for GeoRef. Before 1978, also search ocean waves AND reflection.

reflectivity
No longer a valid term for GeoRef.
use reflectance

reflectograms (1978)
SA reflection
SA reflection methods

reforestation
use revegetation

refraction (1978)
As of 1981, restricted to seismic meaning only. For minerals, use refractive index.
SA diffraction
SA geophysical surveys
SA reflection
SA refractive index
SA seismic surveys
SA seismology

refraction methods (1978)
Before 1978, search refraction AND seismic methods or geophysical methods.
SA geophysical methods
SA seismic methods
SA sonobuoys

refraction waves
As of 1981, no longer a valid index term for GeoRef. Before 1978, also search ocean waves AND refraction.

refractive index (1978)
UF index of refraction
UF refractivity
UF refractometry
SA birefringence
SA optical dispersion
SA optical properties
SA refraction

refractivity
use refractive index

refractometry
use refractive index

refractories
use refractory materials

refractory clay
use fireclay

refractory materials (1978)
Before 1969, also search refractories.
IN As a commodity index with ceramic materials.
UF refractories
SA carborundum
SA ceramic materials
SA clays
SA construction materials
SA fireclay
SA flint clay
SA graphite deposits
SA industrial minerals
SA kaolin deposits
SA magnesite deposits
SA materials
SA sintering

Refugian (1978)
North America. Above Fresnian, below Zemorrian.
BT Eocene
BT Paleogene
BT Tertiary
BT Cenozoic
SA Oligocene

Regensburg
No longer a valid term for GeoRef. See Regensburg Germany.

Regensburg Germany (1993)
City on Danube River in E central Bavaria, SE Germany.
BT Bavaria Germany
BT Germany
BT Central Europe
BT Europe

Reggio
Not a valid term for GeoRef. See Reggio di Calabria Italy.

Reggio Calabria
use Reggio di Calabria Italy

Reggio di Calabria
No longer a valid term for GeoRef as of 1993.
use Reggio di Calabria Italy

Reggio di Calabria Italy (1993)
City and province on Strait of Messina in S Italy. also search Reggio and Reggio Calabria.
UF Reggio Calabria
UF Reggio di Calabria
BT Calabria Italy
BT Italy
BT Southern Europe
BT Europe

regimes, water
use water regimes

Regina
No longer a valid term for GeoRef. As of 1993, see Regina Saskatchewan.

Regina Saskatchewan (1993)
City in S Saskatchewan. Before 1993, also search Regina AND Saskatchewan.
BT Saskatchewan
BT Western Canada
BT Canada

regional (1978)
After 1977, used only as a general term for geography. A valid level 2 term through 1977 used in combination with metamorphism; after 1977, regional metamorphism is used.
SA areal geology
SA geography

regional anomalies (1981)
SA anomalies
SA residual anomalies

Regional Aquifer-System Analysis Program
use RASA

regional metamorphism (1978)
Before 1978, also search meta-morphism AND regional.
BT metamorphism
SA burial metamorphism
SA dynamic metamorphism

regional patterns (1978)
SA geothermal gradient
SA heat flow
SA heat sources
SA patterns

regional planning (1978)
SA environmental geology
SA land use
SA planning
SA urban planning

regolith (1978)
IN May also be used used for the Moon and planets.
SA debris
SA overburden
SA sedimentary cover
SA soils

Regosols (1985)
BT soils

regression (1978)
Term is used in stratigraphy for Pre-Quaternary sedimentation. For Quaternary, use changes of level. For statistical meaning, use regression analysis.
SA changes of level
SA stratigraphy
SA transgression

regression analysis (1978)
Autoposting of this term began in 1978.
BT statistical analysis
NT autoregression
SA analysis
SA least-squares analysis

Reguibat Ridge (1978)
In NW Mauritania and Western Sahara.
BT Africa
SA Mauritania
SA Western Sahara

regulations (1993)
SA decision-making
SA economic geology
SA engineering geology
SA environmental geology
SA government agencies
SA hydrogeology
SA land use
SA legislation
SA management
SA policy
SA pollution
SA preservation
SA protection
SA reclamation
SA resources
SA waste disposal

Rehamma
  use Rehamna

Rehamna (1978)
Tribal area N of Marrakech in W central Morocco.
UF Rehamma
BT Morocco
BT North Africa
BT Africa

Reichenstein
  use Zloty Stok Poland

Reindeer Lake (1981)
NE Saskatchewan and NW Manitoba.
CO N562000N575500
W1013000W1032000
BT Canada
SA Manitoba
SA Saskatchewan

reinforced materials (1993)
Before 1993, also search reinforced soil and reinforced earth as applicable.
SA dams
SA embankments
SA engineering geology
SA foundations
SA highways
SA materials
SA reservoirs
SA rock mechanics
SA slope stability
SA soil mechanics

reinjection wells (1985)
SA artificial recharge
SA fluid injection
SA gas injection
SA injection
SA irrigation
SA waste disposal
SA wells

relation
  A valid general term through 1977.

relative age (1978)
Also search specific methods if applicable.
IN Index specific method or methods.
SA absolute age
SA age
SA biochronology
SA cosmochronology
SA electron paramagnetic resonance
SA epimerization
SA exposure age
SA fission-track dating
SA geochronology
SA hydration of glass
SA isochrons
SA lichenometry
SA meteor craters
SA optical mineralogy
SA overprinting
SA paleomagnetism
SA palynology
SA particle-track dating
SA racemization
SA radiation damage
SA tephrochronology
SA thermoluminescence
SA tree rings
SA varves

relaxation energy (1982)
SA elastic waves
SA energy
SA seismology

release fractures (1978)
Before 1978, also search fractures AND release.
BT fractures
SA joints
SA stress

reliability (1978)
Used as a general term.
SA accuracy
SA errors
SA precision
SA probability
SA sampling
SA statistical analysis

relict materials (1985)
SA materials
SA metamorphic rocks
SA meteorites
SA parent bodies
SA permafrost
SA provenance
SA reworking
SA sediments

relief (1978)
SA digital terrain models
SA geomorphology
SA landform description
SA landform evolution
SA ocean floors
SA paleorelief
SA relief exposure
SA relief inversion
SA terrains
SA topography

relief effects (1981)
For use in relation to the Earth's magnetic field.
SA effects
SA magnetic field

relief exposure (1981)
SA geomorphology
SA landforms
SA relief

relief features
  use landforms

relief inversion (1981)
SA geomorphology
SA landform evolution
SA relief

Relizian (1985)
North America. Miocene, above Saucesian and below Luisian.
BT Miocene
BT Neogene
BT Tertiary
BT Cenozoic
SA lower Miocene
SA middle Miocene

remagnetization (1978)
SA demagnetization
SA magnetization
SA natural remanent magnetization
SA paleomagnetism
SA secondary magnetization

remanent magnetism
  use remanent magnetization

remanent magnetization (1978)
Autoposting of magnetization to this term began in 1989.
UF remanent magnetism
BT magnetization
NT anhysteretic remanent magnetization
NT chemical remanent magnetization
NT depositional remanent magnetization
NT inverse thermoremanent magnetization
NT isothermal remanent magnetization
NT natural remanent magnetization
NT partial thermoremanent magnetization
NT physical remanent magnetization
NT piezoremanent magnetization
NT thermochemical remanent magnetization
NT thermoremanent magnetization
NT viscous remanent magnetization
SA adiabatic demagnetization
SA chemical demagnetization
SA coercivity
SA demagnetization
SA Koenigsberger ratio
SA magnetic field
SA magnetic properties
SA paleomagnetism
SA Thellier Method
SA thermal demagnetization

remediation (1993)
Before 1993, also search reclamation, water treatment or soil treatment if applicable.
NT bioremediation
SA leaking underground storage tanks
SA pollution
SA reclamation
SA soil treatment
SA strip mining
SA waste disposal
SA water treatment

remobilization
  As of 1981, no longer a valid term for GeoRef. Included use in petrology.

remote sensing (1978)
Used for both methods and applications. Before 1972, also search remote-sensing methods; remote-sensing surveys; remote sensing data.
UF remote sensing data
UF remote sensing methods
UF remote-sensing methods
UF remote-sensing surveys
UF sensing, remote
SA aerial photography
SA airborne methods
SA albedo
SA Apollo 8
SA Apollo 9
SA Apollo 11
SA Apollo 12
SA Apollo 13
SA Apollo 14
SA Apollo 15
SA Apollo 16
SA Apollo 17
SA Apollo Program
SA Apollo-Soyuz Program
SA automated analysis
SA color imagery
SA coronae
SA engineering geology
SA environmental geology
SA Explorer 35
SA Explorer Program
SA Fraunhofer line discriminators
SA geobotanical methods
SA geophysical methods
SA geophysical surveys
SA Geosat
SA Geoscan
SA Global Positioning System
SA Gravsat
SA ground truth
SA helicopter methods
SA high-resolution methods
SA image enhancement
SA imagery
SA infrared methods
SA infrared surveys
SA Lageos
SA land use
SA Landsat
SA laser methods
SA laser ranging
SA lineaments
SA Luna 9
SA Luna 16
SA Luna 20
SA Luna 24
SA Luna Program
SA lunar laser ranging

SA Magellan Program
SA Magsat
SA maps
SA Mariner 4
SA Mariner 5
SA Mariner 6
SA Mariner 7
SA Mariner 9
SA Mariner 10
SA Mariner 11
SA Mariner Program
SA microwave methods
SA mineral exploration
SA mosaics
SA multispectral analysis
SA multispectral scanner
SA orbital observations
SA passive methods
SA photogeologic maps
SA photogeologic methods
SA photogeology
SA photogrammetry
SA photography
SA Pioneer 10
SA Pioneer 11
SA Pioneer Program
SA pixels
SA POGO
SA RADAM
SA radar methods
SA sand sheets
SA SAR
SA satellite methods
SA SeaMarc
SA Seasat
SA Shuttle Imaging Radar
SA Skylab
SA SLAR
SA sonar methods
SA space photography
SA Spacelab Program
SA SPOT
SA surface properties
SA Surveyor 3
SA Surveyor 5
SA Surveyor 7
SA Surveyor Program
SA telemetry
SA thematic mapper
SA thermal emission
SA total-field methods
SA vegetation
SA Venera Program
SA vertical-gradient methods
SA very long baseline interferometry
SA video methods
SA Viking Program
SA Voyager 1
SA Voyager 2
SA Voyager Program

remote sensing data
 No longer a valid GeoRef index term. Before 1971, was used on level 2 in subfile N.
 use remote sensing

remote sensing methods
 No longer a valid GeoRef index term. Before 1969, was used on level 1 in subfile E.
 use remote sensing

remote-sensing methods
 Not a valid GeoRef index term after 1971. Was used on level 1 in subfiles G and N.
 use remote sensing

remote-sensing surveys
 Not a valid GeoRef index term after 1971. Was used on level 1 in subfiles G and N.
 use remote sensing

Renalcis (1989)
 Genus. Autoposting of thallophytes and microfossils began in 1990. This term has multiple hierarchies.
 BT1 Cyanophyta
 BT1 algae
 BT1 thallophytes
 BT1 Plantae
 BT2 Cyanophyta
 BT2 algae
 BT2 microfossils

Renault Formation (1989)
 S Illinois, S Indiana, W and central Kentucky, and SE Missouri.
 BT Chesterian
 BT Upper Mississippian
 BT Mississippian
 BT Carboniferous
 BT Paleozoic
 SA Illinois
 SA Indiana
 SA Kentucky
 SA Missouri

Rendzinas (1978)
 BT soils
 SA soil group

Renfrew County
 No longer a valid term for GeoRef. As of 1993, see Renfrew County Ontario.

Renfrew County Ontario (1993)
 SE Ontario. Before 1993, also search Renfrew County AND Ontario.
 CO N450300N462000
   W0762000W0782500
 BT Ontario
 BT Eastern Canada
 BT Canada
 NT Chalk River Ontario
 SA Perch Lake

renierite (1978)
 BT sulfogermanates
 BT sulfosalts

Renmark
 No longer a valid term for GeoRef. As of 1993, see Renmark Australia.

Renmark Australia (1993)
 Town on Murray River in SE South Australia.
 BT South Australia
 BT Australia
 BT Australasia

Rennes
 No longer a valid term for GeoRef. See Rennes France.

Rennes France (1993)
 City in E Brittany, NW France.
 BT Ille-et-Vilaine France
 BT France
 BT Western Europe
 BT Europe

Reno
 Valid through 1988. Search in combination with state term. After 1988, use specific city-state term.

Reno County
 Valid through 1988. Search in combination with state term. After 1988, use specific county-state term.

Reno County Kansas (1989)
 S central Kansas. Before 1989, also search Reno County AND Kansas.
 CO N374500N381000
   W0974400W0983000
 BT Kansas
 BT United States
 NT Hutchinson Kansas

Reno Nevada (1989)
 City on Truckee River in W Nevada. Before 1989, search Reno AND Nevada.
 CO N393200N393200
   W1194900W1194900
 BT Washoe County Nevada
 BT Nevada
 BT United States

Rensselaer County
 Valid through 1988. Search in combination with state term. After 1988, use specific county-state term.

Rensselaer County New York (1989)
 E New York. Before 1989, also search Rensselaer County AND New York.
 CO N422700N425700
   W0731800W0735000
 BT New York
 BT United States

Renville County
 Valid through 1988. Search in combination with state term. After 1988, use specific county-state term.

Renville County Minnesota (1989)
 S Minnesota. Before 1989, also search Renville County AND Minnesota.
 CO N442700N445400
   W0943000W0952900
 BT Minnesota
 BT United States

Renville County North Dakota (1989)
 N North Dakota. Before 1989, also search Renville County AND North Dakota.
 CO N482700N490000
   W1010400W1020200
 BT North Dakota
 BT United States

replacement
 A valid term through 1975. After 1975, use metasomatism or pseudomorphism.

report (1978)
 Autoposting of this term began in 1993.
 UF brief report
 NT annual report
 NT progress report
 SA associations
 SA atlas
 SA current research
 SA symposia

repositories
 Not a valid term for GeoRef. See storage or waste disposal.

representative basins (1981)
 SA drainage basins
 SA ground water
 SA hydrogeology
 SA hydrology
 SA models
 SA water resources

reproduction (1978)
 SA larvae
 SA paleontology
 SA sexual dimorphism

reptiles (1981)
 Common name for Reptilia.
 IN Index for all non-paleontologic studies of fossils.
 BT tetrapods
 BT vertebrates
 NT diapsids
 SA biostratigraphy
 SA dinosaurs
 SA Reptilia

Reptilia (1978)
 Autoposting of this term began in 1978.
 BT Tetrapoda
 BT Vertebrata
 BT Chordata
 NT Anapsida
 NT Diapsida
 NT Synapsida
 SA dinosaurs
 SA eggs
 SA gastroliths
 SA reptiles

Republic County
 Valid through 1988. Search in combination with state term. After 1988, use specific county-state term.

Republic County Kansas (1989)
 N Kansas. Before 1989, also search Republic County AND Kansas.
 CO N394000N400000
   W0972300W0975500
 BT Kansas
 BT United States

research (1978)
 SA associations
 SA bibliography
 SA current research
 SA geology
 SA programs
 SA progress report
 SA survey organizations

research vessels (1983)
 SA CYANA
 SA Deep Sea Drilling Project
 SA expeditions
 SA FAMOUS
 SA geophysical methods
 SA geophysical surveys
 SA GEOSECS
 SA Glomar Challenger
 SA International Decade of Ocean Exploration
 SA IPOD
 SA marine geology
 SA marine methods
 SA Ocean Drilling Program
 SA oceanography
 SA submersibles

reserves (1978)
 IN Index specific commodity as applicable.
 SA ore grade
 SA prediction maps
 SA resources
 SA strategic minerals

reservoir properties (1978)
 Not to be used in relation to surface reservoirs.
 SA confining pressure
 SA Darcy's law
 SA natural gas
 SA naturally fractured reservoirs
 SA oil and gas fields
 SA overpressure
 SA petroleum
 SA properties
 SA tight sands

reservoir rocks (1978)
 SA cap rocks
 SA enhanced recovery

SA geopressure
SA geothermal energy
SA meteoric water
SA natural gas
SA naturally fractured reservoirs
SA oil and gas fields
SA overpressure
SA petroleum
SA petroleum engineering
SA rocks
SA roughness
SA sedimentation
SA source rocks
SA stratigraphic wedges
SA tight sands

reservoirs (1978)
Since 1978, used only for surface reservoirs (e.g. water reservoirs behind dams). For petroleum reservoirs, see reservoir rocks, reservoir properties. See subsurface reservoirs; surface reservoirs.
IN Used only for geological studies on surface reservoirs.
SA Archie's law
SA construction
SA dams
SA deposition
SA design
SA drawdown
SA earthquakes
SA embankments
SA engineering geology
SA feasibility studies
SA floods
SA geologic hazards
SA geomembranes
SA geotextiles
SA hydrogeology
SA hydrology
SA lakes
SA levels
SA maintenance
SA naturally fractured reservoirs
SA ponds
SA reinforced materials
SA retaining walls
SA rock mechanics
SA seepage
SA seismic response
SA siltation
SA site exploration
SA slope stability
SA soil mechanics
SA storage
SA waste disposal
SA water balance
SA water resources
SA water storage
SA water supply
SA waterways

residual anomalies (1981)
SA anomalies
SA gravity anomalies
SA regional anomalies

residual clays (1981)
BT clastic sediments
BT sediments
SA clay
SA clays

residues, insoluble
use insoluble residues

residues, organic
use organic residues

residuum (1978)
BT clastic sediments
BT sediments
SA terrigenous materials
SA weathering

resinite (1978)
BT exinite

BT macerals

resins (1978)
UF fossil resins
UF mineral resins
BT organic materials
SA amber

resistivity (1978)
UF electrical resistivity
SA apparent resistivity
SA Archie's law
SA conductivity
SA crosshole methods
SA electrical conductivity
SA electrical logging
SA electrical methods
SA electrical sounding
SA electrical surveys
SA geophysical methods
SA geophysical surveys
SA Laterolog
SA Schlumberger methods
SA well-logging

Resita
No longer a valid term for GeoRef. See Resita Romania.

Resita Romania (1993)
City in Caras-Severin County in W Romania. Before 1978, also search Recita. This term has multiple hierarchies.
UF Recita Romania
BT1 Banat
BT1 Southern Europe
BT1 Europe
BT2 Romania
BT2 Southern Europe
BT2 Europe

resonance (1978)
SA electron paramagnetic resonance
SA nuclear magnetic resonance
SA oscillations
SA vibration

Resource Conservation and Recovery Act (1989)
United States legislative act enacted in the fall of 1976. Concerns waste disposal management, especially regulation of hazardous waste.
UF RCRA
SA conservation
SA environmental geology
SA ground water
SA liquid waste
SA natural resources
SA pollution
SA protection
SA radioactive waste
SA resources
SA solid waste
SA waste disposal
SA water management
SA water quality
SA water resources

resources (1978)
SA conservation
SA energy sources
SA mineral resources
SA natural resources
SA regulations
SA reserves
SA Resource Conservation and Recovery Act
SA water resources

respiration (1981)
SA geochemistry
SA photosynthesis
SA processes

response
As of 1981, no longer a valid term for GeoRef. Used as a general term.

Restigouche Estuary (1978)
W extremity of Chaleur Bay.
IN Index provinces as applicable.
BT Canada
SA Chaleur Bay
SA New Brunswick
SA Quebec

restorations
No longer a valid GeoRef index term. Before 1971, was used on level 1 in subfiles E and N. See physical models; reclamation; reconstruction; sample preparation.

retaining walls (1993)
Wall designed to resist lateral pressure of material behind it and maintain differences in elevation.
UF walls, retaining
SA berms
SA embankments
SA foundations
SA landslides
SA levees
SA pressure
SA reservoirs
SA rock mechanics
SA slope stability
SA soil mechanics
SA stabilization
SA tailings ponds

retention (1978)
The amount of water from precipitation that has not escaped as runoff or through evapotranspiration.
SA atmospheric precipitation
SA hydrology
SA runoff
SA water

Retezat Mountains (1978)
Group in W Transylvanian Alps in SW Transylvania, W central Romania.
BT Transylvania
BT Romania
BT Southern Europe
BT Europe
SA Transylvanian Alps

Retort Phosphatic Shale Member (1989)
Of Phosphoria Formation. E Idaho, SW Montana, Nevada, NE Utah and W Wyoming.
BT Permian
BT Paleozoic
SA Idaho
SA Montana
SA Nevada
SA Phosphoria Formation
SA Utah
SA Wyoming

retorting (1985)
SA combustion
SA in situ
SA oil shale
SA petroleum engineering
SA recovery

retrieval, data
use data retrieval

retrograde
A valid term through 1977.

retrograde metamorphism (1978)

Before 1978, also search metamorphism AND retrograde; diaphthoresis.
UF diaphthoresis
UF retrogressive metamorphism
BT metamorphism
SA dynamic metamorphism
SA polymetamorphism
SA prograde metamorphism

retrogressive metamorphism
use retrograde metamorphism

Reunion (1978)
One of the Mascarene Islands and a French overseas territory 425 miles E of Madagascar. In 1981, broader term changed from Indian Ocean to Indian Ocean Islands. Piton de la Fournaise, Piton des Neiges and Saint-Louis are narrower terms as of 1981. Before 1972, also search Reunion Island.
CO S212500S205000
    E0560000E0550000
UF Reunion Island
BT Mascarene Islands
BT Indian Ocean Islands
NT Piton de la Fournaise
NT Piton des Neiges
NT Saint-Louis Reunion

Reunion Island
Not a valid GeoRef index term after 1971. Was used on level 1 in subfile G.
use Reunion

revegetation (1993)
Before 1993, also search reforestation.
UF reforestation
SA ecology
SA fires
SA forests
SA grasslands
SA prairies
SA savannas
SA steppes
SA tundra

Revelstoke
No longer a valid term for GeoRef. See Revelstoke British Columbia.

Revelstoke British Columbia (1993)
City in SE British Columbia.
BT British Columbia
BT Western Canada
BT Canada

reversals (1978)
Before 1972, also search self-reversals; self reversal.
UF self reversal
UF self-reversals
SA Brunhes Epoch
SA Earth
SA Gauss Epoch
SA Gilbert Epoch
SA Jaramillo Event
SA magnetic field
SA Matuyama Epoch
SA Olduvai Event
SA paleomagnetism
SA pole positions

reverse
A valid term through 1977. After 1977, use reverse faults.

reverse faults (1978)
Before 1978, also search faults AND reverse.
BT faults
SA displacements
SA imbricate tectonics
SA thrust faults

reverse slip faults
  use thrust faults

revetments (1993)
  Facing of stone, concrete or other material on embankment to prevent scour by weathering or water action.
  SA embankments
  SA engineering geology
  SA seawalls
  SA shorelines
  SA slope stability
  SA waterways

Revett Quartzite (1985)
  In Ravalli Group of Belt Supergroup. NE Idaho and NW Montana.
  BT middle Proterozoic
  BT Proterozoic
  BT upper Precambrian
  BT Precambrian
  SA Belt Supergroup
  SA Idaho
  SA Montana
  SA Ravalli Group
  SA Washington

review (1978)
  Used as a general term for bibliographic reviews, annual reviews and syntheses.
  SA bibliography
  SA book reviews
  SA critical review

revision (1978)
  Used for revision of classification and taxonomy, especially in fossils.
  SA fossils
  SA new taxa
  SA taxonomy

Rewa
  No longer a valid term for GeoRef. Former state in N Madhya Pradesh, central India. See Rewa India.

Rewa India (1993)
  City in N Madhya Pradesh, central India.
  BT Madhya Pradesh India
  BT India
  BT Indian Peninsula
  BT Asia

reworking (1978)
  Used in relation to fossils.
  SA deposition
  SA fossils
  SA relict materials
  SA sedimentation
  SA sediments

Rexroad Formation (1978)
  Underlies Angell Member of Ballard Formation.
  BT upper Pliocene
  BT Pliocene
  BT Neogene
  BT Tertiary
  BT Cenozoic
  SA Kansas
  SA Oklahoma

Reykjanes Peninsula (1978)
  WSW of Reykjavik. Also search Reykjanes.
  CO N633000N643000
    W0220000W0240000
  BT Iceland
  BT Western Europe
  BT Europe

Reykjanes Ridge (1978)
  SW of Iceland.
  BT Atlantic Ocean

  SA DSDP Site 407
  SA DSDP Site 408

Reynolds County
  Valid through 1988. Search in combination with state term. After 1988, use specific county-state term.

Reynolds County Missouri (1989)
  SE Missouri. Before 1989, also search Reynolds County AND Missouri.
  CO N370500N374000
    W0904000W0912000
  BT Missouri
  BT United States

Reynolds Creek (1978)
  Flows into Snake River in extreme SW Idaho.
  IN Index counties or region as applicable.
  SA Idaho

Rh
  use rhodium

rhabdite
  use schreibersite

Rhabdomesidae (1978)
  Family.
  BT Cryptostomata
  BT Bryozoa
  BT Invertebrata

rhabdophane (1978)
  BT phosphates

Rhabdopleurida (1978)
  IN Index Hemichordata or Invertebrata as applicable.
  UF Rhapdopleurida
  BT Pterobranchia
  SA Hemichordata
  SA Invertebrata

Rhaetian (1978)
  Europe. Above Norian, below Hettangian (Jurassic).
  BT Upper Triassic
  BT Triassic
  BT Mesozoic

Rhaetian Alps (1978)
  Division of Central Alps principally in Graubunden Canton, Switzerland.
  IN Index countries as applicable.
  BT Central Alps
  BT Alps
  BT Europe
  NT Adamello Massif
  SA Austria
  SA Graubunden Switzerland
  SA Lombardy Italy

Rhapdopleurida
  use Rhabdopleurida

Rhea Satellite (1985)
  One of the satellites of Saturn. Before 1985, also search Rhea AND Saturn.
  SA icy satellites
  SA satellites
  SA Saturn
  SA Voyager Program

Rhein River
  use Rhine River

Rhein Valley
  use Rhine Valley

Rheingau (1978)
  Region on S slope of Rheingau Mountains on right bank of Rhine E of Rudesheim in extreme W Hesse.
  BT Hesse Germany

  BT Germany
  BT Central Europe
  BT Europe

Rheinland
  use Rhineland

Rhenish Hesse (1978)
  Administrative division along the left bank of the Rhine River with Worms and Mainz the major cities.
  BT Rhineland-Palatinate Germany
  BT Germany
  BT Central Europe
  BT Europe

Rhenish Massif
  use Rhenish Schiefergebirge

Rhenish Schiefergebirge (1978)
  Rhenish Slate Mountains. Extensive plateau with Belgium and Luxembourg on the W, the Lahn River on the E, the Bonn area in the N, and including the Hunsruck mountain region on the S.
  IN Index states as applicable.
  CO N495700N513200
    E0091200E0064500
  UF Rhenish Massif
  UF Rhenish Slate Mountains
  UF Schiefergebirge
  BT Germany
  BT Central Europe
  BT Europe
  SA Ebbe Anticlinorium
  SA Eifel
  SA Hesse Germany
  SA North Rhine-Westphalia Germany
  SA Rhineland-Palatinate Germany
  SA Westerwald

Rhenish Slate Mountains
  use Rhenish Schiefergebirge

rhenium (1978)
  Autoposting of metals to this term began in 1989.
  UF Re
  BT metals
  SA Re/Os

rheology (1978)
  SA constitutive equations
  SA deformation
  SA flow lines
  SA flow mechanism
  SA plastic deformation
  SA viscosity

rheomorphism (1978)
  SA metamorphism

Rhin River
  use Rhine River

Rhin Valley
  use Rhine Valley

Rhine Basin (1978)
  Also search Rhine.
  IN Index Lake Constance and countries as applicable.
  UF Rhine region
  BT Europe
  SA Austria
  SA France
  SA Germany
  SA Lake Constance
  SA Liechtenstein
  SA Lower Rhine Basin
  SA Netherlands
  SA Switzerland

Rhine Graben (1978)
  Rift valley extending N from Basel to Mainz.

  UF Rhine Trough
  BT Central Europe
  BT Europe
  NT Lower Rhine Graben
  NT Swiss Rhine Graben
  NT Upper Rhine Graben
  SA Germany
  SA Switzerland

Rhine region
  use Rhine Basin

Rhine River (1978)
  Formed by confluence of the Hinterrhein and the Vorderrhein rivers in SE Switzerland; it flows through Lake Constance and then W, N, and NW to the North Sea. Also search Rhine.
  IN Index Lake Constance and countries as applicable.
  UF Rhein River
  UF Rhin River
  UF Rijn River
  BT Europe
  SA Austria
  SA Bergisches Land
  SA France
  SA Germany
  SA Lake Constance
  SA Liechtenstein
  SA Netherlands
  SA Switzerland

Rhine Trough
  use Rhine Graben

Rhine Valley (1978)
  Commonly recognized as that part of the river valley extending from Basel, Switzerland, to the North Sea. Also search Rhine.
  IN Index countries as applicable.
  UF Rhein Valley
  UF Rhin Valley
  UF Rijn Valley
  BT Europe
  SA France
  SA Germany
  SA Lower Rhine Basin
  SA Netherlands
  SA Switzerland

Rhine Westphalian Basin (1981)
  CO N493000N522000
    E0085700E0060000
  BT Germany
  BT Central Europe
  BT Europe

Rhineland (1978)
  Region. The part of Germany W of the Rhine River (left bank).
  IN Index states as applicable.
  UF Rheinland
  BT Germany
  BT Central Europe
  BT Europe
  SA North Rhine-Westphalia Germany
  SA Rhineland-Palatinate Germany

Rhineland-Palatinate
  No longer a valid term for GeoRef.
  use Rhineland-Palatinate Germany

Rhineland-Palatinate Germany (1993)
  CO N490000N510000
    E0083000E0061000
  UF Rhineland-Palatinate
  BT Germany
  BT Central Europe
  BT Europe
  NT Eifel
  NT Hunsruck

NT Laacher See
NT Mainz Germany
NT Meerfeld crater lake
NT Rhenish Hesse
SA Hunsruck Shale
SA Lahn River
SA Lahn River valley
SA Mainz Basin
SA Nahe
SA Palatinate
SA Rhenish Schiefergebirge
SA Rhineland
SA Saar Basin
SA Saar-Nahe Basin
SA Upper Rhine Valley
SA Westerwald

Rhinocerotidae (1978)
Family. Autoposting of this term to Coelodonta antiquitatis began in 1989. Autoposting of Ceratomorpha, Eutheria and Theria to this term began in 1989. Before 1981, also search rhinoceros.
BT Ceratomorpha
BT Perissodactyla
BT Eutheria
BT Theria
BT Mammalia
BT Tetrapoda
BT Vertebrata
BT Chordata
NT Coelodonta antiquitatis

Rhipidistia (1978)
Suborder. Autoposting of Crossopterygii to this term ended in 1980 and began again in 1989.
BT Crossopterygii
BT Sarcopterygii
BT Osteichthyes
BT Pisces
BT Vertebrata
BT Chordata

Rhizocorallium (1989)
Genus.
BT ichnofossils

Rhode Island (1978)
Autoposting of this term began in 1978.
CO N411000N420200 W0710700W0715500
BT United States
NT Kent County Rhode Island
NT Narragansett Bay
NT Newport County Rhode Island
NT Providence County Rhode Island
NT Washington County Rhode Island
SA Atlantic Coastal Plain
SA Avalon Terrane
SA Blackstone Series
SA Block Island Sound
SA Lake Char Fault
SA Narragansett Basin
SA Narragansett Pier Granite
SA Westerly Granite

Rhodes (1978)
Largest island of the Dodecanese off SW Turkey. This term has multiple hierarchies.
BT1 Dodecanese
BT1 Greek Aegean Islands
BT1 Greece
BT1 Southern Europe
BT1 Europe
BT2 Dodecanese
BT2 Greek Aegean Islands
BT2 Aegean Islands
BT2 Mediterranean region
SA Hellenic Arc

Rhodesia
A valid term through 1980.

use Zimbabwe

Rhodesian Plateau (1978)
Part of great South African plateau.
BT Zimbabwe
BT Southern Africa
BT Africa

rhodium (1978)
Chemical element. As of 1985, use rhodium ores for rhodium as a commodity. Autoposting of platinum group and metals to this term began in 1989.
UF Rh
BT platinum group
BT metals
SA rhodium ores

rhodium ores (1985)
Before 1985, also search rhodium AND (deposit OR deposits OR ore OR ores OR economic) in the basic index. Autoposting of metal ores to this term began in 1985.
BT metal ores
SA platinum ores
SA rhodium

rhodochrosite (1978)
UF dialogite
BT carbonates
SA manganese minerals
SA manganese ores
SA rhodonite

rhodonite (1978)
BT chain silicates
BT silicates
SA manganese minerals
SA manganese ores
SA ornamental materials
SA rhodochrosite

Rhodope
No longer a valid term for GeoRef. As of 1993 see Rhodope Greece.

Rhodope Greece (1993)
Department in central Greek Thrace, NE Greece. Before 1993 also search Rhodope AND Greece. This term has multiple hierarchies.
BT1 Greek Thrace
BT1 Greece
BT1 Southern Europe
BT1 Europe
BT2 Greek Thrace
BT2 Thrace
BT2 Europe

Rhodope Massif
use Rhodope Mountains

Rhodope Mountains (1978)
Major mountain system primarily in S Bulgaria. Also search Rhodope.
IN Index countries as applicable.
CO N411000N421000 E0253000E0233000
UF Rhodope Massif
BT Southern Europe
BT Europe
NT Bulgarian Rhodope Mountains
SA Bulgaria
SA Greek Macedonia
SA Greek Thrace
SA Rila Mountains

Rhodophyta (1978)
Including red algae. Autoposting of thallophytes and microfossils began in 1990. This term has multiple hierarchies.
UF red algae
BT1 algae
BT1 thallophytes

BT1 Plantae
BT2 algae
BT2 microfossils
NT Corallinaceae
NT Gymnocodiaceae

Rhoen Mountains
use Rhon Mountains

Rhon Mountains (1978)
Also search Rhon.
IN Index states as applicable.
UF Rhoen Mountains
BT Germany
BT Central Europe
BT Europe
SA Bavaria Germany
SA Hesse Germany
SA Thuringia Germany

Rhone
No longer a valid term for GeoRef. As of 1993, see Rhone France.

Rhone Basin
use Rhone Valley

Rhone Delta (1978)
On Gulf of Lion W of Marseilles.
BT Bouches-du-Rhone France
BT France
BT Western Europe
BT Europe
SA Gulf of Lion

Rhone France (1993)
Department in E central France.
CO N453000N462000 E0050300E0041500
BT France
BT Western Europe
BT Europe
NT Lyons France
SA Monts du Lyonnais

Rhone River (1978)
Rises in Rhone Glacier in S central Switzerland; it flows through Lake Geneva and then W and S into the Gulf of Lion.
IN Index Lake Geneva and countries as applicable.
BT Western Europe
BT Europe
SA France
SA Lake Geneva
SA Switzerland

Rhone River valley
use Rhone Valley

Rhone Valley (1978)
IN Index Lake Geneva and countries as applicable.
UF Rhone Basin
UF Rhone River valley
BT Western Europe
BT Europe
SA France
SA Lake Geneva
SA Saone-Rhone Basin
SA Switzerland

rhonite (1978)
BT chain silicates
BT silicates
SA aenigmatite

Rhum (1978)
Isle of Rhum. One of the Inner Hebrides S and SE of the Isle of Skye. This term has multiple hierarchies.
UF Rhum Island
UF Rum
BT1 Inner Hebrides
BT1 Hebrides
BT1 Scotland
BT1 Great Britain
BT1 United Kingdom

BT1 Western Europe
BT1 Europe
BT2 Inverness-shire Scotland
BT2 Highland region Scotland
BT2 Scotland
BT2 Great Britain
BT2 United Kingdom
BT2 Western Europe
BT2 Europe

Rhum Island
use Rhum

rhyacolite
use sanidine

Rhynchocephalia
As of 1993, no longer a valid term for GeoRef.
use Rhynchosauria

Rhynchonellida (1978)
Autoposting of this term began in 1978.
BT Articulata
BT Brachiopoda
BT Invertebrata
NT Rhynchonellidae

Rhynchonellidae (1978)
Family.
BT Rhynchonellida
BT Articulata
BT Brachiopoda
BT Invertebrata

Rhynchosauria (1993)
Order. Before 1993, Lepidosauria was autoposted as a broader term to Rhynchocephalia. Before 1993, also search Rhynchocephalia.
UF Rhynchocephalia
BT Diapsida
BT Reptilia
BT Tetrapoda
BT Vertebrata
BT Chordata
SA Lepidosauria

rhyodacite
As of 1989, no longer a valid term for GeoRef. From 1981 to 1988, rhyodacites and volcanic rocks were autoposted to this term.
use rhyodacites

rhyodacites (1981)
Before 1989, search rhyodacite. Autoposting of this term to rhyodacite ended in 1989.
UF rhyodacite
BT volcanic rocks
BT igneous rocks
NT dellenite
NT quartz latite

rhyolite
As of 1989, no longer a valid term for GeoRef. From 1981 to 1988, rhyolites and volcanic rocks were autoposted to this term.
use rhyolites

rhyolite porphyry (1978)
BT rhyolites
BT volcanic rocks
BT igneous rocks
SA porphyry

rhyolite tuff (1978)
BT pyroclastics
BT volcanic rocks
BT igneous rocks
SA tuff

rhyolites (1981)
Before 1981, search andesite-rhyolite family. From 1978-1980, andesite-rhyolite family was autoposted to the rocks that were classified under that term. Before then, variants of the term may occur in GeoRef.

1989, also search rhyolite. Autoposting of this term to rhyolite ended in 1989.
UF rhyolite
BT volcanic rocks
BT igneous rocks
NT comendite
NT liparite
NT liparite porphyry
NT pantellerite
NT quartz porphyry
NT rhyolite porphyry
SA metarhyolite
SA rhyolitic composition

rhyolitic composition (1978)
Before 1978, search rhyolitic; rhyolitic AND composition.
SA composition
SA igneous rocks
SA rhyolites

rhythmic bedding (1978)
BT planar bedding structures
BT sedimentary structures
SA bedding
SA cyclothems
SA rhythmite

rhythmite (1978)
BT planar bedding structures
BT sedimentary structures
SA cyclothems
SA rhythmic bedding
SA sedimentation

Rians
No longer a valid term for GeoRef. See Rians France.

Rians France (1993)
Village in Var, SE France.
BT Var France
BT France
BT Western Europe
BT Europe

Riasi (1978)
Town and district in SW Jammu and Kashmir. As of 1993, India is no longer a broader term.
IN Index Jammu and Kashmir and India and/or Pakistan as applicable.
UF Reasi
SA India
SA Jammu and Kashmir
SA Pakistan

Ribeira de Iguape River (1978)
Flows into the Atlantic Ocean near Iguape in S Brazil.
UF Ribeira do Iguape River
BT Sao Paulo Brazil
BT Brazil
BT South America

Ribeira do Iguape River
use Ribeira de Iguape River

Rice County
Valid through 1988. Search in combination with state term. After 1988, use specific county-state term.

Rice County Kansas (1989)
Central Kansas. Before 1989, also search Rice County AND Kansas.
CO N381000N383000
W0975500W0982800
BT Kansas
BT United States

Rice County Minnesota (1989)
SE Minnesota. Before 1989, also search Rice County AND Minnesota.
CO N441300N443100
W0930300W0933100
BT Minnesota
BT United States

Rice Lake (1978)
Lake in SE Ontario, and lake in Manitoba.
IN Index provinces as applicable.
BT Canada
SA Manitoba
SA Ontario

Rice Lake Group (1978)
BT Archean
BT Precambrian
SA Canada
SA Manitoba

Rich County
Valid through 1988. Search in combination with state term. After 1988, use specific county-state term.

Rich County Utah (1989)
N Utah. Before 1989, also search Rich County AND Utah.
CO N411900N420000
W1110400W1113200
BT Utah
BT United States

Richards equation (1985)
UF Richards' equation
SA equations
SA ground water
SA hydrodynamics
SA hydrology
SA infiltration
SA mathematical models
SA numerical models
SA soils
SA unsaturated zone
SA water regimes

Richards Island (1978)
Large island in Beaufort Sea at mouth of Mackenzie River, Mackenzie District.
BT Mackenzie District Northwest Territories
BT Northwest Territories
BT Western Canada
BT Canada

Richards' equation
use Richards equation

Richardson Mountains (1978)
IN Index territories as applicable.
BT Canada
SA Northwest Territories
SA Yukon Territory

Richat Mountain (1978)
E of S border of Spanish Sahara. Also search Richat.
BT Mauritania
BT West Africa
BT Africa

Richland
Valid through 1988. Search in combination with state term. After 1988, use specific city-state term.

Richland County
Valid through 1988. Search in combination with state term. After 1988, use specific county-state term.

Richland County Illinois (1989)
SE Illinois. Before 1989, also search Richland County AND Illinois.
CO N383600N385200
W0870600W0881900
BT Illinois
BT United States

Richland County Montana (1989)
NE Montana. Before 1989, also search Richland County AND Montana.
CO N472200N480800
W1040200W1051500
BT Montana
BT United States

Richland County North Dakota (1989)
Extreme SE North Dakota. Before 1989, also search Richland County AND North Dakota.
CO N455700N463800
W0963300W0971800
BT North Dakota
BT United States

Richland County Ohio (1989)
N central Ohio. Before 1989, also search Richland County AND Ohio.
CO N403300N405900
W0822000W0824300
BT Ohio
BT United States

Richland County South Carolina (1989)
Central South Carolina. Before 1989, also search Richland County AND South Carolina.
CO N334500N341700
W0803700W0812100
BT South Carolina
BT United States
NT Columbia South Carolina

Richland County Wisconsin (1989)
S central Wisconsin. Before 1989, also search Richland County AND Wisconsin.
CO N431000N433400
W0901100W0904000
BT Wisconsin
BT United States

Richland Parish Louisiana (1989)
NE Louisiana. Before 1989, also search Richland Parish; Richland County AND Louisiana.
CO N320900N323900
W0912600W0920300
BT Louisiana
BT United States

Richland Washington (1989)
Village on the Columbia River in S Washington. Before 1989, search Richland AND Washington.
CO N461700N461700
W1191700W1191700
BT Benton County Washington
BT Washington
BT United States

Richmond
Valid through 1988. Search in combination with state term. After 1988, use specific city-state term.

Richmond Basin (1989)
Small rift basin near Richmond, Virginia. Search in conjunction with Virginia.
IN Index counties or regions as applicable.
SA Virginia

Richmond Group (1978)
Includes Queenston Shale, Arnheim Formation, Waynesville Formation, Liberty Formation, Whitewater Formation, Sequatchie Formation, Fernvale Limestone, Marnie Shale, Oswego Sandstone, Juniata Formation, Saluda Formation, Elkhorn Formation.
BT Upper Ordovician
BT Ordovician
BT Paleozoic
SA Fernvale Formation
SA Illinois
SA Indiana
SA Juniata Formation
SA Kentucky
SA Missouri
SA Ohio
SA Queenston Shale

Richmond Virginia (1989)
City on the James River in but independent of Henrico County, E central Virginia. Before 1989, search Richmond AND Virginia.
CO N373400N373400
W0772700W0772700
BT Henrico County Virginia
BT Virginia
BT United States

Richmondian (1985)
North America. Upper Ordovician, above Maysvillian and below Lower Silurian.
BT Cincinnatian
BT Upper Ordovician
BT Ordovician
BT Paleozoic

Richter Scale (1981)
Numerical scale of earthquake magnitude.
SA earthquakes
SA magnitude
SA seismology

richterite (1978)
BT clinoamphibole
BT amphibole group
BT chain silicates
BT silicates

Richton Dome (1985)
Salt dome, SE Mississippi.
IN Index county or region as applicable.
SA Gulf Coastal Plain
SA Perry County Mississippi
SA salt domes

Ridder Kazakhstan
use Leninogorsk Kazakhstan

Riddle
Valid through 1988. Search in combination with state term. After 1988, use specific city-state term. Town in Douglas County in SW.

Riddle Oregon (1989)
Town in SW Oregon. Before 1989, search Riddle AND Oregon.
CO N425600N425600
W1232400W1232400
BT Douglas County Oregon
BT Oregon
BT United States

Ridge Basin (1985)
S California.
IN Index counties or region as applicable.
SA California

Ridge Route Formation (1985)
Upper Miocene or lower Pliocene. S California.
BT Neogene
BT Tertiary
BT Cenozoic
SA California
SA lower Pliocene
SA upper Miocene

ridges
As of 1981, no longer a valid term for GeoRef. See geomorphology; ocean floors; mid-ocean ridges; mountains.

Riding Mountain National Park (1978)
SW Manitoba. Also search Riding Mountain.
BT Manitoba
BT Western Canada
BT Canada

riebeckite (1978)
BT clinoamphibole
BT amphibole group
BT chain silicates
BT silicates
SA crocidolite

Ries Basin
use Ries Crater

Ries Crater (1978)
Meteor crater. Before 1978, the terms Ries Basin and Nordlingen were used. Also search Ries.
UF Nordlingen
UF Nordlinger Ries
UF Nordlinger Ries Crater
UF Ries Basin
BT Bavaria Germany
BT Germany
BT Central Europe
BT Europe

Riesengebirge
use Karkonosze Mountains

Rif (1978)
Hilly region constituting part of former Spanish Morocco extending along Mediterranean Sea from E of Melilla on E to Ceuta on W.
CO N340000N360000
    W0020500W0063500
UF Er Rif
UF Er Riff
UF Riff
BT Morocco
BT North Africa
BT Africa
NT Beni Bouchera

Riff
use Rif

Rifle
Valid through 1988. Search in combination with state term. After 1988, use specific city-state term.

Rifle Colorado (1989)
Town on Colorado River in W Colorado. Before 1989, search Rifle AND Colorado.
CO N393200N393200
    W1074700W1074700
BT Garfield County Colorado
BT Colorado
BT United States

Rift Valley
use Great Rift Valley

rift valleys
As of 1981, no longer a valid term for GeoRef.
use rift zones

rift zones (1978)
Before 1981, also search rift valleys.
UF rift valleys
UF rifts
UF zones, rift
SA A-type granites
SA fault zones
SA median valley
SA passive margins
SA plate divergence
SA plate tectonics
SA rifting
SA tectonics
SA tectonophysics

rifting (1978)
As of 1993, also search accreting plate boundaries and plate boundaries if applicable.
SA accreting plate boundary
SA accretion
SA extension tectonics
SA median valley
SA passive margins
SA plate boundaries
SA plate divergence
SA plate tectonics
SA rift zones
SA sea-floor spreading
SA submarine volcanoes
SA suture zones
SA taphrogeny
SA tesserae
SA underplating

rifts
use rift zones

Riga
No longer a valid term for GeoRef. See Riga Latvia.

Riga Latvia (1993)
City on Gulf of Riga.
BT Latvia
BT Baltic region
BT Europe

Riggins Group (1989)
W Idaho and E Oregon.
IN Index ages as applicable.
SA Idaho
SA Mesozoic
SA Oregon
SA Paleozoic

right-lateral faults (1978)
Before 1978, search faults AND right-lateral; faults AND right-slip. In 1989, lateral faults became a broader autoposted term.
UF dextral faults
UF right-lateral slip faults
UF right-slip faults
BT lateral faults
BT faults
SA left-lateral faults

right-lateral slip faults
use right-lateral faults

right-slip faults
use right-lateral faults

rigidities, cutoff
Not a valid term for GeoRef. Usually out-of-scope. Used as cutoff rigidities.

rigidity (1978)
SA brittleness
SA elasticity
SA mechanical properties
SA rock mechanics
SA stiffness
SA stress

rigidity modulus
use shear modulus

Rijn River
use Rhine River

Rijn Valley
use Rhine Valley

Rila
No longer a valid term for GeoRef. See Rila Bulgaria.

Rila Bulgaria (1993)
Village in W Bulgaria.
BT Sofia Bulgaria
BT Bulgaria
BT Southern Europe
BT Europe

Rila Mountains (1978)
Range in SW at W end of the Rhodope Mountains.
BT Bulgaria
BT Southern Europe
BT Europe
SA Bulgarian Rhodope Mountains
SA Rhodope Mountains

Riley County
Valid through 1988. Search in combination with state term. After 1988, use specific county-state term.

Riley County Kansas (1989)
NE Kansas. Before 1989, also search Riley County AND Kansas.
CO N390500N393500
    W0962000W0965600
BT Kansas
BT United States
NT Manhattan Kansas
NT Tuttle Creek Dam

rilles (1978)
SA gorges
SA Moon
SA valleys

rills (1993)
SA channels
SA erosion
SA erosion features
SA fluvial features
SA furrows
SA gullies
SA hydrology
SA shore features
SA soil erosion
SA streams
SA waterways

Rincon Formation (1978)
Tertiary of SW California. Also Cretaceous of Bonaire, Netherlands Antilles.
IN Index ages as applicable.
SA Bonaire
SA California
SA Cretaceous
SA Miocene
SA Netherlands Antilles
SA Oligocene
SA West Indies

Rincon Mountains (1985)
SE Arizona.
BT Arizona
BT United States
SA Basin and Range Province
SA Cochise County Arizona
SA Pima County Arizona

ring
A valid term through 1976. After 1976, see ring dikes. After 1977, also see ring structures and ring complexes.

ring complexes (1978)
BT intrusions
SA complexes
SA dikes
SA ring dikes
SA ring structures

ring dikes (1978)
Before 1976, also search ring.
UF ring-fracture intrusion
BT dikes
BT intrusions
SA dike swarms
SA ring complexes

SA ring structures

ring silicates (1978)
Autoposting of this term began in 1978.
UF cyclosilicates
BT silicates
NT aquamarine
NT axinite
NT barylite
NT bertrandite
NT beryl
NT buergerite
NT cordierite
NT dravite
NT ekanite
NT elbaite
NT emerald
NT eudialyte
NT malayaite
NT milarite
NT osumilite
NT roedderite
NT schorl
NT tourmaline

ring structures (1978)
SA arcuate structures
SA dikes
SA intrusions
SA ring complexes
SA ring dikes

ring-fracture intrusion
use ring dikes

Ringerike (1978)
Region NW of Oslo.
BT Norway
BT Scandinavia
BT Western Europe
BT Europe

Ringerike Sandstone (1978)
Included in Ringerike Series. S Norway.
BT Paleozoic
SA Devonian
SA Norway
SA Silurian

Ringold Formation (1985)
Pliocene. Previously considered middle and upper Pleistocene. NE Oregon and SE Washington.
BT Pliocene
BT Neogene
BT Tertiary
BT Cenozoic
SA Oregon
SA Pleistocene
SA Washington

rings, planetary
use planetary rings

rings, tree
use tree rings

rinneite (1978)
BT chlorides
BT halides

Rio Amazonas
use Amazon River

Rio Arriba County
Valid through 1988. Search in combination with state term. After 1988, use specific county-state term.

Rio Arriba County New Mexico (1989)
NW New Mexico. Before 1989, also search Rio Arriba County AND New Mexico.
CO N355500N370000
    W1053500W1074200
BT New Mexico
BT United States

NT Nacimiento Mountains
NT Tierra Amarilla New Mexico
SA Taos Plateau

Rio Blanco Basin (1978)
River basin in NW Argentina. Also search Rio Blanco.
UF Blanco River basin
SA Salta Argentina

Rio Blanco County
Valid through 1988. Search in combination with state term. After 1988, use specific county-state term.

Rio Blanco County Colorado (1989)
NW Colorado. Before 1989, also search Rio Blanco County AND Colorado.
CO N394000N402000
    W1070000W1090500
BT Colorado
BT United States
NT Rangely Colorado

Rio Bonito Formation (1978)
BT Permian
BT Paleozoic
SA Brazil
SA upper Paleozoic

Rio Bravo
use Rio Grande

Rio Bravo del Norte
use Rio Grande

Rio Claro
No longer a valid term for GeoRef. See Rio Claro Brazil.

Rio Claro Brazil (1993)
City in E central Sao Paulo, S Brazil.
BT Sao Paulo Brazil
BT Brazil
BT South America

Rio de Janeiro
No longer a valid term for GeoRef. As of 1993, see Rio de Janeiro Brazil for the state or Rio de Janeiro City Brazil.

Rio de Janeiro Brazil (1993)
State in SE Brazil. Before 1993, also search Rio de Janeiro. Before 1981, used for both the city and the state. Before 1993, also search Guanabara if applicable.
CO S233000S205000
    W0410000W0450000
BT Brazil
BT South America
NT Angra dos Reis Brazil
NT Guanabara Bay
NT Itaborai Brazil
NT Rio de Janeiro City Brazil
SA Macacu River

Rio de Janeiro City
No longer a valid term for GeoRef.
use Rio de Janeiro City Brazil

Rio de Janeiro City Brazil (1993)
City in S Rio de Janeiro, SE Brazil. Before 1981, also search Rio de Janeiro.
UF Rio de Janeiro City
BT Rio de Janeiro Brazil
BT Brazil
BT South America

Rio de la Plata (1978)
Estuary of the combined Parana River and the Uruguay River.
IN Index countries as applicable.
UF Rio de la Plata Estuary
UF River Plate
BT South America
SA Argentina
SA Uruguay

Rio de la Plata Estuary
use Rio de la Plata

Rio Grande (1978)
River which rises in SW Colorado and flows S and then SE into the Gulf of Mexico.
IN Index Mexican and U.S. states as applicable.
UF Grande River
UF Rio Bravo
UF Rio Bravo del Norte
UF Rio Grande River
SA Chihuahua Mexico
SA Coahuila Mexico
SA Colorado
SA New Mexico
SA Nuevo Leon Mexico
SA Rio Grande Valley
SA Tamaulipas Mexico
SA Texas

Rio Grande County
Valid through 1988. Search in combination with state term. After 1988, use specific county-state term.

Rio Grande County Colorado (1989)
S Colorado. Before 1989, also search Rio Grande County AND Colorado.
CO N372200N375000
    W1060300W1064300
BT Colorado
BT United States

Rio Grande Depression (1978)
Lower Rio Grande Valley.
IN Index Mexican and U. S. states as applicable.
BT North America
SA Tamaulipas Mexico
SA Texas

Rio Grande do Norte
No longer a valid term for GeoRef. See Rio Grande do Norte Brazil.

Rio Grande do Norte Brazil (1993)
State on the Atlantic Ocean in extreme NE Brazil.
CO S070000S045000
    W0345500W0384000
BT Brazil
BT South America
NT Potiguar Basin
SA Borborema

Rio Grande do Sul
No longer a valid term for GeoRef. See Rio Grande do Sul Brazil.

Rio Grande do Sul Brazil (1993)
State in extreme S Brazil.
CO S335000S270000
    W0493000W0574000
BT Brazil
BT South America
NT Encruzilhada do Sul Brazil
NT Sao Gabriel Brazil
SA Itarare Subgroup
SA Pelotas Basin

Rio Grande Rift (1978)
IN Index Mexican and U.S. states as applicable.
CO N313000N370000
    W1053000W1073000
BT North America
SA Chihuahua Mexico
SA Chihuahua tectonic belt
SA Coahuila Mexico
SA Colorado
SA Espanola Basin
SA New Mexico
SA Taos Plateau
SA Texas

Rio Grande Rise (1985)
South American Atlantic. As of 1990, Atlantic Ocean is autoposted to this term.
CO S330000S300000
    W0350000W0410000
BT South American Atlantic
BT South Atlantic
BT Atlantic Ocean
SA DSDP Site 357
SA DSDP Site 516
SA DSDP Site 517
SA Leg 39
SA Leg 72

Rio Grande River
use Rio Grande

Rio Grande Valley (1978)
Several possible locations.
IN Index North America or Mexican and U. S. states as applicable.
SA Chihuahua Mexico
SA Coahuila Mexico
SA Colorado
SA Mexico
SA New Mexico
SA North America
SA Nuevo Leon Mexico
SA Rio Grande
SA Tamaulipas Mexico
SA Texas

Rio Negro (1978)
River in NW South America. Flows into the Amazon River.
IN Index countries or continent as applicable.
UF Negro River
SA Amazonas Venezuela
SA Brazil
SA Colombia
SA South America
SA Venezuela

Rio Negro Argentina (1993)
In central Argentina. Before 1982, also search Rio Negro AND Argentina. Before 1993, also search Rio Negro Province.
CO S420000S373000
    W0630000W0720000
UF Rio Negro Province
BT Argentina
BT South America
SA Neuquen Basin
SA Patagonia

Rio Negro Department
No longer a valid term for GeoRef. As of 1993 see Rio Negro Uruguay.

Rio Negro Meteorite (1981)
Impact in Brazil. Before 1981, also search Rio Negro AND meteorites.
BT L chondrites
BT chondrites
BT stony meteorites
BT meteorites
SA Brazil

Rio Negro Province
No longer a valid term for GeoRef.
use Rio Negro Argentina

Rio Negro Uruguay (1993)
Department in W Uruguay. Before 1993 also search Rio Negro or Rio Negro Department AND Uruguay.
CO S332500S322000
    W0563500W0582500
BT Uruguay
BT South America

Rio Puerco (1989)
River generally flowing N-S through Sandoval, Bernalillo, Valencia and Socorro counties, central New Mexico. Also a river flowing W from McKinley County, NW New Mexico to Apache and Navajo counties, NE Arizona.
IN Index states and counties as applicable.
UF Puerco River
SA Apache County Arizona
SA Arizona
SA Bernalillo County New Mexico
SA McKinley County New Mexico
SA Navajo County Arizona
SA New Mexico
SA Sandoval County New Mexico
SA Socorro County New Mexico
SA Valencia County New Mexico

Rio Tinto
No longer a valid term for GeoRef. As of 1993, see Rio Tinto Spain.

Rio Tinto Spain (1993)
Town in copper mining area of W Betic Cordillera in SW Spain. Before 1993, also search Rio Tinto and Spain.
UF Riotinto
BT Huelva Spain
BT Andalusia Spain
BT Spain
BT Iberian Peninsula
BT Southern Europe
BT Europe

Rioni Basin (1978)
River basin in W Georgian Republic. Also search Rioni; Rioni River.
BT Georgian Republic
BT Europe

Riotinto
use Rio Tinto Spain

Riphean (1978)
Europe. As of 1989, upper Proterozoic and Proterozoic are autoposted to this term.
BT upper Proterozoic
BT Proterozoic
BT upper Precambrian
BT Precambrian
NT Taratash Complex
NT upper Riphean
SA Ediacaran
SA Eocambrian

Ripley County Indiana (1989)
SE Indiana. Before 1989, search Ripley County AND Indiana.
CO N385500N391800
    W0850300W0852700
BT Indiana
BT United States
NT Versailles Indiana

Ripley Formation (1978)
In Selma Group. Includes Coon Creek Tongue, Chiwapa Sandstone Member, McNairy Sand Member, Keownville Limestone Member. Alabama, Georgia, S Illinois, W Kentucky, SE Missouri, N and central Mississippi, and W Tennessee.

BT Upper Cretaceous
BT Cretaceous
BT Mesozoic
SA Alabama
SA Georgia
SA Illinois
SA Kentucky
SA Mississippi
SA Missouri
SA Selma Group
SA Tennessee

ripple drift-cross laminations (1978)
UF ripple-cross-laminations
BT planar bedding structures
BT sedimentary structures
SA cross-laminations
SA laminations

ripple marks (1978)
BT bedding plane irregularities
BT sedimentary structures
SA Bouma sequence
SA primary structures
SA sand ridges
SA sand waves

ripple-cross-laminations
use ripple drift-cross laminations

rise, continental
use continental rise

risk assessment (1989)
SA geologic hazards
SA impact statements
SA safety

risk, seismic
use seismic risk

Riss (1978)
Europe. Above Mindel, below Wurm.
BT Pleistocene
BT Quaternary
BT Cenozoic

Riss-Wurm
use Riss/Wurm Interglacial

Riss-Wurm Interglacial
use Riss/Wurm Interglacial

Riss/Wurm Interglacial (1978)
Europe. A term applied in the Alps to the third interglacial stage of the Pleistocene Epoch, after the Riss glacial stage and before the Wurm.
UF Riss-Wurm
UF Riss-Wurm Interglacial
BT Pleistocene
BT Quaternary
BT Cenozoic
SA Sangamonian

Rita Blanca Lake (1978)
Formed by Rita Blanca Dam on Rita Blanca Creek in Texas panhandle.
BT Texas
BT United States

Ritchie County
Valid through 1988. Search in combination with state term. After 1988, use specific county-state term.

Ritchie County West Virginia (1989)
NW West Virginia. Before 1989, also search Ritchie County AND West Virginia.
CO N390000N392300
W0804800W0812000
BT West Virginia
BT United States

Riukiu Islands
use Ryukyu Islands

river basins
After 1992, use drainage basins.

river capture
use stream capture

River Mountains (1989)
SE Nevada.
IN Index county as applicable.
SA Clark County Nevada

river piracy
use stream capture

river plain
use alluvial plains

River Plate
use Rio de la Plata

Rivera fracture zone (1985)
North American Pacific, off Mexico.
BT North American Pacific
BT Pacific Ocean

Rivera Ocean Seismic Experiment (1985)
UF ROSE
SA East Pacific Rise
SA Equatorial Pacific
SA marine geology
SA marine methods
SA oceanic crust
SA oceanography
SA Pacific Ocean
SA seismic surveys
SA seismology
SA transform faults

rivers (1978)
Autoposting of fluvial features to this term began in 1989.
BT fluvial features
SA channel geometry
SA channels
SA drainage basins
SA drainage patterns
SA estuaries
SA floodplains
SA fluvial currents
SA fluvial environment
SA geomorphology
SA hydrogeology
SA hydrology
SA hydrosphere
SA meanders
SA oxbow lakes
SA rapids
SA rivers and streams
SA runoff
SA stream capture
SA stream placers
SA stream transport
SA streams
SA surface water
SA tributaries
SA water supply
SA waterfalls
SA watersheds
SA waterways

rivers and streams (1978)
SA bedforms
SA channel geometry
SA drainage basins
SA estuaries
SA floods
SA fluvial currents
SA fluvial environment
SA fluvial features
SA hydrogeology
SA hydrology
SA hydrosphere
SA ice streams
SA low water
SA meanders
SA oxbow lakes
SA precipitation
SA rivers
SA runoff
SA stream capture
SA stream gradient
SA stream placers
SA stream transport
SA streams
SA tributaries
SA valleys
SA wadis
SA waterways

Riverside
Valid through 1988. Search in combination with state term. After 1988, use specific city-state term.

Riverside California (1989)
City on Santa Ana River in S California. Before 1989, search Riverside AND California.
CO N335900N335900
W1172200W1172200
BT Riverside County California
BT California
BT United States

Riverside County
Valid through 1988. Search in combination with state term. After 1988, use specific county-state term.

Riverside County California (1989)
S California. Before 1989, also search Riverside County AND California.
CO N332500N340500
W1142500W1174200
BT California
BT United States
NT Big Maria Mountains
NT Little Maria Mountains
NT Mission Creek Fault
NT Palm Springs California
NT Perris California
NT Riverside California
SA Crestmore
SA North Palm Springs earthquake 1986
SA Salton Sea
SA San Bernardino Mountains
SA San Jacinto Fault
SA San Jacinto Mountains
SA Santa Ana Mountains
SA Santa Ana River

Riyadh
No longer a valid term for GeoRef. As of 1993 see Riyadh Saudi Arabia.

Riyadh Saudi Arabia (1993)
CO N243800N243800
E0464300E0464300
BT Saudi Arabia
BT Arabian Peninsula
BT Asia

Rn
use radon

Rn-222 (1978)
Autoposting of broader terms began in 1989. This term has multiple hierarchies.
UF radon-222
UF thoron
BT1 radioactive isotopes
BT1 isotopes
BT2 radon
BT2 noble gases

road log (1978)
UF log, road
SA field trips
SA guidebook
SA maps

road materials
No longer a valid GeoRef index term. Before 1971, was used on level 1 in subfiles E and N. See construction materials; gravel deposits; aggregate; asphalt; cement materials.

Road River Formation (1989)
IN Index ages as applicable.
BT Paleozoic
SA Alaska
SA British Columbia
SA Lower Devonian
SA Northwest Territories
SA Ordovician
SA Silurian
SA Upper Cambrian
SA Yukon Territory

road tests (1981)
SA engineering geology
SA highways
SA testing

Roadless Area Review & Evaluation, Phase II
use RARE II regions

roads
Not a valid term for GeoRef. See highways.

Roan Supergroup (1978)
Interlayered with Carolina Gneiss in Spruce Pine District. N Georgia, W North Carolina, NW South Carolina, and E Tennessee.
BT Precambrian
SA Georgia
SA North Carolina
SA South Carolina
SA Tennessee

Roane County
Valid through 1988. Search in combination with state term. After 1988, use specific county-state term.

Roane County Tennessee (1989)
E Tennessee. Before 1989, also search Roane County AND Tennessee.
CO N354000N360300
W0841800W0844800
BT Tennessee
BT United States

Roane County West Virginia (1989)
W West Virginia. Before 1989, also search Roane County AND West Virginia.
CO N383300N385600
W0810500W0813300
BT West Virginia
BT United States

Roanoke
Valid through 1988. Search in combination with state term. After 1988, use specific city-state term.

Roanoke County Virginia (1989)
SW Virginia. Before 1989, search Roanoke County AND Virginia.
CO N370700N372500
W0795000W0801800
BT Virginia
BT United States
NT Roanoke Virginia

Roanoke Virginia (1989)

City on Roanoke River in but independent of Roanoke County, SW Virginia. Before 1989, search Roanoke AND Virginia.
CO N371500N371500
    W0795800W0795800
BT Roanoke County Virginia
BT Virginia
BT United States

Roaring Brook Valley (1978)
Valley in Hartford County and another in Tolland County.
IN Index counties as applicable.
SA Hartford County Connecticut
SA Tolland County Connecticut

robbery
use stream capture

Robertinacea (1978)
Superfamily. Autoposting of microfossils and Protista to this term began in 1990. This term has multiple hierarchies.
BT1 Rotaliina
BT1 foraminifera
BT1 Protista
BT1 Invertebrata
BT2 Rotaliina
BT2 foraminifera
BT2 Protista
BT2 microfossils

Roberts Mountain Allochthon
use Roberts Mountains Allochthon

Roberts Mountains (1978)
Eureka County in central Nevada.
IN Index county or region as applicable.
SA Eureka County Nevada

Roberts Mountains Allochthon (1989)
Central Nevada.
UF Roberts Mountain Allochthon
BT Eureka County Nevada
BT Nevada
BT United States

Roberts Mountains Formation (1978)
Underlies Rabbit Hill Formation. E Nevada, and W Utah.
BT Middle Silurian
BT Silurian
BT Paleozoic
SA Nevada
SA Utah

Robertson Bay Group (1989)
N Victoria Land. Upper Proterozoic to Ordovician.
IN Index ages as applicable.
SA Antarctica
SA Cambrian
SA Ordovician
SA upper Proterozoic
SA Victoria Land

Robles Formation (1978)
BT Cretaceous
BT Mesozoic
SA Puerto Rico
SA West Indies

Roca Formation (1978)
Underlies Sallyards Limestone Member of Grenola Limestone; overlies Howe Limestone Member of Red Eagle Limestone. NE Kansas, SE Nebraska, and N Oklahoma.
BT Wolfcampian
BT Lower Permian
BT Permian
BT Paleozoic
SA Council Grove Group
SA Kansas

SA Nebraska
SA Oklahoma
SA Red Eagle Limestone

Roccamonfina (1978)
Volcano in NW Campania, S Italy.
BT Campania Italy
BT Italy
BT Southern Europe
BT Europe

Rochechouart
No longer a valid term for GeoRef. See Rochechouart France.

Rochechouart France (1993)
Town in Haute-Vienne, W central France.
BT Haute-Vienne France
BT France
BT Western Europe
BT Europe

Rochester
Valid through 1988. Search in combination with state term. After 1988, use specific city-state term.

Rochester Formation (1978)
Includes Keefer Sandstone Member, Decew Waterlime Bed.
BT Middle Silurian
BT Silurian
BT Paleozoic
SA Clinton Group
SA Keefer Sandstone
SA Maryland
SA New York
SA Ontario
SA Pennsylvania
SA West Virginia

Rochester New York (1989)
City on S Lake Ontario in W New York. Before 1989, search Rochester AND New York.
CO N431200N431200
    W0773700W0773700
BT Monroe County New York
BT New York
BT United States

rock bursts (1978)
UF bursts, rock
UF rockbursts
SA geologic hazards
SA mining geology

Rock Creek Park (1978)
Also search Rock Creek.
IN Index state or region as applicable.
SA District of Columbia
SA Maryland
SA United States

rock creep
Not a valid GeoRef index term after 1970. Was used on level 1 in subfile N.
use creep

rock crystal
use quartz crystal

rock descriptions
No longer a valid GeoRef index term. Before 1971, was used on level 1 in subfiles E and N. See igneous rocks; metamorphic rocks; sedimentary rocks.

Rock Eval
use Rock-Eval

rock failure
use failures

rock glaciers (1978)
Autoposting of glaciers to this term began in 1989.

BT glaciers
SA boulders
SA frost action
SA glacial geology
SA gravity flows
SA ice
SA permafrost

Rock Island
Valid through 1988. Search in combination with state term. After 1988, use specific city-state term.

Rock Island County Illinois (1989)
NW Illinois. Before 1989, search Rock Island County AND Illinois.
CO N412100N414600
    W0901100W0910400
BT Illinois
BT United States
NT Rock Island Illinois

Rock Island Illinois (1989)
City on Mississippi River in NW Illinois. Before 1989, search Rock Island AND Illinois.
CO N413000N413000
    W0903400W0903400
BT Rock Island County Illinois
BT Illinois
BT United States

Rock Lake Shale Member (1978)
Of Stanton Limestone. E Kansas, NW Missouri, and SE Nebraska.
BT Missourian
BT Upper Pennsylvanian
BT Pennsylvanian
BT Carboniferous
BT Paleozoic
SA Kansas
SA Missouri
SA Nebraska
SA Stanton Formation

rock mechanics (1978)
Used for geotechnical studies.
SA acoustical emissions
SA asperities
SA Atterberg limits
SA boundary conditions
SA brittle materials
SA brittleness
SA case studies
SA cohesive materials
SA competent materials
SA compressive strength
SA congelifraction
SA constitutive equations
SA cracks
SA cryoturbation
SA dams
SA deformation
SA dispersivity
SA ductile deformation
SA ductility
SA durability
SA dynamic loading
SA earthquakes
SA elastic constants
SA elastic materials
SA elasticity
SA engineering geology
SA excavations
SA explosions
SA extensometers
SA failures
SA finite strain analysis
SA foundations
SA fractured materials
SA frost action
SA geologic hazards
SA geomembranes
SA geotechnical maps
SA geotextiles
SA highways
SA hydraulic fracturing

SA hysteresis
SA land subsidence
SA load tests
SA marine installations
SA materials
SA mechanics
SA microcracks
SA nuclear facilities
SA penetration
SA permeability
SA pipelines
SA plastic deformation
SA plastic materials
SA plasticity
SA polycrystalline materials
SA reinforced materials
SA reservoirs
SA retaining walls
SA rigidity
SA rupture
SA saturated materials
SA shear strength
SA shear tests
SA site exploration
SA slope stability
SA soil mechanics
SA strain
SA strength
SA stress
SA stressmeters
SA structural analysis
SA testing
SA triaxial tests
SA tunnels
SA underground installations
SA vibration
SA weak rocks
SA weathered materials

rock salt
use halite

rock slides
use rockslides

rock slip
use rockslides

Rock Springs
Valid through 1988. Search in combination with state term. After 1988, use specific city-state term.

Rock Springs Formation (1989)
In Mesaverde Group. Includes McCourt Tongue and Glades coal bed. NE Utah and SW Wyoming.
BT Upper Cretaceous
BT Cretaceous
BT Mesozoic
SA Hilliard Shale
SA Mesaverde Group
SA Utah
SA Wyoming

Rock Springs Uplift (1985)
SW Wyoming.
BT Wyoming
BT United States
SA Rock Springs Wyoming
SA Rocky Mountains
SA Sweetwater County Wyoming

Rock Springs Wyoming (1989)
City in SW Wyoming. Before 1989, search Rock Springs AND Wyoming.
CO N413500N413500
    W1091300W1091300
BT Sweetwater County Wyoming
BT Wyoming
BT United States
SA Rock Springs Uplift

Rock-Eval (1993)
Pyrolysis. Standard screening method in petroleum geochemistry. As of 1993, for discussion of instruments and analysis.

UF Rock Eval
SA chemical analysis
SA hydrocarbons
SA instruments
SA kerogen
SA petroleum
SA pyrolysis

rock-stratigraphy
use lithostratigraphy

rock-water interface (1985)
UF water-rock interaction
UF water-rock interface
SA halmyrolysis
SA interfaces
SA marine geology
SA sediment-water interface

Rockall Bank (1978)
About 225 miles W of the Hebrides.
BT Atlantic Ocean
SA DSDP Site 117

Rockall Plateau (1978)
Rise or submarine plateau just N of Rockall Bank W of the Hebrides.
UF Rockall Rise
BT Atlantic Ocean
SA DSDP Site 116
SA DSDP Site 117
SA DSDP Site 403
SA DSDP Site 404
SA DSDP Site 405
SA DSDP Site 406
SA DSDP Site 552
SA DSDP Site 553
SA DSDP Site 554
SA DSDP Site 555
SA Leg 81

Rockall Rise
use Rockall Plateau

Rockall Trench
use Rockall Trough

Rockall Trough (1978)
SE of Rockall Bank W of the Hebrides.
UF Rockall Trench
BT Atlantic Ocean
SA DSDP Site 610

Rockbridge County
Valid through 1988. Search in combination with state term. After 1988, use specific county-state term.

Rockbridge County Virginia (1989)
W Virginia. Before 1989, also search Rockbridge County AND Virginia.
CO N373300N380600
    W0791000W0794200
BT Virginia
BT United States

rockbridgeite (1978)
BT phosphates

rockbursts
use rock bursts

rockfalls (1978)
Autoposting of mass movements to this term began in 1989.
BT mass movements
SA geomorphology
SA slope stability

rockfill dams (1982)
BT dams

Rockhampton
No longer a valid term for GeoRef. See Rockhampton Australia.

Rockhampton Australia (1993)
City in E Queensland.
BT Queensland Australia
BT Australia
BT Australasia

Rockingham County
Valid through 1988. Search in combination with state term. After 1988, use specific county-state term.

Rockingham County New Hampshire (1989)
SE New Hampshire. Before 1989, also search Rockingham County AND New Hampshire.
CO N424300N431700
    W0704300W0712700
BT New Hampshire
BT United States
SA Great Bay

Rockingham County North Carolina (1989)
N North Carolina. Before 1989, also search Rockingham County AND North Carolina.
CO N361500N363300
    W0793100W0803300
BT North Carolina
BT United States

Rockingham County Virginia (1989)
NW Virginia. Before 1989, also search Rockingham County AND Virginia.
CO N381300N385100
    W0782900W0791400
BT Virginia
BT United States

Rockland County
Valid through 1988. Search in combination with state term. After 1988, use specific county-state term.

Rockland County New York (1989)
SE New York. Before 1989, also search Rockland County AND New York.
CO N410000N412000
    W0735400W0741500
BT New York
BT United States

Rocknest Formation (1989)
Lower Proterozoic of Northwest Territories. May also occur in Western Australia.
IN Index age as applicable.
SA Australia
SA lower Proterozoic
SA Northwest Territories
SA Western Australia

rocks (1978)
SA aggregate
SA bedrock
SA cap rocks
SA carbonate rocks
SA chemically precipitated rocks
SA country rocks
SA crystalline rocks
SA extrusive rocks
SA fluid inclusions
SA host rocks
SA hypabyssal rocks
SA inclusions
SA metaigneous rocks
SA metamorphic rocks
SA metaplutonic rocks
SA metasedimentary rocks
SA metasomatic rocks
SA metavolcanic rocks
SA paragenesis
SA plutonic rocks
SA reservoir rocks
SA sedimentary rocks
SA silicate rocks
SA soft rocks
SA source rocks
SA standard rocks
SA tight sands
SA volcanic rocks
SA wall rocks
SA weak rocks

rocks and ores
No longer a valid GeoRef index term. Before 1972, was used on level 2 in subfile G. See metal ores; rocks; igneous rocks; metamorphic rocks; sedimentary rocks; crystalline rocks; mineral resources.

rockslides (1978)
Autoposting of mass movements to this term began in 1989.
UF rock slides
UF rock slip
BT mass movements
SA geomorphology
SA landslides

Rockwood Formation (1978)
Lower and Middle Silurian. Underlies Hancock Limestone; overlies Sequatchie Formation. E Tennessee and NW Georgia.
BT Silurian
BT Paleozoic
SA Georgia
SA Tennessee

Rocky Mountain Arsenal (1985)
NE central Colorado. Autoposting of broader terms to this term began in 1989.
CO N394000N395000
    W1044500W1045500
BT Adams County Colorado
BT Colorado
BT United States

Rocky Mountain foreland
use Rocky Mountains foreland

Rocky Mountain National Park (1989)
N Colorado, in Boulder, Grand and Larimer counties.
IN Index counties as applicable.
CO N400900N403000
    W1053000W1055500
BT Colorado
BT United States
SA Boulder County Colorado
SA Front Range
SA Grand County Colorado
SA Larimer County Colorado
SA national parks

Rocky Mountain Trench (1978)
A trough in the Northern Rockies, E of the interior plateau of British Columbia, between a series of parallel ranges extending WNW from Montana to the headwaters of the Yukon River.
IN Index British Columbia, Washington, and Yukon Territory as applicable.
BT North America
SA British Columbia
SA Montana
SA Northern Rocky Mountains
SA Rocky Mountains
SA Yukon Territory

Rocky Mountains (1978)
Mountain system in W North America extending from N Alaska to the Mexican frontier. Autoposting of this term began in 1978.
IN Index Alaska, and countries as applicable.
BT North America
NT Canadian Rocky Mountains
NT Central Rocky Mountains
NT Northern Rocky Mountains
NT Southern Rocky Mountains
NT U. S. Rocky Mountains
SA Alaska
SA Big Belt Mountains
SA Bitterroot Range
SA Bridger Range
SA Brooks Range
SA Canada
SA Canadian Cordillera
SA Cariboo Mountains
SA Colorado Lineament
SA Disturbed Belt
SA Franklin Mountains
SA Mackenzie Mountains
SA Madison Range
SA North American Cordillera
SA Rock Springs Uplift
SA Rocky Mountain Trench
SA Rocky Mountains foreland
SA Seminoe Mountains
SA Stikinia Terrane
SA United States
SA Western Overthrust Belt
SA Yellowstone River

Rocky Mountains foreland (1981)
UF Rocky Mountain foreland
BT North America
SA Rocky Mountains

Rocroi
No longer a valid term for GeoRef. See Rocroi France.

Rocroi France (1993)
Town in N France. Before 1993, also search Rocroi. Before 1978, also search Rocroy.
CO N495600N495600
    E0043100E0043100
UF Rocroy France
BT Ardennes France
BT France
BT Western Europe
BT Europe
SA Ardennes

Rocroy France
use Rocroi France

Rodentia (1978)
Order. Autoposting of Eutheria and Theria to this term began in 1989. Before 1981, also search rodents.
BT Eutheria
BT Theria
BT Mammalia
BT Tetrapoda
BT Vertebrata
BT Chordata
NT Castoridae
NT Hystricomorpha
NT Myomorpha
NT Sciuromorpha
SA rodents

rodents (1981)
Common name for Rodentia. Autoposting of mammals to this term began in 1989.
BT mammals
BT tetrapods
BT vertebrates
SA biostratigraphy
SA Rodentia

Rodessa Formation (1989)
In Trinity Group. This term has multiple hierarchies.
BT1 Comanchean
BT1 Cretaceous
BT1 Mesozoic
BT2 Lower Cretaceous
BT2 Cretaceous
BT2 Mesozoic
SA Arkansas
SA Louisiana
SA Mississippi
SA Texas
SA Trinity Group

Rodez Pass
 use Rodez Trough

Rodez Trough (1978)
Between Segala Plateau in S and Causse du Comtal in N. Also search Rodez.
IN Index departments as applicable.
UF Rodez Pass
BT France
BT Western Europe
BT Europe
SA Aveyron France
SA Haute-Garonne France
SA Tarn France

rodingite (1978)
BT gabbros
BT plutonic rocks
BT igneous rocks

Rodna
No longer a valid term for GeoRef. As of 1993 see Rodna Romania.

Rodna Mountains (1978)
Range of the Carpathians along border between Maramures and Cluj Counties in N Transylvania.
UF Rodnei Mountains
BT Transylvania
BT Romania
BT Southern Europe
BT Europe
SA Carpathians
SA Transylvanian Alps

Rodna Romania (1993)
Village in NE Cluj County in NE Transylvania. Before 1993 also search Rodna AND Romania.
BT Transylvania
BT Romania
BT Southern Europe
BT Europe

Rodnei Mountains
 use Rodna Mountains

Rodrigues Island
 use Rodriguez Island

Rodriguez Island (1978)
A dependency of Mauritius in the Mascarene Islands about 500 miles E of the Malagasy Republic. Mascarene Islands is considered a broader term as of 1981. In 1985, broader term changed from Indian Ocean to Indian Ocean Islands.
UF Rodrigues Island
BT Mascarene Islands
BT Indian Ocean Islands

roedderite (1978)
Meteorite mineral.
BT ring silicates
BT silicates
SA meteorites

roemerite (1978)
BT sulfates

roestone
 use oolite

Rogaland
No longer a valid term for GeoRef. As of 1993 see Rogaland Norway.

Rogaland Norway (1993)
County in SW Norway. Before 1993 also search Rogaland AND Norway.
BT Norway
BT Scandinavia
BT Western Europe
BT Europe
NT Sandnes Norway
NT Sogndal Norway
NT Stavanger Norway

roggianite (1978)
Autoposting of sorosilicates began in 1985.
BT sorosilicates
BT orthosilicates
BT silicates

Rogue River (1978)
SW Oregon. Also search Rogue.
BT Oregon
BT United States

Rokko Mountains (1978)
NE of Kobe in Hyogo Prefecture, central Honshu.
SA Honshu
SA Hyogo Japan

roll orebodies
 use roll-type

roll-type (1993)
From 1978-1992, also search roll-type deposits.
UF ore rolls
UF roll orebodies
UF roll-type deposits
SA mineral deposits, genesis
SA uranium ores

roll-type deposits
No longer a valid term for GeoRef.
 use roll-type

Roma Italy
 use Rome Italy

Romagna (1978)
Historical region in N central Italy on the Adriatic Sea.
BT Emilia-Romagna Italy
BT Italy
BT Southern Europe
BT Europe

Romanche Deep
 use Romanche Trench

Romanche fracture zone (1978)
Cuts across the Mid-Atlantic Ridge between the bulge of South America and Liberia.
BT Atlantic Ocean

Romanche Gap
 use Romanche Trench

Romanche Trench (1978)
Extends in a NW-SE direction to the SW of the Romanche fracture zone between bulge of South America and Liberia.
UF Romanche Deep
UF Romanche Gap
BT Atlantic Ocean

Romania (1978)
CO N434000N481000 E0294500E0201500
UF Roumania
UF Rumania
BT Southern Europe
BT Europe
NT Altin-Tepe
NT Apuseni Mountains
NT Arad Romania
NT Arges River
NT Babadag Lake basin
NT Bacau Romania
NT Bicaz Valley
NT Bistrita Mountains
NT Bistrita Valley
NT Bucegi Mountains
NT Constanta Romania
NT Dacian Basin
NT Delinesti Romania
NT Fagaras Mountains
NT Getic Nappe
NT Gutii Mountains
NT Haghimas Syncline
NT Iacobeni Romania
NT Jiu River valley
NT Mehadia Romania
NT Mehedinti Plateau
NT Moldava Valley
NT Moldavian Platform
NT Moldova Noua Romania
NT Olt River
NT Olt River valley
NT Paring Mountains
NT Piatra-Neamt Romania
NT Rarau Massif
NT Resita Romania
NT Romanian Dobruja
NT Romanian Pannonian Basin
NT Romanian Plain
NT Rusca Montana Romania
NT Sasca-Montana Romania
NT Semenic Mountains
NT Severin
NT Suceava Romania
NT Sulina Romania
NT Svinita Romania
NT Teleajen Valley
NT Transylvania
NT Transylvanian Alps
NT Trotus Valley
NT Tulcea Romania
NT Vrancea
NT Vulcan Mountains
NT Walachia
SA Balkan Peninsula
SA Banat
SA Black Sea region
SA Bukovina
SA Carpathian Foredeep
SA Carpathian Foreland
SA Carpathians
SA Crai
SA Danube Delta
SA Danube Plain
SA Danube River
SA Danube Valley
SA Dobruja Basin
SA Eastern Carpathians
SA Moesia
SA Moesian Platform
SA Moldavia
SA Pannonian Basin
SA Prut River
SA Siret River
SA Somes Basin
SA Subcarpathians
SA Tisza River
SA Tulghes Series
SA Vrancea earthquake 1977

Romania Plain
 use Romanian Plain

Romanian (1978)
Provincial series, Europe.
BT upper Pliocene
BT Pliocene
BT Neogene
BT Tertiary
BT Cenozoic

Romanian Dobruja (1982)
Region on the Black Sea. Autoposting of Romania to this term began in 1989. Before 1981, search Dobruja AND Romania. As of 1990, Southern Europe and Europe are autoposted to this term. Before 1993, also search Dobrogea. This term has multiple hierarchies.
UF Dobrogea
BT1 Dobruja Basin
BT1 Southern Europe
BT1 Europe
BT2 Romania
BT2 Southern Europe
BT2 Europe
SA Altin-Tepe
SA Babadag Lake basin
SA Constanta Romania
SA Mangalia Romania
SA Sulina Romania
SA Tulcea Romania

Romanian Pannonian Basin (1982)
Autoposting of Romania to this term began in 1989. As of 1990, Southern Europe and Europe are autoposted to this term. This term has multiple hierarchies.
BT1 Pannonian Basin
BT1 Europe
BT2 Romania
BT2 Southern Europe
BT2 Europe

Romanian Plain (1978)
Area between the Transylvanian Alps on the N, and the Danube River on the S and E. Also search Romanian.
IN Index regions as applicable.
UF Romania Plain
BT Romania
BT Southern Europe
BT Europe
SA Moldavia
SA Siret River
SA Walachia

Romashkino Field (1978)
Orenburg Oblast in Southern Urals. Before 1978, also search Romashkino; Romashkino oil field.
IN Index regions as applicable.
SA oil and gas fields
SA Orenburg Russian Federation

Rome
No longer a valid term for GeoRef. See Rome Italy.

Rome Formation (1978)
Named for exposures south of Rome, Georgia.
BT Lower Cambrian
BT Cambrian
BT Paleozoic
SA Alabama
SA Georgia
SA North Carolina
SA Tennessee
SA Virginia

Rome Italy (1993)
Province including the city on the Tiber River. Before 1993, search Rome and Italy. Before 1978, also search Roma.
UF Roma Italy
BT Latium Italy
BT Italy
BT Southern Europe
BT Europe

Rome Trough (1985)

S Appalachians from Georgia to West Virginia.
IN Index states or region as applicable.
SA Appalachian Basin
SA Georgia
SA Kentucky
SA Southern Appalachians
SA Tennessee
SA United States
SA West Virginia

Ronda Massif
use Serrania de Ronda

Ronda Sierra
use Serrania de Ronda

Ronda, Serrania de
use Serrania de Ronda

Rondonia
No longer a valid term for GeoRef. See Rondonia Brazil.

Rondonia Brazil (1993)
State in W Brazil on Bolivian border. Before 1993, search Rondonia or Guapore.
CO S135000S075500
    W0600000W0664500
UF Guapore Brazil
BT Brazil
BT South America

Rondout Formation (1978)
Upper Silurian and Lower Devonian. New York and Pennsylvania. Includes Fuyk Sandstone Member.
IN Index ages as applicable.
BT Paleozoic
SA Lower Devonian
SA New York
SA Pennsylvania
SA Upper Silurian

Ronne Ice Shelf (1989)
Adjacent to Weddell Sea and Filchner Ice Shelf.
BT Antarctica
SA Filchner Ice Shelf
SA Weddell Sea

roof control (1985)
SA mines
SA mining geology
SA underground mining

Rooks County
Valid through 1988. Search in combination with state term. After 1988, use specific county-state term.

Rooks County Kansas (1989)
NW central Kansas. Before 1989, also search Rooks County AND Kansas.
CO N391000N393500
    W0990400W0993700
BT Kansas
BT United States

Roorkee
No longer a valid term for GeoRef. See Roorkee India.

Roorkee India (1993)
Town in N Uttar Pradesh. Before 1993, also search Roorkee. Before 1978, also search Rurki.
UF Rurki India
BT Uttar Pradesh India
BT India
BT Indian Peninsula
BT Asia

Roosevelt County
Valid through 1988. Search in combination with state term. After 1988, use specific county-state term.

Roosevelt County Montana (1989)
NE Montana. Before 1989, also search Roosevelt County AND Montana.
CO N480000N483300
    W1040200W1055500
BT Montana
BT United States

Roosevelt County New Mexico (1989)
E New Mexico. Before 1989, also search Roosevelt County AND New Mexico.
CO N333500N343700
    W1030300W1035700
BT New Mexico
BT United States

Roosevelt Hot Springs
use Roosevelt Hot Springs KGRA

Roosevelt Hot Springs KGRA (1981)
Known Geothermal Resources Area. SW Utah, in Beaver County. Before 1981, also search Roosevelt Hot Springs.
CO N382500N383300
    W1125000W1130000
UF KGRA, Roosevelt Hot Springs
UF Roosevelt Hot Springs
BT Beaver County Utah
BT Utah
BT United States

root clay
use underclay

Root River (1989)
River in Fillmore and Houston counties, extreme SE Minnesota. Also in Waukesha and Racine counties, extreme SE Wisconsin. Also in SW Mackenzie District, Northwest Territories.
IN Index states and counties; or province as applicable.
SA Fillmore County Minnesota
SA Houston County Minnesota
SA Mackenzie District Northwest Territories
SA Minnesota
SA Northwest Territories
SA Racine County Wisconsin
SA Waukesha County Wisconsin
SA Wisconsin

roots (1985)
SA flowers
SA fossil wood
SA fossils
SA paleobotany
SA Plantae
SA plants
SA vegetation
SA wood

roquesite (1978)
BT sulfides

Roraima
No longer a valid term for GeoRef. See Roraima Brazil.

Roraima Brazil (1993)
Territory in N Brazil.
CO S013000N051000
    W0585000W0644500
BT Brazil
BT South America

Roraima Formation (1978)
Tentatively dated from lower Proterozoic to as young as Miocene. In the Pacaraima Mountains.
SA Brazil
SA Guyana
SA South America
SA Venezuela

Ros River valley (1978)
W central Ukraine. Also search Ros; Ros River. This term has multiple hierarchies.
BT1 Ukraine
BT1 Europe
BT2 Ukraine
BT2 Commonwealth of Independent States

Rosario Formation (1978)
BT Upper Cretaceous
BT Cretaceous
BT Mesozoic
SA Baja California
SA Campanian
SA Maestrichtian
SA Mexico

Roscommon County
Valid through 1988. Search in combination with state term. After 1988, use specific county-state term.

Roscommon County Michigan (1989)
N central Michigan. Before 1989, also search Roscommon County AND Michigan.
CO N441000N443000
    W0842300W0845200
BT Michigan Lower Peninsula
BT Michigan
BT United States

ROSE
use Rivera Ocean Seismic Experiment

Rose Canyon Formation (1978)
Lower and middle Eocene. Overlies Torrey Sand; underlies Poway Conglomerate. S California. Type section in San Diego County.
BT Eocene
BT Paleogene
BT Tertiary
BT Cenozoic
SA California

Rose Hill Formation (1985)
In Clinton Group. Maryland, Pennsylvania and N Virginia.
BT Middle Silurian
BT Silurian
BT Paleozoic
SA Clinton Group
SA Maryland
SA Pennsylvania
SA Virginia

Rosebery
No longer a valid term for GeoRef. See Rosebery Australia.

Rosebery Australia (1993)
Village in W Tasmania.
BT Tasmania Australia
BT Australia
BT Australasia

Rosebud County
Valid through 1988. Search in combination with state term. After 1988, use specific county-state term.

Rosebud County Montana (1989)
E central Montana. Before 1989, also search Rosebud County AND Montana.
CO N451000N465200
    W1060700W1075500
BT Montana
BT United States

Rosetown
No longer a valid term for GeoRef. As of 1993, see Rosetown Saskatchewan.

Rosetown Saskatchewan (1993)
Town in SW Saskatchewan. Before 1993, also search Rosetown AND Saskatchewan.
BT Saskatchewan
BT Western Canada
BT Canada

Rosia
No longer a valid term for GeoRef. As of 1993 see Rosia Romania.

Rosia Romania (1993)
Village 25 miles SE of Oradea in Bihor County in W Transylvania. Before 1993 also search Rosia AND Romania.
BT Bihor Romania
BT Transylvania
BT Romania
BT Southern Europe
BT Europe

Rosidae (1981)
BT Dicotyledoneae
BT angiosperms
BT Spermatophyta
BT Plantae

Ross Barrier
use Ross Ice Shelf

Ross Dependency (1978)
Section lying S of 60° S and between 160° E and 150° W extending to the South Pole. Placed under jurisdiction of New Zealand in 1923 by act of British Parliament. Also search Ross.
BT Antarctica
SA Beacon Supergroup
SA Beardmore Glacier
SA Coalsack Bluff
SA Fremouw Formation
SA Lewis Cliff Meteorites
SA Wright Valley

Ross Formation (1993)
Lower Devonian and Middle Devonian. W Tennessee and Mississippi. Also cited in Antarctica, Tasmania (Ross Sandstone, Triassic), and New Zealand.
IN Index ages and locations as applicable.
SA Antarctica
SA Devonian
SA Lower Devonian
SA Middle Devonian
SA Mississippi
SA New Zealand
SA Tasmania Australia
SA Tennessee
SA Triassic

Ross Ice Shelf (1978)
Covers S part of Ross Sea between Marie Byrd Land and Victoria Land with its S end at the foot of Queen Maud Range on the edge of the Antarctic continent. Also search Ross.
UF Ross Barrier
BT Antarctica
SA Beacon Supergroup
SA Beardmore Glacier

SA Coalsack Bluff
SA Fremouw Formation
SA Lewis Cliff Meteorites
SA McMurdo Ice Shelf
SA Queen Maud Range
SA Ross Sea
SA Shackleton Glacier

Ross Island (1978)
In Ross Sea at W end of Ross Ice Shelf separated from Victoria Land by McMurdo Sound. Also search Ross.
BT Antarctica
SA Hut Point Peninsula
SA Mount Erebus

Ross Sea (1978)
Arm of S Pacific Ocean just N of Ross Ice Shelf between Victoria Land and Edward VII Peninsula. Also search Ross.
CO S783000S713000
    W1580000E1620000
BT Antarctic Ocean
SA DSDP Site 270
SA DSDP Site 271
SA DSDP Site 272
SA DSDP Site 273
SA DSDP Site 274
SA Pacific Ocean
SA Ross Ice Shelf
SA Victoria Land Basin

Ross-shire
No longer a valid term for GeoRef as of 1993.
use Ross-shire Scotland

Ross-shire Scotland (1993)
A separate county until the 17th century but became part of Ross and Cromarty County in N Scotland. In 1974, became part of Highland region.
CO N573000N580000
    W0043000W0053000
UF Ross-shire
BT Highland region Scotland
BT Scotland
BT Great Britain
BT United Kingdom
BT Western Europe
BT Europe

Rosses Granite (1978)
IN Index regions as applicable.
SA Donegal Ireland
SA Ireland

Rossland
No longer a valid term for GeoRef. See Rossland British Columbia.

Rossland British Columbia (1993)
City in SE British Columbia.
BT British Columbia
BT Western Canada
BT Canada

Rostock Bezirk
No longer a valid term for GeoRef. District in N East Germany. See Mecklenburg-Western Pomerania Germany.

Rostov
No longer a valid term for GeoRef. As of 1993, see Rostov Russian Federation.

Rostov Russian Federation (1993)
Oblast in Black Sea region, Russian Federation. This term has multiple hierarchies.
BT1 Russian Federation
BT1 Commonwealth of Independent States
BT2 Europe
NT Azov Russian Federation
NT Rostov-na-Donu Russian Federation

Rostov-na-Donu
As of 1993, no longer a valid term for GeoRef. Valid 1981-1992.
use Rostov-na-Donu Russian Federation

Rostov-na-Donu Russian Federation (1993)
City in Rostov (oblast). Before 1993, also search Rostov-na-Donu. This term has multiple hierarchies.
UF Rostov-na-Donu
BT1 Rostov Russian Federation
BT1 Russian Federation
BT1 Commonwealth of Independent States
BT2 Rostov Russian Federation
BT2 Europe
SA Don River

Rostroconchia (1978)
BT Mollusca
BT Invertebrata

Rotalia (1978)
Genus. Autoposting of microfossils and Protista to this term began in 1990. This term has multiple hierarchies.
BT1 Rotaliacea
BT1 Rotaliina
BT1 foraminifera
BT1 Protista
BT1 Invertebrata
BT2 Rotaliacea
BT2 Rotaliina
BT2 foraminifera
BT2 Protista
BT2 microfossils

Rotaliacea (1978)
Autoposting of microfossils and Protista to this term began in 1990. This term has multiple hierarchies.
BT1 Rotaliina
BT1 foraminifera
BT1 Protista
BT1 Invertebrata
BT2 Rotaliina
BT2 foraminifera
BT2 Protista
BT2 microfossils
NT Ammonia
NT Elphidium
NT Heterostegina
NT Miogypsinidae
NT Nummulitidae
NT Rotalia

Rotaliina (1978)
Autoposting of this term to Carterinacea began in 1981. Autoposting of microfossils and Protista to this term began in 1990. This term has multiple hierarchies.
BT1 foraminifera
BT1 Protista
BT1 Invertebrata
BT2 foraminifera
BT2 Protista
BT2 microfossils
NT Buliminacea
NT Carterinacea
NT Cassidulinacea
NT Discorbacea
NT Globigerinacea
NT Lagenidae
NT Nodosariacea
NT Orbitoidacea
NT Robertinacea
NT Rotaliacea
NT Spirillinacea

Rotalipora (1978)
Genus. Autoposting of microfossils and Protista to this term began in 1990. This term has multiple hierarchies.
BT1 Globigerinacea
BT1 Rotaliina
BT1 foraminifera
BT1 Protista
BT1 Invertebrata
BT2 Globigerinacea
BT2 Rotaliina
BT2 foraminifera
BT2 Protista
BT2 microfossils

rotation (1978)
See plate rotation.
SA Earth
SA length of day
SA motions
SA plate rotation

rotational wave
use S-waves

Rotliegendes (1978)
Europe. Below Zechstein.
BT Permian
BT Paleozoic
SA Autunian
SA Lower Permian
SA Middle Permian
SA Saxonian

Rottleberode
No longer a valid term for GeoRef. See Rottleberode Germany.

Rottleberode Germany (1993)
Village in Saxony-Anhalt, E Germany.
BT Saxony-Anhalt Germany
BT Germany
BT Central Europe
BT Europe

Rouen
No longer a valid term for GeoRef. See Rouen France.

Rouen France (1993)
City on the Seine River in S Seine-Maritime, N France.
BT Seine-Maritime France
BT France
BT Western Europe
BT Europe

Rouergue (1978)
Ancient province. Primarily in Aveyron Department in S central France.
IN Index departments as applicable.
BT France
BT Western Europe
BT Europe
SA Aveyron France
SA Tarn-et-Garonne France

Rough Creek fault zone (1978)
W Kentucky.
BT Kentucky
BT United States

roughness (1978)
SA hydraulics
SA permeability
SA reservoir rocks

Roumania
use Romania

Round Mountain (1978)
On E spur of Great Dividing Range in NE New South Wales.
IN Index state or region as applicable.
SA New South Wales Australia

rounding (1978)
SA particles
SA pebbles
SA roundness
SA sand
SA sedimentary rocks
SA sorting

roundness (1978)
SA particles
SA rounding
SA sediments
SA shape analysis
SA sphericity

Roussillon (1978)
Region and former province roughly coextensive with Pyrenees-Orientales Department on the Spanish border.
BT France
BT Western Europe
BT Europe
SA Pyrenees-Orientales France

Routt County
Valid through 1988. Search in combination with state term. After 1988, use specific county-state term.

Routt County Colorado (1989)
NW Colorado. Before 1989, also search Routt County AND Colorado.
CO N395000N410000
    W1063500W1073000
BT Colorado
BT United States
SA Steamboat Springs
SA Yampa River

Rouyn
No longer a valid term for GeoRef. As of 1993, see Rouyn Quebec.

Rouyn Quebec (1993)
Mining city in SW Quebec near N Ontario border. Before 1993, also search Rouyn AND Quebec.
BT Temiscamingue County Quebec
BT Quebec
BT Eastern Canada
BT Canada

Rowan County
Valid through 1988. Search in combination with state term. After 1988, use specific county-state term.

Rowan County Kentucky (1989)
NE Kentucky. Before 1989, also search Rowan County AND Kentucky.
CO N380300N382300
    W0831100W0834000
BT Kentucky
BT United States

Rowan County North Carolina (1989)
W central North Carolina. Before 1989, also search Rowan County AND North Carolina.
CO N353000N355200
    W0801000W0804700
BT North Carolina
BT United States

Rowne
use Zloty Stok Poland

Roxana Silt (1989)
 SW Illinois and W Kentucky.
 BT Pleistocene
 BT Quaternary
 BT Cenozoic
 SA Illinois
 SA Kentucky

Roxbury Conglomerate (1989)
 In Boston Bay Group. E Massachusetts. Previously thought to be Devonian or Carboniferous in age. Includes Brookline, Squantum and Dorchester members.
 IN Index ages as applicable.
 SA Boston Bay Group
 SA Lower Cambrian
 SA Massachusetts
 SA upper Proterozoic

Roy Group (1989)
 BT Archean
 BT Precambrian
 SA Quebec

Royal Creek (1978)
 BT Yukon Territory
 BT Western Canada
 BT Canada

Rozdol
 No longer a valid term for GeoRef. As of 1993, see Rozdol Ukraine.

Rozdol Ukraine (1993)
 Town in Lvov Oblast in E Ukraine. This term has multiple hierarchies.
 IN Index oblast as applicable.
 BT1 Ukraine
 BT1 Europe
 BT2 Ukraine
 BT2 Commonwealth of Independent States
 SA Lvov Ukraine

rozenite (1978)
 BT sulfates

Roztocze (1978)
 Mountain range.
 IN Index Poland and/or Ukraine as applicable.
 UF Tomaszow-Lvov Ridge
 BT Europe
 SA Lvov Ukraine
 SA Poland
 SA Ukraine

Ru
 use ruthenium

Ru-106 (1985)
 Autoposting of broader terms began in 1989. This term has multiple hierarchies.
 BT1 radioactive isotopes
 BT1 isotopes
 BT2 ruthenium
 BT2 platinum group
 BT2 metals

Ruanda
 use Rwanda

Ruapehu (1978)
 Volcano in Tongariro National Park in S central North Island.
 IN Also index North Island.
 UF Ruapehu Volcano
 BT New Zealand
 BT Australasia
 SA North Island

Ruapehu Volcano
 use Ruapehu

rubblerock
 use breccia

rubidium (1978)
 Autoposting of alkali metals and metals to this term began in 1989.
 UF Rb
 BT alkali metals
 BT metals
 NT Rb-87/Sr-86
 SA isotopes
 SA K/Rb
 SA lithophile elements
 SA Rb/Sr

rubidium-strontium
 Not a valid term for GeoRef. Before 1971, was used in subfiles E, G, N and B.
 use Rb/Sr

rubies (1982)
 Before 1982, also search ruby. Before 1993, included use when of economic value as well as for mineralogical studies. As of 1993, use corundum for the mineral.
 IN Indes with gems for the commodity. See List C.
 SA corundum
 SA gems
 SA sapphire

Ruby Mountains (1978)
 Range in Elko and White Pine counties in NE Nevada. Also search Ruby Range AND Nevada.
 IN Index counties as applicable.
 SA Elko County Nevada
 SA Nevada
 SA Ruby Range
 SA White Pine County Nevada

Ruby Range (1978)
 N extension of Snowcrest Mountains in SW Montana, lies just W of Ruby River.
 IN Index counties or region as applicable.
 SA Montana
 SA Ruby Mountains

Ruda
 No longer a valid term for GeoRef. As of 1993 see Ruda Poland.

Ruda Poland (1993)
 Town in coal mining region in N central Katowice, S Poland. Before 1993 also search Ruda AND Poland.
 BT Katowice Poland
 BT Poland
 BT Central Europe
 BT Europe

Rudistae (1978)
 Before 1981, also search rudists.
 BT Bivalvia
 BT Mollusca
 BT Invertebrata
 SA rudists

rudists (1978)
 As of 1981, restricted to use as the common name for Rudistae. Autoposting of bivalves and mollusks to this term began in 1989.
 BT bivalves
 BT mollusks
 BT invertebrates
 SA biostratigraphy
 SA Rudistae

Rudki
 No longer a valid term for GeoRef. As of 1993, see Rudki Ukraine.

Rudki Ukraine (1993)
 Town in Lvov Oblast in E Ukraine. This term has multiple hierarchies.
 IN Index oblast as applicable.
 BT1 Ukraine
 BT1 Europe
 BT2 Ukraine
 BT2 Commonwealth of Independent States
 SA Lvov Ukraine

Rudnichny
 No longer a valid term for GeoRef. As of 1993, see Rudnichny Russian Federation.

Rudnichny Russian Federation (1993)
 Town in W Chelyabinsk Oblast in Southern Urals of W Siberia. Before 1993, also search Rudnichny or Rudnichnyy.
 BT Chelyabinsk Russian Federation
 BT Russian Federation
 BT Commonwealth of Independent States

Rudny Altai (1978)
 Region in the Altai Mountains in Vostochno Kazakhstan Oblast and in Altai Kray. This term has multiple hierarchies.
 IN Index former Soviet republics as applicable.
 CO N480000N520000 E0873000E0823000
 UF Rudnyy Altai
 UF Rudnyy Altay
 BT1 Commonwealth of Independent States
 BT2 Asia
 SA Altai Mountains
 SA Altai Russian Federation
 SA Eastern Kazakhstan
 SA Kazakhstan
 SA Russian Federation

Rudnyy Altai
 use Rudny Altai

Rudnyy Altay
 use Rudny Altai

Rudolf (Lake)
 use Lake Turkana

Ruegen (Island)
 use Rugen Island

Rugen Island (1978)
 In the Baltic Sea just off Mecklenburg-Western Pomerania Germany. Largest island of Germany. Also search Rugen.
 UF Ruegen (Island)
 BT Mecklenburg-Western Pomerania Germany
 BT Germany
 BT Central Europe
 BT Europe

Rugosa (1978)
 Order. Autoposting of Zoantharia to this term began in 1989. Before 1989, also search Tetracorallia. Also search tetracorals.
 UF Tetracorallia
 BT Zoantharia
 BT Anthozoa
 BT Coelenterata
 BT Invertebrata
 SA Scleractinia
 SA Tabulata

Ruhla
 No longer a valid term for GeoRef. See Ruhla Germany.

Ruhla Germany (1993)
 Town in Thuringian Forest in Thuringia, central Germany.
 BT Thuringia Germany
 BT Germany
 BT Central Europe
 BT Europe
 SA Thuringian Forest

Ruhr (1978)
 Major coal-mining and industrial region. Includes the Ruhr River valley and the Dusseldorf area to the S along the Rhine River.
 UF Ruhr Basin
 UF Ruhr Valley
 BT North Rhine-Westphalia Germany
 BT Germany
 BT Central Europe
 BT Europe

Ruhr Basin
 use Ruhr

Ruhr Valley
 use Ruhr

Rum
 use Rhum

Rumania
 use Romania

Rumilly
 No longer a valid term for GeoRef. See Rumilly France.

Rumilly France (1993)
 Town in Haute-Albanais, E France. Before 1993, search Rumilly. Before 1978, also search Rumilly-Albanais.
 UF Rumilly-Albanais France
 BT Haute-Savoie France
 BT France
 BT Western Europe
 BT Europe

Rumilly-Albanais France
 use Rumilly France

Ruminantia (1978)
 Suborder. Autoposting of this term to Tylopoda began in 1989. Autoposting of Eutheria and Theria to this term began in 1989.
 BT Artiodactyla
 BT Eutheria
 BT Theria
 BT Mammalia
 BT Tetrapoda
 BT Vertebrata
 BT Chordata
 NT Bovidae
 NT Cervidae
 NT Tylopoda

Rundle Group (1993)
 Lower Mississippian and Upper Mississippian. Alberta, British Columbia, Yukon Territory, and Montana. Rundle Formation and Rundle oil shale have also been used for Eocene oil shale deposit in Queensland.
 IN Index ages as applicable.
 SA Alberta
 SA Australia
 SA British Columbia
 SA Eocene
 SA Lower Mississippian
 SA Mississippian
 SA Montana
 SA Queensland Australia
 SA Upper Mississippian
 SA Yukon Territory

Runnels County
 Valid through 1988. Search in combination with state term. After 1988, use specific county-state term.

Runnels County Texas (1989)
 W central Texas. Before 1989, also search Runnels County AND Texas.
 CO N313000N320700

W0994400W1001500
BT Texas
BT United States

runoff (1978)
Before 1993, also search overland flow if applicable.
SA discharge
SA hydrology
SA precipitation
SA retention
SA rivers
SA rivers and streams
SA sediment yield
SA storms
SA streamflow
SA streams
SA surface water
SA Universal Soil Loss Equation
SA water balance
SA water yield
SA watersheds

runout
 use water yield

Rupelian (1978)
Europe. Above Tongrian, below Chattian. Autoposting of lower Oligocene to this term began in 1989.
BT lower Oligocene
BT Oligocene
BT Paleogene
BT Tertiary
BT Cenozoic
SA Krosno Beds
SA Stampian

rupture (1982)
SA deformation
SA failures
SA faults
SA fractures
SA rock mechanics
SA stress
SA tensile strength

Rurki India
 use Roorkee India

Rusca Montana
 No longer a valid term for GeoRef. See Rusca Montana Romania.

Rusca Montana Romania (1993)
Village in the Poiana-Rusca Mountains in Caras-Severin County in E Romania. Before 1993, also search Rusca Montana. Before 1978, also search Rusca-Montana. This term has multiple hierarchies.
UF Rusca-Montana Romania
BT1 Banat
BT1 Southern Europe
BT1 Europe
BT2 Romania
BT2 Southern Europe
BT2 Europe

Rusca-Montana Romania
 use Rusca Montana Romania

Ruse
 No longer a valid term for GeoRef. See Ruse Bulgaria.

Ruse Bulgaria (1993)
Province and city in N Bulgaria.
BT Bulgaria
BT Southern Europe
BT Europe

Rush County
 Valid through 1988. Search in combination with state term. After 1988, use specific county-state term.

Rush County Indiana (1989)
E central Indiana. Before 1989, also search Rush County AND Indiana.
CO N392700N394800
  W0851800W0853800
BT Indiana
BT United States

Rush County Kansas (1989)
Central Kansas. Before 1989, also search Rush County AND Kansas.
CO N382500N384000
  W0990300W0993500
BT Kansas
BT United States

Rusk County
 Valid through 1988. Search in combination with state term. After 1988, use specific county-state term.

Rusk County Texas (1989)
E Texas. Before 1989, also search Rusk County AND Texas.
CO N315100N322200
  W0942400W0945600
BT Texas
BT United States

Rusk County Wisconsin (1989)
N Wisconsin. Before 1989, also search Rusk County AND Wisconsin.
CO N451800N453800
  W0904000W0913200
BT Wisconsin
BT United States

Rusophycus (1978)
BT ichnofossils
SA Trilobita

Russell Cave (1981)
NE Alabama.
IN Index county or region as applicable.
SA Jackson County Alabama

Russell County
 Valid through 1988. Search in combination with state term. After 1988, use specific county-state term.

Russell County Alabama (1989)
E Alabama. Before 1989, also search Russell County AND Alabama.
CO N320400N323200
  W0845300W0852500
BT Alabama
BT United States

Russell County Kansas (1989)
Central Kansas. Before 1989, also search Russell County AND Kansas.
CO N384000N390700
  W0982900W0990300
BT Kansas
BT United States

Russell County Kentucky (1989)
S Kentucky. Before 1989, also search Russell County AND Kentucky.
CO N365000N371000
  W0844900W0851500
BT Kentucky
BT United States

Russell County Virginia (1989)
SW Virginia. Before 1989, also search Russell County AND Virginia.
CO N364300N370900
  W0814700W0822500
BT Virginia

BT United States

Russia
 No longer a valid term for GeoRef. Before 1971, included use in subfiles E, N and B for the USSR.
 use Russian Federation

Russian (1978)
Used to indicate language of a catalog, dictionary, glossary or lexicon.
SA catalogs
SA dictionaries
SA glossaries
SA lexicons

Russian Arctic (1993)
That area N of the Arctic Circle in European Russian Federation and Siberia, and including Russian islands in the Arctic Ocean. In Siberia, arctic climatic conditions prevail far S of the Arctic Circle because of its great continental expanse. Before 1978, also search USSR AND Arctic region. Before 1993, also search Soviet Arctic. This term has multiple hierarchies.
CO N663200N900000
  W1700000E0300000
UF Soviet Arctic
BT1 Russian Federation
BT1 Commonwealth of Independent States
BT2 Arctic region
NT New Siberian Islands
SA Taymyr Dolgan-Nenets Russian Federation

Russian Basin
 use Russian Plain

Russian Far East (1993)
Region occupying easternmost Siberia including Amur Oblast, Khabarovsk Kray, Primorye Kray, and Sakhalin Oblast. Before 1993, also search Soviet Far East. Before 1978, also search USSR AND Far East. This term has multiple hierarchies.
CO N420000N750000
  W1700000E1300000
UF Siberia-Soviet Far East
UF Soviet Far East
BT1 Russian Federation
BT1 Commonwealth of Independent States
BT2 Asia
SA Amur Basin
SA Amur Russian Federation
SA Argun River
SA Baikal-Amur Railroad region
SA Dzhagdy Range
SA Far East
SA Kamchatka Peninsula
SA Karymskaya Sopka
SA Khabarovsk Russian Federation
SA Komsomolsk Russian Federation
SA Kuril Islands
SA Okhotsk-Chukchi
SA Primorye Russian Federation
SA Sakhalin
SA Sakhalin Russian Federation
SA Siberia
SA Sikhote-Alin Range
SA Tancheng-Lujiang fault zone
SA Tatar Strait

Russian Federation (1993)
Formerly the Russian Soviet Federated Socialist Republic (RSFSR) or Russian Republic in Europe and Asia. Before 1993, also search Russian Republic.
IN Index Asia and/or Europe as applicable.
CO N370000N830000
  W1700000E0280000
UF Russia
UF Russian Republic
BT Commonwealth of Independent States
NT Abakan Russian Federation
NT Aldan Plateau
NT Aldan River
NT Allarechenskiy
NT Altai Russian Federation
NT Amur region
NT Amur Russian Federation
NT Anadyr Basin
NT Anadyr Range
NT Angara River
NT Arkhangelsk Russian Federation
NT Astrakhan Russian Federation
NT Baikal Mountains
NT Baikal region
NT Baikal rift zone
NT Baikal-Amur Railroad region
NT Bakal Russian Federation
NT Balei Russian Federation
NT Bashkiria Russian Federation
NT Belgorod Russian Federation
NT Biryusa
NT Blyava
NT Bodaibo Russian Federation
NT Bryansk Russian Federation
NT Buryat Russian Federation
NT Caucasus Foreland
NT Central Urals
NT Chadobets Uplift
NT Chaya Massif
NT Chechen-Ingush Russian Federation
NT Chelyabinsk Russian Federation
NT Chita Russian Federation
NT Chukchi Peninsula
NT Chuya Alps
NT Dagestan Russian Federation
NT Dzhagdy Range
NT Dzhida River
NT Dzhugdzhur region
NT Gornaya Shoriya
NT Imandra
NT Irkutsk Basin
NT Irkutsk Russian Federation
NT Iya River
NT Kabardin-Balkar Russian Federation
NT Kaliningrad Russian Federation
NT Kalmyk Russian Federation
NT Kama River
NT Kamchatka Russian Federation
NT Kamenka River
NT Kansk-Achinsk Basin
NT Karelia Russian Federation
NT Karymskaya Sopka
NT Kemerovo Russian Federation
NT Khabarovsk Russian Federation
NT Khamar-Daban Range
NT Khatanga Basin
NT Khibiny Mountains
NT Kiya River
NT Klichka Russian Federation
NT Kodar Range
NT Kola
NT Kola Peninsula
NT Kolyma River
NT Kolyma River basin
NT Kolyma Uplift
NT Komi Russian Federation
NT Kommunar Russian Federation

NT Komsomolsk Russian Federation
NT Kostroma Russian Federation
NT Kozhim
NT Krasnodar Russian Federation
NT Krasnoyarsk Russian Federation
NT Kuban
NT Kuma Basin
NT Kursk magnetic anomaly
NT Kursk Russian Federation
NT Kuznetsk Alatau
NT Kuznetsk Basin
NT Lake Baikal
NT Lake Ladoga
NT Lake Onega
NT Lena Basin
NT Lena River
NT Lower Tunguska River
NT Magadan Russian Federation
NT Malgobek Russian Federation
NT Mama River
NT Mariinsk Russian Federation
NT Maya River basin
NT Maymecha
NT Maymecha-Kotuy
NT Megion Field
NT Mezen River basin
NT Minusinsk Basin
NT Moneron Island
NT Moscow Basin
NT Moscow Russian Federation
NT Moscow Syneclise
NT Moskva River
NT Murmansk Russian Federation
NT Nayba River basin
NT Nizhniy Novgorod Russian Federation
NT North Ossetia Russian Federation
NT Northern Caucasus
NT Northern Dvina River
NT Northern Urals
NT Novgorod Russian Federation
NT Novosibirsk Russian Federation
NT Ob River
NT Oka River
NT Okhotsk region
NT Okhotsk-Chukchi
NT Okhotsk-Chukchi volcanic belt
NT Olekma
NT Olekma-Vitim Highlands
NT Omolon
NT Omolon Block
NT Omsk Russian Federation
NT Orenburg Russian Federation
NT Orlov Russian Federation
NT Pachelma Russian Federation
NT Patom Plateau
NT Pechora Basin
NT Pechora River
NT Penzhina Bay
NT Perm Russian Federation
NT Polar Urals
NT Primorye Russian Federation
NT Pskov Russian Federation
NT Rostov Russian Federation
NT Russian Arctic
NT Russian Far East
NT Russian Fennoscandia
NT Russian Pacific region
NT Ryazan Russian Federation
NT Sadon Russian Federation
NT Saint Petersburg Russian Federation
NT Sakhalin Russian Federation
NT Sakmara
NT Salair Ridge
NT Salym
NT Samara Russian Federation
NT Saratov Russian Federation
NT Sette-Daban Range
NT Severnaya Zemlya
NT Shoriya Mountains
NT Siberian Platform
NT Sikhote-Alin Range
NT Smolensk Russian Federation
NT Stanovoy Range
NT Stavropol Russian Federation
NT Stony Tunguska River
NT Suchan Basin
NT Tagil Basin
NT Tambov Russian Federation
NT Tatar Arch
NT Tatar Russian Federation
NT Tatar Strait
NT Tataria
NT Timan Ridge
NT Timan-Pechora region
NT Tomsk Russian Federation
NT Transbaikalia
NT Tula Russian Federation
NT Tunguska Basin
NT Tunguska River
NT Tunguska Syneclise
NT Turukhan
NT Tuva Russian Federation
NT Tver Russian Federation
NT Tyrny-Auz Russian Federation
NT Tyumen Russian Federation
NT Uchur River basin
NT Uda River
NT Udmurtia Russian Federation
NT Udokan Mountains
NT Ufaley Russian Federation
NT Ulyanovsk Russian Federation
NT Ural region
NT Ural-Tau
NT Uralian Foreland
NT Valdai
NT Verkhoyansk region
NT Vilyuy River
NT Vilyuy River basin
NT Vitim
NT Vladimir Russian Federation
NT Vodino Russian Federation
NT Volga region
NT Volga River
NT Volga-Don region
NT Volga-Urals
NT Volgograd Russian Federation
NT Vologda Russian Federation
NT Voronezh Russian Federation
NT Vyatka River
NT Vyatka Russian Federation
NT Vyatka-Kama Interfluve
NT Vychegda River
NT Western Transbaikalia
NT Wrangel Island
NT Yakutia Russian Federation
NT Yaroslavl Russian Federation
NT Yekaterinburg Russian Federation
NT Yenisei River
NT Yudoma
NT Zeya
SA Altai-Sayan region
SA Amur Basin
SA Amur River
SA Arctic Coastal Plain
SA Arctic Ocean
SA Argun River
SA Asia
SA Astrakhan Arch
SA Azov region
SA Baikalian Phase
SA Baltic Glaciation
SA Baltic Plain
SA Baltic region
SA Baltic Shield
SA Barabash Suite
SA Baskunchak Series
SA Bazhenov Formation
SA Caspian Basin
SA Caspian Depression
SA Caspian Sea
SA Caucasus
SA Dnieper Basin
SA Dnieper River
SA Dnieper-Donets Basin
SA Don Basin
SA Don River
SA Donets Basin
SA Dvina River
SA Elbrus
SA Eurasia
SA European Platform
SA Fennoscandia
SA Gay
SA Greater Caucasus
SA Ilimaussaq
SA Irtysh River
SA Ishim
SA Kainsaz Meteorite
SA Khanka Lake
SA Korenevskaya Formation
SA Kuban River
SA Kuban Valley
SA Kulunda Steppe
SA Kursk Series
SA Ladoga Series
SA Lake Ladoga region
SA Lapland
SA Leningrad Mining Institute
SA Maikop Series
SA Moscow University
SA Moscow-Pechora Syneclise
SA Moty Formation
SA Ob-Irtysh Interfluve
SA Omulevka
SA Ossetia
SA Poltava Series
SA Protopivskaya Formation
SA Rudny Altai
SA Russian Plain
SA Russian Platform
SA Sarbay
SA Sayan
SA Scythian Platform
SA Selenga River valley
SA Serebryanka Formation
SA Siberia
SA Siberian fold belt
SA Siberian Lowland
SA Southern Urals
SA Steppes region
SA Tancheng-Lujiang fault zone
SA Tannu-Ola Range
SA Terek River
SA Tunguska Series
SA Tura Formation
SA Udokan Series
SA Ural River
SA Urals
SA Urup
SA Usa Series
SA USSR Academy of Sciences
SA Uzen
SA Vetluga Series
SA Volga-Ural region
SA Voronezh-Volga Anteclise
SA West Siberia
SA Western Sayan
SA White Sea
SA Yenisei Basin
SA Yenisei-Khatanga basin
SA Yudoma Series
SA Yudomian

Russian Fennoscandia (1993)
Region of Russian Federation in Fennoscandia. Also search Soviet Fennoscandia. Autoposting of Russian Platform to Soviet Fennoscandia began in 1989. This term has multiple hierarchies.
CO N600000N700000
E0420000E0280000
UF Soviet Fennoscandia
BT1 Russian Federation
BT1 Commonwealth of Independent States
BT2 Fennoscandia
BT2 Europe
BT3 Russian Platform

Russian Pacific region (1993)
Also search Soviet Pacific region. This term has multiple hierarchies.
CO N433000N650000
E1800000E1413000
UF Soviet Pacific region
BT1 Russian Federation
BT1 Commonwealth of Independent States
BT2 Asia
NT Kamchatka Peninsula
NT Koryak Range
NT Kuril Islands
SA Sakhalin Russian Federation
SA Siberia

Russian Plain (1978)
Comprises most of the former European USSR W of the Urals and is bordered on the S by the Carpathians, the Crimean Mountains and the Caucasus. Also search Russian AND plain.
IN Index former Soviet republics as applicable.
UF East European Plain
UF Russian Basin
BT Europe
SA Belarus
SA Estonia
SA Latvia
SA Lithuania
SA Moldavia
SA Moldova
SA Russian Federation
SA Russian Platform
SA Ukraine

Russian Platform (1978)
Ancient platform of Precambrian crystalline rocks overlain by sedimentary deposits of the Russian or East European Plain. Also search East European Platform; Russian AND platform. Autoposting of this term began in 1981. Before 1993, USSR was a broader term.
IN Index Europe or Asia OR former Soviet republics as applicable.
CO N523000N840000
E0660000E0190000
UF East European Platform
NT Baltic Plain
NT Dnieper-Donets Basin
NT Franz Josef Land
NT Moscow-Pechora Syneclise
NT Russian Fennoscandia
NT Timan Ridge
NT Ukrainian Shield
NT Ukrainian Syneclise
NT Voronezh-Volga Anteclise
SA Crimea Ukraine
SA Latvia
SA Moldavia
SA Russian Federation
SA Russian Plain
SA Transcarpathia Ukraine
SA Tsarev Meteorite
SA Ukraine
SA Ukrainian Carpathians

Russian Republic
As of 1993, no longer a valid term for GeoRef.
use Russian Federation

Rustler Formation (1985)
W Texas and SE New Mexico.
BT Permian
BT Paleozoic
SA Culebra Dolomite Member
SA New Mexico

SA Texas

ruthenium (1978)
Chemical element. As of 1985, use ruthenium ores for ruthenium as a commodity. Autoposting of platinum group and metals to this term began in 1989.
UF Ru
BT platinum group
BT metals
NT Ru-106
SA ruthenium ores

ruthenium ores (1985)
Before 1985, also search (ruthenium OR Ru) AND (deposit OR deposits OR ore OR ores OR economic) in the basic index. Autoposting of metal ores to this term began in 1985.
IN Commodity. See List C.
BT metal ores
SA platinum ores
SA ruthenium

rutile (1978)
BT oxides
SA anatase
SA brookite
SA heavy minerals
SA rutile structure

rutile structure (1978)
SA rutile

Rutland
Valid through 1988. Search in combination with state term. After 1988, use specific city-state term.

Rutland County
Valid through 1988. Search in combination with state term. After 1988, use specific county-state term.

Rutland County Vermont (1989)
SW Vermont. Before 1989, also search Rutland County AND Vermont. Before 1974, Rutland County was also a county in England that is presently a part of Leicestershire.
CO N431800N435100
W0724300W0732500
BT Vermont
BT United States
NT Brandon Vermont
NT Castleton Vermont
NT Rutland Vermont

Rutland Vermont (1989)
City in W central Vermont. Before 1989, search Rutland AND Vermont.
CO N433700N433700
W0725900W0725900
BT Rutland County Vermont
BT Vermont
BT United States

Rwanda (1978)
Formerly Ruanda which was part of the Belgian trust territory of Ruanda-Urundi.
CO S025000S010500
E0305000E0285000
UF Ruanda
BT Central Africa
BT Africa
SA East African Rift
SA Lake Kivu

Ryazan
No longer a valid term for GeoRef. As of 1993, see Ryazan Russian Federation.

Ryazan Russian Federation (1993)
Oblast and city located 120 miles SE of Moscow. Before 1993, also search Ryazan. This term has multiple hierarchies.
BT1 Russian Federation
BT1 Commonwealth of Independent States
BT2 Europe

Rybnik
No longer a valid term for GeoRef. As of 1993 see Rybnik Poland.

Rybnik Poland (1993)
City in SW Katowice, S Poland. Before 1993 also search Rybnik AND Poland.
BT Katowice Poland
BT Poland
BT Central Europe
BT Europe

Rye Formation (1989)
SW Maine and SE New Hampshire. May be considered Silurian, Ordovician, or upper Proterozoic.
IN Index age as applicable.
SA Maine
SA New Hampshire

Ryoke
No longer a valid term for GeoRef. See Ryoke Belt. Region midway between Yokohama and Nagoya.

Ryoke Belt (1978)
Region midway between Yokohama and Nagoya. Metamorphic belt. Also search Ryoke.
BT Honshu
BT Japan
BT Far East
BT Asia

Ryujima (1978)
Island in Japan Sea off the Noto Peninsula. Formerly used for a manganese mine in Nagano Prefecture.
SA Honshu
SA Nagano Japan

Ryukyu Group (1989)
SW Japan.
BT Pleistocene
BT Quaternary
BT Cenozoic
SA Japan

Ryukyu Islands (1978)
600 mile chain of islands in W Pacific Ocean between Taiwan and Kyushu. Returned to Japan from U.S. control in 1972. Includes all of Okinawa Prefecture and part of Kagoshima Prefecture.
CO N240000N301000
E1301000E1225000
UF Nansei-shoto
UF Riukiu Islands
BT Japan
BT Far East
BT Asia
NT Okinawa
NT Oshima
NT Tanega-shima
SA Kagoshima Japan
SA Kuroshio
SA Ryukyu Trench

Ryukyu Trench (1989)
NW Philippine Sea.
BT Philippine Sea
BT West Pacific
BT Pacific Ocean
SA Ryukyu Islands

Rzeszow
No longer a valid term for GeoRef. As of 1993 see Rzeszow Poland.

Rzeszow Poland (1993)
Province and industrial commune in SE Poland. Before 1993 also search Rzeszow AND Poland.
BT Poland
BT Central Europe
BT Europe
SA Galitsiya

S

S
use sulfur

S-32 (1978)
Autoposting of broader terms began in 1989. This term has multiple hierarchies.
BT1 stable isotopes
BT1 isotopes
BT2 sulfur
SA S-34/S-32

S-32/S-34
use S-34/S-32

S-34 (1978)
Autoposting of broader terms began in 1989. This term has multiple hierarchies.
BT1 stable isotopes
BT1 isotopes
BT2 sulfur
SA S-34/S-32

S-34/S-32 (1978)
Autoposting of broader terms began in 1989. This term has multiple hierarchies.
UF S-32/S-34
BT1 stable isotopes
BT1 isotopes
BT2 sulfur
SA geologic thermometry
SA isotope ratios
SA paleoclimatology
SA S-32
SA S-34

S-35 (1989)
Autoposting of broader terms began in 1989. This term has multiple hierarchies.
BT1 radioactive isotopes
BT1 isotopes
BT2 sulfur

S-type granites (1989)
Derived mainly from sedimentary source material.
BT granites
BT plutonic rocks
BT igneous rocks
SA A-type granites
SA I-type granites
SA peraluminous composition

S-waves (1978)
Autoposting of this term began in 1978.
UF rotational wave
UF secondary wave
UF shake wave
UF shear wave
UF tangential wave
UF transverse wave
BT body waves
BT elastic waves
NT SH-waves
NT SV-waves
SA P-waves
SA PKS-waves
SA PPS-waves
SA PS-waves
SA ScS-waves
SA seismology
SA SKP-waves
SA SKS-waves
SA Sn-waves
SA waves

Saale glacial stage
use Saalian

Saale Glaciation
use Saalian

Saale River (1978)
Rises in the Fichtelgebirge Range in NE Bavaria and flows N into the Elbe River SE of Magdeburg. Also search Saale.
IN Index states as applicable.
BT Germany
BT Central Europe
BT Europe
SA Bavaria Germany
SA Saxony-Anhalt Germany
SA Thuringia Germany

Saalian (1978)
Term applied in NW Europe to the second latest glacial stage of the Pleistocene Epoch, after the Elster glacial stage and before the Warthe; equivalent to the Riss.
UF Saale glacial stage
UF Saale Glaciation
BT upper Pleistocene
BT Pleistocene
BT Quaternary
BT Cenozoic
SA Central Polish Glaciation

Saalian Orogenic Phase
use Saalian Phase

Saalian Phase (1978)
Orogenic phase. Before 1977 also search Saalian AND orogeny.
UF Saalian Orogenic Phase
BT Permian
BT Paleozoic
SA orogeny

Saanich Inlet (1978)
Arm of Strait of Georgia off SE Vancouver Island, NW of Victoria.
BT British Columbia
BT Western Canada
BT Canada
SA Strait of Georgia

Saar
Not a valid term for GeoRef. See Saarland Germany.

Saar Basin (1978)
River basin. Also search Saar.
IN Index French departments and German states as applicable.
BT Europe
SA Bas-Rhin France
SA Moselle France
SA Rhineland-Palatinate Germany
SA Saarland Germany

Saar Nahe (Basin)
use Saar-Nahe Basin

Saar-Nahe Basin (1978)
As of 1981, refers to a region in Saarland and Rhineland-Palatinate, corresponding roughly to the Saar and Nahe river basins in Germany. Before 1981, part of France was included.
CO N490800N500700

E0081000E0061100
UF Saar Nahe (Basin)
BT Germany
BT Central Europe
BT Europe
SA Nahe
SA Rhineland-Palatinate Germany
SA Saarland Germany

Saarbrucken Anticline (1978)
BT Saarland Germany
BT Germany
BT Central Europe
BT Europe

Saare
use Saaremaa

Saarema Island
use Saaremaa

Saaremaa (1989)
Island in E Baltic Sea, W Estonia. Autoposting of Russian Republic to this term ended in 1990.
CO N575800N584500
E02330E0214500
UF Oesel
UF Saare
UF Saarema Island
UF Sarema
BT Estonia
BT Baltic region
BT Europe
SA Baltic Sea

Saarland
No longer a valid term for GeoRef. See Saarland Germany.

Saarland Germany (1993)
State in SW Germany constituting an industrial region, formerly known as the Saar. Achieved statehood within West Germany in 1957. Before 1978, also search Saar.
CO N490700N493800
E0072100E0062000
BT Germany
BT Central Europe
BT Europe
NT Saarbrucken Anticline
SA Nahe
SA Saar Basin
SA Saar-Nahe Basin

Sabah
No longer a valid term for GeoRef. As of 1993, see Sabah Malaysia.

Sabah Malaysia (1993)
State of East Malaysia on NE Borneo. Formerly British North Borneo. Also called North Borneo. This term has multiple hierarchies.
CO N040000N074000
E1193000E1150000
UF British North Borneo
UF North Borneo
BT1 East Malaysia
BT1 Borneo
BT1 Far East
BT1 Asia
BT2 East Malaysia
BT2 Borneo
BT2 Malay Archipelago
BT3 East Malaysia
BT3 Malaysia
BT3 Far East
BT3 Asia
NT Darvel Bay
NT Sandakan Malaysia

Sabana de Bogota (1978)
Plateau 55 miles long and 25 miles wide with Bogota at its center.
UF Bogota Plateau
BT Colombia
BT South America
SA Bogota Colombia

Sabatini Mountains (1978)
N of Lake Bracciano NW of Rome. Also search Sabatini.
BT Latium Italy
BT Italy
BT Southern Europe
BT Europe

Sabinas Basin (1989)
NE Mexico.
BT Mexico

Sabine Lake (1978)
Formed by expansion of Sabine River which flows through the lake and Sabine Pass to Gulf of Mexico.
IN Index counties or region as applicable.
SA Cameron Parish Louisiana
SA Jefferson County Texas
SA Louisiana
SA Orange County Texas
SA Texas
SA United States

Sabine Uplift (1985)
NW Louisiana and NE Texas.
IN Index states as applicable.
BT United States
SA Gulf Coastal Plain
SA Louisiana
SA Texas

Sabkah
use Sabkha Syria

Sabkha
No longer a valid term for GeoRef. As of 1993 see Sabkha Syria.

sabkha environment
use sebkha environment

Sabkha Syria (1993)
Village on right bank of Euphrates River in N central Syria. Before 1993 also search Sabkha Syria.
UF Sabkah
BT Syria
BT Middle East
BT Asia

Sable Island (1978)
Low, sandy island in North Atlantic 115 miles off mainland.
BT Nova Scotia
BT Maritime Provinces
BT Eastern Canada
BT Canada

Sable Island Bank (1978)
Marine bank in vicinity of Sable Island approximately 115 miles off mainland of Nova Scotia.
BT Atlantic Ocean

sabulous texture
use arenaceous texture

Saco River (1978)
Flows into Atlantic Ocean from S Maine.
IN Index states and counties as applicable.
SA Carroll County New Hampshire
SA Cumberland County Maine
SA Maine
SA New Hampshire
SA Oxford County Maine
SA United States
SA York County Maine

Sacoglossa (1981)
BT Gastropoda
BT Mollusca
BT Invertebrata

Sacramento
Valid through 1988. Search in combination with state term. After 1988, use specific city-state term.

Sacramento Basin (1989)
N central California.
IN Index counties or region as applicable.
SA California

Sacramento California (1989)
City in central California. Before 1989, search Sacramento AND California.
CO N383300N383300
W1213000W1213000
BT Sacramento County California
BT California
BT United States

Sacramento County
Valid through 1988. Search in combination with state term. After 1988, use specific county-state term.

Sacramento County California (1989)
Central California. Before 1989, also search Sacramento County AND California.
CO N380000N384500
W1210200W1215300
BT California
BT United States
NT Sacramento California
SA San Francisco Bay region

Sacramento Mountains (1978)
Range in Otero County NE of El Paso in S New Mexico.
IN Index county or region as applicable.
SA Guadalupe Mountains
SA Otero County New Mexico

Sacramento River (1989)
N California.
IN Index state or region as applicable.
SA California
SA Northern California

Sacramento Valley (1978)
River valley. Northern half of Central Valley. Extends from Lake Shasta to San Joaquin Valley. Also search Sacramento River.
IN Index counties or region as applicable.
SA California
SA Central Valley

Saddle Mountain Basalt
use Saddle Mountains Basalt

Saddle Mountains Basalt (1989)
In Yakima Basalt of the Columbia River Basalt Group. Includes Ice Harbor Member. N Idaho, N Oregon and S Washington.
UF Saddle Mountain Basalt
BT Miocene
BT Neogene
BT Tertiary
BT Cenozoic
SA Columbia River Basalt Group
SA Idaho
SA Oregon
SA Washington
SA Yakima Basalt

Sadlerochit Formation
As of 1989, no longer a valid term for GeoRef.
use Sadlerochit Group

Sadlerochit Group (1989)
N Alaska. Includes Ivishak Formation. Before 1989, also search Sadlerochit Formation.
IN Index ages as applicable.
UF Sadlerochit Formation
SA Alaska
SA Ivishak Formation
SA Lower Triassic
SA Permian

Sadlerochit Mountains (1993)
NE Brooks Range, N Alaska.
BT Brooks Range
BT Northern Alaska
BT Alaska
BT United States
SA Arctic National Wildlife Refuge

Sado Island (1978)
Mountainous island in E Japan Sea off NW coast of Honshu.
BT Honshu
BT Japan
BT Far East
BT Asia
SA Niigata Japan

Sadon
No longer a valid term for GeoRef. As of 1993, see Sadon Russian Federation.

Sadon Russian Federation (1993)
Town in lead-zinc-silver mining area in North Ossetia in Northern Caucasus. Before 1993, also search Sadon.
IN Index regions as applicable.
BT Russian Federation
BT Commonwealth of Independent States
SA North Ossetia Russian Federation

Safaga
No longer a valid term for GeoRef. See Safaga Egypt.

Safaga Egypt (1993)
City on N Red Sea.
BT Egypt
BT North Africa
BT Africa

safety (1993)
Methods or techniques of avoiding accident or disease. See risk assessment for safety assessment.
SA design
SA engineering geology
SA geologic hazards
SA risk assessment
SA seismic risk
SA volcanic risk

safflorite (1978)
Autoposting of sulfides to this term ended in 1989.
BT arsenides

Sag River Sandstone (1989)
Prudhoe Bay, N Alaska.
BT Upper Triassic
BT Triassic
BT Mesozoic
SA Alaska

Saga
No longer a valid term for GeoRef. See Saga Japan.

Saga Japan (1993)
Prefecture in NW Kyushu, S Japan.
BT Kyushu
BT Japan
BT Far East
BT Asia

Sagadahoc County
Valid through 1988. Search in combination with state term. After 1988, use specific county-state term.

Sagadahoc County Maine (1989)
On Atlantic Ocean in S Maine. Before 1989, also search Sagadahoc County AND Maine.
CO N434000N441500
W0694000W0700500
BT Maine
BT United States

Sagami Bay (1978)
Inlet of the Pacific Ocean SW of Yokohama.
UF Sagami Sea
UF Sagami-nada
BT Honshu
BT Japan
BT Far East
BT Asia
SA Izu Peninsula
SA Kanagawa Japan
SA Shizuoka Japan

Sagami Sea
use Sagami Bay

Sagami-nada
use Sagami Bay

Saganaga Lake (1978)
In chain of lakes near Lake Superior.
IN Index Minnesota and/or Ontario as applicable.
BT North America
SA Cook County Minnesota
SA Minnesota
SA Ontario

Sagar
No longer a valid term for GeoRef. See Sagar India.

Sagar India (1993)
Town in N central Madhya Pradesh, central India.
BT Madhya Pradesh India
BT India
BT Indian Peninsula
BT Asia

Saghalin
use Sakhalin

Saginaw Formation (1978)
Includes Verne Limestone Member, Eaton, Ionia, and Woodville Sandstone members. Lower Peninsula.
BT Pennsylvanian
BT Carboniferous
BT Paleozoic
SA Michigan

Saguache County
Valid through 1988. Search in combination with state term. After 1988, use specific county-state term.

Saguache County Colorado (1989)
S Colorado. Before 1989, also search Saguache County AND Colorado.
CO N374000N383000
W1053000W1070000
BT Colorado
BT United States
SA San Luis Valley

Saguenay County Quebec (1993)
Before 1993, also search Saguenay County AND Quebec.
BT Quebec
BT Eastern Canada
BT Canada
NT Schefferville Quebec
NT Sept-Iles Quebec

Saguenay earthquake 1988 (1993)
Saguenay Valley, Quebec.
BT earthquakes
SA Quebec

Saguenay Valley (1978)
River valley between Lake Saint Jean and Saint Lawrence River NE of Quebec City. Also search Saguenay; Saguenay River.
BT Quebec
BT Eastern Canada
BT Canada
SA Saint Lawrence River basin

Sahara (1978)
Vast arid region extending across North Africa from the Atlantic Ocean to the Red Sea. From 1977-80, documents on the Spanish Sahara were indexed under Sahara as the level 1 term. As of 1981, use Western Sahara for the Spanish Sahara.
IN Index countries as applicable.
CO N140000N350000
E0380000E0172000
UF Sahara Desert
BT Africa
SA Adrar des Iforas
SA Ahaggar
SA Algeria
SA Chad
SA Egypt
SA ergs
SA Libya
SA Mali
SA Mauritania
SA Morocco
SA Niger
SA Sudan
SA Tanezrouft
SA Tunisia

Sahara Desert
use Sahara

Sahel (1978)
Region. A transitional steppe belt just S of the Sahara.
IN Index countries as applicable.
BT Africa
SA Burkina Faso
SA Chad
SA Mali
SA Mauritania
SA Niger
SA Senegal

Sahul Shelf (1978)
SW of W Timor Island.
BT Timor Sea
BT North Australian Seas
BT West Pacific
BT Pacific Ocean

Saihun River
use Syr Darya

Saint Bernard Parish Louisiana (1989)
Extreme SE Louisiana, bordering the Gulf of Mexico. Before 1989, search Saint Bernard Parish AND Louisiana.
CO N293700N301100
W0885000W0900200
UF St. Bernard Parish
BT Louisiana
BT United States

Saint Catharines
No longer a valid term for GeoRef. As of 1993, see Saint Catharines Ontario.

Saint Catharines Ontario (1993)
City NW of Niagara Falls on Welland Canal in Niagara regional municipality, S Ontario.
UF St. Catherines Ontario
BT Ontario
BT Eastern Canada
BT Canada

Saint Charles County
Valid through 1988. Search in combination with state term. After 1988, use specific county-state term.

Saint Charles County Missouri (1989)
E Missouri. Before 1989, also search Saint Charles County AND Missouri.
CO N383200N385800
W0900700W0905800
BT Missouri
BT United States

Saint Clair County
Valid through 1988. Search in combination with state term. After 1988, use specific county-state term.

Saint Clair County Alabama (1989)
NE central Alabama. Before 1989, also search Saint Clair County AND Alabama.
CO N332200N335800
W0860200W0863300
BT Alabama
BT United States

Saint Clair County Illinois (1989)
SW Illinois. Before 1989, also search Saint Clair County AND Illinois.
CO N381200N383900
W0894000W0901500
BT Illinois
BT United States

Saint Clair County Michigan (1989)
E Michigan. Before 1989, also search Saint Clair County AND Michigan.
CO N423300N430900
W0822600W0830000
BT Michigan Lower Peninsula
BT Michigan
BT United States
SA Lake Saint Clair
SA Saint Clair River
SA Saint Clair River delta

Saint Clair County Missouri (1989)
W Missouri. Before 1989, also search Saint Clair County AND Missouri.
CO N375000N381300
W0932900W0940400
BT Missouri
BT United States

Saint Clair River (1978)
Connects Lake Huron with Lake Saint Clair.
IN Index Michigan and/or Ontario as applicable.
UF St. Clair River
BT North America
SA Michigan
SA Ontario
SA Saint Clair County Michigan

Saint Clair River delta (1978)
Formed at river mouth in Lake Saint Clair.
IN Index Michigan and/or Ontario as applicable.
UF St. Clair River delta
BT North America
SA Michigan
SA Ontario
SA Saint Clair County Michigan

Saint Croix (1978)
Largest and most populous of the U.S. Virgin Islands which are E of Puerto Rico. Before 1978, also search Saint Croix Island; St. Croix.
CO N173500N175000
W0643500W0645500
UF Saint Croix Island
UF St. Croix
UF St. Croix Island
BT U. S. Virgin Islands
BT Virgin Islands
BT Lesser Antilles
BT Antilles
BT West Indies
BT Caribbean region

Saint Croix Island
use Saint Croix

Saint Elias Mountains (1978)
Range near the Pacific Ocean.
IN Index Alaska and/or Yukon Territory as applicable.
UF St. Elias Mountains
BT North America
SA Alaska
SA Chitistone Pass
SA Coast Ranges
SA Glacier Bay National Park
SA Steele Glacier
SA Variegated Glacier
SA Yukon Territory

Saint Francois County
Valid through 1988. Search in combination with state term. After 1988, use specific county-state term.

Saint Francois County Missouri (1989)
SE Missouri. Before 1989, also search Saint Francois County AND Missouri.
CO N374000N380500
W0901000W0904000
BT Missouri
BT United States

Saint Francois Mountains (1978)
WNW of Cape Girardeau in SE Missouri.
IN Index counties or region as applicable.
CO N372000N380500
W0901000W0904000
UF St. Francois Mountains
BT Missouri
BT United States
SA Ozark Mountains

Saint Gall
No longer a valid term for GeoRef. As of 1993, see Saint Gall Switzerland.

Saint Gall Switzerland (1993)
Canton in NE Switzerland. Before 1993 also search Saint Gall AND Switzerland.
UF Saint Gallen
UF St. Gallen
BT Switzerland
BT Central Europe
BT Europe

Saint Gallen
use Saint Gall Switzerland

Saint George Basin (1985)
Basin in S Bering Sea. Also a basin in Washington County, Utah.
IN Index Bering Sea or Utah as applicable.
SA Bering Sea
SA Utah
SA Washington County Utah

Saint George Batholith (1985)
S New Brunswick.
SA Carboniferous
SA Mississippian
SA New Brunswick
SA Saint George Group

Saint George Group (1989)
W Newfoundland.
UF St. George Group
BT Lower Ordovician
BT Ordovician
BT Paleozoic
SA Newfoundland Island
SA Saint George Batholith

Saint George Island (1989)
In NW Florida, barrier island in Gulf of Mexico. In Alaska, one of the Pribilof Islands in the Bering Sea.
IN Index states and counties as applicable.
SA Alaska
SA Apalachicola Bay
SA Bering Sea
SA Florida
SA Gulf of Mexico

Saint Gotthard
use Gotthard Massif

Saint Helena (1978)
British island in S Atlantic Ocean about 1200 miles from W coast of Africa. In 1981, broader term changed from Atlantic Ocean to Atlantic Ocean Islands.
CO S170000S150000
    W0050000W0070000
UF Saint Helena Island
UF St. Helena
UF St. Helena Island
BT Atlantic Ocean Islands
SA Gough Island
SA Tristan da Cunha

Saint Helena Island
use Saint Helena

Saint John (1993)
No longer a valid term for GeoRef.

Saint John County New Brunswick (1993)
Before 1993, also search Saint John County AND New Brunswick.
UF St. John County
BT New Brunswick
BT Maritime Provinces
BT Eastern Canada
BT Canada
NT Saint John New Brunswick

Saint John Island (1989)
In various locations. One of the U.S. Virgin Islands, Lesser Antilles. Also in Saint John Bay, NW Newfoundland.
IN Index regions as applicable.
SA Leeward Islands
SA Lesser Antilles
SA Newfoundland Island
SA U. S. Virgin Islands
SA Virgin Islands

Saint John New Brunswick (1993)
City on Bay of Fundy. Before 1993, also search Saint John AND New Brunswick.
UF St. John
BT Saint John County New Brunswick
BT New Brunswick
BT Maritime Provinces
BT Eastern Canada
BT Canada

Saint John River (1989)
Somerset and Aroostook counties, N Maine; W and S New Brunswick. Also a river in central Liberia. Also a river on the E Gaspe Peninsula, Quebec.
IN Index regions as applicable.
UF St. John River
SA Aroostook County Maine
SA Gaspe Peninsula
SA Liberia
SA Maine
SA New Brunswick
SA Quebec
SA Somerset County Maine

Saint John's
No longer a valid term for GeoRef. Use Saint John's Newfoundland if applicable.

Saint John's Newfoundland (1993)
City on the Atlantic Ocean in SE Newfoundland.
BT Newfoundland Island
BT Newfoundland
BT Eastern Canada
BT Canada

Saint Johns County
Valid through 1988. Search in combination with state term. After 1988, use specific county-state term.

Saint Johns County Florida (1989)
NE Florida. Before 1989, also search Saint Johns County AND Florida.
CO N293800N301500
    W0811300W0814000
BT Florida
BT United States

Saint Johns River basin (1978)
NE Florida.
IN Index counties or region as applicable.
UF St. Johns River basin
SA Florida

Saint Kilda (1989)
Island in Outer Hebrides of W Scotland. After 1992, use Saint Kilda Australia for the municipality in S Victoria, Australia.
IN Index regions as applicable.
SA Outer Hebrides
SA Scotland
SA Victoria Australia

Saint Laurent Limestone (1978)
Regarded as lower Hamilton (Cazenovia Stage) and below Lingle Limestone. S Illinois and E Missouri.
UF St. Laurent Limestone
BT Middle Devonian
BT Devonian
BT Paleozoic
SA Illinois
SA Missouri

Saint Lawrence County
Valid through 1988. Search in combination with state term. After 1988, use specific county-state term.

Saint Lawrence County New York (1989)
N New York. Before 1989, also search Saint Lawrence County AND New York.
CO N440400N450000
    W0743200W0755000
UF St. Lawrence County
BT New York
BT United States
NT Balmat-Edwards mining district
NT Gouverneur New York
NT Ogdensburg New York
SA Benson Mines
SA Saint Lawrence Lowlands
SA Saint Lawrence River
SA Saint Lawrence Valley

Saint Lawrence Estuary (1978)
The lower, wide part of the river from below Quebec City to its mouth on the Gulf of Saint Lawrence.
UF St. Lawrence Estuary
BT Quebec
BT Eastern Canada
BT Canada
SA Saint Lawrence Valley

Saint Lawrence Lowlands (1978)
Along both sides of river from Lake Ontario to Quebec.
IN Index New York counties and Canadian provinces as applicable.
UF St. Lawrence Lowlands
BT North America
SA Champlain Sea
SA Franklin County New York
SA Great Appalachian Valley
SA Jefferson County New York
SA Laurentide ice sheet
SA Monteregian Hills
SA New York
SA Ontario
SA Quebec
SA Saint Lawrence County New York
SA Saint Lawrence Valley

Saint Lawrence River (1978)
Flows NE out of Lake Ontario into the Gulf of Saint Lawrence. Also search Saint Lawrence AND river.
IN Index New York and Canadian provinces as applicable.
UF St. Lawrence River
BT North America
SA Canada
SA Franklin County New York
SA Jefferson County New York
SA New York
SA Ontario
SA Quebec
SA Saint Lawrence County New York
SA Saint Lawrence River basin
SA Saint Lawrence Valley

Saint Lawrence River basin (1993)
Drainage basin of Saint Lawrence River. E North America. Includes N New York, Vermont, Quebec and Ontario. Principal tributaries of the Saint Lawrence River include the Richelieu River, Ottawa River, Saint Francis River, Saint Maurice River and Saguenay River.
IN Index states and provinces as applicable.
BT North America
SA New York
SA Ontario
SA Ottawa River
SA Quebec
SA Saguenay Valley
SA Saint Lawrence River
SA Saint Lawrence Valley
SA Vermont

Saint Lawrence Valley (1978)
River valley extending to the Gulf of Saint Lawrence. Also search Saint Lawrence AND valley.
IN Index New York counties and Canadian provinces as applicable.
UF St. Lawrence River valley
UF St. Lawrence Valley
BT North America
SA Franklin County New York
SA Great Appalachian Valley
SA Jefferson County New York
SA Laurentide ice sheet
SA New York
SA Ontario
SA Quebec
SA Saint Lawrence County New York
SA Saint Lawrence Estuary
SA Saint Lawrence Lowlands
SA Saint Lawrence River
SA Saint Lawrence River basin

Saint Louis
Valid through 1988. Search in combination with state term. After 1988, use specific city-state term.

Saint Louis County
Valid through 1988. Search in combination with state term. After 1988, use specific county-state term.

Saint Louis County Minnesota (1989)
NE Minnesota. Before 1989, also search Saint Louis County AND Minnesota.
CO N463900N483700
    W0914800W0930300
BT Minnesota
BT United States
NT Duluth Minnesota
NT Ely Minnesota
SA Mesabi Range
SA Rainy Lake
SA Vermilion Range

Saint Louis County Missouri (1989)
E Missouri. Before 1989, also search Saint Louis County AND Missouri.
CO N382300N385300
    W0900800W0904500
BT Missouri
BT United States
NT Saint Louis Missouri

Saint Louis Limestone (1978)
Comprises Croton and Verdi members.
UF St. Louis Limestone
BT Meramecian
BT Upper Mississippian
BT Mississippian
BT Carboniferous
BT Paleozoic
SA Alabama
SA Georgia
SA Illinois
SA Indiana

SA Iowa
SA Kentucky
SA Missouri
SA Oklahoma
SA Tennessee
SA Virginia

Saint Louis Missouri (1989)
City on the Mississippi River in E Missouri. Before 1989, search Saint Louis AND Missouri.
CO N384000N384000
W0901500W0901500
UF St. Louis
BT Saint Louis County Missouri
BT Missouri
BT United States

Saint Lucia (1978)
Island. A self-governing state in association with United Kingdom in Windward Islands of the Lesser Antilles. Before 1981, West Indies was the broader term.
IN Index Windward Islands.
UF St. Lucia
BT Lesser Antilles
BT Antilles
BT West Indies
BT Caribbean region
SA Windward Islands

Saint Lucie County
Valid through 1988. Search in combination with state term. After 1988, use specific county-state term.

Saint Lucie County Florida (1989)
On Atlantic Ocean in SE Florida. Before 1989, also search Saint Lucie County AND Florida.
CO N271000N273500
W0801000W0804000
BT Florida
BT United States

Saint Mary Parish
Valid through 1988. Search in combination with state term. After 1988, use specific county-state term.

Saint Mary Parish Louisiana (1989)
S Louisiana, bordering the Gulf of Mexico. Before 1989, also search Saint Mary Parish AND Louisiana.
CO N292800N295600
W0910500W0915400
UF St. Mary Parish
BT Louisiana
BT United States
SA Atchafalaya Bay
SA Belle Isle

Saint Marys County
Valid through 1988. Search in combination with state term. After 1988, use specific county-state term.

Saint Marys County Maryland (1989)
S Maryland. Before 1989, also search Saint Marys County AND Maryland.
CO N380200N383130
W0761900W0765300
BT Maryland
BT United States

Saint Marys Formation (1978)
In Chesapeake Group. Middle and upper Miocene. Delaware, E Maryland and E Virginia.
UF St. Marys Formation
BT Miocene
BT Neogene
BT Tertiary
BT Cenozoic
SA Chesapeake Group
SA Delaware
SA Maryland
SA Virginia

Saint Paul
Valid through 1988. Search in combination with state term. After 1988, use specific city-state term.

Saint Paul Island (1978)
Most northerly of Pribilof Islands. Also an island in the Indian Ocean.
IN Index Alaska or Indian Ocean Islands as applicable.
SA Alaska
SA Bering Sea
SA Indian Ocean Islands

Saint Paul Minnesota (1989)
City on the Mississippi River E of Minneapolis in E Minnesota. Before 1989, search Saint Paul AND Minnesota.
CO N450000N450000
W0931000W0931000
UF St. Paul
BT Ramsey County Minnesota
BT Minnesota
BT United States

Saint Paul Rocks (1978)
Group of uninhabited volcanic, rocky islets belonging to Brazil about 600 miles NE of Natal. In 1985, broader term changed from Atlantic Ocean to Atlantic Ocean Islands.
UF Saint Paul's Rocks
UF St. Paul's Rock
UF St. Paul's Rocks
BT Atlantic Ocean Islands

Saint Paul's Rocks
use Saint Paul Rocks

Saint Peter Sandstone (1978)
Overlies Prairie du Chien Group and underlies Glenwood Shale Member of Platteville Formation.
UF St. Peter Sandstone
BT Middle Ordovician
BT Ordovician
BT Paleozoic
SA Arkansas
SA Illinois
SA Indiana
SA Iowa
SA Kansas
SA Kentucky
SA Michigan
SA Minnesota
SA Missouri
SA Ohio
SA Oklahoma
SA Platteville Formation
SA Wisconsin

Saint Petersburg Russian Federation (1993)
Oblast and city on the Gulf of Finland. Before 1993, also search Leningrad. This term has multiple hierarchies.
UF Leningrad
BT1 Russian Federation
BT1 Commonwealth of Independent States
BT2 Europe
SA Baltic region

Saint Pierre and Miquelon (1981)
Consists of two islands in the Atlantic Ocean just S of Newfoundland. Became a French department in 1976. Before 1971, also search Saint-Pierre and Miquelon; St. Pierre and Miquelon. This term has multiple hierarchies.
CO N464500N471000
W0560800W0563000
UF Saint-Pierre and Miquelon
UF St. Pierre and Miquelon
BT1 North America
BT2 Atlantic Ocean Islands

Saint Regis Formation (1989)
In Ravalli Group of Belt Supergroup. NE Idaho, NW Montana and Washington.
UF St. Regis Formation
BT middle Proterozoic
BT Proterozoic
BT upper Precambrian
BT Precambrian
SA Belt Supergroup
SA Idaho
SA Montana
SA Ravalli Group
SA Washington

Saint Severin France
use Saint-Severin France

Saint Severin Meteorite
use Saint-Severin Meteorite

Saint Thomas (1978)
Island in the U.S. Virgin Islands E of Puerto Rico.
UF St. Thomas
BT U. S. Virgin Islands
BT Virgin Islands
BT Lesser Antilles
BT Antilles
BT West Indies
BT Caribbean region

Saint Vincent (1978)
Self governing British state in the Windward Islands comprising Saint Vincent Island and the northern Grenadines.
IN Index Windward Islands.
UF St. Vincent
BT Lesser Antilles
BT Antilles
BT West Indies
BT Caribbean region
NT Soufriere

Saint Vincent Bay
use Saint Vincent Gulf

Saint Vincent Gulf (1978)
Inlet of Indian Ocean between Yorke Peninsula and mainland. Also search Saint Vincent.
UF Saint Vincent Bay
BT South Australia
BT Australia
BT Australasia

Saint-Chinian
No longer a valid term for GeoRef. See Saint-Chinian France.

Saint-Chinian France (1993)
Village in S France.
BT Herault France
BT France
BT Western Europe
BT Europe

Saint-Etienne Basin
use Saint-Etienne coal basin

Saint-Etienne coal basin (1978)
In E central France. Also search Saint-Etienne.
UF Saint-Etienne Basin
BT Loire France
BT France
BT Western Europe

BT Europe
SA coal fields

Saint-Girons
No longer a valid term for GeoRef. See Saint-Girons France.

Saint-Girons France (1993)
Town in S France at foot of central Pyrenees.
BT Ariege France
BT France
BT Western Europe
BT Europe

Saint-Louis
No longer a valid term for GeoRef. As of 1993 see Saint-Louis Reunion.

Saint-Louis Reunion (1993)
Town on Reunion Island E of the Madagascar. In 1981, broader terms changed from Mascarene Islands and Indian Ocean to Reunion. Before 1993 also search Saint-Louis AND Reunion.
BT Reunion
BT Mascarene Islands
BT Indian Ocean Islands

Saint-Pierre and Miquelon
Not a valid GeoRef index term after 1970. Was used on level 1 in subfile N.
use Saint Pierre and Miquelon

Saint-Severin
No longer a valid term for GeoRef. See Saint-Severin France.

Saint-Severin France (1993)
Village in E France. Before 1993, also search Saint-Severin. Before 1978, also search St. Severin or Saint Severin.
UF Saint Severin France
UF St. Severin France
BT Charente France
BT France
BT Western Europe
BT Europe
SA Saint-Severin Meteorite

Saint-Severin Meteorite (1985)
Impact at Saint-Severin in Charente, France. Before 1985, also search (Saint-Severin OR Saint Severin) AND meteorites.
UF Saint Severin Meteorite
UF St. Severin Meteorite
BT chondrites
BT stony meteorites
BT meteorites
SA Charente France
SA France
SA Saint-Severin France

Saint-Sylvestre Massif (1978)
In the Monts d'Ambazac in E central France.
BT Haute-Vienne France
BT France
BT Western Europe
BT Europe

Saint-Vallier
No longer a valid term for GeoRef. See Saint-Vallier France.

Saint-Vallier France (1993)
Village in Provence Alps in extreme SE France.
UF Saint-Vallier-de-Thiey
BT Alpes-Maritimes France
BT France
BT Western Europe
BT Europe

Saint-Vallier-de-Thiey
  use Saint-Vallier France

Sainte Genevieve Limestone (1978)
  Includes Fredonia Limestone, Rosiclare Sandstone, and Levias Limestone members. N Alabama, Georgia, S Illinois, Indiana, Iowa, Kentucky, E Missouri, Tennessee.
  UF Ste. Genevieve Limestone
  BT Meramecian
  BT Upper Mississippian
  BT Mississippian
  BT Carboniferous
  BT Paleozoic
  SA Alabama
  SA Georgia
  SA Illinois
  SA Indiana
  SA Iowa
  SA Kentucky
  SA Missouri
  SA Tennessee

Sainte-Baume Massif (1978)
  In lower Provence Alps in SE France. Also search Sainte-Baume. This term has multiple hierarchies.
  BT1 France
  BT1 Western Europe
  BT1 Europe
  BT2 Western Alps
  BT2 Alps
  BT2 Europe
  SA Provence
  SA Provence Alps
  SA Var France

Sainte-Marie-aux-Mines
  No longer a valid term for GeoRef. See Sainte-Marie-aux-Mines France.

Sainte-Marie-aux-Mines France (1993)
  Town near crest of the Vosges Mountains in NE France.
  BT Haut-Rhin France
  BT France
  BT Western Europe
  BT Europe

Sainte-Victoire Massif
  use Sainte-Victoire Mountain

Sainte-Victoire Mountain (1978)
  In S France.
  UF Sainte-Victoire Massif
  BT Bouches-du-Rhone France
  BT France
  BT Western Europe
  BT Europe

Saipan (1978)
  United States island N of Guam.
  BT Mariana Islands
  BT Micronesia
  BT Oceania

Saitama
  No longer a valid term for GeoRef. See Saitama Japan.

Saitama Japan (1993)
  Prefecture N of Tokyo.
  BT Honshu
  BT Japan
  BT Far East
  BT Asia
  NT Chichibu Japan
  NT Chichibu Mine
  SA Fujiyama
  SA Kanto Plain

Sakar Mountains (1978)
  Between Maritsa River and Tundzha River in SE Bulgaria. Also search Sakar; Sakar Mountain.
  BT Bulgaria
  BT Southern Europe
  BT Europe

sakhaite (1978)
  This term has multiple hierarchies.
  BT1 borates
  BT2 carbonates

Sakhalin (1978)
  Island N of Hokkaido in W Okhotsk Sea. This term has multiple hierarchies.
  CO N460000N550000 E1450000E1413000
  UF Karafuto
  UF Saghalin
  UF Sakhalin Island
  BT1 Sakhalin Russian Federation
  BT1 Russian Federation
  BT1 Commonwealth of Independent States
  BT2 Sakhalin Russian Federation
  BT2 Asia
  SA Kuril Islands
  SA Moneron Island
  SA Nayba River basin
  SA Okhotsk region
  SA Russian Far East

Sakhalin Island
  use Sakhalin

Sakhalin Russian Federation (1993)
  Oblast including the island and Kuril Islands. This term has multiple hierarchies.
  BT1 Russian Federation
  BT1 Commonwealth of Independent States
  BT2 Asia
  NT Kuril Islands
  NT Sakhalin
  SA Russian Far East
  SA Russian Pacific region

Sakmara (1978)
  River. Before 1993, also used for village in central Orenburg Oblast in Southern Urals in European Russian Federation. After 1993, see Sakmara Russian Federation for the village. This term has multiple hierarchies.
  CO N520000N523000 E0552000E0522000
  BT1 Russian Federation
  BT1 Commonwealth of Independent States
  BT2 Europe

Sakmara Russian Federation (1993)
  Village in central Orenburg Oblast in Southern Urals in European Russian Federation. Before 1993, also search Sakmara. This term has multiple hierarchies.
  BT1 Orenburg Russian Federation
  BT1 Russian Federation
  BT1 Commonwealth of Independent States
  BT2 Orenburg Russian Federation
  BT2 Europe

Sakmarian (1978)
  Europe. Above Stephanian (Carboniferous), below Artinskian.
  BT Lower Permian
  BT Permian
  BT Paleozoic

Sakoa Basin (1978)
  Main source of coal in SW Madagascar. Also search Sakoa. This term has multiple hierarchies.
  BT1 Madagascar
  BT1 Indian Ocean Islands
  BT2 Madagascar
  BT2 Africa
  SA coal

Saksagan River (1978)
  In Dnepropetrovsk Oblast in E central Ukraine. Also search Saksagan. This term has multiple hierarchies.
  BT1 Dnepropetrovsk Ukraine
  BT1 Ukraine
  BT1 Europe
  BT2 Dnepropetrovsk Ukraine
  BT2 Ukraine
  BT2 Commonwealth of Independent States

Sakura-jima (1978)
  Peninsula and volcano. NW projection of Osumi Peninsula in S Kyushu. An island until 1914. Before 1993, Ontake was sometimes used for the volcano. Before 1993, also search Ontake AND Kyushu for the volcano.
  UF Sakurajima
  BT Kagoshima Japan
  BT Kyushu
  BT Japan
  BT Far East
  BT Asia
  SA Aira Caldera
  SA Ebino Japan
  SA Ontake

Sakurajima
  use Sakura-jima

Salado Basin (1978)
  600 mile-long basin of the combined Desaguadero and Salado river systems. Also search Salado.
  IN Index provinces as applicable.
  BT Argentina
  BT South America
  SA La Pampa Argentina
  SA La Rioja Argentina
  SA Mendoza Argentina
  SA San Luis Argentina

Salado Formation (1989)
  SE New Mexico and W Texas. Includes McNutt Potash Member.
  BT Upper Permian
  BT Permian
  BT Paleozoic
  SA New Mexico
  SA Texas

Salair
  No longer a valid term for GeoRef. As of 1993, see Saliar Russian Federation.

Salair Ridge (1978)
  Along borders of Altai Kray and Kemerovo Oblast in S Siberia. Also search Salair. This term has multiple hierarchies.
  BT1 Russian Federation
  BT1 Commonwealth of Independent States
  BT2 Asia
  SA Altai Russian Federation
  SA Kemerovo Russian Federation

Salair Russian Federation (1993)
  City on Salair Ridge in W Kemerovo Oblast in S Siberia. Before 1993, also search Salair. This term has multiple hierarchies.
  BT1 Kemerovo Russian Federation
  BT1 Russian Federation
  BT1 Commonwealth of Independent States
  BT2 Kemerovo Russian Federation
  BT2 Asia

Salaj
  No longer a valid term for GeoRef. As of 1993 see Salaj Romania.

Salaj Romania (1993)
  County in NW Romania. Before 1993 also search Salaj AND Romania.
  BT Transylvania
  BT Romania
  BT Southern Europe
  BT Europe

Salamanca
  No longer a valid term for GeoRef. As of 1993, see Salamanca City Spain.

Salamanca City Spain (1993)
  Refers to only the city in NE Salamanca Province, W Spain. Before 1993, also search Salamanca and Spain.
  BT Salamanca Spain
  BT Leon region
  BT Spain
  BT Iberian Peninsula
  BT Southern Europe
  BT Europe

Salamanca Province
  No longer a valid term for GeoRef. As of 1993, see Salamanca Spain.

Salamanca Spain (1993)
  Refers only to the province in W Spain on the Portuguese border. From 1981-1992, also search Salamanca Province and Spain. Before 1981, also search Salamanca and Spain. For the city, see Salamanca City Spain.
  CO N401300N411800 W0050800W0065200
  BT Leon region
  BT Spain
  BT Iberian Peninsula
  BT Southern Europe
  BT Europe
  NT Salamanca City Spain

Salat Valley (1978)
  River valley in S France. Also search Salat River.
  IN Index departments as applicable.
  BT France
  BT Western Europe
  BT Europe
  SA Ariege France
  SA Haute-Garonne France

Salem County
  Valid through 1988. Search in combination with state term. After 1988, use specific county-state term.

Salem County New Jersey (1989)
  SW New Jersey. Before 1989, also search Salem County AND New Jersey.
  CO N392200N394700 W0750500W0753500

BT New Jersey
BT United States

Salem Limestone (1978)
Includes Kidd, Fults, Chalfin, and Rocher members. S Illinois, S Indiana, SE Iowa, W and central Kentucky, and E Missouri. Also search Salem.
BT Meramecian
BT Upper Mississippian
BT Mississippian
BT Carboniferous
BT Paleozoic
SA Illinois
SA Indiana
SA Iowa
SA Kentucky
SA Missouri

Salentina Peninsula (1978)
SE Apulia. The heel of the Italian boot. Also search Salentina.
UF Salentine Peninsula
BT Apulia Italy
BT Italy
BT Southern Europe
BT Europe

Salentine Peninsula
use Salentina Peninsula

Salerno
No longer a valid term for GeoRef. See Salerno Italy.

Salerno Italy (1993)
City on Gulf of Salerno SE of Naples in W Campania, S Italy.
BT Campania Italy
BT Italy
BT Southern Europe
BT Europe
SA Lucania

sales (1981)
SA economics

Salida
Valid through 1988. Search in combination with state term. After 1988, use specific city-state term.

Salida Colorado (1989)
City on Arkansas River in central Colorado. Before 1989, search Salida AND Colorado.
CO N383300N383300
  W1060100W1060100
BT Chaffee County Colorado
BT Colorado
BT United States

Salina Group (1978)
Includes Vernon Shale, Camillus Shale, Bertie Formation, Akron Dolomite, Syracuse Formation. Michigan and N Ohio.
BT Upper Silurian
BT Silurian
BT Paleozoic
SA Michigan
SA Ohio

Salinas
Valid through 1988. Search in combination with state term. After 1988, use specific city-state term.

Salinas California (1989)
City in W California. Before 1989, search Salinas AND California.
CO N363900N363900
  W1214000W1214000
BT Monterey County California
BT California
BT United States

Salinas River valley
use Salinas Valley

Salinas Valley (1978)
River valley in San Luis Obispo and Monterey counties in W California. Also search Salinas River.
IN Index counties or region as applicable.
UF Salinas River valley
SA California
SA Monterey County California
SA San Luis Obispo County California

saline composition (1978)
Before 1978, search saline.
SA composition
SA ground water
SA salinity

Saline County
Valid through 1988. Search in combination with state term. After 1988, use specific county-state term.

Saline County Arkansas (1989)
Central Arkansas. Before 1989, also search Saline County AND Arkansas.
CO N342400N345100
  W0921200W0930400
BT Arkansas
BT United States

Saline County Illinois (1989)
SE Illinois. Before 1989, also search Saline County AND Illinois.
CO N373500N375400
  W0882200W0884200
BT Illinois
BT United States

Saline County Kansas (1989)
Central Kansas. Before 1989, also search Saline County AND Kansas.
CO N383500N385700
  W0972300W0975500
BT Kansas
BT United States

Saline County Missouri (1989)
Central Missouri. Before 1989, also search Saline County AND Missouri.
CO N385600N392400
  W0925000W0932900
BT Missouri
BT United States

Saline County Nebraska (1989)
SE Nebraska. Before 1989, also search Saline County AND Nebraska.
CO N402200N404100
  W0965400W0972300
BT Nebraska
BT United States

saline soils
use Solonchak soils

saline water
use salt water

Salinian Block (1981)
Block structure in the California Coast Ranges.
CO N343000N390000
  W1190000W1240000
BT California
BT United States
SA Coast Ranges

salinity (1978)
SA brackish water
SA brackish-water environment
SA desalinization
SA hydrochemistry
SA hypersaline environment
SA paleoecology
SA paleosalinity

SA saline composition
SA salt
SA salt marshes
SA sea water
SA sedimentation
SA soil treatment
SA soils
SA Solonchak soils
SA springs
SA water treatment

Salisbury
No longer a valid term for GeoRef. As of 1993, see Harare Zimbabwe if applicable.

Salisbury Embayment (1985)
Central Atlantic Coastal Plain.
IN Index states as applicable.
BT United States
SA Atlantic Coastal Plain
SA Maryland
SA North Carolina
SA Virginia

salite (1989)
Gray-green to black variety of diopside.
BT clinopyroxene
BT pyroxene group
BT chain silicates
BT silicates
SA diopside

Salix (1985)
Genus.
BT Dicotyledoneae
BT angiosperms
BT Spermatophyta
BT Plantae

Salmo
No longer a valid term for GeoRef. See Salmo British Columbia.

Salmo British Columbia (1993)
Village in S British Columbia.
BT British Columbia
BT Western Canada
BT Canada

Salmon
Valid through 1988. Search in combination with state term. After 1988, use specific city-state term.

Salmon Idaho (1989)
City in E Idaho. Before 1989, search Salmon AND Idaho.
CO N451100N451100
  W1135500W1135500
BT Lemhi County Idaho
BT Idaho
BT United States

Salmon River (1978)
Rises in central Idaho and flows N then W, and again N to empty into Snake River in W Idaho. Also search Salmon.
IN Index counties or regions as applicable.
SA Idaho
SA Salmon River breaks

Salmon River breaks (1978)
Large gorge or canyon in lower course of Salmon River.
BT Idaho
BT United States
SA Salmon River

Salonika
No longer a valid term for GeoRef. As of 1993 see Salonika Greece

Salonika Greece (1993)
City on Gulf of Salonika in S central Greek Macedonia, N Greece. Before 1993 also search Salonika AND Greece. This term has multiple hierarchies.
UF Saloniki
UF Thessaloniki
BT1 Greek Macedonia
BT1 Greece
BT1 Southern Europe
BT1 Europe
BT2 Greek Macedonia
BT2 Macedonia
BT2 Southern Europe
BT2 Europe

Saloniki
use Salonika Greece

Salop
No longer a valid term for GeoRef as of 1993. See Shropshire England.

Salop England
use Shropshire England

Salsigne Mine (1978)
Gold ores. N of Carcassonne in S France. Also search Salsigne.
BT Aude France
BT France
BT Western Europe
BT Europe
SA gold ores
SA mines

salt (1978)
IN See List C for use as a commodity term.
BT evaporites
BT chemically precipitated rocks
BT sedimentary rocks
SA brines
SA bromine
SA bromine deposits
SA chlorine
SA desalinization
SA diapirs
SA evaporite deposits
SA halite
SA iodine deposits
SA paleosalinity
SA salinity
SA salt domes
SA salt tectonics
SA sea water
SA sodium chloride

Salt Creek (1978)
Central Wyoming.
IN Index counties or region as applicable.
SA Wyoming

salt domes (1978)
SA cap rocks
SA diapirism
SA diapirs
SA domes
SA folds
SA Oakwood Dome
SA Richton Dome
SA salt
SA salt tectonics

Salt Lake City
Valid through 1988. Search in combination with state term. After 1988, use specific city-state term.

Salt Lake City Utah (1989)
City on Jordan River in N Utah. Before 1989, search Salt Lake City AND Utah.
CO N404500N404500
  W1115500W1115500
BT Salt Lake County Utah
BT Utah
BT United States

Salt Lake County
Valid through 1988. Search in combination with state term. After 1988, use specific county-state term.

Salt Lake County Utah (1989)
N Utah. Before 1989, also search Salt Lake County AND Utah.
CO N402500N405500
   W1113500W1121500
BT Utah
BT United States
NT Bingham Utah
NT Salt Lake City Utah
SA Great Salt Lake
SA Wasatch fault zone

Salt Lake Formation (1989)
SE Idaho, N Nevada, N Utah and W Wyoming. Also search Salt Lake Group NOT Hawaii. For formation in Hawaii, search Salt Lake Group AND Hawaii.
IN Index ages as applicable.
BT Neogene
BT Tertiary
BT Cenozoic
SA Idaho
SA Miocene
SA Nevada
SA Pliocene
SA Utah
SA Wyoming

salt lakes (1978)
SA brackish water
SA brines
SA lakes
SA salt water

salt marshes (1978)
BT marshes
BT shore features
SA brackish-water environment
SA fluvial features
SA geomorphology
SA lacustrine features
SA paleoecology
SA salinity
SA sebkha environment
SA sedimentation

Salt Range (1978)
Between the Indus River and the Jhelum River.
CO N330000N340000
   E0730000E0720000
BT Punjab Pakistan
BT Pakistan
BT Indian Peninsula
BT Asia

Salt River (1978)
Rises in E Arizona and flows W into Gila River W of Phoenix. Also in other locations.
IN Index counties or region as applicable.
SA Arizona
SA Salt River valley

Salt River valley (1989)
Valley of Salt River in Maricopa County, Arizona. Also in N Kentucky.
IN Index states and counties as applicable.
SA Arizona
SA Kentucky
SA Maricopa County Arizona
SA Salt River

salt structures
No longer a valid GeoRef index term. Before 1969, was used on level 1 in subfile E. See salt tectonics; salt domes.

salt tectonics (1978)
Used for the study of the structure and mechanism of emplacement of salt domes.
UF halokinesis
BT tectonics
SA cap rocks
SA deformation
SA diapirism
SA diapirs
SA emplacement
SA faults
SA folds
SA Louann Salt
SA salt
SA salt domes
SA tectonophysics

Salt Valley (1985)
SW Utah.
IN Index counties or region as applicable.
SA Colorado Plateau
SA Grand County Utah
SA San Juan County Utah
SA Utah

Salt Wash Sandstone Member (1985)
Of Morrison Formation. E Utah, NE Arizona, SW Colorado and NW New Mexico.
BT Upper Jurassic
BT Jurassic
BT Mesozoic
SA Arizona
SA Colorado
SA Morrison Formation
SA New Mexico
SA Utah

salt water (1978)
UF saline water
SA brackish water
SA brines
SA chemical dispersion
SA fresh water
SA ground water
SA hydrochemistry
SA salt lakes
SA salt-water intrusion
SA sea water
SA water

salt water intrusion
Not a valid GeoRef index term after 1970. Was used on level 1 in subfile N.
use salt-water intrusion

salt-water contamination
Not a valid term for GeoRef. Use salt-water intrusion.

salt-water intrusion (1978)
Before 1971, also search salt water intrusion.
UF encroachment (ground water)
UF intrusion (ground water)
UF salt water intrusion
UF sea-water encroachment
UF sea-water intrusion
SA brines
SA chemical dispersion
SA ground water
SA salt water
SA sea water

Salta
No longer a valid term for GeoRef. See Salta Argentina.

Salta Argentina (1993)
Province in NNW Argentina.
UF Salta Province
BT Argentina
BT South America
SA Pampean Mountains
SA Rio Blanco Basin
SA Yacoraite Formation

Salta Province
use Salta Argentina

saltation (1978)
SA sediment transport
SA sedimentation

Salton Sea (1978)
Shallow saline lake just N of Imperial Valley in S California.
CO N331000N333000
   W1153500W1160500
BT California
BT United States
SA Colorado Desert
SA Imperial County California
SA Riverside County California
SA Salton Sea geothermal field

Salton Sea geothermal field (1989)
S California.
BT Imperial County California
BT California
BT United States
SA geothermal fields
SA Salton Sea
SA Salton Trough

Salton Trough (1978)
Depression including from NW to SE, the Coachella Valley, the Salton Sea, and the Imperial Valley in S California.
CO N310000N340000
   W1140000W1170000
BT California
BT United States
SA Salton Sea geothermal field
SA Superstition Hills earthquake 1987

Saltville Fault (1978)
SW Virginia. Also search Saltville.
IN Index counties or region as applicable.
SA Virginia

Salvador
No longer a valid term for GeoRef. See Salvador Brazil.

Salvador Brazil (1993)
City on the Atlantic Ocean in E Bahia, E Brazil. Formerly Sao Salvador or Bahia.
BT Bahia Brazil
BT Brazil
BT South America

Salym (1978)
Stream in NW Novosibirsk Oblast in S Siberia. This term has multiple hierarchies.
BT1 Russian Federation
BT1 Commonwealth of Independent States
BT2 Asia
SA Novosibirsk Russian Federation

Salym Field (1985)
Oil field, W Siberia. This term has multiple hierarchies.
CO N600000N630000
   E0720000E0680000
BT1 Tyumen Russian Federation
BT1 Russian Federation
BT1 Commonwealth of Independent States
BT2 Tyumen Russian Federation
BT2 Asia
SA oil and gas fields
SA West Siberia

Salzach River (1978)
Primarily in Salzburg State. Also search Salzach.
IN Index Austrian and German states as applicable.
BT Central Europe
BT Europe
SA Bavaria Germany
SA Pinzgau
SA Salzburg State Austria
SA Upper Austria

Salzburg
No longer a valid term for GeoRef. See Salzburg Austria.

Salzburg Austria (1993)
City in Salzburg State, NW Austria. As of 1981, Salzburg refers to only the city. Before 1981, Salzburg referred to both the state and the city in central Austria.
BT Salzburg State Austria
BT Austria
BT Central Europe
BT Europe

Salzburg State
No longer a valid term for GeoRef.
use Salzburg State Austria

Salzburg State Austria (1993)
Before 1981, search Salzburg. Also search Salzburg State.
CO N465500N480300
   E0140000E0120400
UF Salzburg State
BT Austria
BT Central Europe
BT Europe
NT Bad Gastein Austria
NT Mitterberg Austria
NT Pinzgau
NT Salzburg Austria
SA Hohe Tauern
SA Salzach River
SA Salzkammergut
SA Venediger Group

Salzkammergut (1978)
Lake and mountain region of Eastern Alps.
IN Index states as applicable.
BT Austria
BT Central Europe
BT Europe
SA Eastern Alps
SA Salzburg State Austria
SA Styria Austria
SA Upper Austria

Samail Ophiolite (1985)
N and E Oman.
SA Cretaceous
SA Oman

Samar (1978)
Island on E side of central part of Philippine Islands.
BT Philippine Islands
BT Far East
BT Asia

Samara Bend (1978)
Region in former Kuibyshev Oblast within oxbow of the middle Volga River where it reaches its easternmost point. Also search Samara. This term has multiple hierarchies.
BT1 Samara Russian Federation
BT1 Russian Federation
BT1 Commonwealth of Independent States
BT2 Samara Russian Federation
BT2 Europe

Samara Russian Federation (1993)
Oblast and city in W Russian Federation. Renamed in 1990s. Before 1993, also search Kuibyshev.

Before 1978, also search Kuybyshev. This term has multiple hierarchies.
    UF Kuibyshev Russian Federation
    UF Kuybyshev Russian Federation
    BT1 Russian Federation
    BT1 Commonwealth of Independent States
    BT2 Europe
    NT Samara Bend
    SA Sarbay
    SA Vodino Russian Federation
Samaria (1978)
    Region extending from the Mediterranean Sea to the Jordan River S of Galilee and N of Judaea. Includes most of the Israeli occupied West Bank of the Jordan River.
    IN Index countries as applicable.
    BT Middle East
    BT Asia
    SA Israel
    SA Jordan
samarium (1978)
    Autoposting of rare earths and metals to this term began in 1989.
    UF Sm
    BT rare earths
    BT metals
    NT Sm-147/Nd-144
    SA Sm/Nd
Samarkand
    No longer a valid term for GeoRef. As of 1993, see Samarkand Uzbekistan.
Samarkand Uzbekistan (1993)
    Oblast and city in SE Uzbekistan. Before 1978, also search Samarqand. This term has multiple hierarchies.
    UF Samarqand Uzbekistan
    BT1 Uzbekistan
    BT1 Asia
    BT2 Uzbekistan
    BT2 Commonwealth of Independent States
    NT Lyangar Uzbekistan
    NT Zirabulak Uzbekistan
    SA Nura-Tau
Samarqand Uzbekistan
    use Samarkand Uzbekistan
samarskite (1978)
    Autoposting of niobotantalates began in 1989. This term has multiple hierarchies.
    BT1 niobotantalates
    BT1 niobates
    BT1 oxides
    BT2 niobotantalates
    BT2 tantalates
    BT2 oxides
Sambagawa Belt (1978)
    Innermost and oldest of three belts of different metamorphic grades of schist in a Paleozoic Group. SW Honshu, Kyushu, and Shikoku. Also search Sambagawa; Sambagawa Schist.
    UF Sanbagawa Belt
    BT Paleozoic
    SA Honshu
    SA Kyushu
    SA Shikoku
Sambalpur
    No longer a valid term for GeoRef. See Sambalpur India.
Sambalpur India (1993)
    Town in N Orissa, E India.
    BT Orissa India

    BT India
    BT Indian Peninsula
    BT Asia
Samoa (1978)
    Group of volcanic islands in SW central Pacific Ocean. American Samoa is in E part of group, and independent Western Samoa comprises the W part.
    UF American Samoa
    UF Samoa Islands
    UF Western Samoa
    BT Polynesia
    BT Oceania
Samoa Islands
    use Samoa
Samos (1978)
    Island in the Aegean Sea off W coast of Turkey. Also a Greek department. This term has multiple hierarchies.
    BT1 Greek Aegean Islands
    BT1 Greece
    BT1 Southern Europe
    BT1 Europe
    BT2 Greek Aegean Islands
    BT2 Aegean Islands
    BT2 Mediterranean region
sample location maps
    use site location maps
sample preparation (1978)
    SA centrifuge methods
    SA chemical analysis
    SA decrepitation
    SA differential thermal analysis
    SA micropaleontology
    SA paleontology
    SA palynology
    SA preparation
    SA samples
    SA sampling
    SA soil sampling
    SA soils
    SA staining
    SA techniques
    SA thermal analysis
samplers (1985)
    SA instruments
    SA sampling
    SA sediment traps
samples (1978)
    Used for physical samples, not for statistical meaning.
    SA dredged samples
    SA lunar samples
    SA sample preparation
    SA sampling
    SA thin sections
sampling (1978)
    SA analysis
    SA dredged samples
    SA field studies
    SA panning
    SA placers
    SA probability
    SA reliability
    SA sample preparation
    SA samplers
    SA samples
    SA sediment traps
    SA soil sampling
    SA techniques
San Andreas Fault (1978)
    A fault system or zone extending for more than 600 miles from the Pacific Ocean at Point Arena through the San Francisco Peninsula and on into S California. Also search San Andreas.
    CO N340000N400000
        W1173000W1243000

    BT California
    BT United States
    SA Hosgri Fault
    SA Parkfield earthquakes
    SA San Gregorio Fault
San Andres Formation (1978)
    In Manzano Group. Central and SE New Mexico. Also search San Andres.
    BT Permian
    BT Paleozoic
    SA Guadalupian
    SA Leonardian
    SA New Mexico
San Antonio
    Valid through 1988. Search in combination with state term. After 1988, use specific city-state term.
San Antonio Mine (1993)
    Gold ores in Bissett area, SE Manitoba and in Sonora Mexico. Polymetallic ores in Santa Eulalia District, Chichuahua, Mexico. Also occurs with numerous other locations and commodities.
    IN Index country and commodities as applicable.
    SA Chihuahua Mexico
    SA gold ores
    SA Manitoba
    SA Mexico
    SA mines
    SA polymetallic ores
    SA Sonora Mexico
San Antonio Texas (1989)
    City on San Antonio River in S central Texas. Before 1989, search San Antonio AND Texas.
    CO N292500N292500
        W0983000W0983000
    BT Bexar County Texas
    BT Texas
    BT United States
San Benito County
    Valid through 1988. Search in combination with state term. After 1988, use specific county-state term.
San Benito County California (1989)
    W California. Before 1989, also search San Benito County AND California.
    CO N361300N365800
        W1203700W1213700
    BT California
    BT United States
    NT Hollister California
    NT Pinnacles National Monument
    NT San Juan Bautista California
    SA Santa Clara Valley
San Bernardino
    Valid through 1988. Search in combination with state term. After 1988, use specific city-state term.
San Bernardino California (1989)
    City in S California. Before 1989, search San Bernardino AND California.
    CO N340700N340700
        W1171800W1171800
    BT San Bernardino County California
    BT California
    BT United States
San Bernardino County
    Valid through 1988. Search in combination with state term. After 1988, use specific county-state term.

San Bernardino County California (1989)
    S California. Before 1989, also search San Bernardino County AND California.
    CO N340000N354500
        W1141500W1174000
    BT California
    BT United States
    NT Avawatz Mountains
    NT Barstow California
    NT Cady Mountains
    NT Cajon Pass
    NT Cima volcanic field
    NT Needles California
    NT Old Woman Mountains
    NT San Bernardino California
    NT San Gorgonio Mountain
    NT San Gorgonio Pass
    NT Trona California
    NT Whipple Mountains
    SA Borax Lake
    SA Coyote Lake
    SA Death Valley
    SA Kingston Range
    SA Mohave Mountains
    SA Mojave Desert
    SA Nopah Range
    SA San Bernardino Mountains
    SA Santa Ana River
    SA Searles Lake
San Bernardino Mountains (1978)
    In San Bernardino and Riverside counties in S California.
    IN Index counties and Coast Ranges as applicable.
    SA California
    SA Coast Ranges
    SA Riverside County California
    SA San Bernardino County California
    SA San Gorgonio Mountain
    SA San Gorgonio Pass
    SA Transverse Ranges
San Buenaventura
    use Ventura California
San Carlos Indian Reservation (1978)
    SE central Arizona. Also search San Carlos.
    BT Arizona
    BT United States
    SA Indian reservations
    SA San Carlos Olivine
San Carlos Olivine (1993)
    Peridotite and pyroxenite inclusions from San Carlos, Arizona, 30 kilometers E of Globe, Arizona. Also search Peridot Mesa, Rice, Gila County and Arizona.
    SA Gila County Arizona
    SA materials
    SA mineral inclusions
    SA San Carlos Indian Reservation
San Clemente Island (1978)
    One of the Channel Islands in the Pacific Ocean W of San Diego. Also search San Clemente. This term has multiple hierarchies.
    BT1 Channel Islands
    BT1 California
    BT1 United States
    BT2 Los Angeles County California
    BT2 California
    BT2 United States
San Diego
    Valid through 1988. Search in combination with state term. After 1988, use specific city-state term.
San Diego California (1989)

City on the Pacific Ocean in S California. Before 1989, search San Diego AND California.
CO N324500N324500
   W1171000W1171000
BT San Diego County California
BT California
BT United States
SA La Jolla California

San Diego County
Valid through 1988. Search in combination with state term. After 1988, use specific county-state term.

San Diego County California (1989)
S California. Before 1989, also search San Diego County AND California.
CO N323000N333000
   W1161000W1174000
BT California
BT United States
NT La Jolla California
NT Point Loma
NT San Diego California
SA Anza Desert State Park
SA San Jacinto Fault

San Diego Formation (1978)
Rests with angular unconformity upon Rose Canyon Formation of the La Jolla, or on Poway Conglomerate, or overlaps them and rests with marked unconformity upon Black Mountain volcanics; unconformably underlies Sweitzer Formation. S California.
BT Pliocene
BT Neogene
BT Tertiary
BT Cenozoic
SA California

San Diego Trough (1978)
Just off San Diego.
BT Pacific Ocean

San Emigdio Mountains (1989)
Part of the southern wall of the San Joaquin Valley, linking Temblor Range with Tehachapi Mountains in S California.
IN Index county or counties as applicable.
BT California
BT United States
SA Kern County California
SA San Joaquin Valley
SA Tehachapi Mountains
SA Temblor Range
SA Ventura County California

San Felipe Formation (1978)
BT Cretaceous
BT Mesozoic
SA Mexico

San Fernando
Valid through 1988. Search in combination with state term. After 1988, use specific city-state term.

San Fernando California (1989)
City in S California. Before 1989, search San Fernando AND California.
CO N341700N341700
   W1182700W1182700
BT Los Angeles County California
BT California
BT United States
SA San Fernando earthquake 1971

San Fernando earthquake 1971 (1985)
Epicenter near San Fernando, California.
UF San Fernando Valley earthquake 1971
BT earthquakes
SA California
SA San Fernando California

San Fernando Valley (1978)
Fertile basin about 20 miles NW of downtown Los Angeles.
IN Index counties or region as applicable.
SA California
SA Sylmar Fault

San Fernando Valley earthquake 1971
use San Fernando earthquake 1971

San Francisco
Valid through 1988. Search in combination with state term. After 1988, use specific city-state term.

San Francisco Bay (1978)
Inlet connected to the Pacific Ocean via the Golden Gate in W California.
BT California
BT United States
SA Berkeley California
SA Greenville Fault
SA San Francisco Bay region

San Francisco Bay region (1978)
Composed of the 12 counties surrounding San Francisco Bay in W California.
IN Index counties as applicable.
CO N365300N385600
   W1205500W1233400
BT California
BT United States
SA Alameda County California
SA Contra Costa County California
SA Loma Prieta earthquake 1989
SA Marin County California
SA Napa County California
SA Sacramento County California
SA San Francisco Bay
SA San Francisco County California
SA San Francisco Peninsula
SA San Joaquin County California
SA San Mateo County California
SA Santa Clara County California
SA Solano County California
SA Sonoma County California
SA Yolo County California

San Francisco California (1989)
City on the Pacific Ocean coextensive with San Francisco County, W California. Before 1989, search San Francisco AND California.
CO N374500N374500
   W1222700W1222700
BT San Francisco County California
BT California
BT United States
SA San Francisco earthquake 1906

San Francisco County
Valid through 1988. Search in combination with state term. After 1988, use specific county-state term.

San Francisco County California (1989)

Coextensive with city of San Francisco. Before 1989, also search San Francisco County AND California.
CO N374200N374800
   W1222500W1223500
BT California
BT United States
NT San Francisco California
SA San Francisco Bay region

San Francisco de la Selva
use Copiapo Chile

San Francisco earthquake 1906 (1985)
Epicenter near San Francisco, California.
BT earthquakes
SA California
SA San Francisco California

San Francisco Mountain (1978)
One of the three peaks N of Flagstaff in Coconino County known as San Francisco Peaks.
IN Index county or region as applicable.
UF Humphreys Peak
SA Coconino County Arizona
SA San Francisco Peaks

San Francisco Mountains (1978)
Range primarily in Catron County, New Mexico.
IN Index counties or regions as applicable.
SA Arizona
SA Catron County New Mexico
SA New Mexico
SA United States

San Francisco Peaks (1978)
An eroded volcano 10 miles N of Flagstaff in Coconino County with Agassiz, Fremont and San Francisco (Humphreys) peaks on its rim.
UF San Francisco Volcanic Field
BT Arizona
BT United States
SA San Francisco Mountain

San Francisco Peninsula (1978)
Extends S from San Francisco the length of San Francisco Bay.
IN Index counties or region as applicable.
SA California
SA San Francisco Bay region

San Francisco Volcanic Field
use San Francisco Peaks

San Gabriel Fault (1985)
S California.
BT California
BT United States
SA San Gabriel Mountains

San Gabriel Mountains (1978)
Range SW of the Mojave Desert primarily in Los Angeles County in S California. Also search San Gabriel.
IN Index counties and Coast Ranges or region as applicable.
BT California
BT United States
SA Coast Ranges
SA Los Angeles County California
SA San Gabriel Fault
SA Transverse Ranges

San Giorgio Mountain (1978)
Ticino, S Switzerland.
BT Ticino Switzerland
BT Switzerland
BT Central Europe
BT Europe

SA Alps
SA Lepontine Alps

San Gorgonio Mountain (1989)
Highest peak in the San Bernardino Mountains, S California.
IN Index Coast Ranges and San Bernardino Mountains as applicable.
CO N340500N340500
   W1165000W1165000
BT San Bernardino County California
BT California
BT United States
SA Coast Ranges
SA San Bernardino Mountains
SA San Gorgonio Pass

San Gorgonio Pass (1989)
Railroad and highway pass located between San Gorgonio and San Jacinto mountains in NW San Bernardino Mountains, S California. Connects San Bernardino Valley with the Coachella Valley.
CO N340300N340700
   W1164800W1165200
BT San Bernardino County California
BT California
BT United States
SA San Bernardino Mountains
SA San Gorgonio Mountain
SA San Jacinto Mountains

San Gregorio Fault (1981)
W central California. Component of San Andreas fault system.
CO N361000N375500
   W1214500W1225000
BT California
BT United States
SA Hosgri Fault
SA San Andreas Fault

San Jacinto Fault (1978)
Imperial, Orange, Riverside and San Diego counties in S California; Sonora, Mexico. Also search San Jacinto.
IN Index North America or counties and/or states as applicable.
SA California
SA Imperial County California
SA Mexico
SA North America
SA Orange County California
SA Riverside County California
SA San Diego County California
SA Sonora Mexico

San Jacinto Mountains (1989)
Part of the Coast Ranges in SW California.
IN Index Coast Ranges as applicable.
BT California
BT United States
SA Coast Ranges
SA Riverside County California
SA San Gorgonio Pass

San Joaquin Basin (1989)
Central California.
IN Index counties or region as applicable.
SA California
SA San Joaquin County California
SA San Joaquin River
SA San Joaquin Valley

San Joaquin County
Valid through 1988. Search in combination with state term. After 1988, use specific county-state term.

San Joaquin County California (1989)
Central California. Before 1989, also search San Joaquin County AND California.
CO N372800N382000
    W1205500W1213500
BT California
BT United States
SA San Francisco Bay region
SA San Joaquin Basin

San Joaquin River (1985)
Central California. Flows into San Pablo Bay.
IN Index counties or region as applicable.
SA California
SA San Joaquin Basin
SA San Joaquin Valley

San Joaquin Valley (1978)
Southern half of Central Valley. Extends from Buena Vista Lake in S to Sacramento Valley in N. Also search San Joaquin.
IN Index counties as applicable.
CO N354000N380500
    W1190000W1214000
BT California
BT United States
SA Central Valley
SA San Emigdio Mountains
SA San Joaquin Basin
SA San Joaquin River
SA White Wolf Fault

San Jose
Valid through 1988. Search in combination with state term. After 1988, use specific city-state term.

San Jose California (1989)
City in W California. Before 1989, search San Jose AND California.
CO N372000N372000
    W1215500W1215500
BT Santa Clara County California
BT California
BT United States

San Jose Formation (1978)
Lithology of formation is highly variable, both vertically and horizontally. N New Mexico, and S Colorado.
BT lower Eocene
BT Eocene
BT Paleogene
BT Tertiary
BT Cenozoic
SA Colorado
SA New Mexico

San Juan
No longer a valid term for GeoRef. From 1981-1989 referred to the city on the Atlantic Ocean in N Puerto Rico. Before 1981, also referred to province in Argentina. See San Juan Puerto Rico.

San Juan Argentina (1993)
In W Argentina. Before 1981, also search San Juan AND Argentina. Before 1993, also search San Juan Province AND Argentina.
CO S320000S283000
    W0670000W0703000
BT Argentina
BT South America
NT Barreal Argentina
NT Talacasto Argentina
SA Paganzo Basin
SA Pampean Mountains
SA San Juan River

San Juan Basin (1978)
IN Index states or region as applicable.
SA Arizona
SA Colorado
SA New Mexico
SA San Juan mining district
SA San Juan Mountains
SA San Juan River
SA San Juan volcanic field
SA United States
SA Utah

San Juan Bautista
No longer a valid term for GeoRef. Valid through 1988. Search in combination with state term. After 1988, use specific city-state term.

San Juan Bautista California (1989)
Town in San Benito County, W California. Before 1989, search San Juan Bautista AND California.
BT San Benito County California
BT California
BT United States

San Juan County
Valid through 1988. Search in combination with state term. After 1988, use specific county-state term.

San Juan County Colorado (1989)
SW Colorado. Before 1989, also search San Juan County AND Colorado.
CO N373800N375700
    W1073000W1075800
BT Colorado
BT United States
NT Silverton Caldera
SA Needle Mountains
SA San Juan mining district
SA Sunlight
SA Sunnyside Mine

San Juan County New Mexico (1989)
Extreme NW New Mexico. Before 1989, also search San Juan County AND New Mexico.
CO N360000N370000
    W1072800W1090400
BT New Mexico
BT United States

San Juan County Utah (1989)
SE Utah. Before 1989, also search San Juan County AND Utah.
CO N370000N383000
    W1090300W1111300
BT Utah
BT United States
NT Monticello Utah
SA Canyonlands National Park
SA Gibson Dome
SA Homestake Mine
SA Lake Powell
SA Monticello Reservoir
SA Salt Valley

San Juan County Washington (1989)
NW Washington. Before 1989, also search San Juan County AND Washington.
CO N481700N485100
    W1224200W1231800
BT Washington
BT United States
SA San Juan Islands

San Juan District
As of 1993, no longer a valid term for GeoRef.
use San Juan mining district

San Juan Formation (1978)
Middle and upper Tertiary in SW Colorado; Pleistocene in Puerto Rico.
IN Index ages as applicable.
BT Cenozoic
SA Colorado
SA Pleistocene
SA Puerto Rico
SA Tertiary

San Juan Islands (1978)
Group of islands off NW Washington and E of Vancouver Island.
IN Index county or region as applicable.
SA Cache Creek Group
SA Chuckanut Formation
SA San Juan County Washington

San Juan mining district (1993)
Mining region in SW Colorado. Also in other locations. Before 1981, also search San Juan AND Colorado. Before 1993, also search San Juan District.
IN Index county or region as applicable.
UF San Juan District
SA San Juan Basin
SA San Juan County Colorado
SA San Juan Mountains
SA San Juan River
SA San Juan volcanic field
SA silver ores

San Juan Mountains (1978)
Range of the Rocky Mountains in SW Colorado and N New Mexico. Autoposting of United States to this term began in 1989. This term has multiple hierarchies.
IN Index states and counties as applicable.
CO N370000N374500
    W1063000W1073000
BT1 U. S. Rocky Mountains
BT1 Rocky Mountains
BT2 U. S. Rocky Mountains
BT2 United States
SA Colorado
SA New Mexico
SA San Juan Basin
SA San Juan mining district
SA San Juan River
SA San Juan volcanic field
SA Silverton Caldera
SA Sunlight

San Juan Province
No longer a valid term for GeoRef. See San Juan Argentina.

San Juan Puerto Rico (1993)
City on the Atlantic Ocean in N Puerto Rico. Before 1993 also search San Juan AND Puerto Rico.
BT Puerto Rico
BT Greater Antilles
BT Antilles
BT West Indies
BT Caribbean region

San Juan River (1978)
Rises in S Colorado and flows SW, bends W, then NW emptying into the Colorado River in SE Utah. Also search San Juan AND (Colorado OR New Mexico OR Utah). Also in S San Juan Province, W Argentina; SW Bolivia; W Colombia; Nuevo Leon and Tamaulipas, NE Mexico; S Nicaragua; Sucre and Monagas, NE Venezuela.
IN Index states and/or countries as applicable.
SA Argentina
SA Bolivia
SA Colombia
SA Colorado
SA Mexico
SA Monagas Venezuela
SA New Mexico
SA Nicaragua
SA Nuevo Leon Mexico
SA San Juan Argentina
SA San Juan Basin
SA San Juan mining district
SA San Juan Mountains
SA San Juan volcanic field
SA Sucre Venezuela
SA Tamaulipas Mexico
SA Utah
SA Venezuela

San Juan volcanic field (1978)
In San Juan Mountains of SW Colorado. Autoposting of broader terms to this term began in 1989.
BT Colorado
BT United States
SA Fish Canyon Tuff
SA San Juan Basin
SA San Juan mining district
SA San Juan Mountains
SA San Juan River

San Lorenzo Formation (1978)
Eocene and Oligocene. Includes Twobar Shale and Rices Mudstone. In Santa Cruz Mountain region of S California. Also search San Lorenzo.
IN Index ages as applicable.
BT Paleogene
BT Tertiary
RT Cenozoic
SA California
SA Eocene
SA Oligocene

San Luis
No longer a valid term for GeoRef. See San Luis Argentina or other country if applicable.

San Luis Argentina (1993)
Province including city in W central Argentina.
BT Argentina
BT South America
SA Pampean Mountains
SA Salado Basin

San Luis Obispo
Valid through 1988. Search in combination with state term. After 1988, use specific city-state term.

San Luis Obispo California (1989)
City in SW California. Before 1989, search San Luis Obispo AND California.
CO N351600N351600
    W1204000W1204000
BT San Luis Obispo County California
BT California
BT United States

San Luis Obispo County
Valid through 1988. Search in combination with state term. After 1988, use specific county-state term.

San Luis Obispo County California (1989)

SW California. Before 1989, also search San Luis Obispo County AND California.
  CO N345300N354800
     W1192800W1212300
  BT California
  BT United States
  NT Cholame California
  NT Pismo Basin
  NT San Luis Obispo California
  SA Hosgri Fault
  SA Salinas Valley
  SA Santa Lucia Range
  SA Santa Maria Basin

San Luis Potosi
  No longer a valid term for GeoRef. As of 1993 see San Luis Potosi Mexico.

San Luis Potosi Mexico (1993)
  State and city in E central Mexico. Before 1993 also search San Luis Potosi AND Mexico.
  BT Mexico
  NT Valles Mexico
  SA Sierra Madre Oriental

San Luis Valley (1978)
  Once bottom of extensive lake. In Saguache, Alamosa, and Conejos counties in S Colorado and in Taos County in New Mexico.
  IN Index states and counties as applicable.
  SA Alamosa County Colorado
  SA Colorado
  SA Conejos County Colorado
  SA New Mexico
  SA Saguache County Colorado
  SA Taos County New Mexico
  SA United States

San Manuel
  Valid through 1988. Search in combination with state term. After 1988, use specific city-state term.

San Manuel Arizona (1989)
  Village in S central Arizona. Before 1989, search San Manuel AND Arizona.
  CO N323559N323559
     W1103749W1103749
  BT Pinal County Arizona
  BT Arizona
  BT United States

San Marcos Arch (1978)
  S central Texas. Also search San Marcos.
  BT Texas
  BT United States

San Marino (1978)
  Republic in the Apennines, N part of Italian peninsula near the Adriatic coast.
  BT Southern Europe
  BT Europe
  SA Apennines
  SA Emilia-Romagna Italy
  SA Italy
  SA Marches Italy

San Mateo County
  Valid through 1988. Search in combination with state term. After 1988, use specific county-state term.

San Mateo County California (1989)
  W California. Before 1989, also search San Mateo County AND California.
  CO N370600N374200
     W1221000W1223500
  BT California

  BT United States
  NT Portola Valley California
  SA San Francisco Bay region
  SA San Pedro Valley

San Mateo Mountains (1978)
  Mountains in SW Socorro County, SW central New Mexico. Also in other location.
  IN Index county or regions as applicable.
  SA New Mexico
  SA Socorro County New Mexico

San Miguel County
  Valid through 1988. Search in combination with state term. After 1988, use specific county-state term.

San Miguel County Colorado (1989)
  SW Colorado. Before 1989, also search San Miguel County AND Colorado.
  CO N374800N380800
     W1074500W1090200
  BT Colorado
  BT United States

San Miguel County New Mexico (1989)
  NE New Mexico. Before 1989, also search San Miguel County AND New Mexico.
  CO N350400N355300
     W1033900W1054300
  BT New Mexico
  BT United States

San Miguel Formation (1985)
  S Texas.
  BT Gulfian
  BT Upper Cretaceous
  BT Cretaceous
  BT Mesozoic
  SA Texas

San Miguel Island (1978)
  Northernmost of Channel Islands in Pacific Ocean SW of Santa Barbara. May occur in other locations. Also search San Miguel AND appropriate area.
  IN Index Channel Islands and county or region as applicable.
  SA Channel Islands
  SA Santa Barbara County California

San Nicolas Basin (1989)
  California continental borderland, S California.
  IN Index county or region as applicable.
  SA California

San Nicolas Island (1978)
  In central Channel Islands in the Pacific Ocean off SW California. Also search San Nicolas.
  IN Index Channel Islands and Ventura County California as applicable.
  SA Channel Islands
  SA Ventura County California

San Onofre Breccia (1978)
  Underlies Monterey Formation; overlies Cozy Dell. San Diego County in S California.
  BT middle Miocene
  BT Miocene
  BT Neogene
  BT Tertiary
  BT Cenozoic
  SA California

San Patricio County
  Valid through 1988. Search in combination with state term. After 1988, use specific county-state term.

San Patricio County Texas (1989)
  S Texas. Before 1989, also search San Patricio County AND Texas.
  CO N275000N281400
     W0971000W0975500
  BT Texas
  BT United States
  SA Corpus Christi Bay

San Pedro (1978)
  Harbor in Los Angeles County. Former city which was annexed to Los Angeles in 1909.
  IN Index county or regions as applicable.
  SA Los Angeles County California
  SA Palos Verdes Peninsula

San Pedro Basin (1985)
  S California continental borderland.
  IN Index county or regions as applicable.
  SA California
  SA North American Pacific

San Pedro Hills
  use Palos Verdes Hills

San Pedro Valley (1978)
  In San Mateo County in W California. Also in SE Arizona. Also search San Pedro AND California or Arizona if applicable.
  IN Index county or region as applicable.
  SA Arizona
  SA San Mateo County California

San Rafael Group (1985)
  Middle and Upper Jurassic. Arizona, New Mexico, E Utah and W Colorado.
  BT Jurassic
  BT Mesozoic
  SA Arizona
  SA Carmel Formation
  SA Colorado
  SA Entrada Sandstone
  SA New Mexico
  SA Todilto Formation
  SA Utah

San Rafael Swell (1978)
  E central Utah. Also search San Rafael.
  BT Utah
  BT United States

San Saba County
  Valid through 1988. Search in combination with state term. After 1988, use specific county-state term.

San Saba County Texas (1989)
  Central Texas. Before 1989, also search San Saba County AND Texas.
  CO N305200N313000
     W0982500W0990800
  BT Texas
  BT United States

San Salvador (1978)
  Island in E central Bahamas. As of 1993, use San Salvador El Salvador for the city.
  IN Index Bahamas if applicable.
  SA Bahamas

San Salvador earthquake 1986 (1993)
  El Salvador. Before 1993, also search El Salvador earthquake 1986.
  UF El Salvador earthquake 1986
  BT earthquakes
  SA El Salvador
  SA San Salvador El Salvador

San Salvador El Salvador (1993)
  City in central El Salvador. Before 1993, search San Salvador AND El Salvador.
  BT El Salvador
  BT Central America
  SA San Salvador earthquake 1986

San San
  use Sansan Panama

San Sebastian
  No longer a valid term for GeoRef. As of 1993, see San Sebastian Spain.

San Sebastian Spain (1993)
  City on Bay of Biscay in Basque country, N Spain. Before 1993, also search San Sebastian and Spain.
  BT Guipuzcoa Spain
  BT Basque Provinces Spain
  BT Spain
  BT Iberian Peninsula
  BT Southern Europe
  BT Europe

Sanbagawa Belt
  use Sambagawa Belt

sand (1978)
  As of 1981, use sands or gravel deposits for sand as a commodity. Also includes use as a material in engineering geology.
  BT clastic sediments
  BT sediments
  SA aggregate
  SA alluvium
  SA arenaceous texture
  SA carbonate sediments
  SA construction materials
  SA ergs
  SA grains
  SA gravel
  SA greensand
  SA loam
  SA Ottawa Sand
  SA quartz sand
  SA quicksand
  SA rounding
  SA sand sheets
  SA sands
  SA sandstone
  SA silica
  SA silt
  SA soils
  SA terrigenous materials

sand bars
  use bars

sand bodies (1978)
  BT planar bedding structures
  BT sedimentary structures
  SA bedding plane irregularities
  SA blowouts

sand boils (1989)
  SA blowouts
  SA discharge
  SA engineering geology
  SA floods
  SA fluvial features
  SA geomorphology
  SA levees
  SA soil mechanics

SA springs

sand dunes
   use dunes

sand equivalent tests (1981)
   SA drilling
   SA petroleum engineering
   SA testing

Sand Hills (1978)
   Belt, 20 to 40 miles wide, of low, sandy hills extending along inner border of coastal plain from central North Carolina to central Georgia. Also large area of stable dunes covered with vegetation in NW central Nebraska.
   IN Index states as applicable.
   SA Georgia
   SA Nebraska
   SA North Carolina
   SA South Carolina
   SA United States

sand ridges (1993)
   Use sand waves if applicable.
   BT bedding plane irregularities
   BT sedimentary structures
   SA barrier beaches
   SA barrier islands
   SA bars
   SA bottom features
   SA dunes
   SA ergs
   SA longshore bars
   SA megaripples
   SA ripple marks
   SA sand waves
   SA shore features

sand seas (1993)
   SA deserts
   SA dunes
   SA eolian features
   SA ergs
   SA periglacial features
   SA sand sheets

sand sheets (1993)
   BT eolian features
   SA deserts
   SA dunes
   SA ergs
   SA remote sensing
   SA sand
   SA sand seas

sand volcanoes
   use mud volcanoes

Sand Wash Basin (1981)
   NW Colorado.
   IN Index county or region as applicable.
   SA Colorado
   SA Green River basin

sand waves (1978)
   BT bedding plane irregularities
   BT sedimentary structures
   SA antidunes
   SA dunes
   SA megaripples
   SA ripple marks
   SA sand ridges

Sandakan
   No longer a valid term for GeoRef. As of 1993, see Sandakan Malaysia.

Sandakan Malaysia (1993)
   City on Sandakan Harbor on Sulu Sea in Sabah in East Malaysia. This term has multiple hierarchies.
   BT1 Sabah Malaysia
   BT1 East Malaysia
   BT1 Borneo
   BT1 Far East
   BT1 Asia
   BT2 Sabah Malaysia
   BT2 East Malaysia
   BT2 Borneo
   BT2 Malay Archipelago
   BT3 Sabah Malaysia
   BT3 East Malaysia
   BT3 Malaysia
   BT3 Far East
   BT3 Asia

sandarac
   use realgar

Sandelzhausen
   No longer a valid term for GeoRef. See Sandelzhausen Germany.

Sandelzhausen Germany (1993)
   Village N of Munich.
   BT Bavaria Germany
   BT Germany
   BT Central Europe
   BT Europe

Sanders County
   Valid through 1988. Search in combination with state term. After 1988, use specific county-state term.

Sanders County Montana (1989)
   NW Montana. Before 1989, also search Sanders County AND Montana.
   CO N471000N482000
        W1141500W1160500
   BT Montana
   BT United States
   SA Coeur d'Alene mining district

Sandia Formation (1989)
   Pennsylvanian; in Magdalena Group of N central New Mexico and S Colorado. Also in Peru.
   IN Index age as applicable.
   SA Colorado
   SA Magdalena Group
   SA New Mexico
   SA Pennsylvanian
   SA Peru

Sandia Granite (1989)
   Central New Mexico.
   BT Precambrian
   SA New Mexico

Sandia Mountains (1978)
   NE of Albuquerque in N central New Mexico.
   BT New Mexico
   BT United States

Sandnes
   No longer a valid term for GeoRef. As of 1993 see Sandnes Norway.

Sandnes Norway (1993)
   Town S of Stavanger in SW Rogaland, SW Norway. Before 1993 also search Sandnes AND Norway.
   BT Rogaland Norway
   BT Norway
   BT Scandinavia
   BT Western Europe
   BT Europe

Sandomierz
   No longer a valid term for GeoRef. As of 1993 see Sandomierz Poland.

Sandomierz Poland (1993)
   Town on Vistula River in central Tarnobrzeg, SE central Poland. Before 1993 also search Sadomierz AND Poland.
   UF Sandomir
   BT Tarnobrzeg Poland
   BT Poland
   BT Central Europe
   BT Europe

Sandomir
   use Sandomierz Poland

Sandoval County
   Valid through 1988. Search in combination with state term. After 1988, use specific county-state term.

Sandoval County New Mexico (1989)
   NW central New Mexico. Before 1989, also search Sandoval County AND New Mexico.
   CO N351200N361200
        W1061500W1074200
   BT New Mexico
   BT United States
   NT Baca Field
   NT Fenton Hill
   SA Rio Puerco

sands (1978)
   Before 1981, search sand AND deposits. For sand used as a construction material, use gravel deposits. Before 1981, included use as a type of material in engineering geology; as of 1981, use sand.
   IN See List C for use as a commodity term for glass, ceramic, chemical use, etc.
   SA aggregate
   SA construction materials
   SA glass materials
   SA gravel deposits
   SA sand
   SA silica

sandstone (1978)
   As of 1981, use sandstone deposits for sandstone as a commodity.
   UF psammite
   BT clastic rocks
   BT sedimentary rocks
   SA arenaceous texture
   SA arkose
   SA gaize
   SA graywacke
   SA greensand
   SA metasandstone
   SA microbreccia
   SA orthoquartzite
   SA red beds
   SA sand
   SA sandstone deposits
   SA sandstone dikes
   SA subgraywacke
   SA terrigenous materials

sandstone deposits (1981)
   Before 1981, search sandstone or psammite AND deposits.
   IN Commodity. See List C.
   SA building stone
   SA construction materials
   SA dimension stone
   SA sandstone

sandstone dikes (1978)
   BT soft sediment deformation
   BT sedimentary structures
   SA clastic dikes
   SA sandstone

sandstone-type (1989)
   SA gold ores
   SA mineral deposits, genesis
   SA uranium ores

Sandur
   No longer a valid term for GeoRef. See Sandur India.

Sandur India (1993)
   Town in central Karnataka, SW India.
   BT Karnataka India
   BT India
   BT Indian Peninsula
   BT Asia

sandurs
   use outwash plains

Sandusky County
   Valid through 1988. Search in combination with state term. After 1988, use specific county-state term.

Sandusky County Ohio (1989)
   N Ohio. Before 1989, also search Sandusky County AND Ohio.
   CO N411500N413000
        W0824900W0832500
   BT Ohio
   BT United States
   SA Sandusky River basin

Sandusky River basin (1981)
   N Ohio. Autoposting of broader terms to this term began in 1989.
   IN Index counties as applicable.
   CO N404000N413000
        W0824200W0832500
   BT Ohio
   BT United States
   SA Crawford County Ohio
   SA Sandusky County Ohio
   SA Seneca County Ohio
   SA Wyandot County Ohio

Sandy Hook (1978)
   Peninsula in NE Monmouth County constituting S side of entrance to lower New York Bay.
   IN Index county or region as applicable.
   SA Monmouth County New Jersey

sandy texture
   use arenaceous texture

Sangamon (1978)
   River which flows into the Illinois River in central Illinois.
   IN Index counties as applicable.
   BT Illinois
   BT United States

Sangamon County Illinois (1989)
   Central Illinois. Before 1989, search Sangamon County AND Illinois.
   CO N393100N395800
        W0891200W0900000
   BT Illinois
   BT United States
   NT Springfield Illinois

Sangamonian (1978)
   Pertaining to third interglacial stage of Pleistocene Epoch in North America.
   BT upper Pleistocene
   BT Pleistocene
   BT Quaternary
   BT Cenozoic
   SA Riss/Wurm Interglacial

Sanganguey (1989)
   Volcano. Nayarit, W Mexico.
   CO N212630N212630
        W1044000W1044000
   BT Nayarit Mexico
   BT Mexico

Sangerville Formation (1989)
   Lower and Middle Silurian.
   BT Silurian
   BT Paleozoic
   SA Maine

Sangilen Mountains (1978)

S Tuva Autonomous Republic just N of Mongolian border. Also search Sangilen. This term has multiple hierarchies.
BT1 Tuva Russian Federation
BT1 Russian Federation
BT1 Commonwealth of Independent States
BT2 Tuva Russian Federation
BT2 Siberian fold belt
BT2 Asia

Sangre de Cristo Mountains (1978)
Range of the Rocky Mountains. This term has multiple hierarchies.
IN Index states as applicable.
BT1 U. S. Rocky Mountains
BT1 Rocky Mountains
BT2 U. S. Rocky Mountains
BT2 United States
NT Spanish Peaks
SA Colorado
SA New Mexico
SA Questa Caldera
SA Sierra Blanca

Sangun Zone (1981)
Region of metamorphic rocks, Honshu, Japan.
BT Honshu
BT Japan
BT Far East
BT Asia

Sanibel Island (1985)
Off W coast of Florida. Autoposting of broader terms to this term began in 1989.
CO N262500N263000
  W0820000W0821500
BT Lee County Florida
BT Florida
BT United States
SA Gulf Coastal Plain
SA Gulf of Mexico

sanidine (1978)
Autoposting of alkali feldspar to this term began in 1981.
UF glassy feldspar
UF rhyacolite
BT alkali feldspar
BT feldspar group
BT framework silicates
BT silicates
SA K-feldspar
SA orthoclase

sanitary landfills (1978)
BT landfills
SA waste disposal

Sannoisian (1978)
Europe.
BT lower Oligocene
BT Oligocene
BT Paleogene
BT Tertiary
BT Cenozoic
SA Lattorfian
SA Stampian
SA Tongrian

Sanpete County
Valid through 1988. Search in combination with state term. After 1988, use specific county-state term.

Sanpete County Utah (1989)
Central Utah. Before 1989, also search Sanpete County AND Utah.
CO N390300N395000
  W1111700W1120500
BT Utah
BT United States

Sansan
No longer a valid term for GeoRef. As of 1993 see Sansan Panama.

Sansan Panama (1993)
Village in NW Panama. Before 1993 also search Sansan AND Panama.
UF San San
BT Panama
BT Central America

Santa Ana
Valid through 1988. Search in combination with state term. After 1988, use specific city-state term.

Santa Ana California (1989)
City at base of Santa Ana Mountains in S California. Before 1989, search Santa Ana AND California.
CO N334400N334400
  W1175400W1175400
BT Orange County California
BT California
BT United States

Santa Ana Mountains (1978)
Range along border between Orange and Riverside counties in S California. Also search Santa Ana AND California.
IN Index counties and mountains as applicable.
BT California
BT United States
SA Coast Ranges
SA Orange County California
SA Riverside County California

Santa Ana River (1985)
S California. Flows into Pacific Ocean.
IN Index counties or regions as applicable.
SA California
SA Orange County California
SA Riverside County California
SA San Bernardino County California

Santa Barbara
Valid through 1988. Search in combination with state term. After 1988, use specific city-state term.

Santa Barbara Basin (1978)
Undersea feature between Santa Barbara County and the N Channel Islands.
BT Pacific Ocean
SA California
SA Channel Islands

Santa Barbara California (1989)
City on Santa Barbara Channel in Santa SW California. Before 1989, search Santa Barbara AND California.
CO N342500N342500
  W1194100W1194100
BT Santa Barbara County California
BT California
BT United States

Santa Barbara Channel (1978)
Between Santa Barbara County and N Channel Islands.
BT California
BT United States
SA Channel Islands

Santa Barbara County
Valid through 1988. Search in combination with state term. After 1988, use specific county-state term.

Santa Barbara County California (1989)
SW California. Before 1989, also search Santa Barbara County AND California.
CO N342500N351000
  W1192500W1204500
BT California
BT United States
NT Lompoc California
NT Point Conception
NT Point Sal
NT Santa Barbara California
NT Santa Maria California
SA Channel Islands
SA Hosgri Fault
SA San Miguel Island
SA Santa Cruz Island
SA Santa Maria Basin
SA Ventura Basin

Santa Barbara Formation (1985)
S California.
BT Pleistocene
BT Quaternary
BT Cenozoic
SA California

Santa Barbara Islands
use Channel Islands

Santa Catalina Island (1978)
Island in central Channel Islands in Pacific Ocean off Long Beach. Also search Santa Catalina AND California. Also in other location.
IN Index Channel Islands and Los Angeles County California as applicable.
SA Channel Islands
SA Los Angeles County California

Santa Catalina Mountains (1978)
Small range in NE Pima County in S Arizona. Also search Santa Catalina AND Arizona.
SA Pima County Arizona

Santa Catarina
No longer a valid term for GeoRef. See Santa Catarina Brazil and Santa Catarina Island.

Santa Catarina Brazil (1993)
CO S292000S255000
  W0483000W0535000
BT Brazil
BT South America
SA Brazilian Shield
SA Estrada Nova Formation
SA Pelotas Basin
SA Santa Catarina Island
SA Serra do Mar

Santa Catarina Island (1978)
Island in Atlantic Ocean just off central Santa Catarina. Also search Santa Catarina.
SA Santa Catarina Brazil

Santa Clara
Valid through 1988. Search in combination with state term. After 1988, use specific city-state term. For Cuba, See Villa Clara Cuba.

Santa Clara California (1989)
City in W California. Before 1989, search Santa Clara AND California.
CO N372100N372100
  W1215800W1215800
BT Santa Clara County California
BT California
BT United States

Santa Clara County
Valid through 1988. Search in combination with state term. After 1988, use specific county-state term.

Santa Clara County California (1989)
W California. Before 1989, also search Santa Clara County AND California.
CO N365200N373000
  W1211200W1221200
BT California
BT United States
NT Palo Alto California
NT San Jose California
NT Santa Clara California
NT Sargent California
SA Coyote Lake earthquake 1979
SA Morgan Hill earthquake 1984
SA San Francisco Bay region
SA Santa Clara Valley

Santa Clara Formation (1985)
W California.
IN Index ages as applicable.
BT Cenozoic
SA California
SA Pleistocene
SA Pliocene

Santa Clara Meteorite (1985)
Impact in Mexico. Before 1985, also search Santa Clara AND Mexico AND meteorites.
BT iron meteorites
BT meteorites
SA Mexico

Santa Clara Valley (1978)
River valley in Los Angeles and Ventura counties. Also the S extension of the San Francisco Bay depression in Santa Clara and San Benito counties is called the Santa Clara Valley.
IN Index counties as applicable.
BT California
BT United States
SA Los Angeles County California
SA San Benito County California
SA Santa Clara County California
SA Ventura County California

Santa Cruz
No longer a valid term for GeoRef. See Santa Cruz Bolivia, Santa Cruz Argentina, or Santa Cruz Island.

Santa Cruz Argentina (1993)
Province including city in S Argentina.
BT Argentina
BT South America
NT Puerto Deseado
SA Baquero Formation
SA Patagonia
SA Santa Cruz River

Santa Cruz Bolivia (1993)
Department including the city in E Bolivia.
BT Bolivia
BT South America

Santa Cruz County
Valid through 1988. Search in combination with state term. After 1988, use specific county-state term.

Santa Cruz County Arizona (1989)
S Arizona. Before 1989, also search Santa Cruz County AND Arizona.
CO N311900N333900

Before then, variants of the term may occur in GeoRef.

   W1090600W1102800
BT Arizona
BT United States
NT Santa Rita Mountains
SA Santa Cruz River

Santa Cruz County California
 (1989)
 W California. Before 1989, also search Santa Cruz County AND California.
 CO N365100N371600
   W1213600W1221800
 BT California
 BT United States
 SA Loma Prieta earthquake 1989
 SA Monterey Bay

Santa Cruz Island (1978)
 Easternmost of northern islands in Channel Islands off Ventura County. Also search Santa Cruz AND California. For the "Santa Cruz Island" in the Galapagos Islands, use Chaves Island. May also occur in other locations.
 IN Index Channel Islands and Santa Barbara County California as applicable.
 SA Channel Islands
 SA Chaves Island
 SA Santa Barbara County California

Santa Cruz Mountains (1978)
 One of the Coast Ranges extending along the W side of Santa Clara Valley just S of San Francisco Bay. Also search Santa Cruz AND California.
 IN Index Coast Ranges and county or region as applicable.
 SA California
 SA Coast Ranges

Santa Cruz River (1989)
 River in Pinal, Pima and Santa Cruz counties, S Arizona. Also in Santa Cruz, S Argentina.
 IN Index Arizona or Argentina as applicable.
 SA Argentina
 SA Arizona
 SA Pima County Arizona
 SA Pinal County Arizona
 SA Santa Cruz Argentina
 SA Santa Cruz County Arizona

Santa Eulalia
 No longer a valid term for GeoRef. As of 1993 see Santa Eulalia Peru.

Santa Eulalia Peru (1993)
 Town on Santa Eulalia River with water reservoir and hydroelectric plant serving Lima. Before 1993 also search Santa Eulalia AND Peru.
 BT Lima Peru
 BT Peru
 BT South America

Santa Fe
 No longer a valid term for GeoRef. As of 1989, use Santa Fe New Mexico for the city in New Mexico. As of 1993, use Santa Fe Argentina for the province and city in Argentina. Search in combination with state or country.

Santa Fe Argentina (1993)
 Province including city in NE central Argentina.
 BT Argentina
 BT South America

Santa Fe County
 Valid through 1988. Search in combination with state term. After 1988, use specific county-state term.

Santa Fe County New Mexico
 (1989)
 N central New Mexico. Before 1989, also search Santa Fe County AND New Mexico.
 CO N350400N360000
   W1054200W1061500
 BT New Mexico
 BT United States
 NT Santa Fe New Mexico

Santa Fe Group (1985)
 N New Mexico and S central Colorado.
 IN Index ages as applicable.
 BT Cenozoic
 SA Colorado
 SA Miocene
 SA New Mexico
 SA Pleistocene
 SA Pliocene

Santa Fe New Mexico (1989)
 City in N central New Mexico. Before 1989, search Santa Fe AND New Mexico.
 CO N354100N354100
   W1055700W1055700
 BT Santa Fe County New Mexico
 BT New Mexico
 BT United States

Santa Isabel (1978)
 Volcanic island of the Solomon Islands in E central Solomon Islands.
 SA Solomon Islands

Santa Lucia Range (1978)
 One of the Coast Ranges in Monterey and San Luis Obispo counties in W California.
 IN Index county and Coast Ranges or region as applicable.
 SA California
 SA Coast Ranges
 SA Monterey County California
 SA San Luis Obispo County California

Santa Margarita Formation
 (1978)
 Includes Quatal Red Clay Member. S California. Also search Santa Margarita.
 BT upper Miocene
 BT Miocene
 BT Neogene
 BT Tertiary
 BT Cenozoic
 SA California

Santa Maria
 Valid through 1988. Search in combination with state term. After 1988, use specific city-state term.

Santa Maria Basin (1985)
 SW California. Autoposting of broader terms to this term began in 1989.
 IN Index counties as applicable.
 BT California
 BT United States
 SA Hosgri Fault
 SA San Luis Obispo County California
 SA Santa Barbara County California

Santa Maria California (1989)
 City in SW California. Before 1989, search Santa Maria AND California.
 CO N345600N345600
   W1202500W1202500
 BT Santa Barbara County California
 BT California
 BT United States

Santa Maria Formation (1978)
 S California.
 IN Index ages as applicable.
 BT Cenozoic
 SA California
 SA Pleistocene
 SA Pliocene

Santa Marta
 No longer a valid term for GeoRef. See Santa Marta Colombia.

Santa Marta Colombia (1993)
 City on Caribbean Sea, Magdalena Department, N Colombia.
 BT Magdalena Colombia
 BT Colombia
 BT South America
 SA Sierra Nevada de Santa Marta

Santa Monica Mountains (1978)
 E-W range paralleling N shore of Santa Monica Bay in S. Also search Santa Monica AND California.
 IN Index mountains as applicable.
 BT California
 BT United States
 SA Coast Ranges

Santa Rita
 Valid through 1988. Search in combination with state term. After 1988, use specific city-state term.

Santa Rita Mountains (1978)
 Santa Cruz County in SE Arizona. Also search Santa Rita.
 BT Santa Cruz County Arizona
 BT Arizona
 BT United States

Santa Rita New Mexico (1989)
 Village in SW New Mexico. Before 1989, search Santa Rita AND New Mexico.
 CO N324800N324800
   W1080400W1080400
 BT Grant County New Mexico
 BT New Mexico
 BT United States

Santa Rosa
 Valid through 1988. Search in combination with state term. After 1988, use specific city-state term.

Santa Rosa California (1989)
 City N of San Francisco in W California. Before 1989, search Santa Rosa AND California.
 CO N382600N382600
   W1224300W1224300
 BT Sonoma County California
 BT California
 BT United States

Santa Rosa Mountains (1978)
 SE California. Range along W side of Coachella Valley.
 IN Index counties and mountains as applicable.
 BT California
 BT United States

Santa Rosa Range (1978)
 In NE Humboldt County in N Nevada.
 BT Humboldt County Nevada
 BT Nevada
 BT United States

Santa Rosa Sandstone (1989)
 In Dockum Group. NE New Mexico and W Texas. Also search Santa Rosa Formation AND United States. For Lower Devonian formation in Argentina, search Santa Rosa Formation AND Argentina.
 BT Upper Triassic
 BT Triassic
 BT Mesozoic
 SA Dockum Group
 SA New Mexico
 SA Texas

Santa Susana Formation (1989)
 S California and Baja California state, Mexico.
 IN Index ages as applicable.
 BT Paleogene
 BT Tertiary
 BT Cenozoic
 SA Baja California Mexico
 SA California
 SA lower Eocene
 SA Mexico
 SA Paleocene

Santa Ynez Mountains (1978)
 E-W coastal range bordering Santa Barbara Channel in SW California. Also search Santa Ynez.
 IN Index counties and mountains as applicable.
 BT California
 BT United States
 SA Coast Ranges

Santana Formation (1978)
 BT Cretaceous
 BT Mesozoic
 SA Brazil

Santander
 As of 1981, no longer a valid index term for GeoRef. After 1992, see Santander Colombia or Santander Spain as applicable. Before 1981, "Santander" referred to both the department in Colombia and the province in Spain.

Santander Colombia (1993)
 In E Colombia. Also search Santander Department AND Colombia. Before 1981, search Santander AND Colombia. When referring to Spain, use Santander Spain.
 UF Santander Department
 BT Colombia
 BT South America
 NT California City Colombia

Santander Department
 No longer a valid term for GeoRef.
 use Santander Colombia

Santander Massif (1978)
 In the E central Cantabrian Mountains in N Spain.
 BT Santander Spain
 BT Old Castile Spain
 BT Castile Spain
 BT Spain
 BT Iberian Peninsula
 BT Southern Europe
 BT Europe

Santander Province
 No longer a valid term for GeoRef. As of 1993, see Santander Spain.

Santander Spain (1993)
 In N Spain on the Bay of Biscay. From 1981-1993, also search Santander Province and Spain.

Before 1981, also search Santander and Spain.
CO N424500N433000
W0031000W0045300
BT Old Castile Spain
BT Castile Spain
BT Spain
BT Iberian Peninsula
BT Southern Europe
BT Europe
NT Santander Massif
SA Cantabrian Basin

Santee Limestone (1985)
E and central South Carolina and North Carolina.
BT middle Eocene
BT Eocene
BT Paleogene
BT Tertiary
BT Cenozoic
SA North Carolina
SA South Carolina

Santee River (1978)
Flows into Atlantic Ocean. SE central South Carolina. Also search Santee.
IN Index counties as applicable.
BT South Carolina
BT United States

Santiago
No longer a valid term for GeoRef. See Santiago Chile.

Santiago Chile (1993)
Province including the city in central Chile.
BT Chile
BT South America
SA Chile earthquake 1985

Santiago de Compostela
No longer a valid term for GeoRef. As of 1993, see Santiago de Compostela Spain.

Santiago de Compostela Spain (1993)
City in S La Coruna Province, NW Spain. Before 1993, also search Santiago de Compostela and Spain.
BT La Coruna Spain
BT Galicia Spain
BT Spain
BT Iberian Peninsula
BT Southern Europe
BT Europe

Santiago de Cuba (1993)
Province including the city, SE Cuba. Before 1993, also search Santiango de Cuba OR Oriente AND Cuba.
UF Oriente Cuba
BT Cuba
BT Greater Antilles
BT Antilles
BT West Indies
BT Caribbean region

Santiago del Estero
No longer a valid term for GeoRef.
use Santiago del Estero Argentina

Santiago del Estero Argentina (1993)
Province including the city in N Argentina.
UF Santiago del Estero
BT Argentina
BT South America

Santiaguito (1978)
Volcano.
BT Guatemala
BT Central America

Santo Domingo
No longer a valid term for GeoRef. See Santo Domingo Dominican Republic.

Santo Domingo Dominican Republic (1993)
City on the Caribbean Sea.
UF Ciudad Trujillo
BT Dominican Republic
BT Hispaniola
BT Greater Antilles
BT Antilles
BT West Indies
BT Caribbean region

Santonian (1978)
Europe. Above Coniacian, below Campanian.
BT Senonian
BT Upper Cretaceous
BT Cretaceous
BT Mesozoic

Santorin (1978)
Volcano. A portion of the crater forms the island of Thera, southernmost of the Cyclades, in the Aegean Sea. This term has multiple hierarchies.
UF Santorin Volcano
UF Santorini Volcano
BT1 Cyclades
BT1 Greek Aegean Islands
BT1 Greece
BT1 Southern Europe
BT1 Europe
BT2 Cyclades
BT2 Greek Aegean Islands
BT2 Aegean Islands
BT2 Mediterranean region
SA Thera

Santorin Volcano
use Santorin

Santorini
use Thera

Santorini Volcano
use Santorin

Sao Francisco Basin (1978)
Drainage basin of Sao Francisco River. Also search Sao Francisco; Sao Francisco River.
IN Index states as applicable.
SA Alagoas Brazil
SA Bahia Brazil
SA Brazil
SA Minas Gerais Brazil
SA Pernambuco Brazil
SA Sergipe Brazil

Sao Francisco Craton (1993)
Archean to lower Proterozoic. Includes parts of Bahia, Minas Gerais, and Espirito Santo in E Brazil.
IN Index states as applicable.
CO S210000S090000
W0370000W0470000
BT Brazil
BT South America
SA Bahia Brazil
SA Espirito Santo Brazil
SA Minas Gerais Brazil

Sao Gabriel
No longer a valid term for GeoRef. See Sao Gabriel Brazil.

Sao Gabriel Brazil (1993)
City in extreme central Rio Grande do Sul, S Brazil.
BT Rio Grande do Sul Brazil
BT Brazil
BT South America

Sao Miguel Island (1978)
Largest island in group. In Portugal's Ponta Delgado District. Also search Sao Miguel.
BT Azores
BT Atlantic Ocean Islands

Sao Paulo
No longer a valid term for GeoRef. See Sao Paulo Brazil.

Sao Paulo Brazil (1993)
State including the city in S Brazil.
CO S252000S194000
W0442000W0530000
BT Brazil
BT South America
NT Amparo Brazil
NT Marilia Brazil
NT Ribeira de Iguape River
NT Rio Claro Brazil
SA Bauru Formation
SA Botucatu Formation
SA Brazilian Shield
SA Estrada Nova Formation
SA Irati Formation
SA Itarare Subgroup
SA Serra do Mar
SA Serra Geral Formation
SA Tubarao Group

Sao Salvador
Not a valid term for GeoRef. See Salvador Brazil.

Sao Tome (1981)
Island in the Gulf of Guinea. Formerly Portuguese territory, now part of independent nation Sao Tome e Principe.
CO S003000N004000
E0070000E0060000
BT Sao Tome e Principe
BT Atlantic Ocean Islands
SA Principe

Sao Tome e Principe (1985)
Island group and nation in the Gulf of Guinea; formerly a Portuguese overseas province.
BT Atlantic Ocean Islands
NT Principe
NT Sao Tome

Saone Valley (1978)
River valley in E central part of country extending from the Vosges Mountains to the Rhone River at Lyons. Also search Saone.
BT France
BT Western Europe
BT Europe
SA Saone-Rhone Basin

Saone-et-Loire
No longer a valid term for GeoRef. See Saone-et-Loire France.

Saone-et-Loire France (1993)
Department in E central France.
CO N461000N471000
E0053000E0033500
BT France
BT Western Europe
BT Europe
NT Autun France
SA Morvan

Saone-Rhone Basin (1981)
BT France
BT Western Europe
BT Europe
SA Rhone Valley
SA Saone Valley

Saoura
No longer a valid term for GeoRef as of 1981. See Saoura region.

Saoura region
No longer a valid term for GeoRef. Used as large department of French Algeria. As of 1993, see Algeria for departments. Before 1981, search Saoura.

Sapelo Island (1978)
In Atlantic Ocean off McIntosh County.
BT McIntosh County Georgia
BT Georgia
BT United States

saponite (1978)
BT clay minerals
BT sheet silicates
BT silicates
SA smectite

Sappada
No longer a valid term for GeoRef. See Sappada Italy.

Sappada Italy (1993)
Village on Piave River in NE Italy.
BT Veneto Italy
BT Italy
BT Southern Europe
BT Europe

sapphire (1978)
Variety of corundum. Includes use when of economic value as well as for mineralogical studies.
BT oxides
SA corundum
SA gems
SA rubies

sapphirine (1978)
Autoposting of nesosilicates began in 1985.
BT nesosilicates
BT orthosilicates
BT silicates
SA aluminosilicates

saprolite (1978)
BT clastic rocks
BT sedimentary rocks

sapropel (1978)
BT organic materials
SA anaerobic environment
SA gyttja
SA hydrocarbons
SA ooze

sapropelic coal
use sapropelite

sapropelite (1978)
This term has multiple hierarchies.
UF sapropelic coal
BT1 coal
BT1 organic residues
BT2 coal
BT2 sedimentary rocks

SAR (1989)
Acronym.
UF synthetic aperture radar
SA airborne methods
SA mosaics
SA multispectral scanner
SA radar methods
SA remote sensing
SA side-scanning methods
SA SLAR

Saragossa
No longer a valid term for GeoRef. As of 1993, see Saragossa City Spain.

Saragossa City Spain (1993)
Refers to only the city in NE Spain. When referring to Mexico, use Zaragoza. Before 1993, also search Saragossa and Spain.

BT Saragossa Spain
BT Aragon Spain
BT Spain
BT Iberian Peninsula
BT Southern Europe
BT Europe
SA Zaragoza Mexico

Saragossa Province
No longer a valid term for GeoRef. As of 1993, see Saragossa Spain.

Saragossa Spain (1993)
Refers only to the province. From 1981-1992, also search Saragossa Province and Spain. Before 1981, also search Saragossa and Spain. For the city, see Saragossa City Spain. In NE Spain.
CO N405500N424500
   E0002300W0021000
BT Aragon Spain
BT Spain
BT Iberian Peninsula
BT Southern Europe
BT Europe
NT Saragossa City Spain
SA Calatayud-Teruel Basin
SA Zaragoza Mexico

Sarajevo (1978)
City in E Bosnia-Herzegovina
UF Serajevo
BT Bosnia
BT Bosnia-Herzegovina
BT Yugoslavia
BT Southern Europe
BT Europe

Sarasota County
Valid through 1988. Search in combination with state term. After 1988, use specific county-state term.

Sarasota County Florida (1989)
On Gulf of Mexico In SW Florida. Before 1989, also search Sarasota County AND Florida.
CO N265500N272500
   W0820500W0824000
BT Florida
BT United States

Saratoga Chalk (1978)
Includes a lower chalk member and an upper argillaceous-arenaceous unit. SW Arkansas.
BT Upper Cretaceous
BT Cretaceous
BT Mesozoic
SA Arkansas

Saratoga County
Valid through 1988. Search in combination with state term. After 1988, use specific county-state term.

Saratoga County New York (1989)
E New York. Before 1989, also search Saratoga County AND New York.
CO N424700N432300
   W0733500W0741200
BT New York
BT United States

Saratov
No longer a valid term for GeoRef. As of 1993, see Saratov Russian Federation.

Saratov Russian Federation (1993)

Oblast in SW Russian Federation including the city on W bank of the Volga River. This term has multiple hierarchies.
BT1 Russian Federation
BT1 Commonwealth of Independent States
BT2 Europe
SA Uzen

Sarawak
No longer a valid term for GeoRef. As of 1993, see Sarawak Malaysia.

Sarawak Malaysia (1993)
State of East Malaysia in NW Borneo. Former British protectorate governed by the Brooke family. Joined Malaysia in 1963. Included use as a first order area term until 1976. This term has multiple hierarchies.
CO N005000N050000
   E1155000E1093000
BT1 East Malaysia
BT1 Borneo
BT1 Far East
BT1 Asia
BT2 East Malaysia
BT2 Borneo
BT2 Malay Archipelago
BT3 East Malaysia
BT3 Malaysia
BT3 Far East
BT3 Asia

Sarbay (1978)
River in former Kuibyshev Oblast in the Middle Volga region.
IN Index regions as applicable.
SA Russian Federation
SA Samara Russian Federation

sarcopside (1978)
From 1978-1980, halides and phosphides were autoposted to this term. Autoposting of phosphates began in 1981.
BT phosphates

Sarcopterygii (1993)
Subclass.
BT Osteichthyes
BT Pisces
BT Vertebrata
BT Chordata
NT Crossopterygii
NT Dipnoi

Sardinia
No longer a valid term for GeoRef. See Sardinia Italy.

Sardinia Italy (1993)
Island and autonomous region in the Mediterranean Sea.
CO N385000N413000
   E0094500E0081500
BT Italy
BT Southern Europe
BT Europe
NT Bosano
NT Gerrei
NT Iglesiente
NT Logudoro
NT Orosei Italy
NT Sarrabus
NT Sulcis
SA Mediterranean region
SA Mediterranean Sea
SA Orosei

Sarema
use Saaremaa

Sargasso Sea (1978)

Large tract of relatively still water between the West Indies and Bermuda. Named after Sargasso weed which floats there.
BT Atlantic Ocean

Sargent
Valid through 1988. Search in combination with state term. After 1988, use specific city-state term.

Sargent California (1989)
Town in W California. Before 1989, search Sargent AND California.
CO N365500N365500
   W1213400W1213400
BT Santa Clara County California
BT California
BT United States

Sarmatian (1978)
Europe.
BT upper Miocene
BT Miocene
BT Neogene
BT Tertiary
BT Cenozoic

Sarrabus (1978)
Region in SE Sardinia.
BT Sardinia Italy
BT Italy
BT Southern Europe
BT Europe

Sartanian (1985)
Term applied in USSR to last Pleistocene glacial stage. Above Karaginskian, below Holocene.
BT upper Pleistocene
BT Pleistocene
BT Quaternary
BT Cenozoic
SA Vistulian
SA Weichselian
SA Wisconsinan
SA Wurm

Sarthe
No longer a valid term for GeoRef. See Sarthe France.

Sarthe France (1993)
Department in NW France.
CO N473000N483000
   E0005000W0003000
BT France
BT Western Europe
BT Europe
NT Le Mans France

Sary-Su
use Sarysu

Sary-Su River
use Sarysu

Sarysu (1978)
River in central Kazakhstan. Flows S into desert but becomes dry before reaching Syr Darya.
IN Index Kazakhstan as applicable.
UF Sary-Su
UF Sary-Su River
SA Chu-Sarysu Depression
SA Kazakhstan
SA Syr Darya

Sasca-Montana
No longer a valid term for GeoRef. See Sasca-Montana Romania.

Sasca-Montana Romania (1993)
Village in Caras-Severin County in SW Romania. This term has multiple hierarchies.
BT1 Banat
BT1 Southern Europe

BT1 Europe
BT2 Romania
BT2 Southern Europe
BT2 Europe

Sasebo
No longer a valid term for GeoRef. See Sasebo Japan.

Sasebo Japan (1993)
City on inlet of East China Sea in Nagasaki, NW Kyushu, S Japan. Coal region.
BT Nagasaki Japan
BT Kyushu
BT Japan
BT Far East
BT Asia

Saskatchewan (1978)
CO N490000N600000
   W1012000W1100000
BT Western Canada
BT Canada
NT Aberfeldy Field
NT Amisk Lake
NT Athabasca District
NT Beaverlodge
NT Cree Lake
NT Delmas Saskatchewan
NT Eldorado Saskatchewan
NT Esterhazy Saskatchewan
NT Estevan Saskatchewan
NT Key Lake Deposits
NT La Ronge Domain
NT La Ronge Saskatchewan
NT Lac La Ronge
NT Leoville Saskatchewan
NT Midwest Lake Deposits
NT Rabbit Lake Deposit
NT Rabbit Lake Saskatchewan
NT Regina Saskatchewan
NT Rosetown Saskatchewan
NT Saskatoon Saskatchewan
NT Tazin Lake
NT Wapawekka Lake
NT Wollaston Lake Belt
NT Wynyard Saskatchewan
SA Assiniboine River
SA Assiniboine River valley
SA Athabasca Formation
SA Bakken Formation
SA Belly River Formation
SA Birdbear Formation
SA Canadian Shield
SA Carswell Structure
SA Charles Formation
SA Churchill Province
SA Clearwater River
SA Cold Lake
SA Coronation Mine
SA Cypress Hills Formation
SA Duperow Formation
SA Edmonton Formation
SA Elk Point Basin
SA Elk Point Group
SA Great Plains
SA Hanson Lake
SA Interlake Formation
SA Kisseynew Complex
SA Lake Athabasca
SA Mannville Group
SA Mazama Ash
SA Midwest Lake
SA Missi Group
SA Missouri River basin
SA Nokomis Group
SA North Saskatchewan River
SA Paskapoo Formation
SA Prairie Evaporite
SA Prince Albert Group
SA Ravenscrag Formation
SA Reindeer Lake
SA Saskatchewan River
SA Shaunavon Formation
SA Sherridon Group

SA Souris River basin
SA South Saskatchewan River
SA Swan Hills Formation
SA Whitemud Formation
SA Williston Basin
SA Winnipeg Formation
SA Wollaston Group
SA Wood Mountain Formation

Saskatchewan River   (1978)
Formed by confluence of the North and South Saskatchewan rivers in central Saskatchewan. Empties into N Lake Winnipeg.
IN   Index provinces as applicable.
BT   Western Canada
BT   Canada
SA   Manitoba
SA   North Saskatchewan River
SA   Saskatchewan
SA   South Saskatchewan River

Saskatoon
No longer a valid term for GeoRef. As of 1993, see Saskatoon Saskatchewan.

Saskatoon Saskatchewan   (1993)
City in S central Saskatchewan. Before 1993, also search Saskatoon AND Saskatchewan.
BT   Saskatchewan
BT   Western Canada
BT   Canada

satellite
A valid term through 1977. After 1977, use satellite methods.

satellite measurements   (1978)
SA   geodesy
SA   geodetic coordinates
SA   measurement
SA   satellite methods
SA   telemetry
SA   very long baseline interferometry

satellite methods   (1978)
Used for artificial satellites. For natural satellites, see satellites. Before 1978, search satellite AND methods. Autoposting of this term began in 1985.
SA   broad-band spectra
SA   Earth Observing System
SA   geophysical methods
SA   geophysical surveys
SA   Geosat
SA   Global Positioning System
SA   Gravsat
SA   ground truth
SA   image enhancement
SA   Lageos
SA   Landsat
SA   laser ranging
SA   Magsat
SA   Mariner 11
SA   methods
SA   multispectral scanner
SA   orbital observations
SA   POGO
SA   remote sensing
SA   satellite measurements
SA   SeaMarc
SA   Seasat
SA   Shuttle Imaging Radar
SA   space photography
SA   Spacelab Program
SA   SPOT
SA   thematic mapper
SA   Venera Program
SA   very long baseline interferometry
SA   Voyager Program

satellites   (1978)
Used for natural satellites other than the Earth's Moon. For artificial satellites, see remote sensing; satellite methods; satellite measurements.
NT   icy satellites
SA   Adrastea Satellite
SA   Amalthea Satellite
SA   Ariel Satellite
SA   Callisto Satellite
SA   Charon Satellite
SA   Deimos Satellite
SA   Dione Satellite
SA   Enceladus Satellite
SA   Europa Satellite
SA   Galilean satellites
SA   Ganymede Satellite
SA   Hyperion Satellite
SA   Iapetus Satellite
SA   Io Satellite
SA   Janus Satellite
SA   Jupiter
SA   Mars
SA   Mimas Satellite
SA   Miranda Satellite
SA   Moon
SA   Neptune
SA   Nereid Satellite
SA   Oberon Satellite
SA   Phobos Satellite
SA   Phoebe Satellite
SA   Pluto
SA   Rhea Satellite
SA   Saturn
SA   Tethys Satellite
SA   Titan Satellite
SA   Titania Satellite
SA   Triton Satellite
SA   Umbriel Satellite
SA   Uranus

Satna
No longer a valid term for GeoRef. See Satna India.

Satna India   (1993)
Town in NE Madhya Pradesh, central India. Before 1993, search Satna. Before 1978, also search Sutna.
UF   Sutna India
BT   Madhya Pradesh India
BT   India
BT   Indian Peninsula
BT   Asia

Satpura Range   (1978)
Line of hills forming N limit of Deccan Plateau. Primarily in Madhya Pradesh. Also search Satpura.
IN   Index states as applicable.
BT   India
BT   Indian Peninsula
BT   Asia
SA   Madhya Pradesh India
SA   Maharashtra India
SA   Uttar Pradesh India

Satsuma Peninsula   (1978)
Between East China Sea and Kagoshima in SW Kyushu. Also search Satsuma.
BT   Kagoshima Japan
BT   Kyushu
BT   Japan
BT   Far East
BT   Asia

Satu Mare
No longer a valid term for GeoRef. As of 1993 see Satu Mare Romania.

Satu Mare Romania   (1993)
County in NW Romania. Before 1993 also search Satu Mare AND Romania.
BT   Transylvania
BT   Romania
BT   Southern Europe
BT   Europe

saturated hydrocarbons   (1989)
Used for hydrocarbons containing only single bonds between carbon atoms.
BT   hydrocarbons
BT   organic materials
SA   alkanes
SA   biomarkers
SA   hopanes
SA   steranes

saturated materials   (1993)
Term valid from 1978 to 1981. Reintroduced in 1993. Between 1978-1992, saturated zone was used. Before 1978, also search saturated.
SA   materials
SA   rock mechanics
SA   soil mechanics

saturated zone   (1981)
Subsurface zone in which all interstices are filled with water under greater than atmospheric pressure.
UF   phreatic zone
SA   capillary water
SA   ground water
SA   soils
SA   unsaturated zone
SA   water table

saturation   (1978)
Used as a general term, e.g. under sea water or springs.
SA   concentration
SA   dilution
SA   geochemistry
SA   hydrochemistry
SA   sea water
SA   soils
SA   solubility
SA   springs
SA   water

saturation magnetization   (1981)
BT   magnetization

Saturn   (1978)
SA   Dione Satellite
SA   Enceladus Satellite
SA   Hyperion Satellite
SA   Iapetus Satellite
SA   Janus Satellite
SA   Mimas Satellite
SA   outer planets
SA   Phoebe Satellite
SA   Pioneer 11
SA   Pioneer Program
SA   planetary rings
SA   planetology
SA   planets
SA   Rhea Satellite
SA   satellites
SA   solar system
SA   Tethys Satellite
SA   Titan Satellite
SA   Voyager 1
SA   Voyager 2
SA   Voyager Program

Sau Alpe
use Sau Alps

Sau Alps   (1978)
Mountains in NW Carinthia. This term has multiple hierarchies.
UF   Sau Alpe
UF   Saualpe
BT1   Carinthia Austria
BT1   Austria
BT1   Central Europe
BT1   Europe
BT2   Alps
BT2   Europe

Saualpe
use Sau Alps

Sauce Grande River valley   (1978)
S Buenos Aires Province Argentina. Also search Sauce Grande; Sauce Grande River.
BT   Buenos Aires Argentina
BT   Argentina
BT   South America

Saucesian   (1978)
North America. Above Zemorrian, below Relizian.
BT   lower Miocene
BT   Miocene
BT   Neogene
BT   Tertiary
BT   Cenozoic
SA   Oligocene

Saudi Arabia   (1978)
From 1926 to 1932, the dual kingdom of Nejd and Hejaz. Before 1932, also search Najd, Nejd, Hejaz, Hedjaz, Hijaz.
CO   N170000N323000
E0570000E0344500
BT   Arabian Peninsula
BT   Asia
NT   Hail Saudi Arabia
NT   Hejaz Saudi Arabia
NT   Jabal Idsas
NT   Riyadh Saudi Arabia
SA   Arab Formation
SA   Arabian Plate
SA   Arabian Shield
SA   Red Sea Basin
SA   Red Sea Rift
SA   Thamama Group

Sauerland   (1978)
Region S and E of the Ruhr River.
BT   North Rhine-Westphalia Germany
BT   Germany
BT   Central Europe
BT   Europe
SA   Ebbe Anticlinorium

Saugeen Peninsula
use Bruce Peninsula

Saugus Formation   (1989)
S California.
IN   Index ages as applicable.
BT   Cenozoic
SA   California
SA   Pleistocene
SA   Pliocene

Sauk County Wisconsin   (1989)
S central Wisconsin. Before 1989, search Sauk County AND Wisconsin.
CO   N431000N433800
W0893700W0901900
BT   Wisconsin
BT   United States
NT   Baraboo Wisconsin

Sauk Sequence   (1985)
Upper Proterozoic to Cambrian. Central and western U.S.
IN   Index ages as applicable.
SA   Cambrian
SA   Midwest
SA   United States
SA   upper Proterozoic
SA   Western U.S.

Sault Sainte Marie
No longer a valid term for GeoRef. As of 1993, see Sault Sainte Marie Ontario or Sault Sainte Marie Michigan.

Sault Sainte Marie Michigan (1989)
City at the falls on Saint Marys River in NE Michigan. Before 1989, also search Sault Sainte Marie AND Michigan.
CO N462900N462900 W0842200W0842200
UF Sault Ste. Marie
BT Chippewa County Michigan
BT Michigan Upper Peninsula
BT Michigan
BT United States

Sault Sainte Marie Ontario (1993)
City in Algoma District, S Ontario. Before 1993, also search Sault Sainte Marie AND Ontario.
UF Sault Ste. Marie
BT Algoma District Ontario
BT Ontario
BT Eastern Canada
BT Canada

Sault Ste. Marie
use Sault Sainte Marie Ontario

Saurashtra (1978)
Region and former state comprising greater part of Kathiawar Peninsula in Gujarat, W India.
BT Gujarat India
BT India
BT Indian Peninsula
BT Asia

Saurischia (1978)
Order. Autoposting of this term began in 1978. This term has multiple hierarchies.
BT1 Archosauria
BT1 Diapsida
BT1 Reptilia
BT1 Tetrapoda
BT1 Vertebrata
BT1 Chordata
BT2 dinosaurs
BT2 Tetrapoda
BT2 Vertebrata
BT2 Chordata
NT Sauropoda
NT Theropoda

Sauropoda (1978)
Infraorder. This term has multiple hierarchies.
BT1 Saurischia
BT1 Archosauria
BT1 Diapsida
BT1 Reptilia
BT1 Tetrapoda
BT1 Vertebrata
BT1 Chordata
BT2 Saurischia
BT2 dinosaurs
BT2 Tetrapoda
BT2 Vertebrata
BT2 Chordata

Sauropterygia (1981)
Before 1993, Euryapsida was autoposted to this term. Autoposting of Diapsida to this term began in 1993.
BT Diapsida
BT Reptilia
BT Tetrapoda
BT Vertebrata
BT Chordata
NT Nothosauria
NT Plesiosauria

sausage structure
use boudinage

Sausar Series (1978)
Indian Shield region of central S and S India.
BT Archean
BT Precambrian
SA India

Savage River (1978)
In Garrett County, extreme W Maryland.
BT Garrett County Maryland
BT Maryland
BT United States

Savanna Formation (1978)
In Krebs Group. Includes Spaniard Limestone Member, Spiro Sandstone, Sam Creek Limestone, Doneley Limestone. W Arkansas and E and S Oklahoma. Also search Savanna.
BT Desmoinesian
BT Middle Pennsylvanian
BT Pennsylvanian
BT Carboniferous
BT Paleozoic
SA Arkansas
SA Krebs Group
SA Oklahoma
SA Spiro Sandstone

Savannah River (1978)
Serves as boundary between Georgia and South Carolina, and flows into Atlantic Ocean.
IN Index counties and states as applicable.
BT United States
SA Georgia
SA South Carolina

Savannah River Plant (1978)
U.S. Atomic Energy Commission plant in Aiken and Barnwell counties on Savannah River.
BT South Carolina
BT United States
SA Aiken County South Carolina
SA Barnwell County South Carolina

savannas (1993)
SA ecology
SA geomorphology
SA grasslands
SA plains
SA revegetation

Savoie
No longer a valid term for GeoRef. See Savoie France.

Savoie France (1993)
Department in SE France.
CO N450000N455000 E0070000E0053000
BT France
BT Western Europe
BT Europe
NT Ambin Massif
NT Tarentaise
SA Graian Alps
SA Savoy

Savona
No longer a valid term for GeoRef. See Savona Italy.

Savona Italy (1993)
City and province near Genoa in central Liguria, NW Italy.
BT Liguria Italy
BT Italy
BT Southern Europe
BT Europe

Savoy (1978)
Historical region in SE France, chiefly in departments of Haute-Savoie and Savoie.
BT France
BT Western Europe
BT Europe
SA Haute-Savoie France
SA Savoie France

Savoy Alps (1978)
NW offshoots of Graian Alps S of Lake Geneva. Also search Savoy. This term has multiple hierarchies.
BT1 Haute-Savoie France
BT1 France
BT1 Western Europe
BT1 Europe
BT2 Western Alps
BT2 Alps
BT2 Europe
NT Mont Blanc
NT Vanoise
SA Chablais
SA Graian Alps
SA Tarentaise

Sawatch Range (1978)
Mountain range of the Rocky Mountains in central Colorado. This term has multiple hierarchies.
BT1 Colorado
BT1 United States
BT2 U. S. Rocky Mountains
BT2 Rocky Mountains
BT3 U. S. Rocky Mountains
BT3 United States
NT Mount Antero

Sawtooth Range (1978)
Large group of mountain ranges in S central Idaho. May also occur elsewhere.
IN Index counties or region and mountains as applicable.
SA Blaine County Idaho
SA Camas County Idaho
SA Custer County Idaho
SA Elmore County Idaho
SA Idaho
SA Lewis and Clark County Montana
SA Montana
SA Teton County Montana

Saxonian (1978)
Europe. Above Autunian, below Thuringian.
BT Permian
BT Paleozoic
SA Rotliegendes

Saxonian Massif (1981)
CO N502200N523000 E0150400E0110000
BT Germany
BT Central Europe
BT Europe
SA Saxony Germany

Saxony
No longer a valid term for GeoRef. See Saxony Germany.

Saxony Germany (1993)
State, NE Germany. Before 1993, also search E German bezirks: Cottbus Bezirk, Dresden Bezirk, Karl-Marx-Stadt Bezirk, and Leipzig Bezirk and Saxony-Thuringia as applicable.
BT Germany
BT Central Europe
BT Europe
NT Altenberg Germany
NT Chemnitz Germany
NT Dresden Germany
NT Freiberg Germany
NT Leipzig Germany
NT Upper Lusatia
NT Weissenberg Germany
NT Wildenfels Germany
NT Zwickau Germany
SA Erzgebirge
SA Frankenberger Germany
SA Harz Foreland
SA Harz region
SA Saxonian Massif
SA Saxony-Thuringia
SA Schwarzenberg Germany
SA Vogtland
SA Weisse Elster Basin

Saxony-Anhalt
No longer a valid term for GeoRef. See Saxony-Anhalt Germany.

Saxony-Anhalt Germany (1993)
State, N central Germany. Before 1993, also search Madgeberg Bezirk and Halle Bezirk if applicable.
BT Germany
BT Central Europe
BT Europe
NT Aschersleben Germany
NT Bitterfeld Germany
NT Elbingerode Germany
NT Halberstadt Germany
NT Halle Germany
NT Magdeburg Germany
NT Mansfeld Syncline
NT Merseburg Germany
NT Rottleberode Germany
NT Stassfurt Germany
SA Brocken Massif
SA Geisel Valley
SA Harz Mountains
SA Harz region
SA Kyffhauser Range
SA North German Plain
SA Saale River
SA Unstrut River
SA Weisse Elster Basin

Saxony-Thuringia (1978)
Region in E Germany including the German states of Saxony and Thuringia.
BT Germany
BT Central Europe
BT Europe
SA Saxony Germany
SA Thuringia Germany

Sayak
No longer a valid term for GeoRef. As of 1993, see Sayak Kazakhstan.

Sayak Kazakhstan (1993)
Village N of W Lake Balkash in W central Kazakhstan. This term has multiple hierarchies.
BT1 Kazakhstan
BT1 Central Asia
BT1 Asia
BT2 Kazakhstan
BT2 Commonwealth of Independent States

Sayan (1978)
E-W range just N of Mongolia extending from S Krasnoyarsk Kray on W across Tuva Autonomous Republic into W Irkutsk Oblast on the E. Sayan Mountains used from 1976-1978. Autoposting of this term began in 1978.
IN Index countries and regions as applicable.
UF Sayan Mountains
UF Sayan Range
BT Siberian fold belt
BT Asia
NT Eastern Sayan
NT Western Sayan

SA Altai-Sayan region
SA Russian Federation

Sayan Mountains
use Sayan

Sayan Range
use Sayan

Sb
use antimony

Sc
use scandium

Scaglia Formation (1978)
Cretaceous to Eocene. In the Apennines and NE Italy.
IN Index ages as applicable.
SA Apennines
SA Cretaceous
SA Eocene
SA Italy
SA K-T boundary
SA Paleocene

scale factor (1993)
Factor by which reading of instrument or solution of problem should be multiplied to give true final value when a corresponding scale factor is used initially to bring the magnitude within the range of the instrument or computer. Not for use with cartographic scales or magnitude of earthquakes.
SA cartographic scales
SA factors
SA magnitude
SA scale models
SA time scales

scale models (1982)
SA models
SA scale factor

scales, cartographic
use cartographic scales

scales, time
use time scales

Scandentia (1981)
Autoposting of Proteutheria, Insectivora, Eutheria and Theria to this term began in 1989.
BT Proteutheria
BT Insectivora
BT Eutheria
BT Theria
BT Mammalia
BT Tetrapoda
BT Vertebrata
BT Chordata

Scandinavia (1978)
Region comprising Denmark, Norway, Sweden, and Finland. Iceland sometimes included. Autoposting of this term began in 1981.
IN Index countries as applicable.
CO N543000N710000
E0320000E0050000
BT Western Europe
BT Europe
NT Denmark
NT Finland
NT Koli Nappe
NT Norway
NT Sulitjelma
NT Sweden
NT Western Gneiss region
SA Baltic ice lake
SA Caledonides
SA NORESS
SA NORSAR

scandium (1978)

Autoposting of rare earths and metals to this term began in 1989.
UF Sc
BT rare earths
BT metals

Scania
use Skane

scanning
Use scanning method.

scanning electron microscopy (1993)
For the method. Before 1978, also search scanning. Before 1993, also search scanning method AND electron microscopy.
SA chemical analysis
SA electron microscopy
SA SEM data
SA TEM data
SA transmission electron microscopy

scanning electron microscopy data
use SEM data

scanning method
No longer a valid term for GeoRef. See scanning electron microscopy for the method.

Scaphites (1978)
Genus.
BT Ammonoidea
BT Tetrabranchiata
BT Cephalopoda
BT Mollusca
BT Invertebrata

Scaphopoda (1978)
Autoposting of this term began in 1978.
BT Mollusca
BT Invertebrata
NT Dentalium

scapolite (1978)
BT scapolite group
BT framework silicates
BT silicates
SA scapolitization

scapolite group (1978)
Autoposting of this term to meionite began in 1981.
BT framework silicates
BT silicates
NT kenyaite
NT magadiite
NT meionite
NT scapolite

scapolitization (1989)
BT metasomatism
SA alteration
SA processes
SA scapolite

Scarborough Formation (1989)
Wisconsinan (upper Pleistocene) of Toronto, Ontario. Also a Bajocian (Middle Jurassic) formation of England.
IN Index ages as applicable.
SA England
SA Middle Jurassic
SA Ontario
SA upper Pleistocene

scarps (1978)
UF escarpments
SA cliffs
SA cuestas
SA erosion features
SA fault scarps
SA faults
SA geomorphology
SA slopes

scattering
No longer a valid term for GeoRef. See wave dispersion.

scawtite (1978)
BT chain silicates
BT silicates

Schaffhausen
No longer a valid term for GeoRef. As of 1993, see Schaffhausen Switzerland.

Schaffhausen Switzerland (1993)
Northernmost canton. Also a city in N central Switzerland. Before 1993 also search Schaffhausen AND Switzerland.
BT Switzerland
BT Central Europe
BT Europe

schapbachite
use matildite

schapbacite
use matildite

scheelite (1978)
BT tungstates
SA tungsten ores

Schefferville
No longer a valid term for GeoRef. As of 1993, see Schefferville Quebec.

Schefferville Quebec (1993)
Town in NE Quebec near Labrador border. Before 1993, also search Schefferville AND Quebec.
BT Saguenay County Quebec
BT Quebec
BT Eastern Canada
BT Canada

Schela
No longer a valid term for GeoRef. As of 1993 see Schela Romania.

Schela Romania (1993)
Village in SW Romania. Before 1993 also search Schela AND Romania.
BT Walachia
BT Romania
BT Southern Europe
BT Europe

Schelde River
use Scheldt River

Scheldt River (1978)
Rises in N France and empties into North Sea through two estuaries, the East and West Scheldt.
IN Index countries as applicable.
UF Schelde River
BT Western Europe
BT Europe
SA Belgium
SA France
SA Netherlands

Schenectady Formation (1978)
Underlies Indian Ladder Beds; overlies Snake Hill Formation. E central New York. Also search Schenectady.
BT Middle Ordovician
BT Ordovician
BT Paleozoic
SA New York

Schiefergebirge
use Rhenish Schiefergebirge

Schimper, Wilhelm-Philippe (1985)
UF Wilhelm-Philippe Schimper
SA biography

schist
As of 1981, no longer a valid term for GeoRef. From 1978-1980, schists was autoposted to this term.
use schists

schistosity (1978)
BT foliation
SA axial-plane structures
SA cleavage
SA flow cleavage
SA metamorphic rocks
SA orientation
SA schists
SA slaty cleavage
SA slip cleavage
SA structural analysis
SA tourmalinite

schists (1978)
Before 1981, also search schist. Autoposting of this term to schist and mica schist ended in 1981.
UF schist
BT metamorphic rocks
NT biotite schist
NT blueschist
NT calc-schist
NT chlorite schist
NT epidiorite
NT glaucophane schist
NT greenschist
NT greenstone
NT hornblende schist
NT listwanite
NT muscovite schist
NT tourmalinite
SA schistosity

schizomycetes
use bacteria

Schizophoria (1978)
Genus.
BT Enteletacea
BT Orthida
BT Articulata
BT Brachiopoda
BT Invertebrata

Schizophyta
use Cyanophyta

Schleswig-Holstein
No longer a valid term for GeoRef.
use Schleswig-Holstein Germany

Schleswig-Holstein Germany (1993)
IN State, N Germany.
CO N531500N550000
E0112500E0081000
UF Schleswig-Holstein
BT Germany
BT Central Europe
BT Europe
NT Helgoland
NT Holstein
NT Kiel Germany
NT Sylt
SA Alster River
SA Hamburg Germany
SA Jutland
SA North German Plain

schlieren (1978)
SA igneous rocks
SA inclusions
SA plutonic rocks

Schlotheim
No longer a valid term for GeoRef. See Schlotheim Germany.

Schlotheim Germany (1993)
Town in E central Germany.
BT Thuringia Germany
BT Germany
BT Central Europe

BT Europe

Schlumberger methods (1989)
UF Schlumberger sounding
SA arrays
SA electrical methods
SA geophysical methods
SA geophysical surveys
SA resistivity
SA well logs
SA well-logging

Schlumberger sounding
use Schlumberger methods

Schneeberg
No longer a valid term for GeoRef. See Schneeberg Germany.

Schneeberg Germany (1993)
Town in Erzgebirge.
BT Thuringia Germany
BT Germany
BT Central Europe
BT Europe

Schoharie County
Valid through 1988. Search in combination with state term. After 1988, use specific county-state term.

Schoharie County New York (1989)
E central New York. Before 1989, also search Schoharie County AND New York.
CO N422200N425000
   W0741000W0744000
BT New York
BT United States

scholzite (1978)
BT phosphates

Schoonover Sequence (1989)
NE Nevada.
BT Carboniferous
BT Paleozoic
SA Nevada

schorl (1993)
Forms a series with dravite.
BT ring silicates
BT silicates
SA buergerite
SA dravite
SA elbaite
SA tourmaline

schorlomite (1978)
BT garnet group
BT nesosilicates
BT orthosilicates
BT silicates

schreibersite (1978)
Meteorite mineral. From 1978-1980, native elements and alloys; sulfides were autoposted to this term. Autoposting of phosphides began in 1981. Autoposting of alloys began in 1989. Autoposting of sulfides ended in 1989.
UF rhabdite
BT phosphides
BT alloys
SA meteorites

Schuler Formation (1978)
In Cotton Valley Group. Includes Morgan Sands Zone, Jones Sand, Dorcheat and Shongaloo members. Subsurface.
BT Upper Jurassic
BT Jurassic
BT Mesozoic
SA Arkansas
SA Cotton Valley Group
SA Louisiana

SA Texas

schuppen texture
use imbricate tectonics

Schuylkill County
Valid through 1988. Search in combination with state term. After 1988, use specific county-state term.

Schuylkill County Pennsylvania (1989)
E Central Pennsylvania. Before 1989, also search Schuylkill County AND Pennsylvania.
CO N402800N405600
   W0754700W0764400
BT Pennsylvania
BT United States

Schwagerina (1978)
Genus. Autoposting of microfossils and Protista to this term began in 1990. This term has multiple hierarchies.
BT1 Fusulinidae
BT1 Fusulinina
BT1 foraminifera
BT1 Protista
BT1 Invertebrata
BT2 Fusulinidae
BT2 Fusulinina
BT2 foraminifera
BT2 Protista
BT2 microfossils

Schwartzwalder Mine (1985)
N central Colorado. Autoposting of broader terms to this term began in 1989.
CO N393500N395000
   W1051000W1052500
BT Jefferson County Colorado
BT Colorado
BT United States
SA mines
SA uranium ores

Schwarzburg Anticlinorium (1978)
In Thuringia.
SA Thuringia Germany

Schwarzenberg
No longer a valid term for GeoRef. See Schwarzenberg Germany.

Schwarzenberg Germany (1993)
City in Saxony, E Germany. Also occurs in Bavaria.
IN Index states as applicable.
BT Germany
BT Central Europe
BT Europe
SA Bavaria Germany
SA Saxony Germany

Schwarzwald
use Black Forest

Schwechat Valley (1978)
River valley SE of Vienna in E Lower Austria.
BT Lower Austria
BT Austria
BT Central Europe
BT Europe

Schwerin
No longer a valid term for GeoRef. See Schwerin Germany.

Schwerin Bezirk
No longer a valid term for GeoRef. District in NW East Germany. Before 1981, search Schwerin. See Mecklenburg-Western Pomerania Germany.

Schwerin Germany (1993)

City in Mecklenburg-Western Pomerania, NE Germany. As of 1981, Schwerin refers to only the city in NW Schwerin Bezirk, NW East Germany. Before 1981, it also included the East German district.
BT Mecklenburg-Western Pomerania Germany
BT Germany
BT Central Europe
BT Europe

scientific creationism
use creationism

scintillations (1978)
SA gamma rays
SA particles
SA radioactivity

Scioto drainage basin
use Scioto River basin

Scioto River basin (1978)
Central and S central Ohio.
UF Scioto drainage basin
BT Ohio
BT United States

Sciuromorpha (1981)
Suborder. Autoposting of Eutheria and Theria to this term began in 1989. Autoposting of this term to Castoridae ended in 1989.
BT Rodentia
BT Eutheria
BT Theria
BT Mammalia
BT Tetrapoda
BT Vertebrata
BT Chordata

Scleractinia (1978)
Order. Autoposting of Zoantharia to this term began in 1989.
UF Hexacorallia
UF Madreporaria
UF madrepores
BT Zoantharia
BT Anthozoa
BT Coelenterata
BT Invertebrata
NT Acropora cervicornis
NT Acropora palmata
NT Montastrea
NT Porites
SA Rugosa
SA Tabulata

sclerites (1978)
SA Echinodermata
SA fossils
SA Holothuroidea

sclerotinite (1981)
BT inertinite
BT macerals

scolecite (1978)
BT zeolite group
BT framework silicates
BT silicates

scolecodonts (1978)
This term has multiple hierarchies.
BT1 worms
BT1 Invertebrata
BT2 microfossils
SA Annelida

Scoresby Land (1978)
Region in E on Greenland Sea between King Oscar Fjord on N and Scoresby Sound on the S.
BT Greenland
BT Arctic region

Scoresby Sound (1978)

Deep inlet and fjord system of Greenland Sea on central E coast.
UF Scoresby Sund
BT Greenland
BT Arctic region

Scoresby Sund
use Scoresby Sound

scoria (1978)
BT pyroclastics
BT volcanic rocks
BT igneous rocks
SA pumice

scorodite (1978)
BT arsenates

Scotia Ridge (1978)
Extends in an arc in the South Atlantic Ocean from S of the Falkland Islands eastward to the South Sandwich Islands and then westward to S of the South Orkney Islands.
BT Atlantic Ocean
SA ODP Site 696

Scotia Sea (1978)
Part of the South Atlantic Ocean within the arc of the Scotia Ridge and E of Drake Passage.
CO S620000S530000
   W0270000W0650000
BT Antarctic Ocean
SA Atlantic Ocean
SA Leg 36

Scotia Sea Islands (1981)
Islands off Antarctica in the Scotia Sea.
CO S640000S530000
   W0250000W0650000
BT Antarctica
NT South Georgia
NT South Orkney Islands
NT South Sandwich Islands
NT South Shetland Islands

Scotian Shelf (1978)
Off Nova Scotia.
UF Nova Scotia Shelf
UF Nova Scotian Shelf
BT Atlantic Ocean

Scotland (1978)
Shetland Islands is considered a narrower term as of 1981.
CO N544000N610000
   W0004500W0083000
BT Great Britain
BT United Kingdom
BT Western Europe
BT Europe
NT Angus Scotland
NT Argyllshire Scotland
NT Berwickshire Scotland
NT Cairngorm Mountains
NT Central region Scotland
NT Dundee Scotland
NT Fife region Scotland
NT Firth of Clyde
NT Galloway Scotland
NT Grampian Highlands
NT Grampian region Scotland
NT Great Glen Fault
NT Hebrides
NT Highland region Scotland
NT Kirkcudbrightshire Scotland
NT Lothian region Scotland
NT Midlothian Scotland
NT Moine thrust zone
NT Moray Firth
NT Orkney Islands
NT Perthshire Scotland
NT Scottish Lowlands
NT Scottish Northern Highlands
NT Shetland Islands

NT Strathclyde region Scotland
NT Tay Estuary
SA Ballantrae Complex
SA Carnmenellis Granite
SA Central Graben
SA Cheviot Hills
SA Dalradian
SA Esk Trough
SA Fair Isle
SA Forth Valley
SA Highland Boundary Fault
SA Kimmeridge Clay
SA Laurentia
SA Lewisian Complex
SA Mercia Mudstone
SA Midland Valley
SA North Sea Coast
SA North Sea region
SA Orcadian Basin
SA Red Hill
SA Saint Kilda
SA Southern Uplands
SA Tyne River

Scott County
Valid through 1988. Search in combination with state term. After 1988, use specific county-state term.

Scott County Arkansas (1989)
W Arkansas. Before 1989, also search Scott County AND Arkansas.
CO N343900N350500
W0931400W0942600
BT Arkansas
BT United States

Scott County Illinois (1989)
W central Illinois. Before 1989, also search Scott County AND Illinois.
CO N393000N394500
W0901700W0904000
BT Illinois
BT United States

Scott County Indiana (1989)
SE Indiana. Before 1989, also search Scott County AND Indiana.
CO N383500N385000
W0853300W0855400
BT Indiana
BT United States

Scott County Iowa (1989)
E Iowa. Before 1989, also search Scott County AND Iowa.
CO N412700N414700
W0902000W0905400
BT Iowa
BT United States
NT Davenport Iowa

Scott County Kansas (1989)
W Kansas. Before 1989, also search Scott County AND Kansas.
CO N381500N384200
W1004300W1010800
BT Kansas
BT United States

Scott County Kentucky (1989)
N Kentucky. Before 1989, also search Scott County AND Kentucky.
CO N380700N382900
W0842500W0844500
BT Kentucky
BT United States

Scott County Minnesota (1989)
S Minnesota. Before 1989, also search Scott County AND Minnesota.
CO N443200N444900
W0931700W0935300
BT Minnesota
BT United States

Scott County Mississippi (1989)
Central Mississippi. Before 1989, also search Scott County AND Mississippi.
CO N321400N323900
W0892000W0894800
BT Mississippi
BT United States

Scott County Missouri (1989)
SE Missouri. Before 1989, also search Scott County AND Missouri.
CO N365200N371500
W0892000W0894600
BT Missouri
BT United States

Scott County Tennessee (1989)
N Tennessee. Before 1989, also search Scott County AND Tennessee.
CO N361000N363400
W0841300W0844700
BT Tennessee
BT United States

Scott County Virginia (1989)
SW Virginia. Before 1989, also search Scott County AND Virginia.
CO N363600N365300
W0821800W0830000
BT Virginia
BT United States

Scottish Lowlands (1981)
Region S of Dumbarton-Stonehaven line to the Southern Uplands. Before 1981, search Lowlands AND Scotland.
BT Scotland
BT Great Britain
BT United Kingdom
BT Western Europe
BT Europe

Scottish Northern Highlands (1993)
Before 1993, also search Northern Highlands and Scotland.
CO N563000N584000
W0030000W0062000
BT Scotland
BT Great Britain
BT United Kingdom
BT Western Europe
BT Europe
SA Grampian Highlands

scour (1989)
UF scouring
BT erosion
SA glacial erosion
SA ice
SA water erosion
SA wind erosion

scour casts (1978)
BT bedding plane irregularities
BT sedimentary structures
SA casts
SA flute casts

scour marks (1978)
BT bedding plane irregularities
BT sedimentary structures
SA current markings
SA flute casts

Scourie
No longer a valid term for GeoRef as of 1993. See Scourie Scotland.

Scourie Scotland (1993)
Village on North Minch in NW Scotland.
BT Sutherland Scotland
BT Highland region Scotland
BT Scotland
BT Great Britain
BT United Kingdom
BT Western Europe
BT Europe

scouring
use scour

Scripps Institution of Oceanography (1978)
In La Jolla, a NW section of San Diego, in San Diego County.
SA academic institutions
SA California

ScS-waves (1981)
BT body waves
BT elastic waves
SA S-waves
SA seismology
SA waves

Sculptured Hills (1978)
BT Moon

Scutelluidae (1978)
BT Ptychopariida
BT Trilobita
BT Trilobitomorpha
BT Arthropoda
BT Invertebrata

Scyphozoa (1978)
Autoposting of this term began in 1978.
BT Coelenterata
BT Invertebrata
NT Conularida

Scythian (1978)
Europe. Above Tatarian (Permian), below Anisian.
UF Skythian
BT Lower Triassic
BT Triassic
BT Mesozoic
SA Werfenian

Scythian Platform (1978)
Broad, sedimentary plain N and NE of the Black Sea. This term has multiple hierarchies.
IN Index former Soviet republics as applicable.
BT1 Commonwealth of Independent States
BT2 Europe
SA Russian Federation
SA Ukraine

Se
use selenium

Sea Beam
use Seabeam

sea fan
use submarine fans

sea floor spreading
use sea-floor spreading

sea floors
use ocean floors

sea ice (1978)
SA ice
SA ice fields
SA icebergs
SA oceanography
SA sea water

sea level
Not a valid term for GeoRef. Use changes of level for recent sea level. Use regression or transgression when referring to older sea levels.

Sea Mapping and Remote Characterization
use SeaMarc

sea mounts
use seamounts

Sea of Azof
use Azov Sea

Sea of Azov
use Azov Sea

Sea of Crises (1978)
Also search Crisium.
UF Mare Crisium
BT Moon

Sea of Fertility (1978)
UF Mare Fecunditatis
BT Moon

Sea of Galilee (1978)
Freshwater lake lying in N part of the Great Rift Valley, Northern Israel.
UF Lake Kinneret
UF Lake Tiberias
BT Israel
BT Middle East
BT Asia
SA Great Rift Valley

Sea of Japan
Not a valid index term for GeoRef since 1973.
use Japan Sea

Sea of Marmara region (1981)
CO N390500N420500
E0310000E0254000
BT Turkey
BT Middle East
BT Asia

Sea of Nectar (1978)
UF Mare Nectaris
BT Moon

Sea of Okhotsk
Not a valid index term for GeoRef since 1973.
use Okhotsk Sea

Sea of Rains (1978)
UF Imbrium Basin
UF Mare Imbrium
BT Moon
SA Aristarchus

Sea of Serenity (1978)
UF Mare Serenitatis
BT Moon

Sea of Tranquillity (1978)
UF Mare Tranquillitatis
BT Moon

Sea of Waves (1978)
UF Mare Undarum
BT Moon

sea water (1978)
Used for studies on the composition and properties of the water of the oceans. For ancient water, see sedimentation or geochemistry.
UF seawater
SA atmosphere
SA brackish water
SA brines
SA carbonate compensation depth
SA chemical dispersion
SA density
SA desalinization
SA geochemistry
SA halmyrolysis
SA ion exchange
SA isotopes
SA marine geology

SA nepheloid layer
SA oceanography
SA paleosalinity
SA pH
SA salinity
SA salt
SA salt water
SA salt-water intrusion
SA saturation
SA sea ice
SA solubility
SA submarine springs
SA suspended materials
SA upwelling

sea-floor spreading   (1978)
Used for topics related to the hypothesis that the oceanic crust is increasing by convective upwelling of magma along the mid-ocean ridges or world rift system.
UF ocean-floor spreading
UF sea floor spreading
UF spreading concept
UF spreading-floor hypothesis
SA accreting plate boundary
SA Benioff zone
SA continental drift
SA crust
SA fracture zones
SA geosynclines
SA geotectonic maps
SA heat flow
SA Iceland Research Drilling Project
SA magnetic anomalies
SA mantle
SA marine geology
SA mid-ocean ridge basalts
SA mid-ocean ridges
SA Mohorovicic discontinuity
SA ocean basins
SA ocean floors
SA paleo-oceanography
SA paleobathymetry
SA paleomagnetism
SA plate divergence
SA plate tectonics
SA rifting
SA spreading centers
SA submarine volcanoes
SA tectonophysics
SA Wilson cycle

sea-floor trench
 use trenches

sea-level changes
 use changes of level

sea-water encroachment
 use salt-water intrusion

sea-water intrusion
 use salt-water intrusion

Seabeam   (1989)
UF Sea Beam
SA acoustical methods
SA acoustical surveys
SA bathymetry
SA data processing
SA geophysical methods
SA geophysical surveys
SA instruments
SA marine methods
SA seismic surveys
SA sonar methods

sealing   (1981)
SA cap rocks
SA engineering geology

SeaMarc   (1989)
Acronym.
UF Sea Mapping and Remote Characterization
SA geophysical methods
SA geophysical surveys

SA image enhancement
SA imagery
SA remote sensing
SA satellite methods

seamounts   (1978)
UF guyots
UF sea mounts
SA bottom features
SA ocean floors

seams, coal
 use coal seams

Searles Lake   (1978)
Dry lake in NW San Bernardino and Inyo counties in S California. Source of potash and borates.
IN Index counties or regions as applicable.
SA California
SA Inyo County California
SA potash
SA San Bernardino County California
SA trona
SA Trona California

seas, epeiric
 use epicontinental seas

seas, epicontinental
 use epicontinental seas

seas, marginal
 use marginal seas

Seasat   (1985)
SA geophysical methods
SA geophysical surveys
SA marine geology
SA oceanography
SA remote sensing
SA satellite methods

seasonal
A valid term through 1977. After 1977, use seasonal variations.

seasonal variations   (1978)
Before 1978, search seasonal AND variations.
SA annual variations
SA magnetic field
SA secular variations
SA variations

seat clay
 use underclay

seat earth
 use underclay

Seattle
Valid through 1988. Search in combination with state term. After 1988, use specific city-state term.

Seattle Washington   (1989)
City on Puget Sound in W central Washington. Before 1989, search Seattle AND Washington.
CO N473500N473500
    W1222000W1222000
BT King County Washington
BT Washington
BT United States

seawalls   (1978)
SA revetments
SA shorelines

seawater
 use sea water

seaweed
 use Phaeophyta

Sebes
No longer a valid term for GeoRef. As of 1993 see Sebes Romania

Sebes Mountains   (1978)

SW Transylvania.
BT Transylvania
BT Romania
BT Southern Europe
BT Europe

Sebes Romania   (1993)
Town in Alba County in SW Transylvania, central Romania. Before 1993 also search Sebes AND Romania.
BT Transylvania
BT Romania
BT Southern Europe
BT Europe

Sebkha de Tindouf
 use Tindouf Basin

Sebkha el Melah   (1978)
BT Tunisia
BT North Africa
BT Africa

sebkha environment   (1978)
Before 1978, search sebkha.
UF sabkha environment
SA depositional environment
SA environment
SA hypersaline environment
SA paleogeography
SA playas
SA salt marshes
SA sedimentation
SA supratidal environment

Seboomook Formation   (1989)
BT Lower Devonian
BT Devonian
BT Paleozoic
SA Maine
SA Quebec

secondary dispersion   (1978)
SA dispersion patterns
SA mineral exploration
SA primary dispersion

secondary ion mass spectroscopy   (1993)
Before 1993, also search secondary ion methods.
UF SIMS
SA chemical analysis
SA ions
SA mass spectroscopy
SA spectroscopy

secondary magnetization   (1981)
BT magnetization
SA remagnetization

secondary migration   (1985)
Not for migration of elastic waves.
SA migration
SA natural gas
SA petroleum

secondary minerals   (1978)
SA amygdules
SA minerals

secondary porosity   (1989)
BT porosity
SA permeability
SA physical properties
SA porous materials
SA sedimentary rocks
SA sediments

secondary recovery   (1978)
BT enhanced recovery
SA fluid injection
SA petroleum
SA petroleum engineering
SA recovery
SA steam injection
SA tertiary recovery
SA waterflooding

secondary structures   (1978)

Autoposting of this term to nodules ended in 1981.
BT sedimentary structures
NT armored mud balls
NT concretions
NT cone-in-cone
NT geodes
NT microstylolites
NT septaria
NT stylolites
SA nodules
SA primary structures

secondary wave
 use S-waves

sections
No longer a valid term for GeoRef. As of 1993, see borehole sections, cross sections, pit sections, polished sections, stratotypes, thin sections, and type sections.

secular variations   (1978)
UF geomagnetic secular variation
SA annual variations
SA magnetic field
SA paleomagnetism
SA seasonal variations
SA variations

Sedan
No longer a valid term for GeoRef. See Sedan France.

Sedan France   (1993)
City near Belgian border in NE Ardennes, NE France.
CO N494200N494200
    E0045700E0045700
BT Ardennes France
BT France
BT Western Europe
BT Europe
SA Ardennes

Sedgwick Basin   (1978)
S central Kansas.
IN Index counties or region as applicable.
SA Kansas

Sedgwick County
Valid through 1988. Search in combination with state term. After 1988, use specific county-state term.

Sedgwick County Colorado   (1989)
Extreme NE Colorado. Before 1989, also search Sedgwick County AND Colorado.
CO N404600N410000
    W1020300W1023700
BT Colorado
BT United States

Sedgwick County Kansas   (1989)
S Kansas. Before 1989, also search Sedgwick County AND Kansas.
CO N373000N375400
    W0970800W0974700
BT Kansas
BT United States
NT Wichita Kansas

sediment budget
 use bedload

sediment load
 use bedload

sediment sampling
As of 1981, no longer a valid term for GeoRef. Use sampling and sediments. Term was introduced in 1978.

sediment supply (1993)
  SA controls
  SA hydrology
  SA sedimentation
  SA supply

sediment transport (1993)
  Before 1993, also search transport AND sedimentation. A valid term before 1976.
  SA erosion rates
  SA flume studies
  SA glacial transport
  SA ice rafting
  SA laminar flow
  SA marine transport
  SA ore transport
  SA saltation
  SA sedimentation
  SA stream transport
  SA suspension
  SA transportation
  SA turbidity currents
  SA wind transport

sediment traps (1985)
  SA hydrology
  SA instruments
  SA oceanography
  SA samplers
  SA sampling
  SA sedimentation
  SA sediments
  SA suspended materials

sediment yield (1978)
  SA erosion
  SA runoff
  SA sediments
  SA streams
  SA yields

sediment-water interface (1978)
  Before 1978, search interface AND sediment-water.
  UF interface, sediment-water
  SA interfaces
  SA rock-water interface
  SA sedimentation
  SA sediments
  SA water

sedimentary basins (1978)
  Term should be used in paleogeographic papers, not for recent basins.
  BT basins
  SA foreland basins
  SA paleogeography

sedimentary bodies (1981)
  SA sedimentary rocks
  SA sedimentation
  SA sediments

sedimentary cover (1978)
  UF cover, sedimentary
  SA basement
  SA overburden
  SA regolith
  SA tectonics

sedimentary environment
  use depositional environment

sedimentary fault
  use growth faults

sedimentary petrology (1978)
  Used for the discipline as a whole.
  SA bibliography
  SA catalogs
  SA clay mineralogy
  SA diagenesis
  SA domains
  SA flume studies
  SA glossaries
  SA lexicons
  SA marine geology
  SA microscope methods
  SA oceanography
  SA petrology
  SA sedimentary rocks
  SA sedimentary structures
  SA sedimentation
  SA sedimentology
  SA sediments
  SA shape analysis

sedimentary processes (1978)
  As of 1993, restricted to economic studies.
  SA deposition
  SA mineral deposits, genesis
  SA processes

sedimentary quartzite
  use orthoquartzite

sedimentary rocks (1978)
  Autoposting of this term began in 1978.
  IN   For description of hierarchy see List I.
  NT bone beds
  NT carbonate rocks
  NT caustobiolith
  NT chemically precipitated rocks
  NT clastic rocks
  NT coal
  NT gas sands
  NT grapestone
  NT oil sands
  NT oil shale
  NT torbanite
  SA agglomerate
  SA anhydrite
  SA arenaceous texture
  SA argillaceous texture
  SA arkosic composition
  SA authigenesis
  SA bauxite
  SA boulders
  SA calcareous composition
  SA catagenesis
  SA cement
  SA clasts
  SA clay mineralogy
  SA coarse-grained materials
  SA concretions
  SA diagenesis
  SA dolomitic composition
  SA dolomitization
  SA environmental analysis
  SA fabric
  SA ferruginous composition
  SA fines
  SA fossiliferous materials
  SA glauconitic composition
  SA grains
  SA greensand
  SA gypsum
  SA hardground
  SA insoluble residues
  SA iron-rich composition
  SA kerogen
  SA lenses
  SA lithofacies
  SA lithogeochemistry
  SA lithostratigraphy
  SA metasedimentary rocks
  SA oolite
  SA oolitic texture
  SA organic materials
  SA organic residues
  SA packing
  SA paragenesis
  SA pelitic texture
  SA petrography
  SA petrology
  SA phi scale
  SA pisolites
  SA pisoliths
  SA pisolitic texture
  SA porosity
  SA provenance
  SA pyroclastics
  SA reefs
  SA rocks
  SA rounding
  SA secondary porosity
  SA sedimentary bodies
  SA sedimentary petrology
  SA sedimentary structures
  SA sedimentation
  SA sedimentology
  SA sediments
  SA shape analysis
  SA shells
  SA siliceous composition
  SA siliciclastics
  SA size distribution
  SA sorting
  SA stylolitization
  SA tempestite
  SA terrigenous materials
  SA turbidite
  SA volcaniclastics
  SA wildflysch

sedimentary structures (1978)
  Used for a structure in a sedimentary rock or sediment, formed either contemporaneously with deposition (primary) or subsequently to deposition (secondary). Autoposting of this term to ichnofossils, nodules and shrinkage cracks ended in 1981. Autoposting of this term to borings, burrows, coprolites, tracks and trails ended in 1985. Autoposting of this term to rhythmite began in 1985.
  IN   See list K for types and names of structures.
  NT bedding plane irregularities
  NT biogenic structures
  NT cylindrical structures
  NT planar bedding structures
  NT primary structures
  NT secondary structures
  NT soft sediment deformation
  NT sole marks
  NT turbidity current structures
  SA borings
  SA burrows
  SA casts
  SA channels
  SA coprolites
  SA deposition
  SA diagenesis
  SA environmental analysis
  SA ergs
  SA fluvial features
  SA furrows
  SA glacial geology
  SA hardground
  SA intraformational structures
  SA nodules
  SA paleogeography
  SA reefs
  SA sedimentary petrology
  SA sedimentary rocks
  SA sedimentation
  SA sediments
  SA solution cavities
  SA spherules
  SA tracks
  SA trails

sedimentation (1978)
  Used for the act or process of forming or accumulating sediment in layers, including all processes from transport through diagenesis. Before 1976, also search aggregation. Autoposting of this term began in 1993.
  IN   Index specific type of sedimentation if applicable.
  NT biochemical sedimentation
  NT bioclastic sedimentation
  NT chemical sedimentation
  NT coastal sedimentation
  NT deep-sea sedimentation
  NT deltaic sedimentation
  NT detrital sedimentation
  NT estuarine sedimentation
  NT fluvial sedimentation
  NT fresh-water sedimentation
  NT glacial sedimentation
  NT intertidal sedimentation
  NT lacustrine sedimentation
  NT lagoonal sedimentation
  NT marine sedimentation
  NT nearshore sedimentation
  NT paludal sedimentation
  NT pelagic sedimentation
  NT terrestrial sedimentation
  SA ablation
  SA accretion
  SA aquatic environment
  SA artificial islands
  SA atectonic processes
  SA base surges
  SA basins
  SA beaches
  SA bedload
  SA biofacies
  SA bioturbation
  SA braided streams
  SA carbonate platforms
  SA cave environment
  SA cementation
  SA changes of level
  SA channels
  SA clastic wedges
  SA climate
  SA climatic controls
  SA coastal environment
  SA continental margin sedimentation
  SA continental shelf
  SA continental slope
  SA current directions
  SA cyclic processes
  SA cyclothems
  SA deep-sea environment
  SA deltaic environment
  SA deltas
  SA deposition
  SA depositional environment
  SA diagenesis
  SA dolomitization
  SA engineering geology
  SA environment
  SA environmental analysis
  SA erosion rates
  SA estuaries
  SA estuarine environment
  SA flow structures
  SA flume studies
  SA fluvial environment
  SA foreland basins
  SA fresh water
  SA fresh-water environment
  SA geochemical controls
  SA geopressure
  SA glacial environment
  SA glacial geology
  SA glacial transport
  SA glaciofluvial environment
  SA glaciolacustrine environment
  SA glaciomarine environment
  SA granulometry
  SA gravity flows
  SA heavy minerals
  SA HEBBLE
  SA hemipelagic environment
  SA high-energy environment
  SA hydrology
  SA hypersaline environment
  SA ice rafting
  SA interglacial environment
  SA intertidal environment
  SA lacustrine environment

SA lagoonal environment
SA lagoons
SA lakes
SA laminar flow
SA lithification
SA localization
SA low-energy environment
SA marine environment
SA marine geology
SA marine transport
SA nearshore environment
SA nodules
SA ocean circulation
SA ocean floors
SA oceanography
SA organic materials
SA packing
SA paleocurrent maps
SA paleocurrents
SA paleoecology
SA paleoenvironment
SA paleogeography
SA paludal environment
SA pelagic environment
SA periglacial environment
SA periodicity
SA postglacial environment
SA precipitation
SA progradation
SA provenance
SA punctuated aggradational cycles
SA reef environment
SA reefs
SA reservoir rocks
SA reworking
SA rhythmite
SA salinity
SA salt marshes
SA saltation
SA sebkha environment
SA sediment supply
SA sediment transport
SA sediment traps
SA sediment-water interface
SA sedimentary bodies
SA sedimentary petrology
SA sedimentary rocks
SA sedimentary structures
SA sedimentation rates
SA sedimentology
SA sediments
SA shallow-water environment
SA shelf environment
SA silicification
SA siltation
SA slope environment
SA slumping
SA sorting
SA storm environment
SA stream transport
SA stylolitization
SA subalpine environment
SA subglacial environment
SA subtidal environment
SA supratidal environment
SA swamps
SA synsedimentary processes
SA taiga environment
SA tectonic controls
SA terrestrial environment
SA tidal flats
SA tombolos
SA turbidity currents
SA uniformitarianism
SA urban environment
SA wind transport

sedimentation rates (1982)
Before 1982, search sedimentation AND rates.
UF rate of sedimentation
SA rates
SA sedimentation

sedimentology (1978)
For the discipline.
SA geomorphology
SA marine geology
SA sedimentary petrology
SA sedimentary rocks
SA sedimentation
SA sediments

sediments (1978)
Used for unconsolidated solid fragmental material, or a mass of such material, that originates from weathering of rocks.
IN See List N.
NT carbonate sediments
NT clastic sediments
NT guano
NT gyttja
NT marine sediments
NT peat
SA amygdules
SA anhydrite
SA arenaceous texture
SA argillaceous texture
SA arkosic composition
SA ash flows
SA bedload
SA biofacies
SA breccia
SA calcareous composition
SA cement
SA clasts
SA clay mineralogy
SA coarse-grained materials
SA compactness
SA concretions
SA dehydration
SA deposition
SA diagenesis
SA dolomitic composition
SA dolomitization
SA engineering properties
SA entropy
SA environmental analysis
SA evaporites
SA fabric
SA ferruginous composition
SA filtration
SA fines
SA fossiliferous materials
SA glacial geology
SA glauconitic composition
SA grain size
SA grains
SA granulometry
SA gypsum
SA halite
SA hardground
SA heavy minerals
SA hydrology
SA iron-rich composition
SA lake sediments
SA liquefaction
SA lithification
SA lithofacies
SA lithostratigraphy
SA littoral drift
SA marine geology
SA ocean floors
SA oolitic texture
SA organic materials
SA organic residues
SA permeability
SA petrography
SA petrology
SA phi scale
SA pisoliths
SA pisolitic texture
SA placers
SA pore water
SA porosity
SA provenance
SA pyroclastic flows
SA pyroclastics
SA relict materials
SA reworking
SA roundness
SA secondary porosity
SA sediment traps
SA sediment yield
SA sediment-water interface
SA sedimentary bodies
SA sedimentary petrology
SA sedimentary rocks
SA sedimentary structures
SA sedimentation
SA sedimentology
SA shape analysis
SA shells
SA siliceous composition
SA siliciclastics
SA size distribution
SA soils
SA sorting
SA sphericity
SA spherules
SA stream sediments
SA stylolitization
SA suspension
SA tempestite
SA terrigenous materials
SA turbidite
SA unconsolidated materials
SA volcanic ash
SA volcaniclastics
SA weathering
SA wildflysch

seeds (1978)
SA angiosperms
SA cones
SA gymnosperms
SA lignin
SA paleobotany
SA plants

Seeland
use Zealand

seepage (1978)
UF leakage
SA aquitards
SA drainage
SA engineering geology
SA far-field
SA foundations
SA geomembranes
SA geotextiles
SA ground water
SA leachate
SA leaking underground storage tanks
SA leaky aquifers
SA near-field
SA nuclear facilities
SA oil seeps
SA percolation
SA petroleum
SA reservoirs
SA soil mechanics
SA soils
SA tailings ponds
SA tunnels
SA underground installations
SA waste disposal
SA water
SA waterways

seeps, oil
use oil seeps

segmentation (1989)
Used as a general term applicable to a variety of subjects including faults, fracture zones, lagoon subdivision, drainage patterns, and arthropod bodies.
SA arthropods
SA drainage patterns
SA faults
SA fracture zones
SA lagoons

Segovia
No longer a valid term for GeoRef. As of 1993, see Segovia Spain.

Segovia Spain (1993)
Province in central Spain. Before 1993, also search Segovia and Spain.
CO N404000N413600 W0031500W0044200
BT Old Castile Spain
BT Castile Spain
BT Spain
BT Iberian Peninsula
BT Southern Europe
BT Europe

Segozero (1978)
Lake in S central Karelian Automomous Republic in NW European Russian Federation. This term has multiple hierarchies.
BT1 Karelia Russian Federation
BT1 Russian Federation
BT1 Commonwealth of Independent States
BT2 Karelia Russian Federation
BT2 Asia

Segre Valley (1978)
River valley in Catalonia, Spain. Also search Segre; Segre River.
IN Index Spain if applicable.
SA Lerida Spain

segregation (1981)
SA crystal fractionation
SA fractional crystallization
SA geochemistry
SA magmas
SA phase equilibria

seiche
Not a valid GeoRef index term after 1970. Was used on level 1 in subfile N.
use seiches

seiches (1978)
Before 1971, also search seiche.
UF seiche
SA lakes
SA limnology
SA ocean waves
SA waves

Seiland (1978)
Island in Arctic Ocean off NW Finnmark.
BT Finnmark Norway
BT Norway
BT Scandinavia
BT Western Europe
BT Europe

Seine (1978)
Former department which constituted Paris proper and a ring of industrial and residential suburbs. Department officially abolished January 1, 1960.
BT France
BT Western Europe
BT Europe
SA Paris France

Seine Estuary (1978)
The tidal mouth of the Seine River emptying into the Bay of the Seine at Le Havre.
IN Index departments as applicable.
BT France
BT Western Europe
BT Europe
SA Calvados France
SA Eure France
SA Seine-Maritime France

Seine River (1978)
Rises in Cote-d'Or Department in E and flows NW through Paris and into the English Channel near Le Havre.
BT France
BT Western Europe
BT Europe

Seine Valley (1978)
River valley in N central and N France.
BT France
BT Western Europe
BT Europe

Seine-et-Marne
No longer a valid term for GeoRef. See Seine-et-Marne France.

Seine-et-Marne France (1993)
Department in N central France.
CO N481000N491000
E0033000E0021500
BT France
BT Western Europe
BT Europe
NT Fontainebleau France
SA Yonne Valley

Seine-et-Oise (1978)
Former department outside Paris suburbs which completely surrounded former Seine Department. Officially abolished January 1, 1968.
BT France
BT Western Europe
BT Europe
SA Paris France

Seine-Maritime
No longer a valid term for GeoRef. See Seine-Maritime France.

Seine-Maritime France (1993)
Department in NW France.
CO N491500N500500
E0015000E0000000
BT France
BT Western Europe
BT Europe
NT Caux
NT Dieppe France
NT Le Havre France
NT Rouen France
SA Normandy
SA Seine Estuary

Seine-Saint-Denis
No longer a valid term for GeoRef. See Seine-Saint-Denis France.

Seine-Saint-Denis France (1993)
Department in N France.
CO N485000N490000
E0023500E0021500
BT France
BT Western Europe
BT Europe

Seislog (1978)
SA instruments
SA seismic methods

seismic data
A valid term through 1977. Use data and earthquakes.

seismic effects
A valid term through 1977. Use effects and earthquakes.

seismic energy (1981)
Energy produced by earthquakes.
SA earthquakes
SA energy
SA seismology

seismic exploration
Not a valid GeoRef index term after 1971. Was used on level 1 in subfiles G and N. Now use seismic surveys, seismic methods, or seismic logging.

seismic focus
use focus

seismic gaps (1985)
SA active faults
SA earthquake prediction
SA earthquakes
SA faults
SA seismicity
SA seismology
SA seismotectonics

seismic intensity (1981)
SA amplitude
SA earthquakes
SA intensity
SA magnitude
SA modified Mercalli scale
SA seismology

seismic logging (1984)
BT well-logging
SA acoustical logging
SA impedance
SA seismic methods
SA seismic surveys
SA tube waves
SA vertical seismic profiles

seismic methods (1978)
Autoposting of this term to reflection methods and refraction methods ended in 1985.
BT geophysical methods
SA air guns
SA AVO methods
SA BIRPS
SA CALCRUST
SA COCORP
SA COCRUST
SA common-depth-point method
SA crosshole methods
SA deep seismic sounding
SA DEKORP
SA dip moveout
SA elastic waves
SA far-field
SA filters
SA frequency domain analysis
SA geophones
SA GEOS
SA Green function
SA hydrocarbon indicators
SA hydrophones
SA impedance
SA Kalman filters
SA Lithoprobe
SA microseismic methods
SA moveout
SA multichannel methods
SA near-field
SA near-field spectra
SA normal moveout
SA ocean bottom seismographs
SA PASSCAL
SA passive methods
SA Radon transforms
SA ray tracing
SA reflection methods
SA refraction methods
SA Seislog
SA seismic logging
SA seismic migration
SA seismic networks
SA seismic profiles
SA seismic stratigraphy
SA seismic surveys
SA signal-to-noise ratio
SA sonobuoys
SA stacking
SA three-dimensional methods
SA tomography
SA velocity analysis
SA vertical seismic profiles
SA Vibroseis

seismic migration (1993)
Inversion operation for rearranging seismic information to plot reflections and diffractions at their true locations. Before 1993, also search migration AND seismic methods.
UF migration (seismic)
SA arrival time
SA diffraction
SA dip moveout
SA finite difference analysis
SA imagery
SA inverse problem
SA Kirchhoff integral
SA migration
SA raypaths
SA reflection
SA seismic methods
SA seismic surveys
SA seismology

seismic moment (1981)
Before 1993, also search moment tensors and seismology.
SA displacements
SA earthquakes
SA faults
SA seismology
SA shear modulus

seismic networks (1993)
Before 1993, also search arrays or networks with seismic methods, seismology or seismic surveys.
SA arrays
SA geophysical methods
SA geophysical surveys
SA GEOS
SA IRIS network
SA LASA
SA networks
SA NORESS
SA NORSAR
SA PASSCAL
SA seismic methods
SA seismology

seismic noise
A valid term through 1978. After 1978, use noise.

seismic profiles (1985)
SA geophysical profiles
SA Lithoprobe
SA seismic methods
SA seismic surveys
SA vertical seismic profiles

seismic reflection profiles
Not a valid term for GeoRef. As of 1989, use seismic profiles in combination with reflection.

seismic response (1981)
Response of a man-made structure to an earthquake.
SA acceleration
SA aseismic design
SA dams
SA earthquakes
SA engineering geology
SA foundations
SA highways
SA marine installations
SA microzonation
SA nuclear facilities
SA railroads
SA reservoirs
SA seismic risk
SA seismology
SA soil mechanics
SA soil-structure interface
SA tunnels
SA underground installations
SA waterways

seismic risk (1978)
UF risk, seismic
SA engineering geology
SA geologic hazards
SA microzonation
SA nuclear facilities
SA safety
SA seismic response
SA seismic zoning

seismic sea waves
use tsunamis

seismic sounding
use deep seismic sounding

seismic sources (1978)
UF sources, seismic
SA earthquakes
SA elastic waves
SA explosions
SA seismology
SA waveforms

seismic stratigraphy (1985)
Before 1993, also search seismostratigraphy.
UF seismostratigraphy
SA hydrocarbon indicators
SA lithofacies
SA lithostratigraphy
SA seismic methods
SA seismic surveys
SA stratigraphy

seismic studies
A valid term through 1976. Use seismic surveys and theoretical studies or experimental studies.

seismic surge
use tsunamis

seismic survey maps (1985)
BT maps
SA geophysical survey maps
SA seismic surveys

seismic surveys (1978)
BT geophysical surveys
BT surveys
SA BIRPS
SA CALCRUST
SA COCORP
SA COCRUST
SA crust
SA DEKORP
SA dip moveout
SA explosions
SA GEOS
SA Lithoprobe
SA microseismic methods
SA moveout
SA multichannel methods
SA normal moveout
SA passive methods
SA ray tracing
SA reflection
SA refraction
SA Rivera Ocean Seismic Experiment
SA Seabeam
SA seismic logging
SA seismic methods
SA seismic migration
SA seismic profiles
SA seismic stratigraphy
SA seismic survey maps
SA seismology
SA three-dimensional methods
SA tube waves
SA vertical seismic profiles
SA Vibroseis

seismic waves
A valid term through 1975.
use elastic waves

seismic zones
use seismic zoning

seismic zoning (1981)
UF seismic zones
SA earthquakes
SA engineering geology
SA environmental geology
SA microzonation
SA seismic risk
SA seismicity maps
SA seismology
SA seismotectonic maps
SA zoning

seismicity (1978)
UF microseismicity
SA coupling
SA intraplate processes
SA paleoseismicity
SA seismic gaps
SA seismicity maps
SA seismology
SA seismotectonics

seismicity maps (1982)
BT maps
SA seismic zoning
SA seismicity
SA seismology
SA seismotectonic maps

seismograms (1978)
UF earthquake record
NT numerical seismograms
NT synthetic seismograms
NT theoretical seismograms
SA coda waves
SA earthquakes
SA seismographs
SA seismology
SA signal-to-noise ratio
SA spectral analysis
SA vertical seismic profiles

seismographs (1978)
Before 1981, also search seismometers.
UF seismometers
NT electromagnetic seismographs
NT horizontal-component seismographs
NT inertial seismographs
NT long-period seismographs
NT ocean bottom seismographs
NT short-period seismographs
NT strain seismographs
NT three-component seismographs
NT vertical-component seismographs
SA earthquakes
SA geophones
SA high-level mode
SA instruments
SA LASA
SA seismograms
SA seismology
SA spheroidal mode
SA toroidal mode

seismology (1978)
Used for the discipline as a whole as well as topics pertaining to seismology.
SA accelerograms
SA accelerometers
SA acoustical emissions
SA acoustical properties
SA acoustical waves
SA aftershocks
SA Airy waves
SA amplitude
SA amplitude distortion

SA anelastic materials
SA anelasticity
SA anisotropy
SA arrival time
SA attenuation
SA b-values
SA Backus-Gilbert analysis
SA Benioff zone
SA body waves
SA coda waves
SA Conrad discontinuity
SA core
SA core-mantle boundary
SA crust
SA curved seismic interface
SA deconvolution
SA deep seismic sounding
SA deformation
SA diffraction
SA dilatancy
SA dilatometers
SA direct problem
SA discontinuities
SA distortion
SA Earth
SA earthquakes
SA elastic properties
SA elastic waves
SA epicenters
SA explosions
SA far-field
SA faults
SA filters
SA focus
SA frequency domain analysis
SA geodesy
SA geophysical methods
SA geophysics
SA Green function
SA ground motion
SA group velocity
SA Gruneisen parameters
SA half-space
SA heterogeneity
SA high-level mode
SA high-resolution methods
SA Hilbert transformations
SA homogeneity
SA horizontal discontinuities
SA impedance
SA inclined seismic interface
SA induced earthquakes
SA interior
SA inverse problem
SA IRIS network
SA isochrons
SA isoseismic maps
SA Kalman filters
SA Lg-waves
SA linear distortion
SA long-period waves
SA Love waves
SA low-velocity zones
SA magnitude-frequency ratio
SA main shocks
SA mantle
SA marker beds
SA mathematical transformations
SA maximum entropy analysis
SA microearthquakes
SA microseisms
SA modified Mercalli scale
SA Mohorovicic discontinuity
SA multichannel methods
SA near-field
SA near-field spectra
SA neotectonics
SA noise
SA non-linear distortion
SA NORESS
SA NORSAR
SA nuclear explosions
SA observatories
SA P-waves
SA paleoseismicity

SA PcP-waves
SA periodicity
SA phase distortion
SA phase transitions
SA phase velocity
SA PKiKP-waves
SA PKP-waves
SA PKS-waves
SA plate tectonics
SA Pn-waves
SA PP-waves
SA PPP-waves
SA PPS-waves
SA precursors
SA propagation
SA PS-waves
SA Q
SA Rayleigh waves
SA raypaths
SA reflection
SA refraction
SA relaxation energy
SA Richter Scale
SA Rivera Ocean Seismic Experiment
SA S-waves
SA ScS-waves
SA seismic energy
SA seismic gaps
SA seismic intensity
SA seismic migration
SA seismic moment
SA seismic networks
SA seismic response
SA seismic sources
SA seismic surveys
SA seismic zoning
SA seismicity
SA seismicity maps
SA seismograms
SA seismographs
SA seismotectonic maps
SA seismotectonics
SA SH-waves
SA shock waves
SA short-period waves
SA signal distortion
SA signal-to-noise ratio
SA signals
SA SKP-waves
SA SKS-waves
SA Sn-waves
SA sounding
SA spectral analysis
SA spherical models
SA spheroidal mode
SA Stoneley waves
SA strain
SA strain relaxation
SA stress
SA stress drops
SA strong motion
SA surface waves
SA SV-waves
SA swarms
SA tangential discontinuities
SA tectonophysics
SA teleseismic signals
SA Thomson-Haskell analysis
SA three-dimensional models
SA tilt
SA tiltmeters
SA time variations
SA tomography
SA toroidal mode
SA traveltime
SA traveltime curves
SA traveltime residuals
SA tsunamis
SA two-dimensional models
SA velocity structure
SA vertical discontinuities
SA vertical seismic profiles
SA vibration
SA volcanology

SA wave absorption
SA wave dispersion
SA waveforms
SA Wiener-Hopf analysis

seismometers
As of 1981, no longer a valid term for GeoRef.
use seismographs

seismostratigraphy
use seismic stratigraphy

seismotectonic maps (1981)
Before 1981, search seismotectonics AND maps.
BT maps
SA seismic zoning
SA seismicity maps
SA seismology
SA seismotectonics
SA tectonics

seismotectonics (1978)
BT tectonics
SA paleoseismicity
SA plate tectonics
SA seismic gaps
SA seismicity
SA seismology
SA seismotectonic maps

selachians (1989)
As of 1981, used as the common name for Selachii.
BT fish
BT vertebrates
SA biostratigraphy
SA Selachii

Selachii (1978)
Order. Before 1981, also search selachians. As of 1981, selachians may be used as a common name for Selachii.
BT Elasmobranchii
BT Chondrichthyes
BT Pisces
BT Vertebrata
BT Chordata
SA selachians

Selangor
No longer a valid term for GeoRef. As of 1993, see Selangor Malaysia.

Selangor Malaysia (1993)
State on W coast of Malay Peninsula in West Malaysia. This term has multiple hierarchies.
BT1 West Malaysia
BT1 Malay Peninsula
BT1 Far East
BT1 Asia
BT2 West Malaysia
BT2 Malaysia
BT2 Far East
BT2 Asia
NT Kuala Lumpur Malaysia

selection, natural
use natural selection

selenates (1981)
Before 1981, search selenates and selenites.
SA minerals

selenates and selenites
As of 1981, no longer a valid term for GeoRef. Use selenates or selenites as applicable.

Selenga
No longer a valid term for GeoRef. As of 1993 see Selenga Mongolia.

Selenga Mongolia (1993)

Province on the Buryat border in N Mongolia. Before 1993 also search Selenga AND Mongolia.
UF Selenge
BT Mongolia
BT Far East
BT Asia

Selenga River valley (1978)
Extends to SE Lake Baikal in Russian Federation. Also search Selenga.
IN Index countries as applicable.
UF Selenge River valley
BT Asia
SA Mongolia
SA Russian Federation

Selenge
use Selenga Mongolia

Selenge River valley
use Selenga River valley

selenides (1978)
Autoposting of sulfides to this term ended in 1989.
NT eskebornite
NT umangite
SA minerals

selenite (1978)
Used for the variety of gypsum. Not related to the group name, selenites.
BT sulfates
SA gypsum

selenites (1981)
Before 1981, search selenates and selenites.
NT onofrite
SA minerals

selenium (1978)
UF Se
SA chalcophile elements
SA chemical elements
SA lithophile elements
SA onofrite

Selennyakh (1978)
River which rises in the N Cherskiy Range and flows into the Indigirka River in NE Yakutia. This term has multiple hierarchies.
UF Selenyak
BT1 Yakutia Russian Federation
BT1 Russian Federation
BT1 Commonwealth of Independent States
BT2 Yakutia Russian Federation
BT2 Asia

selenodesy (1978)
SA geodesy
SA Moon

Selenyak
use Selennyakh

self reversal
use reversals

self-potential methods (1978)
Before 1978, also search self-potential. Before 1985, also search spontaneous potential.
UF spontaneous polarization
UF spontaneous potential
SA electrical methods
SA electrical surveys
SA geophysical methods
SA geophysical surveys
SA well-logging

self-reversals
use reversals

Seligdar
No longer a valid term for GeoRef. As of 1993, see Seligdar Russian Federation.

Seligdar Russian Federation (1993)
Town in SE Yakutia. Before 1993, also search Seligdar. This term has multiple hierarchies.
CO N584300N584300
    E1253000E1253000
BT1 Yakutia Russian Federation
BT1 Russian Federation
BT1 Commonwealth of Independent States
BT2 Yakutia Russian Federation
BT2 Asia

seligmannite (1978)
BT sulfarsenites
BT sulfosalts

Selkirk Mountains (1978)
Range in the Rocky Mountains in SE British Columbia. This term has multiple hierarchies.
UF Selkirks Mountains
BT1 Canadian Rocky Mountains
BT1 Western Canada
BT1 Canada
BT2 Canadian Rocky Mountains
BT2 Rocky Mountains
BT3 British Columbia
BT3 Western Canada
BT3 Canada
NT Esplanade Range
SA Glacier National Park

Selkirks Mountains
use Selkirk Mountains

Selma Group (1985)
Alabama, Mississippi and Tennessee.
BT Upper Cretaceous
BT Cretaceous
BT Mesozoic
SA Alabama
SA Blufftown Formation
SA Eutaw Formation
SA Mississippi
SA Mooreville Chalk
SA Ripley Formation
SA Tennessee

Selukwe
No longer a valid term for GeoRef. As of 1993 see Selukwe Zimbabwe.

Selukwe Zimbabwe (1993)
Town in central Zimbabwe. Before 1993 also search Selukwe AND Zimbabwe.
BT Zimbabwe
BT Southern Africa
BT Africa

Selva
Not a valid term for GeoRef. See Copiapo Chile. Also may be used in other locations.

Selway-Bitterroot Wilderness (1985)
E Idaho and W Montana. Autoposting of broader terms to this term began in 1989.
IN Index states and counties as applicable.
CO N454000N464500
    W1140000W1153000
BT United States
SA Bitterroot Range
SA Idaho
SA Idaho County Idaho
SA Missoula County Montana
SA Montana

SA Ravalli County Montana
SA wilderness areas

Selwyn Basin (1985)
E central Yukon Territory, Canada.
CO N620000N640000
    W1270000W1350000
BT Yukon Territory
BT Western Canada
BT Canada
SA Canadian Rocky Mountains
SA Selwyn Mountains

Selwyn Mountains (1981)
On border of Yukon Territory and Northwest Territories, W of the Mackenzie Mountains. Before 1981, also search Selwyn Range AND Canada.
IN Index provinces as applicable.
BT Canada
SA Northwest Territories
SA Selwyn Basin
SA Selwyn Range
SA Yukon Territory

Selwyn Range (1978)
W central Queensland.
BT Queensland Australia
BT Australia
BT Australasia
SA Selwyn Mountains

SEM data (1978)
For the method, see scanning electron microscopy.
UF scanning electron microscopy data
SA data
SA electron microscopy
SA electron microscopy data
SA scanning electron microscopy
SA TEM data
SA transmission electron microscopy

Semail Ophiolite (1989)
SA Oman
SA United Arab Emirates

Semarkona Meteorite (1989)
Impact in Chhindwara District, Madhya Pradesh, India.
BT LL chondrites
BT chondrites
BT stony meteorites
BT meteorites
SA India
SA Madhya Pradesh India

Semata Formation (1978)
BT Pleistocene
BT Quaternary
BT Cenozoic
SA Japan

Semenic Mountains (1978)
Central Banat. Also search Semenic.
BT Romania
BT Southern Europe
BT Europe
SA Banat

semi-arid environment (1978)
Before 1978, also search semiarid; semiarid regions.
UF semiarid regions
SA arid environment
SA climate
SA depositional environment
SA ecology
SA environment
SA geomorphology

semi-infinite models (1981)
SA infinite models
SA mathematical models

SA models

semi-variograms
use semivariograms

semiarid regions
use semi-arid environment

Seminoe Mountains (1985)
Central Wyoming. Autoposting of broader terms to this term began in 1989.
BT Carbon County Wyoming
BT Wyoming
BT United States
SA Rocky Mountains

Seminole County
Valid through 1988. Search in combination with state term. After 1988, use specific county-state term.

Seminole County Florida (1989)
E central Florida. Before 1989, also search Seminole County AND Florida.
CO N283600N285000
    W0810000W0812900
BT Florida
BT United States

Seminole County Georgia (1989)
Extreme SW Georgia. Before 1989, also search Seminole County AND Georgia.
CO N304600N310600
    W0844200W0850000
BT Georgia
BT United States

Seminole County Oklahoma (1989)
Central Oklahoma. Before 1989, also search Seminole County AND Oklahoma.
CO N345200N352800
    W0962500W0964700
BT Oklahoma
BT United States

Seminole Formation (1978)
In Skiatook Group. Includes Lenapah Limestone, Nowata Shale, Memorial Shale. NE, central, and central S Oklahoma.
BT Missourian
BT Upper Pennsylvanian
BT Pennsylvanian
BT Carboniferous
BT Paleozoic
SA Oklahoma

Semipalatinsk
No longer a valid term for GeoRef. As of 1993, see Semipalatinsk Kazakhstan.

Semipalatinsk Kazakhstan (1993)
Oblast and city in NE Kazakhstan. This term has multiple hierarchies.
BT1 Kazakhstan
BT1 Central Asia
BT1 Asia
BT2 Kazakhstan
BT2 Commonwealth of Independent States
NT Akzhal Kazakhstan
NT Chingis-Tau

semitropical environment
use subtropical environment

semivariograms (1985)
UF semi-variograms
BT variance analysis
BT statistical analysis
SA covariance analysis
SA geostatistics

SA mathematical geology
SA mathematical methods
SA variograms
Semmering (1978)
Resort area in Eastern Alps.
SA Austria
SA Lower Austria
SA Styria Austria
Semri Series (1978)
BT Cambrian
BT Paleozoic
SA Bihar India
SA India
SA Kaimur Sandstone
SA Madhya Pradesh India
SA Uttar Pradesh India
SA Vindhyan
semseyite (1978)
BT sulfantimonites
BT sulfosalts
senarmontite (1978)
BT oxides
Sendai
No longer a valid term for GeoRef. See Sendai Japan.
Sendai Japan (1993)
City near E coast in Miyagi Prefecture, N Honshu.
BT Miyagi Japan
BT Honshu
BT Japan
BT Far East
BT Asia
Seneca County
Valid through 1988. Search in combination with state term. After 1988, use specific county-state term.
Seneca County New York (1989)
W central New York. Before 1989, also search Seneca County AND New York.
CO N423300N430200
W0763700W0765800
BT New York
BT United States
SA Cayuga Lake
SA Seneca Lake
Seneca County Ohio (1989)
N Ohio. Before 1989, also search Seneca County AND Ohio.
CO N405800N411500
W0825000W0832500
BT Ohio
BT United States
SA Sandusky River basin
Seneca Lake (1978)
One of the Finger Lakes in Seneca and Yates counties. Also search Seneca.
IN Index counties as applicable.
CO N422000N424500
W0764500W0770000
BT Finger Lakes
BT New York
BT United States
SA Seneca County New York
SA Yates County New York
Senegal (1978)
Formerly a republic in French Community. Achieved independence in 1960.
CO N123000N164000
W0112000W0173000
BT West Africa
BT Africa
NT Dakar Senegal
SA Mauritanides
SA Sahel
SA Senegal Basin

SA Senegal River
Senegal Basin (1978)
River basin. Also search Senegal.
IN Index countries as applicable.
UF Senegal River basin
BT West Africa
BT Africa
SA Guinea
SA Mali
SA Mauritania
SA Senegal
Senegal River (1978)
Rises in the highlands of Guinea and flows N and then NW into the Atlantic Ocean at Saint-Louis in Senegal. Also search Senegal.
IN Index countries as applicable.
BT Africa
SA Guinea
SA Mali
SA Mauritania
SA Senegal
Senegal River basin
use Senegal Basin
Senonian (1978)
Europe. Above Turonian, below Danian. Autoposting of this term began in 1978.
BT Upper Cretaceous
BT Cretaceous
BT Mesozoic
NT Campanian
NT Coniacian
NT Futaba Group
NT Izumi Group
NT Maestrichtian
NT Santonian
sensing, remote
use remote sensing
sensitive clays (1981)
SA clay
SA clays
sensitivity analysis (1989)
SA mathematical methods
SA mathematical models
SA numerical analysis
sensors
Not a valid term for GeoRef. See geophysical methods, chemical analysis, and instruments for specific description of methods.
Sentinel Butte Formation (1978)
In Fort Union Formation. Overlies Tongue River Member. NE Montana, and SW North Dakota.
BT Paleocene
BT Paleogene
BT Tertiary
BT Cenozoic
SA Fort Union Formation
SA Montana
SA North Dakota
separation (1978)
SA mixing
SA processes
sepiolite (1978)
UF meerschaum
BT clay minerals
BT sheet silicates
BT silicates
Sept-Iles
No longer a valid term for GeoRef. As of 1993, see Sept-Iles Quebec.
Sept-Iles Quebec (1993)
City on the Saint Lawrence River in SE Quebec. Before 1993, also search Sept-Iles AND Quebec.
UF Seven Isles

BT Saguenay County Quebec
BT Quebec
BT Eastern Canada
BT Canada
septa (1978)
UF septae
SA fossils
SA morphology
septae
use septa
septaria (1978)
BT secondary structures
BT sedimentary structures
SA concretions
SA cone-in-cone
Septibranchia (1981)
BT Bivalvia
BT Mollusca
BT Invertebrata
Sequanian (1978)
Europe.
BT Upper Jurassic
BT Jurassic
BT Mesozoic
Sequatchie River valley
use Sequatchie Valley
Sequatchie Valley (1978)
River valley in SE central Tennessee.
IN Index counties as applicable.
UF Sequatchie River valley
BT Tennessee
BT United States
sequence stratigraphy (1993)
SA stratigraphy
SA unconformities
Sequoia (1978)
Genus.
BT Taxodiaceae
BT Coniferales
BT gymnosperms
BT Spermatophyta
BT Plantae
Serajevo
use Sarajevo
Serbia (1978)
Constituent republic in E and NE Yugoslavia. Formerly Servia.
CO N420500N461000
E0230000E0185500
UF Servia
BT Yugoslavia
BT Southern Europe
BT Europe
NT Aleksinac
NT Belgrade
NT Kopaonik
NT Kosovo-Metohija
NT Kostolac
NT Majdanpek
NT Metohija
NT Racha
NT Soko Banja
NT Timok Basin
NT Trepca Mine
NT Vojvodina
NT Zapadna Morava
NT Zlatibor Mountains
SA Banat
SA Bor
SA Moesia
SA Morava River valley
SA Osogovo Mountains
SA Serbo-Macedonian Massif
Serbo-Macedonian Massif (1978)
Macedonia and S Serbia.
IN Index republics as applicable.
BT Yugoslavia

BT Southern Europe
BT Europe
SA Serbia
SA Yugoslav Macedonia
Serebryanka Formation (1978)
Lower and Middle Triassic (?). SE and E Ukraine and adjoining area in Russian Republic.
BT Lower Triassic
BT Triassic
BT Mesozoic
SA Middle Triassic
SA Russian Federation
SA Ukraine
Sereth River
use Siret River
Sergipe
No longer a valid term for GeoRef. See Sergipe Brazil.
Sergipe Basin
use Sergipe-Alagoas Basin
Sergipe Brazil (1993)
State in NE Brazil.
CO S113000S092000
W0362000W0382000
BT Brazil
BT South America
NT Aracaju Brazil
NT Propria Geosyncline
SA Sao Francisco Basin
SA Sergipe-Alagoas Basin
Sergipe-Alagoas Basin (1978)
In NE Brazil.
IN Index states as applicable.
UF Sergipe Basin
SA Alagoas Brazil
SA Sergipe Brazil
sericite (1978)
BT mica group
BT sheet silicates
BT silicates
SA muscovite
sericitization (1978)
BT metasomatism
SA hydrothermal alteration
Serov
No longer a valid term for GeoRef. As of 1993, see Serov Russian Federation.
Serov Russian Federation (1993)
City in Yekaterinburg Oblast E of Urals.
BT Yekaterinburg Russian Federation
BT Russian Federation
BT Commonwealth of Independent States
Serozem
use Sierozems
serpentine (1978)
BT serpentine group
BT sheet silicates
BT silicates
SA cerolite
serpentine group (1978)
Autoposting of this term began in 1981.
BT sheet silicates
BT silicates
NT antigorite
NT chrysotile
NT garnierite
NT lizardite
NT serpentine
serpentine rock
use serpentinite
serpentinite (1978)

Compositional term. This term has multiple hierarchies.
UF serpentine rock
BT1 metaigneous rocks
BT1 metamorphic rocks
BT2 metasomatic rocks
BT2 metamorphic rocks
SA serpentinization

serpentinization (1978)
BT metasomatism
SA alteration
SA hydrothermal alteration
SA processes
SA serpentinite

Serpukhovian (1989)
BT Upper Mississippian
BT Mississippian
BT Carboniferous
BT Paleozoic

Serpulidae (1978)
Family. Before 1981, also search serpulids. As of 1981, serpulids may be used as the common name for Serpulidae.
BT Polychaetia
BT worms
BT Invertebrata

Serra do Espinhaco (1985)
Mountain range, E central Brazil.
BT Brazil
BT South America
SA Bahia Brazil
SA Minas Gerais Brazil

Serra do Mar (1978)
Coastal mountain range in S and SE Brazil.
IN Index states as applicable.
BT Brazil
BT South America
SA Parana Brazil
SA Santa Catarina Brazil
SA Sao Paulo Brazil

Serra do Navio
No longer a valid term for GeoRef. See Serra do Navio Brazil.

Serra do Navio Brazil (1993)
Town in central Amapa, N Brazil.
BT Amapa Brazil
BT Brazil
BT South America

Serra Geral Formation (1978)
BT Mesozoic
SA Brazil
SA Cretaceous
SA Jurassic
SA Paraguay
SA Sao Paulo Brazil

Serrania de Cuenca (1978)
Mountain range in E central Spain.
IN Index provinces as applicable.
BT Spain
BT Iberian Peninsula
BT Southern Europe
BT Europe
SA Cuenca Spain
SA Guadalajara Spain

Serrania de Ronda (1978)
Spur of the Cordillera Penibetica in Andalusia. This term has multiple hierarchies.
IN Index provinces as applicable.
UF Ronda Massif
UF Ronda Sierra
UF Ronda, Serrania de
UF Sierra de Ronda
BT1 Betic Cordillera
BT1 Spain

BT1 Iberian Peninsula
BT1 Southern Europe
BT1 Europe
BT2 Andalusia Spain
BT2 Spain
BT2 Iberian Peninsula
BT2 Southern Europe
BT2 Europe
SA Cadiz Spain
SA Malaga Spain

Serravallian (1978)
Europe.
BT middle Miocene
BT Miocene
BT Neogene
BT Tertiary
BT Cenozoic

Servia
use Serbia

Sesia Valley (1978)
River valley E of Lago Maggiore in N Italy.
BT Italy
BT Southern Europe
BT Europe
SA Piemonte Italy

Sesia Zone (1978)
N Piemonte.
BT Piemonte Italy
BT Italy
BT Southern Europe
BT Europe

Sesia-Lanzo Zone (1978)
N Piemonte and N Lombardy.
IN Index autonomous regions as applicable.
BT Italy
BT Southern Europe
BT Europe
SA Alps
SA Lombardy Italy
SA Piemonte Italy

Sespe Formation (1978)
Upper Eocene and Oligocene. Overlying Coldwater Sandstone Member of Tejon Formation and underlying Vaqueros Formation. In S California.
IN Index ages as applicable.
BT Paleogene
BT Tertiary
BT Cenozoic
SA California
SA Oligocene
SA Tejon Formation
SA upper Eocene

Sete
No longer a valid term for GeoRef. See Sete France.

Sete France (1993)
City on Gulf of Lion in S Herault, S France. Before 1978, also search Cette.
BT Herault France
BT France
BT Western Europe
BT Europe

Seto Inland Sea
use Inland Sea

Seto Sea
use Inland Sea

Seto-chi-umi
use Inland Sea

Seto-Naikai
use Inland Sea

Setouchi
use Inland Sea

Sette Daban (Range)
use Sette-Daban Range

Sette-Daban Range (1978)
NW Khabarovsk Kray and SE Yakutia. Also search Sette-Daban. This term has multiple hierarchies.
UF Sette Daban (Range)
BT1 Russian Federation
BT1 Commonwealth of Independent States
BT2 Asia
SA Khabarovsk Russian Federation
SA Yakutia Russian Federation

Setting Lake (1978)
Central Manitoba.
BT Manitoba
BT Western Canada
BT Canada

settlement (1978)
SA compactness
SA engineering geology
SA foundations
SA land subsidence
SA raft foundations
SA soil mechanics

Setubal
No longer a valid term for GeoRef. As of 1993 see Setubal Portugal.

Setubal Portugal (1993)
District in SW Portugal. Before 1993 also search Setubal AND Portugal.
BT Portugal
BT Iberian Peninsula
BT Southern Europe
BT Europe
SA Estremadura Portugal

Sevan
No longer a valid term for GeoRef. As of 1993, see Sevan Armenia.

Sevan Armenia (1993)
Town on shore of Lake Sevan in N Armenia. This term has multiple hierarchies.
BT1 Armenia
BT1 Commonwealth of Independent States
BT2 Armenia
BT2 Europe
SA Lake Sevan

Seven Devils Mountains (1978)
Range in Adams and Idaho counties in W Idaho.
IN Index counties as applicable.
BT Idaho
BT United States
SA Adams County Idaho
SA Idaho County Idaho

Seven Isles
use Sept-Iles Quebec

Seven Rivers Formation (1985)
SE New Mexico and W Texas.
BT Guadalupian
BT Permian
BT Paleozoic
SA New Mexico
SA Texas

Severin (1978)
Former province whose capital was Caransebes. Much of the province is now Caras-Severin County. This term has multiple hierarchies.
BT1 Banat
BT1 Southern Europe
BT1 Europe
BT2 Romania

BT2 Southern Europe
BT2 Europe

Severn Estuary (1989)
England or Wales. Also near Chesapeake Bay of Maryland and Virginia.
IN Index regions as applicable.
SA England
SA Maryland
SA Virginia
SA Wales

Severn Formation (1989)
Named for exposures in cliffs at Round Bay on the Severn River. E Maryland and Delaware.
IN Index ages as applicable.
SA Delaware
SA Eocene
SA K-T boundary
SA Maryland
SA United States
SA Upper Cretaceous

Severn Valley (1978)
River valley. Also search Severn River.
IN Index regions as applicable.
SA England
SA Wales

Severnaya Zemlya (1981)
Autoposting of Russian Republic to this term began in 1989. This term has multiple hierarchies.
CO N780000N820000
E1060000E0900000
BT1 West Siberia
BT1 Commonwealth of Independent States
BT2 West Siberia
BT2 Asia
BT3 Russian Federation
BT3 Commonwealth of Independent States

Sevier County
Valid through 1988. Search in combination with state term. After 1988, use specific county-state term.

Sevier County Arkansas (1989)
SW Arkansas. Before 1989, also search Sevier County AND Arkansas.
CO N334400N341100
W0940000W0943100
BT Arkansas
BT United States

Sevier County Tennessee (1989)
E Tennessee. Before 1989, also search Sevier County AND Tennessee.
CO N353300N360300
W0831600W0834800
BT Tennessee
BT United States

Sevier County Utah (1989)
Central Utah. Before 1989, also search Sevier County AND Utah.
CO N383000N390400
W1111800W1123500
BT Utah
BT United States

Sevier Desert (1985)
W central Utah.
IN Index counties as applicable.
BT Utah
BT United States
SA Basin and Range Province
SA Juab County Utah
SA Millard County Utah

Sevier orogenic belt (1978)

Central and W Utah, S Idaho and SW Wyoming. Also search Sevier AND belt; Sevier AND orogeny.
UF Sevier Orogeny
UF Sevier Plateau
BT United States
SA Idaho
SA Utah
SA Wyoming

Sevier Orogeny
use Sevier orogenic belt

Sevier Plateau
use Sevier orogenic belt

Sevilla
use Seville City Spain

Seville
No longer a valid term for GeoRef. As of 1993, see Seville City Spain.

Seville City Spain (1993)
Refers to only the city in SW Spain. Before 1993, also search Seville and Spain.
UF Sevilla
BT Seville Spain
BT Andalusia Spain
BT Spain
BT Iberian Peninsula
BT Southern Europe
BT Europe

Seville Province
No longer a valid term for GeoRef. As of 1993, see Seville Spain.

Seville Spain (1993)
Refers only to the province in SW Spain. From 1981-1992, also search Seville Province and Spain. Before 1981, also search Seville and Spain. For the city, see Seville City Spain.
CO N365000N381300
    W0044000W0063300
BT Andalusia Spain
BT Spain
BT Iberian Peninsula
BT Southern Europe
BT Europe
NT Carmona Spain
NT Seville City Spain

sewage (1978)
SA pollution
SA sewage sludge
SA waste disposal
SA water quality

sewage sludge (1993)
Before 1993, also search sewage and sludge.
SA pollution
SA sewage
SA sludge
SA soil treatment
SA solid waste
SA waste disposal
SA water quality

Seward County
Valid through 1988. Search in combination with state term. After 1988, use specific county-state term.

Seward County Kansas (1989)
SW Kansas. Before 1989, also search Seward County AND Kansas.
CO N370000N372200
    W1003800W1010500
BT Kansas
BT United States

Seward County Nebraska (1989) SE Nebraska. Before 1989, also search Seward County AND Nebraska.
CO N404100N410300
    W0965400W0972300
BT Nebraska
BT United States

Seward Peninsula (1978)
On Bering Strait between Kotzebue and Norton sounds in W Alaska. Also search Seward AND Alaska.
BT West-Central Alaska
BT Alaska
BT United States
SA Nome Alaska

sexual dimorphism (1978)
SA reproduction

Seychelles (1978)
Officially designated as the Republic of Seychelles. Island group about 700 miles NE of Malagasy Republic. In 1981, broader term changed from Indian Ocean to Indian Ocean Islands. Mahe Island is considered a narrower term as of 1981.
CO S050000S040000
    E0560000E0550000
BT Indian Ocean Islands
NT Aldabra Island
NT Mahe Island
SA Indian Ocean

Seymour Island (1985)
Island in the Weddell Sea, off the Antarctic Peninsula. Also one of the Galapagos Islands. Before 1993, also search Marambio Island.
IN Index Antarctica or Galapagos Islands as applicable.
UF Marambio Island
UF Vicecomodoro Island
SA Antarctica
SA East Pacific Ocean Islands
SA Galapagos Islands
SA La Meseta Formation
SA Lopez de Bertodano Formation
SA Sobral Formation
SA Weddell Sea

SH waves
use SH-waves

SH-waves (1978)
UF SH waves
BT S-waves
BT body waves
BT elastic waves
SA seismology
SA waves

Shaanxi
No longer a valid term for GeoRef. See Shaanxi China.

Shaanxi China (1993)
Province in E central China. Before 1993, also search Shaanxi or Shensi.
CO N314000N393000
    E1112000E1060000
BT China
BT Far East
BT Asia
NT Qinling Mountains
SA Han River basin
SA Huang Ho
SA Luoquan Formation

Shaba
No longer a valid term for GeoRef. As of 1993 see Shaba Zaire.

Shaba Zaire (1993)
Province in S Zaire. Before 1993 also search Shaba AND Zaire.
CO S130000S050000
    E0310000E0220000
UF Katanga, Zaire
BT Zaire
BT Central Africa
BT Africa
NT Kamoto

Shackleton Glacier (1978)
One of glaciers feeding Ross Ice Shelf from Queen Maud Range. Also Shackleton Ice Shelf off Queen Mary Coast extending into Indian Ocean is sometimes called Shackleton Glacier.
BT Antarctica
SA Queen Maud Range
SA Ross Ice Shelf

Shackleton Range (1978)
SE of the Filchner Ice Shelf which is on the Weddell Sea on the Atlantic Ocean side.
BT Antarctica

shaded relief maps (1981)
BT maps
SA Moon

Shadow Lake Formation (1989)
S Ontario.
BT Cambrian
BT Paleozoic
SA Ontario

Shadow Mountains (1978)
IN Index states as applicable.
SA California
SA Nevada
SA United States

Shady Dolomite (1978)
Equivalent to Tomstown Dolomite. N Alabama, NW Georgia, W North Carolina, E Tennessee, and SW Virginia.
BT Lower Cambrian
BT Cambrian
BT Paleozoic
SA Alabama
SA Chilhowee Group
SA Georgia
SA North Carolina
SA Tennessee
SA Virginia

Shahejie Formation (1989)
Eocene?.
BT Tertiary
BT Cenozoic
SA China

Shakanai Mine (1993)
BT Akita Japan
BT Honshu
BT Japan
BT Far East
BT Asia
SA Hokuroku Japan
SA metal ores
SA mines

shake wave
use S-waves

Shakh-Dag Range
use Shakhdag Range

Shakhdag Range (1978)
In the Lesser Caucasus on N shore of Lake Sevan. This term has multiple hierarchies.
IN Index former Soviet republics as applicable.
UF Shakh-Dag Range
BT1 Commonwealth of Independent States
BT2 Lesser Caucasus
BT2 Caucasus
BT2 Europe
SA Armenia
SA Azerbaidzhan

Shakopee Formation (1978)
In Prairie du Chien Group. Overlies and interbedded with Root Valley Sandstone. N Illinois, Iowa, S Minnesota, and S Wisconsin.
BT Lower Ordovician
BT Ordovician
BT Paleozoic
SA Illinois
SA Iowa
SA Minnesota
SA Prairie du Chien Group
SA Wisconsin

shale (1978)
Does not include oil shale. Use clays for shale as brick clay. Before 1976, also search pelite.
UF bloating shale
UF pelite
BT clastic rocks
BT sedimentary rocks
SA argillaceous texture
SA black shale
SA clays
SA construction materials
SA Eastern Gas Shales Project
SA oil shale
SA slate deposits
SA slates
SA terrigenous materials
SA torbanite

shale oil
use oil shale

shallow
A valid term through 1977. After 1977, use shallow-water environment or shallow-focus earthquakes if applicable. After 1992, use shallow depth if applicable.

shallow aquifers (1993)
Also use unconfined aquifers if applicable.
BT aquifers
SA depth
SA ground water
SA hydrogeology
SA levels
SA springs
SA surficial aquifers
SA surficial geology
SA unconfined aquifers
SA water table

shallow depth (1993)
Use shallow-water environment for paleoecology and shallow-focus earthquakes for earthquakes and upper crust for geophysics and surficial aquifers for aquifers as applicable.
SA aquifers
SA depth
SA ground water
SA hydrogeology
SA levels
SA springs
SA surficial geology
SA unconfined aquifers
SA water table

shallow earthquake
use shallow-focus earthquakes

shallow-focus earthquakes (1978)
Before 1978, search earthquakes AND shallow focus.
UF normal earthquake
UF shallow earthquake
BT earthquakes

SA Chile earthquake 1960
SA deep-focus earthquakes
SA focus
SA intermediate-focus earthquakes
SA Kettleman Hills earthquake 1985
SA Xingtai earthquake 1966

shallow-water environment (1978)
Before 1978, search shallow-water or shallow.
SA carbonate ramps
SA deep-water environment
SA depositional environment
SA ecology
SA environment
SA reef environment
SA sedimentation
SA tempestite

Shamlug
No longer a valid term for GeoRef. As of 1993, see Shamlug Armenia.

Shamlug Armenia (1993)
Town in NE Armenia. This term has multiple hierarchies.
BT1 Armenia
BT1 Commonwealth of Independent States
BT2 Armenia
BT2 Europe

Shan State
No longer a valid term for GeoRef. See Shan State Burma.
use Shan State Burma

Shan State Burma (1993)
E Burma. Before 1993, search Shan State. Before 1978, also search Federated Shan State and Shan States.
UF Federated Shan States
UF Shan State
BT Burma
BT Far East
BT Asia

Shan-tung China
use Shandong China

Shandong
No longer a valid term for GeoRef. See Shandong China.

Shandong China (1993)
Province in E central China on the Yellow Sea. Before 1993, also search Shandong. Before 1981, also search Shan-tung, Shan-tung Province, Shantung, or Shantung Province.
CO N343000N380000
   E1223500E1150000
UF Shan-tung China
UF Shantung China
BT China
BT Far East
BT Asia
NT Jiyang Depression
NT Shandong Peninsula
SA Anshan Group
SA Bohai Bay
SA Dongpu Depression
SA Huang Ho
SA Tancheng-Lujiang Fault
SA Tancheng-Lujiang fault zone
SA Yishu Fault

Shandong Peninsula (1989)
E Shandong, E China. Extends between Bohai Bay and the Yellow Sea. Also search Jiaodong Peninsula.
CO N360000N375500
   E1224000E1200000
UF Jiaodong Peninsula
BT Shandong China
BT China
BT Far East
BT Asia
SA Bohai Bay
SA Yellow Sea

Shanghai
No longer a valid term for GeoRef. See Shanghai China.

Shanghai China (1993)
Independent municipality. E China. Before 1993, also search Shanghai or Shanghai region.
CO N303000N315000
   E1220000E1205000
UF Shanghai region
BT China
BT Far East
BT Asia
SA Jiangsu China

Shanghai region
No longer a valid term for GeoRef.
use Shanghai China

Shannon County
Valid through 1988. Search in combination with state term. After 1988, use specific county-state term.

Shannon County Missouri (1989)
S Missouri. Before 1989, also search Shannon County AND Missouri.
CO N365300N372500
   W0910200W0914000
BT Missouri
BT United States

Shannon County South Dakota (1989)
SW South Dakota. Before 1989, also search Shannon County AND South Dakota.
CO N430000N434200
   W1020500W1030000
BT South Dakota
BT United States

Shannon Sandstone Member (1989)
Also search Shannon Formation NOT Australia. For formation in Western Australia, search Shannon Formation AND Australia.
BT Upper Cretaceous
BT Cretaceous
BT Mesozoic
SA Montana
SA South Dakota
SA Wyoming

Shansi
No longer a valid term for GeoRef. See Shanxi China.

Shansi Sheng
use Shanxi China

Shantung China
use Shandong China

Shanxi
No longer a valid term for GeoRef. See Shanxi China.

Shanxi China (1993)
Province in N central China. Before 1993, also search Shansi or Shanxi or Shanxi Sheng.
UF Shansi Sheng
UF Shanxi Sheng
BT China
BT Far East
BT Asia
NT Taiyuan China
SA Benxi Formation
SA Huang Ho
SA Hutuo Group
SA Taihang Mountains
SA Taiyuan Formation
SA Wufeng Formation

Shanxi Sheng
use Shanxi China

Shap Granite (1978)
The Shap Granite and associated igneous and metamorphic rocks have been called Shap Andesites and Shap Rhyolites. English Lake District in NW England.
UF Shap Rhyolites
BT Ordovician
BT Paleozoic
SA England
SA United Kingdom

Shap Rhyolites
use Shap Granite

shape
As of 1981, no longer a valid term for GeoRef. See shape analysis for grain shape. See figure of Earth and geoid for shape of the Earth.

shape analysis (1978)
SA analysis
SA grain boundaries
SA grains
SA granulometry
SA roundness
SA sedimentary petrology
SA sedimentary rocks
SA sediments
SA sphericity
SA textures

Sharjah (1978)
Emirate. One of federation of 7 states at S end of Persian Gulf.
BT United Arab Emirates
BT Arabian Peninsula
BT Asia

Shark Bay (1978)
Large inlet of Indian Ocean in Western Australia.
IN Index state or region as applicable.
UF Sharks Bay
SA Indian Ocean
SA Western Australia

Sharks Bay
use Shark Bay

Sharon Conglomerate (1978)
In Pottsville Group. Maryland, E Ohio, W Pennsylvania, and N West Virginia.
BT Pennsylvanian
BT Carboniferous
BT Paleozoic
SA Maryland
SA Ohio
SA Pennsylvania
SA Pottsville Group
SA West Virginia

Sharps Meteorite (1981)
Impact near Sharps, in Richmond County, Virginia. Before 1981, also search Sharps AND meteorites.
BT chondrites
BT stony meteorites
BT meteorites
SA Virginia

Shasta County
Valid through 1988. Search in combination with state term. After 1988, use specific county-state term.

Shasta County California (1989)
N California. Before 1989, also search Shasta County AND California.
CO N402000N411000
   W1212000W1230800
BT California
BT United States
NT Lassen Peak
NT West Shasta mining district
SA Lassen Volcanic National Park
SA Trinity Complex

Shatskiy Rise
use Shatsky Rise

Shatsky Rise (1978)
In mid Northwest Pacific Basin E of Tokyo.
UF Shatskiy Rise
BT Pacific Ocean
SA DSDP Site 305
SA DSDP Site 577
SA Leg 132

shatter cones (1978)
Autoposting of cryptoexplosion features to this term began in 1989.
UF cones, shatter
BT cryptoexplosion features
SA cones
SA deformation
SA explosions
SA geomorphology
SA impact features
SA metamorphism
SA meteor craters
SA meteorites
SA nuclear explosions
SA shock waves

shattuckite (1978)
BT chain silicates
BT silicates

Shaunavon Formation (1978)
Subsurface.
BT Jurassic
BT Mesozoic
SA Canada
SA Saskatchewan

Shawangunk Formation (1993)
Upper Ordovician to Middle Silurian. SE New York, N New Jersey, and N Pennsylvania.
IN Index ages as applicable.
BT Paleozoic
SA New Jersey
SA New York
SA Ordovician
SA Pennsylvania
SA Silurian

Shawangunk Mountains (1978)
Part of the Kittatinny Mountains in S New York. Autoposting of broader terms to this term began in 1989. Appalachians is a broader term as of 1993. This term has multiple hierarchies.
BT1 Ulster County New York
BT1 New York
BT1 United States
BT2 Appalachians

Shawnee County
Valid through 1988. Search in combination with state term. After 1988, use specific county-state term.

Shawnee County Kansas (1989)

NE Kansas. Before 1989, also search Shawnee County AND Kansas.
CO N385000N391500
W0953000W0960300
BT Kansas
BT United States
NT Topeka Kansas

Shawnee Group (1978)
Comprises Oread Limestone, Kanwaka Shale, Lecompton Limestone, Tecumseh Shale, Deer Creek Limestone, Calhoun Shale, Topeka Limestone, Severy Shale, Howard Limestone, White Cloud Shale, Happy Hollow Limestone, Cedar Vale Shale, Rulo Limestone, Silver Lake-Auburn Shale Interval, Reading Limestone, Harveyville Shale, Elmont or Preston Limestone, and Williard Shale.
BT Virgilian
BT Upper Pennsylvanian
BT Pennsylvanian
BT Carboniferous
BT Paleozoic
SA Kansas
SA Lecompton Limestone
SA Nebraska
SA Oread Limestone

shear (1978)
As of 1978, used only to indicate the mechanics of deformation. Before 1978, also used to indicate style of structure.
UF shear strain
SA confining pressure
SA deformation
SA mechanical properties
SA mechanics
SA shear strength
SA shear stress
SA shear tests
SA shear zones
SA similar folds
SA strain
SA stress

shear cleavage
 use slip cleavage

shear folds
 use similar folds

shear modulus (1978)
UF Coulomb's modulus
UF modulus of rigidity
UF modulus, shear
UF rigidity modulus
UF torsion modulus
BT elastic constants
SA bulk modulus
SA elasticity
SA seismic moment

shear strain
 use shear

shear strength (1978)
UF cohesion
SA friction angles
SA mechanical properties
SA rock mechanics
SA shear
SA shear tests
SA soil mechanics
SA strain
SA strength
SA stress
SA tensile strength
SA vane tests

shear stress (1978)
UF tangential stress
SA deformation
SA engineering geology
SA shear
SA stress

shear tests (1985)
SA asperities
SA rock mechanics
SA shear
SA shear strength
SA soil mechanics
SA testing

shear wave
 use S-waves

shear zones (1978)
UF zones, shear
SA breccia
SA faults
SA fractures
SA metamorphic core complexes
SA mylonites
SA shear
SA strain
SA structural analysis
SA tectonics

shear-cleavage folds
 use cleavage folds

shearing
A valid term through 1975. After 1975, see shear. See also fractures, foliation, and structural analysis.

sheath folds (1989)
BT folds

Sheep Pass Formation (1989)
Upper Cretaceous? to Eocene. E central Nevada. At the type locality, unconformably overlies Ely Limestone and Chainman Shale; unconformably overlain by Garrett Ranch volcanic group.
IN Index ages as applicable.
SA Eocene
SA Nevada
SA Paleocene
SA Upper Cretaceous

sheet erosion (1981)
BT erosion
SA nivation
SA soil erosion
SA water erosion

sheet silicates (1978)
Autoposting of this term began in 1978.
UF phyllosilicates
BT silicates
NT apophyllite
NT astrophyllite
NT bavenite
NT berthierine
NT cerolite
NT charoite
NT chlorite group
NT chrysocolla
NT clay minerals
NT cymrite
NT gillespite
NT gyrolite
NT hisingerite
NT hydrobiotite
NT hydromica
NT hydromuscovite
NT leucosphenite
NT mica group
NT penninite
NT prehnite
NT pyrophyllite
NT serpentine group
NT stilpnomelane
NT talc
NT zussmanite
SA asbestos
SA bentonite
SA clay mineralogy

sheets
As of 1981, no longer a valid term for GeoRef. See intrusions.

Sheffield
No longer a valid term for GeoRef. As of 1993, see Sheffield England.

Sheffield England (1993)
City in South Yorkshire in N England. Before 1993, also search Sheffield AND England.
BT Yorkshire England
BT England
BT Great Britain
BT United Kingdom
BT Western Europe
BT Europe

Shelby County
Valid through 1988. Search in combination with state term. After 1988, use specific county-state term.

Shelby County Alabama (1989)
Central Alabama. Before 1989, also search Shelby County AND Alabama.
CO N330300N333000
W0862300W0870200
BT Alabama
BT United States

Shelby County Illinois (1989)
Central Illinois. Before 1989, also search Shelby County AND Illinois.
CO N391500N394100
W0882800W0890800
BT Illinois
BT United States

Shelby County Indiana (1989)
Central Indiana. Before 1989, also search Shelby County AND Indiana.
CO N302000N394200
W0853800W0855600
BT Indiana
BT United States

Shelby County Iowa (1989)
W Iowa. Before 1989, also search Shelby County AND Iowa.
CO N413000N415200
W0950300W0953400
BT Iowa
BT United States

Shelby County Kentucky (1989)
N Kentucky. Before 1989, also search Shelby County AND Kentucky.
CO N380300N382100
W0845700W0853000
BT Kentucky
BT United States

Shelby County Missouri (1989)
NE Missouri. Before 1989, also search Shelby County AND Missouri.
CO N393700N395700
W0915000W0921800
BT Missouri
BT United States

Shelby County Ohio (1989)
W Ohio. Before 1989, also search Shelby County AND Ohio.
CO N401200N405800
W0840000W0842500
BT Ohio
BT United States

Shelby County Tennessee (1989)
Extreme SW Tennessee. Before 1989, also search Shelby County AND Tennessee.
CO N345900N352400
W0894000W0901800
BT Tennessee
BT United States
NT Memphis Tennessee

Shelby County Texas (1989)
E Texas. Before 1989, also search Shelby County AND Texas.
CO N313500N315800
W0935000W0942900
BT Texas
BT United States

shelf
A valid term through 1978 used to distinguish older sedimentation from more recent. After 1978, use shelf environment.

shelf environment (1978)
Autoposting of the term environment to this term from 1989-1992. Use this term or slope environment for older sedimentation. For recent sedimentation, use continental shelf or continental slope. Before 1978, search shelf.
SA carbonate platforms
SA continental shelf
SA depositional environment
SA environment
SA insular shelf
SA paleoecology
SA sedimentation
SA slope environment

shelf, continental
 use continental shelf

shelf-slope break (1985)
SA continental shelf
SA continental slope
SA inner slope
SA oceanography
SA outer shelf

Shelikof Strait (1985)
S Alaska, between Alaska Peninsula and Kodiak Island.
BT Southwestern Alaska
BT Alaska
BT United States
SA Gulf of Alaska

shells (1978)
Before 1975, also search shell.
SA biomineralization
SA exoskeletons
SA fossils
SA Invertebrata
SA ornamentation
SA paleontology
SA paleopathology
SA sedimentary rocks
SA sediments
SA tests

shelves, ice
 use ice shelves

Shemakha
No longer a valid term for GeoRef. As of 1993, see Shemakha Azerbaidzhan.

Shemakha Azerbaidzhan (1993)
City on E Caucasus in E Azerbaidzhan. This term has multiple hierarchies.
BT1 Azerbaidzhan
BT1 Europe
BT2 Azerbaidzhan
BT2 Commonwealth of Independent States

Shemya Island (1978)

One of the Semichi Islands at W end of Aleutian Islands. Also search Shemya.
BT Aleutian Islands
BT Southwestern Alaska
BT Alaska
BT United States

Shenandoah Valley (1978)
River valley between the Allegheny and Blue Ridge mountains. Primarily in Virginia.
IN Index states as applicable.
BT United States
SA Allegheny Mountains
SA Blue Ridge Mountains
SA Great Appalachian Valley
SA Virginia
SA West Virginia

Shensi
No longer a valid term for GeoRef. See Shaanxi China.

Shepard Formation (1989)
In Missoula Group of Belt Supergroup. NW Montana.
BT middle Proterozoic
BT Proterozoic
BT upper Precambrian
BT Precambrian
SA Belt Supergroup
SA Missoula Group
SA Montana

Sherburne County
Valid through 1988. Search in combination with state term. After 1988, use specific county-state term.

Sherburne County Minnesota (1989)
Central Minnesota. Before 1989, also search Sherburne County AND Minnesota.
CO N451200N453300
    W0933500W0941000
BT Minnesota
BT United States

Sherghati Meteorite
use Shergotty Meteorite

shergottite (1985)
BT achondrites
BT stony meteorites
BT meteorites
NT Shergotty Meteorite
SA SNC Meteorites

Shergotty Meteorite (1985)
Impact at Sherghati in Bihar, India. Before 1985, also search Shergotty or Sherghati AND meteorites.
UF Sherghati Meteorite
BT shergottite
BT achondrites
BT stony meteorites
BT meteorites
SA Bihar India
SA India
SA SNC Meteorites

Sheridan County
Valid through 1988. Search in combination with state term. After 1988, use specific county-state term.

Sheridan County Kansas (1989)
NW Kansas. Before 1989, also search Sheridan County AND Kansas.
CO N390700N393400
    W1001000W1004500
BT Kansas
BT United States

Sheridan County Montana (1989)
Extreme NE Montana. Before 1989, also search Sheridan County AND Montana.
CO N482200N490000
    W1040200W1050300
BT Montana
BT United States
SA Medicine Lake

Sheridan County Nebraska (1989)
NW Nebraska. Before 1989, also search Sheridan County AND Nebraska.
CO N420000N430000
    W1020000W1024800
BT Nebraska
BT United States

Sheridan County North Dakota (1989)
Central North Dakota. Before 1989, also search Sheridan County AND North Dakota.
CO N472000N475100
    W1000300W1004000
BT North Dakota
BT United States

Sheridan County Wyoming (1989)
N Wyoming. Before 1989, also search Sheridan County AND Wyoming.
CO N443300N450000
    W1060000W1075400
BT Wyoming
BT United States

Sherman County
Valid through 1988. Search in combination with state term. After 1988, use specific county-state term.

Sherman County Kansas (1989)
NW Kansas. Before 1989, also search Sherman County AND Kansas.
CO N390700N393300
    W1012500W1020300
BT Kansas
BT United States

Sherman County Nebraska (1989)
Central Nebraska. Before 1989, also search Sherman County AND Nebraska.
CO N410200N412300
    W0984400W0991300
BT Nebraska
BT United States

Sherman County Oregon (1989)
N Oregon. Before 1989, also search Sherman County AND Oregon.
CO N450500N454300
    W1202200W1210300
BT Oregon
BT United States

Sherman County Texas (1989)
Extreme N Texas. Before 1989, also search Sherman County AND Texas.
CO N360500N363400
    W1013800W1020900
BT Texas
BT United States

Sherridon Group (1989)
In Kisseynew Complex. Above Nokomis Group.
BT Precambrian
SA Kisseynew Complex

SA Manitoba
SA Saskatchewan

Sherwood Sandstone (1989)
BT Triassic
BT Mesozoic
SA England
SA Northern Ireland
SA United Kingdom
SA Wales

Shetland Islands (1978)
Archipelago off N Scotland 50 miles NE of Orkney Islands. Also search Shetland. As of 1981, Scotland is a broader term. Autoposting of Atlantic Ocean Islands to this term began in 1989. Autoposting of Atlantic Ocean ended in 1989. As of 1990, Great Britain, United Kingdom, Western Europe and Europe are autoposted to this term. This term has multiple hierarchies.
CO N595000N610000
    W0004500W0014500
UF Shetlands
UF Zetland
BT1 Scotland
BT1 Great Britain
BT1 United Kingdom
BT1 Western Europe
BT1 Europe
BT2 Atlantic Ocean Islands
NT Unst
SA Fair Isle
SA Highland Boundary Fault
SA South Shetland Islands

Shetlands
use Shetland Islands

Sheveluch (1978)
Volcano on E central Kamchatka Peninsula. This term has multiple hierarchies.
CO N563801N563801
    E1611900E1611900
UF Sheveluch Volcano
UF Shiveluch Volcano
BT1 Kamchatka Russian Federation
BT1 Russian Federation
BT1 Commonwealth of Independent States
BT2 Kamchatka Russian Federation
BT2 Asia
SA Kamchatka Peninsula

Sheveluch Volcano
use Sheveluch

shield volcanoes (1978)
UF basaltic domes
UF lava domes
BT volcanoes
BT volcanic features
SA stratovolcanoes
SA volcanism

shields (1978)
Usage restricted to tectonics.
SA basement
SA cratons
SA crust
SA tectonics

Shiga
No longer a valid term for GeoRef. See Shiga Japan.

Shiga Japan (1993)
Prefecture in S Honshu.
BT Honshu
BT Japan
BT Far East
BT Asia
NT Lake Biwa

Shikoku (1978)
Smallest of the 4 major Japanese islands. In S Japan.
CO N323500N342500
    E1344500E1320000
BT Japan
BT Far East
BT Asia
NT Ehime Japan
NT Kagawa Japan
NT Kochi Japan
NT Kurosegawa Zone
NT Tokushima Japan
NT Tosa Bay
NT Yoshino River
SA Inland Sea
SA Izumi Group
SA Median Tectonic Line
SA Sambagawa Belt
SA Shimanto Belt
SA Shimanto Group

Shikoku Basin (1978)
Undersea feature between Shikoku on the NW and the Bonin and Volcano islands to the SE.
BT Pacific Ocean
SA DSDP Site 442
SA Izu-Bonin Arc

Shikotsu (1978)
Lake in SW Hokkaido.
BT Hokkaido
BT Japan
BT Far East
BT Asia
SA Tarumae

Shilka Valley (1978)
River valley in SW central Chita Oblast in Transbaikalia. Also search Shilka; Shilka River. This term has multiple hierarchies.
BT1 Chita Russian Federation
BT1 Russian Federation
BT1 Commonwealth of Independent States
BT2 Chita Russian Federation
BT2 Asia
SA Argun River

Shillong
No longer a valid term for GeoRef. See Shillong India.

Shillong India (1993)
City in Meghalaya, formerly in Assam, NE India.
BT Meghalaya India
BT Northeastern India
BT India
BT Indian Peninsula
BT Asia
SA Assam India

Shillong Plateau (1978)
Undulating tableland in W Assam, NE India.
SA Assam India
SA Meghalaya India

Shimane
No longer a valid term for GeoRef. See Shimane Japan.

Shimane Japan (1993)
Prefecture in SW Honshu.
BT Honshu
BT Japan
BT Far East
BT Asia
NT Oki Islands
NT Shinji Lake
SA Chugoku
SA Dogo
SA Naka-no-umi

Shimanto Belt (1989)
SW Japan.

UF Shimanto Terrain
BT Japan
BT Far East
BT Asia
SA Honshu
SA Kyushu
SA Shikoku

Shimanto Group (1978)
Includes Terazoma Series, Nishigawa Series, and Higashigawa Series. Triassic-Cretaceous, presumably. Kyushu and S Shikoku. Also search Shimanto.
BT Mesozoic
SA Japan
SA Kyushu
SA Shikoku

Shimanto Terrain
use Shimanto Belt

Shimokita Peninsula (1978)
In Mutsu Bay area in extreme N Honshu.
BT Aomori Japan
BT Honshu
BT Japan
BT Far East
BT Asia

Shimosa Group (1978)
BT Pleistocene
BT Quaternary
BT Cenozoic
SA Japan

Shinano River (1978)
W central Honshu. Flows N into Japan Sea. Longest river in Japan.
BT Niigata Japan
BT Honshu
BT Japan
BT Far East
BT Asia

Shinarump Member (1985)
Of Chinle Formation. Utah, N Arizona, SE Nevada and NW New Mexico.
BT Upper Triassic
BT Triassic
BT Mesozoic
SA Arizona
SA Chinle Formation
SA Nevada
SA New Mexico
SA Utah

shingle (1978)
BT clastic sediments
BT sediments
SA cobbles
SA gravel
SA pebbles

Shinji Lake (1978)
SW Honshu.
BT Shimane Japan
BT Honshu
BT Japan
BT Far East
BT Asia

Shiobara
No longer a valid term for GeoRef. See Shiobara Japan.

Shiobara Japan (1993)
Town in central Honshu.
BT Tochigi Japan
BT Honshu
BT Japan
BT Far East
BT Asia

Shiraki Steppe (1978)
Semidesert plain in E Georgian Republic. Also search Shiraki.
BT Georgian Republic

BT Europe

Shiranish Formation (1978)
BT Cretaceous
BT Mesozoic
SA Iraq

Shirasu
As of 1993, no longer a valid term for GeoRef. See Shirasu River or shirasu.

shirasu (1993)
Volcanic materials.
SA ignimbrite
SA Kyushu
SA pumice
SA pyroclastic flows
SA pyroclastics
SA volcanic ash
SA volcanic glass
SA volcanic rocks

Shirasu River (1993)
Stream NW of Tokyo in central Honshu. Before 1993, also search Shirasu.
BT Honshu
BT Japan
BT Far East
BT Asia

Shiraz
No longer a valid term for GeoRef. See Shiraz Iran.

Shiraz Iran (1993)
City in Fars (province), SW central Iran.
CO N293800N293800 E0523400E0523400
BT Fars Iran
BT Iran
BT Middle East
BT Asia

Shirley Basin (1978)
S central Wyoming. Autoposting of broader terms to this term began in 1989.
BT Carbon County Wyoming
BT Wyoming
BT United States

Shiveluch Volcano
use Sheveluch

Shizukuishi
No longer a valid term for GeoRef.
use Shizukuishi Japan

Shizukuishi Japan (1993)
Town in N Honshu.
UF Shizukuishi
BT Iwate Japan
BT Honshu
BT Japan
BT Far East
BT Asia

Shizuoka
No longer a valid term for GeoRef. See Shizuoka Japan.

Shizuoka Japan (1993)
Prefecture and city in S central Honshu. Before 1993, also search Shuizuoka AND Japan.
BT Honshu
BT Japan
BT Far East
BT Asia
NT Ito Japan
SA Akaishi Mountains
SA Chubu Japan
SA Fuji River
SA Itoigawa-Shizuoka tectonic line
SA Izu Peninsula
SA Sagami Bay

SA Suruga Bay
SA Tokai Japan

Shizuoka-Itoigawa tectonic line
use Itoigawa-Shizuoka tectonic line

shoaling (1978)
SA ocean waves
SA waves

shoals (1978)
SA coastlines
SA fluvial features
SA geomorphology
SA lacustrine features
SA marine terraces
SA oceanography
SA reefs
SA shore features
SA spits

shock
A valid term through 1977.

shock metamorphism (1978)
Before 1978, also search metamorphism AND shock.
BT metamorphism
SA cryptoexplosion features
SA Hugoniot analysis
SA meteor craters

shock waves (1978)
SA b-values
SA deformation
SA foreshocks
SA interplanetary space
SA metamorphism
SA propagation
SA seismology
SA shatter cones
SA waves

Shonai River (1978)
In the Nagoya area of S central Honshu.
BT Aichi Japan
BT Honshu
BT Japan
BT Far East
BT Asia

Shonkin Sag Laccolith (1978)
In Highwood Mountains area of N central Montana.
IN Index counties as applicable.
BT Montana
BT United States

shonkinite (1978)
BT syenites
BT plutonic rocks
BT igneous rocks
SA alkali gabbros

Shoo Fly Complex (1993)
Upper Cambrian, Ordovician, Silurian, Devonian. N and Central California. Metamorphic complex. Before 1993, also search Shoo Fly Formation.
IN Index ages as applicable.
UF Shoo Fly Formation
BT Paleozoic
SA California
SA Cambrian
SA Central California
SA Devonian
SA Northern California
SA Ordovician
SA Silurian

Shoo Fly Formation
No longer a valid term for GeoRef.
use Shoo Fly Complex

shooting star
use meteors

shore drift
use littoral drift

shore environment
use coastal environment

shore features (1978)
Autoposting of this term began in 1978.
UF coastal features
UF features, shore
UF shoreline features
NT barrier beaches
NT bays
NT beach ridges
NT beaches
NT capes
NT chenier plains
NT cheniers
NT coastal dunes
NT coastlines
NT estuaries
NT fjords
NT inlets
NT lagoons
NT mangrove swamps
NT marine terraces
NT marshes
NT spits
NT tidal channels
NT tidal flats
NT tidal inlets
NT tombolos
SA barrier islands
SA bars
SA benches
SA blowouts
SA bluffs
SA caves
SA chimneys
SA cliffs
SA coastal plains
SA continental dunes
SA deltas
SA dunes
SA embayments
SA fan deltas
SA foredunes
SA forelands
SA geomorphology
SA Gilbert-type deltas
SA longshore bars
SA mud banks
SA mud flats
SA mud lumps
SA outliers
SA plains
SA progradation
SA rills
SA sand ridges
SA shoals
SA shorelines
SA wave-cut platforms
SA wetlands

shore lines
No longer a valid GeoRef index term. Before 1971, was used on level 1 in subfiles E and N. See shorelines; shore features; coastlines.

shore lines (abandoned)
No longer a valid GeoRef index term. Before 1971, was used on level 1 in subfiles E and N. See shore features; coastlines; terraces; changes of level.

shore reef
use fringing reefs

shoreline features
use shore features

shorelines (1978)
Used for geological studies on the engineering aspects of shorelines. Use coastlines for geomorpho-

logic aspects of shorelines as of 1983.
 SA artificial islands
 SA barrier islands
 SA bays
 SA beach ridges
 SA beaches
 SA breakwaters
 SA capes
 SA changes of level
 SA cheniers
 SA coastal environment
 SA coastal sedimentation
 SA coastlines
 SA construction
 SA design
 SA dynamics
 SA engineering geology
 SA erosion
 SA geomorphology
 SA harbors
 SA hydraulics
 SA inlets
 SA littoral erosion
 SA management
 SA marine installations
 SA marine terraces
 SA ocean circulation
 SA ocean waves
 SA offshore
 SA paleogeography
 SA polders
 SA revetments
 SA seawalls
 SA shore features
 SA slope stability
 SA stabilization
 SA tidal inlets
 SA waterways

Shoriya Mountains (1978)
 Mountain range at SW end of Kuznetsk Alatau just S of Kuznetsk Basin. Also search Shoriya. This term has multiple hierarchies.
 BT1 Russian Federation
 BT1 Commonwealth of Independent States
 BT2 Asia

short-period seismographs (1981)
 BT seismographs

short-period waves (1981)
 BT elastic waves
 SA Lg-waves
 SA seismology
 SA traveltime
 SA waves

shortening, crustal
 use crustal shortening

shortite (1978)
 BT carbonates

Shorty Crater (1978)
 BT Moon

Shoshone County
 Valid through 1988. Search in combination with state term. After 1988, use specific county-state term.

Shoshone County Idaho (1989)
 N Idaho. Before 1989, also search Shoshone County AND Idaho.
 CO N465500N480500
   W1145500W1162000
 BT Idaho
 BT United States
 NT Bunker Hill Mine
 SA Coeur d'Alene mining district
 SA Idaho Panhandle

Shoshone Mountains (1978)

Mountain range W of Toiyabe Range and Reese River in central Nevada.
 IN Index counties as applicable.
 BT Nevada
 BT United States
 SA Lander County Nevada
 SA Nye County Nevada

shoshonite (1978)
 BT basalts
 BT volcanic rocks
 BT igneous rocks

Showa
 use Syowa Station

Showa Station
 use Syowa Station

shrinkage cracks
 As of 1981, no longer a valid term for GeoRef.
 use mudcracks

Shropshire
 No longer a valid term for GeoRef as of 1993.
 use Shropshire England

Shropshire England (1993)
 County on Welsh border. Before 1993, also search Shropshire or Salop.
 CO N522000N530000
   W0021500W0031500
 UF Salop England
 UF Shropshire
 BT England
 BT Great Britain
 BT United Kingdom
 BT Western Europe
 BT Europe
 NT Church Stretton England
 NT Wenlock Edge

Shublick Formation
 use Shublik Formation

Shublik Formation (1989)
 N Alaska.
 UF Shublick Formation
 BT Triassic
 BT Mesozoic
 SA Alaska

Shubuta Member (1989)
 Of Yazoo Clay.
 BT upper Eocene
 BT Eocene
 BT Paleogene
 BT Tertiary
 BT Cenozoic
 SA Alabama
 SA Mississippi
 SA Yazoo Clay

Shugnan
 No longer a valid term for GeoRef. As of 1993, see Shugnan Tadzhikistan.

Shugnan Tadzhikistan (1993)
 Village in Gorno Badakhshan Autonomous Oblast in Pamirs in E Tadzhikistan. This term has multiple hierarchies.
 BT1 Tadzhikistan
 BT1 Asia
 BT2 Tadzhikistan
 BT2 Commonwealth of Independent States
 SA Badakhshan
 SA Leninabad Tadzhikistan

Shuksan Thrust (1989)
 N Cascades, N Washington; SW British Columbia.
 IN Index British Columbia and/or Washington as applicable.
 BT North America

 SA British Columbia
 SA Cascade Range
 SA Washington

Shumagin Islands (1989)
 Off SE coast of the Alaska Peninsula, SW Alaska.
 CO N545400N552000
   W1591500W1604500
 BT Aleutian Islands
 BT Southwestern Alaska
 BT Alaska
 BT United States

Shungura Formation (1989)
 In Omo Group.
 IN Index ages as applicable.
 BT Cenozoic
 SA Ethiopia
 SA Kenya
 SA Omo Group
 SA Pleistocene
 SA Pliocene
 SA Tanzania

Shuswap Complex (1978)
 Underlies the southern culmination of the Omineca Geanticline. Comprises rocks ranging in age from Proterozoic to early Mesozoic. S British Columbia.
 SA British Columbia
 SA lower Mesozoic
 SA Monashee Complex
 SA Paleozoic
 SA Proterozoic

Shuttle Imaging Radar (1989)
 Also can be abbreviated as SIR. Also search space shuttle and radar methods.
 SA geophysical methods
 SA geophysical surveys
 SA image enhancement
 SA imagery
 SA radar methods
 SA remote sensing
 SA satellite methods

Shventoyi River
 use Sventoji River

Si
 use silicon

Si-32 (1978)
 Autoposting of broader terms began in 1989. This term has multiple hierarchies.
 BT1 radioactive isotopes
 BT1 isotopes
 BT2 silicon
 SA accelerator mass spectroscopy

sial
 use granitic layer

Siam
 use Thailand

Sibay
 No longer a valid term for GeoRef. As of 1993, see Sibay Russian Federation.

Sibay Russian Federation (1993)
 Town in the Southern Urals in Bashkiria in SE European Russian Federation. This term has multiple hierarchies.
 BT1 Bashkiria Russian Federation
 BT1 Russian Federation
 BT1 Commonwealth of Independent States
 BT2 Bashkiria Russian Federation
 BT2 Europe

Siberia (1978)

Region in Asia extending from the Urals to the Pacific Ocean, and from the Arctic Ocean to the Chinese and Mongolian borders and including N Kazakhstan.
 IN Index former Soviet republics as applicable.
 CO N500000N800000
   W1700000E0600000
 BT Asia
 SA Baikal rift zone
 SA Baikalian Phase
 SA Beringia
 SA Central Asia
 SA Kazakhstan
 SA Khatanga Basin
 SA Moty Formation
 SA Ob River
 SA Omolon Block
 SA Russian Far East
 SA Russian Federation
 SA Russian Pacific region
 SA Siberian Platform
 SA Tunguska Basin
 SA Tunguska Syneclise
 SA West Siberia
 SA Yakutia Russian Federation
 SA Yudoma Series

Siberia-Soviet Far East
 use Russian Far East

Siberian fold belt (1981)
 Before 1993, USSR was a broader term.
 CO N490000N610000
   E1210000E0820000
 BT Asia
 NT Baikal region
 NT Sayan
 NT Tuva Russian Federation
 NT Western Transbaikalia
 SA Altai Mountains
 SA Russian Federation

Siberian Lowland (1978)
 Region comprising the West Siberian Plain which extends from the Urals in the W to the Yenisei River in the E, and from the Kara Sea in the N to the Kazakh Hills in the S. Also search Siberian. This term has multiple hierarchies.
 IN Index former Soviet republics as applicable.
 CO N500000N733000
   E0950000E0620000
 UF West Siberian Basin
 UF West Siberian Lowland
 UF West Siberian Plain
 BT1 West Siberia
 BT1 Commonwealth of Independent States
 BT2 West Siberia
 BT2 Asia
 SA Kazakhstan
 SA Ob-Irtysh Interfluve
 SA Russian Federation

Siberian Platform (1978)
 Region comprising the Central Siberian Plateau which lies between the Yenisei River on the W, and the Lena River on the E. Also search Siberian. Autoposting of this term began in 1981. This term has multiple hierarchies.
 CO N523000N750000
   E1390000E0820000
 UF Central Siberian Plateau
 UF West Siberian Platform
 BT1 Russian Federation
 BT1 Commonwealth of Independent States
 BT2 Asia
 NT Aldan Shield
 NT Anabar Shield

NT Angara-Lena Basin
NT North Siberian Plain
NT Tunguska
NT Vilyuy Syneclise
NT Yakutia region
NT Yenisei Ridge
SA Bazhenov Formation
SA Mir Pipe
SA Siberia
SA Udachnaya Pipe

Sibiu
No longer a valid term for GeoRef. As of 1993 see Sibiu Romania.

Sibiu Romania   (1993)
City and county in S Transylvania. Before 1993 also search Sibiu AND Romania.
BT Transylvania
BT Romania
BT Southern Europe
BT Europe

Sichuan
No longer a valid term for GeoRef. See Sichuan China.

Sichuan Basin   (1993)
SW China. Before 1993, also search Szechwan Basin.
UF Szechwan Basin
BT China
BT Far East
BT Asia
SA Sichuan China

Sichuan China   (1993)
Province, SW China. Before 1993, also search Sichuan or Szechwan or Sichuan Sheng, or Szechuan.
CO N260000N341000
   E1100000E0973000
UF Sichuan Sheng
UF Szechuan China
BT China
BT Far East
BT Asia
NT Panxi Rift
SA Luoquan Formation
SA Qinghai-Xizang Plateau
SA Sichuan Basin
SA Wufeng Formation
SA Xianshuihe fault zone
SA Xichang China
SA Yangtze Platform

Sichuan Sheng
 use Sichuan China

Sicilia
Not a valid term for GeoRef. See Sicily Italy.

Sicilian   (1978)
Europe.
BT upper Pleistocene
BT Pleistocene
BT Quaternary
BT Cenozoic

Sicily
No longer a valid term for GeoRef. See Sicily Italy.

Sicily Italy   (1993)
Island in Mediterranean Sea and autonomous region. Also search Sicilia.
CO N364500N381500
   E0150000E0120000
BT Italy
BT Southern Europe
BT Europe
NT Caltanissetta Italy
NT Ciminna Basin
NT Enna Italy
NT Lipari Islands
NT Madonie Mountains
NT Messina Italy
NT Mount Etna
NT Nebrodi Mountains
NT Palermo Italy
NT Pantelleria
NT Peloritani Mountains
NT Strona Valley
NT Syracuse Italy
NT Trapani Italy
NT Ustica Island
SA Mediterranean region
SA Mediterranean Sea

Sicker Group   (1989)
Vancouver Island, British Columbia. Pennsylvanian? to Permian.
BT upper Paleozoic
BT Paleozoic
SA British Columbia
SA Pennsylvanian
SA Permian

side-looking airborne radar
 use SLAR

side-scanning methods   (1981)
SA acoustical methods
SA geophysical methods
SA GLORIA
SA methods
SA radar methods
SA SAR
SA SLAR
SA sonar methods

sideraerolite
 use stony irons

siderite   (1978)
UF chalybite
BT carbonates
SA iron minerals
SA iron ores

siderite (meteorite)
 use iron meteorites

siderolite
 use stony irons

siderophile elements   (1978)
Chemical elements which tend to concentrate in metallic phases of meteorites. Also, an element possessing weak affinity for sulfur and oxygen.
SA chalcophile elements
SA chemical elements
SA cobalt
SA gold
SA iron
SA lithophile elements
SA nickel
SA phosphorus
SA platinum

siderophyre   (1984)
UF siderophyres
BT stony irons
BT meteorites

siderophyres
 use siderophyre

Sidhi
No longer a valid term for GeoRef. See Sidhi India.

Sidhi India   (1993)
Village and district in E Madhya Pradesh, central India.
BT Madhya Pradesh India
BT India
BT Indian Peninsula
BT Asia

Sidobre Massif   (1978)
In S Central Massif NE of Castres. Also search Sidobre. This term has multiple hierarchies.
BT1 Central Massif
BT1 France
BT1 Western Europe
BT1 Europe
BT2 Tarn France
BT2 France
BT2 Western Europe
BT2 Europe

Siebengebirge   (1978)
Hills in the Westerwald on right bank of Rhine River SSE of Bonn.
BT North Rhine-Westphalia Germany
BT Germany
BT Central Europe
BT Europe
SA Westerwald

Siedlce
No longer a valid term for GeoRef. As of 1993 see Siedlce Poland.

Siedlce Poland   (1993)
Province and city in E Poland. Before 1993 also search Siedlce AND Poland.
UF Syedlets Poland
BT Poland
BT Central Europe
BT Europe

Siegenian   (1978)
Europe. Above Gedinnian, below Emsian.
BT Lower Devonian
BT Devonian
BT Paleozoic

siegenite   (1985)
BT sulfides
SA linnaeite

Siena
No longer a valid term for GeoRef. See Siena Italy.

Siena Italy   (1993)
City and province in S central Tuscany.
BT Tuscany Italy
BT Italy
BT Southern Europe
BT Europe

Sierozems   (1978)
UF Cerozem
UF Gray desert soil
UF Gray earth
UF Serozem
BT soils
SA soil group
SA Zonal soils

Sierra Ancha   (1978)
Ridge in E central Arizona. Autoposting of broader terms to this term began in 1989.
BT Gila County Arizona
BT Arizona
BT United States

Sierra Blanca   (1978)
Range in Otero and Lincoln counties in S central New Mexico.
IN Index counties and mountains as applicable.
SA Lincoln County New Mexico
SA New Mexico
SA Otero County New Mexico
SA Sangre de Cristo Mountains

Sierra County
Valid through 1988. Search in combination with state term. After 1988, use specific county-state term.

Sierra County California   (1989)
NE California. Before 1989, also search Sierra County AND California.
CO N392500N394400
   W1200000W1210000
BT California
BT United States

Sierra County New Mexico   (1989)
SW New Mexico. Before 1989, also search Sierra County AND New Mexico.
CO N323500N333000
   W1062200W1080000
BT New Mexico
BT United States
NT Truth or Consequences New Mexico
SA Black Range
SA Caballo Mountains
SA Jornada del Muerto
SA Orogrande Basin

Sierra de Gador   (1978)
Range in SE Spain.
BT Almeria Spain
BT Andalusia Spain
BT Spain
BT Iberian Peninsula
BT Southern Europe
BT Europe

Sierra de Gredos   (1978)
Range W of Madrid in W central Spain.
BT Spain
BT Iberian Peninsula
BT Southern Europe
BT Europe

Sierra de Guadarrama   (1978)
Range N of Madrid separating Old Castile from New Castile.
BT Spain
BT Iberian Peninsula
BT Southern Europe
BT Europe

Sierra de Juarez   (1985)
Mountain range, N Baja California, Mexico.
CO N313000N324000
   W1154000W1162000
BT Baja California Mexico
BT Mexico
SA Baja California

Sierra de la Demanda   (1978)
Range of the Cordillera Iberica in Old Castile in N Spain.
BT Spain
BT Iberian Peninsula
BT Southern Europe
BT Europe

Sierra de los Filabres   (1978)
Range in W central Almeria Province, SE Spain.
BT Almeria Spain
BT Andalusia Spain
BT Spain
BT Iberian Peninsula
BT Southern Europe
BT Europe

Sierra de los Organos   (1978)
Range in W Pinar del Rio, W Cuba.
SA Pinar del Rio Cuba

Sierra de Perija   (1985)
Range of the Andes on the Colombia-Venezuela border.
IN Index countries as applicable.
CO N090000N113000
   W0721500W0733000
BT Andes
BT South America
SA Colombia

SA Sonora Mexico
SA Zacatecas Mexico

Sierra Madre Oriental (1978)
Range running parallel to the Gulf of Mexico.
IN Index Mexican states and Sierra Madre as applicable.
BT Mexico
SA Coahuila Mexico
SA Hidalgo Mexico
SA Nuevo Leon Mexico
SA Puebla Mexico
SA San Luis Potosi Mexico
SA Sierra Madre
SA Tamaulipas Mexico
SA Veracruz Mexico

Sierra Madre Range (1978)
Range in Continental Divide, S Wyoming, just W of North Platte River. Before 1978, also search Sierra Madre AND Wyoming.
BT Wyoming
BT United States
SA Mullen Creek-Nash Fork shear zone
SA Sierra Madre

Sierra Morena
A valid term through mid-1978. See Betic Cordillera.

Sierra Nevada (1978)
Mountain range in E California. Search Sierra Nevada AND California. May also occur in other locations. After 1981, see Spanish Sierra Nevada.
IN Index states or regions and mountains as applicable.
SA California
SA Cascade Range
SA Long Valley Caldera
SA Melones Fault
SA Mount Lyell
SA North American Cordillera
SA Spanish Sierra Nevada
SA Table Mountain
SA White Mountain

Sierra Nevada Batholith (1981)
E central California.
CO N350000N402000
      W1180000W1210000
BT California
BT United States

Sierra Nevada de Santa Marta (1978)
Mountain range in N Colombia on Caribbean coast.
BT Colombia
BT South America
SA Santa Marta Colombia

Sierra Pena Blanca (1989)
Chihuahua, N Mexico.
BT Chihuahua Mexico
BT Mexico

Sierras Pampeanas
use Pampean Mountains

Sierrita Mountains (1978)
IN Index counties or region as applicable.
SA Arizona

Sifton Basin (1978)
N British Columbia. Also search Sifton.
BT British Columbia
BT Western Canada
BT Canada

Sigillaria (1978)
Genus.
BT Lycopsida
BT pteridophytes

SA Eastern Cordillera
SA Venezuela
SA Zulia Venezuela

Sierra de Ronda
use Serrania de Ronda

Sierra Gorda (1978)
E range of S Sierra Madre Occidental in central Mexico.
BT Guanajuato Mexico
BT Mexico

Sierra Leone (1978)
Former British colony and protectorate. Became independent in 1961.
CO N070000N100000
      W0110000W0133000
BT West Africa
BT Africa

Sierra Leone Rise (1981)
SW of West Africa in Cape Verde Atlantic. As of 1990, Atlantic Ocean is autoposted to this term.
CO N023000N093000
      W0180000W0260000
BT Cape Verde Atlantic
BT North Atlantic
BT Atlantic Ocean
SA DSDP Site 366
SA Leg 41
SA ODP Site 665
SA ODP Site 666
SA ODP Site 667
SA ODP Site 668

Sierra Madre (1978)
Chief Mexican mountain system extending 1500 miles SE from U.S. border including 3 separate ranges which enclose the great central plateau. Before 1978, this term also included the Sierra Madre Range of Wyoming. Also used for ranges on Luzon, Philippines and in California. As of 1993, search in combination with country.
IN Index state or country as applicable.
SA California
SA Luzon
SA Mexico
SA North American Cordillera
SA Sierra Madre del Sur
SA Sierra Madre Occidental
SA Sierra Madre Oriental
SA Sierra Madre Range

Sierra Madre del Sur (1978)
Coastal range in SW Mexico.
IN Index Sierra Madre and Mexican states as applicable.
CO N154500N194500
      W0950000W1022000
BT Mexico
SA Guerrero Mexico
SA Mexican volcanic belt
SA Oaxaca Mexico
SA Sierra Madre

Sierra Madre Occidental (1978)
Range running for 700 miles parallel to the Pacific Ocean.
IN Index states and Sierra Madre as applicable.
CO N220000N311500
      W1042000W1120000
BT Mexico
SA Chihuahua Mexico
SA Colima Mexico
SA Durango Mexico
SA Jalisco Mexico
SA Mexican volcanic belt
SA Nayarit Mexico
SA Sierra Madre

BT Plantae

Sigma Mine (1989)
Gold ores. Val d'Or, SW Quebec.
BT Quebec
BT Eastern Canada
BT Canada
SA Abitibi Belt
SA gold ores
SA mines
SA Val d'Or Quebec

signal distortion (1981)
SA distortion
SA elastic waves
SA seismology

Signal Hill Formation (1978)
Unconformably underlies Random Formation and overlies Avalonian Formation.
BT Huronian
BT Proterozoic
BT upper Precambrian
BT Precambrian
SA Canada
SA Newfoundland Island

signal to noise ratios
use signal-to-noise ratio

signal-to-noise ratio (1981)
UF signal to noise ratios
SA geophysical methods
SA geophysical surveys
SA noise
SA seismic methods
SA seismograms
SA seismology
SA signals
SA stacking

signals (1978)
SA filters
SA geophysical methods
SA noise
SA seismology
SA signal-to-noise ratio
SA teleseismic signals

Signy Island (1978)
One of the South Orkney Islands SE of the Falkland Islands.
BT South Orkney Islands
BT Scotia Sea Islands
BT Antarctica
SA Atlantic Ocean

Sigsbee Deep (1978)
In SW central part of Gulf of Mexico. Before 1993, also search Mexico Basin AND Gulf of Mexico.
BT Gulf of Mexico
BT North American Atlantic
BT North Atlantic
BT Atlantic Ocean
SA Campeche Scarp
SA DSDP Site 3
SA DSDP Site 91
SA Mississippi Fan

Sikhote Alin
A valid term through 1976.
use Sikhote-Alin Range

Sikhote-Alin Range (1978)
Mountain range in Primorye and Khabarovsk krays along the Japan Sea and the Tatar Strait. Before 1977, also search Sikhote-Alin. This term has multiple hierarchies.
CO N423000N543000
      E1420000E1310000
UF Sikhote Alin
BT1 Russian Federation
BT1 Commonwealth of Independent States
BT2 Asia

SA Khabarovsk Russian Federation
SA Primorye Russian Federation
SA Russian Far East

Sikkim
No longer a valid term for GeoRef. Former British and Indian protectorate between Bhutan and Nepal. Included use on level 1 as an area term until 1977. See Sikkim India.

Sikkim India (1993)
Became an associated Indian state on Sept. 7, 1974.
BT India
BT Indian Peninsula
BT Asia
SA Himalayas
SA Lesser Himalayas
SA Siwalik Mountains
SA Siwalik System

Sila Massif (1978)
In the Southern Apennines forming central part of toe of the Italian boot.
BT Calabria Italy
BT Italy
BT Southern Europe
BT Europe
SA Southern Apennines

silcrete (1978)
BT chemically precipitated rocks
BT sedimentary rocks
SA duricrust
SA terrigenous materials

Silesia (1978)
Region in E central Europe lying mostly in SW Poland. It comprises both Upper and Lower Silesia plus small areas in E Germany and N Moravia.
IN Index countries as applicable.
BT Central Europe
BT Europe
SA Czechoslovakia
SA Germany
SA Lower Silesia
SA Moravia
SA Poland
SA Upper Silesia

Silesian (1978)
BT Carboniferous
BT Paleozoic

Silesian Basin
use Silesian coal basin

Silesian coal basin (1978)
In Lower Silesia in Wroclaw Province, and in Upper Silesia in Katowice Province.
IN Index provinces as applicable.
UF Silesian Basin
BT Poland
BT Central Europe
BT Europe
SA coal fields
SA Katowice Poland
SA Lower Silesia
SA Upper Silesia
SA Upper Silesian coal basin
SA Wroclaw Poland

silexite
As of 1981, no longer a valid term for GeoRef. Use quartzolite (for the igneous rock) or chert as applicable.

silica (1978)
As of 1985, no longer considered a commodity term (list C). Used for silicon dioxide.
SA abrasives

SA glass materials
SA Ottawa Sand
SA sand
SA sands
SA silica minerals
SA silicates
SA siliceous composition
SA silicon

silica group
  use silica minerals

silica minerals (1978)
  Autoposting of this term began in 1978.
  UF silica group
  BT framework silicates
  BT silicates
  NT agate
  NT amethyst
  NT carnelian
  NT chalcedony
  NT chrysoprase
  NT coesite
  NT cristobalite
  NT hyalite
  NT jasper
  NT keatite
  NT lechatelierite
  NT melanophlogite
  NT onyx
  NT opal
  NT quartz
  NT smoky quartz
  NT stishovite
  NT tridymite
  SA silica

silicate minerals
  use silicates

silicate rocks (1978)
  SA petrology
  SA rocks
  SA siliceous composition
  SA standard materials

silicates (1978)
  As of 1978, the term silicates is autoposted to all individual silicate minerals. Before 1978, also search individual silicate groups. Before 1981, maskelynite was included as a narrower term.
  IN In indexing, use this term only for broad treatments of the entire class of minerals; otherwise, use a narrower term, e.g. orthosilicates, ring silicates, chain silicates, etc.
  UF silicate minerals
  NT aluminosilicates
  NT asbestos
  NT barium silicates
  NT borosilicates
  NT chain silicates
  NT chlorophaeite
  NT framework silicates
  NT orthosilicates
  NT ring silicates
  NT sheet silicates
  SA clay mineralogy
  SA minerals
  SA silica
  SA silicon

siliceous composition (1978)
  Before 1978 also search siliceous in combination with specific sediment type. Also search siliceous rocks.
  SA clastic rocks
  SA composition
  SA insoluble residues
  SA jasperoid
  SA sedimentary rocks
  SA sediments
  SA silica

SA silicate rocks
SA siliceous sinter
SA siliciclastics

siliceous earth
  Not a valid GeoRef index term after 1970. Was used on level 1 in subfile N. Now use diatomaceous earth or diatomite as applicable.

siliceous rocks
  A valid term through 1977. After 1977, use siliceous composition.

siliceous sinter (1978)
  Also search sinter AND siliceous.
  BT chemically precipitated rocks
  BT sedimentary rocks
  SA siliceous composition
  SA sintering

silicic composition
  use acidic composition

siliciclastic rocks
  use siliciclastics

siliciclastics (1989)
  UF siliciclastic rocks
  SA clastic rocks
  SA clastic sediments
  SA sedimentary rocks
  SA sediments
  SA siliceous composition

silicides (1978)
  Autoposting of this term began in 1981.
  BT alloys
  NT ferrosilicon
  SA moissanite

silicification (1978)
  UF silification
  SA chertification
  SA diagenesis
  SA fossilization
  SA metasomatism
  SA processes
  SA sedimentation

silicified wood
  use fossil wood

Silicispongiae (1978)
  UF Noncalcarea
  BT Porifera
  BT Invertebrata

Silicoflagellata (1978)
  Autoposting of this term began in 1978. This term has multiple hierarchies.
  BT1 Protista
  BT1 Invertebrata
  BT2 Protista
  BT2 microfossils
  NT Dictyocha
  NT Distephanus
  SA silicoflagellates

silicoflagellates (1985)
  Common name for Silicoflagellata. This term has multiple hierarchies.
  BT1 Protista
  BT1 Invertebrata
  BT2 Protista
  BT2 microfossils
  BT3 invertebrates
  SA biostratigraphy
  SA Silicoflagellata

silicon (1978)
  UF Si
  NT Si-32
  SA chemical elements
  SA ferrosilicon
  SA moissanite
  SA silica
  SA silicates

silicon carbide
  Not a valid term for GeoRef. As of 1993, see carborundum for economic papers or carbides or moissanite, for minerals. Before 1993, also search carborundum.

silification
  use silicification

Siljan (1978)
  Lake in central Sweden. Also used for possible meteor crater before 1993.
  BT Sweden
  BT Scandinavia
  BT Western Europe
  BT Europe
  SA Siljan Ring

Siljan Ring (1993)
  Impact crater containing the lake, Siljan in central Sweden. Site of several deep drilling boreholes. Before 1993, also search Siljan.
  CO N604500N611300
     E0034000E0025500
  BT Sweden
  BT Scandinavia
  BT Western Europe
  BT Europe
  SA Gravberg Well
  SA Siljan

sillimanite (1978)
  Autoposting of nesosilicates began in 1985. As of 1981, use sillimanite deposits for sillimanite as a commodity.
  UF fibrolite
  BT nesosilicates
  BT orthosilicates
  BT silicates
  SA andalusite
  SA kyanite
  SA sillimanite deposits

sillimanite deposits (1981)
  Before 1981, search sillimanite AND deposits.
  IN May be used as a commodity term with ceramic materials.
  SA ceramic materials
  SA sillimanite

sillimanite gneiss (1978)
  BT gneisses
  BT metamorphic rocks

sills (1978)
  BT intrusions
  SA dikes

silt (1978)
  UF silts
  BT clastic sediments
  BT sediments
  SA alluvium
  SA clay
  SA loam
  SA loess
  SA mud
  SA sand
  SA siltation
  SA siltstone
  SA terrigenous materials

siltation (1982)
  SA deposition
  SA dredging
  SA maintenance
  SA reservoirs
  SA sedimentation
  SA silt
  SA waterways

siltite
  use siltstone

silts
  use silt

siltstone (1978)
  UF siltite
  BT clastic rocks
  BT sedimentary rocks
  SA metasiltstone
  SA mudstone
  SA silt
  SA terrigenous materials

Silurian (1978)
  Above Ordovician, below Devonian. From 1978-1980, Paleozoic was autoposted to this term.
  UF Gothlandian
  UF Gotlandian
  BT Paleozoic
  NT Aberystwyth Grits
  NT Bass Islands Dolomite
  NT Clinch Sandstone
  NT Crab Orchard Formation
  NT Fusselman Dolomite
  NT Henryhouse Formation
  NT Lower Silurian
  NT Middle Silurian
  NT Perry Mountain Formation
  NT Red Mountain Formation
  NT Rockwood Formation
  NT Sangerville Formation
  NT Upper Silurian
  NT Vassalboro Formation
  NT Waterville Formation
  SA Berwick Formation
  SA Broken River Formation
  SA Dalradian
  SA Downtonian
  SA Ellis Bay Formation
  SA Hanson Creek Formation
  SA Henderson Gneiss
  SA Hidden Valley Dolomite
  SA Hunton Group
  SA Kittery Formation
  SA Madrid Formation
  SA Merrimack Group
  SA New York City Group
  SA Peel Sound Formation
  SA Rangeley Formation
  SA Read Bay Formation
  SA Ringerike Sandstone
  SA Road River Formation
  SA Shawangunk Formation
  SA Shoo Fly Complex
  SA Taconic Orogeny
  SA Talladega Group
  SA Waits River Formation

silvanite
  use sylvanite

silver (1978)
  Chemical element. As of 1981, use silver ores for silver as a commodity. Autoposting of metals to this term began in 1989.
  UF Ag
  BT metals
  SA argentite
  SA chalcophile elements
  SA electrum
  SA heavy metals
  SA native elements
  SA silver ores

Silver Bow County
  Valid through 1988. Search in combination with state term. After 1988, use specific county-state term.

Silver Bow County Montana (1989)
  SW Montana. Before 1989, also search Silver Bow County AND Montana.
  CO N453700N461000
     W1121000W1130500
  BT Montana
  BT United States

NT Butte Montana

Silver City
Valid through 1988. Search in combination with state term. After 1988, use specific city-state term.

Silver City Nevada (1989)
Hamlet in W Nevada. Before 1989, search Silver City AND Nevada.
BT Lyon County Nevada
BT Nevada
BT United States

Silver City New Mexico (1989)
Town in SW New Mexico. Before 1989, search Silver City AND New Mexico.
CO N324700N324700
W1081600W1081600
BT Grant County New Mexico
BT New Mexico
BT United States

silver ores (1981)
Before 1981, also search (silver OR Ag) AND (deposit OR deposits OR ore OR ores OR economic) in the basic index. Autoposting of metal ores to this term began in 1985.
IN Commodity. See List C.
BT metal ores
SA acanthite
SA argentite
SA Equity Mine
SA freibergite
SA Kamioka Mine
SA Leadville mining district
SA native elements
SA nuggets
SA San Juan mining district
SA silver
SA stephanite
SA Sullivan Mine
SA Temagami Mine
SA tetrahedrite
SA Toyoha Mine

Silver Peak Mountains (1978)
Small range in W Esmeralda County in SW Nevada. Also search Silver Peak.
BT Esmeralda County Nevada
BT Nevada
BT United States

Silverton Caldera (1978)
Near town of Silverton in San Juan County in San Juan Mountains of SW. Also search Silverton.
BT San Juan County Colorado
BT Colorado
BT United States
SA San Juan Mountains

Silvretta Group (1978)
Mountain group. Also search Silvretta.
UF Silvretta Massif
BT Paleozoic
SA Switzerland
SA Tyrol Austria

Silvretta Massif
use Silvretta Group

sima
use basaltic layer

Simbirsk Russian Federation
use Ulyanovsk Russian Federation

Simferopol
No longer a valid term for GeoRef. As of 1993, see Simferopol

Simferopol Ukraine (1993)
City in the S central Crimea, S Ukraine. This term has multiple hierarchies.
BT1 Crimea Ukraine
BT1 Ukraine
BT1 Europe
BT2 Crimea Ukraine
BT2 Ukraine
BT2 Commonwealth of Independent States

Simi Hills (1978)
In Ventura County NW of Los Angeles.
IN Index county or region as applicable.
BT Ventura County California
BT California
BT United States

Simi Valley California (1989)
City in S California.
CO N341200N341600
W1184500W1185000
BT Ventura County California
BT California
BT United States

simians (1981)
Autoposting of Eutheria and Theria to this term began in 1989. As of 1993, Pongidae and Cercopithecidae are narrower terms.
BT Primates
BT Eutheria
BT Theria
BT Mammalia
BT Tetrapoda
BT Vertebrata
BT Chordata
NT Cercopithecidae
NT Platyrrhina
NT Pongidae

similar folds (1978)
Before 1978, also search shear folds. Before 1993, also search slip folds.
UF shear folds
UF slip folds
BT folds
SA concentric folds
SA shear

Simla
No longer a valid term for GeoRef. See Simla India.

Simla Hills (1978)
Hill and mountain area of the outer Kumaun Himalayas around Simla in Himachal Pradesh. Also search Simla. This term has multiple hierarchies.
BT1 Himachal Pradesh India
BT1 India
BT1 Indian Peninsula
BT1 Asia
BT2 Kumaun Himalayas
BT2 Himalayas
BT2 Asia

Simla India (1993)
Town and hill resort in S Himachal Pradesh, N India.
BT Himachal Pradesh India
BT India
BT Indian Peninsula
BT Asia

Simleu Basin (1978)
In Bihor Mountains near town of Simleu Silvaniei in NW Transylvania. Also search Simleu.
BT Transylvania
BT Romania
BT Southern Europe
BT Europe

Simplon region (1978)
Includes Simplon Pass, Simplon Road, and Simplon Tunnel area. Also search Simplon.
IN Index Italy and/or Switzerland.
BT Europe
SA Italy
SA Piemonte Italy
SA Switzerland

Simpson Desert (1978)
Primarily in SE Northern Territory.
IN Index Northern Territory and states as applicable.
SA Australia
SA Northern Territory Australia
SA Queensland Australia
SA South Australia

Simpson Field (1989)
North Slope, Alaska. Also one in Nebraska. Search in conjunction with appropriate state.
IN Index states as applicable.
SA Alaska
SA Nebraska
SA North Slope
SA oil and gas fields

Simpson Group (1978)
Comprises Joins, Oil Creek, McLish, Tulip Creek, Bromide, and Corbin Ranch formations. Central and S Oklahoma.
BT Middle Ordovician
BT Ordovician
BT Paleozoic
SA Bromide Formation
SA Oklahoma

SIMS
use secondary ion mass spectroscopy

simulation (1978)
SA analog simulation
SA data processing
SA digital simulation
SA mathematical geology
SA mathematical models
SA models

Sinai (1978)
Peninsula extending from the Mediterranean Sea to the Red Sea. It is bounded on the W by the Suez Canal and the Gulf of Suez, and on the E by Israel and the Gulf of Aqaba. As of 1993, see Sinai Egypt for the governorate.
UF Sinai Peninsula
SA Egypt
SA Gavish Sebkha
SA Israel
SA Sinai Egypt
SA Solar Lake

Sinai Egypt (1993)
Governorate, NE Egypt in Sinai.
BT Egypt
BT North Africa
BT Africa
SA Sinai

Sinai Peninsula
use Sinai

Sinaia
No longer a valid term for GeoRef. As of 1993 see Sinaia Romania.

Sinaia Romania (1993)
Town and health resort at SE foot of Bucegi Mountains of the Transylvanian Alps in Prahova County. Before 1993 also search Sinaia AND Romania.
BT Prahova Romania
BT Walachia
BT Romania
BT Southern Europe
BT Europe
SA Bucegi Mountains

Sinaloa
No longer a valid term for GeoRef as of 1993 see Sinaloa Mexico.

Sinaloa Mexico (1993)
State in W Mexico. Before 1993 also search Sinaloa AND Mexico.
BT Mexico
NT Mazatlan Mexico

Sind
No longer a valid term for GeoRef. See Sind Pakistan.

Sind Pakistan (1993)
Province in SE Pakistan. Also search Sindh.
BT Pakistan
BT Indian Peninsula
BT Asia
NT Chor Pakistan
NT Hyderabad Pakistan
SA Nari Series

Sinemurian (1978)
Europe. Above Hettangian, below Pliensbachian.
BT lower Liassic
BT Lower Jurassic
BT Jurassic
BT Mesozoic

Singapore (1978)
Island republic and city off the southern tip of the Malay Peninsula.
CO N010000N014500
E1042000E1032000
BT Far East
BT Asia
SA ASEAN

Singhbhum
No longer a valid term for GeoRef. See Singhbhum India.

Singhbhum Granite (1978)
Also search Singhbhum.
BT Archean
BT Precambrian
SA Bihar India
SA India
SA Orissa India
SA Singhbhum India

Singhbhum India (1993)
District, Chota Nagpur division, S Bihar.
BT Bihar India
BT India
BT Indian Peninsula
BT Asia
NT Sini India
SA Singhbhum Granite

Singhbhum shear zone (1978)
Also search Singhbhum.
IN Index states as applicable.
BT India
BT Indian Peninsula
BT Asia
SA Bihar India
SA Orissa India

single domains (1981)
BT magnetic domains

single-crystal method (1978)
Before 1978, search single crystal.
SA crystal growth
SA crystal structure
SA methods
SA minerals
SA powder method

SA X-ray analysis
Singleton Coal Measures (1978)
E New South Wales.
IN Index age as applicable.
SA Australia
SA Carboniferous
SA New South Wales Australia
SA Permian
SA Triassic

Sini
No longer a valid term for GeoRef. See Sini India.

Sini India (1993)
Village in Singhbhum District. Site of limestone quarries for Jamshedpur iron and steel works.
BT Singhbhum India
BT Bihar India
BT India
BT Indian Peninsula
BT Asia

Sinian (1985)
China. Autoposting of Precambrian to this term ended in 1989. Autoposting of upper Precambrian and Precambrian to this term began in 1990.
BT Proterozoic
BT upper Precambrian
BT Precambrian
NT Dengying Formation
NT Doushantuo Formation
NT Nantuo Formation
NT upper Sinian
NT Wumishan Formation
SA Eocambrian

Sinjar Formation (1978)
Paleocene-lower Eocene. In the Jabel Sinjar Mountains.
IN Index ages as applicable.
BT Paleogene
BT Tertiary
BT Cenozoic
SA Iraq
SA lower Eocene
SA Middle East
SA Paleocene

sink holes
use sinkholes

sinkholes (1978)
Autoposting of solution features to this term began in 1989.
UF sink holes
BT solution features
SA dolines
SA geologic hazards
SA geomorphology
SA karst
SA potholes

Sinkiang
Not a valid term for GeoRef. See Xinjiang China.

Sinkiang Province
use Xinjiang China

Sinkiang Uighur
No longer a valid term for GeoRef.
use Xinjiang China

Sinkiang Weiwu'er Zizhiqu
use Xinjiang China

sinking
Use land subsidence or subsidence as applicable.

sinks (1978)
Substance used to dissipate heat or remove fluids during a chemical reaction. For the solution feature, use sinkholes.
SA chemical reactions

SA geochemistry
sinnerite (1978)
BT sulfides

Sino-Korean Paraplatform
use Sino-Korean Platform

Sino-Korean Platform (1989)
IN Index regions as applicable.
UF Sino-Korean Paraplatform
BT Far East
BT Asia
SA China
SA Korea

sinter
A valid term through 1977. After 1977, use siliceous sinter.

sintering (1985)
SA deformation
SA diffusion
SA meteorites
SA phase equilibria
SA refractory materials
SA siliceous sinter

sinuosity (1978)
SA channels
SA fluvial features
SA meanders
SA streams

Sioux County
Valid through 1988. Search in combination with state term. After 1988, use specific county-state term.

Sioux County Iowa (1989)
NW Iowa. Before 1989, also search Sioux County AND Iowa.
CO N425400N431600 W0955100W0963500
BT Iowa
BT United States

Sioux County Nebraska (1989)
Extreme NW Nebraska. Before 1989, also search Sioux County AND Nebraska.
CO N420000N430000 W1032500W1040500
BT Nebraska
BT United States

Sioux County North Dakota (1989)
S North Dakota. Before 1989, also search Sioux County AND North Dakota.
CO N455600N462300 W1003200W1020000
BT North Dakota
BT United States

Sioux Quartzite (1978)
Has been correlated with Baraboo Quartzite in Wisconsin. NW Iowa, SW Minnesota, NE Nebraska, and SE South Dakota.
BT Precambrian
SA Iowa
SA Minnesota
SA Nebraska
SA South Dakota

Siphonapteroida (1981)
BT Insecta
BT Mandibulata
BT Arthropoda
BT Invertebrata

Sipunculoida (1978)
BT worms
BT Invertebrata

Siqueiros fracture zone (1985)
Near E end of Clipperton fracture zone.
BT Equatorial Pacific

BT Pacific Ocean

SIR
Not a valid term for GeoRef. See Shuttle Imaging Radar, if applicable.

Sir Darya
use Syr Darya

Siracusa
Not a valid term for GeoRef. See Syracuse New York or Syracuse Italy.

Sirenia (1978)
Order. Autoposting of Eutheria and Theria to this term began in 1989.
BT Eutheria
BT Theria
BT Mammalia
BT Tetrapoda
BT Vertebrata
BT Chordata

Siret River (1978)
Rises on the E slopes of the Carpathians in the Ukraine and flows SE into the Danube River just N of Galati. Also search Siret.
IN Index Moldavia and Romania; and/or Ukraine.
UF Sereth River
UF Siretul River
BT Europe
SA Moldavia
SA Moldavian Platform
SA Romania
SA Romanian Plain
SA Ukraine

Siretul River
use Siret River

Sirohi
No longer a valid term for GeoRef. Former Indian state now part of Rajasthan. See Sirohi India.

Sirohi India (1993)
Town in SW Rajasthan.
BT Rajasthan India
BT India
BT Indian Peninsula
BT Asia

Sirte Basin (1978)
On the Gulf of Sidra in N Libya. Also search Sirte.
UF Syrte
BT Libya
BT North Africa
BT Africa

Sisian
No longer a valid term for GeoRef. As of 1993, see Sisian Armenia.

Sisian Armenia (1993)
Village in S Armenia. This term has multiple hierarchies.
BT1 Armenia
BT1 Commonwealth of Independent States
BT2 Armenia
BT2 Europe

Siskiyou County
Valid through 1988. Search in combination with state term. After 1988, use specific county-state term.

Siskiyou County California (1989)
N California. Before 1989, also search Siskiyou County AND California.
CO N410000N420000

W1213000W1214500
BT California
BT United States
NT Medicine Lake Highland
NT Mount Shasta
SA Marble Mountains
SA Medicine Lake
SA Preston Peak
SA Trinity Complex

Sisquoc Formation (1985)
Upper Miocene to Pliocene. S California.
IN Index ages as applicable.
BT Neogene
BT Tertiary
BT Cenozoic
SA California
SA Pliocene
SA upper Miocene

site evaluation
use site exploration

site exploration (1978)
Before 1978, also search site selection. Collection of data about and testing of, surface and subsurface materials (including phyisical properties, distribution, and geologic structure) at a site, to prepare suitable design for an engineering structure or other use.
UF site evaluation
UF site investigation
UF site selection
SA dams
SA engineering geology
SA exploration
SA explosions
SA feasibility studies
SA foundations
SA geologic hazards
SA highways
SA land subsidence
SA marine installations
SA nuclear facilities
SA permafrost
SA railroads
SA reservoirs
SA rock mechanics
SA slope stability
SA soil mechanics
SA tunnels
SA underground installations
SA waste disposal

site investigation
use site exploration

site location maps (1989)
UF sample location maps
BT maps
SA index maps

site selection
use site exploration

sites, archaeological
use archaeological sites

sites, landing
use landing sites

Sitka Sound (1978)
On W side of Baranof Island in SE Alaska. Entrance to Sitka from Gulf of Alaska. Also search Sitka.
BT Southeastern Alaska
BT Alaska
BT United States
SA Alexander Archipelago
SA Gulf of Alaska

Sivamalai (1978)
A hill in the Coimbatore District in E Tamil Nadu, S India.
BT Coimbatore India
BT Tamil Nadu India
BT India

BT Indian Peninsula
BT Asia

Sivapithecus (1985)
Genus. Autoposting of Pongidae, Eutheria and Theria to this term began in 1989. Autoposting of Hominidae to this term ended in 1989.
BT Pongidae
BT simians
BT Primates
BT Eutheria
BT Theria
BT Mammalia
BT Tetrapoda
BT Vertebrata
BT Chordata

Sivash (1978)
Salt lagoons and marshes in N and NE Crimea. This term has multiple hierarchies.
UF Putrid Sea
BT1 Crimea Ukraine
BT1 Ukraine
BT1 Europe
BT2 Crimea Ukraine
BT2 Ukraine
BT2 Commonwealth of Independent States

Siwalik Formation
use Siwalik System

Siwalik Group
use Siwalik System

Siwalik Hills
use Siwalik Range

Siwalik Range (1978)
Range of foothills parallel with the main Himalayan system and extending 1000 miles SE from N Punjab in Pakistan to Sikkim. Also search Siwalik.
IN Index Sikkim and countries as applicable.
UF Siwalik Hills
UF Siwaliks
BT Asia
SA Himalayas
SA India
SA Nepal
SA Pakistan
SA Sikkim India

Siwalik Sandstone
use Siwalik System

Siwalik Series
use Siwalik System

Siwalik System (1978)
Middle Miocene-Pleistocene. Extends beyond limits of Siwalik Range to include parts in Baluchistan and North-West Frontier Province in Pakistan, Tipam Series in Assam, and Irrawaddy Series in Burma. Includes Lower (Kamlial and Chinji), Middle (Nagri and Dhokpathan) and Upper (Tatrot, Pinjor and Boulder conglomerates) Siwalik Sub-groups. Also search Siwalik.
IN Index ages as applicable.
UF Siwalik Formation
UF Siwalik Group
UF Siwalik Sandstone
UF Siwalik Series
BT Cenozoic
SA Burma
SA India
SA Miocene
SA Pakistan
SA Pleistocene
SA Pliocene

SA Sikkim India

Siwaliks
use Siwalik Range

Sixes River (1978)
N Curry County in SW Oregon. Flows into the Pacific Ocean.
BT Curry County Oregon
BT Oregon
BT United States

Siyeh Formation
use Helena Formation

size
As of 1981, no longer a valid term for GeoRef. See grain size. As of 1993, see scale factor.

size analysis
A valid term through mid-1978. Now use granulometry.

size distribution (1978)
SA distribution
SA grain size
SA granulometry
SA sedimentary rocks
SA sediments
SA textures

Sjaeland
use Zealand

sjogrenite (1978)
BT carbonates

Skaergaard Intrusion (1978)
BT Greenland
BT Arctic region

Skagerrak (1978)
Arm of the North Sea between Norway and Denmark.
BT North Sea
BT European Atlantic
BT North Atlantic
BT Atlantic Ocean

Skagit County
Valid through 1988. Search in combination with state term. After 1988, use specific county-state term.

Skagit County Washington (1989)
NW Washington. Before 1989, also search Skagit County AND Washington.
CO N481800N484000
    W1204000W1224500
BT Washington
BT United States
NT South Cascade Glacier

Skagit Valley (1978)
River valley. Primarily in NW Washington. Also search Skagit; Skagit River.
IN Index British Columbia and/or Washington as applicable.
BT North America
SA British Columbia
SA Washington

Skamania County
Valid through 1988. Search in combination with state term. After 1988, use specific county-state term.

Skamania County Washington (1989)
SW Washington. Before 1989, also search Skamania County AND Washington.
CO N453500N462300
    W1213000W1221500
BT Washington
BT United States

NT Mount Saint Helens
SA Spirit Lake
SA Toutle River

Skane (1978)
Region in S Sweden.
IN Index counties as applicable.
UF Scania
BT Sweden
BT Scandinavia
BT Western Europe
BT Europe
SA Kristianstad Sweden
SA Malmohus Sweden

skarn (1978)
Compositional term.
BT metasomatic rocks
BT metamorphic rocks

Skeena Mountains (1978)
NW central British Columbia. E of Coast Mountains.
BT British Columbia
BT Western Canada
BT Canada

Skeidararjokull (1978)
Glacier.
BT Iceland
BT Western Europe
BT Europe

skeletons (1978)
SA anatomy
SA bones
SA fossils
SA ichthyoliths
SA muscles
SA paleontology
SA paleopathology
SA skulls
SA teeth
SA Vertebrata

Skellefte (1978)
River in N Sweden. Flows SE into Gulf of Bothnia.
IN Index counties as applicable.
BT Sweden
BT Scandinavia
BT Western Europe
BT Europe
SA Norrbotten Sweden
SA Skellefte mining district
SA Vasterbotten Sweden

Skellefte District
As of 1993, no longer a valid term for GeoRef.
use Skellefte mining district

Skellefte mining district (1993)
Polymetallic ores. Before 1993, also search Skellefte District.
UF Skellefte District
BT Sweden
BT Scandinavia
BT Western Europe
BT Europe
SA mines
SA polymetallic ores
SA Skellefte

skewness (1978)
BT statistical analysis
SA kurtosis

Skiddaw Slates (1978)
Arenig and Llanvirn Series. Cumberland County in NW England.
BT Ordovician
BT Paleozoic
SA England
SA United Kingdom

Skolithos (1978)
BT ichnofossils

Skomer Island (1978)

In Saint Georges Channel off Dyfed.
BT Pembrokeshire Wales
BT Dyfed Wales
BT Wales
BT Great Britain
BT United Kingdom
BT Western Europe
BT Europe

Skopje (1978)
City on Vardar River in N Macedonia. This term has multiple hierarchies.
UF Skoplje
BT1 Yugoslav Macedonia
BT1 Macedonia
BT1 Southern Europe
BT1 Europe
BT2 Yugoslav Macedonia
BT2 Yugoslavia
BT2 Southern Europe
BT2 Europe

Skoplje
use Skopje

SKP-waves (1981)
BT body waves
BT elastic waves
SA P-waves
SA S-waves
SA seismology
SA waves

SKS-waves (1981)
BT body waves
BT elastic waves
SA P-waves
SA S-waves
SA seismology
SA waves

Skull Creek Shale (1985)
NE Wyoming, Montana, North Dakota and South Dakota.
BT Lower Cretaceous
BT Cretaceous
BT Mesozoic
SA Colorado Group
SA Montana
SA North Dakota
SA South Dakota
SA Wyoming

skulls (1978)
SA anatomy
SA bones
SA fossils
SA jaws
SA paleontology
SA paleopathology
SA skeletons
SA teeth
SA Vertebrata

skutterudite (1978)
Autoposting of sulfides to this term ended in 1989.
BT arsenides

Skye
use Isle of Skye

Skylab (1978)
SA extraterrestrial geology
SA planetology
SA remote sensing

skystones
use meteorites

Skythian
use Scythian

slabs (1978)
SA plate tectonics

slag (1989)

As of 1989, restricted to material resulting from the processing of metal ores.
  UF slags
  SA copper ores
  SA iron ores
  SA metal ores

slags
  use slag

Slak Dolny
  use Wroclaw Poland

Slanic
  No longer a valid term for GeoRef. As of 1993 see Slanic Romania.

Slanic Romania (1993)
Town in Prahova County in N central Walachia. Before 1993 also search Slanic AND Romania.
  BT Prahova Romania
  BT Walachia
  BT Romania
  BT Southern Europe
  BT Europe

slant stacks
  use Radon transforms

Slany (1978)
Town in W central Bohemia, W Czechoslovakia.
  BT Bohemia
  BT Czech Republic
  BT Czechoslovakia
  BT Central Europe
  BT Europe

SLAR (1985)
Acronym.
  UF side-looking airborne radar
  SA airborne methods
  SA multispectral scanner
  SA radar methods
  SA remote sensing
  SA SAR
  SA side-scanning methods

slate
As of 1981, no longer a valid term for GeoRef. Use slate deposits or slates as applicable. From 1978-1980, slates was autoposted to this term.

slate deposits (1981)
Before 1981, search slate AND deposits.
  IN Commodity. See List C.
  SA shale
  SA slates

slates (1978)
Autoposting of this term to slate ended in 1981. Before 1982, also search pelitic schist.
  UF pelitic schist
  BT metamorphic rocks
  SA shale
  SA slate deposits
  SA slaty cleavage

Slatina
  No longer a valid term for GeoRef. As of 1993 see Slatina Romania.

Slatina Romania (1993)
City in Olt County in SW Walachia, S Romania. Before 1993 also search Slatina AND Romania.
  BT Walachia
  BT Romania
  BT Southern Europe
  BT Europe

slaty cleavage (1978)
  BT foliation
  SA axial-plane structures
  SA cleavage
  SA flow cleavage
  SA metamorphism
  SA schistosity
  SA slates
  SA slip cleavage

Slave Point Formation (1989)
N Alberta, NE British Columbia and Northwest Territories.
  BT Devonian
  BT Paleozoic
  SA Alberta
  SA British Columbia
  SA Northwest Territories

Slave Province (1978)
Structural province of the Canadian Shield.
  CO N620000N690000
       W1050000W1170000
  BT Canadian Shield
  BT North America
  SA Northwest Territories

Slick Hills (1989)
SW Oklahoma.
  IN Index counties or region as applicable.
  SA Comanche County Oklahoma
  SA Kiowa County Oklahoma
  SA Oklahoma

slickensides (1978)
  UF polished surface
  SA breccia
  SA faults
  SA lineation

sliding, gravity
  use gravity sliding

Sligo
  No longer a valid term for GeoRef. See Sligo Ireland.

Sligo Formation (1985)
E and S Texas, NW Louisiana and S Arkansas.
  BT Lower Cretaceous
  BT Cretaceous
  BT Mesozoic
  SA Arkansas
  SA Louisiana
  SA Texas

Sligo Ireland (1993)
County and town in NW Ireland.
  BT Ireland
  BT Western Europe
  BT Europe

slip
As of 1993, use specific term. See faults or slip faults, earthquakes, rockslides, similar folds, slip cleavage, slip rates, or slope stability.

slip cleavage (1978)
Before 1978, search slip AND cleavage; shear cleavage; crenulation cleavage. Before 1993, also search crenulation AND cleavage.
  UF crenulation cleavage
  UF shear cleavage
  UF strain-slip cleavage
  BT foliation
  SA cleavage
  SA schistosity
  SA slaty cleavage

slip faults
Before 1993, search faults and displacements. As of 1993, use specific term. See diagonal-slip faults, dip-slip faults, flexural-slip folds, left-lateral faults, normal faults, oblique-slip faults, reverse faults, right-lateral faults, rockslides, similar folds, slip cleavage, slip rates, slope stability, stick-slip, strike-slip faults, or thrust faults.

slip folds
  use similar folds

slip rates (1989)
  SA displacements
  SA faults

Slocan mining camp (1978)
SE British Columbia. Also search Slocan.
  BT British Columbia
  BT Western Canada
  BT Canada
  SA mines

Slope County
Valid through 1988. Search in combination with state term. After 1988, use specific county-state term.

Slope County North Dakota (1989)
SW North Dakota. Before 1989, also search Slope County AND North Dakota.
  CO N461600N463900
       W1025400W1040300
  BT North Dakota
  BT United States

slope environment (1978)
Use this term or shelf environment for older sedimentation. For recent sedimentation, use continental slope or continental shelf. Before 1978, search slope.
  SA continental shelf
  SA continental slope
  SA deep-water environment
  SA depositional environment
  SA environment
  SA hemipelagic environment
  SA sedimentation
  SA shelf environment

slope stability (1978)
Used for geological studies on the engineering aspects of mass movements. For other aspects, see geomorphology.
  SA avalanches
  SA berms
  SA controls
  SA creep
  SA dams
  SA debris avalanches
  SA debris flows
  SA earthflows
  SA earthquakes
  SA embankments
  SA engineering geology
  SA erosion
  SA excavations
  SA explosions
  SA failures
  SA foundations
  SA friction angles
  SA geologic hazards
  SA geomembranes
  SA geomorphology
  SA geotextiles
  SA highways
  SA land subsidence
  SA landslides
  SA liquefaction
  SA liquefaction potential
  SA mass movements
  SA mudflows
  SA mudslides
  SA railroads
  SA reinforced materials
  SA reservoirs
  SA retaining walls
  SA revetments
  SA rock mechanics
  SA rockfalls
  SA shorelines
  SA site exploration
  SA slopes
  SA slumping
  SA soil erosion
  SA soil mechanics
  SA solifluction
  SA stability
  SA stabilization
  SA talus slopes
  SA tunnels
  SA underground installations

slope, continental
  use continental slope

slope, outer
  use outer slope

slopes (1978)
Autoposting of this term began in 1978.
  NT glacis
  SA bluffs
  SA cliffs
  SA continental slope
  SA cuestas
  SA embankments
  SA fault scarps
  SA geomorphology
  SA landslides
  SA scarps
  SA slope stability
  SA talus slopes

Slovakia (1978)
Region in E Czechoslovakia.
  CO N474500N495500
       E0223000E0160000
  UF Slovensko
  BT Czechoslovakia
  BT Central Europe
  BT Europe
  NT Banska Stiavnica
  NT Bojnico
  NT Bratislava
  NT Brezno
  NT Gemer
  NT Hodrusa
  NT Inner Slovakian Carpathians
  NT Korytnica
  NT Kremnica
  NT Kremnica Mountains
  NT Low Tatra Mountains
  NT Novaky
  NT Orava Valley
  NT Slovakian Karst
  NT Spis-Gemer
  NT Sturovo
  NT Vah Valley
  SA Beskid Mountains
  SA Choc Nappe
  SA Krizna Nappe
  SA Morava River valley
  SA Spis
  SA Tokaj-Eperjes Mountains
  SA Veporides

Slovakian Carpathians (1981)
Autoposting of Czechoslovakia to this term began in 1989. As of 1990, Central Europe and Europe are autoposted to this term. This term has multiple hierarchies.
  CO N480500N495500
       E0223000E0165000
  BT1 Carpathians
  BT1 Europe
  BT2 Czechoslovakia
  BT2 Central Europe
  BT2 Europe
  NT Inner Slovakian Carpathians
  SA Choc Nappe
  SA Krizna Nappe

SA Veporides

Slovakian Karst (1978)
Near the Hungarian border in SE Slovakia, E Czechoslovakia.
CO N482200N484400
E0210000E0202200
BT Slovakia
BT Czechoslovakia
BT Central Europe
BT Europe
SA Permian
SA Triassic

Slovakian Pannonian Basin (1981)
Autoposting of Czechoslovakia to this term began in 1989. As of 1990, Central Europe and Europe are autoposted to this term. This term has multiple hierarchies.
BT1 Pannonian Basin
BT1 Europe
BT2 Czechoslovakia
BT2 Central Europe
BT2 Europe

Slovenia (1978)
Constituent republic in NE Yugoslavia.
CO N451500N465000
E0170000E0133000
UF Slovenija
BT Yugoslavia
BT Southern Europe
BT Europe
NT Celje
NT Ljubljana
SA Adriatic region
SA Alps
SA Dinaric Alps
SA Istria
SA Julian Alps
SA Karawanken

Slovenija
use Slovenia

Slovenske Rudohorie
use Spis-Gemer

Slovensko
use Slovakia

sludge (1993)
SA drilling
SA liquid waste
SA mud
SA pollution
SA sewage sludge
SA solid waste
SA waste disposal

sludging
use solifluction

slump faults
Not a valid term for GeoRef. See normal faults or growth faults.

slump structures (1978)
BT soft sediment deformation
BT sedimentary structures
SA collapse structures
SA convoluted beds
SA slumping

slumping (1978)
UF slumps
SA continental shelf
SA continental slope
SA creep
SA erosion
SA geomorphology
SA gravity flows
SA landslides
SA liquefaction
SA listric faults
SA mass movements
SA sedimentation

SA slope stability
SA slump structures
SA soil erosion

slumps
use slumping

Slyudyanka
No longer a valid term for GeoRef. As of 1993, see Slyudyanka Russian Federation.

Slyudyanka Russian Federation (1993)
City at SW end of Lake Baikal in S Irkutsk Oblast. This term has multiple hierarchies.
BT1 Irkutsk Russian Federation
BT1 Russian Federation
BT1 Commonwealth of Independent States
BT2 Irkutsk Russian Federation
BT2 Asia

Sm
use samarium

Sm-147/Nd-144 (1989)
For age determination, use Sm/Nd. This term has multiple hierarchies.
BT1 radioactive isotopes
BT1 isotopes
BT2 stable isotopes
BT2 isotopes
BT3 neodymium
BT3 rare earths
BT3 metals
BT4 samarium
BT4 rare earths
BT4 metals
SA isotope ratios
SA Sm/Nd

Sm/Nd (1985)
Isotopic ratio used in age determination.
UF Nd/Sm
SA absolute age
SA isotope ratios
SA neodymium
SA samarium
SA Sm-147/Nd-144

Smackover Formation (1978)
Overlies Eagle Mills Formation; underlies Buckner Formation. Subsurface.
BT Upper Jurassic
BT Jurassic
BT Mesozoic
SA Alabama
SA Arkansas
SA Mississippi
SA Texas

Smaland (1978)
Plateau region S of Lake Vattern in S Sweden.
IN Index counties as applicable.
BT Sweden
BT Scandinavia
BT Western Europe
BT Europe
SA Kalmar Sweden

small mines (1985)
Refers to size of operation.
UF small-scale mining
BT mines
SA mining geology

small-scale mining
use small mines

smaragd
use emerald

Smartville Complex (1989)
E California.
BT Jurassic

BT Mesozoic
SA California

smectite (1978)
BT clay minerals
BT sheet silicates
BT silicates
SA fuller's earth
SA montmorillonite
SA saponite

Smilodon (1989)
Pliocene-Pleistocene genus in North and South America belonging to Family Felidae, Suborder Fissipeda. Also a Triassic genus of Europe belonging to Suborder Theropoda, Order Saurischia.
IN Index fossil names as applicable.
BT Tetrapoda
BT Vertebrata
BT Chordata
SA Felidae
SA Theropoda

Smith County
Valid through 1988. Search in combination with state term. After 1988, use specific county-state term.

Smith County Kansas (1989)
N Kansas. Before 1989, also search Smith County AND Kansas.
CO N393300N400000
W0983000W0990600
BT Kansas
BT United States

Smith County Mississippi (1989)
S central Mississippi. Before 1989, also search Smith County AND Mississippi.
CO N314700N321300
W0892100W0894400
BT Mississippi
BT United States

Smith County Tennessee (1989)
N central Tennessee. Before 1989, also search Smith County AND Tennessee.
CO N360600N362500
W0854800W0860800
BT Tennessee
BT United States

Smith County Texas (1989)
E Texas. Before 1989, also search Smith County AND Texas.
CO N320800N324200
W0945500W0953400
BT Texas
BT United States

Smithers
No longer a valid term for GeoRef. See Smithers British Columbia.

Smithers British Columbia (1993)
Village in W central British Columbia.
BT British Columbia
BT Western Canada
BT Canada

Smithian (1985)
North America. Lower Triassic, above Dienerian and below Spathian.
BT Lower Triassic
BT Triassic
BT Mesozoic

Smithsonian Institution (1978)
SA District of Columbia

SA museums

smithsonite (1978)
UF zinc spar
BT carbonates
SA azurite

Smithville Formation (1978)
Above Powell Formation and below Black Rock Limestone in N Arkansas, and below Everton Formation in SE Missouri. N Arkansas and SE Missouri.
BT Lower Ordovician
BT Ordovician
BT Paleozoic
SA Arkansas
SA Missouri

Smithwick Shale (1985)
Lower and Middle Pennsylvanian. In Bend Group. Central Texas.
BT Pennsylvanian
BT Carboniferous
BT Paleozoic
SA Texas

Smokies
use Great Smoky Mountains

Smoky Hill Chalk Member (1978)
Of Niobrara Formation. E Colorado, W Kansas, NE New Mexico, and SE South Dakota.
BT Upper Cretaceous
BT Cretaceous
BT Mesozoic
SA Colorado
SA Kansas
SA New Mexico
SA Niobrara Formation
SA South Dakota

Smoky Hill River basin (1978)
W central and central Kansas. Also search Smoky Hill River.
IN Index counties as applicable.
BT Kansas
BT United States

smoky quartz (1989)
BT silica minerals
BT framework silicates
BT silicates
SA quartz

Smoky River (1978)
Rises in Jasper National Park and flows NNE into Peace River.
BT Alberta
BT Western Canada
BT Canada

Smolensk
No longer a valid term for GeoRef. As of 1993, see Smolensk Russian Federation.

Smolensk Russian Federation (1993)
Oblast including the city on left bank of upper Dnieper River SW of Moscow. This term has multiple hierarchies.
BT1 Russian Federation
BT1 Commonwealth of Independent States
BT2 Europe

smythite (1978)
BT sulfides

Sn
use tin

Sn-waves (1989)
Refracted along Mohorovicic discontinuity.
BT body waves
BT elastic waves

SA continental crust
SA Mohorovicic discontinuity
SA Pn-waves
SA S-waves
SA seismology
SA upper mantle
SA waves

Snaefellsnes Peninsula (1978)
Juts into Denmark Strait between Breidi Fjord and Faxa Bay in W Iceland. Also search Snaefellsnes.
BT Iceland
BT Western Europe
BT Europe

Snake Range (1978)
Mountain range in E Nevada. Autoposting of broader terms to this term began in 1989.
BT White Pine County Nevada
BT Nevada
BT United States

Snake River (1978)
Rises in Yellowstone National Park and flows S, then W, N, and again W into the Columbia River in SE Washington. Also occurs in other location.
IN Index states as applicable.
SA Clearwater River
SA Columbia River
SA Idaho
SA Oregon
SA United States
SA Washington
SA Wyoming

Snake River basin (1978)
IN Index states as applicable.
SA Idaho
SA Nevada
SA Oregon
SA United States
SA Utah
SA Washington
SA Wyoming

Snake River canyon (1978)
Grand Canyon of the Snake River. Extends N-S between Wallowa Mountains, Oregon, and Seven Devils Mountains, Idaho.
IN Index counties and states as applicable.
UF Hell's Canyon
SA Grand Canyon
SA Idaho
SA Oregon
SA United States

Snake River plain (1978)
Crescent-shaped lava tableland across S central Idaho.
CO N421500N441500
W1113000W1160000
BT Idaho
BT United States

SNC Meteorites (1989)
Type of meteorite. Shergottites, nakhlites and chassignites.
BT achondrites
BT stony meteorites
BT meteorites
SA chassignite
SA Chassigny Meteorite
SA Nakhla Meteorite
SA nakhlite
SA shergottite
SA Shergotty Meteorite

Snieznik (1978)
Highest peak in Kralicky Sneznik Mountains in NE Bohemia and Lower Silesia.

IN Index countries as applicable.
BT Europe
SA Bohemia
SA Czechoslovakia
SA Lower Silesia
SA Poland

Snohomish County
Valid through 1988. Search in combination with state term. After 1988, use specific county-state term.

Snohomish County Washington (1989)
NW Washington. Before 1989, also search Snohomish County AND Washington.
CO N474700N481700
W1205500W1222500
BT Washington
BT United States
SA Glacier Peak

Snoqualmie Batholith (1978)
Probably intermediate in age between lower part of Keechelus Andesitic Series and Fifes Peak Andesite. W central Washington.
IN Index ages as applicable.
BT Tertiary
BT Cenozoic
SA Miocene
SA Pliocene
SA Washington

snow (1978)
BT atmospheric precipitation
SA firn
SA glacial geology
SA glaciers
SA hydrology
SA hydrosphere
SA ice
SA ice fields
SA meltwater
SA thawing
SA water

Snow Lake
No longer a valid term for GeoRef. See Snow Lake Manitoba.

Snow Lake Manitoba (1993)
Village in gold-mining region 70 miles E of Flin Flon in W Manitoba.
BT Manitoba
BT Western Canada
BT Canada

snow-patch erosion
use nivation

Snowdon
use Snowdonia

Snowdonia (1978)
Highest mountain in Wales, in Caernarvonshire (Gwynedd) 10 miles SE of Caernarvon; consists of 5 peaks, separated by passes.
UF Snowdon
BT Caernarvonshire Wales
BT Wales
BT Great Britain
BT United Kingdom
BT Western Europe
BT Europe

Snowslip Formation (1989)
In Missoula Group of Belt Supergroup. W Montana.
BT middle Proterozoic
BT Proterozoic
BT upper Precambrian
BT Precambrian
SA Belt Supergroup
SA Missoula Group

SA Montana

Snowy Mountains (1978)
Range of the Australian Alps.
IN Index states or region as applicable.
BT Australia
BT Australasia
SA New South Wales Australia
SA Victoria Australia

Snuggedy Swamp (1985)
E South Carolina.
BT South Carolina
BT United States
SA Atlantic Coastal Plain

SO2
use sulfur dioxide

soap clay
use bentonite

soapstone (1978)
IN May be used as a commodity term with talc deposits.
SA steatite
SA talc
SA talc deposits

Sob River basin (1978)
Near border of former Moldavian USSR in SW Ukraine. Also search Sob. This term has multiple hierarchies.
BT1 Ukraine
BT1 Europe
BT2 Ukraine
BT2 Commonwealth of Independent States

Sobotka (1978)
Town in N Bohemia, W Czechoslovakia.
BT Bohemia
BT Czech Republic
BT Czechoslovakia
BT Central Europe
BT Europe

Sobral Formation (1993)
Maestrichtian? and Tertiary. James Ross Island and Seymour Island, Antarctica. Overlies Lopez de Bertodano Formation.
IN Index ages as applicable.
BT Tertiary
BT Cenozoic
SA Antarctica
SA Cretaceous
SA James Ross Island
SA Paleocene
SA Seymour Island

Sochi
No longer a valid term for GeoRef. As of 1993, see Sochi Russian Federation.

Sochi Russian Federation (1993)
City in S Krasnodar Kray on Black Sea near Georgian border. This term has multiple hierarchies.
BT1 Krasnodar Russian Federation
BT1 Russian Federation
BT1 Commonwealth of Independent States
BT2 Krasnodar Russian Federation
BT2 Europe

societies
Not a valid term for GeoRef. As of 1993, see associations.

Society Islands (1978)
Tahiti is a narrower term as of 1981. Autoposting of this term began in 1978.

CO S190000S150000
W1470000W1550000
BT French Polynesia
BT Polynesia
BT Oceania
NT Tahiti
SA Leeward Islands

Socompa (1989)
Volcano. On the eastern border of Antofagasta, N Chile.
CO S243000S242000
W0681000W0682000
BT Antofagasta Chile
BT Chile
BT South America

Socorro
Valid through 1988. Search in combination with state term. After 1988, use specific city-state term.

Socorro County
Valid through 1988. Search in combination with state term. After 1988, use specific county-state term.

Socorro County New Mexico (1989)
Central New Mexico. Before 1989, also search Socorro County AND New Mexico.
CO N332500N343500
W1055500W1074200
BT New Mexico
BT United States
NT Magdalena New Mexico
NT Socorro New Mexico
SA Jornada del Muerto
SA Magdalena Mountains
SA Orogrande Basin
SA Rio Puerco
SA San Mateo Mountains

Socorro New Mexico (1989)
City on the Rio Grande in W central New Mexico. Before 1989, search Socorro AND New Mexico.
CO N340400N340400
W1065500W1065500
BT Socorro County New Mexico
BT New Mexico
BT United States

Sod-podzolic soils (1993)
Also search Sod-Podzols.
BT soils
SA Podzols
SA soil group

soda ash
use sodium carbonate

sodalite (1978)
BT sodalite group
BT framework silicates
BT silicates

sodalite group (1978)
Autoposting of this term to lazurite and hauyne began in 1981.
BT framework silicates
BT silicates
NT hauyne
NT hydrosodalite
NT lazurite
NT nosean
NT sodalite
NT tugtupite
SA feldspathoid rocks

sodium (1978)
Autoposting of alkali metals and metals to this term began in 1989.
UF Na
UF natrium
BT alkali metals
BT metals
NT Na-22

NT Na-24
SA lithophile elements
SA sodium ion

sodium bicarbonate
For minerals, use mineral name e.g. nahcolite or wegscheiderite if applicable. See carbonates. Use reagents if applicable. Also see sodium ion and bicarbonate ion. See trona for economic studies.

sodium carbonate (1978)
IN See list C for use as a commodity term.
UF soda ash
SA carbonates
SA trona

sodium chloride (1978)
SA chlorides
SA evaporites
SA geochemistry
SA halides
SA halite
SA salt

sodium hydroxide (1993)
Reagent. See hydroxides for minerals.
UF NaOH
SA alkalinity
SA geochemistry
SA hydroxyl ion
SA sodium ion

sodium ion (1985)
SA geochemistry
SA ions
SA sodium
SA sodium hydroxide

sodium sulfate (1978)
IN See list C for use as a commodity term.
SA sulfates

Sofia
No longer a valid term for GeoRef. See Sofia Bulgaria.

Sofia Bulgaria (1993)
Province and city in W Bulgaria. Also search Sofiya and Sophia.
BT Bulgaria
BT Southern Europe
BT Europe
NT Rila Bulgaria
NT Yetropole Bulgaria

Sofiya
Not a valid term for GeoRef. See Sofia Bulgaria.

soft clays (1981)
SA clay
SA clays

soft coal
use bituminous coal

soft rocks (1993)
General term. Restricted to engineering geology. See sedimentary and other rock types.
SA engineering geology
SA mining geology
SA rocks
SA weak rocks

soft sediment deformation (1978)
Autoposting of this term began in 1978.
BT sedimentary structures
NT ball-and-pillow
NT boudinage
NT clastic dikes
NT convoluted beds
NT flame structures
NT flow structures
NT olistoliths
NT olistostromes
NT sandstone dikes
NT slump structures
SA casts
SA flute casts
SA load casts
SA sole marks
SA tool marks

soft soils
As of 1993, see strength and soil mechanics.

software
use computer programs

Sogn (1978)
Mountain region around Sogne Fjord in SW Norway.
BT Norway
BT Scandinavia
BT Western Europe
BT Europe

Sogndal
No longer a valid term for GeoRef. As of 1993 see Sogndal Norway.

Sogndal Norway (1993)
Village on North Sea. Before 1993 also search Sogndal AND Norway.
BT Rogaland Norway
BT Norway
BT Scandinavia
BT Western Europe
BT Europe

Sohm abyssal plain (1985)
NW Atlantic Ocean. As of 1990, Atlantic Ocean is autoposted to this term.
BT North American Atlantic
BT North Atlantic
BT Atlantic Ocean

soil dynamics (1981)
SA dynamics
SA soil mechanics
SA soils

soil erosion (1982)
Before 1982, search soils AND erosion.
BT erosion
SA erosion control
SA erosion rates
SA furrows
SA geomorphology
SA landform evolution
SA landslides
SA mudflows
SA piping
SA rills
SA sheet erosion
SA slope stability
SA slumping
SA soils
SA talus slopes
SA Universal Soil Loss Equation
SA water erosion
SA weathering
SA wind erosion

soil flow
use solifluction

soil fluction
use solifluction

soil gases (1993)
Used for soil and water pollution studies as well as petroleum exploration and mineral exploration.
SA chemical analysis
SA earthquake prediction
SA gases
SA geochemical methods
SA pollution
SA radon emanometry
SA soils
SA volatile organic compounds

soil group (1978)
SA Alfisols
SA Alluvial soils
SA Andosols
SA Aridisols
SA Bog soils
SA Brown forest soils
SA Brown soils
SA Calcareous soils
SA Chernozems
SA Chestnut soils
SA Clay soils
SA Desert soils
SA Entisols
SA Ferralites
SA Ferruginous soils
SA Gleys
SA Gray forest soils
SA Halomorphic soils
SA Histosols
SA Hydromorphic soils
SA Inceptisols
SA Intrazonal soils
SA laterites
SA Latosols
SA Luvisols
SA Meadow soils
SA Mediterranean soils
SA Mollisols
SA Oxisols
SA Paleosols
SA Planosols
SA Podzols
SA Pseudogleys
SA Red soils
SA Rendzinas
SA Sierozems
SA Sod-podzolic soils
SA soils
SA Solonchak soils
SA Solonetz soils
SA Spodosols
SA Terra rossa
SA Tundra soils
SA Ultisols
SA Vertisols
SA Zonal soils

soil horizon
use horizons

soil liners
use disposal barriers

soil management (1978)
SA fertilizers
SA management
SA soil treatment
SA soils
SA tillage

soil mechanics (1978)
Used for geotechnical studies.
SA alluvium
SA AMCOR
SA Atterberg limits
SA bearing capacity
SA boundary conditions
SA bulk density
SA California bearing ratio
SA case studies
SA clay
SA coarse-grained materials
SA cohesive materials
SA compaction
SA compactness
SA competent materials
SA compressibility
SA compressive strength
SA cone penetration tests
SA confining pressure
SA congelifraction
SA consolidation
SA consolidometer tests
SA constitutive equations
SA cryoturbation
SA dams
SA deformation
SA durability
SA dynamic loading
SA earth pressure
SA earthquakes
SA elastic constants
SA elasticity
SA engineering geology
SA engineering properties
SA expansive materials
SA explosions
SA finite strain analysis
SA foundations
SA friction angles
SA frost action
SA frozen ground
SA gelifluction
SA geologic hazards
SA geomembranes
SA geotechnical maps
SA geotextiles
SA granular materials
SA highways
SA hydraulics
SA interlaboratory comparison
SA land subsidence
SA lateral loading
SA liquefaction
SA load tests
SA loading
SA loess
SA marine installations
SA materials
SA mechanical properties
SA mechanics
SA nuclear facilities
SA overconsolidated materials
SA penetration
SA penetration tests
SA penetrometers
SA pipelines
SA plastic deformation
SA plastic materials
SA plate-bearing tests
SA pore pressure
SA quick clay
SA quicksand
SA railroads
SA reinforced materials
SA reservoirs
SA retaining walls
SA rock mechanics
SA sand boils
SA saturated materials
SA seepage
SA seismic response
SA settlement
SA shear strength
SA shear tests
SA site exploration
SA slope stability
SA soil dynamics
SA soil-structure interface
SA soils
SA stiff clays
SA testing
SA thixotropy
SA triaxial tests
SA tunnels
SA unconsolidated materials
SA underground installations
SA vane tests
SA water pressure
SA weathered materials

soil profiles (1981)
Before 1981, search soils AND profiles.
SA pedons
SA soils

soil sampling (1978)

Before then, variants of the term may occur in GeoRef.

SA field studies
SA Filicopsida
SA mineral exploration
SA pedons
SA sample preparation
SA sampling
SA soils

soil treatment (1985)
Before 1985, search treatment AND soils.
SA bioremediation
SA desalinization
SA fertilization
SA irrigation
SA reclamation
SA remediation
SA salinity
SA sewage sludge
SA soil management
SA soils

soil ulmin
use humus

soil zone
use horizons

soil-structure interface (1985)
UF structure-soil interface
SA backfill
SA engineering geology
SA foundations
SA interfaces
SA seismic response
SA soil mechanics

soils (1978)
Used for general pedology as well as for specific topics. Autoposting of this term began in 1978.
NT acid sulfate soils
NT Alfisols
NT Alluvial soils
NT Andosols
NT Aridisols
NT Bog soils
NT Brown forest soils
NT Brown soils
NT Calcareous soils
NT Cambisols
NT Chernozems
NT Chestnut soils
NT Clay soils
NT Cryosols
NT Desert soils
NT Entisols
NT Ferralites
NT Ferralsols
NT Ferruginous soils
NT fragipans
NT Gleys
NT Gray forest soils
NT Halomorphic soils
NT Histosols
NT Hydromorphic soils
NT Inceptisols
NT Intrazonal soils
NT laterites
NT Latosols
NT loam
NT Luvisols
NT Meadow soils
NT Mediterranean soils
NT Mollisols
NT Mor
NT Mull
NT Oxisols
NT Paleosols
NT Planosols
NT Podzols
NT Pseudogleys
NT Red soils
NT Regosols
NT Rendzinas
NT Sierozems
NT Sod-podzolic soils

NT Solonchak soils
NT Solonetz soils
NT Spodosols
NT Terra rossa
NT Tundra soils
NT Ultisols
NT Vertisols
NT Zonal soils
SA active layer
SA agriculture
SA alluvium
SA background level
SA bacteria
SA bedrock
SA biorhexistasy
SA biota
SA bulk density
SA calcification
SA calcrete
SA caliche
SA capillarity
SA capillary water
SA catenas
SA characterization
SA chemical composition
SA clay mineralogy
SA climate
SA coarse-grained materials
SA colluvium
SA color
SA compactness
SA conservation
SA creep
SA degradation
SA desalinization
SA desiccation
SA disposal barriers
SA drainage
SA duricrust
SA ecology
SA eluvium
SA engineering geology
SA engineering properties
SA environmental geology
SA erosion control
SA erosion rates
SA expansive materials
SA factors
SA fertilization
SA fertilizers
SA field capacity
SA field studies
SA fines
SA forestry
SA forests
SA foundations
SA frozen ground
SA fruits
SA fuller's earth
SA geomorphology
SA gilgai
SA glacial geology
SA grain size
SA gyttja
SA highways
SA horizon differentiation
SA horizons
SA humus
SA infiltration
SA ion exchange
SA irrigation
SA land use
SA landslides
SA Langmuir equation
SA laterization
SA leachate
SA leaching
SA lunar soils
SA lysimeters
SA micromorphology
SA microorganisms
SA moisture
SA morphology
SA movement
SA nitrogen

SA nutrients
SA organic materials
SA parent materials
SA particles
SA pedogenesis
SA pedons
SA permafrost
SA phosphorus
SA physical methods
SA physical properties
SA pit sections
SA podzolization
SA reclamation
SA recreation
SA regolith
SA Richards equation
SA salinity
SA sample preparation
SA sand
SA saturated zone
SA saturation
SA sediments
SA seepage
SA soil dynamics
SA soil erosion
SA soil gases
SA soil group
SA soil management
SA soil mechanics
SA soil profiles
SA soil sampling
SA soil treatment
SA soils maps
SA solifluction
SA storage
SA surficial geology
SA surveys
SA tillage
SA time factor
SA topography
SA transformations
SA tundra
SA unconsolidated materials
SA Universal Soil Loss Equation
SA unsaturated zone
SA vegetation
SA waste disposal
SA water erosion
SA water regimes
SA waterlogging
SA weathering
SA weathering crust
SA wind erosion
SA yields

soils maps (1981)
BT maps
SA soils

Sokh
No longer a valid term for GeoRef. As of 1993, see Sokh Uzbekistan. Valid 1978-1992.

Sokh Uzbekistan (1993)
Village S of Kokand near Kirghiz border in E Uzbekistan. This term has multiple hierarchies.
BT1 Uzbekistan
BT1 Asia
BT2 Uzbekistan
BT2 Commonwealth of Independent States

Soko Banja (1978)
Village near Nis in E Serbia, E Yugoslavia.
UF Soko Banya
BT Serbia
BT Yugoslavia
BT Southern Europe
BT Europe

Soko Banya
use Soko Banja

Sokolov (1978)
Town on Ohre River in W Bohemia, W Czechoslovakia.
UF Falkenau
UF Falknov
BT Bohemia
BT Czech Republic
BT Czechoslovakia
BT Central Europe
BT Europe

Sokolov Basin (1978)
W Bohemia. Also search Sokolov.
CO N500200N502000
 E0125800E0121500
BT Bohemia
BT Czech Republic
BT Czechoslovakia
BT Central Europe
BT Europe

Sokoman Formation (1978)
In Knob Lake Group.
UF Sokoman Iron Formation
BT Precambrian
SA Canada
SA Labrador
SA Quebec

Sokoman Iron Formation
use Sokoman Formation

Sokoto Basin (1978)
River basin in N Nigeria. Also search Sokoto.
BT Nigeria
BT West Africa
BT Africa

Solano County
Valid through 1988. Search in combination with state term. After 1988, use specific county-state term.

Solano County California (1989)
Central California. Before 1989, also search Solano County AND California.
CO N380200N383200
 W1213500W1222300
BT California
BT United States
SA San Francisco Bay region

solar activity (1978)
Term includes use under a variety of topics as a cause of whatever is being discussed in the document. Before 1993, also search solar flares if applicable. Before 1978, search flares if applicable.
UF solar flares
SA Sun
SA sunspots

solar cycles (1978)
SA cycles
SA intensity
SA interplanetary space
SA solar wind
SA Sun

solar energy (1978)
SA energy
SA energy sources
SA new energy sources
SA Sun

solar flares
As of 1993, no longer a valid term for GeoRef. Usually out-of-scope.
use solar activity

Solar Lake (1985)
Coastal salt lake, SE Sinai, Egypt.
BT Egypt
BT North Africa
BT Africa
SA Sinai

solar nebula (1978)

UF nebula, solar
SA Sun

solar physics, astrophysics and
Not a valid term for GeoRef. Usually out-of-scope. See astrophysics and solar physics.

solar system (1978)
See also names of planets.
SA asteroids
SA comets
SA cosmochronology
SA Earth
SA Earth-Moon couple
SA extraterrestrial geology
SA Halley's Comet
SA interplanetary space
SA Jupiter
SA Mars
SA Mercury Planet
SA meteorites
SA Moon
SA Neptune
SA outer planets
SA planetology
SA planets
SA Pluto
SA Saturn
SA Sun
SA terrestrial planets
SA Uranus
SA Venus

solar wind (1978)
UF wind, solar
SA comae
SA comets
SA electromagnetic radiation
SA interplanetary space
SA magnetic field
SA Moon
SA particles
SA solar cycles
SA Sun

sole faults
   use detachment faults

sole markings
A valid term through 1971.
   use sole marks

sole marks (1978)
Before 1972, also search sole markings.
IN Also index bedding plane irregularities and/or turbidity current structures as applicable.
UF sole markings
BT sedimentary structures
SA bedding plane irregularities
SA flute casts
SA glacial features
SA glacial geology
SA load casts
SA soft sediment deformation
SA turbidity current structures

Solemyida (1981)
BT Bivalvia
BT Mollusca
BT Invertebrata

Solenhofen
Not a valid term for GeoRef. See Solnhofen Germany.

Solenhofen Limestone
   use Solnhofen Limestone

solfataras (1978)
BT fumaroles
SA gases
SA sublimates
SA vents
SA volcanism

solid phase (1978)
UF phase, solid
UF solids
SA fluid phase
SA gaseous phase
SA geochemistry
SA liquid phase
SA solid solution

solid solution (1978)
UF mixed crystals
SA phase equilibria
SA solid phase
SA solution
SA solutions

solid waste (1978)
UF waste, solid
SA engineering geology
SA environmental geology
SA industrial waste
SA liquid waste
SA radioactive waste
SA Resource Conservation and Recovery Act
SA sewage sludge
SA sludge
SA waste disposal
SA waste disposal sites

solidification
   Use consolidation if applicable.

solids
   use solid phase

solifluction (1978)
UF sludging
UF soil flow
UF soil fluction
UF solifluxion
SA creep
SA engineering geology
SA glacial geology
SA periglacial features
SA permafrost
SA slope stability
SA soils

solifluxion
   use solifluction

Solikamsk
No longer a valid term for GeoRef. As of 1993, see Solikamsk Russian Federation.

Solikamsk Russian Federation (1993)
City in Perm Oblast W of the Urals. This term has multiple hierarchies.
BT1 Perm Russian Federation
BT1 Russian Federation
BT1 Commonwealth of Independent States
BT2 Perm Russian Federation
BT2 Europe

solitary waves
As of 1981, no longer a valid term for GeoRef. Term was introduced in 1978. Before 1978, search ocean waves AND solitary.

Solnhofen
No longer a valid term for GeoRef. See Solnhofen Germany.

Solnhofen Germany (1993)
Village in W central Bavaria. Also search Solenhofen.
BT Bavaria Germany
BT Germany
BT Central Europe
BT Europe

Solnhofen Limestone (1978)
In Bavaria Germany. Also search Solenhofen; Solnhofen.
UF Solenhofen Limestone
BT Jurassic
BT Mesozoic
SA Bavaria Germany
SA Germany

Solomon Islands (1978)
Group of islands E of New Guinea in the SW Pacific Ocean. Bougainville, Buka, and Green Islands in the W are part of Papua New Guinea while the remaining 10 large islands and 4 groups of small islands remained a protectorate of the United Kingdom until 1978 when they became independent. Malaita is a narrower term as of 1981. Bougainville is a narrower term beginning in 1993.
CO S120000S063000 E1630000E1550000
BT Melanesia
BT Oceania
NT Bougainville
NT Malaita
SA Australasia
SA Pacific Ocean
SA Papua New Guinea
SA Santa Isabel

Solomon Sea (1978)
Enclosed on the W by New Guinea, on the N by the Bismarck Archipelago, on the E by the Solomon Islands and on the S by the Coral Sea.
BT Pacific Ocean
SA Woodlark Basin

Solonchak soils (1978)
Saline.
UF saline soils
BT soils
SA Intrazonal soils
SA salinity
SA soil group
SA Solonetz soils

Solonetz soils (1978)
BT soils
SA Intrazonal soils
SA soil group
SA Solonchak soils

solongoite (1978)
BT borates

solubility (1978)
SA concentration
SA dissolved materials
SA geochemistry
SA immiscibility
SA insoluble residues
SA properties
SA saturation
SA sea water
SA solution

Solund Islands (1978)
Island group in the North Sea at mouth of Sogne Fjord. Also search Solund.
BT Norway
BT Scandinavia
BT Western Europe
BT Europe

solutes (1978)
SA dissolved materials
SA geochemistry
SA solution

solution (1978)
Before 1978, also search dissolution.
SA adsorption
SA aqueous solutions
SA carbonate compensation depth
SA caves
SA concentration
SA electrolysis
SA embayments
SA exsolution
SA geochemistry
SA karstification
SA leaching
SA lysoclines
SA pits
SA pressure solution
SA solid solution
SA solubility
SA solutes
SA solution features
SA solution transport
SA solutions

solution cavities (1978)
Autoposting of solution features to this term began in 1989.
UF cavities, solution
BT solution features
SA caves
SA geomorphology
SA karst
SA sedimentary structures
SA underground cavities

solution features (1978)
Before 1969, also search solution phenomena. Autoposting of this term began in 1978.
UF features, solution
UF solution phenomena
NT caverns
NT karren
NT karst
NT paleokarst
NT poljes
NT sinkholes
NT solution cavities
NT speleology
NT speleothems
SA canyons
SA cave environment
SA caves
SA geomorphology
SA gorges
SA karst filling
SA karst hydrology
SA land subsidence
SA natural curiosities
SA pits
SA solution
SA stalactites
SA stalagmites
SA tors

solution mining (1978)
BT mining
SA leaching
SA mines
SA mining geology

solution phenomena
   use solution features

solution transport (1993)
Also search transport AND solution.
SA advection
SA concentration
SA exsolution
SA karstification
SA leaching
SA marine transport
SA ore transport
SA solution
SA stream transport

solutions (1978)
Used for material or mixture. Singular form used to indicate process.
SA aqueous solutions
SA geochemistry
SA leachate
SA reagents
SA solid solution
SA solution

Somali Basin (1978)
E of the Somali Republic and N of the Seychelles.
BT Indian Ocean
SA DSDP Site 236

Somali Republic (1978)
Comprises former British Somaliland and Trust Territory of Somalia (formerly Italian Somaliland). Before 1969, also search Somaliland; British Somaliland.
CO S014000N120000
  E0513000E0410000
UF British Somaliland
UF Somalia
UF Somaliland
BT East Africa
BT Africa

Somalia
use Somali Republic

Somaliland
No longer a valid GeoRef index term. Before 1969, was used on level 1 in subfile E.
use Somali Republic

Somasteroidea (1978)
BT Stelleroidea
BT Asterozoa
BT Echinodermata
BT Invertebrata

Somerset
No longer a valid term for GeoRef as of 1993. See Somerset England.

Somerset County
Valid through 1988. Search in combination with state term. After 1988, use specific county-state term.

Somerset County Maine (1989)
Central and W Maine. Before 1989, also search Somerset County AND Maine.
CO N443400N463300
  W0694000W0703300
BT Maine
BT United States
SA Saint John River

Somerset County Maryland
  (1989)
SE Maryland. Before 1989, also search Somerset County AND Maryland.
CO N375300N381800
  W0753300W0760300
BT Maryland
BT United States

Somerset County New Jersey
  (1989)
Bounded by Passaic River in N central New Jersey. Before 1989, also search Somerset County AND New Jersey.
CO N402200N404600
  W0742400W0744800
BT New Jersey
BT United States
SA Watchung Mountains

Somerset County Pennsylvania
  (1989)
SW Pennsylvania. Before 1989, also search Somerset County AND Pennsylvania.
CO N394300N401700
  W0783800W0792500
BT Pennsylvania
BT United States

Somerset England (1993)
County in SW England. In 1974, N part was taken to form Avon County. Before 1993, also search Somerset AND England.
CO N505000N512000
  W0021500W0035000
BT England
BT Great Britain
BT United Kingdom
BT Western Europe
BT Europe
NT Mendip Hills
SA Avon England
SA Bath England

Somerset Island (1978)
In central Franklin District N of Boothia Peninsula.
CO N720000N740000
  W0900000W0960000
BT Franklin District Northwest Territories
BT Northwest Territories
BT Western Canada
BT Canada
NT Fort Ross Northwest Territories
SA Boothia Peninsula

Somes Basin (1978)
River basin in NE Hungary and NW Romania. Also search Somes; Somes River.
IN Index countries as applicable.
UF Somesul Basin
BT Europe
SA Hungary
SA Romania

Somesul Basin
use Somes Basin

Somme
No longer a valid term for GeoRef. See Somme France.

Somme France (1993)
Department in N France.
CO N493000N503000
  E0031000E0011500
BT France
BT Western Europe
BT Europe
SA Somme River valley

Somme River valley (1978)
IN Index departments as applicable.
UF Somme Valley
BT France
BT Western Europe
BT Europe
SA Aisne France
SA Somme France

Somme Valley
use Somme River valley

Son Valley (1978)
River valley. Also search Son River.
IN Index states as applicable.
BT India
BT Indian Peninsula
BT Asia
SA Bihar India
SA Madhya Pradesh India
SA Narmada-Son Lineament

sonar methods (1978)
Before 1978, search sonar.
SA acoustical methods
SA acoustical surveys
SA borehole televiewers
SA echo sounding
SA geophysical methods
SA geophysical surveys
SA GLORIA
SA methods
SA radar methods
SA remote sensing
SA Seabeam
SA side-scanning methods
SA sounding

Sondre Strom Fjord (1978)
Inlet of Davis Strait on Arctic Circle in SW Greenland.
BT Greenland
BT Arctic region

Sondrio
No longer a valid term for GeoRef. See Sondrio Italy.

Sondrio Italy (1993)
City and province in N Lombardy, N Italy.
BT Lombardy Italy
BT Italy
BT Southern Europe
BT Europe

Songhor
No longer a valid term for GeoRef. See Songhor Iran.

Songhor Iran (1993)
Village in Kermanshahan, W Iran. Also search Sunqur and Sonqor.
BT Kerman Iran
BT Iran
BT Middle East
BT Asia

Songliao Basin (1985)
NE China.
IN Index provinces as applicable.
BT China
BT Far East
BT Asia
SA Nenjiang Formation
SA Qingshankou Formation
SA Quantou Formation

sonic logging
Not a valid term for GeoRef. See acoustical logging.

sonic waves
use acoustical waves

sonobuoys (1978)
SA geophysical methods
SA geophysical surveys
SA marine methods
SA refraction methods
SA seismic methods

Sonoma County
Valid through 1988. Search in combination with state term. After 1988, use specific county-state term.

Sonoma County California
  (1989)
W California. Before 1989, also search Sonoma County AND California.
CO N380700N385000
  W1222200W1233500
BT California
BT United States
NT Bodega Head
NT Santa Rosa California
SA San Francisco Bay region
SA The Geysers

Sonora
No longer a valid term for GeoRef. As of 1993 see Sonora Mexico.

Sonora Mexico (1993)
State in NW Mexico. Before 1993 also search Sonora AND Mexico.
CO N263000N323000
  W1083000W1150500
BT Mexico
NT Cananea Mexico
NT Moctezuma Mexico
NT Nacozari de Garcia Mexico
SA Colorado River
SA Colorado River delta
SA Pedregosa Basin
SA San Antonio Mine
SA San Jacinto Fault
SA Sierra Madre Occidental
SA Sonoran Desert

Sonoran Desert (1978)
IN Index Sonora Mexico and U.S. states as applicable.
BT North America
SA Arizona
SA California
SA Mexico
SA Sonora Mexico

Sonqor
Not a valid term for GeoRef. See Sonqhor Iran.

Sonyea Group (1978)
Includes Middlesex Shale, Pultenay Shale, Rock Stream Siltstone, and Cashaqua Shale members. W New York.
BT Upper Devonian
BT Devonian
BT Paleozoic

Sophia
Not a valid term for GeoRef. See Sofia Bulgaria.

Sor Rondane Mountains
use Sor-Rondane Mountains

Sor-Rondane Mountains (1978)
In Queen Maud Land near Princess Ragnhild Coast in Norwegian Sector on Atlantic Ocean side. Also search Sor-Rondane.
UF Sor Rondane Mountains
BT Antarctica
SA Queen Maud Land

Sor-Trondelag
No longer a valid term for GeoRef. As of 1993 see Sor-Trondelag Norway.

Sor-Trondelag Norway (1993)
County S of Trondheim Fjord in central Norway. Before 1993 also search Sor-Trondelag AND Norway.
BT Norway
BT Scandinavia
BT Western Europe
BT Europe
NT Trondheim Norway
SA Trondelag

Sorachi (1978)
River in W central Hokkaido.
BT Hokkaido
BT Japan
BT Far East
BT Asia

Soria
No longer a valid term for GeoRef. As of 1993, see Soria City Spain.

Soria City Spain (1993)
Refers to only the city in N central Spain. Before 1993, also search Soria and Spain.
BT Soria Spain
BT Old Castile Spain
BT Castile Spain
BT Spain
BT Iberian Peninsula
BT Southern Europe
BT Europe

Soria Province
No longer a valid term for GeoRef. As of 1993, see Soria Spain.

Soria Spain (1993)
Refers only to the province in N central Spain. Also search Soria Province and Spain. Before 1981, also search Soria and Spain. For the city, see Soria City Spain.
CO N410400N420900 W0014800W0033200
BT Old Castile Spain
BT Castile Spain
BT Spain
BT Iberian Peninsula
BT Southern Europe
BT Europe
NT Soria City Spain

Soroka Russian Federation
use Belomorsk Russian Federation

sorosilicates (1981)
Autoposting of this term to sorosilicate minerals that are not in the epidote or melilite groups began in 1985.
BT orthosilicates
BT silicates
NT chevkinite
NT epidote group
NT hemimorphite
NT ilvaite
NT innelite
NT joaquinite
NT julgoldite
NT kilchoanite
NT labuntsovite
NT lawsonite
NT melilite group
NT nenadkevichite
NT perrierite
NT pumpellyite
NT roggianite
NT thalenite
NT thortveitite
NT vesuvianite
NT yttrialite
NT zunyite

Soroy (1978)
Island in Norwegian Sea off NW. Finnmark County.
BT Norway
BT Scandinavia
BT Western Europe
BT Europe

sorption (1978)
SA absorption
SA adsorption
SA geochemistry
SA wave absorption

Sorrento Peninsula (1978)
On S side of Bay of Naples. Also search Sorrento.
BT Campania Italy
BT Italy
BT Southern Europe
BT Europe
SA Bay of Naples

sorting (1978)
See also textures under appropriate rock types.
SA grains
SA particles
SA rounding
SA sedimentary rocks
SA sedimentation
SA sediments

Sosnowice
use Sosnowiec Poland

Sosnowiec
No longer a valid term for GeoRef. As of 1993 see Sosnowiec Poland.

Sosnowiec Poland (1993)
City in central Katowice, S Poland. Before 1993 also search Sosnowiec AND Poland.
UF Sosnowice
BT Katowice Poland
BT Poland
BT Central Europe
BT Europe

Soufriere (1978)
As of 1981, refers to volcano on Saint Vincent. Before 1981, also search Soufriere AND Saint Vincent. Before 1981, West Indies was the broader term, and names of islands may also have been indexed: Guadeloupe, Montserrat, Saint Lucia and/or Saint Vincent.
UF Soufriere Volcano
BT Saint Vincent
BT Lesser Antilles
BT Antilles
BT West Indies
BT Caribbean region
SA La Grande Soufriere
SA Windward Islands

Soufriere Volcano
use Soufriere

Souk-el-Arba salt works (1978)
In Souk-el-Arba area in NW Tunisia.
UF Souk-el-Arba Works
BT Tunisia
BT North Africa
BT Africa

Souk-el-Arba Works
use Souk-el-Arba salt works

sound waves
use acoustical waves

sounding (1978)
SA deep magnetic sounding
SA deep seismic sounding
SA deep sounding
SA echo sounding
SA electrical sounding
SA frequency sounding
SA geophysical methods
SA geophysical surveys
SA measurement-while-drilling
SA seismology
SA sonar methods

Sounds National Park
use Fiordland National Park

source areas
use provenance

source regions
use provenance

source rocks (1978)
SA biomarkers
SA natural gas
SA petroleum
SA petroleum exploration
SA reservoir rocks
SA rocks
SA tight sands

source terrains
use provenance

source terranes
use provenance

sourceland
use provenance

sources
A valid general term through 1978.

sources, energy
use energy sources

sources, heat
use heat sources

sources, seismic
use seismic sources

Souris River basin (1978)
Also search Souris; Souris River.
IN Index North Dakota and Canadian provinces as applicable.
SA Alberta
SA North Dakota
SA Saskatchewan
SA Western Canada

South Africa (1978)
CO S350000S220000 E0330000E0160000
BT Southern Africa
BT Africa
NT Cape Province South Africa
NT Klerksdorp Field
NT Natal South Africa
NT Orange Free State South Africa
NT Transvaal South Africa
NT Vaal River
SA Barberton greenstone belt
SA Beaufort Group
SA Black Mountain
SA Broken Hill Mine
SA Central Rand Group
SA Damara Orogeny
SA Dwyka Formation
SA Elliot Formation
SA Fig Tree Series
SA Kaapvaal Craton
SA Kalahari Desert
SA Karroo Supergroup
SA Lebombo Mountains
SA Limpopo Basin
SA Limpopo Belt
SA Malmani Subgroup
SA Malvern Meteorite
SA Nama System
SA Namaqualand metamorphic complex
SA Onverwacht Group
SA Pongola Supergroup
SA Pretoria Group
SA Stormberg Series
SA Swaziland Sequence
SA Swaziland System
SA Table Mountain Group
SA Transvaal Supergroup
SA Ventersdorp Supergroup
SA Vryheid Formation
SA Walvis Bay
SA Waterberg System
SA Witwatersrand Supergroup
SA Wolkberg Group

South America (1978)
To retrieve all documents, individual countries and physiographic regions should also be searched (see list O). As of 1981, Falkland Islands is a narrower term. Autoposting of this term began in 1978.
CO S550000N130000 W0350000W0820000
NT Amazon Basin
NT Amazon River
NT Andes
NT Argentina
NT Bolivia
NT Brazil
NT Chaco
NT Chile
NT Colombia
NT Ecuador
NT Falkland Islands
NT French Guiana
NT Guajira Peninsula
NT Guiana Basin
NT Guianas
NT Guyana
NT Guyana Shield
NT Lake Titicaca
NT Orinoco River
NT Orinoco River basin
NT Paraguay
NT Parana Basin
NT Parana River
NT Patagonia
NT Peru
NT Precordillera
NT Rio de la Plata
NT Surinam
NT Tierra del Fuego
NT Uruguay
NT Venezuela
SA America
SA Botucatu Formation
SA Galapagos Islands
SA Gondwana
SA Latin America
SA Llanos
SA Pacific mobile belt
SA Rio Negro
SA Roraima Formation
SA South American Plate

South American Atlantic (1981)
Region of the Atlantic Ocean. As of 1990, Atlantic Ocean is autoposted to this term.
BT South Atlantic
BT Atlantic Ocean
NT Argentine Basin
NT Brazil Basin
NT Campos Basin
NT Rio Grande Rise
NT Vema Channel
SA DSDP Site 19
SA DSDP Site 20
SA DSDP Site 355
SA DSDP Site 515
SA Guanabara Bay
SA Leg 36
SA Leg 39
SA Leg 72
SA ODP Site 698
SA ODP Site 699
SA ODP Site 700
SA ODP Site 702
SA Southwest Atlantic

South American Pacific (1981)
Region of the Pacific Ocean.
BT Pacific Ocean
NT Carnegie Ridge
NT Chile Ridge
NT Cocos Ridge
NT Nazca Ridge
SA DSDP Site 424
SA El Nino
SA Leg 34
SA Leg 112
SA ODP Site 679
SA ODP Site 680
SA ODP Site 681
SA ODP Site 682
SA ODP Site 683
SA ODP Site 684
SA ODP Site 685
SA ODP Site 686
SA ODP Site 687
SA ODP Site 688

South American Plate (1981)
Includes the South American continent (excluding a small portion in the Caribbean Plate) and part of the Atlantic Ocean W of the Mid-Atlantic Ridge. Extends northward

to the Caribbean region and southward to the Scotia Ridge.
SA plate tectonics
SA plates
SA South America

South Arabia
Not a valid GeoRef index term after 1969. Was used on level 1 in subfiles E and B.
use Yemen

South Arcot (1978)
Region on Coromandel Coast in E Tamil Nadu, S India.
SA Coromandel Coast
SA Tamil Nadu India

South Atlantic (1978)
Before 1978, search Atlantic Ocean AND south. Autoposting of this term began in 1981.
BT Atlantic Ocean
NT Maurice Ewing Bank
NT South American Atlantic
NT South Atlantic Ridge
NT Southeast Atlantic
SA DSDP Site 356
SA DSDP Site 511
SA DSDP Site 512
SA DSDP Site 513
SA DSDP Site 514
SA DSDP Site 516
SA DSDP Site 517
SA DSDP Site 518
SA DSDP Site 519
SA DSDP Site 520
SA DSDP Site 521
SA DSDP Site 522
SA DSDP Site 523
SA DSDP Site 524
SA DSDP Site 525
SA DSDP Site 526
SA DSDP Site 527
SA DSDP Site 528
SA DSDP Site 529
SA DSDP Site 530
SA DSDP Site 531
SA DSDP Site 532
SA East Atlantic
SA Equatorial Atlantic
SA Leg 71
SA Leg 73
SA Leg 114
SA North Atlantic
SA Northeast Atlantic
SA Northwest Atlantic
SA West Atlantic

South Atlantic Ridge (1981)
As of 1990, Atlantic Ocean is autoposted to this term. This term has multiple hierarchies.
BT1 Mid-Atlantic Ridge
BT1 Atlantic Ocean
BT2 South Atlantic
BT2 Atlantic Ocean
SA ODP Site 662
SA ODP Site 663
SA ODP Site 701

South Australia (1978)
CO S380000S260000
    E1410000E1290000
BT Australia
BT Australasia
NT Adelaide Australia
NT Adelaide Geosyncline
NT Balcanoona Australia
NT Beltana Australia
NT Coorong Lagoon
NT Encounter Bay
NT Eyre Peninsula
NT Fleurieu Peninsula
NT Flinders Range
NT Gawler Craton
NT Kangaroo Island
NT Lake Bonney
NT Lake Eyre
NT Lake Frome
NT Lake Torrens
NT Mount Lofty Ranges
NT Olympic Dam
NT Renmark Australia
NT Saint Vincent Gulf
NT Spencer Gulf
NT Spilsby Island
NT Willunga Australia
NT Yorke Peninsula
SA Broken Hill Block
SA Bulldog Shale
SA Burra Group
SA Cooper Basin
SA Discovery Bay
SA Eucla Basin
SA Gambier Embayment
SA Giles Complex
SA Great Artesian Basin
SA Murray Basin
SA Murray River
SA Musgrave Ranges
SA Nullarbor Plain
SA Officer Basin
SA Otway Basin
SA Simpson Desert
SA Stuart Shelf
SA Tapley Hill Formation
SA Tasman orogenic zone
SA Toolebuc Formation
SA Umberatana Group
SA Weekeroo Station Meteorite
SA Willyama Complex
SA Wilpena Group
SA Winton Formation
SA Wonoka Formation

South Austrian Alps (1981)
Autoposting of Alps to this term ended in 1989 and began again in 1993. This term has multiple hierarchies.
CO N462200N470200
    E0145800E0122500
BT1 Alps
DT1 Europe
BT2 Austria
BT2 Central Europe
BT2 Europe
SA Carinthia Austria

South Austrian Molasse (1981)
CO N463500N474000
    E0164300E0151200
BT Austria
BT Central Europe
BT Europe

South Canadian River
use Canadian River

South Carolina (1978)
Autoposting of this term began in 1978.
CO N320400N351200
    W0783200W0831500
BT United States
NT Aiken County South Carolina
NT Anderson County South Carolina
NT Barnwell County South Carolina
NT Beaufort County South Carolina
NT Berkeley County South Carolina
NT Calhoun County South Carolina
NT Charleston County South Carolina
NT Cherokee County South Carolina
NT Chester County South Carolina
NT Chesterfield County South Carolina
NT Dorchester County South Carolina
NT Fairfield County South Carolina
NT Florence County South Carolina
NT Georgetown County South Carolina
NT Greenwood County South Carolina
NT Horry County South Carolina
NT Jasper County South Carolina
NT Lake Jocassee
NT Lancaster County South Carolina
NT Lee County South Carolina
NT Marion County South Carolina
NT Oconee County South Carolina
NT Pickens County South Carolina
NT Richland County South Carolina
NT Santee River
NT Savannah River Plant
NT Snuggedy Swamp
NT Sumter County South Carolina
NT Union County South Carolina
NT York County South Carolina
SA Atlantic Coastal Plain
SA Avalon Terrane
SA Barnwell Formation
SA Black Creek Formation
SA Blue Ridge Province
SA Carolina Bays
SA Carolina slate belt
SA Castle Hayne Limestone
SA Charleston earthquake 1886
SA Charlotte Belt
SA Dry Branch Formation
SA Duplin Formation
SA Hawthorn Formation
SA Henderson Gneiss
SA Huber Formation
SA Jackson Group
SA Kings Mountain
SA Kings Mountain Belt
SA Kiokee Belt
SA Middendorf Formation
SA Monticello Reservoir
SA Peedee Formation
SA Piedmont
SA Roan Supergroup
SA Sand Hills
SA Santee Limestone
SA Savannah River
SA Tallulah Falls Formation
SA Tuscaloosa Formation
SA Waccamaw Formation
SA Wicomico Formation

South Carpathians
use Transylvanian Alps

South Cascade Glacier (1978)
Near Mount Logan in North Cascades National Park, NW Washington. Autoposting of broader terms to this term began in 1989.
BT Skagit County Washington
BT Washington
BT United States

South China Sea (1978)
Bounded on N by China and Taiwan, on the E by the Philippine Islands, on the S by Malaysia and on the W by Vietnam.
CO N000000N250000
    E1220000E0991000
BT West Pacific
BT Pacific Ocean
NT Gulf of Siam
NT Gulf of Tonkin
NT Manila Trench
SA China Sea
SA Hainan China
SA Leg 124E
SA Xisha Islands
SA Zhujiang River

South Dakota (1978)
Autoposting of this term began in 1978.
CO N423000N455500
    W0962700W1040500
BT United States
NT Badlands National Monument
NT Butte County South Dakota
NT Campbell County South Dakota
NT Clark County South Dakota
NT Clay County South Dakota
NT Custer County South Dakota
NT Dewey County South Dakota
NT Douglas County South Dakota
NT Fall River County South Dakota
NT Grant County South Dakota
NT Harding County South Dakota
NT Harney Peak Granite
NT Hughes County South Dakota
NT Jackson County South Dakota
NT Jones County South Dakota
NT Lake County South Dakota
NT Lawrence County South Dakota
NT Lincoln County South Dakota
NT Marshall County South Dakota
NT McPherson County South Dakota
NT Meade County South Dakota
NT Minnehaha County South Dakota
NT Pennington County South Dakota
NT Potter County South Dakota
NT Shannon County South Dakota
NT Union County South Dakota
SA Arikaree Group
SA Ash Hollow Formation
SA Bald Mountain
SA Bear Mountain
SA Benton Formation
SA Black Hills
SA Brule Formation
SA Carlile Shale
SA Cedar Creek Anticline
SA Chadron Arch
SA Chadron Formation
SA Charles Formation
SA Codell Sandstone Member
SA Colorado Group
SA Colorado Lineament
SA Dakota Formation
SA Deadwood Formation
SA Des Moines Lobe
SA Duperow Formation
SA Elk Point Group
SA Fall River Formation
SA Fort Hays Limestone Member
SA Fort Union Formation
SA Fox Hills Formation
SA Gammon Ferruginous Member
SA Grand River
SA Graneros Shale
SA Great Plains
SA Greenhorn Limestone
SA Hell Creek Formation
SA High Plains Aquifer
SA Homestake Mine
SA Inyan Kara Group
SA James River
SA Lakota Formation
SA Lance Formation
SA Laramie Formation
SA Little Missouri River basin
SA Loup Fork Group
SA Ludlow Member

SA Madison Aquifer
SA Madison Group
SA Medicine Lake
SA Midcontinent
SA Minnelusa Formation
SA Missouri Plateau
SA Missouri River
SA Missouri River basin
SA Missouri River valley
SA Montana Group
SA Morrison Formation
SA Mowry Shale
SA Newcastle Sandstone
SA Niobrara Formation
SA Opeche Shale
SA Pierre Shale
SA Red River
SA Red River Formation
SA Red River valley
SA Shannon Sandstone Member
SA Sioux Quartzite
SA Skull Creek Shale
SA Smoky Hill Chalk Member
SA Spearfish Formation
SA Sundance Formation
SA Swift Formation
SA Tongue River Member
SA Tullock Member
SA Valentine Formation
SA Wall Creek Member
SA White River Group
SA Whitewood Creek
SA Williston Basin
SA Winnipeg Formation
SA Winnipegosis Formation
SA Wyoming Province

South Florida Basin (1989)
BT Florida
BT United States
SA Gulf Coastal Plain

South Georgia (1978)
Island 800 miles E of the Falkland Islands of which it is a dependency. Before 1981, broader term was Atlantic Ocean.
CO S550000S540000
    W0350000W0380000
UF South Georgia Island
BT Scotia Sea Islands
BT Antarctica
SA Atlantic Ocean
SA Falkland Islands

South Georgia Island
use South Georgia

South Greenland (1978)
Before 1978, search Greenland AND south or southern. As of 1990, Arctic region and Polar regions are autoposted to this term.
BT Greenland
BT Arctic region
SA East Greenland
SA West Greenland

South Island (1978)
Largest island of New Zealand. S and SW of North Island. Also occurs in other location.
IN Also index county or country as applicable.
SA Alpine Fault
SA Canterbury New Zealand
SA Christchurch New Zealand
SA Dunedin New Zealand
SA Fiordland National Park
SA Fox Glacier
SA Haast River
SA Inangahua New Zealand
SA Kaikoura
SA Kakanui
SA Marlborough New Zealand
SA Nelson New Zealand
SA New Zealand

SA Oamaru New Zealand
SA Otago New Zealand
SA Otago Peninsula
SA Oxford New Zealand
SA Rangitata River
SA Reefton New Zealand
SA Southern Alps
SA Southland New Zealand
SA Taranaki Basin
SA Westland New Zealand

South Kona
use Kona

South Korea (1978)
Officially Republic of Korea. Bounded on N by North Korea, on E by Japan Sea, on S by the Korea Strait, and on W by the Yellow Sea.
CO N330000N383000
    E1293000E1260000
BT Korea
BT Far East
BT Asia
NT Cheju Island
NT Janggun Mine
NT Kangwon South Korea
NT Kyongsang Basin
NT Pohang South Korea
NT Ulsan South Korea
NT Yongyang South Korea
SA North Korea

South Massif
use Causses

South Mountain (1978)
Ridge in W Maryland and S Pennsylvania.
IN Index counties or states and mountain ranges as applicable.
BT United States
SA Appalachians
SA Blue Ridge Mountains
SA Maryland
SA Pennsylvania

South Mountain Batholith (1985)
SW Nova Scotia.
IN Index Nova Scotia if applicable.
SA Carboniferous
SA Devonian
SA Nova Scotia

South Nahanni River (1978)
SW Mackenzie District.
BT Mackenzie District Northwest Territories
BT Northwest Territories
BT Western Canada
BT Canada

South Orkney Islands (1978)
British islands 850 miles NE of Antarctic Peninsula. Autoposting of this term began in 1981.
CO S620000S600000
    W0430000W0470000
UF South Orkneys
BT Scotia Sea Islands
BT Antarctica
NT Signy Island
SA Falkland Islands
SA Orkney Islands

South Orkneys
use South Orkney Islands

South Pacific (1978)
Before 1978, search Pacific Ocean AND south.
BT Pacific Ocean
NT Louisville Ridge
SA DSDP Site 597
SA DSDP Site 598
SA DSDP Site 599

SA DSDP Site 600
SA DSDP Site 601
SA DSDP Site 602
SA East Pacific
SA Eltanin fracture zone
SA Equatorial Pacific
SA Leg 92
SA North Pacific
SA Northeast Pacific
SA Northwest Pacific
SA Pacific Basin
SA Southeast Pacific
SA Southwest Pacific
SA West Pacific

South Pass (1978)
One of the channels at the mouth of the Mississippi River.
IN Index county or region as applicable.
SA Plaquemines Parish Louisiana

South Platte River (1989)
River in central to NE Colorado and SW Nebraska.
IN Index states as applicable.
BT United States
SA Colorado
SA Nebraska
SA Platte River

South Platte River valley (1978)
Also search South Platte; South Platte River.
IN Index states as applicable.
BT United States
SA Colorado
SA Nebraska

South Pole (1978)
The S extremity of the Earth's axis at 90° S latitude, and the point from which all directions are N.
BT Antarctica
SA Antarctic Polar Cap
SA polar caps
SA polar regions

South Polish Glaciation (1978)
Refers to glaciation in S Poland which was roughly the southern limit of Pleistocene glaciers in central Europe.
BT Cromerian
BT upper Pleistocene
BT Pleistocene
BT Quaternary
BT Cenozoic
SA Poland

South Polynesian Pacific (1981)
Used to indicate a region of the Pacific Ocean.
BT Pacific Ocean
SA DSDP Site 209
SA Leg 136

South Ray Crater (1978)
BT Moon

South Sandwich Islands (1978)
Group of small volcanic islands at E end of Scotia Sea about 1350 miles E of Cape Horn. Part of Falkland Island Dependencies. Before 1981, Antarctic Ocean was autoposted to this term.
CO S580000S540000
    W0250000W0300000
BT Scotia Sea Islands
BT Antarctica
SA Antarctic Ocean
SA Atlantic Ocean
SA Falkland Islands
SA Northwest Atlantic

South Saskatchewan River
(1978)

Rises in the Rocky Mountains of W Alberta.
IN Index provinces as applicable.
BT Western Canada
BT Canada
SA Alberta
SA North Saskatchewan River
SA Saskatchewan
SA Saskatchewan River

South Sea Islands
No longer a valid GeoRef index term. Before 1969, was used on level 1 in subfile E. See Oceania; West Pacific Ocean Islands; East Pacific Ocean Islands.

South Shetland Islands (1978)
N of the Antarctic Peninsula and S of Drake Passage. Part of British Antarctic Territory. Autoposting of this term began in 1981.
CO S630000S610000
    W0540000W0630000
UF South Shetlands
BT Scotia Sea Islands
BT Antarctica
NT Deception Island
NT King George Island
NT Livingston Island
SA Shetland Islands

South Shetlands
use South Shetland Islands

South Victoria Land
use Victoria Land

South Wales (1981)
Region in S part of Wales.
IN Index counties as applicable.
CO N512000N522500
    W0024000W0053000
BT Wales
BT Great Britain
BT United Kingdom
BT Western Europe
BT Europe
SA Dyfed Wales
SA Glamorgan Wales

South Wales coal field (1978)
Concentrated industrial area N of Bristol Channel. Also search South Wales.
UF South Wales Coalfield
BT Wales
BT Great Britain
BT United Kingdom
BT Western Europe
BT Europe
SA coal fields

South Wales Coalfield
use South Wales coal field

South West Africa
Not a valid GeoRef term after 1971. Was used on level 1 in subfile G.
use Namibia

South-Central China
No longer a valid term for GeoRef. See provinces: Hubei China, Hunan China, Guangxi China, and Guangdong China. Henan China and Hainan China were considered part of this region for 1992.

South-West Africa
A valid index term through 1980.
use Namibia

South-West England (1981)
CO N495600N513000
    W0023500W0054500
BT England
BT Great Britain

BT United Kingdom
BT Western Europe
BT Europe
Southeast Asia (1981)
Before 1981, also search Asia AND southeast.
BT Asia
Southeast Atlantic (1978)
Before 1978, search Atlantic Ocean AND southeast.
BT South Atlantic
BT Atlantic Ocean
NT Angola Basin
NT Cape Basin
NT Walvis Ridge
SA DSDP Site 360
SA DSDP Site 365
SA East Atlantic
SA Leg 39
SA Leg 40
SA Leg 74
SA Leg 75
SA Leg 108
SA North Atlantic
SA Northeast Atlantic
SA Northwest Atlantic
SA ODP Site 662
SA ODP Site 663
SA ODP Site 703
SA ODP Site 704
SA Southwest Atlantic
SA West Atlantic
Southeast Indian Ridge (1981)
CO S600000S330000
   E1560000E0783000
UF Indian-Antarctic Ridge
UF Indo-Antarctic Ridge
BT Mid-Indian Ridge
BT Indian Ocean
SA DSDP Site 265
SA mid-ocean ridges
Southeast Pacific (1978)
Before 1978, search Pacific Ocean AND southeast.
BT Pacific Ocean
SA East Pacific
SA Equatorial Pacific
SA North Pacific
SA Northeast Pacific
SA Northwest Pacific
SA South Pacific
SA Southwest Pacific
SA West Pacific
Southeastern Alaska (1993)
Artificial region based on U.S. Geological Survey quadrangle designations. Includes Alexander Archipelago. Also search quadrangles if applicable.
IN Index quadrangles as applicable.
CO N544500N600000
   W1300000W1380000
BT Alaska
BT United States
NT Alaska Panhandle
NT Alexander Archipelago
NT Glacier Bay National Park
NT Juneau Alaska
NT Juneau ice field
NT Sitka Sound
NT Variegated Glacier
SA Alexander Terrane
SA Fairweather Fault
SA Glacier Bay
SA Tracy Arm
SA Yakutat Terrane
Southeastern Nigeria
No longer a valid term for GeoRef. Valid 1981-1992.
Southeastern Polish Hills (1981)
CO N494500N514000

   E0241000E0174500
BT Poland
BT Central Europe
BT Europe
Southeastern Transdanubia (1981)
Autoposting of Hungary to this term began in 1989. As of 1990, Central Europe and Europe are autoposted to this term.
CO N454500N462500
   E0185000E0174000
BT Transdanubia
BT Hungary
BT Central Europe
BT Europe
Southeastern U.S. (1985)
As of 1993, term is used only for broad general discussions. From 1978 through 1992, this term was autoposted to Florida, Georgia, North Carolina, South Carolina, and Virginia. As of 1993, see United States or individual states to search those areas.
CO N243000N392800
   W0753000W0873000
BT Eastern U.S.
BT United States
Southeastern Venezuela (1981)
CO N004000N082500
   W0602000W0675500
BT Venezuela
BT South America
Southern Africa (1978)
Before 1978, search Africa AND southern. Autoposting of this term began in 1981.
CO S350000S153000
   E0334000E0120000
BT Africa
NT Barberton greenstone belt
NT Botswana
NT Kaapvaal Craton
NT Lesotho
NT Namaqualand
NT Namaqualand metamorphic complex
NT Namibia
NT Orange River
NT South Africa
NT Swaziland
NT Walvis Bay
NT Zimbabwe
SA Central Africa
SA Damara Orogeny
SA East Africa
SA North Africa
SA West Africa
Southern Alaska (1993)
Artificial region based on U.S. Geological Survey quadrangle designations. Also search quadrangles if applicable.
IN Index quadrangles as applicable.
CO N590000N640000
   W1380000W1530000
BT Alaska
BT United States
NT Anchorage Alaska
NT Chugach Mountains
NT Kenai Peninsula
NT Knik Arm
NT Martin River Glacier
NT Matanuska Valley
NT Prince William Sound
NT Susitna River
NT Susitna River basin
NT Talkeetna Mountains
NT Turnagain Arm
NT Valdez Alaska
NT Wrangell Mountains

NT Yakutat Bay
SA Border Ranges Fault
SA Columbia Glacier
SA Cook Inlet
SA Copper River basin
SA Delta River
SA Icy Bay
SA Middleton Island
SA Montague Island
SA Talkeetna Formation
SA Trans-Alaska Pipeline
SA Yukon-Koyukuk Basin
SA Yukon-Tanana Upland
Southern Alps (1978)
Restricted to central South Island, New Zealand.
IN Also index New Zealand and South Island.
UF Alps, Australian
SA New Zealand
SA South Island
Southern Andes (1978)
Before 1978, also search Andes AND south or southern. As of 1990, South America is autoposted to this term.
BT Andes
BT South America
SA Central Andes
SA Northern Andes
Southern Apennines (1981)
Before 1981, also search Apennines AND South. Northern extent is Sella di Vinchiaturo Pass.
CO N375500N413000
   E0164500E0130000
BT Apennines
BT Italy
BT Southern Europe
BT Europe
SA Sila Massif
Southern Appalachians (1978)
Before 1978, also search Appalachians AND south or southern. As of 1990, North America is are autoposted to this term.
BT Appalachians
BT North America
SA Brevard Zone
SA Central Appalachians
SA Charlotte Belt
SA Kings Mountain Belt
SA Northern Appalachians
SA Rome Trough
Southern Atlantic Coastal Plain (1981)
Before 1981, also search Atlantic Coastal Plain AND South.
CO N250000N340000
   W0780000W0840000
BT Atlantic Coastal Plain
BT United States
Southern California (1978)
Autoposting of broader terms to this term began in 1989. Before 1978, search California AND south.
BT California
BT United States
Southern California Batholith (1978)
BT California
BT United States
Southern Carpathians
use Transylvanian Alps
Southern coal field (1989)
Sydney Basin, E New South Wales.
CO S343000S340000
   E1510000E1500000

UF Southern Coalfield
BT New South Wales Australia
BT Australia
BT Australasia
SA coal fields
SA Sydney Basin
Southern Coalfield
use Southern coal field
Southern Cook Islands
use Cook Islands
Southern Europe (1981)
As of 1993, term is autoposted to Dinaric Alps, Dobruja Basin, Macedonia, and Rhodope Mountains.
BT Europe
NT Albania
NT Banat
NT Bulgaria
NT Danube Plain
NT Dinaric Alps
NT Dobruja Basin
NT Greece
NT Iberian Peninsula
NT Ionian Zone
NT Italy
NT Krajiste
NT Macedonia
NT Malta
NT Moesia
NT Moesian Platform
NT Osogovo Mountains
NT Rhodope Mountains
NT Romania
NT San Marino
NT Struma River valley
NT Vardar Zone
NT Yugoslavia
Southern Great Plains (1978)
Before 1978, search Great Plains AND south or southern; also search Southern High Plains. As of 1990, North America is autoposted to this term.
UF Southern High Plains
BT Great Plains
BT North America
SA Llano Estacado
SA Northern Great Plains
Southern Hemisphere (1978)
Used when discussing many large areas too numerous to mention.
SA Eastern Hemisphere
SA Gondwana
SA Northern Hemisphere
SA Western Hemisphere
Southern High Plains
use Southern Great Plains
Southern Highlands (1978)
Highland region in New South Wales, Australia. and the cratered highland part of the Moon's southern hemisphere. Before 1993, also used as a province in Papua New Guinea. Search in combination with Australia or Moon as applicable.
IN Index regions as applicable.
SA Moon
SA New South Wales Australia
SA Southern Highlands Papua New Guinea
SA Southern Uplands
Southern Highlands Papua New Guinea (1993)
Before 1993, also search Southern Highlands AND Papua New Guinea.
BT Papua New Guinea
BT Australasia
SA Southern Highlands

southern Italy earthquake 1980
use Irpinia earthquake 1980

Southern Norway (1981)
N boundary is Trondheim.
CO N580000N635000
E0125500E0044000
BT Norway
BT Scandinavia
BT Western Europe
BT Europe

Southern Ocean
use Antarctic Ocean

Southern Peninsula, Michigan
use Michigan Lower Peninsula

Southern Province (1978)
Structural province of the Canadian Shield. Primarily in the Lake Superior region.
IN Index Ontario and/or states as applicable.
CO N440000N491500
W0803000W0963000
BT Canadian Shield
BT North America
SA Michigan
SA Minnesota
SA Ontario
SA Wisconsin

Southern Rhodesia
A valid index term through 1972.
use Zimbabwe

Southern Rocky Mountains (1978)
Before 1978, also search Rocky Mountains AND (south OR southern). As of 1990, North America is autoposted to this term.
BT Rocky Mountains
BT North America
SA Central Rocky Mountains
SA Northern Rocky Mountains
SA U. S. Rocky Mountains
SA United States

Southern Swiss Alps (1981)
This term has multiple hierarchies.
BT1 Swiss Alps
BT1 Switzerland
BT1 Central Europe
BT1 Europe
BT2 Swiss Alps
BT2 Alps
BT2 Europe

Southern Transdanubia (1981)
Autoposting of Hungary to this term began in 1989. As of 1990, Central Europe and Europe are autoposted to this term.
CO N454500N472200
E0190200E0160500
BT Transdanubia
BT Hungary
BT Central Europe
BT Europe

Southern U.S. (1981)
As of 1993, term is used only for broad general discussions. From 1978 through 1992, this term was autoposted to Alabama, Arkansas, Kentucky, Louisiana, Mississippi, and Tennessee. As of 1993, for complete search, also search states.
CO N290000N390000
W0814000W0943000
BT United States
SA Mississippi Valley

Southern Uplands (1978)
Between the Scottish Lowlands to the N and the Cheviot Hills on the English border to the S. Also used in Florida and in Nova Scotia.
IN Index regions as applicable.
SA England
SA Florida
SA Nova Scotia
SA Scotland
SA Southern Highlands

Southern Urals (1978)
Section of the Urals N of the Mugodzhar Hills.
CO N510000N560000
E0610000E0550000
BT Urals
BT Commonwealth of Independent States
NT Ural-Tau
SA Bakal Russian Federation
SA Kazakhstan
SA Mugodzhar Hills
SA Russian Federation
SA Taratash Complex
SA Zhetybay Kazakhstan

Southern Yemen
No longer a valid term for GeoRef. Former Peoples Democratic Republic of Yemen. United with Yemen Arab Republic in 1990. Before 1970, also search South Arabia.
use Yemen

Southland
No longer a valid term for GeoRef. See Southland New Zealand.

Southland New Zealand (1993)
A land district in SW South Island.
IN Also index South Island.
BT New Zealand
BT Australasia
NT Fiordland National Park
SA South Island

Southwest Atlantic (1978)
Before 1978, search Atlantic Ocean AND southwest.
BT Atlantic Ocean
SA DSDP Site 21
SA DSDP Site 328
SA East Atlantic
SA North Atlantic
SA Northeast Atlantic
SA Northwest Atlantic
SA South American Atlantic
SA Southeast Atlantic
SA West Atlantic

Southwest Florida Water Management District (1985)
BT Florida
BT United States
SA Floridan Aquifer
SA Gulf Coastal Plain

Southwest Indian Ridge (1981)
CO S413000S223000
E0673000E0440000
BT Indian Ocean
SA Leg 118
SA mid-ocean ridges
SA ODP Site 732
SA ODP Site 733
SA ODP Site 734
SA ODP Site 735

Southwest Pacific (1978)
Before 1978, search Pacific Ocean AND southwest.
BT Pacific Ocean
SA DSDP Site 208
SA DSDP Site 587
SA DSDP Site 588
SA DSDP Site 589
SA DSDP Site 590
SA DSDP Site 591
SA DSDP Site 592
SA DSDP Site 593
SA DSDP Site 594
SA DSDP Site 595
SA DSDP Site 596
SA East Pacific
SA Equatorial Pacific
SA Leg 90
SA Leg 91
SA North Pacific
SA Northeast Pacific
SA Northwest Pacific
SA South Pacific
SA Southeast Pacific
SA West Pacific
SA Woodlark Basin

Southwestern Alaska (1993)
Artificial region based on U.S. Geological Survey quadrangle designations. Includes Aleutian Islands and Kodiak Island. Also search quadrangles if applicable.
IN Also index quadrangles if applicable.
CO N510000N620000
E1710000W1520000
BT Alaska
BT United States
NT Alaska Peninsula
NT Aleutian Islands
NT Katmai
NT Katmai National Monument
NT Kodiak Island
NT Nunivak Island
NT Shelikof Strait
NT Valley of Ten Thousand Smokes
SA Augustine
SA Cook Inlet
SA Pavlof
SA Yukon-Koyukuk Basin

Southwestern China
No longer a valid term for GeoRef. Used from 1985-1992 for the following provinces: Guizhou, Sichuan, Yunnan, and Xizang.

Southwestern German Hills (1981)
CO N473200N504200
E0123700E0080200
BT Germany
BT Central Europe
BT Europe
SA Holzmaden region

Southwestern German Massifs (1993)
Chiefly in S Hesse.
IN Index German states as applicable.
CO N492200N502100
E0092000E0083600
BT Germany
BT Central Europe
BT Europe
SA Hesse Germany

Southwestern U.S. (1978)
As of 1993, term is used only for broad general discussions. Before 1978, also search United States AND (southwest OR southwestern). From 1981 through 1992, autoposted as a regional grouper for Arizona, New Mexico, Oklahoma, and Texas. As of 1993, for complete search, also search states.
CO N260000N370000
W0933000W1144500
BT United States

Southwestern USSR
No longer a valid term for GeoRef. As of 1993, see Commonwealth of Independent States. Before 1978, search USSR AND (southwest OR southwestern).

Soviet Altai Mountains
No longer a valid term for GeoRef. Before 1993, used for the Altai Mountains in the USSR. After 1993, search Altai Mountains with Russian Federation or Kazakhstan. Autoposting of USSR to this term began in 1989. Also see Siberian fold belt.

Soviet Arctic
As of 1993, no longer a valid term for GeoRef.
use Russian Arctic

Soviet Carpathians
As of 1993, no longer a valid term for GeoRef.
use Ukrainian Carpathians

Soviet Central Asia
use Central Asia

Soviet Far East
As of 1993, no longer a valid term for GeoRef.
use Russian Far East

Soviet Fennoscandia
As of 1993, no longer a valid term for GeoRef.
use Russian Fennoscandia

Soviet Pacific region
As of 1993, no longer a valid term for GeoRef. Valid 1981-1992.
use Russian Pacific region

Soviet Union
Not a valid term for GeoRef. See USSR or Commonwealth of Independent States.

sovite (1978)
BT carbonatites
BT plutonic rocks
BT igneous rocks

Sowerbyella (1978)
BT Strophomenida
BT Articulata
BT Brachiopoda
BT Invertebrata

Sowie Mountains (1978)
Range of the Sudeten Mountains in Lower Silesia in SW Poland. This term has multiple hierarchies.
UF Eulengebirge
BT1 Sudeten Mountains
BT1 Central Europe
BT1 Europe
BT2 Poland
BT2 Central Europe
BT2 Europe
SA Klodzko Unit
SA Polish Sudeten Mountains

space
use interplanetary space

space groups (1978)
SA crystal structure

space lattice
use lattice

space photography (1989)
Also search photography AND space shuttle if applicable.
UF spaceborne photography
SA aerial photography
SA mosaics
SA multispectral analysis
SA photogeology

SA photogrammetry
SA photography
SA remote sensing
SA satellite methods

space shuttle
Not a valid term for GeoRef. See space photography or Shuttle Imaging Radar.

space, underground
use underground space

spaceborne photography
use space photography

Spacelab Program (1982)
SA extraterrestrial geology
SA planetology
SA remote sensing
SA satellite methods

Spain (1978)
As of 1981, Balearic Islands is a narrower term.
CO N360000N434500
  E0043000W0093000
BT Iberian Peninsula
BT Southern Europe
BT Europe
NT Alpujarras
NT Andalusia Spain
NT Aragon Spain
NT Asturian Arc
NT Asturias Spain
NT Balearic Islands
NT Basque Provinces Spain
NT Betic Cordillera
NT Betic Zone
NT Calatayud-Teruel Basin
NT Cantabrian Basin
NT Cantabrian Mountains
NT Cantabrian region
NT Castile Spain
NT Catalonia Spain
NT Catalonian Coastal Ranges
NT Ebro Basin
NT Ebro River
NT Extremadura Spain
NT Galicia Spain
NT Guadalquivir
NT Guadalquivir Basin
NT Hercinico Centro
NT Hercinico Sur
NT Iberian Mountains
NT Iberica
NT Leon region
NT Maestrazgo Spain
NT Montes de Toledo
NT Murcia region
NT Navarra Spain
NT Prebetic Zone
NT Serrania de Cuenca
NT Sierra de Gredos
NT Sierra de Guadarrama
NT Sierra de la Demanda
NT Spanish Pyrenees
NT Subbetic Zone
NT Ter River basin
NT Valencia region
SA Ager Formation
SA Altiplano
SA Andalusian
SA Canary Islands
SA Duero Basin
SA Duero River
SA ENADIMSA
SA Galaico Massif
SA Lancara Formation
SA Mediterranean region
SA Meseta
SA North Pyrenean Fault
SA Pyrenees
SA Tagus Basin
SA Tagus River

spallation (1978)
SA isotopes

Spanish Guinea
A valid term through 1975.
use Equatorial Guinea

Spanish Morocco
No longer a valid GeoRef index term. Before 1960, was used on level 1 in subfile E.
use Morocco

Spanish Peaks (1978)
Two mountains in Huerfano and Las Animas counties in S Colorado. This term has multiple hierarchies.
IN Index counties as applicable.
BT1 Colorado
BT1 United States
BT2 Sangre de Cristo Mountains
BT2 U. S. Rocky Mountains
BT2 Rocky Mountains
BT3 Sangre de Cristo Mountains
BT3 U. S. Rocky Mountains
BT3 United States
SA Huerfano County Colorado
SA Las Animas County Colorado

Spanish Pyrenees (1981)
Range in NE Spain. Autoposting of Alps to this term ended in 1989. This term has multiple hierarchies.
BT1 Pyrenees
BT1 Europe
BT2 Spain
BT2 Iberian Peninsula
BT2 Southern Europe
BT2 Europe
SA North Pyrenean Fault

Spanish Sahara
No longer a valid term for GeoRef. A valid level 1 term through 1976. From 1977-80, documents on the Spanish Sahara were indexed using Sahara as the level 1 term and the term west.
use Western Sahara

Spanish Sierra Nevada (1981)
Mountain range in S Spain. Before 1981, also search Sierra Nevada AND Spain.
CO N364500N373500
  W0025000W0043000
BT Betic Cordillera
BT Spain
BT Iberian Peninsula
BT Southern Europe
BT Europe
SA Sierra Nevada

Spanish West Africa
Not a valid GeoRef index term after 1969. Was used on level 1 in subfiles E and B. Now use Africa, Morocco, Mauritania or Equatorial Guinea or West Africa.

sparagmite (1978)
BT clastic rocks
BT sedimentary rocks
SA terrigenous materials

Sparagmite Division
use Sparagmite Group

Sparagmite Group (1978)
Upper Precambrian (Riphean?)-lower Paleozoic.
IN Index ages as applicable.
UF Sparagmite Division
UF Sparagmite Series
SA lower Paleozoic
SA Norway
SA Sweden
SA upper Precambrian

Sparagmite Series
use Sparagmite Group

Sparnacian (1978)
Europe. Above Thanetian, below Ypresian of France.
BT lower Eocene
BT Eocene
BT Paleogene
BT Tertiary
BT Cenozoic
SA Paleocene

Sparta Sand (1989)
In Claiborne Group.
BT middle Eocene
BT Eocene
BT Paleogene
BT Tertiary
BT Cenozoic
SA Arkansas
SA Claiborne Group
SA Louisiana
SA Mississippi
SA Texas

Spartina alterniflora (1993)
BT Gramineae
BT Monocotyledoneae
BT angiosperms
BT Spermatophyta
BT Plantae
SA wetlands

Spathian (1989)
North America. Upper Lower Triassic. Above Smithian; below Middle Triassic.
BT Lower Triassic
BT Triassic
BT Mesozoic

Spathognathodus (1978)
Genus. Autoposting of microfossils to this term began in 1990.
BT Conodonta
BT microfossils

spatial distribution (1981)
SA distribution

spatial frequency filters (1981)
SA filters

spatial variations (1978)
Used as a general term.
SA variations

Spearfish Formation (1978)
Underlies Gypsum Spring Formation, geographically extended into Black Hills area. NW Nebraska, W South Dakota, and E Wyoming.
IN Index ages as applicable.
SA Nebraska
SA Permian
SA South Dakota
SA Triassic
SA Wyoming

speciation (1985)
Restricted to paleontology. For chemical speciation, see chemical fractionation.
SA biologic evolution
SA paleontology
SA phylogeny
SA species diversity

species
As of 1981, no longer a valid term for GeoRef.

species diversity (1978)
SA diversity
SA fossils
SA paleontology
SA speciation

specific conductance
use conductivity

specific gravity (1978)
IN May be used with appropriate material name, e.g. with soils and sediments.
SA buoyancy
SA density

specific heat (1978)
UF heat, specific
SA geochemistry
SA heat capacity
SA physical properties
SA temperature
SA thermal conductivity
SA thermal inertia
SA thermal properties
SA thermodynamic properties

specific surface (1982)
SA mass
SA properties
SA volume

specimens, type
use type specimens

spectra (1978)
Used for data. For methodology, use spectroscopy.
UF spectrum
NT atomic absorption spectra
NT emission spectra
NT EPR spectra
NT gamma-ray spectra
NT infrared spectra
NT mass spectra
NT Mossbauer spectra
NT NMR spectra
NT optical spectra
NT Raman spectra
NT ultraviolet spectra
NT X-ray fluorescence spectra
NT X-ray spectra
SA broad-band spectra
SA geochemistry
SA minerals
SA spectroscopy

spectral analysis (1978)
As of 1981, restricted to seismology, seismic methods and seismic surveys. Before 1981, included use with other geophysical methods and surveys, including remote sensing.
SA analysis
SA near-field spectra
SA seismograms
SA seismology

spectrochemical analysis
Not a valid GeoRef index term after 1970. Was used on level 1 in subfile N.
use spectroscopy

spectrographic analysis
No longer a valid GeoRef index term. Before 1971, was used on level 1 in subfile N.
use spectroscopy

spectrometric analysis
No longer a valid GeoRef index term. Before 1971, was used on level 1 in subfile N.
use spectroscopy

spectrometry
A valid term through 1974.
use spectroscopy

spectrophotometric analysis
Not a valid GeoRef index term after 1970. Was used on level 1 in subfile N.
use spectroscopy

spectroscopic analysis
No longer a valid GeoRef index term. Before 1971, was used on level 1 in subfile N.

use spectroscopy
spectroscopy (1978)
Used only for methodology. For data, see appropriate spectra terms. Before 1973, also search mass spectroscopy; before 1975, also search spectrometry. Before 1971, also search spectrochemical analysis; spectrophotometric analysis; spectrographic analysis; spectrometric analysis; spectroscopic analysis.
  UF spectrochemical analysis
  UF spectrographic analysis
  UF spectrometric analysis
  UF spectrometry
  UF spectrophotometric analysis
  UF spectroscopic analysis
  NT alpha-ray spectroscopy
  NT Auger spectroscopy
  NT electron paramagnetic resonance
  NT emission spectroscopy
  NT gamma-ray spectroscopy
  NT inductively coupled plasma methods
  NT infrared spectroscopy
  NT mass spectroscopy
  NT microwave spectroscopy
  NT Mossbauer spectroscopy
  NT nuclear magnetic resonance
  NT optical spectroscopy
  NT radio-frequency spectroscopy
  NT Raman spectroscopy
  NT ultraviolet spectroscopy
  NT X-ray spectroscopy
  SA absorption
  SA alpha rays
  SA analysis
  SA atomic absorption
  SA atomic absorption spectra
  SA chemical analysis
  SA chromatography
  SA clay mineralogy
  SA colorimetry
  SA crystallography
  SA differential thermal analysis
  SA electron microscopy
  SA electron probe
  SA electron probe data
  SA emission spectra
  SA EPR spectra
  SA flame photometry
  SA gamma-ray spectra
  SA geochemistry
  SA infrared spectra
  SA interlaboratory comparison
  SA ion probe
  SA ion probe data
  SA laser methods
  SA laser ranging
  SA major-element analyses
  SA mass spectra
  SA Mossbauer spectra
  SA neutron activation analysis
  SA neutron probe
  SA neutron probe data
  SA NMR spectra
  SA optical spectra
  SA photometry
  SA plasma emission spectroscopy
  SA Raman spectra
  SA secondary ion mass spectroscopy
  SA spectra
  SA standard materials
  SA surface properties
  SA thermal analysis
  SA trace-element analyses
  SA ultraviolet spectra
  SA wave dispersion
  SA wavelength
  SA X-ray analysis
  SA X-ray diffraction analysis
  SA X-ray fluorescence
  SA X-ray fluorescence spectra
  SA X-ray spectra

spectrum
  use spectra

specularite (1978)
  UF gray hematite
  BT oxides
  SA hematite

Speeton Clay (1978)
Cretaceous: Yorkshire.
  BT Lower Cretaceous
  BT Cretaceous
  BT Mesozoic
  SA Barremian
  SA England
  SA United Kingdom

speleology (1978)
Autoposting of solution features to this term began in 1989.
  BT solution features
  SA cave environment
  SA caves
  SA exploration
  SA geomorphology
  SA speleothems

speleothems (1978)
Autoposting of solution features to this term began in 1989. Before 1993, also search flowstone.
  UF flowstone
  BT solution features
  SA caves
  SA geomorphology
  SA speleology
  SA stalactites
  SA stalagmites

Spencer Gulf (1985)
Inlet of Indian Ocean on S coast of South Australia.
  CO S350000S320000 E1380000E1355000
  BT South Australia
  BT Australia
  BT Australasia
  SA Indian Ocean
  SA Spilsby Island

Spergen Formation (1985)
S Indiana, Illinois, E Iowa, W Kentucky and E Missouri.
  BT Meramecian
  BT Upper Mississippian
  BT Mississippian
  BT Carboniferous
  BT Paleozoic
  SA Illinois
  SA Indiana
  SA Iowa
  SA Kentucky
  SA Missouri

Spermatophyta (1981)
  BT Plantae
  NT angiosperms
  NT gymnosperms
  SA fossil wood

sperrylite (1978)
Autoposting of sulfides to this term ended in 1989.
  BT arsenides
  SA platinum minerals

Spessart (1978)
Low mountain range between the Odenwald and the Rhon Mountains in NW Bavaria.
  UF The Spessart
  BT Bavaria Germany
  BT Germany
  BT Central Europe
  BT Europe

spessartine (1978)
  BT garnet group
  BT nesosilicates
  BT orthosilicates
  BT silicates

spessartite (1978)
  BT lamprophyres
  BT plutonic rocks
  BT igneous rocks

Sphaeroidinella dehiscens (1978)
Autoposting of microfossils and Protista to this term began in 1990. This term has multiple hierarchies.
  BT1 Globigerinidae
  BT1 Globigerinacea
  BT1 Rotaliina
  BT1 foraminifera
  BT1 Protista
  BT1 Invertebrata
  BT2 Globigerinidae
  BT2 Globigerinacea
  BT2 Rotaliina
  BT2 foraminifera
  BT2 Protista
  BT2 microfossils

sphaerolites
  use spherulites

Sphagnum (1993)
Genus coextensive with the order Sphagnales. For modern ecology studies use wetlands if applicable.
  BT Musci
  BT bryophytes
  BT Plantae
  SA peat bogs

sphalerite (1978)
  UF pseudogalena
  UF zinc blende
  BT sulfides
  SA zinc ores

sphene
  use titanite

Sphenolithus (1978)
Genus. Autoposting of thallophytes and microfossils began in 1990. This term has multiple hierarchies.
  BT1 nannofossils
  BT1 algae
  BT1 thallophytes
  BT1 Plantae
  BT2 nannofossils
  BT2 algae
  BT2 microfossils
  BT3 nannofossils
  BT3 algal flora
  BT3 plants
  BT4 nannofossils
  BT4 algal flora
  BT4 microfossils
  SA Discoasteridae

Sphenophyllum (1978)
Genus of Paleozoic fossil plants.
  BT Sphenopsida
  BT pteridophytes
  BT Plantae

Sphenopsida (1978)
Including Articulatae. Autoposting of this term began in 1978.
  BT pteridophytes
  BT Plantae
  NT Articulatae
  NT Equisetales
  NT Sphenophyllum
  NT Sphenopteris
  SA vascular taxa

Sphenopteris (1978)
Genus. This term has multiple hierarchies.
  BT1 Filicopsida
  BT1 pteridophytes
  BT1 Plantae
  BT2 Sphenopsida
  BT2 pteridophytes
  BT2 Plantae
  SA Articulatae
  SA Pteridospermae

spherical harmonic analysis (1978)
  UF spherical harmonics
  SA analysis
  SA Earth
  SA magnetic field
  SA plate tectonics

spherical harmonics
  use spherical harmonic analysis

spherical models (1981)
  SA mathematical models
  SA models
  SA seismology

sphericity (1978)
  SA particles
  SA roundness
  SA sediments
  SA shape analysis

spheroidal mode (1981)
  SA oscillations
  SA seismographs
  SA seismology

spherules (1978)
  SA sedimentary structures
  SA sediments
  SA tektites

spherulites (1978)
  UF sphaerolites
  SA orbicular texture

Sphinctozoa (1978)
  BT Calcispongea
  BT Porifera
  BT Invertebrata

Spice Islands
  use Moluccas

spicules (1978)
  SA Porifera

spilite (1978)
  BT alkali basalts
  BT basalts
  BT volcanic rocks
  BT igneous rocks
  SA pillow lava
  SA spilitization

spilitization (1978)
  BT metasomatism
  SA albitization
  SA spilite

spills, oil
  use oil spills

Spilsby Island (1978)
Largest of Sir Joseph Banks Islands in Spencer Gulf about 5 miles off SE coast of Eyre Peninsula. Also search Spilsby.
  BT South Australia
  BT Australia
  BT Australasia
  SA Spencer Gulf

spinal column (1981)
  SA anatomy
  SA morphology
  SA Vertebrata

spinel (1978)
  BT oxides
  SA chrome spinel
  SA spinel group

spinel group (1978)
　BT oxides
　SA gahnite
　SA hercynite
　SA jacobsite
　SA maghemite
　SA magnesioferrite
　SA magnetite
　SA spinel
　SA ulvospinel

spinel lherzolite (1978)
　BT peridotites
　BT ultramafics
　BT plutonic rocks
　BT igneous rocks
　SA lherzolite

spinel peridotite (1985)
　As of 1990, ultramafics is autoposted to this term.
　BT peridotites
　BT ultramafics
　BT plutonic rocks
　BT igneous rocks

spinifex texture (1981)
　Before 1981, also search spinifex.
　BT textures
　SA igneous rocks

spirals, growth
　use growth spirals

Spiriferida (1978)
　Autoposting of this term to Atrypidae, Atrypa and Cyrtospirifer began in 1981.
　BT Articulata
　BT Brachiopoda
　BT Invertebrata
　NT Atrypidae
　NT Cyrtospirifer
　NT Spiriferidina

Spiriferidae (1978)
　Family.
　BT Spiriferidina
　BT Spiriferida
　BT Articulata
　BT Brachiopoda
　BT Invertebrata

Spiriferidina (1978)
　Suborder. Autoposting of this term began in 1978.
　BT Spiriferida
　BT Articulata
　BT Brachiopoda
　BT Invertebrata
　NT Spiriferidae
　NT Spiriferina

Spiriferina (1978)
　Genus.
　BT Spiriferidina
　BT Spiriferida
　BT Articulata
　BT Brachiopoda
　BT Invertebrata

Spirillinacea (1978)
　Autoposting of microfossils and Protista to this term began in 1990. This term has multiple hierarchies.
　BT1 Rotaliina
　BT1 foraminifera
　BT1 Protista
　BT1 Invertebrata
　BT2 Rotaliina
　BT2 foraminifera
　BT2 Protista
　BT2 microfossils

Spirit Lake (1989)
　Skamania County, SW Washington; Dickinson County, NW Iowa.
　IN Index states and counties as applicable.
　SA Dickinson County Iowa
　SA Iowa
　SA Skamania County Washington
　SA Washington

Spirit River Formation (1989)
　Includes Falher Member.
　BT Lower Cretaceous
　BT Cretaceous
　BT Mesozoic
　SA Alberta

Spiro Sandstone (1993)
　Desmoinesian. Arkoma Basin. E Oklahoma, Arkansas. In Savanna Formation.
　BT Desmoinesian
　BT Middle Pennsylvanian
　BT Pennsylvanian
　BT Carboniferous
　BT Paleozoic
　SA Arkansas
　SA Arkoma Basin
　SA Oklahoma
　SA Savanna Formation

Spis (1978)
　Region. An historic area in dispute before WWI between Hungary, Austria, and Russia. After 1920, area was split between Slovakia and Poland.
　IN Index Poland and/or Slovakia as applicable.
　BT Central Europe
　BT Europe
　SA Czechoslovakia
　SA Poland
　SA Slovakia

Spis-Gemer (1978)
　The Slovak Ore Mountains. A range of the Carpathians in S Slovakia.
　CO N482600N485600
　　　E0212000E0190600
　UF Slovenske Rudohorie
　UF Spis-Gemer Mountains
　BT Slovakia
　BT Czechoslovakia
　BT Central Europe
　BT Europe
　SA Carpathians

Spis-Gemer Mountains
　use Spis-Gemer

Spitak earthquake 1988 (1993)
　Armenia. Before 1993, also search Armenia earthquake 1988.
　UF Armenia earthquake 1988
　BT earthquakes
　SA Armenia

Spiti (1978)
　Region in NE Punjab State, N India.
　BT Punjab India
　BT India
　BT Indian Peninsula
　BT Asia

spits (1978)
　Autoposting of shore features to this term began in 1989.
　BT shore features
　SA beaches
　SA capes
　SA coastal environment
　SA coastlines
　SA geomorphology
　SA littoral erosion
　SA marine environment
　SA reefs
　SA shoals
　SA tombolos

Spitsbergen (1978)
　Norwegian archipelago, 360 miles N of Norway, including the main island of Spitsbergen plus North East Land, Edge Island, and Barents Island. Part of the Svalbard Island group. Before 1972, also search Spitzbergen. As of 1993, use Spitsbergen Island for the island. Autoposting of Polar regions to this term began in 1990.
　CO N760000N810000
　　　E0300000E0100000
　UF Spitzbergen
　BT Svalbard
　BT Arctic region
　NT Nordaustlandet
　NT Spitsbergen Island
　SA Hecla Hoek Formation

Spitsbergen Island (1993)
　Main island, of the Spitsbergen Archipelago of the Svalbard island group. As of 1993, Brogger Peninsula, Hornsund, and Ny Friesland are narrower terms. Before 1993, also search West Spitsbergen and Vestspitsbergen.
　UF Vestspitsbergen
　UF West Spitsbergen
　BT Spitsbergen
　BT Svalbard
　BT Arctic region
　NT Brogger Peninsula
　NT Hornsund
　NT Ny Friesland
　SA Hecla Hoek Formation
　SA Kings Bay

Spitz
　No longer a valid term for GeoRef. See Spitz Austria.

Spitz Austria (1993)
　Village on left bank of Danube River.
　BT Lower Austria
　BT Austria
　BT Central Europe
　BT Europe

Spitzbergen
　Not a valid GeoRef index term after 1971. Was used on level 1 in subfile G.
　use Spitsbergen

spline theory
　use splines

splines (1989)
　UF spline theory
　SA gravity field
　SA mathematical methods

Split (1978)
　City on the central Dalmatian Coast on the Adriatic Sea in W Yugoslavia.
　BT Croatia
　BT Yugoslavia
　BT Southern Europe
　BT Europe

Spodosols (1978)
　BT soils
　SA soil group

spodumene (1978)
　BT clinopyroxene
　BT pyroxene group
　BT chain silicates
　BT silicates
　SA aluminosilicates
　SA ceramic materials
　SA kunzite
　SA lithium ores

spoils (1981)
　SA dredged materials
　SA dredging
　SA environmental geology
　SA mining
　SA tailings
　SA tailings ponds
　SA waste disposal

Spokane
　Valid through 1988. Search in combination with state term. After 1988, use specific city-state term.

Spokane County
　Valid through 1988. Search in combination with state term. After 1988, use specific county-state term.

Spokane County Washington (1989)
　E Washington. Before 1989, also search Spokane County AND Washington.
　CO N471500N480400
　　　W1170300W1175000
　BT Washington
　BT United States
　NT Spokane Washington
　SA Midnite Mine

Spokane Formation (1985)
　In Missoula Group. S central and W Montana.
　BT middle Proterozoic
　BT Proterozoic
　BT upper Precambrian
　BT Precambrian
　SA Belt Supergroup
　SA Missoula Group
　SA Montana

Spokane Washington (1989)
　City in E Washington. Before 1989, search Spokane AND Washington.
　CO N474000N474000
　　　W1172500W1172500
　BT Spokane County Washington
　BT Washington
　BT United States

Spoleto
　No longer a valid term for GeoRef. See Spoleto Italy.

Spoleto Italy (1993)
　City in Perugia Province in SE Umbria, central Italy.
　BT Perugia Italy
　BT Umbria Italy
　BT Italy
　BT Southern Europe
　BT Europe

sponges (1985)
　Common name for Porifera.
　IN Index for all non-paleontologic studies of fossils.
　BT invertebrates
　SA biostratigraphy
　SA Porifera

Spongiae
　use Porifera

spongolite (1978)
　BT clastic rocks
　BT sedimentary rocks

spontaneous fission-track dating
　use fission-track dating

spontaneous magnetization (1981)
　BT magnetization

spontaneous polarization
　use self-potential methods

spontaneous potential
　use self-potential methods

Spoon Formation (1985)
　In Kewanee Group.

BT Middle Pennsylvanian
BT Pennsylvanian
BT Carboniferous
BT Paleozoic
SA Illinois
SA Kewanee Group

Spor Mountain (1978)
IN Index county or region and mountains as applicable.
SA Utah

sporangia (1978)
UF sporangium
SA angiosperms
SA gymnosperms
SA paleobotany
SA pteridophytes
SA spores

sporangium
use sporangia

spores (1978)
SA cones
SA exine
SA megaspores
SA miospores
SA palynomorphs
SA pollen
SA pteridophytes
SA sporangia
SA sporopollenin

sporinite (1978)
BT exinite
BT macerals

sporopollenin (1978)
UF sporopollenine
BT organic materials
SA exine
SA miospores
SA pollen
SA spores

sporopollenine
use sporopollenin

SPOT (1985)
Acronym.
UF Systeme Probatoire d'Observation de la Terre
SA geophysical methods
SA geophysical surveys
SA multispectral analysis
SA remote sensing
SA satellite methods

spreading centers (1978)
UF centers, spreading
SA mid-ocean ridge basalts
SA plate tectonics
SA sea-floor spreading
SA submarine volcanoes

spreading concept
use sea-floor spreading

spreading-floor hypothesis
use sea-floor spreading

spring gravimeters (1981)
BT gravimeters

Spring Mountains (1978)
In W Clark County near California line.
IN Index county or region and mountains as applicable.
SA Clark County Nevada

Springdale
Valid through 1988. Search in combination with state term. After 1988, use specific city-state term.

Springdale Utah (1989)
Village in SW Utah. Gateway to Zion National Park. Before 1989, search Springdale AND Utah.
CO N371000N371000
W1130000W1130000
BT Washington County Utah
BT Utah
BT United States

Springer Formation (1978)
Chester and Morrow Series. Conformably overlies Goddard Shale. Central S Oklahoma.
BT Carboniferous
BT Paleozoic
SA Chesterian
SA Mississippian
SA Morrowan
SA Oklahoma
SA Pennsylvanian

Springfield
Valid through 1988. Search in combination with state term. After 1988, use specific city-state term.

Springfield Coal Member (1978)
Of Carbondale Formation. W and N Illinois.
BT Pennsylvanian
BT Carboniferous
BT Paleozoic
SA Carbondale Formation
SA Illinois

Springfield Illinois (1989)
City on the Sangamon River in central Illinois. Before 1989, search Springfield AND Illinois.
CO N394900N394900
W0893900W0893900
BT Sangamon County Illinois
BT Illinois
BT United States

Springfield Massachusetts (1989)
City in SW Massachusetts. Before 1989, search Springfield AND Massachusetts.
CO N420700N420700
W0723500W0723500
BT Hampden County Massachusetts
BT Massachusetts
BT United States

Springfield Missouri (1989)
City in SW Missouri. Before 1989, search Springfield AND Missouri.
CO N371100N371100
W0931900W0931900
BT Greene County Missouri
BT Missouri
BT United States

Springfield Ohio (1989)
City in W central Ohio. Before 1989, search Springfield AND Ohio.
BT Clark County Ohio
BT Ohio
BT United States

springs (1978)
Used for papers stressing spring hydrology. Autoposting of this term began in 1978.
NT hot springs
NT intermittent springs
NT submarine springs
SA circulation
SA discharge
SA exsurgence
SA fumaroles
SA geysers
SA ground water
SA hydrochemistry
SA hydrogeology
SA hydrology
SA mineral waters
SA salinity
SA sand boils
SA saturation
SA shallow aquifers
SA shallow depth
SA surficial aquifers
SA thermal waters
SA underground streams
SA water
SA water resources

Springsure
No longer a valid term for GeoRef. See Springsure Australia.

Springsure Australia (1993)
Village 165 miles WSW of Rockhampton in E central Queensland.
BT Queensland Australia
BT Australia
BT Australasia
SA Bowen Basin

Springwater
Valid through 1988. Search in combination with state term. After 1988, use specific city-state term.

Springwater New York (1989)
Village in W central New York. Before 1989, search Springwater AND New York.
CO N423800N423800
W0773400W0773400
BT Livingston County New York
BT New York
BT United States

Spumellaria
use Spumellina

Spumellina (1978)
Autoposting of Protista and microfossils to this term began in 1990. This term has multiple hierarchies.
UF Spumellaria
BT1 Radiolaria
BT1 Protista
BT1 Invertebrata
BT2 Radiolaria
BT2 Protista
BT2 microfossils

spurrite (1978)
From 1978-1984, orthosilicates and carbonates were autoposted to this term. In 1985, autoposting of nesosilicates began, while autoposting of orthosilicates ended. As of 1989, orthosilicates and silicates are autoposted to this term. This term has multiple hierarchies.
BT1 nesosilicates
BT1 orthosilicates
BT1 silicates
BT2 carbonates

Squamata (1978)
Order. Autoposting of this term to Lacertilia and Mosasauridae began in 1981.
BT Lepidosauria
BT Diapsida
BT Reptilia
BT Tetrapoda
BT Vertebrata
BT Chordata
NT Lacertilia
NT Mosasauridae

SQUID (1985)
Acronym.
UF superconducting quantum interference devices
SA electromagnetic logging
SA electromagnetic methods
SA instruments
SA magnetic methods
SA magnetic surveys
SA magnetometers
SA magnetotelluric methods

Sr
use strontium

Sr-85 (1985)
Autoposting of broader terms began in 1989. This term has multiple hierarchies.
BT1 radioactive isotopes
BT1 isotopes
BT2 strontium
BT2 alkaline earth metals
BT2 metals

Sr-86 (1978)
Autoposting of broader terms began in 1989. This term has multiple hierarchies.
BT1 stable isotopes
BT1 isotopes
BT2 strontium
BT2 alkaline earth metals
BT2 metals
SA Rb-87/Sr-86
SA Sr-87/Sr-86

Sr-86/Rb-87
use Rb-87/Sr-86

Sr-86/Sr-87
use Sr-87/Sr-86

Sr-87 (1978)
Autoposting of broader terms began in 1989. This term has multiple hierarchies.
BT1 stable isotopes
BT1 isotopes
BT2 strontium
BT2 alkaline earth metals
BT2 metals
SA Sr-87/Sr-86

Sr-87/Sr-86 (1978)
Autoposting of broader terms began in 1989. This term has multiple hierarchies.
UF Sr-86/Sr-87
BT1 stable isotopes
BT1 isotopes
BT2 strontium
BT2 alkaline earth metals
BT2 metals
SA isotope ratios
SA Sr-86
SA Sr-87

Sr-89 (1985)
Autoposting of broader terms began in 1989. This term has multiple hierarchies.
BT1 radioactive isotopes
BT1 isotopes
BT2 strontium
BT2 alkaline earth metals
BT2 metals

Sr-90 (1978)
Autoposting of broader terms began in 1989. This term has multiple hierarchies.
BT1 radioactive isotopes
BT1 isotopes
BT2 strontium
BT2 alkaline earth metals
BT2 metals

Sr/Ca (1989)
This term has multiple hierarchies.
UF Ca/Sr
BT1 calcium
BT1 alkaline earth metals
BT1 metals
BT2 strontium
BT2 alkaline earth metals
BT2 metals
SA chemical ratios

Sr/Rb
As of 1989, no longer a valid term for GeoRef.

Sr/Sr
   use Rb/Sr

Sr/Sr (1989)
   Isotopic ratio used for age determination.
   SA absolute age
   SA isotope ratios
   SA strontium

Sredna Gora (1978)
   Mountain range between the Balkan and Rhodope mountains in central Bulgaria.
   CO N421500N424500
         E0250000E0233000
   UF Sredna Gora Mountains
   BT Bulgaria
   BT Southern Europe
   BT Europe
   SA Panagyurishte Bulgaria

Sredna Gora Mountains
   use Sredna Gora

Sredniy Vasyugan
   No longer a valid term for GeoRef. As of 1993, see Sredniy Vasyugan Russian Republic.

Sredniy Vasyugan Russian Federation (1993)
   Village on the Vasyugan River in NW Tomsk Oblast in S central Siberia. This term has multiple hierarchies.
   BT1 Tomsk Russian Federation
   BT1 Russian Federation
   BT1 Commonwealth of Independent States
   BT2 Tomsk Russian Federation
   BT2 Asia

Sri Lanka (1978)
   Formerly Ceylon. Before 1974, also search Ceylon.
   CO N060000N100000
         E0823000E0790000
   UF Ceylon
   BT Indian Peninsula
   BT Asia
   SA Highland Series
   SA Indian Shield

Srikakulam
   No longer a valid term for GeoRef. See Srikakulam India.

Srikakulam India (1993)
   City on Bay of Bengal in NE Andhra Pradesh, SE India.
   BT Andhra Pradesh India
   BT India
   BT Indian Peninsula
   BT Asia

Srinagar (1978)
   City and district in the Kashmir Valley, Jammu and Kashmir. As of 1993, India is no longer a broader term.
   IN Index Jammu and Kashmir AND India and/or Pakistan as applicable.
   SA India
   SA Jammu and Kashmir
   SA Pakistan

St. Bernard Parish
   use Saint Bernard Parish Louisiana

St. Catherines Ontario
   use Saint Catharines Ontario

St. Clair River
   use Saint Clair River

St. Clair River delta
   use Saint Clair River delta

St. Croix
   use Saint Croix

St. Croix Island
   use Saint Croix

St. Elias Mountains
   use Saint Elias Mountains

St. Francois Mountains
   use Saint Francois Mountains

St. Gallen
   use Saint Gall Switzerland

St. George Group
   use Saint George Group

St. Gotthard
   use Gotthard Massif

St. Helena
   use Saint Helena

St. Helena Island
   use Saint Helena

St. John
   use Saint John New Brunswick

St. John County
   use Saint John County New Brunswick

St. John River
   use Saint John River

St. John's
   Use Saint John's Newfoundland if applicable.

St. Johns River basin
   use Saint Johns River basin

St. Laurent Limestone
   use Saint Laurent Limestone

St. Lawrence County
   use Saint Lawrence County New York

St. Lawrence Estuary
   use Saint Lawrence Estuary

St. Lawrence Lowlands
   use Saint Lawrence Lowlands

St. Lawrence River
   use Saint Lawrence River

St. Lawrence River valley
   use Saint Lawrence Valley

St. Lawrence Valley
   use Saint Lawrence Valley

St. Louis
   use Saint Louis Missouri

St. Louis Limestone
   use Saint Louis Limestone

St. Lucia
   use Saint Lucia

St. Mary Parish
   use Saint Mary Parish Louisiana

St. Marys Formation
   use Saint Marys Formation

St. Paul
   use Saint Paul Minnesota

St. Paul's Rock
   use Saint Paul Rocks

St. Paul's Rocks
   use Saint Paul Rocks

St. Peter Sandstone
   use Saint Peter Sandstone

St. Pierre and Miquelon
   No longer a valid GeoRef index term. Before 1971, was used on level 1 in subfile N.
   use Saint Pierre and Miquelon

St. Regis Formation
   use Saint Regis Formation

St. Severin France
   use Saint-Severin France

St. Severin Meteorite
   use Saint-Severin Meteorite

St. Thomas
   use Saint Thomas

St. Vincent
   use Saint Vincent

stability (1978)
   SA engineering geology
   SA foundations
   SA geochemistry
   SA land subsidence
   SA paleomagnetism
   SA slope stability
   SA stabilization
   SA tunnels
   SA underground installations

stabilization (1978)
   SA artificial islands
   SA retaining walls
   SA shorelines
   SA slope stability
   SA stability

stable isotopes (1978)
   Autoposting of this term to D/H, deuterium, C-12, C-13, C-13/C-12, O-16, O-18, O-18/O-16, S-32, S-34 and S-34/S-32 began in 1985.
   BT isotopes
   NT Al-27
   NT Al-27/Al-26
   NT Ar-36
   NT Ar-38/Ar-36
   NT Ar-40
   NT Ar-40/Ar-36
   NT Ar-40/Ar-39
   NT Be-10/Be-9
   NT C-12
   NT C-13
   NT C-13/C-12
   NT C-14/C-12
   NT Cl-37/Cl-35
   NT Cr-53/Cr-52
   NT D/H
   NT deuterium
   NT Fe-57
   NT He-3
   NT He-4
   NT He-4/He-3
   NT Hf-177/Hf-176
   NT Kr-84
   NT Li-6
   NT Li-7
   NT Mg-25/Mg-24
   NT Mg-26/Mg-24
   NT N-15
   NT N-15/N-14
   NT Nd-144/Nd-143
   NT Ne-20
   NT Ne-21
   NT Ne-22
   NT Ne-22/Ne-20
   NT Ne-22/Ne-21
   NT O-16
   NT O-17/O-16
   NT O-18
   NT O-18/O-16
   NT Os-187/Os-186
   NT Pb-206
   NT Pb-206/Pb-204
   NT Pb-207
   NT Pb-207/Pb-204
   NT Pb-207/Pb-206
   NT Pb-208
   NT Pb-208/Pb-204
   NT Pb-208/Pb-206
   NT Rb-87/Sr-86
   NT S-32
   NT S-34
   NT S-34/S-32
   NT Sm-147/Nd-144
   NT Sr-86
   NT Sr-87
   NT Sr-87/Sr-86
   NT U-238/Pb-206
   NT Xe-124
   NT Xe-126
   NT Xe-128
   NT Xe-129
   NT Xe-130/Xe-129
   NT Xe-131
   NT Xe-132
   NT Xe-132/Xe-129
   NT Xe-136/Xe-130
   NT Xe-136/Xe-132
   SA carbon
   SA hydrogen
   SA oxygen
   SA sulfur

Stablo
   See Stavelot Belgium and Stavelot-Venn Massif.

stacking (1993)
   SA crystal structure
   SA data processing
   SA defects
   SA geophysical methods
   SA normal moveout
   SA seismic methods
   SA signal-to-noise ratio
   SA tectonics
   SA time domain analysis
   SA velocity
   SA velocity analysis

Stafford County
   Valid through 1988. Search in combination with state term. After 1988, use specific county-state term.

Stafford County Kansas (1989)
   S central Kansas. Before 1989, also search Stafford County AND Kansas.
   CO N375000N381500
         W0982300W0990200
   BT Kansas
   BT United States

Stafford County Virginia (1989)
   NE Virginia. Before 1989, also search Stafford County AND Virginia.
   CO N381600N383500
         W0771900W0773800
   BT Virginia
   BT United States

Staffordshire
   No longer a valid term for GeoRef as of 1993.
   use Staffordshire England

Staffordshire England (1993)
   County in W central England. After 1974, part of the territory was taken to form West Midlands County. Before 1993, also search Staffordshire.
   CO N523000N531500
         W0013000W0023000
   UF Staffordshire
   BT England
   BT Great Britain
   BT United Kingdom
   BT Western Europe
   BT Europe
   SA Dudley England
   SA Midlands
   SA Wolverhampton England

stage, universal
   use universal stage

stages
   A valid general term through 1977. Use specific term, e.g. stratigraphic units.

stainierite
use heterogenite

staining (1978)
SA paleobotany
SA paleontology
SA petrology
SA sample preparation

Staked Plain
use Llano Estacado

stalactites (1982)
SA caverns
SA caves
SA geomorphology
SA solution features
SA speleothems
SA stalagmites
SA volcanic features

stalagmites (1982)
SA caverns
SA caves
SA geomorphology
SA solution features
SA speleothems
SA stalactites
SA volcanic features

Stalin
Not a valid term for GeoRef. A former name for various places including the city of Brasov in Romania; seaport of Varna in Bulgaria; city of Donetsk in the Ukraine.

Stalinabad Tadzhikistan
use Dushanbe Tadzhikistan

Stalingrad Russian Federation
use Volgograd Russian Federation

Stampian (1978)
France.
BT lower Oligocene
BT Oligocene
BT Paleogene
BT Tertiary
BT Cenozoic
SA Rupelian
SA Sannoisian

standard deviation (1978)
UF deviation, standard
BT statistical analysis

standard materials (1978)
Used for rocks or minerals or other materials that have been designated as standard by geological laboratories. Before 1976, search standard rocks.
IN Index with specific type of material and name of laboratory where available.
SA absolute age
SA chemical analysis
SA crystallography
SA geochemistry
SA igneous rocks
SA interlaboratory comparison
SA isotopes
SA materials
SA minerals
SA petrology
SA silicate rocks
SA spectroscopy
SA standard rocks
SA standardization
SA thermal analysis

standard rocks (1978)
Was used as a first level term until 1976.
IN Also index standard materials.
SA interlaboratory comparison
SA rocks

SA standard materials

standardization (1978)
Used as a general term.
SA interlaboratory comparison
SA standard materials

Stanislaus County
Valid through 1988. Search in combination with state term. After 1988, use specific county-state term.

Stanislaus County California (1989)
Central California. Before 1989, also search Stanislaus County AND California.
CO N370500N380500
W1202500W1213000
BT California
BT United States

Stanislav Ukraine
use Ivano-Frankovsk Ukraine

Stanislawow Ukraine
use Ivano-Frankovsk Ukraine

Stanley Group (1978)
Includes Hatton Tuff Lentil, Ten Mile Creek Formation, Moyers Formation, Chickasaw Creek Formation. W Arkansas, and central S and SE Oklahoma.
BT Mississippian
BT Carboniferous
BT Paleozoic
SA Arkansas
SA Oklahoma

Stanly County
Valid through 1988. Search in combination with state term. After 1988, use specific county-state term.

Stanly County North Carolina (1989)
S central North Carolina. Before 1989, also search Stanly County AND North Carolina.
CO N351000N353000
W0800500W0803100
BT North Carolina
BT United States

stannite (1978)
UF tin pyrites
BT sulfostannates
BT sulfosalts

stannoidite (1978)
BT sulfides

Stanovoi Range
use Stanovoy Range

Stanovoy Range (1978)
Mountain range between Yakutia and Amur Oblast with Khabarovsk Kray on the E. Also search Stanovoy. This term has multiple hierarchies.
CO N530000N573000
E1350000E1150000
UF Stanovoi Range
UF Stanovoy Upland
BT1 Russian Federation
BT1 Commonwealth of Independent States
BT2 Asia
SA Amur region
SA Khabarovsk Russian Federation
SA Yakutia Russian Federation

Stanovoy Upland
use Stanovoy Range

Stansbury Mountains (1978)

In Wasatch National Forest in NW Utah.
BT Tooele County Utah
BT Utah
BT United States

Stanton County
Valid through 1988. Search in combination with state term. After 1988, use specific county-state term.

Stanton County Kansas (1989)
SW Kansas. Before 1989, also search Stanton County AND Kansas.
CO N372200N374100
W1013100W1020300
BT Kansas
BT United States

Stanton County Nebraska (1989)
NE Nebraska. Before 1989, also search Stanton County AND Nebraska.
CO N414300N420600
W0970200W0972200
BT Nebraska
BT United States

Stanton Formation (1978)
In Lansing Group. Includes Captain Creek Limestone, Eudora Shale, Stoner Limestone, Rock Lake Shale, South Bend Limestone members. SW Iowa, E Kansas, NW Missouri, SE Nebraska, and NE Oklahoma.
BT Missourian
BT Upper Pennsylvanian
BT Pennsylvanian
BT Carboniferous
BT Paleozoic
SA Iowa
SA Kansas
SA Lansing Group
SA Missouri
SA Nebraska
SA Oklahoma
SA Rock Lake Shale Member

Stanton's Cave (1989)
Paleontological and archaeological site in Grand Canyon National Park, NW Arizona.
BT Arizona
BT United States
SA Grand Canyon

Star Point Sandstone (1985)
In Mesaverde Group. E central Utah.
BT Upper Cretaceous
BT Cretaceous
BT Mesozoic
SA Mesaverde Group
SA Utah

Stara Planina
use Balkan Mountains

Stark County
Valid through 1988. Search in combination with state term. After 1988, use specific county-state term.

Stark County Illinois (1989)
N central Illinois. Before 1989, also search Stark County AND Illinois.
CO N405800N411300
W0893800W0900000
BT Illinois
BT United States

Stark County North Dakota (1989)

W North Dakota. Before 1989, also search Stark County AND North Dakota.
CO N463800N470100
W1020600W1031400
BT North Dakota
BT United States

Stark County Ohio (1989)
E central Ohio. Before 1989, also search Stark County AND Ohio.
CO N403800N405800
W0810500W0813900
BT Ohio
BT United States

Starobin
No longer a valid term for GeoRef. As of 1993, see Starobin Belarus.

Starobin Belarus (1993)
Town in S central Belarus. This term has multiple hierarchies.
BT1 Belarus
BT1 Europe
BT2 Belarus
BT2 Commonwealth of Independent States

Stassfurt
No longer a valid term for GeoRef. See Stassfurt Germany.

Stassfurt Germany (1993)
Salt mining city in Saxony-Anhalt, central Germany.
BT Saxony-Anhalt Germany
BT Germany
BT Central Europe
BT Europe

Staten Island (1978)
S of Manhattan and across The Narrows from Long Island on the E. Borough of New York City, co-extensive with Richmond County, SE New York. Also in Argentina (Isla de los Estados).
IN Index county and state or region as applicable.
SA Argentina
SA Magallanes Chile
SA New York
SA New York City New York
SA Raritan Bay

Statfjord Formation (1989)
N North Sea.
BT Mesozoic
SA North Sea

statistical analysis (1978)
Used as a general term for studies that use statistics in analyzing data. As of 1981, includes statistical methods. Before 1981, also search statistical methods. Autoposting of this term began in 1981.
UF statistical methods
NT autocorrelation
NT Bayesian analysis
NT canonical analysis
NT cluster analysis
NT correlation coefficient
NT correspondence analysis
NT covariance analysis
NT crosscorrelation
NT dendrograms
NT discriminant analysis
NT factor analysis
NT finite difference analysis
NT finite element analysis
NT geostatistics
NT histograms
NT kriging
NT kurtosis
NT least-squares analysis
NT Markov chain analysis

NT Monte Carlo analysis
NT multivariate analysis
NT phi scale
NT populations
NT principal components analysis
NT probability
NT regression analysis
NT skewness
NT standard deviation
NT time series analysis
NT trend-surface analysis
NT univariate analysis
NT variance analysis
SA analysis
SA asymmetric distribution
SA biometry
SA cladistics
SA data processing
SA eigenvalues
SA equations
SA geometry
SA iterative methods
SA mathematical geology
SA mathematical methods
SA methods
SA numerical analysis
SA outliers
SA pattern recognition
SA reliability
SA statistical distribution
SA water

statistical distribution (1981)
SA asymmetric distribution
SA distribution
SA statistical analysis

statistical measures
Not a valid index term after 1970. Was used on level 1 in subfile N. Now use statistical analysis or geostatistics.

statistical methods
As of 1981, no longer a valid term for GeoRef. From 1978-1980, was autoposted to many types of statistical analysis.
use statistical analysis

statistics
No longer a valid GeoRef index term. Formerly was used in subfiles B and N. Now use statistical analysis or geostatistics.

Staunton Formation (1978)
Overlies Brazil Formation. SW Indiana.
BT Middle Pennsylvanian
BT Pennsylvanian
BT Carboniferous
BT Paleozoic
SA Indiana

staurolite (1978)
Autoposting of nesosilicates began in 1985.
BT nesosilicates
BT orthosilicates
BT silicates

Stavanger
No longer a valid term for GeoRef. As of 1993 see Stavanger Norway.

Stavanger Norway (1993)
City on Stavanger Fjord on North Sea S of Bergen. Before 1993 also search Stavanger AND Norway
BT Rogaland Norway
BT Norway
BT Scandinavia
BT Western Europe
BT Europe

Stavelot
No longer a valid term for GeoRef. See Stavelot Belgium or Stavelot-Venn Massif.

Stavelot Belgium (1993)
Town in the N Ardennes.
BT Liege Belgium
BT Belgium
BT Western Europe
BT Europe
SA Ardennes

Stavelot-Venn Massif (1978)
Primarily in the N Ardennes of Belgium.
IN Index countries as applicable.
BT Europe
SA Belgium
SA Germany
SA Hohe Venn

Stavers Island
use Vostok

Stavropol region (1978)
In Stavropol Kray (Ordzhonikidze Kray before 1945) in the Northern Caucasus. Before 1978, also search Stavropol. This term has multiple hierarchies.
UF Voroshilovsk (region)
BT1 Stavropol Russian Federation
BT1 Russian Federation
BT1 Commonwealth of Independent States
BT2 Stavropol Russian Federation
BT2 Europe

Stavropol Russian Federation (1993)
Kray (formerly Ordzhonikidze Kray) in the Northern Caucasus. This term has multiple hierarchies.
BT1 Russian Federation
BT1 Commonwealth of Independent States
BT2 Europe
NT Prikumsk Russian Federation
NT Stavropol region
SA Northern Caucasus

Ste. Genevieve Limestone
use Sainte Genevieve Limestone

steady flow (1982)
SA fluid dynamics
SA hydraulics
SA hydrodynamics
SA hydrology
SA unsteady flow

steady-state processes (1985)
Also search steady state.
SA equilibrium
SA expanding universe theory
SA processes
SA transient phenomena

steam
use water vapor

steam coal (1981)
This term has multiple hierarchies.
BT1 coal
BT1 organic residues
BT2 coal
BT2 sedimentary rocks

steam injection (1985)
SA enhanced recovery
SA fluid injection
SA petroleum engineering
SA secondary recovery
SA waterflooding

Steamboat Springs (1989)
Hot springs in Routt County, NW Colorado. Also in Washoe County, NW Nevada.
IN Index states and counties as applicable.
SA Colorado
SA Nevada
SA Routt County Colorado
SA Washoe County Nevada

Stearns County
Valid through 1988. Search in combination with state term. After 1988, use specific county-state term.

Stearns County Minnesota (1989)
Central Minnesota. Before 1989, also search Stearns County AND Minnesota.
CO N451500N454700 W0940500W0951000
BT Minnesota
BT United States

Stearns Shale (1978)
In Council Grove Group. Kansas, and SE Nebraska.
BT Permian
BT Paleozoic
SA Council Grove Group
SA Kansas
SA Nebraska

steatite (1985)
BT metasomatic rocks
BT metamorphic rocks
SA soapstone
SA talc
SA talc deposits

Stebnik
No longer a valid term for GeoRef. As of 1993, see Stebnik Ukraine.

Stebnik Ukraine (1993)
Town in Lvov Oblast in SW Ukraine. This term has multiple hierarchies.
BT1 Lvov Ukraine
BT1 Ukraine
BT1 Europe
BT2 Lvov Ukraine
BT2 Ukraine
BT2 Commonwealth of Independent States

Steele Glacier (1978)
Emanates from Steele Mountain in Saint Elias Mountains near Alaska border in SW Yukon Territory.
BT Yukon Territory
BT Western Canada
BT Canada
SA Saint Elias Mountains

Steens Mountain (1978)
Mountain mass in SW Harney County in SE Oregon.
CO N420000N431000 W1181000W1185000
BT Harney County Oregon
BT Oregon
BT United States

Stegodon (1978)
Genus. Autoposting of Eutheria and Theria to this term began in 1989.
BT Elephantidae
BT Elephantoidea
BT Proboscidea
BT Eutheria
BT Theria
BT Mammalia
BT Tetrapoda
BT Vertebrata
BT Chordata

Steinach
No longer a valid term for GeoRef. See Steinach Germany.

Steinach Germany (1993)
Town in Thuringian Forest in Thuringia.
BT Thuringia Germany
BT Germany
BT Central Europe
BT Europe
SA Thuringian Forest

Steinbach
No longer a valid term for GeoRef. See Steinbach Manitoba.

Steinbach Manitoba (1993)
Town 39 miles SE of Winnipeg.
BT Manitoba
BT Western Canada
BT Canada

Steinheim
No longer a valid term for GeoRef. See Steinheim Germany.

Steinheim Basin (1978)
BT Germany
BT Central Europe
BT Europe

Steinheim Germany (1993)
Towns in two states.
IN Index states as applicable.
BT Germany
BT Central Europe
BT Europe
SA Hesse Germany
SA North Rhine-Westphalia Germany

Stellarton Group (1993)
Carboniferous. Pictou Coalfield, Nova Scotia.
BT Carboniferous
BT Paleozoic
SA Nova Scotia

stellerite (1978)
BT zeolite group
BT framework silicates
BT silicates

Stelleroidea (1978)
Autoposting of this term began in 1978.
BT Asterozoa
BT Echinodermata
BT Invertebrata
NT Asteroidea
NT Ophiuroidea
NT Somasteroidea

step faults (1978)
Before 1978, also search faults AND step.
BT faults
SA thrust faults

Stephanian (1978)
Europe. Above Westphalian, below Sakmarian of Permian.
BT Upper Carboniferous
BT Carboniferous
BT Paleozoic

stephanite (1978)
BT sulfantimonites
BT sulfosalts
SA silver ores

Stephens County
Valid through 1988. Search in combination with state term. After 1988, use specific county-state term.

Stephens County Georgia (1989)

NE Georgia. Before 1989, also search Stephens County AND Georgia.
CO N342700N344200
W0830600W0832800
BT Georgia
BT United States

Stephens County Oklahoma (1989)
S Oklahoma. Before 1989, also search Stephens County AND Oklahoma.
CO N341700N344100
W0973300W0980800
BT Oklahoma
BT United States

Stephens County Texas (1989)
N central Texas. Before 1989, also search Stephens County AND Texas.
CO N323100N325800
W0983100W0990700
BT Texas
BT United States

Stephenson County
Valid through 1988. Search in combination with state term. After 1988, use specific county-state term.

Stephenson County Illinois (1989)
N Illinois. Before 1989, also search Stephenson County AND Illinois.
CO N421200N423000
W0892500W0895500
BT Illinois
BT United States

Stepnyak
No longer a valid term for GeoRef. As of 1993, see Stepnyak Kazakhstan.

Stepnyak Kazakhstan (1993)
City in Kokchetav Oblast in N Kazakhstan. This term has multiple hierarchies.
BT1 Kokchetav Kazakhstan
BT1 Kazakhstan
BT1 Central Asia
BT1 Asia
BT2 Kokchetav Kazakhstan
BT2 Kazakhstan
BT2 Commonwealth of Independent States

Steppe
use Steppes region

steppes (1978)
SA geomorphology
SA grasslands
SA landforms
SA plains
SA prairies
SA revegetation
SA taiga environment

Steppes region (1978)
An extensive, treeless, semi-arid, grassland area of the mid-latitudes extending from the western border of the former USSR to the Altai Mountains in the E. Before 1982, also search Steppes.
IN Index former Soviet republics as applicable.
UF Steppe
BT Commonwealth of Independent States
SA Kazakhstan
SA Russian Federation
SA Ukraine

steranes (1985)
Cycloalkane. From 1985 to 1988, alkanes and aliphatic hydrocarbons were autoposted to this term.
BT hydrocarbons
BT organic materials
SA saturated hydrocarbons

Sterea Ellas (1981)
CO N373500N392000
E0240500E0204000
UF Central Greece
BT Greece
BT Southern Europe
BT Europe
NT Aegina
NT Attica Greece
NT Boeotia Greece
NT Kremasta Greece
NT Laurion Greece
NT Parnassus
SA Othrys
SA Pindus Mountains

stereochemistry
use crystal chemistry

stereographic projection (1989)
SA cartography
SA maps
SA photogrammetry
SA structural geology

Sterling County
Valid through 1988. Search in combination with state term. After 1988, use specific county-state term.

Sterling County Texas (1989)
W central Texas. Before 1989, also search Sterling County AND Texas.
CO N313000N320700
W1005000W1012200
BT Texas
BT United States

Sterling Hill (1985)
Mineral locality and mine site in N New Jersey. After 1993, Sterling Hill Mine is used for the mine.
CO N410000N412000
W0743000W0744000
BT Sussex County New Jersey
BT New Jersey
BT United States

sternbergite (1978)
BT sulfides

steroids (1985)
BT organic materials
NT sterols
SA biochemistry

sterols (1985)
This term has multiple hierarchies.
BT1 alcohols
BT1 organic materials
BT2 steroids
BT2 organic materials

Stettin
use Szczecin Poland

Steubenville Ohio (1989)
City on the Ohio River in E Ohio. Before 1989, search Steubenville AND Ohio.
CO N402200N402200
W0803900W0803900
BT Jefferson County Ohio
BT Ohio
BT United States

Stevens County
Valid through 1988. Search in combination with state term. After 1988, use specific county-state term.

Stevens County Kansas (1989)
SW Kansas. Before 1989, also search Stevens County AND Kansas.
CO N370000N372200
W1010400W1013400
BT Kansas
BT United States
NT Hugoton Kansas

Stevens County Minnesota (1989)
W Minnesota. Before 1989, also search Stevens County AND Minnesota.
CO N452400N454500
W0954500W0961700
BT Minnesota
BT United States

Stevens County Washington (1989)
NE Washington. Before 1989, also search Stevens County AND Washington.
CO N474500N490000
W1172500W1182500
BT Washington
BT United States
SA Colville River
SA Midnite Mine

stevensite (1978)
BT clay minerals
BT sheet silicates
BT silicates
SA cerolite

Stevns Klint (1989)
Promontory in Storstrom and Roskilde, E Denmark, extending into the Baltic Sea.
CO N551300N552500
E0122500E0120600
BT Denmark
BT Scandinavia
BT Western Europe
BT Europe
SA Baltic Sea
SA Zealand

Stewart Valley (1978)
River valley in W central Yukon Territory. Also search Stewart River.
IN Index territory or region as applicable.
SA Yukon Territory

stibiotantalite (1978)
Autoposting of niobotantalates began in 1989. This term has multiple hierarchies.
BT1 niobotantalates
BT1 niobates
BT1 oxides
BT2 niobotantalates
BT2 tantalates
BT2 oxides

stibium
use antimony

stibnite (1978)
UF antimonite
BT sulfides
SA antimony ores
SA metastibnite

stick slip
use stick-slip

stick-slip (1978)
UF stick slip
SA earthquakes
SA faults
SA mechanics

stiff clays (1981)
SA clay
SA clays
SA soil mechanics

stiffness (1993)
SA consistent materials
SA deformation
SA elastic properties
SA engineering properties
SA linear materials
SA properties
SA rigidity

Stigmaria (1978)
Genus.
BT Lycopsida
BT pteridophytes
BT Plantae

Stikine Terrane
use Stikinia Terrane

Stikinia Terrane (1989)
IN Index provinces as applicable.
UF Stikine Terrane
BT Canada
SA British Columbia
SA Canadian Rocky Mountains
SA Rocky Mountains
SA Yukon Territory

stilbite (1978)
UF desmine
BT zeolite group
BT framework silicates
BT silicates

Stillwater
Valid through 1988. Search in combination with state term. After 1988, use specific city-state term.

Stillwater Complex (1978)
Consists of four zones: basal, ultramafic zone, banded zone and upper zone.
BT Precambrian
SA J-M Reef
SA Montana

Stillwater County
Valid through 1988. Search in combination with state term. After 1988, use specific county-state term.

Stillwater County Montana (1989)
S Montana. Before 1989, also search Stillwater County AND Montana.
CO N451000N461000
W1085000W1100500
BT Montana
BT United States

Stillwater Oklahoma (1989)
City in N central Oklahoma. Before 1989, search Stillwater AND Oklahoma.
CO N360700N360700
W0970300W0970300
BT Payne County Oklahoma
BT Oklahoma
BT United States

stillwellite (1978)
Autoposting of nesosilicates began in 1985.
BT nesosilicates
BT orthosilicates
BT silicates

stilpnomelane (1978)
BT sheet silicates
BT silicates
SA aluminosilicates
SA iron ores

stimulation
As of 1993, use well stimulation if applicable. See recovery.

Stirling Quartzite (1978)
Considered synonym for Prospect Mountain Quartzite. E California and SE Nevada.
UF Prospect Mountain Quartzite
BT Precambrian
SA California
SA Nevada

stishovite (1978)
BT silica minerals
BT framework silicates
BT silicates

stochastic processes (1978)
UF random processes
SA kriging
SA probability
SA processes

Stockholm
No longer a valid term for GeoRef. As of 1993 see Stockholm Sweden.

Stockholm Sweden (1993)
County and city on the Baltic Sea in SE Sweden. Before 1993 also search Stockholm AND Sweden.
BT Sweden
BT Scandinavia
BT Western Europe
BT Europe

stocks (1978)
BT intrusions
SA batholiths
SA igneous rocks

Stockton Formation (1989)
Upper Triassic; Newark Supergroup of New Jersey, New York and Pennsylvania. Also a Lower Permian formation of Western Australia.
IN Index ages as applicable.
SA Australia
SA Lower Permian
SA New Jersey
SA New York
SA Newark Supergroup
SA Pennsylvania
SA Upper Triassic
SA Western Australia

stockwork deposits (1978)
Before 1978, search ore deposits AND stockwork, or search deposits AND stockwork.
UF network deposits
SA mineral deposits, genesis

stoichiometry (1978)
SA chemical reactions
SA endothermic reactions
SA geochemistry

Stokes County
Valid through 1988. Search in combination with state term. After 1988, use specific county-state term.

Stokes County North Carolina (1989)
NW North Carolina. Before 1989, also search Stokes County AND North Carolina.
CO N361500N363500 W0800200W0802600
BT North Carolina
BT United States

Stolonoidea (1978)
IN Index Hemichordata or Invertebrata as applicable.
BT Graptolithina

SA Hemichordata
SA Invertebrata

stolzite (1978)
BT tungstates

Stomachorda
use Hemichordata

Stomiosphaera (1978)
This term has multiple hierarchies.
BT1 Tintinnidae
BT1 Protista
BT1 Invertebrata
BT2 Tintinnidae
BT2 Protista
BT2 microfossils

Stone Age (1978)
Archaeological classification. Includes Paleolithic, Mesolithic and Neolithic. As of 1993, this term is autoposted to its narrower terms.
IN Index ages as applicable. See List E.
BT Cenozoic
NT Mesolithic
NT Neolithic
NT Paleolithic
SA archaeology
SA Quaternary
SA Tertiary

Stone Corral Formation (1978)
In Sumner Group. E Kansas.
BT Permian
BT Paleozoic
SA Kansas

Stone Mountain (1978)
Huge gray granite monadnock in De Kalb County near Atlanta. Also in other locations.
IN Index county or region as applicable.
SA De Kalb County Georgia

stone, building
use building stone

stone, dimension
use dimension stone

Stonehenge (1978)
Prehistoric assemblage of stones on the Salisbury Plain 7 miles N of Salisbury in S England.
BT Wiltshire England
BT England
BT Great Britain
BT United Kingdom
BT Western Europe
BT Europe

Stoneley waves (1981)
BT surface waves
BT elastic waves
SA Rayleigh waves
SA seismology
SA waves

Stones River Group (1985)
Tennessee, W Maryland, NE Mississippi, S Pennsylvania, W Virginia and NE West Verginia.
BT Middle Ordovician
BT Ordovician
BT Paleozoic
SA Lenoir Limestone
SA Maryland
SA Mississippi
SA Pennsylvania
SA Tennessee
SA Virginia
SA West Virginia

stony irons (1978)
Autoposting of this term to mesosiderite and pallasite began in 1981. Autoposting of this term to siderophyre began in 1986.

UF iron-stony meteorite
UF lithosiderite
UF sideraerolite
UF siderolite
BT meteorites
NT mesosiderite
NT pallasite
NT siderophyre
SA achondrites
SA chondrites
SA iron meteorites

stony meteorites (1982)
Autoposting of this term began in 1985.
BT meteorites
NT achondrites
NT chondrites

Stony Tunguska River (1978)
Rises in SE Evenk National Okrug and flows WNW into mid course of the Yenisei River. This term has multiple hierarchies.
UF Middle Tunguska River
UF Podkamennaya Tunguska
BT1 Russian Federation
BT1 Commonwealth of Independent States
BT2 Asia
SA Krasnodar Russian Federation
SA Tunguska
SA Tunguska River

storage (1978)
SA data storage
SA gas storage
SA ground water
SA heat storage
SA oil storage
SA reservoirs
SA soils
SA underground installations
SA underground storage
SA waste disposal
SA water regimes
SA water storage

storage coefficient (1985)
SA aquifers
SA ground water
SA hydrodynamics

Storey County
Valid through 1988. Search in combination with state term. After 1988, use specific county-state term.

Storey County Nevada (1989)
W Nevada. Before 1989, also search Storey County AND Nevada.
CO N391500N393800 W1191800W1194300
BT Nevada
BT United States

storm environment (1985)
SA depositional environment
SA environment
SA sedimentation
SA storms

Stormberg Series (1978)
BT Upper Triassic
BT Triassic
BT Mesozoic
SA Karroo Supergroup
SA South Africa

storms (1978)
SA atmosphere
SA climate
SA dust storms
SA geologic hazards
SA hurricanes
SA meteorology
SA monsoons
SA precipitation

SA runoff
SA storm environment
SA winds

Strabo Trench (1989)
E Mediterranean Sea.
CO N333000N360000 E0270500E0230000
BT East Mediterranean
BT Mediterranean Sea
SA DSDP Site 129
SA Leg 13

Straight Creek Fault (1985)
W Washington and SW British Columbia.
IN Index Washington and British Columbia as applicable.
BT North America
SA British Columbia
SA Cascade Range
SA Coast Ranges
SA Washington

strain (1978)
SA brittleness
SA deformation
SA distortion
SA elastic limit
SA elastic strain
SA elasticity
SA engineering geology
SA finite strain analysis
SA Hooke's law
SA hysteresis
SA mechanical properties
SA Poisson's ratio
SA rock mechanics
SA seismology
SA shear
SA shear strength
SA shear zones
SA strainmeters
SA stress
SA structural analysis
SA viscoelasticity

strain relaxation (1981)
SA seismology

strain seismographs (1981)
BT seismographs

strain-slip cleavage
use slip cleavage

strainmeters (1978)
SA deformation
SA extensometers
SA geodesy
SA strain
SA stressmeters

Strait of Georgia (1978)
Channel between Vancouver Island on the W and the mainland of British Columbia and Washington on the E.
IN Index British Columbia and/or Washington as applicable.
BT North America
SA British Columbia
SA Saanich Inlet
SA Washington

Strait of Gibraltar (1978)
Passage connecting Mediterranean Sea and the Atlantic Ocean between Spain and Morocco.
BT Atlantic Ocean
SA Gibraltar

Strait of Juan de Fuca
use Juan de Fuca Strait

Strait of Malacca (1978)
Channel between the S Malay Peninsula and the island of Sumatra connecting the Indian Ocean with the South China Sea.
UF Malacca Strait

UF Malacca Straits
UF Straits of Malacca
BT Asia

Strait of Sicily (1978)
Between Sicily and Tunisia.
BT Mediterranean Sea

Straits of Florida (1978)
Wide channel between Florida Keys and Cuba connecting the Atlantic Ocean with the Gulf of Mexico.
UF Florida Strait
UF Florida Straits
BT Atlantic Ocean
SA Leg 101
SA Northwest Providence Channel
SA ODP Site 626

Straits of Malacca
 use Strait of Malacca

Stramberk (1978)
Town SSW of Ostrava in NE Moravia, central Czechoslovakia.
BT Moravia
BT Czech Republic
BT Czechoslovakia
BT Central Europe
BT Europe

Strandzha Mountains
 use Istranca Mountains

Strasbourg
No longer a valid term for GeoRef. See Strasbourg France.

Strasbourg France (1993)
City on the Rhine River in Alsace, E Bas-Rhin, NE France. Also search Strassburg.
BT Bas-Rhin France
BT France
BT Western Europe
BT Europe

Strassburg
Not a valid term for GeoRef. See Strasbourg France.

strata
 use stratigraphic units

strata-bound deposits
 use stratabound deposits

stratabound deposits (1978)
Before 1978, search ore deposits AND stratabound; deposits AND stratabound; strata-bound.
UF strata-bound deposits
SA besshi-type
SA kuroko-type
SA mantos
SA mineral deposits, genesis
SA stratiform deposits

strategic materials
 use strategic minerals

strategic minerals (1985)
UF strategic materials
SA economic geology
SA economics
SA Exclusive Economic Zone
SA international cooperation
SA legislation
SA mineral resources
SA NURE
SA policy
SA reserves
SA United States Exclusive Economic Zone

Strathclyde region Scotland (1993)
Administrative region in W Scotland created in 1975 from most of Argyllshire including Mull Island, Bute, Ayrshire, Lanarkshire, and Renfrewshire.
BT Scotland
BT Great Britain
BT United Kingdom
BT Western Europe
BT Europe
NT Ayrshire Scotland
NT Gairloch Scotland
NT Glasgow Scotland
NT Islay
NT Kintyre
NT Mull Island
SA Argyllshire Scotland
SA Arran
SA Firth of Clyde
SA Inner Hebrides
SA Loch Lomond

Strathcona Mine (1978)
Copper and nickel ores.
BT Ontario
BT Eastern Canada
BT Canada
SA copper ores
SA mines
SA nickel ores

stratification (1978)
BT planar bedding structures
BT sedimentary structures
SA bedding
SA cross-stratification
SA deposition
SA laminations

stratified volcanic cone
 use stratovolcanoes

stratified volcano
 use stratovolcanoes

stratiform deposits (1978)
Before 1978, search ore deposits AND stratiform; deposits AND stratiform.
UF stratiform ore deposits
SA mineral deposits, genesis
SA stratabound deposits

stratiform ore deposits
 use stratiform deposits

stratigraphic
A valid term through 1977. After 1977, use stratigraphic maps.

stratigraphic boundary (1993)
Before 1993, also search boundary which was used for this concept from 1978-1992.
IN Use in combination with appropriate age terms (list E).
UF boundaries, stratigraphic
SA K-T boundary
SA stratigraphic columns
SA stratigraphy
SA stratotypes

stratigraphic columns (1982)
SA cross sections
SA stratigraphic boundary
SA stratigraphic units
SA stratigraphy
SA stratotypes
SA type sections

stratigraphic controls (1978)
As of 1993, restricted to economic studies.
SA controls
SA lithologic controls
SA mineral deposits, genesis
SA structural controls

stratigraphic correlation
 use correlation

stratigraphic gaps (1981)
SA hardground
SA lithostratigraphy
SA stratigraphy
SA unconformities

stratigraphic geology
 use stratigraphy

stratigraphic maps (1978)
Before 1978, search maps AND stratigraphic.
BT maps
SA structural maps

stratigraphic range
 use range

stratigraphic traps (1978)
Autoposting of this term began in 1978.
UF lithologic traps
UF porosity traps
BT traps
SA natural gas
SA oil and gas fields
SA petroleum
SA petroleum accumulation
SA petroleum exploration
SA stratigraphic wedges
SA structural traps

stratigraphic units (1978)
UF strata
UF units, stratigraphic
SA biozones
SA lithostratigraphy
SA stratigraphic columns
SA stratigraphy
SA tectonostratigraphic units

stratigraphic wedges (1981)
Used for beds that pinch out.
UF pinchouts
SA lithofacies
SA lithostratigraphy
SA natural gas
SA oil and gas fields
SA petroleum
SA petroleum accumulation
SA petroleum exploration
SA reservoir rocks
SA stratigraphic traps
SA stratigraphy

stratigraphy (1978)
Used for the discipline as a whole.
UF stratigraphic geology
SA bibliography
SA biogeography
SA biomarkers
SA biostratigraphy
SA biozones
SA catalogs
SA changes of level
SA chemostratigraphy
SA chronostratigraphy
SA continental drift
SA Continental Offshore Stratigraphic Test
SA correlation
SA cross sections
SA diachronism
SA education
SA encrustations
SA geochronology
SA glacial geology
SA glossaries
SA historical geology
SA index fossils
SA isopach maps
SA key beds
SA lexicons
SA lithostratigraphy
SA magnetostratigraphy
SA micropaleontology
SA outliers
SA paleogeography
SA paleomagnetism
SA paleontology
SA palynology
SA punctuated aggradational cycles
SA range
SA regression
SA seismic stratigraphy
SA sequence stratigraphy
SA stratigraphic boundary
SA stratigraphic columns
SA stratigraphic gaps
SA stratigraphic units
SA stratigraphic wedges
SA stratotypes
SA succession
SA tectonostratigraphic units
SA transgression
SA type localities
SA type sections
SA unconformities
SA uniformitarianism

stratosphere
As of 1993, no longer a valid term for GeoRef. Usually out-of-scope. Used with atmosphere.

stratotypes (1978)
SA stratigraphic boundary
SA stratigraphic columns
SA stratigraphy
SA type sections

stratovolcanoes (1978)
UF bedded volcano
UF composite cone
UF composite volcano
UF stratified volcanic cone
UF stratified volcano
BT volcanoes
BT volcanic features
SA cones
SA shield volcanoes

Strawn Series (1978)
Comprises Millsap Lake and Lone Camp groups.
BT Pennsylvanian
BT Carboniferous
BT Paleozoic
SA New Mexico
SA Texas

stream action
Not a valid index term for GeoRef. See streams and sedimentation.

stream capture (1978)
UF capture
UF piracy
UF river capture
UF river piracy
UF robbery
UF stream piracy
UF stream robbery
SA geomorphology
SA rivers
SA rivers and streams
SA streams
SA tributaries

stream flow
 use streamflow

stream gradient (1978)
Before 1978, search streams AND gradient.
SA rivers and streams
SA streams

stream order (1978)
Autoposting of fluvial features to this term began in 1989.
UF channel order
BT fluvial features
SA drainage patterns
SA geomorphology
SA streams
SA tributaries

stream piracy
 use stream capture

stream placers (1982)
  BT placers
  SA beach placers
  SA heavy mineral deposits
  SA mineral resources
  SA rivers
  SA rivers and streams
  SA stream sediments
  SA streams

stream robbery
  use stream capture

stream sediments (1978)
  Restricted to mineral exploration.
  Also search stream sampling.
  SA fluvial environment
  SA fluvial features
  SA mineral exploration
  SA sediments
  SA stream placers
  SA streams

stream transport (1978)
  UF fluvial transport
  SA bedload
  SA fluvial currents
  SA fluvial sedimentation
  SA rivers
  SA rivers and streams
  SA sediment transport
  SA sedimentation
  SA solution transport
  SA streams
  SA thalwegs

streamflow (1989)
  UF stream flow
  SA channels
  SA discharge
  SA flows
  SA runoff
  SA streams
  SA surface water

streams (1978)
  As of 1981, use fluvial environment for type of environment. Also search alluvial. Autoposting of fluvial features to this term began in 1989.
  BT fluvial features
  NT braided streams
  NT ephemeral streams
  SA bedload
  SA canals
  SA cascades
  SA channel geometry
  SA channelization
  SA channels
  SA drainage patterns
  SA efficiency
  SA exsurgence
  SA fluvial currents
  SA fluvial environment
  SA geomorphology
  SA hydrogeology
  SA hydrosphere
  SA ice streams
  SA levees
  SA meanders
  SA oxbow lakes
  SA rapids
  SA rills
  SA rivers
  SA rivers and streams
  SA runoff
  SA sediment yield
  SA sinuosity
  SA stream capture
  SA stream gradient
  SA stream order
  SA stream placers
  SA stream sediments
  SA stream transport
  SA streamflow
  SA surface water

  SA thalwegs
  SA tributaries
  SA underground streams
  SA watersheds

strengite (1978)
  BT phosphates

strength (1978)
  SA compressive strength
  SA deformation
  SA fracture strength
  SA mechanical properties
  SA rock mechanics
  SA shear strength
  SA stress
  SA stressmeters
  SA tensile strength
  SA thixotropy
  SA weak rocks
  SA yield strength

Streptognathodus (1989)
  Genus. Autoposting of microfossils to this term began in 1990.
  BT Conodonta
  BT microfossils

stress (1978)
  SA bearing capacity
  SA brittleness
  SA creep
  SA deformation
  SA elasticity
  SA extensometers
  SA failures
  SA finite strain analysis
  SA Hooke's law
  SA hysteresis
  SA Poisson's ratio
  SA pore pressure
  SA pressure
  SA release fractures
  SA rigidity
  SA rock mechanics
  SA rupture
  SA seismology
  SA shear
  SA shear strength
  SA shear stress
  SA strain
  SA strength
  SA stress drops
  SA stress fields
  SA stressmeters
  SA structural analysis
  SA tension
  SA torsion
  SA viscoelasticity
  SA yield strength

stress drops (1985)
  SA earthquake prediction
  SA earthquakes
  SA precursors
  SA seismology
  SA stress

stress fields (1985)
  SA deformation
  SA stress

stressmeters (1989)
  SA deformation
  SA engineering geology
  SA instruments
  SA rock mechanics
  SA strainmeters
  SA strength
  SA stress

stretch modulus
  use Young's modulus

striations (1978)
  BT bedding plane irregularities
  BT sedimentary structures
  SA furrows
  SA glacial features
  SA glacial geology

  SA grooves

strike (1978)
  Term used to indicate strike faults through 1977. After 1977, use strike faults when discussing type of faults.
  UF line of strike
  SA bedding
  SA cleats
  SA dip
  SA faults
  SA folds
  SA fractures
  SA joints
  SA orientation
  SA strike faults
  SA strike-slip faults

strike faults (1978)
  Before 1978, also search faults AND longitudinal; faults AND strike.
  UF longitudinal faults
  BT faults
  SA strike

strike-shift faults
  use strike-slip faults

strike-slip faults (1978)
  Before 1978, also search strike-slip.
  UF strike-shift faults
  BT faults
  SA left-lateral faults
  SA oblique-slip faults
  SA pull-apart basins
  SA strike
  SA transcurrent faults
  SA transfer faults
  SA transform faults

strip mines
  use strip mining

strip mining (1978)
  UF strip mines
  BT surface mining
  BT mining
  SA bioremediation
  SA coal
  SA coal mines
  SA environmental geology
  SA land use
  SA mines
  SA mining geology
  SA open-pit mining
  SA quarries
  SA reclamation
  SA remediation

Stripa Mine
  use Stripa region

Stripa region (1985)
  Radioactive waste disposal site, N Orebro, Sweden. Former metals mining site. Before 1993, also search Stripa or Stripa Mine or Guldsmedshyttan in combination with Sweden.
  UF Guldsmedshyttan
  UF Stripa Mine
  BT Orebro Sweden
  BT Sweden
  BT Scandinavia
  BT Western Europe
  BT Europe
  SA radioactive waste
  SA underground storage
  SA waste disposal

stromatactis (1989)
  BT biogenic structures
  BT sedimentary structures

stromatolites (1978)

  As of 1993, Autoposting of this term began in 1978. Search in combination with algae or algal flora as applicable.
  IN Index algae or algal flora as applicable. See list F (fossils).
  BT biogenic structures
  BT sedimentary structures
  NT Collenia
  SA algae
  SA algal flora
  SA algal mats
  SA biostratigraphy
  SA calcareous algae
  SA reef builders
  SA thrombolites

Stromatoporoidea (1978)
  Autoposting of this term began in 1978. This term has also been used with Porifera.
  BT Coelenterata
  BT Invertebrata
  NT Amphipora
  SA Porifera
  SA problematic fossils
  SA stromatoporoids

stromatoporoids (1985)
  Common name for Stromatoporoidea. Autoposting of corals to this term began in 1989.
  BT corals
  BT invertebrates
  SA biostratigraphy
  SA Stromatoporoidea

Stromboli (1978)
  Active volcano on Stromboli Island of Lipari Group in Tyrrhenian Sea.
  BT Lipari Islands
  BT Sicily Italy
  BT Italy
  BT Southern Europe
  BT Europe

strombolian-type eruptions (1982)
  SA eruptions
  SA volcanism
  SA volcanology

stromeyerite (1978)
  BT sulfides

Strona Valley (1978)
  River valley W of Lago Maggiore in N Italy.
  BT Sicily Italy
  BT Italy
  BT Southern Europe
  BT Europe

strong motion (1978)
  UF motion, strong
  SA ground motion
  SA seismology

strontianite (1978)
  BT carbonates

strontium (1978)
  Chemical element. As of 1982, use strontium ores for strontium as a commodity. Autoposting of alkaline earth metals and metals to this term began in 1989.
  UF Sr
  BT alkaline earth metals
  BT metals
  NT Rb-87/Sr-86
  NT Sr-85
  NT Sr-86
  NT Sr-87
  NT Sr-87/Sr-86
  NT Sr-89
  NT Sr-90
  NT Sr/Ca
  SA isotopes

SA Rb/Sr
SA Sr/Sr
SA strontium ores

strontium ores (1982)
Before 1982, also search (strontium OR Sr) AND (deposit OR deposits OR ore OR ores OR economic) in the basic index. Autoposting of metal ores to this term began in 1985.
IN Commodity. See List C.
BT metal ores
SA celestite
SA strontium

Strophomena (1978)
Genus.
BT Strophomenida
BT Articulata
BT Brachiopoda
BT Invertebrata

Strophomenida (1978)
Autoposting of this term began in 1978.
BT Articulata
BT Brachiopoda
BT Invertebrata
NT Sowerbyella
NT Strophomena

structural
A valid term through 1977. After 1977, use structural maps if applicable.

structural analysis (1978)
Used for the analysis of structural features on a relatively small scale (from thin section to outcrop). The analysis may lead to interpretation of larger-scale features such as folds, fractures, and faults, or tectonics. Before 1976, also search fabric analysis; structural petrology; microtectonics.
UF petrogeometry
UF petromorphology
SA analysis
SA axial-plane structures
SA boudinage
SA breccia
SA brittle deformation
SA cleavage
SA conjugate faults
SA deformation
SA dip
SA electron microscopy
SA elongate minerals
SA engineering geology
SA fabric
SA faults
SA flow cleavage
SA fold axes
SA folds
SA foliation
SA fractures
SA geometry
SA geophysics
SA interference patterns
SA joints
SA lineation
SA melange
SA mullions
SA orientation
SA petrofabrics
SA preferred orientation
SA rock mechanics
SA schistosity
SA shear zones
SA strain
SA stress
SA structural geology
SA tectonics
SA tectonophysics
SA transpression

SA universal stage
SA X-ray analysis

structural basins
use basins

structural complexes
No longer a valid term for GeoRef as of 1981. Use complexes.

structural controls (1978)
As of 1981, restricted to economic geology.
SA controls
SA mineral deposits, genesis
SA stratigraphic controls
SA tectonic controls

structural features
No longer a valid term for GeoRef as of mid-1978. Included use in geomorphology, structural geology and petrology (igneous rocks).

structural geology (1978)
Used for the discipline as a whole.
SA allochthons
SA bibliography
SA catalogs
SA faults
SA folds
SA foliation
SA fractures
SA geology
SA glossaries
SA lexicons
SA lineation
SA maps
SA orogeny
SA stereographic projection
SA structural analysis
SA tectonics

structural maps (1978)
Before 1978, search maps AND structural.
UF structure maps
BT maps
SA stratigraphic maps
SA structure contour maps

structural materials
No longer a valid GeoRef index term. Before 1960, was used on level 1 in subfiles E and N. See building stone; clays; construction materials.

structural petrology
A valid term through 1974. After 1974, use structural analysis.

structural traps (1978)
BT traps
SA natural gas
SA petroleum
SA petroleum accumulation
SA petroleum exploration
SA stratigraphic traps

structure
As of 1981, no longer a valid index term.

structure contour maps (1978)
BT maps
SA contour maps
SA structural maps
SA tectonic maps

structure maps
use structural maps

structure-soil interface
use soil-structure interface

structures (1978)
Term is restricted to engineering geology.
SA buildings
SA engineering geology

SA foundations

Struma River valley (1978)
Also search Struma; Struma River.
IN Index countries as applicable.
UF Struma Valley
BT Southern Europe
BT Europe
SA Bulgaria
SA Greece

Struma Valley
use Struma River valley

Strunian (1981)
BT Upper Devonian
BT Devonian
BT Paleozoic

struvite (1978)
BT phosphates

Strzegom
No longer a valid term for GeoRef. As of 1993 see Strzegom Poland.

Strzegom Poland (1993)
City in Lower Silesia in N Walbrzych, SW Poland. Before 1993 also search Strzegom AND Poland.
BT Walbrzych Poland
BT Poland
BT Central Europe
BT Europe

Strzegom-Sobotka Granitoid Massif
use Strzegom-Sobotka Massif

Strzegom-Sobotka Massif (1978)
Granitoid massif in Lower Silesia. Also search Strzegom-Sobotka.
UF Strzegom-Sobotka Granitoid Massif
BT Walbrzych Poland
BT Poland
BT Central Europe
BT Europe
SA Lower Silesia

Strzelin
No longer a valid term for GeoRef. As of 1993 see Strzelin Poland.

Strzelin Poland (1993)
City of Lower Silesia in S Wroclaw, SW Poland. Before 1993 also search Strzelin AND Poland.
BT Wroclaw Poland
BT Poland
BT Central Europe
BT Europe

Stuart Shelf (1989)
IN Index South Australia or region as applicable.
SA Adelaide Geosyncline
SA South Australia

Stubai Alps (1978)
A NE group of Otztal Alps in central Tyrol. Also search Stubai. This term has multiple hierarchies.
UF Stubai Massif
UF Stubai Mountains
BT1 Otztal Alps
BT1 Eastern Alps
BT1 Alps
BT1 Europe
BT2 Tyrol Austria
BT2 Austria
BT2 Central Europe
BT2 Europe

Stubai Massif
use Stubai Alps

Stubai Mountains
use Stubai Alps

studies, areal
use areal studies

studies, case
use case studies

studies, experimental
use experimental studies

studies, faunal
use faunal studies

studies, feasibility
use feasibility studies

studies, field
use field studies

studies, floral
use floral studies

studies, laboratory
use laboratory studies

studies, theoretical
use theoretical studies

study and teaching
No longer a valid GeoRef index term. Before 1971, was used on level 1 in subfiles E and N.
use education

Stump Formation (1978)
Overlies Preuss Sandstone and underlies Bechler and Ephraim conglomerates undifferentiated. SE Idaho and W Wyoming.
BT Upper Jurassic
BT Jurassic
BT Mesozoic
SA Idaho
SA Wyoming

Sturgeon Lake (1989)
NW Ontario, W central Alberta and E Minnesota.
IN Index provinces or states as applicable.
SA Alberta
SA Minnesota
SA Ontario

Sturgis Formation (1985)
W central Kentucky.
BT Upper Pennsylvanian
BT Pennsylvanian
BT Carboniferous
BT Paleozoic
SA Kentucky

Sturovo (1978)
Town on left bank of Danube River in S Slovakia, E Czechoslovakia.
BT Slovakia
BT Czechoslovakia
BT Central Europe
BT Europe

Stutsman County
Valid through 1988. Search in combination with state term. After 1988, use specific county-state term.

Stutsman County North Dakota (1989)
Central North Dakota. Before 1989, also search Stutsman County AND North Dakota.
CO N463800N472000
 W0982500W0992900
BT North Dakota
BT United States

Stuttgart
No longer a valid term for GeoRef. As of 1989, use Stuttgart Arkansas if applicable. After 1992, see Stuttgart Germany if applicable.

Stuttgart Germany (1993)

City in Baden-Wurttemberg, SW Germany. Before 1989, search in combination with Germany.
BT Baden-Wurttemberg Germany
BT Germany
BT Central Europe
BT Europe

style (1978)
SA folds
SA foliation
SA fractures
SA lineation
SA orientation

stylolites (1978)
BT secondary structures
BT sedimentary structures
SA microstylolites
SA stylolitization

stylolitization (1985)
SA diagenesis
SA microstylolites
SA sedimentary rocks
SA sedimentation
SA sediments
SA stylolites

Stylommatophora (1981)
BT Gastropoda
BT Mollusca
BT Invertebrata

Stylophora (1978)
BT Homalozoa
BT Echinodermata
BT Invertebrata

Styria
No longer a valid term for GeoRef. See Styria Austria.

Styria Austria (1993)
State in SE Austria.
CO N463500N474500 E0161500E0133000
BT Austria
BT Central Europe
BT Europe
NT Graz Austria
NT Hartberg Austria
SA Koralpe Range
SA Salzkammergut
SA Semmering
SA Totes Gebirge
SA Wechsel

suanite (1978)
BT borates

sub-alpine environment
 use subalpine environment

sub-bituminous coal
 use subbituminous coal

sub-glacial environment
 use subglacial environment

subaerial environment (1978)
Before 1978 search subaerial.
SA depositional environment
SA environment

subalpine environment (1989)
UF sub-alpine environment
SA alpine environment
SA depositional environment
SA ecology
SA environment
SA sedimentation

subantarctic regions (1978)
Pertaining or relating to the regions immediately outside of the Antarctic circle. Before 1978, also search subantarctic.
SA Antarctic Ocean
SA Antarctica

Subarctic
 use Preboreal

subarctic regions (1978)
Pertaining or relating to the regions immediately outside of the Arctic circle or to areas that have characteristics such as climate, vegetation, and animals similar to these regions. Before 1978, also search subarctic.
SA Arctic region

subarkose (1978)
BT clastic rocks
BT sedimentary rocks
SA arkose

Subathu Formation (1989)
N India.
BT Eocene
BT Paleogene
BT Tertiary
BT Cenozoic
SA Himachal Pradesh India
SA India
SA Jammu and Kashmir
SA Uttar Pradesh India

Subbetic Zone (1978)
A geographic term with stratigraphic-tectonic connotations.
BT Spain
BT Iberian Peninsula
BT Southern Europe
BT Europe
SA Betic Cordillera
SA Prebetic Zone

subbituminous coal (1989)
This term has multiple hierarchies.
UF sub-bituminous coal
BT1 coal
BT1 organic residues
BT2 coal
BT2 sedimentary rocks

Subboreal (1978)
Europe. A term used primarily in Europe for an interval of postglacial time following the Atlantic and preceding the Subatlantic.
BT Holocene
BT Quaternary
BT Cenozoic

Subcarpathians (1978)
Sub-ranges of the Carpathians.
IN Index countries as applicable.
BT Carpathians
BT Europe
SA Carpathian Foreland
SA Czechoslovakia
SA Hungary
SA Poland
SA Romania
SA Ukraine

subcretion
 use underplating

subduction (1978)
SA Benioff zone
SA marginal basins
SA marginal seas
SA obduction
SA plate tectonics
SA subduction zones
SA tectonophysics
SA trenches

subduction zones (1978)
UF zones, subduction
SA active margins
SA back-arc basins
SA Benioff zone
SA fore-arc basins
SA plate tectonics
SA plates
SA subduction
SA tectonophysics
SA trenches
SA underplating

subgelisol
 use taliks

subglacial environment (1989)
UF sub-glacial environment
SA depositional environment
SA environment
SA glacial environment
SA glacial geology
SA glaciers
SA sedimentation

subgraywacke (1978)
BT clastic rocks
BT sedimentary rocks
SA graywacke
SA sandstone
SA terrigenous materials

Sublette County
Valid through 1988. Search in combination with state term. After 1988, use specific county-state term.

Sublette County Wyoming (1989)
W Wyoming. Before 1989, also search Sublette County AND Wyoming.
CO N421500N432700 W1090300W1103800
BT Wyoming
BT United States
NT Pinedale Anticline
SA Moxa Arch

sublimates (1978)
SA fumaroles
SA gases
SA solfataras
SA sublimation
SA volcanism
SA volcanoes

sublimation (1978)
SA evaporation
SA geochemistry
SA mineral deposits, genesis
SA sublimates

sublittoral environment
A valid term for GeoRef from 1978 to 1989.
 use subtidal environment

submarine
A valid term through 1977. After 1977, use submarine environment.

submarine canyons (1978)
Before 1976, also search canyons for submarine canyons.
UF submarine valleys
SA Baltimore Canyon
SA bottom features
SA canyons
SA continental margin
SA continental shelf
SA continental slope
SA levees
SA Lydonia Canyon
SA marine geology
SA Monterey Canyon
SA ocean floors
SA Redondo Canyon
SA submarine fans
SA Trou Sans Fond
SA turbidity currents
SA Wilmington Canyon

submarine cone
 use submarine fans

submarine delta
 use submarine fans

submarine dunes (1981)
BT dunes
SA abyssal hills

SA bottom features
SA ocean floors

submarine environment (1978)
Before 1978, search submarine.
SA abyssal hills
SA bottom features
SA depositional environment
SA environment
SA marine environment
SA reef environment

submarine fans (1978)
UF abyssal cones
UF abyssal fans
UF deep-sea fans
UF sea fan
UF submarine cone
UF submarine delta
SA cones
SA ocean floors
SA submarine canyons
SA turbidity currents

submarine features
 use bottom features

submarine geology
No longer a valid index term for GeoRef. Before 1971, included use in subfiles E, G, N and B.
 use marine geology

submarine hills
 use abyssal hills

submarine installations (1978)
UF installations, submarine
BT marine installations
SA artificial islands
SA engineering geology
SA tunnels

submarine springs (1982)
BT springs
SA ground water
SA marine environment
SA sea water

submarine valleys
 use submarine canyons

submarine volcanoes (1989)
BT volcanoes
BT volcanic features
SA bottom features
SA ocean basins
SA rifting
SA sea-floor spreading
SA spreading centers
SA volcanism
SA volcanology

submarine weathering
 use halmyrolysis

submergence (1978)
SA changes of level
SA embayments
SA transgression

submersibles (1978)
SA CYANA
SA FAMOUS
SA marine geology
SA oceanography
SA research vessels

subsequent folds
 use superposed folds

subsidence (1978)
SA crater lakes
SA land subsidence
SA neotectonics
SA tectonics
SA uplifts
SA vertical movements

subsidies (1981)
SA economics

substitution (1978)

SA crystal chemistry
SA lattice
SA order-disorder

substorms
As of 1993, no longer a valid term for GeoRef. Usually out-of-scope. Used with magnetosphere.

substrates (1978)
SA ecology
SA paleoecology

subsurface
A valid term through mid-1978 under engineering geology. See subsurface reservoirs. Was sometimes used in combination with other terms; e.g. subsurface geology; subsurface geologic methods.

subsurface dams
 use ground-water dams

subsurface mining
 use underground mining

subsurface reservoirs
A valid term from 1978 to 1980. As of 1981, no longer a valid term for GeoRef. Included use for "natural" reservoirs (deposits) of petroleum and natural gas. Also included use for studies on the use of underground formations or cavities for storage or waste disposal. See petroleum; natural gas; geothermal energy; reservoir rocks; aquifers; petroleum engineering; underground installations; waste disposal; storage; gas storage; oil storage; underground storage; heat storage; water storage. Before 1978, search reservoirs AND storage.

subsurface storage
 use underground storage

subtidal environment (1978)
Before 1978, search sublittoral; subtidal. Before 1981, search neritic; neritic environment. Before 1989, also search sublittoral environment.
UF neritic environment
UF sublittoral environment
SA coastal environment
SA depositional environment
SA environment
SA sedimentation

subtropical environment (1978)
Before 1978, search subtropical.
UF semitropical environment
SA climate
SA depositional environment
SA ecology
SA environment
SA paleoclimatology
SA tropical environment

subways (1978)
SA tunnels

succession (1989)
Used in ecologic and stratigraphic sense.
SA communities
SA ecology
SA ecosystems
SA paleoecology
SA stratigraphy

Suceava
No longer a valid term for GeoRef. As of 1993 see Suceava Romania.

Suceava Romania (1993)
Town and county in N Romania. Before 1993 also search Suceava AND Romania. This term has multiple hierarchies.
BT1 Moldavia
BT1 Europe
BT2 Romania
BT2 Southern Europe
BT2 Europe
SA Rarau Massif

Suchan Basin (1978)
Coal basin in Primorye Kray 60 miles ENE of Vladivostok. Also search Suchan. This term has multiple hierarchies.
BT1 Russian Federation
BT1 Commonwealth of Independent States
BT2 Asia
SA coal fields
SA Primorye Russian Federation

Sucre
No longer a valid term for GeoRef. See Sucre Colombia, Sucre Venezuela and Sucre Bolivia.

Sucre Bolivia (1993)
City in Chuquisaca Department, S central Bolivia.
BT Bolivia
BT South America

Sucre Colombia (1993)
Department in N Colombia.
BT Colombia
BT South America

Sucre Venezuela (1993)
State in NE Venezuela.
BT Venezuela
BT South America
NT Araya Peninsula
NT Cumana Venezuela
NT Gulf of Cariaco
NT Paria Peninsula
SA San Juan River

Sudan (1978)
CO N030000N220000
 E0383000E0220000
BT East Africa
BT Africa
NT Darfur Sudan
NT Gezira
NT Kordofan Sudan
NT Malakal Sudan
SA Blue Nile
SA Kapoeta Meteorite
SA Lake Turkana
SA Libyan Desert
SA Near East
SA Nile River
SA Nile Valley
SA Nubia
SA Nubian Sandstone
SA Nubian Shield
SA Red Sea Basin
SA Red Sea Hills
SA Red Sea Rift
SA Sahara
SA Turkana Basin

Sudbury
As of 1985, no longer a valid term for GeoRef. Use Sudbury Basin, Sudbury District Ontario or Sudbury Irruptive as applicable.

Sudbury Basin (1978)
Mining basin N of Georgian Bay in NE Ontario. Also search Sudbury.
BT Ontario
BT Eastern Canada
BT Canada

Sudbury District
No longer a valid term for GeoRef. As of 1993, see Sudbury District Ontario.

Sudbury District Ontario (1993)
Central Ontario. Before 1993, also search Sudbury District, District of Sudbury, or Sudbury in conjunction with Ontario.
CO N455000N483000
 W0802000W0834000
UF District of Sudbury
BT Ontario
BT Eastern Canada
BT Canada

Sudbury Irruptive (1978)
An intrusive region around Sudbury in SE Ontario.
UF Sudbury Nickel Irruptive
BT Ontario
BT Eastern Canada
BT Canada

Sudbury Nickel Irruptive
 use Sudbury Irruptive

sudden commencements
As of 1993, no longer a valid term for GeoRef. Usually out-of-scope. Used with magnetosphere.

Sudeten (1978)
Region. All the borderlands of Bohemia and Moravia formerly inhabited by German speaking people (Sudeten Germans).
IN Index Bohemia and/or Moravia as applicable.
UF Sudetenland
BT Czechoslovakia
BT Central Europe
BT Europe
SA Bohemia
SA Moravia

Sudeten Mountains (1978)
Ranges between NW Czechoslovakia and SW Poland. Also search Sudeten.
IN Index Poland and Czechoslovakian regions as applicable.
CO N495000N510500
 E0180000E0145000
UF Sudetes
UF Sudetes Mountains
UF Sudetic Mountains
UF Sudety Mountains
BT Central Europe
BT Europe
NT Czech Sudeten Mountains
NT Izera Mountains
NT Karkonosze Mountains
NT Polish Sudeten Mountains
NT Sowie Mountains
SA Bohemia
SA Czech Republic
SA Czechoslovakia
SA Moravia
SA North Sudetic Basin
SA Poland

Sudetenland
 use Sudeten

Sudetes
 use Sudeten Mountains

Sudetes Mountains
 use Sudeten Mountains

Sudetic Basin (1978)
In the Sudeten Mountains. Also search Sudetic.
IN Index countries as applicable.
UF Sudetic Depression
BT Central Europe
BT Europe
SA Czechoslovakia
SA Moravia
SA North Sudetic Basin
SA Poland

Sudetic Depression
 use Sudetic Basin

Sudetic Mountains
 use Sudeten Mountains

Sudety Mountains
 use Sudeten Mountains

sudoite (1978)
BT chlorite group
BT sheet silicates
BT silicates
SA aluminosilicates

suevite (1978)
SA breccia
SA cryptoexplosion features
SA impact features
SA impactite
SA metamorphic rocks
SA meteorites

Suez (Canal)
 use Suez Canal

Suez Canal (1978)
Sea level canal crossing the Isthmus of Suez between the E Mediterranean Sea and the Gulf of Suez. Also search Suez.
UF Suez (Canal)
BT Egypt
BT North Africa
BT Africa

Suffield
Valid through 1988. Search in combination with state term. After 1988, use specific city-state term.

Suffield Connecticut (1989)
Town in N Connecticut. Before 1989, search Suffield AND Connecticut.
CO N415800N415800
 W0723800W0723800
BT Hartford County Connecticut
BT Connecticut
BT United States

Suffolk
No longer a valid term for GeoRef. As of 1993, see Suffolk England.

Suffolk County
Valid through 1988. Search in combination with state term. After 1988, use specific county-state term.

Suffolk County Massachusetts (1989)
E Massachusetts. Before 1989, also search Suffolk County AND Massachusetts.
CO N421300N422600
 W0705400W0711100
BT Massachusetts
BT United States
NT Boston Massachusetts

Suffolk County New York (1989)
SE New York. Before 1989, also search Suffolk County AND New York.
CO N403700N411700
 W0715000W0733100
BT New York
BT United States
SA Fire Island
SA Great South Bay
SA Plum Island

Suffolk England (1993)
County in SE England. Before 1993, also search Suffolk AND England.

CO N515500N523500
   E0014500W0003000
BT East Anglia
BT England
BT Great Britain
BT United Kingdom
BT Western Europe
BT Europe

suffosion (1985)
SA glacial geology
SA ground ice
SA ground water
SA permafrost
SA thermokarst
SA water erosion
SA weathering

sugars (1978)
BT carbohydrates
BT organic materials

Suhl Bezirk
No longer a valid term for GeoRef. District in SW East Germany. See Thuringia Germany.

Suidae (1978)
Family. Autoposting of Eutheria and Theria to this term began in 1989.
BT Suiformes
BT Artiodactyla
BT Eutheria
BT Theria
BT Mammalia
BT Tetrapoda
BT Vertebrata
BT Chordata

Suiformes (1981)
Autoposting of Eutheria and Theria to this term began in 1989.
BT Artiodactyla
BT Eutheria
BT Theria
BT Mammalia
BT Tetrapoda
BT Vertebrata
BT Chordata
NT Suidae

Sulaiman Range (1978)
W of the Indus River.
IN Index provinces as applicable.
BT Pakistan
BT Indian Peninsula
BT Asia
SA Baluchistan Pakistan
SA North-West Frontier Pakistan

Sulawesi
  use Celebes

Sulcis (1978)
Region in extreme SW Sardinia.
BT Sardinia Italy
BT Italy
BT Southern Europe
BT Europe

sulfantimonates (1978)
Autoposting of this term began in 1978.
BT sulfosalts
NT cylindrite
NT famatinite

sulfantimonites (1978)
Vrbaite is a narrower term as of 1981. Autoposting of this term began in 1978.
BT sulfosalts
NT boulangerite
NT bournonite
NT chalcostibite
NT freibergite
NT freieslebenite
NT geocronite

NT heteromorphite
NT jamesonite
NT kobellite
NT luzonite
NT meneghinite
NT miargyrite
NT plagionite
NT polybasite
NT pyrargyrite
NT semseyite
NT stephanite
NT tetrahedrite
NT vrbaite
NT zinckenite

sulfarsenates (1981)
BT sulfosalts
NT enargite

sulfarsenites (1978)
Vrbaite is a narrower term as of 1981. Autoposting of this term began in 1978.
BT sulfosalts
NT dufrenoysite
NT geocronite
NT gratonite
NT jordanite
NT luzonite
NT pearceite
NT proustite
NT seligmannite
NT tennantite
NT vrbaite

sulfate ion (1981)
SA geochemistry
SA ions
SA sulfates

sulfates (1978)
Before 1981, hauyne was included as a narrower term. Autoposting of this term began in 1978.
UF sulphates
NT alabaster
NT alum
NT alunite
NT anglesite
NT anhydrite
NT barite
NT bassanite
NT beudantite
NT brochantite
NT celestite
NT chalcanthite
NT copiapite
NT coquimbite
NT epsomite
NT ettringite
NT glauberite
NT gypsum
NT halotrichite
NT hexahydrite
NT hinsdalite
NT jarosite
NT jouravskite
NT kainite
NT kieserite
NT langbeinite
NT lazurite
NT leadhillite
NT melanterite
NT mirabilite
NT natroalunite
NT pickeringite
NT polyhalite
NT posnjakite
NT roemerite
NT rozenite
NT selenite
NT svanbergite
NT thaumasite
NT thenardite
NT voltaite
NT wenkite

SA calcium sulfate
SA hauyne
SA minerals
SA sodium sulfate
SA sulfate ion

sulfides (1978)
Before 1981, voltaite and niggliite were included as narrower terms. Autoposting of this term to oxysulfides began in 1981. Autoposting of this term to antimonides, arsenides, bismuthides, selenides, and tellurides ended in 1989.
UF sulphides
NT acanthite
NT aikinite
NT alabandite
NT alloclasite
NT anilite
NT argentite
NT arsenopyrite
NT arsenosulfides
NT berndtite
NT betekhtinite
NT bismuthinite
NT bohdanowiczite
NT bornite
NT bravoite
NT briartite
NT carrollite
NT cattierite
NT chalcocite
NT chalcopyrite
NT cinnabar
NT cobaltite
NT copper sulfides
NT covellite
NT cubanite
NT digenite
NT djurleite
NT galena
NT genthelvite
NT gersdorffite
NT greenockite
NT greigite
NT gudmundite
NT heazlewoodite
NT helvite
NT herzenbergite
NT idaite
NT iron sulfides
NT jordisite
NT joseite
NT kesterite
NT lazurite
NT linnaeite
NT mackinawite
NT marcasite
NT mawsonite
NT metacinnabar
NT metastibnite
NT millerite
NT molybdenite
NT niningerite
NT oldhamite
NT onofrite
NT orpiment
NT oxysulfides
NT pentlandite
NT polydymite
NT pyrite
NT pyrrhotite
NT realgar
NT roquesite
NT siegenite
NT sinnerite
NT smythite
NT sphalerite
NT stannoidite
NT sternbergite
NT stibnite
NT stromeyerite
NT talnakhite
NT tetradymite

NT tochilinite
NT troilite
NT tungstenite
NT ullmannite
NT ultrabasite
NT vaesite
NT valleriite
NT violarite
NT willyamite
NT wurtzite
NT zinc sulfides
SA black smokers
SA gossan
SA hydrogen sulfide
SA minerals

sulfobismuthites (1978)
Autoposting of this term to matildite began in 1981.
BT sulfosalts
NT berryite
NT cosalite
NT emplectite
NT galenobismutite
NT gustavite
NT hammarite
NT lillianite
NT matildite
NT wittichenite

sulfogermanates (1978)
Autoposting of this term began in 1978.
BT sulfosalts
NT canfieldite
NT germanite
NT renierite

sulfosalts (1978)
Includes sulfantimonates, sulfantimonites, sulfarsenates, sulfarsenites, sulfobismuthites, sulfogermanates, sulfostannates, sulfovanadates. Autoposting of this term began in 1978.
NT sulfantimonates
NT sulfantimonites
NT sulfarsenates
NT sulfarsenites
NT sulfobismuthites
NT sulfogermanates
NT sulfostannates
NT sulfovanadates
SA minerals

sulfostannates (1978)
Autoposting of this term began in 1978.
BT sulfosalts
NT canfieldite
NT stannite

sulfovanadates (1978)
Autoposting of this term began in 1978.
BT sulfosalts
NT sulvanite

sulfur (1978)
Chemical element. As of 1981, use sulfur deposits for sulfur as a commodity.
UF S
UF sulphur
NT S-32
NT S-34
NT S-34/S-32
NT S-35
SA chalcophile elements
SA chemical elements
SA isotopes
SA native elements
SA organic sulfur
SA stable isotopes
SA sulfur deposits

sulfur deposits (1981)
Before 1981, search sulfur or S AND deposits.

IN Commodity. See List C.
SA Grzybow
SA native elements
SA pyrite ores
SA sulfur

sulfur dioxide (1978)
UF SO2
SA atmosphere
SA pollution

sulfuric acid (1978)
UF H2SO4
SA acid mine drainage
SA acids
SA inorganic acids
SA pollution
SA waste disposal

Sulina
No longer a valid term for GeoRef. As of 1993 see Sulina Romania.

Sulina Romania (1993)
Town on the Black Sea in Tulcea County in NE Romanian Dobruja, E Romania. Before 1993 also search Sulina AND Romania.
BT Romania
BT Southern Europe
BT Europe
SA Romanian Dobruja
SA Tulcea Romania

Sulitjelma (1978)
Peak in Kjolen Mountains on Norwegian-Swedish border in N part of Scandinavian Peninsula.
IN Index countries as applicable.
BT Scandinavia
BT Western Europe
BT Europe
SA Norway
SA Sweden

Sullivan County
Valid through 1988. Search in combination with state term. After 1988, use specific county-state term.

Sullivan County Indiana (1989)
SW Indiana. Before 1989, also search Sullivan County AND Indiana.
CO N385400N391500
W0871500W0874000
BT Indiana
BT United States

Sullivan County Missouri (1989)
N Missouri. Before 1989, also search Sullivan County AND Missouri.
CO N400200N402300
W0925200W0932200
BT Missouri
BT United States

Sullivan County New Hampshire (1989)
SW New Hampshire. Before 1989, also search Sullivan County AND New Hampshire.
CO N430800N433700
W0715700W0722600
BT New Hampshire
BT United States

Sullivan County New York (1989)
SE New York. Before 1989, also search Sullivan County AND New York.
CO N412500N420200
W0742300W0750700
BT New York
BT United States
NT Monticello New York
SA Mystery Cave

Sullivan County Pennsylvania (1989)
NE Pennsylvania. Before 1989, also search Sullivan County AND Pennsylvania.
CO N411700N413600
W0761400W0764900
BT Pennsylvania
BT United States

Sullivan County Tennessee (1989)
NE Tennessee. Before 1989, also search Sullivan County AND Tennessee.
CO N362500N363700
W0815100W0824200
BT Tennessee
BT United States

Sullivan Mine (1993)
Use Bunker Hill Mine for the lead-zinc deposits in Idaho. Refers to lead-zinc deposits and silver ores in Kimberley, British Columbia. Also refers to copper ores in Cochise County Arizona and to uranium ores in Carbon County Wyoming and to lead-zinc deposits in Alberta.
IN Index county or province or country AND commodity as applicable.
SA Alberta
SA Arizona
SA British Columbia
SA Carbon County Wyoming
SA Cochise County Arizona
SA copper ores
SA lead-zinc deposits
SA Manitoba
SA mines
SA silver ores
SA uranium ores
SA Wyoming

sulphates
use sulfates

sulphides
use sulfides

sulphur
use sulfur

Sulu Sea (1978)
Large interisland sea between the Philippine Islands and NE Borneo with Palawan Island on the NW and the Sulu Archipelago on the SW. Also search Sulu.
BT West Pacific
BT Pacific Ocean
SA Leg 124
SA ODP Site 768
SA ODP Site 769
SA ODP Site 771

sulvanite (1978)
BT sulfovanadates
BT sulfosalts

Sumatra (1978)
One of the islands of the Malay Archipelago SE and S of the Malay Peninsula. The second largest Indonesian island.
CO S060000N054500
E1061000E0950500
BT Indonesia
BT Far East
BT Asia
NT Toba Lake
SA Bangka
SA Malay Archipelago

Summit County
Valid through 1988. Search in combination with state term. After 1988, use specific county-state term.

Summit County Colorado (1989)
N central Colorado. Before 1989, also search Summit County AND Colorado.
CO N392700N395400
W1054400W1062900
BT Colorado
BT United States
NT Dillon Colorado
SA Vail Pass

Summit County Ohio (1989)
NE Ohio. Before 1989, also search Summit County AND Ohio.
CO N405300N412200
W0812300W0814200
BT Ohio
BT United States
NT Akron Ohio

Summit County Utah (1989)
NE Utah. Before 1989, also search Summit County AND Utah.
CO N403300N411200
W1100000W1113900
BT Utah
BT United States
NT Park City Utah

Sumner County
Valid through 1988. Search in combination with state term. After 1988, use specific county-state term.

Sumner County Kansas (1989)
S Kansas. Before 1989, also search Sumner County AND Kansas.
CO N370000N373000
W0970800W0974800
BT Kansas
BT United States

Sumner County Tennessee (1989)
N Tennessee. Before 1989, also search Sumner County AND Tennessee.
CO N361400N363800
W0861300W0864500
BT Tennessee
BT United States

Sumter County
Valid through 1988. Search in combination with state term. After 1988, use specific county-state term.

Sumter County Alabama (1989)
W Alabama. Before 1989, also search Sumter County AND Alabama.
CO N321700N325800
W0875100W0882700
BT Alabama
BT United States

Sumter County Florida (1989)
Central Florida. Before 1989, also search Sumter County AND Florida.
CO N281600N285800
W0815700W0821800
BT Florida
BT United States

Sumter County Georgia (1989)
SW central Georgia. Before 1989, also search Sumter County AND Georgia.
CO N314300N321300
W0835500W0842600
BT Georgia
BT United States

Sumter County South Carolina (1989)
Central South Carolina. Before 1989, also search Sumter County AND South Carolina.
CO N333800N341000
W0795300W0803800
BT South Carolina
BT United States

Sun (1978)
SA Moon
SA planetology
SA planets
SA solar activity
SA solar cycles
SA solar energy
SA solar nebula
SA solar system
SA solar wind

Sunbury Shale (1985)
Ohio and NE Kentucky.
BT Mississippian
BT Carboniferous
BT Paleozoic
SA Kentucky
SA Ohio

Sunda Arc (1985)
Island arc extending from Sumatra to Timor.
CO S110000N060000
E1280000E0950000
BT Indonesia
BT Far East
BT Asia
SA Java
SA Lesser Sunda Islands
SA Timor

Sunda Shelf (1978)
The continental shelf between Borneo and Java.
UF Borneo-Java Shelf
BT Java Sea
BT Indonesian Seas
BT West Pacific
BT Pacific Ocean

Sundance Formation (1978)
Includes Canyon Springs Sandstone, Stockade Beaver Shale, Hulett Sandstone, Lak, and Redwater Shale members. Central N Colorado, central S Montana, NW Nebraska, SW South Dakota, and SW Wyoming.
BT Upper Jurassic
BT Jurassic
BT Mesozoic
SA Colorado
SA Montana
SA Nebraska
SA South Dakota
SA Wyoming

sundry minerals
No longer a valid term for GeoRef. Before 1971, included use in subfiles N and E. Miscellaneous minerals was used before 1993.
use mineral inventory

Sunlight (1978)
Peak in SW Colorado and mountain in NW Wyoming.
IN Index states, counties, and mountain ranges as applicable.
SA Colorado
SA Park County Wyoming
SA San Juan County Colorado
SA San Juan Mountains
SA Wyoming

Sunniland Limestone (1989)

S Florida. This term has multiple hierarchies.
BT1 Comanchean
BT1 Cretaceous
BT1 Mesozoic
BT2 Lower Cretaceous
BT2 Cretaceous
BT2 Mesozoic
SA Florida

Sunnyside Mine (1989)
Metal ores in Eureka mining district, San Juan County, Colorado. Also gold ores on Thunder Mountain, Valley County, Idaho. Also coal mine and oil sands in the Uinta Basin, Carbon County, E Utah. Also gold ores in Victoria, Australia.
IN Index countries or states and counties and commodities as applicable.
SA Carbon County Utah
SA Colorado
SA gold ores
SA Idaho
SA metal ores
SA mines
SA oil sands
SA San Juan County Colorado
SA Utah
SA Valley County Idaho
SA Victoria Australia

Sunqur
Not a valid term for GeoRef. See Songhor Iran.

Sunrise Formation (1978)
Overlies Gabbs Formation; unconformably underlies Dunlap Formation. SW Nevada.
BT Lower Jurassic
BT Jurassic
BT Mesozoic
SA Nevada

sunspots (1978)
SA geologic hazards
SA solar activity

Sunzha (1978)
Mountain range. N outlier of the central Greater Caucasus in Northern Caucasus. This term has multiple hierarchies.
BT1 Northern Caucasus
BT1 Caucasus
BT1 Europe
BT2 Northern Caucasus
BT2 Russian Federation
BT2 Commonwealth of Independent States
SA Greater Caucasus

Supai Formation (1978)
Pennsylvanian and Permian. In Aubrey Group. Includes Esplanade Sandstone Member, Fort Apache Limestone, Kinishba Beds, Amos Wash Member, Big "A" Member, Apache Member, Corduroy Member, Packard Member, Oak Creek Member. N Arizona, E California, E Nevada, W New Mexico, and S Utah.
IN Index ages as applicable.
BT Paleozoic
SA Arizona
SA California
SA Esplanade Sandstone Member
SA Nevada
SA New Mexico
SA Pennsylvanian
SA Permian
SA Utah

superconducting quantum interference devices
use SQUID

Superfund (1993)
UF CERCLA
UF Comprehensive Environmental Response, Compensation and Liability Act
SA hazardous waste
SA programs
SA reclamation
SA U. S. Environmental Protection Agency
SA waste disposal

supergene processes (1978)
As of 1993, restricted to economic studies.
UF hypergene processes
SA enrichment
SA mineral deposits, genesis
SA processes

superimposed folds
use superposed folds

superimposed metamorphism
use polymetamorphism

Superior Province (1978)
Structural province of the Canadian Shield.
IN Index Minnesota and provinces as applicable.
CO N460000N614000 W0660000W0990000
BT Canadian Shield
BT North America
NT Abitibi Belt
NT English River Belt
NT Kapuskasing Zone
NT Michipicoten Belt
NT Quetico Belt
NT Wabigoon Belt
NT Wawa Belt
SA Eye-Dashwa Lakes Pluton
SA Labrador
SA Manitoba
SA Minnesota
SA Ontario
SA Quebec
SA Timiskaming Group
SA Wyoming Province

superposed folds (1978)
Includes use as orientation of folds relative to spatially associated macroscopic structures such as large folds, fold systems, and orogenic zones. Before 1978, also search folds AND superposed; folds AND cross.
UF cross folds
UF subsequent folds
UF superimposed folds
UF transverse folds
BT folds
SA orientation

Superstition Hills earthquake 1987 (1993)
S. California, Salton Trough area.
BT earthquakes
SA California
SA Salton Trough

Superstition Mountains (1978)
S central Arizona. Also in other location. Also search Superstition.
IN Index county and mountains as applicable.
SA Arizona
SA Pinal County Arizona

superstructure (1978)
SA crystal structure

supply (1981)
SA consumption
SA economics
SA markets
SA sediment supply

supply, water
use water supply

supracrustals (1993)
Sedimentary and volcanic rocks which are sufficiently untransformed for them to be recognizable.
BT metamorphic rocks
SA basement
SA basement tectonics
SA greenstone belts
SA metamorphism
SA metasedimentary rocks
SA metavolcanic rocks
SA tectonics

supralittoral environment
use supratidal environment

supratidal environment (1978)
Before 1978, search supratidal.
UF supralittoral environment
SA depositional environment
SA ecology
SA environment
SA sebkha environment
SA sedimentation
SA tidal flats

Sur fault zone (1978)
In the Point Sur area in W California.
IN Index counties as applicable.
UF Sur-Nacimiento fault zone
BT California
BT United States

Sur-Nacimiento fault zone
use Sur fault zone

Surakhany
No longer a valid term for GeoRef. As of 1993, see Surakhany Azerbaidzhan.

Surakhany Azerbaidzhan (1993)
Town on central Apsheron Peninsula in oil fields near Baku. Also search Surakhany oil field. This term has multiple hierarchies.
BT1 Azerbaidzhan
BT1 Europe
BT2 Azerbaidzhan
BT2 Commonwealth of Independent States
SA Apsheron Peninsula
SA oil and gas fields

Surat Basin (1978)
Sedimentary basin in SE Queensland.
CO S290000S270000 E1510000E1470000
BT Queensland Australia
BT Australia
BT Australasia

surface
A valid term through 1977. Also included use in combination with geologic (i.e. See engineering geology; reservoirs; surface reservoirs; surficial geology; surficial geology maps.

surface defects (1982)
SA crystal dislocations
SA crystal structure
SA crystallography
SA crystals
SA defects
SA minerals

surface features (1978)
Used for features on the surfaces of the planets, their satellites and the Sun.
IN See entries for individual planets (e.g. Mars).
UF features, surface
SA coronae
SA geomorphology
SA Moon
SA planetology
SA planets
SA surface properties
SA surface textures

surface mining (1981)
BT mining
NT open-pit mining
NT strip mining
SA cut-and-fill mining
SA land use
SA mines
SA mining geology
SA quarries

surface phenomena
As of 1993, no longer a valid term for GeoRef. Usually out-of-scope. Used with astrophysics and solar physics.

surface properties (1978)
For photographic characteristics, electromagnetic responses, etc.
SA electrical properties
SA Moon
SA properties
SA remote sensing
SA spectroscopy
SA surface features
SA surface textures

surface reservoirs
As of 1981, no longer a valid term for GeoRef. Use reservoirs. Term was introduced in 1978. Before 1978, also search surface AND reservoirs.

surface textures (1985)
SA surface features
SA surface properties
SA surficial geology
SA textures

surface water (1978)
SA basin management
SA decontamination
SA drinking water
SA dye tracers
SA environmental geology
SA floods
SA fluorimetry
SA ground water
SA hydrochemistry
SA hydrogeology
SA hydrologic cycle
SA hydrology
SA hydrosphere
SA irrigation
SA lakes
SA nonpoint sources
SA polluted water
SA pollution
SA ponds
SA potability
SA radioactive tracers
SA rivers
SA runoff
SA streamflow
SA streams
SA swamps
SA tracers
SA water
SA water hardness
SA water management
SA water resources
SA water rights
SA water supply

SA water treatment
SA watersheds

surface waves (1978)
Autoposting of this term began in 1978.
UF L waves
BT elastic waves
NT coda waves
NT Lg-waves
NT Love waves
NT Rayleigh waves
NT Stoneley waves
NT tube waves
SA Airy waves
SA body waves
SA earthquakes
SA engineering geology
SA ground motion
SA seismology
SA waves

surfaces, erosion
use erosion surfaces

surficial aquifers (1993)
BT aquifers
SA shallow aquifers
SA springs
SA unconfined aquifers
SA water table

surficial geology (1978)
See surficial geology maps.
SA geology
SA geomorphology
SA landform description
SA shallow aquifers
SA shallow depth
SA soils
SA surface textures
SA surficial geology maps

surficial geology maps (1983)
Before 1983, also search maps AND surficial geology. Before 1978, also search maps AND surficial; maps AND surface AND geologic.
BT maps
SA geomorphologic maps
SA geomorphology
SA surficial geology

surges (1978)
As of 1978, term is restricted for use with ocean waves. Before 1978, term was also used for glacier surges.
SA base surges
SA glacier surges
SA ocean waves
SA tidal surges
SA tides
SA waves

Surgut
No longer a valid term for GeoRef. As of 1993, see Surgut Russian Federation.

Surgut Russian Federation (1993)
Town on right bank of Ob River in S Khanty-Mansi National Okrug in W Siberia. This term has multiple hierarchies.
BT1 Tyumen Russian Federation
BT1 Russian Federation
BT1 Commonwealth of Independent States
BT2 Tyumen Russian Federation
BT2 Asia

Surigao del Norte
No longer a valid term for GeoRef. As of 1993 see Surigao del Norte Philippine Islands.

Surigao del Norte Philippine Islands (1993)
Province, S Philippine Islands. Before 1993 also search Surigao del Norte AND Philippine Islands.
CO N092000N103000 E1261500E1252000
BT Philippine Islands
BT Far East
BT Asia
SA Mindanao

Surinam (1978)
CO N015000N060000 W0535900W0580500
UF Dutch Guiana
UF Netherlands Guiana
BT South America
NT Wilhelmina Mountains
SA Guiana Basin
SA Guianas
SA Guyana Shield

Surkhan Darya (1978)
River which rises in the Hissar Range of Tadzhikistan and flows SSW into the Amu Darya near Termez. This term has multiple hierarchies.
IN Index former Soviet republics as applicable.
BT1 Commonwealth of Independent States
BT2 Asia
SA Amu Darya
SA Tadzhikistan
SA Uzbekistan

Surkhan Darya basin (1978)
River basin, primarily in Surkhan Darya Oblast in S Uzbekistan. This term has multiple hierarchies.
IN Index former Soviet republics as applicable.
UF Surkhan-Darya basin
BT1 Commonwealth of Independent States
BT2 Asia
SA Tadzhikistan
SA Uzbekistan

Surkhan-Darya basin
use Surkhan Darya basin

Surma Group (1989)
NE India and Bangladesh.
BT Miocene
BT Neogene
BT Tertiary
BT Cenozoic
SA Bangladesh
SA India

Surrey
No longer a valid term for GeoRef. As of 1993, see Surrey England.

Surrey England (1993)
County SW of London, SE England. Before 1993, also search Surrey AND England.
CO N511000N513000 E0001000W0004500
BT England
BT Great Britain
BT United Kingdom
BT Western Europe
BT Europe
SA The Weald

Surry County
Valid through 1988. Search in combination with state term. After 1988, use specific county-state term.

Surry County North Carolina (1989)
NW North Carolina. Before 1989, also search Surry County AND North Carolina.
CO N361300N363400 W0802500W0805800
BT North Carolina
BT United States

Surry County Virginia (1989)
SE Virginia. Before 1989, also search Surry County AND Virginia.
CO N365700N371400 W0764100W0771000
BT Virginia
BT United States

Surtsey (1978)
Volcanic island off S Iceland.
BT Iceland
BT Western Europe
BT Europe

Suruga Bay (1978)
Inlet of Pacific Ocean SW of Yokohama on SE coast.
BT Honshu
BT Japan
BT Far East
BT Asia
SA Fuji River
SA Izu Peninsula
SA Shizuoka Japan
SA Tokai Japan

survey organizations (1981)
Used for the work of geological surveys, national or local. Before 1981, search surveys as a level 1 term.
IN Index name of survey organization in original language.
UF geologic surveys
BT government agencies
NT ENADIMSA
NT Geological Survey of Canada
NT Illinois State Geological Survey
NT Ontario Geological Survey
NT U. S. Geological Survey
SA annual report
SA associations
SA bibliography
SA catalogs
SA current research
SA history
SA museums
SA organization
SA progress report
SA publications
SA research
SA surveys

Surveyor III
use Surveyor 3

Surveyor Program (1981)
Autoposting of this term to specific Surveyor missions began in 1989.
NT Surveyor 3
NT Surveyor 5
NT Surveyor 7
SA extraterrestrial geology
SA Moon
SA planetology
SA remote sensing

Surveyor V Mission
use Surveyor 5

Surveyor VII
use Surveyor 7

Surveyor 3 (1978)
Autoposting of Surveyor Program to this term began in 1989.
UF Surveyor III
BT Surveyor Program
SA Moon
SA remote sensing

Surveyor 5 (1989)
UF Surveyor V Mission
BT Surveyor Program
SA Moon
SA remote sensing

Surveyor 7 (1989)
UF Surveyor VII
BT Surveyor Program
SA Moon
SA remote sensing

surveys (1978)
As of 1981, restricted to actual surveying and its results. Before 1981, also used for the work of geological surveys. See survey organizations.
NT geophysical surveys
SA exploration
SA geochemistry
SA geodesy
SA geotraverses
SA ground water
SA hydrology
SA maps
SA soils
SA survey organizations
SA three-dimensional methods
SA triangulation

susceptibility
A valid term through 1977. Use magnetic susceptibility.

Susitna River (1989)
S central Alaska. Flows from Mount Hayes in the Alaska Range to Cook Inlet near Anchorage.
BT Southern Alaska
BT Alaska
BT United States
SA Susitna River basin

Susitna River basin (1978)
N of Anchorage in S Alaska. Also search Susitna. As of 1989, use Susitna River for river.
BT Southern Alaska
BT Alaska
BT United States
SA Susitna River

suspect terranes
use terranes

suspended materials (1978)
Before 1978, search materials AND suspended.
SA dissolved materials
SA flocculation
SA hydrochemistry
SA materials
SA sea water
SA sediment traps

suspension (1978)
SA geochemistry
SA hydrology
SA sediment transport
SA sediments

suspension current
use turbidity currents

Susquehanna River (1978)
Rises in Otsego Lake in central New York and flows S emptying into N Chesapeake Bay.
IN Index counties and states as applicable.
BT United States
SA Maryland
SA New York
SA Pennsylvania

Susquehanna River basin (1978)
Primarily in Pennsylvania.

Susquehanna Valley
  IN Index counties and states as applicable.
  UF Susquehanna Valley
  BT United States
  SA Maryland
  SA New York
  SA Pennsylvania
Susquehanna Valley
  use Susquehanna River basin
Sussex
  No longer a valid term for GeoRef as of 1993. See Sussex England.
Sussex County
  Valid through 1988. Search in combination with state term. After 1988, use specific county-state term.
Sussex County Delaware (1989)
  S Delaware. Before 1989, also search Sussex County AND Delaware.
  CO N382700N385700
      W0750500W0754300
  BT Delaware
  BT United States
Sussex County New Jersey (1989)
  Extreme NW New Jersey. Before 1989, also search Sussex County AND New Jersey.
  CO N405300N412300
      W0742300W0745900
  BT New Jersey
  BT United States
  NT Beemerville New Jersey
  NT Sterling Hill
  SA New Jersey Highlands
Sussex County Virginia (1989)
  SE Virginia. Before 1989, also search Sussex County AND Virginia.
  CO N364200N370700
      W0765700W0773700
  BT Virginia
  BT United States
Sussex England (1993)
  Former county on English Channel. Divided into two administrative counties: East Sussex and West Sussex in 1974.
  CO N504500N511500
      E0004500W0010000
  BT England
  BT Great Britain
  BT United Kingdom
  BT Western Europe
  BT Europe
  NT Horsham England
  NT Piltdown England
  SA The Weald
Sussex Sandstone Member (1989)
  BT Upper Cretaceous
  BT Cretaceous
  BT Mesozoic
  SA Wyoming
Sustut Basin (1978)
  River basin in N central British Columbia.
  BT British Columbia
  BT Western Canada
  BT Canada
Sutam River (1978)
  S Yakutia near N border of Amur Oblast in SE Siberia. Also search Sutam. This term has multiple hierarchies.
  BT1 Yakutia Russian Federation
  BT1 Russian Federation
  BT1 Commonwealth of Independent States
  BT2 Yakutia Russian Federation
  BT2 Asia
Sutherland
  No longer a valid term for GeoRef. As of 1993, see Sutherland Scotland or Sutherland South Africa.
Sutherland Scotland (1993)
  Former county in highlands of extreme N Scotland. Before 1993, also search Sutherland AND Scotland.
  CO N575000N583000
      W0033500W0052500
  BT Highland region Scotland
  BT Scotland
  BT Great Britain
  BT United Kingdom
  BT Western Europe
  BT Europe
  NT Assynt
  NT Scourie Scotland
Sutherland South Africa (1993)
  Cape Province, South Africa. Before 1993 also search Sutherland AND South Africa.
  BT Cape Province South Africa
  BT South Africa
  BT Southern Africa
  BT Africa
Sutna India
  use Satna India
Sutter County California (1989)
  N central California. Before 1989, also search Sutter County AND California.
  CO N384200N392000
      W1212600W1215500
  BT California
  BT United States
  NT Nicolaus California
Sutton County
  Valid through 1988. Search in combination with state term. After 1988, use specific county-state term.
Sutton County Texas (1989)
  W Texas. Before 1989, also search Sutton County AND Texas; Sutton AND Texas.
  CO N301500N304000
      W1000800W1010000
  BT Texas
  BT United States
suture zones (1983)
  Used for tectonic sutures. For paleontology, use sutures.
  UF tectonic sutures
  UF zones, suture
  SA plate tectonics
  SA rifting
  SA tectonics
sutures (1978)
  Use suture zones for tectonic sutures.
  SA fossils
  SA morphology
Suva
  No longer a valid term for GeoRef. See Suva Fiji.
Suva Fiji (1993)
  Town with one of best harbors in South Pacific on SE coast of Viti Levu Island. Viti Levu is considered a broader term as of 1985.
  BT Viti Levu
  BT Fiji
  BT Melanesia
  BT Oceania
Suvalkai
  use Suwalki Poland
Suvalki
  use Suwalki Poland
Suwalki
  No longer a valid term for GeoRef. As of 1993 see Suwalki Poland.
Suwalki Poland (1993)
  Region and city E of the Masurian Lakes in NE Suwalki, extreme NE Poland. Before 1993 also search Suwalki AND Poland.
  UF Suvalkai
  UF Suvalki
  BT Poland
  BT Central Europe
  BT Europe
  NT Augustow Poland
Suwannee Limestone (1978)
  Upper part of formation considered to be equivalent to lower Chickasawhay Marl of E Mississippi, and lower part possibly equivalent to Byram Formation and Marianna Limestone of Florida. S central Georgia, and E Florida.
  BT upper Oligocene
  BT Oligocene
  BT Paleogene
  BT Tertiary
  BT Cenozoic
  SA Florida
  SA Georgia
Suzak
  No longer a valid term for GeoRef. As of 1993, see Suzak Kyrgyzstan.
Suzak Kyrgyzstan (1993)
  Village in central Kyrgyzstan. This term has multiple hierarchies.
  BT1 Kyrgyzstan
  BT1 Asia
  BT2 Kyrgyzstan
  BT2 Commonwealth of Independent States
SV-waves (1981)
  BT S-waves
  BT body waves
  BT elastic waves
  SA seismology
  SA waves
Svalbard (1978)
  Norwegian island group including Spitsbergen group and Bear Island. In 1981, broader term changed from Arctic Ocean to Arctic region. Spitsbergen and Bear Island are narrower terms as of 1981. Autoposting of this term began in 1978.
  BT Arctic region
  NT Spitsbergen
  SA Arctic Coastal Plain
  SA Arctic Ocean
  SA Bear Island
  SA Fram Strait
svanbergite (1978)
  BT sulfates
Svanetia (1978)
  Region on S slopes of the Greater Caucasus in NW Georgian Republic.
  BT Georgian Republic
  BT Europe
  SA Greater Caucasus
Svecofennian (1978)
  Europe. Autoposting of middle Precambrian and Precambrian to this term ended in 1989. As of 1989, lower Proterozoic and Proterozoic are autoposted to this term. Autoposting of upper Precambrian and Precambrian to this term began in 1990.
  BT lower Proterozoic
  BT Proterozoic
  BT upper Precambrian
  BT Precambrian
Sventoji River (1978)
  Central Lithuania. Also search Sventoji.
  UF Shventoyi River
  BT Lithuania
  BT Baltic region
  BT Europe
Sverdlovsk
  No longer a valid term for GeoRef. As of 1993, see Yekaterinburg Russian Federation.
Sverdlovsk Russian Federation
  use Yekaterinburg Russian Federation
Sverdrup Basin (1978)
  In the Sverdrup Channel area of the Sverdrup Islands in N Franklin District.
  BT Franklin District Northwest Territories
  BT Northwest Territories
  BT Western Canada
  BT Canada
Svinita
  No longer a valid term for GeoRef. See Svinita Romania.
Svinita Romania (1993)
  Village in S Banat, W Romania. This term has multiple hierarchies.
  BT1 Banat
  BT1 Southern Europe
  BT1 Europe
  BT2 Romania
  BT2 Southern Europe
  BT2 Europe
Svratka (1978)
  Village in W central Moravia.
  BT Moravia
  BT Czech Republic
  BT Czechoslovakia
  BT Central Europe
  BT Europe
Swabia (1978)
  Region. Duchy of medieval Germany. Nearly coextensive with Baden-Wurttemberg, Hesse, and W Bavaria.
  IN Index states as applicable.
  BT Germany
  BT Central Europe
  BT Europe
  SA Baden-Wurttemberg Germany
  SA Bavaria Germany
  SA Hesse Germany
Swabian Alb (1978)
  Mountain range between the Neckar River and the upper Danube River.
  UF Swabian Jura
  BT Baden-Wurttemberg Germany
  BT Germany
  BT Central Europe
  BT Europe
  SA Holzmaden region
Swabian Jura
  use Swabian Alb
swamps (1978)

As of 1981, use paludal environment for type of environment. Autoposting of this term began in 1989.
NT mangrove swamps
SA bogs
SA brackish-water environment
SA ecology
SA fluvial features
SA geomorphology
SA lacustrine features
SA marshes
SA paludal environment
SA peat bogs
SA sedimentation
SA surface water
SA wetlands

Swan Hills (1978)
Hilly region S of Lesser Slave Lake in central Alberta.
SA Alberta

Swan Hills Formation (1993)
Middle-Upper Devonian. Alberta, Northwest Territories, and Saskatchewan. Petroleum-bearing layer.
IN Index ages as applicable.
BT Devonian
BT Paleozoic
SA Alberta
SA Middle Devonian
SA Northwest Territories
SA Saskatchewan
SA Upper Devonian

Swan Peak Formation (1978)
Includes Watson Ranch Tongue. SE Idaho and NE Utah.
BT Middle Ordovician
BT Ordovician
BT Paleozoic
SA Idaho
SA Utah

Swansea
No longer a valid term for GeoRef. See Swansea Wales if applicable. Before 1993, also used for population centers in New South Wales, England, and U. S. states.

Swansea Bay (1985)
Inlet of Bristol Channel, S Wales.
CO N512500N513500
W0034500W0040000
BT Glamorgan Wales
BT Wales
BT Great Britain
BT United Kingdom
BT Western Europe
BT Europe
SA Bristol Channel
SA Swansea Wales

Swansea Wales (1993)
City on Bristol Channel in West Glamorgan County, S Wales. Before 1993, also search Swansea AND Wales.
BT Glamorgan Wales
BT Wales
BT Great Britain
BT United Kingdom
BT Western Europe
BT Europe
SA Swansea Bay

swarms (1978)
As of 1981, restricted to earthquakes.
UF earthquake swarms
SA dike swarms
SA earthquakes
SA intrusions
SA seismology

Swartkrans (1978)
Mountains near Lesotho border in SE Orange Free State.
BT Orange Free State South Africa
BT South Africa
BT Southern Africa
BT Africa

Swat
No longer a valid term for GeoRef. Former princely state. See Swat Pakistan.

Swat Pakistan (1993)
District in the valley of the Swat River NNE of Peshawar.
BT North-West Frontier Pakistan
BT Pakistan
BT Indian Peninsula
BT Asia

Swauk Formation (1978)
Unconformably overlies metamorphic basement of unknown age. Central and N central Washington.
BT Eocene
BT Paleogene
BT Tertiary
BT Cenozoic
SA Washington

Swaziland (1978)
Borders on Mozambique and South Africa. Administered by a British High Commissioner until independence in 1968.
CO S273000S260000
E0321500E0321500
BT Southern Africa
BT Africa
SA Barberton greenstone belt
SA Kaapvaal Craton
SA Pongola Supergroup
SA Swaziland Sequence
SA Swaziland System

Swaziland Sequence (1978)
In South Africa, and Swaziland.
BT Precambrian
SA South Africa
SA Swaziland

Swaziland System (1978)
In South Africa and Swaziland.
BT Precambrian
SA South Africa
SA Swaziland

Sweden (1978)
CO N551500N691500
E0241500E0110000
BT Scandinavia
BT Western Europe
BT Europe
NT Alno
NT Blekinge Sweden
NT Dalarna
NT Gota Valley
NT Goteborg Sweden
NT Gotland Sweden
NT Gravberg Well
NT Halland Sweden
NT Jamtland Sweden
NT Kalmar Sweden
NT Kristianstad Sweden
NT Malmohus Sweden
NT Norrbotten Sweden
NT Nykoping Sweden
NT Orebro Sweden
NT Siljan
NT Siljan Ring
NT Skane
NT Skellefte
NT Skellefte mining district
NT Smaland
NT Stockholm Sweden
NT Uppsala Sweden
NT Varmland Sweden
NT Vasterbotten Sweden
NT Vastergotland
SA Ancylus Lake
SA Arctic region
SA Baltic Glaciation
SA Baltic ice lake
SA Baltic region
SA Baltic Shield
SA Fennoscandia
SA Koli Nappe
SA Lapland
SA Litorina Sea
SA North Sea Coast
SA North Sea region
SA Oresund
SA Sparagmite Group
SA Sulitjelma
SA University of Lund
SA Yoldia Sea

Sweet Grass County
Valid through 1988. Search in combination with state term. After 1988, use specific county-state term.

Sweet Grass County Montana (1989)
S central Montana. Before 1989, also search Sweet Grass County AND Montana.
CO N451000N461500
W1092500W1102000
BT Montana
BT United States
SA Crazy Mountains Basin

Sweetgrass Arch (1978)
IN Index Alberta and/or Montana as applicable.
BT North America
SA Alberta
SA Blackleaf Formation
SA Montana

Sweetwater County
Valid through 1988. Search in combination with state term. After 1988, use specific county-state term.

Sweetwater County Wyoming (1989)
SW Wyoming. Before 1989, also search Sweetwater County AND Wyoming.
CO N410000N422000
W1073000W1101000
BT Wyoming
BT United States
NT Rock Springs Wyoming
SA Great Divide Basin
SA Lake Gosiute
SA Lost Soldier Field
SA Moxa Arch
SA Red Desert
SA Rock Springs Uplift

swelling soils
use expansive materials

Swietokrzyskie Mountains
use Swiety Krzyz Mountains

Swiety Krzyz Mountains (1978)
Mountain group between the Vistula River and Pilica River in SE central Poland. Also search Swiety Krzyz.
CO N503500N510000
E0213000E0202000
UF Holy Cross Mountains
UF Swietokrzyskie Mountains
BT Poland
BT Central Europe
BT Europe
SA Galezice
SA Kielce Poland

Swift Formation (1985)
In Ellis Group. N central Montana, W South Dakota, North Dakota and E Wyoming.
BT Upper Jurassic
BT Jurassic
BT Mesozoic
SA Montana
SA North Dakota
SA South Dakota
SA Wyoming

Swisher County
Valid through 1988. Search in combination with state term. After 1988, use specific county-state term.

Swisher County Texas (1989)
NW Texas. Before 1989, also search Swisher County AND Texas.
CO N342000N344500
W1012500W1020000
BT Texas
BT United States

Swiss Alps (1978)
Those ranges of the Alps mountain system which lie within Switzerland. Autoposting of Alps to this term ended in 1989 and began again in 1993. This term has multiple hierarchies.
BT1 Switzerland
BT1 Central Europe
BT1 Europe
BT2 Alps
BT2 Europe
NT Central Swiss Alps
NT Eastern Swiss Alps
NT Northern Swiss Alps
NT Southern Swiss Alps
SA Glarus Alps
SA Oberalp Pass
SA Tavetsch

Swiss Jura Mountains (1981)
Jura Mountains in Switzerland. Autoposting of Switzerland to this term began in 1989. As of 1990, Central Europe and Europe are autoposted to this term. This term has multiple hierarchies.
BT1 Jura Mountains
BT1 Europe
BT2 Switzerland
BT2 Central Europe
BT2 Europe
SA Jura Switzerland

Swiss Molasse Basin (1981)
Autoposting of Switzerland to this term began in 1989. As of 1990, Central Europe and Europe are autoposted to this term. This term has multiple hierarchies.
CO N460700N474800
E0094100E0055600
BT1 Molasse Basin
BT1 Central Europe
BT1 Europe
BT2 Switzerland
BT2 Central Europe
BT2 Europe

Swiss Rhine Graben (1981)
Autoposting of Switzerland to this term began in 1989. As of 1990, Central Europe and Europe are autoposted to this term. This term has multiple hierarchies.
BT1 Rhine Graben
BT1 Central Europe
BT1 Europe
BT2 Switzerland
BT2 Central Europe
BT2 Europe

Switzerland (1978)
CO N454500N474500
E0103000E0055000
BT Central Europe
BT Europe
NT Aar Massif
NT Aar Valley
NT Aargau Switzerland
NT Basel Switzerland
NT Bergell Massif
NT Bern Switzerland
NT Bernese Alps
NT Geneva Switzerland
NT Glarus Alps
NT Glarus Switzerland
NT Gotthard Massif
NT Graubunden Switzerland
NT Grindelwald
NT Jura Switzerland
NT Lake of Zurich
NT Lucerne Switzerland
NT Neuchatel Switzerland
NT Oberalp Pass
NT Saint Gall Switzerland
NT Schaffhausen Switzerland
NT Swiss Alps
NT Swiss Jura Mountains
NT Swiss Molasse Basin
NT Swiss Rhine Graben
NT Thur Valley
NT Ticino Switzerland
NT Uri Switzerland
NT Valais Switzerland
NT Vaud Switzerland
NT Zurich Switzerland
SA Alps
SA Arve Valley
SA Ceneri Zone
SA Central Alps
SA Lake Constance
SA Molasse Basin
SA Rhine Basin
SA Rhine Graben
SA Rhine River
SA Rhine Valley
SA Rhone River
SA Rhone Valley
SA Silvretta Group
SA Simplon region
SA Ticino
SA Western Alps

sychnodymite
use carrollite

Sydney
No longer a valid term for GeoRef. See Sydney Australia or Sydney Nova Scotia.

Sydney Australia (1993)
City on the Pacific Ocean in E New South Wales.
CO S335500N335500
E1511000E1511000
BT New South Wales Australia
BT Australia
BT Australasia

Sydney Basin (1978)
Including the Sydney area in New South Wales, it extends from NW of Newcastle on the N to W of Wollongong on the south.
IN Index states or region as applicable.
SA Gunnedah Basin
SA New South Wales Australia
SA Newcastle Coal Measures
SA Southern coal field

Sydney coal field (1985)
E Cape Breton Island, Nova Scotia.
IN Index Nova Scotia if applicable.
UF Sydney Coalfield
SA Cape Breton Island
SA coal fields
SA Nova Scotia

Sydney Coalfield
use Sydney coal field

Sydney Nova Scotia (1993)
Before 1993, also search Sydney AND Nova Scotia.
BT Cape Breton County Nova Scotia
BT Nova Scotia
BT Maritime Provinces
BT Eastern Canada
BT Canada

Syedlets Poland
use Siedlce Poland

syenite
As of 1989, no longer a valid term for GeoRef. From 1981 to 1988, syenites was autoposted to this term.
use syenites

syenite family
As of 1981, no longer a valid term for GeoRef. Use syenites, alkali syenites, or monzonites as applicable. From 1978-1980, was autoposted to the rocks that were classified under it.

syenite porphyry (1978)
BT syenites
BT plutonic rocks
BT igneous rocks
SA porphyry

syenites (1981)
Before 1981, search syenite family. From 1978-1980, syenite family was autoposted to the rocks that were classified under it. Before 1989, also search syenite. Autoposting of this term to syenite ended in 1989.
UF syenite
BT plutonic rocks
BT igneous rocks
NT agpaite
NT albitite
NT albitophyre
NT alkali syenites
NT bostonite
NT foyaite
NT granosyenite
NT kakortokite
NT lujavrite
NT miaskite
NT microsyenite
NT monzonites
NT naujaite
NT nepheline syenite
NT pulaskite
NT quartz syenite
NT shonkinite
NT syenite porphyry
SA appinite

syenodiorite (1978)
BT diorites
BT plutonic rocks
BT igneous rocks

Sylmar Fault (1978)
Named after town in San Fernando Valley in Los Angeles County. Also search Sylmar.
IN Index counties as applicable.
BT California
BT United States
SA Los Angeles County California
SA San Fernando Valley

Sylt (1978)
Main island of the North Frisian Islands just off the mainland in the North Sea.
UF Sylt Island
BT Schleswig-Holstein Germany
BT Germany
BT Central Europe
BT Europe

Sylt Island
use Sylt

Sylvania Formation (1978)
Underlies Detroit River Group; overlies Bois Blanc Formation. SE Michigan, NW Ohio, and Ontario.
BT Middle Devonian
BT Devonian
BT Paleozoic
SA Michigan
SA Ohio
SA Ontario

sylvanite (1978)
Autoposting of sulfides to this term ended in 1989.
UF silvanite
BT tellurides

sylvine
use sylvite

sylvinite (1993)
A rock containing chiefly impure potassium chloride.
BT evaporites
BT chemically precipitated rocks
BT sedimentary rocks
SA evaporite deposits
SA halite
SA sylvite

sylvite (1978)
UF leopoldite
UF sylvine
BT chlorides
BT halides
SA sylvinite

symbiosis (1978)
SA epibiotism

symmetric folds (1978)
Before 1978, search folds AND symmetric.
BT folds
SA asymmetric folds
SA normal folds

Symmetrodonta (1978)
Order. Autoposting of Theria to this term began in 1989.
BT Theria
BT Mammalia
BT Tetrapoda
BT Vertebrata
BT Chordata

symmetry (1978)
UF asymmetry
SA asymmetric distribution
SA crystal chemistry
SA crystal field
SA crystal growth
SA crystal structure
SA crystal systems
SA crystallography

symmicton
use diamicton

symplectite (1989)
Intergrowth of two different minerals. Also used for the rock characterized by symplectic texture.
UF symplectites
SA crystal growth
SA crystalline rocks
SA igneous rocks
SA metamorphic rocks
SA minerals

symplectites
use symplectite

symposia (1978)
Before 1972, also search symposium. For 1959, also search symposiums.
IN Index for proceedings volumes.
UF colloquia
UF conferences
UF meetings
UF symposium
UF symposiums
SA annual report
SA associations
SA geology
SA organization
SA progress report
SA report

symposium
No longer a valid term for GeoRef.
use symposia

symposiums
No longer a valid term for GeoRef. In 1959, included use on level 1 in subfile N.
use symposia

Synapsida (1978)
Subclass. "Mammal-like reptiles". Autoposting of this term began in 1978.
BT Reptilia
BT Tetrapoda
BT Vertebrata
BT Chordata
NT Pelycosauria
NT Therapsida

synchisite (1978)
From 1978-1980, carbonates and halides were autoposted to this term. Autoposting of fluorides began in 1981. As of 1989, halides is autoposted to this term. This term has multiple hierarchies.
UF synchysite
BT1 carbonates
BT2 fluorides
BT2 halides

synchrotron radiation (1989)
SA chemical analysis
SA electromagnetic field
SA electromagnetic methods
SA experimental studies
SA radioactivity
SA synchrotrons
SA X-ray analysis
SA X-ray diffraction analysis
SA X-ray fluorescence

synchrotrons (1989)
Used for the instrument.
SA electromagnetic methods
SA electromagnetic radiation
SA instruments
SA particles
SA radioactivity
SA synchrotron radiation
SA X-ray analysis
SA X-ray diffraction analysis
SA X-ray fluorescence

synchysite
use synchisite

synclinal
A valid term through 1977. After 1977, use synclines.

synclines (1978)
Before 1978, also search folds AND synclinal.
BT folds
SA anticlines

SA synform folds

synclinoria (1978)
BT folds
SA anticlinoria
SA geanticlines
SA geosynclines
SA systems

synform folds (1978)
Before 1978, search folds AND synform.
BT folds
SA antiform folds
SA synclines

syngenesis (1978)
BT diagenesis
SA mineral deposits, genesis

synonymy (1978)
SA paleontology
SA taxonomy

synsedimentary processes (1985)
SA processes
SA sedimentation
SA syntectonic processes
SA tectonics

syntectonic processes (1978)
Before 1978, search syntectonic.
SA post-tectonic processes
SA processes
SA synsedimentary processes
SA tectonics

synthesis (1978)
Before 1971, also search mineral synthesis.
SA crystal growth
SA minerals
SA synthetic materials

synthetic
A valid term through 1977 used in combination with the name of the mineral to index artificial minerals. After 1977, use synthetic materials.

synthetic aperture radar
use SAR

synthetic materials (1978)
Used to index artificial minerals. Before 1978, also search synthetic AND name of mineral.
SA carborundum
SA materials
SA minerals
SA natural materials
SA synthesis

synthetic seismograms (1981)
BT seismograms
SA numerical seismograms

Syowa Station (1989)
Japanese research facility on Prince Olav Coast bordering Lutzow-Holm Bay in E Antarctica.
CO S690000S690000 E0393500E0393500
UF Showa
UF Showa Station
BT Antarctica
SA Lutzow-Holm Bay
SA Prince Olav Coast

Syr Darya (1978)
River which rises in the Tien Shan and flows W and NW into the Aral Sea. This term has multiple hierarchies.
IN Index former Soviet republics as applicable.
UF Jaxartes River
UF Saihun River
UF Sir Darya
UF Syr Darya River
UF Syr-Darya
BT1 Commonwealth of Independent States
BT2 Asia
SA Kazakhstan
SA Kyrgyzstan
SA Sarysu
SA Tadzhikistan
SA Uzbekistan

Syr Darya River
use Syr Darya

Syr-Darya
use Syr Darya

Syracuse
No longer a valid term for GeoRef. Before 1989, included use for city in Onondaga County, New York. Now use Syracuse New York for city in New York. As of 1993, see Syracuse Italy.

Syracuse Italy (1993)
City in SE Sicily. Search in conjunction with Italy. Also search Siracusa.
CO N370400N370400 E0151800E0151800
BT Sicily Italy
BT Italy
BT Southern Europe
BT Europe

Syracuse New York (1989)
City in central New York. Before 1989, search Syracuse AND New York.
CO N430300N430300 W0761000W0761000
BT Onondaga County New York
BT New York
BT United States

Syria (1978)
CO N323000N371500 E0423000E0353000
BT Middle East
BT Asia
NT Damascus Syria
NT Golan Heights
NT Sabkha Syria
SA Euphrates River
SA Ghareb Formation
SA Great Rift Valley
SA Jordan River
SA Jordan Valley
SA Judea Group
SA Mediterranean region
SA Menuha Formation
SA Mesopotamia
SA Mount Scopus Group
SA Near East

Syrte
use Sirte Basin

Sysert
No longer a valid term for GeoRef. As of 1993, see Sysert Russian Federation.

Sysert Russian Federation (1993)
City 25 miles SSE of Yekaterinburg in Yekaterinburg Oblast in W Siberia.
BT Yekaterinburg Russian Federation
BT Russian Federation
BT Commonwealth of Independent States

system analysis
use systems analysis

systematics
A valid term through 1977. Use taxonomy.

Systeme Probatoire d'Observation de la Terre
use SPOT

systems (1978)
SA anticlinoria
SA closed systems
SA coordinate systems
SA en echelon faults
SA faults
SA folds
SA fractures
SA geographic information systems
SA geothermal systems
SA grabens
SA half grabens
SA horsts
SA information systems
SA open systems
SA synclinoria
SA systems analogs
SA systems analysis

systems analogs (1978)
SA ground water
SA systems
SA systems analysis

systems analysis (1985)
UF system analysis
SA analysis
SA computer programs
SA data processing
SA information systems
SA systems
SA systems analogs

szaibelyite (1978)
UF ascharite
BT borates

Szczecin
No longer a valid term for GeoRef. As of 1993 see Szczecin Poland.

Szczecin Poland (1993)
Province and city in NW Poland. Before 1993 also search Szczecin AND Poland.
UF Stettin
BT Poland
BT Central Europe
BT Europe
NT Wolin
NT Wolin Poland

Szechuan China
use Sichuan China

Szechwan
No longer a valid term for GeoRef. See Sichuan China.

Szechwan Basin
No longer a valid term for GeoRef. use Sichuan Basin

Szolnok
No longer a valid term for GeoRef. See Szolnok Hungary.

Szolnok Hungary (1993)
County including the city in E central Hungary.
BT Hungary
BT Central Europe
BT Europe

T

T'ai-hang Mountains
use Taihang Mountains

T'ung-hai earthquake 1970
use Tonghai earthquake 1970

Ta
use tantalum

taaffeite (1978)
BT oxides

Taal (1978)
On Volcano Island in center of Lake Taal in Batangas Province in S Luzon.
UF Taal Volcano
BT Luzon
BT Philippine Islands
BT Far East
BT Asia

Taal Volcano
use Taal

Tabasco
No longer a valid term for GeoRef as of 1993 see Tabasco Mexico

Tabasco Mexico (1993)
State in SE Mexico. Before 1993 also search Tabasco AND Mexico.
BT Mexico
SA Gulf Coastal Plain

tabetisol
use taliks

Tabianian (1978)
Europe.
BT lower Pliocene
BT Pliocene
BT Neogene
BT Tertiary
BT Cenozoic

Table Head Group (1989)
W Newfoundland.
BT Middle Ordovician
BT Ordovician
BT Paleozoic
SA Newfoundland Island

Table Mountain (1985)
Mountain near Cape Town, South Africa. Also a mountain in Tuolumne County, California.
IN Index California or South Africa as applicable.
SA California
SA Cape Province South Africa
SA Cape Town South Africa
SA Sierra Nevada
SA Tuolumne County California

Table Mountain Group (1989)
Includes Soom Member, Sardinia Bay Formation. Located in South Africa, Namibia and Agulhas Bank.
IN Index ages as applicable.
SA Agulhas Bank
SA Devonian
SA Indian Ocean
SA Namibia
SA Ordovician
SA South Africa

tables
No longer a valid GeoRef index term. Before 1969, was used on level 2 in subfile E.

Tabulata (1978)

Order. Autoposting of Zoantharia to this term began in 1989.
BT Zoantharia
BT Anthozoa
BT Coelenterata
BT Invertebrata
NT Chaetitidae
NT Favositidae
NT Heliolitidae
SA Rugosa
SA Scleractinia

Tachira
No longer a valid term for GeoRef. As of 1993 see Tachira Venezuela.

Tachira Venezuela (1993)
State in W Venezuela. Before 1993 also search Tachira AND Venezuela.
BT Venezuela
BT South America

Taconian Orogeny
use Taconic Orogeny

Taconic Allochthon (1978)
In Northern Appalachians. Also search Taconic.
IN Index countries as applicable.
SA Appalachians
SA Canada
SA United States

Taconic Orogeny (1978)
An orogeny in the latter part of the Ordovician period, named for the Taconic Range of eastern New York State and well developed through most of the northern Appalachians in U.S. and Canada. Before 1978, search Taconic AND orogeny; Taconian; Taconian Orogeny.
UF Taconian Orogeny
BT Paleozoic
SA Canada
SA Ordovician
SA orogeny
SA Silurian
SA United States

taconite (1978)
BT chemically precipitated rocks
BT sedimentary rocks
SA iron formations

TACT (1989)
Acronym.
UF Trans Alaska Crustal Transect
UF Trans-Alaska Crustal Transect
SA Alaska
SA geophysical surveys
SA geotraverses
SA programs

tactite (1978)
Compositional term.
BT metasomatic rocks
BT metamorphic rocks

Tadjoura
Not a valid term for GeoRef. See Gulf of Tadjoura.

Tadzhik Basin
use Tadzhik Depression

Tadzhik Depression (1978)
North section of Afghan-Tadzhik Depression in S and SW Tadzhikistan. Also search Tadzhik. This term has multiple hierarchies.
UF Tadzhik Basin
UF Tadzhikistan Depression
BT1 Tadzhikistan
BT1 Asia
BT2 Tadzhikistan

BT2 Commonwealth of Independent States
SA Afghan-Tadzhik Depression

Tadzhikhistan
use Tadzhikistan

Tadzhikistan (1978)
Republic. Formerly the Tadzhik Soviet Socialist Republic in Soviet Central Asia. This term has multiple hierarchies.
CO N360000N410000 E0750000E0670000
UF Tadzhikhistan
UF Tadzhkistan
UF Tajikistan
BT1 Asia
BT2 Commonwealth of Independent States
NT Altyn-Topkan Tadzhikistan
NT Badakhshan
NT Darvaz
NT Darvaza Range
NT Dushanbe Tadzhikistan
NT Garm Tadzhikistan
NT Hissar Range
NT Karategin Range
NT Kulyab Tadzhikistan
NT Leninabad Tadzhikistan
NT Nurek Tadzhikistan
NT Shugnan Tadzhikistan
NT Tadzhik Depression
NT Turkestan Range
NT Vakhsh
NT Vanchskiy Range
NT Zeravshan Range
NT Zeravshan-Hissar
SA Afghan-Tadzhik Depression
SA Alai Range
SA Amu Darya
SA Amu Darya Basin
SA Badakhshan Afghanistan
SA Central Asia
SA Fergana Basin
SA Hindu Kush
SA Kurama Range
SA Pamirs
SA Surkhan Darya
SA Surkhan Darya basin
SA Syr Darya
SA Zeravshan River

Tadzhikistan Depression
use Tadzhik Depression

Tadzhikstan
use Tadzhikistan

Tadzhura
Not a valid term for GeoRef. See Gulf of Tadjoura.

taele
use frozen ground

Taeniodonta (1978)
Autoposting of Eutheria and Theria to this term began in 1989.
BT Eutheria
BT Theria
BT Mammalia
BT Tetrapoda
BT Vertebrata
BT Chordata

Taeniopteris (1978)
Genus.
BT Filicopsida
BT pteridophytes
BT Plantae
SA Cycadales
SA Glossopteris
SA gymnosperms
SA Pteridospermae

taenite (1978)
Meteorite mineral.
BT alloys
SA kamacite

SA meteorites
SA plessite
SA tetrataenite

Tafilalet
use Tafilalt

Tafilalt (1978)
Oasis in SE Morocco.
UF Tafilalet
UF Tafilelt
UF Tafilet
BT Morocco
BT North Africa
BT Africa

Tafilelt
use Tafilalt

Tafilet
use Tafilalt

TAG hydrothermal field (1989)
Mid-Atlantic Ridge, North Atlantic.
CO N260800N260800 W0444900W0444900
BT Atlantic Ocean
SA Mid-Atlantic Ridge
SA mid-ocean ridges

Tagil Basin (1978)
River basin in Yekaterinburg, formerly Sverdlovsk Oblast E of the Central Urals. Also search Tagil; Tagil River.
BT Russian Federation
BT Commonwealth of Independent States
SA Yekaterinburg Russian Federation

Tagliamento Valley (1978)
River valley extending from Carnic Alps to the head of the Gulf of Venice. Also search Tagliamento; Tagliamento River.
IN Index autonomous regions as applicable.
BT Italy
BT Southern Europe
BT Europe
SA Friuli-Venezia Giulia Italy
SA Veneto Italy

Tagus Basin (1978)
River basin in central Spain and S central Portugal. Also search Tagus.
IN Index countries as applicable.
UF Tagus River basin
BT Iberian Peninsula
BT Southern Europe
BT Europe
SA Portugal
SA Spain
SA Tagus River

Tagus River (1978)
Longest river in the Iberian Peninsula. Rises in E central Spain and flows W and SW entering the Atlantic Ocean at Lisbon.
IN Index countries as applicable.
BT Iberian Peninsula
BT Southern Europe
BT Europe
SA Portugal
SA Spain
SA Tagus Basin

Tagus River basin
use Tagus Basin

Tahiti (1978)
Island of E group of Society Islands in French Polynesia. Society Islands and French Polynesia are considered broader terms as of 1981.
CO S180000S170000 W1490000W1500000

BT Society Islands
BT French Polynesia
BT Polynesia
BT Oceania
SA DSDP Site 597
SA DSDP Site 598
SA DSDP Site 599
SA DSDP Site 600
SA DSDP Site 601
SA DSDP Site 602

Taiga
No longer a valid term for GeoRef. As of 1993, see Taiga Russian Federation.

taiga environment (1978)
Before 1978, also search taiga.
SA climate
SA depositional environment
SA ecology
SA environment
SA geomorphology
SA paleoclimatology
SA sedimentation
SA steppes
SA tundra

Taiga Russian Federation (1993)
City in NW Kemerovo Oblast. Before 1993, also search Taiga or Tayga and Russian Republic. This term has multiple hierarchies.
UF Tayga Russian Federation
BT1 Kemerovo Russian Federation
BT1 Russian Federation
BT1 Commonwealth of Independent States
BT2 Kemerovo Russian Federation
BT2 Asia

Taigonos Peninsula
use Taygonos Peninsula

Taihang Mountains (1989)
Hebei and Shanxi, N China.
IN Index provinces as applicable.
UF T'ai-hang Mountains
UF Taihang Shan
UF Taihangshan
BT China
BT Far East
BT Asia
SA Hebei China
SA Shanxi China

Taihang Shan
use Taihang Mountains

Taihangshan
use Taihang Mountains

Taihua Group (1989)
Henan.
BT Archean
BT Precambrian
SA China
SA Henan China

tailing dams
use tailings dams

tailing ponds
use tailings ponds

tailings (1978)
SA disposal barriers
SA dredged materials
SA dredging
SA mining
SA reclamation
SA spoils
SA tailings dams
SA tailings ponds
SA waste disposal

tailings dams (1985)

Before 1989, also included use for tailings ponds.
UF tailing dams
UF tailings impoundments
BT dams
SA mining geology
SA tailings
SA tailings ponds

tailings impoundments
use tailings dams

tailings ponds (1989)
Before 1989, see tailings dams.
UF tailing ponds
SA mining
SA mining geology
SA retaining walls
SA seepage
SA spoils
SA tailings
SA tailings dams
SA waste disposal

Taimyr (Peninsula)
use Taymyr Peninsula

Taimyr Peninsula
use Taymyr Peninsula

Taimyr Russian Federation
use Taymyr Dolgan-Nenets Russian Federation

Tainan
No longer a valid term for GeoRef. As of 1993 see Tainan Taiwan.

Tainan Taiwan (1993)
City on SW coast of Taiwan. Before 1993 also search Tainan AND Taiwan.
BT Taiwan
BT Far East
BT Asia

Taipei
No longer a valid term for GeoRef. As of 1993 see Taipei Taiwan.

Taipei Taiwan (1993)
City at N end of island. Before 1993 also search Taipei AND Taiwan.
BT Taiwan
BT Far East
BT Asia

Taishu Group (1978)
Chiefly composed of sandstone and shale in alternation, and intercalating arkose sandstone and red-coloured conglomerates. On Tsushima Island a part of Nagasaki Prefecture, Kyushu.
BT lower Tertiary
BT Tertiary
BT Cenozoic
SA Japan
SA Kyushu
SA Tsushima

Taito
use Taitung Taiwan

Taitung
No longer a valid term for GeoRef. As of 1993, see Taitung Taiwan.

Taitung Taiwan (1993)
City on SE coast of Taiwan.
UF Taito
BT Taiwan
BT Far East
BT Asia

Taiwan (1978)
Island off Fujian Province of China. Seat of Chinese Nationalist government known as the Republic of China. Before 1977, also search Formosa.

CO N220000N253000 E1230000E1200000
UF Formosa
BT Far East
BT Asia
NT Chiayi-Hsinying area
NT Chinkuashih Mine
NT Chinkuashih Taiwan
NT Hengchun Peninsula
NT Hsinchu
NT Miaoli Taiwan
NT Peikang Taiwan
NT Penghu Islands
NT Tainan Taiwan
NT Taipei Taiwan
NT Taitung Taiwan
NT Taiwanese Central Range
NT Taiwanese Coastal Range
NT Taoyuan Taiwan
NT Tatun Shan
NT Tiehchenshan
SA Tananao Schist

Taiwan Strait
use Formosa Strait

Taiwanese Central Range (1989)
Before 1989, also search Central Range AND Taiwan. Before 1977, also search Central Range AND Formosa.
BT Taiwan
BT Far East
BT Asia

Taiwanese Coastal Range (1993)
Also search Coastal Range AND Taiwan OR Formosa.
UF Coastal Range, Taiwan
BT Taiwan
BT Far East
BT Asia

Taiyuan
No longer a valid term for GeoRef. See Taiyuan China.

Taiyuan China (1993)
City in Shanxi, N China.
BT Shanxi China
BT China
BT Far East
BT Asia

Taiyuan Formation (1989)
In Anhui, Henan and Shanxi provinces.
IN Index ages as applicable.
BT Paleozoic
SA Anhui China
SA Carboniferous
SA China
SA Henan China
SA Permian
SA Shanxi China

Tajikistan
use Tadzhikistan

Takanuki
No longer a valid term for GeoRef. See Takanuki Japan.

Takanuki Japan (1993)
Village NE of Utsunomiya in E central Honshu.
BT Honshu
BT Japan
BT Far East
BT Asia

Takla Group (1989)
N central British Columbia.
BT Triassic
BT Mesozoic
SA British Columbia

Tal Formation (1978)

Consists of 2 members: the quartzite member and the limestone member. The formation has been considered of Jurassic-Cretaceous age, but current literature indicates an unconformity between the 2 members so that the quartzite member is Precambrian and the limestone member is Cretaceous.
IN Index ages as applicable.
SA Cretaceous
SA India
SA Jurassic
SA Precambrian
SA Uttar Pradesh India

Talacasto
No longer a valid term for GeoRef. See Talacasto Argentina.

Talacasto Argentina (1993)
Village in central Argentina.
BT San Juan Argentina
BT Argentina
BT South America

Talas Ala-Tau
use Talas Range

Talas Range (1978)
Branch of the Tien Shan in NW Kyrgyzstan. Before 1978, also search Talas. This term has multiple hierarchies.
UF Talas Ala-Tau
BT1 Kyrgyzstan
BT1 Asia
BT2 Kyrgyzstan
BT2 Commonwealth of Independent States
BT3 Tien Shan
BT3 Asia

Talasea
No longer a valid term for GeoRef. As of 1993 see Talasea New Guinea.

Talasea New Guinea (1993)
Settlement on E side of Willaumez Peninsula on N coast of New Britain, Bismark Archipelago. Before 1993 also search Talsea AND New Guinea.
BT Papua New Guinea
BT Australasia

Talbot County
Valid through 1988. Search in combination with state term. After 1988, use specific county-state term.

Talbot County Georgia (1989)
W Georgia. Before 1989, also search Talbot County AND Georgia.
CO N323100N325100 W0841800W0843900
BT Georgia
BT United States

Talbot County Maryland (1989)
E Maryland. Before 1989, also search Talbot County AND Maryland.
CO N383500N385700 W0755400W0762300
BT Maryland
BT United States

talc (1978)
As of 1981, use talc deposits for talc as a commodity.
BT sheet silicates
BT silicates
SA cerolite
SA soapstone
SA steatite

SA talc deposits

talc deposits (1981)
Before 1981, search talc AND deposits.
IN Commodity. See List C.
SA Henderson Mine
SA soapstone
SA steatite
SA talc

talc rock (1978)
Compositional term.
BT metasomatic rocks
BT metamorphic rocks

Talcher
use Talchir

Talcher Coalfield
use Talchir coal field

Talchir (1978)
Region. Former princely state in Orissa States.
UF Talcher
BT Orissa India
BT India
BT Indian Peninsula
BT Asia

Talchir coal field (1978)
E central Orissa. Also search Talchir.
UF Talcher Coalfield
SA coal fields
SA Orissa India

Talchir Formation (1978)
Upper Carboniferous to Permo-Carboniferous. Consists of boulder bed with boulders set in an ill-sorted matrix, often containing angular clastics, and is followed by greenish shales, silty shales, and sandstones. In peninsular India.
BT Paleozoic
SA Carboniferous
SA India
SA Permian
SA Upper Carboniferous

Talchir Series (1978)
Upper Carboniferous. At the base is a glacial boulder bed (Talchir Boulder Bed). Overlain by Damuda Series. In Orissa.
BT Upper Carboniferous
BT Carboniferous
BT Paleozoic
SA Damuda Series
SA India
SA Orissa India

taliks (1993)
Unfrozen layer adjacent to or in permafrost.
UF subgelisol
UF tabetisol
SA active layer
SA frozen ground
SA glacial features
SA glacial geology
SA ground ice
SA hummocks
SA ice wedges
SA palsas
SA periglacial features
SA permafrost
SA pingos
SA thawing

Talimu Basin
use Tarim Basin

Talin Estonia
use Tallinn Estonia

Talkeetna Formation (1993)
S central Alaska.

BT Lower Jurassic
BT Jurassic
BT Mesozoic
SA Alaska
SA Copper River basin
SA Southern Alaska

Talkeetna Mountains (1985)
S Alaska.
CO N614000N625000
　　W1470000W1493000
BT Southern Alaska
BT Alaska
BT United States

Talladega Front (1978)
E Talladega County in E central Alabama.
UF Talladega slate belt
BT Alabama
BT United States
SA Hillabee Chlorite Schist

Talladega Group (1978)
According to the latest published Lexique, the age of Talladega Formation is Precambrian (?) to Carboniferous (?). In NW Georgia, the Talladega Series has been grouped into 10 formations (ascending): Pinelog Quartzite, Hiawassee Slate, Great Smoky Formation, Natahala Schist, Tusquitee Quartzite, Brasstown Schist, Valleytown Schist, Murphy Marble, Andrews Schist and Nottley Quartzite. In E Alabama, North Carolina, NW Georgia, and Tennessee.
IN Index ages as applicable.
BT Paleozoic
SA Alabama
SA Cambrian
SA Devonian
SA Georgia
SA Hillabee Chlorite Schist
SA Mississippi
SA North Carolina
SA Ordovician
SA Pennsylvanian
SA Precambrian
SA Silurian
SA Tennessee

Talladega slate belt
　use Talladega Front

Tallahatta Formation (1985)
In Claiborne Group. S Alabama, W Georgia, and Mississippi.
BT middle Eocene
BT Eocene
BT Paleogene
BT Tertiary
BT Cenozoic
SA Alabama
SA Claiborne Group
SA Georgia
SA Mississippi

Tallapoosa County
Valid through 1988. Search in combination with state term. After 1988, use specific county-state term.

Tallapoosa County Alabama (1989)
E Alabama. Before 1989, also search Tallapoosa County AND Alabama.
CO N323000N331000
　　W0853700W0860100
BT Alabama
BT United States

Tallin Estonia
　use Tallinn Estonia

Tallinn
No longer a valid term for GeoRef. See Tallinn Estonia.

Tallinn Estonia (1993)
City on the Gulf of Finland. Also search Tallinn or Talin or Tallin.
UF Talin Estonia
UF Tallin Estonia
BT Estonia
BT Baltic region
BT Europe

Tallulah Falls Formation (1989)
Proterozoic and/or lower Paleozoic.
IN Index age as applicable.
SA Georgia
SA lower Paleozoic
SA North Carolina
SA Proterozoic
SA South Carolina

talnakhite (1978)
BT sulfides

talus
A valid term through 1978. Use talus slopes.

talus fan
　use alluvial fans

talus slopes (1978)
Before 1978, search talus; debris slopes.
UF debris slopes
BT erosion features
SA cliffs
SA colluvium
SA desert pavement
SA erosion
SA geomorphology
SA landslides
SA mass movements
SA slope stability
SA slopes
SA soil erosion

Talysh Mountains (1978)
NW extremity of the Elburz in NW Iran. Also search Talysh.
BT Iran
BT Middle East
BT Asia
SA Elburz

Tamagawa (1978)
River flowing into Tokyo Bay between Tokyo and Yokohama in Tokyo Prefecture, Honshu. Also used for hot springs in Akita Japan.
BT Honshu
BT Japan
BT Far East
BT Asia
SA Akita Japan
SA Tokyo Japan

Taman
No longer a valid term for GeoRef. As of 1993, see Taman Russian Federation.

Taman Peninsula (1978)
Peninsula of W Krasnodar Kray jutting into Kerch Strait connecting Black Sea and Azov Sea. Also search Taman. This term has multiple hierarchies.
BT1 Krasnodar Russian Federation
BT1 Russian Federation
BT1 Commonwealth of Independent States
BT2 Krasnodar Russian Federation
BT2 Europe

Taman Russian Federation (1993)
Village on N shore of Taman Peninsula near Kerch Strait in W Krasnodar Kray in Northern Caucasus. This term has multiple hierarchies.
BT1 Krasnodar Russian Federation
BT1 Russian Federation
BT1 Commonwealth of Independent States
BT2 Krasnodar Russian Federation
BT2 Europe

Tamar Estuary (1989)
Cornwall and Devonshire, extreme SW England. Empties into the English Channel. Also a river in N Tasmania which empties into Bass Strait.
IN Index Tasmania OR England and counties as applicable.
UF Tamar River
SA Cornwall England
SA Devonshire England
SA England
SA Tasmania Australia

Tamar River
　use Tamar Estuary

Tamar Valley (1978)
River valley in Cornwall and Devonshire in SW England. Also a river valley in Tasmania. Also search Tamar River.
IN Index English counties or Tasmania as applicable.
SA Cornwall England
SA Devonshire England
SA England
SA Tasmania Australia

Tamaulipas
No longer a valid term for GeoRef. As of 1993 see Tamaulipas Mexico.

Tamaulipas Mexico (1993)
State in E Mexico. Before 1993 also search Tamaulipas AND Mexico.
BT Mexico
SA Gulf Coastal Plain
SA Laguna Madre
SA Rio Grande
SA Rio Grande Depression
SA Rio Grande Valley
SA San Juan River
SA Sierra Madre Oriental

Tamayo fracture zone (1985)
North American Pacific, near mouth of Gulf of California. Autoposting of Pacific Ocean to this term began in 1990.
CO N222000N231000
　　W1073000W1084000
BT North American Pacific
BT Pacific Ocean

Tamba Plateau (1978)
In the Kyoto area. Also search Tamba.
BT Honshu
BT Japan
BT Far East
BT Asia
SA Kyoto Japan

Tambov
No longer a valid term for GeoRef. As of 1993, see Tambov Russian Federation.

Tambov Russian Federation (1993)
Oblast including the city 260 miles SE of Moscow. This term has multiple hierarchies.
BT1 Russian Federation
BT1 Commonwealth of Independent States
BT2 Europe

Tamdy-Tau
　use Tamdytau

Tamdytau (1978)
Mountains in the Kyzylkum SE of the Aral Sea. This term has multiple hierarchies.
UF Tamdy-Tau
UF Tamdytau Range
BT1 Uzbekistan
BT1 Asia
BT2 Uzbekistan
BT2 Commonwealth of Independent States
BT3 Kyzylkum
BT3 Commonwealth of Independent States
BT4 Kyzylkum
BT4 Central Asia
BT4 Asia

Tamdytau Range
　use Tamdytau

Tamiami Formation (1985)
BT upper Miocene
BT Miocene
BT Neogene
BT Tertiary
BT Cenozoic
SA Florida

Tamil Nadu
No longer a valid term for GeoRef. See Tamil Nadu India.

Tamil Nadu India (1993)
State in extreme S and SE India. Formerly part of Madras.
CO N080000N135000
　　E0803000E0762000
BT India
BT Indian Peninsula
BT Asia
NT Ariyalur India
NT Coimbatore India
NT Kodaikanal India
NT Kondapalli India
NT Madras India
NT Neyveli India
NT North Arcot India
NT Tiruchirapalli India
NT Vridhachalam India
SA Cauvery Basin
SA Coromandel Coast
SA Cuddalore Series
SA Neyveli Lignite
SA Niniyur Group
SA South Arcot

Tampa Bay (1978)
Inlet of Gulf of Mexico on W coast of Hillsborough County in W Florida.
BT Florida
BT United States
SA Hillsborough County Florida
SA Pinellas County Florida

Tampere
No longer a valid term for GeoRef. See Tampere Finland.

Tampere Finland (1993)
City in SW Finland.
BT Finland
BT Scandinavia
BT Western Europe
BT Europe

Tamworth
No longer a valid term for GeoRef. See Tamworth Australia.

Tamworth Australia (1993)
Town in E central New South Wales.
BT New South Wales Australia
BT Australia
BT Australasia

Tan-Lu Fault
use Tancheng-Lujiang Fault

Tan-Lu fault zone
use Tancheng-Lujiang fault zone

Tana Fjord (1978)
Inlet of Arctic Ocean which receives the Tana River on NE coast of Norway.
BT Finnmark Norway
BT Norway
BT Scandinavia
BT Western Europe
BT Europe

Tanana River (1985)
Source in SW Yukon Territory. Joins Yukon River at Tanana, central Alaska.
BT North America
SA Alaska
SA Delta River
SA Yukon Territory

Tananao Schist (1989)
E Taiwan.
BT Cretaceous
BT Mesozoic
SA Taiwan

Tananarive
No longer a valid term for GeoRef. As of 1993 see Antananarivo Madagascar

Tananarive Madagascar
use Antananarivo Madagascar

Tancheng-Lujiang Fault (1993)
E China. Extends from Anhui through Jiangsu to Shandong. Part of the Tancheng-Lujiang fault zone.
IN Index provinces as applicable.
UF Tan-Lu Fault
UF Tanlu Fault
BT China
BT Far East
BT Asia
SA Anhui China
SA Jiangsu China
SA Shandong China

Tancheng-Lujiang fault zone (1993)
Major fault zone. Extends from South-Central China to Heilongjiang. Some consider it to extend across Russian Federation Far East territory into the Sea of Okhotsk. Tancheng-Lujiang Fault is a narrower concept.
IN Index provinces as applicable.
UF Tan-Lu fault zone
UF Tanlu fault zone
SA Anhui China
SA China
SA Heilongjiang China
SA Hubei China
SA Jiangsu China
SA Jilin China
SA Liaoning China
SA Okhotsk Sea
SA Russian Far East
SA Russian Federation
SA Shandong China
SA Yishu Fault

Tanco Pegmatite (1978)
BT Manitoba
BT Western Canada
BT Canada

Tanega-shima (1978)
Largest of the Osumi Islands off S Kyushu. This term has multiple hierarchies.
UF Tanegashima Island
BT1 Ryukyu Islands
BT1 Japan
BT1 Far East
BT1 Asia
BT2 Kagoshima Japan
BT2 Kyushu
BT2 Japan
BT2 Far East
BT2 Asia

Tanegashima Island
use Tanega-shima

Tanezrouft (1978)
Section of the Sahara primarily in Algeria.
IN Index countries as applicable.
BT Africa
SA Algeria
SA Mali
SA Sahara

Tanganyika
Not a valid index term. Former British U.N. Trust Territory which became independent in 1961. See Tanzania.

tangential discontinuities (1981)
Used for a type of interface in seismology.
BT discontinuities
SA crust
SA mantle
SA seismology

tangential stress
use shear stress

tangential wave
use S-waves

Tanger Morocco
use Tangier Morocco

Tangier
No longer a valid term for GeoRef. As of 1993, see Tangier Morocco.

Tangier Morocco (1993)
City at W end of the Strait of Gibraltar. Also search Tanger or Tangiers.
UF Tanger Morocco
UF Tangiers Morocco
BT Morocco
BT North Africa
BT Africa

Tangiers Morocco
use Tangier Morocco

Tangshan
No longer a valid term for GeoRef. See Tangshan China.

Tangshan China (1993)
Towns in Hebei and Anhui provinces, N and E China.
IN Index provinces as applicable.
BT China
BT Far East
BT Asia
SA Anhui China
SA Hebei China
SA Tangshan earthquake 1976

Tangshan earthquake 1976 (1985)
Epicenter near Tangshan in Hebei, NE China.
BT earthquakes
SA China
SA Hebei China
SA Tangshan China

Tanlu Fault
use Tancheng-Lujiang Fault

Tanlu fault zone
use Tancheng-Lujiang fault zone

Tannu Tuva Russian Federation
use Tuva Russian Federation

Tannu-Ola Range (1978)
On the Mongolian-Tuva Aautonomous Republic border. Also search Tannu-Ola.
IN Index Mongolia and/or Russian Federation as applicable.
BT Asia
SA Mongolia
SA Russian Federation

Tanquary Fiord (1978)
W central Ellesmere Island in Franklin District.
BT Ellesmere Island
BT Franklin District Northwest Territories
BT Northwest Territories
BT Western Canada
BT Canada

Tansill Formation (1989)
SE New Mexico and W Texas.
BT Guadalupian
BT Permian
BT Paleozoic
SA New Mexico
SA Texas

tantalates (1978)
Autoposting of this term began in 1981.
BT oxides
NT niobotantalates

tantalite (1978)
Autoposting of niobotantalates began in 1989. This term has multiple hierarchies.
BT1 niobotantalates
BT1 niobates
BT1 oxides
BT2 niobotantalates
BT2 tantalates
BT2 oxides

tantalum (1978)
Chemical element. As of 1981, use tantalum ores for tantalum as a commodity. Autoposting of metals to this term began in 1989.
UF Ta
BT metals
SA heavy metals
SA niobium
SA tantalum ores

tantalum ores (1981)
Before 1981, also search tantalum AND (deposit OR deposits OR ore OR ores OR economic) in the basic index. Autoposting of metal ores to this term began in 1985.
IN Commodity. See List C.
BT metal ores
SA columbite
SA niobium ores
SA tantalum

Tanzania (1978)
Former British U.N. Trust Territory of Tanganyika, which became independent in 1961. United with the island of Zanzibar to become Tanzania in 1964. Before 1964, also search Tanganyika. Autoposting of this term began in 1978.
CO S114000S010000 E0403000E0293000
BT East Africa
BT Africa
NT Kilimanjaro
NT Laetoli
NT Oldoinyo Lengai
NT Olduvai Gorge
NT Zanzibar
SA East African Rift
SA Gregory Rift
SA Koobi Fora Formation
SA Laetoli Beds
SA Lake Malawi
SA Lake Natron
SA Lake Tanganyika
SA Lake Victoria
SA Shungura Formation
SA Umba

tanzanite (1978)
Variety of zoisite.
BT epidote group
BT sorosilicates
BT orthosilicates
BT silicates
SA zoisite

Tanzawa Massif
use Tanzawa Mountains

Tanzawa Mountainland
use Tanzawa Mountains

Tanzawa Mountains (1978)
Between Fujiyama and Yokohama. Also search Tanzawa.
UF Tanzawa Massif
UF Tanzawa Mountainland
BT Honshu
BT Japan
BT Far East
BT Asia
SA Kanagawa Japan
SA Yamanashi Japan

Taos County
Valid through 1988. Search in combination with state term. After 1988, use specific county-state term.

Taos County New Mexico (1989)
N New Mexico. Before 1989, also search Taos County AND New Mexico.
CO N360000N370000 W1051500W1060500
BT New Mexico
BT United States
NT Questa Caldera
NT Questa Mine
SA San Luis Valley
SA Taos Plateau

Taos Plateau (1985)
N New Mexico.
IN Index counties as applicable.
BT New Mexico
BT United States
SA Colorado Plateau
SA Rio Arriba County New Mexico
SA Rio Grande Rift
SA Taos County New Mexico

Taoudeni
use Taoudenni

Taoudeni Basin
use Taoudenni

Taoudenni (1978)
An oasis in the Sahara in NW Mali.
UF Taoudeni
UF Taoudeni Basin
UF Taoudenni Basin

BT Mali
BT West Africa
BT Africa

Taoudenni Basin
 use Taoudenni

Taourirt
 No longer a valid term for GeoRef. See Taourirt Morocco or Taourirt Algeria.

Taourirt Algeria (1993)
 Village in W central Algeria.
 BT Algeria
 BT North Africa
 BT Africa

Taourirt Morocco (1993)
 Town in NE Morocco.
 BT Morocco
 BT North Africa
 BT Africa

Taoyuan
 No longer a valid term for GeoRef. As of 1993 see Taoyuan Taiwan.

Taoyuan Taiwan (1993)
 Town in N Taiwan.
 BT Taiwan
 BT Far East
 BT Asia

Tapeats Sandstone (1985)
 In Tonto Group. Lower and Middle Cambrian. Central and N Arizona, SE California and S Nevada.
 BT Cambrian
 BT Paleozoic
 SA Arizona
 SA California
 SA Nevada

taphocenoses
 use thanatocenoses

taphocoenosis
 use thanatocenoses

taphonomy (1978)
 SA biomineralization
 SA fossilization
 SA paleoecology
 SA paleopathology
 SA preservation
 SA thanatocenoses

taphrogeny (1978)
 Implies rifting.
 SA orogeny
 SA plate tectonics
 SA rifting
 SA tectonics
 SA tesserae

tapiolite (1978)
 Autoposting of niobotantalates began in 1989. This term has multiple hierarchies.
 BT1 niobotantalates
 BT1 niobates
 BT1 oxides
 BT2 niobotantalates
 BT2 tantalates
 BT2 oxides

Tapley Hill Formation (1989)
 S Australia.
 BT Proterozoic
 BT upper Precambrian
 BT Precambrian
 SA Australia
 SA South Australia

tar sands
 use oil sands

Taranaki
 No longer a valid term for GeoRef. See Taranaki New Zealand.

Taranaki Basin (1993)
 Back-arc basin in West Pacific which extends onshore in N South Island and the Taranaki Peninsula of SW North Island, New Zealand. Includes Taranaki Graben complex and Western Platform areas. Contains petroleum resources.
 IN Index North Island and/or South Island as appicable.
 CO S411500S380000
  E1744000E1723000
 BT West Pacific
 BT Pacific Ocean
 SA New Zealand
 SA North Island
 SA South Island

Taranaki New Zealand (1993)
 Provincial district in W North Island.
 IN Also index North Island.
 BT New Zealand
 BT Australasia
 NT Mokoia New Zealand
 SA North Island

Taranto
 No longer a valid term for GeoRef. See Tarnato Italy.

Taranto Italy (1993)
 City on N Gulf of Taranto on heel of Italian boot.
 BT Apulia Italy
 BT Italy
 BT Southern Europe
 BT Europe

Tarapaca
 No longer a valid term for GeoRef. See Tarapaca Chile.

Tarapaca Chile (1993)
 Province in N Chile.
 BT Chile
 BT South America

Taratash Complex (1985)
 S Urals.
 BT Riphean
 BT upper Proterozoic
 BT Proterozoic
 BT upper Precambrian
 BT Precambrian
 SA Southern Urals

Tarawera volcanic complex (1978)
 N central North Island.
 IN Also index North Island.
 BT New Zealand
 BT Australasia
 SA North Island

Tarbagatai Range
 use Tarbagatay Range

Tarbagatay Range (1978)
 Northern outlier of the Tien Shan. Also search Tarbagatay.
 IN Index Kazakhstan and/or Xinjiang as applicable.
 UF Tarbagatai Range
 BT Asia
 SA Kazakhstan
 SA Tien Shan
 SA Xinjiang China

Tarentaise (1978)
 Alpine valley of upper Isere River in Savoy Alps.
 BT Savoie France
 BT France
 BT Western Europe
 BT Europe
 SA Isere Valley
 SA Savoy Alps

Tareya
 No longer a valid term for GeoRef. As of 1993, see Tareya Russian Federation.

Tareya Russian Federation (1993)
 Village in NE Taymyr National Okrug in NW Siberia. This term has multiple hierarchies.
 BT1 Taymyr Dolgan-Nenets Russian Federation
 BT1 Krasnoyarsk Russian Federation
 BT1 Russian Federation
 BT1 Commonwealth of Independent States
 BT2 Taymyr Dolgan-Nenets Russian Federation
 BT2 Krasnoyarsk Russian Federation
 BT2 Asia

Tarfaia Morocco
 use Tarfaya Morocco

Tarfaya
 No longer a valid term for GeoRef. As of 1993 see Tarfaya Morocco.

Tarfaya Morocco (1993)
 Town in extreme SW Morocco. A former Spanish enclave. Also search Tarfaia.
 UF Tarfaia Morocco
 BT Morocco
 BT North Africa
 BT Africa

Tarim Basin (1989)
 Largest inland basin in China. SE Xinjiang.
 CO N360000N402000
  E0922000E0760000
 UF Talimu Basin
 BT Xinjiang China
 BT China
 BT Far East
 DT Asia

Tarkhankut Peninsula (1978)
 In extreme W Crimea jutting into Black Sea. Also search Tarkhankut. This term has multiple hierarchies.
 BT1 Crimea Ukraine
 BT1 Ukraine
 BT1 Europe
 BT2 Crimea Ukraine
 BT2 Ukraine
 BT2 Commonwealth of Independent States

Tarn
 No longer a valid term for GeoRef. See Tarn France.

Tarn France (1993)
 Department in S France.
 CO N432000N441200
  E0025500E0013000
 BT France
 BT Western Europe
 BT Europe
 NT Sidobre Massif
 SA Montagne Noire
 SA Rodez Trough

Tarn-et-Garonne
 No longer a valid term for GeoRef. See Tarn-et-Garonne France.

Tarn-et-Garonne France (1993)
 Department in SW France.
 CO N434000N443000
  E0020000E0004500
 BT France
 BT Western Europe

BT Europe
SA Orgueil Meteorite
SA Quercy
SA Rouergue

Tarnobrzeg
 No longer a valid term for GeoRef. As of 1993 see Tarnobrzeg Poland.

Tarnobrzeg Poland (1993)
 Province and town in SE Poland. Before 1993 also search Tarnobrzeg AND Poland.
 BT Poland
 BT Central Europe
 BT Europe
 NT Sandomierz Poland

Tarnow
 No longer a valid term for GeoRef. As of 1993 see Tarnow Poland.

Tarnow Poland (1993)
 Province and city in S Poland. Before 1993 also search Tarnow AND Poland.
 BT Poland
 BT Central Europe
 BT Europe
 NT Bochnia Poland

Taro Valley (1978)
 River valley extending in a SW-NE direction from the N Apennines to the Po River. Also search Taro.
 BT Emilia-Romagna Italy
 BT Italy
 BT Southern Europe
 BT Europe

Tarragona
 No longer a valid term for GeoRef. As of 1993, see Tarragona Spain.

Tarragona Spain (1993)
 Province in Catalonia in NE Spain. Before 1993, also search Tarragona and Spain.
 CO N403000N413700
  E0014000E0001000
 BT Catalonia Spain
 BT Spain
 BT Iberian Peninsula
 BT Southern Europe
 BT Europe

Tarrant County
 Valid through 1988. Search in combination with state term. After 1988, use specific county-state term.

Tarrant County Texas (1989)
 N Texas. Before 1989, also search Tarrant County AND Texas.
 CO N323000N330000
  W0970500W0973700
 BT Texas
 BT United States

Tartary
 use Tataria

Tartu
 No longer a valiid term for GeoRef. As of 1993, see Tartu Estonia.

Tartu Estonia (1993)
 City on Ema River W of Lake Peipus in E Estonia. Before 1993, also search Dorpat or Tartu AND Estonia.
 UF Dorpat Estonia
 BT Estonia
 BT Baltic region

BT Europe

Tarumae (1978)
Active volcano. South of Sapporo in S Hokkaido.
UF Tarumae Volcano
BT Hokkaido
BT Japan
BT Far East
BT Asia
SA Shikotsu

Tarumae Volcano
use Tarumae

Tarvisio
No longer a valid term for GeoRef. See Tarvisio Italy.

Tarvisio Italy (1993)
Town near Austrian and Yugoslav borders.
BT Friuli-Venezia Giulia Italy
BT Italy
BT Southern Europe
BT Europe

Tary Ekan Tadzhikistan
use Tary-Ekam Tadzhikistan

Tary-Ekam
No longer a valid term for GeoRef. As of 1993, see Tary-Ekam Tadzhikistan.

Tary-Ekam Tadzhikistan (1993)
Village in Leninabad Oblast in N Tadzhikistan. Before 1978, also search Tary Ekan or Tary-Ekan. This term has multiple hierarchies.
UF Tary Ekan Tadzhikistan
UF Tary-Ekan Tadzhikistan
BT1 Leninabad Tadzhikistan
BT1 Tadzhikistan
BT1 Asia
BT2 Leninabad Tadzhikistan
BT2 Tadzhikistan
BT2 Commonwealth of Independent States

Tary-Ekan Tadzhikistan
use Tary-Ekam Tadzhikistan

Tas-Khayakhtakh Range (1978)
Northern section of Cherski Range N of Okhotsk Sea in NE Yakutia. Also search Tas-Khayakhtakh. This term has multiple hierarchies.
BT1 Yakutia Russian Federation
BT1 Russian Federation
BT1 Commonwealth of Independent States
BT2 Yakutia Russian Federation
BT2 Asia

Tashkent
No longer a valid term for GeoRef. As of 1993, see Tashkent Uzbekistan.

Tashkent Uzbekistan (1993)
Oblast and city in NE Uzbekistan. This term has multiple hierarchies.
BT1 Uzbekistan
BT1 Asia
BT2 Uzbekistan
BT2 Commonwealth of Independent States
NT Angren Uzbekistan
NT Chirchik Uzbekistan

Tasman Basin (1978)
Extends S of Tasman Sea proper between South Tasmanian Rise and Macquarie Ridge SW of New Zealand. Also search Tasman.
UF Tasman-Becken
BT Pacific Ocean

Tasman fold belt
use Tasman orogenic zone

Tasman Geosyncline (1978)
Also search Tasman AND geosynclines.
IN Index states as applicable.
BT Australia
BT Australasia
SA geosynclines
SA New South Wales Australia
SA Queensland Australia
SA Tasmania Australia
SA Victoria Australia

Tasman orogenic zone (1978)
Also search Tasman.
IN Index Australian provinces and/or New Zealand as applicable.
UF Tasman fold belt
BT Australasia
SA New South Wales Australia
SA New Zealand
SA Queensland Australia
SA South Australia
SA Tasmania Australia
SA Victoria Australia

Tasman Sea (1978)
Between SE Australia and Tasmania on W and New Zealand on the E. Also search Tasman.
CO S510000S300000
E1743000E1465500
BT West Pacific
BT Pacific Ocean
NT Challenger Plateau
NT Dampier Ridge
SA DSDP Site 207
SA DSDP Site 284

Tasman-Becken
use Tasman Basin

Tasmania
No longer a valid term for GeoRef. See Tasmania Australia.

Tasmania Australia (1993)
Island and state S of Victoria.
CO S434000S393000
E1483000E1435000
BT Australia
BT Australasia
NT Burnie Australia
NT Hobart Australia
NT Rosebery Australia
NT Tasmania Basin
NT Wynyard Australia
NT Zeehan Australia
SA Coles Bay
SA Flinders Island
SA Forth Valley
SA Great Lake
SA Heemskirk Granite
SA Lachlan fold belt
SA Mersey River
SA Mersey Valley
SA Mount Read Volcanics
SA Perth Australia
SA Ross Formation
SA Tamar Estuary
SA Tamar Valley
SA Tasman Geosyncline
SA Tasman orogenic zone

Tasmania Basin (1981)
BT Tasmania Australia
BT Australia
BT Australasia

Tasmanites (1978)
Autoposting of thallophytes and microfossils began in 1990. This term has multiple hierarchies.
BT1 Chlorophyta
BT1 algae
BT1 thallophytes
BT1 Plantae
BT2 Chlorophyta
BT2 algae
BT2 microfossils
SA palynomorphs

Tassili des Adjer
use Tassili n'Ajjer

Tassili n'Ajjer (1978)
An arid plateau region which is a NE extension of the Ahaggar Mountains in SE Algeria. Also search Tassili.
UF Tassili des Adjer
BT Algeria
BT North Africa
BT Africa

Tatar
No longer a valid term for GeoRef. As of 1993, see Tatar Russian Federation.

Tatar Arch (1978)
E central European Russian Federation centering on former Tatar A.S.S.R. Before 1978, also search Tatar. This term has multiple hierarchies.
BT1 Russian Federation
BT1 Commonwealth of Independent States
BT2 Europe
SA Tatar Russian Federation

Tatar Russian Federation (1993)
The former Tatar Autonomous Soviet Socialist Republic. In area around great bend of Volga River where it heads S at Kazan. Before 1993, also search Tatar. Before 1978, also search Tatarstan. This term has multiple hierarchies.
CO N540000N560000
E0540000E0480000
UF Tatarstan Russian Federation
BT1 Russian Federation
BT1 Commonwealth of Independent States
BT2 Europe
NT Kazan Russian Federation
SA Kainsaz Meteorite
SA Tatar Arch
SA Tataria

Tatar Strait (1978)
Between E coast of Primorye Kray and W coast of Sakhalin Island at the N end of Japan Sea. Also search Tatar. This term has multiple hierarchies.
BT1 Russian Federation
BT1 Commonwealth of Independent States
BT2 Asia
SA Russian Far East

Tataria (1978)
Originally applied to area of Mongol Empire at its height. Now, since displacement of Crimean Tatars, it can be applied to the Tartar A.S.S.R and surrounding areas inhabited by Tatars. This term has multiple hierarchies.
UF Tartary
UF Tatary
BT1 Russian Federation
BT1 Commonwealth of Independent States
BT2 Europe
SA Tatar Russian Federation

Tatarian (1978)
Europe. Above Kazanian, below Scythian (Triassic).
BT Upper Permian
BT Permian
BT Paleozoic

Tatarstan Russian Federation
use Tatar Russian Federation

Tatary
use Tataria

Tatra Mountains (1978)
Chief mountain group of central Carpathians. This term has multiple hierarchies.
IN Index countries as applicable.
CO N490400N492200
E0202000E0193800
UF High Tatra
UF High Tatra Mountains
UF High Tatras
UF Tatras
BT1 Central Europe
BT1 Europe
BT2 Carpathians
BT2 Europe
SA Czechoslovakia
SA Krizna Nappe
SA Low Tatra Mountains
SA Poland

Tatras
use Tatra Mountains

Tatum Basin (1978)
SE New Mexico.
BT New Mexico
BT United States

Tatun Shan (1978)
Mountain 9 miles N of Taipei. Also search Tatun.
BT Taiwan
BT Far East
BT Asia

Taubach
No longer a valid term for GeoRef. See Taubach Germany.

Taubach Germany (1993)
Town in Thuringia, E central Germany.
CO N505600N505600
E0112400E0112400
BT Thuringia Germany
BT Germany
BT Central Europe
BT Europe

Tauern Tunnel (1978)
Railroad tunnel through the Hohe Tauern in SW Austria. Also search Tauern.
BT Austria
BT Central Europe
BT Europe
SA Hohe Tauern

Tauern Window (1985)
SW Austria.
BT Austria
BT Central Europe
BT Europe
SA Eastern Alps
SA Hohe Tauern
SA Tyrol Austria

Taunus (1978)
Mountain range E of the Rhine River and N of the lower Main River.
UF Taunus Mountains
BT Hesse Germany
BT Germany
BT Central Europe
BT Europe

Taunus Mountains
use Taunus

Taupo
No longer a valid term for GeoRef. See Taupo New Zealand.

Taupo New Zealand (1993)
Town on Lake Taupo in central North Island.
IN   Also index North Island.
BT   New Zealand
BT   Australasia
SA   North Island

Taupo volcanic zone (1978)
Mountain region surrounding Lake Taupo in central North Island. Also search Taupo.
IN   Also index North Island.
BT   New Zealand
BT   Australasia
SA   Broadlands
SA   North Island
SA   Tongariro

Taurus Littrow
use Taurus-Littrow

Taurus Mountains (1978)
Mountain chain on Mediterranean coast of Turkey.
CO   N355000N383000
        E0374000E0283000
BT   Turkey
BT   Middle East
BT   Asia
SA   Antalya Turkey

Taurus-Littrow (1978)
Before 1978, also search Tranquillity Base.
UF   Taurus Littrow
UF   Taurus-Littrow Valley
UF   Tranquillity Base
BT   Moon

Taurus-Littrow Valley
use Taurus-Littrow

Tavetsch (1981)
Valley of the upper Vorderrhein River.
CO   N363500N364500
        E0085200E0083900
BT   Graubunden Switzerland
BT   Switzerland
BT   Central Europe
BT   Europe
SA   Alps
SA   Swiss Alps

tavistockite
use carbonate apatite

Tavua
No longer a valid term for GeoRef. See Tavua Fiji.

Tavua Fiji (1993)
Town on Tauva Bay, Fiji. Viti Levu is considered a broader term as of 1985.
BT   Viti Levu
BT   Fiji
BT   Melanesia
BT   Oceania

Tawi River
use Tawi Valley

Tawi Valley (1978)
River valley in SW Jammu and Kashmir, disputed territory between India and Pakistan. As of 1993, India is no longer considered a broader term.
IN   Index Jammu and Kashmir AND India and/or Pakistan as applicable.
UF   Tawi River
SA   India
SA   Jammu and Kashmir
SA   Pakistan

taxa, ahermatypic
use ahermatypic taxa

taxa, anaerobic
use anaerobic taxa

taxa, benthonic
use benthonic taxa

taxa, colonial
use colonial taxa

taxa, endemic
use endemic taxa

taxa, extinct
use extinct taxa

taxa, hermatypic
use hermatypic taxa

taxa, heteromorphic
use heteromorphic taxa

taxa, new
use new taxa

taxa, planktonic
use planktonic taxa

taxes (1981)
SA   economics

Taxodiaceae (1978)
Family. Autoposting of this term began in 1978.
BT   Coniferales
BT   gymnosperms
BT   Spermatophyta
BT   Plantae
NT   Sequoia
SA   Pinaceae
SA   Taxodium

Taxodium (1978)
BT   Coniferales
BT   gymnosperms
BT   Spermatophyta
BT   Plantae
SA   Taxodiaceae

taxonomy (1978)
SA   cladistics
SA   classification
SA   extinct taxa
SA   heteromorphic taxa
SA   holotypes
SA   homonymy
SA   miscellanea
SA   new names
SA   new taxa
SA   paleontology
SA   paratypes
SA   phenotypes
SA   range
SA   revision
SA   synonymy
SA   type specimens

taxonomy review
A valid term through 1973. Use taxonomy.

Tay Estuary (1978)
N of the Firth of Forth on the North Sea.
UF   Firth of Tay
BT   Scotland
BT   Great Britain
BT   United Kingdom
BT   Western Europe
BT   Europe
SA   Fife region Scotland

Tayga Russian Federation
use Taiga Russian Federation

Taygonos Peninsula (1978)
On N Okhotsk Sea between Gizhiga and Penzhina bays. Also search Taygonos. This term has multiple hierarchies.
UF   Taigonos Peninsula
BT1 Magadan Russian Federation
BT1 Russian Federation
BT1 Commonwealth of Independent States
BT2 Magadan Russian Federation
BT2 Asia

Taylor County
Valid term through 1988. Search in combination with state term. After 1988, use specific county-state term.

Taylor County Florida (1989)
On the Gulf of Mexico in N Florida. Before 1989, also search Taylor County AND Florida.
CO   N294100N301800
        W0832200W0840000
BT   Florida
BT   United States

Taylor County Georgia (1989)
W central Georgia. Before 1989, also search Taylor County AND Georgia.
CO   N322200N324400
        W0835900W0842600
BT   Georgia
BT   United States

Taylor County Iowa (1989)
SW Iowa. Before 1989, also search Taylor County AND Iowa.
CO   N403400N405400
        W0942800W0945600
BT   Iowa
BT   United States

Taylor County Kentucky (1989)
Central Kentucky. Before 1989, also search Taylor County AND Kentucky.
CO   N371100N372800
        W0850500W0853500
BT   Kentucky
BT   United States

Taylor County Texas (1989)
W central Texas. Before 1989, also search Taylor County AND Texas.
CO   N320700N323400
        W0994800W1000900
BT   Texas
BT   United States
NT   Abilene Texas

Taylor County West Virginia (1989)
N West Virginia. Before 1989, also search Taylor County AND West Virginia.
CO   N391400N392700
        W0795400W0801300
BT   West Virginia
BT   United States

Taylor County Wisconsin (1989)
N central Wisconsin. Before 1989, also search Taylor County AND Wisconsin.
CO   N450200N452300
        W0900300W0905500
BT   Wisconsin
BT   United States

Taylor Glacier (1978)
In the McMurdo Sound area in S Victoria Land just NW of Ross Ice Shelf. Also search Taylor AND appropriate area.
BT   Antarctica
SA   Taylor Valley

Taylor Marl (1993)
Gulfian. Central and E Texas, Louisiana.
BT   Gulfian
BT   Upper Cretaceous
BT   Cretaceous
BT   Mesozoic
SA   Louisiana
SA   Texas

Taylor Valley (1978)
At McMurdo Sound in S Victoria Land in the Taylor Glacier area NW of Ross Ice Shelf. Also search Taylor.
IN   Index Antarctica or region as applicable.
SA   Antarctica
SA   McMurdo Sound
SA   Taylor Glacier
SA   Victoria Land

Taymyr
No longer a valid term for GeoRef. As of 1993, see Taymyr Dolgan-Nenets Russian Federation.

Taymyr Dolgan-Nenets Russian Federation (1993)
Autonomous Okrug in Arctic Krasnoyarsk Kray. Includes the Taymyr Peninsula. Before 1993, also search Taymyr or Dolgan-Nenets. Before 1978, also search Taimyr. Autoposting of Russian Republic to Taymyr began in 1989. This term has This term has multiple hierarchies.
UF   Dolgan-Nenets Russian Federation
UF   Taimyr Russian Federation
BT1 Krasnoyarsk Russian Federation
BT1 Russian Federation
BT1 Commonwealth of Independent States
BT2 Krasnoyarsk Russian Federation
BT2 Asia
NT   Khatanga Russian Federation
NT   Norilsk region
NT   Norilsk Russian Federation
NT   Popigay Russian Federation
NT   Tareya Russian Federation
NT   Taymyr Peninsula
NT   Ust-Yenisei Basin
SA   Khatanga Basin
SA   Khatanga River
SA   Kheta River
SA   Kotui
SA   Russian Arctic

Taymyr Peninsula (1978)
Between the Yenisei River and the Khatanga River in the Taymyr National Okrug in northernmost Siberia. Also search Taymyr. This term has multiple hierarchies.
CO   N720000N775000
        E1140000E0800000
UF   Taimyr (Peninsula)
UF   Taimyr Peninsula
BT1 Taymyr Dolgan-Nenets Russian Federation
BT1 Krasnoyarsk Russian Federation
BT1 Russian Federation
BT1 Commonwealth of Independent States
BT2 Taymyr Dolgan-Nenets Russian Federation
BT2 Krasnoyarsk Russian Federation
BT2 Asia

Taz Basin (1978)
River basin in E and NE Yamal Nenets National Okrug in NW Siberia. Also search Taz. This term has multiple hierarchies.
BT1 Yamal-Nenets Russian Federation
BT1 Tyumen Russian Federation

BT1 Russian Federation
BT1 Commonwealth of Independent States
BT2 Yamal-Nenets Russian Federation
BT2 Tyumen Russian Federation
BT2 Asia

Taza
No longer a valid term for GeoRef. As of 1993 see Taza Morocco.

Taza Morocco (1993)
Town E of Fez in N Morocco.
BT Morocco
BT North Africa
BT Africa

Tazewell County
Valid through 1988. Search in combination with state term. After 1988, use specific county-state term.

Tazewell County Illinois (1989)
Central Illinois. Before 1989, also search Tazewell County AND Illinois.
CO N402000N404300 W0891500W0895500
BT Illinois
BT United States

Tazewell County Virginia (1989)
SW Virginia. Before 1989, also search Tazewell County AND Virginia.
CO N365600N372000 W0811500W0815400
BT Virginia
BT United States

Tazin Lake (1978)
Extreme NW Saskatchewan.
BT Saskatchewan
BT Western Canada
BT Canada

Tb
  use terbium

Tbilisi
No longer a valid term for GeoRef. As of 1993, see Tbilisi Georgian Republic.

Tbilisi Georgian Republic (1993)
City in SE Georgian Republic. Before 1993, also search Tiflis.
UF Tiflis Georgian Republic
BT Georgian Republic
BT Europe

Tc
  use technetium

Tc-99 (1985)
Autoposting of broader terms began in 1989. This term has multiple hierarchies.
BT1 radioactive isotopes
BT1 isotopes
BT2 technetium
BT2 metals

Tchornozem
  use Chernozems

Te
  use tellurium

Te Aroha
No longer a valid term for GeoRef. See Te Aroha New Zealand.

Te Aroha New Zealand (1993)
Town 70 miles SE of Auckland in N central North Island.
IN Also index North Island.
BT New Zealand
BT Australasia
SA North Island

Teapot Dome (1978)
Near Casper in central Wyoming.
IN Index county as applicable.
BT Wyoming
BT United States
SA Natrona County Wyoming

Teapot Sandstone Member (1989)
E Wyoming.
BT Upper Cretaceous
BT Cretaceous
BT Mesozoic
SA Wyoming

tear faults (1989)
BT faults

technetium (1978)
Autoposting of metals to this term began in 1989.
UF Tc
BT metals
NT Tc-99

technical cooperation (1981)
SA international cooperation

technique
No longer a valid GeoRef index term. Before 1972, was used on levels 1 and 2 in subfiles E, N and G.
  use techniques

technique and apparatus
No longer a valid GeoRef index term. Before 1969, was used on level 1 in subfile E. See instruments; techniques; apparatus.

techniques (1978)
This term deals with sampling. Before 1972, also search technique; technique and apparatus.
UF technique
SA instruments
SA irradiation
SA measurement-while-drilling
SA methods
SA precision
SA sample preparation
SA sampling
SA three-dimensional methods

technology (1978)
SA mining geology
SA production control
SA technology transfer

technology transfer (1981)
SA economics
SA technology

tectites
Not a valid GeoRef index term after 1971. Was used on level 1 in subfile G.
  use tektites

tectogenesis
  use orogeny

tectonic
A valid term through 1977. Used in reference to maps. See tectonic maps.

tectonic breccia (1981)
BT breccia
BT clastic rocks
BT sedimentary rocks
SA tectonics

tectonic controls (1978)
A valid term through 1978. Term reintroduced in 1981. Use structural controls when related to genesis of mineral deposits.
SA controls
SA geomorphologic controls
SA geomorphology
SA sedimentation
SA structural controls

tectonic elements (1981)
SA faults
SA folds
SA fractures
SA tectonic units
SA tectonics

tectonic imbrication
  use imbricate tectonics

tectonic lines
  use lineaments

tectonic maps (1978)
Before 1978, also search maps AND tectonic. Use geotectonic maps for plate tectonics, continental drift and related subjects.
BT maps
SA geotectonic maps
SA structure contour maps
SA tectonic units
SA tectonics

tectonic platforms (1981)
Before 1981, search platforms.
SA continents
SA cratons
SA platforms
SA tectonic units
SA tectonics

tectonic sutures
  use suture zones

tectonic units (1982)
SA cratons
SA tectonic elements
SA tectonic maps
SA tectonic platforms
SA tectonic wedges
SA tectonics

tectonic wedges (1981)
SA faults
SA imbricate tectonics
SA tectonic units
SA tectonics

tectonics (1978)
Used for the structural makeup and structural evolution of regions. See also structural analysis for smaller structures; deformation for microscopic structures. Neotectonics (used for post-Miocene tectonics) is a narrower term as of 1993.
UF geotectonics
UF tectonism
NT basement tectonics
NT compression tectonics
NT extension tectonics
NT imbricate tectonics
NT neotectonics
NT salt tectonics
NT seismotectonics
NT thin-skinned tectonics
SA Acadian Phase
SA accretion
SA accretionary wedges
SA Alleghany Orogeny
SA allochthons
SA Alpine Orogeny
SA alpine-type
SA Andean Orogeny
SA Antler Orogeny
SA Appalachian Phase
SA arches
SA Assyntic Orogeny
SA Asturian Orogeny
SA atectonic processes
SA aulacogens
SA autochthons
SA back-arc basins
SA basement
SA basin range structure
SA basins
SA Brazilian Cycle
SA breccia
SA brittle deformation
SA Cadomian Orogeny
SA Cimmerian Orogeny
SA continental drift
SA continental margin
SA Cordilleran Orogeny
SA cratons
SA crust
SA crustal shortening
SA crustal thinning
SA Damara Orogeny
SA decollement
SA deep-seated structures
SA deformation
SA epeirogeny
SA eugeosynclines
SA exotic terranes
SA faults
SA fold and thrust belts
SA fold belts
SA folds
SA foliation
SA fore-arc basins
SA foreland basins
SA forelands
SA fractures
SA geanticlines
SA geodynamics
SA geosynclines
SA glaciotectonics
SA gravity sliding
SA half grabens
SA Hercynian Orogeny
SA Himalayan Orogeny
SA horizontal movements
SA Indosinian Orogeny
SA intracratonic basins
SA intraplate processes
SA isostasy
SA isostatic rebound
SA klippen
SA Laramide Orogeny
SA lineaments
SA lineation
SA Median Tectonic Line
SA metamorphic core complexes
SA meteor craters
SA microcontinents
SA miogeosynclines
SA mobile belts
SA mobility
SA nappes
SA Nevadan Orogeny
SA ophiolite complexes
SA oroclines
SA orogenic belts
SA orogeny
SA Pan-African Orogeny
SA plate tectonics
SA post-tectonic processes
SA pull-apart basins
SA rift zones
SA sedimentary cover
SA seismotectonic maps
SA shear zones
SA shields
SA stacking
SA structural analysis
SA structural geology
SA subsidence
SA supracrustals
SA suture zones
SA synsedimentary processes
SA syntectonic processes
SA taphrogeny
SA tectonic breccia
SA tectonic elements
SA tectonic maps
SA tectonic platforms
SA tectonic units
SA tectonic wedges

SA tectonophysics
SA tectonosphere
SA tectonostratigraphic units
SA terranes
SA thrust sheets
SA transpression
SA undation
SA uplifts
SA vertical movements
SA volcanology
SA windows

tectonism
use tectonics

tectonite (1978)
SA deformation
SA fabric
SA preferred orientation

tectonophysics (1978)
Used for treatments of the application of physics in tectonics.
IN Use for the discipline as a whole.
SA asthenosphere
SA bibliography
SA continental drift
SA continental margin
SA convection
SA convection currents
SA core
SA core-mantle boundary
SA crust
SA crustal thinning
SA deformation
SA Earth
SA Earth tides
SA faults
SA folds
SA foliation
SA fractures
SA geophysics
SA geosynclines
SA geotectonic maps
SA heat flow
SA island arcs
SA isostasy
SA lithosphere
SA mantle
SA Mohorovicic discontinuity
SA neotectonics
SA ocean basins
SA paleomagnetism
SA plate tectonics
SA rift zones
SA salt tectonics
SA sea-floor spreading
SA seismology
SA structural analysis
SA subduction
SA subduction zones
SA tectonics
SA thin-skinned tectonics

tectonosphere (1978)
SA biosphere
SA hydrosphere
SA lithosphere
SA plate tectonics
SA tectonics

tectonostratigraphic terranes
use terranes

tectonostratigraphic units (1985)
SA lithostratigraphy
SA melange
SA stratigraphic units
SA stratigraphy
SA tectonics
SA terranes

tectosilicates
use framework silicates

teeth (1978)
SA anatomy

SA bones
SA jaws
SA paleopathology
SA skeletons
SA skulls
SA Vertebrata

Tehachapi Mountains (1978)
Range running E-W between S end of Sierra Nevada and the Coast Ranges in Kern County in S California. Also search Tehachapi.
BT California
BT United States
SA Kern County California
SA San Emigdio Mountains

Tehama County
Valid through 1988. Search in combination with state term. After 1988, use specific county-state term.

Tehama County California (1989)
N California. Before 1989, also search Tehama County AND California.
CO N394700N402500 W1212000W1230500
BT California
BT United States
SA Lassen Volcanic National Park

Teheran
No longer a valid term for GeoRef. Former province in NW central Iran. Also search Tehran. See Teheran Iran.

Teheran Iran (1993)
City at S foot of the Elburz in Central province, Iran. Also search Tehran.
BT Iran
BT Middle East
BT Asia

Tehran
Not a valid term for GeoRef. See Teheran Iran.

Tehri
Not a valid term for GeoRef. See Tehri Garhwal India.

Tehri Garhwal
No longer a valid term for GeoRef. See Tehri Garhwal India.

Tehri Garhwal India (1993)
District in Himalayas in NW Uttar Pradesh. Also search Tehri.
BT Uttar Pradesh India
BT India
BT Indian Peninsula
BT Asia

Teichichnus (1985)
BT ichnofossils

Teign Valley (1978)
River valley in Devonshire in SW England. Also search Teign River.
BT Devonshire England
BT England
BT Great Britain
BT United Kingdom
BT Western Europe
BT Europe

Tejon Formation (1978)
At the type section, subdivided into 4 members (ascending): Uvas Conglomerate, Liveoak, Metralla Sandstone, and Reed Canyon Silt. In the Ventura region, includes Matilija, Cozy Dell and Coldwater formations. Underlies Tecuya Formation; overlies basement complex. In W California.
BT upper Eocene
BT Eocene
BT Paleogene
BT Tertiary
BT Cenozoic
SA California
SA Sespe Formation

Tekeli
No longer a valid term for GeoRef. As of 1993, see Tekeli Kazakhstan.

Tekeli Kazakhstan (1993)
Town in Taldy Kurgan Oblast near Xinjiang border in E Kazakhstan. This term has multiple hierarchies.
BT1 Kazakhstan
BT1 Central Asia
BT1 Asia
BT2 Kazakhstan
BT2 Commonwealth of Independent States

tektites (1978)
Used for the descriptions of small, rounded, pitted bodies of silicate glass of nonvolcanic origin. Before 1972, also search tectites. Autoposting of this term began in 1978.
UF tectites
NT australite
NT Darwin glass
NT indochinite
NT irghizite
NT microtektites
NT moldavite
NT muong nong type
NT philippinite
SA asteroids
SA comets
SA glasses
SA isotopes
SA meteorites
SA petrology
SA planetology
SA spherules

tele
use frozen ground

Teleajen River
use Teleajen Valley

Teleajen Valley (1978)
River valley in Prahova County in NE Moldavia. Also search Teleajen. This term has multiple hierarchies.
UF Teleajen River
BT1 Moldavia
BT1 Europe
BT2 Romania
BT2 Southern Europe
BT2 Europe
SA Prahova Romania

Telemark
No longer a valid term for GeoRef. As of 1993 see Telemark Norway.

Telemark Norway (1993)
Before 1993 also search Telemark AND Norway.
BT Norway
BT Scandinavia
BT Western Europe
BT Europe
NT Bamble Norway
NT Langesund Fjord

telemetry (1989)
SA Landsat
SA measurement
SA microwave methods
SA multispectral analysis

SA remote sensing
SA satellite measurements

Teleosauridae (1978)
Family.
BT Crocodilia
BT Archosauria
BT Diapsida
BT Reptilia
BT Tetrapoda
BT Vertebrata
BT Chordata

Teleostei (1978)
Infraclass. Autoposting of this term began in 1981.
BT Actinopterygii
BT Osteichthyes
BT Pisces
BT Vertebrata
BT Chordata
NT Cyprinidae

teleseismic signals (1978)
SA earthquakes
SA seismology
SA signals

telethermal processes (1981)
Also search telethermal.
SA mineral deposits, genesis
SA processes

television logging
Not a valid GeoRef index term after 1970. Was used on level 1 in subfile N. Now use well-logging.

telinite (1981)
BT vitrinite
BT macerals

Tell (1978)
The Mediterranean coastal region of Algeria which is favored by a Mediterranean type climate.
BT Algeria
BT North Africa
BT Africa

Teller County
Valid through 1988. Search in combination with state term. After 1988, use specific county-state term.

Teller County Colorado (1989)
Central Colorado. Before 1989, also search Teller County AND Colorado.
CO N384000N391000 W1045500W1052000
BT Colorado
BT United States
NT Cripple Creek Colorado

tellurates (1981)
Before 1981, search tellurates and tellurites.
SA minerals

tellurates and tellurites
As of 1981, no longer a valid term for GeoRef. Use either tellurates or tellurites as applicable. From 1978-1980, was autoposted to mroseite.

tellurbismuth
use tellurobismuthite

telluric currents
Not a valid index term for GeoRef. See Earth-current methods.

telluric methods
use Earth-current methods

telluric surveys
use Earth-current surveys

tellurides (1978)

Autoposting of sulfides to this term ended in 1989. Hedleyite is a narrower term as of 1981.
NT altaite
NT calaverite
NT coloradoite
NT hedleyite
NT hessite
NT joseite
NT sylvanite
NT tellurobismuthite
NT tetradymite
SA minerals
SA wehrlite

tellurites (1981)
Before 1981, search tellurates and tellurites.
NT mroseite
SA minerals

tellurium (1978)
UF Te
SA chalcophile elements
SA chemical elements
SA heavy metals

tellurobismuthite (1978)
Autoposting of sulfides to this term ended in 1989.
UF tellurbismuth
BT tellurides

TEM data (1978)
Acronym. For the method, see transmission method. For transient electromagnetic (TEM) response, use transient methods and electromagnetic methods, electromagnetic surveys, or electromagnetic logging as applicable.
UF transmission electron microscopy data
SA data
SA electron microscopy
SA electron microscopy data
SA scanning electron microscopy
SA SEM data

Temagami Mine (1978)
Copper, gold and silver ores. Near village of Temagami on E end of Lake Temagami near Quebec border. Also search Temagami; Timagami.
UF Timagami Mine
BT Ontario
BT Eastern Canada
BT Canada
SA copper ores
SA gold ores
SA mines
SA silver ores

Temblor Formation (1989)
Central and Southern California.
BT Miocene
BT Neogene
BT Tertiary
BT Cenozoic
SA California

Temblor Range (1978)
Along SW San Joaquin Valley in W and SW Kern County.
IN Index counties as applicable.
BT California
BT United States
SA Kern County California
SA San Emigdio Mountains

Temiscamingue County
No longer a valid term for GeoRef. As of 1993, see Temiscamingue County Quebec.

Temiscamingue County Quebec (1993)
SW Quebec on Ontario border. Before 1993, also search Temiscamingue County AND Quebec.
UF Timiskaming County
UF Temiskaming County
BT Quebec
BT Eastern Canada
BT Canada
NT Kipawa Lake
NT Rouyn Quebec
SA Lake Timiskaming
SA Timiskaming District Ontario

Temiscamingue District
use Timiskaming District Ontario

Temiscamingue Lake
use Lake Timiskaming

Temiskaming County
use Temiscamingue County Quebec

Temiskaming District
use Timiskaming District Ontario

Temiskaming Lake
use Lake Timiskaming

Temnospondyli (1993)
Order.
BT Labyrinthodontia
BT Amphibia
BT Tetrapoda
BT Vertebrata
BT Chordata

temperate environment (1978)
Before 1978, search temperate.
SA climate
SA depositional environment
SA ecology
SA environment
SA paleoclimatology

temperature (1978)
For temperatures of mineral formation, see phase equilibria. As of 1982, use paleotemperature under paleoclimatology. Before 1974, also search paleotemperature; paleotemperatures.
UF temperatures
NT high temperature
NT low temperature
SA climate
SA cooling
SA fluid inclusions
SA geologic thermometry
SA global warming
SA greenhouse effect
SA heat flow
SA isotherms
SA Kirchhoff integral
SA P-T conditions
SA paleotemperature
SA phase equilibria
SA specific heat
SA temperature logging
SA thermal history
SA thermal properties
SA thermal waters
SA thermocline

temperature anomaly
use thermal anomalies

temperature logging (1985)
BT well-logging
SA heat flow
SA temperature

temperature methods
use heat flow

temperature surveys
use heat flow

temperatures
use temperature

tempestite (1989)
Storm deposit.
SA clastic rocks
SA clastic sediments
SA sedimentary rocks
SA sediments
SA shallow-water environment

temporal distribution (1981)
SA distribution

Tendoy Range (1985)
SW Montana. Autoposting of broader terms to this term began in 1989. As of 1993, U. S. Rocky Mountains is autoposted to this term. This term has multiple hierarchies.
BT1 Beaverhead County Montana
BT1 Montana
BT1 United States
BT2 U. S. Rocky Mountains
BT2 Rocky Mountains
BT3 U. S. Rocky Mountains
BT3 United States

Tenerife (1978)
Largest island in the Spanish controlled Canary Islands off NW Africa.
UF Teneriffe
BT Canary Islands
BT Atlantic Ocean Islands

Teneriffe
use Tenerife

Tengchong (1989)
Volcano-geothermal region in Yunnan, SW China.
BT Yunnan China
BT China
BT Far East
BT Asia

Tengiz (1978)
Salt lake NW of Karaganda in N central Kazakhstan. This term has multiple hierarchies.
UF Teniz
BT1 Kazakhstan
BT1 Central Asia
BT1 Asia
BT2 Kazakhstan
BT2 Commonwealth of Independent States

Teniz
use Tengiz

Tennant Creek
No longer a valid term for GeoRef. See Tennant Creek Australia.

Tennant Creek Australia (1993)
Village 280 miles N of Alice Springs in central Northern Territory.
BT Northern Territory Australia
BT Australia
BT Australasia

tennantite (1978)
BT sulfarsenites
BT sulfosalts
SA copper minerals
SA copper ores

Tennessee (1978)
Autoposting of this term began in 1978.
CO N350000N364500 W0814000W0901500
BT United States
NT Anderson County Tennessee
NT Bedford County Tennessee
NT Benton County Tennessee
NT Blount County Tennessee
NT Campbell County Tennessee
NT Cannon County Tennessee
NT Carroll County Tennessee
NT Carter County Tennessee
NT Chester County Tennessee
NT Claiborne County Tennessee
NT Clay County Tennessee
NT Coffee County Tennessee
NT Cumberland County Tennessee
NT Davidson County Tennessee
NT De Kalb County Tennessee
NT Decatur County Tennessee
NT Fayette County Tennessee
NT Franklin County Tennessee
NT Giles County Tennessee
NT Grainger County Tennessee
NT Greene County Tennessee
NT Grundy County Tennessee
NT Hamilton County Tennessee
NT Hancock County Tennessee
NT Hardeman County Tennessee
NT Hardin County Tennessee
NT Henderson County Tennessee
NT Henry County Tennessee
NT Houston County Tennessee
NT Jackson County Tennessee
NT Jefferson County Tennessee
NT Johnson County Tennessee
NT Knox County Tennessee
NT Lake County Tennessee
NT Lauderdale County Tennessee
NT Lawrence County Tennessee
NT Lewis County Tennessee
NT Lincoln County Tennessee
NT Macon County Tennessee
NT Madison County Tennessee
NT Marion County Tennessee
NT Marshall County Tennessee
NT Maury County Tennessee
NT Monroe County Tennessee
NT Montgomery County Tennessee
NT Moore County Tennessee
NT Morgan County Tennessee
NT Obion County Tennessee
NT Overton County Tennessee
NT Perry County Tennessee
NT Polk County Tennessee
NT Putnam County Tennessee
NT Reelfoot Lake
NT Roane County Tennessee
NT Scott County Tennessee
NT Sequatchie Valley
NT Sevier County Tennessee
NT Shelby County Tennessee
NT Smith County Tennessee
NT Sullivan County Tennessee
NT Sumner County Tennessee
NT Trousdale County Tennessee
NT Union County Tennessee
NT Warren County Tennessee
NT Washington County Tennessee
NT Wayne County Tennessee
NT White County Tennessee
NT Williamson County Tennessee
NT Wilson County Tennessee
SA Appalachian Basin
SA Appalachian Plateau
SA Ashe Formation
SA Bangor Limestone
SA Beekmantown Group
SA Bigby-Cannon Limestone
SA Brassfield Formation
SA Buffalo River
SA Central Basin
SA Chattanooga Shale
SA Chesterian
SA Chickamauga Group
SA Chilhowee Group
SA Clayton Formation
SA Clinch Sandstone
SA Conasauga Group
SA Conococheague Formation
SA Crab Orchard Mountains Group

Before then, variants of the term may occur in GeoRef.

SA Cumberland Plateau
SA Eutaw Formation
SA Fernvale Formation
SA Fort Payne Formation
SA Gizzard Group
SA Golconda Formation
SA Great Smoky Fault
SA Great Smoky Group
SA Great Smoky Mountains
SA Hayesville Fault
SA Holston Formation
SA Juniata Formation
SA Kingsport Formation
SA Knox Group
SA Lee Formation
SA Lenoir Limestone
SA Lincolnshire Limestone
SA Mascot Dolomite
SA Midcontinent
SA Midway Group
SA Mississippi Embayment
SA Mississippi River
SA Mississippi Valley
SA Monteagle Limestone
SA Murphy Marble
SA Nashville Dome
SA New Madrid region
SA Newman Limestone
SA Oak Ridge National Laboratory
SA Ocoee Series
SA Pennington Formation
SA Pine Mountain Window
SA Porters Creek Formation
SA Pottsville Group
SA Pulaski thrust sheet
SA Reedsville Formation
SA Reelfoot Rift
SA Ripley Formation
SA Roan Supergroup
SA Rockwood Formation
SA Rome Formation
SA Rome Trough
SA Ross Formation
SA Saint Louis Limestone
SA Sainte Genevieve Limestone
SA Selma Group
SA Shady Dolomite
SA Stones River Group
SA Talladega Group
SA Tennessee River
SA Tennessee Valley
SA Tuscaloosa Formation
SA Valley and Ridge Province
SA Waldron Shale
SA Warsaw Formation
SA Wilcox Group

Tennessee River (1978)
Formed near Knoxville in E Tennessee. It flows SW, W, and then N entering the Ohio River at Paducah in W Kentucky.
IN   Index states as applicable.
BT   United States
SA   Alabama
SA   Kentucky
SA   Tennessee

Tennessee Sandstone (1978)
Name has either been abandoned by author or rejected for use by U.S. Geological Survey. Typically exposed in Tennessee Ridge, Sebastian County, in W Arkansas.
BT   Pennsylvanian
BT   Carboniferous
BT   Paleozoic
SA   Arkansas

Tennessee Valley (1978)
River valley.
IN   Index states as applicable.
BT   United States
SA   Alabama
SA   Great Appalachian Valley

SA   Kentucky
SA   Tennessee

tensile strength (1978)
SA   deformation
SA   elastic properties
SA   engineering geology
SA   failures
SA   mechanical properties
SA   rupture
SA   shear strength
SA   strength
SA   tensiometers
SA   tension

tensiometers (1985)
SA   instruments
SA   tensile strength
SA   tension

tension (1978)
SA   compression
SA   deformation
SA   extension
SA   extension faults
SA   extension fractures
SA   extension tectonics
SA   fractures
SA   pressure
SA   stress
SA   tensile strength
SA   tensiometers
SA   torsion
SA   yield strength

Tensleep Sandstone (1978)
In Montchauve Group. Pennsylvanian and Lower Permian. Overlies Amsden Formation.
IN   Index ages as applicable.
BT   Paleozoic
SA   Lower Permian
SA   Montana
SA   Pennsylvanian
SA   Wyoming

Tentaculites (1978)
Genus. Autoposting of Mollusca began in 1981.
BT   Tentaculitidae
BT   Tentaculitida
BT   Mollusca
BT   Invertebrata
SA   problematic fossils

Tentaculitida (1978)
Order. Autoposting of Mollusca began in 1981.
BT   Mollusca
BT   Invertebrata
NT   Tentaculitidae
SA   problematic fossils

Tentaculitidae (1978)
Family. Autoposting of Mollusca began in 1981.
BT   Tentaculitida
BT   Mollusca
BT   Invertebrata
NT   Tentaculites
SA   problematic fossils

Tepee Trail Formation (1985)
NW Wyoming.
BT   upper Eocene
BT   Eocene
BT   Paleogene
BT   Tertiary
BT   Cenozoic
SA   Wyoming

tephra
  use pyroclastics

tephrite (1978)
BT   alkali basalts
BT   basalts
BT   volcanic rocks
BT   igneous rocks

tephrochronology (1978)

SA   geochronology
SA   relative age

tephroite (1978)
BT   olivine group
BT   nesosilicates
BT   orthosilicates
BT   silicates
SA   manganese minerals

Teplice (1978)
City in the Erzgebirge, NW Czechoslovakia NNW of Prague.
UF   Teplice-Sanov
BT   Bohemia
BT   Czech Republic
BT   Czechoslovakia
BT   Central Europe
BT   Europe

Teplice-Sanov
  use Teplice

Ter River basin (1978)
In Catalonia, Spain. Also search Ter River.
IN   Index Spain and provinces as applicable.
BT   Spain
BT   Iberian Peninsula
BT   Southern Europe
BT   Europe
SA   Barcelona Spain
SA   Catalonia Spain
SA   Gerona Spain

terbium (1978)
Autoposting of rare earths and metals to this term began in 1989.
UF   Tb
BT   rare earths
BT   metals

Terceira Island (1978)
Central island of Portuguese controlled group. Site of U. S. Lajes Air Force Base. Also search Terceira.
BT   Azores
BT   Atlantic Ocean Islands

Tereblya
No longer a valid term for GeoRef. As of 1993, see Tereblya Ukraine.

Tereblya Ukraine (1993)
Village in SW Ukraine near the Romanian border. This term has multiple hierarchies.
BT1 Ukraine
BT1 Europe
BT2 Ukraine
BT2 Commonwealth of Independent States

Terebratulida (1978)
Order. Autoposting of this term began in 1978.
BT   Articulata
BT   Brachiopoda
BT   Invertebrata
NT   Terebratulidae
NT   Terebratulina

Terebratulidae (1978)
Family.
BT   Terebratulida
BT   Articulata
BT   Brachiopoda
BT   Invertebrata

Terebratulina (1978)
Genus.
BT   Terebratulida
BT   Articulata
BT   Brachiopoda
BT   Invertebrata

Terek River (1978)

Flows into the Caspian Sea N of Makhachkala in the Northern Caucasus. Also search Terek.
IN   Index former Soviet republics as applicable.
BT   Europe
SA   Georgian Republic
SA   Russian Federation

Terlingua
Valid through 1988. Search in combination with state term. After 1988, use specific city-state term.

Terlingua Texas (1989)
Village in W Texas. Before 1989, search Terlingua AND Texas.
CO   N292000N292000
     W1033900W1033900
BT   Brewster County Texas
BT   Texas
BT   United States

terminology
A valid term through mid-1978.
  use nomenclature

termites (1989)
Insects belonging to the order Isoptera.
BT   insects
BT   arthropods
BT   invertebrates

Terni
No longer a valid term for GeoRef. See Terni Italy.

Terni Italy (1993)
City NE of Rome in S Umbria, central Italy.
BT   Umbria Italy
BT   Italy
BT   Southern Europe
BT   Europe

terpanes (1985)
From 1985 to 1989, alkanes and aliphatic hydrocarbons were autoposted to this term.
BT   hydrocarbons
BT   organic materials
SA   diterpanes
SA   triterpanes

Terra rosa
  use Terra rossa

Terra rossa (1978)
UF   Terra rosa
BT   soils
SA   soil group

terraces (1978)
As of 1982, use marine terraces for the shore features.
SA   benches
SA   changes of level
SA   erosion features
SA   floodplains
SA   fluvial features
SA   geomorphology
SA   lacustrine features
SA   marine terraces

terrain classification (1978)
Before 1978, search classification AND terrains.
SA   classification
SA   digital terrain models
SA   geomorphology
SA   terrains

terrains (1978)
Used for tracts of land and their physical features. As of 1983, use terranes for the new concept in tectonics. Before 1983, also search terranes.
SA   digital terrain models
SA   geomorphology

SA glaciated terrains
SA relief
SA terrain classification
SA terranes
SA tesserae
SA topography

terranes (1983)
Used for the new concept in tectonics. Use terrains as a general term for tracts of land.
UF suspect terranes
UF tectonostratigraphic terranes
NT exotic terranes
SA accretionary wedges
SA greenstone belts
SA metamorphic belts
SA metamorphic core complexes
SA plate tectonics
SA tectonics
SA tectonostratigraphic units
SA terrains

Terrebonne Parish Louisiana (1989)
Extreme SE Louisiana, bordering the Gulf of Mexico. Before 1989, search Terrebonne Parish AND Louisiana.
CO N290200N294700
W0902300W0912300
BT Louisiana
BT United States
NT Isles Dernieres
SA Atchafalaya Bay

terrestrial
A valid term through 1977. After 1977, use terrestrial environment for type of environment. See terrestrial materials.

terrestrial comparison (1982)
Used to indicate a comparison between the Earth and an extraterrestrial body or material. Also search interplanetary comparison AND Earth. Before 1981, also search terrestrial materials.
SA Earth
SA extraterrestrial geology
SA interplanetary comparison
SA Moon
SA planetology
SA planets

terrestrial environment (1978)
Before 1978, also search terrestrial. Used to distinguish from marine environment.
UF continental environment
SA cave environment
SA depositional environment
SA ecology
SA environment
SA paleoecology
SA sedimentation
SA terrestrial sedimentation

terrestrial materials
No longer a valid term for GeoRef as of 1981. See terrestrial comparison; interplanetary comparison. Term was introduced in 1978. Included use to indicate a comparison between Earth and Moon (lunar samples).

terrestrial planets (1978)
The four inner planets in the solar system: Mercury Planet, Venus, Earth and Mars.
SA Earth
SA Mars
SA Mercury Planet
SA outer planets
SA planetology
SA planets

SA solar system
SA Venus

terrestrial sedimentation (1981)
BT sedimentation
SA terrestrial environment

terrigenous
No longer a valid term for GeoRef. A valid term through 1977 used in combination with clastic rocks or clastic sediments. After 1977, see terrigenous materials.

terrigenous materials (1978)
Before 1978, also search terrigenous AND clastic rocks or clastic sediments.
SA arenite
SA argillite
SA arkose
SA bentonite
SA black shale
SA boulders
SA breccia
SA cinerite
SA clastic rocks
SA clastic sediments
SA clay
SA claystone
SA cobbles
SA colluvium
SA conglomerate
SA contourite
SA coquina
SA diamictite
SA drift
SA dust
SA eluvium
SA eolianite
SA fanglomerate
SA flint clay
SA flysch
SA gravel
SA graywacke
SA hemipelagic environment
SA loess
SA marl
SA materials
SA microbreccia
SA molasse
SA mud
SA mudstone
SA orthoquartzite
SA outwash
SA pebbles
SA red beds
SA residuum
SA sand
SA sandstone
SA sedimentary rocks
SA sediments
SA shale
SA silcrete
SA silt
SA siltstone
SA sparagmite
SA subgraywacke
SA till
SA tillite
SA tilloid
SA tonstein
SA turbidite
SA volcanic ash
SA volcaniclastics

Tertiary (1978)
Autoposting of this term began in 1978.
BT Cenozoic
NT Anahuac Formation
NT Arikareean
NT Barreiras Formation
NT Catahoula Formation
NT Challis Volcanics
NT Climax Porphyry
NT Esna Shale

NT John Day Formation
NT lower Tertiary
NT Maikop Series
NT middle Tertiary
NT Muddy Creek Formation
NT Neogene
NT Neyveli Lignite
NT Paleogene
NT Poltava Series
NT Shahejie Formation
NT Snoqualmie Batholith
NT Sobral Formation
NT Twin River Formation
NT Twin Sisters Dunite
NT upper Tertiary
NT Vaqueros Formation
NT Warkalli Formation
NT Zambales Ophiolite
SA Brianconnais Zone
SA Cretaceous
SA Hector Formation
SA Ionian Zone
SA Laramide Orogeny
SA Muro Group
SA Paleolithic
SA Quaternary
SA San Juan Formation
SA Stone Age
SA Villafranchian

tertiary recovery (1978)
Use in relation to petroleum.
BT enhanced recovery
SA acidification
SA gas injection
SA hydraulic fracturing
SA mercury injection
SA petroleum
SA petroleum engineering
SA recovery
SA secondary recovery
SA thermal recovery

Teruel
No longer a valid term for GeoRef. As of 1993, see Teruel Spain.

Teruel Spain (1993)
Province in NE central Spain. Before 1993, also search Teruel and Spain.
CO N395200N412200
E0001700W0015000
BT Aragon Spain
BT Spain
BT Iberian Peninsula
BT Southern Europe
BT Europe
SA Calatayud-Teruel Basin
SA Maestrazgo Spain

teschenite (1978)
BT alkali gabbros
BT gabbros
BT plutonic rocks
BT igneous rocks

Tesnus Formation (1989)
Marathon Geosyncline of W Texas.
BT Carboniferous
BT Paleozoic
SA Marathon Geosyncline
SA Texas

tesserae (1993)
Mosaic-like terrain on other planetary bodies.
SA block structures
SA plate tectonics
SA rifting
SA taphrogeny
SA terrains
SA Venus

Tessin
use Ticino Switzerland

testing (1981)

SA consolidometer tests
SA engineering geology
SA lateral loading
SA load tests
SA penetration tests
SA plate-bearing tests
SA road tests
SA rock mechanics
SA sand equivalent tests
SA shear tests
SA soil mechanics
SA triaxial tests
SA uniaxial tests
SA vane tests

tests (1978)
As of 1976, term is only to be used with fossils. For tests used in engineering geology, see testing.
SA chitin
SA Invertebrata
SA shells

Tetagouche Group (1989)
N New Brunswick.
BT Ordovician
BT Paleozoic
SA New Brunswick

Tete
No longer a valid term for GeoRef. As of 1993 see Tete Mozambique.

Tete Mozambique (1993)
Town on the Zambezi River and district in NW Mozambique. Before 1993 also search Tete AND Mozambique.
BT Mozambique
BT East Africa
BT Africa

Teterev River valley (1978)
NW Ukraine. Also search Teterev; Teterev River. This term has multiple hierarchies.
UF Teterev Valley
BT1 Ukraine
BT1 Europe
BT2 Ukraine
BT2 Commonwealth of Independent States

Teterev Valley
use Teterev River valley

Teteven
No longer a valid term for GeoRef. See Teteven Bulgaria.

Teteven Bulgaria (1993)
Town in central Bulgaria.
BT Pleven Bulgaria
BT Bulgaria
BT Southern Europe
BT Europe

Tethys (1978)
An elongated east-west sea, similar to the Mediterranean Sea, that separated Europe and Africa and extended across southern Asia in Pre-Tertiary time.
IN Index continents as applicable.
UF Tethys Sea
SA Africa
SA Asia
SA continental drift
SA Europe
SA Mesogaea
SA Paratethys

Tethys Satellite (1978)
One of the satellites of Saturn.
SA icy satellites
SA satellites
SA Saturn
SA Voyager Program

Tethys Sea
  use Tethys
Teton County
  Valid through 1988. Search in combination with state term. After 1988, use specific county-state term.
Teton County Idaho (1989)
  E Idaho. Before 1989, also search Teton County AND Idaho.
  CO N432900N435800
       W1110200W1112400
  BT Idaho
  BT United States
Teton County Montana (1989)
  N central Montana. Before 1989, also search Teton County AND Montana.
  CO N473000N480900
       W1112500W1130100
  BT Montana
  BT United States
  SA Sawtooth Range
Teton County Wyoming (1989)
  NW Wyoming. Before 1989, also search Teton County AND Wyoming.
  CO N431500N444200
       W1100300W1110300
  BT Wyoming
  BT United States
  NT Grand Teton National Park
  NT Jackson Hole
  NT Old Faithful Geyser
  SA Yellowstone National Park
Teton National Forest (1978)
  IN Index counties as applicable.
  BT Wyoming
  BT United States
  SA Grand Teton National Park
Tetrabranchiata (1981)
  BT Cephalopoda
  BT Mollusca
  BT Invertebrata
  NT Ammonoidea
  NT Nautiloidea
tetrachloroethylene (1989)
  UF perchloroethylene
  BT chlorinated hydrocarbons
  SA ethylene
Tetracorallia
  As of 1989, no longer a valid term for GeoRef. Before 1981, tetracorals was also used.
  use Rugosa
tetradymite (1978)
  This term has multiple hierarchies.
  BT1 tellurides
  BT2 sulfides
tetrahedra (1978)
  BT polyhedra
tetrahedrite (1978)
  BT sulfantimonites
  BT sulfosalts
  SA copper minerals
  SA copper ores
  SA freibergite
  SA silver ores
Tetrapoda (1978)
  BT Vertebrata
  BT Chordata
  NT Amphibia
  NT Aves
  NT dinosaurs
  NT Mammalia
  NT Reptilia
  NT Smilodon
  SA tetrapods
tetrapods (1989)
  Used as the common name for Tetrapoda.
  IN Index for all non-paleontologic studies of fossils.
  BT vertebrates
  NT amphibians
  NT birds
  NT mammals
  NT reptiles
  SA biostratigraphy
  SA Tetrapoda
tetrataenite (1989)
  Meteorite mineral.
  BT alloys
  SA meteorites
  SA taenite
Tetyukhe
  No longer a valid term for GeoRef. As of 1993, see Tetyukhe Russian Federation.
Tetyukhe Russian Federation (1993)
  Town in E Primorye Kray 220 miles NE of Vladivostok. This term has multiple hierarchies.
  BT1 Primorye Russian Federation
  BT1 Russian Federation
  BT1 Commonwealth of Independent States
  BT2 Primorye Russian Federation
  BT2 Asia
Teutoburg Forest (1978)
  Range of hills S of Osnabruck in NW Germany.
  IN Index states as applicable.
  UF Teutoburger Wald
  BT Germany
  BT Central Europe
  BT Europe
  SA Lower Saxony Germany
  SA North Rhine-Westphalia Germany
Teutoburger Wald
  use Teutoburg Forest
Texas (1978)
  Baffin Bay was a narrower term from 1981 through 1992. Autoposting of this term began in 1978.
  CO N254500N363000
       W0933000W1063000
  BT United States
  NT Anderson County Texas
  NT Andrews County Texas
  NT Armstrong County Texas
  NT Atascosa County Texas
  NT Balcones fault zone
  NT Bastrop County Texas
  NT Baylor County Texas
  NT Bell County Texas
  NT Bexar County Texas
  NT Brazoria County Texas
  NT Brazos County Texas
  NT Brazos River
  NT Brewster County Texas
  NT Burleson County Texas
  NT Burnet County Texas
  NT Caldwell County Texas
  NT Calhoun County Texas
  NT Cass County Texas
  NT Chambers County Texas
  NT Cherokee County Texas
  NT Christmas Mountains
  NT Clay County Texas
  NT Comal County Texas
  NT Comanche County Texas
  NT Crane County Texas
  NT Culberson County Texas
  NT Dalhart Basin
  NT Dallas County Texas
  NT Dawson County Texas
  NT Deaf Smith County Texas
  NT Delta County Texas
  NT Dimmit County Texas
  NT Duval County Texas
  NT East Texas
  NT East Texas Field
  NT Ector County Texas
  NT Edwards Aquifer
  NT Edwards County Texas
  NT Edwards Plateau
  NT El Paso County Texas
  NT Ellis County Texas
  NT Falls County Texas
  NT Fayette County Texas
  NT Floyd County Texas
  NT Fort Bend County Texas
  NT Fort Worth Basin
  NT Franklin County Texas
  NT Freestone County Texas
  NT Frio County Texas
  NT Gaines County Texas
  NT Galveston Bay
  NT Galveston County Texas
  NT Garza County Texas
  NT Gonzales County Texas
  NT Guadalupe County Texas
  NT Hall County Texas
  NT Hamilton County Texas
  NT Hardeman County Texas
  NT Hardin County Texas
  NT Harris County Texas
  NT Harrison County Texas
  NT Haskell County Texas
  NT Hays County Texas
  NT Henderson County Texas
  NT Hidalgo County Texas
  NT Hill County Texas
  NT Hockley County Texas
  NT Houston County Texas
  NT Howard County Texas
  NT Hudspeth County Texas
  NT Jackson County Texas
  NT Jasper County Texas
  NT Jeff Davis County Texas
  NT Jefferson County Texas
  NT Johnson County Texas
  NT Jones County Texas
  NT Karnes County Texas
  NT Kenedy County Texas
  NT Kent County Texas
  NT King County Texas
  NT Kinney County Texas
  NT Kleberg County Texas
  NT La Salle County Texas
  NT Lamar County Texas
  NT Lavaca County Texas
  NT Lee County Texas
  NT Leon County Texas
  NT Liberty County Texas
  NT Limestone County Texas
  NT Live Oak County Texas
  NT Llano County Texas
  NT Llano Uplift
  NT Lubbock County Texas
  NT Madison County Texas
  NT Marathon Geosyncline
  NT Marfa Basin
  NT Marion County Texas
  NT Martin County Texas
  NT Mason County Texas
  NT Matagorda Bay
  NT Maverick County Texas
  NT McLennan County Texas
  NT Medina County Texas
  NT Midland County Texas
  NT Milam County Texas
  NT Mills County Texas
  NT Mitchell County Texas
  NT Montgomery County Texas
  NT Moore County Texas
  NT Morris County Texas
  NT Newton County Texas
  NT Nueces County Texas
  NT Nueces River
  NT Oakwood Dome
  NT Orange County Texas
  NT Panola County Texas
  NT Parker County Texas
  NT Pecos County Texas
  NT Polk County Texas
  NT Potter County Texas
  NT Presidio County Texas
  NT Reeves County Texas
  NT Rita Blanca Lake
  NT Runnels County Texas
  NT Rusk County Texas
  NT San Marcos Arch
  NT San Patricio County Texas
  NT San Saba County Texas
  NT Shelby County Texas
  NT Sherman County Texas
  NT Smith County Texas
  NT Stephens County Texas
  NT Sterling County Texas
  NT Sutton County Texas
  NT Swisher County Texas
  NT Tarrant County Texas
  NT Taylor County Texas
  NT Texas Panhandle
  NT Travis County Texas
  NT Trinity County Texas
  NT Upshur County Texas
  NT Upton County Texas
  NT Uvalde County Texas
  NT Val Verde Basin
  NT Val Verde County Texas
  NT Victoria County Texas
  NT Walker County Texas
  NT Ward County Texas
  NT Washington County Texas
  NT Webb County Texas
  NT West Texas
  NT Wheeler County Texas
  NT Wichita County Texas
  NT Williamson County Texas
  NT Wilson County Texas
  NT Winkler County Texas
  NT Wise County Texas
  NT Wood County Texas
  NT Yoakum County Texas
  NT Zavala County Texas
  SA Aguja Formation
  SA Albuquerque Basin
  SA Anadarko Basin
  SA Anahuac Formation
  SA Antlers Sands
  SA Ardmore Basin
  SA Ashmore Meteorite
  SA Austin Chalk
  SA Austin Group
  SA Baffin Bay
  SA Barnett Formation
  SA Barton Springs
  SA Basin and Range Province
  SA Beaumont Clay
  SA Bell Canyon Formation
  SA Blaine Formation
  SA Blancan
  SA Bone Spring Limestone
  SA Bossier Formation
  SA Buckner Formation
  SA Buda Limestone
  SA Calvert Bluff Formation
  SA Canadian River
  SA Canyon Group
  SA Capitan Formation
  SA Carrizo Mountain Formation
  SA Carrizo Sand
  SA Castile Formation
  SA Catahoula Formation
  SA Central Basin Platform
  SA Chappel Limestone
  SA Cherry Canyon Formation
  SA Chicot Aquifer
  SA Chihuahua tectonic belt
  SA Cisco Group
  SA Citronelle Formation
  SA Claiborne Group
  SA Clear Fork Group
  SA Cockfield Formation
  SA Colorado River

SA Comanche Peak Limestone
SA Cook Mountain Formation
SA Corpus Christi Bay
SA Cotton Valley Group
SA Culebra Dolomite Member
SA Cupido Formation
SA Delaware Basin
SA Delaware Mountain Group
SA Diablo Platform
SA Dockum Group
SA Eagle Ford Formation
SA Edwards Formation
SA El Paso Group
SA Ellenburger Group
SA Escondido Formation
SA Evangeline Aquifer
SA Fleming Formation
SA Franklin Mountains
SA Fredericksburg Group
SA Frio Formation
SA Fusselman Dolomite
SA Georgetown Formation
SA Glass Mountains
SA Glen Rose Formation
SA Goliad Sand
SA Grayburg Formation
SA Great Plains
SA Guadalupe Mountains
SA Guadalupe River
SA Gulf Coastal Plain
SA Hackberry Formation
SA Hardeman Basin
SA Haymond Formation
SA Haynesville Formation
SA Hazel Formation
SA High Plains Aquifer
SA Hosston Formation
SA Hueco Limestone
SA Hueco Mountains
SA Jackson Group
SA Laguna Madre
SA Llano Estacado
SA Louann Salt
SA Loup Fork Group
SA Magdalena Group
SA Marathon Basin
SA Maravillas Formation
SA Marble Canyon
SA Marble Falls Group
SA Maverick Basin
SA Mesa Rica Sandstone
SA Midcontinent
SA Midland Basin
SA Midway Group
SA Montoya Group
SA Navarro Group
SA Norphlet Formation
SA Oakville Sandstone
SA Odessa Meteorite
SA Ogallala Aquifer
SA Ogallala Formation
SA Olmos Formation
SA Orogrande Basin
SA Ouachita Belt
SA Padre Island
SA Palo Duro Basin
SA Paluxy Formation
SA Pearsall Formation
SA Pecos River
SA Pecos River valley
SA Permian Basin
SA Queen City Formation
SA Queen Formation
SA Red River
SA Red River valley
SA Rio Grande
SA Rio Grande Depression
SA Rio Grande Rift
SA Rio Grande Valley
SA Rodessa Formation
SA Rustler Formation
SA Sabine Lake
SA Sabine Uplift
SA Salado Formation
SA San Miguel Formation
SA Santa Rosa Sandstone
SA Schuler Formation
SA Seven Rivers Formation
SA Sligo Formation
SA Smackover Formation
SA Smithwick Shale
SA Sparta Sand
SA Strawn Series
SA Tansill Formation
SA Taylor Marl
SA Tesnus Formation
SA Trans-Pecos
SA Travis Peak Formation
SA Trinity Group
SA Trinity River
SA Trinity River basin
SA Uvalde Gravel
SA Vicksburg Group
SA Vinton Canyon
SA Washita Group
SA Washita River valley
SA Whitsett Formation
SA Wichita Group
SA Wilberns Formation
SA Wilcox Group
SA Wolfcampian
SA Woodbine Formation
SA Woods Hollow Shale
SA Yates Formation
SA Yegua Formation
SA Yeso Formation

Texas County
Valid through 1988. Search in combination with state term. After 1988, use specific county-state term.

Texas County Missouri (1989)
S central Missouri. Before 1989, also search Texas County AND Missouri.
CO N370300N373700
W0913800W0921500
BT Missouri
BT United States

Texas County Oklahoma (1989)
Extreme NW Oklahoma. Before 1989, also search Texas County AND Oklahoma.
CO N363000N370000
W1005500W1020200
BT Oklahoma
BT United States
SA Oklahoma Panhandle

Texas Panhandle (1993)
NW Texas. Also search Panhandle AND Texas.
IN Index counties as applicable.
BT Texas
BT United States
SA West Texas

text, explanatory
use explanatory text

textbooks (1978)
Use under disciplines.
SA book reviews
SA education
SA glossaries
SA manuals
SA monographs
SA publications

Textulariina (1978)
Autoposting of microfossils and Protista to this term began in 1990. This term has multiple hierarchies.
BT1 foraminifera
BT1 Protista
BT1 Invertebrata
BT2 foraminifera
BT2 Protista
BT2 microfossils

NT Ammodiscacea
NT Lituolacea

textures (1978)
As of 1993, see grain size or size distribution for soils if applicable.
NT aphanitic texture
NT arenaceous texture
NT argillaceous texture
NT framboidal texture
NT gneissic texture
NT graphic texture
NT oolitic texture
NT orbicular texture
NT pelitic texture
NT pisolitic texture
NT porphyritic texture
NT porphyroblastic texture
NT porphyroclastic texture
NT spinifex texture
NT vesicular texture
SA coarse-grained materials
SA compactness
SA fines
SA grain boundaries
SA grapestone
SA intraclasts
SA megacrysts
SA packing
SA shape analysis
SA size distribution
SA surface textures
SA vugs

TGA data (1981)
SA data
SA thermal analysis
SA thermogravimetric analysis

Th
use thorium

Th-228 (1978)
Autoposting of broader terms began in 1989. This term has multiple hierarchies.
BT1 radioactive isotopes
BT1 isotopes
BT2 thorium
BT2 actinides
BT2 metals

Th-230 (1978)
Autoposting of broader terms began in 1989. Used for ionium as of 1985. Before 1985, also search ionium. This term has multiple hierarchies.
UF ionium
BT1 radioactive isotopes
BT1 isotopes
BT2 thorium
BT2 actinides
BT2 metals
SA Io/Th
SA Io/U
SA Th-232/Th-230
SA U-234/Th-230
SA U-238/Th-230

Th-230/Th-232
use Th-232/Th-230

Th-230/U-234
use U-234/Th-230

Th-230/U-238
use U-238/Th-230

Th-232 (1978)
Autoposting of broader terms began in 1989. This term has multiple hierarchies.
BT1 radioactive isotopes
BT1 isotopes
BT2 thorium
BT2 actinides
BT2 metals
SA Th-232/Th-230

Th-232/Th-230 (1978)
Autoposting of broader terms began in 1989. This term has multiple hierarchies.
UF Th-230/Th-232
BT1 radioactive isotopes
BT1 isotopes
BT2 thorium
BT2 actinides
BT2 metals
SA isotope ratios
SA Th-230
SA Th-232

Th-234 (1978)
Autoposting of broader terms began in 1989. This term has multiple hierarchies.
BT1 radioactive isotopes
BT1 isotopes
BT2 thorium
BT2 actinides
BT2 metals

Th/Th (1978)
Isotopic ratio used in age determination.
SA absolute age
SA isotope ratios
SA Th/U
SA thorium

Th/U (1978)
Isotopic ratio used in age determination. Before 1971, also search thorium-uranium; uranium-thorium.
UF thorium-uranium
UF U/Th
UF uranium-thorium
SA absolute age
SA isotope ratios
SA Th/Th
SA thorium
SA U/Th/Pb
SA uranium

Thailand (1978)
CO N054500N203000
E1060000E0963000
UF Siam
BT Far East
BT Asia
NT Bangkok Thailand
NT Changwat Nakhon Phanom Thailand
NT Chao Phraya River
NT Chiang Mai Thailand
NT Khorat Plateau
NT Lampang Thailand
SA ASEAN
SA Indochina
SA Khorat Group
SA Malay Peninsula
SA Phuket Group

Thalassinoides (1978)
BT ichnofossils

thalenite (1978)
Autoposting of sorosilicates began in 1985.
BT sorosilicates
BT orthosilicates
BT silicates

thallium (1978)
Autoposting of metals to this term began in 1989.
UF Tl
BT metals

thallophytes (1978)
Has narrower terms as of 1981.
BT Plantae
NT algae
NT bacteria
NT fungi
NT lichens

NT prokaryotes
  SA bryophytes
  SA Pteropsida
thalwegs (1989)
  SA channels
  SA fluvial currents
  SA ground water
  SA stream transport
  SA streams
Thamama Group (1989)
  Arabian Peninsula.
  BT Lower Cretaceous
  BT Cretaceous
  BT Mesozoic
  SA Oman
  SA Qatar
  SA Saudi Arabia
  SA United Arab Emirates
Thames Estuary (1978)
  The broad tidal mouth of the Thames River in SE England. Also search Thames AND England. May also occur in other locations.
  SA England
  SA Essex England
  SA Isle of Sheppey
  SA Kent England
  SA London England
Thames River (1978)
  Flows from the E slope of the Cotswold Hills E to the North Sea in SE England. Also in SE Connecticut which flows into Long Island Sound. Also in SE Ontario which flows into Lake Saint Clair. Also search Thames. Search in combination with state or country.
  IN Index regions as applicable.
  UF The Thames
  SA Connecticut
  SA England
  SA Essex England
  SA Isle of Sheppey
  SA Kent England
  SA Lake Saint Clair
  SA London England
  SA New London County Connecticut
  SA Ontario
thanatocenoses (1978)
  UF death assemblages
  UF taphocenoses
  UF taphocoenosis
  UF thanatocoenosis
  SA biocenoses
  SA paleontology
  SA taphonomy

thanatocoenosis
  use thanatocenoses
Thanetian (1978)
  Europe. Above Montian, below Ypresian (Eocene).
  BT upper Paleocene
  BT Paleocene
  BT Paleogene
  BT Tertiary
  BT Cenozoic
Tharsis (1978)
  As of 1978, term is restricted to morphologic features on Mars. Before 1978 term was also used for village in Spain.
  BT Mars
  SA Arsia Mons
  SA Ascraeus Mons
  SA Tharsis Montes
Tharsis Montes (1993)
  In Tharsis region includes Ascraeus Mons, Pavonis Mons, and Arsia Mons.

  BT Mars
  SA Arsia Mons
  SA Ascraeus Mons
  SA Tharsis
thaumasite (1978)
  From 1978-1984, carbonates, orthosilicates and sulfates were autoposted to this term. In 1985, autoposting of orthosilicates ended, while autoposting of nesosilicates began. As of 1989, orthosilicates and silicates are autoposted to this term. This term has multiple hierarchies.
  BT1 carbonates
  BT2 nesosilicates
  BT2 orthosilicates
  BT2 silicates
  BT3 sulfates
thawing (1978)
  SA freezing
  SA frost action
  SA glacial geology
  SA ice
  SA permafrost
  SA snow
  SA taliks
Thaynes Formation (1978)
  Consists of limestone, calcareous sandstone, sandstone, shale, and in the middle, red shale member. In SE Idaho, SW Montana, NE Utah, and SW Wyoming.
  BT Lower Triassic
  BT Triassic
  BT Mesozoic
  SA Beaverhead Formation
  SA Idaho
  SA Montana
  SA Utah
  SA Wyoming

The Banks
  use Outer Banks

The Coorong
  use Coorong Lagoon
The Geysers (1978)
  Eight miles NE of Geyserville in Sonoma County, W California.
  UF Geysers, The
  BT California
  BT United States
  SA geothermal fields
  SA Sonoma County California

The Great Oasis
  use Kharga Oasis

The Himalaya
  use Himalayas

The Rand
  use Witwatersrand

The Spessart
  use Spessart

The Thames
  use Thames River
The Weald (1978)
  Wooded region in Kent, Surrey, and Sussex counties of SE England.
  CO N504500N513000 E0013000W0010000
  UF Weald
  UF Weald region
  BT England
  BT Great Britain
  BT United Kingdom
  BT Western Europe
  BT Europe
  SA Kent England
  SA Surrey England
  SA Sussex England
Thebes Formation (1989)
  E Egypt. Overlies Esna Shale; underlies Minia Formation.
  BT lower Eocene
  BT Eocene
  BT Paleogene
  BT Tertiary
  BT Cenozoic
  SA Egypt
  SA Esna Shale
Thecamoeba (1978)
  This term has multiple hierarchies.
  BT1 Protista
  BT1 Invertebrata
  BT2 Protista
  BT2 microfossils
Thecideidae (1978)
  BT Thecideidina
  BT Articulata
  BT Brachiopoda
  BT Invertebrata
Thecideidina (1978)
  Autoposting of this term began in 1978.
  BT Articulata
  BT Brachiopoda
  BT Invertebrata
  NT Thecideidae
Thecodontia (1981)
  Order.
  BT Archosauria
  BT Diapsida
  BT Reptilia
  BT Tetrapoda
  BT Vertebrata
  BT Chordata

Theiss River
  use Tisza River
Thellier Method (1989)
  SA magnetic field
  SA methods
  SA paleomagnetism
  SA remanent magnetization
  SA thermoremanent magnetization
thematic mapper (1985)
  SA airborne methods
  SA digital cartography
  SA geophysical methods
  SA geophysical surveys
  SA instruments
  SA Landsat
  SA mosaics
  SA multispectral analysis
  SA multispectral scanner
  SA remote sensing
  SA satellite methods
thenardite (1978)
  BT sulfates
theoretical models (1985)
  Used for discussion of a model which is not a mathematical model or a physical model, but which exists in the form of a drawing, diagram, chart or description.
  UF conceptual models
  SA models
theoretical seismograms (1981)
  BT seismograms
theoretical studies (1978)
  UF studies, theoretical
  SA experimental studies
  SA laboratory studies

theory
  A valid term through mid-1978. Use theoretical studies.
Thera (1978)
  Volcanic island. The southernmost of the Cyclades in the Aegean Sea. Also search Santorin, Santorini or Thira. This term has multiple hierarchies.
  UF Santorini
  UF Thira
  BT1 Cyclades
  BT1 Greek Aegean Islands
  BT1 Greece
  BT1 Southern Europe
  BT1 Europe
  BT2 Cyclades
  BT2 Greek Aegean Islands
  BT2 Aegean Islands
  BT2 Mediterranean region
  SA Santorin
Therapsida (1978)
  Order. Autoposting of this term to Lystrosaurus began in 1981.
  BT Synapsida
  BT Reptilia
  BT Tetrapoda
  BT Vertebrata
  BT Chordata
  NT Cynodontia
  NT Lystrosaurus
Theria (1978)
  Subclass. Autoposting of this term to Pantotheria and Symmetrodonta ended in 1980 and began again in 1989. Autoposting of this term to Eutheria and Marsupialia began in 1989.
  BT Mammalia
  BT Tetrapoda
  BT Vertebrata
  BT Chordata
  NT Eutheria
  NT Metatheria
  NT Pantotheria
  NT Symmetrodonta

thermal
  A valid term through 1977.
thermal alteration (1978)
  Before 1978, also search thermal AND alteration.
  SA alteration
  SA hydrothermal alteration
  SA thermal effects
thermal analysis (1978)
  Used for methodology and not for data. For data, use thermal analysis data. Before 1978, thermogravimetric analysis and differential thermal analysis were used as level 1 terms.
  IN Index also the specific type of analysis if possible [e.g. differential thermal analysis, thermogravimetric analysis, thermomagnetic analysis].
  SA analysis
  SA chemical analysis
  SA clay mineralogy
  SA differential thermal analysis
  SA DTA data
  SA electron microscopy
  SA endothermic reactions
  SA interlaboratory comparison
  SA sample preparation
  SA spectroscopy
  SA standard materials
  SA TGA data
  SA thermal analysis data
  SA thermogravimetric analysis
  SA thermomagnetic analysis
  SA X-ray analysis
thermal analysis data (1985)
  Used for data and not for methodology.
  UF thermal data

SA thermal analysis
thermal anomalies (1989)
Used as a general term. Use more specific terms (e.g. hot spots) if applicable.
UF temperature anomaly
SA anomalies
SA geothermal fields
SA heat flow
SA hot spots
SA hot springs

thermal aureole
use aureoles

thermal circulation (1978)
Before 1978, also search oceans AND circulation AND thermal.
SA circulation
SA ocean circulation

thermal conductivity (1978)
Before 1976, also search conductivity.
SA conductivity
SA electrical conductivity
SA geothermal gradient
SA heat flow
SA measurement
SA specific heat
SA thermal diffusivity
SA thermal inertia

thermal data
use thermal analysis data

thermal demagnetization (1981)
BT demagnetization
SA geophysics
SA magnetization
SA paleomagnetism
SA remanent magnetization

thermal diffusivity (1978)
UF diffusivity, thermal
SA diffusivity
SA heat flow
SA thermal conductivity

thermal discharge
use thermal pollution

thermal effects (1981)
SA effects
SA thermal alteration

thermal emission (1978)
UF emission, thermal
SA infrared methods
SA remote sensing

thermal energy storage
use heat storage

thermal evolution
As of 1993, use thermal maturity for petroleum or thermal alteration or thermal history.

thermal gradient
Not a valid term for GeoRef. See geothermal gradient if applicable.

thermal history (1978)
Before 1993, also search thermal evolution.
SA geosynclines
SA paleotemperature
SA temperature
SA thermal maturity
SA thermal regime

thermal inertia (1985)
SA heat flow
SA specific heat
SA thermal conductivity
SA thermal properties
SA thermodynamic properties

thermal maturation
use thermal maturity

thermal maturity (1985)
Before 1993, also search thermal evolution and petroleum.
UF thermal maturation
SA biomarkers
SA maturity
SA petroleum
SA thermal history

thermal metamorphism (1978)
Not a valid index term through 1977. Before 1978, also search metamorphism AND thermal.
UF thermometamorphism
BT metamorphism

thermal methods
Not a valid GeoRef index term after 1971. Was used on level 1 in subfiles G and N. See heat flow or thermal analysis.

thermal pollution (1978)
Before 1978, also search pollution AND thermal; thermal discharge.
UF thermal discharge
SA pollution
SA waste disposal

thermal properties (1978)
SA engineering geology
SA heat capacity
SA permafrost
SA physical properties
SA properties
SA specific heat
SA temperature
SA thermal inertia
SA thermochemical properties
SA thermodynamic properties
SA thermoelastic properties
SA thermomechanical properties

thermal recovery (1985)
UF heat treatment
SA enhanced recovery
SA recovery
SA tertiary recovery

thermal regime (1981)
SA engineering geology
SA geothermal systems
SA heat flow
SA hydrogeology
SA thermal history

thermal remanent magnetization
use thermoremanent magnetization

thermal springs
use hot springs

thermal surveys
use heat flow

thermal waters (1978)
SA brines
SA fractures
SA fumaroles
SA geothermal energy
SA geothermal fields
SA geothermal gradient
SA geothermal systems
SA geysers
SA ground water
SA heat flow
SA heat sources
SA hot springs
SA hydrochemistry
SA hydrogeology
SA hydrothermal alteration
SA medicinal waters
SA migration of elements
SA mineral waters
SA permeability
SA phreatomagmatism
SA springs
SA temperature
SA trace elements
SA wall-rock alteration
SA water
SA water resources

thermochemical properties (1978)
SA calorimetry
SA chemical properties
SA geochemistry
SA properties
SA thermal properties

thermochemical remanent magnetization (1981)
Autoposting of magnetization to this term began in 1989.
BT remanent magnetization
BT magnetization
SA paleomagnetism

thermocline (1989)
SA geothermal gradient
SA lakes
SA ocean circulation
SA temperature

thermodynamic properties (1978)
SA enthalpy
SA entropy
SA free energy
SA fugacity
SA geochemistry
SA isotherms
SA Kirchhoff integral
SA properties
SA specific heat
SA thermal inertia
SA thermal properties
SA thermomechanical properties

thermodynamics
As of 1981, no longer a valid term for GeoRef. See geochemistry; thermodynamic properties.

thermoelastic properties (1989)
SA elastic properties
SA mechanical properties
SA physical properties
SA properties
SA thermal properties

thermography
use differential thermal analysis

thermogravimetric analysis (1978)
A valid level 1 term through 1977 used for methods.
UF thermogravimetry
SA analysis
SA chemical analysis
SA differential thermal analysis
SA TGA data
SA thermal analysis

thermogravimetry
use thermogravimetric analysis

thermohaline circulation (1978)
Before 1978, also search oceans AND circulation AND thermohaline.
SA circulation
SA ocean circulation

thermokarst (1978)
UF cryokarst
SA frost action
SA geomorphology
SA glacial geology
SA ground ice
SA karst
SA periglacial features
SA permafrost
SA pseudokarst
SA suffosion

thermoluminescence (1978)
SA fluorescence
SA geochronology
SA irradiation
SA luminescence
SA relative age

thermomagnetic analysis (1978)
SA analysis
SA thermal analysis

thermomechanical properties (1985)
UF thermophysical properties
SA mechanical properties
SA physical properties
SA thermal properties
SA thermodynamic properties

thermometamorphism
use thermal metamorphism

thermometry
A valid term through mid-1978. Use geologic thermometry.

thermonatrite (1978)
BT carbonates

thermophysical properties
use thermomechanical properties

thermoremanence
use thermoremanent magnetization

thermoremanent magnetization (1978)
Autoposting of magnetization to this term began in 1989.
UF thermal remanent magnetization
UF thermoremanence
UF TRM
BT remanent magnetization
BT magnetization
SA Curie point
SA inverse thermoremanent magnetization
SA magnetic field
SA paleomagnetism
SA partial thermoremanent magnetization
SA Thellier Method

Theropoda (1978)
Suborder. This term has multiple hierarchies.
BT1 Saurischia
BT1 Archosauria
BT1 Diapsida
BT1 Reptilia
BT1 Tetrapoda
BT1 Vertebrata
BT1 Chordata
BT2 Saurischia
BT2 dinosaurs
BT2 Tetrapoda
BT2 Vertebrata
BT2 Chordata
SA Smilodon

theses (1982)
UF dissertations
SA education
SA monographs
SA publications

Thessaloniki
use Salonika Greece

Thessaly
No longer a valid term for GeoRef. As of 1993 see Thessaly Greece.

Thessaly Greece (1993)
Administrative region in E central Greece. Before 1993 also search Thessaly AND Greece.
CO N390000N401200
   E0231500E0210700
BT Greece
BT Southern Europe

BT Europe
NT Olympus
SA Othrys
SA Pindus Mountains

Thetford Mines (1978)
City and site of asbestos mines about 50 miles S of Quebec City in S Quebec. Also search Thetford AND Quebec.
BT Quebec
BT Eastern Canada
BT Canada
SA asbestos deposits
SA mines

thickness (1978)
SA crust
SA crustal thinning

thin sections (1978)
SA cross sections
SA electron microscopy
SA microfacies
SA microscope methods
SA petrography
SA polished sections
SA samples

thin-skin tectonics
use thin-skinned tectonics

thin-skinned tectonics (1989)
UF thin-skin tectonics
BT tectonics
SA crust
SA crustal thinning
SA tectonophysics

Thiobacillus ferrooxidans (1985)
Species.
BT bacteria
BT thallophytes
BT Plantae

Thira
use Thera

Third World (1981)
UF developing countries
SA economics
SA industrialized countries

thixotropy (1982)
SA clay
SA colloidal materials
SA engineering geology
SA gels
SA mechanical properties
SA properties
SA soil mechanics
SA strength

tholeiite (1978)
BT basalts
BT volcanic rocks
BT igneous rocks
SA ocean-island basalts
SA olivine tholeiite
SA tholeiitic composition

tholeiitic basalt (1978)
BT basalts
BT volcanic rocks
BT igneous rocks

tholeiitic composition (1978)
Before 1978, search tholeiitic.
SA basalts
SA composition
SA igneous rocks
SA tholeiite

tholeiitic dolerite (1978)
BT diabase
BT plutonic rocks
BT igneous rocks

Thomar
use Tomar Portugal

Thomas Range (1981)

W Utah.
IN Index counties as applicable.
CO N394000N400000 W1125000W1131500
BT Utah
BT United States
SA Juab County Utah
SA Tooele County Utah

Thompson nickel belt (1985)
Metallogenic province, N Manitoba.
CO N550000N560000 W0973000W0983000
BT Manitoba
BT Western Canada
BT Canada

thomsenolite (1978)
BT fluorides
BT halides

Thomson Formation (1989)
NE Minnesota and NW Wisconsin. Includes the slate which has been referred to in literature as Thomson, Saint Louis, Cloquet, and Carlton slates. And which has usually been considered an equivalent of the Animikie-Virginia Slate of the Mesabi Range. In NE Minnesota and NW Wisconsin. Before 1993, also search Thomson Slate.
UF Thomson Slate
BT lower Proterozoic
BT Proterozoic
BT upper Precambrian
BT Precambrian
SA Minnesota
SA Virginia Formation
SA Wisconsin

Thomson Slate
As of 1993, no longer a valid term for GeoRef.
use Thomson Formation

Thomson-Haskell analysis (1981)
SA analysis
SA seismology

thomsonite (1978)
From 1978-1980, carbonates and framework silicates were autoposted to this term. Autoposting of zeolite group and silicates began in 1981.
BT zeolite group
BT framework silicates
BT silicates

thorianite (1978)
BT oxides

thorite (1978)
Autoposting of nesosilicates began in 1985.
BT nesosilicates
BT orthosilicates
BT silicates

thorium (1978)
Chemical element. As of 1981, use thorium ores for thorium as a commodity. Autoposting of actinides and metals to this term began in 1989.
UF Th
BT actinides
BT metals
NT Th-228
NT Th-230
NT Th-232
NT Th-232/Th-230
NT Th-234
NT U-234/Th-230
NT U-238/Th-230
SA Io/Th

SA Io/U
SA isotopes
SA Pb/Th
SA Th/Th
SA Th/U
SA thorium ores
SA U/Th/Pb

thorium ores (1981)
Before 1981, also search (thorium OR Th) AND (deposit OR deposits OR ore OR ores OR economic) in the basic index. Autoposting of metal ores to this term began in 1985.
BT metal ores
SA monazite deposits
SA thorium

thorium-lead
use Pb/Th

thorium-uranium
Not a valid term for GeoRef. Before 1971, included use in subfiles E, N, T and B.
use Th/U

Thorn
use Torun Poland

thoron
use Rn-222

thortveitite (1978)
Autoposting of sorosilicates began in 1985.
BT sorosilicates
BT orthosilicates
BT silicates

Thrace (1978)
Administrative region in extreme NE Greece (Western Thrace), and all of Turkey in Europe (Eastern Thrace).
IN Index counties as applicable.
BT Europe
NT Greek Thrace
SA Greece
SA Turkey

Three Forks
Valid through 1988. Search in combination with state term. After 1988, use specific city-state term.

Three Forks Montana (1989)
Town in SW Montana. Before 1989, search Three Forks AND Montana.
CO N455400N455400 W1113200W1113200
BT Gallatin County Montana
BT Montana
BT United States

three-component seismographs (1981)
BT seismographs

three-dimensional methods (1993)
Also search three-dimensional with methods or techniques or surveys. From 1978 through 1992, also search three-dimensional surveys if applicable.
UF three-dimensional surveys
SA boreholes
SA geophysical methods
SA geophysical surveys
SA seismic methods
SA seismic surveys
SA surveys
SA techniques

three-dimensional models (1978)
Before 1978, search three-dimensional AND models.

SA asperities
SA four-dimensional models
SA mathematical models
SA models
SA one-dimensional models
SA seismology
SA two-dimensional models

three-dimensional surveys
As of 1993, no longer a valid term for GeoRef.
use three-dimensional methods

thrombolites (1993)
BT biogenic structures
BT sedimentary structures
SA algal structures
SA stromatolites

thrust
A valid term through 1977. After 1977, see thrust faults. After 1988, also see thrust sheets.

thrust faults (1978)
Before 1978, also search faults AND thrust. Before 1971, also search thrusts and thrusting. See overthrust faults; compression tectonics.
UF reverse slip faults
UF thrust slip faults
UF thrusts and thrusting
BT faults
SA compression tectonics
SA crustal shortening
SA duplexes
SA fold and thrust belts
SA foreland basins
SA imbricate tectonics
SA overthrust faults
SA ramps
SA reverse faults
SA step faults
SA thrust sheets

thrust plates
use thrust sheets

thrust sheets (1989)
UF thrust plates
SA faults
SA fold and thrust belts
SA neotectonics
SA tectonics
SA thrust faults

thrust slip faults
use thrust faults

thrusts and thrusting
No longer a valid GeoRef index term. Before 1971, was used on level 1 in subfile N.
use thrust faults

Thule
No longer a valid term for GeoRef. As of 1993 see Thule Greenland.

Thule Greenland (1993)
Settlement on coast of Hayes Peninsula N of Cape York in NW Greenland. U.S. Air Force base nearby. Before 1993 also search Thule AND Greenland.
BT Greenland
BT Arctic region

thulium (1978)
Autoposting of rare earths and metals to this term began in 1989.
UF Tm
BT rare earths
BT metals

Thunder Bay (1978)
Refers only to the inlet of NW Lake Superior. For the city on the bay, see Thunder Bay Ontario.

Also search Thunder AND Ontario.
BT Ontario
BT Eastern Canada
BT Canada
SA Lake Superior
SA Thunder Bay District Ontario

Thunder Bay District
No longer a valid term for GeoRef. As of 1993, see Thunder Bay District Ontario.

Thunder Bay District Ontario (1993)
N Ontario. Before 1993, also search Thunder Bay District AND Ontario.
CO N474000N520000 W0852000W0910000
UF District of Thunder Bay
BT Ontario
BT Eastern Canada
BT Canada
NT Manitouwadge Ontario
NT Marshall Lake
NT Northern Light Lake
NT Thunder Bay Ontario
SA Michipicoten Island
SA Thunder Bay

Thunder Bay Ontario (1993)
Refers only to the city on the bay. For the inlet of NW Lake Superior, see Thunder Bay. Before 1993, also search Thunder or Thunder Bay in conjunction with Ontario.
BT Thunder Bay District Ontario
BT Ontario
BT Eastern Canada
BT Canada

thunder eggs
No longer a valid GeoRef index term. Before 1971, was used on levels 1 and 2 in subfile N.
use geodes

thunderstorms
As of 1993, no longer a valid term for GeoRef. Usually out-of-scope. Used with meteorology. After 1993, see storms.

Thur Valley (1978)
River valley SSW of Lake Constance extending to the Rhine River. Also search Thur River.
BT Switzerland
BT Central Europe
BT Europe

Thuringer Wald
use Thuringian Forest

Thuringia
No longer a valid term for GeoRef. See Thuringia Germany.

Thuringia Basin
use Thuringian Basin

Thuringia Germany (1993)
State in central Germany. Includes the Thuringian Forest. Before 1993, also search Erfurt Bezirk and Gera Bezirk if applicable.
CO N531500N550000 E0112500E0081000
BT Germany
BT Central Europe
BT Europe
NT Erfurt Germany
NT Gera Germany
NT Ruhla Germany
NT Schlotheim Germany
NT Schneeberg Germany
NT Steinach Germany
NT Taubach Germany
NT Weimar Germany
SA Geisel Valley
SA Kyffhauser Range
SA Rhon Mountains
SA Saale River
SA Saxony-Thuringia
SA Schwarzburg Anticlinorium
SA Thuringian Basin
SA Thuringian Forest
SA Thuringian Hills
SA Thuringian Massif
SA Unstrut River
SA Vogtland
SA Weisse Elster Basin
SA Werra River

Thuringian (1981)
BT Permian
BT Paleozoic

Thuringian Basin (1978)
Thuringian Forest region in Thuringia.
IN Index states as applicable.
UF Thuringia Basin
BT Germany
BT Central Europe
BT Europe
SA Thuringia Germany

Thuringian Forest (1978)
Forested mountain range extending NW-SE in Thuringia.
UF Thuringer Wald
BT Germany
BT Central Europe
BT Europe
SA Franconian Forest
SA Ruhla Germany
SA Steinach Germany
SA Thuringia Germany

Thuringian Hills (1981)
Region in Germany.
IN Index states as applicable.
CO N504000N513700 E0123000E0095500
BT Germany
BT Central Europe
BT Europe
SA Thuringia Germany

Thuringian Massif (1981)
Region in Germany.
IN Index states as applicable.
CO N500700N510200 E0130500E0101300
BT Germany
BT Central Europe
BT Europe
SA Thuringia Germany

Thurston County
Valid through 1988. Search in combination with state term. After 1988, use specific county-state term.

Thurston County Nebraska (1989)
NE Nebraska. Before 1989, also search Thurston County AND Nebraska.
CO N420200N421700 W0961600W0965000
BT Nebraska
BT United States

Thurston County Washington (1989)
W Washington. Before 1989, also search Thurston County AND Washington.
CO N464600N471100 W1221500W1231200
BT Washington
BT United States
NT Tono Washington

Thysanopteroida (1981)
BT Insecta
BT Mandibulata
BT Arthropoda
BT Invertebrata

Ti
use titanium

Tian Shan
use Tien Shan

Tian-Shan
use Tien Shan

Tianjin
No longer a valid term for GeoRef. See Tianjin China.

Tianjin China (1993)
Municipality located within Hebei, N China.
CO N383000N401000 E1180000E1164000
BT Hebei China
BT China
BT Far East
BT Asia

Tiber Valley (1978)
River valley. Also search Tiber.
IN Index autonomous regions as applicable.
BT Italy
BT Southern Europe
BT Europe
SA Latium Italy
SA Tuscany Italy
SA Umbria Italy

Tiberias-Dead Sea Rift valley
use Jordan Valley

Tibesti
No longer a valid term for GeoRef. As of 1993 see Tibesti Libya.

Tibesti Libya (1993)
Habitation in S central Libya.
BT Libya
BT North Africa
BT Africa

Tibesti Massif (1978)
Highest mountain group of the Sahara in NW Chad.
UF Tibesti Mountains
BT Chad
BT West Africa
BT Africa
SA Libya
SA Niger

Tibesti Mountains
use Tibesti Massif

Tibet
No longer a valid term for GeoRef. See Xizang China.

Tibet China
use Xizang China

Tibetan Plateau
No longer a valid term for GeoRef. Valid 1985-1992.
use Qinghai-Xizang Plateau

Ticino (1978)
A river in Italy and Switzerland. For the canton use Ticino Switzerland.
BT Europe
SA Italy
SA Switzerland
SA Ticino Switzerland

Ticino Switzerland (1993)
Canton in Lepontine Alps in S Switzerland. Before 1993 also search Ticino AND Switzerland.
UF Tessin
BT Switzerland
BT Central Europe
BT Europe
NT San Giorgio Mountain
SA Lago Maggiore
SA Lepontine Alps
SA Ticino

Ticonderoga
Valid through 1988. Search in combination with state term. After 1988, use specific city-state term.

Ticonderoga New York (1989)
Village on N outlet of Lake George in NE New York. Before 1989, search Ticonderoga AND New York.
CO N435100N435100 W0732600W0732600
BT Essex County New York
BT New York
BT United States

tidal channels (1978)
Autoposting of shore features to this term began in 1989.
BT shore features
SA channels
SA geomorphology
SA tidal flats
SA tidal inlets

tidal currents (1993)
SA longshore currents
SA marine transport
SA nearshore environment
SA ocean currents
SA oceanography
SA tides

tidal energy (1981)
SA energy
SA energy sources
SA new energy sources

tidal environment
As of 1981, no longer a valid term for GeoRef. Use intertidal environment, intertidal sedimentation, estuarine environment, estuarine sedimentation, tidal channels, tidal energy, tidal flats, tidal inlets, or tides as applicable.

tidal flats (1978)
Autoposting of shore features to this term began in 1989.
BT shore features
SA brackish-water environment
SA coastal environment
SA geomorphology
SA marine environment
SA mud flats
SA sedimentation
SA supratidal environment
SA tidal channels
SA tides

tidal glaciers
use tidewater glaciers

tidal inlets (1978)
Autoposting of shore features to this term began in 1989.
UF tidal outlets
BT shore features
SA geomorphology
SA inlets
SA shorelines
SA tidal channels

tidal outlets
use tidal inlets

tidal surges (1993)
Before 1993, also search surges AND tides.
SA surges
SA tides

tidal wave
   use tsunamis

tides (1978)
   SA coastlines
   SA Earth tides
   SA littoral erosion
   SA low water
   SA ocean circulation
   SA surges
   SA tidal currents
   SA tidal flats
   SA tidal surges

tidewater glaciers (1993)
   UF tidal glaciers
   BT glaciers
   SA fjords
   SA glacial geology
   SA hydrology
   SA icebergs

Tieh-Chen Shan
   use Tiehchenshan

Tiehchenshan (1978)
   Hill region in NW Taiwan.
   UF Tieh-Chen Shan
   BT Taiwan
   BT Far East
   BT Asia

Tien Shan (1978)
   Mountain region in Kyrgyzstan and Xinjiang, NW China. From 1981-1992, restricted to portion in Kirghizia.
   IN Index Kyrgyzstan and/or Xinjiang China as applicable.
   CO N400000N445000
      E0960000E0680000
   UF Tian Shan
   UF Tian-Shan
   UF Tien Shan Range
   UF Tien-Shan
   UF Tien-Shan Range
   BT Asia
   NT Alai Range
   NT Chatkal Range
   NT Chu-Ili Mountains
   NT Karatau Range
   NT Talas Range
   SA Central Asia
   SA Kurama Range
   SA Kyrgyzstan
   SA Tarbagatay Range
   SA Xinjiang China
   SA Zeravshan Range
   SA Zeravshan-Hissar

Tien Shan Range
   use Tien Shan

Tien-Shan
   use Tien Shan

Tien-Shan Range
   use Tien Shan

Tientsin
   Not a valid term for GeoRef. See Tianjin China.

Tierra Amarilla
   Valid through 1988. Search in combination with state term. After 1988, use specific city-state term.

Tierra Amarilla New Mexico (1989)
   Village in N New Mexico. Before 1989, search Tierra Amarilla AND New Mexico.
   CO N364200N364200
      W1063400W1063400
   BT Rio Arriba County New Mexico
   BT New Mexico
   BT United States

Tierra del Fuego (1978)
   Archipelago comprising all islands S of Strait of Magellan to Drake Passage. Also name of main island.
   IN Index countries as applicable.
   BT South America
   SA Argentina
   SA Chile
   SA Magallanes Chile

Tieschitz Meteorite (1989)
   Impact at Prerov, Moravia, Czechoslovakia. Olivine-bronzite chondrite.
   BT H chondrites
   BT chondrites
   BT stony meteorites
   BT meteorites
   SA Czechoslovakia
   SA Moravia

Tiffanian (1985)
   North America. Lower Paleocene, above Torrejonian and below Clarkforkian.
   BT lower Paleocene
   BT Paleocene
   BT Paleogene
   BT Tertiary
   BT Cenozoic

Tiflis Georgian Republic
   use Tbilisi Georgian Republic

tight sands (1989)
   SA energy sources
   SA natural gas
   SA oil and gas fields
   SA oil sands
   SA organic materials
   SA permeability
   SA petroleum
   SA reservoir properties
   SA reservoir rocks
   SA rocks
   SA source rocks

Tigre
   No longer a valid term for GeoRef. See Tigre Ethiopia.

Tigre Ethiopia (1993)
   Province in N Ethiopia.
   BT Ethiopia
   BT East Africa
   BT Africa

Tigris River (1985)
   Source in E Turkey. Joins Euphrates River at Basra, Iraq.
   IN Index countries as applicable.
   BT Middle East
   BT Asia
   SA Iraq
   SA Turkey

Tijeras Canyon (1985)
   Central New Mexico. Autoposting of broader terms began in 1989.
   CO N345500N351500
      W1061500W1063000
   BT Bernalillo County New Mexico
   BT New Mexico
   BT United States
   SA Basin and Range Province

Tilia (1985)
   Genus.
   BT Dicotyledoneae
   BT angiosperms
   BT Spermatophyta
   BT Plantae

till (1978)
   Autoposting of this term began in 1978.
   BT clastic sediments
   BT sediments
   NT lodgement till
   SA boulder clay
   SA boulder trains
   SA diamicton
   SA drift
   SA drumlins
   SA glacial features
   SA glacial geology
   SA glacial transport
   SA moraines
   SA terrigenous materials
   SA tillite
   SA tilloid

tillage (1978)
   SA soil management
   SA soils
   SA yields

Tillamook County
   Valid through 1988. Search in combination with state term. After 1988, use specific county-state term.

Tillamook County Oregon (1989)
   NW Oregon. Also search Tillamook AND Oregon. Before 1989, also search Tillamook County AND Oregon.
   CO N450300N454600
      W1232000W1240000
   BT Oregon
   BT United States

tillite (1978)
   Before 1971, also search tillites.
   UF tillites
   BT clastic rocks
   BT sedimentary rocks
   SA glacial features
   SA terrigenous materials
   SA till

tillites
   Not a valid GeoRef index term after 1970. Was used on level 1 in subfile N.
   use tillite

Tillodontia (1978)
   Order. Autoposting of Eutheria and Theria to this term began in 1989.
   BT Eutheria
   BT Theria
   BT Mammalia
   BT Tetrapoda
   BT Vertebrata
   BT Chordata

tilloid (1978)
   BT clastic rocks
   BT sedimentary rocks
   SA terrigenous materials
   SA till

tilt (1978)
   As of 1981, for the tilt of a planet on its axis, use obliquity of the ecliptic.
   UF tilting
   SA geodesy
   SA obliquity of the ecliptic
   SA seismology
   SA tiltmeters
   SA volcanic earthquakes
   SA volcanology

tilting
   use tilt

tiltmeters (1978)
   SA earthquakes
   SA instruments
   SA seismology
   SA tilt
   SA volcanology

Timagami Mine
   use Temagami Mine

Timan Ridge (1978)
   An eroded range of ancient sedimentary rocks in N European Russian Federation dividing the eastern Pechora Basin from the plains to the W. Autoposting of Russian Republic to this term began in 1989. Also search Timan. This term has multiple hierarchies.
   CO N620000N673000
      E0550000E0473000
   BT1 Commonwealth of Independent States
   BT2 Russian Federation
   BT2 Commonwealth of Independent States
   BT3 Russian Platform
   BT4 Europe

Timan-Pechora region (1978)
   Includes the Timan Ridge area; and the Pechora Basin in Komi Autonomous Republic and Nenets National Okrug. In N and NE European Russian Federation. Before 1978, also search Timan-Pechora. This term has multiple hierarchies.
   BT1 Russian Federation
   BT1 Commonwealth of Independent States
   BT2 Europe
   SA Komi Russian Federation
   SA Nenets Russian Federation

Timbered Hills Group (1989)
   S Oklahoma.
   BT Upper Cambrian
   BT Cambrian
   BT Paleozoic
   SA Oklahoma
   SA Reagan Sandstone

time
   As of 1981, no longer a valid term for GeoRef. Use time factor.

time domain analysis (1981)
   Also search time domain.
   SA analysis
   SA frequency domain analysis
   SA geophysical methods
   SA Hilbert transformations
   SA stacking

time factor (1981)
   Before 1981, search time.
   SA factors
   SA pedogenesis
   SA soils

time scales (1978)
   UF scales, time
   SA geochronology
   SA scale factor

time series analysis (1983)
   BT statistical analysis
   SA analysis
   SA mathematical geology

time variations (1978)
   As of 1982, restricted to seismology.
   SA seismology
   SA variations

Timiskaming County
   use Temiscamingue County Quebec

Timiskaming District
   No longer a valid term for GeoRef. As of 1993, see Timiskaming District Ontario.

Timiskaming District Ontario (1993)

E central Ontario, on Quebec border. Before 1993, also search Timiskaming District AND Ontario.
CO N470000N483000
  W0793000W0820000
UF District of Temiscamingue
UF District of Temiscaming
UF District of Timiskaming
UF Temescamingue District
UF Temiskaming District
BT Ontario
BT Eastern Canada
BT Canada
NT Cobalt Ontario
NT Kirkland Lake Ontario
NT New Liskeard Ontario
SA Lake Timiskaming
SA Larder Lake District Ontario
SA Temiscamingue County Quebec

Timiskaming Group (1989)
Abitibi Belt of Superior Province.
BT Archean
BT Precambrian
SA Abitibi Belt
SA Ontario
SA Quebec
SA Superior Province

Timiskaming Lake
use Lake Timiskaming

Timmins
No longer a valid term for GeoRef. As of 1993, see Timmins Ontario.

Timmins Ontario (1993)
Mining town 135 miles N of Sudbury between Lake Huron and James Bay. Before 1993, also search Timmins AND Ontario.
BT Cochrane District Ontario
BT Ontario
BT Eastern Canada
BT Canada

Timna (1978)
Ancient site in Asaylan area in E Arabian Peninsula.
BT Arabian Peninsula
BT Asia

Timok Basin (1978)
River and coal basin near the Bulgarian and Romanian borders in E Serbia. Also search Timok; Timok River.
UF Timok River basin
BT Serbia
BT Yugoslavia
BT Southern Europe
BT Europe
SA coal fields

Timok River basin
use Timok Basin

Timor (1978)
Island in the Lesser Sundas separated from Australia by the Timor Sea. W half of island belongs to Indonesia. The E half was given up by Portugal in 1975 and subsequently occupied by Indonesia. Before 1979, also search Portuguese Timor. As of 1981, Lesser Sunda Islands and Malay Archipelago are broader terms. This term has multiple hierarchies.
IN Index country as applicable.
UF Portuguese Timor
BT1 Lesser Sunda Islands
BT1 Far East
BT1 Asia
BT2 Malay Archipelago
SA Banda Arc
SA Indonesia
SA Portugal

SA Sunda Arc

Timor Sea (1978)
Between the island of Timor and NW Australia. Before 1981, broader term was Indian Ocean. Autoposting of Pacific Ocean to this term began in 1990.
BT North Australian Seas
BT West Pacific
BT Pacific Ocean
NT Bonaparte Gulf basin
NT Sahul Shelf
NT Timor Trough
SA Browse Basin
SA Indian Ocean
SA Joseph Bonaparte Gulf

Timor Trench
use Timor Trough

Timor Trough (1978)
Undersea feature off the S coast of Timor.
UF Timor Trench
BT Timor Sea
BT North Australian Seas
BT West Pacific
BT Pacific Ocean
SA Banda Arc
SA DSDP Site 262

tin (1978)
Chemical element. As of 1981, use tin ores for tin as a commodity. Autoposting of metals to this term began in 1989.
UF Sn
BT metals
SA niggliite
SA tin ores

tin ores (1981)
Before 1981, also search (tin OR Sn) AND (deposit OR deposits OR ore OR ores OR economic) in the basic index. Autoposting of metal ores to this term began in 1985.
BT metal ores
SA cassiterite
SA Dachang Deposit
SA placers
SA tin

tin pyrites
use stannite

Tinaquillo
No longer a valid term for GeoRef. As of 1993 see Tinaquillo Venezuela.

Tinaquillo Venezuela (1993)
Town in N Cojedes, NW Venezuela. Before 1993 also search Tinaquillo AND Venezuela.
BT Cojedes Venezuela
BT Venezuela
BT South America

Tindouf
No longer a valid term for GeoRef. As of 1993 see Tindouf Algeria.

Tindouf Algeria (1993)
Village and oasis in extreme W Algeria.
BT Algeria
BT North Africa
BT Africa

Tindouf Basin (1978)
Salt flats in extreme W Algeria. Also search Tindouf.
UF Sebkha de Tindouf
BT Algeria
BT North Africa
BT Africa

tinguaite (1978)

BT phonolites
BT volcanic rocks
BT igneous rocks

Tintic District
No longer a valid term for GeoRef.
use Tintic mining district

Tintic mining district (1993)
Area around East Tintic Creek in Juab County in W central Utah. Also search Tintic or Tintic District AND Utah.
UF Tintic District
BT Utah
BT United States
SA Juab County Utah
SA metal ores

Tintic Quartzite (1978)
Lower and Middle Cambrian. Underlies Ophir Formation and unconformably overlies crystalline complex. Central and N Utah.
BT Cambrian
BT Paleozoic
SA Utah

Tintina Fault (1989)
Before 1989, also search Tintina fault zone.
IN Index Alaska and/or Yukon Territory as applicable.
UF Tintina fault zone
UF Tintina Trench
BT North America
SA Alaska
SA Yukon Territory
SA Yukon-Tanana Terrane

Tintina fault zone
As of 1989, no longer a valid term for GeoRef. Used in subfiles B, N, and G.
use Tintina Fault

Tintina Trench
use Tintina Fault

Tintinnidae (1978)
Family. Including Calpionellidae. Before 1981, also search tintinnids. As of 1981, tintinnids may be used as a common name for Tintinnidae. Autoposting of this term began in 1978. This term has multiple hierarchies.
BT1 Protista
BT1 Invertebrata
BT2 Protista
BT2 microfossils
NT Calpionella
NT Calpionellidae
NT Calpionellites
NT Stomiosphaera

Tioga Bentonite (1978)
According to the latest published Lexique, Tioga Bentonite Bed is in Seneca Member of Onondaga Limestone. It is subsurface in Pennsylvania.
BT Middle Devonian
BT Devonian
BT Paleozoic
SA New York
SA Ohio
SA Pennsylvania
SA Virginia
SA West Virginia

Tioga County
Valid through 1988. Search in combination with state term. After 1988, use specific county-state term.

Tioga County New York (1989)

S New York. Before 1989, also search Tioga County AND New York.
CO N420000N422400
  W0760600W0763500
BT New York
BT United States

Tioga County Pennsylvania (1989)
N Pennsylvania. Before 1989, also search Tioga County AND Pennsylvania.
CO N413300N420000
  W0765200W0773700
BT Pennsylvania
BT United States

Tippecanoe County
Valid through 1988. Search in combination with state term. After 1988, use specific county-state term.

Tippecanoe County Indiana (1989)
W central Indiana. Before 1989, also search Tippecanoe County AND Indiana.
CO N401300N403400
  W0864300W0870500
BT Indiana
BT United States

Tipperary
No longer a valid term for GeoRef. See Tipperary Ireland.

Tipperary Ireland (1993)
County in S Ireland.
BT Ireland
BT Western Europe
BT Europe
NT Cashel Ireland

Tiraspol
No longer a valid term for GeoRef. See Tiraspol Moldova.

Tiraspol Moldova (1993)
City on the Dniester River in SE Moldova. This term has multiple hierarchies.
BT1 Moldova
BT1 Europe
BT2 Moldova
BT2 Commonwealth of Independent States
SA Moldavia

tirodite (1978)
BT clinoamphibole
BT amphibole group
BT chain silicates
BT silicates

Tirol
Not a valid term for GeoRef. See Tyrol Austria.

Tirol Austria
use Tyrol Austria

Tiruchirapalli
No longer a valid term for GeoRef. See Tiruchirapalli India.

Tiruchirapalli India (1993)
City 200 miles SSW of Madras in Tamil Nadu. Also search Trichinopoly.
UF Trichinopoly India
BT Tamil Nadu India
BT India
BT Indian Peninsula
BT Asia

Tirupati
No longer a valid term for GeoRef. See Tirupati India.

Tirupati India (1993)

City in S Andhra Pradesh, central India.
BT Andhra Pradesh India
BT India
BT Indian Peninsula
BT Asia

Tisa River
use Tisza River

Tishomingo
Valid through 1988. Search in combination with state term. After 1988, use specific city-state term.

Tishomingo County Mississippi (1989)
County in extreme NE Mississippi. Before 1989, also search Tishomingo County AND Mississippi.
CO N342800N350000
  W0880700W0882200
BT Mississippi
BT United States
NT Tishomingo Mississippi

Tishomingo Mississippi (1989)
Village in NE Mississippi. Before 1989, search Tishomingo AND Mississippi.
CO N343700N343700
  W0881500W0881500
BT Tishomingo County Mississippi
BT Mississippi
BT United States

Tismana
No longer a valid term for GeoRef. As of 1993 see Tismana Romania.

Tismana Romania (1993)
Village in S foothills of Transylvanian Alps in NW. Before 1993 also search Tismana AND Romania.
BT Walachia
BT Romania
BT Southern Europe
BT Europe

Tisza River (1978)
Rises in Carpathians in W Ukraine and flows W, SW and then S, into the Danube River N of Belgrade. Also search Tisza.
IN Index countries as applicable.
UF Theiss River
UF Tisa River
BT Europe
SA Hungary
SA Romania
SA Ukraine
SA Vojvodina

Titan Satellite (1989)
One of the satellites of Saturn. Before 1989, also search Titan AND Saturn.
SA icy satellites
SA satellites
SA Saturn
SA Voyager Program

titanaugite (1978)
BT clinopyroxene
BT pyroxene group
BT chain silicates
BT silicates
SA augite

Titania Satellite (1989)
One of the satellites of Uranus. Before 1989, also search Titania AND Uranus.
SA icy satellites
SA satellites
SA Uranus

SA Voyager Program

titanite (1978)
Autoposting of nesosilicates began in 1985. Before 1981, also search sphene.
UF sphene
BT nesosilicates
BT orthosilicates
BT silicates
SA heavy minerals

titanium (1978)
Chemical element. As of 1981, use titanium ores for titanium as a commodity. Autoposting of metals to this term began in 1989.
UF Ti
BT metals
SA titanium ores

titanium ores (1981)
Before 1981, also search (titanium OR Ti) AND (deposit OR deposits OR ore OR ores OR economic) in the basic index. Autoposting of metal ores to this term began in 1985.
BT metal ores
SA ilmenite
SA leucoxene
SA titanium
SA titanomagnetite

titanoclinohumite (1978)
BT humite group
BT nesosilicates
BT orthosilicates
BT silicates
SA clinohumite

titanomaghemite (1978)
BT oxides
SA maghemite

titanomagnetite (1978)
BT oxides
SA iron ores
SA magnetic minerals
SA magnetite
SA titanium ores

Tithonian (1978)
Europe.
BT Portlandian
BT Upper Jurassic
BT Jurassic
BT Mesozoic

titration (1985)
UF volumetric analysis
SA chemical analysis
SA wet methods

tjaele
use frozen ground

Tkibuli
No longer a valid term for GeoRef. As of 1993, see Tkibuli Georgian Republic.

Tkibuli Georgian Republic (1993)
City in W Georgian Republic. Before 1993, also search Tkvibuli.
UF Tkvibuli Georgian Republic
BT Georgian Republic
BT Europe

Tkvibuli Georgian Republic
use Tkibuli Georgian Republic

Tl
use thallium

TM
As of 1993, not a valid term for GeoRef. See thematic mapper if applicable.

Tm
use thulium

Toarcian (1978)
Europe.
BT upper Liassic
BT Lower Jurassic
BT Jurassic
BT Mesozoic

Toba Lake (1978)
In Barisan Mountains in N central Sumatra. Thought to occupy the crater of an extinct volcano. Also search Toba.
BT Sumatra
BT Indonesia
BT Far East
BT Asia

Tobacco Root Mountains (1978)
Range of Rocky Mountains in SW Montana. As of 1993, U. S. Rocky Mountains is a broader term. This term has multiple hierarchies.
BT1 Madison County Montana
BT1 Montana
BT1 United States
BT2 U. S. Rocky Mountains
BT2 Rocky Mountains
BT3 U. S. Rocky Mountains
BT3 United States

Tobago (1978)
Small island NE of Trinidad. A constituent part of Trinidad and Tobago.
BT Trinidad and Tobago
BT Lesser Antilles
BT Antilles
BT West Indies
BT Caribbean region

tobermorite (1978)
BT chain silicates
BT silicates

Tocantins River region (1978)
N central and N Brazil. Also search Tocantins.
IN Index states as applicable.
BT Brazil
BT South America
SA Goias Brazil
SA Maranhao Brazil
SA Para Brazil

Tochigi
No longer a valid term for GeoRef. See Tochigi Japan.

Tochigi Japan (1993)
Prefecture including town in E central Honshu.
BT Honshu
BT Japan
BT Far East
BT Asia
NT Shiobara Japan
SA Ashio Japan
SA Ashio Mine
SA Kanto Plain
SA Nasudake

tochilinite (1993)
BT sulfides
SA iron sulfides
SA valleriite

Todilto Formation (1978)
In San Rafael Group.
BT Upper Jurassic
BT Jurassic
BT Mesozoic
SA Arizona
SA Colorado
SA New Mexico
SA San Rafael Group

todorokite (1978)
BT oxides
SA buserite
SA manganese minerals

Todos Santos Bay (1978)
Just S and SW of Ensenada. Also search Todos Santos.
BT Baja California Mexico
BT Mexico
SA Baja California

Tofino Basin (1978)
SW Vancouver Island. Also search Tofino.
BT British Columbia
BT Western Canada
BT Canada
SA Vancouver Island

Tofua (1978)
Volcanic island. The largest of Haapai group in central Tonga.
BT Tonga
BT Polynesia
BT Oceania

Togo (1978)
Formerly French Togo.
CO N061000N111000
  E0015000W0001000
BT West Africa
BT Africa
SA Ghana
SA Volta Basin

Tohoku (1978)
Literally Northeast. Region, comprising N Honshu, which represents a transitional zone between cold Hokkaido and temperate and subtropical areas farther S. Includes Akita, Fukushima, Aomori, Iwate, Miyagi, and Yamagata prefectures.
IN Index prefectures as applicable.
UF Northeast Japan
BT Honshu
BT Japan
BT Far East
BT Asia
SA Akita Japan
SA Aomori Japan
SA Fukushima Japan
SA Iwate Japan
SA Miyagi Japan
SA Yamagata Japan

Toiyabe Range (1989)
Nye and Lander counties, central Nevada. Extends N-S along E bank of Reese River.
IN Index counties as applicable.
CO N384500N393500
  W1165400W1173000
BT Nevada
BT United States
SA Lander County Nevada
SA Nye County Nevada

Tokachi (1978)
Volcanic peak in central Hokkaido. After 1992, use Tokachi Japan for the sub-prefecture of Hokkaido.
BT Hokkaido
BT Japan
BT Far East
BT Asia
SA Hidaka Mountains

Tokachi earthquake 1968
use Tokachi-Oki earthquake 1968

Tokachi Plain (1978)
Tokachi sub-prefecture, S central Hokkaido.
BT Hokkaido
BT Japan
BT Far East
BT Asia

Tokachi-Oki earthquake 1968 (1993)

Submarine earthquake off Hokkaido, Japan.
UF Tokachi earthquake 1968
BT earthquakes
SA Hokkaido
SA Japan
SA Japan Sea
SA Oki Islands

Tokai
No longer a valid term for GeoRef. See Tokai Japan.

Tokai Japan (1993)
District in E central Honshu.
BT Ibaraki Japan
BT Honshu
BT Japan
BT Far East
BT Asia
SA Chubu Japan
SA Izu Peninsula
SA Shizuoka Japan
SA Suruga Bay

Tokaj
No longer a valid term for GeoRef. See Tokaj Hungary.

Tokaj Hungary (1993)
Town in NE Hungary. Also search Tokaj or Tokay in combination with Hungary.
UF Tokay Hungary
BT Hungary
BT Central Europe
BT Europe

Tokaj Mountains (1978)
In NE Hungary. Hungarian section of Tokaj-Eperjes Mountains. Also search Tokaj. This term has multiple hierarchies.
BT1 Hungary
BT1 Central Europe
BT1 Europe
BT2 Tokaj-Eperjes Mountains
BT2 Europe

Tokaj-Eperjes Mountains (1978)
Outlier of the Carpathians.
IN Index Hungary and/or Slovakia as applicable.
CO N480600N484000
   E0215000E0210600
BT Europe
NT Tokaj Mountains
SA Carpathians
SA Hungary
SA Slovakia

Tokay Hungary
use Tokaj Hungary

Tokelau (1981)
Island group N of Western Samoa.
UF Union Islands
BT Polynesia
BT Oceania

Toki
No longer a valid term for GeoRef. See Toki Japan.

Toki Japan (1993)
City NE of Nagoya in Gifu Prefecture, central Honshu.
BT Gifu Japan
BT Honshu
BT Japan
BT Far East
BT Asia

Tokio
Not a valid term for GeoRef. See Tokyo Japan.

Tokrau Synclinorium (1978)
N of Lake Balkhash. Also search Tokrau. This term has multiple hierarchies.
BT1 Kazakhstan
BT1 Central Asia
BT1 Asia
BT2 Kazakhstan
BT2 Commonwealth of Independent States

Tokushima
No longer a valid term for GeoRef. As of 1993, see Tokushima Japan.

Tokushima Japan (1993)
Prefecture and city on E coast of Shikoku, S Japan.
BT Shikoku
BT Japan
BT Far East
BT Asia
SA Yoshino River

Tokyo
No longer a valid term for GeoRef. See Tokyo Japan.

Tokyo Bay (1981)
SE Honshu.
CO N344000N354200
   E1401000E1393000
BT Honshu
BT Japan
BT Far East
BT Asia
SA Chiba Japan
SA Kanagawa Japan
SA Tokyo Japan

Tokyo Imperial University
use University of Tokyo

Tokyo Japan (1993)
Prefecture including city on Tokyo Bay in SE Honshu. Before 1993, also search Edo, Medo, Tokio, Yeddo and Tokyo AND Honshu.
BT Honshu
BT Japan
BT Far East
BT Asia
NT Musashino Japan
SA Kanto Plain
SA Tamagawa
SA Tokyo Bay

Tolbachik (1981)
Volcano on Kamchatka Peninsula. This term has multiple hierarchies.
CO N554902N554902
   E1602202E1602202
BT1 Kamchatka Peninsula
BT1 Russian Pacific region
BT1 Russian Federation
BT1 Commonwealth of Independent States
BT2 Kamchatka Peninsula
BT2 Russian Pacific region
BT2 Asia

Toledo
Valid through 1988. Search in combination with state term. After 1988, use specific city-state term.

Toledo City
No longer a valid term for GeoRef. As of 1993, see Toledo City Spain.

Toledo City Spain (1993)
Refers only to the city in central Toledo Province, central Spain. Before 1993, also search Toledo City and Spain. Before 1981, also search Toledo and Spain.
BT Toledo Spain
BT New Castile Spain
BT Castile Spain
BT Spain
BT Iberian Peninsula
BT Southern Europe
BT Europe

Toledo Ohio (1989)
City on Maumee River in NW Ohio. Before 1989, search Toledo AND Ohio.
CO N414000N414000
   W0833500W0833500
BT Lucas County Ohio
BT Ohio
BT United States

Toledo Province
No longer a valid term for GeoRef. As of 1993, see Toledo Spain.

Toledo Spain (1993)
Refers only to the province in central Spain. From 1981-1992, also search Toledo Province and Spain. Before 1981, also search Toledo and Spain. For the city, see Toledo City Spain.
CO N391600N401800
   W0030000W0052500
BT New Castile Spain
BT Castile Spain
BT Spain
BT Iberian Peninsula
BT Southern Europe
BT Europe
NT Toledo City Spain
SA Montes de Toledo

Tolfa Hills (1978)
ENE of Civitavechia in NW Latium. Also search Tolfa.
BT Latium Italy
BT Italy
BT Southern Europe
BT Europe

Tolland County
Valid through 1988. Search in combination with state term. After 1988, use specific county-state term.

Tolland County Connecticut (1989)
NE Connecticut. Before 1989, also search Tolland County AND Connecticut.
CO N413500N420300
   W0720500W0723200
BT Connecticut
BT United States
NT Columbia Connecticut
SA Roaring Brook Valley

Toluca Meteorite (1981)
Impact near Xiquipilco in Toluca area W of Mexico City. Before 1981, also search Toluca AND meteorites.
BT octahedrite
BT iron meteorites
BT meteorites
SA Mexico

toluene (1989)
BT aromatic hydrocarbons
BT hydrocarbons
BT organic materials

Tomales Bay (1978)
Inlet of Pacific Ocean on NW coast of Marin County N of San Francisco.
BT Marin County California
BT California
BT United States

Tomar
No longer a valid term for GeoRef. As of 1993 see Tomar Portugal.

Tomar Portugal (1993)
City in W central Portugal. Before 1993 also search Tomar AND Portugal.
UF Thomar
BT Portugal
BT Iberian Peninsula
BT Southern Europe
BT Europe

Tomaszow Lubelski
No longer a valid term for GeoRef. As of 1993 see Tomaszow Lubelski Poland.

Tomaszow Lubelski Poland (1993)
Town in SE Zamosc, E Poland. Before 1993 also search Tomaszow Lubelski AND Poland.
BT Poland
BT Central Europe
BT Europe

Tomaszow Mazowiecki
No longer a valid term for GeoRef. As of 1993 see Tomaszow Mazowiecki Poland.

Tomaszow Mazowiecki Poland (1993)
Town SE of Lodz in NE Piotrkow, central Poland. Before 1993 also search Tomaszow Mazowiecki AND Poland.
BT Poland
BT Central Europe
BT Europe

Tomaszow-Lvov Ridge
use Roztocze

Tombigbee River (1985)
Source in NE Mississippi. Joins Alabama River in SW Alabama, then redivides into Mobile and Tensaw rivers, which flow into Mobile Bay.
IN Index states as applicable.
BT United States
SA Alabama
SA Gulf Coastal Plain
SA Mississippi
SA Mobile Bay

tombolos (1993)
BT shore features
SA barrier islands
SA bars
SA breakwaters
SA coastlines
SA sedimentation
SA spits

Tombstone
Valid through 1988. Search in combination with state term. After 1988, use specific city-state term.

Tombstone Arizona (1989)
City in SE Arizona. Before 1989, search Tombstone AND Arizona.
CO N314400N314400
   W1100400W1100400
BT Cochise County Arizona
BT Arizona
BT United States

Tommotian (1985)
USSR.
BT Lower Cambrian
BT Cambrian
BT Paleozoic

tomography (1985)
UF geotomography
SA crosshole methods
SA geophysical methods
SA Radon transforms
SA seismic methods

SA seismology
SA well-logging
SA X-ray analysis

Tompkins County New York (1989)
W central New York. Before 1989, search Tompkins County AND New York.
CO N421500N423800
    W0761500W0764200
BT New York
BT United States
NT Ithaca New York
SA Cayuga Lake

Tomsk
No longer a valid term for GeoRef. As of 1993, see Tomsk Russian Federation.

Tomsk Russian Federation (1993)
Oblast including the city on right bank of Tom River near its junction with Ob River in S central Siberia. Before 1978, also search Tomsk. This term has multiple hierarchies.
BT1 Russian Federation
BT1 Commonwealth of Independent States
BT2 Asia
NT Narym Russian Federation
NT Sredniy Vasyugan Russian Federation
SA Kiya River
SA Kuznetsk Basin

tonalite (1978)
BT diorites
BT plutonic rocks
BT igneous rocks
SA quartz diorites

tonalite gneiss (1978)
BT gneisses
BT metamorphic rocks

Tonga (1978)
An archipelago of about 150 islands NE of New Zealand. Formerly a British protectorate which became independent in 1970.
CO S230000S150000
    W1710000W1770000
UF Friendly Islands
UF Tonga Arc
UF Tonga Archipelago
UF Tonga island arc
UF Tonga Islands
BT Polynesia
BT Oceania
NT Eua Island
NT Tofua
SA Ata Caldera
SA Pacific Ocean
SA Tonga Trench

Tonga Arc
use Tonga

Tonga Archipelago
use Tonga

Tonga island arc
use Tonga

Tonga Islands
use Tonga

Tonga Trench (1978)
E of the Tongtapu group of the S Tonga Islands.
BT Pacific Ocean
SA DSDP Site 595
SA DSDP Site 596
SA Leg 91
SA Tonga

Tonganoxie Sandstone (1978)
Of Stranger Formation. Underlies Westphalia Limestone Member. In E Kansas and NW Missouri.
UF Tunganoxie Sandstone Member
BT Virgilian
BT Upper Pennsylvanian
BT Pennsylvanian
BT Carboniferous
BT Paleozoic
SA Kansas
SA Missouri

Tongariro (1993)
Volcano in Tongariro National Park, central North Island, New Zealand.
IN Also index North Island.
CO S390800S390800
    E1754200E1754200
BT New Zealand
BT Australasia
SA North Island
SA Taupo volcanic zone

Tonghai earthquake 1970 (1989)
Epicenter near Tonghai, Yunnan, SW China.
UF T'ung-hai earthquake 1970
BT earthquakes
SA China
SA Yunnan China

Tongrian (1978)
Lattorfian. Europe. Above Ludian (Eocene), below Rupelian. Also search Lattorfian.
BT upper Eocene
BT Eocene
BT Paleogene
BT Tertiary
BT Cenozoic
SA Krosno Beds
SA Lattorfian
SA Sannoisian

Tongue of the Ocean (1978)
Strait between Andros Island on the W and New Providence Island and Exuma Cays on the E.
BT Bahamas
BT West Indies
BT Caribbean region

Tongue River (1985)
Tributary of the Yellowstone River, N central Wyoming and SE Montana. Also a tributary of the Pembina River, NE North Dakota.
IN Index counties and states as applicable.
SA Montana
SA North Dakota
SA Wyoming

Tongue River Member (1978)
In Fort Union Formation. In Montana and North Dakota, composed of light-yellow, tan, and gray sandstones and shales; thin lenses of limestone; and numerous beds of lignite. Found in E Montana, SW North Dakota, South Dakota and NW Wyoming.
BT Paleocene
BT Paleogene
BT Tertiary
BT Cenozoic
SA Fort Union Formation
SA Montana
SA North Dakota
SA South Dakota
SA Wyoming

Tonkin Gulf
use Gulf of Tonkin

Tonnerre
No longer a valid term for GeoRef. see Tonnerre France.

Tonnerre France (1993)
Town in E Yonne, central France.
BT Yonne France
BT France
BT Western Europe
BT Europe

Tono (1978)
Village in N Honshu. Before 1989, also used for village in the state of Washington.
BT Honshu
BT Japan
BT Far East
BT Asia

Tono Washington (1989)
Village in SW Washington. Before 1989, search Tono AND Washington.
CO N464600N464600
    W1224800W1224800
BT Thurston County Washington
BT Washington
BT United States

Tonoloway Limestone (1978)
Cayugan. Underlies Keyser Limestone.
BT Cayugan
BT Upper Silurian
BT Silurian
BT Paleozoic
SA Maryland
SA Pennsylvania
SA Virginia
SA West Virginia

Tonopah
Valid through 1988. Search in combination with state term. After 1988, use specific city-state term.

Tonopah Nevada (1989)
Village in S central Nevada. Before 1989, search Tonopah AND Nevada.
CO N380500N380500
    W1171500W1171500
BT Nye County Nevada
BT Nevada
BT United States

Tons Member (1978)
According to the India Lexique, the Tons Series is of Cambrian age. The Tons Series and Son Series are united to make up the Ken Subsystem, or lower main division of the Vindhyan System. In central India.
BT Cambrian
BT Paleozoic
SA India

tonstein (1978)
BT clastic rocks
BT sedimentary rocks
SA terrigenous materials

Tonto Basin (1978)
Valley in N Gila County in central Arizona.
BT Gila County Arizona
BT Arizona
BT United States

Tooele County
Valid through 1988. Search in combination with state term. After 1988, use specific county-state term.

Tooele County Utah (1989)
NW Utah. Before 1989, also search Tooele County AND Utah.
CO N395500N410500
    W1121000W1140500
BT Utah
BT United States
NT Bonneville Salt Flats
NT Stansbury Mountains
SA Great Salt Lake
SA Thomas Range

tool marks (1978)
BT bedding plane irregularities
BT sedimentary structures
SA current markings
SA soft sediment deformation

Toole County
Valid through 1988. Search in combination with state term. After 1988, use specific county-state term.

Toole County Montana (1989)
N Montana. Before 1989, also search Toole County AND Montana.
CO N481500N490000
    W1112000W1121500
BT Montana
BT United States

Toolebuc Formation (1989)
E Australia.
BT Cretaceous
BT Mesozoic
SA Australia
SA New South Wales Australia
SA Northern Territory Australia
SA Queensland Australia
SA South Australia

tools
As of 1993, use artifacts with archaeology if applicable. See instruments for other uses.

topaz (1978)
From 1978-1980, orthosilicates and silicates were autoposted to this term. Autoposting of fluorides began in 1981. In 1985, autoposting of orthosilicates ended, while autoposting of nesosilicates began. As of 1985, halides, orthosilicates and silicates are autoposted to this term. This term has multiple hierarchies.
BT1 fluorides
BT1 halides
BT2 nesosilicates
BT2 orthosilicates
BT2 silicates
SA gems

Topeka
Valid through 1988. Search in combination with state term. After 1988, use specific city-state term.

Topeka Kansas (1989)
City on Kansas River in NE Kansas. Before 1989, search Topeka AND Kansas.
CO N390200N390200
    W0954100W0954100
BT Shawnee County Kansas
BT Kansas
BT United States

tophus
use tufa

Topley Intrusions (1978)
Central British Columbia.
BT British Columbia
BT Western Canada
BT Canada

topographic correction (1982)
SA cartography
SA corrections
SA geophysical methods
SA topographic maps

SA topography

topographic maps (1978)
Before 1978, also search maps AND topographic.
BT maps
SA base maps
SA contour maps
SA topographic correction
SA topography

topography (1978)
SA altimetry
SA cartography
SA geodesy
SA geomorphology
SA hills
SA karst
SA ocean floors
SA physiographic provinces
SA plains
SA plateaus
SA pseudokarst
SA relief
SA soils
SA terrains
SA topographic correction
SA topographic maps

topology (1985)
SA geometry
SA mathematical geology

Topopah Spring Member (1989)
Of Paintbrush Tuff. S Nevada.
BT Miocene
BT Neogene
BT Tertiary
BT Cenozoic
SA Nevada
SA Paintbrush Tuff

Toquima Range (1985)
Central Nevada. Autoposting of broader terms to this term began in 1989.
IN Index counties as applicable.
BT Nevada
BT United States
SA Basin and Range Province
SA Lander County Nevada
SA Mount Jefferson
SA Nye County Nevada

Tor Formation (1978)
Devonian. Overlies Antelope Valley Limestone. In central Nevada. Also an Upper Cretaceous formation in the North Sea oilfields.
IN Index age and region as applicable.
SA Cretaceous
SA Devonian
SA Nevada
SA North Sea
SA Upper Cretaceous

torbanite (1978)
This term has multiple hierarchies.
UF bituminite
BT1 organic residues
BT2 sedimentary rocks
SA oil shale
SA shale

torbernite (1989)
BT phosphates

Torino Italy
 use Turin Italy

Torlesse Supergroup (1978)
BT Upper Triassic
BT Triassic
BT Mesozoic
SA New Zealand

Torngat Mountains (1989)
N Labrador.
CO N580000N602000
 W0623000W0650000
BT Labrador
BT Newfoundland
BT Eastern Canada
BT Canada

toroidal mode (1981)
SA oscillations
SA seismographs
SA seismology

Torok Formation (1985)
N Alaska.
BT Lower Cretaceous
BT Cretaceous
BT Mesozoic
SA Alaska

Toronto
No longer a valid term for GeoRef. As of 1993, see Toronto Ontario.

Toronto Ontario (1993)
City at NW end of Lake Ontario. Before 1993, also search Toronto AND Ontario.
CO N434000N434000
 W0792800W0792800
BT Ontario
BT Eastern Canada
BT Canada
SA Gibraltar Point

Toroweap Formation (1978)
In Aubrey Group. In Arizona can be subdivided into 3 main units (ascending): red to buff sandstone; calcareous sandstone and arenaceous limestone; and alternating red and buff sandstone, siltstone, and some shale. In NW Arizona, SE Nevada, and SW Utah.
BT Permian
BT Paleozoic
SA Arizona
SA Nevada
SA Utah

Torquay
No longer a valid term for GeoRef as of 1993. See Torquay England.

Torquay England (1993)
Resort city on English Channel in SW England. Before 1993, also search Torquay AND England.
BT Devonshire England
BT England
BT Great Britain
BT United Kingdom
BT Western Europe
BT Europe

Torrance County
Valid through 1988. Search in combination with state term. After 1988, use specific county-state term.

Torrance County New Mexico (1989)
Central New Mexico. Before 1989, also search Torrance County AND New Mexico.
CO N341400N350300
 W1051800W1063000
BT New Mexico
BT United States
SA Manzano Mountains

Torrejonian (1985)
North America. Lower Paleocene, above Dragonian and below Tiffanian.
BT lower Paleocene
BT Paleocene
BT Paleogene
BT Tertiary
BT Cenozoic

Torres Strait (1978)
Between the island of New Guinea and the N tip of Cape York Peninsula, Queensland. Connects the Arafura Sea with the Coral Sea. Also search Torres.
BT Pacific Ocean

Torridonian (1978)
Europe. Autoposting of upper Precambrian and Precambrian to this term ended in 1989. As of 1989, upper Proterozoic and Proterozoic are autoposted to this term. Autoposting of upper Precambrian and Precambrian to this term began in 1990.
BT upper Proterozoic
BT Proterozoic
BT upper Precambrian
BT Precambrian

tors (1978)
SA erosion features
SA geomorphology
SA glacial geology
SA hills
SA periglacial features
SA solution features

torsion (1978)
SA deformation
SA extension
SA extension tectonics
SA stress
SA tension

torsion faults
 use wrench faults

torsion modulus
 use shear modulus

Tortonian (1978)
Europe.
BT upper Miocene
BT Miocene
BT Neogene
BT Tertiary
BT Cenozoic
SA Helvetian

tortuosity (1978)
Property of materials.
SA geometry
SA properties

Torun
No longer a valid term for GeoRef. As of 1993 see Torun Poland.

Torun Poland (1993)
Province and city in N central Poland. Before 1993 also search Torun AND Poland.
UF Thorn
BT Poland
BT Central Europe
BT Europe
NT Grudziadz Poland

Tosa Bay (1978)
Inlet of the Pacific Ocean on S coast. Also search Tosa.
BT Shikoku
BT Japan
BT Far East
BT Asia
SA Kochi Japan

tosudite (1978)
BT clay minerals
BT sheet silicates
BT silicates

total field methods
 use total-field methods

total rock
 use whole rock

total-field methods (1985)
UF total field methods
SA airborne methods
SA high-resolution methods
SA magnetic methods
SA magnetic surveys
SA methods
SA remote sensing

Totes Gebirge (1978)
Mountain range in W central Austria.
IN Index states as applicable.
BT Austria
BT Central Europe
BT Europe
SA Styria Austria
SA Upper Austria

Tottori
No longer a valid term for GeoRef. See Tottori Japan.

Tottori Japan (1993)
City and prefecture on Japan Sea NW of Kyoto in S Honshu.
BT Honshu
BT Japan
BT Far East
BT Asia
SA Chugoku
SA Dogo
SA Naka-no-umi

Toulon
No longer a valid term for GeoRef. See Toulon France.

Toulon France (1993)
City on the Mediterranean Sea in S Var, SE France.
BT Var France
BT France
BT Western Europe
BT Europe

Toulouse
No longer a valid term for GeoRef. See Toulouse France.

Toulouse France (1993)
City on the Garonne River in N Haute-Garonne, S France.
BT Haute-Garonne France
BT France
BT Western Europe
BT Europe

Toumodi
No longer a valid term for GeoRef. See Toumodi Ivory Coast.

Toumodi Ivory Coast (1993)
Village NW of Abidjan in S central Ivory Coast.
BT Ivory Coast
BT West Africa
BT Africa

Touraine (1978)
Historical region of NW central France. The area now comprises Indre-et-Loire Department and part of Indre Department.
IN Index departments as applicable.
BT France
BT Western Europe
BT Europe
SA Indre France
SA Indre-et-Loire France

tourmaline (1978)
BT ring silicates
BT silicates
SA buergerite
SA dravite
SA elbaite
SA gems
SA heavy minerals
SA peridot

SA schorl
SA tourmalinization

tourmaline schist
use tourmalinite

tourmalinite   (1993)
UF tourmaline schist
BT schists
BT metamorphic rocks
SA schistosity

tourmalinization   (1985)
BT metasomatism
SA alteration
SA hydrothermal alteration
SA tourmaline

Tournai
No longer a valid term for GeoRef. See Tournai Belgium.

Tournai Belgium   (1993)
City on Scheldt River near the French border in W Hainaut, SW Belgium. Also search Tournay or Tournai or Doornik AND Belgium.
UF Doornik Belgium
UF Tournay Belgium
BT Hainaut Belgium
BT Belgium
BT Western Europe
BT Europe

Tournaisian   (1978)
Europe. Autoposting of this term began in 1978.
BT Dinantian
BT Carboniferous
BT Paleozoic
NT upper Tournaisian

Tournay Belgium
use Tournai Belgium

Tours
No longer a valid term for GeoRef. See Tours France.

Tours France   (1993)
City on the Loire River in central Indre-et-Loire, NW central France.
BT Indre-et-Loire France
BT France
BT Western Europe
BT Europe

Toutle River   (1985)
Tributary of the Cowlitz River, SW Washington. Autoposting of broader terms began in 1989.
IN Index counties as applicable.
BT Washington
BT United States
SA Cowlitz County Washington
SA Cowlitz River
SA Skamania County Washington

Towada   (1978)
Volcano in Towada-Hachimantai National Park in N Honshu.
UF Towada Volcano
BT Honshu
BT Japan
BT Far East
BT Asia

Towada Volcano
use Towada

Towns County
Valid through 1988. Search in combination with state term. After 1988, use specific county-state term.

Towns County Georgia   (1989)
NE Georgia. Before 1989, also search Towns County AND Georgia.
CO N344700N345800
W0833300W0835700

BT Georgia
BT United States
SA Lake Chatuge

Townsville
No longer a valid term for GeoRef. See Townsville Australia.

Townsville Australia   (1993)
City on Halifax Bay on the Coral Sea in E Queensland.
BT Queensland Australia
BT Australia
BT Australasia

toxic materials   (1978)
Before 1978, also search toxic AND materials. Before 1984, also search toxic substances. Toxic substances was used primarily under soils.
SA cyanides
SA environmental geology
SA hazardous waste
SA materials
SA PCBs
SA pollution
SA toxicity
SA waste disposal

toxic waste
Not a valid term for GeoRef. See hazardous waste; radioactive waste; toxic materials; waste disposal.

toxicity   (1978)
SA environmental geology
SA toxic materials
SA waste disposal

Toyama
No longer a valid term for GeoRef. See Toyama Japan.

Toyama Japan   (1993)
Prefecture and city on S shore of Toyama Bay on Japan Sea in central Honshu. Also search Toyama AND Japan.
BT Honshu
BT Japan
BT Far East
BT Asia
SA Chubu Japan
SA Hida Mountains
SA Japanese Alps
SA Noto Peninsula

Toyoha Mine   (1993)
BT Hokkaido
BT Japan
BT Far East
BT Asia
SA lead-zinc deposits
SA mines
SA polymetallic ores
SA silver ores

Toyoma
No longer a valid term for GeoRef.

Toyoma Japan   (1993)
Town SE of Taira on the Pacific Ocean in central Honshu.
BT Honshu
BT Japan
BT Far East
BT Asia

Toyoura Sand   (1989)
SA Japan

trace elements   (1978)
From 1985 to 1988, this term was autoposted to rare earths.
UF guest element
UF microelement
UF trace-elements
SA chemical elements

SA major elements
SA minor elements
SA nutrients
SA rare earths
SA thermal waters
SA trace metals
SA trace-element analyses

trace fossils
A valid term through 1972. After 1972, use ichnofossils. However, as of 1981, a distinction is made between ichnofossils and lebensspuren. Use lebensspuren when the emphasis is sedimentary petrology; use ichnofossils for paleontologic studies.

trace metals   (1978)
IN May be used with commodities (list C).
SA environmental geology
SA geochemistry
SA metals
SA pollution
SA trace elements

trace-element analyses   (1978)
For data, see appropriate material.
SA analysis
SA chemical analysis
SA chemical elements
SA major-element analyses
SA minor-element analyses
SA spectroscopy
SA trace elements
SA voltammetry

trace-elements
use trace elements

tracer experiments
As of 1981, no longer a valid term for GeoRef. Use tracers.

tracers   (1978)
Before 1981, also search tracer experiments.
SA D/H
SA dye tracers
SA fluorimetry
SA ground water
SA hydrogeology
SA hydrology
SA isotopes
SA radioactive tracers
SA radioactivity
SA surface water

tracheophyte
As of 1993, use vascular taxa AND plants.

trachyandesite
As of 1989, no longer a valid term for GeoRef. From 1981 to 1988, trachyandesites and volcanic rocks were autoposted to this term.
use trachyandesites

trachyandesites   (1981)
Before 1989, also search trachyandesite. Autoposting of this term to trachyandesite ended in 1989.
UF trachyandesite
BT volcanic rocks
BT igneous rocks
NT latite

trachybasalt
As of 1989, no longer a valid term for GeoRef. From 1981 to 1988, trachybasalts, alkali basalts, basalts and volcanic rocks were autoposted to this term.
use trachybasalts

trachybasalts   (1981)

Before 1989, also search trachybasalt. Autoposting of this term to trachybasalt ended in 1989.
UF trachybasalt
BT alkali basalts
BT basalts
BT volcanic rocks
BT igneous rocks
SA benmoreite
SA trachydolerite

trachydolerite   (1978)
BT alkali basalts
BT basalts
BT volcanic rocks
BT igneous rocks
SA trachybasalts

Trachyleberididae   (1978)
As of 1990, microfossils, Crustacea, Mandibulata, and Arthropoda are autoposted to this term. This term has multiple hierarchies.
UF Trachyleberidinae
BT1 Cytheracea
BT1 Cytherocopina
BT1 Podocopida
BT1 Ostracoda
BT1 Crustacea
BT1 Mandibulata
BT1 Arthropoda
BT1 Invertebrata
BT2 Cytheracea
BT2 Cytherocopina
BT2 Podocopida
BT2 Ostracoda
BT2 microfossils

Trachyleberidinae
use Trachyleberididae

trachyte
As of 1989, no longer a valid term for GeoRef. From 1981 to 1988, trachytes and volcanic rocks were autoposted to this term.
use trachytes

trachyte-phonolite family
As of 1981, no longer a valid term for GeoRef. Use trachytes or phonolites as applicable. From 1978-1980, was autoposted to the rocks that were classified under it.

trachytes   (1981)
Before 1981, search trachyte-phonolite family. From 1978-1980, trachyte-phonolite family was autoposted to the rocks that were classified under that term. Before 1989, also search trachyte. Autoposting of this term to trachyte ended in 1989.
UF trachyte
BT volcanic rocks
BT igneous rocks
NT keratophyre
NT quartz keratophyre

tracks   (1978)
Before 1977, also search tracks and trails. From 1978-1984, the terms biogenic structures and sedimentary structures were autoposted to this term.
SA biogenic structures
SA ichnofossils
SA lebensspuren
SA sedimentary structures
SA trails

tracks and trails
Not a valid index term for GeoRef. Use tracks or trails as separate terms.

Tracy Arm   (1989)
SE Alaska.

BT Alaska
BT United States
SA Southeastern Alaska

Tradewater Formation (1985)
W Kentucky.
BT Middle Pennsylvanian
BT Pennsylvanian
BT Carboniferous
BT Paleozoic
SA Kentucky

trails (1978)
Before 1977, also search tracks and trails. From 1978-1984, the terms biogenic structures and sedimentary structures were autoposted to this term.
SA biogenic structures
SA ichnofossils
SA lebensspuren
SA sedimentary structures
SA tracks

Tranquillity Base
A valid term through 1977.
use Taurus-Littrow

Trans Alaska Crustal Transect
use TACT

Trans-Alaska Crustal Transect
use TACT

Trans-Alaska Pipeline (1978)
Pipeline constructed for the movement of crude oil from the North Slope at Prudhoe Bay on the Arctic Ocean in the NE to the port of Valdez off Prince William Sound in the SE.
IN Index Northern Alaska, East-Central Alaska and/or Southern Alaska as applicable.
BT Alaska
BT United States
SA East-Central Alaska
SA Northern Alaska
SA Southern Alaska
SA Valdez Alaska

Trans-Aravalli Vindhyan Basin (1978)
IN Index states as applicable.
BT India
BT Indian Peninsula
BT Asia
SA Madhya Pradesh India
SA Rajasthan India

Trans-Hudsonian Orogeny
use Hudsonian Orogeny

Trans-Jordan
No longer a valid GeoRef index term. Before 1969, was used on level 1 in subfile E.
use Jordan

Trans-Pecos (1978)
Region W of the Pecos River and E of the Rio Grande in West Texas and S central New Mexico.
IN Index states as applicable.
CO N290000N320000
W1010000W1063000
BT United States
SA Marathon Geosyncline
SA New Mexico
SA Texas
SA West Texas

Transantarctic Mountains (1978)
A series of mountain ranges extending across the continent from the Filchner Ice Shelf on the Atlantic Ocean side through Victoria Land on the Pacific Ocean side separating East Antarctica from West Antarctica.
BT Antarctica

Transbaikal
use Transbaikalia

Transbaikalia (1978)
Region of SE Siberia E of Lake Baikal including Buryatia; and Chita and Amur oblasts. This term has multiple hierarchies.
CO N490000N573000
E1250000E1023000
UF Transbaikal
BT1 Russian Federation
BT1 Commonwealth of Independent States
BT2 Asia
SA Amur Russian Federation
SA Argun River
SA Baikal region
SA Buryat Russian Federation
SA Chita Russian Federation
SA Western Transbaikalia

Transcarpathia
No longer a valid term for GeoRef. Region in extreme W Ukraine which, prior to World War II, was that part of Czechoslovakia known as Ruthenia or the Carpatho-Ukraine. Now coextensive with the Transcarpathia Oblast. See Transcarpathia Ukraine.

Transcarpathia Ukraine (1993)
Autoposting of Ukraine to Transcarpathia began in 1989. This term has multiple hierarchies.
CO N480000N500000
E0230000E0220000
BT1 Ukraine
BT1 Europe
BT2 Ukraine
BT2 Commonwealth of Independent States
NT Beregovo Ukraine
NT Borkut Deposit
NT Rakhov Ukraine
SA Rakhov Massif
SA Russian Platform

Transcaucasia (1978)
Region including Armenia, Azerbaidzhan and the Georgian Republic.
IN Index former Soviet republics as applicable.
UF Transcaucasus
BT Europe
SA Armenia
SA Azerbaidzhan
SA Caucasus
SA Georgian Republic
SA Greater Caucasus
SA Kabardin-Balkar Russian Federation
SA Lesser Caucasus

Transcaucasus
use Transcaucasia

Transcontinental Arch (1993)
Prominent Middle Ordovician feature extending from the Canadian Shield, southwest across the Great Plains to NW Mexico.
IN Index states, countries, or regions as applicable.
BT North America
SA Canada
SA Great Plains
SA Mexico
SA Midcontinent
SA United States

transcurrent faults (1978)
Before 1978, also search faults AND transcurrent.
BT faults
SA displacements
SA strike-slip faults
SA transverse faults

Transdanubia (1978)
Fertile, hilly region between the Danube River and the Austrian border. Autoposting of this term began in 1978.
UF Dunantul
BT Hungary
BT Central Europe
BT Europe
NT Central Transdanubia
NT Northwestern Transdanubia
NT Southeastern Transdanubia
NT Southern Transdanubia

Transdanubian Central Mountains
use Bakony Mountains

transects
use geotraverses

transfer faults (1993)
BT faults
SA dip-slip faults
SA fault zones
SA strike-slip faults

transfer functions (1985)
SA data processing
SA functions
SA mathematical geology
SA mathematical methods
SA mathematical models

transfer, heat
use heat transfer

transfer, mass
use mass transfer

transform
A valid term through 1977. After 1977, use transform faults.

transform faults (1978)
Before 1978, also search faults AND transform.
BT faults
SA displacements
SA fracture zones
SA mid-ocean ridges
SA plate tectonics
SA Rivera Ocean Seismic Experiment
SA strike-slip faults

transformations (1978)
Before 1978, transformation was used under ocean waves.
SA changes
SA clay mineralogy
SA crystal chemistry
SA minerals
SA ocean waves
SA phase equilibria
SA phase transitions
SA soils

transformations, Laplace
use Laplace transformations

transgression (1978)
Term is used in stratigraphy for Pre-Quaternary sedimentation. For Quaternary, use changes of level.
SA changes of level
SA paleogeography
SA regression
SA stratigraphy
SA submergence

transient electromagnetic logging
Use transient methods and electromagnetic logging.

transient electromagnetic methods
Use transient methods and electromagnetic methods.

transient electromagnetic response
Use transient methods and electromagnetic methods, electromagnetic surveys or electromagnetic logging as applicable.

transient electromagnetic surveys
Use transient methods and electromagnetic surveys.

transient flow
use unsteady flow

transient methods (1981)
Includes use with electromagnetic methods, electromagnetic surveys or electromagnetic logging for transient electromagnetic response.
SA electromagnetic logging
SA electromagnetic methods
SA electromagnetic surveys
SA geophysical methods
SA transient phenomena

transient phenomena (1985)
UF phenomena, transient
SA changes
SA steady-state processes
SA transient methods

transition
A valid general term through 1977.

transition zones (1978)
UF zones, transition
SA continental crust
SA core
SA core-mantle boundary
SA crust
SA mantle
SA oceanic crust
SA phase transitions
SA plate tectonics

translation lattice
use lattice

transmissibility
use transmissivity

transmissibility coefficient
As of 1981, no longer a valid term for GeoRef. Use transmissivity.

transmission
A valid term through 1978. After 1978, use transmission method, e.g. under electron microscopy.

transmission electron microscopy (1993)
For the method. Before 1978 search transmission. Before 1993, also search transmission method AND electron microscopy.
SA chemical analysis
SA electron microscopy
SA scanning electron microscopy
SA SEM data

transmission electron microscopy data
use TEM data

transmission method
No longer a valid term for GeoRef. See transmission electron microscopy.

transmissivity (1978)
Restricted to usage in relation to hydrology. Not to be used for remote sensing.
UF coefficient of transmissibility
UF transmissibility
SA aquifers

    SA ground water
    SA hydraulic conductivity
    SA hydrogeology
    SA hydrology
    SA permeability
transport
    No longer a valid term for GeoRef. See chemical dispersion or wave dispersion, glacial transport, marine transport, ore transport, sediment transport, solution transport, stream transport, transportation or wind transport.
transportation (1978)
    As of 1978, term is restricted to non-geological meaning.
    SA pipelines
    SA sediment transport
transpression (1989)
    SA compression tectonics
    SA crustal shortening
    SA faults
    SA orogeny
    SA plate tectonics
    SA structural analysis
    SA tectonics
Transvaal
    No longer a valid term for GeoRef. As of 1993 see Transvaal South Africa.
Transvaal South Africa (1993)
    Province in NE South Africa. Before 1993 also search Transvaal AND South Africa.
    CO S281000S220500
        E0320300E0244000
    BT South Africa
    BT Southern Africa
    BT Africa
    NT Barberton Mountain Land
    NT Barberton South Africa
    NT Bushveld Complex
    NT Johannesburg South Africa
    NT Merensky Reef
    NT Pretoria South Africa
    NT Ventersdorp South Africa
    NT Witwatersrand
    SA Barberton greenstone belt
    SA Fig Tree Series
    SA Klerksdorp Field
    SA Lebombo Mountains
    SA Limpopo Basin
    SA Vaal River
    SA Waterberg System
    SA Witwatersrand Supergroup
    SA Wolkberg Group
Transvaal Supergroup (1978)
    BT Precambrian
    SA Malmani Subgroup
    SA Pretoria Group
    SA South Africa
transverse faults (1978)
    Before 1978, also search faults AND transverse.
    BT faults
    SA orientation
    SA transcurrent faults
transverse folds
    use superposed folds
Transverse Ranges (1978)
    In S California.
    IN Index counties as applicable.
    CO N333000N350000
        W1153000W1193000
    BT California
    BT United States
    SA San Bernardino Mountains
    SA San Gabriel Mountains
transverse wave
    use S-waves

Transylvania (1978)
    Region bounded by the Carpathians on the E and the Transylvanian Alps on the S, in central and NW Romania.
    BT Romania
    BT Southern Europe
    BT Europe
    NT Alba-Iulia Romania
    NT Almas Valley basin
    NT Aries Valley
    NT Baraolt Basin
    NT Baraolt Romania
    NT Beius Basin
    NT Bihor Mountains
    NT Bihor Romania
    NT Bistrita Romania
    NT Brad Romania
    NT Brasov Romania
    NT Calimani Mountains
    NT Cluj Romania
    NT Covasna Romania
    NT Crisana-Maramures
    NT Gurghiu Mountains
    NT Harghita Mountains
    NT Harghita Romania
    NT Hateg Romania
    NT Maramures Romania
    NT Muntii Metalici
    NT Mures Romania
    NT Padurea Craiului Mountains
    NT Persani Mountains
    NT Petrosani Basin
    NT Poiana-Rusca Mountains
    NT Retezat Mountains
    NT Rodna Mountains
    NT Rodna Romania
    NT Salaj Romania
    NT Satu Mare Romania
    NT Sebes Mountains
    NT Sebes Romania
    NT Sibiu Romania
    NT Simleu Basin
    NT Transylvanian Basin
    SA Crai
    SA Fagaras Mountains
    SA Getic Nappe
    SA Jiu River valley
    SA Olt River
    SA Olt River valley
    SA Transylvanian Alps
Transylvania Basin
    use Transylvanian Basin
Transylvanian Alps (1978)
    Continuation of the Carpathians extending E-W in central Romania.
    IN Index region as applicable.
    UF South Carpathians
    UF Southern Carpathians
    BT Romania
    BT Southern Europe
    BT Europe
    SA Bucegi Mountains
    SA Carpathians
    SA Fagaras Mountains
    SA Getic Nappe
    SA Paring Mountains
    SA Retezat Mountains
    SA Rodna Mountains
    SA Transylvania
    SA Vulcan Mountains
    SA Walachia
Transylvanian Basin (1978)
    Primarily a plateau.
    UF Transylvania Basin
    UF Transylvanian Plain
    BT Transylvania
    BT Romania
    BT Southern Europe
    BT Europe
Transylvanian Plain
    use Transylvanian Basin

trap rock
    No longer a valid term for GeoRef as of 1981.
    use trap rocks
trap rocks (1978)
    Before 1981, also search trap rock. For petroleum and natural gas, use traps if applicable.
    UF trap rock
    BT basalts
    BT volcanic rocks
    BT igneous rocks
    SA construction materials
    SA diabase
Trap Spring Field (1981)
    Oil field in central Nevada. Autoposting of broader terms began in 1989.
    CO N380000N385000
        W1153000W1163000
    BT Nye County Nevada
    BT Nevada
    BT United States
    SA oil and gas fields
Trapani
    No longer a valid term for GeoRef. See Tranpani Italy.
Trapani Italy (1993)
    Province in W Sicily.
    BT Sicily Italy
    BT Italy
    BT Southern Europe
    BT Europe
trapped particles
    As of 1993, no longer a valid term for GeoRef. Usually out-of-scope. Used with magnetosphere.
traps (1978)
    Autoposting of this term began in 1978.
    NT stratigraphic traps
    NT structural traps
    SA natural gas
    SA oil and gas fields
    SA petroleum
    SA petroleum accumulation
Tras-os-Montes (1978)
    Former province in NE Portugal. Now it comprises Braganca and Vila Real districts.
    IN Index districts as applicable.
    BT Portugal
    BT Iberian Peninsula
    BT Southern Europe
    BT Europe
    SA Braganca Portugal
    SA Vila Real Portugal
Travale
    use Travale Field
Travale Field (1989)
    Geothermal field in Tuscany, W Italy.
    UF Travale
    BT Tuscany Italy
    BT Italy
    BT Southern Europe
    BT Europe
    SA geothermal fields
travel time
    use traveltime
traveltime (1978)
    UF travel time
    SA dip moveout
    SA elastic waves
    SA long-period waves
    SA moveout
    SA normal moveout
    SA seismology
    SA short-period waves
    SA traveltime curves

traveltime curves (1978)
    UF curves, traveltime
    UF hodographs
    SA elastic waves
    SA seismology
    SA traveltime
traveltime residuals (1981)
    SA seismology
Traverse Group (1978)
    Middle and Upper Devonian. Comprises (ascending) Belle Shale, Rockport Quarry Limestone, Ferron Point Shale, Genshaw Formation, Koehler Limestone, Gravel Point Formation, and Beebe School Formation. In S Michigan.
    BT Devonian
    BT Paleozoic
    SA Michigan
traverses
    use geotraverses
travertine (1978)
    UF calc-sinter
    UF calcareous sinter
    BT carbonate rocks
    BT sedimentary rocks
    SA calcite
    SA carbonate sediments
    SA chemically precipitated rocks
    SA limestone
    SA tufa
Travis County
    Valid through 1988. Search in combination with state term. After 1988, use specific county-state term.
Travis County Texas (1989)
    S Central Texas. Before 1989, also search Travis County AND Texas.
    CO N300300N304500
        W0972500W0981500
    BT Texas
    BT United States
    NT Austin Texas
    SA Barton Springs
    SA Edwards Aquifer
Travis Peak Formation (1989)
    In Trinity Group. S and E Texas. This term has multiple hierarchies.
    BT1 Comanchean
    BT1 Cretaceous
    BT1 Mesozoic
    BT2 Lower Cretaceous
    BT2 Cretaceous
    BT2 Mesozoic
    SA Texas
    SA Trinity Group
treatment
    As of 1985, no longer a valid term for GeoRef. See soil treatment; water treatment.
tree rings (1978)
    UF annual growth rings
    UF dendrochronology
    UF rings, tree
    SA geochronology
    SA relative age
trees (1993)
    As of 1993, use tree rings, wood or fossil wood for geochronology. Use geobotanical methods for mineral exploration and Plantae or plants for paleobotany or stratigraphy studies, respectively.
    SA drainage patterns
    SA forests
    SA networks

Trego County
Valid through 1988. Search in combination with state term. After 1988, use specific county-state term.

Trego County Kansas (1989)
W central Kansas. Before 1989, also search Trego County AND Kansas.
  CO N384000N391000
       W0993800W1001000
  BT Kansas
  BT United States

Tremadoc Bay (1978)
Inlet of Saint Georges Channel in Gwynedd County in NW Wales. Also search Tremadoc.
  BT Wales
  BT Great Britain
  BT United Kingdom
  BT Western Europe
  BT Europe
  SA Caernarvonshire Wales
  SA Merionethshire Wales

Tremadocian (1978)
Europe. Above Dolgellian (Cambrian), below Arenigian.
  BT Lower Ordovician
  BT Ordovician
  BT Paleozoic
  NT Halifax Formation

tremolite (1978)
  BT clinoamphibole
  BT amphibole group
  BT chain silicates
  BT silicates

Tremp
No longer a valid term for GeoRef. As of 1993, see Tremp Spain.

Tremp Spain (1993)
Town in Catalonia in W Lerida, NE Spain. Before 1993, also search Tremp and Spain.
  BT Lerida Spain
  BT Catalonia Spain
  BT Spain
  BT Iberian Peninsula
  BT Southern Europe
  BT Europe

Trempealeauan (1978)
North America. Upper Cambrian, above Franconian, below Lower Ordovician.
  BT Upper Cambrian
  BT Cambrian
  BT Paleozoic

trenches (1978)
  UF marginal trench
  UF oceanic trench
  UF sea-floor trench
  SA Benioff zone
  SA bottom features
  SA fore-arc basins
  SA furrows
  SA island arcs
  SA morphostructures
  SA ocean floors
  SA plate tectonics
  SA subduction
  SA subduction zones
  SA troughs

trend analysis
  use trend-surface analysis

trend surface analysis
  use trend-surface analysis

trend-surface analysis (1978)
  UF trend analysis
  UF trend surface analysis
  BT statistical analysis

  SA analysis
  SA data processing

Trent
Not a valid term for GeoRef. See Trento Italy and Trent Valley.

Trent Valley (1978)
River valley in central England, SE North Carolina, and SE Ontario. Also search Trent.
  IN Index England, North Carolina, or Ontario as applicable.
  SA England
  SA North Carolina
  SA Ontario

Trentino (1978)
S part of Trentino-Alto Adige autonomous region.
  BT Trentino-Alto Adige Italy
  BT Italy
  BT Southern Europe
  BT Europe
  SA Alto Adige

Trentino-Alto Adige
No longer a valid term for GeoRef. use Trentino-Alto Adige Italy

Trentino-Alto Adige Italy (1993)
Autonomous region in NE Italy. Formerly Venezia-Tridentina.
  CO N454000N470800
       E0123000E0102500
  UF Trentino-Alto Adige
  UF Venezia-Tridentina
  BT Italy
  BT Southern Europe
  BT Europe
  NT Alto Adige
  NT Bolzano Italy
  NT Cima d'Asta
  NT Latemar Massif
  NT Predazzo Italy
  NT Trentino
  NT Trento Italy
  NT Valsugana
  SA Dolomites
  SA Marmolada Glacier
  SA Otztal Alps
  SA Venetia
  SA Zillertal Alps

Trento
No longer a valid term for GeoRef. See Trento Italy.

Trento Italy (1993)
Province and city in S Trentino-Alto Adige. Also search Trent and Italy.
  BT Trentino-Alto Adige Italy
  BT Italy
  BT Southern Europe
  BT Europe

Trenton
Valid through 1988. Search in combination with state term. After 1988, use specific city-state term.

Trenton Group (1978)
Standard section of group comprises (ascending) Rockland Limestone, Hull Formation, Sherman Fall Formation, Cobourg Formation, Collingwood Shale, and Gloucester Shale. In Georgia, Michigan, N Ohio, New York, Vermont, and W Virginia.
  BT Trentonian
  BT Middle Ordovician
  BT Ordovician
  BT Paleozoic
  SA Georgia
  SA Michigan
  SA New York
  SA Ohio

  SA Pennsylvania
  SA Vermont
  SA Virginia

Trenton New Jersey (1989)
City on the Delaware River in W New Jersey. Before 1989, search Trenton AND New Jersey.
  CO N401500N401500
       W0744300W0744300
  BT Mercer County New Jersey
  BT New Jersey
  BT United States

Trentonian (1978)
North America. Upper Mohawkian, above Blackriverian.
  BT Middle Ordovician
  BT Ordovician
  BT Paleozoic
  NT Trenton Group
  SA Champlainian

Trepca Mine (1978)
Lead-zinc deposits. In the Kosovo region of S Serbia. Also search Trepca.
  UF Trepcha Mine
  BT Serbia
  BT Yugoslavia
  BT Southern Europe
  BT Europe
  SA lead-zinc deposits
  SA mines

Trepcha Mine
  use Trepca Mine

Trepostomata (1978)
  BT Bryozoa
  BT Invertebrata

trevorite (1978)
  BT oxides
  SA iron oxides

Trialet Range (1978)
N mountain range of the Lesser Caucasus in central Georgian Republic. Also search Trialet. This term has multiple hierarchies.
  BT1 Georgian Republic
  BT1 Europe
  BT2 Lesser Caucasus
  BT2 Caucasus
  BT2 Europe

triangulation (1978)
  SA cartography
  SA geodesy
  SA geodetic networks
  SA geotraverses
  SA neotectonics
  SA surveys
  SA trilateration
  SA very long baseline interferometry

Triassic (1978)
Above Permian (of Paleozoic), below Jurassic. From 1978-1980, Mesozoic was autoposted to this term.
  BT Mesozoic
  NT Andalusian
  NT Chanares Formation
  NT Cordevolian
  NT Fremouw Formation
  NT Hallstatt Limestone
  NT Hawkesbury Sandstone
  NT Ipswich Coal Measures
  NT Lower Triassic
  NT Middle Triassic
  NT Moenkopi Formation
  NT Narrabeen Group
  NT Nicola Group
  NT Red Peak Formation
  NT Sherwood Sandstone
  NT Shublik Formation

  NT Takla Group
  NT Upper Triassic
  NT Wetterstein Limestone
  SA Alleghany Orogeny
  SA Baskunchak Series
  SA Beacon Supergroup
  SA Beaufort Group
  SA Botucatu Formation
  SA Cache Creek Group
  SA Chugwater Formation
  SA Dwyka Formation
  SA Estrada Nova Formation
  SA Glen Canyon Group
  SA Gondwana System
  SA Indosinian Orogeny
  SA Karroo Supergroup
  SA Khorat Group
  SA Korenevskaya Formation
  SA Louann Salt
  SA lower Gondwana System
  SA New Red Sandstone
  SA Nikolai Greenstone
  SA Passaic Formation
  SA Ross Formation
  SA Singleton Coal Measures
  SA Slovakian Karst
  SA Spearfish Formation
  SA upper Gondwana System

triaxial tests (1978)
  SA confining pressure
  SA deformation
  SA rock mechanics
  SA soil mechanics
  SA testing
  SA uniaxial tests

tributaries (1993)
  SA drainage basins
  SA drainage patterns
  SA fluvial features
  SA geomorphology
  SA glacial geology
  SA glaciers
  SA rivers
  SA rivers and streams
  SA stream capture
  SA stream order
  SA streams
  SA underground streams
  SA valleys
  SA watersheds

Trichinopoly India
  use Tiruchirapalli India

trichloroethane (1989)
Sometimes a parent compound of DDT.
  BT chlorinated hydrocarbons
  SA DDT
  SA ethane

trichloroethylene (1989)
Produced from heating tetrachloroethane with hydrated lime.
  BT chlorinated hydrocarbons
  SA ethylene

triclinic
  use triclinic system

triclinic system (1985)
Before 1985, also search triclinic.
  UF triclinic
  BT crystal systems
  SA crystallography

Triconodonta (1978)
Order.
  BT Mammalia
  BT Tetrapoda
  BT Vertebrata
  BT Chordata

Tridacna (1978)
Genus.
  BT Bivalvia
  BT Mollusca

    BT   Invertebrata

tridymite   (1978)
    BT   silica minerals
    BT   framework silicates
    BT   silicates
    SA   cristobalite
    SA   opal-CT

Trieste
    No longer a valid term for GeoRef. See Trieste Italy.

Trieste Italy   (1993)
    City at head of the Adriatic Sea on Gulf of Trieste in SE Friuli-Venezia Giulia, NE Italy.
    BT   Friuli-Venezia Giulia Italy
    BT   Italy
    BT   Southern Europe
    BT   Europe
    SA   Dinaric Alps

Trigonia   (1978)
    Genus.
    BT   Trigoniidae
    BT   Bivalvia
    BT   Mollusca
    BT   Invertebrata

Trigoniidae   (1978)
    Autoposting of this term began in 1978.
    BT   Bivalvia
    BT   Mollusca
    BT   Invertebrata
    NT   Trigonia

trilateration   (1985)
    SA   cartography
    SA   geodesy
    SA   geodetic coordinates
    SA   leveling
    SA   triangulation

Trilobita   (1978)
    Class. Autoposting of this term began in 1978.
    BT   Trilobitomorpha
    BT   Arthropoda
    BT   Invertebrata
    NT   Agnostida
    NT   Corynexochida
    NT   Lichida
    NT   Odontopleurida
    NT   Phacopida
    NT   Ptychopariida
    NT   Redlichiida
    SA   Crustacea
    SA   Cruziana
    SA   Rusophycus
    SA   trilobites
    SA   Trilobitoidea

trilobites   (1981)
    Common name for Trilobita.
    IN   Index for all non-paleontologic studies of fossils.
    BT   arthropods
    BT   invertebrates
    SA   biostratigraphy
    SA   Trilobita

Trilobitoidea   (1978)
    Class.
    BT   Trilobitomorpha
    BT   Arthropoda
    BT   Invertebrata
    SA   Trilobita

Trilobitomorpha   (1978)
    Autoposting of this term began in 1978.
    BT   Arthropoda
    BT   Invertebrata
    NT   Trilobita
    NT   Trilobitoidea

Trilophosauria   (1981)
    Includes the family Trilophosauridae. As of 1993, considered a narrower term of Archosauromorpha. Before 1993, Araeoscelidia was autoposted to this term.
    BT   Diapsida
    BT   Reptilia
    BT   Tetrapoda
    BT   Vertebrata
    BT   Chordata
    SA   Araeoscelidia

Trinidad   (1978)
    Island off NE coast of Venezuela in the Lesser Antilles. Major part of independent state of Trinidad and Tobago.
    CO   N100200N105100
         W0605300W0615600
    BT   Trinidad and Tobago
    BT   Lesser Antilles
    BT   Antilles
    BT   West Indies
    BT   Caribbean region
    SA   Northern Range

Trinidad and Tobago   (1978)
    Comprises the islands of Trinidad and Tobago, Atlantic Ocean, off NE coast of Venezuela.
    CO   N100000N113000
         W0601500W0620000
    BT   Lesser Antilles
    BT   Antilles
    BT   West Indies
    BT   Caribbean region
    NT   Tobago
    NT   Trinidad
    SA   Northern Range

Trinity Complex   (1993)
    Klamath Mountains, N California. Also search Trinity AND (ophiolite or peridotite) if applicable.
    IN   Index county as applicable.
    CO   N405000N413000
         W1221500W1225500
    BT   California
    BT   United States
    SA   Klamath Mountains
    SA   Shasta County California
    SA   Siskiyou County California
    SA   Trinity County California

Trinity County
    Valid through 1988. Search in combination with state term. After 1988, use specific county-state term.

Trinity County California   (1989)
    N California. Before 1989, also search Trinity County AND California.
    CO   N395800N412000
         W1223000W1233900
    BT   California
    BT   United States
    SA   Eel River
    SA   Trinity Complex
    SA   Trinity River

Trinity County Texas   (1989)
    E Texas. Before 1989, also search Trinity County AND Texas.
    CO   N305300N312200
         W0945100W0952800
    BT   Texas
    BT   United States

Trinity Group   (1978)
    Lower Cretaceous. Subdivided into (ascending) Pearsall, Rodessa, Ferry Lake, and Rusk time-stratigraphic units. In SW Arkansas, NW Louisiana, central S and SE Oklahoma, and Texas. This term has multiple hierarchies.
    BT1 Comanchean
    BT1 Cretaceous
    BT1 Mesozoic
    BT2 Lower Cretaceous
    BT2 Cretaceous
    BT2 Mesozoic
    SA   Antlers Sands
    SA   Arkansas
    SA   Glen Rose Formation
    SA   Louisiana
    SA   Oklahoma
    SA   Paluxy Formation
    SA   Pearsall Formation
    SA   Rodessa Formation
    SA   Texas
    SA   Travis Peak Formation

Trinity River   (1978)
    Flows into Trinity Bay off Galveston Bay on the Gulf of Mexico in E Texas. Also a river in Humboldt and Trinity counties, NW California. Search Trinity River in combination with appropriate state.
    IN   Index states and counties as applicable.
    SA   California
    SA   Humboldt County California
    SA   Texas
    SA   Trinity County California
    SA   Trinity River basin

Trinity River basin   (1989)
    E Texas and NW California. Search in combination with appropriate state.
    IN   Index states as applicable.
    SA   California
    SA   Texas
    SA   Trinity River

triphylite   (1978)
    BT   phosphates

triple junctions   (1978)
    SA   plate geometry
    SA   plate tectonics

triplite   (1978)
    From 1978-1980, phosphates and halides were autoposted to this term. Autoposting of fluorides began in 1981. As of 1989, halides is autoposted to this term. This term has multiple hierarchies.
    BT1 fluorides
    BT1 halides
    BT2 phosphates

Tripoli
    No longer a valid term for GeoRef. As of 1993 see Tripoli Libya.

Tripoli Libya   (1993)
    City on the Mediterranean Sea in NW Libya. Before 1993 also search Tripoli AND Libya.
    BT   Libya
    BT   North Africa
    BT   Africa

Tripolis
    No longer a valid term for GeoRef. As of 1993 see Tripolis Greece.

Tripolis Greece   (1993)
    City in central Peloponnesus, S Greece. Before 1993 also search Tripolis AND Greece.
    BT   Peloponnesus Greece
    BT   Greece
    BT   Southern Europe
    BT   Europe

Tripolitania   (1978)
    NW part of country which was a province until 1963. Under the Italians, it was a major province covering the entire W part of colony.
    BT   Libya
    BT   North Africa
    BT   Africa
    NT   Garian Libya

tripolite
    use diatomaceous earth

Tripura
    No longer a valid term for GeoRef. See Tripura India.

Tripura India   (1993)
    State E of Bangladesh in NE India.
    BT   Northeastern India
    BT   India
    BT   Indian Peninsula
    BT   Asia
    SA   Bengal

Tristan da Cunha   (1978)
    A group of 5 British volcanic islands in the central South Atlantic Ocean. Also the name of the main island of the group. Dependencies of Saint Helena. In 1985, broader term changed from Atlantic Ocean to Atlantic Ocean Islands.
    UF   Tristan da Cunha Islands
    BT   Atlantic Ocean Islands
    NT   Gough Island
    SA   Saint Helena

Tristan da Cunha Islands
    use Tristan da Cunha

triterpanes   (1985)
    From 1985 to 1989, alkanes and aliphatic hydrocarbons were autoposted to this term.
    BT   hydrocarbons
    BT   organic materials
    SA   diterpanes
    SA   terpanes

Triticites   (1978)
    Genus. Autoposting of microfossils and Protista to this term began in 1990. This term has multiple hierarchies.
    BT1 Fusulinidae
    BT1 Fusulinina
    BT1 foraminifera
    BT1 Protista
    BT1 Invertebrata
    BT2 Fusulinidae
    BT2 Fusulinina
    BT2 foraminifera
    BT2 Protista
    BT2 microfossils
    NT   Triticites ventricosus

Triticites ventricosus   (1978)
    Autoposting of microfossils and Protista to this term began in 1990. This term has multiple hierarchies.
    BT1 Triticites
    BT1 Fusulinidae
    BT1 Fusulinina
    BT1 foraminifera
    BT1 Protista
    BT1 Invertebrata
    BT2 Triticites
    BT2 Fusulinidae
    BT2 Fusulinina
    BT2 foraminifera
    BT2 Protista
    BT2 microfossils

tritium   (1978)
    In 1989, hydrogen, radioactive isotopes, and isotopes were autoposted to this term. This term has multiple hierarchies.
    UF   H-3
    BT1 radioactive isotopes

BT1 isotopes
BT2 hydrogen
SA absolute age

Triton Satellite (1989)
One of satellites of Neptune. Before 1989, also search Triton AND Neptune.
SA icy satellites
SA Neptune
SA Nereid Satellite
SA satellites
SA Voyager Program

Trivandrum District
No longer a valid term for GeoRef. See Trivandrum India.

Trivandrum India (1993)
Kerala, SW India. Also search Trivandrum District.
BT Kerala India
BT India
BT Indian Peninsula
BT Asia

TRM
use thermoremanent magnetization

Trochammina (1985)
Genus. Autoposting of microfossils and Protista to this term began in 1990. This term has multiple hierarchies.
BT1 Lituolacea
BT1 Textulariina
BT1 foraminifera
BT1 Protista
BT1 Invertebrata
BT2 Lituolacea
BT2 Textulariina
BT2 foraminifera
BT2 Protista
BT2 microfossils

troctolite (1978)
BT gabbros
BT plutonic rocks
BT igneous rocks

Troia
use Troy

troilite (1978)
BT sulfides
SA meteorites
SA pyrrhotite

Troja
use Troy

Trompia Valley (1978)
Valley of upper Mella River in E Lombardy, N Italy.
BT Lombardy Italy
BT Italy
BT Southern Europe
BT Europe

Troms
No longer a valid term for GeoRef. As of 1993 see Troms Norway.

Troms Norway (1993)
County in far N Norway. Before 1993 also search Troms AND Norway.
BT Norway
BT Scandinavia
BT Western Europe
BT Europe
NT Lyngen Peninsula
SA Vesteralen

trona (1978)
IN May be used as a commodity term with sodium carbonate.
BT carbonates
SA Searles Lake
SA sodium carbonate

Trona California (1989)
Village in Mojave Desert of S California. Before 1977, search Trona AND California. From 1978 to 1989, also search Trona Village AND California.
CO N354600N354600
W1172400W1172400
UF Trona Village
BT San Bernardino County California
BT California
BT United States
SA Searles Lake

Trona Village
Valid through 1988. Search in combination with state term. After 1988, use specific city-state term.
use Trona California

Trondelag (1978)
Region in central part of country between Norwegian Sea and Swedish border around Trondheim Fjord.
IN Index counties as applicable.
BT Norway
BT Scandinavia
BT Western Europe
BT Europe
SA Nord-Trondelag Norway
SA Sor-Trondelag Norway

Trondheim
No longer a valid term for GeoRef. As of 1993 see Trondheim Norway.

Trondheim Norway (1993)
City on the Norweigan sea. Before 1993 also search Trondheim AND Norway.
BT Sor-Trondelag Norway
BT Norway
BT Scandinavia
BT Western Europe
BT Europe

trondhjemite (1978)
BT diorites
BT plutonic rocks
BT igneous rocks
SA plagiogranite

Troodos Massif (1978)
Mountainous mass SW of Nicosia which includes Mount Troodos, the highest peak on the island.
UF Troodos Mountain
UF Troodos Mountains
BT Cyprus
BT Middle East
BT Asia
SA Agrokipia Deposit

Troodos Mountain
use Troodos Massif

Troodos Mountains
use Troodos Massif

Troodos Ophiolite (1989)
SA Agrokipia Deposit
SA Cyprus
SA Middle East

trophic analysis (1985)
SA biomass
SA communities
SA diet
SA ecology
SA ecosystems
SA nutrition
SA paleoecology
SA paleontology
SA predation

tropical
A valid term through 1977. After 1977, use tropical environment.

tropical environment (1978)
Before 1978, search tropical.
SA climate
SA depositional environment
SA ecology
SA environment
SA equatorial region
SA paleoclimatology
SA subtropical environment

troposphere
As of 1993, no longer a valid term for GeoRef. Usually out-of-scope. Used with atmosphere.

Trotus Valley (1978)
River valley SSW of Bacau in W Moldavia. Also search Trotus; Trotus River. This term has multiple hierarchies.
BT1 Moldavia
BT1 Europe
BT2 Romania
BT2 Southern Europe
BT2 Europe

Trou Sans Fond (1978)
Submarine canyon off SW Nigeria in the Bight of Benin.
UF Le Trou Sans Fond
UF Trou-Sans-Fond
BT Atlantic Ocean
SA Ivory Coast
SA submarine canyons

Trou-Sans-Fond
use Trou Sans Fond

troughs (1978)
SA aulacogens
SA bottom features
SA channels
SA furrows
SA gorges
SA ocean floors
SA trenches

Trousdale County
Valid through 1988. Search in combination with state term. After 1988, use specific county-state term.

Trousdale County Tennessee (1989)
N Tennessee. Before 1989, also search Trousdale County AND Tennessee.
CO N361700N363000
W0855800W0861500
BT Tennessee
BT United States

Troy (1978)
Ancient ruined city. An archeological site, named Hissarlik, on the Menderes River S of the Dardanelles in NW Anatolia.
UF Ilion
UF Ilium
UF Troia
UF Troja
BT Turkey
BT Middle East
BT Asia

Trucial Coast
A 350 mile section of S Persian Gulf littoral extending from Qatar to Ruus al Jubal. Often formerly used interchangeably with Trucial Oman and Trucial States. A valid term through 1974.
use United Arab Emirates

Trucial Oman
Former name of 7 British protected states on the Trucial Coast now known as the United Arab Emirates. A valid term through 1972.
use United Arab Emirates

Truckee River (1978)
Rises in Lake Tahoe in E California and flows NE and E into Pyramid Lake in NW Nevada. Also search Truckee.
IN Index counties and states as applicable.
BT United States
SA California
SA Nevada

Trujillo
No longer a valid term for GeoRef. As of 1993 see Trujillo Venezuela.

Trujillo Venezuela (1993)
State in NW Venezuela. Before 1993 also search Trujillo AND Venezuela.
BT Venezuela
BT South America
SA Lake Maracaibo
SA Maracaibo Basin

Trumbull County
Valid through 1988. Search in combination with state term. After 1988, use specific county-state term.

Trumbull County Ohio (1989)
NE Ohio. Before 1989, also search Trumbull County AND Ohio.
CO N410730N413000
W0803200W0810000
BT Ohio
BT United States

Trust Territory of the Pacific Islands
Not a valid term for GeoRef. U.S. trust territory. As of 1993, see Micronesia, or Caroline Islands, Marshall Islands, and Mariana Islands for the Northern Mariana Islands as applicable.

Truth or Consequences New Mexico (1989)
Town in SW New Mexico.
CO N330800N330800
W1071600W1071600
BT Sierra County New Mexico
BT New Mexico
BT United States

Trypanites (1989)
Genus.
BT ichnofossils

Tsaidam Basin
use Qaidam Basin

Tsarev Meteorite (1989)
Impact in Voronezh-Volga Anteclise, Russian Platform, in the former USSR.
BT L chondrites
BT chondrites
BT stony meteorites
BT meteorites
SA Russian Platform
SA Volgograd Russian Federation
SA Voronezh-Volga Anteclise

Tsaritsyn Russian Federation
use Volgograd Russian Federation

tschermakite (1978)
BT clinoamphibole
BT amphibole group
BT chain silicates
BT silicates

Tschernobyl Ukraine
  use Chernobyl Ukraine

Tschernosem
  use Chernozems

Tschernosiom
  use Chernozems

Tsinghai China
  use Qinghai China

Tsuga (1985)
  Genus.
  BT Coniferales
  BT gymnosperms
  BT Spermatophyta
  BT Plantae

Tsugaru Strait (1978)
  Channel between islands of Honshu and Hokkaido connecting the Pacific Ocean with the Japan Sea. Also search Tsugaru.
  IN Index islands as applicable.
  BT Japan
  BT Far East
  BT Asia
  SA Hokkaido
  SA Honshu
  SA Oshima Peninsula

Tsumeb
  No longer a valid term for GeoRef. As of 1993, see Tsumeb Namibia.

Tsumeb Namibia (1993)
  Mineral-collecting locality and mining center in N Namibia. Before 1993 also search Tsumeb AND Namibia.
  CO S191300S191300
       E0174200E0174200
  BT Namibia
  BT Southern Africa
  BT Africa

tsunami
  Not a valid GeoRef index term after 1971. Was used on level 1 in subfile G.
  use tsunamis

tsunamis (1978)
  Before 1972, also search tsunami.
  UF earthquake sea wave
  UF seismic sea waves
  UF seismic surge
  UF tidal wave
  UF tsunami
  SA catastrophic waves
  SA earthquakes
  SA geologic hazards
  SA ocean waves
  SA seismology
  SA waves

Tsushima (1978)
  island off NW Kyushu.
  BT Kyushu
  BT Japan
  BT Far East
  BT Asia
  SA Taishu Group

Tuamotu Islands (1978)
  Group of about 80 small islands in French Polynesia E of Society Islands and S of Marquesas Islands. Also search Tuamotu. Mururoa Atoll is considered a narrower term as of 1981.
  CO S230000S140000
       W1340000W1490000
  UF Dangerous Islands
  UF Low Archipelago
  BT French Polynesia
  BT Polynesia
  BT Oceania
  NT Mururoa Atoll

Tuapse
  No longer a valid term for GeoRef. As of 1993, see Tuapse Russian Federation.

Tuapse Russian Federation (1993)
  City on the Black Sea in S Krasnodar Kray in the Northern Caucasus. This term has multiple hierarchies.
  BT1 Krasnodar Russian Federation
  BT1 Russian Federation
  BT1 Commonwealth of Independent States
  BT2 Krasnodar Russian Federation
  BT2 Europe

Tuar-Kyr
  No longer a valid term for GeoRef. As of 1993, see Tuar-Kyr Turkmenia.

Tuar-Kyr Turkmenia (1993)
  Village near Kara-Bogaz-Gol Gulf in W Turkmenia. Before 1978, also search Tuarkyr. This term has multiple hierarchies.
  UF Tuarkyr Turkmenia
  BT1 Turkmenia
  BT1 Asia
  BT2 Turkmenia
  BT2 Commonwealth of Independent States

Tuarkyr Turkmenia
  use Tuar-Kyr Turkmenia

Tubarao Group (1978)
  BT Carboniferous
  BT Paleozoic
  SA Brazil
  SA Sao Paulo Brazil

tube waves (1989)
  Used in acoustical logging.
  BT surface waves
  BT elastic waves
  SA acoustical logging
  SA boreholes
  SA seismic logging
  SA seismic surveys
  SA waves

tubes, lava
  use lava tubes

Tubiphytes (1989)
  Genus. This term has multiple hierarchies.
  BT1 problematic microfossils
  BT1 microfossils
  BT2 problematic microfossils
  BT2 problematic fossils

Tuboidea (1978)
  IN Index Hemichordata or Invertebrata if applicable.
  BT Graptolithina
  SA Hemichordata
  SA Invertebrata

Tubuai Islands
  use Austral Islands

Tubulidentata (1978)
  Order. Autoposting of Eutheria and Theria to this term began in 1989.
  BT Eutheria
  BT Theria
  BT Mammalia
  BT Tetrapoda
  BT Vertebrata
  BT Chordata

Tucano Basin (1978)
  NE Bahia. Also search Tucano.
  BT Bahia Brazil

  BT Brazil
  BT South America

Tucson
  Valid through 1988. Search in combination with state term. After 1988, use specific city-state term.

Tucson Arizona (1989)
  City in SE Arizona. Before 1989, search Tucson AND Arizona.
  CO N321500N321500
       W1105700W1105700
  BT Pima County Arizona
  BT Arizona
  BT United States

Tucson Basin (1978)
  S Arizona.
  IN Index counties as applicable.
  BT Arizona
  BT United States

Tucson Mountains (1978)
  W of Tucson in Pima County.
  BT Pima County Arizona
  BT Arizona
  BT United States

Tucuman
  No longer a valid term for GeoRef. See Tucuman Argentina.

Tucuman Argentina (1993)
  Province in NW Argentina.
  UF Tucuman Province
  BT Argentina
  BT South America
  SA Pampean Mountains

Tucuman Province
  use Tucuman Argentina

Tucumcari
  Valid through 1988. Search in combination with state term. After 1988, use specific city-state term.

Tucumcari Basin (1989)
  E New Mexico.
  IN Index counties as applicable.
  BT New Mexico
  BT United States

Tucumcari New Mexico (1989)
  City in NE New Mexico. Before 1989, search Tucumcari AND New Mexico.
  CO N351100N351100
       W1034400W1034400
  BT Quay County New Mexico
  BT New Mexico
  BT United States

tufa (1978)
  IN See list I (sedimentary rocks) and list N (sediments).
  UF calc-tufa
  UF calcareous tufa
  UF petrified moss
  UF tophus
  UF tuft
  BT chemically precipitated rocks
  BT sedimentary rocks
  SA calcite
  SA carbonate rocks
  SA carbonate sediments
  SA travertine

tuff (1978)
  BT pyroclastics
  BT volcanic rocks
  BT igneous rocks
  SA andesite tuff
  SA ash flows
  SA ash-flow tuff
  SA bentonite
  SA green tuff
  SA ignimbrite
  SA metatuff
  SA pyroclastic flows

  SA rhyolite tuff
  SA tuffite
  SA volcanic ash
  SA volcaniclastics
  SA welded tuff

tuff lava
  use welded tuff

tuffite (1978)
  BT pyroclastics
  BT volcanic rocks
  BT igneous rocks
  SA tuff

tuft
  use tufa

Tugela Basin (1978)
  River basin in central Natal. Also search Tugela; Tugela River.
  BT Natal South Africa
  BT South Africa
  BT Southern Africa
  BT Africa

tugtupite (1978)
  BT sodalite group
  BT framework silicates
  BT silicates

Tuimazy Russian Federation
  use Tuymazy Russian Federation

Tuktoyaktuk Peninsula (1978)
  Extends into the Beaufort Sea E of the Mackenzie River delta in Mackenzie District. Also search Tuktoyaktuk.
  CO N683000N703000
       W1293000W1350000
  BT Mackenzie District Northwest Territories
  BT Northwest Territories
  BT Western Canada
  BT Canada

Tula
  No longer a valid term for GeoRef. As of 1993, see Tula Russian Federation.

Tula Russian Federation (1993)
  Oblast and city S of Moscow. This term has multiple hierarchies.
  BT1 Russian Federation
  BT1 Commonwealth of Independent States
  BT2 Europe
  NT Chekalin Russian Federation

Tulameen coal area (1978)
  S British Columbia.
  BT British Columbia
  BT Western Canada
  BT Canada
  SA coal fields

Tulare County
  Valid through 1988. Search in combination with state term. After 1988, use specific county-state term.

Tulare County California (1989)
  S central California. Before 1989, also search Tulare County AND California.
  CO N354700N364500
       W1175900W1193500
  BT California
  BT United States
  SA Golden Trout Wilderness

Tulare Formation (1978)
  Pliocene and Pleistocene (?). Consists primarily of sandstone and conglomerate. In Southern California.
  BT Cenozoic
  SA California
  SA Pleistocene

SA Pliocene

Tularosa Basin (1989)
S New Mexico.
IN Index counties as applicable.
BT New Mexico
BT United States

Tulcea
No longer a valid term for GeoRef. As of 1993 see Tulcea Romania.

Tulcea Romania (1993)
County including the city in the Danube Delta area in N Romanian Dobruja, SE Romania. Before 1993 also search Tulcea AND Romania.
BT Romania
BT Southern Europe
BT Europe
SA Romanian Dobruja
SA Sulina Romania

Tulear Basin (1978)
Before 1978, also search Tulear. This term has multiple hierarchies.
BT1 Madagascar
BT1 Indian Ocean Islands
BT2 Madagascar
BT2 Africa

Tulghes Series (1978)
BT Paleozoic
SA Eastern Carpathians
SA Romania

Tullock Member (1989)
In Fort Union Formation. E Montana, SW North Dakota, N South Dakota and Wyoming.
BT Paleocene
BT Paleogene
BT Tertiary
BT Cenozoic
SA Fort Union Formation
SA Montana
SA North Dakota
SA South Dakota
SA Wyoming

Tully Limestone (1978)
In Susquehanna Group. Divided into several members in different areas: Apulia, Laurens, New Lisbon, Tinkers Falls, and West Brook.
BT Middle Devonian
BT Devonian
BT Paleozoic
SA New York
SA Pennsylvania

Tulsa
Valid through 1988. Search in combination with state term. After 1988, use specific city-state term.

Tulsa County
Valid through 1988. Search in combination with state term. After 1988, use specific county-state term.

Tulsa County Oklahoma (1989)
NE central Oklahoma. Before 1989, also search Tulsa County AND Oklahoma.
CO N355000N362500
W0954500W0961500
BT Oklahoma
BT United States
NT Tulsa Oklahoma

Tulsa Oklahoma (1989)
City on Arkansas River in NE Oklahoma. Before 1989, search Tulsa AND Oklahoma.
CO N360700N360700
W0955800W0955800

BT Tulsa County Oklahoma
BT Oklahoma
BT United States

tundra (1978)
SA ecology
SA geomorphology
SA landforms
SA landscapes
SA palsas
SA permafrost
SA plains
SA revegetation
SA soils
SA taiga environment

Tundra soils (1978)
BT soils
SA permafrost
SA soil group
SA Zonal soils

Tunganoxie Sandstone Member
use Tonganoxie Sandstone

tungstates (1978)
Autoposting of this term began in 1978.
NT ferberite
NT scheelite
NT stolzite
NT wolframite
SA minerals
SA molybdates

tungsten (1978)
Chemical element. As of 1981, use tungsten ores for tungsten as a commodity. Also search wolfram. Autoposting of metals to this term began in 1989.
UF W
BT metals
SA tungsten ores

tungsten ores (1981)
Before 1981, also search (tungsten OR wolfram) AND (deposit OR deposits OR ore OR ores OR economic) in the basic index. Autoposting of metal ores to this term began in 1985.
IN Commodity. See List C.
BT metal ores
SA scheelite
SA tungsten
SA wolframite

tungstenite (1978)
BT sulfides

Tunguska (1978)
In GeoRef indexing, this term is used to indicate a general area. Name of three rivers flowing into the Yenisei River in central Siberia: the Lower Tunguska River, Stony Tunguska River, and the lower course of the Angara River which was called the Upper Tunguska. This term has multiple hierarchies.
IN Index rivers as applicable.
CO N570000N710000
E1120000E0840000
BT1 Siberian Platform
BT1 Russian Federation
BT1 Commonwealth of Independent States
BT2 Siberian Platform
BT2 Asia
SA Angara River
SA Lower Tunguska River
SA Stony Tunguska River
SA Tunguska River

Tunguska Basin (1978)

Large coal basin in central Siberia between the Yenisei River and the Lena River in an area drained by the lower Angara River (also called the Upper Tunguska), the Stony Tunguska River, and the Lower Tunguska River. Also search Tunguska. This term has multiple hierarchies.
BT1 Russian Federation
BT1 Commonwealth of Independent States
BT2 Asia
SA coal fields
SA Siberia

Tunguska River (1978)
Tributary of the Amur River in the Soviet Far East near Khabarovsk. Not to be confused with the Lower Tunguska River, Stony Tunguska River, and Angara River (Upper Tunguska) rivers that drain into the Yenisei River in central Siberia. This term has multiple hierarchies.
BT1 Russian Federation
BT1 Commonwealth of Independent States
BT2 Asia
SA Angara River
SA Lower Tunguska River
SA Stony Tunguska River
SA Tunguska

Tunguska Series (1978)
Composed of 2 sections (ascending): productive section with coal beds and a section of tufa containing intercalary beds of normal sedimentary rock. In Tunguska Basin of central Siberia.
BT upper Paleozoic
BT Paleozoic
SA Russian Federation

Tunguska Syncline
use Tunguska Syneclise

Tunguska Syneclise (1978)
The Tunguska Basin area in central Siberia. Also search Tunguska. This term has multiple hierarchies.
UF Tunguska Syncline
BT1 Russian Federation
BT1 Commonwealth of Independent States
BT2 Asia
SA Siberia

Tunis
No longer a valid term for GeoRef. As of 1993 see Tunis Tunisia.

Tunis Tunisia (1993)
City on the Gulf of Tunis in N Tunisia. Before 1993 also search Tunis AND Tunisia.
BT Tunisia
BT North Africa
BT Africa

Tunisia (1978)
CO N303000N373000
E0120000E0073000
BT North Africa
BT Africa
NT Sebkha el Melah
NT Souk-el-Arba salt works
NT Tunis Tunisia
NT Zarzis Tunisia
SA Atlas Mountains
SA Gulf of Gabes
SA Mediterranean region
SA Sahara

tunnel boring machines (1989)
Sometimes abbreviated as TBM.

SA engineering geology
SA instruments
SA tunnels

tunnel erosion
use piping

tunnels (1978)
Used for geological studies on man-made tunnels.
SA construction
SA design
SA engineering geology
SA excavations
SA feasibility studies
SA foundations
SA geologic hazards
SA highways
SA land subsidence
SA mining
SA New Austrian Tunnelling Method
SA Poisson's ratio
SA railroads
SA rock mechanics
SA seepage
SA seismic response
SA site exploration
SA slope stability
SA soil mechanics
SA stability
SA submarine installations
SA subways
SA tunnel boring machines
SA underground installations

Tunsbergdalsbreen (1978)
Glacier emanating from larger Jostedalsbreen Glacier in W Norway.
BT Norway
BT Scandinavia
BT Western Europe
BT Europe

Tuolumne County
Valid through 1988. Search in combination with state term. After 1988, use specific county-state term.

Tuolumne County California (1989)
E central California. Before 1989, also search Tuolumne County AND California.
CO N373800N382500
W1191300W1204000
BT California
BT United States
SA Table Mountain
SA Yosemite National Park

Tura Formation (1978)
According to USSR Lexique, the age of Tura Series is Upper Silurian. Central Urals and West Siberian Plain. Also search Tura.
BT Upper Silurian
BT Silurian
BT Paleozoic
SA Russian Federation

Turan (1978)
Desert lowland N, S, and SE of the Aral Sea. This term has multiple hierarchies.
IN Index former Soviet republics as applicable.
BT1 Commonwealth of Independent States
BT2 Asia
SA Kazakhstan
SA Turkmenia
SA Uzbekistan

Turanian Platform (1978)

Extends from the Caspian Sea on the W to the Tien Shan in the E. This term has multiple hierarchies.
IN   Index former Soviet republics as applicable.
BT1 Commonwealth of Independent States
BT2 Asia
SA   Central Asia
SA   Kazakhstan
SA   Turkmenia
SA   Uzbekistan

turbidite (1978)
IN   See lists I and N. Index clastic sediments or sedimentary rocks if applicable.
SA   Bouma sequence
SA   clastic rocks
SA   clastic sediments
SA   sedimentary rocks
SA   sediments
SA   terrigenous materials
SA   turbidity current structures
SA   turbidity currents

turbidity (1978)
SA   aerosols
SA   meteorology

turbidity current structures (1978)
Autoposting of this term began in 1978. Before 1993, this term was autoposted to sole marks.
BT   sedimentary structures
NT   Bouma sequence
NT   graded bedding
NT   load casts
SA   casts
SA   convoluted beds
SA   flow structures
SA   flute casts
SA   olistoliths
SA   olistostromes
SA   sole marks
SA   turbidite
SA   turbidity currents

turbidity currents (1978)
UF   suspension current
BT   currents
SA   density currents
SA   HEBBLE
SA   sediment transport
SA   sedimentation
SA   submarine canyons
SA   submarine fans
SA   turbidite
SA   turbidity current structures

turbulence (1978)
SA   boundary layer
SA   critical flow
SA   HEBBLE
SA   meteorology
SA   ocean circulation
SA   oceanography
SA   waterways

Turgai Basin
   use Turgay Basin

Turgai Downwarp
   use Turgay Basin

Turgai Gates
   use Turgay Basin

Turgai Kazakhstan
   use Turgay Kazakhstan

Turgay
   No longer a valid term for GeoRef. As of 1993, see Turgay Kazakhstan.

Turgay Basin (1978)
An elongated depression joining the Turan Lowland and West Siberian Plain. Also search Turgai; Turgay.
CO N450000N570000
      E0680000E0573000
UF   Turgai Basin
UF   Turgai Downwarp
UF   Turgai Gates
BT   Central Asia
BT   Asia
SA   Kazakhstan

Turgay Kazakhstan (1993)
Oblast and village NE of Aral Sea in central Kazakhstan. Before 1978, also search Turgai. This term has multiple hierarchies.
UF   Turgai Kazakhstan
BT1 Kazakhstan
BT1 Central Asia
BT1 Asia
BT2 Kazakhstan
BT2 Commonwealth of Independent States

Turin
   No longer a valid term for GeoRef. See Turin Italy.

Turin Italy (1993)
Province and city in W central Piemonte, NW Italy. Also search Torino.
UF   Torino Italy
BT   Piemonte Italy
BT   Italy
BT   Southern Europe
BT   Europe

Turkana Basin (1989)
IN   Index countries as applicable.
BT   East Africa
BT   Africa
SA   Ethiopia
SA   Kenya
SA   Lake Turkana
SA   Sudan

Turkana District (1978)
Administrative district W of Lake Turkana (Rudolf) in NW Kenya.
BT   Kenya
BT   East Africa
BT   Africa

Turkestan (1978)
Region including S Kazakhstan and the rest of former Soviet Central Asia, plus a small section of NE Afghanistan and W and SW Xinjiang.
IN   Index Afghanistan, Xinjiang China, and Central Asian republics as applicable.
UF   Turkistan
BT   Asia
SA   Afghanistan
SA   Central Asia
SA   Kazakhstan
SA   Xinjiang China

Turkestan Range (1978)
W Tadzhikistan. This term has multiple hierarchies.
BT1 Tadzhikistan
BT1 Asia
BT2 Tadzhikistan
BT2 Commonwealth of Independent States

Turkey (1978)
Includes all of Turkey within its political boundaries. Before 1970, also search Asia Minor.
CO N355000N420000
      E0444500E0260000
UF   Asia Minor
BT   Middle East
BT   Asia
NT   Adana Turkey
NT   Alanya Turkey
NT   Amanos Mountains
NT   Amasra Basin
NT   Anatolia
NT   Ankara Turkey
NT   Antalya Turkey
NT   Ararat
NT   Bitlis Turkey
NT   Bosporus
NT   East Anatolian Fault
NT   Elbistan Turkey
NT   Eskisehir Turkey
NT   Hatay Turkey
NT   Istanbul Turkey
NT   Keban Mine
NT   Kocaeli Turkey
NT   Konya Basin
NT   Konya Turkey
NT   Lycian Taurus
NT   Malatya Turkey
NT   North Anatolian Fault
NT   Pontic Mountains
NT   Sea of Marmara region
NT   Taurus Mountains
NT   Troy
NT   Turkish Aegean region
NT   Tuz Golu
NT   Zonguldak Turkey
SA   Antalya Complex
SA   Balkan Peninsula
SA   Black Sea region
SA   Euphrates River
SA   Istranca Mountains
SA   Kura River
SA   Maden Complex
SA   Mardin Formation
SA   Maritsa River
SA   Mediterranean region
SA   Mudurnu earthquake 1967
SA   Near East
SA   Thrace
SA   Tigris River

Turkish Aegean region (1981)
CO N363000N400500
      E0313000E0260500
BT   Turkey
BT   Middle East
BT   Asia
SA   Aegean Islands

Turkistan
   use Turkestan

Turkmenia (1978)
Republic. Formerly the Turkmen Soviet Socialist Republic in Soviet Central Asia. Autoposting of this term began in 1978. This term has multiple hierarchies.
CO N350000N430000
      E0660000E0520000
UF   Turkmenistan
BT1 Asia
BT2 Commonwealth of Independent States
NT   Ashkhabad Turkmenia
NT   Badkhyz
NT   Balkhan
NT   Chardzhou Turkmenia
NT   Cheleken Peninsula
NT   Kara-Bogaz Gulf
NT   Kara-Bogaz-Gol Turkmenia
NT   Karakum
NT   Kizyl-Dere
NT   Kopet-Dag Range
NT   Kurt Turkmenia
NT   Tuar-Kyr Turkmenia
SA   Amu Darya
SA   Amu Darya Basin
SA   Aral region
SA   Caspian Basin
SA   Central Asia
SA   Murgab Basin
SA   Turan
SA   Turanian Platform
SA   Ustyurt

Turkmenistan
   use Turkmenia

Turku
   No longer a valid term for GeoRef. See Abo Finland.

Turku Finland
   use Abo Finland

Turnagain Arm (1989)
Arm of Cook Inlet, S Alaska.
BT   Southern Alaska
BT   Alaska
BT   United States
SA   Cook Inlet

Turnyauz Russian Federation
   use Tyrny-Auz Russian Federation

Turolian (1985)
Europe. Upper Miocene, above Vallesian and below Ruscinian.
BT   upper Miocene
BT   Miocene
BT   Neogene
BT   Tertiary
BT   Cenozoic

Turonian (1978)
Europe. Above Cenomanian, below Coniacian. Autoposting of this term began in 1978.
BT   Upper Cretaceous
BT   Cretaceous
BT   Mesozoic
NT   lower Turonian
SA   Middle Cretaceous

turquoise (1978)
BT   phosphates
SA   copper minerals
SA   gems

Turritella (1978)
BT   Turritellidae
BT   Gastropoda
BT   Mollusca
BT   Invertebrata

Turritellidae (1978)
Autoposting of this term began in 1978.
BT   Gastropoda
BT   Mollusca
BT   Invertebrata
NT   Turritella

Turukhan (1978)
River which flows into Yenisei River just N of Turukhansk in N Krasnoyarsk Kray in central Siberia. This term has multiple hierarchies.
BT1 Russian Federation
BT1 Commonwealth of Independent States
BT2 Asia
SA   Krasnoyarsk Russian Federation

Turukhansk
   No longer a valid term for GeoRef. As of 1993, see Turukhansk Russian Federation.

Turukhansk Russian Federation (1993)
Town on Yenisei River at the mouth of the Lower Tunguska River in N Krasnoyarsk Kray in central Siberia. This term has multiple hierarchies.
BT1 Krasnoyarsk Russian Federation
BT1 Russian Federation
BT1 Commonwealth of Independent States

BT2 Krasnoyarsk Russian Federation
BT2 Asia

Tusas Mountains (1978)
IN Index counties and mountains as applicable.
BT New Mexico
BT United States

Tuscaloosa County
Valid through 1988. Search in combination with state term. After 1988, use specific county-state term.

Tuscaloosa County Alabama (1989)
W central Alabama. Before 1989, also search Tuscaloosa County AND Alabama.
CO N330100N333700
W0870500W0875000
BT Alabama
BT United States

Tuscaloosa Formation (1978)
Irregularly or obscurely bedded quartzitic and miracaceous sands interbedded with heterogeneous clays. Also lenticular pebble beds. Atlantic Coastal Plain from W Tennessee, NE Mississippi, N Louisiana and NW Alabama, across Alabama, Georgia, South Carolina, and North Carolina.
BT Upper Cretaceous
BT Cretaceous
BT Mesozoic
SA Alabama
SA Georgia
SA Louisiana
SA Mississippi
SA North Carolina
SA South Carolina
SA Tennessee

Tuscany
No longer a valid term for GeoRef. See Tuscany Italy.

Tuscany Italy (1993)
Autonomous region in W Italy.
CO N422000N442000
E0122500E0094000
BT Italy
BT Southern Europe
BT Europe
NT Apuane Alps
NT Arno River Basin
NT Florence Italy
NT Grosseto Italy
NT Larderello
NT Livorno Italy
NT Lucca Italy
NT Monte Amiata
NT Monte Capanne
NT Monte Pisano
NT Mount Amiata
NT Pisa Italy
NT Siena Italy
NT Travale Field
SA Elba
SA Tiber Valley

Tuscarawas County
Valid through 1988. Search in combination with state term. After 1988, use specific county-state term.

Tuscarawas County Ohio (1989)
E central Ohio. Before 1989, also search Tuscarawas County AND Ohio.
CO N401200N404000
W0811700W0814300
BT Ohio
BT United States

Tuscarora Formation (1978)
In central Pennsylvania, it is gradational with overlying Rose Hill Formation and transitional with underlying Juniata Formation. In W Maryland, central S and E Pennsylvania, W Virginia, and E West Virginia.
BT Lower Silurian
BT Silurian
BT Paleozoic
SA Maryland
SA Pennsylvania
SA Virginia
SA West Virginia

Tushar Mountains (1989)
Beaver and Piute counties, SE central Utah. Extends S from Pavant Mountains along W bank of Sevier River.
IN Index counties as applicable.
CO N382000N383500
W1122000W1124500
BT Utah
BT United States
SA Beaver County Utah
SA Piute County Utah

Tuttle Creek Dam (1978)
On Tuttle Creek forming reservoir in NE Kansas. Also search Tuttle Creek.
BT Riley County Kansas
BT Kansas
BT United States

Tuva
No longer a valid term for GeoRef. As of 1993, see Tuva Russian Federation.

Tuva Russian Federation (1993)
Former Tuva Autonomous Soviet Socialist Republic. On the Mongolian border between the Altai Mountains on the W and the Sayan Mountains on the NE. Before 1993, also search Tuva or Tannu Tuva. This term has multiple hierarchies.
CO N493000N533000
E0990000E0890000
UF Tannu Tuva Russian Federation
BT1 Russian Federation
BT1 Commonwealth of Independent States
BT2 Siberian fold belt
BT2 Asia
NT Khovu-Aksy Russian Federation
NT Sangilen Mountains

Tuvalu (1981)
Island group NE of Australia, N of Fiji Islands, and SE of Kiribati.
CO S110000S050000
E1800000E1750000
UF Ellice Islands
BT Polynesia
BT Oceania
SA Pacific Ocean

Tuymazy
No longer a valid term for GeoRef. As of 1993, see Tuymazy Russian Federation.

Tuymazy Russian Federation (1993)
Town W of Ufa in W Bashkiria in SE European Russian Federation. Before 1978, also search Tuimazy. This term has multiple hierarchies.
UF Tuimazy Russian Federation
BT1 Bashkiria Russian Federation
BT1 Russian Federation
BT1 Commonwealth of Independent States
BT2 Bashkiria Russian Federation
BT2 Europe

Tuz Golu (1978)
Salt lake in W central Anatolia (Turkey).
UF Tuz Lake
BT Turkey
BT Middle East
BT Asia
SA Anatolia

Tuz Lake
use Tuz Golu

Tver Russian Federation (1993)
Oblast and city in W Russian Federation. Formerly Kirov before 1990s. Before 1993, also search Kirov or Tver. This term has multiple hierarchies.
UF Kalinin Russian Federation
BT1 Russian Federation
BT1 Commonwealth of Independent States
BT2 Europe
NT Vyshkovo Russian Federation

Twente (1978)
Region in SE Overijssel.
UF De Twente
BT Overijssel Netherlands
BT Netherlands
BT Western Europe
BT Europe

Twiggs Clay (1978)
In Dry Branch Formation. Consists of pale green hackly fuller's earth clay, green hackly clay, gray marl, and calcareous sand at base of member. In E Georgia.
BT upper Eocene
BT Eocene
BT Paleogene
BT Tertiary
BT Cenozoic
SA Dry Branch Formation
SA Georgia

Twiggs County
Valid through 1988. Search in combination with state term. After 1988, use specific county-state term.

Twiggs County Georgia (1989)
Central Gorgia. Before 1989, also search Twiggs County AND Georgia.
CO N322600N325300
W0831200W0833700
BT Georgia
BT United States

Twilight Gneiss (1978)
According to latest published Lexique, the age of Twilight Granite is Precambrian. IN SE Colorado.
BT Precambrian
SA Colorado

Twin Creek Limestone (1985)
Middle and Upper Jurassic. SW Wyoming, SE Idaho and NE Utah.
BT Jurassic
BT Mesozoic
SA Idaho
SA Utah
SA Wyoming

Twin Lakes (1978)
N California and central Colorado. Also in other locations.
IN Index counties or regions as applicable.
SA California
SA Colorado

Twin River Formation (1978)
Upper Eocene to Miocene. Three mappable sequences recognized: lower member consisting of thin-bedded sandstone and siltstone; middle member of massive siltstone that grades westward into bedded siltstone and sandstone; and upper member composed chiefly of massive mudstone. In NW Washington.
IN Index ages as applicable.
BT Tertiary
BT Cenozoic
SA Miocene
SA Oligocene
SA upper Eocene
SA Washington

Twin Sisters Dunite (1978)
An ultrabasic intrusive. In NW Washington.
BT Tertiary
BT Cenozoic
SA Washington

twinning (1978)
SA crystal growth
SA epitaxy
SA lattice

Two Harbors
Valid through 1988. Search in combination with state term. After 1988, use specific city-state term.

Two Harbors Minnesota (1989)
City on Lake Superior in NE Minnesota. Before 1989, search Two Harbors AND Minnesota.
CO N470200N470200
W0914000W0914000
BT Lake County Minnesota
BT Minnesota
BT United States

Two Medicine Formation (1993)
Upper Cretaceous. NW Montana. In Montana Group. Overlies Virgelle Sandstone; underlies Bearpaw Formation.
BT Upper Cretaceous
BT Cretaceous
BT Mesozoic
SA Bearpaw Formation
SA Montana
SA Montana Group

two-dimensional models (1978)
Before 1978, search two-dimensional AND models.
SA four-dimensional models
SA mathematical models
SA models
SA one-dimensional models
SA seismology
SA three-dimensional models

two-layer models (1981)
SA models

two-mica granite (1985)
BT granites
BT plutonic rocks
BT igneous rocks

two-phase models (1981)
SA models

Twocreekan (1981)
Substage of Wisconsinan stage. Before 1981, also search Twocreekian.
BT upper Pleistocene
BT Pleistocene

BT Quaternary
BT Cenozoic
SA Wisconsinan

Tyee Formation (1978)
Overlies Siletz River Volcanic Series. It consists of thick series of rhythmically bedded sandstone and intercalated siltstone. In W Oregon.
BT middle Eocene
BT Eocene
BT Paleogene
BT Tertiary
BT Cenozoic
SA Oregon

Tyler Formation (1989)
Pennsylvanian of central Montana and SW North Dakota. Also, lower Proterozoic of N Michigan and N Wisconsin.
IN Index ages as applicable.
SA Bear Gulch Limestone Member
SA lower Proterozoic
SA Michigan
SA Montana
SA North Dakota
SA Pennsylvanian
SA Wisconsin

Tylopoda (1981)
Infraorder. Autoposting of this term to Camelidae began in 1989. Autoposting of Ruminantia, Eutheria and Theria to this term began in 1989.
BT Ruminantia
BT Artiodactyla
BT Eutheria
BT Theria
BT Mammalia
BT Tetrapoda
BT Vertebrata
BT Chordata
NT Camelidae

Tynagh
No longer a valid term for GeoRef. See Tynagh Ireland.

Tynagh Ireland (1993)
Village in S County Galway in W Ireland.
BT Galway Ireland
BT Ireland
BT Western Europe
BT Europe

Tyne River (1978)
Flows by Newcastle and enters the North Sea at Tynemouth in NE England. Also enters North Sea from Lothian in SE Scotland.
IN Index regions as applicable.
BT Great Britain
BT United Kingdom
BT Western Europe
BT Europe
SA East Lothian Scotland
SA England
SA Lothian region Scotland
SA Scotland

Tyonek Formation (1989)
S Alaska.
BT Paleogene
BT Tertiary
BT Cenozoic
SA Alaska

type localities (1985)
SA archaeology
SA mineral exploration
SA mineralogy
SA paleontology
SA stratigraphy

type sections (1978)
SA cross sections
SA stratigraphic columns
SA stratigraphy
SA stratotypes

type specimens (1978)
Used as a general term.
UF specimens, type
SA holotypes
SA neotypes
SA paratypes
SA phenotypes
SA taxonomy

types
A valid general term through 1977.

typomorphism (1978)
SA fossils
SA minerals

Tyrny-Auz
No longer a valid term for GeoRef. As of 1993, see Tyrny-Auz Russian Federation.

Tyrny-Auz Russian Federation (1993)
Town NW of Ordzhonikidze in the Northern Caucasus. Before 1993, also search Turnyauz. This term has multiple hierarchies.
UF Turnyauz Russian Federation
BT1 Russian Federation
BT1 Commonwealth of Independent States
BT2 Europe

Tyrol
No longer a valid term for GeoRef. See Tyrol Austria.

Tyrol Austria (1993)
State between Germany and Italy in W Austria. Italian part of former Tirol is in Alto Adige, in Trentino-Alto Adige. Also search Tirol.
CO N464500N474500 E0124300E0100400
UF Tirol Austria
BT Austria
BT Central Europe
BT Europe
NT Baumkirchen
NT Innsbruck Austria
NT Lanersbach Austria
NT Lechtal Alps
NT Stubai Alps
NT Vernagt Glacier
SA Allgau Alps
SA Bavarian Alps
SA Inn Valley
SA Isar Valley
SA Otztal Alps
SA Silvretta Group
SA Tauern Window
SA Venediger Group
SA Wetterstein Limestone
SA Zillertal Alps

Tyrone
No longer a valid term for GeoRef. After 1992, see Tyrone Northern Ireland.

Tyrone Northern Ireland (1993)
Traditional county in W Northern Ireland. Includes the smaller administrative region of the same name adopted after 1973.
BT Northern Ireland
BT United Kingdom
BT Western Europe
BT Europe

Tyrrhenian (1978)
Europe. Upper Pleistocene. Above Milazzian, below Versilian.
BT upper Pleistocene

BT Pleistocene
BT Quaternary
BT Cenozoic

Tyrrhenian Basin (1981)
Basin of the Tyrrhenian Sea.
CO N380000N430000 E0160000E0093000
BT Mediterranean Sea
SA Leg 107
SA Tyrrhenian Sea

Tyrrhenian Sea (1978)
Between Corsica and Sardinia on the W, the mainland of Italy on the E, and Sicily on the S. Also search Tyrrhenian.
CO N380000N430000 E0160000E0093000
BT West Mediterranean
BT Mediterranean Sea
NT Orosei
SA DSDP Site 132
SA Leg 107
SA ODP Site 652
SA ODP Site 653
SA ODP Site 654
SA ODP Site 655
SA ODP Site 656
SA Tyrrhenian Basin

Tyumen
No longer a valid term for GeoRef. As of 1993, see Tyumen Russian Federation.

Tyumen Russian Federation (1993)
City and oblast E of Sverdlovsk in W Siberia. This term has multiple hierarchies.
BT1 Russian Federation
BT1 Commonwealth of Independent States
BT2 Asia
NT Salym Field
NT Surgut Russian Federation
NT Yamal-Nenets Russian Federation

U

U
use uranium

U-234 (1978)
Autoposting of broader terms began in 1989. This term has multiple hierarchies.
BT1 radioactive isotopes
BT1 isotopes
BT2 uranium
BT2 actinides
BT2 metals
SA U-234/Th-230
SA U-238/U-234

U-234/Th-230 (1985)
Autoposting of broader terms began in 1989. This term has multiple hierarchies.
UF Th-230/U-234
BT1 radioactive isotopes
BT1 isotopes
BT2 thorium
BT2 actinides
BT2 metals
BT3 uranium
BT3 actinides
BT3 metals
SA isotope ratios
SA Th-230

SA U-234

U-234/U-238
use U-238/U-234

U-235 (1978)
Autoposting of broader terms began in 1989. This term has multiple hierarchies.
BT1 radioactive isotopes
BT1 isotopes
BT2 uranium
BT2 actinides
BT2 metals
SA U-238/U-235

U-235/U-238
use U-238/U-235

U-238 (1978)
Autoposting of broader terms began in 1989. This term has multiple hierarchies.
BT1 radioactive isotopes
BT1 isotopes
BT2 uranium
BT2 actinides
BT2 metals
SA U-238/Pb-204
SA U-238/Th-230
SA U-238/U-234
SA U-238/U-235

U-238/Pb-204 (1989)
For isotopic ratio used in age determination, use U/Pb. This term has multiple hierarchies.
BT1 radioactive isotopes
BT1 isotopes
BT2 lead
BT2 metals
BT3 uranium
BT3 actinides
BT3 metals
SA Pb-204
SA U-238

U-238/Pb-206 (1978)
As of 1989, for isotopic ratio used in age determination, use U/Pb. This term has multiple hierarchies.
BT1 radioactive isotopes
BT1 isotopes
BT2 stable isotopes
BT2 isotopes
BT3 lead
BT3 metals
BT4 uranium
BT4 actinides
BT4 metals
SA isotope ratios

U-238/Th-230 (1985)
Autoposting of broader terms began in 1989. This term has multiple hierarchies.
UF Th-230/U-238
BT1 radioactive isotopes
BT1 isotopes
BT2 thorium
BT2 actinides
BT2 metals
BT3 uranium
BT3 actinides
BT3 metals
SA isotope ratios
SA Th-230
SA U-238

U-238/U-234 (1978)
Autoposting of broader terms began in 1989. This term has multiple hierarchies.
UF U-234/U-238
BT1 radioactive isotopes
BT1 isotopes
BT2 uranium
BT2 actinides
BT2 metals

SA isotope ratios
SA U-234
SA U-238

U-238/U-235 (1985)
Autoposting of broader terms began in 1989. This term has multiple hierarchies.
UF U-235/U-238
BT1 radioactive isotopes
BT1 isotopes
BT2 uranium
BT2 actinides
BT2 metals
SA isotope ratios
SA U-235
SA U-238

U-stage
use universal stage

U. S. Bureau of Mines (1985)
UF U.S. Bureau of Mines
UF USBM
BT government agencies

U. S. Department of Energy (1985)
UF Department of Energy, U.S.
UF DOE
UF U.S. Department of Energy
BT government agencies

U. S. Environmental Protection Agency (1989)
UF Environmental Protection Agency, U.S.
UF EPA
UF U.S. Environmental Protection Agency
BT government agencies
SA Superfund

U. S. Geological Survey (1978)
UF U.S. Geological Survey
UF U.S.G.S.
UF United States Geological Survey
BT survey organizations
BT government agencies
SA AMCOR
SA CUSMAP
SA RASA
SA USGS

U. S. Nuclear Regulatory Commission (1985)
UF NRC
UF U.S. Nuclear Regulatory Commission
BT government agencies

U. S. Rocky Mountains (1981)
Before 1981, search Rocky Mountains AND United States. Autoposting of this term began in 1993. This term has multiple hierarchies.
IN Index states as applicable.
BT1 Rocky Mountains
BT2 United States
NT Absaroka Range
NT Bighorn Mountains
NT Gravelly Range
NT Laramie Mountains
NT Little Belt Mountains
NT Medicine Bow Mountains
NT Owl Creek Mountains
NT San Juan Mountains
NT Sangre de Cristo Mountains
NT Sawatch Range
NT Tendoy Range
NT Tobacco Root Mountains
NT Uinta Mountains
NT Wasatch Range
NT Wet Mountains
NT Wind River Range
SA Beaverhead Mountains
SA Bitterroot Range
SA Bridger Range
SA Central Rocky Mountains
SA Colorado
SA Crazy Mountains
SA Elk Mountains
SA Flint Creek Range
SA Front Range
SA Gore Range
SA Great Divide Basin
SA Idaho
SA Madison Range
SA Mahogany Zone
SA Montana
SA Mosquito Range
SA Mullen Creek-Nash Fork shear zone
SA New Mexico
SA North American Cordillera
SA Northern Rocky Mountains
SA Pikes Peak
SA Red Desert
SA Southern Rocky Mountains
SA Utah
SA Washington
SA Wyoming

U. S. S. R.
Not a valid term for GeoRef after 1971. See USSR. Was used on level 1 in subfile G.

U. S. Virgin Islands (1981)
Virgin Islands of the United States. Before 1981, search Virgin Islands.
BT Virgin Islands
BT Lesser Antilles
BT Antilles
BT West Indies
BT Caribbean region
NT Saint Croix
NT Saint Thomas
SA Saint John Island
SA United States

U.S. Bureau of Mines
use U. S. Bureau of Mines

U.S. Department of Energy
use U. S. Department of Energy

U.S. Environmental Protection Agency
use U. S. Environmental Protection Agency

U.S. Exclusive Economic Zone
use United States Exclusive Economic Zone

U.S. Geological Survey
use U. S. Geological Survey

U.S. Nuclear Regulatory Commission
use U. S. Nuclear Regulatory Commission

U.S.G.S.
use U. S. Geological Survey

U/He (1978)
Isotopic ratio used in age determination.
SA absolute age
SA helium
SA isotope ratios
SA uranium

U/Pb (1978)
Isotopic ratio used in age determination. Before 1972, also search uranium-lead.
UF uranium-lead
SA absolute age
SA isotope ratios
SA lead
SA Pb-210
SA Pb/Pb
SA Pb/Th
SA uranium
U/Th
use Th/U

U/Th/Pb (1978)
Isotopic ratio used in age determination.
SA absolute age
SA isotope ratios
SA lead
SA Th/U
SA thorium
SA uranium
SA uranium disequilibrium

Uaua
No longer a valid term for GeoRef. See Uaua Brazil.

Uaua Brazil (1993)
Village in NE Bahia, E Brazil.
BT Bahia Brazil
BT Brazil
BT South America

Ubangi-Shari
As of 1959, no longer a valid term for GeoRef. Used in subfile E.
use Central African Republic

Ube coal field (1978)
E of Shimonoseki in SW Honshu.
UF Ube Coalfield
BT Yamaguchi Japan
BT Honshu
BT Japan
BT Far East
BT Asia
SA coal fields

Ube Coalfield
use Ube coal field

Uchaly
No longer a valid term for GeoRef. As of 1993, see Uchaly Russian Federation.

Uchaly Russian Federation (1993)
Village in E Bashkiria in the Southern Urals in SE European Russian Federation. This term has multiple hierarchies.
BT1 Bashkiria Russian Federation
BT1 Russian Federation
BT1 Commonwealth of Independent States
BT2 Bashkiria Russian Federation
BT2 Europe

Uchi Lake (1978)
Lake in NW Ontario. After 1993, use Uchi Lake Ontario for the village nearby. Also search Uchi.
BT Kenora District Ontario
BT Ontario
BT Eastern Canada
BT Canada

Uchur River basin (1978)
Central W Khabarovsk Kray and SE Yakutia in E Siberia. Before 1978, also search Uchur; Uchur River. This term has multiple hierarchies.
BT1 Russian Federation
BT1 Commonwealth of Independent States
BT2 Asia
SA Khabarovsk Russian Federation
SA Yakutia Russian Federation

Uda River (1978)
Rises in Irkutsk Oblast and flows into Angara River just above its junction with the Yenisei River in S central Siberia. Also search Uda. This term has multiple hierarchies.
UF Chuma River
BT1 Russian Federation
BT1 Commonwealth of Independent States
BT2 Asia
SA Irkutsk Russian Federation

Udachnaya
No longer a valid term for GeoRef. As of 1993, see Udachnaya Russian Federation.

Udachnaya Pipe (1989)
Kimberlite pipe. This term has multiple hierarchies.
BT1 Yakutia Russian Federation
BT1 Russian Federation
BT1 Commonwealth of Independent States
BT2 Yakutia Russian Federation
BT2 Asia
SA Anabar Shield
SA Siberian Platform

Udachnaya Russian Federation (1993)
Village just S of the Arctic Circle in W central Yakutia. This term has multiple hierarchies.
BT1 Yakutia Russian Federation
BT1 Russian Federation
BT1 Commonwealth of Independent States
BT2 Yakutia Russian Federation
BT2 Asia

Udaipur
No longer a valid term for GeoRef. See Udaipur India.

Udaipur District
use Udaipur India

Udaipur India (1993)
City and district in S Rajasthan, NW India. Before 1985, Udaipur was valid for the city only. Also search Udaipur District.
UF Udaipur District
BT Rajasthan India
BT India
BT Indian Peninsula
BT Asia

Udmurt Russian Federation
use Udmurtia Russian Federation

Udmurtia
No longer a valid term for GeoRef. As of 1993, see Udmurtia Russian Federation.

Udmurtia Russian Federation (1993)
Former Udmurt Autonomous Soviet Sovialist Republic. W of the Urals and NE of Kazan. Before 1978, also search Udmurt. This term has multiple hierarchies.
UF Udmurt Russian Federation
BT1 Russian Federation
BT1 Commonwealth of Independent States
BT2 Europe
SA Ural region

Udokan Mountains (1978)
In area where Amur Oblast, Chita Oblast and Yakutia meet NE of Lake Baikal. Also search Udokan. This term has multiple hierarchies.
BT1 Russian Federation
BT1 Commonwealth of Independent States
BT2 Asia

Udokan Series (1978)

Before then, variants of the term may occur in GeoRef.

Lower Proterozoic. An aggregate of metamorphised sedimentary rocks including schists, phyllites, sandstones, and crystalline limestone. In NW Amur Oblast, N Chita Oblast, and S Yakutia NE of Lake Baikal.
  BT Proterozoic
  BT upper Precambrian
  BT Precambrian
  SA Amur region
  SA Chita Russian Federation
  SA Russian Federation
  SA Yakutia Russian Federation

Ufa
  No longer a valid term for GeoRef. As of 1993, see Ufa Russian Federation.

Ufa Russian Federation (1993)
  City in Bashkiria W of the Southern Urals. This term has multiple hierarchies.
  BT1 Bashkiria Russian Federation
  BT1 Russian Federation
  BT1 Commonwealth of Independent States
  BT2 Bashkiria Russian Federation
  BT2 Europe

Ufaley
  No longer a valid term for GeoRef. As of 1993, see Ufaley Russian Federation.

Ufaley Russian Federation (1993)
  Town in NW Chelyabinsk Oblast in the Central Urals of W Siberia. Before 1993, also search Nizhniy Ufaley.
  IN Index regions as applicable.
  UF Nizhniy Ufaley Russian Federation
  BT Russian Federation
  BT Commonwealth of Independent States
  SA Chelyabinsk Russian Federation

Uganda (1978)
  Former British protectorate which became independent in 1962.
  CO S020000N040000
     E0350000E0290000
  BT East Africa
  BT Africa
  NT Bukusu
  NT Kakuto Uganda
  NT Moroto Mountain
  SA East African Rift
  SA Lake Albert
  SA Lake Edward
  SA Lake Victoria
  SA Mount Baker
  SA Nile River

Uinta Basin (1978)
  NW Colorado and NE Utah.
  IN Index states as applicable.
  UF Uintah Basin
  BT United States
  SA Colorado
  SA Utah

Uinta County
  Valid through 1988. Search in combination with state term. After 1988, use specific county-state term.

Uinta County Wyoming (1989)
  Extreme SW Wyoming. Before 1989, also search Uinta County AND Wyoming.
  CO N410000N413500
     W1100400W1110400
  BT Wyoming
  BT United States
  SA Moxa Arch

Uinta Formation (1985)
  Colorado and NE Utah.
  BT upper Eocene
  BT Eocene
  BT Paleogene
  BT Tertiary
  BT Cenozoic
  SA Colorado
  SA Duchesne River Formation
  SA Utah

Uinta Mountain Group (1985)
  NE Utah and NW Colorado.
  BT Precambrian
  SA Colorado
  SA Utah

Uinta Mountains (1978)
  Range chiefly in NE Utah. This term has multiple hierarchies.
  IN Index counties and states as applicable.
  BT1 U. S. Rocky Mountains
  BT1 Rocky Mountains
  BT2 U. S. Rocky Mountains
  BT2 United States
  SA Utah
  SA Wyoming

Uintah Basin
  use Uinta Basin

Uintah County
  Valid through 1988. Search in combination with state term. After 1988, use specific county-state term.

Uintah County Utah (1989)
  NE Utah. Before 1989, also search Uintah County AND Utah.
  CO N392500N405000
     W1090500W1100400
  BT Utah
  BT United States
  SA Dinosaur National Monument

uintahite
  use gilsonite

uintaite
  use gilsonite

Uintan (1989)
  North America. Middle to upper Eocene. Above Bridgerian; below Duchesnian.
  BT Eocene
  BT Paleogene
  BT Tertiary
  BT Cenozoic

Ukraine (1978)
  Republic. Formerly the Ukrainian Soviet Socialist Republic. This term has multiple hierarchies.
  CO N443000N523000
     E0400000E0220000
  BT1 Europe
  BT2 Commonwealth of Independent States
  NT Artemovsk Ukraine
  NT Bar Ukraine
  NT Belozerka Ukraine
  NT Bitkov Ukraine
  NT Boltyshka Depression
  NT Borshchev Ukraine
  NT Cherkassy Ukraine
  NT Chernigov Ukraine
  NT Crimea Ukraine
  NT Dnepropetrovsk Ukraine
  NT Dolina Ukraine
  NT Glinsk Ukraine
  NT Gorlovka Basin
  NT Il'Intsa
  NT Ingul River
  NT Ingulets River
  NT Ivano-Frankovsk Ukraine
  NT Kharkov Ukraine
  NT Kiev Ukraine
  NT Kirovograd Ukraine
  NT Konstantinovka Ukraine
  NT Kosov Ukraine
  NT Krivoy Rog Basin
  NT Lebedin Ukraine
  NT Lvov Ukraine
  NT Lvov-Volyn Basin
  NT Nikitovka Ukraine
  NT Odessa Ukraine
  NT Pervomaisk Ukraine
  NT Podolia
  NT Pokutye
  NT Poltava Ukraine
  NT Ros River valley
  NT Rozdol Ukraine
  NT Rudki Ukraine
  NT Sob River basin
  NT Tereblya Ukraine
  NT Teterev River valley
  NT Transcarpathia Ukraine
  NT Ukrainian Carpathians
  NT Ukrainian Shield
  NT Vodino Ukraine
  NT Volhynia
  NT Volyn Ukraine
  NT Volyn-Podolia
  NT Zaporozhe Ukraine
  NT Zhdanov Ukraine
  NT Zhitomir Ukraine
  SA Azov region
  SA Beshchady Mountains
  SA Black Sea region
  SA Brest Basin
  SA Bug region
  SA Bug River
  SA Bukovina
  SA Carpathian Foredeep
  SA Carpathian Foreland
  SA Carpathians
  SA Danube Delta
  SA Danube River
  SA Danube Valley
  SA Dnieper Basin
  SA Dnieper River
  SA Dnieper-Donets Basin
  SA Dniester River
  SA Dniester-Prut Interfluve
  SA Donets Basin
  SA Eastern Carpathians
  SA European Platform
  SA Galitsiya
  SA Kiev Member
  SA Korenevskaya Formation
  SA Krivoy Rog Series
  SA Maikop Series
  SA Novorayskoe Formation
  SA Odessa Meteorite
  SA Ovruch Series
  SA Pieniny Klippen Belt
  SA Polesye
  SA Poltava Series
  SA Pripet Basin
  SA Protopivskaya Formation
  SA Prut River
  SA Roztocze
  SA Russian Plain
  SA Russian Platform
  SA Scythian Platform
  SA Serebryanka Formation
  SA Siret River
  SA Steppes region
  SA Subcarpathians
  SA Tisza River
  SA Ukrainian Syneclise

Ukrainian Carpathians (1978)
  E extension of the Carpathians in SW Ukraine. Before 1993, also search Soviet Carpathians. As of 1993, Ukraine is autoposted to this term. This term has multiple hierarchies.
  CO N480000N500000
     E0260000E0230000
  UF Soviet Carpathians
  BT1 Ukraine
  BT1 Europe
  BT2 Ukraine
  BT2 Commonwealth of Independent States
  BT3 Carpathians
  BT3 Europe
  NT Rakhov Massif
  SA Eastern Carpathians
  SA Russian Platform

Ukrainian Shield (1978)
  Extends from the S Polesye in the NW to the Azov Sea. Autoposting of Ukraine to this term began in 1989. This term has multiple hierarchies.
  CO N450000N510000
     E0390000E0280000
  UF Volyno-Azov Massif
  BT1 Ukraine
  BT1 Europe
  BT2 Ukraine
  BT2 Commonwealth of Independent States
  BT3 Russian Platform

Ukrainian Syneclise (1981)
  This term has multiple hierarchies.
  IN Index former Soviet republics as applicable.
  CO N450000N530000
     E0320000E0230000
  BT1 Russian Platform
  BT2 Europe
  SA Ukraine

Ulan Bator
  No longer a valid term for GeoRef. As of 1993 see Ulan Bator Mongolia.

Ulan Bator Mongolia (1993)
  City in N central Mongolia. Before 1993 also search Ulan Bator AND Mongolia.
  UF Urga
  BT Mongolia
  BT Far East
  BT Asia

Ulatisian (1978)
  North America. Above Penutian, below Narizian.
  BT middle Eocene
  BT Eocene
  BT Paleogene
  BT Tertiary
  BT Cenozoic

ulexite (1978)
  UF boronatrocalcite
  UF natroborocalcite
  BT borates

Ulianovsk Russian Federation
  use Ulyanovsk Russian Federation

Ulkan
  No longer a valid term for GeoRef. As of 1993, see Ulkan Russian Federation.

Ulkan Russian Federation (1993)
  Village NW of N Lake Baikal in Irkutsk Oblast. This term has multiple hierarchies.
  BT1 Irkutsk Russian Federation
  BT1 Russian Federation
  BT1 Commonwealth of Independent States
  BT2 Irkutsk Russian Federation
  BT2 Asia

ullmannite (1978)
UF nickel-antimony glance
BT sulfides
SA willyamite

Ulmaceae (1978)
BT Dicotyledoneae
BT angiosperms
BT Spermatophyta
BT Plantae

Ulmus (1985)
Genus.
BT Dicotyledoneae
BT angiosperms
BT Spermatophyta
BT Plantae

Ulsan
No longer a valid term for GeoRef. As of 1993 see Ulsan South Korea.

Ulsan South Korea (1993)
Town in S South Korea. Before 1993 also search Ulsan AND South Korea.
BT South Korea
BT Korea
BT Far East
BT Asia

Ulster
Not a valid term for GeoRef as of 1993. See Northern Ireland.

Ulster County
Valid through 1988. Search in combination with state term. After 1988, use specific county-state term.

Ulster County New York (1989)
S New York. Before 1989, also search Ulster County AND New York.
CO N413500N421000
W0735500W0744500
BT New York
BT United States
NT Shawangunk Mountains

Ultisols (1978)
BT soils
SA soil group

ultrabasic composition
As of 1981, no longer a valid term for GeoRef. Term was introduced in 1978.
use ultramafic composition

ultrabasic rocks
Not a valid term for GeoRef. Before 1971, included use in subfiles E, G, N and B. Now use ultramafics or ultramafic composition.

ultrabasite (1978)
For ultramafic rocks, use ultramafics or ultramafic composition.
UF diaphorite
BT sulfides

ultramafic
A valid term through 1977. After 1977, use ultramafic composition.

ultramafic composition (1978)
Before 1978, search ultramafic; ultrabasic. Before 1981, also search ultrabasic composition.
UF ultrabasic composition
SA composition
SA igneous rocks
SA mafic composition

ultramafic family
As of 1981, no longer a valid term for GeoRef. Use ultramafics or peridotites as applicable. From 1978-1980, was autoposted to the rocks classified under it.

ultramafic rocks
A valid term through 1976. Use ultramafics or ultramafic composition.

ultramafics (1981)
Before 1981, search ultramafic family. From 1978-1980, ultramafic family was autoposted to the rocks that were classified under that term.
BT plutonic rocks
BT igneous rocks
NT ariegite
NT augitite
NT bronzitite
NT chromitite
NT clinopyroxenite
NT diallagite
NT eulysite
NT garnet pyroxenite
NT griquaite
NT hornblendite
NT kimberlite
NT komatiite
NT limburgite
NT meimechite
NT olivinite
NT peridotites
NT pyroxenite
NT websterite
SA ankaramite
SA leucitite
SA melilitite
SA nephelinite
SA ophiolite
SA picrite
SA picrite porphyry

ultrametamorphism (1978)
BT metamorphism

ultramylonite (1989)
UF ultramylonites
BT mylonites
BT metamorphic rocks
SA faults

ultramylonites
use ultramylonite

ultrasonic
use ultrasonic methods

ultrasonic methods (1985)
UF ultrasonic
SA acoustical methods
SA acoustical surveys
SA borehole televiewers
SA geophysical methods
SA geophysical surveys
SA methods

ultrastructure (1978)
Before 1978, also search microstructure.
SA paleontology
SA petrography

ultraviolet spectra (1978)
Before 1978, also search ultraviolet.
UF ultraviolet spectrum
UF UV spectra
BT spectra
SA spectroscopy
SA ultraviolet spectroscopy

ultraviolet spectroscopy (1978)
Before 1978, also search spectroscopy AND ultraviolet.
BT spectroscopy
SA analysis
SA chemical analysis
SA ultraviolet spectra

ultraviolet spectrum
use ultraviolet spectra

Ulu-Tau (1978)
Range W of the Kazakh Hills in Dzhezkazgan Oblast in central Kazakhstan. This term has multiple hierarchies.
UF Ulutau
BT1 Kazakhstan
BT1 Central Asia
BT1 Asia
BT2 Kazakhstan
BT2 Commonwealth of Independent States
SA Dzhezkazgan Kazakhstan

Ulutau
use Ulu-Tau

ulvospinel (1978)
BT oxides
SA iron oxides
SA spinel group

Ulyanovsk
No longer a valid term for GeoRef. As of 1993, see Ulyanovsk Russian Federation.

Ulyanovsk Russian Federation (1993)
Oblast and also city on the right bank of the Volga River S of Kazan. Before 1993, also search Simbirsk or Ulianovsk. This term has multiple hierarchies.
UF Simbirsk Russian Federation
UF Ulianovsk Russian Federation
BT1 Russian Federation
BT1 Commonwealth of Independent States
BT2 Europe

Umanak
No longer a valid term for GeoRef. As of 1993 see Umanak Greenland.

Umanak Greenland (1993)
Settlement on S shore of Umanak Fjord midway up W coast. Before 1993 also search Umanak AND Greenland.
BT Greenland
BT Arctic region

umangite (1978)
Autoposting of sulfides to this term ended in 1989.
BT selenides

Umaria
No longer a valid term for GeoRef. See Umaria India.

Umaria India (1993)
Town in NE Madhya Pradesh, central India.
BT Madhya Pradesh India
BT India
BT Indian Peninsula
BT Asia

Umba (1978)
River in Kenya which flows into Indian Ocean just N of Tanzanian border.
IN Index countries as applicable.
BT Africa
SA Kenya
SA Tanzania

Umberatana Group (1989)
E South Australia.
BT Proterozoic
BT upper Precambrian
BT Precambrian
SA Australia
SA South Australia

Umbria
No longer a valid term for GeoRef. See Umbria Italy.

Umbria Italy (1993)
Autonomous region in central Italy.
CO N422000N434000
E0131500E0115000
BT Italy
BT Southern Europe
BT Europe
NT Perugia Italy
NT Terni Italy
SA Tiber Valley

Umbriel
use Umbriel Satellite

Umbriel Satellite (1989)
One of the satellites of Uranus. Before 1989, also search Umbriel AND Uranus.
UF Umbriel
SA satellites
SA Uranus
SA Voyager Program

Umiat Field (1989)
Oil field. North Slope, N Alaska.
CO N692000N693000
W1521500W1523000
BT Northern Alaska
BT Alaska
BT United States
SA North Slope
SA oil and gas fields

Umm al Qaiwain
use Umm al-Qaiwain

Umm al-Qaiwain (1978)
Emirate. One of federation of 7 states at S end of Persian Gulf.
UF Umm al Qaiwain
BT United Arab Emirates
BT Arabian Peninsula
BT Asia

umohoite (1978)
BT molybdates

Umpqua Formation (1978)
Lower to middle Eocene. Predominantly medium-grained sandstone with some shaly and conglomerate layers. In N California and SW Oregon.
BT Eocene
BT Paleogene
BT Tertiary
BT Cenozoic
SA California
SA Oregon

Umrer
No longer a valid term for GeoRef. See Umrer India.

Umrer India (1993)
Town in central Madhya Pradesh, central India.
BT Madhya Pradesh India
BT India
BT Indian Peninsula
BT Asia

Unalaska Island (1989)
In Fox Islands, Aleutian Islands, SW Alaska.
CO N533400N533400
W1665500W1665500
BT Aleutian Islands
BT Southwestern Alaska
BT Alaska
BT United States
NT Makushin

UNCLOS
use Law of the Sea

Uncompahgre Plateau
  use Uncompahgre Uplift
Uncompahgre Uplift (1989)
  IN   Index states as applicable.
  UF   Uncompahgre Plateau
  BT   United States
  SA   Colorado
  SA   New Mexico
  SA   Paradox Basin
  SA   Utah
unconfined aquifers (1978)
  BT   aquifers
  SA   ground water
  SA   shallow aquifers
  SA   shallow depth
  SA   surficial aquifers
  SA   water table
unconformities (1978)
  UF   unconformity
  NT   angular unconformities
  NT   erosional unconformities
  SA   hardground
  SA   lithostratigraphy
  SA   sequence stratigraphy
  SA   stratigraphic gaps
  SA   stratigraphy
unconformity
  use unconformities
unconformity-type (1989)
  SA   mineral deposits, genesis
  SA   uranium ores
unconsolidated materials (1978)
  Before 1978, search unconsolidated.
  SA   coarse-grained materials
  SA   compactness
  SA   consolidation
  SA   fines
  SA   materials
  SA   particulate materials
  SA   sediments
  SA   soil mechanics
  SA   soils
UNCTAD (1981)
  Acronym.
  UF   United Nations Conference on Trade and Development
  SA   associations
  SA   economics
  SA   international cooperation
  SA   United Nations
Unda
  No longer a valid term for GeoRef. As of 1993, see Unda Russian Federation.
Unda Russian Federation (1993)
  Village in S Chita Oblast in Transbaikalia. This term has multiple hierarchies.
  BT1 Chita Russian Federation
  BT1 Russian Federation
  BT1 Commonwealth of Independent States
  BT2 Chita Russian Federation
  BT2 Asia
undation (1978)
  SA   crust
  SA   tectonics
underclay (1978)
  UF   coal clay
  UF   root clay
  UF   seat clay
  UF   seat earth
  UF   underearth
  SA   clay
  SA   coal seams
  SA   fireclay
underearth
  use underclay

underground
  A valid term through 1977. After 1977, use underground space.
underground cavities (1993)
  Use solution cavities if applicable. Also see underground space for engineering geology.
  SA   caverns
  SA   caves
  SA   geomorphology
  SA   solution cavities
  SA   underground space
  SA   volcanic features
underground channels (1981)
  SA   channels
  SA   drainage
  SA   ground water
  SA   infiltration galleries
  SA   irrigation
  SA   underground installations
  SA   waterways
underground disposal (1985)
  UF   underground waste disposal
  SA   far-field
  SA   landfills
  SA   near-field
  SA   radioactive waste
  SA   underground installations
  SA   underground storage
  SA   waste disposal
  SA   waste disposal sites
underground installations (1978)
  Used for geological studies on underground cavities (natural or otherwise), excluding tunnels.
  UF   installations, underground
  SA   construction
  SA   design
  SA   engineering geology
  SA   excavations
  SA   far-field
  SA   feasibility studies
  SA   foundations
  SA   gas storage
  SA   geologic hazards
  SA   land subsidence
  SA   leaking underground storage tanks
  SA   mines
  SA   near-field
  SA   oil storage
  SA   permafrost
  SA   pipelines
  SA   rock mechanics
  SA   seepage
  SA   seismic response
  SA   site exploration
  SA   slope stability
  SA   soil mechanics
  SA   stability
  SA   storage
  SA   tunnels
  SA   underground channels
  SA   underground disposal
  SA   underground mining
  SA   underground space
  SA   underground storage
  SA   waste disposal
underground mines
  use underground mining
underground mining (1982)
  UF   subsurface mining
  UF   underground mines
  BT   mining
  SA   cut-and-fill mining
  SA   longwall mining
  SA   mines
  SA   mining geology
  SA   roof control
  SA   underground installations
underground space (1978)

  UF   space, underground
  SA   land use
  SA   underground cavities
  SA   underground installations
  SA   underground storage
underground storage (1982)
  UF   subsurface storage
  SA   far-field
  SA   gas storage
  SA   high-level waste
  SA   leaking underground storage tanks
  SA   low-level waste
  SA   near-field
  SA   oil storage
  SA   storage
  SA   Stripa region
  SA   underground disposal
  SA   underground installations
  SA   underground space
  SA   waste disposal
  SA   waste disposal sites
  SA   Waste Isolation Pilot Plant
underground streams (1982)
  SA   caverns
  SA   caves
  SA   exsurgence
  SA   geomorphology
  SA   ground water
  SA   hydrogeology
  SA   hydrology
  SA   karst
  SA   springs
  SA   streams
  SA   tributaries
  SA   unsaturated zone
underground waste disposal
  use underground disposal
underground water
  No longer a valid GeoRef index term. Before 1945, was used on levels 1 and 2 in subfiles E and N.
  use ground water
Underhill Formation (1989)
  IN   Index ages as applicable.
  SA   Lower Cambrian
  SA   upper Proterozoic
  SA   Vermont
underplating (1993)
  Used to refer to different concepts in plate tectonics: In areas of rifting and thinning of continental crust, used to describe emplacement of basaltic magma at the base of the continental crust. In oceanic subduction zones, used to refer to accretion of material from the lower plate onto the underside of the upper plate. For underthrusting of fault slices in subduction zones, use underthrust faults.
  UF   subcretion
  SA   accretionary wedges
  SA   active margins
  SA   continental crust
  SA   extension tectonics
  SA   lithosphere
  SA   oceanic crust
  SA   plate tectonics
  SA   rifting
  SA   subduction zones
  SA   underthrust faults
underthrust faults (1978)
  Before 1978, search underthrust AND faults.
  BT   faults
  SA   underplating
UNDP (1981)
  UF   United Nations Development Programme

  SA   associations
  SA   United Nations
undulatory extinction (1981)
  SA   microscope methods
  SA   optical extinction
Ungava (1978)
  Region including Quebec N of Eastmain River and SW Labrador.
  IN   Index Labrador and/or Quebec as applicable.
  CO   N580000N623000
       W0690000W0780000
  UF   New Quebec
  BT   Canada
  SA   Labrador
  SA   Quebec
Ungulata (1978)
  Informal grouper for the hoofed mammals. Includes Amblypoda, Artiodactyla, Condylarthra, Dinocerata, Litopterna, Pantodonta, Perissiodactyla, Pyrotheria, and Xenungulata. Autoposting of Eutheria and Theria to this term began in 1989. Before 1981, also search ungulates. As of 1981, ungulates may be used as the common name for Ungulata.
  BT   Eutheria
  BT   Theria
  BT   Mammalia
  BT   Tetrapoda
  BT   Vertebrata
  BT   Chordata
  SA   Amblypoda
  SA   Artiodactyla
  SA   Condylarthra
  SA   Dinocerata
  SA   Litopterna
  SA   Pantodonta
  SA   Perissodactyla
  SA   Pyrotheria
  SA   Xenungulata
uniaxial tests (1978)
  Before 1978, search uniaxial.
  SA   engineering geology
  SA   testing
  SA   triaxial tests
UNIDO (1981)
  UF   United Nations Industrial Development Organization
  SA   associations
  SA   United Nations
uniformitarianism (1978)
  UF   actualism
  SA   catastrophism
  SA   creationism
  SA   deposition
  SA   geology
  SA   history
  SA   sedimentation
  SA   stratigraphy
Union County
  Valid through 1988. Search in combination with state term. After 1988, use specific county-state term.
Union County Arkansas (1989)
  S Arkansas. Before 1989, also search Union County AND Arkansas.
  CO   N330100N332100
       W0920500W0925900
  BT   Arkansas
  BT   United States
Union County Florida (1989)
  N Florida. Before 1989, also search Union County AND Florida.
  CO   N295500N301300

W0820700W0823100
BT Florida
BT United States

Union County Georgia (1989)
N Georgia. Before 1989, also search Union County AND Georgia.
CO N343800N345900
W0834700W0841100
BT Georgia
BT United States

Union County Illinois (1989)
S Illinois. Before 1989, also search Union County AND Illinois.
CO N372000N373500
W0890200W0893300
BT Illinois
BT United States

Union County Indiana (1989)
E Indiana. Before 1989, also search Union County AND Indiana.
CO N393200N394300
W0845200W0850200
BT Indiana
BT United States

Union County Iowa (1989)
S Iowa. Before 1989, also search Union County AND Iowa.
CO N405300N410900
W0940100W0942800
BT Iowa
BT United States

Union County Kentucky (1989)
W Kentucky. Before 1989, also search Union County AND Kentucky.
CO N373000N375300
W0874000W0881000
BT Kentucky
BT United States

Union County Mississippi (1989)
N Mississippi. Before 1989, also search Union County AND Mississippi.
CO N342300N343600
W0884300W0891500
BT Mississippi
BT United States

Union County New Jersey (1989)
Before 1989, also search Union County AND New Jersey.
BT New Jersey
BT United States
SA Watchung Mountains

Union County New Mexico (1989)
Extreme NE New Mexico. Before 1989, also search Union County AND New Mexico.
CO N354500N370000
W1030000W1040000
BT New Mexico
BT United States
SA Pasamonte Meteorite

Union County North Carolina (1989)
S North Carolina. Before 1989, also search Union County AND North Carolina.
CO N344800N351200
W0801800W0805000
BT North Carolina
BT United States

Union County Ohio (1989)
Central Ohio. Before 1989, also search Union County AND Ohio.
CO N400700N403100
W0831000W0833300
BT Ohio
BT United States

Union County Oregon (1989)
NE Oregon. Before 1989, also search Union County AND Oregon.
CO N445600N455100
W1171700W1184200
BT Oregon
BT United States
SA Wallowa Mountains

Union County Pennsylvania (1989)
Central Pennsylvania. Before 1989, also search Union County AND Pennsylvania.
CO N404800N410800
W0765000W0772200
BT Pennsylvania
BT United States

Union County South Carolina (1989)
N South Carolina. Before 1989, also search Union County AND South Carolina.
CO N342700N345400
W0812500W0815100
BT South Carolina
BT United States

Union County South Dakota (1989)
SE South Dakota. Before 1989, also search Union County AND South Dakota.
CO N422800N430600
W0962700W0964800
BT South Dakota
BT United States

Union County Tennessee (1989)
NE Tennessee. Before 1989, also search Union County AND Tennessee.
CO N360900N362300
W0834200W0840000
BT Tennessee
BT United States

Union Islands
use Tokelau

Union of Soviet Socialist Republics
No longer a valid GeoRef index term. Before 1969, was used on level 1 in subfile E. See USSR.

Unionidae (1978)
BT Bivalvia
BT Mollusca
BT Invertebrata

unit cell (1978)
UF cell, unit
SA crystal structure
SA crystal systems
SA lattice
SA lattice parameters

United Arab Emirates (1978)
Federation of 7 states which was achieved in 1972. Formerly known as Trucial States, Trucial Oman, or Trucial Coast. Not strictly equivalent to Trucial Coast, but replaces Trucial Coast. Before 1973, also search Trucial Oman. Before 1975, also search Trucial Coast.
CO N224000N255000
E0563000E0510000
UF Trucial Coast
UF Trucial Oman
BT Arabian Peninsula
BT Asia
NT Abu Dhabi
NT Ajman
NT Dubai
NT Fujairah
NT Ras al-Khaimah
NT Sharjah
NT Umm al-Qaiwain
SA Arab Formation
SA Middle East
SA Near East
SA Semail Ophiolite
SA Thamama Group

United Kingdom (1978)
Comprising Great Britain and Northern Ireland. Autoposting of this term began in 1981.
IN Index political divisions as applicable.
CO N500000N610000
E0110000E0020000
BT Western Europe
BT Europe
NT English Channel Islands
NT Great Britain
NT Isle of Man
NT Northern Ireland
SA Aberystwyth Grits
SA Ballantrae Complex
SA Barton Beds
SA Bembridge Marls
SA British Virgin Islands
SA London Clay
SA Lower Greensand
SA Oxford Clay
SA Penrhyn Slate
SA Shap Granite
SA Sherwood Sandstone
SA Skiddaw Slates
SA Speeton Clay
SA Upper Old Red Sandstone
SA Weald Clay
SA Wenlock Limestone

United Nations (1978)
SA associations
SA International Monetary Fund
SA UNCTAD
SA UNDP
SA UNIDO
SA World Bank

United Nations Conference on Trade and Development
use UNCTAD

United Nations Convention on the Law of the Sea
use Law of the Sea

United Nations Development Programme
use UNDP

United Nations Industrial Development Organization
use UNIDO

United States (1978)
Autoposting of this term to Atlantic Coastal Plain began in 1981. In 1978, this term began autoposting to Eastern U.S., New England, Southwestern U.S., Southern U.S., Western U.S., Pacific Coast, and Midwest. After 1985, this term is autoposted to Southeastern U.S. As of 1993, this term is autoposted to all states and to most U.S. features larger than a state.
NT Absaroka Fault
NT Alabama
NT Alaska
NT Albuquerque Basin
NT Allegheny Front
NT Allegheny Mountains
NT Allegheny Plateau
NT Amargosa Desert
NT Amite River
NT Anacostia River basin
NT Anadarko Basin
NT Ardmore Basin
NT Arizona
NT Arkansas
NT Arkansas River
NT Arkansas River valley
NT Arkoma Basin
NT Atlantic Coastal Plain
NT Basin and Range Province
NT Battle Mountain High
NT Benton Uplift
NT Black Warrior Basin
NT Blue Ridge Mountains
NT Bronson Hill Anticlinorium
NT California
NT Carolina Bays
NT Cascade Range
NT Catskill Delta
NT Cedar Creek Anticline
NT Chadron Arch
NT Chattahoochee River
NT Chemung River
NT Chesapeake Bay
NT Chicot Aquifer
NT Choptank River
NT Cincinnati Arch
NT Clark's Fork Basin
NT Colorado
NT Colorado Lineament
NT Colorado Plateau
NT Columbia Plateau
NT Columbia River estuary
NT Connecticut
NT Connecticut River
NT Connecticut Valley
NT Crawford Thrust
NT Culpeper Basin
NT Dan River basin
NT Delaware
NT Delaware Basin
NT Delaware Bay
NT Delaware River
NT Delaware River basin
NT Delmarva Peninsula
NT Denver Basin
NT Dinosaur National Monument
NT District of Columbia
NT Dunkard Basin
NT Eastern U.S.
NT Florida
NT Forest City Basin
NT Gallatin Range
NT Genesee River
NT Georgia
NT Gettysburg Basin
NT Gila River
NT Great Smoky Fault
NT Great Smoky Mountains
NT Hawaii
NT Hayfork Terrane
NT High Plains Aquifer
NT Hugoton Embayment
NT Hurricane Ridge Syncline
NT Idaho
NT Idaho Batholith
NT Illinois
NT Indiana
NT Iowa
NT Ismay Zone
NT Kansas
NT Kentucky
NT Kings Mountain Belt
NT Kiokee Belt
NT Klamath Mountains
NT Lake Char Fault
NT Lake Chatuge
NT Lake Powell
NT Lake Tahoe
NT Lewis and Clark Lineament
NT Little Missouri River basin
NT Llano Estacado
NT Louisiana
NT Magothy Aquifer
NT Maine

NT Maryland
NT Massachusetts
NT Maumee River valley
NT Merrimack River valley
NT Merrimack Synclinorium
NT Michigan
NT Midcontinent
NT Midcontinent geophysical anomaly
NT Middle Atlantic Bight
NT Midwest
NT Minnesota
NT Mississippi
NT Mississippi Embayment
NT Mississippi River
NT Mississippi Sound
NT Mississippi Valley
NT Missouri
NT Missouri Plateau
NT Missouri River
NT Missouri River valley
NT Mohave Mountains
NT Montana
NT Monticello Reservoir
NT Mount Adams
NT Moxa Arch
NT Mullen Creek-Nash Fork shear zone
NT Narragansett Basin
NT Nashville Dome
NT Navajo Indian Reservation
NT Nebraska
NT Nemaha Ridge
NT Neosho River valley
NT Nevada
NT New Hampshire
NT New Jersey
NT New Madrid region
NT New Mexico
NT New York
NT Newark Basin
NT Newark-Gettysburg Basin
NT North Carolina
NT North Dakota
NT Ogallala Aquifer
NT Ohio
NT Ohio River
NT Ohio River basin
NT Ohio River valley
NT Okefenokee Swamp
NT Oklahoma
NT Oregon
NT Orogrande Basin
NT Ouachita Belt
NT Ouachita Mountains
NT Owyhee Mountains
NT Ozark Mountains
NT Palo Duro Basin
NT Pecos River
NT Pecos River valley
NT Pennsylvania
NT Permian Basin
NT Pilot Range
NT Pine Mountain Window
NT Potomac River
NT Potomac River basin
NT Powder River basin
NT Pulaski thrust sheet
NT Raft River
NT Ramapo Fault
NT Rangeley Lakes
NT Raritan Bay
NT Reading Prong
NT Reelfoot Rift
NT Rhode Island
NT Sabine Uplift
NT Salisbury Embayment
NT Savannah River
NT Selway-Bitterroot Wilderness
NT Sevier orogenic belt
NT Shenandoah Valley
NT South Carolina
NT South Dakota
NT South Mountain
NT South Platte River

NT South Platte River valley
NT Southern U.S.
NT Southwestern U.S.
NT Susquehanna River
NT Susquehanna River basin
NT Tennessee
NT Tennessee River
NT Tennessee Valley
NT Texas
NT Tombigbee River
NT Trans-Pecos
NT Truckee River
NT U. S. Rocky Mountains
NT Uinta Basin
NT Uncompahgre Uplift
NT Utah
NT Verdigris River valley
NT Vermont
NT Virgin River valley
NT Virginia
NT Wasatch fault zone
NT Wasatch Front
NT Washakie Basin
NT Washington
NT Washita River valley
NT West Virginia
NT Western U.S.
NT Wiggins Arch
NT Wisconsin
NT Wyoming
NT Wyoming Province
NT Yakima fold belt
NT Yellowstone National Park
SA AMCOR
SA Appalachians
SA Bear River Range
SA Beaverhead Mountains
SA Bighorn Basin
SA Black Hills
SA Blue Mountains
SA Brevard Zone
SA California Current
SA Cambridge Arch
SA Canadian River
SA Canyon Group
SA Central Basin Platform
SA Charlotte Belt
SA Clark Fork
SA Clearwater River
SA COCORP
SA conterminous regions
SA Cumberland Plateau
SA CUSMAP
SA Deschutes River
SA Desert Creek Zone
SA Disturbed Belt
SA Eastern Gas Shales Project
SA Eastern Overthrust Belt
SA Elk River
SA Evangeline Aquifer
SA Flint Hills
SA Four Corners
SA Great Bay
SA Great Lakes
SA Great Lakes region
SA Great Plains
SA Green River basin
SA Guadalupe Mountains
SA Gulf Coastal Plain
SA Hardeman Basin
SA Hartford Basin
SA Homestake Mine
SA Hudson River
SA Hudson Valley
SA Hueco Mountains
SA Huronian
SA Illinois Basin
SA Illinois River
SA Indian reservations
SA Kings Mountain
SA Kootenay Arc
SA Lake Bonneville
SA Lake Chicago
SA Lake Lahontan
SA Lake Mead

SA Lake Michigan
SA Lake Missoula
SA Lake Uinta
SA Madison Aquifer
SA Mississippi River basin
SA Missouri River basin
SA Mount Jefferson
SA NASA
SA National Coal Resources Data System
SA NAWDEX
SA New River
SA NOAA
SA North America
SA North American Cordillera
SA NSF
SA NURE
SA Paradox Basin
SA Pearl River
SA Platte River basin
SA Raft River basin
SA RARE II regions
SA RASA
SA Raton Basin
SA Rattlesnake Hills
SA Rock Creek Park
SA Rocky Mountains
SA Rome Trough
SA Sabine Lake
SA Saco River
SA San Francisco Mountains
SA San Juan Basin
SA San Luis Valley
SA Sand Hills
SA Sauk Sequence
SA Severn Formation
SA Shadow Mountains
SA Snake River
SA Snake River basin
SA Snake River canyon
SA Southern Rocky Mountains
SA Taconic Allochthon
SA Taconic Orogeny
SA Transcontinental Arch
SA U. S. Virgin Islands
SA United States Exclusive Economic Zone
SA Western Canada Basin
SA Western Interior
SA Western Interior Seaway
SA Whitewater River valley
SA Yellowstone River

United States Exclusive Economic Zone (1985)
Before 1985, also search EEZ or Exclusive Economic Zone AND United States.
UF U.S. Exclusive Economic Zone
SA economics
SA Exclusive Economic Zone
SA exploration
SA Gulf of Mexico
SA Law of the Sea
SA mineral exploration
SA North American Atlantic
SA North American Pacific
SA ocean floors
SA offshore
SA policy
SA strategic minerals
SA United States

United States Geological Survey
use U. S. Geological Survey

units, stratigraphic
use stratigraphic units

univariate analysis (1989)
BT statistical analysis
SA analysis
SA multivariate analysis

Universal Soil Loss Equation (1989)
UF USLE
SA equations

SA erodibility
SA erosion
SA mathematical methods
SA runoff
SA soil erosion
SA soils

universal stage (1978)
UF Fedorov stage
UF stage, universal
UF U-stage
SA microscope methods
SA ore microscopy
SA structural analysis

universities
use academic institutions

University of Arizona (1978)
In Tucson.
SA academic institutions
SA Arizona

University of Bonn (1978)
In Bonn, North Rhine-Westphalia.
SA academic institutions
SA Germany

University of California (1978)
Has geoscience departments located in Berkeley, Davis, Los Angeles, Riverside, San Diego, Santa Barbara and Santa Cruz.
SA academic institutions
SA California
SA Los Alamos Scientific Laboratory

University of Cambridge
use Cambridge University

University of Louvain (1978)
In Louvain E of Brussels.
SA academic institutions
SA Belgium

University of Lund (1978)
In Lund, Malmohus, extreme S Sweden.
SA academic institutions
SA Sweden

University of Michigan (1978)
In Ann Arbor W of Detroit.
SA academic institutions
SA Michigan

University of Pennsylvania (1978)
In Philadelphia.
SA academic institutions
SA Pennsylvania

University of Rome (1978)
In Rome.
SA academic institutions
SA Italy

University of Tokyo (1978)
In Tokyo.
UF Tokyo Imperial University
SA academic institutions
SA Japan

University of Wisconsin (1978)
Has geoscience departments located in Madison, Eau Claire, Green Bay, Wausau (Marathon County campus), Milwaukee, Oshkosh, Kenosha (Parkside campus), Platteville, River Falls, Stevens Point, Superior and Whitewater.
SA academic institutions
SA Wisconsin

Unkar Group (1978)
Grand Canyon Series. Divided into (descending) Dox Sandstone, Shinumo Quartzite, Hakatai Shale, Bass Limestone, and

Hotuata Conglomerate. In N Arizona.
BT Precambrian
SA Arizona

unmixing
use exsolution

unsaturated zone (1981)
Subsurface zone with pressure less than atmospheric. In GeoRef, it is not considered ground water.
UF vadose zone
UF zone of aeration
SA capillary water
SA ground water
SA hydrogeology
SA infiltration
SA Richards equation
SA saturated zone
SA soils
SA underground streams
SA water regimes
SA water table

Unst (1978)
Northernmost large island of the Shetland Islands. In 1985, broader term changed from Scotland to Shetland Islands. This term has multiple hierarchies.
BT1 Shetland Islands
BT1 Scotland
BT1 Great Britain
BT1 United Kingdom
BT1 Western Europe
BT1 Europe
BT2 Shetland Islands
BT2 Atlantic Ocean Islands

unsteady flow (1982)
UF transient flow
SA fluid dynamics
SA hydraulics
SA hydrodynamics
SA hydrology
SA steady flow

Unstrut River (1978)
Tributary of the Saale River. Also search Unstrut.
IN Index states as applicable.
BT Germany
BT Central Europe
BT Europe
SA Saxony-Anhalt Germany
SA Thuringia Germany

Unzen (1978)
Active volcano. In Unzen National Park, Nagasaki Prefecture on central Shimabar Peninsula E of Nagasaki.
UF Unzen Volcano
BT Nagasaki Japan
BT Kyushu
BT Japan
BT Far East
BT Asia

Unzen Volcano
use Unzen

Uonuma Group (1978)
Upper Pliocene to lower Pleistocene. Composed of Oguni and Tsukayama formations. In NW Honshu.
IN Index ages as applicable.
BT Cenozoic
SA Honshu
SA Japan
SA lower Pleistocene
SA upper Pliocene

uplands (1993)
Use more specific term if applicable.
SA geomorphology
SA highlands
SA mesas
SA plateaus

uplifts (1978)
SA domes
SA epeirogeny
SA glacial rebound
SA isostatic rebound
SA neotectonics
SA subsidence
SA tectonics
SA vertical movements

upper Albian (1985)
BT Albian
BT Lower Cretaceous
BT Cretaceous
BT Mesozoic

upper Archean (1985)
From 1985 to 1989, Precambrian was autoposted to this term. Autoposting of Precambrian to this term began in 1990.
BT Archean
BT Precambrian
NT Amitsoq Gneiss
NT Qorqut Granite
SA Ventersdorp Supergroup

upper atmosphere
As of 1993, no longer a valid term for GeoRef. Usually out-of-scope.

Upper Austria (1978)
CO N472600N484700 E0150000E0124500
BT Austria
BT Central Europe
BT Europe
NT Gosau Austria
NT Linz Austria
NT Weyer Austria
SA Enns Valley
SA North Austrian Crystallines
SA Salzach River
SA Salzkammergut
SA Totes Gebirge

Upper Bavaria
No longer a valid term for GeoRef. See Upper Bavaria Germany.

Upper Bavaria Germany (1993)
An administrative division in the extreme S in which Munich, the Bavarian Alps and Salzburg Alps are located.
BT Bavaria Germany
BT Germany
BT Central Europe
BT Europe
NT Berchtesgaden Germany
NT Kreuth Germany
NT Munich Germany
SA Bavarian Alps

upper boundary
A valid general term through 1976. Now use boundary for stratigraphic meaning.

Upper Cambrian (1978)
Autoposting of this term began in 1978.
BT Cambrian
BT Paleozoic
NT Bonneterre Formation
NT Conococheague Formation
NT Dresbachian
NT Dunderberg Shale
NT Eau Claire Formation
NT Franconia Formation
NT Galesville Sandstone
NT Goldenville Formation
NT Lamotte Sandstone
NT Little Falls Formation
NT Mount Simon Sandstone
NT Nopah Formation
NT Orr Formation
NT Pilgrim Formation
NT Potsdam Sandstone
NT Reagan Sandstone
NT Timbered Hills Group
NT Trempealeauan
NT Wilberns Formation
NT Wonewoc Formation
SA Arbuckle Group
SA Deadwood Formation
SA Knox Group
SA Lower Cambrian
SA Middle Cambrian
SA Road River Formation

upper Campanian (1985)
BT Campanian
BT Senonian
BT Upper Cretaceous
BT Cretaceous
BT Mesozoic

Upper Carboniferous (1978)
Autoposting of this term began in 1978.
BT Carboniferous
BT Paleozoic
NT Bashkirian
NT Coal Measures
NT Gzhelian
NT Karharbari Stage
NT Kasimovian
NT Millstone Grit
NT Moscovian
NT Namurian
NT Stephanian
NT Talchir Series
NT Uralian
NT Westphalian
SA Dinantian
SA Gondwana System
SA Itarare Subgroup
SA lower Gondwana System
SA Middle Carboniferous
SA Pennsylvanian
SA Talchir Formation

upper Cenomanian (1989)
Above Albian; below Turonian.
BT Cenomanian
BT Upper Cretaceous
BT Cretaceous
BT Mesozoic

upper Cenozoic (1978)
BT Cenozoic
SA Quaternary

Upper Cretaceous (1978)
Autoposting of this term began in 1978.
BT Cretaceous
BT Mesozoic
NT Almond Formation
NT Ariyalur Stage
NT Bagh Beds
NT Bearpaw Formation
NT Belly River Formation
NT Black Creek Formation
NT Blackhawk Formation
NT Blair Formation
NT Bluffport Marl Member
NT Blufftown Formation
NT Bridge Creek Limestone Member
NT Buda Limestone
NT Cardium Formation
NT Carlile Shale
NT Cenomanian
NT Codell Sandstone Member
NT Cody Shale
NT Coker Formation
NT Colville Group
NT Crevasse Canyon Formation
NT Cupido Formation
NT Demopolis Chalk
NT Djadokhta Formation
NT Duwi Formation
NT Eagle Sandstone
NT Edmonton Formation
NT Elkhorn Mountains Volcanics
NT Eutaw Formation
NT Ferron Sandstone Member
NT Forbes Formation
NT Fort Hays Limestone Member
NT Fox Hills Formation
NT Frontier Formation
NT Fruitland Formation
NT Gallup Sandstone
NT Gammon Ferruginous Member
NT Ghareb Formation
NT Greenhorn Limestone
NT Gulfian
NT Harebell Formation
NT Hell Creek Formation
NT Hilliard Shale
NT Holz Shale
NT Hornbrook Formation
NT Horseshoe Canyon Formation
NT Judith River Formation
NT K-T boundary
NT Kirtland Shale
NT Kodiak Formation
NT La Luna Formation
NT La Ventana Sandstone
NT Ladd Formation
NT Lance Formation
NT Laramie Formation
NT Lewis Shale
NT Lopez de Bertodano Formation
NT Magothy Formation
NT Marca Shale Member
NT Marshalltown Formation
NT Menefee Formation
NT Menuha Formation
NT Mesaverde Group
NT Middendorf Formation
NT Milk River Formation
NT Monmouth Group
NT Montana Group
NT Mooreville Chalk
NT Moreno Formation
NT Mowry Shale
NT Navesink Formation
NT Niobrara Formation
NT Ojo Alamo Sandstone
NT Oldman Formation
NT Parkman Sandstone
NT Peedee Formation
NT Pictured Cliffs Sandstone
NT Pierre Shale
NT Point Loma Formation
NT Point Lookout Sandstone
NT Price River Formation
NT Raritan Formation
NT Ripley Formation
NT Rock Springs Formation
NT Rosario Formation
NT Saratoga Chalk
NT Selma Group
NT Senonian
NT Shannon Sandstone Member
NT Smoky Hill Chalk Member
NT Star Point Sandstone
NT Sussex Sandstone Member
NT Teapot Sandstone Member
NT Turonian
NT Tuscaloosa Formation
NT Two Medicine Formation
NT Wall Creek Member
NT Williams Fork Formation
NT Williams Formation
NT Woodbury Clay
SA Bauru Formation
SA Beaverhead Formation
SA Comanchean
SA Dakhla Shale
SA Deccan Traps
SA Denver Formation

SA Difunta Group
SA Fort Union Formation
SA Great Valley Sequence
SA Intertrappean Beds
SA Laramide Orogeny
SA Lower Cretaceous
SA Maravillas Formation
SA Middle Cretaceous
SA Mount Scopus Group
SA Nanaimo Group
SA Niniyur Group
SA North Horn Formation
SA Paskapoo Formation
SA Raton Formation
SA Severn Formation
SA Sheep Pass Formation
SA Tor Formation
SA Vraconian
SA Washita Group

upper crust (1978)
BT crust
SA lower crust

Upper Devonian (1978)
Autoposting of this term began in 1978.
BT Devonian
BT Paleozoic
NT Birdbear Formation
NT Brallier Shale
NT Chemung Formation
NT Cleveland Member
NT Duperow Formation
NT Famennian
NT Frasnian
NT Greenland Gap Group
NT Grosmont Formation
NT Hampshire Formation
NT Huron Member
NT Jefferson Group
NT Kanayut Conglomerate
NT Lime Creek Formation
NT New Albany Shale
NT Nisku Formation
NT Ohio Shale
NT Olentangy Shale
NT Palliser Formation
NT Perry Formation
NT Sonyea Group
NT Strunian
NT West Falls Formation
SA Antrim Shale
SA Bakken Formation
SA Chattanooga Shale
SA Endicott Group
SA Lower Devonian
SA Martin Formation
SA Middle Devonian
SA Pilot Shale
SA Swan Hills Formation

upper Eocene (1978)
Autoposting of this term began in 1978.
BT Eocene
BT Paleogene
BT Tertiary
BT Cenozoic
NT Auversian
NT Barnwell Formation
NT Bartonian
NT Cowlitz Formation
NT Dry Branch Formation
NT Jackson Group
NT La Meseta Formation
NT Lattorfian
NT Ludian
NT Mission Valley Formation
NT Moodys Branch Formation
NT Ocala Group
NT Piney Point Formation
NT Poway Conglomerate
NT Priabonian
NT Shubuta Member
NT Tejon Formation

NT Tepee Trail Formation
NT Tongrian
NT Twiggs Clay
NT Uinta Formation
NT Whitsett Formation
NT Yazoo Clay
SA lower Eocene
SA middle Eocene
SA Narizian
SA Sespe Formation
SA Twin River Formation

Upper Franconia
No longer a valid term for GeoRef. As of 1993 see Upper Franconia Germany.

Upper Franconia Germany (1993)
Administration division in NE Bavaria. Part of old historical region of Franconia.
BT Bavaria Germany
BT Germany
BT Central Europe
BT Europe
SA Franconia

upper Gondwana
use upper Gondwana System

upper Gondwana System (1985)
Peninsular India.
UF upper Gondwana
BT Gondwana System
SA Athgarh Sandstone
SA Cretaceous
SA Gondwana
SA Jurassic
SA Lower Cretaceous
SA lower Gondwana System
SA Mesozoic
SA Triassic

upper Holocene (1978)
BT Holocene
BT Quaternary
BT Cenozoic

Upper Jurassic (1978)
Autoposting of this term began in 1978.
BT Jurassic
BT Mesozoic
NT Arab Formation
NT Argovian
NT Bossier Formation
NT Bowser Lake Group
NT Brushy Basin Shale Member
NT Buckner Formation
NT Cotton Valley Group
NT Entrada Sandstone
NT Galice Formation
NT Haynesville Formation
NT Jabalpur Series
NT Josephine Ophiolite
NT Josephine Peridotite
NT Kimmeridge Clay
NT Kimmeridgian
NT Lusitanian
NT Malm
NT Morrison Formation
NT Naknek Formation
NT Oxfordian
NT Portlandian
NT Raghavapuram Shales
NT Rauracian
NT Salt Wash Sandstone Member
NT Schuler Formation
NT Sequanian
NT Smackover Formation
NT Stump Formation
NT Sundance Formation
NT Swift Formation
NT Todilto Formation
NT Volgian
NT Westwater Canyon Sandstone Member

NT Zuloaga Limestone
SA Coast Range Ophiolite
SA Lower Jurassic
SA Middle Jurassic
SA Nicoya Complex

upper Liassic (1981)
BT Lower Jurassic
BT Jurassic
BT Mesozoic
NT Toarcian

Upper Lusatia (1978)
That part of Lusatia in E Germany.
BT Saxony Germany
BT Germany
BT Central Europe
BT Europe
SA Doberlug Germany
SA Lusatia
SA Weissenberg Germany

upper Maestrichtian (1985)
BT Maestrichtian
BT Senonian
BT Upper Cretaceous
BT Cretaceous
BT Mesozoic

upper mantle (1978)
UF outer mantle
BT mantle
SA asthenosphere
SA lateral heterogeneity
SA lower crust
SA lower mantle
SA pyrolite
SA Sn-waves

upper Mesozoic (1978)
BT Mesozoic
SA middle Mesozoic

upper Miocene (1978)
Autoposting of this term began in 1978.
BT Miocene
BT Neogene
BT Tertiary
BT Cenozoic
NT Castaic Formation
NT Duplin Formation
NT Eastover Formation
NT Meotian
NT Messinian
NT Modelo Formation
NT Montesano Formation
NT Pannonian
NT Pontian
NT Puente Formation
NT Punchbowl Formation
NT Santa Margarita Formation
NT Sarmatian
NT Tamiami Formation
NT Tortonian
NT Turolian
NT Yorktown Formation
SA Capistrano Formation
SA Clarendonian
SA Cuddalore Series
SA Hemphillian
SA lower Miocene
SA middle Miocene
SA Mohnian
SA Neyveli Lignite
SA Ridge Route Formation
SA Sisquoc Formation
SA Warkalli Formation
SA Wildcat Group
SA Yakima Basalt

Upper Mississippi Valley (1978)
That part of the Mississippi Valley N of Cairo, Illinois. As of 1990, United States is autoposted to this term.
IN Index states as applicable.
BT Mississippi Valley

BT United States
SA Illinois
SA Iowa
SA Minnesota
SA Missouri
SA Wisconsin

Upper Mississippian (1978)
Autoposting of this term began in 1978.
BT Mississippian
BT Carboniferous
BT Paleozoic
NT Bangor Limestone
NT Chesterian
NT Fayetteville Formation
NT Greenbrier Limestone
NT Hartselle Sandstone
NT Heath Formation
NT Mauch Chunk Formation
NT Meramecian
NT Monteagle Limestone
NT Parkwood Formation
NT Pennington Formation
NT Pitkin Limestone
NT Serpukhovian
SA Bird Spring Formation
SA Ely Limestone
SA Lower Mississippian
SA Manning Canyon Shale
SA Rundle Group
SA Valmeyeran

upper Neogene (1989)
Autoposting of Tertiary to this term began in 1990.
BT Neogene
BT Tertiary
BT Cenozoic
SA Pliocene

Upper Old Red Sandstone (1978)
Represented by Nairn, Boghole, Alves-Scaat Craig, Rosebrae, Plateau, and Portishead beds; by Quartz Conglomerate; and by Farlow Sandstone. In Great Britain.
BT Devonian
BT Paleozoic
SA Great Britain
SA United Kingdom

upper Oligocene (1978)
Autoposting of this term began in 1978.
BT Oligocene
BT Paleogene
BT Tertiary
BT Cenozoic
NT Chattian
NT Chickasawhay Formation
NT Egerian
NT Suwannee Limestone
SA Anahuac Formation
SA Arikareean
SA John Day Formation
SA lower Oligocene
SA middle Oligocene

Upper Ordovician (1978)
Autoposting of this term began in 1978.
BT Ordovician
BT Paleozoic
NT Ashgillian
NT Caradocian
NT Cincinnatian
NT Cortlandt Complex
NT Edenian
NT Fairview Formation
NT Fernvale Formation
NT Fish Haven Dolomite
NT Juniata Formation
NT Kope Formation
NT Maquoketa Formation

NT Neda Formation
NT Queenston Shale
NT Red River Formation
NT Reedsville Formation
NT Richmond Group
NT Utica Shale
NT Wufeng Formation
SA Bala
SA Ellis Bay Formation
SA Lower Ordovician
SA Maravillas Formation
SA Matapedia Group
SA Middle Ordovician

Upper Palatinate (1978)
That part of the historical region of the Palatinate located in E Bavaria, SE Germany. Before 1993, also search Oberpfalz.
UF Oberpfalz
BT Bavaria Germany
BT Germany
BT Central Europe
BT Europe
SA Palatinate

upper Paleocene (1978)
Autoposting of this term began in 1978.
BT Paleocene
BT Paleogene
BT Tertiary
BT Cenozoic
NT Landenian
NT Thanetian
SA Ewekoro Formation
SA lower Paleocene

upper Paleogene (1985)
BT Paleogene
BT Tertiary
BT Cenozoic
SA Eocene
SA Oligocene

upper Paleolithic (1985)
Archaeological classification.
BT Paleolithic
BT Stone Age
BT Cenozoic
SA archaeology
SA Pleistocene
SA Quaternary

upper Paleozoic (1978)
BT Paleozoic
NT Antrim Shale
NT Arkansas Novaculite
NT Bakken Formation
NT Calaveras Formation
NT Dwyka Formation
NT Fountain Formation
NT Kaskaskia Sequence
NT Narragansett Pier Granite
NT Sicker Group
NT Tunguska Series
NT Wood River Formation
SA Monashee Complex
SA Rio Bonito Formation

Upper Peninsula, Michigan
  use Michigan Upper Peninsula

Upper Pennsylvanian (1978)
Autoposting of this term began in 1978.
BT Pennsylvanian
BT Carboniferous
BT Paleozoic
NT Ames Limestone
NT Canyon Group
NT Cisco Group
NT Missourian
NT Sturgis Formation
NT Virgilian
NT Wescogame Formation
SA Laborcita Formation
SA Lower Pennsylvanian

SA Middle Pennsylvanian

Upper Permian (1978)
Autoposting of this term began in 1978.
BT Permian
BT Paleozoic
NT Barabash Suite
NT Cadeby Formation
NT Kazanian
NT Raniganj Formation
NT Salado Formation
NT Tatarian
NT Usuguru Conglomerate
NT Vetluga Series
NT Zechstein
SA Guadalupian
SA Lower Permian
SA Middle Permian

upper Pleistocene (1978)
Autoposting of this term began in 1978.
BT Pleistocene
BT Quaternary
BT Cenozoic
NT Baltic ice lake
NT Cromerian
NT Devensian
NT Eemian
NT Holsteinian
NT Hoxnian
NT Illinoian
NT Ipswichian
NT Old Crow Tephra
NT Rancholabrean
NT Saalian
NT Sangamonian
NT Sartanian
NT Sicilian
NT Twocreekan
NT Tyrrhenian
NT Vistulian
NT Weichselian
NT Wisconsinan
NT Wurm
SA Baltic Glaciation
SA Clovis
SA Glasford Formation
SA Lake Bonneville
SA lower Pleistocene
SA Scarborough Formation

upper Pliocene (1978)
Autoposting of this term began in 1978.
BT Pliocene
BT Neogene
BT Tertiary
BT Cenozoic
NT Akchagylian
NT Astian
NT Chowan River Formation
NT Levantinian
NT Redonian
NT Rexroad Formation
NT Romanian
SA lower Pliocene
SA Matuyama Epoch
SA middle Pliocene
SA Uonuma Group

upper Precambrian (1978)
Autoposting of this term began in 1978.
BT Precambrian
NT Proterozoic
SA Porsanger Dolomite Formation
SA Puncoviscana Formation
SA Sparagmite Group
SA Yudomian

upper Proterozoic (1978)
Autoposting of Precambrian to this term ended in 1989. Autoposting of upper Precambrian and Precambrian to this term began in 1990.
BT Proterozoic
BT upper Precambrian
BT Precambrian
NT Blackstone Series
NT Brioverian
NT Burra Group
NT Hadrynian
NT Horsethief Creek Group
NT Infracambrian
NT Jotnian
NT Lynchburg Formation
NT McCoy Creek Group
NT Moinian
NT Riphean
NT Torridonian
NT Vendian
NT Wilpena Group
NT Wonoka Formation
SA Boston Bay Group
SA Cambridge Argillite
SA Damara Orogeny
SA Robertson Bay Group
SA Roxbury Conglomerate
SA Sauk Sequence
SA Underhill Formation
SA Yudomian

upper Quaternary (1978)
BT Quaternary
BT Cenozoic
NT Baltic Glaciation
NT Bull Lake Glaciation
NT Pinedale Glaciation
NT Yoldia Sea

Upper Rhine
  use Upper Rhine Valley

Upper Rhine Graben (1978)
BT Rhine Graben
BT Central Europe
BT Europe

Upper Rhine Valley (1978)
As of 1981, restricted to that part of the Rhine Valley between Basel and Mainz in W Germany. Before 1981, included part of France.
IN Index states as applicable.
CO N473000N502800
    E0090700E0073200
UF Upper Rhine
BT Germany
BT Central Europe
BT Europe
SA Baden-Wurttemberg Germany
SA Hesse Germany
SA Rhineland-Palatinate Germany

upper Riphean (1985)
Autoposting of upper Precambrian and Precambrian to this term ended in 1989. As of 1989, upper Proterozoic and Proterozoic are autoposted to this term. Autoposting of upper Precambrian and Precambrian to this term began in 1990.
BT Riphean
BT upper Proterozoic
BT Proterozoic
BT upper Precambrian
BT Precambrian

Upper Silesia (1978)
The eastern part of former German Silesia plus Polish Silesia lying SE of Lower Silesia centering on Katowice and Cracow.
UF Upper Silesian area
BT Poland
BT Central Europe
BT Europe
SA Lower Silesia
SA Silesia

SA Silesian coal basin
SA Upper Silesian coal basin

Upper Silesia coal basin
  use Upper Silesian coal basin

Upper Silesian area
  use Upper Silesia

Upper Silesian Basin
  use Upper Silesian coal basin

Upper Silesian coal basin (1978)
N foot of the W Beskid Mountains in Upper Silesia.
UF Upper Silesia coal basin
UF Upper Silesian Basin
UF Upper Silesian coal field
BT Katowice Poland
BT Poland
BT Central Europe
BT Europe
SA coal fields
SA Silesian coal basin
SA Upper Silesia

Upper Silesian coal field
  use Upper Silesian coal basin

Upper Silurian (1978)
Wenlockian was considered a narrower term before 1993. Autoposting of this term began in 1978.
BT Silurian
BT Paleozoic
NT Bloomsburg Formation
NT Cayugan
NT Ludlovian
NT Pridolian
NT Salina Group
NT Tura Formation
SA Downtonian
SA Keyser Limestone
SA Lower Silurian
SA Middle Silurian
SA Rondout Formation
SA Wenlock Limestone

upper Sinian (1989)
China. Autoposting of upper Precambrian and Precambrian to this term began in 1990.
BT Sinian
BT Proterozoic
BT upper Precambrian
BT Precambrian

upper Tertiary (1978)
BT Tertiary
BT Cenozoic
NT Poznan Clays
SA Akchagylian
SA Gauss Epoch
SA Gilbert Epoch
SA lower Tertiary
SA Matuyama Epoch
SA middle Tertiary
SA Neogene

upper Tournaisian (1985)
BT Tournaisian
BT Dinantian
BT Carboniferous
BT Paleozoic

Upper Triassic (1978)
Autoposting of this term began in 1978.
BT Triassic
BT Mesozoic
NT Carnian
NT Chinle Formation
NT Dockum Group
NT Karmutsen Group
NT Kayenta Formation
NT Keuper
NT Lockatong Formation
NT Luning Formation
NT Mercia Mudstone

Before then, variants of the term may occur in GeoRef.

NT Norian
NT Petrified Forest Member
NT Protopivskaya Formation
NT Redonda Formation
NT Rhaetian
NT Sag River Sandstone
NT Santa Rosa Sandstone
NT Shinarump Member
NT Stormberg Series
NT Torlesse Supergroup
NT Wingate Sandstone
SA Brunswick Formation
SA Cimmerian Orogeny
SA Elliot Formation
SA Hallstatt Limestone
SA Kunga Group
SA Lower Triassic
SA Middle Triassic
SA Moenave Formation
SA Navajo Sandstone
SA Passaic Formation
SA Pucara Group
SA Stockton Formation

Upper Tunguska
 use Angara River

upper Visean (1985)
 BT Visean
 BT Dinantian
 BT Carboniferous
 BT Paleozoic

Upper Volta
 As of 1985, no longer a valid term for GeoRef.
 use Burkina Faso

upper Weichselian (1989)
 W Europe.
 BT Weichselian
 BT upper Pleistocene
 BT Pleistocene
 BT Quaternary
 BT Cenozoic

upper Wisconsinan (1985)
 BT Wisconsinan
 BT upper Pleistocene
 BT Pleistocene
 BT Quaternary
 BT Cenozoic

Upper Zaire (1981)
 The province of Haut-Zaire in NW.
 CO S020000N053000
  E0312000E0221000
 UF Haut-Zaire
 BT Zaire
 BT Central Africa
 BT Africa
 SA Lake Albert

Uppsala
 No longer a valid term for GeoRef. As of 1993 see Uppsala Sweden.

Uppsala Sweden (1993)
 County and city in E Sweden. Before 1993 also search Uppsala AND Sweden.
 BT Sweden
 BT Scandinavia
 BT Western Europe
 BT Europe

upright folds (1993)
 BT folds
 SA anticlines
 SA anticlinoria
 SA antiform folds
 SA vertical orientation

Upshur County
 Valid through 1988. Search in combination with state term. After 1988, use specific county-state term.

Upshur County Texas (1989)

NE Texas. Before 1989, also search Upshur County AND Texas.
 CO N323100N325200
  W0944100W0950800
 BT Texas
 BT United States

Upshur County West Virginia (1989)
 Central West Virginia. Before 1989, also search Upshur County AND West Virginia.
 CO N384200N390700
  W0800400W0802400
 BT West Virginia
 BT United States

Upton County
 Valid through 1988. Search in combination with state term. After 1988, use specific county-state term.

Upton County Texas (1989)
 W Texas. Before 1989, also search Upton County AND Texas.
 CO N310000N314000
  W1014700W1022500
 BT Texas
 BT United States

upwelling (1978)
 SA currents
 SA ocean circulation
 SA ocean currents
 SA oceanography
 SA sea water

Urach
 Not a valid term for GeoRef. See Urach Germany.

Urach Germany (1993)
 Central Baden-Wurttemberg, in the Swabian Alb. SE of Stuttgart. Also search Urach region.
 BT Baden-Wurttemberg Germany
 BT Germany
 BT Central Europe
 BT Europe

Urach region
 No longer a valid term for GeoRef. See Urach Germany.

Ural Mountains
 use Urals

Ural Range
 use Urals

Ural region (1978)
 A mining and industrial region on both sides of the Central Urals comprising the Chelyabinsk, Yekaterinburg (formerly Sverdlovsk), Kurgan, Orenburg, and Perm oblasts; and Bashkiria and Udmurtia. Also search Ural.
 UF Urals Region
 BT Russian Federation
 BT Commonwealth of Independent States
 SA Bashkiria Russian Federation
 SA Chelyabinsk Russian Federation
 SA Kurgan Russian Federation
 SA Orenburg Russian Federation
 SA Perm Russian Federation
 SA Udmurtia Russian Federation
 SA Yekaterinburg Russian Federation

Ural River (1978)
 Rises at S end of Urals and flows into the Caspian Sea. Also search Ural.

 IN Index former Soviet republics as applicable.
 BT Commonwealth of Independent States
 SA Kazakhstan
 SA Russian Federation

Ural-Tau (1978)
 Low, water-divide mountain range in E Southern Urals. This term has multiple hierarchies.
 BT1 Russian Federation
 BT1 Commonwealth of Independent States
 BT2 Southern Urals
 BT2 Urals
 BT2 Commonwealth of Independent States

Ural-Volga Interfluve
 use Volga-Ural region

Ural-Volga region
 use Volga-Ural region

Uralian (1978)
 Russia. Above Gzhelian, below Sakmarian of Permian.
 BT Upper Carboniferous
 BT Carboniferous
 BT Paleozoic

Uralian Foreland (1978)
 The western slope and foothills of the Urals. This term has multiple hierarchies.
 BT1 Russian Federation
 BT1 Commonwealth of Independent States
 BT2 Urals
 BT2 Commonwealth of Independent States

uralitization (1978)
 BT metasomatism
 SA hydrothermal alteration
 SA metamorphism

Urals (1978)
 Mountain range extending from the Kara Sea in the N to W Kazakhstan in the S separating Europe from Asia. Autoposting of this term to Mugodzhar Hills, Novaya Zemlya and Polar Urals began in 1981.
 IN Index ranges as applicable.
 CO N473000N773000
  E0700000E0560000
 UF Ural Mountains
 UF Ural Range
 BT Commonwealth of Independent States
 NT Central Urals
 NT Mugodzhar Hills
 NT Northern Urals
 NT Novaya Zemlya
 NT Polar Urals
 NT Southern Urals
 NT Uralian Foreland
 SA Eurasia
 SA Kazakhstan
 SA Russian Federation

Urals Region
 use Ural region

uraninite (1978)
 BT oxides
 SA pitchblende
 SA uranium ores

uranium (1978)
 Chemical element. As of 1981, use uranium ores for uranium as a commodity. Autoposting of actinides and metals to this term began in 1989.
 UF U
 BT actinides

 BT metals
 NT U-234
 NT U-234/Th-230
 NT U-235
 NT U-238
 NT U-238/Pb-204
 NT U-238/Pb-206
 NT U-238/Th-230
 NT U-238/U-234
 NT U-238/U-235
 SA Io/U
 SA isotopes
 SA lithophile elements
 SA radium
 SA Th/U
 SA U/He
 SA U/Pb
 SA U/Th/Pb
 SA uranium disequilibrium
 SA uranium minerals
 SA uranium ores

uranium disequilibrium (1978)
 UF uranium-series method
 SA absolute age
 SA isotopes
 SA U/Th/Pb
 SA uranium

uranium minerals (1978)
 IN May be used under commodities (list C).
 SA minerals
 SA ore minerals
 SA uranium
 SA uranium ores

uranium ores (1981)
 Before 1981, also search uranium AND (deposit OR deposits OR ore OR ores OR economic) in the basic index. Autoposting of metal ores to this term began in 1985.
 IN Commodity. See List C.
 BT metal ores
 SA Alligator Rivers Field
 SA Ambrosia Lake mining district
 SA autunite
 SA brannerite
 SA carnotite
 SA coffinite
 SA Colorado mineral belt
 SA Grants mining district
 SA Jabiluka Australia
 SA Key Lake Deposits
 SA Klerksdorp Field
 SA Midnite Mine
 SA Midwest Lake Deposits
 SA NURE
 SA Oklo
 SA Olympic Dam
 SA pitchblende
 SA Rabbit Lake Deposit
 SA Ranger Mine
 SA roll-type
 SA sandstone-type
 SA Schwartzwalder Mine
 SA Sullivan Mine
 SA unconformity-type
 SA uraninite
 SA uranium
 SA uranium minerals
 SA uranophane

uranium-lead
 Not a valid term for GeoRef. Before 1972, was used in subfiles E, G, N and B.
 use U/Pb

uranium-series method
 use uranium disequilibrium

uranium-thorium
 Not a valid term for GeoRef. Before 1970, included use in subfiles E, N and B.
 use Th/U

uranophane (1978)
Autoposting of nesosilicates began in 1985.
BT nesosilicates
BT orthosilicates
BT silicates
SA uranium ores

Uranus (1978)
SA Ariel Satellite
SA coronae
SA Miranda Satellite
SA Oberon Satellite
SA outer planets
SA planetary rings
SA planetology
SA planets
SA satellites
SA solar system
SA Titania Satellite
SA Umbriel Satellite
SA Voyager 2
SA Voyager Program

urban environment (1989)
SA depositional environment
SA ecology
SA environment
SA environmental geology
SA human ecology
SA pollution
SA sedimentation
SA urbanization

urban planning (1978)
SA environmental geology
SA land use
SA planning
SA regional planning
SA urbanization

Urbana
Valid through 1988. Search in combination with state term. After 1988, use specific city-state term.

Urbana Illinois (1989)
City in E central Illinois. Before 1989, search Urbana AND Illinois.
CO N400700N400700
 W0881200W0881200
BT Champaign County Illinois
BT Illinois
BT United States

Urbana Ohio (1989)
City in W central Ohio. Before 1989, search Urbana AND Ohio.
CO N400600N400600
 W0834500W0834500
BT Champaign County Ohio
BT Ohio
BT United States

urbanization (1978)
SA conservation
SA environmental geology
SA land use
SA urban environment
SA urban planning

ureilite (1981)
Before 1981, also search ureilites.
UF ureilites
BT achondrites
BT stony meteorites
BT meteorites
NT Haveroe Meteorite

ureilites
As of 1981, no longer a valid term for GeoRef.
use ureilite

Urga
use Ulan Bator Mongolia

Urgonian (1978)
Europe.
BT Lower Cretaceous
BT Cretaceous
BT Mesozoic

Uri
No longer a valid term for GeoRef. As of 1993 see Uri Switzerland.

Uri Switzerland (1993)
Canton in central Switzerland. Before 1993 also search Uri AND Switzerland.
BT Switzerland
BT Central Europe
BT Europe
SA Bernese Alps
SA Oberalp Pass

Ursidae (1978)
Family. Autoposting of Fissipeda, Eutheria and Theria to this term began in 1989.
BT Fissipeda
BT Carnivora
BT Eutheria
BT Theria
BT Mammalia
BT Tetrapoda
BT Vertebrata
BT Chordata
NT Ursus

Ursus (1978)
Genus. Autoposting of Fissipeda, Eutheria and Theria to this term began in 1989.
BT Ursidae
BT Fissipeda
BT Carnivora
BT Eutheria
BT Theria
BT Mammalia
BT Tetrapoda
BT Vertebrata
BT Chordata
NT Ursus spelaeus

Ursus spelaeus (1978)
Species. Autoposting of Fissipeda, Eutheria and Theria to this term began in 1989.
BT Ursus
BT Ursidae
BT Fissipeda
BT Carnivora
BT Eutheria
BT Theria
BT Mammalia
BT Tetrapoda
BT Vertebrata
BT Chordata

urtite (1978)
BT alkali gabbros
BT gabbros
BT plutonic rocks
BT igneous rocks
SA ijolite
SA melteigite

Uruguay (1978)
CO S350000S300000
 W0530000W0583000
BT South America
NT Montevideo Uruguay
NT Rio Negro Uruguay
SA Rio de la Plata

Urup (1978)
Volcanic island in S central Kuril Islands NE of Japan. Before 1993, also used for copper mining district in N Caucasus. After 1993, see Urup mining district.
IN Index Kuril Islands as applicable.
UF Urup Island
SA Kuril Islands
SA Russian Federation

Urup Island
use Urup

Urup mining district (1993)
Copper mining district in N Caucasus. Before 1993, also search Urup.
SA copper ores
SA Greater Caucasus
SA Northern Caucasus

Usa Series (1978)
Includes marble plus dolomites interspersed with beds of siliceous schist. In the Kuznetsk Alatau mountain region of Kemerovo Oblast and Khakass Autonomous Oblast E of Novosibirsk.
BT Lower Cambrian
BT Cambrian
BT Paleozoic
SA Russian Federation

USBM
use U. S. Bureau of Mines

uses
A valid term through 1976. Use applications or utilization.

USGS (1985)
Use for documents published by the U. S. Geological Survey. For discussions about the Survey and its activities, use U. S. Geological Survey.
UF U.S.G.S.
SA U. S. Geological Survey

USLE
use Universal Soil Loss Equation

Uspenski Kazakhstan
use Uspenskiy Kazakhstan

Uspenskiy
No longer a valid term for GeoRef. As of 1993, see Uspenskiy Kazakhstan.

Uspenskiy Kazakhstan (1993)
Town in Karaganda Oblast in E central Kazakhstan. Before 1978, also search Uspenski. This term has multiple hierarchies.
UF Uspenski Kazakhstan
BT1 Karaganda Kazakhstan
BT1 Kazakhstan
BT1 Central Asia
BT1 Asia
BT2 Karaganda Kazakhstan
BT2 Kazakhstan
BT2 Commonwealth of Independent States

USSR
No longer a valid term for GeoRef. As of 1993, see Commonwealth of Independent States or former constituent republics now represented by Armenia, Azerbaidzhan, Belarus, Estonia, Georgian Republic, Kazakhstan, Kyrgyzstan, Latvia, Lithuania, Moldova, Russian Federation, Tadzhikistan, Turkmenia, Ukraine, or Uzbekistan. Before 1972, also search Russia, Soviet Union, U. S. S. R.; Union of Soviet Socialist Republics. Autoposting of this term began in 1978 and ended in 1992.

USSR Academy of Sciences (1978)
In Moscow.
BT government agencies
SA Commonwealth of Independent States
SA Russian Federation

Ust-Kut
No longer a valid term for GeoRef. As of 1993, see Ust-Kut Russian Federation.

Ust-Kut Russian Federation (1993)
Town in N central Irkutsk Oblast NW of Lake Baikal. This term has multiple hierarchies.
BT1 Irkutsk Russian Federation
BT1 Russian Federation
BT1 Commonwealth of Independent States
BT2 Irkutsk Russian Federation
BT2 Asia

Ust-Urt
use Ustyurt

Ust-Urt Plateau
use Ustyurt

Ust-Yenisei Basin (1978)
That part of the lower Yenisei Basin from Ust-Port to Yenisei Bay which constitutes a drowned river mouth or estuary. Located in NW Taymyr-Nenets National Okrug. This term has multiple hierarchies.
BT1 Taymyr Dolgan-Nenets Russian Federation
BT1 Krasnoyarsk Russian Federation
BT1 Russian Federation
BT1 Commonwealth of Independent States
BT2 Taymyr Dolgan-Nenets Russian Federation
BT2 Krasnoyarsk Russian Federation
BT2 Asia

Ustica Island (1978)
Small island in the Tyrrhenian Sea NW of Sicily. Also search Ustica.
BT Sicily Italy
BT Italy
BT Southern Europe
BT Europe

Ustron
No longer a valid term for GeoRef. As of 1993, see Ustron Belarus.

Ustron Belarus (1993)
Village in central Belarus. This term has multiple hierarchies.
BT1 Belarus
BT1 Europe
BT2 Belarus
BT2 Commonwealth of Independent States

Ustyurt (1978)
Desert plateau extending from the Caspian Sea to the Aral Sea. This term has multiple hierarchies.
IN Index former Soviet republics as applicable.
CO N394500N450000
 E0590000E0500000
UF Ust-Urt
UF Ust-Urt Plateau
BT1 Commonwealth of Independent States
BT2 Central Asia
BT2 Asia
SA Barsakelmes
SA Kazakhstan
SA Turkmenia
SA Uzbekistan

Usu (1978)
On NE Uchiura Bay in SW Hokkaido.
UF Usu Volcano
BT Hokkaido

BT Japan
BT Far East
BT Asia

Usu Volcano
    use Usu

Usuginu Conglomerate (1978)
    Includes large boulders of limestone, compact slate, sandstone, granite, and porphyrite. In N Honshu.
    BT Upper Permian
    BT Permian
    BT Paleozoic
    SA Honshu
    SA Japan

Usuki
    No longer a valid term for GeoRef. As of 1993 see Usuki Japan.

Usuki Japan (1993)
    Town on Bungo Strait in NE Kyushu, S Japan.
    BT Kyushu
    BT Japan
    BT Far East
    BT Asia

Utah (1978)
    Autoposting of this term began in 1978.
    CO N370000N420000
        W1090500W1140500
    BT United States
    NT Arches National Park
    NT Beaver County Utah
    NT Box Elder County Utah
    NT Cache County Utah
    NT Canyonlands National Park
    NT Carbon County Utah
    NT Daggett County Utah
    NT Davis County Utah
    NT Duchesne County Utah
    NT Emery County Utah
    NT Garfield County Utah
    NT Grand County Utah
    NT Great Salt Lake
    NT Iron County Utah
    NT Juab County Utah
    NT Kaiparowits Plateau
    NT Kane County Utah
    NT Millard County Utah
    NT Morgan County Utah
    NT Oquirrh Mountains
    NT Piute County Utah
    NT Rich County Utah
    NT Salt Lake County Utah
    NT San Juan County Utah
    NT San Rafael Swell
    NT Sanpete County Utah
    NT Sevier County Utah
    NT Sevier Desert
    NT Summit County Utah
    NT Thomas Range
    NT Tintic mining district
    NT Tooele County Utah
    NT Tushar Mountains
    NT Uintah County Utah
    NT Utah County Utah
    NT Wah Wah Mountains
    NT Wasatch County Utah
    NT Wasatch Plateau
    NT Washington County Utah
    NT Wayne County Utah
    NT Weber County Utah
    SA Absaroka Fault
    SA Arapien Shale
    SA Bald Mountain
    SA Basin and Range Province
    SA Bear Lake
    SA Bear River basin
    SA Bear River Formation
    SA Bear River Range
    SA Bird Spring Formation
    SA Black Rock
    SA Blackhawk Formation
    SA Blair Formation
    SA Book Cliffs
    SA Bridger Formation
    SA Brigham Group
    SA Browns Park Formation
    SA Brushy Basin Shale Member
    SA Bull Lake Glaciation
    SA Burro Canyon Formation
    SA Cache Valley
    SA Carmel Formation
    SA Cedar Mountain Formation
    SA Chainman Shale
    SA Chinle Formation
    SA Cloverly Formation
    SA Coconino Sandstone
    SA Colorado Lineament
    SA Colorado Plateau
    SA Colorado River
    SA Colorado River basin
    SA Colton Formation
    SA Columbia River basin
    SA Cutler Formation
    SA Desert Creek Zone
    SA Diamond Peak Formation
    SA Dinosaur National Monument
    SA Dinwoody Formation
    SA Drum Mountains
    SA Duchesne River Formation
    SA Dunderberg Shale
    SA Ely Limestone
    SA Ely Springs Dolomite
    SA Entrada Sandstone
    SA Eureka Quartzite
    SA Ferron Sandstone Member
    SA Fillmore Formation
    SA Fish Haven Dolomite
    SA Flagstaff Formation
    SA Four Corners
    SA Frontier Formation
    SA Glen Canyon Group
    SA Goose Egg Formation
    SA Great Basin
    SA Green River
    SA Green River basin
    SA Green River Formation
    SA Guilmette Formation
    SA Henry Mountains
    SA Hermosa Formation
    SA Hilliard Shale
    SA Homestake Mine
    SA Honaker Trail Formation
    SA Ismay Zone
    SA Jefferson Group
    SA Joana Limestone
    SA Kaibab Formation
    SA Kayenta Formation
    SA Lake Bonneville
    SA Lake Powell
    SA Lake Uinta
    SA Lodgepole Formation
    SA Madison Group
    SA Mancos Shale
    SA Manning Canyon Shale
    SA Marjum Formation
    SA McCoy Creek Group
    SA Meade Peak Member
    SA Mesaverde Group
    SA Mission Canyon Limestone
    SA Moenave Formation
    SA Moenkopi Formation
    SA Montana Group
    SA Monticello Reservoir
    SA Morrison Formation
    SA Moxa Arch
    SA Muddy Creek Formation
    SA Navajo Indian Reservation
    SA Navajo Sandstone
    SA North Horn Formation
    SA Nugget Sandstone
    SA Oquirrh Formation
    SA Orr Formation
    SA Parachute Creek Member
    SA Paradox Basin
    SA Paradox Member
    SA Park City Formation
    SA Petrified Forest Member
    SA Phosphoria Formation
    SA Pilot Range
    SA Pilot Shale
    SA Pioche Shale
    SA Pogonip Group
    SA Price River basin
    SA Price River Formation
    SA Raft River
    SA Raft River basin
    SA Red Hill
    SA Retort Phosphatic Shale Member
    SA Roberts Mountains Formation
    SA Rock Springs Formation
    SA Saint George Basin
    SA Salt Lake Formation
    SA Salt Valley
    SA Salt Wash Sandstone Member
    SA San Juan Basin
    SA San Juan River
    SA San Rafael Group
    SA Sevier orogenic belt
    SA Shinarump Member
    SA Snake River basin
    SA Spor Mountain
    SA Star Point Sandstone
    SA Sunnyside Mine
    SA Supai Formation
    SA Swan Peak Formation
    SA Thaynes Formation
    SA Tintic Quartzite
    SA Toroweap Formation
    SA Twin Creek Limestone
    SA U. S. Rocky Mountains
    SA Uinta Basin
    SA Uinta Formation
    SA Uinta Mountain Group
    SA Uinta Mountains
    SA Uncompahgre Uplift
    SA Virgin River valley
    SA Wasatch fault zone
    SA Wasatch Formation
    SA Wasatch Front
    SA Wasatch Range
    SA Weber Sandstone
    SA Wells Formation
    SA Western Interior
    SA Western Interior Seaway
    SA Western Overthrust Belt
    SA Westwater Canyon Sandstone Member
    SA Wheeler Formation
    SA Williams Fork Formation
    SA Wingate Sandstone
    SA Wyoming Province

Utah County
    Valid through 1988. Search in combination with state term. After 1988, use specific county-state term.

Utah County Utah (1989)
    N central Utah. Before 1989, also search Utah County AND Utah.
    CO N394500N403500
        W1105300W1121500
    BT Utah
    BT United States
    NT Fairfield Utah
    NT Payson Utah
    NT Provo Utah

utahite
    use jarosite

Ute Creek (1978)
    Flows into the Canadian River in NE New Mexico.
    IN Index county or region as applicable.
    SA New Mexico

Utica Shale (1978)
    Consists of Nowadaga, Loyal Creek, Hopkinson, Holland, and Patent members.
    BT Upper Ordovician
    BT Ordovician
    BT Paleozoic
    SA Michigan
    SA New York
    SA Ontario

utilization (1978)
    As of 1982, restricted to economic geology.
    IN May be used with commodity terms (list C).
    SA consumption
    SA economic geology

Utrecht
    No longer a valid term for GeoRef. As of 1993 see Utrecht Netherlands.

Utrecht Netherlands (1993)
    Province. Also a city S of Amsterdam. Before 1993 also search Utrecht AND Netherlands.
    BT Netherlands
    BT Western Europe
    BT Europe

Uttar Pradesh
    No longer a valid term for GeoRef.
    use Uttar Pradesh India

Uttar Pradesh India (1993)
    State in N central India.
    CO N240000N320000
        E0840000E0770000
    UF Uttar Pradesh
    BT India
    BT Indian Peninsula
    BT Asia
    NT Almora India
    NT Dehra Dun India
    NT Garhwal Himalayas
    NT Garhwal India
    NT Jhansi India
    NT Meerut India
    NT Mirzapur India
    NT Mussoorie India
    NT Naini Tal India
    NT Ramnagar India
    NT Roorkee India
    NT Tehri Garhwal India
    SA Bundelkhand
    SA Jumna River
    SA Kaimur Sandstone
    SA Krol Formation
    SA Kumaun Himalayas
    SA Mussoorie Syncline
    SA Muth Quartzite
    SA Satpura Range
    SA Semri Series
    SA Subathu Formation
    SA Tal Formation

UV spectra
    use ultraviolet spectra

Uvalde County
    Valid through 1988. Search in combination with state term. After 1988, use specific county-state term.

Uvalde County Texas (1989)
    SW Texas. Before 1989, also search Uvalde County AND Texas.
    CO N290000N294000
        W0992000W1001000
    BT Texas
    BT United States
    SA Edwards Aquifer

Uvalde Gravel (1978)

Consists almost wholly of rounded flat cobbles, boulders, and occasional limestone pebbles. In Texas.
BT Pliocene
BT Neogene
BT Tertiary
BT Cenozoic
SA Texas

uvarovite (1978)
UF onvarovite
BT garnet group
BT nesosilicates
BT orthosilicates
BT silicates

Uvigerina (1978)
Genus. Autoposting of Uvigerinidae to this term began in 1989. Autoposting of microfossils and Protista to this term began in 1990. This term has multiple hierarchies.
BT1 Uvigerinidae
BT1 Buliminacea
BT1 Rotaliina
BT1 foraminifera
BT1 Protista
BT1 Invertebrata
BT2 Uvigerinidae
BT2 Buliminacea
BT2 Rotaliina
BT2 foraminifera
BT2 Protista
BT2 microfossils
NT Uvigerina peregrina

Uvigerina peregrina (1989)
Species. Autoposting of microfossils and Protista to this term began in 1990. This term has multiple hierarchies.
BT1 Uvigerina
BT1 Uvigerinidae
BT1 Buliminacea
BT1 Rotaliina
BT1 foraminifera
BT1 Protista
BT1 Invertebrata
BT2 Uvigerina
BT2 Uvigerinidae
BT2 Buliminacea
BT2 Rotaliina
BT2 foraminifera
BT2 Protista
BT2 microfossils

Uvigerinidae (1978)
Family. Autoposting of this term to Uvigerina began in 1989. Autoposting of microfossils and Protista to this term began in 1990. This term has multiple hierarchies.
BT1 Buliminacea
BT1 Rotaliina
BT1 foraminifera
BT1 Protista
BT1 Invertebrata
BT2 Buliminacea
BT2 Rotaliina
BT2 foraminifera
BT2 Protista
BT2 microfossils
NT Uvigerina

Uzbekistan (1978)
Republic. Formerly the Uzbek Soviet Socialist Republic. In Soviet Central Asia. This term has multiple hierarchies.
CO N370000N450000 E0720000E0550000
BT1 Asia
BT2 Commonwealth of Independent States
NT Almalyk Uzbekistan

NT Bukhara Uzbekistan
NT Bukhara-Khiva
NT Chadak Uzbekistan
NT Karshi Steppe
NT Karshi Uzbekistan
NT Khiva Uzbekistan
NT Kul'dzhuktau
NT Mubarek Uzbekistan
NT Nura-Tau
NT Oktyabr Uzbekistan
NT Samarkand Uzbekistan
NT Sokh Uzbekistan
NT Tamdytau
NT Tashkent Uzbekistan
NT Ziaetdin Uzbekistan
SA Amu Darya
SA Amu Darya Basin
SA Aral region
SA Bukantau
SA Central Asia
SA Chirchik River
SA Dengizkul
SA Fergana Basin
SA Gazli earthquake 1976
SA Gazli earthquake 1984
SA Kurama Range
SA Kyzylkum
SA Pamirs
SA Surkhan Darya
SA Surkhan Darya basin
SA Syr Darya
SA Turan
SA Turanian Platform
SA Ustyurt
SA Zeravshan River

Uzen (1978)
Two steppe rivers, the Greater Uzen and Lesser Uzen, which flow parallel to one another in Saratov Oblast and NW Kazakhstan.
IN Index former Soviet republics as applicable.
BT Commonwealth of Independent States
SA Kazakhstan
SA Russian Federation
SA Saratov Russian Federation

Uzon (1978)
Volcano. SE Kamchatka Peninsula.
IN Index regions as applicable.
UF Ozun
SA Kamchatka Peninsula
SA Kamchatka Russian Federation

V

V
 use vanadium

Vaal River (1978)
Forms boundary between Transvaal and Orange Free State, and empties into Orange River in N Cape Province.
IN Index provinces as applicable.
BT South Africa
BT Southern Africa
BT Africa
SA Cape Province South Africa
SA Orange Free State South Africa
SA Transvaal South Africa

Vacherie Dome (1985)

Salt dome, Bienville and Webster parishes, NW Louisiana.
IN Index counties as applicable.
CO N321000N324500 W0924500W0932500
BT Louisiana
BT United States
SA Gulf Coastal Plain

vacuum fusion analysis (1978)
SA analysis
SA chemical analysis

Vadia
 use Wadia

Vadito Group (1989)
Central New Mexico.
BT Precambrian
SA New Mexico

vadose zone
 use unsaturated zone

vaesite (1978)
BT sulfides

Vah River valley
 use Vah Valley

Vah Valley (1978)
River valley in W Slovakia. Also search Vah; Vah River.
UF Vah River valley
BT Slovakia
BT Czechoslovakia
BT Central Europe
BT Europe

Vaigach Island
 use Vaygach Island

Vail Pass (1978)
In Summit and Eagle counties W of Denver.
IN Index counties as applicable.
BT Colorado
BT United States
SA Eagle County Colorado
SA Gore Range
SA Summit County Colorado

Vakhsh (1978)
River. A tributary of the Amu Darya in SW Tadzhikistan. This term has multiple hierarchies.
UF Vaksh
BT1 Tadzhikistan
BT1 Asia
BT2 Tadzhikistan
BT2 Commonwealth of Independent States
SA Amu Darya

Vaksh
 use Vakhsh

Val d'Aosta
 Not a valid term for GeoRef. See Valle d'Aosta Italy.

Val d'Or
 No longer a valid term for GeoRef. As of 1993, see Val d'Or Quebec.

Val d'Or Quebec (1993)
Town in SW Quebec. Before 1993, also search Val d'Or AND Quebec.
BT Abitibi County Quebec
BT Quebec
BT Eastern Canada
BT Canada
SA Sigma Mine

Val Gardena Sandstone (1978)
NE Italy.
BT Permian
BT Paleozoic
SA Italy

Val Verde Basin (1978)
SW Texas.

IN Index counties as applicable.
BT Texas
BT United States

Val Verde County
 Valid through 1988. Search in combination with state term. After 1988, use specific county-state term.

Val Verde County Texas (1989)
SW Texas. Before 1989, also search Val Verde County AND Texas.
CO N291000N301500 W1005000W1014500
BT Texas
BT United States
NT Del Rio Texas

Val-d'Oise
 No longer a valid term for GeoRef.
 use Val-d'Oise France

Val-d'Oise France (1993)
Department just NNW of Paris.
CO N485200N491000 E0023000E0013000
UF Val-d'Oise
BT France
BT Western Europe
BT Europe
NT Cormeilles-en-Parisis France
SA Oise River valley

Val-de-Marne
 No longer a valid term for GeoRef.
 use Val-de-Marne France

Val-de-Marne France (1993)
Department in N France.
CO N484000N485500 E0023500E0021500
UF Val-de-Marne
BT France
BT Western Europe
BT Europe
SA Marne Valley

Valais
 No longer a valid term for GeoRef. As of 1993 see Valais Switzerland.

Valais Switzerland (1993)
Canton in SW Switzerland. Before 1993 also search Valais AND Switzerland.
BT Switzerland
BT Central Europe
BT Europe
NT Binnental
NT Zermatt Switzerland
SA Bernese Alps
SA Lake Geneva
SA Monte Rosa
SA Pennine Alps

Valanginian (1978)
Europe. Above Berriasian, below Hauterivian.
BT Lower Cretaceous
BT Cretaceous
BT Mesozoic
SA Neocomian

Valdai (1978)
Moraine region between Moscow and Saint Petersburg. This term has multiple hierarchies.
UF Valdai Hills
UF Valdai Uplands
BT1 Russian Federation
BT1 Commonwealth of Independent States
BT2 Europe

Valdai Hills
 use Valdai

Valdai Uplands
use Valdai

Valdez
Valid through 1988. Search in combination with state term. After 1988, use specific city-state term.

Valdez Alaska (1989)
Town and port on Valdez Arm of Prince William Sound, S Alaska. Terminal point of Trans-Alaska Pipeline. Before 1989, search Valdez AND Alaska.
CO N610700N610700
　　W1461700W1461700
UF　Port Valdez
BT　Southern Alaska
BT　Alaska
BT　United States
SA　Trans-Alaska Pipeline

Valdez Group (1989)
S central Alaska.
BT　Cretaceous
BT　Mesozoic
SA　Alaska

Valdres (1978)
Mountainous region W and SW of Fagernes in S central Norway.
BT　Norway
BT　Scandinavia
BT　Western Europe
BT　Europe

Vale of Glamorgan (1978)
In Brisbane area, SE Queensland.
IN　Index state or region as applicable.
BT　Queensland Australia
BT　Australia
BT　Australasia

Vale of Kashmir
use Kashmir Valley

Valence
No longer a valid term for GeoRef. See Valence France.

Valence France (1993)
City on left bank of Rhone River in NW Drome, SE France.
BT　Drome France
BT　France
BT　Western Europe
BT　Europe

Valencia
No longer a valid term for GeoRef. As of 1993, see Valencia City Spain.

Valencia City Spain (1993)
Refers to only the city in SE Spain. Before 1993, also search Valencia and Spain.
BT　Valencia Spain
BT　Valencia region
BT　Spain
BT　Iberian Peninsula
BT　Southern Europe
BT　Europe

Valencia County
Valid through 1988. Search in combination with state term. After 1988, use specific county-state term.

Valencia County New Mexico (1989)
W New Mexico. Before 1989, also search Valencia County AND New Mexico. In 1981, W part was made into Cibola County.
CO N342500N345800
　　W1062500W1071500
BT　New Mexico
BT　United States

NT　Grants New Mexico
SA　Cibola County New Mexico
SA　Manzano Mountains
SA　Rio Puerco

Valencia Province
No longer a valid term for GeoRef.
use Valencia Spain

Valencia region (1981)
Before 1981, search Valencia.
CO N375100N404700
　　E0003100W0013300
BT　Spain
BT　Iberian Peninsula
BT　Southern Europe
BT　Europe
NT　Alicante Spain
NT　Castellon de la Plana Spain
NT　Valencia Spain

Valencia Spain (1993)
Refers only to the province. From 1981-1992, also search Valencia Province and Spain. Before 1981, also search Valencia and Spain. For the city, see Valencia City Spain.
CO N394000N401300
　　W0000200W0013200
UF　Valencia Province
BT　Valencia region
BT　Spain
BT　Iberian Peninsula
BT　Southern Europe
BT　Europe
NT　Bunol Spain
NT　Valencia City Spain

Valencia Trough (1985)
Off E coast of Spain.
BT　West Mediterranean
BT　Mediterranean Sea

Valenciennes
No longer a valid term for GeoRef. See Valenciennes France.

Valenciennes France (1993)
City near the Belgian frontier in central Nord Department, N France.
BT　Nord France
BT　France
BT　Western Europe
BT　Europe

valency (1978)
SA　chemical properties
SA　geochemistry

Valentian
use Llandoverian

Valentine Formation (1978)
In Ogallala Group. Composed largely of unconsolidated layers of greenish silty marl sand and very few local lenses of diatomaceous marl and light bluish-gray volcanic ash. In W Kansas, NW Nebraska, and SW South Dakota.
BT　Pliocene
BT　Neogene
BT　Tertiary
BT　Cenozoic
SA　Kansas
SA　Nebraska
SA　Ogallala Formation
SA　South Dakota

valentinite (1978)
UF　white antimony
BT　oxides

Valhalla Complex (1993)
Omineca Belt, British Columbia.
IN　Index ages as applicable.
UF　Valhalla gneiss complex

SA　British Columbia
SA　Omineca Belt

Valhalla gneiss complex
use Valhalla Complex

Valladolid Province
No longer a valid term for GeoRef.
use Valladolid Spain

Valladolid Spain (1993)
Refers only to the province in N Spain. From 1981-1992, also search Valladolid Province and Spain. Before 1981, also search Valladolid and Spain.
CO N410500N421800
　　W0035800W0053300
UF　Valladolid Province
BT　Old Castile Spain
BT　Castile Spain
BT　Spain
BT　Iberian Peninsula
BT　Southern Europe
BT　Europe

Valle d'Aosta
No longer a valid term for GeoRef.
use Valle d'Aosta Italy

Valle d'Aosta Italy (1993)
Autonomous region in extreme NW Italy. Also search Aosta Valley and Val d'Aosta.
CO N452800N460000
　　E0075800E0064800
UF　Aosta Valley Italy
UF　Valle d'Aosta
BT　Italy
BT　Southern Europe
BT　Europe
SA　Graian Alps
SA　Mont Blanc

Valle Grande Mountains (1978)
NW of Santa Fe in N New Mexico.
UF　Jemez Mountains
BT　New Mexico
BT　United States
SA　Fenton Hill

valleriite (1978)
BT　sulfides
SA　tochilinite

Valles
No longer a valid term for GeoRef. As of 1993 see Valles Mexico.

Valles Caldera (1978)
Also search Valles AND New Mexico.
BT　New Mexico
BT　United States
SA　Baca Field
SA　Fenton Hill

Valles Marineris (1993)
Large interconnected canyon feature on Mars extending from the Tharsis bulge to the Chryse Planitia-Margaritifer Sinus region.
BT　Mars

Valles Mexico (1993)
City in E central Mexico. Before 1993 also search Valles AND Mexico.
UF　Ciudad de Valles
BT　San Luis Potosi Mexico
BT　Mexico

Vallesian (1978)
Europe.
BT　Miocene
BT　Neogene
BT　Tertiary
BT　Cenozoic

Valley and Ridge Province (1978)
Folded mountain ridges and parallel valleys between the Appalachian Plateau on the W and the Older Appalachians on the E. Extends from along the Hudson River in the N into Alabama in the S. Also search Valley and Ridge. In 1985, broader term changed from United States to Appalachians and North America.
IN　Index states as applicable.
BT　Appalachians
BT　North America
SA　Alabama
SA　Georgia
SA　Great Appalachian Valley
SA　Maryland
SA　New Jersey
SA　New York
SA　Pennsylvania
SA　Tennessee
SA　Virginia
SA　West Virginia

Valley County
Valid through 1988. Search in combination with state term. After 1988, use specific county-state term.

Valley County Idaho (1989)
Central Idaho. Before 1989, also search Valley County AND Idaho.
CO N440800N451100
　　W1144500W1161400
BT　Idaho
BT　United States
SA　Rainbow Mountain
SA　Sunnyside Mine

Valley County Montana (1989)
NE Montana. Before 1989, also search Valley County AND Montana.
CO N474200N490000
　　W1054700W1072300
BT　Montana
BT　United States

Valley County Nebraska (1989)
Central Nebraska. Before 1989, also search Valley County AND Nebraska.
CO N412300N414300
　　W0984400W0991300
BT　Nebraska
BT　United States

Valley of Mexico (1978)
Large oval basin 50 miles by 40 miles which is a subdivision of the plateau of Anahuac in central part of country.
IN　Index Federal District Mexico and/or Mexico Mexico as applicable.
BT　Mexico
SA　Federal District Mexico
SA　Mexico City Mexico
SA　Mexico state

Valley of Ten Thousand Smokes (1978)
Volcanic region in Katmai National Monument W of Mount Katmai in SW Alaska.
BT　Southwestern Alaska
BT　Alaska
BT　United States
SA　Alaska Peninsula
SA　Katmai National Monument

valley, median
use median valley

valleys (1978)
NT　gorges

SA buried valleys
SA canyons
SA drainage basins
SA dry valleys
SA erosion
SA erosion features
SA fluvial features
SA furrows
SA geomorphology
SA glacial features
SA glacial geology
SA landforms
SA poljes
SA rilles
SA rivers and streams
SA tributaries

Valmeyeran (1978)
Provincial series, Illinois. Lower and Upper Mississippian. Equivalent to Osagian and Meramecian elsewhere.
BT Mississippian
BT Carboniferous
BT Paleozoic
SA Lower Mississippian
SA Meramecian
SA Osagian
SA Upper Mississippian

Valpacos
No longer a valid term for GeoRef. As of 1993 see Valpacos Portugal.

Valpacos Portugal (1993)
Town in Vila Real, N central Portugal.
BT Vila Real Portugal
BT Portugal
BT Iberian Peninsula
BT Southern Europe
BT Europe

Valparaiso
No longer a valid term for GeoRef. See Valparaiso Chile.

Valparaiso Chile (1993)
Province including the city on the Pacific in central Chile. Also search Aconcagua or Aconcagua Province which became part of this region in 1975.
UF Aconcagua Chile
BT Chile
BT South America
NT La Ligua
NT La Ligua Chile
SA Chile earthquake 1985

Valsequillo
No longer a valid term for GeoRef. As of 1993, see Valsequillo Spain.

Valsequillo Spain (1993)
Village in NW Cordoba, SE central Spain. Before 1993, also search Valsequillo and Spain.
BT Cordoba Spain
BT Andalusia Spain
BT Spain
BT Iberian Peninsula
BT Southern Europe
BT Europe

Valsugana (1978)
Valley of upper Brenta River in SE Trinto-Alto Adige, N Italy.
BT Trentino-Alto Adige Italy
BT Italy
BT Southern Europe
BT Europe

Valtellina (1978)
Valley of upper Adda River in S Lombardy, N Italy.
BT Lombardy Italy
BT Italy

BT Southern Europe
BT Europe

valves (1978)
IN May be used with appropriate fossil group (list F).
SA morphology

Van Horn
Valid through 1988. Search in combination with state term. After 1988, use specific city-state term.

Van Horn Texas (1989)
Town in extreme W Texas. Before 1989, search Van Horn AND Texas.
CO N310300N310300
    W1045100W1045100
BT Culberson County Texas
BT Texas
BT United States

vanadates (1978)
Autoposting of this term began in 1978.
NT carnotite
NT descloizite
NT vanadinite
NT volborthite
SA minerals

vanadinite (1978)
From 1978-1980, halides and vanadates were autoposted to this term. Autoposting of chlorides began in 1981. As of 1989, halides is autoposted to this term. This term has multiple hierarchies.
BT1 chlorides
BT1 halides
BT2 vanadates

vanadium (1978)
Chemical element. As of 1981, use vanadium ores for vanadium as a commodity. Autoposting of metals to this term began in 1989.
UF V
BT metals
SA vanadium ores

vanadium garnet (1978)
BT garnet group
BT nesosilicates
BT orthosilicates
BT silicates
SA grossular

vanadium ores (1981)
Before 1981, also search vanadium AND (deposit OR deposits OR ore OR ores OR economic) in the basic index. Autoposting of metal ores to this term began in 1985.
IN Commodity. See List C.
BT metal ores
SA carnotite
SA vanadium

Vanch Range
use Vanchskiy Range

Vanchskiy Range (1978)
Mountains near Afghan border in S central Tadzhikistan. Also search Vanch. This term has multiple hierarchies.
UF Vanch Range
BT1 Tadzhikistan
BT1 Asia
BT2 Tadzhikistan
BT2 Commonwealth of Independent States

Vancouver
No longer a valid term for GeoRef. As of 1993, for city at mouth of Burrard Inlet in SW British Colum-

bia, use Vancouver British Columbia. Before 1989, was also used for city in Washington. As of 1989, for city in SW Washington, use Vancouver Washington.

Vancouver British Columbia (1993)
City at mouth of Burrard Inlet in SW British Columbia. Also search Vancouver in combination with British Columbia.
CO N491300N491300
    W1230600W1230600
BT British Columbia
BT Western Canada
BT Canada

Vancouver Island (1978)
Large island off SW coast of mainland British Columbia.
CO N483000N510000
    W1233000W1283000
BT British Columbia
BT Western Canada
BT Canada
SA Bonanza Group
SA Coast Ranges
SA Karmutsen Group
SA Leech River Fault
SA Queen Charlotte Sound
SA Tofino Basin

Vancouver Washington (1989)
City on the Columbia River in SW Washington. Before 1989, search Vancouver AND Washington.
CO N453800N453800
    W1124000W1224000
BT Clark County Washington
BT Washington
BT United States

vane tests (1982)
SA engineering geology
SA shear strength
SA soil mechanics
SA testing

Vanimo Papua New Guinea (1993)
In NW Papua New Guinea. Before 1993, also search Vanimo or Vanimo region AND Papua New Guinea.
CO S024000S024000
    E1411700E1411700
BT Papua New Guinea
BT Australasia

Vanimo region
No longer a valid term for GeoRef. See Vanimo Papua New Guinea.

Vanoise (1978)
High mountain group of the Savoy Alps in Savoie Department. This term has multiple hierarchies.
UF Vanoise Massif
BT1 Savoy Alps
BT1 Haute-Savoie France
BT1 France
BT1 Western Europe
BT1 Europe
BT2 Savoy Alps
BT2 Western Alps
BT2 Alps
BT2 Europe

Vanoise Massif
use Vanoise

Vanua Levu (1985)
Second-largest island in Fiji group.
CO S170500S161000
    W1795000E1783000
BT Fiji
BT Melanesia

BT Oceania

Vanuatu (1985)
Formerly New Hebrides. Group of islands NE of New Caledonia and W of Fiji. Gained independence in 1984. Before 1985, also search New Hebrides.
CO S203000S130000
    E1690000E1660000
UF New Hebrides
BT Melanesia
BT Oceania
NT Malekula
SA DSDP Site 286
SA Pacific Ocean

vapor, water
use water vapor

vaporization
Not a valid term for GeoRef. Use evaporation or sublimation if applicable.

Vaqueros Formation (1985)
Oligocene? and lower Miocene. S California.
BT Tertiary
BT Cenozoic
SA California
SA lower Miocene
SA Oligocene

Var
No longer a valid term for GeoRef. See Var France.

Var France (1993)
Department in SE France.
CO N430000N434500
    E0070000E0053000
BT France
BT Western Europe
BT Europe
NT Maures Massif
NT Plan de la Tour France
NT Rians France
NT Toulon France
SA Esterel
SA Provence
SA Provence Alps
SA Sainte-Baume Massif

Varanger Fjord (1978)
Inlet of the Barents Sea S of the Varanger Peninsula in extreme NE Norway.
UF Varangerfjord
BT Finnmark Norway
BT Norway
BT Scandinavia
BT Western Europe
BT Europe

Varanger Halvoy
use Varanger Peninsula

Varanger Peninsula (1978)
Extreme NE Norway jutting into the Barents Sea.
UF Varanger Halvoy
UF Varangerhalvoeya
UF Varangerhalvoya
BT Finnmark Norway
BT Norway
BT Scandinavia
BT Western Europe
BT Europe

Varangerfjord
use Varanger Fjord

Varangerhalvoeya
use Varanger Peninsula

Varangerhalvoya
use Varanger Peninsula

Vardar River (1978)

Rises in S Yugoslavia and flows into Gulf of Salonika.
IN Index countries as applicable.
BT Europe
SA Greece
SA Yugoslavia

Vardar Zone (1985)
N Greece, S Yugoslavia and S Bulgaria.
IN Index countries as applicable.
BT Southern Europe
BT Europe
SA Bulgaria
SA Greece
SA Yugoslavia

Varese
No longer a valid term for GeoRef. See Varese Italy.

Varese Italy (1993)
Province and city in NW Lombardy, N Italy.
BT Lombardy Italy
BT Italy
BT Southern Europe
BT Europe

variance analysis (1978)
Autoposting of this term began in 1978.
BT statistical analysis
NT semivariograms
NT variograms
SA analysis
SA Bayesian analysis
SA covariance analysis
SA kriging

variations (1978)
UF perturbations
SA annual variations
SA changes
SA diurnal variations
SA fluctuations
SA magnetic field
SA seasonal variations
SA secular variations
SA spatial variations
SA time variations
SA variometers

Variegated Glacier (1989)
SE Alaska, near the Saint Elias Mountains.
IN Index region as applicable.
BT Southeastern Alaska
BT Alaska
BT United States
SA Saint Elias Mountains

varieties (1978)
Used as a general term.
SA minerals

variograms (1985)
BT variance analysis
BT statistical analysis
SA geostatistics
SA mathematical geology
SA mathematical methods
SA semivariograms

variometers (1985)
SA electromagnetic methods
SA gravity methods
SA instruments
SA magnetic methods
SA magnetotelluric methods
SA variations

Variscan
A valid index term through 1976. After 1976, use Hercynian Orogeny.

Variscan Orogeny
use Hercynian Orogeny

Variscides (1978)
Mountain system raised in the latter part of the Paleozoic era, particularly in central Europe; more or less equivalent to Hercynides.
UF Hercynides
BT Europe
SA Hercynian Orogeny

variscite (1978)
BT phosphates

Varmland
No longer a valid term for GeoRef. As of 1993 see Varmland Sweden.

Varmland Sweden (1993)
County in SW Sweden. Before 1993 also search Varmland AND Sweden.
BT Sweden
BT Scandinavia
BT Western Europe
BT Europe

Varna
No longer a valid term for GeoRef. See Varna Bulgaria.

Varna Bulgaria (1993)
Province and city along the Black Sea in NE Bulgaria. From 1949 to 1957, the city was named Stalin; therefore, also search Stalin AND Bulgaria for the city.
BT Bulgaria
BT Southern Europe
BT Europe

Varpalota
No longer a valid term for GeoRef. See Varpalota Hungary.

Varpalota Hungary (1993)
Town N of E Lake Balaton in E Veszprem, W Hungary.
BT Veszprem Hungary
BT Hungary
BT Central Europe
BT Europe

Varta
use Warta

varved clay
A valid term through 1975. After 1975, use varves.

varves (1978)
Before 1976, also search varved clay.
BT planar bedding structures
BT sedimentary structures
SA geochronology
SA glacial features
SA glacial lakes
SA glaciolacustrine sedimentation
SA graded bedding
SA relative age

vascular taxa (1993)
Meaning includes division of plants with vascular systems including the Psilopsida, Sphenopsida, Pteropsida, and Lycopsida. Before 1993, also search vascular plants if applicable.
SA Lycopsida
SA morphology
SA Psilopsida
SA pteridophytes
SA Pteropsida
SA Sphenopsida

Vassalboro Formation (1989)
S Maine.
BT Silurian
BT Paleozoic
SA Maine

Vasterbotten
No longer a valid term for GeoRef. As of 1993 see Vasterbotten Sweden.

Vasterbotten Sweden (1993)
County in N Sweden.
BT Sweden
BT Scandinavia
BT Western Europe
BT Europe
NT Boliden Sweden
NT Nasliden Mine
SA Skellefte

Vastergotland (1985)
Region, SW Sweden. Includes Skaraborg, N Alvsborg and S Goteborg och Bohus counties.
BT Sweden
BT Scandinavia
BT Western Europe
BT Europe
SA Goteborg Sweden

Vastervik
No longer a valid term for GeoRef. As of 1993 see Vastervik Sweden.

Vastervik Sweden (1993)
City on Baltic Sea in NE Kalmar, SE Sweden. Before 1993 also search Vastervik AND Sweden.
BT Kalmar Sweden
BT Sweden
BT Scandinavia
BT Western Europe
BT Europe

vaterite (1978)
Artificial mineral.
BT carbonates

Vatnajokull (1978)
Glacier region and snow field in SE Iceland.
BT Iceland
BT Western Europe
BT Europe
SA Grimsvotn

Vatukoula
No longer a valid term for GeoRef. See Vantukoula Fiji.

Vatukoula Fiji (1993)
Town on N coast of Viti Levu island in S Fiji. Viti Levu is considered a broader term as of 1985.
BT Viti Levu
BT Fiji
BT Melanesia
BT Oceania

Vaucluse
No longer a valid term for GeoRef. See Vaucluse France.

Vaucluse France (1993)
Department in SE France.
CO N434000N442000 E0054500E0044500
BT France
BT Western Europe
BT Europe
NT Apt France
NT Mormoiron France
SA Provence

Vaud
No longer a valid term for GeoRef. As of 1993 see Vaud Switzerland.

Vaud Switzerland (1993)
Canton in W Switzerland. Before 1993 also search Vaud AND Switzerland.
BT Switzerland
BT Central Europe

BT Europe
SA Jura Mountains
SA Lake Geneva

Vaygach Island (1978)
Between NE Nenets National Okrug and island of Novaya Zemlya off N European Russian Federation. Also search Vaigach; Vaygach. This term has multiple hierarchies.
UF Vaigach Island
BT1 Nenets Russian Federation
BT1 Arkhangelsk Russian Federation
BT1 Russian Federation
BT1 Commonwealth of Independent States
BT2 Nenets Russian Federation
BT2 Arkhangelsk Russian Federation
BT2 Europe
SA Polar Urals

vegetation (1978)
After 1992, also search afforestation if applicable.
UF afforestation
SA ecology
SA floral studies
SA flowers
SA leaves
SA paleobotany
SA Plantae
SA remote sensing
SA roots
SA soils
SA wetlands

vein systems
use veins

veins (1978)
General term not restricted to ore deposits.
UF vein systems
NT quartz veins
SA bars
SA fractures
SA host rocks
SA intrusions
SA mineral deposits, genesis
SA wall rocks

Vejer de la Frontera
No longer a valid term for GeoRef. As of 1993, see Vejer de la Frontera Spain.

Vejer de la Frontera Spain (1993)
City near the Atlantic Ocean in SW Cadiz Province, SW Spain. Before 1993, also search Vejer de la Frontera and Spain.
BT Cadiz Spain
BT Andalusia Spain
BT Spain
BT Iberian Peninsula
BT Southern Europe
BT Europe

Velay (1978)
Region of the Central Massif in W and central Haute-Loire.
UF Velay Massif
UF Velay Plateau
BT Haute-Loire France
BT France
BT Western Europe
BT Europe
SA Central Massif

Velay Massif
use Velay

Velay Plateau
use Velay

Velebit Mountains (1978)

Range of Dinaric Alps along Adriatic Sea opposite Pag Island in W Croatia. Also search Velebit. This term has multiple hierarchies.
  BT1 Dinaric Alps
  BT1 Southern Europe
  BT1 Europe
  BT2 Dinaric Alps
  BT2 Eastern Alps
  BT2 Alps
  BT2 Europe
  BT3 Croatia
  BT3 Yugoslavia
  BT3 Southern Europe
  BT3 Europe

Velez Rubio
  No longer a valid term for GeoRef. As of 1993, see Velez Rubio Spain.

Velez Rubio Spain (1993)
  Town in Andalusia, N Almeria Province, SE Spain. Before 1993, also search Velez Rubio and Spain.
  BT Almeria Spain
  BT Andalusia Spain
  BT Spain
  BT Iberian Peninsula
  BT Southern Europe
  BT Europe

velocities
  use velocity

velocity (1978)
  Includes use as a general term, e.g. under seismology.
  UF velocities
  SA acceleration
  SA Conrad discontinuity
  SA flows
  SA half-space
  SA kinematics
  SA kinetics
  SA ray tracing
  SA stacking
  SA velocity structure
  SA vibration

velocity analysis (1993)
  Calculation of stacking or normal moveout velocity from measurements of normal moveout. Can involve common-midpoint data.
  SA analysis
  SA common-depth-point method
  SA normal moveout
  SA seismic methods
  SA stacking

velocity structure (1978)
  SA COCORP
  SA crust
  SA lateral heterogeneity
  SA mantle
  SA seismology
  SA velocity

Vema Channel (1989)
  Located between Sao Paulo Plateau and Rio Grande Plateau, connecting the Argentina and Brazil basins. Main deep passageway for the northerly flowing Antarctic bottom water. As of 1990, Atlantic Ocean is autoposted to this term.
  CO S320000S283000
     W0380500W0394700
  UF Vema Gap
  BT South American Atlantic
  BT South Atlantic
  BT Atlantic Ocean
  SA Brazil Basin
  SA DSDP Site 518
  SA Leg 72

Vema fracture zone (1978)
  Crosses Mid-Atlantic Ridge near 11° N. As of 1990, Atlantic Ocean is autoposted to this term.
  BT North Atlantic
  BT Atlantic Ocean
  SA Leg 39
  SA Mid-Atlantic Ridge
  SA North Atlantic Ridge

Vema Gap
  use Vema Channel

Vendee
  No longer a valid term for GeoRef. See Vendee France.

Vendee France (1993)
  Department on Bay of Biscay in W France.
  CO N462500N471000
     W0004500W0020500
  BT France
  BT Western Europe
  BT Europe
  SA Bay of Bourgneuf
  SA Poitou

Vendian (1978)
  Europe. Autoposting of Precambrian to this term ended in 1989. As of 1989, Proterozoic is autoposted to this term. Autoposting of upper Precambrian and Precambrian to this term began in 1990.
  BT upper Proterozoic
  BT Proterozoic
  BT upper Precambrian
  BT Precambrian
  NT Ediacaran
  SA Eocambrian
  SA Infracambrian

Vendsyssel (1978)
  Region of Jutland Peninsula N of Lim Fjord on the Skagerrak.
  BT Denmark
  BT Scandinavia
  BT Western Europe
  BT Europe
  SA Jutland

Venediger Group (1978)
  Mountain group in SE Tyrol and SW Salzburg State just S of W Hohe Tauern.
  IN Index states as applicable.
  UF Venediger Gruppe
  UF Venediger Massif
  BT Austria
  BT Central Europe
  BT Europe
  SA Salzburg State Austria
  SA Tyrol Austria

Venediger Gruppe
  use Venediger Group

Venediger Massif
  use Venediger Group

Venera Program (1985)
  Before 1985, also search Venera.
  SA extraterrestrial geology
  SA Maxwell Montes
  SA planetology
  SA remote sensing
  SA satellite methods
  SA Venus

Venerida (1981)
  BT Bivalvia
  BT Mollusca
  BT Invertebrata
  NT Calyptogena

Venetia (1978)
  Region E of Lombardy. Also a former autonomous region now known as Veneto.

  IN Index autonomous regions as applicable.
  BT Italy
  BT Southern Europe
  BT Europe
  SA Friuli-Venezia Giulia Italy
  SA Trentino-Alto Adige Italy
  SA Veneto Italy

Veneto
  No longer a valid term for GeoRef. See Veneto Italy.

Veneto Italy (1993)
  Autonomous region in NE Italy. Formerly Venetia autonomous region.
  CO N444800N464100
     E0130500E0104000
  UF Venezia Euganea
  BT Italy
  BT Southern Europe
  BT Europe
  NT Belluno Italy
  NT Cortina D'Ampezzo Italy
  NT Euganean Hills
  NT Monte Baldo
  NT Monti Berici
  NT Piave Valley
  NT Po Delta
  NT Possagno Italy
  NT Sappada Italy
  NT Venice Italy
  NT Verona Italy
  NT Vicenza Italy
  SA Carnic Alps
  SA Dolomites
  SA Lessini Mountains
  SA Marmolada Glacier
  SA Po River
  SA Po Valley
  SA Tagliamento Valley
  SA Venetia

Venezia Euganea
  use Veneto Italy

Venezia Italy
  use Venice Italy

Venezia-Tridentina
  use Trentino-Alto Adige Italy

Venezuela (1978)
  CO N004500N121000
     W0595500W0731500
  BT South America
  NT Amazonas Venezuela
  NT Anzoategui Venezuela
  NT Apure Guarico Plain
  NT Aragua Venezuela
  NT Barinas Venezuela
  NT Bolivar Venezuela
  NT Carabobo Venezuela
  NT Cojedes Venezuela
  NT Cordillera de la Costa
  NT Eastern Venezuela
  NT Falcon Venezuela
  NT Federal District Venezuela
  NT Guarico Venezuela
  NT Lake Maracaibo
  NT Lara Venezuela
  NT Maracaibo Basin
  NT Maracaibo Falcon Plain
  NT Merida Venezuela
  NT Miranda Venezuela
  NT Monagas Venezuela
  NT Nueva Esparta Venezuela
  NT Orinoco Belt
  NT Orinoco Delta
  NT Southeastern Venezuela
  NT Sucre Venezuela
  NT Tachira Venezuela
  NT Trujillo Venezuela
  NT Venezuelan Andes
  NT Venezuelan Islands
  NT Yaracuy Venezuela
  NT Zulia Venezuela

  SA Amazon Basin
  SA Andes
  SA Caracas Group
  SA Caribbean region
  SA Freites Formation
  SA Guajira Peninsula
  SA La Luna Formation
  SA La Quinta Formation
  SA Llanos
  SA Los Testigos
  SA Mirador Formation
  SA Misoa Formation
  SA Oficina Formation
  SA Orinoco River
  SA Orinoco River basin
  SA Punta Mosquito Formation
  SA Rio Negro
  SA Roraima Formation
  SA San Juan River
  SA Sierra de Perija

Venezuela Basin
  use Venezuelan Basin

Venezuelan Andes (1981)
  Autoposting of Venezuela to this term began in 1989. Autoposting of South America to this term began in 1990. This term has multiple hierarchies.
  BT1 Andes
  BT1 South America
  BT2 Venezuela
  BT2 South America

Venezuelan Basin (1978)
  Marine basin between island of Hispaniola to the N and Venezuela to the S.
  UF Venezuela Basin
  BT Caribbean Sea
  BT North American Atlantic
  BT North Atlantic
  BT Atlantic Ocean
  SA Leg 15

Venezuelan Islands (1981)
  CO N103000N120000
     W0630000W0680000
  BT Venezuela
  BT South America

Venice
  No longer a valid term for GeoRef. See Venice Italy.

Venice Italy (1993)
  Province and city on Gulf of Venice off N Adriatic Sea in E Veneto, NE Italy. Also search Venezia.
  UF Venezia Italy
  BT Veneto Italy
  BT Italy
  BT Southern Europe
  BT Europe

Ventersdorp
  No longer a valid term for GeoRef see Ventersdorp South Africa.

Ventersdorp South Africa (1993)
  Town in SW Transvaal.
  BT Transvaal South Africa
  BT South Africa
  BT Southern Africa
  BT Africa

Ventersdorp Supergroup (1989)
  IN Index ages as applicable.
  BT Precambrian
  SA Botswana
  SA lower Proterozoic
  SA South Africa
  SA upper Archean

ventifacts (1985)
  SA abrasion
  SA pebbles
  SA wind transport

vents (1978)
As of 1993, restricted to volcanic vents. Before 1993, also search vents or chimneys and volcanic features. See hydrothermal vents and chimneys.
UF volcanic vents
SA chimneys
SA fumaroles
SA geothermal systems
SA hydrothermal vents
SA solfataras
SA volcanic features
SA volcanic necks
SA volcanology

Ventura
Valid through 1988. Search in combination with state term. After 1988, use specific city-state term.

Ventura Basin (1978)
WNW of Los Angeles in Ventura and Santa Barbara counties.
IN Index counties or region as applicable.
SA California
SA Santa Barbara County California
SA Ventura County California

Ventura California (1989)
City NW of Los Angeles in S California. Before 1989, search Ventura AND California.
CO N341500N341500 W1191800W1191800
UF San Buenaventura
BT Ventura County California
BT California
BT United States

Ventura County
Valid through 1988. Search in combination with state term. After 1988, use specific county-state term.

Ventura County California (1989)
S California. Before 1989, also search Ventura County AND California.
CO N340300N345300 W1184000W1192700
BT California
BT United States
NT Oxnard California
NT Point Mugu
NT Simi Hills
NT Simi Valley California
NT Ventura California
SA Channel Islands
SA San Emigdio Mountains
SA San Nicolas Island
SA Santa Clara Valley
SA Ventura Basin

Venus (1978)
NT Aphrodite Terra
NT Beta Regio
NT Ishtar Terra
NT Lakshmi Planum
NT Maxwell Montes
SA coronae
SA landing sites
SA Magellan Program
SA Mariner 5
SA Mariner 10
SA Mariner Program
SA Pioneer Program
SA planetology
SA planets
SA solar system
SA terrestrial planets
SA tesserae
SA Venera Program

Venus Radar Mapper
No longer a valid term for GeoRef. Valid from 1989-1992.
use Magellan Program

Veporides (1985)
Mountain range, central Slovakia.
BT Czechoslovakia
BT Central Europe
BT Europe
SA Choc Nappe
SA Slovakia
SA Slovakian Carpathians
SA Western Carpathians

Veracruz
No longer a valid term for GeoRef as of 1993 see Veracruz Mexico.

Veracruz Mexico (1993)
State including city on the Gulf of Mexico in E Mexico. Before 1993 also search Veracruz AND Mexico.
CO N170000N221500 W0930000W0990000
BT Mexico
NT Papaloapan River basin
NT Poza Rica Mexico
SA Gulf Coastal Plain
SA Sierra Madre Oriental

Vercelli
No longer a valid term for GeoRef. See Vercelli Italy.

Vercelli Italy (1993)
Province and city in E Piemonte, NW Italy.
BT Piemonte Italy
BT Italy
BT Southern Europe
BT Europe
NT Biella Italy

Vercors (1978)
Limestone massif of the Dauphine Alps SE France. This term has multiple hierarchies.
IN Index departments as applicable.
UF Vercors Massif
BT1 Dauphine Alps
BT1 Western Alps
BT1 Alps
BT1 Europe
BT2 Dauphine Alps
BT2 France
BT2 Western Europe
BT2 Europe
SA Drome France
SA Isere France

Vercors Massif
use Vercors

Verde Formation (1978)
Pliocene (?) or Pleistocene. Unconformably overlies Hickey Formation in Jerome area. Central Arizona.
BT Cenozoic
SA Arizona
SA Pleistocene
SA Pliocene

Verde Valley (1978)
River valley in central Arizona.
IN Index counties or region as applicable.
SA Arizona

Verdigris River valley (1978)
SE Kansas, and NE Oklahoma. Also search Verdigris River.
IN Index counties and states as applicable.
BT United States
SA Kansas
SA Oklahoma

Verdon Valley (1978)
River valley in SE France. Also search Verdon.
SA Alpes-de-Haute Provence France

Verdun
No longer a valid term for GeoRef. See Verdun France.

Verdun France (1993)
City on the Meuse River in central Meuse, NE France. Also search Verdun-sur-Meuse.
UF Verdun-sur-Meuse
BT Meuse France
BT France
BT Western Europe
BT Europe

Verdun-sur-Meuse
use Verdun France

Verkhovtsevo
No longer a valid term for GeoRef. As of 1993, see Verkovtsevo Ukraine.

Verkhovtsevo Ukraine (1993)
Town in W central Dnepropetrovsk Oblast in S central. This term has multiple hierarchies.
BT1 Dnepropetrovsk Ukraine
BT1 Ukraine
BT1 Europe
BT2 Dnepropetrovsk Ukraine
BT2 Ukraine
BT2 Commonwealth of Independent States

Verkhoyansk
No longer a valid term for GeoRef. As of 1993, see Verkhoyansk Russian Federation.

Verkhoyansk Range (1978)
Mountain range extending along the E banks of the Lena River and the lower Aldan River in N central Yakutia. This term has multiple hierarchies.
CO N630000N700000 E1400000E1240000
UF Verkhoyanski Mountains
BT1 Yakutia Russian Federation
BT1 Russian Federation
BT1 Commonwealth of Independent States
BT2 Yakutia Russian Federation
BT2 Asia
SA Verkhoyansk region

Verkhoyansk region (1981)
Region of the Verkhoyansk and Cherski ranges. This term has multiple hierarchies.
CO N570000N730000 E1600000E1240000
BT1 Russian Federation
BT1 Commonwealth of Independent States
BT2 Asia
SA Magadan Russian Federation
SA Verkhoyansk Range
SA Yakutia Russian Federation

Verkhoyansk Russian Federation (1993)
Town N of Arctic Circle on the Yana River in N central Yakutia in N central Siberia. This term has multiple hierarchies.
BT1 Yakutia Russian Federation
BT1 Russian Federation
BT1 Commonwealth of Independent States
BT2 Yakutia Russian Federation
BT2 Asia

Verkhoyanski Mountains
use Verkhoyansk Range

Vermes
Not a valid GeoRef index term except during 1969 and 1974. Was used on level 1 in subfile B.
use worms

vermiculite (1978)
As of 1981, use vermiculite deposits for vermiculite as a commodity.
BT clay minerals
BT sheet silicates
BT silicates
SA clay mineralogy
SA mica group
SA vermiculite deposits

vermiculite deposits (1981)
Before 1981, search vermiculite AND deposits.
IN Commodity. See List C.
SA clays
SA mica deposits
SA vermiculite

Vermilion granitic complex (1989)
N Minnesota and NW Ontario.
BT Archean
BT Precambrian
SA Minnesota
SA Ontario

Vermilion Parish
Valid through 1988. Search in combination with state term. After 1988, use specific county-state term.

Vermilion Parish Louisiana (1989)
On the Gulf of Mexico in S Louisiana. Before 1989, also search Vermilion Parish AND Louisiana; Vermilion AND Louisiana.
CO N293300N300900 W0915700W0924300
BT Louisiana
BT United States

Vermilion Range (1978)
Iron mining area in Saint Louis and Lake counties in NE Minnesota. Also search Vermilion AND Minnesota.
IN Index counties as applicable.
BT Minnesota
BT United States
SA Lake County Minnesota
SA Saint Louis County Minnesota

Vermillion Creek coal bed (1989)
In Niland Tongue of Wasatch Formation. SW Wyoming.
BT Eocene
BT Paleogene
BT Tertiary
BT Cenozoic
SA Wasatch Formation
SA Wyoming

Vermont (1978)
Autoposting of this term began in 1978.
CO N424500N450000 W0713000W0732500
BT United States
NT Addison County Vermont
NT Caledonia County Vermont
NT Chittenden County Vermont
NT Essex County Vermont
NT Franklin County Vermont
NT Orange County Vermont
NT Rutland County Vermont
NT Washington County Vermont
NT Windham County Vermont

NT Windsor County Vermont
SA Ammonoosuc Volcanics
SA Barre Granite
SA Beekmantown Group
SA Champlain Sea
SA Champlain Valley
SA Chazy Group
SA Cheshire Formation
SA Clough Formation
SA Connecticut River
SA Connecticut Valley
SA Dunham Dolomite
SA Gile Mountain Formation
SA Green Mountains
SA Hazens Notch Formation
SA Hoosac Formation
SA Lake Champlain
SA Littleton Formation
SA Normanskill Formation
SA Partridge Formation
SA Pinnacle Formation
SA Pinney Hollow Formation
SA Potsdam Sandstone
SA Saint Lawrence River basin
SA Trenton Group
SA Underhill Formation
SA Waits River Formation
SA Wilcox Formation

vernadite (1993)
BT oxides
SA manganese minerals
SA nsutite
SA pyrolusite
SA ramsdellite

Vernadskii, V. I.
use Vernadskiy, Vladimir Ivanovich

Vernadskiy, Vladimir Ivanovich (1993)
Before 1993, also search Vernadskii, V. I., Vernadsky, V. I., and Vernadskyi, V. I.
UF Vernadskii, V. I.
UF Vernadskyi, V. I.
UF Vladimir Ivanovich Vernadskiy
SA biography

Vernadskyi, V. I.
use Vernadskiy, Vladimir Ivanovich

Vernagt Glacier (1978)
In Otztal Alps S of the Wildspitze in S Tyrol, W Austria.
BT Tyrol Austria
BT Austria
BT Central Europe
BT Europe
SA Otztal Alps

Vernyi Kazakhstan
use Alma-Ata Kazakhstan

Verona
No longer a valid term for GeoRef. See Verona Italy.

Verona Italy (1993)
Province and city in W Veneto, NE Italy.
BT Veneto Italy
BT Italy
BT Southern Europe
BT Europe

Versailles
No longer a valid term for GeoRef. Before 1989, was also used for numerous cities in the U.S. As of 1989, use Versailles in combination with appropriate state term for those. As of 1993, also see Versailles France.

Versailles France (1993)
Before 1989, search Versailles AND France.
BT Yvelines France
BT France

BT Western Europe
BT Europe

Versailles Illinois (1989)
Town in W Illinois. Before 1989, search Versailles AND Illinois.
CO N395306N395306
   W0903921W0903921
BT Brown County Illinois
BT Illinois
BT United States

Versailles Indiana (1989)
Town in SE Indiana. Before 1989, search Versailles AND Indiana.
CO N390400N390400
   W0851600W0851600
BT Ripley County Indiana
BT Indiana
BT United States

Versailles Kentucky (1989)
City in central Kentucky. Before 1989, search Versailles AND Kentucky.
CO N380200N380200
   W0844500W0844500
BT Woodford County Kentucky
BT Kentucky
BT United States

Versailles Missouri (1989)
City in central Missouri. Before 1989, search Versailles AND Missouri.
CO N382500N382500
   W0925100W0925100
BT Morgan County Missouri
BT Missouri
BT United States

Versilian (1978)
Europe.
BT Holocene
BT Quaternary
BT Cenozoic

Vertebrata (1978)
Term is to be used only when more specific terms do not apply, or are too numerous to be recorded.
BT Chordata
NT Agnatha
NT Pisces
NT Tetrapoda
SA affinities
SA bones
SA coprolites
SA extremities
SA fossil man
SA jaws
SA nervous system
SA skeletons
SA skulls
SA spinal column
SA teeth
SA vertebrates

vertebrates (1981)
Common name for Vertebrata.
IN Index for all non-paleontologic studies of fossils.
NT agnathans
NT fish
NT tetrapods
SA biostratigraphy
SA Vertebrata

Vertes
No longer a valid term for GeoRef. See Vertes Hungary.

Vertes Hungary (1993)
Town in E Hungary.
BT Hungary
BT Central Europe
BT Europe

Vertes Mountains (1978)

N central Hungary. Also search Vertes.
BT Hungary
BT Central Europe
BT Europe

vertical
A valid index term through 1977. After 1977, use vertical orientation for folds.

vertical discontinuities (1981)
Used for a type of interface in seismology.
BT discontinuities
SA crust
SA mantle
SA seismology

vertical gradient methods
use vertical-gradient methods

vertical movements (1978)
Before 1981, also search vertical tectonics.
UF movements, vertical
SA horizontal movements
SA isostasy
SA neotectonics
SA subsidence
SA tectonics
SA uplifts

vertical orientation (1978)
Reference to folds defined on the basis of orientation in relation to the geographic horizontal plane. Before 1978, also search folds AND vertical.
SA folds
SA orientation
SA upright folds

vertical seismic profiles (1985)
SA geophysical methods
SA geophysical profiles
SA seismic logging
SA seismic methods
SA seismic profiles
SA seismic surveys
SA seismograms
SA seismology

vertical tectonics
As of 1981, no longer a valid term for GeoRef. Use vertical movements.

vertical-component seismographs (1981)
BT seismographs

vertical-gradient methods (1985)
UF vertical gradient methods
SA airborne methods
SA geophysical methods
SA high-resolution methods
SA magnetic methods
SA magnetic surveys
SA methods
SA remote sensing

Vertisols (1978)
BT soils
SA soil group

very long base interferometry
use very long baseline interferometry

very long baseline interferometry (1982)
UF very long base interferometry
UF VLBI
SA geodesy
SA geodetic coordinates
SA geodetic networks
SA geophysical methods
SA geophysical surveys
SA interferometry

SA remote sensing
SA satellite measurements
SA satellite methods
SA triangulation

Vesdre Valley (1978)
River valley in central Liege, E Belgium.
BT Liege Belgium
BT Belgium
BT Western Europe
BT Europe

vesicular texture (1982)
BT textures
SA igneous rocks
SA lava

Vest Agder Norway
use Vest-Agder Norway

Vest-Agder
No longer a valid term for GeoRef. As of 1993 see Vest-Agder Norway.

Vest-Agder Norway (1993)
County in extreme S Norway. Before 1993 also search Vest-Agder or Vest Agder.
UF Vest Agder Norway
BT Norway
BT Scandinavia
BT Western Europe
BT Europe

Vesteraalen
use Vesteralen

Vesteralen (1978)
Island group N of the Lofoten Islands in the Norwegian Sea off NW mainland.
IN Index counties as applicable.
UF Vesteraalen
BT Norway
BT Scandinavia
BT Western Europe
BT Europe
SA Nordland Norway
SA Troms Norway

Vestfold
No longer a valid term for GeoRef. As of 1993 see Vestfold Norway.

Vestfold Hills (1989)
On Ingrid Christensen Coast bordering Prydz Bay. Near Davis research station (Australian).
CO S684000S682200
   E0783300E0774900
BT Antarctica

Vestfold Norway (1993)
County in SE Norway. Before 1993 also search Vestfold AND Norway.
UF Jarlsberg
BT Norway
BT Scandinavia
BT Western Europe
BT Europe
NT Larvik Norway

Vestman Islands
use Vestmannaeyjar

Vestmann Islands
use Vestmannaeyjar

Vestmannaeyjar (1978)
Small island group of volcanic origin S of Iceland. As of 1993, Heimaey is a narrower term.
UF Vestman Islands
UF Vestmann Islands
UF Westman Islands
BT Iceland
BT Western Europe
BT Europe

NT Heimaey

Vestspitsbergen
As of 1993, no longer a valid term for GeoRef.
use Spitsbergen Island

Vesuvian garnet
use leucite

vesuvian-type eruptions
use vulcanian-type eruptions

vesuvianite (1978)
Autoposting of sorosilicates began in 1985.
UF idocrase
BT sorosilicates
BT orthosilicates
BT silicates

Vesuvio
use Vesuvius

Vesuvius (1978)
Active volcano on E side of the Bay of Naples.
UF Vesuvio
UF Vesuvius Volcano
BT Campania Italy
BT Italy
BT Southern Europe
BT Europe
SA Monte Somma

Vesuvius Volcano
use Vesuvius

Veszprem
No longer a valid term for GeoRef. See Veszprem Hungary.

Veszprem Hungary (1993)
County in W Hungary including the city N of Lake Balaton.
BT Hungary
BT Central Europe
BT Europe
NT Lake Balaton
NT Varpalota Hungary

Vetluga Series (1978)
According to the USSR Lexique, the age of the Vetluga Horizon is upper Permian. It is composed of red spotted clay, and maroon and gray sandstone in an interwinning stratification not characterized paleontologically. NE of Moscow.
BT Upper Permian
BT Permian
BT Paleozoic
SA Russian Federation

Vetrenyy Ridge (1978)
S of Onega Bay in W Arkhangelsk Oblast in NW European Russian Federation. Also search Vetrenyy. This term has multiple hierarchies.
BT1 Arkhangelsk Russian Federation
BT1 Russian Federation
BT1 Commonwealth of Independent States
BT2 Arkhangelsk Russian Federation
BT2 Europe

Viatka River
use Vyatka River

vibrating-string gravimeters (1981)
BT gravimeters

vibration (1978)
UF vibrations
SA earthquakes
SA oscillations
SA resonance
SA rock mechanics
SA seismology

SA velocity

vibrations
use vibration

Vibroseis (1978)
SA elastic waves
SA seismic methods
SA seismic surveys

Viburnum Trend (1985)
Metallogenic province, SE Missouri.
BT Missouri
BT United States

vicariance (1985)
SA biogeography
SA biologic evolution
SA paleobotany
SA paleoecology
SA paleontology

Vicecomodoro Island
use Seymour Island

Vicenza
No longer a valid term for GeoRef. See Vicenza Italy.

Vicenza Italy (1993)
Province and city W of Venice, central Veneto, NE Italy.
BT Veneto Italy
BT Italy
BT Southern Europe
BT Europe
NT Berici Hills

Vich
No longer a valid term for GeoRef. As of 1993, see Vich Spain.

Vich Spain (1993)
City NE of Barcelona City, NE Spain. Before 1993, also search Vich and Spain.
BT Barcelona Spain
BT Catalonia Spain
BT Spain
BT Iberian Peninsula
BT Southern Europe
BT Europe

Vicksburg Group (1978)
Includes Marianna Limestone and Byram Formation. Gulf Coastal Plain. Vicksburg.
BT middle Oligocene
BT Oligocene
BT Paleogene
BT Tertiary
BT Cenozoic
SA Alabama
SA Bucatunna Formation
SA Byram Formation
SA Florida
SA Glendon Limestone
SA Louisiana
SA Marianna Limestone
SA Mississippi
SA Texas

Victoria
No longer a valid term for GeoRef. See Victoria Australia, Lake Victoria, Victoria Land, Victoria Island, and Victoria earthquake 1980.

Victoria Australia (1993)
CO S390000S340000 E1500000E1400000
BT Australia
BT Australasia
NT Gippsland Australia
NT Gippsland Basin
NT Horsham Australia
NT Melbourne Australia
NT Mildura Australia

NT Port Campbell Australia
SA Discovery Bay
SA Gambier Embayment
SA Green Gully
SA Lachlan fold belt
SA Latrobe Group
SA Latrobe Valley
SA Lilydale Limestone
SA Murchison Meteorite
SA Murray Basin
SA Murray River
SA Otway Basin
SA Philip Island
SA Port Phillip Bay
SA Saint Kilda
SA Snowy Mountains
SA Sunnyside Mine
SA Tasman Geosyncline
SA Tasman orogenic zone
SA Western Port

Victoria County
No longer a valid term for GeoRef. As of 1989, for county in S Texas, use Victoria County Texas. As of 1993, see Victoria County New Brunswick, Victoria County Nova Scotia, or Victoria County Ontario.

Victoria County New Brunswick (1993)
NW New Brunswick. Before 1993, search Victoria County AND New Brunswick.
BT New Brunswick
BT Maritime Provinces
BT Eastern Canada
BT Canada

Victoria County Nova Scotia (1993)
NE Nova Scotia. Before 1993, search Victoria County AND Nova Scotia.
BT Nova Scotia
BT Maritime Provinces
BT Eastern Canada
BT Canada

Victoria County Ontario (1993)
S Ontario. Before 1993, search Victoria County AND Ontario.
BT Ontario
BT Eastern Canada
BT Canada

Victoria County Texas (1989)
S Texas. Before 1989, also search Victoria County AND Texas.
CO N283000N290900 W0964300W0972200
BT Texas
BT United States

Victoria earthquake 1980 (1989)
Mexico.
BT earthquakes
SA Mexico

Victoria Island (1989)
Franklin District, Northwest Territories. Also in Lagos, SW Nigeria.
IN Index regions as applicable.
SA Franklin District Northwest Territories
SA Nigeria
SA Northwest Territories

Victoria Land (1978)
W of Ross Sea and W of NW Ross Ice Shelf, primarily on the Ross Dependency on the Pacific Ocean side S of New Zealand.
UF North Victoria Land
UF South Victoria Land
BT Antarctica
NT Allan Hills

NT Lake Vanda
NT Yamato Mountains
SA ALHA 77005
SA ALHA 81005
SA Allan Hills Meteorites
SA Beacon Supergroup
SA Bowers Supergroup
SA Ferrar Group
SA Kirkpatrick Basalt
SA Robertson Bay Group
SA Taylor Valley
SA Victoria Land Basin
SA Wright Valley
SA Yamato Meteorites

Victoria Land Basin (1989)
Sedimentary basin in Ross Sea.
BT Antarctica
SA Ross Sea
SA Victoria Land

Victoria Valley (1978)
River valley in NW Northern Territory.
IN Index state or region as applicable.
SA Northern Territory Australia

Vidda
use Hardangervidda

video methods (1993)
General term relating to use of picture signals or television system technology.
SA borehole televiewers
SA photogeologic methods
SA photography
SA remote sensing
SA well-logging

Vidin
No longer a valid term for GeoRef. See Vidin Bulgaria.

Vidin Bulgaria (1993)
Province in extreme NW Bulgaria. Including the city on the Danube River near the Yugoslav border.
BT Bulgaria
BT Southern Europe
BT Europe
NT Lom Depression

Vienerwald
use Wiener Wald

Vienna
No longer a valid term for GeoRef. See Vienna Austria.

Vienna Austria (1993)
Independent city on the Danube River inside E Lower Austria, NE Austria. Also search Wien.
UF Wien Austria
BT Lower Austria
BT Austria
BT Central Europe
BT Europe

Vienna Basin (1978)
IN Index countries as applicable.
BT Central Europe
BT Europe
NT Austrian Vienna Basin
SA Austria
SA Czechoslovakia

Vienne
No longer a valid term for GeoRef. See Vienne France.

Vienne France (1993)
Department in W central France.
CO N460500N471000 E0011500W0000500
BT France
BT Western Europe
BT Europe
NT Poitiers France

SA Poitou

Viet Nam
Not a valid GeoRef index term after 1971. Was used on level 1 in subfile G.
use Vietnam

Vietnam (1978)
North Vietnam (Democratic Republic of Vietnam) and South Vietnam (Republic of Vietnam) were combined to form a unified nation on June 24, 1976. This nation is known as the Socialist Republic of Vietnam. Before 1977, also search Vietnam AND north; search Vietnam AND south; Viet Nam.
CO N083000N231500
   E1092000E1020000
UF North Vietnam
UF Viet Nam
BT Far East
BT Asia
NT Baoha Vietnam
NT Dalat Vietnam
NT Mekong Delta
NT Nong-Son Vietnam
NT Yenbay Vietnam
SA Indochina
SA Red River Fault

Vigarano Meteorite (1981)
Impact near village of Vigarano in Ferrara Province, NE Emilia-Romagna. Before 1981, also search Vigarano AND meteorites.
BT carbonaceous chondrites
BT chondrites
BT stony meteorites
BT meteorites
SA Emilia-Romagna Italy
SA Italy

Vigo County
Valid through 1988. Search in combination with state term. After 1988, use specific county-state term.

Vigo County Indiana (1989)
W Indiana. Before 1989, also search Vigo County AND Indiana.
CO N391500N393600
   W0871200W0873800
BT Indiana
BT United States

Viking
As of 1981, no longer a valid term for GeoRef. Use Viking Program.

Viking Formation (1978)
Consists of sandstone which grades eastward into siltstone and shale and into the marine shales of the Ashville Formation. In SW Alberta, the sandstones grade into dark marine shales and sandstones of the Bow Island Formation. Oil reservoirs in beds equivalent to the lower Colorado are found in sandstone of the Viking Formation. NE and SW Alberta.
BT Cretaceous
BT Mesozoic
SA Alberta

Viking Graben (1985)
N North Sea.
BT North Sea
BT European Atlantic
BT North Atlantic
BT Atlantic Ocean

Viking Program (1981)
SA extraterrestrial geology

SA landing sites
SA Mars
SA planetology
SA remote sensing

Vila Real
No longer a valid term for GeoRef. As of 1993 see Vila Real Portugal.

Vila Real Portugal (1993)
District in N Portugal. Before 1993 also search Vila Real AND Portugal.
BT Portugal
BT Iberian Peninsula
BT Southern Europe
BT Europe
NT Valpacos Portugal
NT Vilarandelo Portugal
SA Tras-os-Montes

Vilaine Bay (1978)
At the mouth of the Vilaine River on the Bay of Biscay in SE Brittany. Also search Vilaine.
SA Bay of Biscay
SA Morbihan France

Vilarandelo
No longer a valid term for GeoRef. As of 1993 see Vilarandelo Portugal.

Vilarandelo Portugal (1993)
Village in Vila Real, N Portugal. Before 1993 also search Vilarandelo AND Portugal.
BT Vila Real Portugal
BT Portugal
BT Iberian Peninsula
BT Southern Europe
BT Europe

Vilas County
Valid through 1988. Search in combination with state term. After 1988, use specific county-state term.

Vilas County Wisconsin (1989)
N Wisconsin. Before 1989, also search Vilas County AND Wisconsin.
CO N455200N461800
   W0885600W0900400
BT Wisconsin
BT United States

Villa Clara Cuba (1993)
Province, N central Cuba. Also search Las Villas and Santa Clara AND Cuba.
BT Cuba
BT Greater Antilles
BT Antilles
BT West Indies
BT Caribbean region

Villach
No longer a valid term for GeoRef. See Villach Austria.

Villach Austria (1993)
City W of Klagenfurt on the Drava River in S Carinthia, S Austria. Also search Villach or Beljak AND Austria.
UF Beljak Austria
BT Carinthia Austria
BT Austria
BT Central Europe
BT Europe

Villafranca d'Asti
No longer a valid term for GeoRef. See Villafranca d'Asti Italy.

Villafranca d'Asti Italy (1993)
Village E of Turin in central Italy.
BT Piemonte Italy
BT Italy

BT Southern Europe
BT Europe

Villafranchian (1978)
Europe.
BT lower Pleistocene
BT Pleistocene
BT Quaternary
BT Cenozoic
SA Calabrian
SA Tertiary

Villany Mountains (1978)
In Baranya, S Hungary.
BT Hungary
BT Central Europe
BT Europe

villiaumite (1978)
BT fluorides
BT halides

Vilna
No longer a valid term for GeoRef. See Vilna Lithuania.

Vilna Lithuania (1993)
City near the Belarus border in SE Lithuania. Also search Vilna or Vilnius or Vilno or Vilnyus or Wilna or Wilno.
UF Vilnius Lithuania
UF Vilno Lithuania
UF Vilnyus Lithuania
UF Wilna Lithuania
UF Wilno Lithuania
BT Lithuania
BT Baltic region
BT Europe

Vilnius Lithuania
use Vilna Lithuania

Vilno Lithuania
use Vilna Lithuania

Vilnyus Lithuania
use Vilna Lithuania

Vilyui Basin
use Vilyuy River basin

Vilyui River
use Vilyuy River

Vilyui Syneclise
use Vilyuy Syneclise

Vilyuy Basin
use Vilyuy River basin

Vilyuy River (1978)
Rises on the Central Siberian Plateau of E Evenk National Okrug and flows through W and W central Yakutia into the Lena River. Also search Vilyui; Vilyuy. This term has multiple hierarchies.
UF Vilyui River
BT1 Russian Federation
BT1 Commonwealth of Independent States
BT2 Asia

Vilyuy River basin (1978)
In E Evenk National Okrug and W and W central Yakutia in central Siberia. This term has multiple hierarchies.
UF Vilyui Basin
UF Vilyuy Basin
BT1 Russian Federation
BT1 Commonwealth of Independent States
BT2 Asia

Vilyuy Sineclise
use Vilyuy Syneclise

Vilyuy Syneclise (1978)
The lower Vilyuy River basin in the area of the river's flow into Lena River in central Yakutia. This term has multiple hierarchies.
CO N593000N720000
   E1280000E1110000
UF Vilyui Syneclise
UF Vilyuy Sineclise
BT1 Siberian Platform
BT1 Russian Federation
BT1 Commonwealth of Independent States
BT2 Siberian Platform
BT2 Asia
SA Mir Pipe
SA Yakutia Russian Federation

vimsite (1978)
BT borates

Vindhya Basin
use Vindhyan Basin

Vindhyan (1978)
Precambrian-Cambrian. Includes series of sandstones, shales, and limestones extensively exposed in the highlands of central India.
IN Index ages as applicable.
UF Vindhyan System
SA Bhander Group
SA Cambrian
SA India
SA Kaimur Sandstone
SA Precambrian
SA Semri Series

Vindhyan Basin (1978)
IN Index countries as applicable.
UF Vindhya Basin
BT Asia
SA Bangladesh
SA India

Vindhyan System
use Vindhyan

Vindobonian (1978)
Europe.
BT Miocene
BT Neogene
BT Tertiary
BT Cenozoic

Vinini Formation (1989)
BT Ordovician
BT Paleozoic
SA Nevada

Vinton Canyon (1978)
IN Index counties or region as applicable.
SA Texas

Viola Limestone (1978)
Middle and Upper Ordovician. In Patterson Ranch Group. Overlies Bromide Formation and underlies Fernvale Formation. Central S and SW Oklahoma.
BT Ordovician
BT Paleozoic
SA Oklahoma

violarite (1978)
BT sulfides

Virgilian (1978)
Provincial series, North America. Above Missourian, below Wolfcampian (Permian).
UF Virgilian Series
BT Upper Pennsylvanian
BT Pennsylvanian
BT Carboniferous
BT Paleozoic
NT Beil Limestone Member
NT Douglas Group
NT Ervine Creek Limestone
NT Lawrence Formation

NT Lecompton Limestone
NT Oread Limestone
NT Plattsmouth Limestone Member
NT Shawnee Group
NT Tonganoxie Sandstone
NT Wabaunsee Group
SA Wescogame Formation

Virgilian Series
use Virgilian

Virgin Islands (1978)
As of 1981, includes the British Virgin Islands and the U. S. Virgin Islands. From 1978-1980, term referred to only the U. S. Virgin Islands. Before 1978, included both British and U. S. Virgin Islands. Autoposting of this term began in 1981. Before 1981, West Indies was the broader term.
IN Index Leeward Islands.
BT Lesser Antilles
BT Antilles
BT West Indies
BT Caribbean region
NT British Virgin Islands
NT U. S. Virgin Islands
SA Saint John Island

Virgin River valley (1978)
Also search Virgin River.
IN Index states as applicable.
BT United States
SA Arizona
SA Nevada
SA Utah

Virginia (1978)
Autoposting of this term began in 1978.
CO N363500N392800
   W0751500W0833500
BT United States
NT Accomack County Virginia
NT Albemarle County Virginia
NT Alleghany County Virginia
NT Arlington County Virginia
NT Augusta County Virginia
NT Bath County Virginia
NT Bedford County Virginia
NT Botetourt County Virginia
NT Brunswick County Virginia
NT Buchanan County Virginia
NT Buckingham County Virginia
NT Campbell County Virginia
NT Carroll County Virginia
NT Charlotte County Virginia
NT Chesterfield County Virginia
NT Clarke County Virginia
NT Cumberland County Virginia
NT Dickenson County Virginia
NT Dinwiddie County Virginia
NT Essex County Virginia
NT Fairfax County Virginia
NT Floyd County Virginia
NT Franklin County Virginia
NT Frederick County Virginia
NT Giles County Virginia
NT Gloucester County Virginia
NT Greene County Virginia
NT Hanover County Virginia
NT Henrico County Virginia
NT Henry County Virginia
NT Highland County Virginia
NT Lancaster County Virginia
NT Lee County Virginia
NT Madison County Virginia
NT Mecklenburg County Virginia
NT Middlesex County Virginia
NT Montgomery County Virginia
NT Nelson County Virginia
NT Norfolk County Virginia
NT Northampton County Virginia
NT Northumberland County Virginia
NT Orange County Virginia
NT Page County Virginia
NT Pamunkey River
NT Pittsylvania County Virginia
NT Prince William County Virginia
NT Princess Anne County Virginia
NT Pulaski County Virginia
NT Rappahannock River
NT Roanoke County Virginia
NT Rockbridge County Virginia
NT Rockingham County Virginia
NT Russell County Virginia
NT Scott County Virginia
NT Stafford County Virginia
NT Surry County Virginia
NT Sussex County Virginia
NT Tazewell County Virginia
NT Warren County Virginia
NT Washington County Virginia
NT Westmoreland County Virginia
NT Wise County Virginia
NT Wythe County Virginia
NT York County Virginia
SA Allegheny Front
SA Allegheny Group
SA Allegheny Mountains
SA Allegheny Plateau
SA Amelia
SA Antietam Formation
SA Appalachian Basin
SA Appalachian Plateau
SA Aquia Formation
SA Ashe Formation
SA Atlantic Coastal Plain
SA Baltimore Gneiss
SA Beekmantown Group
SA Bloomsburg Formation
SA Blue Ridge Mountains
SA Blue Ridge Province
SA Brallier Shale
SA Calvert Formation
SA Carolina Bays
SA Catoctin Formation
SA Catskill Formation
SA Chemung Formation
SA Chesapeake Bay
SA Chesapeake Group
SA Chickamauga Group
SA Chilhowee Group
SA Chopawamsic Formation
SA Choptank Formation
SA Chowan River Formation
SA Clinch Sandstone
SA Clinton Group
SA Coeymans Formation
SA Conasauga Group
SA Conemaugh Group
SA Conococheague Formation
SA Culpeper Basin
SA Cumberland Plateau
SA Dan River basin
SA Delmarva Peninsula
SA Eastover Formation
SA Genesee Group
SA Glenarm Series
SA Greenbrier Limestone
SA Greenland Gap Group
SA Hampshire Formation
SA Helderberg Group
SA Holston Formation
SA Hurricane Ridge Syncline
SA James River
SA Juniata Formation
SA Kanawha Formation
SA Keyser Limestone
SA Kingsport Formation
SA Kinzers Formation
SA Knox Group
SA Lee Formation
SA Lenoir Limestone
SA Lincolnshire Limestone
SA Lynchburg Formation
SA Marcellus Shale
SA Martinsburg Formation
SA Mascot Dolomite
SA McKenzie Formation
SA Millboro Shale
SA Monongahela Group
SA Nanjemoy Formation
SA New River
SA New River Formation
SA Newark Supergroup
SA Newman Limestone
SA Ocoee Series
SA Onondaga Limestone
SA Oriskany Sandstone
SA Pennington Formation
SA Petersburg Granite
SA Piedmont
SA Pine Mountain Window
SA Piney Point Formation
SA Pocahontas Formation
SA Pocono Formation
SA Potomac Group
SA Potomac River
SA Potomac River basin
SA Pottsville Group
SA Price Formation
SA Pulaski Fault
SA Pulaski thrust sheet
SA Pungo River Formation
SA Reedsville Formation
SA Richmond Basin
SA Rome Formation
SA Rose Hill Formation
SA Saint Louis Limestone
SA Saint Marys Formation
SA Salisbury Embayment
SA Saltville Fault
SA Severn Estuary
SA Shady Dolomite
SA Sharps Meteorite
SA Shenandoah Valley
SA Stones River Group
SA Tioga Bentonite
SA Tonoloway Limestone
SA Trenton Group
SA Tuscarora Formation
SA Valley and Ridge Province
SA Wicomico Formation
SA Williamsport Sandstone
SA Wissahickon Formation
SA York River
SA Yorktown Formation

Virginia Beach
Valid through 1988. Search in combination with state term. After 1988, use specific city-state term.

Virginia Beach Virginia (1989)
City in but independent of Princess Anne County, SE Virginia, on the Atlantic Ocean 18 miles E of Norfolk. Before 1989, search Virginia Beach AND Virginia.
CO N365100N365100
   W0755900W0755900
BT Princess Anne County Virginia
BT Virginia
BT United States

Virginia Formation (1989)
N Minnesota.
BT lower Proterozoic
BT Proterozoic
BT upper Precambrian
BT Precambrian
SA Animikie Group
SA Minnesota
SA Thomson Formation

viridine (1978)
Autoposting of nesosilicates began in 1985.
BT nesosilicates
BT orthosilicates
BT silicates

Viruan (1978)
Europe.
BT Ordovician

BT Paleozoic

Viruddhachalam
Not a valid term for GeoRef. See Vridhachalam India.

Visagapatnam
Not a valid term for GeoRef. See Visakhapatnam India.

Visakhapatnam
No longer a valid term for GeoRef. See Visakhapatnam India.

Visakhapatnam India (1993)
City on the Coromandel Coast of the Bay of Bengal in NE Andhra Pradesh, central India. Also search Visagapatnam, Vishakhapatnam, Vizagapatam, and Vizagapatnam.
BT Andhra Pradesh India
BT India
BT Indian Peninsula
BT Asia

viscoelastic materials (1981)
SA elastic materials
SA elasticity
SA elastoviscoplastic materials
SA engineering geology
SA materials
SA viscosity
SA viscous materials

viscoelasticity (1978)
SA deformation
SA elastic limit
SA elastic materials
SA elastic strain
SA elasticity
SA elastoviscoplastic materials
SA strain
SA stress
SA viscosity
SA viscous materials

viscoplastic materials (1981)
SA elastoviscoplastic materials
SA engineering geology
SA materials
SA plastic deformation
SA plastic materials
SA plasticity
SA viscosity
SA viscous materials

viscosity (1978)
SA deformation
SA effusion
SA elastoviscoplastic materials
SA flow mechanism
SA lava
SA magmas
SA plasticity
SA rheology
SA viscoelastic materials
SA viscoelasticity
SA viscoplastic materials
SA viscous materials

viscous materials (1978)
SA elastoviscoplastic materials
SA materials
SA viscoelastic materials
SA viscoelasticity
SA viscoplastic materials
SA viscosity

viscous remanent magnetization (1978)
Autoposting of magnetization to this term began in 1989.
BT remanent magnetization
BT magnetization
SA paleomagnetism

Visean (1978)

Europe. Above Tournaisian, below Namurian. Autoposting of this term began in 1978.
BT Dinantian
BT Carboniferous
BT Paleozoic
NT Great Scar Limestone
NT upper Visean

Viseu
No longer a valid term for GeoRef. As of 1993 see Viseu Portugal.

Viseu Portugal (1993)
District in N central Portugal. Also a city NE of Coimbra. Before 1993 also search Viseu AND Portugal.
UF Vizeu
BT Portugal
BT Iberian Peninsula
BT Southern Europe
BT Europe

Vishakhapatnam
Not a valid term for GeoRef. See Visakhapatnam India.

Visla River
use Vistula River

Vistula River (1978)
Rises on the N slope of the Carpathian Mountains and flows into the Baltic Sea at Gdansk. Also search Vistula.
UF Visla River
UF Weichsel River
UF Wisla River
BT Poland
BT Central Europe
BT Europe

Vistula River valley (1978)
S, E central, and N Poland. Also search Vistula.
UF Vistula Valley
BT Poland
BT Central Europe
BT Europe

Vistula Valley
use Vistula River valley

Vistulian (1985)
Term applied in Poland, Spitsbergen and the former European USSR to the last Pleistocene glacial stage. Considered by some authors to be equivalent to Wurm.
BT upper Pleistocene
BT Pleistocene
BT Quaternary
BT Cenozoic
SA Sartanian
SA Weichselian
SA Wisconsinan
SA Wurm

Viterbo
No longer a valid term for GeoRef. See Viterbo Italy.

Viterbo Italy (1993)
City and province in NW Latium, W central Italy.
BT Latium Italy
BT Italy
BT Southern Europe
BT Europe

Viti Levu (1978)
Largest island in Fiji group. Has narrower terms as of 1985.
BT Fiji
BT Melanesia
BT Oceania
NT Suva Fiji
NT Tavua Fiji
NT Vatukoula Fiji

Vitim (1978)
River which rises in central Buryatia and flows across NE Irkutsk Oblast into the Lena River on the SW border of Yakutia. This term has multiple hierarchies.
UF Vitim River
BT1 Russian Federation
BT1 Commonwealth of Independent States
BT2 Asia
SA Mama River

Vitim Plateau (1978)
Gold mining area between the Vitim and Barguzin rivers E of Lake Baikal in NE Buryat. Also search Vitim. This term has multiple hierarchies.
BT1 Buryat Russian Federation
BT1 Russian Federation
BT1 Commonwealth of Independent States
BT2 Buryat Russian Federation
BT2 Asia

Vitim River
use Vitim

Vitoria
No longer a valid term for GeoRef. As of 1993, see Vitoria Spain or Vitoria Brazil if applicable.

Vitoria Brazil (1993)
City on the Atlantic Ocean 250 miles NE of Rio de Janeiro in Espirito Santo. Also search Vitoria in combination with Brazil.
BT Espirito Santo Brazil
BT Brazil
BT South America

Vitoria Spain (1993)
City in Alava in N Spain. Before 1993, search Vitoria AND Spain.
BT Alava Spain
BT Basque Provinces Spain
BT Spain
BT Iberian Peninsula
BT Southern Europe
BT Europe

vitrain (1978)
UF pure coal
BT macerals

vitrinite (1978)
Autoposting of this term began in 1978.
BT macerals
NT collinite
NT telinite

vitrophyre (1978)
BT igneous rocks
SA porphyritic texture

Vivarais (1978)
Ancient region now mostly in Ardeche department in SE France.
BT France
BT Western Europe
BT Europe
SA Ardeche France

vivianite (1978)
BT phosphates

Viviparus (1978)
BT Gastropoda
BT Mollusca
BT Invertebrata

Vizagapatam
Not a valid term for GeoRef. See Visakhapatnam India.

Vizagapatnam
Not a valid term for GeoRef. See Visakhapatnam India.

Vizcaino Peninsula (1985)
NW Baja California Sur, Mexico.
CO N265500N275000 W1135000W1151000
BT Baja California Sur Mexico
BT Mexico
SA Baja California

Vizcaya
No longer a valid term for GeoRef. As of 1993, see Vizcaya Spain.

Vizcaya Spain (1993)
One of the Basque Provinces in N Spain. Before 1993, also search Vizcaya and Spain.
CO N425500N432800 W0022200W0033100
BT Basque Provinces Spain
BT Spain
BT Iberian Peninsula
BT Southern Europe
BT Europe
NT Bilbao Spain

Vizeu
use Viseu Portugal

Vladeasa Massif
use Vladeasa Mountain

Vladeasa Mountain (1978)
In the Bihor Mountains of W central Transylvania. Also search Vladeasa.
UF Vladeasa Massif
BT Bihor Mountains
BT Transylvania
BT Romania
BT Southern Europe
BT Europe

Vladimir
Former principality of central Russia. No longer a valid term for GeoRef. As of 1993, see Vladimir Russian Federation for the oblast.

Vladimir Ivanovich Vernadskiy
use Vernadskiy, Vladimir Ivanovich

Vladimir Russian Federation (1993)
Oblast and city 110 miles of Moscow. This term has multiple hierarchies.
BT1 Russian Federation
BT1 Commonwealth of Independent States
BT2 Europe

Vladivostok
No longer a valid term for GeoRef. As of 1993, see Vladivostok Russian Federation.

Vladivostok Russian Federation (1993)
City on the southern tip of a peninsula extending into Peter the Great Bay on the Japan Sea in S Primorye Kray. This term has multiple hierarchies.
BT1 Primorye Russian Federation
BT1 Russian Federation
BT1 Commonwealth of Independent States
BT2 Primorye Russian Federation
BT2 Asia

VLBI
use very long baseline interferometry

VLF
As of 1981, no longer a valid term for GeoRef. See electromagnetic waves.

VOC
See volatile organic compounds.

Vocontian Trough (1978)
BT France
BT Western Europe
BT Europe

Vodino
No longer a valid term for GeoRef. As of 1993, see Vodino Ukraine for village in 3 locations in the Ukraine; or Vodino Russian Federation for villages in Kuybyshev and Volgograd oblasts of the European area of the Russian Federation and a village in Novosibirsk Oblast of the Russian Federation in Siberia.

Vodino Russian Federation (1993)
Villages in Samara and Volgograd oblasts of the European area of the Russian Federation, also a village in Novosibirsk Oblast of the Russian Federation in Siberia. Before 1993, also search Vodino AND Russian Republic.
IN Index oblasts and Europe or Asia as applicable.
BT Russian Federation
BT Commonwealth of Independent States
SA Novosibirsk Russian Federation
SA Samara Russian Federation
SA Volgograd Russian Federation

Vodino Ukraine (1993)
Village in 3 locations in the Ukraine. Before 1993, search Vodino AND Ukraine. This term has multiple hierarchies.
BT1 Ukraine
BT1 Europe
BT2 Ukraine
BT2 Commonwealth of Independent States

Vogels Berg
use Vogelsberg

Vogelsberg (1978)
Mountain range in central Hesse.
UF Vogels Berg
BT Hesse Germany
BT Germany
BT Central Europe
BT Europe

vogesite (1985)
BT lamprophyres
BT plutonic rocks
BT igneous rocks

Vogtland (1978)
Region NW of the W edge of the Erzgebirge around Plauen in E Germany.
IN Index states as applicable.
BT Germany
BT Central Europe
BT Europe
SA Saxony Germany
SA Thuringia Germany

Voivodina
use Vojvodina

Vojvodina (1978)
Autonomous province on the Hungarian and Romanian borders in N Serbia.
UF Voivodina

UF Voyvodina
BT Serbia
BT Yugoslavia
BT Southern Europe
BT Europe
SA Banat
SA Tisza River

volatile combustible
use volatiles

volatile elements (1978)
SA chemical elements
SA volatiles

volatile matter
use volatiles

volatile organic chemicals
use volatile organic compounds

volatile organic compounds (1993)
Hydrocarbon in combination with any other element in a compound which has a low vapor pressure or other reactive organic compounds. This term has multiple hierarchies.
UF volatile organic chemicals
BT1 organic materials
BT2 volatiles
SA chlorinated hydrocarbons
SA compounds
SA crude oil
SA hydrocarbons
SA pollutants
SA soil gases
SA volatilization

volatiles (1978)
For compounds.
UF volatile combustible
UF volatile matter
NT volatile organic compounds
SA coal
SA geochemistry
SA volatile elements
SA volatilization

volatilization (1989)
SA gases
SA geochemistry
SA processes
SA volatile organic compounds
SA volatiles

volborthite (1978)
BT vanadates

volcanic
A valid term through 1977.

volcanic arcs
use island arcs

volcanic ash (1978)
From 1978-1980, pyroclastics and glasses was autoposted to this term. From 1981-1984, pyroclastics and volcanic rocks were autoposted to this term.
SA ash
SA ash falls
SA ash flows
SA ash-flow tuff
SA bentonite
SA clastic sediments
SA dust
SA ejecta
SA ignimbrite
SA nuees ardentes
SA pyroclastic flows
SA pyroclastics
SA sediments
SA shirasu
SA terrigenous materials
SA tuff
SA volcanic rocks
SA volcaniclastics

volcanic belts (1978)
UF belts, volcanic
SA plate tectonics
SA volcanoes

volcanic breccia (1981)
BT breccia
BT clastic rocks
BT sedimentary rocks
SA breccia pipes
SA pyroclastics
SA volcaniclastics

volcanic centers (1993)
Features on scale larger than volcanoes.
BT volcanic features
SA volcanism
SA volcanoes

volcanic clay
use bentonite

volcanic earthquakes (1981)
UF volcanic tremors
BT earthquakes
SA eruptions
SA geologic hazards
SA tilt
SA volcanic risk
SA volcanism
SA volcanoes
SA volcanology

volcanic features (1978)
UF features, volcanic
NT calderas
NT cauldrons
NT cinder cones
NT lava channels
NT lava fields
NT lava lakes
NT lava tubes
NT maars
NT volcanic centers
NT volcanic necks
NT volcanoes
SA black smokers
SA caves
SA crater lakes
SA craters
SA diatremes
SA geomorphology
SA lava
SA lava flows
SA stalactites
SA stalagmites
SA underground cavities
SA vents
SA volcanism
SA volcanology

volcanic fields (1985)
SA lava fields
SA volcanic rocks
SA volcanism
SA volcanology

volcanic glass (1978)
For basalt glass, use volcanic glass and basalt.
UF glass, volcanic
BT glasses
BT volcanic rocks
BT igneous rocks
SA crystallites
SA devitrification
SA ignimbrite
SA obsidian
SA perlite
SA pitchstone
SA shirasu

volcanic necks (1982)
BT volcanic features
SA diatremes
SA erosion features
SA geomorphology
SA intrusions

SA pipes
SA plugs
SA vents
SA volcanism
SA volcanoes
SA volcanology

volcanic processes (1982)
Before 1982, also search volcanism AND mineral deposits, genesis. As of 1993, restricted to economic studies.
UF volcanogenic processes
SA igneous processes
SA mineral deposits, genesis
SA processes
SA volcanism
SA volcanology

volcanic risk (1989)
SA effects
SA engineering geology
SA eruptions
SA geologic hazards
SA safety
SA volcanic earthquakes
SA volcanism

volcanic rocks (1978)
Not a valid term from 1975 through 1977. Before 1978, also search igneous rocks AND volcanic; volcanics; eruptive rocks. Autoposted to diorite porphyry and granophyre from 1978-1980. As of 1981, autoposted to its narrower terms in list H (including their respective narrower terms) as well as granophyre, melilitite and vitrophyre. Autoposting to the term pyroclastics began in 1985. Autoposting to the narrower terms of pyroclastics began in 1981, but this ended for volcanic ash and ash flows in 1985.
UF eruptive rocks
UF volcanics
BT igneous rocks
NT andesites
NT basalts
NT benmoreite
NT dacites
NT glasses
NT granophyre
NT melilitite
NT phonolites
NT pyroclastics
NT rhyodacites
NT rhyolites
NT trachyandesites
NT trachytes
SA ash flows
SA breccia pipes
SA diorite porphyry
SA lava
SA metavolcanic rocks
SA pyroclastic flows
SA rocks
SA shirasu
SA volcanic ash
SA volcanic fields
SA volcanism
SA volcanoes
SA volcanology

volcanic tremors
use volcanic earthquakes

volcanic vents
use vents

volcanicity
use volcanism

volcaniclastic rocks
use volcaniclastics

volcaniclastics (1984)
Used to indicate material, within sediments or sedimentary rocks, which originated from a volcano. Use pyroclastics for the igneous rock.
UF volcaniclastic rocks
SA ash falls
SA clastic rocks
SA clastic sediments
SA ejecta
SA igneous rocks
SA lapilli
SA pyroclastic flows
SA pyroclastics
SA sedimentary rocks
SA sediments
SA terrigenous materials
SA tuff
SA volcanic ash
SA volcanic breccia

volcanics
A valid term through 1976. Included use in subfiles E, G, N, T, and B.
use volcanic rocks

volcanism (1978)
UF volcanicity
UF vulcanism
SA aa lava
SA ash falls
SA ash flows
SA base surges
SA breccia pipes
SA calderas
SA cauldrons
SA cinder cones
SA eruptions
SA explosive eruptions
SA fumaroles
SA hawaiian-type eruptions
SA heat sources
SA intraplate processes
SA lahars
SA lava
SA maars
SA nuees ardentes
SA pahoehoe
SA phreatomagmatism
SA plinian-type eruptions
SA pyroclastic flows
SA shield volcanoes
SA solfataras
SA strombolian-type eruptions
SA sublimates
SA submarine volcanoes
SA volcanic centers
SA volcanic earthquakes
SA volcanic features
SA volcanic fields
SA volcanic necks
SA volcanic processes
SA volcanic risk
SA volcanic rocks
SA volcanoes
SA volcanology
SA vulcanian-type eruptions

Volcano Island
use Vulcano

volcanoes (1978)
Used when discussing specific volcanoes. Autoposting of this term began in 1978. Before 1993, also search volcanic centers.
UF volcanos
BT volcanic features
NT shield volcanoes
NT stratovolcanoes
NT submarine volcanoes
SA ash falls
SA breccia pipes
SA calderas
SA cinder cones
SA domes

SA eruptions
SA geologic hazards
SA geomorphology
SA hot spots
SA intraplate processes
SA lava
SA mud volcanoes
SA nuees ardentes
SA pipes
SA plugs
SA sublimates
SA volcanic belts
SA volcanic centers
SA volcanic earthquakes
SA volcanic necks
SA volcanic rocks
SA volcanism
SA volcanology

volcanogenic processes
use volcanic processes

volcanology (1978)
Used for the discipline as a whole and for discussions of specific volcanoes.
UF vulcanology
SA ash falls
SA ash flows
SA atmosphere
SA bibliography
SA calderas
SA catalogs
SA cauldrons
SA continental drift
SA craters
SA crust
SA ejecta
SA eruptions
SA explosive eruptions
SA extrusive rocks
SA fumaroles
SA hawaiian-type eruptions
SA igneous activity
SA igneous rocks
SA intrusions
SA lava
SA lava flows
SA magmas
SA mud volcanoes
SA orogeny
SA periodicity
SA petrology
SA phreatomagmatism
SA plate tectonics
SA plinian-type eruptions
SA seismology
SA strombolian-type eruptions
SA submarine volcanoes
SA tectonics
SA tilt
SA tiltmeters
SA vents
SA volcanic earthquakes
SA volcanic features
SA volcanic fields
SA volcanic necks
SA volcanic processes
SA volcanic rocks
SA volcanism
SA volcanoes
SA vulcanian-type eruptions

volcanos
use volcanoes

Volga
A valid term through 1976. Use Volga River or Volga region.

Volga region (1978)
The middle and lower courses of the Volga River and its tributaries. It includes the Volga drainage basin with the exception of parts of the NE, NW, and SW. Also search Volga.
Before then, variants of the term may occur in GeoRef.

IN Index Europe if applicable.
BT Russian Federation
BT Commonwealth of Independent States
SA Astrakhan Arch
SA Europe
SA Voronezh-Volga Anteclise

Volga River (1978)
Longest river in Europe. It rises in the Valdai Hills NE of Moscow and flows SE and then S into the N Caspian Sea. Also search Volga. This term has multiple hierarchies.
BT1 Russian Federation
BT1 Commonwealth of Independent States
BT2 Europe
SA Voronezh-Volga Anteclise

Volga-Don Interfluve
use Volga-Don region

Volga-Don region (1978)
Between the Don River and Volga River in the Volga-Don Canal area. This term has multiple hierarchies.
UF Volga-Don Interfluve
BT1 Russian Federation
BT1 Commonwealth of Independent States
BT2 Europe
SA Volgograd Russian Federation

Volga-Ural Basin
use Volga-Ural region

Volga-Ural Interfluve
use Volga-Ural region

Volga-Ural region (1978)
Between the parallel courses of the lower Volga River and lower Ural River. Also search Ural-Volga.
IN Index former Soviet republics as applicable.
CO N470000N550000
    E0550000E0450000
UF Ural-Volga Interfluve
UF Ural-Volga region
UF Volga-Ural Basin
UF Volga-Ural Interfluve
BT Commonwealth of Independent States
SA Kazakhstan
SA Russian Federation

Volga-Urals (1978)
Region between the middle course of the Volga River and the Urals. This term has multiple hierarchies.
BT1 Russian Federation
BT1 Commonwealth of Independent States
BT2 Europe

Volgian (1978)
Europe.
BT Upper Jurassic
BT Jurassic
BT Mesozoic
SA Malm

Volgograd
No longer a valid term for GeoRef. As of 1993, see Volgograd Russian Federation.

Volgograd City Russian Federation (1993)
City in S Volgograd Oblast on the right bank of the Volga River about 280 miles from its mouth. Before 1993, also search Volgograd. Before 1978, also

search Stalingrad or Tsaritsyn. This term has multiple hierarchies.
BT1 Volgograd Russian Federation
BT1 Russian Federation
BT1 Commonwealth of Independent States
BT2 Volgograd Russian Federation
BT2 Europe

Volgograd Russian Federation (1993)
Oblast, SW Russian Federation. Before 1993, also search Volgograd. Before 1978, also search Stalingrad or Tsaritsyn. This term has multiple hierarchies.
UF Stalingrad Russian Federation
UF Tsaritsyn Russian Federation
BT1 Russian Federation
BT1 Commonwealth of Independent States
BT2 Europe
NT Kamyshin Russian Federation
NT Mikhailovka Russian Federation
NT Volgograd City Russian Federation
SA Tsarev Meteorite
SA Vodino Russian Federation
SA Volga-Don region

Volhynia (1978)
Forested region with marshland and many lakes S and SW of the Pripet Marshes in NW Ukraine. This term has multiple hierarchies.
UF Volynia
BT1 Ukraine
BT1 Europe
BT2 Ukraine
BT2 Commonwealth of Independent States
SA Volyn Ukraine

Vologda
No longer a valid term for GeoRef. As of 1993, see Vologda Russian Federation.

Vologda Russian Federation (1993)
Oblast and city on the Vologda River 330 miles E of Saint Petersburg. This term has multiple hierarchies.
BT1 Russian Federation
BT1 Commonwealth of Independent States
BT2 Europe

Volta Basin (1978)
River basin including Lake Volta. Also search Volta.
IN Index countries as applicable.
BT Africa
SA Burkina Faso
SA Ghana
SA Ivory Coast
SA Togo

Voltaic Republic
use Burkina Faso

voltaite (1978)
From 1978-1980, borates, sulfates and sulfides were autoposted to this term.
BT sulfates

voltammetry (1985)
SA chemical analysis
SA electrochemical properties
SA electrolytic analysis
SA trace-element analyses

Voltri Group (1985)
N Italy.

UF Voltri Massif
SA Italy
SA Mesozoic

Voltri Massif
use Voltri Group

Voltzia Sandstone (1978)
BT Bunter
BT Lower Triassic
BT Triassic
BT Mesozoic
SA France
SA Vosges France

volume (1978)
Used as a general term.
SA specific surface

volume elasticity
use bulk modulus

volume susceptibility (magnetic)
use magnetic susceptibility

volumetric analysis
use titration

Volyn
No longer a valid term for GeoRef. As of 1993, see Volyn Ukraine.

Volyn Ukraine (1993)
Oblast in NW Ukraine. A province of Poland prior to WW II. Before 1978, also search Wolyn. This term has multiple hierarchies.
UF Wolyn Ukraine
BT1 Ukraine
BT1 Europe
BT2 Ukraine
BT2 Commonwealth of Independent States
SA Lvov-Volyn Basin
SA Volhynia

Volyn-Podolia (1978)
An upland region between the Dniester River and Dnieper River in W Ukraine. This term has multiple hierarchies.
UF Volyn-Podolian Platform
UF Volyn-Podolian Upland
BT1 Ukraine
BT1 Europe
BT2 Ukraine
BT2 Commonwealth of Independent States
SA Podolia

Volyn-Podolian Platform
use Volyn-Podolia

Volyn-Podolian Upland
use Volyn-Podolia

Volynia
use Volhynia

Volyno-Azov Massif
use Ukrainian Shield

vonsenite (1978)
BT borates

Vorarlberg
No longer a valid term for GeoRef. See Vorarlberg Austria.

Vorarlberg Austria (1993)
State in extreme W Austria.
CO N464800N473500
    E0101500E0093100
BT Austria
BT Central Europe
BT Europe
SA Lake Constance

Voring Plateau (1978)
Underwater feature S of the Lofoten Basin off the central Norwegian coast.
BT Norwegian Sea

BT Arctic Ocean
SA DSDP Site 338
SA DSDP Site 343
SA Leg 104
SA ODP Site 642
SA ODP Site 643
SA ODP Site 644

Vorkuta
No longer a valid term for GeoRef. As of 1993, see Vorkuta Russian Federation.

Vorkuta Russian Federation (1993)
Town on the Vorkuta River in extreme NE Komi at N end of the Urals. This term has multiple hierarchies.
BT1 Komi Russian Federation
BT1 Russian Federation
BT1 Commonwealth of Independent States
BT2 Komi Russian Federation
BT2 Europe

Voronezh
No longer a valid term for GeoRef. As of 1993, see Voronezh Russian Federation.

Voronezh Anteclise (1978)
Near the Ukrainian border in SW Voronezh Oblast. This term has multiple hierarchies.
UF Voronezh Anticline
UF Voronezh Anticlise
UF Voronezh Massif
BT1 Voronezh Russian Federation
BT1 Russian Federation
BT1 Commonwealth of Independent States
BT2 Voronezh Russian Federation
BT2 Europe
SA Voronezh-Volga Anteclise

Voronezh Anticline
use Voronezh Anteclise

Voronezh Anticlise
use Voronezh Anteclise

Voronezh Massif
use Voronezh Anteclise

Voronezh Russian Federation (1993)
Oblast including the city on the Voronezh River near its junction with the Don River. N of E Ukraine. This term has multiple hierarchies.
CO N503000N530000
   E0400000E0384500
BT1 Russian Federation
BT1 Commonwealth of Independent States
BT2 Europe
NT Golovanevsk Russian Federation
NT Voronezh Anteclise

Voronezh-Volga Anteclise (1981)
This term has multiple hierarchies.
IN Index former Soviet republics as applicable.
CO N470000N563000
   E0470000E0300000
BT1 Commonwealth of Independent States
BT2 Europe
BT3 Russian Platform
SA Russian Federation
SA Tsarev Meteorite
SA Volga region
SA Volga River
SA Voronezh Anteclise

Voroshilovsk (region)
use Stavropol region

Vosges
No longer a valid term for GeoRef. See Vosges France.

Vosges France (1993)
Department in NE France.
CO N474500N483000
   E0071500E0053000
BT France
BT Western Europe
BT Europe
SA Lorraine
SA Voltzia Sandstone
SA Vosges Mountains

Vosges Mountains (1978)
Range extending from Belfort Gap to German border. Also search Vosges.
IN Index departments as applicable.
CO N473000N493000
   E0080000E0060000
BT France
BT Western Europe
BT Europe
SA Haut-Rhin France
SA Vosges France

Vostok (1978)
Small uninhabited coral island in S Line Islands. In 1981, broader terms changed from Line Islands and Pacific Ocean to Line Islands and Polynesia.
IN Index Line Islands as applicable.
UF Stavers Island
SA Kiribati
SA Line Islands

Vostok Station (1978)
Two Soviet IGY stations. Vostok 1 Station is S of the Queen Mary Coast and Vostok 2 Station is midway between Queen Mary Coast and the South Pole near the South Geomagnetic Pole. Both stations are on the Indian Ocean side of the continent.
CO S782800S782800
   E1064800E1064800
BT Antarctica

Vourinos (1978)
Mountain SW of Kozane in W Macedonia. This term has multiple hierarchies.
UF Burono
UF Vourinos Massif
BT1 Greek Macedonia
BT1 Macedonia
BT1 Southern Europe
BT1 Europe
BT2 Greek Macedonia
BT2 Greece
BT2 Southern Europe
BT2 Europe

Vourinos Massif
use Vourinos

Voyager Program (1985)
Also search Voyager 1; Voyager 2.
UF Program Voyager
UF Project Voyager
UF Voyager Project
NT Voyager 1
NT Voyager 2
SA Adrastea Satellite
SA Amalthea Satellite
SA Ariel Satellite
SA Callisto Satellite
SA Dione Satellite
SA Enceladus Satellite
SA Europa Satellite
SA extraterrestrial geology
SA Ganymede Satellite
SA Hyperion Satellite
SA Iapetus Satellite
SA Io Satellite
SA Janus Satellite
SA Jupiter
SA Mimas Satellite
SA Miranda Satellite
SA Neptune
SA Nereid Satellite
SA Oberon Satellite
SA outer planets
SA planetology
SA remote sensing
SA Rhea Satellite
SA satellite methods
SA Saturn
SA Tethys Satellite
SA Titan Satellite
SA Titania Satellite
SA Triton Satellite
SA Umbriel Satellite
SA Uranus

Voyager Project
use Voyager Program

Voyager 1 (1989)
Also see Voyager Program.
BT Voyager Program
SA Jupiter
SA remote sensing
SA Saturn
SA Voyager 2

Voyager 2 (1989)
Also see Voyager Program.
BT Voyager Program
SA Jupiter
SA Neptune
SA remote sensing
SA Saturn
SA Uranus
SA Voyager 1

Voyvodina
use Vojvodina

Vraca
Not a valid term for GeoRef. See Vratsa Bulgaria.

Vraconian (1978)
Europe.
BT Cretaceous
BT Mesozoic
SA Lower Cretaceous
SA Upper Cretaceous

Vrancea (1978)
Range of the Carpathians near their junction with the Transylvanian Alps. This term has multiple hierarchies.
UF Vrancei Mountains
BT1 Romania
BT1 Southern Europe
BT1 Europe
BT2 Carpathians
BT2 Europe
SA Vrancea earthquake 1977

Vrancea earthquake 1977 (1985)
Epicenter in the Vrancea mountain range, Romania.
BT earthquakes
SA Romania
SA Vrancea

Vrancei Mountains
use Vrancea

Vratsa
No longer a valid term for GeoRef. See Vratsa Bulgaria.

Vratsa Bulgaria (1993)
Province including the city in NW Bulgaria. Also search Vraca and Vrattsa.
BT Bulgaria
BT Southern Europe
BT Europe

Vrattsa
Not a valid term for GeoRef. See Vratsa Bulgaria.

vrbaite (1978)
Autoposting of sulfantimonites and sulfarsenites began in 1981. This term has multiple hierarchies.
BT1 sulfantimonites
BT1 sulfosalts
BT2 sulfarsenites
BT2 sulfosalts

Vredefort Dome (1978)
N Orange Free State. Also search Vredefort.
BT Orange Free State South Africa
BT South Africa
BT Southern Africa
BT Africa

Vridhachalam
No longer a valid term for GeoRef. See Vridhachalam India.

Vridhachalam India (1993)
Town SW of Cuddalore in NE Tamil Nadu, S India. Also search Viruddhachalam and Viruddhachalam.
BT Tamil Nadu India
BT India
BT Indian Peninsula
BT Asia

Vries Island
use Izu-Oshima

Vryheid Formation (1993)
Permian. South Africa. In Ecca Group of the Karroo Supergroup.
BT Permian
BT Paleozoic
SA Ecca Group
SA South Africa

VSP
Not a valid term for GeoRef. See vertical seismic profiles.

vugs (1985)
SA geodes
SA textures

Vulcan Mountains (1978)
Section of the Transylvanian Alps in W central Romania. Also search Vulcan.
BT Romania
BT Southern Europe
BT Europe
SA Transylvanian Alps

vulcanian-type eruptions (1982)
UF vesuvian-type eruptions
SA eruptions
SA volcanism
SA volcanology

vulcanism
use volcanism

Vulcano (1978)
Southernmost of the Lipari Islands in the Tyrrhenian Sea off NE Sicily.
UF Volcano Island
UF Vulcano Island
BT Lipari Islands
BT Sicily Italy
BT Italy
BT Southern Europe
BT Europe

Vulcano Island
  use Vulcano

vulcanology
  use volcanology

Vulture Mountain (1978)
  Peak in Glacier National Park in NW Montana.
  IN Index county or region and mountains as applicable.
  SA Glacier National Park
  SA Montana

Vuori-Yarvi Russian Federation
  use Vuoriyarvi Russian Federation

Vuoriyarvi
  No longer a valid term for GeoRef. As of 1993, see Vuoriyarvi Russian Federation.

Vuoriyarvi Russian Federation (1993)
  Village near the Finnish border in NW Karelia. Before 1993, also search Vuori-Yarvi. This term has multiple hierarchies.
  UF Vuori-Yarvi Russian Federation
  BT1 Karelia Russian Federation
  BT1 Russian Federation
  BT1 Commonwealth of Independent States
  BT2 Karelia Russian Federation
  BT2 Asia

Vyatka River (1978)
  Rises in W foothills of the Central Urals and flows into the Kama River in N Tatar. Before 1978, also search Vyatka. This term has multiple hierarchies.
  UF Viatka River
  BT1 Russian Federation
  BT1 Commonwealth of Independent States
  BT2 Europe

Vyatka Russian Federation (1993)
  Oblast including the city in E central European Russian Federation. Before 1993, also search Kirov. Before 1978, also search Vyatka for the city before 1940. This term has multiple hierarchies.
  UF Kirov Russian Federation
  BT1 Russian Federation
  BT1 Commonwealth of Independent States
  BT2 Europe
  SA Orlov Russian Federation

Vyatka-Kama Interfluve (1978)
  Region between the lower reaches of the Vyatka and Kama rivers in the Izhevsk area W of the Urals. Also search Vyatka-Kama. This term has multiple hierarchies.
  CO N550000N600000
      E0570000E0480000
  BT1 Russian Federation
  BT1 Commonwealth of Independent States
  BT2 Europe
  SA Kama River

Vychegda River (1978)
  Rises in the Komi autonomous republic and flows W into the Northern Dvina at Kotlas in SE Arkhangelsk Oblast in N European Russian Federation. Before 1978, also search Vychegda. This term has multiple hierarchies.
  BT1 Russian Federation
  BT1 Commonwealth of Independent States

  BT2 Europe

Vyernyi Kazakhstan
  use Alma-Ata Kazakhstan

Vyshkovo
  No longer a valid term for GeoRef. As of 1993, see Vyshkovo Belarus or Vyshkovo Russian Federation.

Vyshkovo Belarus (1993)
  Village N of Minsk in Belarus. Before 1993, also search Vyshkovo AND Byelorussia. This term has multiple hierarchies.
  BT1 Belarus
  BT1 Europe
  BT2 Belarus
  BT2 Commonwealth of Independent States

Vyshkovo Russian Federation (1993)
  Village in Tver formerly Kalinin Oblast NW of Moscow. Before 1993, also search Vyshkovo AND Russian Republic. This term has multiple hierarchies.
  BT1 Tver Russian Federation
  BT1 Russian Federation
  BT1 Commonwealth of Independent States
  BT2 Tver Russian Federation
  BT2 Europe

W

W
  use tungsten

Wabasca Alberta (1993)
  N Alberta. Before 1993, also search Wabasca and Wabasca region.
  CO N555800N555800
      W1135600W1135600
  BT Alberta
  BT Western Canada
  BT Canada

Wabasca region
  No longer a valid term for GeoRef. See Wabasca Alberta.

Wabaunsee County
  Valid through 1988. Search in combination with state term. After 1988, use specific county-state term.

Wabaunsee County Kansas (1989)
  NE central Kansas. Before 1989, also search Wabaunsee County AND Kansas.
  CO N384500N391300
      W0955500W0963000
  BT Kansas
  BT United States

Wabaunsee Group (1978)
  Includes (ascending) Severy Shale, Howard Limestone, Scranton Shale, Bern Limestone, Auburn Shale, Emporia Limestone, Willard Shale, Zeandale Limestone, Pillsbury Shale, Stotler Limestone, Root Shale, and Wood Siding Formation. E Kansas, SW Iowa, NW Missouri, SE Nebraska, and N Oklahoma.
  BT Virgilian
  BT Upper Pennsylvanian
  BT Pennsylvanian
  BT Carboniferous
  BT Paleozoic
  SA Iowa
  SA Kansas
  SA Missouri
  SA Nebraska
  SA Oklahoma

Wabigoon Belt (1985)
  Structural subprovince of the Superior Province, Canadian Shield. W Ontario and E Manitoba.
  CO N484500N504500
      W0850000W0963000
  UF Wabigoon Subprovince
  BT Superior Province
  BT Canadian Shield
  BT North America
  SA Manitoba
  SA Ontario

Wabigoon Subprovince
  use Wabigoon Belt

Wabush (1978)
  Lake near the Quebec border in SW Labrador.
  BT Labrador
  BT Newfoundland
  BT Eastern Canada
  BT Canada

Waccamaw Formation (1978)
  Consists of soft limestones and loose gray to buff fine quartz sands in which occasional small quartz pebbles are present. S North Carolina and S and E South Carolina. Since second edition of Thesaurus, age has been redefined to be lower Pleistocene. Second edition stated lower Pliocene.
  BT lower Pleistocene
  BT Pleistocene
  BT Quaternary
  BT Cenozoic
  SA North Carolina
  SA South Carolina

Wachapreague Inlet (1978)
  Inlet connecting small bay off town of Wachapreague with Atlantic Ocean. In Accomack County on the eastern shore. Also search Wachapreague.
  BT Accomack County Virginia
  BT Virginia
  BT United States

Wachusett-Marlborough Tunnel (1978)
  Rock tunnel between the Wachusett Reservoir and Marlborough in E central Massachusetts. Part of the tunnel system which provides water to Boston.
  IN Index counties as applicable.
  BT Massachusetts
  BT United States

wackestone (1978)
  BT carbonate rocks
  BT sedimentary rocks

Waco
  Valid through 1988. Search in combination with state term. After 1988, use specific city-state term.

Waco Texas (1989)
  City on the Brazos River in E central Texas. Before 1989, search Waco AND Texas.
  CO N313300N313300
      W0971000W0971000
  BT McLennan County Texas
  BT Texas
  BT United States

Wadati zone
  use Benioff zone

Wadati-Benioff Zone
  use Benioff zone

Wadden Sea
  use Wadden Zee

Wadden Zee (1978)
  Outer section of the former Zuider Zee between the West Frisian Islands and the dike enclosing Ijsseelmeer.
  UF Wadden Sea
  UF Waddenzee
  BT Netherlands
  BT Western Europe
  BT Europe

Waddenzee
  use Wadden Zee

Wadi Araba (1978)
  Depression along a valley extending from the Dead Sea to the Gulf of Aqaba. Part of the Great Rift Valley.
  IN Index countries as applicable.
  UF Wadi Arabah
  BT Middle East
  BT Asia
  SA Great Rift Valley
  SA Jordan

Wadi Arabah
  use Wadi Araba

Wadia (1978)
  Former Western Kathiawar state of Western India States Agency; merged in 1948 with Saurashtra. Also search Vadia. Now part of Gujarat.
  UF Vadia
  BT Gujarat India
  BT India
  BT Indian Peninsula
  BT Asia
  SA Kathiawar

wadian
  use wadis

wadies
  use wadis

wadis (1985)
  Autoposting of fluvial features to this term began in 1989.
  UF oueds
  UF wadian
  UF wadies
  UF widan
  BT fluvial features
  SA arid environment
  SA arroyos
  SA deserts
  SA ephemeral streams
  SA geomorphology
  SA hydrology
  SA rivers and streams

Wadowice
  No longer a valid term for GeoRef. As of 1993 see Wadowice Poland.

Wadowice Poland (1993)
  Town on Skawa River in NE Bielsko, S Poland. Before 1993 also search Wadowice AND Poland.
  BT Bielsko Poland
  BT Poland
  BT Central Europe
  BT Europe

Wager (1978)
  Bay off Roes Welcome Sound of N Hudson Bay in NE Keewatin District.

    BT Keewatin District Northwest Territories
    BT Northwest Territories
    BT Western Canada
    BT Canada

Wagga Wagga
    No longer a valid term for GeoRef. See Wagga Wagga Australia.

Wagga Wagga Australia (1993)
    Town on the Murrumbidgee River in S New South Wales.
    BT New South Wales Australia
    BT Australia
    BT Australasia

Wah Wah Mountains (1989)
    Beaver, Iron and Millard counties, SW Utah.
    IN Index counties as applicable.
    BT Utah
    BT United States
    SA Beaver County Utah
    SA Iron County Utah
    SA Millard County Utah

Waianae (1978)
    Mountain range extending along SW side of Oahu Island. Autoposting of broader terms to this term began in 1989. This term has multiple hierarchies.
    UF Waianae Range
    BT1 Oahu
    BT1 Honolulu County Hawaii
    BT1 Hawaii
    BT1 United States
    BT2 Oahu
    BT2 Honolulu County Hawaii
    BT2 Hawaii
    BT2 Polynesia
    BT2 Oceania
    BT3 Oahu
    BT3 Honolulu County Hawaii
    BT3 Hawaii
    BT3 East Pacific Ocean Islands

Waianae Range
    use Waianae

Waikato Basin (1978)
    River basin in central North Island.
    IN Also index North Island.
    BT New Zealand
    BT Australasia
    SA Huntly
    SA North Island

Waimanalo
    Valid through 1988. Search in combination with state term. After 1988, use specific city-state term.

Waimanalo Hawaii (1989)
    Village on the SE coast of Oahu, Hawaii. Before 1989, search Waimanalo AND Hawaii. This term has multiple hierarchies.
    CO N212000N212000
         W1574300W1574300
    BT1 Oahu
    BT1 Honolulu County Hawaii
    BT1 Hawaii
    BT1 United States
    BT2 Oahu
    BT2 Honolulu County Hawaii
    BT2 Hawaii
    BT2 Polynesia
    BT2 Oceania
    BT3 Oahu
    BT3 Honolulu County Hawaii
    BT3 Hawaii
    BT3 East Pacific Ocean Islands

Wairakei (1978)
    Geothermal field N of Lake Taupo in central North Island.
    IN Also index North Island.
    BT New Zealand
    BT Australasia
    SA Broadlands
    SA geothermal fields
    SA North Island

wairakite (1978)
    BT zeolite group
    BT framework silicates
    BT silicates

Wairarapa (1978)
    Lake E of Wellington in S North Island.
    IN Also index North Island.
    BT New Zealand
    BT Australasia
    SA North Island

Waitemata Group (1978)
    Comprises Manukau Breccias, Parnell Grit, Orakei Bay Greensand, Pakaurangi Beds, Albany Conglomerate, and Oneroa Beds all of which are known to be interbedded with the Waitemata Sandstone. In N North Island, New Zealand.
    BT lower Miocene
    BT Miocene
    BT Neogene
    BT Tertiary
    BT Cenozoic
    SA New Zealand
    SA North Island

Waits River Formation (1989)
    E Vermont, W Massachusetts and NW New Hampshire. Includes Standing Pond Volcanic Member.
    IN Index ages as applicable.
    BT Paleozoic
    SA Devonian
    SA Massachusetts
    SA New Hampshire
    SA Silurian
    SA Vermont

Wakayama
    No longer a valid term for GeoRef. See Wakayama Japan.

Wakayama Japan (1993)
    City and prefecture in SW Honshu.
    BT Honshu
    BT Japan
    BT Far East
    BT Asia
    NT Arida-gawa
    SA Kii Peninsula
    SA Kinki Japan

Wake (1978)
    Coral atoll and 3 islets (Wake, Peale, and Wilkes) between Hawaii and Guam in the N Pacific Ocean.
    UF Wake Island
    BT Pacific Ocean

Wake County
    Valid through 1988. Search in combination with state term. After 1988, use specific county-state term.

Wake County North Carolina (1989)
    Central North Carolina. Before 1989, also search Wake County AND North Carolina.
    CO N353200N360500
         W0781500W0785900
    BT North Carolina
    BT United States

Wake Island
    use Wake

Walachia (1978)
    Region between the Danube River and the Transylvania Alps.
    BT Romania
    BT Southern Europe
    BT Europe
    NT Braila Romania
    NT Bucharest Romania
    NT Buzau River
    NT Buzau Romania
    NT Muntenia
    NT Oltenia
    NT Pitesti Romania
    NT Prahova Romania
    NT Schela Romania
    NT Slatina Romania
    NT Tismana Romania
    SA Danube Plain
    SA Fagaras Mountains
    SA Jiu River valley
    SA Mehedinti Plateau
    SA Olt River
    SA Olt River valley
    SA Romanian Plain
    SA Transylvanian Alps

Walbrzych
    No longer a valid term for GeoRef. As of 1993 see Walbrzych Poland.

Walbrzych Poland (1993)
    Province and city on the Bobr River in SW Poland. Before 1993 also search Walbrzych AND Poland.
    UF Waldenburg
    UF Waldenburg in Schlesien
    BT Poland
    BT Central Europe
    BT Europe
    NT Boguszow Poland
    NT Nowa Ruda Poland
    NT Strzegom Poland
    NT Strzegom-Sobotka Massif
    NT Zloty Stok Poland
    SA Polish Sudeten Mountains

Waldeck
    No longer a valid term for GeoRef. Used for former German state in W central Germany and town. See Waldeck Germany for the town.

Waldeck Germany (1993)
    Town in N Hesse, central Germany.
    BT Hesse Germany
    BT Germany
    BT Central Europe
    BT Europe

Waldenburg
    use Walbrzych Poland

Waldenburg in Schlesien
    use Walbrzych Poland

Waldo County
    Valid through 1988. Search in combination with state term. After 1988, use specific county-state term.

Waldo County Maine (1989)
    On Penobscot Bay in S central Maine. Before 1989, also search Waldo County AND Maine.
    CO N441500N444500
         W0684500W0693000
    BT Maine
    BT United States

Waldron Shale (1978)
    A gray or greenish-gray calcareous shale with occasional thin beds of limestone or argillaceous shale. S Indiana, W central Kentucky, and central Tennessee.
    BT Middle Silurian
    BT Silurian
    BT Paleozoic
    SA Indiana
    SA Kentucky
    SA Tennessee

Wales (1978)
    CO N513000N533000
         W0024000W0051500
    BT Great Britain
    BT United Kingdom
    BT Western Europe
    BT Europe
    NT Anglesey Wales
    NT Caernarvonshire Wales
    NT Cardigan Bay
    NT Denbighshire Wales
    NT Dyfed Wales
    NT Glamorgan Wales
    NT Harlech Dome
    NT Merionethshire Wales
    NT Monmouthshire Wales
    NT Pontnewydd Cave
    NT Powys Wales
    NT South Wales
    NT South Wales coal field
    NT Tremadoc Bay
    NT Welsh Basin
    SA Aberystwyth Grits
    SA Bala
    SA Black Mountain
    SA Bristol Channel
    SA Penrhyn Slate
    SA Severn Estuary
    SA Severn Valley
    SA Sherwood Sandstone
    SA Welsh Borderland
    SA Wye Valley

Walfischbai
    use Walvis Bay

Walfish Bay
    use Walvis Bay

Walker Branch watershed (1981)
    NE Tennessee. Autoposting of broader terms to this term began in 1989.
    CO N355200N360000
         W0841000W0842000
    BT Anderson County Tennessee
    BT Tennessee
    BT United States

Walker County
    Valid through 1988. Search in combination with state term. After 1988, use specific county-state term.

Walker County Alabama (1989)
    NW central Alabama. Before 1989, also search Walker County AND Alabama.
    CO N333100N340000
         W0865600W0873700
    BT Alabama
    BT United States

Walker County Georgia (1989)
    NW Georgia. Before 1989, also search Walker County AND Georgia.
    CO N343600N345900
         W0850400W0853400
    BT Georgia
    BT United States

Walker County Texas (1989)
    E central Texas. Before 1989, also search Walker County AND Texas.
    CO N303200N310400
         W0952000W0955300
    BT Texas
    BT United States

Walker Creek Field (1985)
Oil field in Lafayette and Columbia counties, Arkansas. Also an oil field in Niobrara County, Wyoming.
IN Index states and counties as applicable.
SA Arkansas
SA Columbia County Arkansas
SA Lafayette County Arkansas
SA Niobrara County Wyoming
SA oil and gas fields
SA Powder River basin
SA Wyoming

Walker Lake (1981)
Mineral County, W Nevada. Fed by Walker River. Also occurs in Coconino County Arizona.
IN Index county or region as applicable.
SA Coconino County Arizona
SA Mineral County Nevada

Walker River (1978)
Flows through Walker River Indian Reservation into Walker Lake in W central Nevada.
IN Index counties as applicable.
BT Nevada
BT United States

Wall Creek Member (1989)
Of Frontier Formation. E Wyoming and W South Dakota.
BT Upper Cretaceous
BT Cretaceous
BT Mesozoic
SA Frontier Formation
SA South Dakota
SA Wyoming

wall rocks (1993)
Before 1993, also search wallrocks.
UF wallrocks
SA country rocks
SA faults
SA host rocks
SA intrusions
SA rocks
SA veins
SA wall-rock alteration

wall-rock alteration (1993)
Before 1993, also search wallrock alteration.
UF wallrock alteration
BT metasomatism
SA alteration
SA country rocks
SA hydrothermal alteration
SA mineral deposits, genesis
SA thermal waters
SA wall rocks

Walla Walla County
Valid through 1988. Search in combination with state term. After 1988, use specific county-state term.

Walla Walla County Washington (1989)
SE Washington. Before 1989, also search Walla Walla County AND Washington.
CO N460000N463500
    W1180000W1190000
BT Washington
BT United States

Wallace County
Valid through 1988. Search in combination with state term. After 1988, use specific county-state term.

Wallace County Kansas (1989)
W Kansas. Before 1989, also search Wallace County AND Kansas.
CO N383000N390500
    W1014000W1021000
BT Kansas
BT United States

Wallace Formation (1978)
In Piegan Group of Belt Supergroup. Thin-bedded bluish and greenish more or less calcareous shales, underlain by rapidly alternating thin beds of argillite, calcareous sandstone, and impure limestone. NE Idaho, and W Montana.
BT middle Proterozoic
BT Proterozoic
BT upper Precambrian
BT Precambrian
SA Belt Supergroup
SA Idaho
SA Montana
SA Newland Limestone

Wallowa County
Valid through 1988. Search in combination with state term. After 1988, use specific county-state term.

Wallowa County Oregon (1989)
NE Oregon. Before 1989, also search Wallowa County AND Oregon.
CO N450500N460000
    W1163000W1175500
BT Oregon
BT United States

Wallowa Mountains (1978)
Range in NE Oregon.
BT Oregon
BT United States
SA Baker County Oregon
SA Union County Oregon

wallrock alteration
No longer a valid term for GeoRef. Valid 1978-1992.
use wall-rock alteration

wallrocks
use wall rocks

walls, retaining
use retaining walls

Walsh County
Valid through 1988. Search in combination with state term. After 1988, use specific county-state term.

Walsh County North Dakota (1989)
NE Dakota. Before 1989, also search Walsh County AND North Dakota.
CO N481200N483300
    W0971000W0982000
BT North Dakota
BT United States

Walton County
Valid through 1988. Search in combination with state term. After 1988, use specific county-state term.

Walton County Florida (1989)
NW Florida. Before 1989, also search Walton County AND Florida.
CO N301600N310000
    W0852300W0862200
BT Florida
BT United States
SA Choctawhatchee Bay

Walton County Georgia (1989)
N central Georgia. Before 1989, also search Walton County AND Georgia.
CO N333600N335400
    W0832900W0835800
BT Georgia
BT United States

Walvis Basin
use Cape Basin

Walvis Bay (1978)
Inlet of the Atlantic Ocean on the central W coast of Namibia. Before 1993, also a town which, along with the bay and immediate vicinity, constitute an exclave of Cape Province, South Africa. See Walvis Bay South Africa for the town.
IN Index South Africa or Namibia as applicable.
UF Walfischbai
UF Walfish Bay
BT Southern Africa
BT Africa
SA Cape Province South Africa
SA Namibia
SA South Africa

Walvis Bay South Africa (1993)
Town on the Namibian coast which, along with the harbor and immediate vicinity, constitutes an exclave of Cape Province, South Africa.
BT Cape Province South Africa
BT South Africa
BT Southern Africa
BT Africa
SA Namibia

Walvis Ridge (1978)
Extends in a NE-SW direction off Namibia and South Africa in Southeast Atlantic.
UF Walvish Ridge
BT Southeast Atlantic
BT South Atlantic
BT Atlantic Ocean
SA DSDP Site 362
SA DSDP Site 363
SA DSDP Site 523
SA DSDP Site 524
SA DSDP Site 525
SA DSDP Site 526
SA DSDP Site 527
SA DSDP Site 528
SA DSDP Site 529
SA DSDP Site 530
SA DSDP Site 531
SA DSDP Site 532
SA Leg 39
SA Leg 40

Walvish Ridge
use Walvis Ridge

Wanapum Basalt (1989)
In Yakima Basalt of the Columbia River Basalt Group. Includes Frenchman Springs Member.
BT Miocene
BT Neogene
BT Tertiary
BT Cenozoic
SA Columbia River Basalt Group
SA Frenchman Springs Member
SA Idaho
SA middle Miocene
SA Oregon
SA Washington
SA Yakima Basalt

wandering, polar
use polar wandering

Wanganui
No longer a valid term for GeoRef. See Wanganui New Zealand.

Wanganui New Zealand (1993)
City at mouth of the Wanganui River in SW North Island.
IN Also Index North Island.
BT Wellington New Zealand
BT New Zealand
BT Australasia
SA North Island

Wanganui Valley (1978)
River valley in SW central North Island. Also search Wanganui; Wanganui River.
IN Also index North Island.
BT New Zealand
BT Australasia
SA North Island

Wann Formation (1978)
In Ochelata Group. Includes both Clem Creek and Washington Irving sandstones. In NE Oklahoma.
BT Missourian
BT Upper Pennsylvanian
BT Pennsylvanian
BT Carboniferous
BT Paleozoic
SA Oklahoma

Wapanucka Limestone (1978)
Underlies Atoka Formation in McAlestar Basin and Ouachita Front. Overlies Bloyd Formation in McAlestar Basin and Caney Shale and Union Valley (sandstone) in Ouachita Front. In central S and SE Oklahoma.
BT Pennsylvanian
BT Carboniferous
BT Paleozoic
SA Oklahoma

Wapawekka Lake (1978)
At foot of the Wapawekka Hills in central Saskatchewan. Also search Wapawekka.
BT Saskatchewan
BT Western Canada
BT Canada

Wapiti Formation (1985)
In Sunlight Group. Montana and NW Wyoming.
BT Eocene
BT Paleogene
BT Tertiary
BT Cenozoic
SA Montana
SA Wyoming

Warburton Meteorite (1981)
Impact in Warburton Range. Before 1981, also search Warburton AND meteorites.
BT ataxite
BT iron meteorites
BT meteorites
SA Australia
SA Western Australia

Ward County
Valid through 1988. Search in combination with state term. After 1988, use specific county-state term.

Ward County North Dakota (1989)
N central North Dakota. Before 1989, also search Ward County AND North Dakota.
CO N475100N484800
    W1005800W1021500
BT North Dakota

BT United States

Ward County Texas (1989)
Extreme W Texas. Before 1989, also search Ward County AND Texas.
CO N311800N314000
W1024700W1033500
BT Texas
BT United States
SA Central Basin Platform

Wardak
No longer a valid term for GeoRef. See Wardak Afghanistan.

Wardak Afghanistan (1993)
Province in E central Afghanistan. Before 1993, also search Wardak and Afghanistan.
BT Afghanistan
BT Indian Peninsula
BT Asia

Wardha River valley (1978)
In E Maharashtra, central India. Also search Wardha.
BT Maharashtra India
BT India
BT Indian Peninsula
BT Asia

Ware
No longer a valid term for GeoRef. See Ware British Columbia.

Ware British Columbia (1993)
Location in NE British Columbia.
BT British Columbia
BT Western Canada
BT Canada

Warkalli Formation (1978)
According to "Indian Stratigraphical Nomenclature", the age of Warkalli Beds ranges from upper Miocene to Pliocene. They consist of sandy clays and lignite seams. In Kerala in S India.
IN Index ages as applicable.
BT Tertiary
BT Cenozoic
SA India
SA Kerala India
SA Pliocene
SA upper Miocene

Warrawoona Group (1993)
Archean. Pilbara Block, Western Australia.
BT Archean
BT Precambrian
SA Pilbara Block
SA Western Australia

Warren County
Valid through 1988. Search in combination with state term. After 1988, use specific county-state term.

Warren County Georgia (1989)
E Georgia. Before 1989, also search Warren County AND Georgia.
CO N331600N333700
W0822300W0825100
BT Georgia
BT United States

Warren County Illinois (1989)
W Illinois. Before 1989, also search Warren County AND Illinois.
CO N403600N410400
W0902300W0904900
BT Illinois
BT United States

Warren County Indiana (1989)
W Indiana. Before 1989, also search Warren County AND Indiana.
CO N400700N402800
W0870500W0873200
BT Indiana
BT United States

Warren County Iowa (1989)
S central Iowa. Before 1989, also search Warren County AND Iowa.
CO N410900N413200
W0932000W0934200
BT Iowa
BT United States

Warren County Kentucky (1989)
S Kentucky. Before 1989, also search Warren County AND Kentucky.
CO N364600N371200
W0860900W0863900
BT Kentucky
BT United States

Warren County Mississippi (1989)
W Mississippi. Before 1989, also search Warren County AND Mississippi.
CO N320300N323700
W0903500W0911200
BT Mississippi
BT United States

Warren County Missouri (1989)
E central Missouri. Before 1989, also search Warren County AND Missouri.
CO N383300N385900
W0905800W0912500
BT Missouri
BT United States

Warren County New Jersey (1989)
NW New Jersey. Before 1989, also search Warren County AND New Jersey.
CO N403600N410600
W0744700W0751300
BT New Jersey
BT United States
SA New Jersey Highlands

Warren County New York (1989)
E New York. Before 1989, also search Warren County AND New York.
CO N431400N434800
W0732800W0741300
BT New York
BT United States
SA Lake George

Warren County North Carolina (1989)
N North Carolina. Before 1989, also search Warren County AND North Carolina.
CO N361200N363300
W0775400W0782000
BT North Carolina
BT United States

Warren County Ohio (1989)
SW Ohio. Before 1989, also search Warren County AND Ohio.
CO N391500N393600
W0835800W0842300
BT Ohio
BT United States

Warren County Pennsylvania (1989)
NW Pennsylvania. Before 1989, also search Warren County AND Pennsylvania.
CO N413700N420000
W0785400W0793700
BT Pennsylvania
BT United States

Warren County Tennessee (1989)
Central Tennessee. Before 1989, also search Warren County AND Tennessee.
CO N353200N355100
W0853300W0855800
BT Tennessee
BT United States

Warren County Virginia (1989)
N Virginia. Before 1989, also search Warren County AND Virginia.
CO N384600N390400
W0780000W0782300
BT Virginia
BT United States

Warrior coal field (1985)
N Alabama.
IN Index counties as applicable.
BT Alabama
BT United States
SA Black Warrior Basin
SA coal fields

Warsaw
No longer a valid term for GeoRef. As of 1993 see Warsaw Poland.

Warsaw Formation (1978)
In Osage Group. Lower division consists of massive fine-grained earthy geode-bearing limestone below, thin bed of locally brownish dolomite cherty limestone in middle and bluish-gray slightly calcareous geode-bearing shale above. N Alabama, Illinois, Indiana, Iowa, Kentucky, NE Mississippi, E Missouri, and Tennessee.
BT Meramecian
BT Upper Mississippian
BT Mississippian
BT Carboniferous
BT Paleozoic
SA Alabama
SA Illinois
SA Indiana
SA Iowa
SA Kentucky
SA Mississippi
SA Missouri
SA Tennessee

Warsaw Poland (1993)
Province. Also a city on both banks of the Vistula River in E central Poland. Before 1993 also search Warsaw AND Poland.
UF Warschau
UF Warszawa
BT Poland
BT Central Europe
BT Europe

Warschau
use Warsaw Poland

Warszawa
use Warsaw Poland

Warta (1978)
River which rises NW of Krakow and flows NW and W into the Oder River at Kostrzyn on the German border.
UF Varta
UF Warthe
BT Poland
BT Central Europe
BT Europe
SA Oder Valley

Warthe
use Warta

Warwickshire
No longer a valid term for GeoRef as of 1993.
use Warwickshire England

Warwickshire England (1993)
County in central England. After 1974, some territory, including the city of Birmingham, was used to form West Midlands County.
CO N515500N524000
W0011500W0020000
UF Warwickshire
BT England
BT Great Britain
BT United Kingdom
BT Western Europe
BT Europe
SA Birmingham England
SA Brandon
SA Midlands

Wasatch County
Valid through 1988. Search in combination with state term. After 1988, use specific county-state term.

Wasatch County Utah (1989)
N central Utah. Before 1989, also search Wasatch County AND Utah.
CO N395400N404500
W1105500W1113700
BT Utah
BT United States

Wasatch Fault
No longer a valid term for GeoRef. Valid 1978-1992.
use Wasatch fault zone

Wasatch fault zone (1993)
Central Utah to SE Idaho. Forms the E boundary of Basin and Range Province. Also search Wasatch Fault.
IN Index counties as applicable.
UF Wasatch Fault
BT United States
SA Idaho
SA Salt Lake County Utah
SA Utah

Wasatch Formation (1978)
Paleocene and Eocene. In Wasatch Plateau, consists of three members: lower member of sandstone, varicolored shale, conglomerate and small amounts of fresh-water limestone; and an upper member of varicolored shale and sandstone. W Colorado, central S and SE Montana, NE New Mexico, SW North Dakota, Utah, and W Wyoming.
BT Paleogene
BT Tertiary
BT Cenozoic
SA Colorado
SA Eocene
SA Montana
SA New Mexico
SA North Dakota
SA Paleocene
SA Utah
SA Vermillion Creek coal bed
SA Wyoming

Wasatch Front (1978)
Outer slope of the Wasatch Range.
IN Index states and mountains as applicable.
BT United States
SA Idaho

SA Utah
SA Wasatch Range

Wasatch Plateau (1978)
High tableland at S end of Wasatch Range in central Utah.
IN Index counties as applicable.
BT Utah
BT United States

Wasatch Range (1978)
Range of Rocky Mountains extending from SE Idaho to central Utah. This term has multiple hierarchies.
IN Index counties and states as applicable.
BT1 U. S. Rocky Mountains
BT1 Rocky Mountains
BT2 U. S. Rocky Mountains
BT2 United States
SA Idaho
SA Utah
SA Wasatch Front

Wasatchian (1993)
North America. Above Clarkforkian and below Brigerian.
BT lower Eocene
BT Eocene
BT Paleogene
BT Tertiary
BT Cenozoic

Wasekwan Group (1989)
BT Precambrian
SA Manitoba

Washakie Basin (1978)
Primarily in S Wyoming but also just within NW Colorado.
IN Index counties as applicable.
BT United States
SA Colorado
SA Wyoming

Washakie County
Valid through 1988. Search in combination with state term. After 1988, use specific county-state term.

Washakie County Wyoming (1989)
N central Wyoming. Before 1989, also search Washakie County AND Wyoming.
CO N433000N441000
    W1070700W1083800
BT Wyoming
BT United States
SA Bighorn River

Washington (1978)
Autoposting of this term began in 1978.
CO N453000N490000
    W1165500W1244500
BT United States
NT Adams County Washington
NT Benton County Washington
NT Chelan County Washington
NT Clallam County Washington
NT Clark County Washington
NT Columbia County Washington
NT Cowlitz County Washington
NT Cowlitz River
NT Douglas County Washington
NT Ferry County Washington
NT Franklin County Washington
NT Garfield County Washington
NT Golden Horn Batholith
NT Grand Coulee Dam
NT Grant County Washington
NT Grays Harbor County Washington
NT Hanford Reservation
NT Jefferson County Washington
NT King County Washington
NT Kitsap County Washington
NT Kittitas County Washington
NT Klickitat County Washington
NT Lewis County Washington
NT Lincoln County Washington
NT Mason County Washington
NT Midnite Mine
NT Mount Rainier National Park
NT Okanogan County Washington
NT Olympic Mountains
NT Olympic Peninsula
NT Pacific County Washington
NT Pend Oreille County Washington
NT Pierce County Washington
NT Puget Lowland
NT Puget Sound
NT San Juan County Washington
NT Skagit County Washington
NT Skamania County Washington
NT Snohomish County Washington
NT Spokane County Washington
NT Stevens County Washington
NT Thurston County Washington
NT Toutle River
NT Walla Walla County Washington
NT Whatcom County Washington
NT Whitman County Washington
NT Yakima County Washington
NT Yakima Indian Reservation
SA Aldridge Formation
SA Astoria Formation
SA Belt Basin
SA Belt Supergroup
SA Blue Glacier
SA Blue Mountain
SA Blue Mountains
SA Brigham Group
SA Cache Creek Group
SA Cascade Range
SA Chilliwack Group
SA Chuckanut Formation
SA Chumstick Formation
SA Coast plutonic complex
SA Coast Ranges
SA Columbia Plateau
SA Columbia River
SA Columbia River Basalt Group
SA Columbia River basin
SA Columbia River estuary
SA Colville River
SA Cowlitz Formation
SA Crescent Formation
SA Deschutes River
SA Discovery Bay
SA Elk Lake
SA Ellensburg Formation
SA Frenchman Springs Member
SA Grande Ronde Basalt
SA Grays Harbor
SA Juan de Fuca Strait
SA Kootenay Arc
SA Ladner Group
SA Lake Missoula
SA Lake Washington
SA Ledbetter Slate
SA Leech River Fault
SA Lincoln Creek Formation
SA Mazama Ash
SA Metaline Limestone
SA Methow Basin
SA Montesano Formation
SA Mount Adams
SA Mount Baker
SA Nanaimo Group
SA Okanagan Valley
SA Okanogan Range
SA Pasayten Group
SA Pasco Basin
SA Rattlesnake Hills
SA Revett Quartzite
SA Ringold Formation
SA Saddle Mountains Basalt
SA Saint Regis Formation
SA Shuksan Thrust
SA Skagit Valley
SA Snake River
SA Snake River basin
SA Snoqualmie Batholith
SA Spirit Lake
SA Straight Creek Fault
SA Strait of Georgia
SA Swauk Formation
SA Twin River Formation
SA Twin Sisters Dunite
SA U. S. Rocky Mountains
SA Wanapum Basalt
SA Windermere System
SA Wrangellia
SA Yakima Basalt
SA Yakima fold belt
SA Yellow Aster Complex

Washington County
Valid through 1988. Search in combination with state term. After 1988, use specific county-state term.

Washington County Alabama (1989)
SW Alabama. Before 1989, also search Washington County AND Alabama.
CO N310700N314200
    W0875500W0882700
BT Alabama
BT United States

Washington County Arkansas (1989)
NW Arkansas. Before 1989, also search Washington County AND Arkansas.
CO N354300N361300
    W0935500W0943400
BT Arkansas
BT United States
NT Fayetteville Arkansas
SA Fayetteville Meteorite

Washington County Colorado (1989)
NE Colorado. Before 1989, also search Washington County AND Colorado.
CO N393300N403400
    W1023800W1033800
BT Colorado
BT United States

Washington County Florida (1989)
Bounded by Choctashatchee River in NW Florida. Before 1989, also search Washington County AND Florida.
CO N302200N304900
    W0852300W0860000
BT Florida
BT United States

Washington County Georgia (1989)
E central Georgia. Before 1989, also search Washington County AND Georgia.
CO N324700N331400
    W0823000W0830600
BT Georgia
BT United States

Washington County Idaho (1989)
W Idaho. Before 1989, also search Washington County AND Idaho.
CO N440800N445000
    W1162000W1171500
BT Idaho

BT United States

Washington County Illinois (1989)
SW Illinois. Before 1989, also search Washington County AND Illinois.
CO N381400N383200
    W0890800W0894300
BT Illinois
BT United States

Washington County Indiana (1989)
S Indiana. Before 1989, also search Washington County AND Indiana.
CO N382500N384700
    W0855200W0862000
BT Indiana
BT United States

Washington County Iowa (1989)
SE Iowa. Before 1989, also search Washington County AND Iowa.
CO N411000N413100
    W0912800W0914900
BT Iowa
BT United States

Washington County Kansas (1989)
N Kansas. Before 1989, also search Washington County AND Kansas.
CO N393400N400000
    W0965000W0972200
BT Kansas
BT United States

Washington County Kentucky (1989)
Central Kentucky. Before 1989, also search Washington County AND Kentucky.
CO N373700N375300
    W0850100W0852500
BT Kentucky
BT United States

Washington County Maine (1989)
Easternmost county in Maine and U.S. Before 1989, also search Washington County AND Maine.
CO N442500N453800
    W0665600W0680700
BT Maine
BT United States

Washington County Maryland (1989)
W Maryland. Before 1989, also search Washington County AND Maryland.
CO N392000N394330
    W0772800W0782200
BT Maryland
BT United States

Washington County Minnesota (1989)
E Minnesota. Before 1989, also search Washington County AND Minnesota.
CO N444500N451800
    W0924300W0930100
BT Minnesota
BT United States

Washington County Mississippi (1989)
W Mississippi. Before 1989, also search Washington County AND Mississippi.
CO N330100N333200
    W0904200W0911100
BT Mississippi

BT United States

Washington County Missouri (1989)
SE central Missouri. Before 1989, also search Washington County AND Missouri.
CO N374400N381300
W0903800W0910600
BT Missouri
BT United States

Washington County Nebraska (1989)
E Nebraska. Before 1989, also search Washington County AND Nebraska.
CO N412200N414000
W0955400W0962800
BT Nebraska
BT United States

Washington County New York (1989)
E New York. Before 1989, also search Washington County AND New York.
CO N425600N434800
W0731700W0733800
BT New York
BT United States
SA Lake George

Washington County North Carolina (1989)
E North Carolina. Before 1989, also search Washington County AND North Carolina.
CO N354100N355800
W0762200W0765100
BT North Carolina
BT United States

Washington County Ohio (1989)
SE Ohio. Before 1989, also search Washington County AND Ohio.
CO N391200N393800
W0810300W0815200
BT Ohio
BT United States

Washington County Oklahoma (1989)
NE Oklahoma. Before 1989, also search Washington County AND Oklahoma.
CO N362500N370000
W0954700W0960000
BT Oklahoma
BT United States

Washington County Oregon (1989)
NW Oregon. Before 1989, also search Washington County AND Oregon.
CO N452000N454600
W1224500W1233000
BT Oregon
BT United States

Washington County Pennsylvania (1989)
SW Pennsylvania. Before 1989, also search Washington County AND Pennsylvania.
CO N395700N402800
W0795100W0803200
BT Pennsylvania
BT United States

Washington County Rhode Island (1989)
SW Rhode Island. Before 1989, also search Washington County AND Rhode Island.
CO N411800N413700
W0712500W0715300
BT Rhode Island

BT United States

Washington County Tennessee (1989)
NE Tennessee. Before 1989, also search Washington County AND Tennessee.
CO N360700N362800
W0821800W0824200
BT Tennessee
BT United States

Washington County Texas (1989)
S central Texas. Before 1989, also search Washington County AND Texas.
CO N300400N302500
W0960600W0964700
BT Texas
BT United States
NT Brenham Texas

Washington County Utah (1989)
SW Utah. Before 1989, also search Washington County AND Utah.
CO N370000N373700
W1125500W1140300
BT Utah
BT United States
NT Pine Valley Mountains
NT Springdale Utah
SA Saint George Basin

Washington County Vermont (1989)
Central Vermont. Before 1989, also search Washington County AND Vermont.
CO N440200N443000
W0721300W0735200
BT Vermont
BT United States

Washington County Virginia (1989)
SW Virginia. Before 1989, also search Washington County AND Virginia.
CO N363600N365500
W0813700W0822000
BT Virginia
BT United States

Washington County Wisconsin (1989)
E Wisconsin. Before 1989, also search Washington County AND Wisconsin.
CO N431200N433200
W0880200W0882500
BT Wisconsin
BT United States

Washita Group (1978)
Lower and Upper Cretaceous. In Oklahoma, includes (ascending) Duck Creek Formation, Fort Worth Limestone, Denton Clay, Weno Clay, Pawpaw Formation, Main Street Limestone, and Grayson Shale. SW Arkansas, NW Louisiana, S Oklahoma, and Texas.
IN Index Lower Cretaceous and/or Upper Cretaceous as applicable.
BT Comanchean
BT Cretaceous
BT Mesozoic
SA Arkansas
SA Buda Limestone
SA Edwards Formation
SA Georgetown Formation
SA Louisiana
SA Lower Cretaceous
SA Oklahoma
SA Texas

SA Upper Cretaceous

Washita River valley (1978)
E Texas panhandle, and W and S central Oklahoma. Also search Washita River.
IN Index counties or states as applicable.
BT United States
SA Oklahoma
SA Texas

Washoe County
Valid through 1988. Search in combination with state term. After 1988, use specific county-state term.

Washoe County Nevada (1989)
NW Nevada. Before 1989, also search Washoe County AND Nevada.
CO N390700N420000
W1191000W1200000
BT Nevada
BT United States
NT Reno Nevada
SA Lake Tahoe
SA Pyramid Lake
SA Steamboat Springs

Washtenaw County Michigan (1989)
SE Michigan. Before 1989, search Washtenaw County AND Michigan.
CO N420500N422400
W0833200W0840800
BT Michigan Lower Peninsula
BT Michigan
BT United States
NT Ann Arbor Michigan

waste disposal (1978)
IN Index specific type of waste if applicable.
UF disposal, waste
SA agricultural waste
SA Asse Mine
SA biorcmediation
SA conservation
SA controls
SA creosote
SA DDT
SA detergents
SA dispersivity
SA disposal barriers
SA dredged materials
SA effluents
SA engineering geology
SA environmental geology
SA far-field
SA fluid injection
SA geologic hazards
SA geomembranes
SA geotextiles
SA Gorleben
SA ground water
SA hazardous waste
SA herbicides
SA high-level waste
SA human waste
SA impact statements
SA industrial waste
SA injection
SA Konrad Mine
SA land use
SA landfills
SA leachate
SA leaking underground storage tanks
SA liquid waste
SA low-level waste
SA marine installations
SA near-field
SA pesticides
SA pits
SA pollutants

SA pollution
SA radioactive waste
SA radioactivity
SA reclamation
SA regulations
SA reinjection wells
SA remediation
SA reservoirs
SA Resource Conservation and Recovery Act
SA sanitary landfills
SA seepage
SA sewage
SA sewage sludge
SA site exploration
SA sludge
SA soils
SA solid waste
SA spoils
SA storage
SA Stripa region
SA sulfuric acid
SA Superfund
SA tailings
SA tailings ponds
SA thermal pollution
SA toxic materials
SA toxicity
SA underground disposal
SA underground installations
SA underground storage
SA waste disposal sites
SA Waste Isolation Pilot Plant
SA waste water

waste disposal sites (1993)
Before 1993, also search waste disposal.
SA Asse Mine
SA Gorleben
SA hazardous waste
SA human waste
SA industrial waste
SA Konrad Mine
SA landfills
SA liquid waste
SA pollution
SA radioactive waste
SA reclamation
SA solid waste
SA underground disposal
SA underground storage
SA waste disposal
SA Waste Isolation Pilot Plant

Waste Isolation Pilot Plant (1985)
SE New Mexico. Autoposting of broader terms to this term began in 1989.
UF WIPP
UF WIPP site
BT Eddy County New Mexico
BT New Mexico
BT United States
SA radioactive waste
SA underground storage
SA waste disposal
SA waste disposal sites

waste repositories
Not a valid term for GeoRef. After 1993, see storage or underground storage and waste disposal.

waste water (1978)
UF water, waste
SA decontamination
SA detergents
SA effluents
SA environmental geology
SA industrial waste
SA liquid waste
SA polluted water
SA reclamation
SA waste disposal
SA water

SA water treatment
waste, agricultural
  use agricultural waste
waste, human
  use human waste
waste, industrial
  use industrial waste
waste, liquid
  use liquid waste
waste, radioactive
  use radioactive waste
waste, solid
  use solid waste
Watahomigi Formation (1985)
  Lower and Middle Pennsylvanian. NW Arizona.
  BT Pennsylvanian
  BT Carboniferous
  BT Paleozoic
  SA Arizona
  SA Atokan
  SA Morrowan
Watchung Mountains (1989)
  Newark Basin, N central and NE New Jersey. Two long, low ridges in Essex, Union and Somerset counties.
  IN Index counties as applicable.
  CO N403500N405300 W0741200W0744000
  BT New Jersey
  BT United States
  SA Essex County New Jersey
  SA Newark Basin
  SA Somerset County New Jersey
  SA Union County New Jersey
WATEQF (1989)
  Computer program for the calculation of chemical equilibria in waters.
  BT computer programs
  SA Fortran IV
  SA geochemistry
  SA ground water
  SA hydrochemistry
  SA phase equilibria
water (1978)
  SA aqueous solutions
  SA artesian waters
  SA atmospheric precipitation
  SA bottling
  SA bottom water
  SA brackish water
  SA clouds
  SA connate waters
  SA decontamination
  SA drainage
  SA drinking water
  SA droplets
  SA environmental geology
  SA fossil waters
  SA fresh water
  SA gauging
  SA geochemistry
  SA ground water
  SA humidity
  SA hydration
  SA hydrochemistry
  SA hydrologic cycle
  SA hydrological methods
  SA hydrology
  SA hydrosphere
  SA hydrothermal processes
  SA ice
  SA impurities
  SA irrigation
  SA juvenile water
  SA low water
  SA medicinal waters
  SA meltwater
  SA meteoric water
  SA meteorology
  SA mineral waters
  SA moisture
  SA oil-water interface
  SA percolation
  SA polluted water
  SA pollution
  SA pore water
  SA potability
  SA precipitation
  SA purification
  SA raindrops
  SA retention
  SA salt water
  SA saturation
  SA sediment-water interface
  SA seepage
  SA snow
  SA springs
  SA statistical analysis
  SA surface water
  SA thermal waters
  SA waste water
  SA water hardness
  SA water management
  SA water pressure
  SA water regimes
  SA water storage
  SA water supply
  SA water treatment
  SA water vapor
  SA water yield
  SA waterways
  SA wettability
water analysis
  Not a valid term for GeoRef. For methodology, use specific type of water (e.g. ground water) AND specific analytical method. For data, use specific type of water AND specific element or chemical composition, if applicable.
water balance (1978)
  Also search water budget.
  UF balance, water
  UF hydrologic budget
  UF water budget
  SA aquifers
  SA discharge
  SA drainage basins
  SA drought
  SA evapotranspiration
  SA hydrologic cycle
  SA hydrology
  SA lakes
  SA lysimeters
  SA mass balance
  SA recharge
  SA reservoirs
  SA runoff
water biscuits
  use algal biscuits
water budget
  use water balance
water content
  use moisture
water crop
  use water yield
water cycle
  use hydrologic cycle
water erosion (1978)
  BT erosion
  SA denudation
  SA erosion control
  SA erosion features
  SA geomorphology
  SA nivation
  SA piping
  SA scour
  SA sheet erosion
  SA soil erosion
  SA soils
  SA suffosion
  SA weathering
  SA wind erosion
water falls
  Not a valid GeoRef index term after 1970. Was used on level 1 in subfile N.
  use waterfalls
water flooding
  use waterflooding
water hardness (1981)
  Before 1981, also search hardness AND water.
  SA ground water
  SA hardness
  SA hydrochemistry
  SA surface water
  SA water
  SA water management
  SA water quality
  SA water resources
water harnessing (1981)
  SA hydroelectric energy
  SA water resources
water management (1982)
  SA basin management
  SA conservation
  SA decontamination
  SA desalinization
  SA ground water
  SA management
  SA polluted water
  SA pollution
  SA potability
  SA reclamation
  SA Resource Conservation and Recovery Act
  SA surface water
  SA water
  SA water hardness
  SA water quality
  SA water resources
  SA water rights
  SA water supply
  SA water treatment
water of crystallization (1981)
  SA crystallization
  SA geochemistry
  SA lithogeochemistry
  SA minerals
  SA water of dehydration
water of dehydration (1981)
  SA dehydration
  SA geochemistry
  SA lithogeochemistry
  SA minerals
  SA water of crystallization
water opal
  use hyalite
water power
  use hydroelectric energy
water pressure (1978)
  BT pressure
  SA soil mechanics
  SA water
water quality (1978)
  SA bottling
  SA decontamination
  SA desalinization
  SA drinking water
  SA fresh water
  SA ground water
  SA hydrochemistry
  SA hydrology
  SA impurities
  SA irrigation
  SA medicinal waters
  SA nonpoint sources
  SA polluted water
  SA potability
  SA protection
  SA purification
  SA Resource Conservation and Recovery Act
  SA sewage
  SA sewage sludge
  SA water hardness
  SA water management
  SA water resources
  SA water rights
  SA water supply
  SA water treatment
water recovery (1978)
  Use for recovery in relation to ground water. Before 1978 search recovery AND ground water.
  SA discharge
  SA ground water
  SA pumping
  SA recovery
  SA water wells
water regimes (1978)
  UF regimes, water
  SA characterization
  SA drainage
  SA ground water
  SA infiltration
  SA irrigation
  SA lysimeters
  SA moisture
  SA movement
  SA Richards equation
  SA soils
  SA storage
  SA unsaturated zone
  SA water
water resources (1978)
  Used for economically oriented discussions of water.
  IN Commodity. See List C.
  SA aquifers
  SA artesian waters
  SA basin management
  SA bottling
  SA decontamination
  SA desalinization
  SA drinking water
  SA drought
  SA economic geology
  SA energy sources
  SA environmental geology
  SA fresh water
  SA ground water
  SA ground-water provinces
  SA hydroelectric energy
  SA hydrogeology
  SA hydrology
  SA lakes
  SA medicinal waters
  SA NAWDEX
  SA ponds
  SA potability
  SA pumping
  SA purification
  SA representative basins
  SA reservoirs
  SA Resource Conservation and Recovery Act
  SA resources
  SA springs
  SA surface water
  SA thermal waters
  SA water hardness
  SA water harnessing
  SA water management
  SA water quality
  SA water rights
  SA water storage
  SA water supply
  SA water treatment

SA water wells
SA water yield
SA well screens

water rights (1989)
SA ground water
SA legislation
SA surface water
SA water management
SA water quality
SA water resources
SA water supply

water storage (1978)
SA dams
SA engineering geology
SA reservoirs
SA storage
SA water
SA water resources
SA water supply

water supply (1978)
UF supply, water
SA aquifers
SA drinking water
SA drought
SA hydrology
SA irrigation
SA lakes
SA leaking underground storage tanks
SA pollution
SA ponds
SA potability
SA reservoirs
SA rivers
SA surface water
SA water
SA water management
SA water quality
SA water resources
SA water rights
SA water storage

water table (1981)
SA capillary water
SA ground water
SA hydrogeology
SA hydrology
SA levels
SA potentiometric surface
SA saturated zone
SA shallow aquifers
SA shallow depth
SA surficial aquifers
SA unconfined aquifers
SA unsaturated zone

water treatment (1985)
Before 1985, search treatment AND water.
SA bioremediation
SA decontamination
SA desalinization
SA dilution
SA ground water
SA impurities
SA polluted water
SA pollution
SA potability
SA purification
SA reclamation
SA remediation
SA salinity
SA surface water
SA waste water
SA water
SA water management
SA water quality
SA water resources

water use
Generally out of scope for GeoRef. See water supply, water management, pumping or drawdown.

water vapor (1978)
UF steam
UF vapor, water
SA evaporation
SA geochemistry
SA humidity
SA moisture
SA water

water wells (1982)
BT wells
SA drawdown
SA economic geology
SA engineering geology
SA environmental geology
SA ground water
SA hydrogeology
SA levels
SA monitoring
SA observation wells
SA pollution
SA potentiometric surface
SA production
SA pump tests
SA pumping
SA water recovery
SA water resources
SA well screens
SA well-logging

water yield (1978)
UF runout
UF water crop
SA aquifers
SA drainage basins
SA runoff
SA water
SA water resources
SA yields

water, ground and surface
No longer a valid GeoRef index term. Before 1970, was used on level 1 in subfile E. See ground water; surface water; water; hydrogeology; hydrology.

water, waste
use waste water

water-break
use breakwaters

water-rock interaction
use rock-water interface

water-rock interface
use rock-water interface

Waterberg System (1978)
Composed primarily of sedimentary rocks. In Transvaal, South Africa.
BT Precambrian
SA South Africa
SA Transvaal South Africa

waterfalls (1978)
Autoposting of fluvial features to this term began in 1989. Before 1971, also search water falls.
UF water falls
BT fluvial features
SA cascades
SA geomorphology
SA rapids
SA rivers

waterflooding (1985)
UF water flooding
SA enhanced recovery
SA fluid injection
SA petroleum engineering
SA secondary recovery
SA steam injection

waterlogging (1981)
SA hydrology
SA infiltration
SA soils

Waterloo Island
use King George Island

watersheds (1978)
SA drainage basins
SA floods
SA hydrology
SA lakes
SA rainfall
SA rivers
SA runoff
SA streams
SA surface water
SA tributaries
SA waterways

Waterville Formation (1978)
According to the latest published Lexique, the age of Waterville Shale is Silurian. A series of shales, fine grained sandstones, and impure limestones; often pyritiferous. In S central Maine.
BT Silurian
BT Paleozoic
SA Maine

waterways (1978)
Used for geological studies on man-made or man-modified water channels.
SA canals
SA channelization
SA channels
SA controls
SA design
SA dredging
SA engineering geology
SA environmental geology
SA erosion
SA estuaries
SA floods
SA geomembranes
SA geometry
SA geomorphology
SA geotextiles
SA gorges
SA harbors
SA hydraulics
SA hydroelectric energy
SA hydrographic maps
SA hydrology
SA irrigation
SA levees
SA marine installations
SA ocean circulation
SA rapids
SA reservoirs
SA revetments
SA rills
SA rivers
SA rivers and streams
SA seepage
SA seismic response
SA shorelines
SA siltation
SA turbulence
SA underground channels
SA water
SA watersheds

Waterways Formation (1978)
BT Devonian
BT Paleozoic
SA Alberta

Wattenberg Field (1985)
SW Weld County and NW Adams County, Colorado.
IN Index counties as applicable.
CO N395000N402000 W1043000W1050000
BT Colorado
BT United States
SA Adams County Colorado
SA Denver Basin
SA oil and gas fields
SA Weld County Colorado

Waukesha County
Valid through 1988. Search in combination with state term. After 1988, use specific county-state term.

Waukesha County Wisconsin (1989)
Extreme SE Wisconsin. Before 1989, also search Waukesha County AND Wisconsin.
CO N425000N431300 W0880500W0883400
BT Wisconsin
BT United States
SA Root River

Waulsortian facies (1985)
SA Carboniferous
SA Dinantian
SA lithofacies
SA Lower Mississippian
SA Mississippian
SA Osagian

wave absorption (1982)
Before 1982, also search waves AND absorption.
SA absorption
SA distortion
SA elastic waves
SA seismology
SA sorption
SA wave dispersion
SA waves

wave dispersion (1981)
Before 1981, also search waves AND dispersion; waves AND dispersal. Before 1993, also search scattering.
UF dispersion, wave
SA attenuation
SA distortion
SA elastic waves
SA electron microscopy
SA optical properties
SA seismology
SA spectroscopy
SA wave absorption
SA waves

wave forms
use waveforms

wave length
use wavelength

wave propagation
As of 1981, no longer a valid term for GeoRef. Use propagation.

wave-cut platforms (1981)
SA benches
SA erosion features
SA erosion surfaces
SA geomorphology
SA intertidal environment
SA littoral erosion
SA marine terraces
SA shore features

waveforms (1985)
Not to be used for ocean waves.
UF wave forms
SA elastic waves
SA seismic sources
SA seismology

wavelength (1989)
UF wave length
UF wavelengths
SA deformation
SA elastic waves
SA folds
SA spectroscopy
SA waves

wavelengths
  use wavelength

wavellite   (1978)
  BT phosphates

waves   (1978)
  SA acoustical waves
  SA Airy waves
  SA body waves
  SA breaking waves
  SA catastrophic waves
  SA coda waves
  SA elastic waves
  SA electromagnetic waves
  SA Hilbert transformations
  SA ideal waves
  SA internal waves
  SA Lg-waves
  SA long-period waves
  SA Love waves
  SA meteorology
  SA ocean waves
  SA P-waves
  SA PcP-waves
  SA PKiKP-waves
  SA PKP-waves
  SA PKS-waves
  SA Pn-waves
  SA PP-waves
  SA PPP-waves
  SA PPS-waves
  SA propagation
  SA PS-waves
  SA Rayleigh waves
  SA S-waves
  SA ScS-waves
  SA seiches
  SA SH-waves
  SA shoaling
  SA shock waves
  SA short-period waves
  SA SKP-waves
  SA SKS-waves
  SA Sn-waves
  SA Stoneley waves
  SA surface waves
  SA surges
  SA SV-waves
  SA tsunamis
  SA tube waves
  SA wave absorption
  SA wave dispersion
  SA wavelength

Wawa
  No longer a valid term for GeoRef. As of 1993, see Wawa Ontario.

Wawa Belt   (1985)
  Metamorphic belt and structural subprovince in central Ontario.
  CO N462000N494000
        W0824000W0875000
  BT Superior Province
  BT Canadian Shield
  BT North America
  SA Michipicoten Belt
  SA Ontario

Wawa Ontario   (1993)
  Town near Lake Superior in Algoma District, central Ontario. Before 1993, also search Wawa AND Ontario.
  BT Algoma District Ontario
  BT Ontario
  BT Eastern Canada
  BT Canada

Wayne County
  Valid through 1988. Search in combination with state term. After 1988, use specific county-state term.

Wayne County Georgia   (1989)
  SE Georgia. Before 1989, also search Wayne County AND Georgia.
  CO N312000N315000
        W0813700W0820800
  BT Georgia
  BT United States

Wayne County Illinois   (1989)
  SE Illinois. Before 1989, also search Wayne County AND Illinois.
  CO N381600N383600
        W0880900W0884300
  BT Illinois
  BT United States

Wayne County Indiana   (1989)
  E Indiana. Before 1989, also search Wayne County AND Indiana.
  CO N394200N400100
        W0845000W0851400
  BT Indiana
  BT United States

Wayne County Iowa   (1989)
  S Iowa. Before 1989, also search Wayne County AND Iowa.
  CO N403500N405400
        W0930600W0933400
  BT Iowa
  BT United States

Wayne County Kentucky   (1989)
  S Kentucky. Before 1989, also search Wayne County AND Kentucky.
  CO N363400N365700
        W0843400W0850300
  BT Kentucky
  BT United States
  NT Monticello Kentucky

Wayne County Michigan   (1989)
  SE Michigan. Before 1989, also search Wayne County AND Michigan.
  CO N420000N422800
        W0825200W0833100
  BT Michigan Lower Peninsula
  BT Michigan
  BT United States
  SA Detroit River
  SA Lake Saint Clair

Wayne County Mississippi   (1989)
  SE Mississippi. Before 1989, also search Wayne County AND Mississippi.
  CO N312400N315200
        W0882700W0885700
  BT Mississippi
  BT United States
  NT Waynesboro Mississippi

Wayne County Missouri   (1989)
  SE Missouri. Before 1989, also search Wayne County AND Missouri.
  CO N365500N371900
        W0900700W0904700
  BT Missouri
  BT United States

Wayne County Nebraska   (1989)
  NE Nebraska. Before 1989, also search Wayne County AND Nebraska.
  CO N420600N422200
        W0965000W0972200
  BT Nebraska
  BT United States

Wayne County New York   (1989)
  W New York. Before 1989, also search Wayne County AND New York.
  CO N430200N432100
        W0764200W0772200
  BT New York
  BT United States

Wayne County North Carolina   (1989)
  E central North Carolina. Before 1989, also search Wayne County AND North Carolina.
  CO N351000N353600
        W0774800W0781800
  BT North Carolina
  BT United States

Wayne County Ohio   (1989)
  N central Ohio. Before 1989, also search Wayne County AND Ohio.
  CO N404100N405800
        W0813800W0820800
  BT Ohio
  BT United States

Wayne County Pennsylvania   (1989)
  NE Pennsylvania. Before 1989, also search Wayne County AND Pennsylvania.
  CO N411300N420000
        W0750200W0752800
  BT Pennsylvania
  BT United States

Wayne County Tennessee   (1989)
  S Tennessee. Before 1989, also search Wayne County AND Tennessee.
  CO N350000N352800
        W0873500W0880200
  BT Tennessee
  BT United States
  NT Waynesboro Tennessee

Wayne County Utah   (1989)
  S central Utah. Before 1989, also search Wayne County AND Utah.
  CO N381000N383000
        W1095300W1115300
  BT Utah
  BT United States
  SA Canyonlands National Park

Wayne County West Virginia   (1989)
  W West Virginia. Before 1989, also search Wayne County AND West Virginia.
  CO N375200N382400
        W0821200W0823800
  BT West Virginia
  BT United States

Waynesboro
  Valid through 1989. Search in combination with state term. After 1989, use specific city-state term.

Waynesboro Georgia   (1989)
  City in E Georgia. Before 1989, search Waynesboro AND Georgia.
  CO N330400N330400
        W0820100W0820100
  BT Burke County Georgia
  BT Georgia
  BT United States

Waynesboro Mississippi   (1989)
  Town in SE Mississippi. Before 1989, search Waynesboro AND Mississippi.
  CO N314000N314000
        W0884000W0884000
  BT Wayne County Mississippi
  BT Mississippi
  BT United States

Waynesboro Pennsylvania   (1989)
  Borough and resort in S Pennsylvania. Before 1989, search Waynesboro AND Pennsylvania.
  CO N394500N394500
        W0773600W0773600
  BT Franklin County Pennsylvania
  BT Pennsylvania
  BT United States

Waynesboro Tennessee   (1989)
  City in S Tennessee. Before 1989, search Waynesboro AND Tennessee.
  CO N352000N352000
        W0874900W0874900
  BT Wayne County Tennessee
  BT Tennessee
  BT United States

Waynesboro Virginia   (1989)
  City in but independent of Augusta County in N central Virginia. Before 1989, search Waynesboro AND Virginia.
  CO N380400N380400
        W0785400W0785400
  BT Augusta County Virginia
  BT Virginia
  BT United States

weak rocks   (1985)
  SA engineering geology
  SA failures
  SA rock mechanics
  SA rocks
  SA soft rocks
  SA strength

Weald
  use The Weald

Weald Clay   (1978)
  Divided into 3 lithological groups: Group 1- buff grey, Group 2- variegated and Group 3- yellow. In Sussex in SE England.
  BT Cretaceous
  BT Mesozoic
  SA England
  SA United Kingdom

Weald region
  use The Weald

Wealden   (1978)
  Europe. Above Purbeckian, below Gault.
  BT Lower Cretaceous
  BT Cretaceous
  BT Mesozoic

weathered materials   (1985)
  SA materials
  SA rock mechanics
  SA soil mechanics
  SA weathering

weathering   (1978)
  For treatments emphasizing the process.
  IN Index type of rock and type of weathering if applicable.
  NT chemical weathering
  NT differential weathering
  NT mechanical weathering
  NT physical weathering
  SA alteration
  SA anchimetamorphism
  SA bauxitization
  SA biodegradation
  SA biorhexistasy
  SA caliche
  SA clay mineralogy
  SA degradation
  SA denudation
  SA detritus
  SA diagenesis
  SA duricrust
  SA ecology

SA engineering geology
SA enrichment
SA erosion
SA etching
SA exfoliation
SA geochemistry
SA geomorphology
SA gossan
SA hydrolysis
SA insolation
SA kaolinization
SA karstification
SA leaching
SA lithogeochemistry
SA mechanical properties
SA migration of elements
SA mineral deposits, genesis
SA parent materials
SA placers
SA residuum
SA sediments
SA soil erosion
SA soils
SA suffosion
SA water erosion
SA weathered materials
SA weathering crust
SA weathering rinds

weathering crust (1978)
UF crust, weathering
BT chemically precipitated rocks
BT sedimentary rocks
SA calcrete
SA caliche
SA duricrust
SA soils
SA weathering
SA weathering rinds

weathering rinds (1989)
SA chemical weathering
SA exfoliation
SA weathering
SA weathering crust

Webb County
Valid through 1988. Search in combination with state term. After 1988, use specific county-state term.

Webb County Texas (1989)
SW Texas. Before 1989, also search Webb County AND Texas.
CO N271500N281200
W0985000W1001000
BT Texas
BT United States

Weber County
Valid through 1988. Search in combination with state term. After 1988, use specific county-state term.

Weber County Utah (1989)
N Utah. Before 1989, also search Weber County AND Utah.
CO N410500N412500
W1112500W1123000
BT Utah
BT United States
NT Ogden Utah
SA Great Salt Lake

Weber Sandstone (1978)
Pennsylvanian and Permian. In Duchesne River area of Utah, consists mainly of fine grained gray and white sandstone that weathers buff. W Colorado, and NE Utah.
IN Index ages as applicable.
BT Paleozoic
SA Colorado
SA Pennsylvanian
SA Permian
SA Utah

weberite (1978)
BT fluorides
BT halides

Webster County
Valid through 1988. Search in combination with state term. After 1988, use specific county-state term.

Webster County Georgia (1989)
W Georgia. Before 1989, also search Webster County AND Georgia.
CO N315600N321300
W0842400W0843700
BT Georgia
BT United States

Webster County Iowa (1989)
Central Iowa. Before 1989, also search Webster County AND Iowa.
CO N421300N423800
W0935500W0942700
BT Iowa
BT United States

Webster County Kentucky (1989)
W Kentucky. Before 1989, also search Webster County AND Kentucky.
CO N372200N373800
W0872000W0875400
BT Kentucky
BT United States

Webster County Mississippi (1989)
Central Mississippi. Before 1989, also search Webster County AND Mississippi.
CO N332800N334300
W0890200W0892900
BT Mississippi
BT United States

Webster County Missouri (1989)
S central Missouri. Before 1989, also search Webster County AND Missouri.
CO N370500N372800
W0924200W0930500
BT Missouri
BT United States

Webster County Nebraska (1989)
S Nebraska. Before 1989, also search Webster County AND Nebraska.
CO N400000N402200
W0981800W0984400
BT Nebraska
BT United States

Webster County West Virginia (1989)
Central West Virginia. Before 1989, also search Webster County AND West Virginia.
CO N375200N382400
W0821200W0823800
BT West Virginia
BT United States

websterite (1978)
BT ultramafics
BT plutonic rocks
BT igneous rocks
SA ariegite
SA pyroxenite

Wechsel (1978)
An outlier of the Eastern Alps in E Austria.
IN Index states as applicable.
BT Austria
BT Central Europe

BT Europe
SA Eastern Alps
SA Lower Austria
SA Styria Austria

Weddell Sea (1978)
Arm of the S Atlantic Ocean in the Antarctic Ocean between the Antarctic Peninsula on the W and Coats Land on the SE.
CO S780000S630000
W0180000W0600000
BT Antarctic Ocean
SA Atlantic Ocean
SA Bransfield Strait
SA Filchner Ice Shelf
SA James Ross Island
SA Leg 113
SA ODP Site 690
SA ODP Site 691
SA ODP Site 692
SA ODP Site 693
SA ODP Site 694
SA Ronne Ice Shelf
SA Seymour Island

weddellite (1978)
BT carbonates
SA whewellite

wedges, fossil ice
use fossil ice wedges

wedges, ice
use ice wedges

Wedron Formation (1978)
Glacial till. In association with Wadsworth Till Member, Shorewood Till Member, Manitowoc Till Member, and Two Rivers Till Member. Indiana, Illinois, and under Lake Michigan.
BT Pleistocene
BT Quaternary
BT Cenozoic
SA Illinois
SA Indiana
SA Lake Michigan

Weekeroo Station Meteorite (1981)
Impact at Weekeroo Station near Mannahill, Australia. Before 1981, also search Weekeroo Station AND meteorites.
BT octahedrite
BT iron meteorites
BT meteorites
SA Australia
SA South Australia

Wegener hypothesis
use continental drift

Wegener, Alfred (1985)
UF Alfred Wegener
SA biography

wehrlite (1978)
Igneous rock. Also a mineral.
IN Index peridotites or tellurides as applicable.
SA igneous rocks
SA minerals
SA peridotites
SA tellurides

Weichsel
use Weichselian

Weichsel River
use Vistula River

Weichselian (1978)
Term applied in Northern Europe to fourth and last glacial stage of the Pleistocene Epoch. Autoposting of this term began in 1978.
UF Weichsel
BT upper Pleistocene

BT Pleistocene
BT Quaternary
BT Cenozoic
NT Allerod
NT Brandenburg Stade
NT Brandon Stade
NT Frankfurt Stade
NT Loch Lomond Stade
NT Older Dryas
NT Oldest Dryas
NT Perth Stade
NT upper Weichselian
NT Younger Dryas
SA Sartanian
SA Vistulian
SA Wisconsinan
SA Wurm

Weimar
No longer a valid term for GeoRef. See Weimar Germany.

Weimar Germany (1993)
City in Thuringia, central Germany.
BT Thuringia Germany
BT Germany
BT Central Europe
BT Europe

Weinheim
No longer a valid term for GeoRef. See Weinheim Germany.

Weinheim Germany (1993)
City NE of Mannheim in NW Baden-Wurttemberg, SW Germany.
BT Baden-Wurttemberg Germany
BT Germany
BT Central Europe
BT Europe

Weipa
No longer a valid term for GeoRef. See Weipa Australia.

Weipa Australia (1993)
Habitation in an aboriginal reserve on the W coast of Cape York Peninsula in N Queensland.
BT Queensland Australia
BT Australia
BT Australasia
SA Cape York Peninsula

Weisse Elster Basin (1978)
River basin in extreme W Bohemia and E Germany. Also search Weisse Elster.
IN Index Bohemia and/or German states as applicable.
UF White Elster Basin
SA Bohemia
SA Saxony Germany
SA Saxony-Anhalt Germany
SA Thuringia Germany

Weissemburg in Bayern
use Weissenburg Germany

Weissenberg
No longer a valid term for GeoRef. See Weissenberg Germany.

Weissenberg Germany (1993)
Town just N of Lobau near the Polish and Czechoslovak borders in Upper Lusatia.
BT Saxony Germany
BT Germany
BT Central Europe
BT Europe
SA Upper Lusatia

Weissenburg
No longer a valid term for GeoRef. See Weissenburg Germany.

Weissenburg Germany (1993)
Town about 30 miles S of Nuremberg in W central Bavaria. Also search Weissemburg in Bayern and Weissenburg-am-Sand.
UF Weissemburg in Bayern
UF Weissenburg-am-Sand
BT Bavaria Germany
BT Germany
BT Central Europe
BT Europe

Weissenburg-am-Sand
use Weissenburg Germany

Weld County
Valid through 1988. Search in combination with state term. After 1988, use specific county-state term.

Weld County Colorado (1989)
NE Colorado. Before 1989, also search Weld County AND Colorado.
CO N400000N410000 W1033000W1051000
BT Colorado
BT United States
NT Eaton Colorado
NT Greeley Colorado
SA Wattenberg Field

welded tuff (1978)
UF tuff lava
BT pyroclastics
BT volcanic rocks
BT igneous rocks
SA ignimbrite
SA tuff

well and drill-hole logs
No longer a valid term for GeoRef. In 1959, was used in subfile N.
use well-logging

well logging
use well-logging

well logs (1985)
SA Schlumberger methods
SA well-logging

well screens (1981)
SA ground water
SA pumping
SA water resources
SA water wells
SA wells

well stimulation (1993)
Also search stimulation.
SA petroleum engineering
SA recovery
SA wells

well-logging (1978)
For treatments that stress methodology. In 1959, well and drill-hole logs was also used. Autoposting of this term began in 1978.
IN Index specific type of logging when available.
UF geophysical logging
UF well and drill-hole logs
UF well logging
UF wireline logging
NT acoustical logging
NT caliper logging
NT dipmeter logging
NT electrical logging
NT electromagnetic logging
NT gravity logging
NT magnetic logging
NT seismic logging
NT temperature logging
SA Archie's law
SA borehole sections
SA borehole televiewers
SA boreholes
SA cores
SA cuttings
SA data processing
SA density logging
SA downhole methods
SA drilling
SA gamma-gamma methods
SA gamma-ray methods
SA gamma-ray spectroscopy
SA geophysical methods
SA geophysical surveys
SA heat flow
SA horizontal drilling
SA Iceland Research Drilling Project
SA induced polarization
SA induction
SA Laterolog
SA measurement-while-drilling
SA microwave methods
SA mineral exploration
SA neutron methods
SA neutron-gamma methods
SA neutron-neutron methods
SA radioactivity
SA resistivity
SA Schlumberger methods
SA self-potential methods
SA tomography
SA video methods
SA water wells
SA well logs
SA wells

Welland Canal (1978)
Ship canal connecting Lake Erie with Lake Ontario in SE Ontario. Also search Welland.
BT Ontario
BT Eastern Canada
BT Canada

wellbore breakouts
use borehole breakouts

Wellington
No longer a valid term for GeoRef. See Wellington New Zealand.

Wellington County Ontario (1993)
Before 1993, also search Wellington County AND Ontario.
BT Ontario
BT Eastern Canada
BT Canada
NT Guelph Ontario

Wellington Formation (1978)
In Sumner Group. In Kansas, includes Milan Limestone Member at top; Hutchinson Salt Member in middle part but not exposed; Carlton Limestone Member below Hutchinson; Hollenburg Limestone Member near base. Central and S Kansas, and N Oklahoma.
BT Permian
BT Paleozoic
SA Hutchinson Salt Member
SA Kansas
SA Oklahoma

Wellington New Zealand (1993)
District in S North Island including the city on Port Nicholson an inlet of Cook Strait on S North Island.
IN Also index North Island.
BT New Zealand
BT Australasia
NT Wanganui New Zealand
SA North Island

wells (1978)
NT observation wells
NT water wells
SA boreholes
SA cores
SA cuttings
SA drilling
SA engineering geology
SA horizontal drilling
SA pumping
SA reinjection wells
SA well screens
SA well stimulation
SA well-logging

Wells Formation (1989)
SE Idaho, N Utah and W Wyoming.
IN Index ages as applicable.
BT Paleozoic
SA Idaho
SA Pennsylvanian
SA Permian
SA Utah
SA Wyoming

weloganite (1978)
BT carbonates

Welsh Basin (1989)
BT Wales
BT Great Britain
BT United Kingdom
BT Western Europe
BT Europe

Welsh Borderland (1978)
As of 1981, refers to region bordering on Wales that is entirely within England. Before 1981, part of Wales was included, and United Kingdom was the broader term.
CO N513000N524500 W0021000W0031500
BT England
BT Great Britain
BT United Kingdom
BT Western Europe
BT Europe
SA Wales

Welwitschiales (1981)
BT gymnosperms
BT Spermatophyta
BT Plantae

wenkite (1978)
Autoposting of silicates began in 1989. This term has multiple hierarchies.
BT1 framework silicates
BT1 silicates
BT2 sulfates

Wenlock Edge (1978)
A limestone ridge in Shropshire in W England.
BT Shropshire England
BT England
BT Great Britain
BT United Kingdom
BT Western Europe
BT Europe
SA Wenlock Limestone

Wenlock Limestone (1978)
Upper, thin-bedded, lenticular and lower, more massive and crystalline limestones, underlying the "Upper Ludlow Rock" and above the "Lower Ludlow Rock". In Shropshire in W England.
BT Wenlockian
BT Lower Silurian
BT Silurian
BT Paleozoic
SA England
SA United Kingdom
SA Upper Silurian
SA Wenlock Edge

Wenlockian (1978)
Europe. Before 1993, Upper Silurian was a broader term.
BT Lower Silurian
BT Silurian
BT Paleozoic
NT Wenlock Limestone

Werfenian (1978)
Triassic.
BT Lower Triassic
BT Triassic
BT Mesozoic
SA Scythian

Werillup Formation (1978)
BT Eocene
BT Paleogene
BT Tertiary
BT Cenozoic
SA Western Australia

Wernecke Mountains (1989)
E to E central Yukon Territory.
CO N640000N642500 W1312000W1370000
BT Yukon Territory
BT Western Canada
BT Canada

Werra
As of 1989, no longer a valid term for GeoRef. See Werra River or Werra Series if applicable.

Werra River (1978)
River which rises in Thuringia and unites with the Fulda River in Hesse, to form the Weser River. Before 1989, also search Werra AND Germany.
IN Index states as applicable.
BT Germany
BT Central Europe
BT Europe
SA Hesse Germany
SA Thuringia Germany
SA Weser River

Werra Series (1989)
BT Permian
BT Paleozoic
SA Germany
SA Poland

Wescogame Formation (1985)
NW Arizona.
BT Upper Pennsylvanian
BT Pennsylvanian
BT Carboniferous
BT Paleozoic
SA Arizona
SA Virgilian

Weser River (1978)
Formed by the confluence of the Fulda River and Werra River. Flows primarily through Lower Saxony into the North Sea. Also search Weser.
IN Index states as applicable.
BT Germany
BT Central Europe
BT Europe
SA Bremen Germany
SA Hesse Germany
SA Lower Saxony Germany
SA North Rhine-Westphalia Germany
SA Werra River

Weser-Ems (1978)
Region between Weser River and Ems River in W and W central Lower Saxony, NE West Germany.
BT Lower Saxony Germany
BT Germany
BT Central Europe

BT Europe
SA Ems River

Wessex Basin (1989)
S England.
BT England
BT Great Britain
BT United Kingdom
BT Western Europe
BT Europe

West Africa (1978)
Autoposting of this term began in 1981.
IN Index countries as applicable.
CO N020000N250000
  E0240000W0173000
BT Africa
NT Adamawa
NT Benin
NT Benue Valley
NT Burkina Faso
NT Cameroon
NT Chad
NT Gambia
NT Ghana
NT Guinea
NT Guinea-Bissau
NT Ivory Coast
NT Lake Chad
NT Liberia
NT Logone River
NT Mali
NT Mali-Niger Syneclise
NT Mauritania
NT Mauritanides
NT Niger
NT Niger River
NT Niger Valley
NT Nigeria
NT Nimba Mountains
NT Senegal
NT Senegal Basin
NT Sierra Leone
NT Togo
SA Central Africa
SA East Africa
SA North Africa
SA Southern Africa
SA Younger Granites

West African Craton
  use West African Shield

West African Shield (1978)
IN Index countries as applicable.
UF West African Craton
BT Africa
SA Mali
SA Niger

West Antarctica (1985)
W of Transantarctic Mountains.
CO S870000S600000
  W0400000W1650000
BT Antarctica

West Atlantic (1978)
Before 1978, also search Atlantic Ocean AND west or western.
BT Atlantic Ocean
SA East Atlantic
SA North Atlantic
SA Northeast Atlantic
SA Northwest Atlantic
SA South Atlantic
SA Southeast Atlantic
SA Southwest Atlantic

West Bengal
No longer a valid term for GeoRef. See West Bengal India.

West Bengal India (1993)
State bordering Bangladesh in E India.
CO N213000N271000
  E0900000E0854000
BT India

BT Indian Peninsula
BT Asia
NT Bankura India
NT Barakar India
NT Burdwan India
NT Calcutta India
NT Cooch Behar India
NT Darjeeling India
NT Digha
NT Purulia India
NT Raniganj coal field
NT Raniganj India
SA Ajay River
SA Bengal
SA Damodar Valley
SA Raniganj Formation

West Bulgarian Hills (1981)
CO N411900N425200
  E0233000E0222000
BT Bulgaria
BT Southern Europe
BT Europe

West Carpathians
  use Western Carpathians

West Charlevoix
No longer a valid term for GeoRef. As of 1993, see Charlevoix-Ouest County Quebec.

West Coast
  use Pacific Coast

West Falls Formation (1985)
W and W central New York.
BT Upper Devonian
BT Devonian
BT Paleozoic
SA New York

West Germany
No longer a valid term for GeoRef. Officially known as Federal Republic of Germany or Bundesrepublik Deutschland. In W central Europe, bounded on N by North Sea and Denmark, on E by East Germany and Czechoslovakia, on SE by Austria, on S by Austria and Switzerland, on SW by France, and on W by Luxembourg, Belgium and the Netherlands. Introduced as a level 1 area term in 1978. As of 1993, see Germany and also Mecklenburg-Western Pomerania Germany, Brandenburg Germany, Saxony-Anhalt Germany, Thuringia Germany, and Saxony Germany.

West Greenland (1978)
Before 1978, search Greenland AND west. Autoposting of Arctic region and Polar regions to this term began in 1990.
BT Greenland
BT Arctic region
NT Isua Belt
SA Nuk Gneiss
SA South Greenland

West Indian Ocean (1978)
Before 1978, also search Indian Ocean AND west or western.
BT Indian Ocean

West Indies (1978)
Islands between SE North America and N South America enclosing the Caribbean Sea. As of 1981, Netherlands Antilles is a narrower term. Autoposting of this term began in 1978. As of 1993, Caribbean region is a broader term.

IN Index island groups as applicable.
CO N100000N253000
  W0590000W0850000
BT Caribbean region
NT Antilles
NT Bahamas
NT Caribbean Mountain Range
NT Cayman Islands
SA America
SA Rincon Formation
SA Robles Formation

West Irian
  use Irian Jaya Indonesia

West Malaysia (1978)
That part of the Federation of Malaysia which is comprised of eleven states on the Malay Peninsula; Johore, Kedah, Kelantan, Malacca, Negri Sembilan, Pahang, Penang, Perak, Perlis, Selangor, and Trengganu. This term has multiple hierarchies.
BT1 Malay Peninsula
BT1 Far East
BT1 Asia
BT2 Malaysia
BT2 Far East
BT2 Asia
NT Johore Malaysia
NT Kedah Malaysia
NT Kelantan Malaysia
NT Langkawi Islands
NT Pahang Malaysia
NT Perak Malaysia
NT Selangor Malaysia
SA Indochina

West Mediterranean (1978)
Before 1978, search Mediterranean Sea AND west. Autoposting of this term began in 1981.
CO N351000N443000
  E0161500W0053000
BT Mediterranean Sea
NT Alboran Sea
NT Gulf of Lion
NT Tyrrhenian Sea
NT Valencia Trough
SA DSDP Site 121
SA DSDP Site 122
SA DSDP Site 123
SA DSDP Site 124
SA DSDP Site 133
SA DSDP Site 134
SA East Mediterranean

West New Guinea
  use Irian Jaya Indonesia

West Pacific (1978)
Before 1978, search Pacific Ocean AND west. Autoposting of this term began in 1981.
BT Pacific Ocean
NT Banda Arc
NT Bering Sea
NT Bismarck Sea
NT Coral Sea
NT East China Sea
NT Emperor Seamounts
NT Indonesian Seas
NT Izu-Bonin Arc
NT Japan Sea
NT Japan Trench
NT Kashima Seamount
NT Kuroshio
NT North Australian Seas
NT Okhotsk Sea
NT Ontong Java Plateau
NT Osaka Bay
NT Philippine Sea
NT South China Sea
NT Sulu Sea
NT Taranaki Basin

NT Tasman Sea
NT Yellow Sea
SA Bering Strait
SA Chibaken-Toho-Oki earthquake 1987
SA DSDP Site 51
SA DSDP Site 171
SA DSDP Site 183
SA DSDP Site 192
SA DSDP Site 206
SA DSDP Site 207
SA DSDP Site 208
SA DSDP Site 210
SA DSDP Site 285
SA DSDP Site 286
SA DSDP Site 288
SA DSDP Site 305
SA DSDP Site 310
SA DSDP Site 430
SA DSDP Site 431
SA DSDP Site 432
SA DSDP Site 433
SA DSDP Site 577
SA DSDP Site 578
SA DSDP Site 579
SA DSDP Site 580
SA East Pacific
SA Equatorial Pacific
SA Leg 124
SA Leg 124E
SA Leg 125
SA Leg 129
SA Leg 130
SA Leg 131
SA Leg 132
SA Leg 134
SA Leg 135
SA Nauru Basin
SA North Pacific
SA Northeast Pacific
SA Northwest Pacific
SA ODP Site 776
SA ODP Site 778
SA ODP Site 779
SA ODP Site 780
SA ODP Site 781
SA ODP Site 782
SA ODP Site 783
SA ODP Site 785
SA ODP Site 800
SA ODP Site 801
SA ODP Site 802
SA ODP Site 803
SA ODP Site 804
SA ODP Site 805
SA ODP Site 806
SA ODP Site 807
SA ODP Site 808
SA South Pacific
SA Southeast Pacific
SA Southwest Pacific

West Pacific Ocean Islands (1985)
Before 1985, many of these islands were narrower terms of Pacific Ocean.
NT Auckland Islands
NT Bonin Islands
NT Kermadec Islands
NT Lord Howe Island
NT Loyalty Islands
NT Macquarie Island
NT Norfolk Island
SA Campbell Island
SA Kiribati

West Pakistan
The west wing of Pakistan prior to the independence of East Pakistan which became Bangladesh. A valid term through 1972. Also search Pakistan AND west.
  use Pakistan

West Philippine Basin (1989)

Region of W Pacific bordered by Philippine Trench, Ryukyu Trench, and Kyushu-Palau Ridge.
BT Philippine Sea
BT West Pacific
BT Pacific Ocean
SA DSDP Site 291
SA Kyushu-Palau Ridge
SA Leg 31
SA Leg 59

West Shasta District
As of 1993, no longer a valid term for GeoRef.
use West Shasta mining district

West Shasta mining district (1993)
N California.
UF West Shasta District
BT Shasta County California
BT California
BT United States
SA metal ores

West Siberia (1981)
This term has multiple hierarchies.
IN Index former Soviet republics as applicable.
CO N500000N820000
    E1140000E0620000
BT1 Commonwealth of Independent States
BT2 Asia
NT Kuznetsk Alatau
NT Minusinsk Basin
NT Severnaya Zemlya
NT Siberian Lowland
SA Bazhenov Formation
SA Kazakhstan
SA Russian Federation
SA Salym Field
SA Siberia
SA West Siberian Plate

West Siberian Basin
use Siberian Lowland

West Siberian Lowland
use Siberian Lowland

West Siberian Plain
use Siberian Lowland

West Siberian Plate (1978)
SA plate tectonics
SA plates
SA West Siberia

West Siberian Platform
use Siberian Platform

West Spitsbergen
use Spitsbergen Island

West Texas (1985)
Autoposting of United States to this term began in 1989.
UF western Texas
BT Texas
BT United States
SA Diablo Platform
SA Texas Panhandle
SA Trans-Pecos

West Valley
Valid through 1988. Search in combination with state term. After 1988, use specific city-state term.

West Valley New York (1989)
Town in W New York. Before 1989, search West Valley AND New York.
CO N422300N422300
    W0783800W0783800
BT Cattaraugus County New York
BT New York
BT United States

West Virginia (1978)

Autoposting of this term began in 1978.
CO N371500N404000
    W0774500W0823000
BT United States
NT Berkeley County West Virginia
NT Boone County West Virginia
NT Calhoun County West Virginia
NT Clay County West Virginia
NT Fayette County West Virginia
NT Grant County West Virginia
NT Greenbrier County West Virginia
NT Hampshire County West Virginia
NT Hancock County West Virginia
NT Hardy County West Virginia
NT Harrison County West Virginia
NT Jackson County West Virginia
NT Jefferson County West Virginia
NT Kanawha County West Virginia
NT Lewis County West Virginia
NT Lincoln County West Virginia
NT Logan County West Virginia
NT Marion County West Virginia
NT Marshall County West Virginia
NT Mason County West Virginia
NT McDowell County West Virginia
NT Mercer County West Virginia
NT Mineral County West Virginia
NT Monongalia County West Virginia
NT Monroe County West Virginia
NT Morgan County West Virginia
NT Ohio County West Virginia
NT Pendleton County West Virginia
NT Pocahontas County West Virginia
NT Preston County West Virginia
NT Putnam County West Virginia
NT Raleigh County West Virginia
NT Randolph County West Virginia
NT Ritchie County West Virginia
NT Roane County West Virginia
NT Taylor County West Virginia
NT Upshur County West Virginia
NT Wayne County West Virginia
NT Webster County West Virginia
NT West Virginia Panhandle
NT Wood County West Virginia
NT Wyoming County West Virginia
SA Allegheny Front
SA Allegheny Group
SA Allegheny Mountains
SA Allegheny Plateau
SA Ames Limestone
SA Antietam Formation
SA Appalachian Basin
SA Appalachian Plateau
SA Berea Sandstone
SA Blackwater
SA Bloomsburg Formation
SA Blue Ridge Mountains
SA Blue Ridge Province
SA Brallier Shale
SA Catoctin Formation
SA Catskill Delta
SA Chemung Formation
SA Clinch Sandstone
SA Clinton Group
SA Coeymans Formation
SA Conemaugh Group
SA Conococheague Formation
SA Cumberland Plateau
SA Dunkard Basin
SA Dunkard Group
SA Eastern Gas Shales Project
SA Elk River
SA Genesee Group
SA Greenbrier Limestone

SA Greenland Gap Group
SA Hamilton Group
SA Hampshire Formation
SA Helderberg Group
SA Hurricane Ridge Syncline
SA Juniata Formation
SA Kanawha Formation
SA Keefer Sandstone
SA Keyser Limestone
SA Kittanning Formation
SA Lockport Formation
SA Marcellus Shale
SA Martinsburg Formation
SA Mauch Chunk Formation
SA McKenzie Formation
SA Millboro Shale
SA Monongahela Group
SA New River
SA New River Formation
SA Ohio River
SA Ohio River basin
SA Ohio River valley
SA Ohio Shale
SA Onondaga Limestone
SA Oriskany Sandstone
SA Pittsburgh Coal
SA Pocahontas Formation
SA Pocono Formation
SA Potomac River
SA Potomac River basin
SA Pottsville Group
SA Reedsville Formation
SA Rochester Formation
SA Rome Trough
SA Sharon Conglomerate
SA Shenandoah Valley
SA Stones River Group
SA Tioga Bentonite
SA Tonoloway Limestone
SA Tuscarora Formation
SA Valley and Ridge Province
SA Williamsport Sandstone

West Virginia Panhandle (1993)
N West Virginia, E of Ohio River. Also NE West Virginia, the 'Eastern Panhandle' between Maryland and Virginia. Before 1993, also search Panhandle AND West Virginia.
IN Index counties as applicable.
UF Eastern Panhandle, West Virginia
BT West Virginia
BT United States

West-Central Alaska (1993)
Artificial region based on U.S. Geological Survey quadrangle designations. Includes Saint Lawrence Island. Also search quadrangles if applicable.
IN Index quadrangles as applicable.
CO N620000N670000
    W1530000W1740000
BT Alaska
BT United States
NT Nixon Fork Terrane
NT Nome Alaska
NT Seward Peninsula
SA Yukon-Koyukuk Basin

West-Central Nigeria
No longer a valid term for GeoRef. Valid 1981-1992.

Westchester County
Valid through 1988. Search in combination with state term. After 1988, use specific county-state term.

Westchester County New York (1989)

SE New York. Before 1989, also search Westchester County AND New York.
CO N405200N412300
    W0732900W0735800
BT New York
BT United States
NT Peekskill New York
SA Palisades Sill

Westerly Granite (1978)
Pennsylvanian or younger. Finely crystalline gray rock, which shows minor variations in color and texture but which is petrographically the same. SE Connecticut and SW Rhode Island.
BT Pennsylvanian
BT Carboniferous
BT Paleozoic
SA Connecticut
SA Rhode Island

Western Alps (1978)
Ranges of the Alps in SE France, NW Italy, and SW Switzerland.
IN Index countries as applicable.
BT Alps
BT Europe
NT Cottian Alps
NT Dauphine Alps
NT Graian Alps
NT Ligurian Alps
NT Maritime Alps
NT Provence Alps
NT Sainte-Baume Massif
NT Savoy Alps
SA France
SA Italy
SA Switzerland

Western Australia (1978)
CO S350000S140000
    E1290000E1130000
BT Australia
BT Australasia
NT Canning Basin
NT Carnarvon Basin
NT Cue Australia
NT Darling Range
NT Hamersley Basin
NT Hamersley Range
NT Kalgoorlie Australia
NT Kambalda Australia
NT Mundrabilla Australia
NT Norseman Australia
NT Norseman-Wiluna Belt
NT North Pole Deposit
NT Pilbara Block
NT Pilbara gold field
NT Western Gneiss Terrain
NT Widgiemooltha Australia
NT Wiluna Australia
NT Wittenoom Gorge
NT Yilgarn
NT Yilgarn Block
SA Amadeus Basin
SA Barrow Island
SA Brockman Iron Formation
SA Browse Basin
SA Dalgaranga
SA Eastern Goldfields
SA Eucla Basin
SA Exmouth Plateau
SA Fortescue Group
SA Fraser Range
SA Giles Complex
SA Hamersley Group
SA Huntly
SA Joseph Bonaparte Gulf
SA Kalgoorlie System
SA Nullarbor Plain
SA Officer Basin
SA Perth Australia
SA Red Hill
SA Rocknest Formation

SA Shark Bay
SA Stockton Formation
SA Warburton Meteorite
SA Warrawoona Group
SA Werillup Formation
SA Wolf Creek Meteorite

Western Canada (1981)
As of 1993, Northwest Territories is a narrower term.
CO N480000N694000
 W0890000W1410000
BT Canada
NT Alberta
NT Assiniboine River
NT Assiniboine River valley
NT British Columbia
NT Canadian Cordillera
NT Canadian Rocky Mountains
NT Kicking Horse River valley
NT Lake Athabasca
NT Manitoba
NT Northwest Territories
NT Peace River
NT Saskatchewan
NT Saskatchewan River
NT South Saskatchewan River
NT Yukon Territory
SA Souris River basin
SA Western Canada Basin

Western Canada Basin (1993)
Sedimentary basin extending from W-central Canada into N-central United States. Petroleum and coal resources.
IN Index provinces and/or states as applicable.
BT North America
SA United States
SA Western Canada

Western Carpathians (1978)
Ranges of the Carpathians in S and SE Poland, and in Slovakia.
IN Index countries as applicable.
UF West Carpathians
BT Carpathians
BT Europe
SA Choc Nappe
SA Czechoslovakia
SA Krizna Nappe
SA Poland
SA Veporides

Western Cordillera (1981)
Western range of the Andes. Before 1981, also search Cordillera Occidental.
IN Index countries as applicable.
UF Cordillera Occidental
BT Andes
BT South America
SA Chile
SA Colombia
SA Ecuador

Western Desert (1978)
Part of the Libyan Desert in W and W central Egypt.
CO N220000N314000
 E0325000E0244000
BT Egypt
BT North Africa
BT Africa
SA Dakhla Oasis
SA Libyan Desert

Western Europe (1978)
Autoposting of this term began in 1981.
BT Europe
NT Andorra
NT Ardennes
NT Belgium
NT Cottian Alps
NT France
NT Iceland
NT Ireland
NT Luxembourg
NT Maritime Alps
NT Meuse River
NT Meuse Valley
NT Monaco
NT Netherlands
NT North Pyrenean Fault
NT Rhone River
NT Rhone Valley
NT Scandinavia
NT Scheldt River
NT United Kingdom

Western Ghat Mountains
 use Western Ghats

Western Ghats (1978)
Mountain range extending 800 miles along SW and W coast as far N as mouth of Tapti River on the Gulf of Cambay.
UF Western Ghat Mountains
BT Ghats
BT India
BT Indian Peninsula
BT Asia

Western Gneiss region (1989)
Scandinavia. Before 1989, also search Western Gneiss Terrain AND Norway.
BT Scandinavia
BT Western Europe
BT Europe
SA Western Gneiss Terrain

Western Gneiss Terrain (1989)
W Australia.
BT Western Australia
BT Australia
BT Australasia
SA Western Gneiss region

Western gravimeters (1981)
BT gravimeters

Western Hemisphere (1978)
Used when discussing many large areas too numerous to mention.
SA Eastern Hemisphere
SA Northern Hemisphere
SA Southern Hemisphere

Western Interior (1978)
Tremendous region in North America including the Great Plains, the Rocky Mountains, the Basin and Range Province of the U. S., and the interior plateaus of Canada.
IN Index countries as applicable.
BT North America
NT Western Interior Seaway
SA Canada
SA Idaho
SA Montana
SA United States
SA Utah
SA Wyoming

Western Interior Seaway (1989)
IN Index countries, states and provinces as applicable.
BT Western Interior
BT North America
SA Alberta
SA British Columbia
SA Canada
SA Canadian Cordillera
SA Idaho
SA Montana
SA North American Cordillera
SA United States
SA Utah
SA Wyoming

Western Islands
 use Hebrides

Western Morava River
 use Zapadna Morava

western Nagano earthquake 1984
 use Nagano earthquake 1984

Western Nigeria
No longer a valid term for GeoRef. Valid 1981-1992.

Western Overthrust Belt (1981)
Before 1981, search Overthrust Belt; Thrust Belt.
IN Index states and provinces as applicable.
BT North America
SA Absaroka Fault
SA Alberta
SA British Columbia
SA Colorado
SA Eastern Overthrust Belt
SA Idaho
SA Mexico
SA Montana
SA New Mexico
SA Rocky Mountains
SA Utah
SA Wyoming

Western Port (1981)
Inlet of Bass Strait in S Victoria, Australia.
IN Index state or region as applicable.
SA Bass Strait
SA Victoria Australia

Western Sahara (1981)
Presently occupied by Morocco in the north. Also search Sahara AND West; Spanish Sahara. As of 1990, North Africa is autoposted to this term.
IN Index Morocco and/or Mauritania as applicable.
CO N212000N274000
 W0084000W0171000
UF Spanish Sahara
BT North Africa
BT Africa
SA Mauritania
SA Morocco
SA Reguibat Ridge

Western Samoa
 use Samoa

Western Sayan (1978)
Before 1978, search Sayan AND west or western.
IN Index countries as applicable.
CO N500000N540000
 E0950000E0860000
BT Sayan
BT Siberian fold belt
BT Asia
SA Altai-Sayan region
SA Russian Federation

western Texas
 use West Texas

Western Transbaikalia (1981)
Autoposting to Russian Republic to this term began in 1989. This term has multiple hierarchies.
CO N490000N560000
 E1210000E1040000
BT1 Siberian fold belt
BT1 Asia
BT2 Russian Federation
BT2 Commonwealth of Independent States
SA Transbaikalia

Western U.S. (1978)
As of 1993, term is used only for broad general discussions. Before 1978, also search United States AND (west OR western). From 1981 through 1992, this term was autoposted to Alaska, Hawaii, and Nevada. As of 1981, this term is autoposted to Pacific Coast. From 1981 through 1992, term was autoposted to five more states: Colorado, Wyoming, Montana, Idaho, and Utah. As of 1993, for complete search, also search states.
BT United States
NT Pacific Coast
SA Battle Mountain High
SA Sauk Sequence

Western World (1981)
Used for the non-communist "industrialized world".
SA communist countries
SA industrialized countries

Westerwald (1978)
Mountainous region extending NE from near Koblenz for about 70 miles between the Rhine River, Sieg River, and Lahn River. Geologically it is considered part of the Rhenish Schiefergebirge (Rhenish Slate Mountains).
IN Index states as applicable.
BT Germany
BT Central Europe
BT Europe
SA Bundenbach
SA Hesse Germany
SA Rhenish Schiefergebirge
SA Rhineland-Palatinate Germany
SA Siebengebirge

Westland
No longer a valid term for GeoRef. See Westland New Zealand.

Westland New Zealand (1993)
Provincial district along the Tasman Sea in W South Island, New Zealand.
IN Also index South Island.
CO S442000S420000
 E1723000E1680000
BT New Zealand
BT Australasia
SA South Island

Westman Islands
 use Vestmannaeyjar

Westmoreland County
Valid through 1988. Search in combination with state term. After 1988, use specific county-state term.

Westmoreland County Pennsylvania (1989)
SW Pennsylvania. Before 1989, also search Westmoreland County AND Pennsylvania.
CO N400200N404100
 W0785800W0795400
BT Pennsylvania
BT United States

Westmoreland County Virginia (1989)
E Virginia. Before 1989, also search Westmoreland County AND Virginia.
CO N375800N381700
 W0763200W0770400
BT Virginia
BT United States

Westmorland
No longer a valid term for GeoRef as of 1993. See Westmorland England.

Westmorland England (1993)

Former county in NW England. In 1974, became part of Cumbria.
CO N541500N544500 W0021000W0031000
BT Cumbria England
BT England
BT Great Britain
BT United Kingdom
BT Western Europe
BT Europe
SA Lake District
SA Morecambe Bay

Weston County
Valid through 1988. Search in combination with state term. After 1988, use specific county-state term.

Weston County Wyoming (1989)
NE Wyoming. Before 1989, also search Weston County AND Wyoming.
CO N433000N441000 W1040500W1050500
BT Wyoming
BT United States

Weston Meteorite (1985)
Impact at Weston in Fairfield County, Connecticut. Before 1985, also search Weston AND meteorites.
BT chondrites
BT stony meteorites
BT meteorites
SA Connecticut
SA Fairfield County Connecticut

Westphalia (1978)
Region and former Prussian province now in parts of 3 German states.
IN Index states as applicable.
BT Germany
BT Central Europe
BT Europe
SA Hesse Germany
SA Lower Saxony Germany
SA North Rhine-Westphalia Germany

Westphalian (1978)
Europe. Above Namurian, below Stephanian.
BT Upper Carboniferous
BT Carboniferous
BT Paleozoic
SA Essen Beds

Westwater Canyon Sandstone Member (1985)
Of Morrison Formation. SE Utah, NE Arizona, SW Colorado and NW New Mexico. Also search Westwater Canyon Member; Westwater Canyon Sandstone.
BT Upper Jurassic
BT Jurassic
BT Mesozoic
SA Arizona
SA Colorado
SA Morrison Formation
SA New Mexico
SA Utah

wet
A valid term through 1977. After 1977, use wet methods.

wet methods (1978)
Before 1978, also search chemical analysis AND wet.
SA analysis
SA chemical analysis
SA electrolytic analysis
SA methods
SA titration

Wet Mountains (1978)
Range of Rocky Mountains in S central Colorado. This term has multiple hierarchies.
IN Index counties as applicable.
BT1 Colorado
BT1 United States
BT2 U. S. Rocky Mountains
BT2 Rocky Mountains
BT3 U. S. Rocky Mountains
BT3 United States
SA Custer County Colorado
SA Fremont County Colorado
SA Huerfano County Colorado
SA Pueblo County Colorado

wetlands (1978)
SA bogs
SA conservation
SA ecology
SA fluvial features
SA geomorphology
SA grasslands
SA lacustrine features
SA marshes
SA shore features
SA Spartina alterniflora
SA swamps
SA vegetation
SA wilderness areas

wettability (1993)
SA absorption
SA adsorption
SA moisture
SA permeability
SA petroleum engineering
SA porosity
SA properties
SA water

Wetterau (1978)
Region NNE of Frankfurt much of it along the Wetter River.
BT Hesse Germany
BT Germany
BT Central Europe
BT Europe

Wetterstein Limestone (1978)
According to the latest published Austrian Lexique, the age of Wetterstein Dolomite is Triassic. In Tyrol in W Austria.
BT Triassic
BT Mesozoic
SA Austria
SA Tyrol Austria

Wewoka Formation (1978)
In Marmaton Group. Conformably overlies Wetunka Formation and conformably underlies Holdenville Shale.
BT Desmoinesian
BT Middle Pennsylvanian
BT Pennsylvanian
BT Carboniferous
BT Paleozoic
SA Marmaton Group
SA Oklahoma

Wexford
No longer a valid term for GeoRef. See Wexford Ireland.

Wexford Ireland (1993)
County in SE Ireland. Also a city on Wexford Harbour off Saint George's Channel.
CO N521000N524500 W0061000W0070000
BT Ireland
BT Western Europe
BT Europe

Weyer
No longer a valid term for GeoRef. See Weyer Austria.

Weyer Austria (1993)
Village on the Enns River in SE Upper Austria, NW Austria.
BT Upper Austria
BT Austria
BT Central Europe
BT Europe

Wharton Basin (1978)
SSW of the Java Trench in E Indian Ocean.
BT Indian Ocean
SA DSDP Site 212
SA DSDP Site 256
SA DSDP Site 257

Whatcom County
Valid through 1988. Search in combination with state term. After 1988, use specific county-state term.

Whatcom County Washington (1989)
On Puget Sound in NW Washington. Before 1989, also search Whatcom County AND Washington.
CO N483700N490000 W1204000W1230300
BT Washington
BT United States
SA Mount Baker

wheel ore
use bournonite

Wheeler County
Valid through 1988. Search in combination with state term. After 1988, use specific county-state term.

Wheeler County Georgia (1989)
SE central Georgia. Before 1989, also search Wheeler County AND Georgia.
CO N315500N321900 W0823200W0825600
BT Georgia
BT United States

Wheeler County Nebraska (1989)
NE central Nebraska. Before 1989, also search Wheeler County AND Nebraska.
CO N414300N420600 W0981800W0984500
BT Nebraska
BT United States

Wheeler County Oregon (1989)
N central Oregon. Before 1989, also search Wheeler County AND Oregon.
CO N441800N450400 W1194100W1203000
BT Oregon
BT United States
NT Clarno Oregon

Wheeler County Texas (1989)
Extreme N Texas. Before 1989, also search Wheeler County AND Texas.
CO N351300N353800 W1000000W1003000
BT Texas
BT United States

Wheeler Formation (1989)
W Utah.
BT Middle Cambrian
BT Cambrian
BT Paleozoic
SA Utah

Wheeling
Valid through 1988. Search in combination with state term. After 1988, use specific city-state term.

Wheeling West Virginia (1989)
City on the Ohio River in the N panhandle in N West Virginia. Before 1989, search Wheeling AND West Virginia.
CO N400500N400500 W0804300W0804300
BT Ohio County West Virginia
BT West Virginia
BT United States

Whetstone Lake (1978)
BT Ontario
BT Eastern Canada
BT Canada

whewellite (1978)
BT carbonates
SA weddellite

Whipple Mountains (1985)
S California. Autoposting of broader terms to this term began in 1989.
BT San Bernardino County California
BT California
BT United States
SA Basin and Range Province

Whirlpool Sandstone (1985)
In Albion Group. Ontario and W New York.
BT Lower Silurian
BT Silurian
BT Paleozoic
SA New York
SA Ontario

whistlers
As of 1993, no longer a valid term for GeoRef. Usually out-of-scope. Used with magnetosphere.

white antimony
use valentinite

White Bay (1978)
Large inlet of the Atlantic Ocean in N Newfoundland.
IN Index province or region as applicable.
SA Newfoundland
SA Newfoundland Island

White County
Valid through 1988. Search in combination with state term. After 1988, use specific county-state term.

White County Arkansas (1989)
Central Arkansas. Before 1989, also search White County AND Arkansas.
CO N350200N353200 W0911800W0920700
BT Arkansas
BT United States

White County Georgia (1989)
NE Georgia. Before 1989, also search White County AND Georgia.
CO N343000N344900 W0833800W0835300
BT Georgia
BT United States

White County Illinois (1989)
SE Illinois. Before 1989, also search White County AND Illinois.
CO N375300N381600 W0875400W0882300
BT Illinois
BT United States

White County Indiana (1989)
NW central Indiana. Before 1989, also search White County AND Indiana.
CO N403400N405600 W0863500W0870700
BT Indiana
BT United States

White County Tennessee (1989)
Central Tennessee. Before 1989, also search White County AND Tennessee.
CO N354600N360600 W0851200W0854100
BT Tennessee
BT United States

White Elster Basin
use Weisse Elster Basin

White Island (1978)
In Bay of Plenty off N central North Island. An active volcano is located on the island.
IN Index North Island and New Zealand or region as applicable.
SA New Zealand
SA North Island

White Limestone (1978)
Name has either been abandoned by author or rejected for use by the U.S. Geological Survey. It was a color term applied in a titular sense to 160 feet of light-gray to dark-gray dolomitic limestone with shaly layers. Leadville mining district, Colorado.
BT Lower Ordovician
BT Ordovician
BT Paleozoic
SA Colorado

white mica
use muscovite

White Mountain (1978)
Peak in the Sierra Nevada Mountains in E central California, a hill W of Prague. As of 1993, use White Mountain Alaska for the village on S Seward Peninsula in W Alaska.
IN Index Bohemia and states or regions as applicable.
SA Alaska
SA Bohemia
SA California
SA Sierra Nevada

White Mountains (1978)
Range of the Appalachians in N central New Hampshire, range in E California and SW Nevada, and a range in Fort Apache Indian Reservation in E Arizona. Also search White-Inyo Mountains or search White-Inyo Range for California mountains.
IN Index states, counties, and mountain ranges as applicable.
SA Appalachians
SA Arizona
SA California
SA Mount Adams
SA Mount Jefferson
SA Nevada
SA New Hampshire

White Pine
Valid through 1988. Search in combination with state term. After 1988, use specific city-state term.

White Pine County
Valid through 1988. Search in combination with state term. After 1988, use specific county-state term.

White Pine County Nevada (1989)
E Nevada. Before 1989, also search White Pine County AND Nevada.
CO N384000N401000 W1140500W1155500
BT Nevada
BT United States
NT Alligator Ridge Mine
NT Ely Nevada
NT Snake Range
SA Egan Range
SA Long Valley
SA Ruby Mountains

White Pine Michigan (1989)
Village on the Iron River in W Michigan Upper Peninsula, NW Michigan. Before 1989, search White Pine AND Michigan.
BT Ontonagon County Michigan
BT Michigan Upper Peninsula
BT Michigan
BT United States

White Pine Mine (1978)
Copper mine in the Porcupine Mountains near Lake Superior in W Michigan Upper Peninsula. Autoposting of broader terms to this term began in 1989. Also search White Pine AND Michigan.
BT Ontonagon County Michigan
BT Michigan Upper Peninsula
BT Michigan
BT United States
SA copper ores
SA mines

white pyrites
use arsenopyrite

White River (1978)
River in Alaska and SW Yukon Territory; a river in Arkansas and S Missouri; and a river in SW Indiana.
IN Index Yukon Territory and states as applicable.
SA Alaska
SA Arkansas
SA Indiana
SA Missouri
SA North America
SA Yukon Territory

White River Group (1978)
In Wyoming, includes Chadron, Brule, and Arikaree formations. NE Colorado, E Montana, Nebraska, North Dakota, South Dakota, Wyoming.
BT Oligocene
BT Paleogene
BT Tertiary
BT Cenozoic
SA Brule Formation
SA Chadron Formation
SA Colorado
SA Montana
SA Nebraska
SA North Dakota
SA South Dakota
SA Wyoming

White River Plateau (1978)
NW central Colorado.
IN Index counties as applicable.
UF White River Uplift
SA Colorado

White River Uplift
use White River Plateau

White Russia
use Belarus

White Sands (1978)
National Monument constituting a great expanse of white gypsum sand and dunes in Tularosa Basin in S New Mexico. Also a proving ground for testing rockets and guided missiles SW of White Sands National Monument. Also in other locations.
IN Index counties or regions as applicable.
SA Dona Ana County New Mexico
SA national monuments
SA New Mexico
SA Otero County New Mexico

White Sea (1978)
Large inlet of the Barents Sea S of the Kola Peninsula in NW European Russian Federation. In 1981, broader term changed from USSR to Arctic Ocean.
CO N640000N660000 E0440000E0350000
BT Arctic Ocean
SA Russian Federation

White Wolf Fault (1989)
S California.
BT Kern County California
BT California
BT United States
SA San Joaquin Valley

White-Inyo Mountains
use Inyo Mountains

White-Inyo Range
use Inyo Mountains

Whitehorse
No longer a valid term for GeoRef. As of 1993, see Whitehorse Yukon Territory.

Whitehorse Trough (1985)
NW British Columbia and SW Yukon Territory, Canada.
IN Index British Columbia or Yukon Territory as applicable.
BT Canada
SA British Columbia
SA Canadian Rocky Mountains
SA Yukon Territory

Whitehorse Yukon Territory (1993)
City on the Yukon River in S Yukon Territory.
BT Yukon Territory
BT Western Canada
BT Canada

Whitemud Formation (1978)
Clay for china, ball, fire, and stoneware products are obtained from this formation. S and SW Saskatchewan.
BT Cretaceous
BT Mesozoic
SA Saskatchewan

Whitewater River valley (1978)
Primarily E and SE Indiana, but also in extreme SW Ohio. Also search Whitewater River. Also in other location.
IN Index counties or states as applicable.
SA Indiana
SA Ohio
SA United States

Whitewood Creek (1993)
Belle Fourche River tributary of Cheyenne River, W South Dakota.
IN Index counties or regions as applicable.
SA Butte County South Dakota
SA Lawrence County South Dakota
SA Meade County South Dakota
SA South Dakota

Whitley County
Valid through 1988. Search in combination with state term. After 1988, use specific county-state term.

Whitley County Indiana (1989)
NE Indiana. Before 1989, also search Whitley County AND Indiana.
CO N410000N411800 W0851900W0854200
BT Indiana
BT United States

Whitley County Kentucky (1989)
SE Kentucky. Before 1989, also search Whitley County AND Kentucky.
CO N363400N365700 W0835100W0842000
BT Kentucky
BT United States

whitlockite (1978)
Before 1993, if applicable, also search merrillite which was used for this mineral in meteorites.
UF merrillite
BT phosphates

Whitman County
Valid through 1988. Search in combination with state term. After 1988, use specific county-state term.

Whitman County Washington (1989)
SE Washington. Before 1989, also search Whitman County AND Washington.
CO N462500N472000 W1170500W1181500
BT Washington
BT United States

Whitsett Formation (1985)
In Jackson Group. S central Texas.
BT upper Eocene
BT Eocene
BT Paleogene
BT Tertiary
BT Cenozoic
SA Jackson Group
SA Texas

Whittier Fault (1978)
E of Los Angeles in S California.
IN Index counties or regions as applicable.
SA California

Whittier Narrows earthquake 1987 (1993)
S. California, Los Angeles region.
BT earthquakes
SA California
SA Los Angeles Basin

whole rock (1978)
UF total rock
UF whole-rock
SA absolute age

whole-rock
use whole rock

wiborgite
use rapakivi

Wichita
Valid through 1988. Search in combination with state term. After 1988, use specific city-state term.

Wichita County
Valid through 1988. Search in combination with state term. After 1988, use specific county-state term.

Wichita County Kansas (1989)
W Kansas. Before 1989, also search Wichita County AND Kansas.
CO N381600N384200 W1010800W1013500
BT Kansas
BT United States

Wichita County Texas (1989)
N Texas. Before 1989, also search Wichita County AND Texas.
CO N335100N341400 W0982700W0985900
BT Texas
BT United States

Wichita Group (1985)
N and central Texas, and W Oklahoma. Also search Wichita Formation.
BT Lower Permian
BT Permian
BT Paleozoic
SA Oklahoma
SA Texas

Wichita Kansas (1989)
City on the Arkansas River in S Kansas. Before 1989, search Wichita AND Kansas.
CO N374300N374300 W0972000W0972000
BT Sedgwick County Kansas
BT Kansas
BT United States

Wichita Mountains (1978)
Range in SW Oklahoma.
IN Index counties as applicable.
CO N343000N350000 W0983000W0990000
BT Oklahoma
BT United States
SA Comanche County Oklahoma
SA Kiowa County Oklahoma

Wicklow
No longer a valid term for GeoRef. See Wicklow Mountains for the mountain range extending along E coast. See Wicklow Ireland for the county.

Wicklow Ireland (1993)
County in E Ireland. Also a city on the Irish Sea.
CO N524000N531500 W0060000W0064500
BT Ireland
BT Western Europe
BT Europe

Wicklow Mountains (1993)
Mountain range extending along E coast. Also search Wicklow.
IN Index counties as applicable.
BT Ireland
BT Western Europe
BT Europe

Wicomico Formation (1978)
In Columbia Group. In Prince Georges County, Maryland, consists of coarse gravel bed at base and finer sand and silt above; color of silt ranges from yellow to drab to dirty white. Atlantic Coastal Plain from Delaware to Florida.
BT Pleistocene
BT Quaternary

BT Cenozoic
SA Delaware
SA District of Columbia
SA Florida
SA Georgia
SA Maryland
SA North Carolina
SA South Carolina
SA Virginia

widan
 use wadis

Widawka Basin (1978)
River basin in central Poland.
UF Widawka River basin
BT Poland
BT Central Europe
BT Europe
SA Lodz Poland

Widawka River basin
 use Widawka Basin

Widgiemoolatha
No longer a valid term for GeoRef. See Widgiemoolatha Australia.

Widgiemoolatha Australia (1993)
Village in S central Western Australia.
BT Western Australia
BT Australia
BT Australasia

Wiehen Mountains (1978)
Low range of the Weser Mountains in NE North Rhine-Westphalia, W central Germany.
BT North Rhine-Westphalia Germany
BT Germany
BT Central Europe
BT Europe

Wieliczka
No longer a valid term for GeoRef. As of 1993 see Wieliczka Poland.

Wieliczka Poland (1993)
Town in central Cracow, S Poland. Before 1993 also search Wieliczka AND Poland.
BT Cracow Poland
BT Poland
BT Central Europe
BT Europe

Wielun
No longer a valid term for GeoRef. As of 1993 see Wielun Poland.

Wielun Poland (1993)
Town in S Sieradz, central Poland. Before 1993 also search Wielun AND Poland.
BT Poland
BT Central Europe
BT Europe

Wien Austria
 use Vienna Austria

wiener filters (1981)
SA filters
SA Wiener-Hopf analysis

Wiener Wald (1978)
A spur of the Eastern Alps W and NW of Vienna in Lower Austria and S of the Danube River. This term has multiple hierarchies.
CO N480000N482000 E0161500E0154000
UF Vienerwald
BT1 Eastern Alps
BT1 Alps
BT1 Europe
BT2 Lower Austria
BT2 Austria
BT2 Central Europe

BT2 Europe

Wiener-Hopf analysis (1981)
SA analysis
SA seismology
SA wiener filters

Wiesbaden
No longer a valid term for GeoRef. See Wiesbaden Germany.

Wiesbaden Germany (1993)
City on the Rhine River 20 miles W of Frankfurt in SW Hesse, W central Germany.
BT Hesse Germany
BT Germany
BT Central Europe
BT Europe

Wiggins Arch (1989)
IN Index counties or states as applicable.
BT United States
SA Alabama
SA Gulf Coastal Plain
SA Gulf of Mexico
SA Mississippi

Wiggins Formation (1985)
NW Wyoming.
BT Oligocene
BT Paleogene
BT Tertiary
BT Cenozoic
SA Wyoming

Wilberns Formation (1978)
Includes four members: Welge Sandstone, Morgan Creek Limestone, Point Peak Shale, and San Saba Limestone. In central Texas.
BT Upper Cambrian
BT Cambrian
BT Paleozoic
SA Texas

Wilcox Formation (1978)
Predominantly green, white, and black schist enclosing thin dolomite beds near base of schist. Pegmatitic quartzose gneiss occurs near middle of the formation. Above the gneiss, dark schistose grits contrast with strictly argillaceous types below. In W central Vermont.
BT Precambrian
SA Vermont

Wilcox Group (1978)
In Louisiana, comprises (ascending) Converse, Lime Hill, "Hall Summit", Marthaville, Pendleton, Sabinetown, and Carrizo formations. Gulf Coastal Plain from Georgia to S Texas plus SW Illinois, W Kentucky, SE Missouri, and W Tennessee.
BT lower Eocene
BT Eocene
BT Paleogene
BT Tertiary
BT Cenozoic
SA Calvert Bluff Formation
SA Florida
SA Georgia
SA Hatchetigbee Formation
SA Illinois
SA Kentucky
SA Louisiana
SA Mississippi
SA Missouri
SA Nanafalia Formation
SA Tennessee
SA Texas

Wildcat Group (1989)

Upper Miocene to lower Pleistocene. N California and Nevada.
IN Index ages as applicable.
BT Cenozoic
SA California
SA lower Pleistocene
SA Nevada
SA Pliocene
SA upper Miocene

Wildenfels
No longer a valid term for GeoRef. See Wildenfels Germany.

Wildenfels Germany (1993)
Town at the foot of the Erzgebirge in Saxony.
BT Saxony Germany
BT Germany
BT Central Europe
BT Europe

wilderness areas (1985)
Also search names of specific areas. Term is to be used in connection with natural resource assessment.
SA conservation
SA ecology
SA environmental geology
SA fires
SA forests
SA Golden Trout Wilderness
SA habitat
SA land use
SA national monuments
SA national parks
SA natural resources
SA policy
SA RARE II regions
SA Selway-Bitterroot Wilderness
SA wetlands

wildflysch (1978)
SA clastic rocks
SA clastic sediments
SA flysch
SA sedimentary rocks
SA sediments

Wilhelm-Philippe Schimper
 use Schimper, Wilhelm-Philippe

Wilhelmina Mountains (1978)
Range in central Surinam.
BT Surinam
BT South America

Wilkes Land (1978)
Coastal region from the Shackleton Ice Shelf to the George V Coast on the Indian Ocean side.
BT Antarctica

Wilkins Peak Member (1985)
Of Green River Formation. Lower and middle Eocene. SW Wyoming.
BT Eocene
BT Paleogene
BT Tertiary
BT Cenozoic
SA Green River Formation
SA Wyoming

Wilkinson Basin (1978)
Off the coast of Massachusetts.
BT Atlantic Ocean

Wilkinson County
Valid through 1988. Search in combination with state term. After 1988, use specific county-state term.

Wilkinson County Georgia (1989)

Central Georgia. Before 1989, also search Wilkinson County AND Georgia.
CO N323500N330000 W0825500W0832300
BT Georgia
BT United States

Wilkinson County Mississippi (1989)
Extreme SW Mississippi. Before 1989, also search Wilkinson County AND Mississippi.
CO N310000N312300 W0910500W0913800
BT Mississippi
BT United States

Will County
Valid through 1988. Search in combination with state term. After 1988, use specific county-state term.

Will County Illinois (1989)
NE Illinois. Before 1989, also search Will County AND Illinois.
CO N411000N414500 W0873000W0881700
BT Illinois
BT United States

Willamette River (1985)
Tributary of the Columbia River NW Oregon.
IN Index counties as applicable.
BT Oregon
BT United States
SA Columbia River
SA Willamette Valley

Willamette River valley
use Willamette Valley

Willamette Valley (1978)
River valley in NW Oregon.
IN Index counties as applicable.
UF Willamette River valley
BT Oregon
BT United States
SA Willamette River

Willapa Bay (1985)
Inlet of the Pacific Ocean in SW Washington. Autoposting of broader terms to this term began in 1989.
CO N461500N464500 W1234500W1241000
BT Pacific County Washington
BT Washington
BT United States
SA North American Pacific

willemite (1978)
Autoposting of nesosilicates began in 1985.
BT nesosilicates
BT orthosilicates
BT silicates

Williams Fork Formation (1985)
In Mesaverde Group. NW Colorado and NE Utah.
BT Upper Cretaceous
BT Cretaceous
BT Mesozoic
SA Colorado
SA Mesaverde Group
SA Utah

Williams Formation (1989)
S California. Includes Pleasants Sandstone Member.
BT Upper Cretaceous
BT Cretaceous
BT Mesozoic
SA California

Williamson County
Valid through 1988. Search in combination with state term. After 1988, use specific county-state term.

Williamson County Illinois (1989)
S Illinois. Before 1989, also search Williamson County AND Illinois.
CO N373500N375100 W0884200W0890800
BT Illinois
BT United States

Williamson County Tennessee (1989)
Central Tennessee. Before 1989, also search Williamson County AND Tennessee.
CO N354200N360300 W0863700W0871200
BT Tennessee
BT United States

Williamson County Texas (1989)
Central Texas. Before 1989, also search Williamson County AND Texas.
CO N302500N305300 W0970900W0980200
BT Texas
BT United States
SA Edwards Aquifer

Williamsport Sandstone (1978)
In Maryland, includes Cedar Hill Limestone Member (reallocated). Abruptly overlies calcareous shales of Mackenzie Formation. W Maryland, W Virginia, and N West Virginia.
BT Cayugan
BT Upper Silurian
BT Silurian
BT Paleozoic
SA Maryland
SA Virginia
SA West Virginia

Williston Basin (1978)
IN Index states and provinces as applicable.
BT North America
SA Birdbear Formation
SA Cedar Creek Anticline
SA Manitoba
SA Montana
SA North Dakota
SA Saskatchewan
SA South Dakota

Willow Creek (1978)
Rises in Blue Mountains in NE Oregon and flows NW into Columbia River.
IN Index counties or regions as applicable.
SA Oregon

Willunga
No longer a valid term for GeoRef. See Willunga Australia.

Willunga Australia (1993)
Village S of Adelaide in SE South Australia.
BT South Australia
BT Australia
BT Australasia

Willwood Formation (1978)
Variegated shales and hornblende-bearing sandstones conformably overlying Polecat Bench Formation near center of Bighorn Basin. In N Wyoming.
BT lower Eocene
BT Eocene

BT Paleogene
BT Tertiary
BT Cenozoic
SA Wyoming

Willyama Complex (1978)
Great igneous and metamorphic complex comprised of slaty and schistose rocks, gneisses, amphibolites, and granites. In extreme W New South Wales, W Queensland and E South Australia.
BT lower Proterozoic
BT Proterozoic
BT upper Precambrian
BT Precambrian
SA Australia
SA New South Wales Australia
SA Queensland Australia
SA South Australia

willyamite (1978)
BT sulfides
SA ullmannite

Wilmington
Valid through 1988. Search in combination with state term. After 1988, use specific city-state term.

Wilmington California (1989)
City on Los Angeles Harbor in S California. Before 1989, search Wilmington AND California.
CO N334700N334700 W1181600W1181600
BT Los Angeles County California
BT California
BT United States

Wilmington Canyon (1978)
Off the coast of Delaware. Also search Wilmington.
UF Wilmington submarine valley
BT Atlantic Ocean
SA submarine canyons

Wilmington Complex (1989)
BT lower Paleozoic
BT Paleozoic
SA Delaware
SA Maryland
SA Pennsylvania

Wilmington Delaware (1989)
City on the Delaware River in N Delaware. Before 1989, search Wilmington AND Delaware.
CO N394600N394600 W0753100W0753100
BT New Castle County Delaware
BT Delaware
BT United States

Wilmington Field (1978)
Giant field in the Wilmington area around Los Angeles Harbor in the Los Angeles Basin. Also search Wilmington. Before 1989, also search Wilmington oil field.
IN Index county as applicable.
UF Wilmington oil field
SA giant fields
SA Los Angeles Basin
SA Los Angeles County California
SA oil and gas fields

Wilmington North Carolina (1989)
City on Cape Fear River in SE North Carolina. Before 1989, search Wilmington AND North Carolina.
CO N341400N341400 W0775500W0775500
BT New Hanover County North Carolina
BT North Carolina
BT United States

Wilmington oil field
As of 1989, no longer a valid term for GeoRef.
use Wilmington Field

Wilmington submarine valley
use Wilmington Canyon

Wilna Lithuania
use Vilna Lithuania

Wilno Lithuania
use Vilna Lithuania

Wilpena Group (1989)
BT upper Proterozoic
BT Proterozoic
BT upper Precambrian
BT Precambrian
SA Australia
SA South Australia

Wilson County
Valid through 1988. Search in combination with state term. After 1988, use specific county-state term.

Wilson County Kansas (1989)
SE Kansas. Before 1989, also search Wilson County AND Kansas.
CO N372200N374500 W0953000W0955700
BT Kansas
BT United States

Wilson County North Carolina (1989)
E central North Carolina. Before 1989, also search Wilson County AND North Carolina.
CO N353500N355200 W0774000W0781200
BT North Carolina
BT United States

Wilson County Tennessee (1989)
N central Tennessee. Before 1989, also search Wilson County AND Tennessee.
CO N355700N362000 W0860200W0863500
BT Tennessee
BT United States

Wilson County Texas (1989)
S Texas. Before 1989, also search Wilson County AND Texas.
CO N285500N292400 W0974400W0982600
BT Texas
BT United States

Wilson cycle (1989)
SA cycles
SA plate tectonics
SA sea-floor spreading

Wilson Lake (1978)
A reservoir in Jerome County, S Idaho. Also in other locations.
IN Index counties or regions as applicable.
SA Jerome County Idaho

Wilts County England
Not a valid term for GeoRef. See Wiltshire England.

Wiltshire
No longer a valid term for GeoRef as of 1993.
use Wiltshire England

Wiltshire England (1993)
County in S England. Before 1993, also search Wiltshire or Wilts County AND England.
CO N510000N514000

W0013000W0022000
UF Wiltshire
BT England
BT Great Britain
BT United Kingdom
BT Western Europe
BT Europe
NT Stonehenge

Wiluna
No longer a valid term for GeoRef. See Wiluna Australia.

Wiluna Australia (1993)
Town in a gold mining area in W central Western Australia.
BT Western Australia
BT Australia
BT Australasia
SA Norseman-Wiluna Belt

wind action
No longer a valid GeoRef index term. Before 1969, was used on level 1 in subfile E. See wind erosion; wind transport.

wind energy (1981)
SA energy
SA energy sources
SA new energy sources
SA winds

wind erosion (1978)
BT erosion
SA ablation
SA desert pavement
SA eolian features
SA erosion control
SA geomorphology
SA scour
SA soil erosion
SA soils
SA water erosion
SA wind transport
SA winds
SA yardangs

Wind River (1978)
Rises in NW Fremont County and flows SE uniting with the Popo Agie River to form the Big Horn River in W central.
BT Fremont County Wyoming
BT Wyoming
BT United States

Wind River basin (1978)
River basin along E slopes of Wind River Range in Fremont County in W central Wyoming.
BT Fremont County Wyoming
BT Wyoming
BT United States

Wind River Formation (1978)
Divides Lysite and Lost Cabin members, each of which consists of two facies. In W Wyoming.
BT lower Eocene
BT Eocene
BT Paleogene
BT Tertiary
BT Cenozoic
SA Wyoming

Wind River Range (1978)
Range of the Rocky Mountains in W central Wyoming. This term has multiple hierarchies.
CO N424500N433000
W1083000W1093000
BT1 U. S. Rocky Mountains
BT1 Rocky Mountains
BT2 U. S. Rocky Mountains
BT2 United States
BT3 Wyoming
BT3 United States

wind transport (1978)
UF eolian transport
SA ablation
SA barchans
SA eolian features
SA sediment transport
SA sedimentation
SA ventifacts
SA wind erosion
SA winds
SA yardangs

wind work
Not a valid index term after 1970. Was used on level in subfile N. Now use wind erosion or wind transport.

wind, solar
use solar wind

Windermere System (1978)
Unconformably overlays the Purcell System in W and N Purcell Mountains of SE British Columbia. Also may occur as Windermere Group or Windermere Supergroup in NW United States. Also search Windermere Group NOT England to exclude that Silurian formation.
BT Proterozoic
BT upper Precambrian
BT Precambrian
SA British Columbia
SA Idaho
SA Montana
SA Purcell System
SA Washington

Windham County
Valid through 1988. Search in combination with state term. After 1988, use specific county-state term.

Windham County Connecticut (1989)
NE Connecticut. Before 1989, also search Windham County AND Connecticut.
CO N413800N420300
W0714700W0721500
BT Connecticut
BT United States

Windham County Vermont (1989)
SE Vermont. Before 1989, also search Windham County AND Vermont.
CO N424300N431700
W0722800W0730000
BT Vermont
BT United States

Windhoek
No longer a valid term for GeoRef. As of 1993 see Windhoek Namibia.

Windhoek Namibia (1993)
City in central Namibia. Before 1993 also search Windhoek AND Namibia.
BT Namibia
BT Southern Africa
BT Africa

windows (1981)
Before 1981, also search fensters.
UF fensters
SA allochthons
SA autochthons
SA erosion features
SA overthrust faults
SA tectonics

winds (1978)
SA circulation
SA climate
SA eolian features
SA hurricanes
SA meteorology
SA storms
SA wind energy
SA wind erosion
SA wind transport

Windsor
No longer a valid term for GeoRef. As of 1993, see specific city-province name.

Windsor County
Valid through 1988. Search in combination with state term. After 1988, use specific county-state term.

Windsor County Vermont (1989)
E Vermont. Before 1989, also search Windsor County AND Vermont.
CO N431200N435800
W0721200W0725800
BT Vermont
BT United States

Windsor Group (1978)
Consists of thick members of massive to poorly bedded, red or red and greenish gray, mottled siltstone, shale, and sandstone intercalated with gypsum, salt, and thin tabular sequences of limestone and dolomite. New Brunswick and Nova Scotia.
IN Index provinces as applicable.
BT Mississippian
BT Carboniferous
BT Paleozoic
SA Canada
SA New Brunswick
SA Nova Scotia

Windsor Ontario (1993)
City on Detroit River across from Detroit in SE Ontario.
BT Essex County Ontario
BT Ontario
BT Eastern Canada
BT Canada
SA Detroit River

Windward Islands (1978)
The S chain of the Lesser Antilles extending from Martinique in the N to Grenada in the S. This term was autoposted to Lesser Antilles islands from 1981 to 1992.
IN Index Lesser Antilles or Society Islands as applicable.
SA Carriacou
SA Grenville Grenada
SA Lesser Antilles
SA Martinique
SA Saint Lucia
SA Soufriere

Wingate Sandstone (1989)
In Glen Canyon Group. N Arizona, W Colorado, W New Mexico and S Utah.
BT Upper Triassic
BT Triassic
BT Mesozoic
SA Arizona
SA Colorado
SA Glen Canyon Group
SA New Mexico
SA Utah

Winkler County
Valid through 1988. Search in combination with state term. After 1988, use specific county-state term.

Winkler County Texas (1989)
W Texas. Before 1989, also search Winkler County AND Texas.
CO N314000N320700
W1024500W1032000
BT Texas
BT United States
SA Central Basin Platform

Winnebago County
Valid through 1988. Search in combination with state term. After 1988, use specific county-state term.

Winnebago County Illinois (1989)
N Illinois. Before 1989, also search Winnebago County AND Illinois.
CO N421000N423000
W0885600W0892500
BT Illinois
BT United States

Winnebago County Iowa (1989)
N Iowa. Before 1989, also search Winnebago County AND Iowa.
CO N431600N433000
W0933000W0935800
BT Iowa
BT United States

Winnebago County Wisconsin (1989)
E central Wisconsin. Before 1989, also search Winnebago County AND Wisconsin.
CO N435300N441500
W0882400W0885400
BT Wisconsin
BT United States

Winnemucca
Valid through 1988. Search in combination with state term. After 1988, use specific city-state term.

Winnemucca Nevada (1989)
City on the Humboldt River in N Nevada. Before 1989, search Winnemucca AND Nevada.
CO N405800N405800
W1174500W1174500
BT Humboldt County Nevada
BT Nevada
BT United States

Winnfield salt dome (1978)
In Winnfield in Winn Parish, N central Louisiana.
IN Index parish as applicable.
BT Louisiana
BT United States

Winnipeg
No longer a valid term for GeoRef. See Winnipeg Manitoba.

Winnipeg Formation (1978)
Defined as the shale and sandstone section which underlies the limestone of Red River Formation and which rests upon Precambrian basement complex in Manitoba. Surface and subsurface in Manitoba, and subsurface in Saskatchewan; and subsurface in Montana, North Dakota, and South Dakota.
BT Middle Ordovician
BT Ordovician
BT Paleozoic
SA Manitoba
SA Montana
SA North Dakota
SA Saskatchewan
SA South Dakota

Winnipeg Manitoba (1993)

City S of Lake Winnipeg in SE Manitoba.
BT Manitoba
BT Western Canada
BT Canada

Winnipegosis Formation (1978)
In Elk Point Group. In outcrop, overlies Elm Point Formation and underlies Dawson Bay Formation. Redefined in subsurface to include Elm Point Limestone. Surface and subsurface in Manitoba, and subsurface in NE Montana, W North Dakota, and NW South Dakota.
BT Middle Devonian
BT Devonian
BT Paleozoic
SA Elk Point Group
SA Keg River Formation
SA Manitoba
SA Montana
SA North Dakota
SA South Dakota

Winona County
Valid through 1988. Search in combination with state term. After 1988, use specific county-state term.

Winona County Minnesota (1989)
SE Minnesota. Before 1989, also search Winona County AND Minnesota.
CO N435000N441200
   W0911800W0920600
BT Minnesota
BT United States

Winton Formation (1989)
SW Queensland and N South Australia.
BT Middle Cretaceous
BT Cretaceous
BT Mesozoic
SA Australia
SA Queensland Australia
SA South Australia

WIPP
use Waste Isolation Pilot Plant

WIPP site
use Waste Isolation Pilot Plant

wireline logging
use well-logging

Wisconsin (1978)
Autoposting of this term began in 1978.
CO N423000N470000
   W0864500W0925000
BT United States
NT Adams County Wisconsin
NT Ashland County Wisconsin
NT Bayfield County Wisconsin
NT Brown County Wisconsin
NT Clark County Wisconsin
NT Columbia County Wisconsin
NT Crawford County Wisconsin
NT Dane County Wisconsin
NT Dodge County Wisconsin
NT Door County Wisconsin
NT Douglas County Wisconsin
NT Dunn County Wisconsin
NT Florence County Wisconsin
NT Forest County Wisconsin
NT Grant County Wisconsin
NT Green County Wisconsin
NT Iowa County Wisconsin
NT Iron County Wisconsin
NT Jackson County Wisconsin
NT Jefferson County Wisconsin
NT Lafayette County Wisconsin
NT Lincoln County Wisconsin
NT Marathon County Wisconsin
NT Marquette County Wisconsin
NT Milwaukee County Wisconsin
NT Monroe County Wisconsin
NT Oneida County Wisconsin
NT Ozaukee County Wisconsin
NT Pierce County Wisconsin
NT Polk County Wisconsin
NT Portage County Wisconsin
NT Racine County Wisconsin
NT Richland County Wisconsin
NT Rusk County Wisconsin
NT Sauk County Wisconsin
NT Taylor County Wisconsin
NT Vilas County Wisconsin
NT Washington County Wisconsin
NT Waukesha County Wisconsin
NT Winnebago County Wisconsin
NT Wisconsin River
NT Wolf River Batholith
NT Wood County Wisconsin
SA Animikie Group
SA Baraboo Quartzite
SA Canadian Shield
SA Decorah Shale
SA Dubuque Formation
SA Eau Claire Formation
SA Franconia Formation
SA Freda Sandstone
SA Galena Dolomite
SA Galesville Sandstone
SA Glasford Formation
SA Glenwood Shale
SA Great Lakes region
SA Green Bay
SA Green Lake
SA Hemlock Formation
SA Keweenawan
SA Lake Algonquin
SA Lake Chicago
SA Lake Michigan
SA Lake Superior
SA Lake Superior region
SA Maquoketa Formation
SA Marquette Range Supergroup
SA Michigamme Formation
SA Michigan Basin
SA Midcontinent
SA Mississippi River
SA Mississippi Valley
SA Mount Simon Sandstone
SA Neda Formation
SA Niagara Escarpment
SA Nonesuch Shale
SA Oronto Group
SA Platteville Formation
SA Prairie du Chien Group
SA Racine Dolomite
SA Root River
SA Saint Peter Sandstone
SA Shakopee Formation
SA Southern Province
SA Thomson Formation
SA Tyler Formation
SA University of Wisconsin
SA Upper Mississippi Valley
SA Wonewoc Formation

Wisconsin Range (1978)
N of the Horlick Mountains and S of Marie Byrd Land.
BT Antarctica

Wisconsin River (1989)
Flows from Vilas County, N Wisconsin to Mississippi River near Prairie du Chien, SW Wisconsin.
IN Index counties as applicable.
BT Wisconsin
BT United States

Wisconsinan (1978)
Stage. Autoposting of this term began in 1978.
UF Wisconsinian
BT upper Pleistocene
BT Pleistocene
BT Quaternary
BT Cenozoic
NT Lavery Till
NT upper Wisconsinan
NT Woodfordian
SA Sartanian
SA Twocreekan
SA Vistulian
SA Weichselian

Wisconsinian
use Wisconsinan

Wise County
Valid through 1988. Search in combination with state term. After 1988, use specific county-state term.

Wise County Texas (1989)
N Texas. Before 1989, also search Wise County AND Texas.
CO N330000N332700
   W0972400W0975400
BT Texas
BT United States

Wise County Virginia (1989)
SW Virginia. Before 1989, also search Wise County AND Virginia.
CO N364800N371300
   W0821800W0825400
BT Virginia
BT United States

Wisla River
use Vistula River

Wissahickon Formation (1978)
Lower Paleozoic (?). Prevalent structure is schistose. Delaware, N Maryland, SE Pennsylvania, and Virginia.
BT Paleozoic
SA Delaware
SA Glenarm Series
SA lower Paleozoic
SA Maryland
SA Pennsylvania
SA Virginia

witherite (1978)
BT carbonates
SA aragonite

Wittenoom Gorge (1978)
Near Wittenoom, a village N of the SE end of the Hamersley Range in NW Western Australia. As of 1993, use Wittenoom Australia to refer to the village.
BT Western Australia
BT Australia
BT Australasia
SA Brockman Iron Formation

wittichenite (1978)
BT sulfobismuthites
BT sulfosalts

Witwatersrand (1978)
Region on a ridge of auriferous rock about 62 miles long and 23 miles wide with Johannesburg nearly at its center in S Transvaal.
CO S270000S260000
   E0300000E0260000
UF Rand
UF The Rand
UF Witwatersrand Basin
BT Transvaal South Africa
BT South Africa
BT Southern Africa
BT Africa
SA Klerksdorp Field

Witwatersrand Basin
use Witwatersrand

Witwatersrand Supergroup (1993)
Contains detrital gold, uranium, and pyrite concentration. In Transvaal, South Africa. Also search Witwatersrand System.
UF Witwatersrand System
BT Precambrian
SA Central Rand Group
SA South Africa
SA Transvaal South Africa

Witwatersrand System
No longer a valid term for GeoRef as of 1993.
use Witwatersrand Supergroup

wodginite (1978)
Autoposting of niobotantalates began in 1989. This term has multiple hierarchies.
BT1 niobotantalates
BT1 niobates
BT1 oxides
BT2 niobotantalates
BT2 tantalates
BT2 oxides

Wolf Creek Meteorite (1981)
Impact S of Hall's Creek, in NE Western Australia. Before 1981, also search Wolf Creek AND meteorites.
BT octahedrite
BT iron meteorites
BT meteorites
SA Australia
SA Western Australia

Wolf River Batholith (1978)
IN Index counties as applicable.
BT Wisconsin
BT United States

Wolfcamp Formation
use Wolfcampian

Wolfcamp Series
use Wolfcampian

Wolfcampian (1978)
Provincial series, North America. Above Virgilian (Pennsylvanian), below Leonardian.
UF Wolfcamp Formation
UF Wolfcamp Series
BT Lower Permian
BT Permian
BT Paleozoic
NT Hueco Limestone
NT Roca Formation
SA Abo Formation
SA New Mexico
SA Texas

wolfram
Not a valid term for GeoRef. Sometimes used for wolframite. Sometimes used for the chemical element tungsten.

wolframite (1978)
Mineral. Principal ore of tungsten. Also search wolfram.
BT tungstates
SA tungsten ores

Wolfsberg
No longer a valid term for GeoRef. See Wolfsberg Austria.

Wolfsberg Austria (1993)
Town in W Carinthia, S Austria.
BT Carinthia Austria
BT Austria
BT Central Europe
BT Europe

Wolin (1978)

Island in the Baltic Sea just off the extreme NW Poland. Use Wolin Poland for the town on the SE shore of the island.
UF Wolin Island
UF Wollin
BT Szczecin Poland
BT Poland
BT Central Europe
BT Europe
SA Wolin Poland

Wolin Island
use Wolin

Wolin Poland (1993)
Town on the SE shore of Wolin Island. Use Wolin for the island. Before 1993 also search Wolin AND Poland.
BT Szczecin Poland
BT Poland
BT Central Europe
BT Europe
SA Wolin

Wolkberg Group (1978)
Composed of sedimentary rocks. In Transvaal, South Africa.
BT Proterozoic
BT upper Precambrian
BT Precambrian
SA South Africa
SA Transvaal South Africa

Wollaston Group (1989)
BT lower Proterozoic
BT Proterozoic
BT upper Precambrian
BT Precambrian
SA Saskatchewan

Wollaston Lake Belt (1978)
Fold belt in NE Saskatchewan. Also search Wollaston Lake.
BT Saskatchewan
BT Western Canada
BT Canada

wollastonite (1978)
BT chain silicates
BT silicates

Wollin
use Wolin

Wolverhampton
No longer a valid term for GeoRef as of 1993. See Wolverhampton England.

Wolverhampton England (1993)
County borough in West Midlands, NW of Birmingham in W central England.
BT England
BT Great Britain
BT United Kingdom
BT Western Europe
BT Europe
SA Staffordshire England

Wolyn Ukraine
use Volyn Ukraine

Womble Shale (1989)
Lower and Middle Ordovician. W Arkansas and S Oklahoma. Overlies Blakely Formation; underlies Bigfork Formation.
BT Ordovician
BT Paleozoic
SA Arkansas
SA Oklahoma

Wonewoc Formation (1989)
NE Iowa and SW Wisconsin. Includes Ironton Member.
BT Upper Cambrian
BT Cambrian
BT Paleozoic

SA Iowa
SA Wisconsin

Wonoka Formation (1989)
E South Australia. Ediacarian.
BT upper Proterozoic
BT Proterozoic
BT upper Precambrian
BT Precambrian
SA Australia
SA South Australia

wood (1978)
SA fossil wood
SA lignin
SA paleobotany
SA paleontology
SA roots

Wood Canyon Formation (1978)
Precambrian to Lower Cambrian. Overlies Stirling Quartzite; underlies Cadiz Formation.
IN Index ages as applicable.
SA California
SA Lower Cambrian
SA Nevada
SA Precambrian

Wood County
Valid through 1988. Search in combination with state term. After 1988, use specific county-state term.

Wood County Ohio (1989)
NW Ohio. Before 1989, also search Wood County AND Ohio.
CO N411000N413800
  W0832500W0835300
BT Ohio
BT United States
SA Ottawa River

Wood County Texas (1989)
NE Texas. Before 1989, also search Wood County AND Texas.
CO N323500N325800
  W0950700W0954000
BT Texas
BT United States

Wood County West Virginia (1989)
W West Virginia. Before 1989, also search Wood County AND West Virginia.
CO N390200N392500
  W0811500W0814600
BT West Virginia
BT United States

Wood County Wisconsin (1989)
Central Wisconsin. Before 1989, also search Wood County AND Wisconsin.
CO N441500N444200
  W0894300W0901900
BT Wisconsin
BT United States

Wood Mountain Formation (1978)
Fluvial deposits of gravels. In S Saskatchewan.
BT Miocene
BT Neogene
BT Tertiary
BT Cenozoic
SA Saskatchewan

Wood River
Valid through 1988. Search in combination with state term. After 1988, use specific city-state term.

Wood River Formation (1985)
Pennsylvanian and Permian. S central Idaho.

IN Index ages as applicable.
BT upper Paleozoic
BT Paleozoic
SA Idaho
SA Pennsylvanian
SA Permian

Wood River Illinois (1989)
City NE of Saint Louis in SW Illinois. Before 1989, search Wood River AND Illinois.
CO N385400N385400
  W0900400W0900400
BT Madison County Illinois
BT Illinois
BT United States

Woodbine Formation (1978)
In Oklahoma, lower member is principally crossbedded dark tuffaceous sand, red clay, and gravel lentils. Upper member mostly gray to brown crossbedded quartz and sandy gravel. SW Arkansas, W Louisiana, central S and SE Oklahoma, and Texas. Also search Woodbine Group.
BT Gulfian
BT Upper Cretaceous
BT Cretaceous
BT Mesozoic
SA Arkansas
SA Louisiana
SA Oklahoma
SA Texas

Woodbury
Valid through 1988. Search in combination with state term. After 1988, use specific city-state term.

Woodbury Clay (1978)
In Matawan Group. Thick black clay which weathers to dove or light chocolate color. In New Jersey.
BT Upper Cretaceous
BT Cretaceous
BT Mesozoic
SA New Jersey

Woodbury Connecticut (1989)
Town on the Pomperaug River in W Connecticut. Before 1989, search Woodbury AND Connecticut.
CO N413200N413200
  W0731200W0731200
BT Litchfield County Connecticut
BT Connecticut
BT United States

Woodbury New Jersey (1989)
City in SW New Jersey. Before 1989, search Woodbury AND New Jersey.
CO N395000N395000
  W0750900W0750900
BT Gloucester County New Jersey
BT New Jersey
BT United States

Woodbury Pennsylvania (1989)
Borough in S Pennsylvania. Before 1989, search Woodbury AND Pennsylvania.
BT Bedford County Pennsylvania
BT Pennsylvania
BT United States

Woodbury Tennessee (1989)
Town on the Stones River in central Tennessee. Before 1989, search Woodbury AND Tennessee.
CO N354900N354900
  W0860600W0860600
BT Cannon County Tennessee

BT Tennessee
BT United States

Woodford County Illinois (1989)
N central Illinois. Before 1989, also search Woodford County AND Illinois.
CO N403700N405600
  W0885300W0893500
BT Illinois
BT United States
NT Eureka Illinois

Woodford County Kentucky (1989)
Central Kentucky. Before 1989, search Woodford County AND Kentucky.
CO N375000N381100
  W0843900W0845300
BT Kentucky
BT United States
NT Versailles Kentucky

Woodford Shale (1978)
Devonian and Mississippian. Upper part consists of alternating beds of black papery shale that weather light gray, and black chert in beds 1 to 4 inches thick. Central S and SE Oklahoma.
IN Index ages as applicable.
BT Paleozoic
SA Devonian
SA Mississippian
SA Oklahoma

Woodfordian (1978)
Substage of Wisconsinan stage.
BT Wisconsinan
BT upper Pleistocene
BT Pleistocene
BT Quaternary
BT Cenozoic
SA Illinois

Woodlark Basin (1989)
SW Pacific.
BT Pacific Ocean
SA Papua New Guinea
SA Solomon Sea
SA Southwest Pacific

Woods County
Valid through 1988. Search in combination with state term. After 1988, use specific county-state term.

Woods County Oklahoma (1989)
NW Oklahoma. Before 1989, also search Woods County AND Oklahoma.
CO N362300N370000
  W0983200W0992500
BT Oklahoma
BT United States

Woods Hole
Valid through 1988. Search in combination with state term. After 1988, use specific city-state term.

Woods Hole Massachusetts (1989)
Village in Falmouth town in SE Massachusetts. Before 1989, search Woods Hole AND Massachusetts. U. S. Fish and Wildlife station located here. Also site of marine biological institute and Oceanographic Institution.
CO N413135N413135
  W0704025W0704025
BT Barnstable County Massachusetts
BT Massachusetts
BT United States

SA Woods Hole Oceanographic Institution

Woods Hole Oceanographic Institution (1978)
Located in village of Woods Hole in town of Falmouth at SW tip of Cape Cod in Barnstable County, SE Massachusetts.
SA academic institutions
SA Massachusetts
SA Woods Hole Massachusetts

Woods Hollow Shale (1989)
SW Texas. Underlies Maravillas Formation; overlies Fort Pena Formation.
BT Middle Ordovician
BT Ordovician
BT Paleozoic
SA Texas

Woodson County
Valid through 1988. Search in combination with state term. After 1988, use specific county-state term.

Woodson County Kansas (1989)
SE Kansas. Before 1989, also search Woodson County AND Kansas.
CO N374300N380500
  W0953000W0955500
BT Kansas
BT United States

Woodstock
No longer a valid term for GeoRef. As of 1993, see Woodstock New Brunswick.

Woodstock New Brunswick (1993)
Town on the Saint John River in W New Brunswick. Before 1993, also search Woodstock AND New Brunswick.
BT Carleton County New Brunswick
BT New Brunswick
BT Maritime Provinces
BT Eastern Canada
BT Canada

Woodward
Valid through 1988. Search in combination with state term. After 1988, use specific city-state term.

Woodward County Oklahoma (1989)
NW Oklahoma. Before 1989, search Woodward County AND Oklahoma.
CO N361000N364800
  W0985700W0993600
BT Oklahoma
BT United States
NT Woodward Oklahoma

Woodward Oklahoma (1989)
City on the North Canadian River in NW Oklahoma. Before 1989, search Woodward AND Oklahoma.
CO N362600N362600
  W0992500W0992500
BT Woodward County Oklahoma
BT Oklahoma
BT United States

Wopmay Orogen
use Wopmay Orogeny

Wopmay Orogeny (1985)
Lower Proterozoic. Before 1985, also search orogeny AND Wopmay.
UF Wopmay Orogen

BT Proterozoic
BT upper Precambrian
BT Precambrian
SA lower Proterozoic
SA orogeny

Worcester
Valid through 1988. Search in combination with state term. After 1988, use specific city-state term.

Worcester County
Valid through 1988. Search in combination with state term. After 1988, use specific county-state term.

Worcester County Maryland (1989)
SE Maryland. Before 1989, also search Worcester County AND Maryland.
CO N380000N382700
  W0750400W0750400
BT Maryland
BT United States

Worcester County Massachusetts (1989)
Central Massachusetts. Before 1989, also search Worcester County AND Massachusetts.
CO N420200N424300
  W0712800W0721700
BT Massachusetts
BT United States
NT Worcester Massachusetts

Worcester Massachusetts (1989)
City on the Blackstone River in central Massachusetts. Before 1989, search Worcester AND Massachusetts.
CO N421700N421700
  W0714800W0714800
BT Worcester County Massachusetts
BT Massachusetts
BT United States

Worcestershire
No longer a valid term for GeoRef as of 1993.
use Worcestershire England

Worcestershire England (1993)
Former county in W central England. In 1974, split between Hereford and Worcester, and West Midlands counties.
CO N515500N523000
  W0014500W0023000
UF Worcestershire
BT England
BT Great Britain
BT United Kingdom
BT Western Europe
BT Europe
SA Cotswold Hills
SA Lake District
SA Malvern England
SA Malvern Hills
SA Midlands

Worden gravimeters (1981)
BT gravimeters

workstations (1993)
Before 1993, also search microcomputers if applicable.
IN Index microcomputers, if other type not specified.
BT computers
SA data processing
SA microcomputers
SA minicomputers

world
use global

World Bank (1981)
SA associations
SA economics
SA international cooperation
SA International Monetary Fund
SA United Nations

world ocean (1978)
Used as a general term. Also see entries for specific oceans.
UF ocean, world
SA global
SA Law of the Sea

worldwide
No longer a valid GeoRef index term. Before 1972, was used on level 2 in subfile G.
use global

worm borings
use borings

worms (1978)
Annelida, Chaetopoda, Polychaetia, Myzostomia, Oligochaetia and Serpulidae are narrower terms as of 1981. Before 1975, also search Vermes. Autoposting of this term began in 1978.
UF Vermes
BT Invertebrata
NT Annelida
NT Chaetognatha
NT Echiurida
NT Myzostomia
NT Nematoida
NT Nematomorpha
NT Nemerta
NT Oligochaetia
NT Phoronida
NT Pogonophora
NT Polychaetia
NT Priapulida
NT scolecodonts
NT Sipunculoida
SA Arenicolites

Wrangel Island (1978)
About 100 miles off N coast of Chukchi National Okrug in the Arctic Ocean NW of Alaska. This term has multiple hierarchies.
BT1 Russian Federation
BT1 Commonwealth of Independent States
BT2 Asia
SA Chukchi Peninsula

Wrangell Mountains (1978)
Range near the Canadian border at the N end of the Alaskan panhandle.
BT Southern Alaska
BT Alaska
BT United States

Wrangellia (1985)
Tectonostratigraphic terrane in SE Alaska, consisting of allochthonous Triassic rocks. Also used for a hypothetical ancient continent, of which the Alaskan Wrangellia terrane, and other Triassic allochthons along the Pacific Coast of North America, may be remnants.
UF Wrangellia Terrane
SA Alaska
SA British Columbia
SA continental drift
SA Oregon
SA plate tectonics
SA Washington

Wrangellia Terrane
use Wrangellia

Wreford Limestone (1978)
In Chase Group. In Kansas, includes (ascending) Threemile Limestone, Havensville Shale, and Schroyer Limestone members. E Kansas, SE Nebraska, and central Oklahoma. Also search Wreford Megacyclothem.
BT Permian
BT Paleozoic
SA Chase Group
SA Kansas
SA Nebraska
SA Oklahoma

wrench faults (1978)
Before 1978, also search faults AND wrench.
UF basculating faults
UF torsion faults
BT faults
SA displacements

Wright County Minnesota (1989)
S central Minnesota. Before 1989, search Wright County AND Minnesota.
CO N445600N452500
  W0933000W0942000
BT Minnesota
BT United States
NT Monticello Minnesota

Wright Valley (1978)
In the Wright Glacier area W of McMurdo Sound in Victoria Land in Ross Dependency on the Pacific Ocean side. Also search Wright AND appropriate area.
BT Antarctica
SA Ross Dependency
SA Victoria Land

wrinkle ridges (1993)
UF mare ridges
SA highlands
SA lunar craters
SA lunar highlands
SA maria
SA Moon

Wroclaw
No longer a valid term for GeoRef. As of 1993 see Wroclaw Poland.

Wroclaw Poland (1993)
Province in SW Poland. Also a city on the Oder River. Also search Breslau or Slak Dolny.
UF Breslau Poland
UF Slak Dolny
BT Poland
BT Central Europe
BT Europe
NT Bardo Mountains
NT Bystrzyca
NT Lesna Poland
NT Strzelin Poland
SA Silesian coal basin

Wuerttemberg
use Wurttemberg

wuestite (1993)
Before 1993, also search wustite.
UF iozite
UF wustite
BT oxides
SA iron oxides

Wufeng Formation (1989)
Anhui, Guizhou, Hubei, Hunan, Jiangsu, Shanxi, and Sichuan provinces.
BT Upper Ordovician
BT Ordovician
BT Paleozoic
SA Anhui China
SA China
SA Guizhou China

SA Hubei China
SA Hunan China
SA Jiangsu China
SA Shanxi China
SA Sichuan China

wulfenite (1978)
BT molybdates

Wumishan Formation (1989)
Hebei and NE China.
IN Index Chinese provinces as applicable.
BT Sinian
BT Proterozoic
BT upper Precambrian
BT Precambrian
SA China
SA Hebei China

Wurm (1978)
Europe, Above Riss, below Holocene.
BT upper Pleistocene
BT Pleistocene
BT Quaternary
BT Cenozoic
SA Sartanian
SA Vistulian
SA Weichselian

Wurttemberg (1978)
Former German state now part of Baden-Wurttemberg.
UF Wuerttemberg
BT Baden-Wurttemberg Germany
BT Germany
BT Central Europe
BT Europe

wurtzite (1978)
BT sulfides

wustite
As of 1993, no longer a valid term for GeoRef.
use wuestite

Wutach River valley
use Wutach Valley

Wutach Valley (1978)
River valley in SW Baden-Wurttemberg. Also search Wutach; Wutach River.
UF Wutach River valley
BT Baden-Wurttemberg Germany
BT Germany
BT Central Europe
BT Europe

Wyandot County
Valid through 1988. Search in combination with state term. After 1988, use specific county-state term.

Wyandot County Ohio (1989)
N central Ohio. Before 1989, also search Wyandot County AND Ohio.
CO N404000N410000
    W0830700W0833200
BT Ohio
BT United States
SA Sandusky River basin

Wyandotte County
Valid through 1988. Search in combination with state term. After 1988, use specific county-state term.

Wyandotte County Kansas (1989)
NE Kansas. Before 1989, also search Wyandotte County AND Kansas.
CO N390000N391400
    W0943500W0945400
BT Kansas

BT United States
NT Kansas City Kansas

Wyandotte Limestone (1978)
In Kansas City Group. Includes (ascending) Frisbie Limestone, Quindaro Shale, Argentine Limestone, Island Creek Shale, and Farley Limestone members. SW Iowa, E Kansas, NW Missouri, and SE Nebraska. Also search Wyandotte Formation.
BT Missourian
BT Upper Pennsylvanian
BT Pennsylvanian
BT Carboniferous
BT Paleozoic
SA Iowa
SA Kansas
SA Kansas City Group
SA Missouri
SA Nebraska

Wye Valley (1978)
River valley in E Wales, and W England. Also search Wye River. In 1985, Great Britain was added as a broader term.
IN Index political divisions as applicable.
BT Great Britain
BT United Kingdom
BT Western Europe
BT Europe
SA England
SA Wales

Wyman Formation (1978)
Spotted schists and phyllites with a few interbedded dolomite. In S California.
BT Precambrian
SA California

Wynyard
No longer a valid term for GeoRef. As of 1993, see Wynyard Australia or Wynyard Saskatchewan.

Wynyard Australia (1993)
Town on Bass Strait in NW Tasmania. Before 1993, also search Wynyard AND Tasmania.
BT Tasmania Australia
BT Australia
BT Australasia
SA Bass Strait

Wynyard Saskatchewan (1993)
Town in S central Saskatchewan. Before 1993, also search Wynyard AND Saskatchewan.
BT Saskatchewan
BT Western Canada
BT Canada

Wyoming (1978)
Autoposting of this term began in 1978.
CO N410000N450000
    W1040500W1110500
BT United States
NT Albany County Wyoming
NT Big Horn County Wyoming
NT Campbell County Wyoming
NT Carbon County Wyoming
NT Converse County Wyoming
NT Crook County Wyoming
NT Fremont County Wyoming
NT Goshen County Wyoming
NT Great Divide Basin
NT Hartzog Draw Field
NT Heart Mountain Fault
NT Hot Springs County Wyoming
NT Johnson County Wyoming
NT Laramie Basin
NT Laramie County Wyoming
NT Lincoln County Wyoming

NT Lost Soldier Field
NT Natrona County Wyoming
NT Niobrara County Wyoming
NT Owl Creek Mountains
NT Park County Wyoming
NT Platte County Wyoming
NT Rock Springs Uplift
NT Sheridan County Wyoming
NT Sierra Madre Range
NT Sublette County Wyoming
NT Sweetwater County Wyoming
NT Teapot Dome
NT Teton County Wyoming
NT Teton National Forest
NT Uinta County Wyoming
NT Washakie County Wyoming
NT Weston County Wyoming
NT Wind River Range
SA Absaroka Fault
SA Absaroka Range
SA Absaroka Supergroup
SA Almond Formation
SA Amsden Formation
SA Arikaree Group
SA Ash Hollow Formation
SA Aycross Formation
SA Bald Mountain
SA Battle Mountain High
SA Bear River Formation
SA Bearpaw Formation
SA Beartooth Mountains
SA Beaver Creek
SA Benton Formation
SA Bighorn Basin
SA Bighorn Mountains
SA Bighorn River
SA Black Hills
SA Blair Formation
SA Bridger Formation
SA Browns Park Formation
SA Brule Formation
SA Bull Lake Glaciation
SA Cache Creek
SA Carlile Shale
SA Casper Formation
SA Chadron Formation
SA Chugwater Formation
SA Clark's Fork Basin
SA Cloverly Formation
SA Cody Shale
SA Colorado Group
SA Colorado Lineament
SA Colorado River basin
SA Columbia River basin
SA Copper Mountain
SA Crawford Thrust
SA Dakota Formation
SA Deadwood Formation
SA Deer Lake Group
SA Denver Basin
SA Dinwoody Formation
SA Eagle Sandstone
SA Fall River Formation
SA Flathead Sandstone
SA Fort Union Formation
SA Fountain Formation
SA Fox Hills Formation
SA Frontier Formation
SA Gallatin Range
SA Gammon Ferruginous Member
SA Gas Hills
SA Glenns Ferry Formation
SA Goose Egg Formation
SA Graneros Shale
SA Granite Mountains
SA Great Plains
SA Green River
SA Green River basin
SA Green River Formation
SA Greenhorn Limestone
SA Hanna Basin
SA Hanna Formation
SA Harebell Formation
SA High Plains Aquifer

SA Hilliard Shale
SA Inyan Kara Group
SA Jefferson Group
SA Judith River Formation
SA Lake Uinta
SA Lakota Formation
SA Lance Formation
SA Laney Shale Member
SA Laramie anorthosite complex
SA Laramie Formation
SA Laramie Mountains
SA Lewis Shale
SA Little Missouri River basin
SA Lodgepole Formation
SA Loup Fork Group
SA Madison Aquifer
SA Madison Group
SA Mancos Shale
SA Meade Peak Member
SA Medicine Bow Mountains
SA Mesaverde Group
SA Midcontinent
SA Minnelusa Formation
SA Missouri River basin
SA Montana Group
SA Morrison Formation
SA Mowry Shale
SA Moxa Arch
SA Muddy Sandstone
SA Mullen Creek-Nash Fork shear zone
SA Newcastle Sandstone
SA Niobrara Formation
SA Nugget Sandstone
SA Ogallala Aquifer
SA Ogallala Formation
SA Opeche Shale
SA Park City Formation
SA Parkman Sandstone
SA Phosphoria Formation
SA Pierre Shale
SA Pinedale Glaciation
SA Pinyon Conglomerate
SA Platte River basin
SA Powder River basin
SA Rattlesnake Hills
SA Red Hill
SA Red Peak Formation
SA Red River Formation
SA Retort Phosphatic Shale Member
SA Rock Springs Formation
SA Salt Creek
SA Salt Lake Formation
SA Sevier orogenic belt
SA Shannon Sandstone Member
SA Skull Creek Shale
SA Snake River
SA Snake River basin
SA Spearfish Formation
SA Stump Formation
SA Sullivan Mine
SA Sundance Formation
SA Sunlight
SA Sussex Sandstone Member
SA Swift Formation
SA Teapot Sandstone Member
SA Tensleep Sandstone
SA Tepee Trail Formation
SA Thaynes Formation
SA Tongue River
SA Tongue River Member
SA Tullock Member
SA Twin Creek Limestone
SA U. S. Rocky Mountains
SA Uinta Mountains
SA Vermillion Creek coal bed
SA Walker Creek Field
SA Wall Creek Member
SA Wapiti Formation
SA Wasatch Formation
SA Washakie Basin
SA Wells Formation
SA Western Interior
SA Western Interior Seaway

SA Western Overthrust Belt
SA White River Group
SA Wiggins Formation
SA Wilkins Peak Member
SA Willwood Formation
SA Wind River Formation
SA Wyoming Province
SA Yellowstone National Park
SA Yellowstone River

Wyoming County
Valid through 1988. Search in combination with state term. After 1988, use specific county-state term.

Wyoming County New York (1989)
W New York. Before 1989, also search Wyoming County AND New York.
CO N423200N425200 W0775700W0782900
BT New York
BT United States

Wyoming County Pennsylvania (1989)
NE Pennsylvania. Before 1989, also search Wyoming County AND Pennsylvania.
CO N412200N413800 W0754400W0761900
BT Pennsylvania
BT United States

Wyoming County West Virginia (1989)
S West Virginia. Before 1989, also search Wyoming County AND West Virginia.
CO N372500N374700 W0811500W0815300
BT West Virginia
BT United States

Wyoming Province (1993)
Block of Archean crust exposed in cores of Laramide uplifts in Wyoming and adjacent states. Considered by some to be buried extension of the Superior or Churchill provinces of the Canadian Shield.
IN Index states as applicable.
CO N404500N483000 W1040000W1160000
BT United States
SA Churchill Province
SA Hudsonian Orogeny
SA Idaho
SA Montana
SA North Dakota
SA South Dakota
SA Superior Province
SA Utah
SA Wyoming

Wythe County
Valid through 1988. Search in combination with state term. After 1988, use specific county-state term.

Wythe County Virginia (1989)
SW Virginia. Before 1989, also search Wythe County AND Virginia.
CO N364500N370500 W0804500W0812400
BT Virginia
BT United States

X

X-ray
A valid term through 1977. See X-ray analysis.

X-ray analysis (1978)
Used for methodology, not data. For data, use X-ray data. X-ray diffraction analysis was used on level 1 through 1977. Before 1978, also search X-ray.
IN Index specific method where available.
SA analysis
SA chemical analysis
SA clay mineralogy
SA crystal structure
SA crystallography
SA electron microscopy
SA goniometry
SA irradiation
SA powder method
SA single-crystal method
SA spectroscopy
SA structural analysis
SA synchrotron radiation
SA synchrotrons
SA thermal analysis
SA tomography
SA X-ray data
SA X-ray diffraction analysis
SA X-ray fluorescence
SA X-ray radiography
SA X-ray spectroscopy

X-ray data (1978)
For methodology, use X-ray analysis.
SA data
SA X-ray analysis
SA X-ray diffraction data

X-ray diffraction
Not a valid GeoRef index term after 1970. Was used on level 1 in subfile N.
use X-ray diffraction analysis

X-ray diffraction analyses
Not a valid GeoRef index term after 1970. Was used on level 1 in subfile N.
use X-ray diffraction analysis

X-ray diffraction analysis (1978)
Used for methodology. Before 1971, also search X-ray diffraction; X-ray diffraction analyses.
UF X-ray diffraction
UF X-ray diffraction analyses
UF Xray diffraction analysis
SA analysis
SA chemical analysis
SA differential thermal analysis
SA diffractograms
SA electron diffraction analysis
SA neutron diffraction analysis
SA powder method
SA spectroscopy
SA synchrotron radiation
SA synchrotrons
SA X-ray analysis
SA X-ray diffraction data
SA X-ray fluorescence

X-ray diffraction data (1989)
Used only for data. For methodology, use X-ray diffraction analysis.
SA data
SA X-ray data
SA X-ray diffraction analysis
SA X-ray fluorescence spectra
SA X-ray spectra

X-ray fluorescence (1978)
Before 1971, also search X-ray fluorescence analysis.
UF X-ray fluorescence analysis
SA analysis
SA chemical analysis
SA fluorescence
SA fluorimetry
SA luminescence
SA spectroscopy
SA synchrotron radiation
SA synchrotrons
SA X-ray analysis
SA X-ray diffraction analysis
SA X-ray fluorescence spectra
SA X-ray spectra

X-ray fluorescence analysis
Not a valid GeoRef index term after 1970. Was used on level 1 in subfile N.
use X-ray fluorescence

X-ray fluorescence spectra (1978)
Refers to data. For methodology, use X-ray fluorescence.
BT spectra
SA fluorescence
SA fluorimetry
SA spectroscopy
SA X-ray diffraction data
SA X-ray fluorescence
SA X-ray spectra

X-ray radiography (1978)
UF radiography, X-ray
SA X-ray analysis

X-ray spectra (1985)
Used for data. For methodology, use X-ray spectroscopy.
UF Xray spectra
BT spectra
SA spectroscopy
SA X-ray diffraction data
SA X-ray fluorescence
SA X-ray fluorescence spectra
SA X-ray spectroscopy

X-ray spectrographic analysis
Not a valid GeoRef index term after 1970. Was used on level 1 in subfile N.
use X-ray spectroscopy

X-ray spectroscopy (1978)
Before 1978, also search spectroscopy AND X-ray. Before 1971, also search X-ray spectrographic analysis.
UF X-ray spectrographic analysis
BT spectroscopy
SA analysis
SA chemical analysis
SA X-ray analysis
SA X-ray spectra

X-rays (1978)
Use when referring to the rays themselves.
SA alpha rays
SA beta rays
SA electromagnetic radiation
SA gamma rays
SA radioactivity

Xainza China (1993)
County and town in central Xizang, SW China. Fossil locality.
BT Xizang China
BT China
BT Far East
BT Asia

xanthochroite
use greenockite

xanthophyllite
use clintonite

xanthosiderite
use goethite

Xe
use xenon

Xe-124 (1978)
Autoposting of broader terms began in 1989. This term has multiple hierarchies.
BT1 stable isotopes
BT1 isotopes
BT2 xenon
BT2 noble gases

Xe-126 (1978)
Autoposting of broader terms began in 1989. This term has multiple hierarchies.
BT1 stable isotopes
BT1 isotopes
BT2 xenon
BT2 noble gases

Xe-128 (1978)
Autoposting of broader terms began in 1989. This term has multiple hierarchies.
BT1 stable isotopes
BT1 isotopes
BT2 xenon
BT2 noble gases

Xe-129 (1978)
Autoposting of broader terms began in 1989. This term has multiple hierarchies.
BT1 stable isotopes
BT1 isotopes
BT2 xenon
BT2 noble gases
SA Xe-130/Xe-129
SA Xe-132/Xe-129

Xe-129/Xe-132
use Xe-132/Xe-129

Xe-130/Xe-129 (1989)
Autoposting of broader terms began in 1989. This term has multiple hierarchies.
BT1 stable isotopes
BT1 isotopes
BT2 xenon
BT2 noble gases
SA isotope ratios
SA Xe-129

Xe-131 (1978)
Autoposting of broader terms began in 1989. This term has multiple hierarchies.
BT1 stable isotopes
BT1 isotopes
BT2 xenon
BT2 noble gases

Xe-132 (1978)
Autoposting of broader terms began in 1989. This term has multiple hierarchies.
BT1 stable isotopes
BT1 isotopes
BT2 xenon
BT2 noble gases
SA Xe-132/Xe-129
SA Xe-136/Xe-132

Xe-132/Xe-129 (1985)
Autoposting of broader terms began in 1989. This term has multiple hierarchies.
UF Xe-129/Xe-132
BT1 stable isotopes
BT1 isotopes
BT2 xenon
BT2 noble gases
SA isotope ratios
SA Xe-129
SA Xe-132

Xe-136/Xe-130 (1989)
Autoposting of broader terms began in 1989. This term has multiple hierarchies.
BT1 stable isotopes
BT1 isotopes
BT2 xenon
BT2 noble gases
SA isotope ratios

Xe-136/Xe-132 (1989)
Autoposting of broader terms began in 1989. This term has multiple hierarchies.
BT1 stable isotopes
BT1 isotopes
BT2 xenon
BT2 noble gases
SA isotope ratios
SA Xe-132

Xenarthra (1981)
Suborder. Autoposting of Edentata, Eutheria and Theria to this term began in 1989.
BT Edentata
BT Eutheria
BT Theria
BT Mammalia
BT Tetrapoda
BT Vertebrata
BT Chordata

xenocrysts (1985)
SA crystals
SA igneous rocks
SA inclusions
SA mineral inclusions
SA xenoliths

xenoliths (1978)
Autoposting of inclusions to this term began in 1989.
UF exogenous inclusions
BT inclusions
SA fluid inclusions
SA host rocks
SA mineral inclusions
SA xenocrysts

xenon (1978)
Autoposting of noble gases to this term began in 1989.
UF Xe
BT noble gases
NT Xe-124
NT Xe-126
NT Xe-128
NT Xe-129
NT Xe-130/Xe-129
NT Xe-131
NT Xe-132
NT Xe-132/Xe-129
NT Xe-136/Xe-130
NT Xe-136/Xe-132
SA I/Xe
SA isotopes

xenotime (1978)
BT phosphates

Xenungulata (1981)
Suborder. Autoposting of Amblypoda, Eutheria and Theria to this term began in 1989.
BT Amblypoda
BT Eutheria
BT Theria
BT Mammalia
BT Tetrapoda
BT Vertebrata
BT Chordata
SA Ungulata

Xianshuihe fault zone (1993)
From Kangding in W Sichuan extends into Yushu County, Qinghai. Includes Xianshuihe, Moxi, Selaha (Kangding), Zheduotang, and Yalahe faults.
IN Index provinces as applicable.
CO N290000N320000
   E1021000E1000000
BT China
BT Far East
BT Asia
SA Qinghai China
SA Sichuan China
SA Xichang China

Xiao Hinggan Ling (1993)
E Heilongjiang. Separates Amur River from Sungari Valley. Before 1993, also search Lesser Khingan Mountains, Little Khingan Mountains, Xiao Hingan Ling, and Xiao Xingan Ling.
UF Lesser Khingan Mountains
UF Little Khingan Mountains
UF Xiao Xingan Ling
BT Heilongjiang China
BT China
BT Far East
BT Asia

Xiao Xingan Ling
use Xiao Hinggan Ling

Xichang China (1993)
County in S Sichuan and town in S Guangxi, China.
IN Index provinces as applicable.
BT China
BT Far East
BT Asia
SA Guangxi China
SA Sichuan China
SA Xianshuihe fault zone

Xingtai earthquake 1966 (1993)
NW China. Before 1985, also search Xingtai or Hsingtai AND earthquakes.
UF Hsingtai earthquake 1966
BT earthquakes
SA Beijing China
SA China
SA Hebei China
SA shallow-focus earthquakes

Xinjiang
No longer a valid term for GeoRef. See Xinjiang China.

Xinjiang China (1993)
Autonomous region in NW China. Also search Sinkiang, Sinkiang Uighur, Sinkiang Province, Xinjiang, and Xinjiang Weiwu'er Zizhiqu.
CO N340000N490000
   E0970000E0730000
UF Sinkiang Weiwu'er Zizhiqu
UF Sinkiang Province
UF Sinkiang Uighur
UF Xinjiang Weiwu'er Zizhiqu
BT China
BT Far East
BT Asia
NT Junggar
NT Junggar Basin
NT Tarim Basin
SA Altai Mountains
SA Dzhungarian Alatau
SA Ili Basin
SA Irtysh River
SA Tarbagatay Range
SA Tien Shan
SA Turkestan

Xinjiang Weiwu'er Zizhiqu
use Xinjiang China

Xiphosura (1978)
Subclass.
BT Merostomata
BT Chelicerata
BT Arthropoda
BT Invertebrata

Xisha Archipelago
use Xisha Islands

Xisha Islands (1989)
S China Sea.
CO N153000N171500
   E1130000E1110000
UF Paracel Islands
UF Xisha Archipelago
BT China
BT Far East
BT Asia
SA South China Sea

Xizang
No longer a valid term for GeoRef. Valid 1992. As of 1993, see Xizang China.

Xizang China (1993)
Autonomous region in extreme SW China. Also search Tibet.
CO N270000N370000
   E0990000E0790000
UF Tibet China
BT China
BT Far East
BT Asia
NT Lhasa Block
NT Lhasa China
NT Xainza China
SA Brahmaputra River
SA Indian Plate
SA Indo-Australian Plate
SA Indus River
SA Indus-Yarlung Zangbo suture zone
SA Kumaun Himalayas
SA Kunlun Mountains
SA Mount Everest
SA Mount Namjagbarwa
SA Qinghai-Xizang Plateau

Xizang Plateau
use Qinghai-Xizang Plateau

xonotlite (1978)
BT chain silicates
BT silicates

Xray diffraction analysis
use X-ray diffraction analysis

Xray spectra
use X-ray spectra

xylene (1989)
BT aromatic hydrocarbons
BT hydrocarbons
BT organic materials

Y

Y
use yttrium

Yacoraite Formation (1978)
In Jujuy and Salta, Argentina.
BT Cretaceous
BT Mesozoic
SA Argentina
SA Jujuy Argentina
SA Salta Argentina

Yahagi River (1978)
Flows into Chita Bay on the Pacific Ocean S of Nagoya in central Honshu. Also search Yahagi.
BT Honshu
BT Japan
BT Far East
BT Asia
SA Aichi Japan
SA Gifu Japan

Yakataga Formation (1989)
Middleton Island, SE Alaska.
IN Index ages as applicable.
BT Cenozoic
SA Alaska
SA Holocene
SA Middleton Island
SA Miocene
SA Pleistocene
SA Pliocene

Yake-Dake (1978)
Peak NE of Kure in W Honshu.
UF Yakedake
UF Yakeyama
SA Gifu Japan
SA Honshu
SA Nagano Japan

Yakedake
use Yake-Dake

Yakeyama
use Yake-Dake

Yakima Basalt (1978)
Upper Miocene. Formerly also included in lower Pliocene? Typical plateau basalt. In Yakima East Quadrangle, it is exposed in four southeast-trending strips that coincide with crestal portions of four anticlinal axes. In E Washington; also in Oregon and Idaho. Sometimes called the Yakima Basalt Subgroup of the Columbia River Basalt Group, it has been considered to have as members, the Wanapum Basalt, Grande Ronde Basalt, and Saddle Mountains Basalt.
IN Index ages as applicable.
BT Miocene
BT Neogene
BT Tertiary
BT Cenozoic
SA Columbia River Basalt Group
SA Grande Ronde Basalt
SA Idaho
SA lower Pliocene
SA Oregon
SA Saddle Mountains Basalt
SA upper Miocene
SA Wanapum Basalt
SA Washington

Yakima County
Valid through 1988. Search in combination with state term. After 1988, use specific county-state term.

Yakima County Washington (1989)
S Washington. Before 1989, also search Yakima County AND Washington.
CO N460200N470500
   W1195200W1213200
BT Washington
BT United States
SA Mount Adams
SA Rattlesnake Hills
SA Yakima Indian Reservation

Yakima fold belt (1993)
Region of the Columbia Plateau, S-central Washington and N-central Oregon.
IN Index states and counties as applicable.
BT United States
SA Columbia Plateau
SA Columbia River Basalt Group

SA Oregon
SA Washington
Yakima Indian Reservation
(1985)
S central Washington. Autoposting of broader terms to this term began in 1989.
IN Index counties as applicable.
CO N454800N463500
    W1200000W1212500
BT Washington
BT United States
SA Indian reservations
SA Klickitat County Washington
SA Lewis County Washington
SA Yakima County Washington

Yakut A.S.S.R.
use Yakutia Russian Federation

Yakutat Bay (1981)
SE Alaska, inlet of the Gulf of Alaska.
CO N592000N600000
    W1391500W1410000
BT Southern Alaska
BT Alaska
BT United States
SA Gulf of Alaska
SA Yakutat Terrane

Yakutat Block
use Yakutat Terrane

Yakutat Terrane (1989)
SE Alaska.
UF Yakutat Block
BT Alaska
BT United States
SA Southeastern Alaska
SA Yakutat Bay

Yakutia
As of 1993, no longer a valid term for GeoRef.
use Yakutia Russian Federation

Yakutia region (1981)
Region which surrounds and includes the city of Yakutsk, extending past the Aldan River on the E and N. This term has multiple hierarchies.
IN Index oblasts or regions as applicable.
CO N573000N633000
    E1350000E1270000
BT1 Siberian Platform
BT1 Russian Federation
BT1 Commonwealth of Independent States
BT2 Siberian Platform
BT2 Asia
SA Yakutia Russian Federation

Yakutia Russian Federation
(1993)
Former Yakut Autonomous Soviet Socialist Republic. Located in E central Siberia. Before 1978, also search Yakut. Before 1993, also search Yakutia. This term has multiple hierarchies.
CO N550000N730000
    E1620000E1080000
UF Yakut A.S.S.R.
UF Yakutia
BT1 Russian Federation
BT1 Commonwealth of Independent States
BT2 Asia
NT Aldan Russian Federation
NT Anabar Bay
NT Anabar River
NT Anabar Shield
NT Bolshoy Selerikan River
NT Indigirka River
NT Kular Range

NT Mir Pipe
NT Mirnyy Russian Federation
NT New Siberian Islands
NT Olenek River
NT Orulgan Mountains
NT Selennyakh
NT Seligdar Russian Federation
NT Sutam River
NT Tas-Khayakhtakh Range
NT Udachnaya Pipe
NT Udachnaya Russian Federation
NT Verkhoyansk Range
NT Verkhoyansk Russian Federation
NT Yakutsk Russian Federation
NT Yana
NT Yana-Indigirka Lowland
NT Zyryanka Russian Federation
SA Aldan Plateau
SA Aldan River
SA Aldan Shield
SA Arbarastakh
SA Kolyma River
SA Kolyma River basin
SA Kolyma Uplift
SA Maya River basin
SA Olekma
SA Omolon
SA Polousnyy
SA Sette-Daban Range
SA Siberia
SA Stanovoy Range
SA Uchur River basin
SA Udokan Series
SA Verkhoyansk region
SA Vilyuy Syneclise
SA Yakutia region

Yakutsk
No longer a valid term for GeoRef. As of 1993, see Yakutsk Russian Federation.

Yakutsk Russian Federation
(1993)
City on the bend of the Lena River in central Yakutia in E central Siberia. Before 1993, also search Yakutsk. This term has multiple hierarchies.
BT1 Yakutia Russian Federation
BT1 Russian Federation
BT1 Commonwealth of Independent States
BT2 Yakutia Russian Federation
BT2 Asia

Yalta
No longer a valid term for GeoRef. As of 1993, see Yalta Ukraine.

Yalta Ukraine (1993)
City on the Black Sea in S Crimea. This term has multiple hierarchies.
BT1 Crimea Ukraine
BT1 Ukraine
BT1 Europe
BT2 Crimea Ukraine
BT2 Ukraine
BT2 Commonwealth of Independent States

Yamagata
No longer a valid term for GeoRef. See Yamagata Japan.

Yamagata Japan (1993)
Prefecture including city in N Honshu.
BT Honshu
BT Japan
BT Far East
BT Asia
NT Yonezawa Japan
SA Chokai

SA Oguni Japan
SA Tohoku
SA Zao

Yamaguchi
No longer a valid term for GeoRef. See Yamaguchi Japan.

Yamaguchi Japan (1993)
City and prefecture in W Honshu. Before 1993, also search Yamaguchi Prefecture or Yamaguchi AND Japan.
BT Honshu
BT Japan
BT Far East
BT Asia
NT Ube coal field
SA Chugoku

Yamal (1978)
Peninsula between the Kara Sea on the W and the Gulf of Ob on the E in the Yamal-Nenets National Okrug in NW Siberia. This term has multiple hierarchies.
UF Yamal Peninsula
BT1 Yamal-Nenets Russian Federation
BT1 Tyumen Russian Federation
BT1 Russian Federation
BT1 Commonwealth of Independent States
BT2 Yamal-Nenets Russian Federation
BT2 Tyumen Russian Federation
BT2 Asia

Yamal Peninsula
use Yamal

Yamal-Nenets
No longer a valid term for GeoRef. As of 1993, see Yamal-Nenets Russian Federation.

Yamal-Nenets Russian Federation (1993)
National Okrug in N Tyumen Oblast in NW Siberia. Before 1993, also search Yamal-Nenets. Before 1978, also search Yamalo-Nenets.
UF Yamalo-Nenets Russian Federation
BT1 Tyumen Russian Federation
BT1 Russian Federation
BT1 Commonwealth of Independent States
BT2 Tyumen Russian Federation
BT2 Asia
NT Taz Basin
NT Yamal

Yamalo-Nenets Russian Federation
use Yamal-Nenets Russian Federation

Yamanashi
No longer a valid term for GeoRef. See Yamanashi Japan.

Yamanashi Japan (1993)
Prefecture W of Tokyo in central Honshu.
BT Honshu
BT Japan
BT Far East
BT Asia
SA Akaishi Mountains
SA Chiba Japan
SA Chubu Japan
SA Fuji River
SA Fujiyama
SA Kanto Mountains
SA Koma-ga-take
SA Tanzawa Mountains
SA Yatsugatake

Yamasaki Fault (1989)
Hyogo Prefecture, S Honshu, SW Japan.
BT Honshu
BT Japan
BT Far East
BT Asia
SA Hyogo Japan

Yamato Meteorites (1981)
Due to numbering of these meteorites, also truncate when searching. Before 1981, also search Yamato AND meteorites.
BT meteorites
SA Antarctica
SA Victoria Land

Yamato Mountains (1978)
In Victoria Land, Antarctica.
BT Victoria Land
BT Antarctica

Yamhill Formation (1978)
Comprises Mill Creek Beds, of predominantly dark-gray shale and siltstone with occasional beds of lime-cemented sandstone, and an overlying sequence of massive sandstone beds which grades upward into more agrillaceous rock. In W Oregon.
BT Eocene
BT Paleogene
BT Tertiary
BT Cenozoic
SA Oregon

Yampa River (1985)
Tributary of the Green River, NW Colorado.
IN Index counties as applicable.
BT Colorado
BT United States
SA Green River
SA Moffat County Colorado
SA Routt County Colorado

Yamuna River
use Jumna River

Yan Shan
use Yanshan Range

Yana (1978)
River which rises in the Verkhoyansk Range of N central Yakutia and flows N into the Laptev Sea. Also search Yana River; Yana River region; Yana River basin. This term has multiple hierarchies.
UF Yana River
BT1 Yakutia Russian Federation
BT1 Russian Federation
BT1 Commonwealth of Independent States
BT2 Yakutia Russian Federation
BT2 Asia

Yana River
use Yana

Yana-Indigirka Depression
use Yana-Indigirka Lowland

Yana-Indigirka Lowland (1978)
Coastal lowland comprising the area around the lower courses and the lower interfluve of the Indigirka River and the Yana in N central Yakutia. This term has multiple hierarchies.
UF Yana-Indigirka Depression
BT1 Yakutia Russian Federation
BT1 Russian Federation
BT1 Commonwealth of Independent States
BT2 Yakutia Russian Federation
BT2 Asia

Yang-tze River valley
  use Yangtze River valley

Yangtze Plate (1993)
  Central and S China. Also may include parts of Japan. Includes Yangtze Platform and the S China Caledonian folded area.
  SA Asia
  SA China
  SA Japan
  SA plate tectonics
  SA plates

Yangtze Platform (1993)
  SE China. Includes nearly entire drainage basin of Yangtze River and S Yellow Sea. Rocks of Sinian, Cambrian, Silurian and Ordovician age.
  IN Index provinces as applicable.
  BT China
  BT Far East
  BT Asia
  SA Anhui China
  SA Eurasian Plate
  SA Guizhou China
  SA Hubei China
  SA Hunan China
  SA Jiangsu China
  SA Jiangxi China
  SA Sichuan China
  SA Yangtze River
  SA Yellow Sea
  SA Yunnan China

Yangtze River (1985)
  Sources in SW Qinghai. Flows into East China Sea N of Shanghai.
  UF Chang Jiang River
  UF Changjiang Estuary
  UF Changjiang River
  BT China
  BT Far East
  BT Asia
  SA Yangtze Platform
  SA Yangtze River valley

Yangtze River valley (1989)
  Valley of the Yangtze River, China.
  UF Yang-tze River valley
  UF Yangtze Valley
  BT China
  BT Far East
  BT Asia
  SA Yangtze River

Yangtze Valley
  use Yangtze River valley

Yanshan Range (1989)
  Mountain range in Hebei, N China.
  UF Yan Shan
  BT Hebei China
  BT China
  BT Far East
  BT Asia

Yanshanian (1989)
  China.
  IN Index ages as applicable.
  UF Yanshanian Orogeny
  UF Yenshan
  UF Yenshanian
  UF Yenshanian Orogeny
  BT Mesozoic
  SA Cretaceous
  SA Jurassic

Yanshanian Orogeny
  use Yanshanian

Yaquina Bay (1978)
  Inlet of the Pacific Ocean at the mouth of the Yaquina River in W Oregon.
  UF Yaquina Estuary
  BT Lincoln County Oregon
  BT Oregon
  BT United States

Yaquina Estuary
  use Yaquina Bay

Yaracuy
  No longer a valid term for GeoRef. As of 1993 see Yaracuy Venezuela.

Yaracuy Venezuela (1993)
  State in NW Venezuela. Before 1993 also search Yaracuy AND Venezuela.
  BT Venezuela
  BT South America

yardangs (1989)
  SA arid environment
  SA deserts
  SA eolian features
  SA erosion features
  SA geomorphology
  SA wind erosion
  SA wind transport

Yarlung Zangbo suture zone
  No longer a valid term for GeoRef.
  use Indus-Yarlung Zangbo suture zone

Yarmouth
  No longer a valid term for GeoRef. As of 1993, see Yarmouth Nova Scotia.

Yarmouth County
  No longer a valid term for GeoRef. As of 1993, see Yarmouth County Nova Scotia.

Yarmouth County Nova Scotia (1993)
  W Nova Scotia. Before 1993, also search Yarmouth County AND Nova Scotia.
  CO N432200N441600
      W0652500W0661000
  BT Nova Scotia
  BT Maritime Provinces
  BT Eastern Canada
  BT Canada
  NT Yarmouth Nova Scotia

Yarmouth Nova Scotia (1993)
  Town on the Atlantic Ocean in SW Nova Scotia. Before 1993, also search Yarmouth AND Nova Scotia.
  BT Yarmouth County Nova Scotia
  BT Nova Scotia
  BT Maritime Provinces
  BT Eastern Canada
  BT Canada

Yaroslavl
  No longer a valid term for GeoRef. As of 1993, see Yaroslavl Russian Federation.

Yaroslavl Russian Federation (1993)
  Oblast including the city NE of Moscow on the Volga River. This term has multiple hierarchies.
  BT1 Russian Federation
  BT1 Commonwealth of Independent States
  BT2 Europe

Yarrol Basin (1978)
  Queensland.
  BT Queensland Australia
  BT Australia
  BT Australasia

Yass (1978)
  River basin in SE New South Wales. Also search Yass Basin.
  UF Yass Basin
  BT New South Wales Australia
  BT Australia
  BT Australasia

Yass Basin
  use Yass

Yates County New York (1989)
  W central New York. Before 1989, search Yates County AND New York.
  CO N422700N424600
      W0765300W0772200
  BT New York
  BT United States
  SA Seneca Lake

Yates Formation (1978)
  In Artesia Group. A 50-foot sandstone that is subsurface in Texas, and subsurface and surface in New Mexico.
  BT Permian
  BT Paleozoic
  SA New Mexico
  SA Texas

Yatsudake Mountain
  use Yatsugatake

Yatsugatake (1978)
  Mountain peak NW of Tokyo in central Honshu.
  UF Yatsudake Mountain
  BT Honshu
  BT Japan
  BT Far East
  BT Asia
  SA Nagano Japan
  SA Yamanashi Japan

Yavapai County
  Valid through 1988. Search in combination with state term. After 1988, use specific county-state term.

Yavapai County Arizona (1989)
  W Central Arizona. Before 1989, also search Yavapai County AND Arizona.
  CO N335300N353200
      W1113000W1132000
  BT Arizona
  BT United States
  NT Jerome Arizona
  SA Date Creek basin

Yazoo Clay (1978)
  In Jackson Group. Divided into (ascending) North Creek Clay, Cocoa Sand, Pachuta Clay, and Shubuta Clay members. SW Alabama, central N Louisiana and Mississippi.
  BT upper Eocene
  BT Eocene
  BT Paleogene
  BT Tertiary
  BT Cenozoic
  SA Alabama
  SA Jackson Group
  SA Louisiana
  SA Mississippi
  SA Shubuta Member

Yb
  use ytterbium

Ybbs Meteorite
  use Ybbsitz Meteorite

Ybbsitz Meteorite (1993)
  Found in SW Lower Austria. Olivine-bronzite chondrite (H4).
  UF Ybbs Meteorite
  BT H chondrites
  BT chondrites
  BT stony meteorites
  BT meteorites
  SA Lower Austria

Yeddo
  Not a valid term for GeoRef. See Tokyo Japan.

Yefremovka Meteorite
  use Efremovka Meteorite

Yegua Formation (1985)
  In Claiborne Group. E and S Texas, NW Louisiana, and W and S Mississippi.
  BT middle Eocene
  BT Eocene
  BT Paleogene
  BT Tertiary
  BT Cenozoic
  SA Claiborne Group
  SA Louisiana
  SA Mississippi
  SA Texas

Yekaterinburg Russian Federation (1993)
  Oblast and city in foothills of the Central Urals in W Siberia. Renamed after 1990. Before 1993, also search Sverdlovsk. Before 1978, also search Ekaterinburg or Yekaterinburg.
  UF Ekaterinburg Russian Federation
  UF Sverdlovsk Russian Federation
  BT Russian Federation
  BT Commonwealth of Independent States
  NT Kushva Russian Republic
  NT Serov Russian Federation
  NT Sysert Russian Federation
  SA Central Urals
  SA Tagil Basin
  SA Ural region

Yelizavetgrad
  Not a valid term for GeoRef. See Kirovograd Ukraine.

yellow arsenic
  use orpiment

Yellow Aster Complex (1989)
  Proterozoic or Paleozoic?.
  SA Washington

Yellow River, China
  use Huang Ho

Yellow Sea (1978)
  Between NE China and the Korean Peninsula.
  CO N333000N411000
      E1263000E1173000
  UF Huang Hai
  UF Huanghai Sea
  UF Hwang Hae
  UF Hwang Hai
  BT West Pacific
  BT Pacific Ocean
  NT Bohai Bay
  SA Shandong Peninsula
  SA Yangtze Platform

Yellowjacket Formation (1985)
  S central Idaho.
  BT Precambrian
  SA Belt Supergroup
  SA Idaho

Yellowknife
  As of 1993, no longer a valid term for GeoRef. See Yellowknife Northwest Territories if applicable.

Yellowknife Group (1978)

Composed of quartzite, conglomerate, graywacke, meta-basalt, meta-andesite, dacite, rhyolite, tuff, agglomerate; undifferentiated gabbro. In Mackenzie District, Northwest Territories.
BT Archean
BT Precambrian
SA Northwest Territories

Yellowknife Northwest Territories (1993)
Town on NW shore of Great Slave Lake at the mouth of the Yellowknife River in S Mackenzie District. Before 1993, also search Yellowknife.
BT Mackenzie District Northwest Territories
BT Northwest Territories
BT Western Canada
BT Canada

Yellowstone County Montana (1989)
S Montana. Before 1989, search Yellowstone County AND Montana.
CO N452600N463100
   W1072822W1085700
BT Montana
BT United States
NT Billings Montana
SA Bighorn River

Yellowstone National Park (1978)
Largest and oldest of U. S. national parks primarily in extreme NW Wyoming. Also search Yellowstone.
IN Index states and counties as applicable.
CO N441000N450000
   W1095000W1110500
UF Yellowstone Park
BT United States
SA Fremont County Idaho
SA Gallatin Range
SA Idaho
SA Montana
SA national parks
SA Old Faithful Geyser
SA Park County Wyoming
SA Teton County Wyoming
SA Wyoming
SA Yellowstone River

Yellowstone Park
 use Yellowstone National Park

Yellowstone River (1985)
Source is Yellowstone Lake, NW Wyoming. Joins Missouri River at Montana-North Dakota border.
IN Index counties or regions as applicable.
SA Great Plains
SA Montana
SA North Dakota
SA Rocky Mountains
SA United States
SA Wyoming
SA Yellowstone National Park

Yemen (1978)
After 1992, refers to Republic of Yemen which includes the former Yemen Arab Republic and the People's Democratic Republic of Yemen (Southern Yemen) which were united in 1990. Before 1993, used for Arab Republic of Yemen. Before 1993, also search Southern Yemen, Peoples Democratic Republic of Yemen, and South Arabia.
CO N123000N183000
   E0530000E0423000
UF Peoples Democratic Republic of Yemen
UF South Arabia
UF Southern Yemen
BT Arabian Peninsula
BT Asia
NT Aden Yemen
SA Arabian Shield
SA Middle East
SA Near East
SA Red Sea Basin
SA Red Sea Rift

Yen Bai
 use Yenbay Vietnam

Yen Bay
 use Yenbay Vietnam

Yenbay
 No longer a valid term for GeoRef. As of 1993 see Yenbay Vietnam.

Yenbay Vietnam (1993)
Town on the Red River NW of Hanoi in N Vietnam. Before 1993 also search Yenbay AND Vietnam.
UF Yen Bai
UF Yen Bay
BT Vietnam
BT Far East
BT Asia

Yenisei
 Not a valid term for GeoRef. Use Yenisei Basin, Yenisei River, or Yenisei Ridge.

Yenisei Basin (1978)
River basin which extends into N Mongolia in the S and drains the mountain areas W of Lake Baikal and much of the Central Siberian Plateau. Also search Yenisei.
IN Index Mongolia and/or Russian Federation as applicable.
BT Asia
SA Mongolia
SA Russian Federation
SA Yenisei-Khatanga basin

Yenisei Range
 use Yenisei Ridge

Yenisei Ridge (1978)
Upland which extends N-S for 400 miles along the right bank of the Yenisei River between the Trans-Siberian R.R. and the Stony Tunguska River. A major gold-mining region. Also search Yenisei, Yenisei Range, and Yenisey Ridge. This term has multiple hierarchies.
CO N560000N630000
   E0950000E0893000
UF Enisei Range
UF Enisei Ridge
UF Yenisei Range
UF Yenisey Ridge
BT1 Siberian Platform
BT1 Russian Federation
BT1 Commonwealth of Independent States
BT2 Siberian Platform
BT2 Asia

Yenisei River (1978)
Rises in the E Sayan Mountains of E Tuva Autonomous Republic; it flows W for a short distance and then N across central Siberia into the Kara Sea via Yenisei Bay and Yenisei Gulf. Also search Yenisei. This term has multiple hierarchies.
UF Enisei River
UF Yenisey River
BT1 Russian Federation
BT1 Commonwealth of Independent States
BT2 Asia

Yenisei-Khatanga basin (1978)
Combined drainage basin of the Khatanga River and the Yenisei River. The Khatanga Basin drains much of the northern Central Siberian Plateau S of the Taymyr Peninsula.
IN Index Mongolia and/or Russian Federation as applicable.
BT Asia
SA Khatanga Basin
SA Mongolia
SA Russian Federation
SA Yenisei Basin

Yenisey Ridge
 use Yenisei Ridge

Yenisey River
 use Yenisei River

Yenshan
 use Yanshanian

Yenshanian
 use Yanshanian

Yenshanian Orogeny
 use Yanshanian

Yeoval
 No longer a valid term for GeoRef. See Yeoval Australia.

Yeoval Australia (1993)
Village NW of Sydney in E central New South Wales.
BT New South Wales Australia
BT Australia
BT Australasia

Yerevan
 No longer a valid term for GeoRef. As of 1993, see Yerevan Armenia.

Yerevan Armenia (1993)
City on the Razdan River in W Armenia. Before 1978, also search Erevan or Erivan. This term has multiple hierarchies.
UF Erevan Armenia
UF Erivan Armenia
BT1 Armenia
BT1 Commonwealth of Independent States
BT2 Armenia
BT2 Europe

Yerington
 Valid through 1988. Search in combination with state term. After 1988, use specific city-state term.

Yerington Nevada (1989)
City on the Walker River in W Nevada. Before 1989, search Yerington AND Nevada.
CO N390000N390000
   W1191000W1191000
BT Lyon County Nevada
BT Nevada
BT United States

Yeso Formation (1989)
BT Permian
BT Paleozoic
SA New Mexico
SA Texas

Yessei Russian Federation
 use Yessey Russian Federation

Yessey
 No longer a valid term for GeoRef, As of 1993, see Yessey Russian Federation.

Yessey Russian Federation (1993)
Village N of the Arctic Circle in N Evenk National Okrug in N central Siberia. Before 1978, also search Essei or Yessei. This term has multiple hierarchies.
UF Essei Russian Federation
UF Yessei Russian Federation
BT1 Krasnoyarsk Russian Federation
BT1 Russian Federation
BT1 Commonwealth of Independent States
BT2 Krasnoyarsk Russian Federation
BT2 Asia

Yetropole
 No longer a valid term for GeoRef. See Yetropole Bulgaria.

Yetropole Bulgaria (1993)
Town ENE of Sofia in W central Bulgaria.
BT Sofia Bulgaria
BT Bulgaria
BT Southern Europe
BT Europe

yield
 use yields

yield strength (1978)
SA compression
SA deformation
SA elastic limit
SA elasticity
SA fracture strength
SA mechanical properties
SA strength
SA stress
SA tension

yields (1978)
Before 1978, also search yield.
UF yield
SA fertilizers
SA forests
SA fruits
SA sediment yield
SA soils
SA tillage
SA water yield

Yilgarn (1978)
Gold field ENE of Perth in SW central Western Australia.
CO S300000S300000
   E1180000E1180000
UF Yilgarn goldfield
BT Western Australia
BT Australia
BT Australasia
SA gold ores

Yilgarn Block (1978)
Western Australia.
UF Yilgarn Shield
BT Western Australia
BT Australia
BT Australasia

Yilgarn goldfield
 use Yilgarn

Yilgarn Shield
 use Yilgarn Block

Yishu Fault (1993)
Middle part of Tancheng-Lujiang fault zone, E and NE China.
IN Index provinces as applicable.
BT China
BT Far East
BT Asia
SA Anhui China
SA Jilin China
SA Shandong China

SA Tancheng-Lujiang fault zone

Ylojarvi
No longer a valid term for GeoRef.
See Ylojarvi Finland.

Ylojarvi Finland (1993)
Village NW of Tampere in SW Finland. Copper and wolfram mines in area.
BT Finland
BT Scandinavia
BT Western Europe
BT Europe

Yoakum County Texas (1993)
W Texas. Before 1989, also search Yoakum County AND Texas.
CO N325700N332300
 W1023300W1030500
BT Texas
BT United States

Yokohama
No longer a valid term for GeoRef.
See Yokohama Japan.

Yokohama Japan (1993)
City on W shore of Tokyo Bay in Kanagawa Prefecture in SE Honshu.
BT Kanagawa Japan
BT Honshu
BT Japan
BT Far East
BT Asia

Yoldia Sea (1993)
Upper Pleistocene to Lower Holocene. Stage of Baltic Glaciation. Covered E Norway, Sweden, Baltic Sea, Finland, and parts of Kola Peninsula.
IN Index ages as applicable.
BT upper Quaternary
BT Quaternary
BT Cenozoic
SA Baltic Glaciation
SA Baltic region
SA Baltic Sea
SA Finland
SA Kola Peninsula
SA Norway
SA Sweden

Yolla Bolly Terrane (1989)
Tectonostratigraphic terrane in the Klamath Mountains of N California.
IN Index counties as applicable.
BT California
BT United States
SA Klamath Mountains

Yolo County
Valid through 1988. Search in combination with state term. After 1988, use specific county-state term.

Yolo County California (1989)
Central California. Before 1989, also search Yolo County AND California.
CO N382000N385700
 W1213000W1222800
BT California
BT United States
SA San Francisco Bay region

Yonezawa
No longer a valid term for GeoRef.
See Yonezawa Japan.

Yonezawa Japan (1993)
City in Yamagata Prefecture in N Honshu.
BT Yamagata Japan
BT Honshu

BT Japan
BT Far East
BT Asia

Yongyang
No longer a valid term for GeoRef.
As of 1993 see Yongyang South Korea.

Yongyang South Korea (1993)
Village N of Taegu in E central South Korea. North Kyongsang. Before 1993 also search Yongyang AND Korea.
BT South Korea
BT Korea
BT Far East
BT Asia

Yonne
No longer a valid term for GeoRef.
See Yonne France.

Yonne France (1993)
Department in NE central France.
CO N472000N482700
 E0042000E0025000
BT France
BT Western Europe
BT Europe
NT Auxerre France
NT Avallon France
NT Tonnerre France
SA Morvan
SA Yonne Valley

Yonne River valley
use Yonne Valley

Yonne Valley (1978)
River valley in central France.
IN Index departments as applicable.
UF Yonne River valley
BT France
BT Western Europe
BT Europe
SA Nievre France
SA Seine-et-Marne France
SA Yonne France

York
No longer a valid term for GeoRef. As of 1993, see York England; As of 1989, use York Pennsylvania for city in S Pennsylvania. Also a town in Ontario. Before 1993, search in combination with U.S. state, Canadian province, or country.

York County
Valid through 1988. Search in combination with state term. After 1988, use specific county-state term.

York County Maine (1989)
Southernmost county in Maine. Before 1989, also search York County AND Maine.
CO N430500N434700
 W0702100W0705900
BT Maine
BT United States
SA Great Bay
SA Saco River

York County Nebraska (1989)
SE Nebraska. Before 1989, also search York County AND Nebraska.
CO N404200N410500
 W0972300W0975000
BT Nebraska
BT United States

York County Pennsylvania
 (1989)

S Pennsylvania. Before 1989, also search York County AND Pennsylvania.
CO N394300N401300
 W0761400W0770800
BT Pennsylvania
BT United States
NT York Pennsylvania

York County South Carolina
 (1989)
N South Carolina. Before 1989, also search York County AND South Carolina.
CO N344800N351200
 W0805200W0812900
BT South Carolina
BT United States

York County Virginia (1989)
SE Virginia. Before 1989, also search York County AND Virginia.
CO N370600N372200
 W0762000W0764500
BT Virginia
BT United States

York England (1993)
City in North Yorkshire in N England. Before 1993, also search York AND England.
BT North Yorkshire England
BT Yorkshire England
BT England
BT Great Britain
BT United Kingdom
BT Western Europe
BT Europe

York Pennsylvania (1989)
City in S Pennsylvania. Before 1989, search York AND Pennsylvania.
CO N395700N395700
 W0764400W0764400
BT York County Pennsylvania
BT Pennsylvania
BT United States

York River (1978)
An estuary receiving the Pamunkey River and the Mattaponi River at West Point and flowing SE to Chesapeake Bay. Also in other location.
IN Index counties or region as applicable.
SA Virginia

York River Formation (1989)
Gaspe Peninsula. Overlies Gaspe Limestone; overlain by Battery Point Formation.
BT Lower Devonian
BT Devonian
BT Paleozoic
SA Quebec

Yorke Peninsula (1978)
Between Spencer Gulf on W and Gulf of Saint Vincent on E in SE South Australia.
IN Index state or region as applicable.
BT South Australia
BT Australia
BT Australasia

Yorkshire
No longer a valid term for GeoRef as of 1993.
use Yorkshire England

Yorkshire England (1993)
Former county in N England. In 1974, it was split to form North Yorkshire, West Yorkshire, South Yorkshire, and parts of Cleveland and Humberside counties. Also

search York. Before 1993, also search Yorkshire.
CO N532000N544000
 E0001000W0023000
UF Yorkshire
BT England
BT Great Britain
BT United Kingdom
BT Western Europe
BT Europe
NT North Yorkshire England
NT Sheffield England
SA Holderness
SA Humberside England

Yorktown Formation (1978)
In Chesapeake Group. Divisible into 2 major zones. Zone 1 (at base) corresponds to beds exposed at Raysor Bridge, South Carolina; Zone 2 (upper) includes: Uppermost Yorktown beds at Suffolk, Virginia, the beds at Yorktown, Virginia, and the Chama-bearing bed which is correlated with the aluminous clay of Florida. E Maryland, E Virginia, and North Carolina.
BT upper Miocene
BT Miocene
BT Neogene
BT Tertiary
BT Cenozoic
SA Maryland
SA North Carolina
SA Virginia

Yosemite National Park (1978)
On W slope of the Sierra Madre in E central California. Also search Yosemite.
IN Index counties as applicable.
UF Yosemite Park
BT California
BT United States
SA Madera County California
SA Mariposa County California
SA national parks
SA Tuolumne County California

Yosemite Park
use Yosemite National Park

Yoshino River (1978)
Rises in mountains of NW and flows E to Kii Channel at Tokushima. Also search Yoshino.
BT Shikoku
BT Japan
BT Far East
BT Asia
SA Kochi Japan
SA Tokushima Japan

Young's modulus (1978)
UF stretch modulus
BT elastic constants
SA elasticity

Younger Dryas (1978)
Term used primarily in Europe for an interval of late-glacial time following the Allerod and preceding the Preboreal. Before 1978, search Dryas.
BT Weichselian
BT upper Pleistocene
BT Pleistocene
BT Quaternary
BT Cenozoic
SA Dryas

Younger Granites (1978)
In S Niger and N Nigeria.
BT Jurassic
BT Mesozoic
SA Niger
SA Nigeria

SA West Africa

Younginiformes
use Eosuchia

Ypresian (1978)
Europe. Above Thanetian (Paleocene), below Cuisian.
BT lower Eocene
BT Eocene
BT Paleogene
BT Tertiary
BT Cenozoic
NT London Clay

ytterbium (1978)
Autoposting of rare earths and metals to this term began in 1989.
UF Yb
BT rare earths
BT metals

yttrialite (1978)
Autoposting of sorosilicates began in 1985.
BT sorosilicates
BT orthosilicates
BT silicates

yttrium (1978)
Autoposting of rare earths and metals to this term began in 1989.
UF Y
BT rare earths
BT metals

Yuba County
Valid through 1988. Search in combination with state term. After 1988, use specific county-state term.

Yuba County California (1989)
N central California. Before 1989, also search Yuba County AND California.
CO N385500N393800 W1210300W1214000
BT California
BT United States

Yucatan
No longer a valid term for GeoRef. As of 1993 see Yucatan Mexico or Yucatan Peninsula.

Yucatan Basin (1978)
Undersea feature between the Yucatan Peninsula and Belize on the W and S Cuba on the NE and E. Also search Yucatan.
BT Caribbean Sea
BT North American Atlantic
BT North Atlantic
BT Atlantic Ocean

Yucatan Channel (1978)
Strait between W tip of Cuba and NE Yucatan Peninsula connecting the Gulf of Mexico and the Caribbean Sea.
BT Atlantic Ocean

Yucatan Mexico (1993)
BT Mexico
NT Merida Mexico
SA Gulf Coastal Plain
SA Yucatan Peninsula

Yucatan Peninsula (1978)
Separates the Gulf of Mexico from the Caribbean Sea in SE Mexico. Also search Yucatan.
IN Index states as applicable.
BT Mexico
SA Campeche Mexico
SA Caribbean region
SA Quintana Roo Mexico
SA Yucatan Mexico

Yucatan Shelf (1978)
N and W of the Yucatan Peninsula.
BT Gulf of Mexico
BT North American Atlantic
BT North Atlantic
BT Atlantic Ocean
SA Campeche Bank

Yucca Flat (1978)
NNW of Frenchman Flat in the Nellis Air Force Range and AEC Nuclear Testing Site.
IN Index counties as applicable.
BT Nevada
BT United States
SA Nevada Test Site
SA Nye County Nevada
SA Yucca Mountain

Yucca Mountain (1985)
S Nevada. Autoposting of broader terms to this term began in 1989.
BT Nye County Nevada
BT Nevada
BT United States
SA Basin and Range Province
SA Nevada Test Site
SA Yucca Flat

Yudoma (1978)
River which rises in extreme N Khabarovsk Kray in East Siberia and flows S and SW into the Maya River. This term has multiple hierarchies.
BT1 Russian Federation
BT1 Commonwealth of Independent States
BT2 Asia
SA Khabarovsk Russian Federation

Yudoma Series (1978)
Homogeneous bed of gray dolomite and of dolomite limestone. In N Khabarovsk Kray and S Yakutia in E Siberia.
BT Lower Cambrian
BT Cambrian
BT Paleozoic
SA Russian Federation
SA Siberia

Yudomian (1985)
Aldan Shield area, Russian Federation. Used most often to apply to Lower Cambrian, but some have used it to refer to upper Proterozoic.
IN Index additional age terms as applicable.
SA Aldan Shield
SA Cambrian
SA Eocambrian
SA Lower Cambrian
SA Precambrian
SA Proterozoic
SA Russian Federation
SA upper Precambrian
SA upper Proterozoic

yugawaralite (1978)
BT zeolite group
BT framework silicates
BT silicates

Yugoslav Macedonia (1981)
Region in S Yugoslavia. Autoposting of Yugoslavia to this term began in 1989. As of 1990, Southern Europe and Europe are autoposted to this term. This term has multiple hierarchies.
CO N405000N422200 E0230500E0203000
BT1 Macedonia
BT1 Southern Europe
BT1 Europe
BT2 Yugoslavia

BT2 Southern Europe
BT2 Europe
NT Skopje
SA Serbo-Macedonian Massif

Yugoslavia (1978)
Before 1981, Dinaric Alps was included as a narrower term.
CO N410000N470000 E0230000E0133000
BT Southern Europe
BT Europe
NT Bosnia-Herzegovina
NT Croatia
NT Dalmatia
NT Istria
NT Montenegro
NT Neretva Valley
NT Serbia
NT Serbo-Macedonian Massif
NT Slovenia
NT Yugoslav Macedonia
SA Adriatic region
SA Alps
SA Balkan Peninsula
SA Banat
SA Danube River
SA Danube Valley
SA Dinaric Alps
SA Eastern Alps
SA Julian Alps
SA Karst region
SA Krajiste
SA Macedonia
SA Mediterranean region
SA Montenegro earthquake 1979
SA Morava River valley
SA Osogovo Mountains
SA Pannonia
SA Pannonian Basin
SA Vardar River
SA Vardar Zone

Yukon
Not a valid GeoRef index term after 1971. Was used on level 1 in subfiles G, N and T.
use Yukon Territory

Yukon River (1978)
Formed by confluence of Lewes and Pelly rivers in SW Yukon Territory. It flows NW into Alaska and then SE into Bering Sea S of Norton Sound. Also search Yukon.
IN Index Alaska and/or Yukon Territory as applicable.
BT North America
SA Alaska
SA Yukon Territory
SA Yukon-Koyukuk Basin

Yukon Territory (1978)
Before 1972, also search Yukon.
CO N600000N700000 W1250000W1410000
UF Yukon
BT Western Canada
BT Canada
NT Blow River
NT Dawson
NT Donjek Glacier
NT Donjek River
NT Hess Mountains
NT Kaskawulsh Glacier
NT Keno Hill Yukon Territory
NT Klondike
NT Kluane Lake
NT Mount Wood
NT Ogilvie Mountains
NT Old Crow
NT Royal Creek
NT Selwyn Basin
NT Steele Glacier
NT Wernecke Mountains
NT Whitehorse Yukon Territory
SA Beaver River

SA Cache Creek
SA Cassiar Mountains
SA Chitistone Pass
SA Coast Mountains
SA Denali Fault
SA Eagle River
SA Earn Group
SA Indian River
SA Kayak Shale
SA Liard River
SA Mackenzie Mountains
SA Michelle Formation
SA Nahanni Formation
SA Old Crow Tephra
SA Peel River
SA Richardson Mountains
SA Road River Formation
SA Rocky Mountain Trench
SA Rundle Group
SA Saint Elias Mountains
SA Selwyn Mountains
SA Stewart Valley
SA Stikinia Terrane
SA Tanana River
SA Tintina Fault
SA White River
SA Whitehorse Trough
SA Yukon River
SA Yukon-Tanana Terrane

Yukon-Koyukuk Basin (1993)
W Alaska. Extends from the Yukon Delta to the Brooks Range. Wedge-shaped expanse of Jurassic and Cretaceous sedimentary and igneous rocks.
IN Index Northern Alaska, West-Central Alaska, East-Central and/or Southern Alaska as applicable.
CO N610000N674500 W1480000W1640000
UF Koyukuk Basin
UF Yukon-Koyukuk Province
BT Alaska
BT United States
SA East-Central Alaska
SA Kaltag Fault
SA Northern Alaska
SA Southern Alaska
SA Southwestern Alaska
SA West-Central Alaska
SA Yukon River

Yukon-Koyukuk Province
use Yukon-Koyukuk Basin

Yukon-Tanana Terrane (1993)
In North American Cordillera. Chiefly in Alaska and and Yukon Territory.
IN Index provinces and Alaska as applicable.
BT North America
SA Alaska
SA British Columbia
SA Tintina Fault
SA Yukon Territory

Yukon-Tanana Upland (1978)
Mountainous area between Yukon River and Tanana River in SE central Alaska.
IN Index East-Central Alaska or Southern Alaska as applicable.
BT Alaska
BT United States
SA East-Central Alaska
SA Southern Alaska

Yule Marble (1978)
Former trade name for marble quarried from Leadville Formation in Gunnison County, Colorado. In 1928 the trade name became Yule Colorado Marble.
BT Mississippian

BT Carboniferous
BT Paleozoic
SA Colorado
SA Leadville Formation

Yuma
Valid through 1988. Search in combination with state term. After 1988, use specific city-state term.

Yuma Arizona (1989)
City on the Colorado River in extereme SW Arizona. Before 1989, search Yuma AND Arizona.
CO N324000N324000
W1143900W1143900
BT Yuma County Arizona
BT Arizona
BT United States

Yuma County
Valid through 1988. Search in combination with state term. After 1988, use specific county-state term.

Yuma County Arizona (1989)
SW Arizona. Before 1989, also search Yuma County AND Arizona.
CO N320200N332600
W1132000W1145300
BT Arizona
BT United States
NT Yuma Arizona
SA Buckskin Mountains
SA Chocolate Mountains
SA Date Creek basin

Yuma County Colorado (1989)
NE Colorado. Before 1989, also search Yuma County AND Colorado.
CO N393300N402600
W1020200W1024500
BT Colorado
BT United States

Yunnan
No longer a valid term for GeoRef. See Yunnan China.

Yunnan China (1993)
Province in SW China.
CO N214000N290000
E1061000E0973000
BT China
BT Far East
BT Asia
NT Jinning China
NT Kunming China
NT Lufeng China
NT Red River Fault
NT Tengchong
SA Kunyang Group
SA Lancang-Gengma earthquake 1988
SA Luoquan Formation
SA Tonghai earthquake 1970
SA Yangtze Platform

Yuryuzan
No longer a valid term for GeoRef. As of 1993, see Yuryuzan Russian Federation.

Yuryuzan Russian Federation (1993)
City on the Yuryuzan River in W Chelyabinsk Oblast in the Southern Urals.
BT Chelyabinsk Russian Federation
BT Russian Federation
BT Commonwealth of Independent States

Yvelines
No longer a valid term for GeoRef. See Yvelines France.

Yvelines France (1993)
Department in N France.
CO N482500N491000
E0021200E0012500
BT France
BT Western Europe
BT Europe
NT Versailles France

Z

Zabriskie Quartzite (1985)
SE California and W Nevada. Also search Zabriskie Formation; Zabriskie Quartzite Member.
BT Lower Cambrian
BT Cambrian
BT Paleozoic
SA California
SA Nevada

Zabrze
No longer a valid term for GeoRef. As of 1993 see Zabrze Poland.

Zabrze Poland (1993)
City in central Katowice, S Poland. Before 1993 also search Zabrze AND Poland.
UF Hindenburg
UF Hindenburg in Oberschlesien
BT Katowice Poland
BT Poland
BT Central Europe
BT Europe

Zacatecas
No longer a valid term for GeoRef. As of 1993 see Zacatecas Mexico.

Zacatecas Mexico (1993)
State in central Mexico. Before 1993 also search Zacatecas AND Mexico.
BT Mexico
SA Sierra Madre Occidental
SA Zuloaga Limestone

Zagros (1978)
Mountain system primarily in W and SW Iran.
IN Index Iran and/or Iraq as applicable.
UF Zagros Mountains
UF Zagros Range
BT Middle East
BT Asia
SA Iran
SA Iraq

Zagros Mountains
use Zagros

Zagros Range
use Zagros

Zaire (1978)
Formerly Belgian Congo and now officially the Democratic Republic of the Congo. Before 1972, also see Congo.
CO S130000N051500
E0320000E0120000
UF Belgian Congo
BT Central Africa
BT Africa
NT Bambu Zaire
NT Bandundu Zaire
NT Equatorial Zaire
NT Kasai
NT Kinshasa Zaire
NT Kivu Zaire
NT Kundelungu Plateau
NT Lower Zaire
NT Lukuga River valley
NT Shaba Zaire
NT Upper Zaire
SA Congo
SA Congo Basin
SA Congo River
SA Kasai River
SA Katangan Orogeny
SA Lake Albert
SA Lake Edward
SA Lake Kivu
SA Lake Tanganyika

Zaire River
use Congo River

Zaisan (1978)
Lake primarily in Vostochno (East) Kazakhstan Oblast in extreme E Kazakhstan. Before 1978, also search Zaisan Lake or Zaysan. After 1993, see Zaisan Kazakhstan for the town SE of the lake. This term has multiple hierarchies.
IN Index regions as applicable.
UF Zaisan Lake
UF Zaysan Lake
BT1 Kazakhstan
BT1 Central Asia
BT1 Asia
BT2 Kazakhstan
BT2 Commonwealth of Independent States

Zaisan Basin (1978)
Includes Zaisan Lake and surrounding area between Narym Range of the Altai Mountains on the N and Tarbagatai Range on the S in Vostochno Oblast and Semipalatinsk Oblast in extreme E Kazakhstan. Before 1978, also search Zaisan. This term has multiple hierarchies.
UF Zaisan Depression
UF Zaysan Basin
UF Zaysan Depression
BT1 Kazakhstan
BT1 Central Asia
BT1 Asia
BT2 Kazakhstan
BT2 Commonwealth of Independent States

Zaisan Depression
use Zaisan Basin

Zaisan Kazakhstan (1993)
Town SE of Lake Zaisan in East Kazakhstan. Before 1978, also search Zaysan. This term has multiple hierarchies.
UF Zaysan Kazakhstan
BT1 Kazakhstan
BT1 Central Asia
BT1 Asia
BT2 Kazakhstan
BT2 Commonwealth of Independent States

Zaisan Lake
use Zaisan

Zakopane
No longer a valid term for GeoRef. As of 1993 see Zakopane Poland.

Zakopane Poland (1993)
Town at the N foot of the Tatra Mountains in S Poland. The chief summer resort and winter sports center in Poland. Before 1993 also search Zakopane AND Poland.
BT Nowy Sacz Poland
BT Poland

BT Central Europe
BT Europe

Zambales Ophiolite (1989)
Eocene? Luzon Island, N Philippine Islands.
BT Tertiary
BT Cenozoic
SA Luzon
SA Philippine Islands

Zambesi Valley
use Zambezi Valley

Zambeze Valley
use Zambezi Valley

Zambezi Valley (1978)
River valley in S central and SE Africa. Also search Zambezi; Zambezi River.
IN Index countries as applicable.
UF Zambesi Valley
UF Zambeze Valley
BT Africa
SA Angola
SA Botswana
SA Mozambique
SA Namibia
SA Zambia

Zambezia
No longer a valid term for GeoRef. As of 1993 see Zambezia Mozambique.

Zambezia Mozambique (1993)
District on the Mozambique Channel in E Mozambique. Before 1993 also search Zambezia AND Mozambique.
BT Mozambique
BT East Africa
BT Africa

Zambia (1978)
Formerly Northern Rhodesia.
CO S170000S080000
E0333000E0220000
UF Northern Rhodesia
BT East Africa
BT Africa
NT Kafue
NT Lusaka Zambia
NT Nchanga Zambia
SA Broken Hill Mine
SA Congo Basin
SA Copperbelt
SA Lake Kariba
SA Lake Tanganyika
SA Zambezi Valley

Zamora
No longer a valid term for GeoRef. As of 1993, see Zamora City Spain.

Zamora City Spain (1993)
Refers to only the city in NW Spain. Before 1993, also search Zamora and Spain.
BT Zamora Spain
BT Leon region
BT Spain
BT Iberian Peninsula
BT Southern Europe
BT Europe

Zamora Province
No longer a valid term for GeoRef. As of 1993, see Zamora Spain.

Zamora Spain (1993)
Refers only to the province in NW Spain on Portuguese border. From 1981-1992, also search Zamora Province and Spain. Before 1981, also search Zamora and Spain. For the city, see Zamora City Spain.

CO N410700N421300
   W0051800W0070300
BT Leon region
BT Spain
BT Iberian Peninsula
BT Southern Europe
BT Europe
NT Zamora City Spain

Zanga
   use Razdan

Zangezur   (1978)
   Mountain range in the Lesser Caucasus extending from E of Lake Sevan S to near the Iranian border. Contains copper and molybdenum ores. This term has multiple hierarchies.
   IN   Index former Soviet republics as applicable.
   UF Zangezur Range
   UF Zangezurskiy Mountains
   BT1 Commonwealth of Independent States
   BT2 Lesser Caucasus
   BT2 Caucasus
   BT2 Europe
   SA Armenia
   SA Azerbaidzhan

Zangezur Range
   use Zangezur

Zangezurskiy Mountains
   use Zangezur

Zanzibar   (1978)
   Island in the Indian Ocean off NE coast of Tanzania. United with Tanganyika in 1964 to form Tanzania. As of 1990, East Africa and Africa are autoposted to this term.
   BT Tanzania
   BT East Africa
   BT Africa

Zao   (1978)
   Volcanic mountain SW of Sendai on prefectural boundary in N central Honshu. After 1993, use Zao Japan for the place in Yamagata prefecture.
   UF Zao Volcano
   BT Honshu
   BT Japan
   BT Far East
   BT Asia
   SA Miyagi Japan
   SA Yamagata Japan

Zao Volcano
   use Zao

Zapadna Morava   (1978)
   River which rises in SW Serbia and flows N and then SW joining the Southern Morava River near Stalac to form the Morava River. Called the Moravica River in its upper course.
   UF Western Morava River
   BT Serbia
   BT Yugoslavia
   BT Southern Europe
   BT Europe

Zaporozhe
   No longer a valid term for GeoRef. As of 1993, see Zaporozhe Ukraine.

Zaporozhe Ukraine   (1993)
   Oblast including city on the left bank of the Dnieper River in S Ukraine. Before 1993, also search Zaporozhye. This term has multiple hierarchies.
   UF Zaporzhye Ukraine
   BT1 Ukraine
   BT2 Europe
   BT2 Commonwealth of Independent States

Zaporzhye Ukraine
   use Zaporozhe Ukraine

Zaragoza
   No longer a valid term for GeoRef. See Zaragoza Mexico.

Zaragoza Mexico   (1993)
   Municipality in Mexico, 12 miles NNW of Mexico City. Also name of towns in Coahuila, Chihuahua, and Puebla states.
   BT Mexico
   SA Saragossa City Spain
   SA Saragossa Spain

Zarand Basin   (1978)
   In Zarand area W of Dasht-i-Lut in N Kerman, S Iran. Also search Zarand.
   SA Kerman Iran

Zaria
   No longer a valid term for GeoRef. As of 1993 see Zaria Nigeria.

Zaria Nigeria   (1993)
   Town SW of Kano in N central Nigeria. Before 1993 also search Zaria AND Nigeria.
   BT Nigeria
   BT West Africa
   BT Africa

Zarzis
   No longer a valid term for GeoRef. As of 1993 see Zarzis Tunisia.

Zarzis Tunisia   (1993)
   Town and oasis on the Mediterranean Sea in SE Tunisia. Before 1993 also search Zarzis AND Tunisia.
   BT Tunisia
   RT North Africa
   BT Africa

Zavala County
   Valid through 1988. Search in combination with state term. After 1988, use specific county-state term.

Zavala County Texas   (1989)
   SW Texas. Before 1989, also search Zavala County AND Texas.
   CO N284000N290800
      W0992000W1001000
   BT Texas
   BT United States

Zawiercie
   No longer a valid term for GeoRef. As of 1993 see Zawiercie Poland.

Zawiercie Poland   (1993)
   BT Katowice Poland
   BT Poland
   BT Central Europe
   BT Europe
   SA Czestochowa-Zawiercie Basin

Zaysan Basin
   use Zaisan Basin

Zaysan Depression
   use Zaisan Basin

Zaysan Kazakhstan
   use Zaisan Kazakhstan

Zaysan Lake
   use Zaisan

Zealand   (1978)
   Largest and easternmost of the major islands of Denmark. Island on which Copenhagen is located.
   UF Seeland
   UF Sjaeland
   BT Denmark
   BT Scandinavia
   BT Western Europe
   BT Europe
   SA Copenhagen Denmark
   SA Stevns Klint

Zechstein   (1978)
   Composed of evaporites (celestite and rock salt). In Denmark and Germany. Also search Zechstein sedimentary rocks, and Zechstein evaporites.
   BT Upper Permian
   BT Permian
   BT Paleozoic
   SA Denmark
   SA Germany
   SA Kupferschiefer

Zeehan
   No longer a valid term for GeoRef. See Zeehan Australia.

Zeehan Australia   (1993)
   Mining town in W Tasmania.
   BT Tasmania Australia
   BT Australia
   BT Australasia

Zeeland
   No longer a valid term for GeoRef. As of 1993 see Zeeland Netherlands.

Zeeland Netherlands   (1993)
   Province consisting of several islands on the North Sea coast and part of the mainland S of the Western Scheldt Estuary in SW Netherlands. Before 1993 also search Zeeland AND Netherlands.
   BT Netherlands
   BT Western Europe
   BT Europe

Zelten
   No longer a valid term for GeoRef. As of 1993 see Zelten Libya.

Zelten Libya   (1993)
   Village W of Tripoli in extreme NW Libya. Before 1993 also search Zelten AND Libya.
   BT Libya
   BT North Africa
   BT Africa

zeolite
   As of 1981, no longer a valid term for GeoRef. Use zeolite group, zeolite facies or (as of 1982) zeolite deposits.

zeolite deposits   (1982)
   Before 1982, search zeolite (truncated) AND the appropriate economic geology category codes (section headings). Before 1981, also search zeolite (truncated) AND deposits.
   SA industrial minerals
   SA Khekordzula Deposit
   SA zeolite facies
   SA zeolite group

zeolite facies   (1978)
   BT facies
   SA metamorphic rocks
   SA zeolite deposits
   SA zeolite group

zeolite group   (1978)
   Autoposting of this term to analcime and thomsonite began in 1981. Autoposting to zeolite and zeolites began in 1981 and 1985 respectively.
   BT framework silicates
   BT silicates
   NT analcime
   NT chabazite
   NT clinoptilolite
   NT edingtonite
   NT epistilbite
   NT erionite
   NT faujasite
   NT ferrierite
   NT garronite
   NT gmelinite
   NT harmotome
   NT heulandite
   NT laumontite
   NT leonhardite
   NT mesolite
   NT mordenite
   NT natrolite
   NT offretite
   NT phillipsite
   NT scolecite
   NT stellerite
   NT stilbite
   NT thomsonite
   NT wairakite
   NT yugawaralite
   SA clathrates
   SA zeolite deposits
   SA zeolite facies
   SA zeolitization

zeolites
   As of 1982, no longer a valid term for GeoRef. Use zeolite group, zeolite facies or (as of 1982) zeolite deposits.

zeolitization   (1978)
   BT metasomatism
   SA hydrothermal alteration
   SA zeolite group

Zeravshan Range   (1978)
   Branch of the Tien Shan extending from Alai Range W for about 200 miles in W Tadzhikistan. Sometimes considered part of the Pamir-Alai system. Also search Zeravshan. This term has multiple hierarchies.
   UF Zeravshanskiy Range
   BT1 Tadzhikistan
   BT1 Asia
   BT2 Tadzhikistan
   BT2 Commonwealth of Independent States
   SA Alai Range
   SA Pamirs
   SA Tien Shan
   SA Zeravshan-Hissar

Zeravshan River   (1978)
   Rises at W end of Alai Range and flows W disappearing in the desert near Bukhara. Also search Zeravshan. This term has multiple hierarchies.
   IN   Index former Soviet republics as applicable.
   BT1 Commonwealth of Independent States
   BT2 Central Asia
   BT2 Asia
   SA Tadzhikistan
   SA Uzbekistan

Zeravshan-Hissar   (1978)
   Parallel ranges of the Tien Shan in W Tadzhikistan. The Zeravshan Range lies to the N of the Hissar

Range. This term has multiple hierarchies.
BT1 Tadzhikistan
BT1 Asia
BT2 Tadzhikistan
BT2 Commonwealth of Independent States
SA Hissar Range
SA Tien Shan
SA Zeravshan Range

Zeravshanskiy Range
use Zeravshan Range

Zerenda
No longer a valid term for GeoRef. As of 1993, see Zerenda Kazakhstan.

Zerenda Kazakhstan (1993)
Village S of Kokchetav in Kokchetav Oblast in N Kazakhstan. This term has multiple hierarchies.
BT1 Kokchetav Kazakhstan
BT1 Kazakhstan
BT1 Central Asia
BT1 Asia
BT2 Kokchetav Kazakhstan
BT2 Kazakhstan
BT2 Commonwealth of Independent States

Zermatt
No longer a valid term for GeoRef. As of 1993 see Zermatt Switzerland.

Zermatt Switzerland (1993)
Resort village in Pennine Alps in S Valais, SW Switzerland. Before 1993 also search Zermatt AND Switzerland
BT Valais Switzerland
BT Switzerland
BT Central Europe
BT Europe

Zetland
use Shetland Islands

Zeya (1978)
River in Transbaikalia which rises in the Stanovoy Mountains in E Chita Oblast and flows S and SE into the Amur River. This term has multiple hierarchies.
BT1 Russian Federation
BT1 Commonwealth of Independent States
BT2 Asia

Zeya-Bureya Basin (1978)
Fertile plain between the lower courses of the Zeya and Bureya rivers in SE Amur Oblast in Transbaikalia. This term has multiple hierarchies.
UF Zeya-Bureya Depression
UF Zeya-Bureya Plain
BT1 Amur Russian Federation
BT1 Russian Federation
BT1 Commonwealth of Independent States
BT2 Amur Russian Federation
BT2 Asia
SA Amur region

Zeya-Bureya Depression
use Zeya-Bureya Basin

Zeya-Bureya Plain
use Zeya-Bureya Basin

Zhamanshin Crater (1993)
Aktyubinsk Province, NW Kazakhstan. Considered doubtful as a meteor crater. Associated with fused glass and tektites. This term has multiple hierarchies.

BT1 Aktyubinsk Kazakhstan
BT1 Kazakhstan
BT1 Central Asia
BT1 Asia
BT2 Aktyubinsk Kazakhstan
BT2 Kazakhstan
BT2 Commonwealth of Independent States
SA irghizite

zhamanshinite (1993)
SA cryptoexplosion features
SA impactite
SA metamorphic rocks

Zhdanov
No longer a valid term for GeoRef. As of 1993, see Zhdanov Ukraine.

Zhdanov Ukraine (1993)
City on N shore of Sea of Azov in Donetsk Oblast, SE Ukraine. Before 1993, also search Mariupol. This term has multiple hierarchies.
UF Mariupol Ukraine
BT1 Ukraine
BT1 Europe
BT2 Ukraine
BT2 Commonwealth of Independent States

Zhejiang
No longer a valid term for GeoRef. See Zhejiang China.

Zhejiang China (1993)
Province. E China. Also search Chekiang.
CO N271000N311000 E1230000E1180000
UF Chekiang China
BT China
BT Far East
BT Asia
NT Changxing China
SA Chihsia Formation

Zhetybay
No longer a valid term for GeoRef. As of 1993, see Zhetybay Kazakhstan.

Zhetybay Kazakhstan (1993)
Village in Uralsk Oblast in extreme W Kazakhstan. This term has multiple hierarchies.
BT1 Kazakhstan
BT1 Central Asia
BT1 Asia
BT2 Kazakhstan
BT2 Commonwealth of Independent States
SA Southern Urals

Zhitkovichi
No longer a valid term for GeoRef. As of 1993, see Zhitkovichi Belarus.

Zhitkovichi Belarus (1993)
Town WNW of Mozyr in S Belarus. This term has multiple hierarchies.
BT1 Belarus
BT1 Europe
BT2 Belarus
BT2 Commonwealth of Independent States

Zhitomir
No longer a valid term for GeoRef. As of 1993, see Zhitomir Ukraine.

Zhitomir Ukraine (1993)
Oblast and city W of Kiev on the Teterev River in N Ukraine. This term has multiple hierarchies.
BT1 Ukraine
BT1 Europe
BT2 Ukraine

BT2 Commonwealth of Independent States
NT Korosten Ukraine

Zhoukoudian (1985)
China.
UF Choukoutien
BT Quaternary
BT Cenozoic

Zhujiang River (1989)
Flows from Canton to the South China Sea. Also search Pearl River AND China.
UF Canton River
UF Pearl River, China
BT Guangdong China
BT China
BT Far East
BT Asia
SA Pearl River
SA South China Sea

Ziaetdin
No longer a valid term for GeoRef. As of 1993, see Ziaetdin Uzbekistan.

Ziaetdin Uzbekistan (1993)
Station on the Trans-Caspian RR SE of Navoi in S central Uzbekistan. Before 1978, also search Ziatdin. This term has multiple hierarchies.
UF Ziatdin Uzbekistan
BT1 Uzbekistan
BT1 Asia
BT2 Uzbekistan
BT2 Commonwealth of Independent States

Ziatdin Uzbekistan
use Ziaetdin Uzbekistan

Zielona Gora
No longer a valid term for GeoRef. As of 1993 see Zielona Gora Poland.

Zielona Gora Poland (1993)
City and province on the East German border in W Poland. Before 1993 also search Zielona Gora AND Poland.
BT Poland
BT Central Europe
BT Europe
NT Nowa Sol Poland

Zillertal Alps (1978)
Range of Eastern Alps W of the Hohe Tauern in the Tyrol and in Italy's Trentino-Alto Adige. Also search Zillertal.
IN Index countries as applicable.
UF Zillertaler Alps
BT Eastern Alps
BT Alps
BT Europe
SA Trentino-Alto Adige Italy
SA Tyrol Austria

Zillertaler Alps
use Zillertal Alps

Zimbabwe (1981)
Replaces Rhodesia. Before 1981, also search Rhodesia. Before 1973, also search Southern Rhodesia.
CO S223000S153000 E0330000E0251500
UF Rhodesia
UF Southern Rhodesia
BT Southern Africa
BT Africa
NT Great Dyke
NT Harare Zimbabwe
NT Kariba Zimbabwe
NT Rhodesian Plateau

NT Selukwe Zimbabwe
SA Beaufort Group
SA Bulawayan Group
SA Dwyka Formation
SA Ecca Group
SA Lake Kariba
SA Limpopo Basin
SA Limpopo Belt

zinc (1978)
Chemical element. As of 1981, use zinc ores for zinc as a commodity. Autoposting of metals to this term began in 1989.
UF Zn
BT metals
NT Zn-65
SA chalcophile elements
SA heavy metals
SA lead-zinc deposits
SA zinc ores

zinc blende
use sphalerite

zinc ores (1981)
Before 1981, also search (zinc OR Zn) AND (deposit OR deposits OR ore OR ores OR economic) in the basic index. Autoposting of metal ores to this term began in 1985.
IN Commodity. See List C.
BT metal ores
SA Agrokipia Deposit
SA base metals
SA Bunker Hill Mine
SA Coronation Mine
SA hemimorphite
SA lead-zinc deposits
SA mississippi valley-type
SA sphalerite
SA zinc
SA zincite

zinc spar
use smithsonite

zinc spinel
use gahnite

zinc sulfides (1985)
BT sulfides
SA gossan

zincite (1978)
UF red zinc ore
BT oxides
SA zinc ores

zinckenite (1978)
UF zinkenite
BT sulfantimonites
BT sulfosalts

zinkenite
use zinckenite

zinnwaldite (1978)
From 1978-1980, halides and sheet silicates were autoposted to this term. Autoposting of fluorides and mica group began in 1981. As of 1989, halides, sheet silicates and silicates are also autoposted to this term. This term has multiple hierarchies.
UF zinwaldite
BT1 fluorides
BT1 halides
BT2 mica group
BT2 sheet silicates
BT2 silicates

Zinovievsk
Not a valid term for GeoRef. See Kirovograd Ukraine.

zinwaldite
use zinnwaldite

Zirabulak
No longer a valid term for GeoRef. As of 1993, see Zirabulak Uzbekistan.

Zirabulak Uzbekistan (1993)
Station on the Trans-Caspian RR W of Kattakurgan in Samarkand Oblast in S central Uzbekistan. This term has multiple hierarchies.
BT1 Samarkand Uzbekistan
BT1 Uzbekistan
BT1 Asia
BT2 Samarkand Uzbekistan
BT2 Uzbekistan
BT2 Commonwealth of Independent States

zircon (1978)
Mineral. As of 1981, use zircon deposits for zircon as a commodity. Autoposting of nesosilicates began in 1985.
BT nesosilicates
BT orthosilicates
BT silicates
SA cyrtolite
SA gems
SA heavy minerals
SA zircon deposits

zircon deposits (1981)
Before 1981, search zircon AND deposits.
IN Commodity. See List C.
SA heavy mineral deposits
SA zircon

zirconium (1978)
Autoposting of metals to this term began in 1989.
UF Zr
BT metals
NT Zr-95

zirconolite (1978)
As of 1981, used for the mineral zirkelite. Before 1981, also search zirkelite AND minerals.
BT niobates
BT oxides

zirkelite (1978)
As of 1981, restricted to use as the igneous rock; for the mineral, use zirconolite. From 1978-1980, oxides; pyroclastics and glasses were autoposted to this term. Autoposting of glasses and volcanic rocks began in 1981. As of 1993, this term is considered obsolete.
BT glasses
BT volcanic rocks
BT igneous rocks

Zlatibor Massif
use Zlatibor Mountains

Zlatibor Mountains (1978)
Range of the S Dinaric Alps in W Serbia, central Yugoslavia. This term has multiple hierarchies.
UF Zlatibor Massif
BT1 Dinaric Alps
BT1 Southern Europe
BT1 Europe
BT2 Dinaric Alps
BT2 Eastern Alps
BT2 Alps
BT2 Europe
BT3 Serbia
BT3 Yugoslavia
BT3 Southern Europe
BT3 Europe

Zlichovian (1985)
Europe. Devonian, above Pragian and below Eifelian.
BT Devonian
BT Paleozoic
SA Lower Devonian
SA Middle Devonian

Zloty Stok
No longer a valid term for GeoRef. As of 1993 see Zloty Stok Poland.

Zloty Stok Poland (1993)
Town E of Klodzko in Lower Silesia in E Walbrzych, SW Poland. Before 1993 also search Zloty Stok AND Poland.
UF Reichenstein
UF Rowne
BT Walbrzych Poland
BT Poland
BT Central Europe
BT Europe

Zmeinogorsk
No longer a valid term for GeoRef. As of 1993, see Zmeinogorsk Russian Federation.

Zmeinogorsk Russian Federation (1993)
Town SE of Rubtsovsk in S Altai Kray near the E Kazakhstan border. This term has multiple hierarchies.
BT1 Altai Russian Federation
BT1 Russian Federation
BT1 Commonwealth of Independent States
BT2 Altai Russian Federation
BT2 Asia

Zn
use zinc

Zn-65 (1985)
Autoposting of broader terms began in 1989. This term has multiple hierarchies.
BT1 radioactive isotopes
BT1 isotopes
BT2 zinc
BT2 metals

Zoantharia (1978)
Autoposting of this term to Actiniaria, Corallimorpharia, Heterocorallia, Hexactiniaria, Rugosa, Scleractinia, Tabulata and Zoanthiniaria began in 1989.
BT Anthozoa
BT Coelenterata
BT Invertebrata
NT Actiniaria
NT Corallimorpharia
NT Heterocorallia
NT Hexactiniaria
NT Rugosa
NT Scleractinia
NT Tabulata
NT Zoanthiniaria

Zoanthiniaria (1978)
Order. Autoposting of Zoantharia to this term began in 1989.
BT Zoantharia
BT Anthozoa
BT Coelenterata
BT Invertebrata

Zod
No longer a valid term for GeoRef. As of 1993, see Zod Armenia.

Zod Armenia (1993)
Village in E Armenia. This term has multiple hierarchies.
BT1 Armenia
BT1 Commonwealth of Independent States
BT2 Armenia
BT2 Europe

zodiacal dust
use cosmic dust

zoisite (1978)
BT epidote group
BT sorosilicates
BT orthosilicates
BT silicates
SA clinozoisite
SA tanzanite

Zonal soils (1978)
BT soils
SA Brown soils
SA Chernozems
SA Chestnut soils
SA Desert soils
SA podzolization
SA Podzols
SA Red soils
SA Sierozems
SA soil group
SA Tundra soils

zonation
A valid term through 1976. See zoning; crystal zoning.

zone of aeration
use unsaturated zone

zone of mobility
use asthenosphere

zone, auroral
Not a valid term for GeoRef. Usually out-of-scope. See auroral zone.

zone, Benioff
use Benioff zone

zone, oxidation
use oxidation zone

zones
A valid term through 1977. See zoning; crystal zoning.

zones, fault
use fault zones

zones, fracture
use fracture zones

zones, low-velocity
use low-velocity zones

zones, rift
use rift zones

zones, shear
use shear zones

zones, subduction
use subduction zones

zones, suture
use suture zones

zones, transition
use transition zones

Zonguldak
No longer a valid term for GeoRef. As of 1993 see Zonguldak Turkey.

Zonguldak Turkey (1993)
Province and city on the Black Sea in NW Anatolia, N Turkey. Before 1993 also search Zonguldak AND Turkey.
BT Turkey
BT Middle East
BT Asia
NT Amasra Turkey

zoning (1978)
As of 1982, use crystal zoning for zoning within crystals. Also search zonation; zones.
SA aureoles
SA crystal zoning
SA haloes
SA intrusions
SA metamorphism
SA metasomatism
SA seismic zoning

zoogeography
Not a valid term for GeoRef. Before 1971, included use in subfiles E and N.
use biogeography

Zoophycos (1985)
BT ichnofossils

zooplankton (1978)
BT plankton

zooxanthellae (1978)
SA algae
SA Protista

Zr
use zirconium

Zr-95 (1978)
Autoposting of broader terms began in 1989. This term has multiple hierarchies.
BT1 radioactive isotopes
BT1 isotopes
BT2 zirconium
BT2 metals

Zubair Formation (1978)
Hauterivian to lower Aptian. In N, central, and S Iraq. In S, comprises over 1200 feet of sandstones, siltstones, and shales.
BT Lower Cretaceous
BT Cretaceous
BT Mesozoic
SA Aptian
SA Hauterivian
SA Iraq
SA Middle East

Zulia
No longer a valid term for GeoRef. As of 1993 see Zulia Venezuela.

Zulia Venezuela (1993)
State encircling almost all of Lake Maracaibo in NW Venezuela. Before 1993 also search Zulia AND Venezuela.
BT Venezuela
BT South America
SA Guajira Peninsula
SA Lake Maracaibo
SA Maracaibo Basin
SA Sierra de Perija

Zuloaga Limestone (1978)
Coahuila and Zacatecas, Mexico.
BT Upper Jurassic
BT Jurassic
BT Mesozoic
SA Coahuila Mexico
SA Zacatecas Mexico

Zululand (1978)
Region in NE Natal.
BT Natal South Africa
BT South Africa
BT Southern Africa
BT Africa

Zungaria
use Junggar

Zuni Mountains (1978)
Range in McKinley and Cibola counties in W New Mexico.
IN Index counties as applicable.
BT New Mexico
BT United States
SA Cibola County New Mexico
SA McKinley County New Mexico

zunyite (1978)

From 1978-1984, halides and orthosilicates were autoposted to this term. Autoposting of sorosilicates began in 1985. As of 1989, orthosilicates and silicates are autoposted to this term. This term has multiple hierarchies.
BT1 halides
BT2 sorosilicates
BT2 orthosilicates
BT2 silicates

Zurich
No longer a valid term for GeoRef. As of 1993 see Zurich Switzerland.

Zurich Switzerland (1993)
Canton in N Switzerland. Also a city at NW end of Lake of Zurich. Before 1993 also search Zurich AND Switzerland.
BT Switzerland
BT Central Europe
BT Europe
SA Lake of Zurich

zussmanite (1978)
BT sheet silicates
BT silicates

Zwickau
No longer a valid term for GeoRef. See Zwickau Germany.

Zwickau Germany (1993)
City on the Mulde River S of Leipzig in W Saxony, E Germany.
BT Saxony Germany
BT Germany
BT Central Europe
BT Europe

Zyryanka
No longer a valid term for GeoRef. As of 1993, see Zyryanka Russian Federation.

Zyryanka Russian Federation (1993)
Town on the Kolyma River at the mouth of the Zyryanka River in NE Yakutia. A coal mining center. This term has multiple hierarchies.
BT1 Yakutia Russian Federation
BT1 Russian Federation
BT1 Commonwealth of Independent States
BT2 Yakutia Russian Federation
BT2 Asia

Zyryanovsk
No longer a valid term for GeoRef. As of 1993, see Zyryanovsk Kazakhstan.

Zyryanovsk Kazakhstan (1993)
City SE of Ust Kamengorsk in Vostochno (East) Kazakhstan Oblast in extreme E Kazakhstan. This term has multiple hierarchies.
BT1 Kazakhstan
BT1 Central Asia
BT1 Asia
BT2 Kazakhstan
BT2 Commonwealth of Independent States

HIERARCHIES AND OTHER LISTS

These lists, A through R are used by the GeoRef indexers for selecting index terms and was formerly used to arrange them into the three-level index entries found in the printed products made from the GeoRef database.

Before 1993 each complete index entry (set) had three parts: first-, second-, and third-level terms. The three entries below appeared in the 1985 *Bibliography and Index of Geology:*

(1) 1st level, 2nd level
    3rd level

        Illinois,
        geomorphology,
        solution features

(2) 1st level, 2nd level
    3rd level

        Illinois,
        geophysical surveys,
        surveys

(3) 1st level, 2nd level
    3rd level

        Illinois,
        hydrogeology,
        ground water

All of the above documents were indexed with the term Illinois. Then, in addition to Illinois, document (2) also was indexed with the terms geophysical surveys (level 2 term) and surveys (level 3 term).

The special lists which follow are itemized on the contents page. They consist of categories of terms used in indexing. Hierarchical arrangements are presented for fossils (List F), age terms (List E), meteorites (List G), igneous rocks (List H), soils (List M), etc.

For each list there are explanatory notes on indexing, online searching and searching in the *Bibliography and Index of Geology* (B.I.G.).

Under "Indexing" the current indexing practice is given. These should be read along with the instructions under the individual terms in the body of the Thesaurus.

The notes on searching attempt to guide the searcher in the use of the list. Searchers might also read the notes on Indexing for further clues but should be aware that these notes reflect current practice which in some cases differs from past. Further notes on specific terms are in the body of the Thesaurus.

LIST A - LEVEL-ONE TERMS

In 1993 the three-level index was abolished and usage of level-one terms was discontinued. Specific terms became access points in the printed *Bibliography and Index of Geology.*

This list consists of terms used as level 1 index terms in GeoRef before 1993. All List A terms are now also mentioned in the body of the Thesaurus, including those no longer in use. Level 1 terms are main topics in the papers they index. They are also broad terms. There are relatively few level 1 terms and throughout the lifespan of GeoRef they have been the most carefully controlled of the index terms.

Level 1 terms were entry points in the subject index of the five subfiles of GeoRef:

(B) Bibliography and Index of Geology, 1969-1992
(E) Bibliography and Index of Geology Exclusive of North America, 1933-1968
(G) Geophysical Abstracts, 1966-1971
(N) Bibliography and Index of North American Geology, 1785-1968
(T) Bibliography of Theses in Geology

INDEXING

In the index sets used in GeoRef there were three levels. The level 2 terms subdivided the citations on level 1, the level 3 terms those on level 2.

e.g. Brazil   plate tectonics   igneous rocks
        stratigraphy   mechanism   basalts
          Cretaceous   convection   textures

ONLINE SEARCHING

In online searches, a term can be limited to its occurrence on level 1. Such a search will retrieve citations in which the term was used as a main topic, as opposed to papers in which it was used on level 2 or level 3 as a subtopic, or as a supplemental index term. Use of this technique is recommended only for the following situations:

Broad area terms (United States, Canada, Australia, and the following continents, Africa, Asia, Europe, and South America) where the searcher is attempting to obtain references to treatments of the area as a whole.

Disciplines treated as a whole (geology, mineralogy, crystallography, sedimentary petrology, petrology, extraterrestrial geology, economic geology, hydrogeology, paleontology, paleobotany, stratigraphy, geophysics, structural geology, palynology, micropaleontology).

To search on ORBIT and STN in this manner, prefix the term with an asterisk. In online searches on DIALOG, a term can be limited to its occurrence on level 1 by using the limit command and limiting the search to major descriptors (/MAJ). Use of this technique is not recommended in other applications because of the nature of GeoRef's hierarchical index entries. Level 1 terms are very general. Some terms may be used both as level 1 terms and as cross-references. An example of this type of term is gold ores. When associated with a geographic area a cross reference is generated from the geographic entry. Gold ores would not be considered level 1 in this instance. Note that beginning in 1993 Level 1 terms are no longer indicated.

B.I.G. SEARCHING

The letters B, E, G, N, and T, following terms in the list, stand for the subfiles in which the term was used. For subfile B the years during which the term was valid are given as well. For the printed bibliography corresponding to each subfile, see above.

Absolute age B72-92
Absolute age, dates B69-71 E G N T
Absolute age, methods B69-71 E G N T
Absorption spectrophotometry N
Academic institutions B89-92
Acoustical exploration N
Acoustical logging N
Actinium B69-92 N
Aden G
Adriatic Sea B69 E G
Aegean Sea B77-92
Aeronomy B75-92
Afars and Issas G
Afars and Issas Territory B73-77
Afghanistan B69-92 E G
Africa B69-92 E G
Agnatha B81-92
Agnathans B89-92
Alabama B69-92 G N T
Alaska B69-92 G N T
Albania B69-92 E
Alberta B69-92 G N T
Algae B69-92 E N T
Algal flora B81-92
Algeria B69-92
Alkali metals B69 E
Alluvial fans N
Alluvial plains N
Alpha activation analysis N
Alps B69-92 E G
Aluminum B69-92 E N T
Aluminum ores B81-92
American Samoa N
Americium B69-92
Amphibia B69-92 E N
Amphibians B81-92
Amphineura N
Andes B69-92 E G T
Andesite N
Andorra B69-92 E
Angiosperm flora B82-92
Angiosperms B69-92 E N
Angola B69-92 E G
Anhui B92
Anhydrite N
Anion exchange-spectrochemical analysis N

Annelida B69-80 E N T
Antarctic Ocean B69-92 E G T
Antarctic region G
Antarctica B69-92 E G T
Anthozoa B69-71 E N T
Anticlines N
Antimony B69-92 E N T
Antimony ores B81
Apennines B69-92 E
Appalachian Basin N
Appalachians B69 G N T
Arabia B69 E G
Arabian Peninsula B69-92 E
Arabian Sea B77-92
Arachnida B69-71 E N
Archaeocyatha B69-92 E N
Archean B78-92
Arctic G N
Arctic America N
Arctic Ocean B69-92 E G N T
Arctic region B69-92 E T
Argentina B69-92 E G
Argon B69-92 E N T
Arid regions N
Arizona B69-92 G N T
Arkansas B69-92 G N T
Arsenic B69-92 E N
Arsenic ores B81-92
Artesian waters and wells B69-71 E N T
Arthropoda B69-92 E N
Arthropods B85-92
Artifacts B69-71 E N
Asbestos B69-80 E N
Asbestos deposits B81-92
Ascension Island B69-72 E G
Asia B69-92 E G T
Asia Minor B69 E
Asphalt N
Associations B69-92 E N
Astatine N
Asteroidea B69-71 E N T
Asteroids B70-92 G
Asteroyoa B69-71 E N
Astrophysics and solar physics B76-92
Atlantic Coastal Plain B70-92 N
Atlantic Ocean B69-92 E G N T
Atlantic Ocean Islands B81-92
Atlantic region B76-92
Atmosphere B69-92 E G N T
Atomic energy N
Aurora B78-92
Australasia B69-92 E
Australia B69-92 E G
Austria B69-92 E G T
Automatic data processing B69-84 E G N T
Autoradiography B70-92 N
Aves B69-92 E N
Azores B69-92 G
Bacteria B69-92 E N
Bahamas B69-92 G N T
Bahrain B77-92
Balearic Islands B74-92
Balkan Peninsula B69-92 E
Baltic region B69-92 E G
Baltic Sea B69-92 E G
Banda Sea G
Bangladesh B73-92

Barbados B74-92 G N
Barents Sea G
Barite B69-80 E N T
Barite deposits B81-92
Barium B69-92 E N T
Bars N
Basalt N
Base metals B78-92 N
Baselevel N
Basin and Range N
Basin and Range Province B78-92 G
Basin, structural N
Basins, structural B69-71 E N T
Basutoland B69 E
Batholiths B69-71 E N T
Bauxite B69-92 E N T
Bay of Biscay G
Beaches N
Bechuanaland G
Beijing B92
Belgium B69-92 E G
Belize B75-92
Benin B69-92 E
Bentonite B69-80 E N
Bentonite deposits B81-92
Bering Sea B69-92 E G N
Bermuda B69-92 G N T
Bermuda Islands G
Beryl N
Beryllium B69-92 E G N
Beryllium ores B81-92
Bhutan B75-92
Bibliography B69-92 E G N
Biogeochemical prospecting B69-71 E N
Biogeochemical surveys N
Biogeochemistry N
Biogeography B69-92 N
Biography B69-92 E N
Birds B81-92
Bismark Islands G
Bismuth B69-92 E N
Bismuth ores B81-92
Bitumens B69-92 E N
Bituminous sands N
Black Sea B69-92 E G
Blastoidea B69-71 E N
Bogs N
Bohemia B69 E
Bolivia B69-92 E G T
Book reviews B73-92
Borates N
Borneo B69-73 E G
Boron B69-92 E G N T
Boron deposits B81
Botswana B69-92 E G
Boudinage B69 E N
Boulders N
Brachiopoda B69-92 E N T
Brachiopods B81-92
Branchiopoda B69-71 E N
Brazil B69-92 E G T
Breccia B69-71 E N T
Brines B69-92 E N
British Columbia B69-92 G N T
British Guiana G
British Honduras B69-74 N T
British Isles G

Bromine B69-92 E N
Bromine deposits B81-92
Brucite N
Bryophytes B69-92 E N
Bryozoa B69-92 E N T
Bryozoans B81-92
Bulgaria B69-92 E G
Burkina Faso B85-92
Burma B69-92 E G
Burundi B69-92
Cadmium B69-92 E N
Cadmium ores B82-92
Calcite B69-92 E N
Calcite deposits B81-92
Calcium B69-92 E N
Calderas N
Caliche N
California B69-92 G N T
Californium B71-92 N
Cambodia B69-92 E G
Cambrian B69-92 E N T
Cameroon B69-92 E
Cameroun G
Canada B69-92 G N T
Canadian Shield B78
Canal Zone N
Canary Islands B74-92 G
Cape Verde Islands B69-92 G
Carbon B69-92 E G N T
Carbonate rocks N
Carbonates N
Carboniferous B69-92 E N T
Caribbean region B69-92 E G N
Caribbean Sea B69-92 E G N T
Caroline Islands B69 E
Carpathians B69-92 E
Cartography B69-71 E N T
Caspian Sea B69-92
Catalogs B69-92 E N
Caves B69-71 E N T
Cayman Islands B81-92
Celebes Sea B69-92 E
Celtic Sea B77
Cenozoic B69-92 E N T
Central African Republic B69-92 E
Central America B69-92 G N T
Cephalopoda B69-71 E N T
Ceramic materials B69-92 E N T
Cerium B69-92 E N
Cerium ores B82-92
Cesium B69-92 E N T
Ceylon B69-73 E G
Chad B69-92 E G
Chad Republic G
Changes of level B69-92 E G N T
Chelicerata B69-70 E
Chemical analyses N
Chemical analysis B69-92 E N T
Chert N
Chesapeake Bay B69-72 N
Chile B69-92 E G T
China B69-92 E G
Chlorine B69-92 E N T
Chordata B69-92 N
Christmas Island B69-92 E
Chromite B69-80 E N T
Chromite ores B81-92

Chromium B69-92 E N
Cirripedia B69-71 E N
Clay N
Clay mineralogy B69-92 E N T
Clays B69-92 E N T
Coal B69-92 E N T
Coal balls N
Coast Rica N
Cobalt B69-92 E N
Cobalt ores B81-92
Coelenterata B69-92 E N
Collapse structures N
Collections B69-71 E N
Colombia B69-92 E G T
Colorado B69-92 E G N T
Colorado Plateau B78-92 G N
Colorimetric analysis N
Columbia G
Columbia Plateau B78-92
Columbia river B69
Columbium B72 N
Comets G
Comoro Islands B77-92
Compressibility N
Concretions B69-71 E N
Conglomerate N
Congo B69-92 E G
Congo Republic G
Connate water B69-71 N
Connecticut B69-92 G N T
Conodonta B81-92
Conodonts B69-92 E N T
Conservation B78-92
Construction materials B69-92 E N T
Continental drift B69-92 E G N
Continental margin G N
Continental shelf B69-92 E N T
Continental slope B69-92 E G N T
Continents B69-71 E G N
Cook Island G
Copper B69-92 E N T
Copper ores B81-92
Coprolite N
Coprolites B69-92 E N
Coral Sea B77-92 G
Corals B81-92
Core B69-92 E G N
Correlation B69-71 E N
Corsica B69-92 G
Corundum B69-80 E N
Corundum deposits B81-92
Cosmic dust B69-71 N
Cosmic-ray methods G
Cosmogeny G
Costa Rica B69-92 G N
Cratering B69-71 E G N
Craters N
Cretaceous B69-92 E N T
Crete B69
Crinoidea B69-71 E N T
Crust B69-92 E G N T
Crustacea B69-71 E N T
Cryogeology B69
Cryptoexplosion structures B69-71 E G N T
Cryptogam B69
Crystal chemistry B69-92 E N T
Crystal growth B71-92 N

Crystal structure B69-92 E N
Crystallography B69-92 E N T
Cuba B69-92 G N
Curium B69-92 N
Cyprus B69-92 E G
Cystoidea B69-71 E N
Czechoslovakia B69-92 E G
Dacite N
Dahomey B69-76 E G
Dams B78-92
Data processing B85-92
Deception Island G
Deformation B69-92 E G N T
Delaware B69-92 G N
Deltas B69-71 E N T
Denmark B69-92 E G
Density G
Desert pavement N
Deserts N
Deuterium B69-92
Devonian B69-92 E N T
Diabase N
Diagenesis B69-92 E N T
Diamond N
Diamonds B69-92 E N
Diapirs B69-71 E N T
Diatomite B69-92 E N
Diatoms B69-71 E N
Differential thermal analysis B69-77 E N
Dikes B69-71 E N T
District of Columbia B69-92 G N
Djibouti B78-92
Dolomite B69-76 E N
Dolomitization N
Dolostone B77-80
Dolostone deposits B81-92
Domes N
Dominican Republic B72-92 N
Drainage changes N
Drainage patterns N
Dunes N
Dysporium N
Dysprosium B73-92
Earth B69-92 E G N T
Earth current exploration N
Earth currents G N
Earth tides B69-71 E G N
Earth-current exploration G N
Earth-current methods B69-71 E G N
Earth-current surveys B69-71 E G N
Earthquakes B69-92 E G N T
East China Sea B78-92 G
East Germany B78-92
East Pacific Ocean Islands B81-92
Easter Island B69-72 E
Eastern Hemisphere B70-92
Eastern U. S. B78-92
Echinodermata B69-92 E N T
Echinoderms B81-92
Echinoidea B69-71 E N T
Ecology B69-92 E N T
Economic geology B69-92 E N T
Ecuador B69-92 E G T
Education B69-92 E G N
Egypt B69-92 E G T
El Salvador B69-92 G N
Elastic properties B69-71 E G N

Elastic waves N
Elasticity N
Elba G
Electric exploration N
Electrical exploration G N
Electrical logging N
Electrical methods B69-71 E G N T
Electrical properties B69-71 E G N T
Electrical surveys B69-71 E G N T
Electron diffraction analysis B71 N
Electron-diffraction analysis N
Electron microscopy B69-92 E N T
Electron paramagnetic resonance N
Electron probe analysis N
Electron-probe analysis N
Electron spin resonance N
Elements B69-73 E G N
Emission spectroscopy N
Energy sources B69-92 E N
Engineering geology B69-92 E N T
England B69-92 E G
English Channel B69-92 E G
English Channel Islands B81-92
Environmental geology B70-92
Eocene B71-92 N
Epeirogenesis B69-71 E N
Epeirogeny B72-92
Equatorial Guinea B78-92
Erbium B78-92
Eritrea G
Erosion B69-71 E N T
Erosion surfaces N
Estonia B92
Estuaries B69-71 E N
Ethiopia B69-92 E G
Ethiopis G
Eurasia B69-92 E
Europe B69-92 E G
Europium B69-92 E N
Eurypterida B69-71 E N
Evaporite deposits B81-92
Evaporites B69-80 E N T
Evolution B69-71 E N T
Explosion phenomena B69-71 E G N T
Explosions B78-92
Extraterrestrial geology B69-71; 81-92 E T
Faeroe Islands B69-92 E G
Faeroes G
Falkland Islands B69-92 E
Far East B69-92 E
Faroe Islands G
Faults B69-92 E G N T
Fauna N
Feldspar B69-80 E
Feldspar deposits B81-92 N
Fennoscandia B69-71 E
Fernando Poo G
Ferns B81-92
Fiji B69-92 G
Finland B69-92 E G
Fish B81-92
Fjords N
Flame photometric analysis N
Flame photometry N
Flora B69 N
Florida B69-92 G N T

Florida keys B69
Fluid inclusions B69-71; 78-92 E N
Fluorescence N
Fluorine B69-92 E N
Fluorite N
Fluorometric analysis N
Fluorspar B69-92 E N T
Folds B69-92 E G N T
Foliation B69-92 E N T
Foliations N
Foraminifera B69-92 E N T
Foraminifers B81-92
Formation pressure N
Formosa B69-76 E G
Fossil man B78-92
Fossils, problematic B69-77 E N
Fossils, problematical N
Foundations B78-92
Fractures B69-92 E G N T
France B69-92 E G
Francium B85-92 N
Franz Josef Land G
French Guiana B69-92 E
Fuel resources B81-92
Fujian B92
Fulgurites B69 E N
Fuller's earth N
Fumaroles B69-71 E G N
Fungi B69-92 E N
Fusulinidae N
Gabbro N
Gabon B69-92 E G
Gadolinium B69-92 N
Galapagos Islands B69-92 E G
Gallium B69-92 E N T
Gambia B70-73
Gansu B92
Garnet B71 N
Gas chromatographic analyses N
Gas, natural B69-77 E N T
Gastroliths N
Gastropoda B69-71 E N T
Gegenschein G
Gems B69-92 E N
General B69-71 E G N
Geobotanical prospecting N T
Geochemical exploration N
Geochemical methods B70
Geochemical prospecting B69-71 E N
Geochemical surveys B69-71 E N T
Geochemistry B69-92 E N T
Geochronology B69-92 E G N T
Geodes N
Geodesy B69-92 E G N T
Geologic barometry B69-71 E G N
Geologic cryometry B69 E
Geologic exploration N
Geologic hazards B78-92
Geologic mapping N
Geologic thermometry B69-71 E G N
Geological barometry G
Geological exploration B69-71 E N T
Geology B72-92
Geology as a profession N
Geomorphology B69-92 E N T
Geophysical exploration G N
Geophysical logging N

Geophysical methods B69-92 E G N T
Geophysical observations B69
Geophysical research G
Geophysical surveys B69-77 E G N T
Geophysics B69-92 E G N T
Georgia B69-92 G N T
Geosynclines B69-92 E G N
Geotectonics N
Geothermal energy B69-92 E G N
Geothermal gradient G N
Geothermal surveys N
Geothermonetry N
Germanium B69-92 E N
Germany B69-92 E G
Geysers B69-71 E G N
Ghana B69-92 E G
Glacial deposits N
Glacial features N
Glacial geology B72-92 N
Glacial lakes N
Glaciation B69-71 E G N T
Glaciers B69-71 E G N T
Glaciology N
Glauconite B69-80 E N
Glauconite deposits B81-92
Glossaries B69-92 E N
Glossary N
Glossopteris flora B69-71
Goa B69-72 E
Gold B69-92 E N T
Gold ores B81-92
Government agencies B89-92
Grabens N
Granite B69-80 E N
Granite deposits B81-92
Granodiorite N
Graphite B69-80 E N
Graphite deposits B81-92
Graptolites B81-92 N
Graptolithina B69-92 E N T
Gravel B69-80 E N
Gravel deposits B81-92
Gravity anomalies N
Gravity exploration G N
Gravity field, Earth B69-71 E G N T
Gravity field, Moon B70-71 G
Gravity methods B69-71 E G N T
Gravity surveys B69-71 E G N T
Gravity tectonics N
Great Basin B78-92
Great Britain B69-92 E G
Great Lakes B69-92
Great Lakes region B69-92 G N
Great Plains B78-92 N
Greater Antilles B78-92
Greece B69-92 E G
Greenland B69-92 G N T
Greenland Sea G
Groundwater T
Ground water B69-92 E N T
Guadalupe B69 E
Guadeloupe B69-92
Guam G N
Guangdong B92
Guangxi B92
Guatemala B69-92 G N T
Guinea B69-92 E G

Guinea-Bissau B75-92
Guizhou B92
Gulf Coastal Plain B69-92 G N
Gulf of Aden B69-92 E G
Gulf of Alaska G
Gulf of Bothnia G
Gulf of California B69-92 G N
Gulf of Maine N
Gulf of Mexico B69-92 G N T
Gulf of Saint Lawrence G N
Gulf of St. Lawrence G N
Guyana B69-92 E G
Guyots N
Gymnosperm flora B82-92
Gymnosperms B69-92 E N
Gypsum B69-80 E N T
Gypsum deposits B81-92
Hafnium B69-92 E N
Hainan B92
Haiti B69-92 N T
Halogens B69-71 E N
Hawaii B69-92 G N T
Heat flow B69-92 E G N T
Heat transfer G
Heavy metals N
Heavy minerals B69-80 E N T
Heavy mineral deposits B81-92
Hebei B92
Helium B69-92 E N
Helium gas B81-92
Hemichordata B76-92
Henan B92
Heilongjiang B92
High-pressure research N
Highways B78-92
Himalayas B69-92 E
Historical geology N
History B69-71 E N
Holmium B78-92
Holocene B71-92
Holothuroidea B69-71 E N
Honduras B71-92 N
Hong Kong B69-92 E
Hubei B92
Hunan B92
Hungary B69-92 E G
Hydrocarbons N
Hydrogen B69-92 E N
Hydrogeology B69-92 E N T
Hydrology B75-92
Hydrosphere N
Hydrothermal alteration B69-71 E N T
Hydrothermal solutions N
Hydrozoa B69-71 E N
Icarus G
Ice N
Ice ages (ancient) N
Ice ages, ancient B69-71 E N
Ice islands N
Ice, non-glacial B69-71 E N T
Ice, nonglacial G N
Iceland B69-92 E G
Ichnofossils B72-92
Idaho B69-92 G N T
Igneous petrology N
Igneous rocks B69-92 E N T
Illinois B69-92 G N T

Illinois Basin N
Ilmenite N
Impact phenomena G N
Impact statements B78-92
Impactite N
Impactites G N
Inclusions B69-92 E N T
India B69-92 E G T
Indian Ocean B69 E G T
Indian Ocean Islands B81-92
Indiana B69-92 G N T
Indium B69-92 E N
Indochina B69-92 E
Indonesia B69-92 E G T
Industrial Materials B68
Industrial minerals B69-92 E N T
Infrared exploration G N
Infrared methods B69-71 E G N
Infrared spectroscopy N
Infrared surveys B69-71 G N
Inner Mongolia B92
Insecta B69-92 E N
Insects B85-92
Interplanetary space B75-92
Intrusions B69-92 E N T
Invertebrata B69-92 E N T
Invertebrates B81-92
Iodine B69-92 E N
Iodine deposits B81-92
Ionian Sea B69-92 E
Ionosphere B77-92
Iowa B69-92 G N T
Iran B69-92 E G
Iraq B69 E G
Ireland B69-92 E G
Iridium B69-92 E N
Iridium ores B85-92
Irish Sea B69-92 E G
Iron B69-92 E N T
Iron ores B81-92
Island ores N
Isostasy B69-92 E G N
Isotopes B69-92 E G N T
Israel B69-92 E G
Italy B69-92 E G
Ivory Coast B69-92 E G
Jamaica B69-92 G N
Jan Mayen B69-92 E G
Japan B69-92 E G
Japan Sea B69-92 E G
Jiangsu B92
Jiangxi B92
Jilin B92
Johnston Island N
Joints G N
Jordan B69-92 E G
Jupiter B69-92 G
Jurassic B69-92 E N
Kansas B69-92 G N T
Kaolin B69-80 E N
Kaolin deposits B81-92
Karst N
Kashmir B69-71
Katanga B69-73
Kentucky B69-92 G N T
Kenya B69-92 E G
Kerguelen Islands G

Kermadec Islands G
Kiribati B85-92
Korea B69-92 E G
Krypton B69-92 E G N
Kuril Islands B69 E
Kuwait B69-92 E
Labrador B69-92 G N T
Laccoliths N
Lake Erie G
Lake Ontario G
Lake Superior G T
Lake Superior region B69-70 G N
Lakes B69-71 E N T
Lakes, extinct B69-71 E N T
Lamprophyre N
Land subsidence B78-92
Land use B78-92
Landforms N
Landslides N
Lanthanides B69-72 E
Lanthanum B69-92 E N
Laos B69-92 E
Lapland B69
Laser methods G N
Laser surveys N
Laterite N
Laterites B69-71 E N
Latvia B92
Lava B69-92 E N T
Lava flows N
Lead B69-92 E G N T
Lead ores B81-92
Lead-zinc deposits B78-92
Lebanon B69-92 E G T
Lesotho B69-92
Lesser Antilles B77-92
Leveling G
Leveling Networks G
Lexicon N
Lexicons B69-92 E N
Liaoning B92
Liberia B69-92 E G
Libya B69-92 E G
Lichens B70-92
Life G
Lignite B69-92 E N T
Limestone B69-80 E N T
Limestone deposits B81-92
Lineaments N
Lineation B69-92 E N T
Lithium B69-92 E G N
Lithium ores B81-92
Lithofacies N
Lithuania B92
Loess N
Louisiana B69-92 G N T
Low-temperature analysis N
Luminescence B69-71 E G N T
Lutetium B69-92
Luxembourg B69-92 E G
Lycopoda B69-71
Macao G
Macedonia B69
Macquarie Island G
Madagascar B69-74 E G T
Madeira B77-92 G
Magma G

Magmas B69-92 E G N T
Magmas and magmatic differentiation N
Magnesite B69-80 E N
Magnesite deposits B81-92
Magnesium B69-92 E N
Magnesium ores B81-92
Magnetic anomalies N
Magnetic exploration G N
Magnetic field of the Earth N
Magnetic field, Earth B69-71 E G N T
Magnetic field, Jupiter G
Magnetic field, Mars B69 G
Magnetic field, Mercury B69 G
Magnetic field, Moon B69-71 G
Magnetic field, Sun G
Magnetic field, Venus B69 G
Magnetic field, asteroids B69
Magnetic field, interplanetary G
Magnetic field, planets G
Magnetic methods B69-71 E G N T
Magnetic properties B69-71 E G N T
Magnetic surveys B69-71 E G N T
Magnetite B71 N
Magnetosphere B75-92
Magnetotelluric exploration G N
Magnetotelluric methods B69-71 E G N T
Magnetotelluric surveys B69-71 E G N T
Maine B69-92 G N T
Major-element analyses B69-71 E N T
Malacostraca B69-71 E N
Malagasy Republic B74-92
Malawi B69-92 E G
Malay Archipelago B71-92
Malaya B69-72 E G
Malaysia B69-92 E G
Maldive Islands B69-92 E
Mali B69-92 E G
Malta B73-92
Mammalia B69 92 E N T
Mammals B81-92
Man, fossil B69-77 E N
Manganese B69-92 E N T
Manganese ores B81-92
Manitoba B69-92 E G N T
Mantle B69-92 E G N T
Maps B72-92
Marble B69-80 E N
Marble deposits B81
Mariana Islands B69-92 E
Marine geology B69-92 E G N T
Marine installations B78-92
Maritime Provinces B79-92
Mars B69-92 E G
Marshall Islands B69-92 E
Martinique N
Maryland B69-92 G N T
Mass spectroscopy N
Mass wastage N
Massachusetts B69-92 G N T
Mathematical geology B71-92
Mauritania B69-92 E G T
Mauritius B76-92 G
Medical geology B71
Mediterranean region B69-92 E G
Mediterranean Sea B69-92 E G T
Melanesia B74-92

Mercury B69-92 E G N T
Mercury ores B81
Mercury Planet B75-92
Merostomata B69-70 E N
Mesozoic B69-92 E N T
Metal B69 E
Metals B69-92 E N T
Metal ores B81-92
Metamorphic petrology N
Metamorphic rocks B69-92 N T
Metamorphism B69-92 E N T
Metasomatic rocks B78-92
Metasomatism B69-92 N T
Metazoa B69
Meteor craters B69-92 E G N
Meteorites B69-92 E G N T
Meteorology B75-92
Meteors G
Mexico B69-92 G N T
Mica B69-80 E N
Mica deposits B81-92
Michigan B69-92 G N T
Micronesia B69-92 E
Micropaleontology B69-92 E N T
Microscope methods B69-71 E N
Microscope techniques N
Microscopic methods B69 E
Microseisms B69-71 E G N T
Microwave exploration N
Microwave methods G N
Microwave surveys G N
Middle East B69-92 E G
Midway Islands N
Midwest B78-92
Migmatites N
Military geology B69-70 N
Mineragraphy B69-71 E N
Mineral collecting B69-71 E N
Mineral collections N
Mineral data B69-71 E N T
Mineral deposits, genesis B69-92 E N T
Mineral descriptions N
Mineral economics B69-71 E N T
Mineral exploration B69-92 E G N T
Mineral resources B69-92 E N T
Mineral zoning B69-71 E N
Mineralogy B69-92 E N T
Minerals B71
Mining geology B69-92 E N T
Minnesota B69-92 G N T
Minor-element analyses N
Miocene B69-92 E N
Mississippi B69-92 G N T
Mississippi Delta G
Mississippi River N
Mississippi Valley B78-92 G N
Mississippi embayment N
Mississippian B69-92 N T
Missouri B69-92 G N T
Models N
Mohorovicic discontinuity B69-92 E G N
Mollusca B69-92 E N T
Mollusks B81
Molybdenum B69-92 E N T
Molybdenum ores B81-92
Monaco B69-92 E
Monazite B69-80 E N

Monazite deposits B81-92
Mongolia B69-92 E G T
Montana B69-92 G N T
Moon B69-92 E G N T
Morocco B69-92 E G
Mossbauer analysis N
Mossbauer effect G
Mozambique B69-92 E G
Mud volcanoes B69-92 E N
Mudflows N
Museums B69-92 E N
Myriapoda B69-71 E N
Namibia B81-92
Natural bridges N
Natural gas B78-92
Near East 69 E
Nebraska B69-92 G N T
Neodymium B71-92
Neogene B78-92
Neon B69-92 E G N
Neotectonics B78-92
Nepal B69-92 E
Nepheline syenite N
Neptune B69-92 G
Neptunium B72-92
Netherlands B69-92 E G
Netherlands Antilles B69-92 N
Neutron activation analysis N
Neutron diffraction analysis N
Nevada B69-92 G N T
New Britain G
New Britain Island G
New Brunswick B69-92 G N T
New Caledonia B69-85 E G
New England B69-92 G N
New Guinea B69-92 E G T
New Hampshire B69-92 G N T
New Hebrides B69-74 E G
New Jersey B69-92 G N T
New Mexico B69-92 G N T
New South Wales B69-92 E
New York B69-92 G N T
New Zealand B69-92 E G T
Newfoundland B69-92 G N T
Nicaragua B70-92 G N
Nickel B69-92 E N T
Nickel ores B81-92
Niger B69-92 E
Niger Republic G
Nigeria B69-92 E G
Ningxia B92
Niobium B69-92 E N T
Niobium ores B81-92
Nitrate deposits B81-92
Nitrates B68-80 N
Nitrogen B69-92 E N
Noble gases B71-92 N
Nodules B69-92 E N T
Nonmetal deposits B81-92
Nonmetals B74-92
North Africa B69-71 E
North America B69-92 G N T
North Carolina B69-92 G N T
North Dakota B69-92 G N T
North Korea G
North Sea B69-92 E G T
North Vietnam G

Northern Hemisphere B69-92
Northern Ireland B69-92 E G T
Northern Territory B69-92 E
Northwest Territories B69-92 G N T
Norway B69-92 E G
Norwegian Sea G
Nova Scotia B69-92 G N T
Nuclear explosions B69-71 E G N T
Nuclear facilities B78-92
Nuclear magnetic resonance N
Nuclear science N
Nucleosynthesis G
Nutation G
Ocean basins B72-92 G N
Ocean circulation B78-92
Ocean floors B72-92
Ocean waves B72-92
Oceania B69-92 E
Oceanography B69-92 E G N T
Oceans G N
Oceans, circulation B72-77
Ohio B69-92 G N T
Oil and gas fields B69-92 E N T
Oil sand B70-71
Oil sands B69; 72-92 N
Oil shale B69-92 E N T
Okhotsk Sea B69-92 E G
Oklahoma B69-92 G N T
Oligocene B69-92 E
Oman B69-92 E
Ontario B69-92 G N T
Oolites N
Optical data processing N
Optical mineralogy B69-71 N T
Optical properties B69 E
Ordovician B69-92 E N T
Oregon B69-92 G N T
Organic materials B69-92 E N T
Orogeny B69-92 E G N T
Osmium B69-92 E N
Osmium ores B85-92
Ostracoda B69-92 E N T
Ostracods B81-92
Oxygen B69-92 E N T
Ozone B73
Pacific Coast B78-92
Pacific Islands G N
Pacific Ocean B69-92 E G N T
Pacific region B69-92
Pakistan B69-92 E G T
Paladium N
Paleobiogeography N
Paleobotany B69-92 E N T
Paleocene B71-92
Paleoclimatology B69-92 E G N T
Paleoecology B69-92 E N T
Paleogene B78-92
Paleogeography B69-92 E N T
Paleomagnetism B69-92 E G N T
Paleontology B69-92 E N T
Paleosalinity G N
Paleotemperature G
Paleotemperatures N
Paleozoic B69-92 E N T
Palladium B69-92 E N
Palladium ores B85-92
Palynology B69-92 E N

Palynomorphs B69-92 E N T
Panama B69-92 G N T
Pantellerite N
Paper chromatography N
Papua B69-75 E G
Papua New Guinea B75-92
Paragenesis B69-92 E N T
Paraguay B69-92 E
Patterned ground B69-71 E N
Peat B69-92 E N
Pebbles B69-71 E N
Pediments N
Pegmatite B69-92 E N T
Pegmatites N
Pelecypoda B69-71 E N T
Peneplanes N
Pennsylvania B69-92 G N T
Pennsylvanian B69-92 N T
Peridotite N
Periglacial features N
Periglacial phenomena N
Perlite N
Permafrost B69-71; 78-92 E G N
Permeability B69-71 E G N
Permian B69-92 E G N T
Persian Gulf E G
Peru B69-92 E G T
Petrofabrics B69-71 E G N T
Petrogenesis N
Petrography B69 E N
Petroleum B69-92 E N T
Petroleum engineering N
Petrology B69-92 E N T
Phanerozoic B72-92 N
Phase equilibria B69-92 E G N T
Philippine Islands B69-92 E T
Philippine Sea B78-92 G
Philippines G
Phosphate B69-80 E N T
Phosphate deposits B81-92
Phosphorescence N
Phosphorous N
Phosphorus B69-92 E N
Photogeology B69-71 E G N T
Photomicroscopy N
Physical geology N
Physical properties G N
Phytoliths B69
Pisces B69-92 E N T
Placers B69-92 E N
Planetology B71-92
Planets B69-71 E G
Plantae B72-92
Plants B85-92
Plasticity N
Plate tectonics B76-92
Platinum B69-92 E N
Platinum ores B81-92
Playas N
Pleistocene B71-92 N
Pliocene B71-92
Pluto B69-92 G
Plutonium B69-92 E G N
Plutons N
Poland B69-92 E G
Polar wandering G
Polarographic analysis N

Pollution B78-92
Polonium B71-92 N
Polymetallic ores B69-92 E N
Polynesia B74-92
Popular and elementary geology B69-72 E N
Porifera B69-92 E N
Porosity B69-71 E G N T
Portugal B69-92 E G
Portuguese Guinea B69-73 E
Portuguese Timor B76
Potash B69-92 E N
Potassium B69-92 E N T
Praseodymium B78-92
Precambrian B69-92 E N T
Primates N
Prince Edward Island B70-92 G N
Problematic fossils B78-92
Problematical fossils N
Promethium B72-92
Protactinium B69-92 N
Proterozoic B69; 78-92 E
Protista B69-92 E N T
Protozoa B69-71 E N
Pseudomorphs N
Pteridophytes B69-92 E N
Pterobranchia B70-92 N
Pteropoda B69-71 E N
Puerto Rico B69-92 G N T
Pumice B70-80 N
Pumice deposits B81-92
Pyrenees B69-92 E
Pyrite B69-80 E N
Pyrite ores B81-92
Qatar B74-92
Qinghai B92
Quark G
Quartz crystal B69-92 E N
Quartzite N
Quaternary B69-92 E N T
Quebec B69-92 G N T
Queensland B69-92 E
Radar exploration N
Radar methods G N
Radar surveys G N
Radioactive-waste disposal N
Radioactivity B69-71 E G N T
Radioactivity exploration G N
Radioactivity logging N
Radioactivity methods B69-71 E G N T
Radioactivity surveys B69-71 E G N T
Radiochemical analysis N
Radiolaria B69-92 E N T
Radiolarians B81-92
Radiowave methods N
Radiowave surveys G N
Radium B69-92 E N
Radon B69-92 E N
Rare earth deposits B81-92
Rare earths B69-92 E N T
Rare gases B71 N
Reclamation B78-92
Red Sea B69-92 E G T
Red Sea region B76-92
Reefs B69-92 E N T
Reflectivity N
Refractometry N
Refractory materials N

Remote sensing B78-92
Remote-sensing methods G N
Remote-sensing surveys G N
Reptiles B81-92
Reptilia B69-92 E N
Reservoirs B78-92
Reunion B69-92 E
Reunion Island G
Rhenium B69-92 E N
Rhode Island B69-92 G N T
Rhodesia B69-80 E G
Rhodium B69-92 N
Rhodium ores B85-92
Ripple marks N
Rivers B69-71 E N T
Rock creep N
Rock glaciers N
Rock mechanics B78-92 G N
Rocky Mountains B69-92 G N
Romania B69-92 E
Rubidium B69-92 E N T
Rumania G
Ruthenium B69-92 N T
Ruthenium ores B85-92
Rwanda B69-92 E G
Ryukyu Islands B69-68 E
Sahara B69-92 E
Saint Helena G
Saint Pierre and Miquelon B81-92
Saint-Pierre and Miquelon N
Salt B69-92 E N T
Salt tectonics B69-92 E G N T
Salt water intrusion N
Samarium B71-92 N
Samoa B69-92 E T
Samoa Islands B69-92 E
Sampling N
San Marino B81-92
Sand B69-80 E N
Sand volcanoes B69-92 E
Sands B81-92
Sandstone B69-80 E N
Sandstone deposits B81-92
Sao Tome e Principe B85-92
Sarawak B69-75 E
Sardinia B74-92 G
Saskatchewan B69-92 G N T
Saturn B69-92 G
Saudi Arabia B69-92 E G T
Scandinavia B69-92 E G
Scandium B69-92 E N
Scaphopoda B69-71 E N
Scolecodonts N
Scotia Arc G
Scotia Sea G
Scotland B69-92 E G T
Sea mounts N
Sea of Japan B69 E G
Sea water B69-92 E N T
Sea-floor spreading B69-92E
Seamounts B69-71 G N
Sedimentary petrology B69-92 E N
Sedimentary rocks B69-92 E N T
Sedimentary structures B69-92 E N T
Sedimentation B69-92 E N T
Sediments B69-92 E N T
Seiche N

Seismic exploration G N
Seismic methods B69-71 E G N T
Seismic surveys B69-71 E G N T
Seismic waves N
Seismology B69-92 E G N T
Selenium B69-92 E N T
Senegal B69-92 E G
Serpentine N
Seychelles B78-92
Shaanxi B92
Shale N
Shandong B92
Shanghai B92
Shanxi B92
Shatter cones N
Shetland Islands B69-92 E
Shore features N
Shorelines B69-71; 78-92 E N T
Sichuan B92
Sierra Leone B69-92 E
Sikkim B69-77 E
Silica N
Silicate rocks N
Siliceous earth N
Silicification N
Silicon B69-92 E N
Sills N
Silurian B69-92 E N T
Silver B69-92 E N
Silver ores B81-92
Singapore B69-92 E
Slate B69-80 N
Slate deposits B81-92
Slope stability B78-92
Slopes N
Snow B69-71 E N T
Society Islands B69; 75-92 E
Sodium B69-92 E N
Sodium carbonate B71-92 N
Sodium sulfate B69-92 N
Soil mechanics B78-
Soils B69-92 E N T
Solar system B70-71 G
Solomon Islands B69-92 E G
Solubility N
Somali Republic B71-92
Somalia B69 G
South Africa B69-92 E G T
South America B69-92 E G T
South Arabia B69 E
South Australia B69-92 E
South Carolina B69-92 G N
South China Sea B78-92
South Dakota B69-92 G N T
South Korea G
South Sandwich Islands G
South Shetland Islands G
South West Africa G
South-West Africa B69-80 E G
Southeastern U.S. B85-92
Southern Hemisphere B69-92 E
Southern Rhodesia B69-72 E G T
Southern U.S. B81-92
Southern Yemen B77-92
Southwestern U.S. B78-92
Spain B69-92 E G
Spanish Guinea B69-73 E

Spanish Sahara B69-76 E G
Spanish West Africa B69-92 E
specific gravity B69-71 G N
Spectrochemical analysis N
Spectrophotometric analysis N
Spectroscopy B69-92 E G N T
Spermatophyta B81-92 N
Spitsbergen B69-92 E G
Spitzbergen G
Sponges B85-92
Springs B69-92 E N T
Sri Lanka B74-92
Standard materials B76-92
Standard rocks B69-75
Statistical measures N
Statistical methods B69-71 E G N T
Statistics B N
Stocks B69-71 E N T
Stratigraphy B69-92 E N T
Stream capture N
Streams N
Strength G N
Stress N
Stromatolites B69-71 E N
Stromatoporoidea B69-71 E N T
Strontium B69-92 E N T
Strontium ores B82-92
Structural analysis B72-92
Structural geology B69-92 E N T
Styolites N
Submarine geology G N
Subsidence B69-71 E N
Sudan B69-92 E G T
Sulfides B76 N
Sulfur B69-92 E N T
Sulfur deposits B81-92
Sun B69-92 G
Surinam B69-92 E G
Survey organizations B81-92
Surveys B69-80 E N T
Swaziland B69-92 E G
Sweden B69-92 E G
Switzerland B69-92 E G
Symposia B69-92 E N
Syria B69-92 E G
Tahiti B69; 71-92 E
Taiwan B77-92 G
Talc B69-80 E N T
Talc deposits B81-92
Tanganyika B69 E
Tantalum B69-92 E N T
Tantalum ores B81-92
Tanzania B69-92 E G
Tasman Sea B77-92 G
Tasmania B 69-92 E G T
Technetium B76-92
Tectites G
Tectonics B69-92 E G N T
Tectonophysics B71-92
Tektites B69-92 E G N T
Television logging N
Tellurium B69-92 E N
Temperature methods G
Temperature surveys G
Tenerife G
Tennessee B69-92 E G N T
Terbium B70-72; 78-92

Terraces N
Tertiary B69-92 E N T
Tetrapoda B81-92
Tetrapods B81-92
Texas B69-92 G N T
Thailand B69-92 E G
Thallium B69-92 E N
Thallophytes B71-92
Thermal analysis B78-92
Thermal conductivity G N
Thermal methods G N
Thermal properties G N
Thermal springs B69 E G N
Thermal surveys G N
Thermal waters B69-92 E N
Thermodynamic properties B69-73 E G N
Thermogravimetric analysis B73-77 N
Thermoluminescence G N
Thorium B69-92 E N T
Thorium ores B81-92
Thoron B69
Throium B70
Thulium B71-72; 78-92 N
Tianjin B92
Till N
Tillites N
Timor B69-73 E
Timor Sea G
Tin B69-92 E N T
Tin ores B81-92
Titanium B69-92 E N T
Titanium ores B81-92
Togo B69-92 E
Tonga B69-92 E
Tonga Islands B69 E G
Trace elements N
Trace-element analyses B69-71 E N T
Tracks and trails B69-71 E N T
Traps N
Triassic B69-92 E N T
Trieste B69-71 E
Trilobita B69-92 E N T
Trilobites B81-92
Trinidad B69-76 G N
Trindad and Tobago B77-92
Tristan da Cunha G
Tritium B69-92 E N
Trucial Coast B69-73 E
Trucial Oman B69-72
Tsunami G
Tsunamis B69-71 E G N T
Tuff N
Tungsten B69-92 E N T
Tungsten ores B81-92
Tunisia B69-92 E G
Tunnels B78-92
Turbidity currents N
Turkey B69-92 E G
Tyrrhenian Sea B69-92 E
U.S.S.R. G
Uganda B69-92 E G
Ultramafic rocks N
Ultraviolet spectroscopy N
Unconformities B69-71 E N T
Underground installations B78-92
Uniformitarianism N
United Arab Emirates B74-92

United Kingdom B74-92 G
United States B69-92 G N T
Uplifts B69-71 E N
Upper Volta B69-85 E G
Uranium B69-92 E N T
Uranium ores B81-92
Uranus B70-92 G
Uruguay B69-92 E G
USSR B69-92 E
Utah B69-92 G N T
Valleys N
Vanadium B69-92 E N
Vanadium ores B81-92
Vanuatu B85-92
Varves N
Veins B69-71 E N T
Venezuela B69-92 E G T
Venus B69-92 E G
Vermes B69; 74
Vermiculite B69-80 E N
Vermiculite deposits B81-92
Vermont B69-92 G N T
Vertebrata B69-92 E N T
Vertebrates B81-92
Victoria B69-92 E
Viet Nam G
Vietnam B69-92 E G T
Virgin Islands G N T
Virginia B69-92 G N T
Volcanism B69-71 E G N T
Volcanoes B69-71 E G N T
Volcanology B72-92
Wake Island N
Wales B69-92 E G
Washington B69-92 G N T
Waste disposal B78-92
Water B69 E N
Water falls N
Water resources B71-92
Waterways B-78
Weathering B69-92 E N T
Well logging G N
Well-logging B69-92 E T
Wells and drill holes B69-71 E N T
West Africa B69-72 G
West Germany B78-92
West Indies B69-92 G N
West Pacific Ocean Islands B85-92
West Virginia B69-92 G N T
Western Australia B69-92 E T
Western Hemisphere B69-92
Western Interior B78-92
Western Sahara B81-92
Western U.S. B78-92
Williston Basin N
Windward Islands T
Wind work N
Wisconsin B69-92 G N T
World B69 E N
Worms B69-92 E N
Wyoming B69-92 G N T
X-ray analysis B78-92
X-ray diffraction N
X-ray diffraction analyses N
X-ray diffraction analysis B69-77 E N T
X-ray fluorescence analysis N
X-ray radiography N

X-ray spectrographic analysis N
Xenoliths N
Xenon B69-92 E G N
Xinjiang B92
Xizang B92
Yellow Sea B78-92 G
Yemen B69-92 E
Ytterbium B70; 77-92 N
Yttrium B69-92 E N
Yugoslavia B69-92 E G
Yukon G N T
Yukon Territory B69-92
Yunnan B92
Zaire B72-92
Zambia B69-92 E G T
Zhejiang B92
Zimbabwe B81-92
Zinc B69-92 E N T
Zinc ores B81-92
Zircon B69-80 E N
Zircon deposits B81-92
Zirconium B69-92 E N

LIST B - AREA SETS

Before 1993 index entries for areas in the *Bibliography and Index of Geology* consisted of a major area term (see List A), followed by one of the level 2-level 3 combinations in this list.

ONLINE SEARCHING

The level 2 terms here are basically equivalent to the GeoRef categories and can be similarly used for general orientation in searching on a geographic area. Thus a search for iron ores in Nigeria would have:

iron (truncated) AND Nigeria AND economic
    geology

B.I.G. SEARCHING

Before 1993 the level 2 terms here were used to group into broad categories the references treated under level 1 area terms (see List A). Within these categories further subdivision was achieved by the topics on level 3.

areal geology (2)
    Used for entries that might properly be placed under three or more of the level 2 headings such as geomorphology, stratigraphy, structural geology.

    topic (3)
        bibliography*, expeditions, explanatory text (for a map), guidebook*, maps* or locality or regional.

e.g. France
   areal geology
      Narbonne

economic geology (2)
   commodity (List C*) (3)

   e.g. Germany
      economic geology
         coal

engineering geology (2)
   topic (3)
      dams*, earthquakes*, explosions*, foundations*, geologic hazards*, highways*, land subsidence*, marine installations*, nuclear facilities*, permafrost*, reservoirs*, rock mechanics, shorelines*, slope stability*, soil mechanics, tunnels*, underground installations*, waste disposal*, waterways*, etc.

   e.g. California
      engineering geology
         earthquakes

environmental geology (2)
   topic (3)
      conservation*, ecology, geologic hazards*, impact statements*, land use*, pollution*, reclamation*, waste disposal*, etc.

   e.g. Idaho
      environmental geology
         land use

geochemistry (2)
   isotopes or element or other terms from List D or material or topic (Main List) including crust*, mantle*, trace elements*. (3)

   e.g. USSR
      geochemistry
         isotopes

geochronology (2)
   topic (3)
      age (List E*) or absolute age or topic from Main List.

   e.g. United States   e.g. Maine
      geochronology   geochronology
         absolute age   Precambrian

geomorphology (2)
   topic (3)
      changes of level*, glacial geology, meteor craters*, mud volcanoes*, weathering, etc.

   e.g. Sahara
      geomorphology
         eolian features

geophysical surveys (2)
   topic (3)
      Type of survey (acoustical surveys*, Earth-current surveys*, electrical surveys*, electromagnetic surveys*, gravity surveys*, infrared surveys*, magnetic surveys*, magnetotelluric surveys*, seismic surveys*) or geodesy, heat flow*, remote sensing*, well-logging.

   e.g. Oklahoma
      geophysical surveys
         infrared surveys

hydrogeology (2)
   topic (3)
      ground water, hydrology, springs*, thermal waters*.

   e.g. Louisiana
      hydrogeology
         ground water

mineralogy (2)
   Name of mineral group (List L). In the case of miscellaneous minerals, that term suffices here.

   e.g. Italy
      mineralogy
         framework silicates, feldspar group

oceanography (2)
   topic (3)
      continental margin*, continental shelf*, continental slope*, estuaries*, marine geology*, nodules, ocean basins*, ocean circulation*, ocean floors*, ocean waves*, reefs*, sea ice, sea water, sedimentation, sediments, etc.

   e.g. Pacific Ocean
      oceanography
         marine geology

paleobotany (2)

The following terms may be used on level 3 in any of the index entries for areas and generate automatic cross-references whenever they appear: bibliography*, guidebooks*, maps*, road log*, textbooks*

Fossil group (List F) (3)
  Repeat set if more than one equally-emphasized fossil group.

  e.g. Arizona
    paleobotany
      gymnosperms

paleontology (2)
  Also used for studies on both fauna and flora.

  Fossil group (List F) or biogeography*, ecology*. (3)
  Repeat set if more than one equally-emphasized fossil group or biogeography*, ecology*.

  e.g. California
    paleontology
      foraminifera

petrology (2)

  topic (3)
    crystalline rocks, fluid inclusions, igneous rocks, inclusions, intrusions, lava, magmas, metamorphic rocks, metamorphism, metasomatism, meteor craters*, phase equilibria, volcanism, volcanology.

  e.g. Colorado
    petrology
      metamorphism

sedimentary petrology (2)

  topic (3)
    clay mineralogy, diagenesis, heavy minerals*, reefs*, sedimentary rocks, sedimentary structures, sedimentation, sediments, weathering.

  e.g. New South Wales
    sedimentary petrology
      sedimentary rocks

seismology (2)

  topic (3)
    core*, crust*, earthquakes*, explosions*, mantle*, Mohorovicic discontinuity*, etc.

  e.g. Washington
    seismology
      earthquakes

soils (2)

  topic (3)
    Name of soil group, topic, or laterites*, Paleosols*.

  e.g. Alberta
    soils
      name of soil group or Paleosols

stratigraphy (2)

  age (List E*) or archaeology*, biogeography, changes of level*, continental drift, paleogeography, paleomagnetism, paleoclimatology. (3)

  Repeat set if more than one equally-emphasized age.

  e.g. South Carolina
    stratigraphy
      Cretaceous

structural geology (2)

  topic (3)
    deformation, epeirogeny, faults, folds, foliation, fractures, geosynclines, isostasy*, lineation, orogeny, neotectonics*, salt tectonics*, structural analysis, tectonics*.

  e.g. India
    structural geology
      tectonics

tectonophysics (2)

  topic (3)
    continental drift, core*, crust*, heat flow*, isostasy*, mantle*, Mohorovicic discontinuity*, ocean basins*, paleomagnetism, plate tectonics*, sea-floor spreading*, etc.

  e.g. North America
    tectonophysics
      plate tectonics

volcanology (2)
  For Quaternary volcanic activity.
  Name of volcano, or locality for several volcanoes (3), or regional

  e.g. Italy
    volcanology
      Mount Etna

LIST C - COMMODITIES

The terms in this list are used for earth materials of actual or potential commercial value, i.e., from an economic viewpoint.

Beginning in 1981, new terms were adopted for some commodities, to distinguish between geochemical and economic treatments. For example,

---

The following terms may be used on level 3 in any of the following index entries for areas and generate automatic cross-references whenever they appear: bibliography*, guidebook*, maps*, road log*, textbooks*.

beryllium ores was added in 1981, to be used for economic papers. Beryllium, the existing term, was restricted to geochemical use. Where new terms have been added, this is noted in the main body of the Thesaurus.

INDEXING

Except for those few marked otherwise, the commodity terms in this list are indexed independently.

For commodities, use economics for all of the energy sources, and water resources. For petroleum use petroleum exploration. For coal use coal exploration. For all other nonmetals, and for all of the metal ores, use mineral economics and mineral exploration.

In general, the most specific commodity term applicable should be used. Indexers must insure that a grouper term is either manually added or will autopost. In some cases (abrasives) it is necessary to index a group name in order to make retrieval of groups possible. The following terms are general terms to be used as groupers: base metals, ceramic materials, construction materials, industrial minerals, metal ores, mineral resources, energy sources, gems, heavy mineral deposits, nonmetal deposits, polymetallic ores, rare earth deposits. Petroleum is used as a grouper for oil and gas.

ONLINE SEARCHING

For complete retrieval, search both specific and general terms. For example, a search on sillimanite deposits should include a search on ceramic materials, excluding andalusite deposits and kyanite deposits.

To isolate economic papers from non-economic, it is advisable to coordinate pre-1981 commodity terms with the economic geology category codes, or with ore deposits (for metals) or deposits (for nonmetals). See specific terms in the Main List for recommended search strategy. Also search the 1981 (et. seq.) form of the term. As of 1985, the term metal ores is autoposted to all the specific metal ore terms.

B.I.G. SEARCHING

Most papers on commodities emphasize specific geographic locations. Before 1993 commodity papers on an area could be found by looking under the area (1), economic geology (2), and the commodity (3). Under the commodity see also references appeared.

    e.g. Coal see also under economic geology under Alabama; Alberta; Australia; Colorado; England; etc.

For those papers on commodities which were not related to specific areas, look under the commodity. For papers on the genesis of commodities, look under mineral deposits, genesis (1).

Other index entries related to commodities were under sedimentary rocks (for coal), metamorphic rocks (for slate deposits or marble deposits), igneous rocks (for granite deposits), placers, mineral exploration, and mining geology. Economic geology as a discipline was found under economic geology.

In current practice the specific commodity is a direct access point.

---

abrasives
    index with industrial minerals
aggregate
    index with construction materials
aluminum ores
andalusite deposits
    index with ceramic materials
anhydrite deposits
    index with gypsum deposits
antimony ores
arsenic ores
asbestos deposits
asphalt
    index with bitumens
barite deposits
base metals
bauxite
bentonite deposits
beryl
    index with beryllium ores or gems
beryllium ores
    includes beryl
bismuth ores
bitumens
    includes asphalt
borate deposits
    index with boron deposits
boron deposits
    includes borate deposits
brines
    see also salt; bromine deposits; iodine deposits
bromine deposits
    see also brines; salt
building stone
    index with one of the following: construction materials; granite deposits; sandstone deposits; limestone deposits; marble deposits
cadmium ores
calcite deposits
cement materials
    index with one of the following: limestone deposits or construction materials
ceramic materials
    includes andalusite deposits, kyanite deposits, refractory materials, sillimanite deposits
cerium ores
chalk deposits
    index with limestone deposits

chromite ores
clays
    includes fuller's earth, and shale used as brick clay
coal
cobalt ores
construction materials
    includes perlite, gravel deposits, building stone, cement materials, dimension stone, aggregate, ornamental materials, insulation materials
copper ores
corundum deposits
cryolite deposits
    index with fluorspar
diamonds
diatomite
dimension stone
    index with one of the following: construction materials; granite deposits; sandstone deposits; limestone deposits; marble deposits
dolostone deposits
energy sources
    a grouper for petroleum; natural gas; coal; uranium ores; geothermal energy; new energy sources; etc.
evaporite deposits
feldspar deposits
fluorspar
    includes cryolite deposits
fuller's earth
    index with clays
garnet deposits
    index with one of the following: industrial minerals or gems
gems
    includes beryl, garnet deposits
geothermal energy
glass materials
    index with sands
glauconite deposits
gold ores
granite deposits
graphite deposits
gravel deposits
    includes sands when sand is used as a construction material
gypsum deposits
    index with anhydrite deposits
heavy mineral deposits
helium gas
hematite
    index with iron ores
industrial minerals
    includes abrasives, garnet deposits, zeolite deposits
insulation materials
    index with construction materials
iodine deposits
    see also brines, salt
iridium ores
iron ores
    includes hematite, limonite, magnetite
kaolin deposits
    see also clays, bentonite deposits
kyanite deposits
    index with ceramic materials
lead ores
lead-zinc deposits
    see also lead ores, zinc ores

lignite
limestone deposits
    includes chalk deposits
    see also dolostone deposits
limonite
    index with iron ores
lithium ores
magnesite deposits
magnesium ores
magnetite
    index with iron ores
manganese ores
marble deposits
mercury ores
metal ores
mica deposits
mineral resources
    for very general treatments
molybdenum ores
monazite deposits
    see also heavy mineral deposits; rare earth deposits; thorium ores
natural gas
    see also petroleum; oil and gas fields
new energy sources
    index with energy sources
nickel ores
niobium ores
nitrate deposits
nonmetal deposits
    do not use for energy sources, or water resources
oil and gas fields
    use only for detailed discussions of specific fields, otherwise use petroleum or natural gas
    see term set options in body of Thesaurus
oil sands
    see also petroleum
oil shale
    see also petroleum
ornamental materials
    index with one of the following: construction materials; granite deposits; limestone deposits; marble deposits
osmium ores
palladium ores
peat
pegmatite
perlite
    index with construction materials
petroleum
    may include natural gas; see also oil and gas fields; natural gas; oil shale; oil sands
phosphate deposits
platinum ores
polymetallic ores
potash
pumice deposits
pyrite ores
quartz crystal
rare earth deposits
refractory materials
    index with ceramic materials

rhodium ores
ruthenium ores
salt
   see also brines; bromine deposits; iodine deposits
sands
   includes glass materials
   use gravel deposits for sand as a construction material
sandstone deposits
shale
   use clays for shale as brick clay
   index with construction materials
sillimanite deposits
   index with ceramic materials
silver ores
slate deposits
soapstone
   index with talc deposits
sodium carbonate
   includes trona
sodium sulfate
strategic minerals
   index with specific commodity
strontium ores
sulfur deposits
talc deposits
   includes soapstone
tantalum ores
thorium ores
   see also monazite deposits
tin ores
titanium ores
trona
   index with sodium carbonate
tungsten ores
uranium ores
vanadium ores
vermiculite deposits
water resources
zeolite deposits
   index with industrial minerals
zinc ores
zircon deposits

LIST D - ELEMENTS

All elements in the periodic table are index terms. Prior to 1981, element terms were used for both economic and geochemical papers. Starting in 1981, other terms have been added for economic papers (see List C), and the element terms have been restricted to geochemical papers.

INDEXING

Before 1993 elements could be used on levels 1, 2 and 3. The following related terms could be used on levels 1, 2 and 3.

   deuterium
   tritium
   metals
   noble gases
      Used for rare gases and inert gases.
   rare earths
      Used for lanthanide series, inner transition elements, and lanthanoans. Includes cerium, dysprosium, erbium, europium, gadolinium, holmium, lanthanum, lutetium, neodymium, praseodymium, promethium, samarium, scandium, terbium, thulium, ytterbium, and yttrium.

Three additional terms related to elements are:
   major elements
   minor elements
   trace elements

ONLINE SEARCHING

For geochemical references entered through 1980, search the element AND (geochemistry OR the appropriate Category Code--See List P). From 1981 on, search the elements, above.

For economic references, see List C, Commodities.

B.I.G. SEARCHING

For geochemical references, see the element as an access point. Also look under the terms isotopes, chemical analysis, geochemistry, X-ray analysis, spectroscopy, thermal analysis, and electron microscopy. Also consider using the terms related to elements, above.

For economic references, see List C, Commodities.

Below is a list of methods used for chemical analysis. In order to distinguish methodology from data, different terms are used. Methods-type terms are used for studies emphasizing methods, techniques, instruments. Data-type terms are used only for reports of actual analysis.

Methods	Data
activation analysis	
alpha-ray spectroscopy	
atomic absorption	atomic absorption spectra
Auger spectroscopy	
autoradiography	
chemical analysis	
chromatography	chromatograms
colorimetry	
electrolytic analysis	
electron diffraction analysis	

electron paramagnetic resonance	EPR spectra
electron probe	electron probe data
emission spectroscopy	emission spectra
flame photometry	
gamma-ray spectroscopy	gamma-ray spectra
inductively coupled plasma methods	
infrared spectroscopy	infrared spectra
ion probe	ion probe data
laser methods	
mass spectroscopy	mass spectra
microwave spectroscopy	
Mossbauer spectroscopy	Mossbauer spectra
neutron activation analysis	neutron activation analysis data
neutron diffraction analysis	neutron diffraction data
neutron probe	neutron probe data
nuclear magnetic resonance	NMR spectra
optical spectroscopy	optical spectra
polarography	
radio-frequency spectroscopy	
Raman spectroscopy	Raman spectra
spectroscopy	spectra
ultraviolet spectroscopy	ultraviolet spectra
vacuum fusion analysis	
volumetric analysis	
X-ray analysis	X-ray data
X-ray diffraction analysis	X-ray diffraction data
X-ray fluorescence	X-ray fluorescence spectra
X-ray radiography	
X-ray spectroscopy	X-ray spectra

LIST E - GEOLOGIC AGE (STRATIGRAPHIC) TERMS

The authority used in GeoRef for age and stratigraphic terms is the Geological Time Table, compiled by F.W.S. Van Eysinga, 3rd edition, Amsterdam, Elsevier, 1975. For this edition of the Thesaurus, the 4th edition of the Geological Time Table compiled by B. U. Haq, et. al. Amsterdam, Elsevier, 1987 was consulted for new terms and changes. This is followed for nomenclature except in a few cases where previous usage in GeoRef strongly supports alternate terms.

Throughout, the "stratigraphic" upper, middle, and lower rather than the "chronologic" early, middle, and late are used to modify age terms, e.g. Upper Cambrian instead of Late Cambrian. "Uppermost" and "lowermost" are not used.

Other stratigraphic and rock-unit terms, in addition to terms listed below, appear in the main body of the Thesaurus.

INDEXING

Listed below is a general outline of the age terms used in GeoRef as well as an alphabetical list. Indexers must ensure that specific age terms are indexed and their hierarchical age terms are added from specific to general until a term from the alphabetical listing is used. Only the specific term need be indexed, since use of the specific term will result in automatic posting of its broader term(s).

"Upper", "middle", and "lower" are capitalized for Mesozoic and Paleozoic periods (e.g. Upper Cretaceous). This practice does not apply to Precambrian and Cenozoic subdivisions. See list below.

ONLINE SEARCHING

All terms in the list may be searched. A high degree of specificity is often possible.

In searching for rock units such as Dakota Sandstone, it is advisable to search for the common variants of such names. Dakota Group OR Dakota Sandstone will give better results than Dakota Sandstone alone (See List R).

In searching for certain broad terms such as Cretaceous, it suffices to use the term as is and not worry about subdivisions, since those are automatically posted to Cretaceous. This applies to the Precambrian and the Paleozoic, Mesozoic, and Cenozoic subdivisions, (e.g. Tertiary or Paleocene), but not the Paleozoic, Mesozoic, and Cenozoic themselves. To find the latter, you must also search for their major subdivisions.

For the years 1977-1992 the search for the age terms restricted to major descriptor will yield those papers that are not related to a certain geographic area. Thus a search for level 1 occurrences of Precambrian in the September 1980 GeoRef update gives four hits, one title of which illustrates the point: "The problem of the Precambrian-Cambrian boundary". An unrestricted search for Precambrian for the same month gives approximately thirty hits, most of which are related to specific geographic areas, illustrated by the following title: "Precambrian age of Reguibat Ridge dolerites, Mauritania".

B.I.G. SEARCHING

The terms to search in the subject index of the B.I.G. are those marked as possibly occurring on level 1, i.e. those followed by (1) in the list. Under Tertiary, Vol. 44, No. 9, Sept. 1980, p. I-198 is the entry:

Tertiary see also under geochronology under France; USSR; see also under stratigraphy under Algeria; Benin; California; Cuba;

Czechoslovakia; England; France; India;
Indian Ocean; Indonesia....

Tertiary - stratigraphy
    biogeography: Tertiary 33978

In this example, the see also reference is a guide to each area set in which Tertiary occurs on level 3. Papers cited under Tertiary as a level 1 term are those not devoted to a specific area. In the example above the title of the paper happens to be "Tertiary".

Papers dealing with stratigraphy as a discipline, i.e. papers on its practice, philosophy, principles, etc., are found under the heading stratigraphy.

Many papers on fossils contain stratigraphic information. Such papers can be found under appropriate fossil terms (see List F). Note that as of 1981, a number of non-systematic terms have been added to index fossils in biostratigraphic papers, e.g. amphibians, birds, crustaceans, brachiopods. From 1981 on, these non-systematic fossil terms are used for stratigraphic papers while the systematic fossil terms are reserved for paleontologic papers.

Dashed lines separate parallel stratigraphic terminology used in different areas.

Phanerozoic
  Cenozoic
    Quaternary
      Holocene
      Flandrian
      Tyrrhenian   (marine)
      Sicilian
      Calabrian
      - - - - - - - - - - - -
      Holocene
      Pleistocene   (continental)
        upper Pleistocene
        middle Pleistocene
        lower Pleistocene
          Villafranchian
    Tertiary
      Neogene
        Pliocene
          upper Pliocene
          middle Pliocene
          lower Pliocene
        Miocene
          upper Miocene
            Meotian
            Sarmatian
            Tortonian
          middle Miocene
            Helvetian
          lower Miocene
            Burdigalian
            Aquitanian
      Paleogene
        Oligocene
          upper Oligocene
          middle Oligocene
          lower Oligocene
            Stampian
        Eocene
          upper Eocene
          middle Eocene
          lower Eocene
        Paleocene
          Thanetian
          Montian
          Danian
  Mesozoic
    Cretaceous
      Upper Cretaceous
        Senonian
          Maestrichtian
          Campanian
          Santonian
          Coniacian
        Turonian
        Cenomanian
      Lower Cretaceous
        Albian
        Aptian
        Neocomian
        Barremian
        Hauterivian
        Valanginian
        Berriasian
    Jurassic
      Upper Jurassic
        Portlandian
        Tithonian
        Kimmeridgian
        Oxfordian
      Middle Jurassic
        Callovian
        Bathonian
        Bajocian
        Aalenian
      Lower Jurassic (use for Liassic)
        upper Liassic
          Toarcian
        middle Liassic
          Domerian
          Pliensbachian
        lower Liassic
          Sinemurian
          Hettangian
    Triassic
      Upper Triassic
        Rhaetian
        Norian
        Carnian
      Middle Triassic
        Ladinian
        Anisian
      Lower Triassic
      - - - - - - - - - - - -
      Keuper
      Muschelkalk
      Bunter
  Paleozoic
    Permian
      Thuringian
      Saxonian
      Autunian
      - - - - - - - - - - - -

Zechstein
Rotliegendes
- - - - - - - - - - - - - -
Tatarian
Kazanian
Kungurian
Artinskian
Sakmarian
- - - - - - - - - - - - - -
Upper Permian
Lower Permian
Carboniferous
  Upper Carboniferous
    Stephanian
    Westphalian
    Namurian
  Dinantian (use for Lower
               Carboniferous)
    Visean
    Tournaisian
- - - - - - - - - - - - - -
Upper Carboniferous
  Gzhelian
  Kasimovia
  Moscovian
  Bashkirian
  Namurian
Dinantian (use for Lower
            Carboniferous)
  Visean
  Tournaisian
- - - - - - - - - - - - - -
Pennsylvanian
  Upper Pennsylvanian
  Middle Pennsylvanian
  Lower Pennsylvanian
Mississippian
  Upper Mississippian
  Middle Mississippian
  Lower Mississippian
Devonian
  Upper Devonian
    Strunian
    Famennian
    Frasnian
  Middle Devonian
    Givetian
    Eifelian
  Lower Devonian
    Emsian
    Siegenian
    Gedinnian
Silurian
  Upper Silurian
    Pridolian
    Ludlovian
  Lower Silurian
    Wenlockian
    Llandoverian
Ordovician
  Upper Ordovician
    Ashgillian
    Caradocian
  Middle Ordovician
    Llandeilian
    Llanvirnian
  Lower Ordovician
    Arenigian
    Tremadocian
Cambrian
  Upper Cambrian
  Middle Cambrian
  Lower Cambrian
Precambrian
  Proterozoic
    upper Proterozoic
      Vendian
    middle Proterozoic
    lower Proterozoic
  Archean

ALPHABETICAL LIST

Aalenian	Middle Jurassic
Acadian	Middle Cambrian
Acheulian	Paleolithic
Adelaidean	Proterozoic
Aftonian	lower Pleistocene
Akchagylian	upper Pliocene
Albian	Lower Cretaceous
Aldanian	Lower Cambrian
Alexandrian	Lower Silurian
Algonkian	Proterozoic
Allerod	Weichselian
Altonian	lower Miocene
Ancylus Lake	lower Holocene
Andalusian	Triassic
Anglian	Quaternary
Anisian	Middle Triassic
Aphebian	lower Proterozoic
Aptian	Lower Cretaceous
Aquitanian	lower Miocene
Archean	Precambrian
Arenigian	Lower Ordovician
Argovian	Upper Jurassic
Arikareean	Tertiary
Ariyalur Stage	Upper Cretaceous
Artinskian	Lower Permian
Asbian	Dinantian
Ashgillian	Upper Ordovician
Asselian	Permian
Astian	upper Pliocene
Atdabanian	Lower Cambrian
Atlantic	Holocene
Atokan	Middle Pennsylvanian
Aurignacian	Paleolithic
Autunian	Lower Permian
Auversian	upper Eocene
Avonian	Carboniferous
Awamoan	lower Miocene
Badenian	Miocene
Bajocian	Middle Jurassic
Bala	Ordovician
Baltic Glaciation	upper Quaternary
Baltic ice lake	upper Pleistocene
Barakar Stage	Lower Permian
Barrandian	Middle Cambrian
Barremian	Lower Cretaceous
Barstovian	Miocene

Bartonian	upper Eocene	Downtonian	Paleozoic
Bashkirian	Upper Carboniferous	Dresbachian	Upper Cambrian
Bathonian	Middle Jurassic	Edenian	Upper Ordovician
Berriasian	Lower Cretaceous	Ediacaran	Vendian
Biarritzian	Eocene	Eemian	upper Pleistocene
Birrimian	lower Proterozoic	Egerian	upper Oligocene
Blackriverian	Middle Ordovician	Eggenburgian	lower Miocene
Blancan	Cenozoic	Eifelian	Middle Devonian
Boreal	Holocene	Elsterian	middle Pleistocene
Brandenburg Stade	Weichselian	Emsian	Lower Devonian
Brandon Stade	Weichselian	Eocambrian	Precambrian
Bridgerian	Eocene	Eocene	Paleogene
Brioverian	upper Proterozoic	Famennian	Upper Devonian
Bronze Age	Cenozoic	Flandrian	Holocene
Brunhes Epoch	Quaternary	Frankfurt Stade	Weichselian
Bull Lake Glaciation	upper Quaternary	Frasnian	Upper Devonian
Bunter	Lower Triassic	Gargasian	Lower Cretaceous
Burdigalian	lower Miocene	Gault	Lower Cretaceous
Calabrian	lower Pleistocene	Gauss Epoch	Pliocene
Callovian	Middle Jurassic	Gedinnian	Lower Devonian
Cambrian	Paleozoic	Gilbert Epoch	lower Pliocene
Campanian	Senonian	Givetian	Middle Devonian
Canadian Series	Lower Ordovician	Guadalupian	Permian
Caradocian	Upper Ordovician	Gulfian	Upper Cretaceous
Carboniferous	Paleozoic	Gunz	Pleistocene
Carixian	Lower Jurassic	Gzhelian	Upper Carboniferous
Carnian	Upper Triassic	Hadrynian	upper Proterozoic
Carpentarian	lower Proterozoic	Harlech Stage	Lower Cambrian
Cayugan	Upper Silurian	Hauterivian	Lower Cretaceous
Cenomanian	Upper Cretaceous	Helderbergian	Lower Devonian
Cenozoic		Helikian	middle Proterozoic
Central Polish Glaciation		Helvetian	middle Miocene
Chadronian	lower Oligocene	Hemingfordian	lower Miocene
Chalcolithic	Cenozoic	Hemphillian	Neogene
Champlainian	Middle Ordovician	Hettangian	lower Liassic
Chattian	upper Oligocene	Hirnantian	Ashgillian
Chazyan	Middle Ordovician	Holocene	Quaternary
Chesterian	Upper Mississippian	Holsteinian	upper Pleistocene
Cimmerian	Pliocene	Hoxnian	upper Pleistocene
Cincinnatian	Upper Ordovician	Huronian	Proterozoic
Clarendonian	Miocene	Ilerdian	Paleogene
Clovis	Quaternary	Illinoian	upper Pleistocene
Coal Measures	Upper Carboniferous	Infracambrian	upper Proterozoic
Comanchean	Cretaceous	Ipswichian	upper Pleistocene
Coniacian	Senonian	Iron Age	Cenozoic
Cordevolian	Triassic	Irvingtonian	Pleistocene
Couvinian	Middle Devonian	Jacksonian	Eocene
Cretaceous	Mesozoic	Jaramillo Event	lower Pleistocene
Cromerian	upper Pleistocene	Jotnian	upper Proterozoic
Cuisian	lower Eocene	Jurassic	Mesozoic
Culm	Carboniferous	Kansan	lower Pleistocene
Dacian	Pliocene	Karelian	lower Proterozoic
Dalradian		Karharbari Stage	Upper Carboniferous
Danian	lower Paleocene	Kasimovian	Upper Carboniferous
Danube Stade	lower Pleistocene	Kazanian	Upper Permian
Desmoinesian	Middle Pennsylvanian	Keuper	Upper Triassic
Devensian	upper Pleistocene	Keweenawan	Proterozoic
Devonian	Paleozoic	Kimmeridgian	Upper Jurassic
Dinantian	Carboniferous	Kinderhookian	Lower Mississippian
Dittonian	Lower Devonian	Kungurian	Permian
Dogger	Middle Jurassic	Ladinian	Middle Triassic
Domerian	middle Liassic	Landenian	upper Paleocene

Langhian	Miocene	middle Eocene	Eocene
Lattorfian	upper Eocene	middle Holocene	Holocene
Laxfordian	middle Proterozoic	Middle Jurassic	Jurassic
Leonardian	Lower Permian	middle Liassic	Lower Jurassic
Levalloisian	Paleolithic	middle Mesozoic	Mesozoic
Levantinian	upper Pliocene	middle Miocene	Miocene
Lewisian	Proterozoic	Middle Mississippian	Mississippian
Litorina Sea	lower Holocene	middle Oligocene	Oligocene
Llandeilian	Middle Ordovician	Middle Ordovician	Ordovician
Llandoverian	Lower Silurian	middle Paleocene	Paleocene
Llanvirnian	Middle Ordovician	middle Paleozoic	Paleozoic
Loch Lomond Stade	Weichselian	Middle Pennsylvanian	Pennsylvanian
Lochkovian	Lower Devonian	Middle Permian	Permian
Lotharingian	Lower Jurassic	middle Pleistocene	Pleistocene
lower Archean	Archean	middle Pliocene	Pliocene
Lower Cambrian	Cambrian	middle Precambrian	Precambrian
lower Cenomanian	Cenomanian	middle Proterozoic	Proterozoic
lower Cenozoic	Cenozoic	middle Quaternary	Quaternary
Lower Cretaceous	Cretaceous	Middle Silurian	Silurian
Lower Devonian	Devonian	middle Tertiary	Tertiary
lower Eocene	Eocene	Middle Triassic	Triassic
lower Holocene	Holocene	Mindel	Pleistocene
Lower Jurassic	Jurassic	Mindel/Riss Interglacial	Pleistocene
lower Liassic	Lower Jurassic	Miocene	Neogene
lower Mesozoic	Mesozoic	Mississippian	Carboniferous
lower Miocene	Miocene	Missourian	Upper Pennsylvanian
Lower Mississippian	Mississippian	Mohnian	Miocene
lower Oligocene	Oligocene	Moinian	upper Proterozoic
Lower Ordovician	Ordovician	Moldanubian	Precambrian
lower Paleocene	Paleocene	Montian	Paleocene
lower Paleogene	Paleogene	Morrowan	Lower Pennsylvanian
lower Paleozoic	Paleozoic	Moscovian	Upper Carboniferous
Lower Pennsylvanian	Pennsylvanian	Mousterian	Paleolithic
Lower Permian	Permian	Muschelkalk	Middle Triassic
lower Pleistocene	Pleistocene	Namurian	Upper Carboniferous
lower Pliocene	Pliocene	Narizian	Eocene
lower Proterozoic	Proterozoic	Nebraskan	lower Pleistocene
lower Quaternary	Quaternary	Neocomian	Lower Cretaceous
Lower Silurian	Silurian	Neogene	Tertiary
lower Tertiary	Tertiary	Neoglacial	Holocene
Lower Triassic	Triassic	Neohelikian	Helikian
lower Turonian	Turonian	Neolithic	Stone Age
Ludian	upper Eocene	Niagaran	Middle Silurian
Ludlovian	Upper Silurian	Norian	Upper Triassic
Luisian	middle Miocene	Ocoee Series	Precambrian
Lusitanian	Upper Jurassic	Old Red Sandstone	Devonian
Lutetian	Eocene	Older Dryas	Weichselian
Maestrichtian	Senonian	Oldest Dryas	Weichselian
Magdalenian	Paleolithic	Olduvai Event	lower Pleistocene
Malm	Upper Jurassic	Oligocene	Paleogene
Matuyama Epoch	Cenozoic	Ordovician	Paleozoic
Maysvillian	Cincinnatian	Osagian	Lower Mississippian
Meotian	upper Miocene	Ottnangian	lower Miocene
Meramecian	Upper Mississippian	Oxfordian	Upper Jurassic
Mesolithic	Stone Age	Paleocene	Paleogene
Mesozoic		Paleogene	Tertiary
Messinian	upper Miocene	Paleolithic	Stone Age
Middle Cambrian	Cambrian	Paleozoic	
Middle Carboniferous	Carboniferous	Pannonian	upper Miocene
middle Cenozoic	Cenozoic	Pennsylvanian	Carboniferous
Middle Cretaceous	Cretaceous	Permian	Paleozoic
Middle Devonian	Devonian	Perth Stade	Weichselian

Phanerozoic		Tertiary	Cenozoic
Pinedale Glaciation	upper Quaternary	Thanetian	upper Paleocene
Plaisancian	lower Pliocene	Thuringian	Permian
Pleistocene	Quaternary	Tiffanian	lower Paleocene
Pliensbachian	middle Liassic	Tithonian	Portlandian
Pliocene	Neogene	Toarcian	upper Liassic
Pontian	upper Miocene	Tommotian	Lower Cambrian
Portlandian	Upper Jurassic	Tongrian	upper Eocene
Pragian	Lower Devonian	Torrejonian	lower Paleocene
Preboreal	Holocene	Torridonian	upper Proterozoic
Precambrian		Tortonian	upper Miocene
Priabonian	upper Eocene	Tournaisian	Dinantian
Pridolian	Upper Silurian	Tremadocian	Lower Ordovician
Proterozoic	upper Precambrian	Trempealeauan	Upper Cambrian
Puercan	lower Paleocene	Trentonian	Middle Ordovician
Purbeckian	Lower Cretaceous	Triassic	Mesozoic
Quaternary	Cenozoic	Turolian	upper Miocene
Rancholabrean	upper Pleistocene	Turonian	Upper Cretaceous
Rauracian	Upper Jurassic	Twocreekan	upper Pleistocene
Redonian	upper Pliocene	Tyrrhenian	upper Pleistocene
Refugian	Eocene	Uintan	Eocene
Relizian	Miocene	Ulatisian	middle Eocene
Rhaetian	Upper Triassic	upper Albian	Albian
Richmondian	Cincinnatian	upper Archean	Archean
Riphean	upper Proterozoic	Upper Cambrian	Cambrian
Riss	Pleistocene	upper Campanian	Campanian
Riss/Wurm Interglacial	Pleistocene	Upper Carboniferous	Carboniferous
Romanian	upper Pliocene	upper Cenomanian	Cenomanian
Rotliegendes	Permian	upper Cenozoic	Cenozoic
Rupelian	lower Oligocene	Upper Cretaceous	Cretaceous
Saalian	upper Pleistocene	Upper Devonian	Devonian
Sakmarian	Lower Permian	upper Eocene	Eocene
Sangamonian	upper Pleistocene	upper Holocene	Holocene
Sannoisian	lower Oligocene	Upper Jurassic	Jurassic
Santonian	Senonian	upper Liassic	Lower Jurassic
Sarmatian	upper Miocene	upper Maestrichtian	Maestrichtian
Sartanian	upper Pleistocene	upper Mesozoic	Mesozoic
Saucesian	lower Miocene	upper Miocene	Miocene
Saxonian	Permian	Upper Mississippian	Mississippian
Scythian	Lower Triassic	upper Neogene	Neogene
Senonian	Upper Cretaceous	upper Oligocene	Oligocene
Sequanian	Upper Jurassic	Upper Ordovician	Ordovician
Serpukhovian	Upper Mississippian	upper Paleocene	Paleocene
Serravallian	middle Miocene	upper Paleogene	Paleogene
Sicilian	upper Pleistocene	upper Paleolithic	Paleolithic
Siegenian	Lower Devonian	upper Paleozoic	Paleozoic
Silesian	Carboniferous	Upper Pennsylvanian	Pennsylvanian
Silurian	Paleozoic	Upper Permian	Permian
Sinemurian	lower Liassic	upper Pleistocene	Pleistocene
Sinian	Proterozoic	upper Pliocene	Pliocene
Smithian	Lower Triassic	upper Precambrian	Precambrian
South Polish Glaciation	Cromerian	upper Proterozoic	Proterozoic
Sparnacian	lower Eocene	upper Quaternary	Quaternary
Spathian	Lower Triassic	upper Riphean	Riphean
Stampian	lower Oligocene	Upper Silurian	Silurian
Stephanian	Upper Carboniferous	upper Sinian	Sinian
Stone Age	Cenozoic	upper Tertiary	Tertiary
Strunian	Upper Devonian	upper Tournaisian	Tournaisian
Subboreal	Holocene	Upper Triassic	Triassic
Svecofennian	lower Proterozoic	upper Visean	Visean
Tabianian	lower Pliocene	upper Weichselian	Weichselian
Tatarian	Upper Permian	upper Wisconsinan	Wisconsinan

Uralian	Upper Carboniferous
Urgonian	Lower Cretaceous
Valanginian	Lower Cretaceous
Vallesian	Miocene
Valmeyeran	Mississippian
Vendian	upper Proterozoic
Versilian	Holocene
Villafranchian	lower Pleistocene
Vindhyan	
Vindobonian	Miocene
Virgilian	Upper Pennsylvanian
Viruan	Ordovician
Visean	Dinantian
Vistulian	upper Pleistocene
Volgian	Upper Jurassic
Vraconian	Cretaceous
Wasatchian	lower Eocene
Wealden	Lower Cretaceous
Weichselian	upper Pleistocene
Wenlockian	Lower Silurian
Werfenian	Lower Triassic
Westphalian	Upper Carboniferous
Wisconsinan	upper Pleistocene
Wolfcampian	Lower Permian
Woodfordian	Wisconsinan
Wurm	upper Pleistocene
Yanshanian	Mesozoic
Yoldia Sea	upper Quaternary
Younger Dryas	Weichselian
Ypresian	lower Eocene
Yudomian	
Zechstein	Upper Permian
Zhoukoudian	Quaternary
Zlichovian	Devonian

LIST F - FOSSILS

The terminology in this list was developed in consultation with micropaleontologists at the American Museum of Natural History, for whom GeoRef produces the monthly *Bibliography and Index of Micropaleontology*, and with the vertebrate paleontologists at the Museum of Paleontology of the University of California - Berkeley, with whom we produced the annual *Bibliography of Fossil Vertebrates* from 1973-1980.

Terms on the left side of the list are systematic names; those on the right are common names. Common names were first introduced in 1981 for biostratigraphy papers. In 1983, the use of common names was expanded to include papers on paleoecology, biogeography, ecology and biochemistry of fossil groups.

INDEXING

Index the most specific fossil term possible. Index higher level systematic terms step-by-step up the hierarchy until an autoposting term is used.

The numbers 8-11 correspond to the paleontology category codes (section headings) in list P:

08 general paleontology
09 paleobotany
10 invertebrate paleontology
11 vertebrate paleontology

A three-character "M" code, under or beside a term in the list, means that a cited paper, in which that fossil group is an important topic, should have that Special Bibliography code for the *Bibliography and Index of Micropaleontology*.

M01 algae
M02 scolecodonts
M03 nannofossils; Coccolithophoraceae
M04 Conodonta
M05 diatoms
M06 foraminifera
M07 problematic microfossils
M08 Ostracoda
M09 palynomorphs
M10 Protista
M11 Radiolaria
M12 miscellanea (includes use when several microfossil groups are discussed)

Prior to 1993 fossil sets were constructed as follows:

fossil group (1)
  fossil subgroup (2)
    age (List E)

e.g. Bryozoa   foraminifera
    Cheilostomata   Globigerinacea
      Miocene   Cretaceous

If age was not known, topic (e.g., morphology) on level 3 was used. For papers on more than one major subgroup (e.g., Rotaliina and Miliolina), the set was constructed as follows:

fossil group (1)
  topic (2)
  (biocenoses, biologic evolution, distribution, faunal studies, floral studies, fossilization, habitat, miscellanea, morphology, occurrence, ontogeny, taxonomy)

    subtopic
    e.g., benthonic taxa, bibliography, biometry, catalogs, faunal list, floral list, paleoclimatology, phylogeny, planktonic taxa, statistical analysis, taphonomy, thanatocenoses, zoning or type of material (e.g., bones, fossil wood, plankton, shells, skeletons, skulls, teeth, tests), etc. or age (List E)

e.g. Mammalia
    faunal studies
        Cenozoic

Reptilia
    biologic evolution
        phylogeny

Use miscellanea for papers in which the author is uncertain of the group's taxonomic affinities (i.e., incertae sedis). Use faunal studies or floral studies if several groups are discussed, or if the indexer is not sure of the group's relationships to other groups. When indexing newly named fossils include the supplementary term new taxa and list all new taxa up to ten.

Beginning in 1981, common terms from the right side of the list were used on level 1 for biostratigraphy papers. As of 1983, common names could also be used on level 1 for papers emphasizing biogeography, paleoecology, ecology or biochemistry.

Prior to 1993 common names were used in sets as follows:

fossil common name (1)
    fossil common name (2) or topic
    (biochemistry, biogeography, biostratigraphy, ecology, occurrence, or paleoecology)
        subtopic
        e.g., age (List E) or specific environment

e.g. mollusks
    biostratigraphy
        Quaternary

    mammals
        biogeography
            Quaternary

    ostracods
        biochemistry
            trace elements

corals
    paleoecology
        Devonian

fish
    ecology
        brackish-water environment

Common names were also used on level 2 in biogeography, ecology and paleoecology sets.

e.g. paleoecology
        corals
            Devonian

    ecology
        fish
            brackish-water environment

biogeography
    mammals
        Quaternary

A level 2 common name was also used under a level 1 common name:

e.g. mollusks
        cephalopods
            Jurassic

reptiles
    dinosaurs
        Mesozoic

As of 1984, common names were also used for some very general studies, particularly popular articles on fossil collecting.

ONLINE SEARCHING

For paleontology papers on a fossil group, search the systematic name. To exclude stratigraphy papers, add NOT stratigraphy to the search strategy. Broader terms in the list are autoposted to their narrower terms, except for Hemichordata which is not autoposted. More specific fossil group names appear in the main body of the Thesaurus. Their broader terms, which appear in the list, are autoposted to them. Before 1978 GeoRef had no autoposting; broader fossil group names were routinely added to the indexing, however, since they were necessary in the formation of sets.

For stratigraphic papers on a fossil group before 1981, search the systematic name AND biostratigraphy. For paleoecology, biogeography, ecology and biochemistry papers, it is best to search both common and systematic names, in combination with the appropriate topic term.

For searches on specific taxa, the species and/or genus name may be used in searching. Keep in mind that general studies discussing more than 10 taxa will rarely be indexed using the specific terms. A more general strategy is advisable for such searches.

B.I.G. SEARCHING

Prior to 1993, both the systematic names and, since 1981, the higher level common names of fossil groups were entry terms in the subject index.

Related headings include paleontology (for general treatments, discussions of biologic evolution, etc.) biogeography, paleoecology, paleoclimatology, ecology, paleobotany, (for the discipline), palynology (discipline or technique), micropaleontology (discipline or technique).

Beginning in 1993 all fossil names, common and systematic, are direct access points.

Fossil Group	Common Name
Agnatha	agnathans
11	
Heterostraci	
Ostracodermi	ostracoderms
algae	algal flora
09, (M01), (M03), (M05)	
Chlorophyta	
Charophyta	charophytes
Chlorophyceae	
Codiaceae	
Dasycladaceae	
Receptaculitaceae	
Desmidiales	desmids
Tasmanites	
Chrysophyta	
Coccolithophoraceae	coccoliths
Cyanophyta	
diatoms	diatom flora
nannofossils (incl.	nannofossils
Nannoconus	nannoconids
Discoasteridae	discoasters
Phaeophyta	
Pyrrhophyta	
Rhodophyta	
Corallinaceae	
Gymnocodiaceae	

stromatolites   stromatolites   Cheilostomata
Amphibia   amphibians   Cryptostomata
11   Ctenostomata
   Labyrinthodontia   Trepostomata
    (incl. Anthracosauria
    Icythyostegalia   Chordata   (incl. Protochordata,
    Temnospondyli)   11   Vertebrata)
   Lepospondyli   Coelenterata   corals
    (incl. Aistropoda   10
    Microsauria   Anthozoa
    Nectridea)   Ceriantipatharia
   Lissamphibia   Octocorallia
    (incl. Anura   Zoantharia
    Apoda   Actiniaria
    Gymnophiona   Corallimorpharia
    Caecilia   Heterocorallia
    Proanura   Hexactiniaria
    Urodela)   Rugosa
angiosperms   angiosperm flora   Scleractinia
09   Tabulata
   Dicotyledoneae   Zoanthiniaria
    Asteridae   Hydrozoa
    Caryophyllidae   Scyphozoa
    Dilleniidae   Conularida
    Hamamelididae   Stromatoporoidea   stromatoporoids
    Magnoliidae   Conodonta   conodonts
    Rosidae   10 (M04)
   Monocotyledoneae   coprolites
    Alismidae   08, 11
    Arecidae   Echinodermata   echinoderms
    Commelinidae   10
    Liliidae   Asteroidea
Archaeocyatha   Blastoidea
10   Camptostromatoidea
Arthropoda   arthropods   Crinoidea   crinoids
10   Cyclocystoidea
   Chelicerata   Cystoidea
    Arachnida   Echinoidea   echinoids
    Merostomata   Edrioasteroidea
   Mandibulata   Edrioblastoidea
    Crustacea   crustaceans   Eocrinoidea
     Branchiopoda   Helicoplacoidea
     Cirripedia   barnacles   Holothuroidea
     Copepoda   Homoiostelea
     Malacostraca   Homostelea
    Myriapoda   Lepidocystoidea
    Trilobitomorpha   Machaeridia
Aves   birds   Ophiocistioidea
11   Ophiuroidea
   Archaeornithes   Parablastoidea
   Neornithes   Paracrinoidea
bacteria   Somasteroidea
09   Stylophora
Brachiopoda   brachiopods   foraminifera   foraminifers
10   10, (M06)
   Articulata   Allogromiina
    Dictyonellidina   Fusulinina
    Orthida   Fusulinidae   fusulinids
    Pentamerida   Miliolina
    Rhynchonellida   Miliolacea
    Spiriferida   Alveolinellidae
    Strophomenida   Rotaliina
    Terebratulida   Buliminacea
    Thecideidina   Carterinacea
   Inarticulata   Cassidulinacea
bryophytes   Discorbacea
09   Globigerinacea
   Hepaticae   Nodosariacea
   Musci   Orbitoidacea
Bryozoa   bryozoans   Orbitoididae
10   Robertinacea
                                                             Rotaliacea
                                                              Nummulitidae
                                                             Spirillinacea

Textulariina
    Ammodiscacea
    Lituolacea

fossil man   Use only when fossil remains are
11   discussed; otherwise use archaeology
            under stratigraphy (2nd level) under area
            terms.

fungi
09
    Myxomycetes

Graptolithina   graptolites
10
    Camaroidea
    Dendroidea
    Graptoloidea
        Didymograptina
        Diplograptina
        Glossograptina
        Monograptina
    Stolonoidea
    Tuboidea

gymnosperms   gymnosperm flora
09
    Bennettitales
    Caytoniales
    Coniferales
    Cordaitales
    Cycadales
    Cycadofilicales
    Ephedrales
    Ginkgoales
    Glossopteridales
    Gnetales
    Nilssoniales
    Pentoxylales
    Welwitschiales

Hemichordata
10
    Enteropneusta

ichnofossils
08,09,10,11   (Use lebensspuren for sedimentary
              structures)

Insecta   insects
10
    Blattopteroida   (Blattoidea)
    Coleopteroida   (Coleoptera)
    Dermapteroida   (Dermaptera)
    Ectotropha
    Entotropha
    Ephemeropteroida   (Ephemeroptera)
    Hemipteroida   (Hemiptera)
    Hymenopteroida   (Hymenoptera)
    Lepidopteroida   (Lepidoptera)
    Mecopteroida   (Mecoptera)
    Neuropteroida   (Neuroptera)
    Odonatopteroida   (Odonata)
    Orthopteroida   (Orthoptera)
    Palaeodictyopteroida (Palaeodictyoptera)
    Psocopteroida   (Psocoptera)
    Siphonapteroida   (Siphonaptera)
    Thysanopteroida   (Thysanoptera)

Invertebrata   invertebrates
10

lichens   lichens
09

Mammalia   mammals
11
    Docodonta
    Monotremata
    Multituberculata

Paucituberculata
Theria
    Eutheria
        Dinocerata
        Pantodonta
        Pyrotheria
        Xenungulata
        Artiodactyla
            Ruminantia
                Tylopoda
            Suiformes
        Astrapotheria
        Carnivora
            Fissipeda
            Pinnipedia
        Cetacea
        Chiroptera
        Condylarthra
            Acreodi
            Arctocyonia
        Creodonta
        Desmostylia
        Edentata
            Xenarthra
        Embrithopoda
        Hyracoidea
        Insectivora
            Dermoptera
            Macroscelida
            Proteutheria
                Anagalida
                Scandentia
        Lagomorpha
        Litopterna
        Notoungulata
        Perissodactyla
            Ceratomorpha
            Hippomorpha
        Pholidota
        Primates
            Hominidae
                Homo sapiens
                Neanderthal
                pre-Neanderthal
            Prosimii
            simians
                Cercopithecidae
                Platyrrhina
                Pongidae
        Proboscidea
            Bathytherioidea
            Deinotherioidea
            Elephantoidea
            Mastodontoidea
            Moeritherioidea
        Rodentia   rodents
            Hystricomorpha
            Myomorpha
            Sciuromorpha
        Sirenia
        Taeniodonta
        Tillodontia
        Tubulidentata
        Ungulata
    Metatheria
        Marsupialia
            Diprotodonta
            Polyprotodontia
    Pantotheria
    Symmetrodonta
Triconodonta

Mollusca   mollusks
10
    Aplacophora
    Bivalvia   bivalves

Actinodontida		Pisces	fish
Arcina		11	
Astartida		Acanthodii	
Carditida		Chondrichthyes	
Ctenodontida		Elasmobranchii	

Actinodontida
Arcina
Astartida
Carditida
Ctenodontida
Cyrtodontida
Ostreacea
Pteriina
  Inocerami
  Pectinacea
Pholadomyida
Praecardiida
Rudistae   rudists
Septibranchia
Solemyida
Venerida
Cephalopoda   cephalopods
  Dibranchiata
    Belemnoidea
  Tetrabranchiata
    Ammonoidea   ammonoids
      Anarcestida
      Bactritida
      Ceratitida
      Clymeniida
      Desmoceratida
      Goniatitida
      Lytoceratida
      Perisphinctida
      Phylloceratida
      Prolecanitida
      Psiloceratida
    Nautiloidea
Gastropoda   gastropods
  Acoela
  Archaeogastropoda
    Bellerophontina
  Basommatophora
  Entomotaeniata
  Mesogastropoda
  Neogastropoda
  Pteropoda
  Sacoglossa
  Stylommatophora
Hyolithes
Monoplacophora
Polyplacophora
Rostroconchia
Scaphopoda
Tentaculida
Ostracoda   ostracods
10, (M08)
  Archeocopida
  Beyrichicopina
  Eridostraca
  Kirkbyocopina
  Leperditicopida
  Myodocopida
    Cladocopina
    Entomozocopina
    Mydocopina
  Paleocopida
  Podocopida
    Bairdiomorpha
    Cypridocopina
    Cytherocopina
      Cytheracea
  Platycopida
palynomorphs   palynomorphs
09, M09
  acritarchs   acritarch flora
  Chitinozoa   chitinozoans
  Dinoflagellata   dinoflagellates
  megaspores   megaspores
  miospores   miospores

Pisces   fish
11
  Acanthodii
  Chondrichthyes
    Elasmobranchii
    Eubradyodonti
    Holocephali
  Osteichthyes
    Brachiopterygii
    Crossopterygii
      Actinistia
      Rhipidistia
    Chondrostei
    Dipnoi
    Holostei
      Halecostomi
    Teleostei
  Placodermi
Plantae   plants
09
Porifera   sponges
10
  Calcispongea
  Demospongea
  Hyalospongea
problematic fossils
08,09,10,11, (M07)
  problematic microfossils
Protista
08,09,10, (M10)
  ebridians
  Silicoflagellata   silicoflagellates
  Thecamoeba
  Tintinnidae
pteridophytes   ferns
09
  Filicopsida
  Lycopsida
  Noeggerathiales
  Psilopsida
  Pteridophyllen
  Sphenopsida
Pterobranchia
10
  Cephalodiscida
  Rhabdopleurida
Radiolaria   radiolarians
10, (M11)
  Acantharina
  Nassellina
  Phaeodarina
  Spumellina
Reptilia   reptiles
11
  Anapsida
    Chelonia
    Cotylosauria
      Captorhinomorpha
      Pareiasauria
    Mesosauria
  Diapsida
    Araeoscelidia
    Archosauria
      Crocodilia
      dinosaurs
        Ornithischia
        Saurischia
      Pterosauria
      Thecodontia
    Eosuchia
    Ichthyosauria
    Lepidosauria

                    Squamata
                    Placodontia
                    Rhynchosauria
                    Sauropterygia
                        Nothosauria
                        Plesiosauria
                    Trilophosauria
                Synapsida
                    Pelycosauria
                    Therapsida
Spermatophyta (incl. angiosperms
09   and gymnosperms)
Tetrapoda   tetrapods
11
thallophytes   thallophytes
Trilobita   trilobites
10
                Agnostida
                Corynexochida
                Lichida
                Odontopleurida
                Phacopida
                Ptychopariida
                Redlichiida
Vertebrata   vertebrates
11
worms   worms
10
                Annelida
                Chaetognatha
                Echiurida
                Myzostomia
                Nematoida
                Nematomorpha
                Nemerta
                Oligochaetia
                Phoronida
                Pogonophora
                Polychaetia
                Priapulida
    (M02)   ~   ~~ scolecodonts
                Sipunculoida

LIST G - METEORITES

INDEXING

Under this general heading entries are made for specific <u>meteorites</u>. Index the most specific meteorite name. Also index type of meteorite if known. Terms in the meteorite hierarchy listed below should be added using the most specific term appropriate. <u>Meteorites</u> are not indexed with geographic locations.

ONLINE SEARCHING

To search for meteorites use the specific term such as <u>Allende Meteorite</u> or the general term <u>meteorites</u>.

B.I.G. SEARCHING

Prior to 1993, look under under the heading <u>meteorites</u>.

In current practice specific <u>meteorites</u> are direct access points.

HIERARCHICAL LIST

meteorites
  iron meteorites
    ataxite
    hexahedrite
    octahedrite
  stony irons
    mesosiderite
    pallasite
    siderophyre
  stony meteorites
    achondrites
      angrite
      chassignite
      chladnite
      diogenite
      eucrite
      howardite
      nakhlite
      shergottite
      ureilite
    chondrites
      carbonaceous chondrites
      enstatite chondrites
      H chondrites
      HL chondrites
      L chondrites
      LL chondrites
      ordinary chondrites

ALPHABETICAL LIST

<u>Specific name</u>	<u>Group name</u>
Abee Meteorite	enstatite chondrites
achondrites	stony meteorites
ALHA 77005	Allan Hills Meteorites
ALHA 81005	Allan Hills Meteorites
Allan Hills Meteorites	meteorites
Allende Meteorite	carbonaceous chondrites
angrite	achondrites
Ashmore Meteorite	H chondrites
ataxite	iron meteorites
Barwell Meteorite	L chondrites
Bjurbole Meteorite	L chondrites
Bruderheim Meteorite	L chondrites
Campo del Cielo Meteorite	ataxite
Canyon Diablo Meteorites	octahedrite
Cape York Meteorite	iron meteorites
carbonaceous chondrites	chondrites
Chainpur Meteorite	LL chondrites
chassignite	achondrites
Chassigny Meteorite	chassignite

chladnite	achondrites	Toluca Meteorite	octahedrite
chondrites	stony meteorites	Tsarev Meteorite	L chondrites
Coldwater Meteorites	meteorites	ureilite	achondrites
Dhajala Meteorite	H chondrites	Vigarano Meteorite	carbonaceous chondrites
diogenite	achondrites	Warburton Meteorite	ataxite
Eagle Station Meteorite	pallasite	Weekeroo Station Meteorite	octahedrite
EETA 79001	Elephant Moraine Meteorites		
Efremovka Meteorite	carbonaceous chondrites	Weston Meteorite	chondrites
Elephant Moraine Meteorites	meteorites	Wolf Creek Meteorite	octahedrite
		Yamato Meteorites	meteorites
enstatite chondrites	chondrites	Ybbsitz Meteorite	H chondrites
eucrite	achondrites		
Fayetteville Meteorite	H chondrites		
Gibeon Meteorite	octahedrite		
H chondrites	chondrites		
Haveroe Meteorite	ureilite		
Haviland Meteorite	H chondrites		
Hedjaz Meteorite	L chondrites		
Henbury Meteorite	octahedrite		
hexahedrite	iron meteorites		
HL chondrites	chondrites		
howardite	achondrites		
Indarch Meteorite	enstatite chondrites		
Innisfree Meteorite	LL chondrites		
iron meteorites	meteorites		
Jilin Meteorite	chondrites		
Juvinas Meteorite	eucrite		
Kainsaz Meteorite	carbonaceous chondrites		
Kapoeta Meteorite	howardite		
Krymka Meteorite	LL chondrites		
L chondrites	chondrites		
Leoville Meteorite	carbonaceous chondrites		
Lewis Cliff Meteorites	meteorites		
LL chondrites	chondrites		
Malvern Meteorite	howardite		
mesosiderite	stony irons		
micrometeorites	meteorites		
Moore County Meteorite	eucrite		
Murchison Meteorite	carbonaceous chondrites		
Murray Meteorite	carbonaceous chondrites		
Nakhla Meteorite	nakhlite		
nakhlite	achondrites		
Norton County Meteorite	chladnite		
octahedrite	iron meteorites		
Odessa Meteorite	meteorites		
ordinary chondrites	chondrites		
Orgueil Meteorite	carbonaceous chondrites		
pallasite	stony irons		
Pasamonte Meteorite	eucrite		
Qingzhen Meteorite	enstatite chondrites		
Rio Negro Meteorite	L chondrites		
Saint-Severin Meteorite	chondrites		
Santa Clara Meteorite	iron meteorites		
Semarkona Meteorite	LL chondrites		
Sharps Meteorite	chondrites		
shergottite	achondrites		
Shergotty Meteorite	shergottite		
siderophyre	stony irons		
SNC Meteorites	achondrites		
stony irons	meteorites		
stony meteorites	meteorites		
Tieschitz Meteorite	H chondrites		

LIST H - IGNEOUS ROCKS

INDEXING
The hierarchical classification of igneous rocks is presented below as well as a list of all specific terms that are included in the thesaurus. The most specific rock name should be used. Specific rock names which do not appear in the body of the Thesaurus should be indexed using the appropriate group from the hierarchical list.

ONLINE SEARCHING
Search for specific rocks by name, truncated at the the end to retrieve singular and plural forms.

To retrieve papers in which igneous rocks were a major topic, search igneous rocks. Since only specific rock names may have been used in papers in which the rocks are minor topics, it is necessary to also search specific rock names and group names for an exhaustive search for igneous rocks. Autoposting of group names began in 1978. For more information on autoposted terms see specific rock names in the body of the Thesaurus.

Search petrology as a major descriptor for papers dealing with the discipline, such as papers on techniques, practice, history, etc.

Related topics such as intrusions, isotopes, metamorphic rocks, metasomatic rocks, magmas, metamorphism, metasomatism, phase equilibria, etc. can be searched separately.

B.I.G. SEARCHING
Before 1993, at the entry for igneous rocks, look under the appropriate term(s) selected from the hierarchical list below.

Look under related headings such as intrusions, isotopes, magmas, metamorphic rocks, metamorphism, metasomatic rocks, metasomatism, and phase equilibria.

Look under petrology for papers dealing with the discipline, such as papers on techniques, practice of, history, etc.

HIERARCHICAL LIST

igneous rocks
  hypabyssal rocks
  plutonic rocks
    carbonatites
    diabase
    diorites
      alkali diorites
      microdiorite
      quartz diorites
    gabbros
      alkali gabbros
        leucitite
        nephelinite
      anorthosite
      microgabbro
      norite
    granites
      alkali granites
      aplite
      charnockite
      granodiorites
      leucogranite
      microgranite
    lamprophyres
    syenites
      alkali syenites
      microsyenite
      monzonites
    ultramafics
      kimberlite
      peridotites
      pyroxenite
  volcanic rocks
    andesites
    basalts
      alkali basalts
        trachybasalts
    dacites
    glasses
      perlite
    phonolites
    pyroclastics
      ignimbrite
      pumice
      tuff
    rhyodacites
    rhyolites
    trachyandesites
    trachytes
      keratophyre

The following list provides the appropriate group name for all the specific rock names currently listed in the body of the Thesaurus.

Specific rock name	Group name
A-type granites	granites
absarokite	basalts
adamellite	granites
agpaite	syenites
alaskite	granites
albitite	syenites
albitophyre	syenites
alkali olivine basalt	alkali basalts
alnoite	lamprophyres
andesite porphyry	andesites
andesite tuff	pyroclastics
ankaramite	basalts
ankaratrite	alkali basalts
anorthosite	gabbros
anorthositic gabbro	gabbros
aplite	granites
apogranite	granites
appinite	diorites or syenites
ariegite	ultramafics
ash-flow tuff	pyroclastics
augitite	ultramafics
basanite	alkali basalts
benmoreite	volcanic rocks
biotite granite	granites
boninite	andesites
bostonite	syenites
bronzitite	ultramafics
camptonite	lamprophyres
charnockite	granites
chromitite	ultramafaics
clinopyroxenite	ultramafics
columnar basalt	basalts
comendite	rhyolites
crinanite	alkali basalts
dacite porphyry	dacites
dellenite	rhyodacites
diabase porphyry	diabase
diallagite	ultramafics
diorite porphyry	diorites
dunite	peridotites
enderbite	granites
essexite	alkali gabbros
eulysite	ultramafics
farsundite	granites
feldspathoid rocks	
felsite	granites
ferrodiorite	diorites
foyaite	syenites
gabbroic anorthosite	gabbros
garnet lherzolite	peridotites
garnet peridotite	peridotites
garnet pyroxenite	ultramafics
granite porphyry	granodiorites
granophyre	volcanic rocks
granosyenite	granites and syenites
graphic granite	granites
green tuff	pyroclastics
griquaite	ultramafics
harzburgite	peridotites
hawaiite	alkali basalts
hornblendite	ultramafics
hyaloclastite	pyroclastics
ignimbrite	pyroclastics
I-type granites	granites

ijolite	alkali gabbros	quartz keratophyre	trachytes
kakortokite	syenites	quartz latite	rhyodacites
keratophyre	trachytes	quartz monzonite	granites
kersantite	lamprophyres	quartz porphyry	rhyolites
kimberlite	ultramafics	quartz syenite	syenites
komatiite	ultramafics	quartzolite	plutonic rocks
labradoritite	gabbros	rapakivi	granites
lamproite	lamprophyres	rhyodacites	volcanic rocks
latite	trachyandesites	rhyolite porphyry	rhyolites
leucitite	alkali gabbros	rhyolite tuff	pyroclastics
leucogranite	granites	rodingite	gabbros
lherzolite	peridotites	S-type granites	granites
limburgite	ultramafics	scoria	pyroclastics
liparite	rhyolites	shonkinite	syenites
liparite porphyry	rhyolites	shoshonite	basalts
lujavrite	syenites	sovite	carbonatites
mangerite	diorites	spessartite	lamprophyres
meimechite	ultramafics	spilite	alkali basalts
melaphyre	basalts	spinel lherzolite	peridotites
melilitite	volcanic rocks	spinel peridotite	peridotites
melteigite	alkali gabbros	syenite porphyry	syenites
miaskite	syenites	syenodiorite	diorites
microdiorite	diorites	tephrite	alkali basalts
microgabbro	gabbros	teschenite	alkali gabbros
microgranite	granites	tholeiite	basalts
micropegmatite	granites	tholeiitic basalt	basalts
microsyenite	syenites	tholeiitic dolerite	diabase
mid-ocean ridge basalts	basalts	tinguaite	phonolites
minette	lamprophyres	tonalite	diorites
monchiquite	lamprophyres	trachybasalts	alkali basalts
monzodiorite	diorites	trachydolerite	alkali basalts
monzogranite	granites	trap rocks	basalts
mugearite	alkali basalts	troctolite	gabbros
muscovite granite	granites	trondhjemite	diorites
naujaite	syenites	tuff	pyroclastics
nepheline basalt	alkali basalts	tuffite	pyroclastics
nepheline syenite	syenites	two-mica granite	granites
nephelinite	alkali gabbros	urtite	alkali gabbros
norite	gabbros	vitrophyre	volcanic rocks
obsidian	glasses	vogesite	lamprophyres
oceanite	basalts	volcanic ash	
olivine basalt	basalts	volcanic glass	glasses
olivine diabase	diabase	websterite	ultramafics
olivine gabbro	gabbros	welded tuff	pyroclastics
olivine nephelinite	alkali gabbros	zirkelite	glasses
olivine tholeiite	basalts		
olivinite	ultramafics		
palagonite	glasses		
pantellerite	rhyolites		
pegmatite	granites		
perlite	glasses		
picrite			
picrite porphyry			
pitchstone	glasses		
plagiogranite	diorites		
pulaskite	syenites		
pumice	pyroclastics		
pyrolite			
pyroxene andesite	andesites		
pyroxenite	ultramafics		
quartz diabase	diabase		
quartz diorites	diorites		

LIST I - SEDIMENTARY ROCKS

INDEXING

Index the most specific term from the alphabetical list below. If the term does not appear in the list, index a higher level term that does.

ONLINE SEARCHING

To search for a specific rock, enter the rock name, truncated, in order to obtain both singular or plural forms.

To be thorough, include variants.

For an exhaustive search use all major groups and the general term sedimentary rocks.

Search only the general term sedimentary rocks if only papers which have sedimentary rocks as the major topic are desired. Autoposting of group names to specific rock names began in 1978. For more information see the specific rock names in the body of the Thesaurus. For sedimentary rocks which are commodities see List C.

Sedimentary petrology can be searched for papers dealing with the discipline, such as papers on techniques, methods, etc.

Related topics such as sedimentation, sedimentary structures, sediments, and diagenesis can be searched separately.

B.I.G. SEARCHING

Prior to 1993, at the entry for sedimentary rocks, look under the appropriate term(s) selected from the list below.

Look under related headings such as sedimentation, sedimentary structures, sediments, diagenesis, and weathering.

For papers dealing with the discipline, such as papers on techniques, practice of, etc., look under sedimentary petrology.

Look under specific sedimentary rocks which are commodities, e.g. limestone deposits (see List C, Commodities).

In current practice the specific rock type is a direct access point.

HIERARCHICAL LIST

sedimentary rocks
    carbonate rocks
        algal limestone
        beachrock
        biocalcarenite
        calcarenite
        calcrete
        chalk
        coquina
        dolomitic limestone
        dolostone
        limestone
        micrite
        microcrystalline limestone
        oolitic limestone
        travertine
    chemically precipitated rocks
        bauxite
        chert
            jasperoid
        evaporites
            anhydrite
            gypsum
            potash
            salt
        ferricrete
        flint
        iron formations
        phosphate rocks
            bone beds
        silcrete
        siliceous sinter
        weathering crust
    clastic rocks
        argillite
        arkose
        bentonite
        breccia
            tectonic breccia
            volcanic breccia
        conglomerate
        diatomaceous earth
        flysch
        graywacke
        marl
        molasse
        radiolarite
        red beds
        sandstone
        shale
        siltstone
        spongolite
        tillite
        tonstein
        turbidite
    organic residues
        coal
            anthracite
            bituminous coal
            coke coal
            lignite
            sapropelite
            steam coal
    oil sands
    oil shale

The following list provides the appropriate group name for all the specific rock names currently listed in the body of the Thesaurus.

Specific rock name	Group name
algal limestone	carbonate rocks
anthracite	organic residues
arenite	clastic rocks
argillite	clastic rocks
arkose	clastic rocks
beachrock	carbonate rocks
bentonite	clastic rocks
biocalcarenite	carbonate rocks
biomicrite	carbonate rocks
biosparite	carbonate rocks
bituminous coal	organic residues
black shale	clastic rocks
boundstone	carbonate rocks

breccia	clastic rocks	sandstone	clastic rocks
calcarenite	carbonate rocks	saprolite	clastic rocks
calcilutite	carbonate rocks	sapropelite	organic residues
calcrete	carbonate rocks	sclerotinite	organic residues
caliche	carbonate rocks	shale	clastic rocks
caustobiolith	organic residues	silcrete	chemically precipitated rocks
chalk	carbonate rocks	siliceous sinter	chemically precipitated rocks
chert	chemically precipitated rocks	siltstone	clastic rocks
cinerite	clastic rocks	sparagmite	clastic rocks
clarain	organic residues	spongolite	clastic rocks
claystone	clastic rocks	steam coal	organic residues
coal	organic residues	subarkose	clastic rocks
coke coal	organic residues	subbituminous coal	organic residues
collinite	organic residues	subgraywacke	clastic rocks
conglomerate	clastic rocks	taconite	chemically precipitated rocks
contourite	clastic rocks	tectonic breccia	clastic rocks
coquina	carbonate rocks	tillite	clastic rocks
cutinite	organic residues	tilloid	clastic rocks
diamictite	clastic rocks	tonstein	clastic rocks
diatomaceous earth	clastic rocks	torbanite	organic residues
dolomitic limestone	carbonate rocks	travertine	carbonate rocks
dolostone	carbonate rocks	tufa	chemically precipitated rocks
duricrust	chemically precipitated rocks	volcanic breccia	clastic rocks
eolianite	clastic rocks	wackestone	carbonate rocks
evaporites	chemically precipitated rocks	weathering crust	chemically precipitated rocks
fanglomerate	clastic rocks		
ferricrete	chemically precipitated rocks		
flint	chemically precipitated rocks		
flysch	clastic rocks		
gaize	clastic rocks		
gas sands	organic residues		
grainstone	carbonate rocks		
graywacke	clastic rocks		
humodetrinite	organic residues		
iron formations	chemically precipitated rocks		
ironstone	chemically precipitated rocks		
jasperoid	chemically precipitated rocks		
jaspilite	chemically precipitated rocks		
lignite	organic residues		
limestone	carbonate rocks		
marl	clastic rocks		
micrite	carbonate rocks		
microbreccia	clastic rocks		
microcrystalline limestone	carbonate rocks		
mixtite	clastic rocks		
molasse	clastic rocks		
mudstone	clastic rocks		
novaculite	clastic rocks		
oil sands	organic residues		
oil shale	organic residues		
oolitic limestone	carbonate rocks		
orthoquartzite	clastic rocks		
packstone	carbonate rocks		
phosphate rocks	chemically precipitated rocks		
porcellanite	clastic rocks		
potash	chemically precipitated rocks		
quartz arenite	clastic rocks		
quartz-pebble conglomerate	clastic rocks		
radiolarite	clastic rocks		
red beds	clastic rocks		
salt	chemically precipitated rocks		

LIST J - METAMORPHIC ROCKS

INDEXING

Index the most specific term from the alphabetical list below. If the term does not appear in the list, index a higher level term that does.

For documents on petrology as a discipline, index petrology.

Index metamorphism for papers which stress the process.

ONLINE SEARCHING

Papers with major emphasis on these rocks can be retrieved by searching for metamorphic rocks as a major descriptor or access point.

For a thorough search of metamorphic rocks, use all major groups, truncated, as well as the general term metamorphic rocks, truncated. Autoposting of group names to specific rock names began in 1978. For more information see the specific rock names in the body of the Thesaurus. For metamorphic rocks which are commodities see List C.

For specific searches on a certain rock, use the rock name, truncated.

For composite rock names, e.g. biotite gneiss, it is advisable to search both the specific rock name and the separate parts of the name truncated and combined.

B.I.G. SEARCHING

Prior to 1993, at the entry for <u>metamorphic rocks</u>, look under the appropriate term(s) selected from the list below.

Look under related headings such as <u>metamorphism</u>, <u>petrology</u>, <u>metasomatism</u>, <u>phase equilibria</u>, and <u>paragenesis</u>.

For papers dealing with the discipline, such as papers on techniques, practice of, etc., look under <u>petrology</u>.

In current practice the specific rock type is a direct access point.

HIERARCHICAL LIST

metamorphic rocks
    amphibolites
        orthoamphibolite
        para-amphibolite
    cataclasites
    eclogite
    fulgurite
    garnetite
    gneisses
        orthogneiss
        paragneiss
    granulites
    hornfels
    itabirite
    marbles
    metaigneous rocks
    metaplutonic rocks
    metasedimentary rocks
    metasomatic rocks
        fenite
        greisen
        propylite
        skarn
    metavolcanic rocks
    migmatites
        anatexite
        embrechite
    mylonites
    phyllites
    phyllonites
    quartzites
    schists
        blueschist
        greenschist
    slates

The following list provides the appropriate term for all the specific rock names currently listed in the body of the Thesaurus.

Specific rock name	Group name
agmatite	migmatites
anatexite	migmatites
augen gneiss	gneisses
banded gneiss	gneisses
beresite	greisen
biotite gneiss	gneisses
biotite schist	schists
blastomylonite	mylonites
blueschist	schists
brucite marble	marbles
calc-schist	schists
calciphyre	marbles
chlorite schist	schists
ectinite	migmatites
embrechite	migmatites
epidiorite	schists
ferruginous quartzites	quartzites
glaucophane schist	schists
granite gneiss	gneisses
greenschist	schists
greenstone	schists
hornblende schist	schists
khondalite	metasedimentary rocks
kinzigite	granulites
leptite	granulites
leucosomes	migmatites
listwanite	schists
meta-andesite	metaigneous rocks
meta-anorthosite	metaigneous rocks
meta-arkose	metasedimentary rocks
metabasalt	metaigneous rocks
metabasite	metaigneous rocks
metachert	metasedimentary rocks
metaconglomerate	metasedimentary rocks
metadacite	metaigneous rocks
metadiabase	metaigneous rocks
metadiorite	metaigneous rocks
metagabbro	metaigneous rocks
metagranite	metaigneous rocks
metagraywacke	metasedimentary rocks
metakomatiite	metaigneous rocks
metalimestone	metasedimentary rocks
metapelite	metasedimentary rocks
metaperidotite	metaigneous rocks
metapyroxenite	metaigneous rocks
metarhyolite	metaigneous rocks
metasandstone	metasedimentary rocks
metasiltstone	metasedimentary rocks
metatuff	metaigneous rocks
muscovite schist	schists
ophicalcite	marbles
orthoamphibolite	amphibolites
orthogneiss	gneisses
para-amphibolite	amphibolites
paragneiss	gneisses
pseudotachylite	mylonites
pyroxene granulite	granulites
serpentinite	metaigneous rocks and metasomatic rocks
sillimanite gneiss	gneisses
steatite	metasomatic rocks
talc rock	metasomatic rocks
talcite	metasomatic rocks
tonalite gneiss	gneisses
tourmalinite	schists
ultramylonite	mylonites

LIST K - SEDIMENTARY STRUCTURES

INDEXING

Index the specific structure from the alphabetical list below. If the term does not appear in the list, index a higher level term that does.

ONLINE SEARCHING

Search for specific or general structure types as applicable.

Papers with major emphasis on sedimentary structures can be retrieved by searching for <u>sedimentary structures</u> as a major descriptor. See List A for more information.

For a thorough search of sedimentary structures, use all structure types in truncated form, plus the general term <u>sedimentary structures</u>. Autoposting of structure types and the general term <u>sedimentary structures</u> began in 1978.

For specific searches on a structure use the structure name, truncated.

B.I.G. SEARCHING

Prior to 1993, at the entry for <u>sedimentary structures</u>, look under appropriate term(s) from the following list. In current practice the specific rock type is a direct access point.

HIERARCHICAL LIST

sedimentary structures
  bedding plane irregularities
    antidunes
    chevron marks
    dune structures
    flute casts
    fossil ice wedges
    frost features
    grooves
    groove casts
    load casts (<u>also under</u> soft sediment deformation <u>and</u> turbidity current structures)
    megaripples
    mounds
    mudcracks
    mud lumps
    parting lineation
    ripple marks (<u>also under</u> primary structures) (asymmetrical, interference, etc.)
    sand waves
    scour casts
    scour marks
    sole marks (<u>also under</u> turbidity current structures)
    striations
    tool marks
  biogenic structures
    algal banks
    algal biscuits
    algal mats
    algal mounds
    algal structures
    banks
    bioherms
    bioturbation
    borings
    burrows
    coprolites
    girvanella
    lebensspuren
    oncolites
    stromatolites
    tracks
    trails
  cylindrical structures
  planar bedding structures (i.e. bedding structures)
    structures
    bars
    bedding (<u>also under</u> primary structures)
    channels
    cross-bedding
    cross-laminations
    cross-stratification
    cut and fill
    cyclothems
    flaser bedding
    graded bedding (<u>also under</u> turbidity current structures)
    hummocky cross-stratification
    imbrication
    laminations
    massive bedding
    megacyclothems
    rhythmic bedding
    rhythmite
    ripple drift-cross laminations
    sand bodies
    stratification
    varves
  primary structures
    bedding (<u>also under</u> planar bedding structures)
    ripple marks (<u>also under</u> bedding plane irregularities) (asymmetrical, interference, etc.)
  secondary structures
    armored mud balls
    concretions
    cone-in-cone
    geodes
    microstylolites
    septaria
    stylolites
  soft sediment deformation
    ball-and-pillow
    boudinage
    clastic dikes
    convoluted beds (<u>also under</u> turbidity current structures)
    flame structures
    flow structures (<u>also under</u> turbidity current structures)
    load casts (<u>also under</u> bedding plane irregularities <u>and</u> turbidity current structures)
    olisoliths (<u>also under</u> turbidity current structures)
    olistostromes (<u>also under</u> turbidity current

structures)
sandstone dikes
slump structures
turbidity current structures
   Bouma sequence
   convoluted beds (<u>also under</u> soft sediment deformation)
   flow structures (<u>also under</u> soft sediment deformation)
   graded bedding (<u>also under</u> planar bedding structures)
   load casts (<u>also under</u> bedding plane irregularities <u>and</u> soft sediment deformation)
   olistoliths (<u>also under</u> soft sediment deformation)
   olistostromes (<u>also under</u> soft sediment deformation)
   sole marks (<u>also under</u> bedding plane irregularities)

The following list provides the appropriate group term for all the specific structures currently listed in the body of the Thesaurus.

<u>Specific structure</u>	<u>Structure type</u>
algal banks	biogenic structures
algal biscuits	biogenic structures
algal mats	biogenic structures
algal mounds	biogenic structures
algal structures	biogenic structures
antidunes	bedding plane irregularities
armored mud balls	secondary structures
ball-and-pillow	soft sediment deformation
banks	biogenic structures
bedding	planar bedding structures <i>or</i> primary structures
bioturbation	biogenic structures
boudinage	soft sediment deformation
Bouma sequence	turbidity current structures
chevron marks	bedding plane irregularities
clastic dikes	soft sediment deformation
concretions	secondary structures
cone-in-cone	secondary structures
convoluted beds	soft sediment deformation <i>or</i> turbidity current structures
cross-bedding	planar bedding structures
cross-laminations	planar bedding structures
cross-stratification	planar bedding structures
current markings	bedding plane irregularities
cut and fill	planar bedding structures
cyclothems	planar bedding structures
dune structures	bedding plane irregularities
flame structures	soft sediment deformation
flaser bedding	planar bedding structures
flow structures	soft sediment deformation <i>or</i> turbidity current structures
flute casts	bedding plane irregularities
fossil ice wedges	bedding plane irregularities
frost features	bedding plane irregularities
geodes	secondary structures
girvanella	biogenic structures
graded bedding	turbidity current structures <i>or</i> planar bedding structures
groove casts	bedding plane irregularities
grooves	bedding plane irregularities
hummocky cross-stratification	planar bedding structures
imbrication	planar bedding structures
laminations	planar bedding structures
lebensspuren	biogenic structures
load casts	turbidity current structures <i>or</i> bedding plane irregularities <i>or</i> soft sediment deformation
massive bedding	planar bedding structures
megacyclothems	planar bedding structures
megaripples	bedding plane irregularities
microstylolites	secondary structures
mounds	bedding plane irregularities
mud lumps	bedding plane irregularities
mudcracks	bedding plane irregularities
olistoliths	soft sediment deformation <i>or</i> turbidity current structures
olistostromes	soft sediment deformation <i>or</i> turbidity current structures
oncolites	biogenic structures
parting lineation	bedding plane irregularities
rhythmic bedding	planar bedding structures
rhythmite	planar bedding structures
ripple drift-cross laminations	planar bedding structures
ripple marks	bedding plane irregularities <i>or</i> primary structures
sand bodies	planar bedding structures
sand waves	bedding plane irregularities
sandstone dikes	soft sediment deformation
scour casts	bedding plane irregularities
scour marks	bedding plane irregularities
septaria	secondary structures
slump structures	soft sediment deformation
sole marks	turbidity current structures <i>or</i> bedding plane irregularities
stratification	planar bedding structures
striations	bedding plane irregularities
stromatactis	biogenic structures
stromatolites	biogenic structures
stylolites	secondary structures
tool marks	bedding plane irregularities
varves	planar bedding structures

LIST L - MINERALS

INDEXING

The following group names (chemical for the non-silicates and structural for the silicates) are used for <u>minerals</u>. A specific mineral name is used whenever applicable. Select a term from the list below.

For papers dealing with new mineral species, the term <u>new minerals</u> is used as an index term, and the chemical formula is given as an index term if available.

ONLINE SEARCHING

To search for a particular mineral species, the common name of that species must be used keeping in mind possible varieties (agate, chalcedony... or titanite and sphene...) and that the American <u>f</u> is used for the British <u>ph</u>. To be thorough, a search should also be made of the chemical formula (or parts thereof) of the mineral species.

Some mineral species may not be treated from a mineralogical point of view. It is advisable therefore to search for "mineralogic" papers using the mineral species together with the general term <u>mineralogy</u> or the mineralogy category.

To search a group of minerals search for the group name and the major specific minerals in that group. Autoposting of group names to specific minerals began in 1978. For more information see specific minerals in the main body of the Thesaurus.

For papers dealing with gemmological aspects, one should search for a mineral in combination with <u>gems</u>. For mineral collecting search a mineral in combination with (<u>collecting</u> OR <u>popular geology</u>).

For papers dealing with the economic geology of minerals, see List C.

Clay mineralogy as a discipline is treated separately. A search of the term <u>clay mineralogy</u> will retrieve papers dealing with <u>all</u> aspects of clays: mineralogy, structure, heat treatment, occurrence as constituents of rocks, etc.

B.I.G. SEARCHING

Prior to 1993, the headings <u>minerals, mineralogy, crystal structure, crystal chemistry, crystal growth, phase equilibria, clay mineralogy,</u> and <u>crystallography</u> were the main access points for mineralogical papers. Other related headings were <u>igneous rocks, metamorphic rocks, sedimentary rocks, metasomatic rocks, sediments, geochemistry,</u> and <u>paragenesis</u>.

Also cross-references were provided to direct your attention to these groups.

  e.g. sulphur   sulfosalts
   see sulfur   see under minerals

For the economic geology of minerals, look under the appropriate minerals selected from List C, Commodities.

In current practice, the specific mineral is a direct access point.

HIERARCHICAL LIST

minerals
 alloys (including carbides, nitrides, phosphides, silicides)
 antimonates
 antimonides
 antimonites
 arsenates
 arsenides
 arsenites
 bismuthides
 borates
 bromides
 carbides
 carbonates
 chlorides
 chromates
 fluoborates
 fluorides
 fluosilicates
 germanates
 halides (includes bromides, chlorides, fluoborates, fluosilicates, iodides, and fluorides)
 hydrates
 iodates
 iodides
 miscellaneous minerals (used for several groups or for
  minerals of unknown affinity)
 molybdates
 native elements
 niobates
 niobotantalates
 nitrates
 nitrides
 organic compounds
 oxalates
 oxides (including germanates, niobates, niobotantalates, tantalates)
 oxysulfides
 phosphates
 phosphides
 selenates
 selenides
 selenites
 silicates (use a narrower term below if dealing with specific mineral; otherwise larger group)
  aluminosilicates
  orthosilicates
   sorosilicates
    orthosilicates, epidote group
    orthosilicates, melilite group
   nesosilicates
    orthosilicates, garnet group
    orthosilicates, humite group
    orthosilicates, olivine group
  ring silicates
  chain silicates (inosilicates)
   chain silicates, amphibole group
    chain silicates, alkalic amphibole
    chain silicates, clinoamphibole
    chain silicates, orthoamphibole
   chain silicates, pyroxene group
    chain silicates, alkalic pyroxene
    chain silicates, clinopyroxene
    chain silicates, orthopyroxene

sheet silicates (phyllosilicates)
    sheet silicates, chlorite group
    sheet silicates, clay minerals
    sheet silicates. mica group
    sheet silicates, serpentine group
framework silicates (tectosilicates)
    framework silicates, feldspar group
        framework silicates, alkali feldspar
        framework silicates, plagioclase
        framework silicates, barium feldspar
    framework silicates, nepheline group
    framework silicates, scapolite group
    framework silicates, silica minerals
    framework silicates, sodalite group
    framework silicates, zeolite group
silicides
sulfates
sulfosalts (including sulfantimonates, sulfantimonites, sulfarsenates, sulfarsenites, sulfobismuthites, sulfogermanates, sulfostannates, and sulfovanadates)
sulfides (including antimonides, arsenides, bismuthides, oxysulfides, selenides, and tellurides)
tantalates
tellurates
tellurides
tellurites
tungstates
vanadates

NOTES

1. Papers on artifical minerals are usually not treated except when a mineralogic application is clearly indicated. The indexing term <u>synthetic materials</u> is then added.

2. Papers about mineral collecting will have the term <u>collecting</u> in the indexing.

The following list provides the appropriate group name for all the specific minerals currently listed in the Thesaurus.

Specific mineral	Group name
acanthite	sulfides
acmite	chain silicates, clinopyroxene
actinolite	chain silicates, clinoamphibole
adamite	arsenates
adularia	framework silicates, alkali feldspar
aegirine	chain silicates, clinopyroxene
aenigmatite	chain silicates
aeschynite	niobates
agate	framework silicates, silica minerals
aikinite	sulfides
akaganeite	oxides
akermanite	orthosilicates, melilite group
alabandite	sulfides
alabaster	sulfates
albite	framework silicates, plagioclase
alexandrite	oxides
allanite	orthosilicates, epidote group
alloclasite	sulfides
allophane	sheet silicates, clay minerals
almandine	orthosilicates, garnet group
alstonite	carbonates
altaite	tellurides
alum	sulfates
alumohydrocalcite	carbonates
alunite	sulfates
amazonite	framework silicates, alkali feldspar
amber	organic compounds
amblygonite	phosphates
amethyst	framework silicates, silica minerals
amosite	chain silicates, clinoamphibole
analbite	framework silicates, plagioclase
analcime	framework silicates, zeolite group
anatase	oxides
andalusite	nesosilicates
andesine	framework silicates, plagioclase
andradite	orthosilicates, garnet group
anglesite	sulfates
anhydrite	sulfates
anilite	sulfides
ankerite	carbonates
anorthite	framework silicates, plagioclase
anorthoclase	framework silicates, alkali feldspar
anthophyllite	chain silicates, orthoamphibole
antigorite	sheet silicates, serpentine group
antiperthite	framework silicates, alkali feldspar
apatite	phosphates
apophyllite	sheet silicates
aquamarine	ring silicates
aragonite	carbonates
arfvedsonite	chain silicates, clinoamphibole
argentite	sulfides
armalcolite	oxides
arsenopyrite	arsenides *and* sulfides
arsenosulfides	sulfides
artinite	carbonates
asbestos	silicates
asbolite	oxides
astrophyllite	sheet silicates
atacamite	chlorides
augite	chain silicates, clinopyroxene
austinite	arsenates
autunite	phosphates
awaruite	alloys

axinite	ring silicates	carnelian	framework silicates, silica minerals
azurite	carbonates		
babingtonite	chain silicates	carnotite	vanadates
baddeleyite	oxides	carpholite	chain silicates
barbosalite	phosphates	carrollite	sulfides
barite	sulfates	cassiterite	oxides
barium silicates	silicates	cattierite	sulfides
barylite	ring silicates	celadonite	sheet silicates, mica group
bassanite	sulfates	celestite	sulfates
bastnaesite	carbonates *and* fluorides	celsian	framework silicates, barium feldspar
bavenite	sheet silicates		
bayerite	oxides	cerolite	sheet silicates, serpentine group
beidellite	sheet silicates, clay minerals		
beraunite	phosphates	cerussite	carbonates
berndtite	sulfides	chabazite	framework silicates, zeolite group
berryite	sulfosalts		
berthierine	sheet silicates	chalcanthite	sulfates
bertrandite	ring silicates	chalcedony	framework silicates, silica minerals
beryl	ring silicates		
betafite	niobotantalates	chalcocite	sulfides
betekhtinite	sulfides	chalcopyrite	sulfides
beudantite	arsenates *and* sulfates	chalcostibite	sulfosalts
bideauxite	chlorides *and* fluorides	chalybite	carbonates
biotite	sheet silicates, mica group	chamosite	sheet silicates, chlorite group
birnessite	oxides	chaoite	native elements
bischofite	chlorides	charoite	sheet silicates
bismuthinite	sulfides	chevkinite	sorosilicates
bixbyite	oxides	chkalovite	framework silicates
boehmite	oxides	chlorapatite	phosphates
bohdanowiczite	sulfides	chlorite	sheet silicates, chlorite group
boracite	borates *and* chlorides	chloritoid	nesosilicates
borax	borates	chlorophaeite	silicates
bornite	sulfides	chrome diopside	chain silicates, clinopyroxene
borosilicates	silicates	chrome spinel	oxides
boulangerite	sulfosalts	chromite	oxides
bournonite	sulfosalts	chrysoberyl	oxides
brannerite	oxides	chrysocolla	sheet silicates
braunite	nesosilicates	chrysoprase	framework silicates, silica minerals
bravoite	sulfides		
briartite	sulfides	chrysotile	sheet silicates, serpentine group
britholite	phosphates *and* orthosilicates		
		cinnabar	sulfides
brochantite	sulfates	clinochlore	sheet silicates, chlorite group
bromellite	oxides	clinoenstatite	chain silicates, clinopyroxene
bronzite	chain silicates, orthopyroxene	clinohumite	orthosilicates, humite group *and* fluorides
brookite	oxides	clinohypersthene	chain silicates, clinopyroxene
brucite	oxides	clinoptilolite	framework silicates, zeolite group
brushite	phosphates		
buergerite	ring silicates	clinozoisite	orthosilicates, epidote group
bustamite	chain silicates	clintonite	sheet silicates, mica group
bytownite	framework silicates, plagioclase	cobaltite	arsenides *and* sulfides
		coesite	framework silicates, silica minerals
cacoxenite	phosphates		
calaverite	tellurides	coffinite	nesosilicates
calcite	carbonates	cohenite	carbides
cancrinite	carbonates *and* framework silicates	colemanite	borates
		coloradoite	tellurides
canfieldite	sulfosalts	columbite	niobotantalates
carbonado	native elements	conichalcite	arsenates
carbonate apatite	phosphates	cookeite	sheet silicates, chlorite group
carnallite	chlorides	copiapite	sulfates

copper sulfides	sulfides	eskebornite	selenides
coquimbite	sulfates	ettringite	sulfates
cordierite	ring silicates	euclase	nesosilicates
coronadite	oxides	eucryptite	nesosilicates
corrensite	sheet silicates, clay minerals	eudialyte	chlorides *and* ring silicates
corundum	oxides	eulite	chain silicates, orthopyroxene
cosalite	sulfosalts		
covellite	sulfides	euxenite	niobotantalates
crandallite	phosphates	famatinite	sulfosalts
cristobalite	framework silicates, silica minerals	fassaite	chain silicates, clinopyroxene
		faujasite	framework silicates, zeolite group
crocidolite	chain silicates, clinoamphibole		
		fayalite	orthosilicates, olivine group
crocoite	chromates	ferberite	tungstates
crossite	chain silicates, clinoamphibole	fergusonite	niobotantalates
		ferrierite	framework silicates, zeolite group
cryolite	fluorides		
cryptomelane	oxides	ferrihydrite	oxides
cryptoperthite	framework silicates, alkali feldspar	ferrimolybdite	molybdates
		ferrocarpholite	chain silicates
cubanite	sulfides	ferrohastingsite	chain silicates, clinoamphibole
cummingtonite	chain silicates, clinoamphibole		
		ferropseudobrookite	oxides
cuprite	oxides	ferrosilicon	silicides
cylindrite	sulfosalts	ferrosilite	chain silicates
cymrite	sheet silicates	florencite	phosphates
cyrtolite	nesosilicates	fluoborite	borates *and* fluorides
dahllite	carbonates *and* phosphates	fluor-phlogopite	sheet silicates, mica group
danburite	framework silicates	fluorapatite	phosphates
datolite	nesosilicates	fluorite	fluorides
dawsonite	carbonates	forsterite	orthosilicates, olivine group
deerite	chain silicates	francolite	phosphates
delafossite	oxides	franklinite	oxides
descloizite	vanadates	freibergite	sulfosalts
diamond	native elements	freieslebenite	sulfosalts
diaspore	oxides	froodite	bismuthides
dickite	sheet silicates, clay minerals	gadolinite	nesosilicates
digenite	sulfides	gahnite	oxides
diopside	chain silicates, clinopyroxene	galena	sulfides
djurleite	sulfides	galenobismutite	sulfosalts
dolomite	carbonates	garnierite	sheet silicates, serpentine group
dravite	ring silicates		
dufrenoysite	sulfosalts	garrelsite	nesosilicates
dumortierite	nesosilicates	garronite	framework silicates, zeolite group
dyscrasite	antimonides		
edingtonite	framework silicates, zeolite group	gaylussite	carbonates
		gedrite	chain silicates, orthoamphibole
ekanite	ring silicates		
elbaite	ring silicates	gehlenite	orthosilicates, melilite group
ellenbergerite	orthosilicates	geikielite	oxides
emerald	ring silicates	genthelvite	framework silicates *and* sulfides
emplecitite	sulfosalts		
enargite	sulfosalts	geocronite	sulfosalts
enstatite	chain silicates, orthopyroxene	germanates	oxides
		germanite	sulfosalts
eosphorite	phosphates	gersdorffite	arsenides *and* sulfides
epidote	orthosilicates, epidote group	gibbsite	oxides
epistilbite	framework silicates, zeolite group	gillespite	sheet silicates
		glauberite	sulfates
epsomite	sulfates	glauconite	sheet silicates, mica group
erionite	framework silicates, zeolite group	glaucophane	chain silicates, clinoamphibole

gmelinite	framework silicates, zeolite group	hyalite	framework silicates, silica minerals
goethite	oxides	hyalophane	framework silicates, barium feldspar
graftonite	phosphates		
grandidierite	nesosilicates	hydrobiotite	sheet silicates
graphite	native elements	hydroboracite	borates
gratonite	sulfosalts	hydrogoethite	oxides
greenockite	sulfides	hydrogrossular	orthosilicates, garnet group
greigite	sulfides	hydromagnesite	carbonates
grossular	orthosilicates, garnet group	hydromica	sheet silicates
groutite	oxides	hydromuscovite	sheet silicates
grunerite	chain silicates, clinoamphibole	hydrosodalite	framework silicates, sodalite group
gudmundite	sulfides	hydroxides	oxides
gustavite	sulfosalts	hydroxylapatite	phosphates
gypsum	sulfates	hypersthene	chain silicates, orthopyroxene
gyrolite	sheet silicates		
haidingerite	arsenates	Iceland spar	carbonates
halite	chlorides	idaite	sulfides
halloysite	sheet silicates, clay minerals	illite	sheet silicates, clay minerals
halotrichite	sulfates	ilmenite	oxides
hammarite	sulfosalts	ilvaite	sorosilicates
harkerite	nesosilicates	imogolite	sheet silicates, clay minerals
harmotome	framework silicates, zeolite group	innelite	sorosilicates
		iron oxides	oxides
hastingsite	chain silicates, clinoamphibole	iron sulfides	sulfides
		ixiolite	niobotantalates
hausmannite	oxides	jacobsite	oxides
hauyne	framework silicates, sodalite group	jade	chain silicates
		jadeite	chain silicates, clinopyroxene
heazlewoodite	sulfides	jamesonite	sulfosalts
hectorite	sheet silicates, clay minerals	jarosite	sulfates
hedenbergite	chain silicates, clinopyroxene	jasper	framework silicates, silica minerals
hedleyite	tellurides *and* alloys		
helvite	framework silicates *and* sulfides	joaquinite	sorosilicates
		johannsenite	chain silicates, clinopyroxene
hematite	oxides	jordanite	sulfosalts
hemihedrite	chromates	jordisite	sulfides
hemimorphite	sorosilicates	joseite	tellurides *and* sulfides
hercynite	oxides	josephinite	alloys
herderite	phosphates	jouravskite	sulfates
herzenbergite	sulfides	julgoldite	sorosilicates
hessite	tellurides	K-feldspar	framework silicates, alkali feldspar
heterogenite	oxides		
heteromorphite	sulfosalts	kaersutite	chain silicates, clinoamphibole
heulandite	framework silicates, zeolite group		
		kainite	chlorides *and* sulfates
hexahydrite	sulfates	kaliborite	borates
hibonite	oxides	kaliophilite	framework silicates, nepheline group
hinsdalite	phosphates *and* sulfates		
hisingerite	sheet silicates	kalsilite	framework silicates, nepheline group
hollandite	oxides		
holmquistite	chain silicates, orthoamphibole	kamacite	alloys
		kammererite	sheet silicates, chlorite group
hornblende	chain silicates, clinoamphibole	kaolinite	sheet silicates, clay minerals
		karpinskyite	framework silicates
howieite	chain silicates	kasolite	nesosilicates
hulsite	borates	keatite	framework silicates, silica minerals
humite	orthosilicates, humite group *and* fluorides		
		kenyaite	framework silicates, scapolite group
huntite	carbonates		
hureaulite	phosphates	kernite	borates

kesterite	sulfides	malachite	carbonates
kieserite	sulfates	malayaite	ring silicates
kilchoanite	sorosilicates	manganese oxides	oxides
kobellite	sulfosalts	manganite	oxides
kornerupine	nesosilicates	manganosite	oxides
kotoite	borates	marcasite	sulfides
kunzite	chain silicates, clinopyroxene	margarite	sheet silicates, mica group
kurchatovite	borates	martite	oxides
kurnakovite	borates	maskelynite	aluminosilicates
kutnahorite	carbonates	matildite	sulfosalts
kyanite	nesosilicates	maucherite	arsenides
labradorite	framework silicates, plagioclase	mawsonite	sulfides
		meionite	framework silicates, scapolite group *and* carbonates
labuntsovite	sorosilicates		
laihunite	nesosilicates		
langbeinite	sulfates	melanite	orthosilicates, garnet group
larnite	nesosilicates	melanocerite	nesosilicates
laumontite	framework silicates, zeolite group	melanophlogite	framework silicates, silica minerals
lawsonite	sorosilicates	melanterite	sulfates
lazulite	phosphates	melilite	orthosilicates, melilite group
lazurite	framework silicates, sodalite group; sulfates *and* sulfides	meneghinite	sulfosalts
		merwinite	nesosilicates
		mesolite	framework silicates, zeolite group
leadhillite	carbonates *and* sulfates		
lechatelierite	framework silicates, silica minerals	metacinnabar	sulfides
		metahalloysite	sheet silicates, clay minerals
legrandite	arsenates	metastibnite	sulfides
leifite	fluorides *and* framework silicates	miargyrite	sulfosalts
		michenerite	bismuthides
leonhardite	framework silicates	microcline	framework silicates, alkali feldspar
lepidocrocite	oxides		
lepidolite	sheet silicates, mica group *and* fluorides	microlite	fluorides *and* niobotantalates
		microperthite	framework silicates, alkali feldspar
lepidomelane	sheet silicates, mica group		
leucite	framework silicates	milarite	ring silicates
leucosphenite	sheet silicates	millerite	sulfides
leucoxene	oxides	mimetite	arsenates *and* chlorides
lillianite	sulfosalts	minyulite	fluorides *and* phosphates
limonite	oxides	mirabilite	sulfates
linnaeite	sulfides	mixite	arsenites
listvenite	sheet silicates, mica group	moissanite	carbides *and* alloys
lithiophilite	phosphates	molybdenite	sulfides
lithiophorite	oxides	monazite	phosphates
lizardite	sheet silicates, serpentine group	monetite	phosphates
		monohydrocalcite	carbonates
lollingite	arsenides	montebrasite	phosphates
lonsdaleite	native elements	monticellite	orthosilicates, olivine group
loparite	niobates	montmorillonite	sheet silicates, clay minerals
ludwigite	borates	moonstone	framework silicates, alkali feldspar
luzonite	sulfosalts		
mackinawite	sulfides	mordenite	framework silicates, zeolite group
magadiite	framework silicates, scapolite group		
		mroseite	tellurites
maghemite	oxides	mullite	nesosilicates
magnesian calcite	carbonates	muscovite	sheet silicates, mica group
magnesioferrite	oxides	myrmekite	framework silicates, plagioclase
magnesioriebeckite	chain silicates, clinoamphibole		
		nacrite	sheet silicates, clay minerals
magnesite	carbonates	nahcolite	carbonates
magnetite	oxides	natrolite	framework silicates, zeolite group
majorite	orthosilicates, garnet group		

neighborite	fluorides	perthite	framework silicates, alkali feldspar
nenadkevichite	niobates *and* sorosilicates		
nepheline	framework silicates, nepheline group	petalite	framework silicates, nepheline group
nephrite	chain silicates, clinoamphibole	pharmacolite	arsenates
		pharmacosiderite	arsenates
neptunite	chain silicates	phenakite	nesosilicates
nesquehonite	carbonates	phengite	sheet silicates, mica group
newberyite	phosphates	phillipsite	framework silicates, zeolite group
niccolite	arsenides		
nigerite	oxides	phlogopite	sheet silicates, mica group
niggliite	alloys	phosgenite	carbonates *and* chlorides
niningerite	sulfides	pickeringite	sulfates
nontronite	sheet silicates, clay minerals	piemontite	orthosilicates, epidote group
norbergite	orthosilicates, humite group *and* fluorides	pigeonite	chain silicates, clinopyroxene
		pitchblende	oxides
nordstrandite	oxides	plagionite	sulfosalts
norsethite	carbonates	plancheite	chain silicates
nosean	framework silicates, sodalite group	plessite	alloys
		plumbogummite	phosphates
nsutite	oxides	pollucite	framework silicates
offretite	framework silicates, zeolite group	polybasite	sulfosalts
		polydymite	sulfides
oldhamite	sulfides	polyhalite	sulfates
oligoclase	framework silicates, plagioclase	polylithionite	sheet silicates, mica group
		posnjakite	sulfates
olivine	orthosilicates, olivine group	powellite	molybdates
omphacite	chain silicates, clinopyroxene	prehnite	sheet silicates
onofrite	selenites *and* sulfides	preobrazhenskite	borates
onyx	framework silicates, silica minerals	priderite	oxides
		prosopite	fluorides
opal	framework silicates, silica minerals	protodolomite	carbonates
		protoenstatite	chain silicates, orthopyroxene
opal-A	framework silicates, silica minerals		
		proustite	sulfosalts
opal-CT	framework silicates, silica minerals	pseudobrookite	oxides
		pseudoleucite	framework silicates
orpiment	sulfides	psilomelane	oxides
orthoclase	framework silicates, alkali feldspar	pumpellyite	sorosilicates
		pyrargyrite	sulfosalts
orthoenstatite	chain silicates, orthopyroxene	pyrite	sulfides
		pyroaurite	carbonates
osumilite	ring silicates	pyrochlore	fluorides *and* niobotantalates
ozocerite	organic compounds	pyrolusite	oxides
palygorskite	sheet silicates, clay minerals	pyromorphite	chlorides *and* phosphates
paracelsian	framework silicates, barium feldspar	pyrope	orthosilicates, garnet group
		pyrophanite	oxides
paragonite	sheet silicates, mica group	pyrophyllite	sheet silicates
pararammelsbergite	arsenides	pyroxferroite	chain silicates
paratacamite	chlorides	pyroxmangite	chain silicates
pargasite	chain silicates, clinoamphibole	pyrrhotite	sulfides
		quartz	framework silicates, silica minerals
pearceite	sulfosalts		
pectolite	chain silicates	rammelsbergite	arsenides
penninite	sheet silicates	ramsdellite	oxides
pentlandite	sulfides	rancieite	oxides
periclase	oxides	realgar	sulfides
peridot	orthosilicates, olivine group	rectorite	sheet silicates, clay minerals
peristerite	framework silicates, plagioclase	reedmergnerite	framework silicates
		renierite	sulfosalts
perovskite	oxides	rhabdophane	phosphates
perrierite	sorosilicates	rhodochrosite	carbonates

rhodonite	chain silicates	sphalerite	sulfides
rhonite	chain silicates	spinel	oxides
richterite	chain silicates, clinoamphibole	spinel group	oxides
		spodumene	chain silicates, clinopyroxene
riebeckite	chain silicates, clinoamphibole	spurrite	nesosilicates *and* carbonates
		stannite	sulfosalts
rinneite	chlorides	stannoidite	sulfides
rockbridgeite	phosphates	staurolite	nesosilicates
roedderite	ring silicates	stellerite	framework silicates, zeolite group
roemerite	sulfates		
roggianite	sorosilicates	stephanite	sulfosalts
roquesite	sulfides	sternbergite	sulfides
rozenite	sulfates	stevensite	sheet silicates, clay minerals
rutile	oxides	stibiotantalite	niobotantalates
safflorite	arsenides	stibnite	sulfides
sakhaite	borates *and* carbonates	stilbite	framework silicates, zeolite group
salite	chain silicates, clinopyroxene		
samarskite	niobotantalates	stillwellite	nesosilicates
sanidine	framework silicates, alkali feldspar	stilpnomelane	sheet silicates
		stishovite	framework silicates, silica minerals
saponite	sheet silicates, clay minerals		
sapphire	oxides	stolzite	tungstates
sapphirine	nesosilicates	strengite	phosphates
sarcopside	phosphates	stromeyerite	sulfides
scapolite	framework silicates, scapolite group	strontianite	carbonates
		struvite	phosphates
scawtite	chain silicates	suanite	borates
scheelite	tungstates	sudoite	sheet silicates, chlorite group
scholzite	phosphates	sulfantimonates	sulfosalts
schorl	ring silicates	sulfantimonites	sulfosalts
schorlomite	orthosilicates, garnet group	sulfarsenates	sulfosalts
schreibersite	phosphides	sulfarsenites	sulfosalts
scolecite	framework silicates, zeolite group	sulfobismuthites	sulfosalts
		sulfogermanates	sulfosalts
scorodite	arsenates	sulfostannates	sulfosalts
selenite	sulfates	sulfovanadates	sulfosalts
seligmannite	sulfosalts	sulvanite	sulfosalts
semseyite	sulfosalts	svanbergite	sulfates
senarmontite	oxides	sylvanite	tellurides
sepiolite	sheet silicates, clay minerals	synchisite	carbonates *and* fluorides
sericite	sheet silicates, mica group	szaibelyite	borates
serpentine	sheet silicates, serpentine group	taaffeite	oxides
		taenite	alloys
shattuckite	chain silicates	talc	sheet silicates
shortite	carbonates	talnakhite	sulfides
siderite	carbonates	tantalite	niobotantalates
siegenite	sulfides	tanzanite	orthosilicates, epidote group
sillimanite	nesosilicates	tapiolite	niobotantalates
sinnerite	sulfides	tellurobismuthite	tellurides
sjogrenite	carbonates	tennantite	sulfosalts
skutterudite	arsenides	tephroite	orthosilicates, olivine group
smectite	sheet silicates, clay minerals	tetradymite	tellurides *and* sulfides
smithsonite	carbonates	tetrahedrite	sulfosalts
smoky quartz	framework silicates, silica minerals	tetrataenite	alloys
		thalenite	sorosilicates
smythite	sulfides	thaumasite	carbonates *and* nesosilicates; sulfates
sodalite	framework silicates, sodalite group		
		thenardite	sulfates
solongoite	borates	thermonatrite	carbonates
specularite	oxides	thomsenolite	fluorides
sperrylite	arsenides	thomsonite	framework silicates, zeolite group
spessartine	orthosilicates, garnet group		

thorianite	oxides	weddellite	carbonates
thorite	nesosilicates	weloganite	carbonates
thortveitite	sorosilicates	wenkite	framework silicates *and* sulfates
tirodite	chain silicates, clinoamphibole		
titanaugite	chain silicates, clinopyroxene	whewellite	carbonates
titanite	nesosilicates	whitlockite	phosphates
titanoclinohumite	orthosilicates, humite group	willemite	nesosilicates
titanomaghemite	oxides	willyamite	sulfides
titanomagnetite	oxides	witherite	carbonates
tobermorite	chain silicates	wittichenite	sulfosalts
todorokite	oxides	wodginite	niobotantalates
topaz	fluorides *and* nesosilicates	wolframite	tungstates
torbernite	phosphates	wollastonite	chain silicates
tosudite	sheet silicates, clay minerals	wulfenite	molybdates
tourmaline	ring silicates	wurtzite	sulfides
tremolite	chain silicates, clinoamphibole	wustite	oxides
		xenotime	phosphates
trevorite	oxides	xonotlite	chain silicates
tridymite	framework silicates, silica minerals	yttrialite	sorosilicates
		yugawaralite	framework silicates, zeolite group
triphylite	phosphates		
triplite	fluorides *and* phosphates	zinc sulfides	sulfides
troilite	sulfides	zincite	oxides
trona	carbonates	zinckenite	sulfosalts
tschermakite	chain silicates, clinoamphibole	zinnwaldite	sheet silicates, mica group *and* fluorides
tugtupite	framework silicates, sodalite group	zircon	nesosilicates
		zirconolite	niobates
tungstenite	sulfides	zoisite	orthosilicates, epidote group
turquoise	phosphates	zunyite	halides *and* sorosilicates
ulexite	borates	zussmanite	sheet silicates
ullmannite	sulfides		
ultrabasite	sulfides		
ulvospinel	oxides		
umangite	selenides		
umohoite	molybdates		
uraninite	oxides		
uranophane	nesosilicates		
uvarovite	orthosilicates, garnet group		
vaesite	sulfides		
valentinite	oxides		
valleriite	sulfides		
vanadinite	chlorides *and* vanadates		
vanadium garnet	orthosilicates, garnet group		
variscite	phosphates		
vaterite	carbonates		
vermiculite	sheet silicates, clay minerals		
vesuvianite	sorosilicates		
villiaumite	fluorides		
vimsite	borates		
violarite	sulfides		
viridine	nesosilicates		
vivianite	phosphates		
volborthite	vanadates		
voltaite	sulfates		
vonsenite	borates		
vrbaite	sulfosalts		
wairakite	framework silicates, zeolite group		
wavellite	phosphates		
weberite	fluorides		

LIST M - SOILS

INDEXING

The classification of soils has undergone several changes in American practice. What follows are two of the classifications using different criteria. In addition there follows a comparative table giving equivalent French and German terms for some of the basic soil terms in English. We do not attempt to standardize the soils. Use the soil terms given in the source documents, without attempting to interpret types of soil according to one of the available classifications.

ONLINE SEARCHING

To search for a soil group, one must search for all possible varieties, including different names in different classifications and in different languages. For example, to search for Latosols, you must also search for Sol-ferralitique. The American names have been used most commonly in GeoRef indexing.

B.I.G. SEARCHING

Prior to 1993, all the soil entries may be found under the term <u>soils</u>. In current practice specific soil types are direct access points.

AMERICAN SOIL CLASSIFICATION SYSTEM - 1964

Zonal soils
    Tundra soils
    Subarctic brown forest soils
    Desert soils
    Red desert soils
    Sierozems
    Brown soils
    Reddish brown soils
    Chestnut soils
    Reddish chestnut soils
    Chernozems
    Brunizems
    Reddish prairie soils
    Noncalcic brown soils
    Gray wooded soils
    Sols bruns acides
    Gray-brown podzolic soils
    Red-yellow podzolic soils
    Reddish-brown lateritic soils
    Yellowish-brown lateritic soils
    Latosols

Intrazonal soils
    Solonchak soils (saline soils)
    Solonetz soils
    Soloth soils
    Humic-gley soils
    Alpine meadow soils
    Bog soils
    Low-humic gley soils
    Planosols
    Ground-water podzols
    Ground-water laterite soils
    Brown forest soils
    Rendzinas
    Grumusols
    Calcisols

Azonal soils
    Lithosols
    Regosols
    Alluvial soils

AMERICAN SOIL CLASSIFICATION SYSTEM - 1965 et seq.

Alfisols	Mollisols
Udalfs	Albolls
Boralfs	Aquolls
Ustalfs	Borolls
Xeralfs	Redolls
	Udolls
Aridisols	Ustolls
Argids	Xerolls
Orthids	
	Oxisols
Entisols	Aquox
Aquents	Humox
Fluvents	Orthox
Orthents	Torrox
Psamments	Ustox
Histosols	Spodosols
Fibrists	Aquods
Folists	Humods
Hemists	Orthods
Saprists	
	Ultisols
Inceptisols	Aquults
Andepts	Humults
Aquepts	Udults
Ochrepts	Ustults
Tropepts	Xerults
Umbrepts	

AMERICAN	GERMAN	FRENCH
<u>Acrisols</u>	Acrisol	Sol-mediterraneen
<u>Albolls</u>		
<u>Alfisols</u>		
<u>Alluvial soils</u>	Auen-Boden	Sol-d'alluvions
<u>Alpine meadow soils</u>	Alpiner Wiesen-boden	Sol-hydromorphe
<u>Andepts</u>		
<u>Andosols</u>	Andosol	Sol-peu-evolue roche-volcanique
<u>Aqualfs</u>		
<u>Aquents</u>		
<u>Aquepts</u>		
<u>Aquods</u>		
<u>Aquolls</u>		
<u>Aquox</u>		
<u>Aquults</u>		
<u>Arctic tundra soils</u>	Arktische Tundra	Sol-de-toundra Boden
<u>Arenosols</u>	Arenosol	Sol-brut sable
<u>Arents</u>		
<u>Argids</u>		
<u>Aridisols</u>		
<u>Azonal soils</u>	Roh-Boden	Sol-brut
Black earth use Chernozems	Schwarzerde	Chernozem
<u>Bog soils</u>	Moorboden	Tourbe
<u>Boralfs</u>		
<u>Boreal frozen taiga soils</u>		Sol-gele
<u>Boreal taiga and forest soils</u>		Sol
<u>Borolls</u>		Mollisol
<u>Brown desert steppe soils</u>	Burozem	Sierozem
<u>Brown forest soils</u>	Brauner Wald Boden	Sol-brun
<u>Brown loam</u>	Braunlehm	Couche-rouge
<u>Brown podzolic soils</u>	Podsoliger Braun-erde	Podzol
<u>Brown soils</u>	Braunerde	Sol-brun
<u>Brunizems</u>	Brunizem	Brunizem
<u>Calcisols</u>		
<u>Cambisols</u>	Cambisol	Sol-brun
<u>Caar</u>	Uebergangsmoor	Tourbe
<u>Chernozems</u>	Chernozem	Chernozem

Chestnut soils	Kastannozem	Sol-chatain	Plaggenesch	Plaggenesch	Sol-rich-en-humus action-homme
Cryosols	Cryosol	Sol-gele	Plaggepts		
Desert soils	Wuesten-Boden	Sol-subdesertique	Planosols	Planosol	Couche-rouge
Desert raw soils	Wuesten-roh-Boden	Sol-de-desert	Podzols	Podsol	Podzol
			Podzoluvisols	Podzoluvisol	Sol lessivage
Dy	Dy	Vase	Poorly developed soils	Unreife-Boden	Sol-peu-developpe
Entisols					
Fen			Prairie soils	Brunizem	Brunizem
Ferralites	Eisen-Silikat-Boden	Sol-fersialitique	Protopedons	Protopedon	Sol-brut or Sol-d'alluvions
Ferralsols	Ferralsol	Sol-ferralitique	Psamments		
Ferrods			Pseudogleys	Pseudogley	Gley
Ferruginous soils	Eisenhaltiger-Boden	Sol-ferrugineux	Rambla	Rambla	Sol-brut or Sol-d'alluvions
Fibrists			Ranker	Ranker	Ranker
Fluvents			Red desert soils	Roter Wuesten-boden	Sol-subdesertique
Fluviosols		Sol-d'alluvions			
Folists			Reddish brown soils	Roetlich-brauner Halbwueste Boden	Sol-subdesertique
Gleys	Gley	Gley			
Glensols	Glensol	Glensol	Reddish-brown lateritic soils		
Gray-brown podzolic soils	Fahlerde \	Sol-brun lessivage	Reddish chestnut soils	Roetlich-Kastanien-farbiger Boden	Sol-chatain
Gray podzolic soils	Podsolierter grauer Boden	Podzol	Reddish prairie soils	Roetlicher Prairie Boden	Brunizem
Gray warp soils	Paternia	Sol-peu-evolue or Sol-d'alluvions	Red-yellow podzolic soils	Gelbig roterpod-soliger Boden	Sol-ferralitique
Gray wooded soils			Reg	Reg	Sol-de-desert
Ground-water podzols	Gley-Podsol	Podzol	Regosols	Regosol	Sol-peu-evolue
Ground-water laterite soils	Grundwasser-Laterite	Laterite	Regur	Regur	Vertisol
			Rendolls		
Grumosols	Grumosol	Vertisol	Rendzinas	Rendzina	Rendzine
Half bog soils	Anmoor	Tourbe	Rhegosols	Rhegosol	Sol-peu-developpe
Halomorphic soils	Salz-Boden	Sol-halomorphe	Rigosols	Rigosol	Sol-brut
Halosols	Halosols	Sal-halomorphe	Rutmark	Raamark	Sol-brut
Hemists			Saprists		
High moor	Hochmoor	Tourbe	Sapropels	Sapropel	Sierozem
Histosols			Sierozems	Sierozem	Sierozem
Humic gley soils	Humus Gley Boden	Sol-hydromorphe	Solonchak soils	Solonchak	Sol-halomorphe
Humic soils	Humus-reiche-Boden	Sol-riche-en-humus	Solonetz soils	Solonetz	Sol-halomorphe
			Soloth soils	Solod	Sol-halomorphe
Humods			Spodosols		
Humox			Stagnogleys	Stagnogley	Gley
Hydromorphic soils	Hydromorpher-Boden	Sol-riche-en-humus	Subarctic brown forest soils		
Inceptisols			Subboreal desert soils		
Intrazonal soils	Intrazonaler Boden	Sol	Subboreal humid soils		
Kastanozems			Subboreal steppe soils		
Krasnozems	Krasnozem	Krasnozem	Subtropical desert soils		
laterites	Laterit-Boden	Sol-ferralitique			
Latosols	Latosol	Sol-ferralitique	Subtropical dry soils		
Lithosols	Gesteins-roh-Boden	Sol-squelettique	Subtropical humid soils		
Low-humic gley soils					
Luvisols	Luvisols	Sol lessivage	Syrogleys	Syrogley	Gley
Mediterranean soils	Mediterraner Boden	Sol-mediterraneen	Syrozems	Syrozem	Sol-brut
Mollisols			Takyr	Takyr	Sol-de-desert
Mor	Auflagehumus	Humus	Terrae calcis	Terrae calcis	Couche-rouge
Mull	Mull	Humus	Terra rossa	Terra rossa	Couche-rouge
Mull soils	Mull-Boden	Sol-a-mull	Terra fusca	Terra fusca	Couche-rouge
Muck	Niedermoor	Tourbe	Tir	Tir	Vertisol
Nitosols	Nitosol		Torrox		
Noncalcic brown soils			Torrerts		
Ochrepts			Tropepts		
Orthents			Tropical desert soils		
Orthids			Tropical dry soils		
Orthods			Tropical humid soils		
Orthox			Tundra soils	Tundra-Boden	Sol-de-toundra
Oxisols			Udalfs		
Parachernozems	Smonitza	Chernozem	Uderts		
Paramosols	Paramosol	Sol-chatain	Udolls		
Pararendzina	Borowina	Rendzine	Udults		
Parasierozems	Parasierozem	Sierozem	Ultisols		
Peat	Torf	Tourbe	Umbrepts		
Pelosols	Pelosol	Sol-hydromorphe	Ustalfs		
Pergelisols	Pergelisol	Sol-gele	Ustserts		

Ustolls
Ustsults
Vega   Vega   Sol-brun *or*
                                             Sol-d'alluvions
Vertisols
Wet meadow soils   Wiesenboden   Sol-hydromorphe
Xeralfs
Xerests
Xerolls
Xerosols   Xerosol   Sierozem
Xerults
Yeltozems   Yeltozem   Sol-mediterraneen
Yermosols   Ermosol   Sol-de-desert
Zonal soils   Zonaler Boden   Sol

LIST N - SEDIMENTS

INDEXING

Under this general heading, entries are made for unconsolidated (Recent or Quaternary) materials. Index the most specific term available and the appropriate higher level term from the list below.

ONLINE SEARCHING

To search for unconsolidated sediments, use the specific term such as sand or the general term sediments, truncated.

B.I.G. SEARCHING

Prior to 1993, under the heading sediments look under the appropriate terms from the list below.

Other related headings include: sedimentary rocks, sedimentary structures, sedimentation, soils, weathering, and the individual commodities such as kaolin deposits and clays.

In current practice specific sediments are direct access points.

sediments
  clastic sediments
    alluvium
    boulder clay
    boulders
    clay
    colluvium
    eluvium
    erratics
    flint clay
    gravel
    kaolin
    loess
    mud
    ooze
    pebbles
    proluvium
    residual clays
    residuum
    sand
    silt
    till
    turbidite
  carbonate sediments
  marine sediments
  organic residues

LIST O - GEOGRAPHIC TERMS

This list includes terms for large regions, continents, countries, large subdivisions of some countries, and large bodies of water.

Within the list, the terms for land areas are separate from the terms for water bodies, with the water bodies coming at the end of the list. Under each continent, physiographic regions are listed separately from countries, with the physiographic regions first. The island groups are not treated as part of any continent, but are listed alphabetically with the land areas (e.g., East Pacific Ocean Islands precedes Europe). West Indies is listed under America, following South America.

In addition to the terms in the list, thousands of terms exist for smaller geographic areas. The more common of these terms have been added to the Thesaurus, based on frequency of use in GeoRef, and can now be found in the main body of the Thesaurus. Other specific geographic terms, not in the Thesaurus, have also been used in GeoRef indexing when given in the cited document. When one of these specific localities is indexed, a broader term from List O is also indexed.

Most Chinese geographic terms are transliterated according to the Pinyin transliteration. See individual terms for history of usage.

Geographic locations are key factors in most geology papers. Since 1978, in addition to the relevant geographic terms, the coordinates of primary areas of papers have been added to GeoRef citations, when those coordinates have been available from the cited document or other sources. These coordinates are latitudes and longitudes defining the boundaries of the areas. This makes it possible to search for particular areas using both geographic terms and coordinates.

INDEXING

The minimum level of indexing required for geography is the name of the country (State or province in the U.S., Canada and Australia) or the major water body. In addition, specific geographic terms should be used as supplementary terms. In the U.S. counties should be included up to 10. Physiographic regions should be included if they can be determined (see U.S. map, Maps & Charts

Section). For Alaska see subdivisions, Maps & Charts Section.

Autoposting of major area terms began in 1978. From 1971 to 1977 continents and major area terms were included in indexing. It is not necessary to search for all the narrower terms of a continent (or United States, Canada, Australia) or other major area term for the part of the file covering 1971 to the present. For further information on specific terms see the main body of the Thesaurus.

ONLINE SEARCHING

Geographic searching is best achieved by using all variants of an area name including political, physiographic and geologic. It is possible to be as specific as town or county, and as general as country or continent. For oceans and continental margins, it is advisable to use the coordinates as well. For information on searching coordinates, see the *GeoRef Online Workshop Training Manual*.

In searching online for a topic related to a particular geographic region on country or continent scale, search for that region using List O for reference, and combine it with the proper subject from List B to give a general subject orientation. Thus for gold ores in Argentina, search for Argentina AND economic geology AND all terms starting with gold.

B.I.G. SEARCHING

Beginning in 1993, all geographic terms were altered to allow autoposting. In order to create unique geographic terms all political subdivisions now include the name of the country. For the U.S. and Canada state names and provinces were used. It was our practice to make an area set (see List B, Area Sets) for each paper cited in GeoRef, excepting the relatively few geology papers which do not relate to geographic locations.

In each issue of the Bibliography, "see also" references from broad terms to narrower terms used in that issue were included in the Subject Index, such as the following from the May 1981 issue:

Africa see also Algeria; Benin; Botswana; Cape
    Verde Islands; Djibouti; Egypt; Ethiopia;...
    Zambia

Beginning in 1993, all specific geographic terms are direct access points in the B.I.G. "See also" references direct the user from broader terms to narrower terms.

---

global
  Western World

    Eastern Hemisphere
    Northern Hemisphere
    Southern Hemisphere

  Western Hemisphere

    Africa
      Sahara
        Western Sahara
      East African Lakes (Lake Albert, Lake
            Baringo, Lake Edward, Lake
            Kariba, Lake Kivu, Lake Magadi,
            Lake Tanganyika, Lake Turkana,
            Lake Victoria)
      Central Africa
        Angola
        Burundi
        Central African Republic
        Congo
        Equatorial Guinea
        Gabon
        Rwanda
        Zaire
          Bandundu Zaire
          Equatorial Zaire
          Kasai
          Kivu Zaire
          Lower Zaire
          Shaba Zaire
          Upper Zaire
      East Africa
        Djibouti
        Ethiopia
        Kenya
        Malawi
        Mozambique
        Uganda
        Somali Republic
        Sudan
        Tanzania
        Zambia
      North Africa
        Algeria
          Ahaggar
          Algiers Algeria
          Constantine Algeria
          Oran Algeria
        Egypt
          Eastern Desert
          Middle Nile Valley
          Nile Delta
          Sinai Egypt
          Western Desert
        Libya
        Morocco
          Anti-Atlas
          Moroccan Atlas Mountains
          Rif
        Tunisia
      Southern Africa
        Botswana
        Lesotho
        Namibia
        South Africa
          Cape Province South Africa
          Natal South Africa
          Orange Free State South Africa
          Transvaal South Africa
        Swaziland
        Zimbabwe
      West Africa

- Benin
- Burkina Faso
- Cameroon
- Chad
- Gambia
- Ghana
- Guinea
- Guinea-Bissau
- Ivory Coast
- Liberia
- Mali
- Mauritania
- Niger
- Nigeria
- Senegal
- Sierra Leone
- Togo

America
- Caribbean region
- Latin America

- Central America
  - Belize
  - Costa Rica
  - El Salvador
  - Guatemala
  - Honduras
  - Nicaragua
  - Panama
- North America
  - Appalachians
  - Atlantic Coastal Plain
  - Basin and Range Province
  - Canadian Shield
  - Colorado Plateau
  - Columbia Plateau
  - Great Basin
  - Great Lakes
  - Great Lakes region
  - Great Plains
  - Gulf Coastal Plain
  - Mississippi Valley
  - Rocky Mountains
  - Western Interior

    - Canada
      - Eastern Canada
        - Maritime Provinces
          - New Brunswick
          - Nova Scotia
          - Prince Edward Island
        - Labrador
        - Newfoundland
        - Ontario
        - Quebec
      - Western Canada
        - Alberta
        - British Columbia
        - Manitoba
        - Northwest Territories
          - Arctic Archipelago
          - Franklin District Northwest Territories
          - Keewatin District Northwest Territories
          - Mackenzie District Northwest Territories
        - Saskatchewan
        - Yukon Territory
    - Mexico
    - Saint Pierre and Miquelon
    - United States
      - New England
      - Pacific Coast

      - Alabama
      - Alaska
      - Arizona
      - Arkansas
      - California
      - Colorado
      - Connecticut
      - Delaware
      - District of Columbia
      - Florida
      - Georgia
      - Hawaii
      - Idaho
      - Illinois
      - Indiana
      - Iowa
      - Kansas
      - Kentucky
      - Louisiana
      - Maine
      - Maryland
      - Massachusetts
      - Michigan
      - Minnesota
      - Mississippi
      - Missouri
      - Montana
      - Nebraska
      - Nevada
      - New Hampshire
      - New Jersey
      - New Mexico
      - New York
      - North Carolina
      - North Dakota
      - Ohio
      - Oklahoma
      - Oregon
      - Pennsylvania
      - Rhode Island
      - South Carolina
      - South Dakota
      - Tennessee
      - Texas
      - Utah
      - Vermont
      - Virginia
      - Washington
      - West Virginia
      - Wisconsin
      - Wyoming
- South America
  - Andes

    - Argentina
      - Argentine Andes
      - Pampas

Bolivia
Brazil
    Acre Brazil
    Alagoas Brazil
    Amapa Brazil
    Amazonas Brazil
    Bahia Brazil
    Ceara Brazil
    Espirito Santo Brazil
    Goias Brazil
    Maranhao Brazil
    Mato Grosso Brazil
    Minas Gerais Brazil
    Para Brazil
    Paraiba Brazil
    Parana Brazil
    Pernambuco Brazil
    Piaui Brazil
    Rio de Janeiro Brazil
    Rio Grande do Norte Brazil
    Rio Grande do Sul Brazil
    Rondonia Brazil
    Roraima Brazil
    Santa Catarina Brazil
    Sao Paulo Brazil
    Sergipe Brazil
Chile
Colombia
Ecuador
Falkland Islands
Guyana
French Guiana
Paraguay
Peru
Surinam
Uruguay
Venezuela
    Apure Guarico Plain
    Maracaibo Falcon Plain
    Venezuelan Andes
    Venezuelan Islands
    Eastern Venezuela
    Southeastern Venezuela
West Indies
    Bahamas
    Cayman Islands
    Greater Antilles
        Cuba
        Dominican Republic
        Haiti
        Jamaica
        Puerto Rico
    Lesser Antilles
        Barbados
        Dominica
        Leeward Islands
            Guadeloupe
            Virgin Islands
                British Virgin Islands
                U. S. Virgin Islands
        Netherlands Antilles
        Trinidad and Tobago
        Windward Islands
            Grenada
            Martinique
Asia
    Himalayas
    Severnaya Zemlya
    Siberian Lowland

    Arabian Peninsula
        Bahrain
        Kuwait
        Oman
        Qatar
        Saudi Arabia
        Southern Yemen
        United Arab Emirates
        Yemen
    Far East
        Borneo
        Indochina

    Brunei
    Burma
    Cambodia
    China
        Anhui China
        Beijing China
        Fujian China
        Gansu China
        Guangdong China
        Guangxi China
        Guizhou China
        Hainan China
        Hebei China
        Heilongjiang China
        Henan China
        Hubei China
        Hunan China
        Inner Mongolia China
        Jaingsu China
        Jilin China
        Liaoning China
        Ningxia China
        Qinghai China
        Shaanxi China
        Shandong China
        Shanghai China
        Shanxi China
        Sichuan China
        Xinjiang China
        Xizang China
        Yunnan China
        Zhejiang China
    Hong Kong
    Indonesia
        Celebes
        Irian Jaya Indonesia
        Java
        Kalimantan Indonesia
        Lesser Sunda Islands
        Moluccas
        Sumatra
    Japan
        Hokkaido
        Honshu
        Kyushu
        Ryukyu Islands
        Shikoku
    Korea
        North Korea
        South Korea

   Laos
   Malaysia
    Sabah Malaysia
    Sarawak Malaysia
   Mongolia
   Philippine Islands
   Singapore
   Taiwan
   Thailand
   Vietnam
  Indian Peninsula
   Afghanistan
   Bangladesh
   Bhutan
   India
    Andhra Pradesh India
    Bengal Islands
    Bihar India
    Gujarat India
    Haryana India
    Himachal Pradesh India
    Jammu and Kashmir
    Karnataka India
    Kerala India
    Laccadive Islands
    Madhya Pradesh India
    Maharashtra India
    Northeastern India
    Orissa India
    Punjab India
    Rajasthan India
    Tamil Nadu India
    Uttar Pradesh India
    West Bengal India
   Nepal
   Pakistan
   Sri Lanka
  Middle East
   Cyprus
   Iran
   Iraq
   Israel
   Jordan
   Lebanon
   Syria
   Turkey
    Anatolia
    Pontic Mountains
    Sea of Marmara region
    Taurus Mountains
    Turkish Aegean region

 Atlantic Ocean Islands
  Azores
  Bermuda
  Canary Islands
  Cape Verde Islands
  Madeira
  Saint Helena
  Sao Tome e Principe
 Atlantic region

 Australasia
  Australia
   New South Wales Australia
   Northern Territory Australia
   Queensland Australia
   South Australia
   Tasmania Australia
   Victoria Australia
   Western Australia
  New Zealand
  Papua New Guinea
 East Pacific Ocean Islands
  Easter Island
  Galapagos Islands

Eurasia
 Europe
  Alps
   Eastern Alps
   Dinaric Alps
   Western Alps
   Bavarian Alps
  Balkan Peninsula
  Carpathians
  Caucasus
  Jura Mountains
  Pyrenees

  Baltic region
   Estonia
   Latvia
   Lithuania
  Central Europe
   Austria
    Austrian Vienna Basin
    Central Austrian Alps
    North Austrian Alps
    North Austrian Crystallines
    North Austrian Molasse
    South Austrian Alps
    South Austrian Molasse

    Burgenland Austria
    Carinthia Austria
    Lower Austria
    Salzburg State Austria
    Styria Austria
    Tyrol Austria
    Upper Austria
    Vorarlberg Austria
   Czechoslovakia
    Berounka System
    Czech Bohemian Forest
    Czech Erzgebirge
    Czech Sudeten Mountains
    Slovakia
     Slovakian Carpathians
     Slovakian Pannonian Basin
   Germany
    Alpenvorland
    Bavarian Massif
    Harz Mountains
    Hesse Basin
    Northeastern German Plain
    Northern German Hills
    Northwestern German Plain
    Rhenish Schiefergebirge
    Rhine Westphalian Basin
    Saar-Nahe Basin
    Saxonian Massif
    Southwestern German Hills
    Southwestern German Massifs

       Thuringian Hills
       Thuringian Massif
       Upper Rhine Valley

       Baden-Wurttemberg Germany
       Bavaria Germany
       Berlin Germany
       Brandenberg Germany
       Bremen Germany
       Hamburg Germany
       Hesse Germany
       Lower Saxony Germany
       Mecklenburg-Western
          Pomerania Germany
       North Rhine-Westphalia Germany
       Rhineland-Palatinate Germany
       Saarland Germany
       Schleswig-Holstein Germany Saxony
       Germany
       Saxony-Anhalt Germany
   Hungary
       Alfold
       Budapest Hungary
       Central Transdanubia
       Northeastern Hungarian Hills
       Northwestern Transdanubia
       Southeastern Transdanubia
       Southern Transdanubia
   Liechtenstein
   Poland
       Northeastern Polish Plain
       Northwestern Polish Plain
       Polish Carpathians
       Polish Sudeten Mountains
       Southeastern Polish Hills
       Swiety Krzyz Mountains
   Switzerland
       Central Swiss Alps
       Eastern Swiss Alps
       Northern Swiss Alps
       Southern Swiss Alps
       Swiss Jura Mountains
       Swiss Molasse Basin
       Swiss Rhine Graben
Southern Europe
   Albania
   Bulgaria
       Balkan Mountains
       Bulgarian Dobruja
       Bulgarian Rhodope Mountains
       Central Bulgaria
       North Bulgarian Hills
       West Bulgarian Hills
   Georgian Republic
   Greece
       Crete
       Epirus Greece
       Euboea
       Greek Aegean Islands
       Greek Ionian Islands
       Greek Macedonia
       Greek Thrace
       Peloponnesus Greece
       Sterea Ellas
       Thessaly Greece
   Iberian Peninsula
       Portugal
       Spain
          Asturian Arc
          Betic Zone
          Cantabrian region
          Catalonian Coastal Ranges
          Ebro Basin
          Guadalquivir Basin
          Hercinico Centro
          Hercinico Sur
          Iberica
          Spanish Pyrenees

          Andalusia Spain
             Almeria Spain
             Cadiz Spain
             Cordoba Spain
             Granada Spain
             Huelva Spain
             Jaen Spain
             Malaga Spain
             Seville Spain
          Aragon Spain
             Huesca Spain
             Saragossa Spain
             Teruel Spain
          Asturias Spain
          Balearic Islands
          Basque Provinces Spain
             Alava Spain
             Guipuzcoa Spain
             Vizcaya Spain
          Catalonia Spain
             Barcelona Spain
             Gerona Spain
             Lerida Spain
             Tarragona Spain
          Extremadura Spain
             Badajoz Spain
             Caceras Spain
          Galicia Spain
             La Coruna Spain
             Lugo Spain
             Orense Spain
             Pontevedra Spain
          Leon region
             Leon Spain
             Salamanca Spain
             Zamora Spain
          Murcia region
             Albacete Spain
             Murcia Spain
          Navarra Spain
          New Castile Spain
             Ciudad Real Spain
             Cuenca Spain
             Guadalajara Spain
             Madrid Spain
             Toledo Spain
          Old Castile Spain
             Avila Spain
             Burgos Spain
             Logrono Spain
             Palencia Spain
             Santander Spain
             Segovia Spain

                Soria Spain
                Valladolid Spain
            Valencia region
                Alicante Spain
                Castellon de la Plana Spain
                Valencia Spain
    Italy
        Apennines

            Abruzzi Italy
            Apulia Italy
            Basilicata Italy
            Calabria Italy
            Campania Italy
            Emilia-Romagna Italy
            Friuli-Venezia Giulia Italy
            Latium Italy
            Liguria Italy
            Lipari Islands
            Lombardy Italy
            Marches Italy
            Molise Italy
            Piemonte Italy
            Sardinia Italy
            Sicily Italy
            Trentino-Alto Adige Italy
            Tuscany Italy
            Umbria Italy
            Valle d'Aosta Italy
            Veneto Italy
    Malta
    Romania
        Apuseni Mountains
        Getic Nappe
        Moldavian Platform
        Romanian Dobruja
        Romanian Pannonian Basin
        Romanian Plain
        Transylvanian Alps
        Transylvanian Basin
    San Marino
    Yugoslavia
        Bosnia-Herzegovina
        Croatia
        Montenegro
        Serbia
        Slovenia
        Yugoslav Macedonia
Western Europe
    Andorra
    Belgium
    France
        Monaco

        Aquitaine Basin
        Armorican Massif
        Central Massif
        French Alps
        French Pyrenees
        Paris Basin
        Saone-Rhone Basin
        Vosges Mountains

            Ain France
            Aisne France
            Allier France
            Alpes-de-Haute-Provence France
            Alpes-Maritimes France
            Ardeche France
            Ardennes France
            Ariege France
            Aube France
            Aude France
            Aveyron France
            Bas-Rhin France
            Belfort France
            Bouches-du-Rhone France
            Calvados France
            Cantal France
            Charente France
            Charente-Maritime France
            Cher France
            Correze France
            Corsica
            Cote-d'Or France
            Cotes-du-Nord France
            Creuse France
            Deux-Sevres France
            Dordogne France
            Doubs France
            Drome France
            Essonne France
            Eure France
            Eure-et-Loir France
            Finistere France
            Gard France
            Gers France
            Gironde France
            Haut-Rhin France
            Haute-Garonne France
            Haute-Loire France
            Haute-Marne France
            Haute-Saone France
            Haute-Savoie France
            Haute-Vienne France
            Hautes-Alpes France
            Hautes-Pyrenees France
            Hauts-de-Seine France
            Herault France
            Ille-et-Vilaine France
            Indre France
            Indre-et-Loire France
            Isere France
            Jura France
            Landes France
            Loir-et-Cher France
            Loire France
            Loire-Atlantique France
            Loiret France
            Lot France
            Lot-et-Garonne France
            Lozere France
            Maine-et-Loire France
            Manche France
            Marne France
            Mayenne France
            Meurthe-et-Moselle France
            Meuse France
            Morbihan France
            Moselle France
            Nievre France
            Nord France
            Oise France
            Orne France

Paris France
  Pas-de-Calais France
  Puy-de-Dome France
  Pyrenees-Atlantiques France
  Pyrenees-Orientales France
  Rhone France
  Saone-et-Loire France
  Sarthe France
  Savoie France
  Seine-et-Marne France
  Seine-Maritime France
  Seine-Saint-Denis France
  Somme France
  Tarn France
  Tarn-et-Garonne France
  Val-de-Marne France
  Val-d'Oise France
  Var France
  Vaucluse France
  Vendee France
  Vienne France
  Vosges France
  Yonne France
  Yvelines France
 Iceland
 Ireland
 Luxembourg
 Netherlands
Scandinavia
 Denmark
  Faeroe Islands
 Finland
 Norway
  Northern Norway
  Southern Norway
 Sweden
  Gotland
United Kingdom
 English Channel Islands
 Great Britain
  England
   London Basin
   Pennines
   South-West England
   Welsh Borderland
  Scotland
   Grampian Highlands
   Hebrides
   Orkney Islands
   Scottish Northern Highlands
   Shetland Islands
  Wales
 Isle of Man
 Northern Ireland

Indian Ocean Islands
 Comoro Islands
 Madagascar
 Maldive Islands
 Mauritius
 Reunion
 Seychelles

Malay Archipelago
 New Guinea

Mediterranean region

Oceania
 Kiribati
 Melanesia
  Fiji
  New Caledonia
  Solomon Islands
  Vanuatu
 Micronesia
  Caroline Islands
  Mariana Islands
  Marshall Islands
  Nauru Island
 Polynesia
  Cook Islands
  French Polynesia
   Marquesas Islands
   Society Islands
   Tahiti
   Tuamotu Islands
  Samoa
  Tokelau
  Tonga
  Tuvalu
 Pacific region

 Antarctica
  Kerguelen Islands
  Scotia Sea Islands
 Arctic region
  Greenland
  Jan Mayen
  Russian Arctic
  Spitsbergen
 Red Sea region

Commonwealth of Independent States
 Dnieper-Donets Basin
 Fergana Basin
 Kyzylkum
 Moscow-Pechora Syneclise
 Pamirs
 Urals
 Ustyurt
 Voronezh-Volga Antecline

 Armenia
 Azerbaidzhan
 Belarus
 Kazabhstan
 Kyrgyzstan
 Moldova
 Russian Federation
  Amur region
  Baikal region
  Chukchi Peninsula
  Dzhugdzhur region
  Franz Josef Land
  Kolyma Uplift
  New Siberian Islands
  Russian Fennoscandia
  Russian Pacific region
   Kamchatka Peninsula
   Koryak Range
   Kuril Islands
   Sakhalin
  Siberian Platform

            Anabar Shield
            Angara-Lena Basin
            Aldan Shield
            North Siberian Plain
            Tunguska
            Vilyuy Syneclise
            Yakutia region
            Yenesei Ridge
         Sikhote-Alin Range
         Stanovoy Range
         Taymyr Dolgan-Nenets Russian
            Federation
         Timan Ridge
         Tuva
         Verkhoyansk region
      Tadzhikistan
      Turkmenia
      Ukraine
      Uzbekistan
      West Pacific Ocean Islands

world ocean
   Antarctic Ocean
      Bellingshausen Sea
      Mid-Antarctic Ridge
      Ross Sea
      Scotia Sea
      Weddell Sea
   Arctic Ocean
      Barents Sea
      Beaufort Sea
      Chukchi Sea
      East Siberian Sea
      Greenland Sea
      Kara Sea
      Laptev Sea
      Mid-Arctic Ocean Ridge
      Norwegian Sea
      White Sea
   Atlantic Ocean
      North Atlantic
         Azores-Gibraltar Ridge
         Cape Verde Atlantic
         European Atlantic
            Baltic Sea
               Gulf of Bothnia
               Gulf of Finland
               Gulf of Riga
            English Channel
            Bay of Biscay
            Irish Sea
            North Sea
         Gulf of Guinea
         Guyanese Atlantic
         North American Atlantic
            Caribbean Sea
            Gulf of Mexico
            Gulf of Saint Lawrence
            Hudson Bay
            Labrador Sea
         North Atlantic Ridge
      South Atlantic
         South American Atlantic
         South Atlantic Ridge
         Southeast Atlantic
   Caspian Sea
   Indian Ocean
      Andaman Sea
      Arabian Sea
         Gulf of Aden
         Gulf of Oman
         Persian Gulf
      Bay of Bengal
      Carlsberg Ridge
      Central Indian Ridge
      Mozambique Channel
      Ninetyeast Ridge
      Red Sea
         Gulf of Aqaba
         Gulf of Suez
      Southeast Indian Ridge
      Southwest Indian Ridge
   Mediterranean Sea
      East Mediterranean
         Adriatic Sea
         Aegean Sea
         Black Sea
         Ionian Sea
      West Mediterranean
         Alboran Sea
         Gulf of Lion
         Tyrrhenian Sea
   Pacific Ocean
      East Pacific Rise
      North American Pacific
         Gulf of Alaska
         Gulf of California
      South American Pacific
         Carnegie Ridge
         Chile Ridge
         Cocos Ridge
         Nazca Ridge
      South Polynesian Pacific
      West Pacific
         Bering Sea
         Coral Sea
         East China Sea
         Indonesian Seas
            Celebes Sea
            Java Sea
         Japan Sea
         North Australian Seas
            Arafura Sea
            Timor Sea
         Okhotsk Sea
         Philippine Sea
         South China Sea
            Gulf of Siam
            Gulf of Tonkin
         Tasman Sea
         Yellow Sea
      Pacific-Antarctic Ridge

LIST P - FIELDS OF INTEREST

INDEXING

All references selected for inclusion in the *Bibliography and Index of Geology* must be assigned to a specific field of interest. In the monthly issues, references are arranged under these fields to allow the user to scan sections of particular interest.

ONLINE SEARCHING

The fields of interest correspond to the category codes, or section headings, in the online citations. On Orbit and STN, they are designated CC and are searched by number.

On Dialog, the field is SH, and the search statement is formed as follows:

> SH=01
> SH=02
> etc.

On CAN/OLE, the designation is S, and the numbers are prefixed by the letters GR:

> GR01,S
> GR02,S
> etc.

It is usually advisable to search the index term(s) corresponding to the section headings also.

For more information on searching the category codes or section headings, see the *GeoRef Online Workshop Training Manual*.

B.I.G. SEARCHING

There are currently 30 fields of interest. It is important to note that beginning in August 1981, references could be assigned to multiple fields of interest. When this occurs, a "see-also" note appears at the end of the field.

The current list of 30 fields is an expansion of the 21 fields used in the *Bibliography and Index of Geology* from 1969 through 1974. From 1975 to 1991 engineering and environmental geology were combined under field 22. Both current and former fields are listed in this chapter, followed by a chart relating the 21 to the 30 at the end of this section.

It is important to remember that the fields of interest are in the monthly issues of the Bibliography, but do not appear in the annual cumulations.

01-Mineralogy and Crystallography
  physical, optical, and chemical properties of naturally occurring inorganic minerals and related synthetic minerals, mineral crystallography, including crystal structure, determination of lattice parameters and unit cells, the bonding of atoms and molecules, crystal form and symmetry, collecting minerals, as well as non-mineral gems such as amber and jet

02-Geochemistry
  abundance of elements
  organic materials
  water
  trace elements
  isotopes
  geochemical processes and properties
  geochemical cycles
  geochemical surveys
  analytical methods such as chemical, spectroscopic, thermal, and X-ray and electron microscopy
  instruments used for analysis
  Related topics in other fields:
  Field 05--petrology, fluid inclusions, geologic thermometry and barometry, meteorites
  Field 03--geochronology
  Field 06--clay mineralogy
  Field 12--paleoclimatology
  Fields 26 & 28--geochemical prospecting

03-Geochronology
  determination of absolute age methods, including radiometric and radiogenic dating establishing the chronology of events by methods such as lichenometry, racemization, tree rings, hydration of glass, varves, paleomagnetism, tephrochronology, thermoluminescence, radiation damage, fission-track dating and particle-track dating

04-Extraterrestrial geology
  of the planets, asteroids, the Moon, and moons of other planets
  planetary composition, evolution, surface features, structure, motions, gravity, magnetic fields and atmosphere
  Exclusions:
  astrophysics and extraterrestrial physics
  Related topics in other fields:
  Field 05-meteorites and tektites (before 1980 these were in Field 04.)

05-Petrology, igneous and metamorphic
  igneous rocks
  metamorphic rocks
  metasomatism
  metamorphism
  phase equilibria
  magmas
  lava
  intrusions
  inclusions
  ancient volcanology
  Related topics in other fields:
  Field 23-volcanic features
  Field 24-modern volcanology

06-Petrology, sedimentary
  sedimentary rocks
  sediments
  sedimentation
  diagenesis
  sedimentary structures
  genesis of peat, lignite and coal
  clay mineralogy

chemical properties of clay minerals
Related topics in other fields:
Field 07-marine sedimentation
Field 24-Quaternary sediments
Field 29-economic studies of peat, lignite and coal

07-Marine geology and oceanography
  ocean floors
  ocean basins
  ocean waves (sediment transport)
  ocean circulation (sediment transport)
  continental shelf
  continental slope
  Exclusions:
  patterns of ocean circulation
  marine biology
  wave propagation
  Related topics in other fields:
  Field 02-geochemistry of sea water
  Field 18-ocean basin evolution
  Field 21-estuarine studies
  Field 24-Quaternary sediments and Recent changes of level

08-Paleontology, general studies of both fossil plants and animals
  life origin
  paleontological textbooks
  paleontological glossaries
  fossil collecting
  conodonts
  problematic fossils
  ichnofossils (if not related to a specific fossil group)
  Related topics in other fields:
  Field 09-fossil plants
  Field 10-invertebrates
  Field 11-vertebrates
  Field 12-biostratigraphy

09-Paleontology, paleobotany
  algae
  angiosperms
  bacteria
  bryophytes
  fungi
  gymnosperms
  lichens
  palynomorphs
  Plantae
  pteridophytes
  thallophytes
  Related topics in other fields:
  Field 12-biostratigraphy
  Field 24-Quaternary palynology

10-Paleontology, invertebrate
  Coelenterata
  Echinodermata
  foraminifera
  Graptolithina
  Hemichordata
  Insecta
  Mollusca
  Ostracoda
  Porifera
  Pterobranchia
  Radiolaria
  Trilobita
  worms
  Related topics in other fields:
  Field 12-biostratigraphy

11-Paleontology, vertebrate
  Agnatha
  Amphibia
  Aves
  Chordata
  fossil man
  Mammalia
  Pisces
  Reptilia
  Related topics in other fields:
  Field 12-Pre-Quaternary archaeology and artifacts
  Field 24-Quaternary archaeology and artifacts

12-Stratigraphy, historical geology and paleoecology
  lithostratigraphy (age relationships of rock strata)
  biostratigraphy
  evolution of land masses (continental drift)
  paleomagnetism
  paleogeography
  biogeography
  paleoclimatology
  Related topics in other fields:
  Field 06-reefs and sedimentation
  Field 08-11-paleontology
  Field 24-Quaternary

13-Areal geology, general
  area studies dealing with several aspects of geology
  entries that might be placed in three or more fields
  guidebooks
  road logs
  bibliographies of geology of an area
  Related topics in other fields:
  Field 14-geologic maps

14-Areal geology, maps and charts
  separately published geologic maps
  geologic maps with explanatory texts
  separately published geologic charts
  methodology of geologic mapping
  Related topics in other fields:
  specific types of maps are found under the specific field, e.g. geomorphologic maps are found in Field 23, Surficial geology, geomorphology

15-Miscellaneous and mathematical geology
   biography (if not related to a specific field)
   bibliography (if not related to a specific field)
   popular geology
   elementary geology (textbooks)
   general mathematical principles
   annual reports of geologic surveys and associations
   historical accounts
   education-curricula, enrollments
   directories to geology departments
   geology as a profession-career opportunities
   forensic geology-geology applied to crime solving
   Related topics in other fields:
   anything related to a specific field will be found in that field rather than here

16-Structural geology
   classical tectonics (regional and local structures resulting from solid-rock movements)
   faults, fractures, folds, orogeny, geosynclines, deformation, structural analysis, epeirogeny, foliation, lineation
   Related topics in other fields:
   Field 18-plate tectonics, continental drift, sea-floor spreading
   Before 1975 the following subjects were found in Structural geology; they are now found in other fields:
   Field 05-batholiths, dikes, intrusions, stocks and volcanism
   Field 06-breccia, sedimentary structures
   Field 12-changes of level, paleogeography, unconformities
   Field 18-crust, diapirs, geodesy, isostasy, Mohorovicic discontinuity
   Field 22-nuclear explosions
   Field 23-cratering, cryptoexplosion features
   Field 24-glaciers

17-Geophysics, general
   experimental and theoretical studies of physical properties of rocks transition states of various compounds under elevated temperature and pressure (applied to core and mantle composition)
   magnetic and electrical properties of minerals and melts that relate to the Earth's magnetic field
   history, development and education in geophysics (since 1981 in Field 18)
   magnetic and gravity fields of the Earth (since 1981 in Field 18)
   Related topics in other fields:
   Field 04-Extraterrestrial geology
   Exclusions:
   Meteorology, magnetosphere, astrophysics, aeronomy and solar physics

18-Geophysics, solid-Earth
   worldwide structure of the Earth
   plate tectonics
   continental drift
   sea-floor spreading
   paleomagnetism
   structure of core, crust and mantle
   Mohorovicic discontinuity
   Related topics in other fields:
   Field 16-Regional and local structure
   Prior to 1975 the following subjects were found in Solid-Earth Geophysics; they are now found in other fields:
   Field 20-Geophysical surveys
   Field 19-Seismology
   Field 17-Gravity and magnetic fields, Earth's orbit, rotation and internal processes

19-Geophysics, seismology
   Earthquakes and elastic waves, including seismograms, wave velocity, seismic sources, seismicity, microearthquakes, microseisms, mechanism of tsunamis and volcanic earthquakes
   Related topics in other fields:
   Field 18-velocity structure of Earth's interior
   Field 22-geologic hazards, seismic risk

20-Geophysics, applied
   Applied studies not related to a specific subject including: well-logging, remote sensing, magnetotelluric surveys, gravity surveys, electrical surveys, magnetic surveys, seismic surveys, electromagnetic surveys
   Related topics in other fields:
   Anything related to a specific field will be found in that field rather than here.

21-Hydrogeology and hydrology
   ground water-geochemistry, movement, resources, mathematical models
   thermal waters
   springs
   geysers
   fumaroles
   surface water-chemistry, sediment transport, hydrologic cycle (from 1969 to 1974, fewer references to surface water were included)
   Exclusions:
   biology of surface water
   hydraulics
   Related topics in other fields:
   Field 02-geochemistry of water
   Field 22-ground-water pollution was included in Field 22 until 1981 and again after 1989

22-Environmental geology
   (engineering geology was removed from this field in 1992)
   geologic hazards-earthquakes, floods, land subsidence, landslides, debris flows, tsunamis, volcanoes
   conservation
   land use

pollution of surface water
ground-water pollution (before 1981 and after 1989)
With the growth of the environmental sciences, inclusion of environmental topics has increased dramatically in recent years.

23-Surficial geology, geomorphology
genesis and evolution of features on the Earth's surface-meteor craters, cryptoexplosion features, eolian features, erosion features, fluvial features, frost action, lacustrine features, mass movements, shore features, solution features, and volcanic features.
Related topics in other fields:
Field 24-Quaternary glacial features

24-Surficial geology, Quaternary geology
the last 2 million years of Earth's history, including:
glacial geology
stratigraphy
palynology
modern volcanology

25-Surficial geology, soils
soils-genesis, morphology, evolution, water regimes, chemistry, erosion and classification
Exclusions:
agricultural studies
Related topics in other fields:
Field 22-soil pollution, engineering properties of soils
Field 23-erosion processes

26-Economic geology, general and mining geology
commodity studies-more than one type of commodity mining geology
Related topics in other fields:
Field 22-mining engineering
Field 27-metals
Field 28-nonmetals
Field 29-energy sources

27-Economic geology, metals
metal ores-genesis, resources, economics, exploration, production and utilization (includes uranium)
Related topics in other fields:
Field 26-general and mining geology

28-Economic geology, nonmetals
nonmetal deposits-genesis, resources, economics, exploration, production, and utilization
Related topics in other fields:
Field 26-general and mining geology

29-Economic geology, energy sources
petroleum
natural gas
coal (economic studies)
oil shale
geothermal energy
oil sands
Related topics in other fields:
Field 26-general and mining geology
Field 06-genesis of coal, peat, and lignite

30-Engineering geology
(was combined with Field 22 until 1992)
waste disposal, reclamation
structures-dams, foundations, highways, marine installations, nuclear facilities, reservoirs, tunnels, underground installations and waterways when geologic subjects such as rock and soil properties are discussed.

PRE-1975 FIELDS OF INTEREST

Between 1969 and 1974 there were 21 Fields of Interest as follows:

01-Areal Geology
Economic geology, engineering geology, extraterrestrial geology, geochemistry, geochronology, geomorphology, geophysics, hydrogeology, igneous and metamorphic petrology, mineralogy, oceanography, paleontology, paleobotany, sedimentary petrology, stratigraphy, structural geology.

02-Economic Geology
Commodities (see List C for a list of specific terms), mineral deposits, genesis, mineral exploration.

03-Engineering Geology
Clays, dams, earthquakes, experimental studies, foundations, gas storage, highways, landslides, materials, permafrost, rock mechanics, shorelines, soils, subsidence, tunnels, waste disposal, environmental geology.

04-Extraterrestrial Geology
Asteroids, moon, meteorites, Mars, Venus (and other planets), tektites.

05-Geochemistry
Biogeochemical prospecting, chemical analysis, chemical elements, clay mineralogy, crust, crystal chemistry, diffusion, electron microscopy, fluid inclusions, geochemical prospecting, geochemical surveys, geochemistry, geochronology, geologic thermometry, ground water, hydrothermal alteration, isotopes, laterites, magmas, major-element analyses, metasomatism, meteorites, mineral exploration, organic materials, paleoclimatology, petrology, phase equilibria, radioactivity, sea water, soils, spectroscopy, tektites, trace-element analyses, weathering, X-ray-diffraction analysis.

06-Geochronology
Absolute age, correlation, isotopes,

paleoclimatology, paleomagnetism, paleontology, tephrochronology, time scales.

07-Hydrology
Artesian waters, connate water, fumaroles, geysers, glaciers, ground water, infrared surveys, mud volcanoes, permeability, porosity, springs, thermal springs.

08-Geomorphology
Basins, caves, erosion, estuaries, glacial geology, glaciation, glaciers, lakes, maps, mud volcanoes, rivers, shorelines, soils, weathering, permafrost.

09-Igneous and Metamorphic Petrology
Batholiths, breccia, cratering, cryptoexplosion structures, dikes, fluid inclusions, fumaroles, geologic thermometry, geothermal energy, orogeny, paragenesis, petrofabrics, petrology, phase equilibria, boudinage, foliation, geologic thermometry, lineation, metamorphic rocks, metamorphism, metasomatism, mineral zoning, igneous rocks, inclusions, intrusions, lava, magmas, meteorites, stocks, tektites, veins, volcanism, volcanoes, weathering.

10-Oceanography
Continental shelf, continental slope, estuaries, genesis, geophysical methods, marine geology, ocean basins, ocean waves, oceans circulation, ocean floors, reefs, sea water, sedimentation, sediments, seismic surveys.

11-Mineralogy
Crystal structure, crystal growth, crystal chemistry, crystallography, minerals.

12-General
Bibliography, current research, education, maps, philosophy, symposia, textbooks, catalogs, biography, associations.

13, 14, 15, 16-Paleobotany, Paleontology
Artifacts, collections, ecology, evolution, micropaleontology, organic materials, paleobotany, paleoecology, paleontology, palynology, reefs, tracks and trails. (See also List E for names of geologic periods)

17-Sedimentary Petrology
Breccia, concretions, diagenesis, heavy minerals, marine geology, nodules, type of material, sedimentary structures, sediments, sedimentary rocks, sedimentation

18-Soils
Chemistry, engineering properties, laterites, organic materials, permeability, porosity, sedimentation, soil group, weathering.

19-Solid Earth Geophysics
Geophysical surveys, seismology, earthquakes, tectonophysics, deformation.

20-Stratigraphy
Biostratigraphy, changes of level, continental drift, ecology, geochronology, geosynclines, glaciation, ice ages (ancient), lithostratigraphy, orogeny, paleobotany, paleoclimatology, paleoecology, paleogeography, paleomagnetism, paleontology, palynology, reefs, sedimentation, stratigraphy, unconformities.

21-Structural Geology
Basins, batholiths, boudinage, breccia, changes of level, continental drift, cratering, crust, cryptoexplosion structures, deformation, diapirs, dikes, epeirogenesis, faults, folds, foliation, fractures, geodesy, geosynclines, glaciers, intrusions, isostasy, lineation, meteor craters, Mohorovicic discontinuity, nuclear explosions, orogeny, paleogeography, petrofabrics, salt tectonics, sedimentary structures, stocks, tectonics, unconformities, uplifts, volcanism.

MAPPING OF OLD FIELDS TO NEW

Pre-1975	1975 to Date
1	13, 14
2	26, 27, 28, 29
3	22, 30
4	4
5	2
6	3
7	21
8	23, 24
9	5
10	7
11	1
12	15
13	9
14	8
15	10
16	11
17	6
18	25
19	17, 18, 19, 20
20	12
21	16

LIST R - ROCK UNITS

INDEXING

This list includes valid terms used for various rock units, including formations, complexes, intrusions, etc. The form of the term used in indexing should conform to the form found in the following list. In order to make the searching of the data base easier for the user the following terms have been controlled. If no form of the rock unit exists in this list, index using the form provided by the author. In addition, index the appropriate age term where possible. Ages are autoposted to those rock units commonly used for correlation. Note that some formations extend over several ages or may be of indeterminate age. In these instances the age must be added manually if known.

ONLINE SEARCHING

To search for rock units the most effective method is to use the first part of the name and view the index (Expand, Neighbor, etc.) to gather variations on the name. See Introduction, p. vii. Unfortunately there is a great deal of variation in the literature. Until a rock unit name is controlled the term is entered as it appears in the document. Occasionally official changes are made in stratigraphic nomenclature. These are incorporated in the controlled vocabulary as they are identified.

B.I.G. SEARCHING

Prior to 1993 rock units could be found by looking under the appropriate geographic term. Beginning in 1993 all rock units are direct access points.

ALPHABETICAL LIST

Aberystwyth Grits
Abo Formation
Absaroka Supergroup
Ackley Granite
Acungui Group
Adirondack Anorthosite
Ager Formation
Aguacate Group
Aguja Formation
Akiyoshi Limestone
Albert Formation
Aldridge Formation
Alisitos Formation
Allegheny Group
Almond Formation
Altyn Limestone
Ameki Formation
Americus Limestone Member
Ames Limestone
Amisk Group
Amitsoq Gneiss
Ammonoosuc Volcanics
Amsden Formation
Anahuac Formation
Animikie Group
Annot Sandstone
Anshan Group
Antalya Complex
Antelope Valley Limestone
Antietam Formation
Antlers Sands
Antrim Shale
Apache Group
Aquia Formation
Arab Formation
Arapien Shale
Aravalli System
Arbuckle Group
Areado Formation
Arikaree Group
Arkansas Novaculite
Arunta Complex
Ash Hollow Formation
Ashe Formation
Astoria Formation
Asu River Group
Athabasca Formation
Athgarh Sandstone
Atoka Formation
Austin Chalk
Austin Group
Aux Vases Sandstone
Aycross Formation
Aztec Sandstone
Baca Formation
Badami Series
Bagh Beds
Bainbridge Formation
Bakhtiari Formation
Bakken Formation
Balaklala Rhyolite
Ballantrae Complex
Baltimore Gneiss
Bambui Group
Banded Gneissic Complex
Bandelier Tuff
Banff Formation
Bangor Limestone
Baquero Formation
Barabash Suite
Baraboo Quartzite
Baraga Group
Barail Group
Baralaba Coal Measures
Barnett Formation
Barnwell Formation
Barre Granite
Barreiras Formation
Barren Measures
Barstow Formation
Bartlesville Sand
Barton Beds
Baskunchak Series
Bass Islands Dolomite
Battery Point Formation
Bauru Formation
Bay of Islands Ophiolite
Bazhenov Formation
Bear Gulch Limestone Member
Bear River Formation
Bearpaw Formation
Beaufort Formation
Beaufort Group
Beaumont Clay
Beaverhead Formation
Bedford Shale

Beech Creek Limestone Member
Beekmantown Group
Beil Limestone Member
Bell Canyon Formation
Belly River Formation
Belt Supergroup
Bembridge Marls
Benton Formation
Benxi Formation
Berea Sandstone
Berwick Formation
Betts Cove Ophiolite
Bevier Coal
Bhander Group
Bhuj Series
Bidahochi Formation
Big Snowy Group
Bigby-Cannon Limestone
Bijawar System
Bird Spring Formation
Birdbear Formation
Birkhead Formation
Bisbee Group
Bishop Tuff
Bitter Springs Formation
Biwabik Iron Formation
Black Creek Formation
Black River Group
Blackhawk Formation
Blackleaf Formation
Blackstone Series
Blaine Formation
Blair Formation
Blairmore Group
Blake River Group
Blake-Bahama Formation
Blondeau Formation
Bloomsburg Formation
Bloyd Formation
Bluesky Formation
Bluff Formation
Bluffport Marl Member
Blufftown Formation
Boggy Shale
Bonanza Group
Bonanza King Formation
Bone Spring Limestone
Bone Valley Formation
Bonner Formation
Bonneterre Formation
Boom Clay
Boone Formation
Bootlegger Cove Clay
Borden Group
Borrowdale Volcanic Series
Bossier Formation
Boston Bay Group
Botucatu Formation
Bowers Supergroup
Bowser Formation

Bowser Lake Group
Boyle Dolomite
Bracklesham Beds
Brallier Shale
Brassfield Formation
Breathitt Formation
Brent Group
Bridge Creek Limestone Member
Bridger Formation
Brigham Group
Bright Angel Shale
Brockman Iron Formation
Broken River Formation
Bromide Formation
Browns Park Formation
Brule Formation
Brunswick Formation
Brushy Basin Shale Member
Bucatunna Formation
Buchans Group
Buckner Formation
Bucksport Formation
Buda Limestone
Buda Marl
Bulawayan Group
Bulldog Shale
Bullfrog Member
Bulli Seam
Bullion Creek Formation
Burbank Sand
Burgess Shale
Burlington Limestone
Burra Group
Burro Canyon Formation
Byram Formation
Cabaniss Formation
Cache Creek Group
Cadeby Formation
Cadomin Formation
Cahill Formation
Calaveras Formation
Calera Limestone
Calvert Bluff Formation
Calvert Formation
Cambridge Argillite
Canyon Group
Cape Elizabeth Formation
Capistrano Formation
Capitan Formation
Caracas Group
Carbondale Formation
Cardium Formation
Carlile Shale
Carlton Rhyolite
Carmel Formation
Carrara Formation
Carrizo Mountain Formation
Carrizo Sand
Casco Bay Group
Caseyville Formation
Casper Formation
Castaic Formation

Castile Formation
Castle Hayne Limestone
Catahoula Formation
Catalina Schist
Catoche Formation
Catoctin Formation
Catskill Formation
Cedar City Formation
Cedar Mountain Formation
Cedar Valley Formation
Central Rand Group
Chadron Formation
Chainman Shale
Challis Volcanics
Chanares Formation
Changcheng System
Changzhougou Formation
Chappel Limestone
Charles Formation
Chase Group
Chatsworth Formation
Chattanooga Shale
Chazy Group
Chelmsford Formation
Chelmsford Granite
Chemung Formation
Cherokee Group
Cherry Canyon Formation
Chesapeake Group
Cheshire Formation
Cheyenne Sandstone
Chickamauga Group
Chickasawhay Formation
Chihsia Formation
Chilhowee Group
Chilliwack Group
Chinle Formation
Chipola Formation
Chopawamsic Formation
Choptank Formation
Chouteau Limestone
Chowan River Formation
Christopher Formation
Chuanlinggou Formation
Chuar Group
Chuckanut Formation
Chugwater Formation
Chumstick Formation
Cisco Group
Citronelle Formation
Claiborne Group
Clarno Formation
Clays Ferry Formation
Clayton Formation
Clear Fork Group
Clearwater Formation
Cleveland Member
Climax Porphyry
Clinch Sandstone
Clinton Group
Cloridorme Formation
Closepet Granite
Clough Formation
Cloverly Formation
Coast plutonic complex
Coast Range Ophiolite
Cobalt Group
Cockfield Formation
Coconino Sandstone
Codell Sandstone Member
Cody Shale
Coeymans Formation
Cohansey Formation
Coker Formation
Coldbrook Group
Collier Shale
Colorado Group
Colton Formation
Columbia River Basalt Group
Columbus Limestone
Colville Group
Comanche Peak Limestone
Conasauga Group
Condrey Mountain Schist
Conemaugh Group
Conococheague Formation
Cook Mountain Formation
Copley Greenstone
Copper Harbor Conglomerate
Cortlandt Complex
Cotton Valley Group
Cottonwood Limestone
Council Grove Group
Cow Head Group
Cowlitz Formation
Crab Orchard Formation
Crab Orchard Mountains Group
Crater Flat Tuff
Crescent Formation
Crevasse Canyon Formation
Croatan Formation
Crooked Creek Formation
Crystal River Formation
Cuddalore Series
Cuddapah System
Culebra Dolomite Member
Cumberland Group
Cupido Formation
Cutler Formation
Cuyahoga Formation
Cypress Hills Formation
Dakhla Shale
Dakota Formation
Damara System
Damuda Series
Davidsville Group
Deadwood Formation
Deccan Traps
Decorah Shale
Dedham Granodiorite
Deer Lake Group
Delaware Limestone
Delaware Mountain Group
Delhi Supergroup

Demopolis Chalk
Dengying Formation
Denver Formation
Deschutes Formation
Detroit River Group
Dharwars
Diamond Peak Formation
Difunta Group
Dinwoody Formation
Djadokhta Formation
Dockum Group
Douglas Group
Doushantuo Formation
Dry Branch Formation
Dubuque Formation
Duchesne River Formation
Dugger Formation
Dundee Limestone
Dunderberg Shale
Dunham Dolomite
Dunkard Group
Duperow Formation
Duplin Formation
Duwi Formation
Dwyka Formation
Eagle Bay Formation
Eagle Ford Formation
Eagle Sandstone
Earn Group
East Berlin Formation
Eastover Formation
Eau Claire Formation
Ecca Group
Eden Shale
Edmonton Formation
Edwards Formation
Ekofisk Formation
El Paso Group
Elberton Granite
Elk Point Group
Elkhorn Mountains Volcanics
Ellenburger Group
Ellensburg Formation
Elliot Formation
Ellis Bay Formation
Ely Limestone
Ely Springs Dolomite
Empire Formation
Endicott Group
Entrada Sandstone
Ervine Creek Limestone
Escondido Formation
Eskridge Shale
Esna Shale
Espanola Formation
Esplanade Sandstone Member
Essen Beds
Estrada Nova Formation
Eureka Quartzite
Eureka Sound Formation
Eutaw Formation
Everton Formation

Ewekoro Formation
Excello Shale
Fairview Formation
Fall River Formation
Fayetteville Formation
Feltville Formation
Fernie Formation
Fernvale Formation
Ferrar Group
Ferron Sandstone Member
Fig Tree Series
Fillmore Formation
Fish Canyon Tuff
Fish Haven Dolomite
Fisset Brook Formation
Flagstaff Formation
Flathead Sandstone
Fleming Formation
Flinton Group
Flournoy Formation
Forbes Formation
Fordham Gneiss
Fort Hays Limestone Member
Fort Payne Formation
Fort Riley Limestone
Fort Union Formation
Fortescue Group
Fountain Formation
Fourchu Group
Fox Hills Formation
Francis Creek Shale
Franciscan Complex
Franconia Formation
Freda Sandstone
Fredericksburg Group
Freeport Formation
Freites Formation
Fremouw Formation
Frenchman Springs Member
Frio Formation
Frontier Formation
Fruitland Formation
Fusselman Dolomite
Futaba Group
Galena Dolomite
Galesville Sandstone
Galice Formation
Gallup Sandstone
Gammon Ferruginous Member
Gander Lake Group
Gangpur Series
Gates Formation
Geirud Formation
Genesee Group
George River Group
Georgetown Formation
Gething Formation
Ghareb Formation
Ghost Rocks Formation
Gile Mountain Formation
Gilman Formation
Gizzard Group

Glasford Formation
Glen Canyon Group
Glen Mountains Complex
Glen Rose Formation
Glenarm Series
Glendon Limestone
Glenns Ferry Formation
Glenwood Shale
Glorieta Sandstone
Gog Group
Golconda Formation
Golden Valley Formation
Goldenville Formation
Goliad Sand
Goose Egg Formation
Gosau Formation
Gowganda Formation
Graneros Shale
Grayburg Formation
Great Oolite Series
Great Scar Limestone
Great Smoky Group
Great Valley Sequence
Green River Formation
Green Tuff Formation
Greenbrier Limestone
Greenhorn Limestone
Greenland Gap Group
Grimmertingen
Grimsby Sandstone
Grosmont Formation
Gubik Formation
Guelph Formation
Guilmette Formation
Gull River Formation
Gunflint Iron Formation
Hackberry Formation
Hadar Formation
Halifax Formation
Hallstatt Limestone
Hamersley Group
Hamilton Group
Hampden Basalt
Hampshire Formation
Hanna Formation
Hanson Creek Formation
Harbour Main Group
Harebell Formation
Harrodsburg Limestone
Hartland Formation
Hartselle Sandstone
Hartshorne Sandstone
Hat Creek Basalt
Hatchetigbee Formation
Hatteras Formation
Hawkesbury Sandstone
Hawthorn Formation
Hayes River Group
Haymond Formation
Haynesville Formation
Hazel Formation

Hazelton Group
Hazens Notch Formation
Heath Formation
Hecla Hoek Formation
Hector Formation
Heemskirk Granite
Helderberg Group
Helena Formation
Hell Creek Formation
Hemlock Conglomerate
Hemlock Formation
Henderson Gneiss
Hennessey Formation
Henryhouse Formation
Hermosa Formation
Herrin Coal Member
Hidden Valley Dolomite
High Bridge Group
Highland Series
Hillabee Chlorite Schist
Hilliard Shale
Holston Formation
Holyoke Basalt
Holz Shale
Honaker Trail Formation
Hoosac Formation
Horn Plateau Formation
Hornbrook Formation
Horseshoe Canyon Formation
Horsethief Creek Group
Horton Group
Hosston Formation
Huanglong Formation
Huber Formation
Hueco Limestone
Hughes Creek Shale
Hunsruck Shale
Hunton Group
Huron Member
Hurwitz Group
Hutchinson Salt Member
Hutuo Group
Idaho Springs Formation
Illawarra Coal Measures
Imo Formation
Imo Shale
Imperial Formation
Interlake Formation
Inyan Kara Group
Ipswich Coal Measures
Irati Formation
Iron Ore Group
Isachsen Formation
Ishbel Group
Itarare Subgroup
Ivishak Formation
Izumi Group
J-M Reef
Jabalpur Series
Jackfork Group
Jackson Group
Jacobsville Sandstone

Jefferson Group
Jeffersonville Limestone
Joana Limestone
John Day Formation
Johnnie Formation
Johns Valley Formation
Josephine Ophiolite
Josephine Peridotite
Judea Group
Judith River Formation
Juniata Formation
Jupiter Member
Kabuh Formation
Kaibab Formation
Kaimur Sandstone
Kaladgi System
Kalgoorlie System
Kalibeng Formation
Kamthi Formation
Kanawha Formation
Kanayut Conglomerate
Kansas City Group
Karewa Group
Karmutsen Group
Karroo Supergroup
Kasauli Series
Kaskaskia Sequence
Kayak Shale
Kayenta Formation
Kazusa Group
Keefer Sandstone
Keg River Formation
Kekiktuk Conglomerate
Kenai Group
Keokuk Limestone
Kewanee Group
Key Largo Limestone
Keyser Limestone
Khorat Group
Kiev Member
Kimmeridge Clay
Kingak Shale
Kingsdown Formation
Kingsport Formation
Kingston Peak Formation
Kinzers Formation
Kiowa Formation
Kirkpatrick Basalt
Kirkwood Formation
Kirtland Shale
Kisseynew Complex
Kittanning Formation
Kittery Formation
Klodzko Unit
Knife Lake Group
Knox Group
Kobiwako Group
Kodiak Formation
Kombolgie Formation
Koobi Fora Formation
Kootenay Formation
Kope Formation

Korenevskaya Formation
Krebs Group
Krivoy Rog Series
Krol Formation
Krosno Beds
Kunga Group
Kunyang Group
Kursk Series
Kuskokwim Group
La Luna Formation
La Meseta Formation
La Quinta Formation
La Salle Limestone
La Ventana Sandstone
Labette Shale
Laborcita Formation
Ladd Formation
Ladner Group
Ladoga Series
Laetoli Beds
Lake Valley Formation
Lakota Formation
Lamotte Sandstone
Lancara Formation
Lance Formation
Laney Shale Member
Lansing Group
Laramie anorthosite complex
Laramie Formation
Lardeau Group
Lathi Formation
Latrobe Group
Lavery Till
Lawrence Formation
Leadville Formation
Lebo Member
Lecompton Limestone
Leda Clay
Ledbetter Slate
Lee Formation
Leinster Granite
Lenoir Limestone
Levis Shale
Lewis Shale
Lewisian Complex
Lexington Limestone
Lilydale Limestone
Lime Creek Formation
Lincoln Creek Formation
Lincolnshire Limestone
Lisbon Formation
Lisburne Group
Little Falls Formation
Littleton Formation
Llajas Formation
Lockatong Formation
Lockport Formation
Lodgepole Formation
Logan Formation
London Clay
Lookingglass Formation
Lopez de Bertodano Formation

Lorrain Formation
Lost Burro Formation
Louann Salt
Loup Fork Group
Lower Greensand
Lucas Formation
Ludlow Member
Ludlowville Formation
Luning Formation
Luoquan Formation
Lushs Bight Group
Luxembourg Sandstone
Lynchburg Formation
Lyons Sandstone
Maden Complex
Madera Formation
Madison Group
Madrid Formation
Magdalena Group
Magothy Formation
Mahantango Formation
Maikop Series
Maiolica Limestone
Malmani Subgroup
Manakacha Formation
Mancos Shale
Manhattan Formation
Manitou Formation
Manlius Formation
Manning Canyon Shale
Mannville Group
Mansalay Formation
Mansfield Formation
Maokou Formation
Maquoketa Formation
Maravillas Formation
Marble Falls Group
Marca Shale Member
Marcellus Shale
Mardin Formation
Marianna Limestone
Marjum Formation
Marmaton Group
Maroon Formation
Marquette Range Supergroup
Marshalltown Formation
Martin Formation
Martinsburg Formation
Mary Lee Coal
Mascot Dolomite
Matagamon Sandstone
Matapedia Group
Mattoon Formation
Mauch Chunk Formation
Mazama Ash
McCloud Limestone
McCoy Creek Group
McHugh Complex
McKenzie Formation

McMurray Formation
McPherson Formation
Meade Peak Member
Medina Formation
Meguma Group
Menefee Formation
Mentor Beds
Menuha Formation
Mercia Mudstone
Merrimack Group
Mesa Rica Sandstone
Mesaverde Group
Metaline Limestone
Miami Limestone
Michelle Formation
Michigamme Formation
Middendorf Formation
Midway Group
Miette Complex
Miette Group
Mikabu System
Milk River Formation
Millboro Shale
Millstone Grit
Minas Series
Minnelusa Formation
Minturn Formation
Mirador Formation
Mishash Formation
Misoa Formation
Mispec Group
Missi Group
Mission Canyon Limestone
Mission Valley Formation
Missoula Group
Mist Mountain Formation
Mizunami Group
Modelo Formation
Moenave Formation
Moenkopi Formation
Monashee Complex
Monmouth Group
Monongahela Group
Montana Group
Monte Cristo Limestone
Monteagle Limestone
Monterey Formation
Montesano Formation
Montoya Group
Moodys Branch Formation
Mooreville Chalk
Moosebar Formation
Moreno Formation
Morien Group
Morrison Formation
Morrow Formation
Moscow Formation
Motojuku Formation
Moty Formation
Mount Read Volcanics
Mount Scopus Group
Mount Shields Formation

Mount Simon Sandstone
Mowry Shale
Muddy Creek Formation
Muddy Sandstone
Mural Limestone
Muro Group
Murphy Marble
Muth Quartzite
Nacimiento Formation
Nahanni Formation
Naheola Formation
Naknek Formation
Nama System
Namaqualand metamorphic complex
Nanafalia Formation
Nanaimo Group
Nanjemoy Formation
Nantuo Formation
Nanushuk Group
Napier Complex
Nari Series
Narita Formation
Narrabeen Group
Narragansett Pier Granite
Navajo Sandstone
Navarro Group
Navesink Formation
Neda Formation
Negaunee Iron Formation
Nenjiang Formation
New Albany Shale
New Red Sandstone
New River Formation
New York City Group
Newark Supergroup
Newcastle Coal Measures
Newcastle Sandstone
Newland Limestone
Newman Limestone
Neyveli Lignite
Ngorora Formation
Nicola Group
Nicoya Complex
Nihewan Formation
Nikolai Greenstone
Niniyur Group
Niobrara Formation
Nipissing Diabase
Nisku Formation
Nokomis Group
Nonesuch Shale
Noonday Dolomite
Nopah Formation
Normanskill Formation
Norphlet Formation
North Horn Formation
North Shore Volcanics
Novorayskoe Formation
Nubian Sandstone

Nugget Sandstone
Nuk Gneiss
Oak Hill Member
Oakville Sandstone
Ocala Group
Oficina Formation
Ogallala Formation
Ohio Shale
Ojo Alamo Sandstone
Oktemberyan Series
Old Crow Tephra
Oldman Formation
Olentangy Shale
Olmos Formation
Omo Group
Onaping Formation
Onondaga Limestone
Onverwacht Group
Opeche Shale
Opemisca Group
Oquirrh Formation
Orca Group
Oread Limestone
Oriskany Sandstone
Orocopia Schist
Oronto Group
Orr Formation
Ortega Group
Osaka Group
Osler Series
Ovruch Series
Oxford Clay
Paganzo Group
Pagoda Formation
Pahrump Series
Paintbrush Tuff
Palliser Formation
Paluxy Formation
Panchet Series
Parachute Creek Member
Paradox Member
Park City Formation
Parkman Sandstone
Parkwood Formation
Partridge Formation
Pasayten Group
Paskapoo Formation
Passa Dois Group
Passaic Formation
Peace River Formation
Peace Valley Beds
Peach Springs Tuff
Pearlette Volcanic Ash
Pearsall Formation
Pebble Shale
Peedee Formation
Peel Sound Formation
Pelona Schist
Peninsular Gneiss
Pennington Formation
Penobscot Formation
Penrhyn Slate

Peoria Loess
Perry Formation
Perry Mountain Formation
Petersburg Formation
Petersburg Granite
Petrified Forest Member
Phosphoria Formation
Phuket Group
Pico Formation
Pictou Group
Picture Gorge Basalt
Pictured Cliffs Sandstone
Pierre Shale
Pilgrim Formation
Pilot Shale
Piney Point Formation
Pinnacle Formation
Pinney Hollow Formation
Pinyon Conglomerate
Pioche Shale
Pirabas Formation
Pitkin Limestone
Pittsburgh Coal
Platteville Formation
Plattsburg Limestone
Plattsmouth Limestone Member
Pleasanton Group
Pocahontas Formation
Pocatello Formation
Pocono Formation
Pogonip Group
Point Loma Formation
Point Lookout Sandstone
Poltava Series
Pongola Supergroup
Porsanger Dolomite Formation
Portage Lake Lava Series
Porters Creek Formation
Portland Formation
Posidonia Shale
Potomac Group
Potsdam Sandstone
Pottsville Group
Poway Conglomerate
Poznan Clays
Prairie du Chien Group
Prairie Evaporite
Presumpscot Formation
Pretoria Group
Price Formation
Price River Formation
Prichard Formation
Prince Albert Group
Protopivskaya Formation
Pucangan Formation
Pucara Group
Puente Formation
Punchbowl Formation
Puncoviscana Formation
Pungo River Formation
Punta Mosquito Formation
Purcell System

Purisima Formation
Qingshankou Formation
Qixia Formation
Qorqut Granite
Quantou Formation
Queen City Formation
Queen Formation
Queenston Shale
Racine Dolomite
Raghavapuram Shales
Rajmahal Series
Ramp Creek Formation
Rampart Group
Ramparts Formation
Rangeley Formation
Ranger Canyon Formation
Raniganj Formation
Raritan Formation
Raton Formation
Ravalli Group
Ravenscrag Formation
Read Bay Formation
Reagan Sandstone
Red Eagle Limestone
Red Fork Sandstone
Red Mountain Formation
Red Peak Formation
Red River Formation
Redonda Formation
Redwall Limestone
Reedsville Formation
Renault Formation
Retort Phosphatic Shale Member
Revett Quartzite
Rexroad Formation
Rice Lake Group
Richmond Group
Ridge Route Formation
Riggins Group
Rincon Formation
Ringerike Sandstone
Ringold Formation
Rio Bonito Formation
Ripley Formation
Road River Formation
Roan Supergroup
Roberts Mountains Formation
Robertson Bay Group
Robles Formation
Roca Formation
Rochester Formation
Rock Lake Shale Member
Rock Springs Formation
Rocknest Formation
Rockwood Formation
Rodessa Formation
Rome Formation
Rondout Formation
Roraima Formation
Rosario Formation
Rose Canyon Formation
Rose Hill Formation

Ross Formation
Roxana Silt
Roxbury Conglomerate
Roy Group
Rundle Group
Rustler Formation
Rye Formation
Ryukyu Group
Saddle Mountains Basalt
Sadlerochit Group
Sag River Sandstone
Saginaw Formation
Saint George Group
Saint Laurent Limestone
Saint Louis Limestone
Saint Marys Formation
Saint Peter Sandstone
Saint Regis Formation
Sainte Genevieve Limestone
Salado Formation
Salem Limestone
Salina Group
Salt Lake Formation
Salt Wash Sandstone Member
Sambagawa Belt
San Andres Formation
San Diego Formation
San Felipe Formation
San Jose Formation
San Juan Formation
San Lorenzo Formation
San Miguel Formation
San Onofre Breccia
San Rafael Group
Sandia Formation
Sandia Granite
Sangerville Formation
Santa Barbara Formation
Santa Clara Formation
Santa Fe Group
Santa Margarita Formation
Santa Maria Formation
Santa Rosa Sandstone
Santa Susana Formation
Santana Formation
Santee Limestone
Saratoga Chalk
Saugus Formation
Sauk Sequence
Sausar Series
Savanna Formation
Scaglia Formation
Scarborough Formation
Schenectady Formation
Schoonover Sequence
Schuler Formation
Seboomook Formation
Selma Group
Semata Formation
Seminole Formation
Semri Series
Sentinel Butte Formation

Serebryanka Formation
Serra Geral Formation
Sespe Formation
Seven Rivers Formation
Severn Formation
Shadow Lake Formation
Shady Dolomite
Shahejie Formation
Shakopee Formation
Shannon Sandstone Member
Shap Granite
Sharon Conglomerate
Shaunavon Formation
Shawangunk Formation
Shawnee Group
Sheep Pass Formation
Shepard Formation
Sherridon Group
Sherwood Sandstone
Shimanto Group
Shimosa Group
Shinarump Member
Shiranish Formation
Shoo Fly Complex
Shublik Formation
Shubuta Member
Shungura Formation
Sicker Group
Signal Hill Formation
Silvretta Group
Simpson Group
Singhbhum Granite
Singleton Coal Measures
Sinjar Formation
Sioux Quartzite
Sisquoc Formation
Siwalik System
Skiddaw Slates
Skull Creek Shale
Slave Point Formation
Sligo Formation
Smackover Formation
Smartville Complex
Smithville Formation
Smithwick Shale
Smoky Hill Chalk Member
Snoqualmie Batholith
Snowslip Formation
Sobral Formation
Sokoman Formation
Solnhofen Limestone
Sonyea Group
Sparagmite Group
Sparta Sand
Spearfish Formation
Speeton Clay
Spergen Formation
Spirit River Formation
Spiro Sandstone
Spokane Formation
Spoon Formation

Springer Formation
Springfield Coal Member
Stanley Group
Stanton Formation
Star Point Sandstone
Statfjord Formation
Staunton Formation
Stearns Shale
Stellarton Group
Stillwater Complex
Stirling Quartzite
Stockton Formation
Stone Corral Formation
Stones River Group
Stormberg Series
Strawn Series
Stump Formation
Sturgis Formation
Subathu Formation
Sunbury Shale
Sundance Formation
Sunniland Limestone
Sunrise Formation
Supai Formation
Surma Group
Sussex Sandstone Member
Suwannee Limestone
Swan Hills Formation
Swan Peak Formation
Swauk Formation
Swaziland Sequence
Swaziland System
Swift Formation
Sylvania Formation
Table Head Group
Table Mountain Group
Taihua Group
Taishu Group
Taiyuan Formation
Takla Group
Tal Formation
Talchir Formation
Talchir Series
Talladega Group
Tallahatta Formation
Tallulah Falls Formation
Tamiami Formation
Tananao Schist
Tansill Formation
Tapeats Sandstone
Tapley Hill Formation
Taratash Complex
Taylor Marl
Teapot Sandstone Member
Tejon Formation
Temblor Formation
Tennessee Sandstone
Tensleep Sandstone
Tepee Trail Formation
Tesnus Formation
Tetagouche Group
Thamama Group

Thaynes Formation
Thebes Formation
Thomson Formation
Timbered Hills Group
Timiskaming Group
Tintic Quartzite
Tioga Bentonite
Todilto Formation
Tonganoxie Sandstone
Tongue River Member
Tonoloway Limestone
Tons Member
Toolebuc Formation
Topopah Spring Member
Tor Formation
Torlesse Supergroup
Torok Formation
Toroweap Formation
Tradewater Formation
Transvaal Supergroup
Traverse Group
Travis Peak Formation
Trenton Group
Trinity Group
Tubarao Group
Tulare Formation
Tulghes Series
Tullock Member
Tully Limestone
Tunguska Series
Tura Formation
Tuscaloosa Formation
Tuscarora Formation
Twiggs Clay
Twilight Gneiss
Twin Creek Limestone
Twin River Formation
Twin Sisters Dunite
Two Medicine Formation
Tyee Formation
Tyler Formation
Tyonek Formation
Udokan Series
Uinta Formation
Uinta Mountain Group
Umberatana Group
Umpqua Formation
Underhill Formation
Unkar Group
Uonuma Group
Upper Old Red Sandstone
Usa Series
Usuginu Conglomerate
Utica Shale
Uvalde Gravel
Vadito Group
Val Gardena Sandstone
Valdez Group
Valentine Formation
Valhalla Complex
Vaqueros Formation
Vassalboro Formation

Ventersdorp Supergroup
Verde Formation
Vermilion granitic complex
Vermillion Creek coal bed
Vetluga Series
Vicksburg Group
Viking Formation
Vinini Formation
Viola Limestone
Virginia Formation
Voltri Group
Voltzia Sandstone
Vryheid Formation
Wabaunsee Group
Waccamaw Formation
Waitemata Group
Waits River Formation
Waldron Shale
Wall Creek Member
Wallace Formation
Wanapum Basalt
Wann Formation
Wapanucka Limestone
Wapiti Formation
Warkalli Formation
Warrawoona Group
Warsaw Formation
Wasatch Formation
Wasekwan Group
Washita Group
Watahomigi Formation
Waterberg System
Waterville Formation
Waterways Formation
Weald Clay
Weber Sandstone
Wedron Formation
Wellington Formation
Wells Formation
Wenlock Limestone
Werillup Formation
Werra Series
Wescogame Formation
West Falls Formation
Westerly Granite
Westwater Canyon Sandstone Member
Wetterstein Limestone
Wewoka Formation
Wheeler Formation
Whirlpool Sandstone
White Limestone
White River Group
Whitemud Formation
Whitsett Formation
Wichita Group
Wicomico Formation
Wiggins Formation
Wilberns Formation
Wilcox Formation
Wilcox Group
Wildcat Group
Wilkins Peak Member

Williams Fork Formation
Williams Formation
Williamsport Sandstone
Willwood Formation
Willyama Complex
Wilmington Complex
Wilpena Group
Wind River Formation
Windermere System
Windsor Group
Wingate Sandstone
Winnipeg Formation
Winnipegosis Formation
Winton Formation
Wissahickon Formation
Witwatersrand Supergroup
Wolkberg Group
Wollaston Group
Womble Shale
Wonewoc Formation
Wonoka Formation
Wood Canyon Formation
Wood Mountain Formation
Wood River Formation
Woodbine Formation
Woodbury Clay
Woodford Shale
Woods Hollow Shale
Wreford Limestone
Wufeng Formation
Wumishan Formation
Wyandotte Limestone
Wyman Formation
Yacoraite Formation
Yakataga Formation
Yakima Basalt
Yamhill Formation
Yates Formation
Yazoo Clay
Yegua Formation
Yellowjacket Formation
Yellowknife Group
Yeso Formation
York River Formation
Yorktown Formation
Younger Granites
Yudoma Series
Yule Marble
Zabriskie Quartzite
Zambales Ophiolite
Zubair Formation
Zuloaga Limestone

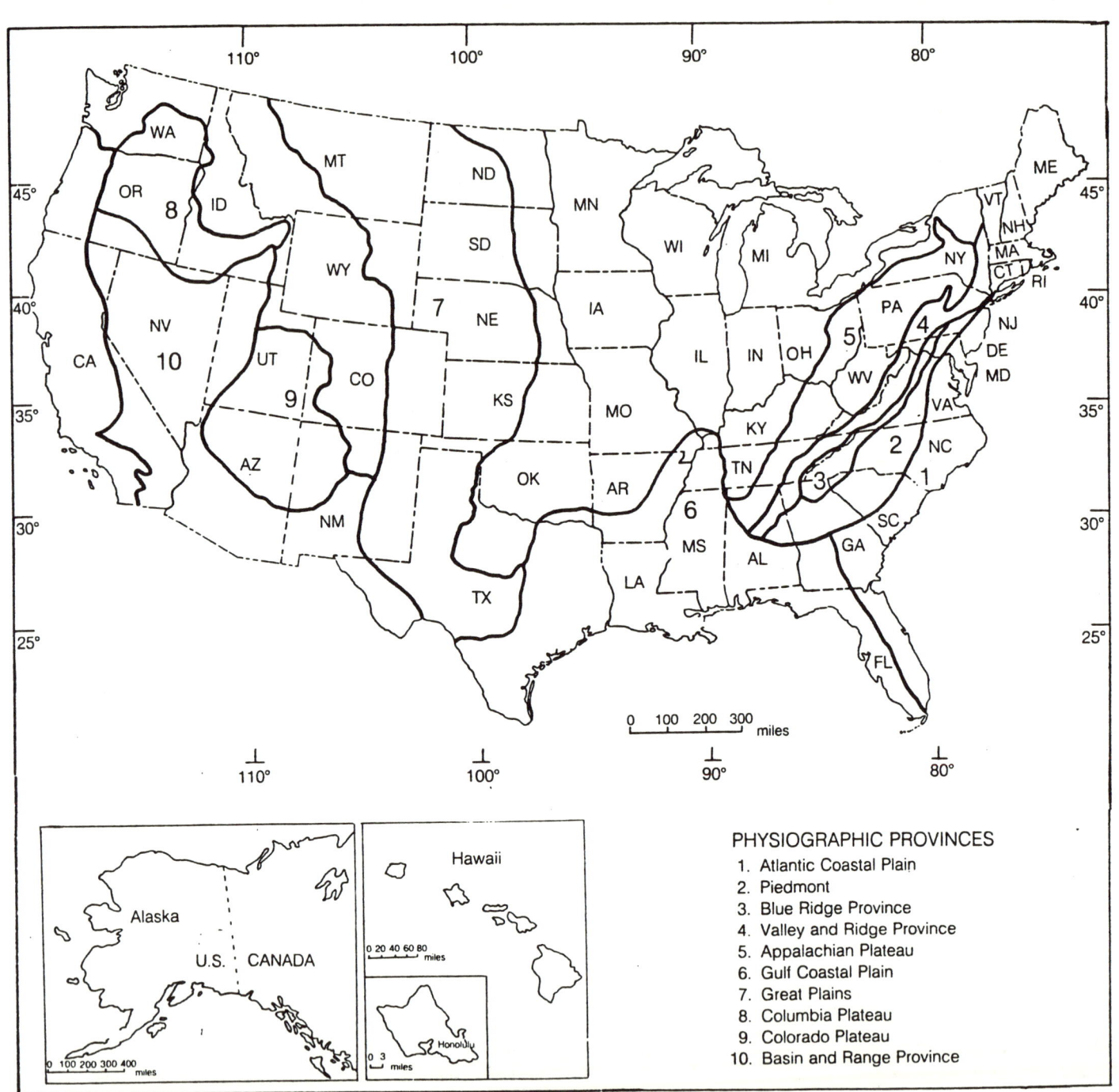

U.S. Physiographic Map

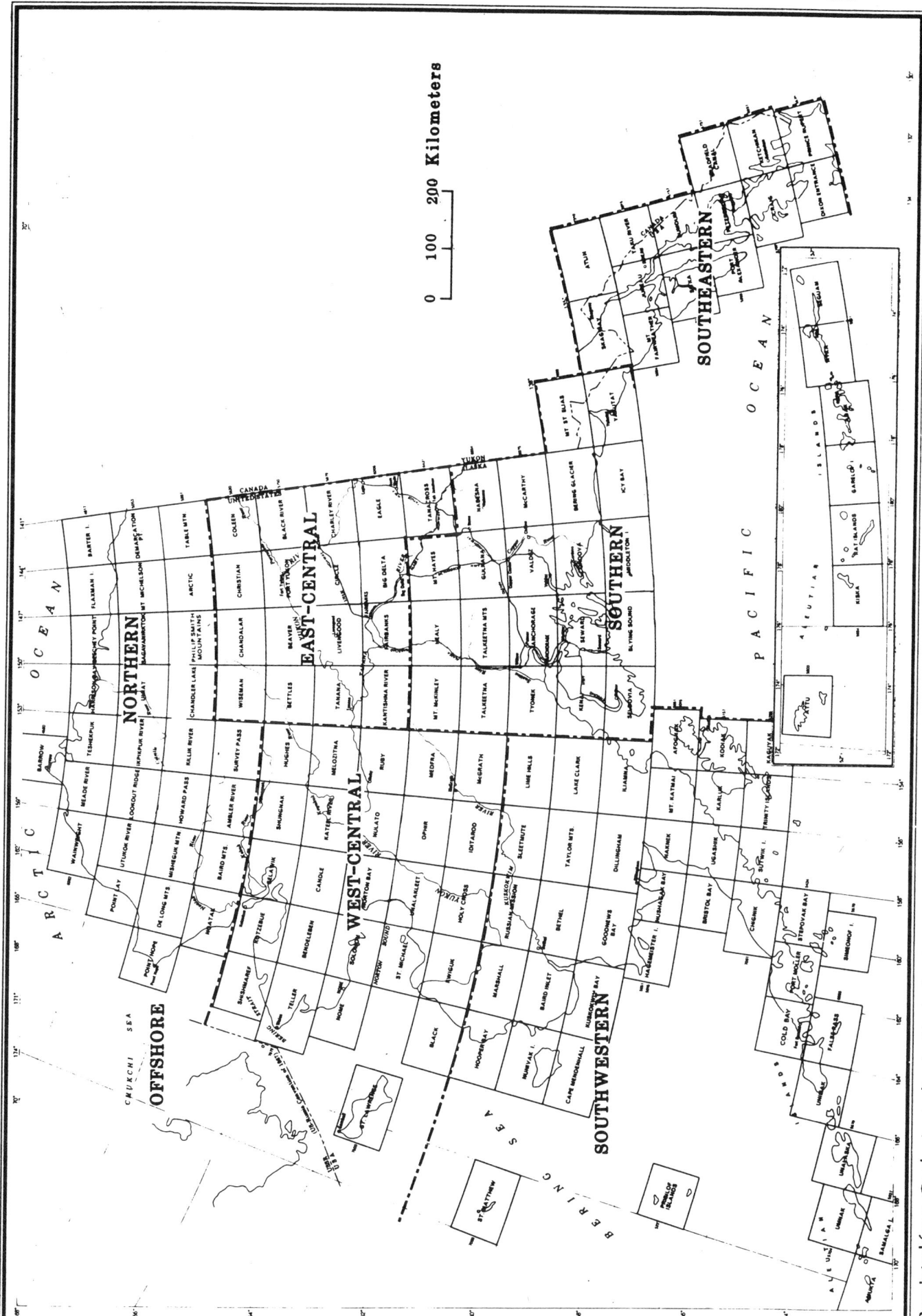

(Reprinted from Geologic studies in Alaska by the U.S. Geological Survey during 1986, U.S. Geological Survey Circular 998, 1987.)

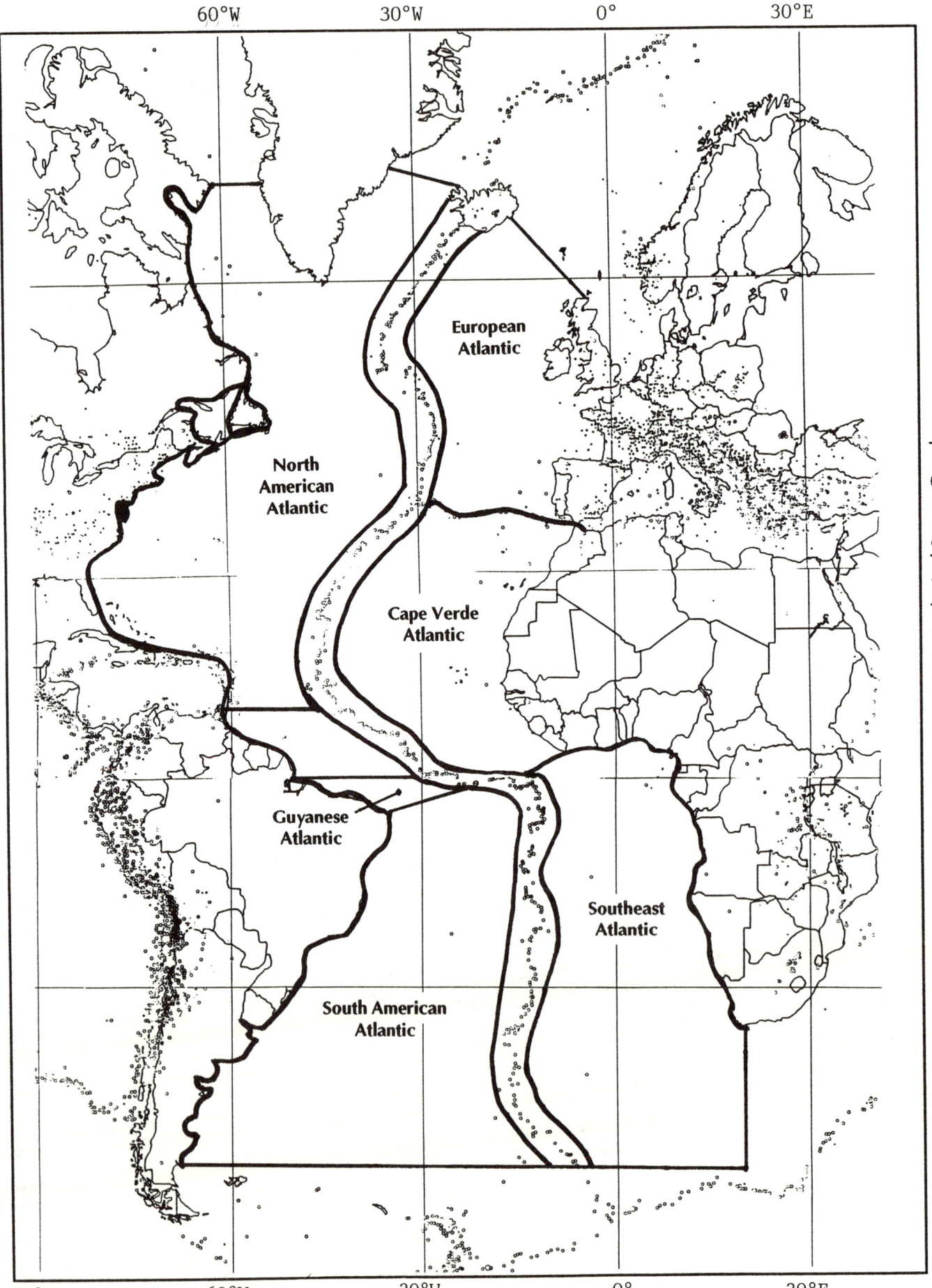

(Adapted from *Map showing global distribution of seismicity, 1977-1986*, U.S. Geological Survey Geophysical Investigations Map GP-989, 1988.)

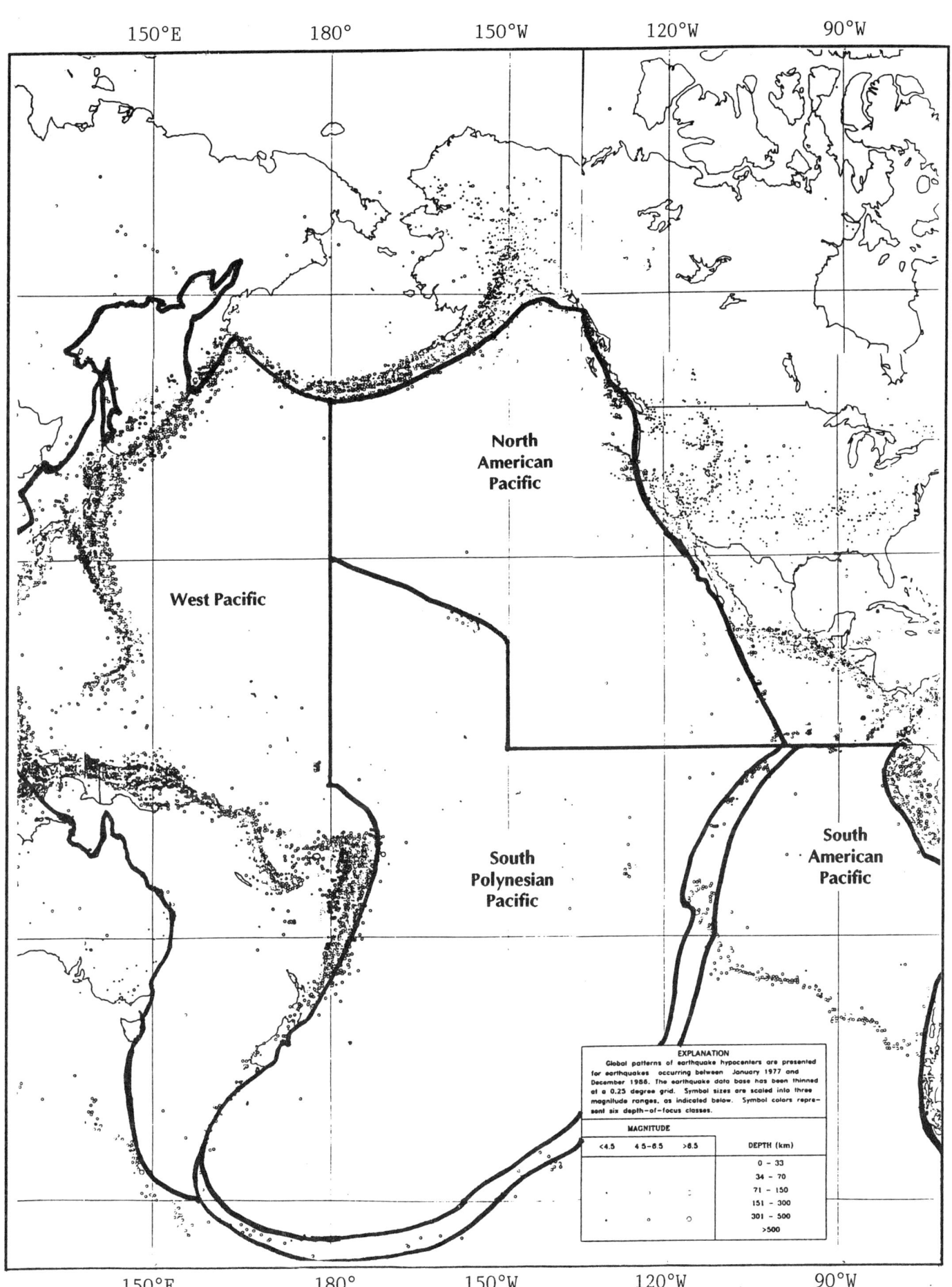

(Adapted from *Map showing global distribution of seismicity, 1977-1986*, U.S. Geological Survey Geophysical Investigations Map GP-989, 1988.)